COMPREHENSIVE ORGANIC SYNTHESIS

IN 9 VOLUMES

EDITORIAL BOARD

COMPREHENSIVE ORGANIC SYNTHESIS

Selectivity, Strategy & Efficiency in Modern Organic Chemistry

Editor-in-Chief
BARRY M. TROST
Stanford University, CA, USA

Deputy Editor-in-Chief
IAN FLEMING
University of Cambridge, UK

Volume 7
OXIDATION

Volume Editor
STEVEN V. LEY, FRS
Imperial College of Science, Technology & Medicine, London, UK

PERGAMON PRESS
OXFORD • NEW YORK • SEOUL • TOKYO

ELSEVIER SCIENCE Ltd
The Boulevard, Langford Lane
Kidlington, Oxford OX5 1GB, UK

First edition 1991
Second impression 1993
Third impression 1999

Library of Congress Cataloging in Publication Data

Comprehensive organic synthesis: selectivity, strategy and efficiency in modern organic chemistry/editor[s] Barry M. Trost, Ian Fleming.
p. cm.
Includes indexes.
Contents: Vol. 1.–2. Additions to C-X[pi]-Bonds — v. 3. Carbon–carbon sigma-Bond formation — v. 4. Additions to and substitutions at C-C[pi]-Bonds — v. 5. Combining C-C[pi]-Bonds — v. 6. Heteroatom manipulation — v. 7. Oxidation — v. 8. Reduction — v. 9. Cumulative indexes.
7. Organic Compounds — Synthesis I. Trost, Barry M. 1941–
II. Fleming, Ian, 1935–
QD262.C535 1991
547.2—dc20 90-26621

British Library Cataloguing in Publication Data

Comprehensive organic synthesis
7. Organic compounds. Synthesis
I. Trost, Barry M. (Barry Martin) 1941–
547.2

ISBN 0-08-040598-3 (Vol. 7)
ISBN 0-08-035929-9 (set)

∞ ™ The paper used in this publication meets the minimum requirements of American National Standard for Information Sciences — Permanence of Paper for Printed Library Materials, ANSI Z39.48-1984.

Contents

Preface

The emergence of organic chemistry as a scientific discipline heralded a new era in human development. Applications of organic chemistry contributed significantly to satisfying the basic needs for food, clothing and shelter. While expanding our ability to cope with our basic needs remained an important goal, we could, for the first time, worry about the quality of life. Indeed, there appears to be an excellent correlation between investment in research and applications of organic chemistry and the standard of living. Such advances arise from the creation of compounds and materials. Continuation of these contributions requires a vigorous effort in research and development, for which information such as that provided by the *Comprehensive* series of Pergamon Press is a valuable resource.

Since the publication in 1979 of *Comprehensive Organic Chemistry*, it has become an important first source of information. However, considering the pace of advancements and the ever-shrinking timeframe in which initial discoveries are rapidly assimilated into the basic fabric of the science, it is clear that a new treatment is needed. It was tempting simply to update a series that had been so successful. However, this new series took a totally different approach. In deciding to embark upon *Comprehensive Organic Synthesis*, the Editors and Publisher recognized that synthesis stands at the heart of organic chemistry.

The construction of molecules and molecular systems transcends many fields of science. Needs in electronics, agriculture, medicine and textiles, to name but a few, provide a powerful driving force for more effective ways to make known materials and for routes to new materials. Physical and theoretical studies, extrapolations from current knowledge, and serendipity all help to identify the direction in which research should be moving. All of these forces help the synthetic chemist in translating vague notions to specific structures, in executing complex multistep sequences, and in seeking new knowledge to develop new reactions and reagents. The increasing degree of sophistication of the types of problems that need to be addressed require increasingly complex molecular architecture to target better the function of the resulting substances. The ability to make such substances available depends upon the sharpening of our sculptors' tools: the reactions and reagents of synthesis.

The Volume Editors have spent great time and effort in considering the format of the work. The intention is to focus on transformations in the way that synthetic chemists think about their problems. In terms of organic molecules, the work divides into the formation of carbon–carbon bonds, the introduction of heteroatoms, and heteroatom interconversions. Thus, Volumes 1–5 focus mainly on carbon–carbon bond formation, but also include many aspects of the introduction of heteroatoms. Volumes 6–8 focus on interconversion of heteroatoms, but also deal with exchange of carbon–carbon bonds for carbon–heteroatom bonds.

The Editors recognize that the assignment of subjects to any particular volume may be arbitrary in part. For example, reactions of enolates can be considered to be additions to C—C π-bonds. However, the vastness of the field leads it to be subdivided into components based upon the nature of the bond-forming process. Some subjects will undoubtedly appear in more than one place.

In attacking a synthetic target, the critical question about the suitability of any method involves selectivity: chemo-, regio-, diastereo- and enantio-selectivity. Both from an educational point-of-view for the reader who wants to learn about a new field, and an experimental viewpoint for the practitioner who seeks a reference source for practical information, an organization of the chapters along the theme of selectivity becomes most informative.

The Editors believe this organization will help emphasize the common threads that underlie many seemingly disparate areas of organic chemistry. The relationships among various transformations becomes clearer and the applicability of transformations across a large number of compound classes becomes apparent. Thus, it is intended that an integration of many specialized areas such as terpenoid, heterocyclic, carbohydrate, nucleic acid chemistry, *etc.* within the more general transformation class will provide an impetus to the consideration of methods to solve problems outside the traditional ones for any specialist.

In general, presentation of topics concentrates on work of the last decade. Reference to earlier work, as necessary and relevant, is made by citing key reviews. All topics in organic synthesis cannot be treated with equal depth within the constraints of any single series. Decisions as to which aspects of a topic require greater depth are guided by the topics covered in other recent *Comprehensive* series. This new treatise focuses on being comprehensive in the context of synthetically useful concepts.

The Editors and Publisher believe that *Comprehensive Organic Synthesis* will serve all those who must face the problem of preparing organic compounds. We intend it to be an essential reference work for the experienced practitioner who seeks information to solve a particular problem. At the same time, we must also serve the chemist whose major interest lies outside organic synthesis and therefore is only an occasional practitioner. In addition, the series has an educational role. We hope to instruct experienced investigators who want to learn the essential facts and concepts of an area new to them. We also hope to teach the novice student by providing an authoritative account of an area and by conveying the excitement of the field.

The need for this series was evident from the enthusiastic response from the scientific community in the most meaningful way — their willingness to devote their time to the task. I am deeply indebted to an exceptional board of editors, beginning with my deputy editor-in-chief Ian Fleming, and extending to the entire board — Clayton H. Heathcock, Ryoji Noyori, Steven V. Ley, Leo A. Paquette, Gerald Pattenden, Martin F. Semmelhack, Stuart L. Schreiber and Ekkehard Winterfeldt.

The substance of the work was created by over 250 authors from 15 countries, illustrating the truly international nature of the effort. I thank each and every one for the magnificent effort put forth. Finally, such a work is impossible without a publisher. The continuing commitment of Pergamon Press to serve the scientific community by providing this *Comprehensive* series is commendable. Specific credit goes to Colin Drayton for the critical role he played in allowing us to realize this work and also to Helen McPherson for guiding it through the publishing maze.

A work of this kind, which obviously summarizes accomplishments, may engender in some the feeling that there is little more to achieve. Quite the opposite is the case. In looking back and seeing how far we have come, it becomes only more obvious how very much more we have yet to achieve. The vastness of the problems and opportunities ensures that research in organic synthesis will be vibrant for a very long time to come.

BARRY M. TROST
Palo Alto, California

Contributors to Volume 7

Professor R. Breslow
Department of Chemistry, University of Columbia, New York, NY 10027, USA

Dr S. M. Brown
ICI Fine Chemicals, PO Box A38, Leeds Road, Huddersfield, Yorkshire HD2 IFF, UK

Dr D. R. Buckle
SmithKline Beecham Pharmaceuticals, Biosciences Research Centre, Great Burgh, Yew Tree Bottom Road, Epsom, Surrey KT18 5XQ, UK

Dr P. C. Bulman Page
Department of Chemistry, University of Liverpool, PO Box 147, Liverpool L69 3BX, UK

Dr T. Chen
Department of Chemistry, Luther College, University of Regina, Regina, Saskatchewan S4S 0A2, Canada

Dr E. W. Colvin
Department of Chemistry, University of Glasgow, Glasgow G12 8QQ, UK

Professor R. H. Crabtree
Department of Chemistry, Yale University, PO Box 6666, New Haven, CT 06511-8118, USA

Dr D. Craig
Department of Chemistry, Imperial College of Science, Technology & Medicine, South Kensington, London SW7 2AY, UK

Dr D. Crich
Department of Chemistry (M/C 111), University of Illinois at Chicago, Box 4348, Chicago, IL 60680, USA

Dr P. J. Dudfield
Schering Agrochemicals Ltd, Chesterford Park Research Station, Saffron Walden, Essex CB10 1XL, UK

Dr T. L. Gilchrist
Department of Chemistry, University of Liverpool, PO Box 147, Liverpool L69 3BX, UK

Dr C. R. A. Godfrey
ICI Agrochemicals, Jealott's Hill Research Station, Bracknell, Berks RG12 6EY, UK

Dr A. Habib
Department of Chemistry, Yale University, PO Box 6666, New Haven, CT 06511-8118, USA

Dr A. H. Haines
School of Chemical Sciences, University of East Anglia, Norwich NR4 7JT, UK

Dr T. Inokuchi
Department of Applied Chemistry, School of Engineering, Okayama University, Tsushima-naka 3, Okayama 700, Japan

Dr R. A. Johnson
Hypersensitivity Diseases Research, The Upjohn Company, Kalamazoo, MI 49001-0199, USA

Dr A. B. Jones
Merck Sharp & Dohme Research Laboratories, PO Box 2000, R80M-119, Rahway, NJ 07065, USA

Dr J. E. G. Kemp
Pfizer Central Research, Sandwich, Kent CT13 9NJ, UK

Dr M. Kennedy
Department of Chemistry, University College Cork, Republic of Ireland

Dr S. N. Kilenyi
Sanofi Research, Avenue de Bejar, B-1120 Brussels, Belgium

Dr W. Kitching
Department of Chemistry, University of Queensland, St Lucia, Queensland 4067, Australia

Professor J. K. Kochi
Department of Chemistry, University of Houston, University Park, 4800 Calhoun Road, Houston, TX 77204-5641, USA

Professor G. R. Krow
Department of Chemistry, Temple University, Philadelphia, PA 19122, USA

Professor P. Laszlo
Laboratoire DCFI, Ecole Polytechnique, F-9128 Palaiseau Cedex, France

Professor D. G. Lee
Department of Chemistry, Luther College, University of Regina, Regina, Saskatchewan S4S 0A2, Canada

Dr T. V. Lee[†]
Department of Chemistry, University of Bristol, Bristol BS8 1TS, UK

Professor S. V. Ley
Department of Chemistry, Imperial College of Science, Technology & Medicine, South Kensington, London SW7 2AY, UK

Professor D. C. Liotta
Department of Chemistry, Emory University, 1515 Pierce Drive, Atlanta, GA 30322, USA

Dr A. Madin
Department of Chemistry, Imperial College of Science, Technology & Medicine, South Kensington, London SW7 2AY, UK

Mr T. J. McCarthy
Department of Chemistry, University of Liverpool, PO Box 147, Liverpool L69 3BX, UK

Professor M. A. McKervey
Department of Chemistry, University College Cork, Republic of Ireland

Professor C. J. Moody
Department of Chemistry, University of Technology, Loughborough, Leicestershire LE11 3TU, UK

Professor A. Pelter
Department of Chemistry, University College Swansea, Singleton Park, Swansea SA2 8PP, UK

Professor H. W. Pinnick
Department of Chemistry, Bucknell University, Lewisburg, PA 17837, USA

Dr I. L. Pinto
SmithKline Beecham Pharmaceuticals, Biosciences Research Centre, Great Burgh, Yew Tree Bottom Road, Epsom, Surrey KT18 5XQ, UK

Professor G. Procter
Department of Chemistry & Applied Chemistry, University of Salford, Salford M5 4WT, UK

Dr A. S. Rao
Emeritus Scientist, Organic Chemistry Division, Indian Institute of Chemical Technology, Hyderabad 500 007, India

Dr M. F. Schlecht
E I Du Pont de Nemours & Co., Inc., Agricultural Products Department, Stine-Haskell Research Center, PO Box 30, Newark, DE 19714, USA

Professor K. B. Sharpless
Department of Chemistry, Massachusetts Institute of Technology, Cambridge, MA 02139, USA

Dr T. K. M. Shing
Department of Chemistry, The Chinese University of Hong Kong, Shatin, N.T., Hong Kong

Professor T. Shono
Institute for Chemical Research, Kyoto University, Uji, Kyoto-fu 611, Japan

Professor K. Smith
Department of Chemistry, University College Swansea, Singleton Park, Swansea SA2 8PP, UK

Dr K. A. Swiss
Department of Chemistry, Emory University, 1515 Pierce Drive, Atlanta, GA 30322, USA

Professor S. Torii
Department of Applied Chemistry, School of Engineering, Okayama University, Tsushima-naka 3, Okayama 700, Japan

Professor J. Tsuji
Department of Applied Chemistry, Okayama University of Science, Ridai-cho, Okayama 700, Japan

Professor S. Uemura
Institute for Chemical Research, Kyoto University, Uji, Kyoto-fu 611, Japan

Abbreviations

The following abbreviations have been used where relevant. All other abbreviations have been defined the first time they occur in a chapter.

Techniques

CD	circular dichroism
CIDNP	chemically induced dynamic nuclear polarization
CNDO	complete neglect of differential overlap
CT	charge transfer
GLC	gas–liquid chromatography
HOMO	highest occupied molecular orbital
HPLC	high-performance liquid chromatography
ICR	ion cyclotron resonance
INDO	incomplete neglect of differential overlap
IR	infrared
LCAO	linear combination of atomic orbitals
LUMO	lowest unoccupied molecular orbital
MS	mass spectrometry
NMR	nuclear magnetic resonance
ORD	optical rotatory dispersion
PE	photoelectron
SCF	self-consistent field
TLC	thin layer chromatography
UV	ultraviolet

***Reagents, solvents,* etc.**

Ac	acetyl
acac	acetylacetonate
AIBN	2,2′-azobisisobutyronitrile
Ar	aryl
ATP	adenosine triphosphate
9-BBN	9-borabicyclo[3.3.1]nonyl
9-BBN-H	9-borabicyclo[3.3.1]nonane
BHT	2,6-di-*t*-butyl-4-methylphenol (butylated hydroxytoluene)
bipy	2,2′-bipyridyl
Bn	benzyl
t-BOC	*t*-butoxycarbonyl
BSA	*N*,*O*-bis(trimethylsilyl)acetamide
BSTFA	*N*,*O*-bis(trimethylsilyl)trifluoroacetamide
BTAF	benzyltrimethylammonium fluoride
Bz	benzoyl
CAN	ceric ammonium nitrate
COD	1,5-cyclooctadiene
COT	cyclooctatetraene
Cp	cyclopentadienyl
Cp*	pentamethylcyclopentadienyl
18-crown-6	1,4,7,10,13,16-hexaoxacyclooctadecane
CSA	camphorsulfonic acid
CSI	chlorosulfonyl isocyanate
DABCO	1,4-diazabicyclo[2.2.2]octane
DBA	dibenzylideneacetone
DBN	1,5-diazabicyclo[4.3.0]non-5-ene
DBU	1,8-diazabicyclo[5.4.0]undec-7-ene

DCC	dicyclohexylcarbodiimide
DDQ	2,3-dichloro-5,6-dicyano-1,4-benzoquinone
DEAC	diethylaluminum chloride
DEAD	diethyl azodicarboxylate
DET	diethyl tartrate (+ or –)
DHP	dihydropyran
DIBAL-H	diisobutylaluminum hydride
diglyme	diethylene glycol dimethyl ether
dimsyl Na	sodium methylsulfinylmethide
DIOP	2,3-*O*-isopropylidene-2,3-dihydroxy-1,4-bis (diphenylphosphino)butane
DIPT	diisopropyl tartrate (+ or –)
DMA	dimethylacetamide
DMAC	dimethylaluminum chloride
DMAD	dimethyl acetylenedicarboxylate
DMAP	4-dimethylaminopyridine
DME	dimethoxyethane
DMF	dimethylformamide
DMI	*N,N'*-dimethylimidazolone
DMSO	dimethyl sulfoxide
DMTSF	dimethyl(methylthio)sulfonium fluoroborate
DPPB	1,4-bis(diphenylphosphino)butane
DPPE	1,2-bis(diphenylphosphino)ethane
DPPF	1,1'-bis(diphenylphosphino)ferrocene
DPPP	1,3-bis(diphenylphosphino)propane
E^+	electrophile
EADC	ethylaluminum dichloride
EDG	electron-donating group
EDTA	ethylenediaminetetraacetic acid
EEDQ	*N*-ethoxycarbonyl-2-ethoxy-1,2-dihydroquinoline
EWG	electron-withdrawing group
HMPA	hexamethylphosphoric triamide
HOBT	hydroxybenzotriazole
$IpcBH_2$	isopinocampheylborane
Ipc_2BH	diisopinocampheylborane
KAPA	potassium 3-aminopropylamide
K-selectride	potassium tri-*s*-butylborohydride
LAH	lithium aluminum hydride
LDA	lithium diisopropylamide
LICA	lithium isopropylcyclohexylamide
LITMP	lithium tetramethylpiperidide
L-selectride	lithium tri-*s*-butylborohydride
LTA	lead tetraacetate
MCPBA	*m*-chloroperbenzoic acid
MEM	methoxyethoxymethyl
MEM-Cl	β-methoxyethoxymethyl chloride
MMA	methyl methacrylate
MMC	methylmagnesium carbonate
MOM	methoxymethyl
Ms	methanesulfonyl
MSA	methanesulfonic acid
MsCl	methanesulfonyl chloride
MVK	methyl vinyl ketone
NBS	*N*-bromosuccinimide
NCS	*N*-chlorosuccinimide

NMO	*N*-methylmorpholine *N*-oxide
NMP	*N*-methyl-2-pyrrolidone
Nu^-	nucleophile
PPA	polyphosphoric acid
PCC	pyridinium chlorochromate
PDC	pyridinium dichromate
phen	1,10-phenanthroline
Phth	phthaloyl
PPE	polyphosphate ester
PPTS	pyridinium *p*-toluenesulfonate
Red-Al	sodium bis(methoxyethoxy)aluminum dihydride
SEM	β-trimethylsilylethoxymethyl
Sia_2BH	disiamylborane
TAS	tris(diethylamino)sulfonium
TBAF	tetra-*n*-butylammonium fluoride
TBDMS	*t*-butyldimethylsilyl
TBDMS-Cl	*t*-butyldimethylsilyl chloride
TBHP	*t*-butyl hydroperoxide
TCE	2,2,2-trichloroethanol
TCNE	tetracyanoethylene
TES	triethylsilyl
Tf	triflyl (trifluoromethanesulfonyl)
TFA	trifluoroacetic acid
TFAA	trifluoroacetic anhydride
THF	tetrahydrofuran
THP	tetrahydropyranyl
TIPBS-Cl	2,4,6-triisopropylbenzenesulfonyl chloride
TIPS-Cl	1,3-dichloro-1,1,3,3-tetraisopropyldisiloxane
TMEDA	tetramethylethylenediamine [1,2-bis(dimethylamino)ethane]
TMS	trimethylsilyl
TMS-Cl	trimethylsilyl chloride
TMS-CN	trimethylsilyl cyanide
Tol	tolyl
TosMIC	tosylmethyl isocyanide
TPP	*meso*-tetraphenylporphyrin
Tr	trityl (triphenylmethyl)
Ts	tosyl (*p*-toluenesulfonyl)
TTFA	thallium trifluoroacetate
TTN	thallium(III) nitrate

Contents of All Volumes

Volume 1 Additions to C—X π-Bonds, Part 1

Nonstabilized Carbanion Equivalents

Heteroatom-stabilized Carbanion Equivalents

Transformation of the Carbonyl Group into Nonhydroxylic Groups

Volume 2 Additions to C—X π-Bonds, Part 2

Uncatalyzed Additions of Nucleophilic Alkenes to C=X

Volume 3 Carbon–Carbon σ-Bond Formation

Volume 5 Combining C—C π-Bonds

Volume 6 Heteroatom Manipulation

Volume 7 Oxidation

Volume 9

1.1
Oxidation by Chemical Methods

ROBERT H. CRABTREE and AFROZE HABIB
Yale University, New Haven, CT, USA

1.1.1 INTRODUCTION

In this chapter, functionalization reactions are emphasized which are of genuine practical utility, but others that are noteworthy and of potential synthetic significance are also discussed. Apart from alkane reactions, some intermolecular functionalization reactions which operate on unactivated C—H bonds, *e.g.* the side chain of a steroid, are examined. Intramolecular cases are covered elsewhere, and general reviews of the area are available.[1,2]

1.1.2 ORGANIZATION OF SECTIONS

Examples of alkane functionalization reactions of the type shown in equation (1) are first considered, in which the atom X to which the new C—X bond is formed comes from metals in Group I, followed by subsequent groups in the Periodic Table. Within each section, radical, electrophilic and carbenoid mechanisms are discussed.

$$\text{R-H} + \text{X} \longrightarrow \text{R-X} \qquad (1)$$

1.1.3 GENERAL PROBLEMS IN ALKANE FUNCTIONALIZATION

Alkanes have no lone pairs nor low-lying empty orbitals, but only the C—H and C—C σ- and σ*-levels. It is therefore relatively hard either to attack the former with an oxidizing agent or to attack the latter with a reducing agent or base. This means that vigorous conditions and reactive reagents often have to be used. It will almost always be the case that the product of an alkane functionalization reaction is more reactive than the starting material and so reacts faster with the functionalizing reagent. This in turn means that overoxidation can be a severe problem. To take a simple case, it is very difficult to stop the air oxidation of methane at the methanol stage. This means that many of the reactions in this section can only be run to low or very low conversion in order to obtain a satisfactory selectivity. This may be tolerable if the substrate is methane, but not so with valuable substrates. Low conversions should be assumed in the cases discussed below unless specifically mentioned.

A second important selectivity issue arises when there are several different types of C—H bond in the molecule, typically, primary, secondary and tertiary C—H bonds. Since tertiary radicals and carbonium ions are more stable than their secondary or primary analogs, many functionalization processes have an intrinsic selectivity pattern: tertiary > secondary > primary. Steric effects favor attack at primary positions, which is seen for very bulky reagents or in reactions in which the C—H bond to be broken is brought side-on to the functionalizing group, and therefore makes the transition state very sensitive to steric effects. The best example is oxidative addition to a transition metal complex.

There are three general classes of mechanism most often encountered in alkane reactions: (i) radical; (ii) electrophilic;[3] and (iii) carbenoid. The C—H bond-breaking steps in (i) and (ii) are shown in equations (2) and (3). Carbenoid reactions can go either by direct insertion into the C—H bond (equation 4), which tends to happen when the carbene in question has singlet character, or by a two-step process (equations 5 and 6), in which H-atom abstraction precedes collapse of the radical pair, a pathway which is characteristic of triplet carbenes.

$$\text{C-H} + \text{Q}\bullet \longrightarrow \text{C}\bullet + \text{Q-H} \qquad (2)$$

$$\text{C-H} + \text{E}^+ \longrightarrow \text{C}^+ + \text{E-H} \qquad (3)$$

$$:\text{Q} + \text{C-H} \longrightarrow \text{C-Q-H} \qquad (4)$$

$$:\text{Q} + \text{C-H} \longrightarrow \text{H-Q}\bullet + \text{C}\bullet \qquad (5)$$

$$\text{H-Q}\bullet + \text{C}\bullet \longrightarrow \text{C-Q-H} \qquad (6)$$

These three mechanistic pathways do not differ very markedly in selectivity. The usual pattern is tertiary > secondary > primary for C—H bonds, because this is the order of increasing stability both of the radical and of the carbonium ion. The concerted carbene route (equation 4) can show the reverse ordering (primary > secondary > tertiary) when the carbenoid in question is very bulky. This is notably the case for transition metal reagents undergoing oxidative addition of a C—H bond. The organometallic literature tends to look at this reaction from the point of view of the metal, as implied by the name 'oxidative addition', but from the point of view of alkane chemistry these reagents are carbenoids, perhaps having singlet character, which insert into an alkane C—H bond.

1.1.4 FORMATION OF R—M BONDS (M = GROUP I AND II METAL)

s-Butylpotassium (formed *in situ* from R_2Hg and K) reacts with alkanes to give the terminally mono- and di-substituted organopotassium compounds. Treatment with CO_2 gives the terminal carboxylic acid and the corresponding *n*-alkylmalonic acid. This selectivity for primary attack is probably the result of deprotonation being the C—H activation step (equation 7).[4]

$$\text{n-}C_5H_{11}\text{—Me} \xrightarrow{Bu^sK} \text{n-}C_5H_{11}CH_2K + \text{n-}C_5H_{11}CHK_2 \xrightarrow{CO_2}$$

$$\text{n-}C_5H_{11}CH_2CO_2H + \text{n-}C_5H_{11}CH(CO_2H)_2 \qquad (7)$$

H/D exchange in alkanes is catalyzed by a number of heterogeneous catalysts, such as Ni/alumina.[5]

1.1.5 FORMATION OF R—M BONDS (M = GROUP III METAL AND THE LANTHANIDES AND ACTINIDES)

Watson[6] was the first to show that methane could be attacked by a Group III metal reagent, in a reaction termed 'σ-bond metathesis'. This reaction probably proceeds *via* electrophilic attack on the C—H bond by the reagent (equation 8). Marks[7] described a nondegenerate example (equation 9) and Wolczanski[8] has shown that a zirconium imidate (Zr=NR) can also activate methane, the basic amine group receiving the proton released from methane (equation 10).

$$Cp^*_2LuMe + {}^{13}CH_4 \longrightarrow Cp^*_2Lu({}^{13}CH_3) + MeH \qquad (8)$$

$$Cp^*_2Th(\text{metallacyclobutane, 3,3-dimethyl}) \xrightarrow{CH_4} Cp^*_2Th(Me)(CH_2CMe_3) \qquad (9)$$

$$Cp^* = \eta^5\text{-}C_5Me_5$$

$$(R_3SiNH)_2Zr{=}NSiR_3 + CH_4 \longrightarrow (R_3SiNH)_2Zr(Me)NHSiR_3 \qquad (10)$$

$$R = Bu^t$$

1.1.6 FORMATION OF R—M BONDS (M = TRANSITION METAL)

The area of alkane activation is of current interest in organometallic chemistry and several examples of electrophilic C—H bond activation and of insertion of a carbenoid metal fragment into an alkane C—H bond have now been observed.[1] The acetone complex (**1**) shown in Scheme 1 is extremely reactive thanks to facile loss of acetone. A number of cyclopentanes react to form cyclopentadienyl (Cp) complexes.[9] For example, cyclopentane itself gives (**2**). The hydrogen removed from the alkane is transferred to the hydrogen acceptor $Bu^tCH{=}CH_2$ to give Bu^tCH_2Me. If a quaternary center is present,[10] as in 1,1-dimethylcyclo-pentane or -hexane, then a diene complex can be formed, which in the cyclopentane case may undergo C—C cleavage by a Green–Eilbracht[11] migration of the methyl group to the metal to give (**4**). Other transformations are shown in Scheme 1. Evanescent intermediates containing M—C bonds are thought to be important in a variety of catalytic alkane conversions, *e.g.* dehydrogenation reactions, and are described in a later section.

Bergman,[12] Graham[13] and Jones[14] and their coworkers observed a series of reactions, of which the examples shown in equations (11) and (12) are typical. The alkyl groups could be functionalized successfully with mercury salts to give the corresponding organomercurial, but most reagents led to elimination of alkane from the metal. The interesting feature of the reaction from a synthetic perspective is that attack at the primary C—H bonds is favored both kinetically and thermodynamically. This is also the case for the catalytic alkane conversions, which are discussed in a later section, which use oxidative addition in the first step. Liquid Xe has been used as a reaction solvent for the iridium system (equation 11;

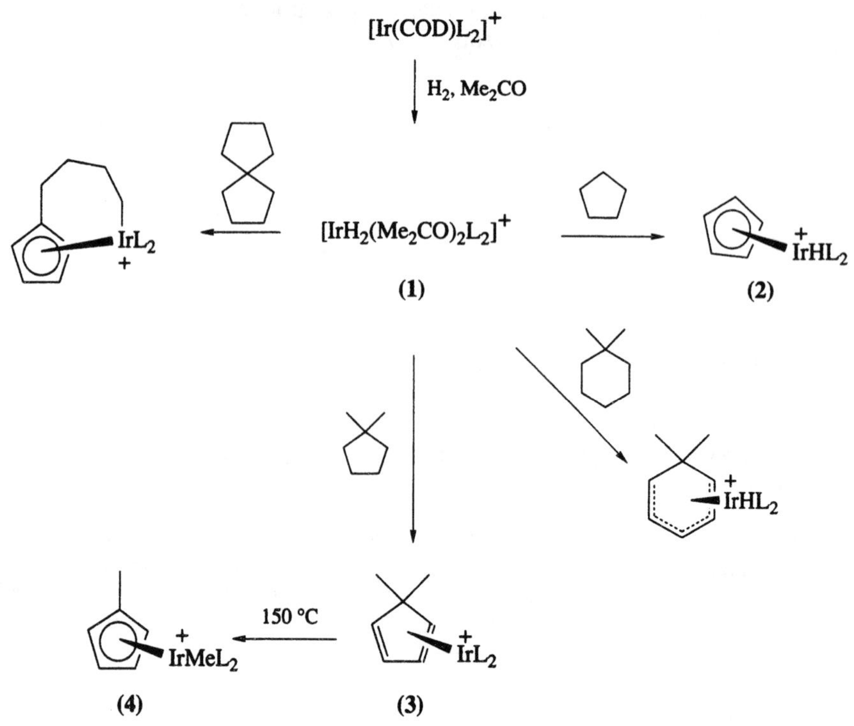

$[Ir(COD)L_2]^+$

H_2, Me_2CO

$[IrH_2(Me_2CO)_2L_2]^+$

(1)

(2)

(3)

(4)

150 °C

Scheme 1

L = PMe_3) to allow the use of rare, gaseous and solid alkanes. These studies show that methane gives the methyl hydride, cubane the cubyl hydride (without C—C bond breaking) and adamantane the secondary adamantyl hydride.[15]

$$CpIrLH_2 \xrightarrow{h\nu,\ RH} CpIrL(R)(H) \quad (11)$$

$$CpReL_3 \xrightarrow{h\nu,\ RH} CpReL_2(R)(H) \quad (12)$$

($Cp = C_5H_5$; $L = PMe_3$, $R = C_6H_{11}$)

Strained alkanes, such as cyclopropane and cubane, react much more easily with transition metal compounds.[16] Bare metal atoms also react with alkanes under metal vapor synthesis conditions to give synthetically useful quantities of alkane derivatives. For example, W atoms, cyclopentane and PMe_3 give $[CpW(PMe_3)H_5]$,[17] while rhenium atoms react with benzene and propane to give $[(\eta^6\text{-}C_6H_6)_2Re_2(\mu\text{-}Me_2C)(\mu\text{-}H)_2]$.[18] Related transformations[1] in mass spectroscopic and ion cyclotron resonance experiments do not have preparative value.

Shilov[19] found that a methylplatinum complex, $[MePtCl_3(PPh_3)_2]$, was formed from methane and H_2PtCl_6 at 120 °C, followed by reaction with PPh_3.

1.1.7 FORMATION OF R—C BONDS

In this section, not only alkane isomerization and dehydrodimerization (equation 13) are considered, but also the dehydrogenation of alkanes to alkenes, as in this case, two adjacent C—H bonds are replaced by a π-type C—C bond. Other C—C bond-forming reactions are also mentioned.

$$2\ R\text{-}H \longrightarrow R\text{-}R + H_2 \quad (13)$$

R R → R R + H_2 (14)
R R R R

1.1.7.1 Alkane Isomerization

Acid catalysts are required for alkane isomerizations and all the reactions of this type probably involve carbonium ions.[20] Optically active tertiary alkanes can be racemized by sulfuric acid at room temperature or by fluorosulfonic acids at –80 °C.[21]

Skeletal isomerization also occurs readily with acid catalysts. If a tertiary C—H is present a less powerful acid (*e.g.* conc. H_2SO_4) is capable of catalyzing the reaction, otherwise something more powerful (*e.g.* aluminum halides or superacids) is used. More recent examples of acid catalysts are: H_2SO_4/TiO_2,[22] SbF_5/Al_2O_3[23] and $H_2SO_4–SbF_5/ZrO_2$.[24] The reactions are driven thermodynamically, so, for example, tetrahydrodicyclopentadiene gives adamantane while higher *n*-alkanes are subject to cracking, and cyclohexane gives both methylcyclopentane and *n*-hexane.[25]

1.1.7.2 Alkane Dehydrodimerization

The mercury-photosensitized dehydrodimerization reaction has been known for many years,[26] but it has only been made preparatively useful very recently.[27] The key feature of the process is that the system is only active in the vapor phase, so that after condensation the product is protected from further conversion. This implies that the reaction can be run to essentially quantitative conversion without a fall-off in yield. In order to run on a gram scale to tens of grams, all that is needed is a quartz flask and a low pressure mercury lamp. Heating the substrate or substrates in the quartz flask with a small drop of mercury leads to smooth formation of the products. Aspects of the process are shown in equations (15) to (18).

$$Hg \xrightarrow{254\ nm} Hg^* \qquad (15)$$

(16): cyclohexane $\xrightarrow{Hg^*}$ cyclohexyl• + H•

(17): cyclohexyl• → bicyclohexyl + cyclohexene + cyclohexane

dimerization disproportionation

(18): H• + cyclohexene → cyclohexyl•

It might have been anticipated that this reaction would not work well because alkyl radicals tend to disproportionate to give alkenes (equation 17). In fact, the H-atoms produced in the initial homolysis rapidly readd to any alkene that forms to produce the alkyl radical (equation 18).

The reaction is also applicable to the dimerization of alcohols to glycols and amines to diamines (equation 19). The alcohol dimerization is important from the point of view of alkane functionalization because the cross dimerization of alkanes (RH) and alcohols (MeOH) gives the carbinol RCH_2OH as the major product. This product is very easy to separate from the glycol by washing with water and from the alkane homodimer, R_2, using column chromatography. By altering the liquid phase ratio of the two reagents there is a corresponding change in the vapor phase ratio. This use of vapor pressure bias can give a different ratio of products. For example, if the alkane were expensive, then an excess of methanol

$$MeOH \xrightarrow{Hg^*} HO\text{—}CH_2CH_2\text{—}OH \qquad (19)$$

could be used in the vapor and only RCH_2OH and R_2 would be formed. The yield based on alkane will exceed 90% and the mixture can be separated by solvent extraction.

Dehydrodimerization is also possible by an electrophilic route using superacids, but a substantial degree of rearrangement of the alkyl skeleton is often observed. For example, using $HSO_3F–SbF_5$ at 140 °C methane gives Me_3C^+, as does neopentane, the former *via* C—C bond formation, the latter by C—C cleavage.[28] The alkylation of alkenes by alkanes can also be brought about in a similar way, but alkene oligomerization is seen as a competing reaction (equations 20 to 23).

$$H_2C{=}CH_2 \quad + \quad H^+ \longrightarrow MeCH_2^+ \tag{20}$$

$$MeCH_2^+ \quad + \quad CH_4 \longrightarrow C_2H_6 \quad + \quad Me^+ \tag{21}$$

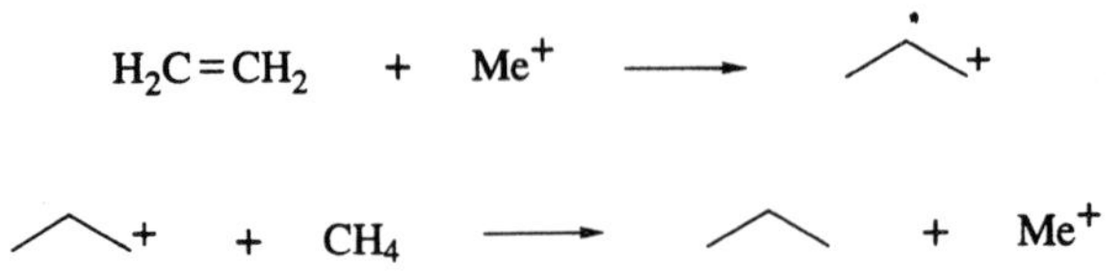

$$H_2C{=}CH_2 \quad + \quad Me^+ \longrightarrow \text{(propyl cation)}^+ \tag{22}$$

$$\text{(propyl cation)}^+ \quad + \quad CH_4 \longrightarrow \text{(propane)} \quad + \quad Me^+ \tag{23}$$

1.1.7.3 Transition Metal Catalyzed Alkane Dehydrogenation and Carbonylation

The groups of Felkin,[29] Crabtree[30] and Tanaka[31] have demonstrated that alkane dehydrogenation *via* oxidative addition is possible (equations 24 and 25). Attack at primary C—H bonds is favored, probably for steric reasons, but the stabilities of the catalysts are not yet sufficient for the reaction to be practically very useful. Tanaka's $[RhCl(CO)(PMe_3)_2]/h\nu$ system also carbonylates alkanes (equations 26 and 27). Lin[32] has applied the iridium system to more complex alkanes (equation 28).

$$C_6H_{12} \xrightarrow{IrH_2(O_2CCF_3)L_2,\ h\nu} C_6H_{10} \quad + \quad H_2 \tag{24}$$

$$n\text{-}C_8H_{18} \xrightarrow{ReH_7L_2} (Bu^nCH{=}CHCH{=}CH_2)ReH_3L_2 \quad + \quad (Pr^nCH{=}CHCH{=}CHMe)ReH_3L_2 \quad +$$
$$(EtCH{=}CHCH{=}CHEt)ReH_3L_2 \xrightarrow{P(OMe)_3} \text{1-Octene} \tag{25}$$

$$C_6H_{12} \xrightarrow{CO,\ RhCl(CO)L'_2,\ h\nu} C_6H_{11}CHO \quad + \quad C_6H_{11}CH_2CH_2OH\ (1.7\ \text{t.o.}) \quad + \quad C_6H_{11}CO_2H \tag{26}$$

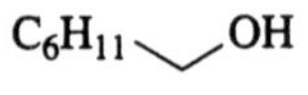

$$n\text{-}C_5H_{12} \xrightarrow{CO,\ RhCl(CO)L'_2,\ h\nu} n\text{-}C_6H_{11}CHO \quad + \quad Pr^n(Me)CHCHO \tag{27}$$

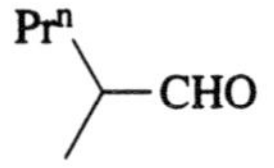

$$L = PPh_3;\ L' = PMe_3$$

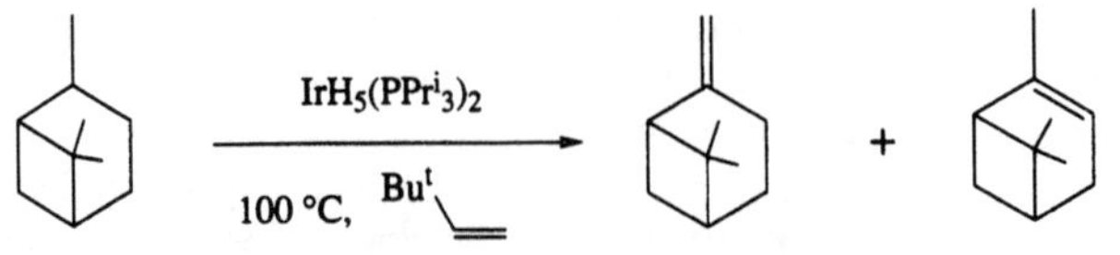

$$\text{(methyl-substituted bicyclic alkane)} \xrightarrow[100\ ^\circ C,\ Bu^tCH{=}CH_2]{IrH_5(PPr^i_3)_2} \text{(exocyclic methylene alkene)} \quad + \quad \text{(endocyclic alkene)} \tag{28}$$

1.1.7.4 Dehydrogenation and Aromatization

The $[IrH_2(Me_2CO)_2(PPh_3)_2]PF_6/Bu^tCH{=}CH_2$ system, mentioned above, has also been used for the aromatization of cyclohexane.[33] Photolysis of pyridine *N*-oxide in an alkane leads to dehydrogenation of the alkane.[34] The formation of by-products in the reaction, as a result of photorearrangement of the *N*-oxide, can be partially suppressed by the addition of BF_3.[35] The classic reaction involving heating with elemental S or Te is a synthetically useful method of aromatization: for example, decalin is converted to naphthalene with Te at 200 °C.[36]

1.1.7.5 Cracking and Reforming

Thermolysis of alkanes at *ca.* 750–800 °C gives dehydrogenation, skeletal rearrangement, cracking and even aromatization in some cases.[37] Catalysts are available which favor one or other of these routes, which are all important in the petroleum-refining industry.[38] Pyrolysis of various alkanes in H_2 can lead to the epitaxial growth of a diamond phase on a suitable substrate.[39] Superacids bring about alkane rearrangements even at 25 °C.[40] Alkylation of alkanes with alkenes under acidic conditions is also well known (equation 29).

$$\text{(isobutene)} + \text{(isobutane)} \xrightarrow{H_2SO_4} \text{Bu}^t\text{CH}_2\text{CH(CH}_3)_2 \quad (29)$$

1.1.7.6 Other Reactions

Oxalyl chloride yields ·COCl radicals on photolysis and this reaction has been used to directly substitute alkanes with the chloroacyl group (equation 30).[41] Similarly, biacetyl reacts with alkanes in a benzoyl peroxide-initiated chain reaction to give ketones in *ca.* 60–70% yield (equation 31).[42,43] Cyanogen chloride affords nitriles under similar conditions with a strong tertiary > secondary > primary selectivity pattern in 50–95% yields (equation 32);[44] MeO_2CCN is also reported to be an alternative reagent for the transformations.[45]

$$C_6H_{12} + (COCl)_2 \xrightarrow[63\%]{h\nu} C_6H_{11}COCl \quad (30)$$

$$C_6H_{12} + (COMe)_2 \xrightarrow[66\%]{\text{peroxide}} C_6H_{11}COMe \quad (31)$$

$$C_6H_{12} + ClCN \xrightarrow[66\%]{\text{peroxide}} C_6H_{11}CN \quad (32)$$

A radical addition reaction has been used to functionalize cyclopentane. The chain carrier, Cl·, was generated and regenerated by a β-elimination process (equation 33).[46] Other activated alkenes, such as maleic anhydride, furanone and acrylonitrile, have also been added to cyclohexane in 15–45% yield in a reaction initiated by Bu^tOOH or light.[47]

$$\text{C}_5\text{H}_{10} \xrightarrow{Q\bullet} \text{C}_5\text{H}_9\bullet \xrightarrow{\text{ClCH=CHCl}} \text{C}_5\text{H}_9\text{CHCl}\dot{\text{C}}\text{HCl} \longrightarrow \text{C}_5\text{H}_9\text{CH=CHCl} + \text{Cl}\bullet \quad (33)$$

Alkane functionalization by electrophilic addition reactions is also possible; for example, the particularly stable tertiary adamantyl cation must be involved in equation (34), a reaction which gives an excellent 75% yield of adduct.[48] In a similar way, a variety of alkenes[49] and arenes[50] can be alkylated by alkanes, or alkanes acylated by $RCOCl/AlBr_3$.[51]

$$\text{AdH} + H_2C{=}CH_2 \xrightarrow{AlBr_3,\ -70\ ^\circ C} \text{AdCH}_2\text{CH}_2\text{Br} \quad (34)$$

In the Koch–Haaf reaction, a superacid/CO mixture leads to carbonylation of the alkane.[52] A variety of products were obtained, *e.g.* Bu^tCO_2H, $Pr^iCHCOMe$ and $Pr^iCH_2COBu^t$, from isobutane using this product. Usually the method is only useful for alkanes containing a tertiary C—H bond,[12] but Sommer has introduced a modification that allows secondary C—H bonds to be functionalized, although only with 4% conversion (equation 35).[53]

$$PrH + HF + SbF_5 + CO \xrightarrow{Br^-,\ -10\ ^\circ C} Pr^iC(=O)F \quad (35)$$

In the Benson process[54] CH_4/Cl_2 mixtures are heated to 700–1700 °C and rapidly quenched. Ethylene, acetylene and benzene are formed by decomposition of the methyl chloride intermediate.

Several electrochemical methods for alkane oxidation have been used by Fleischmann and his coworkers.[55] These proceed *via* carbonium ion intermediates and, as expected, extensive rearrangement can be observed; for example, cyclohexane in FSO_3H gives 1-acetyl-2-methyl-1-cyclopentene as major product![56]

Carbene reagents also functionalize alkanes.[57] Triplet $:CH_2$ adds unselectively to alkane C—H bonds. The product mixture obtained from *n*-pentane was found to be 48% *n*-hexane, 35% 2-methylpentane and 17% 3-methylpentane, so that addition to a primary C—H bond appears to be favored.[58] Monochloromethylcarbene, CHCl, is less reactive and more electrophilic and so the normal tertiary > secondary > primary selectivity pattern was observed.[59] Ethoxycarbonylcarbene, formed on photolysis of the corresponding diazo compound, inserts rather unselectively in to alkane C—H bonds to give the ethoxycarbonylmethyl derivatives in *ca.* 50% yield. Transition metals, such as copper(II)[60] or rhodium(I),[61] also usefully catalyze the insertion of carbenes into alkane C—H bonds.

1.1.8 FORMATION OF R—X BONDS (X = Si, Ge, Sn, Pb)

So far, only the mercury-photosensitized chemistry, discussed above, allows direct functionalization of alkanes with a Si substituent under mild conditions (equation 36).[19]

$$R_3Si{-}H + R{-}H \xrightarrow{Hg^*} R{-}SiR_3 + H_2 \quad (36)$$

1.1.9 FORMATION OF R—N BONDS

Free radical nitration of alkanes has been carried out with nitric acid and related reagents at relatively high temperatures and has been used for the industrial synthesis of $MeNO_2$ from methane. Hydrogen radical atom abstraction from the alkane is thought to be followed by trapping of the radical with NO_2. A mixture of products tends to be formed in these cases, of which the nitroalkane and alkyl nitrite are most prominent.[62] Nitrogen oxides have been used for the same purpose.[63] Aminooxidation of cyclohexane with NH_3/O_2 takes place at 180 °C and 30 atm with copper or cobalt naphthenoate catalysts to give good yields of adiponitrile.[64]

A recent radical-based system, shown in equations (37) to (40), has been developed by Hill *et al.*[65] The catalyst in this process is a manganese(III) porphyrin, which is oxidized by PhIO to give what is believed to be an oxomanganese(V) intermediate. This is thought to abstract an H-atom from the alkane to give the alkyl radical. The resultant R· radical can then abstract either OH or X from the Mn catalyst to give the two chiefly observed products, ROH and RX (X = N_3, Cl, Br, I). In the case of X = N_3, the azide RN_3 is the major product, over 8 catalyst turnovers being observed (*i.e.* 800% yield based on Mn), accompanied by ROH (1.2 turnovers) and ketone (0.4 turnovers). The X = I example also works well, *ca.* 8 turnovers of RI being formed. The X = Cl and Br cases work less well and only *ca.* 1 turnover of RX and *ca.* 2 turnovers of ROH are formed, but an alternate approach is available.[66]

$$Mn(TPP)X + PhIO \longrightarrow O{=}Mn(TPP)X \quad (37)$$

$$RH + O{=}Mn(TPP)X \longrightarrow R\bullet + HO{-}Mn(TPP)X \quad (38)$$

$$R\bullet + HO{-}Mn(TPP)X \longrightarrow ROH + Mn(TPP)X \quad (39)$$

$$R\bullet + HO{-}Mn(TPP)X \longrightarrow RX + Mn(TPP)OH \quad (40)$$

Also developed by Hill[67] is a photochemical system (equations 41 to 48) based on a polyoxoacid, $H_3PW_{12}O_{40}$ (P). The excited state of the acid probably oxidizes the alkane in the first step. The radical can then either attack the solvent to give an iminium radical, which leads to ketone on hydrolysis, or it can be oxidized to the carbonium ion, in which case attack on the solvent leads instead to the *N*-alkylacetamide. If the substrate has two adjacent tertiary C—H bonds, then alkenes tend to be formed. The Barton reaction, normally known as an intramolecular C—H activation, can give some intermolecular reaction in some examples. Thus, when *n*-octyl nitrite is photolyzed in heptane, some nitrosoheptane is observed.[68]

$$P_{ox} + h\nu \longrightarrow P^*_{ox} \qquad (41)$$

$$P^*_{ox} + 2\,RH \longrightarrow P_{red} + 2\,RH^{\bullet+} \qquad (42)$$

$$RH^{\bullet+} \longrightarrow R^{\bullet} + H^+ \qquad (43)$$

$$P_{red} + 2\,H^+ \longrightarrow P_{ox} + H_2 \qquad (44)$$

$$C_6H_{11}^{\bullet} + MeCN \longrightarrow C_6H_{10}(C{=}NH)Me \qquad (45)$$

$$C_6H_{10}(C{=}NH)Me + H_2O \longrightarrow C_6H_{10}COMe \qquad (46)$$

$$C_6H_{11}^+ + MeCN \longrightarrow [C_6H_{11}{-}\overset{+}{N}{\equiv}{-}] \qquad (47)$$

$$[C_6H_{11}{-}\overset{+}{N}{\equiv}{-}] + H_2O \longrightarrow C_6H_{11}NHC(O)CH_3 \qquad (48)$$

A mixture of hydrazine and zinc oxide aminates cyclohexane in *ca.* 40% yield on photolysis.[69] Possibly, the hydrazine is dissociated by ZnO photosensitization and $\cdot NH_2$ radicals both abstract an H-atom from the alkane and quench the resultant carbon radical.

A number of functionalization reactions in which C—N bonds are formed depend on the initial formation of a carbonium ion from the alkane. This cation is quenched by the acetonitrile solvent and an amide or related species is obtained after hydrolysis. In the example shown in equations (49) to (51) Br_2 was used to generate the carbonium ion. Adamantane is a particularly favorable substrate as the carbonium ion is so easily formed and resists elimination. A 92% yield of amide was obtained in this process.[70] In a related reaction, HCN gives amine products (equation 52).[71]

$$Ad{-}H + Br_2 \longrightarrow Ad^+ + HBr + Br^- \qquad (49)$$

$$Ad^+ + MeCN \longrightarrow [Ad{-}\overset{+}{N}{\equiv}{-}] \qquad (50)$$

$$[Ad{-}\overset{+}{N}{\equiv}{-}] + H_2O \longrightarrow AdNHC(O)CH_3 \qquad (51)$$

$$(CH_3)_2CHCH_2CH_3 + HCN \xrightarrow{Bu^tOH,\ H_2SO_4} CH_3CH_2C(CH_3)_2NH_2 \qquad (52)$$

Another electrophilic reaction employs $AlCl_3/NCl_3$, in which case the aluminum reagent generates the carbonium ion which is then quenched by NCl_2^-. The amine is the final product of this reaction after hydrolysis. The system is selective for tertiary C—H bonds, *e.g.* methylcyclohexane gives an 82% yield of the tertiary amine compound. Arenes are also efficiently aminated.[72] NO_2PF_6 reacts with alkanes at 25 °C to give nitroalkanes, but skeletal rearrangements can occur and the yields are often poor.[73,74]

Carbenoid reagents can also introduce C—N bonds into alkanes. For example, cyanogen azide, N_3CN, decomposes at *ca.* 50 °C to give cyanonitrene, the ground state electronic structure of which is believed to be ·N═C═N·. Nevertheless, it reacts as :N—CN and gives insertion products with a variety of alkanes, selectively attacking the tertiary C—H bonds (equation 53). Reduction of the initial product can give the amine. Ethoxycarbonylnitrene reacts similarly.[75]

$$\text{R-H} + \text{:N-CN} \longrightarrow \text{R-N(CN)H} \quad (53)$$

1.1.10 FORMATION OF R—X BONDS (X = P, As, Sb)

In the presence of O_2, PCl_3 reacts readily with alkanes even at 25 °C to give alkylphosphonyl chlorides in yields up to 60%.[76] Surprisingly little use of this reaction has been made in synthesis. No examples of similar reactions have been reported for the analogous As and Sb halides.

1.1.11 FORMATION OF R—O BONDS

1.1.11.1 Autoxidation[77]

Autoxidation, or air oxidation, is one of the simplest functionalization reactions of alkanes. In general, hydroperoxides are the first-formed products, but these can decompose under the conditions of the reaction to give the ketone and alcohol.[78] The reagents used to initiate the reaction are usually *O*-centered radicals or even O_2 itself. These can efficiently start a chain reaction of the sort shown in equations (54) to (56), because the O—H bond energy is usually greater than the C—H bond energy and so both the initiation (equation 54) and the chain-carrying steps (equations 55 and 56) are favorable. The selectivity observed in the liquid phase, tertiary > secondary > primary, is consistent with the radical mechanism proposed. For example, *n*-decane gives ketones formed by attack at all the secondary positions along the chain.[79] Alkyl hydroperoxides have been used as initiators.[80] The RO_2· radical appears to be a more selective abstractor than RO·, and good selectivity for the formation of the tertiary hydroperoxide can be obtained (equation 57).[81]

$$\text{R-H} + \text{Q}\bullet \longrightarrow \text{R}\bullet + \text{Q-H} \quad (54)$$

$$\text{R}\bullet + O_2 \longrightarrow \text{R-O-O}\bullet \quad (55)$$

$$\text{R-O-O}\bullet + \text{R-H} \longrightarrow \text{R}\bullet + \text{R-O-O-H} \quad (56)$$

$$\text{methylcyclohexane} \xrightarrow{O_2,\ \text{peroxide}} \text{1-methylcyclohexyl-O-O-H} \quad (57)$$

A special situation occurs if two tertiary centers are in a 1,3-relationship to each other. In this case, the intermediate peroxy radical tends to abstract an H-atom from the β-C—H bond to give the bis-1,3-hydroperoxide as the final product. The key steps are shown in equation (58).[82]

$$\xrightarrow{O_2} \text{O-O}\bullet \longrightarrow \text{O-OH} \xrightarrow{O_2} \text{O-OH},\ \text{O-O}\bullet \quad (58)$$

Air oxidation of *n*-butane to maleic anhydride is possible over vanadium phosphate and, remarkably, a 60% selectivity is obtained at 85% conversion.[83] In the gas phase oxidation, in contrast to the situation found in the liquid, *n*-alkanes are oxidized more rapidly than branched chain alkanes. This is because secondary radicals are more readily able to sustain a chain; for branched alkanes the relatively stable tertiary radical is preferentially formed but fails to continue the chain process. Vanadium(V)/manganese(II)/AcOH has been used as a catalyst for the autoxidation of cyclohexane to adipic acid, giving 25–30% yields after only 4 h.[84]

1.1.11.2 Hydroxylation and Related Reactions

The classical Fentons reagent,[85] H_2O_2/Fe^{2+}, hydroxylates alkanes by producing hydroxyl radicals in solution.[86] This reagent is relatively unselective and inefficient since much of the peroxide reagent is wasted by catalytic decomposition to O_2. Ferryl radicals, FeO^{2+}, have sometimes been invoked as intermediates by analogy with P-450 chemistry, but conclusive evidence is still lacking. Metal peroxide complexes are also known to hydroxylate alkanes.[87]

Alkane hydroxylation is carried out in nature by a variety of enzymes, but the ones that have attracted most attention are the cytochrome P-450 dependent systems[88] found, for example, in mammalian liver. In the liver they serve to detoxify lipid soluble species, such as drugs, by making them more water soluble and hence more easily eliminated. For some substrates, such as certain arenes, the hydroxylation in fact makes these substrates more toxic, by converting them to epoxides which then alkylate liver DNA. The ultimate source of the O-atom used in the hydroxylation is O_2, but only one of the two O-atoms of the O_2 is incorporated in the substrate, the other is reduced to H_2O. This means two reducing equivalents are also required. Because they introduce only one O-atom from O_2 into the substrate, these enzymes are called monooxygenases (equation 59).

$$\mathrm{R{-}H} + O_2 + 2e^- + 2H^+ \longrightarrow \mathrm{R{-}OH} + H_2O \qquad (59)$$

In cytochrome P-450, an iron(III) coordinated to protoporphyrin IX is bound at the active site of these enzymes by a cysteine thiolato group in the fifth coordination position. In the first step, the iron(III) is reduced to iron(II), which then binds O_2. By a process still not completely elucidated, the distal oxygen is lost as H_2O, leaving the active form of the cofactor, which is believed to be an oxoiron species, probably best described as $O{=}Fe^{IV}(P^{+\cdot})$ (P = porphyrin). The oxo group has the reactivity of an oxene, and can either transfer oxygen to a double bond (*e.g.* forming an arene oxide from an arene) or insert into a C—H bond (*e.g.* to hydroxylate an alkane). In the case of alkane hydroxylation, the oxo group first abstracts an H-atom from the alkane, and the resulting alkyl radical abstracts an OH group from the metal (equations 60 and 61).

$$O{=}Fe^{IV}(P^{+\cdot}) + \mathrm{R{-}H} \longrightarrow \mathrm{HO{-}}Fe^{IV}(P) + \mathrm{R}\cdot \qquad (60)$$

$$\mathrm{HO{-}}Fe^{IV}(P) + \mathrm{R}\cdot \longrightarrow \mathrm{R{-}OH} + Fe^{III}(P) \qquad (61)$$

The early functional models for this oxidation chemistry were rather simple: Udenfriend[89] used iron(II), EDTA, ascorbic acid (as the reducing agent) and O_2 to hydroxylate arenes, while Hamilton[90] showed that the same system hydroxylates unactivated C—H bonds (*e.g.* androsten-3-ol-17-one is converted to androsten-3,7-diol-17-one). Mimoun[91] developed the use of an iron(II)/PhNHNHPh/$PhCO_2H/O_2$ system which is also active for alkane hydroxylation. Curiously, other metals [copper(II), manganese(II), vanadium(II), cobalt(II)] are also active. In the hydroxylation of arenes, an arene oxide is believed to be the intermediate in P-450 dependent systems, because a 1,2-shift of a proton in the arene, the 'NIH shift' is often observed. Neither the Udenfriend nor Mimoun models show such a shift, however.

More physiologically relevant models have been studied by the groups of Groves and of Hill. [Fe(TPP)Cl] (TPP = tetraphenylporphyrin) was used as the catalyst, but instead of the O_2 and reducing agent, iodosobenzene was used as the O-transfer reagent. This reagent is also effective in the enzyme system itself, where it also obviates the need for O_2 and the reducing agent. Conversions are low because the alkane is always used in excess, but yields of 5–25% have been reported with respect to the oxidant.[92] In the presence of the bromine atom donor $BrCCl_3$, the radical intermediates could be converted in part to the corresponding bromide, RBr. The tertiary to secondary selectivity was found to be 10–40:1, depending on the substrate used.

Although less relevant as a model, [Mn(TPP)Cl] is a better reagent for alkane oxidation and up to 70% conversion has been reported for cyclohexane in CH_2Cl_2.[93] Rearrangements were observed for norcarane, which led to 7 products, including ones in which the CH_2Cl_2 solvent had been incorporated. Meunier was the first to show that the far cheaper reagent, hypochlorite, could also be used to oxidize the manganese system.[94]

The chief problem with these systems from the synthetic point of view is the relatively rapid oxidation of the catalyst. Traylor *et al.*[95] have introduced tetraphenylporphyrins bearing chloro substituents at the *ortho* positions, which make the system much more robust, and 440 turnovers have been observed with cyclohexane, for example. Unfortunately, these catalysts are not yet commercially available. A metal-catalyzed acetoxylation of cyclohexane has been reported which utilizes Et_3NO/iron(II)/CF_3CO_2H.[96]

In the chromate oxidation of (+)-3-methylheptane to the corresponding tertiary alcohol, there was 70–80% retention of configuration, which is a useful synthetic reaction.[97] Iridium and ruthenium salts also have been shown to catalyze this reaction.[98]

Hydroxylation is also induced in good yields by the photolysis of alkanes in nitrobenzene. Using a high pressure mercury lamp, the tertiary > secondary > primary selectivities observed have been 300:19:1 (pyrex filter) and 110:7:1 (vycor filter). No retention of configuration was observed in these reactions; consequently, a free radical mechanism was invoked.[99]

An interesting reagent, $CrO_2Cl_2/Me_2C{=}CHMe$, has been described, which is said to oxidize the methyl group of methylcyclohexane to CHO without affecting the tertiary C—H bond (equation 62).[100] This reaction is worth further investigation.

i, CrO_2Cl_2/alkene; ii, H_2O; 25% CHO (62)

Other chromium(VI) reagents are known to attack at tertiary C—H bonds. For example, CrO_3/AcOH appears to be a general reagent for the introduction of an OH group at the 14-position of steroids (equation 63).[101] Yields depend critically upon the amount of water present, 1–2% being best. CrO_2X_2 (X = Cl or OAc) also reacts with alkanes to give oxidized products.[102]

O AcO Br Br CrO_3 AcOH 30% O OH AcO Br Br (63)

Basic $KMnO_4$ has been found to hydroxylate tertiary C—H bonds in certain cases (equation 64).[103] The tertiary alcohol functionality in the starting material seems to be essential in the reaction, and so an intramolecular reaction of a manganate ester is highly likely as an intermediate step. Alternatively, the hydroxy group may simply be required to improve the phase transfer into the aqueous medium, since the organic soluble reagent benzyltrimethylammonium permanganate readily attacks alkanes. *Trans*-decalin affords both the tertiary alcohol (37%) and *trans*-1-decalone (43%).[104] Overoxidation of the products with C—C bond cleavage can be a problem with manganese(VII) reagents, however.[105] $KMnO_4$ in trifluoroacetic acid reacts with alkanes at 25 °C with a tertiary > secondary > primary selectivity ratio of 2100:60:1; k_H/k_D is 4.3 at 25 °C.[106]

OH basic $KMnO_4$ OH OH (64)

Cobalt(III) perchlorate in aqueous MeCN oxidizes alkanes at room temperature with an apparent secondary > primary > tertiary selectivity pattern. This pattern may not be real, however, because the product of tertiary attack may be much more sensitive to further oxidation. 2-Methylpentane was hydroxylated relatively selectively at the 4-position (74%); the minor products have OH groups at the 5- (13%), 1- (6%), 3- (2%) and 2-position (5%).[107] $Co(OCOCF_3)_3$ is a related reagent which has been reported to acetoxylate alkanes.[108]

Lead tetraacetate reacts poorly with acyclic alkanes, even 3-methylpentane,[109] although cyclohexane is readily converted to the corresponding acetate at 80 °C or with irradiation at room temperature.[110] The yield of acetate is increased 10-fold by the addition of Bu^tOH, under which conditions $Bu^tO\cdot$ is thought to act as H-atom abstractor.[111] The more reactive lead(IV) reagent $Pb(OCOCF_3)_4$ has been used to introduce the trifluoroacetate group. Hexafluorobenzene or CF_3CO_2H are satisfactory solvents and hydrolysis to the alcohol is easily accomplished with NaOH, with an overall yield of *ca.* 45%. The secondary C—H bonds are attacked in *n*-alkanes and arenes also react under these conditions.[112]

A similar reagent is thought to be formed from Ag_2O_2 and trifluoroacetic acid. Here, the silver oxide reacts as $Ag^I[Ag^{III}O_2]$ and forms $Ag(OCOCF_3)$ *in situ*, which is believed to be the active oxidant. With adamantane, the normal tertiary substitution product, $AdOCOCF_3$, was obtained in 98% yield. This oxidation could be made catalytic using NH_4NO_3 as cooxidant and $AgOCOCF_3$ as catalyst.[113]

Alkyl peroxycarbonates, which give $\cdot CO_2R$ radicals on thermolysis, function in chain reactions to give good yields of the corresponding carbonates from alkanes (equation 65).[114]

$$C_6H_{12} \quad + \quad (CO_2OR)_2 \xrightarrow{\Delta} C_6H_{11}O\overset{O}{\overset{\|}{C}}OR \quad + \quad HOR \qquad (65)$$

The 'Gif' system, discovered by Barton and coworkers at Gif-sur-Yvette,[115] consisting of air, iron powder, sulfide, organic solvent, acid and water, smoothly hydroxylates alkanes. The sulfide was found to be unnecessary if the reaction temperature exceeded 40 °C,[116] and the basic acetate $[Fe^{II}Fe^{III}_2O(OAc)_6py_3]$ was shown to be active in the presence of a reducing agent. Adamantane afforded up to 11% yield of adamantanone after 18 h. The system does not seem to fall into any of the usual mechanistic categories. Secondary C—H bonds appear to be attacked more readily than either primary or tertiary C—H bonds, but the selectivity is artificially elevated as a result of side reactions which the tertiary products undergo. The intrinsic secondary to tertiary selectivity of *ca.* 1:1 is still much higher than expected for radical or oxometal oxidations,[117] and is therefore of synthetic value. In addition, such species as diphenyl sulfide are not oxidized under the reaction conditions. Ketones, not alcohols, are the major products from cycloalkanes, but the ketones appear to be formed directly, not by oxidation of an alcohol intermediate. Nitrogenous bases, such as pyridine, are essential and Shilov[118] has suggested that the active oxidant may be pyridine-derived, *e.g.* $C_5H_4NO^{+}\cdot$. This suggestion is made more plausible by the fact that the iron can be replaced by other metals.[119] The same paper reports turnovers exceeding 3000 for the most recent version of the Gif system. A reagent which may operate by a related mechanism is iron(II)/Et_3NO/CF_3CO_2H, which gives trifluoroacetates from alkanes also without overoxidation.[120]

Peracids can react with alkanes to give hydroxylated products,[121] as shown in equation (66). This may be an electrophilic reaction because the rate increases with increasing acidity of the peracid. Radical side reactions were thought to be inhibited by added I_2.[122] CF_3CO_3H is also an effective oxidant.[123] The reaction of *trans*-1,2-dimethylcyclohexane with $PhCO_3H$ is reported to be 97% stereoselective (retention) and 97% regioselective for tertiary hydroxylation.[95]

H H — (p-$O_2NC_6H_4CO_3H$), 69% → OH H (66)

98% regioselectivity

Dioxiranes, generated by the oxidation of ketones with $KHSO_5$, insert an oxygen atom into alkane C—H bonds with retention of configuration by an oxenoid mechanism related to that found for peracids. Tertiary C—H bonds are hydroxylated and react faster than secondary CH_2 groups, which are completely oxidized to the ketone. Conversions of up to 50% have been observed.[124] $CF_3(Me)CO_2$ is a more recently developed reagent of the same type.[125] These easily prepared reagents have considerable promise for organic synthesis.

Ozone on silica gel has been used to hydroxylate alkanes and unactivated C—H bonds.[126] The example shown in equation (67) illustrates the application to a steroid, which was achieved in 51% yield at a respectable 11% conversion with regard to ozone.[127]

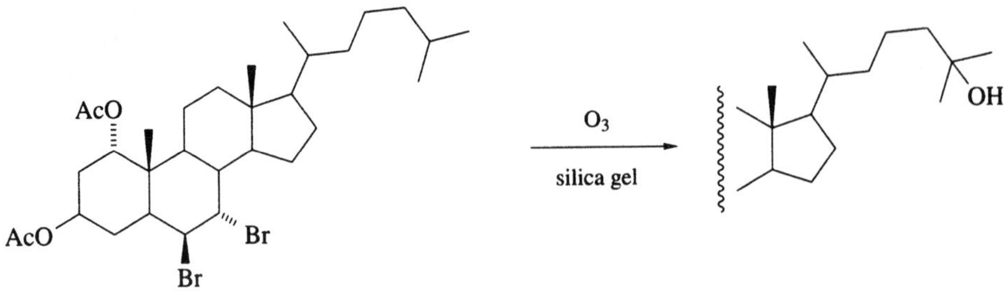

$$\xrightarrow[\text{silica gel}]{O_3} \qquad (67)$$

Ozone on silica gel at –78 °C is also a convenient form of the reagent and is especially useful for tertiary C—H bonds.[128] Ozone and HSO_3F–SbF_6 at –78 °C react with alkanes, but skeletal rearrangement often occurs; for example, methane gives acetone.[129] Oxygen atoms (in the ^3P-state), formed from CO_2 in a microwave discharge at low pressure react with alkanes, to give, for example, the 1,2-epoxides and the tertiary alcohol from 2,3-dimethylbutane; O_3, in contrast, gives the tertiary alcohol and Pr^iCOMe.[130]

Methane may be oxidized to formaldehyde by N_2O at 600 °C over MoO_3. At 5% conversion, 3.5% yield of CH_2O and MeOH are also obtained.[131] These reactions are not especially useful for laboratory scale experiments.

1.1.12 FORMATION OF R—X BONDS (X = S, Se, Te)

Photolysis of C_6F_5SCl in cyclohexane leads to formation of both arylthio- and chloro-cyclohexane by a radical pathway.[132] Sulfuryl chloride in pyridine can chlorosulfonate alkanes by a radical route under photolytic conditions, the chloride being a minor product (equation 68).[133] SO_2 and Cl_2 also gives the sulfonyl chloride by the route shown in equations (69) to (72).[134]

$$R{-}H + SO_2Cl_2 \longrightarrow R{-}SO_2Cl + R{-}Cl \qquad (68)$$

$$Cl_2 \xrightarrow{h\nu} 2Cl\bullet \qquad (69)$$

$$Cl\bullet + R{-}H \longrightarrow R\bullet + H{-}Cl \qquad (70)$$

$$R\bullet + SO_2 \longrightarrow RSO_2\bullet \qquad (71)$$

$$RSO_2\bullet + Cl\bullet \longrightarrow RSO_2Cl \qquad (72)$$

The synthesis of $MeSO_2Cl$ on an industrial scale has been achieved directly from methane by the Elf Aquitaine Company. This is notable not only in being a practical conversion of methane, but also in that it is a photochemical process.[135]

Lead tetraacetate in CF_3CO_2H, followed by RSH, affords the introduction of the SR group (R = Bu^n) into adamantane and bicyclo[3.3.1]nonane in high yields.[136]

Photolysis of alkane/SO_2 mixtures leads to the formation of alkylsulfonic acids,[137] the Hostapon process utilizing $SO_2/O_2/h\nu$.[138] It is curious that this reaction seems to be so efficient, given the low ε for alkane and SO_2 at the wavelengths used, and an efficient chain reaction is presumably involved. The Reed reaction uses $SO_2/Cl_2/h\nu$ to convert alkanes to the corresponding sulfonyl chlorides.[139] Alkanes also react with SO_3 to give alkyl sulfonates, sulfones and sulfates.[140]

Methane reacts with elemental sulfur above 700 °C or at lower temperatures in the presence of a catalyst to give good yields of CS_2, a reaction that has been used for the commercial synthesis of the disulfide.[141] *n*-Butane, however, gives alkenes, dienes and thiophene under similar conditions.

The addition of $(PhSe)_2$ to the Gif system, mentioned above, leads to trapping of the radical intermediates with the formation of products with C—Se bonds, for example, 12% of 2-adamantyl phenyl selenide is formed from adamantane.[142]

1.1.13 FORMATION OF R—F BONDS

Fluorination is one of the few useful methods of preparing fluorocarbons. These materials have important physical and biological properties and are often high value chemicals. The problem with the reaction stems from the very large heat production in the overall process, thanks to the very weak bond strength of F_2, but the very high bond strengths of both H—F and C—F.[143] Nevertheless, by the use of diluted F_2 and appropriate choice of temperatures, a number of organic compounds can be successfully fluorinated (equation 73).[144]

$$\text{bicyclo[2.2.2]octane} \xrightarrow[3.5\%]{F_2/He} \text{bicyclo[2.2.2]octane}\text{—}F_{14} \qquad (73)$$

Barton and coworkers[145] have shown how elemental fluorine can be used in nitrobenzene to obtain a selective fluorination of a steroid, a reaction of importance in drug synthesis (equation 74). A variety of transition metal fluorides, such as CoF_3, are milder fluorinating agents for alkanes.[146]

$$\text{AcO–steroid–OAc} \xrightarrow[\text{PhNO}_2,\ 50\%]{F_2} \text{AcO–steroid(F)–OAc} \qquad (74)$$

Another method that has proved useful involves photolysis of MeOF, which produces a methoxy radical and an F-atom. Substantial amounts of the fluoroalkane are produced *via* H-atom abstraction by MeO· and F·, and quenching the carbon radical with F·. This route was thought to be particularly useful for the synthesis of compounds with useful biological activity (equation 75).[147]

$$C_6H_{12} + MeOF \xrightarrow[44\%]{h\nu} C_6H_{11}F + MeOH \qquad (75)$$

1.1.14 FORMATION OF R—X BONDS (X = Cl, Br, I)

The radical chain halogenation of alkanes is a well-known process and is even commercially practised. The thermodynamics of this process are sufficient to allow the chain to progress for Cl_2 and Br_2, but not for iodine. These halogenations are easy to control and the selectivity of chlorination has been carefully studied.[148]

Photochlorination has been recommended for the preparation of choice of cyclodecene from cyclodecane *via* the intermediate chloride.[149] Bromine is a rather weak brominating agent for alkanes and only unusually favorable substrates, like adamantane, react at a reasonable rate. The mixture HgO/Br_2 is much more reactive than bromine itself, as shown by the facile bromination of 1,1,3,3-tetramethylbutane.[150] Silver hexafluoroantimonate has also been used to activate bromine for this type of reaction.[27] Iodine is normally ineffective in functionalizing alkanes, but use of γ-irradiation of a solution of I_2 in alkanes leads to unselective formation of all possible iodoalkanes.[151]

Rather than using the halogens themselves, other halogen radical donors are more commonly used in laboratory scale synthesis. One of the simplest of these is CCl_4, which can chlorinate alkanes by a free radical chain mechanism.[152] The chain lengths are not very long (equations 76–78), because of their slightly endothermic nature and in part because the reaction is also kinetically rather slow. Elevated temperatures are therefore normally required.[153] Nitrosylchloride at 100 °C has also been used for these reactions.[154]

Trichlorobromomethane appears to be an efficient bromination reagent because of kinetic rather than thermodynamic factors, and fairly long radical chain reactions result.[155] 1,2-Dibromotetrachloroethane is

$$Q\bullet + C_6H_{12} \longrightarrow QH + C_6H_{11}\bullet \qquad (76)$$

$$C_6H_{11}\bullet + CCl_4 \longrightarrow Cl_3C\bullet + C_6H_{11}Cl \qquad (77)$$

$$Cl_3C\bullet \quad + \quad C_6H_{12} \longrightarrow Cl_3CH \quad + \quad C_6H_{11}\bullet \tag{78}$$

Q• = initiator

a useful reagent for the bromination of alkanes. The intermediate halocarbon radical spontaneously β-eliminates to afford a further bromine radical (equation 79).

$$BrCCl_2{-}\dot{C}Cl_2 \longrightarrow Cl_2C{=}CCl_2 \quad + \quad Br\bullet \tag{79}$$

Trichloromethanesulfonyl chloride has also been used as a chlorination reagent for alkanes, but this reaction requires a peroxide initiation step (equation 80).[156]

$$Cl_3CSO_2Cl \quad + \quad R{-}H \longrightarrow R{-}Cl \quad + \quad SO_2 \quad + \quad HCCl_3 \tag{80}$$

N-bromosuccinimide (NBS) is, however, one of the best known bromine donors. For example, it can brominate cyclohexane to give a 30% yield of the corresponding bromide.[157] NBS/dibenzoyl peroxide is not very selective, giving mixtures with methylcyclohexane,[158] although decalin gives tetrabromides in a reasonably well-defined manner (equation 81).[159]

NBS, peroxide (81)

Br Br

Br Br

A useful degree of selectivity for attack at the ω-1 position has been reported in the photochlorination of a variety of linear alkyl compounds, such as *n*-hexyl chloride, with Pr^i_2NCl. The selectivity arises from the fact that many electron-withdrawing groups deactivate adjacent C—H bonds for abstraction by a chlorine radical. By the 'polar' effect, an electronegative atom, such as a chlorine radical, is better able to abstract a hydrogen atom from the most donor C—H bond available. The usual secondary > primary selectivity pattern prevents the terminal methyl group from being the preferred site of attack, hence the next methylene (*i.e.* at the ω-1 position) is attacked preferentially with selectivities of *ca.* 90% being reported.[160] Intramolecular versions of the reaction are also known.[161]

A similar reagent, benzeneiodonium chloride, is also effective under photolytic conditions. A 90% yield of the chloride has been reported for cyclohexane, for example, using this reagent.[162] A recent improvement to the use of $PhICl_2$ employs a trialkylboron as coreagent. *n*-Alkanes are converted to the chlorides in 99% yield under conditions which give no conversion in the absence of the borane catalyst.[163] In all these cases, the reaction is believed to go *via* chlorine radicals and PhICl radicals.

Iodine monochloride, ICl, is another reagent which is useful for the chlorination of alkanes. It was known in the 1950s that ICl was unstable in alkane solution,[164] but the use of irradiation to accelerate the reaction to useful rates was reported later.[165] The chain-carrying step is shown in equation (82).

$$R\bullet \quad + \quad ICl \longrightarrow R{-}Cl \quad + \quad I\bullet \tag{82}$$

Similarly sulfuryl chloride, SO_2Cl_2, has been employed in this type of reaction, using a peroxide-initiated chain reaction to give chlorocarbons from alkanes. The reaction is rather unselective; for example, *n*-heptane gives 15% primary attack and 85% secondary attack. The hydrogen atom abstractor in this chain process is believed to be SOCl·, rather than the chlorine radical. The monochlorinated species is more reactive than the alkane and consequently multiple chlorination takes place. Electron-withdrawing groups destabilize a radical center and so subsequent chlorination events tend to take place at a site remote from the first point of attack.[166] In sulfolane, the same reagent apparently reacts by an electrophilic pathway, and adamantane gives almost exclusively the tertiary halide, compared to the mixture formed during radical reactions. Norbornane gives largely 2-*exo*-chloronorbornane, as the result of the steric bulk of the reagent leading to attack on the least-hindered site.[167] Sulfuryl chloride is also a useful halogenating agent for nonalkane substrates.[168]

t-Butylhypochlorite has been used to chlorinate alkanes with either peroxide or light initiation.[169] Bu^tOI, made from Bu^tOCl and HgI_2, has also been used for iodination in CCl_4 under photolytic conditions. The secondary to primary selectivity in these reactions is good (*ca.* 30:1). Indeed the reaction is only preparatively useful for secondary iodides (yields 30–80%) since the tertiary iodides, although formed, decompose rapidly under the reaction conditions.[170] This process is an example of an apparent secondary > tertiary > primary selectivity pattern resulting from subsequent reaction, and may explain other anomalous selectivities occasionally reported by other workers.

A reagent which may operate by hydrogen atom abstraction from the alkane by the intermediate alkylammonium radical cation is iron(II)/R_2NCl/CF_3CO_2H, which affords secondary chlorides in good yield from *n*-alkanes without overoxidation.[171]

Chemistry reported by Shilov[2,172] allows chlorination of alkanes at the expense of a platinum(IV) halide as the chlorination reagent, but catalyzed by a platinum(II) species. Methane and ethane give the chlorides together with some of the corresponding alcohols. Propane gives a 3:1 mixture of Pr^nCl and Pr^iCl, and *n*-pentane gives a ratio of normal to secondary halides of 56:44, while cyclohexane leads to benzene as the major product. The addition of copper(II) makes the reaction catalytic by reoxidizing the platinum with air as the ultimate oxidant system, although only 5 turnovers were obtained.[173]

$AgSbF_6/Cl_2/CH_2Cl_2$ at –15 to +35 °C is reported to be a convenient and effective reagent for the electrophilic chlorination of tertiary alkanes and cycloalkanes.[174] Adamantane was sufficiently reactive to undergo uncatalyzed electrophilic bromination at 80 °C.[175] Substrates with adjacent tertiary C—H bonds produce α,β,γ,δ-tetrabromides by a series of bromination/dehydrobromination reactions.[176] Alkanes also can be chlorinated by an electrophilic route using heterogeneous catalysts such as $TaOF_3/Al_2O_3$ at 180–250 °C: >90% yield is obtained at 10–50% conversion.[177] $Cl_2/SbF_5/SO_2ClF$ has been used for electrophilic chlorination, but, not surprisingly, skeletal rearrangements are often observed, and less than 20% of products are found from *n*-butane, for example.[178] Schwartz[179] has shown that $[Rh(allyl)_3]$ may be supported in molecular form on alumina, and that the resulting material can be used to chlorinate methane.[180]

1.1.15 REFERENCES

1. R. H. Crabtree, *Chem. Rev.*, 1985, **85**, 245.
2. A. E. Shilov, 'The Activation of Saturated Hydrocarbons by Transition Metal Reagents', Reidel, Dordrecht, 1984.
3. G. A. Olah, *Acc. Chem. Res.*, 1987, **20**, 422.
4. H. Mergard and F. Korte, *Monatsh. Chem.*, 1967, **98**, 763.
5. C. Bosshard, G. Dessauges, T. Gaeumann, H. Oez and O. Piringer, *Chem.-Ing.-Tech.*, 1973, **49**, 493.
6. P. L. Watson, *J. Am. Chem. Soc.*, 1985, **105**, 6491.
7. C. M. Fendrick and T. J. Marks, *J. Am. Chem. Soc.*, 1984, **106**, 2214.
8. C. C. Cummins, S. M. Baxter and P. T. Wolczanski, *J. Am. Chem. Soc.*, 1988, **110**, 8731.
9. R. H. Crabtree, M. F. Mellea, J. M. Mihelcic and J. M. Quirk, *J. Am. Chem. Soc.*, 1982, **104**, 107.
10. R. H. Crabtree, R. P. Dion, D. J. Gibboni, D. V. McGrath and E. M. Holt, *J. Am. Chem. Soc.*, 1986, **108**, 7222.
11. F. W. S. Benfield and M. L. H. Green, *J. Chem. Soc., Dalton Trans.*, 1974, 1325; P. Eilbracht, *Chem. Ber.*, 1980, **113**, 542, 1033, 1420 and 2211.
12. A. H. Janowicz and R. G. Bergman, *J. Am. Chem. Soc.*, 1982, **104**, 352; *J. Am. Chem. Soc.*, 1986, **108**, 4856.
13. J. K. Hoyano and W. A. G. Graham, *J. Am. Chem. Soc.*, 1982, **104**, 3723.
14. W. D. Jones and F. J. Feher, *J. Am. Chem. Soc.*, 1982, **104**, 4240.
15. M. B. Sponsler, B. H. Weiller, P. O. Stoutland and R. G. Bergman, *J. Am. Chem. Soc.*, 1989, **111**, 6841.
16. L. Cassar, P. E. Eaton and J. Halpern, *J. Am. Chem. Soc.*, 1970, **92**, 6366; L. A. Paquette, R. S. Beckley and W. B. Farnham, *J. Am. Chem. Soc.*, 1975, **97**, 1089; L. A. Paquette, R. A. Boggs, W. B. Farnham and R. S. Beckley, *J. Am. Chem. Soc.*, 1975, **97**, 1112.
17. M. L. H. Green and G. Parker, *J. Chem. Soc., Chem. Commun.*, 1984, 1467.
18. J. A. Bandy, F. G. N. Cloke, M. L. H. Green, D. O'Hare and K. Prout, *J. Chem. Soc., Chem. Commun.*, 1984, 240.
19. V. V. Lavrushko, S. A. Lermontov and A. E. Shilov, *Kinet. Katal. Lett.*, 1980, **15**, 269.
20. H. Pines and N. E. Hoffman, in 'Friedel–Crafts and Related Reactions', ed. G. A. Olah, Interscience, New York, vol. II, 1964; C. D. Nenitzescu, in 'Carbonium Ions', ed. G. A. Olah, Wiley, New York, 1970, vol. II.
21. R. L. Burwell, L. C. Maury and R. B. Scott, *J. Am. Chem. Soc.*, 1954, **76**, 5828.
22. M. Hino, S. Kobayashi and K. Arata, *J. Am. Chem. Soc.*, 1979, **101**, 6439.
23. J. Hatton, *J. Catal.*, 1981, **68**, 132.
24. M. Stroecker, *J. Mol. Catal.*, 1985, **29**, 371.
25. C. D. Nenitzescu and I. P. Cantuniari, *Chem. Ber.*, 1932, **66**, 1097.
26. R. J. Cvetanovic, *Prog. React. Kinet.*, 1964, **2**, 77.
27. S. H. Brown and R. H. Crabtree, *J. Chem. Soc., Chem. Commun.*, 1987, 970; *J. Chem. Educ.*, 1988, **65**, 290; *J. Am. Chem. Soc.*, 1989, **111**, 2935, 2946; *J. Am. Chem. Soc.*, 1991, **113**, in press.
28. G. A. Olah and R. H. Schlosberg, *J. Am. Chem. Soc.*, 1968, **90**, 2726; G. A. Olah, G. Klopman and R. H. Schlosberg, *J. Am. Chem. Soc.*, 1969, **91**, 3261.

29. D. Baudry, M. Ephritikhine, H. Felkin and R. Holmes-Smith, *J. Chem. Soc., Chem. Commun.*, 1983, 788; D. Baudry, M. Ephritikhine, H. Felkin and J. Zakrezewski, *J. Chem. Soc., Chem. Commun.*, 1982, 1235; *Tetrahedron Lett.*, 1984, **25**, 1283; H. Felkin, T. Fillebeen-Khan, Y. Gault, R. Holmes-Smith and J. Zakrezewski, *Tetrahedron Lett.*, 1984, **25**, 1279.
30. M. J. Burk and R. H. Crabtree, *J. Am. Chem. Soc.*, 1987, **109**, 8025.
31. T. Sakakura, T. Sodeyama, Y. Tokunaga and M. Tanaka, *Chem. Lett.*, 1987, 263; T. Sakakura and M. Tanaka, *J. Chem. Soc., Chem. Commun.*, 1987, 758; T. Sakakura, T. Sodeyama and M. Tanaka, *Chem. Lett.*, 1988, 683.
32. Y. Lin, D. Ma and X. Lu, *J. Organomet. Chem.*, 1987, **323**, 407.
33. R. H. Crabtree and C. P. Parnell, *Organometallics*, 1985, **4**, 519.
34. D. M. Jerina, D. R. Boyd and J. W. Daly, *Tetrahedron Lett.*, 1970, 457.
35. P. G. Sammes, G. Serra-Errante and A. Tinker, *J. Chem. Soc., Perkin Trans. 1*, 1978, 853.
36. M. Albeck and T. Tamary, *J. Organomet. Chem.*, 1979, **164**, C23.
37. G. Egloff and C. I. Parrish, *Chem. Rev.*, 1936, **19**, 145.
38. B. L. Evering, *Adv. Catal.*, 1954, **6**, 197; R. F. Goldstein, 'The Petroleum Chemicals Industry', Spon, London, 1949; L. Kramer and J. Happel, 'The Chemistry of Petroleum Hydrocarbons', Van Nostrand–Reinhold, New York, 1955; W. A. Gruse and D. R. Stevens, 'Chemical Technology of Petroleum', McGraw–Hill, New York, 1960.
39. K. S. Uspenskaya and D. V. Fedoseev, *Izv. Akad. Nauk. SSSR, Ser. Khim.*, 1979, 2805.
40. G. A. Olah, 'Superacids', Wiley, New York, 1985.
41. M. T. Ahmed, A. J. Swallow, *J. Chem. Soc.*, 1963, 3918.
42. W. G. Bentrude and K. R. Darnell, *J. Am. Chem. Soc.*, 1968, **90**, 3588
43. R. Jaouhari, C. Filliatre, B. Maillard and J. J. Villenave, *Tetrahedron*, 1982, **38**, 3137.
44. D. Tanner and N. J. Bunce, *J. Am. Chem. Soc.*, 1969, **91**, 3028.
45. D. Tanner and P. M. Rahimi, *J. Org. Chem.*, 1979, **44**, 1674.
46. L. P. Schmerling and P. West, *J. Am. Chem. Soc.*, 1949, **71**, 2015.
47. B. H. Toder, S. J. Branca and A. B. Smith, III, *J. Org. Chem.*, 1977, **42**, 904; L. M. Yun, G. G. Galust'yan, A. M. Rushkes and C. S. Kadyrov, *Uzb. Khim. Zh.*, 1972, **16**, 55; 1973, **17**, 28 (*Chem. Abstr.*, 1974, **80**, 20 663; 1978, **73**, 124 227).
48. H. Sletter and P. Goebel, *Chem. Ber.*, 1962, **95**, 1040.
49. L. P. Schmerling, in 'Friedel–Crafts and Related Reactions', ed. G. A. Olah, Interscience, New York, 1964, vol. II.
50. R. Miethchen, S. Steege and C. F. Kroeger, *J. Prakt. Chem.*, 1983, **325**, 823.
51. I. S. Akhrem, A. V. Orlinkov, E. I. Mysov and M. E. Vol'pin, *Tetrahedron Lett.*, 1981, **22**, 3891.
52. K. E. Moller, *Angew. Chem.*, 1963, **73**, 1098.
53. S. Delavarenne, M. Simon, M. Fauconet and J. Sommer, *J. Am. Chem. Soc.*, 1989, **111**, 383; *J. Chem. Soc., Chem. Commun.*, 1989, 1049.
54. S. W. Benson, *US Pat.* 4 199 533 (1980) (*Chem. Abstr.*, 1980, **93**, 70 984t).
55. M. Fleischmann and D. Pletcher, *Tetrahedron Lett.*, 1968, 6255.
56. J. Bertram, M. Fleischmann and D. Pletcher, *Tetrahedron Lett.*, 1971, 349; J. Bertram, J. P. Coleman, M. Fleischmann and D. Pletcher, *J. Chem. Soc., Perkin Trans. 2*, 1973, 374.
57. W. B. DeMore and S. W. Benson, *Adv. Photochem.*, 1964, **2**, 219.
58. W. von E. Doering, R. G. Buttery, R. G. Laughlin and N. Chauduri, *J. Am. Chem. Soc.*, 1956, **78**, 3224.
59. S. Carra and L. Forni, *Catal. Rev.*, 1971, **5**, 159.
60. L. T. Scott and G. J. DeCicco, *J. Am. Chem. Soc.*, 1974, **96**, 322.
61. A. Demonceau, A. F. Noels, A. J. Hubert and P. Teyssie, *J. Chem. Soc., Chem. Commun.*, 1981, 688.
62. A. I. Titov, *Tetrahedron*, 1963, **19**, 557.
63. P. Gray and A. D. Yaffe, *Chem. Rev.*, 1955, **55**, 1102.
64. A. Mee, *Ger. Offen.* 2 413 727 (1973) (*Chem. Abstr.*, 1975, **82**, 30 977j).
65. C. L. Hill and B. C. Schardt, *J. Am. Chem. Soc.*, 1980, **102**, 6374; C. L. Hill, J. A. Smegal and T. J. Henly, *J. Org. Chem.*, 1983, **48**, 3277; J. A. Smegal and C. L. Hill, *J. Am. Chem. Soc.*, 1983, **105**, 3510, 3515.
66. C. L. Hill and J. A. Smegal, *Nouv. J. Chim.*, 1982, **6**, 287.
67. R. F. Renneke and C. L. Hill, *J. Am. Chem. Soc.*, 1986, **108**, 3528.
68. M. Akhtar, *Adv. Photochem.*, 1964, **2**, 263.
69. Y. Ogata, Y. Izawa and H. Tomioka, *Tetrahedron*, 1966, **22**, 483.
70. M. Ohsugi, Y. Inamoto, N. Takeaishi, Y. Fujikura and K. Aigami, *Synthesis*, 1977, 632.
71. L. I. Krimen and D. J. Cota, *Org. React. (N.Y.)*, 1969, **17**, 213.
72. P. Kovacic and S. S. Chaudhary, *Tetrahedron*, 1967, **23**, 3563.
73. G. A. Olah, 'Carbocations and Electrophilic Reagents', Wiley, New York, 1973.
74. G. A. Olah and H. C. Lin, *J. Am. Chem. Soc.*, 1971, **93**, 1259.
75. M. Seno, T. Namba and H. Kise, *J. Org. Chem.*, 1978, **43**, 3345.
76. J. O. Clayton and W. L. Jensen, *J. Am. Chem. Soc.*, 1948, **70**, 3880; R. Graf, *Chem. Ber.*, 1952, **85**, 9.
77. D. J. Hucknall, 'Selective Oxidation of Hydrocarbons', Academic Press, New York, 1974; N. M. Emanuel, E. T. Denisov and Z. Z. Mazius, 'The Oxidation of Hydrocarbons in Liquid Phase', Plenum Press, New York, 1967.
78. A. D. Walsh, *Trans. Faraday Soc.*, 1946, **42**, 269; 1947, **43**, 297, 305.
79. J. L. Benton and M. M. Wirth, *Nature (London)*, 1953, **171**, 269.
80. W. Ester and A. Sommer, *US Pat.* 3 259 661 (*Chem. Abstr.*, 1962, **57**, 1190e); W. Pritzkow and K. H. Grobe, *Chem. Ber.*, 1960, **93**, 2156.
81. K. I. Ivanov and V. K. Savinova, *Dokl. Akad. Nauk SSSR*, 1948, **59**, 493; W. Pritzkow and K. A. Muller, *Justus Liebigs Ann. Chem.*, 1956, **597**, 12.
82. F. F. Rust, *J. Am. Chem. Soc.*, 1960, **93**, 2156.
83. S. C. Arnold, *Hydrocarbon Process.*, 1985, **64**, 123.

84. E. F. Dougherty and C. C. Hobbs, *US Pat.* 3 933 930, (1976) (*Chem. Abstr.*, 1976, **84**, 104 978a; 1976, **85**, 159 415n).
85. H. S. H. Fenton, *J. Chem. Soc.*, 1894, **65**, 899.
86. C. Walling, *Acc. Chem. Res.*, 1975, **8**, 125.
87. H. Mimoun, L. Saussine, E. Daire, M. Postel, J. Fischer and R. Weiss, *J. Am. Chem. Soc.*, 1983, **105**, 3101.
88. R. Sato and T. Omura (ed), 'Cytochrome P-450', Academic Press, New York, 1978.
89. S. Udenfriend, C. T. Clark, J. Axelrod and B. B. Brodie, *J. Biol. Chem.*, 1954, **208**, 731.
90. G. A. Hamilton, *J. Am. Chem. Soc.*, 1964, **86**, 3391.
91. H. Mimoun and I. Seree de Roch, *Tetrahedron*, 1975, **31**, 777.
92. J. T. Groves and T. E. Nemo, *J. Am. Chem. Soc.*, 1983, **105**, 6243.
93. J. T. Groves, W. J. Kruper and R. Haushalter, *J. Am. Chem. Soc.*, 1980, **102**, 6375.
94. E. Guilmet and B. Meunier, *Tetrahedron Lett.*, 1980, **21**, 4449; J. P. Collman, J. I. Brauman, B. Meunier, S. A. Raybush and T. Kodakek, *J. Am. Chem. Soc.*, 1985, **107**, 2000.
95. P. S. Traylor, D. Dolphin and T. G. Traylor, *J. Chem. Soc., Chem. Commun.*, 1984, 279.
96. N. C. Demo and D. G. Pohl, *J. Am. Chem. Soc.*, 1974, **96**, 6680.
97. K. B. Wiberg and G. Foster, *J. Am. Chem. Soc.*, 1961, **83**, 423.
98. E. S. Rudakov, V. P. Tretyakov, L. A. Minto and N. A. Tishchenko, *React. Kinet. Catal. Lett.*, 1981, **16**, 77.
99. J. W. Weller and G. A. Hamilton, *J. Chem. Soc. D*, 1970, 1390.
100. A. Tillotson and B. Houston, *J. Am. Chem. Soc.*, 1951, **73**, 221.
101. C. M. Hol, M. G. J. Bos and H. J. C. Jacobs, *Tetrahedron Lett.*, 1969, 1157.
102. K. B. Wiberg, in 'Oxidation in Organic Chemistry', Academic Press, New York, 1965, p. 69.
103. R. H. Eastman and R. H. Quinn, *J. Am. Chem. Soc.*, 1964, **86**, 3905.
104. H. J. Jager, J. Luetolf and M. W. Meyer, *Angew. Chem., Int. Ed. Engl.*, 1979, **18**, 68, 786, 787.
105. R. Stewart, in 'Oxidation in Organic Chemistry', Academic Press, New York, 1965, p. 1.
106. R. Stewart and U. A. Spitzer, *Can. J. Chem.*, 1978, **56**, 1273.
107. J. Hanotier, P. Camerman, M. Hanotier-Bridoux and P. de Raditsky, *J. Chem. Soc., Perkin Trans. 2*, 1972, 2247.
108. R. Tang and J. K. Kochi, *J. Inorg. Nucl. Chem.*, 1973, **36**, 3845.
109. E. Detilleux and J. Jadot, *Bull. Soc. R. Sci. Liege*, 1955, **24**, 366.
110. J. Ujhazy and E. R. Cole, *Nature (London)*, 1966, **209**, 395.
111. C. Walling and J. Kjellgren, *J. Org. Chem.*, 1969, **34**, 1488.
112. R. E. Partsch, *J. Am. Chem. Soc.*, 1967, **89**, 3662.
113. J. V. Crivello, *Synth. Commun.*, 1976, **6**, 543.
114. J. C. Allen, J. I. G. Cadogan and D. H. Hey, *J. Chem. Soc.*, 1965, 1932.
115. D. H. R. Barton, M. J. Gastiger and W. B. Motherwell, *J. Chem. Soc., Chem. Commun.*, 1983, 41.
116. D. H. R. Barton, M. J. Gastiger and W. B. Motherwell, *J. Chem. Soc., Chem. Commun.*, 1983, 731.
117. D. H. R. Barton, A. K. Gokturk, J. W. Morzycki and W. B. Motherwell, *J. Chem. Soc., Perkin Trans. 1*, 1985, 583.
118. A. E. Shilov, personal communication.
119. D. H. R. Barton, J. Boivin, M. J. Gastiger, J. Monzycki, R. S. Hay-Motherwell, W. B. Motherwell, N. Ozbalik and K. M. Schwartzentrinber, *J. Chem. Soc., Perkin Trans. 1*, 1986, 947.
120. N. C. Deno and D. G. Pohl, *J. Am. Chem. Soc.*, 1974, **96**, 6680.
121. W. Müller and H.-J. Schneider, *Angew. Chem., Int. Ed. Engl.*, 1979, **18**, 407.
122. W. Müller, H.-J. Schneider, *Angew. Chem., Int. Ed. Engl.*, 1979, **18**, 407.
123. G. A. Hamilton, J. R. Giacin, T. M. Hellman, M. E. Snook and J. W. Weller, *Ann. N. Y. Acad. Sci.*, 1973, **212**, 4.
124. R. W. Murray, R. Jeyaraman and L. Mohan, *J. Am. Chem. Soc.*, 1986, **108**, 2470.
125. R. Mello, M. Fiorentino, C. Fusco and R. Curci, *J. Am. Chem. Soc.*, 1989, **111**, 6749.
126. J. R. Durland and H. Adkins, *J. Am. Chem. Soc.*, 1939, **61**, 429.
127. Z. Cohen, E. Keinan, Y. Mazur and A. Ulman, *J. Org. Chem.*, 1976, **41**, 2651.
128. Z. Cohen, E. Keinan, Y. Mazur and T. H. Varonky, *J. Org. Chem.*, 1975, **40**, 2141.
129. G. A. Olah, N. Yoneda and D. G. Parker, *J. Am. Chem. Soc.*, 1976, **98**, 5261.
130. E. Zadok and Y. Mazur, *Angew. Chem., Int. Ed. Engl.*, 1982, **21**, 303; *J. Org. Chem.*, 1982, **47**, 2223.
131. H. F. Liu, R. S. Liu, K. Y. Liew, R. E. Johnson and J. H. Lunsford, *J. Am. Chem. Soc.*, 1984, **106**, 4117.
132. J. F. Harris, *J. Org. Chem.*, 1978, **43**, 1319.
133. M. S. Kharasch and A. T. Read, *J. Am. Chem. Soc.*, 1939, **61**, 3089.
134. M. S. Kharasch, T. H. Chao and H. C. Brown, *J. Am. Chem. Soc.*, 1940, **62**, 2393.
135. J. Ollivier, *La Recherche*, 1990, suppl. vol. 208, p. 40.
136. S. R. Jones and J. M. Mellor, *Synthesis*, 1976, 32.
137. F. S. Dainton and K. J. Ivin, *Trans. Farad. Soc.*, 1950, **46**, 374, 382.
138. W. Hostapon, *Ger. Pat.* 735 096 (1943); B. Bjellquist, *Acta Chem. Scand.*, 1973, **27**, 3180.
139. C. F. Reed, *US Pat.* 2 046 090 (1936) (*Chem. Abstr.*, 1939, **33**, 2911).
140. M. Sveda, *US Pat.* 2 383 752 (1945) (*Chem. Abstr.*, 1945, **40**, 21).
141. C. M. Thacker and E. Miller, *Ind. Eng. Chem.*, 1944, **36**, 182.
142. G. Balavoine, D. H. R. Barton, P. Lecoupanec and P. Lelandais, *Nouv. J. Chim.*, 1989, **13**, 691.
143. L. A. Bigelow, *Chem. Rev.*, 1947, **40**, 68.
144. N. J. Maraschin, B. D. Catskis, L. H. Davis, G. Jarvinen and R. J. Lagow, *J. Am. Chem. Soc.*, 1975, **97**, 513.
145. D. H. R. Barton, R. H. Hesse, R. E. Markwell, M. M. Pechet and S. Rozen, *J. Am. Chem. Soc.*, 1976, **98**, 3036.
146. M. Stacey and J. C. Tatlow, *Adv. Fluorine Chem.*, 1960, **1**, 166.
147. J. Kollonitsch, L. Barash and G. A. Golduras, *J. Am. Chem. Soc.*, 1970, **92**, 7494.
148. G. Egloff, R. E. Schaad and C. D. Lowry, *Chem. Rev.*, 1931, **8**, 1.
149. J. G. Traynham, D. B. Stone and J. L. Couvillon, *J. Org. Chem.*, 1967, **32**, 510.

150. N. J. Bunce, *Can. J. Chem.*, 1972, **50**, 3109.
151. G. Castello and C. D'Amato, *J. Chromatogr.*, 1975, **108**, 194.
152. J. P. West and L. P. Schmerling, *J. Am. Chem. Soc.*, 1950, **72**, 3525.
153. Y. A. Ol'decap, *Dokl. Akad. Nauk SSSR*, 1953, **93**, 75.
154. L. J. Beckham, W. A. Fessler and M. A. Kise, *Chem. Rev.*, 1951, **48**, 354.
155. E. S. Huyser, 'Free Radical Chain Reactions', Wiley, New York, 1970.
156. E. S. Huyser and D. N. DeMott, *Chem. Ind. (London)*, 1963, 54.
157. N. P. Buu-Hoz and P. Demerseman, *J. Org. Chem.*, 1953, **18**, 649.
158. J. R. B. Boocock and W. J. Hickinbottom, *J. Chem. Soc.*, 1963, 1319.
159. R. A. Barnes, *J. Am. Chem. Soc.*, 1948, **70**, 145.
160. N. C. Deno, W. E. Billups, R. Fishbein, C. Pierson, R. Whalen and R. C. Wyckoff, *J. Am. Chem. Soc.*, 1971, **93**, 438; F. Minisci, G. P. Gardini and F. Bertini, *Can. J. Chem.*, 1948, **70**, 544.
161. M. E. Wolff, *Chem. Rev.*, 1962, **62**, 55.
162. D. F. Banks, E. S. Huyser and J. Kleinberg, *J. Org. Chem.*, 1964, **29**, 3692.
163. A. Arase, M. Hoshi and Y. Matsuda, *Chem. Lett.*, 1979, 961.
164. A. I. Popov and W. A. Deskin, *J. Org. Chem.*, 1957, **22**, 1732.
165. C. C. Kelly, W. H. S. Yu and M. H. J. Wijnen, *Can. J. Chem.*, 1970, **48**, 603.
166. M. S. Kharasch and H. C. Brown, *J. Am. Chem. Soc.*, 1939, **61**, 2142.
167. I. Tabushi, J. Hamuro and R. Oda, *Nippon Kagaku Zasshi*, 1968, **89**, 794.
168. I. Tabushi and H. Kitaguchi, *Synth. Reagents*, 1981, **4**, 336.
169. C. Walling and B. B. Jacknow, *J. Am. Chem. Soc.*, 1960, **82**, 6108, 6113.
170. D. Tanner and G. C. Gidley, *J. Am. Chem. Soc.*, 1968, **90**, 808.
171. N. C. Deno, E. J. Gladfelter and D. G. Pohl, *J. Org. Chem.*, 1979, **44**, 3728.
172. A. E. Shilov, *Zh. Fiz. Khim.*, 1972, **46**, 1353.
173. N. F. Goldshleger, V. V. Lavrushko, A. P. Krushch and A. A. Shilov, *Izv. Akad. Nauk SSSR, Ser. Khim.*, 1976, 2174.
174. G. A. Olah and P. Schilling, *J. Am. Chem. Soc.*, 1973, **95**, 7680.
175. H. Stetter, M. Schwartz and A. Hirschhorn, *Chem. Ber.*, 1959, **92**, 1629.
176. H. Stetter and C. Wulff, *Chem. Ber.*, 1960, **93**, 1366.
177. G. A. Olah, B. Gupta, M. Farina, J. D. Felberg, W. M. Ip and A. Husain, *J. Am. Chem. Soc.*, 1985, **107**, 7095.
178. G. A. Olah, R. Renner and Y. K. Mo, *J. Am. Chem. Soc.*, 1973, **95**, 7686.
179. M. D. Ward and J. Schwartz, *Organometallics*, 1982, **1**, 1030; N. Kitajima and J. Schwartz, *J. Mol. Catal.*, 1984, **106**, 2220.
180. J. Schwartz, *Acc. Chem. Res.*, 1985, **18**, 302.

1.2

Oxidation by Nitrene Insertion

CHRISTOPHER J. MOODY

Loughborough University of Technology, UK

1.2.1 INTRODUCTION

Nitrenes are monovalent, neutral, electron deficient nitrogen species. Although a few nitrenes are sufficiently long lived to enable spectroscopic detection and measurements at low temperature, for synthetic purposes, nitrenes can be considered as highly reactive intermediates which have to be generated *in situ* in the presence of the substrate from a suitable stable precursor (see Section 1.2.2.1). The generation and reactivity of nitrenes are discussed in two books and in several reviews, and for general information on specific classes of nitrene, the reader is referred to the appropriate review: alkylnitrenes,[1,2] vinylnitrenes,[3,4] arylnitrenes,[5–7] acylnitrenes,[8–12] cyanonitrenes[13] and sulfonylnitrenes.[14] In addition there are relevant reviews on the nitrenes in heterocyclic synthesis,[15–17] photoaffinity labeling using nitrenes,[18] and on the decomposition and synthetic uses of azides.[19–21]

Although this chapter is primarily concerned with the functionalization of unactivated sp^3 C—H bonds by nitrene insertion, other relevant aspects of nitrene chemistry are included. Thus the brief section on nitrene reactivity highlights the rearrangement reactions which often compete with C—H insertion, and the final section covers insertion into sp^2 C—H bonds, since many of these reactions have found wide use in recent years in the synthesis of natural products.

1.2.2 NITRENE GENERATION AND REACTIVITY: A BRIEF OVERVIEW

1.2.2.1 Generation

The main methods of nitrene generation are summarized in Scheme 1, although it should be borne in mind that the isolation of nitrene-type products does not necessarily imply that the reaction involves a free nitrene. Azides are the most convenient precursors to nitrenes, since, for the most part they are easily prepared, and can be decomposed under 'reagent-free' conditions by the action of heat, light or a suitable catalyst to give the nitrene and molecular nitrogen. The thermal stability of azides varies greatly, with some such as cyanogen azide being notoriously unstable, whilst others require temperatures in excess of 200 °C to initiate decomposition. However, for handling purposes, all azides should be considered potentially explosive. Other reagent-free sources of nitrenes are isocyanates, ylides, small ring heterocycles such as aziridines and oxaziridines, and five-membered ring heterocycles such as 1,3,4-dioxazol-2-ones, *e.g.* **(1)**, and 1,3,2,4-dioxathiazole 2-oxides which can eliminate CO_2 and SO_2 respectively, although none of these are widely used. More commonly used nitrene precursors are nitro compounds, which are deoxygenated by tervalent phosphorus reagents, and compounds which can undergo base-mediated α-elimination reactions such as *N*-chloro compounds and *N*-sulfonyloxy compounds, although of these only *N*-arenesulfonyloxy carbamates, *e.g.* **(2)**, have found wide use as a source of ethoxycarbonylnitrenes. Finally amines can be oxidized using reagents such as lead(IV) acetate to give nitrene intermediates.

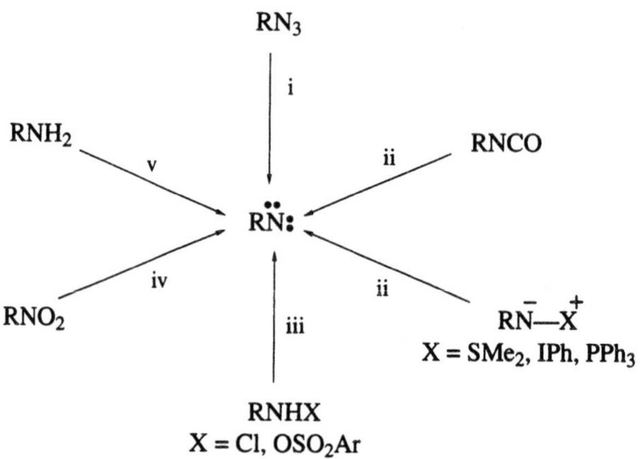

i, heat, *hν*, or catalyst; ii, heat or *hν* ; iii, base; iv, P^{III} reagent; v, oxidant

Scheme 1

N–O Ph O O **(1)**

$EtO_2CNHOSO_2$–C_6H_4–NO_2 **(2)**

$EtO_2C\bar{N}-\overset{+}{S}Me_2$ **(3)**

1.2.2.2 Reactivity

The high reactivity of nitrenes stems from the fact that the nitrogen has only six electrons in its outer shell. Of these, two are bonding electrons, two are the 'normal' nitrogen lone pair, and the remaining two can either be in the same orbital with their spins paired or in separate orbitals with unpaired spins, leading to the possibility of singlet and triplet (diradical) states. Although Hund's rule predicts that nitrenes should have a triplet ground state, many reactions of nitrenes are characteristic of the singlet state.

The four main characteristic reactions of nitrenes are summarized in Scheme 2. All of these reactions have parallels in carbene chemistry; for example, a full discussion of the C—H insertion reaction of carbenes is given in Volume 3, Chapter 4.2. The first reaction, the addition to an alkene to form an aziridine, is covered in detail in Volume 7, Chapter 3.5. The C—H insertion reaction, the subject of this chapter, can, in principle, occur by several mechanisms. However most of the reactions are believed to involve direct insertion of the singlet nitrene into the C—H bond, rather than a stepwise hydrogen abstraction–recombination mechanism involving the triplet, or stepwise ion pair mechanisms. On the other hand, some reactions of nitrenes which lead to products with a new C—N bond that are the result of a formal C—H insertion proceed by an entirely different mechanism. Therefore, although we are largely concerned with the preparative aspects of nitrene C—H insertions, the mechanistic detail cannot always be ignored since it may well impinge on other factors such as the stereoselectivity of the insertion process.

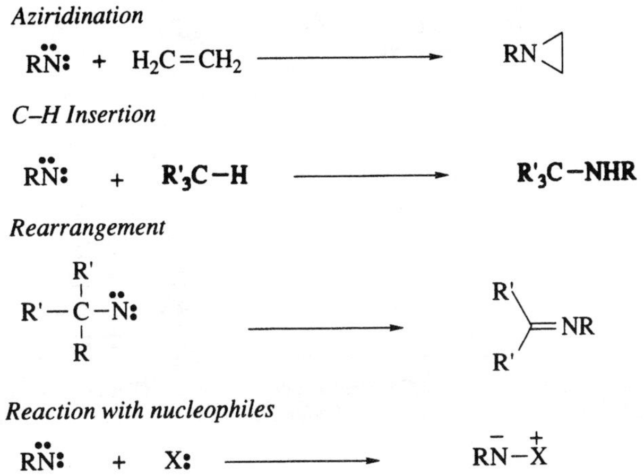

Aziridination

$$\ddot{R N}: + H_2C{=}CH_2 \longrightarrow RN\triangleleft$$

C–H Insertion

$$\ddot{R N}: + R'_3C{-}H \longrightarrow R'_3C{-}NHR$$

Rearrangement

$$R'{-}\overset{R'}{\underset{R}{C}}{-}\ddot{N}: \longrightarrow R'_2C{=}NR$$

Reaction with nucleophiles

$$\ddot{R N}: + X: \longrightarrow R\bar{N}{-}\overset{+}{X}$$

Scheme 2

Nitrenes in common with other electron deficient intermediates are prone to rearrangement by migration of an atom or group from the adjacent carbon to the electron poor center. This type of rearrangement, which often competes with C—H insertion reactions, is particularly favorable in two cases: the decomposition of alkyl azides to give imines, and of acyl azides to give isocyanates (the Curtius rearrangement). Indeed, in the first case, the rearrangement is so facile that the C—H insertion reaction of simple alkylnitrenes is not a synthetically useful process. The final characteristic reaction of nitrenes (Scheme 2) is that with nucleophiles containing a heteroatom lone pair such as sulfides and phosphines to give ylides. Although we are not concerned directly with this reaction, it does have important consequences for C—H insertion reactions that are carried out in heteroatom-containing solvents such as ethers.

1.2.3 INSERTION INTO *sp*[3] C—H BONDS

1.2.3.1 Intermolecular

The C—H insertion reaction of nitrenes is a potentially useful way of functionalizing unactivated C—H bonds, converting hydrocarbons into amine derivatives. In its intermolecular form the synthetic utility of the reaction is highly dependent on the substituents on the nitrene, and on the manner in which it is generated. To exemplify these effects, the results for the functionalization of cyclohexane by insertion of various nitrenes (equation 1) are summarized in Table 1.

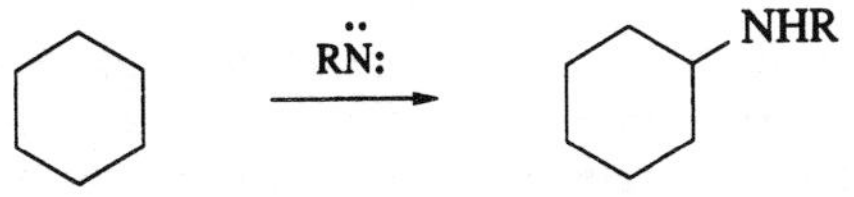

$$C_6H_{12} \xrightarrow{\ddot{R N}:} C_6H_{11}{-}NHR \qquad (1)$$

Several features are immediately apparent. The yields from simple alkylnitrenes are exceedingly poor because of the facile rearrangement by hydrogen migration. Polyfluorinated alkylnitrenes give higher yields, but the reaction is not generally useful. Intermolecular insertion reactions of arylnitrenes are rare, and generally give poor yields, the major reaction being formation of anilines by hydrogen abstraction by the triplet nitrene. Further evidence for triplet involvement in these cases comes from results of phenylnitrene insertion into the tertiary C—H bond of 2-methylpropane and its 2-deuterated analog.[24] The observed isotope effect of 4.1 was considered too large for a concerted C—H insertion reaction of the

Table 1 Functionalization of Cyclohexane by Nitrene Insertion (equation 1)

R	*Conditions of generation*	*Yield (%)*	*Ref.*
Me	Azide, Δ	0.4	22
CF_3CHFCF_2	Azide, *hν*	18[a]	23
Ph	Azide, Δ	*b*	24
Tetrafluoropyrid-4-yl	Azide, Δ	45	25
Bu^tCO	Azide, *hν*	20	26
PhCO	Azide, *hν*	40–46	27, 28
PhCO	(**1**), *hν*	23	28
EtO_2C	Azide, Δ	52–76[b]	29, 30
EtO_2C	Azide, *hν*	51, 78	31, 32
EtO_2C	(**2**), base	13–20	33, 34
EtO_2C	(**3**), *hν*	21	35
$C_{18}H_{37}O_2C$	Azide, Δ	60–73[c]	36
Ms	Azide, *hν*	18	37
$C_5H_{11}SO_2$	Azide, Δ	54	38
Ts	Azide, Δ	58	38
$(EtO)_2PO$	Azide, *hν*	88	39
$(PhO)_2PO$	Azide, *hν*	67	39

[a] After hydrolysis to cyclo-$C_6H_{11}NHCOCHFCF_3$; GLC suggests yield is *ca.* 30%. [b] Gives mainly aniline; yield of cyclo-$C_6H_{11}NHPh$ not quoted. [c] Yield increased in presence of 'additives' such as 1,3-dinitrobenzene; see Section 1.2.3.1.

singlet. Relatively poor yields of C—H insertion products are also obtained from acyl azides; again the problem is a competing reaction, Curtius rearrangement to the isocyanate which in the case of photolysis of benzoyl azide was isolated in 57% yield.[27]

Ethoxycarbonylnitrene, however, gives a synthetically useful yield of the C—H insertion product, *N*-cyclohexylurethane, the carbamate group of which can subsequently be hydrolyzed or modified. The thermal reaction is carried out by simply heating a dilute solution of ethyl azidoformate in cyclohexane, although the yield can be improved by carrying out the reaction in the presence of 'additives' such as 1,3-dinitrobenzene, sulfur or hydroquinone.[29] The C—H insertion process is a singlet nitrene reaction, so the role of these additives, all of which are potential radical traps, is not completely clear. One possibility is that any radicals assist the singlet to triplet nitrene interconversion process, and since the triplet does not insert, yields are higher in the presence of radical traps. Likewise the yield of insertion product is increased in the presence of hexafluorobenzene.[30] Photochemical generation of ethoxycarbonylnitrene from ethyl azidoformate also gives a useful yield of the insertion product,[31,32] but the alternative nitrene precursors such as the *N*-arenesulfonyloxy carbamate (**2**) and the sulfimide (**3**) are less satisfactory,[33–35] the latter precursor giving largely triplet nitrene on irradiation. When the decomposition of the *N*-arenesulfonyloxy carbamate (**2**) was carried out in cyclohexane-d_{12}, an isotope effect of about 1.5 was observed.[33] This small effect is consistent with a concerted C—H insertion of the singlet nitrene. Octadecyl azidoformate also gives a good yield of the insertion product on heating in cyclohexane, although a small amount of intramolecular insertion occurs to give a mixture of five- and six-membered ring products.[36] Again, the yield of the *N*-cyclohexyl carbamate is improved in the presence of 1,3-dinitrobenzene. Finally, some sulfonyl- and phosphonyl-nitrenes give good yields of the corresponding cyclohexylamine derivatives.[37–39]

1.2.3.1.1 Chemoselectivity

The insertion reactions into cyclohexane C–H bonds (Table 1) give some idea of which nitrenes give synthetically useful yields. However, since most other substrates will contain more than one sort of C—H bond, it is important to know the selectivity of nitrenes for different types of C—H bond. Several studies of nitrene selectivity towards tertiary, secondary and primary unactivated C—H bonds have been made, although attempts to study allylic C—H insertion reactions are complicated by the competing nitrene addition to the double bond. In cyclohexene it has been estimated that the allylic C—H bond is only about three times more reactive than the homoallylic C—H bond towards insertion of ethoxycarbonylnitrene.[31,33] However, the reaction is totally unsatisfactory as a means of allylic functionalization since, as shown in Scheme 3, the yields are so low.

The standard hydrocarbon substrate that has been used to determine the relative selectivities of nitrenes for tertiary, secondary and primary unactivated C—H bonds is 2-methylbutane, and the results of

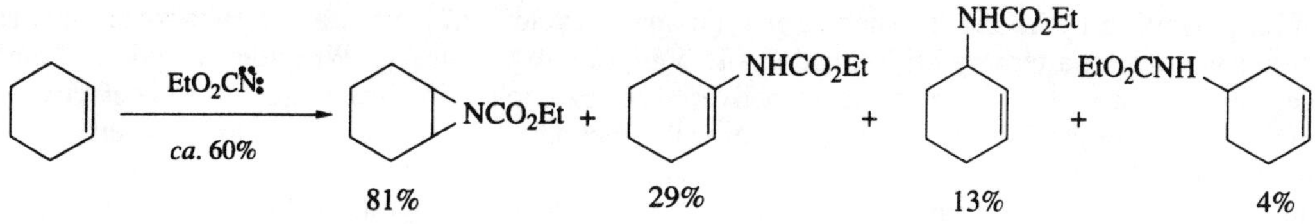

$EtO_2C\ddot{N}$: ca. 60%; NCO_2Et 81%; $NHCO_2Et$ 29%; $NHCO_2Et$ 13%; EtO_2CNH 4%

Scheme 3

several studies are summarized in Table 2. For all nitrenes, the order of reactivity is tertiary > secondary > primary C—H bonds, but there is a considerable variation in the degree of selectivity. Phenylnitrene is the most selective, although as has already been described, the overall yield of C—H insertion is very poor. The selectivity is thought to arise from stabilization of the nitrene by electron donation from the aromatic ring.[40] Pivaloylnitrene is also very selective in its insertion reactions, with benzoylnitrene being less so, although the yields are less than 25% because of the competing isocyanate formation.[26,41] In the latter case, the C—H insertion selectivity is apparently dependent on the nitrene precursor.[28]

Table 2 Relative Reactivity of Tertiary, Secondary and Primary C—H Bonds in 2-Methylbutane towards Nitrene Insertion[a]

R	*Conditions of generation*	*Tertiary*	*Secondary*	*Primary*	*Ref.*
Ph	Azide, Δ	140–280	>7	1	40
Bu^{t}CO	Azide, *hν*	120–200	9	1	26
PhCO	Azide, *hν*	58	7	1	41
EtO_2C	Azide, *hν*	43	6.6	1	42
EtO_2C	Azide, *hν*	34	9	1	33
EtO_2C	Azide, Δ	32	10	1	36
EtO_2C	(**2**), base	27	11	1	33
Ms	Azide, *hν*	9.6	4.2	1	37
Ms	Azide, Δ	5.8	2.2	1[b]	43
$(EtO)_2PO$	Azide, *hν*	6.0	4.3	1	39
$(PhO)_2PO$	Azide, *hν*	3.4	1.2	1	39
NC	Azide, Δ	67		1[c]	46

[a] Statistically corrected for number of C—H bonds. [b] Substrate is 2,4-dimethylpentane. [c] Substrate is 2,3-dimethylbutane.

The much studied ethoxycarbonylnitrene is somewhat less selective, although tertiary C—H bonds are still about 30 times more reactive than primary ones towards the nitrene. The selectivity varies slightly according to which nitrene precursor is used, and is also influenced by the reaction solvent.[33,36] In the presence of dioxane, the selectivity for tertiary C—H insertion over primary decreases with increasing dioxane concentration.[44] The results are explained by formation of a complex (**4**) between dioxane and the singlet nitrene, hence the 'nitrene' is more sterically demanding and exhibits lower selectivity towards tertiary C—H bonds. The relative reactivity of axial and equatorial C—H bonds towards ethoxycarbonylnitrene has been determined using *cis*- and *trans*-1,4-dimethylcyclohexane as substrate. Results show that insertion into equatorial C—H bonds is favored over axial by a factor of about 1.3.[45] Sulfonyl- and phosphonyl-nitrenes are significantly less selective in their insertion reactions, although the reactions involving the phosphonylnitrenes are particularly high yielding.[37,39,43] The relative lack of selectivity compared to acylnitrenes is explained by the fact that both S═O and P═O bonds are considerably less effective than the C═O bond in stabilizing the electron deficient nitrogen. Cyanonitrene, which has been studied in a number of systems, is highly selective and high yielding, although the instability of the precursor, cyanogen azide, detracts from the synthetic utility.[46]

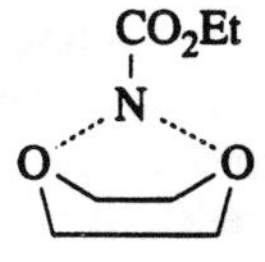

CO_2Et N O O

(**4**)

EtO_2CNH Cl Cl

(**5**)

With polycyclic hydrocarbons such as norbornane, bicyclo[2.2.2]octane and adamantane, nitrene insertion can occur at a tertiary bridgehead C—H bond or at a CH_2 group. With the exception of norbornane, the tertiary C—H is more reactive; for example, in adamantane the selectivity of ethoxycarbonylnitrene for the tertiary C—H over the secondary is about 6:1.[47,48] Similarly, ethoxycarbonylnitrene inserts selectively at the ring junction tertiary C—H bond in bicyclic hydrocarbons such as decalin,[49] although in one experiment a CIDNP effect was observed, suggestive of triplet involvement.[50]

Two groups of insertion substrates that require special mention are ethers and chloroalkanes. In ethers, the nitrene inserts selectively α to the oxygen atom, and although the reaction has only been thoroughly investigated for ethoxycarbonylnitrene and cyclic ethers, the effect does seem to be general.[51–56] The chemoselective insertion α to the ring oxygen of cyclic ethers is synthetically useful, since the resulting insertion products can subsequently be transformed into α,ω-amino alcohol derivatives as shown in Scheme 4.[51]

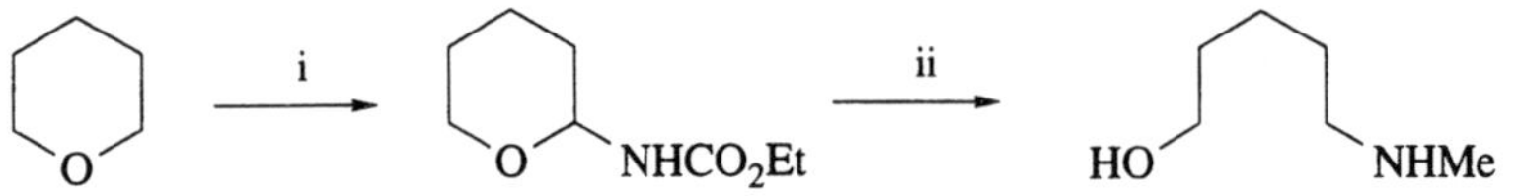

O → (i) O NHCO$_2$Et → (ii) HO NHMe

i, EtO_2CN_3, $h\nu$; ii, $LiAlH_4$

Scheme 4

The relative reactivity of C—H bonds α to ring oxygens have been estimated, and the results are summarized in Scheme 5. The results are rationalized by invoking stabilization of the nitrene by prior coordination to the ring oxygen. Dioxane with two oxygen atoms is particularly effective at stabilizing nitrenes, and the formation of complex (**4**) has been proposed to explain the pronounced solvent effect that dioxane has on a number of insertion reactions of ethoxycarbonylnitrene. Nitrene insertion reactions in chloroalkanes tend to occur away from the chlorine atom; with *trans*-1,2-dichlorocyclohexane as substrate, the insertion product (**5**) is formed in good yield.[57]

1.0 | 0.6 3.0 O | 5.0 O | 11.7 O

Scheme 5

In summary, most nitrenes exhibit some chemoselectivity in their intermolecular C—H insertion reactions, with the order of reactivity being tertiary > secondary > primary C—H bonds. Hence the ease of homolysis of the C—H bond in question would appear to be a good guide to its reactivity towards nitrene insertion, despite the fact that the reaction almost certainly involves a concerted reaction of the singlet nitrene and not a radical process.

1.2.3.1.2 Stereoselectivity

The stereochemistry of nitrene insertion into unactivated C—H bonds has been studied using substituted cyclohexanes as substrates. For arylnitrenes which usually exhibit triplet reactivity, the reaction is nonspecific,[25] but most other nitrenes undergo stereospecific C—H insertion. For example, benzoylnitrene inserts selectively into the tertiary C—H bond of both *cis* and *trans*-1,4-dimethylcyclohexane with retention of configuration.[41,58] Similarly with *cis*- and *trans*-1,2-dimethylcyclohexane as substrate, ethoxycarbonyl-,[59] methanesulfonyl-[43] and cyano-nitrenes[46,60] all insert with retention of configuration at the tertiary C—H bond.

When optically active hydrocarbons have been used as substrates, a similar pattern of insertion reactivity emerges. Phenylnitrene inserts with a maximum of 30% retention into the tertiary C—H bond of optically active 2-phenylbutane implying a high degree of triplet involvement,[24] whereas ethoxycarbonylnitrene inserts stereoselectively with 98–100% retention into the tertiary C—H of (*S*)-(+)-3-methylhexane.[61] The result is independent of the method of nitrene generation, and of concentration, and lends support to the view that only singlet ethoxycarbonylnitrene inserts into unactivated C—H bonds.

1.2.3.2 Intramolecular

Although intermolecular nitrene insertion reactions can be a useful way of functionalizing unactivated C—H bonds, it is the intramolecular version of the reaction that has found the widest use in synthesis. Most types of nitrene will undergo intramolecular C—H insertion, and the following discussion is organized in terms of type of nitrene, scope and selectivity of the reaction, and, finally, specific uses in synthesis.

1.2.3.2.1 Scope and selectivity

Alkylnitrenes are very poor in functionalizing C—H bonds, even if the insertion is intramolecular. Earlier claims that pyrrolidines (**6**) could be formed by insertion of alkylnitrenes derived from simple alkyl azides have subsequently been disproved,[62] and the only apparently genuine example of a reaction of this type involves the azidosteroid 6β-azido-5α-pregnane (**7**), which leads to functionalization of the 19-methyl group upon irradiation, albeit in very poor yield.[63]

N_3 R → (crossed) N H R (2)

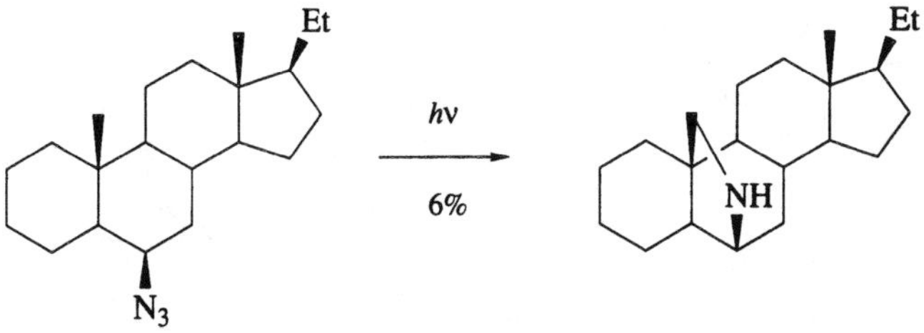

(**6**)

Et Et N_3 hν 6% NH (3)

(**7**)

The decomposition of vinyl azides often leads to products of formal C—H insertion reactions, although in many cases the mechanism does not involve a genuine nitrene insertion. Some examples, which also include insertion into more activated C—H bonds, are shown in Scheme 6.[64–66] The yields of fused pyridines, formed by aromatization of the initial 'insertion' product, are variable, but the reaction is a useful way of constructing polycyclic systems such as the azafluoranthene (**8**) from, in this case, a relatively simple fluorene derivative. The formation of the azepinoindole (**9**) by 'insertion' into the unactivated methyl group of the ethyl substituent rather than into the activated CH_2 group is particularly interesting, although the reaction is solvent dependent with the alternative 'insertion product', ethyl 1-methyl-β-carboline-3-carboxylate being formed in competition in other solvents.[66]

Arylnitrenes, generated by thermolysis of aryl azides or by deoxygenation of the corresponding nitro compounds, readily undergo intramolecular insertion into the C—H bonds of *ortho* alkyl substituents. With azides as precursor, the reaction is often cleaner and higher yielding in the vapor phase than in solution. The reaction is a route to indolines, which may be dehydrogenated to indoles, and, in general, the formation of the five-membered ring indoline is preferred to the six-membered ring tetrahydroquinoline by a factor of about 4:1.[67–70] Some examples are shown in Scheme 7; again the reaction has been applied in the steroid field, the 1-azidoestrone (**10**) giving the C-11 functionalized product in good yield.[71]

The intramolecular insertion reaction of arylnitrenes proceeds with retention of configuration at carbon. For example, heating the (*S*)-aryl azide (**11**; X = N_3) in the vapor phase gives 2-ethyl-2-methylindoline in 50–60% yield in *ca.* 100% optical purity.[72] The optical purity of the product is lower if the azide is heated in solution, or if the nitrene is generated from the corresponding nitro compound (**11**; X = NO_2) with triethyl phosphite.[73]

The intramolecular reaction of acylnitrenes suffers from the same competing Curtius rearrangement as the intermolecular reaction, and therefore the yields of insertion product are often low. In general, the formation of δ-lactams is preferred to γ-lactams by a factor of about 2:1, where the possibility for competing intramolecular insertion into similar C—H bonds exists. Thus irradiation of the azides (**12**; R = Me, Pr) gives a mixture of δ- and γ-lactams in a ratio of 2:1 in lowish overall yield of 30–35% (Scheme 8).[8,74,75] The reaction proceeds stereoselectively at the C—H bond, the nitrene derived from the optically pure (*R*)-azide (**13**) inserting with *ca.* 98% retention of stereochemistry, although in poor chemical yield.[76]

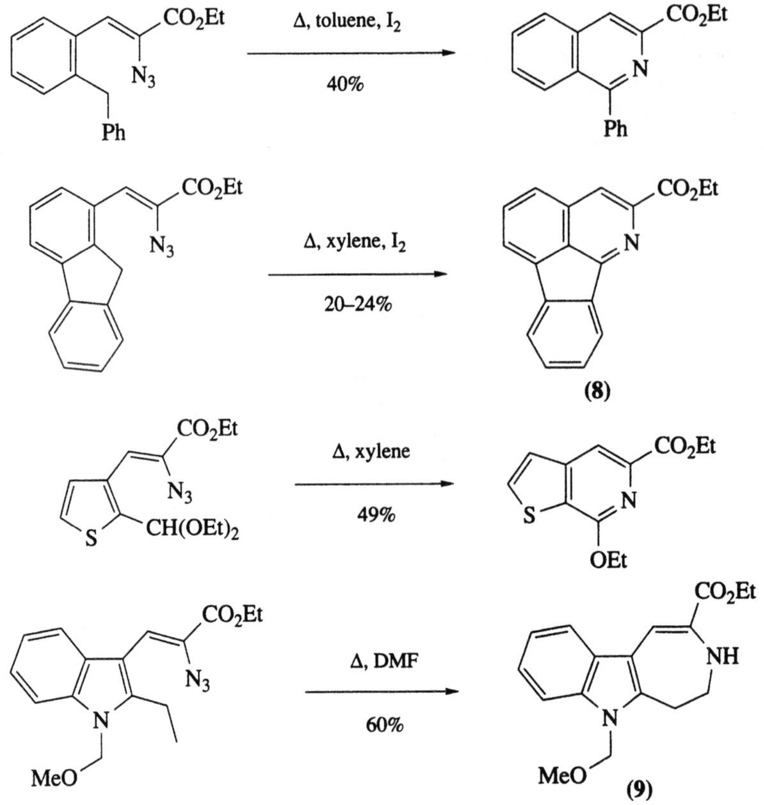

CO_2Et N_3 Ph — Δ, toluene, I_2 / 40% — CO_2Et N Ph

CO_2Et N_3 — Δ, xylene, I_2 / 20–24% — CO_2Et N **(8)**

CO_2Et N_3 S $CH(OEt)_2$ — Δ, xylene / 49% — CO_2Et S N OEt

CO_2Et N_3 N MeO — Δ, DMF / 60% — CO_2Et NH N MeO **(9)**

Scheme 6

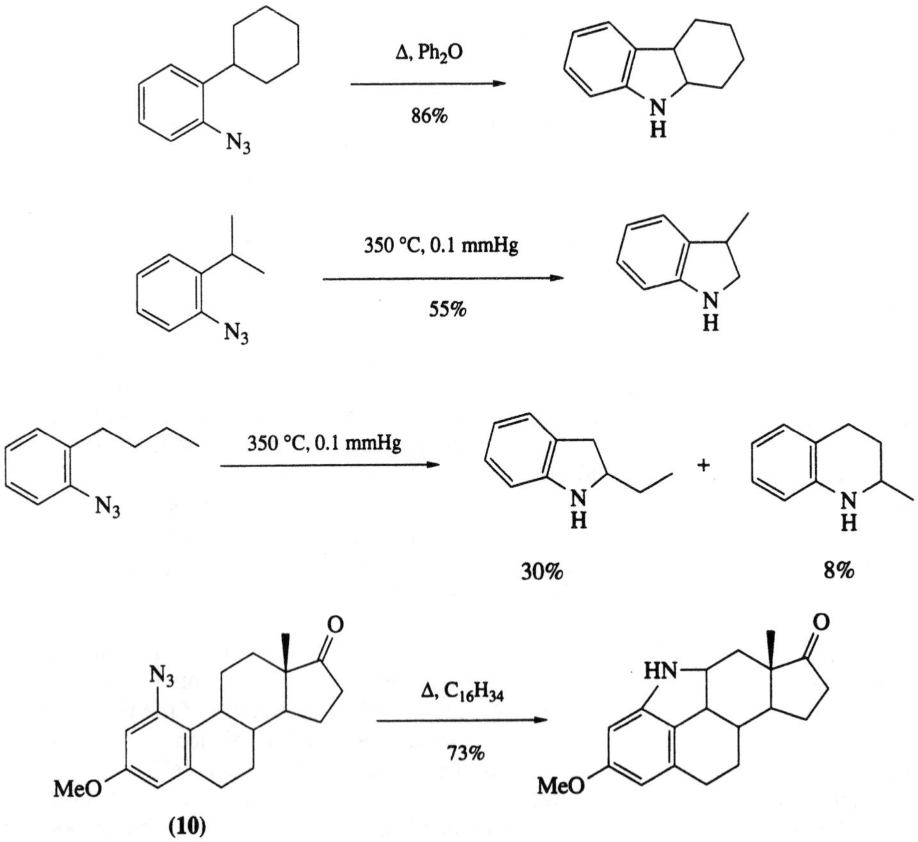

N_3 — Δ, Ph_2O / 86% — N H

N_3 — 350 °C, 0.1 mmHg / 55% — N H

N_3 — 350 °C, 0.1 mmHg — N H 30% + N H 8%

O N_3 MeO **(10)** — Δ, $C_{16}H_{34}$ / 73% — O HN MeO

Scheme 7

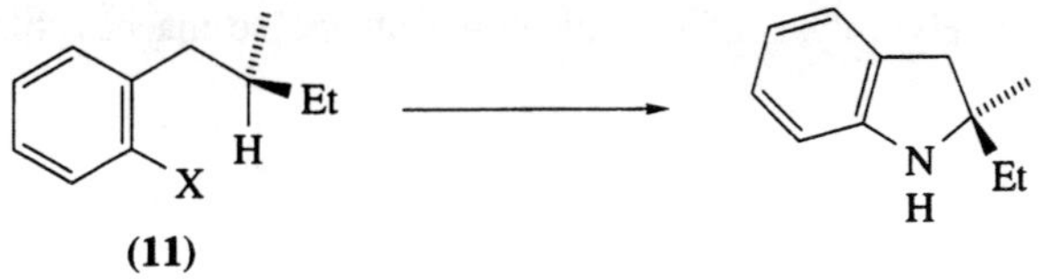

(11) (4)

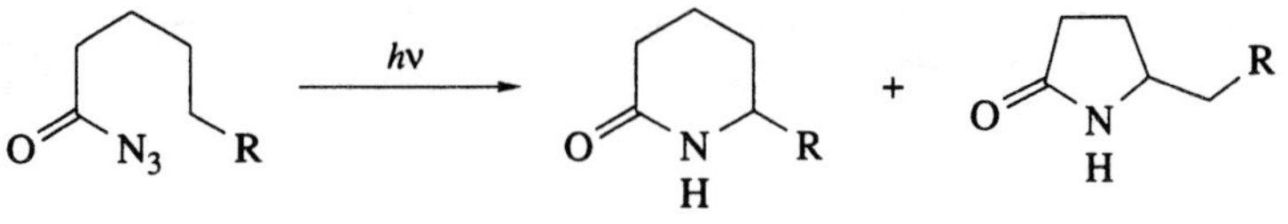

(12) 22% 11%

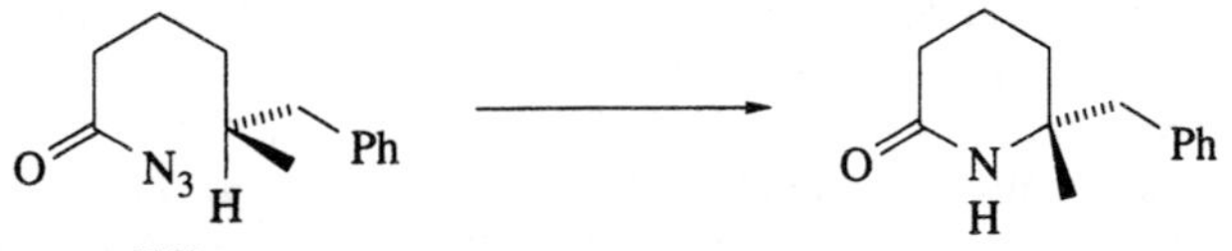

(13)

Scheme 8

Nitrenes derived from azidoformates, however, do not suffer from competing Curtius rearrangement, and undergo intramolecular C—H insertion in good yield. Thus irradiation of *t*-butyl azidoformate results in cyclization by insertion into the unactivated C—H bond of the methyl group to give the oxazolidinone (**14**) in good yield.[77] When the optically active (*S*)-azidoformates (**15**; R = Me, Ph) were irradiated, the corresponding oxazolidinones were formed with >97% retention of stereochemistry,[72,78] confirming once again that high stereoselectivity is a feature of nitrene insertion reactions.

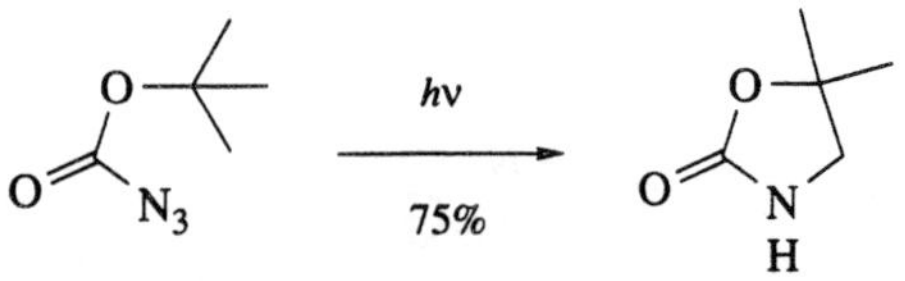

(14) (5)

(15) (6)

Sulfonylnitrenes also undergo intramolecular C—H insertion to give six-membered sultams.[79] In the series of sulfonyl azides (**16**; $n = 0, 1$), the sultams were formed in low yield; no five-membered sultams were observed.

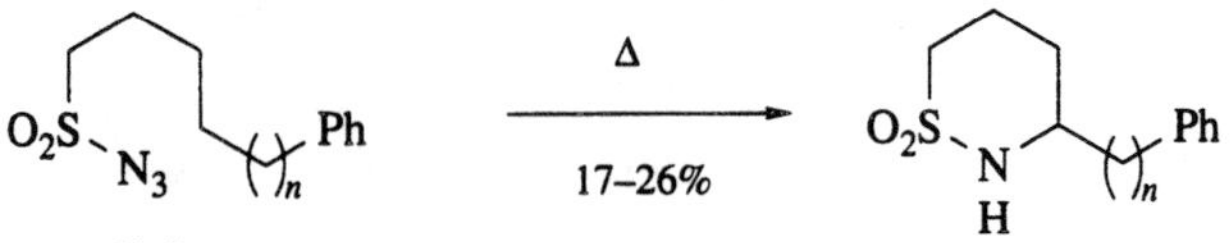

(16) 17–26% (7)

1.2.3.2.2 Use in synthesis

Although the functionalization of unactivated C—H bonds by intramolecular nitrene insertion has been applied to the synthesis of diterpene alkaloids and in the modification of steroids as described below, it has also been used to good effect in simpler systems. For example, 1-adamantyl azidoformate, readily prepared from 1-adamantanol, gives the oxazolidinone (**17**) on irradiation in cyclohexane by intramolecular C—H insertion. Hydrolysis of (**17**) gives the otherwise inaccessible 2-amino-1-adamantanol (Scheme 9).[80]

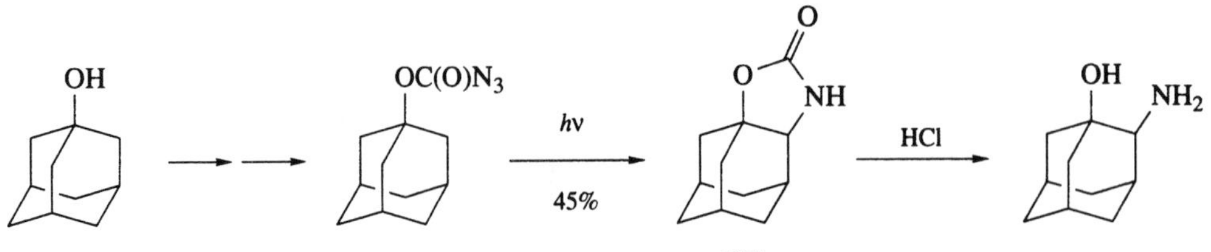

(**17**)

Scheme 9

A similar nitrene insertion reaction was used in the synthesis of 6″-aminogentamycin C_2 by functionalization of the garosamine moiety of the antibiotic.[81] The key steps of the sequence are shown in Scheme 10; heating the azidoformate (**18**) to 130 °C in dichloromethane results in the desired intramolecular nitrene insertion and functionalization of the unactivated methyl group, although some cyclization to the five-position is also observed. The synthesis was completed by hydrogenolysis of the oxazolidinone and removal of the protecting groups.

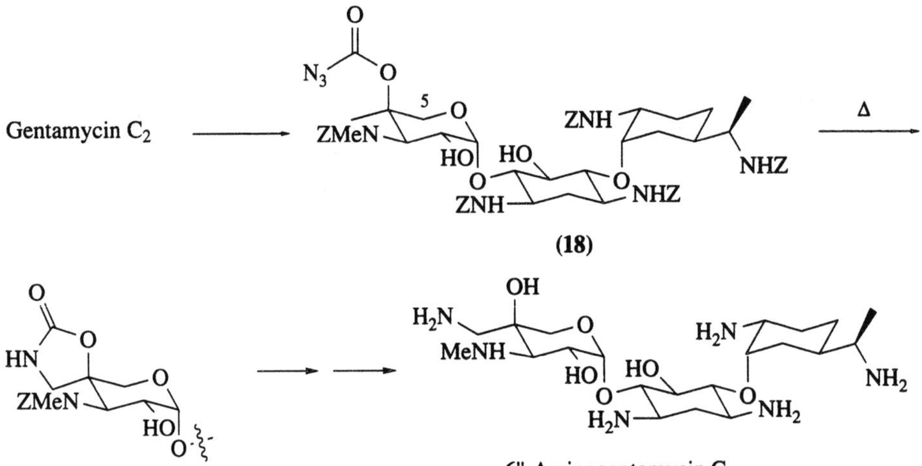

Gentamycin C_2

(**18**)

6″-Aminogentamycin C_2

Scheme 10

The use of acylnitrene cyclizations in the synthesis of diterpene alkaloids goes back to the early 1960s, and although much of the early work has been reviewed,[10] selected examples are included here. The acyl azide (**19**), readily prepared from podocarpic acid, was irradiated to give the δ-lactam (**20**; 20%), which has the azabicyclononane ring system of the diterpene alkaloids such as atisine.[82]

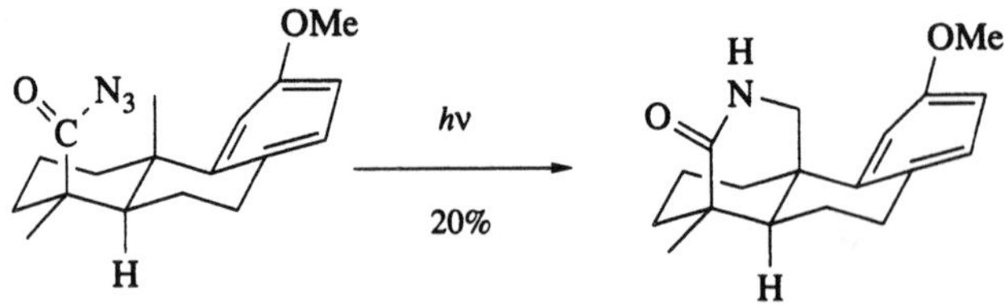

(**19**) (**20**) (8)

The azabicyclononane system is a common structural feature in diterpene alkaloids, and the nitrene insertion route to the ring system has been studied in detail in model decalins as well as in steroids (Scheme 11). Thus irradiation of the *trans*-acyl azide (**21**) gave, in addition to isocyanate (30–35%), a mixture of the γ- and δ-lactams (**22**) and (**23**). The γ-lactam (**22**) predominated, although the overall yield was poor.[83,84] The corresponding *cis*-azide (**24**), however, gave the δ-lactam (**25**) as the major product, again in low yield. One elegant application of this type of intramolecular nitrene insertion reaction was used as a key step (**26**) → (**27**) in Masamune's synthesis of the diterpene alkaloid garryine, although again the reaction was dogged by poor yields.[85]

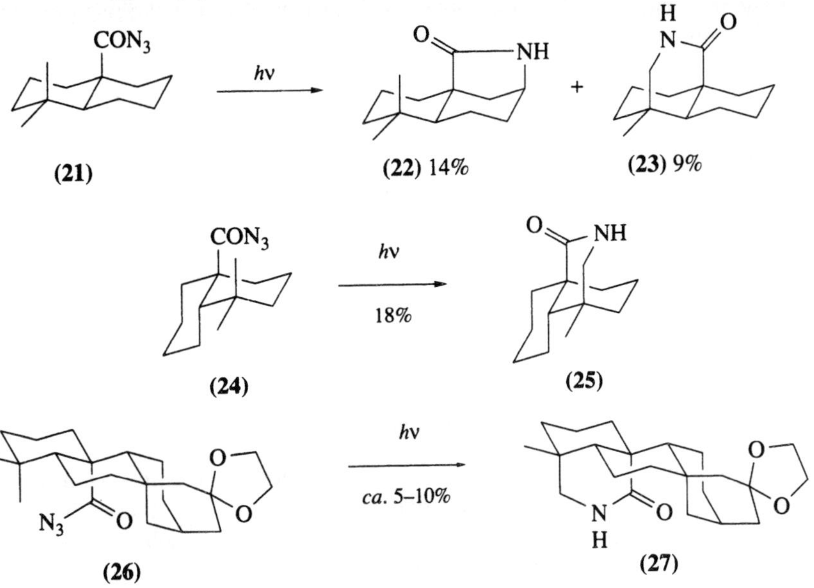

(**21**) (**22**) 14% (**23**) 9%

(**24**) (**25**)

(**26**) (**27**)

Scheme 11

Much of the remaining synthetic work in this area has been concerned with attempts to functionalize the 4,4-dimethyl groups in lanosterol, a process of considerable biosynthetic relevance and importance. For example, in a reaction that has also been studied in model *trans*-decalins,[86,87] thermal decomposition of 3β-lanost-8-enyl azidoformate (**28**) gives rise to a mixture of γ- and δ-lactams (**29**; 55%) and (**30**; 25%) resulting from nitrene insertion into the 2-CH_2 and the 4α-methyl group respectively.[88] Both insertions occur from the α-face of the steroid, and the overall yield of insertion products is excellent. The good yield of insertion products obtained from nitrenes derived from azidoformates is in contrast to the poor yields obtained from acylnitrenes derived from acyl azides. The nitrene derived from the related 7α-azidoformate derivative (**31**) of lanosterol undergoes selective C—H insertion at the 6α-C—H bond to give the modified steroid (**32**) in 52% yield.[89]

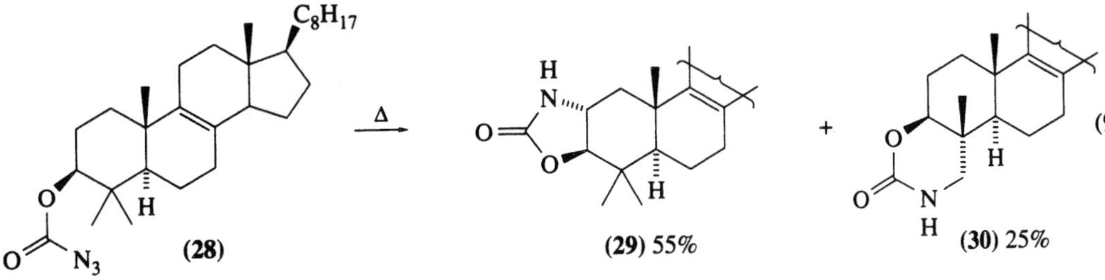

(**28**) (**29**) 55% (**30**) 25% (9)

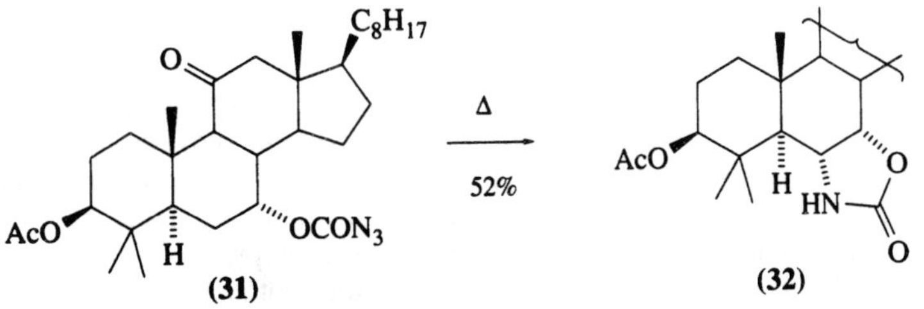

(**31**) (**32**) (10)

1.2.4 INTRAMOLECULAR INSERTION INTO *sp*² C—H BONDS

Intramolecular nitrene insertion reactions into sp^2 C—H bonds have found wide use in recent years in the synthesis of indole alkaloids and related natural products. In general, the reactions are of two types, and involve either vinylnitrenes or arylnitrenes inserting into an aromatic or vinylic sp^2 C—H bond to give an indole or a fused indole (Scheme 12). Although the indole is the product of a 'formal' C—H insertion reaction, the mechanism probably involves a six electron electrocyclic ring closure of the nitrene to give a 7a*H*-indole, followed by an aromatizing hydrogen shift.

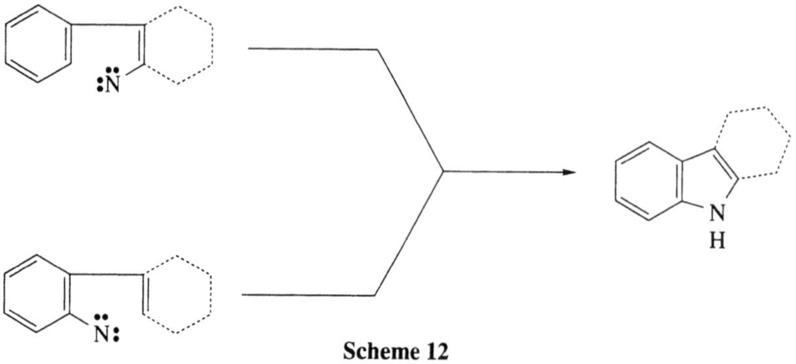

Scheme 12

Simple 3-substituted indoles can be formed in high yield by heating β-azidostyrenes in solution. Thus heating $Ph_2C{=}CHN_3$ in toluene gives 3-phenylindole in 82% yield.[90] The reaction has recently been extended to the preparation of 3,4-bridged indoles (Scheme 13),[91] and since the precursor azides are prepared from readily available cyclic ketones, the nitrene route represents a useful entry to these somewhat inaccessible bridged indoles.

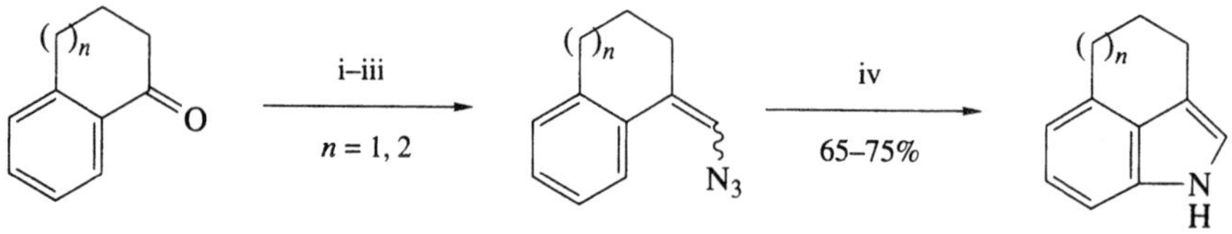

i–iii; $n = 1, 2$; iv; 65–75%

i, $Me_2\overset{+}{S}(O)CH_2^-$; ii, NaN_3; iii, MsCl, py; iv, Δ, mesitylene

Scheme 13

The formation of 2-substituted indoles from β-azidostyrenes can suffer from competing reactions of the azide and/or nitrene. However the discovery in 1970 that azidocinnamates (**33**; R = Me or Et) give indoles in excellent yield on heating in xylene,[92] together with further development in our own laboratories in collaboration with C. W. Rees (see below), has formed the basis of a versatile synthetic method. The azidocinnamates are readily prepared in a single step by base-mediated condensation of benzaldehydes with methyl (or ethyl) azidoacetate, and the reaction has been extended to heteroaromatic aldehydes to give the corresponding fused pyrroles.[93,94]

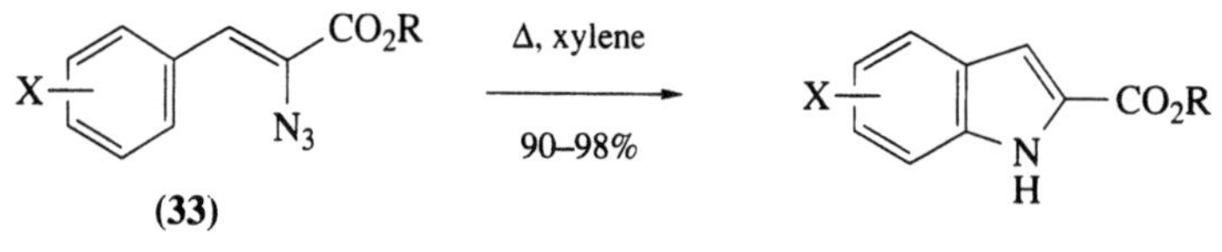

(**33**) CO_2R, N_3 — Δ, xylene, 90–98% — CO_2R (11)

We have used the reaction extensively to prepare the indole moiety of several natural products. For example, the key step in the synthesis of the bacterial coenzyme methoxatin (**36**) is the formation of the indole (**35**) by intramolecular nitrene 'insertion' from the azide (**34**), readily prepared from commercially available 4-aminosalicyclic acid.[95] The third ring was annelated onto the indole (**35**) using conventional chemistry to give, after oxidation to the *ortho*-quinone, the natural product (**36**).

Similar nitrene-mediated cyclizations have been used in the synthesis of the indoles (**38**) and (**40**), key intermediates in the synthesis of the carbazole quinone alkaloid murrayaquinone B (**39**) and the unnatural cyclopropamitosene (**41**), an analog of the aziridinomitosene ring system, respectively.[96,97] In the first example, heating the azidocinnamate (**37**), prepared from 4-hydroxybenzaldehyde, in boiling toluene resulted in sequential indole formation and regioselective Claisen rearrangement to give the 7-isoprenylindole (**38**), the 2-ester group of which was elaborated to the third ring of the natural product.[96] The second example illustrates the value of the nitrene route to indoles, in that the polysubstituted indole (**40**), which

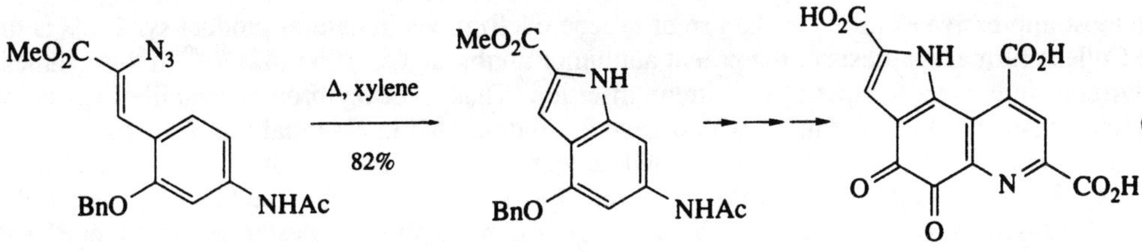

(34) (35) (36) (12)

contains all the functionality for the A-ring of the final product, is constructed from a relatively simple benzaldehyde in just two steps.[97]

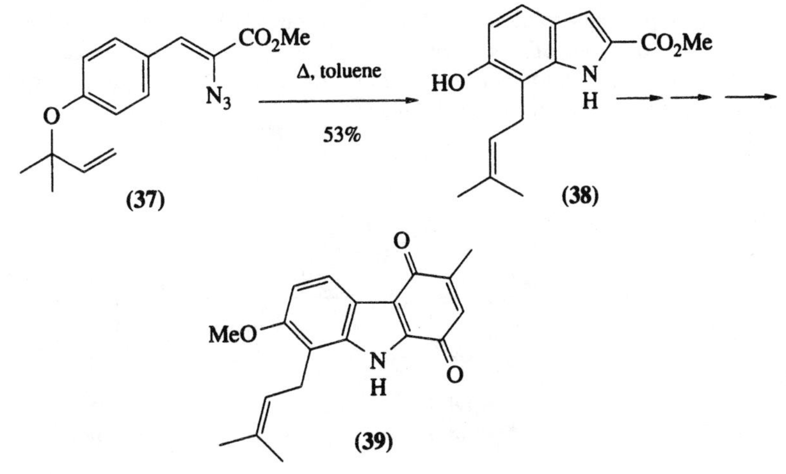

(37) (38)

(39) (13)

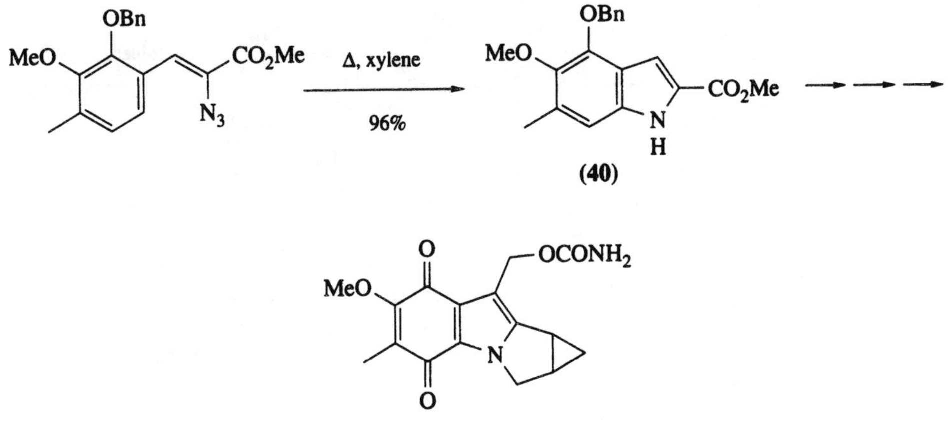

(40)

(41) (14)

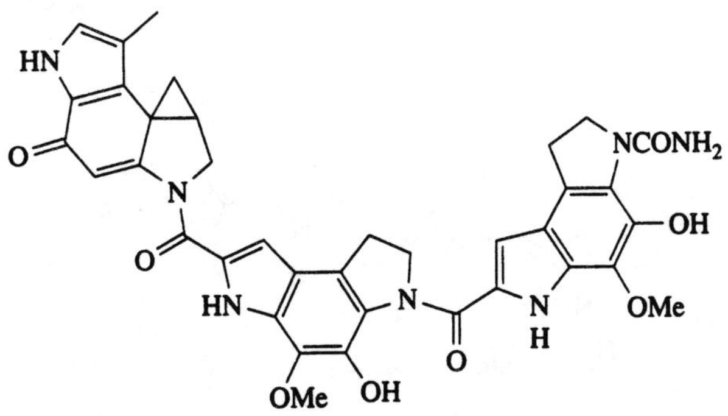

(42)

The most impressive example of the use of nitrene cyclizations in natural product synthesis is the Imperial College formal synthesis of the potent antitumor antibiotic CC-1065 (**42**).[98,99] In this synthesis, all six 'pyrrole' rings were formed using nitrene insertion. Thus *O*-benzylbromoisovanillin was converted into azidocinnamate (**43**), heating of which gave the indole (**44**) in essentially quantitative yield. After removal of the unwanted ester, the 4-bromoindole was converted into the aldehyde (**45**) and hence the azide (**46**). The second nitrene cyclization proceeded in 97% yield to give the key tricyclic indole (**47**), which was subsequently converted into the naturally occurring phosphodiesterase inhibitors PDE-I (**48**) and PDE-II (**49**), and by the coupling together of appropriate pyrroloindoles, into the 'dimer' (**50**), the combined central and right-hand unit of CC-1065. In a separate series of experiments, the left-hand unit of CC-1065 was also assembled using nitrene cyclization reactions. 5-Benzyloxy-2-bromoacetophenone was converted into the azide (**51**), heating of which in mesitylene, followed by reaction with benzenesulfonyl chloride, gave the indole (**52**; 53% over two steps). After introduction of the second azide (**53**), the tricyclic indole (**54**) was formed in 42% yield. Finally, following known chemistry, the indole (**54**) was converted into the cyclopropapyrroloindole (**55**). Since the left-hand unit (**55**) had previously been coupled with the 'dimer' (**50**) by workers at the Upjohn Company, this constituted a formal synthesis of CC-1065. A similar nitrene cyclization was also used by Boger in his total synthesis of CC-1065 to form the second pyrrole ring of the pyrroloindole subunits.[100]

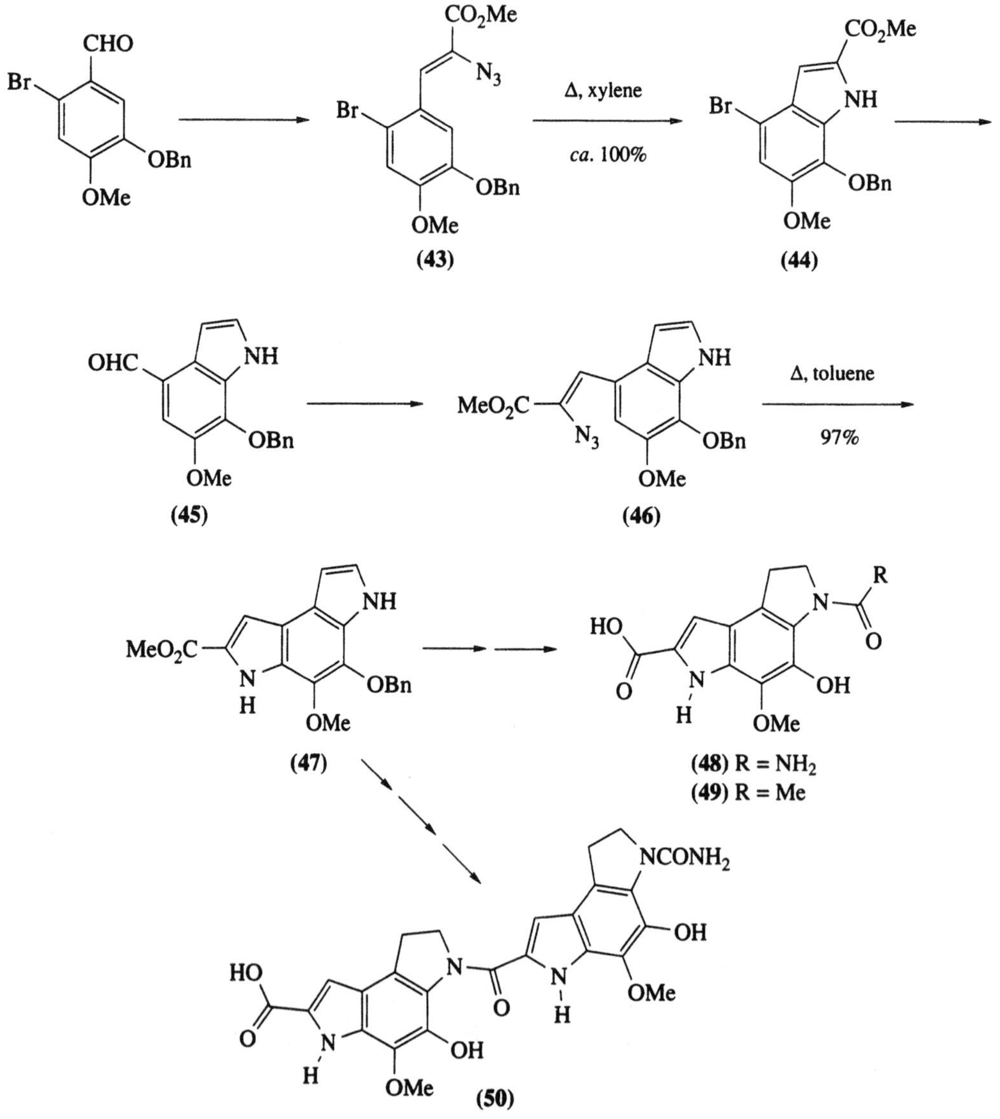

CHO Br OBn OMe → (**43**) CO_2Me N_3 Br OBn OMe — Δ, xylene, *ca.* 100% → (**44**) CO_2Me NH Br OBn OMe →

(**45**) OHC NH OBn OMe → (**46**) MeO_2C N_3 NH OBn OMe — Δ, toluene, 97% →

(**47**) MeO_2C N H NH OBn OMe → → HO O N H N R O OH OMe

(**48**) R = NH_2
(**49**) R = Me

(**50**) HO O N H N O N H $NCONH_2$ OH OMe OH OMe

Scheme 14

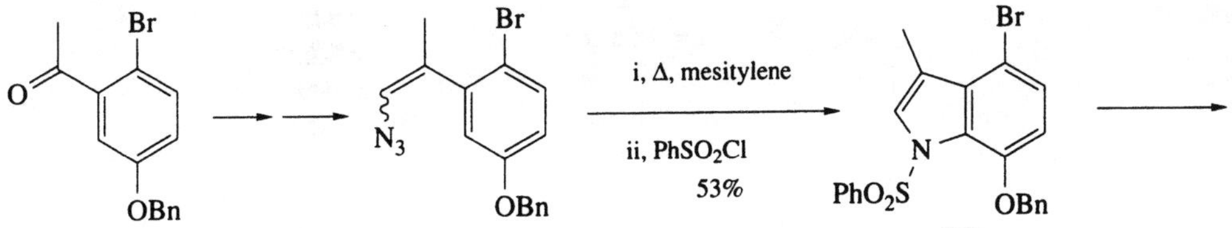

(51) (52)

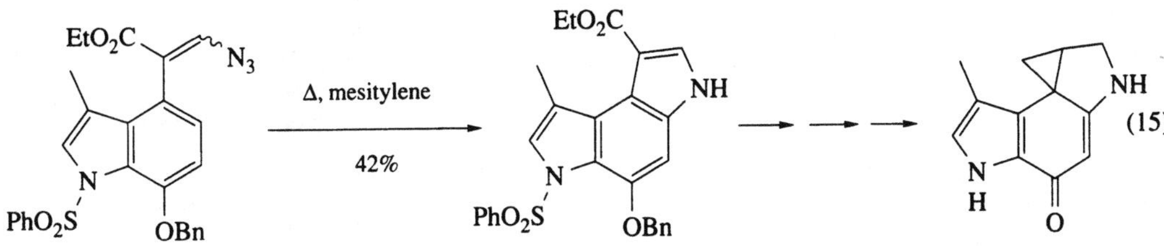

(53) (54) (55) (15)

The intramolecular reaction of vinylnitrenes is not limited to formation of five-membered rings. For example, heating the azide (**56**) results in C—H insertion adjacent to the methoxy group to give (**57**), oxidation of which gave the tetracycle (**58**), a potential precursor to the marine alkaloid amphimedine.[101] Heating the azidocinnamate (**59**) results in formal intramolecular nitrene C—H insertion, followed by hydrogen shift, to give the benzazepine (**60**), a key intermediate in the synthesis of the alkaloid lennoxamine (**61**).[102]

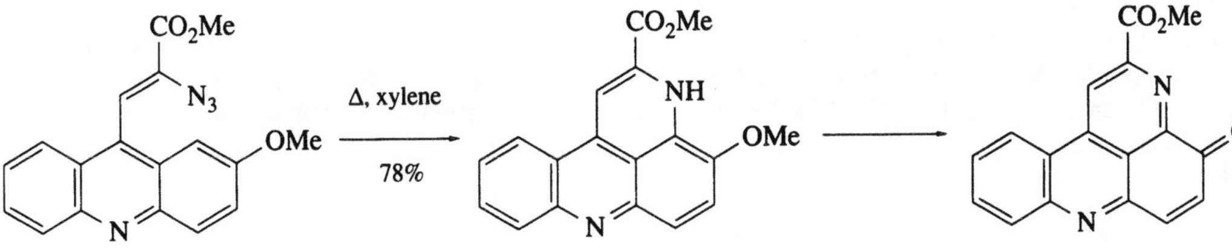

(56) (57) (58) (16)

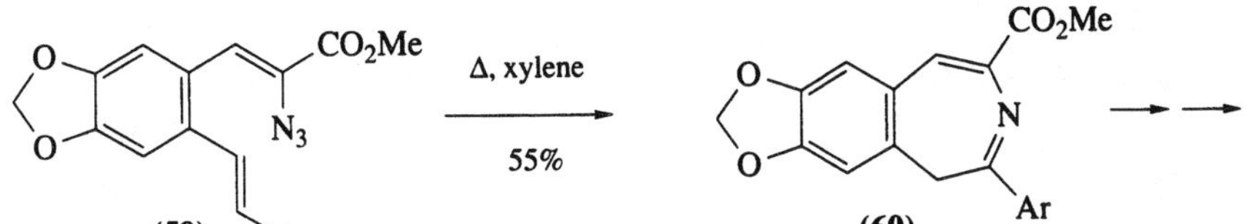

(59) (60)

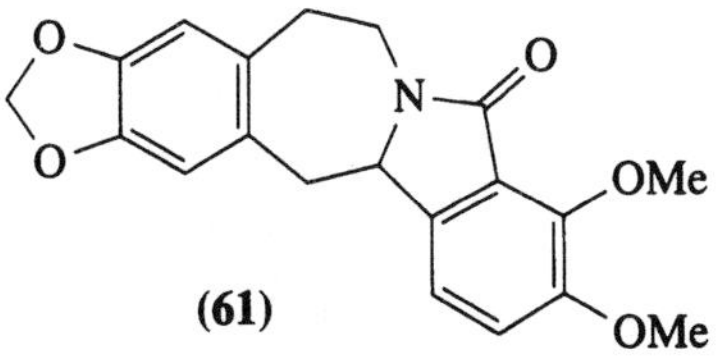

(61) 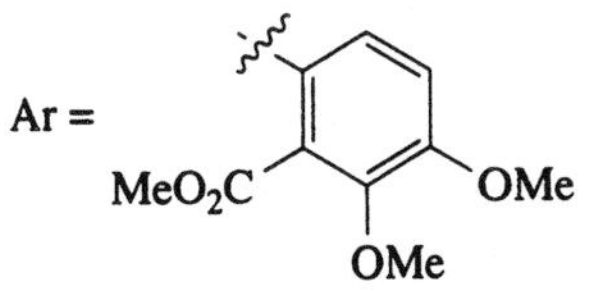 (17)

Indoles can also be formed by arylnitrene cyclizations. Thus nitrenes derived by heating or irradiating 2′-azidostyrenes (**62**; X = N_3) or by deoxygenation of the corresponding nitro compounds (**62**; X = NO_2) cyclize to 2-substituted indoles in moderate to good yield (equation 18),[103,104] although a detailed study of the reaction has confirmed that the mechanism does not involve a genuine C—H insertion.[105] The corresponding reaction of the azidoquinone (**63**) has been used as a key step in the synthesis of the indolequinone (**64**), a precursor to the mitosene analog (**65**).[106]

The intramolecular cyclization of the arylnitrenes derived from azido- or nitro-biphenyls to the adjacent aromatic ring has been well reviewed.[5,7] The reaction is a useful route to carbazoles, two recent examples of which are shown in Scheme 15. In Raphael's elegant approach to the indolocarbazole fam-

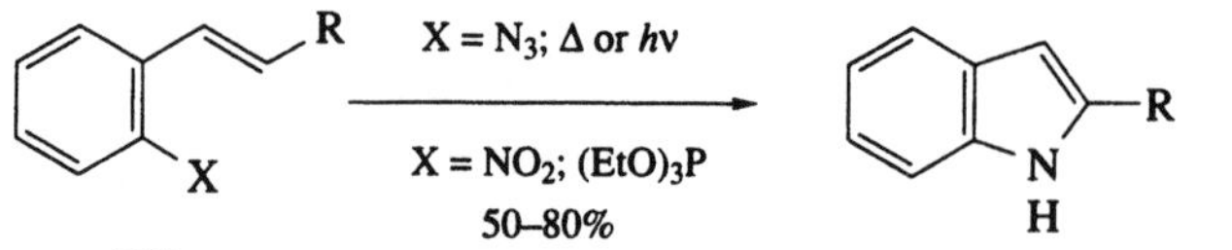

X = N_3; Δ or *hν*

X = NO_2; $(EtO)_3P$

50–80%

(62)

R = Me, Pr, Ph, PhCO

(18)

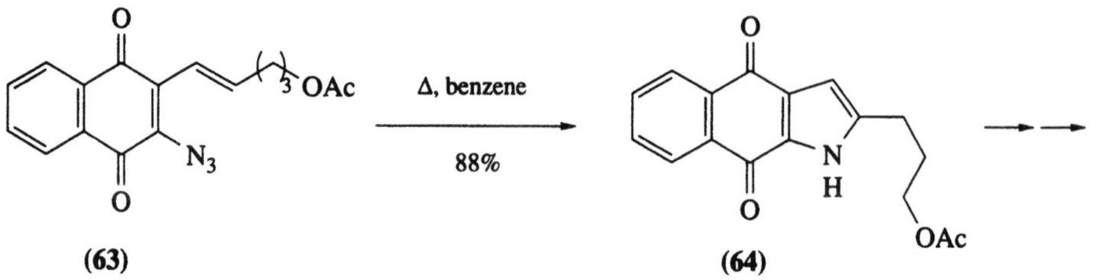

Δ, benzene

88%

(63) **(64)**

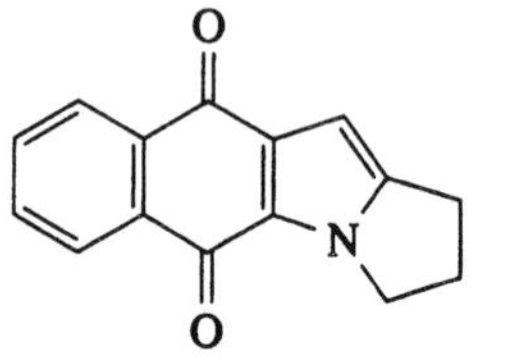

(65)

(19)

ily of antibiotics, the dinitro terphenyl (**66**), readily prepared by dehydrogenation of the Diels–Alder adduct of the appropriate 1,4-diarylbutadiene with maleimide, was deoxygenated with triphenylphosphine in collidine to give the 'double nitrene insertion' product, the indolocarbazole (**67**; 65%), demethylation of which gave arcyriaflavin B (**68**).[107] In the second example, the azide (**69**), constructed in 10 steps from 2,5-dimethylacetanilide, cyclized to the antitumor alkaloid ellipticine (**70**) in excellent yield on heating.[108]

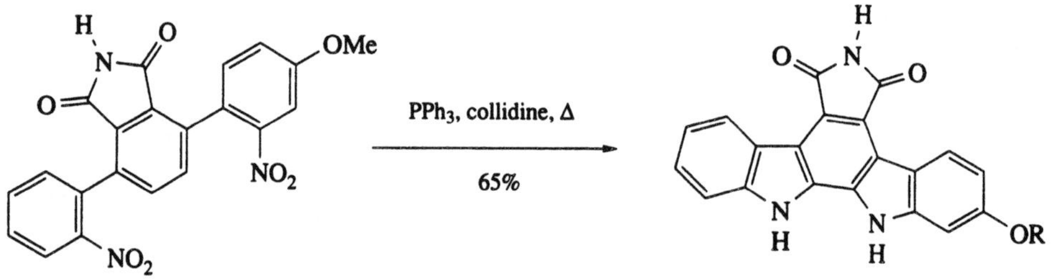

PPh_3, collidine, Δ

65%

(66)

(67) R = Me
(68) R = H

Δ, dodecane

96%

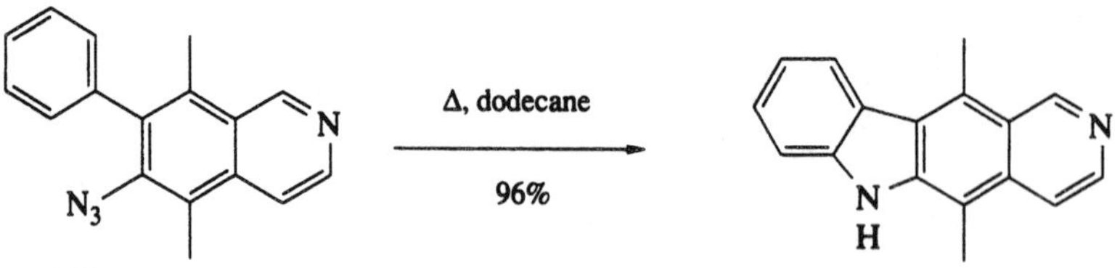

(69) **(70)**

Scheme 15

Thus the intramolecular reaction of nitrenes with sp^2 C—H bonds, although it may not involve a genuine C—H insertion mechanism, is a useful synthetic method, which extends and complements the nitrene insertions into unactivated sp^3 C—H bonds discussed in earlier sections.

1.2.5 REFERENCES

1. F. D. Lewis and W. H. Saunders, in 'Nitrenes', ed. W. Lwowski, Interscience, New York, 1970, chap. 3.
2. E. P. Kyba, in 'Azides and Nitrenes—Reactivity and Utility', ed. E. F. V. Scriven, Academic Press, New York, 1984, chap. 1.
3. G. L'Abbé, *Angew. Chem., Int. Ed. Engl.*, 1975, **14**, 775.
4. A. Hassner, in 'Azides and Nitrenes—Reactivity and Utility', ed. E. F. V. Scriven, Academic Press, New York, 1984, chap. 2.
5. P. A. S. Smith, in 'Nitrenes', ed. W. Lwowski, Interscience, New York, 1970, chap. 4.
6. B. Iddon, O. Meth-Cohn, E. F. V. Scriven, H. Suschitzky and P. T. Gallagher, *Angew. Chem., Int. Ed. Engl.*, 1979, **18**, 900.
7. P. A. S. Smith, in 'Azides and Nitrenes—Reactivity and Utility', ed. E. F. V. Scriven, Academic Press, New York, 1984, chap. 3.
8. W. Lwowski, *Angew. Chem., Int. Ed. Engl.*, 1967, **6**, 897.
9. W. Lwowski, in 'Nitrenes', ed. W. Lwowski, Interscience, New York, 1970, chap. 6.
10. O. E. Edwards, in 'Nitrenes', ed. W. Lwowski, Interscience, New York, 1970, chap. 7.
11. W. Lwowski, in 'Chemistry of the Azido Group', ed. S. Patai, Interscience, New York, 1971, chap. 9.
12. W. Lwowski, in 'Azides and Nitrenes—Reactivity and Utility', ed. E. F. V. Scriven, Academic Press, New York, 1984, chap. 4.
13. A. G. Anastassiou, J. N. Shepelavy, H. E. Simmons and F. D. Marsh, in 'Nitrenes', ed. W. Lwowski, Interscience, New York, 1970, chap. 9.
14. D. S. Breslow, in 'Nitrenes', ed. W. Lwowski, Interscience, New York, 1970, chap. 8.
15. T. Kametani, F. F. Ebetino, T. Yamanaka and K. Nyu, *Heterocycles*, 1974, **2**, 209.
16. O. Meth-Cohn, *Heterocycles*, 1980, **14**, 1497.
17. C. Wentrup, *Adv. Heterocycl. Chem.*, 1981, **28**, 231.
18. H. Bayley and J. V. Staros, in 'Azides and Nitrenes—Reactivity and Utility', ed. E. F. V. Scriven, Academic Press, New York, 1984, chap. 9.
19. G. L'Abbé, *Chem. Rev.*, 1969, **69**, 345.
20. R. A. Abramovitch and E. P. Kyba, in 'Chemistry of the Azido Group', ed. S. Patai, Interscience, New York, 1971, chap. 5.
21. E. F. V. Scriven and K. Turnbull, *Chem. Rev.*, 1988, **88**, 297.
22. W. Pritzkow and D. Timm, *J. Prakt. Chem.*, 1966, **32**, 178.
23. R. E. Banks, D. Berry, M. J. McGlinchey and C. J. Moore, *J. Chem. Soc. (C)*, 1970, 1017.
24. J. H. Hall, J. W. Hill and J. M. Fargher, *J. Am. Chem. Soc.*, 1968, **90**, 5313.
25. R. E. Banks and G. R. Sparkes, *J. Chem. Soc., Perkin Trans. 1*, 1972, 2964.
26. G. T. Tisue, S. Linke and W. Lwowski, *J. Am. Chem. Soc.*, 1967, **89**, 6303.
27. E. Eibler and J. Sauer, *Tetrahedron Lett.*, 1974, 2569.
28. E. Eibler and J. Sauer, *Tetrahedron Lett.*, 1974, 2565.
29. D. S. Breslow and E. I. Edwards, *Tetrahedron Lett.*, 1967, 2123.
30. D. S. Breslow and E. I. Edwards, *Tetrahedron Lett.*, 1972, 2041.
31. W. Lwowski and T. W. Mattingly, *J. Am. Chem. Soc.*, 1965, **87**, 1947.
32. R. Huisgen and H. Blaschke, *Chem. Ber.*, 1965, **98**, 2985.
33. W. Lwowski and T. J. Maricich, *J. Am. Chem. Soc.*, 1965, **87**, 3630.
34. M. Seno, T. Namba and H. Kise, *J. Org. Chem.*, 1978, **43**, 3345.
35. Y. Hayashi and D. Swern, *Tetrahedron Lett.*, 1972, 1921.
36. D. S. Breslow, T. J. Prosser, A. F. Marcantonio and C. A. Genge, *J. Am. Chem. Soc.*, 1967, **89**, 2384.
37. T. Shinaki, M. Inagaki, N. Torimoto and M. Takebayashi, *Chem. Lett.*, 1972, 1181.
38. D. S. Breslow, M. F. Sloan, N. R. Newburg and W. B. Renfrow, *J. Am. Chem. Soc.*, 1969, **91**, 2273.
39. R. Breslow, A. E. Feiring and F. Herman, *J. Am. Chem. Soc.*, 1974, **96**, 5937.
40. J. H. Hall, J. W. Hill and H. Tsai, *Tetrahedron Lett.*, 1965, 2211.
41. M. Inagaki, T. Shingaki and T. Nagai, *Chem. Lett.*, 1981, 1419.
42. T. Shinaki, M. Inagaki, M. Takebayashi and W. Lwowski, *Bull. Chem. Soc. Jpn.*, 1972, **45**, 3567.
43. D. S. Breslow, E. I. Edwards, E. C. Linsay and H. Omura, *J. Am. Chem. Soc.*, 1976, **98**, 4268.
44. H. Takeuchi, Y. Kasamatsu, M. Mitani, T. Tsuchida and K. Koyama, *J. Chem. Soc., Perkin Trans. 2*, 1978, 780.
45. T. Shinaki, M. Inagaki, N. Torimoto and M. Takebayashi, *Chem. Lett.*, 1972, 155.
46. A. G. Anastassiou and H. E. Simmons, *J. Am. Chem. Soc.*, 1967, **89**, 3177.
47. D. S. Breslow, E. I. Edwards, R. Leone and P. von R. Schleyer, *J. Am. Chem. Soc.*, 1968, **90**, 7097.
48. P. Casagrande, L. Pellacani and P. A. Tardella, *J. Org. Chem.*, 1978, **43**, 2725.
49. P. A. Tardella and L. Pellacani, *J. Org. Chem.*, 1976, **41**, 2034.
50. M. R. Brinkman, D. Bethell and J. F. Hayes, *Tetrahedron Lett.*, 1973, 989.
51. H. Nozaki, S. Fujita, H. Takaya and R. Noyori, *Tetrahedron*, 1967, **23**, 45.
52. T. Shingaki, M. Inagaki, N. Torimoto and M. Takebayashi, *Chem. Lett.*, 1972, 297.
53. N. Torimoto, T. Shingaki and T. Nagai, *Bull. Chem. Soc. Jpn.*, 1976, **49**, 2572.
54. H. Takeuchi, K. Kinoshita, S. M. Abdul-Hai, M. Mitani, T. Tsuchida and K. Koyama, *J. Chem. Soc., Perkin Trans. 2*, 1976, 1201.
55. H. Takeuchi, T. Nishiyama, M. Mitani, T. Tsuchida and K. Koyama, *J. Chem. Soc., Perkin Trans. 2*, 1979, 839.
56. H. Takeuchi, N. Murata, Y. Nakagawa, T. Tsuchida and K. Koyama, *J. Chem. Soc., Perkin Trans. 2*, 1977, 80.
57. P. A. Tardella and L. Pellacani, *Tetrahedron Lett.*, 1977, 4451.
58. M. Inagaki, T. Shingaki and T. Nagai, *Chem. Lett.*, 1982, 9.
59. R. C. Belloli, M. A. Whitehead, R. H. Wollenberg and V. A. LaBahn, *J. Org. Chem.*, 1974, **39**, 2128.
60. A. G. Anastassiou and J. N. Shepelavy, *J. Am. Chem. Soc.*, 1968, **90**, 492.

61. J. M. Simson and W. Lwowski, *J. Am. Chem. Soc.*, 1969, **91**, 5107.
62. D. H. R. Barton and L. R. Morgan, *J. Chem. Soc.*, 1962, 622; D. H. R. Barton and A. N. Starratt, *J. Chem. Soc.*, 1965, 2444.
63. A. Pancrazi and Q. Khuong-Huu, *Tetrahedron*, 1975, **31**, 2041.
64. L. Henn, D. M. B. Hickey, C. J. Moody and C. W. Rees, *J. Chem. Soc., Perkin Trans. 1*, 1984, 2189.
65. M. Farnier, S. Soth and P. Fournari, *Can. J. Chem.*, 1976, **54**, 1066.
66. C. J. Moody and J. G. Ward, *J. Chem. Soc., Perkin Trans. 1*, 1984, 2895.
67. G. Smolinsky, *J. Am. Chem. Soc.*, 1961, **83**, 2489.
68. G. Smolinsky, *J. Org. Chem.*, 1961, **26**, 4108.
69. G. Smolinsky and B. I. Feuer, *J. Org. Chem.*, 1964, **29**, 3097.
70. R. J. Sundberg, *J. Am. Chem. Soc.*, 1966, **88**, 3781.
71. E. W. Cantrall, R. B. Conrow and S. Bernstein, *J. Org. Chem.*, 1967, **32**, 3445.
72. G. Smolinsky and B. I. Feuer, *J. Am. Chem. Soc.*, 1964, **86**, 3085.
73. G. Smolinsky and B. I. Feuer, *J. Org. Chem.*, 1966, **31**, 3882.
74. I. Brown and O. E. Edwards, *Can. J. Chem.*, 1967, **45**, 2599.
75. W. Lwowski and S. Linke, *Justus Liebigs Ann. Chem.*, 1977, 8.
76. S. Terashima, M. Nara and S. Yamada, *Chem. Pharm. Bull.*, 1970, **18**, 1124.
77. R. Kreher and G. H. Bockhorn, *Angew. Chem., Int. Ed. Engl.*, 1964, **3**, 589; R. Puttner and K. Hafner, *Tetrahedron Lett.*, 1964, 3119.
78. S. Terashima and S. Yamada, *Chem. Pharm. Bull.*, 1968, **16**, 1953.
79. R. A. Abramovitch, S. B. Hendi and A. O. Kress, *J. Chem. Soc., Chem. Commun.*, 1981, 1087.
80. W. V. Curran and R. B. Angier, *J. Org. Chem.*, 1969, **34**, 3668.
81. J. J. K. Wright, J. A. Albarello and P. Lee, *J. Org. Chem.*, 1982, **47**, 523.
82. J. W. ApSimon and O. E. Edwards, *Can. J. Chem.*, 1962, **40**, 896.
83. W. L. Meyer and A. S. Levinson, *J. Org. Chem.*, 1963, **28**, 2859.
84. R. F. C. Brown, *Aust. J. Chem.*, 1964, **17**, 47.
85. S. Masamune, *J. Am. Chem. Soc.*, 1964, **86**, 290.
86. M. R. Czarny, B. W. Benson and T. A. Spencer, *J. Org. Chem.*, 1977, **42**, 556.
87. P. F. Alewood, M. Benn, J. Wong and A. J. Jones, *Can. J. Chem.*, 1977, **55**, 2510.
88. J. J. K. Wright and J. B. Morton, *J. Chem. Soc., Chem. Commun.*, 1976, 668.
89. O. E. Edwards and Z. Paryzek, *Can. J. Chem.*, 1973, **51**, 3866.
90. G. Smolinsky and C. A. Pryde, *J. Org. Chem.*, 1968, **33**, 2411.
91. A. L. Beck, W. J. Coates and C. J. Moody, *J. Chem. Soc., Perkin Trans. 1*, 1990, 689.
92. H. Hemetsberger, D. Knittel and H. Weidmann, *Monatsh. Chem.*, 1970, **101**, 161.
93. M. Farnier, S. Soth and P. Fournari, *Can. J. Chem.*, 1976, **54**, 1074.
94. C. J. Moody and J. G. Ward, *J. Chem. Soc., Perkin Trans. 1*, 1984, 2903.
95. A. R. MacKenzie, C. J. Moody and C. W. Rees, *Tetrahedron*, 1986, **42**, 3259.
96. T. Martin and C. J. Moody, *J. Chem. Soc., Perkin Trans. 1*, 1988, 241.
97. G. B. Jones and C. J. Moody, *J. Chem. Soc., Chem. Commun.*, 1989, 186.
98. R. E. Bolton, C. J. Moody, C. W. Rees and G. Tojo, *J. Chem. Soc., Perkin Trans. 1*, 1987, 931.
99. R. E. Bolton, C. J. Moody, M. Pass, C. W. Rees and G. Tojo, *J. Chem. Soc., Perkin Trans. 1*, 1988, 2491.
100. D. L. Boger and R. S. Coleman, *J. Am. Chem. Soc.*, 1987, **109**, 2717; 1988, **110**, 1321.
101. C. V. Labarca, A. R. MacKenzie, C. J. Moody, C. W. Rees and J. J. Vaquero, *J. Chem. Soc., Perkin Trans. 1*, 1987, 927.
102. C. J. Moody and G. J. Warrellow, *Tetrahedron Lett.*, 1987, **28**, 6089; *J. Chem. Soc., Perkin Trans. 1*, 1990, 2929.
103. R. J. Sundberg, L.-S. Lin and D. E. Blackburn, *J. Heterocycl. Chem.*, 1969, **6**, 441.
104. R. J. Sundberg, *J. Org. Chem.*, 1965, **30**, 3604.
105. P. A. S. Smith, C. D. Rowe and D. W. Hansen, Jr., *Tetrahedron Lett.*, 1983, **24**, 5169.
106. P. Germeraad and H. W. Moore, *J. Org. Chem.*, 1974, **39**, 774.
107. I. Hughes and R. A. Raphael, *Tetrahedron Lett.*, 1983, **24**, 1441; I. Hughes, W. P. Nolan and R. A. Raphael, *J. Chem. Soc., Perkin Trans. 1*, 1990, 2475.
108. R. B. Miller and S. Dugar, *Tetrahedron Lett.*, 1989, **30**, 297.

1.3
Oxidation by Remote Functionalization Methods

RONALD BRESLOW
University of Columbia, New York, NY, USA

1.3.1 INTRODUCTION

1.3.1.1 Topics Covered

'Oxidation' is not a well-defined concept in organic chemistry. It hardly ever involves simply the removal of electrons, and when covalent bonds are being made and broken some arbitrary choices must be made. When a C—H bond is converted to a C—OH bond, we say that the C—H bond has been oxidized. Similarly, the conversion of ethane to ethylene is generally considered to be an oxidation, and hydrogenation of an alkene is generally considered a reduction. However, if these conversions are indirect we must specify in which step the oxidation or reduction occurred.

As a simple case, the chlorination of a C—H bond, converting it to C—Cl, is certainly an oxidation. The chloride can be solvolyzed to an alcohol, which is an oxidation product of the original C—H bond, and no one would think of the solvolysis as the oxidation step. Thus we will consider as an oxidation any conversion of a C—H bond to a derivative in which the hydrogen has been replaced by a more electronegative element. Furthermore, conversion of a saturated carbon to an unsaturated one will be considered an oxidation of that carbon. In principle the double bond could be hydrated so as to leave the hydroxy group on that carbon, and addition of water to an alkene is not an oxidation step.

The insertion of a carbonyl group into a C—H bond is also an 'oxidation' by this reasoning. The resulting C—C—OH unit can in principle be dehydrated towards the original carbon, converting it to an unsaturated carbon, and dehydration of an alcohol is not an oxidation step. If the insertion was at a tertiary carbon such a simple dehydration would of course be impossible, but it is hard to imagine that carbonyl insertion into the C—H bond of a secondary carbon be considered an oxidation, but insertion at a tertiary carbon not be. Thus this chapter will discuss reactions that functionalize a C—H bond such that it is either directly oxidized or can be converted, by further nonoxidative steps, into an oxidized carbon.

The focus will largely be on cases in which the functionalization or oxidation is performed with geometric control, and in which that control permits selective attack at various predetermined positions remote from any functional groups of the substrate. Those methods in which oxidations, using geometric direction, occur in the near neighborhood of substrate functional groups will also be discussed briefly. Such methods were an important source of inspiration for the more remote functionalization procedures, and they perform many useful transformations.

Controlled functionalization of unactivated positions can also be achieved by putting the substrate into an inhomogeneous environment and then attacking it with an otherwise random reagent.[1] Good examples include hydroxylation with ozone on silica,[2–5] halogenation of substrates bound in zeolite cavities,[6] and reactions in solid inclusion complexes.[7] Some chemically selective reagents can also perform useful conversions of steroids, for instance, to more or less single products.[8–11] Methods that do not involve rationalizable geometric control will not be included in this review.

1.3.1.2 Biomimetic Chemistry

An important inspiration for this field also comes from a consideration of the selective oxidations commonly performed by enzymes. It is commonplace for an oxidative enzyme to convert an isolated methyl group to a carboxy group while leaving double bonds and carbinols alone;[12] this is done by geometric control within the enzyme–substrate complex. The outstanding regioselectivity and stereoselectivity of enzymatic processes led to the coining of the term 'biomimetic' to describe selective chemical functionalizations of unactivated positions similarly directed by geometric constraints.[13] In this chemistry the intrinsic reactivity of the substrate is overridden by the geometric preference of the reagent–substrate combination. The term 'biomimetic' has now been enlarged in scope, to refer to chemistry that mimics biochemical processes or pathways in any aspect.

1.3.2 INTRAMOLECULAR FUNCTIONALIZATIONS IN THE VICINITY OF EXISTING SUBSTRATE FUNCTIONAL GROUPS

Free radicals can undergo 1,5-hydrogen shifts. When the initial radical site is a heteroatom, the 1,5-shift leads to functionalization of the carbon four bonds away. There are many examples of such processes.

An early version was the Hofmann–Loffler–Freytag reaction (Scheme 1).[14] Irradiation of a chloramine in acid leads to formation of the aminium radical, which can abstract a hydrogen to generate a carbon radical. Then the resulting carbon radical abstracts chlorine from another protonated chloramine, producing a chlorinated carbon and regenerating the chain-carrying radical. On treatment with base, the product

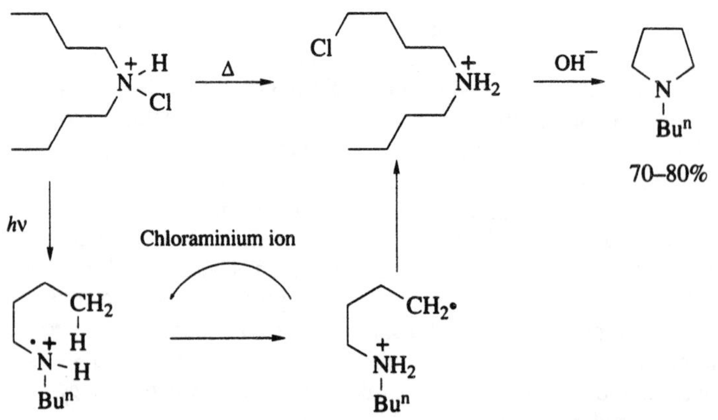

Δ
Cl
$\overset{+}{N}H_2$
OH^-
N
Bu^n
70–80%
$h\nu$
Chloraminium ion
CH_2
$\cdot+$ H
N–H
Bu^n
$CH_2\cdot$
$\overset{+}{N}H_2$
Bu^n

Scheme 1

δ-chloroamine can cyclize to form a pyrrolidine. More recently phosphoramidate radicals have been used to perform similar hydrogen abstractions.[15]

A related reaction occurs on irradiation of a hypochlorite (Scheme 2).[16] The alkoxyl radical can again abstract a hydrogen atom in a 1,5-shift, and the final chloro alcohol can be cyclized to form a tetrahydrofuran. Some processes lead directly to the cyclic product. For instance, an alcohol with an accessible δ-hydrogen can be directly converted to a tetrahydrofuran on refluxing with lead tetraacetate.[17] In a related reaction, treatment of an alcohol with silver carbonate and bromine can lead to the cyclic ether by initial formation of a hypohalite.[18] The cyclization occurs when the rearranged radical is converted to a cation, either by oxidation with $Pb(OAc)_4$ or by silver-assisted loss of halide ion. Hypoiodites are also frequently used, generated *in situ*.[19]

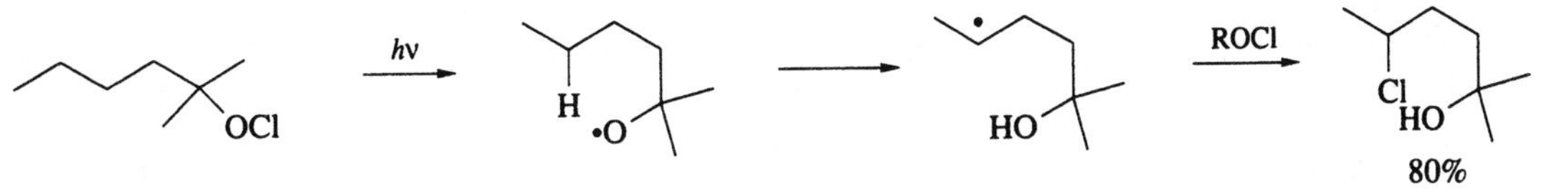

OCl $\xrightarrow{h\nu}$ H •O ⟶ • HO $\xrightarrow{ROCl}$ Cl HO

80%

Scheme 2

A particularly nice conversion is the reaction of a cyanohydrin with I_2 and $Pb(OAc)_4$ (the Heusler–Kalvoda reaction; Scheme 3).[20] After the abstraction of a δ-hydrogen the cyano group migrates to the resulting radical. The final product has a ketone in place of the original cyanohydrin, which was of course formed from that ketone, and a cyano group on the carbon γ to the ketone.

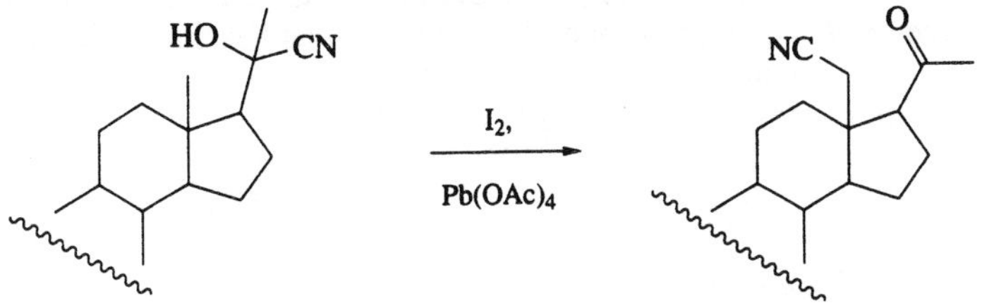

HO CN $\xrightarrow[Pb(OAc)_4]{I_2,}$ NC O

Scheme 3

A functionalization that converts C—H bonds to C—NO bonds occurs when nitrite esters are photolyzed (the Barton reaction; Scheme 4).[21] Again an alkoxyl radical abstracts a δ-hydrogen, and the resulting carbon radical picks up NO. The product nitroso compounds convert easily to oximes. Particularly valuable examples have been studied in the steroid field.[22] If the photolysis is performed in the presence of copper(II) acetate the intermediate carbon radical can be oxidized to an alkene, rather than capture NO.[23] If the alcohol whose nitrite ester is photolyzed is part of a cyanohydrin, then the Heusler–Kalvoda reaction occurs, and the product is a ketone with a migrated cyano group (Scheme 5).[24]

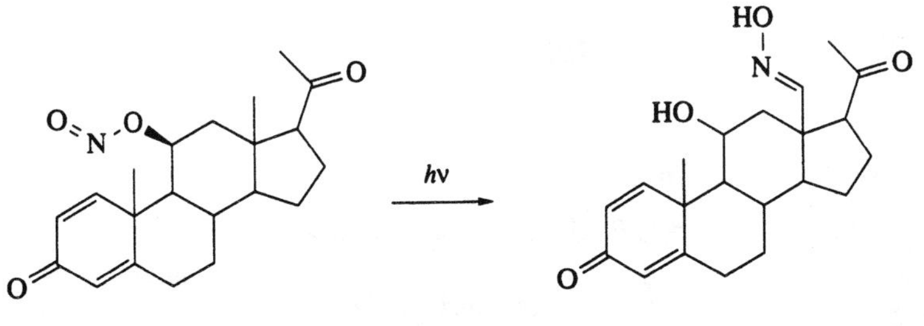

O=N-O O O $\xrightarrow{h\nu}$ HO N HO O O

Scheme 4

Photolysis of ketones can also lead to 1,5-hydrogen shifts, resulting in functionalization of the γ-carbon. The resulting 1,4-diradical can then fragment or cyclize to form a cyclobutanol (Scheme 6).[25] Examples are also known in which hydrogen abstraction involves a seven-membered[26] or a five-membered ring transition state.[27] This photochemical Type II process has been shown to involve the intermediacy of an excited triplet state, with an electron promoted from an unshared pair into the π-system. Hydrogen abstraction occurs by attack of the half-vacant nonbonding orbital on the electrons of the nearby C—H

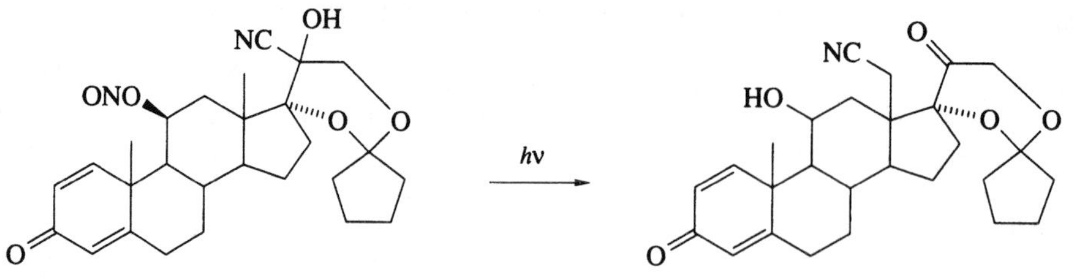

Scheme 5

bond. Thus the geometric requirement is accessibility of the hydrogen to the plane of the carbonyl group, not the π-system.

Scheme 6

Related photochemistry has also been examined with other functional groups such as phthalimides, which also abstract nearby hydrogens with the photoexcited carbonyl group.[28,29] Furthermore, since the hydrogen abstraction is performed by the half-vacant nonbonding orbital of a photoexcited ketone carbonyl, related chemistry is observed if the electron is removed electrochemically, not just photoexcited into a π^*-orbital. Electrochemical functionalization of nearby carbons has been reported in which, after hydrogen atom abstraction by an oxidized ketone, the resulting radical is electrochemically oxidized further to the carbon cation, which reacts with solvent (Scheme 7).[30,31]

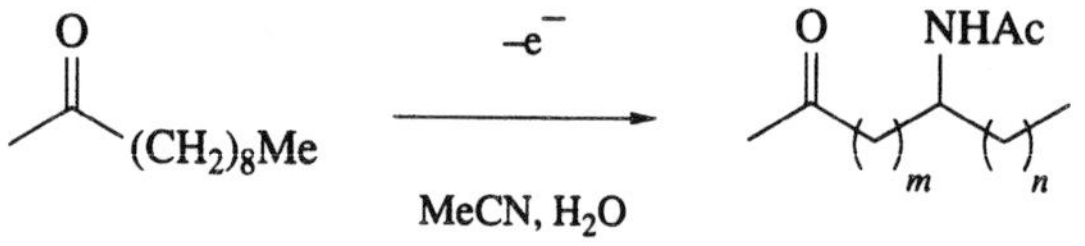

m	2	3	4	5	6
%	45	25	15	10	5

Scheme 7

In all these examples functionalization of unactivated carbons occurred, but at positions only a few carbons removed from a substrate functional group. The rest of this chapter shall consider cases in which this restriction is removed.

1.3.3 OXIDATIONS REMOTE FROM EXISTING SUBSTRATE FUNCTIONAL GROUPS

1.3.3.1 Remote Photochemical Functionalization

In 1969 the general principle of this field was enunciated in a paper[32] reporting the photochemical insertion of a benzophenone carbonyl group into the CH_2 groups of long alkyl chains (Scheme 8). It was pointed out that, as in biochemical reactions, the intrinsic reactivity of a substrate can be overridden by the geometric preferences imposed by a suitably oriented reagent. Specifically, the dodecyl ester of benzophenone-4-carboxylic acid (**1**) underwent photoinsertion into carbons 10 and 11 on irradiation; only minimal insertion occurred at carbon 9, and essentially none in carbons 1 to 8. The results were as expected for the geometry of (**1**) but the distribution of attack sites means that this is not a useful preparative method for a single product. Work with related compounds also gave a distribution of products, reflecting the flexibility of an alkyl chain.[33]

Good selectivity was seen with a flexible substrate immobilized by double ion-pair binding to a benzophenone dication (Scheme 9).[34] The insertion product could be dehydrated, and the resulting alkene fur-

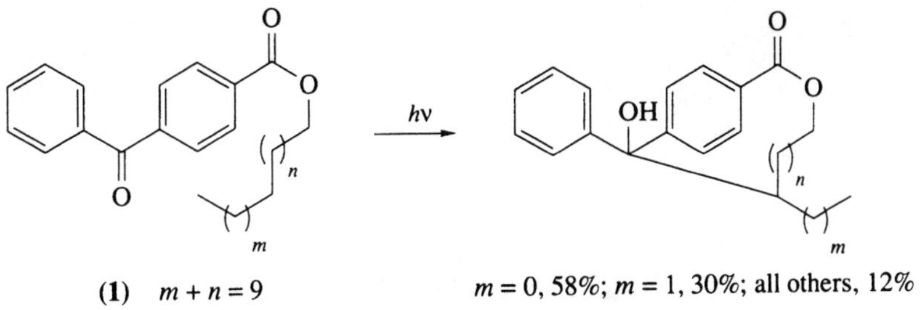

(1) $m + n = 9$ $\quad$ $m = 0$, 58%; $m = 1$, 30%; all others, 12%

Scheme 8

ther oxidized to furnish a keto diacid product with excellent selectivity. The high selectivity is restricted to substrates that are of the correct length to stretch out along the benzophenone reagent.

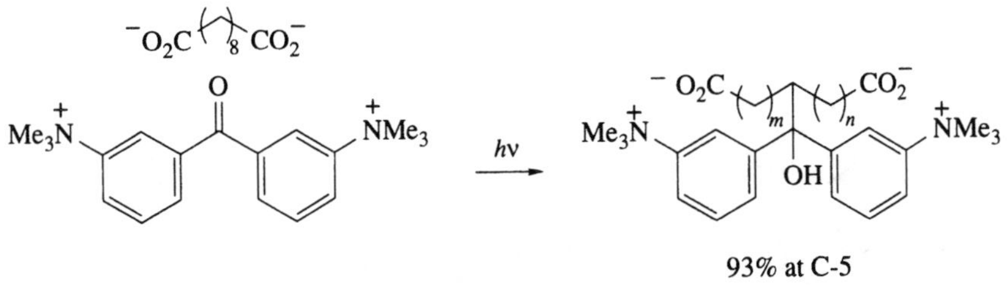

93% at C-5

Scheme 9

Photochemical functionalizations with synthetic potential have been achieved using benzophenone esters of steroids (Scheme 10). In some cases attack occurs on several hydrogens; for instance, a mixture of Δ^{14}- and Δ^{16}-alkenes is produced on irradiation of (**2**).[35] However, with compound (**3**) photolysis produces only (**4**) as a new steroid.[36] The yield of 55% involves some photoreduction of the benzophenone unit by solvent, so the other significant product is starting material with a reduced ketone group. Many other photolyses of benzophenone steroid esters have been studied;[37–39] they lead to useful information about conformations, but not the directed single-site functionalizations that would make them synthetically useful.

Irradiation of nitro aromatics produces excited states in which the nitro group oxygen can remove an accessible hydrogen. The resulting diradical can then undergo hydroxyl transfer to the substrate carbon. This process has been used to hydroxylate dammarane terpenes related to steroids, by preparing appropriate nitrophenyl esters of the substrates and then photolyzing (Scheme 11).[40,41] In the steroid series a related reaction led to remote dehydrogenation, not hydroxylation.[42] Oxygenation of C—H bonds can also be achieved by photolysis of nitroxides (Scheme 12),[43,44] but so far only at nearby carbons, not remote ones.[44]

1.3.3.2 Template-directed Epoxidation

Although epoxidation reactions are treated in detail elsewhere in these volumes, it should be mentioned here that a template ester attached to a steroid alkene can direct epoxidation to remote double bonds using the general concepts of remote functionalization.[45] Steroidal diene (**5**) underwent the epoxidation shown (Scheme 13) with excellent regiochemical and stereochemical control.[46] The product was formed in quantitative yield, although the reaction was carried through to only 25% conversion.

1.3.3.3 Directed Chlorinations

The major work to date on synthetic applications of remote functionalization has involved free radical chlorination. The earliest studies[8,47] involved the direct attachment of aryliodine dichloride units to the steroid substrates, then intramolecular free radical chain chlorination in benzene or chlorobenzene solution (Scheme 14). Yields were only in the 50% region, but fairly good selectivities were observed; compound (**6**) afforded chiefly the 9-chloro derivative, while compound (**7**) produced the 14-chloro steroid. The yields and selectivities were considerably improved when it was realized that aromatic solvents promote intermolecular random processes by forming complexes with Cl·, and when the radical relay method was developed.

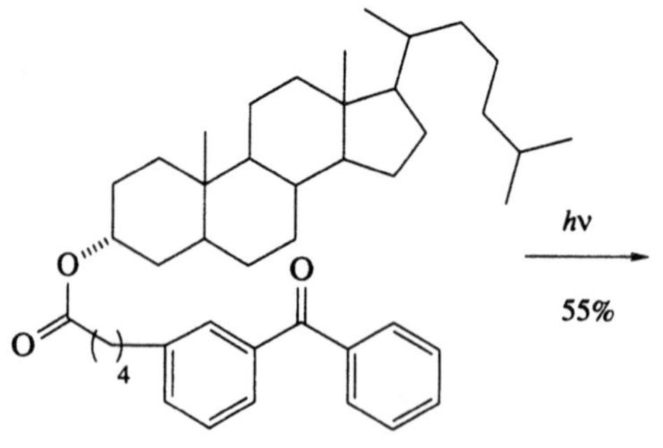

hν
55%

(2)

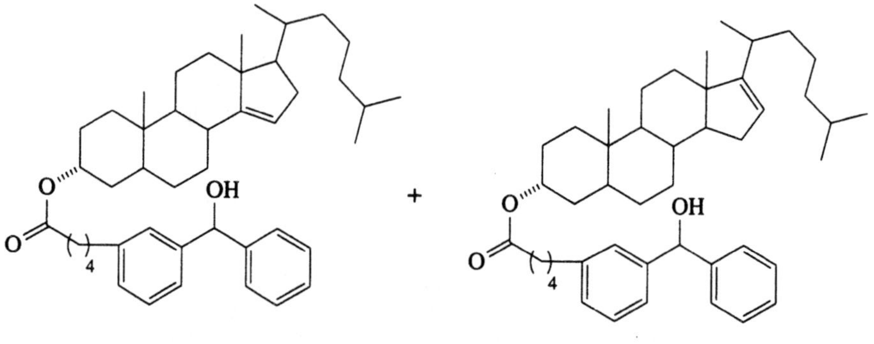

OH

+

OH

43%

57%

hν
55%

OH

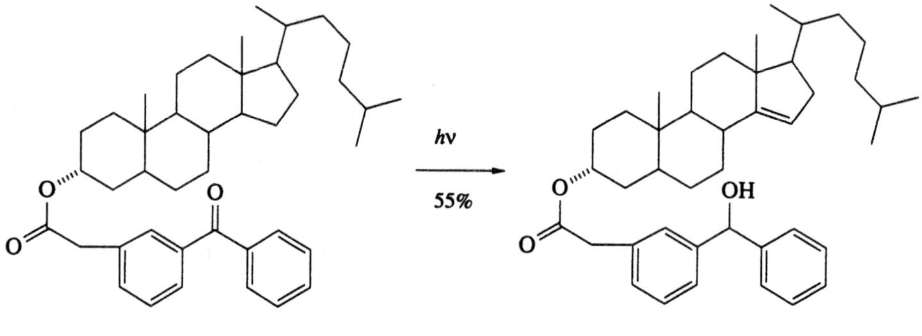

(3)

(4)

only functionalized steroid

Scheme 10

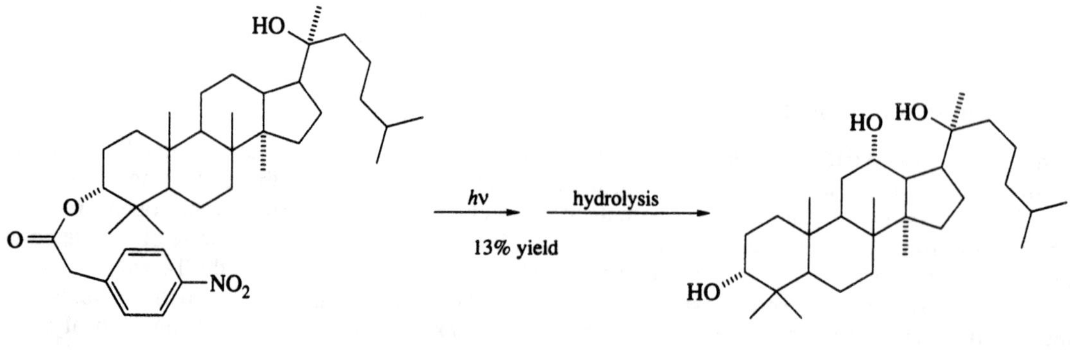

HO

HO HO

NO_2

hν hydrolysis
13% yield

HO

Scheme 11

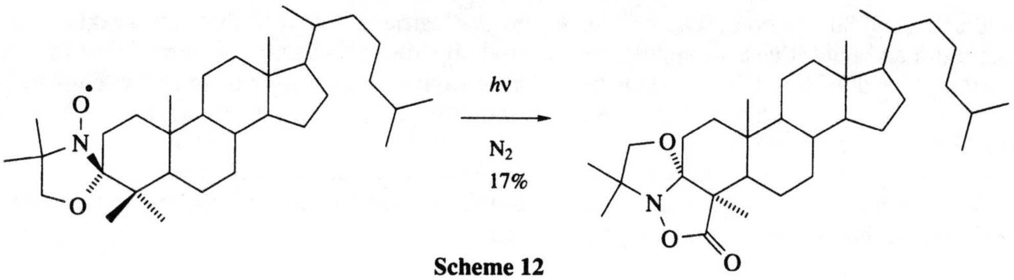

$h\nu$

N_2

17%

Scheme 12

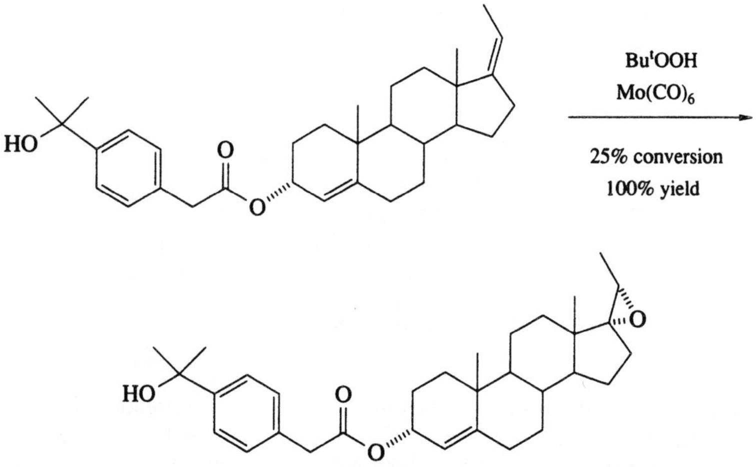

Bu^tOOH

$Mo(CO)_6$

25% conversion

100% yield

Scheme 13

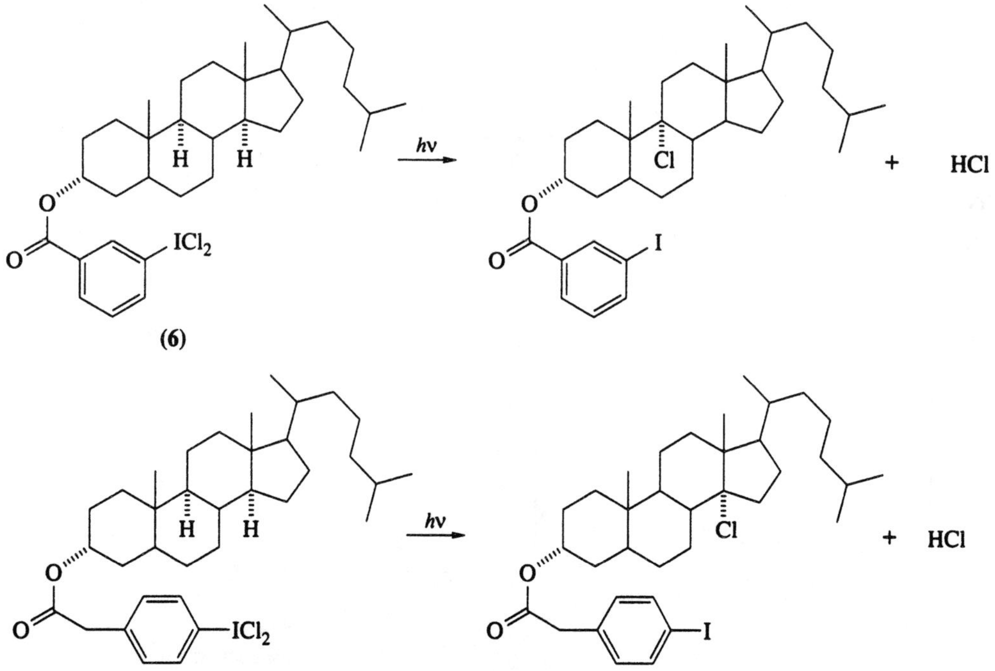

$h\nu$

+ HCl

(6)

$h\nu$

+ HCl

(7)

Scheme 14

In radical relay chlorination (Scheme 15), a substrate carries a template that can weakly bond to a chlorine atom and hold it near an appropriate substrate hydrogen.[48] A chlorine atom donor in solution, such as PhICl· or SO_2Cl· or Cl· itself, puts the chlorine atom on the bonding atom of the template. After the hydrogen atom is removed, under geometric control, the resulting substrate radical picks up a chlorine from the reagent ($PhICl_2$, SO_2Cl_2, or Cl_2) to produce a chlorinated substrate and regenerate the chlorine donor species. Under some conditions the resulting free radical chain process can have a chain length of 20 or so. With substrate concentrations of the order of 10^{-3} to 10^{-2} M there is normally little competition from intermolecular nondirected processes.

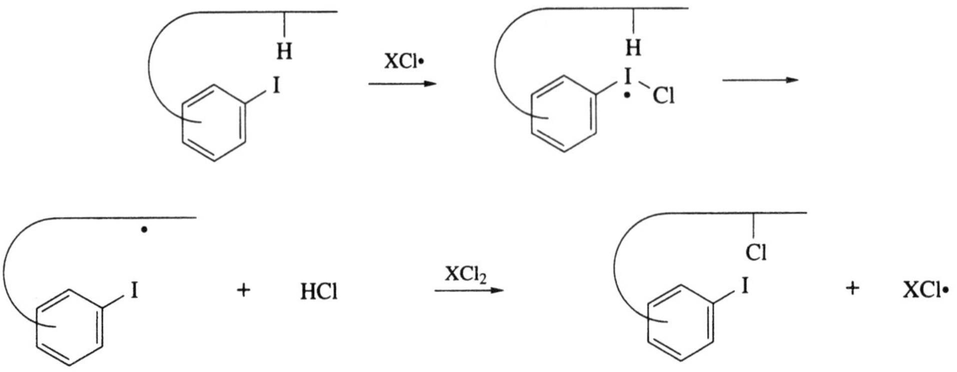

Scheme 15 Radical relay chlorination

Iodoaryl esters of steroids can serve as templates for radical relay chlorination. For instance (Scheme 16), refluxing 10^{-2} M (**8**) in CCl_4 with 1.2 equiv. SO_2Cl_2 and 10 mol % benzoyl peroxide for 5 h, then basic hydrolysis and dehydrochlorination, afforded the $\Delta^{9(11)}$-alkene product (**9**) in 75% yield along with 15% of recovered cholestanol and only 10% of other products.[49,50] These are polar materials derived from further chlorination of alkenes formed *in situ*, and easily separated from the desired product. No isomeric alkenes were detected. As another example, the single template in compound (**10**) catalyzes and

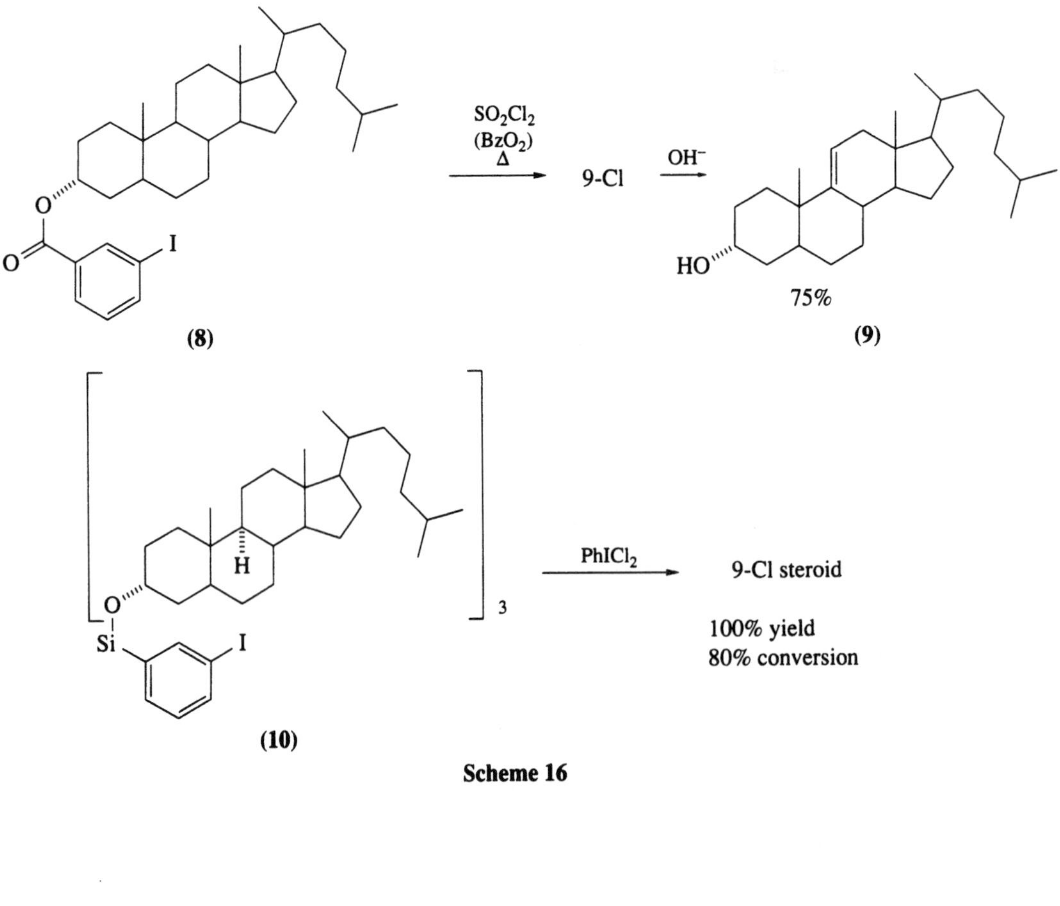

Scheme 16

directs the chlorination of three steroid substrates at C-9 under similar conditions, in quantitative yield and *ca.* 80% conversion.[51] This is possible since the template is regenerated after each functionalization, and the selectivity is so good that once a steroid is chlorinated it does not get attacked in another position.

Longer templates promote chlorination over greater distances. In compound (**11**; Scheme 17) the biphenyl template promotes chlorination of C-17 with good selectivity.[50,52] Dehydrochlorination forms the Δ^{16}-alkene (**12**) in reasonable (66%) yield; this has been used in an indirect scheme to remove the side chain of cholesterol and of sitosterol to afford the 17-keto steroid. Under other conditions the 17-chloro steroids can be dehydrochlorinated toward C-20, and the resulting $\Delta^{17(20)}$-alkenes directly oxidized to afford the 17-keto steroid.[53,54]

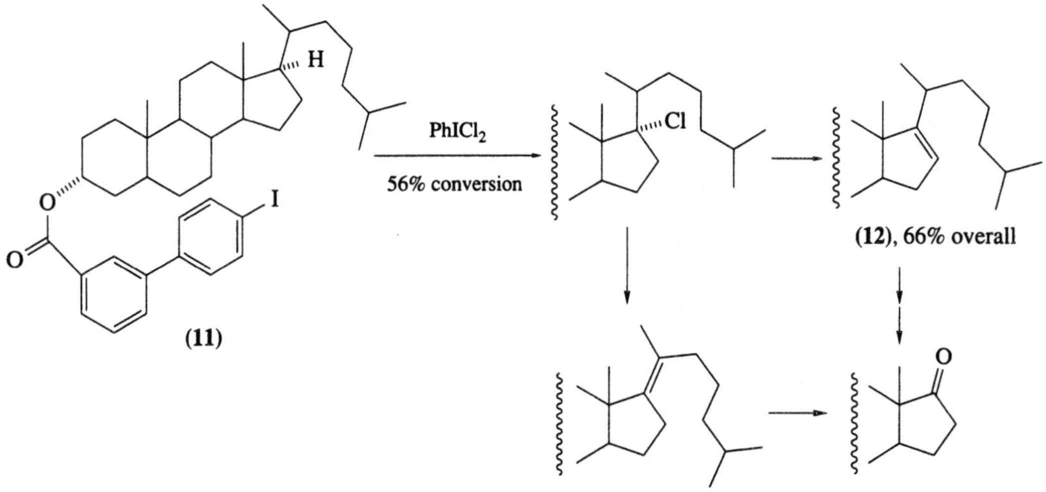

PhICl$_2$
56% conversion
(11)
(12), 66% overall

Scheme 17

The selectivity of the radical relay chlorination is striking. In the case of the enone (**13**; Scheme 18), and in related compounds with A-ring dienones, the *m*-iodobenzoate template at C-17 directs chlorination to C-9 and not to the preexisting functional groups of (**13**).[50] The selective chlorination of C-9 seems to be quantitative, although in the first report[50] the $\Delta^{9(11)}$-alkene (**14**) was isolated in only 77% yield. Later work has shown that the overall introduction of this double bond can have yields in the 90–95% range, and good yields for this reaction have also been reported from another laboratory.[55] Template-directed radical relay chlorination on the α-face of steroids has also been successful in the A/B *cis*-coprostanol steroid series,[56] and in the cholestanol series with iodophenyl templates linked by amide, ether, or sulfonate functions rather than carboxylic esters.[57]

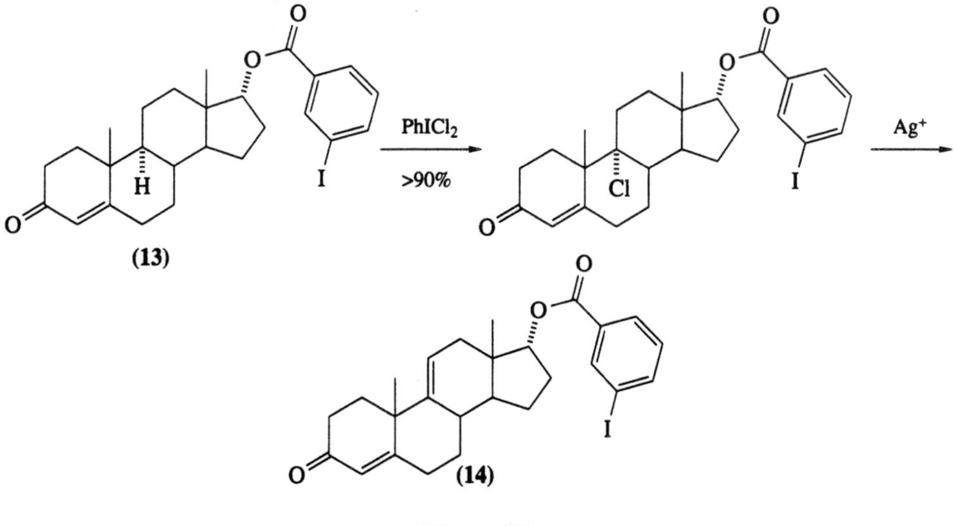

PhICl$_2$
>90%
Ag$^+$
(13)
(14)

Scheme 18

Limited studies have been done on template-directed chlorination on the β-face of steroids. Compound (**15**; Scheme 19) was designed so that the template could curve around the angular C-18 methyl group and direct chlorination to C-20.[58] Reaction with an excess of $PhICl_2$ led to *ca.* 40% chlorination of C-20 with 25% unfunctionalized steroid. The 20-chloro steroid was converted in part to the $\Delta^{20(22)}$-alkene, which was ozonized to form the 17-acetyl steroid (**16**). A similar result was observed with the i-steroid derivative (**17**).[58] The selectivities and yields are not yet up to those of other examples of the radical relay reaction.

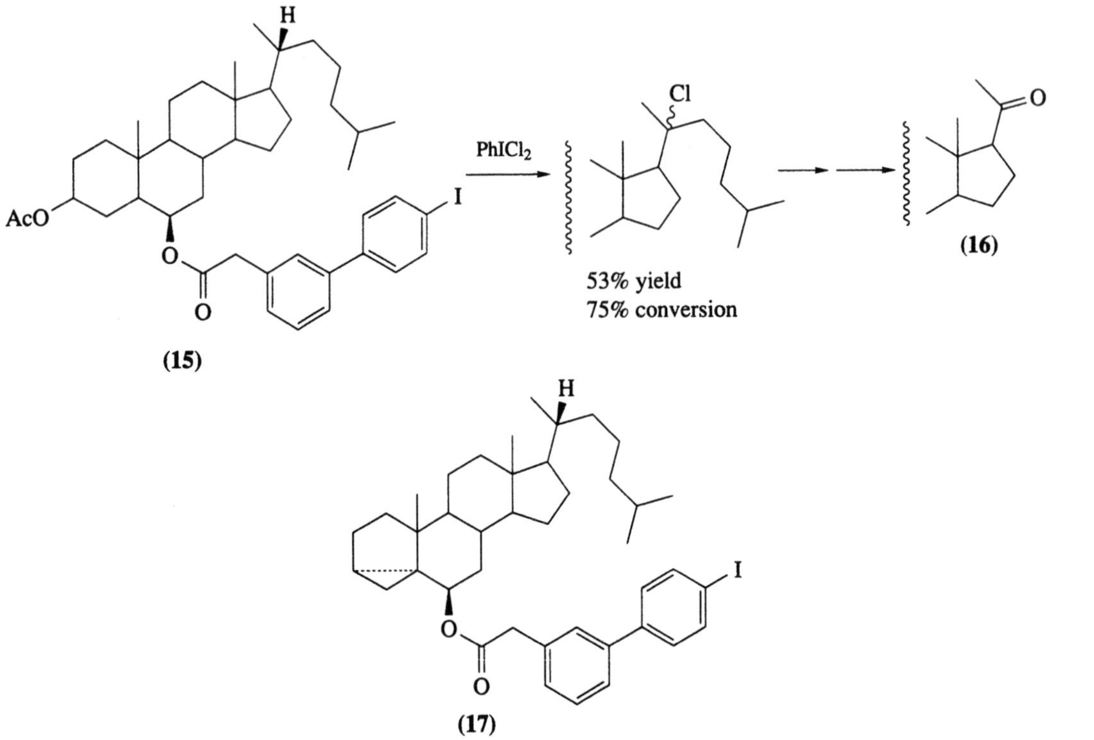

(**15**) → $PhICl_2$ → 53% yield, 75% conversion → (**16**)

(**17**)

Scheme 19

Diaryl sulfide templates have also been used to direct chlorinations.[59] The selectivities indicate that the chlorine atom is bound to the sulfur, but the yields are not as good as those with aryl iodide templates. The problem is that the sulfur gets oxidized under the reaction conditions. As expected, a thiophene ring is more stable to oxidation and its sulfur atom can still bind chlorine in a radical relay process.[60] The best sulfur template so far examined is the thioxanthone system (Scheme 20).[61] Thus with 3 equiv. $PhICl_2$ compound (**18**) undergoes directed C-9 chlorination in 100% conversion, affording a 71% yield of the $\Delta^{9(11)}$-alkene after base treatment, along with some polar products from excessive chlorination. The thioxanthone template can be recovered unchanged.

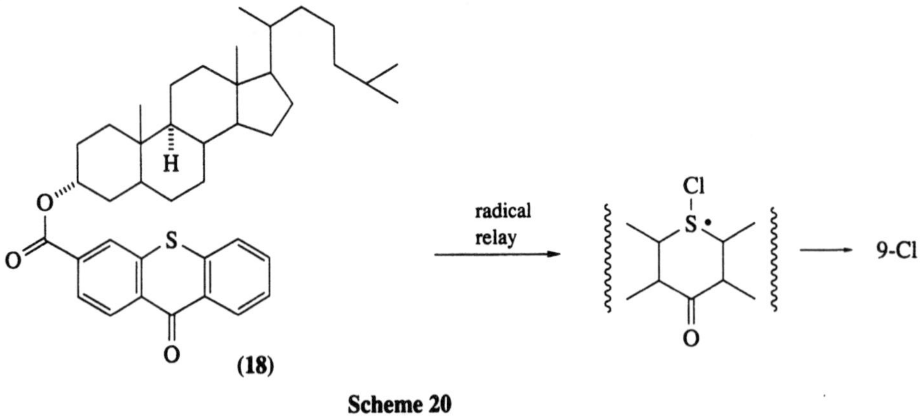

(**18**) → radical relay → 9-Cl

Scheme 20

Binding of Cl· to an aryl iodide may well involve sp^3d hybridization at iodine to accommodate the ninth electron, but the involvement of a *d*-orbital in bonding at sulfur is more controversial. Recently it was discovered that even first row elements can form Cl· complexes; the evidence indicates that these complexes utilize three-electron bonds, not *d*-orbitals.[62] Best explored are templates for radical relay chlorination using nitrogen atoms.

As a striking example, photo-initiated chlorination (Scheme 21) of 3 mM (**19**) with 1.5 equiv. $PhICl_2$ led to the 9-chloro derivative (**20**) in >98% yield; with Ag^+ this was converted to the $\Delta^{9(11)}$-alkene.[63] Again the template-directed reaction overcomes the normal reactivity of the substrate, but at 21 mM (**19**) undirected reactions start to compete and some 6-chloro steroid is also formed. A pyridine *N*-oxide template, that can use three-electron bonding to complex a chlorine to the oxygen atom, seems to be almost as effective.[64] Furthermore, an imidazole template in compound (**21**) directs chlorination at C-9 with similar efficiency to the templates previously examined,[65] and (**21**) is particularly easily prepared using carbonyldiimidazole.

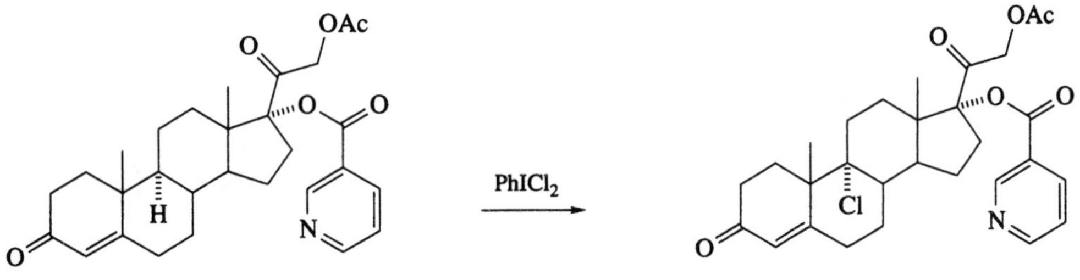

PhICl₂

(**19**)

(**20**) >98% yield

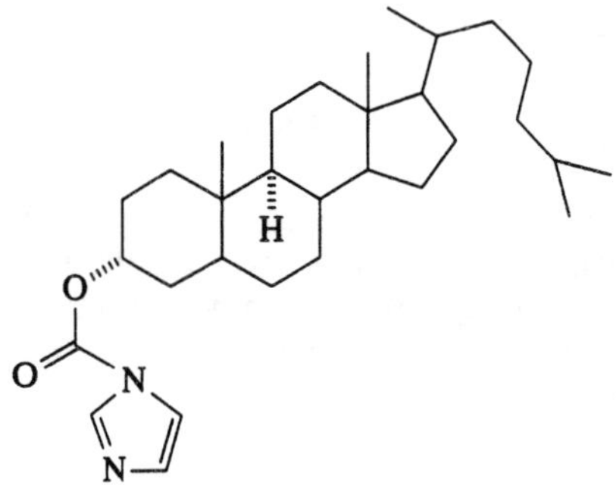

(**21**)

Scheme 21

1.3.4 SELECTIVE REACTIONS IN MOLECULAR COMPLEXES

There is no good reason that a catalytic template, which directs remote functionalization reactions, need be covalently attached to the substrate. Indeed, it would be preferable to use catalytic amounts of such a template that could bind temporarily to a substrate, perform its reaction, and then move on.

A good example of such a process is the template-directed chlorination of an aromatic ring by β-cyclodextrin (Scheme 22).[66–68] Hydrophobic forces hold the complex together temporarily, and within the complex the chlorination is catalyzed and directed by a hydroxy group of the cyclodextrin. An electro-

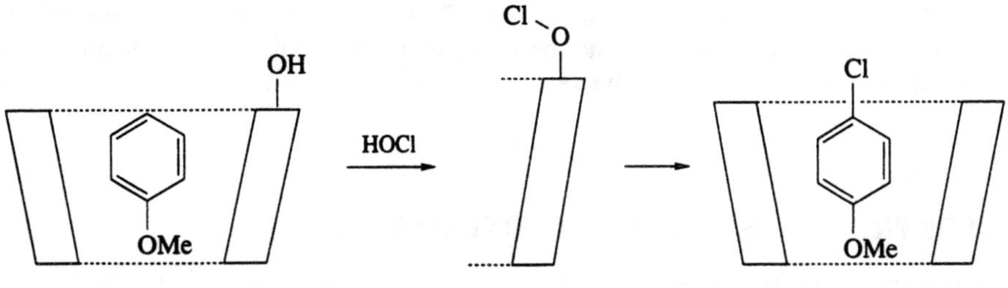

HOCl

Scheme 22

chemical version of this has been devised, in which the cyclodextrin is chemically linked to an electrode and the chlorinating species is generated by anodic oxidation of chloride ion.[69] Other related reactions have also been observed in which cyclodextrin binding is used to direct functionalization chemistry.[70]

The shape-selective metalloporphyrin-catalyzed oxidations of hydrocarbons studied by Suslick are also relevant,[71,72] although the binding forces and geometry are less obvious. Groves' recent hydroxylation of steroids in a bilayer containing a metalloporphyrin (Scheme 23) is also clearly in the spirit of biomimetic chemistry.[73] In this case hydrophobic binding produces a complex with predictable geometry.

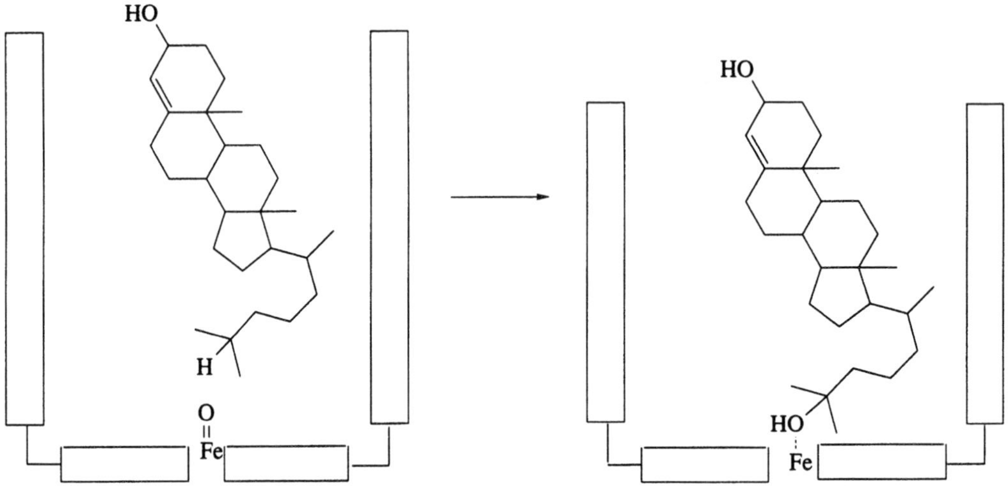

HO H O Fe → HO HO Fe

Scheme 23

Template-catalyzed remote chlorination reactions have also been examined in molecular complexes. In one early study (Scheme 24), ion pairing was used to hold a charged template near a charged substrate.[74] Selective catalyzed radical relay chlorination was observed, but the selectivity was not as good as has been seen when the template is covalently attached to the substrate. In more recent work better selectivity and some catalytic turnover has been observed.

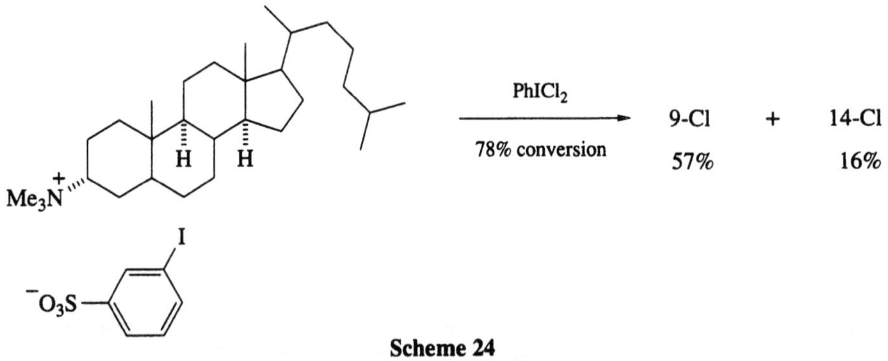

$Me_3\overset{+}{N}$ H H I $^{-}O_3S$ —($PhCl_2$, 78% conversion)→ 9-Cl (57%) + 14-Cl (16%)

Scheme 24

Catalytic turnover has also been seen in radical relay chlorinations in which the template is temporarily linked to the substrate in a mixed metal complex. The steroid phosphate (**22**) and catalytic ligand (**23**) both bind to zinc in a mixed complex, and the iodine atom of (**23**) directs chlorination of (**22**) at C-9 with reasonable selectivity (Scheme 25). Five or more turnovers are seen, when only 10% of the catalyst (**23**) is used.[75]

1.3.5 FUTURE PROSPECTS FOR REMOTE OXIDATIONS

Most of the examples so far have utilized a single covalent bond, or an ionic or ligand bond, to hold the substrate to a catalytic template that can direct chlorination of a remote position. To achieve highly selective reactions several interactions are needed to impose strong geometric constraints, as in the

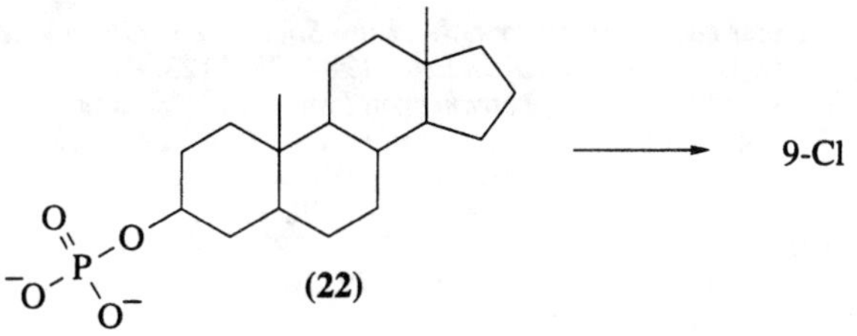

→ 9-Cl

(22)

Zn^{2+}

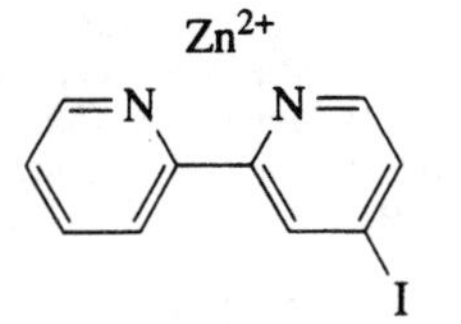

(23)

Scheme 25

double ion pair of Scheme 9. To justify the complex template catalyst that this implies, multiple turnovers are needed so the catalyst can be used in truly catalytic amounts. Furthermore, functionalizations other than radical relay chlorinations are of interest. As these techniques develop, the methods outlined here may become ever more useful in synthesis.

1.3.6 REFERENCES

1. V. Ramamurthy, *Tetrahedron*, 1986, **42**, 5753.
2. Z. Cohen, E. Keinan, Y. Mazur and T. H. Varkony, *J. Org. Chem.*, 1975, **40**, 2141.
3. A. L. J. Beckwith and T. Duong, *J. Chem. Soc., Chem. Commun.*, 1978, 413.
4. R. L. Wife, D. Kyle, L. J. Mulhern and H. C. Volger, *J. Chem. Soc., Chem. Commun.*, 1982, 306.
5. Z. Cohen, E. Keinan, Y. Mazur and A. Ulman, *J. Org. Chem.*, 1976, **41**, 2651.
6. J. van Dijk, J. J. van Daalen and G. B. Paerels, *Recl. Trav. Chim. Pays-Bas*, 1974, **93**, 72.
7. N. Friedman, M. Lahav, L. Leiserowitz, K. Popovitcz-Biro, C.-P. Tang and Z. Zaretzkii, *J. Chem. Soc., Chem. Commun.*, 1975, 864.
8. R. Breslow, J. A. Dale, P. Kalicky, S. P. Liu and W. N. Washburn, *J. Am. Chem. Soc.*, 1972, **94**, 3276.
9. A. Rotman and Y. Mazur, *J. Chem. Soc., Chem. Commun.*, 1974, 15.
10. Y. Mazur and Z. Cohen, *Angew. Chem., Int. Ed. Engl.*, 1978, **17**, 281.
11. D. H. R. Barton, R. S. Hay-Motherwell and W. B. Motherwell, *Tetrahedron Lett.*, 1979, 24.
12. C. J. Sih, *Science (Washington, D.C.)*, 1969, **163**, 1297.
13. R. Breslow, *Chem. Soc. Rev.*, 1972, **1**, 553.
14. Y. L. Chow, W. C. Danen, S. F. Nelsen and D. H. Rosenblatt, *Chem. Rev.*, 1978, **78**, 243.
15. C. Betancor, J. I. Concepcion, R. Hernandez, J. A. Salazar and E. Suárez, *J. Org. Chem.*, 1983, **48**, 4430.
16. C. Walling and D. Bristol, *J. Org. Chem.*, 1972, **37**, 3514.
17. R. Criegee, in 'Oxidation in Organic Chemistry', ed. K. B. Wiberg, Academic Press, New York, 1965, part A, chap. 5.
18. S. G. Wilkinson, in 'Comprehensive Organic Chemistry', ed. D. H. R. Barton and W. D. Ollis, Pergamon Press, Oxford, 1979, vol. 1, p. 653.
19. J. Kalvoda and K. Heusler, *Synthesis*, 1971, 501.
20. C. Meystre, K. Heusler, J. Kalvoda, P. Wieland, G. Anner and A. Wettstein, *Experientia*, 1961, **17**, 475.
21. R. H. Hesse, *Adv. Free-Radical Chem.*, 1969, **3**, 83.
22. D. H. R. Barton, N. K. Basu, M. J. Day, R. H. Hesse, M. M. Pechet and A. N. Starratt, *J. Chem. Soc., Perkin Trans. 1*, 1975, 2243.
23. Z. Cekovic and T. Srnic, *Tetrahedron Lett.*, 1976, 561.
24. J. Kalvoda, *J. Chem. Soc., Chem. Commun.*, 1970, 1002.
25. N. J. Turro, in 'Modern Molecular Photochemistry', Benjamin/Cummings, Menlo Park, CA, 1978, p. 386.
26. P. J. Wagner, P. A. Kelso, A. E. Kemppainen and R. G. Zepp, *J. Am. Chem. Soc.*, 1972, **94**, 7500.
27. H. J. Roth and M. H. El Raie, *Tetrahedron Lett.*, 1970, 2445.
28. Y. Kanaoka, Y. Migita and K. Koyama, *Tetrahedron Lett.*, 1973, 1193.
29. Y. Sato, H. Nakai, H. Ogiwara, T. Mizoguchi, Y. Migita and Y. Kanaoka, *Tetrahedron Lett.*, 1973, 4565.
30. L. L. Miller and V. Ramachandran, *J. Org. Chem.*, 1974, **39**, 369.
31. S. Hammerum and O. Hammerich, *Tetrahedron Lett.*, 1979, 5027.
32. R. Breslow and M. Winnik, *J. Am. Chem. Soc.*, 1969, **91**, 3083.
33. B. Dors, H. Luftmann and H. J. Schäfer, *Chem. Ber.*, 1983, **116**, 761.
34. R. Breslow, R. Rajagopalan and J. Schwarz, *J. Am. Chem. Soc.*, 1981, **103**, 2905.
35. R. Breslow and S. W. Baldwin, *J. Am. Chem. Soc.*, 1970, **92**, 732.
36. R. Breslow, S. W. Baldwin, T. Flechtner, P. Kalicky, S. Liu and W. N. Washburn, *J. Am. Chem. Soc.*, 1973, **95**, 3251.

37. J. F. Baldwin, A. D. Bhatnagar and R. W. Harper, *J. Chem. Soc., Chem. Commun.*, 1970, 659.
38. R. Breslow, U. Maitra and D. Heyer, *Tetrahedron Lett.*, 1984, **25**, 1123.
39. E. Lee, H. Lee, H. K. Chang and D. Y. Lim, *Tetrahedron Lett.*, 1988, **29**, 339.
40. R. Kasai, K. Shinzo, O. Tanaka and K.-I. Kawai, *Chem. Pharm. Bull.*, 1974, **22**, 1213.
41. R. Kasai, K. Shinzo and O. Tanaka, *Chem. Pharm. Bull.*, 1976, **24**, 400.
42. P. C. Scholl and M. R. Van De Mark, *J. Org. Chem.*, 1973, **38**, 2376.
43. J. F. W. Keana, R. J. Dinerstein and F. Batis, *J. Org. Chem.*, 1971, **36**, 209.
44. D. L. Smith and T. A. Spencer, *J. Heterocycl. Chem.*, 1979, **16**, 807.
45. J. R. Knowles, *Acc. Chem. Res.*, 1972, **5**, 155.
46. R. Breslow and L. M. Maresca, *Tetrahedron Lett.*, 1977, 623.
47. R. Breslow, R. J. Corcoran, J. A. Dale, S. Liu and P. Kalicky, *J. Am. Chem. Soc.*, 1974, **96**, 1973.
48. R. Breslow, R. J. Corcoran and B. B. Snider, *J. Am. Chem. Soc.*, 1974, **96**, 6791.
49. R. Breslow, B. B. Snider and R. J. Corcoran, *J. Am. Chem. Soc.*, 1974, **96**, 6792.
50. R. Breslow, R. J. Corcoran, B. B. Snider, R. J. Doll, P. L. Khanna and R. Kaleya, *J. Am. Chem. Soc.*, 1977, **99**, 905.
51. R. Breslow and D. Heyer, *J. Am. Chem. Soc.*, 1982, **104**, 2045.
52. B. B. Snider, R. J. Corcoran and R. Breslow, *J. Am. Chem. Soc.*, 1975, **97**, 6580.
53. P. Welzel, K. Hobart, A. Ponty and T. Milkova, *Tetrahedron Lett.*, 1983, **24**, 3199.
54. R. Breslow and U. Maitra, *Tetrahedron Lett.*, 1984, **25**, 5843.
55. U. Kerb, M. Stahnke, P-E. Schulze and R. Wiechert, *Angew. Chem., Int. Ed. Engl.*, 1981, **20**, 88.
56. R. J. Corcoran, *Tetrahedron Lett.*, 1976, 317.
57. D. Wolner, *Tetrahedron Lett.*, 1979, 4613.
58. U. Maitra and R. Breslow, *Tetrahedron Lett.*, 1986, **27**, 3087.
59. R. Breslow, R. L. Wife and D. Prezant, *Tetrahedron Lett.*, 1976, 1925.
60. R. Breslow and D. Heyer, *J. Am. Chem. Soc.*, 1982, **104**, 2045.
61. R. Breslow and T. Guo, *Tetrahedron Lett.*, 1987, **28**, 3187.
62. R. Breslow, M. Brandl, J. Hunger, N. J. Turro, K. C. Cassidy, K. Krogh-Jespersen and K. Westbrook, *J. Am. Chem. Soc.*, 1987, **109**, 7204.
63. R. Breslow, M. Brandl, J. Hunger and A. D. Adams, *J. Am. Chem. Soc.*, 1987, **109**, 3799.
64. R. Breslow, A. Adams, M. Brandl, T. Guo and J. Hunger, *Lect. Heterocycl. Chem.*, 1987, **9**, 43.
65. R. Batra and R. Breslow, *Heterocycles*, 1989, **28**, 23.
66. R. Breslow and P. Campbell, *J. Am. Chem. Soc.*, 1969, **91**, 3085.
67. R. Breslow and P. Campbell, *Bioorg. Chem.*, 1971, **1**, 140.
68. R. Breslow, H. Kohn and B. Siegel, *Tetrahedron Lett.*, 1976, 1645.
69. T. Matsue, M. Fujihira and T. Osa, *J. Electrochem. Soc.*, 1979, **126**, 500.
70. H.-J. Schneider and K. Philippi, *Chem. Ber.*, 1984, **117**, 3056.
71. B. R. Cook, T. J. Reinert and K. S. Suslick, *J. Am. Chem. Soc.*, 1986, **108**, 7281.
72. K. S. Suslick and B. R. Cook, *J. Chem. Soc., Chem. Commun.*, 1987, 200.
73. J. T. Groves and R. Neumann, *J. Org. Chem.*, 1988, **53**, 3891.
74. R. Breslow and D. Heyer, *Tetrahedron Lett.*, 1983, **24**, 5039.
75. M. Sprecher and M. Kotera, unpublished work.

1.4
Oxidation by Microbial Methods

STEPHEN M. BROWN
ICI Specialties Research Centre, Manchester, UK

1.4.1 INTRODUCTION

Microorganisms have the ability to effect chemical oxidations on a wide variety of substrates, many of which occur with chemo-, regio- or stereo-selectivities unattainable by conventional chemical methods. This is particularly true in the case of unactivated C—H oxidation. What chemical oxidation system, for example, would be capable of the stereoselective hydroxylation of the diol (**1**) to the 3β-hydroxy derivative (**2**)? This transformation has been reported to occur in 80% yield when a microbial oxidation is employed.[1]

Preparative microbial oxidations have long been practiced in organic synthesis, perhaps most prominently in the steroidal field, and a number of comprehensive and specialized reviews have appeared. The most recent review,[2] published in 1981, covers most aspects of biochemical oxidations, and gives an ex-

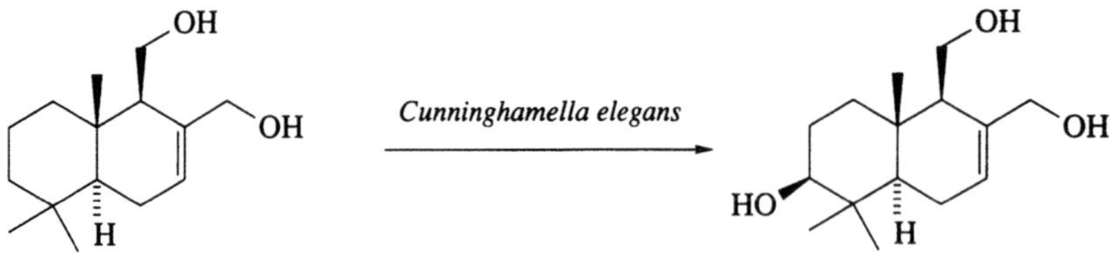

Cunninghamella elegans (1)

(1) (2)

cellent bibliography of previous work and of books relevant to fermentation and microbial transformations in general. Other treatise giving comprehensive coverage of the scope and practical aspects of microbial oxidations (bacteria, fungi, yeasts and spores) alone have also appeared.[3–6]

Despite their well-researched and impressive capabilities, the use of microbial oxidations has not yet become commonplace in organic synthesis. Unfamiliarity and the difficulty in predicting when the use of a microorganism would be advantageous (except of course when there is no clear choice of chemical oxidant), and, if so, which one(s) should be used, are all impediments to wider use. However, with the advent of genetic engineering and increasing emphasis being placed upon the need for single enantiomers of chiral materials, microbial methods are likely to play an increasingly important role in organic synthesis.

A fundamental difficulty with microbial oxidations of unactivated C—H groups is that the dominant factors in controlling the site and extent of the reaction are often steric in nature, whereas in chemical oxidations electronic factors are often more important. Thus microbial and chemical oxidations can rarely be equated, and microbial oxidations cannot be treated directly on a functional group basis. Also, following from this, microbes cannot be considered as reagents; although some classes do show certain characteristics as to the type of hydroxylation effected, in many cases the site of hydroxylation, and even if hydroxylation will take place at all, is dependent upon the nature of the substrate and the strain of microbe used. Thus, unlike chemical oxidations where tens of oxidants are available (although fewer than this are useful for oxidation of unactivated C—H), literally hundreds of microorganisms have been employed, and many thousands are available.

However, the selection of a microorganism capable of bringing about a specific reaction on a new substrate is performed initially by seeking literature analogies. For example the conversion of cinerone (**3**) to the cinerolone (**4**) by *Aspergillus niger*[7] was used as the precedent which led to the stereospecific hydroxylation of the cyclopentenone (**5**) to the prostaglandin synthon (**6**) by the same microorganism (Scheme 1).[8]

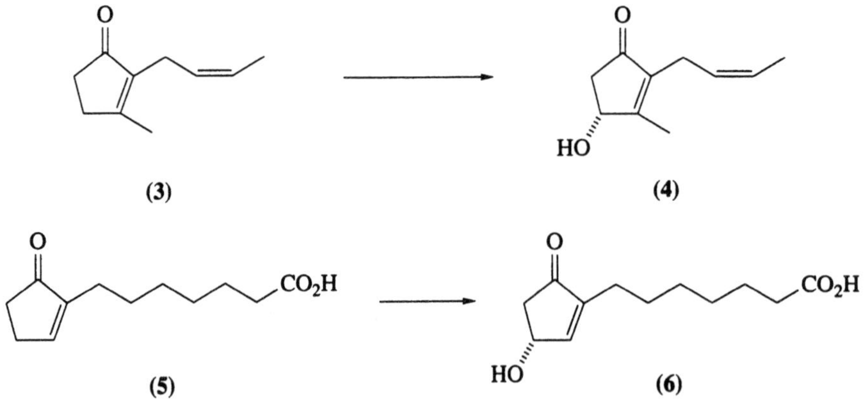

(**3**) (**4**)

(**5**) (**6**)

Scheme 1

In the following sections some of the general characteristics of microbes, and of oxidation of C—H bonds that can be effected by them, are discussed. Because microbes cannot be treated strictly as reagents, the organization of sections is largely on a substrate type basis, with the main division being between nonsteroidal and steroidal substrates.

1.4.2 GENERAL ASPECTS OF THE USE OF MICROORGANISMS

Few synthetic organic chemists are acquainted with the nomenclature or the use of microorganisms, and consequently there is a need to provide some guidance on where to start.

1.4.2.1 Taxonomy of Microorganisms

Microorganisms are classified according to genus and species. The first name of the microorganism is the genus, and the second is the species of that genus to which it belongs. Comprehensive lists of microorganisms have been published[5,9,10] and the detailed organization of genera into families, orders and classes has also been reviewed.[11]

1.4.2.2 Sources of Cultures

Microorganisms can be acquired from various sources and in many cases several strains or mutants will be available for each species of microorganism. In the more recent literature the precise identity (name, source and number) of the microorganisms used are normally stated, and Table 1 lists the more common reference collections. More extensive listings are available.[9] There may be considerable differences in the nature and yield of products from a given microbial oxidation if two different strains of the same species are used, and for this reason older work may be difficult to reproduce.

Table 1 Sources of Supply of Microorganisms

Abbreviation	*Source*
ATCC	American Type Culture Collection, Rockville, MD, USA
CBS	Centraalbureau voor Schimmelcultures, Baarn, The Netherlands.
CMI	Commonwealth Mycological Institute, Kew, Surrey, UK
DSM (or GCM)	German Collection of Microorganisms, Soc. Invest. Biotechnol., Gottingen, Germany
FERM	Fermentation Research Institute, Ibaraki, Japan
IAM	Institute of Applied Microbiology, University of Tokyo, Japan
IFO	Institute for Fermentation, Osaka, Japan
NCIB	National Collection of Industrial Bacteria, Torry Research Station, Aberdeen, UK
NCTC	National Collection of Type Cultures, Central Public Health Laboratory, London, UK
NCYC	National Collection of Yeast Cultures, Brewing Industry Research Foundation, Nutfield, Surrey, UK
NRRL	Culture Collection Unit, Northern Utilization Research Branch, (formerly Northern Regional Research Laboratories), US Department of Agriculture, Peoria 5, IL, USA

In many cases microorganisms that will carry out a required transformation have been selected after screening many naturally occurring colonies. This approach is tedious and lengthy and needs to go hand in hand with the classification of those microorganisms giving the best results. For these reasons this approach is not to be recommended unless no precedents for the required transformation exist.

1.4.2.3 Use of Microorganisms

From many experiments with many different types of substrate and oxidation reaction, it is possible to compile a list of microorganisms that have shown considerable versatility and reliability, and which provide a good starting point for attempting a new oxidation, and these are given in Table 2. More extensive listings are available.[5,9] From this starting point, optimization of the yield and selectivity can be achieved by screening an ever broader range of strains and alternative microorganisms, and in the past it has not been unusual for many hundreds of these to have been assessed for a particularly rare or difficult transformation (one project reportedly screened 2000 strains of microorganisms). The ultimate objective is to identify a particular microorganism that can accomplish the desired oxidation in reasonable yield. If a low yield or selectivity is obtained (even at a few percent the yield may be far superior to that from the alternative chemical method) it may be possible to enhance this by selective mutation (for example by exposure of the organism to radiation or a chemical mutagenic agent). The mutation and selection of improved strains of microorganisms is a science in itself, and is beyond the scope of this work and has been discussed elsewhere.[12]

Table 2 Some Microorganisms Commonly Used for Oxidation of Unactivated C—H Bonds

Microorganism	*Class*	*Some applications*
Beauveria sulfurescens[a] (ATCC 7159)	Fungi imperfecti	11α-Hydroxylation of steroids, hydroxylation of cyclic and acyclic amides, tricyclic terpenes and aromatics
Rhizopus nigricans[b] (ATCC 6227b)	Phycomycetes	11α-Hydroxylation of steroids, hydroxylation of diterpenes
Calonectria decora (ATCC 14767, NRRL 2380)	Ascomycetes	Steroid hydroxylation and dehydrogenation, hydroxylation of cyclic hydrocarbons
Curvularia lunata (ATCC 12017)	Fungi imperfecti	Steroid 11β-hydroxylation
Aspergillus niger	Fungi imperfecti	Very general use
Cunninghamella elegans (NRRL 1393)	Phycomycetes	Steroid hydroxylation and allylic hydroxylation
Pseudomonas putida	Schizomycetes	Unactivated C—H hydroxylation

[a] This is the most commonly used name for this microorganism, but it has recently been reclassified as *Beauveria bassiana*. In earlier literature it is referred to as *Sporotrichum sulfurescens*. [b]This is the most commonly used name for this microorganism, but it has recently been reclassified as *Rhizopus stolonifer*.

By their very nature microbial oxidations require mild, usually near physiological, conditions; this can be both an advantage and a disadvantage. For example, the mild conditions often enable sensitive functional groups to remain unaltered during the oxidation (*e.g.* **1** above), but at the same time the use of mainly aqueous conditions precludes water sensitive substrates and can limit both the concentration of substrate and complicate the recovery of products. Experimental aspects of the use of microorganisms in oxidations have been discussed previously,[4,11–13] and these sources of information will provide all that is necessary to be able to carry out a laboratory microbial oxidation.

1.4.3 MICROBIAL OXIDATION OF NONSTEROIDAL SUBSTRATES

1.4.3.1 Acyclic Hydrocarbons and Their Functionalized Derivatives

Microbial oxidation of alkanes can take place at the terminal carbon, in which case an alcohol is the initial product, or at a subterminal position (often the β-position) to give either the secondary alcohol or a ketone. In both cases further oxidation[14] can take place to give carboxylic acids, themselves liable to β-oxidation and shortening of the carbon chain by successive two-carbon units (Scheme 2).

The mechanisms of these oxidations and the nature of the various enzyme systems involved have been discussed previously,[14] as have some examples of these types of microbial reactions.[15]

Mutation of the microorganism can lead to the blocking of undesirable secondary metabolism (a problem most serious for simple *n*-alkanes). For example a mutant of *Candida cloacae* (M-1) converted *n*-alkanes to α,ω-dicarboxylic acids, and up to 30 g l^{-1} of the products could be accumulated in high yield.[16]

The fungi *Torulopsis gropengiesseri* and *Torulopsis apicola* have the ability to incorporate the primary oxidation products of alkanes and their derivatives into extracellular glycolipids, thus reducing secondary oxidation. Thus hydroxylation of long chain esters or amides can result in preparatively useful product yields (equation 2).[3] However simple alkanes, alcohols, halides or ethers as substrates can result in significant competitive α-oxidation at one or both ends of the carbon chain, to give primary alcohols and/or carboxylic acids.

β-Hydroxylation of short chain aliphatic carboxylic acids can be accomplished by a number of microorganisms[17] (see also Scheme 3) including *Endomyces reessii, Trichosporum fermentans, Torulopsis candida* and *Micrococcus flavus*. Longer chain carboxylic acids can also be satisfactorily hydroxylated.[18]

Many other microorganisms have been studied in relation to the degradation of environmental hydrocarbon pollutants; however the identification of metabolites and elucidation of metabolic pathways has often been the prime concern of these studies, rather than the development of synthetic methods. Methylotropic bacteria are noted for their ability to degrade hydrocarbons and in some cases intermediate hydroxylated products can be recovered. Patel and coworkers have done much of the work in this area[19] and systems capable of converting *n*-alkanes to secondary alcohols[20] or ketones[21] have been developed.

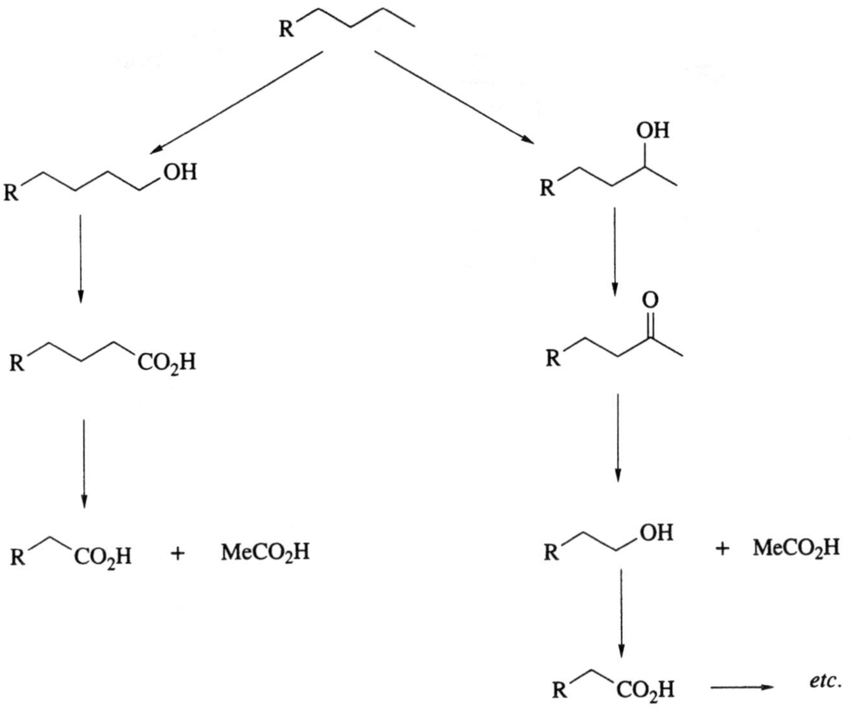

Scheme 2

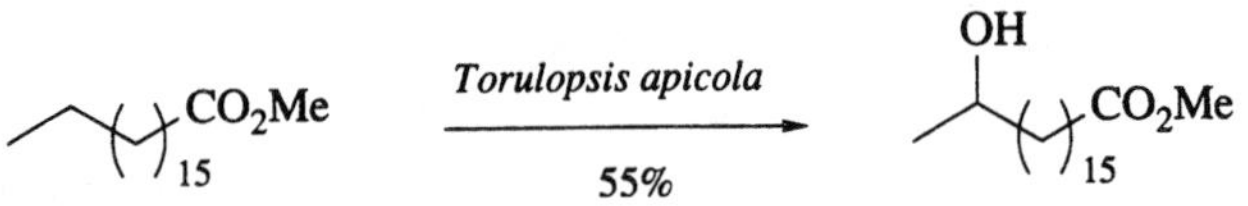

$$\text{CH}_3\text{CH}_2(\text{CH}_2)_{15}\text{CO}_2\text{Me} \xrightarrow[\text{55\%}]{\textit{Torulopsis apicola}} \text{CH}_3\text{CH(OH)}(\text{CH}_2)_{15}\text{CO}_2\text{Me} \quad (2)$$

Species of the genera *Methylosinus* or *Methylococcus* are commonly used, and optically active derivatives can be produced in some cases.[22]

In many cases enantiospecific or enantioselective oxidation of acyclic hydrocarbons or their derivatives is possible using microorganisms, although in few cases has the extent of optical induction been accurately quantified. This is clearly an area where more work is required. In those cases studied so far hydroxylation has been found to occur with retention of configuration at the reacting carbon.

The enantiotopic discrimination of hydrogens during oxidation of unactivated C—H bonds by microorganisms is synthetically extremely useful, and some examples are shown in Scheme 3.[28–31] The resultant products are valuable chiral synthons. For example (*R*)-3-hydroxybutanoic acid (**7**) a versatile homochiral synthon, can be used in the synthesis of antibacterials.[23,24] (*S*)-2-Methyl-3-hydroxypropanoic acid (**8**) has been widely employed as a source of chirality, for example in the synthesis of maysine,[25] macrolide antibiotics[26] and both (*R*)- and (*S*)-muscone.[27] A variety of other optically active 3-hydroxy aliphatic carboxylic acids can be prepared by analogous methods.[18]

In the case of 2-arylpropanoic acids, although the (*S*)-enantiomer (**9**) is available by a terminal oxidation, the alternative (*R*)-enantiomer (**11**) can be prepared by the more extensive oxidative degradation of the alkylbenzene (**10**; equation 3) by *Rhodococcus* spp. (BPM 1613).[32] In this case the optical induction is due to oxidative kinetic resolution of intermediates; the recovered substrate is racemic.

This type of progressive chain shortening can be of general use, as higher homologs of hydrocarbon substrates are often more readily accepted by microbes than lower ones. The progress of the reaction needs to be carefully monitored, of course, to avoid overreaction. Another example of this approach is the synthesis of the homochiral antiulcer agent (**12**; equation 4) in near quantitative yield.[33]

In some cases the regioselectivity of a microbial oxidation is governed by the absolute configuration of the substrate. For example, the racemic amide (**13**; Scheme 4) gives two major products upon culturing with *Beauveria sulfurescens*, one from hydroxylation at C-5 (**14**; 29% yield, 44% *ee*) and derived from the (1*S*)-enantiomer of the substrate, and one from hydroxylation at C-4 (**15**; 22% yield, 53% *ee*) and derived from the (1*R*)-enantiomer.[34]

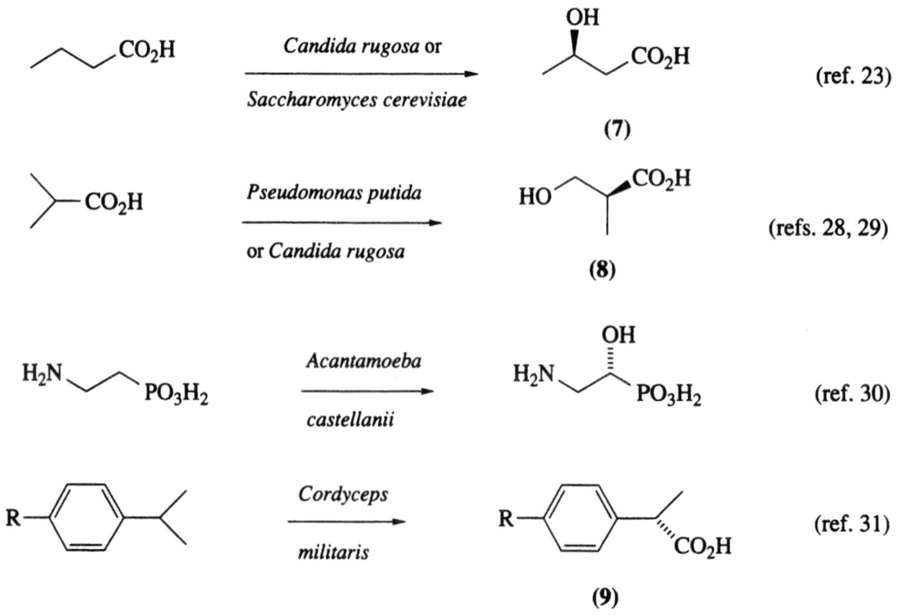

CO_2H —*Candida rugosa* or *Saccharomyces cerevisiae*→ OH CO_2H (7) (ref. 23)

CO_2H —*Pseudomonas putida* or *Candida rugosa*→ HO CO_2H (8) (refs. 28, 29)

H_2N PO_3H_2 —*Acantamoeba castellanii*→ OH H_2N PO_3H_2 (ref. 30)

R —*Cordyceps militaris*→ R CO_2H (9) (ref. 31)

Scheme 3

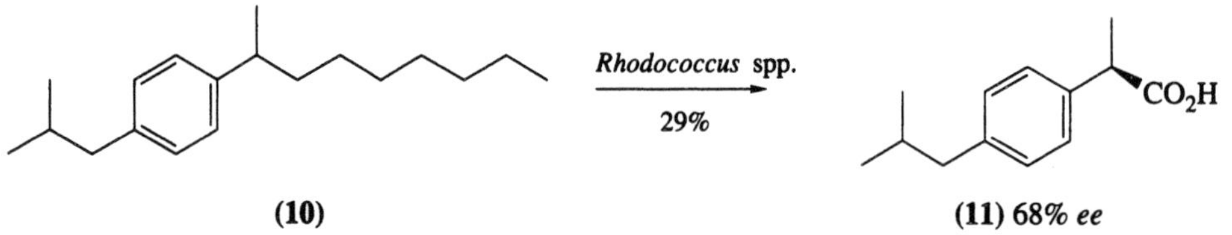

(10) —*Rhodococcus* spp. 29%→ CO_2H (11) 68% *ee* (3)

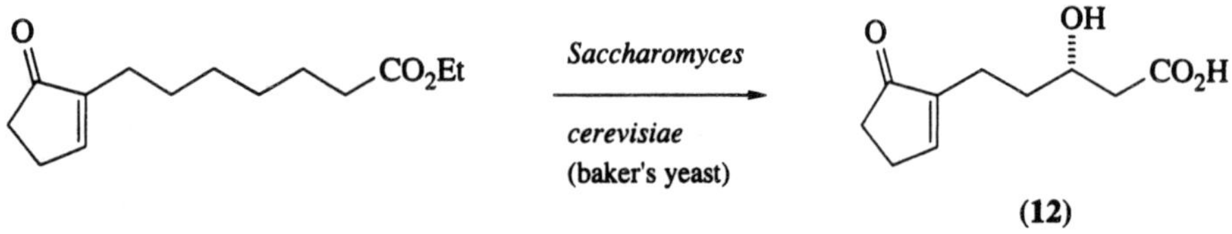

O CO_2Et —*Saccharomyces cerevisiae* (baker's yeast)→ O OH CO_2H (12) (4)

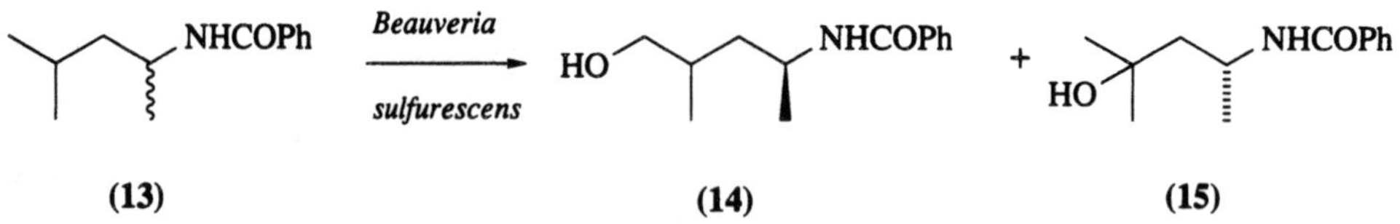

NHCOPh (13) —*Beauveria sulfurescens*→ HO NHCOPh (14) + HO NHCOPh (15)

Scheme 4

1.4.3.2 Cyclic Hydrocarbons and Their Functionalized Derivatives

The literature in this area up until 1975 has been reviewed by Kieslich.[5] In common with acyclic hydrocarbons, simple unsubstituted cyclic hydrocarbons give poor yields of hydroxylated products with most microorganisms, either due to cascade degradation or volatilization of the product during the reaction.

Beauveria sulfurescens has repeatedly been shown to be particularly suitable for hydroxylations of a wide range of cyclic substrates, and much pioneering work was done by Fonken and coworkers, and is recorded in a number of patents held by Upjohn Co., two of which are particularly relevant.[35,36] Cyclohexylcyclohexane (**16**), for example, can be 4,4′-bishydroxylated by *Beauveria sulfurescens* to give a 30% yield of the 4,4′-dihydroxy derivative (**17**). More recent work employing *Beauveria sulfurescens* or *Cun-*

ninghamella blakesleeana gave the 4(e),4′(e) isomer, with some unspecified optical activity residing in the 3(e),4′(e) isomer also isolated.[37]

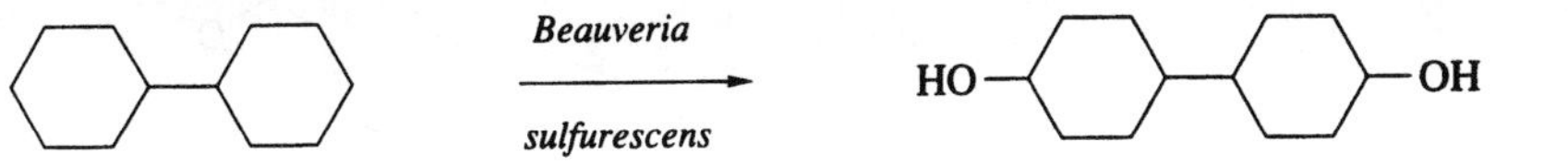

(16) —*Beauveria sulfurescens*→ HO–(17)–OH (5)

Polar groups on the substrate molecule can facilitate hydroxylation of nonactivated positions, and the cyclohexane derivative (**18**) is regio- and stereo-selectively hydroxylated in 71% yield by *Penicillium concavo-rugulosum* (equation 6).[38]

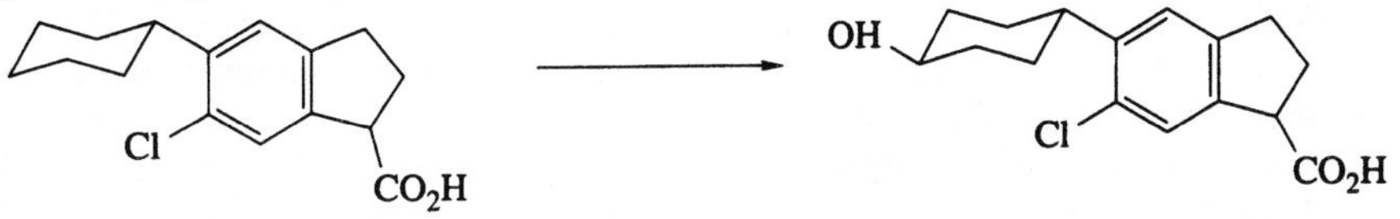

(18) → (6)

Hydroxylation of 7-carboxybicyclo[2.2.1]heptane (**19**) and the unsaturated analog (**20**) provide a good example of what can be achieved with microorganisms. Examination of 119 types of microorganism showed that most gave little or no regio- or stereo-selectivity, however *Aspergillus awamori* (FERM P-8052) showed excellent regio-, diastereo- and enantio-selectivity, resulting in the conversion of the acid (**19**) into the *endo*-alcohol (**21**) (84.7% *ee*)[39] This can then be oxidized to give the ketone (**22**; 92.2% *ee*; Scheme 5).

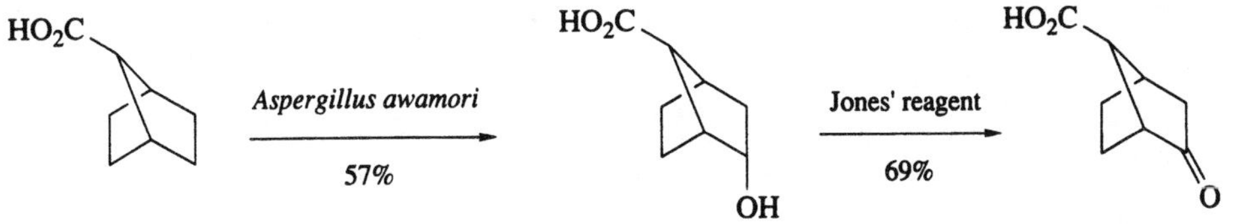

(19) —*Aspergillus awamori*, 57%→ (21) —Jones' reagent, 69%→ (22)

Scheme 5

The unsaturated ester (**20**) can be similarly converted to the *endo*-alcohol (**23**), which can be oxidized to the ketone (**24**; 81.9% ee) with an overall yield of 8% (Scheme 6). These products are potential intermediates for (–)-methyl jasmonate and natural prostaglandins.[40] Here the microorganism is showing good discrimination between the two enantiotopic *endo* hydrogens on C-2 and C-3.

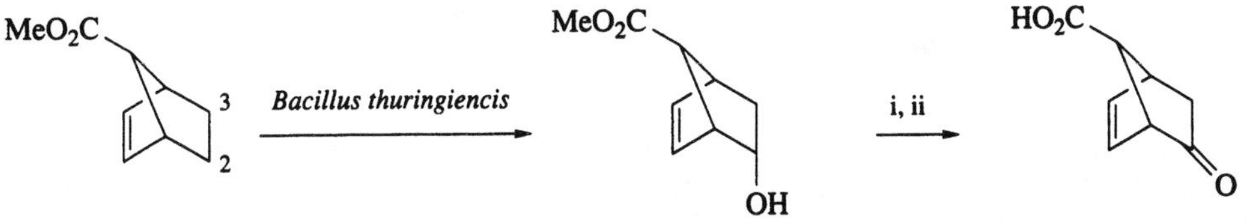

(20) —*Bacillus thuringiencis*→ (23) —i, ii→ (24)

i, hydrolysis; ii, Jones' reagent

Scheme 6

Diastereoselective hydroxylations are more common, for example *Streptomyces rimosus* (NRRL 2234) will hydroxylate zearalenone (**25**) to give the (*S*)-8′-hydroxy derivative (**26**; equation 7).[41] Other microorganisms gave reduction of the 6′-ketone group in (**25**).

The presence of an amine or amide group in cyclic substrates greatly facilitates the hydroxylation by *Beauveria sulfurescens*. Numerous mono-, di- and tri-cyclic amides and saturated nitrogen heterocycles have been studied[5] and a rational basis for the position at which the hydroxy group is introduced into the substrate molecule has been put forward;[42,43] however, yet more work is required to define all the factors controlling the selectivity of hydroxylation. Nevertheless useful regio-, stereo- and in some cases

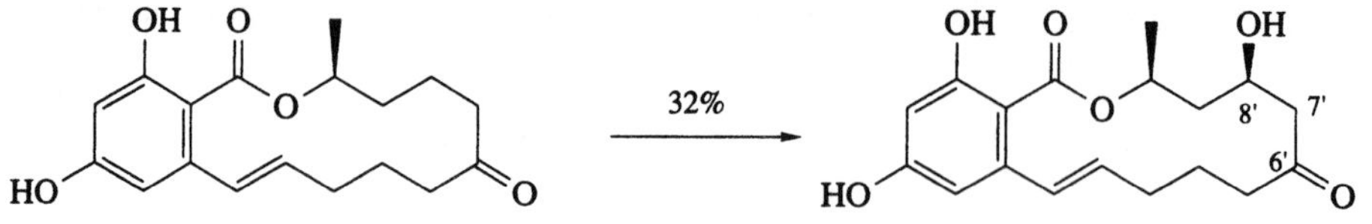

(25) —32%→ (26) (7)

enantio-selectivities are possible, although as with other areas there is still a lack of detailed quantitative data, particularly on the optical purities of products.

Hydroxylation of the bridged piperid-2-one (**27**) can be accomplished with total stereocontrol[44] to give the *exo*-alcohol (**28**), a useful precursor to novel 2′-desoxynucleosides (equation 8). In other cases the regio- and stereo-chemical outcome of the reaction may be highly dependent upon the nature of the substrate; this seems particularly true for lactams such as (**29**). Here mixtures of regioisomers are obtained when using *Beauveria sulfurescens* and optical activity of these is usually low or absent.[45] For example the phenylacetyl amide of pyrollidin-3-one (**30**) gave the (*S*)-3-hydroxy derivative (**31**) with a 30% ee (equation 9).

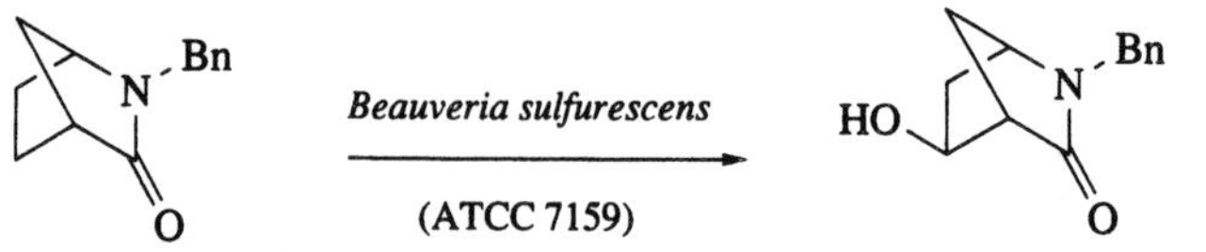

(±)-(**27**) —*Beauveria sulfurescens* (ATCC 7159)→ (2*S*)-(**28**) (8)

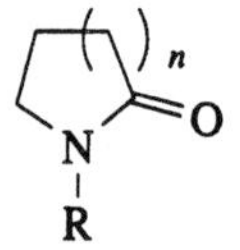

(**29**) $n = 1–3$

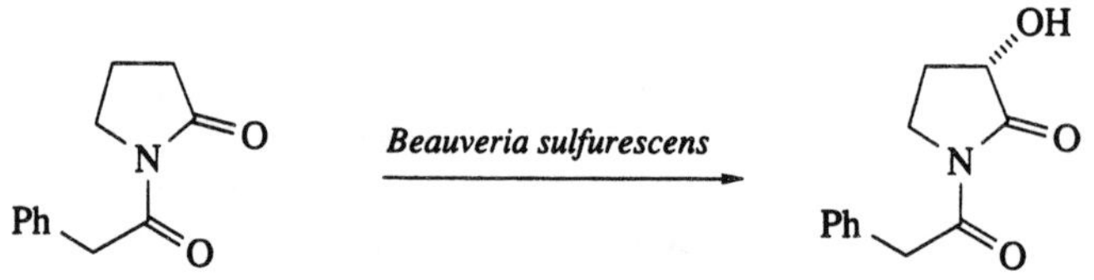

(**30**) —*Beauveria sulfurescens*→ (**31**) (9)

As with acyclic amides, enantiospecific discrimination of the enantiomers of a racemic substrate sometimes occurs. This phenomenon could prove useful in preparing new homochiral synthons. For example the two C-1 enantiomeric amides (**32**) and (**33**) are hydroxylated on different methyls (relative to the *t*-amide nitrogen) of the C-8 *gem*-dimethyl bridging group,[46a] as shown in Scheme 7. Hydroxylation of one methyl in a *gem*-dimethyl grouping by microorganisms is fairly common, and other examples are given later and elsewhere.[46b]

Thus the active site of the hydroxylating enzyme appears insensitive to the relative position of the amide group. This has also been observed for bicyclic amides, for example both *exo* and *endo* isomers of the amide (**34**) are hydroxylated to the *exo*-alcohol (**35**) with the same regio- and stereo-selectivity (equation 10).[43] Further, the hydroxylation can also be insensitive to the position of the carbonyl group. For example, bi-, tri- and tetra-cyclic amides, and the equivalent lactams, of which the amide (**36a**) and lactam (**36b**) are representative cases respectively, are both hydroxylated at the same position (Scheme 8a).[47a]

These observations have led to the development of a trajectory-based model to rationalize the selectivity of the monooxygenase enzyme in *Beauveria sulfurescens*,[47a] and this may allow the regio- and stereochemistry of transformations on new (related) substrates to be predicted. Also of some assistance in predicting the stereochemical outcome of the reactions is the study of the mechanism, and this has been investigated and shown, at least in some cases, to proceed with inversion of stereochemistry.[47b]

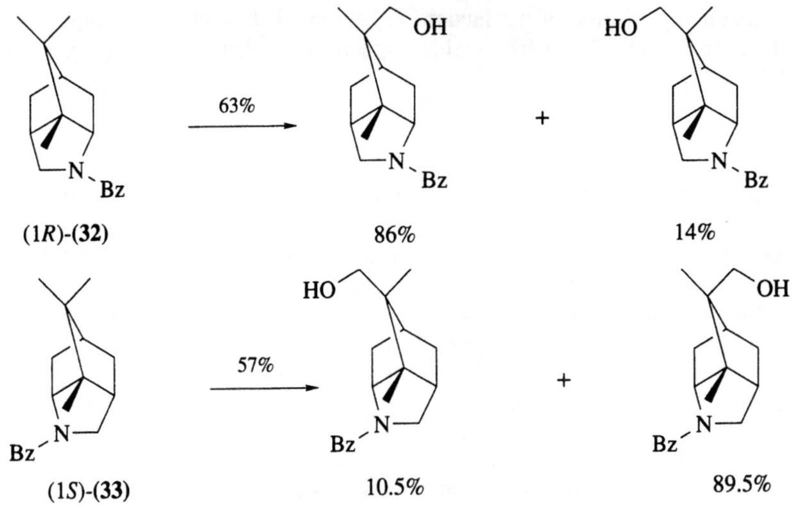

(1*R*)-(**32**) 63% → 86% + 14%

(1*S*)-(**33**) 57% → 10.5% + 89.5%

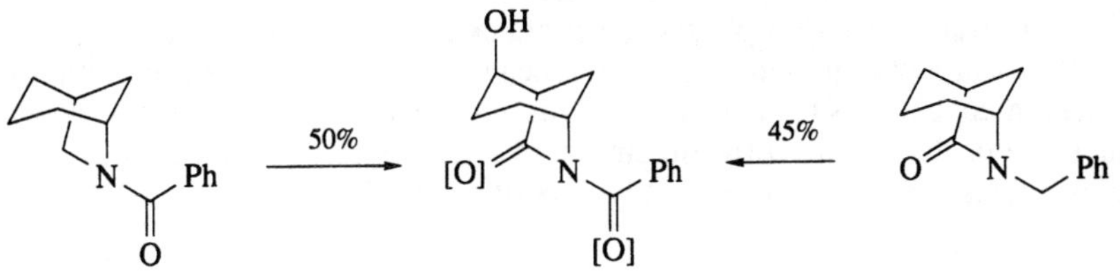

Scheme 7

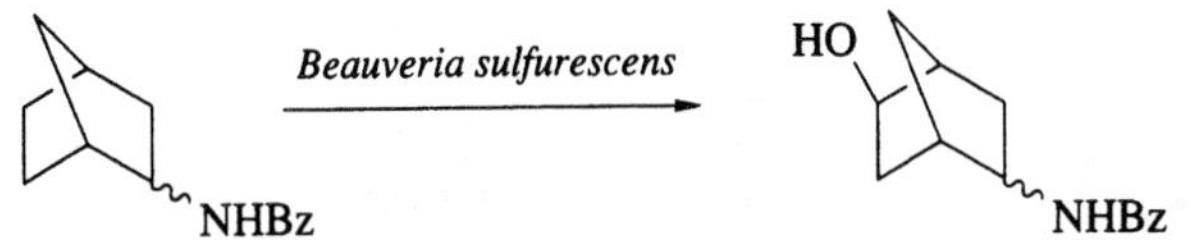

(**34**) *Beauveria sulfurescens* → (**35**) (10)

(**36a**) 50% → ← 45% (**36b**)

Scheme 8a

Although much of the work on the microbial hydroxylation of amides has been directed at active-site mapping of the enzyme responsible, the products themselves are valuable building blocks for further synthesis, for example, for various optically active sesquiterpenes[48] or β-lactams. In this latter context regioselective hydroxylation of unactivated positions is particularly attractive as several β-lactam antibiotics, *e.g.* the carbapenem derivative thienamycin, have a free hydroxy group in their structure.

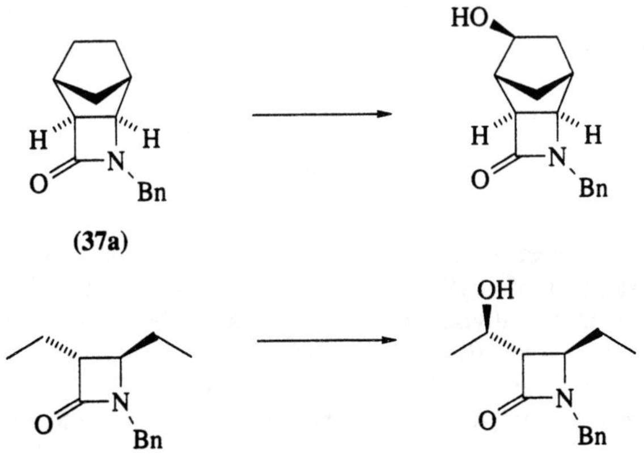

(**37a**)

(**37b**)

Scheme 8b

For example, monohydroxylation of the lactams (**37a**) and (**37b**) can be accomplished by *Beauveria sulfurescens* (ATCC 7159) in 65% and 10% yields respectively (Scheme 8b).[47c]

1.4.3.3 Isoprenoids

The hydroxylation of terpenes by microorganisms is of interest in the preparation of flavor and fragrance compounds

There have been a number of previous reviews on microbial oxidations of terpenes.[5,49,50a] Monoterpenes are often degraded progressively after an initial hydroxylation step, but di-, tri- and sesqui-terpenes can be converted more selectively, to accumulate useful quantities of hydroxylated products. Less systematic work on the microbial oxidation of terpenoids has been carried out than in the case of steroids, and therefore prediction of the regio- and stereo-chemistry is scarcely possible.

Acyclic triterpenes can be considered as aliphatic hydrocarbons and are α-hydroxylated by a number of microorganisms.[50b] The microbial oxidation of a variety of acyclic terpenoid hydrocarbons has been investigated by Nakajima,[51] and although terminal alcohols can be obtained, for example pristanol (**39**) from pristane (**38**; equation 11), further oxidation can also occur.

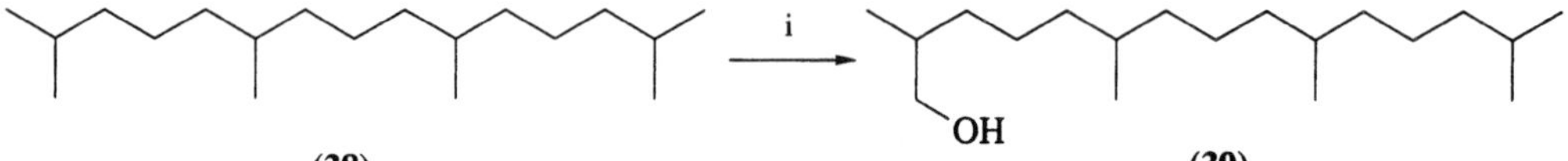

i

OH

(11)

(**38**) (**39**)

i, *Rhodococcus* spp. (BPM 1613)

Citronellol, geraniol and linalool (as their acetates) can be regiospecifically hydroxylated at the terminal allylic carbon by a strain of *Aspergillus niger* (equation 12).[52a] Concurrent hydrolysis of the acetate groupings in these substrates also takes place to a certain extent. This hydroxylation is particularly interesting as previously a strain of *Aspergillus niger* has been reported[5] that rearranged geraniol (**40**) to linalool before oxidation to citral (**41**; Scheme 9). This difference could be due to acetylated *versus* unacetylated substrate or due to two different strains of microorganism being employed. Longer chain acyclic terpenoids can also be hydroxylated in allylic positions by *Aspergillus niger* (ATCC 9142).[52b]

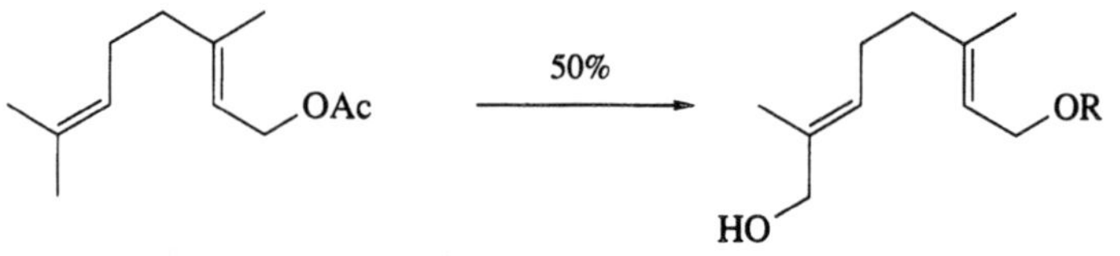

OAc

50%

OR

HO

(12)

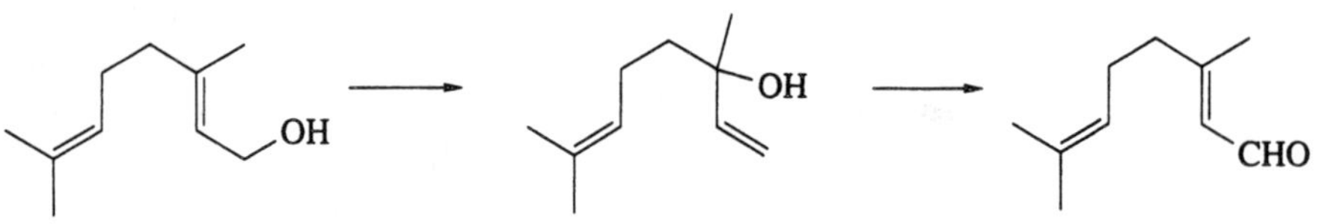

OH

OH

CHO

(**40**) (**41**)

Scheme 9

Mono- and bi-cyclic monoterpenes containing sites of unsaturation tend to be hydroxylated at the allylic position,[5] (more examples of allylic hydroxylation are discussed in Section 1.4.5.2) with regioisomers occurring if more than one allylic position is accessible. Some illustrative examples of reported hydroxylations are shown in Scheme 10.[53–57] The hydroxylation of 1,4-cineole (**42**)[53a] is illustrative of the enantioselectivity that may be achieved in such transformations. *Bacillus cereus* gives a 1:7 mixture of (2*R*)-*exo*- and (2*S*)-*endo*-monohydroxy-1,4-cineole, both with essentially 100% enantiomeric purity.[53b,53c]

In the case of the transformation of bornyl acetate (**43**) by *Fusarium culmorum* it is interesting to note that both enantiomers of the substrate yielded only the 5-*exo*-hydroxybornyl acetate as the major product.

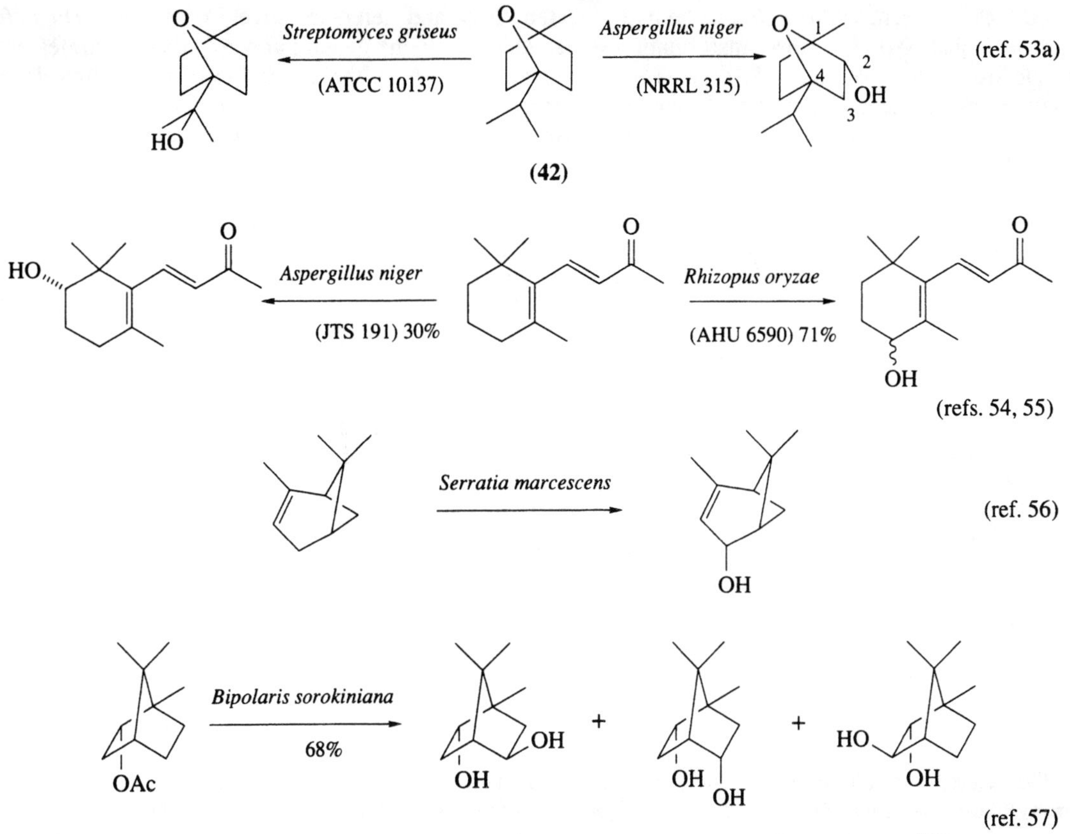

Streptomyces griseus (ATCC 10137)

Aspergillus niger (NRRL 315)

(ref. 53a)

(42)

Aspergillus niger (JTS 191) 30%

Rhizopus oryzae (AHU 6590) 71%

(refs. 54, 55)

Serratia marcescens

(ref. 56)

Bipolaris sorokiniana 68%

OAc

(43)

82% 15% 3%

(ref. 57)

Scheme 10

Also in this case the major product from microbial oxidation corresponded to that from chemical oxidation.[57]

Chemically difficult (or currently impossible) hydroxylations on more complex terpenes can be accomplished by microbial methods, but there is as yet no clear understanding of the factors affecting selectivity. As is often the case, microorganisms suitable for a required transformation have been selected largely on precedent or on an empirical basis. In some cases respectable yields of one stereoisomer may result.

Sesquiterpenes can also be hydroxylated and species of the genera *Aspergillus, Cunninghamella* and *Streptomyces* have all been used to accomplish miscellaneous hydroxylations.[5] Unfortunately prediction of the site of attack is not yet possible. To illustrate the difficulty in understanding the factors controlling the site of attack the sesquiterpene lactones (**44**) and (**45**) and the diol (**1**) serve as examples. Although the lactone (**44**) is hydroxylated predominantly at the reactive 8α-position (*Cunninghamella echinulata*, NRRL 3655),[58] the product from the hydroxylation of the lactone (**45**) (*Aspergillus niger*, MIL 5024) at the equivalent position is not obtained,[59] and the diol (**1**) is not hydroxylated at the expected (allylic) position (*Cunninghamella elegans*).[1]

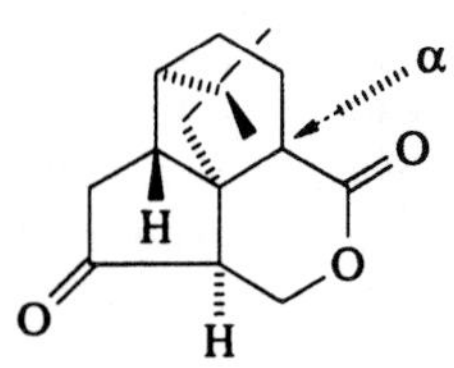

(44) 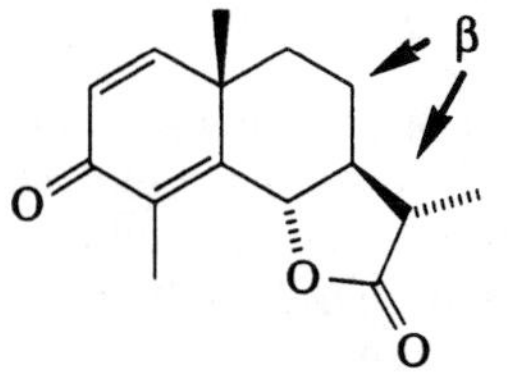 **(45)** 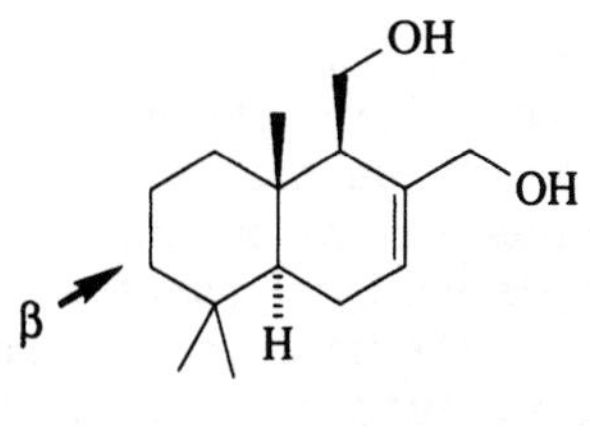 **(1)**

The sesquiterpene cedrol (**46**) can be hydroxylated regio- and stereo-selectively with *Beauveria sulfurescens* (equation 13).[60] This transformation serves to illustrate the general principle that substrates with an electron rich substituent, to serve as an anchor at, or close to, the active site of the hydroxylating enzyme system, generally are transformed with improved selectivity over those with no such anchor. For example, in the above system the unsaturated substrate cedrene (**47**) gives low yields of a mixture of products.[60]

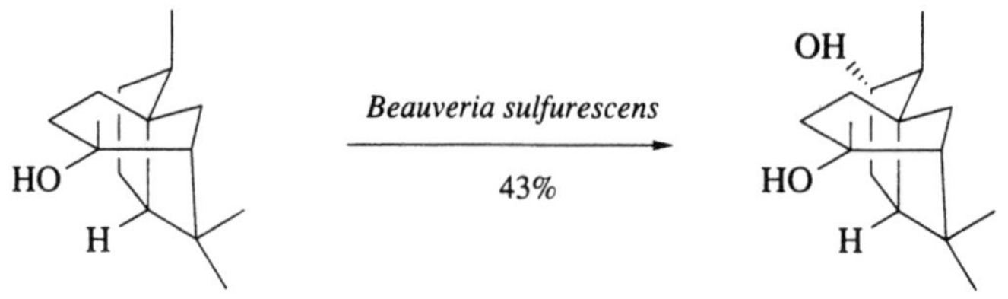

Beauveria sulfurescens 43% (13)

(**46**)

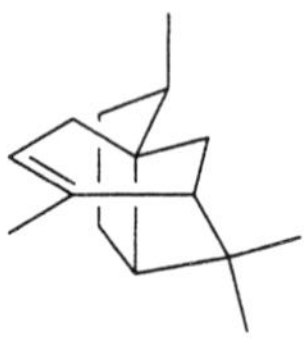

(**47**)

The related tricyclic sesquiterpene patchoulol (**48a**) is a good substrate for microorganisms[48] and hydroxylation at various sites can be accomplished, for example at C-5 by *Choanephora circinana* in 74% yield (equation 14). Hydroxylation of the methyl group on C-4 can be accomplished by *Penicillium rubrum* FX-318 in 75% yield,[61a] and of the C-14 methyl group in (1*R*)-caryolan-1-ol (**48b**) by *Aspergillus niger* (MMP 521) in 26% yield.[61b]

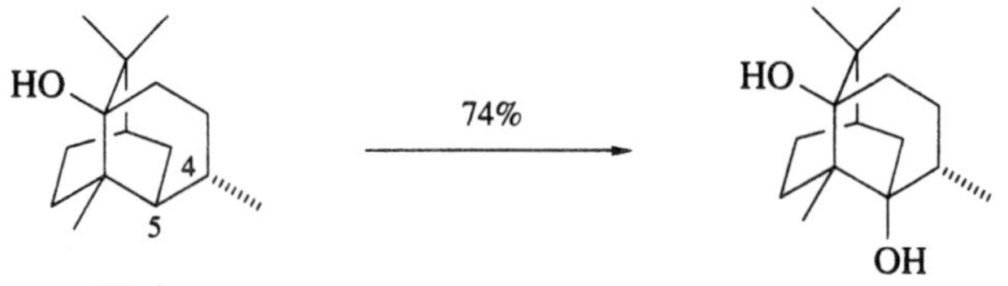

74% (14)

(**48a**)

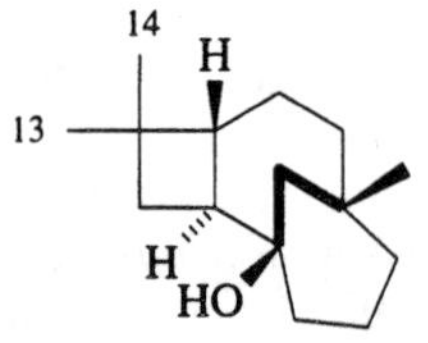

(**48b**) 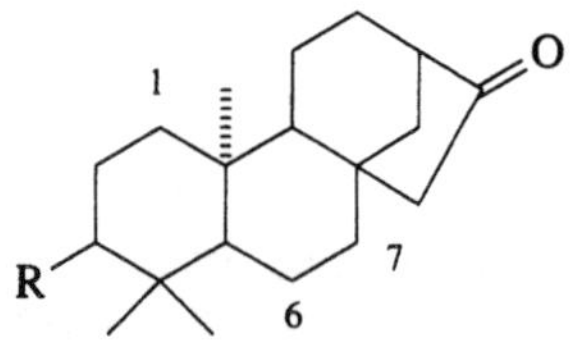 (**49**)

Because of their importance as precursors to gibberellins, diterpenes with the *ent*-kaurane skeleton have been subjected to microbial hydroxylation.[4] Favored microorganisms for these transformations have been *Calonectria decora, Rhizopus nigricans* and *Aspergillus ochraceus*. The hydroxylations are sometimes selective but often mixtures are obtained, made even more complex by di- as well as monohydroxylation. Some representative results illustrating regio- and diastereo-selectivities are shown in Table 3.[62] The same type of binding-site model can be used to rationalize the sites of hydroxylation in the ent-kaurene series as in the steroidal series *vide infra*.[63] Also, as in the case of steroids, better selectivity is obtained if two binding groups are present in the substrate rather than one.

Forskolin derivatives, for example (**50**), can be hydroxylated by, for example, *Neurospora crassa* (ATCC 10336), *Mortierella isabellina* (ATCC 160074) and *Aspergillus niger* (DSM 3210); 2α-, 2β-, 3α- or 3β-hydroxy derivatives can be obtained in this way (*e.g.* equation 15).[64]

Table 3 Hydroxylation of *ent*-17-Norkauranones (**49**)

Substrate (**49**)	*Microorganism*	*Yields (%)* 1α	$7a$	6β
R = —OH	*Rhizopus nigricans*	25	35	
	Calonectria decora		40	
	Aspergillus ochraceus		25	30
R = ═O	*Calonectria decora*		40	

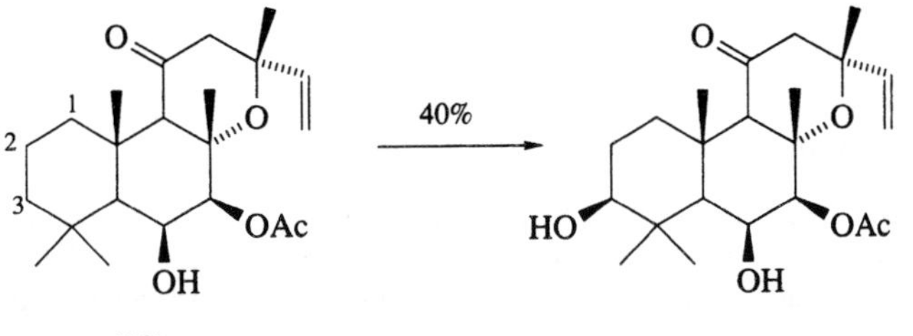

(**50**) —40%→ (15)

1.4.3.4 Alkaloids

Microbial oxidations are relatively common in the alkaloid field and a number of excellent reviews covering the literature up until 1984 have appeared.[13,65–67] Hydroxylation at both aliphatic and aromatic positions of alkaloid molecules can be accomplished, however the latter is favored and the site of hydroxylation in these cases is usually predictable based on the rules governing electrophilic aromatic substitution.[13]

The heteroyohimbine alkaloids can be effectively hydroxylated at the (aromatic) C-10 and C-11 positions.[68] For example ajmalcine (**51**) is 10-hydroxylated in 92% yield by *Cunninghamella elegans* (ATCC 9245) and tetrahydroalstonine (**52**) is 11-hydroxylated in 72% yield by a plant-derived mold. This latter transformation is particularly difficult to achieve by chemical methods.

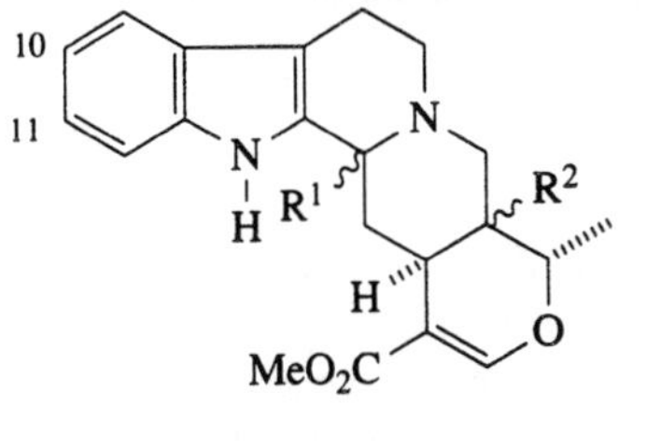

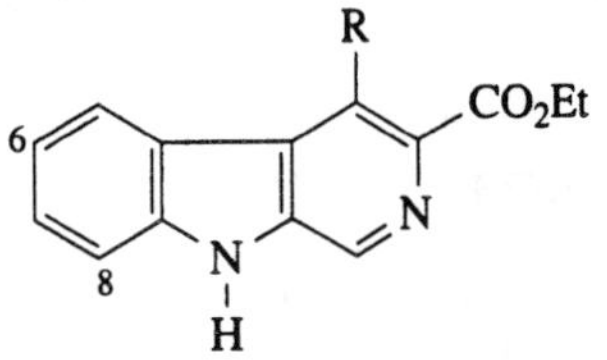

(**51**) $R^1 = \alpha$-H, $R^2 = \beta$-H

(**52**) $R^1 = \alpha$-H, $R^2 = \alpha$-H

(**53**) R = H or alkyl

Carboline alkaloids (**53**) can be hydroxylated by *Beauveria sulfurescens* (ATCC 7195) at either the 6- or 8-position depending on the substituents on the substrate.[69] When R in the alkaloid (**53**) is methyl, then a mixture of products is obtained (20%, 6-hydroxy; 18%, 8-hydroxy) but if R is ethyl then only the 8-hydroxy product is obtained (70% yield).

Stereoselective dehydrogenation of some alkaloids can be carried out using microorganisms (see also steroids, Section 1.4.4.1). Glaucine derivatives (**54**) are dehydrogenated, in very high yield,[70] to dehydro derivatives by *Aspergillus flavipes* (ATCC 1030) and *Fusarium solani* (ATCC 12823) (equation 16). *Aspergillus flavipes* selectively dehydrogenates the *cis*-6a-(*R*)-enantiomer, whilst *Fusarium solani* selectively dehydrogenates only the *cis*-6a-(*S*)-enantiomer.

Hydroxylation of aliphatic carbon has rarely been accomplished in high yield in alkaloids, but some attempts have been made to produce an active site map for the hydroxylase of *Streptomyces roseochromogenes* which may be of predictive value.[71]

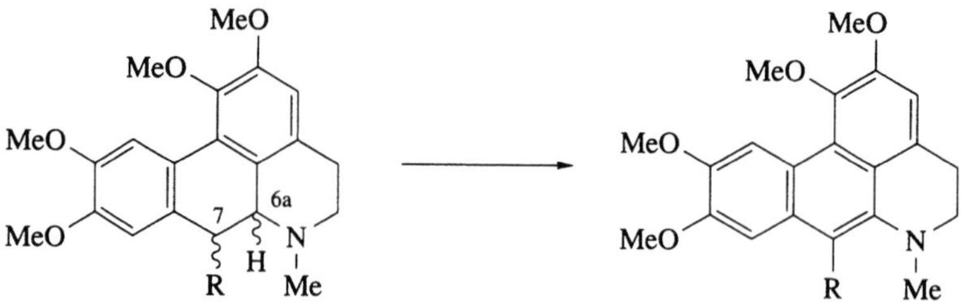

(16)

(54) R = H or Me

1.4.3.5 Prostaglandins and Cannabinoids

Hydroxylation of the side chain of prostaglandins is possible using microorganisms, but is often accompanied by reduction of double bonds, hydrolysis of esters or further oxidation of alcohols. Hydroxylation usually takes place at C-18, C-19 or C-20 [numbering with respect to prostanoic acid (**55**)]. *Streptomyces ruber* (NRRL B-1268), *Aspergillus niger* (ATCC 9142), *Cunninghamella blakesleeana* (ATCC 9245) or *Microascus trigonosporus* (NRRL 1199) can be employed with a wide range of prostaglandin substrates.[11,13] Little is known of the stereochemistry of these reactions.

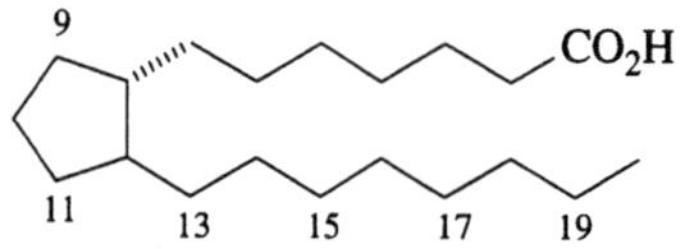

(55)

In cannabinoids, side chain hydroxylation is the most common step in microbial conversions, followed by hydroxylation of the terpenoid ring. Generally mixtures of products are obtained and yields are poor.[11] The following microorganisms can be employed for hydroxylation of unactivated C—H bonds in cannabinoids:[72] *Aspergillus niger* (ATCC 9142), *Botrytis allii* (ATCC 9435), *Cunninghamella elegans* (ATCC 9245), *Mycobacterium rhodochrous* (ATCC 19067) and *Streptomyces aureus* (ATCC 15437).

1.4.4 MICROBIAL OXIDATION OF STEROIDS

Of all the considerable research efforts expended on microbial oxidations, steroids have received the greatest share. The vast amount of work in this area has been stimulated by the medical importance of steroids, and the desire to develop new drugs with new or improved pharmacological properties.[73] The volume of work is so great that only an overview, with reference to the more interesting or higher yielding transformations, can be presented here. Many previous reviews on the subject have appeared, with a handbook by Charney and Herzog[10] being one of the most useful. Other publications of special importance as reviews and sources of references to the voluminous literature in this area have also appeared.[2–4,13,74] A discussion of practical aspects is also of great value to those unfamiliar with this area.[75a] The commercial relevance of steroid hydroxylation can also be appreciated from a recent extensive review on the subject.[75b]

Microbial oxidative transformations of steroids can be divided into a number of key categories:[13] (i) hydroxylation, (a) at all nuclear sites and angular methyl groups, mono-, poly-, carbonyl-activated, allylic, (b) at some side chain sites; (ii) alcohol dehydrogenation, (a) saturated alcohol to ketone, (b) allylic or homoallylic alcohol to α,β-unsaturated ketone (Δ^1- and/or Δ^4-3-keto steroid formation); (iii) double bond formation, (a) introduction of a Δ^1 double bond in 3-keto steroids, (b) introduction of a Δ^4 double bond in 3-keto steroids, (c) A-ring aromatization; (iv) carbon–carbon bond oxidation, (a) side chain degradation, (b) ring cleavage; and (v) epoxidation.

Carbon–carbon bond oxidation is beyond the scope of this work, but is a consequence of initial hydroxylation followed by further metabolism involving other enzyme systems. Many examples of alcohol dehydrogenation and double bond epoxidation have been published previously,[76,77] and these will not be considered further here.

1.4.4.1 Dehydrogenation

The dehydrogenation of CH—CH, particularly to give α,β-unsaturation in 3-keto steroids can be readily accomplished using microorganisms, and an excellent review on this aspect of steroid oxidations has appeared.[2] Although many microorganisms possess the sterol 1-dehydrogenase enzyme, *Mycobacterium smegmatis* and *Arthrobacter simplex* (also known as *Corynebacterium simplex*), and various mutants thereof, have proved especially useful for 1,2-dehydrogenation. A more complete list is as follows: *Arthrobacter simplex* (ATCC 6946 and NRRL B-8055), *Bacillus lentus* (ATCC 13805), *Glomerella cingulata* (ATCC 10534), *Nocardia asteroides* (ATCC 3308), *Nocardia corallina* (ATCC 999) and *Nocardia restrictus* (ATCC 14887)

As an example, cortisone (**56**) can be converted to prednisone (**57**; equation 17) in up to 90% yield by *Arthrobacter simplex*.[2]

(56) → **(57)** (17)

In these transformations the microorganism will often accept a wide range of steroidal substrates and the conversion, or yield, can often be improved by the addition of an electron carrier such as 1,4-naphthoquinone.[78] The 1,2-dehydrogenation is reversible, but the forward reaction can be encouraged to go to completion by efficient aeration of the fermentation.[13] Existing alcohol groups present in the molecule generally remain unaltered, for example in the conversion of cortisol (**58**) to prednisolone (**59**; equation 18), although 3-ols can be oxidized to 3-keto steroids concomitant with 1,2-dehydrogenation. If the substrate contains a 20-keto group this may be reduced during the 1-dehydrogenation.

(58) —*Arthrobacter simplex*→ **(59)** (18)

The 1,2-dehydrogenation occurs by stereospecific removal of the 1α- and 2β-hydrogens, and, in addition, some microorganisms will differentiate between enantiomers of the substrate. For example, *Arthrobacter simplex* will usually dehydrogenate only the (*R*)-enantiomer of 10-substituted steroids; however, other microorganisms (*e.g. Corynebacterium hoagii*) having 1,2-dehydrogenase activity will accept both enantiomers.[2]

Double dehydrogenation and even complete aromatization of the A-ring of 3-hydroxy or 3-keto steroids can also be effected; the latter is of special interest in the preparation of estrogens, for example the conversion of the diene (**60**) to the α-estradiol derivative (**61**; equation 19) by *Proactinomyces globerula*.[79]

(60) —*Proactinomyces globerula*, 78%→ **(61)** (19)

Immobilized microorganisms, in particular *Arthrobacter simplex* and *Nocardia rhodocrous*, and fungal spores, in particular those from *Septomyxa affinis*, can also be used to effect dehydrogenations.[2,80] Dehydrogenations may also be advantageously carried out in the presence of hydrocarbon solvents, for example the conversion of 6α-methylhydrocortisone (**62**) to 6α-methylpredisolone (**63**; equation 20) with *Arthrobacter simplex*.[81]

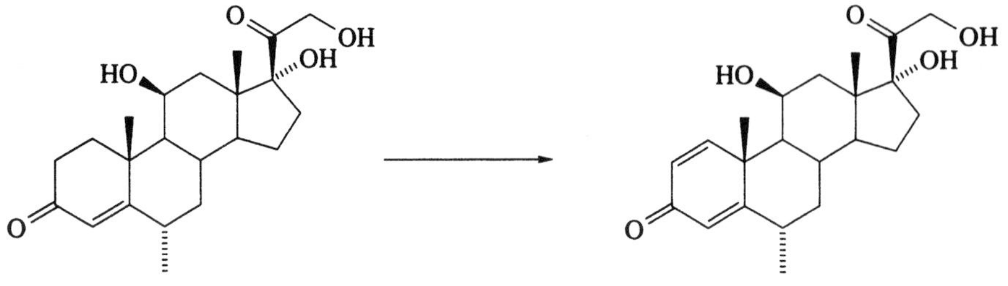

(20)

(62) **(63)**

1.4.4.2 Hydroxylation

Virtually every site in the steroid molecule is accessible for microbial hydroxylation, and almost all positions have been hydroxylated by various microbial strains. Of particular importance are hydroxylated products with the unnatural α-configuration, and of those the derived 11α- and the 16α-alcohols are of greatest synthetic interest.

Much of the fundamental work on the microbial hydroxylation of steroids, including selection of microorganisms and determination of specificities, was carried out in the 1950s and 1960s, following Peterson's original discovery of the 11α-hydroxylation of progesterone (**64**) by *Rhizopus nigricans* in 1952 (equation 21). A mutant strain of *Aspergillus ochraceus* has subsequently been found that will carry out this transformation in 91% yield at an initial substrate concentration of 40 g L^{-1}.[85]

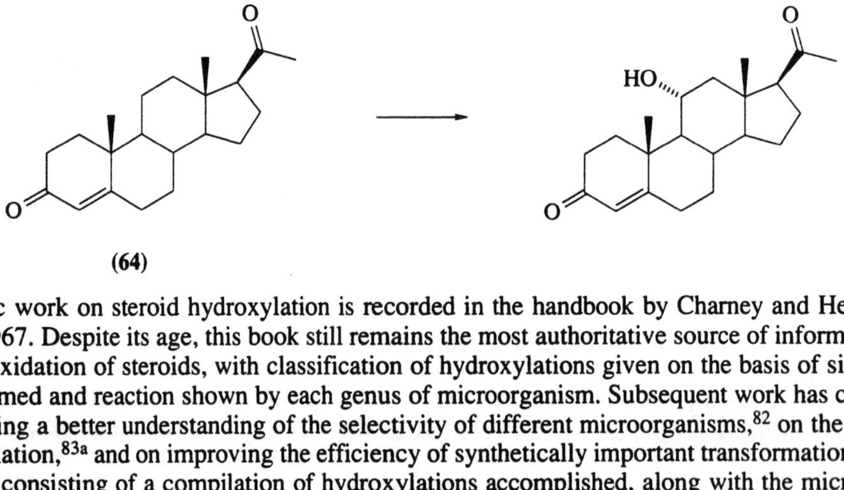

(21)

(64)

The basic work on steroid hydroxylation is recorded in the handbook by Charney and Herzog[10] published in 1967. Despite its age, this book still remains the most authoritative source of information on the microbial oxidation of steroids, with classification of hydroxylations given on the basis of site of attack, product formed and reaction shown by each genus of microorganism. Subsequent work has concentrated on developing a better understanding of the selectivity of different microorganisms,[82] on the mechanism of hydroxylation,[83a] and on improving the efficiency of synthetically important transformations.[83b]

Reviews consisting of a compilation of hydroxylations accomplished, along with the microorganisms employed, have been published to cover the period 1979 to 1988.[76,77,83b]

The greatest contribution towards understanding the structural features in the substrate that effect, or indeed control, the selectivity in microbial hydroxylation of steroids was carried out by Jones and Meakins and their work is recorded in a series of papers entitled 'Microbiological Hydroxylation'. Their last report was published in 1980, and serves as a source of references to earlier material.[84]

Generally speaking fungi are of greatest use in the hydroxylation of steroids, and five genera in particular have been found to be extremely useful; these are *Rhizopus*, *Calonectria*, *Aspergillus*, *Curvularia* and *Cunninghamella*.

As with other classes of substrate, hydroxylase enzymes are responsible for the hydroxylation reactions. A given microorganism may be capable of producing more than one steroid hydroxylase enzyme in response to a substrate, with certain features of the substrate inducing each enzyme. For example progesterone induces a 11α-hydroxylase enzyme in *Aspergillus ochraceus*, whilst a 11α-hydroxypregn-4-en-3-one structural feature induces an independently operating 6β-hydroxylase. For this reason whilst some generalizations are possible in discussing steroid hydroxylations, there is still an element of experimentation involved in developing new transformations.

Practical aspects are much as discussed earlier, and a good laboratory steroid hydroxylation procedure has been published.[86] Numerous additives (*e.g.* surfactants, antibiotics and fungicides) have been used to improve microbial hydroxylation of steroids, and some of these have been discussed previously.[87]

1.4.4.2.1 Chemoselectivity[13]

The value of microbial hydroxylation of steroids has long been that they allow functionalization of positions not easily accessible by normal chemical methods. For this reason yields of a few percent have often been tolerated, but yields approaching quantitative can be achieved in many instances. The objective of much current work is to improve the selectivity of microbial hydroxylations. This can be achieved in a number of ways, for example by structural modifications of the substrate, optimization of fermentation conditions and by strain improvement. The latter is most likely to yield the most significant improvements in selectivity, and has previously been discussed at length.[88]

As in the case of nonsteroidal substrates, selectivity tends to be better with substrates containing more than one polar functional group, and functional group modification can be used to take advantage of this effect. Previous work has shown that steroid hydroxylation and epoxidation can be performed by the same enzyme,[89] and as a result those substrates containing double bonds may be epoxidized as well as hydroxylated. Similarly, steroid dehydrogenase enzymes may also be present or induced by some substrates, and as a consequence alcohol groups (already present in the substrate, or introduced by the microorganism) may be converted to ketones, and double bonds may also be introduced. The Δ^1-dehydrogenation of Δ^4-steroids by microorganisms can however be inhibited by the addition of metabolic poisons such as hydrazine, ammonia or some antibiotics.[90]

In some cases it is desirable to carry out a hydroxylation and a dehydrogenation concurrently, and the use of two microorganisms to take effect of the unique selectivity of each can be of great value. For example the fermentation of the steroid (**65**) with *Pellicularia filamentosa* (TFO 6675) and *Bacillus lentus* (ATCC 13805) gives simultaneous 11β-hydroxylation and Δ^1-dehydrogenation.[91]

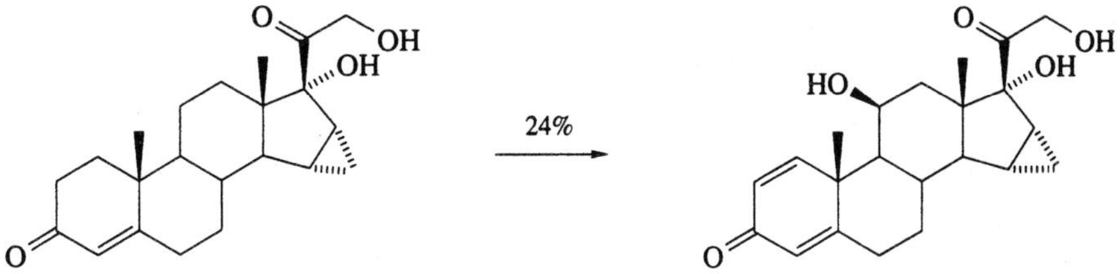

O OH OH → 24% → HO O OH OH (22)

O (**65**) O

Microbial reduction of carbonyl groups is another possible side reaction, with a 3-keto group seemingly most susceptible to this reaction, but again the nature of the substrate and microorganism can have a major effect on the extent of this side reaction. In the hydroxylation of 5α-androstan-3-one (**66**) the results shown in Table 4 were obtained.[92]

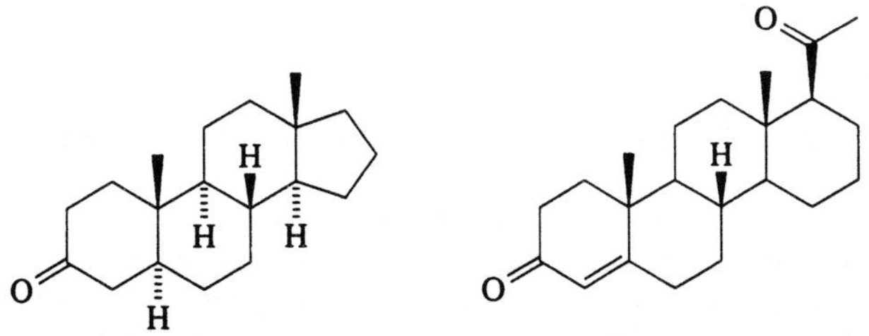

(**66**) (**67**)

Multiple hydroxylation is one of the most frequently encountered problems in steroid hydroxylation, but this can be controlled to a certain extent by appropriate choice of fermentation conditions, usually involving low substrate concentrations and minimization of fermentation time. Some microorganisms are

Table 4 Hydroxylation of 5α-Androstan-3-one (**66**)

Microorganism	*Hydroxylation product*	*Yield (%)*	*Reduction of 3-C=O (%)*
Calonectria decora	12β,15α	47	65
Rhizopus nigricans	11α,16β	35	30
Aspergillus ochraceus	6β,11α	84	0

of value because they will accomplish di- rather than mono-hydroxylation, and again by appropriate selection of strain, isomeric mono- and/or di- hydroxylated products may be synthesized. For example D-homoprogesterone (**67**) can be converted to various mono- and di-hydroxylated products (Table 5).[91]

Table 5 Hydroxylation of D-Homoprogesterone (**67**)

Microorganism	*Main product*	*Yield (%)*
Aspergillus ochraceus	11α-OH	30
Calonectria decora	12β,15α-$(OH)_2$	23
Fusarium lini	15α-OH	13
Glomeralla cingulata	11α,16β-$(OH)_2$	45
Pellicularia filamentosa	11β,17α-$(OH)_2$	44
Rhizopus arrhizus	6β,11α-$(OH)_2$	51

After hydroxylation of a steroidal substrate has taken place, fission of carbon–carbon bonds may occur. Thus 9α-hydroxylation is the first stage in the fission of the B-ring. This process can be suppressed by restriction of metals (such as iron).[2] This has been accomplished using 2,2-bipyridyl,[93] either alone or in combination with an absorbent such as Amberlite XAD-7. Very substantial enhancements in yield are possible. For example 3-ketobisnorcholenol (**68**) gave <10% yield of the 7α-hydroxy derivative with *Botryodiploida theobromae* in the absence of these materials, but up to 45% when they were added together (equation 23).[94]

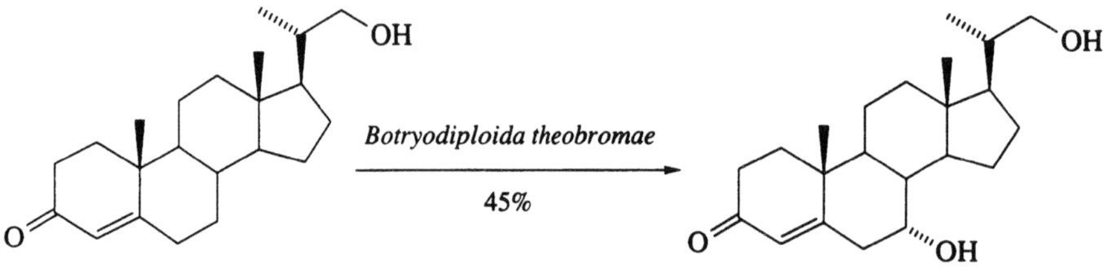

(**68**) —*Botryodiploida theobromae*, 45%→ (23)

Side chain degradation is also a common problem, and those substrates with long chains at C-17 (*e.g.* cholesterol) are difficult to hydroxylate without loss or truncation of the hydrocarbon chain.

Chemoselectivity is indeed a very complex issue in the hydroxylation of steroids, but microorganisms capable of chemoselective transformations can often be selected on precedent.[10,13,74]

1.4.4.2.2 Regioselectivity

Aspects of regioselectivity in the microbial hydroxylation of steroids have been reviewed most recently by Kieslich.[13]

The regioselectivity of hydroxylation with a given microorganism is largely dictated by the nature and position of substituents on the steroid substrate; however, a number of microorganisms do show a tendency to hydroxylate in certain positions irrespective of substituent patterns. For example *Rhizopus nigricans* and *Aspergillus ochraceus* have become known as efficient 11α-hydroxylators, *Curvularia lunata* as a 11β-hydroxylator (not as selective), *Calonectria decora* as a 12β,15α-dihydroxylator and *Rhizopus arrhizus* as a 6β-hydroxylator of 3-keto Δ^4-steroids. *Cunninghamella elegans* is not as easy to categorize in this way, although it has been extensively used, but most frequently causes 7α- and 7β-hydroxylation.

Thus the site of hydroxylation may be influenced by the structure of the substrate or the type of microorganism used. The latter effect is evident in the hydroxylation of progesterone (**64**; Table 6).

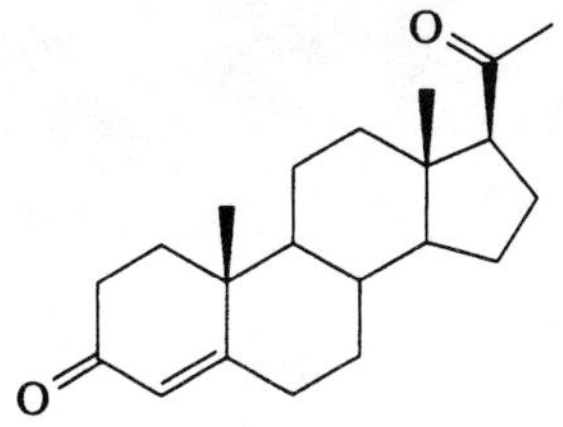

(64)

Table 6 Hydroxylation of Progesterone (**64**)

Microorganism	*Product(s)*	*Yield(s)(%)*	*Ref.*
Aspergillus ochraceus	11α-OH	91	85
Aspergillus ochraceus (spores)	11α-OH		95
Aspergillus phoenicis	11α-OH		10
Rhizopus nigricans	11α-OH		10
Cunninghamella echinulata	11α-OH	29	96
	17α-OH	54	96
Mucor spp.	14α-OH		98
Streptomyces coriofaciens	16α-OH		97

As in the case of nonsteroidal substrates, the regioselectivity can also be influenced by the stereochemistry of the reactant, for example hydroxylation of the 3α- and 3β-stereoisomers of the androstan-17-one (**69**) by *Calonectria decora* gave different regioisomeric products (Scheme 11).[4] In many cases mixtures of regioisomers are obtained.

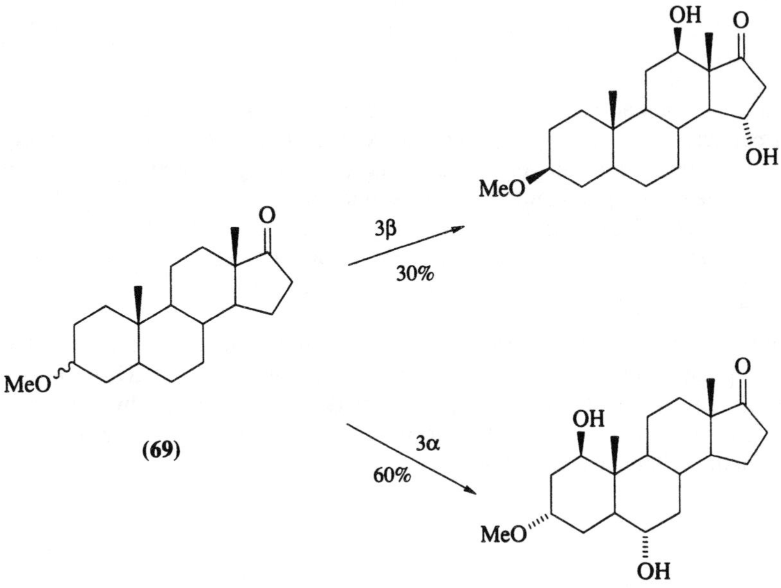

(**69**) 3β 30% 3α 60%

Scheme 11

In the case of keto androstanes some attempts have been made to rationalize the site of hydroxylation, particularly with *Calonectria decora* and *Rhizopus nigricans*[82] and a précis is presented elsewhere.[4] As with other substrates, binding of polar groups to the active site of a cytochrome *P*-450 dependent monooxygenase is thought, in most cases, to control both the regio- and stereo-chemistry of hydroxylation (although see 6β-hydroxylation of Δ^4-steroids in Section 1.4.4.2.3). Thus binding of 16β-hydroxy-5α-androstan-3-one (**70**) to the hydroxylase of *Rhizopus nigricans* results in 11α-hydroxylation, but if the keto and alcohol groups are transposed, reversal of the binding orientation results in the 7α-hydroxylated product (**71**).[99] Similarly with *Rhizopus arrhizus*, whilst 9α,10β-androst-4-ene-3,17-dione (**72**) undergoes 11α- and 6β-hydroxylation, 9β,10α-androst-4-ene-3,17-dione (**73**) is 9β-hydroxylated, again due to the reversed binding orientation (Scheme 12).[100]

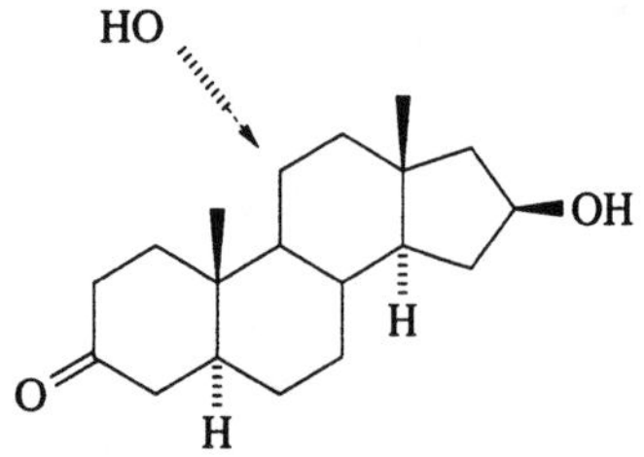

(70) 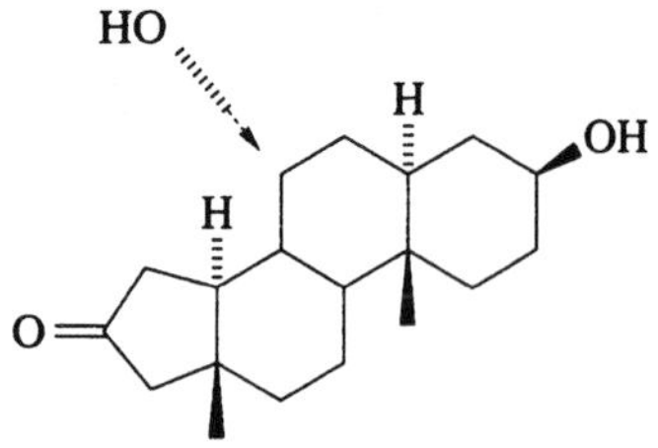 (71)

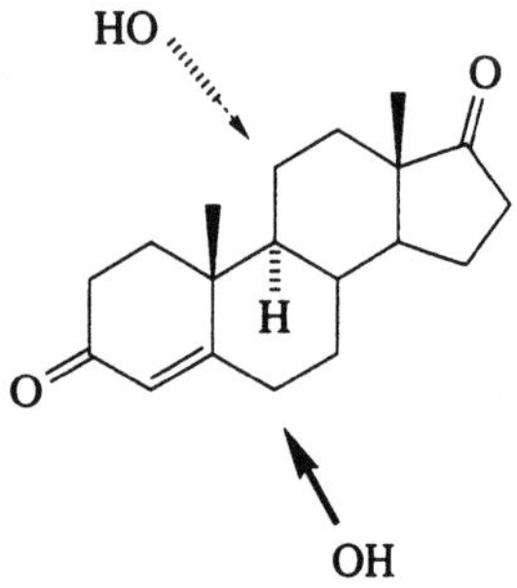

(72) 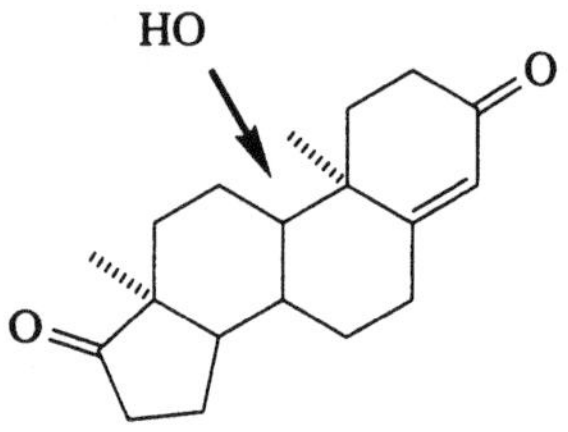 (73)

Scheme 12

1.4.4.2.3 Stereoselectivity

Hydroxylation of steroids at unactivated positions occurs exclusively with net retention of configuration, and this is believed to be the case for all cytochrome *P*-450 dependent steroid hydroxylations, irrespective of the microorganism employed.[83a] As in the case of regioselectivity, certain microorganisms have become associated with introducing hydroxy groups with a preferred stereoselectivity irrespective of position hydroxylated. Thus *Calonectria decora* is associated with the introduction of equatorial —OH groups, *Curvularia lunata* an axial —OH group and *Rhizopus nigricans* largely an equatorial —OH group. *Cunninghamella* spp. are less stereospecific, giving both axial- and equatorial-substituted products.

Stereoselectivity will be dictated in most cases by the binding of the substrate to the hydroxylating enzyme. One exception to this occurs in the 6β-hydroxylation of 3-keto Δ^4-steroids. In this case the stereochemistry of substitution at C-6 of the product is determined largely by conventional stereoelectronic processes, as the mechanism is believed to involve axial addition of oxidant to a conjugate of the substrate and the hydroxylating enzyme (equation 24).[83a]

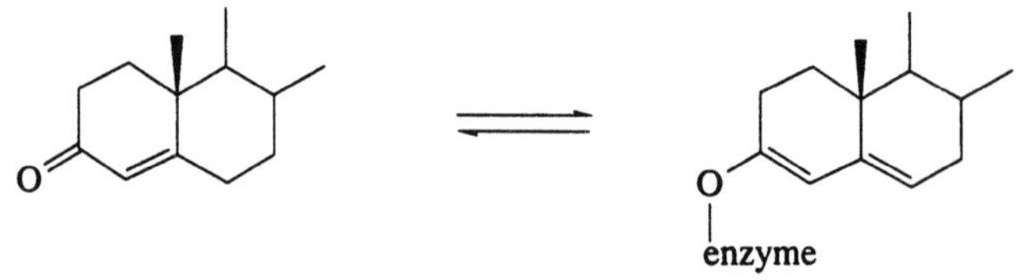

(24)

The principles alluded to above are exemplified by some further steroid hydroxylations in Sections 1.4.4.2.4 and 1.4.4.2.5.

1.4.4.2.4 Saturated substrates

The hydroxylation of 5α-androstanes has previously been discussed on a microorganism basis,[4] and yields of from a few percent up to 50% or better can be achieved. The work has since been extended to cover the microorganism *Leptoporus fissilis* with oxygenated 5α-androstanes[84] and the microbial oxidation of A-nor- and A-homo-5α-androstanes by *Cunninghamella elegans*.[101] The chemically modified steroid (**74**) can be hydroxylated at the 11α-position by *Rhizopus nigricans* when R = H, but when R = Me hydroxylation occurs predominantly at the 1β-position[102] to give the steroid (**75**). Acid-catalyzed rearrangement of the steroid (**75**) can be used to give 1β,3β-dihydroxyandrost-5-en-17-one (**76**; Scheme 13).

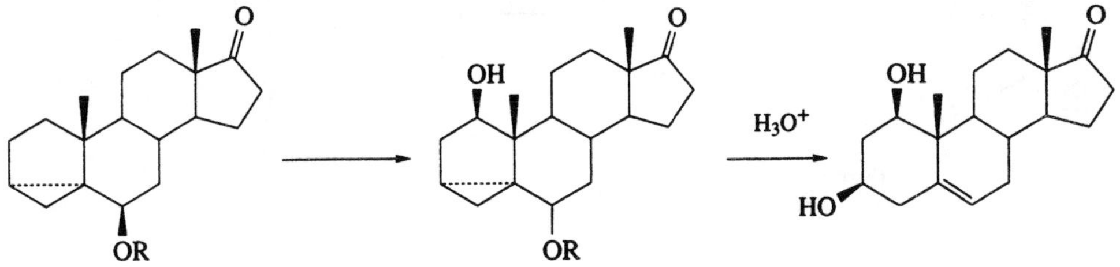

(**74**) (**75**) (**76**)

Scheme 13

Substitution on the steroid nucleus by fluorine leads to complicated changes in selectivity, but usually results in hydroxylation at sites remote from the fluorine substituent. Thus 5α-androstan-17-one (**77**) is primarily 7β,11α-dihydroxylated by *Aspergillus ochraceus*, but the 12,12-difluoro derivative (**78**) is 7β-monohydroxylated (Scheme 14).[92]

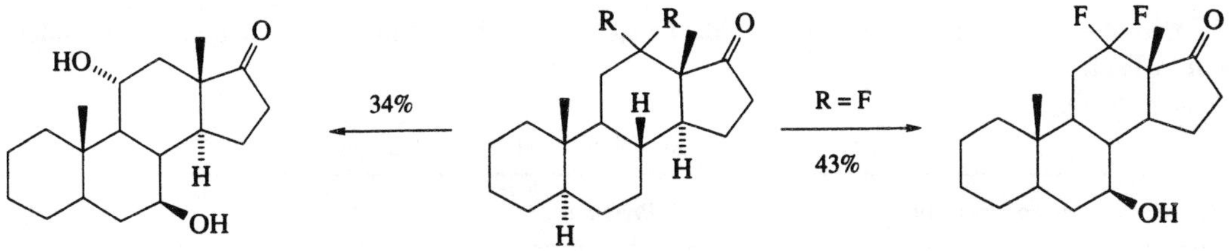

(**77**) R = H
(**78**) R = F

Scheme 14

1.4.4.2.5 Unsaturated substrates

Aspergillus giganteus (ATCC 10059) will dihydroxylate progesterone (**64**) to give the 11α,15β-dihydroxy derivative (**79**; equation 25), which is a precursor to the oogonial steroids.[103]

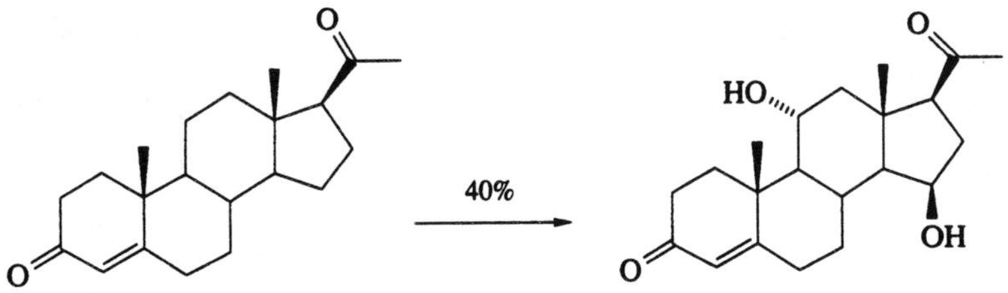

(25)

(**64**) (**79**)

Bile acids have historically received much less attention than other steroids, for example the corticosteroids; however, now that useful therapeutic effects are being observed from some bile acids, there is fresh interest in this area. A strain of *Cunninghamella blakesleeana* has been isolated that will 15β-hydroxylate lithocholic acid (**80**; equation 26) in 31% yield.[104] Further reaction is possible[105] to give 3α,11β,15β-trihydroxy-5β-cholanic acid (15%), 3α,15β,18α-trihydroxy-5β-cholanic acid (4%) and 3α,11α,15β-trihydroxycholanic acid (9%). Taurolithocholic acid (**81**; equation 27) can be 7β-hydroxylated in virtually quantitative yield by *Mortierella ramanniana*.[106]

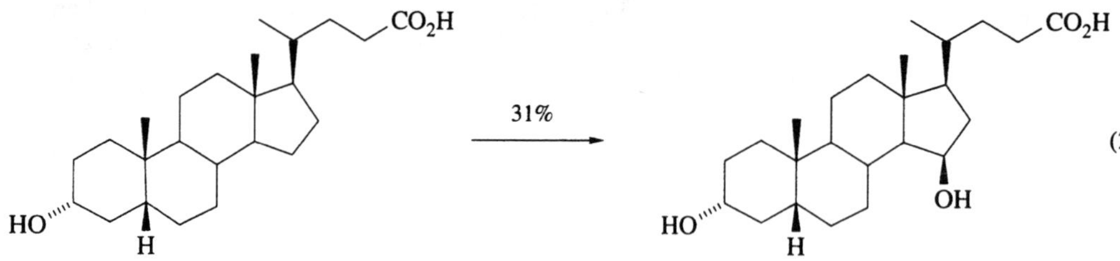

(80)

31%

(26)

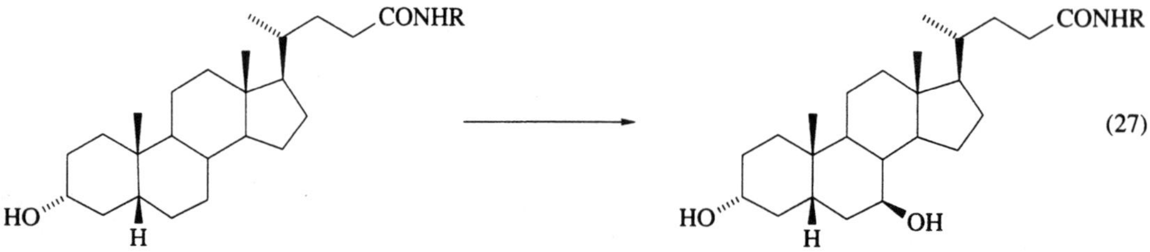

(81) $R = CH_2CH_2SO_3H$

(27)

Androst-4-ene-3,17-dione (**72**) can be hydroxylated readily and in high yield, and some examples are given in Table 7.

Table 7 Hydroxylation of Androst-4-ene-3,17-dione (**72**)

Microorganism	*Product*	*Yield (%)*	*Ref.*
Neosaltria fisheri (IFO-5866)	11α-OH	40	108
Nocardia canicruria (ATCC 31548)	9-OH	45	90
Penicillium stoloniferum (CBS P 102)	15α-OH	77	107

19-Hydroxy steroids are important as direct precursors to 19-norsteroids, but are not readily obtainable by microbial hydroxylation without additional nuclear substitution. However a strain of *Pellicularia filamentosa*[109] will 19-hydroxylate cortexolone (**82**; equation 28) successfully. 11β-Hydroxylation of cortexolone derivatives can be accomplished in up to 86% yield using a mutant strain of *Curvularia lunata* (FERM P-8515).[110]

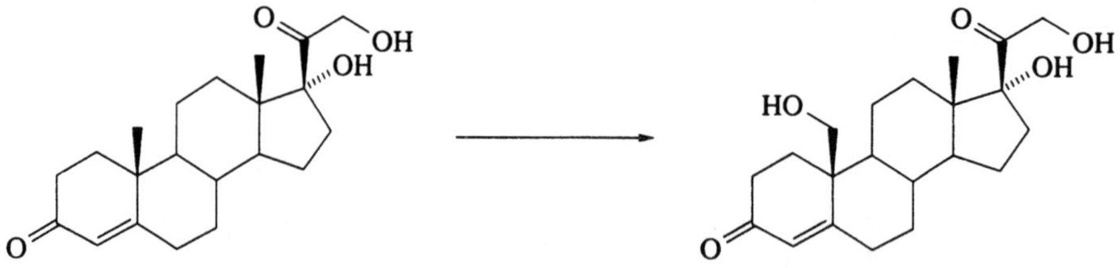

(28)

(82)

As a final example, hydroxylation of the Δ^5-steroids (**83**) and (**84**) can be accomplished microbially, and the products are useful in the synthesis of the aldosterone antagonist, spirorenone (**85**; Scheme 15).[111,112]

1.4.5 SPECIAL METHODS

In the preceding sections oxidation of unactivated C—H bonds has been discussed, and this is the most useful category of oxidation facilitated by microorganisms. However, although not strictly in this ca-

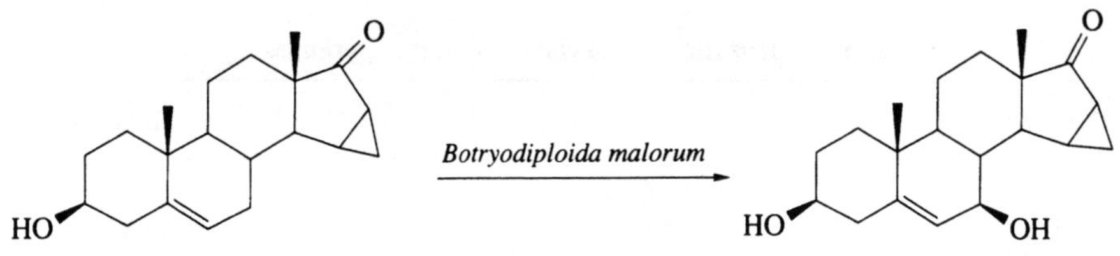

Botryodiploida malorum

(ref. 111)

(83)

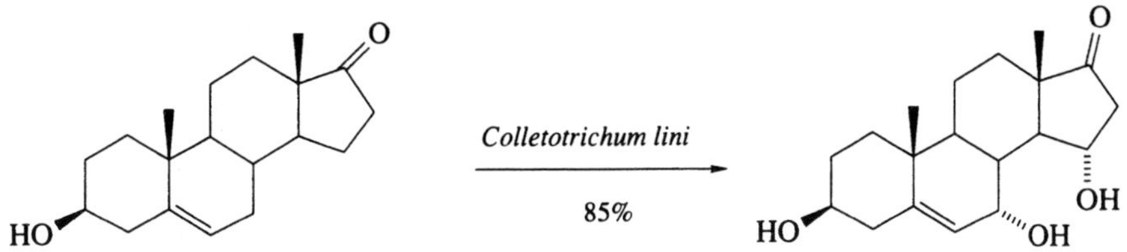

Colletotrichum lini

85%

(ref. 112)

(84)

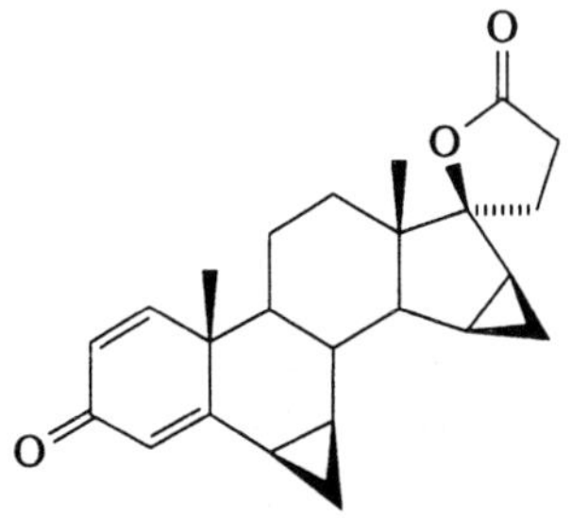

(85)

Scheme 15

tegory, some useful transformations of allylic, benzylic and aromatic substrates can also be accomplished in this way, and these are therefore considered in Sections 1.4.5.1 to 1.4.5.3.

1.4.5.1 Benzylic Substrates

In many cases benzylic oxidation by microorganisms does not stop at the alcohol but proceeds *via* the aldehyde through to the benzoic acid.[2] With some highly substituted substrates, however, good yields of the benzyl alcohol may be obtained. For example *Aspergillus selerotiorum* will convert the toluene derivative (**86**) into the benzyl alcohol (**87**; equation 29) in 66% yield.[113] The same microorganism (IMI 56673), along with *Aspergillus unifer* (NRRL 3228) and *Beauveria sulfurescens*, will convert methylpyridines to the corresponding alcohols, and some results are shown in Table 8.[114] *Penicillium adametzi* and a range of other microorganisms are also useful in benzylic hydroxylations in certain specific cases.[4]

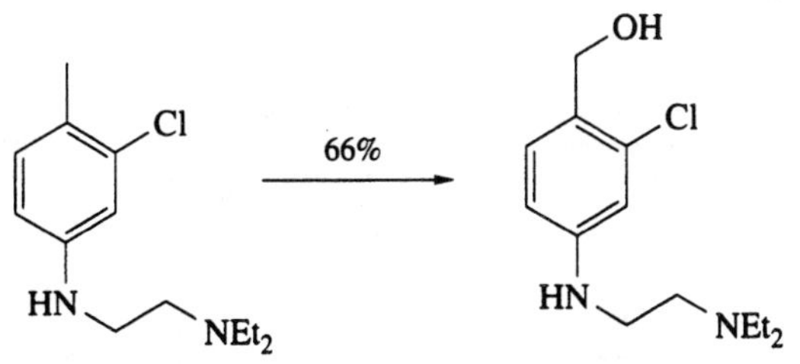

66% (29)

(86) **(87)**

Cunninghamella elegans has shown a degree of generality in the oxidation of benzylic substrates. For example the tetrahydroquinoline (**88**) is converted to the derivative (**89**; equation 30), but the degree of enantioselectivity is not known.[115] Similarly triprolidine (**90**) is converted to the alcohol (**91**; equation 31) in respectable yield.[116]

Table 8 Benzylic Hydroxylation of Methylpyridines

Substrate	*Products*	*Yield (%)*
2-Me	2-CH_2OH	42.7
4-Me	4-CH_2OH	44.4
2,6-Me_2	2-Me,6-CH_2OH	86.7
4,5-Me_2	4-Me,5-CH_2OH	81.7

OH

N CO-*p*-tolyl → N CO-*p*-tolyl (30)

(88) **(89)**

OH

N N *Cunninghamella elegans* (ATCC 9245) 55% N N (31)

(90) **(91)**

The fungus *Mortierella isabellina* (NRRL 1757) hydroxylates the ethylbenzenes (**92**) to the 1-arylethanols (**93**; equation 32) with a degree of enantioselectivity as shown in Table 9. By-products resulting from terminal carbon hydroxylation and overoxidation (acetophenones) are also obtained.[117]

HO

R → R (32)

(92) **(93)**

Table 9 Benzylic Hydroxylation of Ethylbenzenes by *Mortierella isabellina*

R	*Yield (%)*	*ee (%)*	*Configuration*
H	10	33	(*R*)
CN	60	39	(*R*)
Cl	30	34	(*S*)
Pr	45	25	(*R*)
NO_2	40	9	(*R*)
Me	4	20	(*R*)

Cunninghamella echinulata (ATCC 26269) and *Helminthosporium* spp. (NRRL 4671) are also useful in the benzylic oxidation of ethylbenzenes, both exhibiting enantioselectivities similar to *Mortierella isabellina*.[117] The enzyme responsible for these transformations has the characteristics of a cytochrome *P*-

450 dependent monooxygenase and no contribution from an alcohol–carbonyl interconversion at C-1 is observed.

Benzylic oxidations to give aldehydes (rare) and carboxylic acids have been considered previously.[2]

1.4.5.2 Allylic Substrates

There are few examples of synthetically useful allylic hydroxylations except in the steroidal field (Section 1.4.4.2), although some others are shown in Sections 1.4.3.3 and 1.4.1. Previous reviews on microbial oxidation[3,4] have also included some examples of allylic oxidations.

Good regioselectivity is obtained in the allylic hydroxylation of β-damascone (**94**) by *Botryosphaeria rhodina* (equation 33),[118] with the allylic hydroxylation product (**95**) being preferred over the homoallylic one (**96**). Little work seems to have been done on comparing microbial and chemical allylic oxidation methods, although in at least some cases microbial oxidation shows better selectivity to the alcohol and less over oxidation to carbonyl compounds than chemical methods. For example oxidation of the terminal allylic methyl group in novobiocin (**97**; equation 34) and related antibiotics by *Sebekia benihana* (NRRL 11111) is superior to chemical methods.[119] A similar terminal methyl allylic oxidation on an acyclic terpene has also been affected by a *Nocardia* microorganism (FERM-P 1609).[120a]

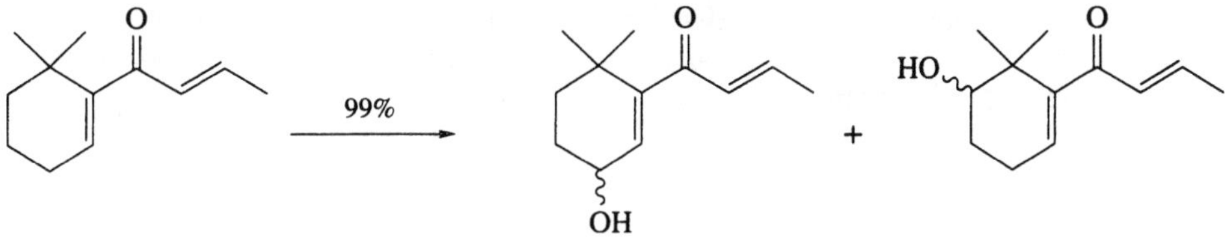

(**94**) → 99% → (**95**) 98% + (**96**) 2% (33)

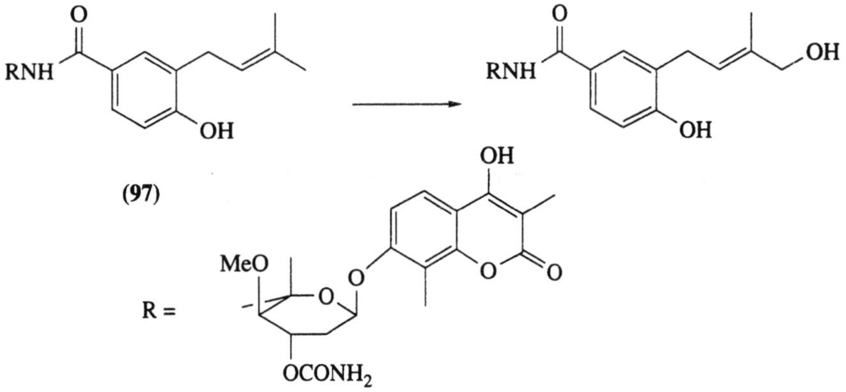

(**97**) → (34)

R =

Allylic hydroxylation of Δ^4-steroids at the 6β-position and Δ^5-steroids at the 7-position is possible using a wide range of microorganisms,[2,10] including *Cunninghamella elegans* and *Rhizopus nigricans*.

Milbemycin derivatives can be stereoselectively hydroxylated at the allylic 13β-position by a variety of microorganisms including *Streptomyces violascens* (ATCC 31560), *Streptomyces carbophilus* (FERM BP-1145) or *Streptomyces diastatochromogenes* (ATCC 31561), and in preparatively useful yields.[120b]

In the general case of unsaturated substrates, double bond attack or migration may take place during microbial oxidation; numerous examples of the latter are to be found in the steroidal field,[2] and less commonly with other classes of substrate. Thus the antibiotics compactin (**98**) and monacolin-K (**99**) can be hydroxylated at the allylic 3- and 8a-positions by *Syncephalastrum nigricans* (SANK 42372) [3α, 26% from (**98**)],[121] *Mucor hiemaliis* (SANK 36372) [3β, 72% from (**98**)][122] or *Schizophyllum commune* [8a, 80% from (**98**)].[123] However hydroxylation of (**98**) with a *Nocardia* species gave rearrangement as well as hydroxylation to give the allyl alcohol[124a] (**100**; R = H), and the use of *Mucor heimaliis* with (**99**) gave the corresponding 6α-hydroxy derivative (**100**; R = Me) as shown in equation (35).[122]

In other examples of allylic hydroxylation, *Streptomyces roseosporus* A-5797 (FERM BP-1574), *Streptomyces sclerotialus* (FERM BP-1370) or *Nocardia autotrophica* (FERM BP-1573)[124b] and *Botrytic cincerea*[124c] have proved useful.

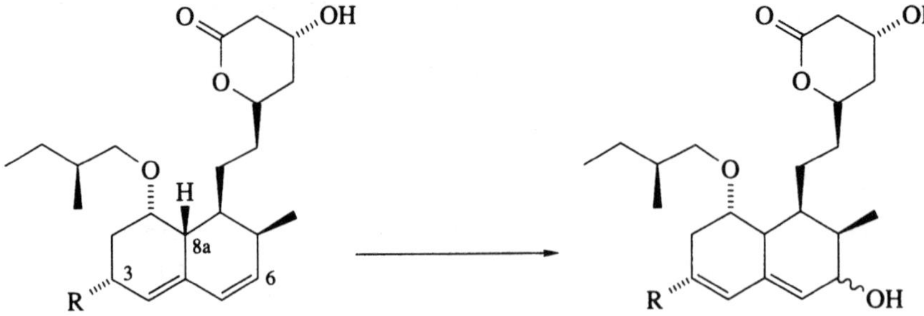

(35)

(98) R = H; **(99)** R = Me **(100)**

1.4.5.3 Nuclear Hydroxylation of Aromatic Hydrocarbons

Much work has been devoted to this area but a large part of this has been on metabolic studies[5] and little can be truly said to be synthetically useful. As with chemical oxidations on aromatic rings, a fundamental problem with microbial oxidations is that once hydroxylated the aromatic ring becomes more susceptible to further degradation. Nevertheless some useful transformations have been accomplished, and many of these have been reviewed previously.[3,4] An impressive example is the conversion of L-tyrosine (**101**) into L-DOPA (**102**; equation 36).

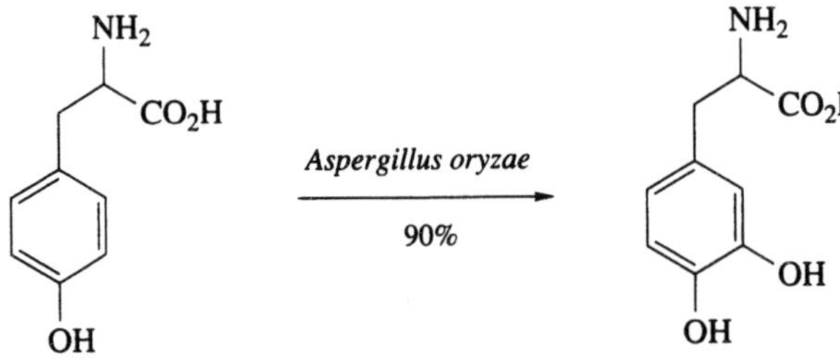

Aspergillus oryzae, 90% (36)

(101) **(102)**

Microbial *ortho*- and *para*-hydroxylation reactions are considered to proceed *via* the arene oxide,[125] and a phenomenon known as the NIH shift is commonly found to occur (Scheme 16) with di- or polysubstituted substrates.

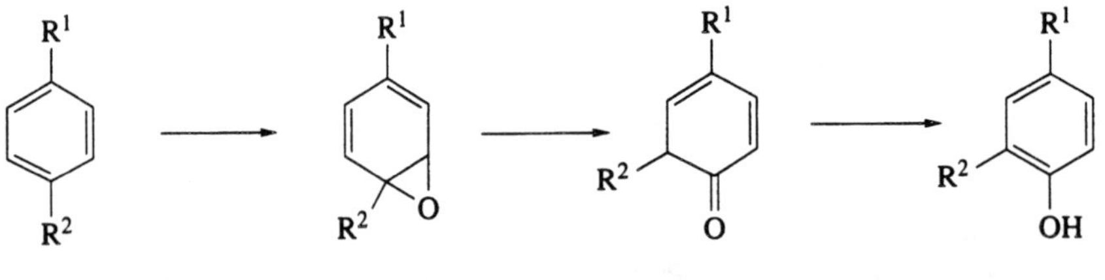

Scheme 16

Fungi appear to preferentially *ortho*-hydroxylate monosubstituted arenes,[125] however there are some exceptions. For example, the near ubiquitous fungus *Beauveria sulfurescens* (ATCC 7159) will hydroxylate the herbicide Propham (**103**) to give a 49% yield of *para*-substituted products (**104**; equation 37),[126] about half of which were *O*-glycosated.

Biphenyl (**105**) can be hydroxylated to either the mono- or di-hydroxylated derivatives. This is perhaps surprising considering the fungicidal properties of these materials; however, acclimatization of the microbe with monohydroxylated products allows a satisfactory rate of reaction. The 4,4′-dihydroxy isomer (**106**; equation 38) may be produced using *Absidia pseudocylindrospora* (NRRL 2770)[127] whilst *Pseudomonas* SG 1043 (FERM-P 4471) gives a mixture of 2- and 3-monohydroxy derivatives, in a ratio of 3:1.[128a]

Species of the *Aspergillus* genus (particularly various strains of *Aspergillus niger*) will also *p*-hydroxylate aromatics and accumulation of products can be good. Other substrates can be *p*-hydroxylated in

Beauveria sulfurescens, 49% (37)

(103) (104)

Absidia pseudocylindrospora (38)

(105) (106) X = OH

good yield by *Cunninghamella echinulata*,[128b] *Streptomyces rimosus* (ATCC 10970) or *Nocardia lyena* (ATCC 21430).[128c]

Methylotropic bacteria (for example *Methylomonas methanica*) will hydroxylate aromatics when methane or the like is also provided as a carbon source,[129] and as in the case of aliphatic hydrocarbons further developments are likely as new strains of these bacteria are isolated.

Little is known about the hydroxylation of heterocyclic substrates, and more experimental work is needed in this area. *Achromobacter xylosoxydans* (DSM 2783) and *Pseudomonas putida* (NCIB 8176 or 10521) are particularly effective in the hydroxylation of nicotinic acid (**107**) to 6-hydroxynicotinic acid (**108**; equation 39), and commercial processes capable of giving up to 97% yield have been developed.[130] It is interesting to note that in this case the product is precipitated as it forms by formation of its magnesium or barium salt, and this could serve to protect the product from further degradation as well as facilitating its recovery.

Achromobacter xylosoxydans, 97% (39)

(107) (108)

Benzimidazole can be hydroxylated in the 5-position by spores of *Absidia spinosa* and by microorganisms of the class micromycetes.[131]

1.4.6 ALTERNATIVES TO MICROORGANISMS

Microorganisms contain a multitude of enzyme systems, some of which are not directly involved in their use in chemical oxidations, but which are necessary for the cell to function in total. As discussed previously these ancillary enzymes can be harmful, and ways of reducing or avoiding their effects have been sought; two of these are considered in the following sections.

1.4.6.1 Oxidations with Isolated Enzymes

Isolated single enzymes have the potential advantage of totally specific transformations.[132,133] However this is often offset by disadvantages such as instability, the need to add or recycle cofactors, and the difficulty in obtaining suitable enzymes for oxidation of unactivated C—H systems.[132]

The nuclear hydroxylation of aromatics can be catalyzed by horseradish peroxidase, and with suitably activated substrates synthetically useful yields can be obtained.[134] The reactions are usually carried out at 0 °C in the presence of dihydroxyfumaric acid cofactor and a source of oxygen. Thus L-DOPA (**102**) has been prepared using this system.[135]

A general review on oxygenase-catalyzed hydroxylation of aromatic compounds has appeared, which discusses this area in greater detail.[136]

A bacterial methane monooxygenase that catalyzes the hydroxylation of saturated hydrocarbons[137] has been isolated, but there are severe limitations in practical use.[132] Similarly the enzymes involved in steroid hydroxylation have been isolated and used *in vitro*,[138,139] but this practice is not sufficiently well refined for general use in organic synthesis.

The cytochrome *P*-450 responsible for the hydroxylation of camphor by *Pseudomonas putida* has been isolated, and is, in contrast to most others, stable.[140] However this enzyme will only hydroxylate substrates closely related to camphor, and the site of hydroxylation may vary with substrate. Thus camphor gives the *exo*-5-hydroxy derivative but 5,5-difluorocamphor gives the 9-hydroxy derivative.[141]

1.4.6.2 Oxidation with Spores

In some cases improved selectivity in a microbial oxidation may be achieved by using spores rather than bacteria or fungal mycellium.[4] The technique has been largely restricted to the steroid area, with Δ^1-dehydrogenation being the most common use.[2] In hydroxylation reactions the spores of *Aspergillus ochraceus* and *Cunninghamella elegans* have proved particularly useful,[142] and in general the spores produce the same hydroxylation product as the derived microorganism. A limitation of the technique is that not all microorganisms produce spores.

1.4.7 REFERENCES

1. D. M. Hollinshead, S. C. Howell, S. V. Ley, M. Mahon, N. M. Ratcliffe and P. A. Worthington, *J. Chem. Soc., Perkin Trans. 1*, 1983, 1579.
2. P. Hartter, in 'Methoden der Organischen Chemie (Houben-Weyl)', Thieme, Stuttgart, 1981, vol. 4/1a, p. 963.
3. G. S. Fonken and R. A. Johnson, in 'Oxidation in Organic Chemistry', ed. J. S. Belew, Dekker, New York, 1972, vol. 2.
4. R. A. Johnson, in 'Oxidation in Organic Chemistry', ed. W. S. Trahanovsky, Academic Press, New York, 1978, vol. 5, p. 131.
5. K. Kieslich, 'Microbial Transformations of Non-Steroid Cyclic Hydrocarbons', Thieme, Stuttgart, 1976.
6. C. J. Suckling and K. E. Suckling, *Chem. Soc. Rev.*, 1974, **3**, 387.
7. B. Tabenkin, R. A. LeMahieu, J. Berger and R. W. Kierstead, *Appl. Microbiol.*, 1969, **17**, 714.
8. S. Kurozumi, T. Toru and S. Ishimoto, *Tetrahedron Lett.*, 1973, 4959.
9. American Type Culture Collection catalog, ATCC, Rockville, Maryland, USA.
10. W. Charney and H. L. Herzog, 'Microbiol Transformations of Steroids', Academic Press, New York, 1967.
11. J. P. Rosazza (ed.), 'Microbial Transformations of Bioactive Compounds', CRC Press, Boca Raton, Florida, 1982, vol. 1.
12. R. P. Elander and L. T. Chang, in 'Microbial Technology', ed. H. J. Peppler and D. Perlman, Academic Press, New York, 1979, p. 244.
13. H.-J. Rehm and G. Reed (eds.), 'Biotechnology', Verlag Chemie, 1984, vol. 6a.
14. C. Ratledge, *J. Am. Oil Chem. Soc.*, 1984, **61**, 447.
15. J. J. Perry, *Microbiol. Rev.*, 1979, **43**, 59.
16. R. Uchio and I. Shiio, *Agric. Biol. Chem.*, 1972, **36**, 426.
17. J. Hasegawa, M. Ogura, S. Hamaguchi, M. Shimazaki, H. Kawaharada and K. Watanabe, *J. Ferment. Technol.*, 1981, **59**, 203.
18. J. Hasegawa, S. Hamaguchi, M. Ogura and K. Watanabe, *J. Ferment. Technol.*, 1981, **59**, 257.
19. R. N. Patel, *Catal. Org. React.*, 1985, **22**, 327.
20. R. N. Patel, C. T. Hou, A. I. Laskin, A. Felix and P. Derelanko, *Appl. Environ. Microbiol.*, 1980, **39**, 720.
21. R. N. Patel, C. T. Hou, A. I. Laskin, A. Felix and P. Derelanko, *Appl. Environ. Microbiol.*, 1980, **39**, 727.
22. T. Imai, H. Takigawa, S. Nakagawa, G. J. Shen, T. Kodama and Y. Minoda, *Appl. Environ. Microbiol.*, 1986, **52**, 1403.
23. J. Hasegawa, *Hakko Kogaku Kaishi* 1984, **62**, 341 (*Chem. Abstr.*, 1984, **101**, 226 524q).
24. T. Timori and M. Shibasah, *Tetrahedron Lett.*, 1985, **26**, 1523.
25. A. I. Meyers and J. P. Hudspath, *Tetrahedron Lett.*, 1981, **22**, 3925.
26. M. R. Johnson, T. Nakata and Y. Kishi, *Tetrahedron Lett.*, 1979, 4347.
27. Q. Branca and A. Fishli, *Helv. Chim. Acta*, 1977, **60**, 925.
28. T. Goodhue and J. R. Schaeffer, *Biotechnol. Bioeng.*, 1971, **13**, 203.
29. J. Hasegawa (Kanegafuchi Chem. Ind.), *Ger. Pat.* 3 041 224 (1981) (*Chem. Abstr.*, 1981, **95**, 59 935y).
30. F. Hammerschmidt, *Oesterr. Chem. -Ztg.*, 1986, **87**, 321 (*Chem. Abstr.*, 1987, **106**, 154 684v).
31. B. W. Robertson (Gist-Brocades NV, Shell Int. Res. Mij. BV), *Eur. Pat.* 205 215 (1986) (*Chem. Abstr.*, 1987, **107**, 132 619w).
32. T. Sugai and K. Mori, *Agric. Biol. Chem.*, 1984, **48**, 2501.
33. I. A. Mil'man (Acad. Sci. Latvian SSR), *Ger. Pat.* 3 531 713 (1987) (*Chem Abstr.*, 1987, **106**, 154 809q).
34. R. A. Johnson, H. C. Murray and L. M. Reineke, *J. Am. Chem. Soc.*, 1971, **93**, 4872.
35. G. S. Fonken (Upjohn Co.), *Br. Pat.* 1 104 944 (1964) (*Chem. Abstr.*, 1967, **66**, 9974r).

36. G. S. Fonken (Upjohn Co.), *Br. Pat.* 1 036 084 (1966) (*Chem. Abstr.*, 1966, **65**, 11 307g).
37. H. G. Davies, M. J. Dawson, G. C. Lawrence, J. Mayall, D. Noble and S. M. Roberts, *Tetrahedron Lett.*, 1986, **27**, 1089.
38. S. Kishimota, H. Sugino, K. Tanaka, A. Kakinuma and S. Noguchi, *Chem. Pharm. Bull.*, 1976, **24**, 584.
39. Y. Yamazaki and H. Maeda, *Agric. Biol. Chem.*, 1985, **49**, 3201.
40. Y. Yamazaki and H. Maeda, *Tetrahedron Lett.*, 1985, **26**, 4775.
41. S. H. El-Sharkaway and Y. J. Abul-Hajj, *J. Org. Chem.*, 1988, **53**, 515.
42. A. Archelas, R. Furstoss, B. Waegell, J. Petit and L. Deveze, *Tetrahedron*, 1984, **40**, 355.
43. J. D. Fourneron, A. Archelas, B. Vigne and R. Furstoss, *Tetrahedron*, 1987, **43**, 2273.
44. A. Archelas and C. Morin, *Tetrahedron Lett.*, 1984, **25**, 1277.
45. A. Archelas, R. Furstoss, D. Srairi and G. Maury, *Bull. Soc. Chim. Fr.*, 1986, 234.
46. (a) A. Archelas, J. D. Fourneron and R. Furstoss, *J. Org. Chem.*, 1988, **53**, 1797; (b) G. Dauphin, A. Kergomard, C. Sainz and H. Veschambre, *Agric. Biol. Chem.*, 1989, **53**, 1433.
47. (a) R. Furstoss, A. Archelas, J. D. Fourneron and B. Vigne, *NATO Adv. Study Inst. Ser., Ser. C*, 1986, **178**, 361; (b) J. D. Fourneron, A. Archelas and R. Furstoss, *J. Org. Chem.*, 1989, **54**, 2478; (c) A. Archelas, J. D. Fourneron and R. Furstoss, *Tetrahedron Lett.*, 1988, **29**, 6611.
48. P. Teisseire, *Bull. Soc. Chim. Fr.*, 1980, II-66.
49. A. Ciegler, 'CRC Handbook of Microbiology', CRC Press, Boca Raton Florida, 1974, vol. 4, p. 449.
50. (a) B. A. Marples, in 'Terpenoids and Steroids', The Royal Society of Chemistry, London, 1980, p. 214; (b) W. R. Abraham, H. A. Arfmann, B. Stumpf, P. Washausen and K. Kieslich, *Bioflavor '87*, 1988, 399.
51. K. Nakajima, in 'Bioreactions and Biotransformations' ed. G. W. Moody and P. B. Baker, Elsevier, London, 1987, p. 219.
52. (a) K. M. Madyastha and M. S. R. KrishnaMurthy, *Tetrahedron Lett.*, 1988, **29**, 579; (b) H. A. Arfmann, W. R. Abraham and K. Kieslich, *Biocatalysis*, 1988, **2**, 59.
53. (a) J. P. N. Rosazza, J. J. Steffens, F. S. Sariaslani, A. Goswami, J. M. Beale, Jr. and S. Reeg, *Appl. Environ. Microbiol.*, 1987, **53**, 2482; (b) W. G. Liu, A. Goswami, R. P. Steffek, R. L. Chapman, F. S. Sariaslani and J. P. N. Rossazza, *J. Org. Chem.*, 1988, **53**, 5700; (c) J. P. N. Rossazza, A. Goswami, W. G. Liu, F. S. Sariaslani, J. J. Steffens and R. P. Steffek, *Dev. Ind. Microbiol.*, 1988, **29**, 181.
54. Jap. Tab. and Salt Co., *Jpn. Pat.* 81 42 591 (1979) (*Chem. Abstr.*, 1981, **95**, 59 934x).
55. Y. Mikami, E. Watanabe, Y. Fukunaga and T. Kisaki, *Agric. Biol. Chem.*, 1978, **42**, 1075.
56. S. J. Wright, P. Caunt, D. Carter and P. B. Baker, *Appl. Microbiol. Biotechnol.*, 1986, **23**, 224.
57. M. S. Allen, N. Darby, P. Salisbury, E. R. Sigurdson and T. Money, *Can. J. Chem.*, 1979, **57**, 733.
58. A. Goswami, J. M. Beale, Jr., R. L. Chapman, D. W. Miller and J. P. Rosazza, *J. Nat. Prod.*, 1987, **50**, 49.
59. K. Yamakawa, K. Nishitani, M. Iida and A. Mikami, *J. Ferment. Technol.*, 1988, **66**, 51.
60. V. Lamare, J. D. Fourneron, R. Furstoss, C. Ehret and B. Corbier, *Tetrahedron Lett.*, 1987, **28**, 6269.
61. (a) A. Fujwara, M. Tazoe, Y. Shiomi and M. Fujiwara, *Adv. Biotechnol.*, 1981, **3**, 519; (b) V. Lamare, A. Archelas, R. France, M. Cesario, C. Pascard and R. Furstoss, *Tetrahedron*, 1989, **45**, 3761.
62. E. L. Ghisalberti, P. R. Jefferies and M. A. Sefton, *J. Chem. Res. (M)*, 1982, 166.
63. R. McCrindle, B. N. Ganguli, J. K. Turnbull and A. B. Anderson, *J. Chem. Soc., Perkin Trans. 1*, 1975, 1202.
64. Y. Khandelwal, P. K. Inamdar, N. J. de Souza, R. H. Rupp, S. Chatterjee and B. N. Ganguli, *Tetrahedron*, 1988, **44**, 1661.
65. P. J. Davis, *Microb. Transform. Bioact. Compd.*, 1982, **2**, 167.
66. J. P. Rosazza (ed.), 'Microbial Transformations of Bioactive Compounds', CRC Press, Boca Raton, Florida, 1982, vol. 2.
67. H. L. Holland, in 'The Alkaloids', ed. R. G. A. Rodrigo, Academic Press, New York, 1981, vol. 18, p. 323.
68. G. Lesma, G. Palmisano, S. Tllari, P. Ame, P. Martelli and U. Valcavi, *J. Org. Chem.*, 1983, **48**, 3825.
69. G. Neef, U. Eder, K. Petzoldt, A. Seeger and H. Wieglepp, *J. Chem. Soc., Chem. Commun.*, 1982, 366.
70. K. M. Kerr and P. J. Davis, *J. Org. Chem.*, 1983, **48**, 928.
71. B. J. Abbott, *Dev. Ind. Microbiol.*, 1979, **20**, chap. 32.
72. L. W. Robertson, R. J. McClanahan, J. A. Schrier and M. M. Tsai, *Abstr. Annu. Meet. Am. Soc. Microbiol.*, 1983, 247.
73. G. Nomine, *Bull. Soc. Chim. Fr.*, 1980, II-18.
74. A. H. Rose (ed.), 'Economic Microbiology', Academic Press, New York, 1980, vol. 5.
75. (a) W. J. Marsheck, Jr., *Prog. Ind. Microbiol.*, 1971, **10**, 49; (b) L. Sedlaczek, *Crit. Rev. Biotechnol.*, 1988, **7**, 187.
76. S. B. Mahato and S. Banerjee, *Phytochemistry*, 1984, **23**, 2131.
77. S. B. Mahato and S. Banerjee, *Phytochemistry*, 1985, **24**, 1403.
78. H. J. Wolf and L. A. Kominek, *Ann. N.Y. Acad. Sci.*, 1984, **434**, 106.
79. J. Protiva and V. Schwartz, *Folia Microbiol. (Prague)*, 1974, **19**, 151.
80. Y. Shen, *Yiyao Gongye*, 1985, **16**, 514 (*Chem. Abstr.*, 1986, **104**, 67 372g).
81. L. A. Kominek, H. J. Wolf and T. W. Evans (Upjohn Co.), *US Pat.* 4 524 134 (1984) (*Chem. Abstr.*, 1984, **100**, 190 328s).
82. E. R. H. Jones, *Pure Appl. Chem.*, 1973, **33**, 39.
83. (a) H. L. Holland, *Chem. Soc. Rev.*, 1982, **11**, 371; (b) S. B. Mahato, S. Banerjee and S. Podder, *Phytochemistry*, 1989, **28**, 7.
84. W. A. Denny, P. M. Fredericks, I. Ghilezan, E. R. H. Jones, G. D. Meakins and J. O. Miners, *J. Chem. Res. (M)*, 1980, 345.
85. P. Somal and C. L. Chopra, *Appl. Microbiol. Biotechnol.*, 1985, **21**, 267.
86. E. J. Volker, *J. Chem. Educ.*, 1977, **54**, 65.
87. A.-M. H. El-Refai, L. A. R. Sallam and S. Sabry, *Rev. Assoc. Bioquim. Argent.*, 1975, **40**, 129 (*Chem. Abstr.*, 1976, **85**, 61 353u).
88. R. T. Rowlands, *Enzyme Microb. Technol.*, 1984, **6**, 3.
89. H. L. Holland and E. Riemland, *Can. J. Chem.*, 1985, **63**, 1121.

90. W. J. Marsheck, Jr. (G. D. Searle and Co.), *US Pat.* 4 397 947 (1983) (*Chem. Abstr.*, 1983, **99**, 174 223j).
91. K. Kieslich, *Bull. Soc. Chim. Fr.*, 1980, II-9.
92. T. G. C. Bird, P. M. Fredericks, E. R. H. Jones and G. D. Meakins, *J. Chem. Soc., Perkin Trans. 1*, 1980, 750.
93. G. Wix, K. G. Buki, E. Tomorkeny and G. Ambrus, *Steroids*, 1968, **11**, 401.
94. C. W. Despreaux, K. R. Rittweger and N. J. Palleroni, *Appl. Environ. Microbiol.*, 1986, **51**, 946.
95. V. Bihari, P. P. Goswani, S. H. M. Rizvi, A. W. Khan, S. K. Basu and V. C. Vora, *Biotechnol. Bioeng.*, 1984, **26**, 1403.
96. I. A. El-Kady and R. M. Metwali, *Qatar Univ. Sci. Bull.*, 1985, **5**, 145 (*Chem. Abstr.*, 1987, **106**, 3752q).
97. K. M. Ghanem, A.-M. H. El-Refai and E. R. El-Helow, *Microbiol. Esp.*, 1985, **38**, 107.
98. K. M. Madyastha, K. Madhava and J. Srivatsas, *Can. J. Microbiol.*, 1987, **33**, 361.
99. V. E. M. Chambers, W. A. Denny, J. M. Evans, E. R. H. Jones, A. Kasal, G. D. Meakins and J. Pragnell, *J. Chem. Soc., Chem. Commun.*, 1973, 1500.
100. J. Favero, T. Ton That and F. Winternitz, *Bull. Soc. Chim. Fr.*, 1979, II-56.
101. T. A. Crabb and N. M. Ratcliffe, *J. Chem. Res. (M)*, 1985, 650.
102. H. Kim Thoa, Z. Prochazka, M. Budesiinsky and P. Kocovsky, *Collect. Czech. Chem. Commun.*, 1978, **43**, 2305.
103. T. C. McMorris, P. H. Le, M. W. Preus, S. R. Schow and G. R. Weihe, *J. Org. Chem.*, 1983, **48**, 3370.
104. S. Kulprecha, T. Hihira, C. Shimomura, K. Yanada, N. Nilubol, T. Yoshida and H. Taguchi, *Tetrahedron*, 1984, **40**, 2843.
105. Y. Jodoi, T. Nihira, H. Naoki, Y. Yamada and H. Taguchi, *Tetrahedron*, 1987, **43**, 492.
106. H. Sawada and M. Watanuki (Yakult Honsha Co. Ltd.), *Eur. Pat.* 119 040 (1984) (*Chem. Abstr.*, 1985, **102**, 4461y).
107. K. Petzoldt (Schering AG), *Eur. Pat.* 150 782 (1985) (*Chem. Abstr.*, 1986, **104**, 166 905q).
108. Dainippon Ink Chem. KK, *Jap. Pat.* 58 179 498 (1983) (*Chem. Abstr.*, 1984, **100**, 84 228z).
109. T. A. Clark, R. Chang and I. S. Maddox, *Appl. Microbiol. Biotechnol.*, 1985, **21**, 132.
110. A. Kamata and Y. Imada (Mitsubishi Chem. Co. KK), *Jap. Pat.* 62 118 898 (1985) (*Chem. Abstr.*, 1987, **107**, 132 637a).
111. D. Bittler, H. Hofmeister, H. Laurent, K. Nickisch, R. Nickolson, K. Petzoldt and R. Wiechert, *Angew. Chem.*, 1982, **94**, 718.
112. K. Petzoldt, H. Laurent and R. Wiechert, *Angew. Chem.*, 1983, **95**, 413.
113. S. Archer and D. Rosi, *Br. Pat.* 1 136 483 (1968) (*Chem. Abstr.*, 1968, **69**, 36 175x).
114. A. M. Kost, L. I. Vorobeva and O. K. Shibilkina (Moscow Lomonosov Univ.), *Russ. Pat.* 803 369 (1987) (*Chem. Abstr.*, 1987, **107**, 96 603a).
115. T. A. Crabb and S. L. Soilleux, *J. Chem. Soc., Perkin Trans. 1*, 1985, 1381.
116. E. B. Hansen, Jr., R. H. Heflich, W. A. Korfmacher, D. W. Miller and C. E. Cerniglia, *J. Pharm. Sci.*, 1988, **77**, 259.
117. H. L. Holland, E. J. Bergen, P. C. Chenchaiah, S. H. Khan, B. Munoz, R. W. Ninniss and D. Richards, *Can. J. Chem.*, 1987, **65**, 502.
118. V. Krasnobajew (Givaudan), *Swiss Pat.* 654 567 (1986) (*Chem. Abstr.*, 1986, **105**, 113 607h).
119. O. K. Sebek and L. A. Dolak, *J. Antibiot.*, 1984, **37**, 136.
120. (a) S. Ishizaka (Eisai Co. Ltd), *US Pat.* 4 338 251 (1982) (*Chem. Abstr.*, 1981, **95**, 187 467j); (b) G. M. Ramos Tombo, O. Ghisalba, H. P. Schaer, B. Frei, P. Maienfisch and A. C. O'Sullivan, *Agric. Biol. Chem.*, 1989, **53**, 1531.
121. N. Serizawa, K. Nakagawa, J. Tsujita, A. Terahara and H. Kuwano, *J. Antibiot.*, 1983, **36**, 608.
122. N. Serizawa, K. Nakagawa, K. Hamano, Y. Tsujita, A. Terahara and H. Kuwano, *J. Antibiot.*, 1983, **36**, 604.
123. H. Yamashita, S. Tsubokawa and A. Endo, *J. Antibiot.*, 1985, **38**, 605.
124. (a) T. Okazaki and A. Naito, *Sankyo Kenkyusho Nempo*, 1986, **38**, 80, (*Chem. Abstr.*, 1987, **107**, 36 265); (b) Taisho-Pharm., *Eur. Pat.* 298 757 (1989); (c) J. Gladysz-Dmochowska, *J. Basic Microbiol.*, 1986, **26**, 577.
125. B. J. Auret, S. K. Balani, D. R. Boyd, R. M. E. Green and G. A. Berchtold, *J. Chem. Soc., Perkin Trans. 1*, 1984, 2659.
126. B. Vigne, A. Archelas, J. D. Fourneron and R. Furstoss, *Tetrahedron*, 1986, **42**, 2452.
127. R. D. Schwartz, A. L. Williams and D. B. Hutchinson, *Appl. Environ. Microbiol.*, 1980, **39**, 702.
128. (a) N. Murakami (Idemitsu Kosan Co. Ltd.), *Jpn. Pat.* 54 151 195 (1979) (*Chem. Abstr.*, 1980, **93**, 43 946a); (b) M. F. Pasulto, N. N. Singh, F. Jamali, R. T. Coutts and S. Abuzar, *J. Pharm. Sci.*, 1987, **76**, 177; (c) Kyowa-Hakko, *Jpn. Pat.* 63 251 088 (1988) (WPI Ac. No. 88-335827).
129. T. Uchiumi (Idemitsu Kosan KK), *Jpn. Pat.* 61 166 398 (1986) (*Chem. Abstr.*, 1987, **107**, 57 469s).
130. P. Lehky, H. Kulla and S. Mischler (Lonza AG), *Eur. Pat.* 152 948 (1985) (*Chem. Abstr.*, 1985, **103**, 176 998r).
131. F. Chapelle, R. Steiman, F. Seigle-Murandi and C. L. Duc, *Microbiol. Biotechnol.*, 1986, **23**, 430.
132. G. M. Whitesides and C.-H. Wong, *Angew. Chem., Int. Ed. Engl.*, 1985, **24**, 617.
133. J. B. Jones, *Tetrahedron*, 1986, **42**, 3351.
134. J. S. Dordick, M. A. Marletta and A. M. Klibanov, *Abstr. Pap. Am. Chem. Soc. 188th*, 1984, MBTD47.
135. A. M. Klibanov, Z. Berman and B. N. Alberti, *J. Am. Chem. Soc.*, 1981, **103**, 6263.
136. J. R. Mason, *Int. Ind. Biotechnol.*, 1988, **8**, 19.
137. R. N. Patel, C. T. Hou, A. I. Laskin and A. Felix, *Appl. Environ. Microbiol.*, 1982, **44**, 1130.
138. M. J. Coon and K. Inouye, *Ann. N.Y. Acad. Sci.*, 1985, **458**, 216.
139. Y. Fujimoto, C. S. Chen, Z. Szeleczky, D. D. Tullio and C. J. Sih, *J. Am. Chem. Soc.*, 1982, **104**, 4718.
140. C.-A. Yu and I. C. Gunsalus, *Biochem. Biophys. Res. Commun.*, 1970, **40**, 1431.
141. K. S. Eble and J. H. Dawson, *J. Biol. Chem.*, 1984, **259**, 14 389.
142. A. Jaworski, L. Sedlaczek and J. Dlugonski, *Appl. Microbiol. Biotechnol.*, 1984, **20**, 313.

2.1 Oxidation Adjacent to C═C Bonds

PHILIP C. BULMAN PAGE and TIMOTHY J. McCARTHY
University of Liverpool, UK

2.1.1 INTRODUCTION

Allylic oxidation remains a reaction of considerable value in organic synthesis. Oxidation reactions in this section are divided into two types: reactions which produce allylic alcohols (equation 1) and those which produce α,β-unsaturated aldehydes or ketones directly (equation 2). Examples from the recent literature fall approximately equally into each type. Examples of allylic oxidations which occur with rearrangement (equation 3 and 4) or give mixtures are discussed individually within each reagent type; however, it is fair to say that many reagents and reaction conditions can give either direct oxidation or oxidative rearrangement, and almost no allylic oxidation system exclusively gives one course of reaction in all cases. In many examples the course of oxidation depends mostly upon substrate structure.

(1)

(2)

(3)

(4)

For most reagents this pattern is multiplied in cases where several similar possible sites of oxidation exist around a particular double bond. While chemoselectivity and stereoselectivity are often good, poor regioselectivity is a weakness afflicting many allylic oxidation methods.

Among oxidations producing allylic alcohols or their derivatives the modern variants of selenium dioxide oxidations are by far the most popular. Systems based on metal acetates, particularly palladium trifluoroacetate, can be very useful and are receiving increasing attention; but the Kharasch–Sosnovsky reaction, once very common for allylic oxidation, is now rarely used. Sensitized photooxidation with singlet oxygen, a very well-known procedure, is still somewhat unpredictable and has perhaps received less consideration than it deserves.

Fewer methods exist for direct allylic/benzylic oxidation to give α,β-unsaturated carbonyl compounds. Some of these occur by initial oxidation to the allylic alcohol, followed by a second oxidation or oxidative rearrangement step, and reagent systems often give mixtures of alcoholic and carbonylic products. The most valuable methods for direct oxidation to enones involve chromium(VI), palladium or selenium reagents, but none had proved particularly satisfactory in terms of predictability, selectivity and generality until the development of the chromium trioxide–3,5-dimethylpyrazole complex, no doubt the best system currently available.

It is interesting to note that few examples of propargylic oxidation were found for either type of oxidation reaction.

2.1.2 ALLYLIC OXIDATION REACTIONS WHICH PRODUCE ALLYLIC ALCOHOLS, ESTERS OR ETHERS

2.1.2.1 Selenium Dioxide Based Reagents

2.1.2.1.1 Stereochemistry and mechanism

Selenium dioxide is still regarded as the most reliable and predictable reagent for this transformation, particularly for more substituted alkenes. A number of reviews of the reaction are available in the literature.[1] Selenium dioxide generally produces unrearranged (*E*)-allylic alcohols, an ene reaction–2,3-sigmatropic rearrangement sequence being the probable major pathway, and while several methods for the oxidation of allylic C—H bonds to produce alcohols are now available, the majority of recent synthetic examples have involved selenium methodology. Several studies of the stereochemistry and mechanism of selenium dioxide allylic oxidations have been carried out over the years and indeed a set of rules for predicting the outcome of such reactions appeared as early as 1939.[2] These studies serve to indicate the complexity of this reaction and the difficulties often observed in predicting its regio- and stereo-chemical outcome. Apart from allylic alcohols, other possible products are dienes, esters, ethers, enones, α-diketones and glycols. Following a number of less conclusive studies,[3] in 1970 Trachtenberg[4] examined the oxidation of a variety of substituted cyclohexene systems. Under conditions of refluxing wet dioxane solution and a deficiency of oxidant they obtained mostly unrearranged allylic alcohols, although enones were also observed in some cases. In contrast to earlier work they suggested that tertiary positions are preferentially oxidized over secondary, and secondary positions over primary ($CH > CH_2 > CH_3$). They further found the reactions to be stereoselective, pseudoaxial alcoholic products being favored in cyclic systems (*e.g.* equation 5).

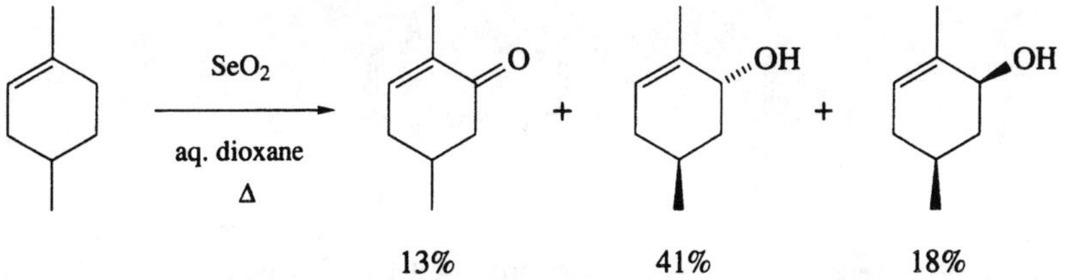

SeO_2, aq. dioxane, Δ (5)

13% 41% 18%

Early proposals concerning the mechanism of selenium dioxide allylic oxidations involved solvolysis of allylseleninic acid intermediates or free radical processes.[1,2,5,7,8] Trachtenberg[4] and Schaefer[7] discounted the involvement of allylseleninic acids on the grounds of inertness towards solvolysis, and subsequent work showed that radicals are not involved.[6,7] In order to explain his findings Trachtenberg[6] suggested a mechanism proceeding through an oxaselenocyclobutane intermediate (Scheme 1). This mechanism was preferred to those of Schaefer (Scheme 2)[7] and Wiberg[8] on stereochemical and other grounds.

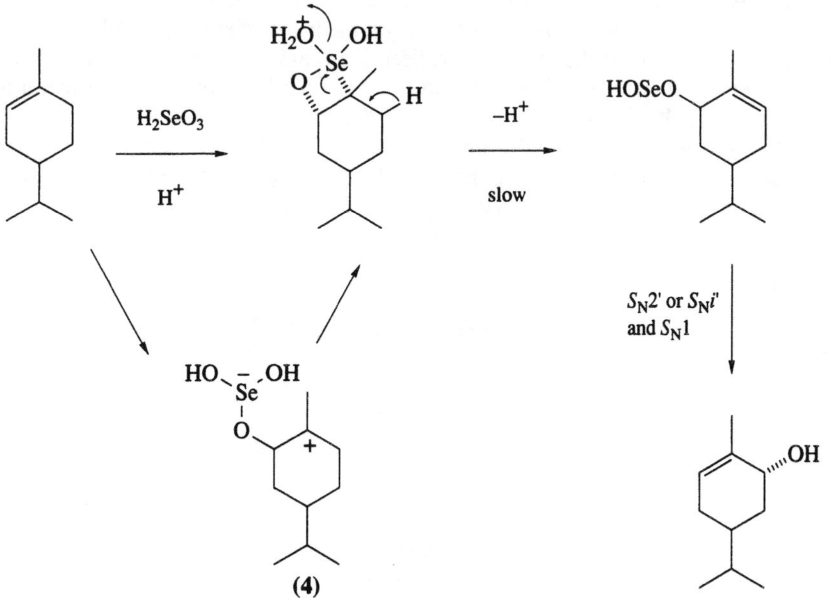

H_2SeO_3, H^+; $-H^+$, slow; HOSeO; S_N2' or S_Ni' and S_N1; (4)

Scheme 1

slow; fast; H_2O

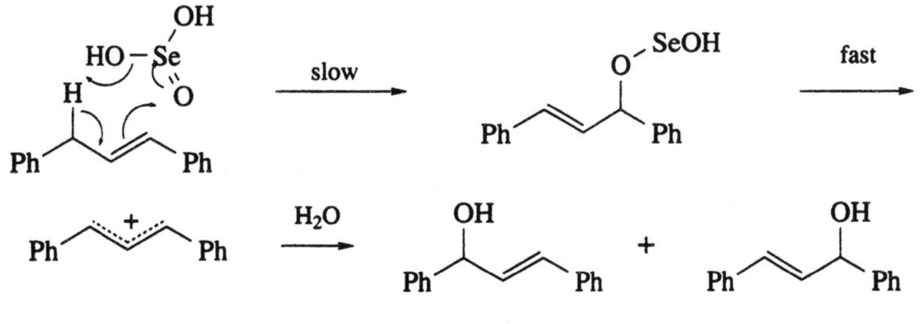

Scheme 2

A more complex picture was painted in a further study by Rapoport,[9] which indicated that both the mechanism and reactivity sequence are dependent upon the alkene structure and reaction conditions:[10] 1,2-disubstituted alkenes (**1**) reacting *via* an oxaselenocyclobutane intermediate with a reactivity sequence CH > CH_2 > CH_3; geminally disubstituted alkenes (**2**) with a reactivity sequence CH > CH_2 > CH_3; and trisubstituted alkenes (**3**) with a reactivity sequence CH_2 > CH_3 > CH, (*E*)-allylic alcohols being the preferred products as established by Büchi;[11] types (**2**) and (**3**) reacting *via* carbenium ion intermediates (**4**) without four-membered ring closure or by unspecified cyclic transition states. Rapoport's evidence also showed the final step to occur by S_Ni' or S_N1 processes and not by S_N2'. Monosubstituted alkenes, particularly arylpropenes, commonly react with rearrangement.[2,12]

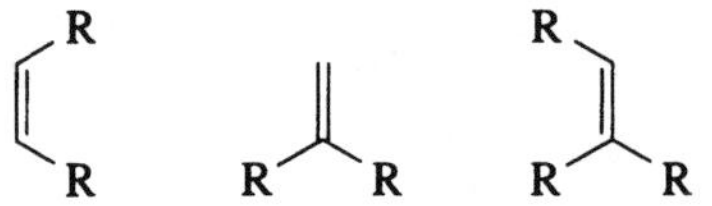

(1) (2) (3)

A new mechanism, now generally accepted, was proposed by Sharpless in 1972.[13] Following some elegant experimental work, Sharpless suggested that allylseleninic acids are indeed intermediates, but that they react by 2,3-sigmatropic rearrangement (well known for the analogous allylic sulfinates and sulfoxides) rather than by solvolysis. The fully regio- and stereo-specific sequence consists of two consecutive pericyclic reactions (an ene reaction followed by the sigmatropic rearrangement), and subsequent selenite ester hydrolysis (Scheme 3). Evidence for an initial ene reaction was provided by isolation from appropriate substrates of selenino ketones (**5**), presumed to be trapped forms of the allylseleninic acid intermediates (*e.g.* equation 6). However, while this mechanism does neatly explain the preference for (*E*)-allylic alcohol formation by invoking steric effects in the six-membered ring chair transition state of the sigmatropic rearrangement, it also implies a much higher level of stereo- and regio-control than is commonly observed. Stephenson[14] rationalized these facts on the basis of some careful kinetic isotope effect studies by proposing a mixture of the stereocontrolled Sharpless mechanism and a stereorandom ionic ene reaction equivalent, perhaps involving a carbon-bonded selenium-containing carbenium ion such as that shown in equation 7. Stephenson also comments that the ionic processes are suppressed in basic media, the Sharpless mechanism then being predominant. A similar ene reaction/2,3-sigmatropic rearrangement mechanism incorporating an alternative explanation for reduced regio- and stereo-selectivity was put forward in 1980 by Woggon[15] following a ^{13}C-labeling study (Scheme 4).

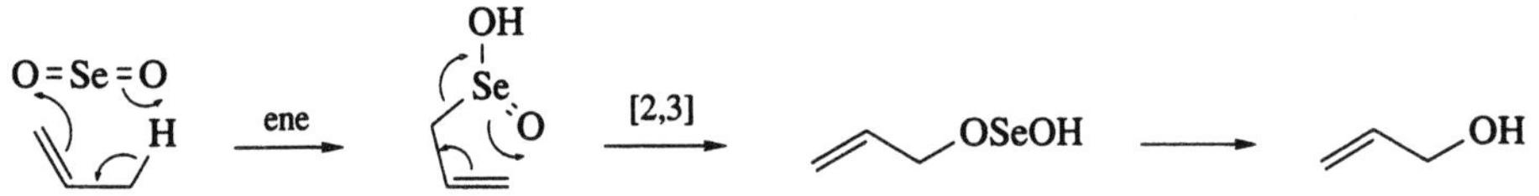

Scheme 3

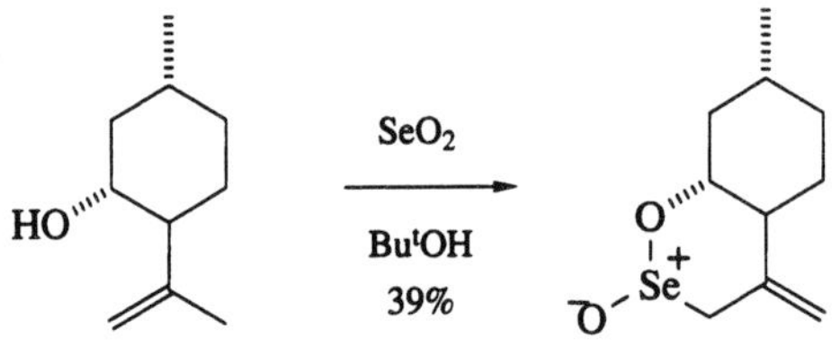

SeO_2, Bu^{t}OH, 39% (6)

(**5**)

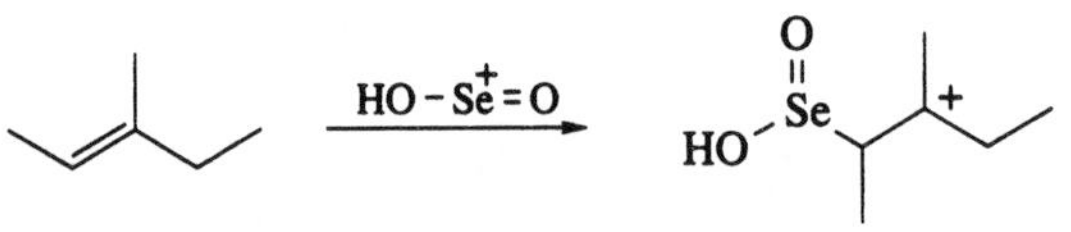

(7)

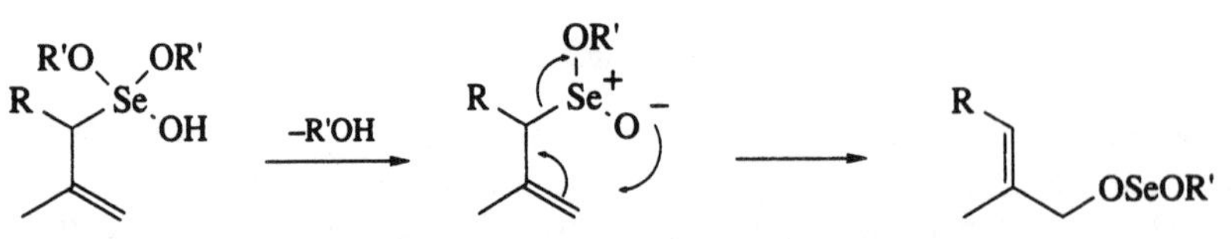

Scheme 4

2.1.2.1.2 Synthetic examples

Among early synthetic examples of selenium dioxide allylic oxidation[16a] is Rapoport's synthesis of sirenin (**6**).[16b] Oxidation of (**7**) with selenium dioxide in ethanol at 90 °C for 13 h gave a mixture of allylic alcohol (**8**) and aldehyde (**9**; equation 7a). The reaction was apparently regiospecific and it is valuable to note the survival of the three-membered ring under these conditions. Rapoport has also reported the oxidation of *cis–trans* mixtures of allylic alcohols to give all-*trans* α,β-unsaturated aldehydes using the same reagent.[17] A mechanism proceeding through allylic selenite esters was proposed (Scheme 5) and the reaction applied to a synthesis of squalene.

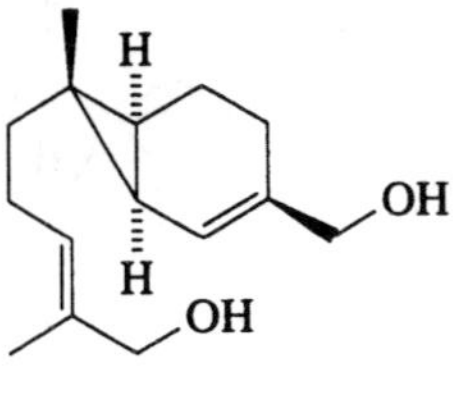

H H OH OH

(6)

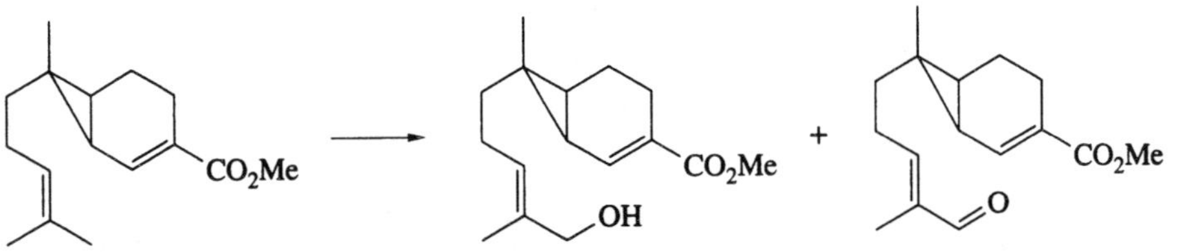

CO₂Me CO₂Me + CO₂Me (7a)

OH O

(7) **(8)** **(9)**

R OH R O Se OH OH O H R CHO

R O OH Se OH H O

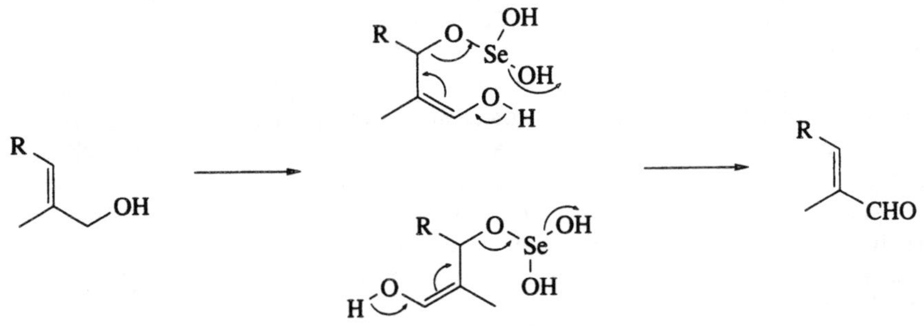

Scheme 5

While most synthetic examples of selenium dioxide allylic oxidation now involve more sophisticated systems (see below), use of the simple stoichiometric reagent alone is still popular.[12,18] One recent example which well illustrates the mildness and possible selectivity of the reaction is shown in equation (8).[18a]

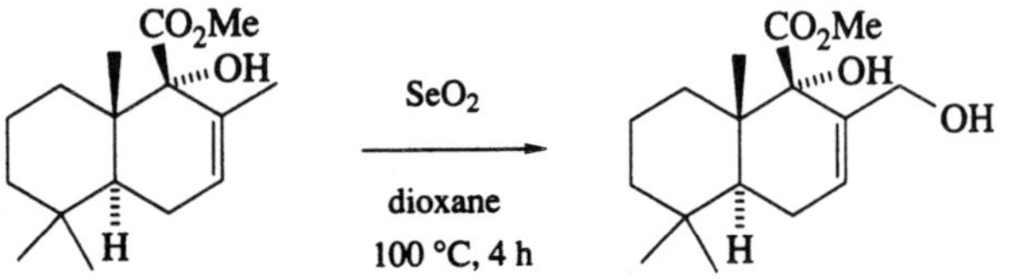

CO_2Me OH H $\xrightarrow[\text{dioxane, 100 °C, 4 h}]{SeO_2}$ CO_2Me OH OH H (8)

In this regiospecific oxidation, taken from a synthesis of (–)-warburganal (**10**) from glycyrrhetinic acid (**11**), only allylic methyl group oxidation occurred, no overoxidation or competing reactions at the tertiary alcohol or ester groups being reported. The reaction conditions, selenium dioxide in dioxane solution at 100 °C for 4 h, are typical, although one very recent example used *t*-butyl alcohol as solvent to produce the sensitive diol (**13**; equation 9).[19]

Selenium dioxide has also been used in combination with pyridine for the preparation of acid-sensitive materials.[20]

A useful modification of selenium dioxide allylic oxidation was introduced by Sharpless in 1977.[21] An inevitable complication of the normal reaction is the production of odoriferous low-valent selenium species which may be difficult to remove from the product mixtures and which can give rise to organoselenium by-products.[22] Sharpless provided a solution to this problem by the introduction into the re-

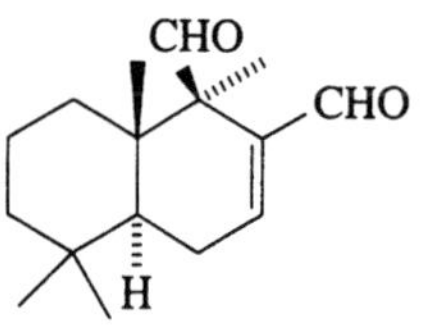

(10) 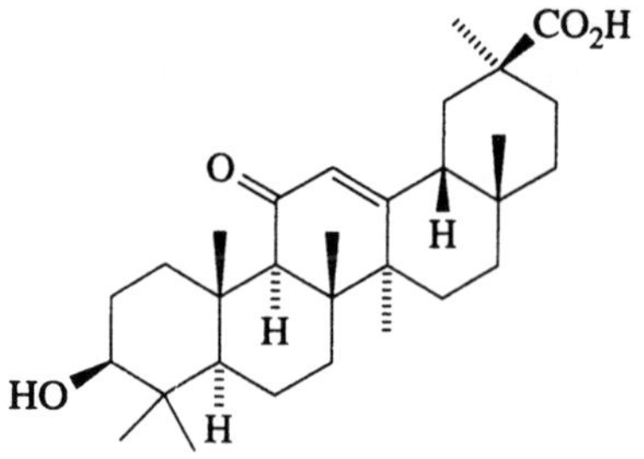 **(11)**

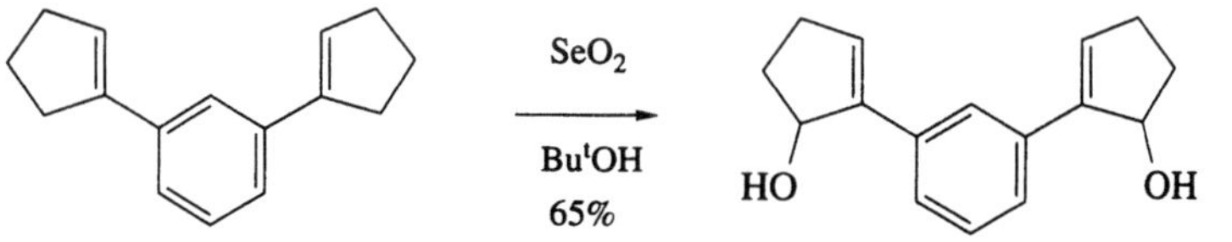

(12) $\xrightarrow[\text{Bu}^t\text{OH, 65\%}]{\text{SeO}_2}$ **(13)** (9)

action of a selective reoxidant for selenium. While the use of hydrogen peroxide[23] tends to give rise to epoxide products, *t*-butyl hydroperoxide is successful and often results in higher yields and a cleaner reaction, although further oxidation of allylic alcohol competes in some cases. Propargyl substrates generally undergo α,α′-dioxygenation.[24] Oxidation of more substituted alkenes can also be carried out at room temperature under catalytic conditions (1.5–2% SeO_2) using *t*-butyl hydroperoxide as reoxidant, and is then most efficient in methylene chloride solution using 3 to 4 equiv. of reoxidant. Some alkenes require water or carboxylic acid catalysis, and indeed in the presence of a hydroxylic solvent the reagent may be selenious acid or an alkyl selenite. Less substituted and unreactive alkenes are not efficiently oxidized using the catalytic system or in the absence of reoxidant; but use of stoichiometric selenium dioxide in combination with 2 equiv. of *t*-butyl hydroperoxide in methylene chloride solution provides a very mild and general allylic and propargylic oxidation reagent. Indeed, this reagent combination is probably still the best currently available for selective allylic oxidation without rearrangement.

Under these conditions for 2 h at room temperature germacrane-type sesquiterpene lactones **(14)**, **(15)** and **(16)** were all regio- and stereo-selectively oxidized at C-14 to give the corresponding allylic alcohols **(17)**, **(18)** and **(19)**, in which the double bond geometry within the 10-membered ring has been inverted.[25] Compound **(17)** was formed in 90% yield, the only observed by-product being aldehyde (**20**; 5%). The authors proposed an ene reaction–2,3-sigmatropic rearrangement pathway similar to those of Sharpless and Woggon but in which the initial reaction is the formation of an activated selenium *t*-butyl hydroperoxide species which undergoes the ene reaction (Scheme 6). The authors also comment that catalytic amounts of selenium dioxide suffice for the reaction.

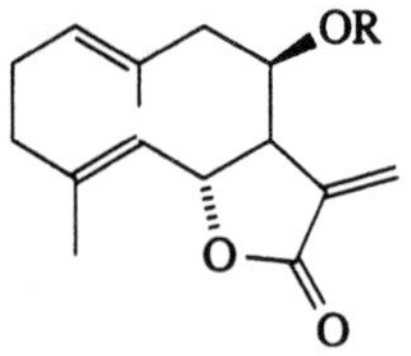

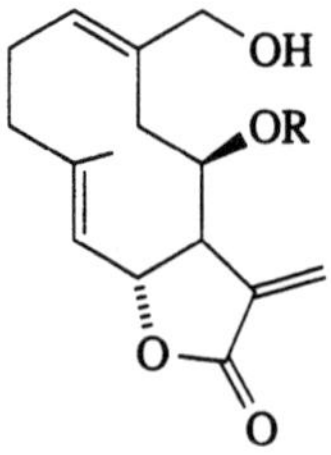

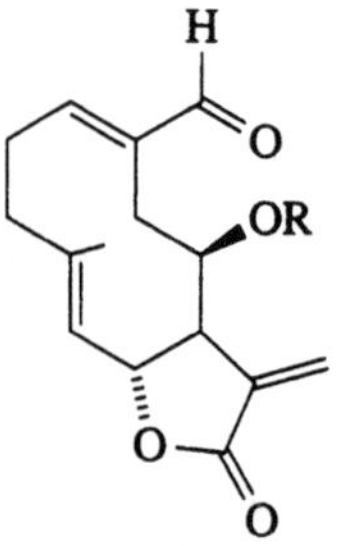

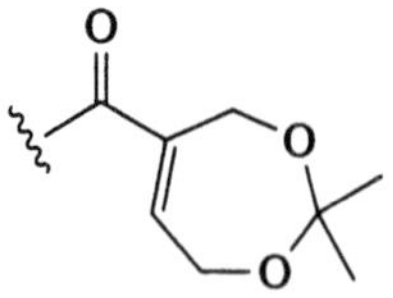

(14) R = Ac
(15) R = X
(16) R = SO_2Me

(17) R = Ac
(18) R = X
(19) R = SO_2Me

(20)

The selectivity of the reaction is nicely illustrated by the oxidation under similar conditions of 6-*N*-(3,3-dimethylallyl)adenosine **(21)**, which regiospecifically gave (*E*)-zeatine β-D-ribofuranoside **(22)**, none of the other functionality within the molecule being affected.[26]

The isomeric (*Z*)-zeatine riboside was obtained from **(22)** by UV irradiation induced equilibration.

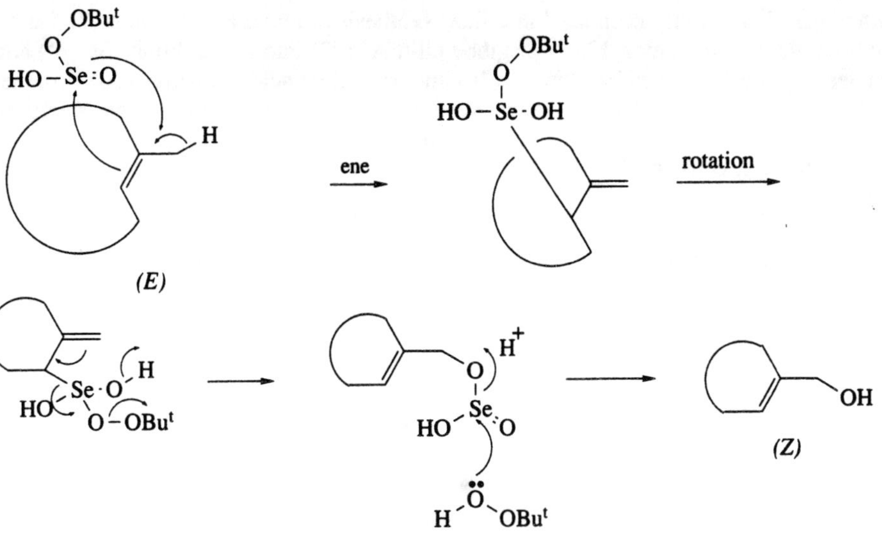

Scheme 6

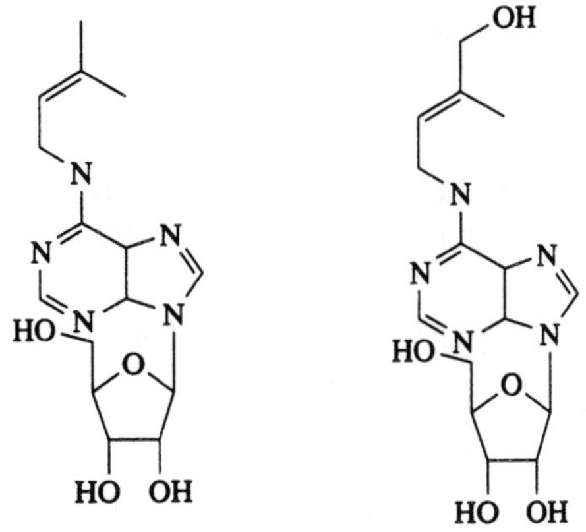

(21) **(22)**

The possible regioselectivity and the preference for (*E*)-allylic alcohol production using the reoxidative modification are demonstrated in a projected synthesis of cembranolides (equation 10).[27] The substrate in this case contains two double bonds and several allylic positions. This oxidation is reported to be even more selective than the analogous oxidation of geranyl acetate originally reported by Sharpless.

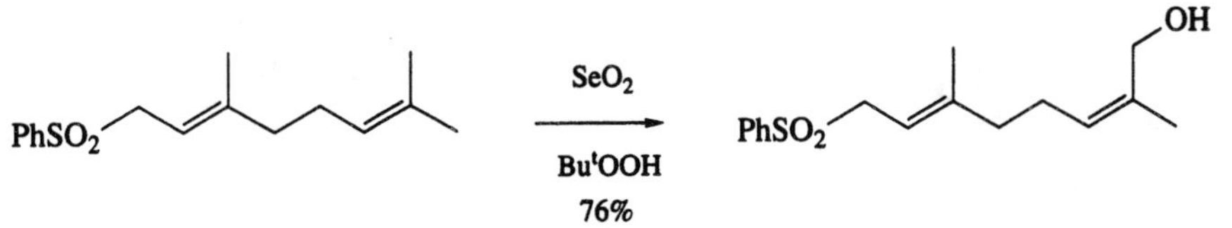

(10)

An allylic oxidation of neryl acetate, apparently using partially catalytic conditions (0.25 equiv. SeO_2, 2 equiv. TBHP), has recently been reported (equation 11).[28] While this reaction is described as highly selective, it should perhaps be noted that the product was obtained in only 45% yield.

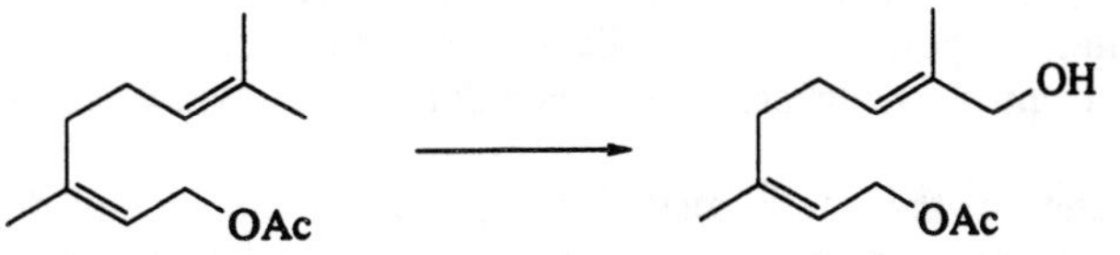

(11)

The method has also recently been used in a short synthesis of lipoic acid (Scheme 7)[29] and featured in MacMillan's conversion (equation 12a)[30] of gibberellin A_3 (**23**) into gibberellins A_{67} (**24**) and A_{68} (**25**), and in Mander's conversion of gibberellin A_7 (**26**) into antheridic acid (equation 12b).[31] In these last two examples the terminally disubstituted alkenes were oxidized with almost complete stereoselectivity to give the 15-hydroxy products consistent with sterically controlled approach of the reagent. The trisubstituted double bond in equation (12b) was unaffected, as was the remaining functionality in both substrate molecules.

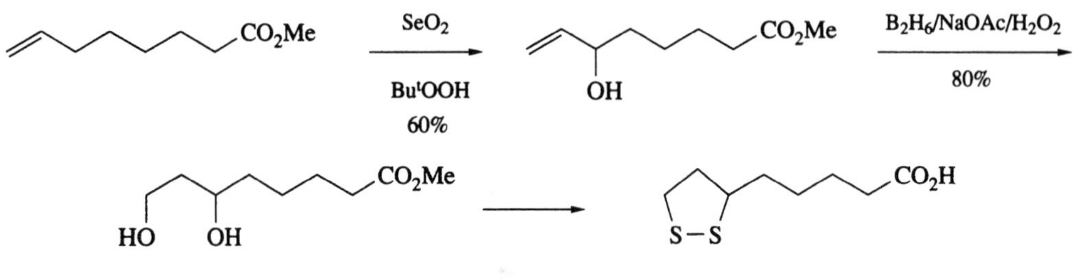

CO_2Me — SeO_2, Bu^tOOH, 60% → OH, CO_2Me — $B_2H_6/NaOAc/H_2O_2$, 80% → HO, OH, CO_2Me → S–S, CO_2H

Scheme 7

H, R^1, O, H, R^3, R^2, O, CO_2H

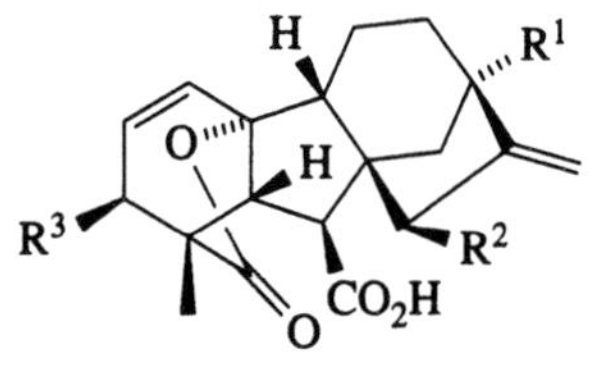

(**23**) R^1 = OH, R^2 = H, R^3 = OH

(**24**) R^1 = OH, R^2 = OH, R^3 = H

(**25**) R^1 = H, R^2 = OH, R^3 = OH

(**26**) R^1 = H, R^2 = H, R^3 = OH

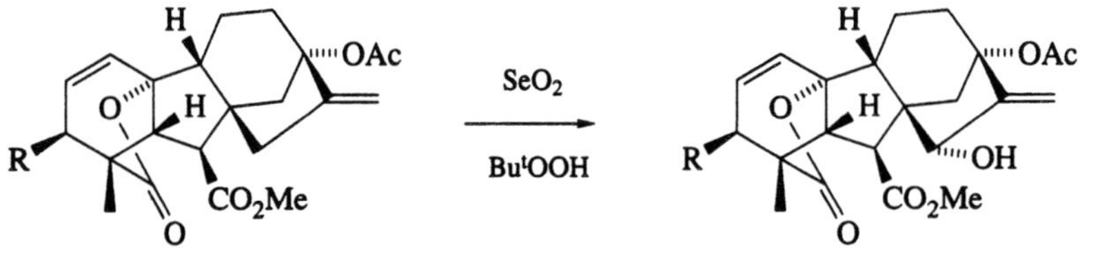

H, OAc, O, H, R, CO_2Me, O — SeO_2, Bu^tOOH → H, OAc, O, H, R, OH, CO_2Me, O (12a)

R = H, OAc

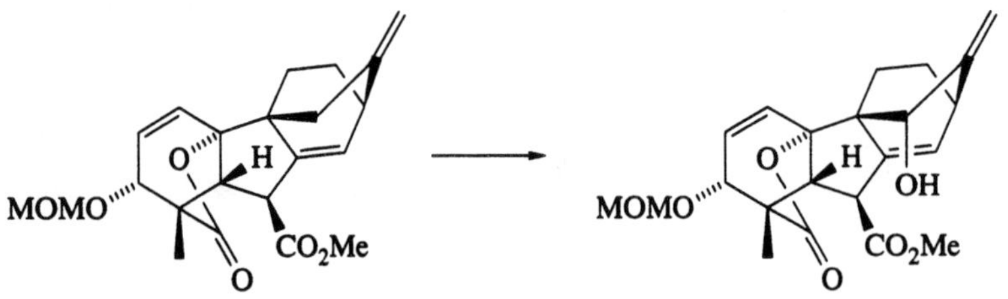

O, H, MOMO, CO_2Me, O → O, H, OH, MOMO, CO_2Me, O (12b)

A number of reoxidants for selenium dioxide have been examined. For example, while hydrogen peroxide is sometimes successful,[23] oxidation of cholecalciferol (**27**) or derivatives with selenium dioxide alone gave poor results not improved by addition of hydrogen peroxide. In this case, use of sodium periodate or tetra-*n*-butylammonium periodate gave increased yields.[32] The reaction was much improved when carried out under reflux in methanol or solvent mixtures containing methanol, and indeed selenous acid and dialkyl selenites, suggested as intermediates in the reaction, both accomplished a similar oxidation in nonalcoholic solvents in the presence of a reoxidant, *N*-methylmorpholine *N*-oxide proving superior.

Double allylic oxidation at the allylic methyl and methinyl positions of drimenyl acetate (**28**) was achieved[33] in boiling dioxane solution using 'catalytic amounts' of selenium dioxide in the presence of

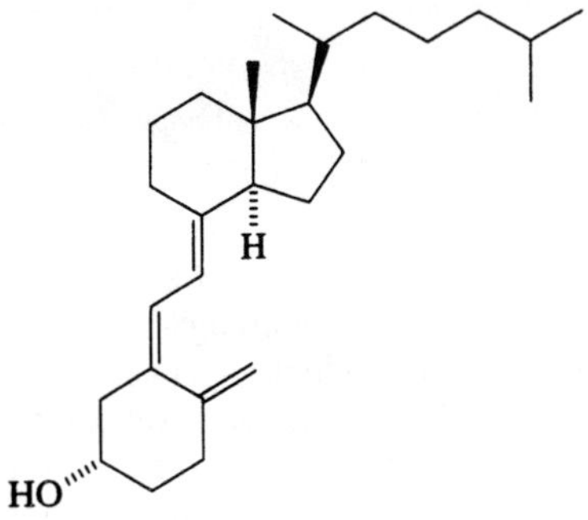

(27) 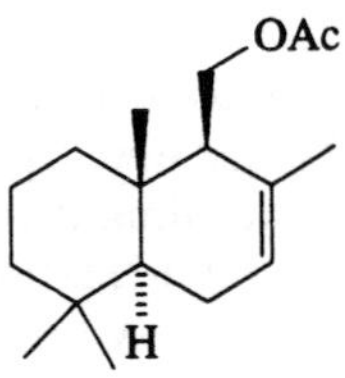 (28) 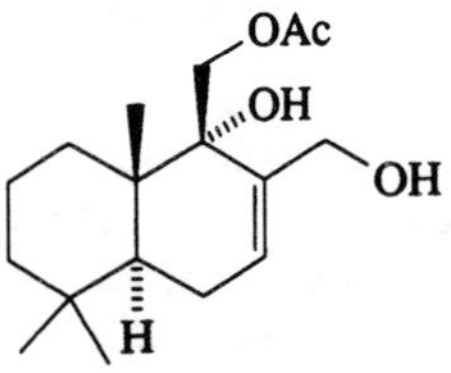 (29)

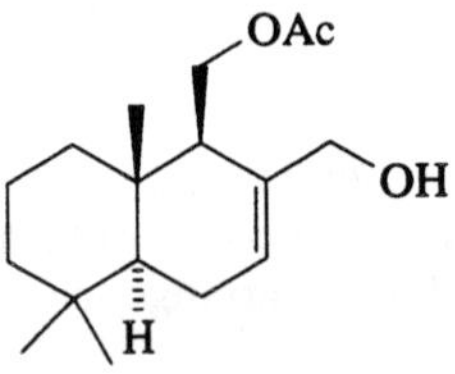

(30) 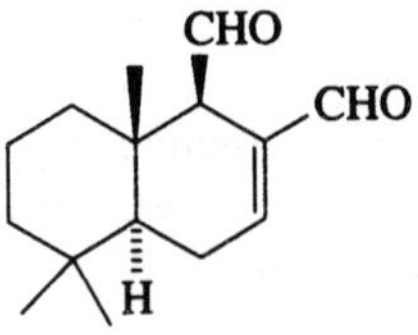 (31)

bis(4-methoxyphenyl) selenoxide (1.5 equiv.) as reoxidant.[34] Compounds (**29**; 60%) and (**30**; 30%) were both isolated from the reaction mixture. Compound (**30**) was used for a synthesis of (–)-polygodial (**31**).

Ethers (**32**) and peroxides (**33**) are seen as by-products in the catalytic selenium dioxide oxidation of cycloalkenes, and these materials can predominate in the case of small rings.[35] Addition of hydroquinone to the reaction mixtures suppresses their formation and consequently a free radical pathway has been proposed (Scheme 8).

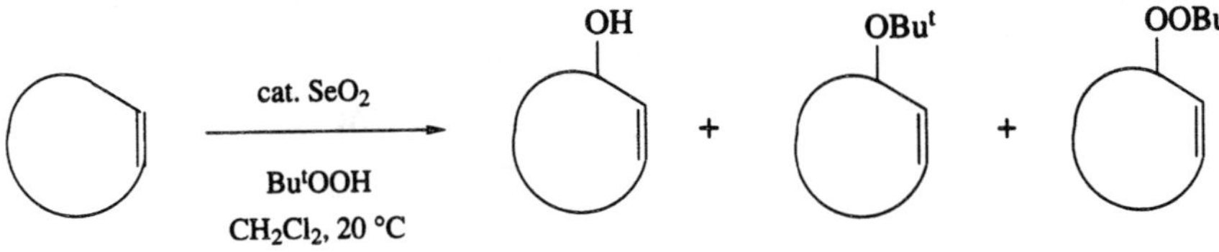

cat. SeO_2, Bu^tOOH, CH_2Cl_2, 20 °C → OH + OBu^t (32) + $OOBu^t$ (33)

$O{=}Se{\cdot}O{-}OBu^t$

+ SeO_2 + $Bu^tO{\bullet}$ → OBu^t

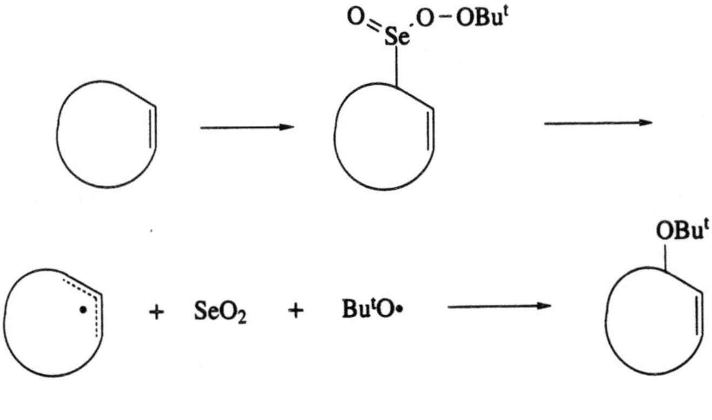

Scheme 8

Selenium dioxide has been used supported on silica gel in combination with *t*-butyl hydroperoxide in hexane or dichloromethane solution.[36]

Sharpless has achieved the allylic oxidation of alkenes using arylselenenic acids, generated *in situ* from the diselenide and *t*-butyl hydroperoxide, a reaction claimed to occur with exclusive allylic rearrangement.[37]

2.1.2.2 Metal Acetates and Related Reagents

2.1.2.2.1 Mercury and related metals

Rappoport[38] and Muzart[39] have noted that the reaction of alkenes with metal acetates and related derivatives,[40,41] including those of mercury,[42] palladium (Section 2.1.2.2.2), thallium, manganese, silver and lead, can give rise to a variety of products including π-complexes, addition products, acetates, diacetates, rearrangement products and allylic esters. Lead tetraacetate in general gives mixtures of substitution, rearrangement and addition products, and is usually not a synthetically useful reagent for allylic oxidation.[40,41a] Thallium triacetate, while a more powerful oxidizing agent than mercury(II) acetate, normally gives allylic oxidation only as a minor side reaction.[40,41b] Manganese triacetate catalyzed by potassium bromide has been reported to convert toluene into benzyl acetate and alkenes into allylic acetates *via* a radical mechanism,[43] and cobalt triacetate can induce allylic oxidation or dihydroxylation of a double bond, depending upon substrate structure.[44]

While allylic oxidation products may arise by elimination of a metal hydride from an intermediate adduct or metal–alkene complex,[45] allylmercury species (**34**) are thought to be intermediates in the case of mercury(II) acetate.[42,46] A number of pathways have been suggested, for example involving radical[46] and carbenium ion[47] intermediates, and addition–elimination[48] and rearrangement processes.[49]

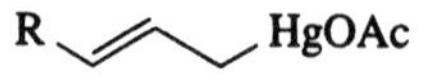

R HgOAc

(34)

Winstein showed that the solvolysis of crotylmercury(II) acetate under kinetically controlled conditions gives >99.5% of α-methylallyl acetate (equation 13).[50] Subsequent work indicated that both the solvolysis of cinnamylmercury(II) acetate and the mercury(II) acetate oxidation of allylbenzene give *ca.* 60% cinnamyl acetate (**35**) and 40% α-phenylallyl acetate (**36**; equation 14).[38] An equilibrium exists between (**35**) and (**36**) favoring the primary ester which constitutes >99.5% of the equilibrium mixture at 75 °C. Oxidation of a range of both 1- and 2-alkenes under kinetically controlled conditions exclusively gave the secondary allylic esters.

HgOAc ⟶ OAc (13)

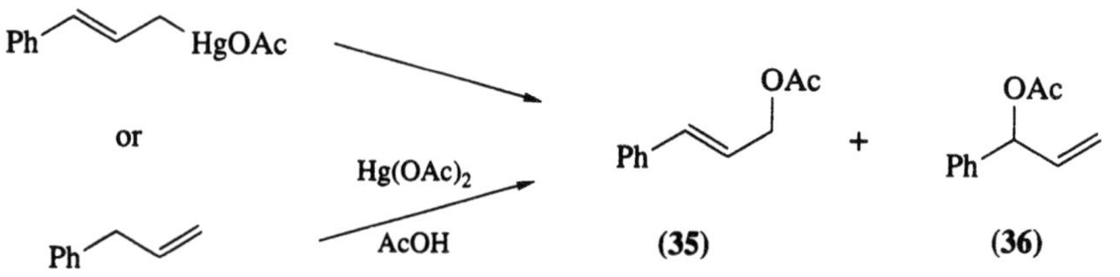

Ph HgOAc or Ph $\xrightarrow[\text{AcOH}]{\text{Hg(OAc)}_2}$ Ph OAc (**35**) + Ph OAc (**36**) (14)

Similarly, both α- and β-pinene give the acetates of both myrtenol and *trans*-pinocarveol (and also the ring cleavage product) upon mercury(II) acetate oxidation (equation 15).[51]

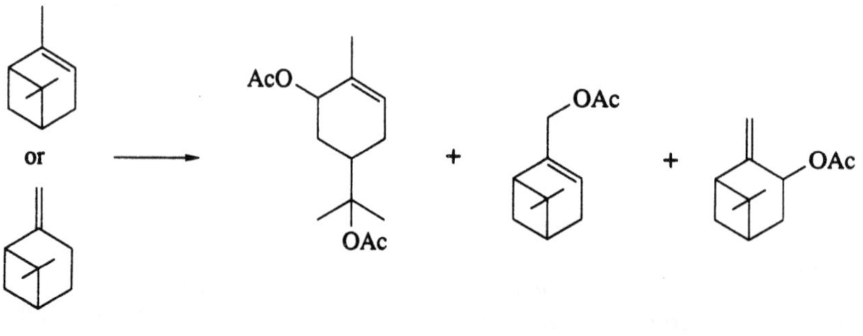

or ⟶ AcO OAc + OAc + OAc (15)

On the basis of this and other evidence allylic mercury(II) acetates are believed to be the reaction intermediates in all allylic oxidations with mercury(II) acetate. The isolation of the less stable secondary acetates from both 1- and 2-alkenes indicates that in the oxidation of 2-alkenes rearrangement of the secondary allylic mercury(II) acetate to the primary isomer is faster than solvolysis of the organomercurial (Scheme 9).

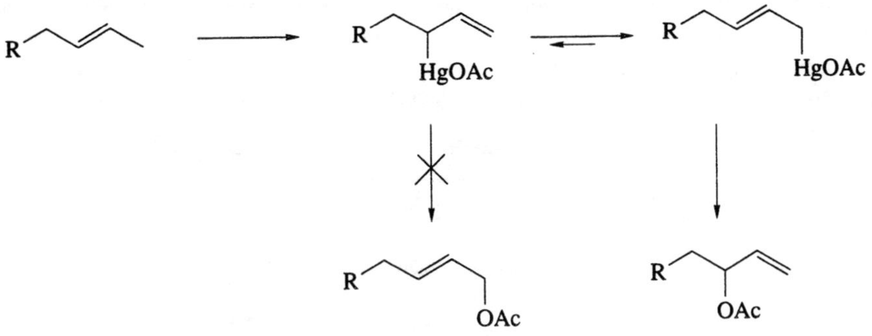

R HgOAc HgOAc R OAc R OAc

Scheme 9

In contrast to lead tetraacetate, simple addition to the double bond does not occur as a side reaction.[41b,42] While allylic rearrangement is common and mixtures of products are frequently obtained, the reaction often proceeds in very high yield and is simple to carry out; the alkene is simply heated in an appropriate solvent with mercury(II) acetate until reaction is complete. Mercury(II) acetate has also been used for dehydrogenation, particularly in the steroid field. One interesting example incorporating simultaneous dehydrogenation and allylic oxidative rearrangement is seen in the reaction of abietic acid (**37**; equation 16).[52]

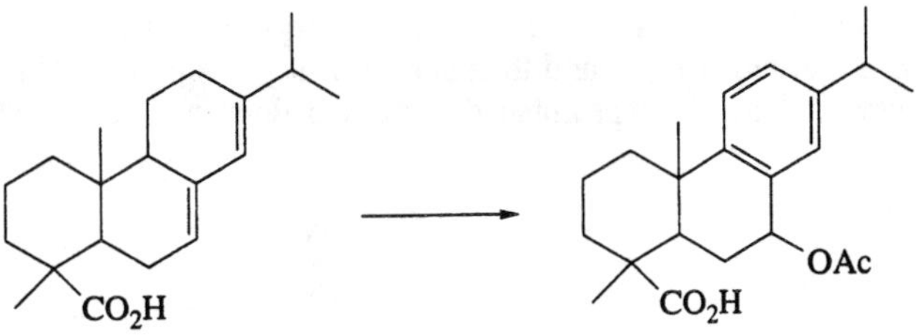

CO_2H OAc CO_2H (16)

(37)

That mercury(II) acetate allylic oxidation can be a useful reaction in the case of complex and sensitive substrates is demonstrated by the oxidation of avermectin A_{2a} (**38**).[53] The reaction, carried out in anhydrous toluene at 100 °C for 40 min, was remarkably selective, allylic oxidation occurring exclusively at the 3,4-double bond with rearrangement to give (**39**) in up to 73% yield (equation 17).

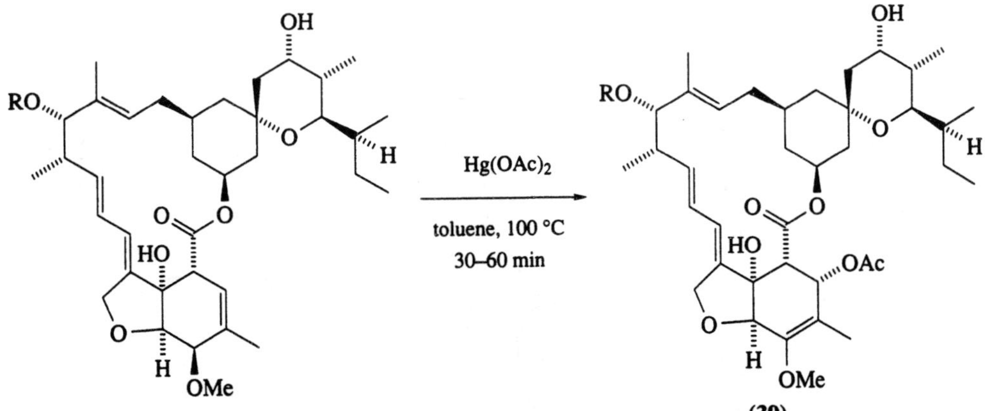

$Hg(OAc)_2$, toluene, 100 °C, 30–60 min (17)

(38) **(39)**

Finally, mercury(II) oxide in combination with fluoroboric acid and ethanol in THF solution has been shown to convert allylbenzenes into the rearranged allylic ethyl ethers.[54]

2.1.2.2.2 Palladium and related metals

Nucleophilic addition to π-allylpalladium complexes is known to take place with a range of nucleophiles, and the mechanism and stereochemistry of these reactions have been thoroughly investigated over the last few years.[39,41,55,60] For example, reaction with acetate anion occurs in the presence of benzoquinone at room temperature in acetic acid solution by initial *cis* attack at the metal atom and subsequent migration.[56] Alternatively, in the presence of chloride ions, a *trans* attack takes place to give the product of opposite stereochemistry (Scheme 10).[57] Intramolecular versions of the reaction are known.[58]

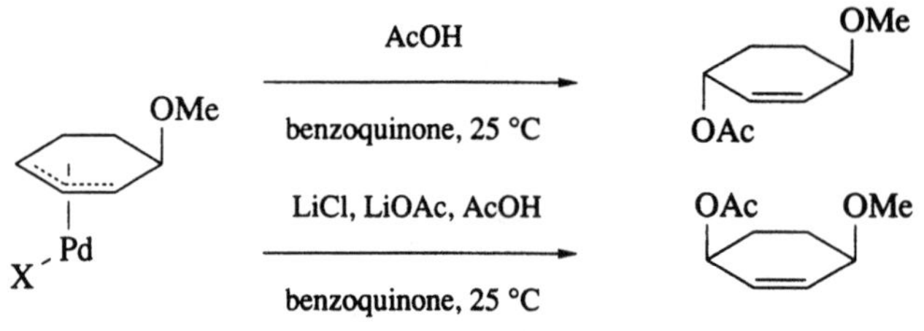

OMe; X Pd; AcOH, benzoquinone, 25 °C → OAc, OMe; LiCl, LiOAc, AcOH, benzoquinone, 25 °C → OAc, OMe

Scheme 10

A new and very highly selective catalytic method for allylic oxidation based on palladium acetate chemistry was discovered in 1984.[59] The alkene, palladium bis(trifluoroacetate) (5 mol %), 2-methoxyacetophenone (added ligand; 20 mol %), and benzoquinone (reoxidant; 1 equiv.) are dissolved in acetic acid and stirred for *ca.* 2 d at room temperature. Geranylacetone (**40**), which contains six allylic carbon atoms plus two carbon atoms adjacent to a carbonyl group, was oxidized using this procedure to give (**41**) and (**42**) in 85% yield and about 2:1 ratio (equation 18). No other available procedure proved as selective. The reaction presumably proceeds *via* formation of a π-allylpalladium complex[60] and attack by an oxygen nucleophile with expulsion of Pd^0, subsequently reoxidized to Pd^{2+} by the oxidizing agent. A range of reoxidants and added ligands was examined. The reaction was used in a synthesis of helminthogermacrene (**43**) and β-elemene (**44**),[61] and then subsequently to prepare (**45**) in a synthetic approach to casbene (**46**);[62] however, despite its promise the method does not seem yet to have been widely adopted.

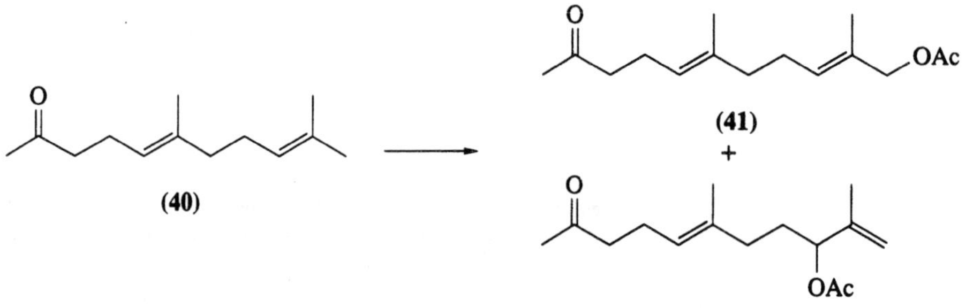

O (**40**) → O, OAc (**41**) + O, OAc (**42**) (18)

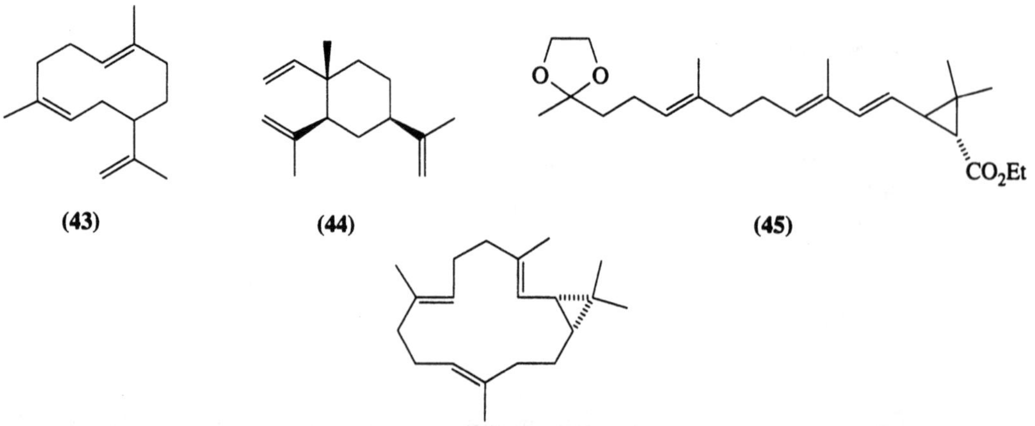

(**43**) (**44**) O O, CO_2Et (**45**)

(**46**)

Palladium chloride in combination with potassium acetate, pentyl nitrite and oxygen,[63] and in combination with silver acetate, *t*-butyl hydroperoxide and tellurium dioxide in a 1:2:10:1 ratio,[64] has been shown to accomplish allylic oxidation, although in somewhat disappointing yields. Other reagent combinations with palladium chloride have also been used,[65] as has a palladium acetate–manganese dioxide–benzoquinone mixture.[66] Several other metal species,[39] among them rhodium(II) acetates,[67,172] rhodium(III)–copper(II) in combination,[68] vanadyl acetylacetonate,[69] rhodium(I)[70a] and iridium(I) chloride complexes,[70] and various iron-,[71,72] manganese-,[72] cobalt-[73a] and chromium-(III)[73b] complexes, catalyze allylic oxidation and epoxidation of alkenes with very limited selectivity in the presence of reoxidants such as molecular oxygen, *t*-butyl hydroperoxide, hydrogen peroxide and iodosylbenzene.

2.1.2.3 Miscellaneous Reagents

2.1.2.3.1 Peroxy esters and peroxides

Oxidation of C—H bonds by copper ion catalyzed reaction with an organic peroxy ester (the Kharasch–Sosnovsky reaction)[74] was at one time very popular for allylic oxidation and has been thoroughly reviewed.[39,75] The reaction is usually carried out by dropwise addition of peroxy ester (commonly *t*-butyl peracetate or *t*-butyl perbenzoate) to a stirred mixture of substrate and copper salt (0.1 mol %; commonly copper(I) chloride or bromide) in an inert solvent at mildly elevated temperature (60–120 °C). The mechanism involves three steps: (i) generation of an alkoxy radical; (ii) hydrogen atom abstraction; and (iii) radical oxidation and reaction with carboxylate anion (Scheme 11).

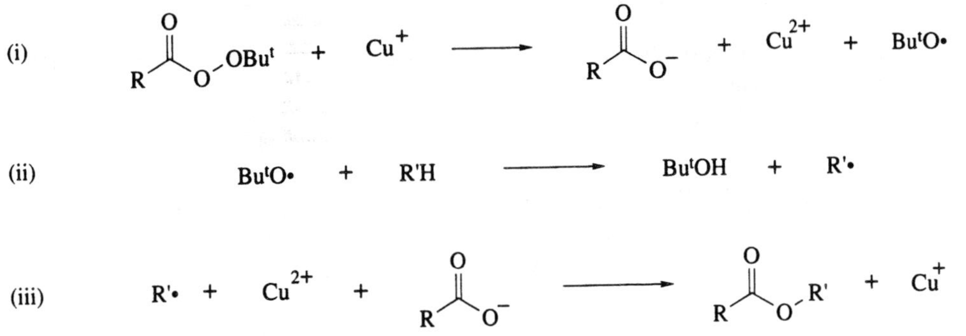

(i) $RCO_2Bu^t + Cu^+ \longrightarrow RCO_2^- + Cu^{2+} + Bu^tO\bullet$

(ii) $Bu^tO\bullet + R'H \longrightarrow Bu^tOH + R'\bullet$

(iii) $R'\bullet + Cu^{2+} + RCO_2^- \longrightarrow RCO_2R' + Cu^+$

Scheme 11

With allylic substrates the intermediate is an allylic radical, and allylic oxidative rearrangement is therefore common; reaction of optically active bicyclo[3.2.1]oct-2-ene gave racemic *exo*-(**47**) after hydrolysis.[76] Benzylic oxidation is slow and proceeds in only moderate yield, but allylic oxidation is usually clean and can occur in high yield (*e.g.* equation 19).[75,77,78]

OH

(**47**)

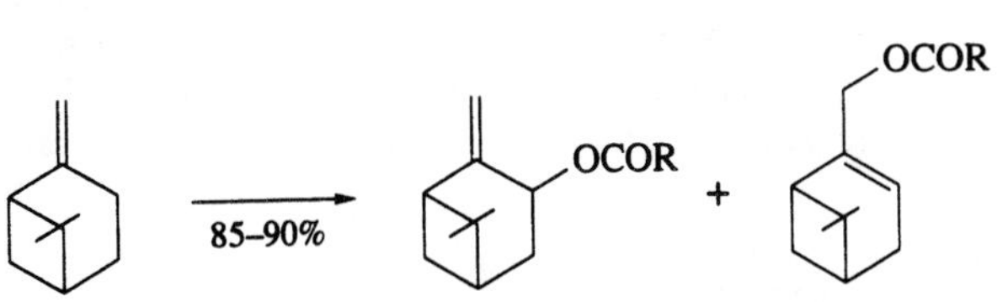

85–90%

OCOR + OCOR (19)

Terminal alkenes are oxidized with negligible rearrangement to give the 3-acyloxy species.[75,79] However, oxidation of internal alkenes, including cycloalkenes, may result in substantial or exclusive rearrangement.[75,80] Most notable is the reaction of 1-phenylpropene exclusively to give 3-acetoxy-3-phenylpropene, in which deconjugation of the double bond takes place (equation 20).[81]

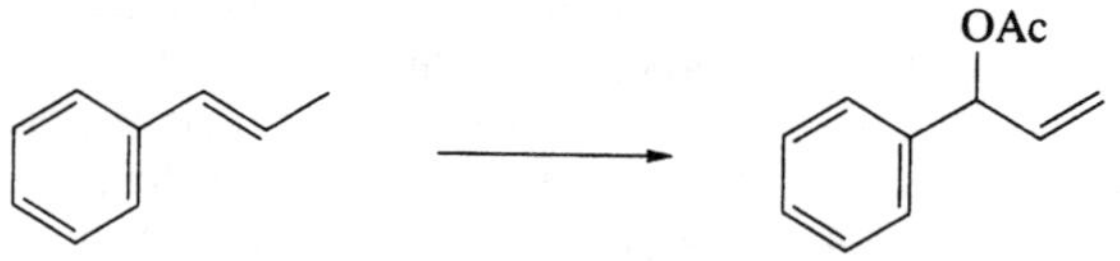

OAc

(20)

Unlike most other allylic oxidation systems, extensive rearrangement can take place with higher alkenes, and this limits the synthetic utility of the reaction. Nevertheless, some very selective allylic oxidations have been achieved in the steroid series;[82] for example treatment of progesterone with *t*-butyl perbenzoate in the presence of copper carbonate gave after hydrolysis and isomerization the 6α-hydroxylated product (**48**),[83] neither rearrangement nor competing reaction α to either carbonyl group being observed.

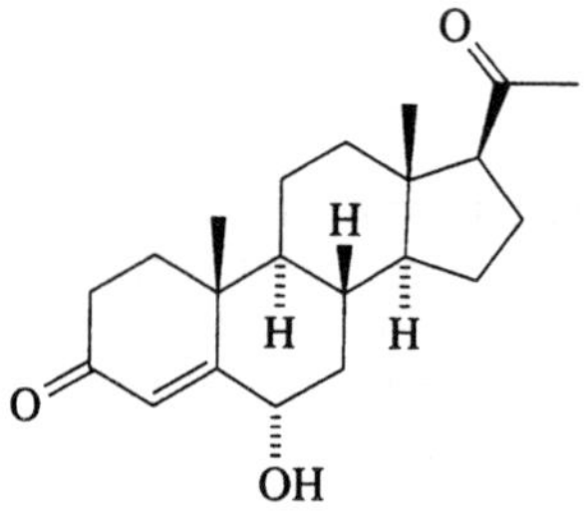

(48) 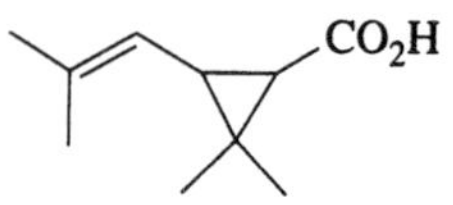 **(49)**

The reaction has also featured in a synthesis of chrysanthemic acid (**49**), exclusive rearrangement occurring to give the secondary allylic benzoate (equation 21).[84]

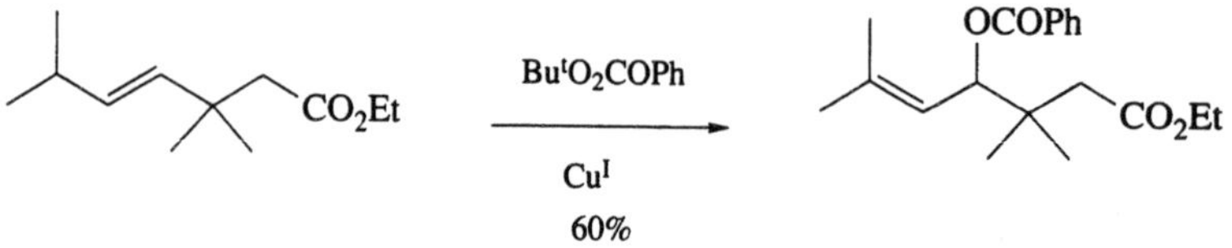

CO_2Et — Bu^tO_2COPh, Cu^I, 60% → OCOPh, CO_2Et (21)

The Kharasch–Sosnovsky reaction may be carried out in the presence of carboxylic acids to introduce the acyloxy moiety of the acid used, and may also be conducted photochemically at room temperature using UV irradiation. Peroxy acids,[75] diacyl peroxides,[75] and peroxyphosphates and peroxyphosphonates[85] are alternative oxidants. *t*-Butyl hydroperoxide may also be used in place of peroxy esters with broadly similar results, although formations of mixed peroxides[75] and *t*-butyl ethers[86] can then compete with allyl ester production.

2.1.2.3.2 Singlet oxygen

Photosensitized oxygenation of alkenes using singlet oxygen is a well-known reaction[87,199] and several comprehensive reviews have appeared.[39,88,89] The normal course of singlet oxygen reactions with monoalkenes is *via* an ene reaction to produce an allylically rearranged hydroperoxide, often in excellent yield, and this may be reduced to give allylic alcohol (Scheme 12). Other mechanisms, which may compete, have also been proposed, involving dioxetane (**50**) and perepoxide (**51**) intermediates,[88] and indeed perepoxides have been proposed as common intermediates for the formation of dioxetanes and other by-products.[90] Typical photosensitizers for singlet oxygen production in these reactions are rose bengal,

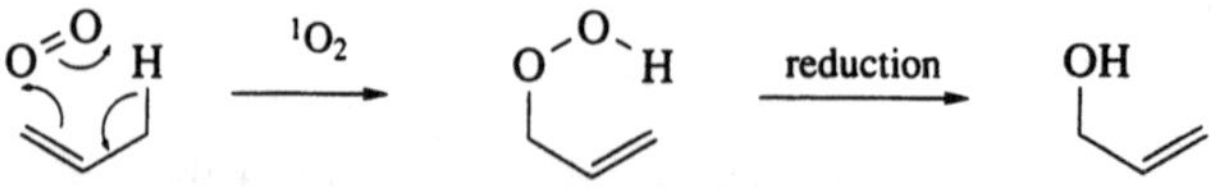

$^1O_2$ — reduction — OH

Scheme 12

hematoporphyrin, eosin, methylene blue, chlorophyll and fluorescein. 18-Crown-6 has been used to solubilize rose bengal and eosin in aprotic solvents.[91] Solvent effects are small, typical solvents being pyridine, methylene chloride, ether and methanol, although the lifetime of singlet oxygen in this last solvent is very much reduced.

(50) 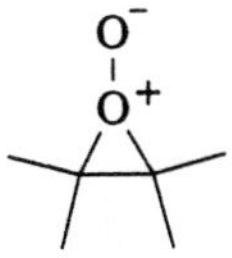 (51)

Oxidation with singlet oxygen is subject to steric effects[92] but can show poor regioselectivity.[88,89,93] For example, (+)-3-carene (**52**) is oxidized to produce a mixture of all three possible regioisomeric hydroperoxides with oxygenation occurring at the face of the allyl system opposite the *gem*-dimethylcyclopropane unit in each case (equation 22).[89,94]

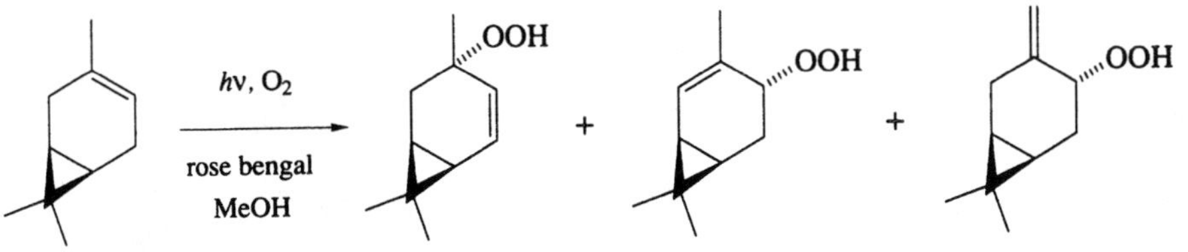

*h*ν, O_2, rose bengal, MeOH (22)

(52)

The sensitivity to steric effects is nicely illustrated in the oxidation of dienes (**53**) and (**54**), where addition of a methyl group at a remote site is sufficient to change completely the course of the reaction (Scheme 13).[95]

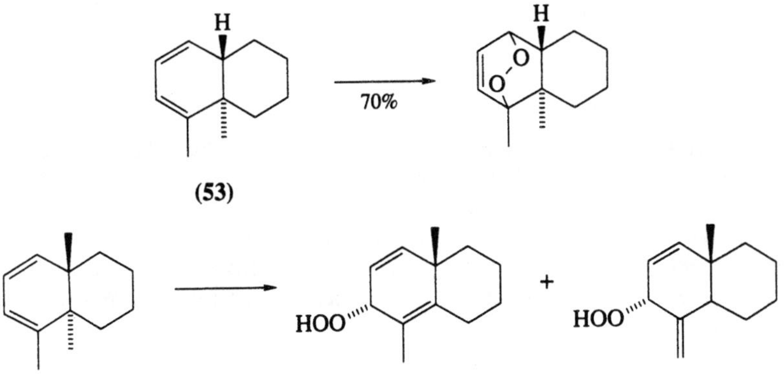

70%

(53)

(54) 85–90% 10–15%

Scheme 13

Singlet oxygen is an electrophilic reagent and increasing substitution around a double bond therefore increases reactivity. Tetrasubstituted alkenes are around 20 times as reactive as trisubstituted alkenes, which are in turn about 150 times as reactive as disubstituted alkenes, which are some 15 times as reactive as monosubstituted alkenes.[88,89] This effect is illustrated in the oxidation of diene (**55**), in which oxidation occurred exclusively at the more substituted double bond.[96]

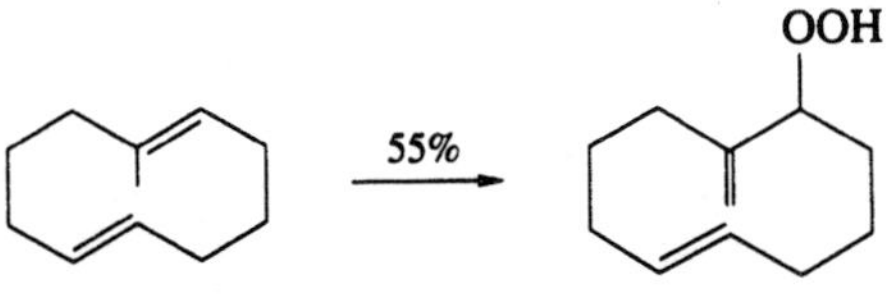

55%

(55)

Furthermore, the presence of electron-withdrawing substituents, such as an allylic hydroxy group, is deactivating, an effect intensified by esterification to a degree such that acylation of an allylic alcohol may be sufficient to protect the double bond during photooxidation at another site.[88]

In some instances the primary product of alkene photooxidation is not the allylically rearranged hydroperoxide, but the dioxetane addition product, *e.g.* (**56**), which may or may not be formed by concerted [2 + 2] cycloaddition.[90] Some of these dioxetanes, *e.g.* (**57**), are relatively stable, although most suffer cleavage to produce carbonyl compounds or other materials.[97] For example, photooxidation of indene gives homophthaldehyde (**58**) which was not produced under identical reaction conditions from hydroperoxide (**59**).[98] Isomeric hydroperoxides (**60**) and (**61**) were also isolated when the oxidation was carried out in methanolic solution (Scheme 14).

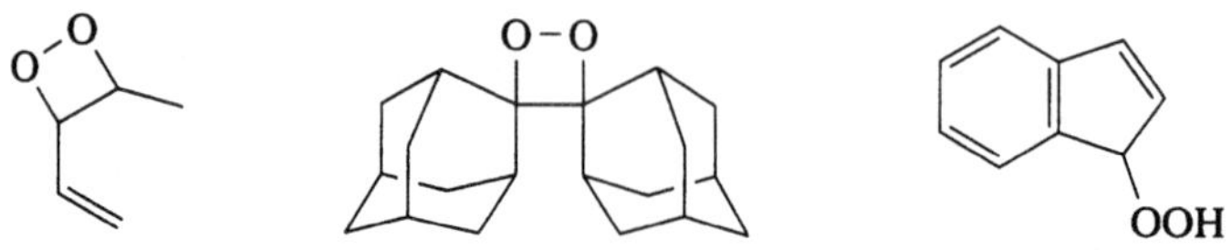

(**56**) (**57**) (**59**)

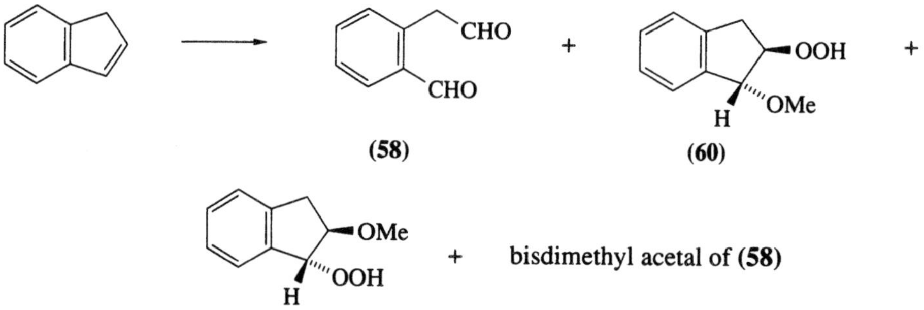

(**58**) (**60**)

(**61**) + bisdimethyl acetal of (**58**)

Scheme 14

The factors which control the reaction pathway followed in individual alkene photooxidations do not seem to be well understood,[99] although dipolar solvents such as acetonitrile seem to favor the dioxetane mode, while less polar solvents such as benzene favor the ene mode. A number of chemical methods for the production of singlet oxygen are known, including generation from triphenyl phosphite ozonide,[100] bromine–alkaline hydrogen peroxide,[101] sodium hypochlorite–hydrogen peroxide,[102] and anthracene endoperoxide;[103] however the potential for greater selectivity offered by such reagents has not yet been thoroughly investigated.

2.1.2.3.3 Electrochemical methods

Torii has reported the electrochemical oxyselenation–deselenation of alkenes to give allylically rearranged allyloxy products.[104] A typical example is given below (equation 23).[105] Benzyl- or acetyl-(*S*)-citronellol was mixed in acetonitrile/water solution with diphenyl diselenide (0.5 mol equiv.) and a catalytic amount of tetraethylammonium bromide, and was electrolyzed at room temperature in an undivided cell using platinum electrodes and a constant current density of 10 mA cm^{-2}. The corresponding allylic alcohols were isolated in excellent yields. In methanolic solution, using a reduced amount of diphenyl diselenide (20 mol %), the methoxy compounds were obtained in slightly reduced yields.

OCH_2Ph ⟶ OH OCH_2Ph (23)

2.1.2.3.4 *Allylic oxidation* via *metallation*

Allylic alcohols may be derived from alkenes by metallation to give the allylpotassium species, followed by treatment with fluorodimethoxyborane. Oxidation of the resultant boronic ester with hydrogen peroxide gives the allylic alcohol (Scheme 15).[106,107] Some allylic rearrangement may be observed; for example, metallation of α-pinene with potassium *t*-butoxide in petroleum ether solution and subsequent boration and oxidation gave myrtenol (42%) and *trans*-pinocarveol (1%) (equation 24), while treatment of the allylpotassium with oxirane gave the alkylated products in a ratio of *ca.* 2:1.[106]

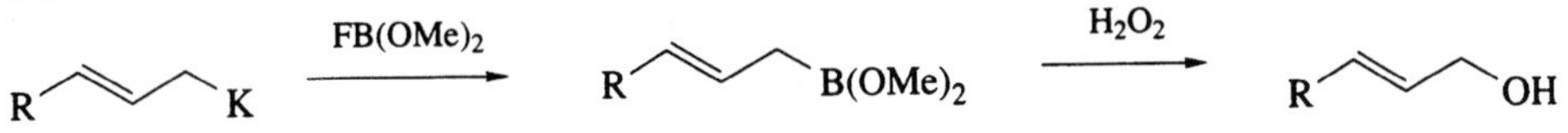

$FB(OMe)_2$; $B(OMe)_2$; H_2O_2 ; R ; K ; OH

Scheme 15

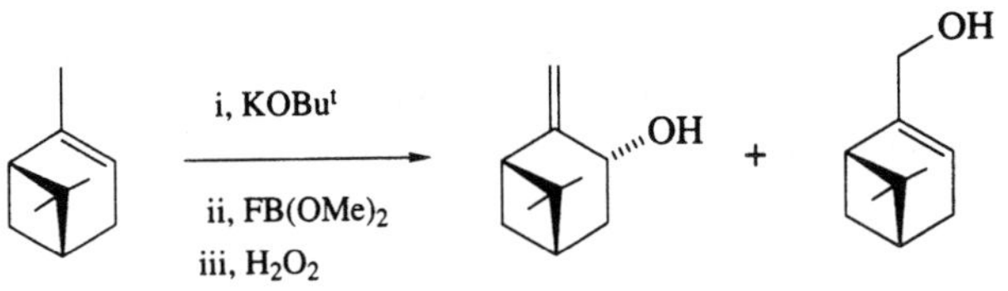

i, $KOBu^t$; ii, $FB(OMe)_2$; iii, H_2O_2 ; OH ; + ; OH (24)

2.1.2.3.5 *Enzymatic methods*

The copper–protein dopamine β-monooxygenase (DBM), which catalyzes hydroxylation at the *pro-(R)* hydrogen atom of dopamine to form norepinephrine in mammalian tissues, has been used for enantioselective benzylic hydroxylation[108] and sulfoxidation.[109] Very recently enantioselective allylic oxygenation by DBM has also been reported.[110] 2-(1-Cyclohexenyl)ethylamine (**62**) was subjected to 'preparative scale' enzymatic reaction and the product (**63**) characterized as having the (*R*)-configuration by HPLC and ^1H NMR analysis of a derivative using the Mosher model. No trace of the (*S*)-enantiomer was observed and neither allylic rearrangement nor epoxidation took place. The authors comment that DBM also catalyzes the allylic hydroxylation of (*Z*)-hex-2-enylamine with no detectable allylic rearrangement, and suggest that the reaction involves interaction of copper with the alkene moiety during catalysis, thus precluding double bond rearrangement.

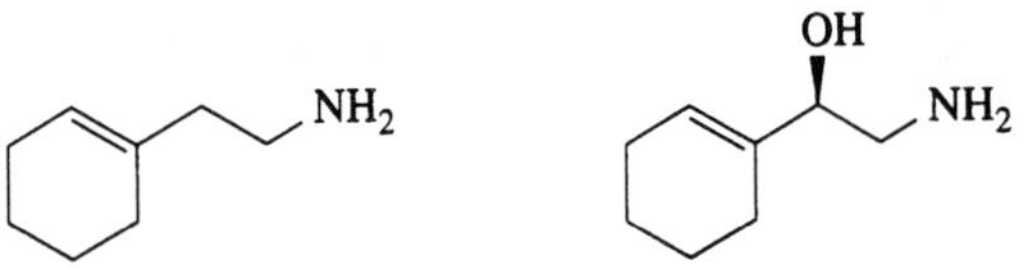

NH_2 ; OH ; NH_2

(62) **(63)**

2.1.3 ALLYLIC OXIDATION REACTIONS WHICH PRODUCE α,β-UNSATURATED CARBONYL COMPOUNDS

2.1.3.1 Chromium(VI)-based Reagents

2.1.3.1.1 *Chromic acid and simple chromate esters*

Examples of the use of chromium(VI) reagents to effect the allylic oxidation of alkenes to give α,β-unsaturated carbonyl compounds are very common in the literature.[111,112] The reaction was first reported by Treibs and Schmidt[113] for the allylic oxidations of α-pinene to verbenone and verbenol, of dipentene to carvone and carveol, and of cyclohexene to cyclohexenol and cyclohexenone, using a solution of chromium trioxide in a mixture of acetic anhydride and carbon tetrachloride. However, yields were low and no synthetic use of this observation was made.

Chromic acid itself has been used in the oxidation of alkenes and in some cases allylic oxidation products were observed; for example, cyclohexene was converted to cyclohexenone in 37% yield and 1-methylcyclohexene was oxidized to a mixture of enones (Scheme 16).[114]

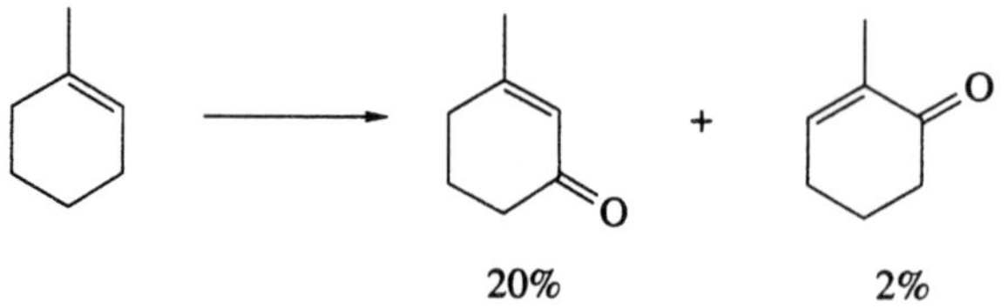

20% + 2%

Scheme 16

Extension of this type of reagent[115] to the use of sodium dichromate in acetic acid[116] was found to furnish the allylic oxidation of 4,4,10-trimethyl-Δ^5-octalin into 7-keto-4,4,10-trimethyl-Δ^5-octalin (**64**)[117] in 65% yield (equation 25); this reaction was employed in the total syntheses of (±)-widdrol (**65**) and of (±)-thujopsene (**66**).[118]

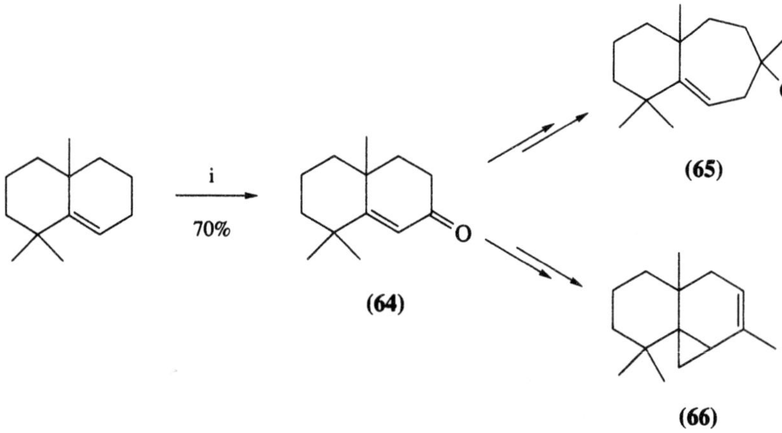

i 70% (**64**) OH (**65**) (**66**) (25)

i, 1.4 equiv. $Na_2Cr_2O_7$, acetic acid, r.t./overnight, then 100 °C/2.5 h

Allylic oxidation of steroids, particularly at the 7-position, has evoked interest over many years. For example, chromium trioxide–acetic acid,[119] sodium dichromate,[120] and *t*-butyl chromate[121–128] have all been used in the oxidation of the 5-α-pregnane series (*e.g.* equation 26).

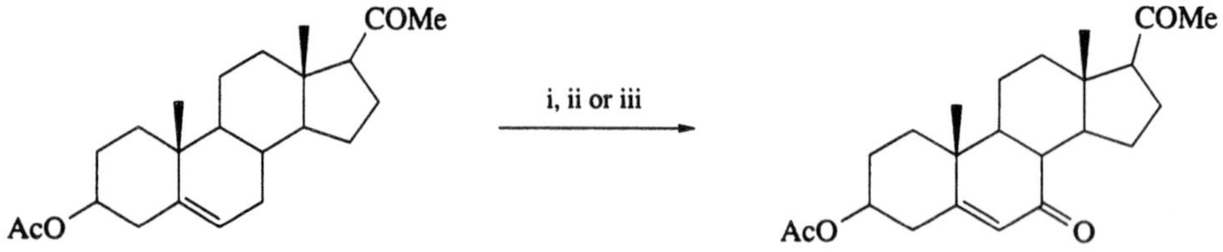

COMe AcO i, ii or iii COMe AcO O (26)

i, CrO_3, AcOH; 50%; ii, $Na_2Cr_2O_7$, AcOH, Ac_2O; 79%; iii, $(Bu^tO)_2CrO_2$; 62%

The overall mechanism of chromium(VI) allylic oxidation appears to consist of removal of a hydrogen atom or hydride ion from the alkene, forming a resonance-stabilized allylic radical or carbocation, which is ultimately converted into the unsaturated ketone (Scheme 17).[8]

An alternative mechanism has also been proposed in which oxidation at the double bond leads to a ketol derivative, elimination of water from which then gives the unsaturated ketone (Scheme 18a).[112] Limited kinetic data are available and suggest that Scheme 17 is obtained for chromic acid oxidations.[129]

The discovery of the chromium trioxide–pyridine complex led to the accessibility of allylic oxidation under much less harsh conditions, typically room temperature reaction in dichloromethane solution[130,131]

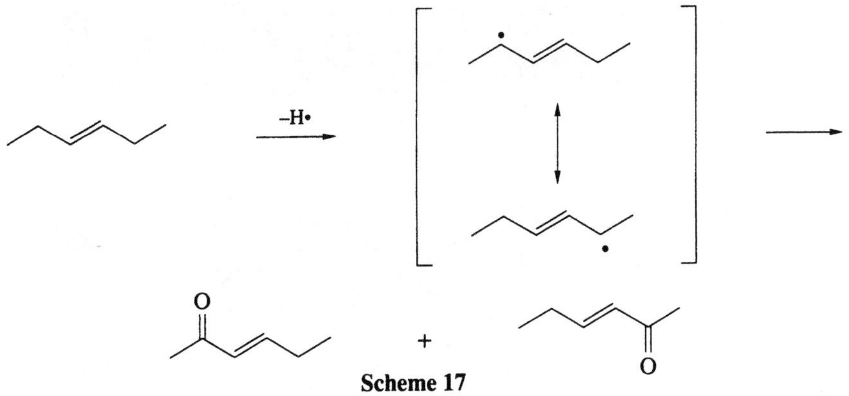

–H•

Scheme 17

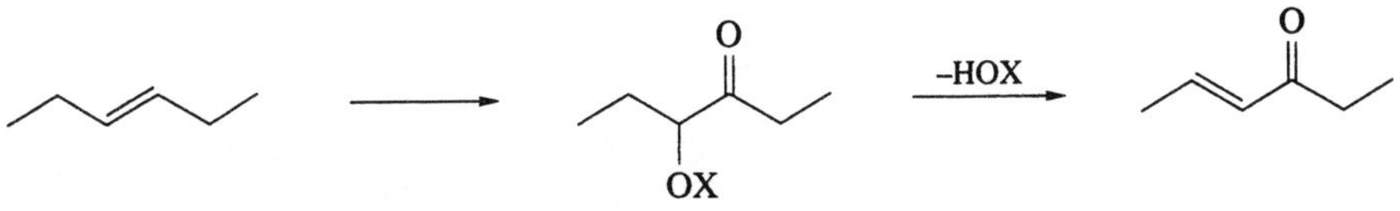

–HOX

OX

Scheme 18a

for a number of days, rather than high temperature conditions for extended periods. Dauben[132] proposed that the products of allylic oxidation using this reagent (according to Scheme 17) would be governed by a large number of factors, for example the steric accessibility of the allylic hydrogen atom towards abstraction, the relative stabilities of possible allylic intermediates, and the stereoelectronic control of the oxygen transfer step at competing sites.

Allylic oxidation in steroid systems provides a good illustration of the factors controlling the reaction pathway. In the oxidation of cholest-5-ene, which has a rigid structure with two allylic hydrogen atoms at C-4 and C-7, assuming axial preference of hydrogen atoms for abstraction,[133] an incoming chromium species should encounter steric hindrance from the methyl group above the plane and a less crowded approach from beneath. Indeed, allylic oxidation proceeds to yield only cholest-5-en-7-one in 52% yield (equation 27).

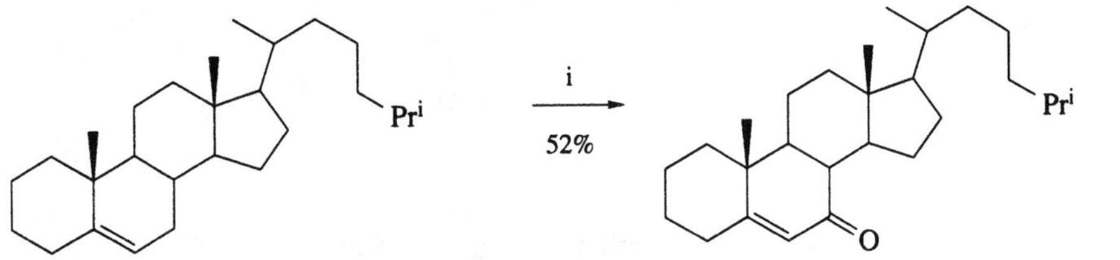

Pr^i i 52% Pr^i (27)

i, 20 equiv. CrO_3•2Py, CH_2Cl_2, r.t./24 h

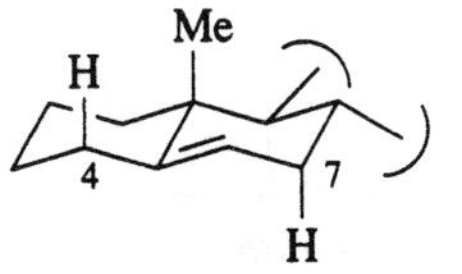

Me H 4 7 H

Dauben postulated that there is a preference for the abstraction of a tertiary allylic hydrogen atom and that this stems from the relative stability of the intermediate radical (or ionic) species rather than relative C—H bond strengths (Scheme 18b; the allylic oxidation of 1-methylcyclohex-2-ene).

From an extensive survey of this reagent system the following guidelines have been proposed: (i) allylic methyl groups are not readily oxidized; (ii) if more than one allylic methylene group is present in a conformationally flexible molecule, enones resulting from attack at all positions are formed, while if the molecule is conformationally rigid, as in a steroid, selectivity is observed; and (iii) attack at an allylic methinyl position yields a rearranged enone wherever possible; similar rearrangements may also occur in methylene systems possessing steric hindrance towards hydrogen atom abstraction.

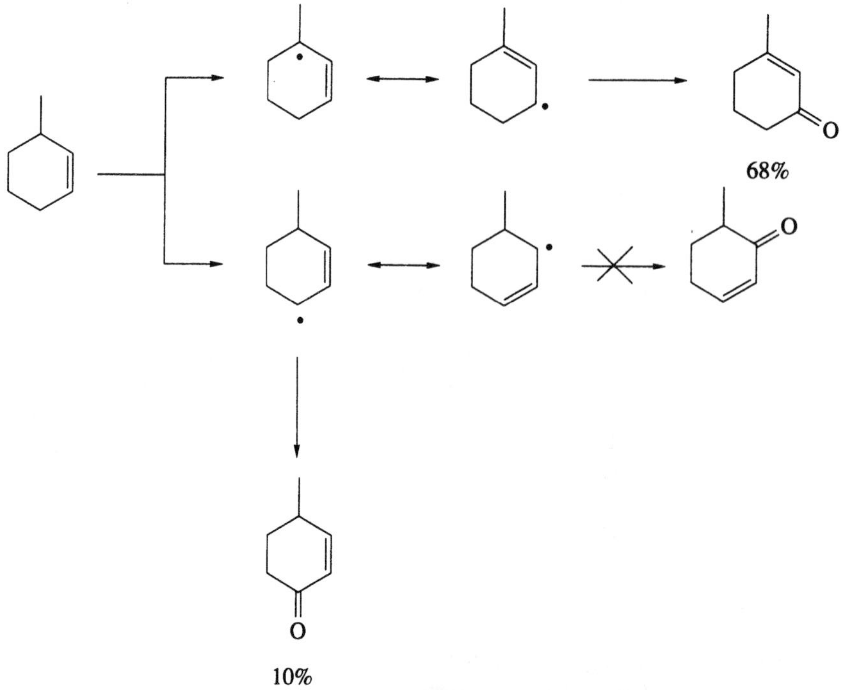

68%

10%

Scheme 18b

Allylic oxidation reactions employing chromium(VI) reagents therefore appear to be very much dependent upon the intrinsic nature of the substrate as to their regiochemical outcome. This is exemplified by the *t*-butyl chromate allylic oxidation of (+)-3-carene (**67**; equation 28)[134] where no great preference for either product exists.

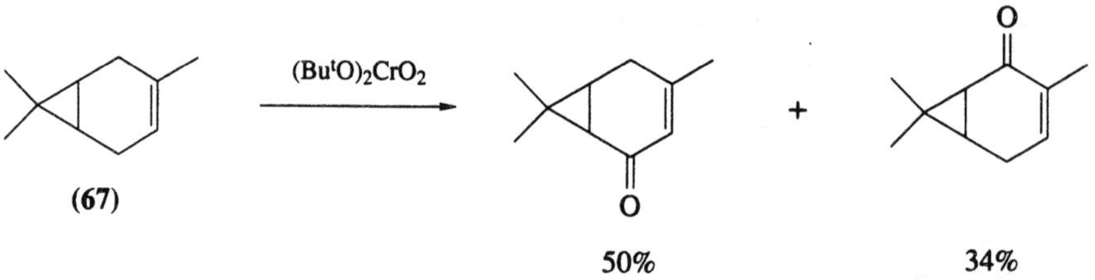

$(Bu^tO)_2CrO_2$ (28)

(67)

50% 34%

An attempted allylic oxidation of (**68**; equation 29) was found by Paquette[135] to be difficult to achieve using a range of reagents due to problems with polymerization and rearrangement. The chromium trioxide–pyridine complex was the only reagent combination found to be successful, albeit in low yield.

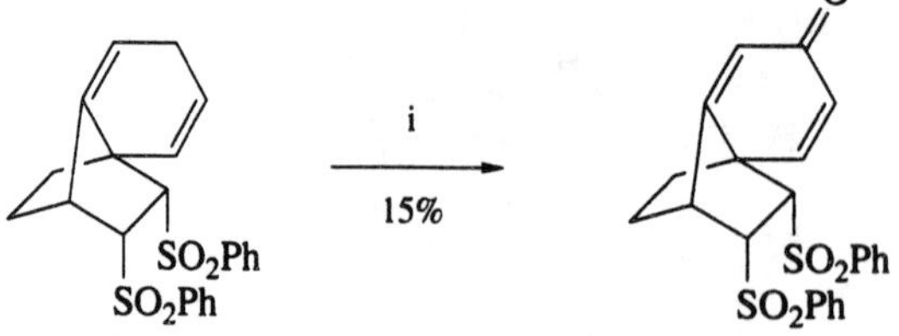

i, 15% (29)

SO_2Ph SO_2Ph SO_2Ph SO_2Ph

(68)

i, 20 equiv. CrO_3•2Py, 25 °C/24h

The chromium trioxide–pyridine complex was also found to be the reagent of choice in a synthesis of α-methylene-γ-butyrolactones (equation 30).[136]

This method was found to be superior to others tested, including selenium dioxide, chromium trioxide–acetic acid, and *t*-butyl chromate.

i
33–62%
(30)

R^1, R^2 = H, Ph or alkyl
R^3 = H or alkyl
i, 20 equiv. $CrO_3{\bullet}2Py$, CH_2Cl_2, reflux, 1 h

2.1.3.1.2 Pyridinium chlorochromate and dichromate

An alternative to the chromium trioxide–pyridine complex is provided by pyridinium chlorochromate (PCC) and pyridinium dichromate (PDC).[137] These reagents, now ubiquitous for chromate-based oxidation of alcohols, overcome the hygroscopic nature of the chromium trioxide–pyridine complex[138] and are prepared by a less hazardous procedure;[139] both are commercially available as are several other derivative reagents.

Pyridinium chlorochromate has been shown to be of particular value in the allylic oxidation of compounds containing an activated methylene group, such as 5,6-dihydropyrans (**69a** and **69b**; equations 31a and 31b).[140]

i
85%
(31a)

(69a) i, 1 equiv. PCC, 60–70°C/9 h

i
70%
(31b)

(69b)
i, 1 equiv. PCC, 60–70 °C/12 h

Indeed, Parish[141] claims that PCC is the reagent of choice in the allylic oxidation of Δ^5-steroids (*e.g.* equation 32). The reactions were carried out using PDC in pyridine solution at 100 °C, PCC in refluxing benzene solution, and PCC in DMSO solution at 100 °C. These solvent systems are claimed to be superior to the more usual methylene chloride.[138,142]

i, ii or iii
(32)

i, PCC, benzene, reflux, 89%; ii, PCC, DMSO, 100 °C, 78%; PDC, pyridine, 100 °C, 64%

One drawback associated with this type of chromium species is the frequent requirement for a large excess of reagent. Recent attempts to combat this problem have involved the use of a PCC–celite mixture in benzene solution under reflux[143] and more successfully a *t*-butyl hydroperoxide–pyridinium dichromate mixture (equation 33).[144]

While this latter modification appears to be efficient and highly regioselective for steroidal substrates which contain a rigid structure and provide a sterically crowded environment for the reacting double bond, the method can be rather poor in less hindered situations (*e.g.* equation 34).[144]

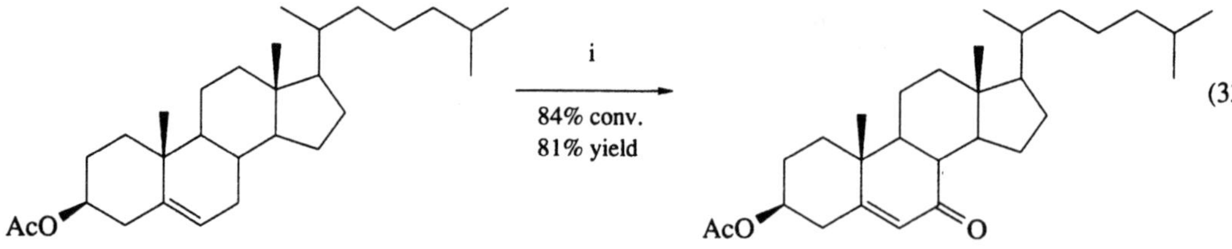
i
84% conv.
81% yield
AcO
AcO
O

(33)

i, 4 equiv. PDC, Bu^{t}OOH, celite, anhydr. benzene, r.t.

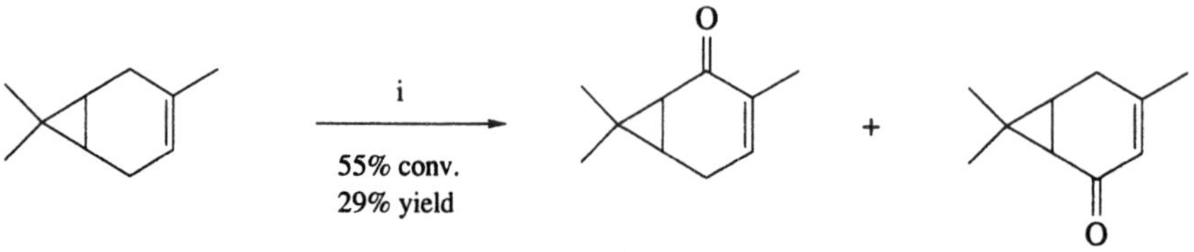
i
55% conv.
29% yield
O
+
O
4:1

(34)

i, 2 equiv. PDC, Bu^{t}OOH, celite, anhydr. benzene, r.t.

2.1.3.1.3 Chromium trioxide–3,5-dimethylpyrazole

With the advent of the chromium trioxide–3,5-dimethylpyrazole complex as an oxidant[145,146] allylic oxidation has become far more valuable as a synthetic transformation. The reagent was applied by Salamond[129] to the allylic oxidation of cholesteryl benzoate to give the corresponding Δ^5-7-ketone (equation 35). However, a 20 molar excess of reagent was still required to effect the reaction in less than 30 min at room temperature.

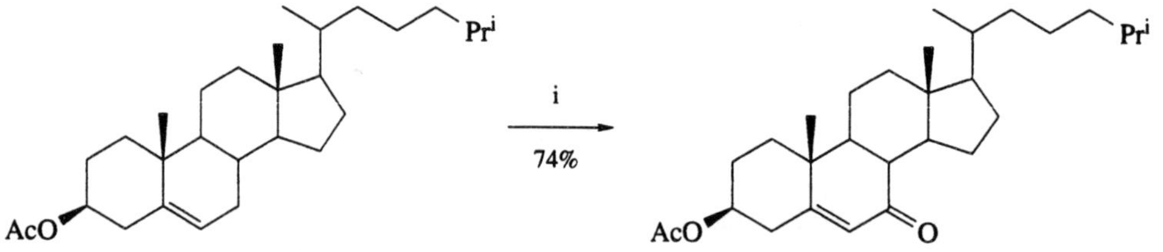
Pr^i
i
74%
AcO
Pr^i
AcO
O

(35)

i, 12 equiv. CrO_3•3,5-DMP, CH_2Cl_2, –15 °C, 4 h

The observed rate enhancement for this reagent over other chromium(VI) species has been rationalized as an effect of increased reagent solubility and by invoking the potential for acceleration by intramolecular participation by the pyrazole nucleus. Two mechanistic pathways have been proposed (Schemes 19a and 19b). The salient feature of both these pathways is that the chromium complex attacks first at the

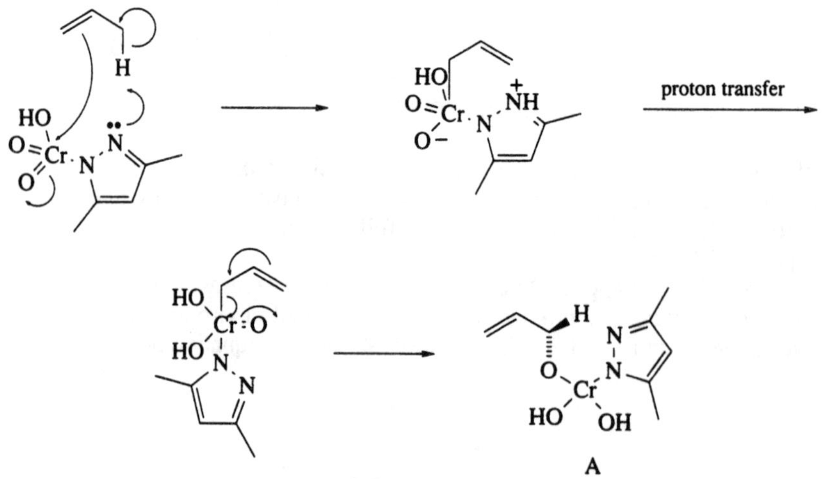
proton transfer
A

Scheme 19a

double bond and not at the allylic methylene group. This type of mechanism is not possible for chromium trioxide–pyridine, or for pyridinium chlorochromate or dichromate because: (i) no ligand sites are available for complexation with π-electrons unless pyridine is first displaced; and (ii) no basic nitrogen atom is available to assist in the removal of the allylic proton, other than by intermolecular deprotonation by a displaced pyridine molecule.

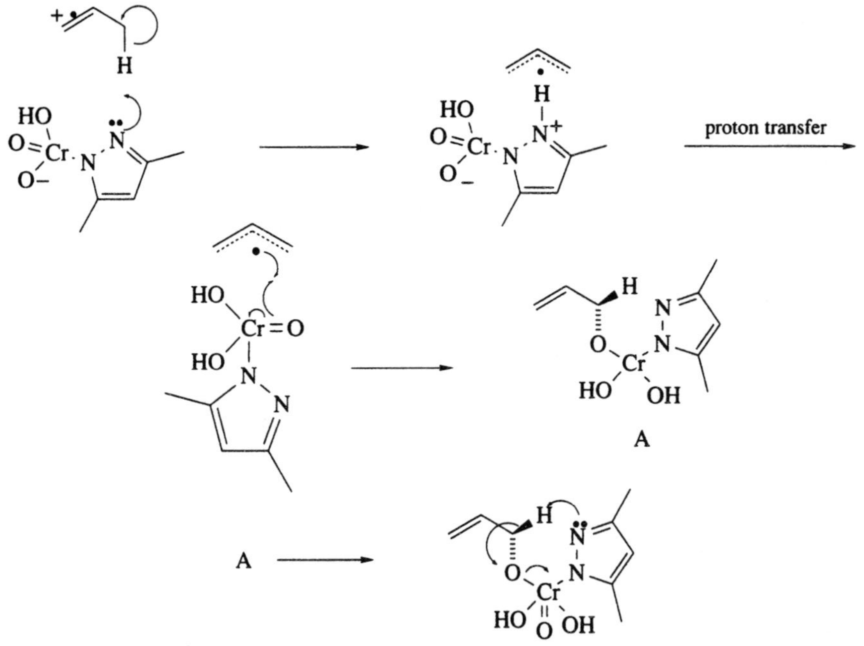

proton transfer

A

A

Scheme 19b

From a practical viewpoint the reagent is simple to make and use. It is prepared *in situ* at low temperature (*ca.* 20–25 °C), requiring about 15 min to form prior to the addition of substrate. It is important to note that the chromium trioxide should be thoroughly dried over phosphorus pentoxide before use.

Some fine examples of the synthetic use of this reagent are available in the literature;[147] for example in a total synthesis of vernolepin,[148] intermediate (**70**), containing a fairly sensitive lactol ether unit, was selectively prepared by the use of chromium trioxide–dimethylpyrazole with formation of only 5% of the allylically rearranged product (equation 36).

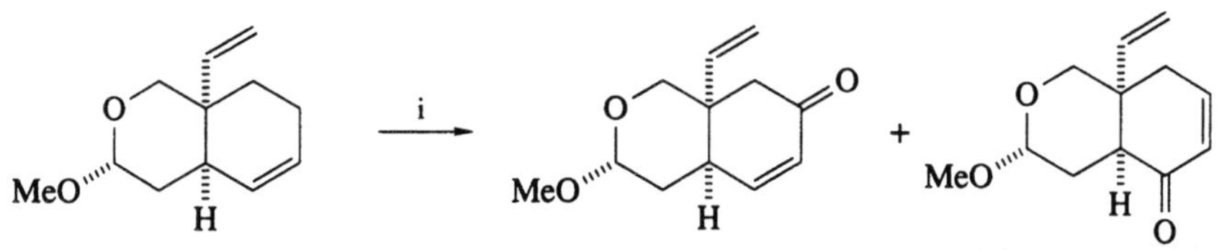

(36)

(**70**) 48% 5%

i, 20 equiv CrO_3•3,5-DMP, CH_2Cl_2, –20 °C/1 h then 0 °C/4 h

The reagent has been used by Magnus[149] in studies directed towards a synthesis of bachrachotoxin (**71**). A *cis*-decalin was oxidized selectively at the 7-position without the acetal or triple bond moieties present being affected (equation 37). The reagent has also found use in an approach to forskolin;[150] cyclohexadienone (**72**) was prepared by exclusive oxidation at the 7-position of (**73**; equation 38).

An interesting example of the incorporation of this oxidation into a synthetic strategy is seen in a route to quadrone (**74**) based on an intramolecular Diels–Alder reaction.[151] In this scheme highly selective allylic oxidative rearrangement of a *trans*-decalin (**75**) occurs to give a product (**76**) containing the double bond at a ring fusion position, allowing subsequent conversion to the desired *cis*-decalin system (Scheme 20). Neither of the other two possible ketonic oxidation products were observed.

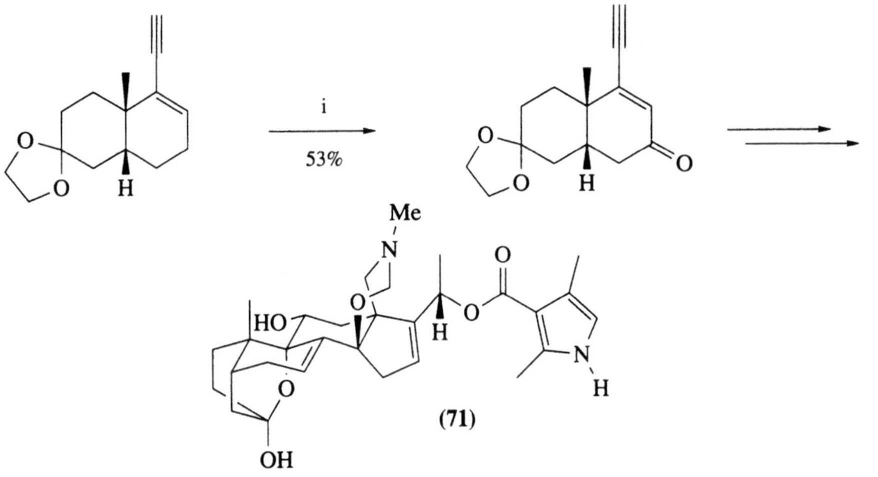

i

53%

Me

N

O

HO

O

H

O

O

N

H

(71)

OH

(37)

i, 30 equiv. CrO_3•3,5-DMP, CH_2Cl_2, r.t.

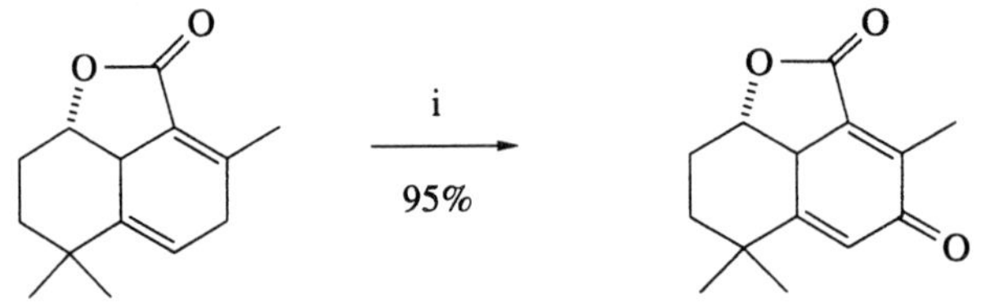

i

95%

(73)

(72)

(38)

i, 30 equiv. CrO_3•3,5-DMP, CH_2Cl_2, r.t.

i

75%

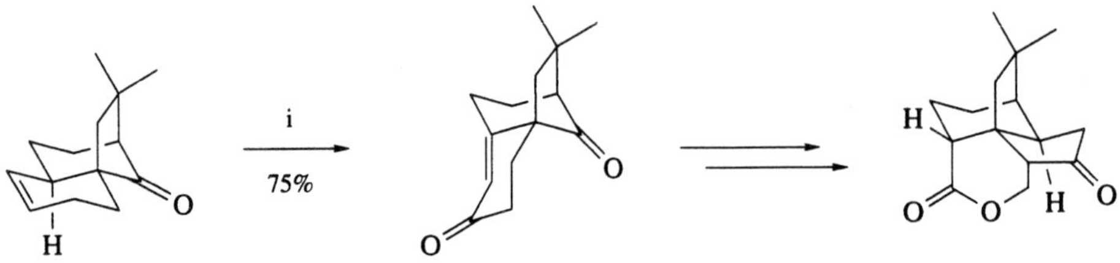

(75)

(76)

(74)

i, CrO_3•3,5-DMP, CH_2Cl_2

Scheme 20

2.1.3.1.4 Other chromium-based reagents

A number of other chromium-based reagents have been developed for allylic oxidation; for example that of steroids by *t*-butyl hydroperoxide in the presence of a catalytic amount (0.05–0.5 mol equiv.) of chromium trioxide[152] in dichloromethane solution at room temperature (equation 39). Yields vary from 32 to 69%. This modification is useful in terms of cost, operational simplicity and yields.

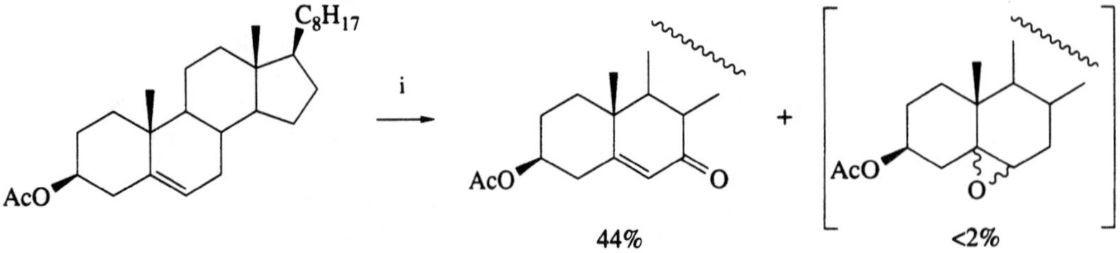

C_8H_{17}

i

AcO

AcO

44%

AcO

<2%

(39)

i, 0.4 equiv. CrO_3, Bu^tOOH, CH_2Cl_2, 5.5 h/r.t.

A chromium hexacarbonyl–*t*-butyl hydroperoxide system has also been developed with the remarkable chemoselective ability to effect allylic oxidation even in the presence of some secondary alcohols (equations 40 and 41).[153–155]

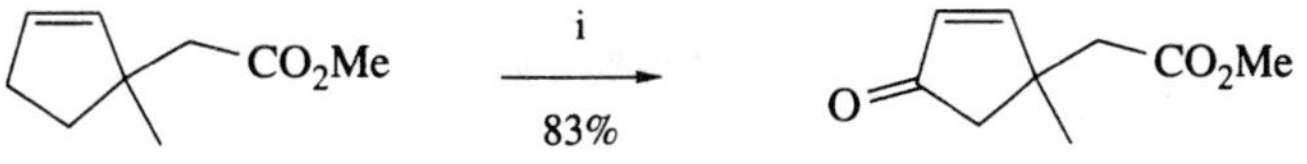

CO$_2$Me —i, 83%→ O= ... CO$_2$Me (40)

i, 0.5 equiv. $Cr(CO)_6$, 1.2 equiv. Bu^{t}OOH, MeCN, reflux/18 h

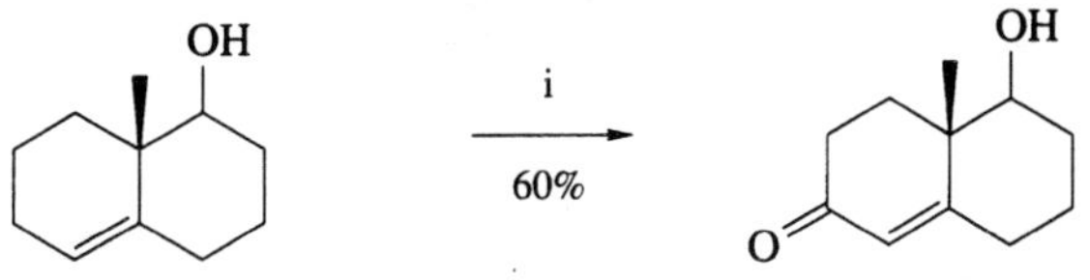

OH —i, 60%→ OH, O (41)

i, 0.25 equiv. $Cr(CO)_6$, 1.2 equiv. Bu^{t}OOH, MeCN, reflux/30 h

2.1.3.2 Other Transition Metal Catalyzed Allylic Oxidations

2.1.3.2.1 Palladium

Palladium catalysts are best known for oxidizing alkenes to ketones or vinyl derivatives.[156] However, formation of α,β-unsaturated carbonyl compounds by UV irradiation of oxygenated solutions of alkenes in the presence of catalytic amounts of palladium salts has been observed by Muzart.[39,157,158] This reaction is believed to proceed through a π-allylpalladium trifluoroacetate complex, *e.g.* (**77**).

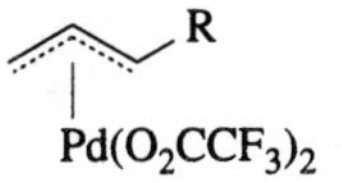

R

$Pd(O_2CCF_3)_2$

(**77**) R = alkyl

The process was later improved by the use of a *p*-toluenesulfonyl substituent[159] at the allylic carbon atom (equation 42). The authors claim that this modification has a powerful influence on both the selectivity and mechanism of the oxidation, exclusive oxidative rearrangement then being observed. Several other methods of achieving allylic oxidation using palladium catalysts have also been reported,[160–165] although these are generally of less importance.

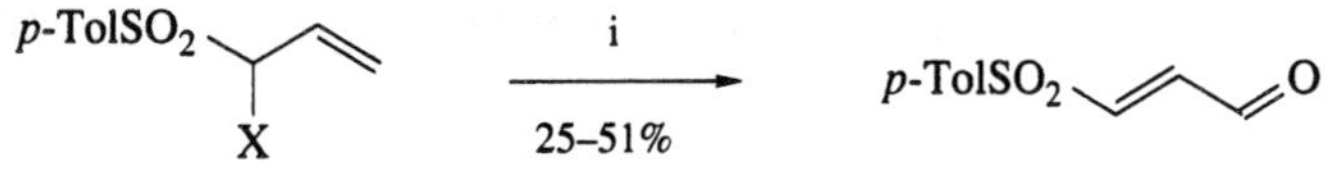

p-TolSO$_2$, X —i, 25–51%→ *p*-TolSO$_2$ ... O (42)

i, $Pd(O_2CCF_3)_2$, $h\nu$, O_2, acetone

2.1.3.2.2 Rhodium

Rhodium catalysis for effecting allylic oxidation has been developed and has led to considerable controversy over the operative mechanistic pathway.[166]

The first example of rhodium catalysis for this purpose utilized chlorotris(triphenylphosphine)rhodium(I) to catalyze the allylic oxidation of a range of alkenes.[167,168] This catalyst has also been shown to successfully oxidize cyclic allylsilanes[169] to afford β-silyl-2-cycloalkenones in very good yields and with exclusive rearrangement (equation 43).

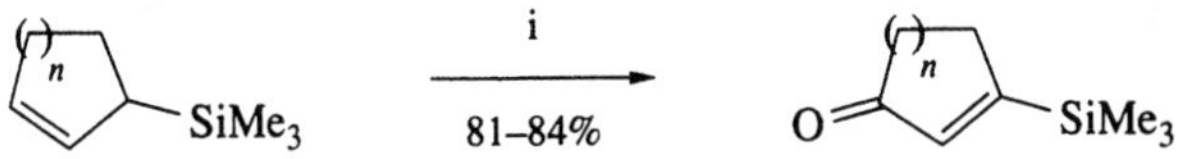

(43)

i, O_2, 0.01 equiv. $RhCl(PPh_3)_2$, 97 °C

81–84%

A combination of rhodium(III) chloride with silver acetate,[170] and treatment of rhodium(II) acetate in acetic acid solution with ozone,[171] are two methods for generation of the μ_3-oxotrimetal–acetato complex of rhodium $[Rh_3O(OAc)_6(H_2O)_3]OAc$. This '$Rh_3O$' complex was found to effect catalytic allylic oxidation of alkenes efficiently to give the corresponding α,β-unsaturated carbonyl compounds[172] in the presence of a reoxidant such as *t*-butyl hydroperoxide, although in disappointing yield (equation 44).

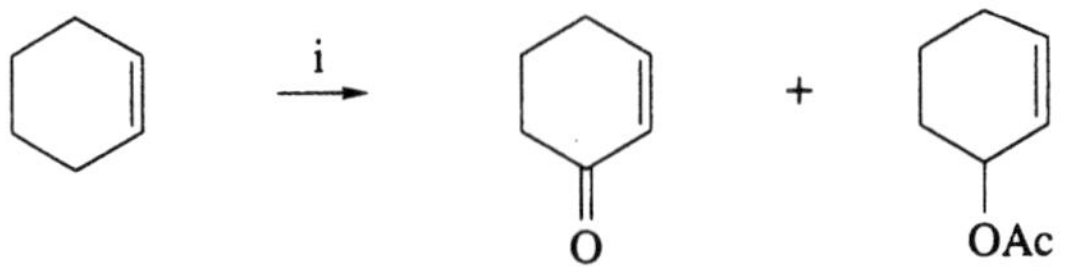

(44)

30% *ca.* 2%

i, cat. $[Rh_3O(OAc)_6(H_2O)_3]OAc$, Bu^tOOH, AcOH

2.1.3.2.3 *Iridium*

Iridium catalysts have not been widely developed for allylic oxidation; however a small number of examples of such use have been reported.[173,174] One example is given below (equation 45).

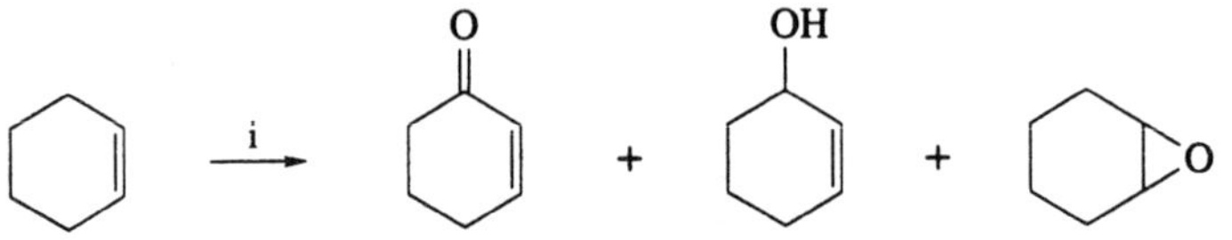

(45)

74% 26% traces

i, cat. $IrCO(PPh_3)_2Cl$, O_2

2.1.3.2.4 *Other transition metals*

Overall, many transition metal complexes have been investigated. Among those not mentioned above which may carry out catalytic allylic oxidation to give enones under certain circumstances are $Co(PPh_3)Cl/O_2$,[175] $Mn(TPP)Cl/O_2$,[176] $[Fe(PPh_3)]_2O$/UV,[177] Ni(phthalocyanine)/O_2[178] and an unusual mercury(II) acetate example[179] in which the enone is formed rather than the expected acetate.

2.1.3.3 Selenium-based Reagents

2.1.3.3.1 *Selenium dioxide*

Selenium-mediated allylic oxidations producing allylic alcohols have been discussed above; however, in some cases oxidation proceeds further to give the α,β-unsaturated carbonyl compounds directly, or mixtures of alcoholic and ketonic products.[180] That the regioselectivity observed in these allylic oxidation reactions closely resembles that found in classical selenium dioxide oxidations is in accord with initial formation of the intermediate allylic alcohol before *in situ* oxidation to the carbonyl compound.[1] This process was studied by Rapoport[17] and was explained mechanistically as an elimination of the intermediate allylic selenite ester *via* a cyclic transition state, analogous to S_Ni' (rather than S_N2') solvolysis (Scheme 21). Of the two possible transition states (**78**) and (**79**), the cyclic alternative (**78**) was preferred because oxidation exclusively yields *trans* aldehydes.

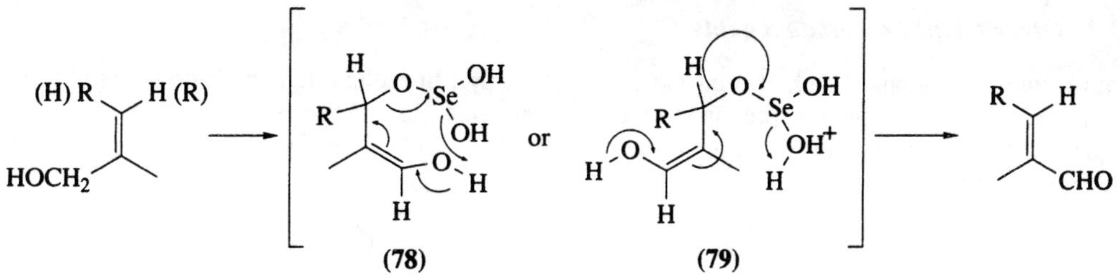

(78) (79)

Scheme 21

Oxidation reactions of this nature are common in the literature.[181] For example, selenium dioxide in refluxing ethanolic solution brought about the allylic oxidative rearrangement of geranyl acetate, which was further functionalized in a synthesis of the norsesquiterpenoid gyrinidal (equation 46).[182] This transformation was also used in a total synthesis of phytol.[183] Similarly, an α,β-unsaturated aldehyde was obtained under similar conditions in studies of a synthesis of pentalenic acid derivatives (equation 47).[184]

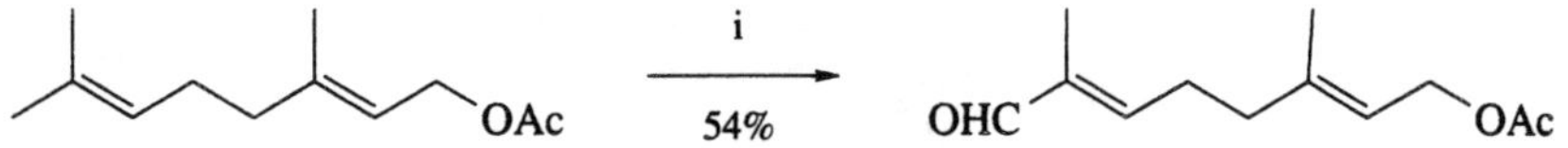

(46)

i, SeO_2, EtOH, reflux

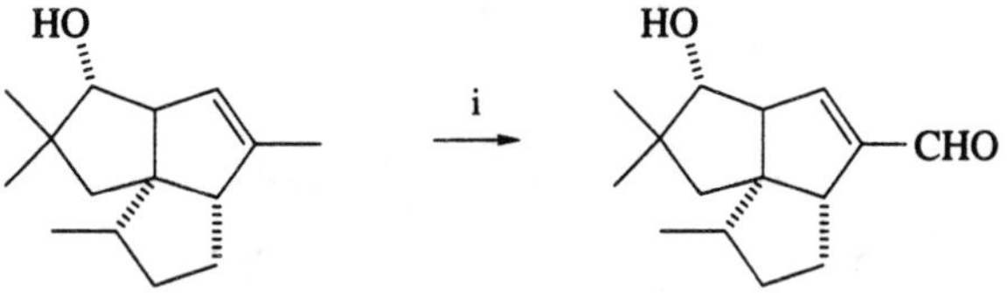

(47)

i, SeO_2, EtOH, reflux overnight

Evidence for the preferential formation of the *trans*-substituted product of selenium dioxide allylic oxidation[185] is seen in the synthesis of part of (13*Z*)-retinoic acid (equation 48). Reaction took place exclusively at the exocyclic double bond without rearrangement. Allylic oxidation of this nature has also been used in the synthesis of 6-conjugated 2-pyrones (equation 49).[186] This intermediate was employed in the total synthesis of natural pyrones such as yangonin.

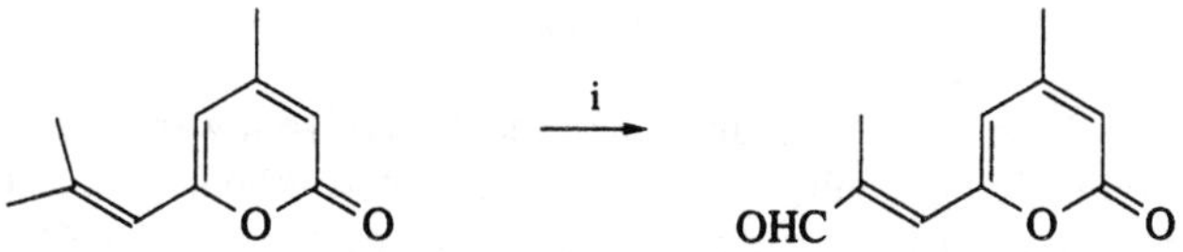

(48)

i, SeO_2, aq. dioxane, reflux/12 h

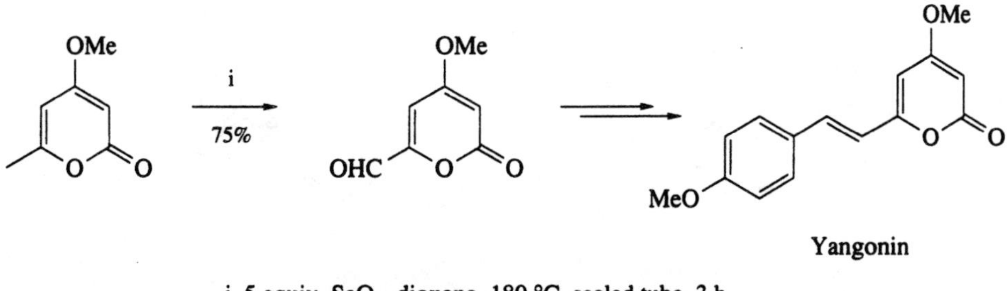

(49)

Yangonin

i, 5 equiv. SeO_2, dioxane, 180 °C, sealed tube, 3 h

2.1.3.3.2 Other selenium-based reagents

More recently Barton and Crich[187] reported the use of 2-pyridineseleninic anhydride in the allylic oxidation of alkenes. This reagent is prepared *in situ* by the oxidation of the corresponding diselenide by iodylbenzene. It effects oxidation to α,β-unsaturated ketones with retention of the double bond regiochemistry (*e.g.* equation 50).

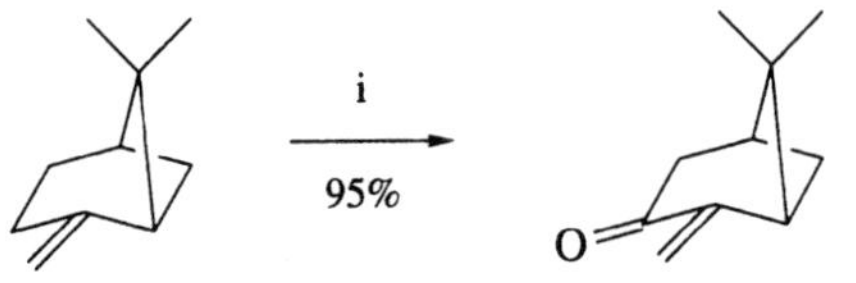

i

95%

(50)

i, 0.1 equiv. 2,2'-dipyridyl diselenide, 3 equiv. iodylbenzene, chlorobenzene, 80 °C/1.5 h

Observation of the reaction by TLC indicated initial formation of the allylic alcohols which were oxidized *in situ* to give the enone. The following mechanism for allylic oxidation was proposed (Scheme 22).

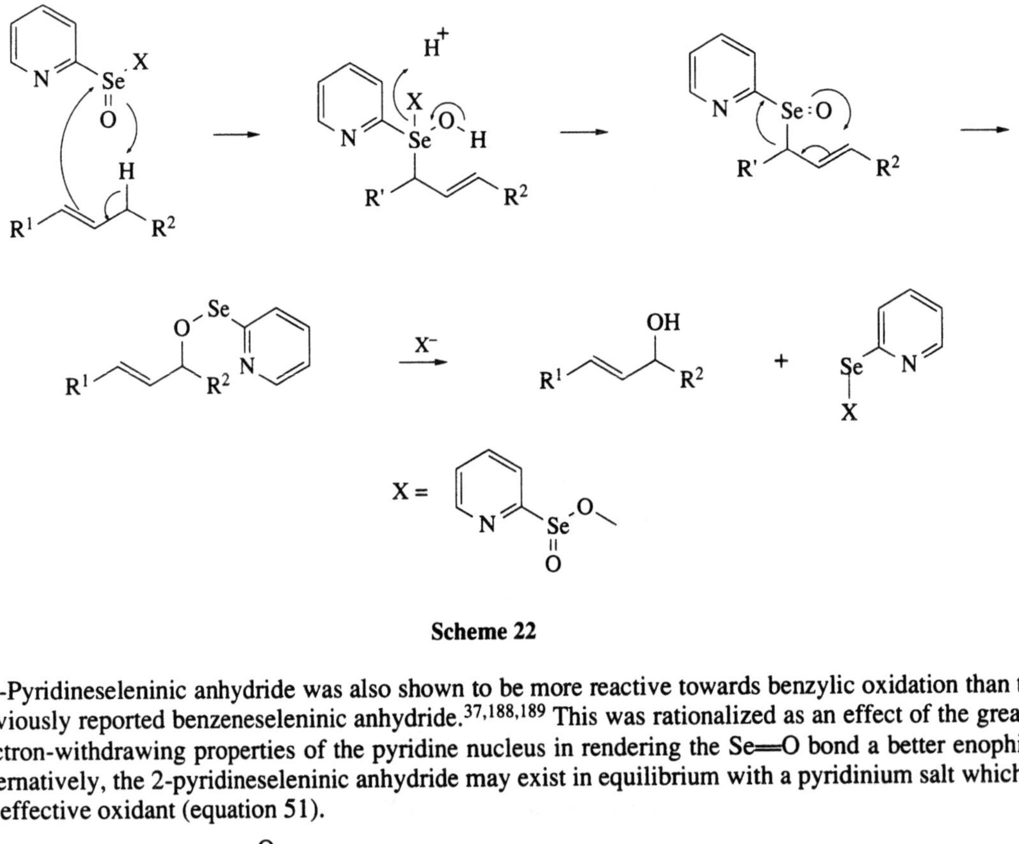

Scheme 22

2-Pyridineseleninic anhydride was also shown to be more reactive towards benzylic oxidation than the previously reported benzeneseleninic anhydride.[37,188,189] This was rationalized as an effect of the greater electron-withdrawing properties of the pyridine nucleus in rendering the Se═O bond a better enophile. Alternatively, the 2-pyridineseleninic anhydride may exist in equilibrium with a pyridinium salt which is the effective oxidant (equation 51).

(51)

2.1.3.4 Singlet Oxygen

The ene reaction is by far the most widely investigated reaction of singlet oxygen,[88] involving the formation of an allylic hydroperoxide from an alkene by a process involving abstraction of an allylic proton with migration of the carbon–carbon double bond. Reduction of the resulting hydroperoxide, as discussed above, provides the corresponding allylic alcohol. Thus subsequent oxidation is required for formation of an α,β-unsaturated carbonyl compound.[190–192] A more direct route to the enone can also occur *via* β-elimination of water from the allylic hydroperoxide. Ireland,[190] in his investigation of α-methylene ketones, studied the photooxidation of some model systems and the subsequent reactions of the intermediate allylic hydroperoxides with acetic anhydride (Scheme 23).[88]

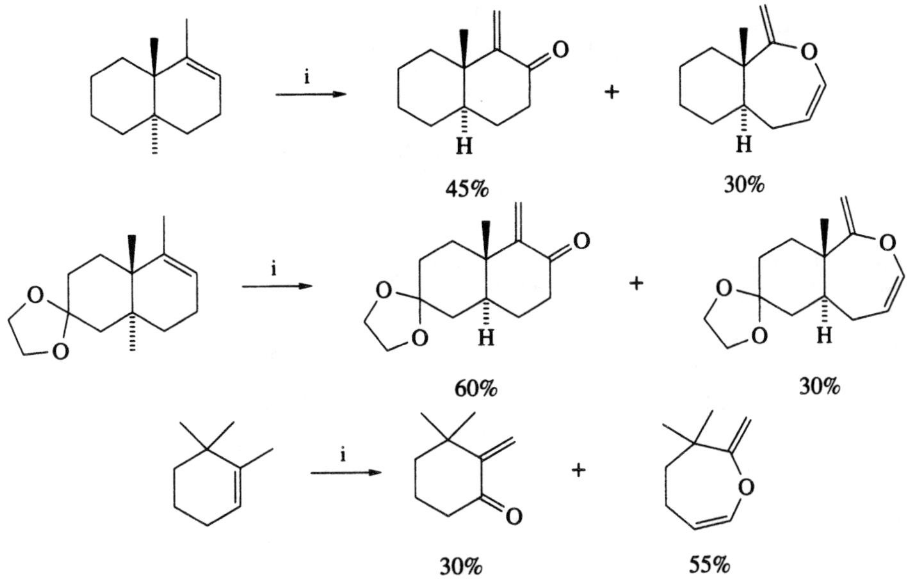

45% 30%

60% 30%

30% 55%

i, $^1O_2$, Ac_2O

Scheme 23

Decomposition of the intermediate allylic peracetates yielded the desired α-methylene ketones along with ring-expanded divinyl ethers, formed *via* a Hock fragmentation.[193] Direct formation of enones has also been reported by Mihelich[194] under similar conditions of photooxidation in the presence of acetic acid and a catalytic amount of base (Scheme 24).

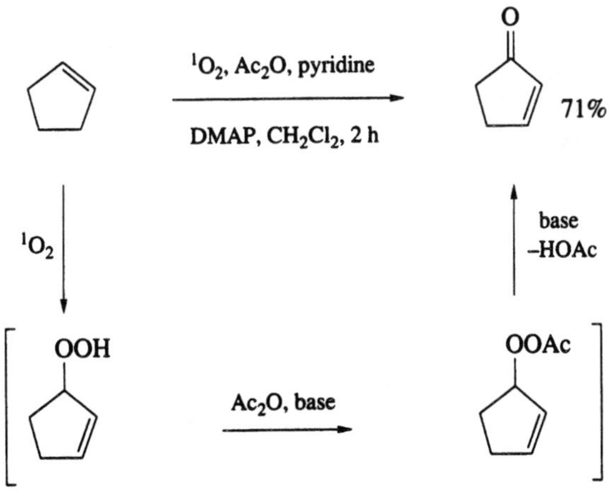

$^1O_2$, Ac_2O, pyridine
DMAP, CH_2Cl_2, 2 h
71%
$^1O_2$
base
–HOAc
OOH
OOAc
Ac_2O, base

Scheme 24

For example, photooxidation of α-pinene led to formation of the desired product in 97% yield, whereas similar reaction of β-pinene was accomplished in only 58% yield (Scheme 25).

Interestingly, an anomalous result was obtained in the photooxidation of 1,3-cholestadiene and related compounds (equation 52).[195] Thus 1,4-cholestadien-3-one was obtained rather than the expected Diels–

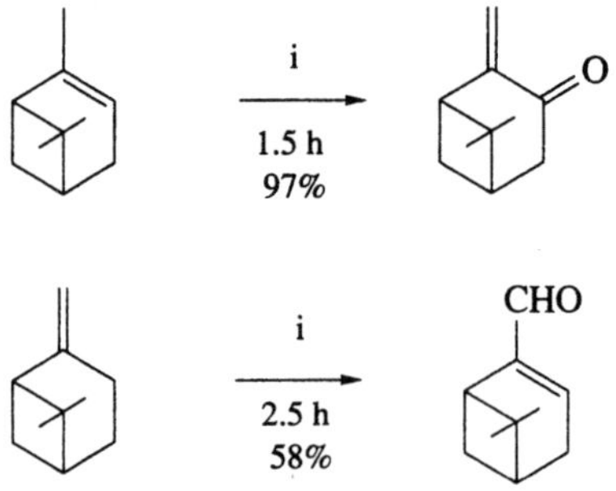

i
1.5 h
97%

CHO
i
2.5 h
58%

i, $^1O_2$, Ac_2O, pyridine, DMAP, CH_2Cl_2

Scheme 25

Alder endoperoxide product. These results are discussed by the authors in terms of failure to meet the steric requirement for endoperoxide formation.[95,195]

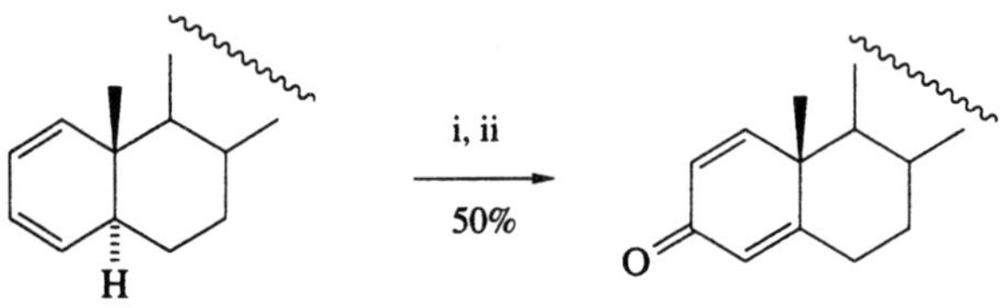

i, ii
50%
(52)

i, $^1O_2$, EtOH, 0 °C, *h*ν, 18 h; ii, Al_2O_3

2.1.3.5 Miscellaneous Reagents

A few reagents have been reported for the allylic oxidation of particular substrates. These include *N*-bromosuccinimide oxidation of α-amyrin acetate in moist dioxane (equation 53),[196] a method later modified by Thomson.[197]

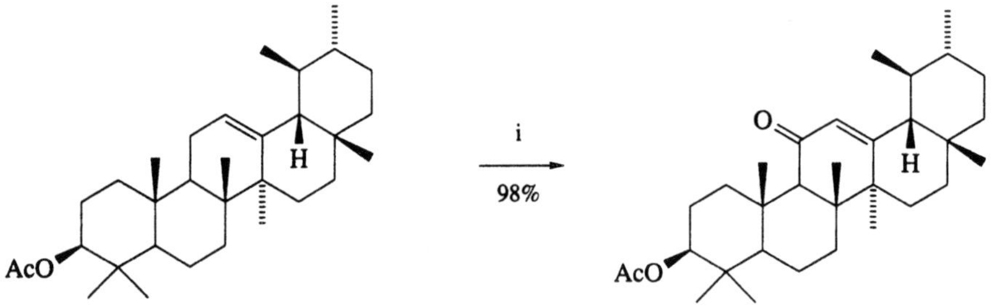

i
98%
(53)

i, *N*-bromosuccinimide, aq. dioxane

A catalytic amount of palladium on charcoal (5 mol %) has been shown by Stoodley to effect allylic oxidation of cephem dioxides (**80**) and (**81**) in yields of about 60% in each case.[198]

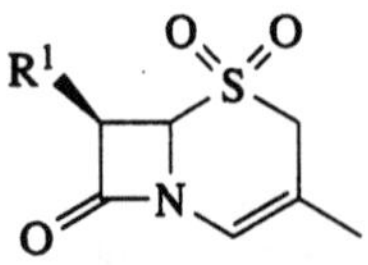

(**80**) **a**: R^1 = H
b: R^1 = $PhOCH_2CONH$

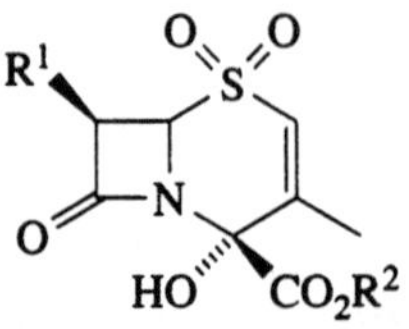

(**81**) **a**: R^1 = H, R^2 = $PhCH_2$
b: R^1 = $PhOCH_2CONH$, R^2 = Ph_2CH

The conversion of (**80b**) to (**81b**) and conversions of other compounds containing acidic allylic methine or methylene protons was found to proceed in the presence of activated carbon (Darco G-60; equations 54 and 55).

MeO_2C ... CO_2Me —i, 53%→ MeO_2C ... CO_2Me (54)

i, 4 equiv. Darco G-60, NEt_3, EtOAc, 24 h

H ... CO_2Et —i, 50%→ HO ... CO_2Et (55)

i, 4 equiv. Darco G-60, NEt_3, EtOAc, 24 h

Clearly, regio- and chemo-selectivities of this reagent are highly dependent on the substrate structure.

Allylic oxidation to give enones has also been reported at the 11-position of steroids upon treatment with nitrosyl fluoride solutions.[200]

2.1.4 REFERENCES

1. N. Rabjohn, *Org. React. (N.Y.)*, 1949, **5**, 331; G. R. Waitkins and C. W. Clark, *Chem. Rev.*, 1945, **36**, 235; T. W. Campbell, H. G. Walker and G. M. Coppinger, *Chem. Rev.*, 1952, **50**, 279; R. A. Jerussi, in 'Selective Organic Transformations', ed. B. S. Thyagarajan, Wiley, New York, 1970, vol. 1; N. Rabjohn, *Org. React. (N.Y.)*, 1978, **24**, 261; S. G. Wilkinson, in 'Comprehensive Organic Chemistry', ed. D. H. R. Barton and W. D. Ollis, Pergamon Press, Oxford, 1979, vol. 1, p. 579; C. Paulmier, 'Selenium Reagents and Intermediates in Organic Synthesis', Pergamon Press, Oxford, 1986.
2. A. Guillemonat, *Ann. Chim. (Rome)*, 1939, **11**, 143.
3. W. Zacharewicz, *Rocz. Chem.*, 1948, **22**, 68; L. F. Fieser and G. Ourisson, *J. Am. Chem. Soc.*, 1953, **75**, 4404; A. K. Macbeth, B. Milligan and J. S. Shannon, *J. Chem. Soc.*, 1953, 2574; C. H. Issidorides, M. Fieser and L. F. Fieser, *J. Am. Chem. Soc.*, 1960, **82**, 2002; Y. Sakuda, *Bull. Chem. Soc. Jpn.*, 1961, **34**, 514; Y. Sakuda, *Nippon Kagaku Zasshi*, 1961, **82**, 117; T. Suga, M. Sugimoto and T. Matsuura, *Bull. Chem. Soc. Jpn.*, 1963, **36**, 1363; G. Ohloff and G. Uhde, *Helv. Chim. Acta*, 1965, **48**, 10; S. Pathak and G. H. Kulkarni, *Chem. Ind. (London)*, 1968, 913.
4. E. N. Trachtenberg and J. R. Carver, *J. Org. Chem.*, 1970, **35**, 1646.
5. W. A. Waters, *J. Chem. Soc.*, 1939, 1805.
6. E. N. Trachtenberg, C. H. Nelson and J. R. Carver, *J. Org. Chem.*, 1970, **35**, 1653.
7. J. P. Schaefer, B. Horvath and H. P. Klein, *J. Org. Chem.*, 1968, **33**, 2647; J. P. Schaefer and B. Horvath, *Tetrahedron Lett.*, 1964, 2023.
8. K. B. Wiberg and S. D. Nielsen, *J. Org. Chem.*, 1964, **29**, 3353.
9. U. T. Bhalerao and H. Rapoport, *J. Am. Chem. Soc.*, 1971, **93**, 4835.
10. A. F. Thomas and W. Bucher, *Helv. Chim. Acta*, 1970, **53**, 770.
11. G. Büchi and H. Wüest, *J. Org. Chem.*, 1969, **34**, 857.
12. H.-L. Pan, C. A. Cole and T. L. Fletcher, *Synthesis*, 1980, 813.
13. K. B. Sharpless and R. F. Lauer, *J. Am. Chem. Soc.*, 1972, **94**, 7154; D. Arigoni, A. Vasella, K. B. Sharpless and H. P. Jensen, *J. Am. Chem. Soc.*, 1973, **95**, 7917; H. P. Jensen and K. B. Sharpless, *J. Org. Chem.*, 1975, **40**, 264.
14. L. M. Stephenson and D. R. Speth, *J. Org. Chem.*, 1979, **44**, 4683.
15. W.-D. Woggon, F. Ruther and M. Egli, *J. Chem. Soc., Chem. Commun.*, 1980, 706.
16. (a) O. Rosenheim and W. W. Starling, *J. Chem. Soc.*, 1937, 377; L. F. Fieser, *J. Am. Chem. Soc.*, 1953, **75**, 4395; Y. Abe, T. Harukawa, H. Ishikawa, T. Miki, M. Sumi and T. Toga, *J. Am. Chem. Soc.*, 1956, **78**, 1422; N. Danieli, Y. Mazur and F. Sondheimer, *Tetrahedron Lett.*, 1961, 310; 1962, 1281; N. Danieli, Y. Mazur and F. Sondheimer, *J. Am. Chem. Soc.*, 1962, **84**, 875; N. Danieli, Y. Mazur and F. Sondheimer, *Tetrahedron*, 1967, **23**, 509; J. E. Pike, F. H. Lincoln, G. B. Spero, R. W. Jackson and J. L. Thompson, *Steroids*, 1968, **11**, 755; A. Furlenmeier, A. Fürst, A. Langemann, G. Waldvogel, U. Kerb, P. Hocks and R. Wiechert, *Helv. Chim. Acta*, 1966, **49**, 1591; J. B. Siddall, J. P. Marshall, A. Bowers, A. D. Cross, J. A. Edwards and J. H. Fried, *J. Am. Chem. Soc.*, 1966, **88**, 379; J. Meinwald, K. Opheim and T. Eisner, *Tetrahedron Lett.*, 1973, 281; (b) J. J. Plattner, U. T. Bhalerao and H. Rapoport, *J. Am. Chem. Soc.*, 1969, **91**, 4933; U. T. Bhalerao, J. J. Plattner and H. Rapoport, *J. Am. Chem. Soc.*, 1970, **92**, 3429.
17. U. T. Bhalerao and H. Rapoport, *J. Am. Chem. Soc.*, 1971, **93**, 5311.

18. (a) S. Manna, P. Yadagiri and J. R. Falck, *J. Chem. Soc., Chem. Commun.*, 1987, 1324; (b) C. H. Miller, J. A. Katzenellenbogen and S. B. Bowlus, *Tetrahedron Lett.*, 1973, 285; M. Somei, Y. Makita and F. Yamada, *Chem. Pharm. Bull.*, 1986, **34**, 948.
19. H. Hart and T. Ghosh, *Tetrahedron Lett.*, 1988, **29**, 881.
20. F. Camps, J. C. Coll and A. Parente, *Synthesis*, 1978, 215.
21. M. A. Umbreit and K. B. Sharpless, *J. Am. Chem. Soc.*, 1977, **99**, 5526.
22. K. B. Sharpless, K. M. Gordon, R. F. Lauer, S. P. Singer and M. W. Young, *Chem. Scr.*, 1975, **8A**, 9.
23. J. M. Coxon, E. Dansted and M. P. Hartshorn, *Org. Synth.*, 1977, **56**, 25.
24. B. Chabaud and K. B. Sharpless, *J. Org. Chem.*, 1979, **44**, 4202.
25. M. Haruna and K. Ito, *J. Chem. Soc., Chem. Commun.*, 1981, 483.
26. S. David, G. de Sennyey and B. Sotta, *Tetrahedron Lett.*, 1982, **23**, 1817.
27. J. A. Marshall and R. C. Andrews, *J. Org. Chem.*, 1985, **50**, 1602.
28. J. A. Marshall and J. Lebreton, *J. Org. Chem.*, 1988, **53**, 4108.
29. A. V. Rama Rao, S. V. Mysorekar and J. S. Yadav, *Synth. Commun.*, 1987, **17**, 1339.
30. S. C. Dolan and J. MacMillan, *J. Chem. Soc., Perkin Trans. 1*, 1985, 2741.
31. M. Furber and L. N. Mander, *J. Am. Chem. Soc.*, 1987, **109**, 6389.
32. D. R. Andrews, D. H. R. Barton, K. P. Cheng, J.-P. Finet, R. H. Hesse, G. Johnson and M. M. Pechet, *J. Org. Chem.*, 1986, **51**, 1635.
33. I. Razmilic, J. López, J. Sierra and M. Cortés, *Synth. Commun.*, 1987, **17**, 95.
34. F. Ogura, T. Otsubo, K. Ariyoshi and H. Yamaguchi, *Chem. Lett.*, 1983, 1833.
35. M. A. Warpehoski, B. Chabaud and K. B. Sharpless, *J. Org. Chem.*, 1982, **47**, 2897.
36. B. R. Chhabra, K. Hayano, T. Ohtsuka, H. Shirahama and T. Matsumoto, *Chem. Lett.*, 1981, 1703.
37. T. Hori and K. B. Sharpless, *J. Org. Chem.*, 1978, **43**, 1689.
38. Z. Rappoport, S. Winstein and W. G. Young, *J. Am. Chem. Soc.*, 1972, **94**, 2320.
39. J. Muzart, *Bull. Soc. Chim. Fr.*, 1986, 65.
40. R. Criegee, in 'Oxidation in Organic Chemistry', ed. K. B. Wiberg, Academic Press, New York, 1965; R. M. Moriarty, in 'Selective Organic Transformations', ed. B. S. Thyagarajan, Wiley, New York, 1972, vol. 2; W. Kitching, *Organomet. Chem. Rev.*, 1968, **3**, 35; W. Kitching, in 'Organometallic Reactions', ed. E. I. Becker and M. Tsutsui, Wiley, New York, 1972, vol. 3; L. H. Briggs, B. F. Cain, R. C. Cambie, B. R. Davis and P. S. Rutledge, *J. Chem. Soc.*, 1962, 1850; M. P. Bertrand, H. Oumar-Mahamet and J.-M. Surzur, *Tetrahedron Lett.*, 1985, **26**, 1209; D. J. Rawlinson and G. Sosnovsky, *Synthesis*, 1973, 567.
41. (a) R. Criegee, *Angew. Chem.*, 1958, **70**, 173; D. G. Lee, in 'Oxidation', ed. R. L. Augustine, Dekker, New York, 1969, vol. 1; O. Dimroth and R. Schweitzer, *Ber. Dtsch. Chem. Ges.*, 1923, **56**, 1375; R. Criegee, *Justus Liebigs Ann. Chem.*, 1930, **481**, 263; L. F. Fieser and J. Cason, *J. Am. Chem. Soc.*, 1940, **62**, 432; J. Cason, *Org. Synth., Coll. Vol.*, 1965, **3**, 3; D. H. R. Barton, E. F. Lier and J. F. McGhie, *J. Chem. Soc. C*, 1968, 1031; M. Stefanovic, A. Jokić, Z. Maksimović, L. Lorenc and M. L. Mihailovic, *Helv. Chim. Acta*, 1970, **53**, 1895; B. M. Trost and Y. Tanigawa, *J. Am. Chem. Soc.*, 1979, **101**, 4413; I. Alkonyi, *Chem. Ber.*, 1963, **96**, 1873; E. Dane and K. Eder, *Justus Liebigs Ann. Chem.*, 1939, **539**, 207; (b) G. Ortar and A. Romeo, *J. Chem. Soc., Perkin Trans. 1*, 1976, 111; M. Ochiai, E. Fujita, M. Arimoto and H. Yamaguchi, *Chem. Pharm. Bull.*, 1984, **32**, 5027.
42. W. Treibs, *Naturwissenschaften*, 1948, **35**, 125; J. Chatt, *Chem. Rev.*, 1951, **48**, 7; G. F. Wright, *Ann. N. Y. Acad. Sci.*, 1957, **65**, 436; A. C. Cope, M. R. Kinter and R. T. Keller, *J. Am. Chem. Soc.*, 1954, **76**, 2757; P. Karrer and C. H. Eugster, *Helv. Chim. Acta*, 1951, **34**, 1400; G. Rosenkrantz, J. Romo, E. Batres and C. Djerassi, *J. Org. Chem.*, 1951, **16**, 298; C. Djerassi, J. Romo and G. Rosenkrantz, *J. Org. Chem.*, 1951, **16**, 754; T. Matsuda, M. Moritake, J. Ihara and S. Akiyoshi, *Nippon Kagaku Zasshi*, 1961, **82**, 1570; K. Takaoka and Y. Izumisawa, *Kogyo Kagaku Zasshi*, 1964, **67**, 1244; I. Alkonyi, *Chem. Ber.*, 1962, **95**, 279; R. C. Larock, *Tetrahedron*, 1982, **38**, 1713; H. Arzoumanian and J. Metzger, *Synthesis*, 1971, 527.
43. M. S. Ahmad, S. Z. Ahmad and J. A. Ansari, *J. Chem. Res. (S)*, 1984, 374; J. R. Gilmore and J. M. Mellor, *J. Chem. Soc. C*, 1971, 2355; W. E. Fristad and J. R. Peterson, *J. Org. Chem.*, 1985, **50**, 10; J. A. Norman, C. B. Thomas and M. J. Burrow, *J. Chem. Soc., Perkin Trans. 1*, 1985, 1087.
44. M. Hirano and T. Morimoto, *J. Chem. Res. (S)*, 1979, 104; M. Hirano and T. Morimoto, *J. Chem. Soc., Perkin Trans. 2*, 1984, 1033; D. A. Baines and W. Cocker, *J. Chem. Soc., Perkin Trans. 1*, 1975, 2232; E. Bordier, *Bull. Soc. Chim. Fr.*, 1973, 2621, 3291.
45. G. H. Whitman, *J. Chem. Soc.*, 1961, 2232.
46. W. Treibs and H. Bast, *Justus Liebigs Ann. Chem.*, 1949, **561**, 165; W. Treibs, G. Lucius, K. Kögler and H. Breslauer, *Justus Liebigs Ann. Chem.*, 1953, **581**, 59; W. Treibs and M. Weissenfels, *Chem. Ber.*, 1960, **93**, 1374.
47. K. B. Wiberg and S. W. Nielsen, *J. Org. Chem.*, 1964, **29**, 3353; S. Wolfe and P. G. C. Campbell, *Can. J. Chem.*, 1965, **43**, 1184.
48. A. Kergomard, *Ann. Chim. (Rome)*, 1953, **8**, 153; W. V. Ruyle, T. A. Jacob, J. M. Chemerda, E. M. Chamberlin, D. W. Rosenberg, G. E. Sita, R. L. Erickson, L. M. Aliminosa and M. Tishler, *J. Am. Chem. Soc.*, 1953, **75**, 2604.
49. D. H. R. Barton and W. J. Rosenfelder, *J. Chem. Soc.*, 1951, 2161.
50. P. D. Sleezer, S. Winstein and W. G. Young, *J. Am. Chem. Soc.*, 1963, **85**, 1890; Z. Rappoport, P. D. Sleezer, S. Winstein and W. G. Young, *Tetrahedron Lett.*, 1965, 3719; C. B. Anderson and S. Winstein, *J. Org. Chem.*, 1963, **28**, 605; Z. Rappoport, L. K. Dyall, S. Winstein and W. G. Young, *Tetrahedron Lett.*, 1970, 3483.
51. A. Kergomard, J.-C. Tardivat and J.-P. Vuillerme, *Bull. Soc. Chim. Fr.*, 1974, 2572.
52. M. P. Irismetov, G. A. Tolstikov, R. S. Irirmetura and M. I. Goryaev, *Izv. Akad. Nauk Kaz. SSR, Ser. Khim.*, 1968, **18**, 73.
53. H. Mrozik, P. Eskola and M. H. Fisher, *J. Org. Chem.*, 1986, **51**, 3058.
54. J. Barluenga, L. Alonso-Cires, P. J. Compos and G. Asensio, *Tetrahedron*, 1984, **40**, 2563.
55. J.-E. Bäckvall, *Acc. Chem. Res.*, 1983, **16**, 335, and refs. contained therein; J. Tsuji, 'Organic Synthesis with Palladium Compounds', Springer-Verlag, Berlin, 1980; P. M. Henry, *Adv. Organomet. Chem.*, 1975, **13**, 363;

R. O. C. Norman, C. B. Thomas and G. Watson, *J. Chem. Soc., Perkin Trans. 2*, 1980, 1099; M. Green, R. N. Haszeldine and J. Lindley, *J. Organomet. Chem.*, 1966, **6**, 107; C. H. Bushweller, *Tetrahedron Lett.*, 1968, 6123; P. M. Maitlis, 'The Organic Chemistry of Palladium', Academic Press, New York, 1971, vol. 2; R. Jira and W. Freiesleben, in 'Organometallic Reactions', ed. E. I. Becker and M. Tsutsui, Wiley, New York, 1972, vol. 3; P. M. Henry, in 'Palladium Catalysed Oxidation of Hydrocarbons', Reidel, Dordrecht, 1979.

56. S. Wolfe and P. G. C. Campbell, *J. Am. Chem. Soc.*, 1971, **93**, 1497, 1499; W. Kitching, T. Sakakiyama, Z. Rappoport, P. D. Sleezer, S. Winstein and W. G. Young, *J. Am. Chem. Soc.*, 1972, **94**, 2329; W. Kitching, Z. Rappoport, S. Winstein and W. G. Young, *J. Am. Chem. Soc.*, 1966, **88**, 2054.
57. J.-E. Bäckvall and R. E. Nordberg, *J. Am. Chem. Soc.*, 1981, **103**, 4959; J.-E. Bäckvall, *Pure Appl. Chem.*, 1983, **55**, 1669; J.-E. Bäckvall, R. E. Nordberg and J. E. Nyström, *Tetrahedron Lett.*, 1982, **23**, 1617; J.-E. Bäckvall, R. E. Nordberg and S. E. Byström, *J. Org. Chem.*, 1984, **49**, 4619; J.-E. Bäckvall, J. Vagberg and R. E. Nordberg, *Tetrahedron Lett.*, 1984, **25**, 2717.
58. T. Hosokawa, J. Miyagi, S.-I. Murahashi and A. Sonada, *J. Org. Chem.*, 1978, **43**, 2752; T. Hosokawa, J. Miyagi, S.-I. Murahashi and A. Sonada, *J. Chem. Soc., Chem. Commun.*, 1978, 687.
59. J. E. McMurry and P. Kocovsky, *Tetrahedron Lett.*, 1984, **25**, 4187.
60. B. M. Trost and P. J. Metzner, *J. Am. Chem. Soc.*, 1980, **102**, 3572.
61. J. E. McMurry and P. Kocovsky, *Tetrahedron Lett.*, 1985, **26**, 2171.
62. J. E. McMurry and G. K. Bosch, *J. Org. Chem.*, 1987, **52**, 4885.
63. J. Tsuji, K. Sakai, H. Nagashima and I. Shimizu, *Tetrahedron Lett.*, 1981, **22**, 131.
64. S. Uemura, S. Fukuzawa, A. Toshimitsu and M. Okano, *Tetrahedron Lett.*, 1982, **23**, 87.
65. A. Heumann, M. Reglier and B. Waegell, *Angew. Chem., Int. Ed. Engl.*, 1982, **21**, 366; C. A. Horiuchi and J. Y. Satoh, *J. Chem. Soc., Perkin Trans. 1*, 1982, 2595; K. Ogura, N. Shibuya and H. Iida, *Tetrahedron Lett.*, 1981, **22**, 1519; K. Ogura, N. Shibuya, K. Takahashi and H. Iida, *Bull. Chem. Soc. Jpn.*, 1984, **57**, 1092.
66. A. Heumann and B. Åkermark, *Angew. Chem., Int. Ed. Engl.*, 1984, **23**, 453.
67. S. Bien and Y. Segal, *J. Org. Chem.*, 1977, **42**, 1685; S. Uemura and S. R. Patil, *Chem. Lett.*, 1982, 1743.
68. H. Mimoun, M. M. Perez Machirant and I. Sérée de Roch, *J. Am. Chem. Soc.*, 1978, **100**, 5437.
69. O. Bortolini, F. Di Furia, P. Scrimin and G. Modena, *J. Mol. Catal.*, 1980, **7**, 59.
70. (a) J. E. Baldwin and J. C. Swallow, *Angew. Chem., Int. Ed. Engl.*, 1969, **8**, 601; B. H. van Vugt and W. Drenth, *Rec. Trav. Chim. Pays-Bas*, 1977, **96**, 225; G. Read and J. Shaw, *J. Chem. Soc., Chem. Commun.*, 1984, 1313; (b) J. E. Lyons and J. O. Turner, *J. Org. Chem.*, 1972, **37**, 2881.
71. M. W. Peterson, D. S. Rivers and R. M. Richman, *J. Am. Chem. Soc.*, 1985, **107**, 2907; H. Arzoumanian, H. Bitar and J. Metzger, *J. Mol. Catal.*, 1980, **7**, 373; S. Ito, K. Inoue and M. Matsumoto, *J. Am. Chem. Soc.*, 1982, **104**, 6450.
72. D. Mansuy, J. Leclaire, M. Fontecave and P. Dansette, *Tetrahedron*, 1984, **40**, 2847.
73. (a) C. S. Sharma, S. C. Sethi and S. Dev, *Synthesis*, 1974, 45; W. K. Rybak and J. J. Ziólkowsky, *Inorg. Chim. Acta*, 1977, **24**, L69; I. M. Reibel and A. F. Sandu, *Russ. J. Phys. Chem. (Engl. Transl.)*, 1976, **50**, 632; I. M. Reibel and A. F. Sandu, *Russ. J. Phys. Chem. (Engl. Transl.)*, 1975, **49**, 946; J. H. Fuhrhop, M. Baccouche, H. Grabow and H. Arzoumanian, *J. Mol. Catal.*, 1980, **7**, 245; H. Kropf, S. K. Ivanov and P. Diercks, *Justus Liebigs Ann. Chem.*, 1974, 2046; (b) J. T. Groves and W. J. Kruper, *J. Am. Chem. Soc.*, 1979, **101**, 7613.
74. M. S. Kharasch and G. Sosnovsky, *J. Am. Chem. Soc.*, 1958, **80**, 756; M. S. Kharasch, G. Sosnovsky and N. C. Yang, *J. Am. Chem. Soc.*, 1959, **81**, 5819; G. Sosnovsky and N. C. Yang, *J. Org. Chem.*, 1960, **25**, 899; G. Sosnovsky and H. J. O'Neill, *C. R. Hebd. Seances Acad. Sci.*, 1962, **254**, 704.
75. G. Sosnovsky and D. J. Rawlinson, in 'Organic Peroxides', ed. D. Swern, Wiley, New York, 1970, vol. 1; D. J. Rawlinson and G. Sosnovsky, *Synthesis*, 1972, 1; G. Sosnovsky, *Tetrahedron*, 1961, **13**, 241.
76. S.-O. Lawesson and C. Frisell, *Ark. Kemi*, 1961, **17**, 409; H. L. Goering and U. Mayer, *J. Am. Chem. Soc.*, 1964, **86**, 3753.
77. J. J. Villenave, H. Francois and R. Lalande, *Bull. Soc. Chim. Fr.*, 1970, 599; R. Lalande and J. J. Villenave, *C. R. Hebd. Seances Acad. Sci., Ser. C*, 1971, **272**, 1825, 1892.
78. F. Bohlmann, W. Klose and K. Nickisch, *Tetrahedron Lett.*, 1979, 3699.
79. B. Cross and G. H. Whitham, *J. Chem. Soc.*, 1961, 1650.
80. J. K. Kochi, *J. Am. Chem. Soc.*, 1962, **84**, 774; D. Z. Denney, A. Appelbaum and D. B. Denney, *J. Am. Chem. Soc.*, 1962, **84**, 4969; D. B. Denney, R. Napier and A. Cammarata, *J. Org. Chem.*, 1965, **30**, 3151; K. Pedersen, P. Jakobsen and S.-O. Lawesson, *Org. Synth.*, 1968, **48**, 18.
81. C. Walling and A. A. Zavitsas, *J. Am. Chem. Soc.*, 1963, **85**, 2084.
82. A. L. J. Beckwith and G. W. Evans, *Proc. Chem. Soc., London*, 1962, 63; L. Stárka, *Collect. Czech. Chem. Commun.*, 1961, **26**, 2452.
83. L. Stárka and K. Syhora, *Czech. Pat.* 100 325 (1961) (*Chem. Abstr.*, 1962, **57**, 11 270f).
84. J. Ficini and J. d'Angelo, *Tetrahedron Lett.*, 1976, 2441.
85. G. Sosnovsky and G. A. Karas, *Z. Naturforsch., Teil A*, 1978, **336**, 1165.
86. P. R. Story, *J. Org. Chem.*, 1961, **26**, 287; P. R. Story and S. Fahrenholtz, *Org. Synth.*, 1964, **44**, 12.
87. A. Nickon, J. B. DiGiorgio and P. J. L. Daniels, *J. Org. Chem.*, 1973, **38**, 533; G. O. Schenck, K. Gollnick and O. A. Neumüller, *Justus Liebigs Ann. Chem.*, 1957, **603**, 46; 1958, **618**, 194; M. J. Kulig and L. L. Smith, *J. Org. Chem.*, 1973, **38**, 3639; H. E. Ensley, R. V. C. Carr, R. S. Martin and T. E. Pierce, *J. Am. Chem. Soc.*, 1980, **102**, 2836; M. Nakagawa, H. Okajima and T. Hino, *J. Am. Chem. Soc.*, 1976, **98**, 635; M. Nakagawa, H. Okajima and T. Hino, *J. Am. Chem. Soc.*, 1977, **99**, 4424.
88. R. W. Denny and A. Nickon, *Org. React. (N.Y.)*, 1973, **20**, 133; H. H. Wasserman and J. L. Ives, *Tetrahedron*, 1981, **21**, 1825.
89. W. R. Adams, in 'Oxidation', ed. R. L. Augustine, Dekker, New York, 1971, vol. 2.
90. A. P. Schaap and G. R. Faler, *J. Am. Chem. Soc.*, 1973, **95**, 3381; N. M. Hasty and D. R. Kearns, *J. Am. Chem. Soc.*, 1973, **95**, 3380; S. Inagaki and K. Fukui, *J. Am. Chem. Soc.*, 1975, **97**, 7480.
91. R. M. Boden, *Synthesis*, 1975, 783.
92. L. A. Paquette, C. C. Liao, D. Liotta and W. E. Fristad, *J. Am. Chem. Soc.*, 1976, **98**, 6412.
93. K. Kondo and M. Matsumoto, *Tetrahedron Lett.*, 1976, 391.

94. K. Gollnick, S. Schroeter, G. Ohloff, G. Schade and G. O. Schenck, *Justus Liebigs Ann. Chem.*, 1965, **687**, 14.
95. I. Sasson and J. Labovitz, *J. Org. Chem.*, 1975, **40**, 3670.
96. S.-K. Chung and A. I. Scott, *J. Org. Chem.*, 1975, **40**, 1652.
97. J. H. Wieringa, J. Strating, H. Wynberg and W. Adam, *Tetrahedron Lett.*, 1972, 169; C. W. Jefford and A. F. Boschung, *Tetrahedron Lett.*, 1976, 4771; C. W. Jefford and C. G. Rimbault, *Tetrahedron Lett.*, 1976, 2479.
98. W. Fenical, D. R. Kearns and P. Radlick, *J. Am. Chem. Soc.*, 1969, **91**, 3396.
99. K. E. O'Shea and C. S. Foote, *J. Am. Chem. Soc.*, 1988, **110**, 7167.
100. E. Friedrich, W. Lutz, H. Eichenauer and D. Enders, *Synthesis*, 1977, 893; P. D. Bartlett and H.-K. Chu, *J. Org. Chem.*, 1980, **45**, 3000.
101. E. McKeown and W. A. Waters, *J. Chem. Soc. B*, 1966, 1040.
102. A. U. Khan and M. Kasha, *J. Am. Chem. Soc.*, 1966, **88**, 1574.
103. H. H. Wasserman and J. R. Scheffer, *J. Am. Chem. Soc.*, 1967, **89**, 3073.
104. S. Torii, K. Uneyama and M. Ono, *Tetrahedron Lett.*, 1980, **21**, 2653; S. Torii, K. Uneyama, M. Ono and T. Bannou, *J. Am. Chem. Soc.*, 1981, **103**, 4606.
105. K. Uneyama, H. Matsuda and S. Torii, *J. Org. Chem.*, 1984, **49**, 4315.
106. G. Rauchschwalbe and M. Schlosser, *Helv. Chim. Acta*, 1975, **58**, 1094.
107. M. Schlosser, H. Bosshardt, A. Walde and M. Stahle, *Angew. Chem., Int. Ed. Engl.*, 1980, **19**, 303; M. Schlosser and G. Rauchschwalbe, *J. Am. Chem. Soc.*, 1978, **100**, 3258.
108. K. Wimalasena and S. W. May, *J. Am. Chem. Soc.*, 1987, **109**, 4036; K. B. Taylor, *J. Biol. Chem.*, 1974, **249**, 454.
109. S. W. May and R. S. Phillips, *J. Am. Chem. Soc.*, 1980, **102**, 5981.
110. S. R. Sirrimanne and S. W. May, *J. Am. Chem. Soc.*, 1988, **110**, 7560.
111. F. Freeman, in 'Organic Synthesis by Oxidation with Metal Compounds', ed. W. J. Mils and C. R. H. I. de Jonge, Plenum Press, New York, 1986.
112. K. B. Wiberg, in 'Oxidation in Organic Chemistry', ed. K. B. Wiberg, Academic Press, New York, 1965.
113. W. Treibs and H. Schmidt, *Ber. Dtsch. Chem. Ges.*, 1928, **61**, 459.
114. F. C. Whitmore and G. W. Pedlow, *J. Am. Chem. Soc.*, 1941, **63**, 758.
115. M. Nakayama, S. Shinke, Y. Matsushita, S. Ohira and S. Hayashi, *Bull. Chem. Soc. Jpn.*, 1979, **52**, 184.
116. L. A. Paquette, R. A. Roberts and G. J. Drtina, *J. Am. Chem. Soc.*, 1984, **106**, 6690.
117. C. Enzell, *Tetrahedron Lett.*, 1962, 185.
118. W. G. Dauben and A. C. Ashcroft, *J. Am. Chem. Soc.*, 1963, **85**, 3673.
119. W. Klyne, *J. Chem. Soc.*, 1951, 3449.
120. C. W. Marshall, R. E. Ray, I. Laos and B. Riegel, *J. Am. Chem. Soc.*, 1957, **79**, 6308.
121. W. Logerman and P. Giraldi, *Gazz. Chim. Ital.*, 1951, **81**, 548.
122. R. V. Oppenauer and H. Oberrauch, *An. Asoc. Quim. Argent.*, 1949, **37**, 246.
123. P. N. Rao and P. Kurath, *J. Am. Chem. Soc.*, 1956, **78**, 5660.
124. R. E. Beyler, A. E. Oberster, F. Hoffman and L. H. Sarett, *J. Am. Chem. Soc.*, 1960, **82**, 170.
125. C. Sannié, J. J. Panouse and S. Vertalier, *Bull. Soc. Chim. Fr.*, 1955, 1039.
126. N. B. Haynes, D. Redmore and C. J. Timmons, *J. Chem. Soc.*, 1963, 2420.
127. A. Katz, *Helv. Chim. Acta*, 1957, **40**, 487.
128. E. Menini and J. K. Norymberski, *Biochem. J.*, 1962, **84**, 195.
129. W. G. Salamond, M. A. Barta and J. L. Havens, *J. Org. Chem.*, 1978, **43**, 2057.
130. H. Sisler and C. E. Accountius, *J. Am. Chem. Soc.*, 1948, **70**, 3827.
131. J. C. Collins, W. W. Hess and F. J. Frank, *Tetrahedron Lett.*, 1968, 3363.
132. W. G. Dauben, M. Lorber and D. S. Fullerton, *J. Org. Chem.*, 1969, **34**, 3587.
133. E. J. Corey and R. H. Sneen *J. Am. Chem. Soc.*, 1956, **78**, 6269; A. B. Turner and H. J. Ringold, *J. Chem. Soc. C*, 1967, 1720.
134. P. H. Boyle, W. Cocker, D. H. Grayson and P. V. R. Shannon, *J. Chem. Soc. C*, 1971, 1073.
135. L. A. Paquette and U. S. Rocherla, *J. Org. Chem.*, 1987, **52**, 3250.
136. M. Okabe, M. Abe and M. Tada, *J. Org. Chem.*, 1982, **47**, 1775.
137. G. Piancatelli, A. Scettri and M. D'Auria, *Synthesis*, 1982, 245.
138. R. Ratcliffe and R. Rodehorst, *J. Org. Chem.*, 1970, **35**, 4000.
139. E. J. Corey and J. W. Suggs, *Tetrahedron Lett.*, 1975, 2647.
140. F. Bonadies and R. D. Fabio, *J. Org. Chem.*, 1984, **49**, 1647.
141. E. J. Parish and T.-Y. Wei, *Synth. Commun.*, 1987, **17**, 1227.
142. E. J. Parish, S. Chitrakorn and T.-Y. Wei, *Synth. Commun.*, 1986, **16**, 1371.
143. R. Rathore, N. Saxena and S. Chadrasekaran, *Synth. Commun.*, 1986, **16**, 1493.
144. N. Chidambaram and S. Chandrasekaran, *J. Org. Chem.*, 1987, **52**, 5048.
145. E. J. Corey and G. W. J. Fleet, *Tetrahedron Lett.*, 1973, 4499.
146. For an example of the use of this reagent in benzylic oxidation, see E. McDonald and A. Suksamrarn, *Tetrahedron Lett.*, 1975, 4425.
147. J. Vanmaele, P. J. de Clercq and M. Vandewalle, *Tetrahedron Lett.*, 1982, **23**, 995; P. Kok, P. J. de Clercq and M. Vandewalle, *J. Org. Chem.*, 1979, **44**, 4553.
148. K. Yoshida and P. A. Grieco, *J. Org. Chem.*, 1984, **49**, 5257.
149. P. D. Magnus, T. Leapheart and C. Walker, *J. Chem. Soc., Chem. Commun.*, 1985, 1185.
150. Z.-Y. Liu, X.-R. Zhou and Z.-M. Wu, *J. Chem. Soc., Chem. Commun.*, 1987, 1868.
151. R. H. Schlessinger, J. L. Wood, A. J. Poss, R. A. Nugent and W. H. Parsons, *J. Org. Chem.*, 1983, **48**, 1146.
152. J. Muzart, *Tetrahedron Lett.*, 1987, **28**, 4665.
153. A. J. Pearson, Y.-S. Chen, G. R. Han, S.-Y. Hsu and T. Ray, *J. Chem. Soc., Perkin Trans. 1*, 1985, 267.
154. A. J. Pearson and G. R. Han, *J. Org. Chem.*, 1985, **50**, 2791.
155. A. J. Pearson, Y.-S. Chen, S.-Y. Hsu and T. Ray, *Tetrahedron Lett.*, 1984, **25**, 1235.
156. J. Tsuji, *Synthesis*, 1984, 369.
157. J. Muzart, P. Pale and J. P. Pete, *Tetrahedron Lett.*, 1982, **23**, 3577.

158. J. Muzart, A. Riahi and J. P. Pete, *J. Organomet. Chem.*, 1985, **280**, 269.
159. J. Muzart, J. P. Pete and A. Riahi, *J. Organomet. Chem.*, 1987, **331**, 113.
160. R. Hüttel and H. Christ, *Chem. Ber.*, 1964, **97**, 1439.
161. G. A. Gray, W. R. Jackson and J. J. Rooney, *J. Chem. Soc. C*, 1970, 1788.
162. H. Mimoun, R. Charpentier, A. Mitschler, J. Fischer and R. Weiss, *J. Am. Chem. Soc.*, 1980, **102**, 1047.
163. M. A. Andrews and C.-W. F. Cheng, *J. Am. Chem. Soc.*, 1982, **104**, 4268.
164. T. Hosokawa, S. Inui and S.-I. Murahashi, *Chem. Lett.*, 1983, 1081.
165. J. Y. Satoh and C. A. Horiuchi, *Bull. Chem. Soc. Jpn.*, 1981, **54**, 625.
166. P. A. Chaloner, in 'Handbook of Coordination Catalysts in Organic Chemistry,' Butterworths, Guildford, 1986.
167. J. P. Collman, M. Kubota and J. W. Hosking, *J. Am. Chem. Soc.*, 1967, **89**, 4809; S. Bien and Y. Segal, *J. Org. Chem.*, 1977, **42**, 1685.
168. K. Takao, M. Wayaku, Y. Fujiwara, T. Imanaka and S. Teranishi, *Bull. Chem. Soc. Jpn.*, 1970, **43**, 3898; C. Dudley and G. Read, *Tetrahedron Lett.*, 1972, 5273; J. Blum, H. Rosenman and E. D. Bergmann, *Tetrahedron Lett.*, 1967, 3665; J. Blum, J. Y. Becker, H. Rosenman and E. D. Bergmann, *J. Chem. Soc. B*, 1969, 1000.
169. J. M. Reuter, A. Sinha and R. G. Salomon, *J. Org. Chem.*, 1978, **43**, 2438.
170. I. B. Baranovskii, G. Y. Mazo and L. M. Dikareva, *Russ. J. Inorg. Chem. (Engl. Transl.)*, 1971, **16**, 1388.
171. S. Uemura, A. Spencer and G. Wilkinson, *J. Chem. Soc., Dalton Trans.*, 1973, 2565.
172. S. Uemura and S. R. Patil, *Tetrahedron Lett.*, 1982, **23**, 4353.
173. A. Fusi, R. Ugo, F. Fox, A. Pasini and S. Cenini, *J. Organomet. Chem.*, 1971, **26**, 417.
174. M. T. Atlay, M. Preece, G. Strukul and B. R. James, *Can. J. Chem.*, 1983, **61**, 1332.
175. Y. Ohkatsu and T. Tsurata, *Bull. Chem. Soc. Jpn.*, 1978, **51**, 188.
176. N. Koga and I. Tabushi, *J. Am. Chem. Soc.*, 1979, **101**, 6456; D. Mansuy, M. Fontecave and J. F. Bartoli, *J. Chem. Soc., Chem. Commun.*, 1983, 253.
177. M. W. Peterson, D. S. Rivers and R. M. Richman, *J. Am. Chem. Soc.*, 1985, **107**, 2907.
178. C. Paquot, *C. R. Hebd. Seances Acad. Sci.*, 1939, **209**, 171.
179. E. C. Blossey and P. Kucinski, *J. Chem. Soc., Chem. Commun.*, 1973, 56.
180. A. J. Waring, in 'Comprehensive Organic Chemistry', ed. D. H. R. Barton and W. D. Ollis, Pergamon Press, Oxford, 1979, vol. 1, p. 1017.
181. K.-Y. Zee-Cheng and C. Cheng, *J. Heterocycl. Chem.*, 1967, **4**, 163.
182. J. Meinwald, K. Opheim and T. Eisner, *Tetrahedron Lett.*, 1973, 281.
183. P. Gramatica, P. Manitto, D. Monti and G. Speranza, *Tetrahedron*, 1987, **43**, 4481.
184. K. Sakai, T. Ohtsuka, S. Misumi, H. Shirahama and T. Matsumoto, *Chem. Lett.*, 1981, 355.
185. H. J. Bestmann and P. Ermann, *Liebigs Ann. Chem.*, 1984, 1740.
186. E. Suzuki, R. Hamajima and S. Inoue, *Synthesis*, 1975, 192.
187. D. H. R. Barton and D. Crich, *Tetrahedron*, 1985, **41**, 4359.
188. D. H. R. Barton, A. G. Brewster, R. A. H. F. Hui, D. J. Lester, S. V. Ley and T. G. Back, *J. Chem. Soc. D*, 1978, 952.
189. D. Laber and A. Krief, *Bull. Soc. Chim. Belg.*, 1984, **93**, 1005.
190. R. E. Ireland, S. W. Baldwin, D. J. Dawson, M. I. Dawson, J. E. Dolfini, J. Newbould, W. S. Johnson, M. Brown, R. J. Crawford, P. F. Hudrlik, G. H. Ramussen and K. K. Schmiegel, *J. Am. Chem. Soc.*, 1970, **92**, 5743.
191. R. E. Ireland, S. W. Baldwin and S. C. Welch, *J. Am. Chem. Soc.*, 1972, **94**, 2056.
192. J. S. Dutcher, J. G. Macmillan and C. H. Heathcock, *J. Org. Chem.*, 1976, **41**, 2663.
193. H. Hock and S. Lang, *Ber. Dtsch. Chem. Ges.*, 1942, **75**, 300; 1944, **77**, 257; H. Hock and H. Kropf, *Angew. Chem.*, 1957, **69**, 313; R. Criegee, *Justus Liebigs Ann. Chem.*, 1948, **560**, 127.
194. E. D. Mihelich and D. J. Eickhoff, *J. Org. Chem.*, 1983, **48**, 4135.
195. J. W. Huffman, *J. Org. Chem.*, 1976, **41**, 3847.
196. S. Corsano and G. Piancatelli, *Ann. Chim. (Rome)*, 1965, **55**, 742.
197. B. W. Finucane and J. B. Thomson, *J. Chem. Soc., Chem. Commun.*, 1969, 1220.
198. G. D. S. Ananda, A. M. Z. Slawin, R. J. Stoodley and D. J. Williams, *J. Chem. Soc D.*, 1986, 165.
199. G. D. S. Ananda, P. J. Cremins and R. J. Stoodley, *J. Chem. Soc., Chem. Commun.*, 1987, 882.
200. D. Rosenthal and J. P. Gratz, *J. Org. Chem.*, 1969, **34**, 409.

2.2
Oxidation Adjacent to C=X Bonds by Dehydrogenation

DEREK R. BUCKLE and IVAN L. PINTO
SmithKline Beecham Pharmaceuticals, Epsom, UK

2.2.1 INTRODUCTION

The synthetic versatility of α,β-unsaturated carbonyl compounds has resulted in the development of a wide variety of methods for their synthesis. Many such procedures rely on the construction of the basic carbon framework from simpler fragments, and are typified by reactions of the Wittig, Knoevenagel, aldol and Reformatsky type.[1] To be able to introduce regioselective unsaturation into a previously established carbon skeleton is, however, an additional tool in the chemist's armamentarium. In this review we have attempted to bring together the main literature relating to dehydrogenation methodology. No attempt has been made to include similar reactions that would generate alkynes or reactions that would result in the formation of carbon atoms doubly bonded to heteroatoms. Several of the intermediates described, and especially those involving α-selenenyl or α-thio moieties, offer the opportunity for further elaboration prior to elimination, since such species are able to stabilize adjacent carbanions.[2–4] The synthetic applications arising from such intermediates are left to the ingenuity of the reader.

2.2.2 HALOGENATION–DEHYDROHALOGENATION REACTIONS

2.2.2.1 Halogenation

The traditional method for introducing α,β-unsaturation into compounds containing C=X groups is by a halogenation–dehydrohalogenation sequence. The halogen leaving group can be either α or β to C=X (Scheme 1), although base-mediated elimination is more facile in the latter instance. Since selective functionalization of the β-position is difficult, except where it is activated by other groups,[5,6] it is usual to introduce the halogen at the α-position.

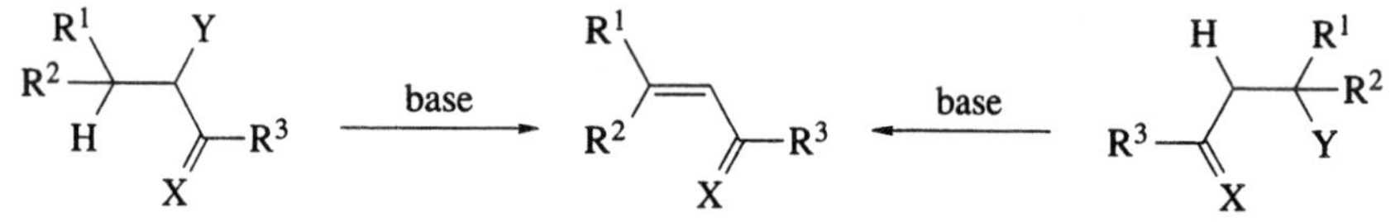

Y = halogen

Scheme 1

A variety of methods are available for the halogenation of aldehydes and ketones, and rely on the ease of enolization of such compounds. Copper(II) chloride[7] or bromide[8] in ethyl acetate at reflux have been shown to be effective reagents and rely on the promotion of enolization by the copper ion prior to the transfer of halogen. Since these conditions tend to favor the thermodynamic enol, unsymmetrical ketones preferentially halogenate at the more highly substituted α-carbon atom.[9] Similar selectivity is observed with NBS.[10,11]

While bromine itself can be used to effect α-bromination of ketones,[12,13] the hydrogen bromide produced can be detrimental.[10] The addition of acid scavengers such as 1,2-epoxycyclohexane (equation 1)[14] or potassium perchlorate[15] can, however, lead to good yields in the more difficult cases. As with copper(II) salts, the conditions for elemental bromine also favor substitution at the more highly substituted carbon atom.

Br_2, *hν*

75%

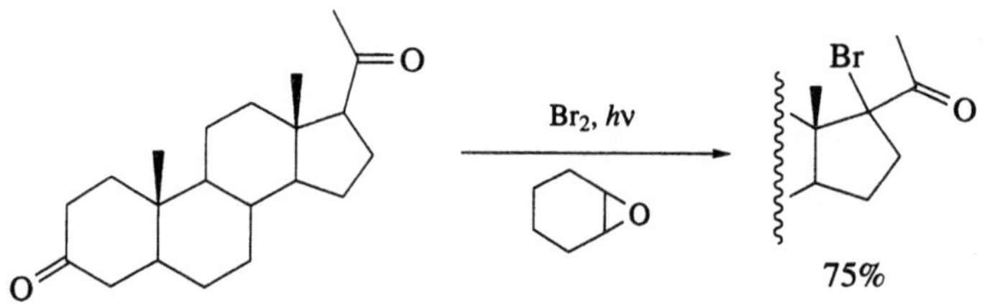

(1)

An alternative procedure to bromine itself is the use of complexed derivatives such as 2-pyrrolidone hydrotribromide (PHT), which is easier to handle. This reagent has been shown to brominate flavanones in THF, while the more vigorous conditions of hot DMSO result in concommitant dehydrohalogenation to give the flavone (Scheme 2).[16]

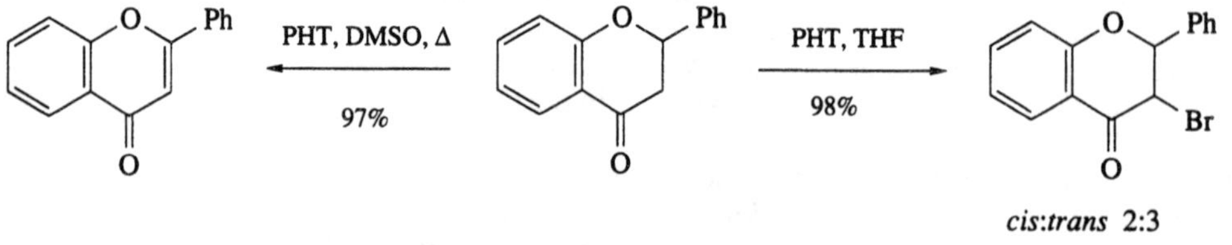

PHT, DMSO, Δ 97%

PHT, THF 98%

cis:*trans* 2:3

Scheme 2

A greater degree of regiocontrol over the above methods can be achieved by quenching the enolate of carbonyl compounds with either bromine[17] or iodine.[18,19] Thus, in the case of unsymmetrical ketones (Scheme 3), low temperature formation of the enolate allows exclusive bromination of the kinetic enolate to afford the haloketone (**1**), which on elimination gives the enone (**2**).[17] A similar procedure allows esters to be iodinated in excellent yields (Scheme 4).[18,19] Lactones (Scheme 5) also undergo bromination, and good yields have been obtained using 1,2-dibromoethane as the halogen source.[20,21]

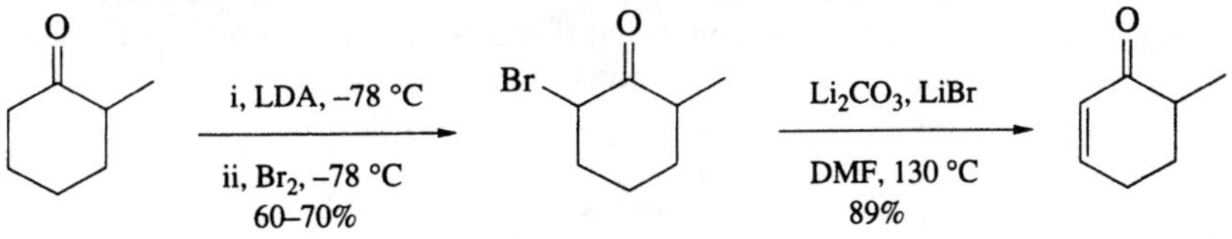

i, LDA, –78 °C; ii, Br_2, –78 °C, 60–70% → (1); Li_2CO_3, LiBr, DMF, 130 °C, 89% → (2)

Scheme 3

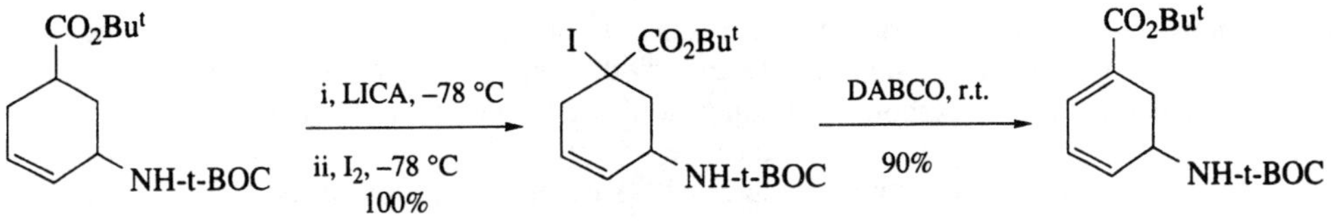

i, LICA, –78 °C; ii, I_2, –78 °C, 100%; DABCO, r.t., 90% (CO_2Bu^t, NH-t-BOC)

Scheme 4

i, Ph_3CLi; ii, $BrCH_2CH_2Br$; DBU, toluene, Δ

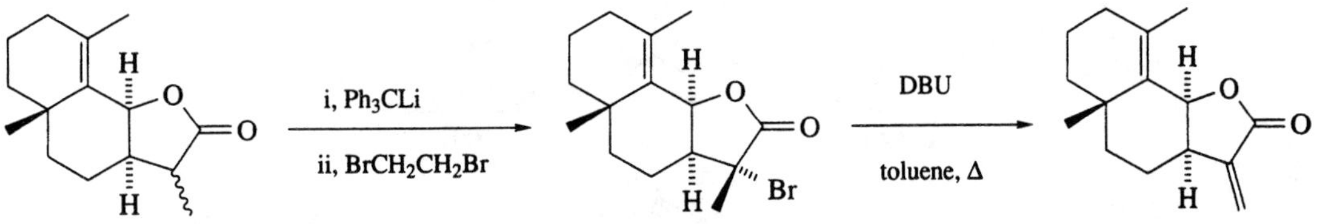

Scheme 5

Since enolate formation requires the use of strong base, compounds that are unstable to such conditions may be α-halogenated *via* the corresponding enol ether.[22,23] Thus, the iodination of enol acetates in the presence of thallium(I) acetate has been reported (entry 1, Table 1),[24] although the toxicity of this reagent limits its use to all but difficult cases. Silver acetate, although expensive, is an effective substitute (entry 2, Table 1).[25] The enol acetates of aldehydes undergo facile reaction with NBS and furnish good yields of the dehydrogenated derivatives on subsequent base treatment.[22,23] Fortunately, silyl enol ethers react well with elemental bromine or NBS under mild conditions[22,23,26,27] which makes this procedure one of the most synthetically useful alternatives (entry 3, Table 1).

Table 1 Reaction of Enol Ethers and Acetates with Halogens

Entry	*Substrate*	*Product*	*Conditions*	*Yield* (%)	*Ref.*
1	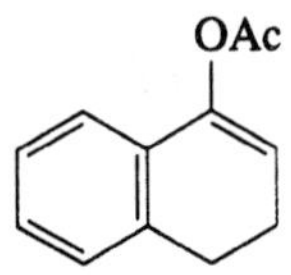 OAc	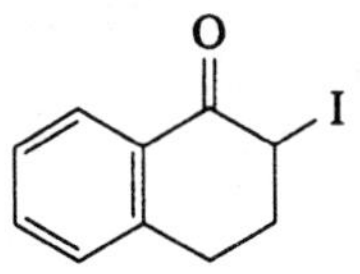 O, I	TlOAc, I_2, CH_2Cl_2	75	24
2	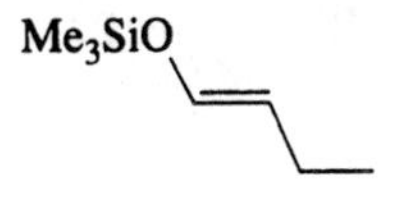 Me_3SiO	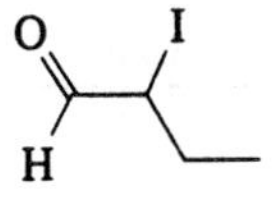 O, I, H	i, AgOAc, I_2, CH_2Cl_2; ii, $Et_3N{\bullet}HF$	64	25
3	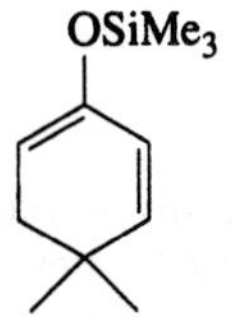 $OSiMe_3$	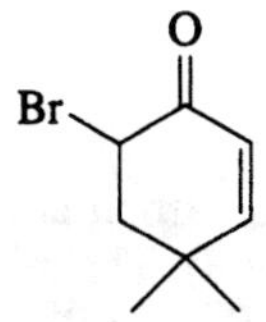 O, Br	Br_2, CCl_4, –20 °C	90	26

The methods discussed so far are applicable to aldehydes, ketones, esters and lactones. The α-halogenation of acids has received relatively little attention, although the traditional Hell–Vollard–Zelinski conditions are adequate in most instances (equation 2).[28] Alternative conditions have been developed, however, in which the acyl halide may be halogenated using NBS.[29] Quenching the reaction with alcohols or amines offers the opportunity of forming carboxylate derivatives.

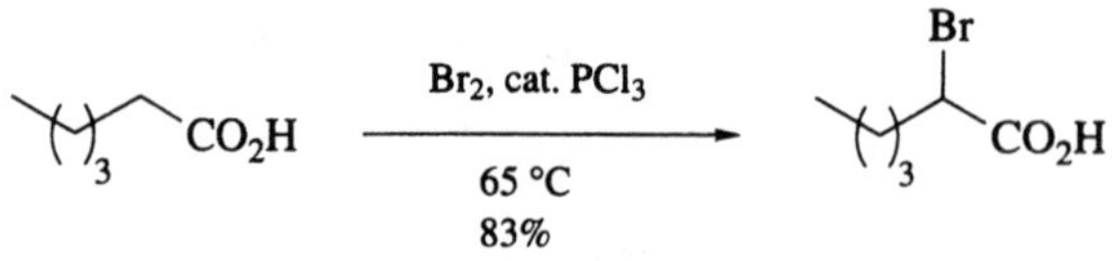

Br
CO_2H —Br_2, cat. PCl_3 / 65 °C / 83%→ CO_2H (2)

Tertiary amides can be converted into their α,β-unsaturated derivatives in good yield by the sequential treatment with phosgene then pyridine *N*-oxide and triethylamine (Scheme 6),[30] provided that only one hydrogen atom is present on the carbon atom adjacent to the amide group. In this instance, the intermediate chloroiminium salt (**3**) undergoes oxidation to the unsaturated amide. This method has the advantage that it does not involve the use of strong base and, with suitable protection of the amino group, furnishes a potential route to dehydro amino acids.[30]

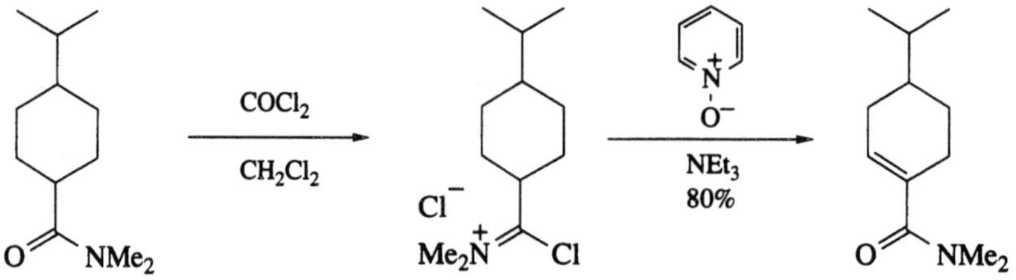

$COCl_2$ / CH_2Cl_2; Cl^- Me_2N^+ Cl (**3**); pyridine N-oxide, NEt_3, 80%; O NMe$_2$

Scheme 6

2.2.2.2 Dehydrohalogenation

The second step in the dehydrogenation sequence involves the base-induced elimination of the α-halide. Depending on the nature of the substrate, it is possible to obtain both (*E*)- and (*Z*)-isomers since elimination usually proceeds *via* an antiperiplanar loss of the halogen acid (Scheme 7). For chiral compounds having only one β-hydrogen atom, the geometry of the resulting product is defined in the transition state (Scheme 7), although the strong thermodynamic preference for the formation of (*E*)-isomers may result in mixtures.

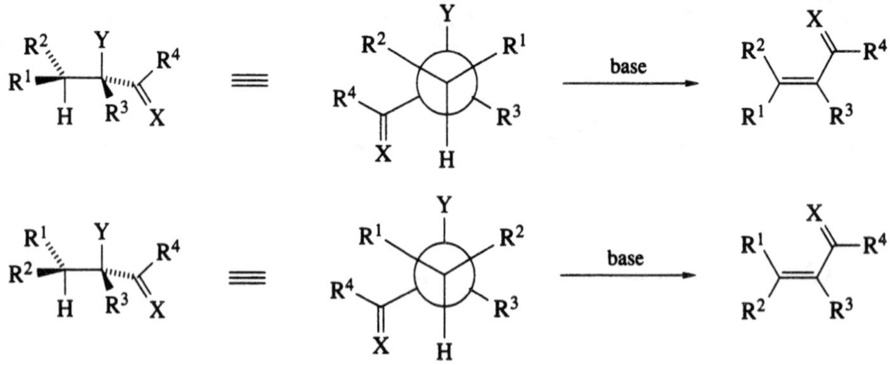

Scheme 7

In those cases where there are two hydrogen atoms on the β-carbon atom, two conformational preferences exist and, consequently, two isomers can result (Scheme 8). The relative ratio of isomers will be dependent on steric interactions in the transition state. Thus, if R^1 and R^2 are large formation of the alkene in which R^1 and R^2 are *trans* to one another will be preferred.

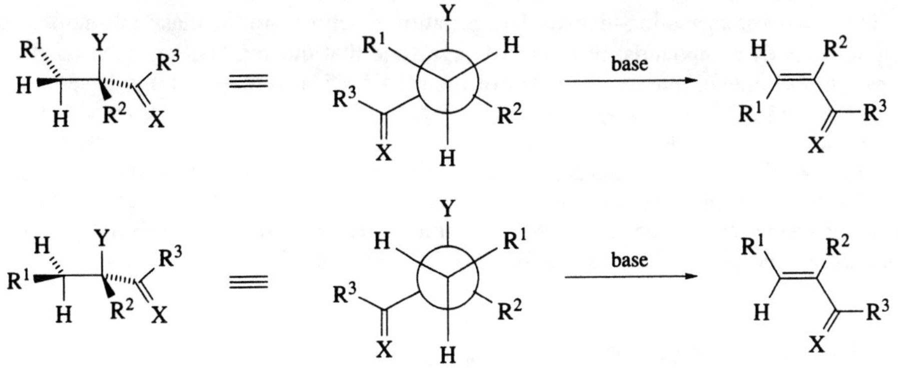

Scheme 8

A variety of bases have been used to effect the elimination of halogen acids from α-halo carbonyl compounds (Table 2). Among the more commonly used organic bases are DBN,[20] DABCO,[19] collidine[31] and triethylamine.[32] Yields tend to be variable with these reagents and reductive dehalogenation or double bond migration have been observed.[31] Inorganic bases such as lithium carbonate[12,17,27,33] or calcium carbonate[13] (Table 2) are usually preferred to organic bases since fewer side reactions are encountered. Stronger bases such as potassium hydroxide[34] and potassium *t*-butoxide[35] have been successfully employed, although with ketones the incidence of Favorskii rearrangement is increased.[36]

Table 2 Base Elimination of α-Halides

Entry	*Substrate*	*Product*	*Conditions*	*Yield (%)*	*Ref.*
1	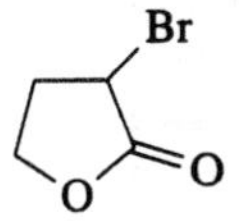	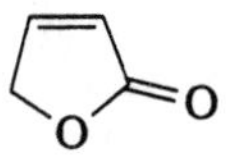	Et_3N, ether, Δ	60	32
2	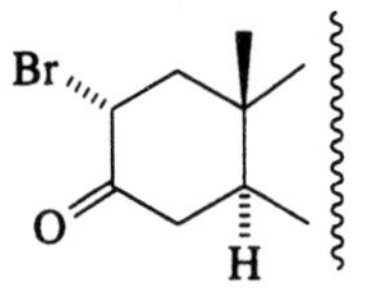 2α-Bromocholestan-3-one	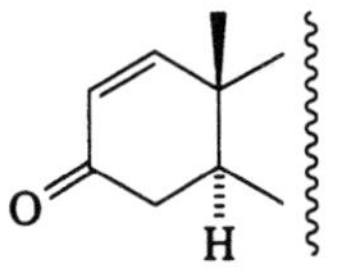	γ-Collidine, Δ	41	31
3	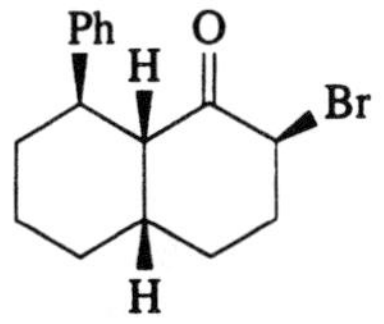	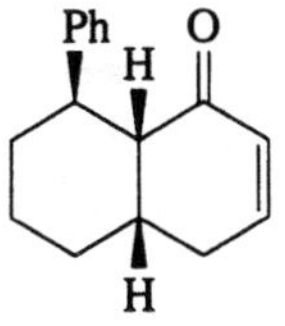	Li_2CO_3, DMF, Δ	51	12, 33
4	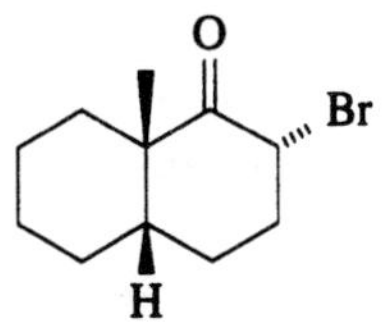	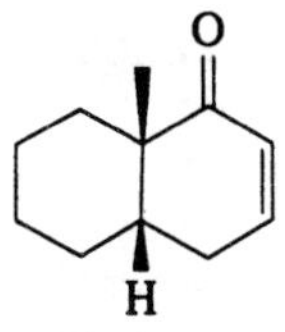	$CaCO_3$, $MeCONMe_2$, Δ	95	13
5	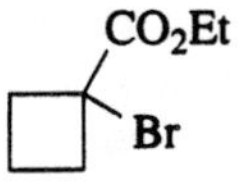	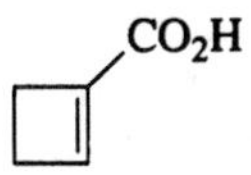	KOH, toluene, Δ	>85	34

In general, although halogenation–dehydrohalogenation reactions are the classical method for the dehydrogenation of C═X compounds, there is little evidence that this method has been used where X is other than oxygen. Moreover, the development of more sophisticated methodology based on sulfur and selenium (see Sections 2.2.3 and 2.2.4) has placed halogenation–dehydrohalogenation low in the order of preferred alternatives. Nonetheless, it is possible to replace one α-substituent by another under nucleophilic conditions. While such processes appear to offer little advantage over the direct insertion methods, it is conceivable that the replacement of halogen atoms by thio[4] and selenenyl[37–39] groups may allow difficult eliminations to proceed under comparatively mild conditions. Such methods may be of particular value when the corresponding halo compounds are readily available.

2.2.3 SULFUR-BASED REAGENTS

Elemental sulfur has been used for many years to effect dehydrogenation reactions[40] and, provided that the substrate is thermally stable, it has the advantages of cheapness and simplicity. In a typical reaction, the carbonyl compound (**4**) and powdered sulfur are heated together to around 200 °C either neat[41] or using a high boiling solvent such as *p*-cymene,[42] as illustrated in equation (3).[41]

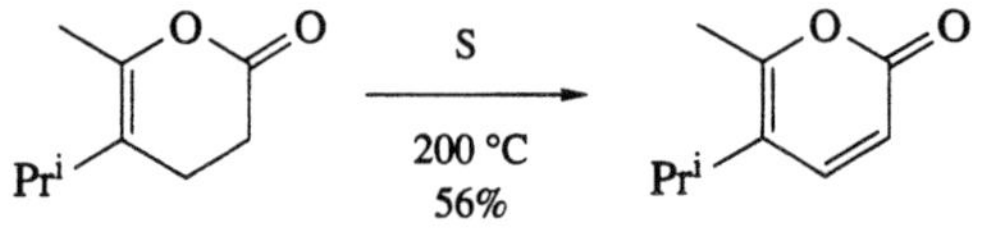

S; 200 °C; 56% (3)

Pr^i (**4**) → Pr^i

Although in some instances elemental sulfur is favorable to halogenation–dehydrohalogenation and quinone dehydrogenation reactions,[41] it has largely been superseded by organosulfur reagents. Like selenium (Section 2.2.4) the success of these reagents is dependent on the ease of the thermolytic *syn* elimination of sulfoxide from compounds bearing a suitably orientated hydrogen atom on the β-carbon atom (equation 4).[4] The temperature at which this elimination ensues varies with the nature of the substituent on sulfur, but aryl sulfoxides usually require temperatures of 25–80 °C compared to 110–130 °C for alkyl sulfoxides.[4] In general, exclusive formation of the (*E*)-geometric isomers is observed with the exception of those compounds having similar β,β-disubstitution or in those cases where this geometry is unattainable or otherwise disfavored.[4,43] Thus, in acyclic systems the regiochemistry for hydrogen abstraction is usually governed by the order C═CCH$_2$ ≈ C≡CCH$_2$ > ArCH$_2$ ≈ CH$_3$ > CH$_2$ >> C—H, whereas in cyclic systems preference is for the formation of endocyclic alkenes.[4]

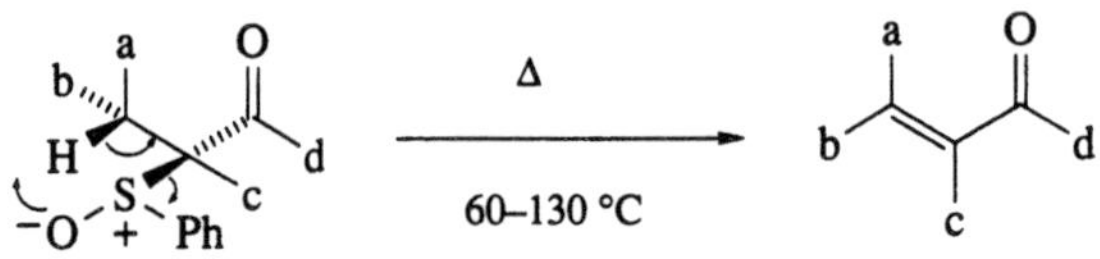

Δ; 60–130 °C (4)

When the temperature needs to be kept as low as possible, elimination may be facilitated by incorporation of electron-withdrawing substituents in the *p*-position of the aromatic ring.[44] The 2-pyridylthio moiety has also been used to good effect in the synthesis of methyl dehydrojasmonate and tuberolactone.[45] Elimination may also be enhanced by conversion of the sulfide to the *N*-*p*-toluenesulfonylsulfilamine with Chloramine T prior to pyrolysis,[46] but the advantages, if any, of this modification have still to be shown. A further approach that assists the elimination of methylthio groups has been described by Vedejs and Engler.[47] Thus, alkylation of the thio compound (**5**) with ethyl trifluoromethylsulfonyloxyacetate generates the ylide (**6**) which spontaneously decomposes under the reaction conditions to give the corresponding α,β-unsaturated carbonyl compound (**7**; Scheme 9).[47]

Several methods are available for the introduction of sulfenyl groups α to carbonyl derivatives and these have been reviewed.[4,43] The most versatile procedure involves reaction of the enolate with an appropriate thiol derivative,[43,48] but the preferred method is largely dependent on the nature of the substrate employed (see below). In most instances, sulfur has been introduced in the divalent state and subsequently oxidized, although the oxidative step has been avoided by the direct introduction of sulfur at the S^{IV} oxidation level.[49–51] The oxidation of sulfides to sulfoxides is a trivial procedure that can be effected by a variety of reagents. Sodium metaperiodate, *m*-chloroperbenzoic acid and hydrogen peroxide are the most common oxidants, but *t*-butyl hydroperoxide, *t*-butyl hypochlorite, *N*-chlorobenzotriazole,

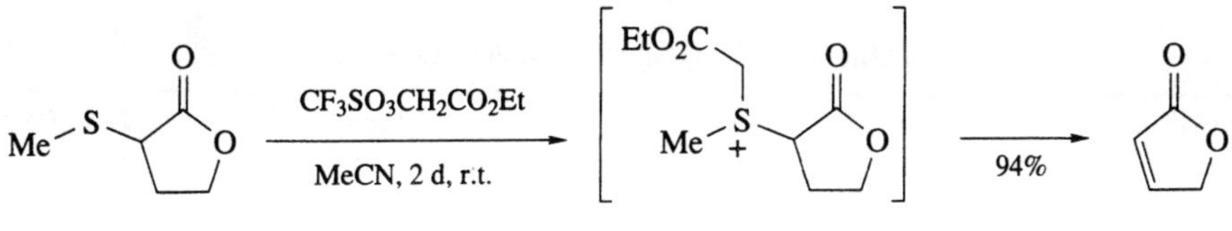

Scheme 9

chromic acid, dinitrogen tetroxide, iodosylbenzene, nitric acid, ozone and a host of other oxidizing agents have also been used.[43] Sodium metaperiodate is preferred in the absence of other determining factors and typical reaction conditions involve treatment of the sulfenyl compound in methanol at room temperature with 1 equiv. of the oxidant.[43]

Dehydrogenation reactions using sulfur reagents have been shown to tolerate a variety of other functional groups, including acetals, alkenes, epoxides and silyl ethers,[43,52] but the milder procedures available using selenenyl moieties may offer advantages in more sensitive molecules (see Section 2.2.4).

2.2.3.1 Sulfur(II) Reagents

The sulfenylation of esters,[43,52] lactones,[48,52,53] carboxylic acids,[48,54] amides[55] and lactams[55,56] may be effected by reaction of the corresponding lithium enolates in THF at –78 to 0 °C with dimethyl or diphenyl disulfides, or, less commonly, with methyl or phenyl sulfenyl halides.[48] The enolates of ketones, however, are insufficiently nucleophilic to react with dialkyl sulfides unless HMPA is added to the reaction mixture,[43] although they do react smoothly with diaryl sulfides.[43,48] This difference allows the selective sulfenylation of esters in the presence of ketones (entry 5, Table 3).[43]

In those instances where sulfenylation of ester enolates results in poor yields, use of the more stable *t*-butyl or silyl esters can offer advantages.[4] The sulfenylation of aldehyde enolates is only possible at –100 °C due to competing aldol condensation reactions.[48] Typical products made in this manner are illustrated in Table 3. The reaction of 2-methylcyclohexanone (**8**; entry 2) is of particular interest and clearly demonstrates the improved yield of kinetic enolate-derived product obtainable with more reactive sulfenylating agents.[43] Conformational studies indicate a slight axial preference for thiolate substitution in cyclohexanones, which is similar to that found on halogenation, although in condensed ring systems 1,3-diaxial interactions result in exclusive equatorial substitution.[43]

Dehydrogenation reactions involving sulfur have proved important in the formation of α-methylenelactones such as (**9**; Scheme 10), but sulfenylation prior to alkylation is necessary for *cis*-fused systems in order to establish the correct geometry for exocyclic double bond formation.[53]

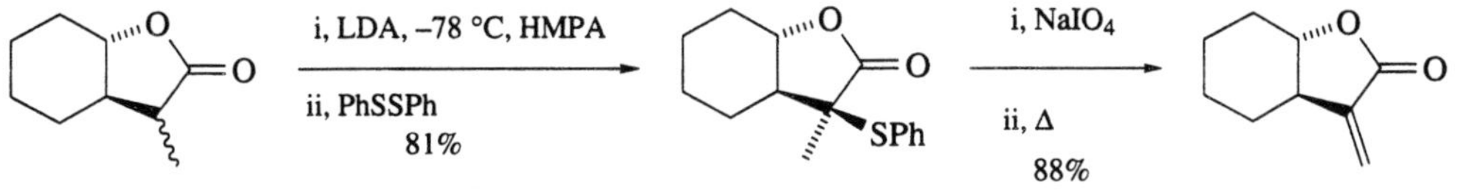

Scheme 10

Because of their tendency to undergo aldol reactions, various conditions have been investigated to develop methods for the sulfenylation of aldehydes. Indirect methods involving metallation of the corresponding imines (**10**; Scheme 11)[4] offer a preferred alternative to the low temperature direct sulfenylation described above, but better methods are still required. One possibility may be to exploit the rapid room temperature enolization of aldehydes observed on treatment with potassium hydride in THF.[57]

The sulfenylation of metalloimines is equally applicable to ketones, although using more reactive sulfur electrophiles it is possible to bring about reaction on the unmetallated enamine.[58,59] Sulfenylation of ketone enol silyl ethers also proceeds well with the more reactive sulfur species.[60] Sulfenamides and their derivatives (*e.g.* **11**) are particularly suited to the direct sulfenylation of ketones and active methylene compounds such as β-diketones, β-keto esters and malonates, which undergo facile reaction at room temperature (equation 5).[59] This procedure, however, does not appear to have been exploited for the dehydrogenation of active methylene compounds (*cf.* Section 2.2.4.1). By preparing the dianion (**13**)

Table 3 Sulfenylation of Lithium Enolates

Entry	*Substrate*	*Product*	*Reagent*	*Yield (%)*[a]	*Ref.*
1	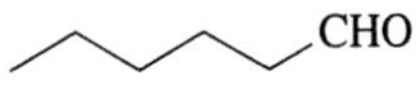	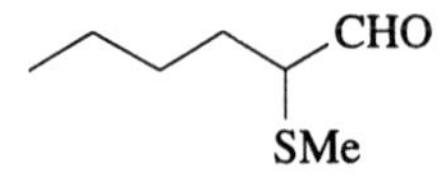	MeSCl	70[b]	48
2	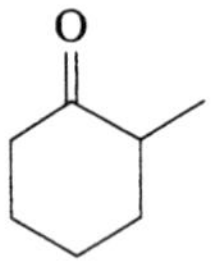 (8)	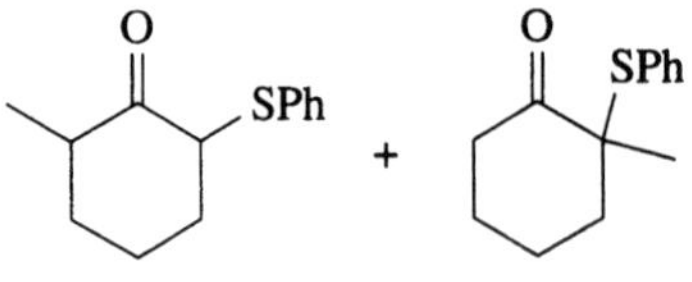 4:1	PhSSPh	87	43
		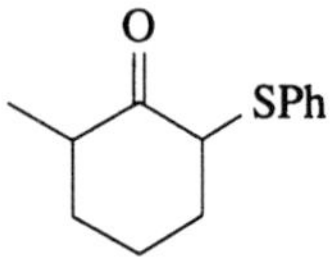	$PhSSO_2Ph$	85	43
3	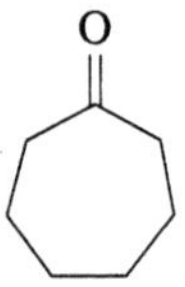	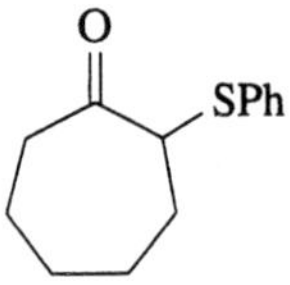	PhSSPh	87	43
4	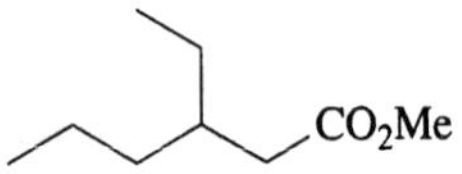	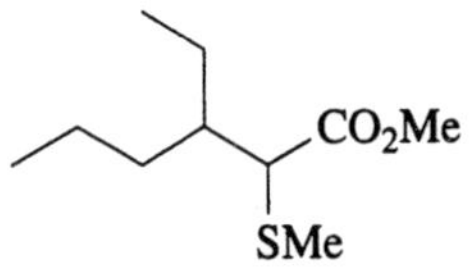	MeSSMe	88	52
5	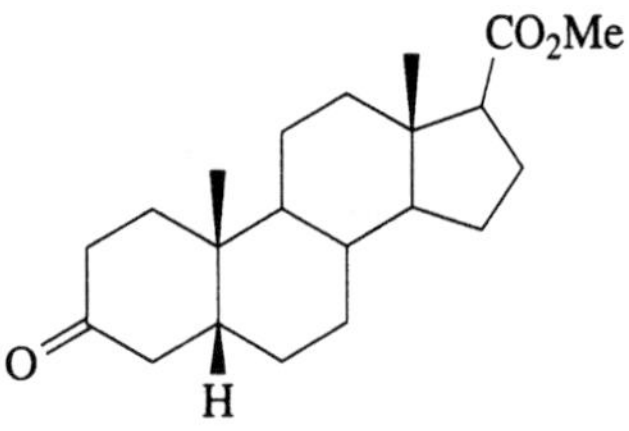	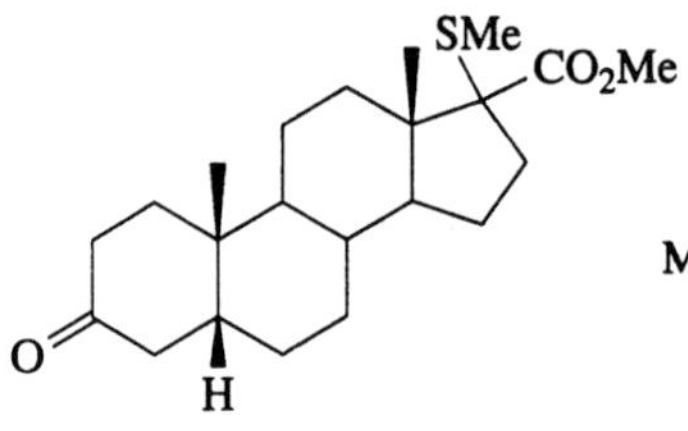	MeSSMe	100	52
6	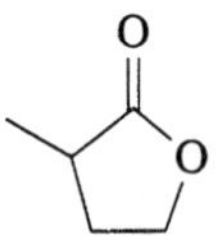	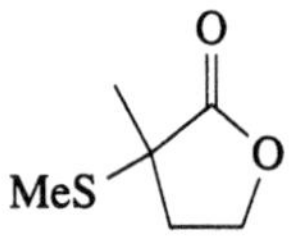	MeSSMe	79	53
7	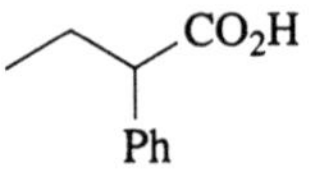	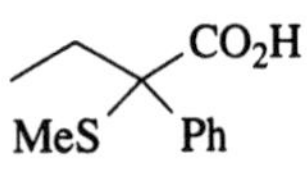	MeSSMe	96	48
8	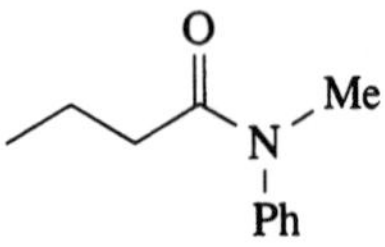	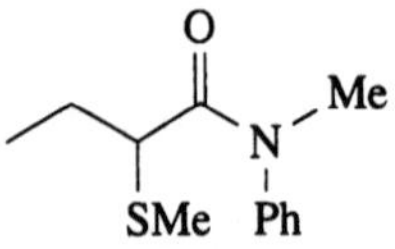	MeSSMe	80	55
9	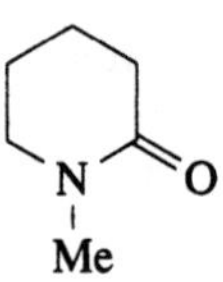	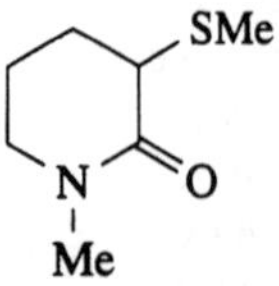	MeSSMe	69	55

[a] Reactions carried out at –78 to 0 °C in THF. [b] Reaction carried out at –100 °C.

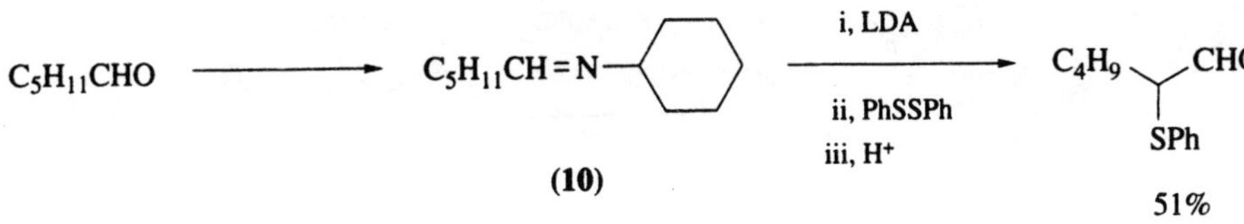

$C_5H_{11}CHO$ ⟶ $C_5H_{11}CH{=}N$–cyclohexyl **(10)** ⟶ (i, LDA; ii, PhSSPh; iii, H^+) $C_4H_9CH(SPh)CHO$ 51%

Scheme 11

of the β-keto ester (**12**; Scheme 12) sulfur was incorporated at the more nucleophilic C-3 position on quenching with diphenyl disulfide.[61] This offers the opportunity to introduce unsaturation out of conjugation with the ester moiety.

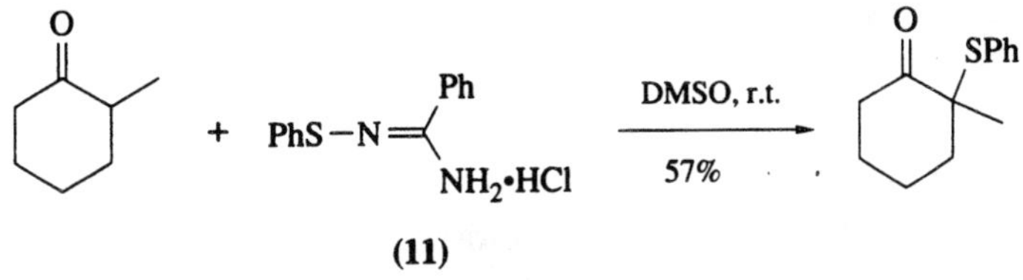

2-Methylcyclohexanone + PhS–N=C(Ph)NH_2•HCl **(11)** ⟶ (DMSO, r.t., 57%) 2-methyl-2-(phenylthio)cyclohexanone (5)

(12) CO$_2$Et, =O ⟶ (NaH, BuLi) **(13)** ⟶ (PhSSPh, THF, r.t., 4 h, 85%) CO$_2$Et, =O, SPh

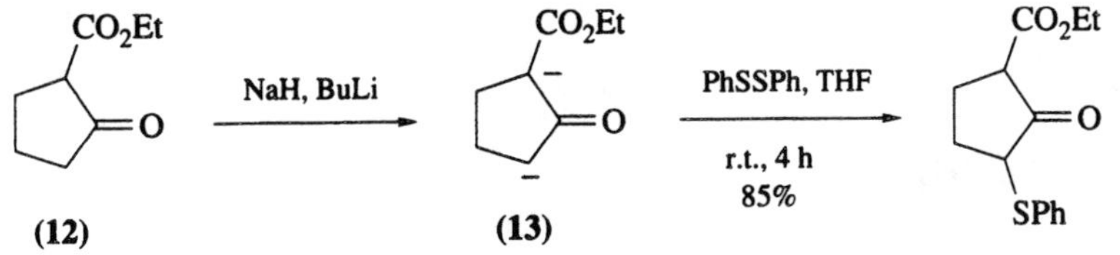

Scheme 12

The direct sulfenylation of *N,N*-dimethylhydrazones *via* the reaction of the α-lithio derivative (**14**) with dimethyl disulfide (Scheme 13) has been reported, and the initially formed product (**15**) shown to isomerize to the more stable (*E*)-isomer (**16**).[62] While further transformations have been carried out on compound (**16**), attempts do not appear to have been made to introduce unsaturation by the elimination of the thiol group.

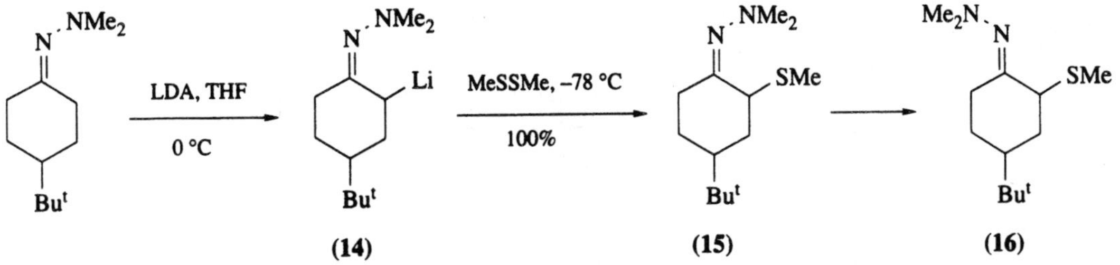

N–NMe$_2$, Bu^t ⟶ (LDA, THF, 0 °C) **(14)** Li ⟶ (MeSSMe, –78 °C, 100%) **(15)** SMe ⟶ **(16)** Me$_2$N–N, SMe

Scheme 13

For a more comprehensive account of the methods available for the sulfenylation of carbonyl compounds the review by Trost is recommended.[4]

2.2.3.2 Sulfur(IV) Reagents

Methods for the sulfinylation of carbonyl species have not been extensively investigated, but are complementary to those used for sulfenylation and have the advantage of avoiding the oxidative step prior to elimination. Typically, ketones and esters have been sulfinylated in good yields by reaction of their enolates in ethereal solvents with methyl-, phenyl- or *p*-toluene-sulfinate at room temperature to reflux[49,50] (equation 6).[49] These bulky reagents have been used to good effect in distinguishing between two otherwise similar methylene groups in a complex asymmetric ketone.[51]

NaH, Et_2O, $PhSO_2Me$

60%

(6)

Little work has been carried out on sulfinylation reactions on those systems having thiocarbonyl and imino moieties. However, hydrazones are converted to α-sulfinyl derivatives on reaction of their anions (prepared from LDA in THF) at –78 °C with sulfinate esters,[63] although the full utility of this reaction remains to be explored. Furthermore, in an unusual reaction, *p*-toluenesulfinyl chloride has been shown to effect a facile one-step dehydrogenation of the thiolactam (**17**; equation 7) in good yield.[64] These reactions contrast with the oxidative removal of thiocarbonyl, hydrazonyl and similar functionalities with Se^{IV} species (see Section 2.2.4.2).

i, *p*-TolSOCl, NPr^i_2Et
MDC, 0 °C

ii, AcOH
75%

(**17**)

(7)

2.2.4 SELENIUM-BASED REAGENTS

Of all the methods currently used for the dehydrogenation of carbonyl and similar compounds, those utilizing selenium-based reagents have possibly received the greatest attention. Historically, selenium first found utility as its dioxide for the dehydrogenation of steroidal ketones (equation 8),[65,171] but the reagent lacks selectivity[37] and has proved problematical with more sensitive compounds.[66] As a consequence therefore, organoselenium reagents have virtually replaced selenium dioxide for effecting this transformation.

SeO_2, Bu^tOH

AcOH

(8)

These newer reagents rely on the extremely facile *syn* elimination of selenoxides in which the β-carbon atom bears at least one hydrogen atom (equation 9).[37,67] In general, the elimination of selenoxides takes place at temperatures between 0 and 25 °C, except in those cases in which some factor renders the *syn* elimination unfavorable.[67] This contrasts with the stability of sulfoxides, which generally require heating to temperatures around 60 to 120 °C in order for elimination to occur.[4] As with sulfoxide eliminations, in those instances where geometric isomers are possible only the (*E*)-isomer is formed.[39] However, not all selenoxides collapse readily and difficulties have been found with primary alkyl selenoxides[38] and some ketones and aldehydes.[68] Since the rate of selenoxide elimination is enhanced by electron-withdrawing groups on the aromatic ring, the introduction of *o*- or *p*-nitro groups is particularly beneficial.[69] The use of 2-pyridylselenenyl bromide was found to be a useful alternative for the dehy-

+ [RSeOH]

(9)

drogenation of ketones and aldehydes,[70] and Chloramine T under phase transfer conditions has also been found to facilitate the elimination of selenium in some difficult cases.[38]

The most versatile method for introduction of the selenenyl moiety is by low temperature reaction of the enolate anion or an enolic derivative with a suitable selenium species, the precise conditions being dependent on the reactivity of both the carbonyl compound and the selenium species.[37,71] Like sulfur, selenium may be introduced either in the divalent state and subsequently oxidized[37,71] or, more recently, as the selenoxide directly (Scheme 14).[72] The choice of method is determined by the subsequent reactions that need to be carried out.

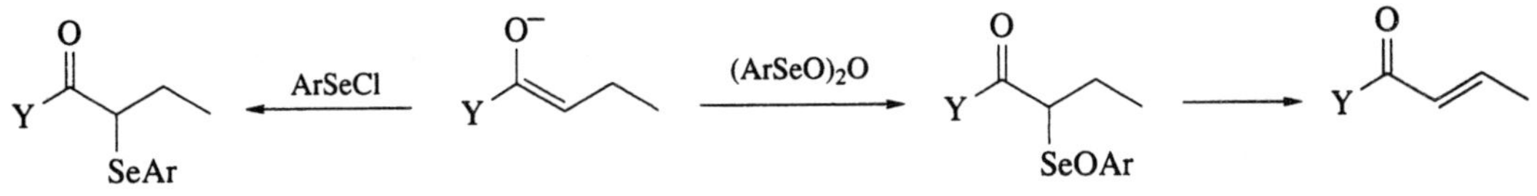

Scheme 14

Selenides are more readily oxidized to Se^{IV} than the corresponding sulfur compounds and most oxidizing agents will effect this transformation.[3] The most commonly used reagents are hydrogen peroxide, sodium periodate, peracids and ozone,[37,67] although a number of more exotic reagents have also been used.[67] Ozone offers several advantages over the use of other oxidants. In particular, it reacts quantitatively with selenides in a variety of solvents at –10 to –50 °C and excess reagent is easy to remove.[37] Furthermore, whereas sulfides are oxidized more slowly than alkenes by ozone, selenides are oxidized considerably faster,[37,67] suggesting that selenium can be selectively oxidized in the presence of both sulfur and alkenes.

The drawbacks to the use of selenium-based reagents are their inherent toxicity, relative expense and the unpleasant odors frequently formed as a result of their use. The development of catalytic processes and polymer-bound systems should ultimately overcome these disadvantages.

2.2.4.1 Selenium(II) Reagents

Selenenylations of ketones,[71] esters,[2] lactones[2,71] and lactams[56] are usually effected by the reaction of the corresponding lithium enolates with PhSeCl, PhSeBr and PhSeSePh (with the exception of ketones) at low temperature.[71] Aldehydes have not been selenenylated in this manner. Table 4 illustrates some typical products that have been made in this way. Selenenylation has been especially useful in natural product synthesis for the formation of α-methylenelactones from the parent α-methyl compounds (Scheme 15 and Table 4),[73,74] and has significant advantages over the more traditional methods for ef-

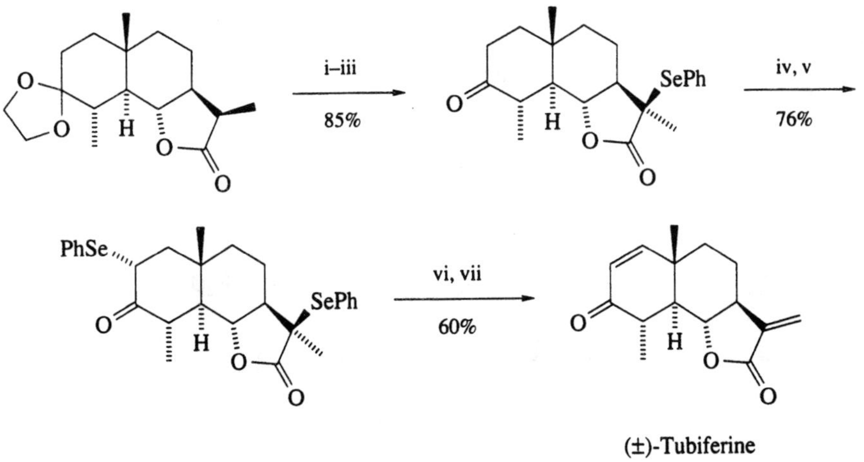

(±)-Tubiferine

i, LDA, THF; ii, $(PhSe)_2$, –20 °C; iii, HCl; iv, LDA, THF, –78 °C; v, PhSeCl, –78 °C; vi, O_3, –78 °C, CH_2Cl_2; vii, 25 °C

Scheme 15

fecting this transformation.[75] In order to ensure the formation of reasonable amounts of the *exo* isomers, however, it has often been found necessary to introduce the selenenyl moiety prior to alkylation, rather than the other way around.[73,74]

Table 4 Dehydrogenation of Ketone, Ester, Lactone and Lactam Enolates[a]

Entry	*Substrate*	*Product*	*Oxidant*	*Yield (%)*	*Ref.*
1	PhCOEt	PhCOCH=CH$_2$	$NaIO_4$	78	71
2	2-methylcyclohexanone	6-methylcyclohex-2-enone	O_3	65	71
3	bicyclic ketone	bicyclic enone	H_2O_2	58	71
4	PhCO, PhSO	PhCO, PhSO	H_2O_2	55	71
5	n-C_9H_{19}, CO_2Et	n-C_9H_{19}, CO_2Et	$MeCO_3H$	79	39
6	CO_2Me	CO_2Me	H_2O_2	96	71
7	lactone	unsaturated lactone	O_3	46	71
8	H	H	$MeCO_3H$	82[b]	74
9	N–Me	N–Me	H_2O_2	31	56

[a] Selenenylation of enolate, usually generated with LDA in THF, with PhSeCl or PhSeBr. [b] Reaction of enolate with $(PhSe)_2$ in HMPA.

Selenenylation of lithium enolates is particularly important in the case of unsymmetrical ketones, when the product of kinetic control is preferentially formed. The more-substituted isomeric derivative is prepared by the selenenylation of the corresponding enol acetate.[76] An interesting base-catalyzed transselenenylation reaction of α-alkyl-α-phenylseleneno ketones to the less-substituted α′-position has recently been reported[77] for which steric crowding at the α-position appears to be an essential requirement.

The selenenylation of aldehydes may be carried out in several ways, but most of the earlier methods involve prior formation of either the enol ether (**18**)[78] or the corresponding enamine (**19**; Scheme 16).[39,79] These stepwise procedures overcome the problems of slow incorporation and low yields of selenenyl moieties observed under acid-catalyzed conditions.[39] Selenium can be efficiently introduced in one step, however, using a combination of PhSeSePh and SeO_2 in the presence of a catalytic amount of sulfuric acid.[80]

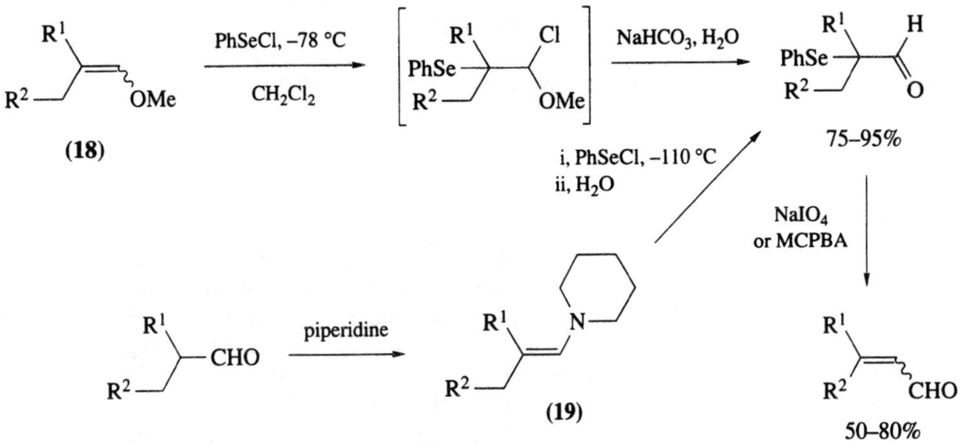

(18) —PhSeCl, –78 °C, CH_2Cl_2→ [intermediate] —$NaHCO_3$, H_2O→ 75–95%

(19) —i, PhSeCl, –110 °C; ii, H_2O→

piperidine; $NaIO_4$ or MCPBA; 50–80%

Scheme 16

An alternative one-step procedure using *N,N*-diethylbenzeneselenamide has been developed[81,82] and, as illustrated in equation (10), this reagent is particularly suitable for differentiating between aldehydic and ketonic moieties in the same molecule.[81] The analogous morpholinoselenamide has been shown to selenenylate the α-keto ester (**20**; equation 11) but no other examples have been reported.[67]

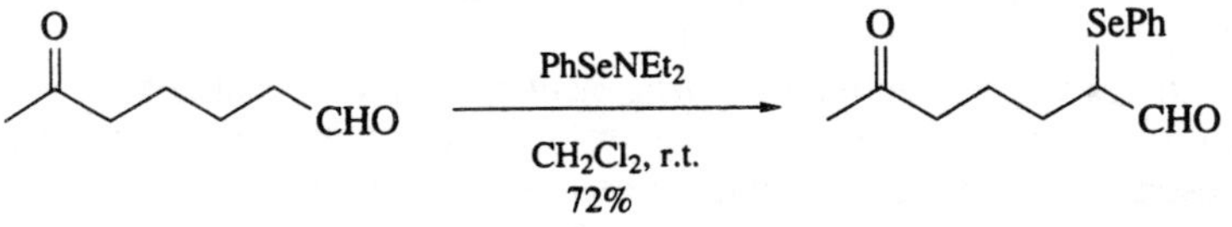

CHO —$PhSeNEt_2$, CH_2Cl_2, r.t., 72%→ SePh, CHO (10)

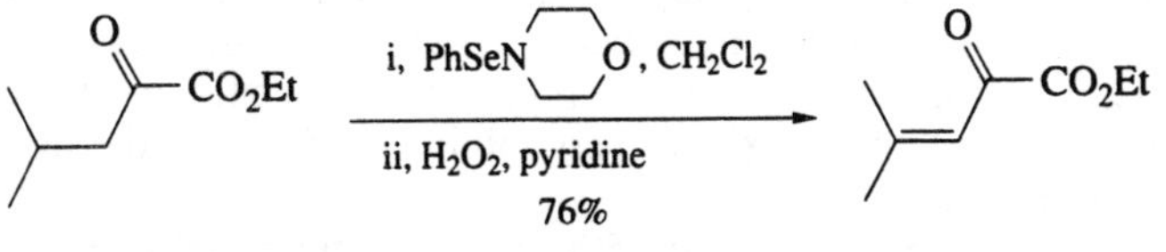

(20) CO_2Et —i, PhSeN(morpholine)O, CH_2Cl_2; ii, H_2O_2, pyridine; 76%→ CO_2Et (11)

Using $PhSeNEt_2$, it is also possible to selenenylate β-dicarbonyl compounds, but this method has received little attention.[82] The enolates of β-dicarbonyl compounds are also selenenylated on treatment with PhSeCl or PhSeBr,[71] although such highly enolized compounds may be converted into their selenenylated derivatives in good yield under milder conditions using a 1:1 complex of PhSeCl and pyridine.[83] This latter method has the advantage that, by the avoidance of strong base, the reaction is compatible with a wide variety of other functional groups, without the need for prior protection. Nonenolized carbonyl compounds were shown not to react under these conditions.[83]

An interesting variation for the introduction of selenenyl species, which has the advantage of using cheaper elemental selenium, has been described by Liotta and coworkers (Scheme 17).[84–86] This reaction involves conversion of the lithium enolate (**21**) to the intermediate selenolate (**22**) which may be directly alkylated to give the selenenyl derivative (**23**) in high yield.[84–86] The reaction works well with ketones, esters and β-dicarbonyl compounds, but has the disadvantage of requiring the use of HMPA.[84,85]

The aromatization of cyclohexenones is an important process that can be easily accomplished by the use of selenium-based reagents using similar techniques to those previously discussed for other carbonyl species. Thus, enolates derived from α,β-enones readily undergo selenenylation at the α′-position and on oxidation and elimination afford the corresponding phenols.[87,88]

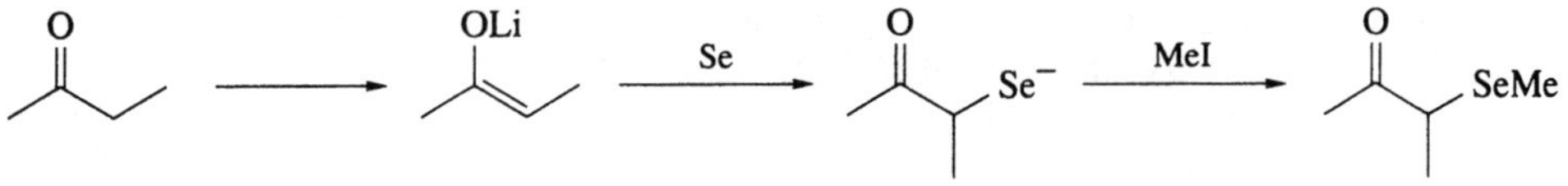

(21) (22) (23)

Scheme 17

A comprehensive review of the methods available for the introduction of selenenyl moieties has recently been published.[89]

2.2.4.2 Selenium(IV) Reagents

For those reactions in which the insertion of α,β-unsaturation is the immediate objective, the introduction of the selenenyl species as Se^{IV}, rather than as Se^{II}, may be expeditious since this obviates the need for a subsequent oxidative step. Selenium dioxide, benzeneseleninic acid and its anhydride act principally as oxidizing agents in their reaction with organic substrates and selenium tetrahalides are powerful halogenating agents.[89] Nevertheless, selenium dioxide effectively dehydrogenates 1,4-dicarbonyl compounds and has been useful for the dehydrogenation of steroidal and terpenoid ketones.[90] Benzeneseleninic anhydride has proved to be particularly suited to this transformation and has been used successfully for the dehydrogenation of steroidal ketones,[72,91] lactones[92] and lactams (Table 5).[93] Where comparisons have been made, this reagent is superior to selenium dioxide.[94] Typically the carbonyl compound is heated at 95 to 130 °C with the anhydride, in solvents such as chlorobenzene, to give high yields of the oxidized products.[72] At the higher temperatures (>120 °C) benzeneseleninic acid is converted to the anhydride and so forms a useful alternative reagent.[72] A feature of the reaction is that the PhSeSePh formed in the oxidation may be isolated and reoxidized to the anhydride with nitric acid if required.[72] The catalytic use of PhSeSePh with *t*-butyl peroxide,[72] iodosylbenzene,[95] or better *m*-iodosylbenzoic acid,[95] has also been described. It is interesting that all attempts to oxidize the γ-lactone (**24**; entry 3, Table 5) resulted only in dehydrogenation of the A-ring ketone.[92] Care must be taken to ensure minimum reaction times when effecting dehydrogenations with benzeneseleninic anhydride in order to avoid angular hydroxylation reactions resulting from further oxidation.[94] The addition of aluminum chloride to troublesome reactions appears to favor dehydrogenation.[96]

As a result of the powerful oxidizing potential of benzeneseleninic anhydride, it is incompatible with the presence of a number of functional groups, although many common moieties are well tolerated. Thus, it has been shown to convert thiocarbonyl compounds such as xanthates, thiocarbonates, thioamides and thiones,[97] and hydrazones, oximes, thiosemicarbazones and hydroxylamines,[98] into the corresponding carbonyl compounds under relatively mild conditions. Furthermore, hydrazo derivatives are converted to the azo compounds.[98]

Little work appears to have been carried out with benzeneseleninic anhydride on substrates other than steroidal or triterpenoid compounds, but it seems likely that the stronger conditions required to effect oxidation with this reagent makes it less attractive than the two-step procedure described above. Indeed, in the few instances reported it failed to convert hydrocinnamamide into cinnamamide,[93] and is said to be of no value for the dehydrogenation of acyclic esters.[99] There are, however, several reports in which either catalytic or stoichiometric benzeneseleninic anhydride has effectively dehydrogenated cyclic ketones in high yield,[94] a typical example being illustrated in equation (12).[95]

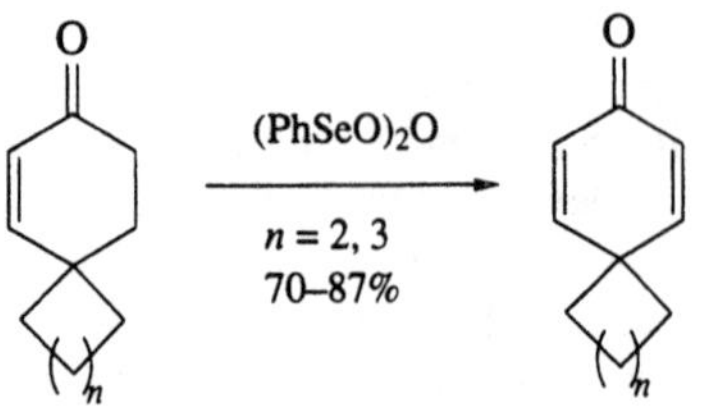

$(PhSeO)_2O$; $n = 2, 3$; 70–87% (12)

In an interesting extension of the use of benzeneseleninic anhydride, Barton and coworkers[100] have dehydrogenated steroidal and other oxazolines (*e.g.* **25**) in high yield (equation 13). This type of reaction has considerable potential for a wide variety of heterocyclic systems, due to the acidity of exocyclic

Table 5 Dehydrogenation of Steroidal Ketones, Lactones and Lactams[a]

Entry	*Substrate*	*Product*	*Yield (%)*	*Ref.*
1	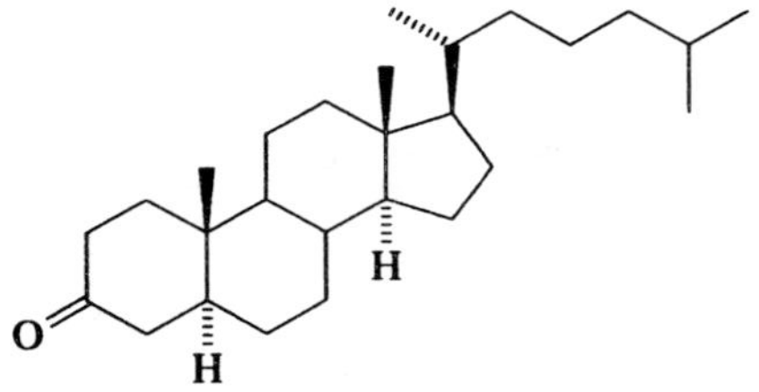 O; H; H	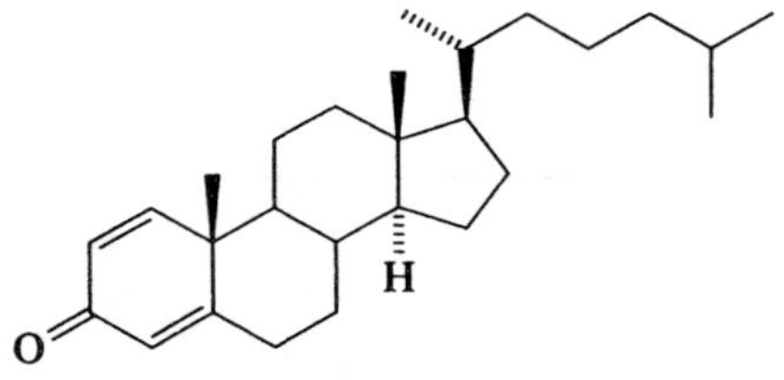 O; H	83[b]	72
2	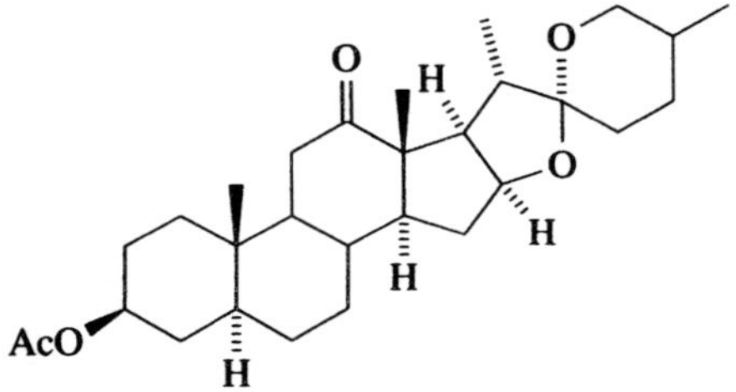 AcO; H; H; O; H; O; O; H	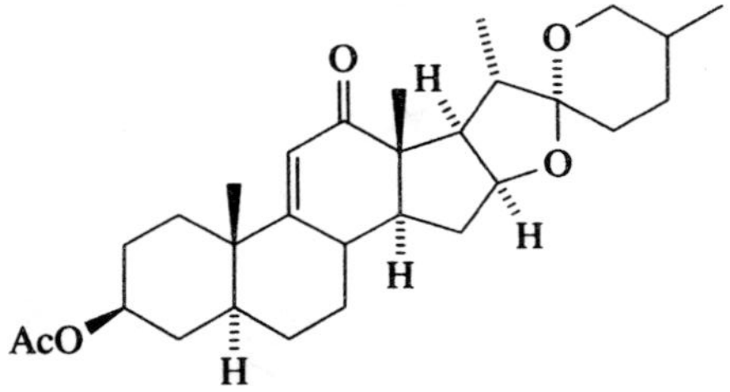 AcO; H; H; O; H; O; O; H	91	72
3	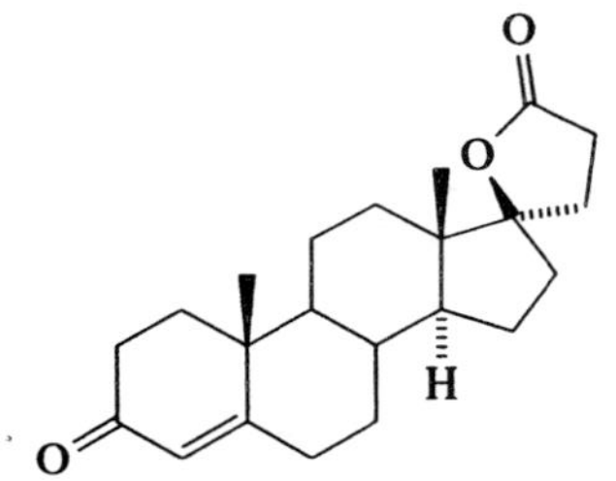 O; H; O; O; **(24)**	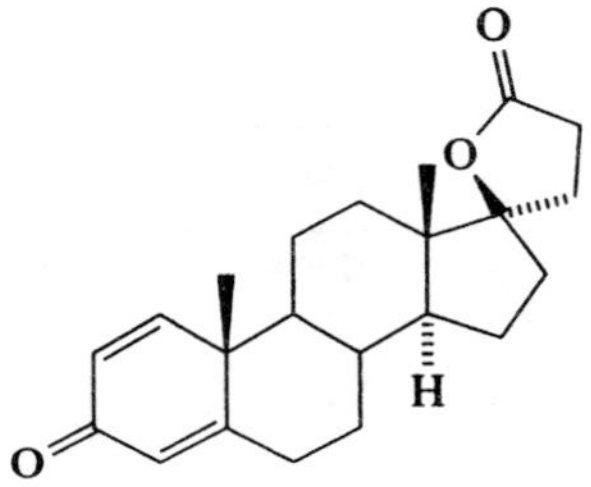 O; H; O; O	84	92

Table 5 *(continued)*

Entry	*Substrate*	*Product*	*Yield (%)*	*Ref.*
4	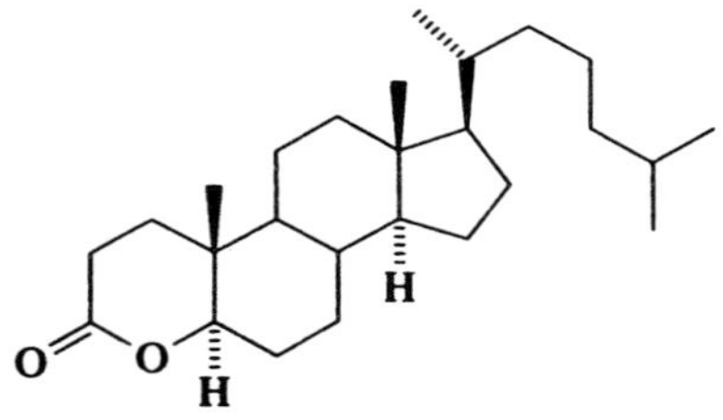 O, O, H, H	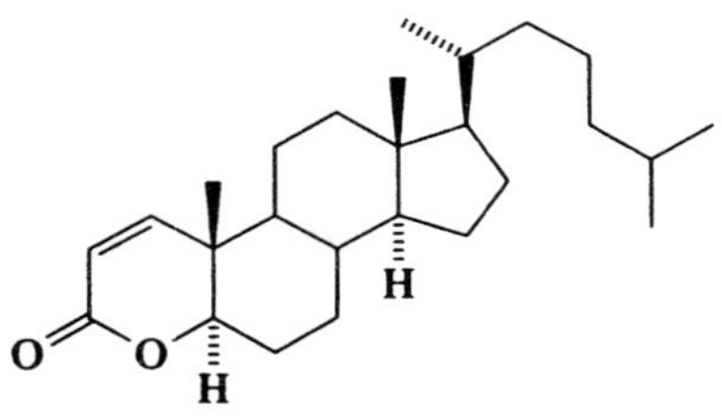 O, O, H, H	90	92
5	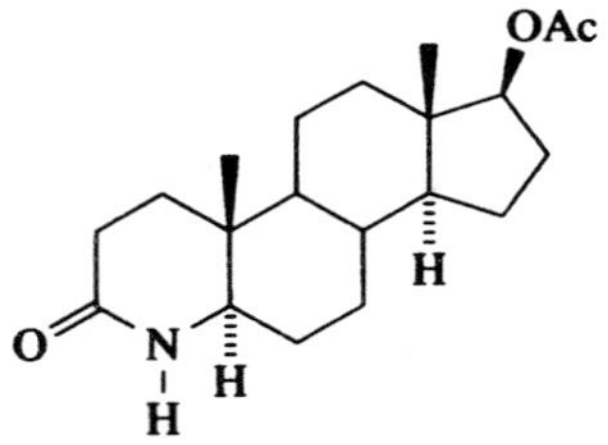 OAc, O, N, H, H, H	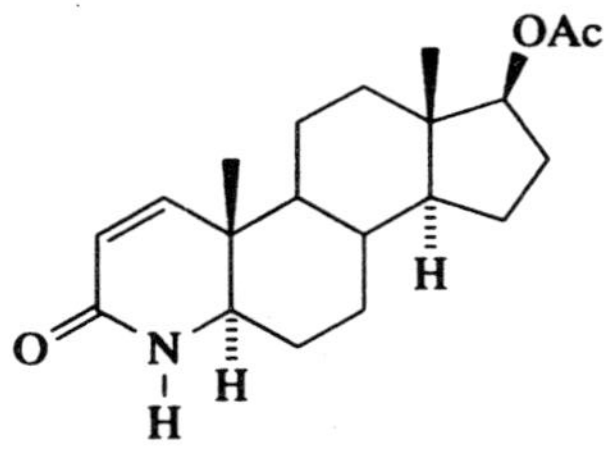 OAc, O, N, H, H, H	64	93

[a] Reaction with benzeneseleninic anhydride at 120–130 °C. [b] With 2 equiv. of anhydride.

methylene groups, but, apart from these few examples, it does not seem to have been explored. Possibly the greatest scope would result from the two-stage procedure using Se^{II}.

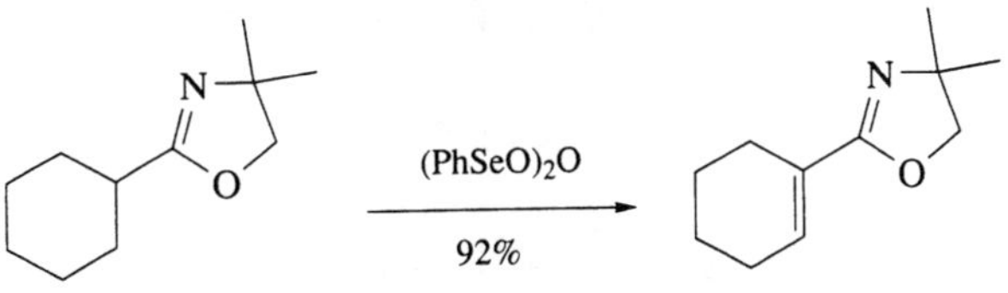

$(PhSeO)_2O$, 92% (13)

(25)

Benzeneseleninyl chloride is another example of a Se^{IV} electrophile, but has found limited use for the dehydrogenation of ketones and esters[71] due to its hygroscopic nature. Thus, although a crystalline solid it is considerably more difficult to handle than the Se^{II} halides. An alternative reagent, phenylselenium trichloride, offers a milder approach for the direct introduction of Se^{IV}, although its utility appears to be limited to ketones.[101] In a typical reaction cyclopentanone (**26**; Scheme 18) may be dehydrogenated *via* the intermediate (**27**) by reaction with phenylselenium trichloride in diethyl ether at 5 °C followed by mild aqueous hydrolysis.[101] The lower reactivity of aldehydes, acids, lactones and esters suggests that ketonic substrates may be selectively dehydrogenated in the presence of these functional groups (*cf.* equation 10).

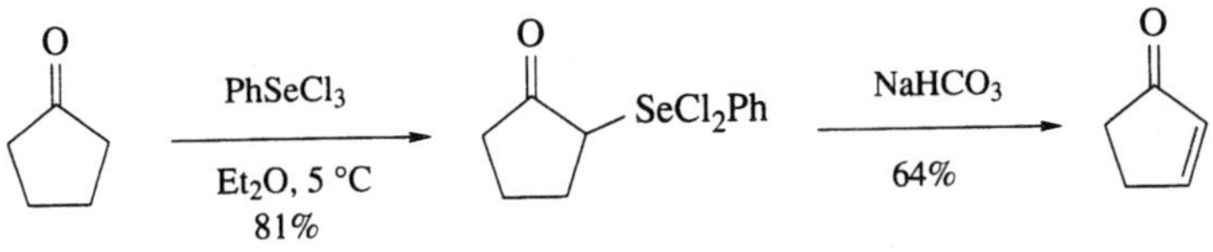

$PhSeCl_3$, Et_2O, 5 °C, 81%; $SeCl_2Ph$; $NaHCO_3$, 64%

(26) (27)

Scheme 18

2.2.5 DICHLORODICYANOQUINONE AND RELATED REAGENTS

Although quinones have been recognized since the turn of the century, it was not until the mid 1950s that Braude and coworkers demonstrated their full potential as dehydrogenation reagents.[102,103] Those quinones bearing electron-withrawing groups showed the highest oxidation potentials, and therefore represented the most effective reagents,[104] which led to the development of 2,3-dichloro-5,6-dicyanobenzoquinone (DDQ; **28**)[105] and chloranil (**29**)[106] as the most commonly used reagents.

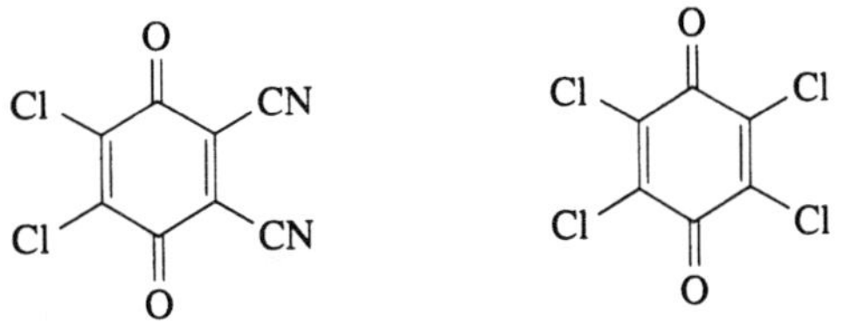

(28) (29)

The dehydrogenation reaction is generally first order in both quinone and substrate and is enhanced in polar solvents. Together with other findings, these observations have suggested an ionic mechanism involving the initial formation of a charge-transfer complex (**30**) followed by hydride abstraction and rapid loss of a proton (Scheme 19).[102]

An alternative mechanism has been proposed to explain the rate enhancement seen in the presence of acids and which is particularly evident with quinones of low oxidation potential (E_0 = 600 mV, *cf.* DDQ, $E_0 \approx$ 1000 mV).[102] In this instance, formation of the quinone conjugate acid (**31**; Scheme 20) has been proposed, which might be expected to be a considerably more powerful hydride abstractor than the parent quinone.

$$\mathrm{RH_2 + Q \xrightarrow{\text{complex}} \underset{(30)}{[RH_2Q]} \xrightarrow{\text{slow}} \overset{+}{R}H + \overset{-}{Q}H \xrightarrow{\text{fast}} R + QH_2}$$

Scheme 19

$$\mathrm{Q + HX \underset{}{\overset{\text{fast}}{\rightleftharpoons}} \underset{(31)}{\overset{+}{Q}H} + X^-}$$

$$\mathrm{RH_2 + \overset{+}{Q}H \xrightarrow{\text{slow}} \overset{+}{R}H + QH_2}$$

$$\mathrm{\overset{+}{R}H + X^- \xrightarrow{\text{fast}} R + HX}$$

Scheme 20

Quinones have been extensively used for aromatization reactions[107] in addition to the dehydrogenation of steroidal ketones and lactones.[105] Interestingly, whereas chloranil (**29**) and a number of other quinones oxidize steroidal 4-ene-3-ones (**32**) selectively to 4,6-dienones (**34**),[108] DDQ (**28**) results only in the formation of the 1,4-dienone (**36**; Scheme 21).[109] This divergent behavior is best explained by the intermediacy of the kinetic enolate (**35**) in the case of the higher potential DDQ, but of the thermodynamic enolate (**33**) in the case of the less reactive quinones.[110] Acidic conditions need to be avoided if the cross-conjugated ketone (**36**) is the desired product since under these conditions the 3,5-dienol (**33**) becomes both the kinetic and the thermodynamic enol, resulting only in the formation of the linear dienone.[110]

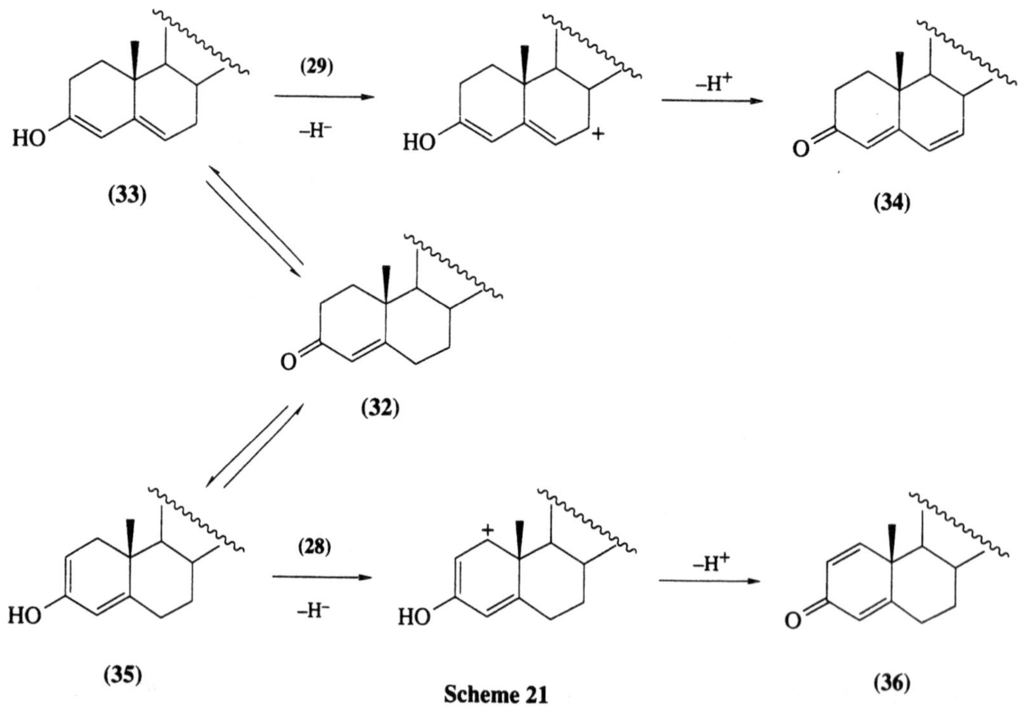

Scheme 21

In general, the dehydrogenation of steroidal ketones is carried out in dry benzene or dioxane at reflux with 1.1 to 2 equiv. of the quinone.[105] Similar conditions have also been used to prepare flavones,[111] chromones[112] and spirodienones[113] in good yields (Table 6). Consistent with the apparent requirement that enolization is a prerequisite to dehydrogenation with quinones,[114] reactants such as α-formyl ketones, *e.g.* (**37**) and (**38**), that have a high enol content, dehydrogenate rapidly at room temperature (Table 6).[115,116]

Steroidal enol ethers have also been shown to undergo facile dehydrogenation with DDQ, but the products formed are dependent on the reaction conditions.[117] Thus, whereas under anhydrous conditions the

Table 6 Dehydrogenation of Ketones using DDQ

Entry	*Substrate*	*Product*	*Conditions*	*Yield (%)*	*Ref.*
1	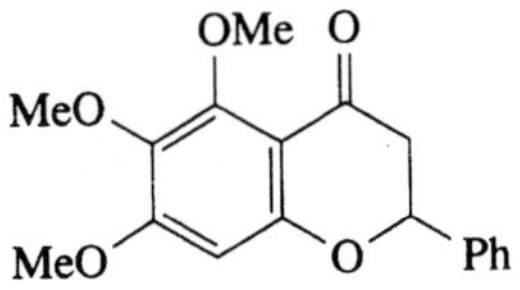	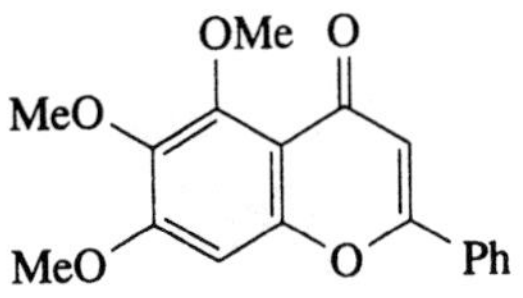	DDQ, dioxane, Δ, 12 h	100	111
2	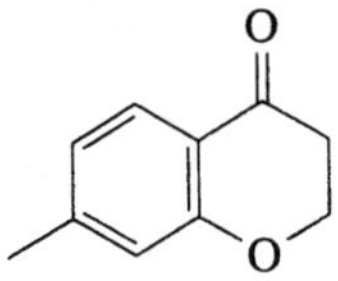	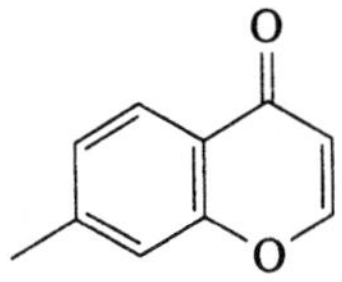	DDQ, dioxane, Δ, 8 h	75	112
3	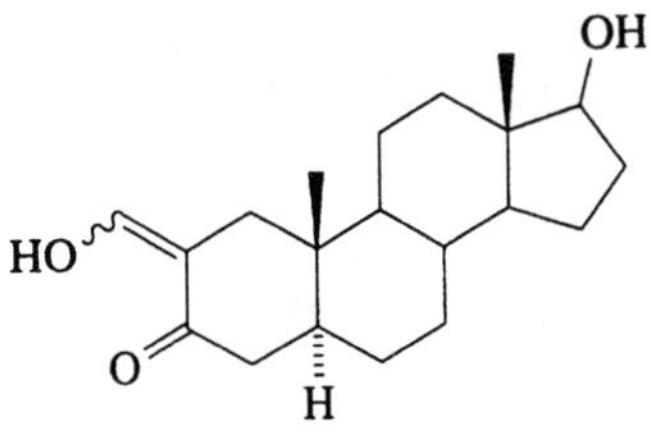 (37)	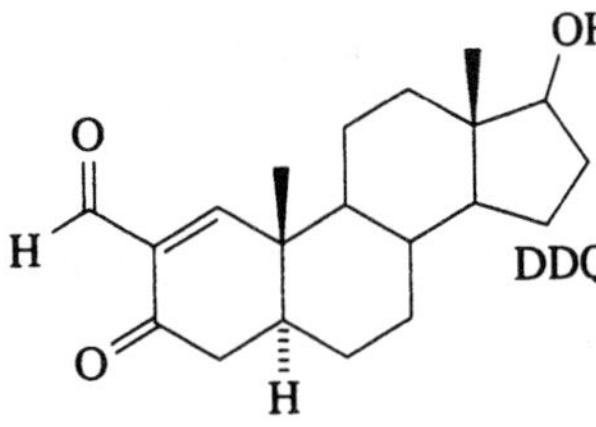	DDQ, dioxane, r.t., 2 min	50	116
4	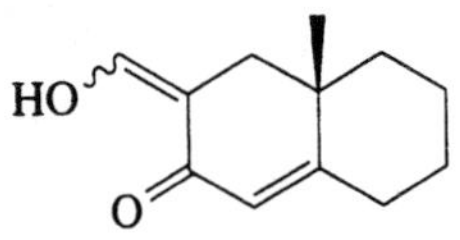 (38)	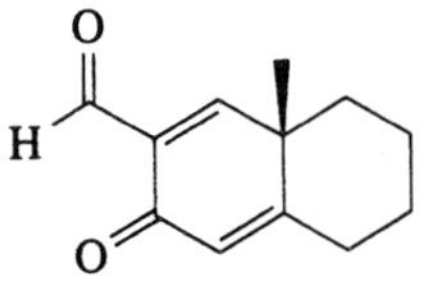	DDQ, dioxane, r.t.	76	115
5			DDQ, dioxane	—	113

dienyl ether (**39**) furnished the trienone (**41**), in the presence of moist acetone the intermediate oxonium ion (**40**) was hydrolyzed to the dienone (**42**; Scheme 22).[117]

A more general approach involves the oxidation of silyl enol ethers with DDQ,[118–120] although for good yields care is needed to ensure removal of the acidic by-product $DDQH_2$. Usually this problem is overcome by the addition of bis(trimethylsilyl)acetamide[118] or bases such as collidine[119] or 2,6-lutidine (Table 7).[120] As with selenium, sulfur and palladium reagents (see Sections 2.2.3, 2.2.4 and 2.2.6) the use of silyl enol ethers allows the regioselective introduction of unsaturation (Table 7).[118] Typically, the dehydrogenation of silyl enol ethers is effected at ambient temperature using 1–1.5 equiv. of quinone in hydrocarbon solvents.

Most examples of quinone dehydrogenations adjacent to C═X have been carried out on steroidal ketones and are essentially limited to readily enolizable species. Reactions on esters and amides (Table 8) are far less common and, because of their relatively low ease of enolization, require harsh conditions.[121] Thus, unless stabilization of the intermediate carbonium ion is possible,[122,123] elevated temperatures and prolonged reaction times are required (Table 8), which increases the incidence of unwanted side reactions. Frequent by-products are those arising as a result of Diels–Alder reactions or Michael addition to the quinone.[105] Allylic alcohols may be rapidly oxidized to aldehydes or ketones under these conditions[105] and require prior protection.

The conversion of carboxylic acids to α,β-unsaturated acids is not a trivial transformation, although it can be effected by treatment of the α-anion of the carboxylate salt (**43**) with DDQ in THF containing HMPA at reflux (Scheme 23).[124,125] Using this procedure, a number of fatty acids have been successfully dehydrogenated, albeit only in around 30% yield. Only the (*E*)-isomers are isolated.

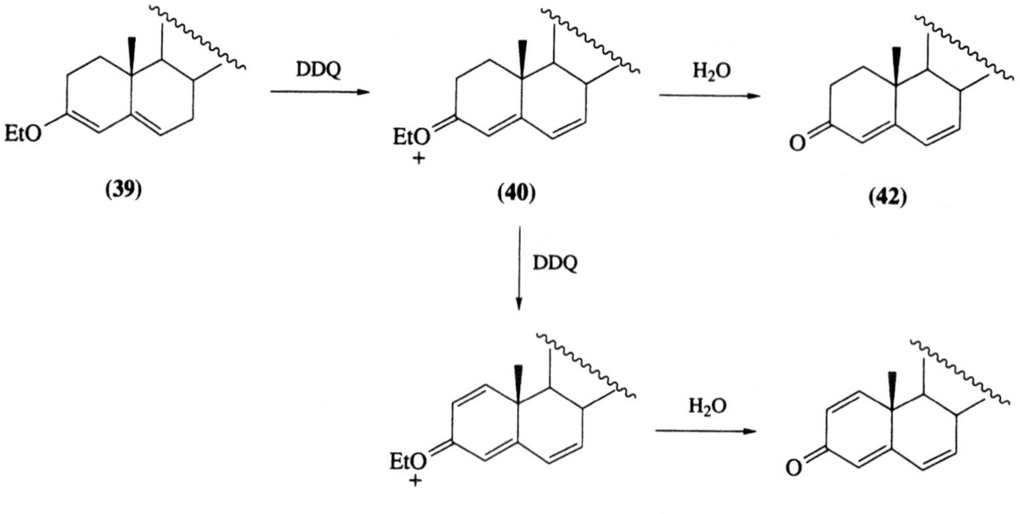

(39) DDQ (40) H_2O (42)

DDQ

H_2O (41)

Scheme 22

Table 7 Dehydrogenation of Trimethylsilyl Enol Ethers using DDQ

Entry	*Substrate*	*Product*	*Conditions* [a]	*Yield (%)*	*Ref.*
1	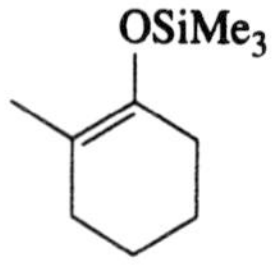 $OSiMe_3$	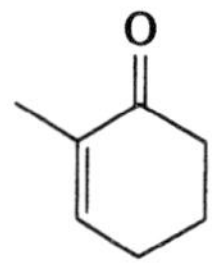 O	BSA, 1 h	50	118
2	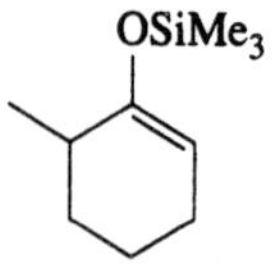 $OSiMe_3$	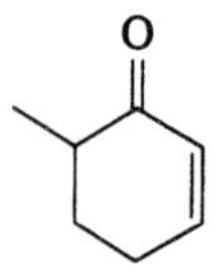 O	BSA, 1 h	53	118
3	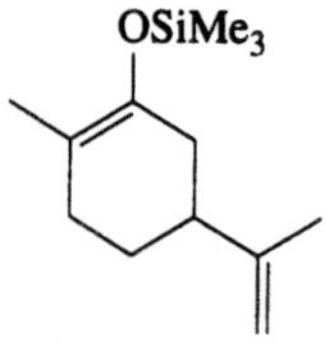 $OSiMe_3$	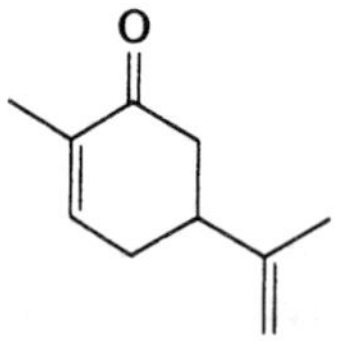 O	Collidine, 1.5 h	52	119

[a] Reactions carried out with DDQ in benzene at room temperature, BSA = *N*,*O*-bis(trimethylsilyl)acetamide.

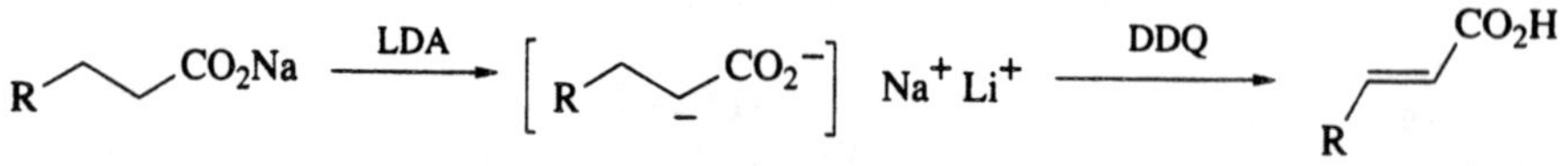

$R\text{-}CH_2CH_2\text{-}CO_2Na \xrightarrow{LDA} [R\text{-}CH_2\text{-}\overset{-}{C}H\text{-}CO_2^-]\ Na^+\ Li^+ \xrightarrow{DDQ} R\text{-}CH{=}CH\text{-}CO_2H$

(43)

Scheme 23

Quinones have also been used to dehydrogenate adjacent to C═N in a variety of substituted nitrogen heterocycles[125–127] with the ultimate generation of aromatic species (equation 14). In some instances DDQ has been claimed to be preferable to other reagents generally used for this purpose[125] but the eventual choice must be determined by the substituents present and the intrinsic stability of both product and starting material.

Table 8 Dehydrogenation of Lactones and Lactams using DDQ

Entry	*Substrate*	*Product*	*Conditions*	*Yield (%)*	*Ref.*
1	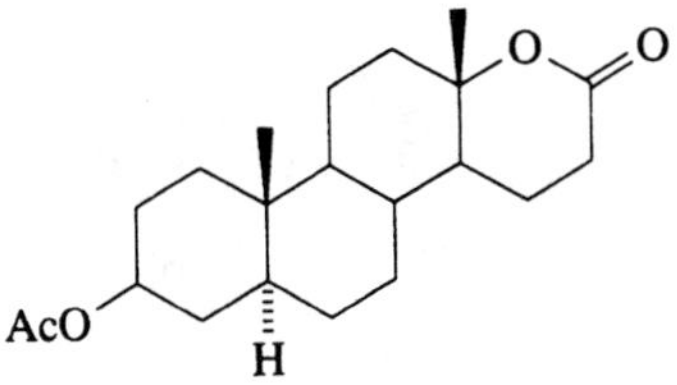 AcO, H, O, O	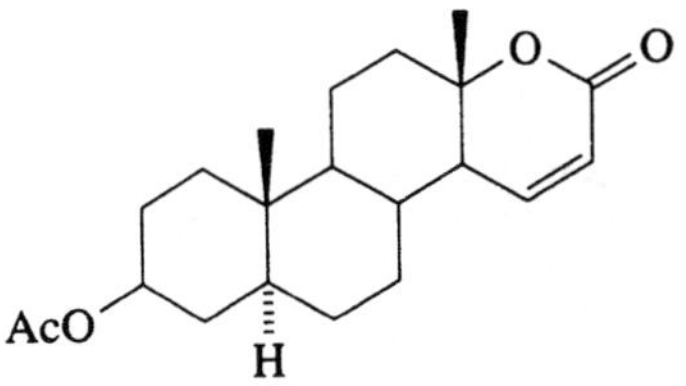 AcO, H, O, O	240 h[a]	25	121
2	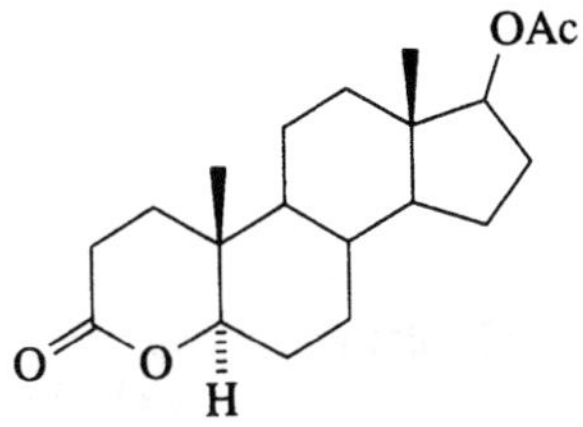 OAc, O, O, H	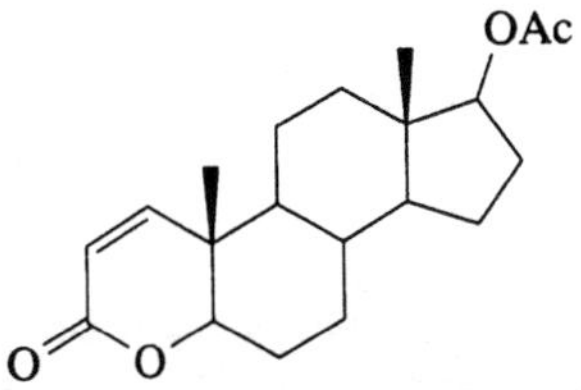 OAc, O, O	130 h[a]	55	121
3	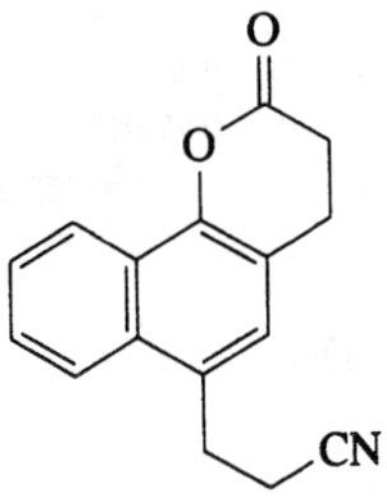 O, O, CN	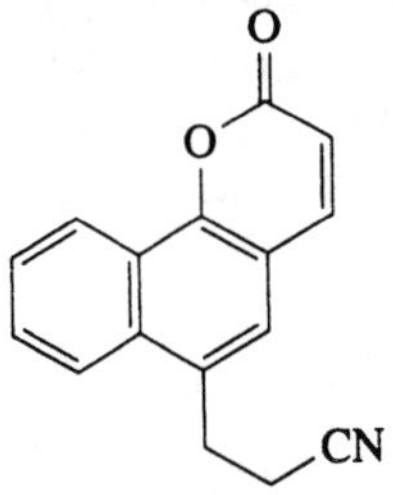 O, O, CN	14 h[a]	70	122
4	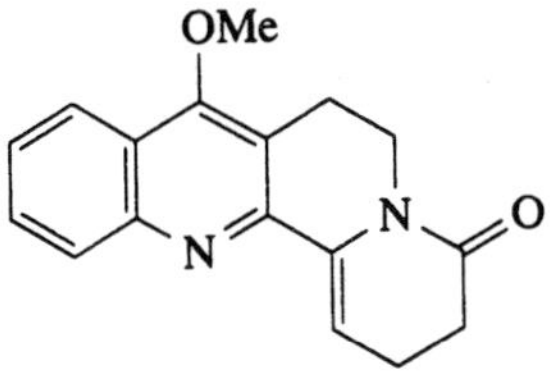 OMe, N, N, O	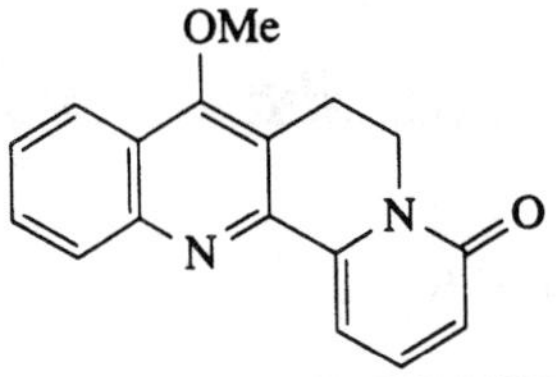 OMe, N, N, O	Benzene	—	123

[a] Reaction carried out in dioxane at reflux.

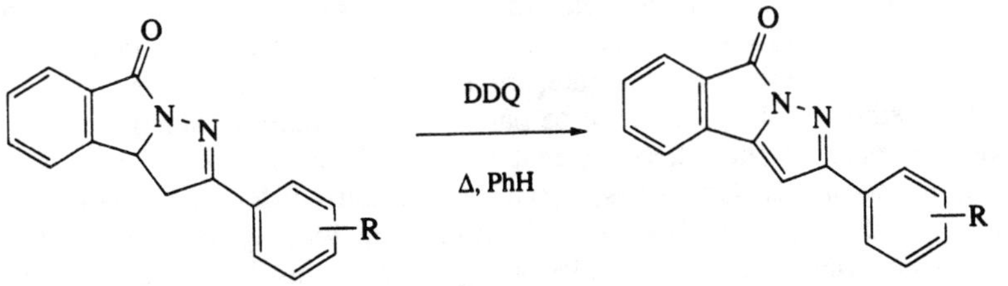

DDQ; Δ, PhH (14)

2.2.6 NOBLE METALS AND THEIR SALTS

The application of transition metals and their salts or complexes to dehydrogenation reactions adjacent to C═X compounds has received relatively scant attention compared to reagents such as selenium. In an early isolated example, however, 10% palladium on charcoal in refluxing *p*-cymene was shown to dehydrogenate the thermally stable steroidal lactone (**44**; equation 15) in good yield in instances where classical reagents such as selenium dioxide and halogenation–dehydrohalogenation failed.[128] Whereas these reaction conditions have proved to be successful for the aromatization of hydroaromatic compounds, newer transition metal based methodology has largely superseded this approach and led to increasing use of palladium salts and their complexes.

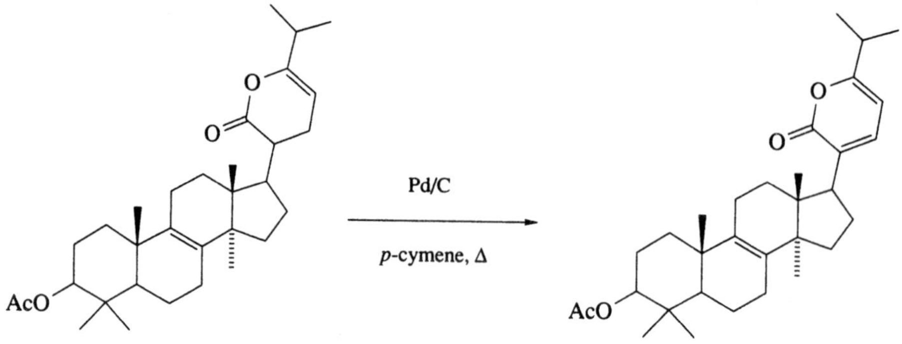

Pd/C

p-cymene, Δ

AcO

(44)

(15)

A detailed study of over 45 catalysts, primarily from Group VIII metal salts and complexes, showed palladium(II) compounds to be the most effective in the dehydrogenation of a variety of aldehydes and ketones.[129] Soluble palladium(II) salts and complexes such as dichloro(triphenylphosphine)palladium(II) and palladium(II) acetylacetonate have been shown to be optimal, with the salts of rhodium, osmium, iridium and platinum having reduced efficacy.[129] Since the dehydrogenation reaction is accompanied by reduction of the palladium(II) catalyst to palladium(0), oxygen and a cooxidant are required to effect reoxidation. Copper(II) salts are favored cooxidants, but quinones, and especially *p*-benzoquinone, are also effective (Scheme 24).[129,130]

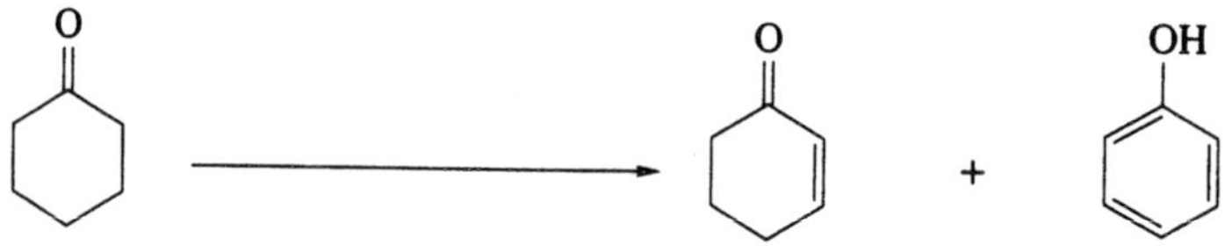

$Pd(acac)_2$, $Cu(OAc)_2$, O_2, AcOH, 100 °C 8%	95%	5%
$PdCl_2$, Bu^{t}OH, conc. HCl, 80 °C 90%	90%	<5%

Scheme 24

Improved conditions for dehydrogenation reactions have been developed using palladium(II) chloride in a mixture of *t*-butyl alcohol and concentrated hydrochloric acid,[130] although these conditions limit the utility of the method to compounds without acid sensitive groups. The yields of enones from readily enolizable carbonyl compounds are usually moderate, and reaction rates generally reflect the ease of enolization.[129,130] Thus, the method has been found to be particularly suitable for aldehydes and cyclic ketones, but acyclic ketones are less efficiently dehydrogenated.[129–131] Carboxylic acids, esters and amides are not dehydrogenated by palladium(II) derivatives,[129] which potentially offers the opportunity to selectively introduce double bonds into compounds possessing mixed functionality. In contrast to sulfur- and selenium-based dehydrogenations, unsymmetrical ketones generally afford a mixture of isomers (equations 16 and 17).[129,130] Like most other methods, however, acyclic aldehydes and ketones furnish *trans* enones exclusively.[129,130]

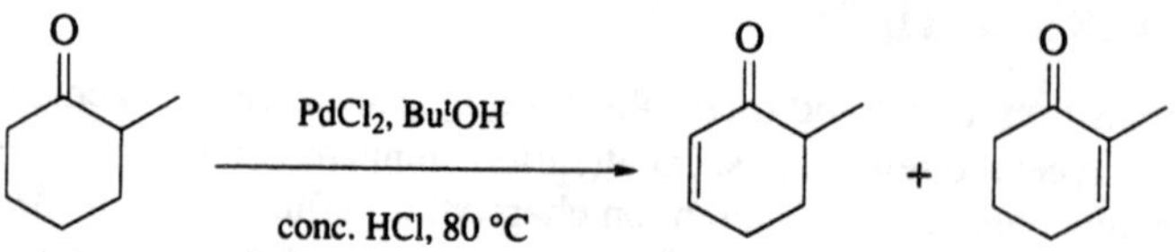

$PdCl_2$, Bu^{t}OH

conc. HCl, 80 °C

45% 46%

(16)

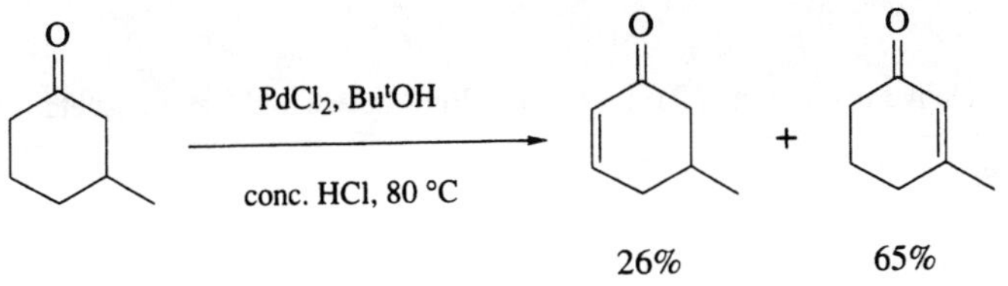

PdCl$_2$, Bu^{t}OH; conc. HCl, 80 °C

26% 65%

(17)

Mechanistically, palladium-catalyzed dehydrogenations have been shown to proceed according to Scheme 25, in which the initially formed π-complex (**45**) rearranges to a σ-complex (**46**) prior to the elimination of palladium hydride.[129,130]

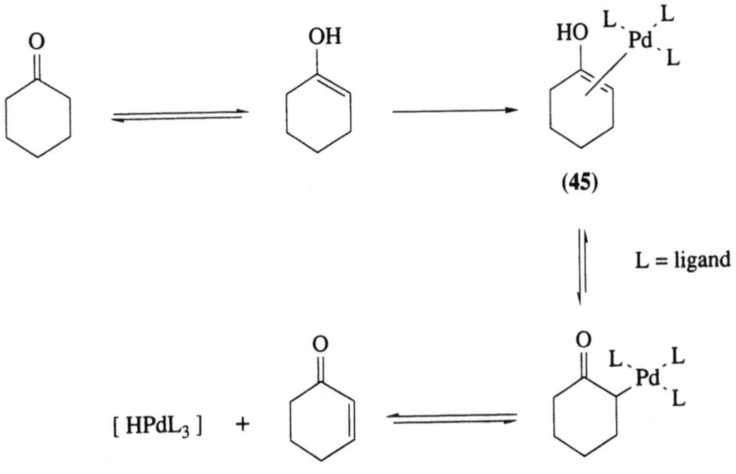

(**45**)

L = ligand

[HPdL$_3$] +

(**46**)

Scheme 25

Improved understanding of the mechanism of palladium-induced dehydrogenations has led to the development of significantly better catalysts and reaction conditions. In particular, mixtures of PdCl$_2$(PhCN)$_2$ and silver triflate in the presence of *N*-methylmorpholine have allowed the efficient dehydrogenation of aldehydes under ambient conditions and in nonacidic media (equation 18).[132] Ketones undergo a similar reaction, affording enones in 60–78% yield, but require prior formation of the tin enolate with tin(II) triflate.[132] Under these conditions, however, 2 equiv. of palladium(II) chloride were used to effect conversion, which severely limits the usefulness of the method.

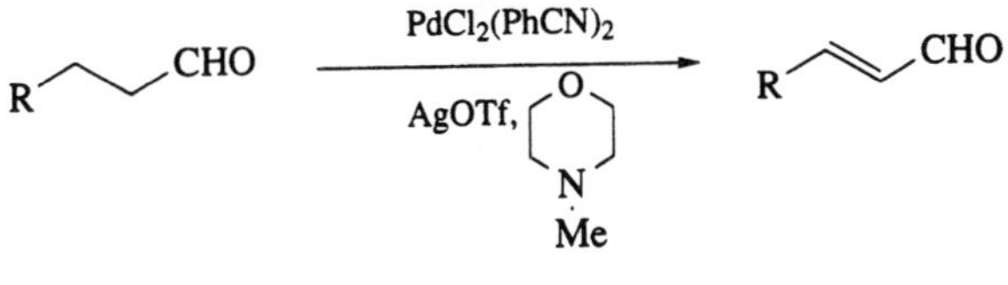

PdCl$_2$(PhCN)$_2$; AgOTf, N-methylmorpholine

(18)

Probably the most widely applicable conditions developed for palladium catalysts utilize silyl enol ethers.[133,134] In one instance,[133] an excellent yield of enone was obtained using 0.5 equiv. each of palladium(II) acetate and *p*-benzoquinone in acetonitrile. The method has the advantage that the position of the double bond is determined by the geometry of the precursor silyl enol ether (Scheme 26). Palla-

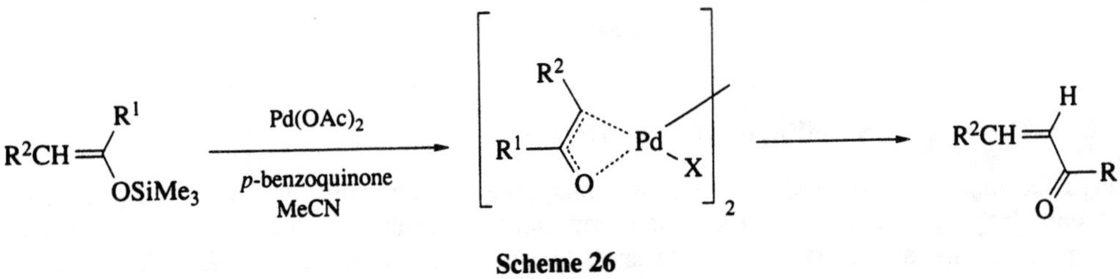

Pd(OAc)$_2$; *p*-benzoquinone, MeCN

Scheme 26

Table 9 Dehydrogenation of Trimethylsilyl Enol Ethers and Ketene Acetals with Pd^0

Entry	*Substrate*	*Product*	*Conditions* [a]	*Yield (%)*	*Ref.*
1	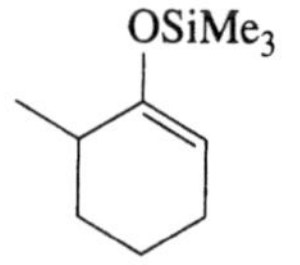 $OSiMe_3$	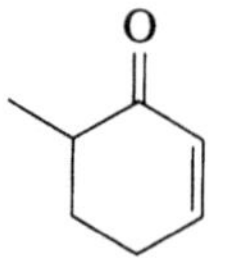 O	A	87	135
2	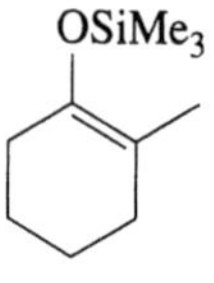 $OSiMe_3$	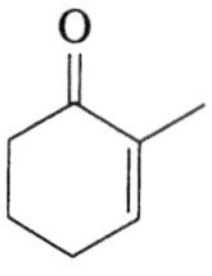 O	A	100	135
3	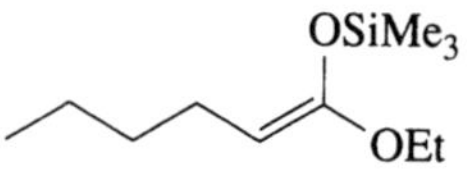 $OSiMe_3$, OEt	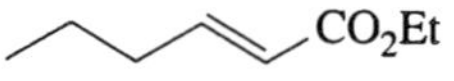 CO_2Et	B	79	137
4	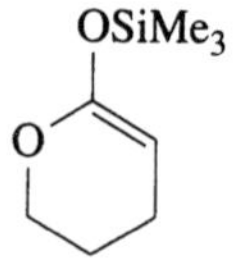 $OSiMe_3$, O	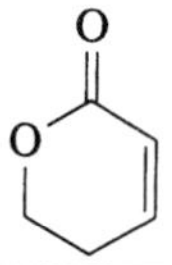 O, O	B	70	137

[a] Conditions. A: 5 mol % $Pd(OAc)_2$, 5 mol % DPPE, 2 equiv. dimethyl carbonate in acetonitrile at reflux. B: 10 mol % $Pd(OAc)_2$, 2 equiv. allyl methyl carbonate in acetonitrile at reflux.

dium(0) has also been used to effect a similar reaction with silyl enol ethers[134,135] and enol carbonates.[134,136] In these cases the palladium is present in truely catalytic quantities and the reaction proceeds with as little as 1 mol % of palladium(II) acetate (DPPE) in acetonitrile. It is also possible to oxidize esters *via* their corresponding silyl ketene acetals with Pd^0, although in this instance the yields are better in the absence of phosphine ligands.[137] Palladium(0) chemistry offers a mild, high yielding entry to α,β-unsaturated ketones and esters (Table 9), and should find a wide application as an alternative to selenium-based dehydrogenations.

A particularly interesting extension of this work is offered by the observed enantioselective hydrogen abstraction from the prochiral cyclohexanone (**47**) on treatment with chiral lithium amide bases (Scheme 27).[138] Thus, quenching the initially formed enolate afforded the asymmetric trimethylsilyl ether (**48**) which gave the chiral enone (**49**) in 65% enantiomeric excess on dehydrogenation.[138] Further work in this area should provide valuable methodology for the formation of chiral α,β-unsaturated carbonyl systems.

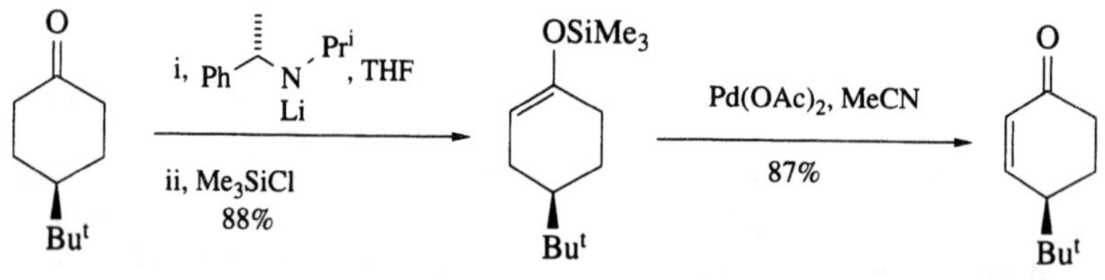

O, Bu^t (**47**) — i, Ph–CH(Me)–N(Li)–Pr^i, THF; ii, Me_3SiCl; 88% → $OSiMe_3$, Bu^t (**48**) — $Pd(OAc)_2$, MeCN; 87% → O, Bu^t (**49**)

Scheme 27

2.2.7 MISCELLANEOUS CHEMICAL METHODS

Although the methods discussed in earlier sections generally constitute the preferred procedures by which to dehydrogenate carbonyl and similar compounds, a variety of other reagents will effect this transformation, and in some instances may offer certain advantages. Manganese dioxide is one reagent that has been extensively utilized, and is particularly suited to the dehydrogenation of heterocyclic compounds and the formation of quinones (Table 10, entries 1 and 2).[139–141] Where manganese dioxide oxidations have been compared with other methods they have frequently been found to give similar or better yields.[139,142] Typically, manganese dioxide oxidations are effected in aprotic solvents such as benzene or dioxane at reflux using approximately 5 equiv. of oxidant for each double bond.[139,141] The quality of the oxidant is important, with activated manganese dioxide[143] affording greatest efficiency.[139,141] Potassium nitrosodisulfonate (Fremy's salt) will also effect the oxidation of dihydroquinones to quinones,[144] in addition to effecting a wide variety of other oxidations.[145]

Table 10 Miscellaneous Oxidations of Carbonyl Compounds and Heterocycles

Entry	*Substrate*	*Product*	*Conditions*	*Yield (%)*	*Ref.*
1	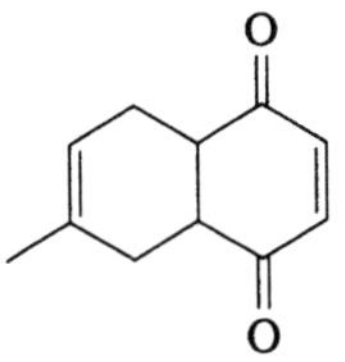	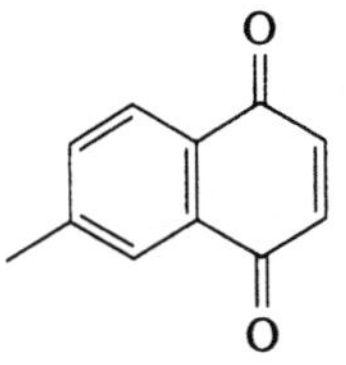	MnO_2, Δ, PhH, 24 h	62	139
2	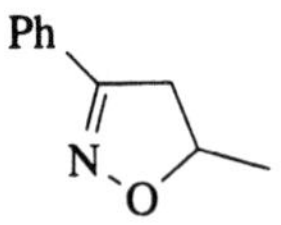	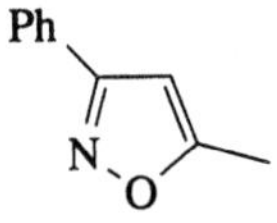	MnO_2, Δ, PhH, 7 h	98	141
3	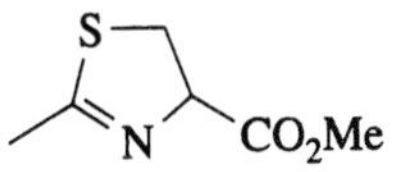	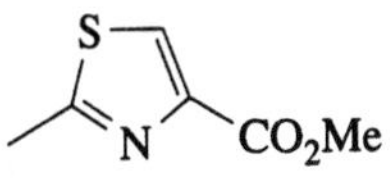	NiO_2, $CHCl_3$, 25 °C, 3 d	81	146
4	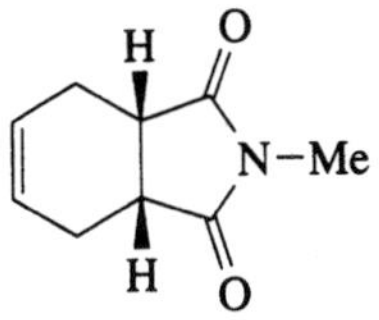	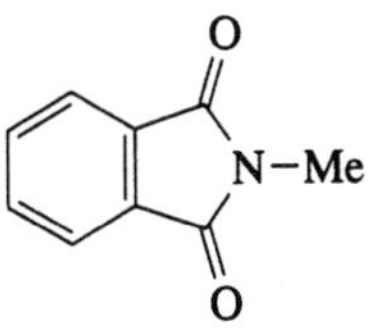	NiO_2, PhH, Δ, 7 h	62	146
5	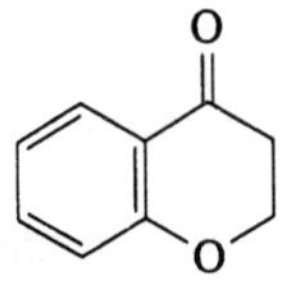	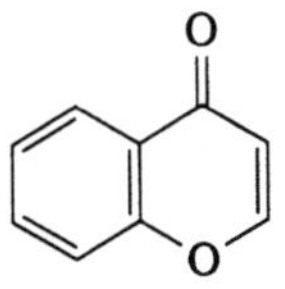	$Tl(NO_3)_3$, MeOH, $HClO_3$, 5 h	76	148
6	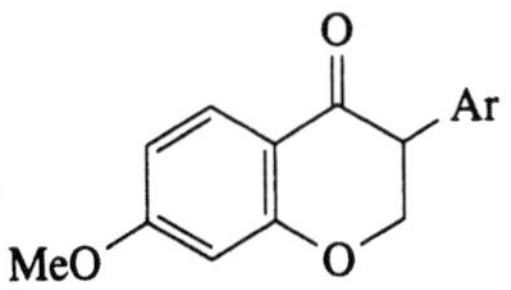	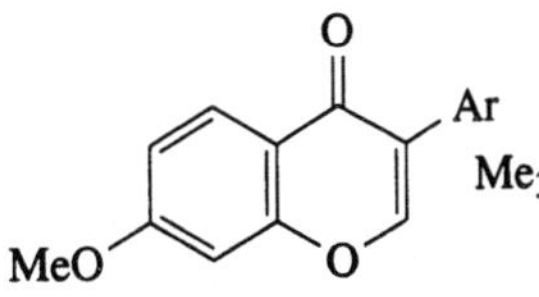	Me_3SiCl, Ac_2O, 60–65 °C, 168 h	55	150
7	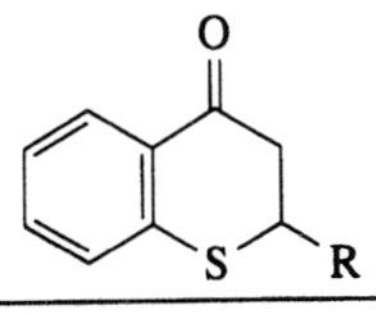	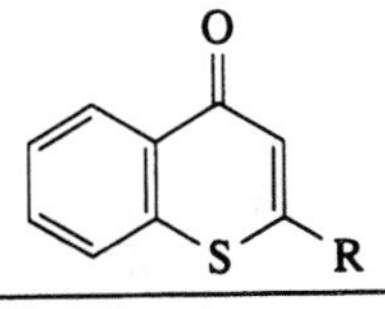	*hν*, MeOH, 48 h	50–55	151

Heterocyclic systems may also be conveniently dehydrogenated using nickel peroxide in aprotic solvents, and good yields may be obtained even in the presence of sensitive functional groups (Table 10, entries 3 and 4).[146] This reagent is not specific for the dehydrogenation of C═X compounds, however,[146] and may not be suitable for reactions requiring selective oxidation. A variety of other oxidants[147] have been shown to effect similar oxidations.

Thallium trinitrate has been shown to be an efficient reagent for the dehydrogenation of chromanones[148] and flavanones (Table 10, entry 5),[149] the reaction being carried out in methanol at room temperature. The addition of perchloric acid to chromanone oxidations enhances the yields and reaction rates by promotion of enolization,[148] but apparently was without effect on flavanones.[149] The ease with which chromanones and similar compounds may be dehydrogenated has also permitted unusual procedures such as trimethylsilyl chloride/acetic anhydride[150] and photolysis[151] to be used (Table 10, entries 6 and 7), but their generality is suspect.

The trityl carbonium ion has proved to be an interesting reagent for the dehydrogenation of ketones and esters *via* their silyl enol ethers (*e.g.* **50**; equation 19),[119,152] although major side reactions involving α-tritylation have been reported and yields are variable.[153] Nonetheless, this is a particularly suitable way to convert tetralones into naphthols (equation 20).[153] Both the perchlorate and tetrafluoroborate counterions are effective. Whether this procedure offers any advantages over the use of DDQ or chloranil, which effect the same transformations, is doubtful.[119,153] Palladium(II) acetate behaves similarly, but is expensive on catalyst.[133] Trityl tetrafluoroborate has also been shown to abstract hydrogen from enamines,[154] but whether this offers a useful alternative for the dehydrogenation of ketones remains to be proven. In common with many other reagents, trityl perchlorate will oxidize 4-chromanones and 4-thiochromanones to their corresponding α,β-unsaturated derivatives in excellent yields.[155]

OTMS, Pr^i — $Ph_3C^+\ BF_4^-$, CH_2Cl_2, 18 h → O, Pr^i (19)

(50) 74%

OTMS — $Ph_3C^+\ BF_4^-$ or $Ph_3C^+\ ClO_4^-$, CH_2Cl_2, 3–5 h → OH (20)

93%

Pyridine *N*-oxide will dehydrogenate carboxylic acids in the presence of acetic anhydride,[156] but this does not represent an efficient method. Tertiary amides, however, may be smoothly oxidized in a two-step procedure *via* an intermediate chloriminium ion (see Section 2.2.2).[30]

Copper(II) bromide is another reagent that has been used successfully for the dehydrogenation of ketones and amides (equation 21).[157,158] This procedure, which presumably proceeds *via* the α-bromo compounds, (*cf.* Section 2.2.2) was found to have particular advantages over a number of alternative methods for the dehydrogenation of some dihydrouracils.[158]

O, $(\ )_6CO_2Me$, R — $CuBr_2$, $CHCl_3$, Δ → O, $(\ )_6CO_2Me$, R (21)

R = H, 66%
R = CHO, 38%

An interesting dehydrogenation of hydrazones (**51**) has been reported by Barton[35] which relies on the available oxygen of aromatic nitro groups (equation 22). In a detailed study, quantitative yields were obtained using 4-nitrobenzoic acid as the oxidant.[35] Whilst this unusual reaction affords some advantages over earlier methods it is unlikely to be the method of choice in most instances.

The introduction of hydroxy groups α to carbonyl-type functions is the subject of another chapter (Chapter 2.3, this volume), but clearly this represents an alternative, though seldom used, procedure for the dehydrogenation of such species. The direct insertion of other oxygen moieties is, however, com-

i, $KOBu^t$, $(CH_2OMe)_2$
4-$NO_2C_6H_4CO_2H$, r.t.
ii, AcOH
100%

(51)

(22)

plementary to those methods already discussed. One method that has found use in terpenoid chemistry involves the incorporation of an α-benzoyloxy group *via* the enolate anion and thermolytic elimination at 450–550 °C to afford the enone.[20] Enolizable ketones will also react with lead tetraacetate and mercury(II) acetate[159] to give α-acetoxy derivatives that can be subsequently eliminated, but this method is unpopular.

More recently, work has been reported showing that silyl enol ethers of ketones, esters and lactones can be efficiently converted to α-sulfonyloxy carbonyl compounds on treatment with either [hydroxy(tosyloxy)iodo]benzene or [hydroxy(mesyloxy)iodo]benzene (equation 23).[160] This method is similar to that used by the same workers to introduce the trifluoromethylsulfonyloxy group α to ketone carbonyl groups *via* their silyl enol ethers.[161] While the major interest in these developments is the further functionalization of the α-position of carbonyl compounds the method clearly offers a route to α,β-unsaturated species.

OTMS → PhI(OH)OTs, CH_2Cl_2, r.t., 80% → O, OTs

(23)

2.2.8 MICROBIAL AND ENZYMATIC METHODS

In addition to the chemically based methods described above, fermentation and enzymatic procedures are also available for the dehydrogenation of C=X compounds,[105,162–164] although this approach has found greatest favor for stereospecific reduction and regiospecific oxidation reactions.[165,166] The class of enzymes that effect dehydrogenation reactions are of the redox type and have been classified by the International Union of Biochemistry as oxidoreductases.[163] A number of such enzymes are now known.[163] Particular advantages of microbial and enzymatic methods are the versatility, efficiency and selectivity with which these reactions are carried out and the mild conditions that are employed. Thus, in contrast to most chemical methods, enzymes are often able to discriminate between enantiomers of racemic mixtures and to generate chiral products from prochiral substrates.[166] As a general rule, however, it seems unlikely that this approach would be favored over chemically based methods for dehydrogenation reactions, unless it was necessary to circumvent a particular synthetic problem.

The nature of oxidative processes requires the removal of electrons from the substrate and many enzymes of the redox class contain transition metals which act as an electron sink.[167] Those enzymes which do not satisfy this requirement need organic cofactors such as nicotinamide adenine dinucleotide or nicotinamide adenine dinucleotide 2′-phosphate to act as electron acceptors,[167] although simple quinones have been shown to suffice.[168]

Much of the work with microorganisms capable of effecting dehydrogenation reactions has been carried out using steroids, particularly those reactions introducing unsaturation at C-1 of 3-keto steroids. Those organisms most frequently used are *Bacillus sphaericus* and *Arthrobacter simplex*, but many others are claimed to be effective.[105] Yields of dehydro compounds formed by microbiological methods are somewhat variable, ranging from very little to 95%.[105] In a typical experiment, good yields of androstene-3,17-dione (**52**) have been obtained using *Bacillus sphaericus* and the mechanism established as proceeding *via* a *trans* diaxial (1α,2β) elimination (equation 24).[168] This result is consistent with the general observation that enzymatic oxidation of 3-keto steroids shows a preference for ring A.[164] Several other enzymes have been shown to dehydrogenate steroids,[163,166,169] but particularly interesting is the stereochemical preference shown by cortisone β-reductase for the 5β-hydrogen atom compared to the 5α-reductase, which shows a preference for the opposite enantiomer.[163]

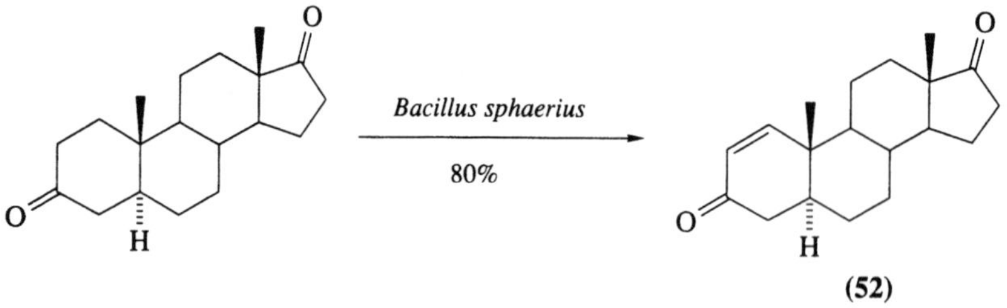

Bacillus sphaerius

80%

(24)

(52)

Specific enzymes have also been identified which convert 5,6-dihydrouracil into uracil, succinate into fumarate and acylated coenzyme A into 2,3-dehydroacyl derivatives.[163] While these are important biological processes, it is doubtful whether they will have general synthetic value. Of greater potential interest are those enzymes capable of converting 3-nitropropanoate to the corresponding acrylate and hexadecanal to its 2,3-unsaturated derivative,[163] although there is little evidence that these reactions can be advantageously exploited relative to alternative chemical methods.

2.2.9 SUMMARY

A wide diversity of reagents exists for effecting the dehydrogenation of C═X compounds. With a few exceptions, the most versatile methods are those based on selenium and sulfur, and there is little to choose between these two elements in most instances. Particular advantages of selenium pertain to the weaker σ-bonds that it forms with carbon, which results in the *syn* elimination of selenoxides being some 1000 times faster than that of sulfoxides.[3] Disadvantages of selenium, on the other hand, relate to its greater toxicity and expense. It is possible that dehydrogenations based on palladium chemistry will offer some advantages, especially in the light of newly developed methodology.

Throughout this review, elimination reactions have been restricted to the loss of an appropriate leaving group from the α-carbon atom, but both β-thio and β-selenenyl groups can be eliminated with ease from C═X compounds following oxidation.[170] As a rule, such derivatives are prepared by conjugate addition to α,β-unsaturated carbonyl compounds,[37] and therefore formation and elimination constitutes a formal protection of these compounds.

Most of the examples reviewed concentrate on instances in which the C═X heteroatom is oxygen, and this reflects the dearth of work that has been reported on other heteroatoms. Thus, although numerous examples of the aromatization of nitrogen heterocycles exist, there is very little pertaining to other systems. This is an area where more exploratory work is needed, especially on oxazolines and related heterocycles.[100] Reactions with C═S compounds are even more rare, presumably because of the ease with which such systems are oxidized under dehydrogenation conditions. Opportunities exist to develop the dehydrogenation of such systems, however, as demonstrated with thioamides which have served as suitable intermediates for the dehydrogenation of otherwise difficult amides.[64]

2.2.10 REFERENCES

1. J. March, 'Advanced Organic Chemistry: Reaction Mechanisms and Structure', 3rd edn., Wiley, New York, 1985.
2. D. L. Clive, *Aldrichimica Acta*, 1978, **11**, 43.
3. H. J. Reich, *Acc. Chem. Res.*, 1979, **12**, 22.
4. B. M. Trost, *Chem. Rev.*, 1978, **78**, 363.
5. A. G. Schultz and J. P. Dittami, *Tetrahedron Lett.*, 1983, **24**, 1369.
6. P. Bakuzis and M. L. F. Bakuzis, *J. Org. Chem.*, 1981, **46**, 235.
7. W. G. Nigh, in 'Oxidation in Organic Chemistry', ed. W. S. Trahanovsky, Academic Press, New York, 1973, p. 67.
8. E. R. Glazier, *J. Org. Chem.*, 1962, **23**, 2937, 4397.
9. C. Rappe, *Acta Chem. Scand.*, 1968, **219**, 1359.
10. H. O. House, 'Modern Synthetic Reactions', 2nd edn., Benjamin, New York, 1972, p. 459.
11. E. J. Corey, *J. Am. Chem. Soc.*, 1953, **75**, 2301, 3297, 4832.
12. G. Jones, Jr., *J. Chem. Soc. C*, 1970, 1230.
13. R. Zurflüh and C. Tamm, *Helv. Chim. Acta*, 1972, **55**, 2495.
14. V. Caló, L. Lopez and G. Pesce, *J. Chem. Soc., Perkin Trans. 1*, 1977, 501.
15. J. R. Catch, D. H. Hey, E. R. H. Jones and W. Wilson, *J. Chem. Soc.*, 1948, 276.
16. N. Takeno, T. Fukushima, S. Takeda and K. Kishimoto, *Bull. Chem. Soc. Jpn.*, 1985, **58**, 1599.
17. P. L. Stotter and K. A. Hill, *J. Org. Chem.*, 1973, **38**, 2576.

18. M. W. Rathke and A. Lindert, *Tetrahedron Lett.*, 1971, 3995.
19. S. P. Singer and K. B. Sharpless, *J. Org. Chem.*, 1978, **43**, 1448.
20. A. E. Greene, J.-C. Muller and G. Ourisson, *J. Org. Chem.*, 1974, **39**, 186.
21. A. E. Greene, J.-C. Muller and G. Ourisson, *Tetrahedron Lett.*, 1972, 2489.
22. D. Ladjama and J.-J. Riehl, *Synthesis*, 1979, 504.
23. F. Jung, D. Ladjama and J.-J. Riehl, *Synthesis*, 1979, 507.
24. R. C. Cambie, R. C. Hayward, J. L. Jurlina, P. S. Rutledge and P. D. Woodgate, *J. Chem. Soc., Perkin Trans. 1*, 1978, 126.
25. G. M. Rubottom and R. C. Mott, *J. Org. Chem.*, 1979, **44**, 1732.
26. L. Blanco, P. Amice and J. M. Conia, *Synthesis*, 1976, 194.
27. U. Ravid and R. Ikan, *Synth. Commun.*, 1975, **5**, 137.
28. H. T. Clarke and E. R. Taylor, *Org. Synth., Coll. Vol.*, 1956, **1**, 115.
29. J. G. Gleason and D. N. Harpp, *Tetrahedron Lett.*, 1970, 3431.
30. R. Da Costa, M. Gillard, J. B. Falmagne and L. Ghosez, *J. Am. Chem. Soc.*, 1979, **101**, 4381.
31. E. W. Warnhoff and D. R. Marshall, *J. Org. Chem.*, 1967, **32**, 2000.
32. C. C. Price and J. M. Judge, *Org. Synth., Coll. Vol.*, 1973, **5**, 255.
33. H. O. House and R. W. Bashe, II, *J. Org. Chem.*, 1965, **30**, 2942.
34. W. G. Dauben and J. R. Wiseman, *J. Am. Chem. Soc.*, 1967, **89**, 3545.
35. D. H. R. Barton, J. Coll, J. F. McGarrity and D. A. Widdowson, *J. Chem. Soc., Perkin Trans. 1*, 1973, 1565.
36. H. O. House and F. A. Richley, Jr., *J. Org. Chem.*, 1967, **32**, 2151.
37. D. L. J. Clive, *Tetrahedron*, 1978, **34**, 1049.
38. K. B. Sharpless, K. M. Gordon, R. F. Lauer, D. W. Patrick, S. P. Singer and M. W. Young, *Chem. Scr.*, 1975, **8A**, 9.
39. K. B. Sharpless, R. F. Lauer and A. Y. Teranishi, *J. Am. Chem. Soc.*, 1973, **95**, 6137.
40. G. R. Pettit, D. C. Fessler, K. D. Paull, P. Hofer and J. C. Knight, *J. Org. Chem.*, 1970, **35**, 1398.
41. M. Trolliet, R. Longeray and J. Dreux, *Bull. Soc. Chim. Fr.*, 1974, 1484.
42. W. Haede, W. Fritsch, K. Radscheit and U. Stache, *Justus Liebigs Ann. Chem.*, 1973, 5.
43. B. M. Trost, T. N. Salzmann and K. Hiroi, *J. Am. Chem. Soc.*, 1976, **98**, 4887.
44. D. W. Emerson and T. J. Korniski, *J. Org. Chem.*, 1969, **34**, 4115.
45. P. Dubs and R. Stüssi, *Helv. Chim. Acta*, 1978, **61**, 998.
46. K. Tsujihara, K. Harada, N. Furukawa and S. Oae, *Tetrahedron*, 1971, **27**, 6101.
47. E. Vedejs and D. A. Engler, *Tetrahedron Lett.*, 1976, 3487.
48. D. Seebach and M. Teschner, *Chem. Ber.*, 1976, **109**, 1601.
49. H. J. Monteiro and J. P. de Souza, *Tetrahedron Lett.*, 1975, 921.
50. R. M. Coates and H. D. Pigott, *Synthesis*, 1975, 319.
51. B. M. Trost and L. H. Latimer, *J. Org. Chem.*, 1978, **43**, 1031.
52. B. M. Trost and T. N. Salzmann, *J. Am. Chem. Soc.*, 1973, **95**, 6840.
53. P. A. Grieco and J. J. Reap, *Tetrahedron Lett.*, 1974, 1097.
54. B. M. Trost and Y. Tamaru, *J. Am. Chem. Soc.*, 1977, **99**, 3101.
55. P. G. Gassman and R. J. Balchunis, *J. Org. Chem.*, 1977, **42**, 3236.
56. P. A. Zoretic and P. Soja, *J. Heterocycl. Chem.*, 1977, **14**, 681.
57. P. Groenewegen, H. Kallenberg and A. van der Gen, *Tetrahedron Lett.*, 1978, 491.
58. M. E. Kuehne, *J. Org. Chem.*, 1963, **28**, 2124.
59. T. Kumamoto, S. Kobayashi and T. Mukaiyama, *Bull. Chem. Soc. Jpn.*, 1972, **45**, 866.
60. S. Murai, Y. Kuroki, K. Hasegawa and S. Tsutsumi, *J. Chem. Soc., Chem. Commun.*, 1972, 946.
61. K. Hiroi, H. Miura, K. Kotsuji and S. Sato, *Chem. Lett.*, 1981, 559.
62. E. J. Corey and S. Knapp, *Tetrahedron Lett.*, 1976, 4687.
63. L. Banfi, L. Colombo and C. Gennari, *Synthesis*, 1982, 829.
64. P. D. Magnus and P. Pappalardo, *J. Am. Chem. Soc.*, 1983, **105**, 6525.
65. L. F. Fieser and M. Fieser, 'Reagents for Organic Synthesis', Wiley, New York, 1967, vol. 1, p. 996.
66. J. N. Marx and G. Minaskanian, *Tetrahedron Lett.*, 1979, 4175.
67. C. Paulmier, 'Selenium Reagents and Intermediates in Organic Synthesis', Pergamon Press, Oxford, 1986, chap. 5.
68. A. Toshimitsu, H. Owada, S. Uemura and M. Okano, *Tetrahedron Lett.*, 1982, **23**, 2105.
69. K. B. Sharpless and M. W. Young, *J. Org. Chem.*, 1975, **40**, 947.
70. A. Toshimitsu, H. Owada, K. Terao, S. Uemura and M. Okano, *J. Org. Chem.*, 1984, **49**, 3796.
71. H. J. Reich, J. M. Renga and I. L. Reich, *J. Am. Chem. Soc.*, 1975, **97**, 5434.
72. D. H. R. Barton, D. J. Lester and S. V. Ley, *J. Chem. Soc., Perkin Trans. 1*, 1980, 2209.
73. P. A. Grieco and M. Nishizawa, *J. Chem. Soc., Chem. Commun.*, 1976, 582.
74. P. A. Grieco and M. Miyashita, *J. Org. Chem.*, 1974, **39**, 120.
75. P. A. Grieco, *Synthesis*, 1975, 67.
76. H. O. House, M. Gall and H. D. Olmstead, *J. Org. Chem.*, 1971, **36**, 2361.
77. D. Liotta, M. Saindane and D. Brothers, *J. Org. Chem.*, 1982, **47**, 1598.
78. K. C. Nicolaou, R. L. Magolda and W. J. Sipio, *Synthesis*, 1979, 982.
79. D. R. Williams and K. Nishitani, *Tetrahedron Lett.*, 1980, **21**, 4417.
80. N. Miyoshi, T. Yamamoto, N. Kambe, S. Murai and N. Sonoda, *Tetrahedron Lett.*, 1982, **23**, 4813.
81. M. Jefson and J. Meinwald, *Tetrahedron Lett.*, 1981, **22**, 3561.
82. H. J. Reich and J. M. Renga, *J. Org. Chem.*, 1975, **40**, 3313.
83. D. Liotta, C. Barnum, R. Puleo, G. Zima, C. Bayer and H. S. Kezar, III, *J. Org. Chem.*, 1981, **46**, 2920.
84. D. Liotta, M. Saindane, C. Barnum, H. E. Ensley and P. Balakrishnan, *Tetrahedron Lett.*, 1981, **22**, 3043.
85. D. Liotta, G. Zima, C. Barnum and M. Saindane, *Tetrahedron Lett.*, 1980, **21**, 3643.
86. R. Monahan, III, D. Brown, L. Waykole and D. Liotta, in 'Organoselenium Chemistry', ed. D. Liotta, Wiley, New York, 1987, chap. 4.

87. L.-F. Tietze, G. V. Kiedrowski and B. Berger, *Tetrahedron Lett.*, 1982, **23**, 51.
88. J. Blumbach, D. A. Hammond and D. A. Whiting, *Tetrahedron Lett.*, 1982, **23**, 3949.
89. T. G. Bach, in 'Organoselenium Chemistry', ed. D. Liotta, Wiley, New York, 1987, chap. 1.
90. N. Rabjohn, *Org. React. (N.Y.)*, 1976, **24**, chap. 4.
91. D. H. R. Barton, D. J. Lester and S. V. Ley, *J. Chem. Soc., Chem. Commun.*, 1978, 130.
92. D. H. R. Barton, R. A. H. F. Hui, S. V. Ley and D. J. Williams, *J. Chem. Soc., Perkin Trans. 1*, 1982, 1919.
93. T. G. Bach, *J. Org. Chem.*, 1981, **46**, 1442.
94. S. V. Ley, in 'Organoselenium Chemistry', ed. D. Liotta, Wiley, New York, 1987, chap. 3.
95. D. H. R. Barton, C. R. A. Godfrey, J. W. Morzycki, W. B. Motherwell and S. V. Ley, *J. Chem. Soc., Perkin Trans. 1*, 1982, 1947.
96. K. Yamakawa, T. Satoh, N. Ohba, R. Sakaguchi, S. Takita and N. Tamura, *Tetrahedron*, 1981, **37**, 473.
97. D. H. R. Barton, N. J. Cussans and S. V. Ley, *J. Chem. Soc., Chem. Commun.*, 1978, 393.
98. D. H. R. Barton, D. J. Lester and S. V. Ley, *J. Chem. Soc., Perkin Trans. 1*, 1980, 1212.
99. D. H. R. Barton, R. A. H. F. Hui, D. J. Lester and S. V. Ley, *Tetrahedron Lett.*, 1979, 3331.
100. D. H. R. Barton, W. B. Motherwell, J. Wozniak and S. Z. Zard, *J. Chem. Soc., Perkin Trans. 1*, 1985, 1865.
101. L. Engman, *Tetrahedron Lett.*, 1985, **26**, 6385.
102. E. A. Braude, L. M. Jackson and R. P. Linstead, *J. Chem. Soc.*, 1954, 3548.
103. L. M. Jackman, *Adv. Org. Chem.*, 1960, **2**, 329.
104. W. M. Clark, in 'Oxidation–Reduction Potential of Organic Systems', Bailliere, Tindell and Cox, London, 1960.
105. P. Walker and J. D. Hiebert, *Chem. Rev.*, 1967, **67**, 153.
106. H.-D. Becker, in 'The Chemistry of the Quinonoid Compounds', ed. S. Patai, Wiley, New York, 1974, part 1, p. 335.
107. P. P. Fu and R. G. Harvey, *Chem. Rev.*, 1978, **78**, 334.
108. E. J. Agnello and G. D. Laubach, *J. Am. Chem. Soc.*, 1960, **82**, 4293.
109. D. Burn, D. N. Kirk and V. Petrov, *Proc. Chem. Soc. London*, 1960, 14.
110. A. B. Turner and H. J. Ringold, *J. Chem. Soc. C*, 1967, 1720.
111. S. Matsuura, M. Iinuma, K. Ishikawa and K. Kagei, *Chem. Pharm. Bull.*, 1978, **26**, 305.
112. C. G. Shanka, B. V. Mallaiah and G. Srimannarayana, *Synthesis*, 1983, 310.
113. V. V. Kane, *Synth. Commun.*, 1976, **6**, 237.
114. H. J. Ringold and A. B. Turner, *Chem. Ind. (London)*, 1962, 211.
115. D. Caine and J. F. DeBardeleben, Jr., *Tetrahedron Lett.*, 1965, 4585.
116. J. A. Edwards, J. C. Orr and A. Bowers, *J. Org. Chem.*, 1962, **27**, 3378.
117. S. K. Pradhan and H. J. Ringold, *J. Org. Chem.*, 1964, **29**, 601.
118. I. Ryu, S. Murai, Y. Hatayama and N. Sonoda, *Tetrahedron Lett.*, 1978, 3455.
119. I. Fleming and I. Paterson, *Synthesis*, 1979, 736.
120. T. L. Fevig, R. L. Elliott and D. P. Curran, *J. Am. Chem. Soc.*, 1988, **110**, 5064.
121. A. D. Cross, (Syntex Corp.), *Neth. Pat.* 6 503 543 (1965) (*Chem. Abstr.*, 1966, **64**, 5177).
122. A. K. Das Gupta, R. M. Chatterje and M. Paul, *J. Chem. Soc. C*, 1971, 3367.
123. T. Tanaka, K. Mashimo and M. Wagatsuma, *Tetrahedron Lett.*, 1971, 2803.
124. G. Cainelli, G. Cardillo and A. U. Ronchi, *J. Chem. Soc., Chem. Commum.*, 1973, 94.
125. N. Latif, N. Mishriki and N. S. Girgis, *Chem. Ind. (London)*, 1976, 28.
126. E. W. Bousquet, M. D. Moran, J. Harmon, A. L. Johnson and J. C. Summers, *J. Org. Chem.*, 1975, **40**, 2208.
127. M. P. Sammes, M. W. L. Chung and A. R. Katritzky, *J. Chem. Soc., Perkin Trans. 1*, 1985, 1773.
128. D. Rosenthal, P. Grabowich, E. F. Sabo and J. H. Fried, *J. Am. Chem. Soc.*, 1963, **85**, 3971.
129. R. J. Theissen, *J. Org. Chem.*, 1971, **36**, 752.
130. B. Bierling, K. Kirschke, H. Oberender and M. Schulz, *J. Prakt. Chem.*, 1972, **314**, 170.
131. S. Wolff and W. C. Agosta, *Synthesis*, 1976, 240.
132. T. Mukaiyama, M. Ohshima and T. Nakatsuka, *Chem. Lett.*, 1983, 1207.
133. Y. Ito, T. Hirao and T. Saegusa, *J. Org. Chem.*, 1978, **43**, 1011.
134. J. Tsuji and I. Minami, *Acc. Chem. Res.*, 1987, **20**, 140.
135. J. Tsuji, I. Minami and I. Shimizu, *Tetrahedron Lett.*, 1983, **24**, 5635.
136. I. Shimizu, I. Minami and J. Tsuji, *Tetrahedron Lett.*, 1983, **24**, 1797.
137. J. Tsuji, K. Takahashi, I. Minami and I. Shimizu, *Tetrahedron Lett.*, 1984, **25**, 4783.
138. R. Shirai, M. Tanaka and K. Koga, *J. Am. Chem. Soc.*, 1986, **108**, 543.
139. S. Mashraqui and P. M. Keehn, *Synth. Commun.*, 1982, **12**, 637.
140. A. Barco, S. Benetti, G. P. Pollini, P. G. Baraldi, M. Guarneri and C. B. Vicentini, *J. Org. Chem.*, 1979, **44**, 105.
141. A. Barco, S. Benetti, G. P. Pollini and P. G. Baraldi, *Synthesis*, 1976, 133.
142. G. Bianchi and M. De Amici, *J. Chem. Res. (S)*, 1979, 311.
143. A. J. Fatiadi, *Synthesis*, 1976, 133.
144. V. H. Powell, *Tetrahedron Lett.*, 1970, 3463.
145. H. Zimmer, D. C. Lankin and S. W. Horgan, *Chem. Rev.*, 1971, **71**, 229.
146. D. L. Evans, D. K. Minster, U. Jordis, S. M. Hecht, A. L. Mazzu, Jr. and A. I. Meyers, *J. Org. Chem.*, 1979, **44**, 497.
147. D. D. Berge and A. V. Kale, *Chem. Ind. (London)*, 1979, 662.
148. P. G. Ciattini, E. Morera and G. Ortar, *J. Heterocycl. Chem.*, 1982, **19**, 395.
149. R. S. Varma and M. Varma, *Synth. Commun.*, 1982, **12**, 927.
150. S. Mohanty and S. K. Grover, *Chem. Ind. (London)*, 1987, **20**, 721.
151. A. Couture, A. Lablanche-Combier and T. Q. Minh, *Tetrahedron Lett.*, 1977, 2873.
152. M. E. Jung, Y.-G. Pan, M. W. Rathke, D. F. Sullivan and R. P. Woodbury, *J. Org. Chem.*, 1977, **42**, 3961.
153. M. T. Reetz and W. Stephan, *Liebigs Ann. Chem.*, 1980, 533.
154. M. T. Reetz, W. Stephan and W. F. Maier, *Synth. Commun.*, 1980, **10**, 867.

155. A. Schönberg and G. Schütz, *Chem. Ber.*, 1960, **93**, 1466.
156. T. Cohen, C. K. Shaw and J. A. Jenkins, *J. Org. Chem.*, 1973, **38**, 3737.
157. D. D. Miller, K. B. Moorthy and A. Hamada, *Tetrahedron Lett.*, 1983, **24**, 555.
158. T. Fuchikami and A. Yamanouchi, *Chem. Lett.*, 1984, 1595.
159. H. O. House, in 'Modern Synthetic Reactions', 2nd edn., Benjamin, Menlo Park, 1972, chap. 7.
160. R. M. Moriarty, R. Penmasta, A. K. Awasthi, W. R. Epa and I. Prakash, *J. Org. Chem.*, 1989, **54**, 1101.
161. R. M. Moriarty, W. R. Epa, R. Penmasta and A. K. Awasthi, *Tetrahedron Lett.*, 1989, **30**, 667.
162. P. O. Larsson, S. Ohlson and K. Mosbach, *Nature (London)*, 1976, **263**, 796.
163. 'Enzyme Nomenclature, Recommendations (1984) of the Nomenclature Committee of the International Union of Biochemistry', 5th edn., Academic Press, New York, 1984, p. 68.
164. C. J. Suckling and K. E. Suckling, *Chem. Soc. Rev.*, 1974, 387.
165. H. Yamada and S. Shimizu, *Angew. Chem., Int. Ed. Engl.*, 1988, **27**, 622.
166. J. B. Jones, *Tetrahedron*, 1986, **42**, 3351.
167. R. Bowen and S. Pugh, *Chem. Ind. (London)*, 1985, 323.
168. H. J. Ringold, M. Gut, M. Hayano and A. B. Turner, *Tetrahedron Lett.*, 1962, 835.
169. C. J. Sih, *J. Pharm. Sci.*, 1961, **50**, 712.
170. K. B. Sharpless, M. W. Young and R. F. Lauer, *Tetrahedron Lett.*, 1973, 1979.
171. C. Meystre, H. Frey, W. Voser and A. Wettstein, *Helv. Chim. Acta*, 1956, **39**, 734.

2.3
Oxidation Adjacent to C=X Bonds by Hydroxylation Methods

A. BRIAN JONES
Merck Sharp & Dohme Research Laboratories, Rahway, NJ, USA

2.3.1 INTRODUCTION

Many of the natural products of current biological importance and synthetic interest consist of highly oxygenated carbon skeletons. The desire to prepare these compounds and their analogs has led to many impressive advances in synthetic technology. The strategy of constructing the carbon skeleton in simplified form and subsequently installing the remainder of the functionality has many desirable attributes. Clearly the insertion of hydroxy groups (or protected hydroxy groups) α to preexisting functionality is valuable in this sense.

The earliest observations of α-hydroxylation relied on simple autoxidation of particularly susceptible compounds. In recent years the act of deliberate α-hydroxylation has been the subject of much attention and the area has grown to provide an abundance of expedient, rational approaches. This in turn has led to an increase in the use of these technologies in synthesis and a subsequent acceptance into basic synthetic planning.

It is the emergence and use of these techniques that this review is intended to cover and in such a way as to aid the selection of a successful procedure for any particular use.

2.3.2 HYDROXYLATION α TO C=O

2.3.2.1 Saturated Ketones

2.3.2.1.1 Directly from ketone/enol

(i) Transition metal salts

One of the oldest methods for effecting the α-hydroxylation of ketones utilizes transition metal salts,[1] the most widely employed being lead tetraacetate (LTA).[2] Treatment of enolizable ketones with LTA (usually at reflux in acetic acid or benzene) affords the corresponding α-acetoxy derivatives. Originally a radical mechanism was proposed (Scheme 1),[3] but elsewhere it has been suggested that an incipient organolead species is involved prior to conversion to the α-acetoxy derivative by inter- or intra-molecular nucleophilic attack (Scheme 2).

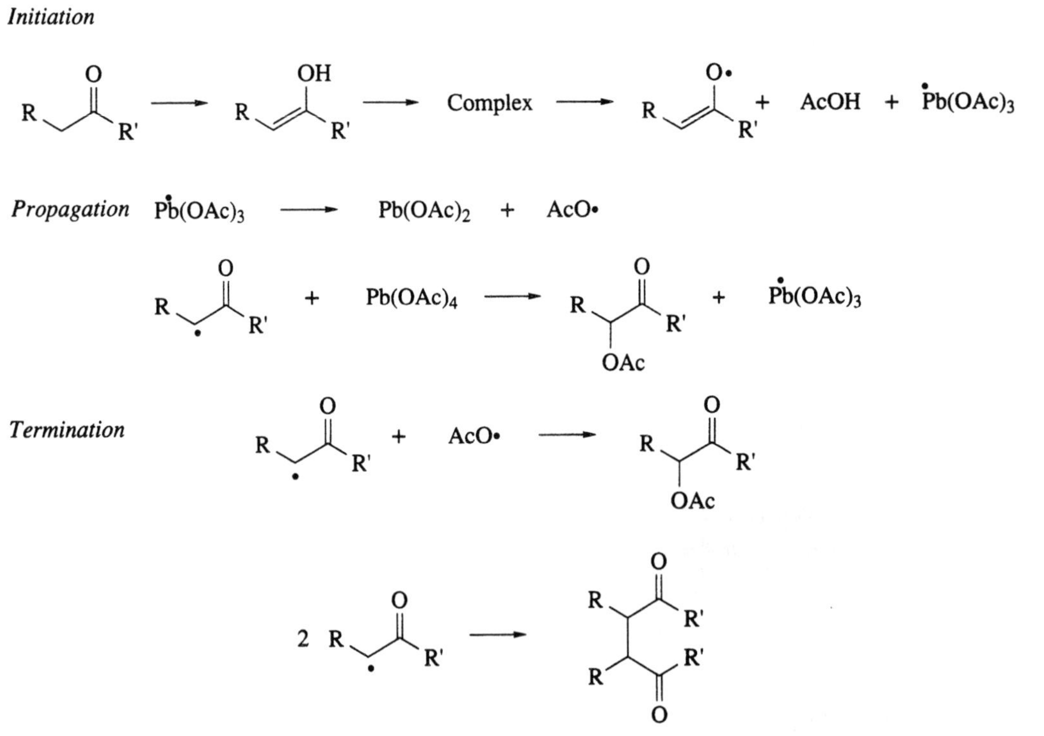

Initiation

RCH₂C(O)R' ⟶ RCH=C(OH)R' ⟶ Complex ⟶ RCH=C(O•)R' + AcOH + $\dot{P}b(OAc)_3$

Propagation $\dot{P}b(OAc)_3$ ⟶ $Pb(OAc)_2$ + AcO•

R$\dot{C}$HC(O)R' + $Pb(OAc)_4$ ⟶ RCH(OAc)C(O)R' + $\dot{P}b(OAc)_3$

Termination

R$\dot{C}$HC(O)R' + AcO• ⟶ RCH(OAc)C(O)R'

2 R$\dot{C}$HC(O)R' ⟶ R'C(O)CH(R)CH(R)C(O)R'

Scheme 1

Alternatively, the reaction may proceed through formation of a lead enolate derived from the enol[4] followed by intramolecular rearrangement[5] to the desired product (Scheme 2). It appears that the first and last mechanisms are operative, the product distribution reflecting a balance between the two dependent on temperature, solvent and substrate. In all three cases however the rate-determining step is the formation of the enol, a factor which heavily influences the choice of reaction conditions.

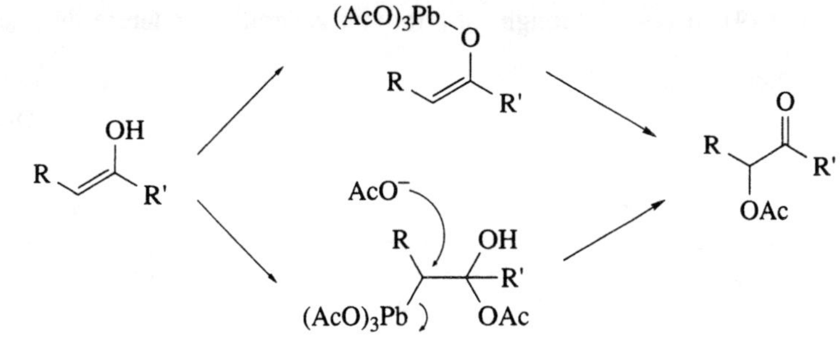
(AcO)3Pb
O
R
R'
OH
R
R'
O
R
R'
OAc
AcO−
R
OH
R'
(AcO)3Pb
OAc

Scheme 2

Numerous examples are available that demonstrate the utility of the process. Sasaki and Eguchi[6] utilized LTA to effect α-acetoxylation of isodihydro-*O*-acetylisophotosantonic lactone (**1**). This oxidation is both chemoselective towards the ketone and regioselective towards the less substituted position (**2**). The opposite regioselectivity has been reported[2] and it does not seem that it can be reliably predicted. In the case of ketone (**1**) the oxidation was stereorandom. This is not an intrinsic problem with the process but rather a reflection of the steric similarity of the enol faces in these systems. Other examples do display stereoselectivity.[2] A recent report described angular acetoxylation of a highly functionalized octahydrobenzofuran[7] producing the stereoisomer indicated in good yield (**3** to **4**). β-Dicarbonyl substrates, α-aryl ketones and β,γ-unsaturated ketones may also be usefully oxidized with LTA.[8] α-Dicarbonyl compounds do not yield simple oxidation products.[9] It is quite possible to effect bisacetoxylation, the second residue being introduced regiospecifically at the α′-position (**5** to **6**).[10] The approach is successful for oxidation of primary, secondary and tertiary centers, but in most cases yields are only moderate.

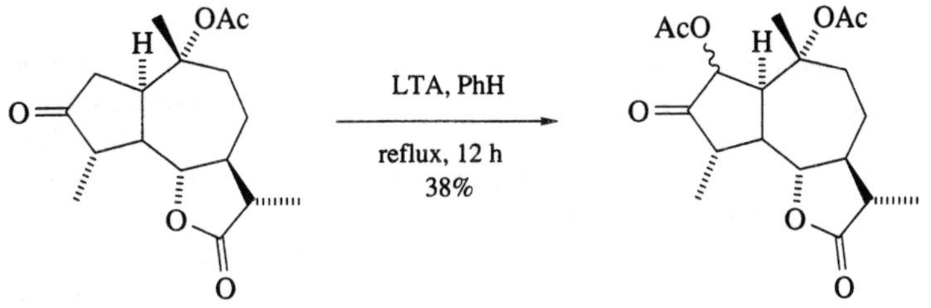
OAc
H
O
O
O
LTA, PhH
reflux, 12 h
38%
AcO
H
OAc
O
O
O

(1) **(2)**

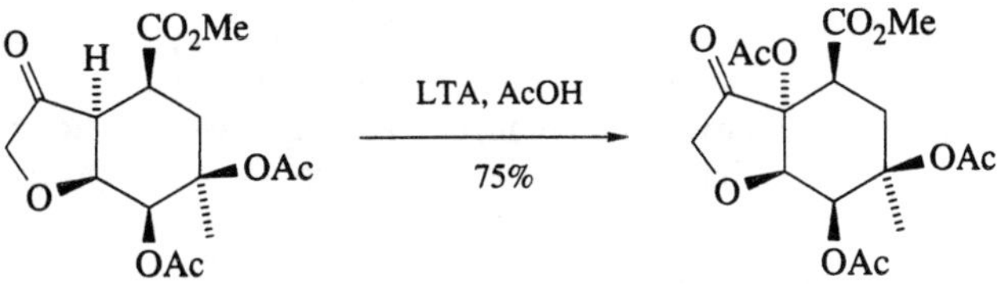
O
H
CO2Me
OAc
O
OAc
LTA, AcOH
75%
O
AcO
CO2Me
OAc
O
OAc

(3) **(4)**

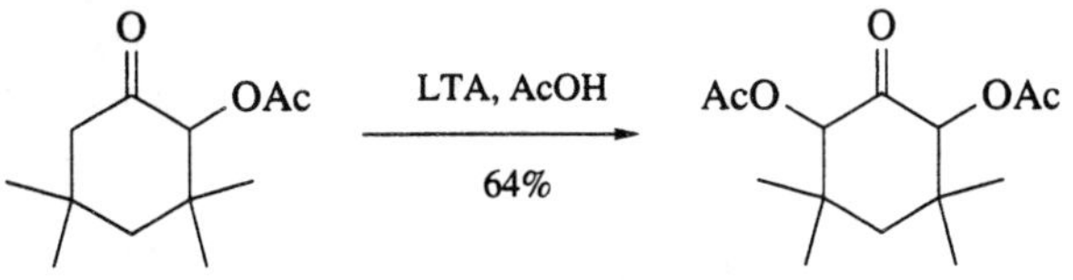
O
OAc
LTA, AcOH
64%
AcO
O
OAc

(5) **(6)**

In an attempt to ameliorate this situation Henbest reasoned[4] that introduction of a Lewis acid would facilitate the rate-determining enol formation. The whole reaction sequence would then be accelerated allowing the use of lower temperatures. This in turn would improve the chemo-, regio- and stereo-selectivity of the process. This proved to be a valid hypothesis.

Reaction of ketones with LTA in benzene in the presence of boron trifluoride etherate at or below room temperature effects rapid α-acetoxylation.[11] The yields are indeed improved in most cases, for example (**7**) to (**8**) and (**9**) to (**10**), although they remain generally moderate despite the remarkable example shown.[11b]

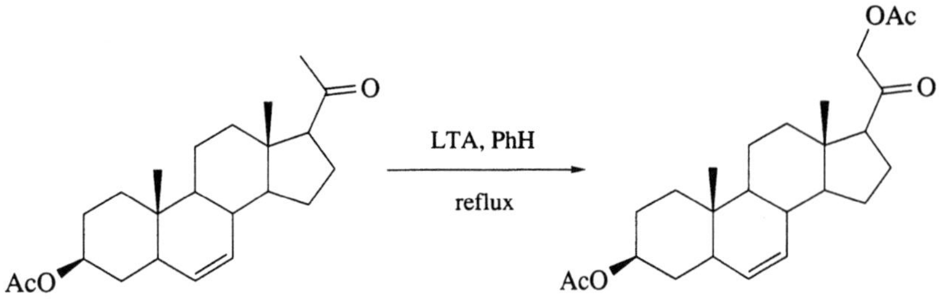

LTA, PhH, reflux

(**7**) → (**8**)

with $BF_3{\bullet}OEt_2$ 71%
without $BF_3{\bullet}OEt_2$ 53%

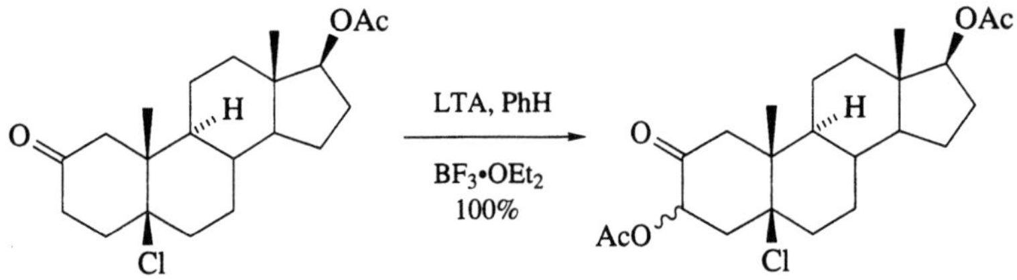

LTA, PhH, $BF_3{\bullet}OEt_2$, 100%

(**9**) → (**10**)

The Lewis acid catalyzed process is not applicable to aryl ketones, where ester products have been observed through aryl migration.[12] Furthermore, anchimeric and solvent effects have been noted[13] during oxidation of a β-carboxy steroidal ketone. α-Oxygenation, although not inhibited, was altered. Despite the frequently disappointing yields the reagent has been widely used, particularly in the steroid field, and remains a useful, if not 'first choice', procedure.

Other transition metal salts mediate in similar oxidations. For example, mercury(II) acetate, a milder reagent than LTA, effects α-acetoxylation[2] through a comparable mechanism. However the corresponding yields for these processes are poor.[14] 3,3-Dimethylcyclohexanone, for example, is oxidized to the α-acetoxy derivative in only 14% yield.[15] The β,γ-unsaturated ketone, isopugelone, exhibits no oxidation at the α- or α′-positions, but affords a product derived from isomerization of the alkene and allylic oxidation.[16] Not surprisingly therefore the reagent has found little synthetic application for this transformation.

Thallium(III) salts also provide a means of α-oxidation. Thallium triacetate, which lies between LTA and mercury(II) acetate in oxidizing power, can induce α-acetoxylation of ketones[2] in hot acetic acid, although again the yields are low.[17] Thallium trinitrate has been shown to produce 2-hydroxycyclohexanone from cyclohexanone in 84% yield. In this case the sequence is thought to involve the intermediacy of the epoxy enol derivative generated through 'oxythallation' of the enol double bond.[19] Subsequent basic hydrolysis produces the required product. Despite the favorable yield, the process does not appear to be widely applicable.[18] Use of the reagent in acetonitrile produces α-nitrato ketones in high yield[20] for both aromatic and aliphatic ketone substrates, although with little or no regioselectivity. The process involves α-thallation of the enol and subsequent intramolecular rearrangement. The α-nitrato ketones can be readily reduced to the α-hydroxy ketones. This apparently efficient procedure has received little attention.

Thallium(III) sulfate can effect a similar α-hydroxylation of straight chain saturated ketones.[21] The vigorous conditions employed together with the apparently limited substrate effectiveness suggests that the procedure will find little synthetic application.

Salts of other transition metals including vanadium, cerium, chromium[22] and manganese[23] have been used for α-oxygenation, although rarely applied in synthesis. Manganese triacetate has been used for the efficient α′-oxidation of enones (Section 2.3.2.2.1.i), but appears not to have been used for the α-hydroxylation of saturated ketones despite its known ability to form the corresponding α-keto radicals.[24] Similarly the use of Lewis acid assisted enolization in the oxidative process appears to have been limited to the LTA-mediated examples.

(ii) Hypervalent iodine reagents

In 1978 Mizukami and coworkers[25] showed that treatment of α-aryl ketones and β-dicarbonyl ketones with phenyliodosyldiacetate in strongly acidic media resulted in their α-acetoxylation (Scheme 3). The process was considered to rest on the coupling of the enol with the iodonium cation generated *in situ*, subsequent nucleophilic attack at the α-position effecting oxidation. The yields for the process were unremarkable. Later Moriarty introduced the use of iodosylbenzene or phenyliodosyldiacetate in basic media (KOH/MeOH) to effect an analogous transformation, yielding the α-hydroxydimethyl acetal.[26] Under these conditions oxidation involves nucleophilic attack of the enol on the iodosyl species.[27] Methoxide addition to the re-formed carbonyl unit results in generation of an epoxide, which is solvolyzed to the observed product (Scheme 4). Since the intermediacy of the reactive iodonium cation is avoided, the reaction can provide improved yields and stereoselectivity. A number of examples are displayed in Scheme 5.[28,29] Noteworthy is the compatibility with the tertiary amine (**11**) and the sulfide (**12**),[28] which are frequently difficult substrates for peroxide-based reagent systems (*vide infra*). Primary and secondary amines are similarly compatible.[28] 1,3-Dicarbonyl substrates are not oxygenated but produce stable iodonium ylides.[26] In some cases work-up is facilitated by the use of 2-iodosylbenzoic acid.[30] Yields are moderate or good.

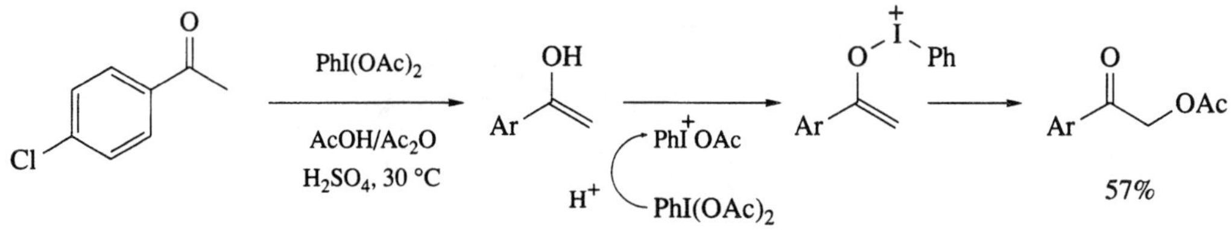

PhI(OAc)$_2$, AcOH/Ac$_2$O, H_2SO_4, 30 °C; $PhI^{+}OAc$; H^+; PhI(OAc)$_2$; Ar; OH; O; Ph; OAc; Cl; 57%

Scheme 3

R; R'; O^-; OH; I; Ar; MeO^-; MeO

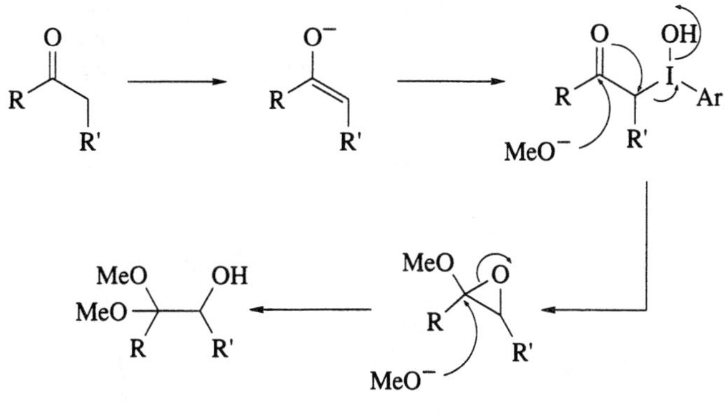

Scheme 4

Stereocontrolled oxidation with these reagents is possible. Thus α-hydroxydimethoxy acetal (**13**) was stereoselectively prepared from the precursor ketone in high yield during the synthesis of (±)-cephalotaxine.[29c] This also demonstrates a useful functional differentiation between ketone and amide carbonyl groups and an interesting although unobvious regioselectivity.

A similar reagent, [hydroxy(tosyloxy)iodo]benzene, has been used to prepare α-tosyloxy ketones, *e.g.* (**14**), from the corresponding ketones.[31a] A similar mechanism is thought to operate except that here the initially formed α-iodo species decomposes to the α-phenyliodonio ketone (the tosyloxy salt has been isolated in one case), which is displaced directly by tosyloxy anion. The yields are generally good for a range of substrates, including β-dicarbonyl systems.

The equivalent reagent for α-mesyloxylation has been reported.[31b] Again yields are high but little or no regioselectivity was observed. The question of stereoselection was not addressed. Most recently an analogous reagent for generating the α-ketophosphate has been reported.[31c]

These reagents provide efficient α-oxidation and their relatively recent emergence will, no doubt, be followed by the expansion of their use in synthesis.

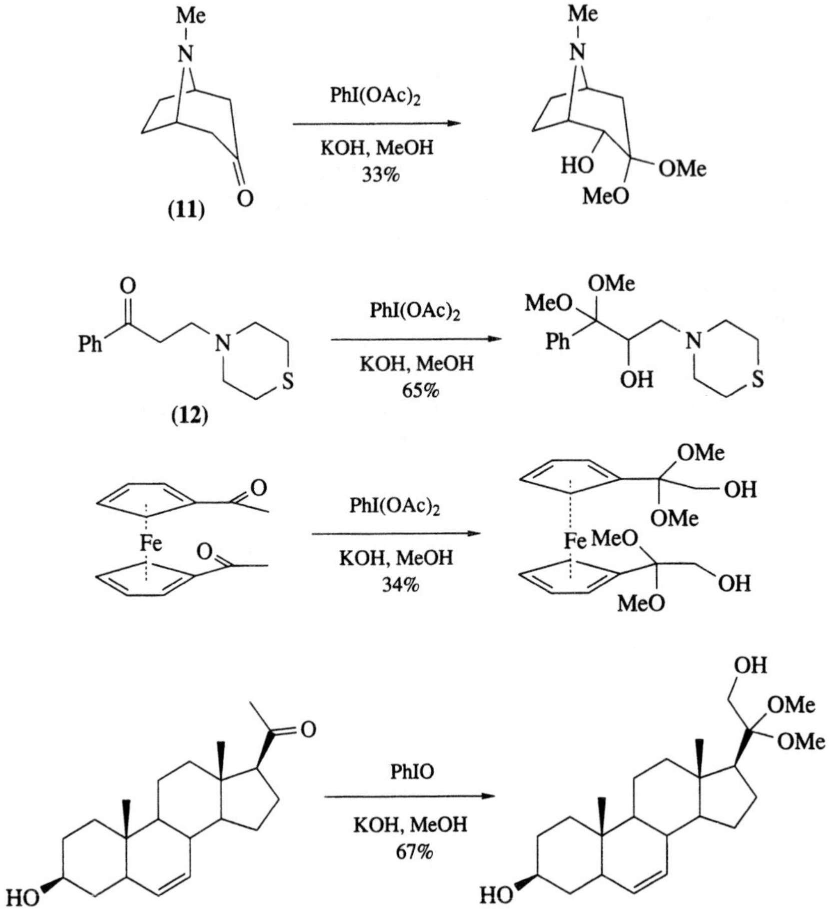

(11)

PhI(OAc)$_2$

KOH, MeOH

33%

(12)

PhI(OAc)$_2$

KOH, MeOH

65%

PhI(OAc)$_2$

KOH, MeOH

34%

PhIO

KOH, MeOH

67%

Scheme 5

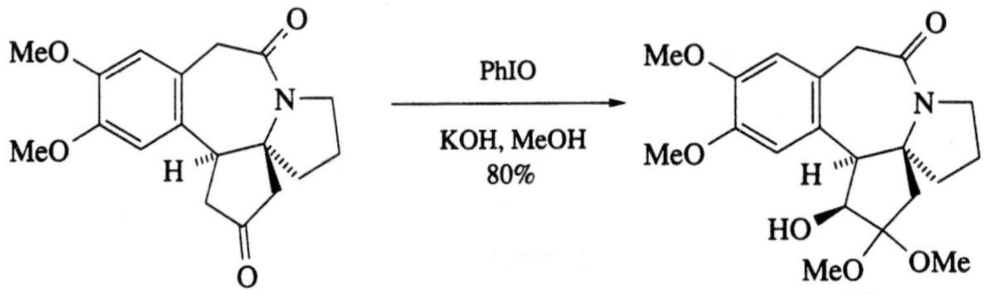

PhIO

KOH, MeOH

80%

(13)

TsO-I(Ph)-OH

MeCN or CH_2Cl_2

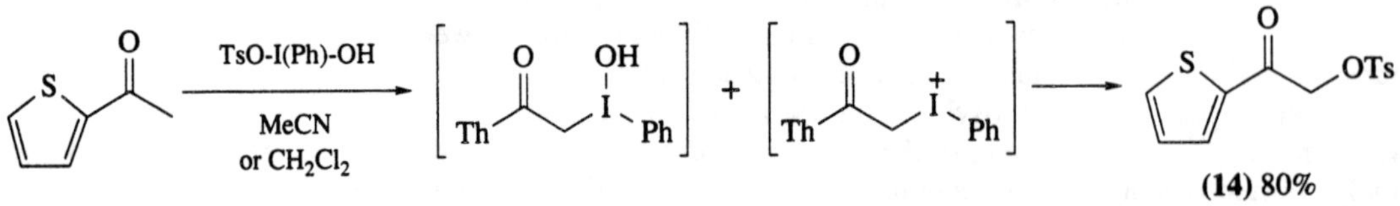

(14) 80%

(iii) Molecular oxygen

It has long been realized that the enol form of ketones can react with molecular oxygen to generate the α-hydroperoxy ketone[32] from which α-hydroxy ketones are readily obtained by reductive work-up. The oxidation was thought to arise either from direct attack of the enol on molecular oxygen or through a radical-mediated process (*vide infra*). Necessarily the oxidation is most efficient where the proportion of enol is enhanced, either as a result of substrate stabilization (for example in β-dicarbonyl or α-aryl ketones), or where the enol is generated as the product of a preceding reaction.

Enslin found that exposure to air of a crude mixture from the hydrogenation of a steroidal enone (**15**) provided an 80% yield of the corresponding α-hydroperoxy ketone (**16**).[32a] A relatively stable enol was formed in this case by the 1,4-addition of hydrogen across the enone. Similarly Crombie demonstrated that α-hydroxylation of an α-aryl ketone, (±)-isorotenone (**17**), could be achieved by simply passing air through an alkaline solution of the ketone.[32e] A number of similar oxygenations have been observed in the tetracycline system[33] involving highly enolized β-diketones. Thus, for example, exposure of ketone (**18**) to oxygen in the presence of platinum or palladium oxide resulted in the formation of the derived hydroxy ketone[33a] as a single stereoisomer. A recently described procedure[34] utilizing potassium superoxide/18-crown-6 and oxygen provides only low yields of the α-hydroperoxy ketones.

H_2/Pt; air

(**15**) (**16**) 80%

KOH, EtOH; air; ~65%

(**17**)

Pt_2O; O_2; DMF

(**18**)

O_2, 50% aq. NaOH; PhMe; $(EtO)_3P$, 5 mol % (**19**); 95%

(S)-(**19a**) 79% *ee*

(**19**)

An excellent extension to these processes is the enantioselective, molecular oxygen mediated α-hydroxylation reported by Shiori.[35] Oxidation in a two-phase system using a chiral phase transfer catalyst (**19**) allowed preparation of α-hydroxy ketones, for example (**19a**), in high yield and with good enantioselectivity. This is the only currently available *catalytic* enantioselective α-hydroxylation process.

It is clear that such mild and efficient techniques can provide a synthetically economic procedure in appropriate cases.

(iv) Miscellaneous

A number of additional methods are available. Among the most useful is the angular hydroxylation by benzeneseleninic anhydride. Oxidation of primary or secondary α-centers produces α,β-unsaturated ketones,[36a] but where the α-center is tertiary, stereoselective hydroxylation is possible, *e.g.* (**20**).[36b] The initially formed 'seleno enolate' undergoes 2,3-sigmatropic rearrangement and subsequent hydrolysis reveals the α-hydroxy ketone in good yield.

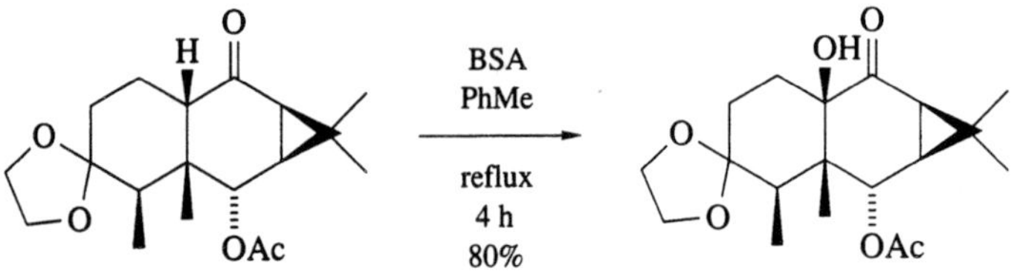

BSA, PhMe, reflux, 4 h, 80%

(**20**)

Treatment of a β-keto ester directly with peracid has been shown in one case (**21**)[37] to effect quantitative α-hydroxylation. Presumably this arises through epoxidation of the enol. Peracid reactions of this kind will be discussed in more detail in Section 2.3.2.1.3.i. Oxidations of the enols of β-keto esters to the α-hydroxy derivatives using singlet oxygen in the presence of fluoride ion occurs in moderate yield through an ene process (Section 2.3.2.1.3.ii).

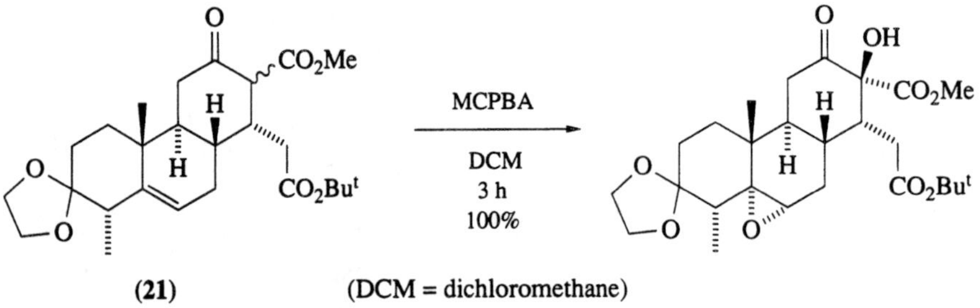

MCPBA, DCM, 3 h, 100%

(**21**) (DCM = dichloromethane)

Hydroxylation using alkali metal based oxidants[32e,33b] (for example $KMnO_4$, $K_2Cr_2O_7$, *etc.*), is possible, although these somewhat harsh reagents frequently give rise to products of overoxidation and are limited with respect to substrate compatibility, particularly when one considers the complex nature of many natural products of current interest.

The peroxy ester reaction[38] provides a method for α-oxygenation, although it is of little synthetic value. The process hinges on the thermal or copper-catalyzed decomposition of a peroxy ester to initiate a radical sequence which ultimately generates and traps an α-keto radical. Yields are very low except for some β-dicarbonyl substrates where relatively efficient conversion is possible. DDQ has been used to effect α-oxygenation in a specific α-aryl case,[39] although the transformation is a reflection of the benzylic nature of the oxidation site. A ruthenium-based electrocatalytic system has been shown to cause α-hydroxylation of cyclohexanone in low yield[40] through a two-electron redox pathway involving hydride transfer. Finally microbial hydroxylation,[41] although usually effecting initial hydroxylation independent of the position of the ketone, has, in some cases, generated α-hydroxy ketones, either as the primary product[43] or through multiple hydroxylation.[42] Neither of these last two methods, although areas of expanding interest, are synthetically useful at the current time.

2.3.2.1.2 Via *preformed enolate*

All the procedures outlined in this section present no dilemma in regioselection, since they may take advantage of the well-documented regiocontrol of enolate formation.

(i) Molecular oxygen

Although autoxidation of enols can effect α-hydroxylation (*vide supra*), attempted oxygenation of ketones in basic media can result in skeletal fragmentation.[44] However the observation that even under strongly basic conditions oxygenation without skeletal alteration could be achieved in some cases[45] provided the basis for what has become a widely used procedure.

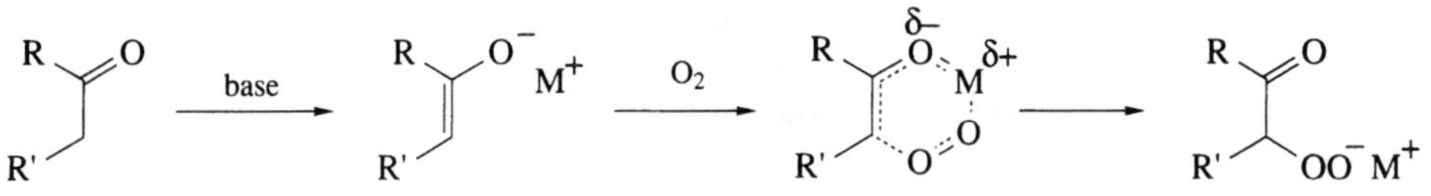

Scheme 6

Two mechanistic rationales have been proposed for this reaction. Electrophilic addition of molecular oxygen to the enolate, activated by counterion complexation in a six-membered transition state, could effect direct oxygenation (Scheme 6).[45] Alternatively oxidation may be thought to proceed through a radical chain mechanism involving single-electron transfer from the enolate to oxygen generating an α-keto radical (Scheme 7).[45] Presumably this process would only require an initiating quantity of the enolate.

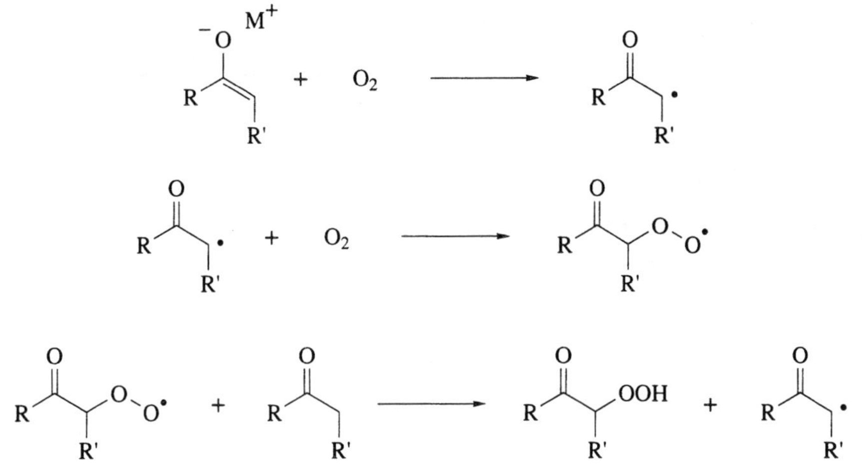

Scheme 7

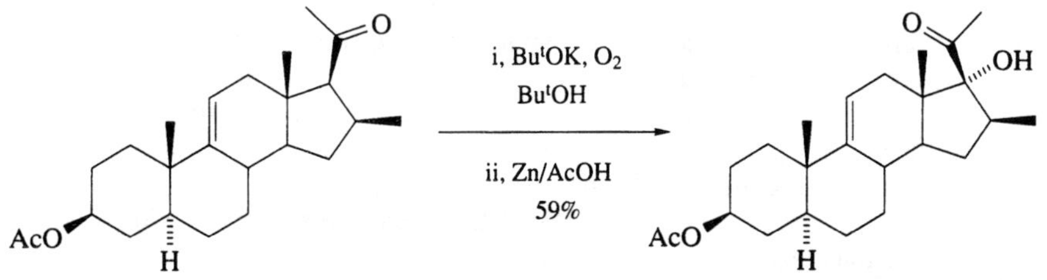

i, Bu^tOK, O_2 Bu^tOH; ii, Zn/AcOH; 59%

(22) **(23)**

The primary product is the α-hydroperoxy ketone. The corresponding alcohol is obtained after reductive work-up. Initially this was achieved using zinc dust in acetic acid, *e.g.* (**22**) to (**23**).[45] Potassium *t*-butoxide was used to generate the enolate in this case[46] and indeed is frequently the preferred base (*vide infra*). Subsequently it was found that the presence of triethyl phosphite in the reaction mixture provided an improved *in situ* reduction.[47] The combination of potassium *t*-butoxide, triethyl phosphite and oxygen in either DMF, *t*-butyl alcohol or monoglyme at temperatures between –30 °C and ambient may be regarded as standard conditions for the process. Where the α-center is primary or secondary, dehydrative overoxidation may occur and the method is generally only viable for the oxidation of tertiary centers, *e.g.* (**24**) to (**25**).[48] Other potential sites of anion formation may also be susceptible, *e.g.* (**26**) to (**27**) and (**28**).[49] Use of alternative combinations of reagents and solvents can promote efficient oxygenation, *e.g.* (**29**) to (**30**), in this case producing a highly enolized β-dicarbonyl substrate used in the total synthesis of (±)-terramycin.[50] Clearly this method is of synthetic value.

Ph N NH2 O O (**24**) — Bu^{t}OK, O_2 / Bu^{t}OH/monoglyme / $(EtO)_3P$, –20 °C / 96% → Ph N HO NH2 O O (**25**)

MeO O Br MeO (**26**) — i, Bu^{t}OK, O_2 DMF / ii, $(EtO)_3P$ → MeO O OH Br MeO (**27**) + MeO O OH Br MeO OH (**28**); (**27**) → (**28**): repeat procedure

(**29**) — i, NaH, O_2 THF, DMF $(EtO)_3P$ / ii, H^+ / 47% → (**30**)

(ii) Molybdenum peroxy complexes

The first report of enolate hydroxylation by reaction with molybdenum peroxy complexes came in 1974 when Vedejs disclosed the use of MoOPH (MoO_5·py·HMPA complex).[51] A later more detailed publication[52] delineated the scope and limitations of the procedure and the advantages of the reagent over other molybdenum peroxy complexes. Molybdenum peroxy complexes, including MoOPH, had previously been prepared and studied with respect to their epoxidation of alkenes.[53–55] MoOPH (**31**) contains two electrophilic bridged peroxy ligands and a single oxo unit. α-Hydroxylation is effected by nucleophilic attack of the enolate at a peroxy oxygen atom. Two modes of attack are possible but the lack of α-hydroperoxy products suggests that the pathway involves only O—O cleavage (Scheme 8). The oxygenation has, in some cases, occurred using less than stoichiometric amounts of MoOPH, indicating that both peroxy bridges may be available for reaction.

In general the ketone enolate is formed and reacted at low temperature (between –50 °C and –30 °C). The preferred base is LDA and gives rise to the kinetic enolate under these conditions. The hydroxylation is frequently found to be sensitive to reaction variables (temperature, stoichiometry, concentration, *etc.*). This contrasts with the less sentient and more reactive ester enolates (Section 2.3.2.4.2.ii. The only noticeable competing reactions are overoxidation and aldol condensation of the product with unconsumed enolate. These processes rarely become noticeable and, where they do, are often significantly diminished by lowering the reaction temperature and/or increasing dilution. Aldol condensation is more

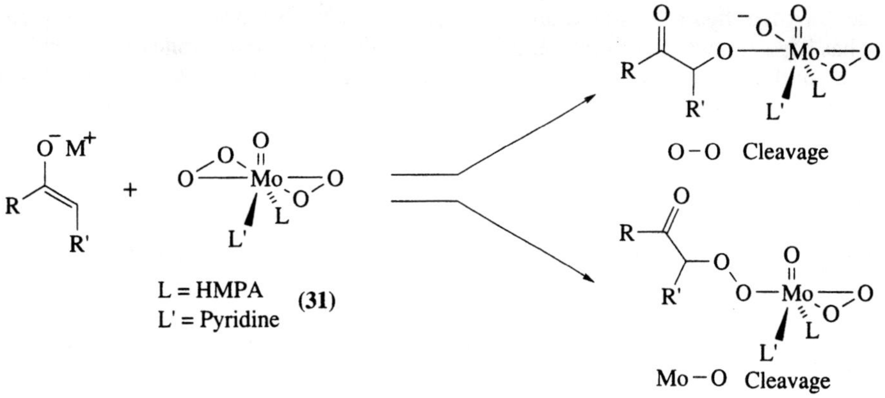

L = HMPA
L' = Pyridine (31)

O–O Cleavage

Mo–O Cleavage

Scheme 8

of a problem where the enolate is unhindered and the process is generally inefficient for methyl ketones, although some improvement is possible by inverse addition of the enolate to MoOPH. β-Dicarbonyl compounds are not hydroxylated.[52]

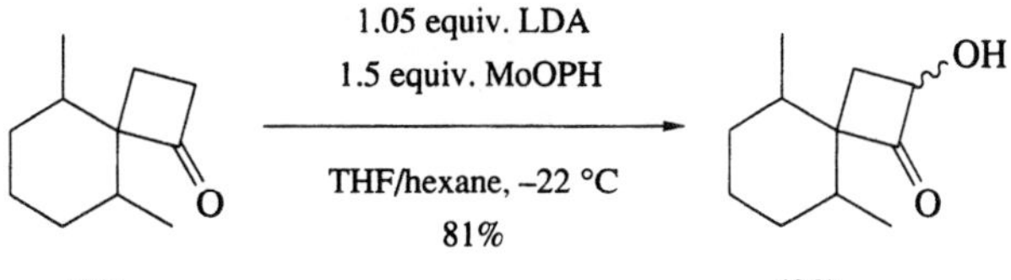

1.05 equiv. LDA
1.5 equiv. MoOPH
THF/hexane, –22 °C
81%

(32) (33)

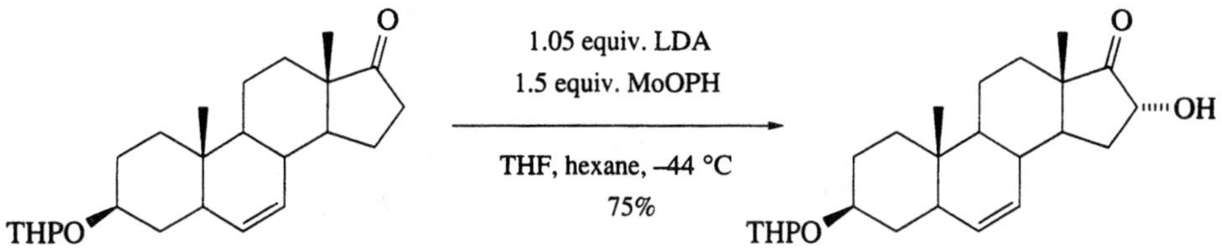

1.05 equiv. LDA
1.5 equiv. MoOPH
THF, hexane, –44 °C
75%

(34) (35)

Simple ketone substrates served to demonstrate the process. Thus bicyclic ketone (**32**) was oxidized at –22 °C to generate a mixture of diastereomers (**33**) in good yield. Good stereoselectivity was observed in the oxidation of steroidal ketone (**34**) to the hydroxy ketone (**35**). Application of the procedure in syn-

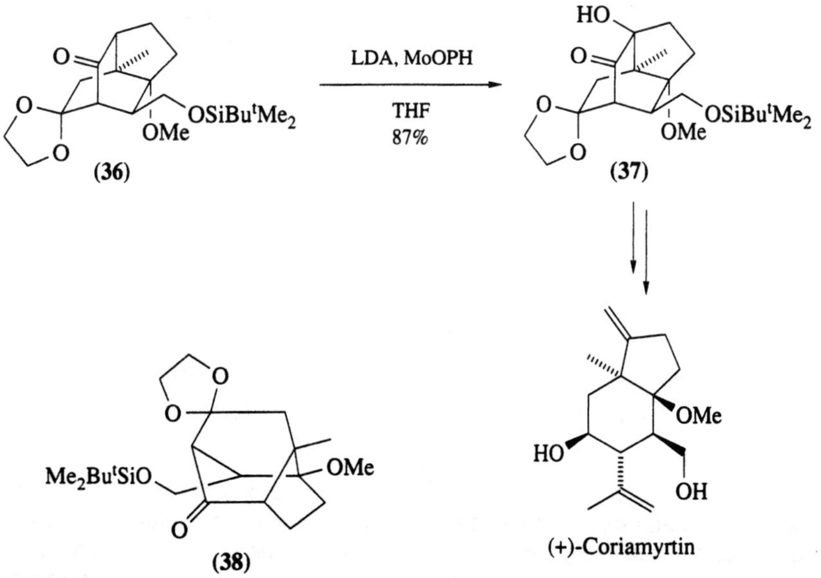

LDA, MoOPH
THF
87%

(36) (37)

(38) (+)-Coriamyrtin

thesis most often involves hydroxylation α to esters[56] or lactones[57] (Section 2.3.2.4.2.ii), although ketone hydroxylation has been applied. For example, bridgehead hydroxylation of ketone (**36**) provided (**37**), an intermediate used in the synthesis of (+)-coriamyrtin and (–)-picrotoxinin.[58] Interestingly, enolization in an anti-Bredt fashion is possible because the cyclohexane is locked in a boat conformation (**38**), with the result that the transoid enolate is effectively generated in a cycloheptane ring.

(iii) 2-Sulfonyloxaziridines

In 1977 Davis reported the synthesis of 2-arylsulfonyl-3-phenyloxaziridines (**39**), the first stable oxaziridines heterosubstituted at nitrogen.[59] The highly electrophilic nature of the ring oxygen in these compounds was soon established.[60–63] That this was due at least in part to the powerfully electron-withdrawing phenylsulfonyl group was equally clear.[64] Reaction of ketone enolates with the reagents produces the α-hydroxy ketone by direct nucleophilic attack on the ring oxygen[61,65] and subsequent β-elimination (Scheme 9). The enolates are generated at –78 °C in THF by treatment with potassium hexamethyldisilazide (lithio bases are less successful) and are reacted and quenched at this temperature.[65] Products from condensation of the enolate with the generated sulfonimine (**40**) have been observed only where the corresponding base was potassium *t*-butoxide. Overoxidation is barely noticeable. The stereoselectivity of the oxidations is generally good and the yields of α-hydroxy ketones are frequently better than those available using MoOPH or molecular oxygen (*vide supra*); for example (**41**) to (**42**), and (**43**) to (**44**).[65] Oxidation of β-dicarbonyl compounds is possible (see Section 2.3.2.2.2.iii).

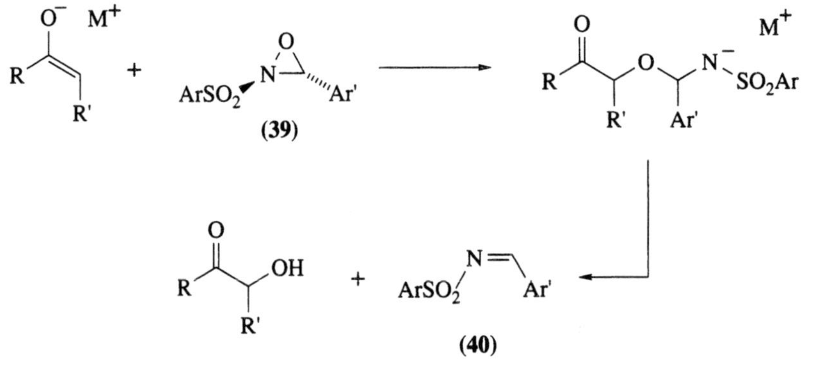

Scheme 9

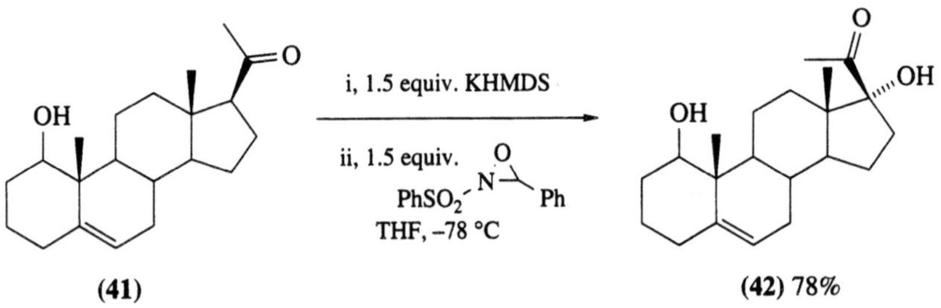

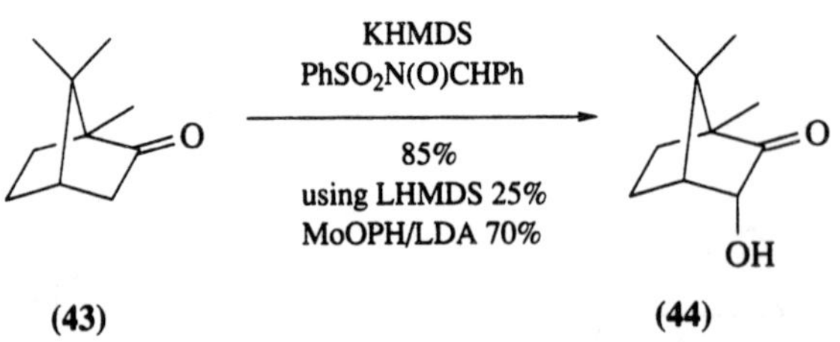

i, 1.5 equiv. KHMDS
ii, 1.5 equiv. $PhSO_2$-oxaziridine-Ph, THF, –78 °C

(**41**) (**42**) 78%

KHMDS
$PhSO_2N(O)CHPh$
85%
using LHMDS 25%
MoOPH/LDA 70%

(**43**) (**44**)

Extension of this procedure to provide a means of asymmetric hydroxylation has been the subject of more recent attention. Initially oxaziridines bearing a camphor-derived residue at nitrogen, for example (**45**), were considered.[66,67] Relatively low levels of chiral induction were achieved[68] and a more rigid compressed system was sought. Camphorsulfonyloxaziridines (**46a** and **b**) were subsequently shown to provide variable although promising enantioselectivities, *e.g.* (**47**) to (**48**).[69] No convincing rationale is available, but it has been suggested that the principle determinant lies with the nonbonded, steric interactions in an open transition state.[69]

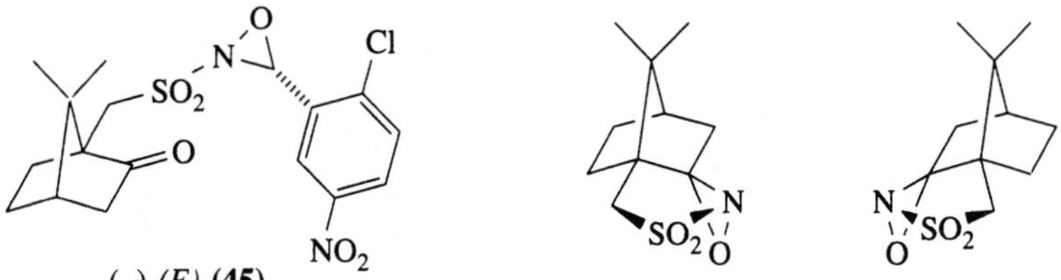

(–)-(*E*)-(**45**)
(also antipode)

(**46a**)

(**46b**)

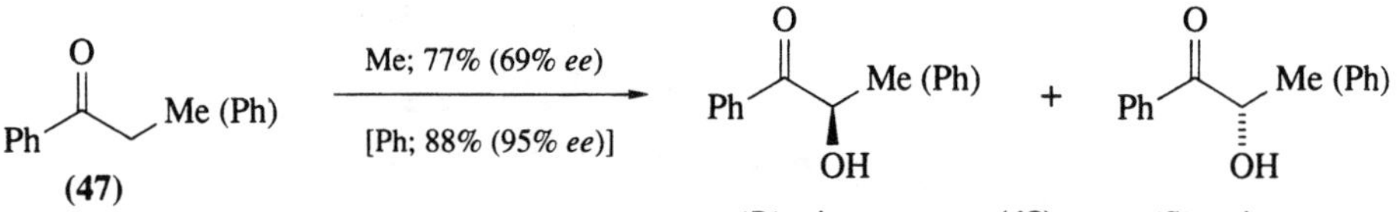

Me; 77% (69% *ee*)
[Ph; 88% (95% *ee*)]

(**47**)

(*R*) minor (**48**) (*S*) major

The relatively recent emergence of this approach to hydroxylation will, no doubt, mature into a well-used facility.

(iv) Miscellaneous

Preformed enolates are susceptible to further methods of oxygenation. For example treatment with LTA in benzene effects α-acetoxylation[70] at lower temperature and more rapidly than the corresponding enol examples. Similarly α-benzoylation using benzoyl peroxide is possible for both lactones[71] and β-keto esters[72] and presumably could be used for less-activated ketones.

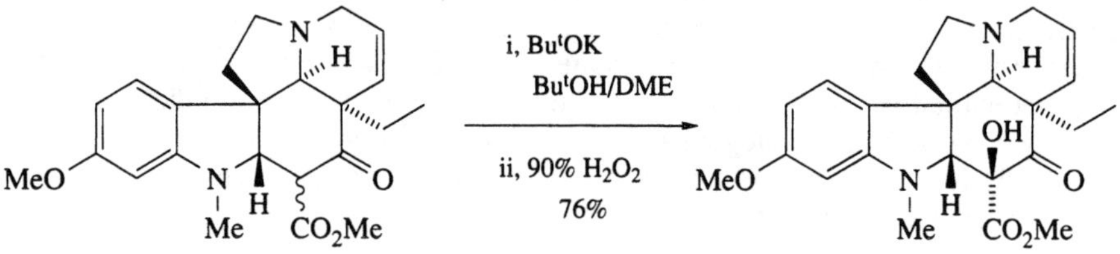

i, Bu^{t}OK Bu^{t}OH/DME
ii, 90% H_2O_2 76%

(**49**) (**50**)

In some cases, where enolate oxygenation with molecular oxygen failed, it has been reported that quenching with 90% hydrogen peroxide allows efficient conversion to the hydroxy ketone, *e.g.* (**49**) to (**50**).[73] Similarly enolate oxidation with organic peracids is possible (*vide infra*). α-Hydroxylation *via* preformed enolates comprises one of most synthetically expedient approaches for achieving this transformation.

2.3.2.1.3 Via *silyl enol ether*

(i) Peracid

Perhaps the most convenient and reliable α-hydroxylation procedure involves treatment of the silyl enol ether[74] of a ketone with organic peracids.[75] Initial epoxidation is followed by silyl migration and generation of the α-silyloxy ketone, which usually forms the hydroxy ketone directly, by rapid hydrolysis (Scheme 10). That epoxides are indeed the intermediates has been demonstrated by their isolation and X-ray characterization.[76,77] The silyl migration may occur by one of two processes involving either an oxacarbenium ion (**51**)[77] or a tight ion pair (**52**).[75] Recent work implicates the former.[76] The process is successful for a range of enol ethers derived from reaction of the enolate with various silylating agents[78] and for various peracid reagents. The synthetic usefulness is enhanced by the efficient regiocontrol arising from selective formation of the kinetic or thermodynamic enol ether.[74]

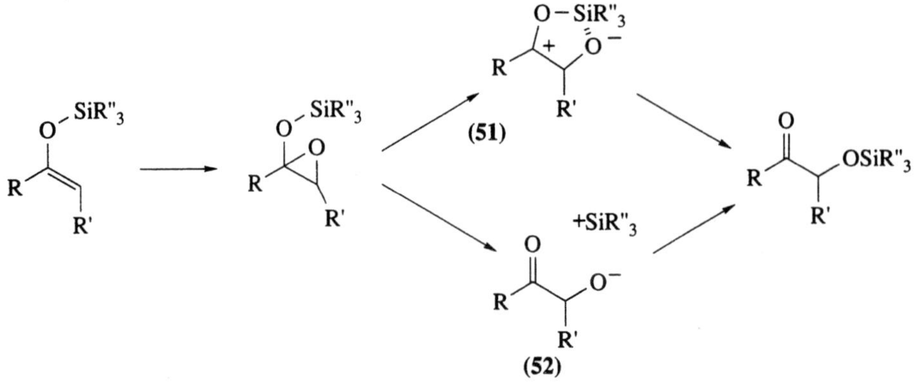

Scheme 10

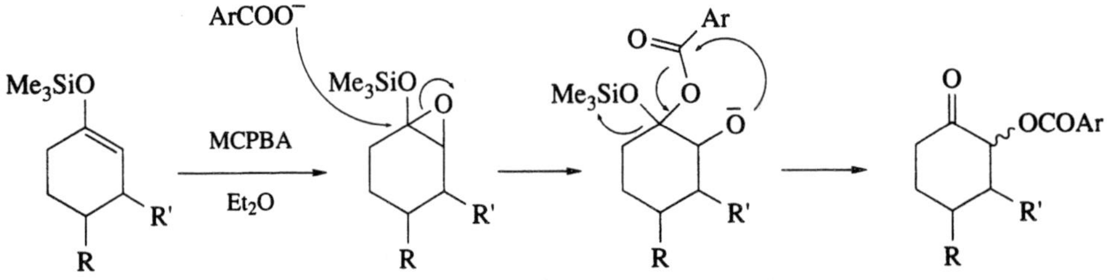

Most frequently the reactions are performed by treating the crude silyl enol ether with MCPBA at 0–25 °C in dichloromethane. Solvent effects have been observed. Thus treatment of enol ether (**53**) with MCPBA in ether resulted in isolation of the benzoate (**54**).[79] This was considered to arise as a result of the increased nucleophilicity of the residual carboxylic acid in ether over that in dichloromethane. Isolation of the silyloxy epoxide by an analogous ethereal oxidation[77] suggests perhaps that the 1,4-silyl migration is intrinsically less facile in this solvent. Generally however the process is efficient and simple substrates are readily oxygenated (Scheme 11).

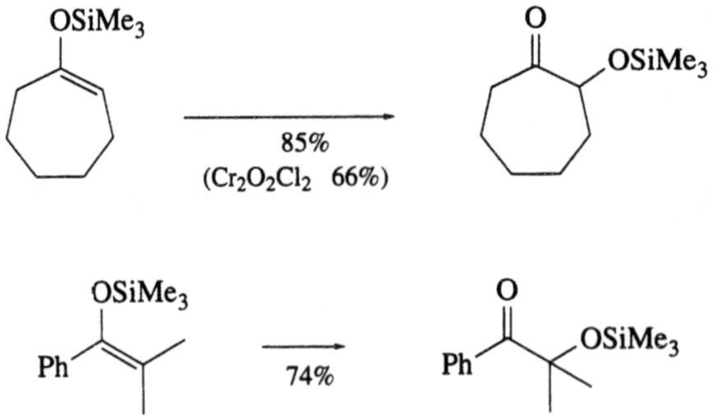

Scheme 11

Synthetic application includes Paquette's recent application in work directed toward the total synthesis of sterpuric acid.[76] Exposure of enol ether (**55**) to peracid provided a single diastereomer of the silyloxy compound (**56**) in good yield. It was from this substrate (**55**) that the first stable trimethylsilyloxy epoxide was obtained (**57**) and examined by X-ray crystallography. Similarly stereoselective oxygenation of β-keto ester (**49**) *via* the corresponding silyl enol ether provided (**50**), also in 76% yield.[80] Lastly efficient and highly stereoselective α-hydroxylation by this method was employed during studies towards the synthesis of helenanolides (**58** to **59**).[81]

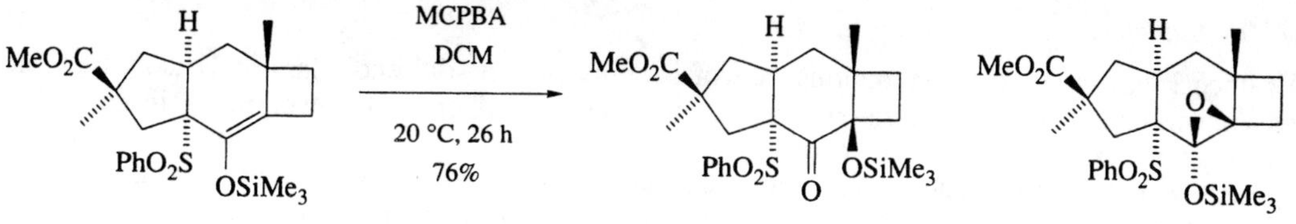

MCPBA DCM, 20 °C, 26 h, 76%

(55) (56) (57)

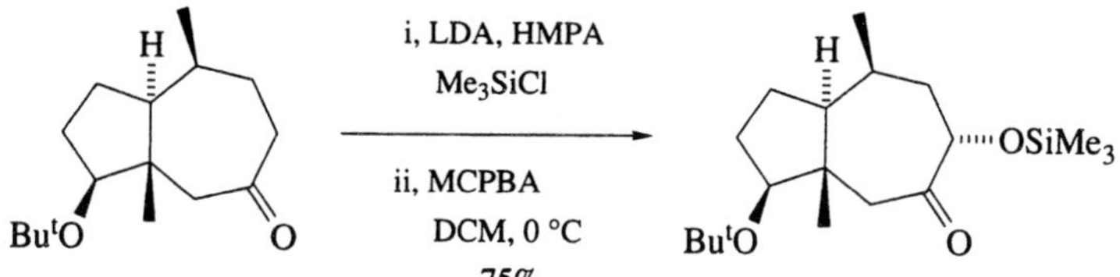

i, LDA, HMPA, Me_3SiCl; ii, MCPBA, DCM, 0 °C, 75%

(58) (59)

(ii) Singlet oxygen

Singlet oxygen has been shown to react with silyl enol ethers in two ways (Scheme 12). Firstly a normal prototropic ene process may occur in a manner analogous to that with isolated alkenes.[82] Secondly a silatropic process cleaving the Si—O bond in a comparable fashion may occur. The latter process gives rise to α-oxygenated products. In general, however, where β-protons are available, the prototropic ene reaction takes precedence.[83,84] Clearly where β-protons are absent the silatropic process is free to run its course. Alternatively, where β-protons are present one could conceive of two situations where the silatropic mode would dominate. The ene process requires the reacting allylic proton to be orthogonal to the plane of the carbon–carbon double bond (coplanar with the π-system).[82] Consequently where allylic protons are not so arranged and are conformationally restricted from attaining such alignment, the silatropic process may be favored. Similarly, where the prototropic ene reaction is inhibited through an increase in strain associated with migration of the alkene the alternative process will again become favorable. That such restrictions are important is apparent from the outcome of the sensitized photooxygenation of silyl enol ether (**60**).[85] The silatropic ene reaction dominates in the presence of an 'unavailable' β-proton.

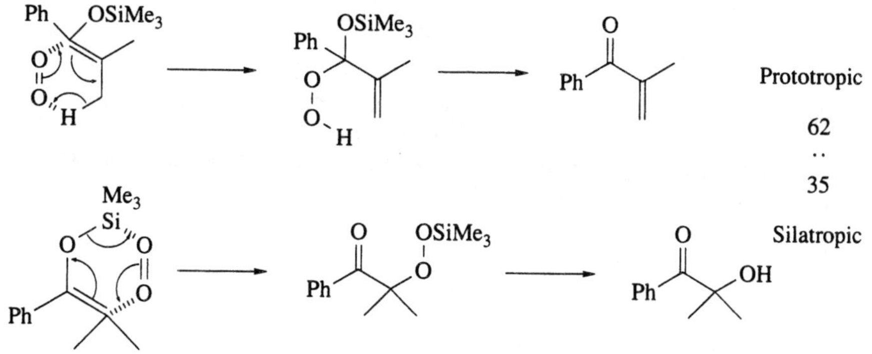

Prototropic 62 : 35 Silatropic

Scheme 12

$^1O_2$, –20 °C, DCM

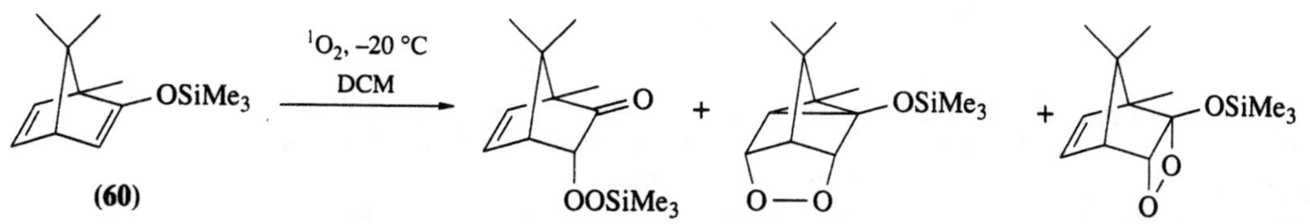

(60) 64% 30% <5%

Clearly then, where examination of substrate conformation suggests poor alignment (or the absence) of allylic protons this mild process may well be viable.

(iii) Miscellaneous

Heathcock has reported an anomalous case of ozonolysis of a silyl enol ether.[86a] Usually these substrates undergo facile oxidative cleavage in the same manner as alkenes. However, in this instance the α-silyloxy ketone (**61**) was obtained in quantitative yield. The intermediacy of a silyloxy epoxide was suggested. A more recent report[86b] has indicated that a similar process is competitive with the simple cleavage reaction, (**63a**) *versus* (**63b**), in the ozonolysis of the steroidal enol ether (**62**).

O_3, MeOH/DCM, −78 °C, 100%; OSiBu^{t}Me$_2$ → OSiBu^{t}Me$_2$

(**61**)

i, O_3, CH_2Cl_2, MeOH; ii, Me_2S

OSiMe$_3$ — OH, O — CHO, CO_2H

(**62**) (**63a**) (**63b**)

Osmium tetroxide-mediated *cis* hydroxylation of a silyl enol ether has been demonstrated to produce the corresponding α-hydroxy ketone in moderate yield after exposure to an acidic work-up,[87] *e.g.* (**64**) to (**65**). The success of the catalytic procedure[88] bodes well for future application and furthermore bears some possibilities for asymmetric hydroxylation.[89]

OSiMe$_3$ — i, OsO_4 cat., NMO, $(Me)_2CO/H_2O$; ii, H^+ (or Bu^{t}OH/py/OsO_4) — OH, O

(**64**) (**65**) 59%

Lee has reported α-hydroxylation through the action of chromyl chloride, *e.g.* cycloheptanone (Scheme 11).[90] The yields were moderately good and no overoxidation was apparent, although the reagent may be of less synthetic value than more mild procedures.

The use of Moriarty's hypervalent iodine system (*vide supra*) has been extended to reaction with silyl enol ethers.[91] In this case a more activated electrophile is required and the reactions are carried out with iodosylbenzene in the presence of boron trifluoride etherate. However, yields are only moderate and the process seems less useful than the corresponding ketone/enol application.

One case has been reported[92] where simple photolysis of a crude silyl enol ether has generated the α-hydroxy derivative (**66** to **67**). This was considered to arise through coupling of the enol ether with photolytically generated silyloxy radicals derived from residual silicon-containing impurities. The presence of benzoyl peroxide, however, failed to provide useful quantities of the α-benzoyloxy ketone.

i, Me_3SiCl, 90 °C; N-(trimethylsilyl)imidazole ($NSiMe_3$); ii, *hν*, 24 h; 60%

(66) → **(67)**

Lead(IV) salts will α-oxygenate enol ethers as they do enols (*vide supra*), although in this case the process involves bisoxygenation of the unsaturated linkage and subsequent hydrolysis. For example, the combination of lead tetrabenzoate and triethylammonium fluoride at 0–25 °C effects efficient α-benzoyloxylation, *e.g.* (**68**) to (**69**).[93] β,γ-Unsaturated ketones are also successfully oxidized, *e.g.* (**70**) to (**71**).[94] The corresponding LTA α-acetoxylations are possible, but the benzoate salt remains the transition metal reagent of choice for these substrates.[74] These reactions appear to be uniformly efficient and perhaps deserve wider synthetic application.

Me_3SiO … $Pb(OBz)_4$, DCM, 0 °C, 30 min → Me_3Si–O, OBz, OBz (F^-) → Et_3NHF → O, OBz

(68) **(69)** 92%

$OSiMe_3$ → i, LTB/DCM; ii, Et_3NHF; 91% → O, OBz

(70) **(71)**

Reagents which effect epoxidation of the enol ether unsaturation effect α-hydroxylation comparable to the peracid approach. Thus a combination of molybdenum hexacarbonyl and *t*-butyl hydroperoxide[95] converts the substrates to α-silyloxy derivatives.[78] The peroxide generated *in situ* from benzonitrile, potassium carbonate and hydrogen peroxide[96] can also perform the oxidation.[78] Molybdenum–peroxy complexes, including MoOPH, could presumably also effect this transformation. Lastly, dimethyldioxirane has been used to epoxidize alkenes and it is likely that application of this useful, debris free, organic peroxide to these reactions will soon emerge.[186]

2.3.2.1.4 Via *enol acetates and alkyl enol ethers*

(i) Peracid

Enol acetates and alkyl enol ethers can be α-hydroxylated through peracid epoxidation in a process analogous to that for silyl enol ethers (*vide supra*). In these cases however the epoxide intermediates are more readily isolable. Acetoxy epoxides, from enol acetates, may be rearranged by the action of heat or acid.[97] Where acid catalyzed, intramolecular rearrangement occurs with retention of configuration at the α-center of the acetoxy epoxide (Scheme 13).[97,98] The thermal rearrangement is thought to involve a slightly different mechanism.[97] Thus enol acetate (**72**) produces the α-acetoxy derivative (**73**) on treatment with perbenzoic acid in benzene.[99] Peracetic acid was less efficient. Numerous other examples are available.[100] Where dienol acetates are utilized, the product is derived from epoxidation of the more nucleophilic, remote unsaturation.[101] Both thermodynamic and kinetic enol acetates may be prepared[102] (although with less precision than the silyl equivalent), allowing a useful degree of regiocontrol.

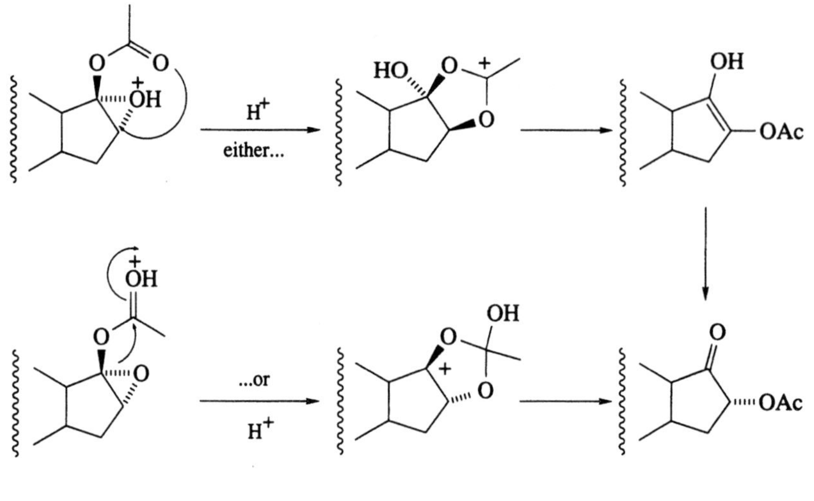

Scheme 13

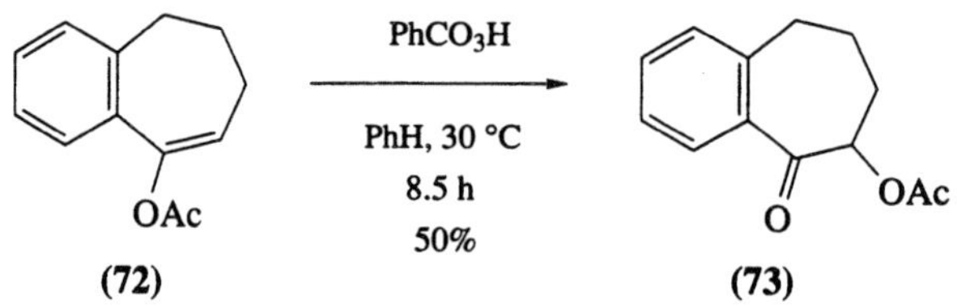

The comparable process for alkyl enol ethers involves participation of solvent,[103] residual peracid[104] or water[105] in cleaving the initially formed epoxide.[106] Thus vinyl ether (**74**) produces the α-hydroxy derivative directly,[105] while (**75**) provides the dimethoxy acetal.[103b]

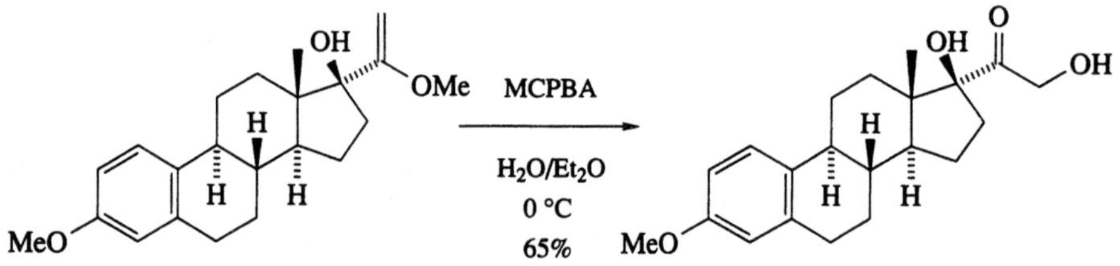

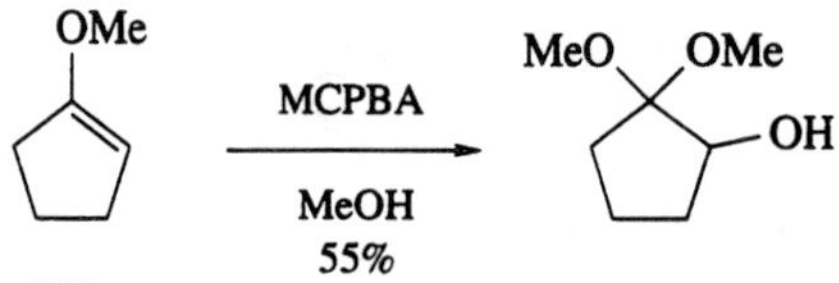

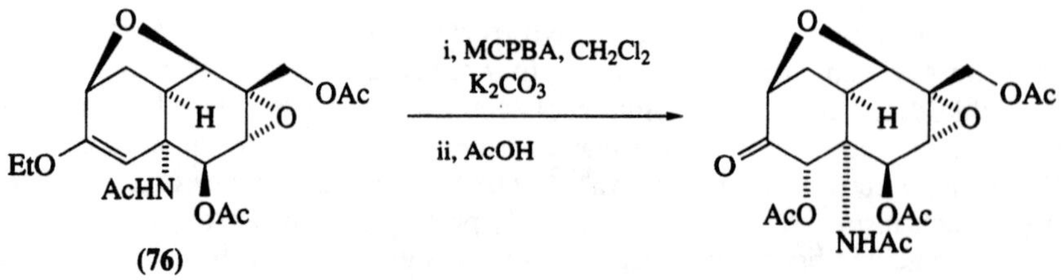

A more complex example is seen in Kishi's tetrodotoxin synthesis.[107] Enol ether (**76**) provided the precursor to α-acetoxy ketone (**77**), which was obtained as a single stereoisomer by acetic acid opening of the initial ethoxy epoxide.

(ii) Singlet oxygen

The discussion outlined above (Section 2.3.2.1.3.ii) for the interaction of singlet oxygen with silyl enol ethers is equally relevant here. Thus the oxygenation pathway competes with the normal ene reaction. The primary work with enol acetates[108] displayed solely the prototropic ene reaction generating the corresponding enones. Subsequent investigations into the reactions of enol acetates bearing less readily available allylic protons revealed the production of α-peroxy ester products derived from a novel acyl migration process (in this case an aldehyde-derived enol acetate was used).[109] Once again, where possible the normal ene process is dominant, although this in itself could provide an indirect oxygenation procedure. For example, if the dominant ene process pivoted on the acetoxy-bearing carbon (Scheme 14), the so-formed enol acetate (**78**)[109] would allow regeneration of the ketone.

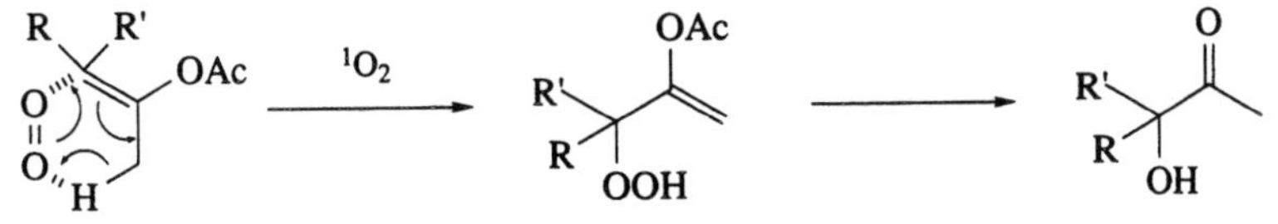

(**78**)

Scheme 14

In the case of alkyl enol ethers the normal ene process competes with solvent incorporated and 1,2-dioxetane products. Here however the ene process seems to be less inevitable when allylic protons are available and the product distribution may be effectively controlled by manipulation of solvent and temperature combinations.[110] Best results are nonetheless achieved where the competitive processes are restricted.[111] Thus enol ether (**79**) produces hydroxylated dimethoxy acetal (**80**) *via* direct incorporation of methanol or through reduction of the 1,2-dioxetane (**81**).

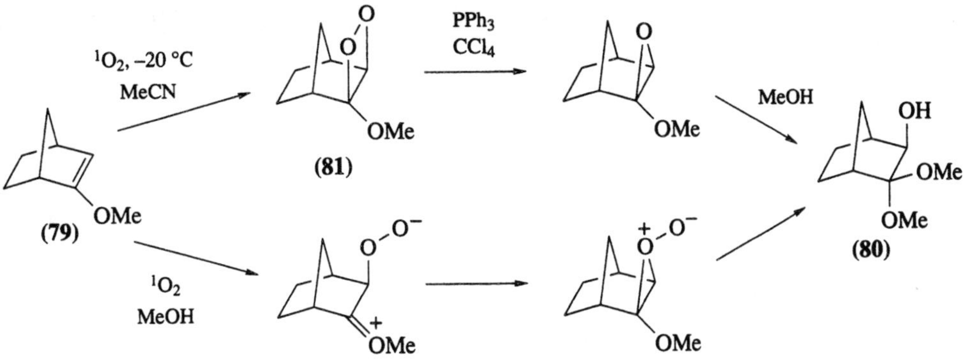

Although perhaps not widely applicable, these processes could find useful application in some instances.

(iii) Miscellaneous

Treatment of enol acetates with LTA in acetic acid affords α-acetoxy ketones.[112,113] For example the tetracyclic substrate (**82**) is converted to the α-acetoxy derivative (**83**) in 95% yield and provides a step in the total synthesis of cycloneosamandione.[114] Vinyl ethers react similarly, suggesting that alkyl enol ethers should follow suit.

α-Arylsulfonyloxy ketones are formed from enol esters (and to a lesser extent from silyl enol ethers) by reaction with arylsulfonyl peroxides[115,116] in methanol at 0 °C. The process involves direct attack on the electrophilic peroxy oxygen atoms and the yields are high.

Reaction of enol acetates with hexamethyldisilyl peroxide in the presence of a Lewis acid (*e.g.* $FeCl_3$, $SnCl_4$ or $BF_3{\cdot}OEt_2$) gives moderate yields of α-acetoxy and α-hydroxy ketones.[117] A similar transforma-

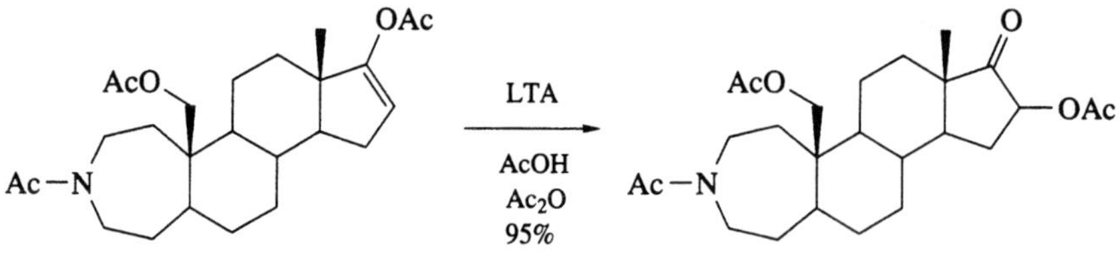

LTA, AcOH, Ac_2O, 95%

(82) **(83)**

tion is possible for some large ring enols. Osmium tetroxide in pyridine converts alkyl enol ethers to the corresponding α-hydroxy ketones[106] although the poor yield may be synthetically restrictive. Finally, electrochemical acetoxylation of enol acetates occurs in moderate yield.[118]

Although some of the procedures encountered in this section are efficient, in general they are less attractive than those using the silyl enol ethers.

2.3.2.1.5 Via *alternative derivatives*

(i) Enamines and enamides

Enamines are readily available ketone derivatives.[119] Exposure of these compounds to certain transition metal salts has been shown to produce α-oxygenated imines which are rapidly hydrolyzed to their ketone counterparts. Thus, for example, morpholino enamines, prepared *in situ*, are α-acetoxylated on treatment with thallium triacetate.[120] The process is thought to involve either direct nucleophilic extraction of an acetate unit or the intermediacy of an organothallium species which subsequently undergoes anchimerically assisted intramolecular acetoxy migration to generate the α-acetoxyimine (Scheme 15).

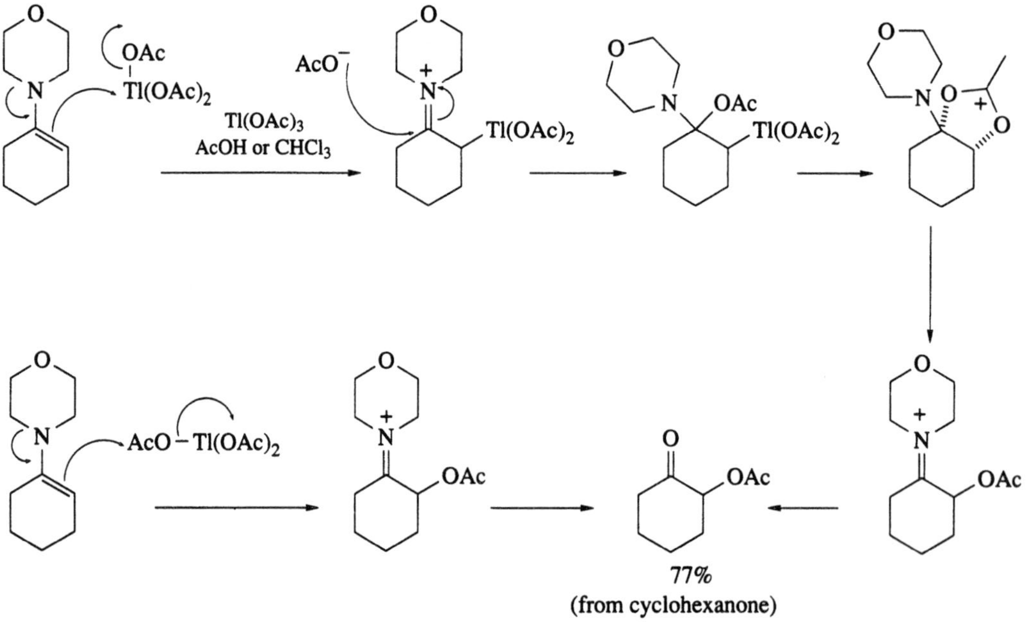

$Tl(OAc)_3$, AcOH or $CHCl_3$

77% (from cyclohexanone)

Scheme 15

Reaction with LTA in benzene generates the bisacetoxy derivative (**84**) analogous to the reaction with silyl enol ethers. Subsequent collapse of the intermediate is however somewhat dendritic (Scheme 16) and consequently of little synthetic value.[121] However it is clear from the nature of the alternative products that enamines possessing no protons at the a-positions could prove to be operable substrates.

Enamides, derived from ketoximes, provide more useful substrates for this procedure.[122] Thus exposure of the relatively stable enamide (**86**) to LTA in benzene affords the α-acetoxyimine (**87**), which can be used, if required, to regenerate the enamide and repeat the process. Alternatively, hydrolysis would re-

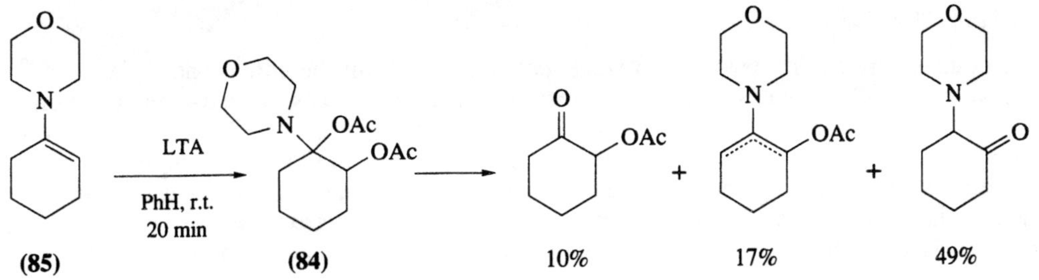

(85) LTA PhH, r.t. 20 min (84) 10% 17% 49%

Scheme 16

veal the monoacetoxy ketone. The overall sequence from the ketone, although efficient, is dissuasively long.

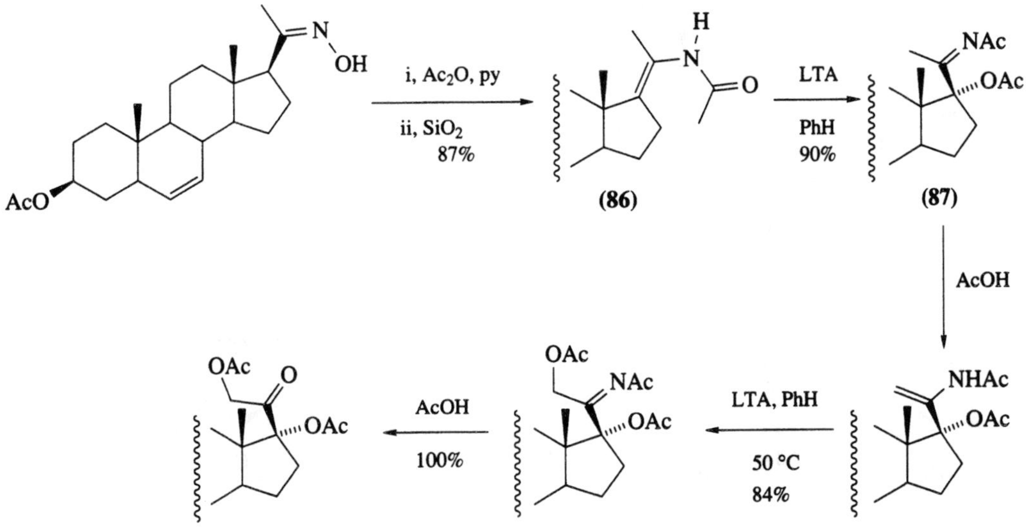

i, Ac_2O, py ii, SiO_2 87% (86) LTA PhH 90% (87) AcOH LTA, PhH 50 °C 84% AcOH 100%

Enamines are susceptible to peracid oxidation, presumably through the epoxide, producing the α-hydroxy ketone after hydrolysis. Thus steroidal ketone (**88**) is converted *via* the pyrollidino enamine to the α-hydroxy derivative (**89**) in approximately 50% overall yield by treatment of the enamine with MCPBA followed by basic work-up.[123] Similar conversion of a steroidal enamide to the α-hydroxy ketone using monoperphthalic acid has been reported.

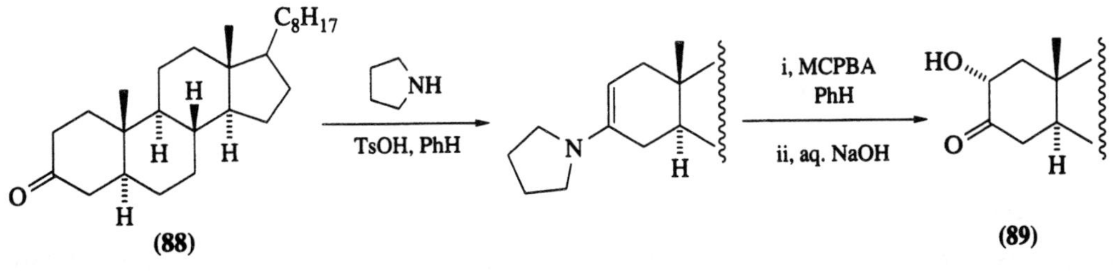

(88) TsOH, PhH i, MCPBA PhH ii, aq. NaOH (89)

The process used in the α-sulfonyloxylation of enol esters (Section 2.3.2.1.4.iii) may be used to regioselectively insert a sulfonyloxy group on morpholino and pyrrolidino enamines.[115,116] Yields of the corresponding ketones are high. Similarly reaction of morpholino enamines with benzoyl peroxide generates the α-benzoyloxy ketones after acidic work-up, although in variable yield (25–82%).[38] A single example describes α-hydroxylation of an enamide by ozonolysis and reductive work-up;[122] the steroidal conversion was achieved in remarkable 94% yield. The generality of the approach is unclear.

(ii) Vinyl cyanides

Vinyl cyanides are readily prepared in single-pot reactions from the corresponding ketones.[124] The conversion is generally very efficient. Transformation of these substrates to the α-hydroxy ketones may be effected in a number of ways. Oxidation with potassium permanganate, although vigorous, provides the required hydroxy ketones, even in the presence of further unsaturation, *e.g.* (**90**) to (**91**).[125] A more recent permanganate-based approach utilizes triphenylmethylphosphonium permanganate at low temperature.[126] The yields are acceptable and conditions are such that a reasonable level of functional group compatibility is achieved, *e.g.* (**92**) to (**93**). Both processes involve initial formation of a cyclic permanganic ester analogous to that involved in the oxidation of isolated alkenes by the reagent. Osmium tetroxide reacts similarly to effect oxygenation.[125] Stoichiometric techniques work reasonably well, while the catalytic process requires the addition of a cyanide trap, in the form of a second transition metal salt (*e.g.* zinc(II) nitrate) to realize the same efficiency, *e.g.* (**94**) to (**95**).[130] Nonetheless the overall sequence provides a useful method.

OAc, CN, O, O, O — $KMnO_4$ → OAc, O, OH, O, O, O

(**90**) (**91**)

CN, O, O, OMe, O — $Ph_3\overset{+}{P}Me\ \overset{-}{MnO_4}$, DCM, −78 °C, 66% → O, OH, O, O, OMe, O

(**92**) (**93**)

CN, OMe — cat. OsO_4, 65% → O, OH, OMe

(**94**) (**95**)

(iii) Vinylsilanes

Vinylsilanes may be prepared from the corresponding ketones by formation of the hydrazone followed by Shapiro reaction, quenching the vinylic anion with chlorosilanes.[127,128] An equally effective process derives the vinylsilane from the vinyl chloride, in turn prepared simply from the ketone.[128] The crucial oxidative transformation may be achieved in two ways. Firstly, ozonolysis in dichloromethane/methanol at approximately 0 °C followed by reductive work-up affords the α-hydroxy ketone, *e.g.* (**96**) to (**97**), *via* the intermediates (**98**) and (**99**). The outcome of the reaction varies with the solvent and work-up conditions, but using the combination indicated, good yields of the desired products are available.

Alternatively if an alkoxysilane is used, a second oxidation method is applicable.[129] Thus epoxidation of the vinylsilane (**100**) and oxidative cleavage of the crude silyl epoxide (**101**) provides a good yield of the product (**102**). The conversion (**101**) to (**102**) involves peroxidation of the silylalkoxy group and coni, O_3, MeOH/DCM −25 °C; ii, Me_2S

(96) → **(98)** → **(99)** → **(97)** 73%

sequent oxiranyl migration. Fluoride-induced fragmentation then reveals **(102)**. The process is ineffective for trialkylsilanes.

i, MCPBA, DCM, 5 h, 0 °C; ii, H_2O_2, KHF_2, $KHCO_3$, MeOH/THF, 3 h, r.t.

(100) → **(101)** → **(102)** 74%

(iv) Vinyl sulfides

Vinyl sulfides are readily prepared directly from the ketone[130,131] in good yield. Exposure of these derivatives to a single equivalent of ozone provides the hydroxylated, alkene-migrated vinyl sulfide, **(103)** to **(104)**, which may be hydrolyzed to the α-hydroxy ketone. Epoxy sulfide **(105)** is instrumental in the conversion, although alternative epoxidation methods fail to epoxidize without oxidizing at sulfur. It could prove possible however to α-oxygenate through the derived epoxy sulfones in a manner reported to occur to generate α-hydroxy aldehydes.[132]

O_3, DCM, Py, −70 °C

(103) → **(105)** → **(104)** 60%

LTA reacts with enol sulfides,[133] to produce thionium ions, *e.g.* **(106)**, and thence allylic acetates **(107)** or bisacetoxylated products **(108)**, in good yields. Presumably either of these compounds could be hydrolyzed to the α-acetoxy ketone.

LTA

(106) → **(108)** and/or **(107)**

2.3.2.2 α,β-Unsaturated Ketones: sp^3 Center

2.3.2.2.1 *Directly from ketone/enol*

(i) Transition metal salts

The majority of procedures outlined in Section 2.3.2.1 are applicable to these substrates through comparable mechanisms. Thus LTA played a dual role in the total synthesis of pyroangolensolide;[134] firstly α′-acetoxylation (**109**) and subsequently carbon–carbon bond cleavage (**110**). Numerous other examples of this process are available, *e.g.* (**111**)[81b] and (**112**).[135,140a] In some cases, however, α-oxidation (rather than α′-oxidation), occurs with alkene deconjugation, although this may in some cases be circumvented by using an alternative reagent (compare refs. 136 and 137). The α′-oxidation product is generally formed regardless of enol distribution.[138]

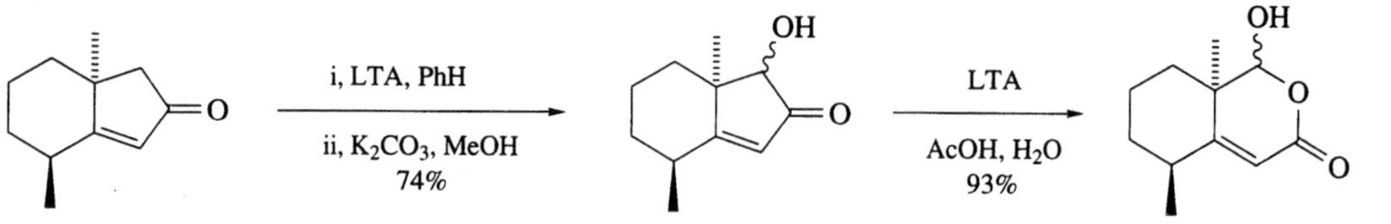
i, LTA, PhH
ii, K_2CO_3, MeOH
74%
OH
O
LTA
AcOH, H_2O
93%

(**109**) (**110**)

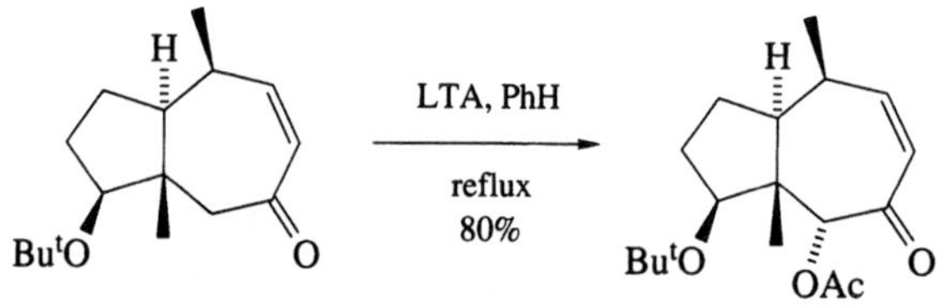
H
Bu^{t}O
O
LTA, PhH
reflux
80%
OAc

(**111**)

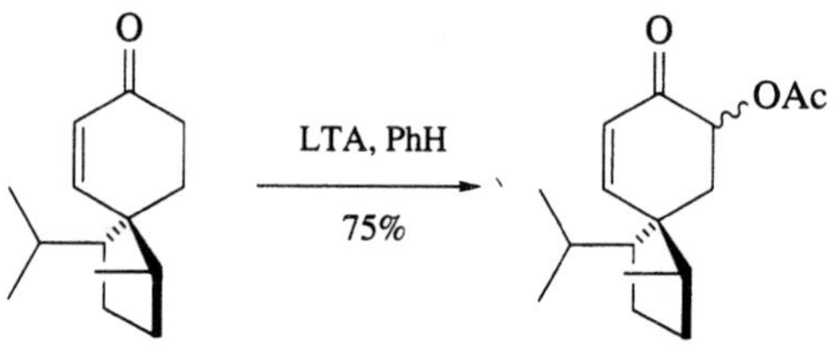
O
LTA, PhH
75%
OAc

(**112**)

Manganese triacetate has been specifically reported as a reagent for α′-oxidation.[139] Mechanistic dualism analogous to LTA (*vide supra*) is observed, although the radical process may be more dominant. Watt and coworkers used this technique during the synthesis of quassinoids.[140] Enone (**113**) was con-

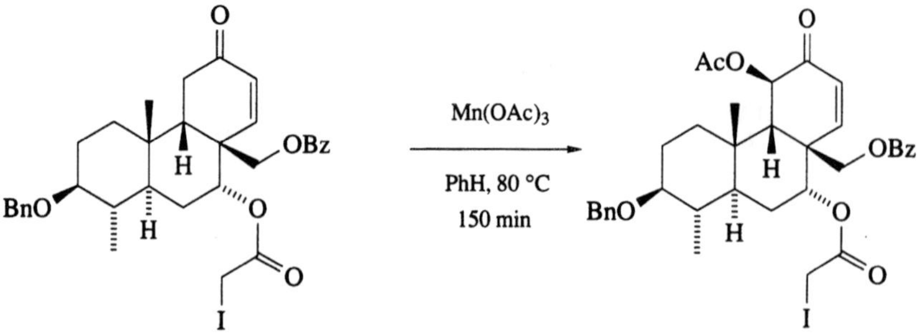
O
OBz
H
BnO
H
I
$Mn(OAc)_3$
PhH, 80 °C
150 min
AcO

(**113**) (**114**)

verted in high yield to the acetoxy derivative (**114**). Despite the potentially radical nature of the reaction, the alkyl iodide (primary, but α-keto) remained intact. That radicals were indeed involved was demonstrated in this system by intramolecular α′-oxygenation; a direct result of trapping of the α′-keto radical by a suitably positioned alcohol (**115** to **116**).

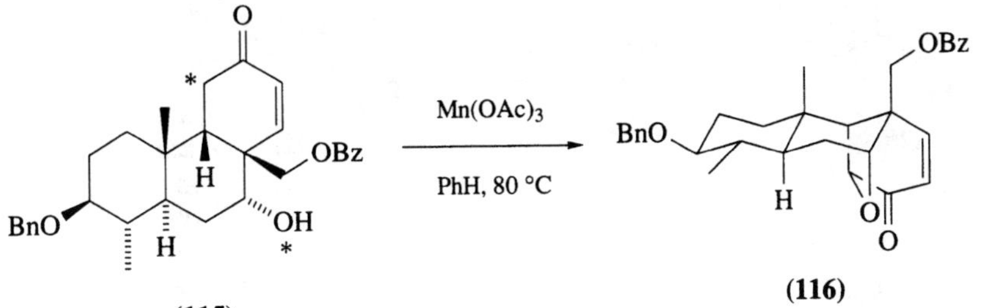

Mn(OAc)$_3$, PhH, 80 °C

(**115**) (**116**)

α′-Acetoxylation of 2,3-dihydro-4-pyrones, *e.g.* (**117**), with this reagent proceeded in moderate yield under similar conditions to give stereochemically pure products.[141]

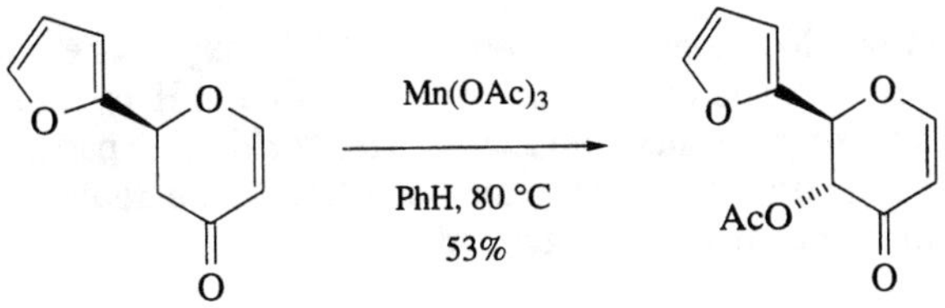

Mn(OAc)$_3$, PhH, 80 °C, 53%

(**117**)

(ii) Miscellaneous

Where the dienol form of an α,β-unsaturated ketone is available, autoxidation giving the α′-hydroxy ketone through the α′-hydroperoxide is possible, as seen for saturated ketones (see Section 2.3.2.1.1.iii). Benzeneseleninic anhydride (see Section 2.3.2.1.1.iv) effects α′-hydroxylation at tertiary centers, *e.g.* (**118**) to (**119**), again in the same manner as for saturated ketones.[36b]

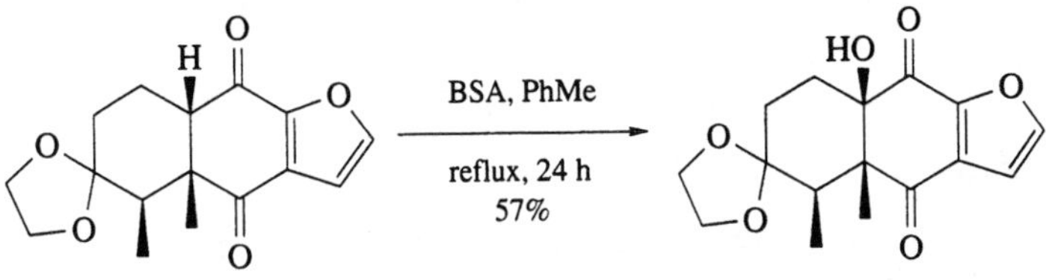

BSA, PhMe, reflux, 24 h, 57%

(**118**) (**119**)

2.3.2.2.2 Via *preformed enolate*

(i) Molecular oxygen

These substrates are as readily susceptible to hydroxylation as saturated ketones (Section 2.3.2.1.2.i). Thus enone (**120**) was oxygenated *via* the corresponding sodium enolate.[142] Exposure to oxygen and *in situ* reductive work-up provided (**121**) in moderate yield during the synthesis of (±)-deoxyaspidodispermine. Similarly (±)-kjellmanianone (**123**) was prepared through oxygenation of the potassium enolate of the corresponding deoxy substrate (**122**). The enolate was derived by the action of a less orthodox base.[143]

(ii) Molybdenum peroxy complexes

MoOPH may be used to α′-hydroxylate lithium enolates of α,β-unsaturated ketones (*cf.* Section 2.3.2.1.2.ii), although the conversion is less efficient than the equivalent process with saturated sub-

(120) → (121) + epimer 5%
$Bu^tOH/NaOH$, DMF, O_2, $P(OEt)_3$, 51%

(122) → (123)
KF, 18-crown-6, O_2, DMSO, $P(OEt)_3$, 3 h, r.t., 60%

strates. By-products formed by aldol condensation of the α′-hydroxy enone with unreacted enolate are more significant here and inverse addition of the enolate to MoOPH is required in order to achieve usable yields.[52] Thus enone (**124**) gives alcohol (**125**) in 52% yield. This particular example also suffers the handicap of being a less favorable methyl ketone substrate (*vide supra*) and the reasonable yield bears testimony to the ameliorating effects of inverse addition.

(124) → (125)
LDA, MoOPH, THF, –20 °C, 52%

(iii) 2-Sulfonyloxaziridines

Although few examples of the oxidation of α,β-unsaturated ketones with these recently established reagents (Section 2.3.2.1.2.iii) have been reported, the application is clearly plausible. The use of a chiral camphor-derived oxaziridine to effect this process has been reported.[67] Thus (+)-kjellmanionone (**123**) was prepared by treatment of the precursor (**122**) with oxaziridine (**45**) in THF at –78 °C. Although the yield in this case was only moderate, it would be unwise to generalize at this stage.

(iv) Miscellaneous

Racemic kjellmanionone (**123**) has also been prepared by direct oxidation of the potassium enolate of the ketone (**122**) with MCPBA[67] and, if general, this would represent a convenient procedure. Oxidation of ketones with benzoyl peroxide is of no synthetic value (*vide supra*), but the process becomes useful if enolates are employed. Treatment of enone (**126**) with LDA in THF at 0 °C followed by quenching with

(126) → (127)
i, LDA, THF; ii, $(PhCOO)_2$, –10 °C, 50%

benzoyl peroxide at −10 °C produced hydroxy ketone (**127**).[144] The initial product, the α′-benzoyloxy ketone, is rapidly hydrolyzed. In this case the use of MoOPH was unsuccessful.

It has not been established whether enone-derived enolates are oxidized by LTA (*cf.* Section 2.3.2.1.2.iv), but the success of such a process seems likely.

2.3.2.2.3 Via *silyl dienol ethers*

(i) Peracid

Silyl dienol ethers are readily available.[74] α′-Hydroxylation through the action of peracids is facile (see Section 2.3.2.1.3.i).[145,146] Hence treatment of simple dienol ether (**128**) with MCPBA followed by fluoride-induced fragmentation of the unisolated epoxides gives α′-hydroxy ketone (**130**) or, in the case of (**129**), in the presence of acetylating agents, α′-acetoxy ketone (**131**). Oxidation of more adventurous substrates has been achieved. Thus, for example, dienol ether (**132**) gives the α′-hydroxy derivatives in 75% yield through the same process, although in this case with poor stereocontrol.[147] The tricyclic substrate (**133**) is similarly oxidized, albeit in surprisingly low yield.

OSiMe$_3$ (**128**) —i, MCPBA; ii, Et$_3$NHF, DCM; 87%→ (**130**) O, OH

OSiMe$_3$ (**129**) —i, MCPBA; ii, Et$_3$NHF, Et$_3$N, Ac$_2$O; 89%→ (**131**) O, OAc

Me$_3$SiO, Bu^{t}Me$_2$SiO, O, O, OSiBu^{t}Me$_2$ (**132**) —MCPBA; 75%→ O, HO, Bu^{t}Me$_2$SiO, O, O, OSiBu^{t}Me$_2$

2:1 mixture of diastereoisomers; major isomer isolated as cyclic ether

Me$_3$SiO, OBz, O (**133**) —i, MCPBA; ii, HCl, THF; 50%→ O, OH, OBz, O

(ii) Singlet oxygen

While the reaction of singlet oxygen with silyl enol ethers was governed by competing prototropic and silatropic ene processes (see Section 2.3.2.1.3.ii), the interaction with dienol ethers displays a different mode of reactivity. Singlet oxygen generated from triphenyl phosphite ozonide at low temperature undergoes a [4 + 2] cycloaddition with the diene unit.[82] Reductive cleavage of the so-formed endoperoxide and dehydrative reconstitution of the enone generates the α′-hydroxy enone (Scheme 17).[148]

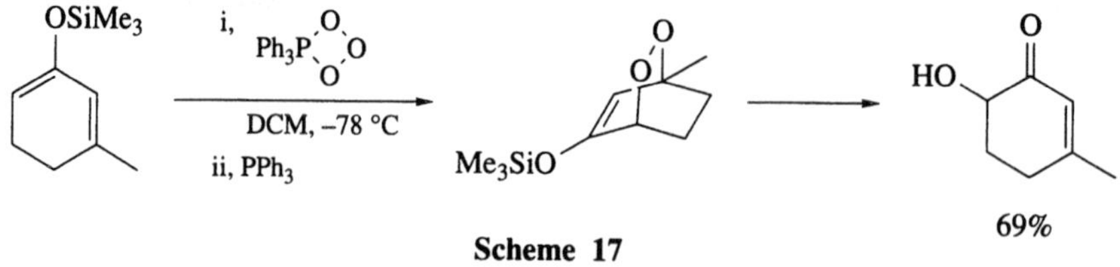

OSiMe3 · i, Ph3P(O3), DCM, −78 °C; ii, PPh3 → Me3SiO (endoperoxide) → HO (α′-hydroxy enone) 69%

Scheme 17

The technique has been applied to the total synthesis of (±)-oxylubimin (**134**) through hydroxylation of dienol ether (**135**).[148] The stereoselectivity is good, particularly when compared to the equivalent selectivities obtained using MCPBA, MoOPH or manganese triacetate.

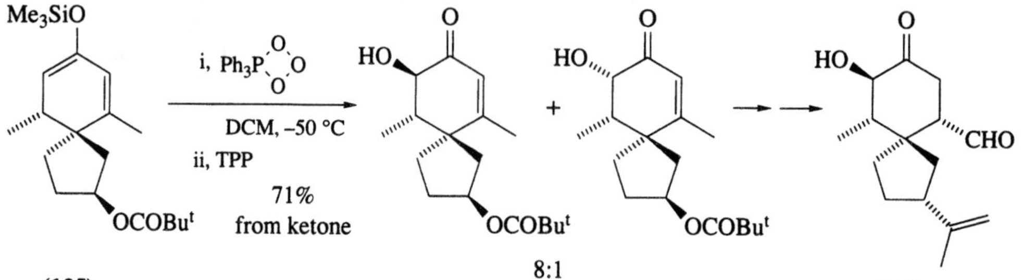

Me3SiO, OCOBu^t (**135**) · i, Ph3P(O3); DCM, −50 °C; ii, TPP; 71% from ketone → HO, OCOBu^t + HO, OCOBu^t 8:1 → → HO, CHO (**134**)

The procedure is mild and reasonably efficient and may therefore find further synthetic use. One might also expect that photosensitized preparation of the singlet oxygen would be equally effective,[82] although this remains to be shown.

(iii) Miscellaneous

Other reagents have been utilized for this transformation. For example, lead tetrabenzoate (*cf.* Section 2.3.2.1.3.iii) provides the corresponding α′-benzoyloxy enones when combined with fluoride-induced hydrolysis of (**136**) and (**137**).[149] However, success is restricted to acyclic ketones. Where cyclic dienol ethers are employed, products derived from α-oxidation are obtained.[149]

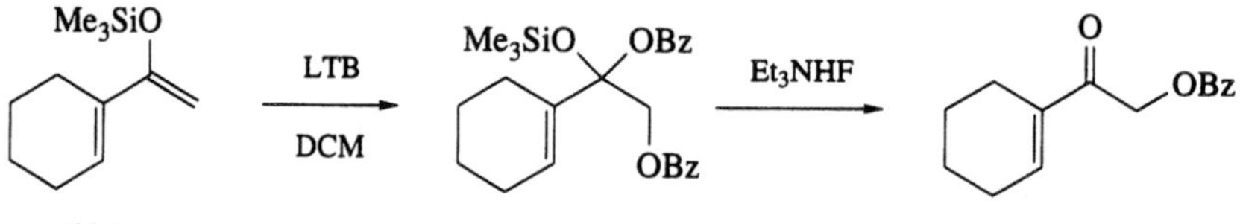

Me3SiO (**136**) · LTB, DCM → Me3SiO, OBz, OBz · Et_3NHF → O, OBz 54%

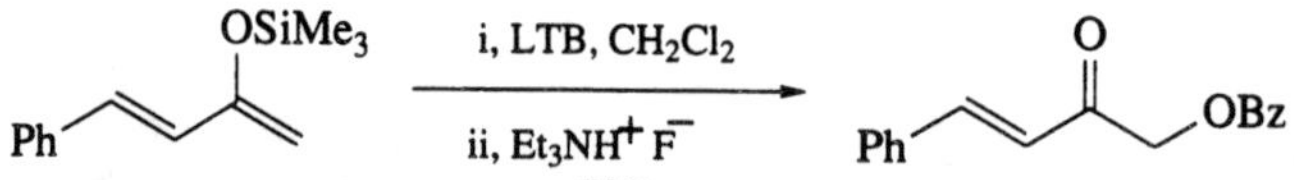

OSiMe3, Ph (**137**) · i, LTB, CH_2Cl_2; ii, $Et_3NH^+F^-$; 78% → Ph, O, OBz

Finally, a radical-induced α′-hydroxylation has been achieved using anhydrous *t*-butyl hydroperoxide in the presence of a copper(I) chloride catalyst at 50 °C in benzene,[150] but it is a relatively low yielding process.

2.3.2.2.4 Via *other derivatives*

While enol acetates from saturated ketones were useful α-oxygenation substrates, the corresponding dienol acetates are not. The relatively electron-deficient alkene bearing the acetoxy group is less attractive to electrophilic oxygenating agents than the unsubstituted double bond. Thus, for example, peracid treatment leads to epoxidation of the unfunctionalized alkene.[151] However, it would seem likely that reactions in basic media, which could initially hydrolyze the enol acetate, could be effective. Conversely the electron-rich alkyl enol ether should allow α′-oxidation of the dienol equivalent by electrophilic reagents (see Section 2.3.2.1.4 and *cf.* Section 2.3.2.1.3), although possibly complicated by skeletal cleavage.[151] By the same token enamine and ketoxime substrates should be effective (see Section 2.3.2.1.5).

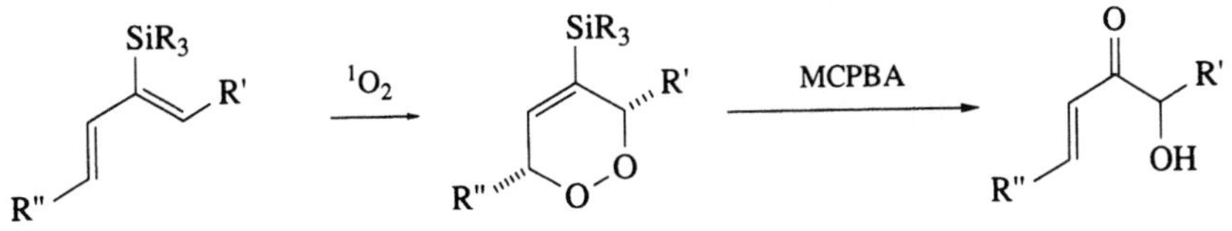

SiR_3 R′ R″ $^1O_2$ → SiR_3 R′ R″ O O MCPBA → O R′ OH R″

Scheme 18

Vinyl cyanides may be useful derivatives since permanganate-induced α-hydroxylation in the presence of alkenes has been demonstrated, *e.g.* (**90**) to (**91**). Oxygenation of α,β-unsaturated vinylsilanes and sulfides using the previously described procedures would not be successful (Sections 2.3.2.1.5.iii and 2.3.2.1.5.iv). However a singlet oxygen cycloaddition process (*cf.* Section 2.3.2.2.3.ii), followed by eliminative hydrolysis, could provide a usable, although lengthy, approach (Scheme 18).

2.3.2.3 α,β-Unsaturated Ketones: *sp*² Center

The overall sequence could be realized by 1,4-addition of a nucleophile to the enone and subsequent quenching at the α-position followed by β-elimination of the initial nucleophilic component.[152] Such multistep processes will not be discussed here. However direct hydroxylation methods are scarce. Moriarty has reported that α,β-unsaturated ketones are oxidized by phenyliodosyldiacetate at the α-site in preference to the α′-position, *e.g.* (**138**) to (**139**),[153] although no yield has been indicated. There is no available mechanistic rationale, although the intermediacy of the α,β-epoxide is precluded.

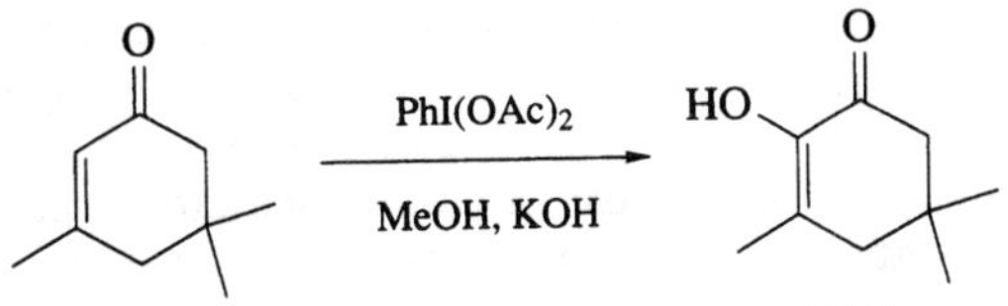

O → $PhI(OAc)_2$ / MeOH, KOH → HO O

(**138**) (**139**)

Clearly there is little precedence for this *direct* procedure and the development of an efficient and reliable method is required.

2.3.2.4 Esters and Lactones

Much of the preceding discussion concerning the α-hydroxylation of ketones is relevant for ester and lactone substrates. Many examples have featured β-keto esters and these are clearly relevant. Reference should be made to these sections.

2.3.2.4.1 Directly from enol form

(i) Hypervalent iodine reagents

The procedure developed by Moriarty for the α-hydroxylation of ketones using iodosylbenzene or its diacetate has been extended for use with esters.[154] Thus treatment of methyl or ethyl esters with iodosylbenzene diacetate in a two-phase system (benzene/aqueous KOH) generates the α-hydroxy acid, while reaction in methanol in the presence of sodium methoxide provides the α-methoxy ester, (Scheme 19). Oxidation of the free acid was unsuccessful (Section 2.3.2.6). Both variations proceed in similar, moderately good yields, in a fashion mechanistically analogous to the reaction with ketones.

Similarly the α-mesyloxy esters are available by reaction with Koser's reagent.[31b] Neither of these studies addressed the question of stereoselectivity.

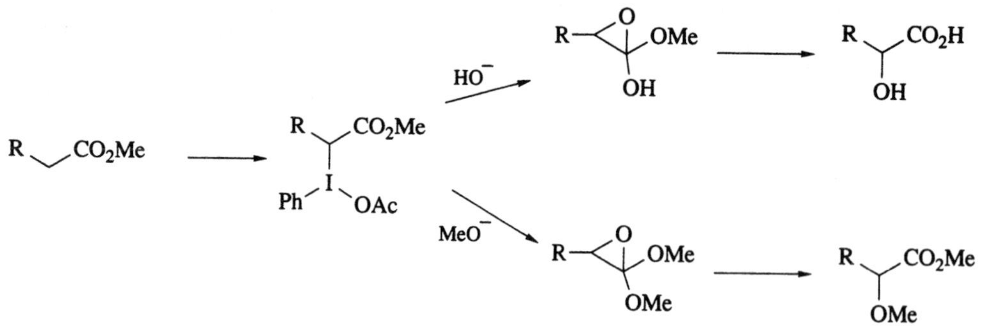

Scheme 19

(ii) Miscellaneous

A number of enol oxidations of β-keto esters utilizing, for example, peracid, singlet oxygen or the peroxy ester reaction have been recorded (see Section 2.3.2.1.1.iv).

2.3.2.4.2 Via *preformed enolate*

(i) Molecular oxygen

Electrophilic consumption of the enolates of esters and lactones is the most widely used process for effecting the α-hydroxylation. In analogy with the ketone series some of the earliest procedures employed molecular oxygen in combination with a suitable reducing agent. In this manner Corey produced hydroxy ester (**140**) during studies in the prostaglandin area.[155] Treatment of a lithio enolate with molecular oxygen at –78 °C and *in situ* reduction with triethyl phosphite gave a good yield of the required product, although with only minimal stereoselectivity. The oxidation of this substrate demonstrates a workable procedure but, as other studies have shown,[156] oxidation at nontertiary centers by a similar procedure are generally not useful. α-Hydroxylation of the extended enolates of enoate substrates, reducing with tin(II) chloride, has also been demonstrated.[157]

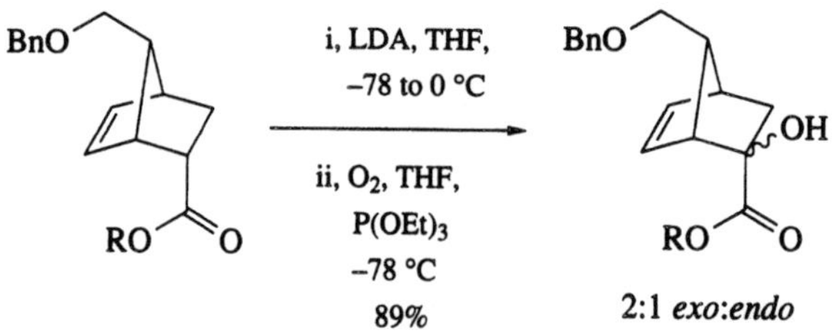

i, LDA, THF, –78 to 0 °C; ii, O_2, THF, $P(OEt)_3$, –78 °C, 89%

2:1 *exo:endo*

(**140**)

These methods have been somewhat superseded by the introduction of specific hydroxylating reagents (*vide infra*).

(ii) Molybdenum peroxy complexes

Reaction of lactone or ester enolates with MoOPH (see Section 2.3.2.1.2.ii) produces the α-hydroxy derivative in high yield.[52] They appear to be better substrates than ketones, reacting at lower temperatures and with generally greater efficiency. The use of lower temperatures suggests an ameliorating effect on stereoselectivity. An interesting example (**141**) to (**142**)[158] cites interaction of the enolate counterion with a pendant trifluoromethyl group as providing the source of the stereochemical bias in a typically efficient oxidation. The benefit of this procedure over the simple molecular oxygen approach, in terms of stereoselectivity, is illustrated in the oxidation of the lactone (**143**).[159] A similar comparison of lactone enolate oxidation between MoOPH and 2-sulfonyloxaziridines has, however, suggested that better selectivity is achieved with the latter reagent (*vide infra*).

Taking stereoselectivity a stage further, MoOPH has been successfully employed in the enantioselective oxidation of esters bearing an enolate face discriminating chiral auxillary.[160] Thus exposure of the

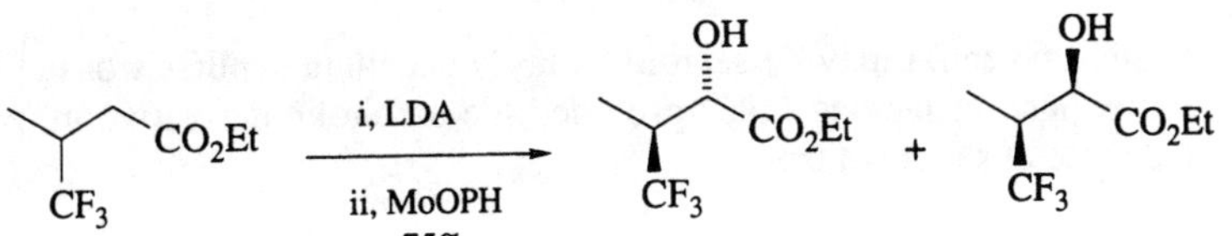

(141) i, LDA; ii, MoOPH; 75% → (142) 97:3

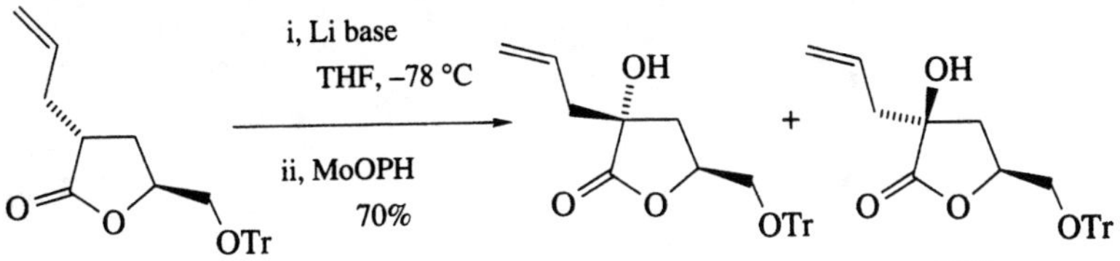

(143) i, Li base, THF, –78 °C; ii, MoOPH; 70% → 90% : 10% (ii, O_2 50% : 50%)

potassium enolate of ester (**144**) to MoOPH at *ca.* –50 °C delivers the α-hydroxy ester with a high level of diastereoselection. The process is independent of asymmetry at the adjacent carbon atom, but fails for oxidation of tertiary positions. Either diastereomer is available by appropriate selection of the (*E*)- or (*Z*)-enolate. Basic hydrolysis of the diastereomeric products produces the enantiomeric α-hydroxy acids without racemization.

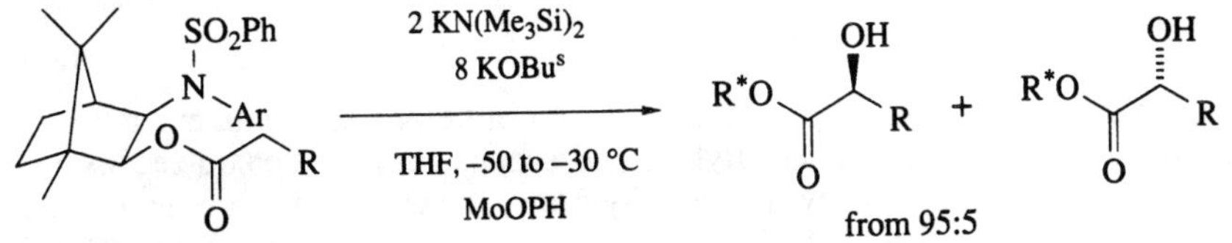

(**144**) Ar = 2,4,6-$Me_3C_6H_2$; 2 $KN(Me_3Si)_2$, 8 $KOBu^s$, THF, –50 to –30 °C, MoOPH → from 95:5 to 99:1

(iii) 2-Sulfonyloxaziridines

Recent introduction of these reagents as a source of 'electrophilic oxygen' for a variety of oxidative processes was extended to the α-hydroxylation of ketone enolates (Section 2.4.2.1.2.iii), and, at the same time, to the analogous ester/lactone oxidations.

A comparative study,[65] contrasting this oxidation with that of MoOPH indicated a markedly improved yield and stereoselectivity (Scheme 20). Although undoubtedly a valid example, the distinction is unlikely to be so universally clear cut and both of these valuable reagents will be useful in synthetic application.

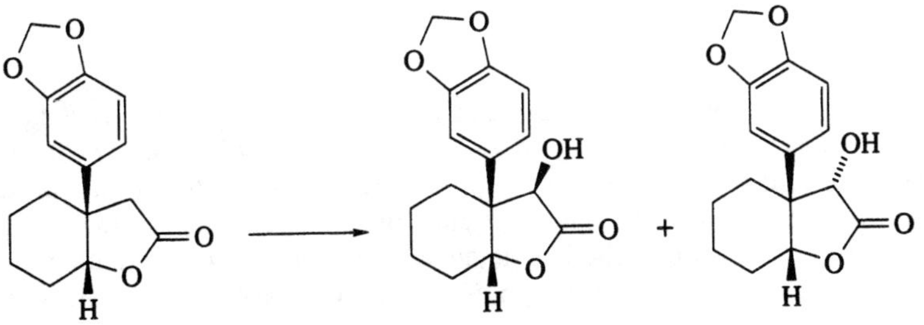

LDA/MoOPH	15%	75	:	25
LHMDS/ oxaziridine	62%	100	:	0
KHMDS/ oxaziridine	91%	100	:	0

Scheme 20

A recent display of such potential may be seen in Corey's excellent synthesis of (±)-Ginkgolide B.[161] Oxidation of the lithio enolate of lactone (**145**) provided the alcoholic derivative in, presumably, good yield and with a high degree of stereocontrol.

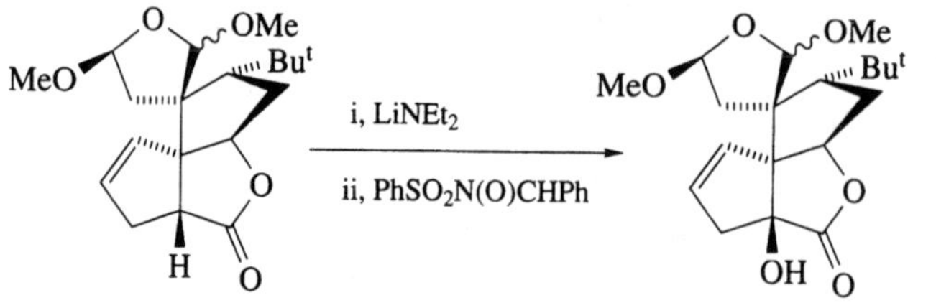

i, $LiNEt_2$

ii, $PhSO_2N(O)CHPh$

(**145**)

(iv) Miscellaneous

α-Oxidation of the dienolate of dienoate systems through the intervention of benzoyl peroxide at low temperature has been demonstrated to be an efficient method for introducing an α-benzoyloxy substituent.[157] The procedure appears to have drawn little attention despite its obvious effectiveness.

2.3.2.4.3 Via *silyl ketene acetals*

(i) Peracid

In a manner exactly analogous to the α-hydroxylation of ketone silyl enol ethers (Sections 2.3.2.1.3.i and 2.3.2.2.3.i) the corresponding ester silyl ketene acetals may be epoxidized by peracid and subsequently cleaved with fluoride to reveal the α-hydroxy ester.[162] Yields are good if hexanes are employed as solvent, while competing hydrolysis hampers the process in other media. The equivalent lactone hydroxylations are, however, not possible since hydrolysis is the dominant process even in hexane. This solvent limitation may prove restrictive to the widespread use of this technique.

(ii) Singlet oxygen

One report has indicated the potential of this mild reagent for α-oxidation of silyl ketene acetals.[163] Once again the usefulness is restricted by competition between the required silatropic ene process and the prototropic ene reaction giving rise to the enoate. For substrates lacking β-protons the α-hydroperoxy ester is readily obtained in good yield. In one case where β-protons were present (methyl group) an 80% yield of the silatropic product was obtained by carefully optimizing the reaction conditions. This may not however always be possible.

(iii) Transition metal salts

α-Acetoxylation and α-benzoylation of esters and lactones *via* their ketene acetals may be achieved using the appropriate lead(IV) salt.[164] Thus good yields of the acyloxy lactones (**146**) were obtained after fluoride treatment of the crude 'ortho ester' (*cf.* Section 2.3.2.1.3.iii). Although no information pertinent to the stereoselectivity was available here, subsequent studies have demonstrated that introduction of a steric bias can be fruitful. Oxidation of the ketene acetals of esters bearing camphor-derived auxiliaries produced the acetoxylated derivatives with a high degree of diastereoselectivity (Scheme 21).[165] Chemi-

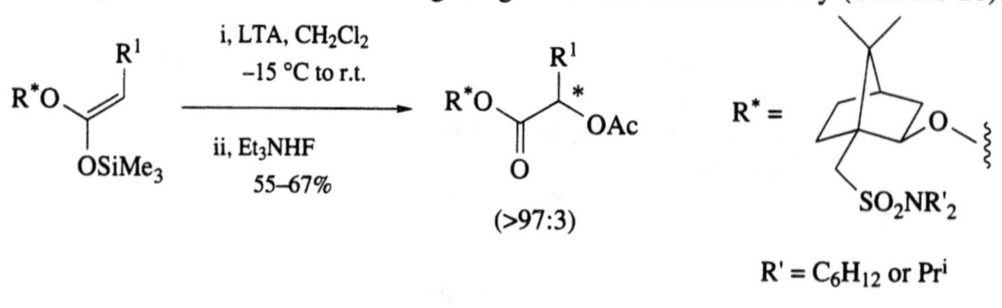

i, LTA, CH_2Cl_2, −15 °C to r.t.; ii, Et_3NHF, 55–67%

(>97:3)

$R^* =$ ... $SO_2NR'_2$

$R' = C_6H_{12}$ or Pr^i

Scheme 21

OSiMe3 Pb(OCOR)4 CH2Cl2 or PhH OSiMe3 OCOR OCOR Et3NHF OCOR

R = Me 82% **(146)**
R = Ph 78%

cal yields are somewhat lower than for simple substrates. Either antipode of the auxiliary may be used with equally good results and effective, nonracemizing hydrolysis procedures have been secured.

2.3.2.5 Amides and Lactams

2.3.2.5.1 Via *preformed enolates*

(i) Molecular oxygen

Just as ketones and esters may be derivatized by the action of molecular oxygen on the corresponding enolates, the same is true of amides. Generation of the lithio enolate (clearly not as facile as the previous substrates) and quenching with oxygen produce the α-hydroperoxyamide, which after *in situ* reduction, ($P(OEt)_3$),[166] or reductive work-up, ($NaHSO_3$),[156] gives good yields of the required material.

BnO H O N–Et(Bn) H Ph i, LHMDS THF, 0 °C ii, O_2, $P(OEt)_3$ 0 °C to r.t. 60% BnO OH O N–Et(Bn) H Ph

(147)

Stereoselective oxidation is possible. A notable synthetic example involved hydroxylation of lactam (**147**) to produce a single diastereomer of the functionalized isoindoline — an intermediate required during work related to the cytochalasins.[166] It should be noted that in all the amide enolate examples the tertiary amide is utilized.

(ii) Molybdenum peroxy complexes

MoOPH (Section 2.3.2.1.2.ii)[52] may be a suitable reagent for lactam enolate hydroxylation. This is suggested by the oxidation of lactam (**148**).[167] Clearly the label 'enolate' is not strictly applicable to the bridgehead carbanion and it is likely that more forcing conditions would be necessary for genuine enolate hydroxylations. It is not clear whether *N*-oxidation would then emerge as a source of problems.

Me N O H O O N–Me OSiBu^{t}Me$_2$ i, Bu^{t}Li, THF, –78 °C ii, MoOPH 62% Me N O OH O O N–Me OSiBu^{t}Me$_2$

(148)

(iii) 2-Sulfonyloxaziridines

Hydroxylation of amide and lactam enolates using Davis' reagents (Section 2.3.2.1.2.iii) occurs readily at low temperature. The process can be highly stereoselective and is viable for oxidation of both secondary and tertiary positions (presumably also primary). Thus hydroxy lactam (**150**) is obtained as the sole diastereomer[168] after exposure of the lithio enolate of the lactam (**149**) to the oxidant at –78 °C.

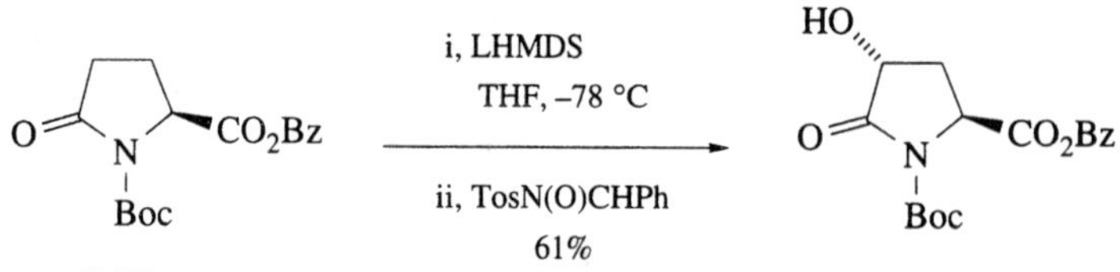

i, LHMDS THF, –78 °C; ii, TosN(O)CHPh; 61%

(**149**) → (**150**)

Futhermore the reagent has proven valuable for asymmetric hydroxylation of face-discriminating amide and oxazolidinone enolates. Davis has shown[169] that oxidation of secondary centers *via* the enolate derived from chiral pyrrolidinoamide (**151a**) provides the hydroxylated derivative in high yield (93–96%) and with excellent diastereoselection in either direction. The equivalent processes with methoxymethylpyrrolidine (**151b**) were of little value. This latter point contrasts with later studies concerning similar oxidations at tertiary centers.[170] Utilizing the chiral oxaziridines (**46a**) and (**46b**) and the methoxymethylpyrrolidinoamide (**152**), the corresponding tertiary alcohols were generated with good diastereoselection, although a little less efficiently than secondary alcohol products. Nonetheless the process represents the highest level of asymmetric induction of any electrophile from chiral, acyclic, tetrasubstituted enolates.

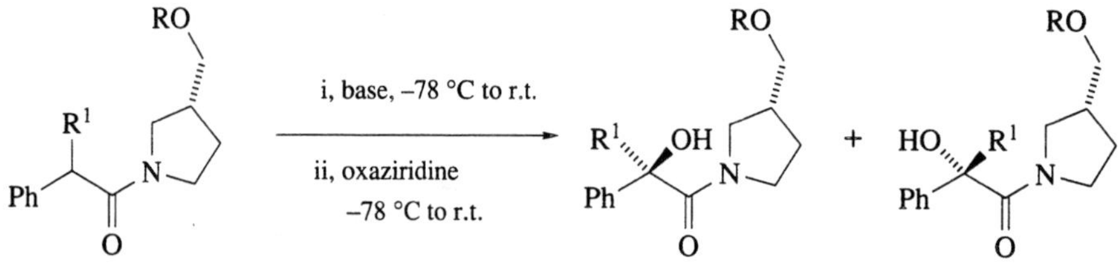

i, base, –78 °C to r.t.; ii, oxaziridine, –78 °C to r.t.

(**151**) **a**: R = H; R^1 = H
b: R = Me; R^1 = H
(**152**) R = Me; R^1 = Me

(*R*) (**151a**) (*S*)

		(*R*):(*S*)
LDA/THF	96%	2:98
NaHMDS/THF	93%	96:4

Similar high levels of asymmetric induction using the oxazolidinone methodology developed by Evans have been observed.[171] In contrast to Davis' observations with amides[169,170] the sodium enolates were preferred in these cases, allowing diastereomeric ratios in the range 90:10–99:1 to be realized (Scheme 22).

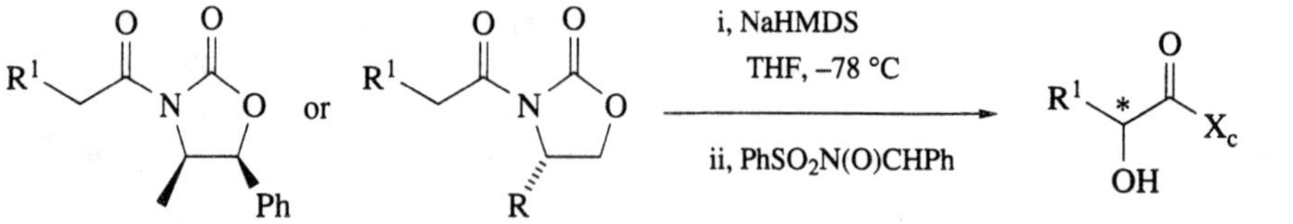

i, NaHMDS THF, –78 °C; ii, $PhSO_2N(O)CHPh$

R = Pr^i or $PhCH_2$–

diastereomeric ratio 90:10 to 99:1

Scheme 22

The methodology was extended to show chemoselective hydroxylation (imide over ester) and α-hydroxylation of extended enolates. It also provided support for a stepwise mechanism (see Section 2.3.2.1.2.iii) involving a counterion dependent equilibrium.

(iv) Miscellaneous

Hydroxylation of face-differentiated oxazolidinone enolates (*vide supra*) has also been enacted using dibenzoyl peroxydicarbonate as the electrophilic spouse.[172] Initial production of the α-benzyl carbonate (with analogously impressive diasereoselectivity) may be followed by facile hydrolysis to give essentially enantiomerically pure α-hydroxy esters.

2.3.2.6 Carboxylic Acids

2.3.2.6.1 Via *enol form*

Direct treatment of a free acid with thallium triacetate provides the α-acetoxy acids *via* intramolecular reductive rearrangement of the derived thallium enolate.[173] Only simple acids have been used and the necessity to use a large excess of the substrate acid limits the synthetic usefulness of the procedure.

2.3.2.6.2 Via *preformed enolate dianion*

A number of related procedures for this oxidation exist. Initially they were performed by bubbling air through solutions of the enolate,[174] but were subsequently improved by using oxygen or ethereal solutions of oxygen.[175] Reduction of the primary products (α-hydroperoxides) may be effected *in situ* or during work-up (see Section 2.3.2.5.1.i).

2.3.2.6.3 Via *bis-silyl ketene acetals*

(i) Peracid

These derivatives of carboxylic acids behave analogously to ester silyl ketene acetals (Section 2.3.2.4.3.i). Thus treatment with peracid in hexane followed by acidic work-up allows isolation of good yields of the α-hydroxy acids.[176]

(ii) Singlet oxygen

Bis-silyl ketene acetals devoid of β-protons undergo a clean silatropic ene reaction with singlet oxygen (see Sections 2.3.2.1.3.ii and 2.3.2.4.3.ii) to generate the α-silylperoxy silyl ester quantitatively.[177] Treatment with methanol affords the α-hydroperoxy acid, also quantitatively (Scheme 23). Hydrogenation over platinum reveals the α-hydroxy acid, once again, quantitatively. Despite this encouragement the substrate limitation is severe.

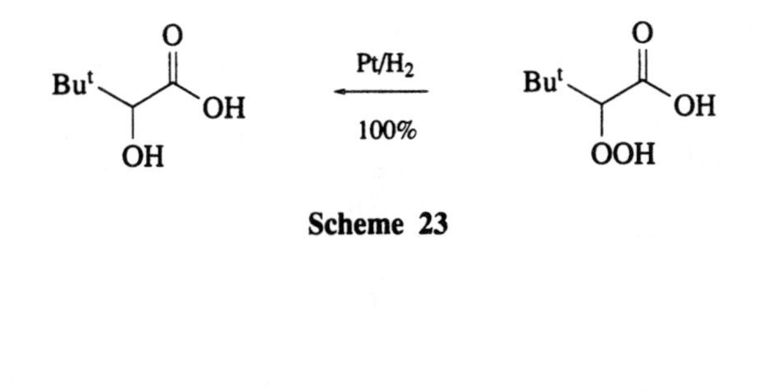

$Bu^tCH=C(OSiMe_3)_2 \xrightarrow[100\%]{^1O_2} Bu^tCH(OOSiMe_3)CO_2SiMe_3 \xrightarrow[100\%]{MeOH} Bu^tCH(OOH)CO_2H \xrightarrow[100\%]{Pt/H_2} Bu^tCH(OH)CO_2H$

Scheme 23

2.3.2.7 Aldehydes

Few methods for the α-hydroxylation of aldehydes are available. This is a reflection of their notorious instability towards polymerization and rearrangement to α-hydroxy ketones. All the useful procedures generate protected α-hydroxy aldehydes.

Exposure of the silyl enol ethers of aldehydes to peracid in dichloromethane, followed by treatment of the intermediate masked hydroxy aldehyde (*cf.* (**53**) to (**54**); Section 2.3.2.1.3.i) with acetic anhydride and triethylamine, allows isolation of the product α-acetoxy aldehydes in moderate yield.[178] Similarly treatment of the silyl enol ethers with LTA in acetic acid containing potassium acetate effects the same transformation.[179]

An indirect, although very valuable method provides a means of preparing optically active α-hydroxy aldehydes from the aldehyde-derived hydrazones (see Section 2.3.3.2).

2.3.3 HYDROXYLATION α TO C=N

2.3.3.1 Oxime Acetates and Nitrones

House has reported[180] that α-acetoxy ketones are available through 3,3-sigmatropic rearrangement of enamines derived from oxime acetates. Thus treatment of an oxime acetate with trimethyloxonium tetrafluoroborate generates the corresponding iminium salt (**153**) which, in the presence of triethylamine, rapidly isomerizes to the enamine. This in turn equally rapidly rearranges to the α-acetoxyimine (**154**) from which the corresponding ketone is recovered by acidic hydrolysis. The regioselectivity of the acetoxylation is dependent on the position of the enamine unsaturation rather than the original stereochemistry of the oxime.

$Me_3O^+ BF_4^-$, $MeNO_2$; Et_3N; H_3O^+

(**153**) (**154**)

major product 50%

An identical rearrangement can be arrived at through acylation of *N*-methyl nitrones,[181] conveniently prepared from the ketone (**155** to **156**), although the process is apparently restricted to cyclic substrates. Finally a similar one-pot procedure for conversion of ketoximes to α-acetoxy ketones under the conditions shown (**157** to **158**), allows the transformation to be carried out simply and efficiently.[182] In this case rearrangement produces α-acyloxyenimides, whose hydrolysis provides the keto equivalent.

i, $MeNHOH{\cdot}HCl$, 70 °C; ii, NH_3; i, RCOCl, Et_3N; ii, H_2O, HOAc

(**155**)

R = Bu[t], *n* = 1 } 59% (**156**)

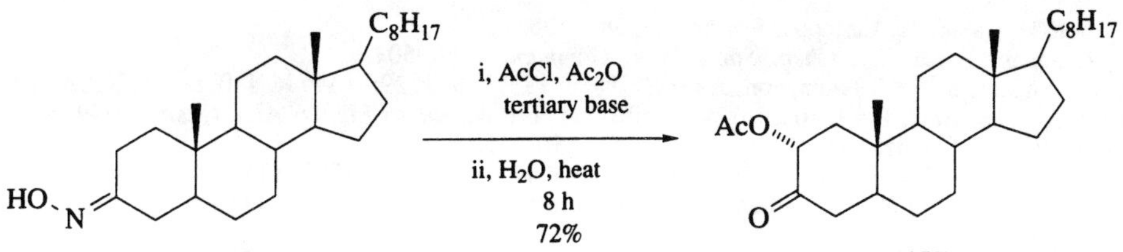

i, AcCl, Ac_2O tertiary base; ii, H_2O, heat 8 h 72%

(157) (158)

2.3.3.2 Ketoximes and Hydrazones

Oxygenation of the dianion of a ketoxime[183,184] by MoOPH (*cf.* Section 2.3.2.1.2) provides α-hydroxy ketones after hydrolysis, albeit in relatively low yield. The equivalent process for hydrazones was unsuccessful,[52] although it appears that this was a problem associated with manipulation of the products rather than an intrinsic failure of reaction. Indeed subsequent work with hydrazone anions has shown that the process is viable.

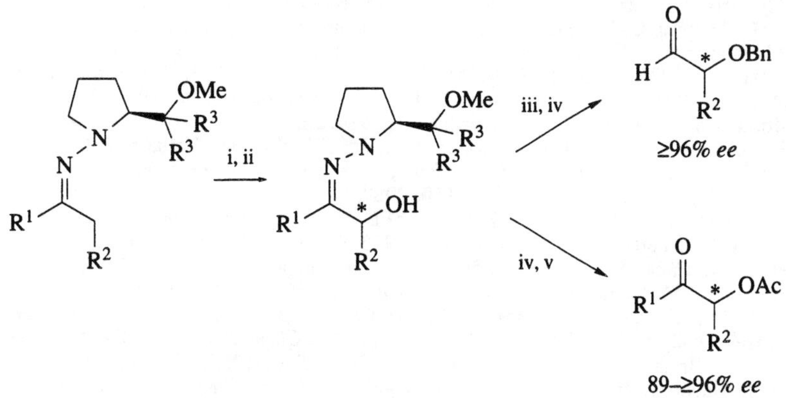

≥96% *ee*

89–≥96% *ee*

i, LDA, THF, 0 °C; ii. oxaziridine; iii, NaH, BnCl; iv, O_3 ; v, Ac_2O, DMAP

Scheme 24

Enders and coworkers have shown that deprotonation of chiral SAMP/RAMP hydrazones (or their substituted analogs) derived from ketones or aldehydes, followed by reaction with Davis' oxaziridine reagent provides the α-hydroxy hydrazones in moderate yield but with high diastereoselectivity.[185] Direct unmasking or protection followed by unmasking provides the corresponding α-hydroxy ketones or aldehydes respectively (Scheme 24). Both antipodes of the hydroxylated compounds are available by appropriate choice of (*S*)- or (*R*)-proline-derived auxiliaries. The direction of induction is predictable, if not wholly uniform (R^3 substitution alters the α-stereochemistry for aldehyde hydrazones). The process clearly provides a valuable approach to both systems.

2.3.4 REFERENCES

1. O. Dimroth and R. Schweitzer, *Ber. Dtsch. Chem. Ges.*, 1923, **56**, 1375.
2. D. J. Rawlinson and G. Sosnovsky, *Synthesis*, 1973, 567.
3. G. W. K. Cavill and D. H. Solomon, *J. Chem. Soc.*, 1955, 4426.
4. H. B. Henbest, D. N. Jones and G. P. Slater, *J. Chem. Soc.*, 1961, 4472.
5. J. A. Marshall and G. L. Bundy, *J. Chem. Soc.*, 1966, 500.
6. T. Sasaki and S. Eguchi, *Nippon Kagaku Zasshi*, 1968, **48**, 2453; see also H. Miura, Y. Fujimoto and T. Tatsumo, *Synthesis*, 1979, 898.
7. S. Hanessian, P. Beaulieu and D. Dube, *Tetrahedron Lett.*, 1986, **27**, 5071.
8. L. O. Krampitz, *Arch. Biochem.*, 1948, **17**, 81.
9. L. Canonica, B. Danieli, P. Manitto, G. Russo, S. Maroni and T. Salvatori, *Gazz. Chim. Ital.*, 1967, **97**, 1370.
10. G. W. K. Cavill and D. H. Solomon, *Aust. J. Chem.*, 1960, **13**, 121.
11. (a) G. R. Pettit, C. L. Herald and Y. P. Yardley, *J. Org. Chem.*, 1970, **35**, 1389; (b) G. Habermahl and A. Haaf, *Z. Naturforsch., Teil B*, 1968, **23**, 880; (c) E. Yoshii, T. Koizumi, H. Ikeshima, K. Ozaki and I. Hayashi, *Chem. Pharm. Bull.*, 1975, **23**, 2496.

12. B. Myrboh, H. Ila and H. Junjappa, *Synthesis*, 1981, 126.
13. D. M. Piatak and E. Caspi, *J. Chem. Soc., Chem. Commun.*, 1966, 501.
14. (a) H. O. House and H. W. Thompson, *J. Org. Chem.*, 1961, **26**, 3729; (b) P. R. Jeffries, A. K. Macbeth and B. Milligan, *J. Chem. Soc.*, 1954, 705; (c) W. Treibs and H. Bast, *Justus Liebigs Ann. Chem.*, 1949, **561**, 165.
15. A. Brenner and H. Schinz, *Helv. Chim. Acta*, 1952, **35**, 1615.
16. L. H. Zalkow, J. W. Ellis and M. R. Brennan, *J. Org. Chem.*, 1963, **28**, 1705, and refs. cited therein.
17. S. Uemura, T. Nakano and K. Ichikawa, *Nippon Kaguku Zasshi*, 1967, **88**, 1111 (*Chem. Abstr.*, 1968, **69**, 43 582).
18. A. McKillop and J. D. Hunt, *J. Org. Chem.*, 1972, **37**, 3381, and refs. cited for the use of thallium(III) perchlorate.
19. A. McKillop, B. P. Swann and E. C. Taylor, *J. Am. Chem. Soc.*, 1971, **93**, 4919.
20. A. McKillop, D. W. Young, M. Edwards, R. P. Hug and E. C. Taylor, *J. Org. Chem.*, 1978, **43**, 3773.
21. M. Pechal, Z. Cvengrosova, M. Hrusovsky, J. Vojtko and A. Kasonyi, *Czech. Pat.* 202 345 B (1982) (*Chem. Abstr.*, 1983, **98**, 71 491v).
22. J. S. Littler, *J. Chem. Soc.*, 1962, 832.
23. A. V. Drummond and W. A. Waters, *J. Chem. Soc.*, 1955, 497.
24. E. I. Heiba and R. M. Dessau, *J. Am. Chem. Soc.*, 1971, **93**, 524, and refs. cited therein.
25. F. Mizukami, M. Ando, T. Tanaka and J. Imamura, *Bull. Chem. Soc. Jpn.*, 1978, 335.
26. R. M. Moriarty, H. Hu and S. C. Gupta, *Tetrahedron Lett.*, 1981, **22**, 1283.
27. R. M. Moriarty, K.-C. Hu and I. Prakash, *Org. Synth.*, 1986, **64**, 184.
28. R. M. Moriarty, O. Prakash, P. Karalis and I. Prakash, *Tetrahedron Lett.*, 1984, **25**, 4745; see also R. M. Moriarty, O. Prakash, C. T. Thachet and H. A. Musallam, *Heterocycles*, 1985, **23**, 633.
29. (a) R. M. Moriarty, S. G. Engerer, O. Prakash, I. Prakash, U. S. Gill and W. A. Freeman, *J. Chem. Soc., Chem. Commun.*, 1985, 1715; (b) R. M. Moriarty, L. S. John and P. C. Du, *J. Chem. Soc., Chem. Commun.*, 1981, 641; (c) S. Yasuda, T. Yamada and M. Hanaoka, *Tetrahedron Lett.*, 1986, **27**, 2023.
30. R. M. Moriarty and K. C. Hou, *Tetrahedron Lett.*, 1984, **25**, 691.
31. (a) G. F. Koser, A. G. Relenyi, A. N. Kalos, L. Rebovic and R. M. Wettach, *J. Org. Chem.*, 1982, **47**, 2487; (b) J. S. Lodaya and G. F. Koser, *J. Org. Chem.*, 1988, **53**, 210; (c) G. F. Koser, J. S. Lodaya, D. G. Ray, III and P. D. Kokil, *J. Am. Chem. Soc.*, 1988, **110**, 2987.
32. (a) P. R. Enslin, *Tetrahedron*, 1971, **27**, 1909; (b) E. P. Kohler, *Am. Chem. J.*, 1906, **36**, 177; (c) E. P. Kohler, M. Tishler and H. Potter, *J. Am. Chem. Soc.*, 1935, **57**, 2517; (d) R. C. Fuson, B. C. McKusick and F. W. Spangler, *J. Am. Chem. Soc.*, 1945, **67**, 597; (e) L. Crombie and P. J. Godin, *J. Chem. Soc.*, 1961, 2861; (f) H. Irie, J. Katakawa, M. Tomita and Y. Mizuno, *Chem. Lett.*, 1981, 637, and refs. cited in (a) to (f).
33. (a) H. Muxfeldt, G. Buhr and R. Bangert, *Angew. Chem., Int. Ed. Engl.*, 1962, **1**, 157, and refs. cited therein; (b) C. E. Holmlund, W. W. Andres and A. J. Shay, *J. Am. Chem. Soc.*, 1959, **81**, 4748, and refs. cited therein; see also A. V. Drummond and W. A. Waters, *J. Chem. Soc.*, 1955, 497; (c) J. J. Korst, J. D. Johnston, K. Butler, E. J. Bianco, L. H. Conover and R. B. Woodward, *J. Am. Chem. Soc.*, 1968, **90**, 439.
34. C. Betancor, C. G. Francisco, R. Freire and E. Suárez, *J. Chem. Soc., Chem. Commun.*, 1988, 947.
35. M. Masui, A. Ando and T. Shioiri, *Tetrahedron Lett.*, 1988, **29**, 2835.
36. (a) K. Yamakawa, T. Satoh, N. Obha, R. Sakaguchi, S. Takita and N. Tamura, *Tetrahedron*, 1981, **37**, 473; (b) K. Yamakawa, T. Satoh, N. Obha and R. Sakaguchi, *Chem. Lett.*, 1979, 763.
37. C. H. Heathcock, C. Mahain, M. F. Schlecht and T. Utawanit, *J. Org. Chem.*, 1984, **49**, 3264.
38. D. J. Rawlinson and G. Sosnovsky, *Synthesis*, 1972, 1.
39. G. M. Buchan, J. W. A. Findlay and A. B. Tumer, *J. Chem. Soc., Chem. Commun.*, 1975, 126.
40. M. S. Thompson, W. F. DeGiovani, B. A. Moyer and T. J. Meyer, *J. Org. Chem.*, 1984, **49**, 4972.
41. J. B. Jones, *Tetrahedron*, 1986, **42**, 3351, and refs. cited therein.
42. A. S. Bailey, M. L. Gilpinand and E. R. H. Jones, *J. Chem. Soc., Perkin Trans. 1*, 1977, 265.
43. C. E. Holmlund, W. W. Andres and A. J. Shay, *J. Am. Chem. Soc.*, 1959, **81**, 4750.
44. W. von E. Doering and R. M. Haines, *J. Am. Chem. Soc.*, 1954, **76**, 482.
45. E. J. Bailey, D. H. R. Barton, J. Elks and J. F. Templeton, *J. Chem. Soc.*, 1962, 1578 and refs. cited therein.
46. Alternatively potassium *t*-pentoxide or their sodium equivalents.[147] A trace of water was found to be advantageous although presumably a discrete enol/enolate is still present.
47. J. N. Gardner, F. E. Carlton and O. Gnoj, *J. Org. Chem.*, 1968, **33**, 3294.
48. G. Buchi, P. Kulsa, K. Ogasawara and R. L. Rosati, *J. Am. Chem. Soc.*, 1970, **92**, 999.
49. C. E. Coburn, K. Anderson and J. S. Swenton, *J. Org. Chem.*, 1983, **48**, 1455; see also S. D. Kimball, D. R. Watt and F. Johnson, *J. Am. Chem. Soc.*, 1981, **103**, 1563.
50. H. Muxfeldt, G. Haas, G. Hardtmann, F. Kathawala, J. B. Mooberry and E. Vedejs, *J. Am. Chem. Soc.*, 1979, **101**, 689; see also B. Glatz, G. Helmchen, H. Muxfeldt, H. Porcher, R. Prewo, J. Senn, J. J. Stezowski, R. J. Stojda and D. R. White, *J. Am. Chem. Soc.*, 1979, **101**, 2171.
51. E. Vedejs, *J. Am. Chem. Soc.*, 1974, **96**, 5944.
52. E. Vedejs, D. A. Engler and J. E. Telshow, *J. Org. Chem.*, 1978, **43**, 188.
53. H. Mimoun, I. Seree de Roch and L. Sajus, *Bull. Soc. Chim. Fr.*, 1969, 1481; H. Mimoum, I. Seree de Roch and L. Sajus, *Tetrahedron*, 1970, **26**, 37; see also A. A. Achrem, A. Timoschtschuk and D. I. Metelitza, *Tetrahedron*, 1974, **30**, 3165.
54. K. B. Sharpless, J. M. Townsend and D. R. Williams, *J. Am. Chem. Soc.*, 1972, **94**, 295.
55. H. Arakawa, Y. Moro-Oka and A. Ozaki, *Bull. Chem. Soc. Jpn.*, 1974, **47**, 2958.
56. R. Gamboni and C. Tamm, *Tetrahedron Lett.*, 1986, **27**, 3999 and refs. cited therein.
57. S. Hanessian, S. P. Sahoo and P. J. Murray, *Tetrahedron Lett.*, 1985, **26**, 5631.
58. H. Niwa, K. Wakamatsu, T. Hida, K. Niiyama, H. Kigoshi, M. Yamada, H. Nagase, M. Suzuki and X. Yamada, *J. Am. Chem. Soc.*, 1984, **106**, 4547.
59. F. A. Davis, U. K. Nadir and E. W. Kluger, *J. Chem. Soc., Chem. Commun.*, 1977, 25.
60. F. A. Davis, R. H. Jenkins, Jr. and S. G. Yocklovidi, *Tetrahedron Lett.*, 1978, 5171, for oxidation of sulfides.

61. F. A. Davis, P. A. Mancinelli, K. Balasubramanian and U. K. Nadir, *J. Am. Chem. Soc.*, 1979, **101**, 1044, for oxygenation of organometallics.
62. F. A. Davis, N. F. Abdul-Malik, S. B. Awad and M. E. Harakal, *Teterahedron Lett.*, 1981, **22**, 917, for epoxidation of alkenes.
63. F. A. Davis, O. D. Stringer and J. M. Billmers, *Tetrahedron Lett.*, 1983, **24**, 1213, for oxidation of selenides.
64. F. A. Davis, J. Lamendola, Jr., U. K. Nadir, E. W. Kluger, T. C. Sedergran, T. W. Panunto, R. Billmers, R. H. Jenkins, Jr., I. J. Turchi, W. H. Watson, J. S. Chen and M. Kimura, *J. Am. Chem. Soc.*, 1980, **102**, 2000.
65. F. A. Davis, L. C. Vishwakarma, J. M. Billmers and J. Finn, *J. Org. Chem.*, 1984, **49**, 3241.
66. F. A. Davis, R. H. Jenkins, Jr., S. Q. R. Rizvi and T. W. Panunto, *J. Chem. Soc., Chem. Commun.*, 1979, 600.
67. D. Boschelli, A. B. Smith, III, O. D. Stringer, R. H. Jenkins, Jr. and F. A. Davis, *Tetrahedron Lett.*, 1981, **22**, 4385.
68. F. A. Davis and L. C. Vishwakarma, *Tetrahedron Lett.*, 1985, **26**, 3539; D. A. Evans, M. M. Morrissey and R. L. Dorow, *J. Am. Chem. Soc.*, 1985, **107**, 4346, for oxaziridine-mediated asymmetric hydroxylation of chiral ester enolates.
69. F. A. Davis and M. S. Haque, *J. Org. Chem.*, 1986, **51**, 4083, and refs. cited therein.
70. J. W. Ellis, *J. Chem. Soc., Chem. Commun.*, 1970, 406.
71. A. E. Greene, J.-C. Muller and G. Ourisson, *Tetrahedron Lett.*, 1972, 3375.
72. S.-O. Lawesson and S. Gronwall, *Acta Chem. Scand.*, 1960, **14**, 144.
73. M. Ando, G. Buchi and T. Ohuma, *J. Am. Chem. Soc.*, 1975, **97**, 6880; see also G. Buchi, K. E. Matsumoto and H. Nishimura, *J. Am. Chem. Soc.*, 1971, **93**, 3299.
74. P. Brownbridge, *Synthesis*, 1983, 1.
75. G. M. Rubottom, M. A. Vazquez and D. R. Pelegrina, *Tetrahedron Lett.*, 1974, 4319.
76. L. A. Paquette, H. S. Lin and J. C. Gallucci, *Tetrahedron Lett.*, 1987, **128**, 1363.
77. A. G. Brook and D. M. McCrae, *J. Organomet. Chem.*, 1974, **77**, C19.
78. R. A. Amos and J. A. Katzenellenbogen, *J. Org. Chem.*, 1977, **42**, 2537.
79. R. K. Boeckman, Jr. and M. Ramaiah, *J. Org. Chem.*, 1977, **42**, 1581.
80. R. Z. Andriamialisoa, N. Langlois and Y. Langlois, *Tetrahedron Lett.*, 1985, **26**, 3563.
81. P. T. Lansbury and J. P. Vacca, *Tetrahedron Lett.*, 1982, **23**, 2623.
82. R. W. Denny and A. Nickon, *Org. React. (N.Y.)*, 1973, **20**, 133.
83. G. M. Rubottom and M. I. Lopez Nieves, *Tetrahedron Lett.*, 1972, 2423.
84. E. Friedrich and W. Lutz, *Angew. Chem., Int. Ed. Engl.*, 1977, **16**, 413.
85. C. W. Jefford and C. G. Rimbault, *J. Am. Chem. Soc.*, 1978, **100**, 6515; see also *Tetrahedron Lett.*, 1977, 2375.
86. (a) R. D. Clark and C. H. Heathcock, *Tetrahedron Lett.*, 1974, 2027; (b) W. S. Zhou, B. Jiang and X.-F. Pan, *J. Chem. Soc., Chem. Commun.*, 1988, 791.
87. J. P. McCormick, W. Tamasik and M. W. Johnson, *Tetrahedron Lett.*, 1981, **22**, 607.
88. M. Schröder, *Chem. Rev.*, 1980, **80**, 187.
89. T. Yamada and K. Narasaka, *Chem. Lett.*, 1986, 131; see also M. Tokles and J. K. Snyder, *Tetrahedron Lett.*, 1986, **27**, 3951.
90. T. V. Lee and J. Toczek, *Tetrahedron Lett.*, 1982, **23**, 2917.
91. R. M. Moriarty, O. Prakash and M. P. Duncan, *Synthesis*, 1985, 943.
92. G. M. Maume and E. C. Horning, *Tetrahedron Lett.*, 1969, 343.
93. G. M. Rubottom, J. M. Gruber and G. M. Mong, *J. Org. Chem.*, 1976, **41**, 1673.
94. G. M. Rubottom, in 'Oxidation in Organic Chemistry', ed. W. S. Trahanovsky, Academic Press, New York, 1982, p. 46.
95. K. B. Sharpless and R. C. Michaelson, *J. Am. Chem. Soc.*, 1973, **95**, 6136.
96. G. B. Payne, *Tetrahedron*, 1962, **18**, 763.
97. K. L. Williamson, J. I. Coburn and M. F. Herr, *J. Org. Chem.*, 1967, **32**, 3934.
98. A. L. Draper, W. J. Heilman, W. E. Schaefer, H. J. Shine and J. N. Shoolery, *J. Org. Chem.*, 1962, **27**, 2727.
99. P. D. Gardner, *J. Am. Chem. Soc.*, 1956, **78**, 3421.
100. P. G. Carlson and J. K. Pierce, *J. Org. Chem.*, 1971, **36**, 2319; K. L. Williamson and W. S. Johnson, *J. Org. Chem.*, 1961, **26**, 4563; H. J. Shine and G. E. Hunt, *J. Am. Chem. Soc.*, 1958, **80**, 2434.
101. T. Ito, N. Tomiyoshi, K. Nakamura, S. Azuma, M. Izawa, F. Maruyama, M. Yanagiya, H. Shirahama and T. Matsumoto, *Tetrahedron Lett.*, 1982, **23**, 1721; D. N. Kirk and M. F. Herr, *J. Chem. Soc., Chem. Commun.*, 1970, 518.
102. H. O. House, in 'Modern Synthetic Reactions', Benjamin, Menlo Park, CA, 1972, 499, 763.
103. (a) A. A. Frimer, *Synthesis*, 1977, 578; (b) F. Huet, A. Lechevallier and J. M. Conia, *Synth. Commun.*, 1980, **10**, 83.
104. C. L. Stevens and S. J. Dykstra, *J. Am. Chem. Soc.*, 1953, **75**, 5975.
105. J. E. Baldwin, O. W. Lever, Jr. and N. R. Tzodikov, *J. Org. Chem.*, 1976, **41**, 2312.
106. C. L. Stevens and J. Tazuma, *J. Am. Chem. Soc.*, 1954, **76**, 715.
107. Y. Kishi, M. Aratani, T. Fukuyama, F. Nakatsubo and T. Goto, *J. Am. Chem. Soc.*, 1972, **94**, 9217.
108. J. Pusset, D. Guenard and R. Beugelmans, *Tetrahedron*, 1971, **27**, 2939.
109. S. L. Wilson and G. B. Schuster, *J. Am. Chem. Soc.*, 1983, **105**, 679; S. L. Wilson and G. B. Schuster, *J. Org. Chem.*, 1986, **51**, 2056.
110. E. W. H. Asveld and R. M. Kellogg, *J. Am. Chem. Soc.*, 1980, **102**, 3644, aldehyde-derived enol ethers.
111. C. W. Jefford and C. G. Rimbault, *J. Am. Chem. Soc.*, 1978, **100**, 294.
112. S. Moon and M. Bohm, *J. Org. Chem.*, 1972, **37**, 4338.
113. W. S. Johnson, B. Gastambide and R. Pappo, *J. Am. Chem. Soc.*, 1957, **79**, 1991.
114. K. Oka and S. Hara, *J. Am. Chem. Soc.*, 1977, **99**, 3859.
115. R. V. Hoffman, *Synthesis*, 1985, 760.
116. R. V. Hoffman and C. S. Carr, *Tetrahedron Lett.*, 1986, **27**, 5811.
117. S. Matsubara, K. Takai and H. Nozaki, *Bull. Chem. Soc. Jpn.*, 1983, **56**, 2029.

118. T. Shono, M. Okawa and I. Nishiguchi, *J. Am. Chem. Soc.*, 1975, **97**, 6144.
119. H. O. House, in 'Modern Synthetic Reactions' Benjamin, Menlo Park, CA, 1972, p. 570.
120. M. E. Kuehne and T. J. Giacobbe, *J. Org. Chem.*, 1968, **33**, 3359.
121. F. Corbani, B. Rindone and C. Scolastico, *Tetrahedron Lett.*, 1972, 2597.
122. R. B. Boar, J. F. McGhie, M. Robinson and D. H. R. Barton, *J. Chem. Soc., Perkin Trans. 1*, 1975, 1242, and the preceding paper.
123. M. Otaka and Y. Lizuka, *Jpn. Pat.* 50 32 156 (*Chem. Abstr.*, 1975, **83**, 114 751q).
124. S. Harusawa, R. Yoneda, T. Kuritiara, Y. Hamada and T. Shiara, *Tetrahedron Lett.*, 1984, **25**, 427; M. Oda, A. Yamamuro and T. Watabe, *Chem. Lett.*, 1979, 1427.
125. R. W. Freerken, M. L. Raggio, C. A. Thomas and D. S. Watt, *J. Org. Chem.*, 1979, **44**, 702, and refs. cited therein.
126. R. W. Frank and T. V. John, *J. Org. Chem.*, 1980, **45**, 1170, and refs. cited therein.
127. R. T. Taylor, C. R. Degenhardt, W. P. Melega and L. A. Paquette, *Tetrahedron Lett.*, 1977, 159.
128. T. M. Chan, A. Baldassarre and D. Massuda, *Synthesis*, 1976, 801.
129. K. Tamao and K. Maeda, *Tetrahedron Lett.*, 1986, **27** 65.
130. B. M. Trost and A. C. Lavoie, *J. Am. Chem. Soc.*, 1983, **105**, 5075, and refs. cited therein.
131. F. Akiyama, *Bull. Chem. Soc. Jpn.*, 1977, **50**, 936, and refs. cited therein.
132. E. Nagashima, K. Suzuki, M. Ishikawa and M. Sekiya, *Heterocycles*, 1985, **23**, 1873.
133. B. M. Trost and Y. Tanigawa, *J. Am. Chem. Soc.*, 1979, **101**, 4413.
134. Y. Fukuyama and T. Tokoroyama, *Tetrahedron Lett.*, 1973, 4869.
135. (a) W. Oppolzer and K. K. Mahalanabis, *Tetrahedron Lett.*, 1975, 3411; (b) T. Sone, S. Terashima and S. Yamada, *Synthesis*, 1974, 725; (c) J. A. Marshall and G. L. Bundy, *J. Chem. Soc., Chem. Commun.*, 1966, 500; (d) S. Ohnishi and Y. Osawa, *J. Chem. Soc., Chem. Commun.*, 1986, 1089.
136. L. H. Zalkow and J. W. Ellis, *J. Org. Chem.*, 1964, **29**, 2626.
137. L. H. Zalkow, J. W. Ellis and M. R. Brennan, *J. Org. Chem.*, 1963, **28**, 1705.
138. J. W. Ellis, *J. Org. Chem.*, 1969, **34**, 1154.
139. G. J. Williams and N. R. Hunter, *Can. J. Chem.*, 1976, **54**, 3830, and refs. cited therein.
140. N. K. Dunlap, M. R. Sabol and D. S. Watt, *Tetrahedron Lett.*, 1984, **25**, 5839.
141. S. J. Danishefsky and M. Bednarski, *Tetrahedron Lett.*, 1985, **26**, 3411.
142. T. Ohnuma, K. Seki, T. Oishi and Y. Ban, *J. Chem. Soc., Chem. Commun.*, 1974, 296.
143. H. Irie, J. Katakawa, M. Tomita and Y. Mizuno, *Chem. Lett.*, 1981, 637.
144. J. W. Huffman, R. C. Desai and G. F. Hillenbrand, *J. Org. Chem.*, 1984, **49**, 982.
145. G. M. Rubottom, J. M. Gruber, H. D. Juve, Jr. and D. A. Charleson, *Org. Synth.*, 1985, **64**, 118.
146. G. M. Rubottom and J. M. Gruber, *J. Org. Chem.*, 1978, **43**, 1599.
147. E. Colvin and S. Cameron, *Tetrahedron Lett.*, 1988, **29**, 493.
148. C. Iwata, Y. Takemoto, A. Nakamura and T. Imanishi, *Tetrahedron Lett.*, 1985, **26**, 3227, and following paper.
149. G. M. Rubottom and J. M. Gruber, *J. Org. Chem.*, 1977, **42**, 1051.
150. T. Hosakawa, S. Inui and S.-I. Murahashi, *Chem. Lett.*, 1983, 1081.
151. H. O. House, in 'Modern Synthetic Reactions', Benjamin, Menlo Park, CA, 1972, p. 316.
152. H. O. House and R. L. Wasson, *J. Am. Chem. Soc.*, 1957, **79**, 1489.
153. R. M. Moriarty, S. C. Gupta, H. Hu, D. R. Berenschot and K. B. White, *J. Am. Chem. Soc.*, 1981, **103**, 686.
154. R. M. Moriarty and H. Hu, *Tetrahedron Lett.*, 1981, **22**, 2747.
155. E. J. Corey and H. E. Ensley, *J. Am. Chem. Soc.*, 1975, **97**, 6908.
156. H. H. Wasserman and B. H. Lipshutz, *Tetrahedron Lett.*, 1975, 1731.
157. P. R. Ortiz de Montellano and C. K. Hsu, *Tetrahedron Lett.*, 1976, 4215.
158. Y. Morizawa, A. Yasuda and K. Uchida, *Tetrahedron Lett.*, 1986, **27**, 1833.
159. S. Takano, M. Morimoto and K. Ogasawara, *J. Chem. Soc., Chem. Commun.*, 1984, 82.
160. R. Gamboni and C. Tamm, *Tetrahedron Lett.*, 1988, **29**, 3999, and refs. cited therein.
161. E. J. Corey, M. Kang, M. C. Desai, A. K. Ghosh and I. N. Houpis, *J. Am. Chem. Soc.*, 1988, **110**, 649.
162. G. M. Rubottom and R. Marrero, *Synth. Commun.*, 1981, **11**, 505.
163. W. Adam and J. del Fierro, *J. Org. Chem.*, 1978, **43**, 1159.
164. G. M. Rubottom, J. M. Gruber, R. Marrero, H. D. Juve, Jr. and C.-W. Kim, *J. Org. Chem.*, 1983, **48**, 4940.
165. W. Oppolzer and P. Dudfield, *Helv. Chim. Acta*, 1985, **68**, 216.
166. M. Y. Kim, J. E. Starrett and S. M. Weinreb, *J. Org. Chem.*, 1981, **46**, 5383.
167. R. M. Williams, *Tetrahedron Lett.*, 1981, **22**, 2341.
168. T. Ohta, A. Hosoi and S. Nozoe, *Tetrahedron Lett.*, 1988, **29**, 329.
169. F. A. Davis and L. C. Vishwakarma, *Tetrahedron Lett.*, 1985, **26**, 3539.
170. F. A. Davis, *J. Org. Chem.*, 1987, **52**, 5288.
171. D. A. Evans, M. M. Morrissey and R. L. Dorow, *J. Am. Chem. Soc.*, 1985, **107**, 4346.
172. M. P. Gore and J. C. Vederas, *J. Org. Chem.*, 1986, **51**, 3700.
173. E. C. Taylor, H. W. Atland and G. McGillivray, *Tetrahedron Lett.*, 1970, 5285.
174. G. W. Moersch and M. L. Zwiesler, *Synthesis*, 1971, 647.
175. D. A. Konen, L. S. Silbert and P. E. Pfeffer, *J. Org. Chem.*, 1975, **40**, 3253; W. Adam and O. Cueto, *J. Org. Chem.*, 1977, **42**, 38.
176. G. Rubottom and R. Marrero, *J. Org. Chem.*, 1975, **40**, 3783.
177. W. Adam and J. C. Liu, *J. Am. Chem. Soc.*, 1972, **94**, 2894.
178. A. Hassner, R. H. Reuss and H. W. Pinnick, *J. Org. Chem.*, 1975, **40**, 3427.
179. G. M. Rubottom, R. Marrero and J. M. Gruber, *Tetrahedron*, 1983, **39**, 861.
180. H. O. House and F. A. Richey, Jr., *J. Org. Chem.*, 1969, **34**, 1430.
181. C. H. Cumins and R. M. Coates, *J. Org. Chem.*, 1983, **48**, 2070.
182. G. S. Reddy and M. V. Bhatt, *Synthesis*, 1981, 223.
183. M. E. Jung, P. A. Blair and J. A. Lowe, *Tetrahedron Lett.*, 1976, 1439.

184. F. E. Henoch, K. G. Hampton and G. R. Hauser, *J. Am. Chem. Soc.*, 1968, **91**, 676.
185. D. Enders and V. Bhushan, *Tetrahedron Lett.*, 1988, **29**, 2437.
186. L. Troisi, L. Cassidei, L. Lopez, R. Mello and R. Curci, *Tetrahedron Lett.*, 1989, **30**, 257.

2.4
Oxidation Adjacent to Sulfur

MICHAEL KENNEDY and M. ANTHONY McKERVEY
University College Cork, Republic of Ireland

2.4.1 INTRODUCTION

In this chapter oxidation of an activated C—H bond adjacent to a sulfur atom refers to any process whereby a C—H bond at the α-position of an alkyl sulfide is replaced by a C—X bond, where X is a halogen atom or an oxygen-, nitrogen-, carbon-, or sulfur-based substituent (equation 1). Processes in which a stabilized anion is generated adjacent to the sulfur atom of a sulfide, sulfoxide or sulfone and subsequently used as a nucleophile in addition or substitution reactions are excluded from this section. The use of such anions in organic synthesis is dealt with in Volume 1, Chapter 2.3.

$$R^1S(R^2)(R^3)C\text{—}H \longrightarrow R^1S(R^2)(R^3)C\text{—}X \quad (1)$$

Conventional oxidants are little used for bringing about the functionalization reactions encompassed by equation (1), largely because of the ease with which oxidation occurs at the sulfur atom producing sulfoxides and sulfones. Nevertheless, the oxidation of sulfide to sulfoxide is very significant within the context of equation (1) since it provides the first stage of a valuable two-stage route from sulfides to α-functionalized sulfides. The second stage, involving reduction of the sulfoxide group with concomitant oxidation of the α-carbon atom, is often referred to as the Pummerer rearrangement. It is in fact one of two closely related processes by which the oxidation in equation (1) can most readily be accomplished. The second is ionic halogenation, in practice predominantly chlorination (Scheme 1). These two functionalization procedures have been in widespread use for many years, reflecting the very considerable utility of the reaction products as intermediates in organic synthesis (*vide infra*).

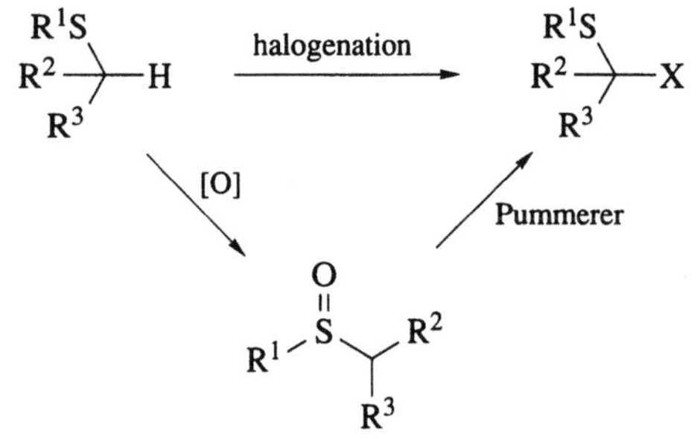

$$R^1S(R^2)(R^3)C{-}H \xrightarrow{\text{halogenation}} R^1S(R^2)(R^3)C{-}X$$

$$R^1S(R^2)(R^3)C{-}H \xrightarrow{[O]} R^1S(=O)C(R^2)(R^3)H \xrightarrow{\text{Pummerer}} R^1S(R^2)(R^3)C{-}X$$

Scheme 1

2.4.2 THE PUMMERER REARRANGEMENT

In the early 1900s Pummerer observed that exposure of phenylsulfinylacetic acid (**1**) to either acetic acid or dilute sulfuric acid led to the formation of benzenethiol, glyoxylic acid and bis(thiophenoxy)acetic acid (equation 2).[1] Smythe, in the same year, found that dibenzyl sulfoxide in hot hydrochloric acid behaved rather similarly, producing phenylmethanethiol and benzaldehyde (equation 3).[2] Pummerer's study also contained the observation that exposure of the ethyl ester of acid (**1**) to hot acetic anhydride furnished the α-acetoxy sulfide shown in equation (4). These three reactions all contain the elements of a transformation which later was to become known as the Pummerer rearrangement. In equations (2) and (3) rearrangement produces unstable intermediates from which the products isolated are derived, whereas the rearrangement in equation (4) yields a stable product whose structure more clearly typifies the essential features of the reaction, namely reduction of the sulfoxide to sulfide with concomitant oxidation of the α-carbon atom. Originally, the term 'Pummerer rearrangement' was used to describe oxidation in the literal sense of replacement of an α-C—H bond by an oxygenated group. However, many later applications were to involve functionalization with carbon-, nitrogen-, sulfur- and halogen-based groups. Furthermore, the realization that Pummerer rearrangements can be brought about intramolecularly has proved especially effective in carbocyclic and heterocyclic synthesis.

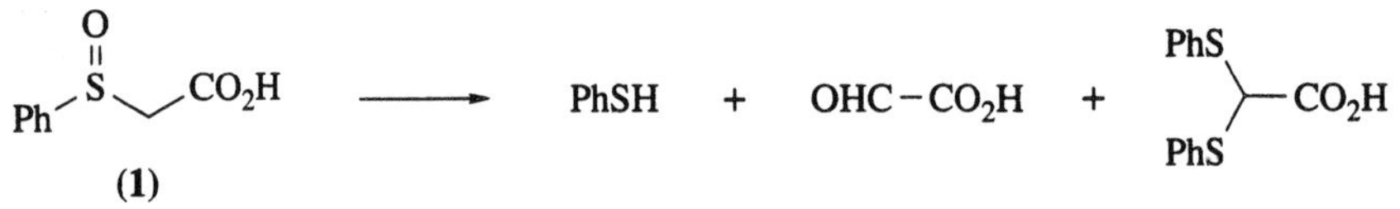

$$\underset{(\mathbf{1})}{PhS(O)CH_2CO_2H} \longrightarrow PhSH + OHC-CO_2H + (PhS)_2CH-CO_2H \tag{2}$$

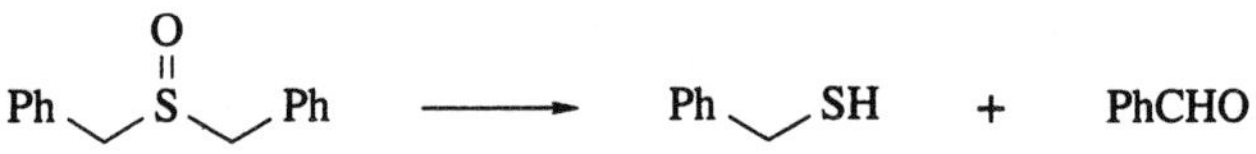

$$PhCH_2S(O)CH_2Ph \longrightarrow PhCH_2SH + PhCHO \tag{3}$$

$$PhS(O)CH_2CO_2Et \longrightarrow PhSCH(OAc)CO_2Et \tag{4}$$

Use of the Pummerer rearrangement to produce α-functionalized sulfides presupposes access to sulfoxides from sulfides. Fortunately, many alkyl sulfides can be synthesized from readily available compounds and several reliable methods exist for their efficient conversion to sulfoxides. The latter include the use of (with respective examples in equations 5–9) *m*-chloroperbenzoic acid,[3] monoperphthalic acid magnesium salt,[4] sodium periodate,[5] hydrogen peroxide[6] and *t*-butyl hypochlorite.[7] Details of the use of these and other less common oxidants are available in Volume 7, Chapter 6.2. Madesclaire's recent extensive compilation of oxidants is also available.[8] Enzymic methods include the use of chloroperoxidase to catalyze oxygen transfer to sulfides.[9] β-Keto sulfoxides are particularly useful substrates for Pummerer rearrangement. Reaction can be brought about under very mild conditions and the presence of the β-keto group ensures the regioselectivity of the group transfer. β-Keto sulfoxides are readily available from methyl esters and dimethyl sulfoxide in the presence of base.[10]

Electrophilic reagents capable of bringing about Pummerer rearrangement include inorganic and organic acids, carboxylic anhydrides, acyl halides, isocyanates, carbodiimides, trimethylsilyl halides and triflate, sulfonyl and sulfenyl halides, phosphorus pentoxide and many typical Lewis acids such as boron

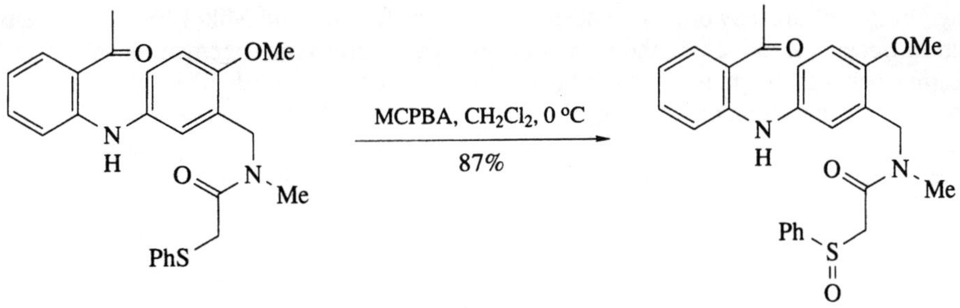

MCPBA, CH_2Cl_2, 0 °C
87% (5)

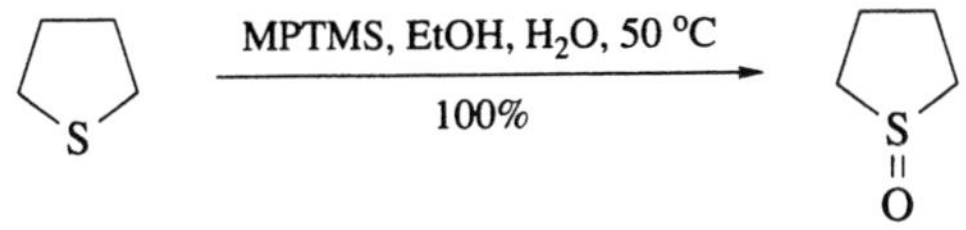

MPTMS, EtOH, H_2O, 50 °C
100% (6)

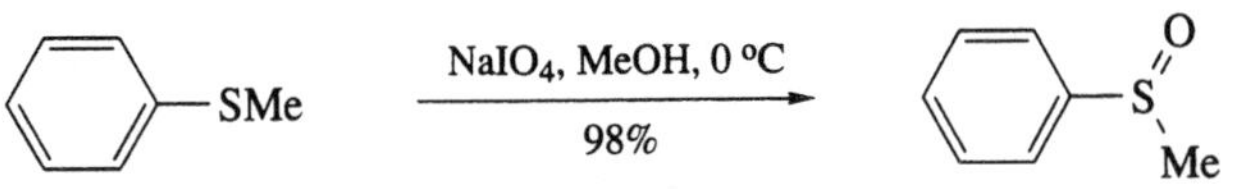

$NaIO_4$, MeOH, 0 °C
98% (7)

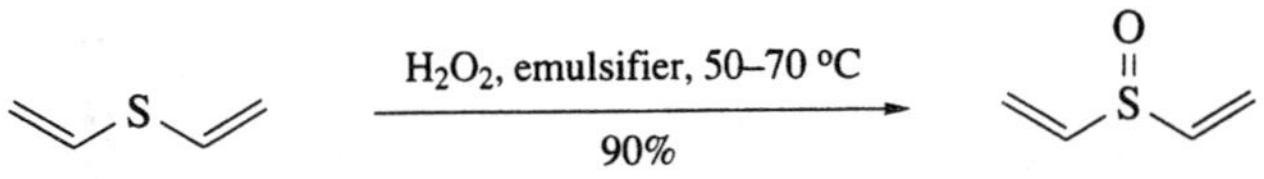

H_2O_2, emulsifier, 50–70 °C
90% (8)

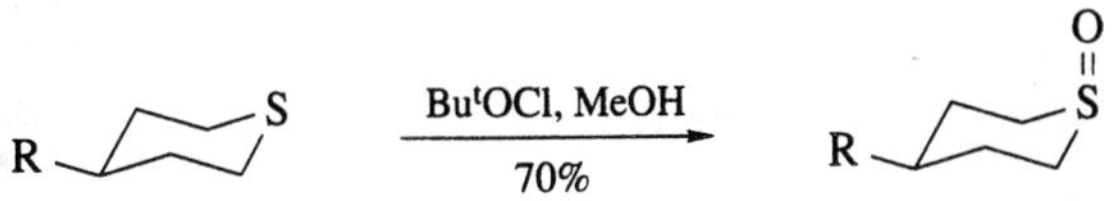

Bu^tOCl, MeOH
70% (9)

trifluoride, boron trichloride, silicon tetrachloride, phosphorus pentachloride and phosphorus oxychloride. Of these, carboxylic anhydrides are probably the most serviceable in synthetic procedures.

The generally accepted mechanism of the Pummerer rearrangement is one in which there is an initial attack on the sulfoxide oxygen atom by an electrophilic species, *e.g.* protonation or acylation. The latter process with methyl phenyl sulfoxide as the substrate is used in Scheme 2 to illustrate the likely mechanistic details. Acylation is followed by proton abstraction by a base (which in this case is the conjugate of the electrophile, though it need not be) from the α-carbon atom of the sulfoxide to form an ylide, which rapidly eliminates an acetate ion to form the α-thiocarbocation. Addition of acetate ion to the α-thiocarbocation completes the formation of the α-functionalized sulfide. The mechanism of ionic halogenation of alkyl sulfides can similarly be understood (*vide infra*). Ylide formation from sulfoxonium salts is well recognized and this aspect of the mechanism has received considerable experimental support. Although not definitive, there is also evidence to suggest that transfer of the acetoxy group from sulfur to carbon may be either intramolecular or may involve an intimate ion pair. Some of the evidence arises from the results of a study of Pummerer rearrangement of optically active sulfoxide where the asymmetry is transferred from sulfur to carbon with, in some instances, high enantioselectivity. For example, ethyl *p*-tolylsulfinylacetate with dicyclocarbodiimide in acetic anhydride furnishes the α-acetoxy sulfide shown in equation (10) with an *ee* of about 70%. Other examples of asymmetric synthesis in Pummerer rearrangement will be discussed later. ^{18}O-Labeling studies have been interpreted in terms of both inter- and intramolecular mechanisms. For a much fuller account of mechanistic aspects of the rearrangement see the

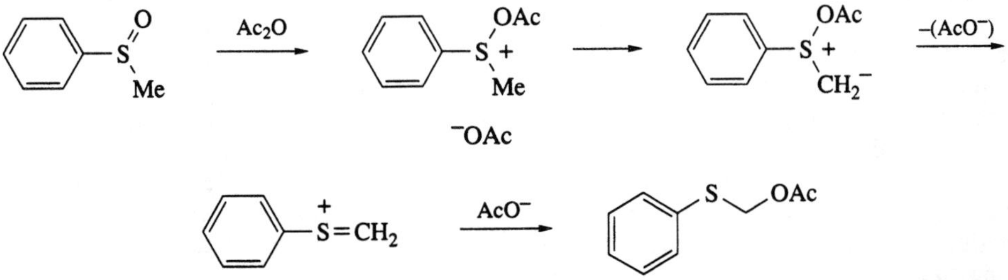

Ac_2O → $-(AcO^-)$ → AcO^-

Scheme 2

analysis by Wolfe and his coworkers[11] and the review by Russell and Mikol.[12,13] The extension of the mechanism in Scheme 2 to include the intramolecular Pummerer rearrangement is shown in Scheme 3, the key feature being the capture of the α-thiocarbocation (or its equivalent) by an internal nucleophile thus completing the formation of a carbocyclic or heterocyclic ring. The discussion of the applications of the Pummerer rearrangement that follows has been organized according to the nature of the nucleophilic addition step which completes the functionalization sequence shown in Scheme 2.

Ac_2O, DCC (10)

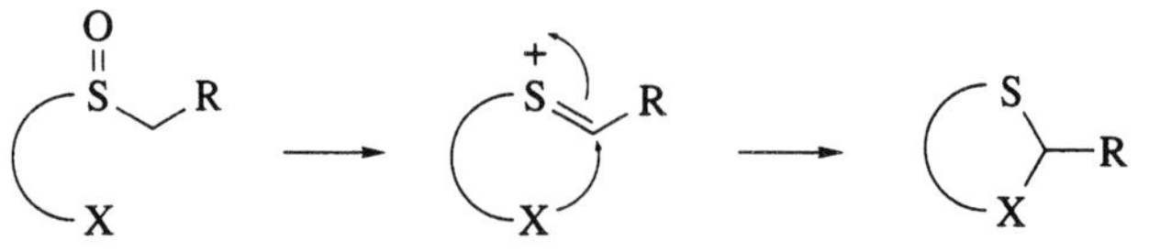

Scheme 3

2.4.3 THE PUMMERER REARRANGEMENT EXEMPLIFIED

2.4.3.1 Reaction with Carboxylic Anhydrides

α-Acetoxylation is probably the most commonly encountered form of the rearrangement. Acetic anhydride alone or accompanied by sodium acetate are popular reagents, though reaction usually requires elevated temperatures, often at reflux for several hours. Acetic anhydride containing dicyclohexylcarbodiimide has also been used. Convenient rearrangement rates can be realized at much lower temperatures (0–25 °C) if trifluoroacetic anhydride is employed as the reagent. Examples of the use of mixtures of acetic and trifluoroacetic anhydrides, with and without sodium acetate, are also available. Table 1 contains a selection of representative examples of the α-acetoxylation of alkyl sulfides *via* Pummerer rearrangement of sulfoxides.

Dimethyl sulfoxide in hot acetic anhydride (entry 1) furnishes acetoxymethyl methyl sulfide. Reaction of acetic anhydride with methyl alkyl sulfoxides not containing a β-keto or other acid strengthening substituent, *e.g.* entry 2, leads to functionalization of the methyl group. The reactivity series for alkyl substituents in the sulfoxide appears to be methyl > *n*-alkyl > vinyl. Entry 3 provides an example of a vinyl sulfoxide rearrangement in acetic anhydride. Entry 4 illustrates rearrangement in an amino acid substrate. The examples shown in entries 5 and 6 are part of an aldehyde synthesis; the yields quoted refer to the product obtained after hydrolysis of the α-acetoxy sulfide. Entry 7 contains a comparison of the use of acetic acid alone and admixed with trifluoroacetic anhydride. Entries 8 and 9 contain β-keto sulfoxides which are particularly popular substrates; where there is a choice, as in entry 9, rearrangement invariably occurs so as to replace the more acidic hydrogen atom between the keto and sulfide moieties. Other activated systems which undergo ready rearrangement are those containing the phosphoryl and cyano groups shown in entries 10 and 11. An additional feature of the former is that use of the sulfoxide in homochiral form yields α-acetoxy-α-(dimethylphosphoryl)methyl *p*-tolyl sulfide of 24% optical purity, a result used as evidence for an acetoxy migration *via* an intramolecular mechanism. Cyclic sulfoxides are also amenable to α-acetoxylation as shown by the examples in entries 12 and 13. Entry 14 is an example of the Pummerer rearrangement in the Sharpless–Masamune carbohydrate synthesis, the α-acetoxy sulfide being converted into a primary hydroxyl group by reduction with lithium aluminum hydride. Two other uses of the rearrangement are summarized in entries 15 and 16, the latter as part of side chain elaboration for cardenolide analogs.

These various intermolecular α-acetoxylation reactions have intramolecular counterparts. For example, treatment of the sulfinylbutanoic acid shown in equation (11) with acetic anhydride containing *p*-toluenesulfonic acid yields a sulfenylated butanolide, the carboxylic acid function having intercepted the α-thiocarbocation intermediate.[29] Yet another demonstration of the intramolecular process, due to Allenmark,[30] is the cyclization of *o*-carboxyphenyl benzyl sulfoxide with acetic anhydride to form the 1,3-benzoxathian-4-one shown in equation (12). This reaction was also conducted with one of the

Table 1 Acetoxylation of Sulfides from Sulfoxides *via* the Pummerer Rearrangement

Entry	*Substrate*	*Reagents*	*Product*	*Yield (%)*	*Ref.*
1	MeS(O)Me	Ac_2O, heat	$MeSCH_2OAc$	—	14
2	$Bu^nS(O)Me$	Ac_2O, heat	Bu^nSCH_2OAc	—	15
3	PhCH=CHS(O)Me	Ac_2O, heat	$PhCH=CHSCH_2OAc$	—	16
4	$HO_2CCH(NH_2)CH_2CH_2S(O)Me$	Ac_2O, heat	$HO_2CCH(NHAc)CH_2CH_2SCH_2OAc$	—	15
5	$PhCH_2S(O)Ph$	$(CF_3CO)_2O$, 0 °C	$PhCH(OCOCF_3)SPh$	—	17
6	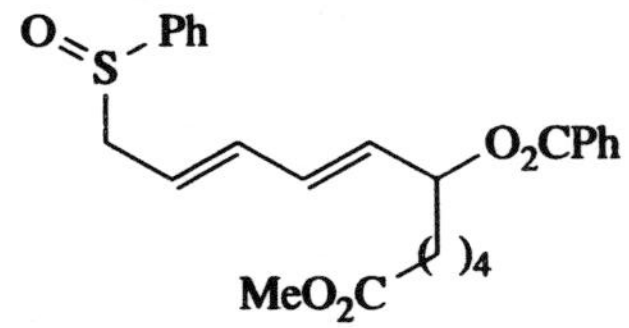	Ac_2O, $(CF_3CO)_2O$, 2,6-lutidine	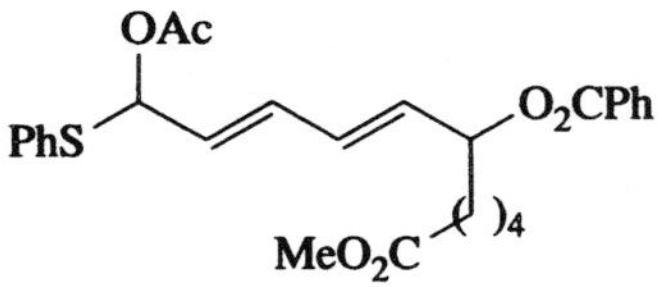	>65	18
7a	$Me(CH_2)_4CH_2S(O)Ph$	Ac_2O, heat	$Me(CH_2)_4CH(OAc)SPh$	62	19
7b	$Me(CH_2)_4CH_2S(O)Ph$	Ac_2O, $(CF_3CO)_2O$, 0.5 h, 20 °C	$Me(CH_2)_4CH(OAc)SPh$	84	19
8	$MeC_6H_4S(O)CH_2CO_2Et$	Ac_2O, DCC, heat	$MeC_6H_4SCH(OAc)CO_2Et$	43	20
9	$PhCOCH_2S(O)Me$	Ac_2O, heat	PhCOCH(OAc)SMe	95	21
10	$(EtO)_2P(O)CH_2S(O)Me$	$(CF_3CO)_2O$, –78 °C, 15 min	$(EtO)_2P(O)CH(OCOCF_3)SMe$	76	22
11	$PhS(O)CH_2CN$	Ac_2O, heat	PhSCH(OAc)CN	85–90	23
12	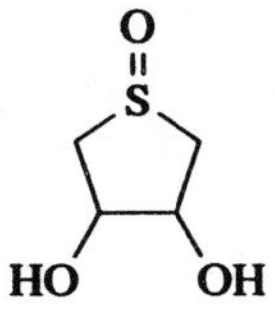	Ac_2O, NaOAc, heat	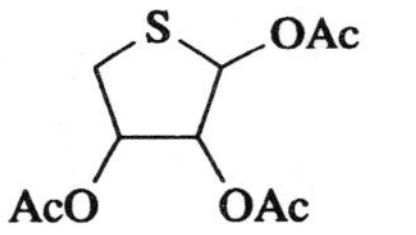	61	24
13a		Ac_2O, 100 °C	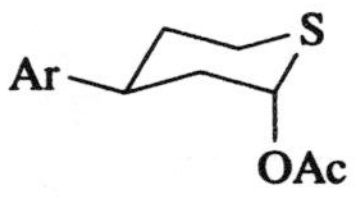	70 (from *cis* and *trans*)	25
13b		Ac_2O, DCC	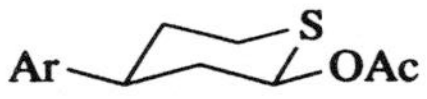	83	

Table 1 *(continued)*

Entry	*Substrate*	*Reagents*	*Product*	*Yield (%)*	*Ref.*
14	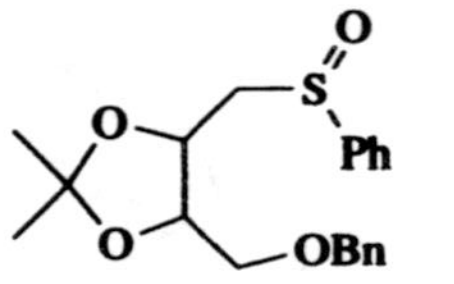	Ac_2O, heat	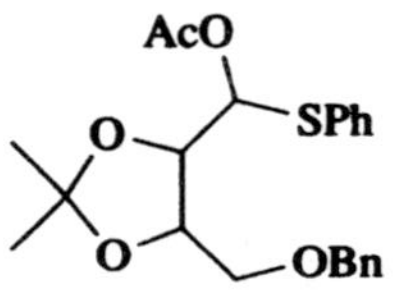	81	26
15	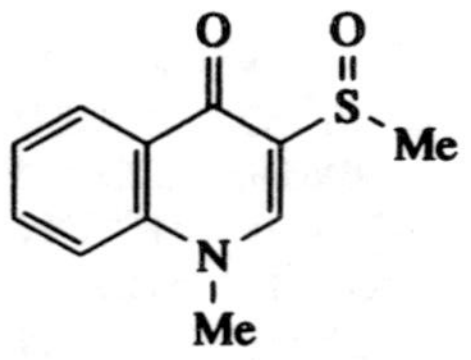	Ac_2O, heat	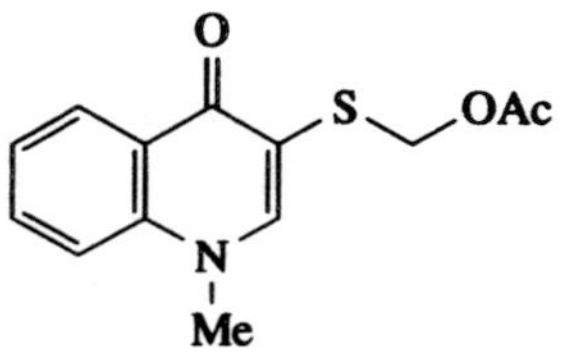	92	27
16	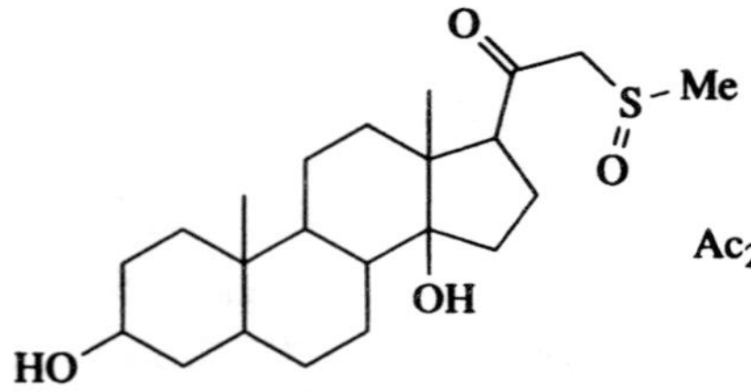	Ac_2O, pyridine, 20 °C	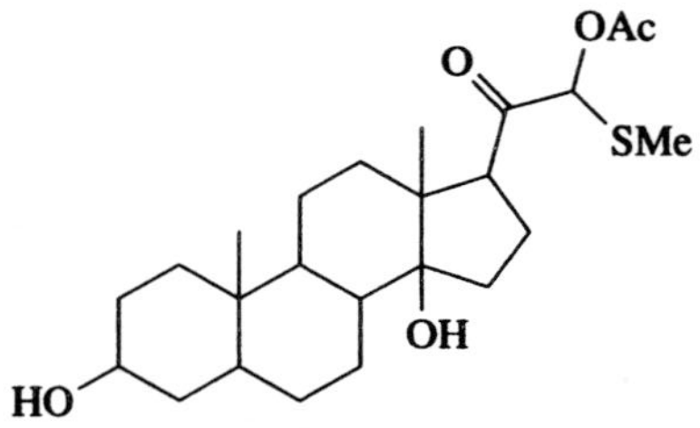	55 33 (3-OAc)	28

enantiomers of the sulfoxide and asymmetry transfer to the extent of 11% was observed.[30] When the reaction was brought about by dicyclohexylcarbodiimide in dichloroethane, the optical purity increased to 30%.[11] This asymmetric synthesis has been interpreted as providing evidence for the formation of diastereoisomeric ylides as reaction intermediates.

Ac_2O, *p*-TsOH, PhMe; 63% (11)

Ac_2O (12)

2.4.3.2 Pummerer Rearrangement with α-Alkylation and Arylation

Various kinds of α-alkyl and α-aryl substituted sulfides are accessible by Pummerer rearrangement where the α-thiocarbocation intermediate, or its equivalent, is intercepted, either intermolecularly or intramolecularly, by a nucleophilic carbon species. The most useful versions of this process involve interception by a carbon–carbon double bond. The transformation of the sulfoxide in equation (13) into the sulfide shown using 1-pentene in the presence of trifluoroacetic anhydride provides an illustrative example of the intermolecular process.[31] The nucleophilicity of silyl enol ethers has also been exploited for carbon–carbon bond formation with Pummerer intermediates, the example in equation (14) having been brought about at low temperature using trimethylsilyl triflate and diisopropylethylamine as the reagents.[32] A combination of trimethylsilyl triflate and Hunig's base in dichloromethane at –78 °C has been used to combine the silyl enol ether of cyclohexanone with the bifunctional reagent 3-phenylsulfinyl-2-(trimethylsilylmethyl) propene as shown in equation (15).[33] The intramolecular mode, which is in effect a π-route cyclization, has extended the usefulness of the Pummerer rearrangement very considerably. For example, the acyclic sulfoxide in equation (16) has been converted into the cyclic products shown by the action of trifluoroacetic anhydride.[34] The construction of a five-membered ring by the same approach is shown in equation (17) with the additional significant feature that use of the sulfoxide precursor in homochiral form produced the bicyclic sulfide with an enantiomeric excess of 75%.[35] Bridged-ring structures are also accessible, two notable successes in this area being the cyclization leading to bicyclo[3.2.1]octane derivatives in equation (18) reported by Mander and Mundill[36] and that in equation (19) due to Magnus and his coworkers, where the participating double bond is provided by the indole system.[37]

i, TFAA, 20 °C; ii, 1-pentene, TFA, 0 °C; 86% (13)

Me_3SiOTf, Pr^i_2NEt, –78 °C; 63% (14)

Me_3SiOTf, Pr^i_2NEt; 75% (15)

TFAA, CH_2Cl_2, r.t. 43% 35% (16)

TFAA, CH_2Cl_2, 20 °C 87% (17)

TFAA, PhH, 5 °C (18)

TFAA, Bu^t N Bu^t, 0–110 °C 65% (19)

Participation by aromatic rings is also possible and there are now several examples of electrophilic aromatic substitution involving Pummerer intermediates. Equation (20), the alkylation of benzene with dimethyl sulfoxide in trifluoroacetic anhydride, illustrates the process in its simplest form.[38] As with alkenes, reaction with aromatics has been more widely exploited in intramolecular versions for the construction of carbocycles and heterocycles. In many cases the sulfoxide precursor is of the β-keto variety, thus ensuring regiospecificity in the point of cyclization. Equation (21) (formation of a six-membered carbocycle),[39] equation (22) (formation of a six-membered sulfur heterocycle),[41] equation (23) (formation of a six-membered nitrogen heterocycle)[40] and equation (24) (formation of a seven-membered nitrogen, sulfur heterocycle)[42] provide illustrations of the versatility of this form of intramolecular aromatic alkylation.

i, TFAA, 0 °C; ii, $SnCl_4$ 62% (20)

TFAA, PhH 85% (21)

p-TsOH, benzene, reflux; 86% (22)

H_2SO_4, AcOH, reflux; 37% (23)

p-TsOH, PhH, reflux; 76% (24)

2.4.3.3 Pummerer Rearrangement with Participation by Nitrogen

A combination of the Pummerer rearrangement and the Ritter reaction occurs in the reaction of acetonitrile with methyl phenyl sulfoxide (equation 25) in a mixture of trifluoroacetic acid and its anhydride, although a substantial amount of the normal α-acetoxylation also occurs.[43] Participation by amido groups is also possible, the interest here being largely in the construction of lactams *via* the intramolecular cyclization mode. Whereas Wolfe and his coworkers were unable to find conditions for the cyclization of *S*-phenylcysteinamide sulfoxides under Pummerer conditions, Kaneko found that variously substituted 3-phenylsulfinylpropionamides (equation 26) were converted into 4-(phenylthio)-2-azetidinones in 14–50% yields by the combined action of trimethylsilyl triflate and triethylamine.[44] Kaneko has suggested that the sulfonium ion intermediate in this rearrangement may be considered as a chemical equivalent of an intermediate believed to be involved in the biosynthesis of β-lactam antibiotics. Keneko and his coworkers have extended this study to include the enantioselective Pummerer cyclization of homochiral

TFAA/TFA, MeCN, 0 °C; 45% (25)

Me_3SiOTf, NEt_3; CH_2Cl_2, –20 °C (26)

R¹	*R²*	*Yield (%)*
R^1 = Me	R^2 = OBz	51
R^1 = H	R^2 = OBz	14
R^1 = H	R^2 = H	41

3-phenylsulfinylpropionamide to produce the sulfenylated β-lactam with 67% *ee*. In this case the reagents employed were trimethylsilyl triflate and diisopropylethylamine.[45] An example of δ-lactam formation from an amide is shown in equation (27), the reaction being initiated by a ketone silyl acetal in the presence of zinc iodide in acetonitrile.[46] Equation (28) shows a variant of amide participation operating through the oxygen atom leading to the construction of a seven-membered heterocycle.

MeO–C(=CH$_2$)–OSiBu^{t}Me$_2$, MeCN, ZnI$_2$ (27)

Ac$_2$O (28)

Use of trimethylsilyl triflate to bring about Pummerer rearrangement requires the presence of a base such as a tertiary amine (*vide supra* equations 15 and 26). In some instances, involving attempts to alkylate Pummerer intermediates with silyl enol ethers under such conditions, the base has been found to compete as a nucleophile.[33] In the absence of the silyl enol ether, amine addition can be very efficient. For example, treatment of methallyl phenyl sulfoxide with diisopropylethylamine and trimethylsilyl triflate in dichloromethane (equation 29) at 0 °C yields the ammonium triflate indicated in 91% yield.[33] Other tertiary amines which undergo this reaction include triethylamine and *N,N*-diethyltrimethylsilamine. In the latter case with allyl phenyl sulfoxide as the substrate and a mildly acidic work-up, the Mannich derivative shown in equation (30) can be obtained in 90% yield.[33]

Pr^i_2NEt, Me_3SiOTf, CH_2Cl_2, 0 °C; 91% (29)

Et_2NSiMe_3, Me_3SiOTf; 90% (30)

2.4.3.4 Pummerer Rearrangement in Hydroxylic Solvents

Pummerer[1] and Smythe[2] were the first to note that aqueous acids cause sulfoxides with an α-hydrogen atom to decompose to aldehydes, thiols and other products. In some cases, however, particularly those involving β-keto sulfoxides, stable mixed hemithioacetals, the primary Pummerer rearrangement products, may be isolated. Such a case is summarized in equation (31), where exposure of the phenothiazine sulfoxide shown to *p*-toluenesulfonic acid hydrate in tetrahydrofuran at 65 °C furnished the mixed hemithioacetal in 95% yield;[47] in this particular example aromatic participation in the Pummerer process did not occur. An indirect route to mixed hemithioacetals, shown in equation (32), involves Pummerer rearrangement in trifluoroacetic anhydride to bring about α-acetoxylation (*cf.* Section 2.4.3.1) followed by hydrolysis or solvolysis of the product. In this case an α-acetoxy intermediate is produced, which is cleaved to the hemithioacetal on treatment with sodium methoxide.[48] In another example addition of ethanol to the formyl group leads intramolecularly to attack on the α-acetoxy group with formation of the bridged product (equation 33).[49] Both intermolecular and intramolecular participation by hydroxy groups in Pummerer processes are known. Methylation of cyclopropyl phenyl sulfoxide (equation 34) with trimethyloxonium tetrafluoroborate, followed by treatment with methoxide ion in methanol produces the mixed thioacetal.[50] The conversion of 2-hydroxy-3-methoxy-l-methylsulfinylacetylbenzene (equation 35) into a benzofuranone on exposure to phosgene in pyridine illustrates the intramolecular version of hydroxy group participation.[51] Treatment of the vinyl sulfide shown in equation (36) with toluenesulfonic acid produces a tetrahydrofuran derivative whose formation probably also results from a thiocarbocation intermediate.[52]

p-TsOH, THF, 65 °C
90–95% (31)

i, TFAA, –60 °C
ii, NaOMe/MeOH
54% (32)

i, TFAA, C_6H_6, 20 °C
ii, EtOH/H_2O
43% (33)

i, Me_3OBF_4
ii, NaOMe/MeOH, 20 °C
54% (34)

$(COCl)_2$, py, 80 °C
48% (35)

p-TsOH (36)

2.4.3.5 Miscellaneous and Abnormal Pummerer Rearrangements

Among other electrophilic reagents capable of bringing about the Pummerer rearrangement are halides of organic and inorganic acids. As these halides transform sulfoxides into α-chlorosulfides they complement the sulfide chlorination route to these compounds. Thionyl chloride reacts readily with sulfoxides and β-keto sulfoxides; methyl phenyl sulfoxide furnishes chloromethyl phenyl sulfide (equation 37).[53] Benzoyl chloride and acetyl chloride behave similarly.[53] Cyanuric chloride is transformed into cyanuric acid by dimethyl sulfoxide, which in turn is transformed into methyl chloromethyl sulfide (equation 38).[54,55]

In the examples of the Pummerer rearrangement presented above, the common feature has been the formation of an α-thiocarbocation, which is then captured either intermolecularly or intramolecularly by a nucleophilic species. There are of course other outlets through which the α-thiocarbocation can pro-

$PhS(O)Me \xrightarrow{SOCl_2} PhSCH_2Cl$ (37)

Me_2SO + cyanuric chloride ⟶ $MeSCH_2Cl$ + cyanuric acid (38)

gress to a stable neutral product, such as β-elimination. The reaction of cyclohexyl phenyl sulfoxide (equation 39) with trifluoroacetic anhydride and triethylamine to form 1-phenylthiocyclohexene is an illustrative example.[56] β-Elimination is also the main pathway of the Pummerer rearrangements summarized in equations (40) and (41). The reagents in these cases were trimethylsilyl chloride[7] and trifluoroacetic anhydride–lutidine,[57] respectively. The Pummerer process shown in equation (42) culminates in a δ-proton elimination.[58] A final example of the eliminative process is provided by the reaction shown in equation (43), which is brought about by heat alone and has been referred to as the 'sila-Pummerer' rearrangement.[59] A vinylogous Pummerer rearrangement has been observed in the reaction of the vinyl sulfoxide shown in equation (44) with acetic anhydride.[60]

TFAA, Et_3N, CH_2Cl_2; 97% (39)

Me_3SiCl, CCl_4, reflux; 88% — 82% + 18% (40)

TFAA, lutidine, 25 °C; 37% — CO_2Me, $CO_2C_6H_4(NO_2)$-*p* (41)

Ac_2O, 70 °C; 99% (42)

PhH, reflux; 88% — $R—≡—R$ 37% + SPh 19% + $OSiMe_3$, SPh 32% (43)

R = 2-(succinimido)ethyl

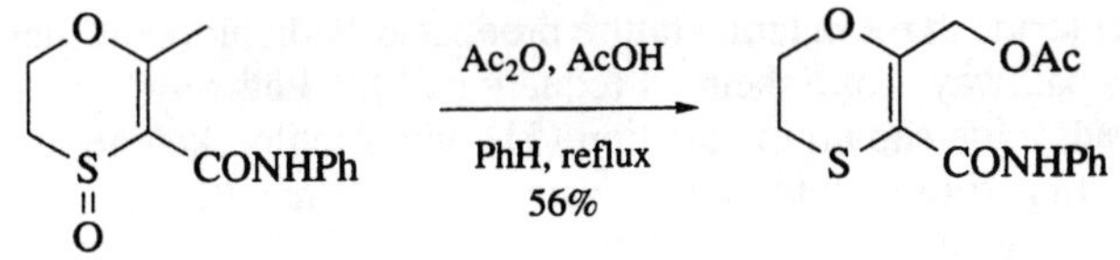

Ac$_2$O, AcOH; PhH, reflux; 56% (44)

This section concludes with a selection of reactions which are believed to follow the Pummerer pathway, but which lead to abnormal or unexpected products. For example, treatment of the sulfoxide shown in equation (45) with acetic anhydride produces two sulfides in a 5:1 ratio.[7] The major product is that of a ring contraction process which could involve a bicyclic episulfonium ion as an intermediate. Attack by acetate can occur at two sites yielding the products shown. C—S bond cleavage is also a feature of the Pummerer rearrangement of penicillin sulfoxide derivatives (equation 46)[61] and of the sulfoxide shown in equation (47).[61] Despite the absence of an α-hydrogen atom the sulfoxide shown in equation (48) does undergo a Pummerer rearrangement without C—S bond cleavage when exposed to hydrogen chloride in methanol.[62] A likely interpretation is that addition of methanol across the double bond precedes rearrangement during which a second molecule of methanol is added. A transannular Pummerer rearrangement involving sulfide participation and the formation of a dication intermediate has been proposed for the α-acetoxylation process shown in equation (49).[63]

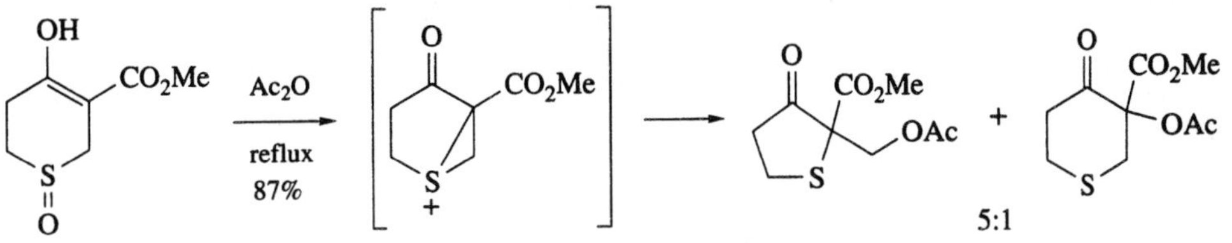

Ac$_2$O; reflux; 87%; 5:1 (45)

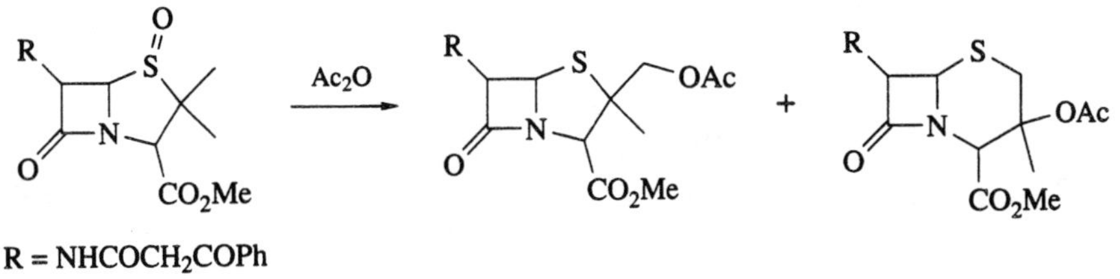

R = NHCOCH$_2$COPh; Ac$_2$O (46)

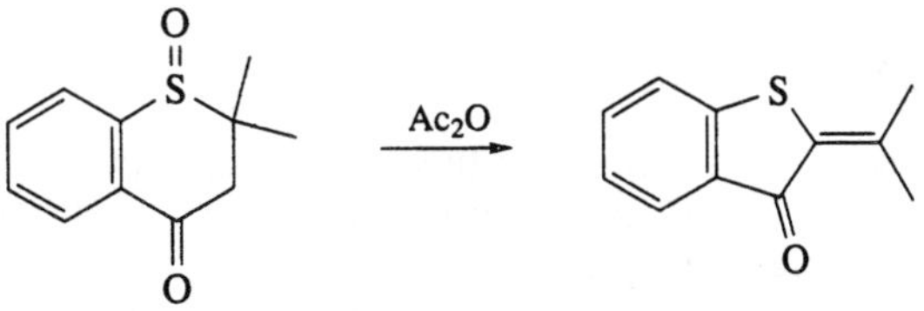

Ac$_2$O (47)

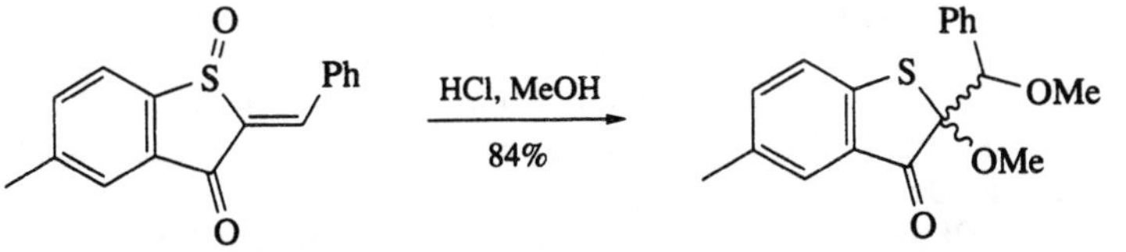

HCl, MeOH; 84% (48)

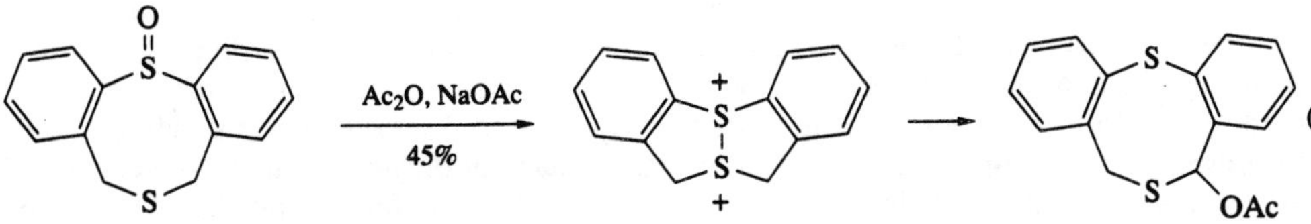

Ac$_2$O, NaOAc; 45% (49)

A few examples of additive Pummerer rearrangements are known involving direct conversion of α,β-unsaturated sulfoxides into α,β-disubstituted sulfides.[61,64,65] Scheme 4 illustrates two general pathways for such processes and specific examples of each are known. For example, alk-l-enyl phenyl sulfoxides on exposure to thionyl chloride at room temperature produce α,β-dichlorosulfides in excellent yield *via* a route which exemplifies pathway 1 of Scheme 4 (equation 50).[66] Pathway 2 is illustrated by the reaction of the cyclopentenone sulfoxide shown in equation (51) with dichloroketene to form a bicyclic product which was subsequently transformed into methyl jasmonate.[66] When the sulfoxide used in this sequence was enantiomerically pure, the product was assessed as being 20% optically pure.

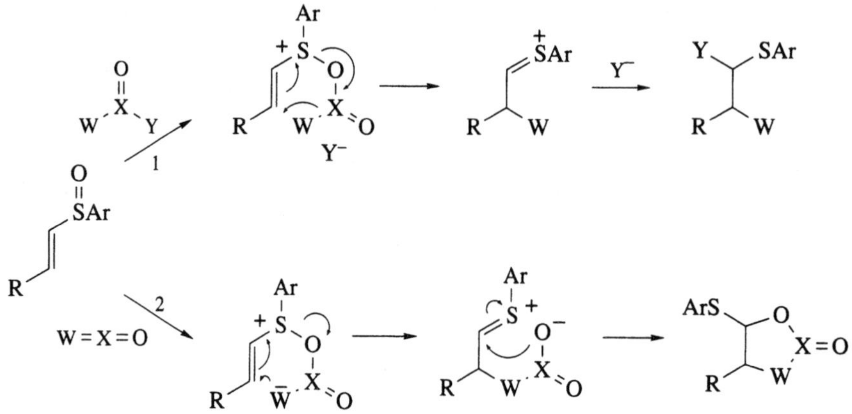

Scheme 4

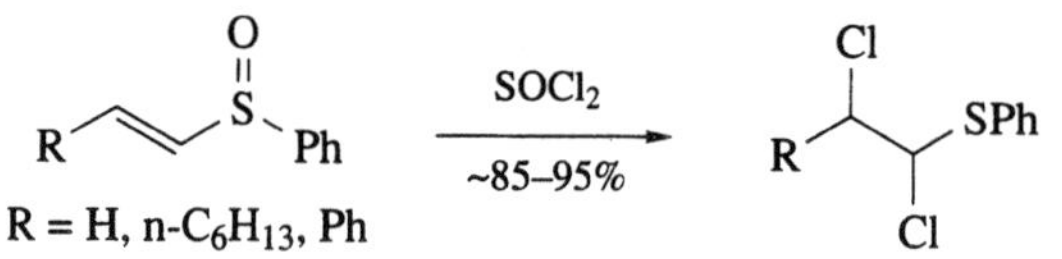

$$\text{RCH=CH-S(O)Ph} \xrightarrow[\sim 85\text{–}95\%]{SOCl_2} \text{RCHCl-CHCl-SPh} \quad (50)$$

R = H, n-C_6H_{13}, Ph

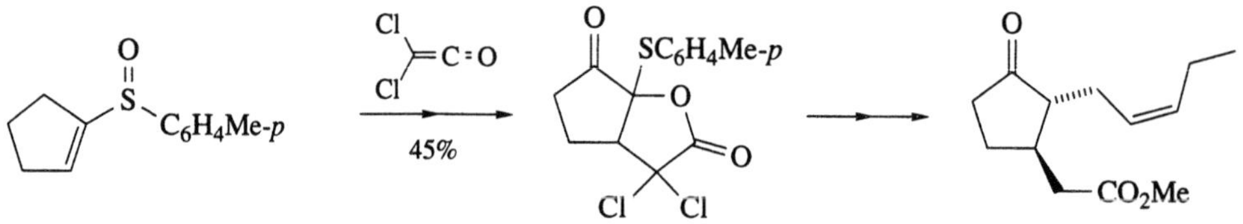

$$(51)$$

2.4.4 α-HALOGENATION OF SULFIDES

Whereas use of the Pummerer rearrangement to functionalize an alkyl sulfide normally requires prior preparation and isolation of the intermediate sulfoxide, α-halogenation may be accomplished in a single operation employing one of several readily available halogenation agents. The structural requirements for the two processes are very similar and their mechanisms undoubtedly have much in common. Although halogenation by all the halogens, with the exception of fluorine, does occur, chlorination is the most important. The earliest studies in this area were conducted using chlorine alone or in an inert solvent such as carbon tetrachloride.[67,68] However, molecular chlorine is no longer the reagent of choice for most sulfides, having been replaced by more convenient alternatives such as sulfuryl chloride, thionyl chloride, *N*-chlorosuccinimide (NCS), trichloroisocyanuric acid (chloreal), iodobenzene dichloride and benzenesulfenyl chloride. Of these, sulfuryl chloride and NCS are by far the most commonly used.

Sulfuryl chloride reacts vigorously with dimethyl sulfide at ambient temperatures, although at –15 °C it is possible to obtain a 45% yield of chloromethyl methyl sulfide (equation 52).[69] Bohme and Gran preferred the use of sulfuryl chloride in carbon tetrachloride for chlorination of dibenzyl sulfide (equation 53) and obtained the monochloride in 79% yield.[70] The same combination of reagent and solvent has been used to transform the isothiazolidine shown in equation (54) into the corresponding α-chlorosulfide.[71] Bordwell and Pitt employed sulfuryl chloride in pentane or dichloromethane to prepare α-chlorosulfides from several alkyl methyl sulfides and aryl methyl sulfides (equation 55).[72] However, complications often arise when sulfuryl chloride is used to chlorinate alkyl sulfides containing β-hydrogen atoms due to the ease of elimination of the products and subsequent further reactions. This is apparent during the chlorination of thiane and thiolane with sulfuryl chloride (equations 56 and 57), where only trace amounts of the normal products are obtained, the major products being 3,4-dihydro-2*H*-thiin and 2,3-dichlorothiolane, respectively.[72] On the other hand, Wilson and Albert have reported that addition of an equivalent amount of pyridine or triethylamine to solutions of sulfuryl chloride suppressed the elimination process to the extent that good yields of 2-chlorothiolane can be obtained from thiolane.[73] Controlled introduction of two chlorine atoms at the α-position of alkyl phenyl sulfides has been achieved using sulfuryl chloride–pyridine (1:1 molar ratio) in carbon tetrachloride at –5 °C.[74]

Me S Me + SO_2Cl_2 ⟶ Me S Cl + HCl + SO_2 (52)

Ph S Ph + SO_2Cl_2 ⟶ Ph S Ph (Cl) + HCl + SO_2 (53)

Ph Ph O N–S Bu^t + SO_2Cl_2 ⟶ Ph Ph O Cl N–S Bu^t + HCl + SO_2 (54)

R S Me + SO_2Cl_2 —pentane or CH_2Cl_2⟶ R S Cl (55)

R = alkyl, aryl

S + SO_2Cl_2 ⟶ S (56)

S + SO_2Cl_2 ⟶ Cl S Cl (57)

The introduction of NCS by Tuleen and Stevens, as a reagent for sulfide chlorination led to a major improvement in the preparation of α-chlorosulfides.[75] This crystalline reagent is easily handled and its reactivity is such that chlorination can be controlled to afford the monochlorination product selectively. Furthermore, NCS can be used for chlorination of acid-sensitive substrates. NCS is soluble in carbon tetrachloride at room temperature at ordinary concentrations, whereas its conjugate product, succinimide, is not. Solutions of α-chlorosulfides are therefore often prepared with NCS in CCl_4 and simply filtered prior to use without further purification. Other nonpolar solvents that have been used with NCS include chloroform, dichloromethane and benzene. Some of the very many examples of the use of NCS for sulfide chlorination from the recent literature are summarized in Table 2. For several of the entries yields were not recorded. This is almost always due to the fact that α-chlorosulfides are produced and treated as unstable reaction intermediates *en route* to more stable products.

The examples in Table 2 have been chosen so as to highlight significant features of the use of NCS. The isolation of α-chlorosulfides from substrates of diverse structural type in which β-elimination is possible is particularly noteworthy (entries 2–15). It is also apparent that functional groups such as alkenes, anhydrides, imides, esters, trimethylsilyls, acyl chlorides, amides, NBOC, acetals, β-lactams and ethers are all unaffected during the reaction (entries 12–23). However, allylic rearrangement does occur in some cases with allylic sulfides (entry 13 and equation 59). Small ring systems may be prone to ring-opening rearrangement as exemplified by the behavior of the bicyclo[3.2.0] system (entry 24). Highly reactive chlorosulfides such as 2-chloro-1,3-dithiane are also accessible (entry 25). This compound was also obtained using sulfuryl chloride as the reagent, provided very low temperatures were employed.

Although NCS continues to be the reagent of choice for sulfide chlorination, Cohen and his coworkers have advocated the use of trichloroisocyanuric acid as a less expensive alternative.[83] This substance, which is available commercially as an industrial deodorant and household cleaner under the trade name of Chloreal^R, is also useful for sulfide chlorination. Two examples of its use with sulfides are shown in

Table 2 NCS Chlorination of Representative Sulfides

Entry	*Substrate*	*Product*	*Yield (%)*	*Ref.*
1	PhSMe	$PhSCH_2Cl$	—	75
2	PhSEt	PhSCHClMe	—	75
3	$PhS(CH_2)_4Me$	$PhSCHClBu^n$	>90	76
4	$Me_3SiO(CH_2)_2SPh$	$Me_3SiOCH_2CHClSPh$	>95	77
5	$Me_3Sn(CH_2)_4SPh$	$Me_3Sn(CH_2)_3CHClSPh$	100	78
6	$PhS(CH_2)_4SPh$	$PhSCHCl(CH_2)_2CHClSPh$	>65	79
7	S	S Cl	—	80
8	S	S Cl	—	80
9	O S	O S Cl	—	80
10	SPh SPh	Cl SPh SPh Cl	>62	79
11	H S	Cl H S	100	81
12	SPh	Cl SPh	>95	82
13	SPh	SPh Cl	—	83
14	SPh O O O	Cl SPh O O O	95	84
15	O Bz N N Bz S	O Bz N N Bz S Cl	100	85
16	$MeSCH_2CO_2Me$	$MeSCHClCO_2Me$	78	86
17	$PhSCH_2SiMe_3$	$PhSCHClSiMe_3$	100	87
18	$PhSCH_2COCl$	PhSCHClCOCl	86	88

Table 2 *(continued)*

Entry	*Substrate*	*Product*	*Yield (%)*	*Ref.*
19	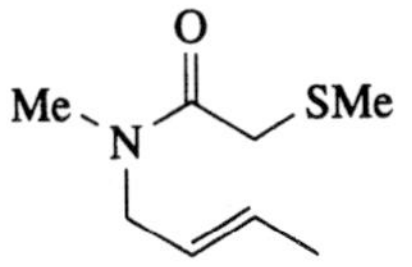 [structure]	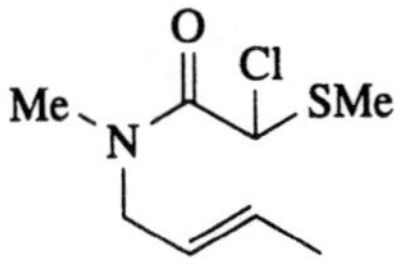 [structure]	~100	89
20	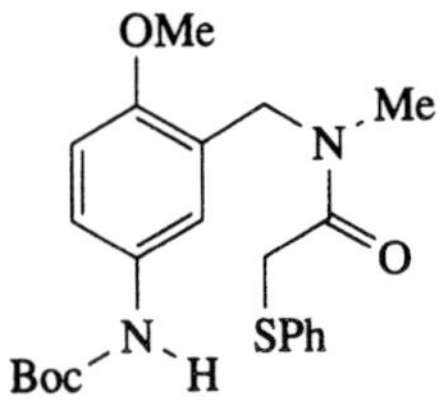 [structure]	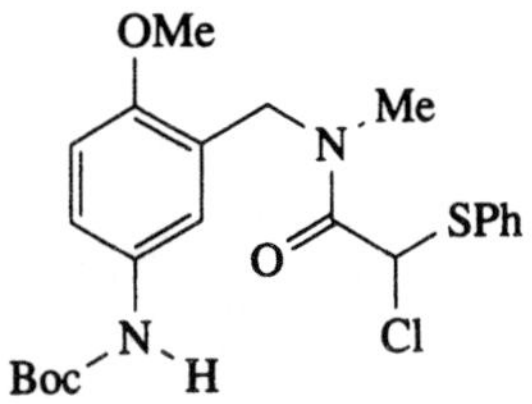 [structure]	100	90
21	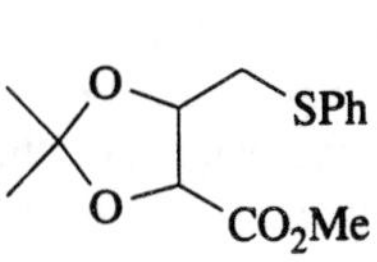 [structure]	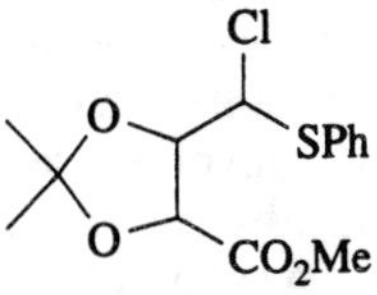 [structure]	—	90
22	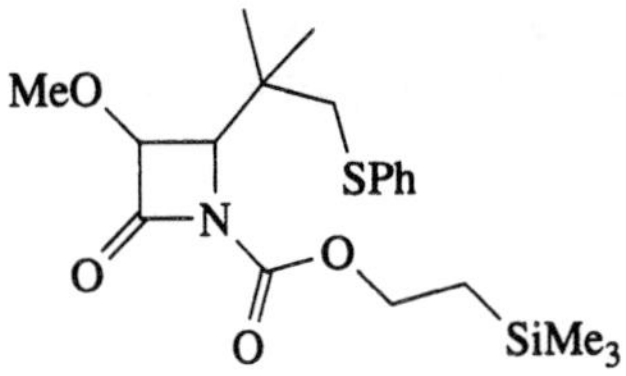 [structure]	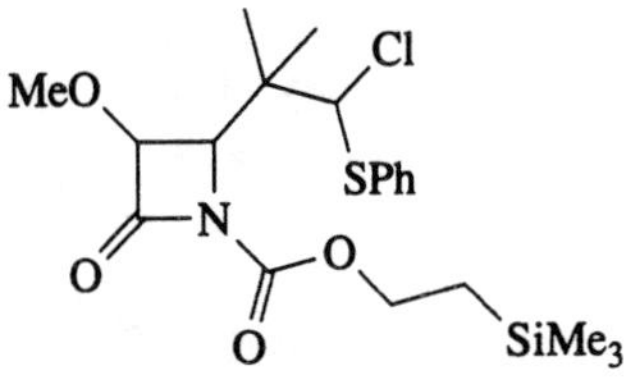 [structure]	—	91
23	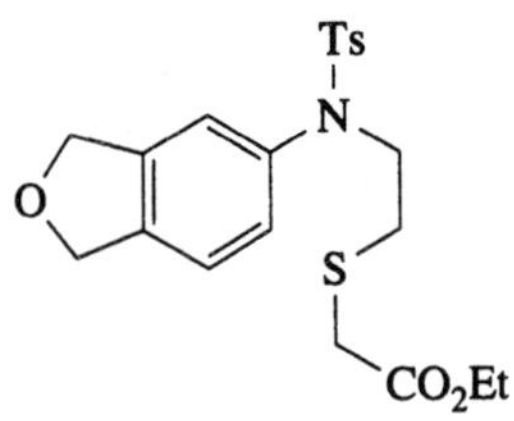 [structure]	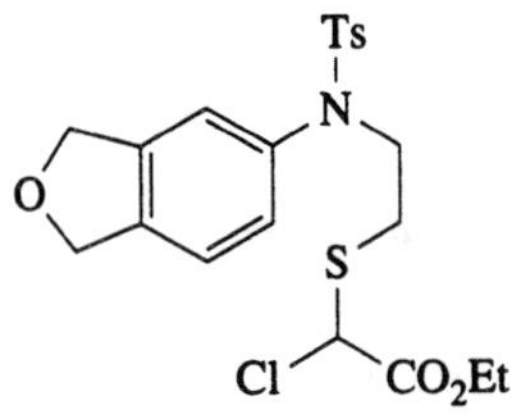 [structure]	—	92
24	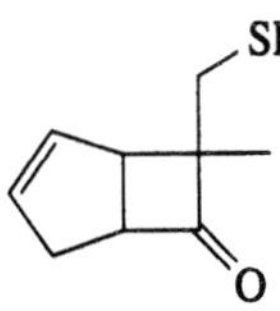 [structure]	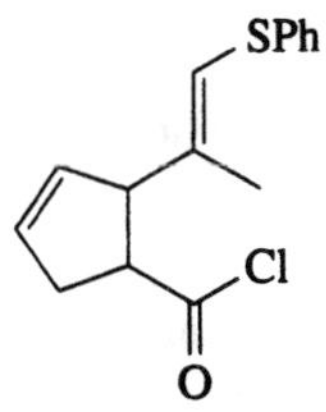 [structure]	—	93
25	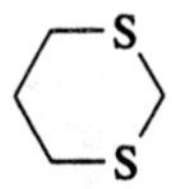 [structure]	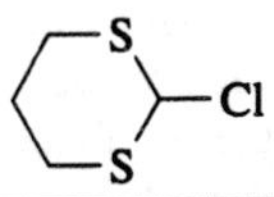 [structure]	—	94

equations (58) and (59), the latter revealing that allylic rearrangement, as with NCS, is a feature of its reactivity. In a comparative study of the chlorination of allylic sulfides (equation 59) Cohen found that for substrates with R^1 = alkyl, Chloreal produced a much faster reaction at room temperature than did NCS.[83] Furthermore, chlorination of primary allylic phenyl sulfides (equation 59; R^1 = H) was also significantly faster with Chloreal than with NCS. Thus, phenyl crotyl sulfide (equation 59; R^1 = H, R^2 = Me) gave a 36% yield of chlorosulfide after 24 h at 5 °C with NCS, whereas the yield was quantitative with Chloreal; however, the amount of (*E*)-isomer in the product was greater with NCS.[83] Chloreal has also been used by Cohen to prepare the cyclopropyl chlorosulfide shown in equation (58).[95]

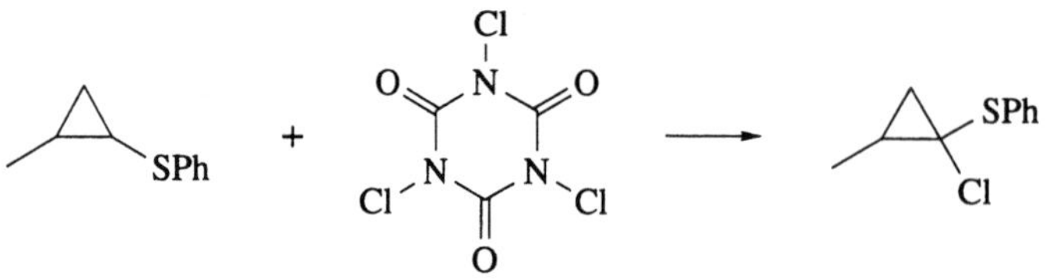

(58)

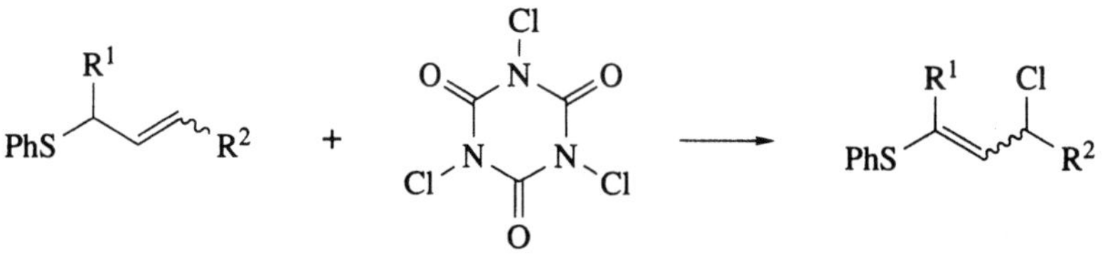

(59)

These various sulfide chlorinations are believed to follow ionic pathways similar to those summarized earlier for the Pummerer rearrangement of sulfoxides, the net result being the transfer of a chlorine atom from sulfur to carbon.[72] The broad outlines of the mechanism of action of NCS and of sulfuryl chloride are summarized in Scheme 5. The initiating step involves electrophilic attack by the reagent on the sulfur atom to form a sulfonium salt which, in some instances, can be isolated and converted into α-chlorosulfide on heating. Pathways A and B in Scheme 5 suggest two mechanistic extremes as to how this transformation may be brought about. The pathway favored for any particular substrate will reflect structural features, *e.g.* the acidity of the α-hydrogen atom(s) undergoing substitution and the extent of hyperconjugation. The choice of chlorinating agent may also be significant since the basicity of its conjugate anion is also implicated in the mechanism.

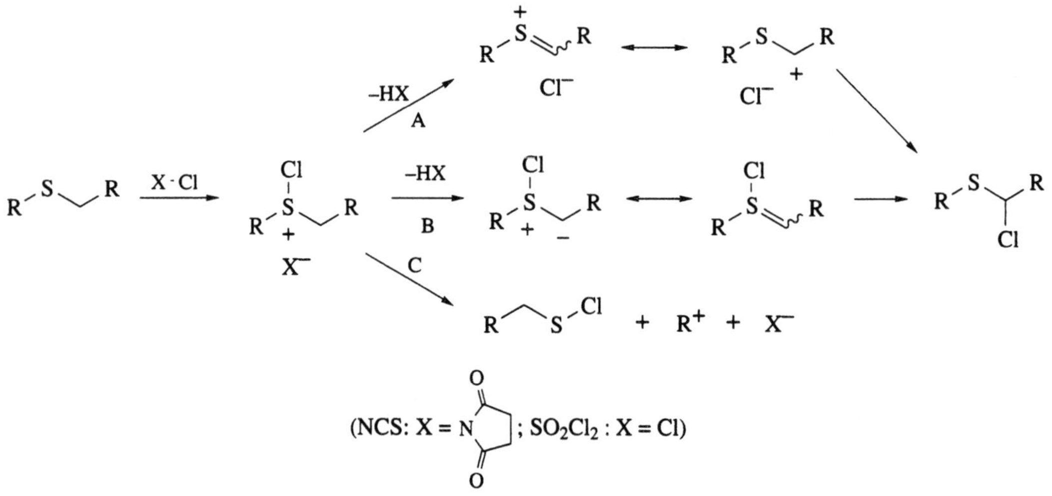

(NCS: X = N(succinimide); SO_2Cl_2 : X = Cl)

Scheme 5 Mechanisms of chlorination by NCS and SO_2Cl_2

The work of Tuleen and Stevens on the regioselectivity of chlorination of a series of unsymmetrically substituted dialkyl sulfides with NCS provides clues to the directive effects implicit in the mechanisms encompassed by Scheme 5.[96] These observations are collected in Scheme 6 where the preferred site of chlorination in each case is indicated with an arrow, the number over the arrow indicating the major:minor product ratio (minor = 1). In the first example, chlorination of benzyl methyl sulfide (**2**) produces chlorobenzyl methyl sulfide exclusively.[70] Secondly, chlorination of benzyl ethyl sulfide (**3**) and benzyl isopropyl sulfide (**4**) also shows a marked, though not exclusive, preference for the benzylic position.[96] In the latter case the extent of benzylic chlorination can be modulated by ring substitution. The directive effects in these internal competitions for *p*-methyl and *p*-chloro substituents are correlated by the Hammett relationship with a value of $\rho = 1.0$, which is consistent with a mechanism involving abstraction of the more acidic proton in the chlorosulfonium ion intermediate.[96] Further indications of the importance of proton acidity are shown by the behavior of alkyl sulfides (**5**)–(**8**) in Scheme 6, where the preferred site for chlorination is consistently that adjacent to the electron-withdrawing substituent.[96] The α-chlorosulfide carboxylate derived from (**6**) is a particularly notable (and useful) example of very high regioselectivity for the internal position.[86]

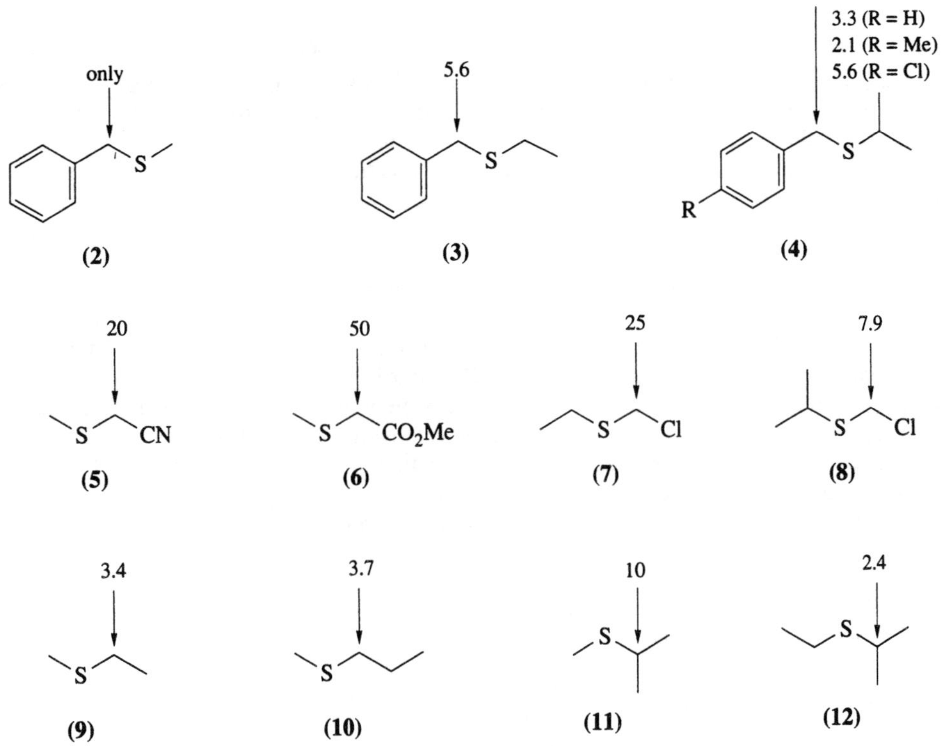

only
(**2**)
5.6
(**3**)
3.3 (R = H)
2.1 (R = Me)
5.6 (R = Cl)
(**4**)
20
(**5**)
50
(**6**)
25
(**7**)
7.9
(**8**)
3.4
(**9**)
3.7
(**10**)
10
(**11**)
2.4
(**12**)

Scheme 6

While consideration of the relative acidities of the α-hydrogen atoms accounts satisfactorily for the dominant direction of chlorination of many alkyl sulfides, it does not explain the directive effects exhibited by the simple dialkyl sulfides (**9**)–(**12**) of which (**11**), with a 10:1 preference for the internal position, is the most notable. These substrates reveal an increasing susceptibility to chlorination of alkyl groups in the order methyl < ethyl ≈ *n*-propyl < isopropyl, which is also the order expected if in the course of the reaction the α-carbon atom assumes some degree of carbocation character in the transition state for α-hydrogen atom abstraction. This is consistent with pathway A in Scheme 5 in which concerted removal of hydrogen chloride from the chlorosulfonium ion generates a delocalized thiocarbocation. Pathways A and B are in fact variations on the *E*2 and *E*1cb mechanisms for 1,2-elimination, the latter leading to an ylide intermediate (*cf.* the Pummerer rearrangement), which should therefore become significant in sulfonium salts having an α-hydrogen atom of pronounced acidity.[97] The regiochemistry of further chlorination of chlorosulfides (**7**) and (**8**) demonstrates that a chloro substituent exerts a powerful directive effect in these systems. This effect is also apparent in the double chlorination of the tricyclic sulfide shown in equation (60), where formation of the geminal dichloride is favored over the 1,3-dichloride by a factor of 4:1 when the chlorinating agent is sulfuryl chloride.[98] On the other hand, NCS chlorination of the symmetrical bis-sulfide series in equation (61) produces symmetrical 1,X-dichlorides rather than geminal dichlorides.[79]

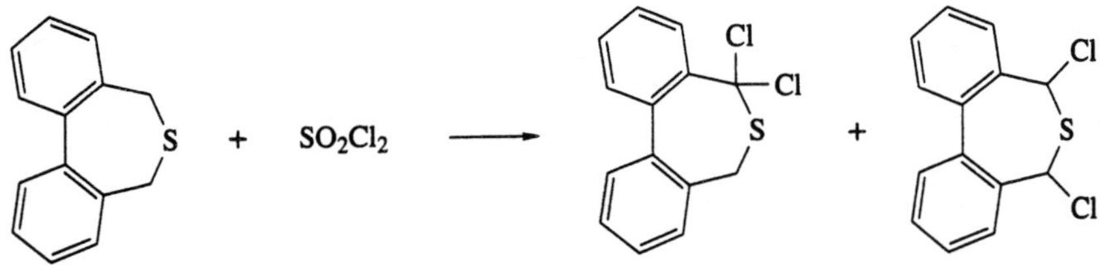

$$ + SO_2Cl_2 \longrightarrow + \quad (60)$$

PhS–(CH2)n–SPh + 2 NCS ⟶ PhS–CH(Cl)–(CH2)n–CH(Cl)–SPh (61)

$n = 0$, 1 and 2

Tuleen and Stephens noted a regioselectivity difference during the chlorination of ethyl methyl sulfide by NCS and sulfuryl chloride, the latter being the more selective towards the internal position (ratio: 4.9 *versus.* 3.4).[96] In chlorination with sulfuryl chloride, proton abstraction from the chlorosulfonium ion is brought about by chloride ion, whereas with NCS the more basic succinimidyl ion is responsible. This difference suggests that of the two, the sulfuryl chloride reaction is more likely to follow the mechanism of pathway A in Scheme 5. Consequently, carbocation relative stabilities should have a greater influence in determining product composition with sulfuryl chloride than with NCS. Although pathway C in Scheme 5 does not lead to chlorosulfide, it is included here to highlight further differences between the reactivity of sulfuryl chloride and NCS. For example, while benzyl *t*-butyl sulfide and NCS give the expected chlorosulfide shown in equation (62), use of sulfuryl chloride affords predominantly the fragmentation products *t*-butyl chloride and dibenzyl disulfide (equation 63).[96] Similarly, benzyl *p*-methoxybenzyl sulfide behaves normally with NCS, but is cleaved to *p*-methoxybenzyl chloride and disulfide by sulfuryl chloride. Carbon–sulfur bond cleavage with sulfuryl chloride may be interpreted in terms of fragmentation of the chlorosulfonium cation (pathway C in Scheme 5), a process which should be facilitated by the release of a relatively stable carbocation, in this case *t*-butyl. That this does not happen with NCS suggests that the competition between pathway C leading to fragmentation and those leading to chlorosulfide is controlled by the relative basicities of chloride and succinimidyl ion.

PhCH2–S–Bu^t —NCS→ PhCH(Cl)–S–Bu^t (62)

PhCH2–S–Bu^t —SO_2Cl_2→ PhCH2–S–S–CH2Ph + Bu^{t}Cl (63)

2.4.4.1 Miscellaneous Routes to α-Chloro Sulfides

Although this chapter is devoted primarily to methods of functionalizing alkyl sulfides at the α-position, it is appropriate to mention alternative routes to the same synthetic objective which do not involve direct functionalization of preformed sulfides. One such method, devised by Bohme and his coworkers, is based on the condensation of an aldehyde with a thiol in the presence of hydrogen chloride.[68,99] Use of formaldehyde in this way produces a primary chlorosulfide (equation 64). Higher aldehydes such as propanal yield secondary chlorosulfides (equation 65). The process is particularly useful in the synthesis of regiochemically pure chlorosulfides without danger of contamination by isomers in situations where the alternative of direct chlorination of a dialkyl sulfide would be nonregioselective. For example, NCS chlorination of benzyl *p*-methylbenzyl sulfide gives both possible monochlorides as shown in equation (66), whereas the condensation alternative of aldehyde, thiol and hydrogen chloride shown in equations (67) and (68) gives one or other of the individual monochlorides.[100]

PhSH + CH_2O + HCl ⟶ Ph–S–CH2–Cl (64)

PhSH + CH3CH2CHO + HCl ⟶ Ph–S–CH(Cl)–CH2CH3 (65)

PhCH2–S–CH2–C6H4-*p*-Me + NCS ⟶ PhCH(Cl)–S–CH2–C6H4-*p*-Me + PhCH2–S–CH(Cl)–C6H4-*p*-Me (66)

Ph⌒SH + (4-methylbenzaldehyde, CHO) + HCl ⟶ Ph⌒S–CH(Cl)–(4-methylphenyl) (67)

PhCHO + (4-methylbenzenethiol, SH) + HCl ⟶ Ph–CH(Cl)–S–(4-methylphenyl) (68)

Yet another route to α-chlorosulfides which does not require the availability of sulfide is based on the reaction of diazo compounds, especially diazocarbonyls, with sulfenyl halides. Weygand and Bestmann found that α-diazo ketones and benzenesulfenyl chloride react smoothly together with loss of nitrogen at room temperature to furnish α-chloro-α-phenylthio ketones in excellent yield (equation 69).[101,102] This route, which features the simultaneous introduction of both chlorine and sulfur moieties, is particularly useful when one wishes to place the chlorosulfide unit adjacent to the carbonyl group of a ketone in a regiospecific manner. Since terminal α-diazo ketones may be obtained efficiently from acyl chloride and diazomethane, this route to α-chloro-α-phenylthio ketones is especially useful since it depends neither on the availability of the parent ketone nor on prior regiospecific introduction of the phenylthio group. More recent applications of this method to α-chlorosulfides include the reaction of benzenesulfenyl chloride with cyclic α-diazo ketones as summarized in equation (70).[103] The process is also applicable to α-diazo esters (equation 71) and 2-diazo-1,3-dicarbonyl compounds (equation 72).[102] Dimedone can be converted directly into an α-chlorosulfide without prior activation *via* the diazo intermediate (equation 73).[104]

PhC(O)C(H)=N_2 + PhSCl ⟶ PhC(O)CH(SPh)Cl (69)

cyclic α-diazo ketone (ring size ( )$_n$) + PhSCl —(>90%)⟶ α-chloro-α-phenylthio cyclic ketone (70)

n = 1, 2, 3 and 4

EtOC(O)C(H)=N_2 + PhSCl ⟶ EtOC(O)CH(SPh)Cl (71)

MeC(O)C(=N_2)C(O)Me + PhSCl ⟶ MeC(O)C(Cl)(SPh)C(O)Me (72)

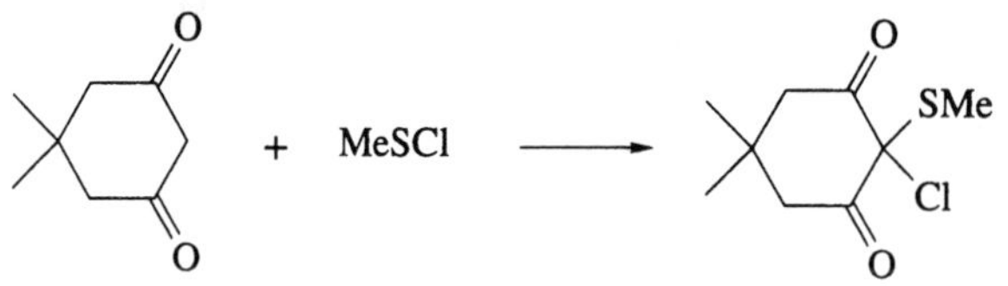

(73)

2.4.4.2 Summary of Uses of α-Chloro Sulfides in Organic Synthesis

α-Chlorosulfides exhibit a range of reactivity which makes them very useful as intermediates in organic synthesis.[105] We conclude this chapter with a schematic survey of their more important uses (see Scheme 7). In many ways they complement the uses of Pummerer products and intermediates.

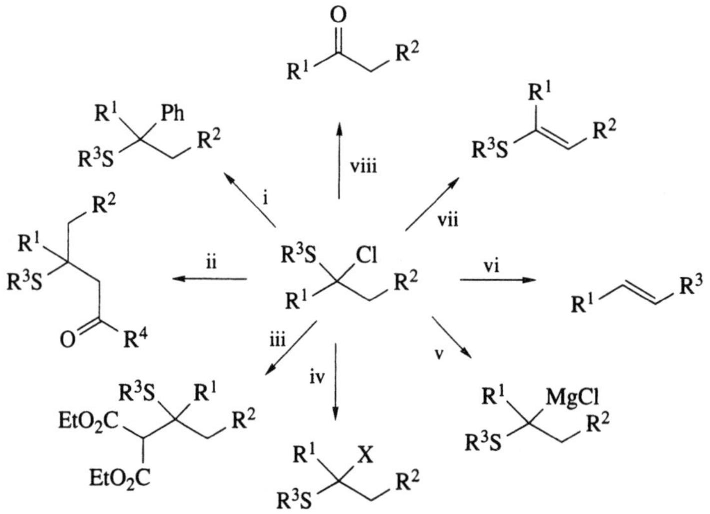

X = OR^4, $OCOR^4$, SR^4

i, aromatic alkylation; ii, enol ether alkylation; iii, active methylene alkylation; iv, alcohol, thiol, carboxylate; v, Grignard formation; vi, Ramberg–Bäcklund rearrangement; vii, eliminate HCl; viii, hydrolysis

Scheme 7 Summary of α-chlorosulfide uses

α-Chlorosulfides are prone to solvolysis; in aqueous media they serve as aldehyde and ketone precursors. Exposure to alcohols or thiols leads to hemithioacetals or dithioacetals, which can serve as protecting groups. Under anhydrous conditions α-chlorosulfides are oxidized to α-chlorosulfones, which have been exploited extensively as alkene precursors in the Ramberg–Bäcklund synthesis. α-Chlorosulfides form and couple with Grignard reagents. Elimination of hydrogen chloride provides access to vinyl sulfides and thiocarbenoids. α-Chlorosulfides are also important sources of reactive electrophiles for alkylation reactions of aromatics, alkenes, alkynes, enolates and silyl enol ether derivatives of aldehydes, ketones, esters and lactones. As with Pummerer intermediates, intramolecular versions of α-chlorosulfide uses have also been developed.

2.4.5 REFERENCES

1. R. Pummerer, *Chem. Ber.*, 1909, **42**, 2282; 1910, **43**, 1401.
2. J. A. Smythe, *J. Chem. Soc.*, 1909, **95**, 349.
3. M. Kennedy, C. J. Moody and C. W. Rees, unpublished results.
4. P. Brougham, M. S. Cooper, D. A. Cummerson, H. Heaney and N. Thompson, *Synthesis*, 1987, 1015.
5. N. J. Leonard and C. R. Johnson, *J. Org. Chem.*, 1962, **27**, 282.
6. B. A. Trofimov, N. K. Gusarova, S. V. Amosova and G. G. Efremova, *Chem. Abstr.*, 1980, **92**, 22 043t.
7. C. R. Johnson and D. McCants, Jr., *J. Am. Chem. Soc.*, 1965, **87**, 1109.
8. M. Madesclaire, *Tetrahedron*, 1986, **42**, 5459.

9. S. Colonna, N. Gaggero, A. Manfredi, L. Casella and M. Gullotti, *J. Chem. Soc., Chem. Commun.*, 1988, 1451.
10. M. Chaykovsky and E. J. Corey, *J. Am. Chem. Soc.*, 1965, **87**, 1345.
11. S. Wolfe, P. M. Kazmaier and H. Auksi, *Can. J. Chem.*, 1979, **57**, 2404.
12. G. A. Russell and G. J. Mikol, 'Mechanisms of Molecular Migrations', ed. B. S. Thyagarajan, Interscience, New York, 1968, vol. 1, p. 157.
13. For more specialist accounts of the Pummerer rearrangement see S. Oae and T. Numata, *Isot. Org. Chem.*, 1980, **5**, 45; S. Oae, T. Numata and Y. Yoshimura, 'The Chemistry of the Sulphonyl Group', ed. C. J. M. Stirling and S. Patai, Wiley, New York, 1981, part 2, chap. 15; S. Oae, 'Topics in Organic Sulphur Chemistry', ed. T. Tisler, University Press, Ljubljana, 1978, p. 289; T. Durst, *Adv. Org. Chem.*, 1969, **6**, 358.
14. S. Oae, T. Kitao, J. Kawamura and Y. Kitaoka, *Tetrahedron*, 1963, **19**, 817.
15. S. Oae, T. Kitao and S. Kawamura, *Tetrahedron*, 1963, **19**, 1783.
16. E. Sabourin, unpublished results quoted in ref. 12.
17. H. Sugihara, R. Tanikaga and A. Kaji, *Synthesis*, 1978, **12**, 881.
18. E. J. Corey and D. J. Hoover, *Tetrahedron Lett.*, 1982, **23**, 3463.
19. R. Tanikaga, Y. Yabuki, N. Ono and A. Kaji, *Tetrahedron Lett.*, 1976, 2257.
20. T. Numata, O. Itoh and S. Oae, *Tetrahedron Lett.*, 1979, 1869.
21. A. Dossena, R. Marchelli and G. Casnati, *J. Chem. Soc., Perkin Trans. 1*, 1981, 2737.
22. M. Mikolajczyk, A. Zatorski, S. Grzejszczak, B. Costisella and W. Midura, *J. Org. Chem.*, 1978, **43**, 2518.
23. T. Numata and S. Oae, *Tetrahedron Lett.*, 1977, 1337.
24. J. E. McCormick and R. S. McElhinney, *J. Chem. Soc., Perkin Trans. 1*, 1976, 2533.
25. T. Numata, O. Itoh and S. Oae, *Tetrahedron Lett.*, 1979, 161.
26. S. Y. Ko, A. W. M. Lee, S. Masamune, L. A. Reed, III, K. B. Sharpless and F. J. Walker, *Science (Washington, D.C.)*, 1983, **220**, 949; T. Katsuki, A. W. M. Lee, P. Ma, V. S. Martin, S. Masamune, K. B. Sharpless, D. Tuddenham and F. J. Walker, *J. Org. Chem.*, 1982, **47**, 1373.
27. D. T. Connor and M. von Strandtmann, *J. Heterocycl. Chem.*, 1978, **15**, 113; see also D. T. Connor, P. A. Young and M. von Strandtmann, *J. Heterocycl. Chem.*, 1978, **15**, 115.
28. F. Theil, C. Lindig and K. Repke, *Z. Chem.*, 1980, **11**, 414.
29. M. Watanabe, S. Nakomori, H. Hasegawa, K. Shirai and T. Kumamoto, *Bull. Chem. Soc. Jpn.*, 1981, **54**, 817.
30. B. Stridsberg and S. Allenmark, *Acta Chem. Scand., Ser. B*, 1974, **28**, 591.
31. H. Ishibashi, H. Komatsu and M. Ikeda, *J. Chem. Res. (S)*, 1987, 296.
32. R. Hunter and C. D. Simon, *Tetrahedron Lett.*, 1986, **27**, 1385.
33. R. Hunter and C. D. Simon, *Tetrahedron Lett.*, 1988, **29**, 2257.
34. Y. Tamura, H. Maeda, S. Akai, K. Ishiyama and H. Ishibashi, *Tetrahedron Lett.*, 1981, **22**, 4301.
35. H. Ishibashi, H. Ozeki and M. Ikeda, *J. Chem. Soc., Chem. Commun.*, 1986, 654.
36. L. N. Mander and P. H. C. Mundill, *Synthesis*, 1981, 620.
37. K. Cardwell, B. Hewitt and P. D. Magnus, *Tetrahedron Lett.*, 1987, **28**, 3303; T. Gallagher and P. D. Magnus, *J. Am. Chem. Soc.*, 1983, **105**, 2086; T. Gallagher, P. D. Magnus and J. C. Huffman, *J. Am. Chem. Soc.*, 1982, **104**, 1140.
38. I. K. Stamos, *Tetrahedron Lett.*, 1985, **26**, 477.
39. I. A. Blair, L. N. Mander and P. H. C. Mundill, *Aust. J. Chem.*, 1981, **34**, 1235.
40. C. J. Moody, C. W. Rees and M. Kennedy, unpublished results.
41. Y. Tamura, H. D. Choi, H. Shindo, J. Uenishi and H. Ishibashi, *Tetrahedron Lett.*, 1981, **22**, 81.
42. H. Ishibashi, M. Okada, A. Akiyama, K. Nomura and M. Ikeda, *J. Heterocycl. Chem.*, 1986, **23**, 1163.
43. Y. Vankar and C. T. Rao, *Tetrahedron*, 1985, **41**, 3405.
44. T. Kaneko, *J. Am. Chem. Soc.*, 1985, **107**, 5490.
45. T. Kaneko, Y. Okamoto and K. Hatada, *J. Chem. Soc., Chem. Commun.*, 1987, 1511.
46. Y. Kita, O. Tamura, T. Miki and Y. Tamura, *Tetrahedron Lett.*, 1987, **28**, 6479.
47. L. C. Vishwakarma and A. R. Martin, *J. Heterocycl. Chem.*, 1982, **19**, 103.
48. R. Lett and Y. Kuroki, *Tetrahedron Lett.*, 1982, **23**, 5541.
49. H. L. Yale, *J. Heterocycl. Chem.*, 1978, **15**, 331.
50. M. Bhupathy and T. Cohen, *Tetrahedron Lett.*, 1987, **28**, 4797.
51. D. T. O'Connor and R. J. Sorenson, *J. Heterocycl. Chem.*, 1981, **18**, 587.
52. T. Naga, T. Miki and I. Kumadaki, *Chem. Pharm. Bull.*, 1987, **35**, 3620.
53. F. G. Bordwell and B. M. Pitt, *J. Am. Chem. Soc.*, 1955, **77**, 572.
54. S. K. Ray, R. A. Shaw and B. C. Smith, *Nature (London)*, 1962, **196**, 372.
55. R. Ratz and O. J. Sweeting, *Tetrahedron Lett.*, 1963, 529.
56. P. Bakuzis and M. L. F. Bakuzis, *J. Org. Chem.*, 1985, **50**, 2569.
57. C. U. Kim, P. F. Misco and D. N. McGregor, *J. Org. Chem.*, 1982, **47**, 170.
58. K. Praefcke and Ch. Weichsel, *Tetrahedron Lett.*, 1976, 2229.
59. D. J. Hart and Y.-M. Tsai, *Tetrahedron Lett.*, 1983, **24**, 4387.
60. R. R. King, *J. Org. Chem.*, 1980, **45**, 5347.
61. R. B. Morin, B. G. Jackson, R. A. Mueller, E. R. Lavagnino, W. B. Scanlon and S. L. Andrews, *J. Am. Chem. Soc.*, 1963, **85**, 1896.
62. L. S. S. Réamonn and W. I. O'Sullivan, *J. Chem. Soc., Chem. Commun.*, 1976, 642.
63. H. Fujihara, J. J. Chiu and N. Furukawa, *J. Chem. Res. (S)*, 1987, 204.
64. H. Kosugi, H. Uda and S. Yamagiwa, *J. Chem. Soc., Chem. Commun.*, 1976, 71; O. De Lucchi, G. Marchioro and G. Modena, *J. Chem. Soc., Chem. Commun.*, 1984, 513.
65. R. Malherbe and D. Bellus, *Helv. Chim. Acta*, 1978, **61**, 3096; J. P. Marino and M. Neisser, *J. Am. Chem. Soc.*, 1981, **103**, 7687.
66. G. H. Posner, E. Asirvatham and S. F. Ali, *J. Chem. Soc., Chem. Commun.*, 1985, 542.
67. M. A. Riche, *Ann. Chim. Phys.*, 1855, **43**, 283.
68. H. Böhme, H. Fischer and R. Frank, *Justus Liebigs Ann. Chem.*, 1949, **563**, 54.

69. W. E. Truce, G. H. Birum and E. T. McBee, *J. Am. Chem. Soc.*, 1952, **74**, 3594.
70. H. Böhme and H. J. Gran, *Justus Liebigs Ann. Chem.*, 1953, **581**, 133.
71. C. J. Easton, *J. Chem. Soc., Chem. Commun.*, 1983, 1349.
72. F. G. Bordwell and B. M. Pitt, *J. Am. Chem. Soc.*, 1955, **77**, 572.
73. G. E. Wilson, Jr. and R. Albert, *J. Org. Chem.*, 1973, **38**, 2156; 2160.
74. C. C. Fortes, H. C. Fortes and D. C. R. G. Concalves, *J. Chem. Soc., Chem. Commun.*, 1982, 857.
75. D. L. Tuleen and T. B. Stevens, *Chem. Ind. (London)*, 1966, 1555.
76. I. Paterson and I. Fleming, *Tetrahedron Lett.*, 1979, 993, 995, 2179.
77. H. A. Khan and I. Paterson, *Tetrahedron Lett.*, 1982, **23**, 5083.
78. E. Murayama, M. Uematsu, H. Nishio and T. Sato, *Tetrahedron Lett.*, 1984, **25**, 313.
79. J. P. Cronin, B. M. Dilworth and M. A. McKervey, *Tetrahedron Lett.*, 1986, **27**, 757.
80. D. J. Tuleen and R. H. Bennett, *J. Heterocycl. Chem.*, 1969, **6**, 115.
81. P. G. Gassman and S. M. Bonser, *J. Am. Chem. Soc.*, 1983, **105**, 667.
82. H. A. Khan and I. Paterson, *Tetrahedron Lett.*, 1982, **23**, 2399.
83. A. J. Mura, Jr., D. A. Bennett and T. Cohen, *Tetrahedron Lett.*, 1975, 4433.
84. J. A. Kaydos and D. L. Smith, *J. Org. Chem.*, 1983, **48**, 1096.
85. H. A. Bates and S. B. Rosenblum, *J. Org. Chem.*, 1986, **51**, 3447.
86. Y. Tamura, H. Shindo, J. Uenishi and H. Ishibashi, *Tetrahedron Lett.*, 1980, **21**, 2547.
87. H. Ishibashi, H. Nakatana, Y. Umi, W. Yamamoto and M. Ikeda, *J. Chem. Soc., Perkin Trans. 1*, 1987, 589.
88. C. W. Rees, C. J. Moody and M. Kennedy, unpublished results.
89. H. Ishibashi, M. Ikeda, H. Maeda, K. Ishiyama, M. Yoshida, S. Akai and Y. Tamura, *J. Chem. Soc., Perkin Trans. 1*, 1987, 1099.
90. T. Mukaiyama, T. Sugaya, S. Marui and T. Nakatsuka, *Chem. Lett.*, 1982, **10**, 1555.
91. D. T. W. Chu, J. E. Henegeveld and D. Lester, *Tetrahedron Lett.*, 1983, **24**, 139.
92. H. Ishibashi, M. Okada, A. Akiyama, K. Nomura and M. Ikeda, *J. Heterocycl. Chem.*, 1986, **23**, 1163.
93. Y. Ohshiro, H. Ishida, J. Shibata, T. Minami and T. Agawa, *Chem. Lett.*, 1982, 587.
94. K. Arai and M. Oki, *Bull. Chem. Soc. Jpn.*, 1976, **49**, 553.
95. T. Cohen and J. R. Matz, *Tetrahedron Lett.*, 1981, **22**, 2455.
96. D. L. Tuleen and T. B. Stephens, *J. Org. Chem.*, 1969, **34**, 31.
97. For a more detailed analysis of the mechanism of sulfide chlorination, including a reassessment of product controlling factors, see T. P. Ahern, D. G. Kay and R. F. Langler, *Can. J. Chem.*, 1978, **56**, 2422.
98. L. A. Paquette, *J. Am. Chem. Soc.*, 1964, **86**, 4089.
99. H. Böhme and H. J. Gran, *Justus Liebigs Ann. Chem.*, 1952, **577**, 68.
100. L. A. Paquette and L. S. Wittenbrook, *J. Am. Chem. Soc.*, 1967, **89**, 4483.
101. F. Weygand and H. J. Bestmann, *Z. Naturforsch., Teil B*, 1955, **10B**, 296.
102. F. Weygand, H. J. Bestmann and H. Fritzsche, *Chem. Ber.*, 1960, **93**, 2340.
103. M. A. McKervey and P. Ratananukul, *Tetrahedron Lett.*, 1983, **24**, 117.
104. I. W. J. Still and G. W. Kutney, *J. Org. Chem.*, 1981, **46**, 4911.
105. For a detailed survey of these reactions see B. M. Dilworth and M. A. McKervey, *Tetrahedron*, 1986, **42**, 3731.

2.5
Oxidation Adjacent to Nitrogen

HAROLD W. PINNICK
Bucknell University, Lewisburg, PA, USA

2.5.1 INTRODUCTION

Oxidation of carbon adjacent to nitrogen refers to the introduction of either an oxygen atom or a carbon atom next to a nitrogen atom, since either raises the oxidation state of the carbon atom attached to the nitrogen group. Sometimes, the nitrogen functionality is then lost on work-up or in a further hydrolysis step.

2.5.2 NITRO COMPOUNDS

2.5.2.1 The Nef Reaction

2.5.2.1.1 Traditional Nef reactions

Perhaps the most well-known example of this process is the Nef reaction.[1] This reaction was reported by the Swiss chemist Nef in 1894[2] and converts a primary or secondary nitro compound into an aldehyde or ketone (equation 1). The traditional base used in the reaction is sodium hydroxide and the acid is sulfuric acid. Side reactions sometimes are very significant. A study of pH dependence of the second step showed that the best results are obtained at a pH of 0.5.[3] At lower acidity, the starting nitro compound was recovered, in addition to some oxime and nitroso nitro compound (pseudonitrole).

$$R^1R^2CH{-}NO_2 \xrightarrow[\text{ii, } H^+]{\text{i, base}} R^1R^2C{=}O \quad (1)$$

The popularity of the Nef reaction is due in part to the ready availability of nitro compounds.[4–7] Primary and secondary halides react with sodium nitrite in dimethyl sulfoxide (DMSO) or dimethylformamide (DMF) to give useful yields of nitro compounds.[4] Primary amines can be oxidized to nitro compounds with potassium permanganate, *m*-chloroperbenzoic acid[4] or ozone.[5] Chlorination of oximes with hypochlorous acid and reduction with magnesium, zinc or hydrogen/palladium gives secondary nitro compounds.[6] Stabilized carbanions can be nitrated by treatment with a nitrate ester,[4] and enol acetates are nitrated by acetyl nitrate to give nitro ketones.[7]

A wide array of compounds undergo the Nef reaction.[1] These include simple nitro compounds, as well as complex multifunctional species. For example, the carbohydrate in equation (2) was isolated as its hydrazone derivative in 72% yield.[8] Nitronate anions generated *in situ* can be acidified to give good yields of aldehydes and ketones. As an illustration, Michael addition of the dianion of hexanoic acid to 2-nitropropene gives the expected keto acid after acidification (equation 3).[9]

$$CH_2NO_2{-}[CH(OAc)]_3{-}CH_2OAc \xrightarrow[\text{ii, } H_2SO_4\text{; 72\%}]{\text{i, NaOH}} CHO{-}[CH(OH)]_3{-}CH_2OH \quad (2)$$

$$CH_3(CH_2)_3CH(Li)CO_2Li \xrightarrow[\text{ii, HCl; 65\%}]{\text{i, } CH_2{=}C(NO_2)CH_3\text{, }-100\ ^\circ C} CH_3COCH_2CH(CO_2H)(CH_2)_3CH_3 \quad (3)$$

2.5.2.1.2 Problems

There are several problems and side reactions with the Nef reaction. The nitro compound or the anion is often insoluble in water; however, this shortcoming has been circumvented by running the reaction in a solvent such as methanol (equation 4).[10] An acetal may be formed if methanolic acid is used for the acidification stage (equation 5).[11] Interaction of neighboring groups leads to unexpected products in some cases such as with the nitro lactone in equation (6).[12] Systems which are prone to form carbocations may undergo rearrangement under Nef conditions. For example, attempts to prepare norcamphor derivatives led to an *N*-hydroxylactam as the product (equation 7).[13]

i, NaOH, MeOH; ii, H_2SO_4; 95% (4)

i, NaOMe; ii, H_2SO_4, MeOH; 95% (5)

i, NaOMe; ii, H_2SO_4, MeOH; 98% (6)

i, KOH, MeOH; ii, HCl; 40% (7)

Many cyclic α-nitro ketones undergo ring opening when exposed to nucleophilic reagents. For example, 2-nitrocyclohexanone gives an 85% yield of 6-nitrohexanoic acid when stirred with aqueous sodium hydrogencarbonate at room temperature.[14] Interestingly, this strategy has been useful in the construction of macrocyclic nitro compounds (equation 8).[15]

cat. NaH, DME, Δ; 93% (8)

2.5.2.1.3 *Modified Nef reactions*

The power of the Nef reaction as a tool in synthetic organic chemistry has been enhanced greatly by the development of modified reaction conditions. Many more polyfunctional compounds now can undergo one or more of these modified Nef reactions so that the scope of the original reaction is much greater. Many of these newer methods rely on oxidizing or reducing agents. Several of the more useful ones are outlined here, but a more comprehensive discussion is also available.[1]

Although discovered in the early 1900s, the oxidation of nitro compounds with potassium permanganate has become a truly useful reaction only within the last 25 years. An alkoxide base is used to form the nitronate salt, and aqueous potassium permanganate gives the oxidized product in excellent yield. Even aldehydes can be obtained as in equation (9) without loss of the pivaloyl group.[16] Cetyltrimethylammonium permanganate in dichloromethane also converts nitro compounds into aldehydes and ketones in synthetically useful yields.[17]

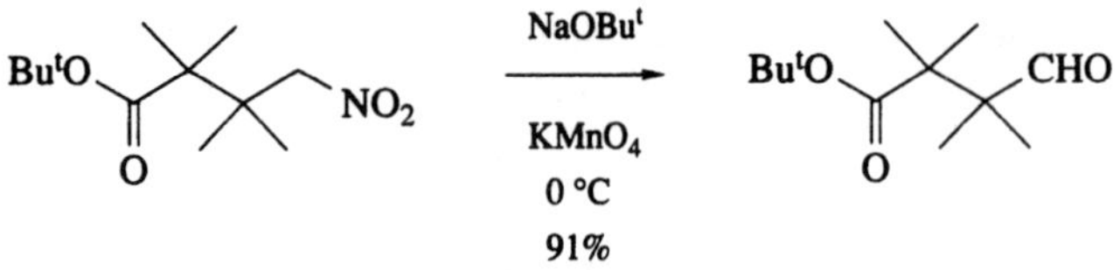

(9)

Titanium trichloride functions as an excellent reductive Nef alternative reagent. This aqueous reagent is very acidic, so that acid sensitive groups such as ketals and esters do not survive unless an acetate buffer is used.[18] Systems prone to acid-catalyzed rearrangements may then successfully undergo the reaction (equation 10).[19] Some very sensitive multifunctional compounds have been obtained using this modified Nef procedure (equation 11).[20] A related process is the formation of 1,4-diketones via *in situ* generation of a nitronate anion by the Lewis acid catalyzed addition of an enol silyl ether to a nitroalkene (equation 12).[21]

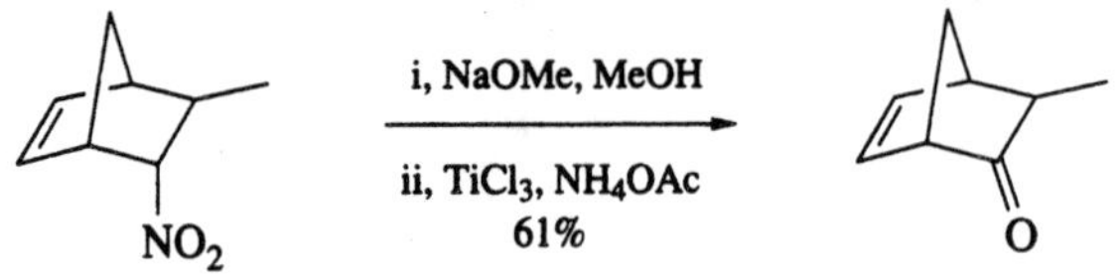

(10)

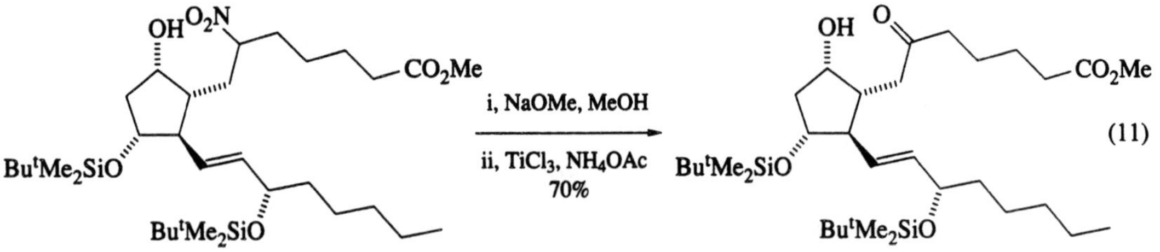

(11)

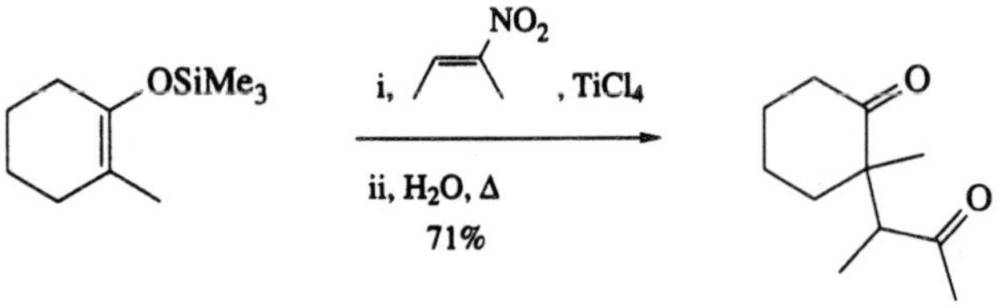

(12)

Ozonolysis of intermediate nitronate anions also yields carbonyl compounds,[22] and, while unsaturation and acetal groups cannot be tolerated, other sensitive molecules have been prepared using this reaction (equation 13).[23]

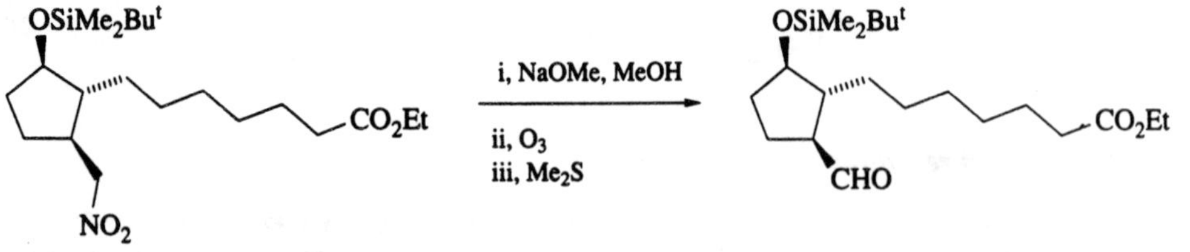

(13)

2.5.2.2 The Meyer Reaction

The Meyer reaction is generally not of major synthetic significance. It is observed when a nitro compound is exposed to strong acid. In this way, carboxylic acids are obtained from primary nitro compounds. The reaction is thought to involve nitrile oxides and hydroxamic acids (RCONHOH) as intermediates. The latter can be isolated by avoiding heat,[24] and the former have been trapped by 1,3-dipolar cycloaddition to alkenes and alkynes.[25]

2.5.3 TERTIARY AMINES

2.5.3.1 Mercury(II) Ion Oxidation

Tertiary amines form complexes with mercury(II) ion, which then give iminium ions by loss of a proton. Addition of perchloric acid permits isolation of the iminium ion as the perchlorate salt and generally the more-substituted ion is favored (equation 14).[26] Intramolecular trapping by a hydroxyalkyl group is also possible to form aminals (equation 15).[27] The lactam products result from over-oxidation, which is promoted by heat. Basification on the other hand usually allows the isolation of enamines[28] although hydroxyenamines have been obtained by reaction of enamines with mercury(II) acetate (equation 16),[29] while dihydroaromatic systems undergo aromatization (equation 17).[30]

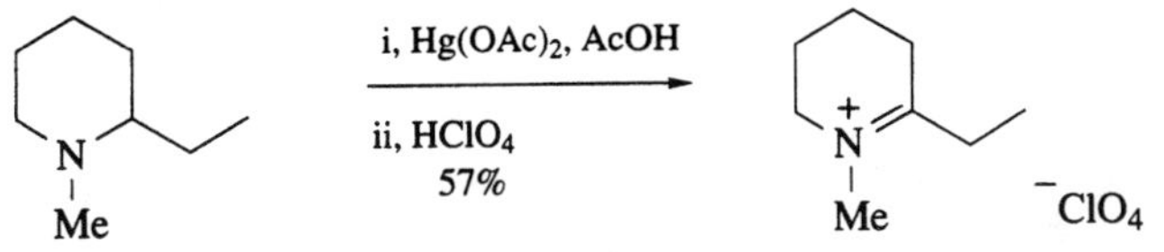

(14)

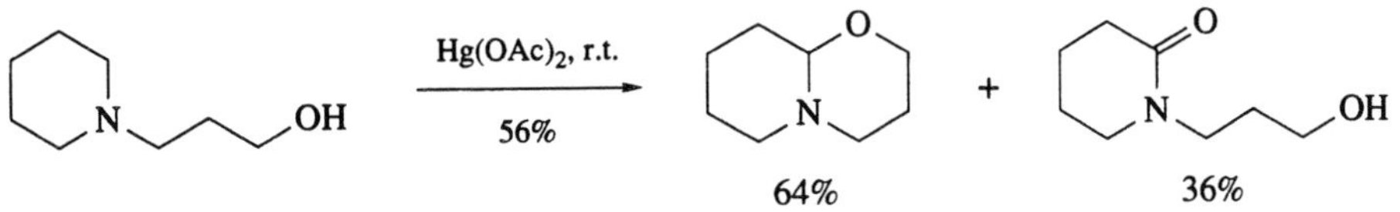

(15)

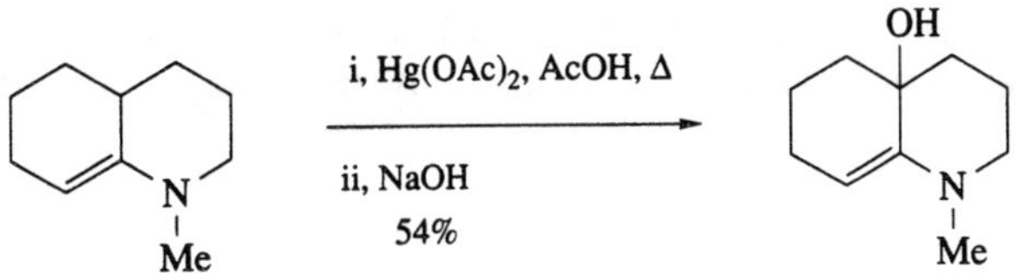

(16)

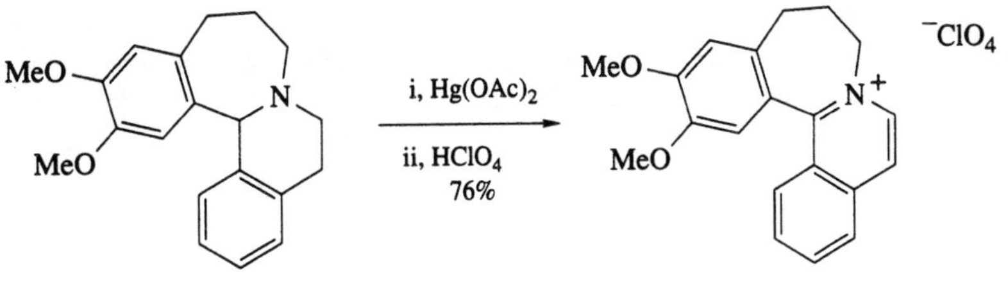

(17)

Lactams are obtained at higher pH by using Hg^{II}–ethylenediaminetetraacetic acid (EDTA) as the oxidizing agent (equation 18).[31] Nicotine gives an 88% yield of cotinine under these reaction conditions.

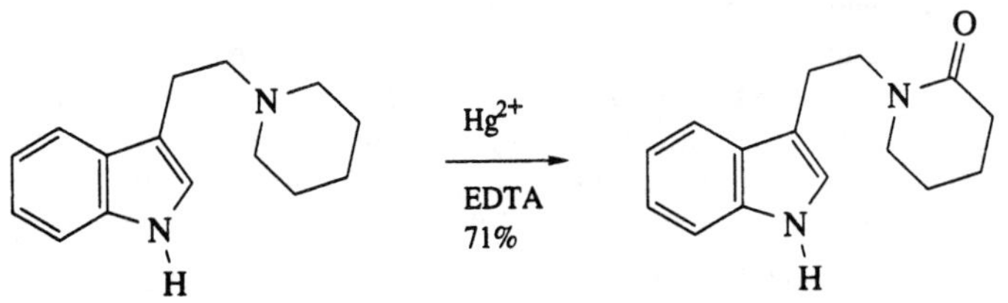

(18)

2.5.3.2 Chromium and Manganese Reagents

Chromium trioxide and pyridine have been used to form amides and lactams from tertiary amines. The yields are however only impressive for the preparation of formamides (equation 19).[32]

Manganese dioxide converts tertiary amines into amides. For example, *N,N*-dimethylaniline gives *N*-methyl-*N*-phenylformamide in 78% yield[33] and *N*-phenylpyrrolidine yields *N*-phenylformamide in 48% yield.[34] Dimethylaminocyclohexane on oxidation leads to cyclohexanone in 85% yield.[35] There is also

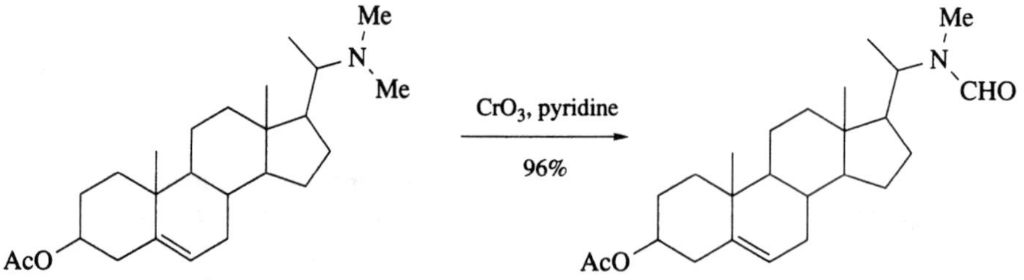

CrO_3, pyridine

96%

(19)

an isolated example of a lactam being isolated in about 5% yield using potassium permanganate as the oxidant (equation 20).[36]

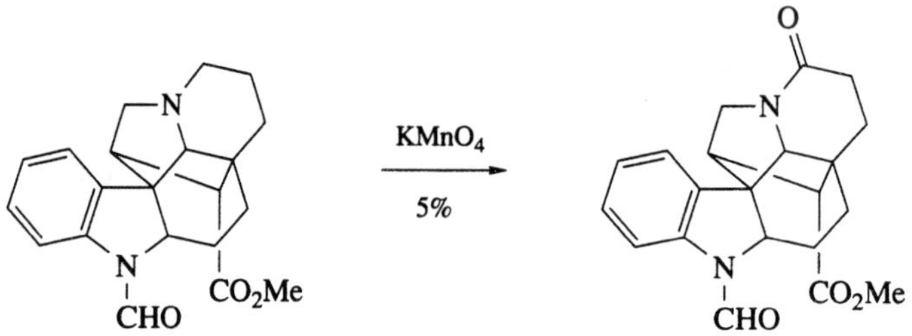

$KMnO_4$

5%

(20)

2.5.3.3 Hypervalent Iodine

Iodosylbenzene converts cyclic tertiary amines into lactams,[37] as indicated by nicotine affording a 20% yield of cotinine and *N*-methylpyrrolidine being converted into the corresponding lactam in 55% yield.

2.5.3.4 Platinum/Oxygen

Tertiary amines react with oxygen in the presence of platinum to give amides[38] showing a strong preference for reaction at methyl groups. For example, oxidation of trimethylamine gives *N,N*-dimethylformamide in 74% yield, and *N*-methylcyclohexylamine yields *N*-formylcyclohexylamine in quantitative yield.

2.5.3.5 Amine Oxides

2.5.3.5.1 The Polonovski reaction

The Polonovski reaction occurs when an amine oxide reacts with an acylating agent.[39] The accepted mechanism involves proton removal to give a nitrogen ylide which loses acetate (using acetic anhydride) which attacks the carbon adjacent to the nitrogen atom giving an α-acetoxyamine. These intermediates can be hydrolyzed to give aldehydes or eliminated to give enamines (equation 21).[40] This latter case was used to help determine the structure of the natural product nupharidine. A recent variant of the procedure occurred by silylation of amine *N*-oxides to give α-silyloxyamines which subsequently underwent elimination to give iminium ion intermediates which reacted with nucleophiles (equation 22).[41]

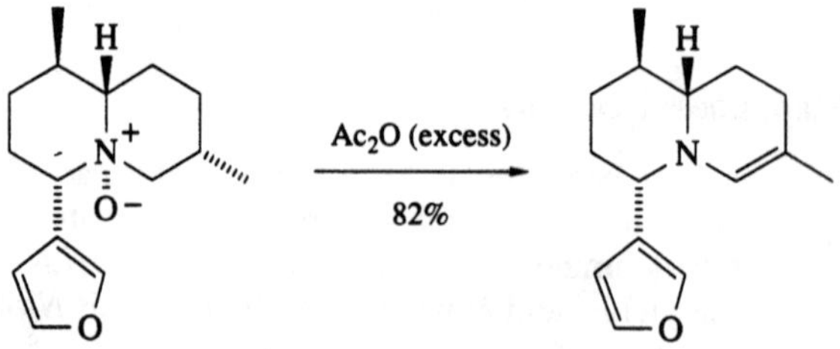

Ac_2O (excess)

82%

(21)

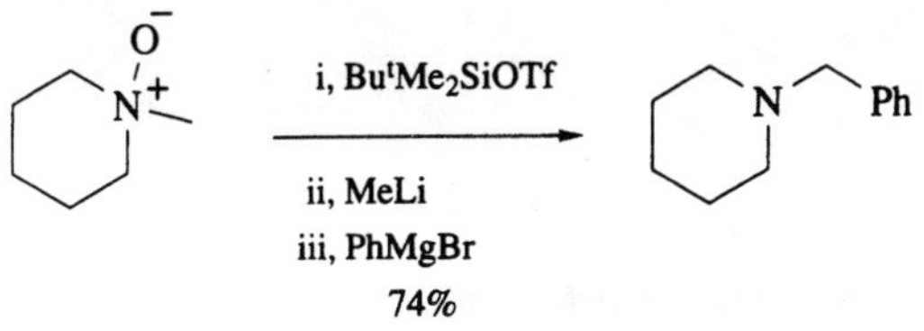

i, Bu^tMe_2SiOTf; ii, MeLi; iii, PhMgBr; 74% (22)

2.5.3.5.2 *Pyridine* N-*oxides*

Pyridine *N*-oxides may be deprotonated to give ylides which react with electrophiles such as carbon dioxide and ketones. For example, 4-chloropyridine *N*-oxide reacts with butyllithium at –65 °C followed by quenching with carbon dioxide to give 4-chloropyridine *N*-oxide 2-carboxylic acid in 49% yield. Quinuclidine *N*-oxide can be deprotonated with *t*-butyllithium to give the anion which can be trapped with deuterium oxide or benzaldehyde.[42]

2.5.3.6 Miscellaneous Oxidation Reactions of Tertiary Amines

Irradiation of some amines in the presence of 1,4-dicyanonaphthalene causes the formation of radical cations, which give iminium ions by loss of a proton. Intramolecular addition of a hydroxylic nucleophile yields aminals (equation 23).[43]

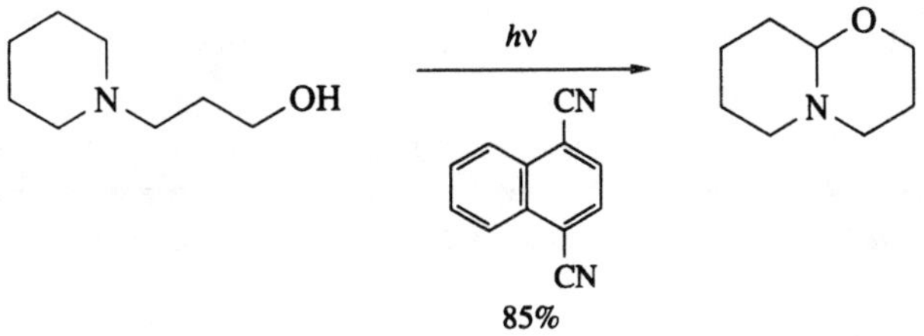

hν, 1,4-dicyanonaphthalene; 85% (23)

Benzylic quaternary salts react with hot dimethyl sulfoxide (DMSO) to give benzaldehyde.[44] For example, benzylammonium chloride gives 60% benzaldehyde plus 24% benzylmethylamine and 5% benzyldimethylamine.

Hydride abstraction from tertiary amines by arylmethyl cations leads to iminium ions which can be hydrolyzed or trapped with nucleophiles.[45] For example, *t*-butyldimethylamine reacts with triphenylmethyl perchlorate to give a 93% yield of the iminium salt. This can be trapped with acetophenone to give the Mannich product (Scheme 1).

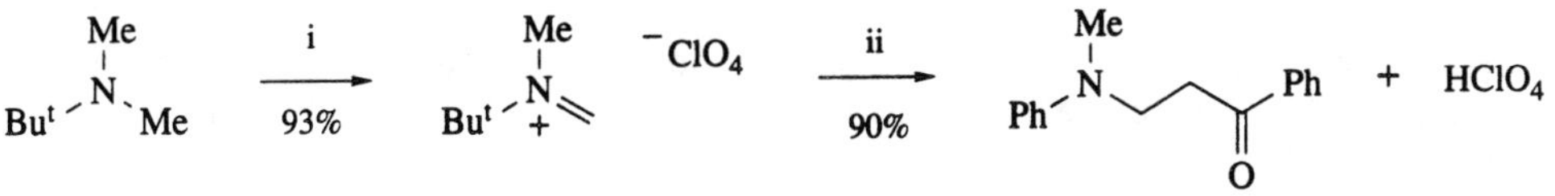

i (93%); $^-ClO_4$; ii (90%); + $HClO_4$

i, $Ph_3C^+ClO_4^-$; ii, PhCOMe

Scheme 1

Benzylic amines react with hydrogen peroxide to yield benzaldehydes.[46] For example, 4-bromobenzyldimethylamine gives a 60% yield of 4-bromobenzaldehyde. Only six cases were reported, with both dimethylamino and diethylamino groups undergoing the reaction.

2.5.4 SECONDARY AMINES

2.5.4.1 Formamidines

Secondary amines are easily converted into formamidines by reaction with amidines (equation 24).[47] Deprotonation and addition of electrophiles, including alkyl halides, acyl halides, aldehydes and diselenides, permit the introduction of various groups adjacent to the amino center, and therefore can be considered as an oxidation. Alkylation with dihalides gives intermediates which lead to new carbocyclic rings after removal of the formamidine group (equation 25).[48]

Me_2N⌒NBu^t; H^+, Δ; 90% (24)

i, Bu^tLi; ii, Cl⌒⌒Br; iii, NH_2NH_2, AcOH; 68% (25)

High enantioselectivity is also possible in these reactions by using an optically active amidine, readily available from (*S*)-valinol.[49] A recent synthesis of (+)-reticuline made elegant use of this chiral intermediate (equation 26).[50] Likewise, (–)-yohimbone was prepared in 98% enantiomeric excess using these reactions.[51]

i, Bu^tLi; ii, (Br, MeO, O⌒Ph); iii, NH_2NH_2, AcOH; 70% (26)

Reticuline, 99% *ee*

Another formamidine which allows facile removal of hindered tertiary hydrogens has been recently introduced (equation 27).[52] The bridgehead position of a bicyclic amine has also been alkylated in good yields via the *t*-butylformamidine.[53]

i, Bu^tLi; ii, EtI; iii, NH_2NH_2; 81% (27)

2.5.4.2 *N*-Nitrosamines

The acidity of hydrogen atoms adjacent to the nitrogen substituent of *N*-nitrosamines has been known for many years,[54] although Seebach and coworkers played a major role in developing this into a useful synthetic process. For example, the anion of *N*-nitrosodimethylamine is formed by using LDA, and can be alkylated or condensed with carbonyl compounds or nitriles (Scheme 2).[55] 1-Bromo-3-iodopropane reacts with this anion, and formation of the kinetic anion of this intermediate gives N-nitrosopiperidine in 54% yield.[56] The key step in the total synthesis of macrostomine was accomplished using a nitrosamine alkylation (equation 28).[57] Many additional examples using N-nitrosamines are also available.[58,59]

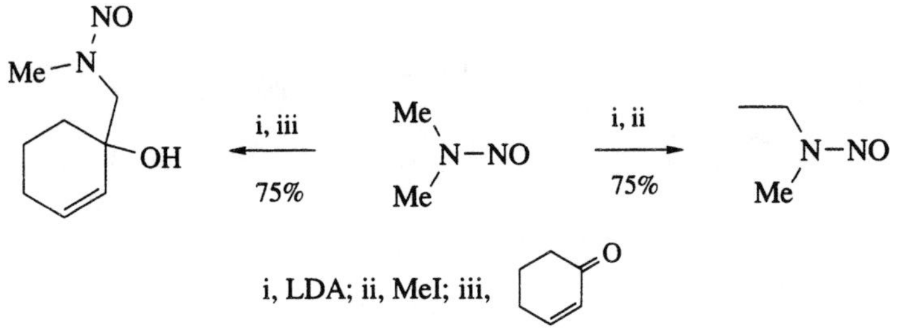

i, iii 75% i, ii 75%

i, LDA; ii, MeI; iii,

Scheme 2

i, 2LDA

ii,

iii, Raney Ni/H_2

80%

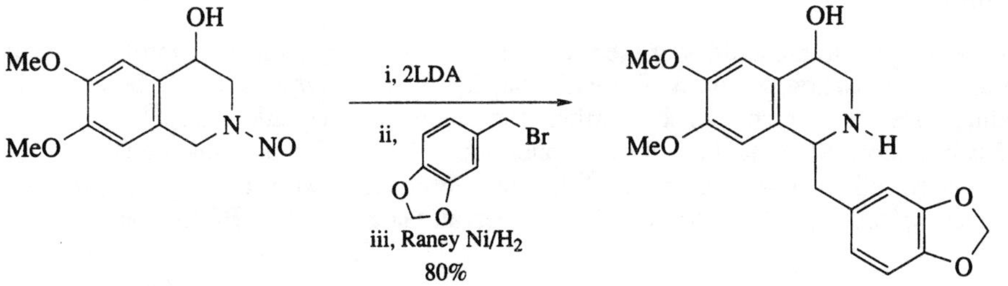

(28)

N-Nitroso-α-amino acids may be converted into α-acetoxynitrosamines.[60] For example, N-nitrosoproline gives a 40% yield of 2-acetoxy-1-nitrosopyrrolidine using lead tetraacetate in pyridine.

2.5.4.3 Amides

Secondary amines can be acylated with acyl groups bearing no α-hydrogens. Deprotonation next to the nitrogen atom and introduction of electrophiles allow the oxidation of that position. Early work showed that these anions underwent self-condensation. For example, N,N-dimethylbenzamide can be deprotonated with lithium 2,2,6,6-tetramethylpiperidide (LITMP) to give N-methyl-N-phenacylbenzamide in 60% yield.[61] It is clear that lithium ion is crucial to the success of this reaction. Crown ethers prevent the reaction,[62] and recent kinetic studies show intermediate lithium complex formation.[63]

Simple benzamides undergo slow deprotonation with amide bases. Stronger base systems such as s-butyllithium in tetramethylethylenediamine (TMEDA) are too nucleophilic for simple benzamides but N,N-dimethyl-2,4,6-triisopropylbenzamide is deprotonated in only 5 min at –78 °C, and this anion is alkylated with methyl iodide to give a 77% yield of the expected product.[64]

Other amides have been alkylated by this method. For example, the system in equation (29) gives clean products even with secondary iodides.[65] Vinylogous amides behave in a similar manner (equation 30).[66] Thioamides also undergo alkylation without complications.[67]

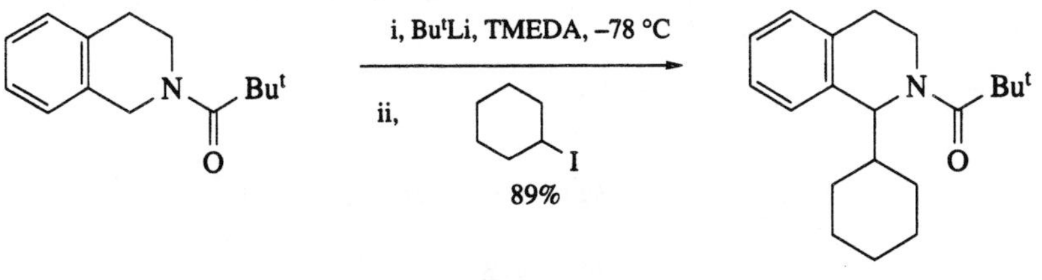

i, Bu^tLi, TMEDA, –78 °C

ii,

89%

(29)

i, LDA

ii, BuBr

58%

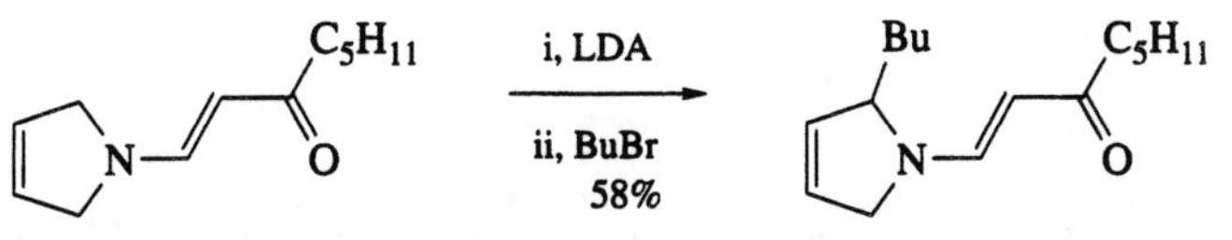

(30)

Trapping of these so-called dipole-stabilized anions with either oxygen or the complex of oxodiperoxymolybdenum with HMPA and pyridine gives hemiaminals which provide secondary amides.[68] The procedure can therefore be used as a method of benzyl deprotection of amides (equation 31).

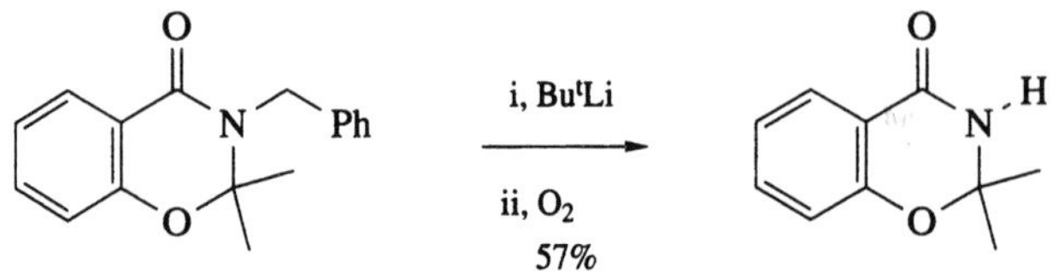
O
N
Ph
O
i, Bu^{t}Li
ii, O_2
57%
O
N
H
O

(31)

2.5.4.4 Urethanes

2.5.4.4.1 Anions

Urethanes analogous to the amides of the previous section undergo similar deprotonation followed by alkylation and condensation reactions. For example, 2,4,6-tri-*t*-butylphenol may be converted into the corresponding urethane which can be further functionalized (equation 32).[69] *N*-Carbomethoxy-3-pyrroline has been converted into both the trail pheromone for the Pharaoh ant and gephyrotoxin 223 by using regiospecific alkylations (Scheme 3).[70] Similar approaches were used in the preparation of the natural product supinidine.[71] Piperidines also have been alkylated via the *t*-BOC-protected amines.[72]

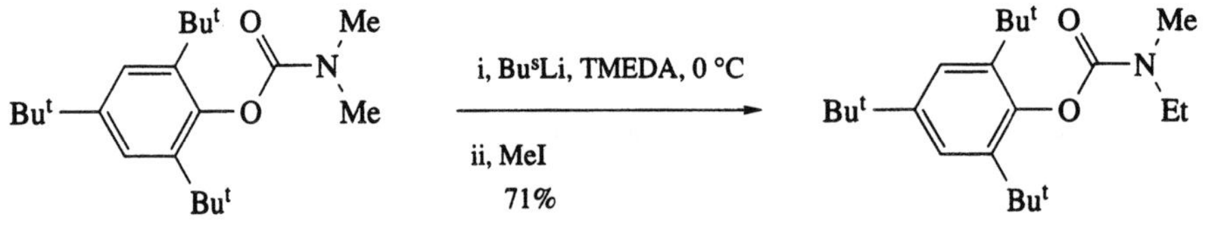
Bu^t
O
Me
N
Me
Bu^t
O
Bu^t
i, Bu^{s}Li, TMEDA, 0 °C
ii, MeI
71%
Bu^t
O
Me
N
Et
Bu^t
O
Bu^t

(32)

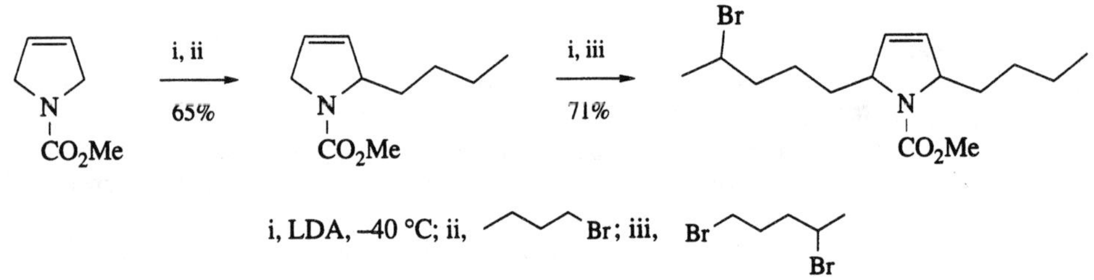
N
CO_2Me
i, ii
65%
N
CO_2Me
i, iii
71%
Br
N
CO_2Me
i, LDA, –40 °C; ii,
Br; iii,
Br
Br

Scheme 3

Carbamic acid dianions have also been reported and behave in the analogous manner (Scheme 4).[73]

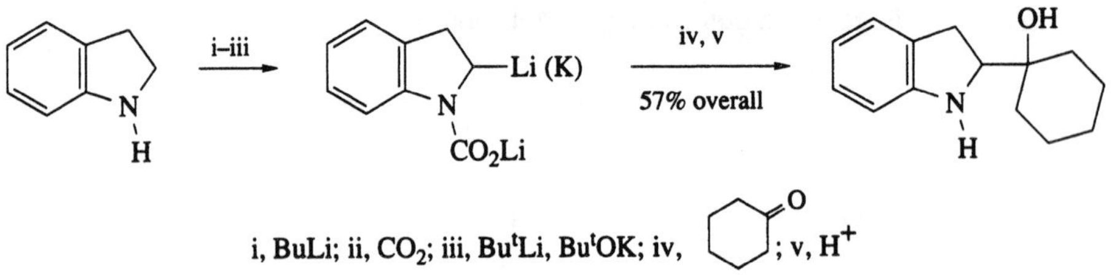
N
H
i–iii
–Li (K)
N
CO_2Li
iv, v
57% overall
OH
N
H
i, BuLi; ii, CO_2; iii, Bu^{t}Li, Bu^{t}OK; iv,
O
; v, H^+

Scheme 4

2.5.4.4.2 Electrolysis

Anodic electrolysis of urethanes in methanol gives α-methoxyurethanes, which have been used in reactions with nucleophiles or eliminated to give enamine derivatives. Yields from the electrolysis step are usually acceptable (typically 50%).[74] Several natural products have been prepared using this process as the key step.[75,76] For example, a simple synthesis of hygrine is possible (Scheme 5).[77] Only simple urethanes have been electrolyzed, which suggests a lack of functional group compatibility.

i, electrolysis, MeOH; ii, $TiCl_4$,

Scheme 5

2.5.4.5 Miscellaneous Methods

Secondary amines react with 2-ethoxy-4-isopropyloxazoline and acid to give oxazoline derivatives which may also be deprotonated. High enantiomeric excess is possible during alkylation by using chiral oxazolines,[78] as in the preparation of (+)-salsolidine (equation 33).[79] Likewise, isoindoline also has been alkylated by using a similar approach.[80]

i, Bu^tLi, –78 °C; ii, MeI, –100 °C; iii, NH_2NH_2; 93% (+)-Salsolidine (33)

Iodosylbenzene oxidizes secondary amines to lactams in reasonable yields, much like the tertiary amines discussed in Section 2.5.3.3.[37] If only 1 equiv. of iodosylbenzene is used, the imine intermediates are isolated. Thus, 1,2,3,4-tetrahydroisoquinoline gives 3,4-dihydroisoquinoline in 61% yield.[81] This general reaction also has been reported with either iodosylbenzene or iodosylbenzene/$RuCl_2(PPh_3)_3$.[82]

Mercury(II) ion oxidizes secondary amines to imines when reacted in the presence of acetic acid.[83] For example, 2-*t*-butylpiperidine gives a 75% yield of the more substituted imine. Mercury(II) ion in EDTA gives lactams.[31] For example, 4-*t*-butylpiperidone is isolated in 81% yield from 4-*t*-butylpiperidine using mercury(II) ion in EDTA.

Photochemically induced oxidation of one secondary amine in the presence of 1,4-dicyanonaphthalene allows the introduction of oxygen into the position adjacent to the nitrogen atom (equation 34).[43]

Secondary amines have been oxidized to imines by several routes. The nitrogen atom can be halogenated and the *N*-chloramine eliminated with base. *N*-Chlorosuccinimide and potassium hydroxide are popular reagents[84] although *t*-butyl hypochlorite followed by potassium superoxide has also been used with some success.[85] The overall yields in these reactions are usually good. Diphenyl selenoxide and trifluoroacetic anhydride convert tetrahydroisoquinolines into dihydroisoquinolines.[86] Benzeneseleninic anhydride converts secondary amines into imines, which can be trapped by cyanide ion to give α-cyanoamines.[87] Tristriphenylphosphineruthenium dichloride and *t*-butyl hydroperoxide convert secondary amines into imines, although the products in all cases are conjugated with aromatic rings.[88] Finally, dehydrogenation of secondary amines into imines can be accomplished with a cobalt catalyst in the presence of oxygen at 60 °C.[89]

$h\nu$, 82% (34)

2.5.5 PRIMARY AMINES

2.5.5.1 Conversion into Aldehydes and Ketones

Primary amines have been transformed into imines which when metallated react with carbonyl compounds. Treatment with butyllithium, alkylation with allyl bromide and hydrolysis gives highly substituted aldehydes (Scheme 6).[90] Thus, in this example, the carbon adjacent to the original amino nitrogen atom becomes the carbonyl carbon. The technique has been used several times in the course of total synthesis, as with a recent approach to crinine[91] and as the key step in a recent highly regioselective preparation of α,β-unsaturated aldehydes (equation 35).[92]

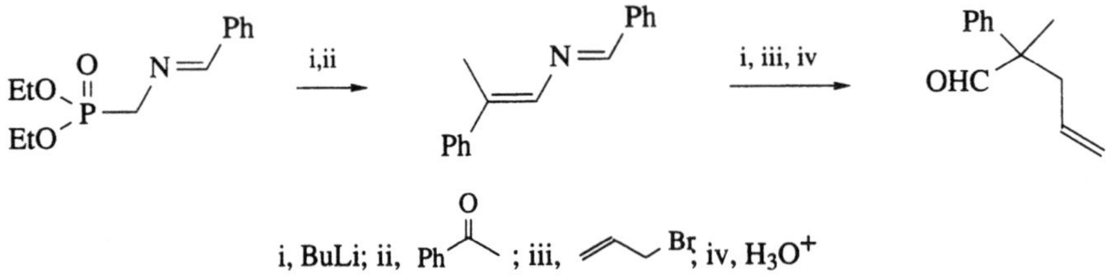

EtO, EtO, P=O, N=, Ph —i,ii→ Ph, N=, Ph —i, iii, iv→ Ph, OHC

i, BuLi; ii, PhC(O)Me; iii, allyl Br; iv, H_3O^+

Scheme 6

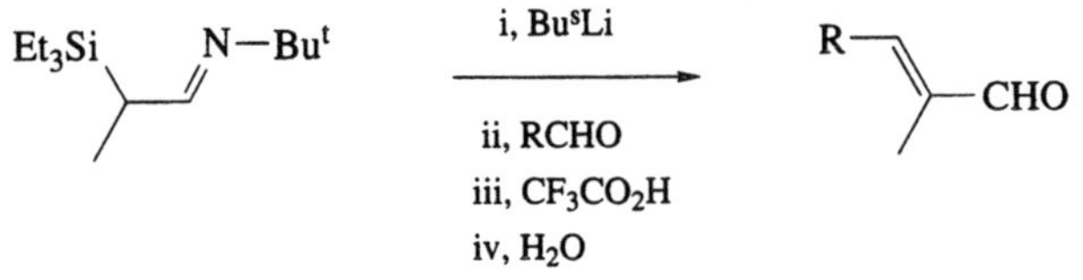

Et_3Si, N—Bu^t —i, Bu^sLi; ii, RCHO; iii, CF_3CO_2H; iv, H_2O→ R, CHO (35)

Many other methods have been developed for the conversion of primary amines into carbonyl groups *via* imines. Prostaglandin E_1 was prepared in over 25% yield from the corresponding amine with hydroxy groups protected as THP acetals. This was achieved by *N*-bromination (NBS) followed by loss of HBr and hydrolysis of the resulting imine.[93] The approach has been applied successfully to several primary amines having only one α-hydrogen atom using sodium hypochlorite under phase transfer conditions (nitriles are isolated from amines having two α-hydrogens). Cyclohexanone is obtained in 98% yield and norcamphor in 84% yield.[94] Another clever method for amine oxidation uses 3,5-di-*t*-butyl-1,2-benzoquinone to give an intermediate imine which is hydrolyzed to the carbonyl compound.[95] Camphor is isolated in 69% yield and cyclohexanone in 97% yield. The mechanism of these reactions has been examined in some detail recently.[96] The imines from reaction of primary amines with 2,6-di-*t*-butyl-1,4-benzoquinone react with oxygen and potassium *t*-butoxide or potassium hydroxide to give amides.[97] For example, benzylamine gives benzamide in 50% overall yield. Another method uses aromatic aldehydes to give imines which can be isomerized and then hydrolyzed, as with 2-pyridinecarboxaldehyde which converts undecylamine into the corresponding aldehyde in 94% yield.[98] In a similar manner, *N*-methyl-4-pyridinecarbaldehyde has been used for the oxidative deamination of primary amines (even lysine and lysine esters with the α-amino group protected).[99] Phenylselenyl chloride converts primary amines into imines.[100] 2-Adamantylamine gives adamantanone in 91% isolated yield using this method. Di-*t*-butyliminoxyl radical, from cerium(IV) ammonium nitrate oxidation of the oxime of di-*t*-butyl ketone, also oxidizes amines into imines.[101]

Primary amines *via* pyrilium salts react with 2-pyridone *N*-oxides and on decomposition give aldehydes in fair yields (equation 36).[102]

Various reagents will oxidize primary amines to intermediates, which are often hydrolyzed in acceptable yields *in situ* to give aldehydes and ketones. Hot potassium permanganate in a buffered (calcium sulfate) medium gives reasonable yields of carbonyl compounds, although of course double bonds and many other functional groups would not be compatible with this method.[103] Sodium persulfate and catalytic silver nitrate convert primary amines into aldehydes (43–96%).[104] Similarly, palladium(II) chloride or gold(III) chloride[105] as well as silver picolinate[106] have been used for this purpose also, although the latter reagent gives considerable amounts of nitrile as a side product and can be the major product in many cases depending upon the substituents present.

i, [2,4,6-triphenylpyrylium BF_4^-]; ii, [sodium salt of N-oxido-6-phenyl-4-phenylpyridin-2-one]; iii, Δ; 47% (36)

Aromatic aldehydes are obtained by reaction of primary amines with *p*-nitrobenzenesulfonyl peroxide followed by mild hydrolysis with dilute acid.[107]

N,N-Disulfonamides are decomposed with sodium hydrogencarbonate in DMSO to give ketones.[108]

Imines formed from the reaction of primary amines with carbonyl compounds can be oxidized to oxaziridines with MCPBA which hydrolyze to aldehydes or ketones with acid. When acetone is used, the final by-products are ammonia and acetone (equation 37).[109] The use of 2-pyridinecarbaldehyde is preferred since it gives an acid-soluble by-product which aids work-up (equation 38).[110]

i, acetone; ii, MCPBA; iii, H_3O^+; 40% (37)

i, 2-pyridinecarbaldehyde; ii, MCPBA; iii, ^-OH; iv, H_3O^+; 40% (38)

2.5.5.2 Conversion of Primary Amines into Nitriles

Many reagents convert primary amines into nitriles. Some of these have been mentioned above and represent serious limitations on methods for generating carbonyl compounds. Other ways of oxidizing amines to nitriles are the use of nickel peroxide,[111] lead tetraacetate,[112] copper(I) chloride plus oxygen and pyridine,[113] iodine pentafluoride[114] and benzeneseleninic anhydride.[115] A double bromination–dehydrobromination can be effected for the preparation of nitriles with 2 equiv. of NBS and trimethylamine.[116] Likewise, fluorination and elimination of HF gives nitriles.[117]

2.5.6 AMINO ACIDS

2.5.6.1 Alkylation

The introduction of alkyl groups at the α-carbon of amino acids has been accomplished most efficiently by formation of imine esters. For example, the benzaldehyde imine of ethyl glycinate can be deprotonated and alkylated (equation 39).[118] Other imines also have been used.[119] Optical activity has been introduced by using chiral palladium ligands during the alkylation step,[120] chiral alcohols to form the ester,[121] and chiral ketones to form the imine.[122] Alkylation of 2-pyrrole acetate esters has been accomplished in a similar fashion.[123]

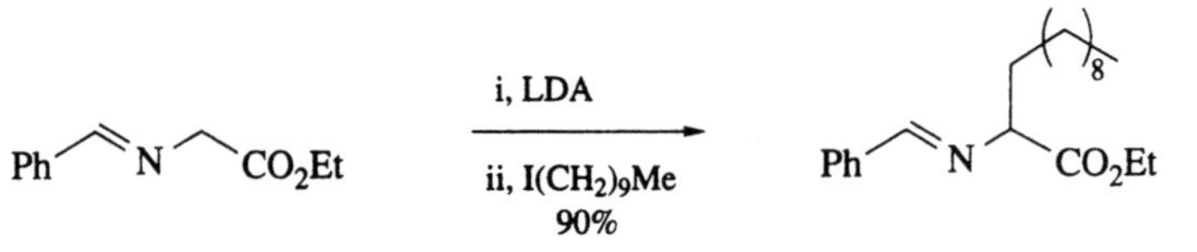

$$\text{PhCH=NCH}_2\text{CO}_2\text{Et} \xrightarrow[\text{ii, I(CH}_2)_9\text{Me} \quad 90\%]{\text{i, LDA}} \text{PhCH=NCH(CH}_2(\text{CH}_2)_8\text{Me)CO}_2\text{Et} \quad (39)$$

Alkylation of amino acids has also been achieved by first forming heterocyclic derivatives. For example, the oxazolidinone from CBZ-phenylalanine and formaldehyde reacts with potassium hexamethyldisilazide followed by allyl bromide to give the expected product in 76% yield.[124] Chiral tetrahydro-1,4-oxazin-2-ones have been used in a similar manner to give amino acids with a degree of diastereoselectivity.[125,126]

2.5.6.2 Degradation of Amino Acids

Amino acids are converted into aldehydes, the most popular method being the Strecker degradation.[127] The amino acid is simply mixed with reagents such as ninhydrin and heated to form ammonia and carbon dioxide as by-products. Sodium hypochlorite can be used in a process accelerated by UV irradiation.[128] A similar reaction is the Akabori reaction where the amino acid is heated with compounds such as glucose to give aldehydes, as in the Strecker process.[129]

N-Alkyl amino acids are decarboxylated to iminium salts by brief treatment with hot $POCl_3$.[130] These may be intercepted by internal nucleophiles to form cyclic compounds (equation 40).[131]

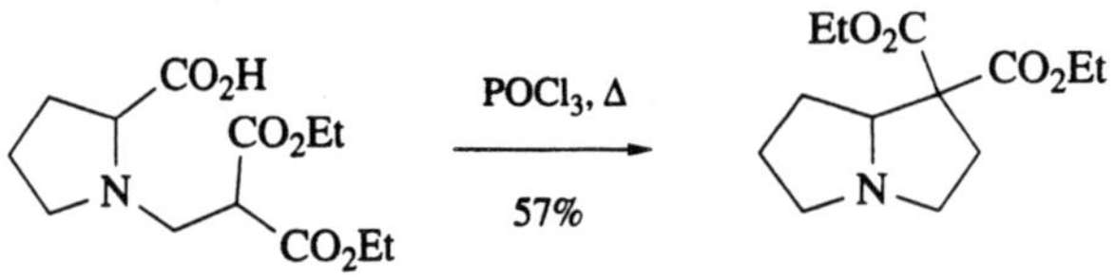

$$\text{(40)} \quad \xrightarrow[57\%]{\text{POCl}_3,\ \Delta}$$

2.5.6.3 Miscellaneous

Oxidation with excess iodosylbenzene converts proline into 2-pyrrolidinone in up to 70% yield.[78] Use of only 1 equiv. of iodosylbenzene in this reaction gives 1-pyrroline, plus the corresponding trimer.

Oxygen substituents have been introduced into the α-position of amino acids by *N*-halogenation and elimination/addition in the presence of an alkoxide. For example, the methyl ester of BOC-phenylalanine reacts with *t*-butylhypochlorite followed by methoxide to give the α-methoxy amino acid derivative in 76% yield.[132] The intermediate halo compound reacts with Grignard reagents to give higher amino acids. *N*-BOC-glycinates react with NBS followed by 2 equiv. of a methyl Grignard reagent to give the protected alanine ester by a process which must involve an imino ester.[133] Similar intermediates may be obtained by alternative methods involving addition of alkoxides to imidates (equation 41).[134]

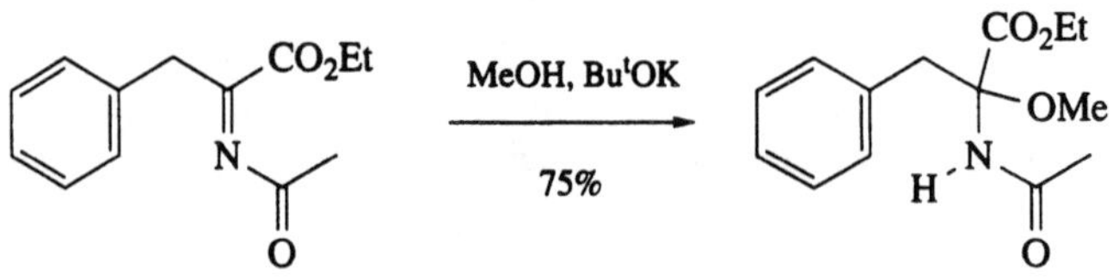

$$\text{(41)} \quad \xrightarrow[75\%]{\text{MeOH, Bu}^t\text{OK}}$$

2.5.7 MISCELLANEOUS OXIDATIONS

2.5.7.1 Lactams/Amides

Lactams and amides are easily oxidized to imides with hydroperoxides in the presence of $Mn(acac)_2$ or Co^{II} naphthenate.[135,136] For example, 2-piperidone is oxidized to glutarimide in 72% yield using $Mn(acac)_2$.

2.5.7.2 α-Cyanoamines

α-Cyanoamines can be deprotonated to allow the introduction of electrophiles at the α-position. The resulting adducts can be hydrolyzed to give ketones. These intermediate anions are therefore acyl anion equivalents. For example, aromatic ketones are efficiently prepared as illustrated in equation (42).[137] More recently, 1-benzyl-2,6-dicyanopiperidine was bisalkylated using sequential treatment with LDA and alkyl halides and the resulting product was hydrolyzed to give 1,5-diketones.[138]

MeO, MeO, NEt_2, CN —— i, NaH, DMF; ii, $PhCH_2Br$; iii, H^+; 68% ——> MeO, MeO, O, Ph (42)

2.5.7.3 Pyridinium Salts

The classic Krohnke aldehyde synthesis results from the displacement of pyridinium salts by aromatic nitroso compounds to give nitrones which are hydrolyzed to aldehydes.[139] Phenacyl bromide reacts with pyridine and then nitrosobenzene to give phenylglyoxal in 76% yield after acid hydrolysis.[140] The pyridinium salts in these reactions must be activated in some way toward displacement to effect efficient conversions.

2.5.7.4 Hydrazones/Oximes

Lead tetraacetate converts hydrazones to carbonyl compounds plus the corresponding alcohols. For example, benzophenone hydrazone is converted into benzophenone in 36–67% yields plus lesser amounts of benzhydrol.[141,142] Substituted ketone hydrazones give α-acetoxyazo compounds with 1 equiv. of lead tetraacetate,[143] although 2 equiv. of lead tetraacetate results in ketones. This also has been observed with tosylhydrazones.[144] Arylhydrazones of aldehydes give diacylhydrazines (equation 43).[145,146]

H, NO_2, N, N, NO_2 —— $Pb(OAc)_4$; 72% ——> O, Ac, NO_2, N, N, H, NO_2 (43)

Lead tetraacetate oxidizes oximes to various products. For example, the oxime of cyclohexanone gives α-acetoxynitrosocyclohexane in 35% yield.[147] Aldoximes react with lead tetraacetate to give nitrile oxides which then yield acetyl hydroxamates by reaction with acetic acid.[148]

Miscellaneous derivatives are hydrolyzed back to the starting carbonyl compounds. Phenylhydrazones of ketones are converted into the parent ketones upon treatment with manganese dioxide.[149] Tosylhydrazones react with molybdenyl chloride[150] or tungsten tetrafluoride[151] to give both aldehydes and ketones. Sodium peroxide converts aldoximes into carboxylic acids.[152]

The anions of hindered hydrazones allow the introduction of electrophiles at the α-position, thus functioning as acyl anion equivalents after isomerization back to a hydrazone and hydrolysis.[153,154] For example, the *t*-butylhydrazone of acetaldehyde gives phenylacetone following the sequence of reactions shown in equation 44.

H, N, N, Bu^t —— i, BuLi; ii, $PhCH_2Br$; iii, CF_3CO_2H; iv, HO_2CCO_2H, H_2O; 71% ——> O, Ph (44)

2.5.7.5 Isocyanides

Isocyanides similarly upon deprotonation and alkylation give amines or ketones after hydrolysis. For example, methyl isocyanide reacts with butyllithium, allyl bromide, and aqueous acid to give 3-butenamine.[155] Tosylmethyl isocyanide (TOSMIC) can be deprotonated with excess sodium hydride in DMSO, alkylated twice with benzyl bromide and hydrolyzed with acid and then base to give 1,3-diphenylpropanone.[156]

2.5.7.6 Nitrones

Certain nitrones react with acylating agents and then rearrange to give products having an oxygen group adjacent to the nitrogen of the original nitrone, somewhat reminiscent of the Polonovski reaction reported earlier.[157,158] An example is the reaction shown in equation (45).

Cl Ph $^{-}$O−N+ N−Me O $\xrightarrow{Ac_2O}$ Cl Ph N N−Me AcO O (45)

2.5.8 REFERENCES

1. For a recent review of this reaction, see H. W. Pinnick, *Org. React. (N.Y.)*, 1990, **38**, 655.
2. J. U. Nef, *Justus Liebigs Ann. Chem.*, 1894, **280**, 263.
3. J. Armand, *Bull. Soc. Chim. Fr.*, 1965, 3246.
4. N. Kornblum, *Org. React. (N.Y.)*, 1962, **12**, 101.
5. E. Keinan and Y. Mazur, *J. Org. Chem.*, 1977, **42**, 844.
6. E. J. Corey and H. Estreicher, *Tetrahedron Lett.*, 1980, **21**, 1117.
7. P. Dampawan and W. W. Zajac, Jr., *J. Org. Chem.*, 1982, **47**, 1176.
8. J. C. Sowden, *J. Am. Chem. Soc.*, 1950, **72**, 808.
9. M. Miyashita, R. Yamaguchi and A. Yoshikoshi, *J. Org. Chem.*, 1984, **49**, 2857.
10. S. Hirano, T. Hiyama, S. Fujita, T. Kawaguti, Y. Hayashi and H. Nozaki, *Tetrahedron*, 1974, **30**, 2633.
11. J. G. Gleason, T. F. Buckley, K. G. Holden, D. B. Bryan and P. Siler, *J. Am. Chem. Soc.*, 1979, **101**, 4730.
12. M. A. Adams, A. J. Duggin, J. Smolanoff and J. Meinwald, *J. Am. Chem. Soc.*, 1979, **101**, 5364.
13. W. E. Noland, R. B. Hart, W. A. Joern and R. G. Simon, *J. Org. Chem.*, 1969, **34**, 2058.
14. A. S. Matlack and D. S. Breslow, *J. Org. Chem.*, 1967, **32**, 1995.
15. R. C. Cookson and P. S. Ray, *Tetrahedron Lett.*, 1982, **23**, 3521.
16. N. Kornblum, A. S. Erickson, W. J. Kelly and B. Henggeler, *J. Org. Chem.*, 1982, **47**, 4534.
17. P. S. Vankar, R. Rathore and S. Chandrasekaran, *Synth. Commun.*, 1987, **17**, 195.
18. J. E. McMurry, *Acc. Chem. Res.*, 1974, **7**, 281.
19. G. Mehta, S. K. Kapoor and P. N. Panday, *Indian J. Chem., Sect. B*, 1976, **14**, 252.
20. R. Noyori and M. Suzuki, *Angew. Chem., Int. Ed. Engl.*, 1984, **23**, 847.
21. M. Miyashita, T. Yanami, T. Kumazawa and A. Yoshikoshi, *J. Am. Chem. Soc.*, 1984, **106**, 2149.
22. J. E. McMurry, J. Melton and H. C. Padgett, *J. Org. Chem.*, 1974, **39**, 259.
23. B. P. Cho, V. K. Chadha, G. C. Le Breton and D. L. Venton, *J. Org. Chem.*, 1986, **51**, 4279.
24. N. Kornblum and R. A. Brown, *J. Am. Chem. Soc.*, 1965, **87**, 1742.
25. P. A. Wade, N. V. Amin, H.-K. Yen, D. T. Price and G. F. Huhn, *J. Org. Chem.*, 1984, **49**, 4595.
26. N. J. Leonard and F. P. Hauck, Jr., *J. Am. Chem. Soc.*, 1957, **79**, 5279.
27. N. J. Leonard and W. K. Musker, *J. Am. Chem. Soc.*, 1960, **82**, 5148.
28. M. G. Reinecke and L. R. Kray, *J. Org. Chem.*, 1966, **31**, 4215.
29. N. J. Leonard, L. A. Miller and P. D. Thomas, *J. Am. Chem. Soc.*, 1956, **76**, 3463.
30. G. Schluter and W. Meise, *Liebigs Ann. Chem.*, 1989, 1.
31. E. Wenkert and E. C. Angell, *Synth. Commun.*, 1988, **18**, 1331.
32. T. R. Govindachari, B. R. Pai, S. Rajappa, N. Yiswanathan, W. G. Kump, K. Nagarajan and H. Schmidt, *Helv. Chim. Acta*, 1962, **45**, 1146.
33. E. F. Curragh, H. B. Henbest and A. Thomas, *J. Chem. Soc.*, 1960, 3559.
34. H. B. Henbest and A. Thomas, *J. Chem. Soc.*, 1957, 3032.
35. H. B. Henbest and M. J. W. Stratford, *J. Chem. Soc. C*, 1966, 995.
36. C. Djerassi, M. Cereghetti, H. Budzikiewicz, M.-M. Janot, M. Plat and J. Le Men, *Helv. Chim. Acta*, 1964, **47**, 827.
37. R. M. Moriarty, R. K. Vaid and M. P. Duncan, *Tetrahedron Lett.*, 1988, **29**, 6913.
38. G. T. Davis and D. H. Rosenblatt, *Tetrahedron Lett.*, 1968, 4085.

39. G. A. Russell and G. J. Mikol, *Mech. Mol. Migr.*, 1968, **1**, 176.
40. R. T. LaLonde, E. Auer, C. F. Wong and V. P. Muralidharan, *J. Am. Chem. Soc.*, 1971, **93**, 2501.
41. N. Tokitoh and R. Okazaki, *Bull. Chem. Soc. Jpn.*, 1988, **61**, 735.
42. D. H. R. Barton, R. Beugelmans and R. N. Young, *Nouv. J. Chim.*, 1978, **2**, 363.
43. G. Pandey and G. Kumaraswamy, *Tetrahedron Lett.*, 1988, **29**, 4153.
44. V. J. Traynelis and R. H. Ode, *J. Org. Chem.*, 1970, **35**, 2207.
45. H. Volz and H.-H. Kiltz, *Justus Liebigs Ann. Chem.*, 1971, **752**, 86.
46. K. Takabe and T. Yamada, *Chem. Ind. (London)*, 1982, 959.
47. A. I. Meyers, *Aldrichimica Acta*, 1985, **18**, 59.
48. A. I. Meyers and S. Hellring, *J. Org. Chem.*, 1982, **47**, 2229.
49. D. A. Dickman, M. Boes and A. I. Meyers, *Org. Synth.*, 1988, **67**, 52.
50. A. I. Meyers and J. Gulles, *Heterocycles*, 1989, **28**, 295.
51. A. I. Meyers, D. B. Miller and F. H. White, *J. Am. Chem. Soc.*, 1988, **110**, 4778.
52. M. A. Gonzalez and A. I. Meyers, *Tetrahedron Lett.*, 1989, **30**, 43, 47.
53. J. A. Monn and K. C. Rice, *Tetrahedron Lett.*, 1989, **30**, 911.
54. L. K. Keefer and C. H. Fodor, *J. Am. Chem. Soc.*, 1970, **92**, 5747.
55. D. Seebach and D. Enders, *Angew. Chem., Int. Ed. Engl.*, 1972, **11**, 301, 1101, 1102.
56. D. Seebach, D. Enders and B. Renger, *Chem. Ber.*, 1977, **110**, 1852.
57. W. Wykypiel and D. Seebach, *Tetrahedron Lett.*, 1980, **21**, 1927.
58. J. E. Saavedra, G.-K. Pei and D. W. Farnsworth, *Org. Prep. Proced. Int.*, 1988, **20**, 385.
59. K. Yoshida and Y. Yano, *J. Chem. Soc., Perkin Trans. 2*, 1988, 2011.
60. J. E. Saavedra, *Tetrahedron Lett.*, 1978, 1923.
61. P. Beak and R. F. Farney, *J. Am. Chem. Soc.*, 1973, **95**, 4771.
62. P. Beak, G. R. Brubraker and R. F. Farney, *J. Am. Chem. Soc.*, 1976, **98**, 3621.
63. D. R. Hay, Z. Song, S. G. Smith and P. Beak, *J. Am. Chem. Soc.*, 1988, **110**, 8145.
64. P. Beak, B. G. McKinnie and D. B. Reitz, *Tetrahedron Lett.*, 1977, 1839.
65. J.-J. Lohmann, D. Seebach, M. A. Syfrig and M. Yoshifuji, *Angew. Chem., Int. Ed. Engl.*, 1981, **20**, 128.
66. J. C. L. Armande and U. K. Pandit, *Tetrahedron Lett.*, 1977, 897.
67. D. Seebach and W. Lubosch, *Angew. Chem., Int. Ed. Engl.*, 1976, **15**, 313.
68. R. M. Williams and E. Kwast, *Tetrahedron Lett.*, 1989, **30**, 451.
69. D. Seebach and T. Hassel, *Angew. Chem., Int. Ed. Engl.*, 1978, **17**, 274.
70. Y. Yamamoto, H. Yatagai and K. Maruyama, *J. Org. Chem.*, 1980, **45**, 193.
71. T. L. Macdonald and B. A. Narayanan, *J. Org. Chem.*, 1983, **48**, 1129.
72. P. Beak and W.-K. Lee, *Tetrahedron Lett.*, 1989, **30**, 1197.
73. A. R. Katritzky and S. Sengupta, *J. Chem. Soc., Perkin Trans. 1*, 1989, 17.
74. T. Shono, H. Hamaguchi and Y. Matsumura, *J. Am. Chem. Soc.*, 1975, **97**, 4264.
75. T. Shono, Y. Matsumura, M. Ogaki and O. Onomura, *Chem. Lett.*, 1987, 1447.
76. T. Nagasaka, H. Hayashi and F. Hamaguchi, *Heterocycles*, 1988, **27**, 1685.
77. T. Shono, Y. Matsumura and K. Tsubata, *J. Am. Chem. Soc.*, 1981, **103**, 1172.
78. R. E. Gawley, G. C. Hart and L. J. Bartolotti, *J. Org. Chem.*, 1989, **54**, 175.
79. R. E. Gawley, G. C. Hart, M. Goicoechea-Pappas and A. L. Smith, *J. Org. Chem.*, 1986, **51**, 3076.
80. R. E. Gawley, S. R. Chemburkar, A. L. Smith and T. V. Anklekar, *J. Org. Chem.*, 1988, **53**, 5381.
81. M. Ochiai, M. Inenaga, Y. Nagao, R. M. Moriarty, R. K. Vaid and M. P. Duncan, *Tetrahedron Lett.*, 1988, **29**, 6917.
82. P. Müller and D. M. Gilabert, *Tetrahedron*, 1988, **44**, 7171.
83. M. F. Grundon and B. E. Reynolds, *J. Chem. Soc.*, 1964, 2445.
84. J. Quick and R. Oterson, *Synthesis*, 1976, 745.
85. F. E. Scully, Jr. and R. C. Davis, *J. Org. Chem.*, 1978, **43**, 1467.
86. J. P. Marino and R. D. Larsen, Jr., *J. Am. Chem. Soc.*, 1981, **103**, 4642.
87. D. H. R. Barton, A. Billion and J. Boivin, *Tetrahedron Lett.*, 1985, **26**, 1229.
88. S.-I. Murahashi, T. Naota and H. Taki, *J. Chem. Soc., Chem. Commun.*, 1985, 613.
89. A. Nishinaga, S. Yamazaki and T. Matsuura, *Tetrahedron Lett.*, 1988, **29**, 4115.
90. S. K. Davidsen, G. W. Phillips and S. F. Martin, *Org. Synth.*, 1987, **65**, 119.
91. S. F. Martin and C. L. Campbell, *J. Org. Chem.*, 1988, **53**, 3184.
92. R. Desmond, S. G. Mills, R. P. Volante and I. Shinkai, *Tetrahedron Lett.*, 1988, **29**, 3895.
93. E. J. Corey, N. H. Anderson, R. M. Carlson, J. Paust, E. Vedejs, I. Vlattas and R. E. K. Winter, *J. Am. Chem. Soc.*, 1968, **90**, 3247.
94. G. A. Lee and H. H. Freedman, *Tetrahedron Lett.*, 1976, 1641.
95. E. J. Corey and K. Achiwa, *J. Am. Chem. Soc.*, 1969, **91**, 1429.
96. R. F. X. Klein, L. M. Bargas and V. Horak, *J. Org. Chem.*, 1988, **53**, 5994.
97. A. Nishinaga, T. Shimizu and T. Matsuura, *J. Chem. Soc., Chem. Commun.*, 1979, 970.
98. J. Babler and B. J. Invergo, *J. Org. Chem.*, 1981, **46**, 1937.
99. T. F. Buckley and H. Rapoport, *J. Am. Chem. Soc.*, 1982, **104**, 4446.
100. M. R. Czarny, *Synth. Commun.*, 1976, **6**, 285.
101. J. Cornejo, K. D. Larson and G. D. Mendenhall, *J. Org. Chem.*, 1985, **50**, 5382.
102. A. R. Katritzky, M. J. Cook, A. Ikizler and G. H. Millet, *J. Chem. Soc., Perkin Trans. 1*, 1979, 2500.
103. S. S. Rawalay and H. Shechter, *J. Org. Chem.*, 1967, **32**, 3129.
104. R. G. R. Bacon and D. Stewart, *J. Chem. Soc. C*, 1966, 1384.
105. M. E. Kuehne and T. C. Hall, *J. Org. Chem.*, 1976, **41**, 2742.
106. J. B. Lee, C. Parkin, M. J. Shaw, N. A. Hampson and K. I. MacDonald, *Tetrahedron*, 1973, **29**, 751.
107. R. V. Hoffman and R. Cadena, *J. Am. Chem. Soc.*, 1977, **99**, 8226.
108. V. A. Curtis, A. Raheja, J. E. Rejowski, R. W. Majewski and R. J. Baumgarten, *Tetrahedron Lett.*, 1975, 3107.

109. D. St. C. Black and N. A. Blackman, *Aust. J. Chem.*, 1975, **28**, 2547.
110. S. E. Dinizo and D. S. Watt, *J. Am. Chem. Soc.*, 1975, **97**, 6900.
111. K. Nakagawa and T. Tsuji, *Chem. Pharm. Bull.*, 1963, **11**, 296.
112. A. Stojiljkovic, V. Andrejevic and M. L. Mihailovic, *Tetrahedron*, 1967, **23**, 721.
113. T. Kametani, K. Takahashi, T. Ohsawa and M. Ihara, *Synthesis*, 1977, 245.
114. T. E. Stevens, *J. Org. Chem.*, 1961, **26**, 2531.
115. M. R. Czarny, *J. Chem. Soc., Chem. Commun.*, 1976, 81.
116. W. Gottardi, *Monatsh. Chem.*, 1973, **104**, 1690.
117. C. M. Sharts, *J. Org. Chem.*, 1968, **33**, 1008.
118. G. Stork, A. Y. W. Leong and A. M. Touzin, *J. Org. Chem.*, 1976, **41**, 3491.
119. M. J. O'Donnell, W. D. Bennett, W. A. Bruder, W. N. Jacobsen, K. Knuth, B. LeClef, R. L. Polt, F. G. Bordwell, S. R. Mrozack and T. A. Cripe, *J. Am. Chem. Soc.*, 1988, **110**, 8520.
120. J.-P. Genet, S. Juge, J. R. Montes and J.-M. Gaudin, *J. Chem. Soc., Chem. Commun.*, 1988, 718.
121. K.-J. Fasth, G. Antoni and B. Langstrom, *J. Chem. Soc., Perkin Trans. I*, 1988, 3081.
122. J. M. McIntosh and K. C. Cassidy, *Can. J. Chem.*, 1988, **66**, 3116.
123. C. Kashima and T. Maruyama, *Heterocycles*, 1988, **27**, 1727.
124. T. M. Zydowsky, J. F. Dellaria, Jr. and H. N. Nellans, *J. Org. Chem.*, 1988, **53**, 5607.
125. R. M. Williams and M.-N. Im, *Tetrahedron Lett.*, 1988, **29**, 6075.
126. J. F. Dellaria, Jr. and B. D. Santarsiero, *Tetrahedron Lett.*, 1988, **29**, 6079.
127. A. Schonber and R. Moubacher, *Chem. Rev.*, 1952, **50**, 261.
128. Y. Ogata, M. Kimura and Y. Kondo, *Bull. Chem. Soc. Jpn.*, 1981, **54**, 2057.
129. S. Akabori, *Chem. Ber.*, 1933, **66**, 143.
130. R. T. Dean, H. C. Padgett and H. Rapoport, *J. Am. Chem. Soc.*, 1976, **98**, 7448.
131. I. G. Csendes, Y. Y. Lee, H. C. Padgett and H. Rapoport, *J. Org. Chem.*, 1979, **44**, 4173.
132. H. Poisel, *Chem. Ber.*, 1977, **110**, 942.
133. P. Ernert, J. Neyer, C. Stucki, J. Schneebeli, and J.-P. Obrecht, *Tetrahedron Lett.*, 1988, **29**, 1265.
134. J. D. M. Herscheid, R. J. F. Nivard, M. W. Tijhuis, H. P. H. Scholten and H. C. J. Ottenheijm, *J. Org. Chem.*, 1980, **45**, 1880.
135. A. R. Doumaux, Jr., J. E. McKeon and D. J. Trecker, *J. Am. Chem. Soc.*, 1969, **91**, 3992.
136. A. R. Doumaux, Jr. and D. J. Trecker, *J. Org. Chem.*, 1970, **35**, 2121.
137. S. F. Dyke, E. P. Tiley, A. W. C. White and D. P. Gale, *Tetrahedron*, 1975, **31**, 1219.
138. K. Takahashi, M. Asakawa and K. Ogura, *Chem. Lett.*, 1988, 1109.
139. S. J. Angyal, *Org. React. (N.Y.)*, 1954, **8**, 203.
140. F. Kröhnke and E. Borner, *Chem. Ber.*, 1936, **69**, 2006.
141. D. H. R. Barton, J. F. McGhie and P. L. Batten, *J. Chem. Soc. C*, 1970, 1033.
142. A. Stojiljkovic, N. Orbovic, S. Sredojevic and M. L. Mihailovic, *Tetrahedron*, 1970, **26**, 1101.
143. D. C. Iffland, L. Salisbury and W. R. Schafer, *J. Am. Chem. Soc.*, 1961, **83**, 747.
144. A. Bhati, *J. Chem. Soc., Chem. Commun.*, 1965, 476.
145. W. A. F. Gladstone, J. B. Aylward and R. O. C. Norman, *J. Chem. Soc. C*, 1969, 2587.
146. J. B. Aylward, *J. Chem. Soc. C*, 1970, 1494.
147. J. W. Lown, *J. Chem. Soc. B*, 1966, 441.
148. G. Just and K. Dahl, *Tetrahedron*, 1968, **24**, 5251.
149. I. Bhatnagar and M. V. George, *J. Org. Chem.*, 1967, **32**, 2252.
150. G. A. Olah, J. Welch, G. K. S. Prakash and T.-L. Ho, *Synthesis*, 1976, 808.
151. G. A. Olah and J. Welch, *Synthesis*, 1976, 809.
152. T.-L. Ho and G. A. Olah, *Synthesis*, 1976, 807.
153. R. M. Adlington, J. E. Baldwin, J. C. Bottaro and M. W. D. Perry, *J. Chem. Soc., Chem. Commun.*, 1983, 1040.
154. J. E. Baldwin, J. C. Bottaro, J. N. Kolhe and R. M. Adlington, *J. Chem. Soc., Chem. Commun.*, 1984, 22.
155. U. Schollkopf, K.-W. Henneke, K. Madawinata and R. Harms, *Justus Liebigs Ann. Chem.*, 1977, 40.
156. O. Possel and A. M. van Leusen, *Tetrahedron Lett.*, 1977, 4229.
157. N. Castagnoli, Jr. and W. Sadee, *J. Med. Chem.*, 1972, **15**, 1076.
158. V. Sunjic, F. Kajfez, D. Kolbah, H. Hofman and M. Stromar, *Tetrahedron Lett.*, 1973, 3209.

2.6
Oxidation Adjacent to Oxygen of Ethers

CHRISTOPHER R. A. GODFREY
ICI Agrochemicals, Bracknell, UK

2.6.1 INTRODUCTION

In the presence of certain oxidants, ethers possessing at least one C—H bond at the position adjacent to oxygen are susceptible to oxidation. The light-induced oxidation of diethyl ether in air is a well-known example of this phenomenon. In most instances, oxidative attack leads, *via* the intermediacy of carbonium ions or free radicals, to either esters (or lactones) or α-substituted ethers. Subsequent breakdown under the reaction conditions may then lead to mixtures of cleavage products, which can in turn suffer further oxidation. In principle, therefore, a wide range of products can result during the oxidation of an ether, depending on the nature of the substrate, the oxidant and the experimental conditions employed.

Although a number of different reagents have been discovered for the selective oxidation of ethers, *e.g.* halogens,[1] iodine tris(trifluoroacetate),[2] trichloroisocyanuric acid,[3] UF_6,[4] *N,N*-dibromobenzenesulfonamide[5] and lead tetraacetate,[6] few have assumed any synthetic importance. Of these, the most significant are the metallic oxidants chromic acid and ruthenium tetroxide. DDQ has also been widely used for the oxidative deprotection of benzyl ethers. It is the aim of this chapter to review the latest developments in ether oxidation by these, and other reagents, with particular emphasis on chemo- and regio-selectivity. Several reviews on the subject have appeared previously.[7–11] The related oxidation of acetals has been reviewed recently[11] and will not be dealt with here.

2.6.2 METALLIC OXIDANTS

2.6.2.1 Introduction

Most of the metallic oxidants which have been used for the oxidation of ethers have been based on oxides of the transition metals chromium, manganese and ruthenium, the latter being of greatest synthetic importance. The first reported example of the application of ruthenium tetroxide in the oxidation of ethers appeared over 30 years ago in 1958,[12] although an indication of its reactivity towards ethers had been obtained some years before.[13] In a systematic study which revealed the powerful oxidizing properties of the reagent, Berkowitz and Rylander demonstrated the quantitative conversion of tetrahydrofuran and *n*-butyl ether into γ-butyrolactone and butyl butyrate, respectively.[12] Significantly, no overoxidation was observed. Apart from an unsuccessful attempt to oxidize ethylene oxide, no further attempts were made by the authors to examine further the scope of this novel transformation. In a series of subsequent publications[14,15] and a patent,[16] Wolf and his coworkers went on to exploit the reaction in the preparation of aldosterone and related steroids (equation 1).

OAc OAc O OAc OAc O O O O

RuO_4 / CCl_4, r.t / 30% (1)

In 1959, Henbest and Nicholls reported the conversion of the ether (**1**) into the corresponding lactone (**2**), on treatment with chromic acid in acetone at room temperature (equation 2).[17] Mechanistic studies on the oxidation of ethers by both ruthenium tetroxide[18] and chromic acid[19] were duly carried out. A more detailed analysis of the mechanistic aspects of these reagents appears in a review published by Müller.[7]

AcO O AcO O O

chromic acid / acetone, 20 °C / 89% (2)

(**1**) (**2**)

Despite these early successes, few synthetic chemists appeared to recognize the potential of metallic oxidants for the oxidation of ethers to esters and lactones, and only a few further developments were published prior to 1980.[20–24] Since that time, as the chemistry of these reagents has become better understood, their use has increased dramatically to the stage where they are now the reagents of choice for many applications, including complex natural product synthesis.

Although most attention has been focused on ruthenium tetroxide and chromium trioxide, some other variants have appeared in the literature which are worthy of note. For example, a Merck group reported that *t*-butyl chromate was effective in the transformation of the spiro ether (**3**) into the corresponding lactone (**4**; equation 3).[25] The use of ruthenium tetroxide, on the other hand, led to extensive side reactions. Benzyltriethylammonium permanganate,[26] zinc dichromate[27] and zinc permanganate supported on silica gel[28] are all potent oxidants, which react readily with a wide range of substrates, including simple ethers. The full scope of these reagents, however, remains essentially unexplored. *cis*-$[Ru^{VI}(6,6'\text{-}Cl_2bpy)_2O_2][ClO_4]_2$ has recently been reported to oxidize a wide variety of substrates, including tetrahydrofuran, to butyrolactone.[29] Trömel and Russ have described the preparation and reactions of dimanganese heptoxide (Mn_2O_7) as dilute solutions in carbon tetrachloride or Freon 113.[30] This potentially explosive oxidant reacts within minutes with ethers at –45 °C to afford lactones or esters in high yield. However, even at this temperature, other functional groups, such as alkenes, alcohols and aromatic systems, are attacked.

t-butyl chromate

HOAc–Ac$_2$O, Δ, 2–4 h 40%

(3)

Cyclic ethers are oxidized to lactones in the presence of cerium(IV) salts.[31] Treatment of tetrahydrofuran with cerium(IV) ammonium nitrate in the presence of primary, secondary or tertiary alcohols leads to the formation of the corresponding tetrahydrofuranyl ethers in quantitative yield.[31] Furthermore, 4-methoxybenzyl ether derivatives of carbohydrates are selectively deprotected to the parent alcohols on reaction with cerium(IV) ammonium nitrate in aqueous acetonitrile.[32–34]

Finally, oxidative cyclization (HgO, I_2, $h\nu$) of appropriately substituted alcoholic ethers formed the basis of Kay's stereoselective syntheses of both 4-hydroxy-1,7-dioxaspiro[5.5]undecane, an olive fly pheromone component,[35] and (±)-talaromycin B (equations 4 and 5).[36] More recently, Danishefsky *et al.* have further extended the scope of this spiroketal-forming reaction in their elegant total synthesis of avermectin A_{1a} (equation 6).[37]

HgO, I_2, $h\nu$

(4)

HgO, I_2, $h\nu$

CCl_4, 20 °C 24 h 55%

3:1

(5)

HgO, I_2, $h\nu$

CCl_4 53%

(6)

2.6.2.2 Reaction Conditions

The chromium-based reagents, particularly chromium trioxide, are commonly used in solution in acetic acid and/or acetic anhydride, or occasionally in a less polar organic solvent, such as dichloromethane. The high reactivity of zinc permanganate was effectively controlled by supporting the reagent on silica gel and carrying out its reactions in dichloromethane.[28] The reagents are used in stoichiometric quantities, or in excess, and reactions are generally carried out at room temperature or above.

Much of the early work with ruthenium tetroxide also made use of stoichiometric amounts of a solution of the reagent in an inert solvent, such as carbon tetrachloride. Reactions were carried out at room temperature. The general acceptance of the reagent as a powerful wide-ranging oxidant,[38] coupled with the expense of ruthenium metal, however, later provided the incentive to develop alternative catalytic procedures. In 1980, Smith and Scarborough published details of a systematic study of ether oxidation by ruthenium tetroxide in which they compared the relative merits of both the stoichiometric and the two-phase catalytic methods.[39] In the latter case, catalytic quantities of the reagent were generated by vigorously stirring a mixture of ruthenium dioxide in carbon tetrachloride with an aqueous solution of sodium periodate at room temperature. Good yields of esters and lactones were obtained from a series of simple ethers using the stoichiometric method (equations 7 and 8). When the two-phase catalytic method was employed, however, reaction times were long, and products unstable to aqueous conditions tended to undergo further oxidation to the corresponding carboxylic acids (equations 9 and 10). Similar results were obtained in a later study using catalytic ruthenium chloride and sodium or calcium hypochlorite as the stoichiometric oxidant under phase transfer conditions.[40] The use of sodium hypochlorite as cooxidant in the oxidation of the bicyclic ketone (**5**) led preferentially to the formation of the Baeyer–Villiger product (**6**), whereas sodium periodate gave the lactone (**7**; Scheme 1).[41]

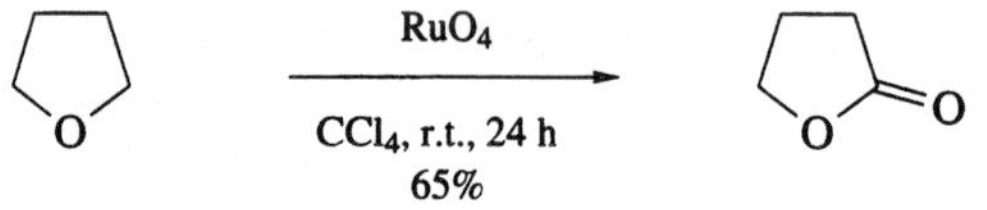

RuO_4; CCl_4, r.t., 24 h; 65% (7)

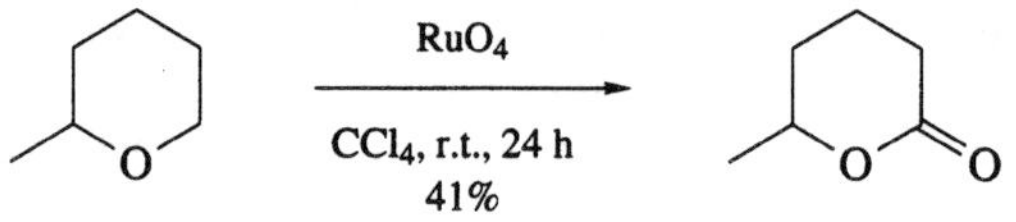

RuO_4; CCl_4, r.t., 24 h; 41% (8)

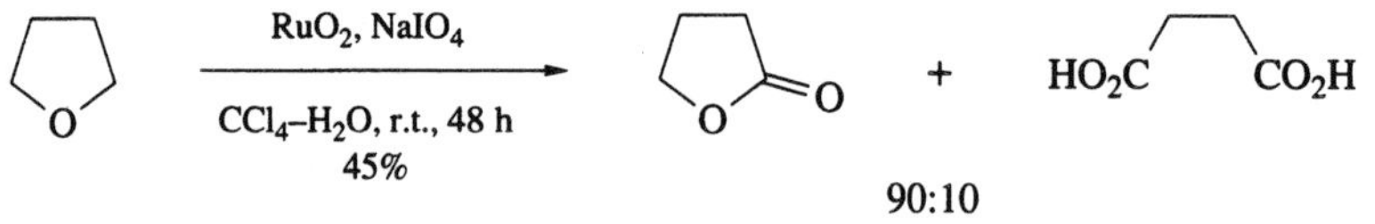

RuO_2, $NaIO_4$; CCl_4–H_2O, r.t., 48 h; 45% + HO_2C CO_2H 90:10 (9)

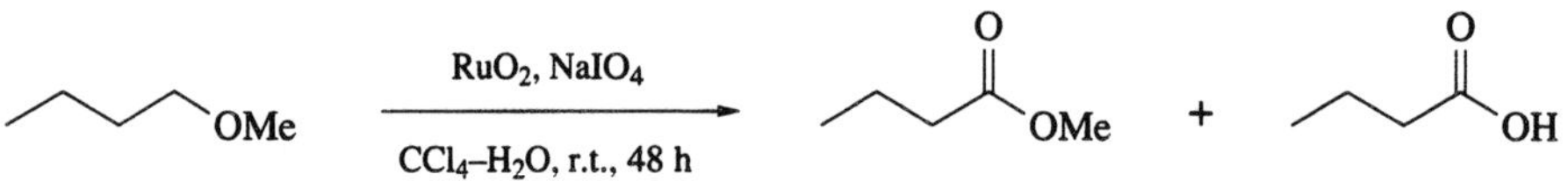

OMe; RuO_2, $NaIO_4$; CCl_4–H_2O, r.t., 48 h; OMe + OH (10)

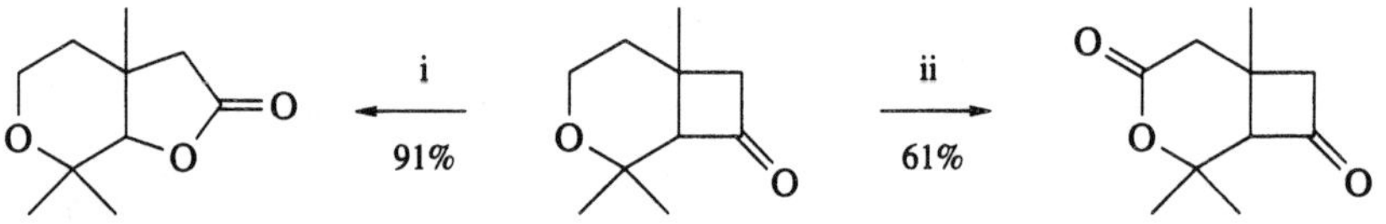

(**6**) ← i, 91% — (**5**) — ii, 61% → (**7**)

i, $RuCl_3$, NaOCl, H_2O–CCl_4, r.t., 5 h; ii, $RuCl_3$, $NaIO_4$, H_2O–CCl_4, r.t., 18 h

Scheme 1

In the acetonitrile modification reported by Sharpless and coworkers, hydrolysis apparently does not take place to any appreciable extent.[42] Consequently the yield of ester can be significantly increased (equation 11). This improved procedure, along with some minor variants,[43] therefore appears to be the method of choice for effecting the oxidation of ethers with ruthenium tetroxide, and has been widely adopted.

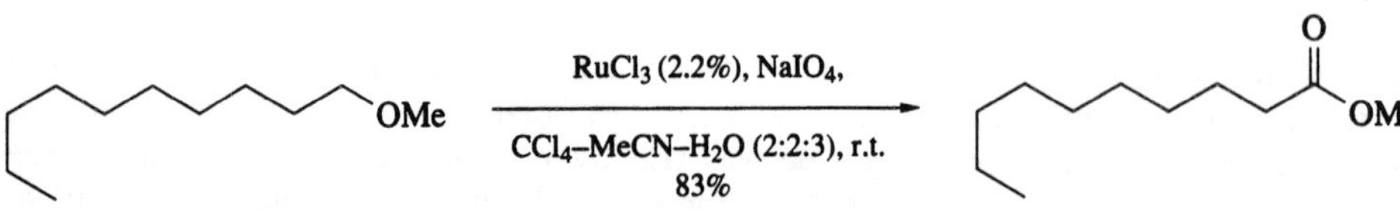

OMe; $RuCl_3$ (2.2%), $NaIO_4$, CCl_4–MeCN–H_2O (2:2:3), r.t.; 83%; OMe (11)

2.6.2.3 Selectivity of Reaction and Application in Synthesis

Since most of the synthetic applications described to date have involved the oxides of chromium and ruthenium, the discussion on reaction selectivity will be limited to these reagents. Even so, much work still remains to be done before the outcome of reactions on complex substrates can be predicted with absolute certainty.

Simple symmetrical ethers, such as tetrahydrofuran or *n*-butyl ether, and ethers possessing only one unsubstituted α-carbon atom and no other functional groups, clearly present no problems in terms of selectivity of reaction (*e.g.* equation 12).[44] In the absence of overriding steric effects, the reaction of simple unsymmetrical ethers with ruthenium and chromium oxidants generally proceeds with high regioselectivity. With complex substrates, however, mixtures of products are often produced.

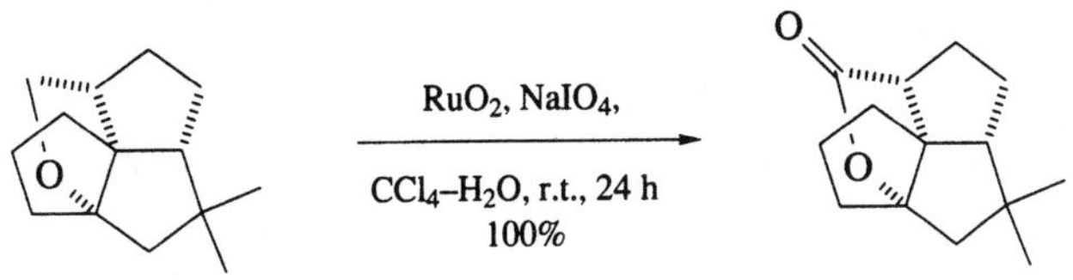

RuO_2, $NaIO_4$, CCl_4–H_2O, r.t., 24 h, 100% (12)

The relative reactivity of primary and secondary positions adjacent to oxygen can be strongly dependent on the nature of the oxidant. For example, treatment of the methyl ethers (**8**)[45] and (**10**)[46] with chromium trioxide in acetic acid leads to the formation of the formates (**9**) and (**11**), respectively (equations 13 and 14). In direct contrast, *n*-decyl methyl ether is oxidized exclusively to methyl *n*-decanoate (83% yield) by ruthenium tetroxide (equation 11).[42] Under similar reaction conditions, 3β-cholestanol methyl ether gives cholestan-3-one as the major product, together with traces of the corresponding formate.[42] Therefore, at least in the case of ruthenium tetroxide, primary positions appear to be more reactive than tertiary.

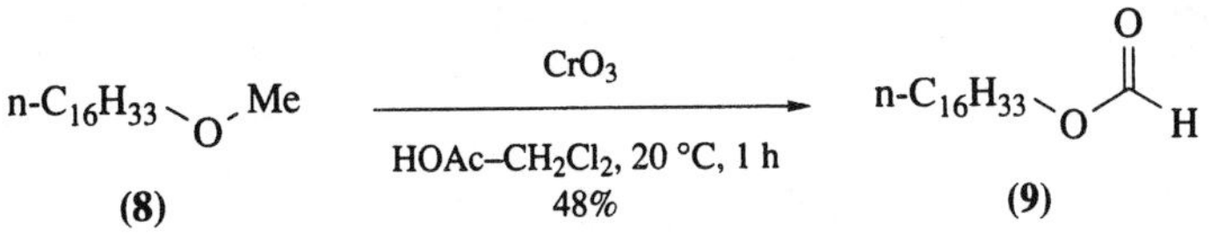

n-$C_{16}H_{33}$–O–Me (**8**) → CrO_3, HOAc–CH_2Cl_2, 20 °C, 1 h, 48% → n-$C_{16}H_{33}$–O–CHO (**9**) (13)

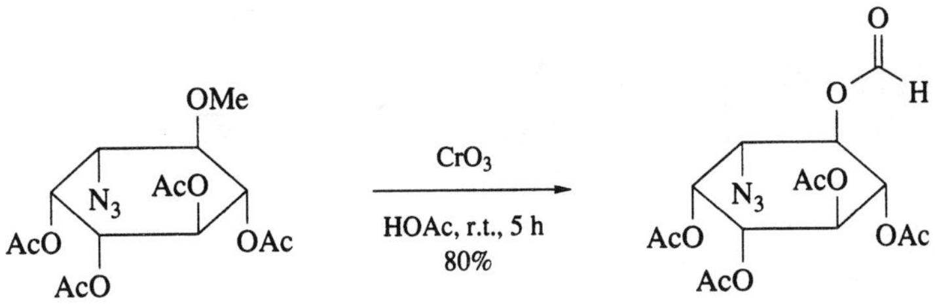

(**10**) (OMe, N_3, AcO, AcO, OAc, AcO) → CrO_3, HOAc, r.t., 5 h, 80% → (**11**) (OCHO, N_3, AcO, AcO, OAc, AcO) (14)

Secondary positions tend to be more reactive towards oxidation than tertiary positions, unless steric hindrance dictates otherwise. Good examples of this are the chromic acid oxidation of the ether (**12**)[47] and the ruthenium tetroxide oxidation of the ether (**13**),[48] both of which lead to lactone formation (Scheme 2). Oxidation of the quassinoid intermediate (**14**), on the other hand, is completely nonselective (equation 15).[49]

Salomon and coworkers observed high levels of selectivity in the oxidation of a series of polycyclic tetrahydrofuran derivatives.[50] Thus, reaction of the tricyclic ethers (**15**; *n* = 2, 5) with ruthenium tetroxide gave the corresponding lactones (Scheme 3). Conversely, the ether (**16**) afforded the lactol (**17**) as the major product of oxidation, together with small amounts of the corresponding lactone (**18**). Similarly, during the synthesis of both enantiomers of grandisol, Mori observed that the regioselectivity of ruthenium tetroxide oxidations was strongly dependent on steric factors (Scheme 4).[51] A further example is provided by Mori's work on the synthesis of both enantiomers of the spiroketal (**19**), a key intermediate in the synthesis of talaromycins A and B (equation 16).[52]

Predictably, when two secondary positions are available for reaction, mixtures of products ensue, with the reaction favoring the less hindered position.[50]

Allylic and benzylic ethers appear to be particularly susceptible to oxidation, and very high selectivities are often observed. For example, selective oxidation of the allylic ethers (**20**)[53] and (**21**)[54] has provided new methods for the synthesis of α-methylene-γ-butyrolactones (Scheme 5). Similarly, on

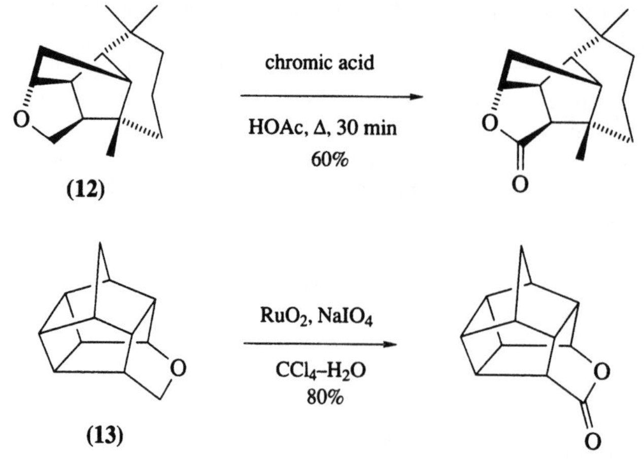

chromic acid

HOAc, Δ, 30 min

60%

(12)

RuO_2, $NaIO_4$

CCl_4–H_2O

80%

(13)

Scheme 2

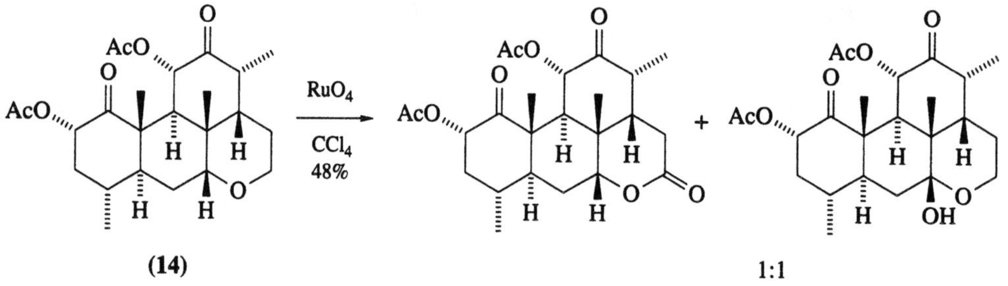

RuO_4

CCl_4

48%

(14)

1:1

(15)

RuO_2, $NaIO_4$

CCl_4–H_2O

(15) n = 2, 5

RuO_2, $NaIO_4$

CCl_4–H_2O

24%

(16) (17) 83:17 (18)

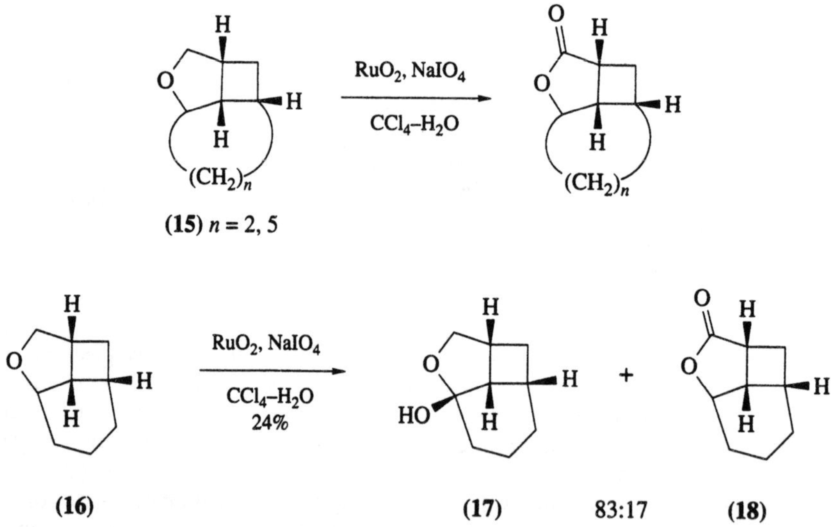

Scheme 3

treatment with pyridinium chlorochromate, the ether (**22**) afforded anhydromevalonolactone, an intermediate in the synthesis of pheromones and verrucarinic acid (equation 17).[55]

Benzylic oxidation using chromium reagents has been reviewed previously.[56] More recently, Pinnick's group has demonstrated that the Jones reagent readily oxidizes benzyl ethers.[57] However, selectivity is poor and mixtures of products result (equation 18). Collins reagent behaved similarly, whereas under the same conditions, pyridinium dichromate was ineffective. In contrast, catalytic oxidation of benzyl methyl ether with ruthenium tetroxide afforded methyl benzoate in excellent yield and none of the corresponding formate was detected.[42] Benzyl ethyl ether reacts similarly under these conditions.[58] During their work on the synthesis of coriolin, Schuda and coworkers carried out a systematic study of the ruthenium tetroxide oxidation of benzyl ethers of primary, secondary and tertiary alcohols to the corresponding ben-

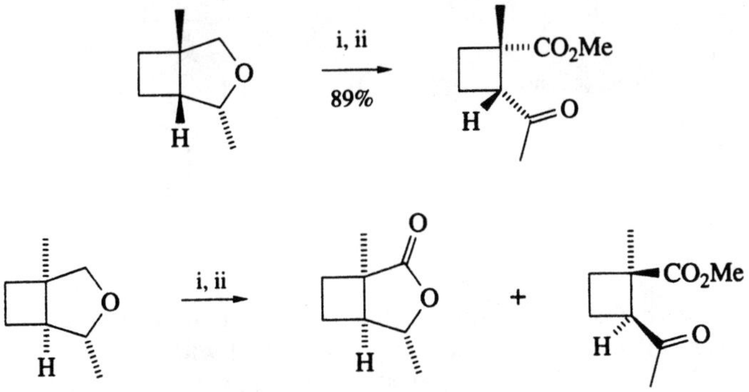

i, ii
89%

i, ii

67%
33%

i, RuO_2, $NaIO_4$, CCl_4–MeCN–phosphate buffer (pH 7), r.t., 12 h; ii, CH_2N_2, Et_2O

Scheme 4

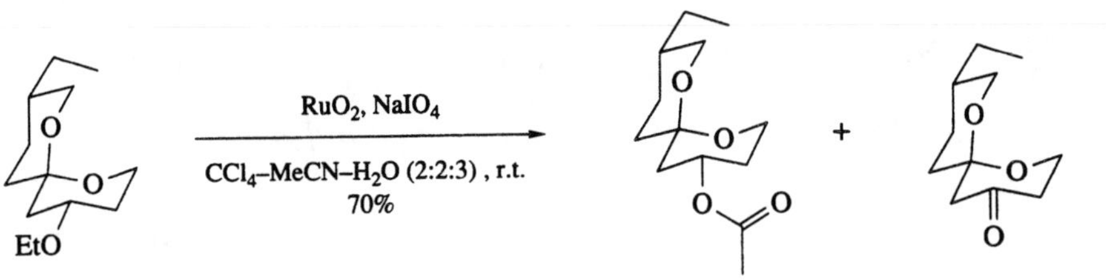

RuO_2, $NaIO_4$
CCl_4–MeCN–H_2O (2:2:3), r.t.
70%

63%
(19) 37%

(16)

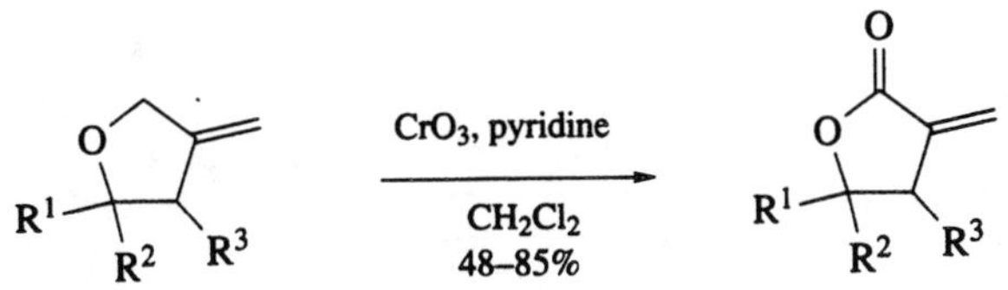

(20)
CrO_3, pyridine
CH_2Cl_2
48–85%

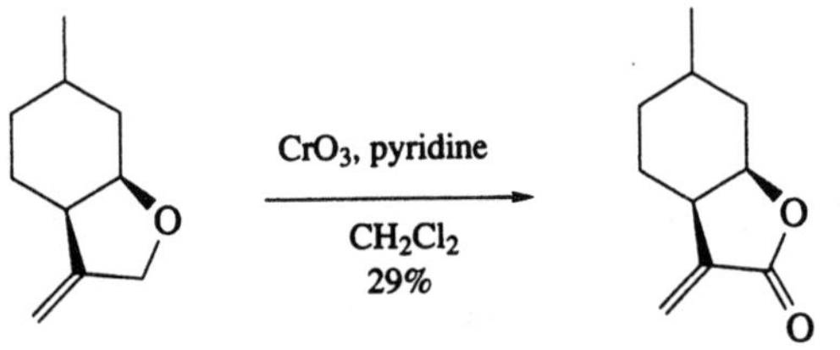

(21)
CrO_3, pyridine
CH_2Cl_2
29%

Scheme 5

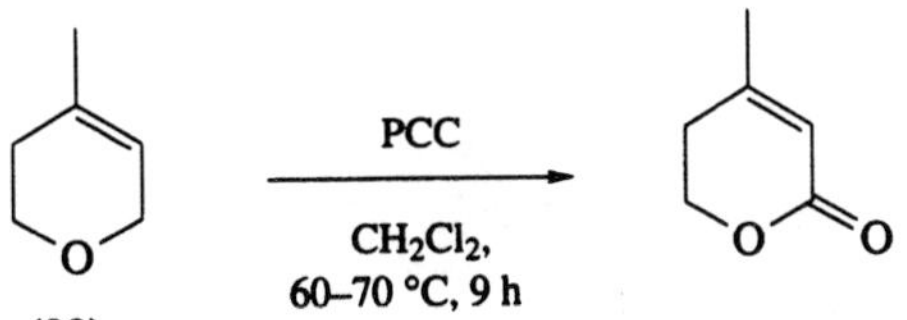

(22)
PCC
CH_2Cl_2,
60–70 °C, 9 h

(17)

zoates, in the presence of other functional groups.[59] They found that ethyleneketals, acetonides, benzoates and aromatic rings (including pyridine) were stable under the reaction conditions, but that other functional groups, such as alkenes and alcohols, were also oxidized. Interestingly, benzyl ethers of phenols appeared to resist oxidation.

Jones (4 equiv.)
0 °C, 12 h

57% 28% + $PhCO_2H$ (18)
53%

Further examples of ether oxidation in natural product synthesis have been published recently which serve to illustrate the levels of selectivity which can be achieved with these reagents. For example, the high yielding oxidation of the ether (**23**) was a key step in model studies directed at the synthesis of taxane (equation 19).[60] In this case, stoichiometric amounts of ruthenium tetroxide were employed. In an alternative photochemical approach to the taxane skeleton, Berkowitz and coworkers achieved a 92% yield in the oxidation of the intermediate (**24**) with catalytic ruthenium tetroxide (equation 20).[61] During their work on the synthesis of the highly oxygenated sesquiterpene anisatin, Niwa's group was able to obtain the lactone (**25**) in 93% yield (equation 21).[62] As part of a programme aimed at the synthesis of known

(**23**) RuO_4, CCl_4, r.t., 10 min, 86% (19)

(**24**) RuO_4, $NaIO_4$, CCl_4–MeCN–H_2O, r.t., 92% (20)

$RuCl_3$, $NaIO_4$, CCl_4–MeCN–phosphate buffer (pH 7), r.t., 14 h, 93% (**25**) (21)

(**26**) RuO_2, $NaIO_4$, CCl_4–MeCN–H_2O (2:2:3), 71% (**27**) (22)

intermediates to leukotrienes B_4 and A_4 from D-arabinose, the intermediate benzyl ether (**26**) was selectively converted into the corresponding benzoate (**27**) in 71% yield (equation 22).[63] Finally, on treatment with chromium trioxide in acetic acid–acetic anhydride, the enone (**28**) furnished the lactone (**29**), a key intermediate in the total synthesis of (±)-desepoxy-4,5-didehydromethylenomycin A (equation 23).[64]

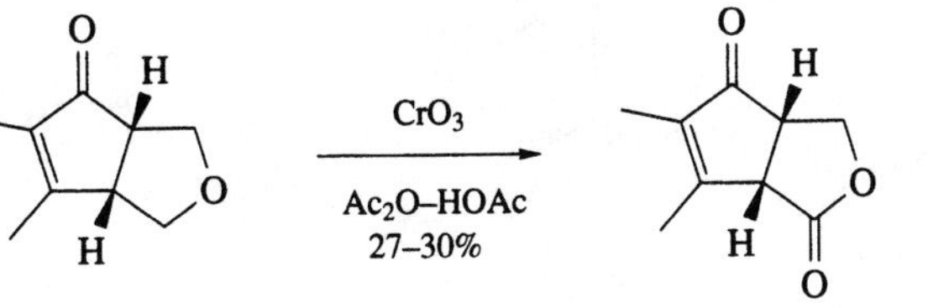

(**28**) (**29**) (23)

Where functional groups are present which are more readily oxidized than the ether group, multiple reactions can occur. For example, in their total synthesis of (+)-tutin and (+)-asteromurin A, Yamada *et al.* observed concomitant oxidation of a secondary alcohol function in the oxidation of the ether (**30**) with ruthenium tetroxide (equation 24).[65] The same group successfully achieved the simultaneous oxidation of both ether functions of the intermediate (**31**) in their related stereocontrolled syntheses of (–)-picrotoxinin and (+)-coriomyrtin (equation 25).[66] Treatment of karahana ether (**32**) with excess ruthenium tetroxide resulted in the formation of the ketonic lactone (**33**) *via* oxidation of both the methylene group adjacent to the ether function and the exocyclic alkenic group (equation 26).[67] In contrast, ruthenium tetroxide oxidation of the steroidal tetrahydrofuran (**34**) gave as a major product the lactone (**35**) in which the alkenic bond had been epoxidized.[68] A small amount of the 5,6-deoxylactone (17%) was also isolated (equation 27). This transformation formed the basis of a facile introduction of the ecdysone side chain into C-20 keto steroids.

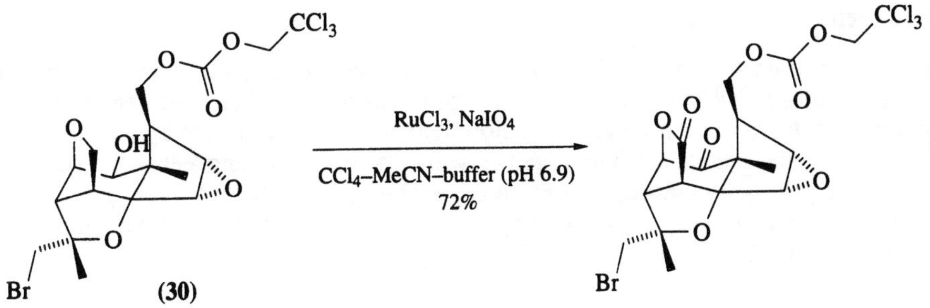

(**30**) (24)

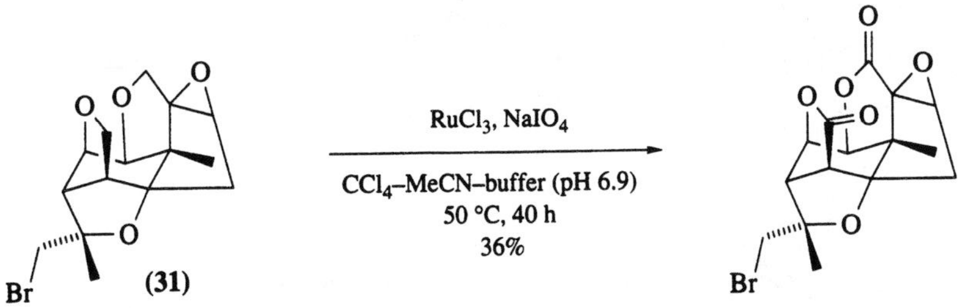

(**31**) (25)

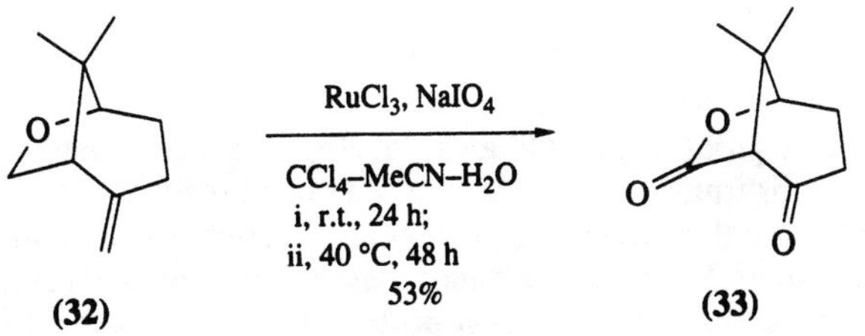

(**32**) (**33**) (26)

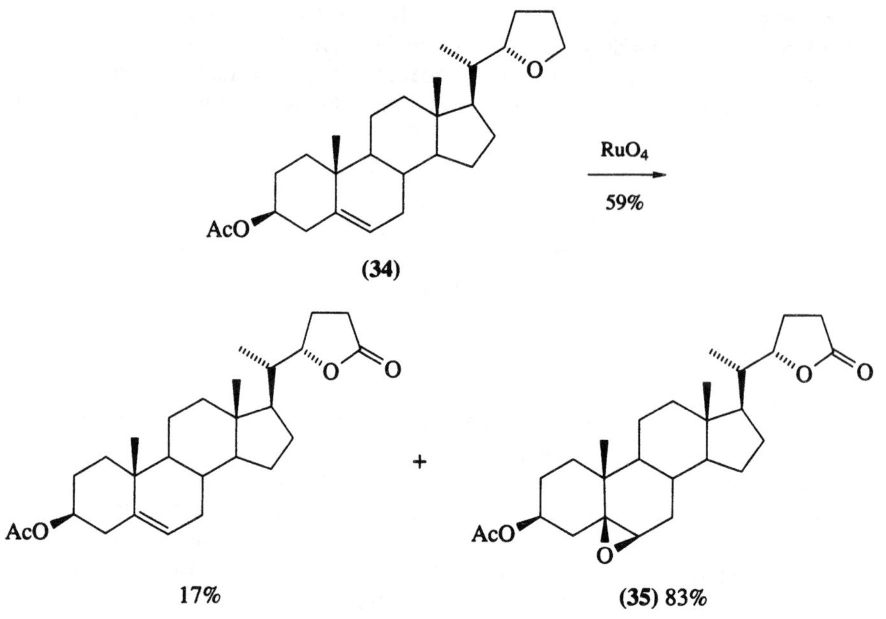

AcO

(34)

RuO_4

59%

AcO

17%

AcO

(35) 83%

(27)

2.6.3 HYDRIDE TRANSFER REAGENTS

2.6.3.1 Introduction

In the early 1970s, Barton *et al.* published the results of their work on the oxidation of acetals and ethers by hydride transfer.[69,70] They observed that substituted benzyl ethers and benzyloxy carbonates, on brief exposure to trityl tetrafluoroborate in dichloromethane at 0 °C followed by aqueous work-up, afforded good yields of the parent alcohols together with the corresponding benzaldehydes. Under the same conditions, the tetrahydropyranyl ether of cholesterol was also efficiently deprotected. A mechanism was proposed which involved an initial hydrogen abstraction, followed by quenching of the resulting stabilized cation by water (Scheme 6).

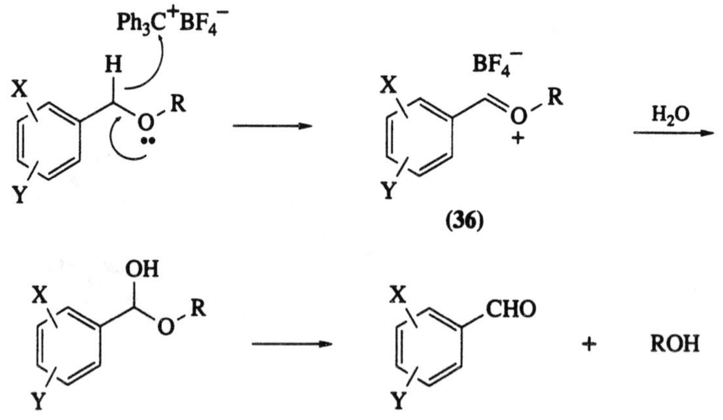

$Ph_3C^+BF_4^-$

BF_4^-

H_2O

(36)

CHO

+ ROH

Scheme 6

Ten years later, a Japanese group led by Oikawa developed a mechanistically related method for the selective debenzylation of substituted benzyl ethers based on the reagent 2,3-dichloro-5,6-dicyanobenzoquinone (DDQ).[71] In contrast to the trityl tetrafluoroborate reaction, the oxidation proceeds at room temperature in the presence of water. Furthermore, under these convenient and essentially neutral conditions, many functional groups, including other common protecting groups, such as isopropylidine, methoxymethyl, benzyloxymethyl, tetrahydropyranyl, acetyl, *t*-butyldimethylsilyl, benzyl, benzoyl and tosyl, are unaffected. As a result of the high levels of selectivity which can be achieved, this method for the deprotection of other derivatives of alcohols and related functional groups[72] has been widely used in the synthesis of complex molecules. In view of its importance, a more detailed analysis of this reaction will follow.

2.6.3.2 Reaction Conditions

Typically, the deprotection sequence involves treatment of the substrate with a 10% excess of DDQ at room temperature in an organic solvent containing traces of water.[73] Although the reaction can take place in aqueous methanol or tetrahydrofuran, reaction rates are slow, and dichloromethane containing approximately 5% water is vastly superior. Aqueous toluene is also occasionally employed. Under these conditions, the weakly acidic 2,3-dichloro-5,6-dicyanohydroquinone by-product is precipitated from solution as the reaction proceeds, and consequently the reaction medium is kept essentially neutral. Reaction times vary according to the reactivity of the benzyl ether function, unsubstituted benzyl groups requiring much longer reaction times than either 4-methoxybenzyl (MPM) or 3,4-dimethoxybenzyl (DMPM) groups. Aliphatic ethers are essentially unreactive.

2.6.3.3 Selectivity of Reaction and Application in Synthesis

The greater reactivity of MPM ethers with respect to unsubstituted benzyl ethers and aliphatic ethers can be attributed to the increased stabilization by the 4-methoxy substituent of a cationic intermediate of type (**36**; Scheme 6). The selectivity between these groups has been exploited in the synthesis of a variety of natural products, including octosyl acid A,[74] oligosaccharides,[75,76] inositol phosphates,[77,78] and the polyether antibiotics X-206[79] and salinomycin (equation 28).[80]

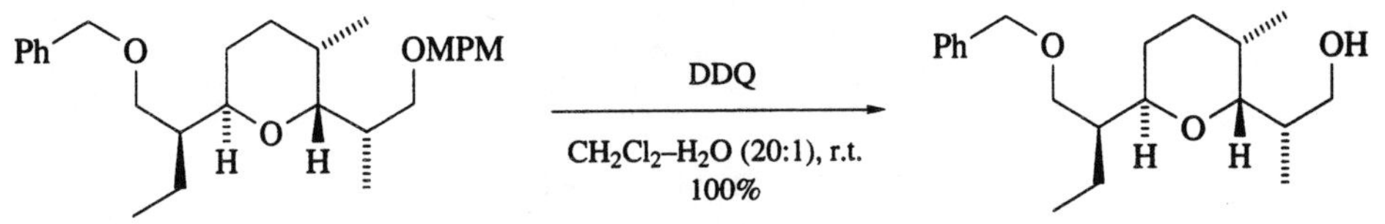

Ph O OMPM H H — DDQ, CH_2Cl_2–H_2O (20:1), r.t., 100% → Ph O OH H H (28)

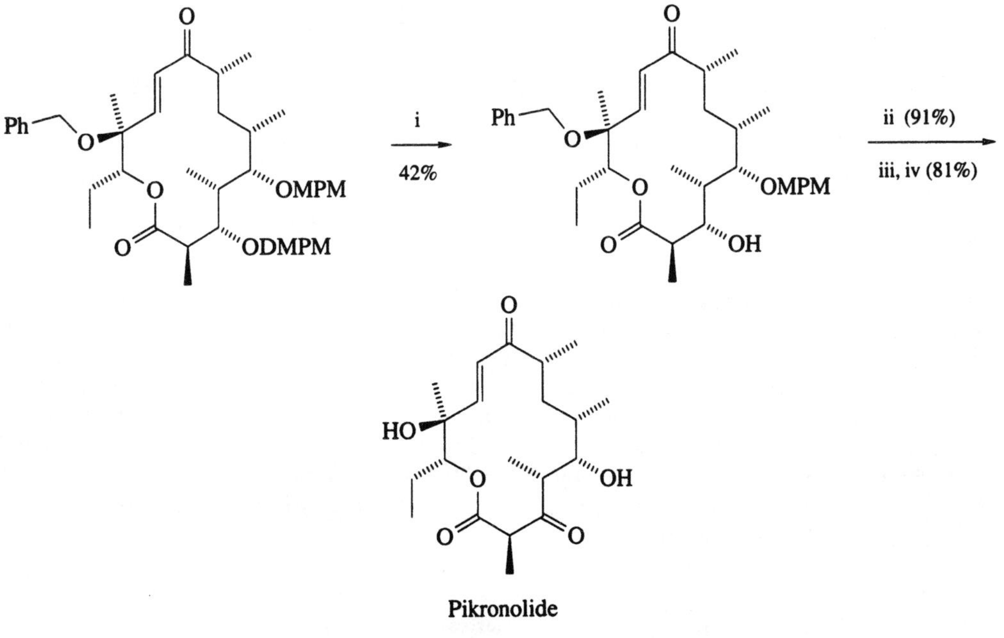

Ph O O OMPM O ODMPM — i, 42% → Ph O O OMPM O OH — ii (91%), iii, iv (81%) → HO O OH O O

Pikronolide

i, DDQ, toluene–H_2O (20:1), 0 °C, 4.5 h; ii, Swern; iii, DDQ, CH_2Cl_2–H_2O (20:1), r.t., 5 min; iv, DDQ, CH_2Cl_2–H_2O (20:1), r.t., 19 h

Scheme 7

The additional presence of a 3-methoxy substituent on the benzyl group confers greater stability on the intermediate cation, and consequently oxidation of DMPM ethers by DDQ is even more facile.[81,82] Yonemitsu and coworkers have used this differential reactivity of substituted benzyl ethers to great effect in the total synthesis of the macrolide antibiotics methynolide,[83] tylonolide,[84] (9*S*)-9-dihydroerythronolide A[85] and pikronolide.[86] The pikronolide synthesis provides an excellent example of the selective, sequential deprotection of DMPM, MPM and benzyl ether protecting groups (Scheme 7).

Recently, as part of their work on the biosynthesis of the ergot alkaloids, Kozikowski and Wu developed the use of (4-methoxybenzyloxy)methyl ethers as alcohol-protecting groups which can be removed oxidatively.[87] MPM ethers, however, are normally much more reactive towards oxidation than benzyloxymethyl (BOM) ethers. Thus, in the synthesis of aplysiatoxin and debromoaplysiatoxin, Kishi's group obtained the unstable diol (**38**) in 70% yield on treatment of the intermediate (**37**) with DDQ in dichloromethane–water at room temperature.[88] Following macrolactonization, the benzyloxymethyl groups were deprotected by hydrogenolysis. DDQ oxidation of a DMPM ether in the presence of an unprotected secondary alcohol and a benxyloxymethyl ether has also been used to selectively unmask the anomeric center of the zincophorin intermediate (**39**; equation 29).[89] Other examples of the selective deprotection at the anomeric center of carbohydrates have also been reported.[90,91] Allyl disaccharides containing MPM protecting groups are selectively deprotected on treatment with DDQ, whereas cerium(IV) ammonium nitrate leads to overoxidation.[92]

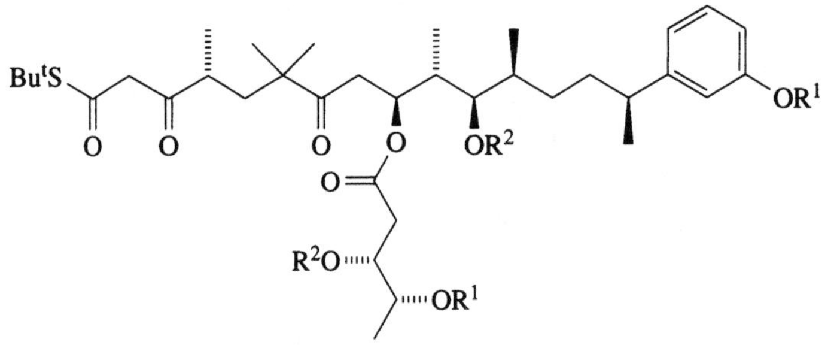

Bu^{t}S, O, O, O, O, OR^2, O, R^{2}O, OR^1, OR^1

(**37**) R^1 = BOM, R^2 = MPM

(**38**) R^1 = BOM, R^2 = H

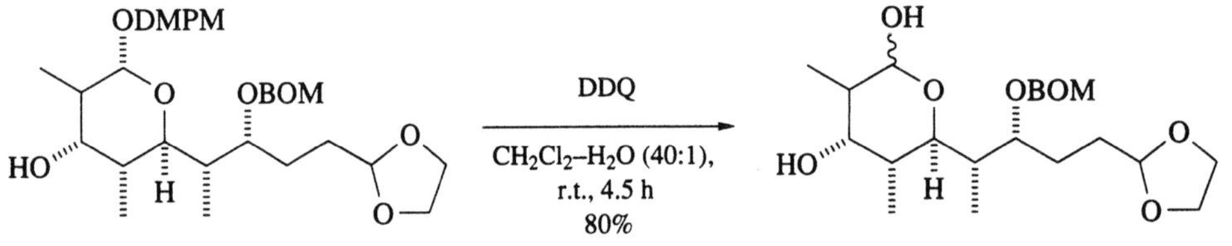

ODMPM, O, OBOM, O, HO, H, O

(**39**)

DDQ

CH_2Cl_2–H_2O (40:1), r.t., 4.5 h

80%

OH, O, OBOM, O, HO, H, O

3:2 mixture of anomers

(29)

In cases where readily oxidized functional groups are also present in either the substrate or the product of the reaction, overoxidation can occur. However, this can sometimes be advantageous. For example, in the total synthesis of (±)-sterepolide Trost and Chung effected the deprotection of the MPM ether (**40**) with DDQ in moist dichloromethane.[93] Under the conditions of the reaction, further oxidation of the allylic alcohol took place to afford the final product (equation 30). However, separate treatment of the allylic alcohol with PDC in dichloromethane was found to be more effective. Similarly, treatment of the ether (**41**) with pyridinium tosylate followed by excess DDQ afforded the Ireland alcohol (**42**), a key intermediate for the synthesis of tirandamycin A (equation 31).[94] Propargylic alcohols, however, would appear to be less susceptible to oxidation.[95]

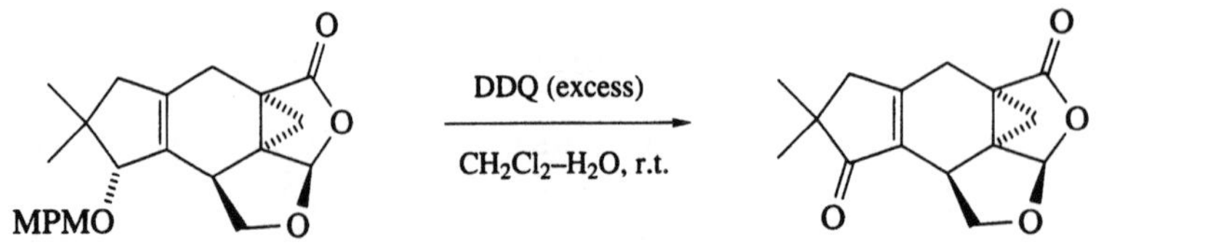

O, O, MPMO, O

(**40**)

DDQ (excess)

CH_2Cl_2–H_2O, r.t.

O, O, O, O

(±)-Sterepolide

(30)

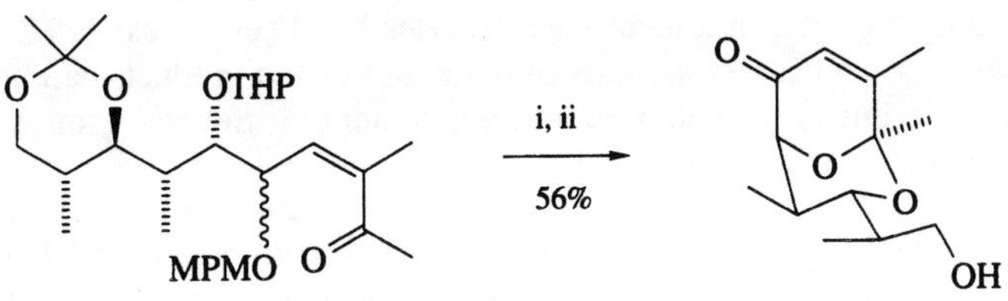

(31)

(41) (42)

i, pyridine–TsOH, MeOH; ii, DDQ, CH_2Cl_2–H_2O

2.6.4 OTHER METHODS

2.6.4.1 Oxygen and Ozone

Few synthetically useful examples of the oxidation of ethers by oxygen or ozone have been published.[7,96–100] In 1978, Ourisson and coworkers reported that ozonization of the natural product cedrane oxide (**43**) on silica gel at –78 °C led to the formation of the corresponding lactone (**44**) in 30% yield (equation 32).[101] A small amount of the tertiary alcohol (**45**) was also produced. Later, in the course of a chiral total synthesis of compactin, Hirama examined the ozonolysis of the alkene (**46**; equation 33).[102] Under carefully controlled conditions, selective ozonolysis of the double bond could be achieved in 88% yield. However, when excess ozone was employed, significant amounts of the benzoate (**47**) were obtained, even at –78 °C. In subsequent studies, benzyl ethers of primary and secondary alcohols,[103] and carbohydrates[104] were oxidized to the corresponding benzoates in excellent yields. Surprisingly, no further synthetic applications of this reaction have been reported.

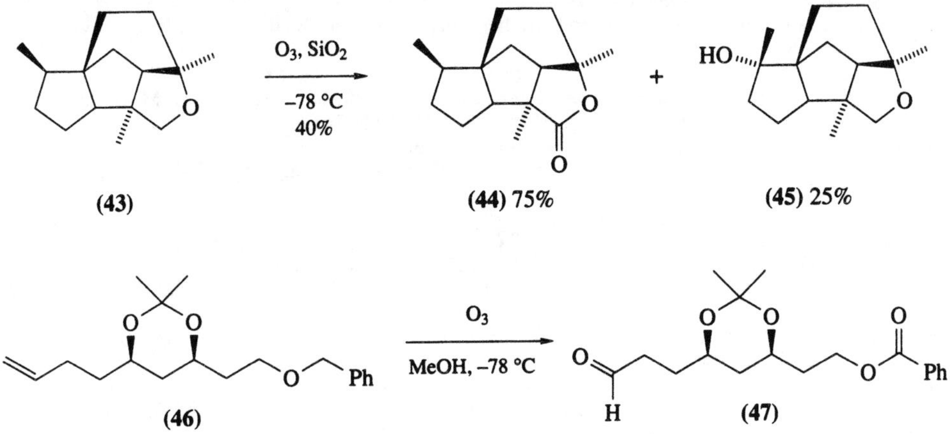

(32)

(43) (44) 75% (45) 25%

(33)

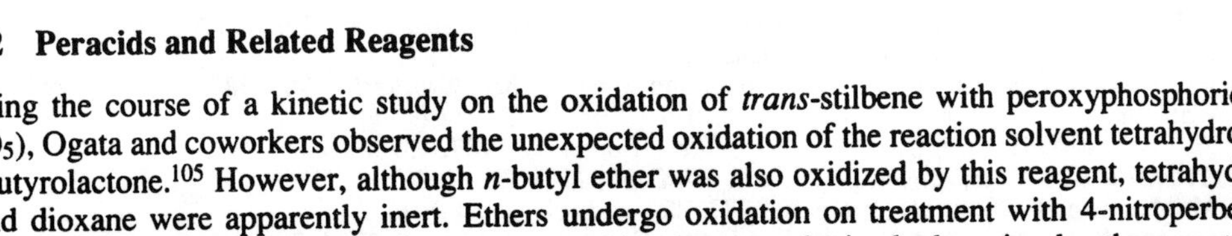

(46) (47)

2.6.4.2 Peracids and Related Reagents

During the course of a kinetic study on the oxidation of *trans*-stilbene with peroxyphosphoric acid (H_3PO_5), Ogata and coworkers observed the unexpected oxidation of the reaction solvent tetrahydrofuran to γ-butyrolactone.[105] However, although *n*-butyl ether was also oxidized by this reagent, tetrahydropyran and dioxane were apparently inert. Ethers undergo oxidation on treatment with 4-nitroperbenzoic acid in chloroform.[106] Moderate yields of esters and lactones are obtained when simple ethers are treated with calcium hypochlorite.[107] At room temperature, reaction times of 4–16 h are necessary. However, primary and secondary alcohols are readily oxidized under these conditions.

2.6.4.3 Electrochemical Oxidation

Several reports on the electrochemical oxidation of ethers have appeared in the literature within the last 10 years, although few have been of direct relevance to the synthetic chemist. The electrochemical α-hydroxylation of tetrahydrofuran in aqueous electrolytes has been investigated recently in some detail.[108] Shono *et al.* have shown that good yields of α-methoxylated products can be obtained *via* the anodic oxidation of aliphatic ethers in methanol and acetic acid.[109] Several groups have worked on the electrochemical oxidation of benzyl ethers.[110–112]

A novel procedure for the oxidative removal of benzyl protecting groups by catalytic homogeneous electron transfer has been developed by Schmidt and Steckhan (equation 34).[113] The selectivity of the reaction can be adjusted by altering the substitution on the aromatic rings of the cation radicals (**48**). Finally, a recent publication describes a photoinduced single electron transfer initiated oxidative cleavage of benzylic ethers.[114]

$$\mathrm{RO\diagdown\!\diagup Ph} \xrightarrow[\mathrm{H_2O,\ -2H^+}]{2\ \mathrm{Ar_3N^{+\bullet}}\ (\mathbf{48})} \mathrm{ROH} + \mathrm{PhCHO} + 2\ \mathrm{Ar_3N} \qquad (34)$$

2.6.5 REFERENCES

1. M. Kratochvil, J. Jonas, O. Bartes and H. Gross, *Chem. Ber.*, 1966, **99**, 1218; N. C. Deno and N. H. Potter, *J. Am. Chem. Soc.*, 1967, **89**, 3550; J. N. Bemiller, R. E. Wing and C. Y. Meyers, *J. Org. Chem.*, 1968, **33**, 4292; C. G. Kruse, N. L. J. M. Broekhof and A. van der Gen, *Tetrahedron Lett.*, 1976, 1725.
2. J. Buddrus and H. Herzog, *Chem. Ber.*, 1979, **112**, 1260; *Justus Liebigs Ann. Chem.*, 1983, 123.
3. E. C. Jeuenge and D. A. Beal, *Tetrahedron Lett.*, 1968, 5819; E. C. Jeuenge, M. D. Corey and D. A. Beal, *Tetrahedron*, 1971, **27**, 2671.
4. G. A. Olah, J. Welch and T.-L. Ho, *J. Am. Chem. Soc.*, 1976, **98**, 6717.
5. Y. Kamiya and S. Takemura, *Chem. Pharm. Bull.*, 1973, **21**, 1401.
6. R. Criegee, in 'Oxidation in Organic Chemistry', ed. K. B. Wiberg, Academic Press, New York, 1965, part A, p. 277.
7. P. Müller, in 'Chemistry of Functional Groups', ed. S. Patai, Wiley, New York, 1980, suppl. E, part 1, p. 469.
8. M. V. Bhatt and S. U. Kulkarni, *Synthesis*, 1983, 249.
9. O. C. Musgrave, *Chem. Rev.*, 1966, **69**, 499.
10. D. G. Lee, in 'Oxidation — Techniques and Applications in Organic Synthesis', ed. R. L. Augustine, Dekker, New York, 1969, vol. 1, p. 53.
11. A. H. Haines, 'Methods for the Oxidation of Organic Compounds', Academic Press, New York, 1988.
12. L. M. Berkowitz and P. N. Rylander, *J. Am. Chem. Soc.*, 1958, **80**, 6682; see also L. M. Berkowitz and P. N. Rylander, *US Pat.* 3 278 558 (1966) (*Chem. Abstr.* 1967, **66**, 10 597g).
13. C. Djerassi and R. R. Engle, *J. Am. Chem. Soc.*, 1953, **75**, 3838.
14. M. E. Wolff, J. F. Kerwin, F. F. Owings, B. B. Lewis, B. Blank, A. Magnani and V. Georgian, *J. Am. Chem. Soc.*, 1960, **82**, 4117.
15. M. E. Wolff, J. F. Kerwin, F. F. Owings, B. B. Lewis and B. Blank, *J. Org. Chem.*, 1963, **28**, 2729.
16. J. Weinstock and M. E. Wolff, *US Pat.* 2 960 503 (1960) (*Chem. Abstr.*, 1961, **55**, 7483e).
17. H. B. Henbest and B. Nicholls, *J. Chem. Soc.*, 1959, 227; and references therein.
18. D. G. Lee and M. van den Engh, *Can. J. Chem.*, 1972, **50**, 3129; M. van den Engh, *Diss. Abstr. Int. B*, 1973, **34**, 602.
19. R. Brownell, A. Leo, Y. W. Chang and F. W. Westheimer, *J. Am. Chem. Soc.*, 1960, **82**, 406.
20. R. M. Moriarty, H. Gopal and T. Adams, *Tetrahedron Lett.*, 1970, 4003.
21. W. J. Rodewald and J. W. Morzycki, *Pol. J. Chem.*, 1979, **53**, 1373.
22. G. F. Reynolds, G. H. Rasmussen, L. Birladeanu and G. E. Arth, *Tetrahedron Lett.*, 1970, 5057.
23. W. J. Rodewald and J. W. Morzycki, *Pol. J. Chem.*, 1978, **52**, 2361.
24. G. Cainelli, B. Kamber, J. Keller, M. L. Mihailovic, D. Arigoni and O. Jeger, *Helv. Chim. Acta*, 1961, **44**, 518.
25. G. F. Reynolds, G. H. Rasmussen, L. Birladeanu and G. E. Arth, *Tetrahedron Lett.*, 1970, 5057.
26. H. J. Schmidt and H. J. Schäfer, *Angew. Chem., Int. Ed. Engl.*, 1979, **1**, 68, 69.
27. H. Firouzabadi, A. Sardarian, H. Moosavipour and G. M. Afshari, *Synthesis*, 1986, 285.
28. S. Wolfe and C. F. Ingold, *J. Am. Chem. Soc.*, 1983, **105**, 7755.
29. C. M. Che and W. H. Leung, *J. Chem. Soc., Chem. Commun.*, 1987, 1376.
30. M. Trömel and M. Russ, *Angew. Chem., Int. Ed. Engl.*, 1987, **26**, 1007.
31. A. M. Maione and A. Romeo, *Synthesis*, 1987, 250, and references therein.
32. B. Classon, P. J. Garegg and B. Samuelsson, *Acta Chem. Scand., Ser. B*, 1984, **38**, 419.
33. R. Johansson and B. Samuelsson, *J. Chem. Soc., Perkin Trans. 1*, 1984, 2371.
34. Y. Ito and T. Ogawa, *Agric. Biol. Chem.*, 1986, **50**, 3231.
35. I. T. Kay and E. G. Williams, *Tetrahedron Lett.*, 1983, **24**, 5915.
36. I. T. Kay and D. Bartholomew, *Tetrahedron Lett.*, 1984, **25**, 2035.
37. S. J. Danishefsky, D. M. Armistead, F. E. Wincott, H. G. Selnick and R. Hungate, *J. Am. Chem. Soc.*, 1989, **111**, 2967; F. E. Wincott, S. J. Danishefsky and G. Schulte, *Tetrahedron Lett.*, 1987, **28**, 4951.
38. J. L. Courtney, *Org. Synth. Oxid. Met. Compd.*, 1986, 445; D. G. Lee and M. van den Engh, in 'Oxidation in Organic Chemistry', ed. W. S. Trahanovsky, Academic Press, New York, 1973, part B, chap. 4, p. 177; R. L. Augustine, in 'Oxidation, Techniques and Application in Organic Synthesis', Dekker, New York, 1969, vol. 1; J. L. Courtney and K. F. Swandborough, *Rev. Pure Appl. Chem.*, 1972, **22**, 47, and references therein.

39. A. B. Smith, III and R. M. Scarborough, Jr., *Synth. Commun*, 1980, **10**, 205.
40. G. Balavoine, C. Eskenazi and F. Meunier, *J. Mol. Catal.*, 1985, **30**, 125.
41. B. D. Johnston, K. N. Slessor and A. C. Oehlschlager, *J. Org. Chem.*, 1985, **50**, 114.
42. P. H. J. Carlsen, T. Katsuki, V. S. Martin and K. B. Sharpless, *J. Org. Chem.*, 1981, **46**, 3936.
43. R. Takeda, A. Zask, K. Nakanishi and M. H. Park, *J. Am. Chem. Soc.*, 1987, **109**, 914; M. Bressan and A. Morvillo, *J. Chem. Soc., Chem. Commun.*, 1989, 421.
44. A. B. Smith, III, B. A. Wexter, C. Y. Tu and J. P. Konepelski, *J. Am. Chem. Soc.*, 1985, **107**, 1308.
45. I. T. Harrison and S. Harrison, *J. Chem. Soc., Chem. Commun.*, 1966, 752.
46. A. Olesker, D. Mercier, S. D. Gero, C. J. Pearce and J. E. G. Barnett, *J. Antibiot.*, 1975, **28**, 490.
47. S. G. Patnekar and S. C. Bhattacharyya, *Tetraherdon*, 1967, **23**, 919.
48. W. G. Dauben and A. F. Cunningham, Jr., *J. Org. Chem.*, 1983, **48**, 2842.
49. H. Hirota, A. Yokoyama, K. Miyaji, T. Nakamura and T. Takahashi, *Tetrahedron Lett.*, 1987, **28**, 435.
50. S. Ghosh, S. R. Raychaudhuri and R. G. Salomon, *J. Am. Chem. Soc.*, 1982, **104**, 6841; *J. Org. Chem.*, 1987, **52**, 83.
51. K. Mori and M. Miyake, *Tetrahedron*, 1987, **43**, 2229.
52. M. Ikunaka and K. Mori, *Agric. Biol. Chem.*, 1987, **51**, 565.
53. M. Okabe, M. Abe and M. Tada, *J. Org. Chem.*, 1982, **47**, 1775.
54. T. J. Brocksom and J. T. B. Ferreira, *Synth. Commun.*, 1981, **11**, 105.
55. F. Bonadies, R. Di Fabio and C. Bonini, *J. Org. Chem.*, 1984, **49**, 1647.
56. K. B. Wiberg, in 'Oxidation in Organic Chemistry', ed. K. B. Wiberg, Academic Press, New York, 1965, vol. 5A, p. 83.
57. B. S. Bal, K. S. Kochhar and H. W. Pinnick, *J. Org. Chem.*, 1981, **46**, 1492.
58. P. H. J. Carlsen and M. Ystenes, *Acta Chem. Scand., Ser. B*, 1986, **40**, 757.
59. P. Schuda, M. B. Cichowicz and M. R. Heimann, *Tetrahedron Lett.*, 1983, **24**, 3829; P. F. Schuda and M. R. Heimann, *Tetrahedron*, 1984, **40**, 2365; *Tetrahedron Lett.*, 1983, **24**, 4267.
60. T. Kojima, Y. Inouye and H. Kakisawa, *Bull. Chem. Soc. Jpn.*, 1985, **58**, 1738.
61. W. F. Berkowitz, A. S. Amarasekara and J. J. Perumattam, *J. Org. Chem.*, 1987, **52**, 1119; *Tetrahedron Lett.*, 1985, **26**, 3665.
62. H. Niwa, T. Mori, T. Hasegawa and K. Yamada, *J. Org. Chem.*, 1986, **51**, 1015.
63. H. Maehr, A. Perrotta and J. Smallheer, *J. Org. Chem.*, 1988, **53**, 832.
64. D. H. Boschelli, R. M. Scarborough, Jr. and A. B. Smith, III, *Tetrahedron Lett.*, 1981, **22**, 19.
65. K. Wakamatsu, H. Kigoshi, K. Niiyama, H. Niwa and K. Yamada, *Tetrahedron*, 1986, **42**, 5551; K. Wakamatsu, K. Niiyama, H. Niwa and K. Yamada, *Chem. Lett.*, 1984, 1763; K. Wakamatsu, H. Kigoshi, K. Niiyama, H. Niwa and K. Yamada, *Tetrahedron Lett.*, 1984, **25**, 3873.
66. H. Niwa, K. Wakamatsu, T. Hida, K. Niiyama, H. Kigoshi, M. Yamada, H. Nagase, M. Suzuki and K. Yamada, *J. Am. Chem. Soc.*, 1984, **106**, 4547.
67. K. Mori and H. Mori, *Tetrahedron*, 1985, **41**, 5487.
68. T. Kametani, T. Katoh, M. Tsubuki and T. Honda, *Chem. Lett.*, 1985, 485.
69. D. H. R. Barton, P. D. Magnus, G. Smith, G. Streckert and D. Zurr, *J. Chem. Soc., Perkin Trans. 1*, 1972, 542.
70. D. H. R. Barton, P. D. Magnus, G. Streckert and D. Zurr, *J. Chem. Soc., Chem. Commun.*, 1971, 1109.
71. Y. Oikawa, T. Yoshioka and O. Yonemitsu, *Tetrahedron Lett.*, 1982, **23**, 885, 889.
72. R. E. Dolle and K. C. Nicolaou, *J. Am. Chem. Soc.*, 1985, **107**, 1695.
73. K. Horita, T. Yoshioka, T. Tanaka, Y. Oikawa and O. Yonemitsu, *Tetrahedron*, 1986, **42**, 3021.
74. S. J. Danishefsky, R. Hungate and G. Schulte, *J. Am. Chem. Soc.*, 1988, **110**, 7434.
75. J. Arnap, M. Haraldsson and J. Lönngren, *J. Chem. Soc., Perkin Trans. 1*, 1985, 535.
76. G. Catelani, A. Marra, F. Paquet and P. Sinay, *Carbohydr. Res.*, 1986, 131.
77. Y. Watanabe, T. Ogosawara, N. Shiotani and S. Ozaki, *Tetrahedron Lett.*, 1987, **28**, 2607.
78. S. Ozaki, M. Kohno, H. Nakahira, M. Bunya and Y. Watanabe, *Chem. Lett.*, 1988, 77.
79. D. A. Evans, S. L. Bender and J. Morris, *J. Am. Chem. Soc.*, 1988, **110**, 2506.
80. K. Horita, Y. Oikawa, S. Nagato and O. Yonemitsu, *Tetrahedron Lett.*, 1988, **29**, 5143; K. Horita, S. Nagato, Y. Oikawa and O. Yonemitsu, *Tetrahedron Lett.*, 1987, **28**, 3253; Y. Oikawa, K. Horita and O. Yonemitsu, *Tetrahedron Lett.*, 1985, **26**, 1541.
81. Y. Oikawa, T. Tanaka, K. Horita, T. Yoshioka and O. Yonemitsu, *Tetrahedron Lett.*, 1984, **25**, 5393, 5397.
82. T. Fukuyama and L. Yang, *J. Am. Chem. Soc.*, 1987, **109**, 7881.
83. Y. Oikawa, T. Tanaka, T. Hamada and O. Yonemitsu, *Chem. Pharm. Bull.*, 1987, **35**, 2184, 2196, 2203; Y. Oikawa, T. Tanaka and O. Yonemitsu, *Tetrahedron Lett.*, 1986, **27**, 3647.
84. T. Tanaka, Y. Oikawa, T. Hamada and O. Yonemitsu, *Chem. Pharm. Bull.*, 1987, **35**, 2209, 2219; *Tetrahedron Lett.*, 1986, **27**, 3651.
85. H. Tone, T. Nishi, Y. Oikawa, M. Hikota and O. Yonemitsu, *Tetrahedron Lett.*, 1987, **28**, 4569, and references therein.
86. N. Nakajima, T. Tanaka, T. Hamada, Y. Oikawa and O. Yonemitsu, *Chem. Pharm. Bull.*, 1987, **35**, 2228; *J. Am. Chem. Soc.*, 1986, **108**, 4645.
87. A. P. Kozikowski and J. P. Wu, *Tetrahedron Lett.*, 1987, **28**, 5125.
88. P. Park, C. A. Broka, B. F. Johnson and Y. Kishi, *J. Am. Chem. Soc.*, 1987, **109**, 6205.
89. S. J. Danishefsky, H. G. Selnick, R. E. Zelle and M. P. DeNinno, *J. Am. Chem. Soc.*, 1988, **110**, 4368.
90. B. F. Molino and B. Fraser-Reid, *Can. J. Chem.*, 1987, **65**, 2834.
91. D. R. Mootoo and B. Fraser-Reid, *J. Org. Chem.*, 1987, **52**, 4511.
92. J. Gigg, R. Gigg, S. Payne and R. Conant, *J. Chem. Soc., Perkin Trans. 1*, 1987, 1165.
93. B. M. Trost and J. Y. L. Chung, *J. Am. Chem. Soc.*, 1985, **107**, 4586.
94. S. Ikegami, T. Katsuki and M. Yamaguchi, *Tetrahedron Lett.*, 1988, **29**, 5285.
95. J. S. Yadav, M. C. Chander and B. V. Joshi, *Tetrahedron Lett.*, 1988, **29**, 2737.
96. R. D. Srivastava, A. B. Stiles and G. A. Jones, *J. Catal.*, 1982, **77**, 192.
97. P. Wan, *Tetrahedron Lett.*, 1985, **26**, 2387.

98. J. A. Pincock, A. L. Pincock and M. A. Fox, *Tetrahedron*, 1985, **41**, 4107.
99. K. A. Pollart and R. E. Miller, *J. Org. Chem.*, 1962, **27**, 2392.
100. R. E. Erickson, R. T. Hansen and J. Harkins, *J. Am. Chem. Soc.*, 1968, **90**, 6777, and references therein.
101. E. Trifilieff, L. Bang, A. S. Narula and G. Ourisson, *J. Chem. Res. (M)*, 1978, 601.
102. M. Hirama and M. Uei, *J. Am. Chem. Soc.*, 1982, **104**, 4251.
103. M. Hirama and M. Shimizu, *Synth. Commun.*, 1983, **13**, 781.
104. P. Angibeaud, J. Defaye, A. Gadelle and J. P. Utille, *Synthesis*, 1985, 1123.
105. Y. Ogata, K. Tomizawa and T. Ikeda, *J. Org. Chem.*, 1979, **44**, 2362; 1980, **45**, 1320.
106. H. J. Schneider, A. Ahlhelm and W. Müller, *Chem. Ber.*, 1984, **117**, 3297.
107. S. O. Nwaukwa and P. M. Keehn, *Tetrahedron Lett.*, 1982, **23**, 35.
108. B. Wermeckes, F. Beck and H. Schultz, *Tetrahedron*, 1987, **43**, 577.
109. T. Shono, Y. Matsumura, O. Onomura and Y. Yamada, *Synthesis*, 1987, 1099.
110. E. A. Mayeda, L. L. Miller and J. F. Wolf, *J. Am. Chem. Soc.*, 1972, **94**, 6812.
111. S. M. Weinreb, G. A. Epling, R. Comi and M. Reitano, *J. Org. Chem.*, 1975, **40**, 1356.
112. M. Masui, S. Hara, T. Ueshima, T. Kawaguchi and S. Ozaki, *Chem. Pharm. Bull.*, 1983, **31**, 4209.
113. W. Schmidt and E. Steckhan, *Angew. Chem., Int. Ed. Engl.*, 1978, **17**, 673; 1979, **18**, 801, 802.
114. G. Pandey and A. Krishna, *Synth. Commun.*, 1988, **18**, 2309.

2.7
Oxidation Adjacent to Oxygen of Alcohols by Chromium Reagents

STEVEN V. LEY and ANDREW MADIN
Imperial College of Science, Technology & Medicine, London, UK

2.7.1 INTRODUCTION

Chromium-based oxidants are probably the most widely used of all oxidizing agents. Over the years they have been continually developed and modified to overcome the typical problems that occur during oxidation and to accept wider ranges of substrates with improved selectivities. They have been accepted readily by synthesis chemists since they are easy to handle and are often 'off the shelf reagents'. However, they are not without their problems: work-up can be problematical; overoxidation can occur; and, at all times, removal of the product from toxic chromium contaminants is a concern, especially with respect to large scale preparations. In an attempt to circumvent these problems the trend has been to develop the use of catalytic and/or supported reagents. This review is concerned for the most part with the applications and limitations of more recent chromium(VI) oxidants. Several other comprehensive reviews have appeared in this area and should be consulted for more detailed descriptions of older methods, chromium(V) oxidants, mechanism of oxidation and for typical experimental procedures.[1–5]

2.7.2 CHROMIUM(VI) IN ACIDIC MEDIA

Many of the early methods devised around chromium oxidants have employed strongly acidic media. Although high yielding for some specific transformations, most methods require the use of harsh conditions that are often unsuited to modern approaches to the synthesis of complex, sensitive natural products and bioactive molecules.

2.7.2.1 In Aqueous Sulfuric Acid

Sodium dichromate in aqueous sulfuric acid has been used since the turn of the century.[6] It is a very strong oxidant; the use of this system to oxidize primary alcohols is severely limited by overoxidation, *via* the aldehyde hydrate, to the corresponding acid. This problem can be partially circumvented in the preparation of volatile aldehydes (in particular aromatic aldehydes[7]), by slow addition to excess alcohol and continuous removal of the aldehyde by distillation.[8] Oxidation of secondary alcohols that are reasonably soluble is acceptable, but milder methods are now available, and are discussed in detail later.

As may be expected, a totally aqueous system is also restricted by the low solubility of many potential substrates. Hence several variations have subsequently been developed from this basic technique.

2.7.2.2 In Aqueous Acetic Acid

The use of acetic acid increases the solubility of organic substrates and also increases the rate of oxidation *via* acid catalysis. Some water, however, is necessary to solubilize the chromium(VI) oxide.

Once again the selective preparation of aldehydes is problematical — oxidation generally proceeds rapidly to give the acid. Secondary alcohols give good yields of the corresponding ketones, but the use of excess oxidant should be avoided since yields may be lowered by cleavage of the desired products *via* the enol form; this becomes more prominent with increasing substitution α to the ketone. In addition to the use of minimum amounts of oxidant, degradation may also be retarded by the addition of manganese(II) or cerium(III) salts to the reaction.[9]

Substrates containing 1,2-diols will normally undergo oxidative cleavage with this oxidant, but again this may be inhibited by the addition of manganese(II) or cerium(III) salts.[10]

2.7.2.3 In DMF/DMSO with Catalytic Sulfuric Acid

Chromium(VI) oxide in DMF is a very poor oxidant. However, the reaction is catalyzed by the addition of concentrated sulfuric acid.[11] Acid sensitive functionalities such as acetonides and acetals were found to be stable to these milder conditions.

Similarly sodium dichromate dihydrate in DMSO is a poor oxidant, but the addition of catalytic quantities of sulfuric acid leads to the effective oxidation of a variety of alcohols to give aldehydes and ketones in good yield (80–90%).[12] Oxidations are normally complete within 90 min at 70 °C, and may be performed in commercial DMSO without the need for further purification.

The suitability of these methods for large scale preparations may be restricted by the problems of removal of large quantities of the polar solvents.

2.7.2.4 In Two-phase Systems

Two-phase oxidations have been developed to protect sensitive substrates from prolonged exposure to strongly acidic conditions, thus avoiding degradation of the product and/or epimerization of compounds containing α-chiral centres. Benzene and dichloromethane are commonly used for the organic phase, but diethyl ether is often found to be superior since it is less likely to form emulsions.

Brown *et al.* initially reported the use of two-phase oxidation for the preparation of a range of ketones;[13] the method has also recently been reported for the oxidation of a selection of primary alcohols.[14]

Phase transfer conditions can be used for substrates with only limited aqueous solubility. Phase transfer agents also facilitate the preparation of aldehydes under biphasic conditions.[15,16]

2.7.2.5 In Aqueous Sulfuric Acid/Acetone (Jones Oxidation[17])

This is one of the best-known and most widely used methods of oxidation using chromium(VI). The procedure, which is amenable to large scale preparations, uses a standard chromic acid/sulfuric acid solution which is simply titrated against a solution of the alcohol in acetone. Acetone performs a dual role: (i) it is an excellent solvent for a wide range of organic molecules; and (ii) it protects the substrate from overoxidation or undesired side reactions by reacting with the excess oxidant itself. Hence it is uncommon to observe substantial epimerization of α-chiral centres.[18,19]

Secondary alcohols give good yields of ketones, but the foremost use of the Jones oxidation has been for the conversion of saturated primary alcohols to the corresponding acid. In direct contrast, primary allylic and benzylic alcohols can be selectively oxidized to aldehydes; acids are obtained only after prolonged exposure to the oxidant. In rigid systems it has been found that axial alcohols are oxidized faster than equatorial alcohols.[20]

Isolated carbon–carbon multiple bonds are not normally attacked by Jones' reagent, but some double-bond isomerization may occur during the preparation of α,β-unsaturated aldehydes. Hydroxy-directed epoxidation (presumably *via* chromate ester formation, followed by oxygen transfer to the double bond) has also been observed in steroidal substrates for axial alcohols (equation 1).[21] Equatorial alcohols undergo oxidation to give the expected enone.

OH C_8H_{17} Jones → 1h, 0–5 °C, 60%

C_8H_{17} (62%) + C_8H_{17} (38%) (1)

Care should be taken to avoid substrates containing either *cis* or *trans* 1,2-diols or α-hydroxy ketones since these groups are liable to be cleaved under the reaction conditions.[22]

Although acetonides and secondary THP groups are sometimes compatible with the Jones oxidation (Table 1), acid labile protecting groups are often removed during the reaction, thus effecting a convenient 'one-pot' deprotection–oxidation protocol. In a similar way nitrile groups may be 'deprotected' to reveal acids during Jones oxidations.[23]

Table 1 Oxidation of Alcohols in Aqueous Sulfuric Acid/Acetone (Jones Oxidation)

Entry	*Substrate*	*Product*	*Yield (%)*	*Ref.*
1	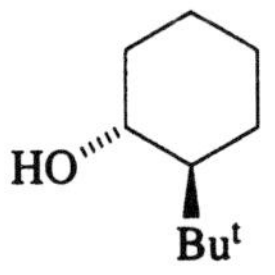 HO, Bu^t	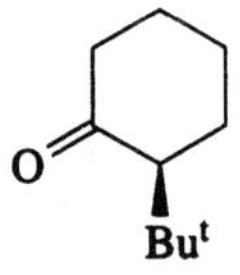 O, Bu^t	40	19
2	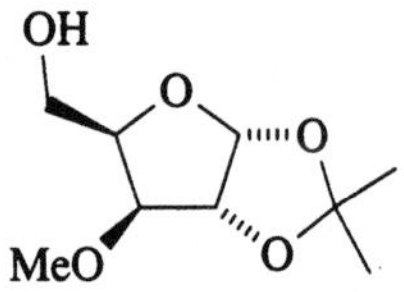 OH, O, O, MeO, O	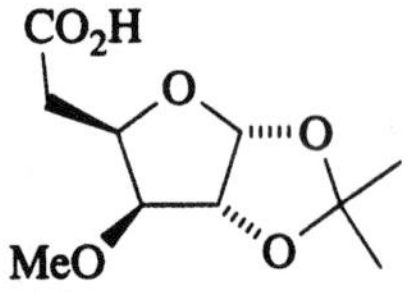 CO_2H, O, O, MeO, O	82	26
3	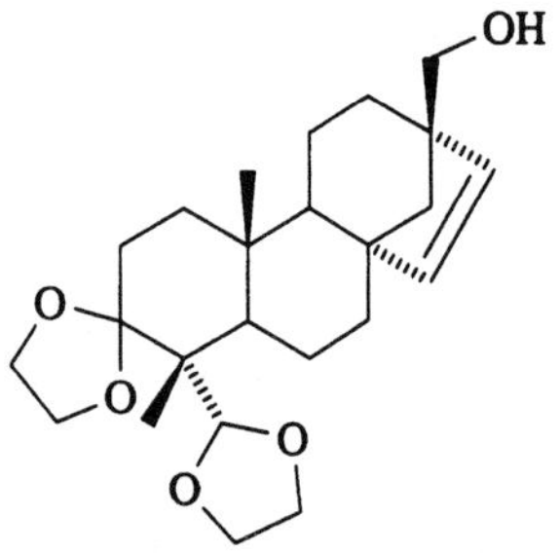 OH, O, O, O, O	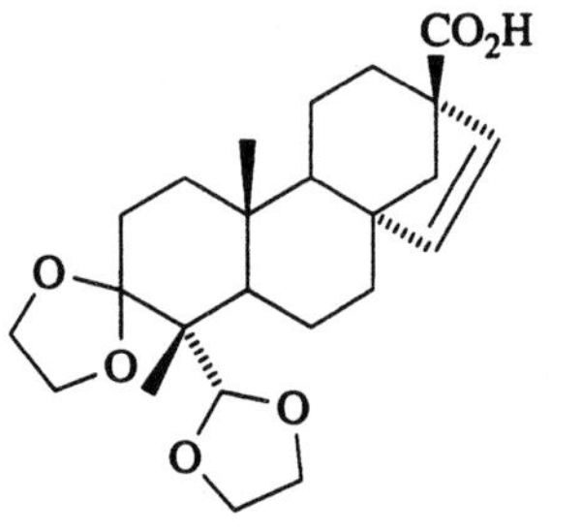 CO_2H, O, O, O, O	82	27
4	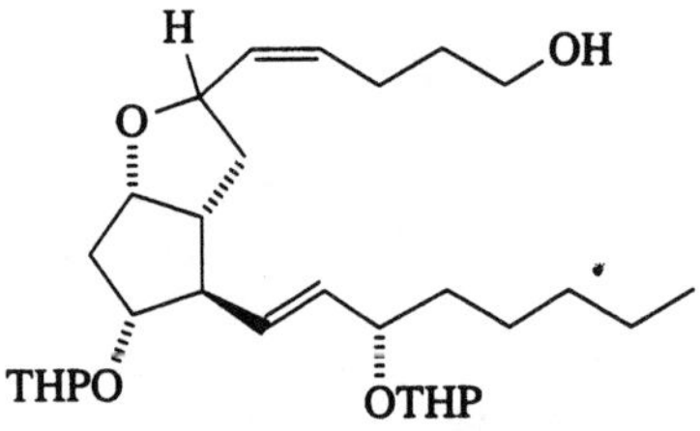 H, O, OH, THPO, OTHP	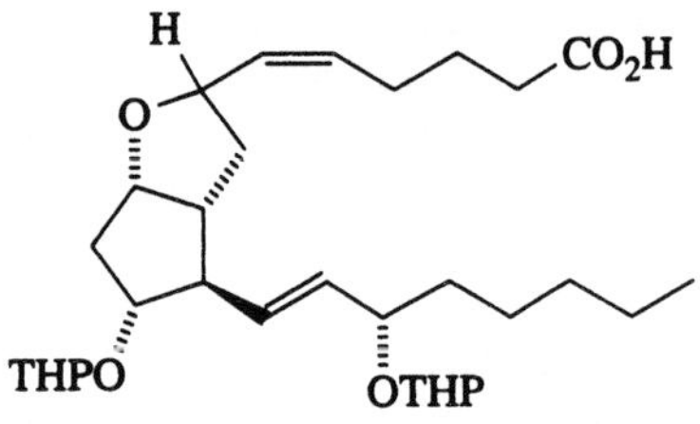 H, O, CO_2H, THPO, OTHP	59	28
5	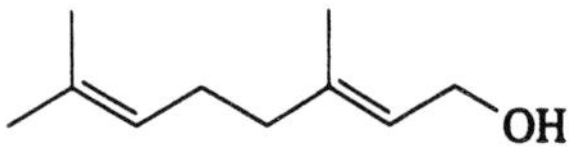 OH	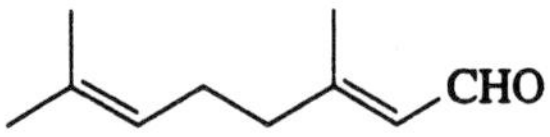 CHO	91 96:4 *(E):(Z)*	29
6	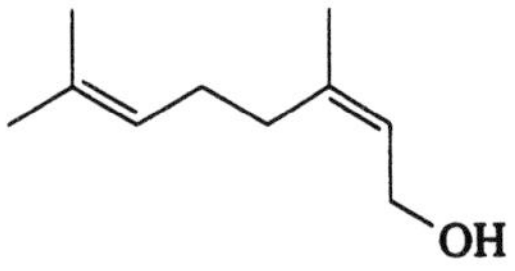 OH	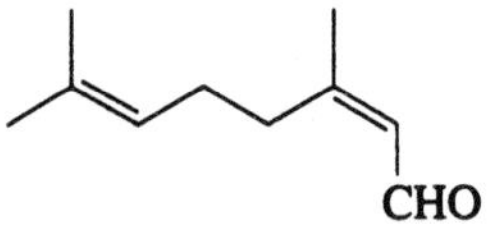 CHO	84 92:8 *(Z):(E)*	29
7	 Ph OH	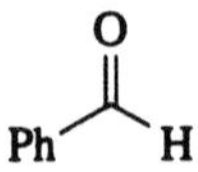 O, Ph H	76	29
8	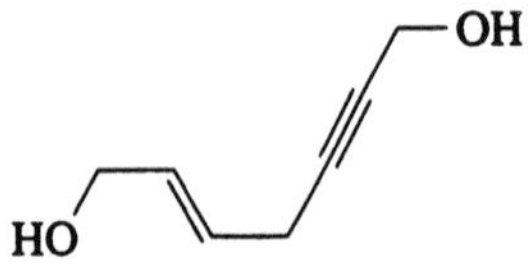 OH, HO	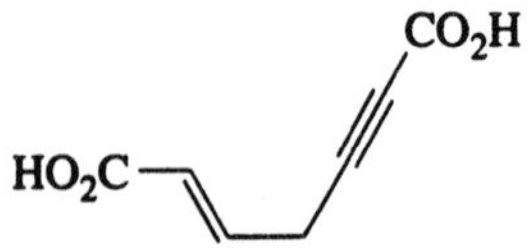 CO_2H, HO_2C	40	30
9	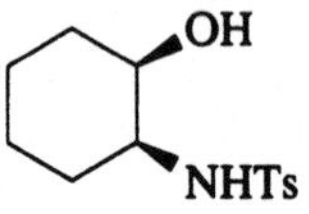 OH, NHTs	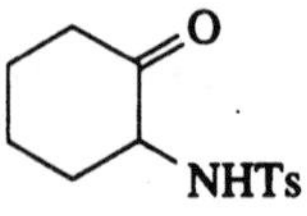 O, NHTs	98	31

Table 1 *(continued)*

Entry	*Substrate*	*Product*	*Yield (%)*	*Ref.*
10	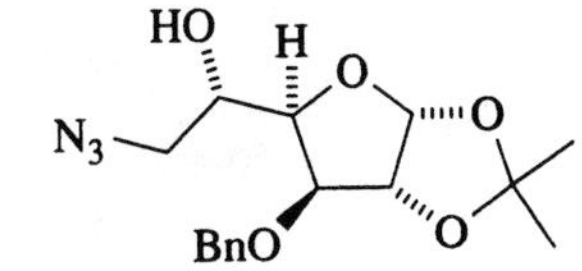	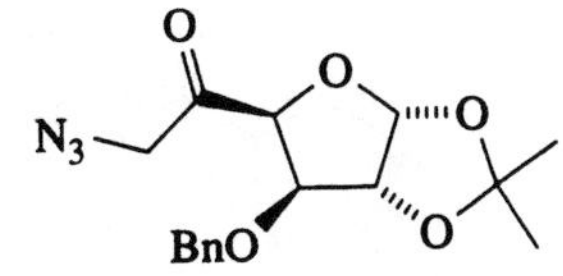	67	32
11	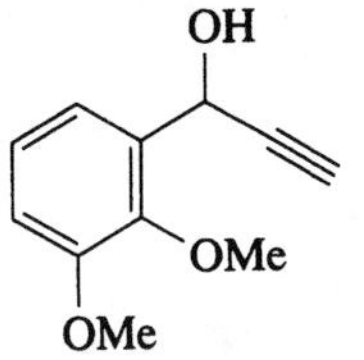	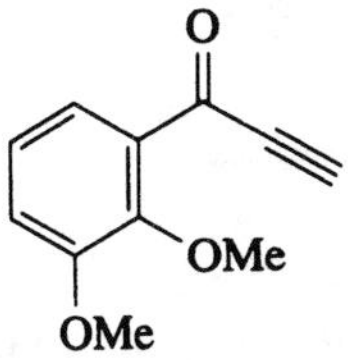	71	33
12	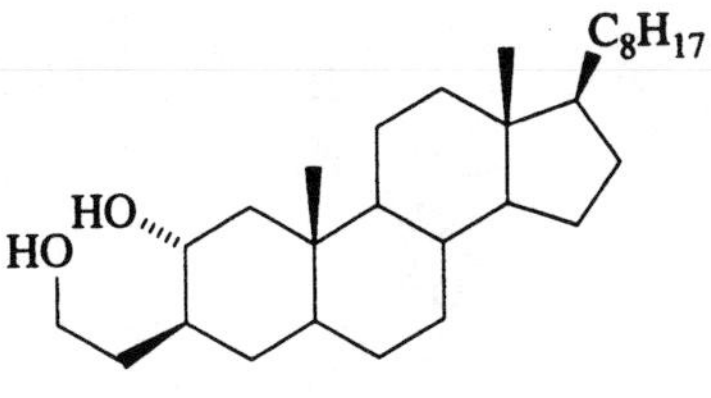	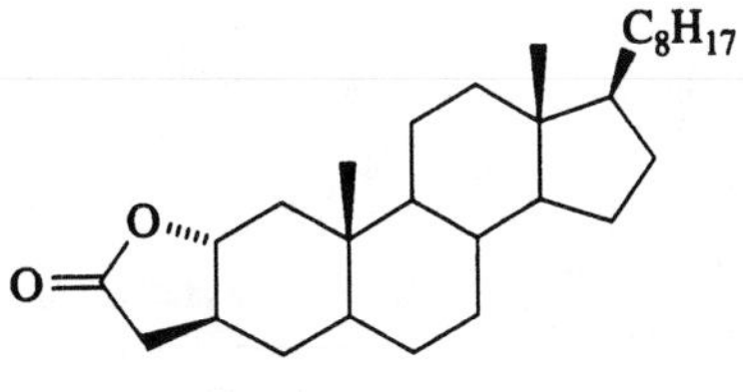	94	34
13	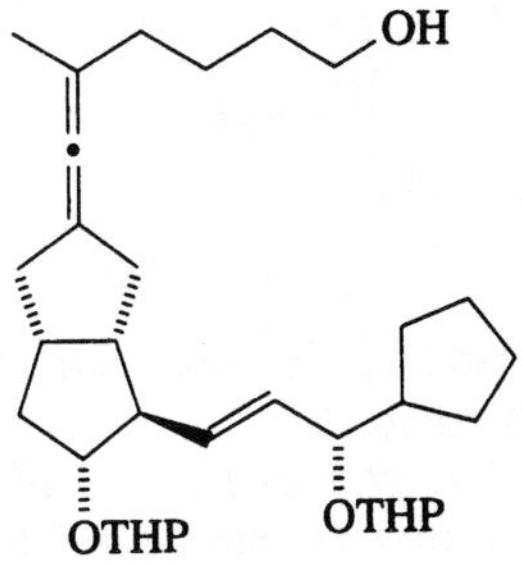	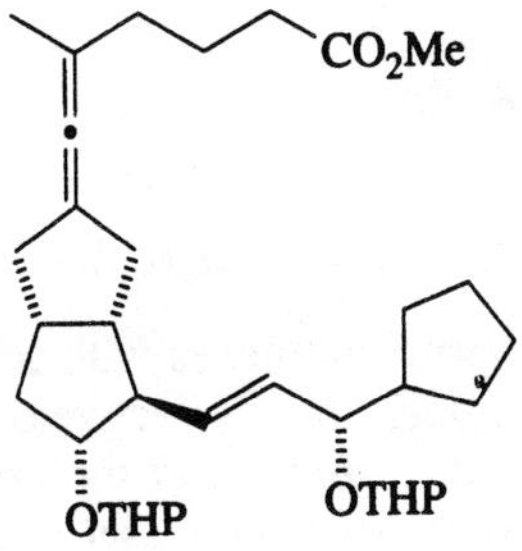	60 (over 2 steps)	35
14	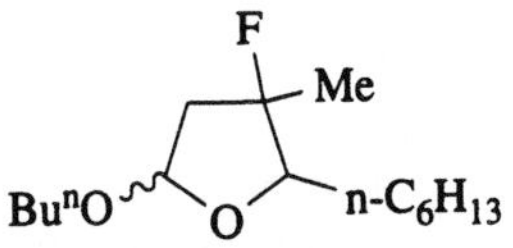	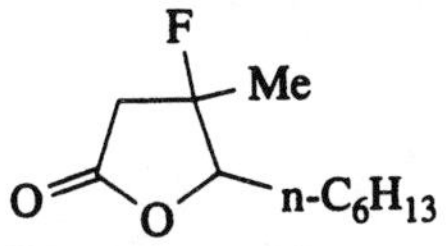	82	36
15	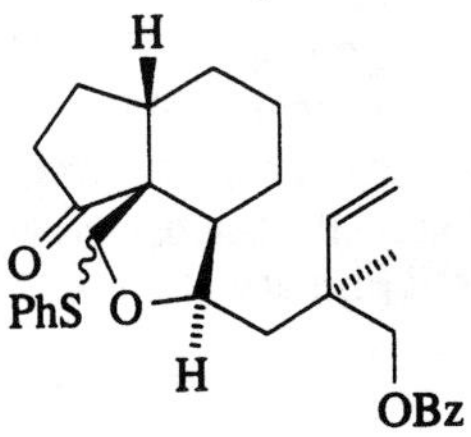	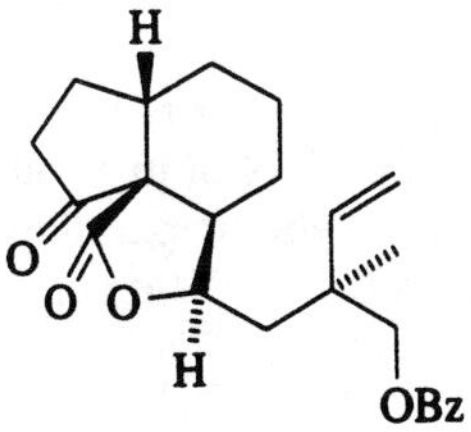	88	37
16	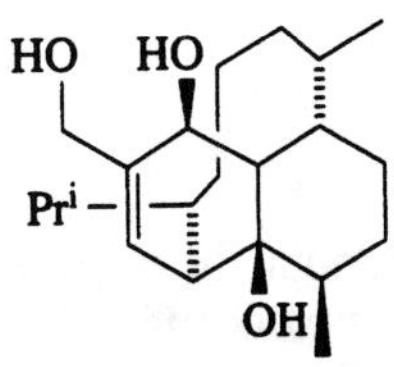	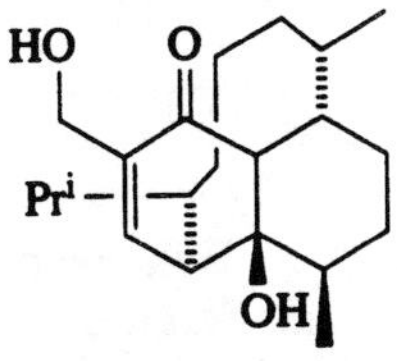	54	38

Table 1 *(continued)*

Entry	Substrate	Product	Yield (%)	Ref.
17	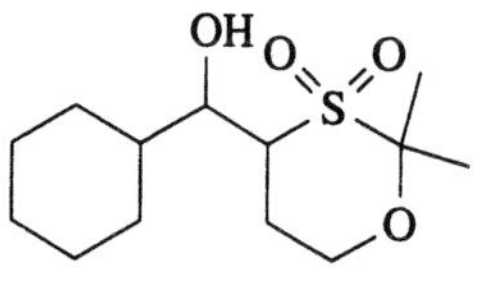	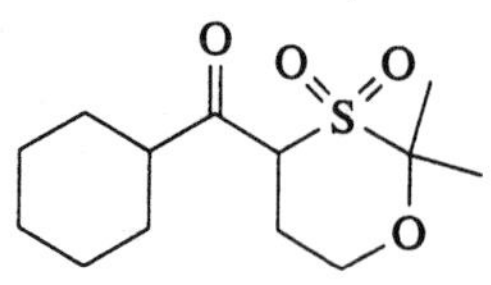	50–90	39
18	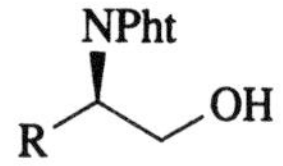	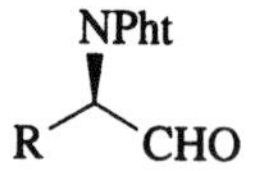	70–80	18

The oxidation of compounds with free N—H groups can be complicated by difficult product isolation.[24] However, conversion to the perchlorate salt prior to oxidation has been reported to alleviate this problem (equation 2).[25]

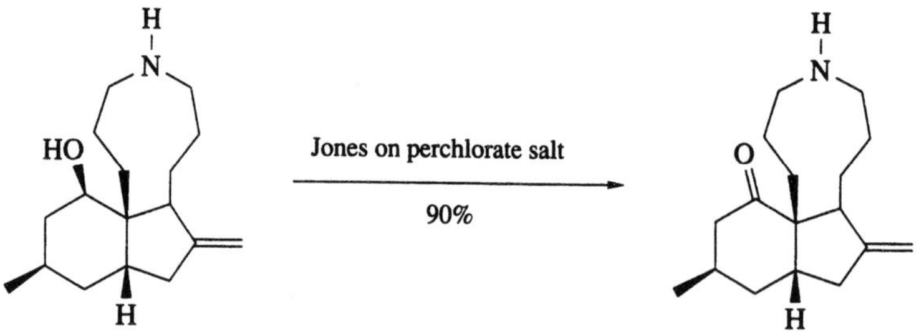

Jones on perchlorate salt
90%
(2)

2.7.3 CHROMIUM(VI) WITH HETEROCYCLIC BASES

Chromium(VI) oxide is known to form complexes with several nitrogen heterocycles, many of which show oxidizing properties. They are milder, more selective oxidants than the acid-based reagent systems. Acid sensitive groups are tolerated much more, and the preparation of aldehydes is generally easier. However, oxidation may be difficult on substrates that contain basic nitrogens, since exchange may occur with the oxidant to give substrate–chromium complexes.

2.7.3.1 Chromium(VI) Oxide·(Pyridine)$_2$

Sarett and coworkers discovered that the complex (**1**) prepared by the addition of chromium(VI) oxide to pyridine (CAUTION—reverse order of addition may cause the mixture to inflame) is an efficient oxidizing agent for the preparation of ketones from secondary alcohols.[40] The reagent, as prepared by Sarett, is moderately soluble in pyridine, but is only sparingly soluble in standard organic solvents. Thus the normal procedure is to add a solution of the alcohol in pyridine to three equivalents of the complex, also in pyridine. This procedure is also useful for the preparation of aromatic and α,β-unsaturated aldehydes, but the use of pyridine as solvent prohibits the oxidation of volatile, saturated primary alcohols.[41]

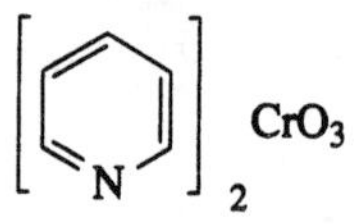

CrO_3

(**1**)

The practical problems encountered during the isolation of products from pyridine led Collins and coworkers to suggest an improved version of the Sarett oxidation, which has subsequently been widely adopted.[42] Collins and coworkers found that during the preparation of the dipyridine complex prolonged stirring gave a deep red macrocrystalline form, which could be isolated and stored (*n.b.* this reagent is very hygroscopic). This form of the complex exhibits moderate solubility in chlorinated solvents. Thus in

Table 2 Oxidation of Alcohols with $CrO_3 \cdot py_2$ (Collins Oxidation)

Entry	*Substrate*	*Product*	*Yield (%)*	*Ref.*
1	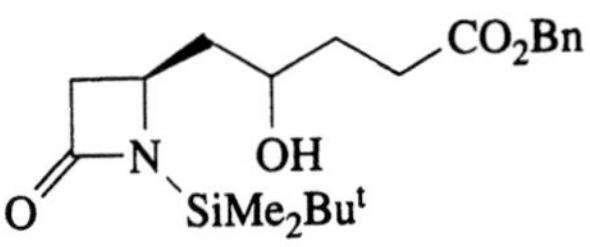 CO_2Bn, N, OH, O, $SiMe_2Bu^t$	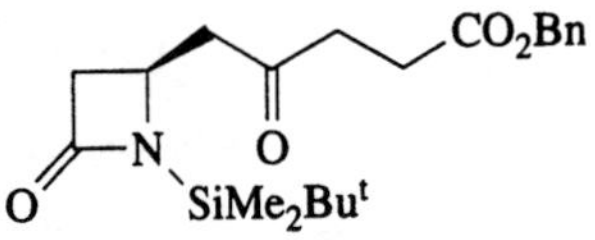 CO_2Bn, N, O, O, $SiMe_2Bu^t$	73	46
2	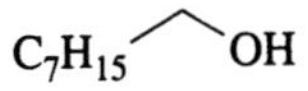 C_7H_{15} OH	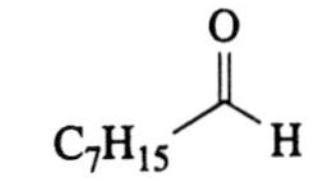 O, C_7H_{15} H	90	45
3	 Ph OH	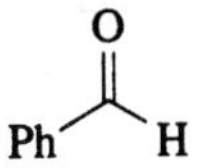 O, Ph H	89	45
4	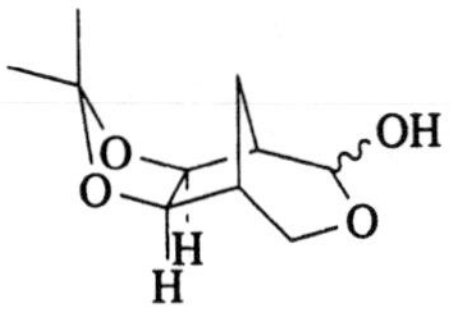 OH, O, O, O, H, H	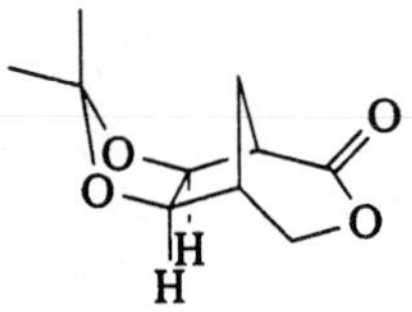 O, O, O, O, H, H	48 (over 3 steps)	47
5	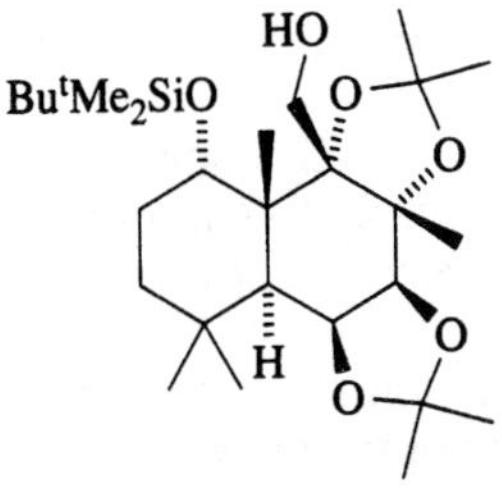 HO, Bu^tMe_2SiO, O, O, H, O, O	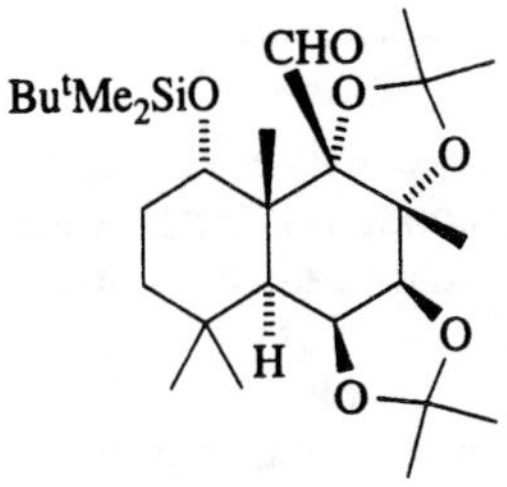 CHO, Bu^tMe_2SiO, O, O, H, O, O	88	48
6	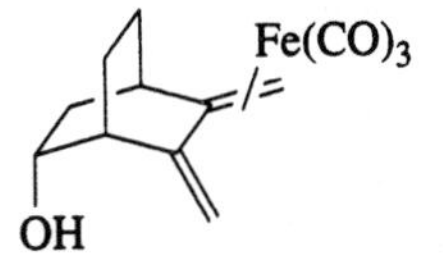 $Fe(CO)_3$, OH	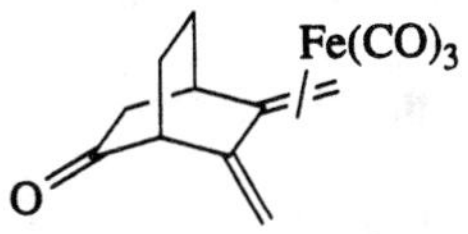 $Fe(CO)_3$, O	71	49
7	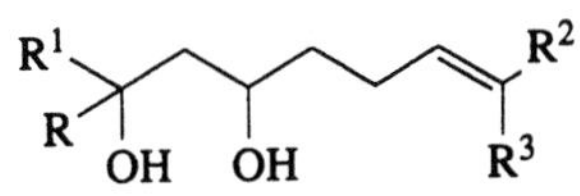 R^1, R, OH, OH, R^2, R^3	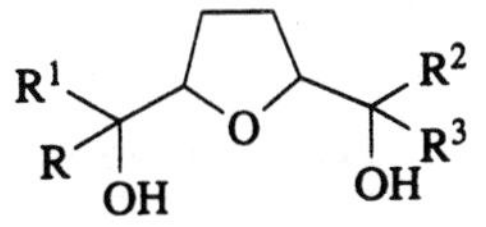 R^1, R, OH, O, R^2, R^3, OH	~50	50
8	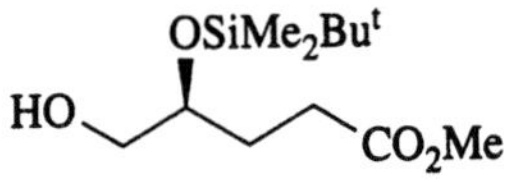 $OSiMe_2Bu^t$, HO, CO_2Me	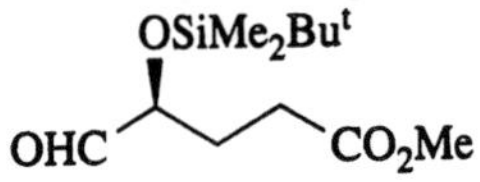 $OSiMe_2Bu^t$, OHC, CO_2Me	79 (over 2 steps)	51
9	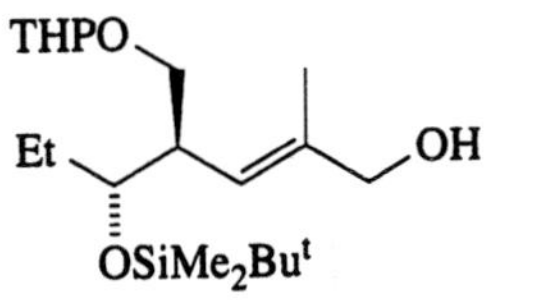 THPO, Et, OH, $OSiMe_2Bu^t$	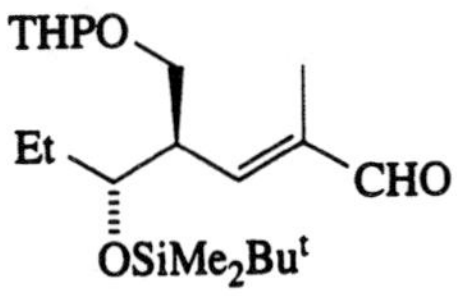 THPO, Et, CHO, $OSiMe_2Bu^t$	94	52
10	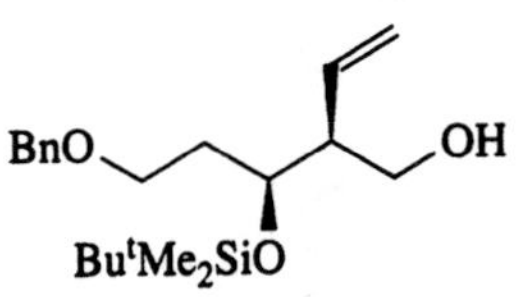 BnO, OH, Bu^tMe_2SiO	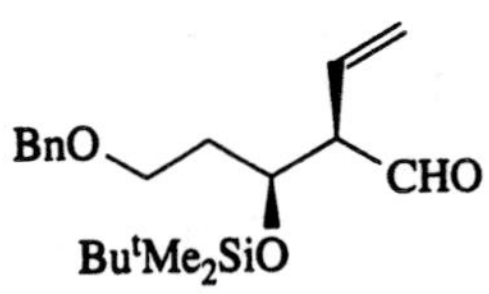 BnO, CHO, Bu^tMe_2SiO	94	52

Table 2 *(continued)*

Entry	*Substrate*	*Product*	*Yield (%)*	*Ref.*
11	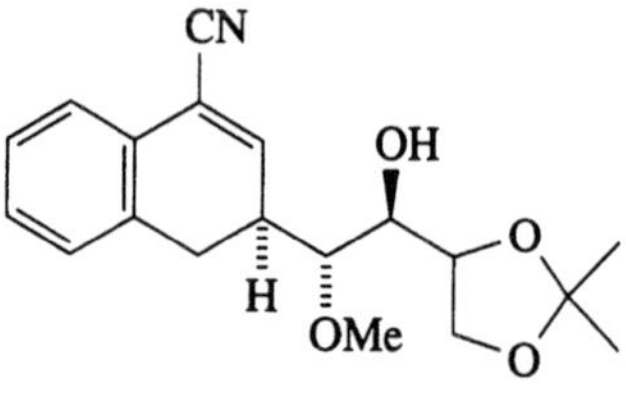 CN, OH, H, OMe, O, O	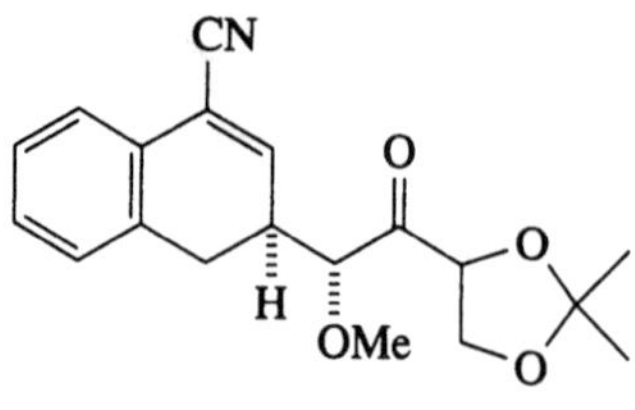 CN, O, H, OMe, O, O	89	53
12	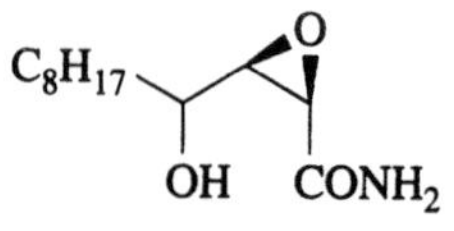 C_8H_{17}, O, OH, $CONH_2$	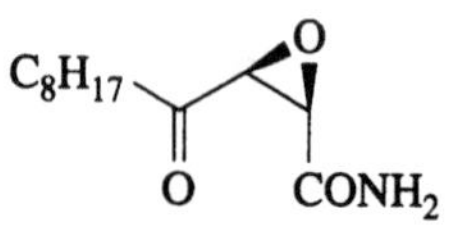 C_8H_{17}, O, O, $CONH_2$	92	54
13	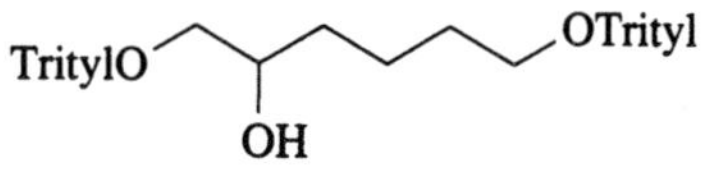 TritylO, OTrityl, OH	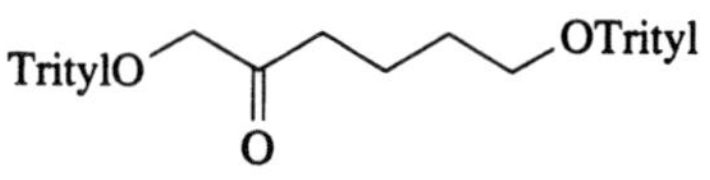 TritylO, OTrityl, O	74	55
14	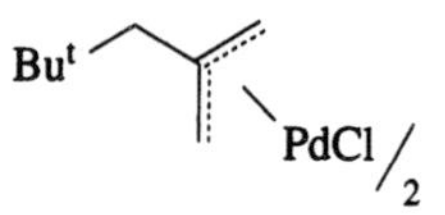 Bu^t, PdCl /2	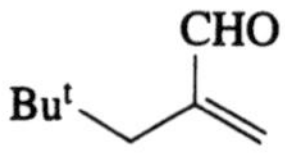 CHO, Bu^t	40	56

dichloromethane, the solvent of choice, oxidations of simple substrates are usually complete within 5–15 min at room temperature. Standard conditions employ six equivalents of oxidant in order to achieve quick, clean reactions. The excess oxidant can cause problems during work-up, but these may be eased by the addition of celite which adsorbs some of the chromium salts.[44]

This procedure for the oxidation of alcohols was further improved by Ratcliffe and Rodehorst:[45] *in situ* preparation of a dichloromethane solution of the complex immediately prior to use avoided the problems associated with the hygroscopic nature of the complex and the fire hazard endured during its preparation and isolation. (If desired, the dichloromethane solution may be stored for up to one month with little loss in activity.)

The Collins oxidation is efficient for the preparation of carbonyl compounds in the presence of a wide range of functionalities (Table 2); however, Dauben and coworkers observed that extended exposure to the complex (24 h at room temperature) can give moderate to excellent yields of allylic C—H oxidation products (equations 3 to 5).[57]

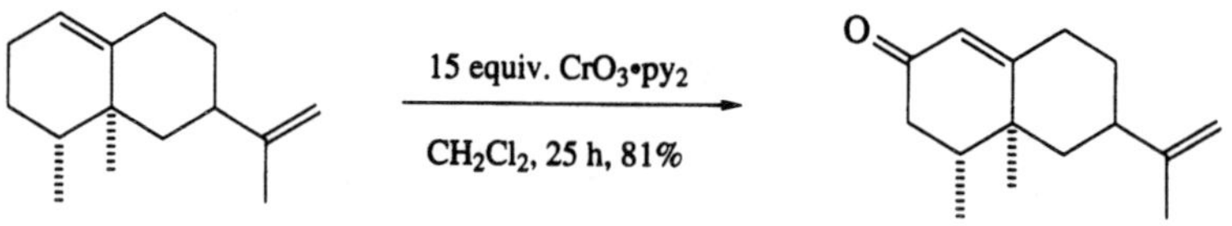

15 equiv. $CrO_3{\bullet}py_2$ / CH_2Cl_2, 25 h, 81% (3)

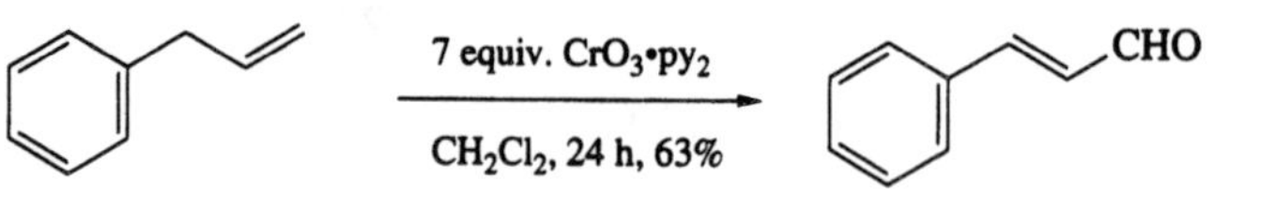

7 equiv. $CrO_3{\bullet}py_2$ / CH_2Cl_2, 24 h, 63% CHO (4)

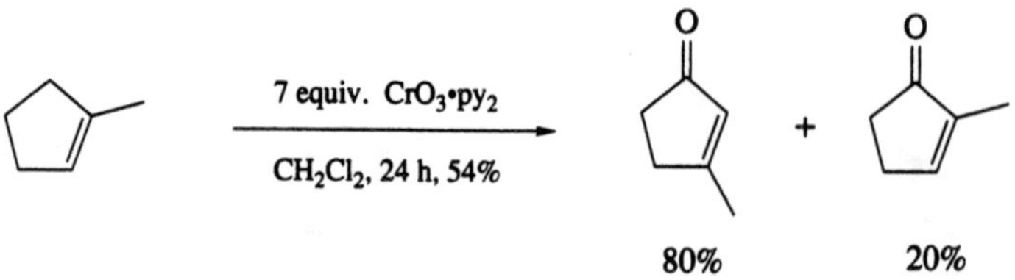

7 equiv. $CrO_3{\bullet}py_2$ / CH_2Cl_2, 24 h, 54% O + O (5)

80% 20%

The chromium(VI) oxide–dipyridine complex also has been found to cause oxidative rearrangement of tertiary allylic alcohols to α,β-epoxy aldehydes and small amounts of α,β-unsaturated aldehydes (equation 6 and Table 3).[58] This is potentially useful as a homologation sequence since the starting materials are readily available from vinyl metal addition to ketones. Use of pyridinium chlorochromate (PCC) for this transformation gives mostly α,β-unsaturated aldehydes.

Table 3 Rearrangement of Tertiary Allylic Alcohols with the Collins Reagent

(6)

Entry	*Substrate*	*Product(s)*		*Yield (%)*	*Ref.*
1	OH	CHO, 76%	CHO, 24%	90	58
2	OH	O, CHO, 89%	CHO, 11%	96	58
3	OH	CHO, O, *ca.* 100%	CHO, trace	50	58
4	$SnPh_3$, HO, OTHP	CHO, OTHP		54	62

Addition of acetic anhydride to the Collins reagent (CrO_3:pyridine:acetic anhydride 1:2:1) has been reported to be suitable for the oxidation of carbohydrates[59] and nucleoside substrates;[60] for example, 5′-*O*-acetylthymidine gives spontaneous elimination of thymine under Pfitzner–Moffat conditions. Similarly, 5′-*O*-tritylthymidine loses thymine when treated with $CrO_3{\cdot}py_2$. However, under modified conditions (CrO_3:pyridine:acetic anhydride 1:2:1) good yields of the corresponding 3′-carbonyl compounds are regularly obtained (equation 7).[60]

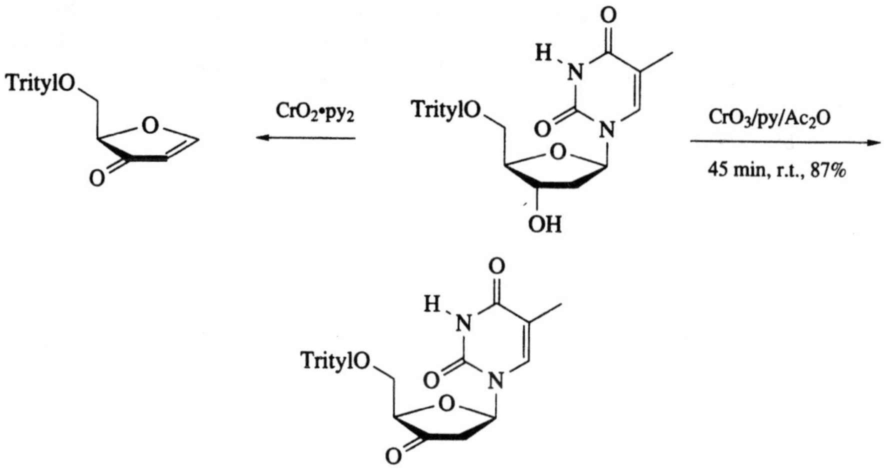

TritylO O O ← $CrO_2{\cdot}py_2$ — TritylO (thymidine, OH) — CrO_3/py/Ac_2O, 45 min, r.t., 87% → TritylO (3'-keto nucleoside) (7)

The complex formed by 2,2′-bipyridine with chromium(VI) oxide is a milder oxidant than the Collins reagent,[61] as indicated by the need for long reaction times (up to 48 h) and a larger excess of oxidant (8 equiv.).

2.7.3.2 Chromium(VI) Oxide·(3,5-Dimethylpyrazole)

In contrast to the Collins reagent, the complex formed by 3,5-dimethylpyrazole with chromium(VI) oxide (**2**) is very soluble in dichloromethane.[63] Hence, reactions (up to 0.1 mol scale) can be carried out in the minimum amount of solvent. Generally, 2.5 equiv. of complex, generated *in situ*, gives good yields of aldehydes and ketones. In addition, upon work-up most of the chromium salts may be precipitated by dilution with diethyl ether.

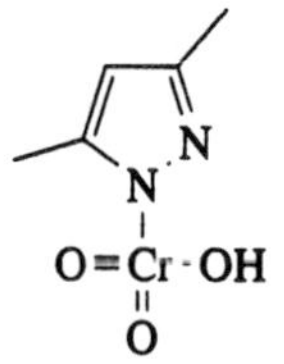

O=Cr(=O)-OH (**2**) 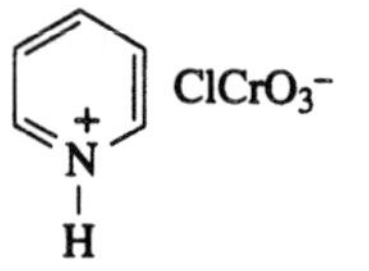 $ClCrO_3^-$ (**3**)

Although chromium(VI) oxide·(3,5-dimethylpyrazole) is also a good oxidant for allylic C—H bonds,[43] it is surprising that this reagent has not been more widely adopted.

2.7.3.3 Pyridinium Chlorochromate (PCC)

Pyridinium chlorochromate (**3**),[64] first developed by Corey and coworkers in 1977, is a commercially available, stable yellow solid, which may be stored in air.[65] With simple substrates, oxidations are normally performed in dichloromethane at room temperature with 1.5 equiv. of oxidant, and are usually complete within 2 h. (More polar solvents, in which PCC has higher solubility, unfortunately lead to prohibitively long reaction times.) Good yields of ketones and aldehydes are regularly obtained, but slight (*E*)/(*Z*)-isomerization is observed in allylic alcohol oxidations (Table 4). Significant overoxidation is rare, but if desired, acids may be prepared from aldehydes with stoichiometric sodium cyanide and PCC in THF at 45 °C.[84] The reaction is selective for aliphatic aldehydes; conjugated and α-oxygenated substrates are recovered unchanged. Oxidation direct from the alcohol is not as efficient.

The acidic properties of PCC, which if necessary may be buffered by the use of powdered sodium acetate, can be used to good advantage to effect oxidative cationic cyclizations.[85,86] Efficient cyclizations are limited to substrates that give tertiary carbonium ions as the initial cyclization product and to the formation of six-membered rings.

Table 4 Oxidation of Alcohols with PCC

Entry	*Substrate*	*Product*	*Conditions*	*Yield (%)*	*Ref.*
1	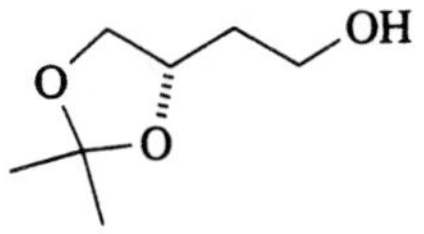 OH	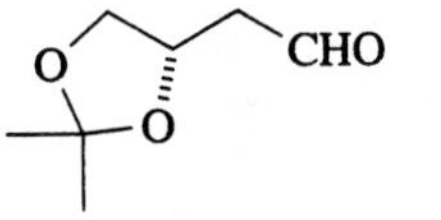 CHO	3 equiv., DCM, 1 h, r.t., 3Å sieves	88	66
2	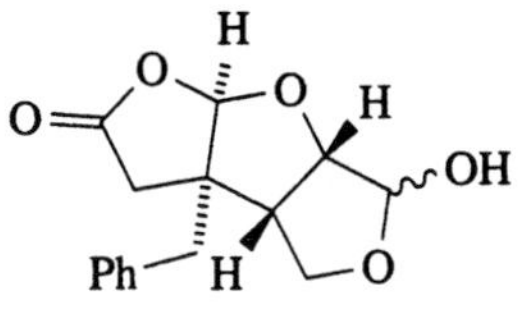 OH	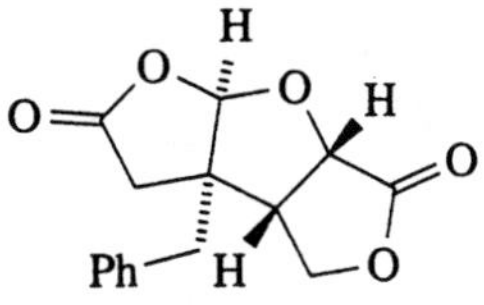	3 equiv., DCM, 4.5 h, r.t.	65	67
3	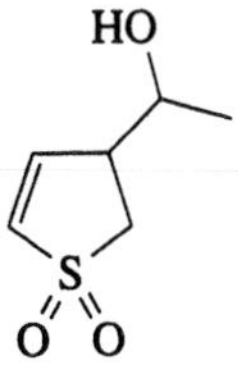 HO	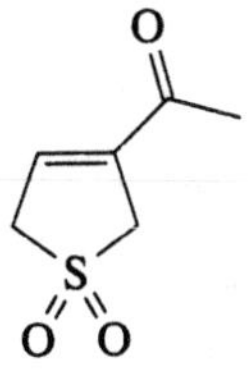	6.5 equiv., THF, 35 °C, 3Å sieves	51	68
4	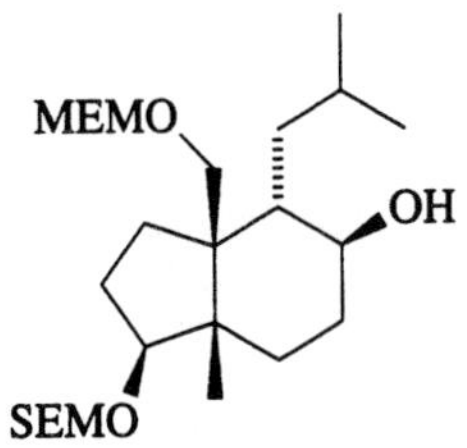 MEMO, OH, SEMO	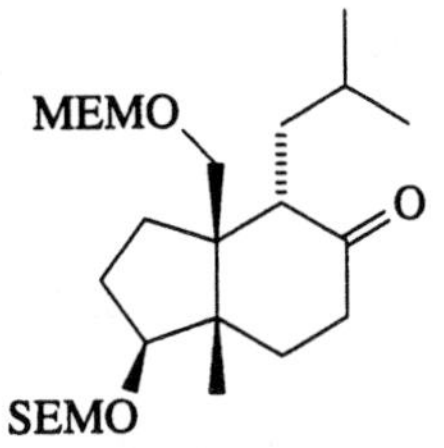 MEMO, SEMO	2 equiv., DCM, 3 h, r.t.	74	69
5	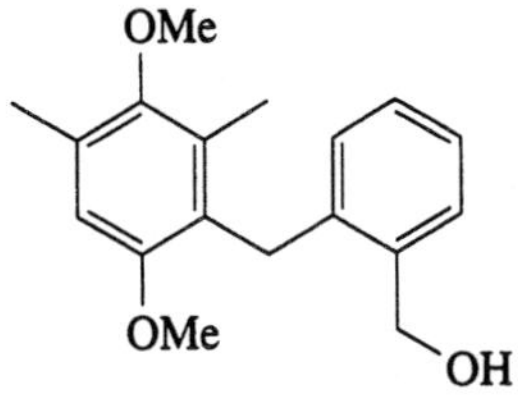 OMe, OMe, OH	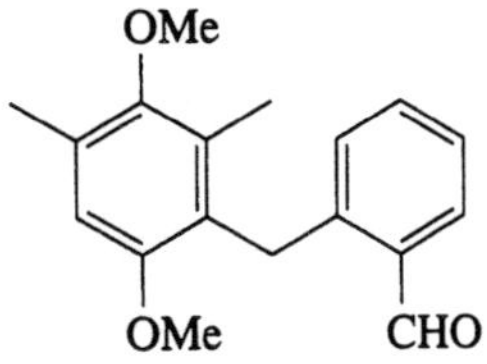 OMe, OMe, CHO	DCM, 1 h, r.t.	96	70
6	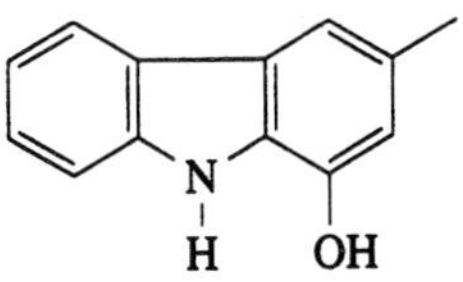 OH	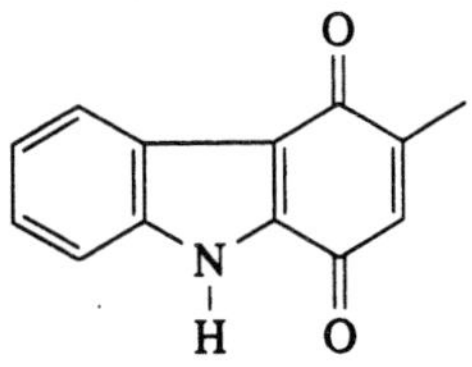	DCM, r.t.	40	71
7	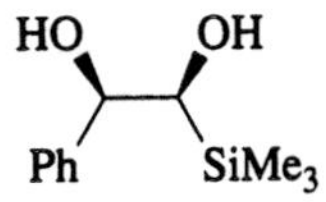 HO, OH, Ph, $SiMe_3$	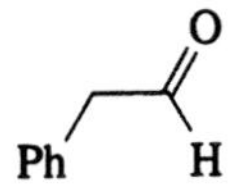 Ph, H	DCM, r.t.	90	72
8	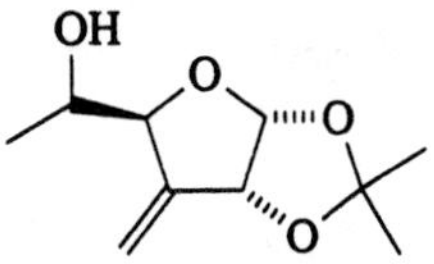 OH	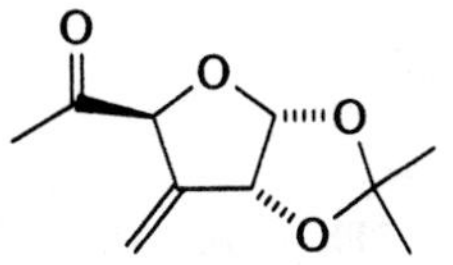	2.5 equiv., DCM, 3Å sieves, 3 h, reflux	72	73

Table 4 *(continued)*

Entry	*Substrate*	*Product*	*Conditions*	*Yield (%)*	*Ref.*
9	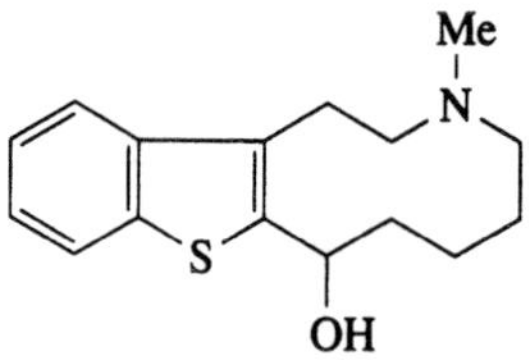	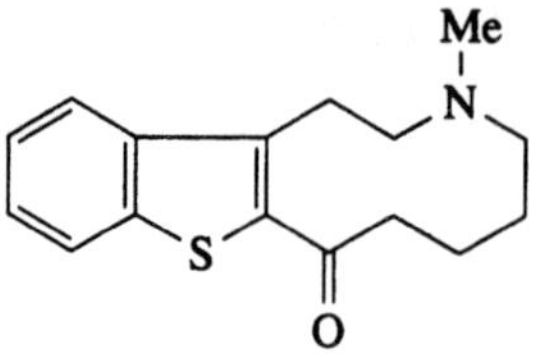	1.5 equiv., DCM, NaOAc, 17 h, r.t.	71	74
10	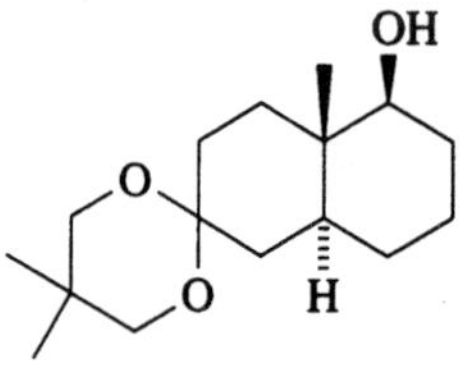	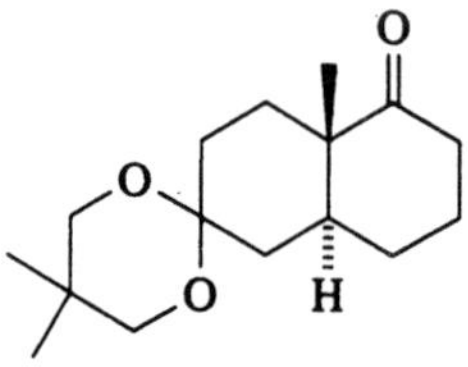	2 equiv., DCM, NaOAc, 3 h, r.t.	88	75
11	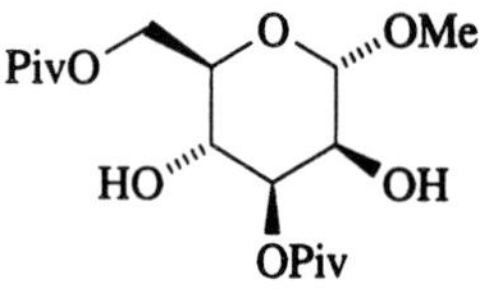	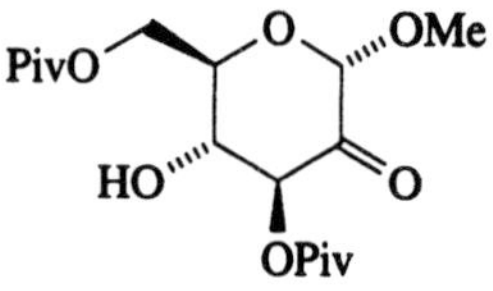	2 equiv., DCM, NaOAc, 24 h, reflux	76	76
12	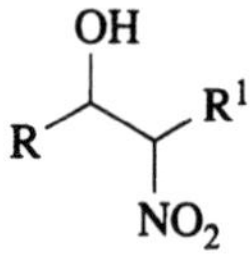	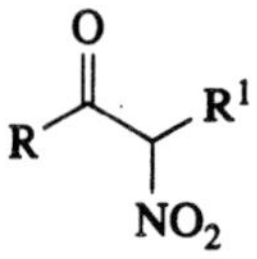	DCM, 3Å sieves, 36 h, r.t.	61–78	77
13	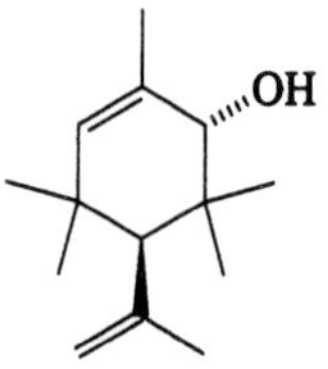	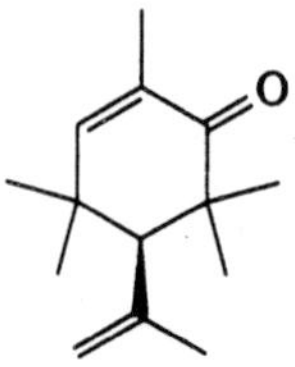	1.1 equiv., DCM, 2 h, r.t.	85	78
14	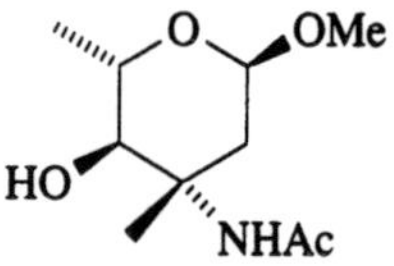	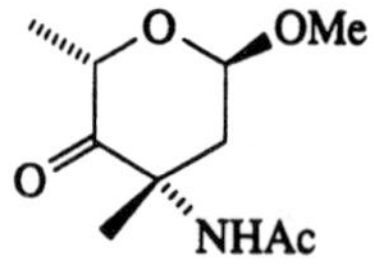	3 equiv., DCM, 3Å sieves, 3 h, r.t.	66	79
15	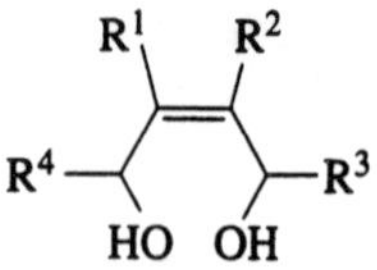	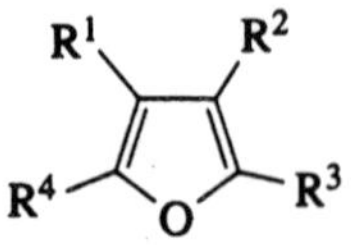	1.5 equiv., DCM, 0.5 h, r.t.	35–89	80
16	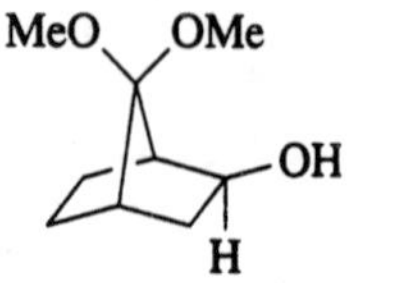	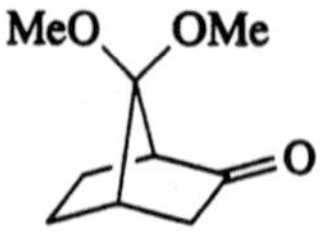	1.4 equiv., DCM, NaOAc, 2 h, r.t.	95	81
17	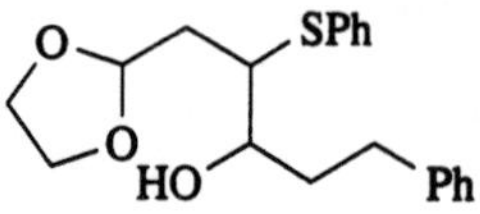	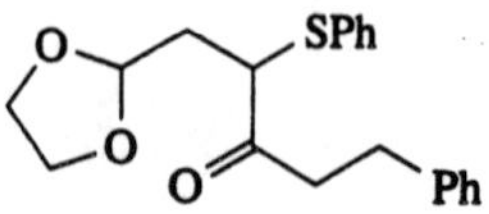	1.7 equiv., DCM, 40 h, reflux	68	82

Table 4 *(continued)*

Entry	*Substrate*	*Product*	*Conditions*	*Yield (%)*	*Ref.*
18	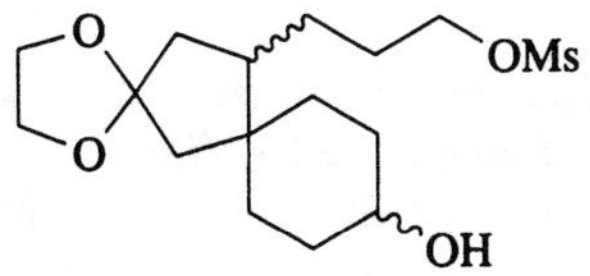 OMs, O, O, OH	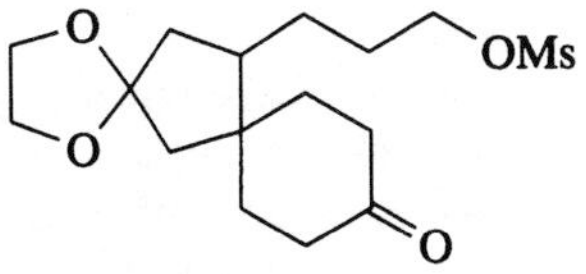 OMs, O, O, O	2 equiv., DCM, NaOAc, 2.5 h, r.t.	98	83

In a similar fashion to the Collins reagent, PCC will also induce oxidative rearrangement of tertiary allylic alcohols (Table 5).[87,88] PCC, and several other chromium oxidants, will also cause tertiary cyclopropyl alcohols to rearrange to give β,γ-unsaturated carbonyl compounds (equation 8).[88]

Table 5 Rearrangements of Tertiary Allylic Alcohols with PCC

Entry	*Substrate*	*Product(s)*	*Conditions*	*Yield (%)*	*Ref.*
1	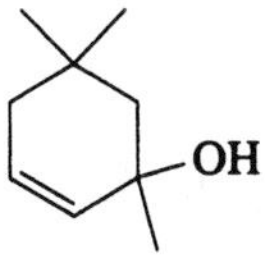 OH	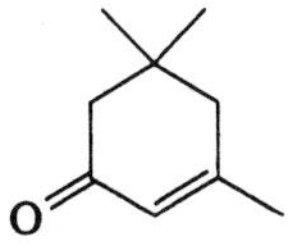 O	3 equiv. PCC, DCM, 16 h	94	87
2	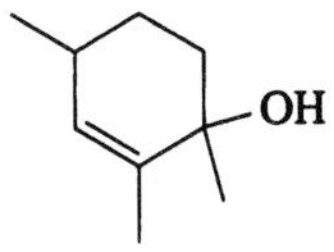 OH	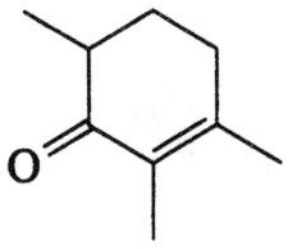 O	3 equiv. PCC, DCM, 16 h	88	87
3	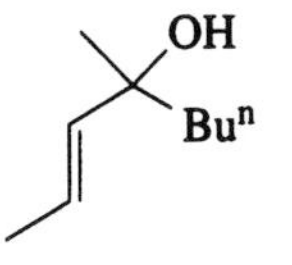 OH, Bu^n	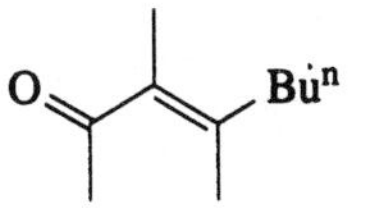 O, Bu^n ~1:1 (*E*):(*Z*)	3 equiv. PCC, DCM, 16 h	50	87
4	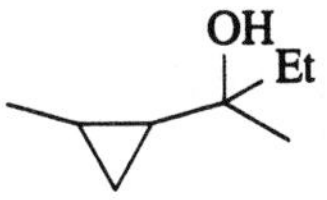 OH, Et	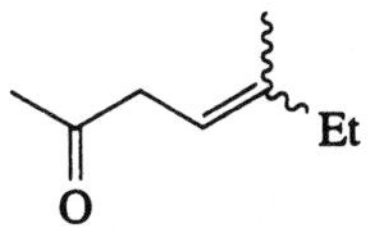 Et, O	8 equiv. PCC, DCM, 6 h	62	88
5	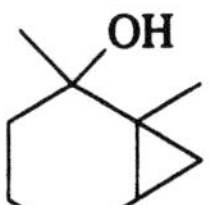 OH	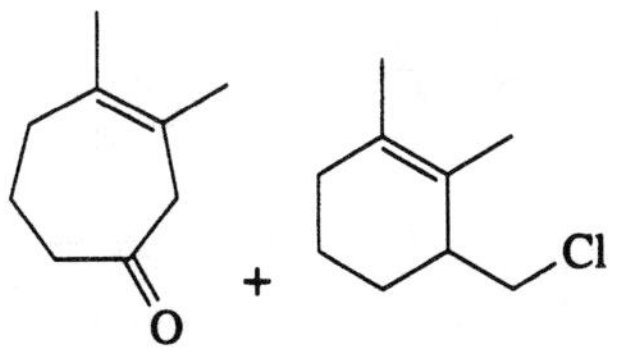 O (85%) + Cl (15%)	8 equiv. PCC, DCM, 6 h	36	88

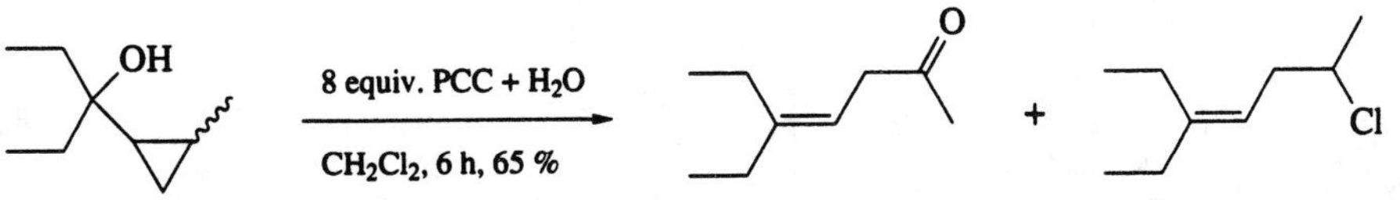

$$\text{OH} \xrightarrow[\text{CH}_2\text{Cl}_2\text{, 6 h, 65 \%}]{\text{8 equiv. PCC + H}_2\text{O}} \text{O} + \text{Cl} \qquad (8)$$

PCC can used for the oxidation of silyl-protected hydroquinones to quinones, except where there are electron-withdrawing substituents on the aromatic ring.[89] Interestingly, there was no evidence for cleavage of the silicon–oxygen bond as the first step, which might be expected under the acidic conditions of a PCC oxidation.

Organoboranes from hydroboration reactions can be oxidized directly to ketones with PCC, thus eliminating the need to isolate the intermediate alcohol.[90]

PCC can be modified to show selectivity for the oxidation of allylic alcohols in steroidal systems. A solution of PCC in dichloromethane with 2% pyridine at *ca.* 2 °C was found to be an effective and selective oxidant (Table 6).[91] In contrast to chromate oxidations of saturated alcohols in rigid systems, Parish and coworkers found that quasiequatorial allylic alcohols were oxidized faster than axial ones. Similar properties were also found for solutions of PCC and 3,5-dimethylpyrazole (2%) in dichloromethane.[92] In addition, Parish and coworkers also examined several other aromatic amines for the ability to promote allylic selectivity.[93] 2,2′-Bipyridine, pyrazine, pyridazine, *s*-triazine and 2,4,6-triphenylpyridine all had some effect, but their efficacy appeared to be substrate dependent. Most recently the combination of PCC

Table 6 Oxidation of Allylic Alcohols with PCC

Entry	*Substrate*	*Product*	*Conditions*	*Yield (%)*	*Ref.*
1	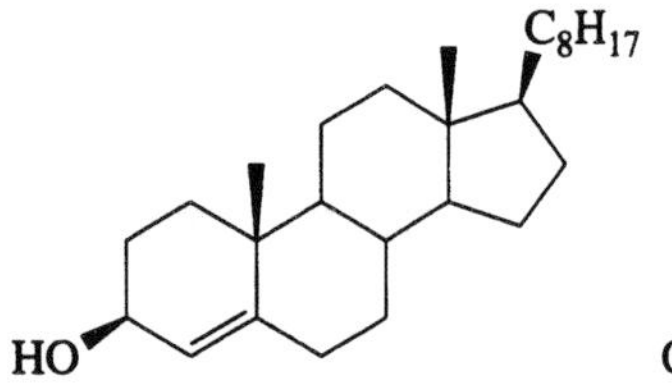	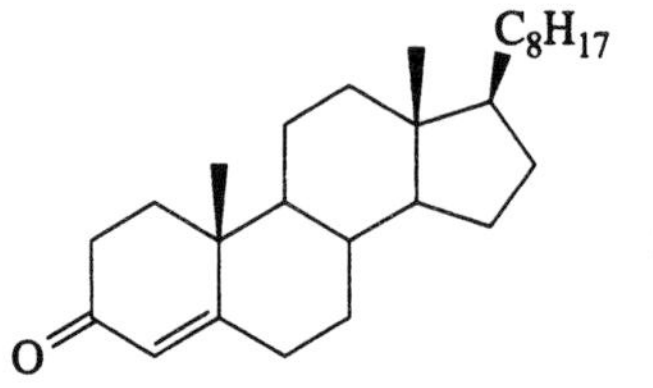	3.5 equiv. PCC, 2% BZT,[a] DCM, 2–3 °C, 30 min	91	94
2	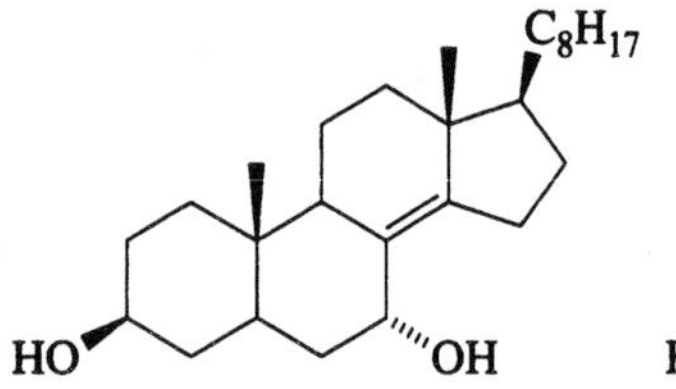	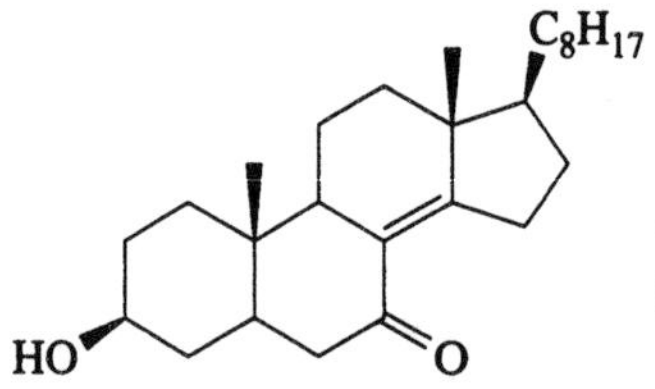	7 equiv. PCC, 2% Py, DCM, 2 °C, 30 min	89	91
3	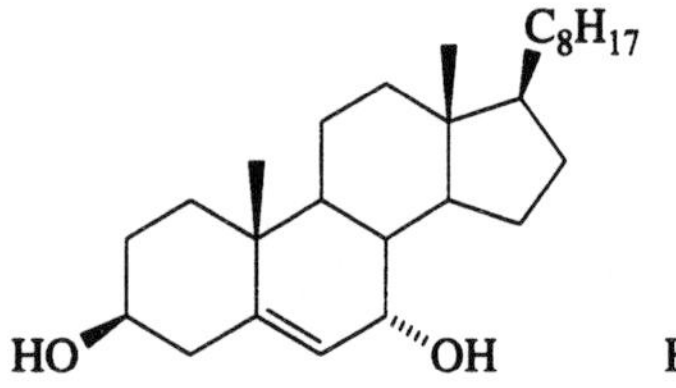	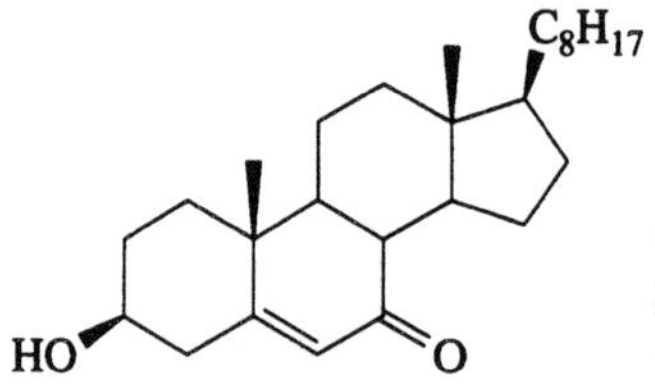	3 equiv. PCC, 2% DMP,[a] DCM, 2–3 °C, 30 min	89	92
4	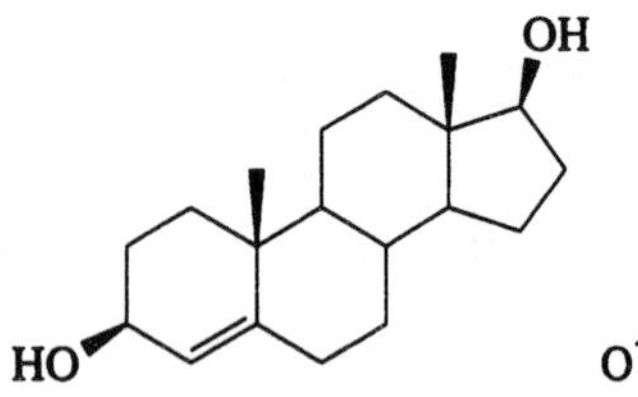	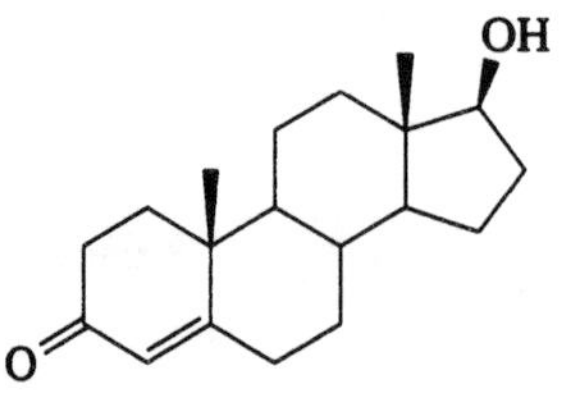	3 equiv. PCC, 2% DMP,[a] DCM, 2–3 °C, 30 min	87	92
			3.5 equiv. PCC, 2% BZT,[a] DCM, 2–3 °C, 30 min	92	94
5	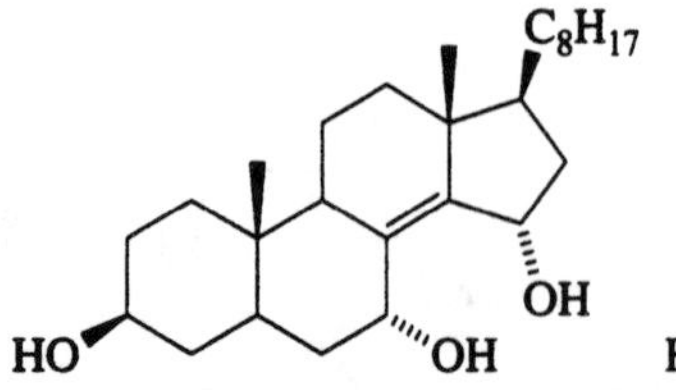	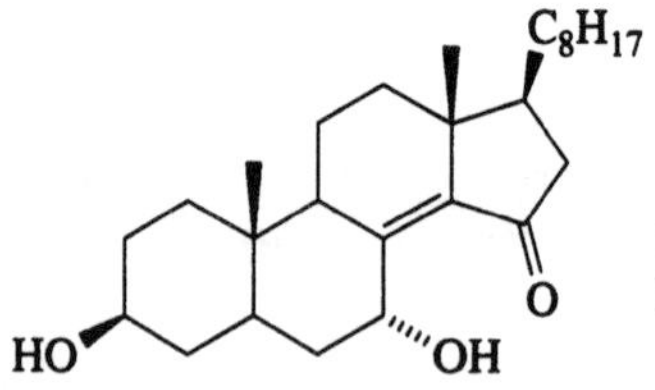	7 equiv. PCC, 2% Py, DCM, 2 °C, 30 min	82	91

[a] BZT = benzotriazole; DMP = 3,5-dimethylpyrazole.

with benzotriazole (2%)[94] has been found to exhibit excellent selectivity for allylic alcohols (~90% allylic oxidation) (equation 9).

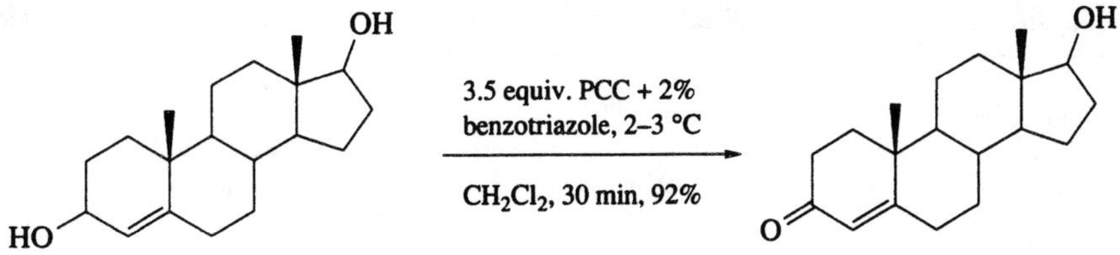

3.5 equiv. PCC + 2% benzotriazole, 2–3 °C; CH_2Cl_2, 30 min, 92% (9)

PCC becomes a much milder oxidant when it is adsorbed on alumina, e.g. added buffer is not required to prevent oxidative cationic cyclizations (see Section 2.7.5.1).[95]

Oxidation of carbohydrates with PCC *via* the standard procedure has been found to be slow, even with large excesses of oxidant. Use of boiling benzene instead of dichloromethane at room temperature can cause drastic reductions in reaction times.[96] Addition of molecular sieves to the oxidation of carbohydrates with PCC (or PDC) also gives dramatic rate enhancement.[97,98] (The use of molecular sieves in

Table 7 Oxidation of Furans with PCC

Entry	*Substrate*	*Product*	*Conditions*	*Yield (%)*	*Ref.*
1	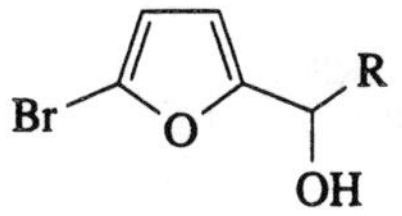 Br, R, OH	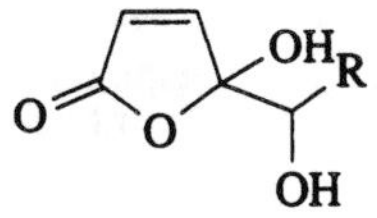 OH, R, OH	2 equiv. PCC, r.t., 1.5 h	60–75	99
2	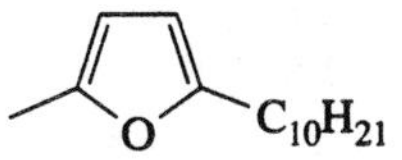 $C_{10}H_{21}$	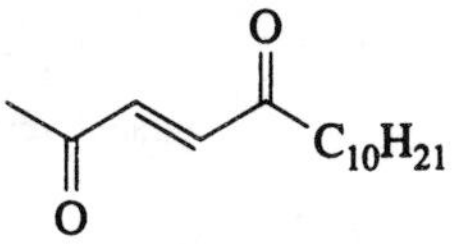 $C_{10}H_{21}$	PCC, DCM, reflux, 24 h,	90	100
3	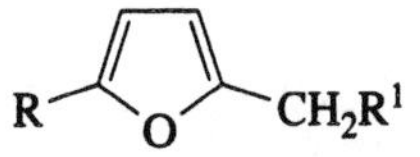 R, CH_2R^1	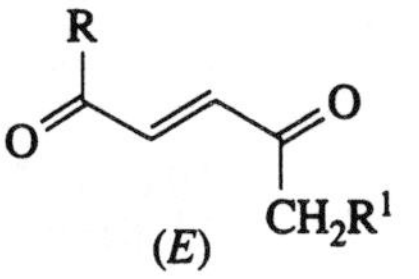 R, CH_2R^1 (*E*)	1.5 equiv. PCC, DCM, r.t., ~24 h, reflux, ~9 h	50–80	101
4	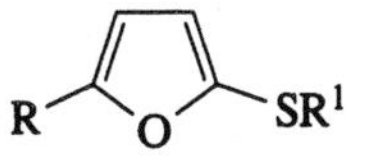 R, SR^1	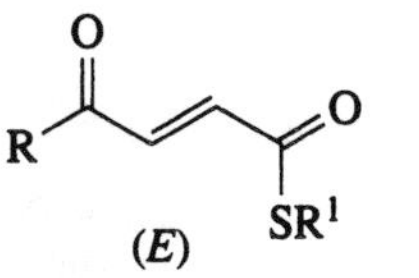 R, SR^1 (*E*)	5 equiv. PCC, r.t., DCM, 2–24 h	60–90	102
5	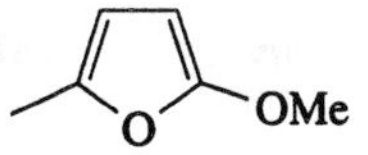 OMe	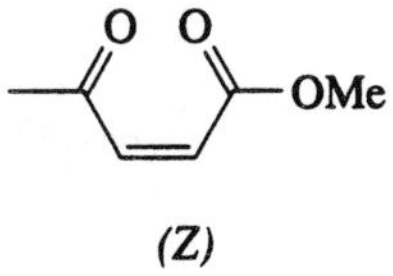 OMe (*Z*)	5 equiv. PCC, DCM, r.t., 10 min	70	102
6	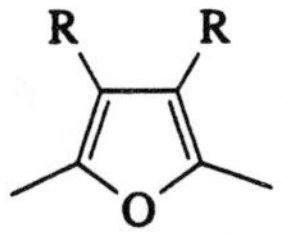 R, R	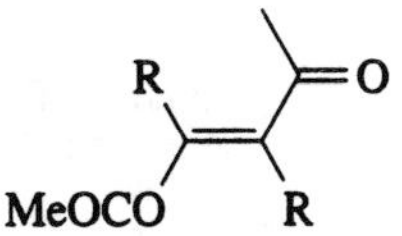 R, MeOCO, R	4 equiv. PCC, DCM, r.t., 10 h	50–70	103
7	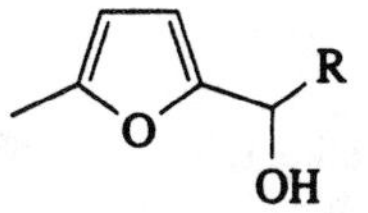 R, OH	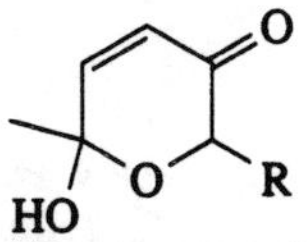 HO, R	1.5 equiv. PCC, DCM, r.t., 1 h	90	104

Table 8 Alternative Oxidation Reactions of PCC

Entry	*Substrate*	*Product*	*Conditions*	*Yield (%)*	*Ref.*
1	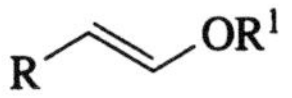 R, OR^1	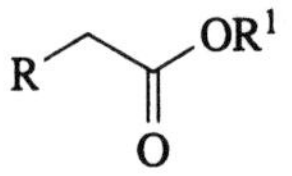 R, OR^1, O	2 equiv. PCC, DCM, r.t., 1 h	70–95	105
2	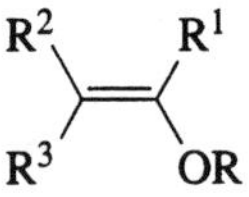 R^2, R^1, R^3, OR	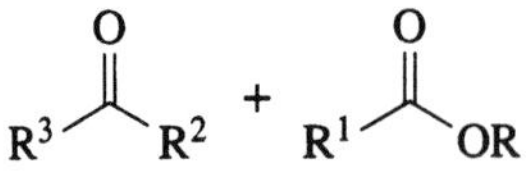 O, R^3, R^2 + O, R^1, OR	4 equiv. PCC/PDC, celite, r.t., 1 h, DCM or benzene	60–90	106
3	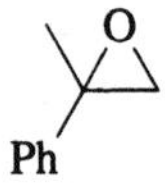 O, Ph	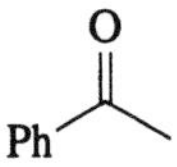 O, Ph	2 equiv. PCC, 4Å sieves, r.t., 1.5 h	65	107
4	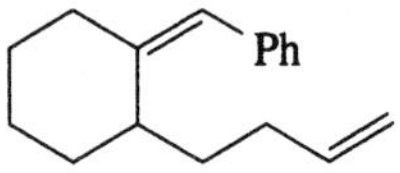 Ph	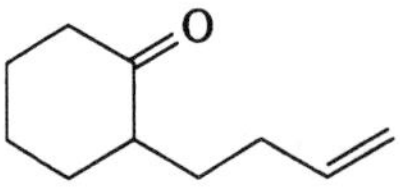 O	5 equiv. PCC, celite, reflux, 25 h	82	108
5	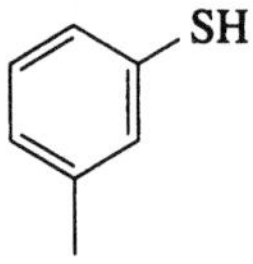 SH	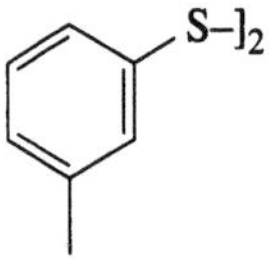 S–]$_2$	1 equiv. PCC, DCM, r.t., 2 min	67	109
6	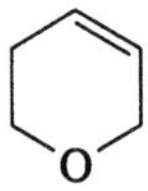 O	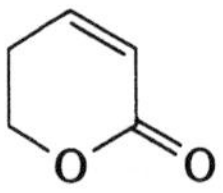 O, O	3 equiv. PCC, DCM, sealed vessel, 60 °C	70	110
7	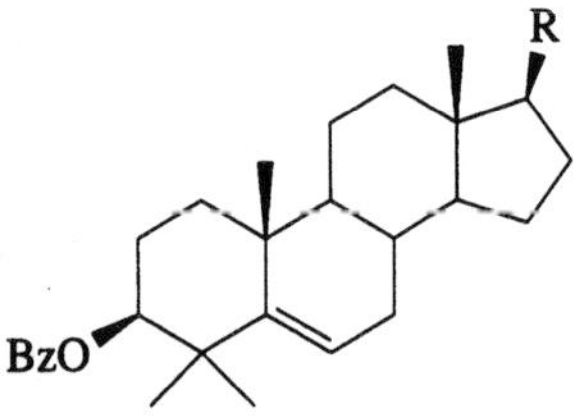 R, BzO	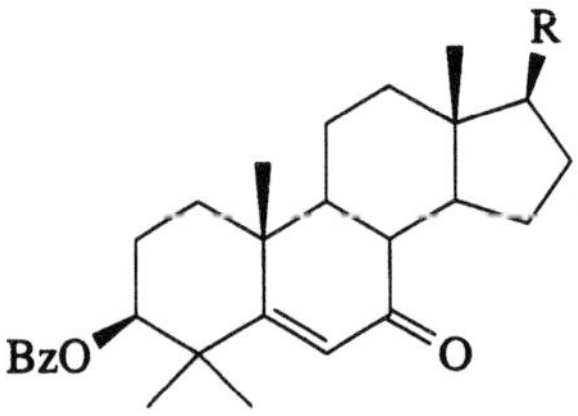 R, BzO, O	60 equiv. PCC, DCM, 3Å sieves, reflux, 48 h	84	111
8	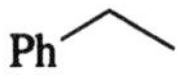 Ph	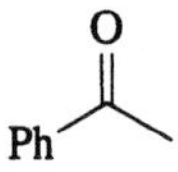 O, Ph	5 equiv. PCC, benzene, celite, r.t., 15 h	71	112
9	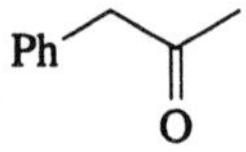 Ph, O	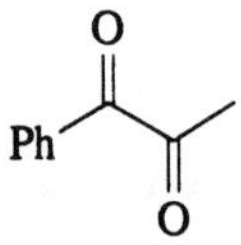 O, Ph, O	3 equiv. PCC, 3 equiv. Py, DCM, reflux, 25 h	81	113
10	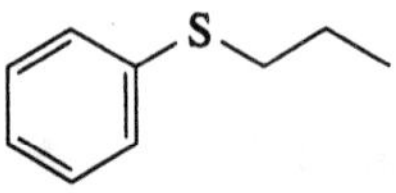 S	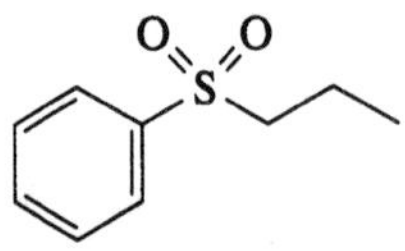 O, O, S	3 equiv. PCC, DCM, r.t., 3.5 h	60	114
11	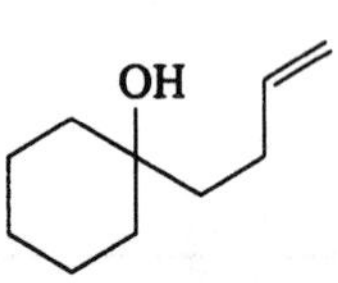 OH	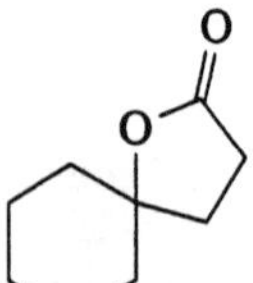 O, O	5 equiv. PCC, celite, DCM, reflux, 48 h	53	115

many types of oxidations is now commonplace.) Comparative studies indicated that 3 Å sieves gave the best results [3 Å > 4 Å > 10 Å > 5 Å]. Celite, alumina and silica were found to have little or no effect upon the rate of oxidation of carbohydrates.

Mechanistic studies imply that there are specific sites in these zeolites that promote hydride transfer.

PCC is an excellent oxidant for the transformation of alcohol to carbonyl, but it is also a good general oxidant. Therefore it should be noted that with unreactive substrates that may require forcing conditions, there are a number of other possible oxidative (and degradative) pathways available: these are outlined below.

Furans appear readily to undergo a variety of oxidation reactions with PCC (Table 7).[99–104]

Simple mono- or di-substituted enol ethers may be oxidized to esters and lactones (Table 8, entries 1 and 2).[105] However, fully substituted enol ethers undergo oxidative cleavage, since the proposed mechanism for oxidation to an ester involves a hydride shift which can no longer take place.[106]

Phenyloxiranes are cleaved by PCC. The phenyl group appears to be essential for carbon–carbon bond scission (Table 8, entry 3).[107]

In general, PCC is inert towards isolated carbon–carbon multiple bonds, but it is possible to cleave aryl-substituted double bonds.[108] Once again, the aromatic group is necessary to impart sufficient reactivity for cleavage to occur (Table 8, entry 4). An exception has been reported by Chakraborty and Chandrasekaran.[115] It was found that γ- and δ-tertiary hydroxyalkenes give good yields of the corresponding spirolactones (Table 8, entry 11) upon treatment with $(bipyH_2)CrOCl_5$ and with PCC. Double-bond cleavage is well known with chromium(V) oxidants, but the use of PCC for this transformation is much less common.

It has been observed that PCC will rapidily dimerize aromatic, but not aliphatic, thiols to their corresponding disulfides in good yield (Table 8, entry 5).[109] Sulfides may undergo oxidation to give sulfones (Table 8, entry 10).[114]

Several chromium oxidants, including PCC, will oxidize activated methylene groups to carbonyl compounds, but much stronger conditions are usually required than for alcohol oxidation.

5,6-Dihydropyrans need only a moderate excess of PCC to be converted into unsaturated lactones (Table 8, entry 6).[110] However, the oxidation of normal allylic and benzylic C—H groups requires a large excess of PCC (Table 8, entries 7 to 9).[111] The amount of PCC needed is lower if benzene[112,116] or DMSO[116] are used as solvents for oxidation, but the reactions still need to be heated to obtain reasonable conversion.

2.7.3.4 Other Chromates

A number of different chromates have been developed for the oxidation of alcohols. Most changes have been made to create oxidants that are milder and more selective in comparison to PCC.

2.7.3.4.1 Pyridinium fluorochromate (PFC)

Pyridinium fluorochromate is a stable solid, which can be stored for long periods.[117] It is as reactive as PCC, but slightly less acidic (pH of a 0.01 M solution = 2.45 compared to 1.75 for PCC), and thus substrates with acid labile groups can be oxidized without the need to add a buffer.[118] With 1.5 equiv. of oxidant in dichloromethane at room temperature, primary and secondary alcohols are oxidized to aldehydes and ketones in high yield (Table 9). Unfortunately, (*E*)/(*Z*)-isomerization has been observed during the oxidation of allylic alcohols.[118]

PFC is also a reasonable oxidant for activated C—H bonds. Allylic oxidation to a variety of ketonic products was observed upon treatment of Δ^3-carene with PFC.[119] Benzylic C—H oxidation with PFC is also known.[117,120]

2.7.3.4.2 Bipyridinium chlorochromate (BPCC)

Bipyridinium chlorochromate (**4**) is a mild, air stable nonhygroscopic oxidant.[61] It is weaker than PCC,[121] and thus 2–4 equiv. are required to obtain good yields of carbonyl compounds. The bipyridyl system acts as an internal buffer, permitting the ready oxidation of alcohols in substrates with acid labile groups. Bipyridinium chlorochromate will also oxidize sulfides to sulfoxides and sulfones.[114]

Table 9 Oxidation of Alcohols with PFC

Entry	Substrate	Product	Conditions	Yield (%)	Ref.
1	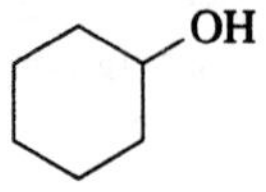 OH	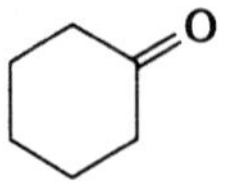 O	1.5 equiv., DCM, 3.5 h, r.t.	89	117
2	n-$C_7H_{15}OH$	n-$C_6H_{13}CHO$	1.5 equiv., DCM, 1 h, r.t.	84	117
3	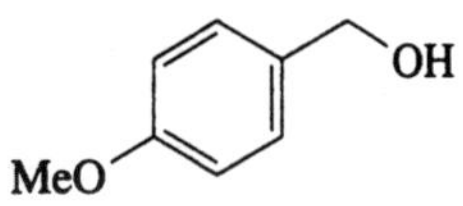 MeO, OH	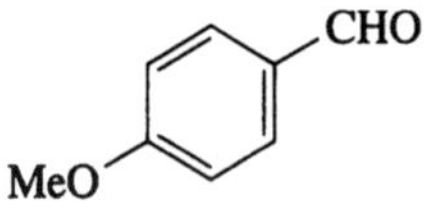 MeO, CHO	1.5 equiv., DCM, 1 h, r.t.	90	117
4	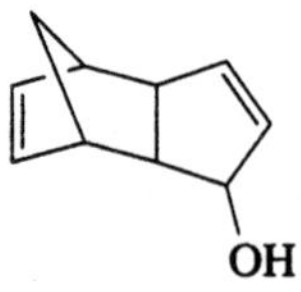 OH	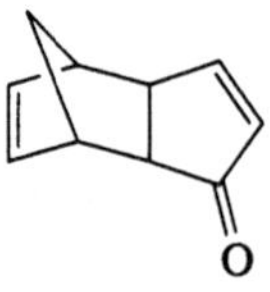 O	1.5 equiv., DCM, 1.5 h, r.t.	92	117
5	 HO, OTHP	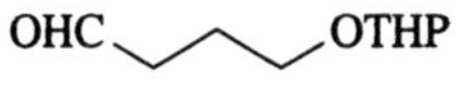 OHC, OTHP	2 equiv., DCM, 4 h, r.t.	78	118
6	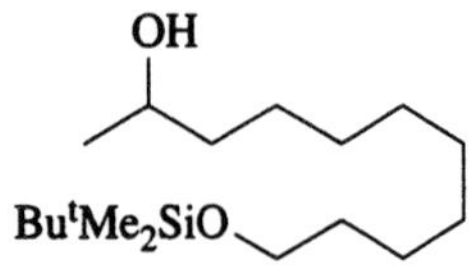 OH, Bu^tMe_2SiO	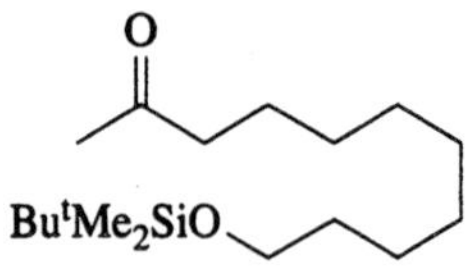 O, Bu^tMe_2SiO	2 equiv., DCM, 3 h, r.t.	89	118
7	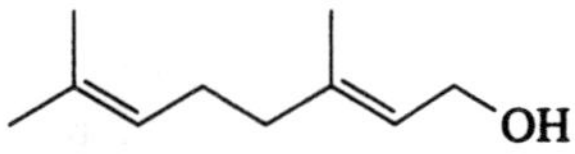 OH	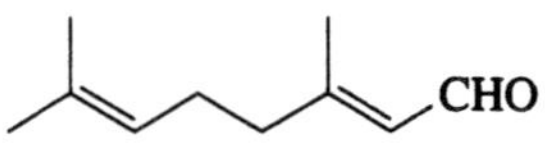 CHO	2 equiv., DCM, 2 h, r.t.	80 85:15 *(E)*:*(Z)*	118
8	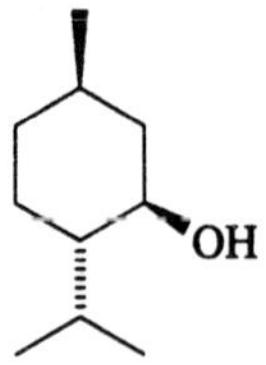 OH	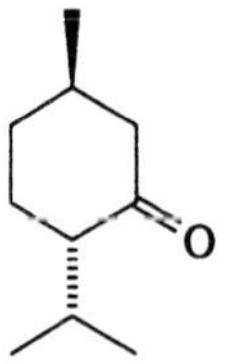 O	2 equiv., DCM, 2 h, r.t.	97	118
9	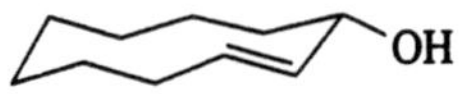 OH	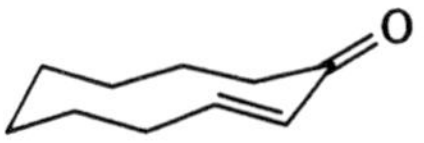 O	2 equiv., DCM, 4 h, r.t.	87	118

N, N, H, +, $ClCrO_3^-$

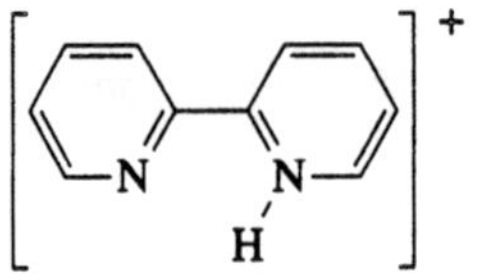

(4)

If necessary, bipyridinium chlorochromate may also be buffered with sodium acetate. In a system especially prone to epimerization (equation 10), bipyridinium chlorochromate/sodium acetate gave 90% of the ketone (**5**) with little isomerization,[122] whereas use of unbuffered PCC on a closely related substrate gave significant epimerization.[123]

Alumina (activity III) may also be used to create an even milder oxidant. Diol (**6**; equation 11) underwent cleavage to the aldehyde when treated with bipyridinium chlorochromate alone,[124] but bipyridinium chlorochromate/alumina (1:1) gave a good yield of the α-hydroxy ketone.[125]

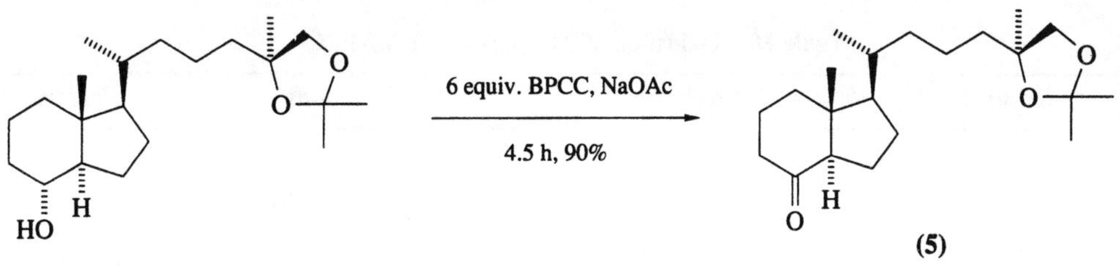

6 equiv. BPCC, NaOAc

4.5 h, 90%

(5)

(10)

BPCC

BPCC/Al_2O_3 (1:1)

(6)

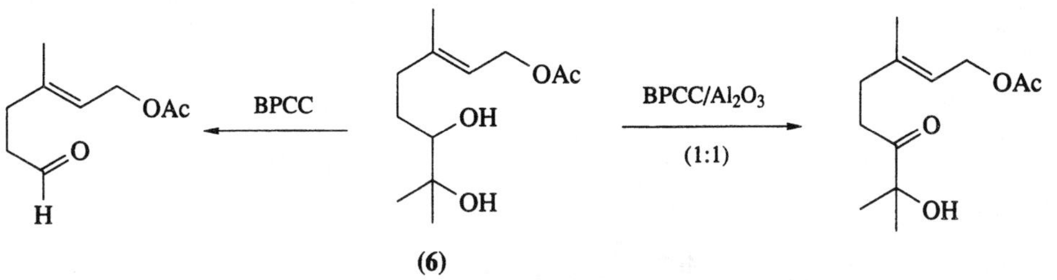

(11)

Other bipyridyl-related systems (**7** and **8**) have been found to be less effective for the oxidation of alcohols.[126] 2,2′:6′,2″-Terpyridinium hydrochloride chlorochromate (**9**) has also been prepared and found to be inert under standard conditions. It has been suggested that the steric bulk prevents effective electron transfer.[121]

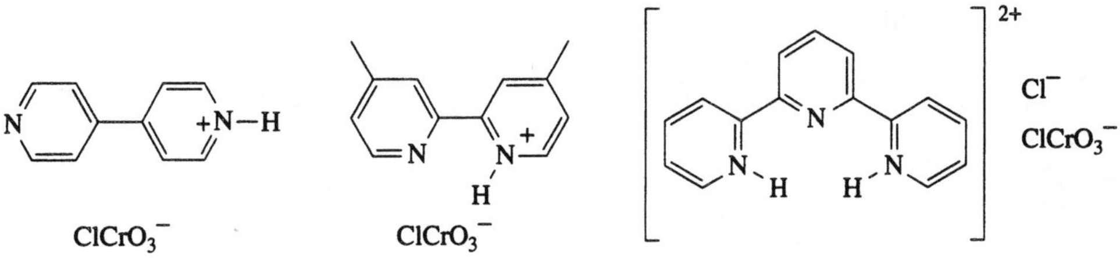

$ClCrO_3^-$ **(7)**

$ClCrO_3^-$ **(8)**

Cl^- $ClCrO_3^-$ **(9)**

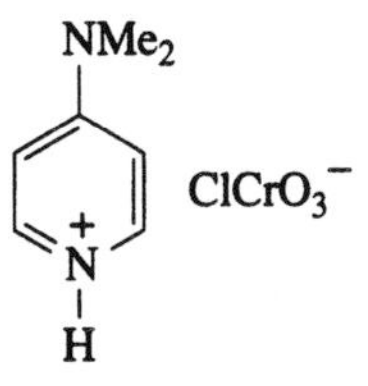

$ClCrO_3^-$ **(10)**

2.7.3.4.3 4-(Dimethylamino)pyridinium chlorochromate

4-(Dimethylamino)pyridinium chlorochromate (**10**) is a commercially available, stable and nonhygroscopic solid, but it is light sensitive.[127] It is selective for allylic and benzylic alcohols. *n*-Heptanol gives only 5% aldehyde after 3 h with 4 equiv. The selectivity is greater over primary than secondary saturated alcohols. Between 4 and 6 equiv. of oxidant are normally required for good conversion, but this is often less than the amount of manganese(IV) oxide required for the same transformation, and reaction times are usually shorter. Unfortunately, (*Z*)-unsaturated aldehydes will isomerize to the (*E*)-isomers under the reaction conditions due to the acidic nature of this reagent (Table 10).

It should be noted that treatment of di-*n*-propylsulfide with 4-(dimethylamino)pyridinium chlorochromate (1 equiv.) in dichloromethane for 20 h gave only 5% oxidation, suggesting that this may be a useful oxidant for sulfur-containing compounds.[114]

Table 10 Oxidation of Alcohols with DMAPCC

Entry	*Substrate*	*Product*	*Conditions*	*Yield (%)*	*Ref.*
1	2-thienylmethanol (OH)	thiophene-2-CHO	3 equiv., DCM, 4 h, r.t.	73	127
2	4-O_2N-benzyl alcohol (OH)	4-O_2N-benzaldehyde (CHO)	6 equiv., DCM, 20 h, r.t.	43	127
3	2-methylbenzyl alcohol (OH)	2-methylbenzaldehyde (CHO)	4 equiv., DCM, 15 h, r.t.	78	127
4	3,4-(MeO)₂-benzyl alcohol (MeO, MeO, OH)	3,4-(MeO)₂-benzaldehyde (MeO, MeO, CHO)	4 equiv., DCM, 14 h, r.t.	91	127
5	4-Cl-benzyl alcohol (Cl, OH)	4-Cl-benzaldehyde (Cl, CHO)	5 equiv., DCM, 21 h, r.t.	91	127
6	cinnamyl alcohol (OH)	cinnamaldehyde (CHO)	6 equiv., DCM, 15 h, r.t.	62	127
7	cyclohex-2-en-1-ol (OH)	cyclohex-2-en-1-one (O)	6 equiv., DCM, 7 h, r.t.	74	127
8	THPO–CH_2CH=CHCH₂–OH (*Z*)	THPO–CH_2CH=CHCHO (*E*)	5 equiv., DCM, 24 h, r.t.	55	127
9	2-pyridylmethanol (N, OH)	pyridine-2-CHO (N, CHO)	4 equiv., DCM, 6 h, r.t.[a]	0	127
10	4-(2-hydroxyethyl)benzyl alcohol (OH, OH)	4-(2-hydroxyethyl)benzaldehyde (CHO, OH)	4 equiv., DCM, 2 h, r.t.	62	127

[a] Note no reaction with this pyridine derivative.

2.7.3.4.4 *1,8-Naphthyridinium chlorochromate (NapCC)*

NapCC[128] is much milder than PCC, and hence is less likely to cause overoxidation. It shows moderate selectivity for allylic and benzylic alcohols (Table 11). It has low solubility in solvents such as dichloromethane and diethyl ether, but is soluble in solvents with higher dielectric constants, such as DMSO or DMF.

Table 11 Oxidation of Alcohols with Naphthyridinium Chlorochromate[128]

Entry	*Substrate*	*Product*	T_{50} (h)[a]
1	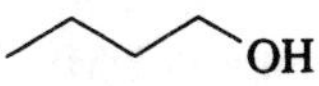 OH	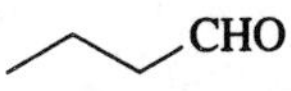 CHO	5.7
2	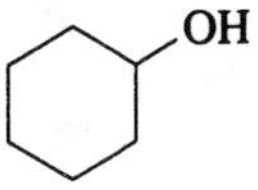 OH	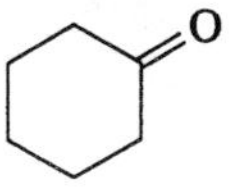 O	9.8
3	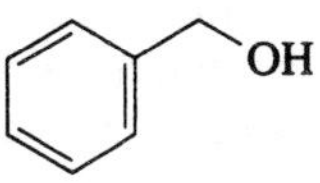 OH	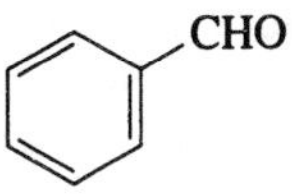 CHO	1.8
4	 OH	 CHO	2.9

[a] T_{50} = time taken for reaction to reach 50% completion.

2.7.3.4.5 *Quinolinium chlorochromate*

In a brief communication, it was reported that quinolinium chlorochromate is a selective oxidant for primary alcohols;[129] for example, this reagent is reported to oxidize the diol (**11**) selectively to the aldehyde (**12**; equation 12).

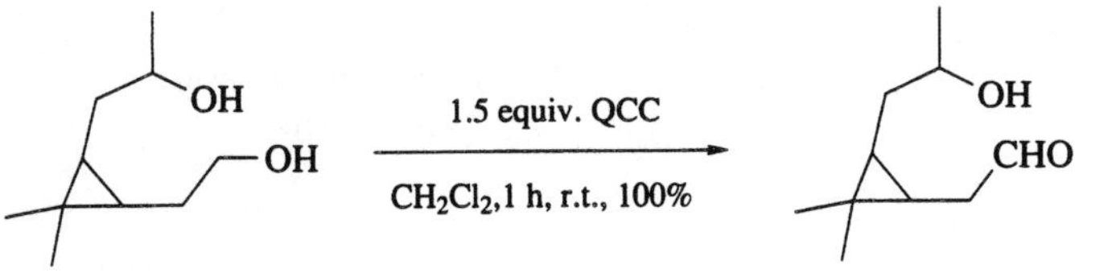

OH OH → 1.5 equiv. QCC, CH_2Cl_2,1 h, r.t., 100% → OH CHO (12)

(**11**) (**12**)

Secondary alcohols can be oxidized but require prolonged reaction times. The reagent is also reported to show sensitivity to substitution β to primary alcohols.

2.7.3.4.6 *Pyrazinium chlorochromate (PzCC)*

PzCC (**13**) is a much milder oxidant than PCC,[128,130] but is only moderately soluble in dichloromethane, carbon tetrachloride and diethyl ether. It is soluble in water and acetonitrile. In pyridine, exchange occurs freely to give PCC and pyrazine. Aldehydes and ketones are easily prepared, with no sign of overoxidation. However, PzCC displays no marked selectivity for any class of alcohols, so it is difficult to predict specific instances where it would be better to use this oxidant in place of the others available.

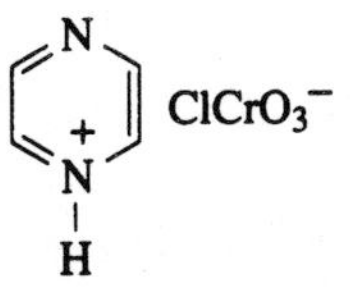

N, +N–H $ClCrO_3^-$ (**13**)

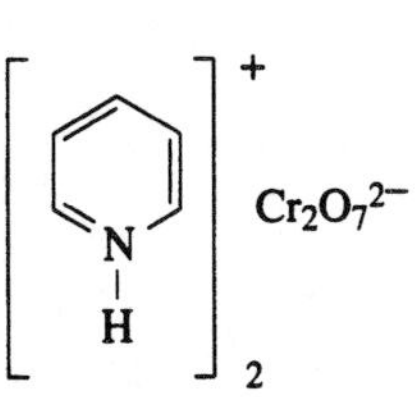

[N–H]$_2^+$ $Cr_2O_7^{2-}$ (**14**)

Preparation of pyrazinium *N*-oxide chlorochromate gives a reagent which is as reactive as PCC, but much less stable.[121] Again there is no apparent selectivity for a particular type of substrate.

2.7.3.5 Pyridinium Dichromate (PDC)

Pyridinium dichromate (**14**) is an isolable, stable orange solid that can be simply and safely prepared.[131] PDC had been used previously,[132,133] but it was Corey and coworkers who demonstrated the wide applicability of this mild and selective oxidant in organic synthesis. PDC is very soluble in solvents such as DMF, water and DMSO, but sparingly soluble in chlorinated hydrocarbons and acetone. It is normally used either as a solution in DMF or as a suspension in dichloromethane (Table 12).

Primary and secondary allylic alcohols and saturated secondary alcohols are oxidized to the corresponding carbonyl compounds quickly and in high yield at room temperature in DMF. There is no appreciable overoxidation of allylic alcohols in DMF, but primary saturated alcohols are readily oxidized to their corresponding acids. Recently, it has been reported that aldehydes may be converted to methyl esters by oxidation with PDC in the presence of methanol.[141] Preparation of other esters, or methyl esters direct from the alcohol, proved to be less efficient.

By using PDC as a suspension in dichloromethane it becomes a selective oxidant for the preparation of aldehydes, saturated or unsaturated. Allylic alcohols are oxidized faster than saturated alcohols, but some (*E*)/(*Z*)-isomerization has been observed during the preparation of α,β-unsaturated aldehydes with PDC in dichloromethane.

Initially, Corey and Schmidt found that by addition of pyridinium trifluoroacetate (0.4 equiv.) to their reactions,[131] there was an increase in rate and the amount of PDC needed for complete oxidation diminished. Subsequently, several other techniques have been devised to improve the rate and efficacy of PDC oxidations (most frequently in the field of carbohydrate research).

As mentioned for PCC,[98] the addition of molecular sieves causes a dramatic increase in the rate of oxidation of carbohydrates. Once again, molecular sieves have been used to improve the oxidations of a wide variety of substrates with PDC.

Addition of small quantities of anhydrous acetic acid and freshly activated sieves[142] to oxidations of carbohydrates has also been found to increase the rate of oxidation. In comparison to the addition of pyridinium trifluoroacetate, reaction times were reduced from days to minutes (Scheme 1). The acetic acid and sieves appear to have a synergistic effect, since both are required to give the dramatic rate enhancement.

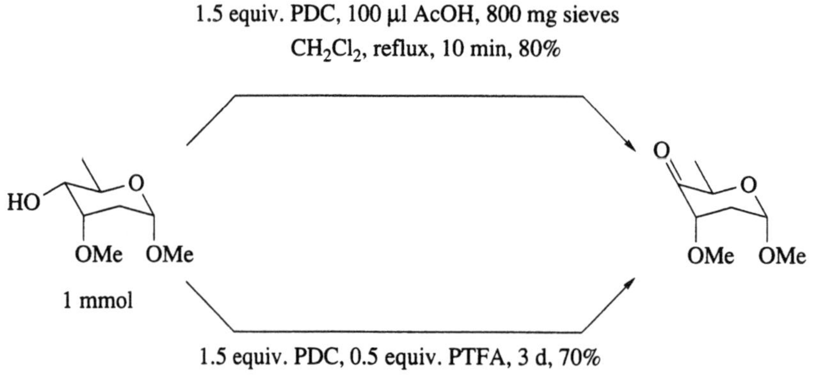

1.5 equiv. PDC, 100 μl AcOH, 800 mg sieves
CH_2Cl_2, reflux, 10 min, 80%

HO O OMe OMe
1 mmol

O O OMe OMe

1.5 equiv. PDC, 0.5 equiv. PTFA, 3 d, 70%

Scheme 1

A combination of acetic acid and PDC has also been used to effect selective allylic oxidations on unprotected, unsaturated carbohydrates.[143] Interestingly, ethyl acetate was found to be the solvent of choice, and the reaction proceeded better if molecular sieves were omitted, since they appeared to cause unselective oxidation in this case (equation 13).

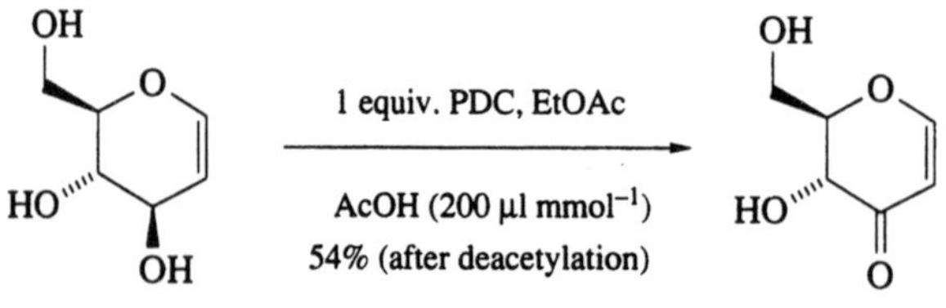

OH O HO OH

1 equiv. PDC, EtOAc
AcOH (200 μl $mmol^{-1}$)
54% (after deacetylation)

OH O HO O

(13)

PDC with acetic anhydride gives a strong but mild, neutral oxidant.[144] It is normally used in dichloromethane at room temperature. Under these conditions secondary alcohols are oxidized to ketones in high yield with 0.6 equiv. of PDC and 3 equiv. of acetic anhydride. With only slightly more PDC (0.7 equiv.) primary alcohols are rapidly and selectively oxidized to aldehydes. (The extra oxidant serves to increase

Table 12 Oxidation of Alcohols with PDC

Entry	*Substrate*	*Product*	*Conditions*	*Yield (%)*	*Ref.*
1	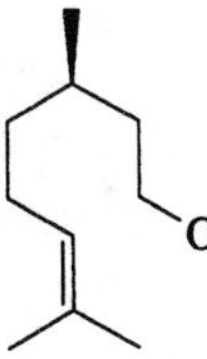 OH	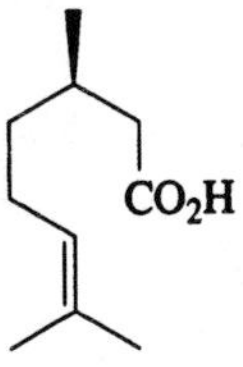 CO_2H	3.5 equiv. PDC, DMF, 7–9 h, r.t.	83	131
2	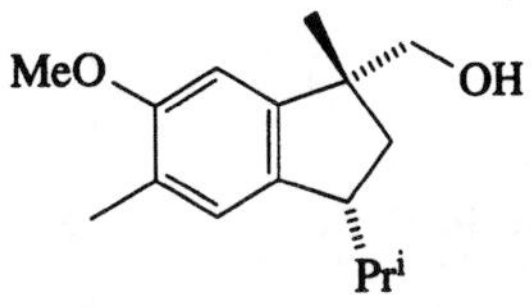 MeO, OH, Pr^i	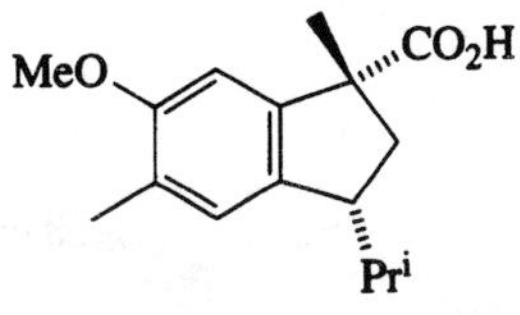 MeO, CO_2H, Pr^i	3.5 equiv. PDC, DMF, 7–9 h, r.t.	85	131
3	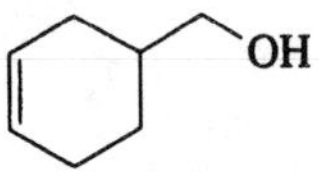 OH	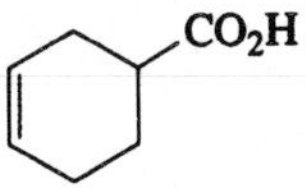 CO_2H	3.5 equiv. PDC, DMF, 7–9 h, r.t.	92	131
4	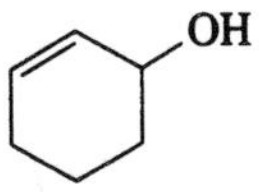 OH	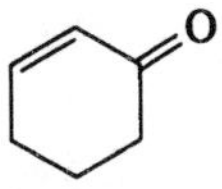 O	1.25 equiv. PDC, DMF, 4–5 h, 0 °C	86	131
5	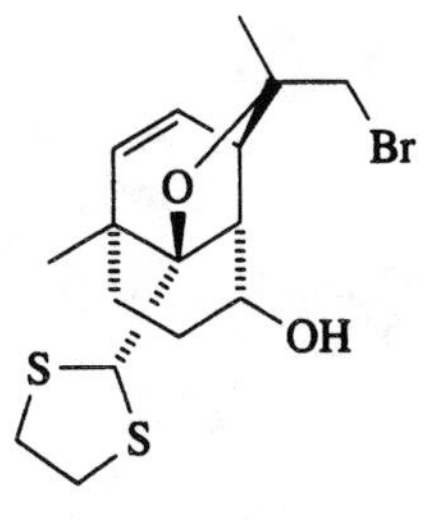 Br, O, OH, S, S	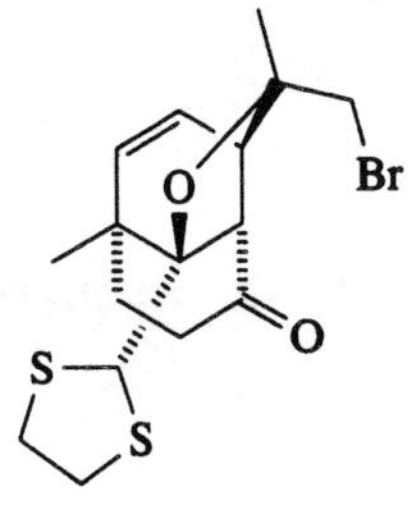 Br, O, O, S, S	7 equiv. PDC, DMF, 6 h, 0 °C	95	131
6	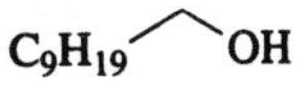 C_9H_{19} OH	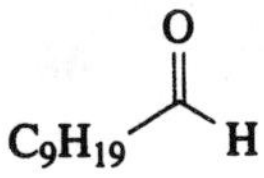 O, C_9H_{19} H	1.5 equiv. PDC, DCM, r.t., 20 h	98	131
7	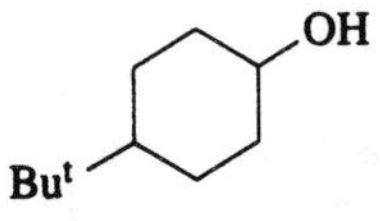 OH, Bu^t	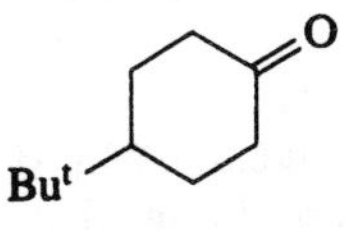 O, Bu^t	1.5 equiv. PDC, 0.4 equiv. PTFA, DCM, r.t., 3 h	97	131
8	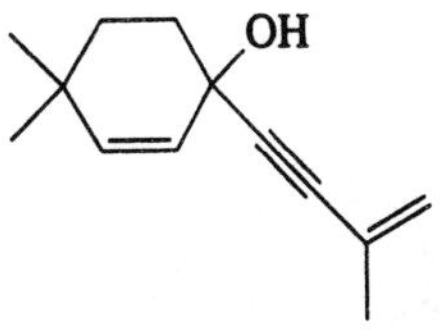 OH	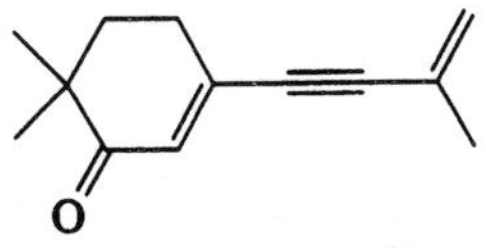 O	1.5 equiv. PDC, DCM, r.t., 16 h	88	134
9	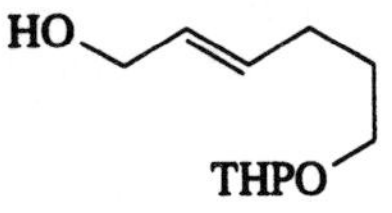 HO, THPO	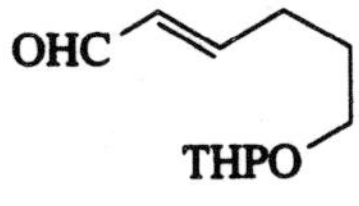 OHC, THPO	1.5 equiv. PDC, DMF, 2 h, 0 °C	85	135

Table 12 *(continued)*

Entry	*Substrate*	*Product*	*Conditions*	*Yield (%)*	*Ref.*
10	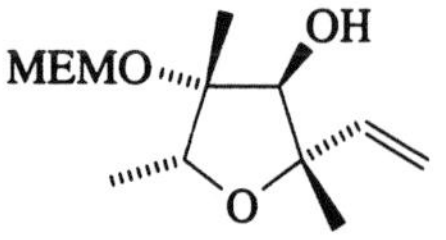 MEMO, OH, O	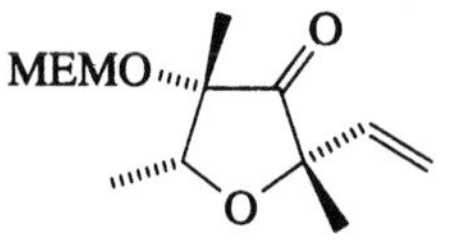 MEMO, O, O	5 equiv. PDC, DCM, r.t., 16 h	85	136
11	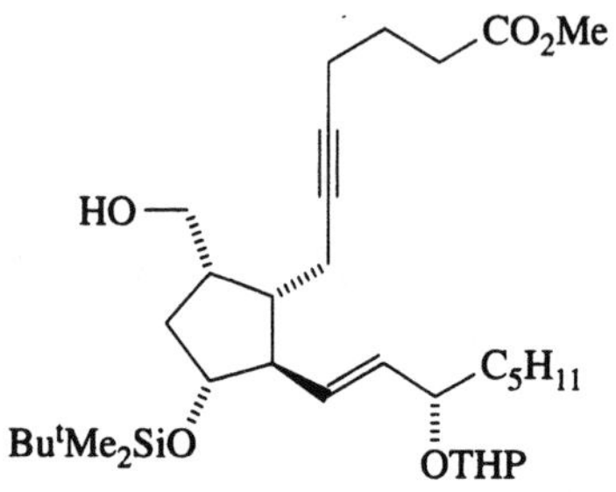 CO_2Me, HO, C_5H_{11}, Bu^tMe_2SiO, OTHP	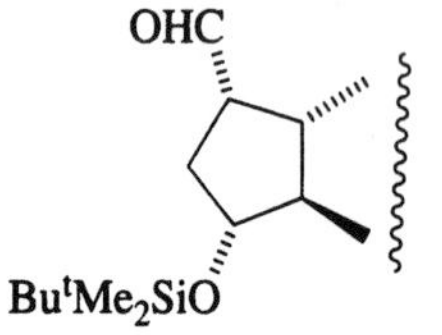 OHC, Bu^tMe_2SiO	20 equiv. PDC, DCM, r.t., 11 h	85	137
12	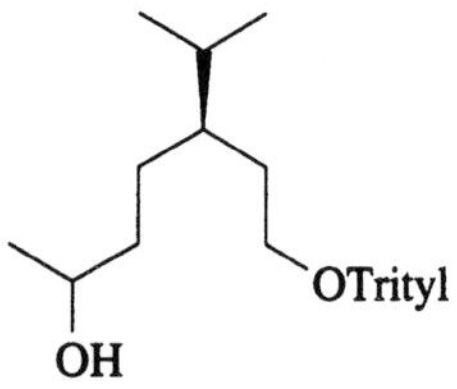 OTrityl, OH	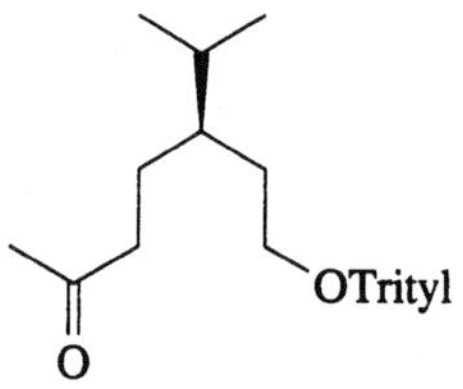 OTrityl, O	3 equiv. PDC, DCM, r.t., 48 h	97	138
13	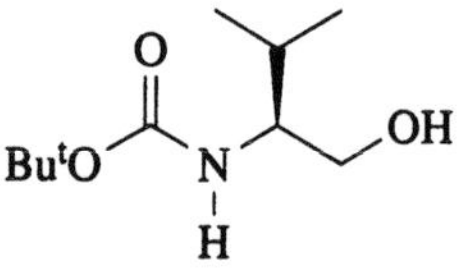 O, Bu^tO, N, H, OH	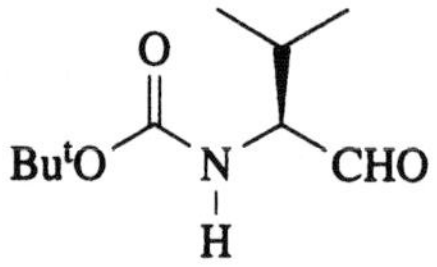 O, Bu^tO, N, H, CHO	1.5 equiv. PDC, DCM, r.t., 16 h	95	139
14	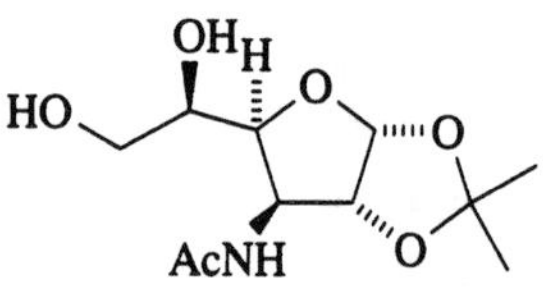 OH, H, HO, O, O, AcNH, O	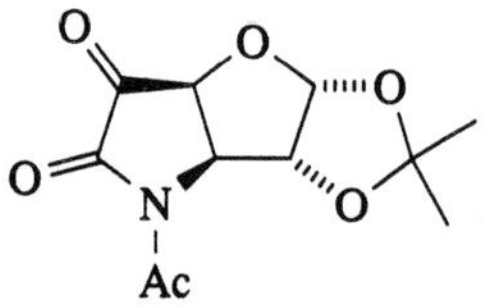 O, O, O, O, N, Ac, O	10 equiv. PDC, DMF, r.t., 24 h	68	140

the rate and thus reduce side reactions.) Overoxidation can be a problem, but is conveniently avoided by the inclusion of *ca.* 20% DMF as cosolvent (equations 14 and 15).[144] In the absence of DMF, and with excess oxidant, this method may also be used to prepare acids in good yields.

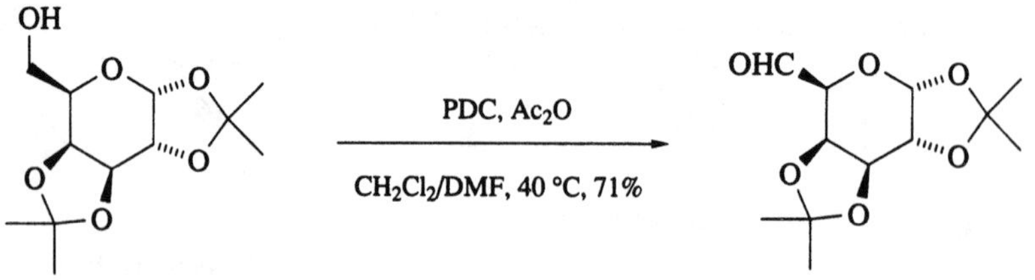

$$\xrightarrow[\text{CH}_2\text{Cl}_2\text{/DMF, 40 °C, 71\%}]{\text{PDC, Ac}_2\text{O}} \qquad (14)$$

$$\xrightarrow[\text{CH}_2\text{Cl}_2\text{/DMF, 40 °C, 93\%}]{\text{PDC, Ac}_2\text{O}} \qquad (15)$$

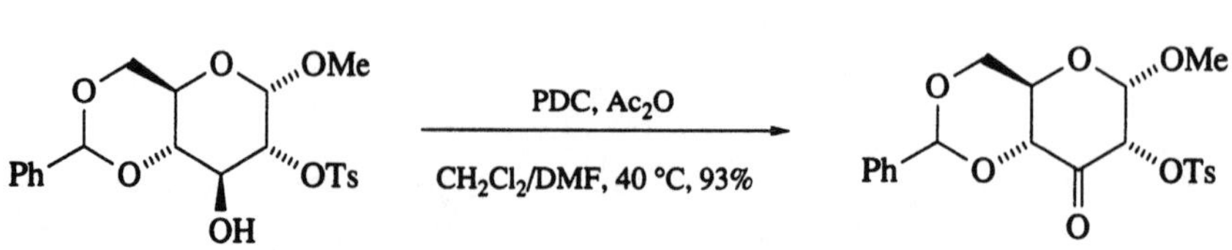

PDC with trimethylsilyl chloride[145] is not only a rapid oxidizing agent for alcohols, but will also effect a deprotection–oxidation sequence for silyl ethers. Both trimethylsilyl and *t*-butyldimethylsilyl ethers, which are normally stable to PDC, can be transformed directly into the corresponding carbonyl compounds in good yield (Table 13).

Table 13 Oxidation of Alcohols and Silyl Ethers with PDC/Me_3SiCl

Entry	*Substrate*	*Product*	*Conditions*	*Yield (%)*	*Ref.*
1	O_2N, OH	O_2N, CHO	1.5 equiv., DCM, 2 h, r.t.	90	145
2	Ph, $OSiMe_2Bu^t$	Ph, CHO	1.5 equiv., DCM, 1 h, r.t.[a]	95	145
3	$OSiMe_2Bu^t$	O	1.5 equiv., DCM, 1 h, r.t.	95	145
4	PhO, Ph, OH, N, O, Ph	PhO, Ph, O, N, O, Ph	1.5 equiv., DCM, 25 min, r.t.	88	145
5	PhO, $OSiMe_2Bu^t$, N, O, Ph	PhO, O, N, O, Ph	1.5 equiv., DCM, 1 h, r.t.	85	145
6	$OSiMe_2Bu^t$, Cl, $OSiMe_2Bu^t$	O, Cl, O	3 equiv., DCM, 20 min, r.t.	70	145
7	$OSiMe_2Bu^t$, MeO, $OSiMe_2Bu^t$	O, MeO, O	3 equiv., DCM, 15 min, r.t.[a]	74	145

[a] Oxidation with preformed reagent.

This deprotection–oxidation can also be applied to the preparation of quinones from trialkylsilyl-protected hydroquinones. This method has wider applicability than that reported employing PCC,[89] since substrates with electron-releasing *and* those with electron-withdrawing groups are oxidized. Use of the reagent prepared *in situ* appears to be preferable to the preformed reagent.

It is possible to use PDC as a catalytic oxidant (10 mol %), with bis(trimethylsilyl) peroxide as the cooxidant, for the preparation of carbonyl compounds.[146,147] It is necessary to add the cooxidant slowly *via* syringe pump since the actual oxidizing agent (**15**) is unstable in solution. A range of primary and secondary alcohols were oxidized in good yields by this method (Table 14). Fortunately, isolated double bonds are inert under these conditions.

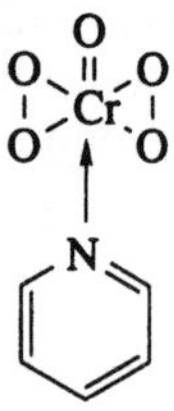

(15)

Table 14 Oxidation of Alcohols and Silyl Ethers with PDC and $(Me_3SiO)_2$

Entry	Substrate	Product	Conditions	Yield (%)	Ref.
1	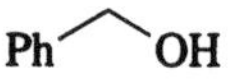 Ph OH	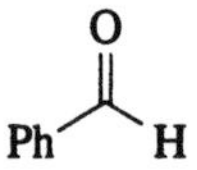 Ph H (O)	3 equiv. $(Me_3SiO)_2$, 10 mol % PDC	91	147
2	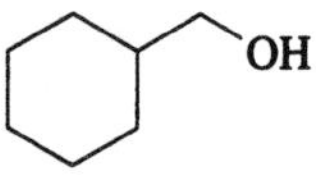 OH	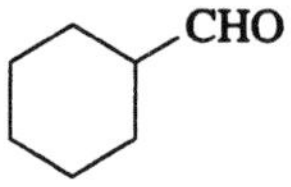 CHO	3 equiv. $(Me_3SiO)_2$, 10 mol % PDC	71	147
3	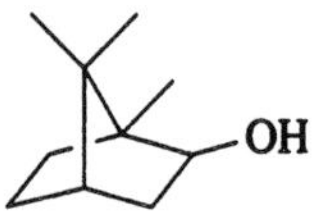 OH	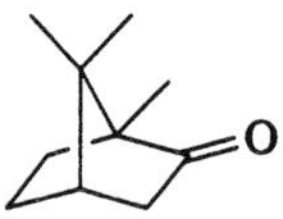 O	3 equiv. $(Me_3SiO)_2$, 10 mol % PDC	98	147
4	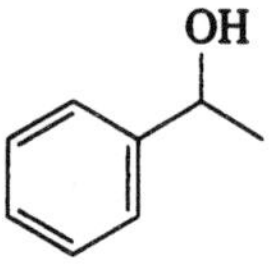 OH	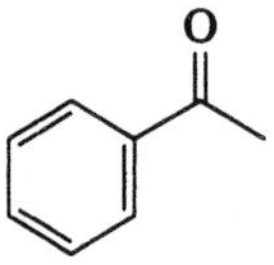 O	3 equiv. $(Me_3SiO)_2$, 10 mol % PDC	97	147
5	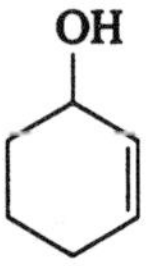 OH	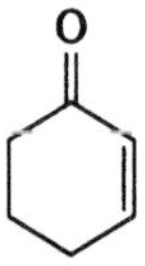 O	3 equiv. $(Me_3SiO)_2$, 10 mol % PDC	90	147

Under similar conditions chromium(VI) oxide and PCC were found to be less effective. Other cooxidants were also examined: hydrogen peroxide gives unacceptable amounts of overoxidation of aldehydes, whilst *t*-butyl hydroperoxide, di-*t*-butyl peroxide and benzoyl peroxide all failed as cooxidants.

For substrates that will withstand it, Czernecki and coworkers have developed a work-up procedure for PDC reactions that should be applicable to other reagents.[143,148] Oxalate is known to be a very good ligand for chromium(III). Thus treatment of a reaction mixture with an aqueous solution of oxalic acid dihydrate and ammonium oxalate monohydrate readily removes residual chromium(III) salts. This procedure is particularly useful for large scale reactions.

Pyridinium dichromate will also undergo several other oxidation reactions, though not as many as PCC.

Aromatic oximes, for example, can be converted back to carbonyl compounds with only 2 equiv. of PDC in 1 h at room temperature.[149] PDC will also cleave enol ethers.[106,150]

Allylic C—H oxidation is known to occur if 1,4-dienes are treated with PDC in boiling chloroform[15] (equation 16).[151] In the presence of *t*-butyl hydroperoxide, PDC becomes an effective allylic and

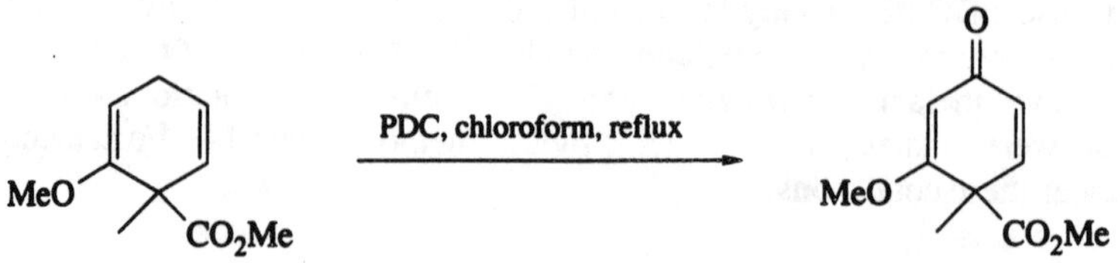

MeO CO$_2$Me —PDC, chloroform, reflux→ O MeO CO$_2$Me (16)

benzylic oxidant.[152] In steroidal systems reasonable amounts of allylic oxidation have been observed simply with excess PDC in benzene, DMSO or pyridine.[116]

2.7.3.6 Other Dichromates

2.7.3.6.1 3-Carboxypyridinium dichromate (nicotinium dichromate)

Nicotinium dichromate (**16**) is a stable, nonhygroscopic, nonphotosensitive, mild oxidant.[153,154] Nicotinium dichromate alone in dichloromethane gives moderate selectivity for benzylic and allylic over saturated alcohols. This selectivity is further increased by the use of benzene as solvent.

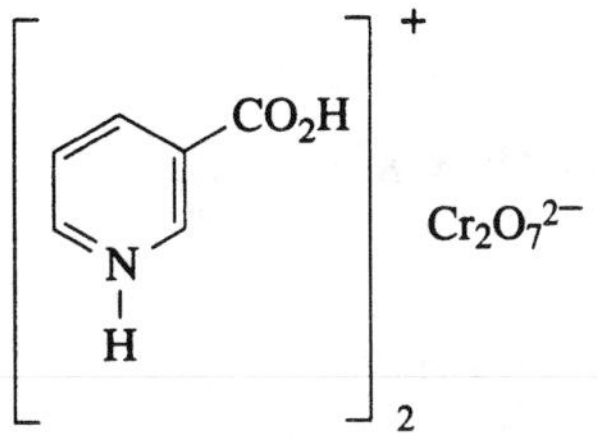

$$\left[\text{3-}CO_2H\text{-}C_5H_4N\text{-}H\right]_2^+ \quad Cr_2O_7^{2-}$$

(16)

It was noted earlier that electron-withdrawing groups on aromatic rings tend to retard benzylic oxidation. However, addition of pyridine was found to improve these sluggish reactions. Indeed, pyridine appears to give good improvements in rates for the oxidation of many substrates. Use of 1:2.5:20 alcohol:nicotinium dichromate:pyridine generally gives quick oxidations, but there is an accompanying loss of selectivity with these faster reactions. Little (*E*)/(*Z*)-isomerization has been detected, but carbon–carbon bond cleavage is observed, especially in situations where there are electron-withdrawing groups β to the hydroxy group. This implies that the excess pyridine may promote formation of the enol, which may then be attacked by the remaining oxidant.

Nicotinium dichromate has been reported to be particularly useful for large scale carbohydrate oxidations where other modified chromium oxidants have failed.[155]

In the presence of pyridine, nicotinium dichromate is also a sufficiently strong oxidant for the preparation of quinones from hydroquinones.

Arenethiols may be dimerized to disulfides by nicotinium dichromate but aliphatic thiols are virtually inert.

4-Carboxypyridinium dichromate (isonicotinium dichromate) has also been prepared by Palomo and coworkers. It appears to have very similar properties to nicotinium dichromate,[154] and offers no further advantage for the oxidation of alcohols.

2.7.3.6.2 Quinolinium dichromate

Quinolinium dichromate (**17**)[156] shows the solubility profile common to most chromium based oxidants—sparingly soluble in chlorinated solvents, but more soluble in more polar solvents. It has been mainly used in dichloromethane at reflux, or in DMF at 30 °C to oxidize primary alcohols. Secondary alcohols are also oxidized reasonably well.

Quinolinium dichromate displays reverse selectivity to PDC: in dichloromethane primary alcohols can be oxidized directly to acids, but in DMF the oxidation is selective for the preparation of aldehydes.

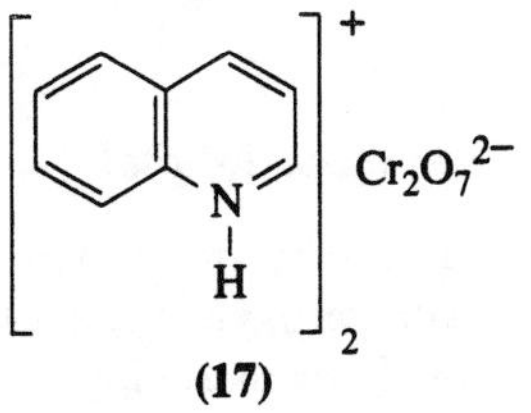

$$\left[C_9H_7N\text{-}H\right]_2^+ \quad Cr_2O_7^{2-}$$

(17)

2.7.3.6.3 Imidazolium Dichromate (IDC)

Imidazolium dichromate is a selective oxidant for allylic and benzylic hydroxy groups.[157] (Allylic alcohols are oxidized faster than benzylic alcohols.) The selectivity over saturated alcohols is similar to that of 4-(dimethylamino)pyridinium chlorochromate. DMF is recommended as the solvent for oxidations, since it appears that the choice of solvent is critical to obtaining high yields. This reagent has also been observed to cause some (*E*)/(*Z*)-isomerization during the oxidation of allylic alcohols.

2.7.3.6.4 1,8-Naphthyridinium dichromate

This mild oxidant is stable in the absence of light, and is a weaker oxidant than PCC or BPCC; it is also less acidic, and displays good selectivity for benzylic alcohols.[121]

2.7.4 OTHER CHROMIUM(VI) OXIDE OXIDANTS

2.7.4.1 Chromium(VI) Oxide in Diethyl Ether

Fleet and coworkers discovered that chromium(VI) oxide in 3:1 dichloromethane/diethylether, in the presence of celite, acts as an efficient oxidant for a range of alcohols, and is tolerant of a wide range of acid labile functionalities.[158] In dichloromethane alone chromium(VI) oxide was reported to be inert due to low solubility. This oxidant is particularly effective for the preparation of ketones, but aldehydes, especially α,β-unsaturated, are found to be prone to overoxidation. The oxidizing species is unstable, and therefore it is better to add chromium(VI) oxide to a solution of the alcohol in 3:1 dichloromethane/diethylether. (*N.b.* Diethyl ether/chlorinated solvent mixtures have been reported elsewhere to inflame spontaneously in the presence of chromium(VI).[159])

2.7.4.2 Chromium(VI) Oxide and Crown Ethers

In contrast to earlier reports, Palomo and coworkers found that chromium(VI) oxide will effect the oxidation of alcohols in one day at room temperature.[159] They also found that the addition of semicatalytic amounts (0.3 equiv.) of crown ethers (either 18-crown-6 or 12-crown-4) led to significant rate enhancements. The crown ethers are thought to generate a soluble oxidizing agent, similar to the alkyl ammonium salts used for solid–liquid phase transfer with chromium(VI) oxide in dichloromethane (*vide infra*).

Chromium(VI) oxide will also oxidize hydroquinones to quinones, and thiols to disulfides.[160]

2.7.4.3 Catalytic Chromium(VI) Oxide Oxidations

2.7.4.3.1 CrO_3 with t-butyl hydroperoxide

Chromium(VI) oxide can be used as a catalytic oxidant for alcohols with *t*-butyl hydroperoxide as the cooxidant.[161] This reagent appears to be selective for allylic and benzylic over saturated alcohols, though (*E*)/(*Z*)-isomerization has been observed during the preparation of α,β-unsaturated aldehydes. This reagent is also a good oxidant for allylic and benzylic C—H bonds; these may be competing pathways in more sophisticated substrates.[162–164]

2.7.4.3.2 CrO_3 complex with 2,4-dimethylpentane-2,4-diol

Corey *et al.*[165] reported that the complex formed by chromium(VI) oxide and 2,4-dimethylpentane-2,4-diol can be used as a catalytic oxidant with peroxyacetic acid as a cooxidant. When used stoichiometrically, secondary alcohols are oxidized quickly even at –20 °C, but the oxidation of primary alcohols is slow and large amounts of ester coupling are observed.

With the catalytic system secondary alcohols are still oxidized quickly (Scheme 2) and in excellent yield with 2 equiv. of peracetic acid at 0 °C. Primary alcohols are also reported to be oxidized to aldehydes in good yields, but details have not been given.[165]

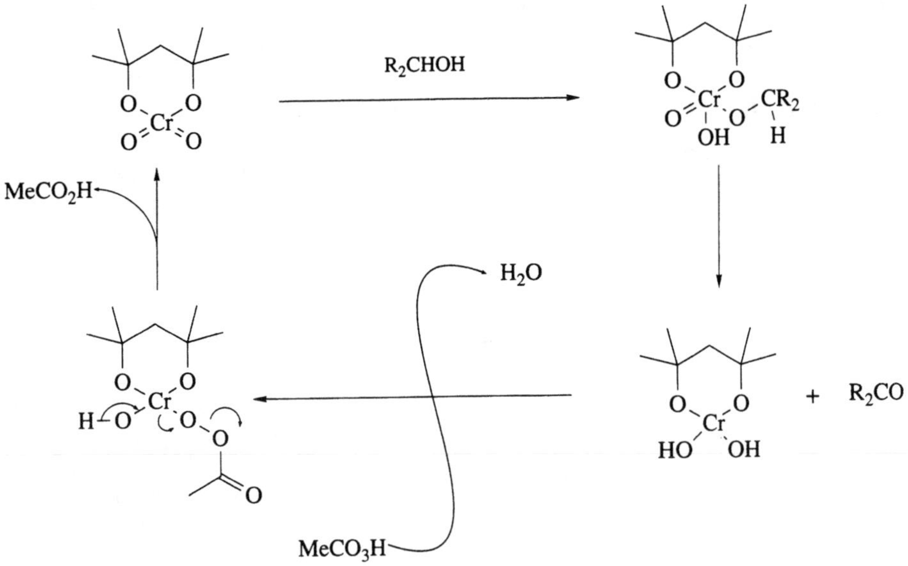

R$_2$CHOH

MeCO$_2$H

H$_2$O

+ R$_2$CO

MeCO$_3$H

Scheme 2

The use of this oxidant is restricted by the sensitivity of potential substrates to the cooxidant (peracetic acid). *t*-Butyl hydroperoxide and hydrogen peroxide were found to be ineffective for the regeneration of the complex.

2.7.5 SUPPORTED OXIDANTS

Many oxidants, especially the older and stronger chromium oxidizing agents, may have their reactivity and selectivity modified by adsorption on to inert supports.[166] Reactions utilizing supported oxidants have the advantage that the residual chromium salts remain bound to the support and thus work-up often becomes reduced to a mere filtration. Many of these systems are discussed in detail later (Volume 7, Chapter 7.3).

Supported chromium oxidants fall in to three main categories: (i) adsorbed on alumina, silica or celite (Section 2.7.5.1); (ii) adsorbed on a polymer or resin (Section 2.7.5.2); and (iii) adsorbed on carbon (Section 2.7.5.3).

2.7.5.1 On Inert Inorganic Supports

Adsorbing PCC onto alumina not only eliminates the need to buffer reactions with sodium acetate, it also enhances its reactivity.[95] In addition, work-up requires only a filtration and then concentration. The reagent is stable for several weeks when stored under vacuum and in the absence of light (Table 15).

Pyridinium chromate on silica[170] is also a good general oxidant, even in the presence of acid labile groups. In contrast to PCC on alumina this oxidant may be stored at room temperature for one year, with no loss of activity. It is convenient to use it in the form of a column when performing small scale oxidations; the substrate is allowed to stand on the column for a few hours, and then the carbonyl compound is obtained simply by eluting the column with a suitable solvent.

Chromic acid is a very strong oxidizing agent, but its reactivity may be tempered by adsorption onto a support. Chromic acid on alumina was found to be inactive,[171] but on silica it gives instantaneous oxidation, in diethyl ether at room temperature, of primary and secondary alcohols to aldehydes and ketones in good to excellent yield. This reagent can also be conveniently used in the form of a column.[168–170]

In a similar fashion, chromyl chloride, normally a very vigorous oxidant, can be used to selectively prepare aldehydes and ketones once it has been adsorbed onto silica–alumina.[172] Unfortunately, double bonds are still cleaved under the reaction conditions.

Table 15 Oxidation of Alcohols with Chromium Reagents on Inert Inorganic Supports

Entry	*Substrate*	*Product*	*Conditions*	*Yield (%)*	*Ref.*
1	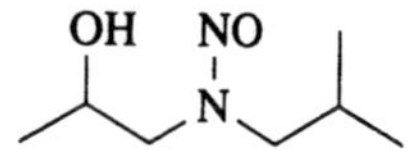 OH NO N	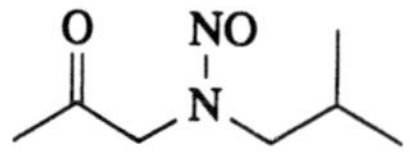 O NO N	3 equiv. CrO_3 on celite, EtOAc, r.t., 15 min	76	167
2	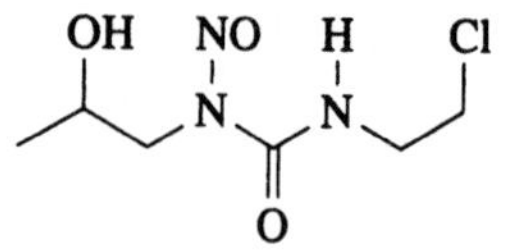 OH NO H Cl N N O	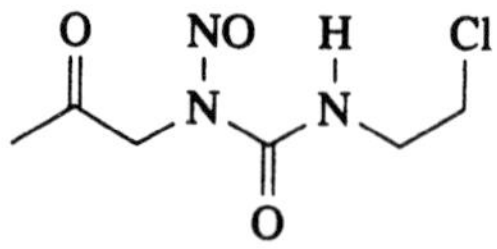 O NO H Cl N N O	3 equiv. CrO_3 on florisil, EtOAc, r.t., 10 min	79	167
3	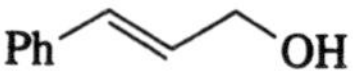 Ph OH	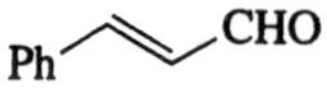 Ph CHO	H_2CrO_4, silica gel, CCl_4, r.t., 20 min	78	168
4	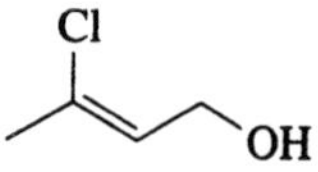 Cl OH	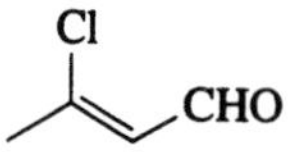 Cl CHO	H_2CrO_4, silica gel, CCl_4, r.t., 20 min	69	168
5	 OH	 CHO	H_2CrO_4, aluminum silicate, petroleum ether, 48 h	74	169
6	 Ph OH	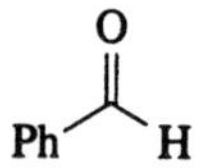 O Ph H	H_2CrO_4, aluminum silicate, petroleum ether, 48 h	90	169
7	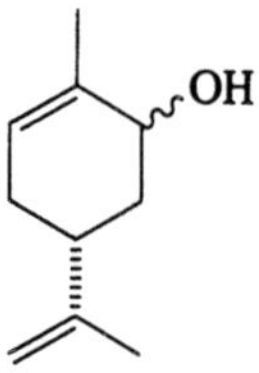 OH	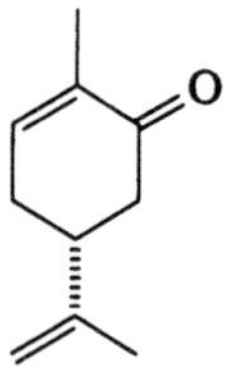 O	1.6 equiv. PCC/alumina, *n*-hexane, 2 h	93	95
8	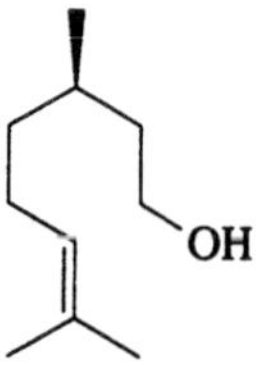 OH	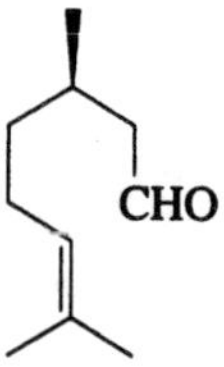 CHO	3 equiv. PCC/alumina, *n*-hexane, 4 h	82	95
9	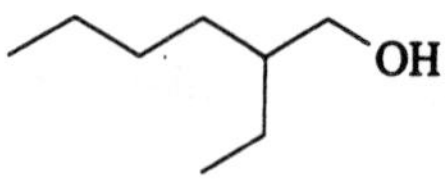 OH	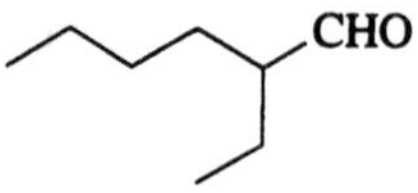 CHO	2.5 equiv. PCC/alumina, *n*-hexane, 2 h	87	95

Chromium trioxide oxidations have also consistently been found to be enhanced by the addition of Florisil or celite.[167]

2.7.5.2 On Resins/Polymers

Chromic acid also becomes a selective oxidant for the preparation of aldehydes and ketones when it is supported on an anion exchange resin (Amberlyst A-26; Table 16).[177] The reaction appears to be general and highly tolerant of a wide range of solvents, unlike many resin-based oxidations where the availability of the oxidant is critically dependent upon the nature of the solvent.

Gelbard *et al.*[173] have prepared a number of neutral and acidic supported ammonium chromates. Neutral resins were found to be generally more effective than acidic resins. The large differences in reactivity between different unbound, soluble alkylammonium chromates (Section 2.7.6.1) are not so pronounced with the polymer-supported oxidants. Some comments about the reactivity of these supported oxidants were made: (i) quaternary ammonium salts are more reactive than the tertiary pyridinium chromates; (ii) the reactivity of a complex chromate $XCrO_3^-$ was found to increase as the basicity of X^- decreases. Best

Table 16 Oxidation of Alcohols with Chromium Reagents on Polymers/Resins

Entry	*Substrate*	*Product*	*Conditions*	*Yield(%)*	*Ref.*
1	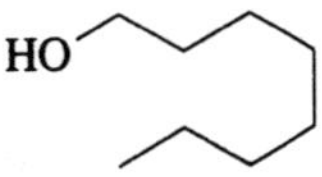 HO	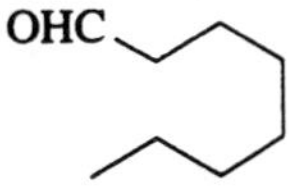 OHC	7 equiv. Cr on Amberlyst A-26, 12h[a]	80	173
2	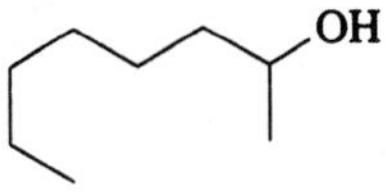 OH	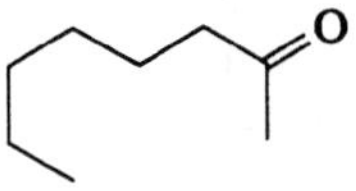 O	7 equiv. Cr on Amberlyst A-26, 12h[a]	92	173
3	 Ph OH	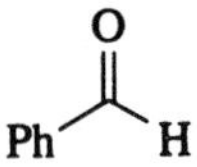 O, Ph H	10 equiv. Cr on Amberlyst A-26, 12 h[a]	73	173
			4 equiv. PVPCC, 0.25 h[b]	95	174
			1.1 equiv. PVPDC, (wet), 2 h[c]	96	175
4	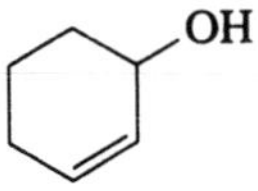 OH	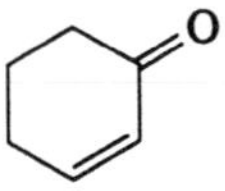 O	8 equiv. Cr on Amberlyst A-26, 8 h[a]	2	173
5	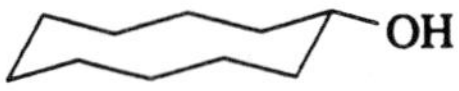 OH	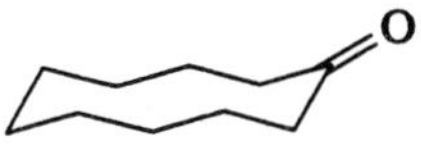 O	Cr^{III}/NAFK with Bu^tOOH[d]	86	176
6	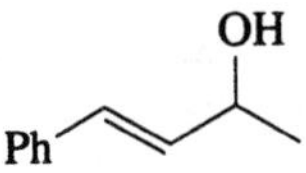 OH, Ph	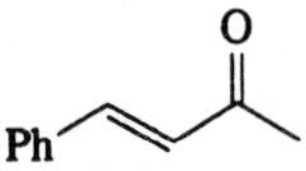 O, Ph	Cr^{III}/NAFK with Bu^tOOH[d]	81	176
7	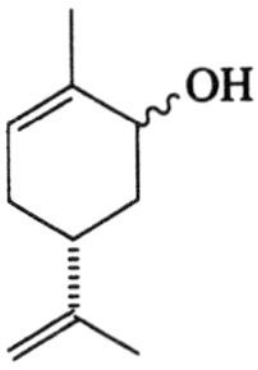 OH	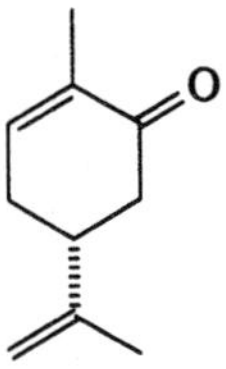 O	Cr^{III}/NAFK with Bu^tOOH[d]	82	176
8	 OH	 CHO	4 equiv. PVPCC, 4.5 h[b]	90	174
			1.1 equiv. PVPDC, (wet), 68 h[c]	81	175
9	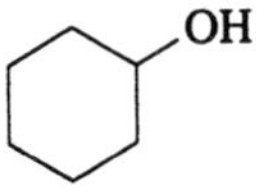 OH	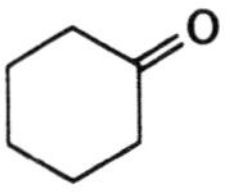 O	4 equiv. PVPCC, 24 h[b]	94	174
			1.1 equiv. PVPDC, (wet), 68 h[c]	66	175
10	 Ph OH	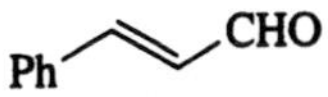 Ph CHO	4 equiv. PVPCC, 0.5 h[b]	100	174
			1.1 equiv. PVPDC, (wet), 4 h[c]	98	175

[a] Reactions performed in cyclohexane at 70 °C. [b] Reaction performed at 70–80 °C in cyclohexane. [c] Reaction performed at 60 °C in cyclohexane. [d] Reactions performed in chlorobenzene at 85 °C for 6–8 h using 4 equiv. Bu^tOOH and 4 mol % of Cr^{III} -impregnated NAFK.

results were obtained with trifluoroacetatochromates; and (iii) nonpolar solvents, such as cyclohexane, enhance reactivity by promoting the diffusion of substrates into the resin. For similar reasons, macroporous resins are preferred to gel types, even when well swollen.

The reagents prepared by Gelbard *et al.* were found to be selective for saturated alcohols, since the oxidation of allylic or benzylic alcohols requires either a vast excess of oxidant or prolonged reaction times.

To overcome the problems of toxicity and work-up associated with many inorganic oxidants, it would be advantageous to develop a catalytic supported oxidant. Towards this aim, chromium(III)-impregnated Nafion 511 (NAFK) has been used as a catalytic oxidant in the presence of *t*-butyl hydroperoxide.[176] This reagent gives good yields of ketones (80–100%), but unfortunately oxidation of primary alcohols leads to the formation of complex mixtures.

Poly(vinylpyridinium chlorochromate) (PVPCC)[174] is a mild oxidant for primary, secondary, allylic and benzylic alcohols. Unfortunately, optimum conditions require the use of very nonpolar solvents (best is cyclohexane) at 80 °C. More polar solvents (that would be more generally useful in synthesis) severely retard the rate of oxidation, thus necessitating an increase in the amount of oxidant used. Oxidations were found to have high inital rates, but were very slow to go to completion due to the inaccessibility of the chromium. This can be overcome by using a lower loading of oxidant or by an alternative preparation of the polymer,[175] where the addition of 1–5% divinylbenzene gives a much more porous resin.

Poly(vinylpyridinium dichromate) can be prepared in a similar way to PVPCC.[175] To be effective it must be used in the presence of water and in the most nonpolar solvent possible. It gives moderate to good yields, but long reaction times may be required for the reaction to proceed to completion. Even with the long reaction times, very little overoxidation is observed. Up to five oxidation–regeneration cycles may be completed without significant loss of activity.

2.7.5.3 On Carbon

Lalancette *et al.* described the preparation of chromium(VI) oxide intercalated in graphite[178] and its use as a selective oxidant for primary alcohols to aldehydes (Table 17). Secondary and tertiary alcohols are inert under the reaction conditions, but 1,2-diols are cleaved. However, Ebert and coworkers demonstrated that Lalancette probably used Cr_2O_8 rather than chromium(VI) oxide intercalated graphite for these oxidations.[179] This reagent does appear to show the properties described by Lalancette *et al.* Indeed, when Kagan and coworkers prepared fully characterized chromium(VI)-intercalated graphite it appeared to lack any oxidative properties.[180]

Table 17 Oxidation of Alcohols with CrO_3 and Graphite

Entry	*Substrate*	*Product*	*Conditions*[a]	*Yield (%)*	*Ref.*
1	$C_{15}H_{31}CH_2CH_2OH$	$C_{15}H_{31}CH_2CHO$	24 h	95	178
2	$PhCH_2CH_2OH$	$PhCH_2CHO$	24 h	98	178
3	$PhCH(OH)CH_2CH_2OH$	$PhCH_2CHO$	24 h	80	178
4	cyclohexyl-CH_2CH_2OH	cyclohexyl-CH_2CHO	48 h	52	178
5	cyclohexyl-$CH(OH)CH_3$	cyclohexyl-$COCH_3$	24 h	2	178
6	$PhCH=CHCH_2OH$	$PhCH=CHCHO$	96 h	100	178
7	2-furyl-CH_2OH	2-furyl-CHO	48 h	72	178

[a] Oxidations in toluene at reflux with 4–10 equiv. of CrO_3.

2.7.6 MISCELLANEOUS CHROMATES AND DICHROMATES

2.7.6.1 Alkylammonium Chromates

Gelbard *et al.*[173,184] reported the use of tetraalkylammonium chromates (Table 18) for the oxidation of alcohols under mild, neutral conditions. They have the advantage of being much more soluble than PCC (and PDC) in standard organic solvents, and are readily prepared from chromium(VI) oxide and the appropriate tetraalkylammonium salt (equation 17).[188] They may be used stoichiometrically or catalytically. In the stoichiometric mode the oxidations are quick, but yields are low due to overoxidation. When used catalytically (Scheme 3 and Table 19), the reactions are much more efficient than the stoichiometric oxidations or oxidation with PCC. This intriguing observation led to the suggestion that there is actually a different oxidizing agent in solution—a complex polychromate (**18**). Overoxidation is not a problem, but the catalytic systems need longer reaction times and thus give rise to some (*E*)/(*Z*)-isomerization.[173]

$$CrO_3 \quad + \quad X^-NR_4^+ \longrightarrow XCrO_3^-NR_4^+ \qquad (17)$$

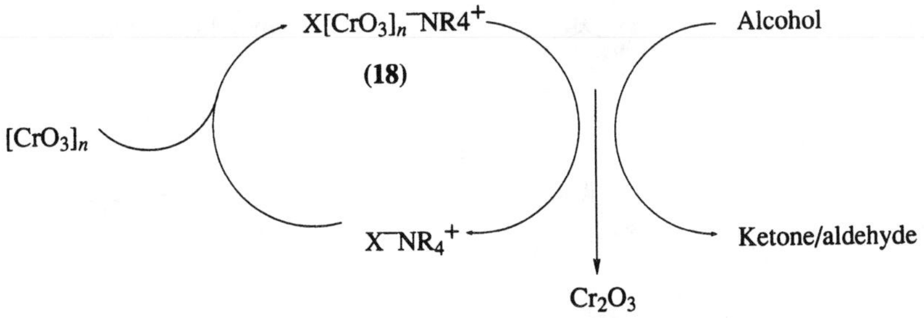

$X[CrO_3]_n^-NR4^+$

(**18**)

$[CrO_3]_n$

Alcohol

$X^-NR_4^+$

Ketone/aldehyde

Cr_2O_3

Scheme 3

The tetraalkylammonium chromates prepared by Gelbard and coworkers appeared to be equally effective for all types of alcohols, but under the conditions of Santaniello *et al.*,[181] tetra-*n*-butylammonium chlorochromate was found to be a selective benzylic and allylic oxidant. It also efficiently converts aliphatic and aromatic thiols to disulfides. Trimethylammonium chlorochromate has also been reported for the oxidation of allylic alcohols in DMF.[185]

Benzyltrimethylammonium chlorochromate is a neutral oxidant with similar selectivity to PFC.[118] However, it normally requires longer reaction times and higher temperatures (at reflux in 1,2-dichloroethane).

With only minor modifications it is possible to prepare a selective oxidant for benzylic alcohols: benzyltriethylammonium chlorochromate under phase transfer conditions exhibits such a preference.[182] The preparation of benzyltriethylammonium chlorochromate had been reported previously,[183] but was initially assigned as the dichromate. It was demonstrated that this reagent (chromate or dichromate) shows good selectivity for benzylic and allylic alcohols, but unfortunately it was necessary to perform the oxidation in HMPT.

2.7.6.2 Trimethylsilyl Chlorochromate (TMSCC)

Trimethylsilyl chlorochromate (**19**) must be formed *in situ* from trimethylsilyl chloride and moist, powdered chromium(VI) oxide,[186,187] since attempted isolation results in explosions.

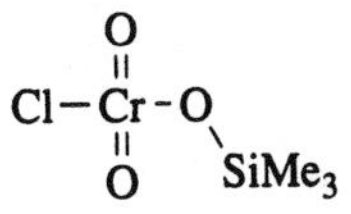

$$Cl-\overset{\overset{O}{\|}}{\underset{\underset{O}{\|}}{Cr}}-O-SiMe_3$$

(**19**)

Trimethylsilyl chlorochromate gives good yields for secondary alcohols and benzylic alcohols, but saturated primary alcohols give complex mixtures. It will also oxidise thiols to disulfides, and cleave

Table 18 Oxidation of Alcohols with Tetraalkylammonium Chromates

Entry	*Substrate*	*Product*	*Conditions*	*Yield (%)*	*Ref.*
1	OH	CHO	3 equiv., Bu_4N^+, 4 h, reflux[a]	65	181
			1.5 equiv., $BnEt_3N^+$, 24 h, r.t.[b]	72	182
			0.7 equiv., $BnEt_3N^+$, 4–6 h, HMPT[c]	90	183
2	MeO, OH	MeO, CHO	3 equiv., Bu_4N^+, 7 h, reflux[a]	85	181
			0.7 equiv., $BnEt_3N^+$, 4–6 h, HMPT[c]	95	183
3	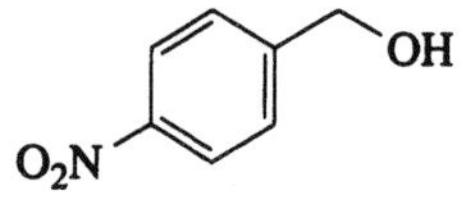 O_2N, OH	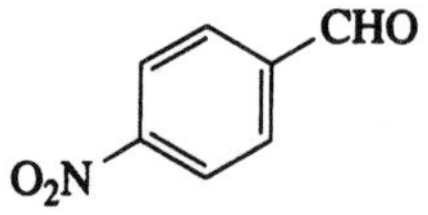 O_2N, CHO	3 equiv., Bu_4N^+, 3 h, reflux[a]	82	181
4	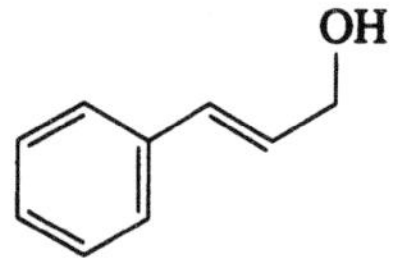 OH	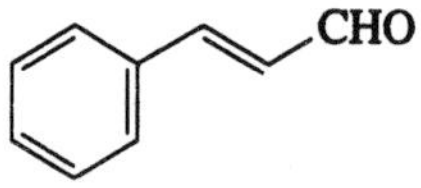 CHO	3 equiv., Bu_4N^+, 3 h, reflux[a]	82	181
5	OH	CHO	3 equiv., Bu_4N^+, 7 h, reflux[a]	72	181
			3 equiv., $BnMe_3N^+$, 6 h, reflux[d]	83 71:29 *(E):(Z)*	118
6	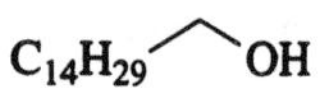 $C_{14}H_{29}$ OH	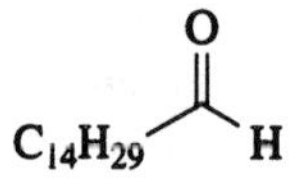 O, $C_{14}H_{29}$ H	3 equiv., Bu_4N^+, 24 h, reflux[a]	10	181
7	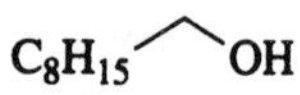 C_8H_{15} OH	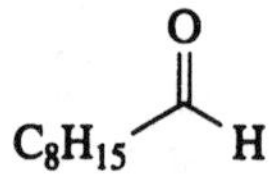 O, C_8H_{15} H	0.7 equiv., $BnEt_3N^+$, 4–6 h, HMPT[c]	30	183
8	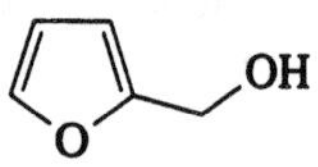 O, OH	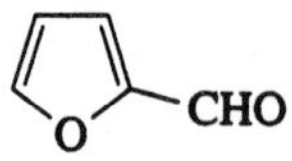 O, CHO	0.7 equiv., $BnEt_3N^+$, 4–6 h, HMPT[c]	71	183
9	Ph OH	Ph CHO	1.5 equiv., $BnEt_3N^+$, 24 h, r.t.[b]	72	182
			0.7 equiv., $BnEt_3N^+$, 4–6 h, HMPT[c]	90	183
10	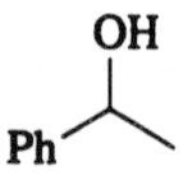 OH, Ph	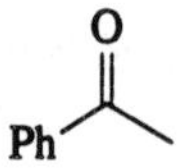 O, Ph	1.5 equiv., $BnEt_3N^+$, 24 h, r.t.[b]	50	182
11	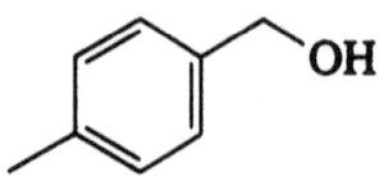 OH	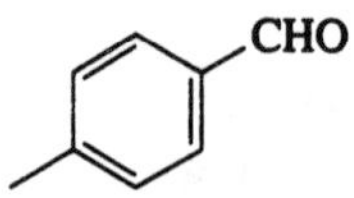 CHO	1.5 equiv., $BnEt_3N^+$, 24 h, r.t.[b]	75	182
12	 C_9H_{17} OH	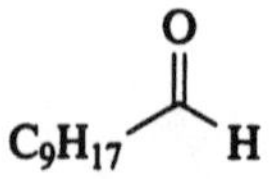 O, C_9H_{17} H	3 equiv., $BnMe_3N^+$, 10 h, reflux[d]	92	118

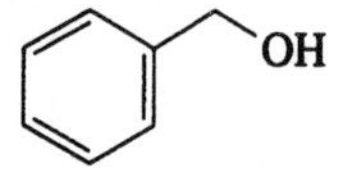

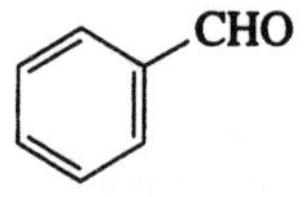

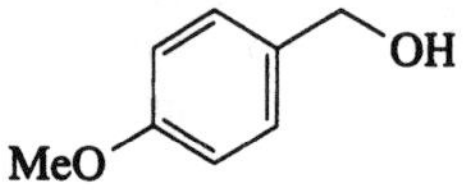

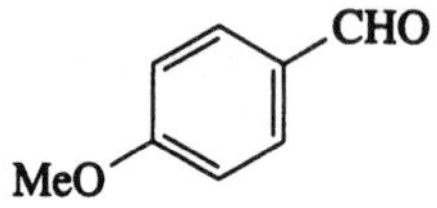

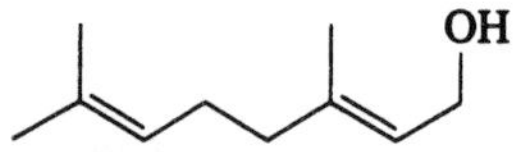

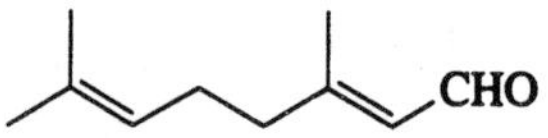

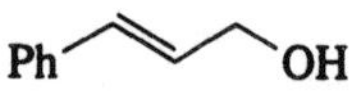

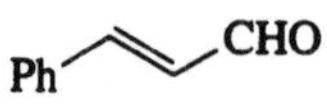

Table 18 *(continued)*

Entry	*Substrate*	*Product*	*Conditions*	*Yield (%)*	*Ref.*
13	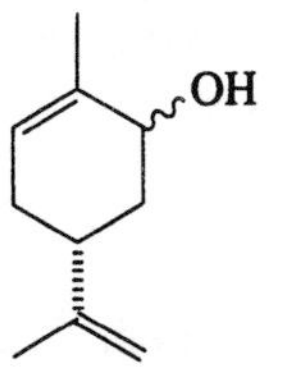	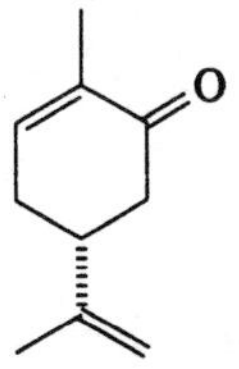	3 equiv., $BnMe_3N^+$, 4 h, reflux[d]	87	118

[a] Reaction in chloroform. [b] Under phase transfer conditions. [c] Reactions performed at 60–80 °C. [d] Reaction in 1,2-dichloroethane.

Table 19 Oxidation of Alcohols with CrO_3 and Catalytic *n*-Butylammonium Chloride

Entry	*Substrate*	*Product*	*Conditions*[a]	*Yield (%)*	*Ref.*
1	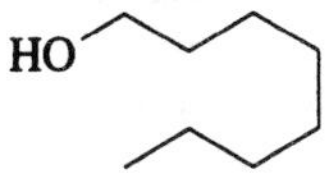 HO	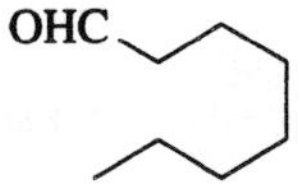 OHC	90 min	60	173
2	 Ph OH	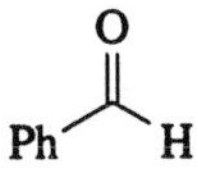 Ph(C=O)H	120 min	70	173
3	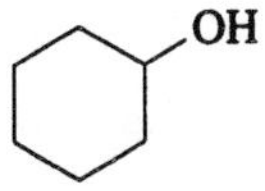 OH	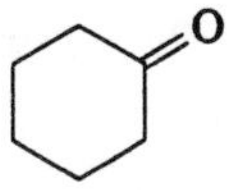 O	90 min	63	173
4	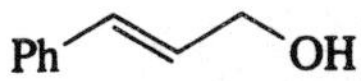 Ph OH	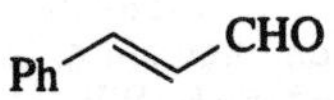 Ph CHO	15 min	85	173

[a] Reactions were carried out on a 50 mmol scale in DCM at r.t. using CrO_3 with 5 mol % of Bu_4NCl.

oximes and benzyl esters. Hence, it appears that trimethylsilyl chlorochromate does not offer any obvious advantage over other reagents available for the oxidation of alcohols.

2.7.6.3 Alkyl Metal Chromates

Oxidations with stoichiometric inorganic oxidants can generate large quantities of toxic waste and thus present potentially serious environmental problems. In part answer, several catalytic systems have been developed around different elements, but there still remains the problem of disposal of the cooxidant. Recently, Shapley and coworkers have reported the preparation of two alkyl metal complexes; *cis*-$[Bu^n{}_4N][Os(N)(CH_2SiMe_3)_2(CrO_4)]$ (**20**) and *cis*-$[PPh_4][Os(N)Me_2(CrO_4)]$ (**21**) (equation 18),[188] that are thermally stable and inert to most standard organic solvents. They act as catalysts (5 mol %) for the oxidation of primary and secondary alcohols, using air as cooxidant. Allylic and benzylic alcohols are oxidized faster than saturated secondary alcohols, which in turn are oxidized faster than saturated primary alcohols. No products of overoxidation could be detected. The chromates also appear to be inert towards

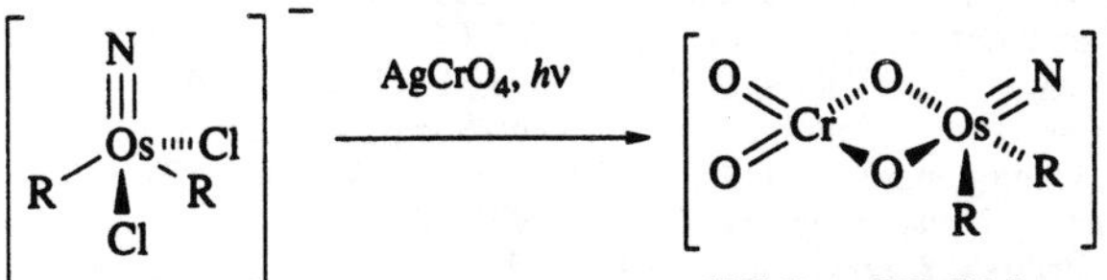

$$\left[R_2Os(N)Cl_2 \right]^- \xrightarrow{AgCrO_4,\ h\nu} \left[O_2Cr(\mu\text{-}O)_2Os(N)R_2 \right]^- \qquad (18)$$

(**20**) R = CH_2SiMe_3
(**21**) R = Me

carbon–carbon double bonds. Chromate (**21**) was found to be more reactive than the more bulky chromate (**20**).

Turnovers for these oxidations range from moderate (57) to poor (2) but can be significantly improved by the addition of a copper(II) salt (10 mol %).

Even though the reactions are slow and the catalyst is degraded under the reaction conditions, the possibility of a ruthenium analog makes this a particularly promising area for further development.

2.7.6.4 Bis(tetrabutylammonium) Dichromate

Bis(tetrabutylammonium) dichromate is a neutral oxidant which at reflux in dichloromethane acts as a selective oxidant for allylic and benzylic alcohols.[189] Only 10% of oxidation products are obtained after treatment of *n*-decanol with bis(tetrabutylammonium) dichromate for 24 h.

2.7.6.5 Tetrakis(pyridine)silver Dichromate

Many chromium oxidants suffer from problems of stability, light sensitivity or acidity, but tetrakis(pyridine)silver dichromate[190] is stable, nonphotosensitive, nonhygroscopic and a neutral oxidant. It can be used to oxidize allylic and benzylic alcohols selectively in benzene. Unfortunately, it cannot be used in chlorinated solvents because it decomposes in these solvents.

A possible drawback for this oxidant is that the use of silver in a stoichiometric oxidant may render the reagent too expensive for large scale oxidations.

2.7.7 REFERENCES

1. G. Cainelli and G. Cardillo, 'Chromium Oxidations in Organic Chemistry', Springer–Verlag, Berlin, 1984, p. 118.
2. F. Freeman, in 'Organic Syntheses by Oxidation with Metal Compounds', ed. W. J. Mijs and C. R. H. I. de Jonge, Plenum Press, New York, 1986, p. 41.
3. F. A. Luzzio and F. S. Guziec, Jr., *Org. Prep. Proced. Int.*, 1988, **20**, 533.
4. H. O. House, in 'Modern Synthetic Reactions', 2nd edn., Benjamin, Menlo Park, CA, 1972, p. 257.
5. K. B. Wiberg, in 'Oxidation in Organic Chemistry', ed. K. B. Wiberg, Academic Press, New York, 1965, vol. 5A, p. 69.
6. H. Kiliani and B. Merk, *Chem. Ber.*, 1901, **34**, 3562.
7. L. Gindraux, *Helv. Chim. Acta*, 1929, **12**, 921.
8. J. Sauer, *Org. Synth., Coll. Vol.*, 1963, **4**, 813.
9. J. Hampton, A. Leo and F. H. Westheimer, *J. Am. Chem. Soc.*, 1956, **78**, 306.
10. B. H. Walker, *J. Org. Chem.*, 1967, **32**, 1098.
11. G. Snatzke, *Chem. Ber.*, 1961, **94**, 729.
12. Y. Shyamsunder Rao and R. Filler, *J. Org. Chem.*, 1974, **39**, 3304.
13. H. C. Brown, C. P. Garg and K.-T. Liu, *J. Org. Chem.*, 1971, **36**, 387.
14. J.-D. Lou, *Chem. Ind. (London)*, 1989, 312.
15. D. Pletcher and S. J. D. Tait, *Tetrahedron Lett.*, 1978, 1601.
16. D. Landini, F. Montanari and F. Rolla, *Synthesis*, 1979, 134.
17. A. Bowers, T. G. Halsall, E. R. H. Jones and A. J. Lemin, *J. Chem. Soc.*, 1953, 2548.
18. J. Mulzer, A. Angermann, B. Schubert and C. Seilz, *J. Org. Chem.*, 1986, **51**, 5294.
19. C. Djerassi, P. A. Hart and E. J. Warawa, *J. Am. Chem. Soc.*, 1964, **86**, 78.
20. S. H. Burstein and H. J. Ringold, *J. Am. Chem. Soc.*, 1967, **89**, 4722.
21. E. Glotter, S. Greenfield and D. Lavie, *J. Chem. Soc. C*, 1968, 1646.
22. R de A. Epifanio, W. Camargo and A. C. Pinto, *Tetrahedron Lett.*, 1988, **29**, 6403.
23. A. Murai, H. Miyazaki, K. Watanabe and T. Masamune, *Chem. Lett.*, 1987, 651.
24. J. K. Thottathil, J. L. Moniot, R. H. Mueller, M. K. Y. Wong and T. P. Kissick, *J. Org. Chem.*, 1986, **51**, 3140.
25. C. H. Heathcock, K. M. Smith and T. A. Blumenkopf, *J. Am. Chem. Soc.*, 1986, **108**, 5022.
26. K. C. Nicolaou, M. R. Pavia and S. P. Seitz, *Tetrahedron Lett.*, 1979, 2327.
27. P. R. Jefferies, R. S. Rosich, D. E. White and M. C. Woods, *Aust. J. Chem.*, 1962, **15**, 521.
28. C. H. Lin and D. L. Alexander, *J. Org. Chem.*, 1982, **47**, 615.
29. K. E. Harding, L. M. May and K. F. Dick, *J. Org. Chem.*, 1975, **40**, 1664.
30. I. Heilbron, E. R. H. Jones and F. Sondheimer, *J. Chem. Soc.*, 1947, 1586.
31. K. B. Sharpless, A. O. Chong and K. Oshima, *J. Org. Chem.*, 1976, **41**, 177.
32. D. E. Kiely and L. Benzing-Nguyen, *J. Org. Chem.*, 1975, **40**, 2630.
33. E. V. Vedejs and J. R. Pribish, *J. Org. Chem.*, 1988, **53**, 1593.
34. Z. Paryzek and K. Blaszczyk, *Can. J. Chem.*, 1987, **65**, 229.
35. S. W. Djuric, M. Miyano, M. Clare and R. M. Rydzewski, *Tetrahedron Lett.*, 1987, **28**, 299.
36. T. Morikawa, T. Nishiwaki, Y. Iitaka and Y. Kobayashi, *Tetrahedron Lett.*, 1987, **28**, 671.

37. L. A. Paquette, P. D. Pansegrau, P. E. Wiedeman and J. P. Springer, *J. Org. Chem.*, 1988, **53**, 1461.
38. I. Uchida, T. Ando, N. Fukami, K. Yoshida, M. Hashimoto, T. Tada, S. Koda and Y. Morimoto, *J. Org. Chem.*, 1987, **52**, 5292.
39. K. Fuji, M. Node and Y. Usami, *Chem Lett.*, 1986, 961.
40. G. I. Poos, G. E. Arth, R. E. Beyler and L. H. Sarett, *J. Am. Chem. Soc.*, 1953, **75**, 422.
41. J. R. Holum, *J. Org. Chem.*, 1961, **26**, 4814.
42. J. C. Collins, W. W. Hess and F. J. Frank, *Tetrahedron Lett.*, 1968, 3363; J. C. Collins and W. W. Hess, *Org. Synth.*, 1972, **52**, 5.
43. W. G. Salmond, M. A. Barta and J. L. Havens, *J. Org. Chem.*, 1978, **43**, 2057.
44. N. H. Anderson and H.-S. Uh, *Synth. Commun.*, 1973, **3**, 115.
45. R. Ratcliffe and R. Rodehorst, *J. Org. Chem.*, 1970, **35**, 4000.
46. R. W. Ratcliffe, T. N. Salzmann and B. G. Christensen, *Tetrahedron Lett.*, 1980, 31.
47. Y. Arai, Y. Hayashi, M. Yamamoto, H. Takayama and T. Koizumi, *Chem Lett.*, 1987, 185.
48. F. E. Ziegler, B. H. Jaynes and M. T. Saindane, *J. Am. Chem. Soc.*, 1987, **109**, 8115.
49. C. A. Barras, R. Roulet, P.-A. Carrupt, F. Berchier and P. Vogel, *Helv. Chim. Acta*, 1984, **67**, 986.
50. D. M. Walba and G. S. Stoudt, *Tetrahedron Lett.*, 1982, **23**, 727.
51. M. Hirama and M. Uei, *J. Am. Chem. Soc.*, 1982, **104**, 4251.
52. S. Masamune, T. Kaiho and D. S. Garvey, *J. Am. Chem. Soc.*, 1982, **104**, 5521.
53. R. W. Franck and T. V. John, *J. Org. Chem.*, 1980, **45**, 1170.
54. A. A. Jakubowski, F. S. Guziec, Jr. and M. Tishler, *Tetrahedron Lett.*, 1977, 2399.
55. W. A. Szarek, O. R. Martin, R. J. Rafka and T. S. Cameron, *Can. J. Chem.*, 1985, **63**, 1222.
56. E. Vedejs, M. F. Salomon and P. D. Weeks, *J. Organomet. Chem.*, 1972, **40**, 221.
57. W. G. Dauben, M. Lorber and D. S. Fullerton, *J. Org. Chem.*, 1969, **34**, 3587.
58. P. Sundararaman and W. Herz, *J. Org. Chem.*, 1977, **42**, 813.
59. P. J. Garegg and B. Samuelsson, *Carbohydr. Res.*, 1978, **67**, 267.
60. F. Hansske and M. J. Robins, *Tetrahedron Lett.*, 1983, **24**, 1589.
61. F. S. Guziec, Jr. and F. A. Luzzio, *Synthesis*, 1980, 691.
62. K. Nozaki, K. Oshima and K. Utimoto, *J. Am. Chem. Soc.*, 1987, **109**, 2547.
63. E. J. Corey and G. W. J. Fleet, *Tetrahedron Lett.*, 1973, 4499.
64. G. Piancatelli, A. Scettri and M. D'Auria, *Synthesis*, 1982, 245.
65. E. J. Corey and J. W. Suggs, *Tetrahedron Lett.*, 1975, 2647.
66. S. Hanessian and B. Vanasse, *Can. J. Chem.*, 1987, **65**, 195.
67. E. B. Villhauer and R. C. Anderson, *J. Org. Chem.*, 1987, **52**, 1186.
68. H.-H. Tso, T.-S. Chou and S. C. Hung, *J. Chem. Soc., Chem. Commun.*, 1987, 1552.
69. T. Sugimura and L. A. Paquette, *J. Am. Chem. Soc.*, 1987, **109**, 3017.
70. A. G. Schultz, W. G. McMahon and R. K. Kullnig, *J. Org. Chem.*, 1987, **52**, 3905.
71. K. Ramesh and R. S. Kapil, *Chem. Ind. (London)*, 1986, 614.
72. P. C. B. Page and S. Rosenthal, *Tetrahedron Lett.*, 1986, **27**, 2527.
73. Y. Oikawa, T. Nishi and O. Yonemitsu, *J. Chem. Soc., Perkin Trans. 1*, 1985, 1.
74. E. J. Browne, *Aust. J. Chem.*, 1985, **38**, 765.
75. E. Piers, B. F. Abeysekera, D. J. Herbert and I. D. Suckling, *Can. J. Chem.*, 1985, **63**, 3418.
76. H.-D. Scharf, J. Runsink and A. Klausener, *Liebigs Ann. Chem.*, 1984, 783.
77. G. Rosini and R. Ballini, *Synthesis*, 1983, 543.
78. R. J. Giguere, G. von Ilsemann and H. M. R. Hoffmann, *J. Org. Chem.*, 1982, **47**, 4948.
79. H. I. Ahmad, J. S. Brimacombe, A. S. Mengech and L. C. N. Tucker, *Carbohydr. Res.*, 1981, **93**, 288.
80. H. Nishiyama, M. Sasaki and K. Itoh, *Chem. Lett.*, 1981, 1363.
81. D. A. Lightner, J. K. Gawronski and T. D. Bouman, *J. Am. Chem. Soc.*, 1980, **102**, 1983.
82. K. Inomata, M. Sumita and H. Kotake, *Chem. Lett.*, 1979, 709.
83. E. J. Corey and J. G. Smith, *J. Am. Chem. Soc.*, 1979, **101**, 1038.
84. P. S. Reddy, P. Yadagiri, S. Lumin, D.-S. Shin and J. R. Falck, *Synth. Commun.*, 1988, **18**, 545.
85. E. J. Corey, H. E. Ensley and J. W. Suggs, *J. Org. Chem.*, 1976, **41**, 380.
86. E. J. Corey and D. L. Boger, *Tetrahedron Lett.*, 1978, 2461.
87. W. G. Dauben and D. M. Michno, *J. Org. Chem.*, 1977, **42**, 682.
88. E. Wada, M. Okawara and T. Nakai, *J. Org. Chem.*, 1979, **44**, 2952.
89. J. P. Willis, K. A. Z. Gogins and L. L. Miller, *J. Org. Chem.*, 1981, **46**, 3215.
90. H. C. Brown, S. U. Kulkarni, C. G. Rao and V. D. Patil, *Tetrahedron*, 1986, **42**, 5515 and refs. therein.
91. E. J. Parish and G. J. Schroepfer, Jr., *Chem. Phys. Lipids*, 1980, **27**, 281.
92. E. J. Parish and A. D. Scott, *J. Org. Chem.*, 1983, **48**, 4766.
93. E. J. Parish, A. D. Scott, J. R. Dickerson and W. Dykes, *Chem. Phys. Lipids*, 1984, **35**, 315.
94. E. J. Parish, S. Chitrakorn, *Synth. Commun.*, 1985, **15**, 393.
95. Y.-S. Cheng, W.-L. Liu and S.-H. Chen, *Synthesis*, 1980, 223.
96. D. H. Hollenberg, R. S. Klein and J. J. Fox, *Carbohydr. Res.*, 1978, **67**, 491.
97. J. Herscovici and K. Antonakis, *J. Chem. Soc., Chem. Commun.*, 1980, 561.
98. J. Herscovici, M.-J. Egron and K. Antonakis, *J. Chem. Soc., Perkin Trans. 1*, 1982, 1967.
99. G. Piancatelli, A. Scettri and M. D'Auria, *Tetrahedron Lett.*, 1979, 1507.
100. R. D'Ascoli, M. D'Auria, A. De Mico, G. Piancatelli and A. Scettri, *J. Org. Chem.*, 1980, **45**, 4500.
101. G. Piancatelli, A. Scettri and M. D'Auria, *Tetrahedron*, 1980, **36**, 661.
102. R. Antonioletti, M. D'Auria, A. De Mico, A. Scettri and G. Piancatelli, *Synthesis*, 1984, 280.
103. M. Seth, M. S. Akhtar and A. P. Bhaduri, *J. Heterocycl. Chem.*, 1985, **22**, 1323.
104. G. Piancatelli, A. Scettri and M. D'Auria, *Tetrahedron Lett.*, 1977, 2199.
105. G. Piancatelli, A. Scettri and M. D'Auria, *Tetrahedron Lett.*, 1977, 3483.
106. S. Baskaran, I. Islam, M. Raghavan and S. Chandrasekaran, *Chem. Lett.*, 1987, 1175.
107. R. Antonioletti, M. D'Auria, A. De Mico, G. Piancatelli and A. Scettri, *Synthesis*, 1983, 890.

108. V. Narasimhan, R. Rathore and S. Chandrasekaran, *Synth. Commun.*, 1985, **15**, 769.
109. H. Firouzabadi, N. Iranpoor, H. Parham, A. Sardarian and J. Toofan, *Synth. Commun.*, 1984, **14**, 717.
110. F. Bonadies and R. Di Fabio, *J. Org. Chem.*, 1984, **49**, 1647.
111. E. J. Parish, S. Chitrakorn and T.-Y. Wei, *Synth. Commun.*, 1986, **16**, 1371.
112. R. Rathore, N. Saxena and S. Chandrasekaran, *Synth. Commun.*, 1986, **16**, 1493.
113. F. Bonadies and C. Bonini, *Synth. Commun.*, 1988, **18**, 1573.
114. Ref. 3, p. 573.
115. T. K. Chakraborty and S. Chandrasekaran, *Chem. Lett.*, 1985, 551.
116. E. J. Parish and T.-Y. Wei, *Synth. Commun.*, 1987, **17**, 1227.
117. M. N. Bhattacharjee, M. K. Chaudhuri, H. S. Dasgupta, N. Roy and D. T. Khathing, *Synthesis*, 1982, 588. Pyridinium Bromochromate (PBC) has also been prepared. (N. Narayanan and T. R. Balasubramanian, *Indian J. Chem., Sect. B*, 1986, **25**, 228.) It appears to be a selective oxidant for benzylic alcohols, *e.g.* treatment of 2-chlorobenzyl alcohol with PBC for 3 h in chloroform at reflux gives the corresponding aldehyde in 95% yield. However, it has not been widely adopted.
118. T. Nonaka, S. Kanemoto, K. Oshima and H. Nozaki, *Bull. Chem. Soc. Jpn.*, 1984, **57**, 2019.
119. R. Varadarajan and R. K. Dhar, *Indian J. Chem., Sect. B*, 1986, **25**, 971.
120. R. Varadarajan and R. K. Dhar, *Indian J. Chem., Sect. B*, 1986, **25**, 746.
121. H. B. Davis, R. M. Sheets, W. W. Paudler and G. L. Gard, *Heterocycles*, 1984, **22**, 2029.
122. P. M. Wovkulich, F. Barcelos, A. D. Batcho, J. F. Sereno, E. G. Baggiolini, B. M. Hennessy and M. R. Uskokovic, *Tetrahedron*, 1984, **40**, 2283.
123. P. E. Peterson, R. L. B. Leffew and B. L. Jensen, *J. Org. Chem.*, 1986, **51**, 1948.
124. D. M. Walba and G. S. Stoudt, *Tetrahedron Lett.*, 1982, **23**, 727.
125. M. Kamber, H. Pfander and K. Noack, *Helv. Chim. Acta*, 1984, **67**, 968.
126. Ref. 3, p. 546.
127. F. S. Guziec, Jr. and F. A. Luzzio, *J. Org. Chem.*, 1982, **47**, 1787.
128. H. B. Davis, R. M. Sheets, J. M. Brannfors, W. W. Paudler and G. L. Gard, *Heterocycles*, 1983, **20**, 2029.
129. J. Singh, P. S. Kalsi, G. S. Jawanda and B. R. Chabra, *Chem. Ind. (London)*, 1986, 751.
130. G. J. S. Doad, *J. Chem. Res. (S)*, 1988, 270.
131. E. J. Corey and G. Schmidt, *Tetrahedron Lett.*, 1979, 399.
132. R. H. Cornforth, J. W. Cornforth and G. Popjak, *Tetrahedron*, 1962, **18**, 1351.
133. W. M. Coates and J. R. Corrigan, *Chem. Ind. (London)*, 1969, 1594.
134. D. Liotta, D. Brown, W. Hoekstra and R. Monahan, III, *Tetrahedron Lett.*, 1987, **28**, 1069.
135. J. A. Marshall, J. Grote and J. E. Audia, *J. Am. Chem. Soc.*, 1987, **109**, 1186.
136. D. R. Williams and F. H. White, *J. Org. Chem.*, 1987, **52**, 5067.
137. M. Suzuki, H. Koyano and R. Noyori, *J. Org. Chem.*, 1987, **52**, 5583.
138. F. Nicotra, L. Panza, F. Ronchetti, G. Russo and L. Toma, *J. Org. Chem.*, 1986, **51**, 1272.
139. C. F. Stanfield, J. E. Parker and P. Kanellis, *J. Org. Chem.*, 1981, **46**, 4797.
140. K. S. Kim and W. A. Szarek, *Carbohydr. Res.*, 1982, **104**, 328.
141. B. O'Connor and G. Just, *Tetrahedron Lett.*, 1987, **28**, 3235.
142. S. Czernecki, C. Georgoulis, C. L. Stevens and K. Vijayakumaran, *Tetrahedron Lett.*, 1985, **26**, 1699.
143. S. Czernecki, K. Vijayakumaran and G. Ville, *J. Org. Chem.*, 1986, **51**, 5472.
144. F. Andersson and B. Samuelsson, *Carbohydr. Res.*, 1984, **129**, C1.
145. F. P. Cossío, J. M. Aizpurua and C. Palomo, *Can. J. Chem.*, 1986, **64**, 225.
146. S. Kanemoto, K. Oshima, S. Matsubara, K. Takai and H. Nozaki, *Tetrahedron Lett.*, 1983, **24**, 2185.
147. S. Kanemoto, S. Matsubara, K. Takai, K. Oshima, K. Utimoto and H. Nozaki, *Bull. Chem. Soc. Jpn.*, 1988, **61**, 3607.
148. S. Czernecki, C. Georgoulis, C. L. Stevens and K. Vijayakumaran, *Synth. Commun.*, 1986, **16**, 11.
149. S. Satish and N. Kalyanam, *Chem. Ind. (London)*, 1981, 809.
150. M. Fetizon, P. Goulaouic and I. Hanna, *Tetrahedron Lett.*, 1988, **29**, 6261.
151. A. G. Schultz, F. P. Lavieri and M. Macielag, *Tetrahedron Lett.*, 1986, **27**, 1481.
152. N. Chidambaram and S. Chandrasekaran, *J. Org. Chem.*, 1987, **52**, 5048.
153. F. P. Cossío, M. C. López and C. Palomo, *Tetrahedron*, 1987, **43**, 3963.
154. C. López, A. González, F. P. Cossío and C. Palomo, *Synth. Commun.*, 1985, **15**, 1197.
155. F. Roldan, A. González and C. Palomo, *Carbohydr. Res.*, 1986, **149**, C1.
156. K. Balasubramanian and V. Prathiba, *Indian J. Chem., Sect. B*, 1986, **25**, 326.
157. S. Kim and D. C. Lhim, *Bull. Chem. Soc. Jpn.*, 1986, **59**, 3297.
158. S. J. Flatt, G. W. J. Fleet and B. J. Taylor, *Synthesis*, 1979, 815.
159. I. Ganboa, J. M. Aizpurua and C. Palomo, *J. Chem. Res. (S)*, 1984, 92.
160. M. Juaristi, J. M. Aizpurua, B. Lecea and C. Palomo, *Can. J. Chem.*, 1984, **62**, 2941.
161. J. Muzart, *Tetrahedron Lett.*, 1987, **28**, 2133.
162. J. Muzart, *Tetrahedron Lett.*, 1987, **28**, 4665.
163. J. Muzart, *Tetrahedron Lett.*, 1987, **28**, 2131.
164. J. Muzart, *Tetrahedron Lett.*, 1986, **27**, 3139.
165. E. J. Corey, E.-P. Barrette and P. A. Magriotis, *Tetrahedron Lett.*, 1985, **26**, 5855.
166. A. McKillop and D. W. Young, *Synthesis*, 1979, 401 and 481.
167. J. E. Saavedra, D. W. Farnsworth and G.-K. Pei, *Synth. Commun.*, 1988, **18**, 313.
168. J.-D. Lou and Y.-Y. Wu, *Chem. Ind. (London)*, 1987, 531.
169. J.-D. Lou and Y.-Y. Wu, *Synth. Commun.*, 1987, **17**, 1717.
170. R. P. Singh, H. N. Subbarao and S. Dev, *Tetrahedron*, 1979, **35**, 1789.
171. E. Santaniello, F. Ponti and A. Manzocchi, *Synthesis*, 1978, 534.
172. J. San Filippo, Jr. and C.-I. Chern, *J. Org. Chem.*, 1977, **42**, 2182.
173. T. Brunelet, C. Jouitteau and G. Gelbard, *J. Org. Chem.*, 1986, **51**, 4016.
174. J. M. J. Fréchet, J. Warnock and M. J. Farall, *J. Org. Chem.*, 1978, **43**, 2618.

175. J. M. J. Fréchet, P. Darling and M. J. Farrall, *J. Org. Chem.*, 1981, **46**, 1728.
176. S. Kanemoto, H. Saimoto, K. Oshima and H. Nozaki, *Tetrahedron Lett.*, 1984, **25**, 3317.
177. G. Cainelli, G. Cardillo, M. Orena and S. Sandri, *J. Am. Chem. Soc.*, 1976, **98**, 6737.
178. J.-M. Lalancette, G. Rollin and P. Dumas, *Can. J. Chem.*, 1972, **50**, 3058.
179. L. B. Ebert, R. A. Huggins and J. I. Brauman, *Carbon*, 1974, **12**, 199.
180. H. B. Kagan, *CHEMTECH*, 1976, **6**, 510.
181. E. Santaniello, F. Milani and R. Casati, *Synthesis*, 1983, 749.
182. C. Someswara Rao, A. A. Deshmukh, M. R. Thakor and P. S. Srinivasan, *Indian. J. Chem., Sect. B*, 1986, **25**, 324.
183. X. Huang and C.-C. Chan, *Synthesis*, 1982, 1091.
184. G. Gelbard, T. Brunelet and C. Jouitteau, *Tetrahedron Lett.*, 1980, **21**, 4653.
185. Q.-J. Ding and K. Cai, *Youji Huaxue*, 1988, **8**, 457.
186. J. M. Aizpurua and C. Palomo, *Tetrahedron Lett.*, 1983, **24**, 4367.
187. J. M. Aizpurua, M. Juaristi, B. Lecea and C. Palomo, *Tetrahedron*, 1985, **41**, 2903.
188. N. Zhang, C. M. Mann and P. A. Shapley, *J. Am. Chem. Soc.*, 1988, **110**, 6591.
189. E. Santaniello and P. Ferraboschi, *Synth. Commun.*, 1980, **10**, 75.
190. H. Firouzabadi, A. Sardarian and H. Gharibi, *Synth. Commun.*, 1984, **14**, 89.

2.8 Oxidation Adjacent to Oxygen of Alcohols by Activated DMSO Methods

THOMAS V. LEE

University of Bristol, UK

2.8.1 INTRODUCTION

The nucleophilic nature of the sulfinyl oxygen of sulfoxides has been used to good effect in synthetic chemistry, most notably in the oxidation of primary and secondary alcohols to aldehydes and ketones using dimethyl sulfoxide (DMSO). The first report of the use of dimethyl sulfoxide in such an oxidation was due to Pfitzner and Moffatt,[1] although Kornblum had previously demonstrated the use of dimethyl sulfoxide as an oxidant for the conversion of alkyl halides to aldehydes.[2] There are now many examples of how advantageous the use of dimethyl sulfoxide can be, most notably in the oxidation of primary alcohols to aldehydes without overoxidation to carboxylic acids. Despite much activity in this area, resulting in a plethora of activated dimethyl sulfoxide reagents which behave as an oxidant, the method is still basically that originally described by Pfitzner and Moffatt. However, considerable practical improvements have been made by a number of groups, most notably that of Swern, who utilized oxalyl chloride as the activator, and the 'Swern oxidation' is now a well-established synthetic method.

To appreciate this important reaction requires a mechanistic understanding of how dimethyl sulfoxide can be activated for use as an oxidant. The following discussion therefore begins with the general mechanism of the reaction, followed by a description of the use of the more important activation methods, many of which are complementary. A description of the process from the viewpoint of the substrate then follows, emphasizing the advantages of using the activated dimethyl sulfoxide oxidation method. Earlier examples of the use of the method have been reviewed previously,[3–5] hence the present discussion concentrates on more recent applications.

2.8.2 GENERAL MECHANISM OF ACTIVATED DMSO OXIDATION

The key to successfully using dimethyl sulfoxide as an oxidant for alcohols is to activate the sulfur atom prior to reaction with a nucleophilic alcohol function. This activation involves electrophilic attack upon the sulfinyl oxygen by a variety of electrophiles. The initial product formed when an alcohol does attack the activated dimethyl sulfoxide is known to be the sulfonium salt (**1**; Scheme 1).

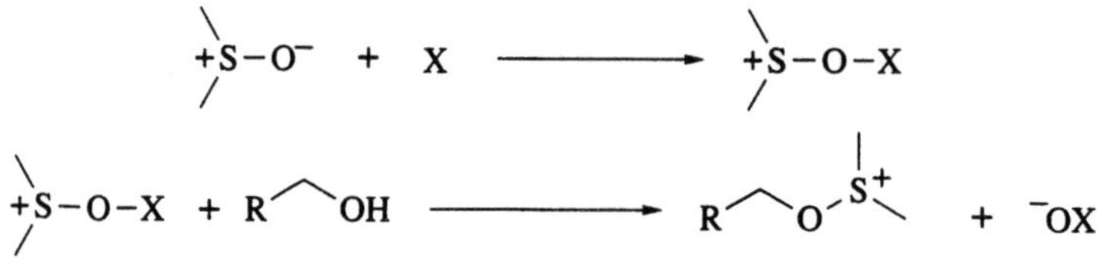

$$\mathrm{{>}S^{+}{-}O^{-}} + \mathrm{X} \longrightarrow \mathrm{{>}S^{+}{-}O{-}X}$$

$$\mathrm{{>}S^{+}{-}O{-}X} + \mathrm{RCH_2OH} \longrightarrow \mathrm{RCH_2O{-}S^{+}Me_2} + \mathrm{^{-}OX}$$

(1)

Scheme 1

Studies using ^{18}O-labeled dimethyl sulfoxide have confirmed this pathway as opposed to an alternative in which the alcohol initially attacks the activator (Scheme 2) to form an ester which is subsequently attacked by dimethyl sulfoxide to give the same sulfonium species as above.[6]

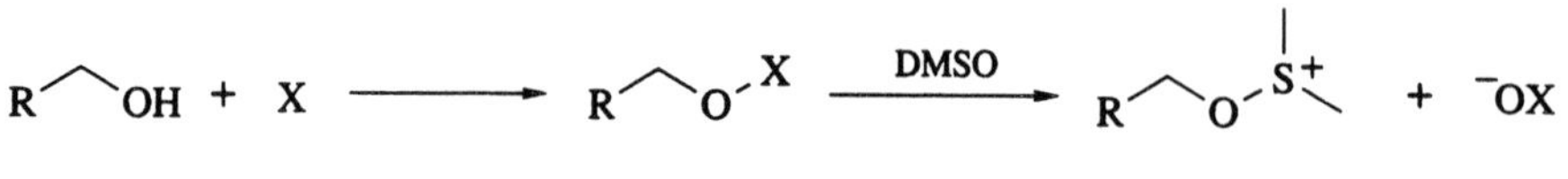

$$\mathrm{RCH_2OH} + \mathrm{X} \longrightarrow \mathrm{RCH_2O{-}X} \xrightarrow{\mathrm{DMSO}} \mathrm{RCH_2O{-}S^{+}Me_2} + \mathrm{^{-}OX}$$

(1)

Scheme 2

The reaction requires the addition of base (commonly triethylamine) and results in the formation of an ylide (**2**; Scheme 3), which collapses intramolecularly to the carbonyl compound.[6,7–9] Further supportive evidence for ylide formation lies in the observation that sulfonium salts lacking a hydrogen α to the sulfur do not break down to form the carbonyl compound.[10]

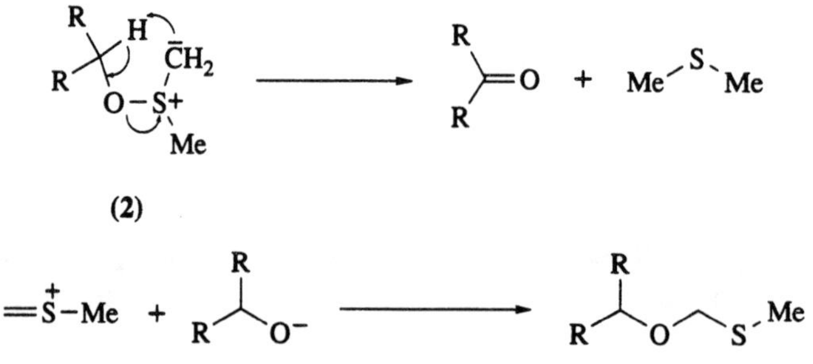

$$\mathrm{R_2CH{-}O{-}S^{+}(Me){-}\bar{C}H_2} \longrightarrow \mathrm{R_2C{=}O} + \mathrm{MeSMe}$$

(2)

$$\mathrm{CH_2{=}\overset{+}{S}{-}Me} + \mathrm{R_2CH{-}O^{-}} \longrightarrow \mathrm{R_2CH{-}O{-}CH_2{-}S{-}Me}$$

(3)

Scheme 3

Common by-products in these reactions are (methylthio)methyl ethers (**3**) formed by a Pummerer rearrangement which occurs *via* an alternative breakdown of (**2**), as shown in Scheme 3. The proportion of the Pummerer rearrangement derived product varies with the electrophilic activator used.

One of the major considerations in the choice of activator is the temperature at which the reaction can be performed. Most of the electrophilic activators used react rapidly with dimethyl sulfoxide, some of them violently; consequently temperatures of less than –30 °C are typically required. Another important consequence of these oxidations is that the initially acidic mixture becomes basic on completion of the oxidation, and highly sensitive substrates may undergo side reactions. An example of this is seen in the oxidation of the carbohydrate (**4**; equation 1) using sulfur trioxide/pyridine complex as the dimethyl sulfoxide activator, which causes spontaneous elimination of a β-acetoxy group.[11] As seen later, the use of the Swern variation overcomes these problems.

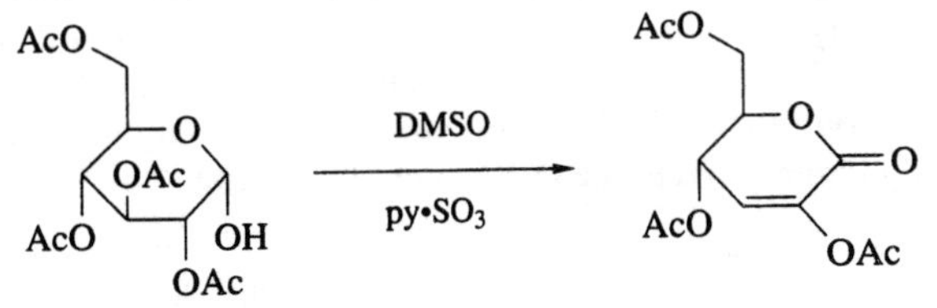

AcO, O, OAc, AcO, OH, OAc — DMSO, py•SO_3 → AcO, O, =O, AcO, OAc (1)

(4)

Oxidation of primary alcohols to aldehydes using these oxidants does not result in overoxidation to carboxylic acids, in contrast to many other oxidants. This arises since the aldehydes are formed under anhydrous conditions and are not capable of hydration, which is a necessary requirement for conversion to a carboxylic acid.

2.8.3 ACTIVATED DMSO REAGENTS

2.8.3.1 DMSO–Dicyclohexylcarbodiimide

The original Pfitzner–Moffatt procedure for alcohol oxidation by activated dimethyl sulfoxide utilized dicyclohexylcarbodiimide (DCC) and a source of protons such as polyphosphoric acid or pyridinium trifluoroacetate.[1] The use of strong acids such as the common mineral acids must be avoided since, although acidic conditions are initially required, the reaction must readily become basic in the later stages of the process. Mechanistically it is reasonable to suggest that the activation follows the pattern whereby initial attack of the nucleophilic sulfinyl oxygen of dimethyl sulfoxide, with the protonated carbodiimide, forms a sulfonium isourea. This is followed by displacement of dicyclohexylurea by the alcohol to form an alkoxysulfonium salt. Base treatment of this salt forms an ylide, which collapses *via* the proven cyclic mechanism to the carbonyl compound and dimethyl sulfide (Scheme 4).

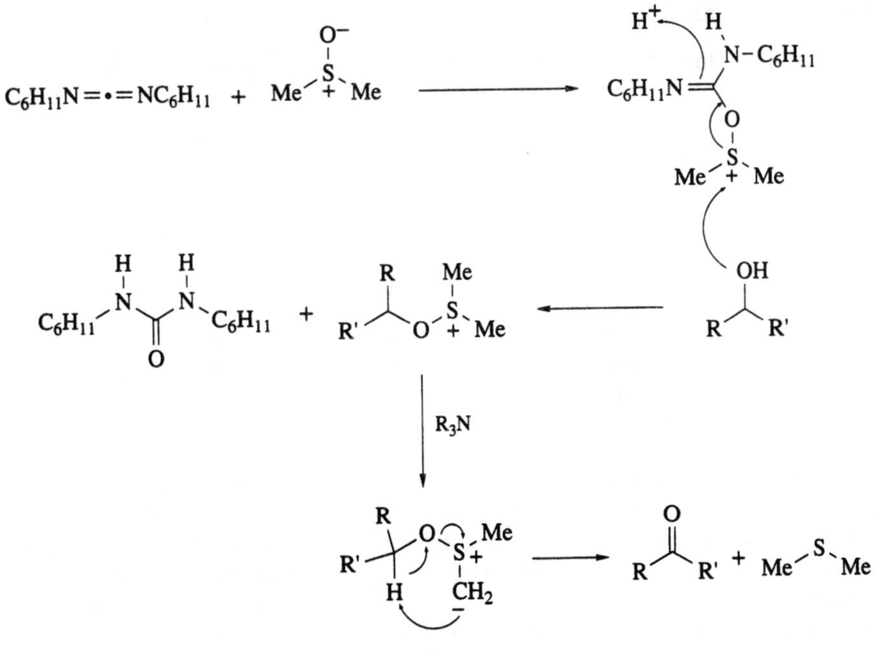

$C_6H_{11}N{=}\bullet{=}NC_6H_{11}$ + $Me_2S^+{-}O^-$ → H^+, H, $N{-}C_6H_{11}$, $C_6H_{11}N{=}$, O, $Me{-}S^+{-}Me$; OH, R, R' → $C_6H_{11}NH{-}C(O){-}NHC_6H_{11}$ + R, R', Me, O, S^+, Me; R_3N ↓ R, R', O, Me, S+, H, CH_2^- → R, R', O + $Me{-}S{-}Me$

Scheme 4

One of the disadvantages of this procedure is that a large excess of DCC is required, the residue of which is difficult to remove during work-up, as is the dicyclohexylurea formed during oxidation. These problems provided the impetus behind the development of alternative activating agents. For example the water-soluble diimide 1-(3-dimethylaminopropyl)-3-ethylcarbodiimide[12] should partially solve the work-up problem, and the use of polymer-bound DCC should permit the efficient removal of the urea by-product.[13] However, despite these limitations, much use has been made of the Pfitzner–Moffatt oxidation, especially during the oxidation of carbohydrate derivatives. For example the amino sugar derivative (**5**; equation 2) was converted in 86% yield to the corresponding ketone.[14]

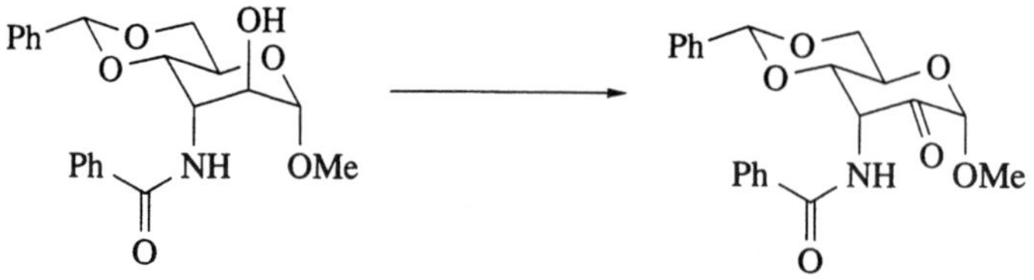

Ph O O OH O Ph NH OMe O (**5**) ⟶ Ph O O O Ph NH O OMe O (2)

This combination was also suitable for the clean conversion of the alcohol (**6**; equation 3)[15] to the corresponding aldehyde, whereas all other attempts resulted in concurrent protiodesilylation or in the formation of the conjugated aldehyde (**7**).

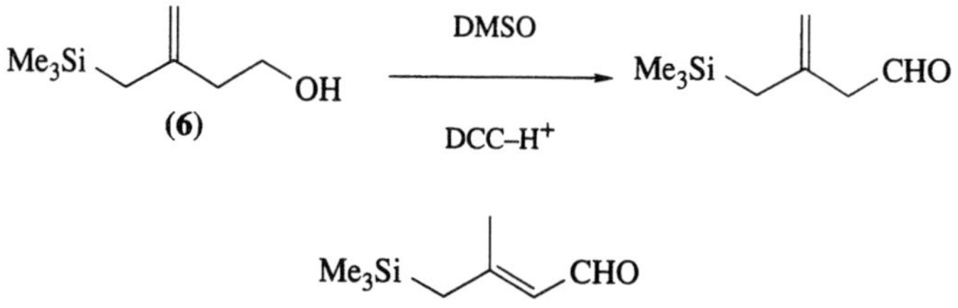

Me_3Si OH (**6**) —DMSO, DCC–H^+⟶ Me_3Si CHO (3)

Me_3Si CHO (**7**)

2.8.3.2 DMSO–Acetic Anhydride

Upon the addition of dimethyl sulfoxide to acetic anhydride the species (**8**; equation 4) is formed, which reacts with alcohols in the usual manner to give carbonyl products.

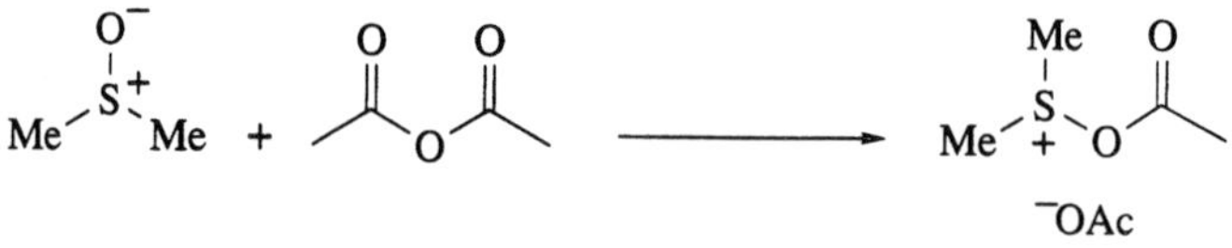

O^- S^+ Me Me + O O O ⟶ Me O S Me + O $^-$OAc (**8**) (4)

One of the advantages of using the acetic anhydride system is that the reaction can be performed at room temperature, which is in contrast to a report that trifluoroacetic anhydride is an unsatisfactory activator at this temperature.[16] Attempts have been made to use acylating activators such as benzoic anhydride, phosphorus pentoxide or polyphosphoric acid, although these tend to give lower yields of carbonyl compounds.[16]

In comparison to some of the other activation methods however, the dimethyl sulfoxide–acetic anhydride procedure has certain disadvantages. The method often requires the use of long reaction times (18–24 h), which can result in many side reactions, especially with sensitive substrates. Notable in this respect is that it is not uncommon for this procedure to result in the formation of substantial yields of the thiomethyl ethers obtained from the Pummerer rearrangement product as described above. In fact upon attempted oxidation of cholesterol with this system, the major product obtained was the corresponding (methylthio)methyl ether.[17] Acetates may also be formed if the alcohol is unhindered. For example the sugar derivative (**9**) reacts under these conditions to form an enol acetate (derived from the required carbonyl compound) in 40% yield contaminated with 30% of the acetate (**10**; equation 5).[18]

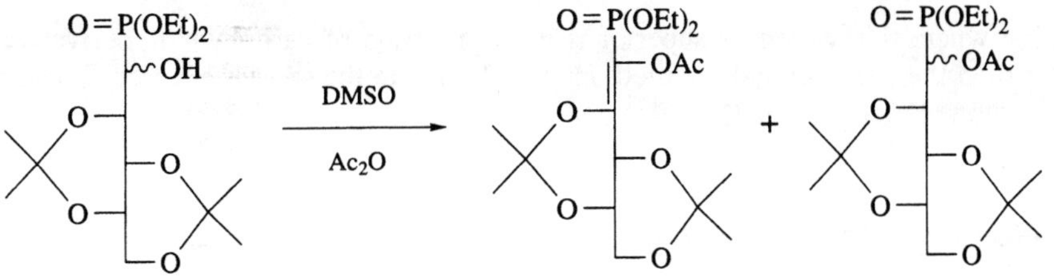

(5)

(9) **(10)**

Despite this, some useful oxidations have been achieved using the method, such as that of yohimbine to yohimbinone in 85% yield, which compares well with that achieved by using dimethyl sulfoxide activated with dicylohexylcarbodiimide.[19] The method has also been successfully applied to the oxidation of carbohydrates,[20] as shown by the formation of (**11**; equation 6), and aromatic α-diketones can be efficiently prepared using this method by oxidation of the corresponding acyloin products. Unfortunately this methodology cannot be extended to the more useful aliphatic diketones.[21]

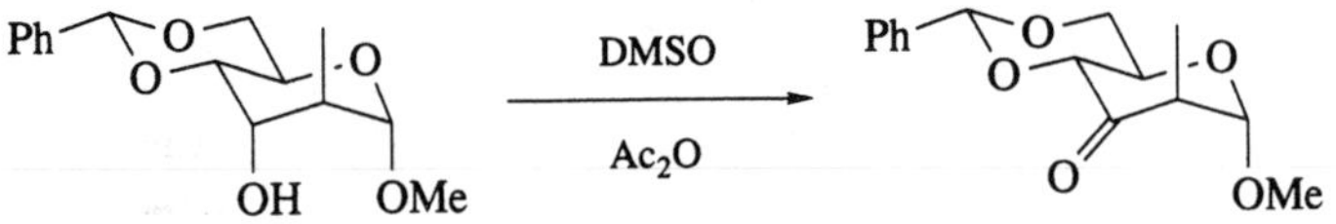

(6)

(11)

Amongst other less successful applications, however, are the oxidations of some steroidal systems such as testosterone, which afforded Δ^4-androsterone-3,17-dione in only 34% yield and 11α-hydroxyprogesterone, in which the equatorial hydroxy group was oxidized to afford a poor yield of 11-oxoprogesterone (13%).[16]

2.8.3.3 DMSO–Trifluoroacetic Anhydride

Trifluoroacetic anhydride upon mixing with dimethyl sulfoxide can undergo a violent reaction at or just below room temperature. However, it is possible to moderate this behavior by working at temperatures below –60 °C in an inert solvent such as dichloromethane,[22] when dimethyl sulfoxide and trifluoroacetic anhydride react exothermically and instantly to form a white precipitate, which is most probably the species (**12**).

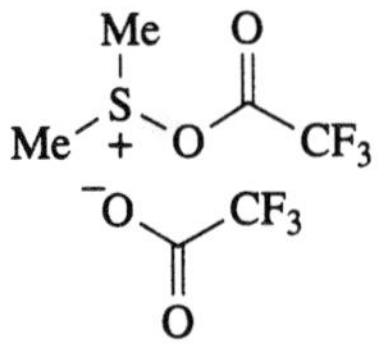

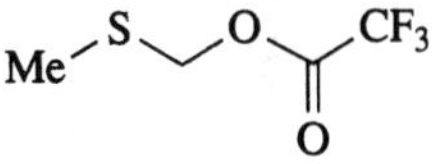

(12) **(13)**

On warming above –30 °C the mixture clears and a Pummerer rearrangement occurs to form (methylthio)methyl trifluoracetate (**13**). However the extent of this by-product formation is minimized at the lower temperature, and the reaction with alcohols gives high yields of carbonyl products over short reaction times. This makes trifluoroacetic anhydride one of the better activators for dimethyl sulfoxide oxidations.

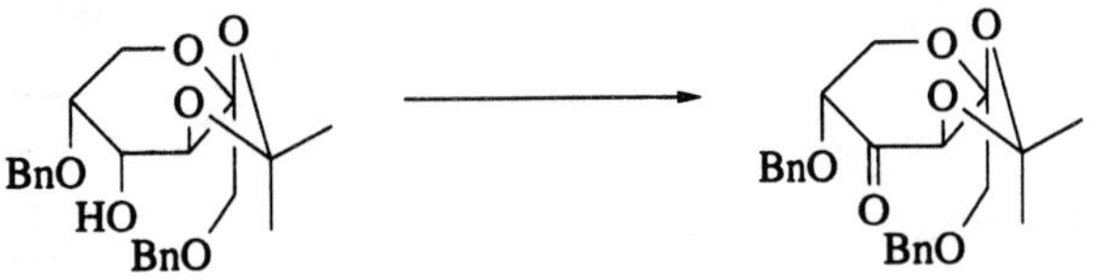

(7)

(14)

An oxidation where this system is superior is in the reaction of carbohydrate derivatives where it is notably better than the acetic anhydride method, mainly due to the shorter reaction times, *e.g.* the preparation of the ketone (**14**; equation 7).[23]

2.8.3.4 DMSO–Sulfur Trioxide/Pyridine

One of the best activators for dimethyl sulfoxide is the complex of sulfur trioxide/pyridine, which in the presence of triethylamine rapidly oxidizes primary and secondary alcohols to aldehydes and ketones in very good yields at ambient temperature.[24] This reagent also allows the very useful conversion of allylic alcohols to the corresponding α,β-unsaturated carbonyl compounds. A further advantage of this procedure over many of the others is the ease of work-up, especially over the dimethyl sulfoxide–dicyclohexylcarbodiimide method.

The precise intermediates formed in this version of the oxidation are not known, but it is possible that the necessary alkoxysulfonium intermediate is derived from a breakdown of the zwitterion (**15**; Scheme 5).

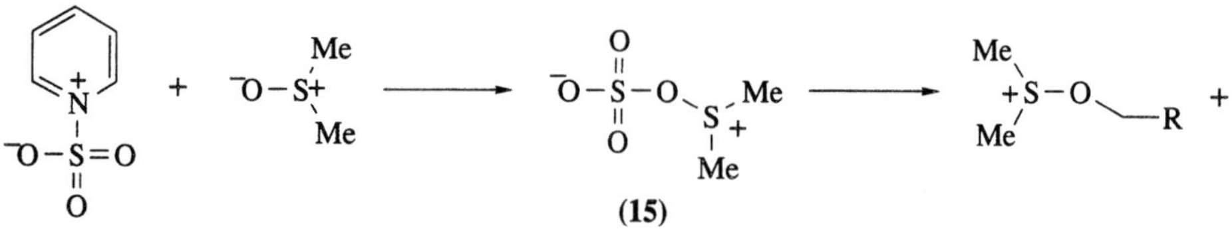

(**15**)

Scheme 5

In an example of the use of this activation method testosterone, with a 17β-hydroxy group, was oxidized to Δ^4-androstene-3,17-dione very rapidly in high yield, in contrast to the use of DMSO–acetic anhydride. During a reaction, when other oxidizing agents were found to be ineffective, sulfur trioxide/dimethyl sulfoxide led to smooth oxidation of the *cis*-diol (**16**; equation 8) to an *o*-quinone in 49% yield and the *cis*-diol (**17**) to (**18**; equation 9) in 98% yield.[25,26] The use of dimethyl sulfoxide–acetic anhydride for this oxidation gave large amounts of the diacetate as the by-product.

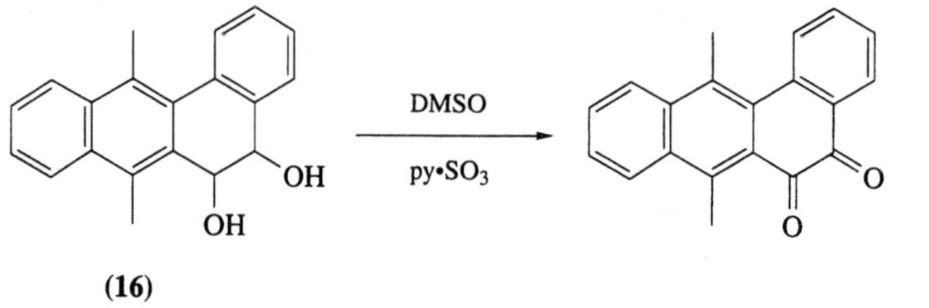

DMSO, py•SO_3 (8)

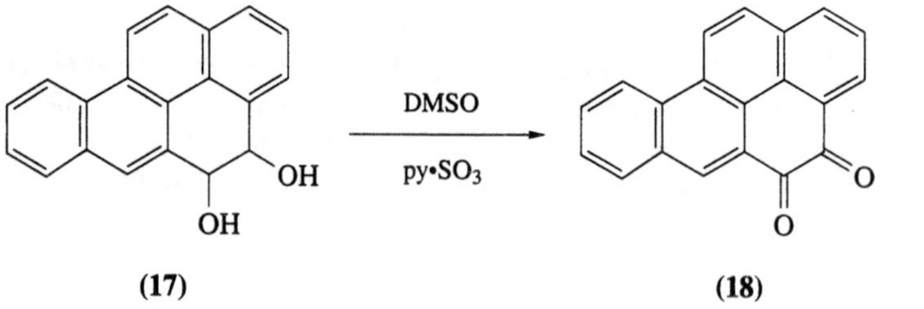

(**16**)

DMSO, py•SO_3 (9)

(**17**) (**18**)

2.8.3.5 DMSO–Oxalyl Chloride

Undoubtedly the most popular variation of these oxidations is the use of oxalyl chloride to activate the dimethyl sulfoxide, which is commonly referred to as the Swern oxidation. The advantages of the method are: the mild conditions; the ease of work-up, due to two of the main by-products being carbon monoxide and carbon dioxide; the low yields, if any, of Pummerer rearrangement products; and the fact that the reaction is usually very rapid. It is advantageous that freshly distilled oxalyl chloride is used in these reactions, and generally yields are excellent.[27,28]

Activation temperatures of about –60 °C are typically used to form the activated dimethyl sulfoxide intermediate (**19**), which arises by spontaneous loss of carbon dioxide and carbon monoxide from an initially formed salt (Scheme 6).

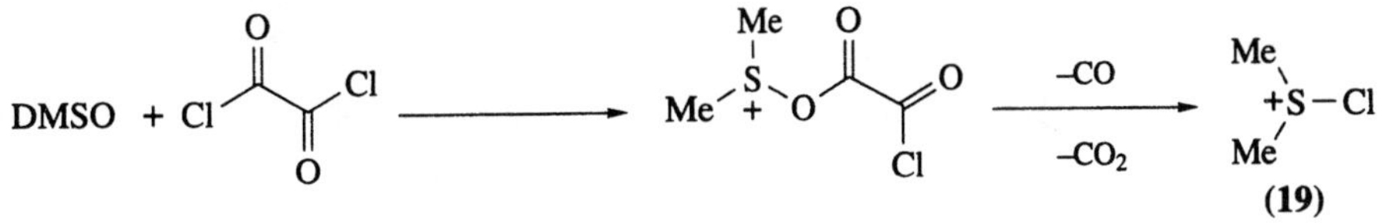

DMSO + Cl–CO–CO–Cl ⟶ Me_2S^+–O–CO–CO–Cl $\xrightarrow[-CO_2]{-CO}$ Me_2S^+–Cl

(**19**)

Scheme 6

Interestingly, this intermediate is identical to that formed in the reaction of dimethyl sulfide with chlorine, a mixture well known to be useful in the oxidation of alcohols to carbonyl compounds.[29]

The relatively mild conditions of this dimethyl sulfoxide activation method have been used to good effect to oxidize many sensitive substrates. For instance the alcohol (**20**; equation 10) was smoothly converted to the aldehyde without any racemization occurring, which can be a problem with other oxidizing systems.[30] Similar advantages were noted in the oxidation of the alcohol (**21**), thus demonstrating the compatibility of some sensitive protecting groups to the reaction conditions.[31]

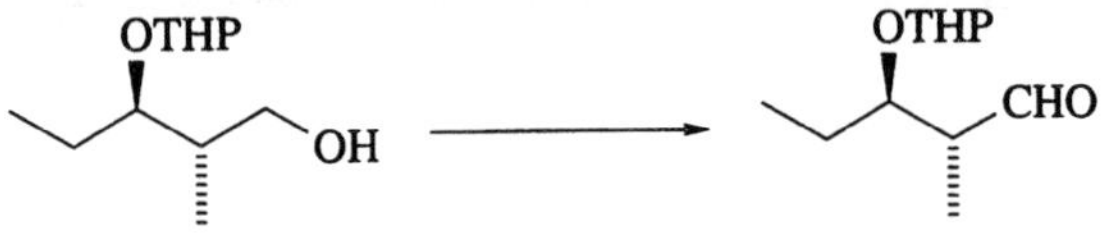

OTHP, OH ⟶ OTHP, CHO (10)

(**20**)

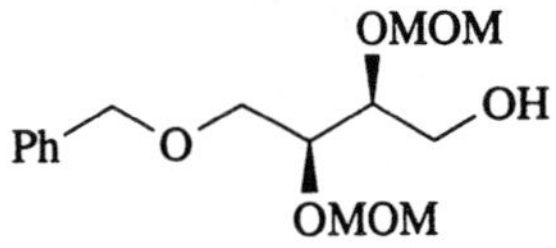

Ph, O, OMOM, OH, OMOM

(**21**)

The unsaturated branched chain sugar (**22**; equation 11) was successfully converted to an enone under the Swern conditions[32] and the allylic alcohol (**23**), containing an α,β-unsaturated amide, was smoothly oxidized by this activated dimethyl sulfoxide reaction.[33]

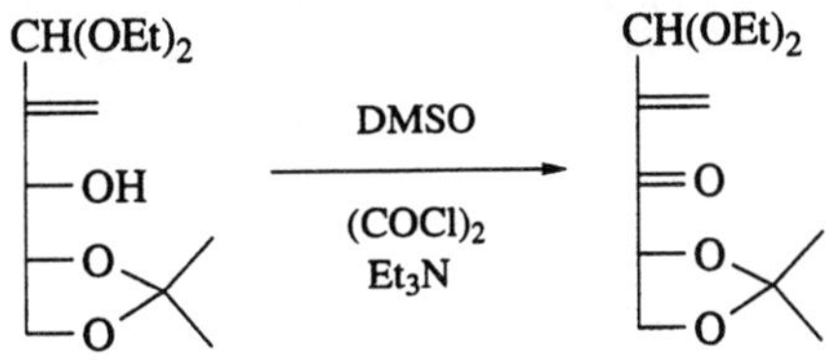

$CH(OEt)_2$, OH, O, O $\xrightarrow[\text{(COCl)}_2,\ Et_3N]{DMSO}$ $CH(OEt)_2$, O, O, O (11)

(**22**)

Ph, N, O, OH

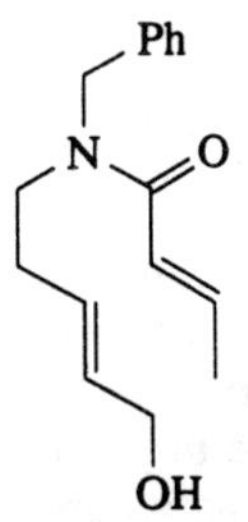

(**23**)

Amongst the more sensitive substrates which can be tolerated during the Swern oxidation is the formation of the aldehyde (**24**),[34] which normally undergoes very rapid epimerization in the presence of a trace of acid. Another example which serves to demonstrate the advantages of the Swern oxidation over other methods is the formation of the lactol (**25**), which is prepared in the presence of a sensitive vinylsilane group.[35]

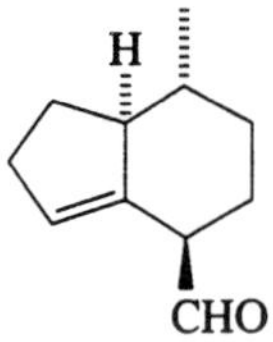

(24) 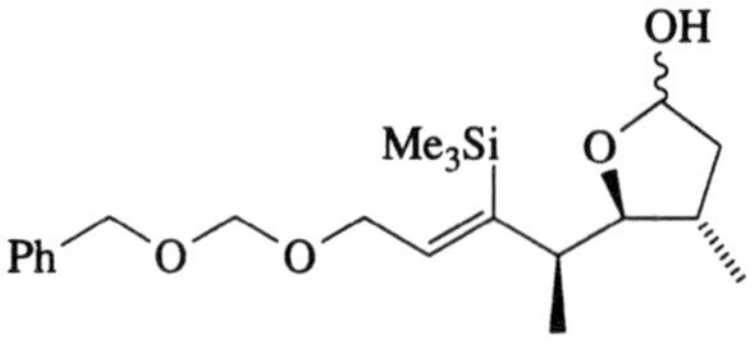 **(25)**

A very useful development of this reaction is the demonstration that trimethylsilyl-protected primary and secondary alcohols can be directly oxidized without prior deprotection,[36] but *t*-butyldimethylsilyl ethers do not similarly react.

The Swern procedure, however, is not without its problems; for instance, it has been shown that electrophilic chlorination can occur as a significant side reaction.[37] In these cases the use of trifluoroacetic or acetic anhydride as activators of dimethyl sulfoxide has been recommended.

2.8.3.6 DMSO–Thionyl Chloride

Thionyl chloride appears to be superior to trifluoroacetic anhydride as an activator of dimethyl sulfoxide during the oxidation of alcohols in terms of yields of carbonyl compounds,[27] although it has not been as widely used. The active species in this process, leading to an alkoxysulfonium species, is probably the ion pair (**26**).

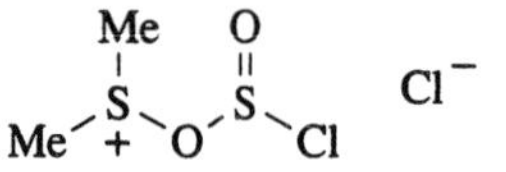

(26)

As with trifluoroacetic anhydride, activation of dimethyl sulfoxide with thionyl chloride must be carried out at low temperatures as the reaction is highly exothermic. Besides the higher yields, a further advantage of thionyl chloride to activate dimethyl sulfoxide over anhydrides is the lack of Pummerer rearrangement products or of esters formed as by-products (as long as the reactions are carried out below –60 °C). This is amply demonstrated by the oxidation of (–)-borneol which proceeds in an excellent 99% yield (equation 12).[27]

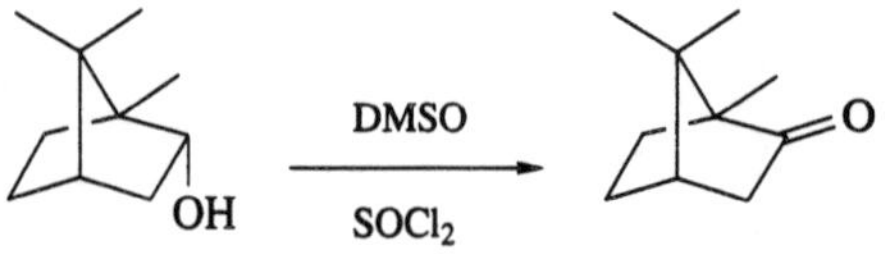

DMSO, $SOCl_2$ (12)

2.8.3.7 DMSO–Chlorine and Halogen Derivatives

Dimethyl sulfoxide and chlorine form highly reactive intermediates which are of some limited use as oxidants for alcohols. These intermediates are related to those derived from the reaction of the halogens with dimethyl sulfide and probably have a structure such as (**27**). When formed at –45 °C they allow the oxidation of primary and secondary alcohols to aldehydes and ketones[38] when used in a two-fold excess. For very simple alcohols the reaction proceeds in yields of greater than 90%, but there are considerable drawbacks if some types of additional functionality are present in the molecule, *e.g.* alkenes react very rapidly to form vicinal dichlorides.

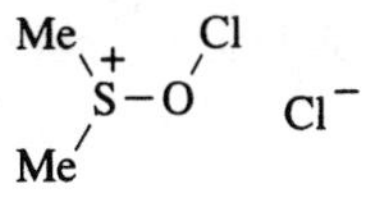

(27)

Attempted use of *N*-chlorosuccinimide or *N*-bromosuccinimide to activate dimethyl sulfoxide is limited, owing to the preferential formation of methylene acetals in good yields,[39] as illustrated in the preparation of the acetal (**28**; equation 13).

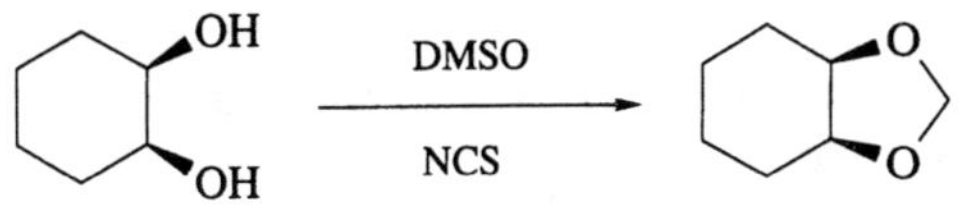

DMSO / NCS (13)

(28)

2.8.3.8 DMSO–Miscellaneous Activators

A large number of other materials have been used to activate dimethyl sulfoxide for the oxidation of primary and secondary alcohols and new methods are still being introduced. The vast majority of these reactions proceed *via* an alkoxysulfonium salt and consequently are variants of the original Pfitzner–Moffatt procedure. Very few of these methods have been exhaustively tested and their advantages are often not apparent.

Despite this they should be considered as possible alternatives in cases where more familiar methods fail. Amongst this group of activators are *p*-toluenesulfonyl chloride,[40] trifluoromethanesulfonic anhydride,[40,41] silver tetrafluoroborate,[42] molybdenum oxide,[43] phosphorous pentoxide,[44] trichloromethyl chloroformate,[45] 2-fluoro-1-methylpyridiniumsulfonate,[46] chlorosulfonyl isocyanate,[47] antimony pentachloride (for which an X-ray structure of the DMSO–$SbCl_5$ complex was obtained)[48] and phenyl dichlorophosphate.[49]

One potential activator for dimethyl sulfoxide that in practice turns out to be very poor, is phosgene.[5] However, a related oxidation of alcohols using dimethyl sulfoxide does use phosgene for the preparation of a chloroformate (or carbonochloridate) such as (**29**).[50,51] This reacts with dimethyl sulfoxide to give, after spontaneous loss of carbon dioxide, an alkoxysulfonium salt (**30**) which upon treatment with triethylamine forms the carbonyl compound (Scheme 7). Relatively little use appears to have been made of this method.

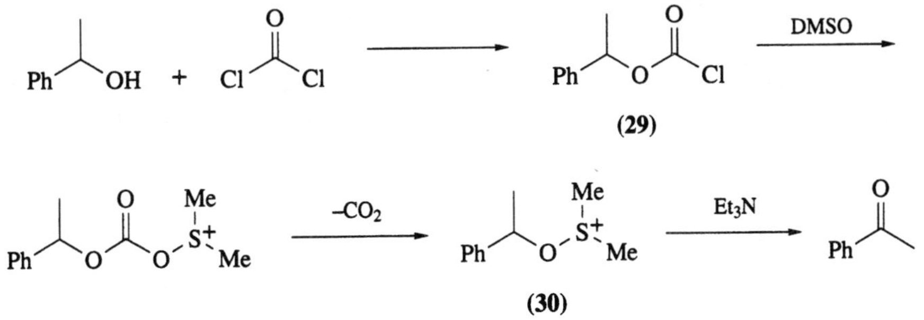

DMSO

(29)

$-CO_2$ Et_3N

(30)

Scheme 7

2.8.4 OXIDATION OF ALCOHOLS

The above discussion has concentrated upon the reagents used, but it is equally of value to comment on the substrate, particularly in reactions for which other oxidation methods have been reported to fail. A good example is the oxidation of the iron–carbonyl complex (**31**) to the ketone (**32**; equation 14). The use of dimethyl sulfoxide activated with sulfur trioxide–pyridine complex gave a 70% yield of the product, in contrast to the use of the Pfitzner–Moffatt procedure (dimethyl sulfoxide–DCC) or the chromium trioxide/pyridine complex, both of which caused the alcohol (**31**) to fragment to benzaldehyde and benzene.[52]

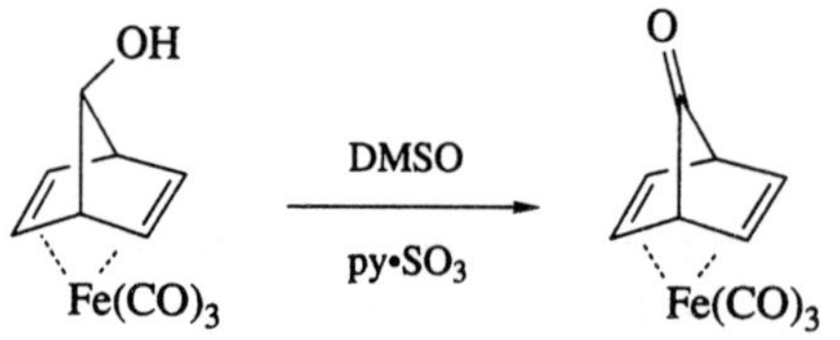

DMSO, py•SO_3 (14)

$Fe(CO)_3$ $Fe(CO)_3$

(**31**) (**32**)

The sulfur trioxide–pyridine activated dimethyl sulfoxide oxidation was also a key step in an excellent synthesis of the Prelog–Djerassi lactonic acid, being highly recommended as the best method to avoid epimerization of the C-2 center in the aldehyde (**33**).[53]

CHO

(**33**)

An area of recent intense synthetic endeavor has been in the synthesis of the avermectins and milbemycins, which contain a range of highly reactive functionalities such as the spiroacetal group, double bonds, and epimerizable centers. However, dimethyl sulfoxide activated oxidations, most notably the Swern variation, have been useful in this area of chemistry. For example, the sensitive spiroacetal (**34**) gave 92% of the derived aldehyde using dimethyl sulfoxide–oxalyl chloride–triethylamine at –50 °C (equation 15).[54]

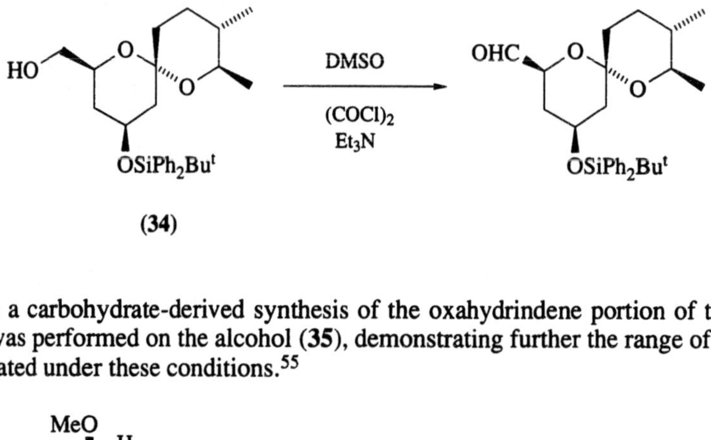

HO ... $OSiPh_2Bu^t$ DMSO, $(COCl)_2$, Et_3N OHC ... $OSiPh_2Bu^t$ (15)

(**34**)

Additionally, in a carbohydrate-derived synthesis of the oxahydrindene portion of the avermectins, a Swern oxidation was performed on the alcohol (**35**), demonstrating further the range of functionality that can be accommodated under these conditions.[55]

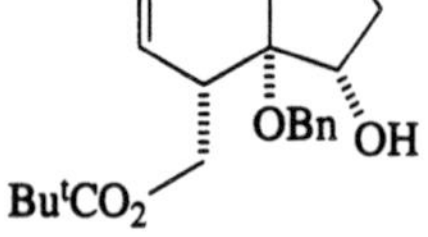

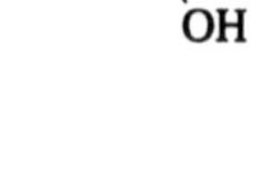

MeO H O OBn OH Bu^tCO_2 Bu OH Bu OH

(**35**) (**36**) (**37**)

One class of compounds that does not react particularly well in activated dimethyl sulfoxide oxidations, however, is the alkynic alcohols, and only a few successful examples are known, *e.g.* (**36**) and (**37**).[5]

α-Diketones are also an important class of compounds which may be obtained very easily by using Swern oxidation of the sensitive α-hydroxy ketone precursors,[56] or vicinal diols.[57] Amongst such oxidations are the preparations of the α-diketones (**38**) and (**39**).

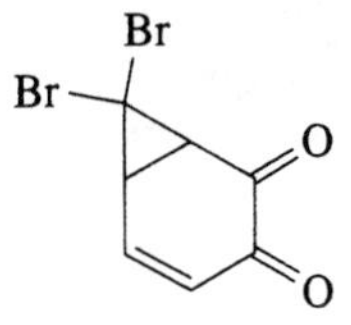

(38)

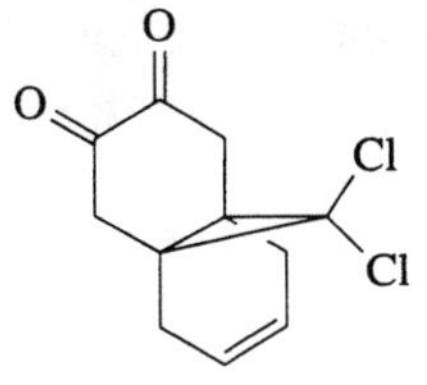

(39)

Activated dimethyl sulfoxide oxidations have been fairly well used in the synthesis of monoterpenes, as seen in the oxidation to an aldehyde of the alcohols (**40**), *en route* to loganin aglycone,[58] and (**41**), a precursor in a synthesis of specionin (equation 16).[59]

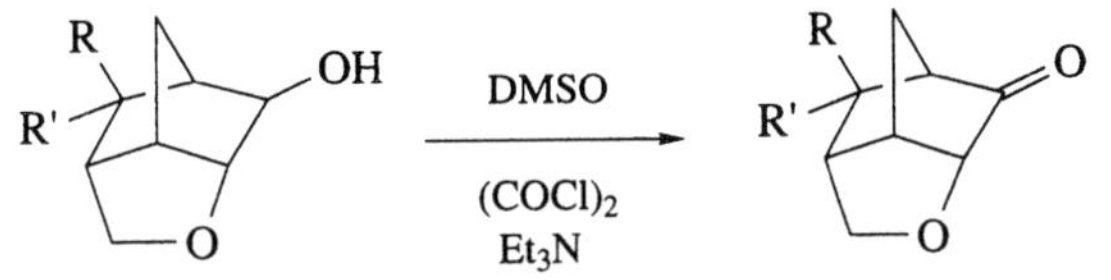

DMSO, $(COCl)_2$, Et_3N (16)

(**40**) R = H, R' = Me

(**41**) R = $CH_2OSiMe_2Bu^t$, R' = H

However, relatively less use of the reaction appears to have been made in the preparation of the higher terpenes, but some notable exceptions to this are the use of a Moffatt oxidation as a key step in an unusual approach to hydrazulene-based sesquiterpenes, (**42**) to (**43**; equation 17),[60] and the use of the Swern variation in the gibberellin synthesis (**44**) to (**45**; equation 18).

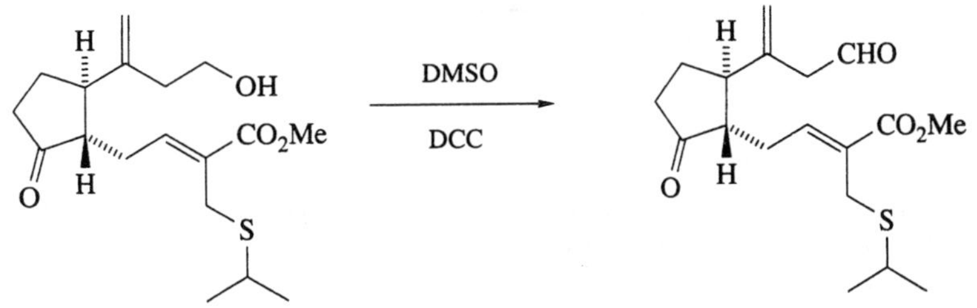

DMSO, DCC (17)

(**42**) (**43**)

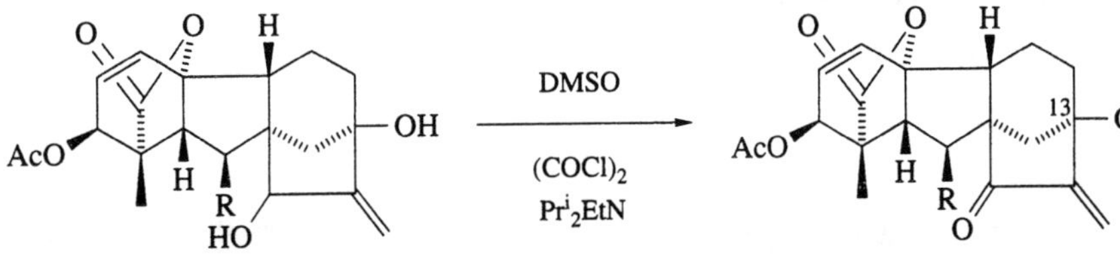

DMSO, $(COCl)_2$, $Pr^i{}_2EtN$ (18)

(**44**) R = $CO_2CH_2CO_2Ph$ (**45**)

Interestingly, this last case is one of the few reports which notes a dependence upon the base used in the oxidation. The use of diisopropylethylamine gives the ketone shown, whereas the usual triethylamine base forms the ketone containing the 13-chloro group, for reasons which are unclear.[61]

As seen above, activated dimethyl sulfoxide oxidations are often used to prepare carbonyl compounds which are highly sensitive and which do not withstand the more vigorous conditions of other, more traditional oxidants such as the Cr^{VI}-based reagents. A striking example of this comes from a series of studies aimed at preparing sensitive carbonyl compounds *in situ*, and subjecting them to further reaction such as a Wittig condensation.[62] The best oxidant for these systems is the dimethyl sulfoxide–oxalyl chloride–triethylamine mixture, *i.e.* the Swern oxidation. The fact that the carbonyl compounds could be used without purification to give high yields of products serves to emphasize that the by-products in these oxidations, carbon monoxide, carbon dioxide, dimethyl sulfide and triethylamine hydrochloride are relatively innocuous. In this study, for which other oxidation systems failed to give clean products, the ketone

(**46**) and the aldehydes (**47**) to (**49**) were prepared and used crude without the need for further purification.

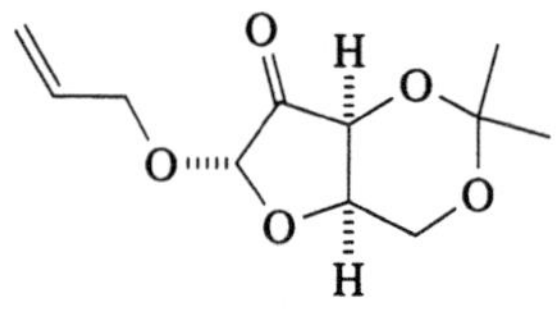

(**46**) 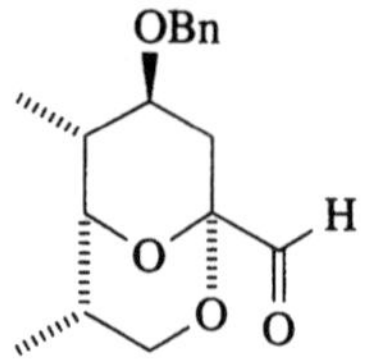 (**47**)

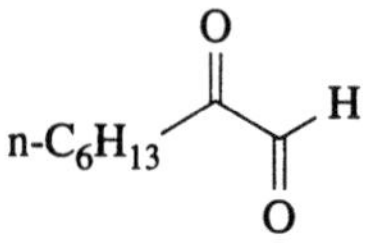

(**48**) 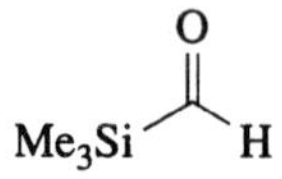 (**49**)

A further, extremely good demonstration of the advantages offered by these reactions is in the oxidation of the alcohol (**50**; equation 19), a precursor to some unusual prostaglandin analogs, which proceeds in 65% yield (with concurrent epimerization) using the Moffatt procedure (DMSO–DCC–CF_3CO_2H). In contrast, the use of Ac_2O–DMSO, DMS–Cl_2, PDC, Ag_2CO_3–celite, Collins reagent or Jones oxidation all failed.[63]

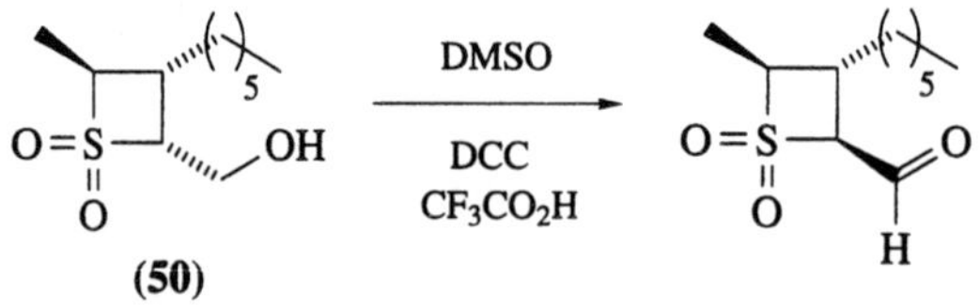

DMSO, DCC, CF_3CO_2H (19)

(**50**)

Very few reports of competitive reactions of alcohols have appeared in activated dimethyl sulfoxide oxidations, but the results obtained from these limited studies are interesting. It is known that dimethyl sulfoxide–trifluoroacetic anhydride oxidation is selective for primary or secondary alcohols in the presence of benzylic or allylic alcohols, due to the latter alcohols being preferentially converted to trifluoroacetates.[64] Additionally, secondary alcohols are known to be more reactive than primary alcohols in these oxidations, although the measured rate differences are not very high[65] and selectivity cannot be guaranteed.

2.8.5 CONCLUSIONS

The above discussion highlights the great synthetic utility of activated dimethyl sulfoxide oxidations in organic chemistry. The enormous amount of effort put into developing these procedures has resulted in a clear picture of their relative value, so enabling one to easily assess the method of choice for a particular oxidation. The popularity of the Swern oxidation reflects the very real advantages that it offers in terms of the mild conditions and high yields. However, there are many instances where alternative activators of dimethyl sulfoxide are better and it is wise to assess these in any synthetic scheme.

The range of substrates which have been oxidized by activated dimethyl sulfoxide covers a wide range of tolerant functional groups, and there are many reports of how these oxidations proved superior to others that were investigated.

Thus activated dimethyl sulfoxide oxidations are now well established as standard synthetic methods and are familiar to all organic chemists, and will continue into the future as new variations are developed.

2.8.6 REFERENCES

1. K. E. Pfitzner and J. G. Moffatt, *J. Am. Chem. Soc.*, 1963, **85**, 3027.
2. N. Kornblum, J. W. Powers, G. J. Anderson, W. J. Jones, H. O. Larson, O. Levand and W. W. Weaver, *J. Am. Chem. Soc.*, 1957, **79**, 6562.
3. W. W. Epstein and F. W. Sweat, *Chem. Rev.*, 1967, **67**, 247.
4. T. Durst, in 'Advances in Organic Chemistry: Methods and Results', ed. E. C. Taylor and H. Wynberg, Wiley, New York, 1969, vol. 6, p. 285.
5. A. J. Mancuso and D. Swern, *Synthesis*, 1981, 165.
6. A. H. Fenselau and J. G. Moffatt, *J. Am. Chem. Soc.*, 1966, **88**, 1762.
7. C. R. Johnson and W. G. Phillips, *J. Org. Chem.*, 1967, **32**, 1926.
8. K. Torssell, *Tetrahedron Lett.*, 1966, 4445.
9. F. W. Sweat and W. W. Epstein, *J. Org. Chem.*, 1967, **32**, 835.
10. E. J. Corey and C. U. Kim, *Tetrahedron Lett.*, 1974, 287.
11. G. M. Cree, D. M. Mackie and A. Perlin, *Can. J. Chem.*, 1969, **47**, 511.
12. J. C. Sheehan, P. A. Cruickshank and G. L. Boshart, *J. Org. Chem.*, 1961, **26**, 2525.
13. N. M. Weinshenker and C.-U. Shen, *Tetrahedron Lett.*, 1972, 3285.
14. B. R. Baker and D. H. Buss, *J. Org. Chem.*, 1965, **30**, 2308.
15. T. V. Lee, R. J. Boucher and C. J. M. Rockell, *Tetrahedron Lett.*, 1988, **29**, 689.
16. J. D. Albright and L. Goldman, *J. Am. Chem. Soc.*, 1967, **89**, 2416.
17. S. M. Ifzal and D. A. Wilson, *Tetrahedron Lett.*, 1967, 1577.
18. Z. I. Glebova, L. A. Uzlova and Yu. A. Zhdanov, *Zh. Obshch. Khim.*, 1985, **55**, 1435 (*Chem. Abstr.*, 1986, **104**, 69 072).
19. J. D. Albright and L. Goldman, *J. Org. Chem.*, 1965, **30**, 1107.
20. S. Hanessian and G. Rancourt, *Can. J. Chem.*, 1977, **55**, 1111.
21. M. S. Newman and C. C. Davis, *J. Org. Chem.*, 1967, **32**, 66.
22. A. K. Sharma, T. Ku, A. D. Dawson and D. Swern, *J. Org. Chem.*, 1975, **40**, 2758.
23. I. I. Cubero and M. T. P. Lopez-Espinosa, *J. Carbohydr. Chem.*, 1985, **5**, 229.
24. J. R. Parikh and W. von E. Doering, *J. Am. Chem. Soc.*, 1967, **89**, 5505.
25. S. H. Goh and R. G. Harvey, *J. Am. Chem. Soc.*, 1973, **95**, 242.
26. R. G. Harvey, S. H. Goh and C. Cortez, *J. Am. Chem. Soc.*, 1975, **97**, 3468.
27. K. Omura and D. Swern, *Tetrahedron*, 1978, **34**, 1651.
28. A. J. Mancuso, S.-L. Huang and D. Swern, *J. Org. Chem.*, 1978, **43**, 2480.
29. E. J. Corey and C. U. Kim, *J. Am. Chem. Soc.*, 1972, **94**, 7586.
30. M. Hondo, T. Katsuki and H. Yamaguchi, *Tetrahedron Lett.*, 1984, **25**, 3857.
31. H. Iida, N. Yamazaki and C. Kibayashi, *Tetrahedron Lett.*, 1985, **26**, 3255.
32. J.-C. Depezay, M. Saniere and D. Mansuy, *Carbohydr. Res.*, 1983, **117**, 313.
33. S. F. Martin and B. Benage, *Tetrahedron Lett.*, 1984, **25**, 4863.
34. R. Baudouy, J. Sartoretti and F. Choplin, *Tetrahedron*, 1983, **39**, 3293.
35. K. Suzuki, K. Tomooka, T. Matsumoto, E. Katayama and G. Tsuchihashi, *Tetrahedron Lett.*, 1985, **26**, 3711.
36. C. M. Alfonso, M. T. Barros and C. D. Maycock, *J. Chem. Soc., Perkin Trans. 1*, 1987, 1221.
37. A. B. Smith, III and T. L. Leenay, *Tetrahedron Lett.*, 1988, **29**, 49.
38. E. J. Corey and C. U. Kim, *Tetrahedron Lett.*, 1973, 919.
39. S. Hanessian, G. Yang-Chung, P. Lavallee and A. Pernet, *J. Am. Chem. Soc.*, 1972, **94**, 8929.
40. J. D. Albright, *J. Org. Chem.*, 1974, **39**, 1977.
41. J. B. Hendrickson and S. M. Schwartzmann, *Tetrahedron Lett.*, 1975, 273.
42. B. Ganem and R. K. Boeckman, Jr., *Tetrahedron Lett.*, 1974, 917.
43. Y. Masuyama, A. Tsuhako and Y. Kurusu, *Tetrahedron Lett.*, 1981, **22**, 3973.
44. K. Onodera, S. Hirano and N. Kashimura, *J. Am. Chem. Soc.*, 1965, **87**, 4651.
45. S. Takano, K. Inomata, S. Tomita, M. Yanase, K. Samizu and K. Ogasawara, *Tetrahedron Lett.*, 1988, **29**, 6619.
46. K. Hojo and T. Mukaiyama, *Chem. Lett.*, 1978, 369.
47. G. A. Olah, Y. D. Vankar and M. Arvanaghi, *Synthesis*, 1980, 141.
48. J. Yamamoto, S. Ito, T. Tsuboi and K. Tsukihara, *Bull. Chem. Soc. Jpn.*, 1985, **58**, 470.
49. H. S. Lui and M. Nyarguhi, *Tetrahedron Lett.*, 1988, **29**, 3167.
50. D. H. R. Barton, B. J. Garner and R. H. Wightman, *J. Chem. Soc.*, 1964, 1855.
51. D. H. R. Barton and C. P. Forbes, *J. Chem. Soc., Perkin Trans. 1*, 1975, 1614.
52. J. M. Landesberg and J. Sieczkowski, *J. Am. Chem. Soc.*, 1971, **93**, 972.
53. D. A. Evans and J. Bartroli, *Tetrahedron Lett.*, 1982, **23**, 807.
54. D. R. Williams, B. A. Barner, K. Nishitani and J. G. Phillips, *J. Am. Chem. Soc.*, 1982, **104**, 4708.
55. M. Prashad and B. Fraser-Reid, *J. Org. Chem.*, 1985, **50**, 1564.
56. A. J. Mancuso, D. S. Brownfain and D. Swern, *J. Org. Chem.*, 1979, **44**, 4148.
57. C. M. Amon, M. G. Banwell and G. L. Gravatt, *J. Org. Chem.*, 1987, **52**, 4851.
58. P. Callant, P. Storme, E. Van der Eycken and M. Vandewalle, *Tetrahedron Lett.*, 1983, **24**, 5797.
59. E. Van der Eycken, J. Van der Eycken and M. Vandewalle, *J. Chem. Soc., Chem. Commun.*, 1985, 1719.
60. M. F. Semmelhack, A. Yamashita, J. C. Tomesch and K. Hirotsu, *J. Am. Chem. Soc.*, 1978, **100**, 5565.
61. S. C. Dolan and J. MacMillan, *J. Chem. Soc., Perkin Trans. 1*, 1985, 2741.
62. R. E. Ireland and D. W. Norbeck, *J. Org. Chem.*, 1985, **50**, 2198.
63. D. N. Jones, T. P. Kogan and R. F. Newton, *J. Chem. Soc., Perkin Trans. 1*, 1982, 1333.
64. K. Omura, A. K. Sharma and D. Swern, *J. Org. Chem.*, 1976, **41**, 957.
65. M. Marx and T. T. Tidwell, *J. Org. Chem.*, 1984, **49**, 788.

2.9
Oxidation Adjacent to Oxygen of Alcohols by Other Methods

GARRY PROCTER
University of Salford, UK

2.9.1 INTRODUCTION

The oxidation of an alcohol to a carbonyl compound is a fundamental reaction which is encountered at all levels of organic synthesis. As might be expected there are numerous methods and reagents which are available for carrying out this transformation and two of the most important general types have been dealt with in the previous two chapters (2.7 and 2.8) of this volume. This chapter is concerned with the oxidation of alcohols using methods which are not based on chromium reagents or 'activated DMSO'. An attempt has been made to include reagents whose worth has already been demonstrated by their use in particular syntheses, or which appear to have some potential for use in organic synthesis by virtue of their selectivity. The usefulness of a reagent for organic synthesis is critically dependent upon its selectivity, mildness and availability *inter alia*. The organization of this chapter into specific synthetic transformations (rather than a classification based on reagent type) was chosen with the practising synthetic chemist in mind. In the author's experience the need to consider a range of possible reagents for a specific selective transformation arises more often than the need to know all about one specific reagent. This latter information is usually easy to obtain from reference sources or original literature.[1] In keeping with the aims of this work the main emphasis will be on reagents which exhibit selectivity of some kind.

2.9.2 OXIDATION OF PRIMARY ALCOHOLS

2.9.2.1 Selective Preparation of Aldehydes

The preparation of aldehydes by the oxidation of primary alcohols is often possible using reagents which also oxidize secondary alcohols to ketones, and some overlap with the section of this chapter dealing with the preparation of ketones is inevitable. Moreover some reagents which oxidize primary alcohols to aldehydes will also cleave 1,2-diols, and this will be pointed out where it is known. Notwithstanding these comments it is possible to carry out the oxidation of primary alcohols to aldehydes in the presence of other groups which themselves are easily oxidized. The extent of this chemoselectivity depends upon the reagent, for example some reagents will oxidize primary alcohols in the presence of other oxidizable hydroxy groups. On the whole, benzylic and allylic alcohols (and correspondingly 'activated' alcohols) are more easily oxidized than alcohols in which the hydroxy group is not 'activated' in this way. Not surprisingly there are many examples of chemoselective oxidation in which such an 'activated' alcohol is oxidized in the presence of other hydroxy groups. The first part of this section will cover some of the more recent examples of reagents which exhibit chemoselectivity.

As stated above the oxidation of allylic or benzylic alcohols is usually particularly easy, and the chemoselective oxidation of such an alcohol in the presence of other oxidizable hydroxy groups is one of the 'standard' examples of reagent selectivity in oxidation. The classical reagent is manganese dioxide in one form or another, and several reviews have appeared on this reagent.[2] Examples of the use of 'active' manganese dioxide are given in Scheme 1.[3–8] It must be emphasized that the success or otherwise of this reagent can depend critically on the method of preparation, and a comparison has been made of various samples of manganese dioxide for the oxidation of the allylic alcohols geraniol and nerol (**1**) to the corresponding aldehydes.[9]

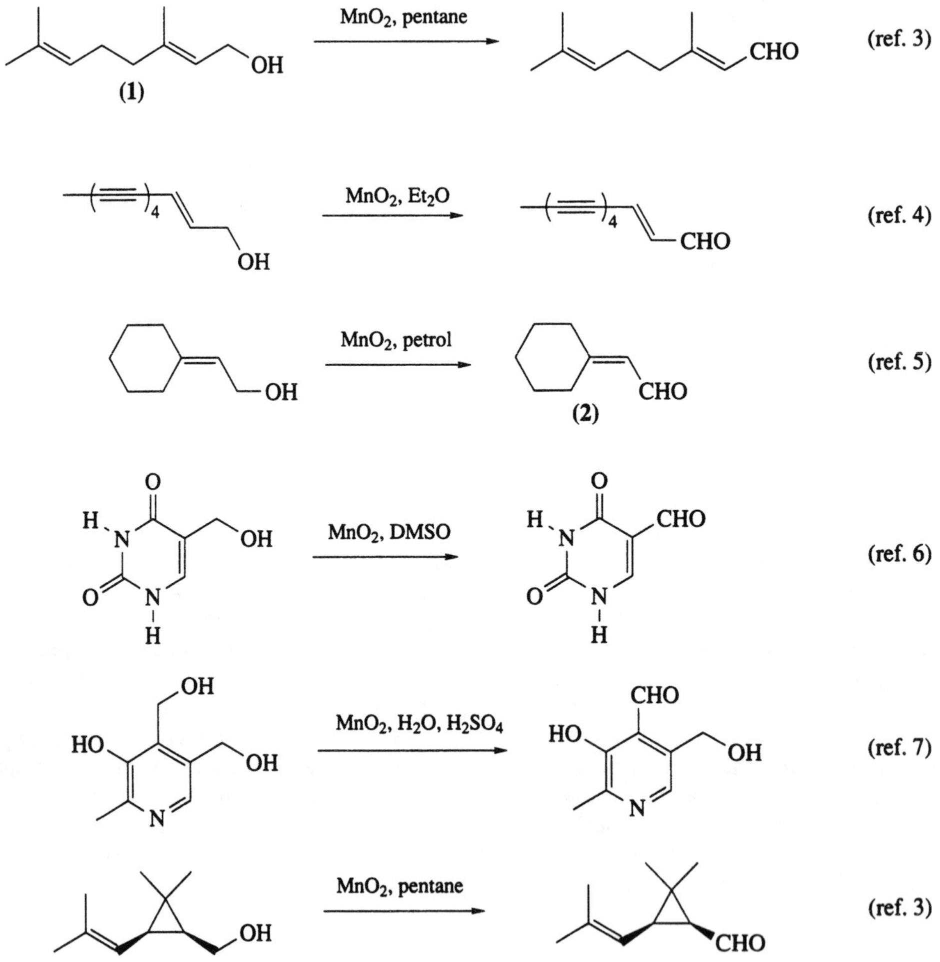

(**1**) OH — MnO_2, pentane → CHO (ref. 3)

OH — MnO_2, Et_2O → CHO (ref. 4)

OH — MnO_2, petrol → CHO (**2**) (ref. 5)

OH — MnO_2, DMSO → CHO (ref. 6)

OH — MnO_2, H_2O, H_2SO_4 → CHO (ref. 7)

OH — MnO_2, pentane → CHO (ref. 3)

Scheme 1

This reagent can be of value not only for its inherent chemoselectivity, but also because of the mild conditions under which oxidation occurs. For example the cyclohexylideneacetaldehydes (**2**) can be produced by manganese dioxide oxidation of the allylic alcohols despite the instability of (**2**) to air, acids and bases.[5] Manganese dioxide is known to cleave 1,2-diols,[10] and can cause oxidative rearrangement to take place in some situations, for example the inositol derivative (**3**) is converted into the lactone (**4**; equation 1).[11]

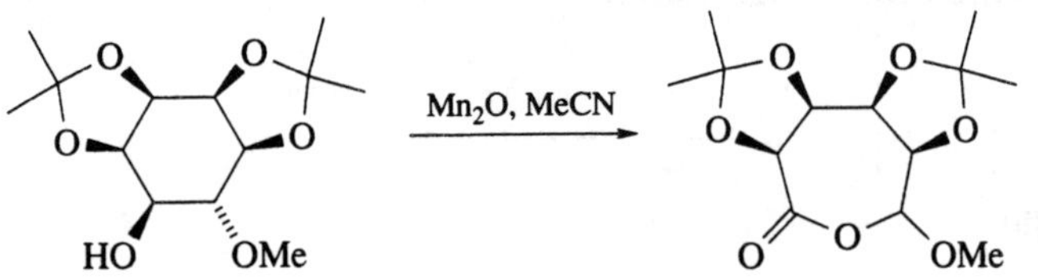

$$\xrightarrow{Mn_2O,\ MeCN} \quad (1)$$

(**3**) (**4**)

Several other reagents will carry out similar chemoselective oxidations of allylic and benzylic alcohols, some of which are outlined here. Of the other manganese-based oxidants which will selectively oxidize allylic primary alcohols to aldehydes, barium manganate ($BaMnO_4$) is possibly the most useful.[12] Primary allylic alcohols can be oxidized by this reagent in dichloromethane at room temperature, and yields are equivalent to, or better than those obtained with manganese dioxide. Being easy to handle, and needing no activation, barium manganate can be particularly useful for large scale reactions.[12] It is of particular interest that barium manganate will give good yields of *vic*-dials (equation 2),[13] since these can be difficult to obtain by direct oxidation of the corresponding diol (however, if the hydroxy groups have sufficiently different reactivities, lactones can be produced in this reaction, see later). The selectivity of this reagent was used in studies on the synthesis of cinnamolide and polygodiol, in which the diol (**5**; equation 3) was selectively oxidized to the lactol (**6**), excess reagent giving the corresponding lactone.[14]

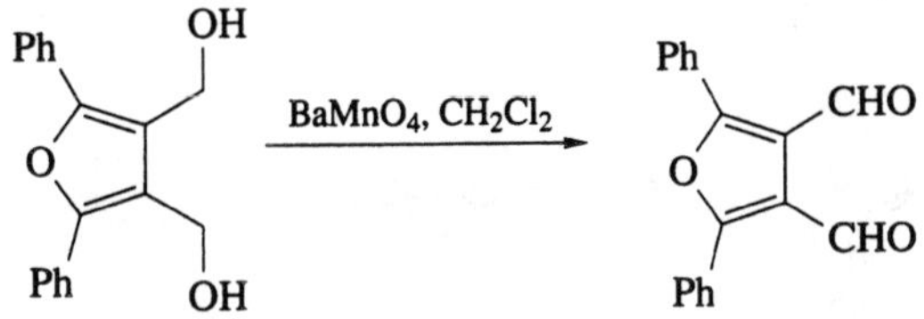

$$\xrightarrow{BaMnO_4,\ CH_2Cl_2} \quad (2)$$

$$\xrightarrow{BaMnO_4,\ CH_2Cl_2} \quad (3)$$

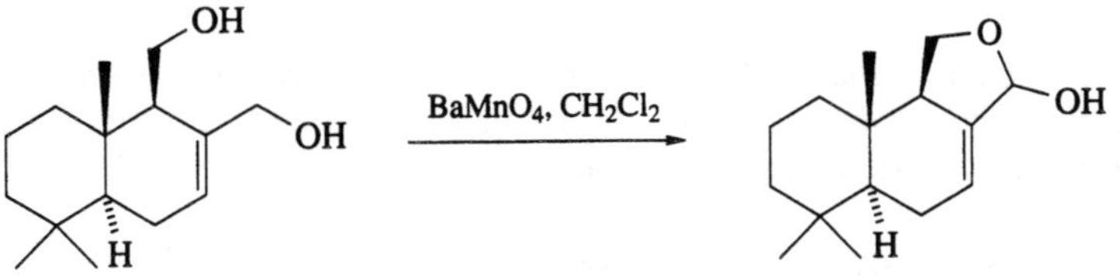

(**5**) (**6**)

Organoselenium reagents have been observed to exhibit selectivity for the oxidation of allylic alcohols, for example a catalytic amount of dimesityl diselenide with *t*-butyl hydroperoxide as cooxidant will oxidize benzylic and allylic alcohols in the presence of saturated alcohols, as in the case of the diol (**7**; equation 4).[15]

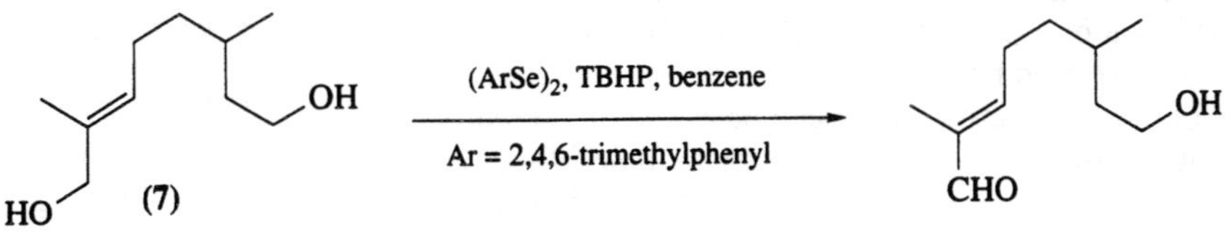

$$\xrightarrow[Ar\,=\,2,4,6\text{-trimethylphenyl}]{(ArSe)_2,\ TBHP,\ benzene} \quad (4)$$

(**7**)

μ-Oxo-bis(chlorotriphenylbismuth) (**8**; equation 5) has also been used for this type of oxidation, and can exhibit very useful levels of chemoselectivity.[16]

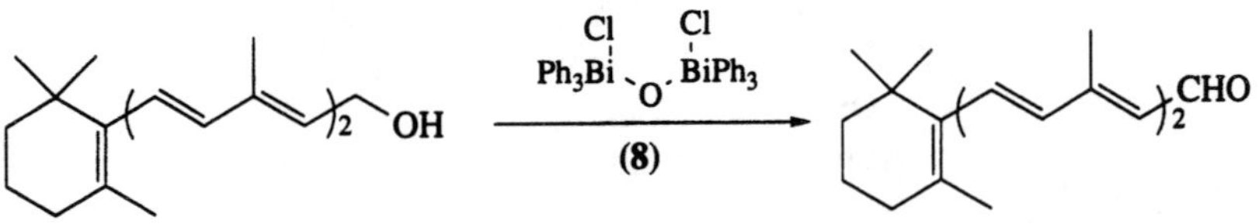

$$\xrightarrow[(\mathbf{8})]{Ph_3Bi(Cl)\text{–}O\text{–}BiPh_3(Cl)} \quad (5)$$

The selective oxidation of a primary allylic (**9**; equation 6) or benzylic (**10**; equation 7) alcohol in the presence of a secondary alcohol has been carried out using bis(trimethylsilyl) peroxide in the presence of bis(triphenylphosphine)ruthenium(II) chloride.[17]

$(Me_3SiO)_2$, $RuCl_3(PPh_3)_3$, CH_2Cl_2 (6)

(**9**)

$(Me_3SiO)_2$, $RuCl_3(PPh_3)_3$, CH_2Cl_2 (7)

(**10**)

Hydrated ruthenium dioxide will act as a catalyst for the oxidation of primary allylic alcohols (equations 8 and 9) in an oxygen atmosphere (a trace of the antioxidant 2,6-di-*t*-butyl-4-methylphenol is required to prevent autoxidation of the aldehyde to the acid).[18] The oxidation is not accompanied by any loss in double bond stereochemistry, secondary allylic alcohols are oxidized but at a decreased rate, and saturated alcohols are scarcely oxidized at all. However, α-hydroxy ketones and α-hydroxylactones will oxidize under forcing conditions, so there is clearly likely to be some degree of substrate dependence.[18]

RuO_2, O_2, $ClCH_2CH_2Cl$ (8)

RuO_2, O_2, $ClCH_2CH_2Cl$ (9)

Primary benzylic alcohols (equation 10) can be oxidized in the presence of saturated primary alcohols using a catalyst derived from ammonium cerium(IV) nitrate supported on charcoal with air as the cooxidant (under these conditions α-hydroxy ketones are oxidized to α-diketones).[19]

$(NH_4)_2Ce(NO_3)_6$/charcoal, O_2, toluene (10)

The chemoselective oxidation of a primary alcohol in the presence of a secondary alcohol is a somewhat more difficult task. Not only is the inherent difference in reactivity less than in the case of the selective oxidation of allylic alcohols discussed above, but most reagents will oxidize secondary alcohols somewhat more rapidly than primary alcohols. Nevertheless there are reagents which will carry out the selective oxidation of a primary alcohol to an aldehyde without oxidizing a secondary alcohol, some of which will be considered here.

This type of chemoselectivity has been observed in the oxidations of alcohols using copper(II) chloride and a catalytic quantity of 2,2,6,6-tetramethylpiperidinyl-1-oxyl (**11**; equation 11).[20] The oxidizing species which is generated, in for example the oxidation of (**12**) to (**13**), is the cation (**14**), which may also be produced using electrochemical oxidation (in the presence of the weak base 2,6-lutidine). Allylic and benzylic alcohols are easily oxidized by this reagent, whereas secondary alcohols react slowly.

(**11**) oxidant (**14**) (11)

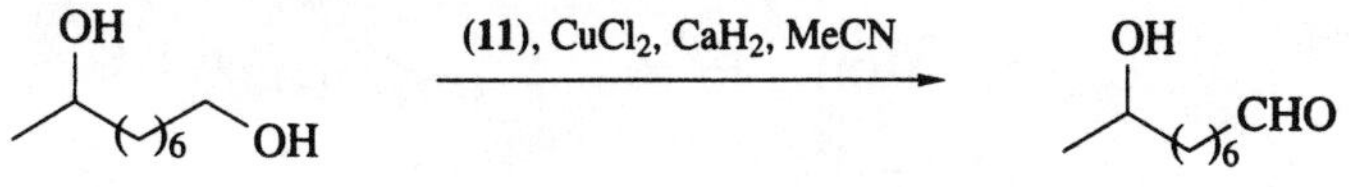

OH (12) → (11), $CuCl_2$, CaH_2, MeCN → OH, CHO (13) (12)

An interesting example of this type of chemoselective oxidation has been reported with the reagent mixture derived from diisopropyl sulfide and *N*-chlorosuccinimide.[21] This reagent will oxidize selectively a primary alcohol to an aldehyde at 0 °C. Surprisingly, this same reagent at –78 °C will oxidize selectively a secondary alcohol to the corresponding ketone (Scheme 2). Allylic and benzylic alcohols are oxidized at both temperatures.

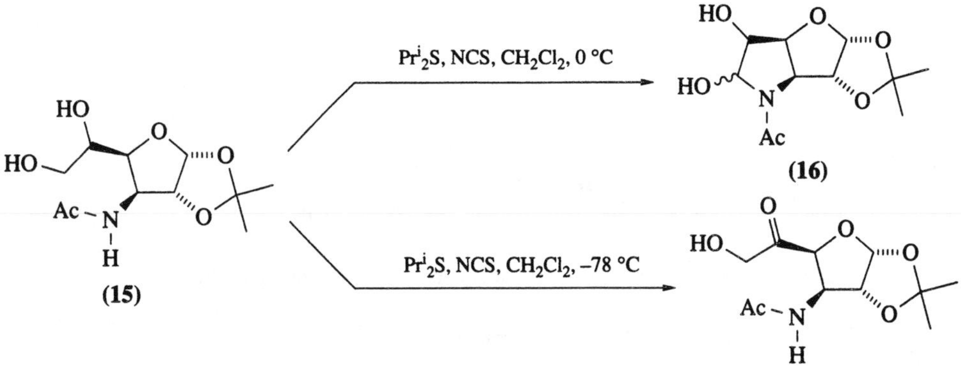

(15) → Pr^i_2S, NCS, CH_2Cl_2, 0 °C → (16)

(15) → Pr^i_2S, NCS, CH_2Cl_2, –78 °C → (17)

Further examples	0 °C	–78 °C
Decan-1-ol	90%	n.r.
Cyclohexylmethanol	84%	n.r.
Heptan-2-ol	n.r.	88%
4-*t*-Butylcyclohexanol	n.r.	83%
(15) → (16)	70%	—
(15) → (17)	—	65%

Scheme 2

A number of derivatives of ruthenium(II) have the potential to oxidize a primary alcohol in the presence of a secondary alcohol; the original report of Sharpless *et al.*[22] has been followed by a number of modifications.[23] The ruthenium complex can be used as a catalyst in conjunction with a cooxidant, which in the original work was *N*-methylmorpholine *N*-oxide. In general benzylic and allylic alcohols react more readily than their saturated counterparts, and primary alcohols react more readily than secondary alcohols. Alkenes can interfere with this oxidation, probably by binding to the metal and inhibiting the catalytic process. The stoichiometric use of tris(triphenylphosphine)ruthenium(II) chloride will oxidize a primary/secondary diol to the corresponding hydroxy aldehyde in excellent yield (equation 13).[24]

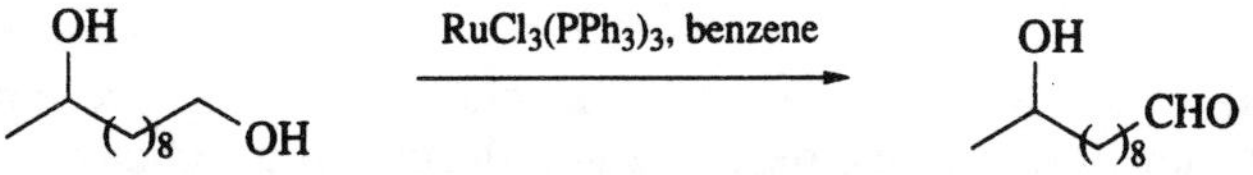

OH → $RuCl_3(PPh_3)_3$, benzene → OH, CHO (13)

Zirconyl acetate [$ZrO(OAc)_2$] has been used as a catalyst for the oxidation of primary aliphatic alcohols to aldehydes, with *t*-butyl hydroperoxide as cooxidant. Under the reaction conditions benzylic and allylic alcohols are also oxidized, but the oxidation of saturated secondary alcohols is slow, and C—C double bonds are unaffected and some degree of chemoselectivity would appear to be feasible.[25]

A zirconium complex, bis(cyclopentadienyl)zirconium(IV) hydride will function as a catalyst for the chemoselective Oppenauer oxidation of primary alcohols in the presence of a hydrogen acceptor (cyclohexanone, benzaldehyde or benzophenone).[26] This method appears to be of some value, since it also allows for the selective monooxidation of primary (and secondary) diols (Scheme 3). 1,2-Diols are not cleaved under these conditions and retro-aldol reactions appear not to be a problem.

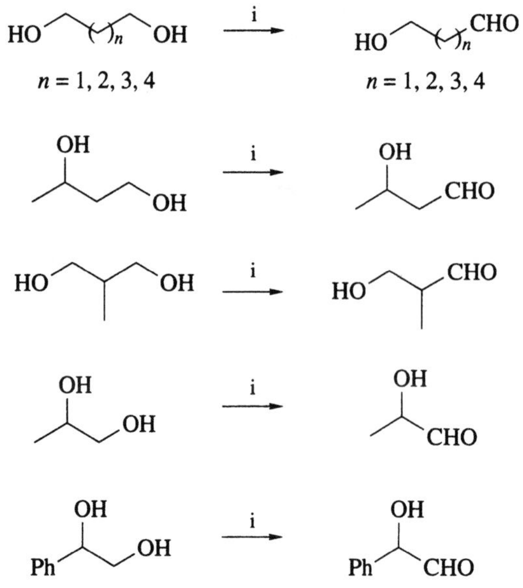

i, Cp_2ZrH_2 + H-acceptor (cyclohexanone, benzaldehyde or benzophenone)

Scheme 3

Under neutral or acidic conditions osmium tetroxide has been reported to exhibit selectivity for the oxidation of primary alcohols, although in the examples shown (equations 14 and 15) the yields appear to be lower than might be desirable (40–50%).[27]

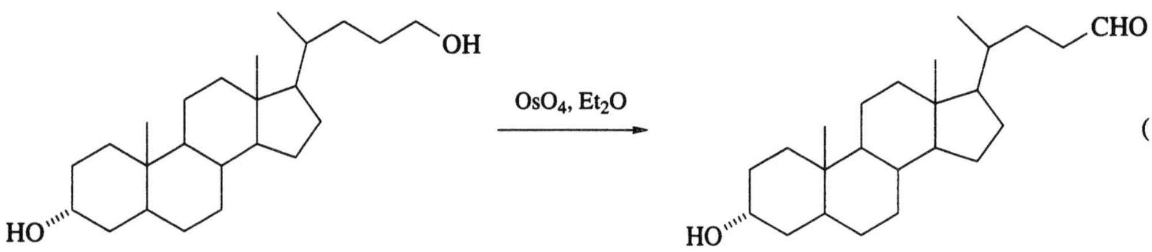

Clearly there are easily oxidizable groups other than alcohols which might be found within a particular structure, and which can interfere with the oxidative introduction of a carbonyl group. Not surprisingly reagents have been developed which will show chemoselectivity for the oxidation of the alcohol function in such systems. Triphenylbismuth carbonate and μ-oxo-bis(chlorotriphenylbismuth) have been reported as such a chemoselective reagent, and will oxidize alcohols in the presence of other easily oxidized species such as benzenethiol, indole and pyrrole (Section 2.9.3).[16] This reagent also cleaves 1,2-diols.

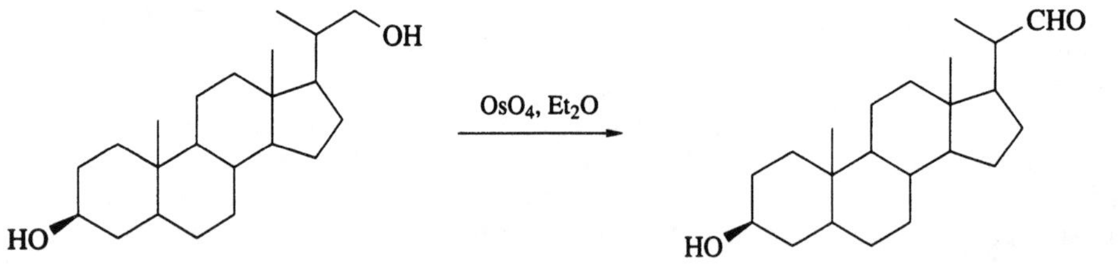

It is possible to oxidize an alcohol in the presence of sulfur- or selenium-containing groups (equation 16) using *t*-butyl hydroperoxide and a diselenide as the oxidizing system (this also oxidizes secondary alcohols, see later).[15] Selenium chemistry can also be used to oxidize benzylic and related primary alcohols to the aldehydes without oxidizing pyridyl (**18**; equation 17) or thiophenyl (**19**; equation 18) groups.[28]

It is possible to oxidize sensitive allylic alcohols to aldehydes using catalytic tris(triphenylphosphine)ruthenium(II) chloride in an oxygen atmosphere, a thiophenyl group survives under these condi-

PhX(R)C=CH–CH2OH —[(ArSe)$_2$, TBHP, benzene; Ar = 2,4,6-trimethylphenyl]→ PhX(R)C=CH–CHO (16)

X = S, Se; R = H, Me, Ph

(18) —[Ar_2SeO, SeO_2, dioxane; Ar = 4-methoxyphenyl]→ pyridine-2-CHO (17)

(19) —[Ar_2SeO, SeO_2, dioxane; Ar = 4-methoxyphenyl]→ thiophene-2-CHO (18)

tions (equation 19), and the polyene retinal is produced from retinol (equation 20) without loss of double bond geometry.[29]

PhS…OH —[$RuCl_2(PPh_3)_3$, O_2, $ClCH_2CH_2Cl$]→ PhS…CHO (19)

retinol —[$RuCl_2(PPh_3)_3$, O_2, 2,6-lutidine]→ retinal (20)

A catalytic method which promises to find wide application in view of its mildness and ease of execution uses a catalytic amount of tetra-*n*-propylammonium perruthenate (TPAP) with *N*-methylmorpholine *N*-oxide (NMO) as the cooxidant.[30] Primary (and secondary) alcohols which contain a range of functional groups (alkenes, tetrahydropyran ethers, epoxides, lactones, silyl ethers and indoles *inter alia*) can be oxidized without interference by the other functional group (equations 21–23). The performance of the reagent is improved further by including molecular sieves in the reaction mixture.[31]

THPO…OH —[$Pr^n_4N(RuO_4)$; NMO, CH_2Cl_2]→ THPO…CHO (no racemization) (21)

HO…OSiPh$_2$Bu^t —[$Pr^n_4N(RuO_4)$; NMO, CH_2Cl_2]→ OHC…OSiPh$_2$Bu^t (22)

(indole ketone with OH) —[$Pr^n_4N(RuO_4)$; NMO, CH_2Cl_2]→ (N–H, O product) (23)

Another reagent which undoubtedly will prove to be of real synthetic value is periodinane, which is an excellent, mild oxidant for primary (and secondary) alcohols.[32] This reagent is discussed further in Section 2.9.3.

The enzyme D-galactose oxidase has been reported to oxidize some racemic diols with kinetic resolution to provide the corresponding hydroxy aldehydes with high enantiomeric excesses.[33] A list of successful and unsuccessful substrates is given in Scheme 4.

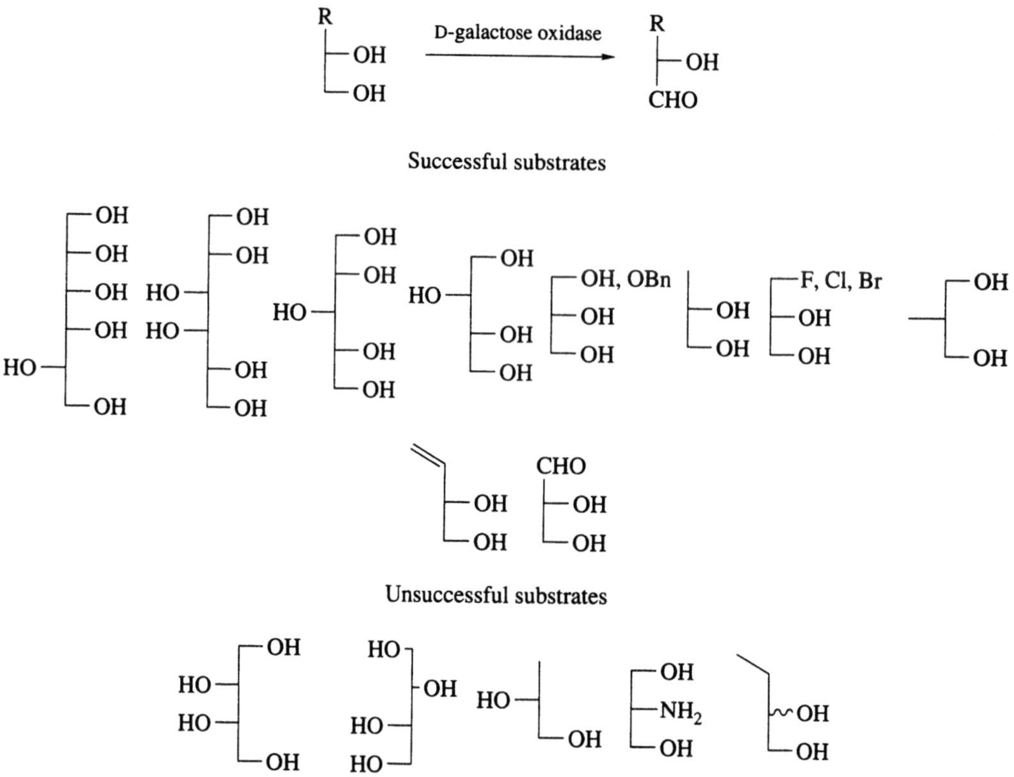

R OH OH D-galactose oxidase R OH CHO

Successful substrates

OH, OBn OH OH F, Cl, Br OH OH CHO OH OH

Unsuccessful substrates

OH NH2 OH

Scheme 4

2.9.2.2 Selective Preparation of Lactones from Diols

The oxidation of primary alcohols to the corresponding carboxylic acid or ester generally requires fairly powerful oxidants, and in most cases the issue of selectivity is dealt with by protection of other oxidizable functionality within the molecule. One important area in which this need not be the case is the oxidation of symmetrical and unsymmetrical diols to the corresponding lactone. The general scheme is presented in Scheme 5, and relies on an initial chemoselective oxidation to the hydroxy aldehyde, which is in equilibrium with the lactol. This lactol is then oxidized to the lactone. In some cases it is possible to halt the reaction at the lactol stage, but usually the lactone is the product. Most of this section will be concerned with this type of selective oxidation.

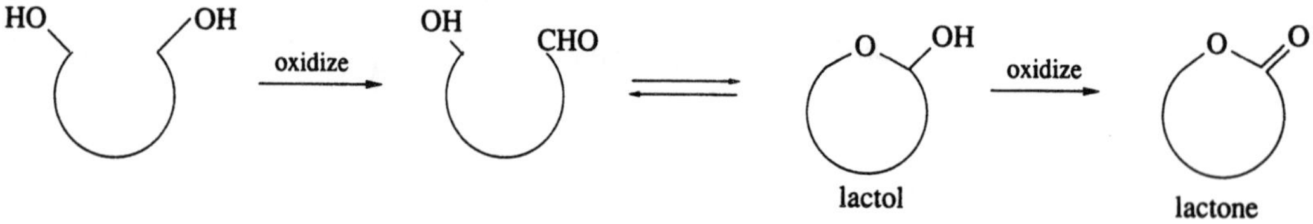

HO OH oxidize OH CHO O OH oxidize O O

lactol lactone

Scheme 5

The most widely known reagent for the oxidation of α,ω-diols to lactones is silver carbonate on Celite.[34] This reagent will oxidize primary 1,4-, 1,5- and 1,6-diols, to lactones. Primary–secondary diols can give the corresponding keto alcohol as a serious side product, but this can be overcome by a change in solvent.[30] As would be expected, primary–tertiary diols give the corresponding lactones in excellent yield. This oxidation was used as the key step in a simple synthesis of racemic mevalonolactone (**20**; Scheme 6).[34] Although this is often the first reagent to be tried in a given situation, this reagent is not free of problems, as often a large excess is required (10–26 equiv. in the original paper), which compounds the problem of the cost of silver salts, and sulfide groups appear to interfere.[34]

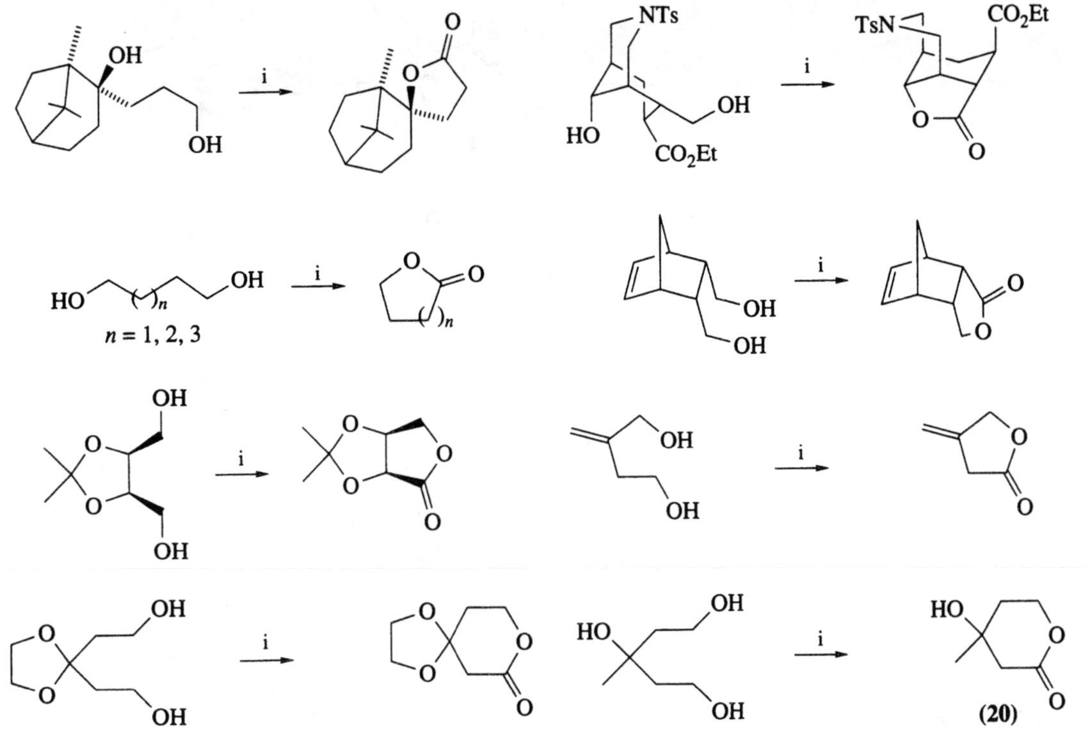

HO OH
$n = 1, 2, 3$

(20)

i, Ag_2CO_3/celite, benzene

Scheme 6

A mixture of potassium permanganate and copper sulfate will oxidize simple 1,4- and 1,5-diols to the corresponding lactones, and will oxidize selectively primary–secondary diols (**21**; equation 24) and (**22**; equation 25) to the corresponding lactones.[35]

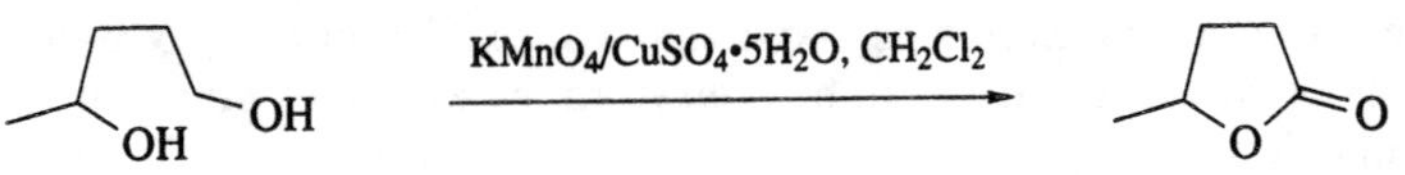

$KMnO_4/CuSO_4{\cdot}5H_2O$, CH_2Cl_2 (24)

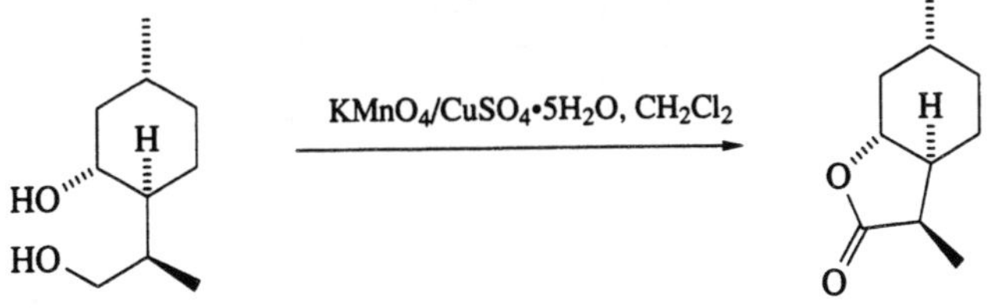

(21)

$KMnO_4/CuSO_4{\cdot}5H_2O$, CH_2Cl_2 (25)

(22)

In work related to natural product synthesis the efficacy of silver carbonate on Celite was compared with a platinum-catalyzed oxidation using an oxygen atmosphere for the oxidation of (**23**; equation 26).[36] In some cases the Pt/O_2 system was superior, but in others the situation was reversed, with no obvious rationale. The Pt/O_2 reagent has been used in the total synthesis of the hydroazulene natural products damsin (**24**; Scheme 7),[37] aromatin (**25**)[38] and aromaticin (**26**; Scheme 8).[38]

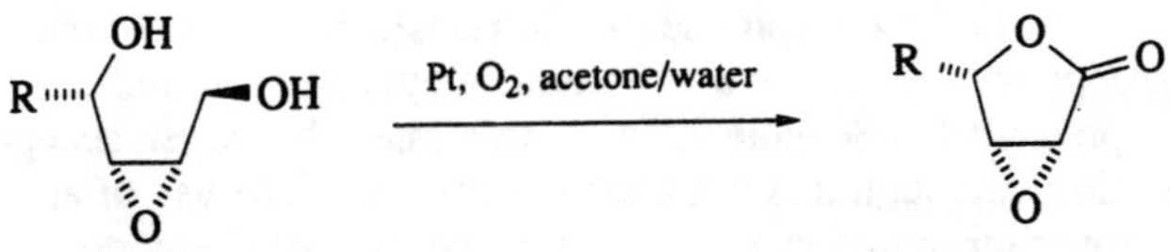

Pt, O_2, acetone/water (26)

(23)

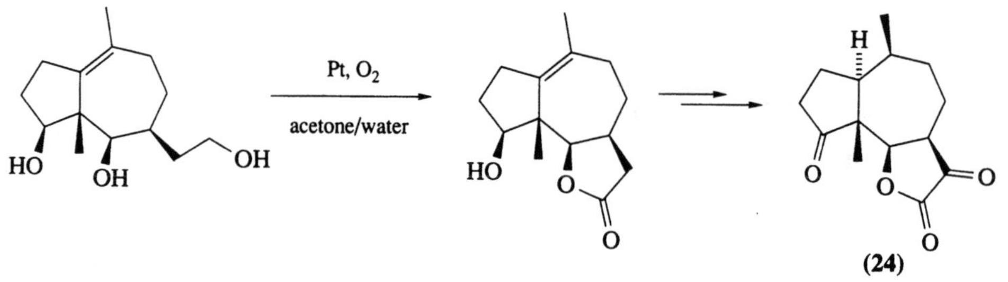

Pt, O_2
acetone/water

(24)

Scheme 7

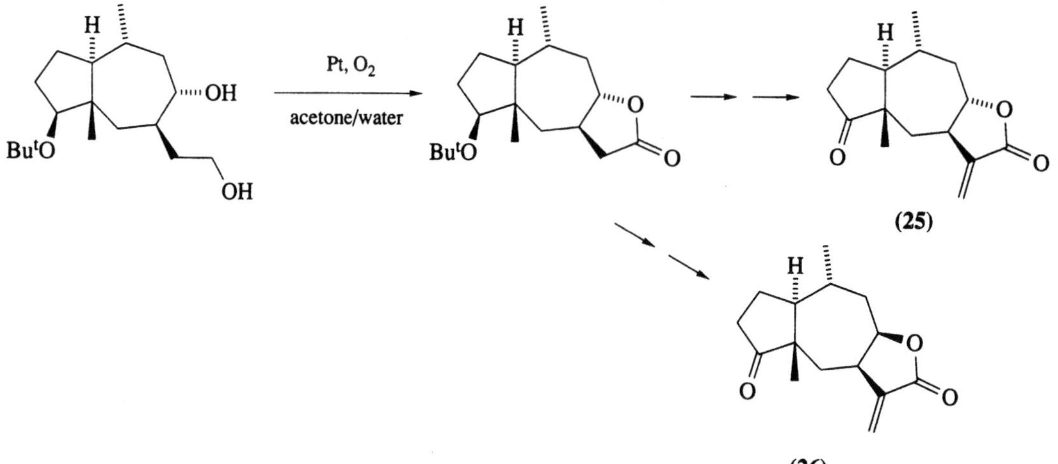

Pt, O_2
acetone/water

(25)

(26)

Scheme 8

Reagent mixtures which utilize a catalytic metal complex, usually with a cooxidant (analogous to the Pt/O_2 reagent) for the selective oxidation of diols to lactones have been studied recently, with some interesting results.

Palladium acetate with bromobenzene as cooxidant will oxidize primary α,ω-diols (**27**; equation 27) and (**28**; equation 28) to lactones, but as can be seen from the examples given, the selectivity in unsymmetrical cases is rather low.[39]

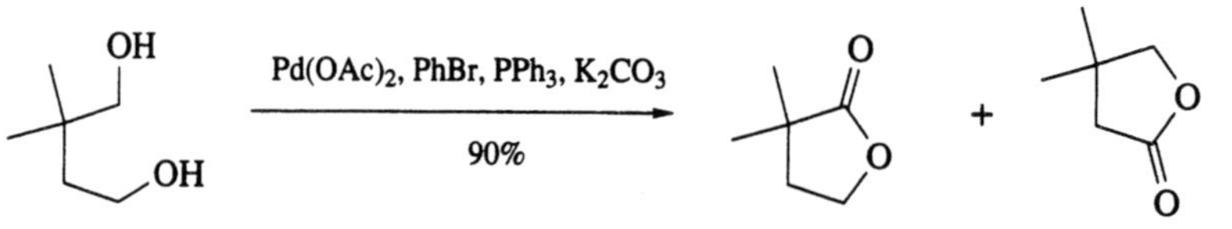

$Pd(OAc)_2$, PhBr, PPh_3, K_2CO_3, 90% (27)

(**27**) 70%:30%

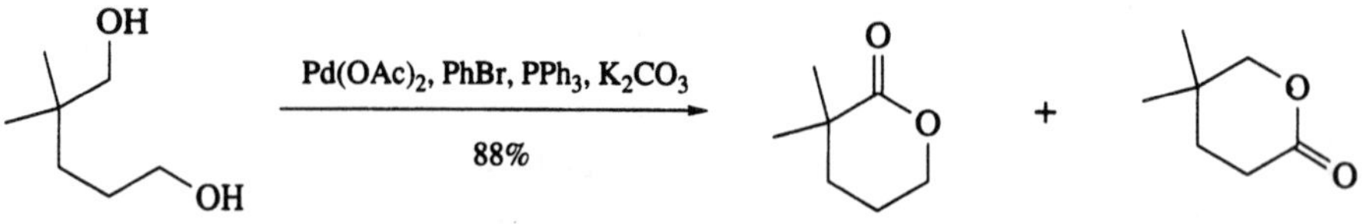

$Pd(OAc)_2$, PhBr, PPh_3, K_2CO_3, 88% (28)

(**28**) 62%:38%

In contrast to this palladium reagent, several ruthenium complexes will catalyze this type of oxidation with high regioselectivity, but in the opposite sense. Tetrakis(triphenylphosphine)ruthenium(II) hydride has been used by two groups for the selective oxidation of symmetrical and unsymmetrical diols; some representative results are presented in Scheme 9.[40,41] In this case a hydrogen acceptor is useful in increasing the yield of the oxidation, although it is not always necessary. The use of an α,β-unsaturated ketone as hydrogen acceptor allows the reaction to be performed at room temperature, whereas the use of acetone for this purpose requires high temperatures (180°C).

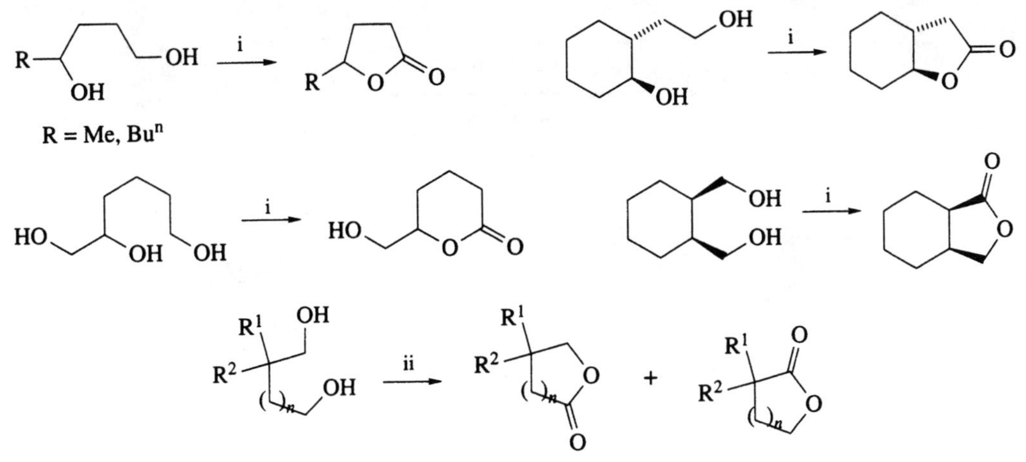

R = Me, Bu^n

(29) (30)

i, $RuH_2(PPh_3)_4$, toluene, acetone, 180 °C (ref. 40); ii, $RuH_2(PPh_3)_4$, (*E*)-4-phenylbut-3-en-2-one, toluene, 20 °C (ref. 41)

R^1	R^2	*n*	*Yield* (%)	**(29):(30)**
Me	H	1	94	93:7
Pr^i	H	1	100	98:2
Ph	H	1	90	97:3
MeO	H	1	100	98:2
Bu^{n}O	H	1	100	98:2
BnO	H	1	100	96:4
MOMO	H	1	91	99:1
Me	Me	1	100	99.6:0.4
Me	H	2	100	84:16
Ph	H	2	93	98:2
Me	H	2	100	99.5:0.5

Scheme 9

A similar selective oxidation can be carried out with tetrakis(triphenylphosphine)rhodium(I) hydride and an α,β-unsaturated ketone as hydrogen acceptor, in this case the use of an optically active phosphine provided an enantioselective synthesis, although the levels of asymmetric induction were rather low (Scheme 10).[42]

An oxidizing system which uses bromine in the presence of carboxylate salts of nickel(II) has been studied in some detail,[43] and supersedes the analogous procedure which uses dibenzoyl peroxide in the presence of nickel(II) bromide. In some unsymmetrical cases the selectivity is somewhat dependent on

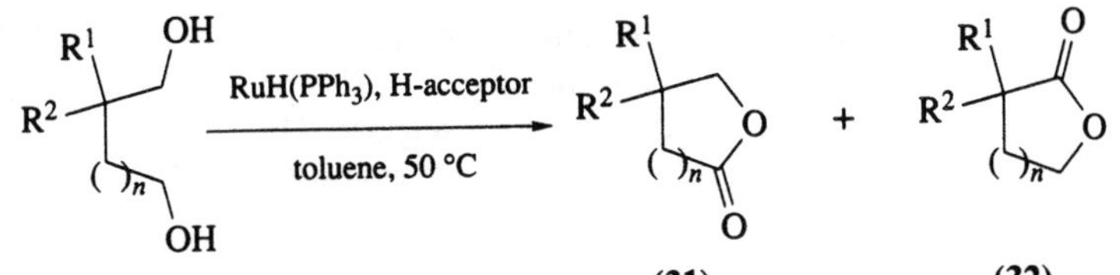

$RuH(PPh_3)$, H-acceptor

toluene, 50 °C

(31) (32)

R^1	R^2	*n*	*Yield* (%)	*(31):(32)*
Me	H	1	82	86:14
Pr^i	H	1	100	85:15
Ph	H	1	80	89:11
MeO	H	1	94	92:8
Bu^{n}O	H	1	90	90:10
Me	Me	1	95	98:2
Me	H	2	86	73:27
Ph	H	2	57	79:21
Me	Me	2	95	91:9

Scheme 10

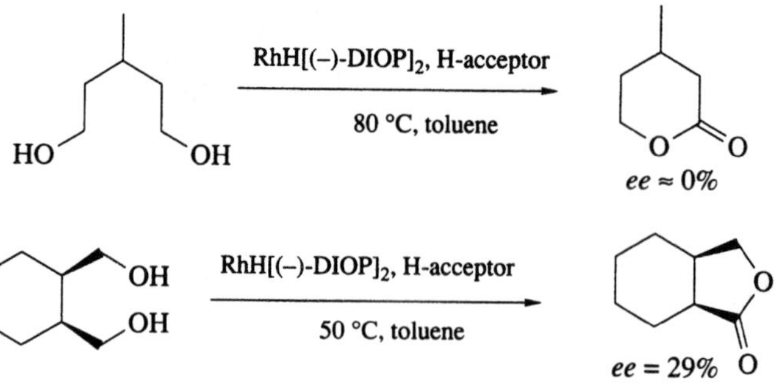

H-acceptor = (*E*)-4-phenylbut-3-en-2-one

Scheme 10 *(continued)*

the carboxylate salt used, the catalyst of choice being nickel(II) 2-ethylhexanoate, although nickel(II) benzoate is often satisfactory. In Scheme 11 a limited comparison is made with triphenylmethyl tetrafluoroborate, a reagent which has also been used for the selective oxidation of unsymmetrical diols.[44]

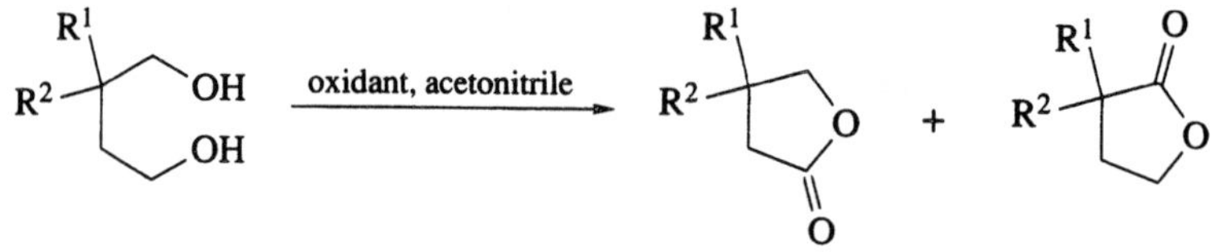

Oxidant	$R^1, R^2 = Ph_2$ Yield[a]	(33):(34)	$R^1, R^2 = Me_2$ Yield[a]	(33):(34)	R^1, R^2 = Me, Et Yield[a]	(33):(34)
Br_2, $Ni(OBz)_2$	99	24:1	87	6.5:1	99	19:1
Br_2, $Ni(Piv)_2$[b]	99	>100:1	99	6.6:1	88	23:1
Br_2, NiR_2[c]	99	>100:1	82	14:1	93	35:1
$Ph_3C^+BF_4^-$	77	>100:1	47	24:1	59	24:1

[a] % Yield. [b] Piv = $Bu^tCO_2^-$. [c] R = 2–ethylhexanoate

Scheme 11

The use of enzymes for the enantioselective oxidation of prochiral (or racemic) diols has proved to be of significant synthetic interest. A range of simple racemic 1,2-diols proved to be good substrates for a system involving coimmobilized horse liver alcohol dehydrogenase (HLADH) and aldehyde dehydrogenase (AldDH) with NAD cofactor recycling.[45] This produced 'enantiomerically pure' α-hydroxycarboxylic acids (Scheme 12).

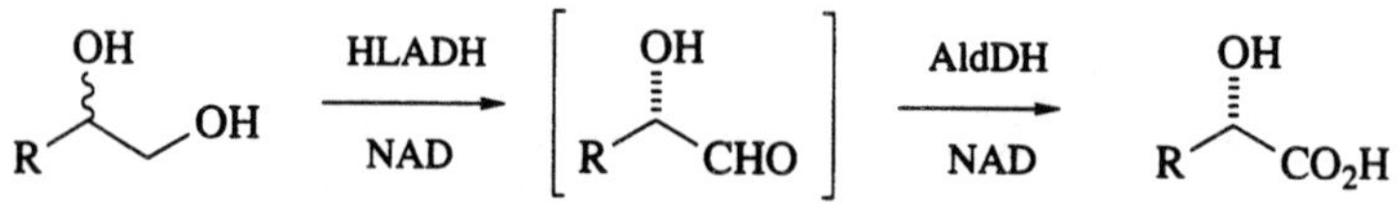

R = $HOCH_2$, FCH_2, $ClCH_2$, $BrCH_2$, CH_2=CH, Me, Et, H_2NCH_2

Scheme 12

Enzymes from *Gluconobacter roseus* organisms proved somewhat less versatile, but did allow a simple enantioselective synthesis of the unnatural enantiomer of mevalonolactone (**35**; Scheme 13) in reasonable optical purity.[46]

A thorough study of the use of HLADH for the enantioselective oxidation of *meso*-diols to lactones has provided a versatile and synthetically useful route to enantiomerically enriched lactones.[46–48] There are two major advantages of this system in that it appears to accept a fair amount of structural variation and full experimental details are available for preparative scale oxidations. A selection of results obtained with this enzyme system is presented in Scheme 14.

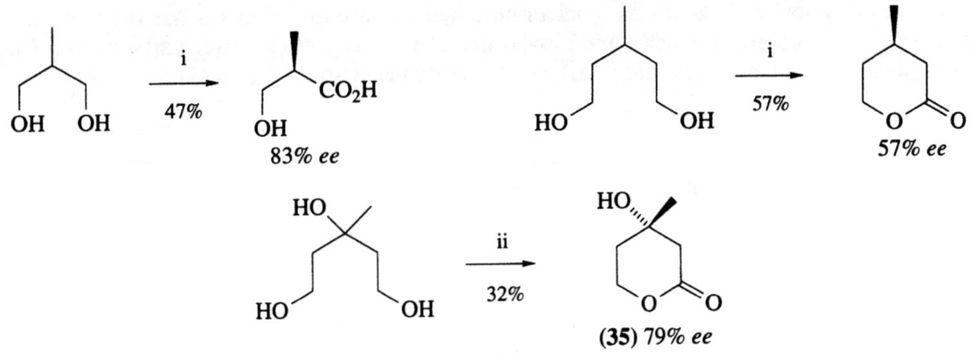

OH OH → i, 47% → OH, CO_2H, 83% *ee*

HO, OH → i, 57% → O, O, 57% *ee*

HO, HO, OH → ii, 32% → HO, O, O, (**35**) 79% *ee*

i, *Gluconobacter roseus* IAM 1841; ii, *Gluconobacter scleroideus* IAM 1842

Scheme 13

General reaction (ref. 46)

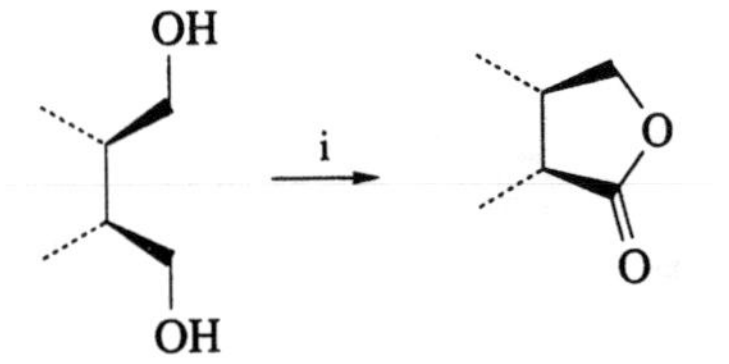

OH, OH → i → O, O

lactones prepared (100% *ee*):

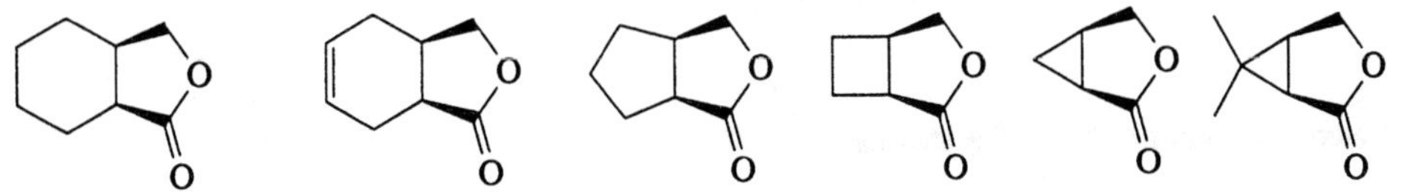

General reaction 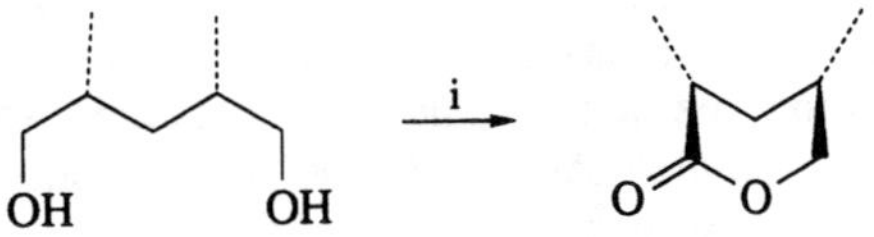 (ref. 47)

OH, OH → i → O, O

lactones prepared (100% *ee*):

OMe

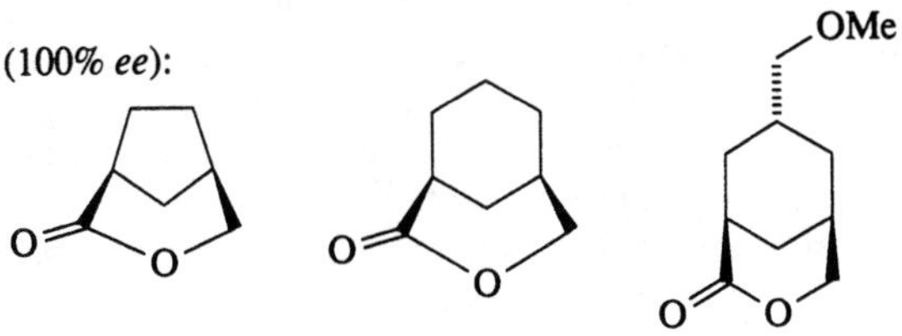

General reaction 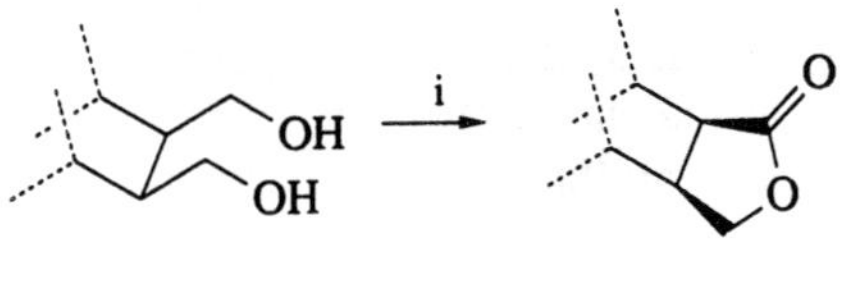 (ref. 48)

OH, OH → i → O, O

lactones prepared (>97% *ee*):

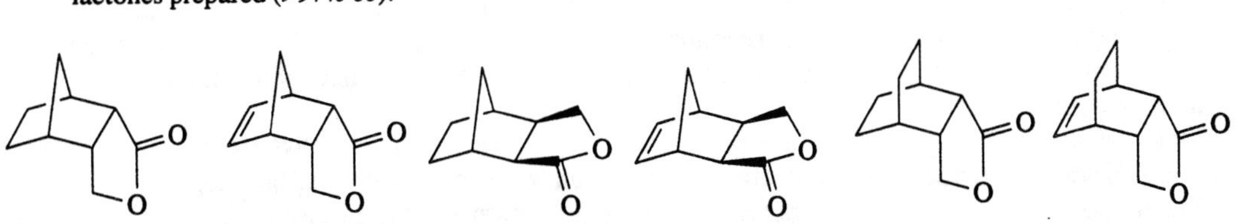

i, HLADH, pH 9, NAD^+, FMN, H_2O

Scheme 14

The selective oxidation of diols in which one or both hydroxy groups are allylic has been reported on a number of occasions. Reagents which have proved useful for this include silver carbonate on Celite,[48] barium manganate,[49] and manganese dioxide,[50] as illustrated in equations (29)–(31).

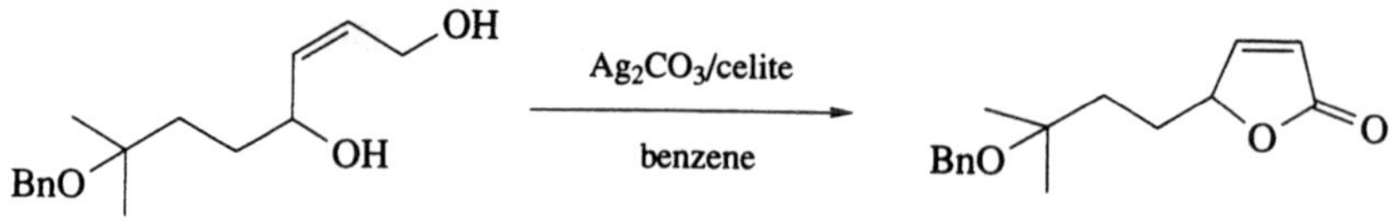

Ag_2CO_3/celite, benzene (29)

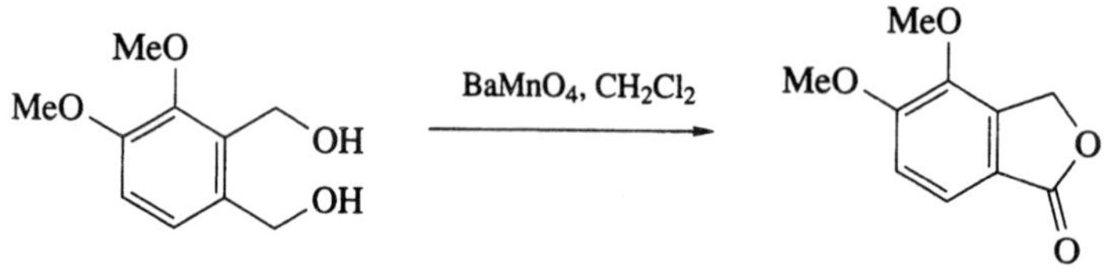

$BaMnO_4$, CH_2Cl_2 (30)

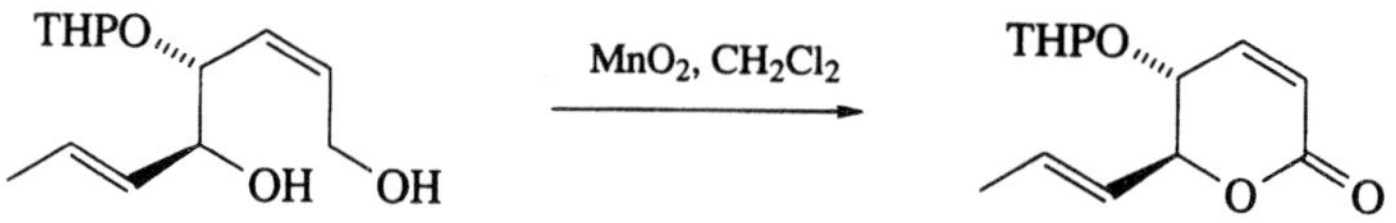

MnO_2, CH_2Cl_2 (31)

2.9.3 OXIDATION OF SECONDARY ALCOHOLS

2.9.3.1 Selective Preparation of Ketones

The oxidation of a secondary alcohol to the corresponding ketone is often a relatively straightforward task. There are several reasons for this. The product ketone is usually stable to the oxidating conditions and moreover with many reagents a secondary alcohol is oxidized more rapidly than a similar primary alcohol. Consequently there are more methods available for the chemoselective oxidation of secondary alcohols in the presence of primary alcohols. As might be expected from the preceding comments the selective oxidation of an allylic or benzylic secondary alcohol to the corresponding α,β-unsaturated ketone is possible with a number of reagents. Essentially the same range of reagents can be used for this as is used for the equivalent oxidation of primary allylic alcohols (see Section 2.9.2.1). A selection of potentially useful reagents is given in Scheme 15.[16,51–58] The bismuth reagent is of particular interest in that it will also oxidize secondary allylic alcohols in preference to the corresponding saturated alcohols, and will not attack other sensitive functional groups (see later).[16]

The chemoselective oxidation of a saturated secondary alcohol in the presence of a saturated primary alcohol is possible with a number of reagents. *N*-Bromosuccinimide in an aqueous organic solvent has been used to carry out this type of selective oxidation and has found use in synthesis.[59,60] The value of this reagent is exemplified by its use in the synthesis of isocyanopupukeanane[59] and in work towards a total synthesis of gelsemine (equations (32) and (33) respectively).[60] Clearly this reagent would not be compatible with all functional groups, given the well-known reactivity of *N*-bromosuccinimide towards unsaturated compounds.

The use of *N*-chlorosuccinimide/diisopropyl sulfide for the selective oxidation of primary/secondary diols was outlined earlier in Section 2.9.2.1, where it was used for the selective oxidation of the primary alcohol. Remarkably, by carrying out the reaction at –78°C (as compared to 0°C in the previous case) this reagent system becomes selective for secondary alcohols in the presence of primary alcohols (see Scheme 2; Section 2.9.2.1).

Possibly the simplest reagent which has been reported to carry out the selective oxidation of a secondary alcohol is sodium hypochlorite in acetic acid.[61] Given the very low cost of the reagents, this system has obvious potential for large-scale operation. A modification of this procedure uses calcium hypochlorite, which has the advantage of being a stable solid and exhibits much the same reactivity as the original system.[62] Examples of these oxidations are given in equations (34) and (35).

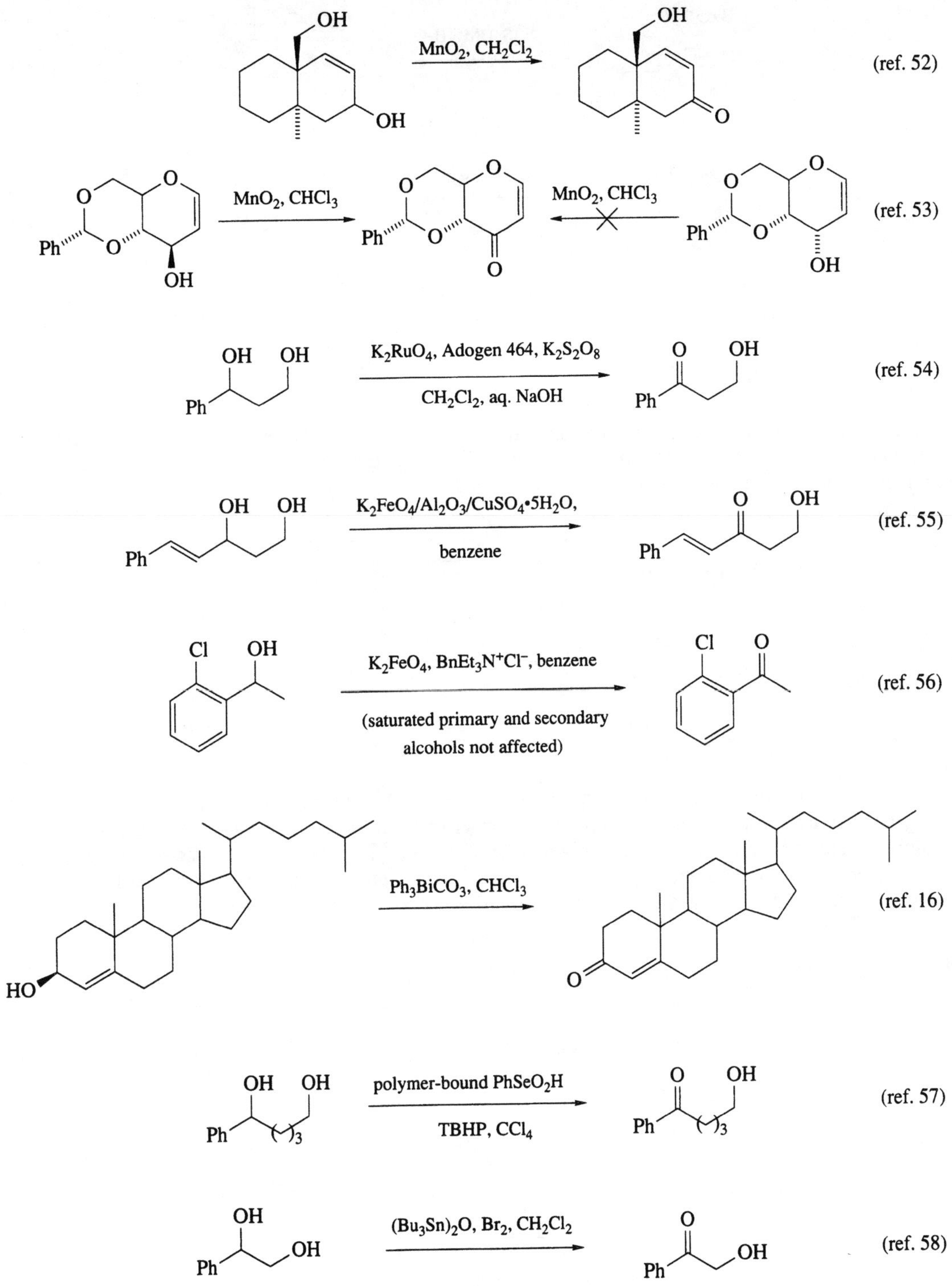

MnO_2, CH_2Cl_2 (ref. 52)

MnO_2, $CHCl_3$; MnO_2, $CHCl_3$ (ref. 53)

K_2RuO_4, Adogen 464, $K_2S_2O_8$, CH_2Cl_2, aq. NaOH (ref. 54)

$K_2FeO_4/Al_2O_3/CuSO_4{\cdot}5H_2O$, benzene (ref. 55)

K_2FeO_4, $BnEt_3N^+Cl^-$, benzene (saturated primary and secondary alcohols not affected) (ref. 56)

Ph_3BiCO_3, $CHCl_3$ (ref. 16)

polymer-bound $PhSeO_2H$, TBHP, CCl_4 (ref. 57)

$(Bu_3Sn)_2O$, Br_2, CH_2Cl_2 (ref. 58)

Scheme 15

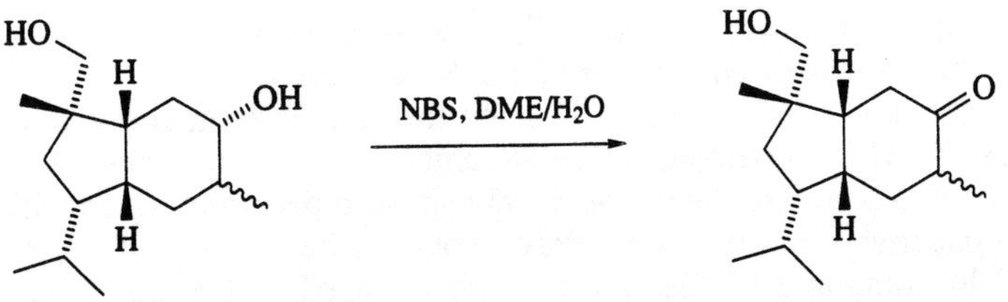

NBS, DME/H_2O (32)

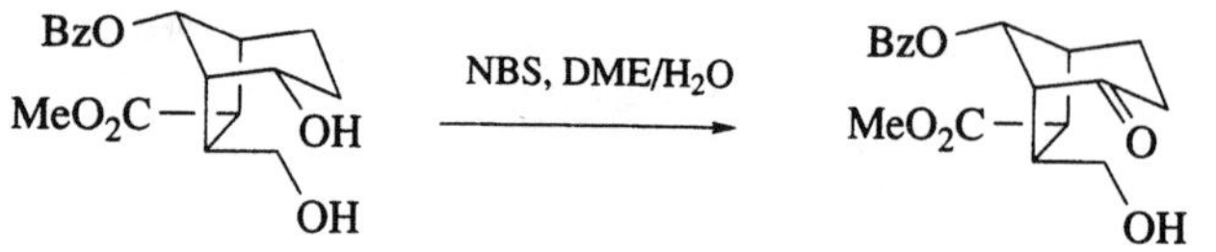

(33)

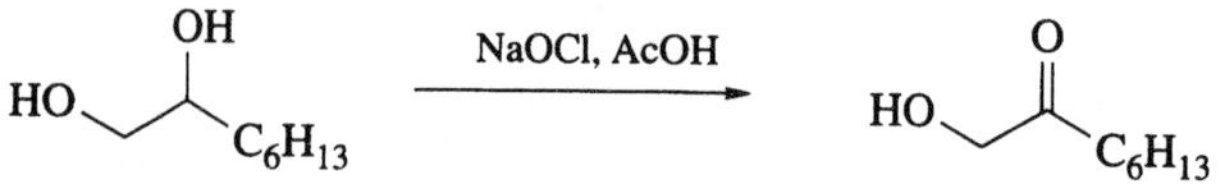

(34)

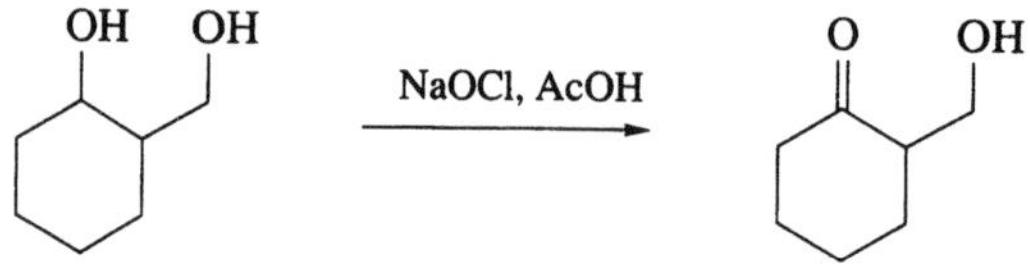

(35)

Examples of the highly chemoselective oxidation of a secondary hydroxy group have been reported using bromine in the presence of bis(tri-*n*-butyltin) oxide (equations 36 and 37), primary alcohols being essentially inert to this reagent mixture.[58]

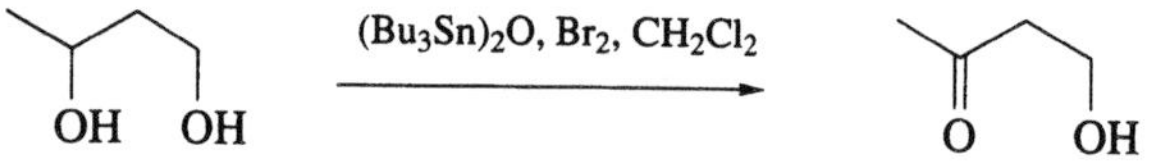

$(Bu_3Sn)_2O$, Br_2, CH_2Cl_2 (36)

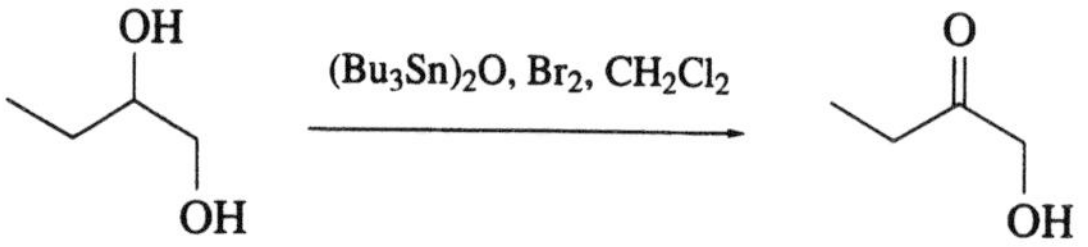

$(Bu_3Sn)_2O$, Br_2, CH_2Cl_2 (37)

A modification of the Oppenauer oxidation which uses trichloroacetaldehyde on alumina provides a good, general oxidation of alcohols, and in particular will oxidize a secondary alcohol in the presence of a primary one, as shown by the examples given in Scheme 16.[63] This method is claimed to be superior to silver carbonate on Celite and much cheaper. Other advantages of this method are that it is neutral, nonaqueous, and halide, ester and lactone functionalities survive the reaction conditions.[63]

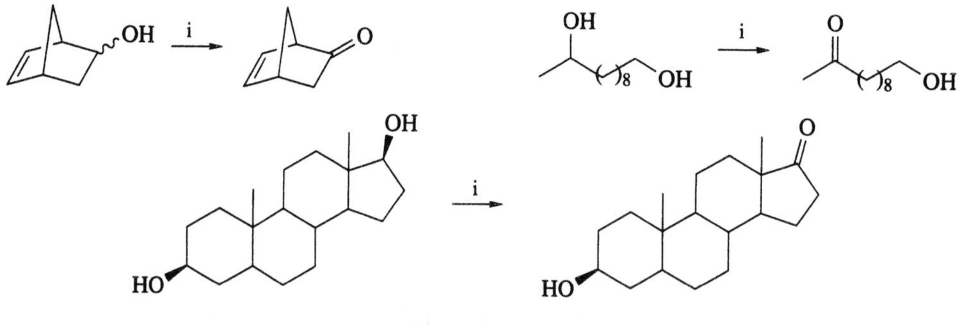

i, Cl_3CCHO/Al_2O_3, CCl_4

Scheme 16

Silver carbonate on Celite itself is a highly selective reagent for this type of chemoselective oxidation as can be seen from the examples in Scheme 17;[34] secondary diols can be oxidized to the hydroxy ketones, and primary diols are oxidized to lactones (see Section 2.9.2.2).

Several procedures for this chemoselective oxidation utilize molybdenum-based catalysts, with either hydrogen peroxide or *t*-butyl hydroperoxide as the stoichiometric oxidant. These include ammonium molybdate in the presence of a phase transfer reagent and hydrogen peroxide, which with pH control (potassium carbonate) will selectively oxidize a secondary alcohol in the presence of a primary alcohol without oxidizing alkenes.[64] In addition hindered alcohols are oxidized in preference to less hindered ones (Scheme 18).

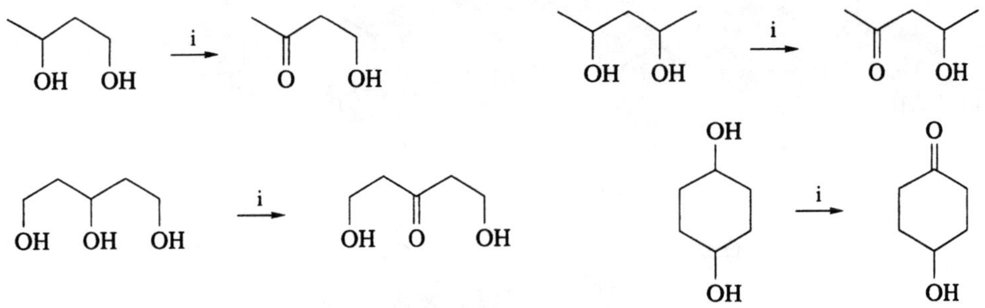

i, Ag_2CO_3/celite, benzene

Scheme 17

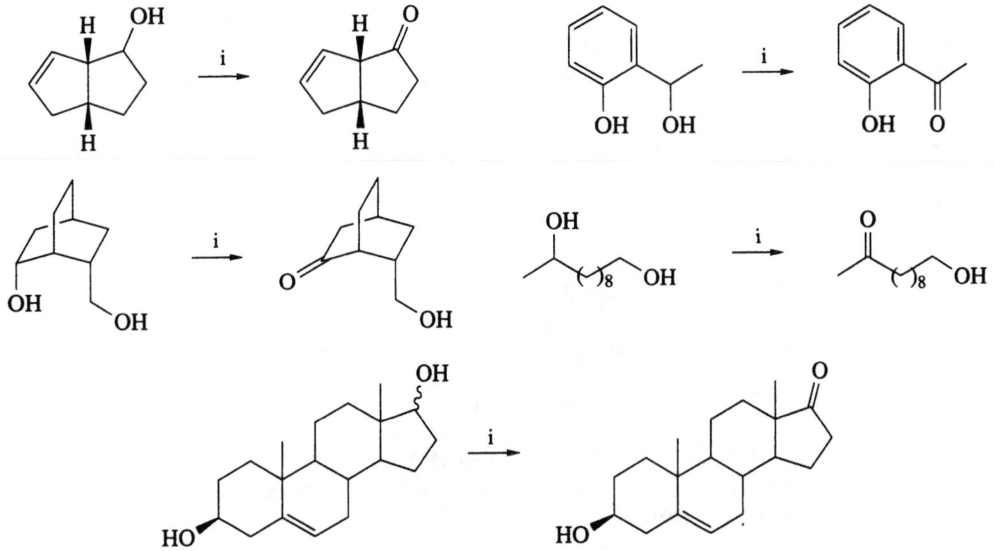

i, $(NH_4)_6Mo_7O_{24}{\bullet}4H_2O$, H_2O_2, K_2CO_3, $Bu_4N^+Cl^-$, THF

Scheme 18

Benzyltrimethylammonium tetrabromooxomolybdate will catalyze the chemoselective oxidation of secondary alcohols with *t*-butyl hydroperoxide as cooxidant.[65] Remote double bonds can interfere with this oxidation, and 1,2-diols are converted into 1,2-diketones (Scheme 19).

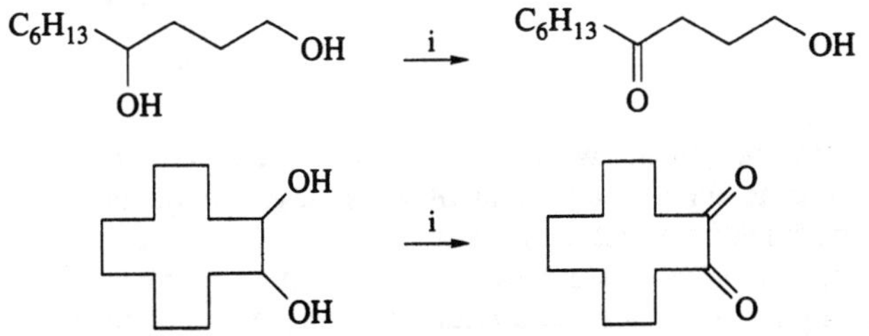

i, $BnMe_3N^+OMoBr_4^-$, TBHP, benzene

Scheme 19

The readily available catalyst vanadyl bisacetylacetonate when used with *t*-butyl hydroperoxide in benzene will oxidize secondary alcohols (Scheme 20) much more rapidly than primary ones (rate ratio > 100:1), but the other oxidizing properties of this system, in particular the epoxidation of allylic alcohols and the cleavage of 1,2-diols might well limit its uses somewhat.[66]

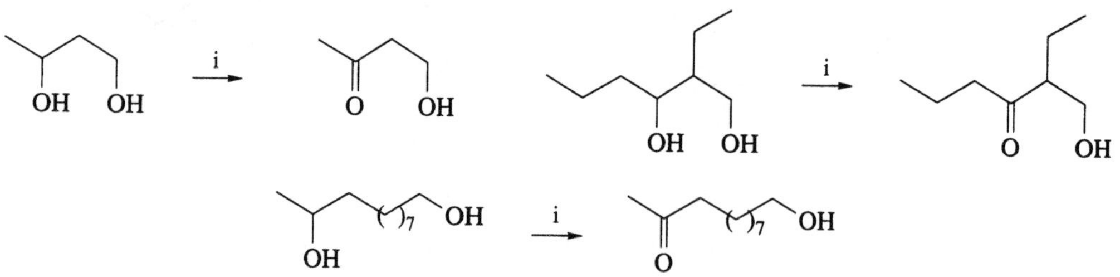

i, $VO(acac)_2$, TBHP, benzene

Scheme 20

Ammonium cerium(IV) nitrate or cerium(IV) sulfate will catalyze the selective oxidation of secondary alcohols with sodium bromate as cooxidant, in this case remote C—C double bonds interfere, but 1,2-diols are not cleaved.[67] It has been found that sodium bromite in aqueous acetic acid will act as a selective oxidant for secondary/primary diols without the need for other catalysts (Scheme 21).[68]

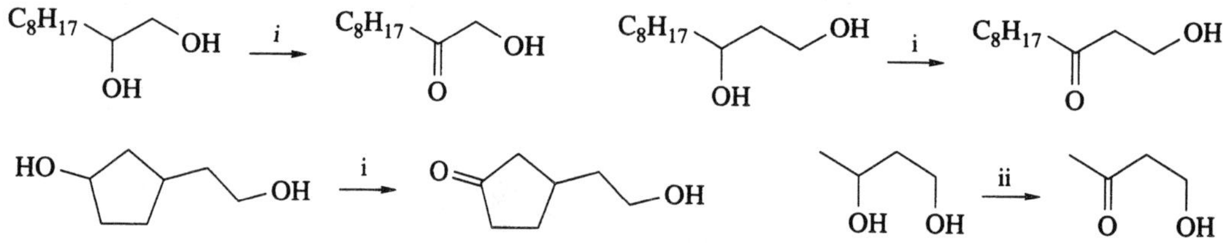

i, $(NH_4)_2Ce(NO_3)_6$, $NaBrO_3$, $MeCN/H_2O$; ii, $NaBrO_2$, $AcOH/H_2O$

Scheme 21

As referred to elsewhere μ-oxobis(chlorotriphenylbismuth) will carry out a number of interesting selective oxidations, amongst which is the selective oxidation of a secondary alcohol (equation 38), although this reagent will cleave 1,2-diols.[16]

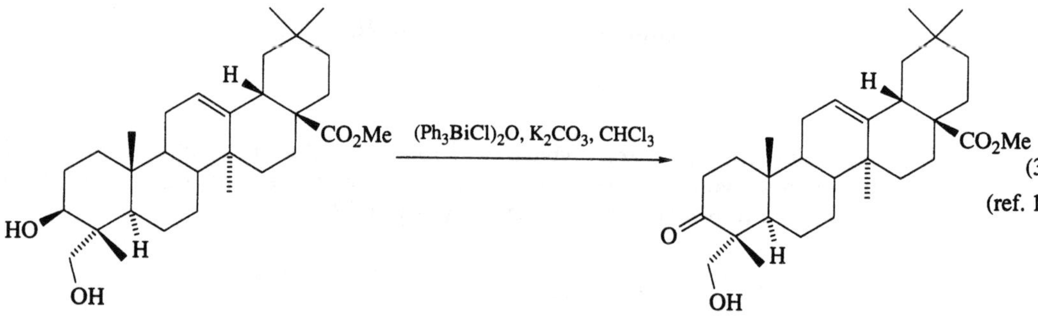

$(Ph_3BiCl)_2O$, K_2CO_3, $CHCl_3$ (38) (ref. 16)

The preceding section dealt specifically with the chemoselective oxidation of secondary/primary diols. There is a clear interest in chemoselective oxidation of secondary alcohols in the presence of other sensitive functional groups, and some of the methods available will be described briefly in this section.

Remarkable chemoselectivity is exhibited by the pentavalent organobismuth reagents μ-oxo bis(triphenylbismuth) and triphenylbismuth carbonate referred to in the preceding sections.[16] The former reagent will oxidize a secondary alcohol without affecting a spiroacetal or an unsaturated carbonyl function. Triphenylbismuth carbonate is a highly selective, nonelectrophilic oxidant, which will oxidize α,β-unsaturated secondary alcohols without oxidizing thiols, pyrrolidine or indole. The selective oxidation of (**36**; Scheme 22) without oxidation of the selenium is also possible with this reagent. Other compounds/functional groups which are unaffected by this reagent include a range of carbonyl derivatives (dinitrophenylhydrazone, phenylhydrazone, semicarbazone, tosylhydrazone, excluding oximes), *O*-acetates, a vinyl ether, aniline, *N*,*N*-dimethylaniline, a dienamine, a thione, and a steroidal xanthate and *N*,*N*-diethyl thionocarbonate. Both these bismuth reagents will cleave 1,2-diols.

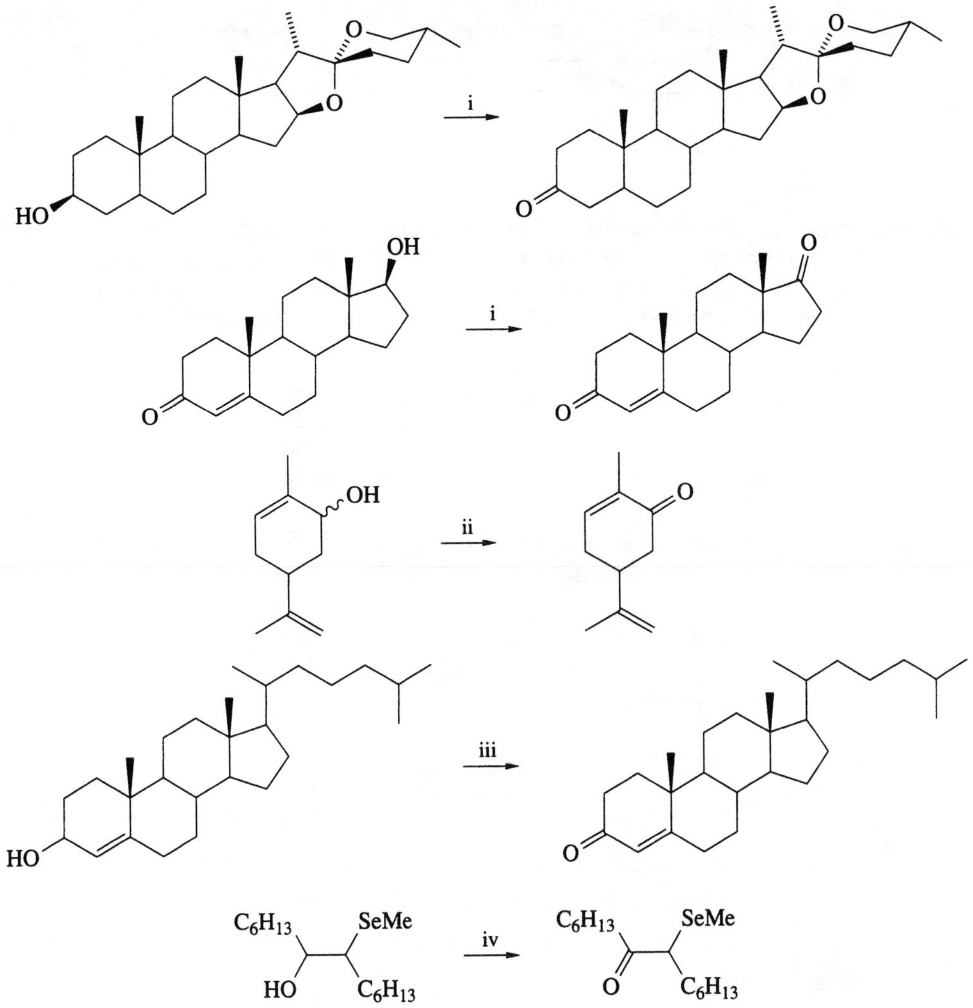

HO O i O O O O

OH O i O O

OH O ii

iii HO O

C_6H_{13} SeMe HO C_6H_{13} (36) iv C_6H_{13} SeMe O C_6H_{13}

i, $(Ph_3BiCl)_2O$, CH_2Cl_2 or $CHCl_3$, $NaHCO_3$ or K_2CO_3; ii, Ph_3BiCO_3, CH_2Cl_2, oxidation proceeds in the presence of either PhSH or Bu^{i}SH without oxidation of the thiol; iii Ph_3BiCO_3, CH_2Cl_2, oxidation proceeds in the presence of either indole or pyrrolidine without oxidation of either; iv Ph_3BiCO_3, CH_2Cl_2

Scheme 22

The selective oxidation of secondary alcohols which contain either a sulfide or selenide is also possible using either selenium-based oxidizing agents (equations 39 and 40),[69] or with a modified Oppenauer system involving trichloroacetaldehyde (equation 41).[15]

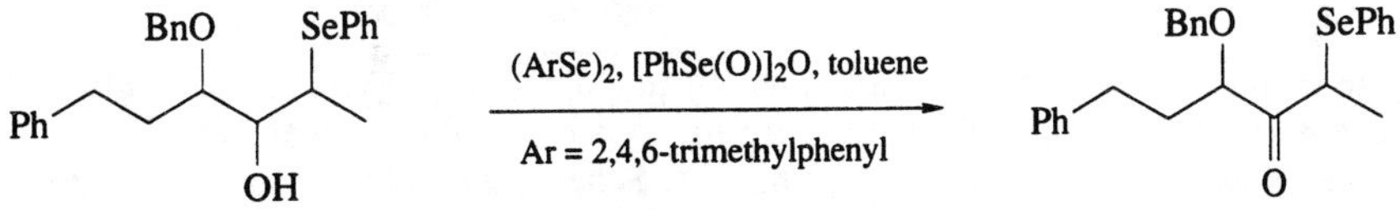

BnO SePh Ph OH $\xrightarrow[\text{Ar = 2,4,6-trimethylphenyl}]{(ArSe)_2,\ [PhSe(O)]_2O,\ toluene}$ BnO SePh Ph O (39)

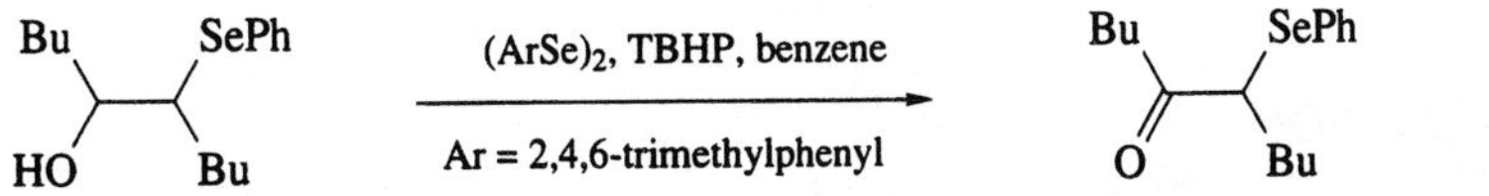

Bu SePh HO Bu $\xrightarrow[\text{Ar = 2,4,6-trimethylphenyl}]{(ArSe)_2,\ TBHP,\ benzene}$ Bu SePh O Bu (40)

t-Butyl hydroperoxide finds several applications as a stoichiometric oxidant in this area, when used with another reagent. Combination with aluminum tri-(*t*-butoxide) (equation 42) produces a mild, selec-

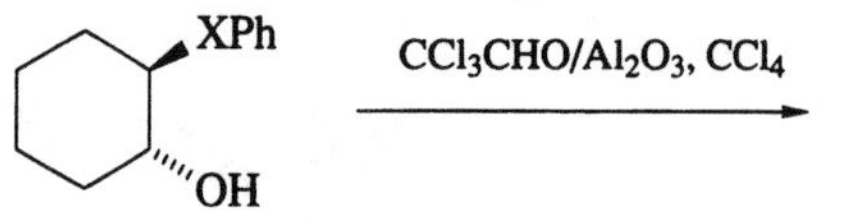

$$\text{(2-XPh-cyclohexanol)} \xrightarrow{CCl_3CHO/Al_2O_3,\ CCl_4} \text{(2-XPh-cyclohexanone)} \quad (41)$$

X = S, Se

tive oxidizing agent which tolerates other functional groups including, iodide, ester, terminal alkyne, aromatic ether and 1,3-dioxolane.[70] Various ruthenium complexes, including $RuCl_3$, $Ru(acac)_3$, $RuCl_2(PPh)_3$ and $[RuCl_2(CO)_3]_2$ in the presence of *t*-butyl hydroperoxide, will catalyze the selective oxidation of secondary alcohols containing styryl, furyl and thienyl groups (Scheme 23).[71]

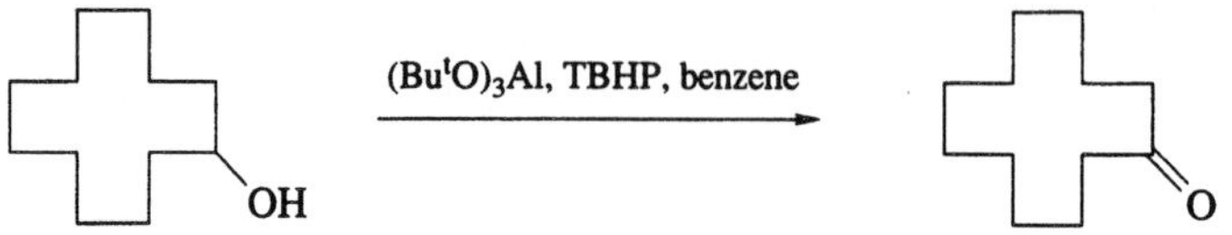

$$\text{(cyclododecanol)} \xrightarrow{(Bu^tO)_3Al,\ TBHP,\ benzene} \text{(cyclododecanone)} \quad (42)$$

1-iodododecane, methyl dodecanoate, phenylacetylene, anisole and dodecanal ethylene acetal were recovered unchanged under the reaction conditions

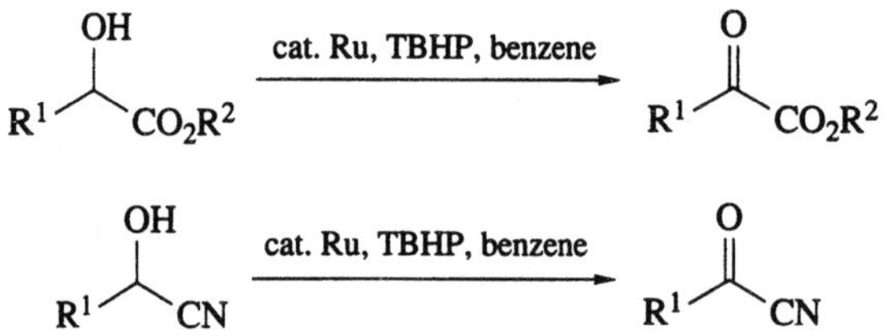

$$R^1CH(OH)CO_2R^2 \xrightarrow{cat.\ Ru,\ TBHP,\ benzene} R^1COCO_2R^2$$

$$R^1CH(OH)CN \xrightarrow{cat.\ Ru,\ TBHP,\ benzene} R^1COCN$$

Ru catalysts include $RuCl_3$, $Ru(acac)_3$, $RuCl_2(PPh_3)_3$, $RuCl_2(CO)_3$

R^1 = methyl, styryl, phenyl, 2-furyl, 2-thienyl; R^2 = methyl, ethyl

Scheme 23

The selective oxidation of a secondary alcohol in the presence of a tertiary amine function has been carried out with manganese dioxide (equations 43 and 44) in the context of an alkaloid synthesis.[72]

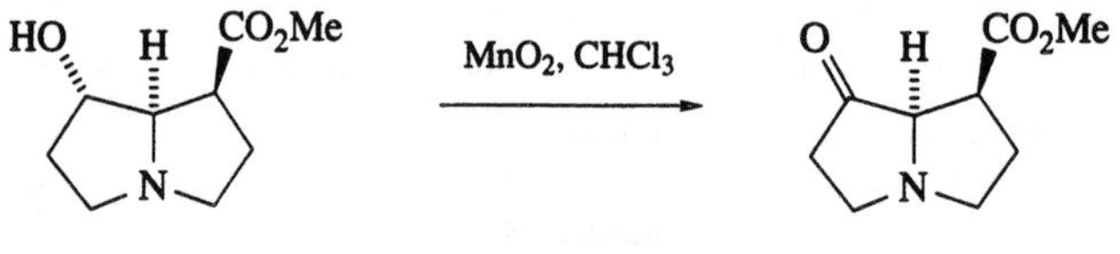

$$\text{(hydroxy pyrrolizidine, } CO_2Me\text{)} \xrightarrow{MnO_2,\ CHCl_3} \text{(keto pyrrolizidine, } CO_2Me\text{)} \quad (43)$$

$$\text{(hydroxy bicyclic amine, } CH=CHCO_2Me\text{)} \xrightarrow{MnO_2,\ CHCl_3} \text{(keto bicyclic amine, } CH=CHCO_2Me\text{)} \quad (44)$$

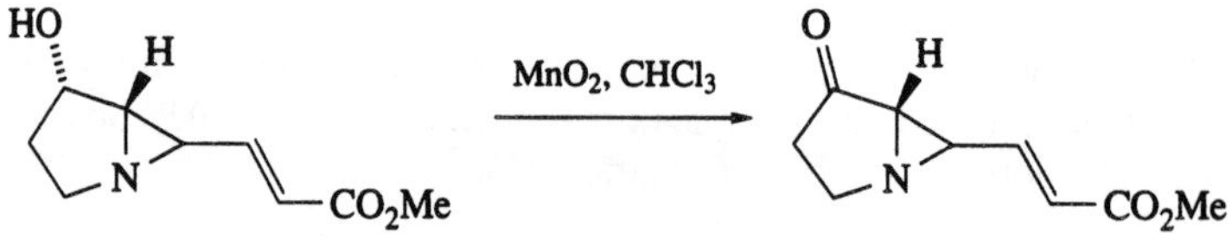

The versatile oxidizing agent 'periodinane', which functions as an excellent reagent for the oxidation of alcohols (see Section 2.9.2.1),[32] will oxidize a functionalized α-hydroxy ester to the corresponding α-keto ester in the amino acid derivatives (**37**) and (**38**; Scheme 24) without loss of stereochemical integrity or interference from the other highly polar groups in these molecules.[73]

2.9.4 CONCLUSION

There are numerous reagents available for the chemoselective oxidation of polyfunctional alcohols. The most promising general type of oxidant must be that in which a mild, clean oxidizing agent (*e.g. t*-butyl hydroperoxide, bromine, air or *N*-methylmorpholine *N*-oxide) is used in conjunction with a reagent which will catalyze the desired selective oxidation. Mild, stoichiometric oxidants (such as periodinane),

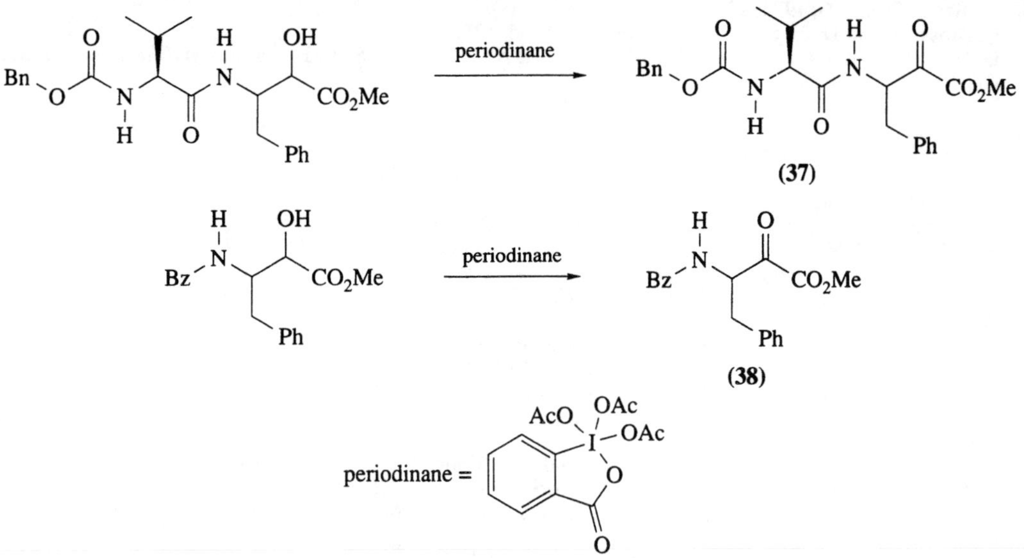

(37)

(38)

Scheme 24

which avoid the use of possibly toxic or expensive metals, and do not produce toxic or noxious by-products, provide an alternative of great value.

It is unlikely that any single reagent will prove to be a universal oxidant for a particular type of chemoselective oxidation. Many of the reagents considered in this chapter would need to be tested further, on more complicated substrates, before a 'reagent of choice' could be arrived at. Nevertheless, reagents which fall into the two general categories delineated in the preceding paragraph would be high on the list of potential reagents for a given selective oxidation.

Given the above it is rather surprising that chemoselective oxidation is encountered relatively rarely in the synthesis of complex molecules. The problem of selective oxidation is usually dealt with by careful choice of strategy, or by protection of the functional group which is not to be oxidized.

Why not use selective oxidants instead of protection or limiting the strategies considered? The usual response to this from those (including the author) involved in such complex syntheses is that the reagents are not proven in such a synthetic context. It is too risky, why take the chance? This is a perfectly reasonable response given the time and effort required to carry out such syntheses, but perhaps a little negative. No reagent is perfect, and in the area of selective oxidation much remains to be achieved, and no doubt new reagents will be discovered and developed in the future. Avoiding the use of relatively untried selective oxidants by protection of functional groups is far from perfect itself. While protecting group chemistry is certainly a more advanced art than selective oxidation, it can have its own problems. Selective deprotection, often a consequence of such protection, can prove far from trivial.

Perhaps organic chemists involved in the synthesis of complex organic compounds should have the courage to try this type of selective oxidant in new situations, especially where it might avoid difficult protection/deprotection, or where it would lead to a more direct synthetic route. Protection/deprotection is often cumbersome and far from elegant, and new or untried reagents can only be evaluated properly within the context of the synthesis of polyfunctional organic compounds. In this way the limitations of existing reagents are uncovered, and the requirements for new reagents may be determined, and the discovery and development of such new selective reagents is a worthy challenge for modern organic chemistry to meet.

2.9.5 REFERENCES

1. A. H. Haines, 'Methods for the Oxidation of Organic Compounds; Alcohols, Alcohol Deriatives, Alkyl Halides, Nitroalkanes, Alkyl Azides, Carbonyl Compounds, Hydroxyarenes, and Aminoarenes', in the series 'Best Synthetic Methods', ed. A. R. Katritzky, O. Meth-Cohn and C. W. Rees, Academic Press, New York, 1988.
2. A. J. Fatiadi, *Synthesis*, 1976, 65, 133; J. S. Pizey, 'Synthetic Reagents,' Halstead Press, New York, 1974, vol. 2, p. 143.
3. M. Harnfeinst, A. Bavley and W. A. Lazier, *J. Org. Chem.*, 1954, **19**, 1608.
4. R. K. Bentley, E. R. H. Jones and V. Thaller, *J. Chem. Soc. (C)*, 1969, 1096.

5. M. Brink, *Synthesis*, 1975, 253.
6. R. Brossmer and D. Ziegler, *Tetrahedron Lett.*, 1966, 5253.
7. D. Heyland, *J. Am. Chem. Soc.*, 1948, **70**, 3434; A. N. Wilson and S. A. Harris, *J. Am. Chem. Soc.*, 1951, **73**, 4693.
8. L. Crombie and J. Crossley, *J. Chem. Soc.*, 1963, 4983.
9. J. W. K. Burrell, R. F. Garwood, L. M. Jackman, E. Oskay and B. C. L. Weedon, *J. Chem. Soc.*, 1966, 2144.
10. G. Ohloff and W. Giersch, *Angew. Chem., Int. Ed. Engl.*, 1973, **12**, 401.
11. H. Fujami, H. S. Koh, T. Sakata and M. Nakajima, *Tetrahedron Lett.*, 1967, 4771.
12. A. J. Fatiadi, *Synthesis*, 1987, 85.
13. H. Firouzabadi and E. Ghaderi, *Tetrahedron Lett.*, 1978, 839.
14. S. C. Howell, S. V. Ley and M. Mahon, *J. Chem. Soc., Chem. Commun.*, 1981, 507.
15. M. Shimizu, H. Urabe and I. Kuwajima, *J. Org. Chem.*, 1982, **47**, 837.
16. D. H. R. Barton, J. P. Kitchin, D. J. Lester, W. B. Motherwell and M. T. B. Papoula, *Tetrahedron*, 1981, **37**, Suppl. 9, 73.
17. S. Kanemoto, K. Oshima, S. Matsubara, K. Takai and H. Nozaki, *Tetrahedron Lett.*, 1983, **24**, 2185.
18. M. Matsumoto and N. Watanabe, *J. Org. Chem.*, 1984, **49**, 3435.
19. T. Imamoto, Y. Hatanaka and M. Tokoyama, *Tetrahedron Lett.*, 1983, **24**, 2399.
20. M. F. Semmelhack, C. R. Schmid, D. A. Cortes and C. S. Chan, *J. Am. Chem. Soc.*, 1984, **106**, 3374, and refs. cited therein.
21. K. S. Kim, I. H. Cho, B. K. Yoo, Y. H. Song and C. S. Hahn, *J. Chem. Soc., Chem. Commun.*, 1984, 762.
22. K. B. Sharpless, K. Akashi and K. Oshima, *Tetrahedron Lett.*, 1975, 2503.
23. P. Müller and J. Godoy, *Tetrahedron Lett.*, 1981, **22**, 2361; P. Muller and J. Godoy, *Helv. Chim. Acta*, 1981, **64**, 2531.
24. H. Tomioka, K. Takai, K. Oshima and H. Nozaki, *Tetrahedron Lett.*, 1981, **22**, 1605.
25. K. Kaneda, Y. Kawanishi and S. Teranishi, *Chem. Lett.*, 1984, **9**, 1481.
26. T. Nakano, T. Teradd, Y. Ishii and M. Ogawa, *Synthesis*, 1986, 774.
27. A. M. Maione and A. Romeo, *Synthesis*, 1984, 955.
28. F. Ogura, T. Otsubo, K. Ariyoshi and H. Yamaguchi, *Chem. Lett.*, 1983, 1833.
29. M. Matsumoto and S. Ito, *J. Chem. Soc., Chem. Commun.*, 1981, 907.
30. W. P. Griffith, S. V. Ley, G. P. Whitcombe and A. D. White, *J. Chem. Soc., Chem. Commun.*, 1987, 1625.
31. S. V. Ley, Tilden Lecture, London, 1989.
32. D. B. Dess and J. C. Martin, *J. Org. Chem.*, 1983, **48**, 4155.
33. R. L. Root, J. R. Durrwachter and C.-H. Wong, *J. Am. Chem. Soc.*, 1985, **107**, 2997.
34. M. Fetizon, M. Golfier and J. M. Louis, *Tetrahedron*, 1975, **31**, 171.
35. C. W. Jefford and Y. Wang, *J. Chem. Soc., Chem. Commun.*, 1988, 634.
36. R. K. Boeckman, Jr. and E. W. Thomas, *Tetrahedron Lett.*, 1976, 4045.
37. R. A. Kretchmer and W. J. Thompson, *J. Am. Chem. Soc.*, 1976, **98**, 3379.
38. R. T. Lansbury, D. G. Hangauer, Jr. and J. P. Vacca, *J. Am. Chem. Soc.*, 1980, **102**, 3964.
39. Y. Tamara, Y. Yamada, K. Inoue, Y. Yamamoto and Z. Yoshida, *J. Org. Chem.*, 1983, **48**, 1286.
40. S.-I. Murahashi, T. Naota, K. Ito, Y. Maeda and H. Taki, *J. Org. Chem.*, 1987, **52**, 4319.
41. Y. Ishii, K. Osakada, T. Ikariya, M. Sabari and M. Yoshikawa, *J. Org. Chem.*, 1986, **51**, 2034.
42. Y. Ishii, K. Suzuki, T. Ikariya, M. Sabari and S. Yoshikawa, *J. Org. Chem.*, 1986, **51**, 2822.
43. M. P. Doyle, R. L. Dow, V. Bagheri, W. J. Patrie and Z. Yoshida, *J. Org. Chem.*, 1983, **48**, 476, and refs. cited therein.
44. M. E. Jung and R. W. Brown, *Tetrahedron Lett.*, 1978, 2771.
45. C.-H. Wong and J. R. Matos, *J. Org. Chem.*, 1985, **50**, 1992.
46. I. J. Jakovac, H. B. Goodbrand, K. P. Lok and J. B. Jones, *J. Am. Chem. Soc.*, 1982, **104**, 4659.
47. A. J. Bridges, P. S. Raman, G. S. Y. Ng and J. B. Jones, *J. Am. Chem. Soc.*, 1984, **106**, 1461.
48. K. P. Lok, I. J. Jakovac and J. B. Jones, *J. Am. Chem. Soc.*, 1985, 2521.
49. T. Chakraborty and S. Chandrasekaran, *Tetrahedron Lett.*, 1984, **25**, 2891.
50. D. Bhattacharjee and F. D. Popp, *J. Heterocycl. Chem.*, 1980, **17**, 315.
51. H. Hiraoka, K. Furuta, N. Ikeda and H. Yamamoto, *Bull. Chem. Soc. Jpn.*, 1984, **57**, 2777.
52. P. C. Mukherjee and A. N. Ganguli, *Tetrahedron*, 1969, **25**, 5281.
53. B. Fraser-Reid, D. L. Walker, S. Y. K. Tam and N. L. Holder, *Can. J. Chem.*, 1973, **51**, 3950; see also B. Fraser-Reid, *Acc. Chem. Res.*, 1975, **8**, 192.
54. K. S. Kim, S. J. Kim, Y. H. Song and C. S. Hahn, *Synthesis*, 1987, 1017.
55. K. S. Kim, Y. H. Song, N. H. Lee and C. S. Hahn, *Tetrahedron Lett.*, 1986, **27**, 2875.
56. K. S. Kim, Y. K. Chang, S. K. Bae and C. S. Hahn, *Synthesis*, 1984, 867.
57. R. T. Taylor and L. A. Flood, *J. Org. Chem.*, 1983, **48**, 5160.
58. Y. Ueno and M. Okawara, *Tetrahedron Lett.*, 1976, 4597; see also K. Saigo, A. Morikawa and T. Mukaiyama, *Bull. Chem. Soc. Jpn.*, 1976, **49**, 1656.
59. E. J. Corey and M. Ishiguro, *Tetrahedron Lett.*, 1979, 2745.
60. C. Clarke, I. Fleming, J. M. D. Fortunak, P. T. Gallagher, M. C. Honan, A. Mann, C. O. Nubling, P. R. Raithby and J. J. Wolff, *Tetrahedron*, 1988, **44**, 3931.
61. R. V. Stevens, K. T. Chapman and H. N. Weller, *J. Org. Chem.*, 1980, **45**, 2030.
62. S. O. Nwaukwa and M. Keehn, *Tetrahedron Lett.*, 1982, **23**, 35.
63. G. H. Posner, R. B. Perfetti and A. W. Runquist, *Tetrahedron Lett.*, 1976, 3499.
64. B. M. Trost and Y. Masuyama, *Tetrahedron Lett.*, 1984, **25**, 173.
65. Y. Masuyama, M. Takahashi and Y. Kurusu, *Tetrahedron Lett.*, 1984, **25**, 4417.
66. K. Kaneda, Y. Kawanishi, K. Jitsukawa and S. Teranishi, *Tetrahedron Lett.*, 1983, **24**, 5009.
67. H. Tomioka, K. Oshima and H. Nozaki, *Tetrahedron Lett.*, 1982, **23**, 539.
68. Y. Ueno, T. Kageyama and M. Okawara, *Synthesis*, 1983, 815.
69. M. Shiiza and I. Kuwajima, *Tetrahedron Lett.*, 1979, 2801.

70. K. Takai, K. Oshima and H. Nozaki, *Tetrahedron Lett.*, 1980, **21**, 1657.
71. M. Tanaka, T. Kobayashi and T. Sakakura, *Angew. Chem., Int. Ed. Engl.*, 1984, **23**, 518.
72. T. Hudlicky, G. Seoane and T. C. Lovelace, *J. Org. Chem.*, 1988, **53**, 2094.
73. J. P. Burkhart, N. P. Peet and P. Bey, *Tetrahedron Lett.*, 1988, **29**, 3435.

2.10
Vinylic and Arylic C—H Oxidation

PHILIP J. DUDFIELD
Schering Agrochemicals, Saffron Walden, UK

2.10.1 INTRODUCTION

A comprehensive introduction to this subject will be found by consulting the excellent books by Haines[1] and Wakefield[2] and the reviews by Harvey[3] and Whiting.[4]

The replacement of an aromatic or vinylic hydrogen by any other element can formally be considered as an oxidative process.[5] In this review only replacement of hydrogen by oxygen and reactions that proceed by such a process have been considered. The subject matter has been divided into several sections covering oxidations that proceed *via* an organometallic species, and finally those that do not. In order to make comparisons or to highlight the uses of a particular reagent, this division of material has not been rigidly held to. With organometallic reagents, transmetallation is often used to obtain the desired intermediate, and these reactions have generally been classified according to the first metallation, as it is this process which controls the regioselectivity of the oxidation.

The bulk of material considered covers arylic C—H oxidation, since vinylic C—H oxidation is essentially covered elsewhere (Chapter 3.4, this volume).

The reaction of an arylorganometallic reagent with an oxygen species is a popular process for introducing oxygen functionality in organic synthesis. The main advantage of such a reaction is its regiospecificity; the oxygen is invariably introduced at the carbon–metal bond. The main difficulty with such a process is regioselective formation of the carbon–metal bond. This can be achieved by one of the following methods: (i) functional group directed metallation, such functionality must be either present in the product or subject to appropriate modification; (ii) halogen–metal exchange, this requires the presence of an appropriately positioned halogen; (iii) electronically directed metallation *e.g.* thallation; or (iv) incorporation of the metal in the synthesis, particularly applicable to silicon and tin. This process, along with (ii), is not formally an oxidation of a C—H bond, but is considered here because it helps to complete the picture for aryl C—O bond formation.

2.10.2 LITHIUM, MAGNESIUM AND BORON INTERMEDIATES

In addition to molecular oxygen, [6] a number of reagents have been developed for introduction of oxygen into a carbon–lithium or carbon–magnesium bond. These include peroxides,[7,8] molybdenum peroxide–pyridine–hexamethylphosphoramide (MoOPH),[9] sulfonyloxaziridines,[10] nitrobenzene[11] and oxidation of the boronate[12] after transmetallation.

2.10.2.1 Imidazolones

Taddei and Ricci[7] reported the use of bis(trimethylsilyl) peroxide for the electrophilic hydroxylation of aryllithiums. The aryllithiums were formed regiochemically pure by either halogen–metal exchange or *ortho*-directed lithiation (Scheme 1). The same authors also noted that when the carbanion is α to a heteroatom, as in the 2-lithiobenzthiazolone, the reaction afforded the trimethylsilyl derivative (**1**), with no trace of the siloxy product (Scheme 2). This is not the case with *N*-protected imidazoles, as shown by Lipshutz *et al.*[13] who reacted the 2-position and 5-position anions of *N*-SEM-protected imidazoles with either bis(trimethylsilyl) peroxide or benzoyl peroxide to give good yields of the imidazolone (**2**; Scheme 3) and the TMS-protected imidazolone (**3**; Scheme 3). The reason why an *N*-protected imidazole gives the siloxy derivative, while a benzthiazolone gives the trimethylsilyl adduct, is not immediately obvious.

$$RX \xrightarrow{Bu^nLi} RLi \xrightarrow{(Me_3SiO)_2} ROSiMe_3 \xrightarrow{H^+} ROH$$

X = Br or H

Scheme 1

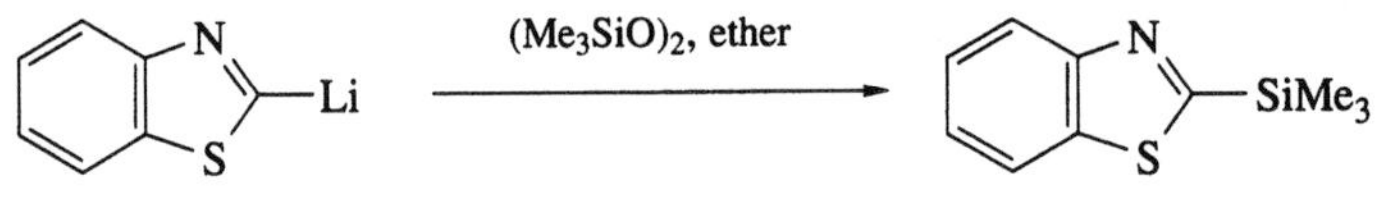

2-Lithiobenzothiazole $\xrightarrow{(Me_3SiO)_2,\ ether}$ 2-(SiMe$_3$)benzothiazole (**1**)

Scheme 2

N-SEM imidazole $\xrightarrow[\text{ii, }(Me_3SiO)_2\text{; 98\%}]{\text{i, }Bu^nLi}$ N-SEM imidazolone (**2**)

N-SEM imidazole $\xrightarrow[\text{ii, }Ph_2S_2]{\text{i, }Bu^nLi}$ 2-SPh N-SEM imidazole $\xrightarrow[\text{ii, }(Me_3SiO)_2\text{; 79\%}]{\text{i, }Bu^nLi}$ 5-Me$_3$SiO-2-SPh N-SEM imidazole (**3**)

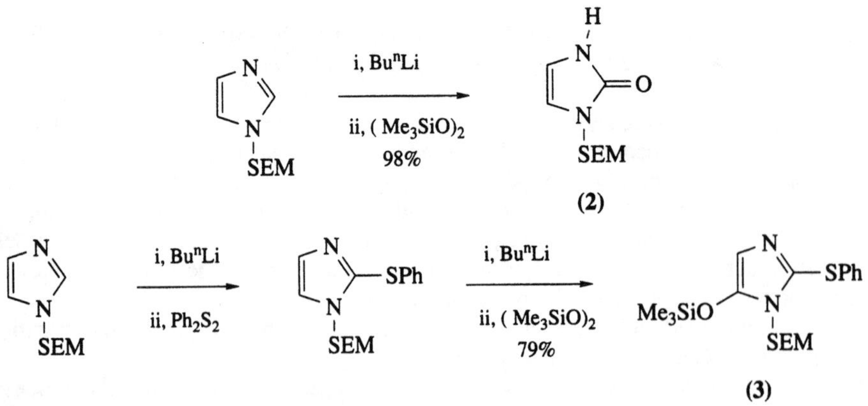

Scheme 3

2.10.2.2 Totarol Metabolites

In order to investigate the production of nagilactones from totarol by a route modeled on the suggested biogenetic pathway, Cambie *et al.*[14] needed to convert totarol (**4**) into 12-hydroxytotarol (**5**). Bromination of 13-methoxytotara-8,11,13-triene (**6**) with bromine and thallium(I) acetate in refluxing carbon tetrachloride gave the bromide (**7**) in 94% isolated yield. All attempts to form the Grignard reagent were unsuccessful, but treatment of the bromide with 1 equiv. of *n*-butyllithium resulted in a clean conversion to the aryllithium compound. Reaction of this intermediate with *t*-butyl perbenzoate gave back only starting material (**6**), but reaction with dry oxygen, MoOPH or lithium *t*-butyl hydroperoxide gave the natural product dispermol (**8**) in 47%, 40% and 36% yields respectively. It should be noted that higher yields of phenols were obtained with all these reagents in model studies. Dispermol (**8**) was then converted into the desired catechol (**5**) by pyridinium hydrochloride.

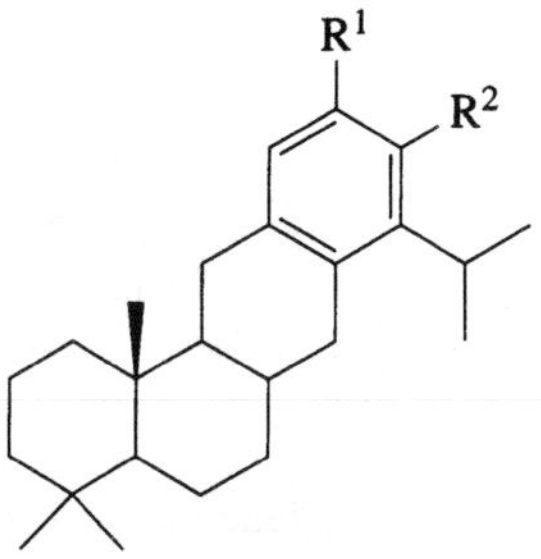

R^1, R^2

(**4**) R^1 =H, R^2 = OH
(**5**) R^1 = OH, R^2 = OH
(**6**) R^1 = H, R^2 = OMe
(**7**) R^1 = Br, R^2 = OMe
(**8**) R^1 = OH, R^2 = OMe
(**9**) R^1 = COMe, R^2 = OMe
(**10**) R^1 = OCOMe, R^2 = OMe

Alternative methods for the formation of the catechol (**5**) were examined, and these serve to show other methods for introduction of the hydroxy group. Nitration of triene (**6**) followed by reduction to the amine and diazotization in the presence of methanol gave the phenol (**8**), but only in poor yield. The best method developed appears to be acetylation of the triene (**6**) with titanium tetrachloride/acetyl chloride to give the ketone (**9**), followed by Baeyer–Villiger oxidation to the acetate (**10**), which on hydrolysis afforded the catechol (**8**) in 70% overall yield.

2.10.2.3 Tetralins

Whilst studying the metabolism of dopaminergic and serotonergic agonists, Wikström *et al.*[15] required the 5- and 8-hydroxy-2-(di-*n*-propylamino)tetralins. The synthesis of the 5-isomer (**11**) is shown in Scheme 4. The yield with nitrobenzene was poor, but this was the best reagent for the required conversion.

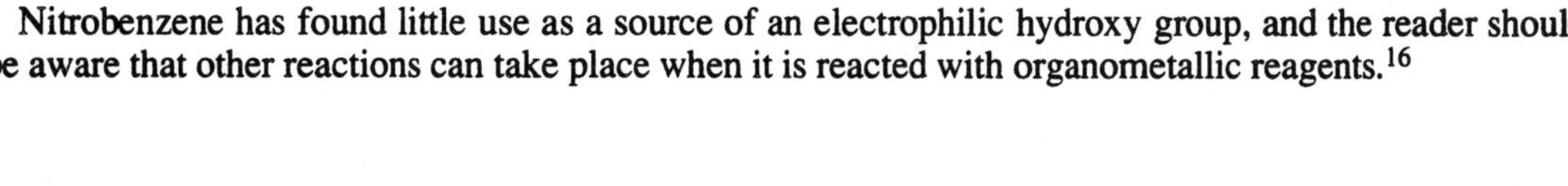

OMe, NPr^n_2 —(i, Bu^nLi; ii, $PhNO_2$)→ HO, OMe, NPr^n_2 —(aq. HBr)→ HO, OH, NPr^n_2 (**11**)

Scheme 4

Nitrobenzene has found little use as a source of an electrophilic hydroxy group, and the reader should be aware that other reactions can take place when it is reacted with organometallic reagents.[16]

2.10.2.4 Estrogens

A fine example demonstrating the use of an arylboron intermediate was reported by Santaniello *et al.*[17] As part of a project aimed at evaluating the biological activity of 2-and 4-substituted estrogens a convenient synthesis of 2-hydroxyestradiol was needed. Classical electrophilic oxidations usually lead to equimolar amounts of 2- and 4-isomers which are not easy to separate, and thus a method for regioselective hydroxylation was required. 3-Methoxyestra-1,3,5(10)-trien-17β-yl acetate (**12**) reacted with mercury(II) acetate in dry acetonitrile, and the reaction mixture was then treated with saturated aqueous sodium chloride to give the 2-chloromercurio derivative (**13**) in 80% yield. This intermediate was then exposed to diborane and the resulting organoboron complex directly oxidized to the 2-hydroxy estrogen which, after acetylation, gave the diacetate (**14**) in average yields of 48–50%. The methyl protecting group was removed and the triol stored as the triacetate (**15**; Scheme 5).

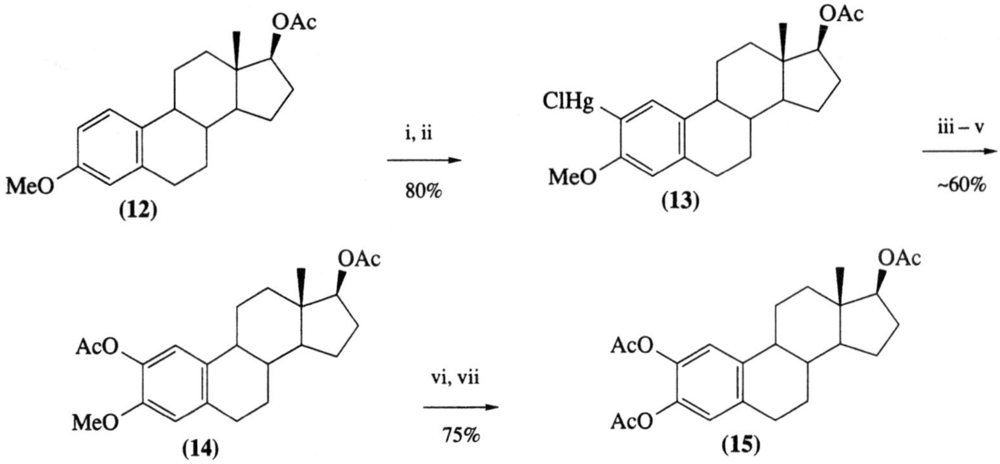

i, $Hg(OAc)_2$, MeCN; ii, aq. NaCl; iii, B_2H_6; iv, H_2O_2; v, Ac_2O; vi, pyridinium hydrobromide; vii, Ac_2O

Scheme 5

It is worthwhile here to put this work in context with related work in the field.[18] If the acetal-protected estrogen (**16**) was treated with *n*-butyllithium/TMEDA then metallation occurred at both the 2- and 4-positions. Conversion to the hydroxylated products (**17**) and (**18**) was then achieved *via* the boron complex using similar methodology to the above (Scheme 6). The same authors, Kirk and Slade,[19] did however report an efficient, regioselective synthesis of the 4-acetate isomer (**19**) by employing lead tetraacetate (LTA) as the oxidizing agent after initially forming the 4-mercurio intermediate (Scheme 7).

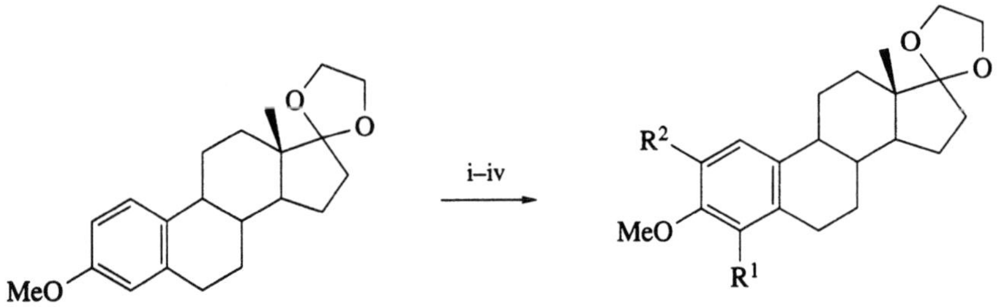

(**17**) R^1 = H, R^2 = OH; 30%
(**18**) R^1 = OH, R^2 = H; 15%

i, Bu^nLi; ii, B_2H_6, $B(OMe)_3$; iii, H_2O_2, NaOH; iv, H^+

Scheme 6

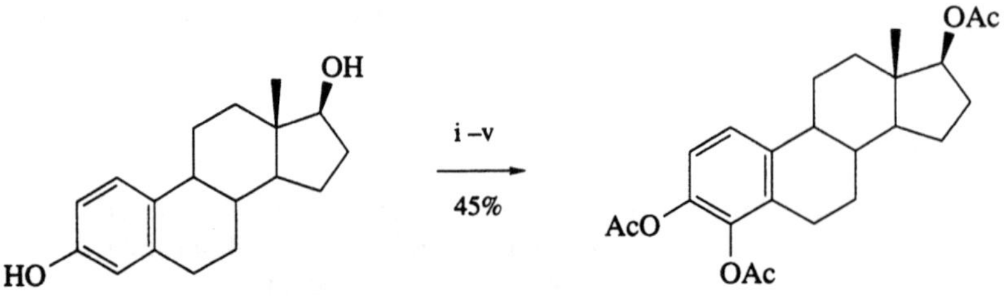

i, $Hg(OAc)_2$, AcOH, cat. $HClO_4$; ii, Ac_2O; iii, LTA, CF_3CO_2H; iv, Ac_2O; v, H^+

Scheme 7

In complete contrast to the above results is the oxidation of the estrone (**20**) by hydrogen peroxide in hydrogen fluoride/antimony tetrafluoride.[20] When subjected to hydroxylating conditions using this reagent, the estrone (**20**) gave a poor-yielding mixture of the oxidized products (**21–23**; Scheme 8).

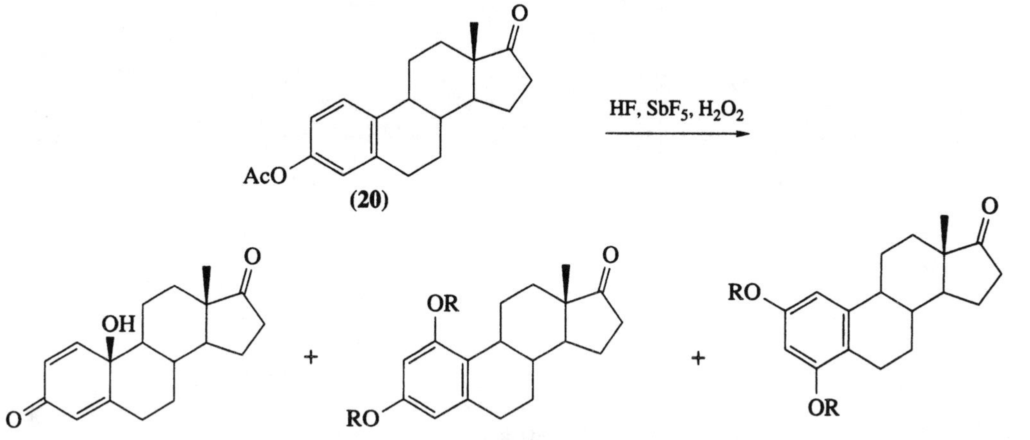

HF, SbF_5, H_2O_2

AcO (**20**)

OH O (**21**) + OR RO O (**22**) + RO O OR (**23**)

R = H and/or Ac

Scheme 8

These examples demonstrate that metal-directed hydroxylation is a powerful tool in organic synthesis.

2.10.2.5 Miscellaneous

A new method for the *ortho* hydroxylation of aromatic aldehydes *via ortho*-lithiated aromatic aminoalkoxides has recently been reported by Einhorn *et al.*[21] Formation of the aminoalkoxide serves two purposes. Firstly, the aldehyde group is protected and, secondly, the aminoalkoxide directs lithiation to the *ortho* position. Oxidation of the lithio species was effected by either MoOPH or molecular oxygen, albeit in poor yield. Alternatively, a two-step, one-pot condensation of the lithio intermediate with tributyl borate followed by oxidation with hydrogen peroxide gave the *ortho*-hydroxy aldehydes (**24**) in slightly better yields (Scheme 9).

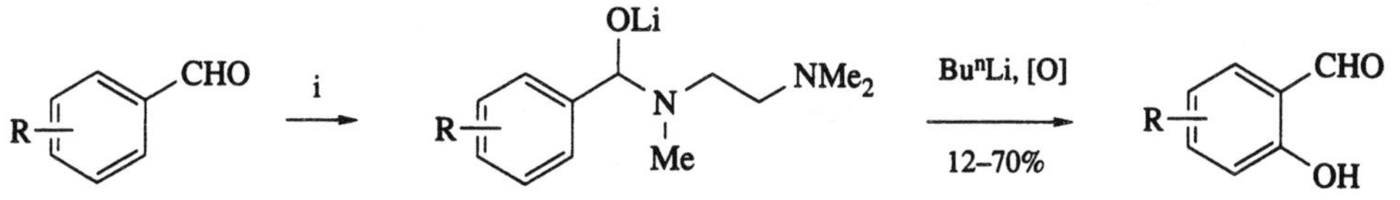

R CHO —i→ R OLi N Me NMe_2 —Bu^nLi, [O]; 12–70%→ R CHO OH (**24**)

i, Bu^nLi, *N,N,N'*-trimethylethylenediamine

Scheme 9

Snieckus and coworkers[22] used the combination of a tertiary amide with a methylenedioxy group to direct lithiation to the *ortho* position, and then converted the lithiated intermediate to the 2-hydroxybenzamide (**25**), which was used as a starting material for the synthesis of acridones (Scheme 10).

A similar methodology was used by Borchardt and coworkers[23] in the synthesis of the phenol (**26**; Scheme 11). Here, the fact that 4-fluorophenol metallates next to fluorine when the phenol is protected as a TBDMS ether was put to good use for introduction of the hydroxy group. Likewise, oxidation of a boron complex was a key step in the synthesis of phenolic dihydrodiols of benzo[a]pyrene, (**27**) and (**28**),[24] and also in the synthesis of phenolic crown ethers (**29**).[25]

The use of thallium is covered later, but it is worthwhile mentioning here that in the hydroxylation of polystyrenes reported by Bullen *et al.*,[26] the mercury or thallium intermediates were converted to the boronic acid residue using diborane and the boronate then transformed into the phenol (**30**) using hydrogen peroxide or trimethylamine *N*-oxide.

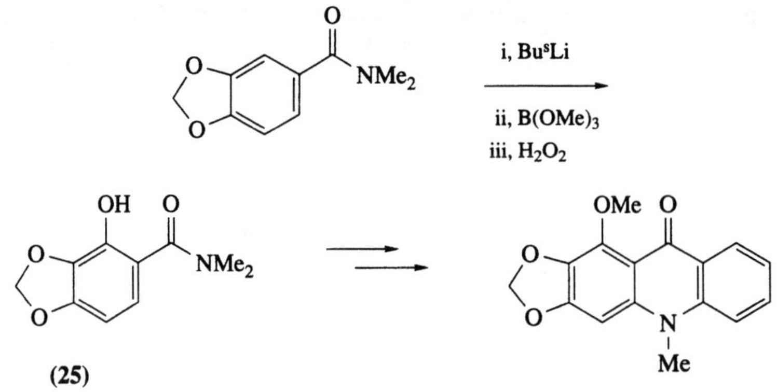

i, Bu^sLi

ii, $B(OMe)_3$

iii, H_2O_2

(25)

Scheme 10

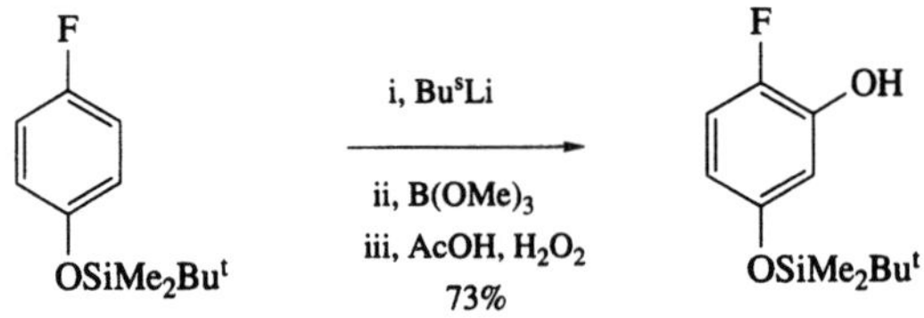

i, Bu^sLi

ii, $B(OMe)_3$

iii, AcOH, H_2O_2

73%

(26)

Scheme 11

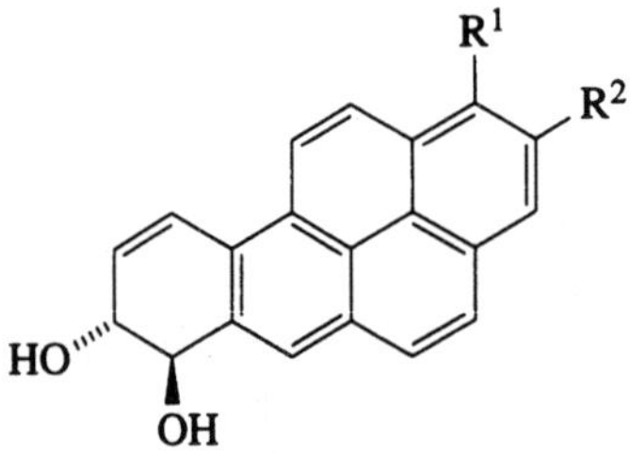

(27) $R^1 = H, R^2 = OH$

(28) $R^1 = OH, R^2 = H$

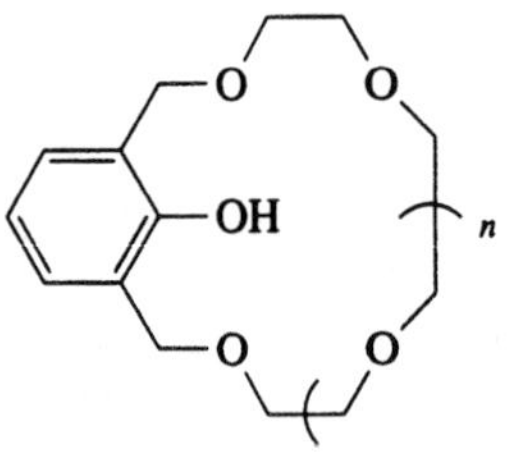

(29) $n = 1–7$

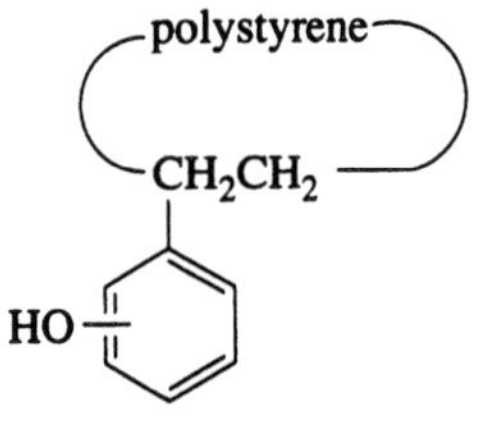

(30)

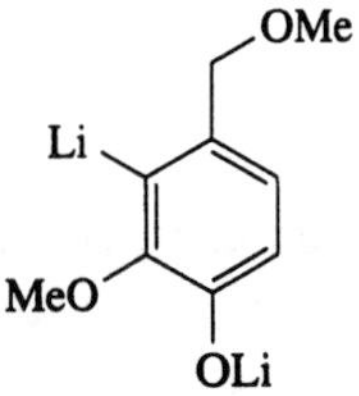

(31)

It is only fair when reviewing syntheses to mention unsuccessful reactions so as to outline any shortcomings of a particular reagent, and this has been done throughout this review. Saà *et al.*[27] have recently reported that reaction of the metallated lithium phenolate (**31**) with either nitrobenzene or bis(trimethylsilyl) peroxide gave only starting material, while reaction with trimethyl borate followed by oxidation furnished a complex mixture.

2.10.3 THALLIUM INTERMEDIATES

The reactions of aromatic substrates with thallium reagents is a fascinating subject which has been reviewed by McKillop and Taylor,[28] two of the prime contributors to this field of chemistry. Two types of reaction are possible, both of which are important for the introduction of oxygen functionality. The first is electrophilic aromatic thallation, whilst the second involves one-electron oxidation.

Arylthallium compounds are very interesting species and so obviously different from their lithium/magnesium/boron counterparts. Formation of an aryllithium/magnesium/boron reagent involves either proton abstraction, halogen–metal exchange or metal–metal exchange, whereas arylthallium compounds are produced by a reversible electrophilic process. Subsequent reactions are also very different in that the thallium species can undergo electrophilic, nucleophilic or free radical reactions depending upon the choice of reagent. For an understanding of the factors controlling the regiochemistry of aromatic thallation the reader should consult McKillop's review; however, aromatic thallation will in general follow the basic rules governing electrophilic aromatic substitution. The following points, however, should be remembered: (i) the process is reversible and can lead in time to the thermodynamic product, (ii) any groups present capable of coordinating to thallium will direct the position of metallation, and (iii) aromatic moieties activated towards electrophilic substitution are likely to undergo one-electron oxidation.

Whilst direct electrophilic hydroxylation of the arylthallium species can be effected using peroxytrifluoroacetic acid, further oxidation of the phenol to a quinone accompanies this process. This over-oxidation can be avoided by initial transmetallation to a lead species with concomitant reduction of the thallium trifluoroacetate (TTFA) by triphenylphosphine, followed by displacement of the lead by trifluoroacetate to give the aryl trifluoroacetate.[28] This hydroxylation method has yet to find use in the synthesis of molecules which are more complex than simple arenes.

2.10.3.1 Indoles

The synthesis of compounds bearing the indole nucleus has received a lot of attention from Somei and coworkers. As part of this program they have developed methods for regioselectively introducing oxygen functionality into the 4- and 7-positions of indoles. Thus, subjecting indole-3-carbaldehyde to thallation with thallium trifluoroacetate, followed by reaction of the thallium intermediate with iodine/copper(I) iodide in DMF, gave the iodide (**32**), which was readily converted to the corresponding methoxyindole (**33**) with sodium methoxide. The overall yield was an excellent 86% (Scheme 12).[29]

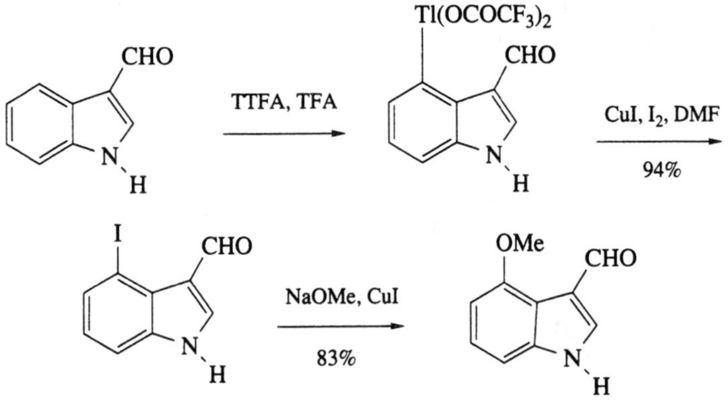

$Tl(OCOCF_3)_2$ CHO CHO N H TTFA, TFA → N H CuI, I_2, DMF → 94%

I CHO N H NaOMe, CuI → 83% OMe CHO N H

(**32**) (**33**)

Scheme 12

The synthesis of 7-methoxyindole was accomplished starting from 1-acetylindoline (**34**). Regioselective introduction of iodine was achieved using thallium trifluoroacetate, then potassium iodide. Deacetylation and oxidation to the indole (**35**), followed by reaction with sodium methoxide in DMF, gave the 7-methoxyindole (**36**) in 48% overall yield (Scheme 13).[30] More recently, Somei *et al.*[31] have reported that treating the intermediate thallium species with copper(II) sulfate pentahydrate gives directly the 1-acetyl-2,3-dihydro-7-hydroxyindole (**37**) in 42% yield (Scheme 14). It remains to be seen whether this is a general process.

A completely different approach to hydroxylation at the 5-position (indole numbering) in tryptophans was reported by Hino and coworkers.[32] Their synthesis involved either LTA or Fremy's salt oxidation of

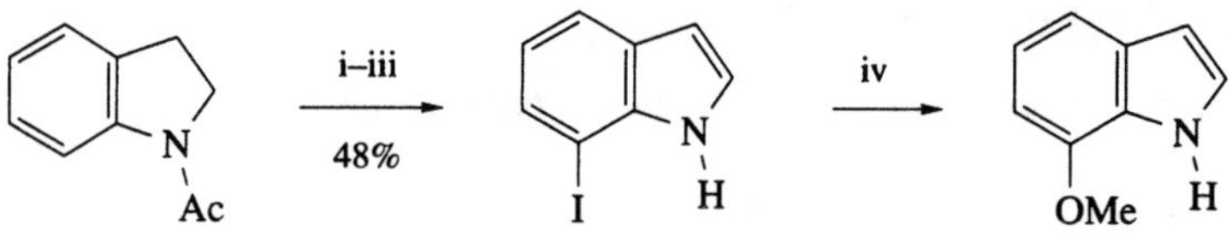

(34) (35) (36)

i, TTFA, TFA; ii, KI; iii, O_2, salcomine; iv, NaOMe, CuI, DMF

Scheme 13

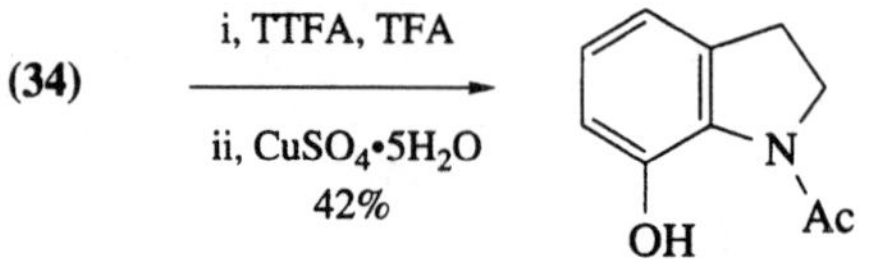

(34) (37)

i, TTFA, TFA; ii, $CuSO_4•5H_2O$; 42%

Scheme 14

the cyclic tautomer derived from tryptophan (**38**) to the *p*-quinoneimine (**39**), followed by sodium borohydride reduction and treatment with acetic acid, to give the hydroxytryptophan (**40**; Scheme 15).

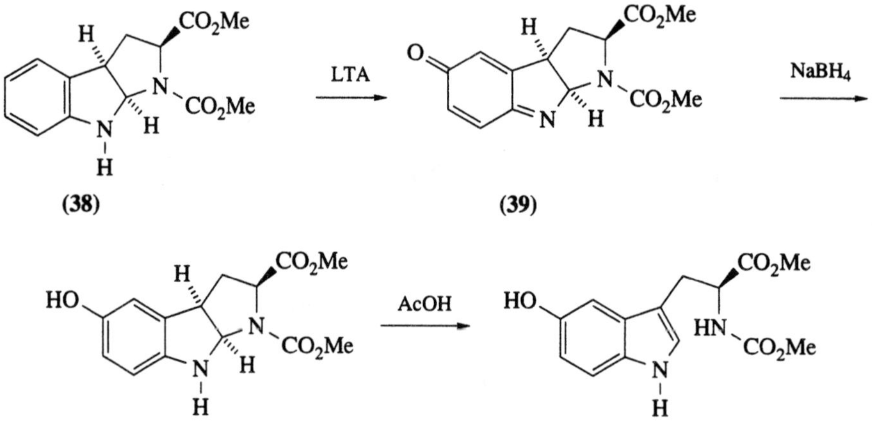

(38) LTA (39) $NaBH_4$ AcOH (40) 44% overall

Scheme 15

2.10.3.2 Towards Lycorine Alkaloids

Although one-electron oxidation of arenes by thallium trifluoroacetate presumably does not proceed *via* formation of a discrete arylthallium bond, some examples involving oxidative cyclization mediated by thallium trifluoroacetate will be considered here. Schwartz and Hudec[33] employed thallium trifluoroacetate to effect an intramolecular cyclization of the amide (**41**) to give a key intermediate (**42**) in their projected synthesis of lycorine alkaloids (Scheme 16). Interestingly, when the bromine was replaced by a hydrogen the yield was poorer.

2.10.3.3 Dihydrocoumarins

Taylor and coworkers[34] have studied the intramolecular capture of radical cations from the thallium trifluoroacetate oxidation of arylalkanoic acids and arylalkanols. For example, 3-(3,4-dimethoxyphenyl)propionic acid (**43**) on treatment with thallium trifluoroacetate in trifluoroacetic acid containing a small amount of boron trifluoride etherate for a few seconds gave the oxidized products (**44–46**; Scheme 17), the exact yields dependent upon the reaction conditions and work-up. In analogous fashion, oxida-

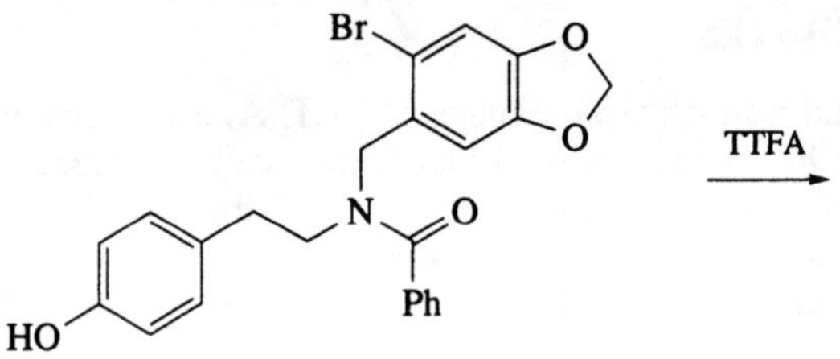

(41)

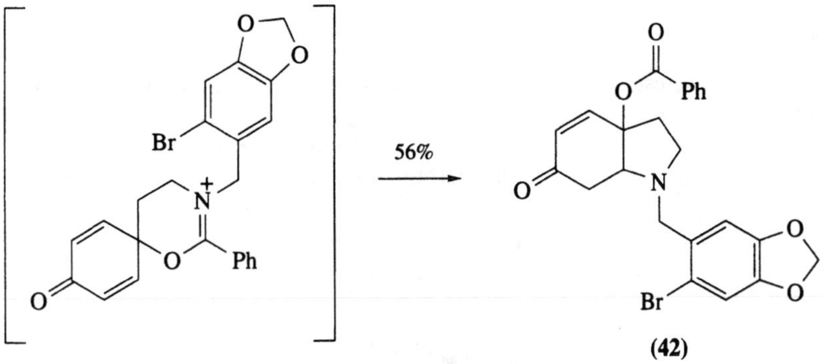

(42)

Scheme 16

tion of 3-(2,3,4-trimethoxyphenyl)propionic acid (**47**) gave the dihydrocoumarin (**48**) in 53% yield (Scheme 18).

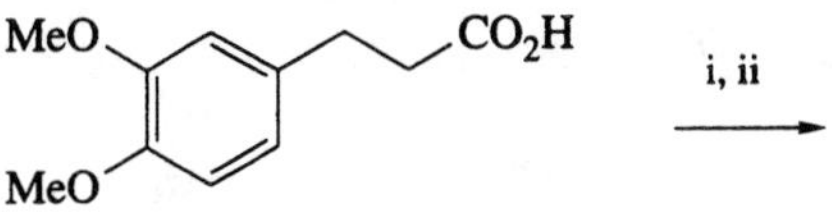

(43)

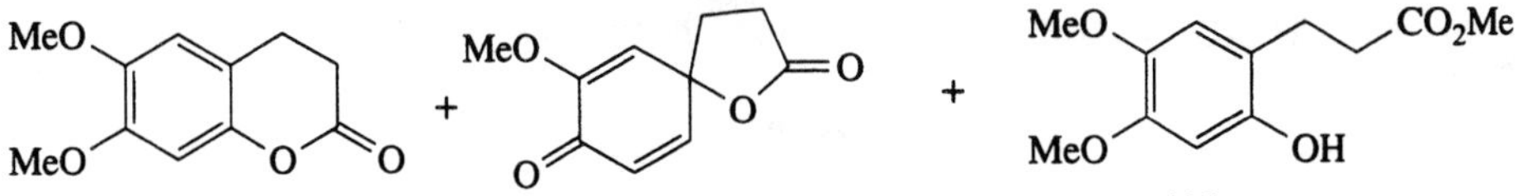

(44) (45) (46)

i, TTFA, TFA, cat. $BF_3•OEt_2$; ii, Bu[t]OH

Scheme 17

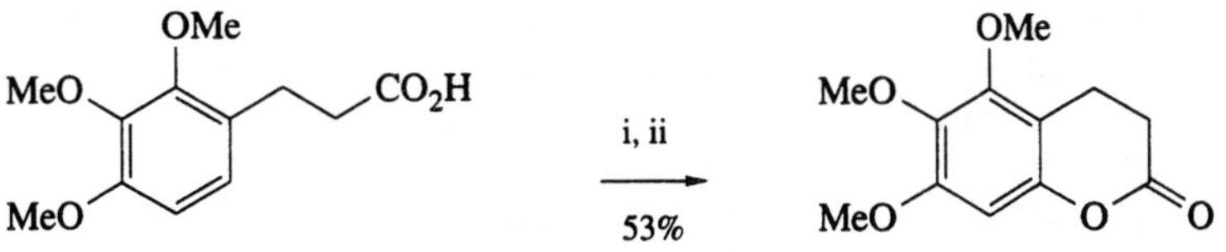

(47) (48)

i, TTFA, TFA, cat. $BF_3•OEt_2$; ii, Bu[t]OH

Scheme 18

Yamamura and coworkers have used a thallium trifluoroacetate mediated oxidative cyclization in their synthesis of aerothionin and related products[35] and also in the synthesis of bastadin-6,[36] a 28-membered ring lactone.

2.10.4 LEAD INTERMEDIATES

Lead tetraacetate (LTA) and lead tetrakisfluoroacetate (LTFA) are common oxidants for the introduction of the hydroxy group. It has been suggested that the oxidation proceeds *via* electrophilic attack directly onto oxygen,[37] but this seems unlikely in view of the fact that aryllead species are well characterized compounds,[38] and they are known to give the corresponding acetates on treatment with trifluoroacetic acid or acetic acid. Some examples of the use of LTA have already been described (see Sections 2.10.2.4 and 2.10.3).

2.10.4.1 Estrone

An impressive synthesis of estrone (**51**) was reported by Vollhardt and coworkers.[39] Cobalt-catalyzed cooligomerization of the diyne (**49**) with bis(trimethylsilyl)acetylene gave the estratrienone (**50**) in 71% yield. Introduction of the hydroxy substituent at C-3 was then cleverly achieved by selective protodesilylation at C-2, followed by oxidation of the carbon–silicon bond using LTFA (Scheme 19).

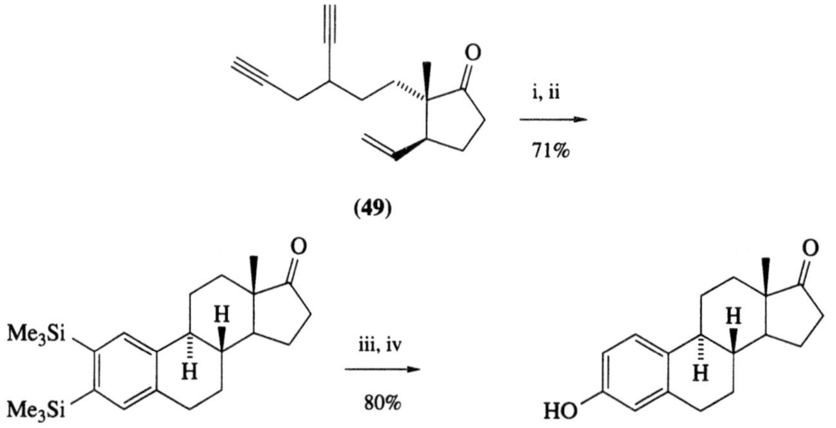

i, ii
71%

(49)

Me_3Si
Me_3Si
H
H
O
iii, iv
80%
HO
H
H
O

(50) **(51)**

i, $CpCo(CO)_2$, BTMSA; ii, xylene, reflux; iii, TFA, –30 °C; iv, LTFA

Scheme 19

2.10.4.2 Vitamin B_6

In a synthesis of vitamin B_6, which also employed a cobalt-mediated diyne cyclization, Vollhardt and coworkers[40] were unsuccessful in converting either the tin compound (**52**) or the corresponding silicon derivative to the hydroxylated material using LTA and trifluoroacetic acid. In fact, attempted hydroxylation of the tin compound (**52**) by transmetallation with *n*-butyllithium followed by reaction with molecular oxygen, nitrobenzene or trimethyl borate/hydrogen peroxide gave complex mixtures containing primarily destannylated material. A solution to the problem was eventually found by introduction of iodine, followed by displacement of iodide using a modification of the method developed by Tiecco (Scheme 20).[41]

2.10.4.3 Quinol Acetates

Pattenden and coworkers[42] have recently evaluated the relative merits of LTA and electrochemical oxidation of phenolic compounds with particular reference to synthesis of the antiallergic compounds sodium chromoglycate (INTAL; **53**) and proxicromil (**54**), which are used for the prophylactic treatment of asthma. The 2-carboxychromone moieties in the compounds (**53**) and (**54**) are synthesized from the appropriate 2′,6′-dihydroxyacetophenones. Oxidation of the 2′-hydroxyacetophenone (**55**) by LTA in dichloromethane gave almost exclusively the quinol acetate (**56**), which was subsequently converted to the 2′,6′-dihydroxyacetophenone (**57**), a precursor to proxicromil (**54**; Scheme 21). By contrast, electro-

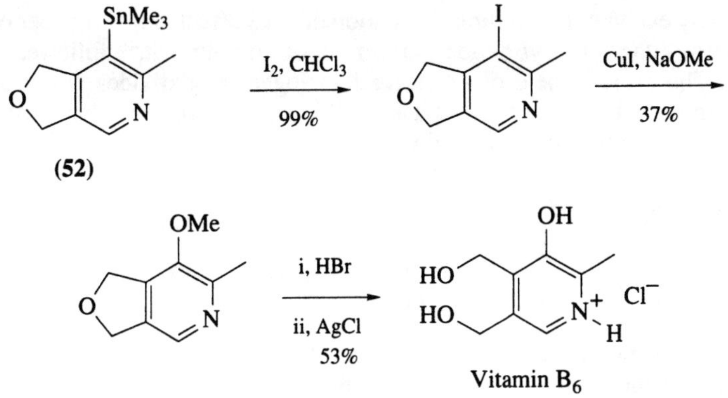

(52)

Vitamin B_6

Scheme 20

chemical oxidation of the acetophenone (**55**) gave four principal products, with the quinol acetate (**58**) as the major one. The acetate (**58**) was then converted to the 2′,6′-dihydroxyacetophenone (**57**; Scheme 22).

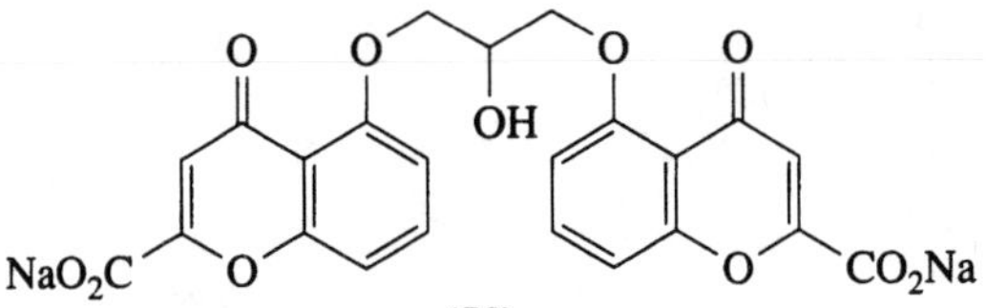

(53) 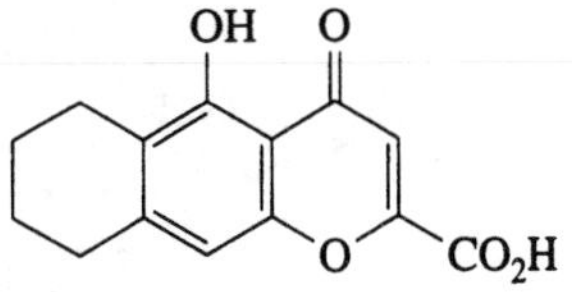 (54)

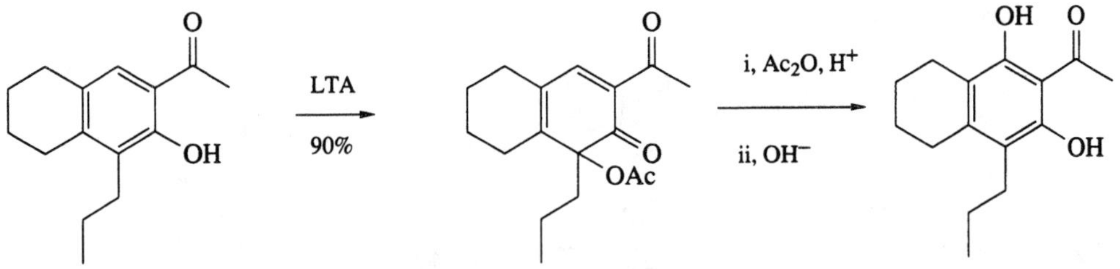

(55) (56) (57)

Scheme 21

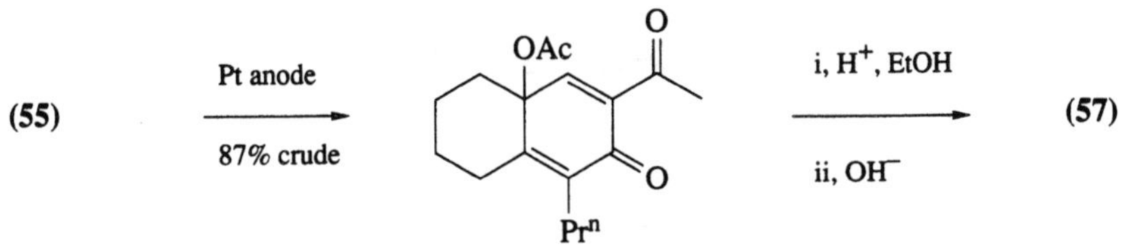

(55) (58) (57)

Scheme 22

Umezawa and coworkers[43] have extensively studied the LTA oxidation of hydroxytetrahydroisoquinolines, highlighting the range of products that can be obtained depending upon the reaction conditions and subsequent transformations.

2.10.5 NONMETALLATED INTERMEDIATES

Methods for the oxidation of arylic and vinylic C—H bonds which do not proceed *via* organometallic intermediates have already been referred to in the preceeding sections when it was appropriate to make comparisons.

Direct introduction of oxygen may be performed by a variety of reagents, but this type of oxidation does not normally extend beyond the preparation of simple phenols.[44] The oxidation of phenols to hydroquinones is occasionally encountered. Indirect oxidation may be effected in a number of ways. A reliable method is formation of either an aryl halide[45] or an aryldiazonium salt,[46] followed by attack with an oxygen nucleophile. The displacement of an aromatic halogen by alkoxides is a procedure which has been developed in particular by Testaferri and Tiecco.[41,47] Alternative procedures for indirect oxidation involve intramolecular delivery of oxygen, and encompass such reactions as the Baeyer–Villiger oxidation and the photochemical rearrangement of aromatic azoxy derivatives,[48] the latter process obviously being restricted to quite specific systems.

2.10.5.1 Hydroquinones

The Elbs persulfate oxidation procedure was used by Bach and coworkers[49] at an early stage of their synthesis of model compounds related to fredericamycin A. Thus the phenol (**59**) gave 2,5-dihydroxy-4-methoxybenzoic acid (**60**) in modest yield (30%). This was then transformed in several steps to the isobenzofuranone (**61**). Generation of the isobenzofuran (**62**) *in situ* and Diels–Alder reaction of this with the enedione (**63**) gave, after loss of the trimethylsilyl group, the desired compound (**64**; Scheme 23) in 62% yield from the isobenzofuranone (**61**).

(59) —i, ii; 30%→ (60) → → (61) —iii→ (62) —iv, (63), v; 62%→ (64)

i, $K_2S_2O_8$, NaOH; ii, H^+; iii, Bu^nLi then Me_3SiCl; iv, (**63**); v, silica gel

Scheme 23

(65) —$K_2S_2O_8$, Et_4NOH; 50%→ (66) —LTA, AcOH; 68%→ (67)

Scheme 24

Looker *et al.*[50] used a modification of the Elbs persulfate oxidation procedure in which tetraethylammonium hydroxide was used as the base in their synthesis of 5,8-quinoflavone. The phenol (**65**) was oxidized to the hydroquinone, primetin (**66**), which upon further oxidation with LTA afforded 5,8-flavoquinone (**67**) in 34% overall yield (Scheme 24).

2.10.5.2 Daunomycin and Adriamycin Analogs

One good example of the use of the Baeyer–Villager oxidation was reported by Mitscher and coworkers[51] in their approach to analogs of adriamycin and daunomycin, two important antitumor antibiotics. The anthracenone (**68**) was cyclized with hydrogen fluoride to give the unstable product (**69**), which upon treatment with hydrogen peroxide in acetic acid/sulfonic acid underwent oxidation to the quinone and also Baeyer–Villiger cleavage of the ketone moiety to give, after esterification, the anthraquinone (**70**), a key intermediate in their projected synthesis (Scheme 25). Unfortunately, problems in the latter stages of the synthesis and attractive routes to similar analogs from other workers led to this approach being abandoned. It is interesting to note that the same authors[52] had great difficulty in introducing a hydroxy group by displacement of a bromide in a similar system (Scheme 26). The desired transformation was finally effected in 65% yield using calcium hydroxide and copper powder in a sealed tube at 200 °C.

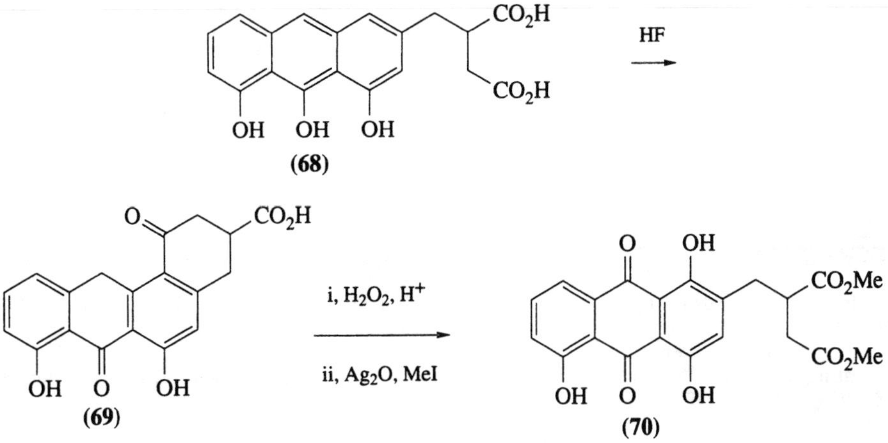

Scheme 25

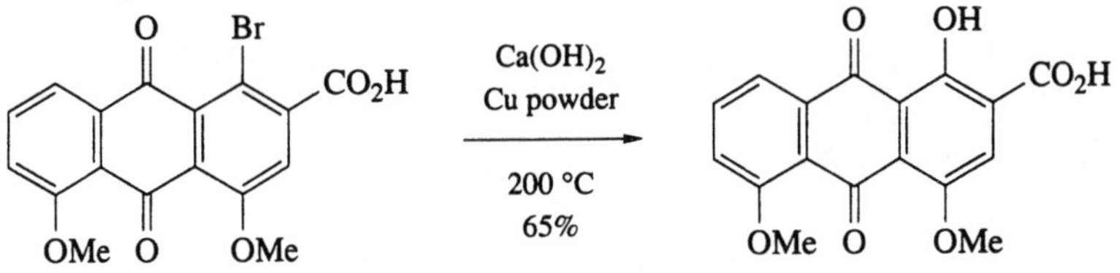

Scheme 26

2.10.5.3 Anthracycline Relatives

Lown and Sondhi[53] were interested in the synthesis of chromophores of the anthracycline antibiotics in which the quinone ring C was replaced by a γ-pyrone. The γ-pyrone ring was prepared by an oxidative cyclization mediated by DDQ, presumably proceeding *via* the quinone (Scheme 27).

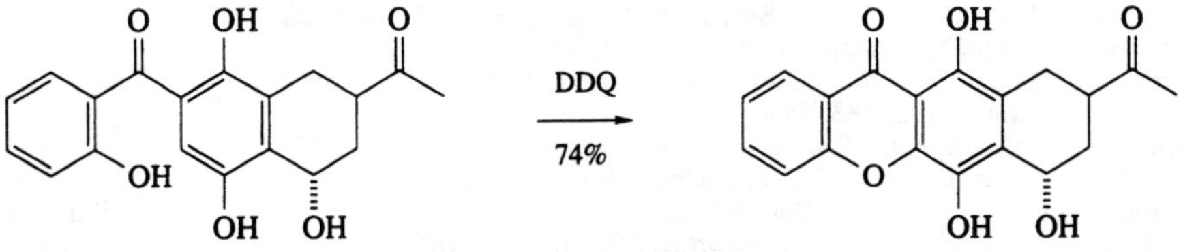

Scheme 27

2.10.5.4 Fervenulin Analogs

The synthesis of analogs of the antibiotic 2-methylfervenulone (MSD-92)[54] highlights an example of the displacement of a chloride by an oxygen nucleophile, although perhaps more interesting is the regioselective introduction of the chlorine by reacting the 4-deazafervenulin 2-oxide (**71**) with the Vilsmeier–Haack reagent to give the 3-chloro-4-deazafervenulin (**72**) in 45% yield. Senga and coworkers then converted this chloro compound (**72**) to the 4-hydroxy analog (**73**), which upon methylation gave 4-deaza-MSD-92 (**74**; Scheme 28).

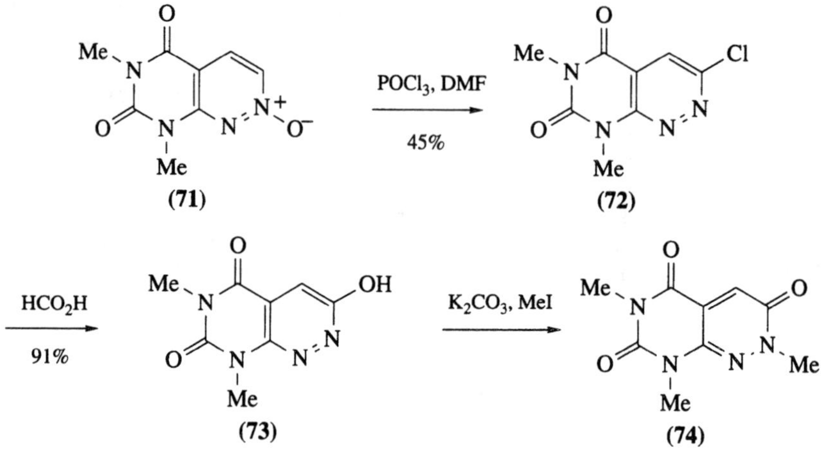

Scheme 28

2.10.6 REFERENCES

1. A. H. Haines, in 'Methods for the Oxidation of Organic Compounds', Academic Press, London, 1985, p. 173.
2. B. J. Wakefield, in 'Organolithium Methods', Academic Press, London, 1988, p. 129.
3. R. G. Harvey, *Synthesis*, 1986, 601.
4. D. A. Whiting, in 'Comprehensive Organic Chemistry', ed. D. H. R. Barton and W. D. Ollis, Pergamon Press, Oxford, 1979, vol. 1, p. 707.
5. Ref. 1, p. 2.
6. K. A. Parker and K. A. Koziski, *J. Org. Chem.*, 1987, **52**, 674.
7. M. Taddei and A. Ricci, *Synthesis*, 1986, 633; L. Camici, A. Ricci and M. Taddei, *Tetrahedron Lett.*, 1986, **27**, 5155.
8. H. J. Jakobsen, E. H. Larsen and S.-O. Lawesson, *Recl. Trav. Chim. Pays-Bas.*, 1963, **82**, 791, and references cited therein.
9. N. J. Lewis, S. Y. Gabhe and M. R. DeLaMater, *J. Org. Chem.*, 1977, **42**, 1479.
10. F. A. Davis, J. Wei, A. C. Sheppard and S. Gubernick, *Tetrahedron Lett.*, 1987, **28**, 5115, and references cited therein.
11. P. Wiriyachitra and M. P. Cava, *J. Org. Chem.*, 1977, **42**, 2274.
12. M. F. Hawthorne, *J. Org. Chem.*, 1957, **22**, 1001; S.-O. Lawesson and N. C. Yang, *J. Am. Chem. Soc.*, 1959, **81**, 4230.
13. B. H. Lipshutz, B. Huff and W. Hagen, *Tetrahedron Lett.*, 1988, **29**, 3411.
14. R. C. Cambie, R. C. Hayward and B. D. Palmer, *Aust. J. Chem.*, 1982, **35**, 1679.
15. H. Wikström, T. Elebring, G. Hallnemo, B. Andersson, K. Svensson, A. Carlsson and H. Rollema, *J. Med. Chem.*, 1988, **31**, 1080.
16. G. Bartoli, E. Mercantoni, M. Bosco and R. Dalpozzo, *Tetrahedron Lett.*, 1988, **29**, 2251.
17. E. Santaniello, A. Fiecchi, P. Ferraboschi and M. Ravasi, *J. Chem. Soc., Perkin Trans. 1*, 1983, 2765.
18. D. N. Kirk and C. J. Slade, *J. Chem. Res. (S)*, 1983, 228.
19. D. N. Kirk and C. J. Slade, *J. Chem. Soc., Chem. Commun.*, 1982, 563.
20. C. Berrier, J. C. Jacquesy and M. P. Jouannetaud, *Tetrahedron*, 1984, **40**, 5135.
21. J. Einhorn, J.-L. Luche and P. Demerseman, *J. Chem. Soc., Chem. Commun.*, 1988, 1350.
22. M. Iwao, J. N. Reed and V. Snieckus, *J. Am. Chem. Soc.*, 1982, **104**, 5531.
23. A. K. Sinhababu, M. Kawase and R. T. Borchardt, *Tetrahedron Lett.*, 1987, **28**, 4139.
24. S. Kumar and P. L. Kole, *Tetrahedron Lett.*, 1988, **29**, 3513.
25. M. Skowronska-Ptasinka, V. M. L. J. Aarts, R. J. M. Egberink, J. Van Eerden, S. Harkema and D. N. Reinhoudt, *J. Org. Chem.*, 1988, **53**, 5484.
26. N. P. Bullen, P. Hodge and F. G. Thorpe, *J. Chem. Soc., Perkin Trans. 1*, 1981, 1863.
27. G. Coll, J. Morey, A. Costa and J. M. Saà, *J. Org. Chem.*, 1988, **53**, 5345.
28. A. McKillop and E. C. Taylor, in 'Comprehensive Organometallic Chemistry', ed. G. Wilkinson, F. G. A. Stone and E. W. Abel, Pergamon Press, Oxford, 1982, vol. 7, p. 465.
29. M. Somei, F. Yamada, M. Kunimoto and C. Kaneko, *Heterocycles*, 1984, **22**, 797.
30. M. Somei and Y. Saida, *Heterocycles*, 1985, **23**, 3113.

31. M. Somei, T. Kawasaki and T. Ohta, *Heterocycles*, 1988, **27**, 2363.
32. M. Tanaguchi, T. Anjik, M. Nakagawa and T. Hino, *Chem. Pharm. Bull.*, 1984, **32**, 2544.
33. M. A. Schwartz and T. M. Hudec, *Synth. Commun.*, 1986, **16**, 1599.
34. E. C. Taylor, J. G. Andrade, G. J. H. Rall, I. J. Turchi, K. Steliou, G. E. Jagdmann, Jr. and A. McKillop, *J. Am. Chem. Soc.*, 1981, **103**, 6856.
35. S. Nishiyama and S. Yamamura, *Bull. Chem. Soc. Jpn.*, 1985, **58**, 3453.
36. S. Nishiyama, T. Suzuki and S. Yamamura, *Chem. Lett.*, 1982, 1851.
37. W. A. Waters, 'Mechanisms of Oxidation of Organic Compounds', Methuen, London, 1964; ref. 4, p. 744.
38. E. R. Cole, G. Crank and H. T. H. Minh, *Aust. J. Chem.*, 1980, **33**, 527, and references cited therein.
39. R. L. Funk and K. P. C. Vollhardt, *J. Am. Chem. Soc.*, 1980, **102**, 5253.
40. C. A. Parnell and K. P. C. Vollhardt, *Tetrahedron*, 1985, **41**, 5791.
41. L. Testaferri, M. Tiecco, M. Tingoli, D. Chianelli and M. Montanucci, *Tetrahedron*, 1983, **39**, 193; see also ref. 47.
42. M. J. Begley, G. B. Gill, G. Pattenden, A. Stapleton and R. A. Raphael, *J. Chem. Soc., Perkin Trans. 1*, 1988, 1677.
43. H. Hara, M. Hosaka, O. Hoshino and B. Umezawa, *J. Chem. Soc., Perkin Trans. 1*, 1980, 1169; H. Hara, H. Shinoki, O. Hoshino and B. Umezawa, *Heterocycles*, 1983, **20**, 2149.
44. Ref. 1, p. 173.
45. H. Yamaguchi, S. Nakajima, M. Arimoto, M. Tanoguchi, T. Ishida and M. Inoue, *Chem. Pharm. Bull.*, 1984, **32**, 1754; L. Testaferri, M. Tiecco, M. Tingoli, D. Bartoli and A. Massoli, *Tetrahedron*, 1985, **41**, 1373.
46. R. A. Johnson, E. G. Nidy, J. W. Aiken, N. J. Crittenden and R. R. Gorman, *J. Med. Chem.*, 1986, **29**, 1461; M. S. Allen, T. J. Hagen, M. L. Trudell, P. W. Codding, P. Skolnick and J. M. Cook, *J. Med. Chem.*, 1988, **31**, 1854.
47. M. Tiecco, *Synthesis*, 1988, 749.
48. A. Albini and M. Alpegiani, *Chem. Rev.*, 1984, **84**, 43.
49. J. C. Evans, R. C. Klix and R. D. Bach, *J. Org. Chem.*, 1988, **53**, 5519.
50. J. H. Looker, J. R. Edman and C. A. Kingsbury, *J. Org. Chem.*, 1984, **49**, 645.
51. J. Alexander, A. V. Bhatia, G. W. Clark, III, A. Leutzow, L. A. Mitscher, S. Omoto and T. Suzuki, *J. Org. Chem.*, 1980, **45**, 24.
52. J. Alexander, A. V. Bhatia, L. A. Mitscher, S. Omoto and T. Suzuki, *J. Org. Chem.*, 1980, **45**, 20.
53. J. W. Lown and S. M. Sondhi, *J. Org. Chem.*, 1984, **49**, 2844.
54. S. Nishigaki, M. Ichiba and K. Senga, *J. Org. Chem.*, 1983, **48**, 1629.

2.11
Synthesis of Quinones

PHILIP J. DUDFIELD
Schering Agrochemicals, Saffron Walden, UK

2.11.1 INTRODUCTION

The synthesis of quinones,[1] including anthracyclinone antibiotics[2] and oxidized metabolites of polynuclear aromatic hydrocarbons,[3] has been reviewed.

The synthesis of quinones from arenes is an area which demands further research, despite the number of reagents presently available for this transformation. This is highlighted by the synthesis of the naphthoquinone (**3**).[4] Direct oxidation of the dibromoarene (**1**) was unsatisfactory, and therefore Bruce and coworkers had to resort to a multistep sequence involving nitration, reduction, diazotization, displacement by hydroxide and finally oxidation of the phenol (**2**) with Fremy's salt (Scheme 1). Although there are examples of the oxidation of polynuclear aromatic hydrocarbons to quinones, the direct oxidation of an arene to a quinone is a process not encountered in the synthesis of more complex molecules.

On the contrary, there are many examples of the synthesis of quinones from activated arenes, the activation normally arising from a hydroxy, alkoxy or amino group. It would be impossible here to cite every example from the last decade; however, it is hoped that the references expanded upon will give the reader a feel for the types of reagent which are most popular, the type of system that has employed this chemistry, and some of the problems involved. A number of reagents[5] are available for the oxidation of activated arenes to quinones, but there are only a few tried and tested reagents which are used again and again. The question posed by the reader who is looking to effect a particular oxidative transformation is

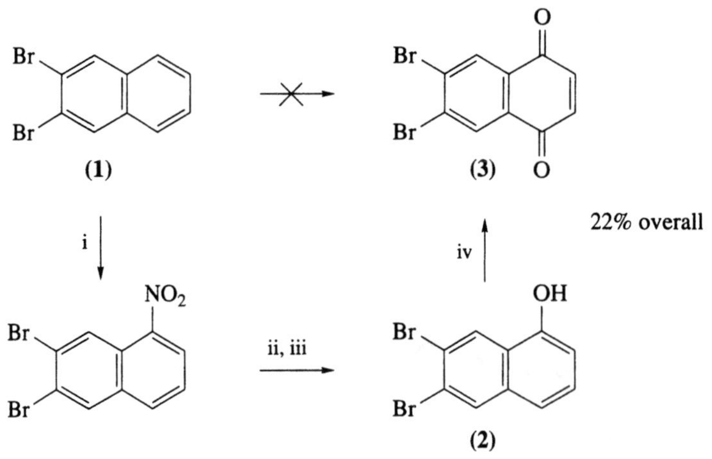

22% overall

i, dil. HNO_3, CH_2Cl_2, conc. H_2SO_4; ii, H_2NNH_2, Pd–C; iii, $NaNO_2$, H_2SO_4 then H_3O^+; iv, Fremy's salt

Scheme 1

which reagent? Unfortunately in this area there is no simple answer. Although it could possibly be said that Fremy's salt is the reagent of choice for the synthesis of *p*-quinones from phenols, while cerium(IV) ammonium nitrate is often preferred for the oxidative demethylation of methoxyarenes, the following examples will serve to emphasize that experimentation is often necessary to find the best reagent for the transformation in question. Although this review is primarily concerned with synthesis it is most convenient to loosely categorize the syntheses according to the reagents that were utilized.

2.11.2 FREMY'S SALT

2.11.2.1 Towards Rubradirin

Fremy's salt (potassium nitrodisulfonate)[6] is a widely used reagent,[7–12] particularly in the synthesis of quinone antibiotics. One example has emerged from Kozikowski's group in their approach to the rubradirin antibiotics.[13] The MEM-protected 2-nitroresorcinol (**4**) was converted in several steps to the phenol

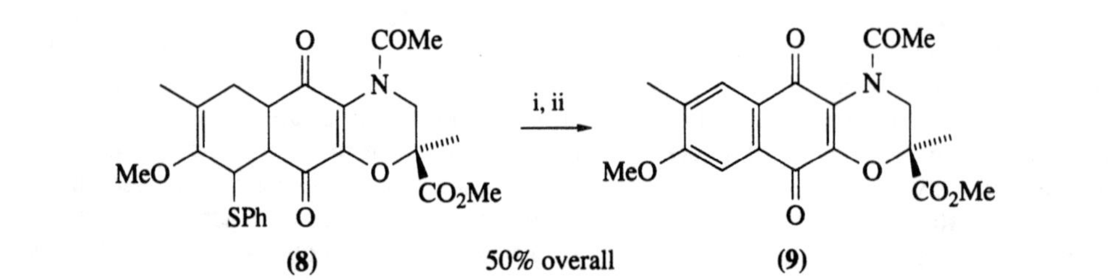

50% overall

i, $KHCO_3$; ii, Fremy's salt

Scheme 2

(**5**), which was then oxidized with Fremy's salt to the quinone (**6**). It was imperative that the nitrogen was protected by an acyl group, otherwise none of the desired quinone was formed. Diels–Alder reaction with the diene (**7**) then gave the cyclohexenedione (**8**) which upon treatment with potassium hydrogencarbonate to eliminate thiophenol and oxidation with Fremy's salt gave the key intermediate quinone (**9**; Scheme 2).

2.11.2.2 Defucogilvocarcin V

A more recent example from McGee and Confalone[14] highlights the use of Fremy's salt in the oxidation of a latent phenol (**10**) to the quinone (**11**). The oxidation was initially problematic because Fremy's salt requires the free phenol and, although the lactone could be opened under basic conditions, the salt readily relactonized on acidification. The oxidation was finally accomplished by adding the initial hydrolysis mixture to a solution of Fremy's salt buffered to pH 7. The quinone (**11**) was then converted to the aglycone defucogilvocarcin V (**12**), an important compound for studying the mechanism of DNA damage by the antitumor gilvocarcins (Scheme 3).

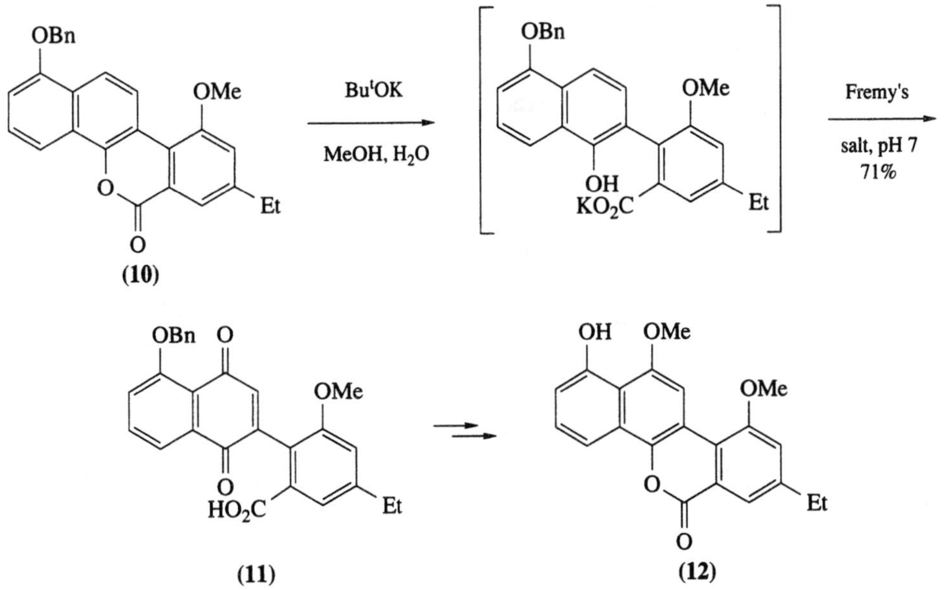

OBn OMe Et O O (**10**) Bu^tOK MeOH, H_2O [OBn OMe OH KO_2C Et] Fremy's salt, pH 7 71%

OBn O OMe O HO_2C Et (**11**) → OH OMe OMe O O Et (**12**)

Scheme 3

2.11.2.3 α-Tocopherol

One complication that can arise when using Fremy's salt is that the substrate may be insoluble in the aqueous reaction medium, despite the use of organic cosolvents. Olson *et al.*[15] introduced an organic-soluble version of Fremy's salt in their synthesis of α-tocopherol. Hexahydrofarnesolacetone (**13**) was transformed to the phenol (**14**), which was oxidized in nearly quantitative yield to the tocopheryl quinone (**15**) using a mixture of Fremy's salt and tricaprylylmethylammonium chloride in a two-phase water/benzene solvent system. The oxidizing agent was presumed to be the bis(tricaprylylmethyl)ammonium nitrodisulfonate. The quinone (**15**) is a known precursor of α-tocopherol (**16**; Scheme 4).

2.11.2.4 Lavendamycin Pharmacophores

The idea of using a phase transfer procedure with Fremy's salt was further developed by Kende and Ebetino[16] who advanced tetra-*n*-butylammonium bisulfate as the phase transfer reagent in a dichloromethane/aq. bicarbonate two-phase system. Although Kende obtained disappointing results with this system in the synthesis of lavendamycin (**17**), Boger *et al.*[17] put these reagent conditions to good use in their synthesis of a similar quinoline-5,8-diene (**18**) which was designed to investigate the potential minimum, potent pharmacophores of lavendamycin (**17**) and the related streptonigrin, two powerful antitumor anti-

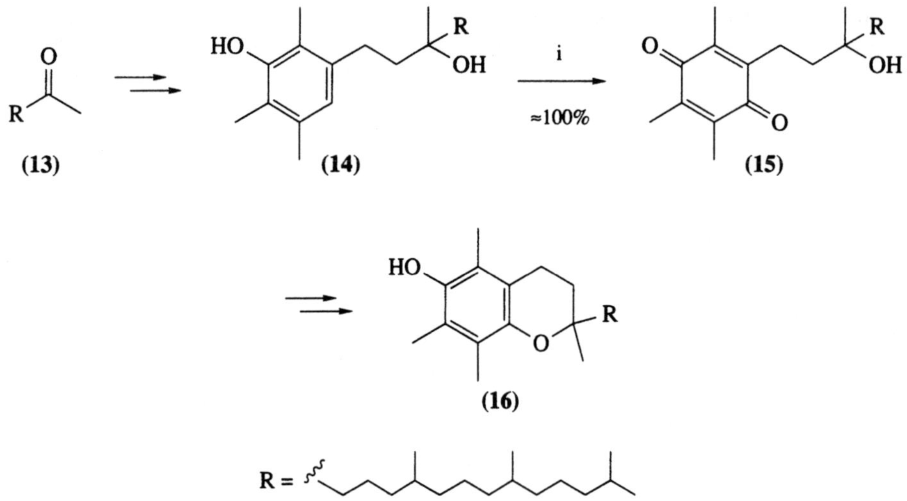

(13) (14) (15)

i

≈100%

(16)

R =

i, Fremy's salt, tricaprylylmethylammonium chloride

Scheme 4

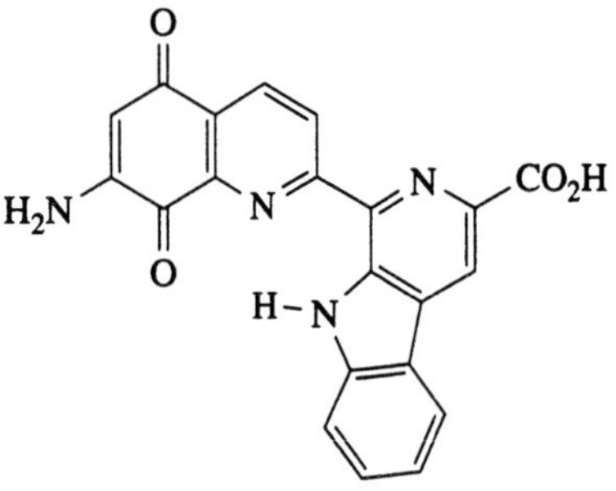

(17) (18)

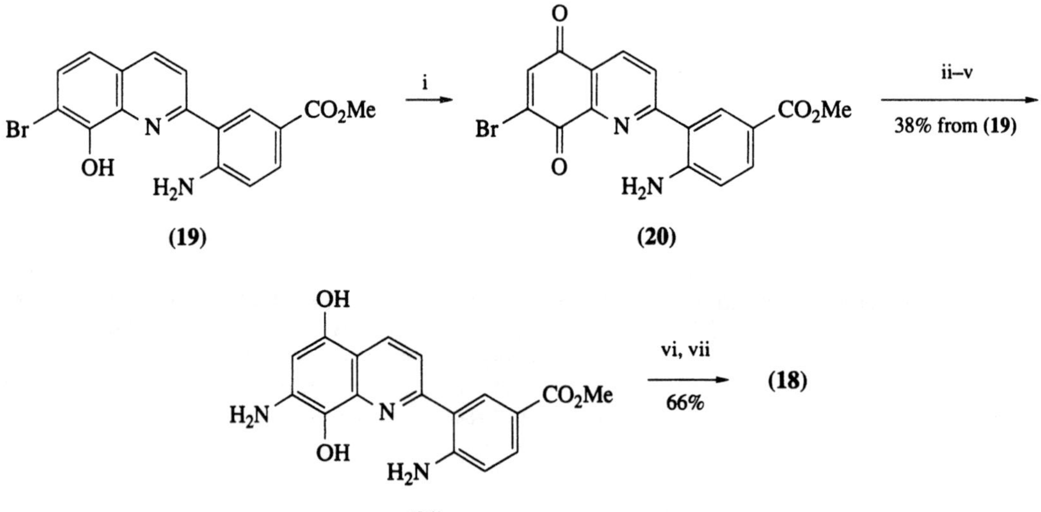

(19) (20)

i

ii–v

38% from **(19)**

(21)

vi, vii

66%

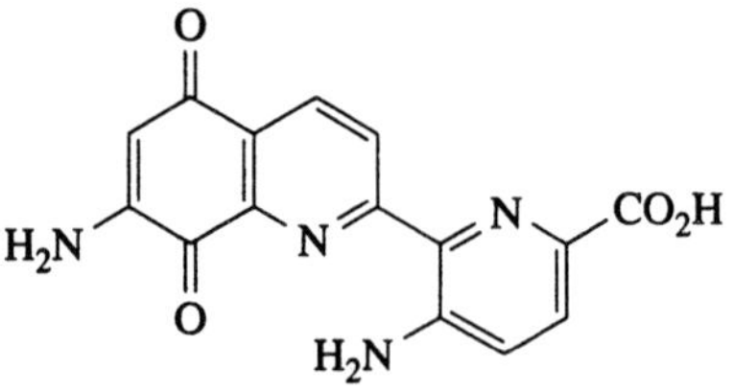

(18)

i, Fremy's salt, $Bu^n_4NHSO_4$; ii, NaN_3; iii, Ph_3P; iv, H^+; v, $Na_2S_2O_4$; vi, LiOH; vii, air

Scheme 5

biotics. In Boger's synthesis, the phenol (**19**) was oxidized using the Kende conditions to give the quinone (**20**), which upon further manipulation afforded the hydroquinone (**21**). This upon exposure to lithium hydroxide, followed by air gave the desired quinone (**18**) (Scheme 5). Interestingly in the synthesis of a related quinone (**25**) by the Boger group, the phenol (**22**) gave only poor erratic yields of the quinone (**24**) using the above oxidation procedure. Cerium(IV) ammonium nitrate and ammonium nitrate/acetic anhydride/trifluoroacetic anhydride were equally unsuccessful. A satisfactory synthesis of the quinone (**24**) was accomplished by a three-step procedure involving nitration, reduction to the amine (**23**) and finally oxidation of the *p*-aminophenol to the quinone (**24**) using manganese dioxide.

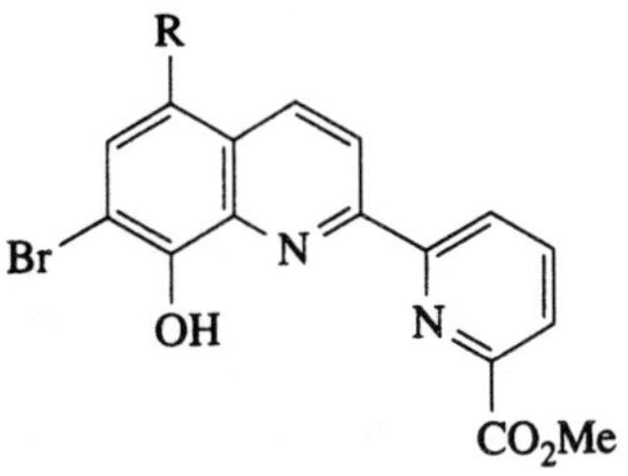

R, Br, OH, N, N, CO_2Me

(**22**) R = H
(**23**) R = NH_2

O, R, O, N, N, CO_2R'

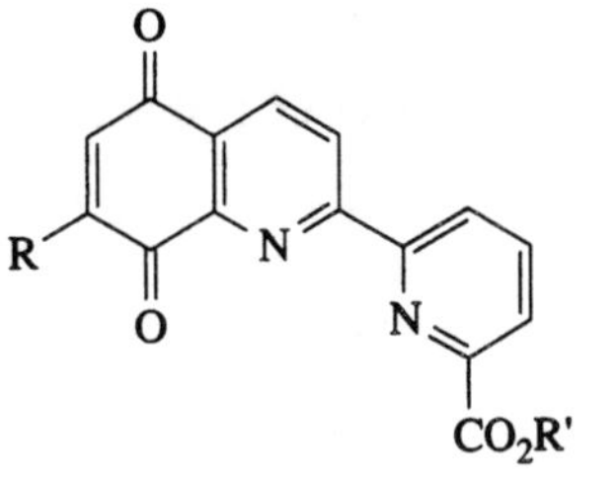

(**24**) R = Br, R' = Me
(**25**) R = NH_2, R' = H

2.11.2.5 Methoxatin

The insolubility of their substrate in the Fremy's salt reaction medium was a problem encountered by Rees and coworkers in their synthesis of the bacterial coenzyme methoxatin (**31**).[18] The methoxatin skeleton was rapidly assembled from the substituted benzaldehyde (**26**), using the nitrene insertion reaction to give the indole nucleus. Introduction of the *o*-quinone moiety, however, proved particularly troublesome. The methoxypyrroloquinoline (**27**) could not be oxidized directly to the quinone (**30**) and, although the oxidation could be effected indirectly by nitration, reduction and finally oxidation with manganese dioxide, the overall yield (29%) was disappointing. The problem was overcome by engaging an organic-soluble nitroxide equivalent of Fremy's salt, namely, benzoyl *t*-butyl nitroxide. Hydrogenolysis of the 4-benzyloxypyrroloquinoline (**28**) gave the phenol (**29**) in 89% yield. Attempted oxidation of this phenol with Fremy's salt was unsatisfactory, but using benzoyl *t*-butyl nitroxide in dichloromethane/methanol the required quinone (**30**) was prepared in excellent yield. Conversion to methoxatin (**31**) then followed a literature procedure (Scheme 6).

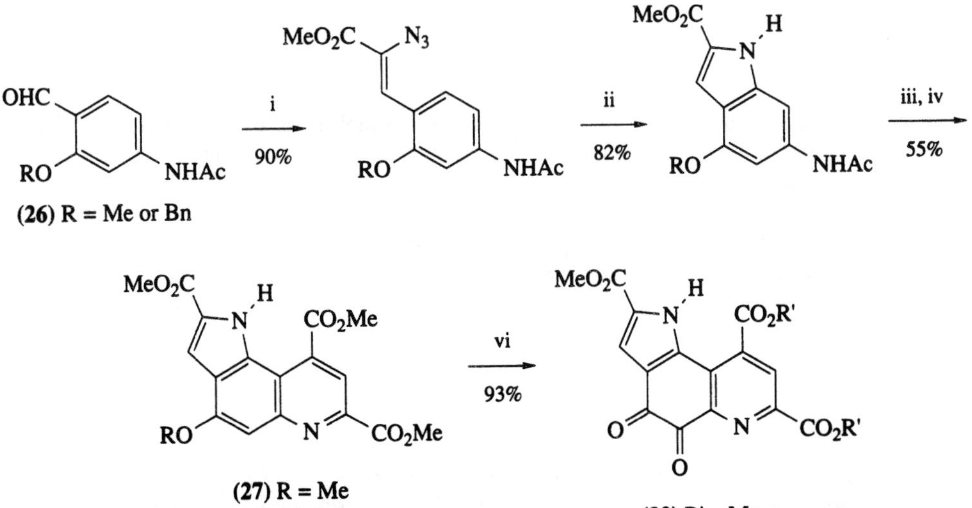

OHC, RO, NHAc — i, 90% → MeO_2C, N_3, RO, NHAc — ii, 82% → MeO_2C, H, N, RO, NHAc — iii, iv, 55% →

(**26**) R = Me or Bn

MeO_2C, H, N, CO_2Me, RO, N, CO_2Me — vi, 93% → MeO_2C, H, N, CO_2R', O, O, N, CO_2R'

(**27**) R = Me
(**28**) R = Bn
v → (**29**) R = H 89%

(**30**) R' = Me
(**31**) R' = H

i, $MeO_2CCH_2N_3$, NaOMe, MeOH; ii, xylene, reflux; iii, H^+; iv, $MeO_2CCOCH{=}CHCO_2Me$ then H^+; v, H_2, Pd–C; vi, $Bu^t(COPh)NO\bullet$; yields quoted for R = Bn

Scheme 6

These 'Fremy's salt variants' have so far found limited use in synthesis, but it is anticipated that they will become more prominent in the years to come.

2.11.3 CERIUM(IV) AMMONIUM NITRATE (CAN)

Cerium(IV) ammonium nitrate has been used increasingly during the last decade in the synthesis of quinones,[19–24] particularly as a means of effecting oxidative demethylation of methoxyarenes.

2.11.3.1 Quinone Derivatives of 2′-Deoxyuridine

The inhibition of thymidylate synthase is recognized as a viable approach to the control of cancer and DNA viral infections. Mertes and coworkers[25] proposed 5-quinone derivatives of 2′-deoxyuridine 5′-phosphate as potential irreversible inhibitors of this enzyme and prepared the quinones (**32a**) by oxidative demethylation of the appropriate dimethoxyarenes using CAN. Interestingly, the quinones (**32b**) were also prepared from the corresponding dimethoxyarenes but using silver(II) oxide and nitric acid in aqueous dioxane to effect this transformation. Cerium(IV) ammonium nitrate could not be used, due to precipitation.

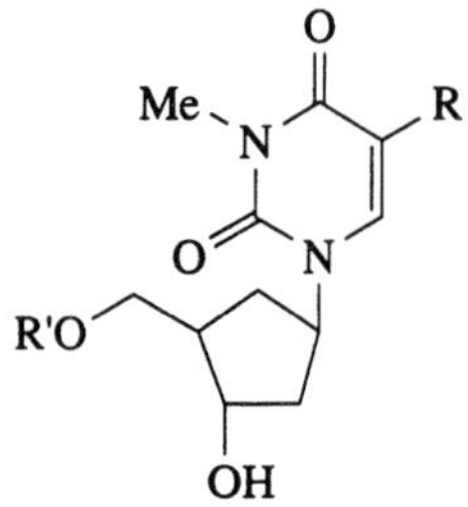

(**32a**) R' = H
(**32b**) R' = PO_3H_2

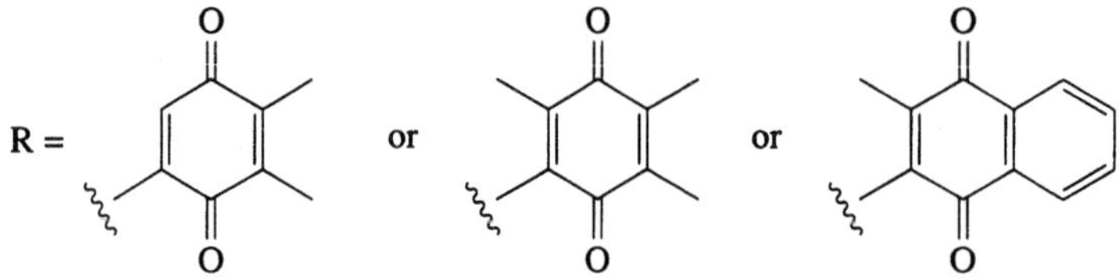

R = or or

2.11.3.2 Pleurotin

Hart and Huang[26] employed CAN in the penultimate step in their synthesis of pleurotin (**38**), an antitumor antibiotic. Treatment of the β-keto ester (**33**) with the organometallic reagent derived from 2,5-dimethoxybenzylmagnesium chloride and cerium trichloride gave the alcohol (**34**; 92% yield), which was then transformed into the pentacycle (**35**). A carboxy group was then attached at C-8 and the dimethoxyarene (**36**) oxidized with CAN to give dihydropleurotin acid (**37**). The final step of the synthesis was accomplished using manganese dioxide and was based upon the known behavior of tetraalkyl-*p*-benzoquinones in the presence of nucleophiles. Although this synthesis was rather long, it represented the first total synthesis of the structurally complex pleurotin (Scheme 7).

2.11.3.3 Saframycin B

In the total synthesis of saframycin B (**40**) recently reported by Kubo *et al.*,[27] the final step, in which the polymethoxyarene (**39**) was converted to saframycin B, provided quite a challenge in itself and emphasizes how a seemingly straightforward transformation can prove difficult when incorporated into a natural product synthesis. The direct oxidative demethylation of the arene (**39**) could only be achieved using 10 M nitric acid, and this gave an unacceptable 1.5% yield of saframycin B. Other commonly employed reagents gave only starting material. Partial demethylation with boron tribromide followed by oxidation with CAN gave saframycin B (**40**) and the monoquinone (**41**) in 17% and 45% yields

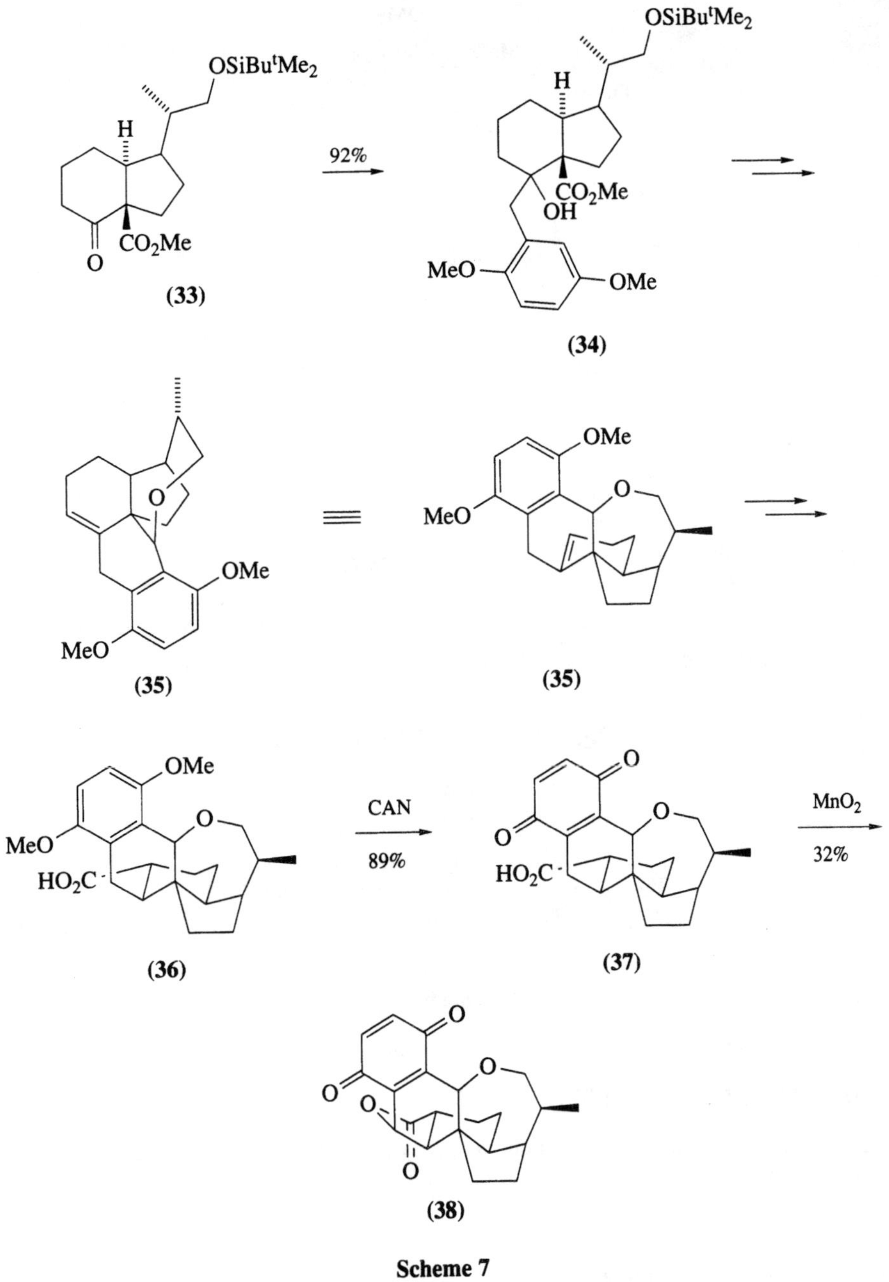

(33) 92% (34)

(35) ≡ (35)

(36) CAN 89% (37) MnO_2 32%

(38)

Scheme 7

respectively (Scheme 8). Finally, partial demethylation followed by oxidation with 10 M nitric acid gave saframycin B (**40**) in an acceptable 41% yield (Scheme 9).

2.11.3.4 Demethoxydaunomycinone

One final interesting application of CAN which exemplifies one of the many possible reactions of quinones comes from Hassall and coworkers[28] in their synthesis of 4-demethoxydaunomycinone (**47**). Thus, oxidation of the boronate (**42**) with CAN gave the crude quinone (**43**) which was reacted with *trans*-1,2-bis(acetoxy)-1,2-dihydrobenzocyclobutene (**44**) to give the tetracyclic quinone (**45**) in an impressive 79% overall yield. Deacetalization and reductive acetylation to the naphthacene (**46**), followed by oxidation with anhydrous chromium trioxide and deprotection with boron trichloride afforded the target compound (**47**; Scheme 10).

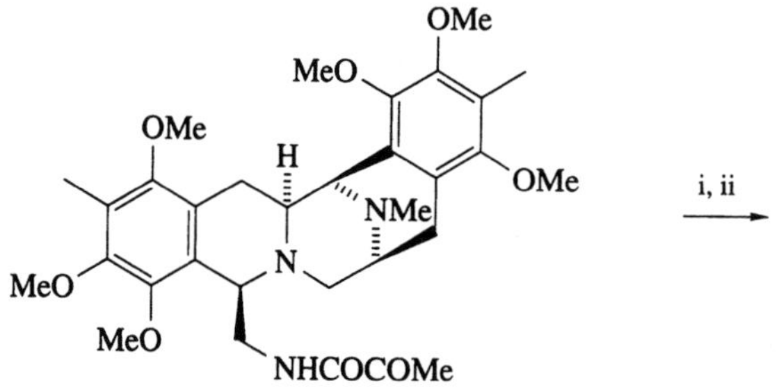

i, BBr_3, CH_2Cl_2, –78 to 0 °C; ii, CAN, MeCN, H_2O, 0 °C

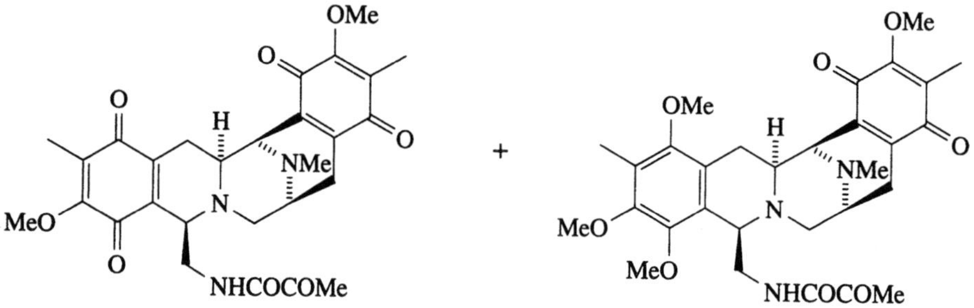

Scheme 8

(39) $\xrightarrow[\text{ii, 10 M HNO}_3\text{, r.t.}]{\text{i, BBr}_3\text{, CH}_2\text{Cl}_2\text{, –78 to 0 °C}}$ **(40)**

41%

Scheme 9

2.11.4 LEAD TETRAACETATE (LTA)

Lead tetraacetate is a versatile reagent which has many applications in organic synthesis, not least in the synthesis of quinones.[29–32]

2.11.4.1 Demethoxydaunomycinone

Lead tetraacetate was employed by Stoodley and coworkers[33] for an oxidative isomerization in their synthesis of 4-demethoxydaunomycinone (**47**). The diene (**48**) reacted with the oxirane dienophile (**49**) *via* the least hindered *endo* transition state to give the cycloadduct (**50**) in 86% yield. Hydrolysis of the silyl enol ether followed by reduction of the oxirane and introduction of the acetylene moiety gave the compound (**51**), which was oxidatively isomerized with LTA in acetic acid to give the quinone (**52**). All that remained now to complete the synthesis was conversion of the acetylene to a methyl ketone and dealkylation of the ether. The last two steps were accomplished in an overall yield of 38%, the low yield attributable to problems in formation of the hydroxy group from the ether (Scheme 11). Bulman-Page and Ley[34] employed LTA for a similar transformation in their synthesis of demethoxydaunomycinone and related anthracyclinones.

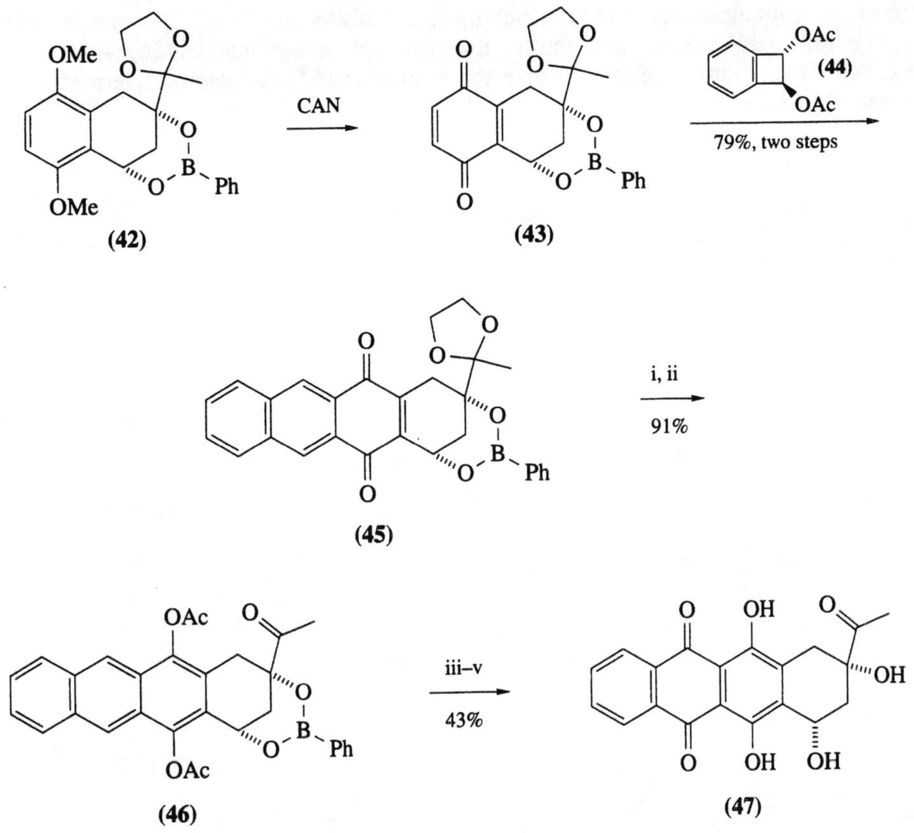

(42) (43) (44) (45) (46) (47)

CAN

79%, two steps

i, ii 91%

iii–v 43%

i, H^+; ii, H_2, Pd–C, Ac_2O, pyridine; iii, CrO_3; iv, BCl_3; v, 2-methylpentane-2,4-diol

Scheme 10

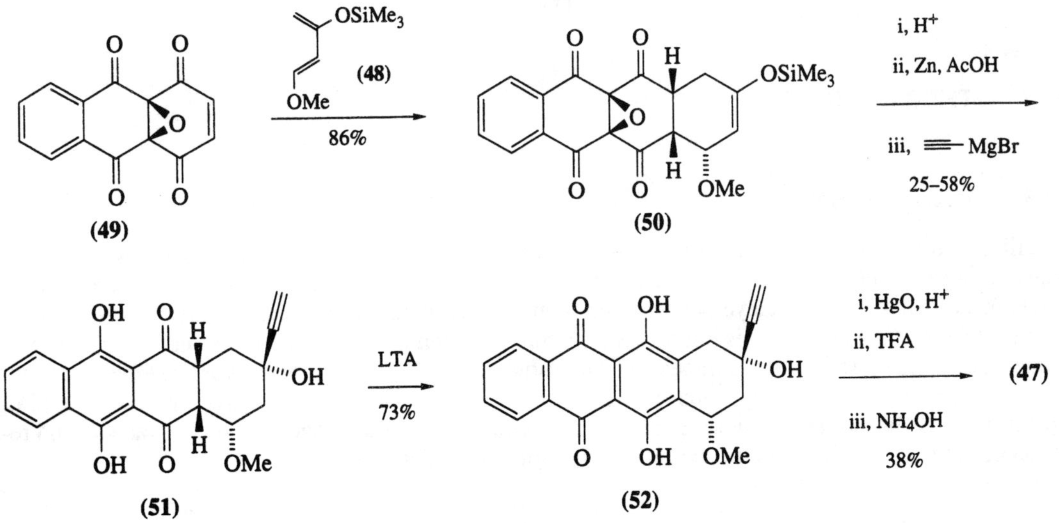

(48) (49) (50) (51) (52)

86%

i, H^+ ii, Zn, AcOH iii, ≡—MgBr 25–58%

LTA 73%

i, HgO, H^+ ii, TFA iii, NH_4OH 38% (47)

Scheme 11

2.11.4.2 Towards Mitomycin C

Another use of LTA was shown by Yoshida and coworkers[35] in studies directed towards the mitomycin family of antibiotics, *e.g.* mitomycin C (**58**). The aniline (**53**) was treated with LTA in dichloromethane to give the *o*-quinoneimide (**54**), which upon hydrolysis and then hydrogenation afforded the catechol (**55**) in reasonable overall yield. Further manipulation gave the phenol (**56**) which was oxidized using the cobalt(II) complex salcomine to give the *p*-quinone (**57**), a potential intermediate for mitomycin synthesis (Scheme 12).

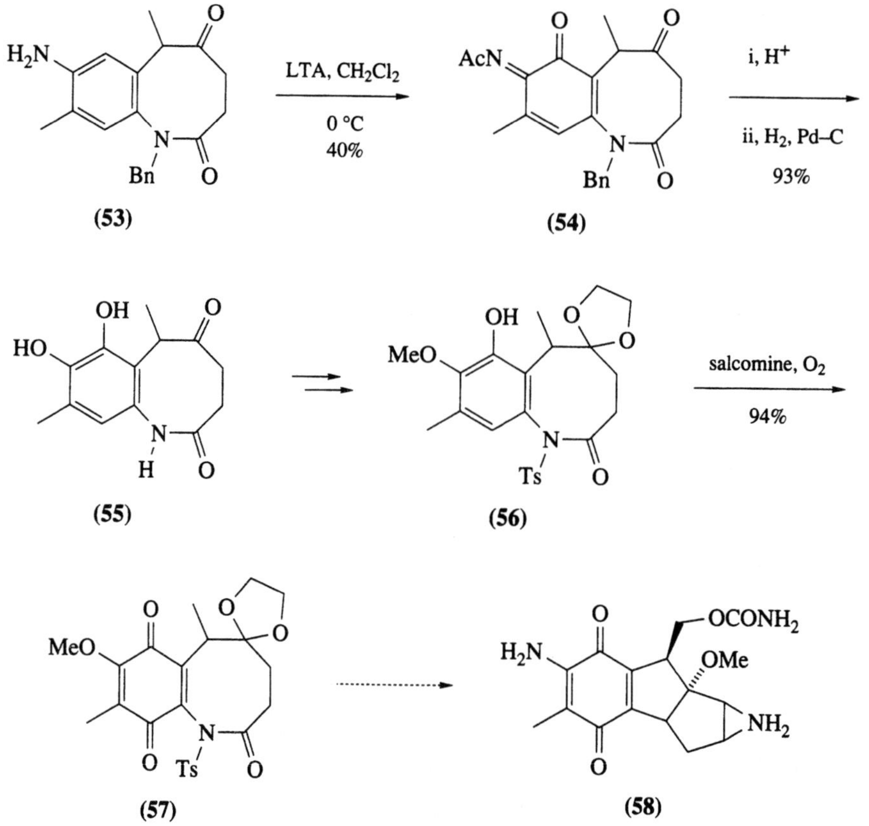

(53) → LTA, CH_2Cl_2, 0 °C, 40% → (54) → i, H^+; ii, H_2, Pd–C, 93% → (55) → (56) → salcomine, O_2, 94% → (57) ⇢ (58)

Scheme 12

2.11.5 OTHER REAGENTS

2.11.5.1 Metacyclophanes

Thallium trifluoroacetate has not enjoyed widespread use as a reagent for quinone synthesis, possibly because it is still a relatively new reagent but more probably because of its toxicity. One example of its use lies in the synthesis of metacyclophanes and related compounds as reported by Tashiro *et al.*[36] Thus the *t*-butylphenol (**59**) gave the bisquinone (**61**), while the phenol (**60**) afforded the monoquinone (**62**). An alternative and more practical synthesis of the bisquinone (**61**) for large scale work involved dealkylation to afford the bisphenol (**63**) which was then treated with sodium nitrite to give the bisoxime (**64**). Hydrolysis of the bisoxime did not give the quinone (**61**), but it could be obtained by zinc/acetic acid reduction of the bisoxime followed by oxidation with nitric acid (Scheme 13).

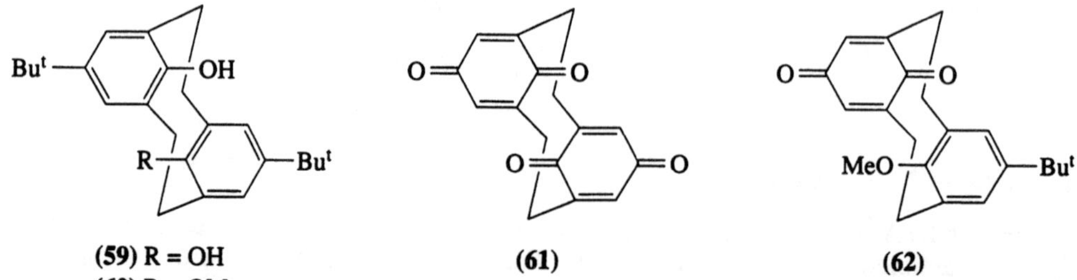

(**59**) R = OH
(**60**) R = OMe

(**61**)

(**62**)

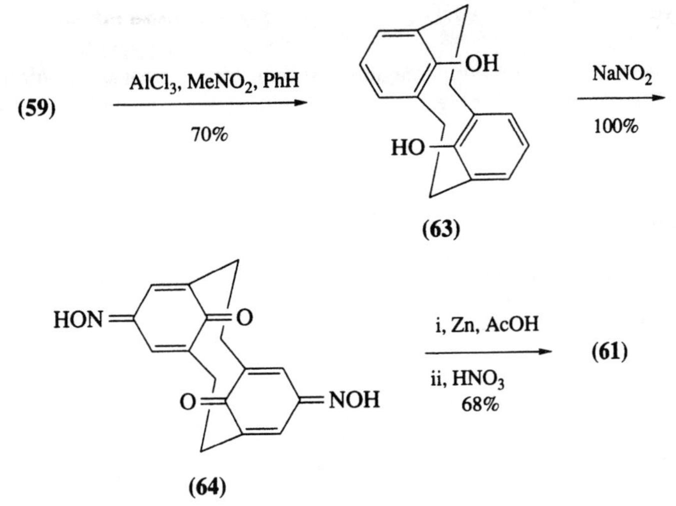

(59) $\xrightarrow[70\%]{AlCl_3, MeNO_2, PhH}$ (63) $\xrightarrow[100\%]{NaNO_2}$

(64) $\xrightarrow[\text{ii, } HNO_3 \text{ 68\%}]{\text{i, Zn, AcOH}}$ (61)

Scheme 13

2.11.5.2 Diazaquinomycin A

The reader should always bear in mind that spontaneous oxidation in air can occur if a hydroquinone is sufficiently activated towards oxidation.[37] An example of this is exemplified by Kelly *et al.*[38] in a short synthesis of diazaquinomycin A (**68**). The synthesis incorporates the first reported use of a double Knorr cyclization. The key intermediate (**66**) was prepared in just two steps from the MOM-protected hydroquinone (**65**). Cyclization of the compound (**66**) then gave diazaquinomycin B (**67**), which either under the reaction conditions or by careful isolation prior to simply stirring the solution in an open flask afforded the antibiotic diazaquinomycin A (**68**), thereby confirming the structure of the only recorded example of the tricyclic 1,8-diazaanthraquinone ring system (Scheme 14).

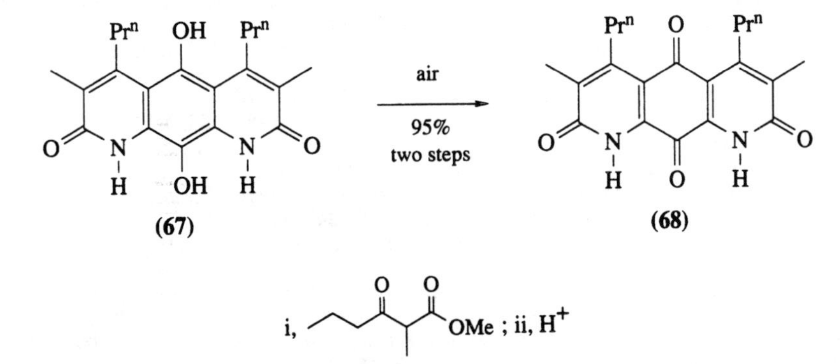

(65) $\xrightarrow[75\%]{\text{i, ii}}$ (66) $\xrightarrow{H_2SO_4}$

(67) $\xrightarrow[\text{95\% two steps}]{\text{air}}$ (68)

i, [methyl 2-methyl-3-oxohexanoate] OMe ; ii, H^+

Scheme 14

2.11.5.3 Miscellaneous

There are numerous other examples of the synthesis of quinones employing reagents such as nitric acid,[27,36] manganese dioxide,[17,18] salcomine/O_2,[24,35,39,40] silver oxide,[25,41–43] chromium oxidants,[16,28,44–47] benzene selenic anhydride[10,30,48–50] and DDQ,[51,52] some of which have already been referred to in the text.

No doubt, the next decade will see other reagents brought forward for the oxidative synthesis of quinones. The practicability of these and more recently introduced reagents will be shown by their efficacy in the synthesis of complex molecules.

2.11.6 REFERENCES

1. T. Laird, in 'Comprehensive Organic Chemistry', ed. D. H. R. Barton and W. D. Ollis, Pergamon Press, Oxford, 1979, vol. 1, p. 1213; A. H. Haines, in 'Methods for the Oxidation of Organic Compounds', Academic Press, London, 1985, p. 182.
2. K. Krohn, *Angew. Chem., Int. Ed. Engl.*, 1986, **25**, 790.
3. R. G. Harvey, *Synthesis*, 1986, 605.
4. A. Ashnagar, J. M. Bruce and P. Lloyd-Williams, *J. Chem. Soc., Perkin Trans. 1*, 1988, 559.
5. A. Fischer and G. N. Henderson, *Synthesis*, 1985, 641, and references cited therein.
6. H. Zimmer, D. C. Larkin and S. W. Horgan, *Chem. Rev.*, 1971, **71**, 229.
7. R. G. Harvey, C. Cortez, T. W. Sawyer and J. DiGiovanni, *J. Med. Chem.*, 1988, **31**, 1308.
8. G. Coll, J. Morey, A. Costa and J. M. Saa, *J. Org. Chem.*, 1988, **53**, 5345; J. M. Saa, A. Llobera, A. Garcia-Raso, A. Costa and P. M. Deya, *J. Org. Chem.*, 1988, **53**, 4263.
9. S. Kumar, A. Czech and E. J. La Voie, *J. Org. Chem.*, 1988, **53**, 1329.
10. J. Lee, J. Lang and J. K. Synder, *Tetrahedron Lett.*, 1987, **28**, 3427.
11. N. Cohen, R. J. Lopresti and G. Saucy, *J. Am. Chem. Soc.*, 1979, **101**, 6710.
12. G. Palmisano, B. Danieli, G. Lesma and F. Trupiano, *J. Org. Chem.*, 1988, **53**, 1056.
13. A. P. Kozikowski, K. Sugiyama and J. P. Springer, *J. Org. Chem.*, 1981, **46**, 2426.
14. L. R. McGee and P. N. Confalone, *J. Org. Chem.*, 1988, **53**, 3695.
15. G. L. Olson, H.-C. Cheung, K. Morgan and G. Saucy, *J. Org. Chem.*, 1980, **45**, 803.
16. A. S. Kende and F. H. Ebetino, *Tetrahedron Lett.*, 1984, **25**, 923.
17. D. L. Boger, M. Yasuda, L. A. Mitscher, S. D. Drake, P. A. Kitos and S. Collins Thompson, *J. Med. Chem.*, 1987, **30**, 1918; M. Yasuda and D. L. Boger, *J. Heterocycl. Chem.*, 1987, **24**, 1253.
18. A. R. MacKenzie, C. J. Moody and C. W. Rees, *Tetrahedron*, 1986, **42**, 3259.
19. M. A. Brimble, R. Hodges and S. J. Stuart, *Tetrahedron Lett.*, 1988, **29**, 5987.
20. W. D. Wulff, J. S. McCallum and F. A. Kunng, *J. Am. Chem. Soc.*, 1988, **110**, 7419.
21. M. Croisy-Delcy, M. Rautureau, C. Huel and E. Bisagni, *J. Org. Chem.*, 1988, **53**, 5301.
22. G. Balanikas, N. Hussain, S. Amin and S. S. Hecht, *J. Org. Chem.*, 1988, **53**, 1007.
23. M. Nakata, M. Kinoshita, S. Ohba and Y. Saito, *Tetrahedron Lett.*, 1984, **25**, 1373.
24. K. A. Parker and J. J. Petraitis, *Tetrahedron Lett.*, 1981, **22**, 397.
25. L. A. Al-Razzak, D. Schwepler, C. J. Decedue, J. Balzarini, E. De Clercq and M. P. Mertes, *J. Med. Chem.*, 1987, **30**, 409.
26. D. J. Hart and H.-C. Huang, *J. Am. Chem. Soc.*, 1988, **110**, 1634.
27. A. Kubo, H. Yamato, K. Masubuchi and M. Nakamura, *J. Org. Chem.*, 1988, **53**, 4295.
28. M. J. Broadhurst, C. H. Hassall and G. J. Thomas, *J. Chem. Soc., Perkin Trans. 1*, 1982, 2239.
29. D. W. Hansen, Jr., R. Pappo and R. B. Garland, *J. Org. Chem.*, 1988, **53**, 4244.
30. R. M. Carman, S. Owsia and J. M. A. M. van Dongen, *Aust. J. Chem.*, 1987, **40**, 333.
31. H. Greenland, J. T. Pinhey and S. Sternhell, *Aust. J. Chem.*, 1987, **40**, 325.
32. M. Tanaguchi, T. Anjik, M. Nakagawa and T. Hino, *Chem. Pharm. Bull.*, 1984, **32**, 2544.
33. R. C. Gupta, D. A. Jackson, R. J. Stoodley and D. J. Williams, *J. Chem. Soc., Perkin Trans. 1*, 1985, 525.
34. P. C. Bulman-Page and S. V. Ley, *J. Chem. Soc., Perkin Trans. 1*, 1984, 1847.
35. K. Yoshida, S. Nakajima, T. Ohnuma, Y. Ban, M. Shibasaki, K. Aoe and T. Date, *J. Org. Chem.*, 1988, **53**, 5355.
36. M. Tashiro, K. Koya and T. Yamato, *J. Am. Chem. Soc.*, 1982, **104**, 3707.
37. J. S. Calderon and R. H. Thomson, *J. Chem. Soc., Perkin Trans. 1*, 1988, 563; M. K. Manthey, *J. Org. Chem.*, 1988, **53**, 1486; ref. 31.
38. T. R. Kelly, J. A. Field and Q. Li, *Tetrahedron Lett.*, 1988, **29**, 3545.
39. E. R. Dockal, Q. B. Cass, T. J. Brocksom, U. Brocksom and A. G. Correa, *Synth. Commun.*, 1985, **15**, 1033.
40. I. Kubo, T. Kamikawa and I. Miura, *Tetrahedron Lett.*, 1983, **24**, 3825.
41. C. B. de Koning, R. G. F. Giles, L. S. Knight, M. L. Niven and S. C. Yorke, *J. Chem. Soc., Perkin Trans. 1*, 1988, 2477.
42. T. R. Kelly, M. Behforouz, A. Echavarren and J. Vaya, *Tetrahedron Lett.*, 1983, **24**, 2331.
43. R. Cassis, M. Scholz, R. Tapia and J. A. Valderrama, *Tetrahedron Lett.*, 1985, **26**, 6281.
44. C. A. Townsend, S. B. Christensen and S. G. Davies, *J. Chem. Soc., Perkin Trans. 1*, 1988, 839.
45. B. M. Trost and W. H. Pearson, *Tetrahedron Lett.*, 1983, **24**, 269.
46. J. R. Wiseman, J. J. Pendery, C. A. Otto and K. G. Chiong, *J. Org. Chem.*, 1980, **45**, 516.
47. S. O. de Silva, M. Watanabe and V. Snieckus, *J. Org. Chem.*, 1979, **44**, 4802.
48. D. H. R. Barton, J.-P. Finet and M. Thomas, *Tetrahedron*, 1988, **44**, 6397, and references cited therein.
49. Y. Miyahara, T. Inazu and T. Yoshino, *Tetrahedron Lett.*, 1982, **23**, 2189.
50. H. Tanaka, M. Ishihara, K. Ichino and K. Ito, *Heterocycles*, 1988, **27**, 2651.
51. P. N. Preston, S. G. Will, T. Winwick and J. O. Morley, *J. Chem. Soc., Perkin Trans. 1*, 1983, 1001.
52. M. Sato, N. Katsumata and S. Ebine, *Synthesis*, 1984, 685.

3.1
Addition Reactions with Formation of Carbon–Oxygen Bonds: (i) General Methods of Epoxidation

A. SOMASEKAR RAO
National Chemical Laboratory, Pune, India

3.1.1 INTRODUCTION

Oxiranes (epoxides) are compounds which contain a saturated three-membered ring having one oxygen atom and two carbon atoms.[1] They are widely distributed in nature and are of industrial, mechanistic and biochemical interest.[1] Squalene 2,3-oxide is the biogenetic precursor of sterols. Leukotriene A (LTA) is the biogenetic precursor of the leukotrienes LTC, LTD and LTE which are important natural mediators of allergic asthma.[2] The ultimate carcinogenic metabolites of polycyclic aromatic hydrocarbons are the tetrahydrodiol epoxides.[3] An allene oxide is regarded as the precursor of preclavunone A.[4]

In their reports on oxiranes Lewars[1] and Rao *et al.*[2] have cited many of the earlier reviews. Recent work on oxiranes has been reviewed by Harvey,[3] Sharpless and Verhoeven,[5] Bartok and Lang,[6] Plesnicar,[7] Mimoun[8a] and Jorgensen.[8b] The various nomenclature systems are illustrated in the naming of (**1**) and (**2**). Cyclohexene oxide, 1,2-epoxycyclohexane and 7-oxabicyclo[4.1.0]heptane are the names used for (**1**) while (**2**) is 2,2-dimethyloxirane.

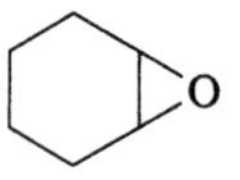

(**1**)

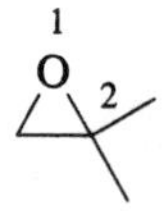

(**2**)

The ease of preparation of epoxides and their facile ring opening have made them important intermediates in organic synthesis for the past several decades. In the present decade the main objective in organic synthesis is to develop reactions which are enantio-, diastereo-, regio- and chemo-selective. With the discovery of enantioselective epoxidation of prochiral acyclic allyl alcohols by Katsuki and Sharpless[8a] and observation of high and predictable diastereoselectivity during (i) the epoxidation of several types of acyclic unsaturated chiral alcohols with MCPBA or TBHP/VO(acac)$_2$, and (ii) the preparation of epoxy alcohols *via* halolactonization, coupled with elegant routes for highly regioselective intramolecular ring opening of epoxides,[9] it may be noted that epoxides are versatile intermediates for organic synthesis in the present decade also. A number of complex compounds such as monensin,[10] maytansine[11] and prostaglandins[12] have been synthesized using epoxides as intermediates. Some of these developments are reviewed in this chapter, and enantioselective methods of epoxidation are presented in Chapter 3.2, this volume. There are also other chapters and sections which deal with the synthesis and reactions of epoxides.

3.1.2 EPOXIDATIONS WITH ORGANIC PEROXY ACIDS

3.1.2.1 General Survey of Reactivity

A large number of organic peroxycarboxylic acids having the general formula (**3**) readily epoxidize alkenes (equation 1).[7] The stereochemistry of the alkene is retained in the epoxide. There are a few apparent exceptions which will be presented later (see equation 38). The reaction is believed to take place *via* the transition state (**5**)[13] and involves the nucleophilic attack on the O—O bond by the π-electrons of the double bond.[5] The reaction rate increases when the groups R^1, R^2, R^3 and R^4, in (**4**), are electron releasing and also when R, in (**3**), is an electron-withdrawing group. In the case of polyunsaturated hydrocarbons, if the double bonds differ in their degree of substitution the regioselectivity can be easily predicted; the more substituted double bond is more reactive. The peroxy acid has a relatively low steric requirement. For example, epoxidation is the only addition reaction which has been carried out on the sterically hindered alkene (**6**).[14]

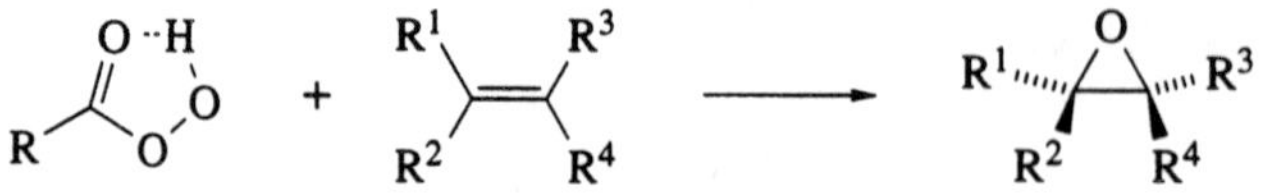

(**3**) (**4**) (1)

Epoxidation of an alkene containing one or more chiral centers can furnish two diastereoisomeric epoxides, depending on the face from which the reagent approaches the π-bond (equation 2). If the two faces of the π-bond are unequally shielded, and if polar and stereoelectronic factors are also involved, the two expected diastereomers will not be formed to the same extent, thus resulting in diastereoselectivity. An inspection of the molecular model may often reveal the face of the alkene which is more shielded, especially when the alkene has a rigid structure.

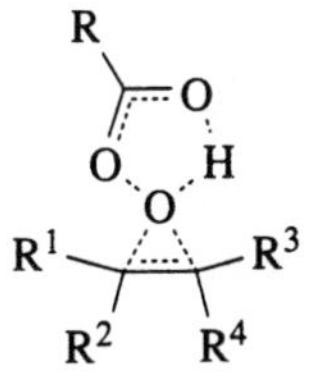

(5)

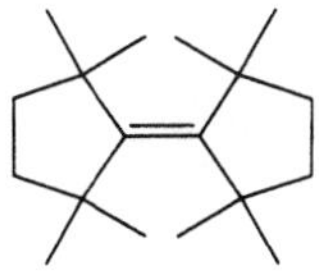

(6) 59%

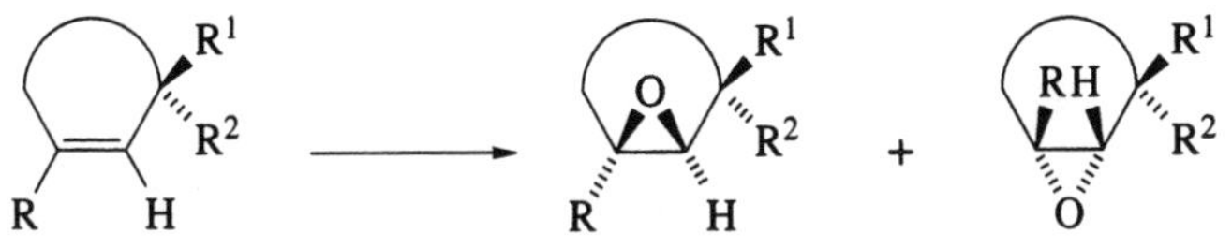

(2)

Data on diastereoselectivity are presented in two ways. When experimental conditions are different from the examples given earlier then an equation is given in full. When experimental conditions employed are similar to those used for earlier examples then two figures are given in brackets below the chemical structure, and after the structure number. The first figure gives the combined yield of both the diastereomers, the second figure gives the percentage of the major diastereomer. A solid arrow indicates that the major isomer was formed by attack on the β-face; a dashed arrow indicates that the preferred direction of attack is on the α-face, as shown in structures (**33**) and (**34**).

Since peroxides and peroxy acids are potentially explosive care is required while carrying out reactions and also during work-up of the reaction mixture.

3.1.2.2 Epoxidations with *m*-Chloroperbenzoic Acid (MCPBA)

3.1.2.2.1 General comments

MCPBA is a relatively stable solid which is soluble in many common organic solvents. It is the reagent of choice for laboratory scale experiments and is commercially available. Though reactions are carried out normally at 0–25 °C in CH_2Cl_2 or $CHCl_3$, one can use elevated temperatures (95 °C, ethylene dichloride) in conjunction with radical inhibitors if the alkene has low reactivity. The experimental conditions for the reaction and subsequent work-up depend on the stability of the epoxides. In the case of acid-sensitive epoxides the usual practice is to control the pH using $NaHCO_3$, aqueous Na_2CO_3, Na_2HPO_4 or KF.[15]

3.1.2.2.2 MCPBA epoxidations of acyclic alkenes lacking directing groups

The reactivity of an alkene depends on the degree of substitution; the least reactive are the monosubstituted (terminal) alkenes. Monosubstituted alkenes can be epoxidized by MCPBA (equation 3).[16] The diene (**7**) has been epoxidized regioselectively at the more electron-rich disubstituted double bond.[17] In the triene (**8**) all the double bonds are trialkylsubstituted; however the C(2)—C(3) double bond is strongly deactivated due to conjugation with electron-withdrawing CO_2Me. The C(10)—C(11) double bond is deactivated, but to a lesser extent due to the location of fluorine. Consequently the predicted reactivities are C(6)—C(7) > C(10)—C(11) > C(2)—C(3). Epoxidation of the triene (**8**) is highly regioselective at –20 °C (equation 4);[18] 30% of the starting material is recovered. However, the regioselectivity is very poor if epoxidation is carried out at 45 °C, when a 1:1 mixture of epoxides (**9**) and (**10**) is obtained.

The preparation of allylic epoxides from conjugated dienes requires careful control of pH, since allylic epoxides undergo ring opening by nucleophiles at acidic pH. The allylic epoxide (**12**) has been prepared from the diene (**11**; equation 5);[19] however, the ring opening of (**12**) cannot be totally suppressed.

(3)

i, MCPBA, CH_2Cl_2, 20 °C, 8 h

MeO 75% Pr

(7)

10 3 CO_2Me MCPBA, –20 °C 6 2 F 70%

(8)

O CO_2Me + O (4)

7 6 11 10 F

(9) 100% (10) 0%

i OH (5)

OAr

(11) (12) 42% (13) 28%

i, MCPBA, aq. $NaHCO_3$, CH_2Cl_2 Ar = 3-chlorobenzoyl

Epoxidation of the allylsilane (**14**) is diastereospecific (equation 6).[20] The favored conformation of (**14**) is (**14a**); the peroxy acid approaches the double bond from the face *anti* to the bulky silyl group. The epoxides (**16**) and (**17**) obtained from the acetonide (**15**; equation 7) can be readily separated in gram quantities using standard chromatographic techniques.[21] The presence of the conformationally rigid acetonide moiety in the epoxides (**16**) and (**17**) facilitates their separation; the corresponding epoxy diols cannot be separated by chromatography. The racemic epoxide (**19**), an intermediate for the synthesis of maytansine has been synthesized from (**18**; equation 8).[22]

$PhMe_2Si$ $PhMe_2Si$ i $PhMe_2Si$ O $PhMe_2Si$ O + (6)

(14a) (14) 95% 5%

i, MCPBA, Na_2HPO_4, CH_2Cl_2, 0 °C, 2 h

MCPBA 88% (7)

(15) (16) 33% (17) 67%

(8)

(18) R = $SiMe_2Bu^t$ (19) 53% (20) 47%

i, MCPBA, CH_2Cl_2, 0 °C

3.1.2.2.3 MCPBA epoxidations of cyclic alkenes, methylenecycloalkanes and unsaturated macrocyclic lactones

The conformational preferences and stereoselective reactions of a number of macrocyclic systems have been studied. The stereochemical results have been explained on the basis of the model of local conformer control. The epoxidation of a macrocyclic alkene containing the substitution pattern (**21**) provides a single epoxide having the stereochemistry (**22**).[23] A macrocycle containing a 1,5-diene system adopts the local conformation (**23**) that is free of torsional strain; epoxidation of (**23**) from the less hindered side furnishes the *syn*-diepoxide (**24**).[23] The MCPBA epoxidations of the unsaturated macrocyclic lactones (**25**) and (**26**) are stereoselective (equations 9 and 10).[23,24] In the epoxidation of (**26**) six new chiral centres are introduced; the reaction product is a 20:1:1 mixture of triepoxides. The triepoxide (**27**) is closely related to the C(9)–C(23) segment of monensin B.

(21) (22) (23) (24)

MCPBA, –78 °C (9)

(25) 90% 10%

i 74% (10)

(26) (27)

i, MCPBA, $NaHCO_3$, CH_2Cl_2

In the peroxy acid epoxidation of cyclohexenes a substituent is more effective in blocking the approach of the reagent when it is pseudoaxial allylic or axial homoallylic than when it is pseudoequatorial allylic or equatorial homoallylic.[13] MCPBA epoxidation of tetramethyllimonene (**28**) is regio- and stereo-selective (equation 11);[25] approach of the reagent from the β-face is blocked by the pseudoaxial methyl at C-3 and the axial methyl at C-5. In contrast, epoxidation of limonene (**28**; R = H) furnishes an approximately 50:50 mixture of epoxides (**29a**; R = H) and (**29b**; R = H); the reaction is not stereoselective since there are no axially oriented bulky groups to selectively shield one face. The sensitive allylic epoxide (**31**) has been prepared stereoselectively employing a biphasic medium (equation 12);[26] the reaction proceeds from the α-face since the β-face is blocked by axial methyl at C-10. The epoxidation of (**32**) is stereoselective (equation 13);[27] (**32**) does not prefer the conformation with CO_2Me equatorial since this leads to steric interference between CO_2Me and hydrogen at C-11 ($A^{1,3}$ strain). In the preferred conformation, (**32**) has the CO_2Me axial which leads to blocking of the β-face. MCPBA epoxidation (ethylene dichloride, 90 °C, 4 h) of (**33**) takes place selectively from the β-face since the α-face is blocked by the axial chlorovinyl group.[28] The more electron-rich trisubstituted double bond is selectively attacked. In the alkene (**34**), the β-methyl at C-10 being axial is more effective in shielding the double bond than the α-methyl at C-4 which is pseudoequatorial.[29] The diastereoselectivities in the epoxidation of the dienes (**35**), (**36**) and (**37**) have been studied (equations 14–16).[30] The selectivity depends on the nature of the groups at C-10 and also on the functional groups present on the double bond at C-8 and C-9. Epoxidation of vitamin D_3 (**38**) is regio- as well as stereo-selective (equation 17).[31] Though both C(5)—C(6) and C(7)—C(8) are trisubstituted double bonds reaction takes place selectively at C(7)—C(8), since only this route leads to the thermodynamically more stable conjugated diene derivative; attack is selectively from the α-face since the β-face is shielded by the axial methyl at C-13. Epoxidation of (**39a**) is stereoselective (equation 18).[32] The selectivity is higher in (**39a**) than in the reaction of (**39b**; R = MOM) due to stereoelectronic repulsive effects involving acetate and peroxy acid in the transition state leading to (**41**).

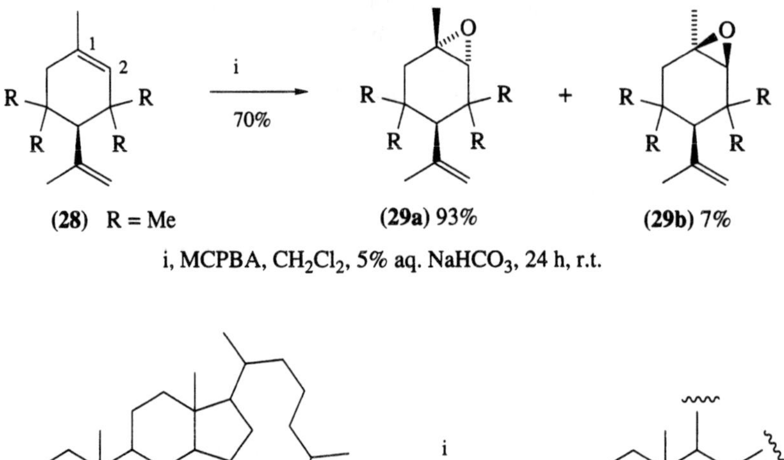

(**28**) R = Me → (**29a**) 93% + (**29b**) 7% (i, 70%) (11)

i, MCPBA, CH_2Cl_2, 5% aq. $NaHCO_3$, 24 h, r.t.

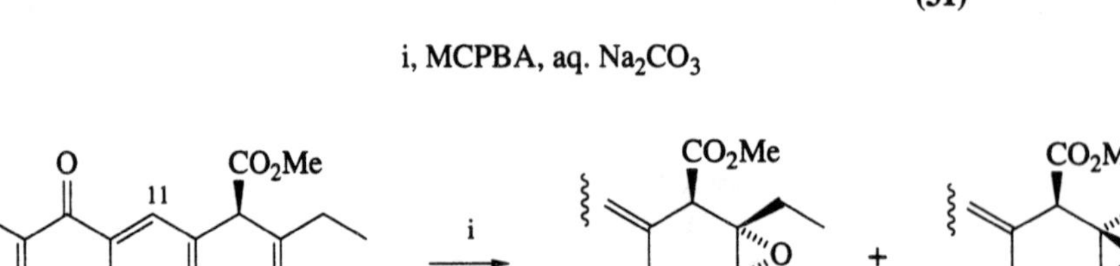

(**30**) → (**31**) (i, 89%) (12)

i, MCPBA, aq. Na_2CO_3

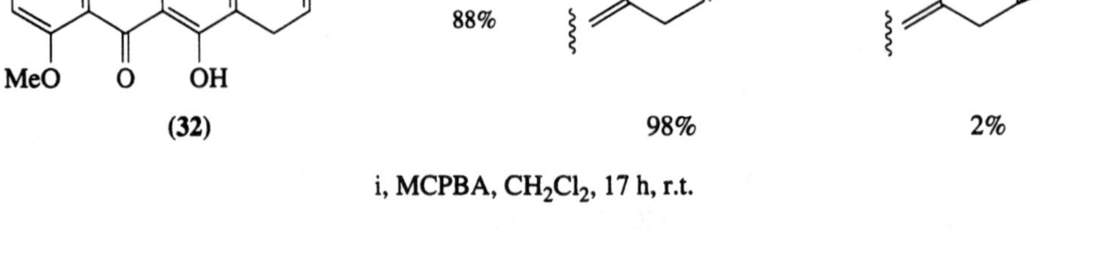

(**32**) → 98% + 2% (i, 88%) (13)

i, MCPBA, CH_2Cl_2, 17 h, r.t.

(33) (83, ≈100) X = Br

(34) (84)

→ indicates attack on β-face forms major isomer ····► indicates attack on α-face

MCPBA

		Yield		
(35)	R^1 = Me, R^2 = H, R^3 = CO_2Et	90%	95:5	(14)
(36)	R^1 = H, R^2 = H, R^3 = CO_2Et	77%	33:67	(15)
(37)	R^1 = H, R^2 = CO_2Et, R^3 = H	82%	10:90	(16)

(38) i, 81% (17)

i, MCPBA, CH_2Cl_2 −70 °C to 0 °C

R = Ac (39a) i, 91% (40) 95% + (41) 5% (18)

i, MCPBA, $CHCl_3$, CCl_4, 32 h, 25 °C

Folded molecules are epoxidized selectively from the less hindered convex side. This is illustrated by the reactions of (**42**; equation 19),[33] (**43**) and (**44**).[34,35] The electron-rich trisubstituted double bond of (**42**) is selectively attacked. The peroxy acid used for the epoxidation of (**44**) is peracetic acid. The compound (**45**) is attacked selectively from the more hindered concave side.[36]

Unhindered methylenecyclohexane derivatives undergo preferential axial epoxidation.[13] During the epoxidation of alkenes (**46**) and (**47**) the reagent approaches selectively from the axial side.[37,38] Probably in the epoxidation of (**47**) the selectivity is mainly due to the shielding of the α-face by the axial CH_2OAc at C-5. The percentage of axial attack during the epoxidation of several 3-substituted methylenecyclohexanes (**48**) has been studied.[39] There is an increase in the proportion of axial attack with an increase in electronegativity of the remote 3-equatorial substituent; when R = *p*-$CF_3C_6H_4$ in (**48**), axial

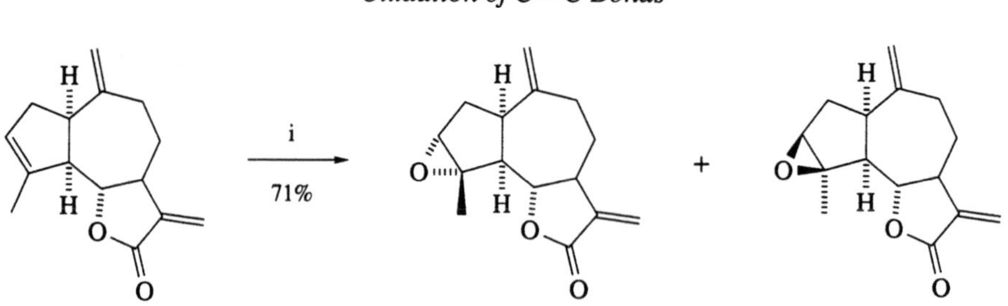

i, 71%

(42) 97% 3% (19)

i, MCPBA, 3 h, 5 °C

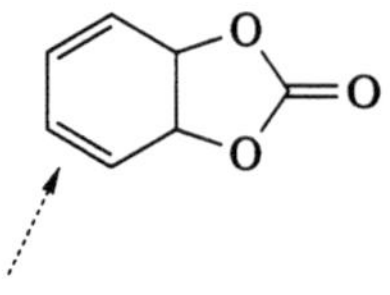

(43) (95, 82) 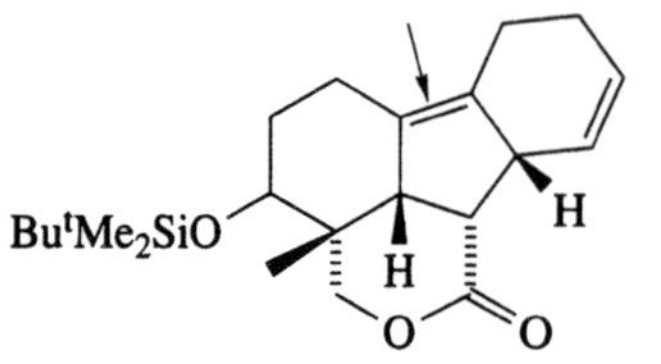 (44) (72, 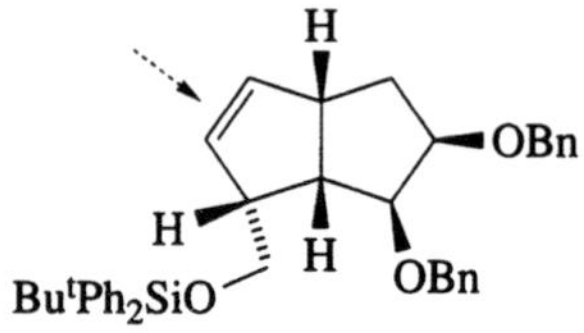 –) (45) (–, 75)

attack is 75%. In the alkene (**49**) the β-oriented siloxy group does not block the β-face since the C—O bond is in the plane of the π-bond; the epoxidation of (**49**) is not stereoselective.[40] The [10.10] betweenanene (**50a**) cannot be epoxidized since both faces of the π-bond are effectively blocked.[41a] Epoxidation of (**50b**) is highly stereoselective.[41b] The β-face of (**50b**) is sterically more shielded than the α-face, since the β-axial hydrogen at C-3 is closer to the π-system than the α-axial hydrogen at C-2a.

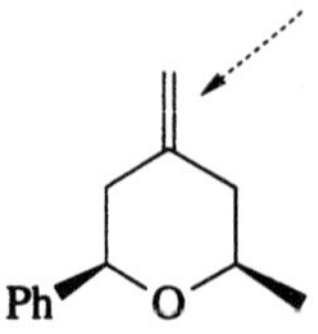

(46) (100, 86) 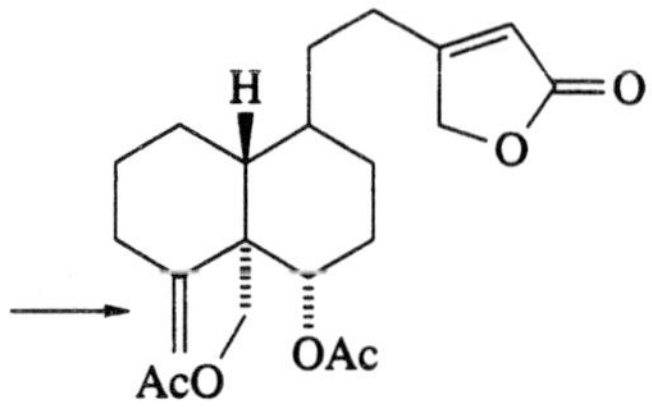 (47) (–, 100)  (48)

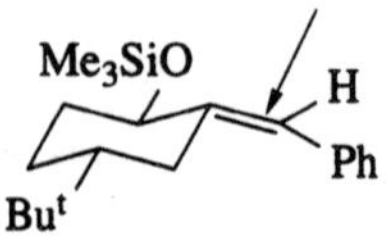

(49) (95, 50) 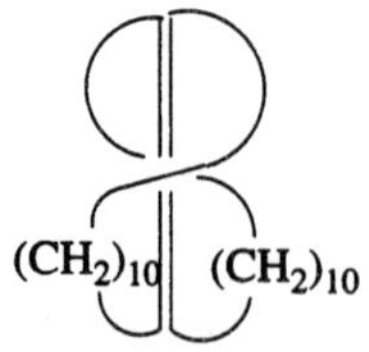 (50a) 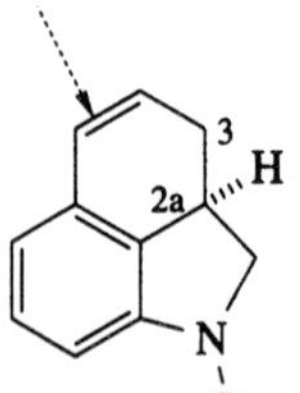 (50b) (97, 98)

3.1.2.2.4 MCPBA epoxidations of cyclic alkenes and methylenecycloalkanes having directing groups

Henbest *et al.* have shown that allylic cyclohexenols undergo epoxidation selectively *cis* to hydroxy if there is no severe steric interference.[13] It has been suggested that in the transition state for epoxidation, the hydroxy group is associated with the peroxy acid through hydrogen bonding. The ideal situation for the transition state is attained when the O—C—C═C dihedral angle is 120°.[5,42] The pseudoequatorial hydroxy is more effective than the pseudoaxial hydroxy in directing epoxidation. The rate of epoxidation of allylic cyclohexenol is about 10 times that of the corresponding allylic acetate and about one half that of cyclohexene.[13] The directive effect is rather weak and the stereoselectivity can be poor if the peroxy acid encounters severe steric interference. Homoallylic hydroxy can direct epoxidation if it can approach the double bond. In the cyclohexane system axial homoallylic hydroxy directs epoxidation. Besides the hydroxy, carbamates, ethers and ketones direct epoxidation.

In the *trans*-diol (**51**) both the hydroxys are equatorial. The homoallylic hydroxy, being equatorial, cannot direct epoxidation. Since only the allylic hydroxy can direct epoxidation (**51**) reacts with MCPBA stereoselectively (equation 20).[43] In the *trans*-diol (**52**) the hydroxys are diaxial. The β-oriented homoallylic hydroxy being axial can direct epoxidation from the β-face; the α-oriented allylic hydroxy also directs epoxidation but from the α-face. Hence the epoxidation of (**52**) is not stereoselective (equation 21).[44] The major diastereoisomer in the epoxidation of (**52**) is the β-epoxide showing that in this diol the directing effect of homoallylic hydroxy is stronger than that of allylic hydroxy. The allylic alcohols (**53**; equation 22),[45] and (**54**)–(**57**)[46–49] undergo stereoselective epoxidation due to the directing influence of hydroxy. Epoxidation of (**58**) furnishes almost exclusively the *trans*-epoxide (**59**; equation 23).[50] The moderate assistance provided by the allylic hydroxy for epoxidation from the β-face is not large enough to overcome steric interference with one of the geminal methyl groups.

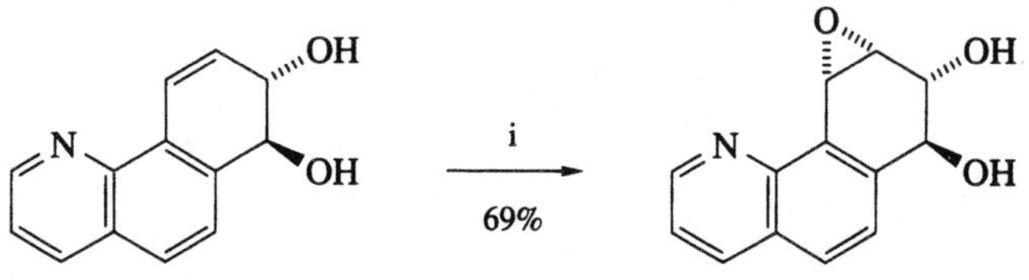

(20)

(**51**) i, MCPBA, dry THF, 48 h, r.t.; 69%

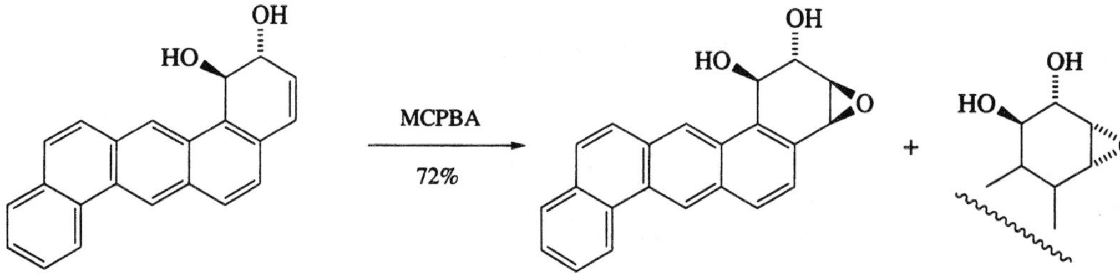

(21)

(**52**) MCPBA, 72%; 75% + 25%

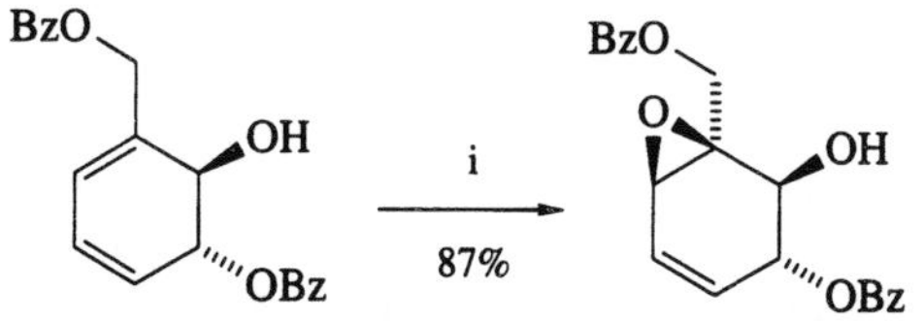

(22)

(**53**) i, 87%

i, MCPBA, CH_2Cl_2, phosphate buffer, pH 8, 0 °C, 1 h

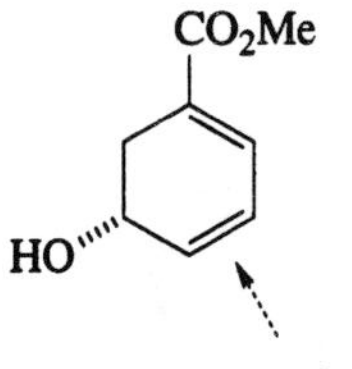

(**54**) (65, 95) 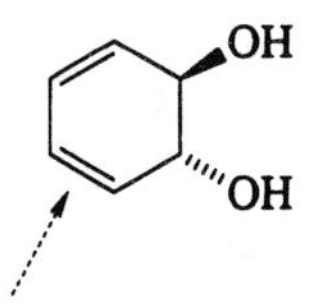 (**55**) (61, 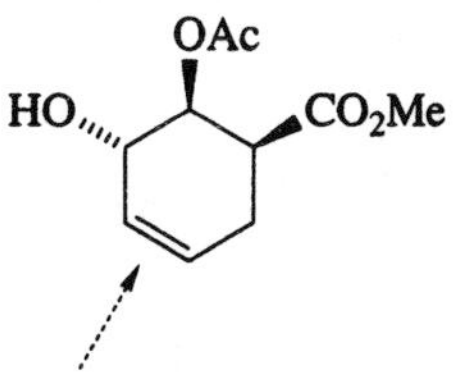 –) (**56**) (91, ~100) 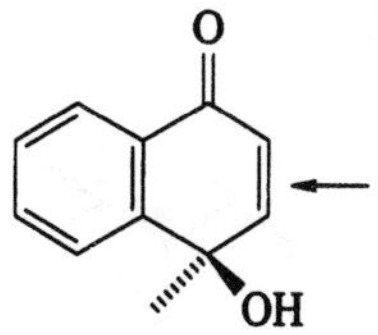 (**57**) (64, –)

When the allylic hydroxy in a methylenecyclohexane, *e.g.* (**61**), is equatorial then the C—O bond is in the plane of the π-bond and it does not direct epoxidation when it is reacted with peroxy acids (equation

i, MCPBA, $CHCl_3$

24);[2,51] the stereochemistry of the product is determined by the steric hindrance exhibited by the axial methyl group at C-10. In the allylic alcohol (**62**) the secondary hydroxy is axial and is not in the plane of the C(5)—C(6) double bond. Hence it is able to direct epoxidation from the β-face; the directing effect is strong enough to overcome steric hindrance due to methyl at C-10, leading to the stereoselective formation of the product (**63**).[52]

i, MCPBA, CH_2Cl_2, r.t., 1.5 h

Moderate selectivity has been observed in the epoxidation of the homoallylic alcohol (**64**; equation 25).[53] In the nonsteroidal conformation of (**64**), the axial hydroxy is suitably oriented for forming a hydrogen bond with the peroxy acid in the transition state. That the epoxidation of (**64**; R = OH) is indeed hydroxy-directed is supported by the observation that the epoxidation of the acetate (**64**; R = OAc) furnishes exclusively the α-epoxide (**66**; R = OAc).

The epoxidation of the acid (**67a**; R = OH) furnishes exclusively the α-epoxide (**67b**; equation 26);[54] the corresponding methyl ester (**67a**; R = OMe) furnishes a 2:1 mixture of 9α,10α- and 9β,10β-epoxides. This shows that the hydroxy group of the acid (**67a**), which is attached to the bishomoallylic carbon, directs the epoxidation.

(26)

(**67a**) R = OH (**67b**) R = OH

i, MCPBA, CH_2Cl_2, reflux, 50 min

Alkene (**68**) undergoes carbamate-directed epoxidation (equation 27).[55] Epoxidation of the diketone (**69**) furnishes exclusively the *syn*-epoxide, probably due to hydrogen bonding between MCPBA and the carbonyl group (equation 28).[56] The epoxidation of the ketal (**70**) furnishes exclusively one epoxide (**71**), but the closely related ketone (**72**) furnishes a 70:30 mixture of 1α,2α- and 1β,2β-epoxides.[57] Comparison of the selectivities suggests that the α-oriented ketal oxygen of (**70**) directs epoxidation through hydrogen bonding to the peroxy acid. Ether-directed epoxidation has been observed in the reaction of alkene (**73**) with trifluoroperacetic acid (equation 29);[58] the selectivity is due to hydrogen bonding involving the hydroxy of the peroxy acid and the oxygen of the allyl ether.

(27)

(**68**) 91% 9%

i, MCPBA, CH_2Cl_2, 0 °C

(28)

(**69**) MCPBA 72% 100% 0%

(**70**) (72, ~100) (**71**) (**72**) (78, 70)

(29)

i, CF_3CO_3H/CH_2Cl_2 buffered with Na_2HPO_4, –40 °C

3.1.2.2.5 MCPBA epoxidations of acyclic alkenes having directing groups

Several intermediates for the synthesis of macrolides and ionophore antibiotics have been prepared from epoxy alcohols, obtained through stereoselective epoxidation of acyclic allyl alcohols. One of the reagents studied extensively to effect stereoselective epoxidations is MCPBA. α,β-Epoxy alcohol (**77**) cannot be prepared through MCPBA epoxidation of geraniol (**76**) in an organic solvent. However, epoxide (**77**) has been prepared regioselectively from geraniol using an emulsion system (equation 30).[59] The emulsion system is prepared by stirring a mixture of geraniol, *n*-hexane, *n*-octanol, water, NaOH and dioctadecyldimethylammonium chloride.

(30)

i, MCPBA in emulsion, 25 °C, 15 h

Epoxidation of the secondary allylic alcohol (**79**) can furnish the diastereoisomers, *threo*-epoxy alcohol (**80**) and *erythro*-epoxy alcohol (**81**). Epoxidation *via* the rotamer (**82**) leads to *threo*-(**80**) and epoxidation *via* the rotamer (**83**) leads to *erythro*-(**81**).[5,42] The steric interactions in the *threo* transition state[60] can be obtained by examining the rotamer (**82**) and the steric interactions in the *erythro* transition state can be obtained by examining the rotamer (**83**). During the epoxidation of (**79**), when both R^c and R^t are alkyl and also when R^c is alkyl and R^t is H, in the *erythro* transition state (see **83**) there is severe steric interference between R^c and R which destabilizes the *erythro* transition state. High *threo* selectivity has been observed during the epoxidation of secondary allylic alcohols having trisubstituted (R^t and R^c are alkyl in **79**) or *cis*-disubstituted double bonds. During the epoxidation of secondary allylic alcohols having a *trans*-disubstituted double bond or a monosubstituted double bond (R^t and R^c are H in **79**) the *erythro* transition state is not sufficiently destabilized, since R interferes only with H; hence diastereoselectivity is poor.

High stereoselectivity has been observed during the epoxidation of the allylic alcohol (**84**) having a trisubstituted double bond and the allylic alcohol (**85**) having a *cis*-disubstituted double bond (equations

31a and 31b).[61,62] In contrast, selectivity is poor during the epoxidation of the allylic alcohols (**86**) and (**87**) having *trans*-disubstituted double bonds.[61,62]

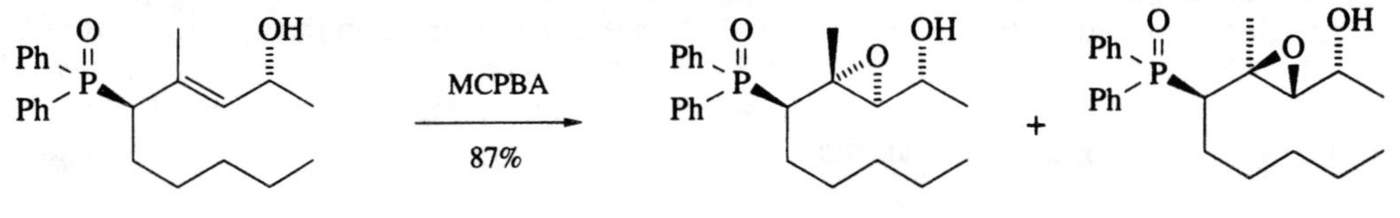

MCPBA, 87%

(**84**) 93% 7% (31a)

MCPBA, CH_2Cl_2, 80–92%

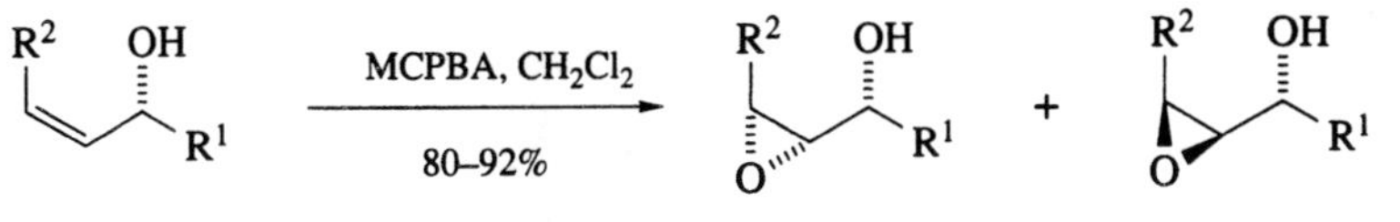

(**85**) 93% 7% (31b)

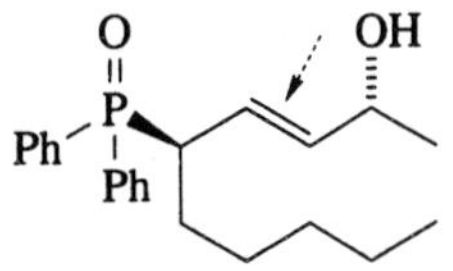

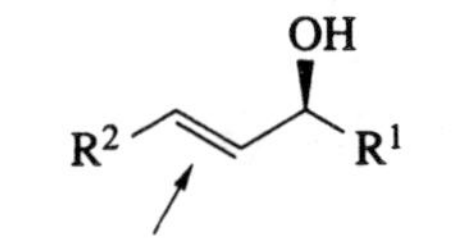

$R^1 = (CH_2)_6CO_2Me$, $R^2 = (CH_2)_5Me$

(**86**) (95, 73) (**87**) (86, 50)

$R^1 = (CH_2)_6 CO_2Me$, $R^2 = (CH_2)_5Me$

MCPBA epoxidation of the allylic alcohol (**88b**; R = H) is not a satisfactory route for the preparation of the *threo*-epoxy alcohol (**91**), as the stereoselectivity in this reaction is poor and a 61:39 mixture of (**91**) and (**92**) is obtained.[63] A convenient route has been developed for the synthesis of (**91**) employing reactions which are highly stereoselective (Scheme 1).[63] The epoxidation of (**88a**; R = TMS) is stereoselective since the *erythro* transition state is destabilized due to 1,3-steric interference between the bulky trimethylsilyl group and *n*-butyl group. The trimethylsilyl group of epoxide (**89**) can be replaced by hydrogen with retention of configuration by reacting it with F^-. Several epoxy alcohols have been prepared stereoselectively starting from alkenes having a trimethylsilyl group on the double bond, following the route given in Scheme 1.

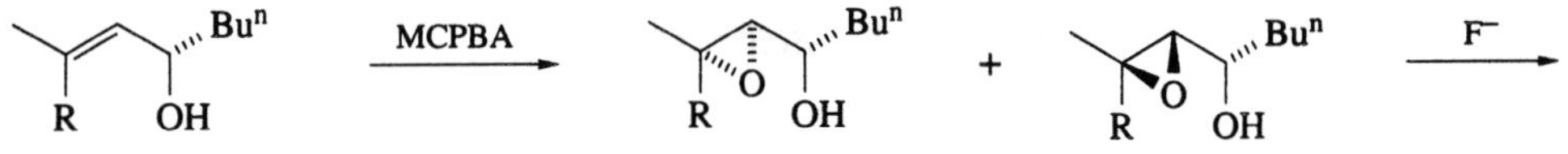

MCPBA F^-

(**88a**) $R = SiMe_3$
(**88b**) R = H

(**89**) (**90**)

(**91**) 99% (**92**) 1%

Combined yield in two steps = 81%

Scheme 1

Kishi *et al.* have observed that the epoxidation of the allylic alcohol (**93**) which has a suitably located ether oxygen is stereoselective (equation 32).[64] In the transition state (**94**), which delivers oxygen from the β-face, MCPBA is complexed by two hydrogen bonds involving participation of ether oxygen as well as hydroxy. In the transition state (**94**) there is steric interference between allylic hydrogen at C-4 and methyl at C-2. In contrast, in the transition state which can deliver oxygen from the α-face, there is steric interaction between methyl at C-4 and methyl at C-2; this steric interaction is more severe than the corresponding interaction in (**94**). Hence epoxidation from the α-face of the alcohol (**93**) is not favored. The epoxidation of (**93**) presented above is an example of the cooperative effect of hydroxy and ether oxygen in directing epoxidation. High stereoselectivities have been observed in the epoxidations of the allylic alcohols (**95**; equation 33) and (**98**; equation 34);[65,66] the selectivities are due to a cooperative effect. Epoxidation of the allyl alcohol (**99**) is stereoselective (equation 35);[67] it is postulated that in the transition state the hydroxy of the allylic alcohol and the carbonyl oxygen form hydrogen bonds with the peroxy acid.

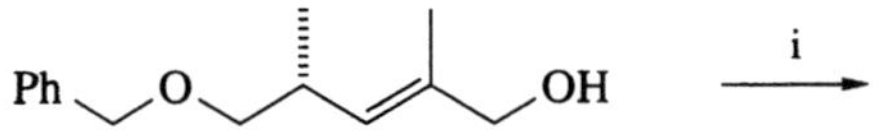

(**93**) —i→ (32)

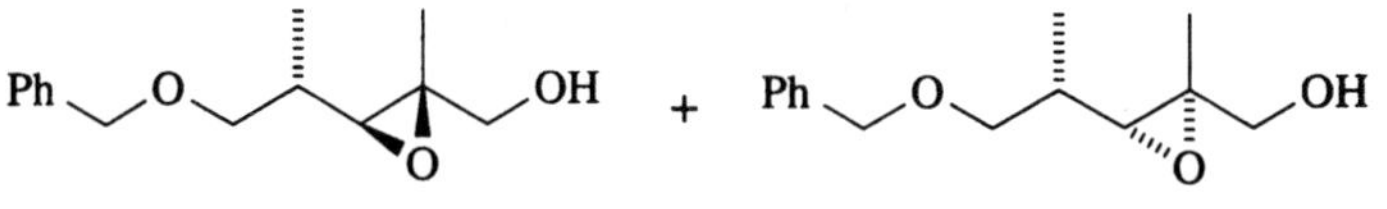

96% + 4%

i, MCPBA, CH_2Cl_2, 0 °C

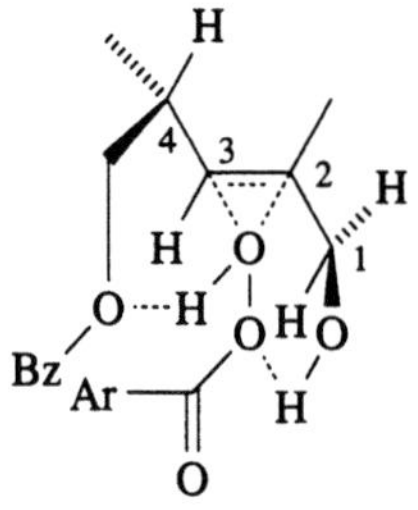

(**94**)

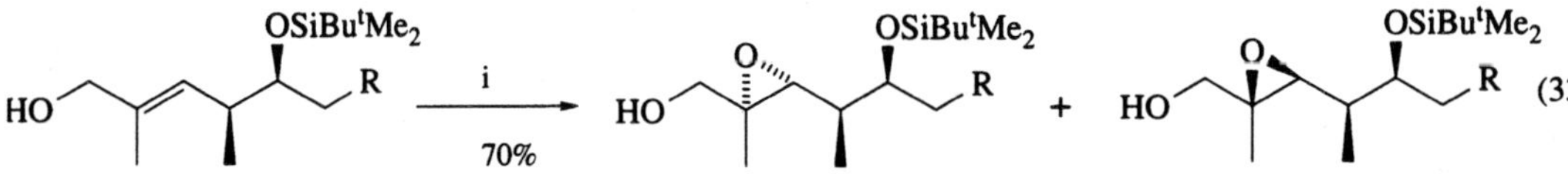

(**95**) —i (70%)→ (**96**) 98% + (**97**) 2% (33)

R = $C(OMe)_2CH(OMe)CH_2OAc$

i, MCPBA, CH_2Cl_2, –20 °C

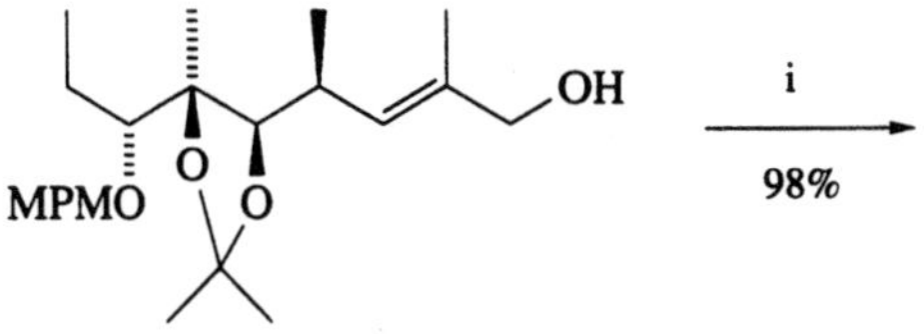

(**98**) —i (98%)→

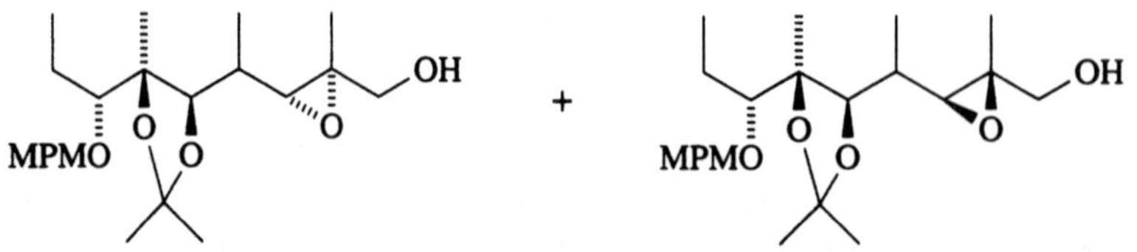

100% + 0% (34)

i, MCPBA, CH_2Cl_2, –15 to –10 °C

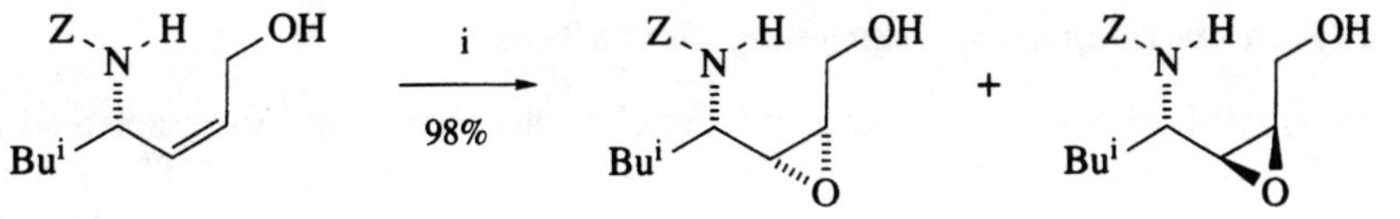

Z = $PhCH_2OCO$

(99) → 95% + 5% (35)

i, MCPBA, CH_2Cl_2, –10 °C

The epoxidation of the homoallylic alcohol (**100**) is regio- and stereo-selective (equation 36).[10] Epoxidation of (**100**) from the β-face involves a transition state which can be approximated by the conformer (**102**) complexed with MCPBA; in this conformation there is steric interference between the tertiary allylic hydrogen and ethyl group. Inspection of conformation (**103**) reveals that in the transition state leading to the α-epoxide there is steric interaction between the ethyl and allyl groups; the steric interaction in (**103**) is much larger than the interaction in (**102**).

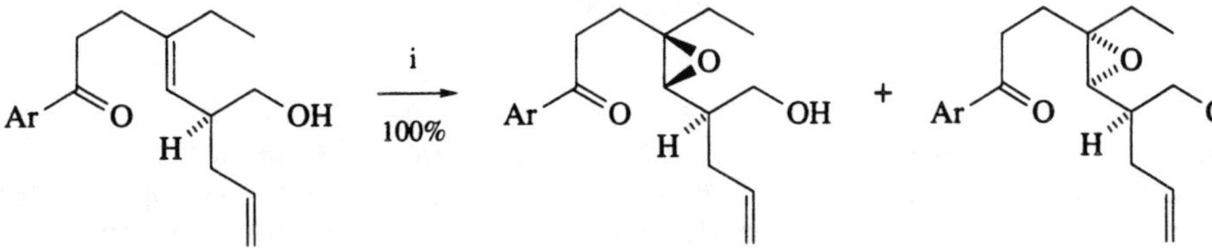

(**100**) → (**101**) 95% + <5% (36)

Ar = 4-$MeOC_6H_4$

i, MCPBA, CH_2Cl_2, aq. $NaHCO_3$, r.t.

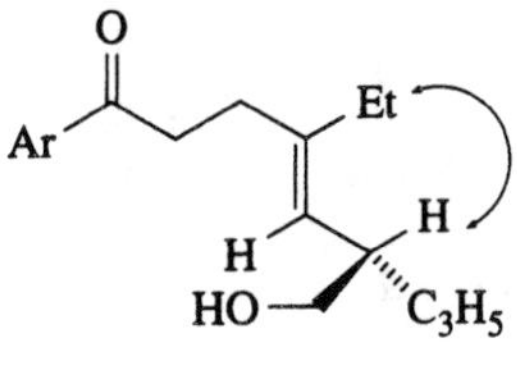

(**102**) 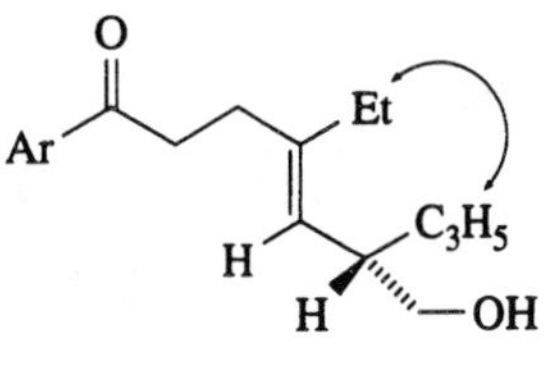 (**103**)

The stereoselectivities of the epoxidations of the homoallylic alcohols (**104**) and (**105**) and their benzoates (**106**) and (**107**) have been studied.[68] The amide-directed epoxidation of the *cis*-disubstituted alkene (**108**) is stereoselective (equation 37).[69]

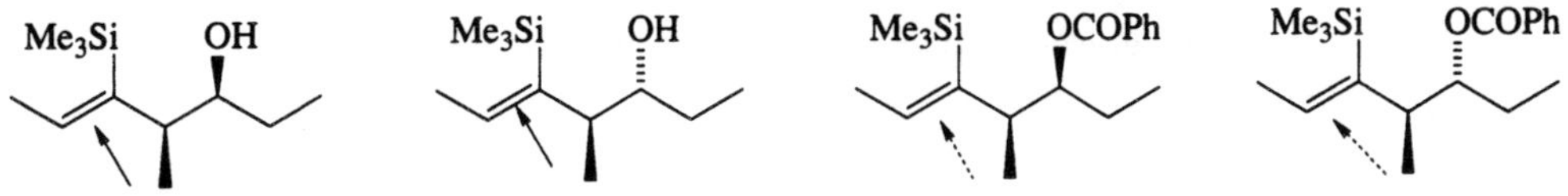

(**104**) (90, 80) (**105**) (88, 75) (**106**) (96, 86) (**107**) (89, 75)

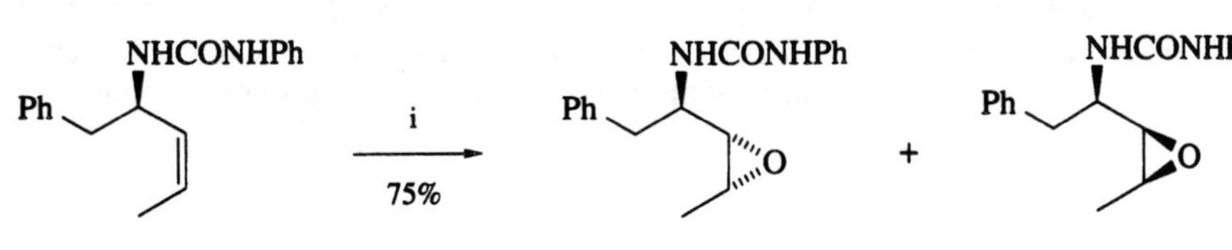

(**108**) → 95% + 5% (37)

i, MCPBA, CH_2Cl_2, 0 °C

3.1.2.2.6 MCPBA epoxidations of electron-deficient alkenes

Alkenes conjugated with C=O are electron-deficient and hence do not react readily with organic peroxy acids. The observation that the stereochemistry of the pyrazolinone (**109**) is not retained in the epoxidation product (**110**) is interesting (equation 38).[70] The epoxide (**110**) is not formed directly from (**109**). Since the double bond in (**109**) is electron-deficient its peroxy acid epoxidation to furnish (**111**) is a slow process. The isomerization of the (*Z*)-pyrazolinone (**109**) to the corresponding (*E*)-isomer is a comparatively fast process. MCPBA epoxidation of the (*E*)-isomer derived from (**109**) furnishes (**110**).

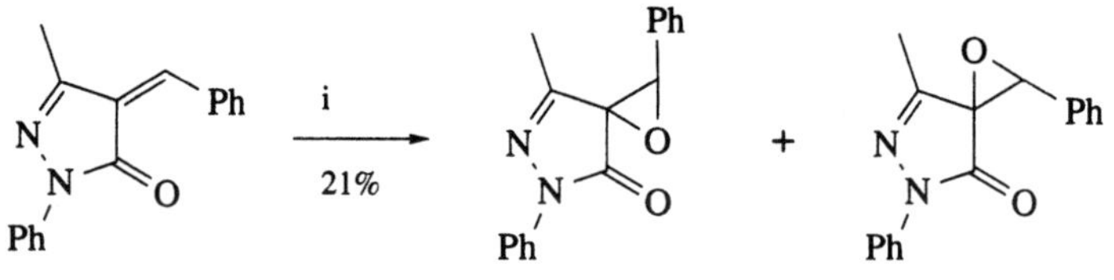

(38)

(**109**) → (**110**) 85% + (**111**) 15%

i, 21%

i, MCPBA, CH_2Cl_2, r.t., 7 d

Dienones with extended conjugation undergo peroxy acid epoxidation regioselectively at the γ,δ-double bond, even if it is less substituted than the α,β-double bond. The epoxidation of (**112**) is regio- and stereo-selective (equation 39).[71] Attempted epoxidation of (**113a**) using nucleophilic reagents furnishes polymeric materials. The epoxidation has been carried out with MCPBA (equation 40).[72a]

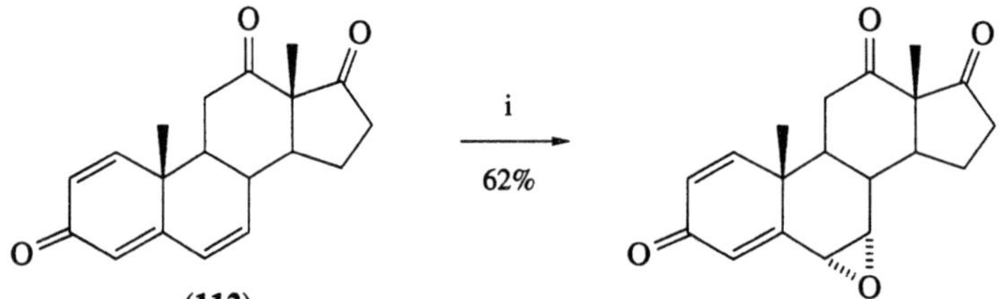

(39)

(**112**)

i, 62%

i, MCPBA, CH_2Cl_2, r.t., 5 d

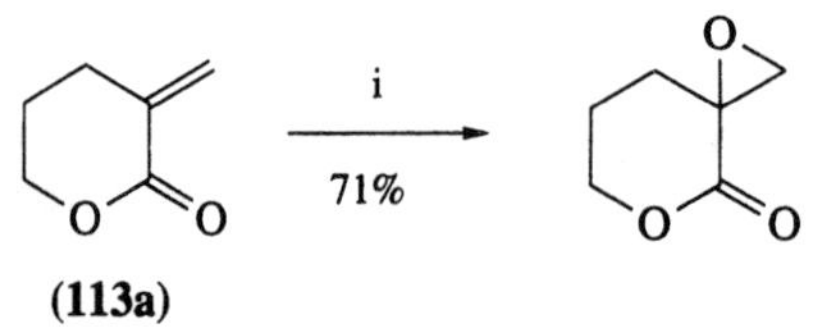

(40)

(**113a**)

i, 71%

i, MCPBA, dichloroethane, 2,6-di-t-butylphenol, reflux in dark, 5 h

3.1.2.3 Some of the Commonly Used Peroxy Acids and Related Reagents

For the large scale preparation of epoxides, reagents which are cheaper than MCPBA are available. Though many electron-deficient alkenes have been epoxidized with MCPBA at elevated temperatures, reagents which are more reactive than MCPBA have been used; when these reagents are employed the reactions can be carried out under comparatively mild conditions, leading to improvements in yields and selectivities.

3.1.2.3.1 Peracetic acid and performic acid

Peracetic acid is available commercially as a 40% solution in acetic acid. Both peracetic acid and performic acid prepared *in situ* are used industrially.[6,7]

3.1.2.3.2 *Peroxycarboximidic acids*

Alkenes have been epoxidized with H_2O_2 in the presence of nitriles such as acetonitrile and benzonitrile. The actual epoxidizing agent is a peroxycarboximidic acid, RC(=NH)CO_3H, generated *in situ*.[13] The reagents are inexpensive and the method is convenient and safe for large scale preparations. Perbenzimidic acid epoxidation of (**113b**) takes place from the more hindered α-face.[72b] It has been suggested that the epoxidizing reagent complexes with the carbonyl group as well as the ether oxygen at C-3.

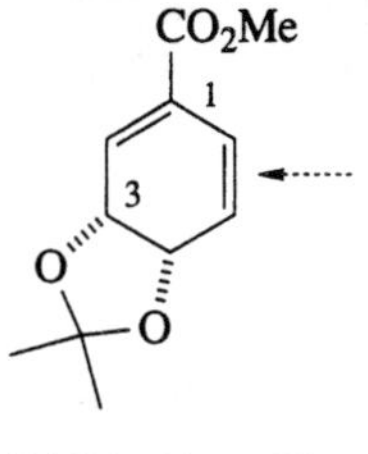

(**113b**) (90, ≈100)

3.1.2.3.3 *Trifluoroperacetic acid*

The unsaturated ester (**114**) has been epoxidized with CF_3CO_3H (equation 41).[73] Only one diastereoisomer is produced in the reaction (see also equation 29).

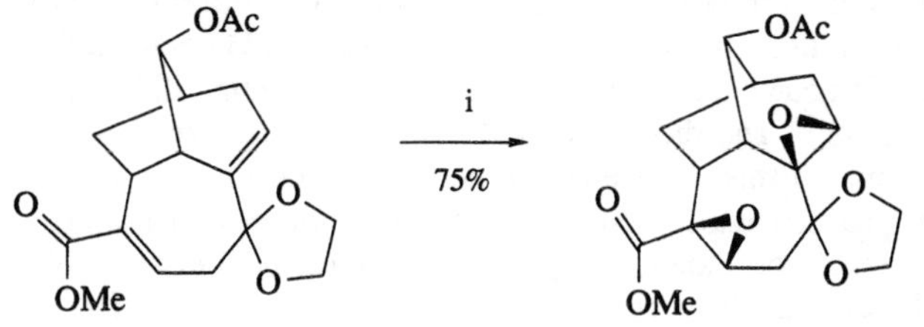

(**114**) —i, 75%→ (41)

i, K_2HPO_4, CF_3CO_3H, CH_2Cl_2, 40 °C

3.1.2.3.4 *4-Nitroperbenzoic acid*

4-Nitroperbenzoic acid has been used for the preparation of the epoxide (**116**; equation 42).[74] In the aldehyde (**115**), the tetrasubstituted C(1)—C(2) double bond is not epoxidized since it is deactivated by conjugation with the aldehyde group. The disubstituted double bond is not sufficiently reactive due to the inductive effect of the allyl ether moieties. The epoxidation takes place from the α-face since the β-face is blocked by the allylic substituents. The epoxide (**116**) cannot be prepared in satisfactory yields using MCPBA.

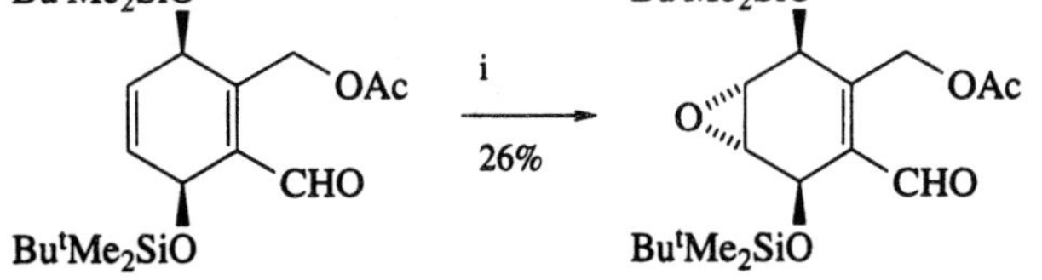

(**115**) —i, 26%→ (**116**) (42)

i, 4-nitroperbenzoic acid, 4,4'-thiobis(6-*t*-butyl-3-methylphenol), $CHCl_3$, reflux, 3 d

3.1.2.3.5 *3,5-Dinitroperbenzoic acid*

The diene (**117**) has been epoxidized with 3,5-dinitroperbenzoic acid (equation 43).[75] Attack at both the double bonds is stereoselective. Epoxides of enol ethers are normally difficult to isolate, but the epoxide (**118**) is quite stable. This stability is due to the attachment of the electron-withdrawing carboxylate substituent to the furan ring. When (**117**) is treated with the comparatively less reactive MCPBA, epoxidation takes place regioselectively at the trisubstituted double bond to furnish a monoepoxide.

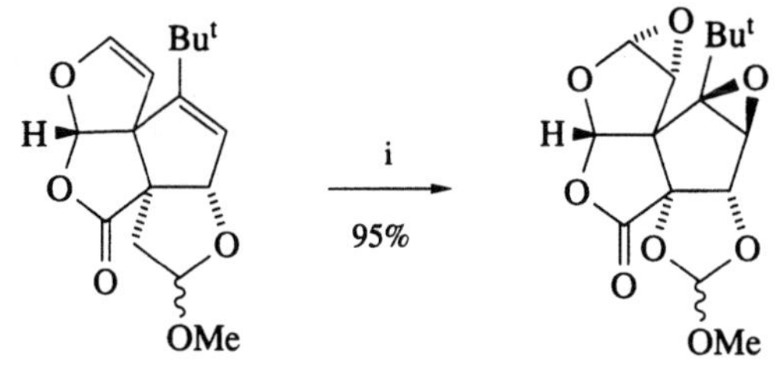

(**117**) → (**118**) i, 95% (43)

i, 3,5-dinitroperbenzoic acid, CH_2Cl_2, $NaHCO_3$, 23 °C, 36 h

3.1.2.3.6 Dialkyldioxirane

Murray *et al.* have shown that (≤0.1 M) solutions of dimethyldioxirane (**119**) in acetone can be obtained through low temperature distillation of caroate–acetone reaction mixtures. More recently it has been shown that solutions of methyl(trifluoromethyl)dioxirane (**120**), a reagent which is more reactive than (**119**), can be prepared starting from trifluoroacetone. A solution of dioxirane (**119**) in dry acetone reacts readily with disubstituted alkenes such as *cis*-3-hexene (**121**) and *trans*-3-hexene (**122**) to furnish corresponding epoxides in nearly quantitative yield. The *cis*-alkene (**121**) reacts about eight times faster than the *trans*-alkene (**122**).[76] The dioxirane (**120**) is highly reactive. Reaction of (**120**) with phenanthrene at –20 °C for 5 min furnishes phenanthrene 9,10-oxide in 93% yield.[77a] Epoxidations with dimethyldioxirane (**119**) proceed under neutral and mild conditions in the absence of nucleophiles and electrophiles. It is the reagent of choice for synthesizing sensitive epoxides of enol esters, enol lactones and enol ethers.[77b] The epoxidations are carried out at –40 to 20 °C. The alkenes (**123a**),[77c] (**123b**), (**123c**), (**123d**)[77d] and aflatoxin B_1[77e] have been epoxidized with the reagent (**119**). For a recent review see Murray.[77f]

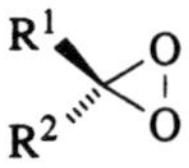

(**119**) $R^1 = R^2 = Me$

(**120**) $R^1 = CF_3$, $R^2 = Me$

(**121**)

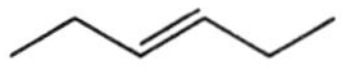

(**122**)

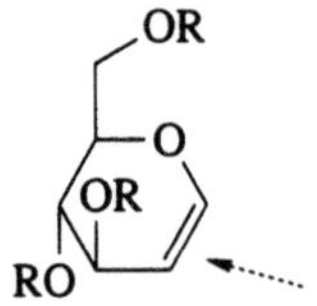

$R = SiMe_2Bu^t$

(**123a**) (100, 100)

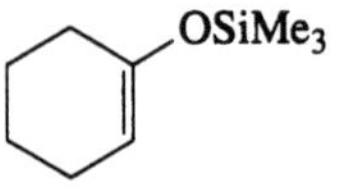

(**123b**) (99)

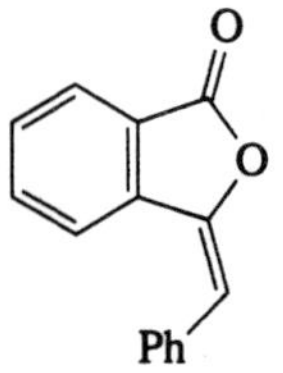

(**123c**) (85)

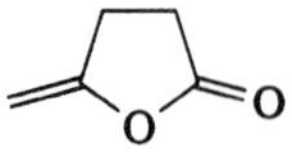

(**123d**) (96)

3.1.2.3.7 Magnesium monoperphthalate hexahydrate (MMPP)

Pure MCPBA is shock sensitive and can deflagrate. Magnesium monoperphthalate is not shock sensitive and does not deflagrate. MMPP is cheaper than MCPBA and loses available oxygen at a slower rate than MCPBA. The reagent is water soluble and hence epoxidations are carried out in a water–isopropanol mixture. Epoxidation of cyclohexene with MMPP in isopropanol–water at 25 °C for 7 h furnishes the epoxidation product in 85% yield.[78]

3.1.2.4 Intramolecular Epoxidations with Peroxy Acids

Treatment of the acid (**67a**) with carbonyldiimidazole and 90% H_2O_2 furnishes the epoxide (**67b**; equation 44).[54] This reaction is more than 100 times faster than the epoxidation of (**67a**) with MCPBA (see equation 26). It has been suggested that under the experimental conditions of equation (44) the acid (**67a**; R = OH) is transformed to the peroxy acid (**67a**; R = O_2H), which reacts regio- and stereo-selectively through an intramolecular reaction.

$$(\mathbf{67a}) \xrightarrow[75\%]{\text{i–iii}} (\mathbf{67b}) \qquad (44)$$

i, CH_2Cl_2, *N,N'*-carbonyldiimidazole, r.t., 30 min; ii, H_2O_2 (90%); iii, 0 °C, 5 min

3.1.3 EPOXIDATIONS WITH ALKYL HYDROPEROXIDES

3.1.3.1 Epoxidations of Alkenes Lacking Directing Groups

One of the important developments in oxirane chemistry during the past 25 years is the use of alkyl hydroperoxides for the preparation of epoxides from alkenes, in the presence of high-valent d^0 transition metal complexes. *t*-Butyl hydroperoxide (TBHP) and ethylbenzene hydroperoxide (EBHP), in the presence of soluble compounds of Mo, are used for the manufacture of propylene oxide.[8] Alkenes can be epoxidized with TBHP using Mo, V, W or Ti complexes as catalyst; however, Mo is the catalyst of choice when the substrates lack directing groups such as hydroxy. Electron-rich alkenes react rapidly, but the reaction is sluggish when electron-deficient alkenes such as 1-decene are the substrates. Polar solvents, particularly alcohols and water, retard the epoxidation by competing with the hydroperoxide for coordination sites on the metal. Water reduces the selectivity by reacting with the epoxide to furnish the corresponding diol. Convenient procedures for the preparation of anhydrous solutions of TBHP in organic solvents and also the precautions that have to be taken while handling TBHP are given in a recent review.[5] The use of a solution of TBHP in toluene is preferred.[79] TBHP is one of the most stable organic peroxides known.[5]

In the synthesis of an epoxide from an alkene with TBHP/Mo, the stereochemistry of the alkene is retained in the epoxide.[8] It has been suggested that the reaction proceeds through the transition state (**124**).[80] 1-Decene and the alkene (**115**) (see equation 42) have been epoxidized with TBHP/Mo (equations 45 and 46).[5,74] The epoxidation of (**115**) is regio- as well as stereo-selective; the reagent approaches the electron-rich double bond from the less hindered face.

(**124**) L = ligand; M = Mo

$$\text{1-decene} \xrightarrow[86\%]{\text{i}} \text{n-}C_8H_{17}\text{-oxirane} \qquad (45)$$

i, TBHP, 1,2-dichloroethane, $Mo(CO)_6$, Na_2HPO_4, reflux, 10 h

$$(\mathbf{115}) \xrightarrow[20\%]{\text{i}} (\mathbf{116}) \qquad (46)$$

i, TBHP, $Mo(CO)_6$, 4,4'-thiobis(6-*t*-butyl-3-methylphenol), benzene, reflux, 6 h

3.1.3.2 Epoxidations of Cyclic Alkenes Having Directing Groups

The rate of epoxidation of an allylic alcohol with TBHP in the presence of a vanadium catalyst is more than 1000 times the rate of epoxidation of the parent alkene.[5] An increase in reaction rate has also been observed in the vanadium-catalyzed epoxidation of homoallylic and bishomoallylic alcohols due to the location of the hydroxy group. Hence the epoxidation of polyunsaturated allylic and homoallylic alcohols with TBHP/V^{5+} is regioselective; only those double bonds which come under the directing influence of hydroxy are epoxidized. For examples see compounds (**38**), (**131**), (**150**) and (**151**). The epoxidation of the cyclic allylic alcohol (**125**) is stereoselective and takes place from the face *cis* to hydroxy (equation 47).[5] Hydroxy-directed epoxidation of cyclic allylic alcohols employing TBHP/V has been used extensively in organic synthesis; high stereoselectivities have been observed during the epoxidations of the cyclic allylic alcohols (**128**)–(**131**).[81–84] The epoxidation of (**131**) is also regioselective; the disubstituted double bond, but not the trisubstituted double bond, is suitably located for hydroxy-directed epoxidation. The *cis*-directing effect of allylic hydroxy in the metal-catalyzed epoxidation is much stronger than the *cis*-directing effect observed in MCPBA epoxidation. When there is severe steric interference to the approach of the reagent *cis* to hydroxy then MCPBA epoxidations exhibit poor *cis* selectivity (see equation 23). In contrast high selectivity has been observed in the hydroxy-directed epoxidation of (**58**) with TBHP/V, even though one of the geminal methyl groups sterically interferes with the approach of the reagent from the face *cis* to the hydroxy (equation 48).[50] The primary allylic alcohol (**132**) undergoes hydroxy-assisted epoxidation, with high stereoselectivity from the less hindered side to furnish in 96% yield the epoxide (**133**), when it is reacted with trityl hydroperoxide/VO(acac)$_2$; there is a decrease in stereoselectivity if TBHP is used instead of trityl hydroperoxide.[85] TBHP epoxidation of the allylic alcohol (**134**) in which the C—O bond is in the plane of the double bond is not stereoselective.[86] The homoallylic alcohols (**135**), (**136**) and the bishomoallylic alcohol (**137**) undergo hydroxy-directed epoxidation stereoselectively.[87–89] Due to the directing effect of homoallylic hydroxy, vitamin D_3 (**38**) is epoxidized regio- and stereo-selectively at room temperature to furnish in 90% yield the (5*S*)-5,6-

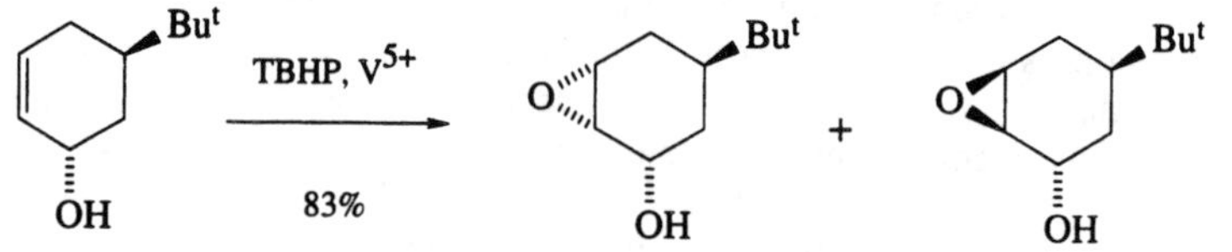

Bu^t / OH — TBHP, V^{5+} / 83% → Bu^t / O / OH + Bu^t / O / OH (47)

(**125**) (**126**) 100% (**127**) 0%

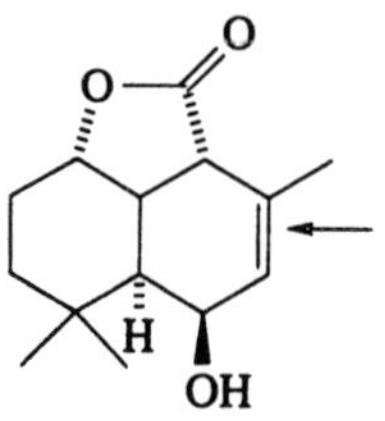

O / O / H / OH

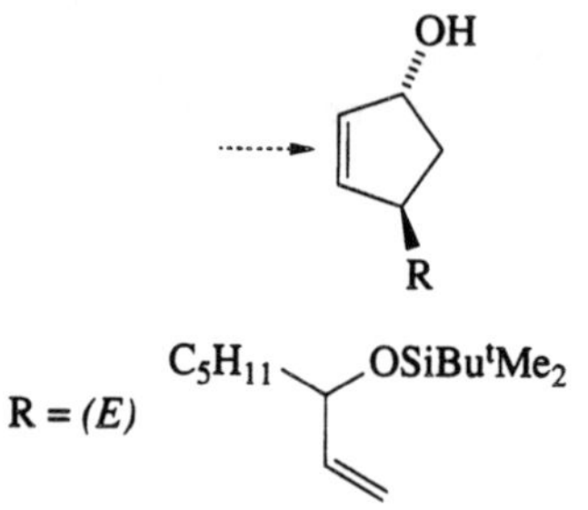

OH / R

R = (*E*) C_5H_{11} OSiBu^{t}Me$_2$

(**128**) (95, ≈100) (**129**) (98, ≈100)

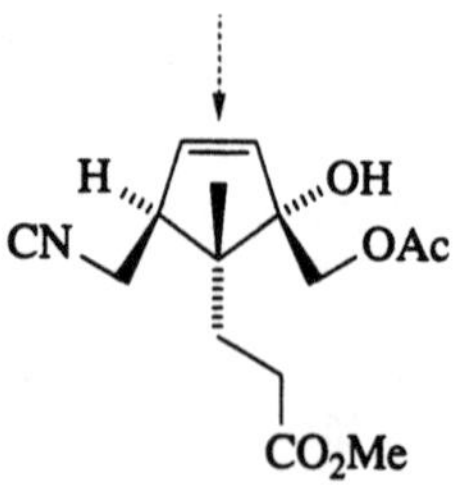

H / CN / OH / OAc / CO_2Me

HO / H / H / Bu^{t}Ph$_2$SiO / O

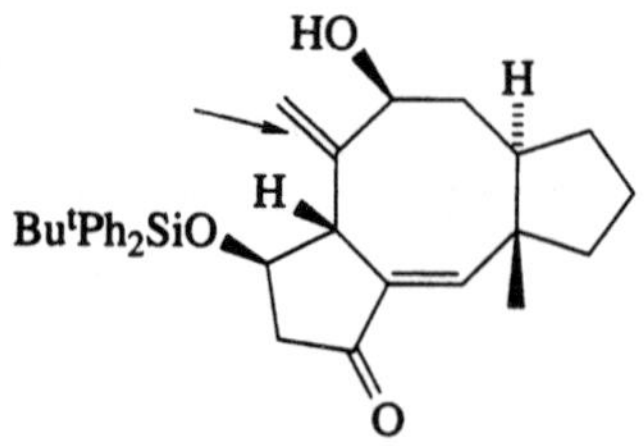

(**130**) (85, ≈100) (**131**) (67)

monoepoxide when TBHP/$VO(acac)_2$ is employed.[31] The epoxidation of (**38**) does not take place at room temperature when the reaction is carried out with TBHP/Mo; when the reaction is carried out at higher temperature a complex mixture is obtained due to the instability of the substrate. Oxidations of some hindered allylic alcohols (*e.g.* **138a**) to the corresponding α,β-unsaturated ketones during attempted preparation of epoxy alcohols employing TBHP/V are reported.[90]

$$\textbf{(58)} \xrightarrow[100\%]{\text{TBHP, VO(acac)}_2} \underset{>99\%}{\textbf{(60)}} + \underset{<1\%}{\textbf{(59)}} \quad (48)$$

(132) (96, ≈100) **(133)** **(134)** (100, 55)

(135) (77, ≈100) **(136)** (95, ≈100) **(137)** (87, ≈100)

(138a)

Ester-directed epoxidations have been observed when $Mo(CO)_6$ is used as catalyst (equation 49).[91]

$$\textbf{(138b)} \xrightarrow[66\%]{\text{TBHP, Mo(CO)}_6} 94\% + 6\% \quad (49)$$

(138c) R = H

(138d) R = $MeOCH_2$

Hydroxy-directed epoxidation of (**138d**) with TBHP/Ti(OPr^i)$_4$ is complete in 2 h at –35 °C and gives in 87% yield the corresponding α-epoxide. In contrast, the epoxidation of the diol (**138c**) is extremely sluggish, probably due to the strong coordination of the Ti cation to the diol.[91b]

3.1.3.3 Epoxidations of Acyclic Alkenes Having Directing Groups

In the vanadium-catalyzed epoxidation of allylic alcohols the ideal geometry for the transition state is reached when the O—C—C═C dihedral angle is 50°.[5] To predict the stereochemistry of epoxidation of allylic alcohol (**139**) two conformers (**140**) and (**141**) have to be considered.[63] In the conformer (**140**) there is steric interference between TMS and H; in the conformer (**141**) there is steric interference between TMS and Me. Since the steric interference is less in (**140**), this conformer determines the stereochemistry of epoxidation and the epoxy alcohol (**142**) is anticipated as the major product. The direction of asymmetric induction during epoxidation of acyclic homoallylic alcohols can be predicted by considering steric interference in the transition state (**143**);[92] stereoselectivity is high when alkyl groups can be equatorially oriented (R^2 and R^3) in the transition state (**143**). The stereoselectivities observed during the epoxidation of several acyclic alcohols with TBHP/V have been given in a recent review.[2]

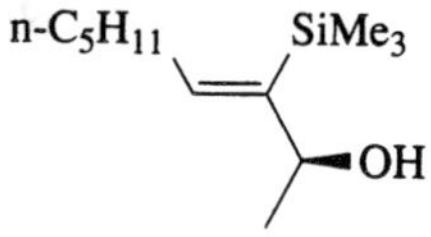

n-C_5H_{11} SiMe$_3$ OH

(139)

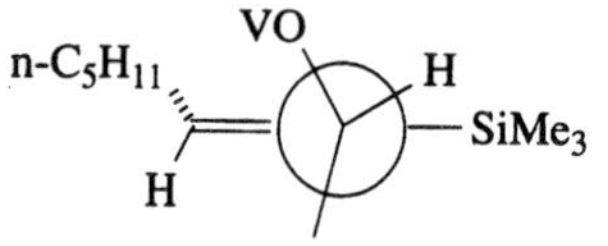

n-C_5H_{11} VO H SiMe$_3$ H

(140)

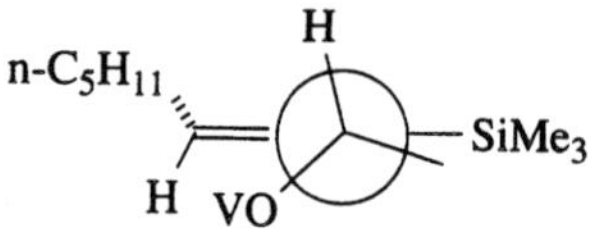

n-C_5H_{11} H SiMe$_3$ H VO

(141)

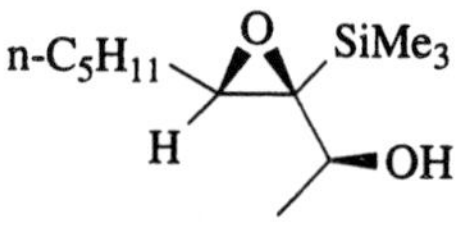

n-C_5H_{11} O SiMe$_3$ H OH

(142)

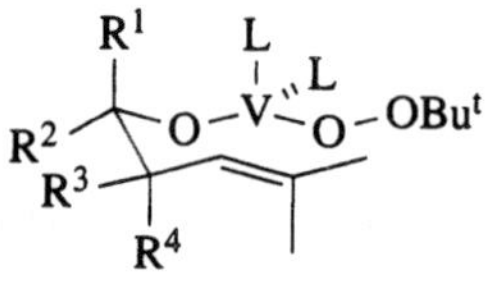

R^1 L L R^2 O V O OBu^t R^3 R^4

(143)

Epoxidations of the allylic alcohols (**144**),[93] (**145**),[94] (**146**; equation 50)[95] and (**148**; equation 51)[96] are highly stereoselective. The transformation of (**148**) to (**149**) is an example of the use of silyloxyalkenes in the stereoselective synthesis of *trans* α,β-epoxy alcohols. The epoxidation of the ester (**150**), which has a hydroxy allylic to a *trans*-disubstituted double bond, does not exhibit high stereoselectivity (equation 52);[97] the epoxidation is regioselective, involving only the C(13)—C(14) double bond.

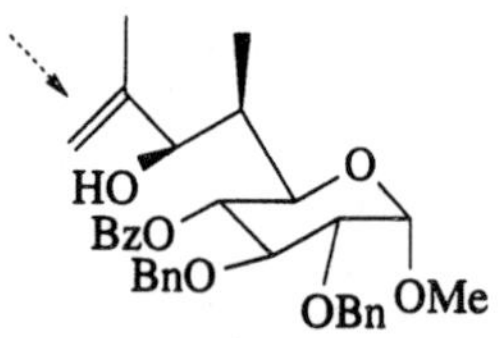

HO BzO BnO O OBn OMe

(144) (90, ≈100)

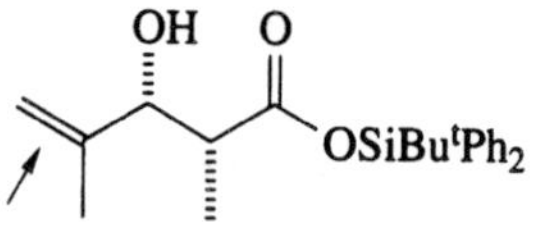

OH O OSiBu^{t}Ph$_2$

(145) (80, ≈100)

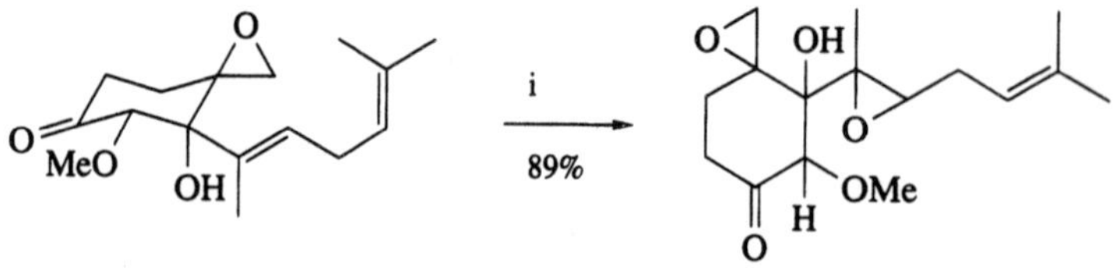

O O MeO OH —i, 89%→ O OH O OMe H O (50)

(146) **(147)**

i, TBHP, VO(acac)$_2$, PhMe, r.t., 2 h

OH SiMe$_2$Ph

i, ii
65%

OH

O

(51)

OMe

(148) **(149)**

i, TBHP, VO(acac)$_2$; ii, Bu$^n{}_4$NF

CO$_2$Me

i
82%

OH
(150)

CO$_2$Me

+

(52)

O

OH

O

OH

70% 30%

i, TBHP, VO(acac)$_2$

The acyclic homoallylic alcohol (**151**) is epoxidized regio- and stereo-selectively (equation 53).[98] High stereoselectivities have been observed in the epoxidations of the homoallylic alcohols (**152**) and (**153**) which have a *cis*-disubstituted double bond.[99] The stereoselectivity is excellent when the reactant is (**154**) but poor when the reactant is (**155**).[100]

H
OH

i
100%

(151)

H H

OH + OH (53)

O O

100% 0%

i, TBHP, VO(acac)$_2$, benzene, reflux, 45 min

Vanadium-catalyzed epoxidation of the diene (**156**) having hydroxy allylic to one double bond and homoallylic to the other double bond does not furnish exclusively (**157**; equation 54).[101] The epoxidation of (**156**) with dibutyltin oxyperoxide is regioselective, furnishing exclusively the regioisomer (**157**) as a 95:5 mixture of *erythro* and *threo* diastereoisomers.[101] Epoxidations of the diol (**159**) and its epimer (**160**) take place selectively from the α-face. On the basis of this observation and further studies it has been concluded that in (**159**) and (**160**) the epoxidation is directed by the homoallylic, but not the allylic,

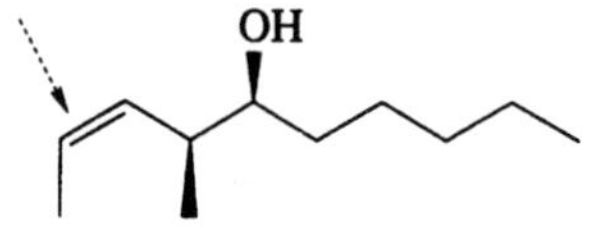

(152) (81, 97)

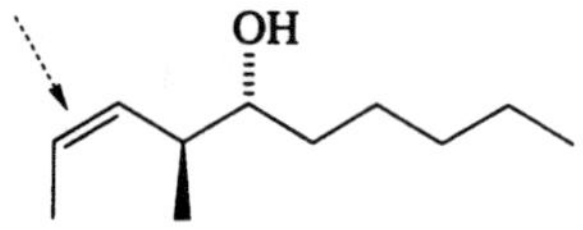

(153) (95, 97)

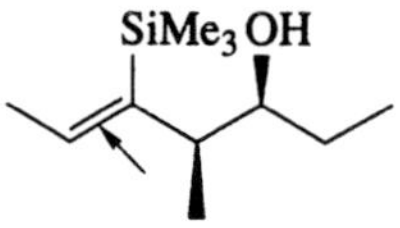

SiMe3 OH

(154) (88, 99)

SiMe3 OH

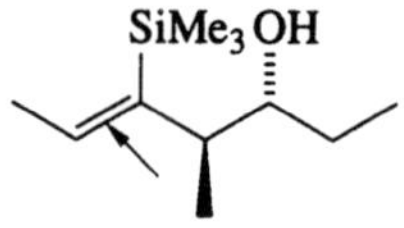

(155) (91, 67)

hydroxy.[102] Stereoselective epoxidations of some acyclic bishomoallylic alcohols have been reported (equation 55).[103]

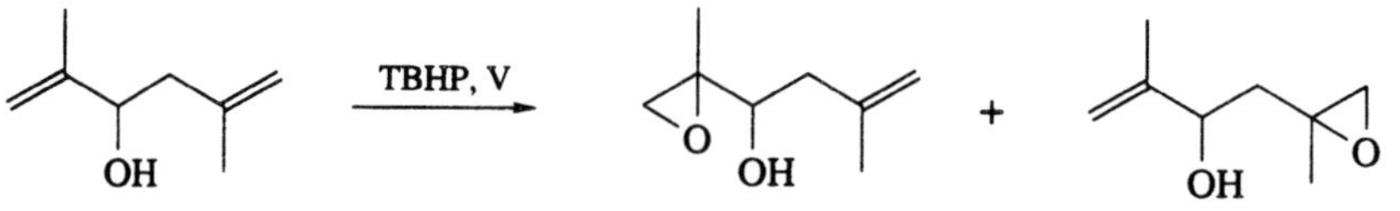

TBHP, V

(156) → (157) 87% + (158) 13% (54)

erythro:*threo* = 98:2

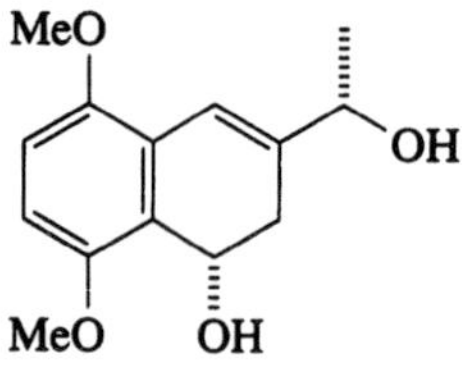

(159)

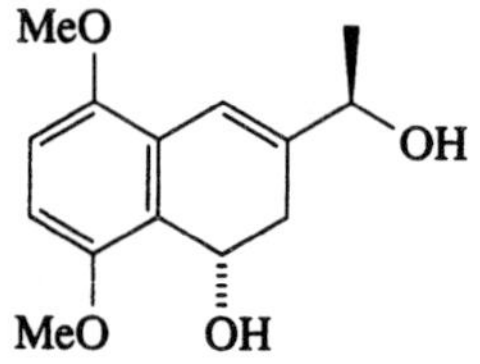

(160)

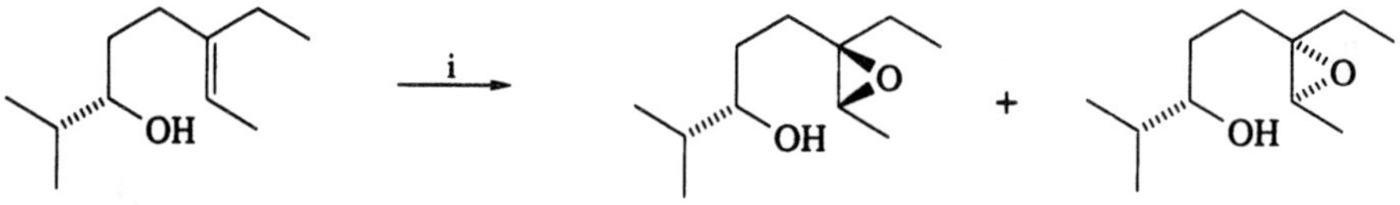

(161) →i 90% + 10% (55)

i, TBHP, $VO(acac)_2$, benzene, r.t.

The epoxy alcohol (**97**), a key intermediate in the synthesis of maytansine, has been prepared through Ti-catalyzed epoxidation of (**95**; equation 56).[65] The alcohol (**95**) exists predominantly in conformation (**162**), with the allylic hydrogen at C-4 and the π-bond very nearly eclipsed. The oxygens of the alcohol and silyl ether which are located below the plane of the π-bond complex with Ti; this complex blocks the approach of the epoxidizing reagent from the α-face and hence the β-epoxide is formed. It is of interest to note that the π-facial selectivity resulting from this route is the opposite of the π-facial selectivity observed in MCPBA epoxidation (see equation 33).

$$(\mathbf{95}) \xrightarrow[80\%]{\text{i}} \underset{99\%}{(\mathbf{97})} + \underset{1\%}{(\mathbf{96})} \qquad (56)$$

i, TBHP, $Ti(OPr^i)_4$, –20 °C, 11 h

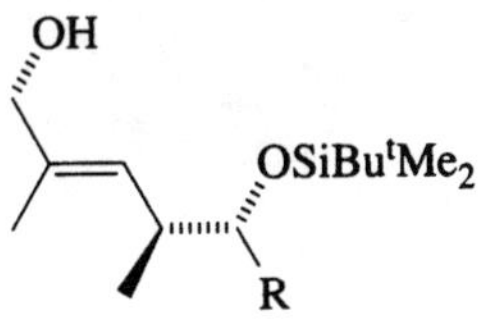

OH

OSiBu^{t}Me$_2$

R

(162)

3.1.4 EPOXIDATIONS UTILIZING SILYL-PROTECTED PEROXY ESTERS

The regioselective intramolecular epoxidation of the peroxy ester (**163**), which can be prepared from farnesol, has been effected by treating it with $Cu(OCOCF_3)_2$ (equation 57).[104] This reaction provides a convenient route for the preparation of the 6,7-epoxide (**164**), which cannot be synthesized from farnesol by conventional methods or even by template-directed epoxidation using $Mo(CO)_6$/TBHP.

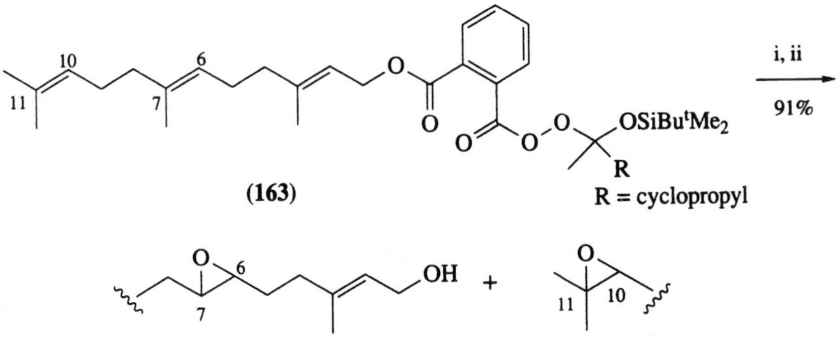

(**163**) R = cyclopropyl; i, ii, 91% → (**164**) 80% + (**165**) 20% (57)

i, $Cu(OCOCF_3)_2$; ii, LiOH/aq. THF

3.1.5 EPOXIDATIONS WITH HYDROGEN PEROXIDE

Several acidic oxides such as MoO_3, WO_3 and compounds of selenium, arsenic and boron are effective catalysts for the epoxidation of alkenes by H_2O_2 through generation of inorganic peroxo acids, such as peroxoselenic and peroxoarsonic acids.[8]

When compared with Mo and V ions, tungsten ion induced H_2O_2 decomposition is very slow. Hence when peroxytungstic acid or peroxytungstates are employed as catalysts it is possible to carry out reactions on a large scale, employing strong H_2O_2 solutions and temperatures up to 70 °C. Allylic and *cis*-homoallylic alcohols can be efficiently epoxidized with H_2O_2/WO_3 as shown for the alcohol (**166**; equation 58).[105] The reaction is carried out in aqueous methanol at pH 4.5. The stereoselectivity of epoxidation is similar to that observed with TBHP/V^{5+}.

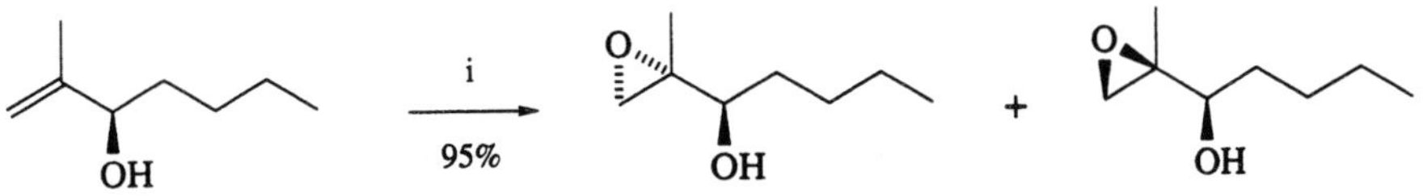

(**166**) —i, 95%→ 95% + 5% (58)

i, 30% H_2O_2, H_2WO_4, Me_3NO, r.t., 8 h

Under the experimental conditions used in the earlier epoxidation studies, H_2O_2 and the alkene to be oxidized are in the same phase. Since water reduces the rate of reaction and lowers the yield, it has to be removed continuously. Recently, it has been observed that epoxidations can be carried out, even with dilute H_2O_2, by employing a biphasic system. A quaternary phosphonium peroxotungstate catalyst has been used for the epoxidation of alkenes and allylic alcohols with 30% H_2O_2 (equation 59).[106] It has been observed that the pH of the reaction mixture remains close to 5–6 and there is no need to buffer the aqueous solution. The epoxidation proceeds in the organic phase where the phosphonium peroxotungstate enters because of the lipophilicity of the phosphonium moiety. Epoxidations have been carried out efficiently with 16% w/v H_2O_2 employing the (diperoxotungsto)phosphate catalyst (**168**) in a biphasic system (equation 60).[107]

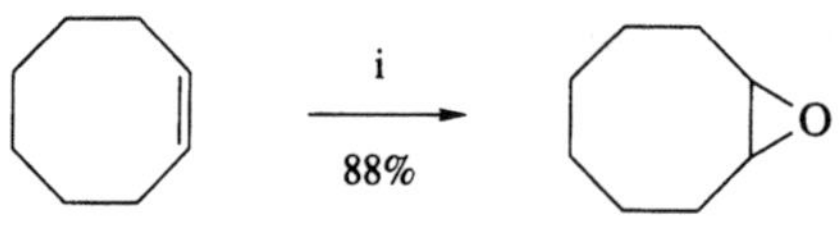

(167) (59)

i, $(Ph_3\overset{+}{P}CH_2Ph)_2\ (W_2O_{11})^{2-}$, 30% H_2O_2, 1,2-dichloroethane, 50 °C, 15 h

$[(C_8H_{17})_3\overset{+}{N}Me]_3\ [PO_4\{W(O)(O_2)_2\}_4]^{3-}$
(168)

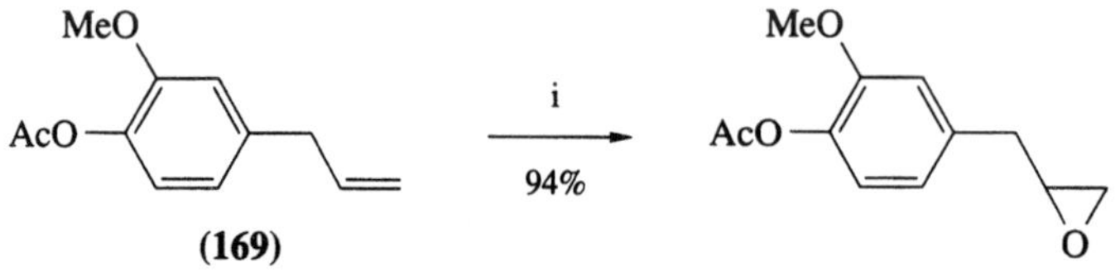

(169) (60)

i, 1,2-dichloroethane, **(168)**, 16% H_2O_2, 60 °C, 150 min

3.1.6 STOICHIOMETRIC EPOXIDATIONS WITH Mo AND W PEROXO COMPLEXES

Epoxidations of triisopropylsilyl (TIS) ethers (**170**) and (**171**) with WO_5·HMPA in dichloroethane take place stereoselectively to furnish *syn*-epoxides;[99] these epoxidations, along with the vanadium-catalyzed epoxidations of (**152**) and (**153**) (described in Section 3.1.3.3), make available a group of all the four possible diastereoisomeric epoxides having four consecutive chiral centers in an acyclic carbon framework.

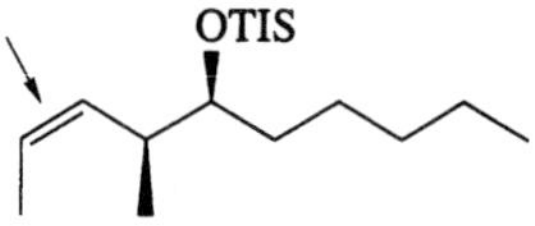

(**170**) (89, 94) 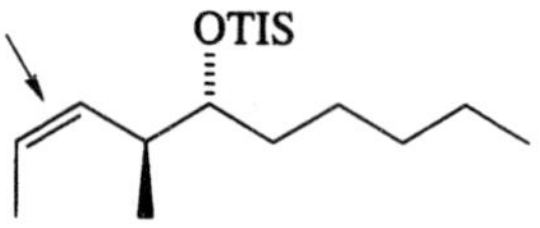 (**171**) (67, 88)

cis-2-Butene-1,4-diones are epoxidized by $MoO_5·H_2O$·HMPA. The diketone (**172**) furnishes the corresponding *cis*-epoxide when it is reacted with the molybdenum reagent in CH_2Cl_2 at room temperature for a week.[108] This is a rare example of epoxidation of an electron-deficient alkene by a MoO_5 complex.

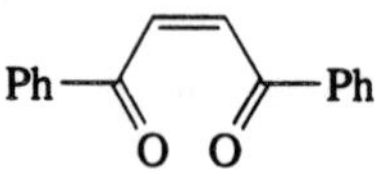

(**172**) (92)

3.1.7 EPOXIDATIONS *VIA* CATALYSIS BY FIRST-ROW TRANSITION METAL COMPLEXES

The observation that iron porphyrins can catalyze, under mild conditions, epoxidations of alkenes when iodosylbenzene is used as the oxidant has been followed up by a number of studies on metalloporphyrins as models for cytochrome *P*-450 enzymes. Cytochrome *P*-450 enzymes catalyze epoxidation of alkenes by molecular oxygen in the presence of a hydrogen donor, NADPH cofactor.[8] This has led to the study of a number of systems based on a metalloporphyrin/O_2/reducing agent, to bring about epoxidation of alkenes.

Cyclooctene has been epoxidized with oxygen using a manganese porphyrin as catalyst (equation 61).[109]

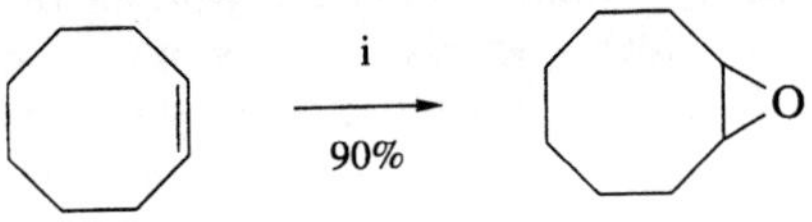

(61)

i, Mn(TPP)Cl, 1-methylimidazole, O_2, Zn, AcOH, 20 °C, 1 h
TPP = *meso*-tetraphenylporphyrin

Cobalt-catalyzed epoxidation of alkenes has been carried out with the cobalt derivative of (**174**), employing iodosylbenzene as the oxidant. Epoxidation of *cis*-β-methylstyrene furnishes exclusively the *cis*-epoxide (equation 62).[110] The reaction proceeds through an active oxo-cobalt(IV) species, and is more selective than reactions proceeding through oxo-chromium or oxo-manganese species. The catalyst can be recovered unchanged by simple filtration.

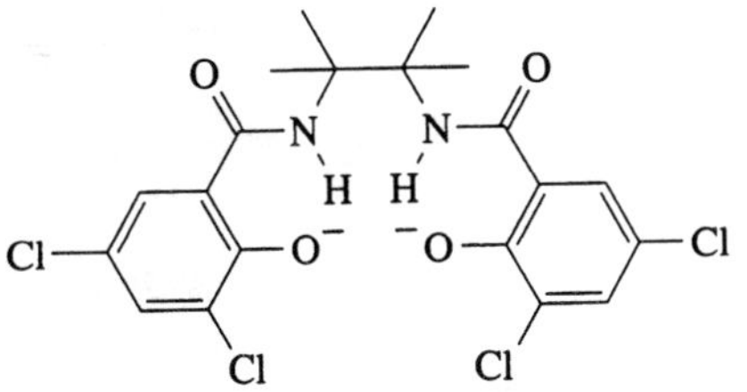

(**174**)

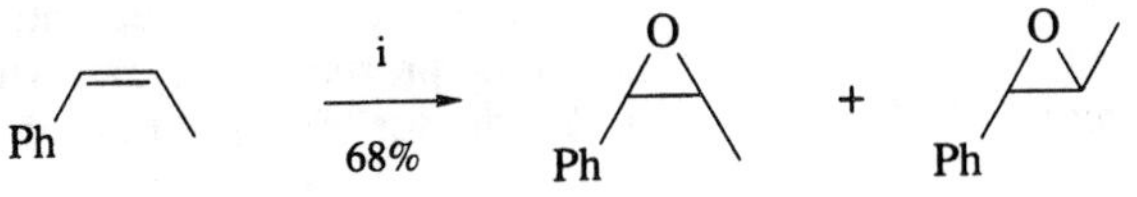

(62)

(**173**) 100% 0%

i, Co^{II}(**174**)•H_2O, PhIO, CH_2Cl_2, r.t., 5 h

Epoxidation of the diene (**175**) with iodosylbenzene/Fe^{II}phthalocyanine is regio- and stereo-selective (equation 63).[111]

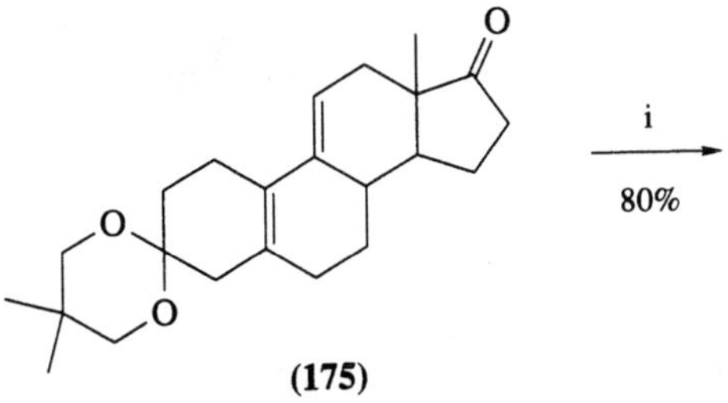

(**175**)

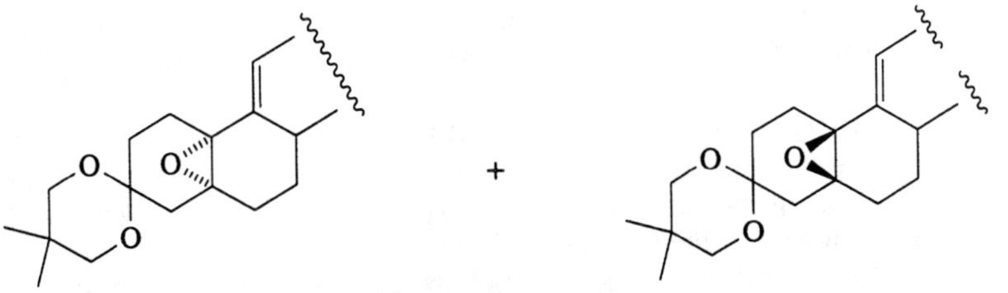

(63)

92% 8%

i, Fe^{II} phthalocyanine, PhIO, MeCN

The applications of a wide variety of metal complexes in catalyzing the epoxidations of alkenes have been reviewed recently.[8]

3.1.8 EPOXIDATIONS USING OXYGEN

Ethylene oxide is manufactured by oxidizing ethylene with air or oxygen in the presence of a silver catalyst.[1] Alkenes furnish hydroperoxides when oxidized by oxygen in the presence of catalysts like salts of cobalt and manganese; the hydroperoxides are transformed to a number of products, including epoxides. Only in a few cases, such as oxidation of 1-phenylcyclooctene, have moderate yields of epoxides been obtained during autoxidation.[112]

Irradiation of an alkene in the presence of molecular oxygen and an α-diketone furnishes the corresponding oxirane in high yields. The reaction proceeds in the complete absence of nucleophiles, and thus can avoid formation of by-products arising from the reaction of nucleophiles with sensitive oxiranes.[2] The photoepoxidation proceeds *via* addition of an acylperoxy radical to the alkene.[113] Photochemical epoxidation of cholesteryl acetate (**176**) has been carried out (equation 64a);[114a] the major epoxidation product is the 5β,6β-epoxide (**177a**). In MCPBA epoxidation of (**176**) the major product is (**177b**).

(**176**) —i, 94%→ (**177a**) 70% + (**177b**) 30% (64a)

i, biacetyl, O_2, $h\nu$

The thermal rearrangement of unsaturated bicyclic 1,4-peroxides, readily available from the reaction of conjugated dienes with singlet oxygen, is a convenient route for the preparation of bisepoxides.[1]

Epoxidation of cholesteryl acetate (**176**) with air in the presence of a catalytic amount of dioxo(tetramesitylporphyrinato)ruthenium(VI) furnishes in 85% yield a 99:1 mixture of the epoxides (**177a**) and (**177b**).[114b]

Epoxy alcohols have been synthesized by carrying out the photooxygenation of alkenes in the presence of the transition metal complexes derived from Ti, V and Mo (for an example see equation 64b).[114c] The hydroperoxides formed from alkenes during the photooxygenation function as oxygen transfer reagents and precursors for the allylic alcohol intermediates.

(**178**) —$^1O_2$, Ti(OPr^i)$_4$, 84%→ 5% + 95% (64b)

3.1.9 CHEMOSELECTIVE EPOXIDATIONS

Chemoselective epoxidations of many alkenes carrying functional groups such as hydroxy, ether, ester, amide and ketone have been presented in this chapter. Chemoselective epoxidations of a few functionalized alkenes have proved difficult, but by using appropriate reagents and strategies the difficulties have been overcome. Some examples are given below.

When the unsaturated tertiary amine, pirprofen (**179**; R = H) is treated with MCPBA the reaction takes place selectively at the more nucleophilic nitrogen to furnish the corresponding amine oxide with the alkene moiety intact. In contrast, peroxycarboximidic acid, prepared *in situ* from acetonitrile/H_2O_2, reacts selectively with the alkene moiety of the ester (**179**; R = Me; equation 65).[115] The sterically hindered nitrogen of (**179**) is able to react with peroxy acids which have a low steric demand, but not with peroxycarboximidic acids which have a large steric demand.

To prevent *N*-oxide formation the tertiary nitrogen of (**181**) is blocked by protonation. The salt prepared from trifluoroacetic acid and (**181**) is epoxidized with CF_3CO_3H; work-up of the reaction mixture furnishes the epoxide (**182**; equation 66).[116]

Sulfides are readily oxidized by peroxy acids as well as TBHP/Mo. Hence the chemoselective epoxidation of the unsaturated sulfide (**183**) has been effected by an indirect method; the alkene is first trans-

(65)

(179) R = Me **(180)**

i, H_2O_2, MeCN, MeOH, $KHCO_3$, pH 7.5–8.0

(66)

(181) **(182)**

i, CF_3CO_2H; ii, CF_3CO_3H

formed to a bromohydrin which is then treated with a base to furnish **(184)**.[117] It has been suggested that unsaturated sulfides can be epoxidized chemoselectively using the photochemical epoxidation route, since under these experimental conditions sulfides remain unchanged.[113]

(183) **(184)**

Cyclobutanones are susceptible to Baeyer–Villiger oxidation. The epoxide **(186)** cannot be prepared by reacting the ketoalkene (**185**; equation 67) with MCPBA. Moderate, chemoselective epoxidation has been observed in the reaction of (**185**; equation 68) with *O*-trichloroethylperoxycarbonic acid **(190)** prepared *in situ* from the triazole **(189)** and H_2O_2.[118].

(185)	**(186)**	**(187)**	**(188)**	
MCPBA	1%	96%	3%	(67)
(189)/H_2O_2	31%	28%	41%	(68)

(189) **(190)**

3.1.10 REFERENCES

1. E. G. Lewars, in 'Comprehensive Heterocyclic Chemistry', ed. A. R. Katritzky, Pergamon Press, Oxford, 1984, vol. 7, p. 95.
2. A. S. Rao, S. K. Paknikar and J. G. Kirtane, *Tetrahedron*, 1983, **39**, 2323.
3. R. G. Harvey, *Synthesis*, 1986, 605.
4. E. J. Corey and S. P. T. Matsuda, *Tetrahedron Lett.*, 1987, **28**, 4247.
5. K. B. Sharpless and T. R. Verhoeven, *Aldrichimica Acta*, 1979, **12**, 63.
6. M. Bartok and K. L. Lang, in 'The Chemistry of Heterocyclic Compounds', ed. A. Weissberger and E. C. Taylor, Wiley, New York, 1985, vol. 42, part 3, p. 1.
7. B. Plesnicar, *Org. Chem. (N.Y.)*, 1978, **5** (C), 211.
8. (a) H. Mimoun, in 'Comprehensive Coordination Chemistry', ed. G. Wilkinson, Pergamon Press, Oxford, 1987, vol. 6, p. 317; (b) K. A. Jorgensen, *Chem. Rev.*, 1989, **89**, 431.
9. W. R. Roush and M. A. Adam, *J. Org. Chem.*, 1985, **50**, 3752.
10. T. Fukuyama, C.-L. J. Wang and Y. Kishi, *J. Am. Chem. Soc.*, 1979, **101**, 260.
11. E. J. Corey, L. O. Weigel, A. R. Chamberlin, H. Cho and D. H. Hua, *J. Am. Chem. Soc.*, 1980, **102**, 6613.
12. J. P. Marino, R. Fernandez de la Pradilla and E. Laborde, *J. Org. Chem.*, 1984, **49**, 5279.
13. G. Berti, *Top. Stereochem.*, 1973, **7**, 93, and refs. cited therein.
14. A. Krebs, W. Rüger, W.-U. Nickel, M. Wilke and U. Burkert, *Chem. Ber.*, 1984, **117**, 310.
15. A. Amann, G. Ourisson and B. Luu, *Synthesis*, 1987, 696.
16. J. Eustache, J.-M. Bernardon and B. Shroot, *Tetrahedron Lett.*, 1987, **28**, 4681.
17. M. F. Semmelhack and A. Zask, *J. Am. Chem. Soc.*, 1983, **105**, 2034.
18. F. Camps, A. Messeguer and F.-J. Sanchez, *Tetrahedron*, 1988, **44**, 5161.
19. M. A. Hashem, E. Manteuffel and P. Weyerstahl, *Chem. Ber.*, 1985, **118**, 1267.
20. I. Fleming, A. K. Sarkar and A. P. Thomas, *J. Chem. Soc., Chem. Commun.*, 1987, 157.
21. J. Rokach, R. N. Young, M. Kakushima, C.-K. Lau, R. Seguin, R. Frenette and Y. Guindon, *Tetrahedron Lett.*, 1981, **22**, 979.
22. A. I. Meyers, D. L. Comins, D. M. Roland, R. Henning and K. Shimizu, *J. Am. Chem. Soc.*, 1979, **101**, 7104.
23. S. L. Schreiber, T. Sammakia, B. Hulin and G. Schulte, *J. Am. Chem. Soc.*, 1986, **108**, 2106.
24. W. C. Still and A. G. Romero, *J. Am. Chem. Soc.*, 1986, **108**, 2105.
25. R. J. Giguere, G. von Ilsemann and H. M. R. Hoffmann, *J. Org. Chem.*, 1982, **47**, 4948.
26. D. P. Michaud, N. T. Nashed and D. M. Jerina, *J. Org. Chem.*, 1985, **50**, 1835.
27. T.-T. Li, Y. L. Wu and T. C. Walsgrove, *Tetrahedron*, 1984, **40**, 4701.
28. P. G. Williard and S. E. de Laszlo, *J. Org. Chem.*, 1985, **50**, 3738.
29. S. V. Govindan and P. L. Fuchs, *J. Org. Chem.*, 1988, **53**, 2593.
30. S. P. Tanis, Y. M. Abdallah and P. G. Williard, *Tetrahedron Lett.*, 1985, **26**, 3651.
31. K. Nakayama, S. Yamada, H. Takayama, Y. Nawata and Y. Iitaka, *J. Org. Chem.*, 1984, **49**, 1537.
32. B. Fraser-Reid, L. Magdzinski, B. F. Molino and D. R. Mootoo, *J. Org. Chem.*, 1987, **52**, 4495.
33. A. A. Devreese, M. Demuynck, P. J. de Clercq and M. Vandewalle, *Tetrahedron*, 1983, **39**, 3049.
34. S. V. Ley and F. Sternfeld, *Tetrahedron Lett.*, 1988, **29**, 5305.
35. E. J. Corey and A. G. Myers, *J. Am. Chem. Soc.*, 1985, **107**, 5574.
36. N. Hussain and J. Leonard, *Tetrahedron Lett.*, 1987, **28**, 4871.
37. A. Schneider and U. Sequin, *Tetrahedron*, 1985, **41**, 949.
38. J. M. Luteijn and A. de Groot, *Tetrahedron Lett.*, 1982, **23**, 3421.
39. C. R. Johnson, B. D. Tait and A. S. Cieplak, *J. Am. Chem. Soc.*, 1987, **109**, 5875.
40. Y. S. Sanghvi and A. S. Rao, *Indian J. Chem., Sect. B*, 1987, **26**, 671.
41. (a) J. A. Marshall and S. D. Rothenberger, *Tetrahedron Lett.*, 1986, **27**, 4845; (b) M. R. Leanna, M. J. Martinelli, D. L. Varie and T. J. Kress, *Tetrahedron Lett.*, 1989, **30**, 3935.
42. B. E. Rossiter, T. R. Verhoeven and K. B. Sharpless, *Tetrahedron Lett.*, 1979, 4733.
43. S. Kumar, A. Czech and E. J. La Voie, *J. Org. Chem.*, 1988, **53**, 1329.
44. H. Lee and R. G. Harvey, *Tetrahedron Lett.*, 1981, **22**, 1657.
45. S. Ogawa and T. Takagaki, *J. Org. Chem.*, 1985, **50**, 2356.
46. D. A. McGowan and G. A. Berchtold, *J. Am. Chem. Soc.*, 1982, **104**, 7037.
47. R. A. Aleksejczyk, G. A. Berchtold and A. G. Braun, *J. Am. Chem. Soc.*, 1985, **107**, 2554.
48. M. Koreeda and M. A. Ciufolini, *J. Am. Chem. Soc.*, 1982, **104**, 2308.
49. J. Carnduff, M. Hafiz, R. Hendric and F. Monaghan, *Tetrahedron Lett.*, 1984, **25**, 6033.
50. Y. S. Sanghvi and A. S. Rao, *J. Heterocycl. Chem.*, 1984, **21**, 317.
51. I. V. Ekhato, J. V. Silverton and C. H. Robinson, *J. Org. Chem.*, 1988, **53**, 2180.
52. M. Hirayama, K. Gamoh and N. Ikekawa, *Tetrahedron Lett.*, 1982, **23**, 4725.
53. S. Manabe and C. Nishino, *Tetrahedron*, 1986, **42**, 3461.
54. M. Furber and L. N. Mander, *J. Am. Chem. Soc.*, 1987, **109**, 6389.
55. P. Kocovsky, *Tetrahedron Lett.*, 1988, **29**, 2475.
56. C. C. Liao, H. S. Lin, T. H. Hseu, C.-P. Tang and J. L. Wang, *J. Am. Chem. Soc.*, 1982, **104**, 292.
57. C. H. Heathcock, C. M. Tice and T. C. Gemroth, *J. Am. Chem. Soc.*, 1982, **104**, 6085.
58. B. A. McKittrick and B. Ganem, *Tetrahedron Lett.*, 1985, **26**, 4895.
59. M. Nakamura, N. Tsutsui, T. Takeda and T. Tokoroyama, *Tetrahedron Lett.*, 1984, **25**, 3231.
60. A. S. Narula, *Tetrahedron Lett.*, 1981, **22**, 2017.
61. A. B. McElroy and S. Warren, *Tetrahedron Lett.*, 1985, **26**, 2119.
62. M. D. Lewis and R. Menes, *Tetrahedron Lett.*, 1987, **28**, 5129.
63. H. Tomioka, T. Suzuki, K. Oshima and H. Nozaki, *Tetrahedron Lett.*, 1982, **23**, 3387.
64. M. R. Johnson and Y. Kishi, *Tetrahedron Lett.*, 1979, 4347.
65. M. Kitamura, M. Isobe, Y. Ichikawa and T. Goto, *J. Org. Chem.*, 1984, **49**, 3517.
66. Y. Oikawa, T. Nishi and O. Yonemitsu, *J. Chem. Soc., Perkin Trans. 1*, 1985, 7.

67. H. Kogen and T. Nishi, *J. Chem. Soc., Chem. Commun.*, 1987, 311.
68. Y. Kobayashi, H. Uchiyama, H. Kanbara and F. Sato, *J. Am. Chem. Soc.*, 1985, **107**, 5541.
69. W. R. Roush, J. A. Straub and R. J. Brown, *J. Org. Chem.*, 1987, **52**, 5127.
70. S. N. Ege, A. D. Adams, E. J. Gess, K. S. Ragone, B. J. Kober, M. B. Lampert, P. Umrigar, D. C. Lankin and G. W. Griffin, *J. Chem. Soc., Perkin Trans. 1*, 1983, 325.
71. J. E. Burdett, Jr., P. N. Rao, H. K. Kim, M. T. Karten and R. P. Blye, *J. Chem. Soc., Perkin Trans. 1*, 1982, 2877.
72. (a) A. W. Murray and R. G. Reid, *Synthesis*, 1985, 35; (b) S. Bowles, M. M. Campbell, M. Sainsbury and G. M. Davies, *Tetrahedron Lett.*, 1989, **30**, 3711.
73. G. Van Beek, J. L. Van Der Baan, G. W. Klumpp and F. Bickelhaupt, *Tetrahedron*, 1986, **42**, 5111.
74. R. K. Duke and R. W. Rickards, *J. Org. Chem.*, 1984, **49**, 1898.
75. E. J. Corey and W.-G. Su, *J. Am. Chem. Soc.*, 1987, **109**, 7534.
76. A. L. Baumstark and P. C. Vasquez, *J. Org. Chem.*, 1988, **53**, 3437.
77. (a) R. Mello, M. Fiorentino, O. Sciacovelli and R. Curci, *J. Org. Chem.*, 1988, **53**, 3890; (b) W. Adam, L. Hadjiarapoglou and X. Wang, *Tetrahedron Lett.*, 1989, **30**, 6497; (c) R. L. Halcomb and S. J. Danishefsky, *J. Am. Chem. Soc.*, 1989, **111**, 6661; (d) W. Adam, L. Hadjiarapoglou, V. Jager and B. Seidel, *Tetrahedron Lett.*, 1989, **30**, 4223; (e) S. Gopalakrishnan, M. P. Stone and T. M. Harris, *J. Am. Chem. Soc.*, 1989, **111**, 7232; (f) R. W. Murray, *Chem. Rev.*, 1989, **89**, 1187.
78. P. Brougham, M. S. Cooper, D. A. Cummerson, H. Heaney and N. Thompson, *Synthesis*, 1987, 1015.
79. J. G. Hill, B. E. Rossiter and K. B. Sharpless, *J. Org. Chem.*, 1983, **48**, 3607.
80. V. P. Rajan, S. N. Bannore, H. N. Subbarao and S. Dev, *Tetrahedron*, 1984, **40**, 983.
81. F. E. Ziegler, B. H. Jaynes and M. Saindane, *Tetrahedron Lett.*, 1985, **26**, 3307.
82. J. P. Marino, R. F. de la Pradilla and E. Laborde, *J. Org. Chem.*, 1987, **52**, 4898.
83. R. V. Stevens, J. H. Chang, R. Lapalme, S. Schow, M. G. Schlageter, R. Shapiro and H. N. Weller, *J. Am. Chem. Soc.*, 1983, **105**, 7719.
84. M. Rowley and Y. Kishi, *Tetrahedron Lett.*, 1988, **29**, 4909.
85. A. K. Singh, R. K. Bakshi and E. J. Corey, *J. Am. Chem. Soc.*, 1987, **109**, 6187.
86. Y. S. Sanghvi and A. S. Rao, *Indian J. Chem., Sect. B*, 1980, **19**, 608.
87. J. M. Luteijn and A. de Groot, *J. Org. Chem.*, 1981, **46**, 3448.
88. M. C. Pirrung and S. A. Thomson, *J. Org. Chem.*, 1988, **53**, 227.
89. E. J. Corey and B. De, *J. Am. Chem. Soc.*, 1984, **106**, 2735.
90. P. A. Grieco, Y. Ohfune, G. F. Majetich and C.-L. J. Wang, *J. Am. Chem. Soc.*, 1982, **104**, 4233.
91. (a) A. J. Pearson and S.-Y. Hsu, *J. Org. Chem.*, 1986, **51**, 2505; (b) L. A. Paquette, R. J. Ross and J. P. Springer, *J. Am. Chem. Soc.*, 1988, **110**, 6192.
92. E. D. Mihelich, K. Daniels and D. J. Eickhoff, *J. Am. Chem. Soc.*, 1981, **103**, 7690.
93. D. R. Mootoo and B. Fraser-Reid, *J. Org. Chem.*, 1987, **52**, 4511.
94. P. A. McCarthy, *Tetrahedron Lett.*, 1982, **23**, 4199.
95. E. J. Corey and J. P. Dittami, *J. Am. Chem. Soc.*, 1985, **107**, 256.
96. H. Hayami, M. Sato, S. Kanemoto, Y. Morizawa, K. Oshima and H. Nozaki, *J. Am. Chem. Soc.*, 1983, **105**, 4491.
97. J. R. Falck, S. Manna, A. K. Siddhanta, J. Capdevila and J. D. Buynak, *Tetrahedron Lett.*, 1983, **24**, 5715.
98. E. E. van Tamelen and E. J. Leopold, *Tetrahedron Lett.*, 1985, **26**, 3303.
99. T. Hanamoto, T. Katsuki and M. Yamaguchi, *Tetrahedron Lett.*, 1987, **28**, 6191.
100. Y. Kobayashi, H. Uchiyama, H. Kanbara and F. Sato, *J. Am. Chem. Soc.*, 1985, **107**, 5541.
101. S. Kanemoto, T. Nonaka, K. Oshima, K. Utimoto and H. Nozaki, *Tetrahedron Lett.*, 1986, **27**, 3387.
102. R. W. Irvine, R. A. Russell and R. N. Warrener, *Tetrahedron Lett.*, 1985, **26**, 6117.
103. T. Fukuyama, B. Vranesic, D. P. Negri and Y. Kishi, *Tetrahedron Lett.*, 1978, 2741.
104. I. Saito, T. Mano, R. Nagata and T. Matsuura, *Tetrahedron Lett.*, 1987, **28**, 1909.
105. D. Prat, B. Delpech and R. Lett, *Tetrahedron Lett.*, 1986, **27**, 711.
106. J. Prandi, H. B. Kagan and H. Mimoun, *Tetrahedron Lett.*, 1986, **27**, 2617.
107. C. Venturello and R. D'Aloisio, *J. Org. Chem.*, 1988, **53**, 1553.
108. C.-S. Chien, T. Kawasaki, M. Sakamoto, Y. Tamura and Y. Kita, *Chem. Pharm. Bull.*, 1985, **33**, 2743.
109. P. Battioni, J. F. Bartoli, P. Leduc, M. Fontecave and D. Mansuy, *J. Chem. Soc., Chem. Commun.*, 1987, 791.
110. J. D. Koola and J. K. Kochi, *J. Org. Chem.*, 1987, **52**, 4545.
111. R. Rohde, G. Neef, G. Sauer and R. Wiechert, *Tetrahedron Lett.*, 1985, **26**, 2069.
112. S. C. Sethi, A. D. Natu and M. S. Wadia, *Heterocycles*, 1982, **18**, 221.
113. Y. Sawaki and Y. Ogata, *J. Org. Chem.*, 1984, **49**, 3344.
114. (a) T. Muto, J. Umehara, H. Masumori, T. Miura and M. Kimura, *Chem. Pharm. Bull.*, 1985, **33**, 4749; (b) J.-C. Marchon and R. Ramasseul, *Synthesis*, 1989, 389; (c) W. Adam, M. Braun, A. Griesbeck, V. Lucchini, E. Staab and B. Will, *J. Am. Chem. Soc.*, 1989, **111**, 203.
115. N. K. Chaudhuri and T. J. Ball, *J. Org. Chem.*, 1982, **47**, 5196.
116. J. Quick, Y. Khandelwal, P. C. Meltzer and J. S. Weinberg, *J. Org. Chem.*, 1983, **48**, 5199.
117. V. Rautenstrauch, *J. Chem. Soc., Chem. Commun.*, 1978, 519.
118. Y. Tsunokawa, S. Iwasaki and S. Okuda, *Chem. Pharm. Bull.*, 1983, **31**, 4578.

3.2
Addition Reactions with Formation of Carbon–Oxygen Bonds: (ii) Asymmetric Methods of Epoxidation

ROY A. JOHNSON
The Upjohn Company, Kalamazoo, MI, USA

and

K. BARRY SHARPLESS
Massachusetts Institute of Technology, Cambridge, MA, USA

3.2.1 INTRODUCTION

Alkenes are found in abundance in the realm of organic molecules, either derived from natural sources or generated as products of the chemical industry. Epoxidation is one of the most useful oxidative transformations of these alkenes and the reagents that have been developed for this process have a high degree of selectivity for the alkenic bond. Epoxidation functionalizes two adjacent carbon atoms while simultaneously activating either of these carbons towards attack by nucleophiles. If the epoxide is unsymmetrically substituted, regioselectivity in the attack of a nucleophile on the oxirane ring will be observed. Only when very similar substituents are present on the epoxide will selectivity of nucleophilic attack be difficult to achieve. A further advantage of an epoxide as an electrophilic intermediate is the fact that competing elimination reactions are rendered stereoelectronically unfavorable by the constraints of the cyclic structure. Whereas nucleophilic substitution at secondary carbon in larger cyclic or in acyclic systems is accompanied by significant elimination and is impossible at tertiary carbon, substitution at secondary or tertiary centers of epoxides is relatively free of competing elimination processes.

With chemoselectivity available in epoxidation reagents and regioselectivity inherent in the opening reactions of many epoxides, there remains the challenge of achieving epoxidation with asymmetric induction. The development of peracids as a standard method for epoxidation (see Volume 7, Chapter 3.1) led to an initial attempt in 1965 by Henbest to achieve asymmetric epoxidation using homochiral (enantiomerically pure) percamphoric acid.[1] Asymmetric induction was observed but the enantiomeric excess (*ee*) was a disappointing 8%. In retrospect, one can see that the stereogenic center of the peracid is far removed from the electrophilic peroxygen and that a low degree of asymmetric induction should not be surprising for this reaction. A brief, but thorough, review of the fitful progress over the following 15 years towards the goal of a synthetically useful asymmetric epoxidation has been recorded elsewhere,[2] while selected highlights of newer methods are described in a later section of this chapter.

In 1980, Katsuki and Sharpless reported that with the unique combination of a titanium(IV) alkoxide, an optically active tartrate ester, and *t*-butyl hydroperoxide, they were able to carry out the epoxidation of a variety of allylic alcohols in good yield and with an enantiomeric excess generally greater than 90%.[3] Subsequent improvements in the reaction have been described[4] and the frequent use of the process as reported in the literature attest to its wide generality and utility. Since, to date, this method provides the most successful general solution to the problem of asymmetric epoxidation, the present chapter deals primarily with this titanium-catalyzed process. A variety of other reagents which yield asymmetric epoxides are summarized briefly. A new approach to asymmetric epoxides proceeding *via* diols, currently at an early stage of development, is also outlined.

The literature has been reviewed through 1989 for the purposes of preparing this chapter but the documentation herein is not intended to be comprehensive. Other reviews have covered various aspects of asymmetric epoxidation including synthetic applications through 1984,[5,6] a thorough compilation of uses through early 1987[7] and an extensive discussion of the mechanism of the reaction.[2] Use of homochiral epoxy alcohols in the synthesis of polyhydroxylated compounds, *e.g.* sugars,[8,9] and for the preparation of various synthetic intermediates has been reviewed.[10] A personal account of the discovery of titanium-catalyzed asymmetric epoxidation has been recorded.[11] A comprehensive review of titanium-catalyzed asymmetric epoxidation is planned.[12]

3.2.2 FUNDAMENTAL ELEMENTS OF TITANIUM TARTRATE CATALYZED ASYMMETRIC EPOXIDATION

The essence of titanium-catalyzed asymmetric epoxidation is illustrated in Figure 1. As shown there, the four essential components of the reaction are the allylic alcohol substrate, a titanium(IV) alkoxide, a chiral tartrate ester and an alkyl hydroperoxide. The asymmetric complex formed from these reagents delivers the peroxy oxygen to one face or the other of the allylic alcohol depending on the absolute configuration of the tartrate used. If D-(–)-tartrate is used, oxygen delivery will be from the top face of the allylic alcohol, when drawn in the orientation shown in Figure 1, and if L-(+)-tartrate is used, oxygen delivery will be from the bottom face. The enantioselectivity of this reaction approaches 100% as measured by the optical purities (% *ee*) of the epoxy alcohol products. An enantiomeric excess of 94%, a degree of optical purity attained in many of the epoxide products, reflects an enantioselectivity of 97:3 for epoxidation of one face of the allylic alcohol over the other.

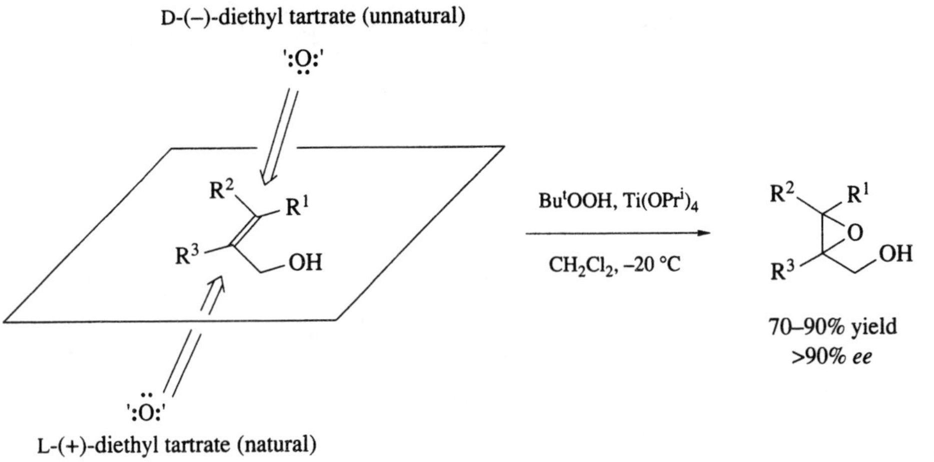

D-(–)-diethyl tartrate (unnatural)

Bu^tOOH, $Ti(OPr^i)_4$

CH_2Cl_2, –20 °C

70–90% yield
>90% *ee*

L-(+)-diethyl tartrate (natural)

Figure 1 Enantiofacial selectivity in the epoxidation of prochiral allylic alcohols with titanium/tartrate/TBHP

The enantioselectivity principles portrayed in Figure 1 have been followed without exception in all epoxidations of prochiral allylic alcohols reported to date. More than 300 prochiral allylic alcohols had been subjected to asymmetric epoxidation by the end of 1989. From this experience and a better understanding of the reaction mechanism, it is now safe to use the enantioselectivity principles portrayed in Figure 1 to assign absolute configurations to the epoxy alcohols prepared by the method. On the other hand, epoxidation of allylic alcohols with chiral substituents at C-1, C-2 and/or C-3 does not always follow these principles and assignment of absolute configuration to the products must be made with care. Even in the latter cases, reliable assignments can usually be made if the outcome (diastereomeric ratio) of epoxidation with both the (+)- and (–)-tartrate ester ligands is compared.

A structural variant of the allylic alcohol not shown in Figure 1 is encountered when a substituent is placed on the C-1 carbon, as illustrated in Figure 2. Such an allylic alcohol is a racemate (unless it has been previously resolved) in which one enantiomer will have the R group oriented in the direction of oxygen delivery, while the other enantiomer will have the R group oriented away from the direction of oxygen delivery. The enantioselective principles of asymmetric epoxidation remain in force for epoxidation of this type of substrate, but now oxygen is delivered at different rates to the two enantiomers depending on the orientation of the R group. Experimental results have shown that the difference in these rates is of sufficient magnitude that one enantiomer of the allylic alcohol will remain largely unoxidized while the other undergoes complete epoxidation, the net result being that a kinetic resolution of the enantiomers is achieved.[13] Experience has further shown that the slow-reacting enantiomer will always be the one having the R group oriented in the direction of oxygen delivery. For the example illustrated in Figure 2, the titanium/D-(–)-diethyl tartrate complex will deliver oxygen to the top face in preference to the bottom face of the substrate, in accordance with the rules implied in Figure 1, and this delivery will be more rapid when the R group is oriented toward the bottom face of the molecule. Opposite results will be obtained with the titanium/L-(+)-diethyl tartrate complex. Additional details for using this reaction in the kinetic resolution mode may be found in Section 3.2.5.9.

An important aspect of asymmetric epoxidation which is not apparent from Figures 1 and 2 is the fact that the allylic alcohol is coordinated to titanium as the alkoxide during the epoxidation process (see Section 3.2.6). Not only does this coordination play a key role in orientation of the allylic alcohol during the epoxidation process, but it also accounts for the selectivity of the process for allylic and homoallylic alcohols in preference to nearly all other alkenes. This effect is most clearly seen in comparison of allylic alcohols with the analogous allylic ethers. The latter are essentially unchanged by the $Ti(OR)_4$/tartrate/TBHP system during the same time required for epoxidation of the allylic alcohol. The $Ti(OR)_4$/tartrate/TBHP reagent thereby exhibits selectivity for allylic and homoallylic alcohols while being

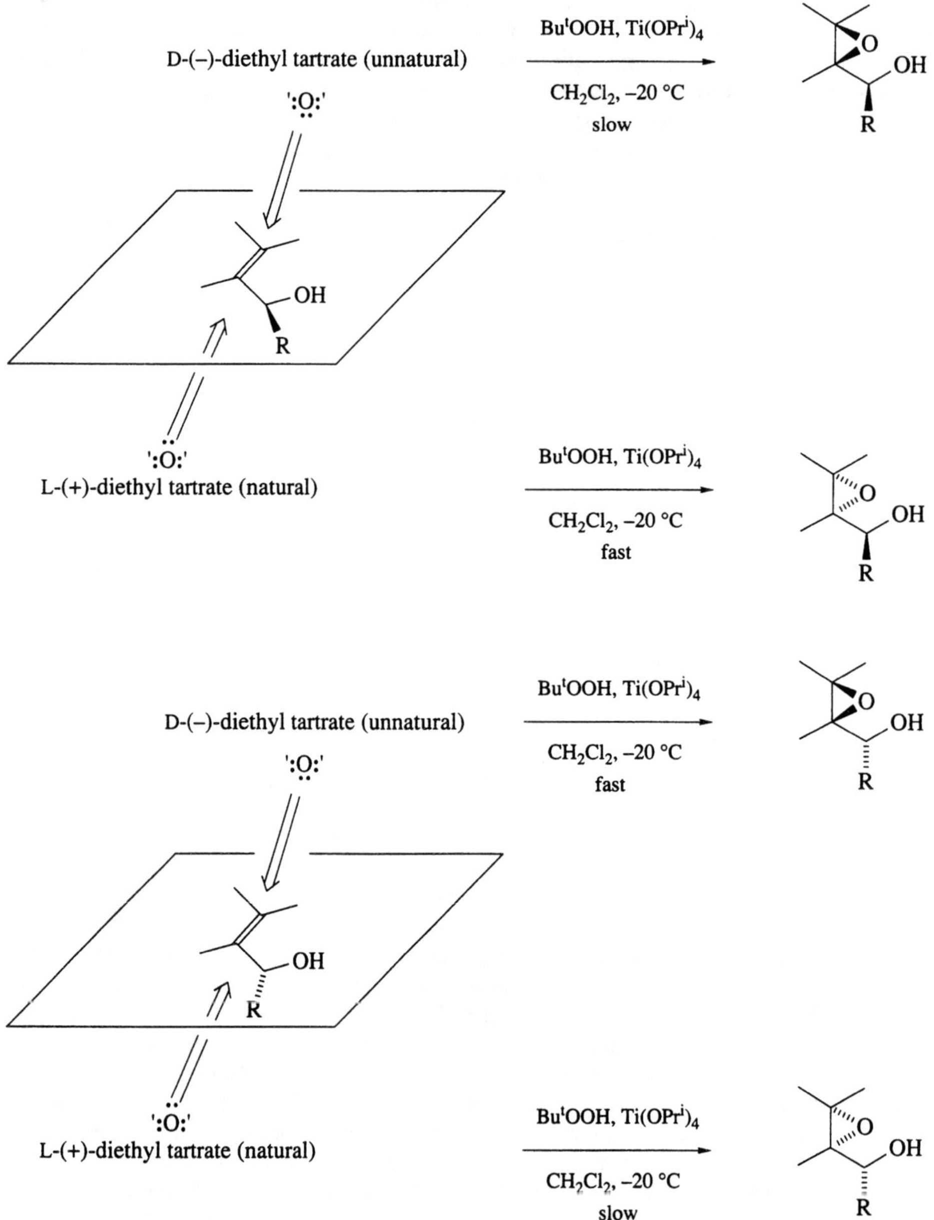

Figure 2 Diastereofacial selectivity in the epoxidation of C-1—substituted allylic alcohols with titanium/tartrate/TBNP

Table 1 Compatibility of Functional Groups with the Asymmetric Epoxidation Reaction

Compatible functional groups		*Incompatible groups*
Acetals, ketals	Ketones	Amines (most)
Alcohols (remote)	Nitriles	Carboxylic Acids
Aldehydes	Nitro	Thiols
Alkenes	Pyridines	Phenols (most)
Alkynes	Silyl ethers	Phosphines
Amides	Sulfones	
Azides	Sulfoxides	
Carboxylic esters	Tetrazoles	
Epoxides	Ureas	
Ethers	Urethanes	
Hydrazides		

compatible with other alkenic groups. Use of the reagent is compatible with many other functional groups as well (see Table 1).

An important improvement in the asymmetric epoxidation process is the finding, reported in 1986, that by adding molecular sieves to the reaction medium virtually all reactions can be performed with a catalytic amount (5–10 mol%) of the titanium tartrate complex.[14] Previously, only a few structural classes of allylic alcohols were efficiently epoxidized by less than stoichiometric amounts of the complex and most reactions were routinely performed with stoichiometric quantities of the reagent. The advantages of using a catalytic amount of complex include reagent economy, mildness of conditions, ease of isolation, increased yields and the potential for *in situ* derivatization of the product. However, there may be occasions where the use of stoichiometric quantities of the catalytic complex is necessary.

In situ derivatization of the crude epoxy alcohol product becomes a viable alternative to isolation when 5–10 mol % of catalyst is used for the epoxidation. This procedure is especially useful in those cases where the product is reactive or is difficult to isolate because of solubility in an aqueous extraction phase.[15,16] Low molecular weight epoxy alcohols, such as glycidol (see Section 3.2.5.1), are readily extracted from the reaction mixture after conversion to ester derivatives such as the 4-nitrobenzoate or 3-nitrobenzenesulfonate.[4,17] This derivatization not only facilitates isolation of the product but also preserves the epoxide in a synthetically useful form.

3.2.3 REACTION VARIABLES FOR TITANIUM TARTRATE CATALYZED ASYMMETRIC EPOXIDATION

This section presents a summary of the currently preferred conditions for performing titanium-catalyzed asymmetric epoxidations and is derived primarily from the detailed account of Gao *et al.*[4] We wish to draw the reader's attention to several aspects of the terminology used here and throughout this chapter. The terms titanium tartrate *complex* and titanium tartrate *catalyst* are used interchangeably. The term *stoichiometric reaction* refers to the use of the titanium tartrate complex in a stoichiometric ratio (100 mol %) relative to the substrate (allylic alcohol). The term *catalytic reaction* (or *quantity*) refers to the use of the titanium tartrate complex in a catalytic ratio (usually 5–10 mol %) relative to the substrate.

3.2.3.1 Stoichiometry

Two aspects of stoichiometry are important in an asymmetric epoxidation: one is the ratio of titanium to tartrate used for the catalyst and the other is the ratio of catalyst to substrate. With regard to the catalyst, it is crucial to obtaining the highest possible enantiomeric excess that at least a 10% excess of tartrate ester to titanium(IV) alkoxide be used in all asymmetric epoxidations. This is important when the reaction is being done with either a stoichiometric or a catalytic quantity of the complex. There appears to be no need to increase the excess of tartrate ester beyond 10–20% and, in fact, a larger excess has been shown to slow the epoxidation reaction unnecessarily.[4]

The second stoichiometry consideration is the ratio of catalyst to substrate. As noted in the previous section, virtually all asymmetric epoxidations can be performed with a catalytic amount of titanium tartrate complex if molecular sieves are added to the reaction milieu. A study of catalyst–substrate ratios in the epoxidation of cinnamyl alcohol revealed a significant loss in enantiomeric excess (Table 2) below the level of 5 mol % catalyst. At this catalyst level, the reaction rate also decreases with the consequence that incomplete epoxidation of the substrate may occur. Presently, the recommended catalyst stoichiometry is from 5% Ti/6% tartrate ester to 10% Ti/12% tartrate ester.[4]

Table 2 Dependence of Enantiomeric Excess on Catalyst Stoichiometry

Ph, OH; 3 Å sieves, TBHP; Ph, O, OH

Entry	*Ti(OPr^i)$_4$* (mol %)	*(+)-DIPT* (mol %)	*Enantiomeric excess* (%)
1	5.0	6.0	92
2	4.0	5.2	87
3	2.0	2.5	69

3.2.3.2 Concentration

The concentration of substrate used in the asymmetric epoxidation must be given consideration because competing side reactions may increase with increased reagent concentration. The use of catalytic quantities of the titanium tartrate complex has greatly reduced this problem. The epoxidation of most substrates under catalytic conditions may be performed at a substrate concentration up to 1 M. By contrast, epoxidations using stoichiometric amounts of complex are best run at substrate concentrations of 0.1 M or lower. Even with catalytic amounts of the complex, a concentration of 0.1 M may be maximal for substrates, such as cinnamyl alcohol, which produce sensitive epoxy alcohol products.[4]

3.2.3.3 Preparation and Aging of the Catalyst

Proper preparation of the catalyst is essential for optimal reaction rates and enantioselectivity. The preparation and storage of stock solutions of the titanium tartrate catalyst should not be attempted as the complex is not sufficiently stable for long term storage. Best results are obtained when the catalyst is prepared by mixing the titanium(IV) alkoxide and the tartrate in a solvent at –20 °C, adding either TBHP *or* the allylic alcohol, and aging the system at this temperature for 20–30 min. This aging period is critical to the success of the reaction and must not be eliminated. On the rare occasion that a bulky titanium(IV) alkoxide such as the *t*-butoxide is used, the aging period should be increased to 1 h.[18] After the aging period, the temperature is adjusted to the desired level and the last reagent, either the allylic alcohol or the hydroperoxide, is added.

3.2.3.4 Oxidant and Epoxidation Solvent

t-Butyl hydroperoxide (TBHP) is used as the oxidant for nearly all titanium-catalyzed asymmetric epoxidations. Exceptions are for allyl alcohol and methallyl alcohol, where cumyl hydroperoxide is used to advantage for the epoxidation.[4] Cumyl hydroperoxide can be used for other epoxidations and is reported to result in slightly faster reaction rates than are observed with TBHP.[4] Trityl hydroperoxide also can serve as an effective replacement for TBHP.[2] TBHP is generally preferred, however, since product isolation is significantly easier when this oxidant is used. The most economical source of TBHP is the commercially available 70% solution in water, in which case steps must be taken to obtain anhydrous material. The detailed instructions for obtaining dry solutions of TBHP have been published elsewhere.[4,14] For smaller laboratory scale reactions, anhydrous solutions of TBHP in 2,2,4-trimethylpentane (isooctane) are available commercially. Storage of TBHP solutions over molecular sieves is not recommended, but brief drying over sieves (*ca.* 30 min) of the required amount of the solution just before use is good practice.

Since the preparation and storage of stock quantities of TBHP is a convenient way in which to deal with this reagent, compatibility with solvent is essential. Much care has gone into finding the optimum solvent for storage of TBHP and recommendations have changed as additional experience has been gained. The current solvent of choice is isooctane with the favored alternatives being dichloromethane or toluene.[4] Dichloroethane should not be used.[19] Dichloromethane solutions of TBHP require storage at 0 °C and toluene solutions occasionally develop a contaminant which inhibits the catalytic reaction. Due to safety considerations (chance of slight pressurization), high density polyethylene bottles are preferred over glass bottles for storage of TBHP solutions. However, both dichloromethane and toluene, but not isooctane, permeate through such bottles with the result that the concentration of the contents slowly changes with time. If the published instructions[4,14] for preparation of anhydrous TBHP in isooctane are followed, a relatively concentrated solution (5–6 M) is obtained. Aliquots of this solution are briefly dried over sieves and added directly to the epoxidation reaction without concern for removal of the isooctane. The use of dilute solutions of TBHP in isooctane (*e.g.* 3 M TBHP is too dilute) should be avoided since the additional isooctane involved in transfer will have an inhibitory effect on the rate of epoxidation and can lead to solubility problems with some substrates. Solutions of 5.5 M TBHP in isooctane now are available commercially.

For the asymmetric epoxidation reaction, dry, alcohol-free dichloromethane (the use of dichloromethane stabilized with methanol must be avoided) is usually the solvent of choice since it is inert to the reagents, has good solvent power for the components of the reaction and supports good epoxidation rates. A fortunate consequence of the asymmetric epoxidation process is that ligation of the allylic alcohol to the titanium center aids in solubilization of the substrate. Substrates that normally may only be modestly soluble in the above-mentioned solvents will be brought into solution as they complex with the titanium tartrate catalyst.

3.2.3.5 Tartrate Esters

Optically active tartrate esters are the source of chirality for the asymmetric epoxidation process. The esters used conventionally are dimethyl (DMT), diethyl (DET) and diisopropyl tartrate (DIPT), and, with a few subtle exceptions, all are equally effective at inducing asymmetry during the crucial epoxidation event. The minor exceptions that have been noted include: (i) a slight improvement in enantioselectivity (from 93% to 95% *ee*) when changing from DIPT to DET in the epoxidation of (*E*)-monosubstituted allylic alcohols such as (*E*)-2-hexen-1-ol (having only a primary alkyl chain at C-3); and (ii) a higher product yield (but no change in enantiomeric excess) when changing from DET to DIPT in the epoxidation of allyl alcohol.[4] Other subtle variations such as these may exist but their discovery awaits execution of the appropriate comparative experiments. If optimal conditions are desired for a specific asymmetric epoxidation, then variation of the tartrate ester is likely to be a useful exercise.

In the kinetic resolution of chiral 1-substituted allylic alcohols, there clearly is benefit to be gained in the choice of tartrate ester used for the reaction. In these reactions (see Section 3.2.5.9), the efficiency of kinetic resolution increases as the size of the tartrate alkyl ester group increases. Data for DMT, DET and DIPT are summarized in Table 8,[2] and the trend shown there continues with the use of the crystalline dicyclohexyl and dicyclododecyl tartrates.[4]

The nonconventional tartrate esters (**1**) to (**3**) have been used to probe the mechanism of the asymmetric epoxidation process.[20a] These chain-linked bis(tartrate) molecules when complexed with 2 equiv. of $Ti(OBu^t)_4$ catalyze asymmetric epoxidation with good enantiofacial selectivity. A number of tartrate-like ligands have been studied as potential chiral auxiliaries in the asymmetric epoxidation and kinetic resolution processes.[2,20b] Although on occasion a ligand has been found that has the capability to induce high enantioselectivity into selected substrates (see Section 3.2.7.3), none has exhibited the broad scope of effectiveness seen with the tartrate esters.

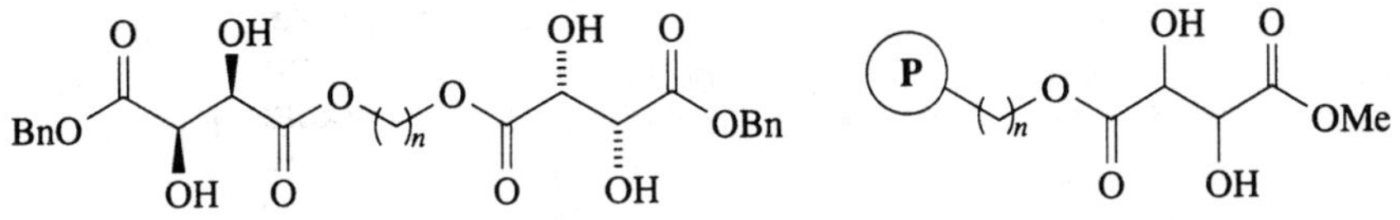

(**1**) $n = 3$
(**2**) $n = 4$
(**3**) $n = 5$

(**4**) $n = 1$
(**5**) $n = 2$

Polymer-linked tartrate esters have been prepared and used for asymmetric epoxidation in efforts to simplify reaction work-up procedures and to allow recycling of the chiral tartrate.[21] The tartrates are linked through an ester bond to either a hydroxymethyl or a hydroxyethyl group on the polymer backbone to form (**4**) or (**5**), respectively. Epoxidation catalysts were prepared from these polymer-linked tartrates by combination with 0.5 equiv. of $Ti(OPr^i)_4$, based on the weight of tartrate ester which had been added to the polymer. Epoxidation of geraniol with (**4**) or (**5**) gave the epoxy alcohol with enantiomeric excesses of 49% and 65%, respectively. Recycling of the polymer-linked tartrate was possible but the subsequent epoxidation suffered from significant loss in enantiomeric excess.[21]

3.2.3.6 Titanium Alkoxides

Titanium(IV) isopropoxide (*Chemical Abstracts* nomenclature: 2-propanol, titanium(4+) salt) is the titanium species of choice for preparation of the titanium tartrate complex in the asymmetric epoxidation process. The use of titanium(IV) *t*-butoxide has been recommended for those reactions in which the epoxy alcohol product is particularily sensitive to ring opening by the alkoxide.[18] The 2-substituted epoxy alcohols (Section 3.2.5.2) are one such class of compounds. Ring opening by *t*-butoxide is much slower than by isopropoxide. With the reduced amount of catalyst that now is needed for all asymmetric epoxidations, the use of $Ti(OBu^t)_4$ appears to be unnecessary in most cases, but the concept is worth noting.

3.2.3.7 Molecular Sieves

The addition of activated molecular sieves (zeolites) to the asymmetric epoxidation milieu has the beneficial effect that virtually all reactions can be carried out with only 5–10 mol % of the titanium tartrate catalyst.[4,14] Without molecular sieves, only a few of the more reactive allylic alcohols are epoxidized efficiently with less than an equivalent of the catalyst. The role of the molecular sieves is thought to be protection of the catalyst from adventitious water and water that may be generated in small amounts by side reactions during the epoxidation process.

There are several important guidelines to be followed in using activated molecular sieves for the asymmetric epoxidation process.[4] Stock solutions of TBHP should not be stored over molecular sieves (the sieves catalyze slow decomposition of TBHP), but the amount of TBHP solution required for a reaction should be placed over sieves briefly (10–60 min) before use. Likewise, neither the tartrate ester nor the titanium(IV) isopropoxide should be stored over sieves. Addition of the sieves at the time of mixing the tartrate ester with the $Ti(OPr^i)_4$ followed by the normal aging of the catalyst is sufficient to dry these reagents, provided they initially are of good quality (see ref. 4, p. 5771). The use of powdered, activated 4 Å molecular sieves is preferred and they are commercially available in preactivated form. Also effective are 3 Å, 4 Å and 5 Å molecular sieves in pellet form. In the case of allyl alcohol only 3 Å sieves are effective since this substrate is small enough to be sequestered by 4 Å or 5 Å sieves. Unactivated sieves can be activated by heating at 200 °C under high vacuum for at least 3 h.

3.2.4 SOURCES OF ALLYLIC ALCOHOLS

One of the amenities of present day organic synthesis is the availability of intermediates from the many chemical supply companies. Over 100 allylic alcohols (excluding extensive listings of phorbol esters and prostaglandin structures) are offered for sale from these sources. Two concerns about such supplies should be noted, the first being the desirability to check the (*E*)/(*Z*) composition of acyclic allylic alcohols when this is not specified, and the second is to check the optical purity of those allylic alcohols offered in optically active form.

When the allylic alcohol needed for asymmetric epoxidation is unavailable from a commercial source, reasonably general synthetic routes have been developed to allylic alcohols of several different substitution patterns. Good methods are available for the preparation of 3-substituted allylic alcohols, whereas synthesis of 2-substituted allylic alcohols is more problematic. 1-Substituted allylic alcohols, the substrates for kinetic resolution, frequently can be derived by addition of alkenyl or alkynyl organometallic reagents to aldehydes followed by modification of the resulting product as required.

The Horner–Emmons addition of dialkyl alkoxycarbonylmethylenephosphonates to aldehydes[22] has been widely used to generate α,β-unsaturated esters which, in turn, can be reduced to allylic alcohols. Under the original conditions of the Horner–Emmons reaction, the stereochemistry of the α,β-unsaturated ester is predominantly *trans*, and therefore the *trans*-allylic alcohol is obtained upon reduction. Still and Gennari have introduced an important modification of the Horner–Emmons reaction which shifts the stereochemistry of the α,β-unsaturated ester to predominantly *cis*.[23] Diisobutylaluminum hydride (DIBAL) has frequently been used for the reduction of the alkoxycarbonyl group to the primary alcohol functionality. The aldehyde needed for reaction with the Horner–Emmons reagent may be derived *via* Swern oxidation[24] of a primary alcohol. The net result is that one frequently sees the reaction sequence shown in equation (1) used for the preparation of (3*E*)- and (3*Z*)-allylic alcohols.

R⌄OH —[$(COCl)_2$, DMSO, Et_3N]→ RCHO → (*E*)-R–CH=CH–CO_2R^1 —[DIBAL-H]→ (*E*)-R–CH=CH–CH_2OH; RCHO → (*Z*)-R–CH=CH–CO_2R —[DIBAL-H]→ (*Z*)-R–CH=CH–CH_2OH (1)

The propargylic alcohol group may be exploited as an allylic alcohol precursor (equation 2) and may be generated by nucleophilic addition to an electrophile[25] or by addition of a formaldehyde equivalent to a preexisting terminal alkyne group.[26] Once in place, reduction of the propargylic alcohol with lithium aluminum hydride or, preferably, with sodium bis(2-methoxyethoxy)aluminum hydride (Red-Al)[27] will produce the *trans*-allylic alcohol. Alternately, catalytic reduction over Lindlar catalyst can be used to obtain the *cis*-allylic alcohol.[28] The addition of other lithium alkynides to ketones produces chiral secondary alcohols which also can be reduced by the preceding methods to the *cis*- or *trans*-allylic alcohols. Additional synthetic approaches to allylic alcohols may be found in the various references cited in this chapter.

RX + ≡—OR → R—≡—OH ← R—≡ + 'H$_2$CO'

Red-Al → R—CH=CH—CH$_2$OH (*trans*)

H$_2$, Lindlar → R—CH=CH—CH$_2$OH (*cis*) (2)

3.2.5 ASYMMETRIC EPOXIDATIONS BY SUBSTRATE STRUCTURE

The scope of allylic alcohol structures which are subject to asymmetric epoxidation was foreshadowed in the first report of this reaction. Examples of nearly all of the possible substitution patterns have been shown to be epoxidized in good yield and with high enantiofacial selectivity.[3] The numerous results that have appeared since the initial report have confirmed and extended the scope of the structures that have been epoxidized. This section of the chapter is intended to illustrate the structural scope without being exhaustive in coverage of the literature. Examples have been chosen as much as possible from those reports in the literature which provide experimentally determined yield and enantiomeric excess data. When there are limitations to the structural scope, as reflected by lower enantiofacial selectivity, these cases are noted. The results presented in this section are divided according to the substitution patterns of the allylic alcohol substrates. This organization is intended to provide easy access to precedent when the synthetic chemist is contemplating asymmetric epoxidation of a new substrate.

Before commencing, the attention of the reader is drawn to our usage of the terms enantiofacial selectivity and diastereoselectivity. The usage in this chapter does not conform to the strictest possible definitions of these terms. In particular, enantiofacial selectivity is used with reference to the selection and delivery of oxygen by the epoxidation catalyst to one face of the alkene in preference to the other. This usage extends to chiral allylic alcohols (primarily the 1-substituted allylic alcohols) when the focus of the discussion is on face selection in the epoxidation process. Diastereoselectivity is used in the discussion of kinetic resolution, when the generation of diastereomeric compounds is emphasized.

3.2.5.1 Allyl Alcohol

Glyceraldehyde derivatives,[29] asymmetrically substituted glycerol[30] and glycidol[31] are three-carbon molecules which, especially in their optically active forms, find widespread use in organic synthesis. In the past, the source of these compounds in optically active form has been almost exclusively from the degradation of natural products such as mannitol. Efficient, multistep routes from the natural products provide access to either enantiomer of these three-carbon compounds. Since the discovery of asymmetric epoxidation in 1980, the potential has existed for a convenient one-step synthesis of optically active glycidol (**7**) from allyl alcohol (**6**).[3] However, because glycidol is one of the more sensitive epoxy alcohols to ring-opening reactions and also is a water soluble molecule, isolation from the stoichiometric asymmetric epoxidation is difficult and very little glycidol has been prepared in this way. Now with the use of catalytic epoxidation in the presence of molecular sieves, it is possible to isolate optically active glycidol of 88–92% *ee* in yields of 50–60%.[4] As a result of these improvements both enantiomers of glycidol are now available commercially.

An attractive alternative to isolation of glycidol is *in situ* derivatization of the crude product during work-up.[15] Two distinct applications of this method have been described. In the first, ring opening of glycidol (*R*)-(**7**) with a nucleophile such as sodium 1-naphthoxide produces an intermediate (**8**) that can be carried on to useful products, *e.g.* for the synthesis of β-adrenergic blocking agents,[15a] antidepress-

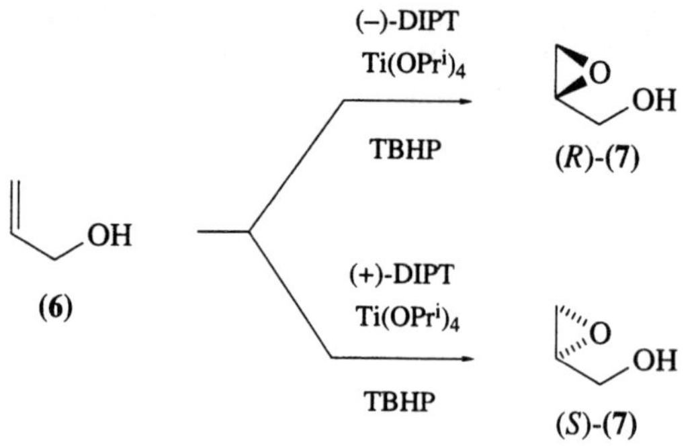

(–)-DIPT
$Ti(OPr^i)_4$
TBHP
(*R*)-(**7**)

(**6**)

(+)-DIPT
$Ti(OPr^i)_4$
TBHP
(*S*)-(**7**)

ants[32] and so on. In the second, esterification of the hydroxy group of glycidol improves the extraction of the glycidol moiety from the reaction mixture and at the same time generates a synthon in which all three carbon centers are differentiated for further reaction. Another benefit is that with certain derivatives, such as the 3-nitrobenzenesulfonate ester (**9**), recrystallization can be used to upgrade the optical purity to >99% *ee*.[4,16a]

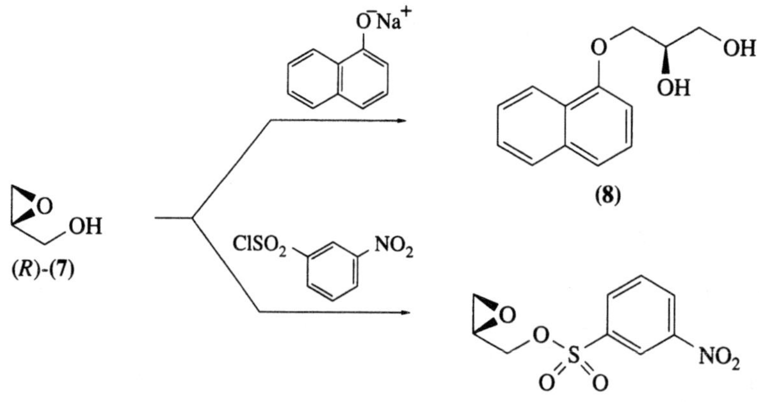

(*R*)-(**7**)

(**8**)

(**9**)

As an industrial process, production of optically active glycidol is at an early stage of development with additional improvements and economies certain to occur. As a chemical intermediate, optically active glycidol is the most versatile epoxy alcohol prepared by asymmetric epoxidation and is poised for exploitation in organic synthesis.[16]

3.2.5.2 2-Substituted Allyl Alcohols

The epoxides (**11**) derived from 2-substituted allylic alcohols (**10**) are particularly susceptible to nucleophilic attack at C-3, a reaction that is promoted by titanium(IV) species.[18] When stoichiometric amounts of titanium tartrate complex are used in these epoxidations considerable product is lost *via* opening of the epoxide before it can be isolated from the reaction. The primary nucleophilic culprit is the isopropoxide ligand of the $Ti(OPr^i)_4$. The use of $Ti(OBu^t)_4$ in place of $Ti(OPr^i)_4$ has been prescribed as a means to reduce this problem (the *t*-butoxide being a poorer nucleophile).[18] Fortunately, a better solution now exists in the form of the catalytic version of the reaction which uses only 5–10 mol % of titanium tartrate complex and greatly reduces the amount of epoxide ring opening. Some comparisons of results from reactions run under the two sets of conditions are possible from the epoxidations summarized in Table 3.

The prototype for this structural class is 2-methyl-2-propen-1-ol (methallyl alcohol) from which asymmetric epoxidation generates optically active 2-methyloxiranemethanol. Like glycidol, 2-methyloxiranemethanol has been difficult to obtain by stoichiometric asymmetric epoxidation, but with the use of the catalytic version reasonable quantities now are produced[4] and the compound has become commercially available. *In situ* derivatization also can be used to recover this epoxy alcohol from the epoxidation reaction. Progress in the isolation of 2-methyloxiranemethanol is reflected in entries 1–3 of Table 3, and

(–)-DIPT
$Ti(OPr^i)_4$
TBHP
(R)-(11)

(10)

(+)-DIPT
$Ti(OPr^i)_4$
TBHP
(S)-(11)

Table 3 Epoxides from 2-Substituted Allylic Alcohols

Entry	Epoxide R^1	R^2	Catalyst (% Ti/% tartrate)	Tartrate	Yield (%)	Enantiomeric excess (%)	Ref.
1	H	Me	100/100[a]	(+)-DET	—	85	33
2	H	Me	27/27	(–)-DET	32	94	34
3	H	Me	7.6/10[a]	(–)-DET	47	>95	35
4	PNB	Me	5/6	(+)-DIPT	78	92 (98)[b]	4
5	Tos	Me	5/6	(+)-DIPT	69	95	4
6	Nps[c]	Me	5/6	(+)-DIPT	60	(92)[b]	4
7	H	Pr^n	4.7/5.9	(+)-DET	88	95	4
8	H	*n*-Nonyl	100/110	(+)-DET	53	>96	36
9	H	*n*-Tetradecyl	100/110[a]	(+)-DET	51	95	18
10	H	*n*-Tetradecyl	10/13	(+)-DET	91	96	4
11	H	Pr^i	65/120	(+)-DET	56	86	37
12	H	Bu^t	120/150[a]	(+)-DET	42	86	38
13	H	Cyclohexyl	100/100	(+)-DET	81	>95	3
14	H	$—CH_2OBn$	7.6/10[a]	(–)-DET	74	>95	35

[a]$Ti(OBu^t)_4$ used in this reaction. [b]Enantiomeric excess in parentheses is after recrystallization. [c]Nps = 2-Naphthalenesulfonyl.

the results of *in situ* derivatization are revealed by entries 4–6. The optical purity of 2-methyloxiranemethanol produced in this way is good (92–95%) and improvement to 98% *ee* is observed after recrystallization of the 4-nitrobenzoate derivative.

Several other allylic alcohols with primary C-2 substituents have been epoxidized with good results (Table 3, entries 7–10 and 14). Epoxy alcohols have been obtained with 95–96% *ee* and when the catalytic version of the reaction is used, as in Table 3, entry 10, the yield is excellent. When the C-2 substituent is more highly branched, as in entries 11–13, there may be some interference to high enantiofacial selectivity by the bulky group, since the *ee* in two cases (entries 11 and 12) is 86%. Another example which supports this possibility of steric interference to selective epoxidation is summarized in equation (3).[39] In this case, the optically active allylic alcohol (**12**) was subjected to epoxidation with both antipodes of the titanium tartrate catalyst. With (+)-DIPT enantiofacial selectivity was 96:4 ('matched pair'),[40a] but with (–)-DIPT selectivity fell to only 1:3 ('mismatched pair'), a further indication that a secondary C-2 substituent can perturb the fit of the substrate to the active catalyst species. In the epoxidation of the allylic alcohol shown in equation (4), the epoxy alcohol is obtained in 96% yield and with a 14:1 ratio of enantiofacial selectivity.[40b] An interesting alternate route to the epoxide of entry 12 (Table 3) has been described, in which 2-*t*-butylpropene is first converted to an allylic hydroperoxide *via* photooxygenation and then, in the presence of the titanium tartrate catalyst, undergoes asymmetric epoxidation (79%

$Ti(OPr^i)_4$ / TBHP

(**12**) (+)-DIPT 94:4
(–)-DIPT 1:3 (3)

yield, 72% *ee*).[38b] The intermediate hydroperoxide serves as the source of oxygen for the epoxidation step.

(−)-DET, $Ti(OPr^i)_4$, TBHP (4)

3.2.5.3 (3*E*)-Substituted Allyl Alcohols

Several factors contribute to the frequent use of (3*E*)-substituted allylic alcohols (**13**) for asymmetric epoxidation. The allylic alcohols are easily prepared, conversion to epoxy alcohol normally proceeding with good chemical yields and with >95% *ee*, and a large variety of functionality in the (3*E*)-position is tolerated by the epoxidation catalyst. Representative epoxy alcohols (**14**) are summarized in Table 4 and Figure 3, with results divided arbitrarily according to whether the (3*E*) substituent is a hydrocarbon

(**13**) — (−)-DIPT, $Ti(OPr^i)_4$, TBHP → (2*R*,3*R*)-(**14**); (+)-DIPT, $Ti(OPr^i)_4$, TBHP → (2*S*,3*S*)-(**14**)

Table 4 Epoxides from (3*E*)-Substituted Allylic Alcohols (Hydrocarbon Substituents)

Entry	Epoxide R	Catalyst (mol % Ti/% tartrate)	Tartrate	Epoxide configuration (2,3)	Yield (%)	Enantiomeric excess (%)	Ref.
1	Me	100/100	(−)-DIPT	(*R,R*)	40–58	95	41–43
2	Me	5/6	(+)-DIPT	(*S,S*)	70	92	4
3	Et	stoichiometric	(−)-DIPT	(*R,R*)	80	>95	44
4	Pr^n	100/104	(+)-DET	(*S,S*)	64	93	45
5	Pr^i	5/6	(+)-DET	(*S,S*)	85	94	4
6	Pr^i	100/104	(+)-DET	(*S,S*)	66	98	45
7	Bu^s	—	(−)-DET	(*R,R*)	a	a	46
8	Bu^t	120/150	(+)-DET	(*S,S*)	52	>95	38
9	CH_2=CH	5/6	(+)-DIPT	(*S,S*)	56	>91	47
10	MeCH=$CHCH_2$	8/10	(+)-DET	(*S,S*)	81	a	48
11	n-C_5H_{11}	a	(+)-DET	(*S,S*)	78	95	49
12	CH_2=$CH(CH_2)_3$	100/100	(+)-DET	(*S,S*)	80	>95	41
13	n-C_7H_{15}	5/7.3	(+)-DET	(*S,S*)	99	96	4
14	n-C_8H_{17}	5/6	(+)-DET	(*S,S*)	78	94	4
15	EtCH=$CHCH_2$CH=$CHCH_2$	a	(+)-DET	(*S,S*)	82	>95	46
16	EtC≡CCH_2C≡CCH_2	a	(−)-DET	(*R,R*)	76	a	50
17	n-$C_{10}H_{21}$	100/100	(+)-DET	(*S,S*)	79	>95	3
18	n-$C_{12}H_{25}$	a	(+)-DET	(*S,S*)	a	a	51
19	$C_{14}H_{29}$	a	(+)-DIPT	(*S,S*)	77	a	52
20	$C_{15}H_{31}$	120/160	(−)-DET	(*R,R*)	88	>95	53

[a]Not reported.

(Table 4) or otherwise (Figure 3). The versatility of these and other 3-substituted epoxy alcohols for organic synthesis is illustrated with several examples in the following discussion.

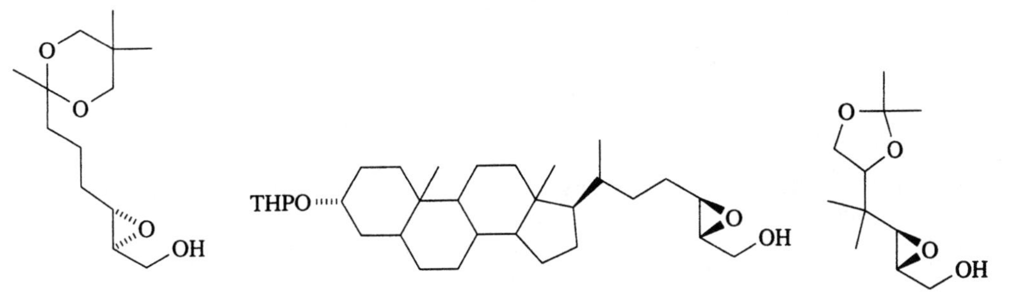

70%, 92% *ee* (ref. 54) 72%, >95% *ee* (ref. 55) 70%, >95% *ee* (ref. 56)

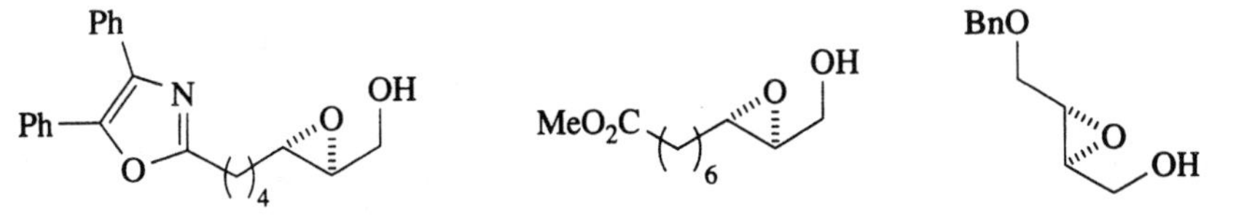

66%, >95% *ee* (ref. 57) 69%, 93% *ee* (ref. 58) 85%, 98% *ee* (ref. 59)

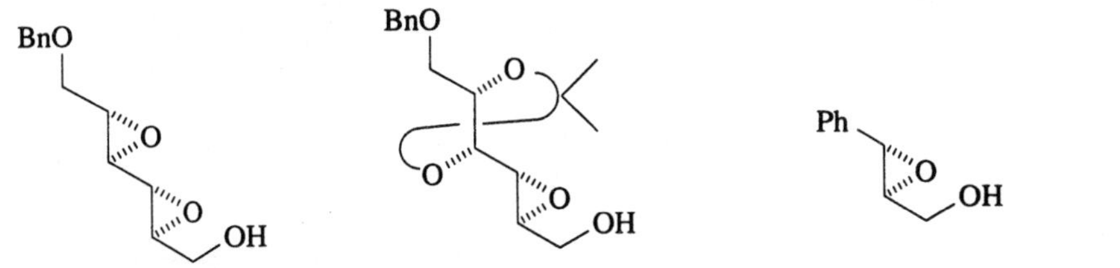

76%, 98% *ee* (ref. 60) 70%, >95% *ee* (ref. 59) 98% *ee* (refs. 4, 61a and 61b)

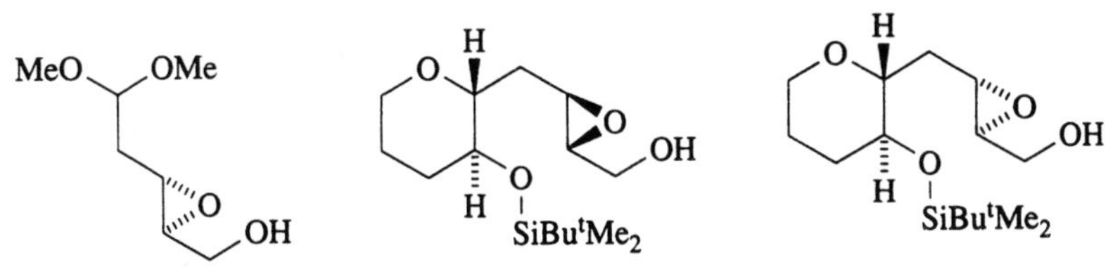

81%, >97% *ee* (ref. 61c) 92%, 95% *ee* (ref. 61d) 78%, 83% *ee* (ref. 61d)

Figure 3 Epoxy alcohols from asymmetric epoxidation of (3*E*)-monosubstituted allylic alcohols

Compatibility of asymmetric epoxidation with acetals, ketals, ethers and esters has led to extensive use of allylic alcohols containing these groups in the synthesis of polyoxygenated natural products. An example which illustrates one such synthetic approach is the asymmetric epoxidation of (**15**), an allylic alcohol derived from (*S*)-glyceraldehyde acetonide.[59,62] In the epoxy alcohol (**16**) obtained from (**15**), each carbon of the five-carbon chain is oxygenated and all stereochemistry has been controlled. The structural relationship of (**16**) to the pentoses is evident, and methods leading to these carbohydrates have been described.[59,62a] This synthetic methodology has been extended by the development of an efficient series of reactions that can transform one allylic alcohol into a second which is two carbons longer than the first. Repetition of the reaction sequence can, in principle, be continued to any desired chain length.

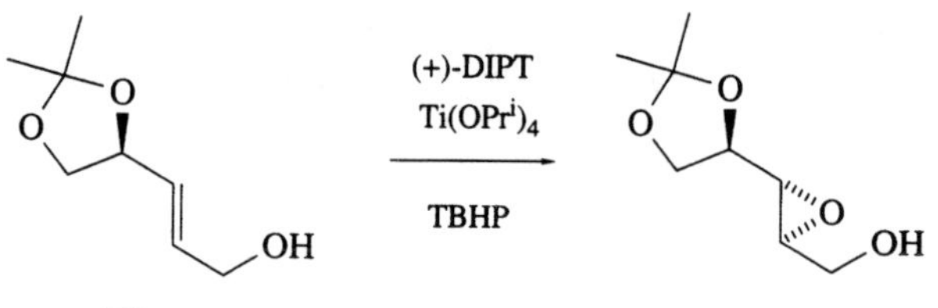

(+)-DIPT, $Ti(OPr^i)_4$, TBHP: (15) → (16)

The key steps in this 'reiterative two-carbon extension cycle' are illustrated in Scheme 1, which shows a synthetic route leading from an achiral alkoxyacetaldehyde (**17**) to L-allose, one of the eight possible L-hexoses.[63] In practice, all eight L-hexoses were synthesized by taking advantage of branch points in the sequence and by using both antipodes of the titanium tartrate catalyst to generate epimeric epoxides. The sequence of reactions begins with the two-carbon, benzyloxyacetaldeyde (**17**), which can be converted to the four-carbon intermediate (**18**) by means of a Wittig reaction. In the actual synthesis, intermediate (**18**) was the starting point and was obtained by an alternative method. The carboxylic acid ester of (**18**) is reduced with DIBAL to the (3*E*)-allylic alcohol (**19**) which, by asymmetric epoxidation, is converted to epoxy alcohol (**20**). Base-catalyzed rearrangement (Payne rearrangement) of (**20**) establishes the

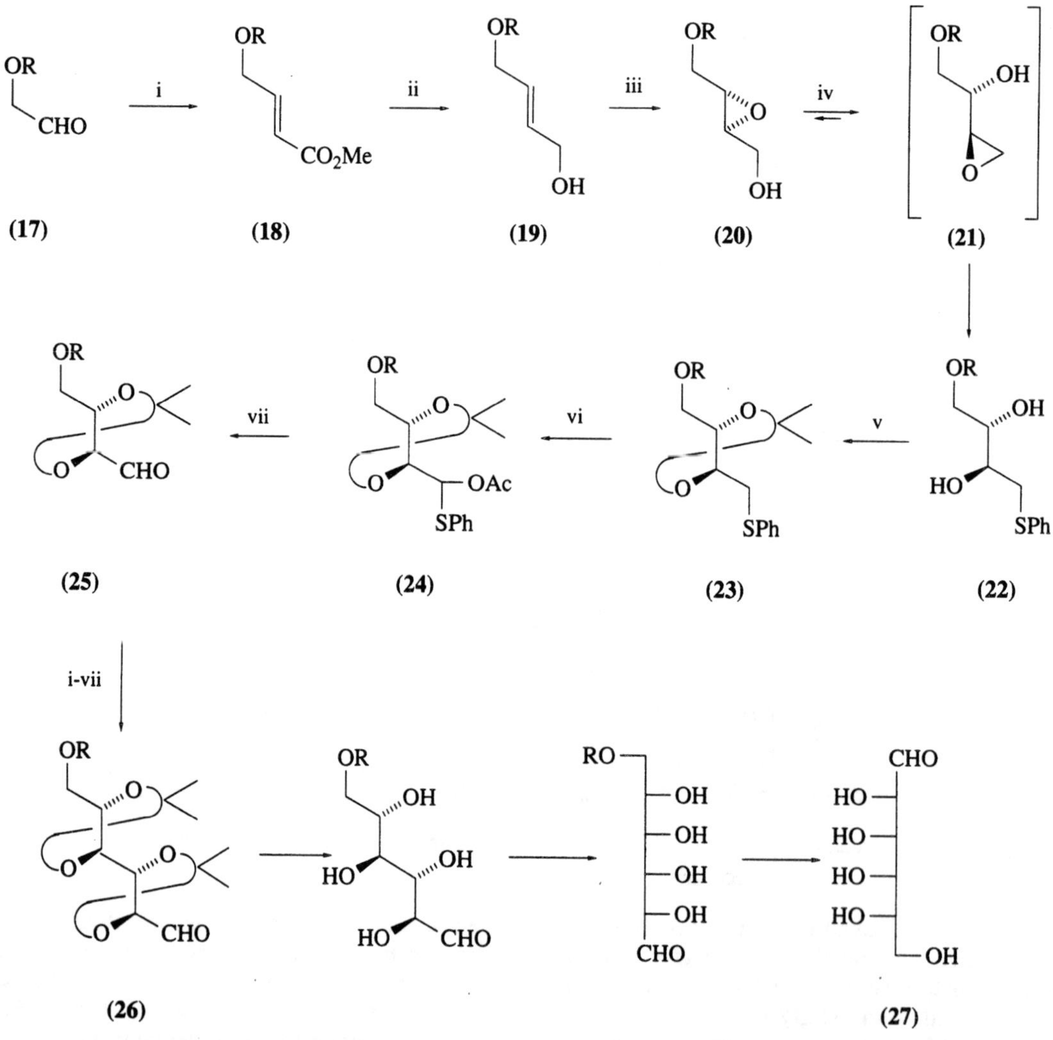

(17) (18) (19) (20) (21)

(25) (24) (23) (22)

(26) (27)

i, $(MeO)_3P(O)CHCO_2Me$; ii DIBAL; iii, (+)-DET/$Ti(OPr^i)_4$; iv, $PhSH/OH^-$; v, $MeC(=CH_2)OMe/H^+$; vi, MCPBA, Ac_2O/AcO^-; vii, DIBAL (1 equiv.), H_2O

Scheme 1

equilibrium shown between (**20**) and the 1,2-epoxy alcohol (**21**). Benzenethiolate reacts regioselectively to open the 1,2-epoxide leading to the dihydroxy sulfide (**22**). The diol is protected by conversion to the acetonide (**23**) which, upon oxidation of the sulfide to a sulfoxide followed by Pummerer rearrangement, is converted to the acetoxythioacetal (**24**). Reduction of the latter (**24**) with one equivalent of DIBAL produces aldehyde (**25**). At this point the synthetic sequence can be branched by converting a portion of the aldehyde (**25**) to the epimeric aldehyde (not shown) by epimerization with potassium carbonate in methanol. Both of these new aldehydes can now be chain extended by repeating steps (i–vii), which in the case of (**25**) leads to the hexose derivative (**26**). In order to obtain all eight hexoses, a further branching during the second cycle is initiated at step (iii), with part of the material (an allylic alcohol) being subjected to asymmetric epoxidation with (–)-DET. Both of these branches are carried on through step (vi) or step (vii), thereby producing all eight L-hexose derivatives. Deprotection of the derivatives completes the synthesis, as shown for L-allose (**27**) in Scheme 1.

1,2-Epoxy-3-alcohols can be derived from 2,3-epoxy-1-alcohols by the base-catalyzed Payne rearrangement, as illustrated in step (iv) of Scheme 1.[59,64] The rearrangement is completely stereospecific but, since it is reversible, it usually results in an equilibrium mixture of the two epoxy alcohols for which the relative proportions are structure dependent. Practical synthetic applications of this rearrangement therefore depend on methods that will shift the equilibrium completely in the direction desired. Nucleophiles such as thiolates and amines are sufficiently selective to react preferentially at C-1 of the 1,2-epoxy-3-alcohol and thereby shift the equilibrium completely in that direction. However, many other nucleophiles are incompatible with the reaction conditions required for the Payne rearrangement and the approach of trapping the 1,2-epoxide cannot be used. To circumvent this problem and increase the scope of the Payne rearrangement/opening process, methods have been developed that lead to isolation of the terminal 1,2-epoxy-3-alcohols.[10,65]

One method uses the 2,3-diol-1-sulfide (**30**) produced by thiolate trapping of the 1,2-epoxide from the Payne rearrangement equilibrium between (**28**) and (**29**).[10,65] The sulfide is alkylated with Me_3OBF_4 in order to produce a good leaving group in (**31**). Then, base-promoted ring closure gives the 1,2-epoxide (**32**) in complete preference to formation of any 1,3-oxetane. The *erythro*-epoxy alcohol precursor (**31**) requires sodium hydride as the base in order to avoid reversal of the Payne rearrangement back to the starting 2,3-epoxy alcohol (**28**). The analogous *threo*-epoxy alcohol precursor can be closed with sodium hydroxide.

R, O, OH (**28**) ⇌ R, OH, O (**29**) —Bu^tS^-→ R, OH, HO, SBu^t (**30**)

(**30**) —$Me_3\overset{+}{O}\ BF_4^-$→ R, OH, HO, $\overset{+}{S}$, Bu^t (**31**) —NaH, CH_2Cl_2→ R, OH, O (**32**)

In a second method, the 2,3-epoxy-1-alcohol (**28**) is first converted to a mesylate (or a tosylate) and then the epoxide is opened hydrolytically with inversion at C-3 to give the dihydroxymesylate (**33**). A slight loss of optical purity has been observed in this process and is due to lack of complete regioselectivity for C-3 opening. Mild base is sufficient to effect ring closure of the dihydroxymesylate (**33**) to give the 1,2-epoxide (**34**).[10,65] The two methods are complementary in terms of stereochemistry such that if a 2,3-epoxy alcohol of the same absolute configuration is used to start each sequence, then the *erythro*-1,2-epoxy-3-ols produced will have opposite configurations at C-2 and C-3. This results from the fact that during the Payne rearrangement inversion occurs at C-2, while in the epoxymesylate opening, inversion occurs at C-3. Detailed discussions of these Payne rearrangement processes, as well as of further synthetic transformations of the 1,2-epoxy alcohols, have been presented elsewhere.[10,65]

When two allylic alcohols are contained in a symmetrical molecule, asymmetric epoxidation proceeds with interesting consequences for stereochemical purity. The results were first described for the asymme-

(28) (33) (34)

tric epoxidation of (2*Z*,6*E*,10*Z*)-dodeca-2,6,10-trien-1,12-diol (**35**).[66] The first epoxidation of (**35**) produces the major and minor enantiomers (**36**) and (**37**). Since the stereogenic centers in these compounds are remote from the second allylic alcohol, each enantiomer undergoes a second epoxidation with essentially the same enantiofacial selectivity as in the first epoxidation. Three bis(epoxide) products result; (**38a**), a *meso* compound (**39**) and (**38b**) (mirror image of **38a**). The overall consequence is that most of the epoxidation resulting from the undesired enantiofacial attack leads to the *meso* compound (**39**) which is in principle separable from the major product. Very little of the mirror image compound (**38b**) is formed and therefore the enantiomeric purity of the major product will be very high. In the example cited, enantiomeric purity could not be determined directly but was calculated according to the expression $(A_1 + B_1)(A_2 + B_2)$, where A_1 and B_1 are the enantiofacial selectivities of the first epoxidation and A_2 and B_2 are the enantiofacial selectivities of the second epoxidation. In the example being discussed an enantiofacial selectivity of 19:1 (90% *ee*) was assumed for both steps. The ratio of the three products therefore should be (19 + 1)(19 + 1), or 361:38:1, and the enantiomeric excess of (**38a**) should be 99.45%.

(35) (36) (37)

(38a) (39) (38b)

Fortunately, a wide variety of functionality is compatible with the titanium tartrate catalyst (see Table 1), but the judicious placement of functional groups relative to the allylic alcohol can lead to further desirable reactions following epoxidation. For example, in (**40**), asymmetric epoxidation of the allylic alcohol is followed by intramolecular cyclization under the reaction conditions to give the tetrahydrofuran (**41**).[67] Likewise, in the epoxidation of (**42**) cyclization of the intermediate epoxy alcohol occurs under the reaction conditions and leads to the cyclic urethane (**43**).[68] Titanium(IV) isopropoxide is an effective reagent for promoting the regioselective attack by nucleophiles at the 3-position of 2,3-epoxy alcohols,[69] 2,3-epoxy acids[70] and 2,3-epoxyamides.[70] This process has been proposed to involve coordination to the metal center in the bidentate manner shown for a 2,3-epoxy alcohol in structure (**44**). Such titanium-assisted nucleophilic opening of epoxides is thought to play a role in the *in situ* reactions leading to (**41**) and (**43**).

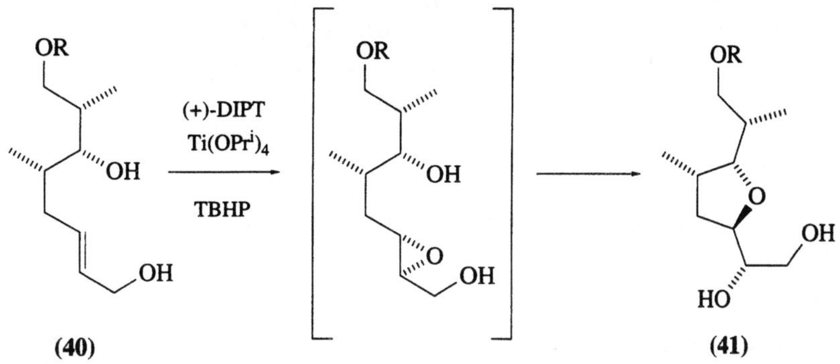

(+)-DIPT, $Ti(OPr^i)_4$, TBHP

(**40**) (**41**)

(+)-DIPT, $Ti(OPr^i)_4$, TBHP

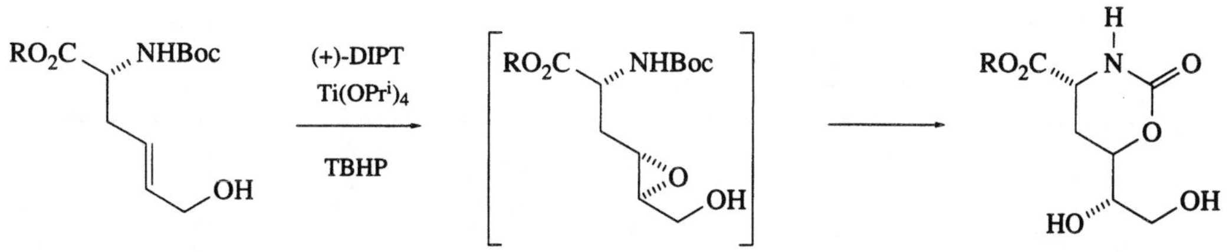

(**42**) (**43**)

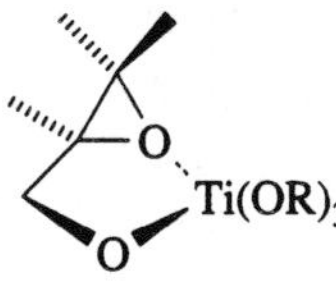

(**44**)

3.2.5.4 (3*Z*)-Monosubstituted Allyl Alcohols

Allylic alcohols having a *cis*-3-substituent (**45**) are the slowest to be epoxidized and give the most variable enantiofacial selectivities. Both of these characteristics suggest that allylic alcohols of this structure have the poorest fit to the requirements of the active epoxidation catalyst. Nevertheless, asymmetric epoxidation of these substrates is still effective and in most cases gives an optical purity of at least 80% *ee* and often as high as 95% *ee*. Patience with the slower reaction rate usually is rewarded with chemical yields of epoxy alcohols comparable to those obtained with other allylic alcohols. A number of representative examples are collected in Table 5.

There is a rough correlation between the enantiomeric excess observed for these epoxy alcohols and the steric complexity at the α-carbon of the C-3 substituent. When the C-3 substituent is a primary group (Table 5, entries 1, 2, 4, 6–12 and 19–21), enantiofacial selectivity is highest and enantiomeric excesses of 80–95% are observed for these compounds. When the substituent is secondary (entries 3 and 15–18) or tertiary (entry 5), enantiofacial selectivity is much more variable. When the substituent is asymmetric, enantiofacial selectivity depends on the absolute configuration, as is evident in comparison of entry 15 with 16 and of 17 with 18 in Table 5. Epoxidation of these chiral allylic alcohols with one antipode of catalyst yields moderate to good diastereoselectivity, while with the other antipode diastereoselectivity is virtually lacking.

(−)-DIPT
$Ti(OPr^i)_4$
TBHP
(2*R*,3*S*)-**(46)**

(45)

(+)-DIPT
$Ti(OPr^i)_4$
TBHP
(2*S*,3*R*)-**(46)**

Table 5 Epoxides from (3*Z*)-Substituted Allylic Alcohols

3*R* 2*S* 3*S* 2*R*

Entry	Epoxide R	Catalyst (mol % Ti/% tartrate)	Tartrate	Epoxide configuration (2,3)	Yield (%)	Enantiomeric excess (%)	Ref.
1	Me	5/6	(+)-DIPT	(*S*,*R*)	68	92	4
2	Et	a	(+)-DET	(*S*,*R*)	60	80	71
3	Pr^i	a	(+)-DET	(*S*,*R*)	54	66	72
4	Bu^i	a	(+)-DET	(*S*,*R*)	80	95	73
5	Bu^t	120/150	(+)-DET	(*S*,*R*)	77	25	38
6	n-C_9H_{19}	100/100	(−)-DET	(*R*,*S*)	80	91	3
7	n-C_7H_{15}	10/14	(+)-DET	(*S*,*R*)	74	86	4
8	n-C_8H_{17}	5/7.4	(+)-DIPT	(*S*,*R*)	63	>80	4
9	$CH_2CH{=}CHC_5H_{11}$	110/110	(+)-DMT	(*S*,*R*)	70	94	74
10	$(CH_2)_3CO_2Me$	a	(+)-DET	(*S*,*R*)	57	95	75
11	CH_2OBn	100/100	(−)-DET	(*R*,*S*)	84	92	59
12	CH_2OBn	14/14	b	—	—	95	4
13	Ph	100/120	(+)-DET	(*S*,*R*)	61	78	76
14	CHMePh	a	(+)-DIPT	(*S*,*R*)	a	a	77
15	(2,2-dimethyl-1,3-dioxolan-4-yl)	100/100	(+)-DET	(*S*,*R*)	55, 57	93, 84	59, 62a
16	as entry 15	100/100	(−)-DET	(*R*,*S*)	a	20	62a
17	$CHMeCH_2OBn$	100/100	(+)-DET	(*S*,*R*)	a	66	78a
18	as entry 17	100/100	(−)-DET	(*R*,*S*)	a	0	78a
19	CH_2CH_2OBn	a	(+)-DIPT	(*S*,*R*)	75	92	78b
20	n-$C_{11}H_{23}$	120/130	(+)-DIPT	(*S*,*R*)	83	92	78c
21	$(CH_2CH{=}CH)_2CH{=}CH_2$	100/148	(+)-DET	(*S*,*R*)	59	89	78d

[a]Not reported. [b]See ref. 4, footnote 9, for this entry.

3.2.5.5 (2,3*E*)-Disubstituted Allyl Alcohols

Extensive use in synthesis has been made of the asymmetric epoxidation of (2,3*E*)-disubstituted allylic alcohols. With few exceptions enantiofacial selectivity is excellent as reflected by enantiomeric excesses in the range of 90–95%. The results for a number of epoxidations of allylic alcohols with smaller substituents are collected in Table 6, while a variety of other compounds with larger groups are illustrated by structures **(47)** to **(60)**.

The epoxy alcohol **(47)** is a squalene oxide analog which has been used to examine substrate specificity in enzymatic cyclizations by baker's yeast.[85] The epoxy alcohol **(48)** provided an optically active intermediate used in the synthesis of 3,6-epoxyauraptene and marmine,[86] and epoxy alcohol **(49)** served as an intermediate in the synthesis of the antibiotic virantmycin.[87] In the synthesis of the three stilbene

Table 6 Epoxides from (2,3*E*)-Disubstituted Allylic Alcohols

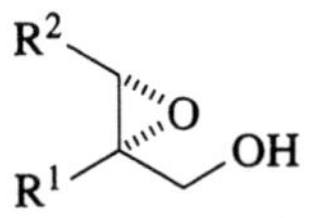

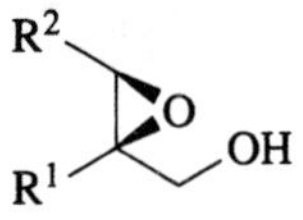

Entry	Epoxide R^1	R^2	Tartrate	Epoxide configuration	Yield (%)	Enantiomeric excess (%)	Ref.
1	Me	Me	(+)-DET	(2*S*,3*S*)	77	94	61b
2	Me	Et	(+)-DMT	(2*S*,3*S*)	79	95	41, 78e
3	—$(CH_2)_3$—		(+)-DET	(2*S*,3*S*)	38	>95	79
4	—$(CH_2)_4$—		(+)-DET	(2*S*,3*S*)	77	93	4
5	Me	CH_2OBn	(–)-DIPT	(2*R*,3*R*)	87	90	80
6	Me	$CHMeCH_2OBn$	—	—	93	>95	81
7	Me	Ph	(+)-DIPT	(2*S*,3*S*)	79	>98[a]	4
8	Ph	Ph	(+)-DET	(2*S*,3*S*)	70	>95	3, 4
9	Me	$CH_2CH_2CH{=}CH_2$	(+)-DET	(2*S*,3*S*)	71	96	82
10	Me	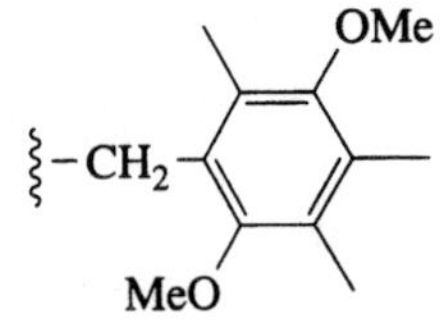 CH_2-(2,5-dimethoxy-3,4,6-trimethylphenyl)	(–)-DET	(2*R*,3*R*)	87	>95	83
11	Me	$CH_2CH{=}CMe_2$	(+)-DET	(2*S*,3*S*)	64	>90	84a
12	Me	$CH_2CH{=}CMe_2$	(–)-DET	(2*R*,3*R*)	59	>91	84a
13	Me	$(CH_2)_3OSiMe_2Bu^t$	(–)-DET	(2*R*,3*R*)	89	93	84b

[a]Enantiomeric excess after crystallization.

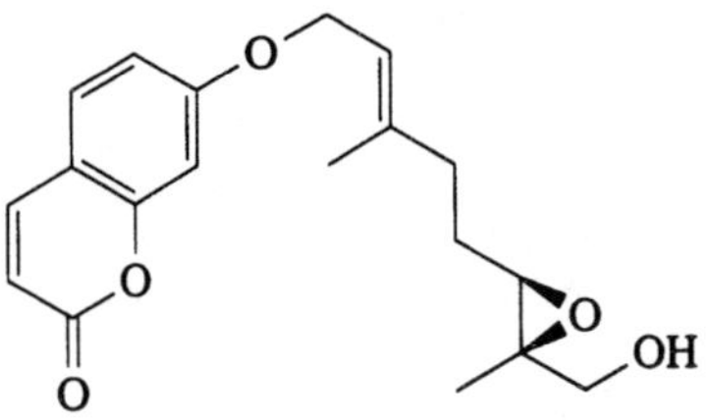

(48)

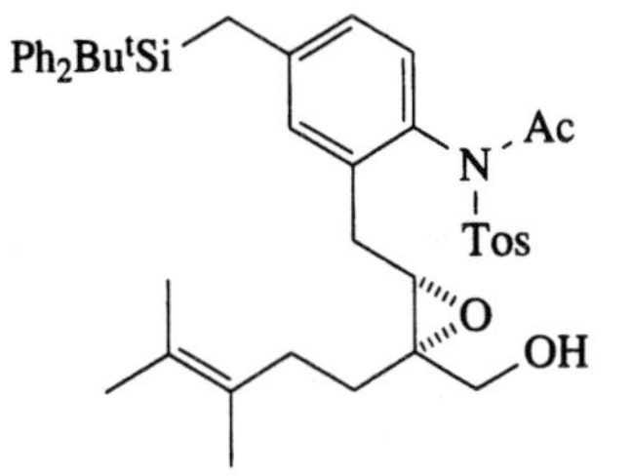

(49)

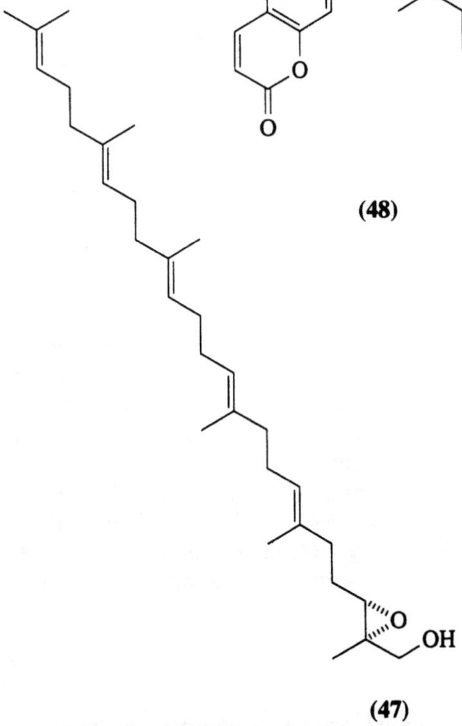

(47)

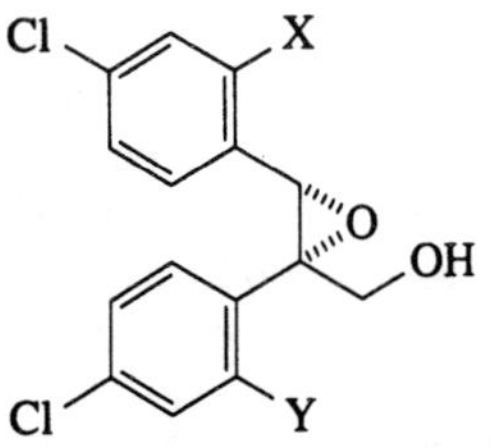

	X	Y	Enantiomeric excess (%)
(50)	H	H	>90
(51)	Cl	H	>90
(52)	Cl	Cl	70

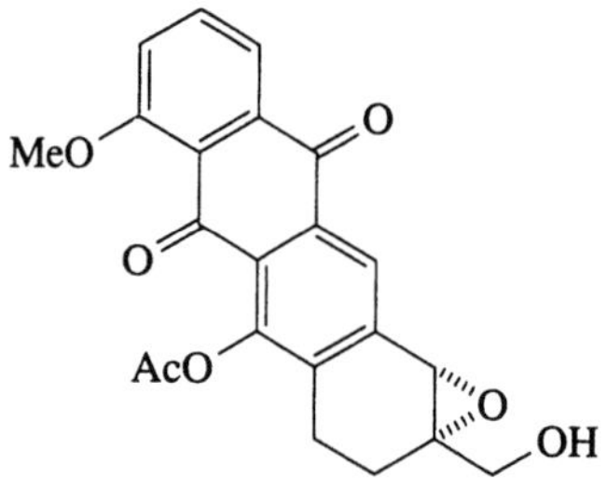

(52a) 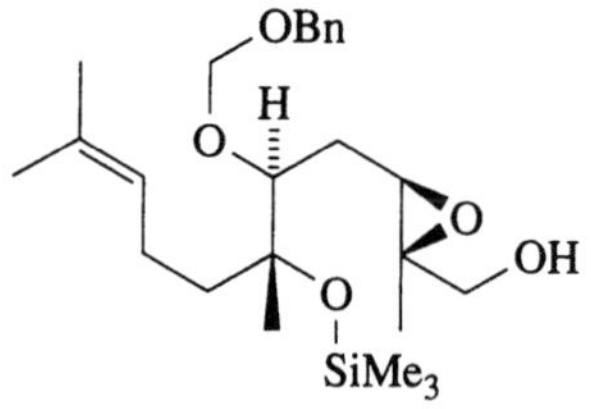 (52b)

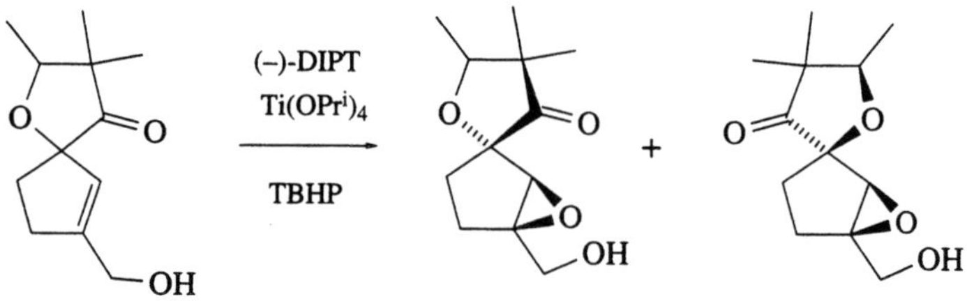

(–)-DIPT, Ti(OPr^i)$_4$, TBHP

(53) (54) (55)

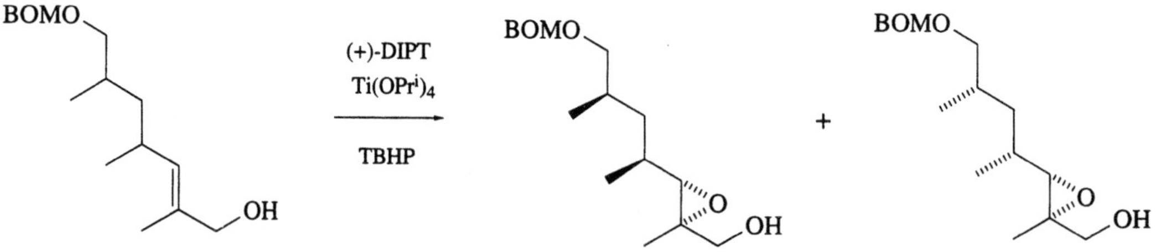

(+)-DIPT, Ti(OPr^i)$_4$, TBHP

(56) (57) (58)

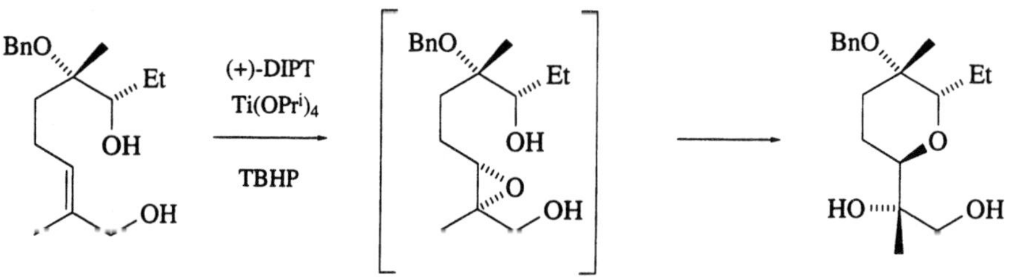

(+)-DIPT, Ti(OPr^i)$_4$, TBHP

(59) (60)

oxides (**50**) to (**52**), the presence of an *ortho* chloro group in the 2-phenyl ring resulted in a lower enantiomeric excess (70%) when compared to the analogs without this chlorine substituent.[88a] The very efficient (80% yield, 96% *ee*) formation of (**52a**) by asymmetric epoxidation of the allylic alcohol precursor offers a synthetic entry to optically active 11-deoxyanthracyclinones,[88b] while epoxy alcohol (**52b**) is one of several examples of asymmetric epoxidation used in the synthesis of brevitoxin precursors.[88c] Diastereomeric epoxy alcohols (**54**) and (**55**) are obtained in combined 90% yield (>95% *ee* each) from epoxidation of the racemic alcohol (**53**).[89] Diastereomeric epoxy alcohols (**57**) and (**58**) also are obtained with high optical purity in the epoxidation of (**56**).[44] The epoxy alcohol obtained from substrate (**59**) undergoes further intramolecular cyclization with stereospecific formation of the cyclic ether (**60**).[90]

3.2.5.6 (2,3*Z*)-Disubstituted Allyl Alcohols

A limited number of allylic alcohols of this type have been subjected to asymmetric epoxidation. With one exception, the C-2 substituent in these substrates has been a methyl group, the exception being a *t*-butyl group.[38] The (3*Z*)-substituents have been more varied and are illustrated by structures (**61**) to (**64**), which show the epoxy alcohols derived from the corresponding allylic alcohol substrates. Epoxidation of (*Z*)-2-methyl-2-hepten-1-ol gave epoxy alcohol (**61**) in 80% yield, 89% *ee*,[3] while (*Z*)-2-methyl-4-phenyl-2-buten-1-ol gave (**62**) in 90% yield, 91% *ee*,[77] and (*Z*)-1-hydroxysqualene gave (**63**) in 93% yield, 78% *ee*.[85] The epoxy alcohol (**64**) was obtained with >95% *ee* after recrystallization.[91] In the epoxidation of (*Z*)-2-*t*-butyl-2-buten-1-ol, the allylic alcohol with a C-2 *t*-butyl group, the epoxy alcohol was obtained in 43% yield and with 60% *ee*.[38] These results lead one to expect that other (2,3*Z*)-disubstituted allylic alcohols will be epoxidized in good yield and with enantioselectivity similar to that observed for the (3*Z*)-monosubstituted allylic alcohols (*i.e.* 80–95% *ee*).

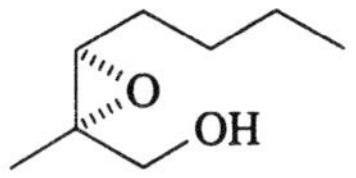

O OH

(61)

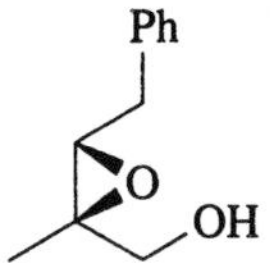

Ph O OH

(62)

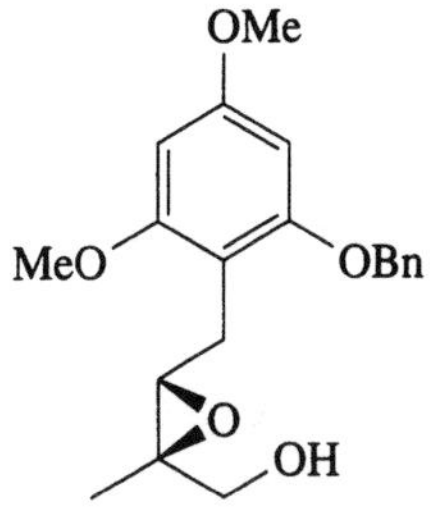

OMe MeO OBn O OH

(64)

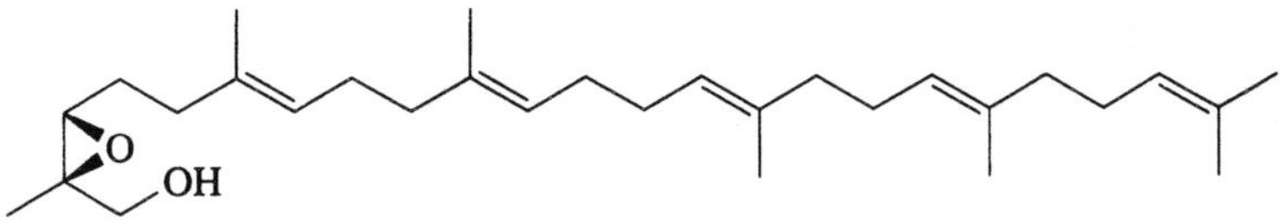

O OH

(63)

3.2.5.7 3,3-Disubstituted Allyl Alcohols

These substrates combine a (3*E*)-substituent with a (3*Z*)-substituent in the same molecule. Allylic alcohols with only a (3*E*)-substituent generally are epoxidized with excellent enantioselectivity, whereas those with only a (3*Z*)-substituent are epoxidized with enantioselectivity in the range of 80–95% *ee*. In the combination many of the reported examples have a methyl substituent at the (3*Z*)-position and all of these are epoxidized with an enantiomeric excess of 90–95% (Table 7, entries 1–4 and 6). Only a limited number of examples with larger groups at the (3*Z*)-position have been reported (entries 5 and 7–12) and in these the enantiomeric excesses are in the range 84–94%.

3,3-Dimethylallyl alcohol was epoxidized with >90% *ee* (Table 7, entry 1) but in low yield when a stoichiometric amount of the titanium tartrate complex was used. However, when a catalytic amount of the complex was used and *in situ* derivatization employed, the *p*-nitrobenzoate (>98% *ee* after recrystallization) and *p*-toluenesulfonate (93% *ee*) were isolated in yields of 70% and 55%, respectively. Likewise, the epoxidation of geraniol with a stoichiometric amount of the complex gave the epoxide (Table 7, entry 3) in 77% yield (95% *ee*) which was improved to 95% yield (91% *ee*) when a catalytic amount of complex was used (entry 4).

3.2.5.8 2,3,3-Trisubstituted Allyl Alcohols

Interesting structural diversity is present in the limited examples of trisubstituted allyl alcohols (equivalent to tetrasubstituted alkenes) to which asymmetric epoxidation has been applied. The epoxides (**65**) to (**70**) have been obtained from the corresponding allylic alcohols with yield and enantiomeric excess as indicated when such data have been reported. The lower enantiomeric excess observed for epoxy alcohol (**69**) may result from disruption of the catalyst structure by the phenolic groups or from alternate modes of binding of substrate to catalyst, again because of the phenolic groups.[102] Phenols bind strongly to titanium(IV), which may account for the fact that a large excess (six equivalents) of the titanium tartrate complex was required to achieve the yield and enantiomeric excess reported in the case of (**69**).

Table 7 Epoxides from 3,3-Disubstituted Allylic Alcohols

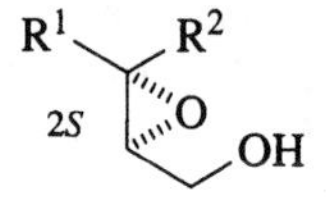

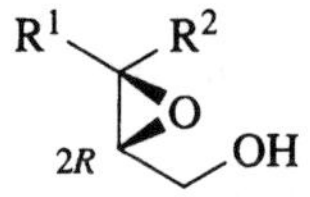

Entry	Epoxide R^1	R^2	Catalyst (mol% Ti/% tartrate)	Tartrate	Epoxide configuration	Yield (%)	ee (%)	Ref.
1	Me	Me	100/100	(–)-DBT	(2*R*)	25	>90	92
2	$CH_2CH{=}CMe_2$	Me	200/200	(+)-DET	(2*S*,3*S*)	67	95	93
3	$(CH_2)_2CH{=}CMe_2$	Me	100/100	(+)-DET	(2*S*,3*S*)	77	95	3
4	$(CH_2)_2CH{=}CMe_2$	Me	5/7.4	(+)-DET	(2*S*,3*S*)	95	91	4
5	Me	$(CH_2)_2CH{=}CMe_2$	100/100	(+)-DET	(2*S*,3*R*)	79	94	3
6	$(CH_2)_2OSiMe_2Bu^t$	Me	105/157	(–)-DET	(2*R*,3*R*)	81	>95	94
7	Me	$(CH_2)_2OSiMe_2Bu^t$	a	(+)-DET	(2*S*,3*R*)	98	90	95
8	Me	$(CH_2)_2OSiMe_2Bu^t$	a	(–)-DET	(2*R*,3*S*)	98	86	95
9	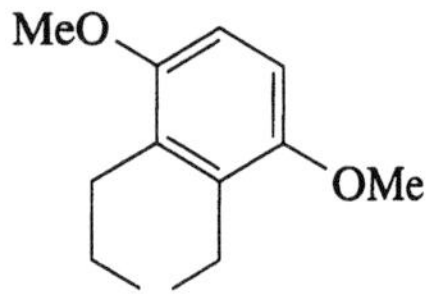 MeO, OMe (structure)		10/15	(+)-DET	(2*S*,3*R*)	97	93	96
10	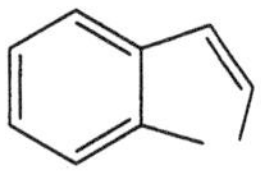 (structure)		100/110	(+)-DET	(2*S*,3*S*)	a	84	97a
11	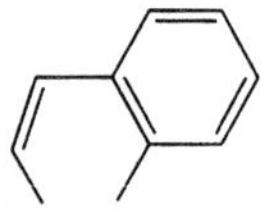 (structure)		100/110	(+)-DET	(2*S*,3*R*)	a	88	97a
12	$(CH_2)_4OBn$	$CH_2CH(Me)CH_2OMEM$	10/12	(–)-DIPT	(2*R*,3*S*)	83	91	97b

[a] Not reported.

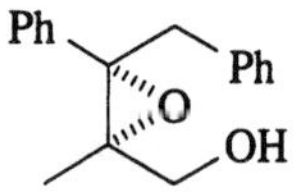

(65)
90%, 94% *ee* (ref. 98)

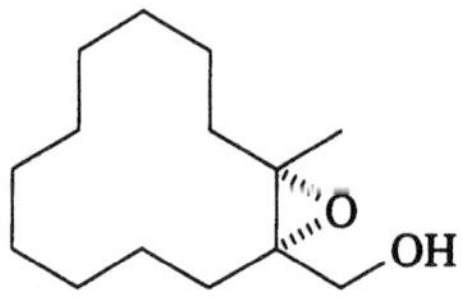

(66)
72%, 94% *ee* (ref. 99)

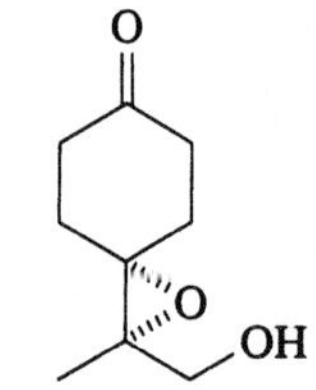

(67)
>90% *ee* (ref. 100)

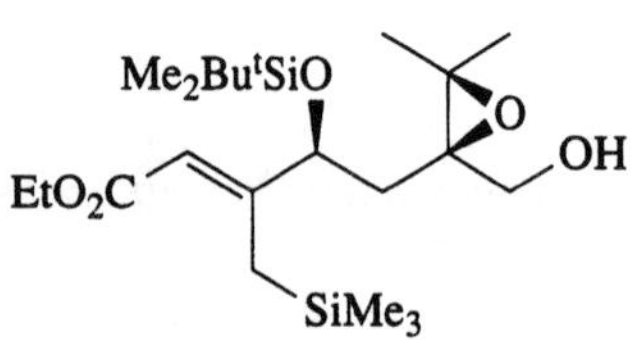

(68)
>90% *ee* (ref. 101)

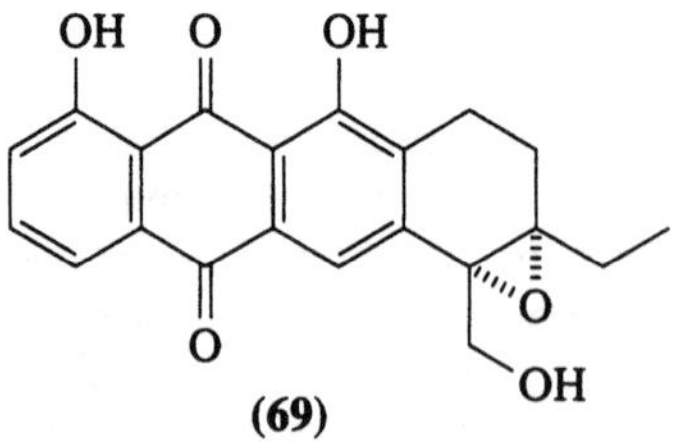

(69)
85%, 53% *ee* (ref. 102)

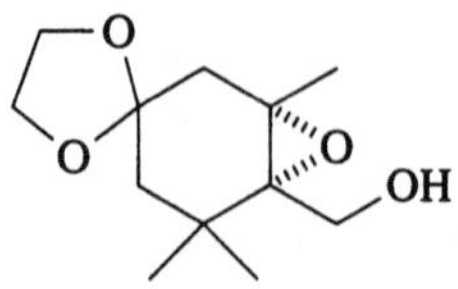

(70)
95%, 95% *ee* (ref. 103)

3.2.5.9 1-Substituted Allyl Alcohols: Kinetic Resolution

The presence of a stereogenic center at C-1 of an allylic alcohol introduces an additional factor into the asymmetric epoxidation process in that now both enantiofacial selectivity and diastereoselectivity must be considered. It is helpful in these cases to examine epoxidation of each enantiomer of the allylic alcohol separately. Epoxidation of one enantiomer proceeds normally and produces an *erythro*-epoxy alcohol in accord with the rules shown in Figure 1. Epoxidation of the other enantiomer proceeds at a reduced rate because contact between the C-1 substituent and the catalyst seriously impedes the necessary approach of alkene to oxidant (see Figure 2). The difference in epoxidation rates for the two enantiomers is usually of sufficient magnitude that either the epoxy alcohol or the recovered allylic alcohol can be produced with high optical purity. The net result is that a kinetic resolution is achieved.[13] In the case of a homochiral C-1-substituted allylic alcohol, asymmetric epoxidation will be fast and highly diastereoselective with one antipode of the titanium tartrate catalyst but not with the other, according to the guidelines of Figure 2. Although kinetic resolution is most frequently encountered and applied to chiral C-1-substituted allylic alcohols, the rationale also is applicable to allylic alcohols with chiral substituents at other positions, examples of which have been given in several preceding subsections.

The ratio of the rates of epoxidation of the two enantiomers, k_{fast}/k_{slow}, has been defined as the relative rate (k_{rel}) and is related to both the percentage conversion of allylic alcohol to epoxy alcohol and the enantiomeric excess of the remaining allylic alcohol. A mathematical relationship between these variables exists and can be represented graphically, as shown in Figure 4.[13] If values are known for two of the three variables, then the third can be predicted by use of this graph. Inspection of the graph reveals that relative rates of 25 or more are very effective for achieving kinetic resolution of 1-substituted allylic alcohols. With a relative rate of 25, the epoxidation need be carried to less than 60% conversion to achieve an enantiomeric excess of essentially 100% for the unreacted alcohol. A convenient method for limiting the extent of epoxidation to 60% is simply by controlling the amount of oxidant used in the reaction. However, for some substrates (see Table 8, entries 1, 9 or 10) even k_{fast} is extremely slow and several days are needed for the epoxidation. To shorten the time needed for such reactions, an alternate practice is to use an excess of oxidant and to monitor the extent of epoxidation by an appropriate analytical method. If the optically active epoxy alcohol is the desired reaction product, then high enantiomeric excess can be insured by running the reaction to approximately 45% completion.

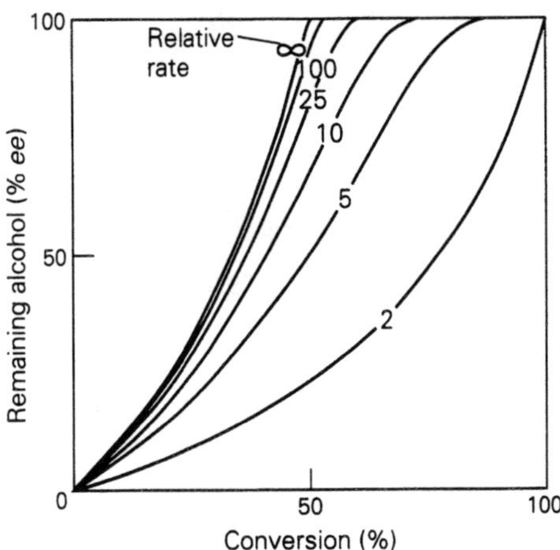

Figure 4 Dependence of enantiomeric excess on relative rate in the epoxidation of C-1-substituted allylic alcohols

Relative rate data for the kinetic resolution/epoxidation of 1-substituted allylic alcohols of varying structure are summarized in Table 8. The k_{rel} values at –20 °C for all entries in Table 8 were determined using DIPT as the chiral ligand. Additionally, for several entries (1–3, 10 and 11) the dependence of k_{rel} on temperature, 0 °C *versus* –20 °C, and on steric bulk of the tartrate ester, DIPT *versus* DET *versus* DMT, has been measured. Lower reaction temperature and larger tartrate ester groups both are factors that clearly increase the magnitude of k_{rel} and, therefore, improve the efficiency of the kinetic resolution process. While the results summarized in Table 8 are all from experiments in which stoichiometric quantities of titanium tartrate complex were used, the catalytic version of the reaction also may be used for kinetic resolution.[4] When comparing results with the same tartrate ester, a slight loss in enantioselectivity is seen in the catalytic mode relative to the stoichiometric reaction. The trend toward higher enantioselectivity with bulkier tartrate esters can be used to advantage in the catalytic reaction by using dicyclo-

Table 8 Relative Rate (k_{rel}) Data for Kinetic Resolution of 1-Substituted Allylic Alcohols

Entry	*Allylic alcohol*	*Reaction time*	*Relative rate at −20 °C*	*ee (%)*	*Relative rates at 0 °C* DIPT	DET	DMT	*Ref.*
1	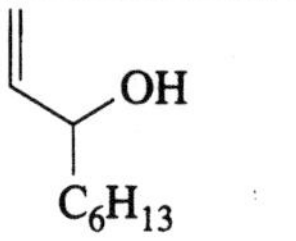 OH, C_6H_{13}	12 d	83	>96	60			2, 13
2	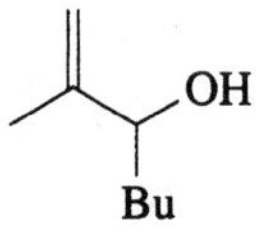 OH, Bu	15 h	138	>96	96	52		2, 13
3	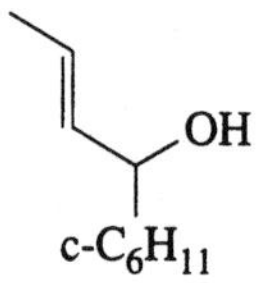 OH, c-C_6H_{11}	15 h	104	>96	74	28	15	2, 13
4	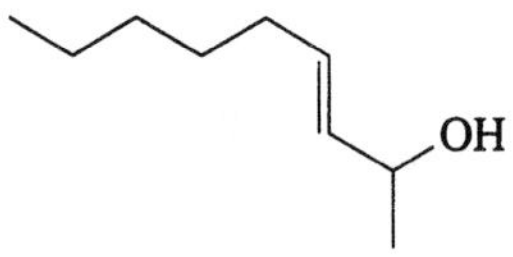 OH		160					104
5	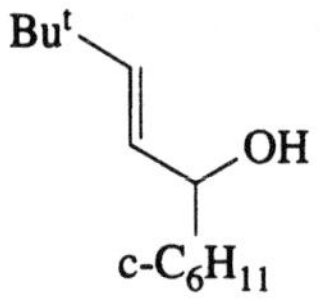 Bu^t, OH, c-C_6H_{11}		300					104
6	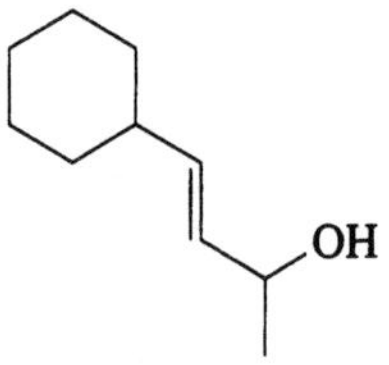 OH		330					104
7	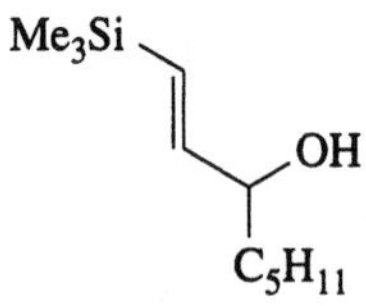 Me_3Si, OH, C_5H_{11}		700					104–106
8	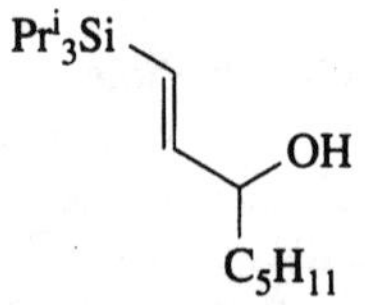 Pr^i_3Si, OH, C_5H_{11}		300					104
9	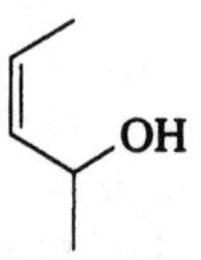 OH	6 d	20	91				2, 13

Table 8 *(continued)*

Entry	*Allylic alcohol*	*Reaction time*	*Relative rate at –20 °C*	*ee (%)*	*Relative rates at 0 °C DIPT*	*DET*	*DMT*	*Ref.*
10	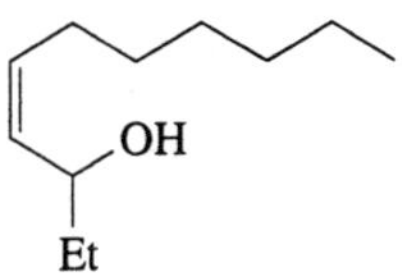 OH, Et	2 d	16	82	13			2, 13
11	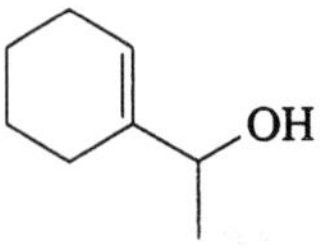 OH	15 h	83	>96	60	38		2, 13

hexyltartrate (DCT), which gives higher selectivity than DIPT, or dicyclododecyl tartrate, which gives yet higher selectivity than DCT.[4]

The efficiency of kinetic resolution is even greater when there is a silicon or iodo substituent in the (3*E*)-position of the C-1 chiral allylic alcohols. The compatibility of silyl substituents with asymmetric epoxidation conditions was first shown by the conversion of (3*E*)-3-trimethylsilylallyl alcohol into (2*R*,3*R*)-3-trimethylsilyloxiranemethanol in 60% yield with >95% *ee*,[107a] and further exploited by the conversion of (*E*)-3-(triphenylsilyl)-2-[2,3-$^2H_2$]propenol into (2*R*,3*R*)-3-triphenylsilyl[2,3-$^2H_2$]oxiranemethanol in 96% yield and with 94% *ee*.[107b,107c] With an *n*-pentyl group at C-1, the k_{rel} for asymmetric epoxidation of the enantiomeric allylic alcohols is 700 (Table 8, entry 7), and both epoxy alcohol and optically active recovered allylic alcohol are obtained in 42% yield with >99% *ee* (see Table 9, entry 1). Equally good yields and optical purities are observed with other substituents in the C-1 position, as is shown by entries 2 to 9 in Table 9. Good yields with high enantiomeric excess also are reported in the kinetic resolution of (3*E*)-iodo analogs (entries 10–14) and of a (3*E*)-chloro analog (entry 15). (3*E*)-Stannyl substituents (entries 16–18) appear similar to carbon substituents in their effect on kinetic resolution.

The influence of both the steric and the electronic properties of the silyl group on the rate of epoxidation have been examined experimentally.[104] Two different rate effects were considered. First, the overall rate of epoxidation of the silyl allylic alcohols was found to be one-fifth to one-sixth that of the similar carbon analogs. This rate difference was attributed to electronic differences between the silicon and carbon substituents. Second, the increase in k_{rel} to 700 for silyl allylic alcohols compared to carbon analogs (*e.g.* 104 for entry 3, Table 8) was attributed to the steric effect of the large trimethylsilyl group. As expected, when a bulky *t*-butyl group was placed at C-3, k_{rel} increased to 300.[104]

At the end of 1989, over 75 1-substituted allylic alcohols had been used in kinetic resolution/asymmetric epoxidation experiments. In slightly over half of these experiments, the desired product was the kinetically resolved allylic alcohol, while in the remainder the epoxy alcohol was desired. In addition to the compounds in Table 8, experimental results for other kinetically resolved alcohols are summarized in Table 10. From these results, it appears that kinetic resolution is successful regardless of the nature of the (3*E*)-substituent and is successful with any except the most bulky substituents at C-2.

In those cases where the allylic alcohol is the desired product of the kinetic resolution process, the accompanying epoxy alcohol also may be converted to the desired allylic alcohol by the two-step sequence shown in Scheme 2. The epoxy alcohol, after separation from the allyl alcohol, is mesylated and then subjected to reaction with sodium telluride, which effects the transformation of epoxymesylate to the allylic alcohol with inversion at the asymmetric carbinol center.[115e] Preliminary results suggest that the rearrangement follows this pathway only when the epoxy alcohol is unsubstituted at the 3-position.

A small, structurally distinct class of 1-substituted allylic alcohols are those which are conformationally restricted by incorporation into a ring system. These allylic alcohols may be further subdivided into two types depending on whether the double bond is endocyclic or exocyclic. For allylic alcohols with endocyclic double bonds, kinetic resolution gives 2-cyclohexen-1-ol (**71**) with 30% *ee*,[13] (4a*S*,2*R*)-4a-methyl-2,3,4,4a,5,6,7,8-octahydronaphthalen-2-ol (**72**) with 55% *ee*[116] and (*R*)-2-cyclohepten-1-ol (**73**) with 80% *ee*.[13] The epoxy alcohols, (1*S*,2*S*,3*R*)-2,3-epoxycyclopenten-1-ol (**74**),[117] (1*S*,2*S*,4a*R*)-4a-decahydronaphthalen-2-ol (**75**)[116] and (1*R*,2*S*,3*R*)-2,3-epoxy-6-cyclononen-1-ol (**76**)[118] are obtained with

Table 9 Kinetic Resolution of 3-Silyl-, Halo- and Stannyl-substituted Allylic Alcohols

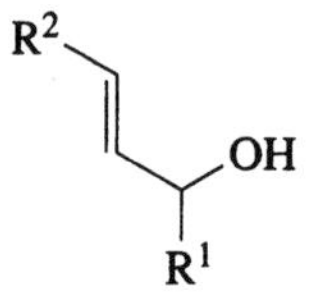

Entry	*Allylic alcohol* R^1	R^2	*Allylic alcohol* Yield (%)[a]	Enantiomeric excess (%)	*Epoxy alcohol* Yield (%)[a]	Enantiomeric excess (%)	*Ref.*
1	n-C_5H_{11}	$SiMe_3$	42	>99	42	>99	105
2	Pr^i	$SiMe_3$	40	>99	41	99	105b
3	Ph	$SiMe_3$	44	>99	42	97	105b
4	CH_2OPh	$SiMe_3$	47	>99	46	>99	105b
5	CH_2OCH_2Ph	$SiMe_3$	43	>99	48	>99	105b
6	$CH_2CH{=}CHC_5H_{11}$	$SiMe_3$	44	>99	43	>99	105b
7	$CH_2CH_2OCH_2Ph$	$SiMe_3$	43	>99	45	>99	105b
8	$CH_2CO_2Bu^n$	$SiMe_3$	44	>99			105c
9	$(CH_2)_3CO_2Me$	$SiMe_3$	43	>99	45	>99	105b
10	n-C_5H_{11}	I	49	>99	49	>99	108
11	Et	I	40	>98			108
12	CH_2-c-C_6H_{11}	I	42	>99			108
13	c-C_5H_9	I	44	>99			108
14	Ph	I	43	>98			108
15	n-C_5H_{11}	Cl	43	>99			108
16	n-C_5H_{11}	$SnBu_3$	40	>99	b	84	109
17	c-C_6H_{11}	$SnBu_3$	41	>99			109
18	CH_2OPh	$SnBu_3$	40	>99			109

[a]Maximum yield is 50%. [b]Not reported.

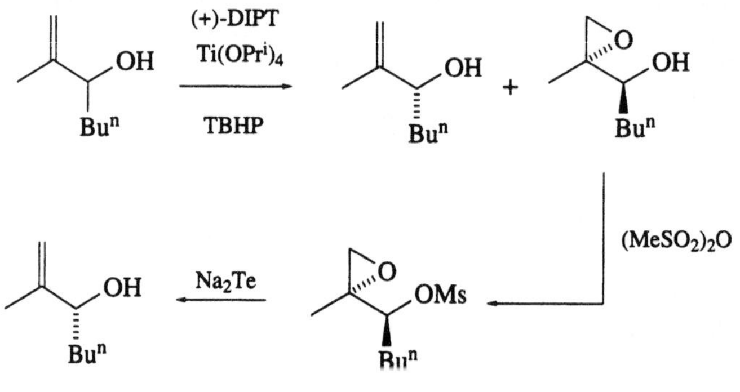

(+)-DIPT, $Ti(OPr^i)_4$, TBHP; $(MeSO_2)_2O$; Na_2Te; OH; OMs; Bu^n

Scheme 2

60% *ee*, 61% *ee* and 90% *ee*, respectively. (*R*)-*trans*-Verbenol (**77**) is epoxidized five times as fast as is (*S*)-*trans*-verbenol when (+)-DIPT is used in the catalyst.[77] For allylic alcohols with an exocyclic double bond, kinetic resolution gives 2-methylenecyclohexanol (**78**) with 80% *ee* and a 46% yield when (–)-DIPT is used.[119] Epoxidation of the homochiral 4-methylene-5α-cholestan-3β-ol (**79**) is reported to be much faster with catalyst derived from (+)-DET than from (–)-DET.[120] The variable enantioselectivities seen in these results likely stem from conformational restraints imposed by the cyclic structures which prevent the allylic alcohols from attaining an ideal conformation for the epoxidation process (see Section 3.2.6 and Figure 5 for the proposed ideal conformation).

One especially interesting kinetic resolution/asymmetric epoxidation substrate is (*R*,*S*)-2,4-hexadien-3-ol (**80**).[77] The racemic diene has eight different alkene faces at which epoxidation can occur and thereby presents an interesting challenge to the selectivity of the epoxidation catalyst. The selectivity can be tested by using slightly less than 0.5 equiv. of oxidant (because the substrate is a racemate, the maximum yield of any one product is 50%). When the reaction was run under these conditions, the only product that was formed was the (1*R*,2*R*,3*R*)-epoxy alcohol (**81**). Three different principles of selectivity are required to achieve this result. First, the difference in rate of epoxidation by the catalyst of a disubstituted

Table 10 Representative Kinetic Resolutions of 1-Substituted Allylic Alcohols

Entry	R^1	Allylic alcohol R^2	R^3	R^4	Yield (%)[a]	Enantiomeric excess (%)	Ref.
1	Me	H	$CH_2CH{=}CH_2$	H	39	90	110
2	c-C_6H_{11}	H	H	H	32	>98	111
3	Et	Me	H	H	b	>98	112
4	Bu^n	H	H	H	43	>90	113
5	2,4-$Cl_2C_6H_3$	2,4-$Cl_2C_6H_3$	H	H	42	90	114
6	Et	H	Ph	H	b	99	77
7	CH_2CH_2Ph	H	H	H	b	99	77
8	$-(CH_2)_3$-(2-methyl-1,3-dioxolan-2-yl)	H	H	H	b	99	77
9	$CH_2C({=}CH_2)CH{=}CH_2$	H	Me	Me	10	>99	77
10	n-$C_{12}H_{25}$	H	Me	H	44	97	115a
11	Me	Bu^t	H	H	b	30	38
12	Bu^t	H	Me	H	b	5	38
13	Bu^n	H	$CH{=}CH_2$	H	40	90	115b
14	CH_2CO_2Et	Me	H	H	11	>95	115c
15	$C{\equiv}C$-n-C_6H_{13}	H	H	H	35	95	115d

[a]Maximum yield is 50%. [b]Not reported.

(71) (72) (73) (74) (75) (76)

(77) (78) (79)

versus a monosubstituted alkene must be such that the propenyl group is epoxidized in complete preference to the vinyl group. The effect of this selectivity is to reduce the choice of alkene faces to the four in the propenyl groups. Second, the inherent enantiofacial selectivity of the catalyst as represented in Figure 1 narrows the choice of propenyl faces from four to two. Finally, the steric factor responsible for kinetic resolution of 1-substituted allylic alcohols (Figure 2) determines the choice between the propenyl groups in the enantiomers of (**80**). The net result is the formation of epoxy alcohol (**81**) and enrichment of the unreacted allylic alcohol in the (3*S*)-enantiomer.

(*R*) OH + (*S*) OH

(–)-DIPT
$Ti(OPr^i)_4$
TBHP

(*R*,*S*)-(**80**)

(**81**)
35%, >95% *ee*

trans-1,2-Dialkylcycloalkenes (**82**) have helical chirality and can be resolved if flipping of the ring from one face of the alkene to the other is restricted. These compounds, when appropriately substituted, also serve as synthetic precursors to the betweenanenes. The asymmetric epoxidation approach to kinetic resolution is ideally suited for the resolution of the cycloalkenes when a hydroxymethyl group is one of the substituents on the double bond, as shown for (**82**). The epoxidation of (**82**) with $Ti(OPr^i)_4$/(+)-DET and 0.6 equiv. of TBHP was complete within 10 min and gave resolved allylic alcohol (**83**) in 41% isolated yield with no detectable enantiomeric impurity and epoxy alcohol (**84**) in 50% yield (the maximum yield possible for both **83** and **84** is 50%).[121a] A variety of analogs of (**82**) including different ring sizes have been resolved by this method and have been used for the synthesis of optically active betweenanenes.[121b]

C_5H_9 $(CH_2)_{10}$ OH

(+)-DIPT
$Ti(OPr^i)_4$
TBHP

C_5H_9 $(CH_2)_{10}$ OH + C_5H_9 $(CH_2)_{10}$ O OH

(**82**) (**83**) (**84**)

A final subclass of 1-substituted allylic alcohols is made up of carbinol derivatives having two identical alkenic substituents, the simplest example being 1,4-pentadien-3-ol or divinylcarbinol (**85**). Although these compounds *per se* are achiral, once they bind to the chiral titanium complex the two vinyl groups become stereochemically nonequivalent (diastereotopic). Asymmetric epoxidation now will occur selectively at one of the two vinyl groups, the choice being controlled by factors identical to those in effect during the kinetic resolution process. The similarity can be seen by comparison of the titanium–allylic alcohol complex portrayed in Scheme 3 with the kinetic resolution process depicted in Figure 2. The *pro-S* and *pro-R* conformations shown will be sterically favored and disfavored, respectively, for the same reasons that the enantiomers of chiral C-1 allylic alcohols are distinguished during kinetic resolution. Therefore, epoxidation of (**85**) produces (2*R*,3*S*)-epoxy alcohol (**86**).[122]

Further analysis of the asymmetric epoxidation of divinylcarbinol (**85**), including the minor products, has led to recognition of a second factor that influences the optical purity of the major product (**86**).[26,123] One of the minor epoxy alcohols is enantiomeric to (**86**) and therefore is responsible for lowering the optical purity of (**86**). However, in this minor isomer the configuration of the remaining allylic alcohol group favors a rapid second epoxidation and this isomer is quickly converted to a diepoxide. As a consequence, the optical purity of the major epoxy alcohol (**86**) increases as the reaction progresses. A mathematical equation relating optical purity to the various rates of epoxidation for these divinylcarbinols has been derived. This analysis can also be applied to asymmetric epoxidation of prochiral compounds such as (**87**).[124]

As noted earlier in this section on C-1-substituted compounds, preparation of the epoxy alcohol has been the synthetic objective nearly as often as has been the optically active allylic alcohol. The principles

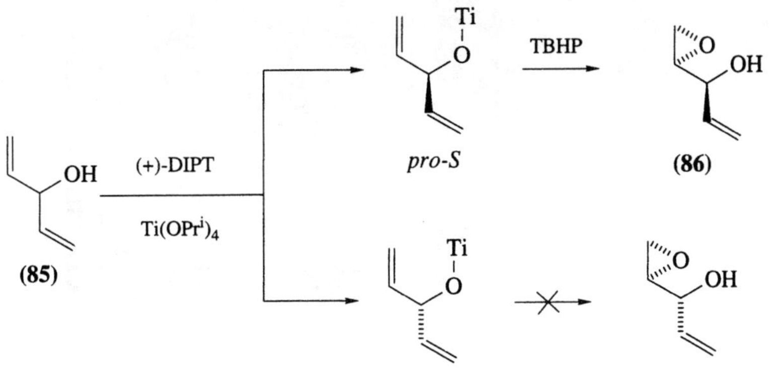
Ti
O
TBHP
O
OH
OH
(+)-DIPT
Ti(OPr^i)$_4$
Ti
O
O
OH

(85) *pro-S* (86) *pro-R*

Scheme 3

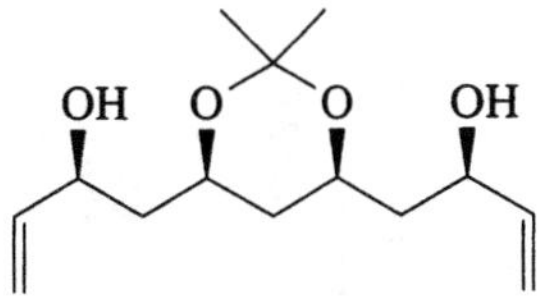
OH
O
O
OH

(87)

outlined in Figure 2 can again be used to guide the choice of tartrate ester needed in order to obtain the *erythro*-epoxy alcohol of desired absolute configuration. By limiting the amount of oxidant (TBHP) used for the epoxidation to 0.4 equiv. (relative to substrate), optimum optical purity of the epoxy alcohol can be assured and, in most cases, will be excellent. A few representative examples of epoxides prepared in this way are summarized in Table 11. In the special case where the substrate is already homochiral (as in Table 11, entry 5), it should be clear from Figure 2 that asymmetric epoxidation will be successful (with regard to diastereomeric purity) only when the choice of catalyst directs delivery of oxygen to the face of the alkene opposite that of the C-1 substituent. Such choice of catalyst is further illustrated in Scheme 4, wherein the two sequential epoxidations each proceed with >97% diastereoselectivity. The bis(epoxide) is obtained in an overall yield of 80%.[130c]

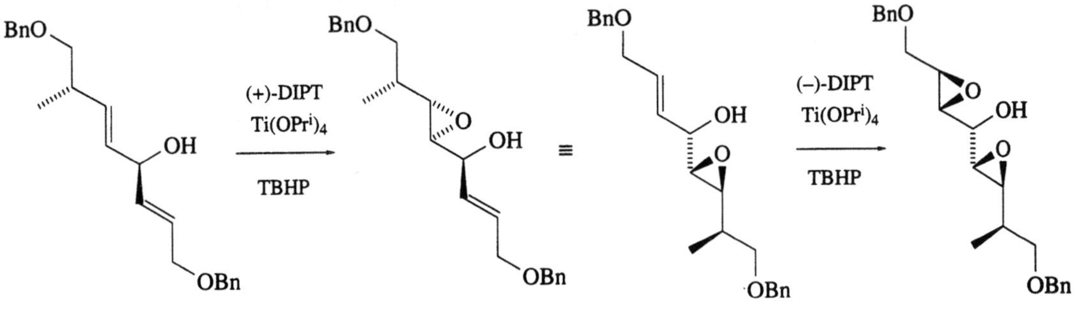
BnO
OH
OBn
(+)-DIPT
Ti(OPr^i)$_4$
TBHP
BnO
O
OH
OBn
≡
BnO
OH
O
OBn
(−)-DIPT
Ti(OPr^i)$_4$
TBHP
BnO
O
OH
O
OBn

Scheme 4

3.2.5.10 1,1-Disubstituted Allyl Alcohols

The rationale that explains the kinetic resolution of the 1-monosubstituted allylic alcohols predicts that a 1,1-disubstituted allylic alcohol will be difficult to epoxidize with the titanium tartrate catalyst. In practice, the epoxidation of 1,1-dimethylallyl alcohol (**88**) with a stoichiometric quantity of the titanium tartrate complex is very slow and no epoxy alcohol is isolated.[131] Clearly, the rate of epoxidation of this substrate is slower than the subsequent reaction(s) of the epoxide.

Table 11 Epoxides from 1-Substituted Allylic Alcohols

Entry	R^1	Epoxide R^2	R^3	R^4	Tartrate	Configuration[a]	Yield (%)[b]	Enantiomeric excess (%)	Ref.
1	Et	H	H	H	(+)-DIPT	(2*R*,3*S*)	c	d	125
2	n-C_5H_{11}	H	H	H	(–)-DIPT	(2*S*,3*R*)	47	91	126
3	$(CH_2)_6CO_2Me$	H	H	H	(+)-DIPT	(2*R*,3*S*)	36	>95	127
4	$CH_2C{\equiv}CSiPr^i_3$	H	H	H	(–)-DIPT	(2*S*,3*R*)	40	>90	128
5[e]	Et	Me	H	H	(+)-DIPT	(2*R*,3*R*)	82	92	90
6	$CH_2CH{=}CH_2$	H	Me	H	(–)-DIPT	(1*R*,2*R*,3*R*)	27	>95	129a–129c
7	$CH(OBn)CH{=}CHMe$	H	Me	H	(+)-DIPT	(1*S*,2*S*,3*S*)	35	>95	130a
8	Me	H	$CH_2CH{=}CH_2$	H	(–)-DIPT	(1*R*,2*R*,3*R*)	40	90	129b
9	C_5H_{11}	H	$SiMe_3$	H	(+)-DIPT	(1*S*,2*S*,3*S*)	40	99	105b
10	CH_2CO_2Et	Me	Et	H	(–)-DET	(1*S*,2*S*,3*R*)	d	>95	115c
11	Me	$—CH_2CH{=}C(Me)CH_2—$		H	(+)-DIPT	(1*S*,2*S*,3*S*)	37	95	130b

[a]Note that the arbitrary numbering used here may not coincide in all cases (*e.g.* entries 7, 10, 11) with correct *Chemical Abstracts* numbering. [b]Maximum yield is 50%, except for entry 5. [c]The epoxy alcohol was converted without isolation to the ethoxyethyl derivative. [d]Not reported. [e](3*S*)-2-Methylpent-1-en-3-ol was used as the substrate for this epoxidation.

(88)

3.2.5.11 Homoallylic, Bis(homoallylic) and Tris(homoallylic) Alcohols

In contrast to allylic alcohols, the asymmetric epoxidation of homoallylic alcohols shows the following three general characteristics:[132] (i) the rates of epoxidation are slower; (ii) enantiofacial selectivity is reversed, *i.e.* oxygen is delivered to the opposite face of the alkene when the same tartrate ester is used; and (iii) the degree of enantiofacial selectivity is lower with enantiomeric excesses of the epoxy alcohols in the range 20–55%. A series of seven model homoallylic alcohols, including all but one of the possible substitution patterns, has been subjected to epoxidation using the stoichiometric version of the reaction with the results providing the basis for the preceding generalizations. An analogous complex composed of zirconium(IV) isopropoxide ($Zr(OPr^i)_4$) and (+)-dicyclohexyltartramide has been found to catalyze asymmetric epoxidation of homoallylic alcohols with the same sense of enantiofacial selectivity as the titanium tartrate ester complex. An improvement in enantiomeric excess was noted for epoxy alcohols derived from (*Z*)-homoallylic alcohols (to 77%), while other epoxy alcohols were obtained with enantiomeric excesses comparable to those achieved with titanium.[133]

The tris(homoallylic) alcohol (**89**) undergoes asymmetric epoxidation in a yield of 74% and with 'high' diastereofacial selectivity to give (**90**). Trityl hydroperoxide, which had been shown to be effective in the asymmetric epoxidation of allylic alcohols,[2] was required in order to attain enantiofacial selectivity in the epoxidation of (**89**).[134a] The titanium/tartrate/TBHP-catalyzed conversions of the bis(homoallylic) phenol (**90a**, $n = 1$, R = H) and the tris(homoallylic) analog (**90a**, $n = 2$, R = Me) into dihydrobenzofuran (**90b**, 22% yield, 29% *ee*) and dihydrobenzopyran (**90c**, 49%, 56% *ee*), respectively, is assumed to occur *via* the intermediate epoxides.[134b] The dihydrofuran (**90b**) is assigned the (2*S*,1′*R*)-configuration, whereas the configuration of the dihydropyran (**90c**) is unspecified.

(89) **(90)**

(90a)

(2*S*,1′*R*)-(**90b**)

(–)-(**90c**)

3.2.6 MECHANISM OF THE TITANIUM TARTRATE CATALYZED ASYMMETRIC EPOXIDATION

The hallmark of titanium tartrate catalyzed asymmetric epoxidation is the high degree of enantiofacial selectivity seen for a wide range of allylic alcohols. The question naturally arises as to what is the mechanism of this reaction and what are the structural features of the catalyst that produce these desirable results. These questions have been studied extensively and the results have been the subject of considerable previous discussion.[2,135,136] For the purpose of this chapter, we wish to review those aspects of the mechanistic–structural studies that may be helpful in devising synthetic applications of this reaction.

$$\text{Ti(OR)}_4 \;+\; \text{tartrate} \;\rightleftharpoons\; [\text{Ti(tartrate)(OR)}_2] \;+\; 2\,\text{ROH} \qquad (5)$$

Of fundamental importance to an understanding of the reaction and its mechanism is the fact that in solution there is rapid exchange of titanium ligands.[2] Thus, when equimolar solutions of a titanium alkoxide and a dialkyl tartrate are mixed, the equilibrium represented by equation (5) will be quickly reached with all but the most sterically demanding alkoxides. This equilibrium is shifted far to the right by virtue of the fact that a chelating diol (*i.e.* the tartrate) has a much higher binding constant for titanium than do monodentate alcohols. The binding of tartrate is also enhanced by the increased acidity of its hydroxy groups (due to the inductive effect of the esters). Spectroscopic evidence clearly reveals that two moles of free monodentate alcohol are present at equilibrium. Rapid ligand exchange continues as the hydroperoxide oxidant and the allylic alcohol substrate are added to the reaction medium. Pseudo-first-order kinetic experiments have shown a first order rate dependence on the titanium tartrate complex, the hydroperoxide and the allylic alcohol and an inverse second order dependence on the nonalkenic alcohol ligands (*i.e.* the isopropyl alcohol). The rate law derived from these results is expressed in equation (6).

$$\text{Rate} = k\,\frac{[\text{Ti(tartrate)(OR)}_2][\text{TBHP}][\text{allylic alcohol}]}{[\text{inhibitor alcohol}]^2} \qquad (6)$$

The mechanistic pathway outlined in Scheme 5 is consistent with equation (6) and clearly illustrates the ligand exchange processes essential for catalytic epoxidation. After formation of the [Ti(tartrate)(OR)$_2$] complex, the two remaining alkoxide ligands are replaced in reversible exchange reactions by the hydroperoxide (TBHP) and the allylic alcohol to give the 'loaded' complex [Ti(tartrate)(TBHP)(allylic alcohol)]. Now, in the rate-controlling step of the process, oxygen transfer from the coordinated hydroperoxide to the allylic alcohol gives the complex [Ti(tartrate)(OBu^t)(epoxy alcohol)]. The product alkoxides are replaced by more allylic alcohol and TBHP to regenerate the 'loaded' complex and complete the catalytic cycle.

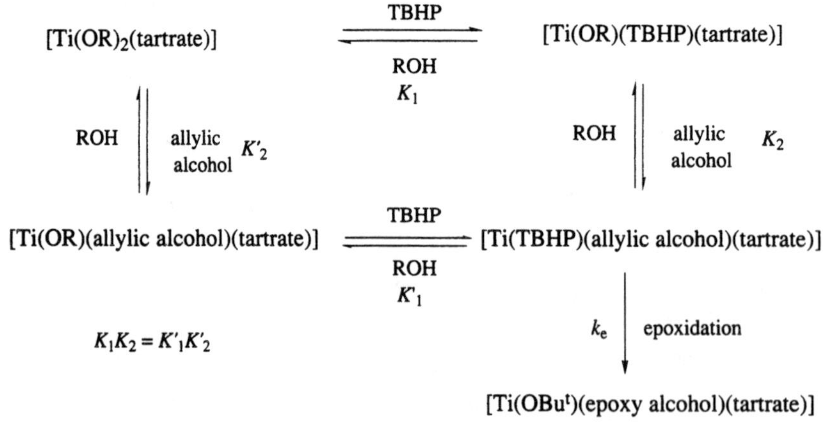

[Ti(OR)$_2$(tartrate)] ⇌ (TBHP / ROH, K_1) [Ti(OR)(TBHP)(tartrate)]

[Ti(OR)$_2$(tartrate)] ⇌ (allylic alcohol / ROH, K'_2) [Ti(OR)(allylic alcohol)(tartrate)]

[Ti(OR)(TBHP)(tartrate)] ⇌ (allylic alcohol / ROH, K_2) [Ti(TBHP)(allylic alcohol)(tartrate)]

[Ti(OR)(allylic alcohol)(tartrate)] ⇌ (TBHP / ROH, K'_1) [Ti(TBHP)(allylic alcohol)(tartrate)]

[Ti(TBHP)(allylic alcohol)(tartrate)] → (k_e, epoxidation) [Ti(OBu^t)(epoxy alcohol)(tartrate)]

$K_1K_2 = K'_1K'_2$

Scheme 5

An alternative mechanism invoking an ion-pair transition state assembly has been proposed to account for the enantioselectivity of the asymmetric epoxidation process.[137] In this proposal, two additional alcohol species are required in the transition state complex. This requirement is inconsistent with the kinetic studies of this reaction which have led to the rate law expressed in equation (5) and, therefore, this proposal must be considered incorrect.

Much of the experimental success of asymmetric epoxidation lies in exercising proper control of equation (5).[2] Both $Ti(OR)_4$ and $[Ti(tartrate)(OR)_2]$ are active epoxidation catalysts and since the former is achiral, any contribution by that species to the epoxidation will result in loss of enantioselectivity. The addition to the reaction of more than one equivalent of tartrate, relative to titanium, will have the effect of minimizing the leftward component of the equilibrium and will suppress the amount of $Ti(OR)_4$ present in the reaction. The excess tartrate, however, forms $Ti(tartrate)_2$ which has been shown to be a catalytically inactive species and which will cause a decrease in reaction rate that is proportional to the excess tartrate added. The need to minimize $Ti(OR)_4$ concentration and, at the same time, to avoid a drastic reduction in rate of epoxidation is the basis for the recommendation of a 10–20 mol % excess of tartrate over titanium for formation of the catalytic complex. After the addition of hydroperoxide and allylic alcohol to the reaction, the concentration of ROH will increase accordingly and this will increase the leftward pressure on the equilibrium. Fortunately, in most situations this shift apparently is extremely slight and is effectively suppressed by the use of excess tartrate. One situation in which a shift in the equilibrium does begin to occur is when the reaction is run in the catalytic mode and the amount of catalyst used is less than about 5 mol % relative to allylic alcohol substrate. Loss in enantioselectivity then may be observed. This factor is the basis of the recommendation for use of 5–10 mol % of titanium tartrate complex when using the catalytic version of asymmetric epoxidation.

Comparison of the epoxidation rates of several *para*-substituted cinnamyl alcohols reveals that the alkene acts as a nucleophile towards the activated peroxide oxygen in the epoxidation reaction.[136] Relative to unsubstituted cinnamyl alcohol (relative rate = 1), an electron-withdrawing *p*-nitro group decreases the rate of epoxidation (0.42), while an electron-releasing group such as *p*-methoxy increases the rate (4.39). These results are consistent with the alkene acting as a nucleophile. Additional support for this conclusion arises from comparison of the rates for epoxidation of less-substituted allylic alcohols with those for more highly substituted analogs. A clear example of this substituent effect is seen in the epoxidation of (*R*,*S*)-2,4-hexadien-3-ol (**80**), described in the preceding section, where the propenyl group is epoxidized in nearly complete preference to the vinyl group.[77] Another example is seen with the allylic–homoallylic alcohol (**91**), where epoxidation occurs preferentially at the tetrasubstituted homoallylic alkene to give (**92**).[99] The preferential epoxidation of the more highly substituted alkene in these compounds is consistent with a nucleophilic role for the alkene.

OH $(CH_2)_{10}$ —(+)-DIPT, $Ti(OPr^i)_4$, TBHP→ OH O $(CH_2)_{10}$

(**91**) (**92**)

While the mechanistic scheme portrayed in Scheme 5 provides important insight into the experimental aspects of asymmetric epoxidation, it sheds little light on the structure of the catalyst and on the features of the catalyst responsible for the concurrent high stereoselectivity and broad generality. The rapid ligand exchange, so crucial to the success of the reaction, makes characterization of the catalyst structure extremely difficult. Some reliable structural information has been obtained from spectroscopic measurements on the complex in solution.[2,136b] These data clearly support the conclusion that the major molecular species formed in solution is the dimeric composite $[Ti_2(tartrate)_2(OR)_4]$. Efforts to isolate this complex, ideally as a crystalline solid, have so far been fruitless. Therefore, assignment of a structure to the dimeric complex has depended on information provided by the X-ray crystallographic structure obtained for the closely related complex $[Ti_2(dibenzyltartramide)_2(OR)_4]$.[138] The assumption of a similarity of structure for these two complexes receives some support from the fact that both catalyze the epoxidation of α-phenylcinnamyl alcohol with the same enantiofacial selectivity. From this analogy, the structure shown in equation (7) has been proposed for the $[Ti_2(tartrate)_2(OR)_4]$ complex. This structure has a C_2 axis of symmetry with the two titanium atoms in identical stereochemical environments. To account for the fact that the tartrate ester groups all are identical in the room temperature NMR spectrum, a fluxional equilibrium between the two structurally degenerate complexes shown in equation (7) has been proposed. Catalysis of the epoxidation process is thought to involve only one of the two titanium atoms but the possibility that both are required has not yet been ruled out.

'Loading' of the catalyst with hydroperoxide and substrate can now be considered in terms of the proposed structure.[2] Orientation of these two ligands on the catalyst becomes a crucial issue. Three coordi-

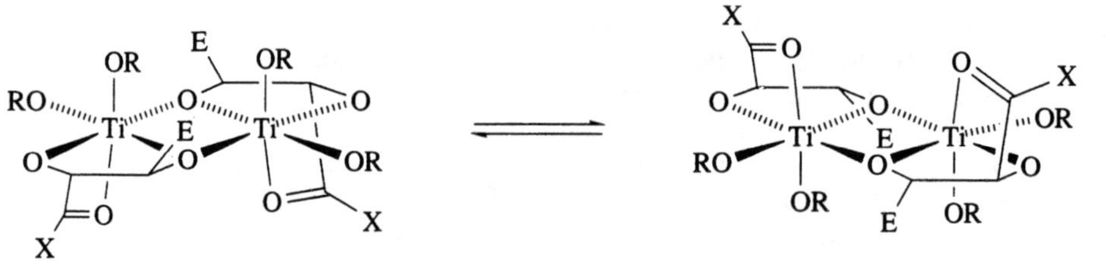

(7)

nation sites, two axial and one equatorial, become available by exchange of two isopropoxides and dissociation of the coordinated ester carbonyl group. These processes can occur with minimal perturbation of the remaining catalyst structure. The three coordination sites are in a semicircular (*i.e.* meridional) array around one edge of the catalyst surface. In the reactive mode, coordination of the hydroperoxide is assumed to be bidentate by analogy to the precedent of bidentate TBHP coordination to vanadium.[2,139] The hydroperoxide must occupy the equatorial and one of the two available axial coordination sites with the allylic alcohol in the remaining axial site. In order to achieve the necessary proximity for transfer of oxygen (the distal peroxide oxygen is assumed to be transferred) to the alkene, the distal oxygen is placed in the equatorial site (Figure 5) and the proximal oxygen is placed in the axial site. The axial site on the lower face of the complex (as drawn in Figure 5) is chosen for the peroxide because of the larger steric demands of the *t*-butyl group, or especially of the trityl group when trityl hydroperoxide is used, in comparison to the allylic alcohol.

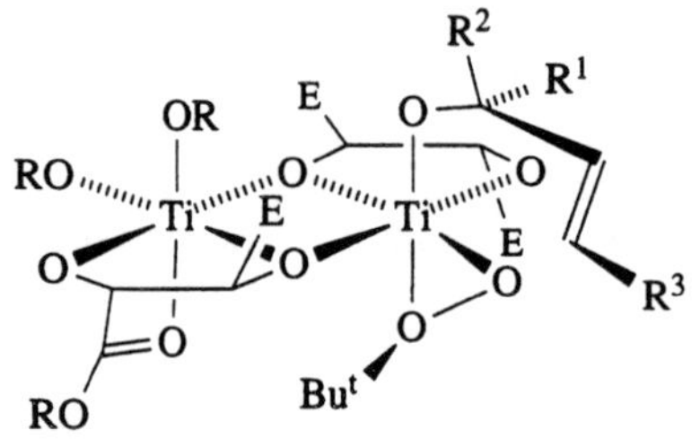

Figure 5 Proposed structure of 'loaded' catalyst at the time of oxygen transfer

The allylic alcohol binds to the remaining axial coordination site where stereochemical and stereoelectronic effects dictate the conformation shown in Figure 5.[2] The structural model of catalyst, oxidant and substrate shown in Figure 5 illustrates a detailed version of the formalized rule presented in Figure 1. Ideally, all the observed stereochemistry of epoxy alcohol and kinetic resolution products can be rationalized according to the compatibility of their binding with the stereochemistry and stereoelectronic requirements imposed by this site.[2] A transition state model for the asymmetric epoxidation complex has been calculated by a frontier orbital approach and is consistent with the formulation portrayed in Figure 5.[140]

3.2.7 OTHER ASYMMETRIC EPOXIDATIONS AND OXIDATIONS CATALYZED BY TITANIUM TARTRATE COMPLEXES

3.2.7.1 Ti_2(tartrate)$_2$ Complex

The discussion to this point has focused entirely on the epoxidation of allylic (and homoallylic) alcohols catalyzed by the [Ti(OR)$_2$(tartrate)] complex. The role of the alkene as a nucleophile towards the activated peroxide oxygen in this reaction has been established (see Section 3.2.6). If the alkene of the allylic alcohol is replaced by another nucleophilic group then, in principle, oxidation of that group may occur (equation 8).[141] In practice, oxidations of this type have been observed and generally have been carried out with a substrate bearing a racemic secondary alcohol so that kinetic resolution is achieved. While these oxidations are not strictly within the scope of this chapter, they are summarized briefly in equations (9) to (11) in order to acquaint the reader with other potential uses for the titanium tartrate catalytic complex. In the kinetic resolutions shown in equations (9) and (10), the oxidations are controlled by limiting the amount of oxidant used to 0.6 equiv. Only modest resolution was attained for the alkynic alcohol (equation 9, 21% *ee*)[77] and the allenic alcohol (equation 10, 40% *ee*).[77] Resolutions of the furanols[142] or the thiophene alcohols[143] of equation (11) generally are excellent (*ca.* 90–98% *ee*, except when R^1 is a *t*-butyl group). Only in the kinetic resolution of the furanols has the oxidation product been identified and, in that case, is a dihydropyranone.

G-(C)$_n$-C-OH ⟶ O←G-(C)$_n$-C-OH (8)

n-C_7H_{15}—≡—CH(OH)Et —[(+)-DIPT, Ti(OPr^i)$_4$, TBHP]→ n-C_7H_{15}—≡—CH(OH)Et (9)

n-C_7H_{15}—CH=C=CH—CH$_2$OH —[(+)-DIPT, Ti(OPr^i)$_4$, TBHP]→ n-C_7H_{15}—CH=C=CH—CH$_2$OH (10)

R^2-(heteroaryl)-CH(OH)R^1 —[(+)-DIPT, Ti(OPr^i)$_4$, TBHP]→ R^2-(heteroaryl)-CH(OH)R^1 (11)

X = O, S

The asymmetric epoxidation of an allylic alcohol in which the carbinol has been replaced by a silanol has been described.[144] As shown in equation (12), (3*E*)-phenylethenyldimethylsilanol is converted to an epoxy silanol in 50% yield with 85–95% *ee*. Note that here the longer Si—C bonds appear to overcome the restriction to epoxidation associated with a fully substituted C-1 atom in the allylic alcohol series. Fluoride cleavage of the silanol group gives (*S*)-styrene oxide.

PhCH=CH—Si(Me)$_2$OH —[(+)-DIPT, Ti(OPr^i)$_4$, TBHP]→ Ph-epoxide—Si(Me)$_2$OH (12)

3.2.7.2 Ti$_2$(tartrate) Complex

The β-hydroxyamines are a class of compounds which fall within the generic definition of equation (8). When the alcohol is secondary, the possibility for kinetic resolution exists if the titanium tartrate complex is capable of catalyzing the enantioselective oxidation of the amine to an amine oxide (or other oxidation product). The use of the 'standard' asymmetric epoxidation complex, *i.e.* Ti$_2$(tartrate)$_2$, to achieve such an enantioselective oxidation was unsuccessful. However, modification of the complex so that the stoichiometry lies between Ti$_2$(tartrate)$_1$ and Ti$_2$(tartrate)$_{1.5}$ leads to very successful kinetic resolutions of β-hydroxyamines. A representative example is shown in equation (13).[141b,141c] The oxidation and kinetic resolution of more than 20 secondary β-hydroxyamines[141,145] provides an indication of the scope of the reaction and of some structural limitations to good kinetic resolution. These results also show a consistent correlation of absolute configuration of the resolved hydroxyamine with the configuration of tartrate used in the catalyst. This correlation is as shown in equation (13), where use of (+)-DIPT results in oxidation of the (*S*)-β-hydroxyamine and leaves unoxidized the (*R*)-enantiomer.

i, 1.2 equiv. (+)-DIPT; 2.0 equiv. $Ti(OPr^i)_4$, 30 min; ii, 0.6 equiv. TBHP

37%, 95% *ee* + 59%, 63% *ee* (13)

3.2.7.3 Ti(tartramide) Complexes

A number of derivatives of the tartaric acid structure have been examined as substitutes for the tartrate ester in the asymmetric epoxidation catalyst. These have included a variety of tartramides, some of which are effective in catalyzing asymmetric epoxidation (although none display the broad consistency of results typical of the esters). One notable example is the dibenzyltartramide which in a 1:1 ratio (in reality, a 2:2 complex as shown by an X-ray crystallographic structure determination[138]) with $Ti(OPr^i)_4$ catalyzes the epoxidation of allylic alcohols with the same enantiofacial selectivity as does the titanium tartrate ester complex.[18] Remarkably, when the ratio of dibenzyltartramide to titanium is changed to 1:2, epoxidation is catalyzed with *reversed* enantiofacial selectivity. These results are illustrated for the epoxidation of α-phenylcinnamyl alcohol (equation 14). α-Phenylcinnamyl alcohol is a particularly felicitous substrate for asymmetric epoxidation; epoxidation of other allylic alcohols with the 1:2 dibenzyltartramide–titanium complex does not give as high enantioselectivities but the reversed selectivity is consistent throughout.[18] An extensive listing of tartramides used in the epoxidation of α-phenylcinnamyl alcohol with both 1:1 and 1:2 catalysts has been tabulated elsewhere.[2]

1 equiv. (*R*,*R*)-di-Bn-tartramide, 1 equiv $Ti(OPr^i)_4$, TBHP → 96% *ee*

1 equiv. (*R*,*R*)-di-Bn-tartramide, 2 equiv. $Ti(OPr^i)_4$, TBHP → 82% *ee* (14)

3.2.7.4 $[Ti(OPr^i)_2Cl_2(tartrate)]$ Complexes

As described in earlier sections of this chapter, certain epoxy alcohols, *e.g.* the 2-monosubstituted epoxy alcohols, are particularly susceptible to ring-opening processes. With the intent of controlling the ring-opening reaction, the epoxidation catalyst was modified by the use of $[Ti(OPr^i)_2Cl_2]$ in place of $Ti(OPr^i)_4$, the idea being to open the ring with chloride to produce a chlorodiol.[18] This modification was successful with 3-chloro-1,2-diols being formed in yields of 60–80% with good regioselectivity. Epoxy alcohols were assumed to be intermediates in these reactions and can be regenerated from the chlorodiols by base-promoted ring closure. Unfortunately, the enantioselectivity of the process varies from 20–70% *ee*. A point of interest concerning the chlorohydroxylation process is the fact that the enantiofacial selectivity is reversed from that of the normal asymmetric epoxidation process and is not altered by changing the $[Ti(OPr^i)_2Cl_2]$:tartrate ratio from 1:1 to 2:1. Chlorohydroxylation of 2-(6-chloropyridin-2-yl)-2-propen-1-ol (shown in equation 15) followed by closure of the epoxide ring has provided a useful route to the optically active epoxy alcohol in 50% yield and with 90% *ee*.[145b]

(–)-DET
$[TiCl_2(OPr^i)_2]$
TBHP

aq. NaOH
Et_2O

(15)

3.2.7.5 $[Ti(tartrate)_2(H_2O)]$ Complex

A complex of $Ti(OPr^i)_4$ and tartrate ester in a 1:2 ratio to which one equivalent of water is added has been found to oxidize prochiral sulfides to optically active sulfoxides.[146,147a] Yields of sulfoxides range from 50–95% and asymmetric inductions are in the range 75–95% *ee* for alkyl aryl sulfoxides and in the range 50–71% *ee* for dialkyl sulfoxides. The correlation between tartrate configuration and sulfoxide configuration shown in equation (16) has been found consistently when one of the sulfide substituents is an aryl group. As substrates, these sulfides differ from previously discussed substrates in one important respect: there is no adjacent hydroxy group by which the molecule may coordinate to titanium during the oxidation. This system appears to provide one of the first examples of effective asymmetric catalysis without the need for prior binding or tethering of the substrate to the catalyst (see Section 3.2.9 for another example of this presently rare phenomenon).[146]

(*R*,*R*)-(+)-tartrate

(16)

3.2.8 OTHER ASYMMETRIC EPOXIDATION METHODS

A review of nonenzymatic asymmetric epoxidations covering the literature through 1983 has been published elsewhere.[2] Improved enantioselectivity (to as high as 64% *ee*) for epoxidations of some alkenes with chiral oxaziridines has been described and results are included in a review of synthetic applications of oxaziridines.[147b] A summary of *catalytic* asymmetric epoxidations of alkenes is presented in Table 12, together with brief comments on each method.

Preliminary results for asymmetric epoxidations of (*E*)-cinnamyl alcohol and geraniol using (1*S*,2*S*)-1,2-di(2-methoxyphenyl)ethane-1,2-diol or (1*S*,2*S*)-1,2-di(4-methoxyphenyl)ethane-1,2-diol as chiral auxiliaries with titanium(IV) isopropoxide and TBHP have been described. High enantioselectivity (95% *ee*) is observed when the 2-methoxyphenyl compound is used, while somewhat lower enantioselectivity (64% *ee*) and *opposite face selectivity* is described for the catalyst comprised of the 4-methoxyphenyl analog.[149a] Further elaboration of the scope and generality of these observations will be of interest.

Sulfur ylides, derived from benzyl bromides and an optically active alkyl sulfide, undergo base-promoted reactions with aryl aldehydes to produce optically active 1,2-diaryl epoxides.[149b] The reaction is illustrated by equation (17) and produces epoxides with optical purities in the range of 28–47% *ee*. The bicyclic sulfide shown in equation (17) was derived from (+)-camphorsulfonic acid and produces the (*R*,*R*)-enantiomer of the epoxide in excess.

KOH
THF

(17)

Asymmetric induction occurs during the alkylation of ketones with the α-sulfinyl carbanion derived from optically active 1-chloroalkyl-*p*-tolyl sulfoxides (equation 18).[149c] The resulting chloro alcohol may be converted to an optically active epoxide under alkaline conditions and the sulfinyl group is removed with n-butyllithium. While the process benefits from high asymmetric induction in the alkylation reaction, it must be recognized that, when either R ≠ H and/or $R^1 \neq R^2$, diastereomeric compounds are formed and require separation.

Asymmetric epoxidation of 2-alkylnaphthoquinones is achieved with TBHP in a buffered (pH 9) medium containing bovine serum albumin, but enantioselectivity is extremely sensitive to reaction condi-

Table 12 Catalytic Asymmetric Epoxidation of Unfunctionalized Alkenes

Entry	*Catalyst*[a]	*Oxidant*	*Substrate*	*ee (%)*	*Features and drawbacks*	*Ref.*
1	$[MoO_2(acac)_2]$/DIPT	TBHP	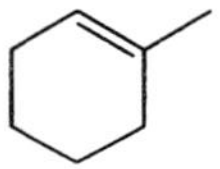	10	50 turnovers All components commercially available and inexpensive Very low selectivity No subsequent promising results in catalysis with similar systems	148a
2	A[a], M = FeCl, R 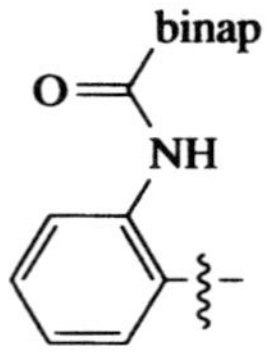 = (binap, O, NH) (α,β,α,β)	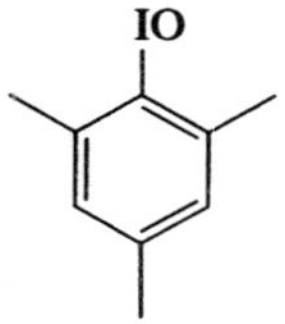 IO	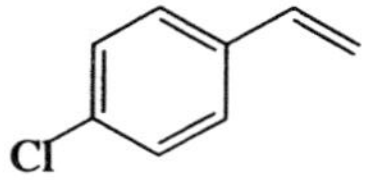 Cl	51	First successful use of a chiral metal oxo-based system for catalytic asymmetric epoxidation Ligand prepared in <<1% yield Reactions taken to 30% conversion of alkene Limited to terminal alkenes	148b
3	B[a], M = FeCl RR 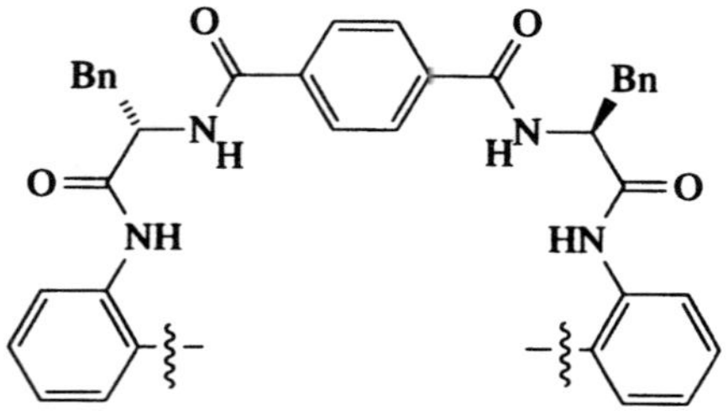 = (Bn, O, NH, H)	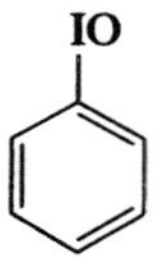 IO	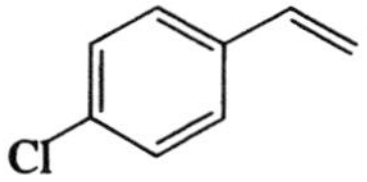 Cl	50	First use of peptide/porphyrin complex 0.3% conversion of alkene Difficult ligand synthesis (overall yield not reported)	148c
4	$[(chiraphos)Pt(CF_3)(CH_2Cl_2)]BF_4$	H_2O_2	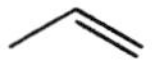 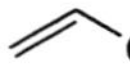 C_6H_{13}	41 36	High turnover numbers Cheap oxidant Tartrate-derived ligand Reasonable success with difficult substrates Only two substrates examined Expensive metal	148d

Table 12 *(continued)*

Entry	*Catalyst*[a]	*Oxidant*	*Substrate*	*ee* (%)	*Features and drawbacks*	*Ref.*
5	C[a], M = FeCl; R,R 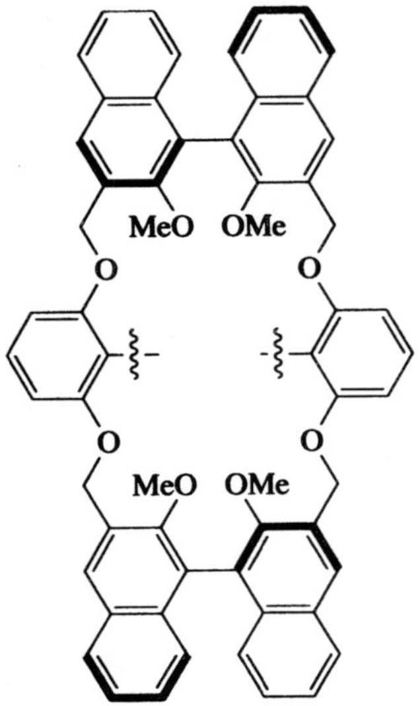 = MeO OMe O O MeO OMe O O	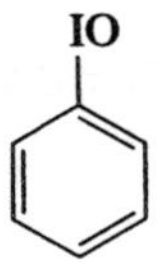 IO	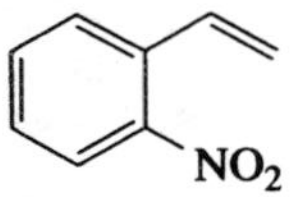 NO_2	80	High % *ee* reported for electron deficient styrenes Ligand prepared in extremely low yield Styrene gave only 14% *ee* Reactions taken to only 20% conversion of alkene	148e
6	A[a], M = MnCl (α,β,α,β); R 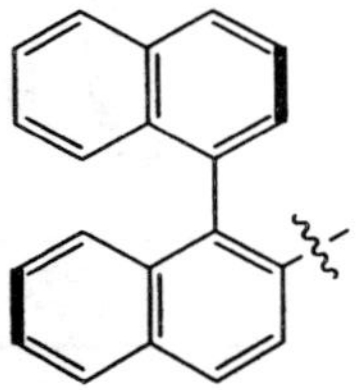 =	NaOCl	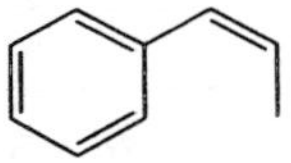	40	Extremely sturdy catalyst (86% recovery after 2800 turnovers) Good turnover rates Cheap oxidant Ligand prepared in low yield Relatively low selectivity	148f

Table 12 *(continued)*

Entry	*Catalyst*[a]	*Oxidant*	*Substrate*	*ee (%)*	*Features and drawbacks*	*Ref.*
7	M = $MnPF_6$ 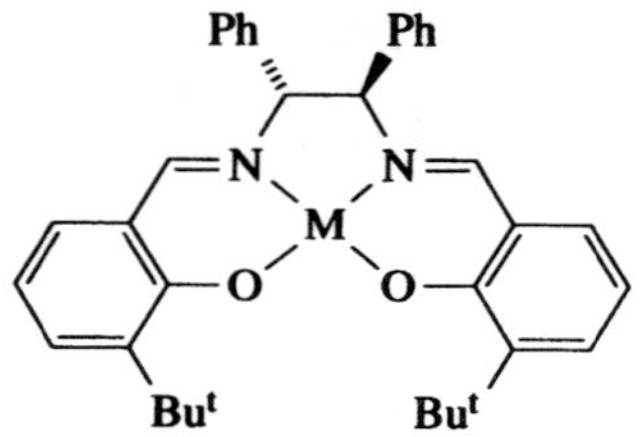 (Ph, Ph, N, M, N, O, O, Bu^t, Bu^t)	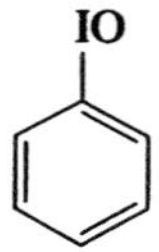 PhIO	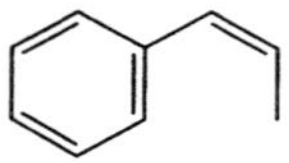	84	Ligand synthesized in simple high yield steps 50 turnovers with NaOCl Complete conversion of alkene Structurally characterized catalyst precursor *trans*-Disubstituted alkenes give only 20–34% *ee*	148g
		or	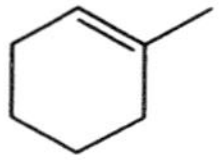	59		
		NaOCl	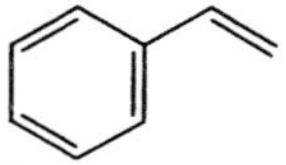	57		

[a] A = (porphyrin, M, N, R)   B = (strapped porphyrin, M, N, R)   C = (strapped porphyrin, M, N, R)

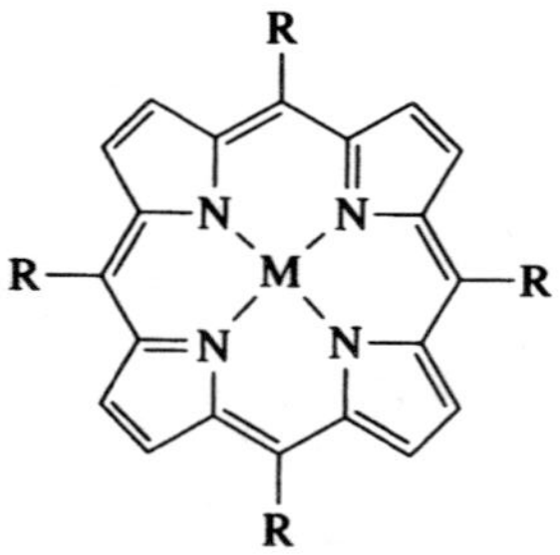

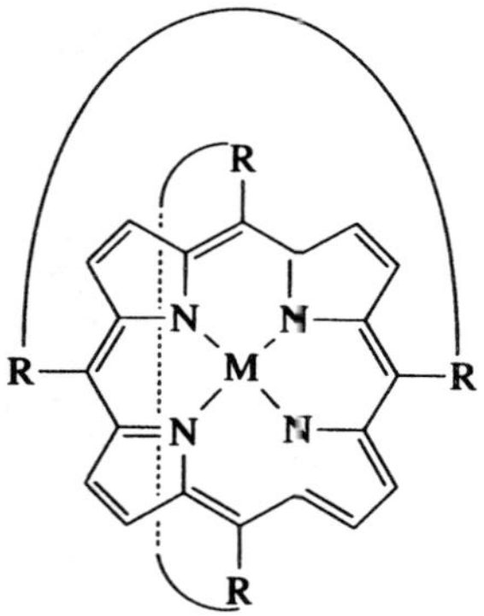

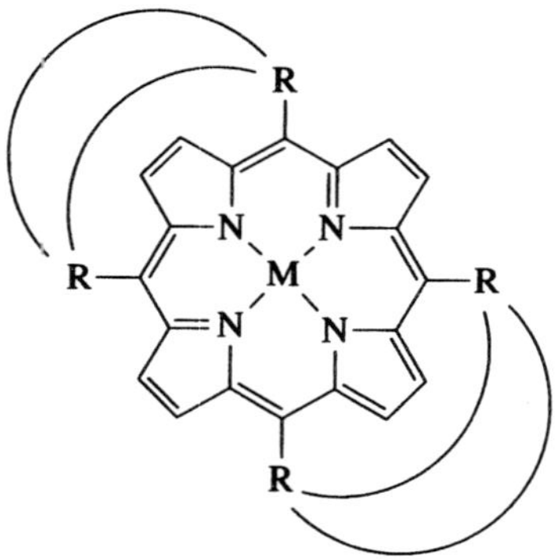

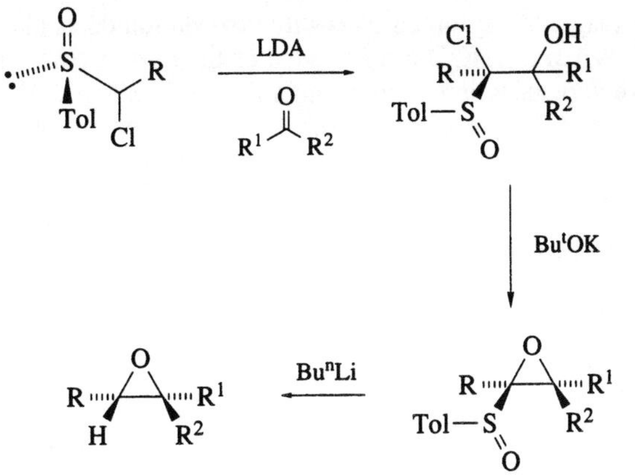

(18)

tions as well as to the nature of the alkyl substituent.[150a] Other routes to asymmetric epoxides have been described which employ enzymatic catalysis. The bacterium *Xanthobacter* Py2 has been used to resolve several 2,3-epoxyalkanes, such as *trans*-2,3-epoxypentane, by selective metabolism of one enantiomer of the substrate.[150b] In a similar vein, both hog pancreatic lipase and hog liver esterase have been used to selectively hydrolyze one enantiomer of 2,3-epoxy alcohol esters and thereby produce the optically active epoxy alcohol.[151]

A final method which has the potential for producing epoxides with high enantiomeric excess is biological epoxidation of alkenes. Microbial epoxidations have been known for at least 30 years[152] and find value as solutions to specific problems. Several examples illustrate the application of this method. Epoxidation of 1,7-octadiene by *Pseudomonas oleovorans* gives (*R*)-7,8-epoxy-1-octene (**93**) in 25% yield and with >80% *ee*.[153] 1-Hexadecene is epoxidized with >95% *ee* and in 41% yield to (*R*)-1,2-epoxyhexadecane (**94**) by *Corynebacterium equi*.[154] Similar epoxidations of terminal alkenes by *Nocardia corallina* yield epoxides with 76–90% *ee*.[155] The antibiotic fosfomycin (**95**) is produced in 90% yield by epoxidation of (*Z*)-2-methyl-1-vinylphosphonic acid with *Penicillium spinulosum*.[156] The optical purity of the epoxide was claimed to be high, based on an optical rotary dispersion measurement. Finally, several organisms useful for dihydroxylation of terpenes, presumably *via* intermediate epoxides, have been described and are the fungi *Corynespora cassicola*,[157a] *Diploda gossypina*[157a] and *Aspergillus niger*.[157b] The first of these is reported to oxidize 1300 g of (*R*)-(+)-limonene to 900 g of the (1*S*,2*S*)-diol (**96**) in 96 h at a level of 20 g l^{-1} of fermentation medium.

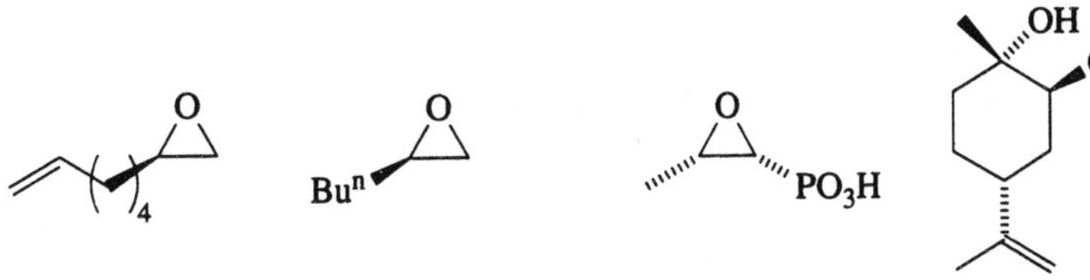

(**93**) (**94**) (**95**) (**96**)

3.2.9 HOMOCHIRAL EPOXIDES *VIA* ASYMMETRIC DIHYDROXYLATION

Allylic alcohols represent a small fraction of the total population of alkenes found in organic molecules. Asymmetric epoxidation of allylic alcohols therefore taps only a small portion of the synthetic potential inherent in a completely general asymmetric epoxidation of isolated (nonfunctionalized) alkenes. A partial solution to this problem now exists. The recent development of a catalytic asymmetric process for the dihydroxylation of alkenes[158] provides an indirect route to epoxides or epoxide-like functionalization of alkenes. The stereochemistry of the process, the scope of enantioselectivity and chemical yield and a summary of key chemical transformations are presented in this section. Since this approach to alkene functionalization is at an early stage of development, the results summarized here are certain to benefit from extensions and improvements as research in this area progresses.

The essential components of the catalyst for the asymmetric dihydroxylation process are osmium tetroxide (OsO_4) and an ester of one or the other of the pseudoenantiomeric cinchona alkaloids dihydroquinidine (DHQD) and dihydroquinine (DHQ). An amine oxide, generally *N*-methylmorpholine *N*-oxide, serves as the oxidant for the reaction.[158] When an alkenic substrate is added very slowly to a mixture of the preceding reagents, asymmetric *cis*-dihydroxylation takes place in accord with the scheme shown in Figure 6.[158,159] When a DHQD ester is used as the asymmetric ligand, oxygen 'delivery' will occur from above the alkene plane when oriented as in Figure 6. When a DHQ ester is the ligand, oxygen delivery will be from below the plane of the alkene. A final ingredient of the reaction is water which is required for hydrolysis of the intermediate osmate ester to regenerate the catalyst and yield the vicinol diol product. The isolated yield of diol usually is in the range of 80–95%.

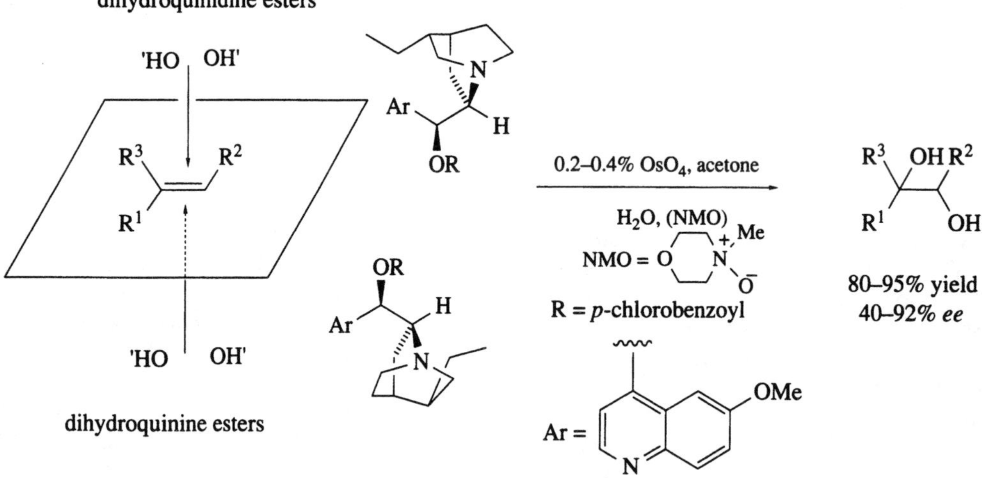

Figure 6 Enantiofacial selectivity in the dihydroxylation of alkenes with osmium tetroxide/alkaloid ester/NMO

The enantioselectivity of the process, as reflected by enantiomeric excesses, shows a rough correlation with the substitution pattern of the alkene, as shown in Figure 7. Highest enantiomeric excesses (70–90%) are obtained with *trans*-1,2-disubstituted alkenes. Trisubstituted (35–80% *ee*) and monosubstituted (25–60% *ee*) alkenes show greater variation in enantioselectivity while poorest enantioselectivity is observed with *cis*-1,2-disubstituted alkenes (<25% *ee*).

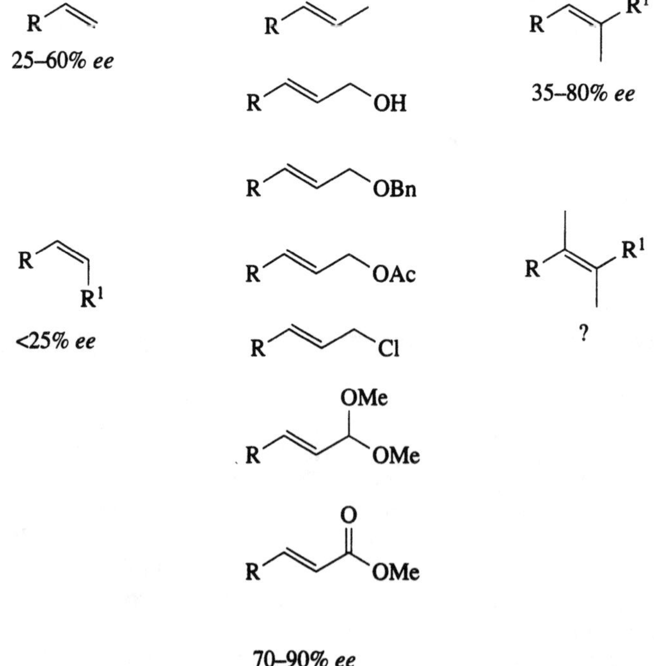

Figure 7 Range of enantioselectivity (expressed as % *ee*) observed in the asymmetric dihydroxylation of alkenes with different substitution patterns

The asymmetric dihydroxylation is one of the simplest catalytic asymmetric processes to perform. The reaction is completely insensitive to water and oxygen and is performed in an open vessel in the range 0–25 °C. The activity of the catalyst is very good so that only 0.2–0.4% is commonly used. There is only weak product inhibition of the reaction so that high substrate concentrations (*e.g.* 2 M) give excellent results. The cinchona alkaloid ligand is easily recovered and reused.

The diols (**97**) from asymmetric dihydroxylation are easily converted to cyclic *sulfite* esters (**98**) and thence to cyclic *sulfate* esters (**99**).[160] This two-step process, reaction of the diol (**97**) with thionyl chloride followed by ruthenium tetroxide catalyzed oxidation, can be done in one pot if desired and transforms the relatively unreactive diol into an epoxide mimic, *i.e.* the 1,2-cyclic sulfate (**99**), which is an excellent electrophile. A survey of reactions shows that cyclic sulfates can be opened by hydride, azide, fluoride, thiocyanide, carboxylate and nitrate ions.[160] Benzylmagnesium chloride and the anion of dimethyl malonate can also be used to open the cyclic sulfates.[160] Opening by a nucleophile leads to formation of an intermediate β-sulfate anion (**100**) which is easily hydrolyzed to a β-hydroxy compound (**101**).[160] Conditions for catalytic acid hydrolysis have been developed that allow for selective removal of the sulfate ester in the presence of other acid sensitive groups such as acetals, ketals and silyl ethers.[161]

$SOCl_2$, CCl_4; $RuCl_3$, $NaIO_4$, MeCN; Nu:; H_3O^+

(**97**) (**98**) (**99**) (**101**) (**100**)

The β-sulfate need not be hydrolyzed and can instead be used as a leaving group for a second nucleophilic displacement reaction. For example, when the first-added nucleophile retains a nucleophilic capability, an intramolecular cyclization can be achieved *via* displacement of the sulfate group. The consequence of this sequence is illustrated by the reaction of cyclic sulfate (**102**) with malonate anion to generate the cyclopropane derivative (**103**).[160] This approach can also be used to prepare aziridines.[162] Opening of the cyclic sulfate (**99**) with azide gives an intermediate azidosulfate (**104**) which, after reduction of the azide to an amino group, undergoes intramolecular cyclization to form an aziridine (**105**). Alternately, opening with a primary amine leads to an aminosulfate (**106**) which can undergo hydrolysis to an amino alcohol (**107**) or cyclization to the *N*-substituted aziridine (**108**).

NaH, $CH_2(CO_2Me)_2$

(**102**) (**103**)

Not surprisingly, the regioselectivity in cyclic sulfate openings is strongly influenced by the nature of groups R^1 and R^2. Notable in this regard is the virtually complete (>100:1) regioselective attack by nucleophiles at the α-carbon of cyclic sulfates adjacent to carboxylic esters such as (**109**).[160] By contrast, analogous glycidic (α,β-epoxy) esters show no clear preference for C-2 *versus* C-3 opening by nucleophiles.

If desired, glycidic esters can be derived from α,β-dihydroxy esters, such as (**110**), by either of two methods. In one method, reaction of the diol with an arenesulfonyl chloride is regioselective, producing the α-arenesulfonate (**111**) in preference to the β-sulfonate. Treatment of (**111**) with an equivalent of base produces the *erythro*-glycidic ester (**112**) in good yield. In the second method, the diol is converted to a bromohydrin (**114**) *via* the acetoxy bromide (**113**). The bromohydrin (**114**) affords the *threo*-glycidic ester (**115**) on exposure to potassium carbonate in methanol.[163]

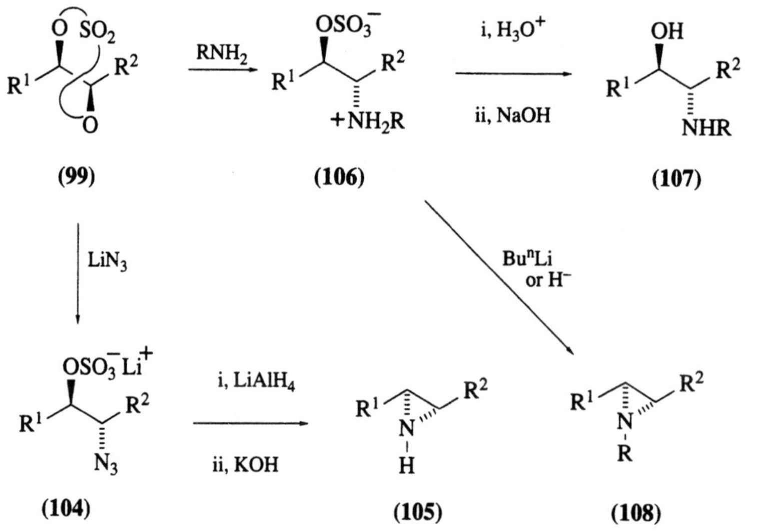

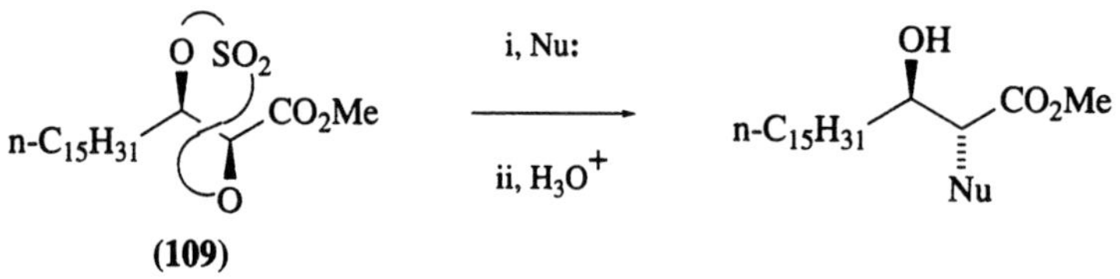

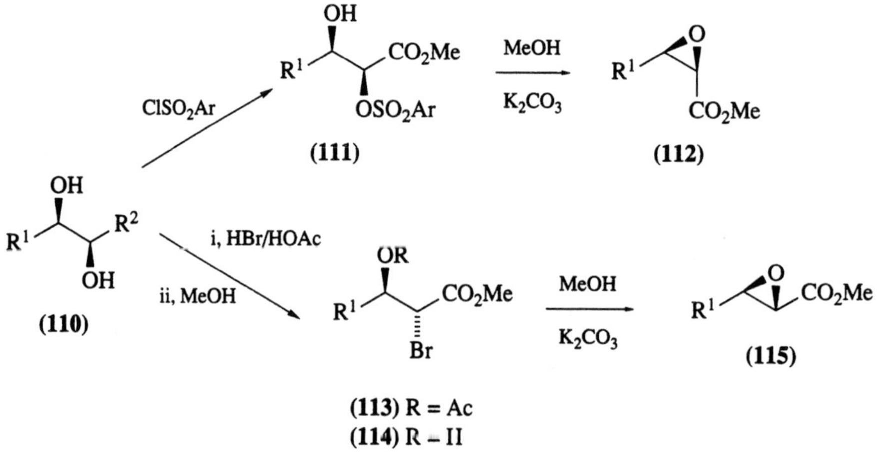

(99) (106) (107)

(104) (105) (108)

(109)

(111) (112)

(110) (113) R = Ac (114) R = H (115)

3.2.10 REFERENCES

1. (a) H. B. Henbest, *Chem. Soc., Spec. Publ.* 1965, **19**, 83; (b) R. C. Ewins, H. B. Henbest and M. A. McKervey, *J. Chem. Soc., Chem. Commun.*, 1967, 1085.
2. M. G. Finn and K. B. Sharpless, in 'Asymmetric Synthesis', ed. J. D. Morrison, Academic Press, New York, 1985, vol. 5, p. 247.
3. T. Katsuki and K. B. Sharpless, *J. Am. Chem. Soc.*, 1980, **102**, 5974.
4. Y. Gao, R. M. Hanson, J. M. Klunder, S. Y. Ko, H. Masamune and K. B. Sharpless, *J. Am. Chem. Soc.*, 1987, **109**, 5765.
5. B. E. Rossiter, in 'Asymmetric Synthesis', ed. J. D. Morrison, Academic Press, New York, 1985, vol. 5, p. 193.
6. A. Pfenniger, *Synthesis*, 1986, 89.
7. K.-P. Zeller, *Methoden Org. Chem. (Houben–Weyl)*, 1988, **E13** (2), 1210.
8. S. Masamune and W. Choy, *Aldrichimica Acta*, 1982, **15**, 47.
9. G. J. McGarvey, M. Kimura, T. Oh and J. M. Williams, *Carbohydr. Chem.*, 1984, **3**, 125.
10. C. H. Behrens and K. B. Sharpless, *Aldrichimica Acta*, 1983, **16**, 67.
11. (a) K. B. Sharpless, *Proc. Robert A. Welch Found. Conf. Chem. Res.*, 1984, **27**, 59; (b) K. B. Sharpless, *CHEMTECH*, 1985, **15**, 692; (c) K. B. Sharpless, *Chem. Br.*, 1986, **22**, 38.
12. T. Katsuki and V. S. Martin, *Org. React. (N.Y.)*, in preparation.

13. V. S. Martin, S. S. Woodard, T. Katsuki, Y. Yamada, M. Ikeda and K. B. Sharpless, *J. Am. Chem. Soc.*, 1981, **103**, 6237.
14. R. M. Hanson and K. B. Sharpless, *J. Org. Chem.*, 1986, **51**, 1922.
15. (a) J. M. Klunder, S. Y. Ko and K. B. Sharpless, *J. Org. Chem.*, 1986, **51**, 3710; (b) S. Y. Ko and K. B. Sharpless, *J. Org. Chem.*, 1986, **51**, 5413.
16. (a) J. M. Klunder, T. Onami and K. B. Sharpless, *J. Org. Chem.*, 1989, **54**, 1295; (b) C. E. Burgos, D. E. Ayer and R. A. Johnson, *J. Org. Chem.*, 1988, **53**, 4973; (c) P. N. Guivisdalsky and R. Bittman, *Tetrahedron Lett.*, 1988, **29**, 4393; (d) R. A. Johnson, C. E. Burgos and E. G. Nidy, *Chem. Phys. Lipids*, 1989, **50**, 119; (e) C. E. Burgos, E. G. Nidy and R. A. Johnson, *Tetrahedron Lett.*, 1989, **30**, 5081; (f) P. N. Guivisdalsky and R. Bittman, *J. Am. Chem. Soc.*, 1989, **111**, 3077; (g) H.-S. Byun and R. Bittman, *Tetrahedron Lett.*, 1989, **30**, 2751; (h) P. N. Guivisdalsky and R. Bittman, *J. Org. Chem.*, 1989, **54**, 4637; (i) P. N. Guivisdalsky and R. Bittman, *J. Org. Chem.*, 1989, **54**, 4643.
17. S. Y. Ko, H. Masamune and K. B. Sharpless, *J. Org. Chem.*, 1987, **52**, 667.
18. L. D.-L. Lu, R. A. Johnson, M. G. Finn and K. B. Sharpless, *J. Org. Chem.*, 1984, **49**, 728.
19. J. G. Hill, B. E. Rossiter and K. B. Sharpless, *J. Org. Chem.*, 1983, **48**, 3607.
20. (a) P. R. Carlier and K. B. Sharpless, *J. Org. Chem.*, 1989, **54**, 4016; (b) C. J. Burns, C. A. Martin and K. B. Sharpless, *J. Org. Chem.*, 1989, **54**, 2826.
21. M. J. Farrall, M. Alexis and M. Trecarten, *Nouv. J. Chim.*, 1983, **7**, 449.
22. (a) W. S. Wadsworth, *Org. React. (N.Y.)*, 1977, **25**, 73; (b) B. J. Walker, in 'Organophosphorus Reagents in Organic Synthesis', ed. J. I. G. Cadogan, Academic Press, New York, 1979, p. 155.
23. W. C. Still and C. Gennari, *Tetrahedron Lett.*, 1983, **24**, 4405.
24. A. J. Mancuso, S.-L. Huang and D. Swern, *J. Org. Chem.*, 1978, **43**, 2480.
25. K. C. Nicolaou, R. A. Daines, J. Uenishi, W. S. Li, D. P. Papahatjia and T. K. Chakraborty, *J. Am. Chem. Soc.*, 1987, **109**, 2205; 1988, **110**, 4672.
26. S. L. Schreiber, T. S. Schreiber and D. B. Smith, *J. Am. Chem. Soc.*, 1987, **109**, 1525.
27. (a) S. E. Denmark and T. K. Jones, *J. Org. Chem.*, 1982, **47**, 4595; (b) T. K. Jones and S. E. Denmark, *Org. Synth.*, 1986, **64**, 182.
28. E. N. Marvell and T. Li, *Synthesis*, 1973, 457.
29. J. Jurczak, S. Pikul and T. Bauer, *Tetrahedron*, 1986, **42**, 447.
30. D. Breitgoff, K. Laumen and M. P. Schneider, *J. Chem. Soc., Chem. Commun.*, 1986, 1523.
31. A. Kleemann and R. Wagner, 'Glycidol', Huethig, New York, 1981.
32. Y. Gao and K. B. Sharpless, *J. Org. Chem.*, 1988, **53**, 4081.
33. J. S. Dung, R. W. Armstrong, O. P. Anderson and R. M. Williams, *J. Org. Chem.*, 1983, **48**, 3592.
34. C. Meister and H.-D. Scharf, *Liebigs Ann. Chem.*, 1983, 913.
35. D. Tanner and P. Somfai, *Tetrahedron*, 1986, **42**, 5985.
36. B. Giese and R. Rupaner, *Liebigs Ann. Chem.*, 1987, 231.
37. K. Mori, T. Ebata and S. Takechi, *Tetrahedron*, 1984, **40**, 1761.
38. (a) M. J. Schweiter and K. B. Sharpless, *Tetrahedron Lett.*, 1985, **26**, 2543; (b) W. Adam, A. Griesbeck and E. Staab, *Tetrahedron Lett.*, 1986, **27**, 2839; (c) W. Adam, M. Braun, A. Griesbeck, V. Lucchini, E. Staab and B. Will, *J. Am. Chem. Soc.*, 1989, **111**, 203.
39. (a) J. D. White and L. R. Jayasinghe, *Tetrahedron Lett.*, 1988, **29**, 2138; (b) J. D. White, J. C. Amedio, Jr., S. Gut and L. R. Jayasinghe, *J. Org. Chem.*, 1989, **54**, 4268.
40. (a) S. Masamune, W. Choy, J. S. Petersen and L. R. Sita, *Angew. Chem., Int. Ed. Engl.*, 1985, **24**, 1; (b) Y. Nishikimi, T. Iimori, M. Sodeoka and M. Shibasaki, *J. Org. Chem.*, 1989, **54**, 3354.
41. B. E. Rossiter, T. Katsuki and K. B. Sharpless, *J. Am. Chem. Soc.*, 1981, **103**, 464.
42. R. Baker, W. J. Cummings, J. F. Hayes and A. Kumar, *J. Chem. Soc., Chem. Commun.*, 1986, 1237.
43. C. Kuroda, P. Theramongkol, J. R. Engebrecht and J. D. White, *J. Org. Chem.*, 1986, **51**, 956.
44. M. Honda, T. Katsuki and M. Yamaguchi, *Tetrahedron Lett.*, 1984, **25**, 3857.
45. L. A. Gorthy, M. Vairamani and C. Djerassi, *J. Org. Chem.*, 1984, **49**, 1511.
46. S. Hanessian, A. Ugolini, D. Dube, P. J. Hodges and C. Andre, *J. Am. Chem. Soc.*, 1986, **108**, 2776.
47. S. Wershofen and H.-D. Scharf, *Synthesis*, 1988, 854.
48. R. D. Tung and D. H. Rich, *Tetrahedron Lett.*, 1987, **28**, 1139.
49. G. A. Molander and G. Hahn, *J. Org. Chem.*, 1986, **51**, 2596.
50. E. J. Corey, S. G. Pyne and W.-G. Su, *Tetrahedron Lett.*, 1983, **24**, 4883.
51. J. Furukawa, S. Iwasaki and S. Okuda, *Tetrahedron Lett.*, 1983, **24**, 5257.
52. Y. Kitano, Y. Kobayashi and F. Sato, *J. Chem. Soc., Chem. Commun.*, 1985, 498.
53. W. R. Roush and M. A. Adam, *J. Org. Chem.*, 1985, **50**, 3752.
54. A. C. Oehlschlager and B. D. Johnston, *J. Org. Chem.*, 1987, **52**, 940.
55. C. K. Lai and M. Gut, *J. Org. Chem.*, 1987, **52**, 685.
56. R. E. Dolle and K. C. Nicolaou, *J. Am. Chem. Soc.*, 1985, **107**, 1691.
57. L. N. Pridgen, S. C. Shilcrat and I. Lantos, *Tetrahedron Lett.*, 1984, **25**, 2835.
58. S. R. Baker, J. R. Boot, S. E. Morgan, D. J. Osborne, W. J. Ross and P. R. Shrubsall, *Tetrahedron Lett.*, 1983, **24**, 4469.
59. T. Katsuki, A. W. M. Lee, P. Ma, V. S. Martin, S. Masamune, K. B. Sharpless, D. Tuddenham and F. J. Walker, *J. Org. Chem.*, 1982, **47**, 1373.
60. P. Ma, V. S. Martin, S. Masamune, K. B. Sharpless and S. M. Viti, *J. Org. Chem.*, 1982, **47**, 1378.
61. (a) H. Brunner and A. Sicheneder, *Angew. Chem., Int. Ed. Engl.*, 1988, **27**, 718; (b) D. A. Evans and J. M. Williams, *Tetrahedron Lett.*, 1988, **29**, 5065; (c) P. Hughes and J. Clardy, *J. Org. Chem.*, 1989, **54**, 3260; (d) K. C. Nicolaou, C. V. C. Prasad, P. K. Somers and C.-K. Hwang, *J. Am. Chem. Soc.*, 1989, **111**, 5330.
62. (a) N. Minami, S. S. Ko and Y. Kishi, *J. Am. Chem. Soc.*, 1982, **104**, 1109; (b) W. R. Roush, M. A. Adam, A. E. Walts and D. J. Harris, *J. Am. Chem. Soc.*, 1986, **108**, 3422.
63. (a) S. Y. Ko, A. W. M. Lee, S. Masamune, L. A. Reed, III, K. B. Sharpless and F. J. Walker, *Science (Washington, D.C.)*, 1983, **220**, 949; (b) S. Y. Ko, A. W. M. Lee, S. Masamune, L. A. Reed, III, K. B.

Sharpless and F. J. Walker, *Tetrahedron*, 1990, **46**, 245; (c) L. A. Reed, III, Y. Ito, S. Masamune and K. B. Sharpless, *J. Am. Chem. Soc.*, 1982, **104**, 6468.
64. J. E. Wrobel and B. Ganem, *J. Org. Chem.*, 1983, **48**, 3761.
65. C. H. Behrens, S. Y. Ko, K. B. Sharpless and F. J. Walker, *J. Org. Chem.*, 1985, **50**, 5687.
66. (a) T. R. Hoye and J. C. Suhadolnik, *J. Am. Chem. Soc.*, 1985, **107**, 5312; (b) T. R. Hoye and J. C. Suhadolnik, *Tetrahedron*, 1986, **42**, 2855.
67. A. M. Doherty and S. V. Ley, *Tetrahedron Lett.*, 1986, **27**, 105.
68. J. E. Baldwin and A. Flinn, *Tetrahedron Lett.*, 1987, **28**, 3605.
69. M. Caron and K. B. Sharpless, *J. Org. Chem.*, 1985, **50**, 1557.
70. J. M. Chong and K. B. Sharpless, *J. Org. Chem.*, 1985, **50**, 1560.
71. R. Baker, C. J. Swain and J. C. Head, *J. Chem. Soc., Chem. Commun.*, 1986, 874.
72. R. D. Wood and B. Ganem, *Tetrahedron Lett.*, 1982, **23**, 707.
73. T. Takahashi, M. Miyazawa, H. Veno and J. Tsuji, *Tetrahedron Lett.*, 1986, **27**, 3881.
74. L. S. Mills and P. C. North, *Tetrahedron Lett.*, 1983, **24**, 409.
75. (a) M. Suzuki, Y. Morita, A. Yanagisawa, R. Noyori, B. J. Baker and P. J. Scheuer, *J. Am. Chem. Soc.*, 1986, **108**, 5021; (b) M. Suzuki, Y. Morita, A. Yanagisawa, B. J. Baker, P. J. Scheuer and R. Noyori, *J. Org. Chem.*, 1988, **53**, 286.
76. J. N. Denis, A. E. Greene, A. A. Serra and M.-J. Luche, *J. Org. Chem.*, 1986, **51**, 46.
77. K. B. Sharpless, C. H. Behrens, T. Katsuki, A. W. M. Lee, V. S. Martin, M. Takatani, S. M. Viti, F. J. Walker and S. S. Woodard, *Pure Appl. Chem.*, 1983, **55**, 589.
78. (a) H. Nagaoka and Y. Kishi, *Tetrahedron*, 1981, **37**, 3873; (b) Y. Hirai, M. Chintani, T. Yamazaki and T. Momose, *Chem. Lett.*, 1989, 1449; (c) T. Ebata and K. Mori, *Agric. Biol. Chem.*, 1989, **53**, 801; (d) K. Mori and T. Takeuchi, *Liebigs Ann. Chem.*, 1989, 453; (e) R. K. Boeckman, Jr. and J. R. Pruitt, *J. Am. Chem. Soc.*, 1989, **111**, 8286.
79. W. C. Still and H. Ohmizu, *J. Org. Chem.*, 1981, **46**, 5242.
80. P. Garner, J. M. Park and V. Rotello, *Tetrahedron Lett.*, 1985, **26**, 3299.
81. A. I. Meyers and J. P. Hudspath, *Tetrahedron Lett.*, 1981, **22**, 3925.
82. N. Niwa, Y. Miyachi, Y. Uosaki and K. Yamada, *Tetrahedron Lett.*, 1986, **27**, 4601.
83. K. Takabe, K. Okisaki, Y. Uchiyama, T. Katagiri and H. Yoda, *Chem. Lett.*, 1985, 561.
84. (a) K. Mori and H. Ueda, *Tetrahedron*, 1981, **37**, 2581; (b) K. C. Nicolaou, C. V. C. Prasad, C.-K. Hwang, M. E. Duggan and C. A. Veale, *J. Am. Chem. Soc.*, 1989, **111**, 5321.
85. J. C. Medina and K. S. Kyler, *J. Am. Chem. Soc.*, 1988, **110**, 4818.
86. M. Aziz and F. Rouessac, *Tetrahedron*, 1988, **44**, 101.
87. Y. Morimoto, K. Oda, H. Shirahama, T. Matsumoto and S. Omura, *Chem. Lett.*, 1988, 909.
88. (a) K. Takahashi and M. Ogata, *J. Org. Chem.*, 1987, **52**, 1877; (b) Y. Naruta, Y. Nishigaichi and K. Maruyama, *Tetrahedron Lett.*, 1989, **30**, 3319; (c) K. C. Nicolaou, M. E. Duggan and C.-K. Hwang, *J. Am. Chem. Soc.*, 1989, **111**, 6676.
89. W. H. Rastetter and J. Adams, *Tetrahedron Lett.*, 1982, **23**, 1319.
90. D. A. Evans, S. L. Bender and J. Morris, *J. Am. Chem. Soc.*, 1988, **110**, 2506.
91. K. S. Reddy, O. H. Ko, D. Ho, P. E. Persons and J. M. Cassidy, *Tetrahedron Lett.*, 1987, **28**, 3075.
92. S. Yamada, M. Shiraishi, M. Ohmori and H. Takayama, *Tetrahedron Lett.*, 1984, **25**, 3347.
93. K. Mori and K. Okada, *Tetrahedron*, 1985, **41**, 557.
94. W. R. Roush and T. A. Blizzard, *J. Org. Chem.*, 1984, **49**, 4332.
95. F. Bonadies, G. Rossi and C. Bonini, *Tetrahedron Lett.*, 1984, **25**, 5431.
96. M. Sodeoka, T. Iimori and M. Shibasaki, *Tetrahedron Lett.*, 1985, **26**, 6497.
97. (a) A. G. Myers, P. J. Porteau and T. M. Handel, *J. Am. Chem. Soc.*, 1988, **110**, 7212; (b) D. R. Williams, D. L. Brown and J. W. Benbow, *J. Am. Chem. Soc.*, 1989, **111**, 1923.
98. T. J. Erickson, *J. Org. Chem.*, 1986, **51**, 934.
99. J. A. Marshall and T. M. Jenson, *J. Org. Chem.*, 1984, **49**, 1707.
100. D. P. G. Hamon and N. J. Shirley, *J. Chem. Soc., Chem. Commun.*, 1988, 425.
101. L. Pettersson, T. Frejd and G. Magnusson, *Tetrahedron Lett.*, 1987, **28**, 2753.
102. J. P. Rizzi and A. S. Kende, *Tetrahedron*, 1984, **40**, 4693.
103. M. Acemoglu, P. Uebelhart, M. Rey and C. H. Eugster, *Helv. Chim. Acta*, 1988, **71**, 931.
104. P. R. Carlier, W. S. Mungall, G. Schröder and K. B. Sharpless, *J. Am. Chem. Soc.*, 1988, **110**, 2978.
105. (a) Y. Kitano, T. Matsumoto, Y. Takeda and F. Sato, *J. Chem. Soc., Chem. Commun.*, 1986, 1323; (b) Y. Kitano, T. Matsumoto and F. Sato, *Tetrahedron*, 1988, **44**, 4073; (c) Y. Kitano, S. Okamoto and F. Sato, *Chem. Lett.*, 1989, 2163.
106. (a) A. T. Russell and G. Procter, *Tetrahedron Lett.*, 1987, **28**, 2041; (b) G. Procter, A. T. Russell, P. J. Murphy, T. S. Tan and A. N. Mather, *Tetrahedron*, 1988, **44**, 3953.
107. (a) T. Katsuki, *Tetrahedron Lett.*, 1984, **25**, 2821; (b) J. M. Schwab and C.-K. Ho, *J. Chem. Soc., Chem. Commun.*, 1986, 872; (c) J. M. Schwab, T. Ray and C.-K. Ho, *J. Am. Chem. Soc.*, 1989, **111** 1057.
108. Y. Kitano, T. Matsumoto, T. Wakasa, S. Okamoto, T. Shimazaki, Y. Kobayashi, F. Sato, K. Miyaji and K. Arai, *Tetrahedron Lett.*, 1987, **28**, 6351.
109. Y. Kitano, T. Matsumoto, S. Okamoto, T. Shimazaki, Y. Kobayashi and F. Sato, *Chem. Lett.*, 1987, 1523.
110. W. R. Roush and A. P. Spada, *Tetrahedron Lett.*, 1983, **24**, 3693.
111. P. A. Aristoff, P. D. Johnson and A. W. Harrison, *J. Am. Chem. Soc.*, 1985, **107**, 7967.
112. L. E. Overman and N.-H. Lin, *J. Org. Chem.*, 1985, **50**, 3670.
113. S. K. Aggarwal, J. S. Bradshaw, M. Eguchi, S. Parry, B. E. Rossiter, K. E. Markides and M. L. Lee, *Tetrahedron*, 1987, **43**, 451.
114. M. Ogata, H. Matsumoto, K. Takahashi, S. Shimizu, S. Kida, A. Murahayashi, M. Shiro and K. Tawara, *J. Med. Chem.*, 1987, **30**, 1054.
115. (a) S. Sugiyama, M. Honda and T. Komori, *Liebigs Ann. Chem.*, 1988, 619; (b) A. Tanaka, H. Suzuki and K. Yamashita, *Agric. Biol. Chem.*, 1989, **53**, 2253; (c) A. R. Chamberlin, M. Dezube, S. H. Reich and D. J. Sall,

J. Am. Chem. Soc., 1989, **111**, 6247; (d) A. V. Rama Rao, A. P. Khrimian, P. Radha Krishna, P. Yagadiri and J. S. Yadav, *Synth. Commun.*, 1989, **18**, 2325; (e) R. P. Discordia and D. C. Dittmer, *J. Org. Chem.*, 1990, **55**, 1414.

116. J. A. Marshall and K. E. Flynn, *J. Am. Chem. Soc.*, 1982, **104**, 7430.
117. E. D. Mihelich, unpublished results.
118. E. Alvarez, E. Manta, J. D. Martin, M. L. Rodriquez and C. Ruiz-Perez, *Tetrahedron Lett.*, 1988, **29**, 2093.
119. R. C. Ronald, S. M. Ruder and T. S. Lillie, *Tetrahedron Lett.*, 1987, **28**, 131.
120. I. V. Ekhato, J. V. Silverton and C. H. Robinson, *J. Org. Chem.*, 1988, **53**, 2180.
121. (a) J. A. Marshall and K. E. Flynn, *J. Am. Chem. Soc.*, 1984, **106**, 723; (b) J. A. Marshall and V. H. Audia, *J. Org. Chem.*, 1987, **52**, 1106.
122. (a) S. Hatakeyama, K. Sakurai and S. Takano, *J. Chem. Soc., Chem. Commun.*, 1985, 1759; (b) B. Hafele, D. Schroter and V. Jager, *Angew. Chem., Int. Ed. Engl.*, 1986, **25**, 87; (c) R. E. Babine, *Tetrahedron Lett.*, 1986, **27**, 5791; (d) D. Askin, R. P. Volante, R. A. Reamer, K. M. Ryan and I. Shinkai, *Tetrahedron Lett.*, 1988, **29**, 277.
123. S. Bergens and B. Bosnich, *Comments Inorg. Chem.*, 1987, **6**, 85.
124. S. L. Schreiber, M. T. Goulet and G. Schule, *J. Am. Chem. Soc.*, 1987, **109**, 4718.
125. K. Mori and Y. B. Seu, *Tetrahedron*, 1985, **41**, 3429.
126. K. Mori and T. Otsuka, *Tetrahedron*, 1985, **41**, 553.
127. M. D. Lewis, J. P. Duffy, B. E. Blough and T. D. Crute, *Tetrahedron Lett.*, 1988, **29**, 2279.
128. E. J. Corey and A. Tramontano, *J. Am. Chem. Soc.*, 1984, **106**, 462.
129. (a) W. R. Roush and R. J. Brown, *J. Org. Chem.*, 1982, **47**, 1373; (b) W. R. Roush and R. J. Brown, *J. Org. Chem.*, 1983, **48**, 5093; (c) P. C. Bulman-Page, J. F. Carefull, L. H. Powell and I. O. Sutherland, *J. Chem. Soc., Chem. Commun.*, 1985, 822.
130. (a) U. Kufner and R. R. Schmidt, *Angew. Chem., Int. Ed. Engl.*, 1986, **25**, 89; (b) G. Frater and J. Muller, *Helv. Chim. Acta*, 1989, **72**, 653; (c) T. Ibuka, M. Tanaka and Y. Yamamoto, *J. Chem. Soc., Chem. Commun.*, 1989, 967.
131. T. Katsuki and K. B. Sharpless, unpublished results.
132. B. E. Rossiter and K. B. Sharpless, *J. Org. Chem.*, 1984, **49**, 3707.
133. S. Ikegami, T. Katsuki and M. Yamaguchi, *Chem. Lett.*, 1987, 83.
134. (a) E. J. Corey and D.-C. Ha, *Tetrahedron Lett.*, 1988, **29**, 3171; (b) T. Hosokawa, T. Kono, T. Shinohara and S.-I. Murahashi, *J. Organomet. Chem.*, 1989, **370**, C13.
135. K. B. Sharpless, S. S. Woodard and M. G. Finn, *Pure Appl. Chem.*, 1983, **55**, 1823.
136. (a) S. S. Woodard, M. G. Finn and K. B. Sharpless, *J. Am. Chem. Soc.*, 1991, **113**, 106; (b) M. G. Finn and K. B. Sharpless, *J. Am. Chem. Soc.*, 1991, **113**, 113.
137. E. J. Corey, *J. Org. Chem.*, 1990, **55**, 1693.
138. I. D. Williams, S. F. Pedersen, K. B. Sharpless and S. J. Lippard, *J. Am. Chem. Soc.*, 1984, **106**, 6430.
139. H. Mimoun, P. Chaumette, M. Mignard, L. Saussine, J. Fischer and R. Weiss, *Nouv. J. Chim.*, 1983, **7**, 467.
140. (a) K. A. Jorgensen, R. A. Wheeler and R. Hoffmann, *J. Am. Chem. Soc.*, 1987, **109**, 3240; (b) K. A. Jorgensen, *Chem. Rev.*, 1989, **89**, 431.
141. (a) T. Katsuki and K. B. Sharpless, *US Pat.* 4 471 130 (1984) (*Chem. Abstr.*, 1985, **102**, 24 872m); (b) S. Miyano, L. D.-L. Lu, S. M. Viti and K. B. Sharpless, *J. Org. Chem.*, 1983, **48**, 3611; (c) S. Miyano, L. D.-L. Lu, S. M. Viti and K. B. Sharpless, *J. Org. Chem.*, 1985, **50**, 4350.
142. (a) Y. Kobayashi, M. Kusakabe, Y. Kitano and F. Sato, *J. Org. Chem.*, 1988, **53**, 1586; (b) T. Kamatani, M. Tsubuki, Y. Tatsuzaki and T. Honda, *Heterocycles*, 1988, **27**, 2107; (c) M. Kusakabe, Y. Kitano, Y. Kobayashi and F. Sato, *J. Org. Chem.*, 1989, **54**, 2085; (d) M. Kusakabe and F. Sato, *J. Org. Chem.*, 1989, **54**, 3486; (e) T. Kamatani, Y. Tatsuzaki, M. Tsubuki and T. Honda, *Heterocycles*, 1989, **29**, 1247.
143. Y. Kitano, M. Kusakabe, Y. Kobayashi and F. Sato, *J. Org. Chem.*, 1989, **54**, 994.
144. T. H. Chan, L. M. Chen and D. Wang, *J. Chem. Soc., Chem. Commun.*, 1988, 1280.
145. (a) M. Kihara, K. Ohnishi and S. Kobayashi, *Heterocycles*, 1988, **25**, 161; (b) T. Jikihara, M. Katsurada, O. Ikeda, K. Yoneyama and T. Takematsu, *Natl. Meet. Am. Chem. Soc., 198th*, 1989, ORGN 220.
146. (a) P. Pitchen, E. Duñach, M. N. Deshmukh and H. B. Kagan, *J. Am. Chem. Soc.*, 1984, **106**, 8188; (b) E. Duñach and H. B. Kagan, *Nouv. J. Chim.*, 1985, **9**, 1; (c) S. H. Zhao, O. Samuel and H. B. Kagan, *Tetrahedron*, 1987, **43**, 5144.
147. (a) O. Bortolini, F. Di Furia, G. Licini and G. Modena, *Phosphorus Sulfur*, 1988, **37**, 171; (b) F. A. Davis and A. C. Sheppard, *Tetrahedron*, 1989, **45**, 5703.
148. (a) K. Tani, M. Hanafusa and S. Otsuka, *Tetrahedron Lett.*, 1979, 3017; (b) J. T. Groves and R. S. Myers, *J. Am. Chem. Soc.*, 1983, **105**, 5791; (c) D. Mansuy, P. Battioni, J.-P. Renaud and P. Guerin, *J. Chem. Soc., Chem. Commun.*, 1985, 155; (d) R. Sinigalia, R. A. Michelin, F. Pinna and G. Strukul, *Organometallics*, 1987, **6**, 728; (e) Y. Naruta, F. Tani and K. Maruyama, *Chem. Lett.*, 1989, 1269; (f) S. O'Malley and T. Kodadek, *J. Am. Chem. Soc.*, 1989, **111**, 9116; (g) W. Zhang, J. L. Loebach, S. R. Wilson and E. N. Jacobsen, *J. Am. Chem. Soc.*, 1990, **112**, 2801
149. (a) K. Yamamoto, H. Ando, T. Shuetake and H. Chikamatsu, *J. Chem. Soc., Chem. Commun.*, 1989, 754; (b) N. Furukawa, Y. Shugihara and H. Fujihara, *J. Org. Chem.*, 1989, **54**, 4222; (c) T. Satoh, T. Oohara, Y. Ueda and K. Yamakawa, *J. Org. Chem.*, 1989, **54**, 3130.
150. (a) S. Colonna, N. Gaggero, A. Manfredi and M. Spadoni, *Tetrahedron*, 1988, **44**, 5169; (b) C. A. G. M. Weijers, A. de Haan and J. A. M. de Bont, *Appl. Microbiol. Biotechnol.*, 1988, **27**, 337.
151. (a) W. E. Ladner and G. M. Whitesides, *J. Am. Chem. Soc.*, 1984, **106**, 7250; (b) J. L. Pawlak and G. A. Berchtold, *J. Org. Chem.*, 1987, **52**, 1765; (c) A. Classen, S. Wershofen, A. Yusufoglu and H.-D. Scharf, *Liebigs Ann. Chem.*, 1987, 629; (d) D. Bianchi, W. Cabri, P. Cesti, F. Francalanci and F. Fama, *Tetrahedron Lett.*, 1988, **29**, 2455; (e) B. A. Marples and M. Rogers-Evans, *Tetrahedrom Lett.*, 1989, **30**, 261; (f) P. Mohr, L. Rosslein and C. Tamm, *Tetrahedron Lett.*, 1989, **30**, 2513.

152. (a) G. S. Fonken and R. A. Johnson, 'Chemical Oxidations with Microorganisms', Dekker, New York, 1972, p. 113; (b) R. A. Johnson, in 'Oxidation in Organic Chemistry', ed. W. S. Trahanovsky, Academic Press, New York, 1978, part C, p. 131.
153. S. W. May and R. D. Schwartz, *J. Am. Chem. Soc.*, 1974, **96**, 4031.
154. H. Ohta and H. Tetsukawa, *J. Chem. Soc., Chem. Commun.*, 1978, 849.
155. (a) O. Takahashi, J. Umezawa, K. Furuhashi and M. Takagi, *Tetrahedron Lett.*, 1989, **30**, 1583; (b) K. Furuhashi, A. Taoka, S. Uchida, I. Karube and S. Suzuki, *Eur. J. Appl. Microbiol. Biotechnol.*, 1981, **12**, 39.
156. R. F. White, J. Birnbaum, R. T. Meyer, J. ten Brocke, J. M. Chemerda and A. L. Demain, *Appl. Microbiol.*, 1971, **22**, 55.
157. (a) W. R. Abraham, H. M. R. Hoffmann, K. Kieslich, G. Reng and B. Stumpf, in 'Enzymes in Organic Synthesis', Pitman, London, 1985, p. 146; (b) J. D. Fourneron, A. Archelas and R. Furstoss, *J. Org. Chem.*, 1989, **54**, 4686.
158. E. N. Jacobsen, I. Marko, W. S. Mungall, G. Schröder and K. B. Sharpless, *J. Am. Chem. Soc.*, 1988, **110**, 1968.
159. J. S. M. Wai, I. Marko, J. S. Svendsen, M. G. Finn, E. N. Jacobsen and K. B. Sharpless, *J. Am. Chem. Soc.*, 1989, **111**, 1123.
160. Y. Gao and K. B. Sharpless, *J. Am. Chem. Soc.*, 1988, **110**, 7538.
161. B. M. Kim and K. B. Sharpless, *Tetrahedron Lett.*, 1989, **30**, 655.
162. B. B. Lohray, Y. Gao and K. B. Sharpless, *Tetrahedron Lett.*, 1989, **30**, 2623.
163. P. Fleming and K. B. Sharpless, *J. Org. Chem.*, in press.

3.3
Addition Reactions with Formation of Carbon–Oxygen Bonds: (iii) Glycol Forming Reactions

ALAN H. HAINES

University of East Anglia, Norwich, UK

3.3.1 INTRODUCTION

The addition of hydroxy groups to the carbon–carbon double bond of an alkene (equation 1) is classed under the IUPAC nomenclature for transformations[1] as a dihydroxy addition. Many reagents can bring about this transformation,[2–9] which can proceed in a *syn* or *anti* manner, as shown in equation (2), and the type of addition which occurs depends on the reagent.

HO OH

(1)

R[1] R[4] R[1] R[4] R[1] OH

R[2] R[3] R[2] R[3] or R[2] R[3] (2)

HO OH HO R[4]

syn addition *anti* addition

Syn hydroxylation is most commonly performed using osmium tetroxide or potassium permanganate, and addition usually occurs from the less hindered side of the double bond in the absence of other directing effects. With these reagents, the stereospecificity of *syn* addition results from formation of cyclic ester intermediates. Support for these arises from isolation of solid addition complexes (**1**) on reaction of alkenes with osmium tetroxide in the presence of tertiary amines (L),[6] and from spectral measurements during permanganate hydroxylations, which have been interpreted[9] in terms of the intermediate (**2**).

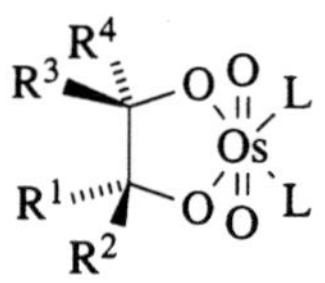

(1) 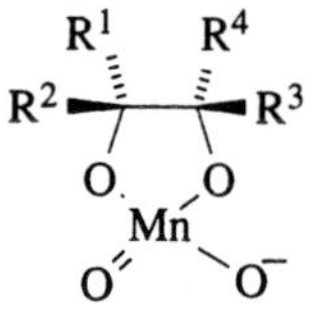 **(2)**

The oxidation of alkenes by osmium tetroxide has been thought, in general, to proceed by a direct oxygen attack at the unsaturated centers in a concerted [3 + 2] cycloaddition step, affording the six-electron transition state (**3**) in equation (3a). This type of transition state is supported by molecular orbital calculations.[10] An alternative mechanism[11,12] involves the intermediacy of the metallocycle (**4**; equation 3b), which might arise from an initially formed complex containing osmium π-bonded to the alkene or, possibly, by a direct [2 + 2] cycloaddition.[12] (A modified mechanism in which the initial π-complex reacts with a ligand to afford a complex like (**4**), but with the ligand attached to osmium, has been suggested.[12] This latter complex then reacts with a second molecule of the ligand, inducing formation of the complex **1**.) Evidence that (**4**) is a possible intermediate in alkene hydroxylation with the tetroxide comes from the enantioselective oxidation of (*E*)-stilbene in the presence of a chiral diamine[13] (see Section 3.3.2.1).

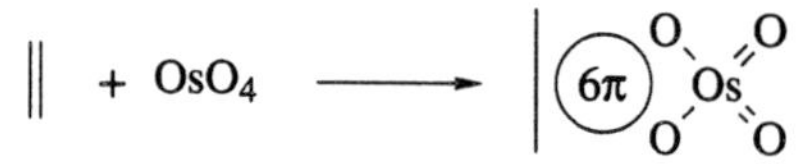

$$\| + OsO_4 \longrightarrow \textbf{(3)} \tag{3a}$$

$$\| + OsO_4 \longrightarrow \textbf{(4)} \tag{3b}$$

Syn hydroxylation from the more hindered face of a π-system can be effected using Woodward's procedure[2,4,5,14] in which an alkene is treated with iodine–silver acetate in acetic acid containing water. Variants of this method avoid the use of silver salts.[15–19] A versatile procedure by which *syn* hydroxylation can be performed on either the more hindered or less hindered face of an alkene relies on stereoselective formation of the appropriate *trans*-bromohydrin from the alkene.[20]

Anti hydroxylation of an alkene is readily achieved with peroxycarboxylic acids.[2,4,5] Acid-catalyzed ring opening of the initial product, an oxirane (epoxide), forms the monoester of a 1,2-diol, hydrolysis of which affords the parent diol. Alternative reagents which are often used for *anti* hydroxylation of alkenes are hydrogen peroxide with oxides of tungsten[2,4,5,21] or selenium,[2,4,5,21] and iodine–silver benzoate (Prévost reaction).[2,4,5]

With careful choice of reagent and reaction conditions, alkenes containing other functionalities can be selectively hydroxylated without complicating side reactions. For example, the oxidation may be carried out in the presence of ester, ether, sulfide, carboxylic acid, acetal, carbonyl, halo, alcohol and aryl groups. Regioselective hydroxylation is also possible in dienes in which one center is electron poor, and some selectivity is also found between isolated double bonds. For example, *syn* hydroxylation of diene (**5**) with a catalytic amount of osmium tetroxide and *N*-methylmorpholine *N*-oxide as the secondary oxidant gives diol (**6**)[22] in 46% yield, and phase transfer catalyzed permanganate oxidation of diene (**7**) affords diol (**8**)[23] in 83% yield.

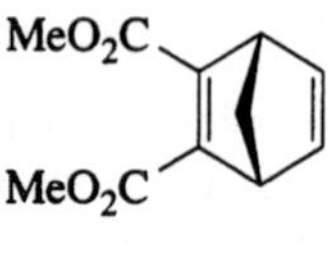

(5) 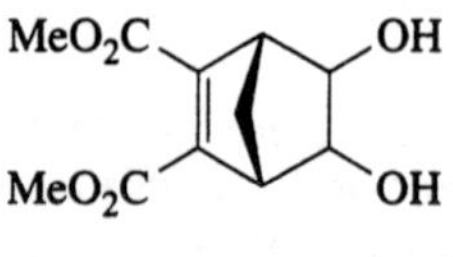 **(6)** 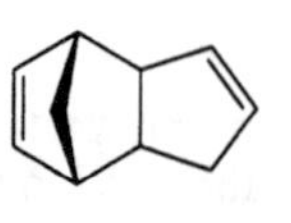 **(7)** 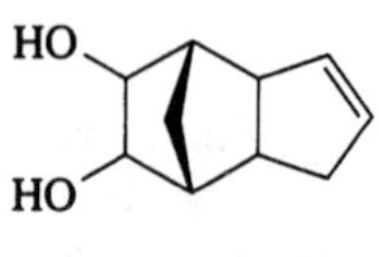 **(8)**

The diastereoselective and enantioselective oxidation of alkenes with osmium tetroxide is considered in Section 3.3.2.1.

3.3.2 *SYN* HYDROXYLATION METHODS

3.3.2.1 Osmium Tetroxide

Despite the development in recent years of new reagents for organic synthesis, osmium tetroxide, used either stoichiometrically or catalytically, remains the reagent of choice for *syn* hydroxylation of alkenes.[5–7] Surprisingly, the hydroxylation of alkynes to give[5–7,24,25] 1,2-dicarbonyl compounds has been much less studied, but it has considerable potential for use in the synthesis of complex molecules.

For the oxidation of alkenes, osmium tetroxide is used either stoichiometrically, when the alkene is precious or only small scale operation is required, or catalytically with a range of secondary oxidants which include metal chlorates, hydrogen peroxide, *t*-butyl hydroperoxide and *N*-methylmorpholine *N*-oxide. The osmium tetroxide/*N*-methylmorpholine *N*-oxide combination is probably the most general and effective procedure which is currently available for the *syn* hydroxylation of alkenes,[26,27] although tetrasubstituted alkenes may be resistant to oxidation.[28] For hindered alkenes, use of the related oxidant trimethylamine *N*-oxide in the presence of pyridine appears advantageous.[29] When *t*-butyl hydroperoxide is used as a cooxidant, problems of overoxidation are avoided which occasionally occur with the catalytic procedures using metal chlorates or hydrogen peroxide. Further, in the presence of tetraethylammonium hydroxide[30] hydroxylation of tetrasubstituted alkenes is possible, but the alkaline conditions clearly limit the application.

With osmium tetroxide, considerable diastereoselectivity may be achieved in the stoichiometric and catalytic hydroxylation of allylic alcohol systems in which the oxygen-containing group forms part of a chiral center (equation 4).[31–34]

$$R^2O\text{–CH}(R^1)\text{–CH=C}R^3R^4 \xrightarrow{OsO_4} \text{(HO, OH; }R^2O, R^1, R^3, R^4\text{)} + \text{(}R^1, R^2O, R^3, R^4\text{; HO, OH)} \quad (4)$$

From a study embracing a large number of substrates, the following general observations were formulated: (i) the stoichiometric hydroxylation procedure provides a slightly higher diastereoselectivity than the catalytic procedure; (ii) protecting groups of the hydroxy at the chiral center, except acyl groups, have only a limited effect in determining the stereochemical course of the oxidation, while with acyl derivatives stereoselectivity is noticeably diminished or is absent; (iii) a hydroxy or alkoxy oxygen seems to play the important role in governing the high degree of stereoselectivity; (iv) the degree of selectivity observed with (*Z*)-alkenes is higher than that for the corresponding (*E*)-alkenes; (v) the relative stereochemistry between the preexisting hydroxy or alkoxy group and the adjacent, newly introduced hydroxy group of the major product is in all cases *erythro*.

A rationalization[31,34] of observations (iv) and (v) is based on conformational analysis of sp^3–sp^2 single bond systems, for which an eclipsed conformation seems to be preferred.[35] Of the three eclipsed conformations (**9a**), (**9b**) and (**9c**) of an allylic alcohol system $R^1(R^2O)CHCH{=}CR^3R^4$, conformation (**9a**) appears to be least sterically hindered and, therefore, most preferred. If this conformational preference is reflected in the transition state for hydroxylation, the major product can be seen to arise from the preferential approach of osmium tetroxide from the π-face opposite to that of the preexisting hydroxy or alkoxy group. An alternative rationalization, based[33] on observations on the hydroxylation of γ-hydroxy-α,β-unsaturated esters, suggests that conformation (**9b**) might be the preferred one, as a result of a favorable interaction between *p*-orbitals of the double bond and an unshared electron pair on the oxygen, and that attack would be directed from the side of the π-system opposite to that occupied by R^1. Interestingly, and in support of this idea, force-field calculations indicate[36] that the reactive conformer in the transition state for addition of a nitrile oxide to the allyl ether, 3-methoxy-1-butene, is one in which the alkoxy group is almost in the plane of the double bond in the 'inside' conformation shown in (**9b**), and a similar preference may apply in osmium tetroxide hydroxylations. A rationalization which focuses on the role of stereoelectronic factors in the transition state has also been proposed.[37] In the case of a (*Z*)-alkene (**9a–c**; R^3 = H) conformational preference for (**9a**) will be greater than that for the corresponding

(E)-alkene (**9a–c**; R^4 = H), leading to an increased steric differentiation of the two π-faces in the case of the former alkene compared to the latter one.

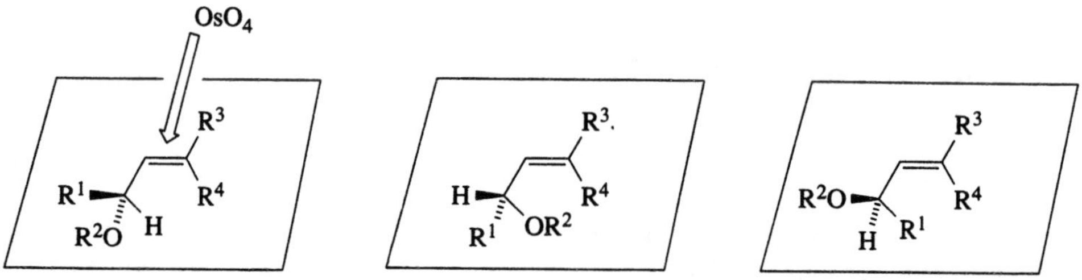

(9a) **(9b)** **(9c)**

Support for the empirical rule has been obtained, subsequently, during synthetic studies in carbohydrate chemistry.[37–39a,39b] For example, catalytic hydroxylation with osmium tetroxide of (E)-allylic alcohol (**10**) and (E)-allylic ether (**11**) gave the octose derivatives (**12**) and (**13**) as the major products, whereas the (Z)-allylic alcohol (**14**) gave (**15**) as the predominant product.[38] Similar steric control was apparent in the synthesis of decose derivatives in a related sequence of reactions,[39a] but some exceptions to the empirical rule are known for conjugated carbonyl compounds,[34,38,40] and it should be applied with caution to such compounds. It appears that the empirical rule may also predict, in general, the outcome of similar oxidations with potassium permanganate.[34]

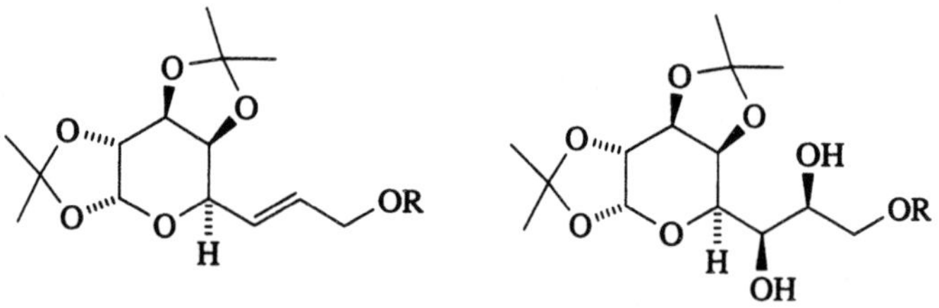

(10) R = H
(11) R = CH_2Ph

(12) R = H
(13) R = CH_2Ph

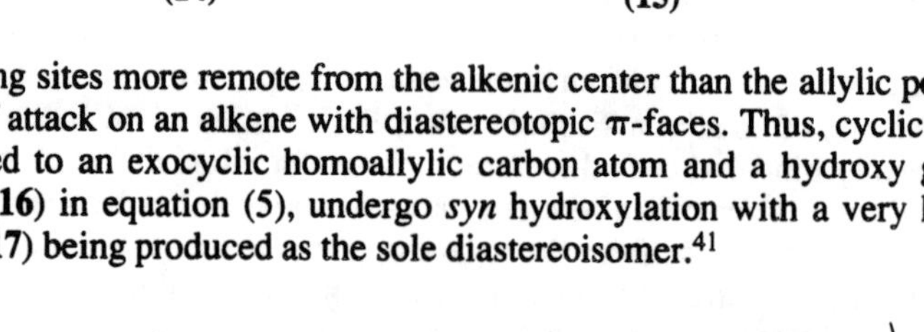

(14) **(15)**

Substituents occupying sites more remote from the alkenic center than the allylic position may also influence the direction of attack on an alkene with diastereotopic π-faces. Thus, cyclic alkenes with a sulfoximine group attached to an exocyclic homoallylic carbon atom and a hydroxy group at the allylic position, for example (**16**) in equation (5), undergo *syn* hydroxylation with a very high degree of diastereoselectivity, diol (**17**) being produced as the sole diastereoisomer.[41]

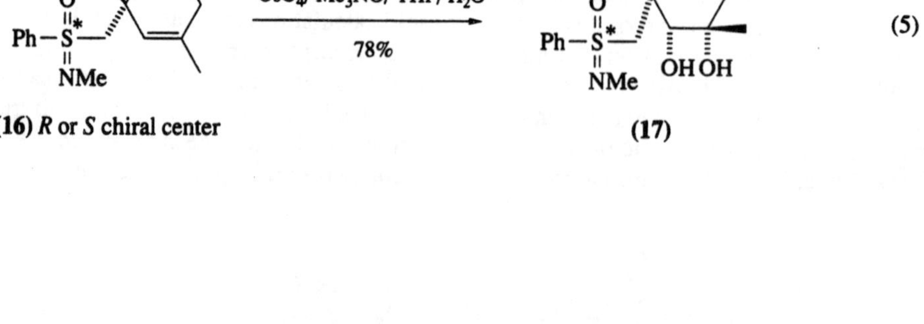

OsO_4/ Me_3NO/ THF/ H_2O, 78% (5)

(16) *R* or *S* chiral center **(17)**

In contrast, the corresponding sulfone (**16**; O replacing NMe) gives, on similar reaction, a 2:1 mixture of diastereoisomeric *cis*-diols with (**17**) predominating.[41]

Diastereoselective hydroxylation has also been observed at an alkenic center in an acyclic system that is guided by a sulfoxide group that is more remote than the homoallylic position. Alkenes (**18**; equation 6) and (**20**; equation 7), were converted[42] by treatment with a catalytic amount of osmium tetroxide and trimethylamine *N*-oxide into diols, which on acetylation gave diacetates (**19**) and (**21**), respectively as the sole products of the individual reactions. Apparently, complexation occurs between the oxygen of the sulfoxide and the osmium reagent prior to hydroxylation of the alkenic center. Hydroxylation of the alkenic sulfone corresponding to (**18**) and (**20**) afforded, after acetylation of the product, diacetates (**19**) and (**21**) in a 3:2 ratio, indicating that it is not an intermediate in the oxidation and that the amide group exerts a relatively small influence in favor of the diol corresponding to (**19**).

(6)

NHCOCCl$_3$ (**18**) — i, OsO_4/Me_3NO; ii, acetylation → OAc, AcO, SO_2Ph, NHCOCCl$_3$ (**19**)

(7)

NHCOCCl$_3$ (**20**) — i, OsO_4/ Me_3NO; ii, acetylation → OAc, AcO, SO_2Ph, NHCOCCl$_3$ (**21**)

Enantioselective *syn* hydroxylation of alkenes with enantiotopic faces may be achieved if addition is performed under the influence of a chiral control element. In practice, the latter may take the form of a chiral grouping temporarily attached to the alkene if suitable functionality is present. (This description distinguishes this type of reaction from diastereoselective hydroxylations which are achieved when the chiral center is an integral part of the substrate such as in **9**.) This renders the π-faces diastereotopic and thus distinguishable by the nonchiral tetroxide. Alternatively, the hydroxylation may be conducted in the presence of a suitable chiral substance capable of coordination with osmium tetroxide, rendering the reagent chiral and able to differentiate, therefore, between the enantiotopic faces of the alkene. (This rationalization supposes a [3 + 2] cycloaddition mechanism.[10] In an alternative mechanism proposed by Sharpless,[11,12] enantioselectivity arises by steric differentiation of the initially formed enantiomeric alkene–osmium π-complexes upon reaction with a chiral reagent such as a chiral amine.[43]) An example of the first approach is the conversion of an ester derived from (*E*)-2-methylbut-2-enoic acid and a chiral alcohol to the corresponding ester of (2*S*,3*R*)- and (2*R*,3*S*)-2,3-dihydroxy-2-methylbut-2-enoic acid in an isomer ratio of 83:17, respectively (equation 8).[44] Chiral oxazolidines obtained by reaction of α,β-unsaturated aldehydes with L-*N*-benzyloxycarbonylnorephedrine have been used in a similar manner to prepare derivatives of chiral α,β-dihydroxy aldehydes.[45]

(8)

CO_2R — OsO_4/ THF/ pyridine/ –78 °C → OH, CO_2R, OH (2*S*,3*R*) + OH, CO_2R, OH (2*R*,3*S*) 5:1

R = O, Ph

In the second approach, a chiral nitrogen-containing compound has most often been used as the ligand to achieve enantioselectivity. Thus, oxidation of (*E*)-stilbene (**22**; equation 9) with a stoichiometric quantity of osmium tetroxide in toluene at room temperature, in the presence of dihydroquinine acetate (**23**), yielded[12] *threo*-hydrobenzoins (**24**) after reductive hydrolysis, with an enantiomeric excess of 83.2% in favor of the (1*S*,2*S*)-(–)-isomer; performing the reaction at –78 °C increased the enantiomeric excess to 89.7%.

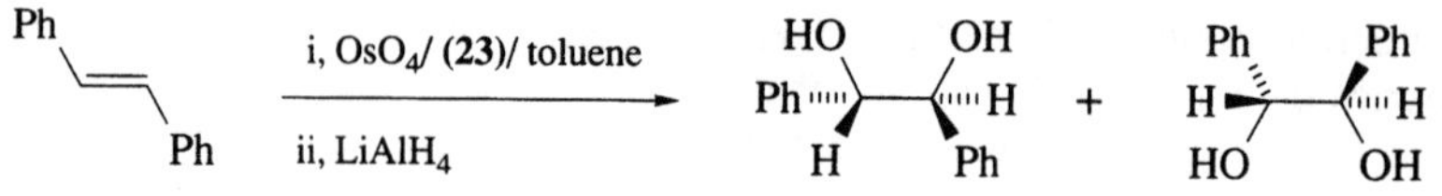

i, OsO_4/ (**23**)/ toluene; ii, $LiAlH_4$ (9)

(**22**) → (1*S*,2*S*) major + (1*R*,2*R*) minor (**24**)

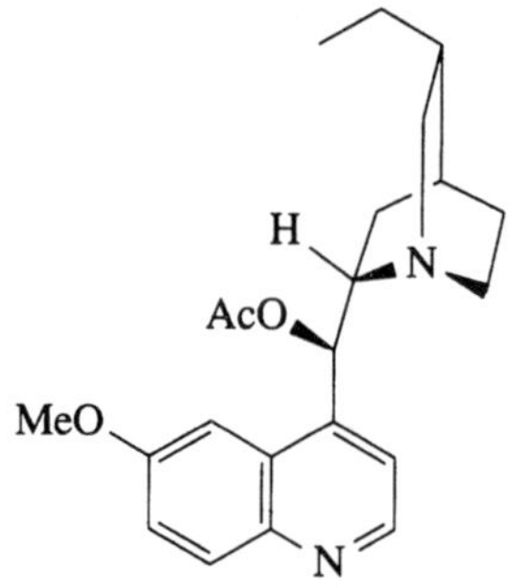

(**23**)

This procedure has been modified[46a] to become an effective catalytic procedure in which *N*-methylmorpholine *N*-oxide is used as the secondary oxidant. In this manner, (*E*)-stilbene has been converted[46a] into (+)-*threo*-hydrobenzoin (55% yield after two recrystallizations, >99% *ee*) on a one molar scale, by treatment with osmium tetroxide (0.002 mol equiv.) and *N*-methylmorpholine *N*-oxide (1.2 mol equiv.) in aqueous acetone in the presence of dihydroquinidine *p*-chlorobenzoate (0.134 mol equiv.). The latter compound can be recovered in 91% yield.

By the seemingly minor modification[46b,46c] to the original catalytic procedure[46a] of adding the alkene slowly to a stirred mixture of the alkaloid derivative, *N*-methylmorpholine *N*-oxide, and osmium tetroxide, nearly all alkenes react faster and give higher enantiomeric excesses in the product diols than with the earlier procedure[46a] in which all reactants, including the alkene, are present from the start of the reaction. This significant enhancement in enantioselectivity has been rationalized[46b] in terms of the existence of at least two diol-producing cycles as summarized in Scheme 1.[46b] The first cycle, which appears to give a high enantiomeric excess in the diol product, consists of reaction of the alkene with the alkaloid–osmium complex (**24a**) to give the monoglycolate ester (**24b**). This compound is oxidized to the key osmium(VIII) trioxoglycolate complex (**24c**), which is hydrolyzed to afford the 1,2-diol and complex (**24a**). By slow addition of the alkene, hydrolysis of (**24c**) can be made to dominate an alternative reaction with a second alkene molecule to give a bisglycolate ester (**24d**). Intrusion of the second cycle leads to reduced enantiomeric excess in the product since diol formation *via* the osmium(VIII) dioxobisglycolate complex (**24e**) proceeds with low enantioselectivity. Further, the rate of turnover in the secondary cycle is generally slower and involvement of this cycle in the reaction binds the catalyst in a relatively unproductive form.

Protected α,β-dihydroxy aldehydes have been prepared[47] by oxidation of acetals of α,β-unsaturated aldehydes with osmium tetroxide in the presence of (**23**), and a remarkable level of enantioselection (*ee* ≥ 90%) thereby achieved. Oxidation of chiral acetals of α,β-unsaturated aldehydes in which chirality resides in the noncarbonyl moiety with osmium tetroxide–dihydroquinine acetate (or dihydroquinidine acetate)[47] may be regarded as a process in which double stereoselection[48] is at work and a high diastereoisomeric ratio of products may be obtained.

Chiral diamines capable of chelating to a metal center, such as (–)-(*R*,*R*)–*N*,*N*,*N′*,*N′*-tetramethyl(*trans*-1,2-cyclohexanediamine (**25**),[43] the tartaric acid derived (–)-diamine (**26**),[49] and the (–)-1,2-dipyrrolidinylethane (**27**),[50] also lead to a high degree of asymmetric induction when alkene hydroxylation with osmium tetroxide is conducted in their presence.

With diamine (**25**), 1-heptene afforded[43] (*R*)-1,2-heptanediol as the major product (86% *ee*) in 75% yield by this procedure but, curiously, oxidation of (*E*)-stilbene proceeded with lower optical yield (34% *ee*). Particularly efficient enantioface differentiation was achieved in the reaction[50] of (*E*)-1-phenylpropene with a stoichiometric amount of osmium tetroxide in the presence of 1 mol equiv. of (–)-(**27**) when essentially optical pure (>99% *ee*) (1*S*,2*S*)-1-phenylpropane-1,2-diol was obtained in 73% yield. This procedure is effective for mono-, (*E*)-di- and tri-substituted alkenes, with enantioface selection being as shown in Scheme 2 but, notably, the oxidation of (*Z*)-alkenes does not give satisfactory optical yields.

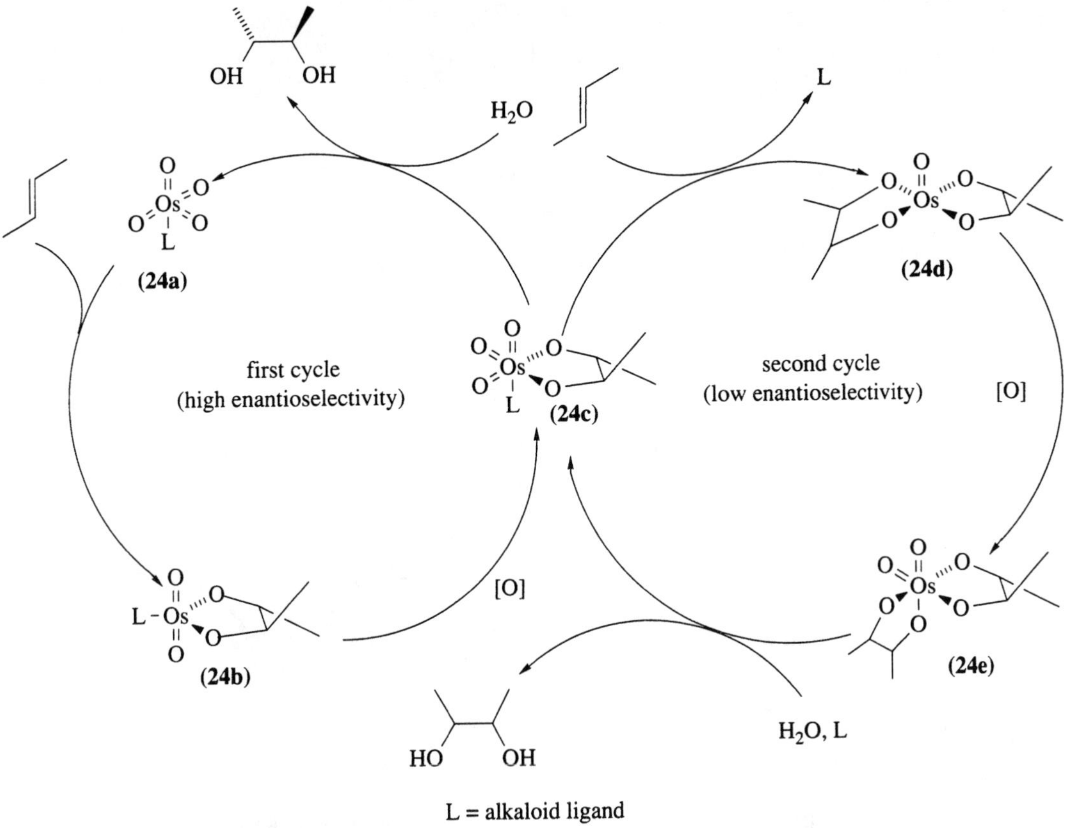

Scheme 1 Proposed mechanism of the osmium-catalyzed asymmetric dihydroxylation of alkenes

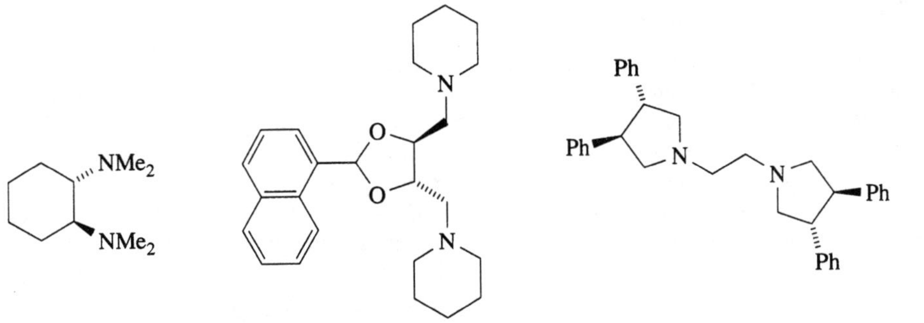

The structure of the osmate(VI) ester–(–)-(**27**) complex, which may be isolated from the reaction with (*E*)-stilbene, has been determined[13] by X-ray crystallography to be (**28**). Although (**28**) would reasonably seem to arise by a [3 + 2] cycloaddition pathway, it does not seem to account for the observed stereochemical outcome of the reaction since, on steric grounds, complex (**29**) would appear to be favored. On the other hand, the alternative pathway[11,12] *via* organometallocycle (**30**) seems to account more satisfactorily for the stereochemical result, since intramolecular attack by the nitrogen of the second pyrrolidine ring moiety places substituent R^1 in the least sterically demanding region and affords osmate ester (**28**) in accord with the observed enantioface differentiation. In the stereoisomeric organometallocycle (**31**), steric interactions between the phenyl group on the coordinated pyrrolidine ring and that on the four-membered metallocycle would disfavor formation of the osmate ester.

The 1:1 complex between bovine serum and an osmate ester is an enantioselective catalyst in the *syn* hydroxylation of certain alkenes,[51a] although synthetic applications appear to be limited. Asymmetric dihydroxylation of alkenes is considered in a review on catalytic asymmetric reactions.[51b]

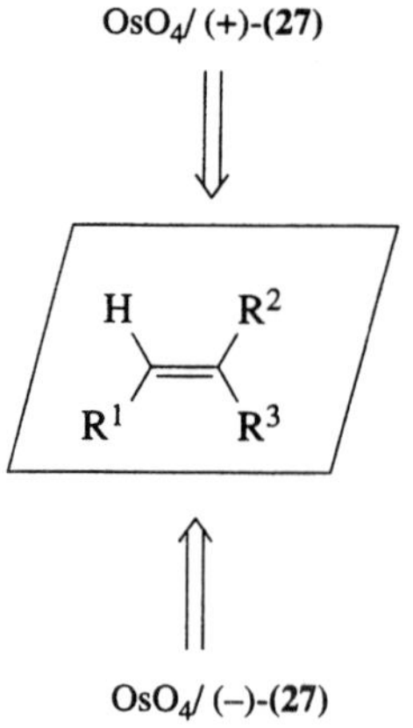

OsO$_4$/ (+)-(**27**)

OsO$_4$/ (–)-(**27**)

Scheme 2 Enantioface selection on hydroxylation of an alkene with osmium tetroxide in the presence of (+)- and (–)-(**27**)

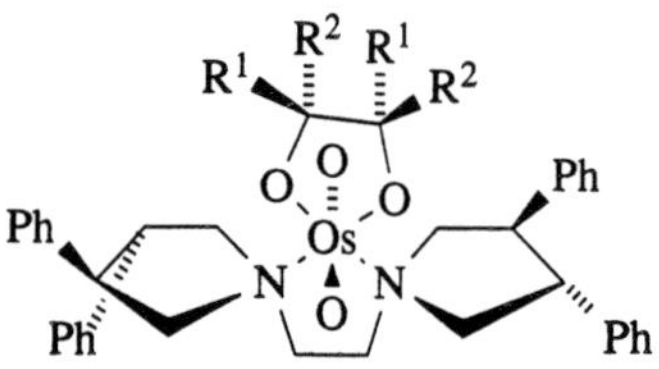

(**28**) R^1 = Ph, R^2 = H
(**29**) R^1 = H, R^2 = Ph

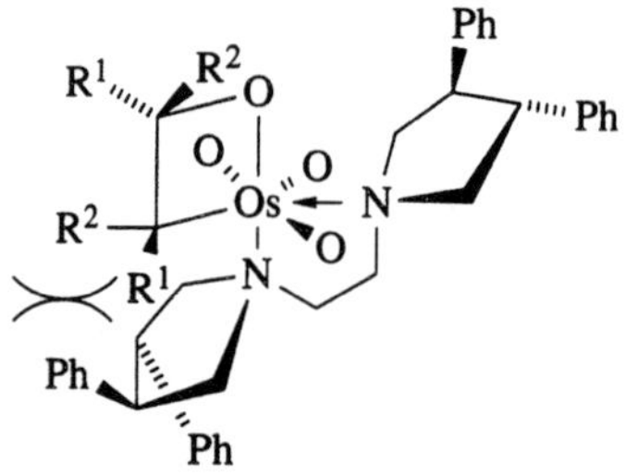

(**30**) R^1 = Ph, R^2 = H
(**31**) R^1 = H, R^2 = Ph

3.3.2.2 Potassium Permanganate

Alkaline aqueous potassium permanganate has long been used to achieve the *syn* hydroxylation of alkenes,[9] but overoxidation and alternative oxidation pathways often pose problems, and yields are rarely as high as those obtained with osmium tetroxide. Nevertheless, permanganate oxidation is less hazardous to perform and is much less expensive for large-scale operations. Improved yields of diols may sometimes be obtained by using phase transfer catalysis.[52–54] Typically, a solution of the alkene in dichloromethane is stirred vigorously with aqueous sodium hydroxide in the presence of a phase transfer agent such as benzyltriethylammonium chloride, while potassium permanganate is added portionwise. *cis*-1,2-Cyclooctanediol is prepared[53] from *cis*-cyclooctene in much higher yield by this method than with the conventional procedure employing basic, aqueous potassium permanganate. Solid–liquid phase transfer may be brought about under nonaqueous conditions, to bring potassium permanganate into solution in an organic solvent which contains the dissolved substrate.[54]

Turbulent stirring and the presence of low concentrations of sodium hydroxide are very beneficial in improving the yield of *cis*-1,2-cyclohexanediol from the hydroxylation of cyclohexene with potassium permanganate.[55–57] Presumably, hydroxylation of other alkenes with this reagent would also benefit by attention to these factors.

Recent evidence on the mechanism of permanganate oxidation of alkenes has been summarized;[9] the initial step probably involves a [3 + 2] cycloaddition between permanganate ion and the alkene to give a cyclic manganese(V) ester (**2**; see Section 3.3.1).

3.3.2.3 Methods Involving Halohydrin Esters as Intermediates

Several related procedures for *syn* hydroxylation of alkenes involve a halohydrin ester (**32**) as the key intermediate. In Woodward's procedure[14] an alkene in glacial acetic acid is treated with iodine and silver acetate. Acetyl hypoiodite, $MeCO_2I$, formed by reaction of the latter two reagents attacks the alkene, $R^1R^2C{=}CR^3R^4$, in an electrophilic manner, from the less hindered side to give, by overall *anti* addition, an iodoacetate (**32**; R^5 = Me, Hal = I), in which the acetoxy group is attached to the more hindered face of the alkene. After addition of water, the reaction mixture is heated, inducing silver(I)-assisted anchimeric displacement of the iodo group with formation of a 1,3-dioxolan-2-ylium ion (**33**). Addition of water to the cation affords orthoacetate (**34**) that rearranges to give a mixture of two diol monoacetates (**35**a) and (**35**b) that may be hydrolyzed to afford a diol having the hydroxy groups bonded to the more hindered side of the alkene. Some alkenes, however, particularly trisubstituted alkenes, can give other products in addition to the expected diol,[58] suggesting that the foregoing mechanism may be oversimplified. Further, hydroxylation of the tetrasubstituted alkene 1,2-dimethylcyclohexene by the Woodward procedure is not stereospecific, giving an approximately 3:2 ratio of *cis*- to *trans*-diol.[59]

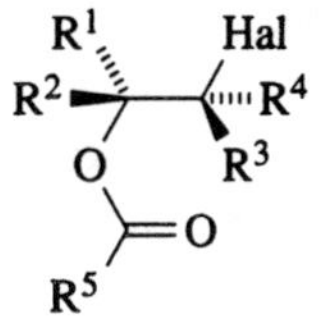

(**32**) 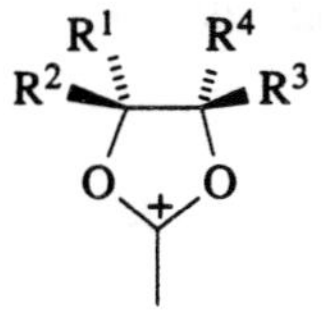 (**33**) 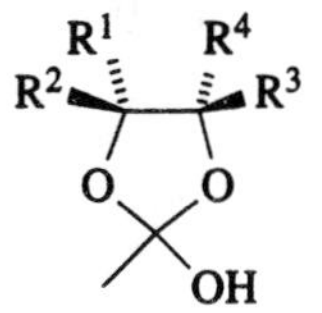 (**34**)

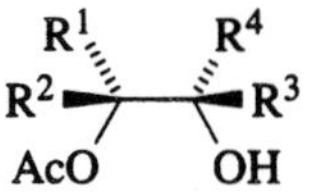

(**35a**) 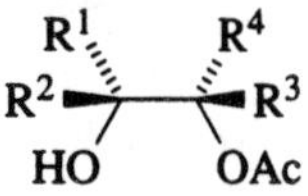 (**35b**)

Since silver salts are expensive, other cheaper reagents have been sought which can bring about the same type of conversion, and the iodine–potassium iodate–potassium acetate combination has been found to provide a useful alternative.[17–19] Thallium(I) acetate may be used[15,16] in place of silver acetate in the Woodward procedure, and *syn* hydroxylation of steroidal alkenes with thallium(III) acetate in acetic acid has been performed.[60]

In the 2-halocyanoacetate procedure,[20] a *trans*-bromohydrin (**36**; Scheme 3), prepared from an alkene, is esterified with cyanoacetic acid to give the 2-halocyanoacetate (**37**), and the latter is treated with sodium hydride to give, *via* the enolate anion, a cyanoketene acetal (**38**). Acid hydrolysis of the latter, followed by deesterification with base of the diol monocyanoacetate so-formed gives diol (**39**). Since the starting *trans*-bromohydrin may be prepared from a cyclohexene by two stereochemically complementary routes using hypobromous acid (oxygen added to the more hindered π-face) or peroxy acid epoxidation followed by cleavage with hydrogen bromide (oxygen added to the less hindered π-face) it is possible to achieve *syn* hydroxylation on either the more or less hindered π-face.

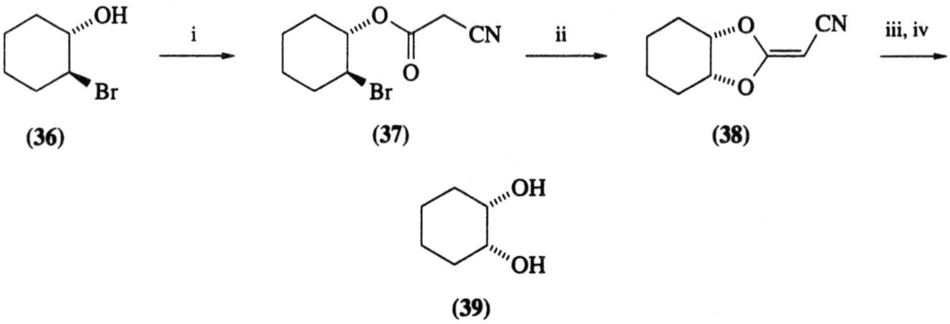

(**36**) (**37**) (**38**) (**39**)

i, $NCCH_2CO_2H$/TsCl; ii, NaH; iii, H_3O^+; iv, base

Scheme 3

3.3.3 *ANTI* HYDROXYLATION METHODS

3.3.3.1 Peroxycarboxylic Acids

The hydroxylation of an alkene may be achieved by treatment with a suitable peroxycarboxylic acid, RCO_3H, the reaction proceeding (Scheme 4) by initial *syn* addition to give an epoxide (oxirane) (**40**), which undergoes acid-catalyzed ring scission in an *anti* manner through attack by the corresponding carboxylic acid, RCO_2H, normally present in the reaction medium, to give the monoesters (**41**) and (**42**). Hydrolysis of the ester mixture then affords a racemic mixture of the enantiomeric diols (**43a**) and (**43b**) with stereochemistry resulting from overall *anti* addition to the alkene.

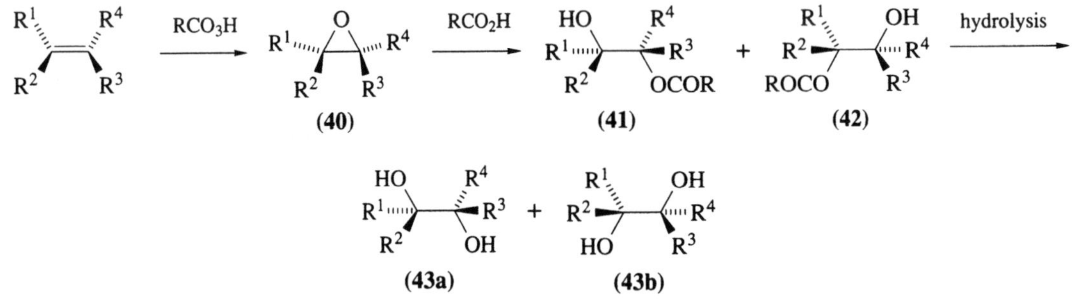

RCO_3H → (**40**) → RCO_2H → (**41**) + (**42**) → hydrolysis → (**43a**) + (**43b**)

Scheme 4

Peroxyformic, peroxyacetic and peroxytrifluoroacetic acids are most commonly used to bring about this type of oxidation, but 2-sulfoperoxybenzoic acid,[61] monoperoxysuccinic acid[62] and disuccinoyl peroxide[63] (which is converted to monoperoxysuccinic acid by hydrolysis), also oxidize alkenes to diols with the advantage that the free diols are obtained directly.

In general, attack of the peroxycarboxylic acid on an alkene will occur from the less hindered π-face and ring opening of the oxirane usually occurs to place the acyloxy group on the more substituted carbon atom.

Sodium perborate ($NaBO_3{\cdot}nH_2O$; $n = 1$–4) is a cheap and widely used industrial chemical. If sulfuric acid is added to a mixture of the perborate and an alkene in acetic anhydride, an exothermic reaction occurs leading to *anti* addition to the double bond with formation of the corresponding 1-hydroxy-2-acetoxy derivative in moderate yield.[64] Peroxybis(diacetoxy)borane, $(AcO)_2BOOB(OAc)_2$, may be the reactive species in this oxidation and it seems likely that the epoxide is an intermediate.

3.3.3.2 Hydrogen Peroxide with an Oxide Catalyst

The oxidation of alkenes by hydrogen peroxide catalyzed by certain oxide catalysts, such as tungsten(VI) oxide (WO_3),[65] tungsten(VI) acid (H_2WO_4)[66] and selenium dioxide (SeO_2),[64,67–69] brings about *anti* hydroxylation with formation of 1,2-diols (equation 10). Tungsten(VI) oxide is a particularly effective catalyst and functions best at elevated temperatures (50–70 °C) in a purely aqueous medium. For oxidation of alkenes which are insoluble in water, a mixture of 30% hydrogen peroxide and acetic acid forms a suitable medium, and with the selenium dioxide–hydrogen peroxide system, *t*-butyl alcohol has been used as a solvent.

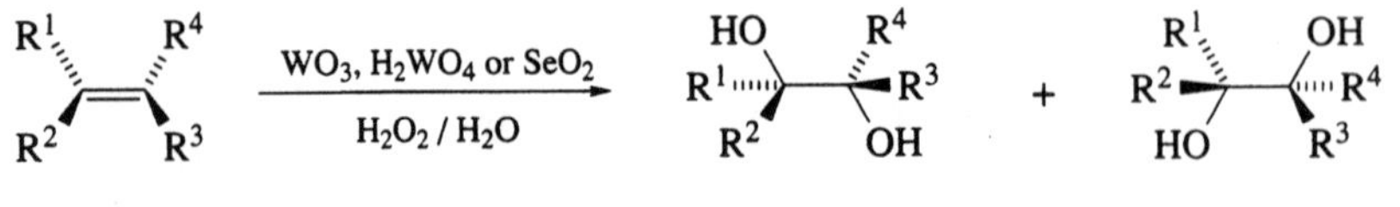

WO_3, H_2WO_4 or SeO_2; H_2O_2 / H_2O (10)

The observed stereochemistry of addition suggests that oxiranes (epoxides) may be intermediates. Oxiranes may, indeed, be isolated from the reaction of certain alkenes with hydrogen peroxide in the presence of sodium tungstate.[70]

Polystyrene-bound phenylseleninic acid has been used in catalytic amounts in a triphasic system consisting of the polymer, aqueous hydrogen peroxide, and dichloromethane, to catalyze the oxidation of alkenes to 1,2-diols;[71] cyclohexene gave *trans*-1,2-cyclohexanediol in high yield with no detectable amounts of the *cis*-diol.

3.3.3.3 Iodine–Silver Benzoate — the Prévost Reaction

In the Prévost reaction, an alkene is treated with 1 mol equiv. of iodine and 2 mol equiv. of silver carboxylate (most often silver benzoate) in an inert solvent (for example, benzene), leading to formation of a racemic mixture of 1,2-diesters (**46a** and **46b**; Scheme 5). Hydrolysis of the latter yields a mixture of the corresponding 1,2-diols with stereochemistry corresponding to *anti* hydroxylation of the alkene. The reaction proceeds through initial formation of a complex RCO_2Ag/RCO_2I (the Simonini complex), which can itself be prepared and isolated separately, prior to use in such a reaction. *Anti* addition of the acyl hypoiodite (RCO_2I) to the alkene through a cyclic iodonium ion gives the 1-acyloxy-2-iodo compound (**44**), and displacement of iodide ion with anchimeric assistance by the neighboring acyloxy group affords the 1,3-dioxolan-2-ylium ion (**45**). Nucleophilic attack by carboxylate anion on (**45**) then leads to diesters (**46a**) and (**46b**).

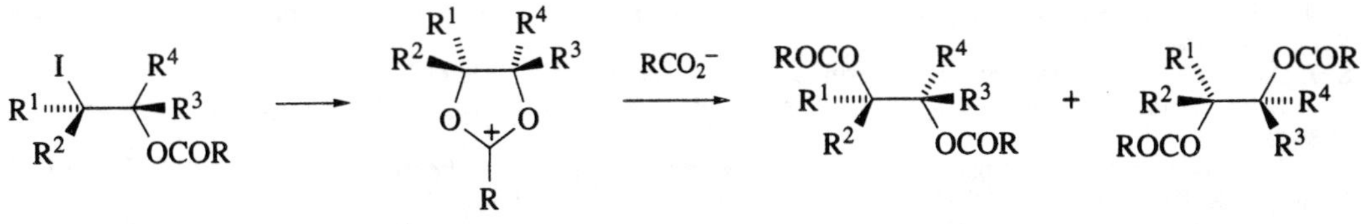

I R⁴ R¹ R³ R² OCOR (**44**) ⟶ R¹ R⁴ R² R³ O + O R (**45**) $\xrightarrow{RCO_2^-}$ ROCO R⁴ R¹ R³ R² OCOR (**46a**) + R¹ OCOR R² R⁴ ROCO R³ (**46b**)

Scheme 5

Thallium(I) acetate has been used in place of a silver carboxylate in a related procedure, for the preparation of *trans*-1,2-cyclohexanediol.[16]

3.3.3.4 Miscellaneous Procedures

Some mono- and di-substituted alkenes have been converted to 1,2-diacetoxy compounds by heating them in acetic acid solution with ammonium persulfate and a catalytic amount of iron(II) sulfate.[72] *Anti* addition is observed with 1,2-disubstituted alkenes; with trisubstituted alkenes complex mixtures are obtained.

Thallium(III) sulfate in water brings about the *anti* hydroxylation of 3-*t*-butyl- and 4-*t*-butyl-cyclohexene.[73] The reagent has been recommended[73] for the one-step preparation of *trans*-diols from conformationally rigid cycloalkenes.

3.3.4 REFERENCES

1. J. F. Bunnett, *Pure Appl. Chem.*, 1981, **53**, 305.
2. F. D. Gunstone, *Adv. Org. Chem.*, 1960, **1**, 103.
3. H. E. House, 'Modern Synthetic Reactions', 2nd edn., Benjamin, Menlo Park, CA, 1972, pp. 275, 296.
4. R. Criegee and H. Kropf, *Methoden Org. Chem. (Houben-Weyl)*, 1979, **6/1a** (1), 592.
5. A. H. Haines, 'Methods for the Oxidation of Organic Compounds. Alkanes, Alkenes, Alkynes, and Arenes', Academic Press, London, 1985, p. 73.
6. M. Schröder, *Chem. Rev.*, 1980, **80**, 187.
7. H. S. Singh, in 'Organic Synthesis by Oxidation with Metal Compounds', ed. W. J. Mijs and C. R. H. I. de Jonge, Plenum Press, New York, 1986, p. 633.
8. R. A. Sheldon and J. K. Kochi, 'Metal-Catalyzed Oxidations of Organic Compounds', Academic Press, New York, 1981, pp. 162, 294.
9. A. J. Fatiadi, *Synthesis*, 1987, 85; see especially pp. 86–87.
10. K. A. Jørgensen and R. Hoffmann, *J. Am. Chem. Soc.*, 1986, **108**, 1867.
11. K. B. Sharpless, A. Y. Teranishi and J.-E. Bäckvall, *J. Am. Chem. Soc.*, 1977, **99**, 3120.
12. S. G. Hentges and K. B. Sharpless, *J. Am. Chem. Soc.*, 1980, **102**, 4263.
13. K. Tomioka, M. Nakajima, Y. Iitaka and K. Koga, *Tetrahedron Lett.*, 1988, **29**, 573.
14. R. B. Woodward and F. V. Brutcher, *J. Am. Chem. Soc.*, 1958, **80**, 209.
15. R. C. Cambie, R. C. Hayward, J. L. Roberts and P. S. Rutledge, *J. Chem. Soc., Perkin Trans. 1*, 1974, 1858.
16. R. C. Cambie and P. S. Rutledge, *Org. Synth.*, 1980, **59**, 169.

17. L. Mangoni, M. Adinolfi, G. Barone and M. Parrilli, *Tetrahedron Lett.*, 1973, 4485.
18. M. Parrilli, G. Barone, M. Adinolfi and L. Mangoni, *Gazz. Chim. Ital.*, 1974, **104**, 835.
19. L. Mangoni, M. Adinolfi, G. Barone and M. Parrilli, *Gazz. Chim. Ital.*, 1975, **105**, 377.
20. E. J. Corey and J. Das, *Tetrahedron Lett.*, 1982, **23**, 4217.
21. Y. Ogata and Y. Sawaki, in 'Organic Syntheses by Oxidation with Metal Compounds', ed. W. J. Mijs and C. R. H. I. de Jonge, Plenum Press, New York, 1986, p. 839.
22. S. Danishefsky, M. Hirama, K. Gombatz, T. Harayama, E. Berman and P. Schuda, *J. Am. Chem. Soc.*, 1978, **100**, 6536.
23. T. Ogino and K. Mochizuki, *Chem. Lett.*, 1979, 443.
24. M. Schröder and W. P. Griffith, *J. Chem. Soc., Dalton Trans.*, 1978, 1599.
25. L. Bassignani, A. Brandt, V. Caciagli and L. Re, *J. Org. Chem.*, 1978, **43**, 4245.
26. V. VanRheenen, R. C. Kelly and D. Y. Cha, *Tetrahedron Lett.*, 1976, 1973.
27. V. VanRheenen, D. Y. Cha and W. M. Hartley, *Org. Synth.*, 1979, **58**, 43.
28. K. Akashi, R. E. Palermo and K. B. Sharpless, *J. Org. Chem.*, 1978, **43**, 2063.
29. R. Ray and D. S. Matteson, *Tetrahedron Lett.*, 1980, **21**, 449.
30. K. B. Sharpless and K. Akashi, *J. Am. Chem. Soc.*, 1976, **98**, 1986.
31. J. K. Cha, W. J. Christ and Y. Kishi, *Tetrahedron Lett.*, 1983, **24**, 3943.
32. W. J. Christ, J. K. Cha and Y. Kishi, *Tetrahedron Lett.*, 1983, **24**, 3947.
33. G. Stork and M. Kahn, *Tetrahedron Lett.*, 1983, **24**, 3951.
34. J. K. Cha, W. J. Christ and Y. Kishi, *Tetrahedron*, 1984, **40**, 2247.
35. G. J. Karabatsos and D. J. Fenoglio, *Top. Stereochem.*, 1970, **5**, 167.
36. K. N. Houk, S. R. Moses, Y.-D. Wu, N. G. Rondan, V. Jäger, R. Schohe and F. R. Fronczek, *J. Am. Chem. Soc.*, 1984, **106**, 3880.
37. S. J. Danishefsky, E. Larson and J. P. Springer, *J. Am. Chem. Soc.*, 1985, **107**, 1274.
38. J. S. Brimacombe, R. Hanna, A. K. M. S. Kabir, F. Bennett and I. D. Taylor, *J. Chem. Soc., Perkin Trans. 1*, 1986, 815.
39. (a) J. S. Brimacombe, R. Hanna and A. K. M. S. Kabir, *J. Chem. Soc., Perkin Trans. 1*, 1986, 823; (b) J. C. Barnes, J. S. Brimacombe and G. McDonald, *J. Chem. Soc., Perkin Trans. 1*, 1989, 1483.
40. H. Fujioka, W. J. Christ, J. K. Cha, J. Leder, Y. Kishi, D. Uemura and Y. Hirata, *J. Am. Chem. Soc.*, 1982, **104**, 7367.
41. C. R. Johnson and M. R. Barbachyn, *J. Am. Chem. Soc.*, 1984, **106**, 2459.
42. F. M. Hauser, S. R. Ellenberger, J. C. Clardy and L. S. Bass, *J. Am. Chem. Soc.*, 1984, **106**, 2458.
43. M. Tokles and J. K. Snyder, *Tetrahedron Lett.*, 1986, **27**, 3951.
44. S. Hatakeyama, Y. Matsui, M. Suzuki, K. Sakurai and S. Takano, *Tetrahedron Lett.*, 1985, **26**, 6485.
45. L. Colombo, C. Gennari, G. Poli and C. Scolastico, *Tetrahedron Lett.*, 1985, **26**, 5459.
46. (a) E. N. Jacobsen, I. Markó, W. S. Mungall, G. Schröder and K. B. Sharpless, *J. Am. Chem. Soc.*, 1988, **110**, 1968; (b) J. S. M. Wai, I. Markó, J. S. Svendsen, M. G. Finn, E. N. Jacobsen and K. B. Sharpless, *J. Am. Chem. Soc.*, 1989, **111**, 1123; (c) B. B. Lohray, T. H. Kalantar, B. M. Kim, C. Y. Park, T. Shibata, J. S. M. Wai and K. B. Sharpless, *Tetrahedron Lett.*, 1989, **30**, 2041.
47. R. Annunziata, M. Cinquini, F. Cozzi, L. Raimondi and S. Stefanelli, *Tetrahedron Lett.*, 1987, **28**, 3139.
48. S. Masamune, W. Choy, J. S. Petersen and L. R. Sita, *Angew. Chem., Int. Ed. Engl.*, 1985, **24**, 1.
49. T. Yamada and K. Narasaka, *Chem. Lett.*, 1986, 131.
50. K. Tomioka, M. Nakajima and K. Koga, *J. Am. Chem. Soc.*, 1987, **109**, 6213.
51. (a) T. Kokubo, T. Sugimoto, T. Uchida, S. Tanimoto and M. Okano, *J. Chem. Soc., Chem. Commun.*, 1983, 769; (b) I. Ojima, N. Clos and C. Bastos, *Tetrahedron*, 1989, **45**, 6901.
52. D. G. Lee, in 'Oxidation in Organic Chemistry', ed. W. S. Trahanovsky, Academic Press, New York, 1982, part D, p. 147; see especially pp. 168–181.
53. W. P. Weber and J. P. Shepherd, *Tetrahedron Lett.*, 1972, 4907.
54. T. Ogino and K. Mochizuki, *Chem. Lett.*, 1979, 443.
55. J. E. Taylor, D. Williams, K. Edwards, D. Otonnaa and D. Samanich, *Can. J. Chem.*, 1984, **62**, 11.
56. J. E. Taylor, *Can. J. Chem.*, 1984, **62**, 2641.
57. J. E. Taylor and R. Green, *Can. J. Chem.*, 1985, **63**, 2777.
58. M. Parrilli, M. Adinolfi, V. Dovinola and L. Mangoni, *Gazz. Chim. Ital.*, 1974, **104**, 819.
59. C. A. Bunton and M. D. Carr, *J. Chem. Soc.*, 1963, 770.
60. E. Glotter and A. Schwartz, *J. Chem. Soc., Perkin Trans. 1*, 1976, 1660.
61. J. M. Bachhawat and N. K. Mathur, *Tetrahedron Lett.*, 1971, 691.
62. R. Lombard and G. Schröder, *Bull. Soc. Chim. Fr.*, 1963, 2800.
63. C. Wilkins, *Synthesis*, 1973, 156.
64. G. Xie, L. Xu, J. Hu, S. Ma, W. Hou and F. Tao, *Tetrahedron Lett.*, 1988, **29**, 2967.
65. M. Mugdan and D. P. Young, *J. Chem. Soc.*, 1949, 2988.
66. G. B. Payne and C. W. Smith, *J. Org. Chem.*, 1957, **22**, 1682.
67. P. Seguin, *C. R. Hebd. Seances Acad. Sci.*, 1943, **216**, 667.
68. A. Stoll, A. Lindenmann and E. Jucker, *Helv. Chim. Acta*, 1953, **36**, 268.
69. N. Sonoda and S. Tsutsumi, *Bull. Chem. Soc. Jpn.*, 1965, **38**, 958.
70. G. B. Payne and P. H. Williams, *J. Org. Chem.*, 1959, **24**, 54.
71. R. T. Taylor and L. A. Flood, *J. Org. Chem.*, 1983, **48**, 5160.
72. W. E. Fristad and J. R. Peterson, *Tetrahedron Lett.*, 1983, **24**, 4547.
73. C. Freppel, R. Favier, J.-C. Richer and M. Zador, *Can. J. Chem.*, 1971, **49**, 2586.

3.4

Addition Reactions with Formation of Carbon–Oxygen Bonds: (iv) The Wacker Oxidation and Related Reactions

JIRO TSUJI

Okayama University of Science, Japan

3.4.1 INTRODUCTION

The oxidation of ethylene to acetaldehyde using $PdCl_2$ and $CuCl_2$ as catalysts under an oxygen atmosphere is well known as the Wacker process (Scheme 1), and is one of the most important industrial processes employing transition metal catalysts.[1,2] This industrial oxidation reaction of ethylene involves the following three stoichiometric reactions. These sequential oxidation and reduction reactions constitute a catalytic cycle.

The Wacker process is carried out in an aqueous medium containing hydrochloric acid. In addition to ethylene, Smidt and coworkers carried out the oxidation of other alkenes in an acidic aqueous solution of $PdCl_2$ to prepare carbonyl compounds.[1,2] After this report, a few studies on the oxidation of higher alkenes were carried out in organic media.[3–6] In general, terminal alkenes are converted to methyl ketones rather than aldehydes (equation 1).

$$\begin{array}{rcl} H_2C{=}CH_2 + H_2O + PdCl_2 & \longrightarrow & MeCHO + Pd^0 + 2\,HCl \\ Pd^0 + 2\,CuCl_2 & \longrightarrow & PdCl_2 + 2\,CuCl \\ 2\,CuCl + 2\,HCl + 0.5\,O_2 & \longrightarrow & 2\,CuCl_2 + H_2O \\ \hline H_2C{=}CH_2 + 1/2\,O_2 & \longrightarrow & MeCHO \end{array}$$

Scheme 1 Wacker process

$$R\text{-}CH{=}CH_2 \longrightarrow R\text{-}C(=O)\text{-}CH_3 \qquad (1)$$

This reaction is a unique method for the one-step synthesis of ketones from alkenes. As such, alkenes can be regarded as masked ketones which are stable to acids, bases and nucleophiles. Thus, the reaction is potentially useful for organic synthesis, and has attracted the attention of synthetic organic chemists.[7] A considerable number of studies have been carried out on this oxidation and its synthetic applications. A number of review articles and books have been published on various palladium-catalyzed oxidative reactions of alkenes.[8–13] One comprehensive review on synthetic applications of the Wacker reaction was published in 1984.[14] In this chapter, the Pd^{II}-catalyzed oxidation of alkenes to carbonyl compounds is surveyed from a standpoint of organic synthesis by citing pertinent examples. Complete coverage of all related references is not the purpose of this review. Emphasis is placed on the oxidation of higher and functionalized alkenes, and many examples of the oxidation of lower simple alkenes are excluded. Oxidative conversion of alkenes to enol or allyl acetates with $Pd(OAc)_2$, although a closely related reaction, is not discussed. Repetition of material which has appeared in other reviews is avoided as far as possible. Also no mechanistic or kinetic discussion is given due to limited space.

3.4.2 REACTION CONDITIONS AND SCOPE OF THE REACTION

3.4.2.1 General Procedure

The oxidation on a laboratory scale can be carried out easily in a way similar to the hydrogenation of alkenes under atmospheric pressure of hydrogen using palladium black as a catalyst. Instead of palladium black and hydrogen, the oxidation is carried out with $PdCl_2$ and a copper salt under an oxygen atmosphere at room temperature using a similar apparatus. However, rates and yields of the oxidation are heavily dependent on the structure of alkenes. Also, the proper selection of solvents and reoxidants is crucial; this is surveyed in the following sections.

3.4.2.2 Solvents

The industrial Wacker process is carried out in aqueous hydrochloric acid using $PdCl_2/CuCl_2$ as the catalyst under oxygen pressure.[1,2] The oxidation of higher terminal alkenes under the same conditions is slow and sometimes accompanied by undesired by-products formed by the chlorination of carbonyl compounds by $CuCl_2$, and isomerization of double bonds. Earlier examples of oxidation of various alkenes, mainly in aqueous solutions, have been tabulated.[1,8] The pseudo-first-order rate constants for oxidation of various alkenes, relative to the value for cycloheptene, with $PdCl_2$ in the presence of benzoquinone in aqueous solution have been reported.[15] An accelerating effect of surfactants such as sodium lauryl sulfate on the stoichiometric oxidation of higher alkenes in an aqueous solution has been reported.[16]

The low rate of the reaction in the aqueous medium can be partly improved by the addition of suitable organic solvents which can mix alkenes with water. Several solvents have been tested, but the results are sometimes conflicting. At first, 1-dodecene and undecenoic acid were oxidized to the corresponding methyl ketones in good yields in aqueous DMF using $PdCl_2$ and $CuCl_2$ or benzoquinone.[3] However, the use of 3-methylsulfolane and NMP gave better results than DMF for the oxidation of 3,3-dimethyl-1-butene at 70–80 °C under 40–99 p.s.i. (1 p.s.i. = 6.9×10^{-2} bar) of oxygen using 10 mol % of $PdCl_2$.[5] The following yields of the methyl ketone were obtained: 91% (in 3-methylsulfolane), 79% (in NMP) and 33% (in DMF). The oxidation in alcoholic solutions was carried out with terminal and internal alkenes and some cyclic alkenes.[4] The reaction in alcohols is faster than in DMF. Parallel oxidations of cyclohexene at 50 °C showed conversion of 30% in ethanol, 1.2% in 1,4-dioxane and less than 0.5% in DMF, DMSO, acetic acid and carbon tetrachloride.[4] Polyethylene glycol (PEG 400) is a good solvent for the oxidation of terminal and internal alkenes with $PdCl_2/CuCl_2$. *cis*-2-Butene was oxidized to 2-butanone in 82% yield.[17]

When the reaction is carried out in pure alcohol, the corresponding acetal is formed. Styrene was converted to the cyclic acetal of phenylacetaldehyde in 90% yield in ethylene glycol, whereas a mixture of products was obtained in ethanol. Interestingly, acrylonitrile was oxidized to 1,3-dioxolan-2-ylacetonitrile (**1**) in ethylene glycol (equation 2).[4,18] 2,2-Dimethoxypropionitrile was obtained in methanol.[4,19,20] Methyl acrylate behaves similarly.

CN + HO⁀⁀OH —($PdCl_2$ / $CuCl_2$)→ (**1**) (2)

Another interesting example of the acetal formation is the synthesis of brevicomin (**2**), a cyclic acetal, by the palladium-catalyzed intramolecular oxidation and acetal formation of 6,7-dihydroxy-1-nonene (equation 3).[21,22]

—($PdCl_2$ / $CuCl_2$ / DMF, 45%)→ (**2**) (3)

Similarly, frontalin,[23] and 2,9-dioxabicyclo[3.3.1]nonane[24] were prepared by the bicyclic acetal formation from terminal alkenes. γ-Butyrolactone is another solvent of choice.[25] A two-phase reaction was carried out using carbon tetrachloride and benzene.[26]

The oxidation and double bond isomerization are competitive reactions, and the extents of these reactions are influenced by solvents. DMF is good for the oxidation, whereas use of acetic acid facilitates the isomerization. Both reactions proceed in alcoholic solvents. Acetonitrile and DMSO retard oxidation by complex formation with the catalysts.[27] The double bond migration is facilitated by high temperatures. The oxidation of 1-octene in *n*-propyl alcohol yielded 2-octanone to the extent of 62% at 90 °C, 85% at 60 °C, and more than 97% at 30 °C. 4-Methyl-1-pentene isomerized to the 2- and 3-alkenes, which formed π-allylic complexes in ethanol, but normal oxidation to the methyl ketone took place in DMF and γ-butyrolactone.[25]

In some cases, ketones are obtained in high yields by the oxymercuration[28] or oxythallation[29] of alkenes, followed by treatment with $PdCl_2$ in aqueous THF.

Oxidation of terminal alkenes may be carried out in benzene–water in the presence of cetyltrimethylammonium bromide at 80 °C[30] although cyclodextrins are better phase-transfer agents. In the presence of a catalytic amount of β-cyclodextrin, 1-decene and *cis*-2-butene were oxidized at 65 °C to 2-decanone (61%) and 2-butanone (76%) respectively.[31] Selective oxidation of linear C_8–C_{10} terminal alkenes took place at 75 °C in the presence of α-cyclodextrin in water,[32] but a low yield was obtained with 1-dodecene.

3.4.2.3 Reoxidants

The essence of the Wacker process is the invention of the reoxidation process for Pd^0 by using $CuCl_2$ as a cocatalyst. Cu^{II} salts are good reoxidants, but chlorination of carbonyl compounds takes place with $CuCl_2$. For example, chloroacetaldehyde is a by-product of the Wacker process. Chlorohydrin is another by-product from the reaction of ethylene with $PdCl_2$ and $CuCl_2$.[33,34] Thus, a number of other reoxidants were introduced. When CuCl, pretreated with oxygen, is used, no chlorination of ketones takes place and the rate of the reaction is higher.[6,7] Also $Cu(NO_3)_2$[4,35] and $Cu(OAc)_2$[36] have been used. Oxidation of cyclopentene with $PdCl_2/Fe(ClO_4)_3$ combined with electrochemical oxidation was carried out.[37] Benzoquinone was used at first by Moiseev *et al.*[38] and later by many other researchers as a good reoxidant, but a stoichiometric amount is necessary. The oxidation of alkenes can be carried out smoothly with catalytic amounts of both $Pd(OAc)_2$ and benzoquinone by means of efficient electrochemical reoxidation of hydroquinone to benzoquinone.[39] A combination of $Pd(OAc)_2$/benzoquinone/Fe phthalocyanine in the presence of $HClO_4$ (5%) in aqueous DMF is an active catalyst.[40] A heterogeneous catalyst system of a Pd^{II} salt and polymers containing quinone and sulfonic acid groups was used for ethylene oxidation.[41]

Alkyl nitrites (**3**) are good and unique reoxidants and used in industrial processes. 2,2-Dimethoxypropionitrile (**4**), an important intermediate for vitamin B_1 synthesis, is produced commercially by the oxidation of acrylonitrile in methanol (equation 4).[19] 3-Phenylpropene is oxidized to 1-phenyl-2,2-dialkoxypropane in a similar fashion.[20] Alkyl nitrites (**3**) are formed *in situ* from O_2, NO, and ROH, and can be recycled (equation 5).

$$\text{CH}_2\text{=CHCN} + 2\ \text{MeONO}\ \textbf{(3)} \xrightarrow{\text{Pd}^{II}} (\text{MeO})_2\text{CHCH}_2\text{CN}\ \textbf{(4)} + 2\ \text{NO} \qquad (4)$$

$$2\ \text{NO} + 1/2\ \text{O}_2 + 2\ \text{MeOH} \longrightarrow 2\ \text{MeONO}\ \textbf{(3)} + \text{H}_2\text{O} \qquad (5)$$

Efficient catalytic oxidation of 1-octene (90–95% yields) was carried out in acetic acid or *t*-butyl alcohol at 80 °C in 3 h by using 5 equiv. of 30% hydrogen peroxide and 1/500 equiv. of $Pd(OAc)_2$.[42] Hydrogen peroxide was used for the styrene oxidation with Na_2PdCl_4 in NMP. Compared with $PdCl_2/CuCl$, the rate of the oxidation was very high.[43] However, extensive double bond migration of terminal alkenes occurred in the presence of hydrogen peroxide. *t*-Butyl hydroperoxide and hydrogen peroxide are efficient reoxidants for the oxidation of internal double bonds conjugated to carbonyl groups.[43] An *endo*-peroxide/$Pd(OAc)_2$[44] and *t*-butyl hydroperoxide/palladium trifluoroacetate[11,45] were used for the oxidation of 1-alkenes.

$PdCl_2$–cobalt–nitro complexes were found to be efficient catalysts for alkene oxidation.[46] Pd–nitro complexes catalyze the oxidation of l-alkenes under oxygen without using other reoxidants to give methyl ketones.[47–51] They are mechanistically different from $PdCl_2/CuCl_2$ catalysts.[50] Heteropolyacids such as $H_3PMo_6O_{40}$ are water soluble and good reoxidants when $PdSO_4$ is used rather than $PdCl_2$ to give methyl ketones with high selectivity.[52,53] The catalyst system $PdSO_4/H_3PMo_6W_6O_{40}$ was used for the oxidation of 1-butene, cyclohexene and cyclopentene in aqueous DMF.[54–57] Cyclohexanone was obtained in 85% yield.

Deactivation of the palladium catalyst is a serious problem. Sometimes, the reaction is stopped by the precipitation of black palladium metal. Furthermore, the deactivation occurs during the reaction even when no precipitation is observed. The formation of a bis(dimethylamine)–$PdCl_2$ complex deactivates the catalyst when DMF is used as a solvent.[5,6] Also, the formation of rather stable π-allylpalladium complexes (**5**) from alkenes may account for the deactivation of the catalyst (equation 6).[58–61]

$$\text{RCH}_2\text{CH=CH}_2 + \text{PdCl}_2 \longrightarrow (\pi\text{-RC}_3\text{H}_4)\text{–PdCl}\ \textbf{(5)} + \text{HCl} \qquad (6)$$

3.4.3 OXIDATION OF TERMINAL ALKENES TO METHYL KETONES

3.4.3.1 Oxidation of Terminal Alkenes Bearing Various Functional Groups

Terminal double bonds are selectively oxidized to methyl ketones. A typical procedure for the oxidation of l-decene to 2-decanone in 65–73% yield with $PdCl_2/CuCl$ is given in *Organic Synthesis*.[62] Although there are several known synthetic methods for methyl ketone preparation, the $PdCl_2$-catalyzed oxidation of terminal alkenes seems to be one of the best. In other words, terminal alkenes can be regarded as precursors of methyl ketones, or as masked methyl ketones based on this reaction. This reaction is useful because terminal alkenes are easily available, stable under acidic and basic conditions and inert to nucleophiles.

Various terminal alkenes with functional groups are oxidized to the corresponding methyl ketones. Since the oxidation proceeds under mild conditions, various functional groups, such as an aldehyde,[63] carboxylic acid,[3,64] ester,[65–68] alcohol,[69–71] MOM ether,[72] acetal,[73] bromide,[74] selenide,[75] sulfonyl ester[73] and amines,[76,77] which are located at suitable positions, remain intact. Though it is known that alcohols are oxidized to aldehydes or ketones with $PdCl_2$,[78–80] the oxidation of terminal alkenes is faster than that of alcohols under these conditions.

Since the rate of oxidation of terminal alkenes is much higher than that of internal alkenes, selective oxidation of terminal alkenic bonds is possible without attacking the internal alkenic bonds in various dienes (equation 7).[6,77,81–93]

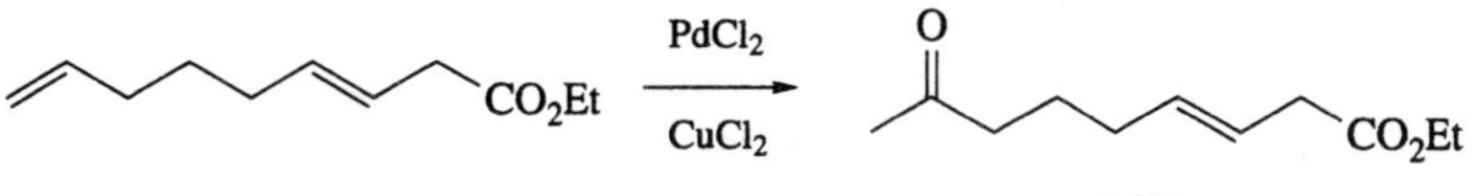

CO_2Et $\xrightarrow[CuCl_2]{PdCl_2}$ O CO_2Et (7)

86%

Steric hindrance considerably affects the rate of the oxidation. For example, one of the terminal double bonds in (**6**) was oxidized in 3 h, but it took 36 h to oxidize the other one (equation 8).[69]

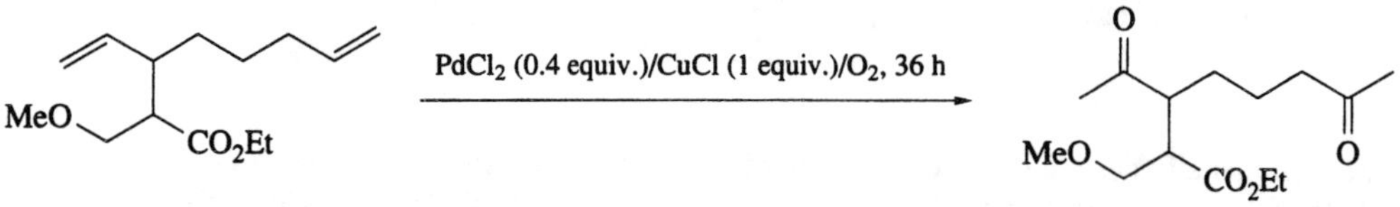

MeO CO_2Et (**6**) $\xrightarrow{PdCl_2\text{ (0.4 equiv.)/CuCl (1 equiv.)/}O_2\text{, 36 h}}$ O MeO O CO_2Et 71% (8)

Some allylic alcohols or acetates with terminal double bonds do not give methyl ketones cleanly. Oxidation of l-undecen-3-ol (**7**) at room temperature gave the methyl ketone (**8**) in 60% yield and l-hydroxy-3-undecanone (**9**) in 14% yield (equation 9).[69]

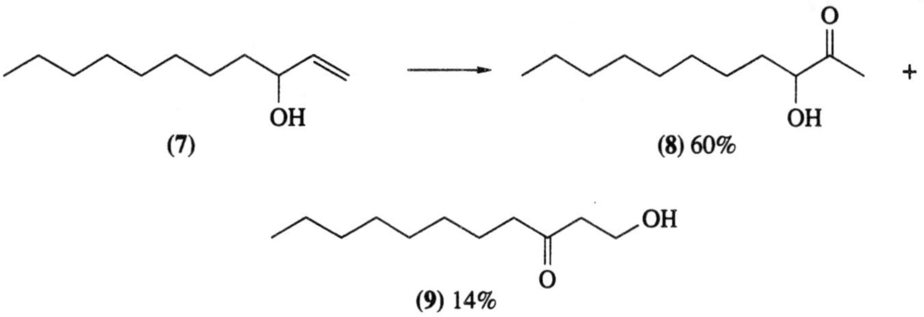

OH (**7**) ⟶ O OH (**8**) 60% +

OH O (**9**) 14% (9)

The acid sensitive 2-methyl-3-buten-2-ol (**10**) was oxidized to the corresponding methyl ketone (**11**) with a palladium–nitro complex (90% conversion, 90% selectivity) or $PdCl_2$–benzoquinone (98% conversion, 90% selectivity) (equation 10).[51]

OH (**10**) $\xrightarrow{81\%}$ O OH (**11**) (10)

1-Vinyl-1-cyclobutanols (**12**) undergo oxidative ring expansion to give cyclopentenones (**13**; Scheme 2).[94]

Oxidation of 3-acetoxy-1-nonene (**14**) at 50 °C gave a mixture of the methyl ketone (**15**) and 1-acetoxy-3-nonanone (**16**) in 33% and 17% yields.[69] The latter was formed by regioselective oxidation of 1-acetoxy-2-nonene, itself formed by the allylic rearrangement of (**14**) promoted by Pd^{II} ions (equation 11).

2,2-Disubstituted ethylenes undergo oxidation with a skeletal rearrangement. Methylenecyclobutane (**17**) was oxidized to cyclopentanone *via* ring expansion (Scheme 3).[95]

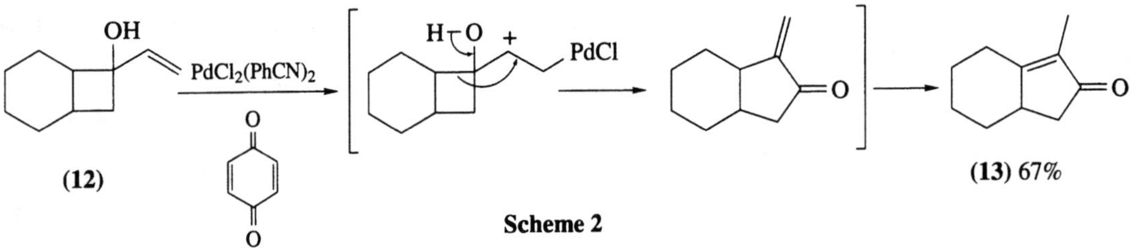

OH PdCl$_2$(PhCN)$_2$ H–O + PdCl O O

(12) (13) 67%

Scheme 2

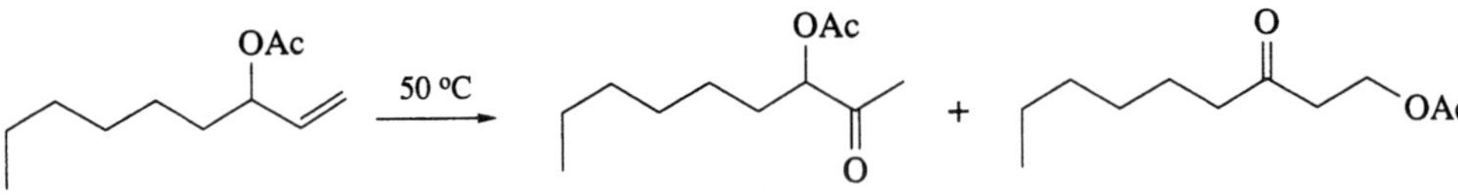

OAc 50 °C OAc O OAc (11)

(14) (15) 33% (16) 17%

PdCl$_2$/CuCl$_2$/ O_2/H_2O/C_6H_6 75% Pd–Cl O–H O

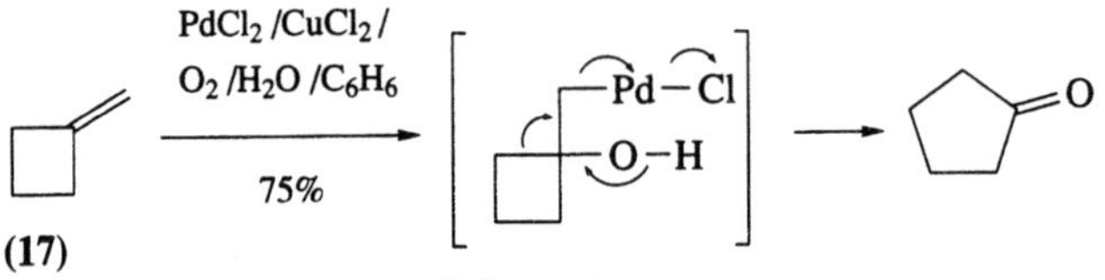

(17)

Scheme 3

Some terminal alkenes are oxidized to aldehydes depending on their structure. As described before, acrylonitrile and acrylate are oxidized to acetals of aldehydes in alcohols or ethylene glycol.[18–20] Selective oxidation of terminal carbons in 4-hydroxy-1-alkenes (**18**) gave the five-membered hemiacetals (**19**), which can be converted to γ-butyrolactones by PCC oxidation (Scheme 4).[96] Formation of a tricyclic six-membered hemiacetal (62%) from a 5-hydroxy-1-alkene system was used for the synthesis of rosaramicin.[97] Formation of aldehydes as a major product from terminal alkenes using $(MeCN)_2Pd(Cl)(NO_2)$ and $CuCl_2$ in *t*-butyl alcohol under selected conditions was reported.[98] The vinyl group in the β-lactam was oxidized mainly to the aldehyde as shown below (equation 12).[99]

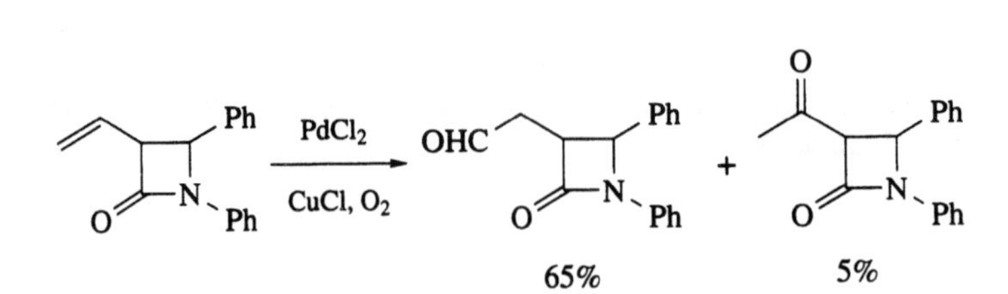

OH C_6H_{13} PdCl$_2$ O O, 62% HO O C_6H_{13} PCC O O C_6H_{13}

(18) (19)

Scheme 4

Ph N Ph O PdCl$_2$ CuCl, O_2 OHC Ph N Ph O + O Ph N Ph O (12)

65% 5%

3.4.3.2 Synthetic Applications

3.4.3.2.1 Synthesis of natural products

Oxidation of terminal alkenes to methyl ketones is useful for the syntheses of natural products.[100–102] Based on this method, simple and efficient syntheses of prostaglandin intermediates,[64] Queen bee substance,[92] zearalenone,[73] dihydrojasmone, jasmone,[82] diplodialide B,[83] lasiodiplodin methyl ether,[84] curvularin,[86] resorcylide,[87] pyrethrolone,[88] muscone,[89,90] recifeiolide[93] and α-vetispirene[63] have been carried out. Syntheses of steroids based on this methodology are treated in Section 3.4.3.2.4.

3.4.3.2.2 Synthesis of 1,4-dicarbonyl compounds

(i) Synthesis via *allylation of carbonyl compounds*

A simple synthetic method for 1,4-dicarbonyl compounds was introduced, based on the allylation of carbonyl compounds with allyl halide as a C_3 component, followed by the palladium-catalyzed oxidation of the terminal alkenes (**20**) to methyl ketones (**21**).[7] In this method, the allyl group is a synthetic equivalent of the 2-oxopropyl group (Scheme 5). This is a good anellation method for cyclopentenones.

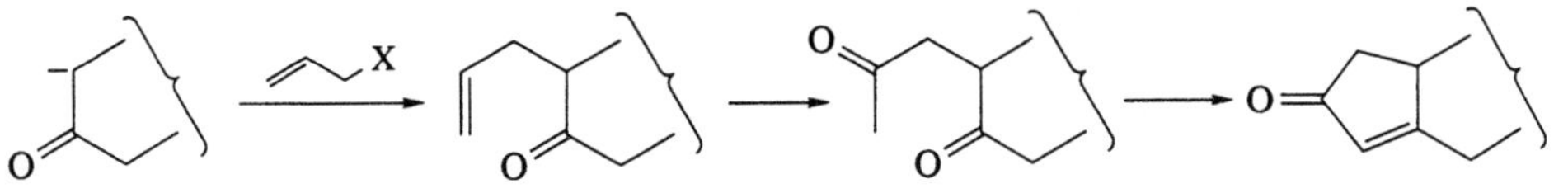

(**20**) (**21**)

Scheme 5

As a typical example, allylation of cyclohexanone *via* the pyrrolidine enamine with allyl bromide gave 2-allylcyclohexanone. Its terminal alkenic bond was oxidized with $PdCl_2/CuCl/O_2$ in aqueous DMF to give the 1,4-diketone (**21**) in 68% yield. The base-catalyzed cyclization of (**21**) gave the indenone (**22**) in 85% yield (Scheme 6).[7] This methodology was applied to the syntheses of pentalenene,[103] laurenene,[74] decarboxyquadrone,[104] and coriolin.[105]

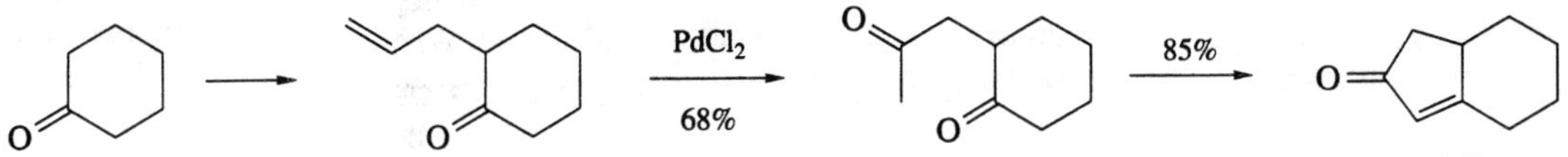

$PdCl_2$, 68%; 85%

(**20**) (**21**) (**22**)

Scheme 6

Bicyclo[10.3.0]-$\Delta^{1,15}$-pentadecen-14-one (**25**) was prepared from cyclododecanone. Allylation of the β-keto ester (**23**) and the oxidation of the terminal alkenic bond afforded the 1,4-diketone (**24**) in 72% yield from the β-keto ester. Base-catalyzed cyclization and deethoxycarbonylation gave the bicycloketone (**25**), which was converted to muscone (**26**; Scheme 7).[106] Thus, this is a method for three-carbon ring expansion.

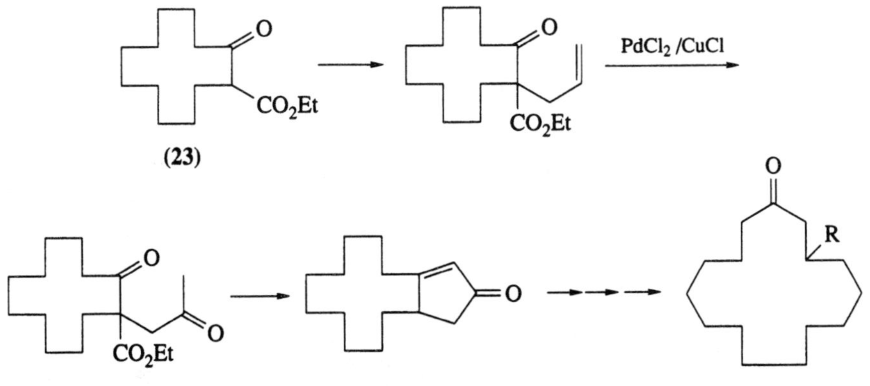

$PdCl_2/CuCl$

(**23**)

(**24**) 72% from (**23**) (**25**) (**26**)

Scheme 7

1,4-Keto aldehydes are prepared by the allylation of aldehydes. Reaction of 2-*p*-tolylpropanal (**27**) with allyl bromide gave 2-*p*-tolyl-2-methyl-4-pentenal (**28**) in 63% yield. The oxidation of the terminal alkenic bond to a methyl ketone gave the 4-oxopentanal (**29**) in 68% yield.[69] The keto aldehyde (**29**) was an intermediate for cuparanone (**30**) synthesis (Scheme 8).

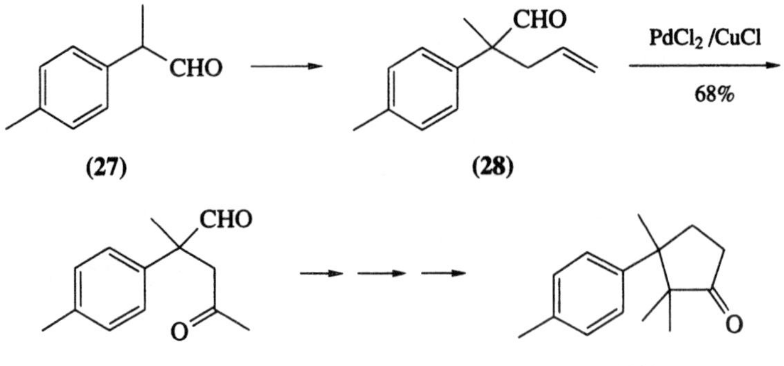

(27) (28) $PdCl_2$/CuCl 68%

(29) (30)

Scheme 8

The facile synthesis of 4,4-dimethyl-2-cyclopentenone by the allylation of isobutanal, followed by the oxidation and aldol condensation of the keto aldehyde is another example.[107]

(ii) Synthesis of 4-oxopentanals via *Claisen rearrangement and oxidation*

4-Oxopentanals may be synthesized from allylic alcohols by 3,3-sigmatropic rearrangement of their vinyl ethers, and subsequent oxidation of the terminal double bond.[108] Cinnamyl alcohol (**31**) was converted to the allyl vinyl ether (**32**), which was subjected to Claisen rearrangement to give 3-phenyl-4-pentenal (**33**) in 50% yield. Oxidation of the terminal double bond of (**33**) gave 3-phenyl-4-oxopentanal (**34**) in 76% yield, which was converted to 2-methyl-3-phenylfuran (**35**) in quantitative yield (Scheme 9).

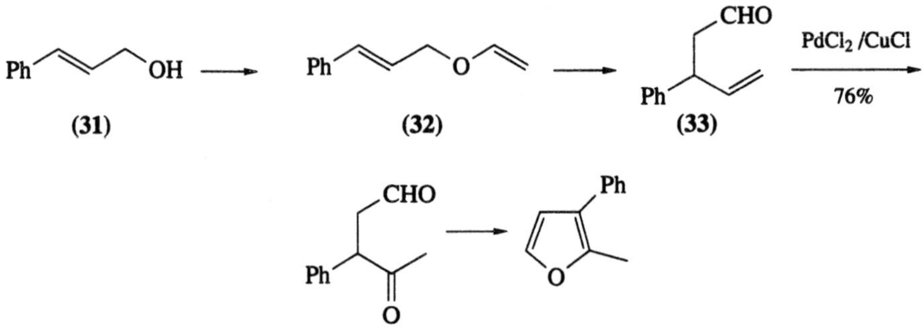

(31) (32) (33) $PdCl_2$/CuCl 76%

(34) (35)

Scheme 9

3,3-Cyclohexano-4-oxopentanal (**37**) was synthesized from the allylic alcohol (**36**) by the same procedure, and the spiro compound (**38**) was obtained by intramolecular aldol condensation of (**37**; Scheme 10).

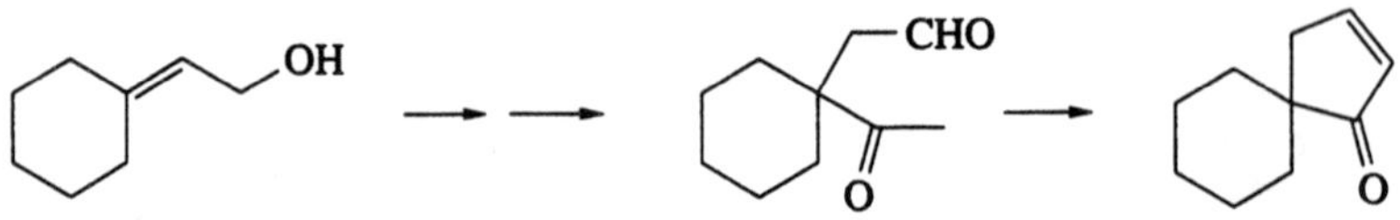

(36) (37) (38)

Scheme 10

The 3,3-sigmatropic rearrangement of 2-octenyl vinyl ether (**39**) afforded 3-pentyl-4-pentenal (**40**) in 79% yield. The terminal double bond was converted to the methyl ketone (**41**) in 90% yield. The keto aldehyde (**41**) was converted to 5-pentyl-2-cyclopentenone (**42**). The double bond migration gave 2-pentyl-2-cyclopentenone (dihydronorjasmone), which was converted to methyl dihydrojasmonate (**43**; Scheme 11).[108,109] Similarly, dihydrojasmone was synthesized from 2-octenyl allyl ether.

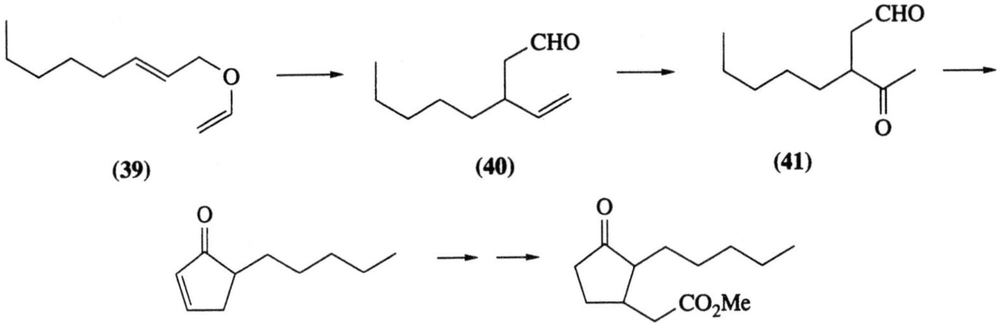

CHO CHO O
(**39**) (**40**) (**41**)
O O CO$_2$Me
(**42**) (**43**)

Scheme 11

(iii) Other methods

1,4-Addition of an acyl anion or its equivalent to α,β-unsaturated ketones is an important synthetic method for 1,4-dicarbonyl compounds. In the palladium method for 1,4-dicarbonyl compounds, a vinyl Grignard or vinyllithium reagent is used as a synthetic equivalent of the acetyl anion. Reaction of lithium divinylcuprate with 2-cyclohexenone (**44**) afforded 3-vinylcyclohexanone (**45**), which was oxidized to the 1,4-diketone (**46**; Scheme 12).[108]

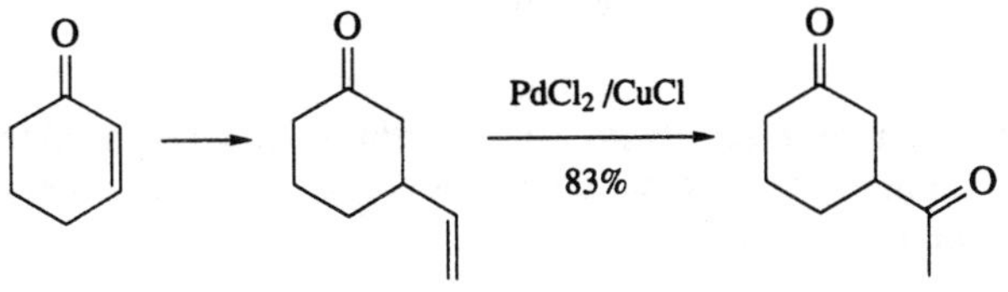

$PdCl_2$/CuCl, 83%
(**44**) (**45**) (**46**)

Scheme 12

The Lewis acid promoted addition of allylsilane (**48**) to nitroalkene (**47**) gave the unsaturated ketone (**49**) after hydrolysis of the nitro group. The palladium-catalyzed oxidation affords the 1,4-diketone (**50**; Scheme 13).[110]

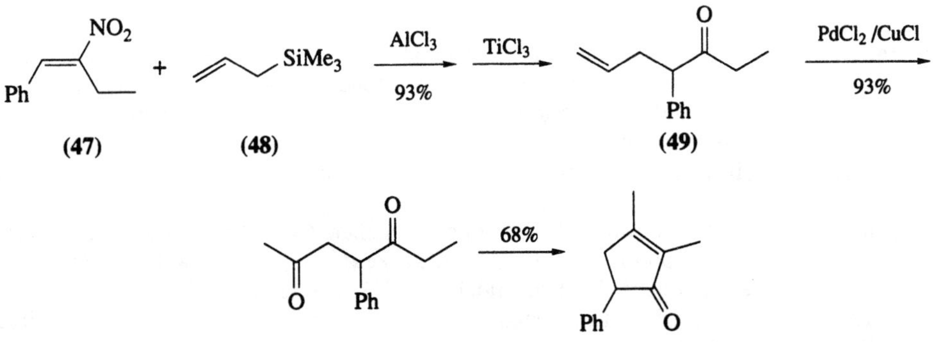

NO_2 Ph + $SiMe_3$ — $AlCl_3$, 93% — $TiCl_3$ — Ph — $PdCl_2$/CuCl, 93%
(**47**) (**48**) (**49**)
O Ph — 68% — Ph O
(**50**)

Scheme 13

The 1,4-diketone (**54**) was prepared by the titanium-catalyzed butenylation of the silylacetylene (**51**), followed by the oxidations of the terminal double bond in (**52**) to give (**53**) and the silylated double bond (Scheme 14).[111]

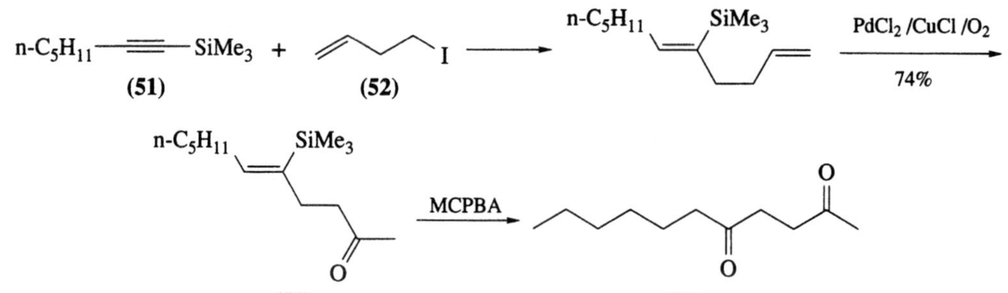

Scheme 14

3.4.3.2.3 Synthesis of 1,5-dicarbonyl compounds

1,5-Dicarbonyl compounds can be prepared by the reaction of ketones with 3-butenyl halide as a C_4 component, following oxidation of the terminal double bond.[7] A modified method for 3-butenylation of ketones by the palladium-catalyzed reaction of 4-acetoxy-2-butenylmethyl carbonate with ketones, followed by the palladium-catalyzed reaction of ammonium formate was reported (Scheme 15).[112]

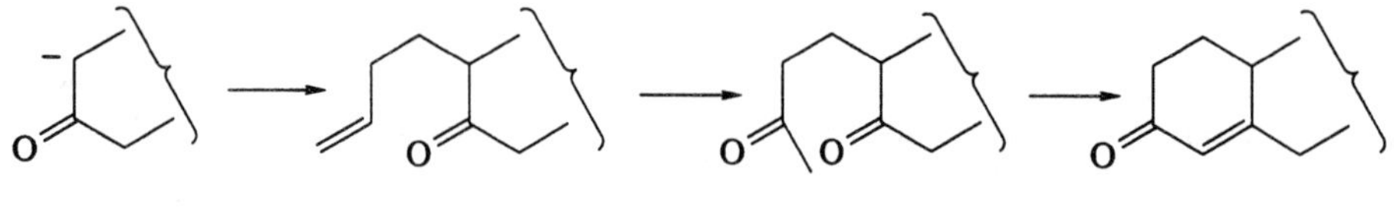

Scheme 15

Reaction of β-keto esters (**55a**) and (**55b**) with 3-butenyl bromide gave the alkenes (**56a**) and (**56b**) in 63% and 57% yields. Oxidation of the terminal double bonds gave the 1,5-diketones (**57a**) and (**57b**) in 61% and 58% yields (Scheme 16).

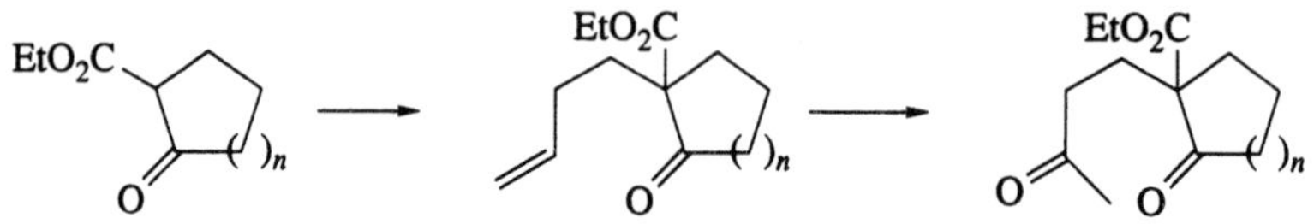

Scheme 16

In this synthesis of 1,5-dicarbonyl compounds, 3-butenyl halide is behaving as a masked 3-oxobutyl reagent, and can be used as an equivalent of methyl vinyl ketone. These reactions offer new anellation methods. Also 1,4-addition of the allyl group to enones, followed by oxidation, offers a convenient synthetic method for 1,5-diketone preparation. Lewis acid promoted Michael addition of allylsilane (**48**) to α,β-unsaturated ketones, followed by the palladium-catalyzed oxidation, affords 1,5-diketones (Scheme 17).[113]

The 1,5-diketone formation by the Michael addition of allylsilane (**48**) to α,β-unsaturated ketones was applied to the synthesis of (+)-nootkatone.[114] Reaction of the keto group of keto aldehyde (**58**) with allyl Grignard reagent and dehydration gave the diene aldehyde (**59**). The selective oxidation of the terminal double bond afforded the 1,5-dicarbonyl compound (**60**), which is not stable and converted directly to pyridines and phenols (Scheme 18).[115]

Synthesis of useful 5-oxohexanals can be carried out *via* the following three reactions: (1) 1,2-addition of allylmagnesium bromide to α,β-unsaturated aldehydes (**61**) to give 3-hydroxy-1,5-dienes (**62**); (2) conversion to 5-hexenals (**63**) by the oxy-Cope rearrangement; (3) the palladium-catalyzed oxidation of

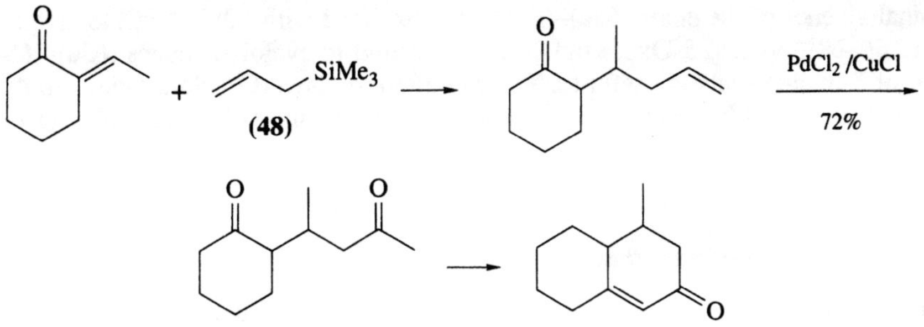

SiMe3
(48)
PdCl2 /CuCl
72%

Scheme 17

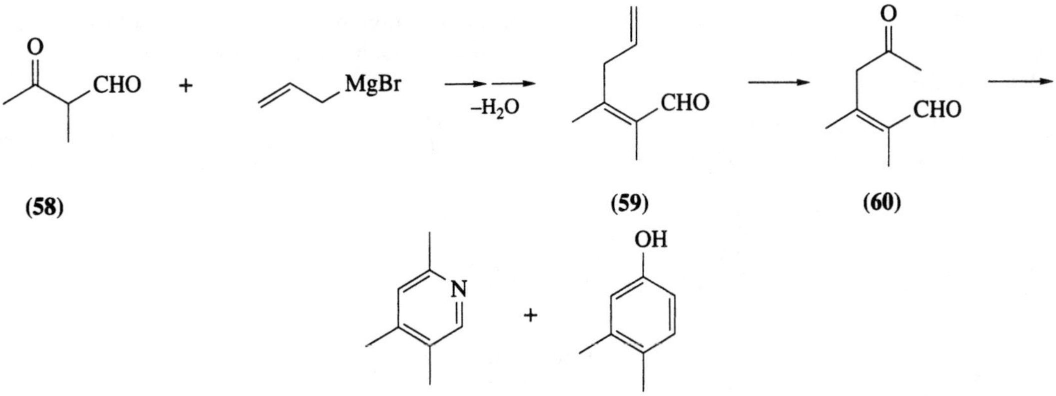

CHO
MgBr
–H2O
CHO
CHO
N
OH
(58)
(59)
(60)

Scheme 18

the terminal alkene to give the 5-oxohexanals (**64**). In this method, α,β-unsaturated aldehydes are used directly without protection. Based on this process, formal Michael addition of the 2-oxopropyl anion to α,β-unsaturated aldehydes can be achieved (Scheme 19).[108]

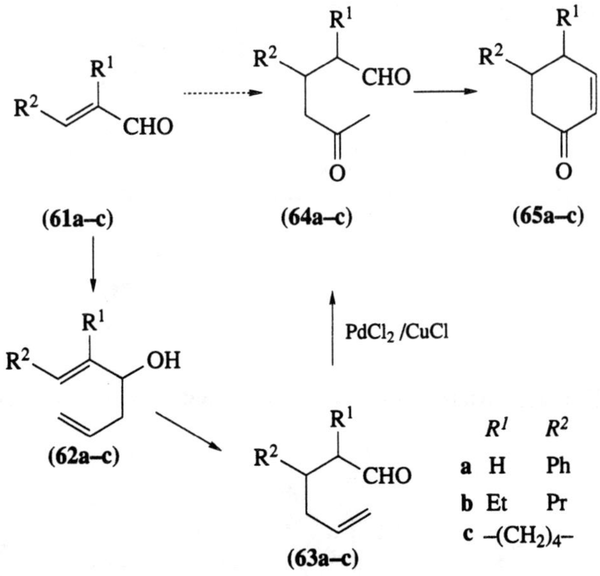

R1
R2
CHO
CHO
O
(61a–c)
(64a–c)
(65a–c)
OH
(62a–c)
PdCl2 /CuCl
(63a–c)
R1 R2
a H Ph
b Et Pr
c –(CH2)4–

Scheme 19

Reaction of allylmagnesium bromide with enals (**61a**)–(**61c**) gave allylvinyl methanols (**62a**)–(**62c**) in good yields. Oxy-Cope rearrangement using potassium hydride gave 5-hexenals (**63a**)–(**63c**) in 60–70%

yields. Terminal alkenes of the enals (**63a**)–(**63c**) were oxidized with $PdCl_2$/CuCl to give 5-oxohexanals (**64a**)–(**64c**) in 66–68% yields. 5-Oxohexanals are converted to cyclohexenones (**65a**)–(**65c**). This synthetic method of 5-oxohexanals is useful for the anellation to convert cyclohexanone to $\Delta^{3,4}$-2-octalone (**65c**), which is difficult to obtain by common Robinson anellation. Acid-catalyzed cyclization of (**64c**) gave (**65c**) in 68% yield.

3.4.3.2.4 Application to steroid synthesis

A new synthetic method for steroids has been developed using a butadiene dimer (**66**) as a building block and the palladium-catalyzed oxidation as the key reaction.[102,116] 3-Acetoxy-1,7-octadiene (**66**), prepared by the palladium-catalyzed reaction of butadiene with acetic acid, is hydrolyzed and oxidized to 1,7-octadien-3-one (**67**) in high yield. The enone (**67**) is a very useful reagent for bisanellation because its terminal double bond can be regarded as a masked ketone which can be readily unmasked by the palladium catalyst to form the 1,5-diketone (**68**) after Michael addition at the enone moiety of (**67**; Scheme 20). Thus, the enone (**67**) is the cheapest and most readily available bisanellation reagent, permitting a simple total synthesis of steroids.

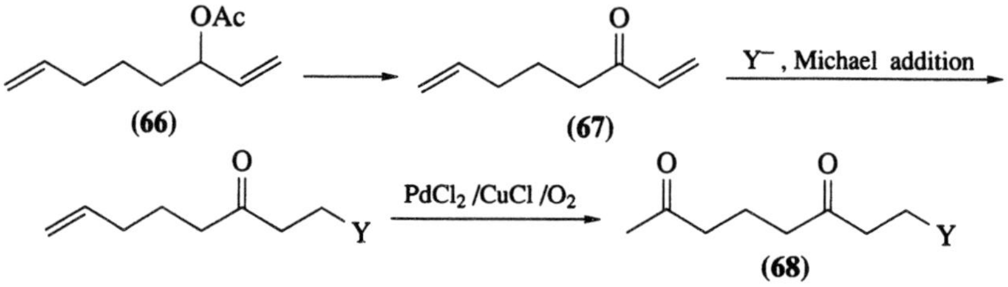

OAc (**66**) → O (**67**) —Y^-, Michael addition→ O ... Y —$PdCl_2$/CuCl/O_2→ O O ... Y (**68**)

Scheme 20

In the simplest example, Michael addition to the enone (**67**) of the cyclohexanone enamine and aldol condensation yielded 4-(3-butenyl)-3-oxo-Δ^4-octalin (**69**). The terminal double bond was oxidized to the ketone (**70**) by $PdCl_2$/CuCl/O_2, and subsequent aldol condensation leads to the tricyclic ketone (**71**; Scheme 21).

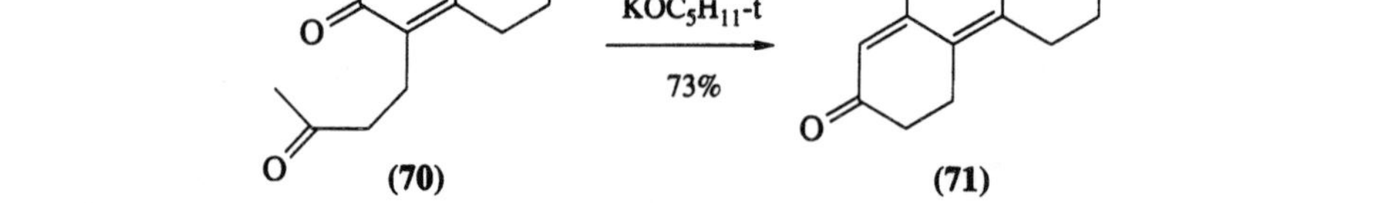

(**67**) + N-enamine —64%→ (**69**) —$PdCl_2$/CuCl/O_2, 83%→ (**70**) —KOC_5H_{11}-t, 73%→ (**71**)

Scheme 21

The synthesis of (+)-19-nortestosterone (**73**) was carried out starting from the optically active keto ester (**72**; equation 13).

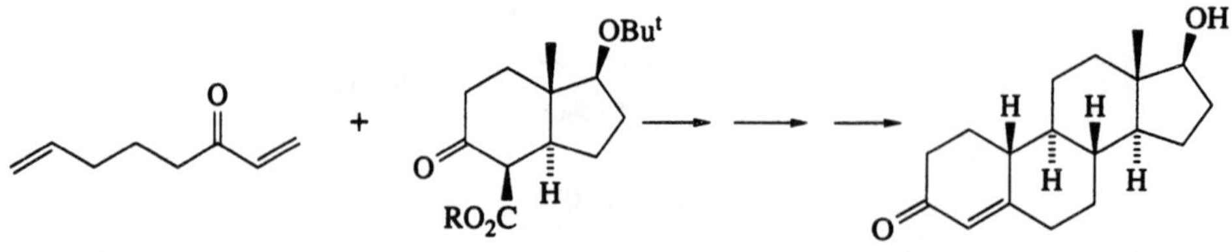

(**67**) + (**72**) $[\alpha]^{25}_D$: +35.1° (OBu^t, RO_2C, H) → → → (**73**) $[\alpha]^{25}_D$: +57.8° (OH, H, H, H, H) (13)

Another route to (+)-19-nortestosterone (**73**) started from 2-methyl-1,3-cyclopentanedione (**74**). The asymmetric aldol condensation of the Michael adduct using L-phenylalanine produced the optically active enone (**75**). The $PdCl_2$-catalyzed oxidation yielded crystalline trione (**76**) in 77% *ee*, which was recrystallized as an optically pure form. Reduction of the double bond and aldol condensation afforded the desired CD *trans*-fused ketone (**77**). The construction of the A-ring was carried out by alkylation with 4-bromo-1-butene to give (**78**), the palladium-catalyzed oxidation, and aldol condensation to give the optically active (+)-19-nortestosterone (**73**; Scheme 22).[117]

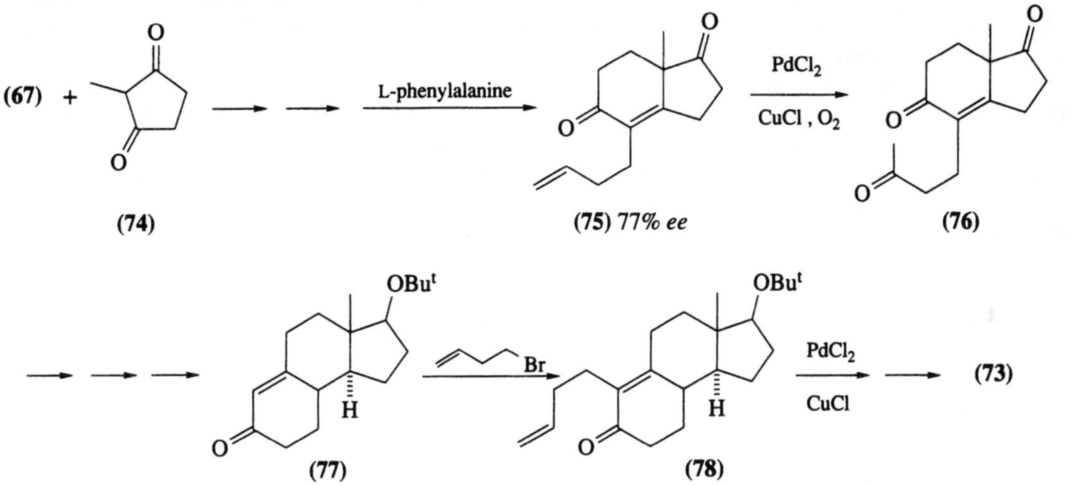

(**67**) + (**74**) → → L-phenylalanine → (**75**) 77% *ee* → $PdCl_2$, CuCl, O_2 → (**76**) → → → (**77**) → 4-bromo-1-butene → (**78**) → $PdCl_2$, CuCl → → (**73**)

Scheme 22

Subsequently the trisanellation reagent, 7-acetoxy-1,11-dodecadien-3-one (**80**) was prepared from the bisanellation reagent (**67**), and the synthesis of D-homo-19-norandrosta-4-en-3-one (**82**) was carried out from (**79**) as shown below.[118] For the A-ring formation, the unmasking of the terminal double bond and hydrogenation afforded the 1,5-diketone (**81**), which was subjected to intramolecular aldol condensation to give D-homo-4-androstene-3,17a-dione (**82**; Scheme 23).

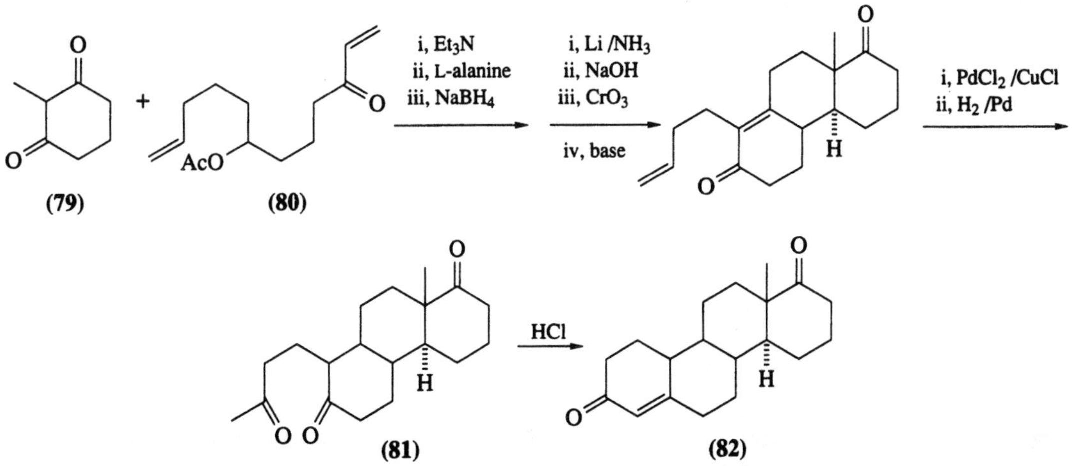

(**79**) + (**80**) → i, Et_3N; ii, L-alanine; iii, $NaBH_4$ → i, Li/NH_3; ii, NaOH; iii, CrO_3; iv, base → → i, $PdCl_2$/CuCl; ii, H_2/Pd → (**81**) → HCl → (**82**)

Scheme 23

3.4.4 OXIDATION OF INTERNAL ALKENES

3.4.4.1 General Remarks

Compared with the facile oxidation of terminal alkenes, the oxidation of internal alkenes is extremely slow under the usual conditions. In addition, the reaction is not regioselective. The reaction would be very useful if cyclic alkenes could be oxidized to cyclic ketones. However, cyclic alkenes such as cyclopentene, cyclohexene and cyclooctene are not oxidized efficiently under the usual conditions. Good results so far reported are those with the use of heteropolyacids as the reoxidant.[54,55] Cyclohexene was oxidized by using $PdSO_4$ and heteropolyacid ($H_3PMo_6W_6O_{40}$) in aqueous DMF, giving cyclohexanone in 85% yield; the turnover of the catalyst was 90. For cyclopentanone, the turnover number was 30. Cyclopentene was oxidized smoothly by electrooxidation catalyzed by Pd^{II} and benzoquinone.[39] Ethanol is a good solvent for the oxidation of cyclohexene and cyclopentene.[119–121] *cis*-2-Butene was oxidized to 2-butanone in 82% yield in polyethylene glycol.[17] The stoichiometric oxidation of 1-methylcyclobutene (**83**) gave cyclopropyl methyl ketone (Scheme 24).[122]

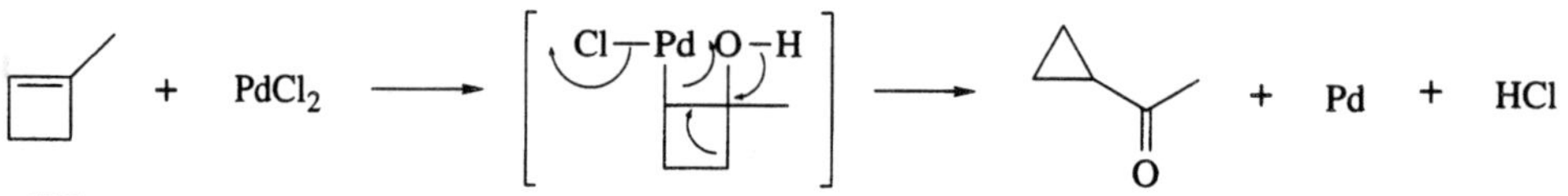

(83)

Scheme 24

The furan ring in khellin and cyclic ethers in sugars were oxidized to esters in methanol (Scheme 25).[35,123] Tricyclo [4.2.2.0^{2,5}]deca-3,7-diene-9,10-dicarboxylate (**84**) was oxidized to the monoketone in 7 h as a primary product, which was further oxidized to the diketone in 10 h in high yield with a stoichiometric amount of $Pd(NO_3)_2$ in refluxing methanol (equation 14).[124]

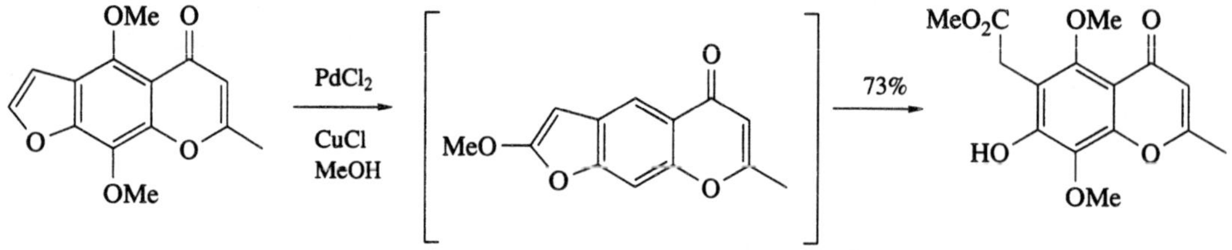

Scheme 25

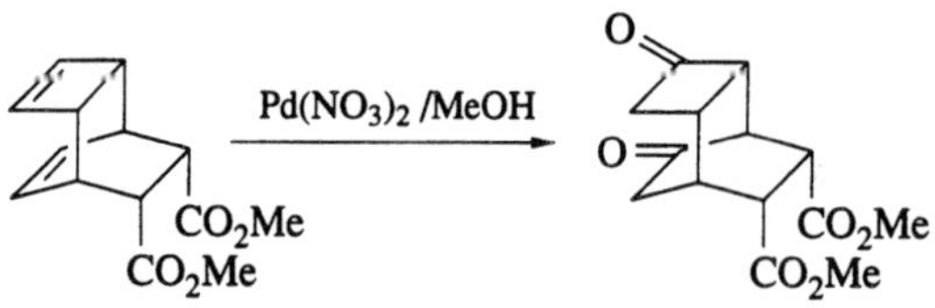

(14)

(84)

Thus, more improvement is still necessary for the oxidation of internal alkenes. However, internal alkenes with some functional groups at suitable positions are oxidized regioselectively by participation of the functional groups under proper conditions as surveyed in the following section.

3.4.4.2 Regioselective Oxidation of α,β- and β,γ-Unsaturated Carbonyl Compounds

The oxidation of α,β-unsaturated carbonyl compounds under the usual conditions in DMF using $PdCl_2/CuCl/O_2$ is very slow. However, regioselective oxidation of α,β-unsaturated esters to β-keto esters (equation 15), and α,β-unsaturated ketones to 1,3-diketones (equation 16) proceeds with Na_2PdCl_4 in solvents such as 50% acetic acid, isopropyl alcohol, and NMP.[43] *t*-Butyl hydroperoxide and hydrogen peroxide are used as the reoxidants of the reduced palladium. The reaction proceeds slowly at room temperature but smoothly between 50 and 80°C. Some typical examples of this process are shown in Table 1.

$$R^1\text{-CH=CH-}CO_2R^2 \longrightarrow R^1\text{-C(=O)-CH}_2\text{-}CO_2R^2 \qquad (15)$$

$$R^3\text{-CH=CH-C(=O)-}R^4 \longrightarrow R^3\text{-C(=O)-CH}_2\text{-C(=O)-}R^4 \qquad (16)$$

Table 1 Oxidation of Various α,β-Unsaturated Esters and Ketones

Alkene	*Product*	*Yield (%)*
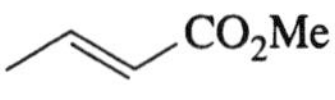 CO_2Me	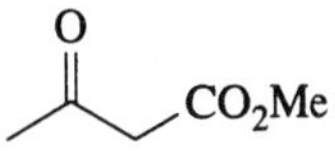 O, CO_2Me	83
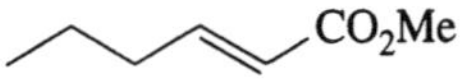 CO_2Me	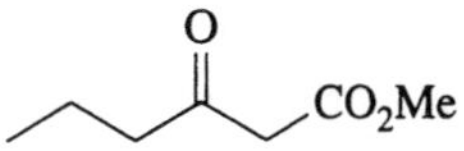 O, CO_2Me	78
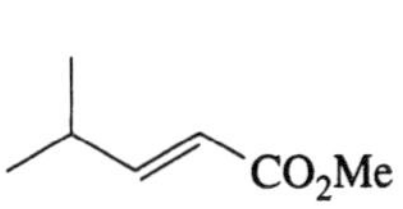 CO_2Me	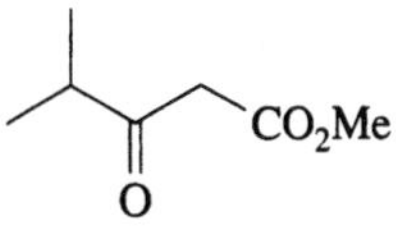 CO_2Me, O	64
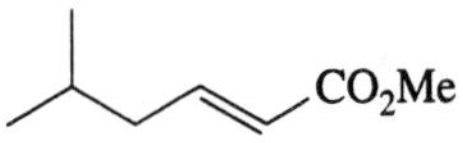 CO_2Me	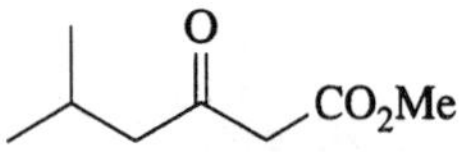 O, CO_2Me	68
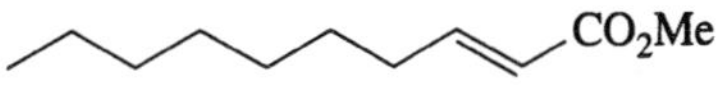 CO_2Me	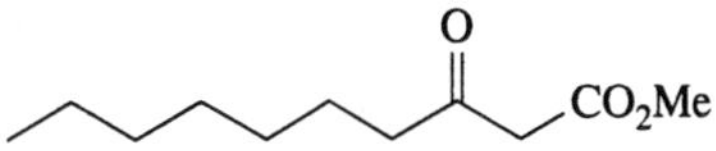 O, CO_2Me	75
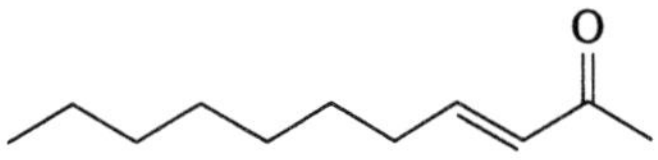 O	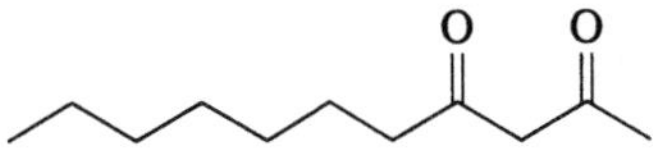 O O	59
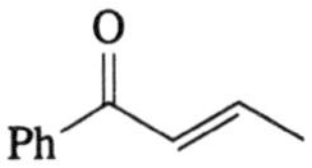 Ph, O	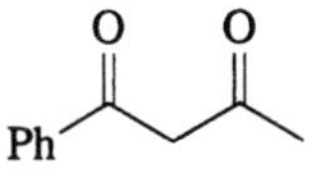 Ph, O O	59

It is well known that π-allylpalladium complexes (**86**) are easily formed by the reaction of $PdCl_2$ with β,γ-unsaturated esters or ketones (**85**).[125] An attempted oxidation of β,γ-unsaturated esters and ketones with the $PdCl_2/CuCl/O_2$ catalyst system in aqueous DMF led to π-allylpalladium complex formation as the main reaction, and the oxidation of the alkenic bond was hardly observed to a significant extent. However, in aqueous dioxane or THF, the oxidation became the main reaction, giving γ-keto esters and 1,4-diketones (**87**), respectively, with high regioselectivity (Scheme 26).[126] Some results are shown in Table 2. In all cases, no β-keto ester or 1,3-diketone was detected. At the end of the reaction, formation of a considerable amount of the π-allylpalladium complex (**86**) was observed. γ-Keto esters and 1,4-diketones are useful intermediates for the preparation of cyclopentanedione and cyclopentenone, respectively, by base-catalyzed cyclization. This regioselective oxidation provides a unique and efficient synthetic method for γ-keto ester and 1,4-diketone synthesis.

1,4-Diketones were obtained mainly by the oxidation of γ,δ-unsaturated ketones and used for cyclopentenone anellation.[127,128] One example is shown below (equation 17). However, the following γ,δ-unsaturated lactone was oxidized regioselectively to give 1,5-ketolactone in 73% yield in one week (equation 18).[129]

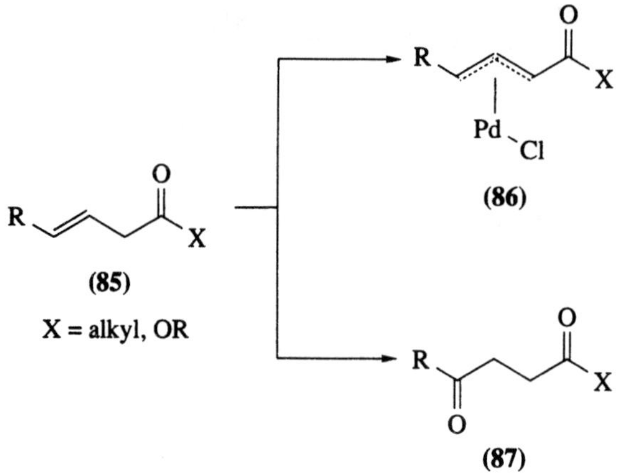

R, X, O, (85), X = alkyl, OR, Pd, Cl, (86), (87)

Scheme 26

Table 2 Oxidation of Various β,γ-Unsaturated Esters and Ketones

Alkene	*Product*	*Yield* (%)
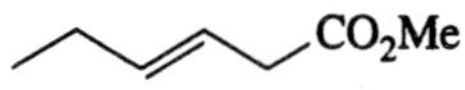 CO_2Me	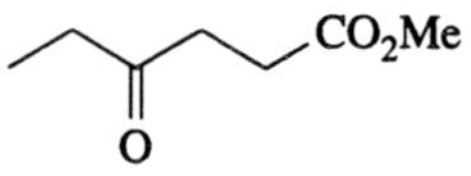 CO_2Me, O	52
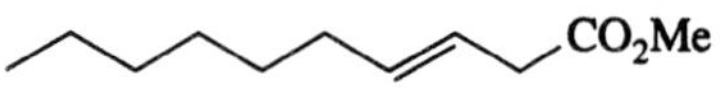 CO_2Me	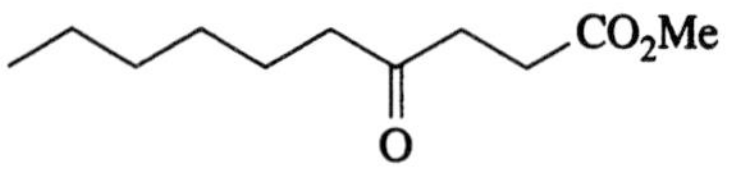 CO_2Me, O	61
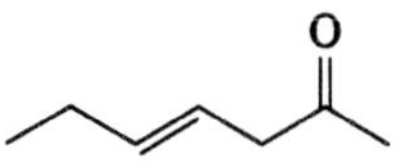 O	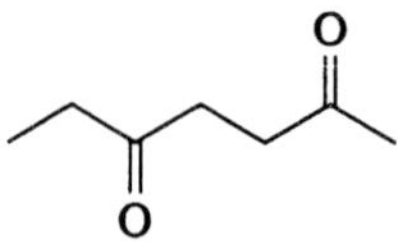 O, O	45
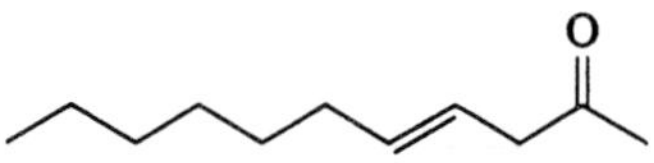 O	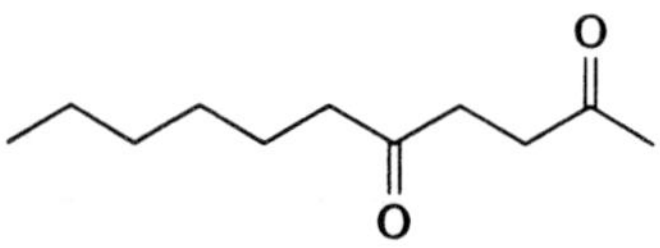 O, O	61

$PdCl_2$, $CuCl_2$ / DMF / 57% → O + O, O (17)

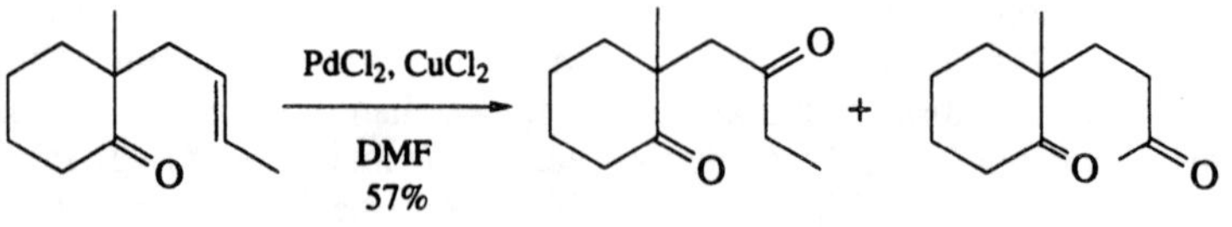

91:9

OMe, OMe, O, O — $PdCl_2$ /CuCl /O_2 → OMe, OMe, O, O, O (18)

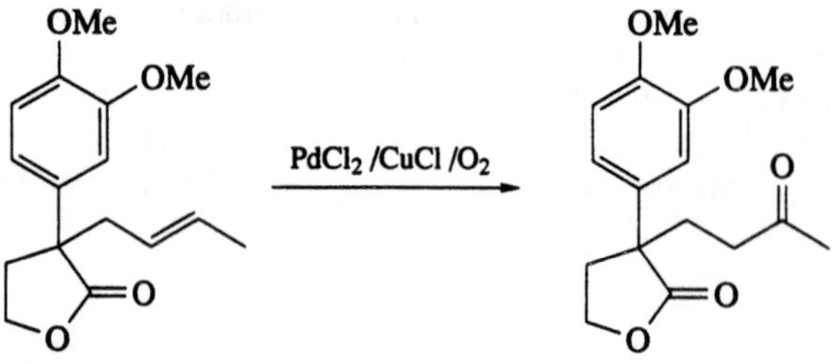

3.4.4.3 Regioselective Oxidation of Allyl and Homoallyl Ethers and Esters

The oxidation of various allyl ethers and acetates with internal alkenes using $PdCl_2/CuCl/O_2$ or $PdCl_2$/*p*-benzoquinone catalyst systems gave the corresponding β-alkoxy ketones regioselectively (equation 19).[130] No α-alkoxy or α-acetoxy ketone was detected.

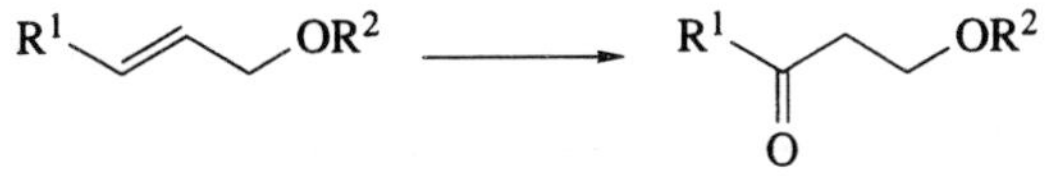

R^1 OR^2 ⟶ R^1 OR^2 O (19)

R^1 = alkyl; R^2 = alkyl, acetyl

In the oxidation of 2-octenyl acetate, in addition to the normal oxidation, palladium-catalyzed allylic rearrangement and subsequent oxidation took place to give a small amount of 3-acetoxy-2-octanone as a byproduct. Ethers of secondary allylic alcohols also underwent the regioselective oxidation to give the corresponding β-alkoxy ketones in 30–40% yields. But in this case too, by-products derived from the allylic rearrangement and subsequent oxidation were also detected. Results of the oxidation of some allyl ethers are shown in Table 3.[130]

Table 3 Oxidation of Various Allyl Ethers

Substrate	*Product*	*Yield (%)*
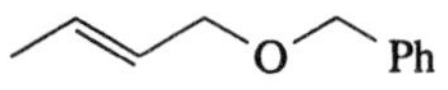 O Ph	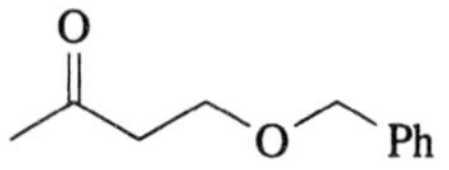 O O Ph	67
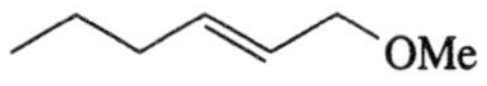 OMe	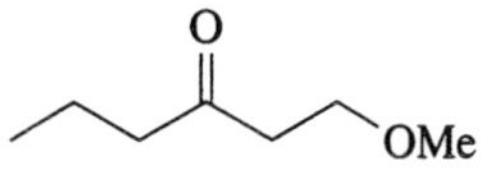 O OMe	65
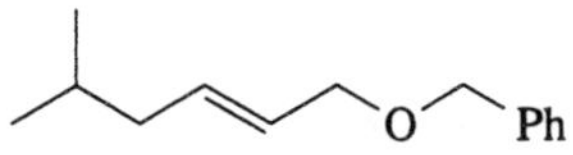 O Ph	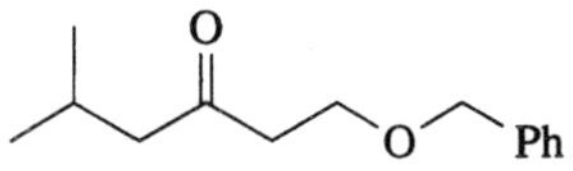 O O Ph	76
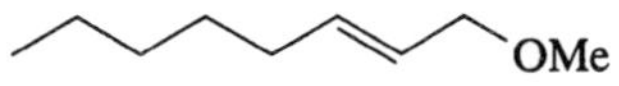 OMe	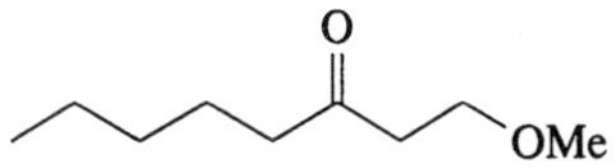 O OMe	50–60
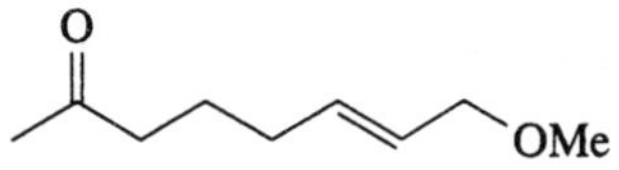 O OMe	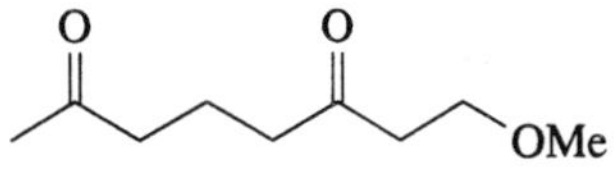 O O OMe	64

Homoallyl acetates were oxidized to form the corresponding γ-acetoxy ketones with high regioselectivity. The results are shown in Table 4. In this oxidation, small amounts of β-acetoxy ketones were sometimes formed (<10%).

In these oxidations of the internal alkenes, an oxygen atom was introduced at the alkenic carbon atom remote from the neighboring alkoxy or acetoxy group. The results suggest that there is definite influence of the alkoxy or acetoxy group, which may be explained by coordination of palladium with the oxygen function, to control the regioselection. As another example of this regioselection, β-tetrahydropyranylstyrene (**88**) was oxidized regioselectively to the phenyl ketone (**89**; equation 20). On the other hand, β-methylstyrene (**90**) was oxidized to give phenylacetone (**91**) and propiophenone (**92**) in a 3:1 ratio (equation 21).[131] The participation effect of the ether oxygen is clear.

The control of the regioselectivity by the neighboring oxygen function is effective only in the oxidation of internal alkenes. Terminal alkenes bearing a neighboring alkoxy or acetoxy group were oxidized to the corresponding methyl ketones, though the effect of the alkoxy or acetoxy group predicts the formation of the corresponding aldehydes (see equation 11).

β-Alkoxy ketones and γ-acetoxy ketones prepared by the oxidation of allyl ethers and homoallyl acetates, respectively, are synthetically useful intermediates. The reaction of (**93**) in the presence of excess sodium methoxide with 2-methylcyclohexanone afforded methyloctalone (**94**) in 42% yield (equation

Table 4 Oxidation of Homoallyl Acetates

(A) (B)

Substrate	*Catalyst* (%)	*Time* (h)	*Yield* (%)	*Ratio* (**A**):(**B**)
OAc	10	12	80	91:9
OAc	10	24	73	88:12
OAc	10	24	72	90:10
OAc	20	12	42	100:0

$PdCl_2$, CuCl; O_2, DMF; 44% (20)

(88) **(89)** 100:0

$PdCl_2$, CuCl; O_2, DMF (21)

(90) **(91)** 75:25 **(92)**

22). Thus, allyl ethers can be regarded as 'masked vinyl ketones' by means of this palladium-catalyzed oxidation.

NaOMe /MeOH; 42% (22)

(93) **(94)**

1,4-Diketones or γ-keto aldehydes (**95**) were prepared from γ-acetoxy ketones by saponification of the acetate with aqueous sodium hydroxide, followed by oxidation with PCC (equation 23).

i, aqueous NaOH; ii, pyridinium $ClCrO_3^-$ (23)

R = H, Me

(95)

3.4.5 REFERENCES

1. J. Smidt, W. Hafner, R. Jira, J. Sedlmeier, R. Sieber, R. Ruttinger and H. Kojer, *Angew. Chem., Int. Ed. Engl.*, 1959, **71**, 176.
2. J. Smidt, W. Hafner, R. Jira, R. Sieber, J. Sedlmeier and A. Sabel, *Angew. Chem., Int. Ed. Engl.*, 1962, **1**, 80.
3. W. H. Clement and C. M. Selwitz, *J. Org. Chem.*, 1964, **29**, 241.
4. W. G. Lloyd and B. J. Luberoff, *J. Org. Chem.*, 1969, **34**, 3949.
5. D. R. Fahey and E. A. Zuech, *J. Org. Chem.*, 1974, **39**, 3276.
6. F. J. McQuillin and D. G. Parker, *J. Chem. Soc., Perkin Trans., 1*, 1974, 809.
7. J. Tsuji, I. Shimizu and K. Yamamoto, *Tetrahedron Lett.*, 1976, 2975.
8. R. Jira and W. Freiesleben, *Organomet. React.*, 1972, **3**, 1.
9. P. M. Maitlis, 'The Organic Chemistry of Palladium', Academic Press, New York, 1971, vol. 2.
10. P. M. Henry, 'Palladium Catalyzed Oxidation of Hydrocarbons', Reidel, Dordrecht, 1980.
11. H. Mimoun, *J. Mol. Catal.*, 1980, **7**, 1.
12. R. F. Heck, in 'Palladium Reagents in Organic Syntheses', Academic Press, New York, 1985, p. 59.
13. J. Tsuji, in 'Organic Syntheses with Palladium Compounds', Springer, Berlin, 1980.
14. J. Tsuji, *Synthesis*, 1984, 369.
15. M. Kolb, E. Bratz and K. Dialer, *J. Mol. Catal.*, 1977, **2**, 399.
16. C. Lapinte and H. Rivière, *Tetrahedron Lett.*, 1977, 3817.
17. H. Alper, K. Januszkiewicz and D. J. H. Smith, *Tetrahedron Lett.*, 1985, **26**, 2263.
18. T. Hosokawa, T. Ohta, S. Kanayama and S. Murahashi, *J. Org. Chem.*, 1987, **52**, 1758.
19. K. Matsui, S. Uchiumi, A. Iwayama and T. Umezu (Ube Industries Ltd.) *Jpn. Pat.* 106 634, 106 635 (1982) (*Chem. Abstr.* 1982, **97**, 162 364).
20. M. Nakai and T. Enomiya (Ube Industries Ltd.) *Jpn Pat.* 128 343 (1984) (*Chem. Abstr.*, 1984, **101**, 54 715).
21. N. T. Byrom, R. Grigg and B. Kongkathip, *J. Chem. Soc., Chem. Commun*, 1976, 216.
22. K. Mori and Y. B. Seu, *Tetrahedron*, 1985, **41**, 3429.
23. B. Kongkathip, R. Sookkho and N. Kongkathip, *Chem. Lett.*, 1985, 1849.
24. (a) P. C. B. Page, C. M. Bayner and I. O. Sutherland, *Tetrahedron Lett.*, 1986, **27**, 3535; (b) Y. Mori, M. Kuhara, A. Takeuchi and M. Suzuki, *Tetrahedron Lett.*, 1988, **29**, 5419.
25. J. Urbanec, M. Hrusovsky and M. Strasak, *Petrochemia*, 1976, **16**, 132 (*Chem. Abstr.*, 1977, **87**, 5123).
26. Asahi Chem. Ind., *Jpn. Pat.* 82 07 439 (1982) (*Chem. Abstr.*, 1982, **96**, 162 158).
27. A. Kaszonyi, J. Vojtko and M. Hrusovsky, *Collect. Czech. Chem. Commun.*, 1982, **47**, 2128.
28. G. T. Rodeheaver and D. F. Hunt, *J. Chem. Soc., Chem. Commun.*, 1971, 818.
29. S. Uemura, K. Zushi, M. Okano and K. Ichikawa, *J. Chem. Soc., Chem. Commun.*, 1972, 234.
30. K. Januszkiewicz and H. Alper, *Tetrahedron Lett.*, 1983, **24**, 5159.
31. H. A. Zahalka, K. Januszkiewicz and H. Alper, *J. Mol. Catal.*, 1986, **35**, 249.
32. A. Harada, Y. Hu and S. Takahashi, *Chem. Lett.*, 1986, 2083.
33. P. M. Henry, *J. Org. Chem.*, 1967, **32**, 2575.
34. H. Stangl and R. Jira, *Tetrahedron Lett.*, 1970, 3589.
35. M. Gouedard, F. Gaudemer and A. Gaudemer, *Bull. Soc. Chim. Fr.*, 1973, 577.
36. M. Hrusovsky, J. Vojtko and M. Cihova, *Hung. J. Ind. Chem.*, 1974, **2**, 137 (*Chem. Abstr.* 1975, **82**, 139 205).
37. H. H. Horowitz, *J. Appl. Electrochem.*, 1984, **14**, 779.
38. I. I. Moiseev, M. N. Vargaftik and Y. K. Syrkin, *Dokl. Akad. Nauk SSSR*, 1960, **133**, 377 (*Chem. Abstr.*, 1960, **54**, 24 350).
39. J. Tsuji and M. Minato, *Tetrahedron Lett.*, 1987, **28**, 3683.
40. J.-E. Bäckvall and R. B. Hopkins, *Tetrahedron Lett.*, 1988, **29**, 2885.
41. H. Arai and M. Yashiro, *J. Mol. Catal.*, 1977/78, **3**, 427.
42. M. Roussel and H. Mimoun, *J. Org. Chem.*, 1980, **45**, 5387.
43. J. Tsuji, H. Nagashima and K. Hori, *Chem. Lett.*, 1980, 257.
44. A. Matsuura, Y. Ito and T. Matsuura, *J. Org. Chem.*, 1985, **50**, 5002.
45. H. Mimoun, R. Charpentier, A. Mitschler, J. Fischer and R. Weiss, *J. Am. Chem. Soc.*, 1980, **102**, 1047.
46. B. S. Tovrog, F. Mares and S. E. Diamond, *J. Am. Chem. Soc.*, 1980, **102**, 6616.
47. A. Heumann, M. Reglier and B. Waegell, *Angew. Chem., Int. Ed. Engl.*, 1979, **18**, 866, 867.
48. A. Heumann, F. Chauvet and B. Waegell, *Tetrahedron Lett.*, 1982, **23**, 2767.
49. A. Andrews and K. P. Kelly, *J. Am. Chem. Soc.*, 1981, **103**, 2894.
50. T. Hosokawa, T. Takahashi, T. Ohta and S. Murahashi, *J. Organomet. Chem.*, 1987, **334**, C5.
51. F. Derdar, J. Martin, C. Martin, J. M. Bregeault and J. Mercier, *J. Organomet. Chem.*, 1988, **338**, C21.
52. Phillips Petrol. Co., *US Pat.* 4 434 082 (1982) (*Chem. Abstr.*, 1984, **100**, 174 277).
53. B. El Ali, J. M. Bregeault and J. Martin, *J. Organomet. Chem.*, 1987, **327**, C9.
54. H. Ogawa, H. Fujinami and K. Taya, *J. Chem. Soc., Chem. Commun.*, 1981, 1274.
55. H. Ogawa, H. Fujinami, K. Taya and S. Teratani, *Bull. Chem. Soc. Jpn.*, 1984, **57**, 1908.
56. S. F. Davison, B. E. Mann and P. M. Maitlis, *J. Chem. Soc., Dalton Trans.*, 1984, 1223.
57. R. B. Gammill and S. A. Nash, *Tetrahedron Lett.*, 1984, **25**, 2953.
58. R. Hüttel and H. Christ, *Chem. Ber.*, 1963, **96**, 3101; 1964, **97**, 1439.
59. A. D. Ketley and J. Braatz, *Chem. Commun.*, 1968, 169.
60. H. C. Volger, *Recl. Trav. Chim. Pays-Bas*, 1969, **88**, 225.
61. F. Conti, M. Donati, G. F. Pregaglia and R. Ugo, *J. Organomet. Chem.*, 1971, **30**, 421.
62. J. Tsuji, H. Nagashima and H. Nemoto, *Org. Synth.*, 1984, **62**, 9.
63. G. Balme, *Tetrahedron Lett.*, 1985, **26**, 2309.
64. C. S. Subramaniam, P. J. Thomas, V. R. Mamdapur and M. S. Chadha, *Synthesis*, 1978, 468.
65. T. Hirao, K. Hayashi, Y. Fujihara, Y. Ohshiro and T. Agawa, *J. Org. Chem.*, 1985, **50**, 279.
66. C. C. Lenznoff, C. R. McArthur and M. Whittaker, *Synth. Commun.*, 1986, **16**, 225.
67. P. Knochel and J. F. Normant, *Tetrahedron Lett.*, 1986, **27**, 4433.

68. R. J. Capon and J. K. MacLeod, *Tetrahedron*, 1988, **44**, 1637.
69. J. Tsuji and coworkers, unpublished results.
70. H. Yatagai, Y. Yamamoto and K. Maruyama, *J. Am. Chem. Soc.*, 1980, **102**, 4548.
71. A. J. Poss and R. K. Belter, *Synth. Commun.*, 1988, **18**, 417.
72. A. J. Pearson and T. Ray, *Tetrahedron Lett.*, 1985, **26**, 2981.
73. T. Takahashi, K. Kasuga, M. Takahashi and J. Tsuji, *J. Am. Chem. Soc.*, 1979, **101**, 5072.
74. G. Mehta and K. S. Rao, *J. Org. Chem.*, 1988, **53**, 425.
75. T. A. Hase and K. McCoy, *Synth. Commun.*, 1979, **9**, 63.
76. T. Kametani, K. Kigasawa, M. Hiiragi, N. Wagatsuma, T. Uryu and H. Inoue, *Yakugaku Zasshi*, 1980, **100**, 192 (*Chem. Abstr.*, 1980, **93**, 150 202).
77. T. Antonsson, S. Hansson and C. Moberg, *Acta Chem. Scand., Ser. B*, 1985, **39**, 593.
78. W. G. Lloyd, *J. Org. Chem.*, 1967, **32**, 2816.
79. T. E. Nalesnik and N. L. Holy, *J. Org. Chem.*, 1977, **42**, 372.
80. T. F. Blackburn and J. Schwartz, *J. Chem. Soc., Chem. Commun.*, 1977, 157.
81. R. Rossi and A. Carpita, *Tetrahedron Lett.*, 1986, **27**, 2529.
82. C. S. Subramaniam, P. J. Thomas, V. R. Mamdapur and M. S. Chadha, *J. Chem. Soc., Perkin Trans. 1*, 1970, 2346.
83. J. Tsuji and T. Mandai, *Tetrahedron Lett.*, 1978, 1817.
84. T. Takahashi, K. Kasuga and J. Tsuji, *Tetrahedron Lett.*, 1978, 4917.
85. J. Tsuji, M. Kaito and T. Takahashi, *Bull. Chem. Soc. Jpn.*, 1978, **51**, 547.
86. T. Takahashi, H. Ikeda and J. Tsuji, *Tetrahedron Lett.*, 1980, **21**, 3885.
87. T. Takahashi, I. Minami and J. Tsuji, *Tetrahedron Lett.*, 1981, **22**, 2651.
88. J. Tsuji, T. Yamakawa and T. Mandai, *Tetrahedron Lett.*, 1979, 3741.
89. J. Tsuji, M. Kaito, T. Yamada and T. Mandai, *Bull. Chem. Soc. Jpn.*, 1978, **51**, 1915.
90. J. Tsuji, T. Yamada, M. Kaito and T. Mandai, *Tetrahedron Lett.*, 1979, 2257.
91. J. Tsuji, M. Masaoka and T. Takahashi, *Tetrahedron Lett.*, 1977, 2267.
92. J. Tsuji and H. Yasuda, *Synth. Commun.*, 1978, **8**, 103.
93. J. Yoshida, K. Tamao, M. Takahashi and M. Kumada, *Tetrahedron Lett.*, 1978, 2161.
94. G. R. Clark and S. Thiensathit, *Tetrahedron Lett.*, 1985, **26**, 2503.
95. P. Boontanonda and R. Grigg, *J. Chem. Soc., Chem. Commun.*, 1977, 583.
96. J. Nokami, H. Ogawa, S. Miyamoto, T. Mandai, S. Wakabayashi and J. Tsuji, *Tetrahedron Lett.*, 1988, **29**, 5181.
97. U. Sunay and B. Fraser-Reid, *Tetrahedron Lett.*, 1986, **27**, 5335.
98. B. L. Feringa, *J. Chem. Soc., Chem. Commun.*, 1986, 909.
99. A. K. Bose, L. Krishnan, D. R. Wagle and M. S. Manhas, *Tetrahedron Lett.*, 1986, **27**, 5955.
100. J. Tsuji, *Ann. N. Y. Acad. Sci.*, 1980, **333**, 250.
101. J. Tsuji, *Top. Curr. Chem.*, 1980, **91**, 30.
102. J. Tsuji, *Pure Appl. Chem.*, 1981, **53**, 2371.
103. G. Mehta and K. S. Rao, *J. Am. Chem. Soc.*, 1986, **108**, 8015.
104. T. Imanishi, M. Matsui, M. Yamashita and C. Iwata, *Tetrahedron Lett.*, 1986, **27**, 3163.
105. K. Iseki, M. Yamazaki, M. Shibasaki and S. Ikegami, *Tetrahedron*, 1981, **37**, 4411.
106. J. Tsuji, T. Yamada and I. Shimizu, *J. Org. Chem.*, 1980, **45**, 5209.
107. P. D. Magnus and M. S. Nobbs, *Synth. Commun.*, 1980, **10**, 273.
108. J. Tsuji, I. Shimizu and Y. Kobayashi, *Isr. J. Chem.*, 1984, **24**, 153.
109. J. Tsuji, Y. Kobayashi and I. Shimizu, *Tetrahedron Lett.*, 1979, 39.
110. M. Ochiai, M. Arimoto and E. Fujita, *Tetrahedron Lett.*, 1981, **22**, 1115.
111. F. Sato, H. Watanabe, Y. Tanaka, T. Yamaji and M. Sato, *Tetrahedron Lett.*, 1983, **24**, 1041.
112. T. Takahashi, M. Miyazawa, H. Ueno and J. Tsuji, *Tetrahedron Lett.*, 1986, **27**, 3881.
113. A. Hosomi, H. Kobayashi and H. Sakurai, *Tetrahedron Lett.*, 1980, **21**, 955.
114. T. Yanami, M. Miyashita and A. Yoshikoshi, *J. Org. Chem.*, 1980, **45**, 607.
115. M. A. Tius, A. Thurkauf and J. W. Truesdell, *Tetrahedron Lett.*, 1982, **23**, 2819.
116. J. Tsuji, I. Shimizu, H. Suzuki and Y. Naito, *J. Am. Chem. Soc.*, 1979, **101**, 5070.
117. I. Shimizu, Y. Naito and J. Tsuji, *Tetrahedron Lett.*, 1980, **21**, 487.
118. J. Tsuji, Y. Kobayashi and T. Takahashi, *Tetrahedron Lett.*, 1980, **21**, 483.
119. K. Takehira, I. H. Ho, V. C. Martinez, R. S. Chavira, T. Hayakawa, H. Orita, M. Shimadzu and T. Ishikawa, *J. Mol. Catal.*, 1987, **42**, 237.
120. K. Takehira, H. Orita, I. H. Ho, C. R. Leobardo, G. C. Martinez, M. Shimadzu, T. Hayakawa and T. Ishikawa, *J. Mol. Catal.*, 1987, **42**, 247.
121. K. Takehira, T. Hayakawa and H. Orita, *Chem. Lett.*, 1985, 1835.
122. J. E. Byrd, L. Cassar, P. E. Eaton and J. Halpern, *J. Chem. Soc. (D)*, 1971, 40.
123. R. B. Gammill and S. A. Nash, *Tetrahedron Lett.*, 1984, **25**, 2953; *J. Org. Chem.*, 1986, **51**, 3116.
124. T. Sasaki, K. Kanematsu and A. Kondo, *J. Chem. Soc., Perkin Trans. 1*, 1976, 2516.
125. J. Tsuji and S. Imamura, *Bull. Chem. Soc. Jpn.*, 1967, **40**, 197.
126. H. Nagashima, K. Sakai and J. Tsuji, *Chem. Lett.*, 1982, 859.
127. T. J. Grattan and J. S. Whitehurst, *J. Chem. Soc., Chem. Commun.*, 1988, 43.
128. A. J. Poss and M. S. Smyth, *Synth. Commun.*, 1987, **17**, 1735.
129. S. Takano, Y. Imamura and K. Ogasawara, *Tetrahedron Lett.*, 1981, **22**, 4479.
130. J. Tsuji, H. Nagashima and K. Hori, *Tetrahedron Lett.*, 1982, **23**, 2679.
131. E. Keinan, K. K. Seth and R. Lamed, *J. Am. Chem. Soc.*, 1986, **108**, 3474.

3.5
Addition Reactions with Formation of Carbon–Nitrogen Bonds

JOHN E. G. KEMP
Pfizer Central Research, Sandwich, UK

3.5.1 INTRODUCTION

This review is written to cover the needs of synthetic chemists with interests in oxidizing alkenes by addition of nitrogenous substituents. Whilst some aspects have been covered in previous reviews (noted in the text), most notably in the Tetrahedron Report No. 144, 'Amination of Alkenes'[1] and prior reviews on aziridines and nitrenes, the present review is the first compilation of references to the whole range of these particular bond-forming processes. A review by Whitham provides a useful general introduction to reaction mechanisms of additions to alkenes in greater detail than can be covered here.[2] The 'oxidation' requirement excludes from the scope the additions of N + H and most additions of N + Metal or N + C. Hence, unmodified Michael and Ritter reactions are excluded. These topics are mostly covered in Volume 4 of the present series.

Half this review is concerned with aziridination, but other cycloadditions are generally excluded, being reviewed elsewhere in this series (Volume 4, Part 4 and Volume 5, Part 4). The second half of this review covers additions of nitrogen and a chalcogen or halogen, or a second nitrogen. No examples of addition of N + P were found: this transformation can be achieved indirectly by opening aziridines with phosphorus nucleophiles.[3]

Surprisingly few of the reactions herein have been used widely, yet some have much to offer for synthesis. Many aziridinations, for example, occur under extremely mild conditions. The potential utility of aziridines as synthetic intermediates is enormous; for example, to make regiospecifically substituted diamines, but published examples to date are few. There is also much scope for wider usage of iodolactamizations and related cyclizations.

3.5.2 AZIRIDINE FORMATION FROM ALKENES

Aziridines are readily accessible, yet under used as synthetic intermediates.[4–7] There is an even greater variety of methods of preparation than for epoxides. Nitrene additions to congested alkenes (Section 3.5.2.3) often show stereoselectivity, and indirect syntheses, *e.g.* from amino alcohols, are stereospecific. Some success has already been achieved in chiral aziridine preparation (Section 3.5.2.5), though the generality, and enantiomeric excesses (except in a few specific cases) do not yet match those of, for example, the Sharpless epoxidation. The main utility of aziridines in synthesis lies in their ring-opening reactions analogous to those of epoxides, but more versatile owing to the varied exocyclic *N*-substituent which modulates the properties and reactivity of the three-membered ring.[4–9] It should be noted, however, that aziridines are toxic, and that the lower molecular weight compounds, particularly, can sometimes polymerize with a very vigorous exotherm, generally initiated by acid catalysis.[7]

3.5.2.1 Aziridines with Nitrogen Unsubstituted

Several reagents will effect the aziridination of alkenes directly, for example, electrophilic alkenes react with diphenylsulfilimine (**1**) in good yield, with the elimination of diphenyl sulfide (Scheme 1).[10–12] A similar reaction using the chiral sulfilimine (*R*)-(+)-(**3**) gives product (**2**; 96%) as the (2*R*,3*S*) form shown (64%), accompanied by the enantiomer (32%).[13] Cephalosporin substrates (**4**) react with achiral (**1**) to give products (**5**), as single diastereoisomers in 52–63% yields.[14]

Pentamethyleneoxaziridine (**6**) also acts as an aminating agent, giving quite good yield of aziridines (Scheme 2).[15,16] Remarkably, none of the corresponding epoxide was reported in this reaction. Cyclohexene yields instead the hydrazino alcohol (**8**), probably *via* the *N*-aminoaziridine (**7**). In view of the recent use of chiral oxaziridines for chiral oxidations, it will be interesting to see whether chiral aziridination is possible when chiral *N*-unsubstituted oxaziridines become available.[17]

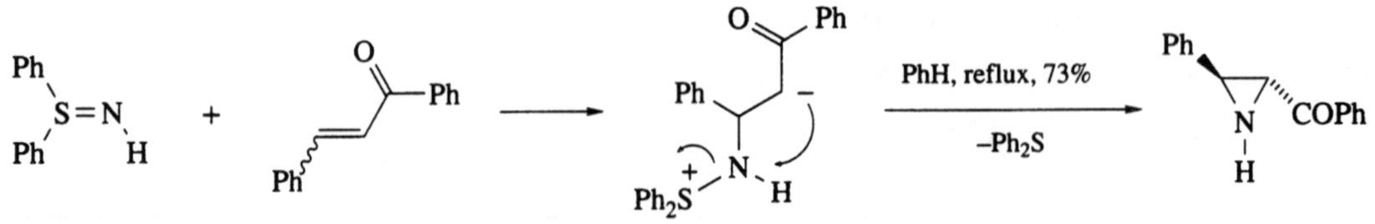

Ph Ph S=N H (**1**) + Ph–CH=CH–COPh → [Ph₂S⁺–NH⁻ adduct] —PhH, reflux, 73%; –Ph₂S→ (**2**)

Scheme 1

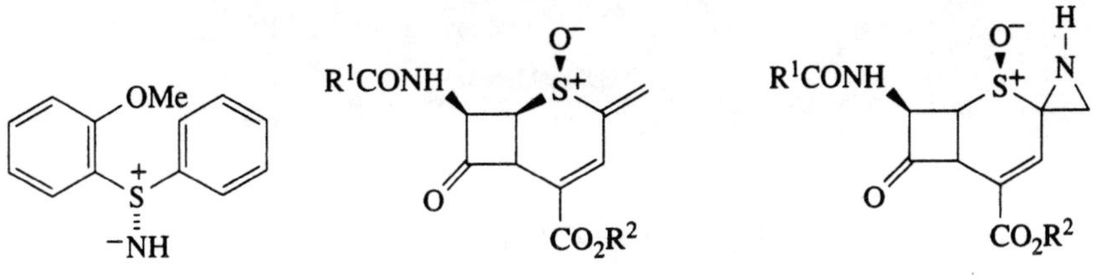

OMe
$\overset{+}{S}$
–NH
(3)

R^1CONH
O^-
S+
O
CO_2R^2
(4)

R^1CONH
O^-
S+
H
N
O
CO_2R^2
(5)

100 mL
i
(6) 74%
reflux 3 h
46%

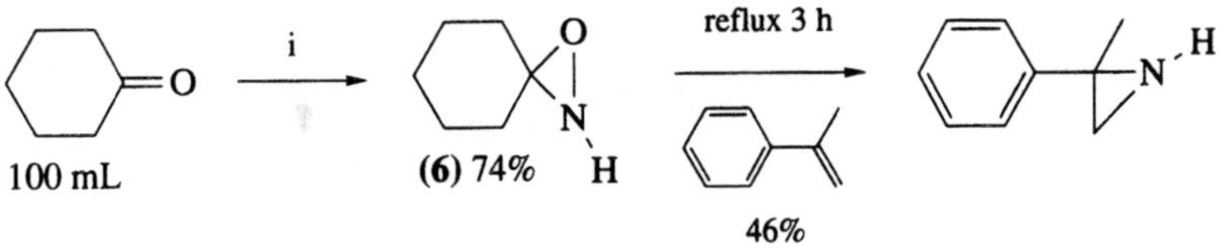

i, NaOCl (200 mL 0.98 M), NH_4OH (100 mL 2 M), PhMe (500 mL) ice/water (400 mL), shake 30 s, separate, wash H_2O, dry ($MgSO_4$)

Scheme 2

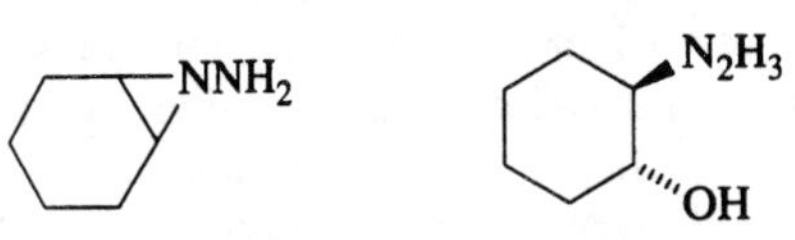

NNH_2
(7)

N_2H_3
OH
(8)

Electron deficient alkenes, *e.g.* NCCH=C(CO_2Me)$_2$, can be aziridinated with *O*-(arylsulfonyl)hydroxylamines. The reaction is believed to involve a Michael addition followed by cyclization with expulsion of a sulfonate anion (yields 30–90%).[18] Less electrophilic alkenes react in lower yield but with a high stereospecificity: a high degree of concertedness would appear to be present in this case (Scheme 3).[19] Related syntheses, where the leaving group is halide, alkoxide or trimethylamine, are discussed in Deyrup's[5] and Červinka's[20] reviews; some examples are given in Schemes 4[21] and 5;[22] reactions of haloamines, particularly with unactivated alkenes, are generally radical processes (Scheme 6).[23] An elegant one-pot aziridination of chalcones has been devised (Scheme 7).[24]

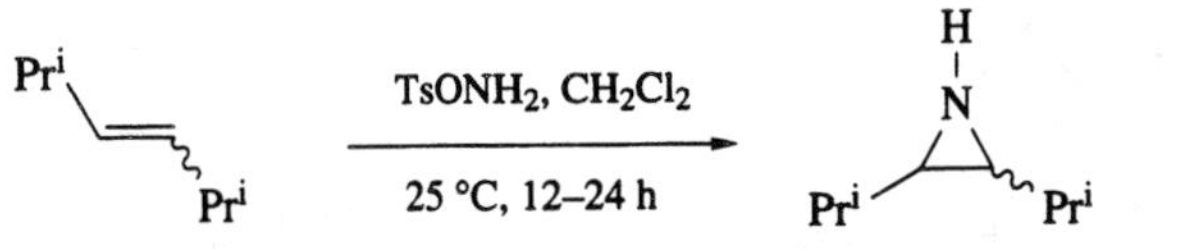

Pr^i Pr^i
$TsONH_2$, CH_2Cl_2
25 °C, 12–24 h
(*cis* or *trans*)
(*cis* or *trans* respectively)

Scheme 3

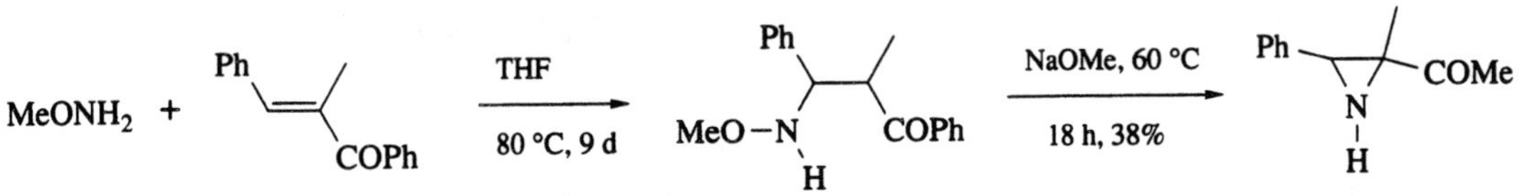

$MeONH_2$ + Ph COPh
THF
80 °C, 9 d
MeO–N(H) Ph COPh
NaOMe, 60 °C
18 h, 38%
Ph COMe N H

Scheme 4

O Me N Ph
i, Me_2NNH_2, 5 h, reflux
ii, MeI
78%
$Me_3\overset{+}{N}$–NH ... N(Me)Ph
NaOMe
reflux, 16 h
35%
CONMePh

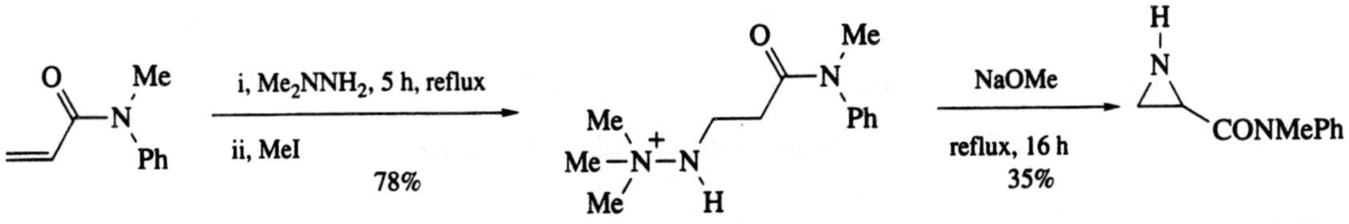

Scheme 5

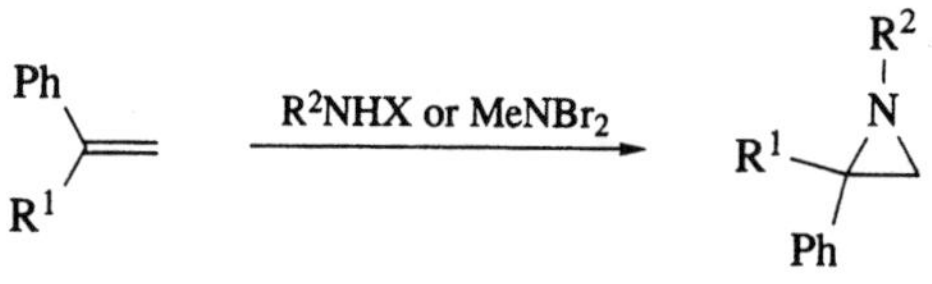

R^2NHX or $MeNBr_2$

$R^1 = H, Me; R^2 = H$, alkyl; X = Cl, Br

Scheme 6

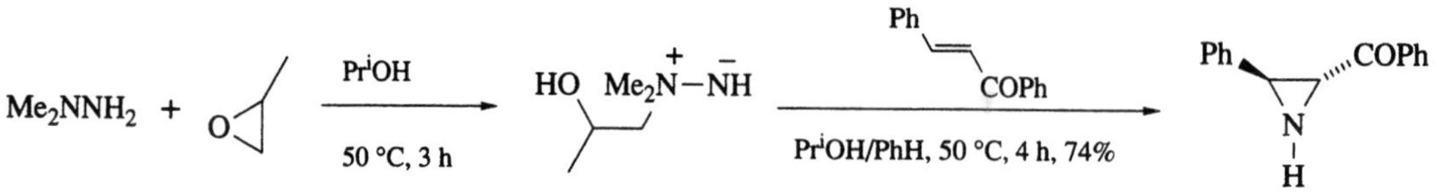

Me_2NNH_2 + (methyloxirane) $\xrightarrow[50\ °C,\ 3\ h]{Pr^iOH}$ HO $Me_2\overset{+}{N}-\overset{-}{N}H$ $\xrightarrow[Pr^iOH/PhH,\ 50\ °C,\ 4\ h,\ 74\%]{PhCH=CHCOPh}$ (Ph, COPh, N–H aziridine)

Scheme 7

Of the numerous other aziridine syntheses, there are several multistep procedures from alkenes. Though not strictly within the scope of this review, the practising chemist will wish to consider their merits alongside the direct syntheses, and the main possibilities are summarized in Scheme 8. There are several good recent reviews,[4,5] and two older compilations remain very useful.[6,7] Syntheses of those intermediates of Scheme 8 accessible from alkenes are described in later sections of the present review, and syntheses of epoxides (Volume 6, Chapter 1.1 and Volume 7, Chapters 3.1 and 3.2) and triazolines (Volume 5, Chapter 3.1) are described elsewhere in 'Comprehensive Organic Synthesis'. It is important to note that by careful choice of route one can either commence with alkene (**14**) and retain the *cis/trans* stereochemistry in the resulting aziridine (**16**), or start with alkene (**13**) and change the *cis/trans* relationships of the substituents.

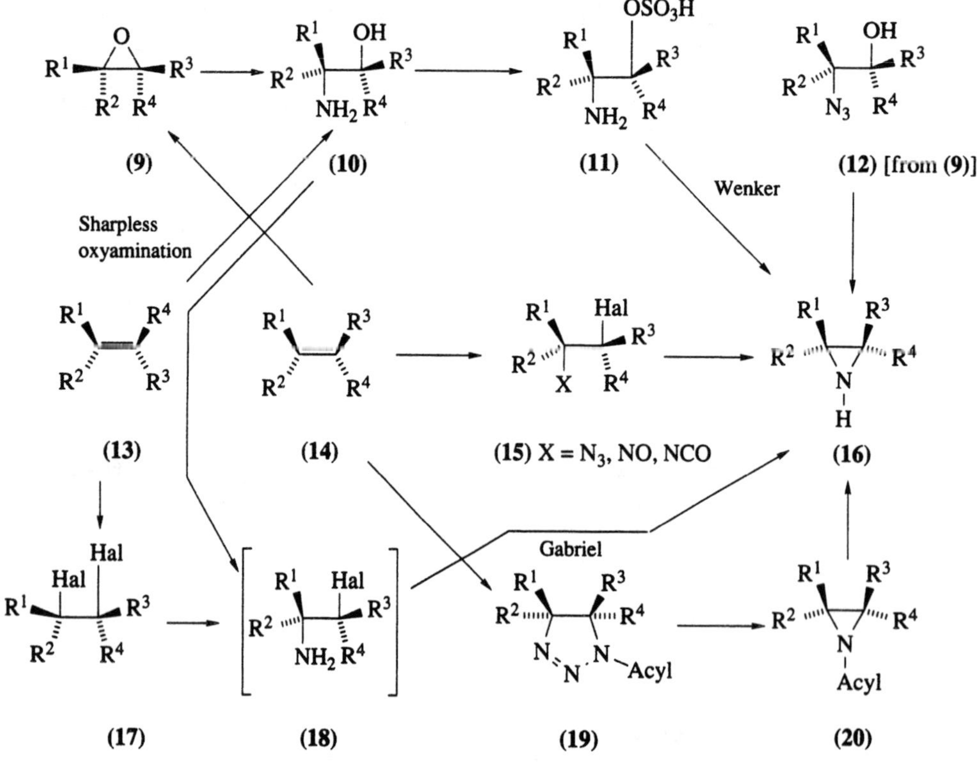

Scheme 8 The major indirect aziridination routes

The Wenker and Gabriel syntheses have great scope and usually give good yields, however the routes *via* triazolines (*e.g.* **19**) are not good for *N*-unsubstituted aziridines or for *N*-alkylaziridines. The Gabriel synthesis is stereospecific, and by choice of appropriate precursor can yield stereospecific syntheses of the desired aziridines. The Wenker synthesis is also *generally* similarly stereospecific: a rare exception being when the sulfate leaving group is attached to a benzylic center.[6] In such systems the Gabriel synthesis remains stereospecific. Both syntheses require that the amine and leaving groups in (**11**) or (**18**) can adopt a *trans* coplanar, or nearly so, orientation. Several resolved amino alcohols are readily available by reduction of amino acids and can be utilized to make chiral aziridines, the Wenker synthesis of (*S*)-(–)-2-methylaziridine from L-alaninol being typical.[25] In certain cases, a resolved epoxide precursor may be available by the Sharpless procedure (Volume 7, Chapter 3.2) as in the PS5 synthesis (Scheme 9). This synthesis illustrates two further features of this utilization of epoxides: clean regiochemistry of the initial ring opening is not required to yield pure, resolved aziridines and, though *cis/trans* relationships are retained, this is by virtue of an inversion at each ring carbon atom.[26] This allows, in principle, the preparation of an *endo* aziridine *via* an *exo*-epoxide of, for example, a norbornene type of alkene, whereas direct aziridination, *e.g.* by nitrene addition to the double bond, would give the *exo*-aziridine.

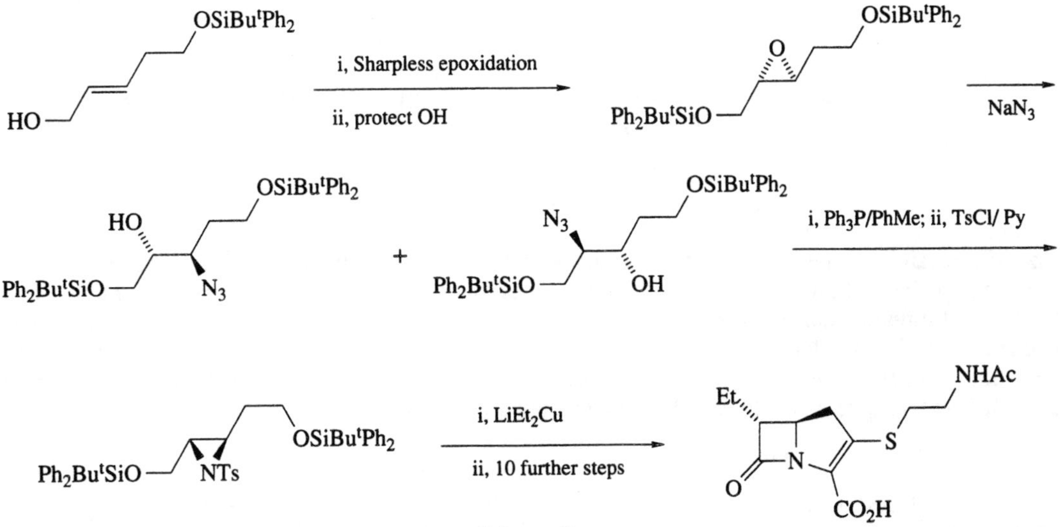

Scheme 9

The alkene pseudohalogen adducts (**15**) of Scheme 8 are also useful intermediates for aziridine synthesis. These adducts are discussed later in Sections 3.5.6.2–4. The iodine azide[27–29] and bromine azide[30] adducts may be reduced to aziridines with many reagents; recent references report use of lithium aluminum hydride[28,29] and dimethylamineborane.[27] The iodine isocyanate aziridination continues to prove useful, as in Scheme 10.[31–33] Since the recent reviews,[4,5] the mechanism of the triphenylphosphine-based cyclization of azido alcohols has appeared (Scheme 11); there are clear steric consequences.[34] Alkenes can be chlorinated in acetonitrile to give intermediates which can be worked up to yield aziridines (Scheme 12).[35]

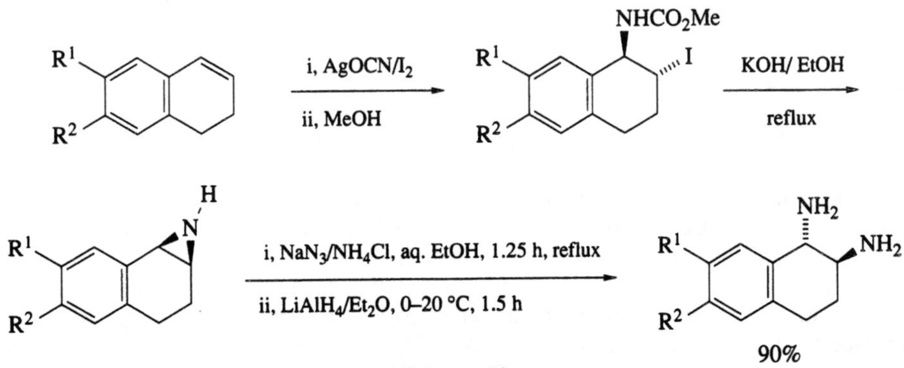

Scheme 10

The regio- and stereo-chemical consequences of preparing an aziridine *via* IN_3, or INCO, or the above chlorination procedure are, in so far as they have been investigated, analogous to those of proceeding *via* an epoxidation, for the simple reason that all these last three reaction types appear to proceed *via* cyclic halonium ions, isosteric with epoxides.

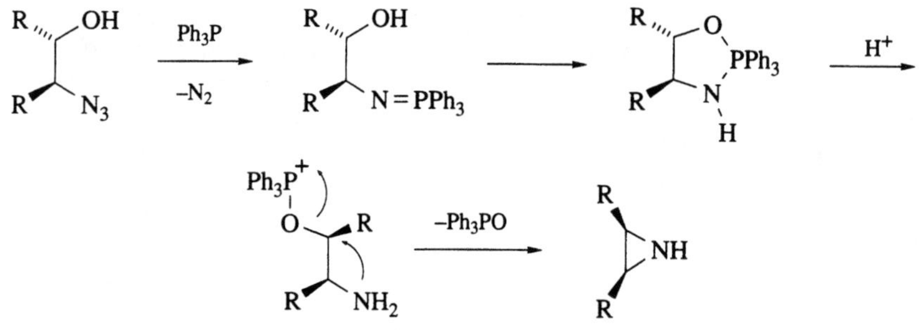

Scheme 11

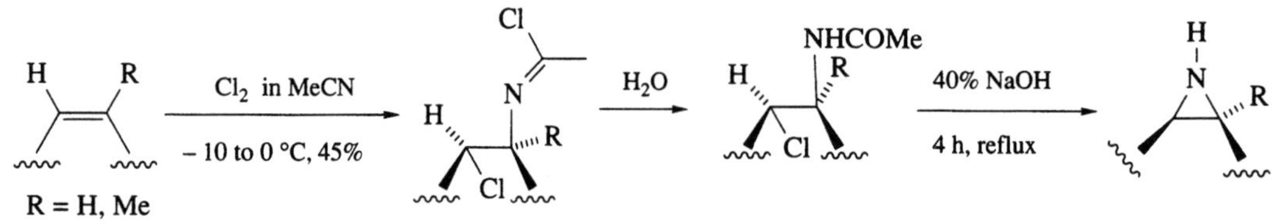

Scheme 12

Aziridines can be made in a sequence starting with the addition of NOCl to alkenes, Scheme 13. The sequence is probably limited to tetrasubstituted alkenes, owing to instability of the nitroso intermediates from less-substituted alkenes (Section 3.5.6.2).[36–38] Since the reagent reacts as NO^+Cl^- and delivers the nitrogen, rather than the leaving group, to the exposed face of an alkene, one would expect an appropriate norbornene derivative to yield an *exo*-aziridine in contrast to the *endo* products obtained *via* epoxides, although this appears not to have been investigated yet.

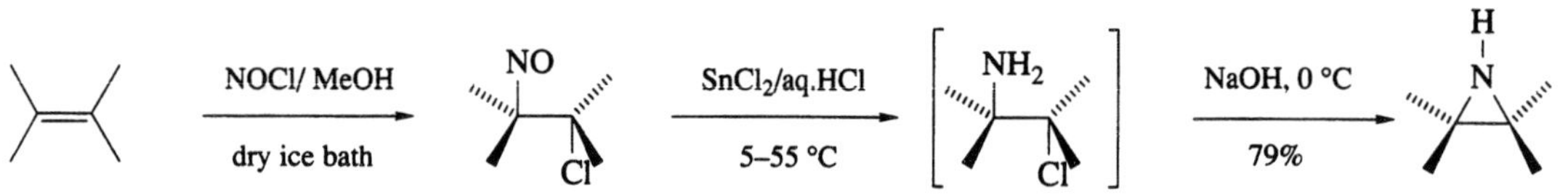

Scheme 13

N-Unsubstituted aziridines may be prepared by removal of *N*-substituents. Surprisingly vigorous reagents or conditions may be tolerated in these reactions: *N*-carbethoxy is cleavable with KOH as in Scheme 10, *N*-cyano[39] or acyl[40] by lithium aluminohydride, this last normally being considered an aldehyde synthesis.[40] An *N*-sulfenyl group (of under-used potential, Section 3.5.2.6) may be removed with sodium borohydride.[41] 1,2,3-Triphenylaziridine is quantitatively ozonolyzed to 2,3-diphenylaziridine.[42]

3.5.2.2 *N*-Alkyl- and *N*-Alkenyl-aziridines

There are no one-step syntheses of *N*-alkylaziridines direct from alkenes of proven generality, the most thoroughly investigated being the reactions of *N*-halo- and *N,N*-dihalo-alkylamines with alkenes referred to earlier (Scheme 6).[23] They are probably radical reactions and are not stereospecific; the *N*-haloamine reagents are difficult to prepare cleanly, and are rather dangerous. Representative examples (6 of 17) prepared from halogenated alkylamines are listed in Scheme 14.

A one-pot conversion of alkenes to *N*-methylaziridines has been achieved by aminopalladation followed by an oxidative work-up with bromine (Scheme 15).[43] The reaction is stereospecific, as shown by deuteration experiments, but has been little exploited.[43,44] 2-Aminopalladium compounds such as (**21**) are extremely versatile; depending on starting compounds and work-up reagents, they can be made to yield a wide variety of products (Scheme 16).[45]

Gabriel and Wenker type syntheses are thoroughly viable for *N*-alkylaziridines,[4–7] and there are examples of aziridination of alkenes, *via vic*-dihalides *etc*., where by use of a chiral amine a kinetic induction of chirality in the aziridine nucleus has been achieved.[46,47] 2-Iodoalkyl azides react with alkyl and aryl dichloroboranes, forming aziridines (Scheme 17).[48] Sequences resembling the Wittig[49] and Wittig–Hor-

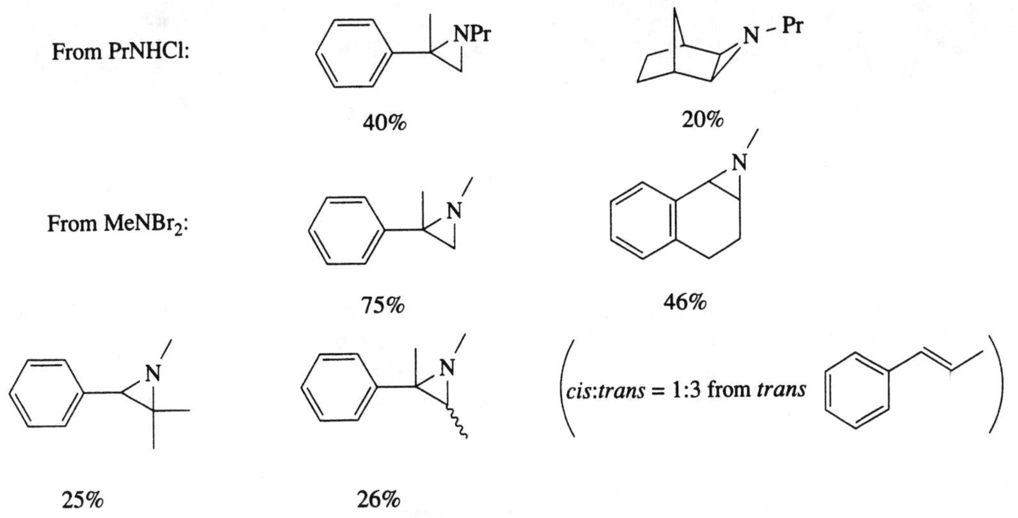

From PrNHCl: NPr 40% N - Pr 20%

From $MeNBr_2$: N 75% N 46%

N 25% N 26% (*cis*:*trans* = 1:3 from *trans*)

Conditions: PhH, dioxane, PhMe, CCl_4 *etc.*; 20–110 °C; 10 min–3 d

Scheme 14 *N*-Alkylaziridines from *N*-haloamines and alkenes

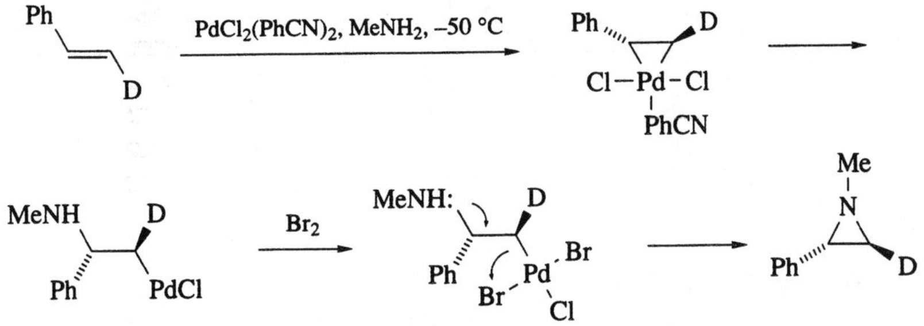

Ph D $PdCl_2(PhCN)_2$, $MeNH_2$, –50 °C Ph D Cl–Pd–Cl PhCN

MeNH D Ph PdCl Br_2 MeNH: D Ph Br Pd Br Cl Me N Ph D

Scheme 15

$-Pd^{II}-$ R_2NH $R_2N-Pd^{II}-$ **(21)** CO, [O], MeOH R_2N CO_2Me

[O] (one R = H) R_2NH, [O] AcO^-, [O] H^-

R N R_2N R_2N R_2N AcO R_2N R_2N H

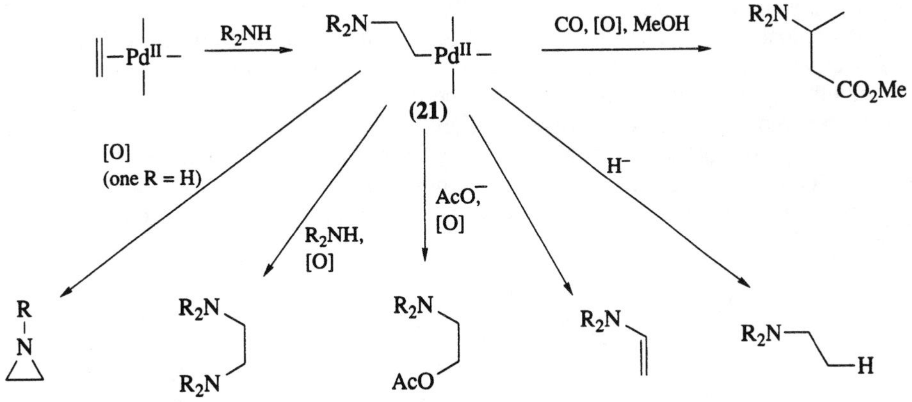

Scheme 16

ner[50] reactions can be carried out on epoxides, resulting in *N*-alkyl- or *N*-aryl-aziridines instead of imines.

Covalent azides add to alkenes in a [2 + 3] cycloaddition, giving triazolines (Volume 4, Chapter 4.10), and these in turn can be thermolyzed or photolyzed with loss of nitrogen to give aziridines. The addition obeys the complementarity principle for cycloadditions, which in this case means that electron-rich azides prefer to react with electron-poor alkenes and *vice versa*.[51] Thus, a few *N*-alkylaziridines have been prepared from electron-poor alkenes *via* triazolines.[52] The thermal stability of triazolines with respect to loss of nitrogen and aziridine formation varies with the electronegativity of the exocyclic *N*-substituent. Thus, *N*-vinyltriazolines decompose more readily than *N*-alkyltriazolines to give *N*-vinylaziridines in good yield, as in Scheme 18.[53–57] Several examples of intramolecular formation of

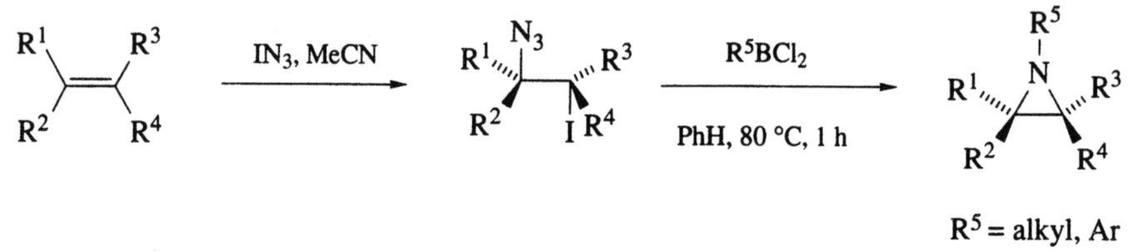

Scheme 17

vinyltriazolines, and subsequent pyrolysis to give fused-ring aziridines are known.[53] *N*-Alkylaziridines are accessible in good yield by intramolecular cyclization of suitable alkenic nitrenes, giving bridged ring systems (analogous intermolecular reactions are not useful).[58–60] This was put to good use in the synthesis of (**22**), a key intermediate in the synthesis of a tetracyclic alkaloid, ibogamine.[61]

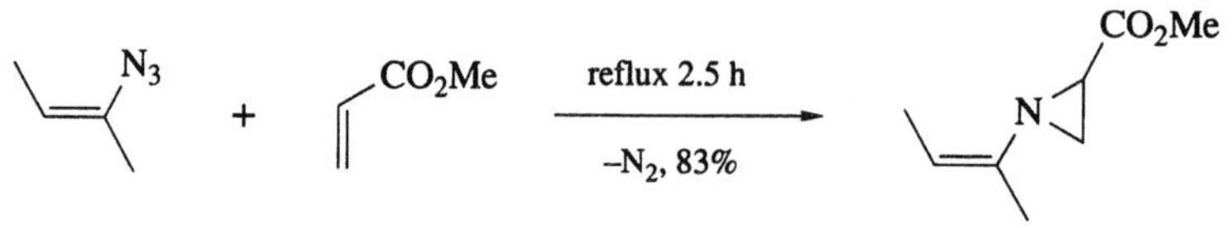

Scheme 18

(**22**)

3.5.2.3 *N*-Aryl- and *N*-Heteroaryl-aziridines

The indirect routes: Gabriel and Wenker syntheses (Scheme 8),[4–7] the borane route (Scheme 17),[48] and Wittig-like syntheses[49,50] are all highly satisfactory for *N*-arylaziridines. It is, however, with the *N*-aryl compounds that the azide/triazoline route begins to be useful, though not, in general, preferable to the other routes: the explosive potential of azides limits their desirability. The preparation, thermolysis and photolysis of triazolines, including many *N*-aryl compounds, has been well reviewed recently.[53] A few key points of selectivity are worth noting. Norbornene-type systems form overwhelmingly the *exo*-triazolines, which on photolysis give *exo*-aziridines. Thermolysis of these systems occurs *via* a ring-opened intermediate and hence is not stereospecific (Scheme 19).[62] Complications are encountered when the thermal sequence is applied to very electron-poor alkenes: the thermolysis is slow, and the resulting aziridine ring opens with C—C cleavage to give an ylide, which then adds to the alkene to give a pyrrolidine (Scheme 20).[53,63]

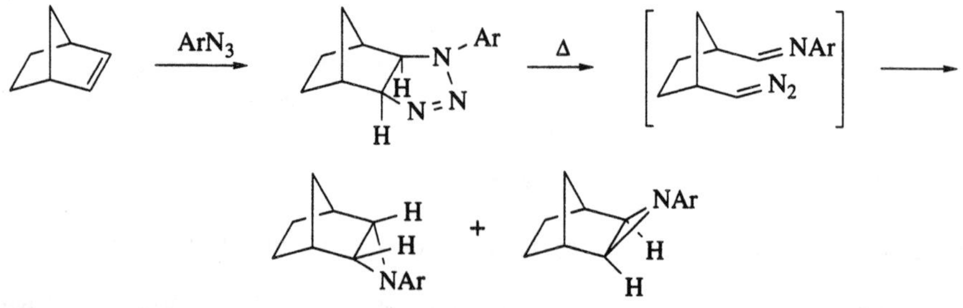

Scheme 19

Aziridine formation from arylnitrenes, rather than *via* triazolines, is known for highly fluorinated arenes.[64,65] Phenyl azide with trifluoroacetic acid generates a nitrenium ion which adds stereospecifically to alkenes to give aziridines. Yields are rather low, partly due to concurrent ring opening of the aziridine by addition of trifluoroacetic acid.[66] Similar reactions can be achieved with Lewis acids such as $AlCl_3$.[67] Enamines with aryl azides can yield either 2-aminoaziridines[68] or amidines.[53]

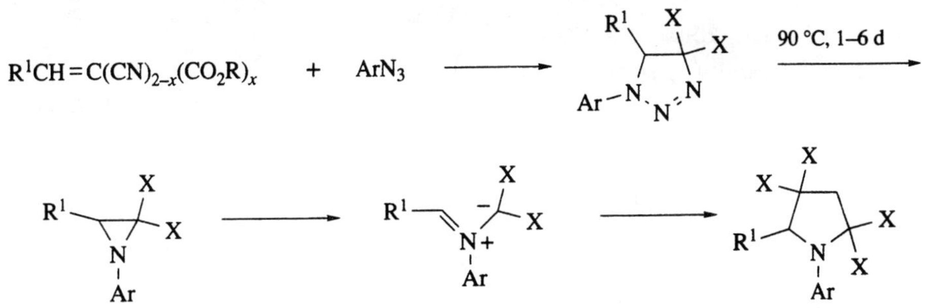

$R^1CH{=}C(CN)_{2-x}(CO_2R)_x$ + ArN_3 → [triazoline] —90 °C, 1–6 d→ [aziridine] → [azomethine ylide] → [pyrrolidine]

R^1 = alkyl, Ar; X = CO_2R, CN

Scheme 20

Heterocyclic azides can give good yields of aziridines (Scheme 21), but the reactions can give other products, depending on the precise combination of starting materials.[69] The reaction shown in Scheme 21 is believed to occur *via* a nitrene, not a triazoline, and is favored by electron-poor alkenes (in this case, electron-rich alkenes give much isomeric thiochroman).

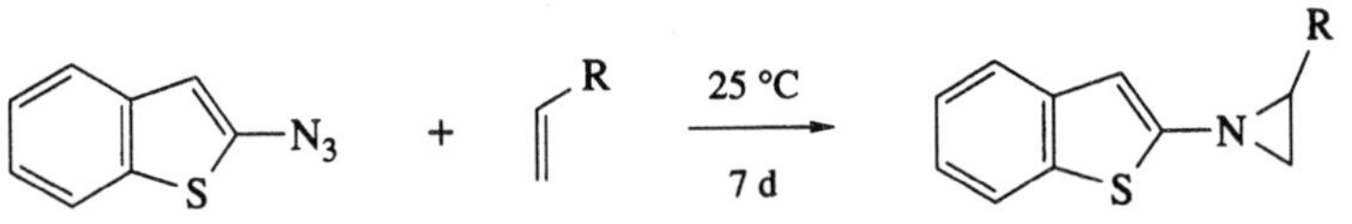

25 °C, 7 d

R = CN, 85%
R = CO_2Me, 90%

Scheme 21

3.5.2.4 *N*-Acyl- and *N*-Cyano-aziridines and Related Compounds

Acylaziridines are not freely available by the Gabriel or Wenker routes, oxazolines being the normal products of such reactions,[70] though *threo*-β-acylamino alcohols (from *trans* internal alkenes) can give aziridines.[71] Acyl azides undergo the Curtius rearrangement to isocyanates,[72] and additions to alkenes to yield triazolines and hence aziridines are rare.[73] Photolysis of acyl azides can yield acylaziridines,[74] though the reaction is only of potential preparative value when a strained ring increases alkene reactivity.[75] Pivaloyl azide, in particular, on photolysis yields a nitrene (50%) which is unusually slow to rearrange to the isocyanate. It adds to alkenes stereospecifically (as singlet) and stereoselectively (as triplet);[76,77] by use of as long a wavelength as possible (300–350 nm), and very pure solvents, especially dichloromethane which stabilizes singlet nitrenes, yields up to 45% are obtainable.

In a reaction modeled on the use of cytochrome *P*-450 to catalyze oxidations with iodosylbenzene, iron or manganese porphyrins have been used to catalyze aziridinations with iodinanes (Scheme 22).[78] In this early report *cis*- or *trans*-stilbene each gave the *trans*-aziridine, but stereoselectivity has since been achieved for the sulfonylaziridines (Section 3.5.2.6).

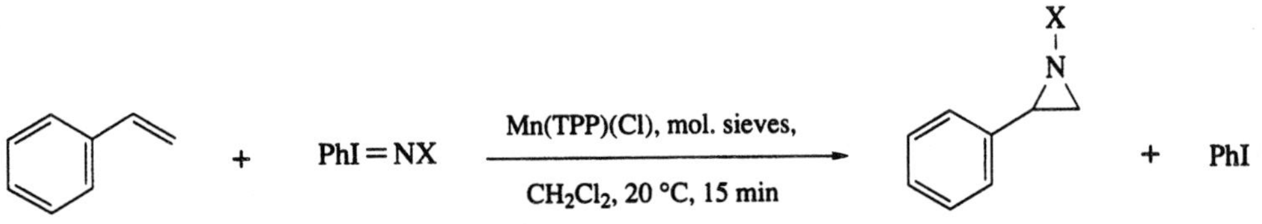

Styrene + PhI=NX —Mn(TPP)(Cl), mol. sieves, CH_2Cl_2, 20 °C, 15 min→ N-X-2-phenylaziridine + PhI

X = COF_3, 50%; $SO_2C_6H_4Me$-*p*, 80%

Scheme 22

Alkoxycarbonyl azides, which do not readily undergo the Curtius rearrangement, have a much richer triazoline and nitrene chemistry, the scope and theory of which are fully discussed by Lwowski.[72] Carbethoxynitrene may be generated by α-eliminations,[79,80] optionally assisted by ultrasound[79] or PTC (Scheme 23).[79,81]

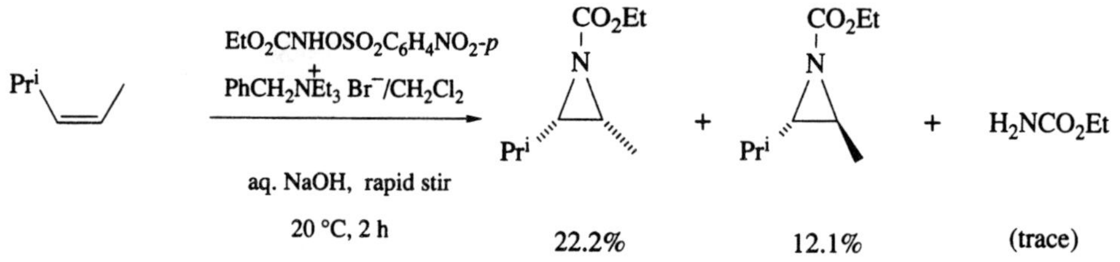

Pr^i (cis-alkene) → [$EtO_2CNHOSO_2C_6H_4NO_2$-p, $PhCH_2\overset{+}{N}Et_3$ Br^-/CH_2Cl_2, aq. NaOH, rapid stir, 20 °C, 2 h] → cis-aziridine (N-CO_2Et) 22.2% + trans-aziridine (N-CO_2Et) 12.1% + H_2NCO_2Et (trace)

Scheme 23

Results from *cis*- and *trans*-alkenes show that all the nitrene is generated in the singlet state (which adds stereospecifically) but concurrent decay to the triplet (1/30 as fast as addition) leads to loss of stereospecificity (Scheme 24).[80,82,83] However, by optimizing conditions, stereospecificities of 97.4% (from *trans*-alkene to *trans*-aziridine) and 98.2% (*cis* to *cis*) are obtainable by the α-elimination route. Thermolysis of ethoxycarbonyl azide also produces singlet nitrene,[84] however photolysis produces two-thirds singlet and one-third triplet, which sets an upper limit to the stereospecificity of the addition. Both reactions can be optimized to give about 70% yields of aziridines. The thermal reaction is complicated in those cases where the triazoline route competes with nitrene formation. This is the case with tetraalkyl-alkenes leading to increased proportions of, chiefly, imine by-products (Scheme 25).[85] Ethoxycarbonyl-nitrene can also be generated and trapped as an aziridine by thermolysis of a bis-silylated hydroxylamine (Scheme 26).[86]

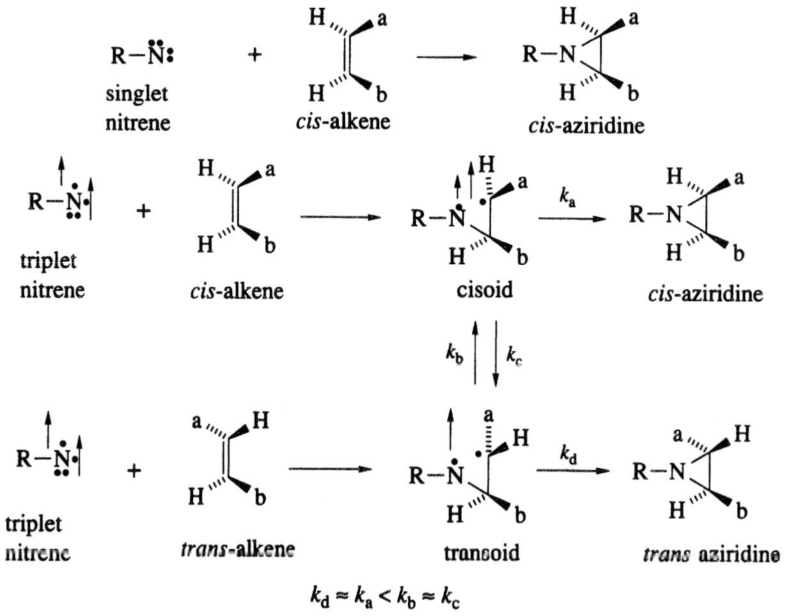

R–N: (singlet nitrene) + *cis*-alkene → *cis*-aziridine

R–N (triplet nitrene) + *cis*-alkene → cisoid —k_a→ *cis*-aziridine

cisoid $\underset{k_c}{\overset{k_b}{\rightleftharpoons}}$ transoid

R–N (triplet nitrene) + *trans*-alkene → transoid —k_d→ *trans* aziridine

$k_d \approx k_a < k_b \approx k_c$

Scheme 24

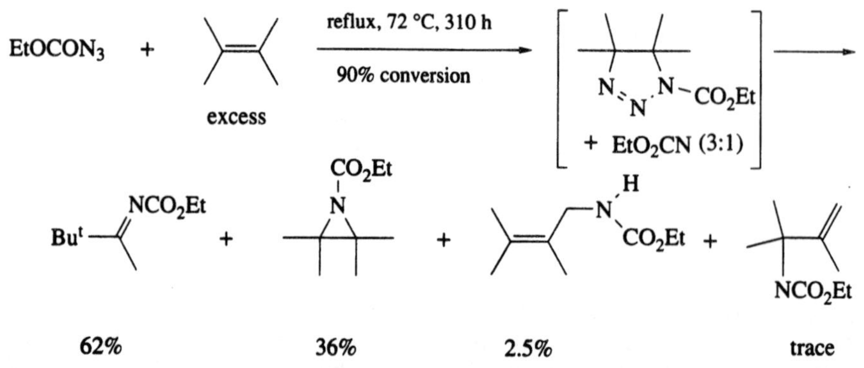

$EtOCON_3$ + (tetramethylethylene, excess) → [reflux, 72 °C, 310 h; 90% conversion] → [triazoline (N-CO_2Et) + EtO_2CN (3:1)] →

Bu^{t}C(=NCO_2Et)Me 62% + aziridine (N-CO_2Et) 36% + allylic $NHCO_2Et$ 2.5% + NCO_2Et alkene trace

Scheme 25

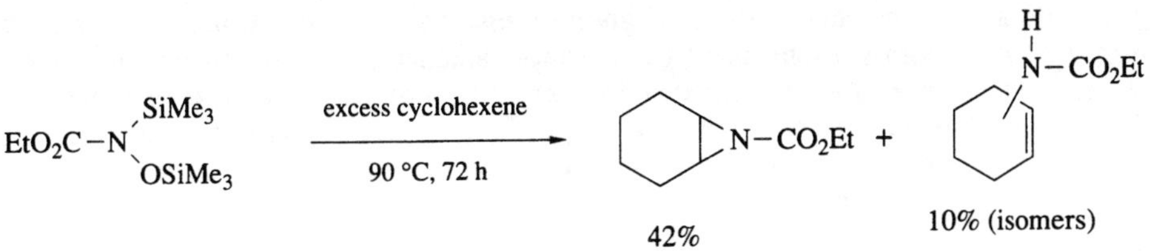

Scheme 26

Norbornene adds to photolytically produced ethoxycarbonylnitrene specifically at the *exo* face;[87] the same aziridine is produced in the thermal addition of ethoxycarbonyl azide, but *via* the triazoline rather than the nitrene, with much imine by-product.[88] There can be problems of selectivity and rearrangements when one reacts ethoxycarbonylnitrene with more complex substrates, *e.g.* alkenic steroids.[89] Ethoxycarbonylnitrene (*via* α-elimination) adds to vinyl chlorides to give 2-chloroaziridines,[90] which can be rearranged thermally to yield 2-chloroallyl carbamates. This nitrene also adds to enamines, giving an array of rearranged products.[91] A modern discussion of the reactivities of ethoxycarbonylnitrene (electrophilic) in comparison with phthalimidonitrene (nucleophilic) towards alkenes of different electronic properties has appeared.[92]

A procedure, primarily intended to yield diamines,[93] can be diverted to yield either imidoylaziridines or imidazolines, though rarely as well controlled as in the following example (Scheme 27). The initial step is regioselective (Markovnikov) and stereospecific. In a related procedure the β-bromocyanamide was reduced to a formamidine (H_2, 1% Pd/C, MeOH–AcOH, 20 °C, 1 atm) as an alternative intermediate.[94]

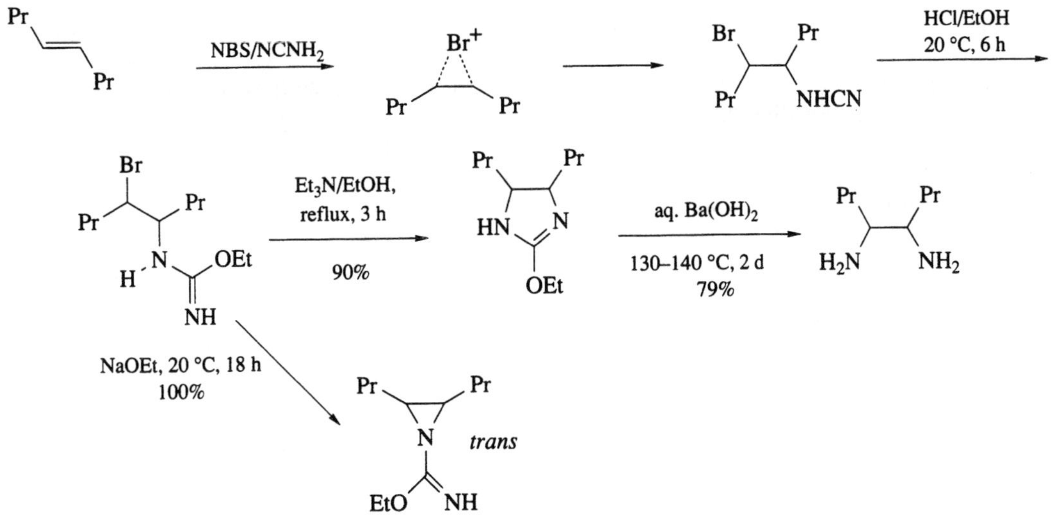

Scheme 27

N-Cyanoaziridines can be made from alkenes *via* either cyanogen azide or cyanonitrene. Cyanonitrene can be generated thermally and photolytically from cyanogen azide,[95] but the azide itself adds so readily to most alkenes to give a transient triazoline that this has prevented much study of the nitrene-mediated reactions from the azide. More recently, our own group oxidized cyanamide in the presence of various nitrene traps: alkenes, such as dicyclopentadiene (Scheme 28), and also phosphines, sulfoxides and sulfides (including penicillins), and obtained the products of formal cyanonitrene addition in each case, though this array of products remains the sole evidence of a nitrene mechanism for these reactions.[96] Cyanonitrene has also been generated by oxidation of sodium hydrogen cyanamide with Bu^{t}OCl, but was not, apparently, added to alkenes.[97]

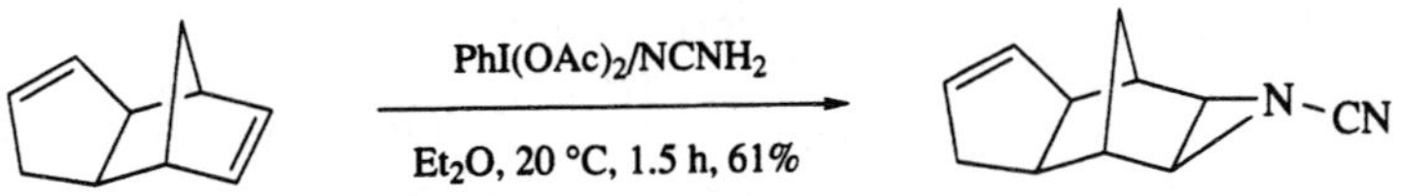

Scheme 28

Cyanogen azide adds to alkenes at 0–35 °C giving transient triazolines, which decompose to alkylidenecyanamides, often with rearrangement (from simple alkenes), or to aziridines (cleanly only from polycyclic, strained or more highly alkylated alkenes).[98] A typically complex example is given in Scheme 29. The NBS/cyanamide addition to alkenes (Scheme 27) can also be modified to yield *N*-cyanoaziridines, by submitting the initial β-bromocyanamide adduct to a Gabriel cyclization (with NaOH/aq. acetone, reflux 30 min).[99]

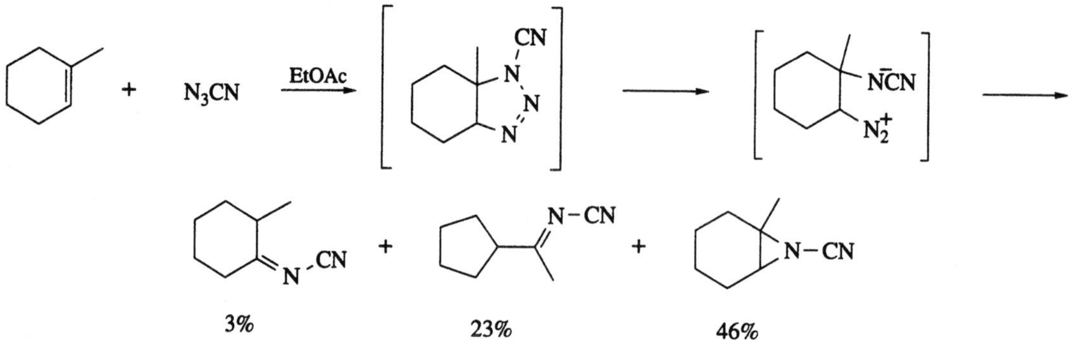

+ N_3CN —EtOAc→ [ ] → [NCN N_2^+] →

N-CN + + N-CN

3% 23% 46%

Scheme 29

3.5.2.5 *N*-Aminoaziridines, *N*-Phosphonylaziridines

Oxidation of the quinazoline (**23**) with lead tetraacetate gives an unstable intermediate (**24**) which shows a first order decomposition in CH_2Cl_2, $k = 3.68 \times 10^{-4}\ s^{-1}$ at 10 °C.[100] The rate of disappearance of (**24**) increases on adding an alkene such as styrene (Scheme 30), with the formation of a metastable *cis* invertomer (**25**) of an aziridine (**26**), which last is the product eventually isolated.[101] It is suggested that the mechanism is analogous to the Bartlett mechanism for epoxidation.[100,102] Quinazoline (**23**) has an interesting chemistry, on oxidation adding to heterosubstituted alkenes to generate novel aziridines which show some previously unrecognized chemistry (Scheme 31).[103]

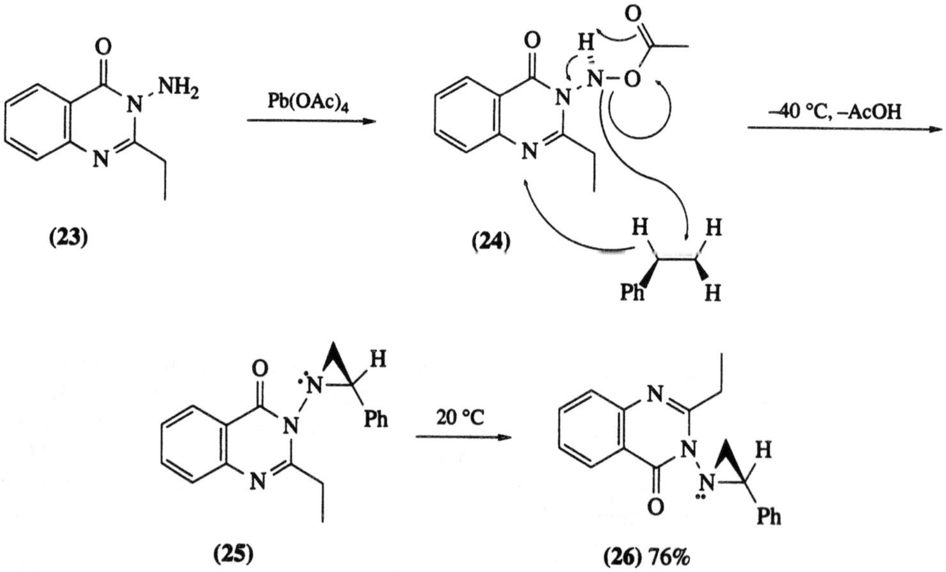

$Pb(OAc)_4$ → –40 °C, –AcOH →

(**23**) (**24**)

20 °C →

(**25**) (**26**) 76%

Scheme 30

It would appear that a similar mechanism probably applies to the many aziridinations of alkenes with lead tetraacetate and a wide variety of other *N*-aminoheterocycles. In any case, such reactions are now thought unlikely to occur *via* a nitrene: in competition experiments, *N*-aminophthalimide/lead tetraacetate reacts with styrene in preference to methyl acrylate (1.5:1), whereas with the supposed genuine nitrene (**27**) (whose nature is also in doubt[103]) prepared by pyrolysis (Scheme 32), the ratio is reversed (1:3).[100] Other sources of supposed (**27**) by pyrolysis include (**28**)[104] and (**29**).[105] Very recently, strong

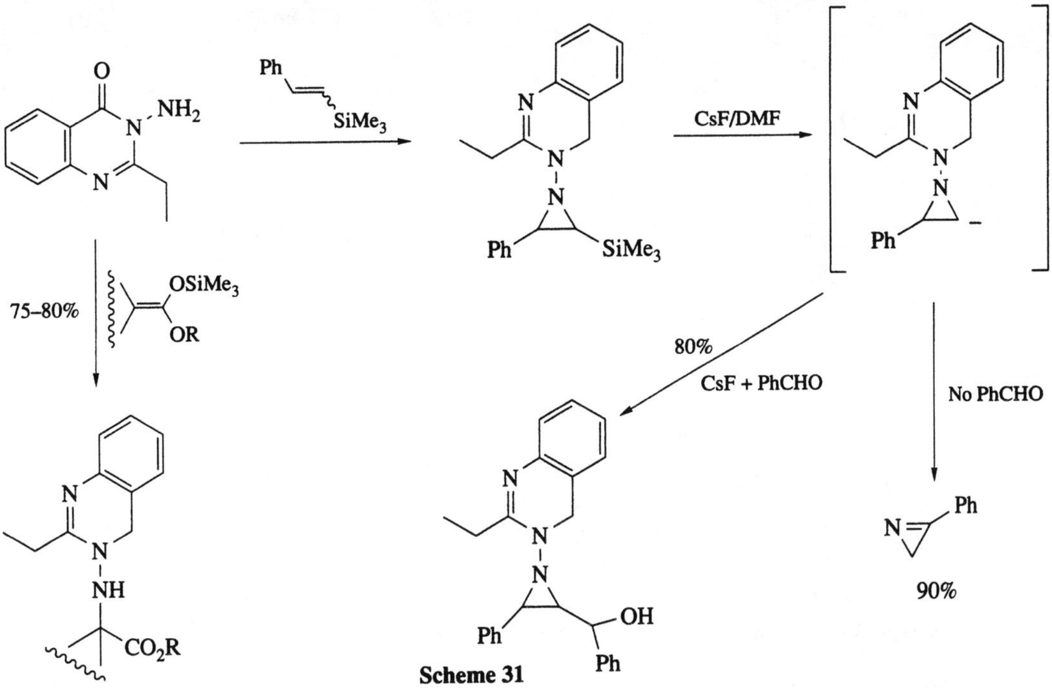

Scheme 31

evidence for an N-acetoxy intermediate has been obtained for the N-aminophthalimide/$Pb(OAc)_4$ reactions.[103,106]

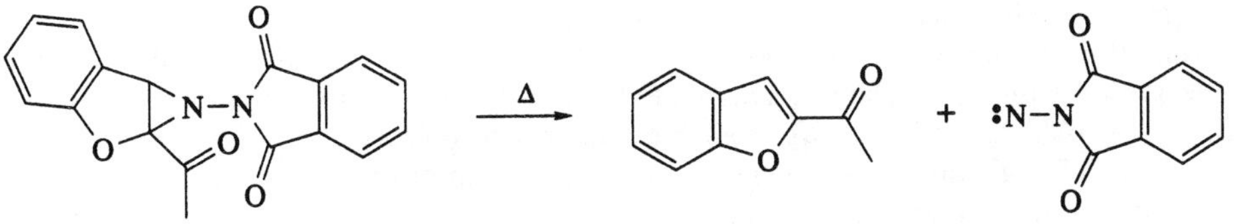

Scheme 32

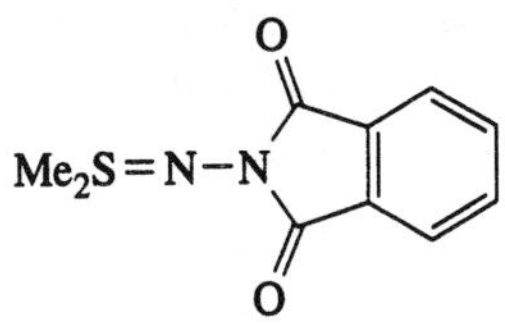

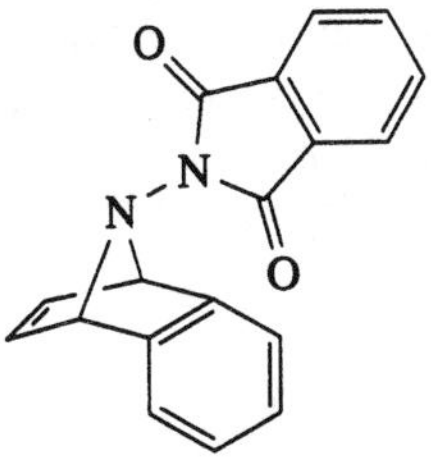

(28) **(29)**

2-Cyclohexen-1-ol is aziridinated *syn* by **(24)**; geraniol ($Me_2C{=}CHCH_2CH_2C(Me){=}CHCH_2OH$) is aziridinated at the 2,3-position due to hydrogen bonding, and geranyl chloride reacts at the 6,7-position.[107] These recent results also generally support the analogy with peroxy acid epoxidation, but there are some differences. Thus 3-chlorohexen-1-ol is also aziridinated *syn*, yet the corresponding epoxidation in this case is not stereoselective.[103]

The many N-aminoheterocycles convertible to aziridines by oxidation with lead tetraacetate, generally at ambient temperature in dichloromethane, are listed in Scheme 33.[5,58,108–111] The most frequently used is N-aminophthalimide, which gives good yields with a wide variety of alkenes, both electron-rich and electron-poor, most recently steroidal alkenes.[112] Prior to the appearance of ref. 100, these reactions were all assumed to occur *via* (singlet) nitrenes, and indeed all the early evidence was consistent with this suggestion, except perhaps, in retrospect, the absence of products of C—H insertion.

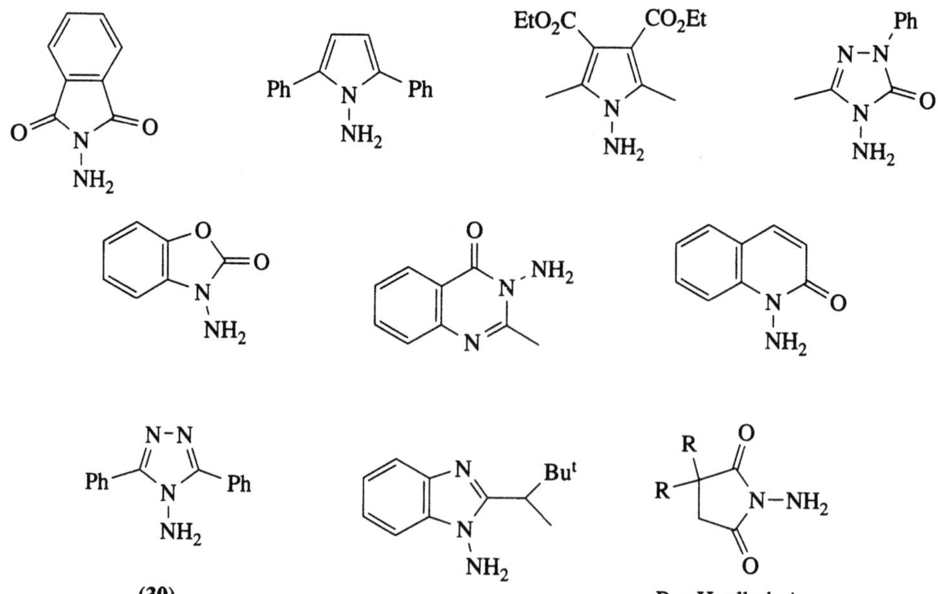

(30)

R = H, alkyl, Ar

Scheme 33 *N*-Amino heterocycles which yield aziridines from alkenes and $Pb(OAc)_4$

Some *N*-aminoheterocycles undergo other reaction modes on oxidation (perhaps *via* even less stable *N*-acetoxyamino compounds) and aziridines are not available from them, a well-known example being 1-aminobenzotriazole, which on oxidation yields benzyne and nitrogen.[113] An interesting case is (**30**; Scheme 33) which on oxidation in the presence of alkenes gives aziridines; in their absence, the intermediate fragments to nitrogen and two molecules of benzonitrile.[114]

A promising start has been made in applying the lead tetraacetate oxidation to chiral *N*-aminoheterocycles in the presence of achiral alkenes, leading to chiral aziridines *via* asymmetric induction, with compound (**31**) being obtained in 100% *de* from the appropriate precursors.[115] Similar reactions with the chiral quinazoline (**32**) give only modest face selectivities (stereoisomer ratios 1.2:1 to 2.4:1); however inclusion of trifluoroacetic acid in the reaction mixtures gives greatly improved purities (stereoisomer ratios 5.2:1 to 23:1) *via* a stabilized protonated intermediate originally formulated as the nitrene (**33**),[116] but which, in the light of later work would seem to be an *N*-acetoxyamino compound.[100] For wide utility of this potentially very useful reaction, a method for cleaving the N—N bond without destroying the aziridine ring would be needed. It would be most interesting to try some oxidations of achiral *N*-aminoheterocycles with lead(IV) salts of chiral acids: it could lead to a new chiral aziridination, but even in the absence of a preparatively useful *ee*, any significant *ee* would provide yet further evidence for (Het)—NHOCOR intermediates. Some resolved aziridines are also available in high optical purity by a kinetic resolution *via* carbonyl insertion with CO and a chiral rhodium complex.[117]

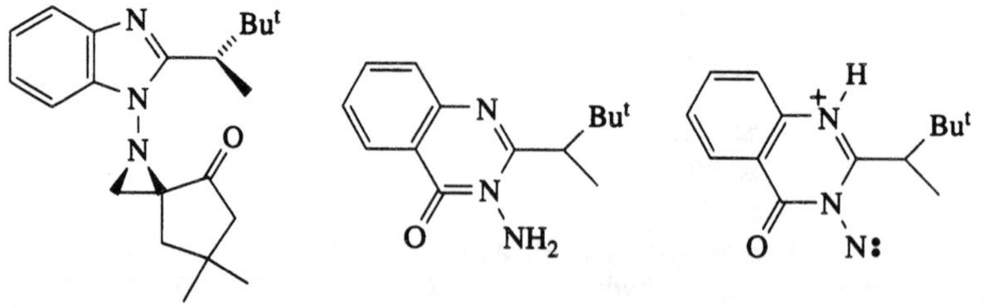

(**31**) racemic (**32**) (**33**)

Phthalimidoaziridines can be cleaved by hydrazinolysis to give 1-aminoaziridines,[118,119] which decompose slowly at room temperature, and rapidly above 48 °C, regenerating the starting alkene with stereochemistry intact. There would appear to be some unexploited potential for phthalimidoaziridines to be used for alkene protection. Phthalimidoaziridines are stable to reflux in chlorobenzene (132 °C) for 24 h.[119] *N*-Aminoaziridines can be acylated to yield *N*-acylaminoaziridines of varying thermal stability:[105] (**34**; $t_{1/2}$ = 15 min at 37 °C) is a ready thermal source of benzamidonitrene from which new aziridines can be made by reaction with other alkenes, whereas the isomeric *cis*-aziridine is stable at 20 °C for at least 15 years. Two unusual intramolecular formations of *N*-aminoaziridines are given in Schemes 34 and 35.[120,121]

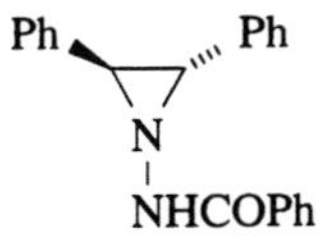

Ph, Ph, N, NHCOPh

(34)

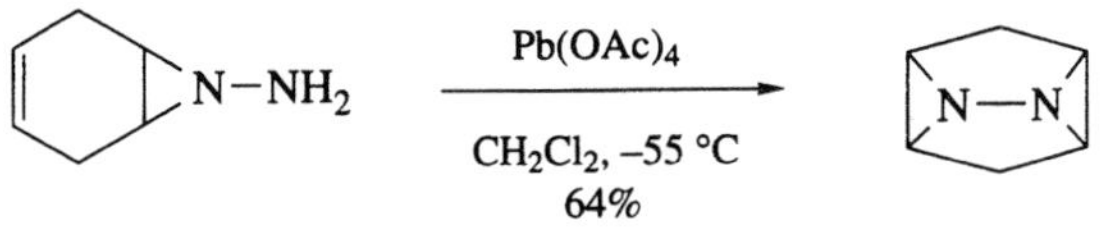

N–NH2 — $Pb(OAc)_4$, CH_2Cl_2, –55 °C, 64% → N—N

Scheme 34

N–NHTs, Ph, Ph — i, NaH, 80 °C, 20 min; ii, 20 °C, 2 d → N–N, Ph, Ph

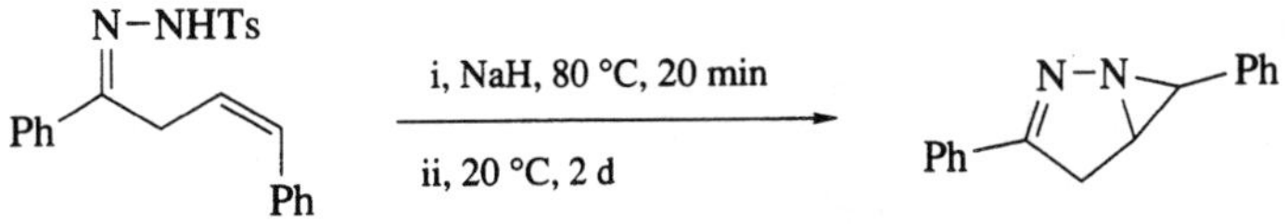

Scheme 35

Diethylphosphoryl azide gives an *exo*-triazoline with norbornene;[122] subsequent photolysis gives the phosphorylated aziridine, but pyrolysis gives the imine isomer. Markovnikov addition of diethyl dibromophosphoramidate to alkenes (BF_3 catalysis) occurs in high yield;[123] the initial adducts were converted to *N*-unsubstituted products but would be readily convertible to phosphorylated aziridines by the Gabriel method.[124] In any case, *N*-unsubstituted aziridines are readily phosphorylated.[5,125]

3.5.2.6 Aziridines *N*-Substituted with O or S

Treatment of *O*-alkylhydroxylamines with a mixture of an alkene and lead tetraacetate in dichloromethane gives aziridines in about 30% yields.[5,108] The products are mixtures of stable invertomers.[126] The reactions are not quite stereospecific.[127] Oxidation of certain nitroarylsulfenamides with lead tetraacetate in dichloromethane in the presence of alkenes gives *N*-sulfenylaziridines, 18–64%, together with some (identified) by-products.[41,108] The addition to *cis*-1-phenylpropene gives a mixture of *cis*- and *trans*-aziridines (3:1), and the unreacted alkene was partly isomerized (to *trans*) also.[128] Kinetic and ESR investigations have enabled these results to be rationalized in terms of a disproportionation of the (singlet) nitrene–sulfenamide mixture yielding amidyl radicals, which add reversibly to the alkene causing isomerization. The resulting *trans*-alkene is then, of course, the precursor of the *trans*-aziridine (Scheme 36). The investigation also concluded that a second aziridine-forming intermediate was involved, but that this was not the triplet nitrene. Mild thermolysis of *N*-arenesulfenylimino-1,4-dihydronaphthalenes also gives the sulfenylnitrenes which, in the presence of excess alkene, can be trapped quantitatively as the aziridines.[129] Though the thermal reaction is still not quite stereospecific, it has advantages over the lead tetraacetate procedure in giving higher yields, and also in working with electron-deficient alkenes. (It is also not dependent on nitro-substitution on the aromatic ring.) An *N*-sulfenyl group is readily removed by reduction ($NaBH_4$),[41] thus providing another overall route to *N*-unsubstituted aziridines from alkenes.

Fluorothiazyne (F—S≡N, not a nitrene) adds to perfluoropropene under photolysis[130] or *via* cesium fluoride catalysis[131] to give various abstruse derivatives of perfluoropropyleneimine. *N*-Arylsulfinylaziridines are available by acylation of the NH compounds.[125] Arenesulfonyl azides add to alkenes, forming unstable triazolines, which decompose spontaneously in a variety of ways, only rarely (norbornene) yielding aziridines as the main product.[72] *N*,*N*-Dibromosulfonamides were early on added to alkenes, a reaction that can give good yields, but has been little used.[132] The regiochemistry is interesting, and seems fairly well established (Scheme 37).

Compound (**35**) is also available in high yield by the porphyrin-catalyzed reaction described earlier (Scheme 22).[78,133] This reaction, originally reported as nonstereospecific, has recently been made stereo-

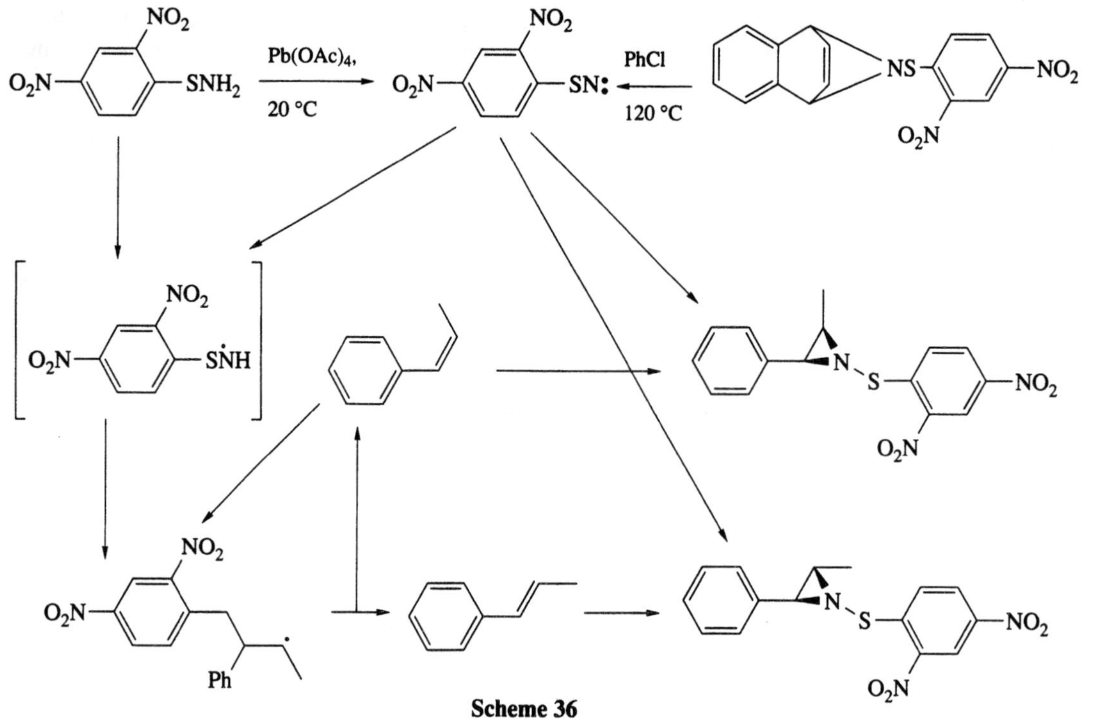

Scheme 36

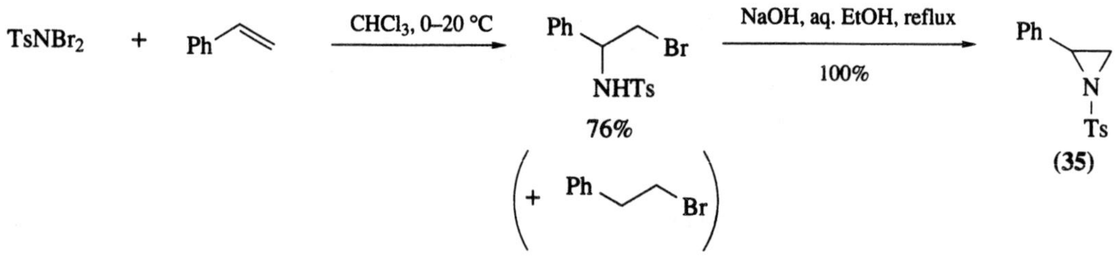

Scheme 37

specific for the tosylaziridination (the trifluoroacetylaziridination[78] appears not to have been followed up) by the adoption of tetrakis-2,6-dichlorophenylporphyriniron(III) perchlorate as catalyst.[133] Manganese porphyrins can also be used with terminal alkenes, but give allylic tosylamination with nonterminal alkenes.[134] Aryloxysulfonylaziridines can be made from the unsubstituted compounds or, in the case of norbornene derivatives, from norbornene and aryloxysulfonyl azides (65–98%).[135]

3.5.3 ADDITION OF TWO NITROGEN ATOMS TO ALKENES

3.5.3.1 Formation of Diamines

There is one direct method of preparing primary diamines from alkenes, in which the alkene is treated with nitric oxide and a cobalt complex, and the intermediate worked up reductively. The two-stage reduction gives better stereoselectivity (Scheme 38).[136] Typical yields from alkenes listed and *cis:trans* selectivities (expected product first) are: cyclopentene, 70%, 70:30; *trans*-3-hexene, 61%, 90:10; *cis*-3-hexene, 43%, 66:34; cyclohexene, 47%, 68:32; *trans*-1-phenylpropene, 90%, 85:15; *cis*-1-phenylpropene, 74%, 72:28. An alternative oxidative work-up with iodine gives, at least from the norbornene adduct, a dioxime.

The palladium-promoted vicinal diamination of alkenes can be achieved according to Scheme 39; the example chosen illustrates how the mechanism and stereochemistry were determined.[137] The independent synthesis of (36) from a quaternized aziridine is also noteworthy. Yields are 60–87% from ter-

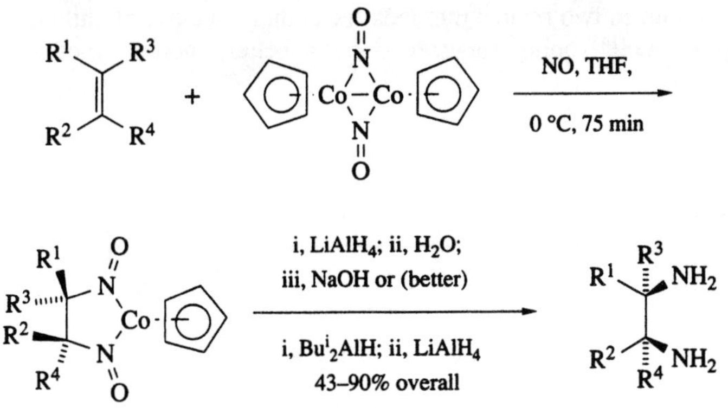

NO, THF,

0 °C, 75 min

i, $LiAlH_4$; ii, H_2O; iii, NaOH or (better)

i, $Bu^i{}_2AlH$; ii, $LiAlH_4$
43–90% overall

Scheme 38

minal alkenes, 35–45% from internal alkenes. Tri- and tetra-substituted alkenes have not been reported, and the only amine used appears to be dimethylamine.

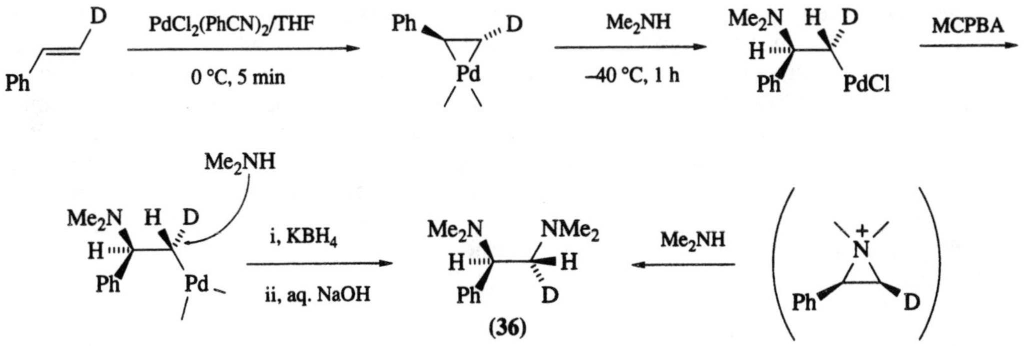

$PdCl_2(PhCN)_2$/THF, 0 °C, 5 min

Me_2NH, –40 °C, 1 h

MCPBA

i, KBH_4; ii, aq. NaOH

Me_2NH

(36)

Scheme 39

Osmium *t*-alkylimides react with alkenes to give *cis*-diamines.[138] The reaction is limited to *t*-alkylamines, and further limited by the lack, so far, of a catalytic version. The problem would appear to be that whilst monoimides are readily made in aqueous solution, the bis- or tris-imides required to achieve diamination have to be prepared from phosphinimines (Scheme 40). The reactivities are modulated by steric and electronic factors; in being favored by electron-withdrawing substituents on the alkenes, the reactions resemble those of $KMnO_4$ rather than OsO_4. Unusually, *trans* double bonds are more reactive than *cis*, though this selectivity is shared by OsO_4, $KMnO_4$, RuO_4 and Pd(II).

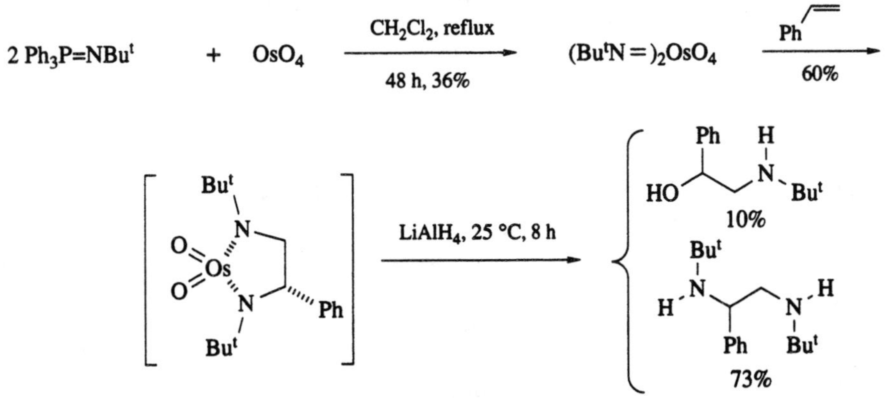

2 $Ph_3P{=}NBu^t$ + OsO_4 → (CH_2Cl_2, reflux, 48 h, 36%) → $(Bu^tN{=})_2OsO_4$ → (styrene, 60%)

$LiAlH_4$, 25 °C, 8 h

10%

73%

Scheme 40

Two older direct methods are of limited scope. Tetrafluorohydrazine reacts thermally with alkenes at 50 °C, 6 h to give α,β-bis(difluoroamino) compounds in good yields.[139] Tetramethyltetrazene–zinc chloride complex, a source of dimethylamino radicals, adds to α-methylstyrene in 30–40% yield.[140] Barluenga's group have introduced two related methods for adding 2 equiv. of aniline across alkenic bonds, an earlier thallium procedure[141] being superseded by a better, mercury-mediated reaction (Scheme 41).[142]

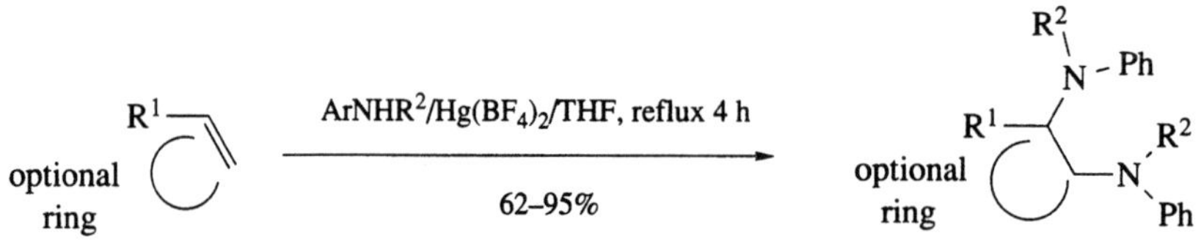

R^1 = H, Me, Ph, $PhCH_2$; R^2 = H, Me

Scheme 41

Bisimides of sulfur dioxide undergo Diels–Alder additions to dienes (Scheme 42). The adducts can be processed to give the products of net 1,2-addition of amides to one of the double bonds.[143] The initial adducts (**37**) and (**38**) are mixtures of two diastereoisomers at sulfur. The S—C(6) *trans* and S—C(6) *cis* isomers have to be processed differently as shown; the *trans* isomers are cleaved with Grignard reagents to give sulfilimines (**39**) which undergo [2,3] sigmatropic rearrangement to sulfenylated diamides, *e.g.* (**40**), requiring only cleavage of the S—N bond [$(MeO)_3P$] to give the desired products (**41**). A similar sequence can be carried out on some of the S—C(6) *cis* isomers (**38**) but requires less sterically demanding organometallics (MeLi). However, these S—C(6) *cis* isomers are well set up for a novel [2,3] sigmatropic rearrangement to yield thiadiazolines (**42**), and these, on cleavage, ($NaBH_4$) happen to yield the same end product diamines. The full scope of this methodology is not yet known. Clearly, saturated diamides are available by reduction of the remaining double bond, and the methoxycarbonyl group (and, less conveniently, the tosyl group) can be modified or removed to give diamines.

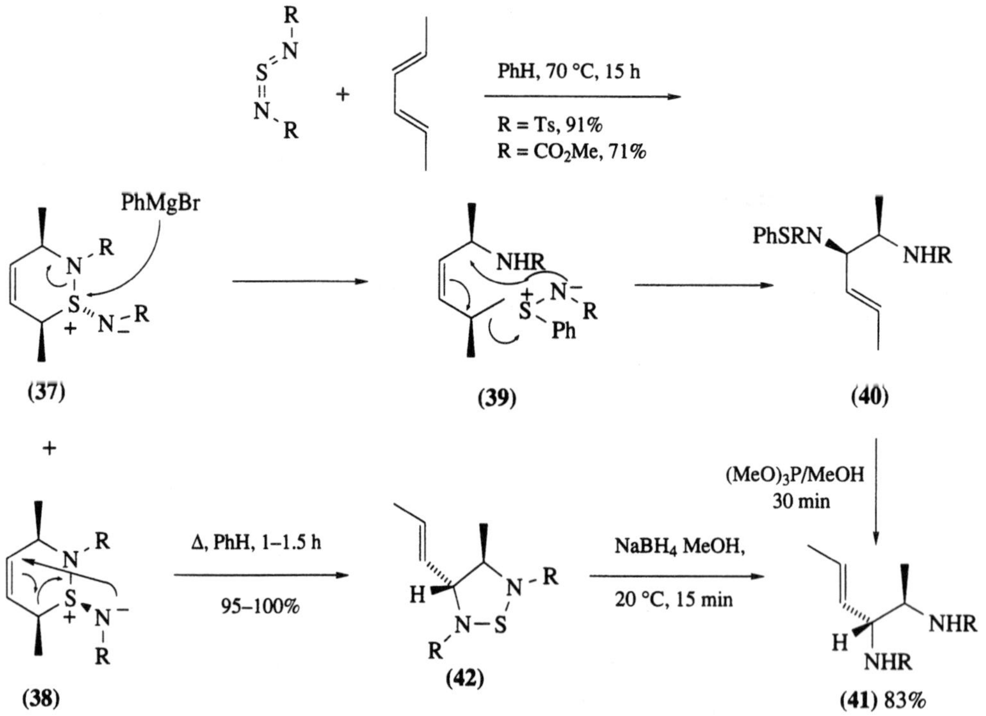

Scheme 42

This sulfur work was preceded by a conceptually related selenium reaction (Scheme 43).[144] In accord with this mechanism, 1-vinylcyclohexene gives predominantly the isomer substituted at the more hindered endocyclic double bond.

A modified work up of the alkene–iodine isocyanate adducts (Section 3.5.6.4), yields imidazolinones (Scheme 44).[145]

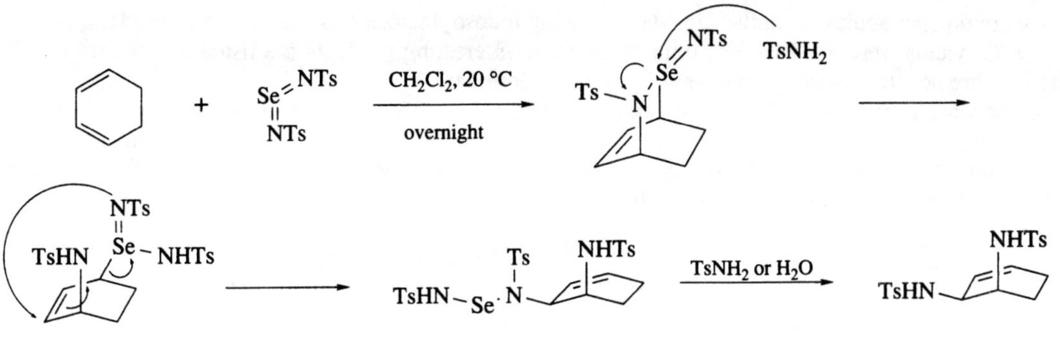

Scheme 43

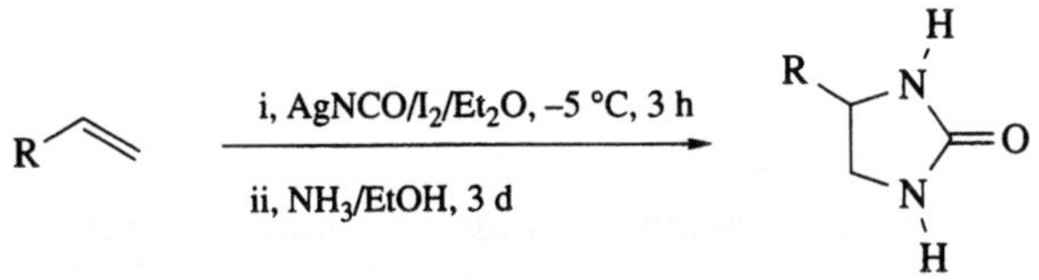

Scheme 44

Indirect methods of making diamines from alkenes include ring opening of aziridines or oxazolines with amines, amides or azides. Aziridines have received limited usage as yet, though there are many parallels with the ring opening of epoxides. In general, aziridines are less readily opened than epoxides unless strongly electron-withdrawing groups are present on the nitrogen. Examples include sulfonyl, but activation can also be achieved by quaternization (Scheme 39), by protonation or by Lewis acids.[9] Recent examples of ring openings with nitrogenous nucleophiles include chiral bisaziridines[146] and *N*-phthalimidoaziridines.[147] Oxazolines have been opened with primary and secondary amines to yield unsymmetrically substituted diamines.[148]

3.5.3.2 Formation of Diazides

Two recent high yielding processes for diazidation of alkenes have been reported.[60] In the first, $Mn(OAc)_3$, NaN_3 and the alkene are heated in glacial acetic acid until the brown colour is discharged (indicating complete conversion to Mn^{II}) (Scheme 45).[149] The paper also reviews the reduction of diazides, favoring Lindlar catalyst.

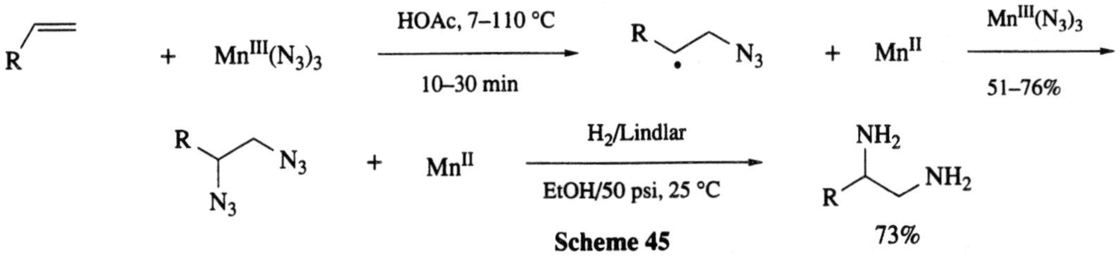

Scheme 45

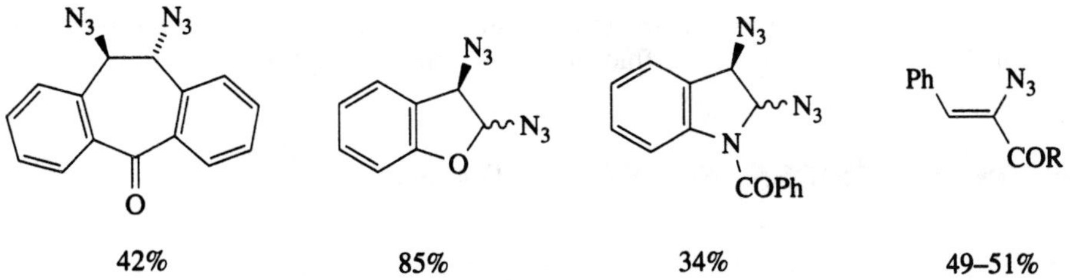

Scheme 46 Diazides and related compounds

A second paper achieves similar diazidations using iodosoylbenzene in acetic acid as oxidant, 2–3 h at 20–50 °C, yields were 34–70%.[150] Some of the more interesting products are listed in Scheme 46. The reactions are not stereospecific, but *trans* isomers predominate.

Earlier work using lead(IV) azide or persulfate mediated by iron as oxidant has been reviewed;[151,152] this last procedure, a radical redox one, can work very well (Scheme 47).[153] Dienes can undergo 1,4-addition with lead(IV) azide;[60] reactions with steroid alkenes are varied (refs. 152,154 and references cited therein, and Section 3.5.11) but can yield the 1,2-adducts.[154]

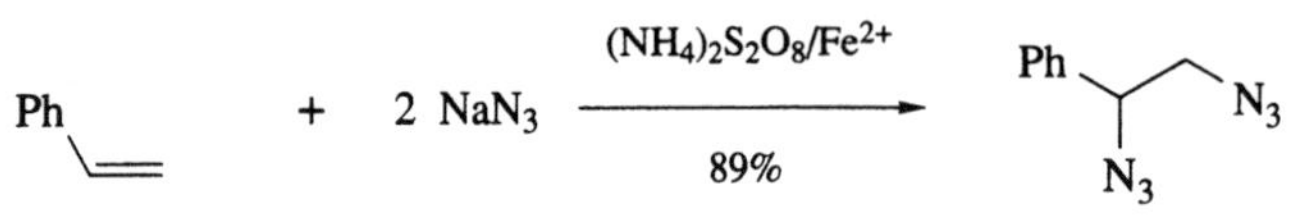

Ph + 2 NaN_3 —($(NH_4)_2S_2O_8/Fe^{2+}$, 89%)→ Ph, N_3, N_3

Scheme 47

3.5.3.3 Addition of Two Nitrogen Atoms, including a Nitroso or Nitro Group

Photoaddition of nitrosamines (carcinogens) to alkenes in the presence of oxygen gives quite good yields of 1:1 adducts as their oxime isomers.[155] Nitronium tetrafluoroborate in acetonitrile adds to alkenes in Markovnikov manner, giving acetamidonitro compounds, 38–84% (from arylalkenes) or 13–20% (from nonaromatic alkenes) (Scheme 48).[156]

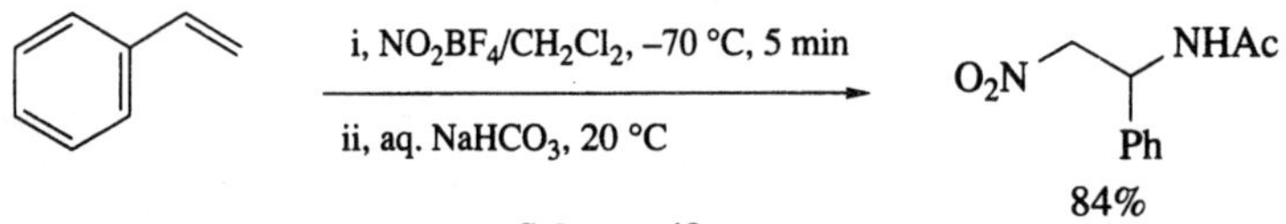

i, NO_2BF_4/CH_2Cl_2, –70 °C, 5 min; ii, aq. $NaHCO_3$, 20 °C → O_2N, NHAc, Ph; 84%

Scheme 48

The reaction is akin to the Ritter reaction, with activation achieved by nitration, rather than protonation, and the products accordingly retain the nitro group. Additions to 1-phenylcyclohexene (59%) and to *trans*-stilbene (72%) are stereospecific (*trans*); *cis*-stilbene gives the expected *threo* product (39%) plus some *erythro* (6%). Reactions of nitrogen dioxide with alkenes are very complex and rarely useful.[157] A recent mechanistic paper gives many key references.[158] Addition of N_2O_3 is occasionally useful, as with dicyclopentadiene (Scheme 49).[159]

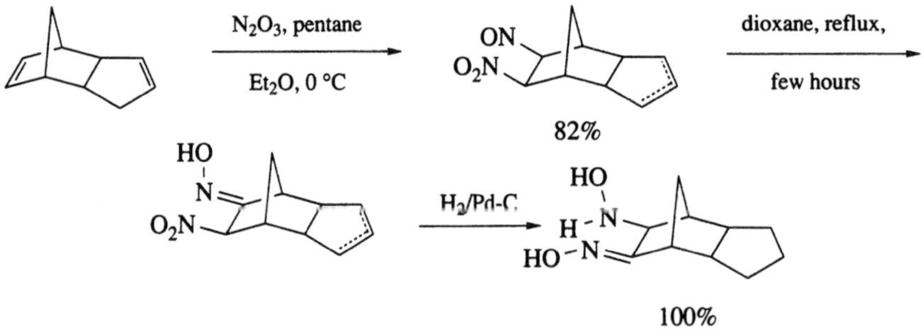

N_2O_3, pentane, Et_2O, 0 °C → ON, O_2N; 82% → dioxane, reflux, few hours → HO, N, O_2N → H_2/Pd-C → HO, H-N, HO-N; 100%

Scheme 49

Preparation of nitronitroso dimers from a variety of straight chain alkenes has been patented.[160] The reactions of nitric oxide with alkenes are extremely complex (*e.g.* isobutylene) and are rarely useful.[161] Perfluoroalkenes add nitric oxide at room temperature in the dark; tetrafluoroethylene gives $ONCF_2CF_2NO$ (68%);[162] the reaction with perfluoropropene is more complex.[163]

3.5.4 ADDITION OF NITROGEN AND OXYGEN TO ALKENES

3.5.4.1 Oxyamination and Oxyamidation

In an extension of the *cis* dihydroxylation of alkenes with osmium tetroxide, Sharpless has developed a series of reactions, reviewed by Gasc *et al.*,[1] in which an osmium imine species is added to an alkene, forming a cyclic amide ester, which on reductive work-up gives a *cis*-amino alcohol or derivative. The nitrogen is delivered to the least substituted end of the C=C bond. The original procedures were confined to *t*-alkylamines, and to stoichiometric usage of osmium tetroxide. In the definitive procedure, it was recommended that the imine be complexed with quinuclidine (Scheme 50).[164] It will be interesting to see if use of an optically active complexing base can yield asymmetric induction, in view of the high *ee* achieved recently in the related osmium-mediated dihydroxylations.[165] The imine reagent is a little milder than osmium tetroxide and appears to have as wide a functional group compatibility. Thus, the *N*-allylaniline yields an amino alcohol in 55% yield, whereas *cis* hydroxylation is achieved in only 10% yield.[166]

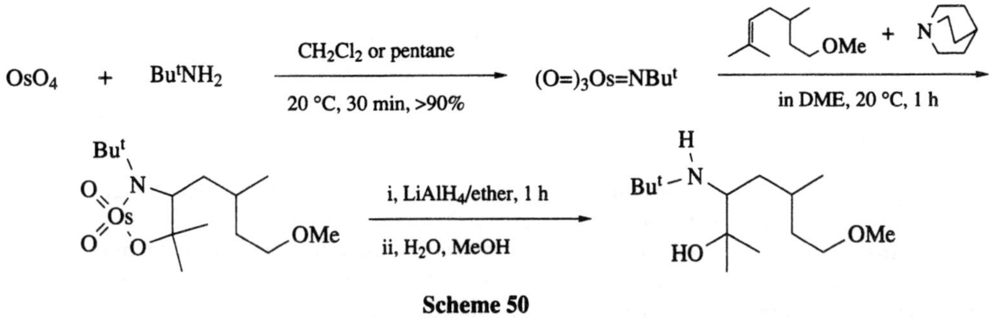

OsO_4 + Bu^tNH_2 —(CH_2Cl_2 or pentane, 20 °C, 30 min, >90%)→ $(O=)_3Os=NBu^t$ —(OMe + N, in DME, 20 °C, 1 h)→

Bu^t N O Os O O O OMe —(i, $LiAlH_4$/ether, 1 h; ii, H_2O, MeOH)→ H Bu^t–N HO OMe

Scheme 50

Two related procedures employing osmium in catalytic quantities, and providing, unlike the above, a removable nitrogen substituent, have been developed.[167,168] Scheme 51 represents a procedure suitable for mono- and 1,2-di-substituted alkenes,[167] but which is too vigorous for diethyl fumarate and enones; Scheme 52 is suitable for 1,1-di- and tri-substituted alkenes;[167] Scheme 53 is applicable to mono- and 1,2-di-substituted alkenes, especially electron-deficient alkenes,[168] and can be extended to trisubstituted alkenes by incorporation of Et_4NOAc in the reaction mixture.[169]

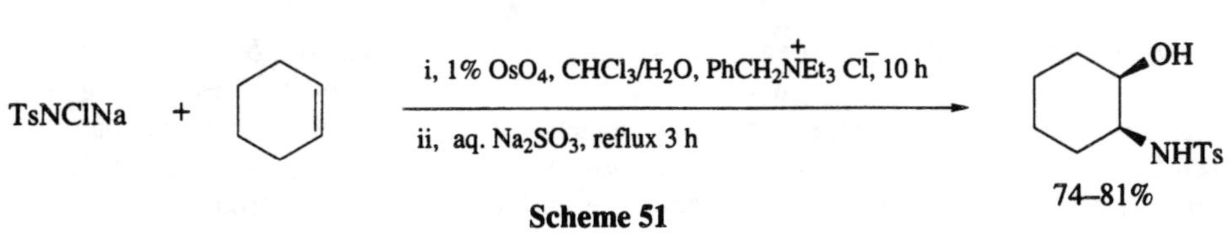

TsNClNa + —(i, 1% OsO_4, $CHCl_3/H_2O$, $PhCH_2\overset{+}{N}Et_3$ $\overset{-}{Cl}$, 10 h; ii, aq. Na_2SO_3, reflux 3 h)→ OH NHTs

74–81%

Scheme 51

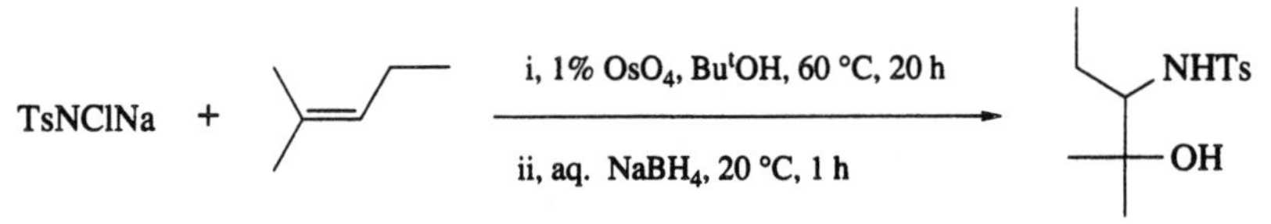

TsNClNa + —(i, 1% OsO_4, Bu^tOH, 60 °C, 20 h; ii, aq. $NaBH_4$, 20 °C, 1 h)→ NHTs OH

Scheme 52

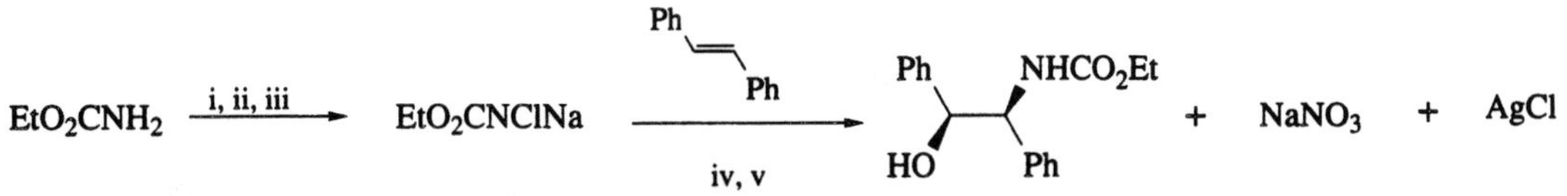

EtO_2CNH_2 —(i, ii, iii)→ $EtO_2CNClNa$ —(Ph–CH=CH–Ph; iv, v)→ Ph NHCO$_2$Et HO Ph + $NaNO_3$ + AgCl

i, Bu^tOCl/MeOH, 0 °C, 15 min; ii, NaOH/MeOH, 0 °C, 10 min; iii, evaporate to dryness;
iv, MeCN, $AgNO_3$, 20 °C, 5 min; v, 1% OsO_4, aq. Bu^tOH, 18 h, 20 °C, then aq. Na_2SO_3, reflux, 3 h

Scheme 53

Recent work has shown that the carbamate reaction shows no regioselectivity in the rather demanding case of 1,2,3,6-tetrahydropyridines; 1,2-dihydropyridines react at the 5,6 double bond, but with no selectivity in the N,O orientation.[170] The regioselectivities of the Sharpless oxyaminations have been rationalized,[171] and the react'on has recently been studied from the point of view of the inorganic chemist.[172] These procedures and ɔther *cis* hydroxyaminations below produce stereochemistry complementary to that provided by ring oƫ ening of the epoxidized alkene.[173]

Alkenes can be palladated to yield a complex which can be opened *trans* by an amine nucleophile.[1,174] The resulting σ-palladium species can be worked up oxidatively to yield an amino alcohol (as its acetate ester). This depalladation occurs with inversion, yielding overall *cis* stereochemistry (Scheme 54). If the acetic acid in step iii is replaced with phenol, a β-phenoxyamine is produced.

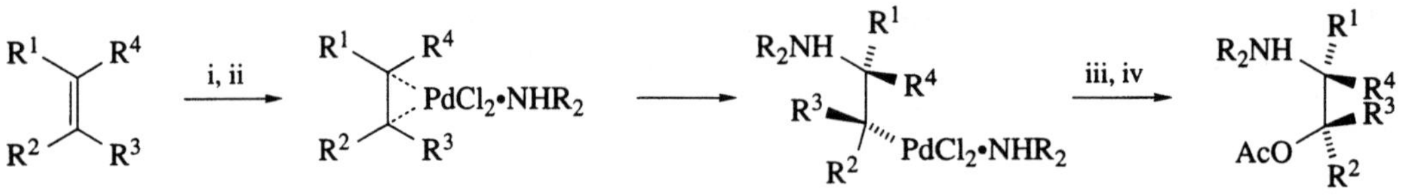

i, $(PhCN)_2PdCl_2$, THF, N_2, 0 °C, 10 min; ii, R_2NH, –50 °C, 50 min;
iii, $Pb(OAc)_4/AcOH$, –50 °C, 5 min, then 20 °C, 2 h; iv, KBH_4, 20 °C, 20 min

Scheme 54

Use of resolved PhCHMeNHMe as complexing agent and nucleophile in this reaction causes asymmetric induction in the amino acetate, with *de* 20–60%;[175] alternatively, resolved $PhCHMeNMe_2$ can be used as complexing agent, followed by an achiral amine as nucleophile, but *ee* of only 3–12% was achieved.[175] Related transformations, but this time *trans* stereospecific, can be achieved using mercury activation; in these cases the mercury itself is the oxidant (Scheme 55).[176]

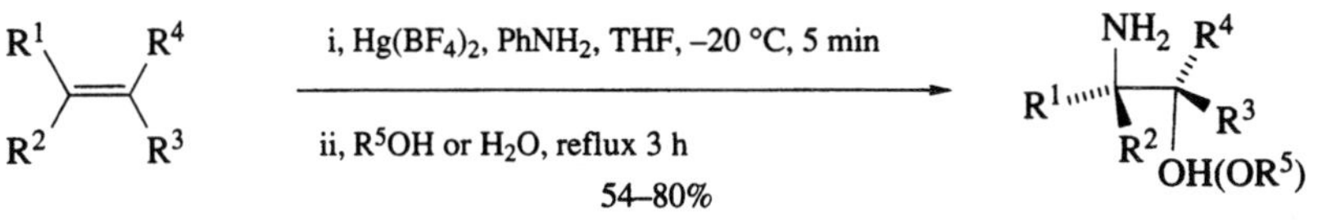

i, $Hg(BF_4)_2$, $PhNH_2$, THF, –20 °C, 5 min
ii, R^5OH or H_2O, reflux 3 h
54–80%

Scheme 55

In its utilization of acetonitrile, the oxazoline synthesis shown in Scheme 56 resembles a Ritter reaction.[177] The procedure is convenient, but yields are variable; the pyrolysis gives starting alkene plus acetamide as by-products. Another oxazoline synthesis and subsequent conversion to a *cis*-amino alcohol is discussed later (Scheme 85). A recent γ-hydroxy-α-amino acid synthesis incorporates the following type of transformation (Scheme 57).[178] If a three-day equilibration with anhydrous HBr was introduced between stages i and ii, almost pure *trans* product was obtained. The paper has many useful references.[178] Yet another modified Ritter reaction is shown in Scheme 58.[179]

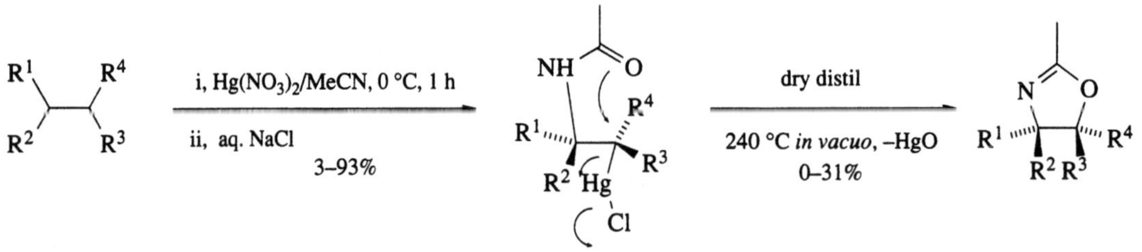

i, $Hg(NO_3)_2$/MeCN, 0 °C, 1 h
ii, aq. NaCl
3–93%

dry distil
240 °C *in vacuo*, –HgO
0–31%

Scheme 56

i, $Hg(NO_3)_2$/MeCN, 1 h
ii, aq. KBr, 2 h
iii, O_2, $NaBH_4$

60% 20%

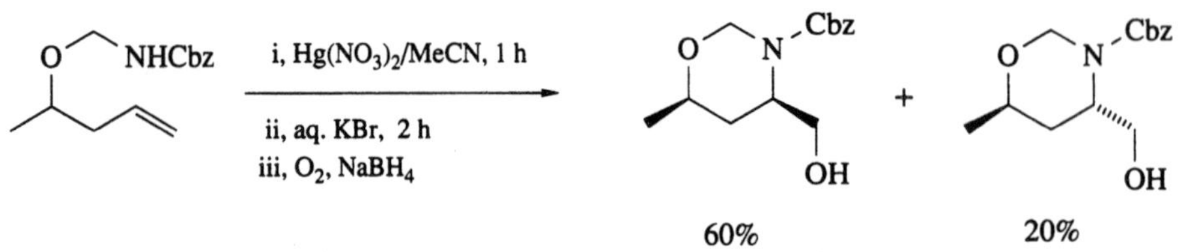

Scheme 57

N-Nitrosopiperidine (carcinogen) with oxygen adds photolytically to alkenes to give *N*-(2-nitroalkyl)piperidines.[155] Indirect methods of preparing amino alcohols from alkenes include the well-known *trans* opening of epoxides with nitrogen nucleophiles and a recent, complementary, *cis* opening of

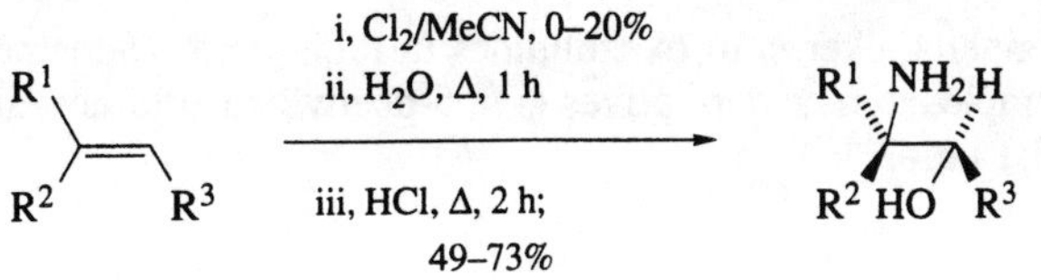

i, Cl_2/MeCN, 0–20%
ii, H_2O, Δ, 1 h
iii, HCl, Δ, 2 h;
49–73%

Scheme 58

acyclic or cyclic vinyl epoxides with tosyl or aryl isocyanates (Scheme 59).[180] Another indirect procedure, also requiring an auxiliary conjugated double bond, is shown in Scheme 60.[181] It is conceptually closely related to the diamination procedure of Scheme 42.

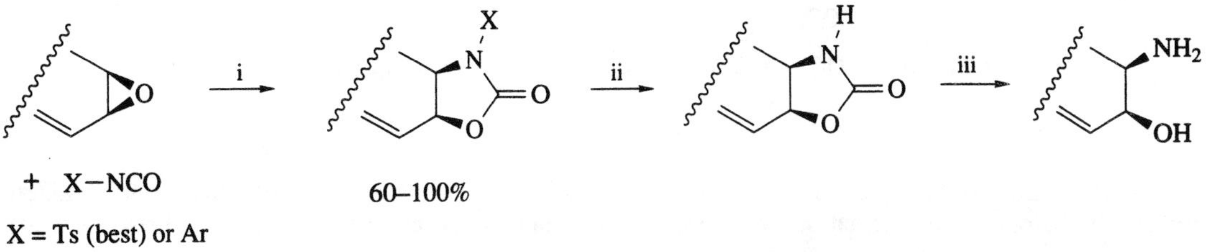

+ X–NCO

X = Ts (best) or Ar

60–100%

i, $[(PhCH{=}CH)_2C{=}O]_3Pd_2$, $CHCl_3$, THF, $(Pr^iO)_3P$, 0–20 °C; ii, $C_{10}H_8^-Na^+$ (for X = Ts) or $Ce(NH_4)_2(NO_3)_6$, MeCN (for X = 4-$MeOC_6H_4$); iii, NaOH/aq. EtOH, reflux

Scheme 59

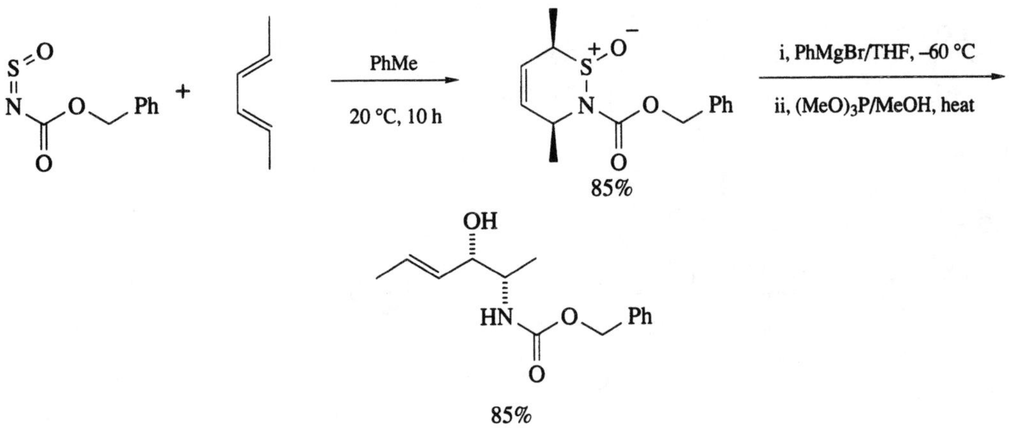

PhMe, 20 °C, 10 h

85%

i, PhMgBr/THF, –60 °C
ii, $(MeO)_3P$/MeOH, heat

85%

Scheme 60

Thallated aziridines can be opened with TFAA in high yield, giving *trans*-azido esters (Scheme 61).[182] A similar sequence with cyclohexene yields *trans*-1-azido-2-trifluoroacetoxycyclohexane (80%). Steroid alkenes yield azido alcohols with chromyl azide.[60]

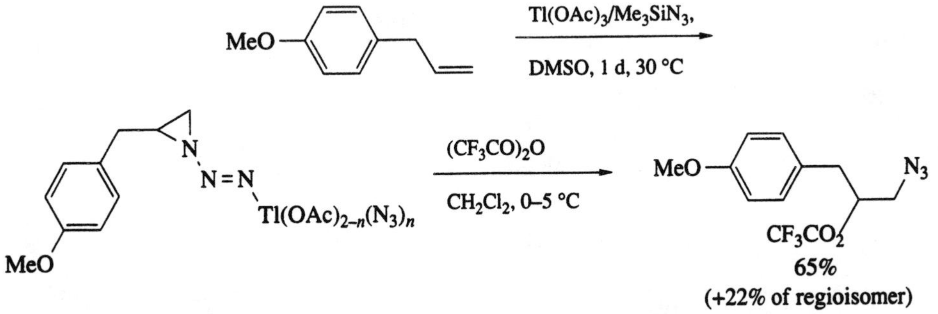

$Tl(OAc)_3/Me_3SiN_3$, DMSO, 1 d, 30 °C

$(CF_3CO)_2O$, CH_2Cl_2, 0–5 °C

$Tl(OAc)_{2-n}(N_3)_n$

65%
(+22% of regioisomer)

Scheme 61

A tellurium reagent converts alkenes to oxazolidines in high yields (Scheme 62) with predictable regiospecificity.[183] For example, styrene derivatives give 5-phenyloxazolidines, and 1-hexene the 5- and 4-propyloxazolidines in a 3:1 ratio.

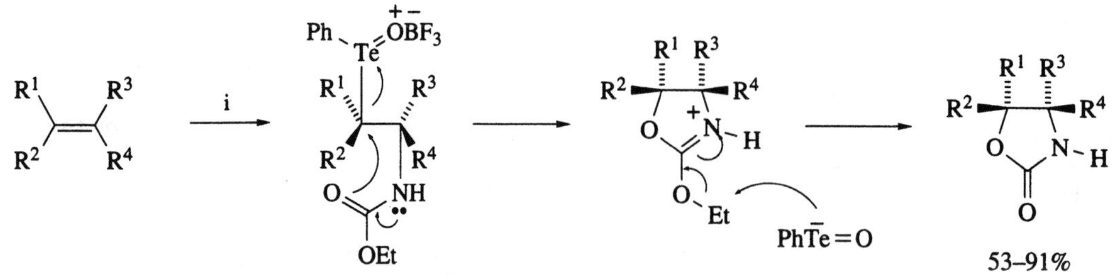

53–91%

i, $PhTe(O)OCOCF_3$, $EtOCONH_2$, $BF_3{\bullet}OEt_2$, $CHCl_2CHCl_2$, reflux 6–20 h

Scheme 62

There are numerous examples of the intramolecular delivery of amide or imidate nitrogen to an alkene activated by halonium ion formation or by epoxidation leading, eventually, to stereo- and regio-specific

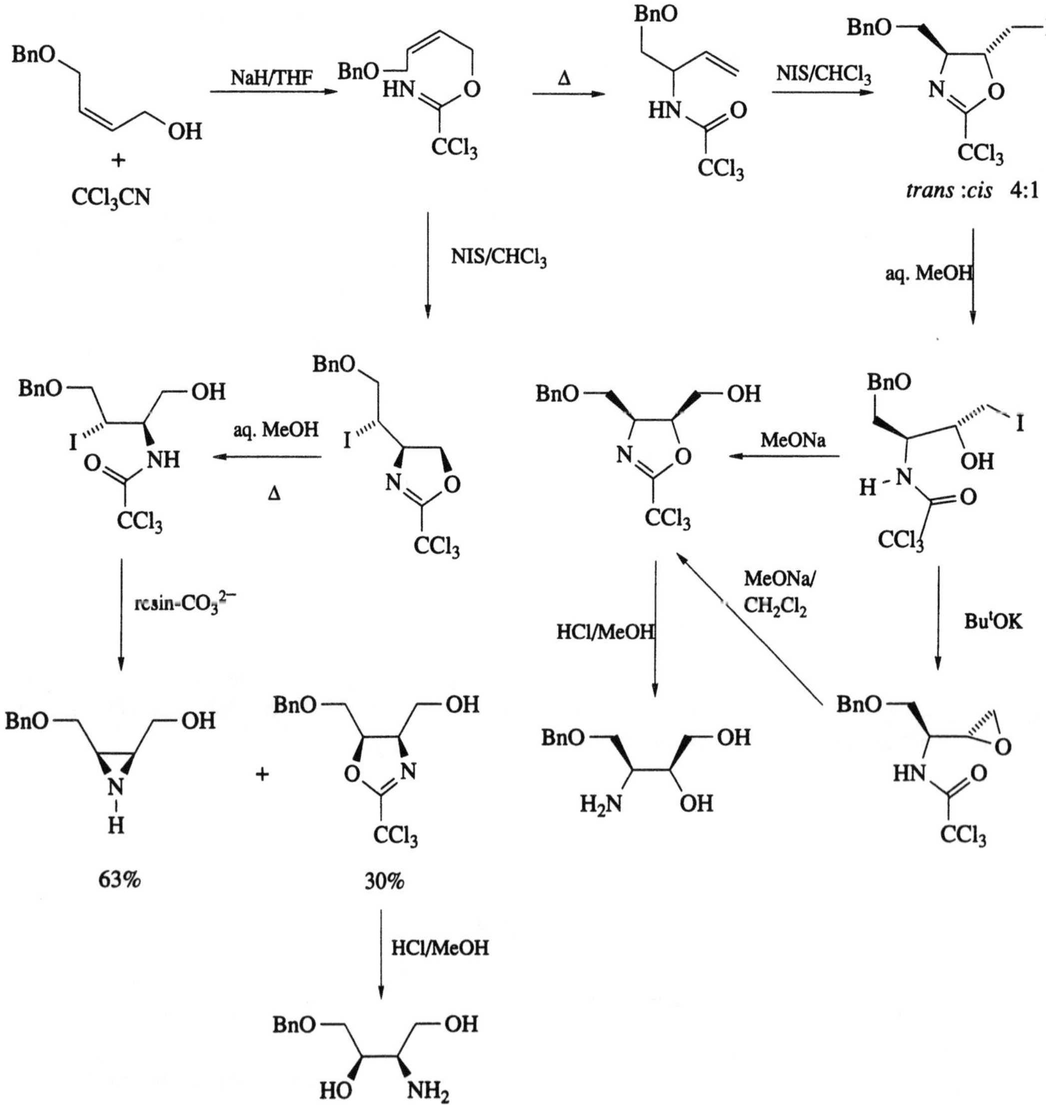

trans :*cis* 4:1

63%

30%

Scheme 63

amino alcohols. An elegant series of reactions, including a Cope rearrangement, is given in Scheme 63.[184] Note that imidates give clean oxazolidinones, whereas with amides there is a tendency for aziridine formation to compete. Other examples are given in refs.185–188.

Whilst ring opening of epoxides (Volume 6, Chapter 1.3) is really beyond the scope of this review, two recent papers are noteworthy: poorly nucleophilic amines can be reacted very cleanly as their diethylaluminum derivatives,[189] and a start has been made on chiral induction of opening of epoxides (*e.g.* cyclohexene oxides).[190] Amino alcohols have been resolved by enantioselective enzymatic hydrolysis of their acetates.[191] Ring opening of phthalimidoaziridines has been achieved with water, phenol and tosic acid, amongst other nucleophiles,[147] giving products of formal N—O addition to the double bond.

3.5.4.2 Additions of Oxygen and a Nitroso or Nitro Group to Alkenes

Nitrosyl hydrogen sulfate adds to alkenes at –40 °C in liquid SO_2, in Markovnikov fashion giving 2-sulfato-oximes (37–84%).[192] Reactions of nitrogen oxides with alkenes are usually complex, yet some have limited industrial, rather than laboratory, importance.[193,194] Nitryl tetrafluoroborate in acetic anhydride at –65 to –45 °C adds Markovnikov fashion to alkenes, with only fair stereoselectivity, giving β-nitroacetates in 36–40% yields.[195] Nitryl fluorosulfonate adds to perfluoroalkenes in fluorosulfonic acid/freon-113 giving 63–92% β-nitroperfluoroalkyl fluorosulfonates.[196]

3.5.5 ADDITION OF NITROGEN AND SULFUR, SELENIUM OR TELLURIUM TO ALKENES

3.5.5.1 Addition of Nitrogen and Sulfur

There are several related reactions involving probable formation of episulfonium ions from alkenes, with subsequent addition of various nitrogen nucleophiles giving products of net *trans*-1,2-(N + S) addition (ref.197 and references cited therein). In a recent example (Scheme 64) the sulfur reagent also provides the nitrogen nucleophile; yields were best with styrene.[197] Related cyclizations are known in the β-lactam area,[198] *e.g.* Scheme 65; compounds (**43**) and (**44**) are also available by related but base-catalyzed cyclizations.[199]

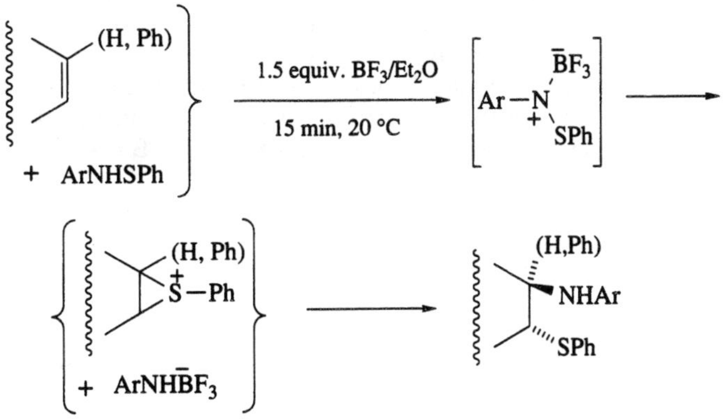

(H, Ph) + ArNHSPh

1.5 equiv. BF_3/Et_2O, 15 min, 20 °C

$\bar{B}F_3$, Ar—N⁺, SPh

(H, Ph), $\overset{+}{S}$—Ph + $ArNH\bar{B}F_3$

(H,Ph), NHAr, SPh

Ar = Ph, 4-ClC_6H_4, 3- or 4-$O_2NC_6H_4$, 4-MeC_6H_4, yields 25–88%;
Ar = 3- or 4-$MeOC_6H_4$, yields 4–9%

Scheme 64

Regiochemistry is generally Markovnikov in aminosulfenylations. However, *N*-phenylsulfenylpyrrolidine reacts with 1-octene ($TfOH/CH_2Cl_2$), to give products chiefly of anti-Markovnikov additions, probably for steric reasons.[200] Using dimethyl(methylthio)sulfonium fluoroborate as the source of MeS^+ permits a wide range of other nitrogen nucleophiles to be used (Scheme 66).[201] The authors draw attention to the versatility of these products, *e.g.* β-nitrosulfides are precursors of nitroalkenes, and oxazolines and *cis* amino alcohols are also available from *trans*-1-acetamido-2-methylthioalkenes.

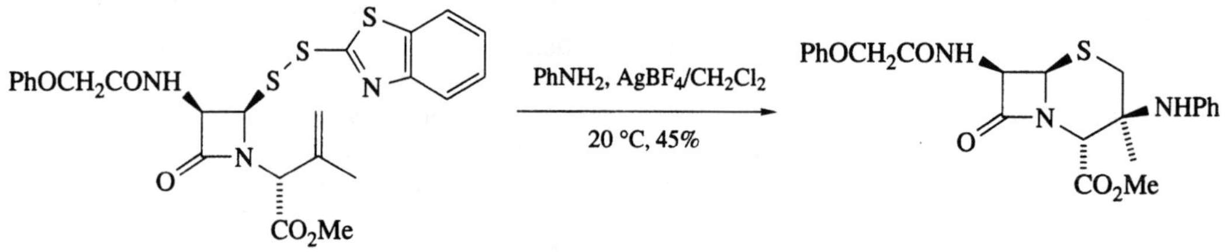
PhOCH2CONH
PhNH2, AgBF4/CH2Cl2
20 °C, 45%
NHPh
CO2Me

Scheme 65

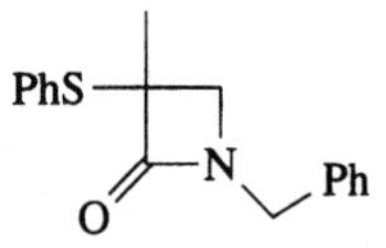

(43) 93%

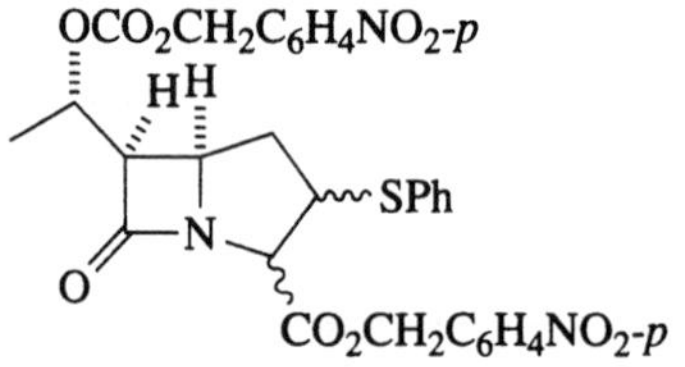

(44) 7%

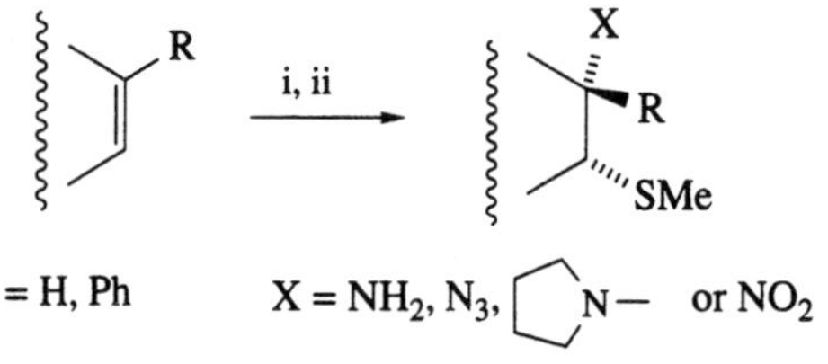
i, ii
R = H, Ph
X = NH2, N3, or NO2
SMe

54–98%

i, $Me_2\overset{+}{S}SMeB\overset{-}{F_4}/CH_2Cl_2$ or MeCN or $MeNO_2$, 0–20 °C, 1 h
ii, NH_4OH or NaN_3 or piperidine or $NaNO_2$, 1–4 d, 20 °C

Scheme 66

If the reactions are carried out in a nitrile as solvent, rather than dichloromethane, using triflic acid as catalyst, a modified Ritter reaction takes place, and the intermediate nitrilium ion traps the liberated amine, forming an amidine (Scheme 67).[200] In an earlier reaction (*cf.* Scheme 67) the lithium perchlorate catalyzed reaction of sulfenyl chlorides with alkenes in the presence of nitriles had also given 1-amido-2-sulfenyl adducts.[202] Ritter products are also obtained in good yields by anodic oxidation (Pt or C, 1.2–1.4 V) of disulfides in acetonitrile, in the presence of excess alkene, using Bu_4NBF_4 as supporting electrolyte (Scheme 68).[203]

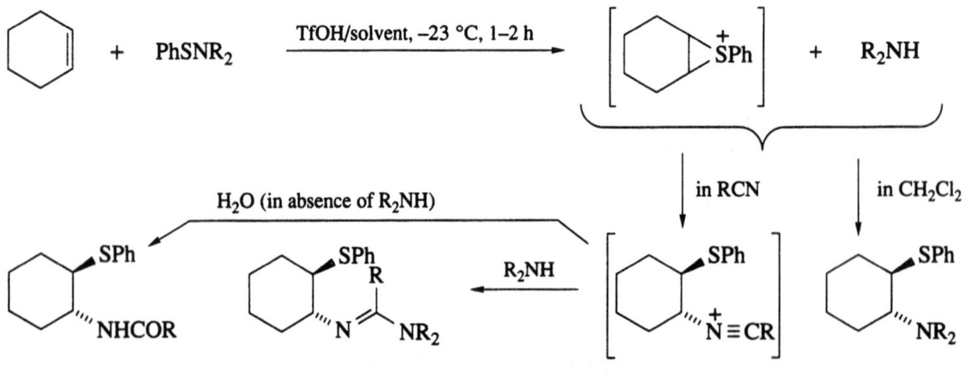
PhSNR2
TfOH/solvent, –23 °C, 1–2 h
SPh
R2NH
in RCN
in CH2Cl2
H2O (in absence of R2NH)
NHCOR
NR2
N≡CR

Scheme 67

Similar products can be obtained in a two-step sequence using lead(IV) or manganese(III) acetate in TFA/CH_2Cl_2, 0 °C, giving an intermediate *trans*-ester sulfide which, after a simple work-up, is treated, crude, with sulfuric acid in acetonitrile, giving the acetamido sulfide product. Both the intermediate ester

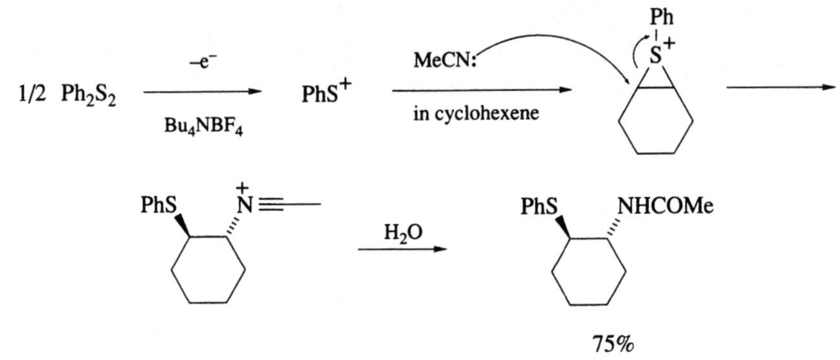

1/2 Ph_2S_2 —($-e^-$, Bu_4NBF_4)→ PhS^+ —(MeCN:, in cyclohexene)→ [episulfonium ion, PhS^+] → [PhS / $\overset{+}{N}$≡— cyclohexane] —(H_2O)→ PhS / NHCOMe cyclohexane 75%

Scheme 68

and the final amide have *trans* Markovnikov orientation, suggesting that each is formed *via* the episulfonium ion (Scheme 69). The amides could be hydrolyzed to amino sulfides (KOH/glycol, reflux 4 h) and, in products from dibenzyl disulfide, the benzyl group removed (Na/NH_3) to give acetamidothiols, sometimes with partial epimerization.[203] Unfortunately, the requisite benzylthio compounds were among the worst yields, a major by-product in these cases[204] and in the electrochemical alternative[203] being benzylacetamide.

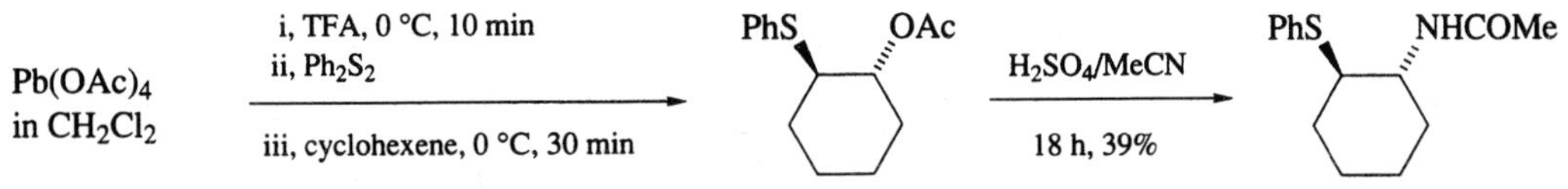

$Pb(OAc)_4$ in CH_2Cl_2 —(i, TFA, 0 °C, 10 min; ii, Ph_2S_2; iii, cyclohexene, 0 °C, 30 min)→ PhS / OAc cyclohexane —(H_2SO_4/MeCN, 18 h, 39%)→ PhS / NHCOMe cyclohexane

Scheme 69

Aziridines can be opened by thiols to give 2-amino sulfides.[146,147] The expected complete Walden inversion in this reaction has been confirmed recently, in the case of *cis*- and *trans*-stilbeneimine, whether the requisite prior activation of the aziridine is achieved by *N*-protonation or sulfonation or benzoylation.[205] Hexavalent sulfur can be introduced by yet another modified Ritter reaction (Scheme 70).[206]

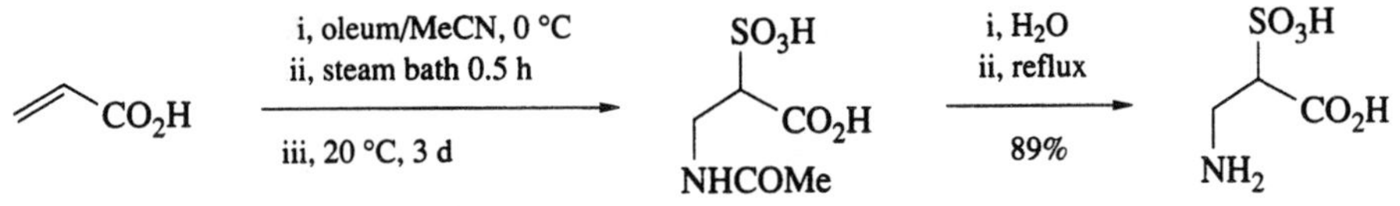

$CH_2{=}CHCO_2H$ —(i, oleum/MeCN, 0 °C; ii, steam bath 0.5 h; iii, 20 °C, 3 d)→ SO_3H / CO_2H / NHCOMe —(i, H_2O; ii, reflux; 89%)→ SO_3H / CO_2H / NH_2

Scheme 70

3.5.5.2 Addition of Nitrogen and Selenium

In a modified Ritter reaction, terminal and 1,2-disubstituted alkenes are amidoselenated by treating with equimolar amounts of PhSeCl in a liquid nitrile solvent containing 5 equiv. of water, at 76–90 °C, for 1 h.[207] Yields are 72–98% with cyclohexene, poor (36%) with styrene. The selenides can be oxidized with H_2O_2 to give selenoxides which spontaneously eliminate PhSeOH to give allylic amides, though the selenoxide from cyclohexene is stabilized by an internal hydrogen bond and requires heating. Alternatively, the selenium can be removed reductively (Ph_3SnH) to give amides (Scheme 71). Terminal alkenes give predominantly (≈7:1) the terminal selenides, which yield stable selenoxides. Internal alkenes give products with the expected stereochemistry: *trans* to *erythro*, *cis* to *threo*. This chemistry has recently been extended to previously unusable alkenes by use of 2,2′-dipyridyl diselenide.[208] Anodic oxidation of diphenyl diselenide in acetonitrile in the presence of an alkene[209] gives acetamido selenides, in a procedure related to the acetamidosulfenylation of Scheme 68.[203]

β-Phenylselenocarbamates (20–95%) are obtained from alkenes with PhSeCl and carbamates with $AgBF_4$ in CH_2Cl_2 (4 h, 25 °C under Ar).[210] The intramolecular form of this reaction had been known previously;[211] indeed, selenocyclizations have been pursued rather more successfully than the related sulfur reactions. The reaction, introduced by Clive *et al.*,[211] has been developed by Toshimitsu's group, whose recent papers give many lead references;[212,213] typical products of selenocyclization include lactams

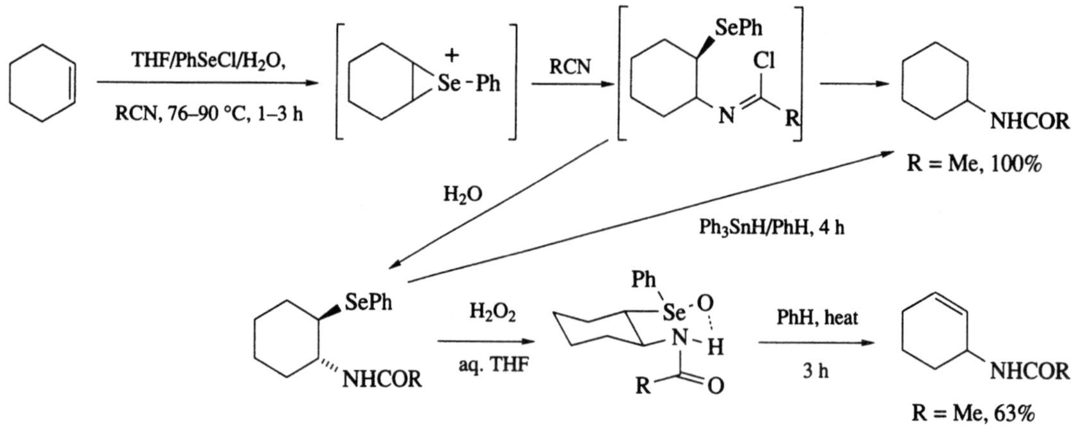

Scheme 71

(47)–(50) and imidates (53) and (54). The reaction with amides can be very sensitive to ring size and substitution patterns: whereas compounds (45) and (46) give lactams, the closely related (51)[213] and also (52)[212] give cyclic imidates (Scheme 72). In some cases the problem is overcome by use of an imidate starting material.[213]

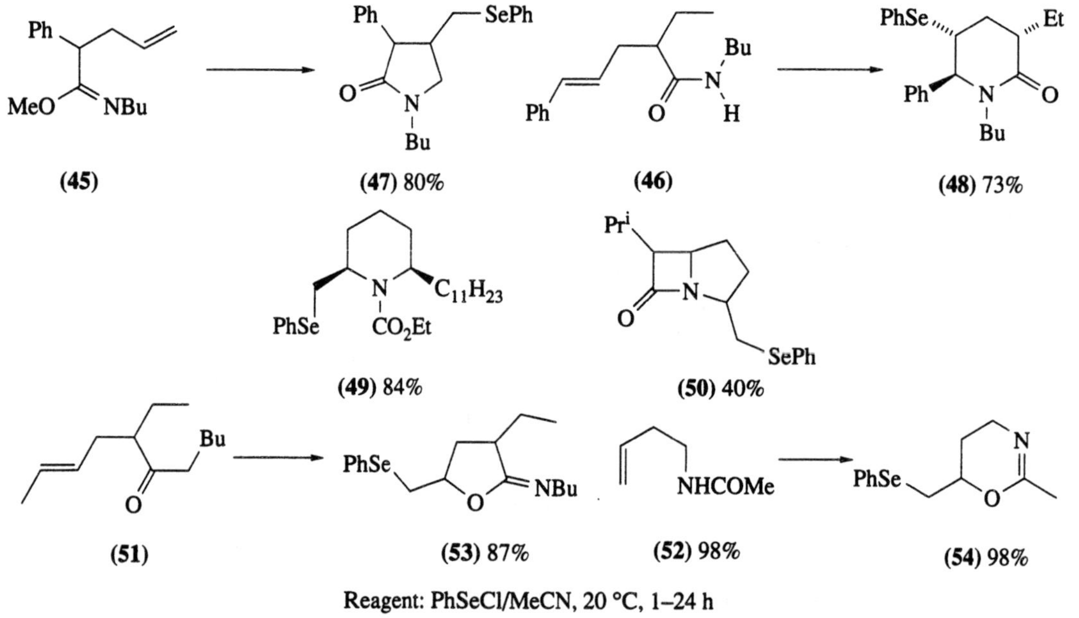

Reagent: PhSeCl/MeCN, 20 °C, 1–24 h

Scheme 72

Mono- and 1,2-di-substituted alkenes react with PhSeCl/Hg(SCN)$_2$ in benzene (0.5–96 h, at 20 °C), giving β-*trans*-phenylselenoalkyl isothiocyanates in 70–94% yields.[214] Terminal alkenes generally give the product with the selenium terminal (an exception is the product from Bu^{t}CH═CH$_2$); internal alkenes show the expected stereochemistry (*cis* to *threo, trans* to *erythro*). Oxidation to selenoxides could be achieved cleanly only with ozone, and the products *cis* eliminate in the usual manner to give predominantly the vinylic isothiocyanates (Scheme 73).

A similar two-step azidoselenation *via* (i) MeSeBr or PhSeBr, (ii) NaN$_3$/CF$_3$CH$_2$OH, and subsequent elimination (iii) (O$_3$) of selenium has been reported; unfortunately, selectivities in the second and third steps are poor.[215] Phenylselenyl azide (made *in situ* from PhSeCl/NaN$_3$/DMSO, room temperature) adds to alkenes (20 °C, overnight, 86–98%) stereospecifically;[216] regiospecificity is poor with terminal alkenes but good with highly polarized alkenes, typical products being (55) to (57); cyclohexadiene gives (58).

β-Nitroselenation of alkenes has been achieved using (i) PhSeBr, (ii) AgNO$_2$/HgCl$_2$ in MeCN/THF, –78 to +22 °C under argon,[217] and the products oxidized with hydrogen peroxide to give vinylic nitro

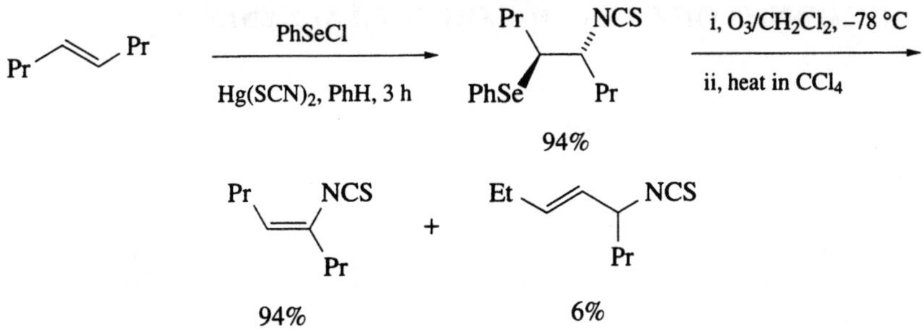

Pr–CH=CH–Pr → (PhSeCl; $Hg(SCN)_2$, PhH, 3 h) → PhSe/Pr/NCS/Pr adduct, 94% → (i, O_3/CH_2Cl_2, –78 °C; ii, heat in CCl_4) → Pr/NCS/Pr alkene, 94% + Et/NCS/Pr, 6%

Scheme 73

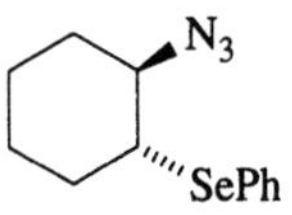

(55) 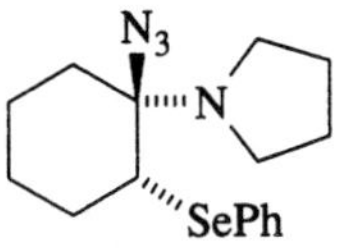 (56) (57) $CH(SePh)(CO_2Et)_2$ / $CH(N_3)OEt$ (58)

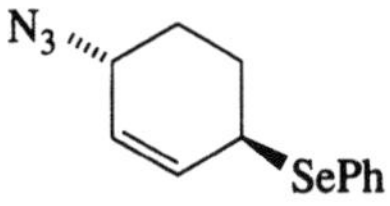

compounds in good yield. The reaction applied to 1-hexene again has only moderate regioselectivity, giving 78% (**59**) and 22% of the anti-Markovnikov product. However, alkenylsilanes react regiospecifically, giving (**60**), convertible in high yield on oxidation to (**61**).[218] Vinylsilanes (*i.e.* R = H) give (**62**), which are useful Diels–Alder dienophiles.[218]

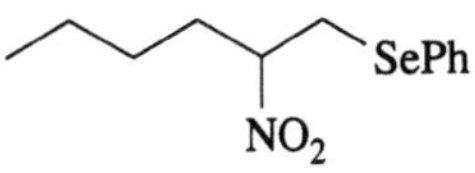

(59) 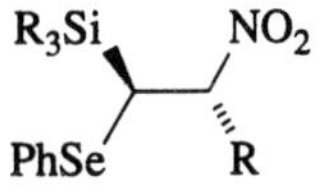 (60) 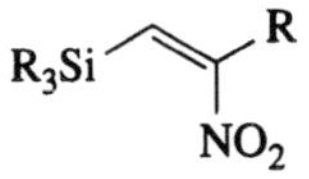 (61) 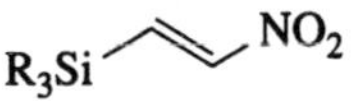 (62)

3.5.5.3 Addition of Nitrogen and Tellurium

Benzenetellurinyl acetate (or trifluoroacetate) reacts with alkenes and carbamates under BF_3 catalysis to give *trans*, predominantly Markovnikov, adducts, conveniently worked up *via* hydrazine reduction to give tellurides (Scheme 74).[219] The *trans* nature of (**63**) was confirmed by independent synthesis from PhTeNa and the appropriate aziridine. Alkenic carbamates undergo a much faster intramolecular telluro-lactamization (Scheme 75); yields are 49–97% over 15 varied examples.

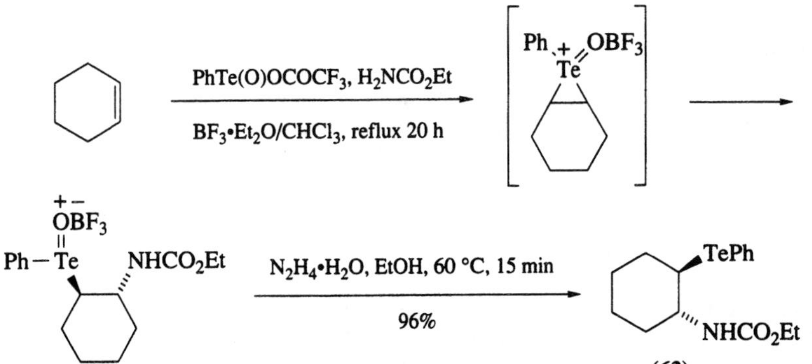

Cyclohexene → ($PhTe(O)OCOCF_3$, H_2NCO_2Et; BF_3•$Et_2O/CHCl_3$, reflux 20 h) → [Ph–Te^+(=OBF_3) bridged intermediate] → Ph–Te(=$\overset{+-}{OBF_3}$)/$NHCO_2Et$ cyclohexane → (N_2H_4•H_2O, EtOH, 60 °C, 15 min; 96%) → TePh/$NHCO_2Et$ cyclohexane (63)

Scheme 74

$NHCO_2Et$ alkene → (i, $PhTe(O)OCOCF_3$, BF_3•$Et_2O/CHCl_3$, reflux 30 min; ii, N_2H_4 (as Scheme 74)) → N-CO_2Et pyrrolidine–CH_2TePh, 96%

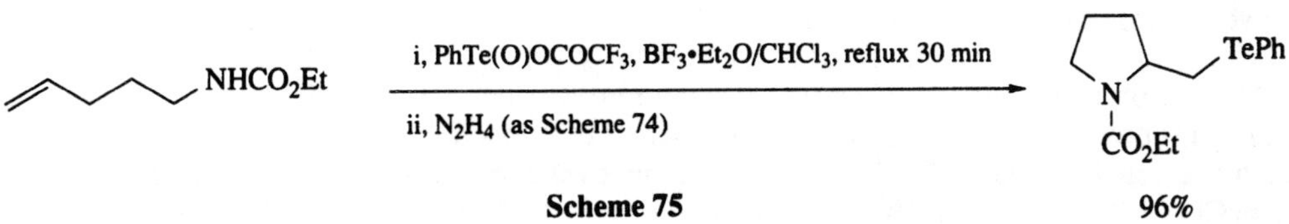

Scheme 75

3.5.6 ADDITION OF NITROGEN AND HALOGEN TO ALKENES

3.5.6.1 Addition of Nitrogen and Fluorine

There are few reports of these reactions. Anodic oxidation of styrene derivatives $PhCHR^1CHR^2$ in $HF/Et_4NF/MeCN$ gives 20–45% yields of fluorinated acetamides $PhC(NHAc)R^1CFR^2$ (R^1 = H, Me; R^2 = H, Me, Ph), generally with substantial quantities of by-products, chiefly isomers and 1,2-difluoro compounds.[220] Allylbenzene gives $PhCH(NHAc)CHFCH_2F$ (20%). Regioselectivity is high, but stereoselectivity low. 1-Fluoro-2-aminoalkanes are also obtained by the reaction of aziridines with pyridine–HF/benzene, generally at room temperature.[221,222] Yields are high, but again regio-, but not stereo-specific. 2-Phenylaziridine gives $PhCHFCH_2NH_2$ (78%) consistent with an S_N1 mechanism. Terminal alkenes react with $Tl(OAc)_3/TMSN_3$ to give thallated aziridines which, on treatment with 40% HF, give fluoro azides, sometimes with rearrangement and often with concomitant formation of hydroxy azides (Scheme 76).[223] Iodide or thiocyanate give the opposite regiochemistry $[ArCH(N_3)CH(I,SCN)]$; chloride or bromide give regioisomeric mixtures. Industrially satisfactory fluoronitrations have been reported, including a 93% conversion of vinylidene fluoride into $CF_3CH_2NO_2$ with HF and HNO_3 in fluorosulfonic acid at 0 °C.[224]

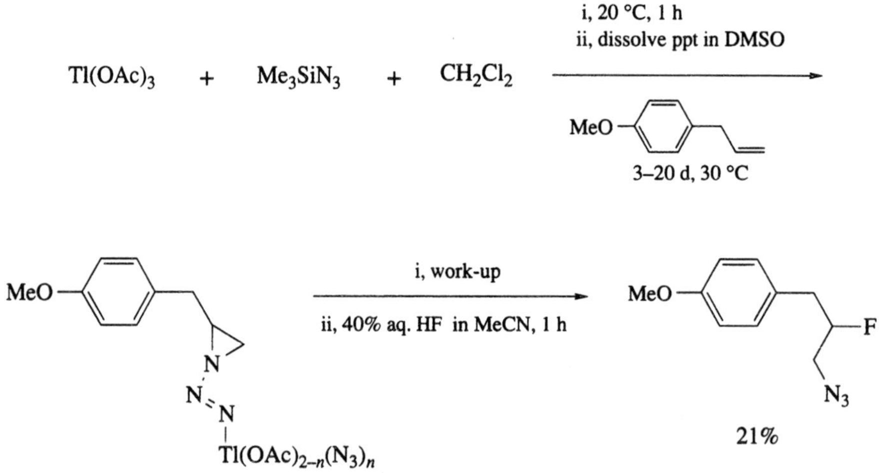

$Tl(OAc)_3$ + Me_3SiN_3 + CH_2Cl_2 —(i, 20 °C, 1 h; ii, dissolve ppt in DMSO; MeO-C6H4-CH2CH=CH2, 3–20 d, 30 °C)→ aziridine (MeO-C6H4-CH2-aziridine, N–N=N–$Tl(OAc)_{2-n}(N_3)_n$) —(i, work-up; ii, 40% aq. HF in MeCN, 1 h)→ MeO-C6H4-CH2CHF-CH2N3 (F, N_3) 21%

Scheme 76

3.5.6.2 Addition of Nitrogen and Chlorine

Surprisingly few heterolytic additions of nitrogen and chlorine to alkenes have been reported, and they are not (yet) synthetically useful. Chloramine-T in acetic acid, 50–60 °C, 2–3 h, gives low (6–25%) yields of adducts with alkenes[225] with regio- and stereo-chemistry consistent with a cyclic chloronium ion intermediate.[226] Chloramine and its alkyl derivatives react with alkenes in CH_2Cl_2/Et_2O at –50 to –10 °C in the presence of $AlCl_3$, to give chloroamines non-regiospecifically.[227] Alkenes add $ClN(SO_2F)_2$ to give products consistent with formation from positive chlorine.[228]

Radical chloroaminations are known, using radical, transition metal ion or photochemical initiation. They also occur without overt initiation, thus anti-Markovnikov additions to terminal alkenes occur with *N,N*-dichlorourethane in benzene at 5–40 °C (yields ≈ 60%).[229] Similar reactions occur with *N,N*-dichloroarenesulfonamides in CH_2Cl_2 at or below room temperature (yields mostly 53–91%; 10% with isobutylene).[230a] The remaining N—Cl bond is readily reduced if desired with sodium sulfite. *N*-Halosulfoximines also add to alkenes thermally or photolytically.[230b]

Free radical additions to alkenes in the presence of redox systems, especially Fe^{II}/Fe^{III}, have been much investigated and reviewed by Minisci.[151] The redox approach may have many beneficial effects, *e.g.* in improving yields: the reaction of *N*-chloroamines with Fe^{II} is faster than the competing electrophilic chlorination of the alkene; it favors 1,2-addition over polymerizations; it increases the scope of both adding groups, which can be provided by different reagents if necessary. The sequences of most relevance are given in Scheme 77, which may be an oversimplification (Minisci favors a complexed species $FeCl^{2+}(R_2N^+HCl)$ as the chlorinating agent in step 4). The reactions of Scheme 77 when applied to *N*-chloropiperidine and cyclohexene give a mixture of *cis* and *trans* adducts. However, an analogous sequence[151,231] can be achieved in a nonacidic medium also, providing in this case predominantly the *cis* isomer, ascribed to an intramolecular delivery of the chlorine atom (Scheme 78).

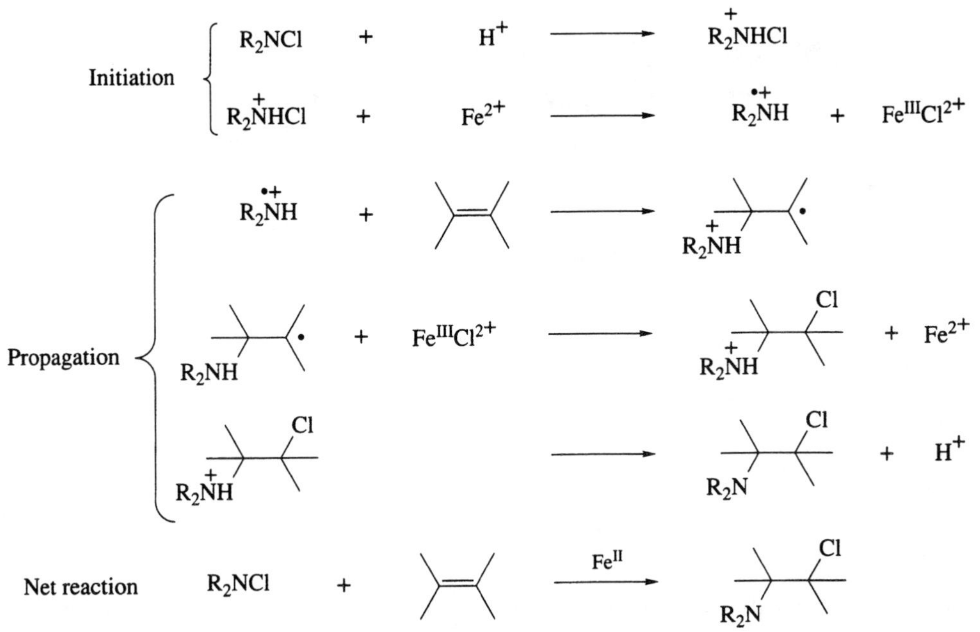

Initiation

$R_2NCl + H^+ \longrightarrow R_2\overset{+}{N}HCl$

$R_2\overset{+}{N}HCl + Fe^{2+} \longrightarrow R_2\overset{\bullet+}{N}H + Fe^{III}Cl^{2+}$

Propagation

$R_2\overset{\bullet+}{N}H$ + alkene ⟶ $R_2\overset{+}{N}H$-substituted radical

radical + $Fe^{III}Cl^{2+}$ ⟶ $R_2\overset{+}{N}H$, Cl adduct + Fe^{2+}

$R_2\overset{+}{N}H$, Cl adduct ⟶ R_2N, Cl adduct + H^+

Net reaction

R_2NCl + alkene $\xrightarrow{Fe^{II}}$ R_2N, Cl adduct

Scheme 77

N–Cl $\xrightarrow{Fe^{2+}}$ $\overset{\bullet}{N}$ → $FeCl^{2+}$ ⟶ (cyclohexene)

N, Cl–Fe^{2+} ⟶ 45% (+22% *trans* isomer) ⟶ N, Cl

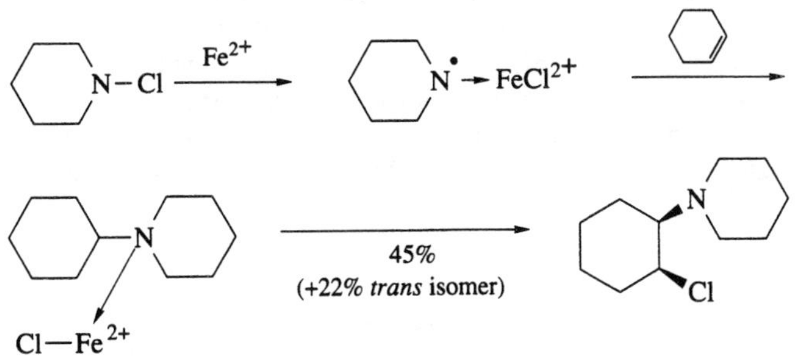

Scheme 78

This last reaction, and related sequences using $TiCl_3$ and $CrCl_2$ have been studied comparatively.[231] The overall yields are 75% with titanium, 14% with chromium. However, the chromium-based procedure can be applied to *N*-haloamides and to urethanes, when the yields are excellent, producing also a higher proportion of *cis* isomers from cyclohexenes. Furthermore, the amidic *N*-substituents are also more versatile. An example, illustrating also the anti-Markovnikov regiochemistry, is given in Scheme 79.[232] Such *cis*-haloamines are relatively inaccessible by other routes, most of which give *trans* products arising from cyclic intermediates (*e.g.* chloronium ions) or starting materials (aziridines), or from neighbouring group effects on attempted inversions (*e.g.* transient aziridinium ions). Intramolecular cyclizations have also been achieved with several different[151,231] metals, a recent example being given in Scheme 80.[233]

Additions of *N*-haloamides to alkenes can also be achieved photolytically, generally in CH_2Cl_2 (ref. 234 and references cited therein). Such additions are anti-Markovnikov. Both *cis* and *trans* internal alkenes give about 2:1 *threo* to *erythro* products. Yields vary widely but can be preparatively useful. Cyclohexene is substituted with predominantly *cis* orientation (2:1); this ratio is reversed in MeOH due to promotion of the competing electrophilic addition. Certain haloamides of electronegatively substituted acids (*e.g.* CCl_3CONCl_2) can give very high *cis* stereoselectivity. It would appear that as the electronegativity of the RCONH group is increased to approach that of halogen, homolytic cleavage is favored over heterolytic, thus decreasing the proportion of *trans* isomer in the product. In the case of *N,N*-dihalosulfonamides also the mechanism and hence the regiochemistry of the product is finely balanced, dibromo

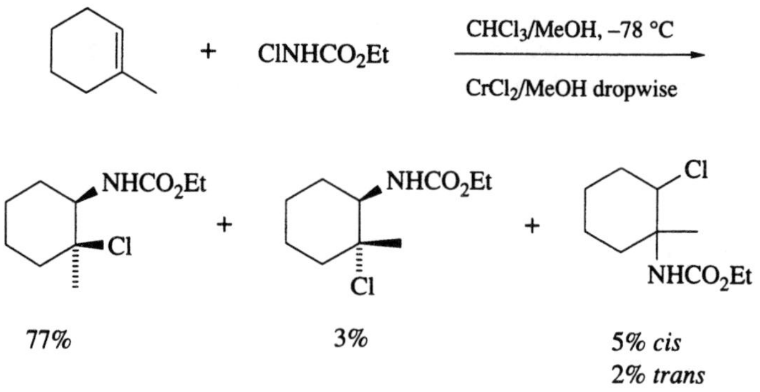

Scheme 79

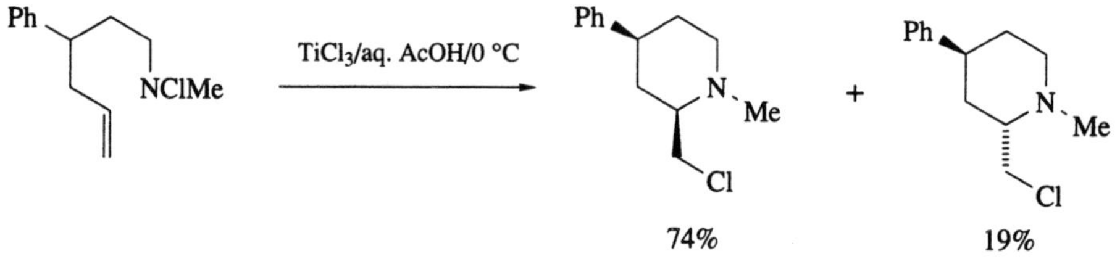

Scheme 80

reacting anti-Markovnikov (radical) and dichloro Markovnikov (ionic), and can switch between radical and ionic mechanisms according to catalyst and conditions.[235,236] A similar switching of mechanism occurs with chlorine azide[237] and with bromine azide additions (Section 3.5.6.3).

Nitrosyl halides add to alkenes: references are scattered through the literature back to 1875 (ref. 194 and references cited therein). The adducts vary enormously in their stability, but when their structures allow they, like nonhalogenated nitroso compounds, isomerize to oximes or dimerize. The orientation of the reaction is consistent with an electrophilic mechanism, in which the reagent is polarized as NO^+Hal^-. Bicyclic substrates and reaction media of low polarity favor *syn* addition, suggesting a four-center transition state (Scheme 81). Aziridine synthesis *via* NOCl/alkene adducts is discussed in Section 3.5.2.1.

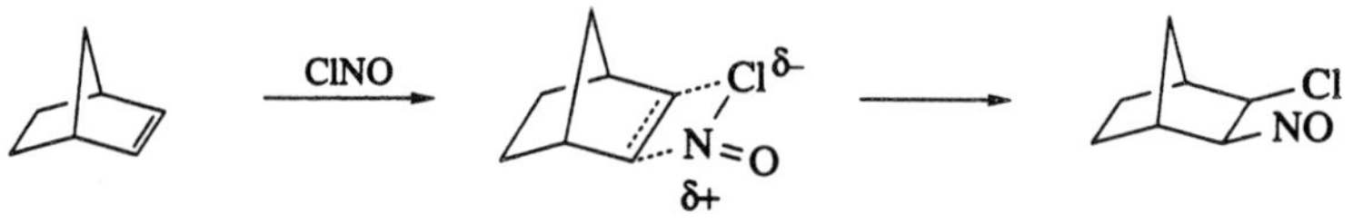

Scheme 81

With excess nitrosyl chloride, a chloronitro product is obtained in certain cases, a wide variety of steroid-5-enes giving 5α-chloro-6β-nitro derivatives in good yield (CH_2Cl_2/CCl_4, –60 to 0 °C, 2–24 h).[238] Nitryl chloride adds to terminal alkenes (56–80%) and to acrylic acid derivatives (refs. 239, 240 and references cited therein) at temperatures close to ambient. The reaction appears to be a radical one, the NO_2 entering the terminal position whatever the electronic requirement of the alkene.

3.5.6.3 Addition of Nitrogen and Bromine

There are several reports of homolytic addition of NBS[241] or *N*-bromophthalimide[242] to alkenes. Yields are moderate. The reaction can be promoted by conversion to a heterolytic mechanism with BF_3 catalysis: concurrent formation of the BrF adduct is of interest also (Scheme 82).[243]

Diethyl *N,N*-dibromophosphoramidate undergoes ready Markovnikov addition to styrene or cyclohexene (92%) in CCl_4 at –20 to +20 °C with $BF_3·Et_2O$ catalysis.[133,244] Alternatively, such reactions can be carried out in refluxing CH_2Cl_2 in the absence of catalyst, when anti-Markovnikov products (47–97%) are formed.[245] All these compounds are versatile intermediates (Scheme 83). A few *N*-bromoperfluoroamines have been added to alkenes, thermally or photolytically.[246]

Bromine azide reacts with alkenes by ionic or radical mechanisms according to conditions, adding to styrene in $MeNO_2/CH_2Cl_2$ to give $PhCH(N_3)CH_2Br$ (95%) or in $CHCl_3$ purged with N_2 (to remove O_2, a

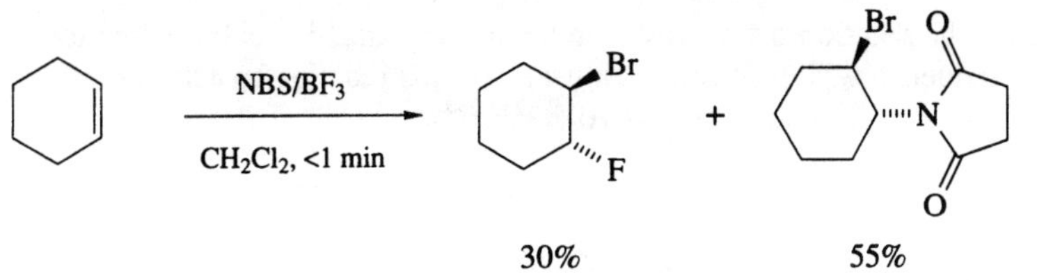

Scheme 82

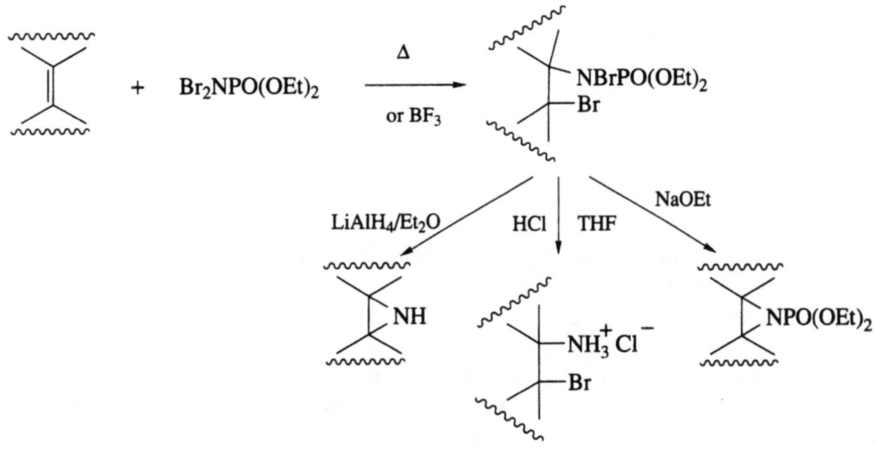

Scheme 83

radical inhibitor), $PhCHBrCH_2N_3$ (100%).[247] The two 2-butenes react stereospecifically in $MeNO_2/CH_2Cl_2$, *trans* to *erythro, cis* to *threo.*[248] The ionic reaction with deuterated styrene is regio- but not stereo-specific, indicating involvement of a benzylic cation.[248] If the reactions are done in acetonitrile, the initial product adds solvent, thus cyclohexene gives *trans*-1-(2-bromocyclohexyl)-2-methyltetrazole.[249]

Alkenes can be nitromercurated ($HgCl_2$/2 $NaNO_2/H_2O$, 30 h, 25 °C) and the products can be bromodemercurated, giving bromonitro compounds. An alternative work-up with base provides a valuable nitroalkene synthesis (Scheme 84).[250]

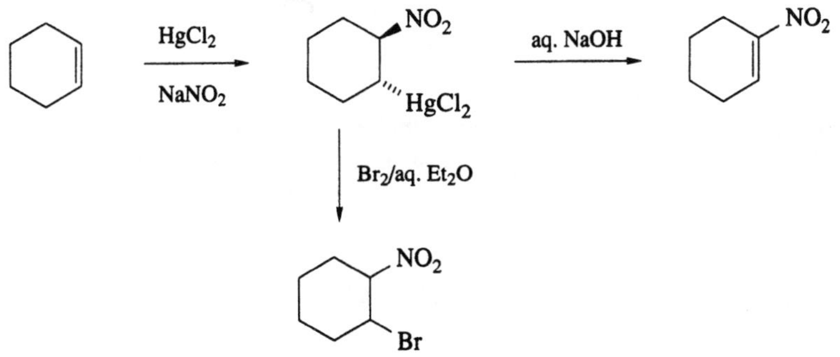

Scheme 84

3.5.6.4 Addition of Nitrogen and Iodine

Iodine isocyanate, preformed or made *in situ* from AgNCO and I_2, adds to alkenes[31–33,251–255] with the regio- and stereo-chemistry expected of reactions proceeding *via* cyclic iodonium ions.[251] When the INCO is made *in situ*, a competing mechanism also occurs (except with the most reactive alkenes) in which the alkene complexes with the iodine, and the complex then reacts with the isocyanate ion to generate the same β-iodoisocyanate as obtained from INCO direct.[34] The reaction can be carried out at –35 to +20 °C in Et_2O, CH_2Cl_2, THF, pentane or excess alkene as solvent. Dichloromethane or ether are the solvents of choice: the reaction occurs faster in the more polar MeCN but with more side reactions.[253] Scheme 85 shows a sequence which illustrates a usage of this reaction to achieve an overall *cis* substitution (compare the *trans* substitution in Scheme 10).[33,253,254]

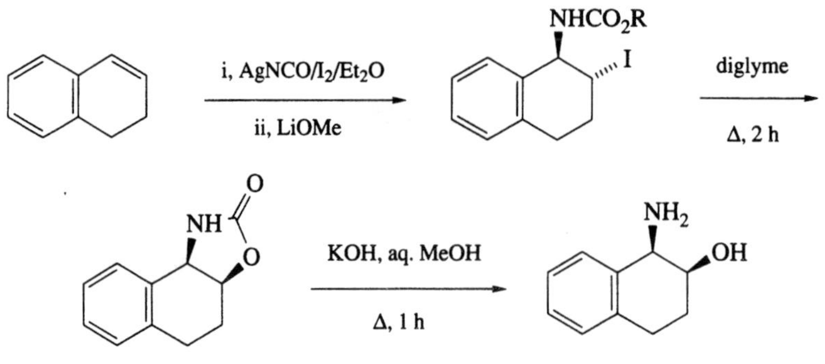

i, $AgNCO/I_2/Et_2O$; ii, LiOMe → $NHCO_2R$, I; diglyme, Δ, 2 h → NH, O, O; KOH, aq. MeOH, Δ, 1 h → NH_2, OH

Scheme 85

Iodine azide, generally made *in situ* from ICl and NaN_3 in MeCN at 0 °C, adds readily to alkenes by a similar heterolytic mechanism to INCO. Whereas the *trans* stereochemistry is generally well established,[256,257] the regiochemistry of the adduct with 1-phenylcyclohexene[28] has been queried recently;[258] it was originally formulated as the 1-azido-2-iodo compound (Scheme 86), but base treatment was subsequently shown to yield what appeared to be 6-azido-1-phenylcyclohexene,[258] which would have arisen from the 1-iodo-2-azido isomer. However, it has very recently been shown by 300 MHz NMR that the elimination product is in fact 3-azido-1-phenylcyclohexene,[259] derived ultimately (Scheme 86) from the originally proposed 1-azido-2-iodo- structure.

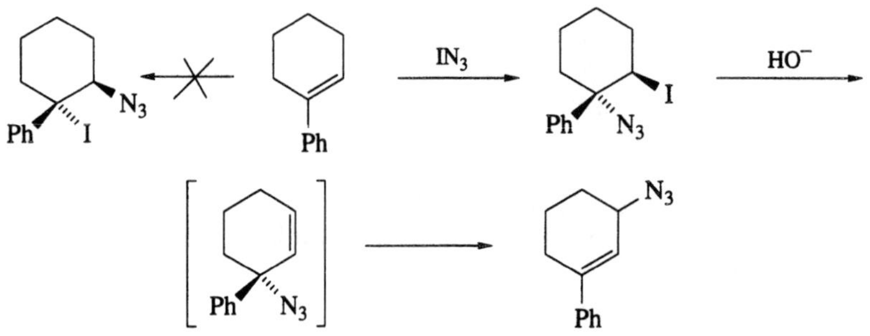

Ph, I, N_3 ← ✕ Ph; IN_3 → Ph, N_3, I; HO^- → [Ph, N_3] → N_3, Ph

Scheme 86

There is one known exception to the *trans* nature of the addition: a hindered polycyclic alkene where a *cis* addition occurs.[260] More recently, small proportions of *cis* adducts have been detected as minor components in additions to cyclohexenes.[261] An interesting recent usage of the IN_3–alkene reaction is in the azirine synthesis of Scheme 87.[262]

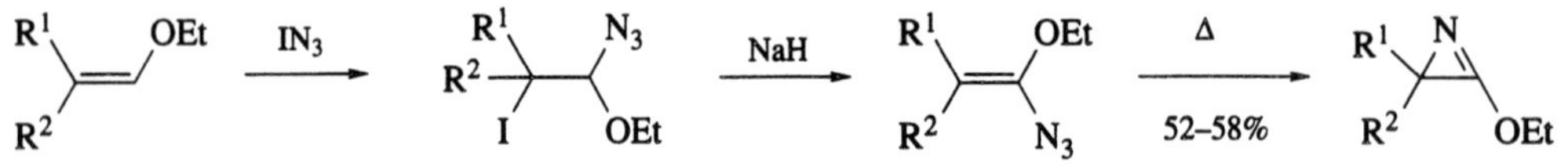

R^1, R^2, OEt; IN_3 → R^1, R^2, N_3, I, OEt; NaH → R^1, R^2, OEt, N_3; Δ, 52–58% → R^1, R^2, N, OEt

Scheme 87

Nitryl iodide, from iodine with silver nitrite[263,264] or from NO_2 and iodine,[265,266] adds to alkenes, generally in ether at room temperature, by a radical mechanism. Yields are 50–90%. The products yield nitroalkenes on treatment with base, a sequence used recently in the preparation of intermediates ($ArCH{=}CMeNO_2$) for amphetamine analogs.[264] The reaction is also a key step in the α-methylenebutyrolactone synthesis shown in Scheme 88.[266,267]

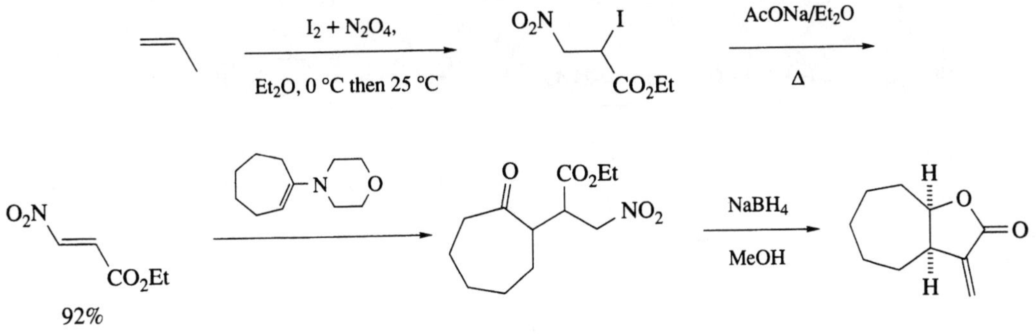

Scheme 88

3.5.6.5 Iodolactamizations and Related Reactions

Early attempts to extend the halolactonization procedure to yield lactams gave cyclic imidates instead, but several approaches favor lactam products. These include: working with silyl imidates,[268] imidate esters[269] or oxazolines,[270] using sulfonylcarbamates[271] or other acidic amides;[272,273] or by using hydroxylamine derivatives with increased nucleophilicity due to the α-effect.[274] Lactams can also be favored as a consequence of steric requirements.[275] In a few cases, amines can be cyclized to cyclic amines: many lead references are given in a recent report on cyclic hydroxylamines such as (**64**).[276] Very recent work has provided a fairly general iodolactamization procedure from unsaturated amides, trimethylsilyl triflate and iodine (Scheme 89).[268]

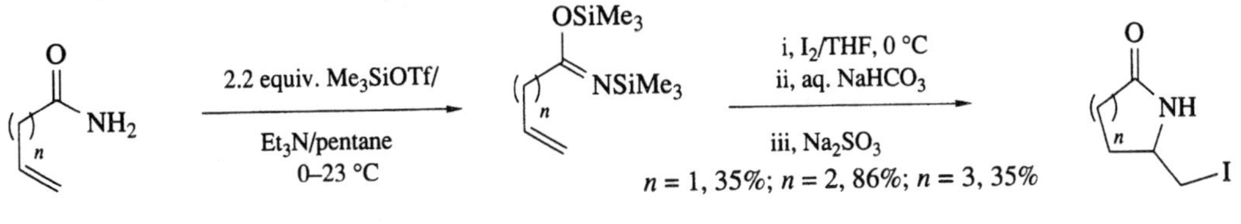

Scheme 89

These products have a varied and useful chemistry, for example, the stereochemistry of substitution with azide can be manipulated by optional formation of an intermediate aziridine (Scheme 90).[268]

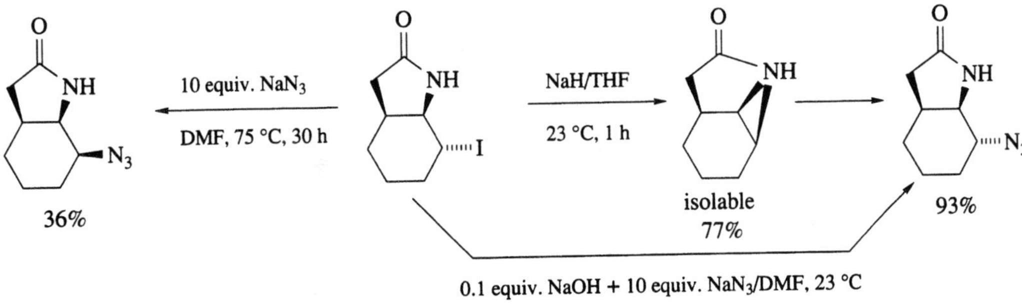

Scheme 90

Alkenic oxazolines[270] undergo kinetically controlled[226] iodolactamizations in 52–91% yields (I_2 and $NaHCO_3$, aq. THF, 0–25 °C, 12 h). The extent and direction of steric induction vary widely and are not fully understood (Scheme 91). A 1989 paper contains many useful references.[270]

Other compounds prepared by halolactamization of alkenes (and preceding the above work[268]) include (**64**),[276] (**65**),[269] (**66**),[271] (**67**),[273] (**68**),[274] (**69**)[275] and (**70**).[277] Note that related sulfeno-, seleno- and telluro-cyclizations have been discussed earlier (Section 3.5.5). Finally, related products may be obtained by radical reactions of *N*-haloamides, giving the following (**71–73**) representative products in reactions promoted thermally (**71**),[278] photochemically (**72** and **73**),[279,280] with dibenzoyl peroxide (**73**)[280] or chromous chloride (**73**).[281]

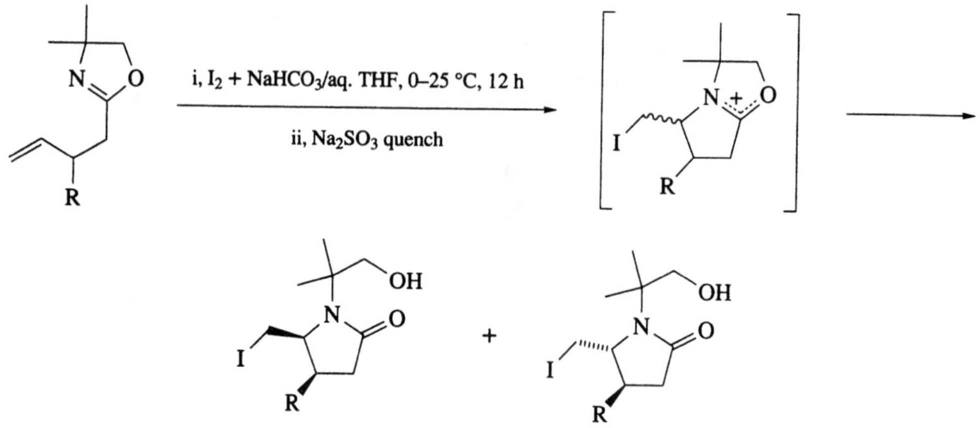

i, I_2 + $NaHCO_3$/aq. THF, 0–25 °C, 12 h

ii, Na_2SO_3 quench

R	*cis*:*trans*
Me	2:1
Bu[t]	1:10

Scheme 91

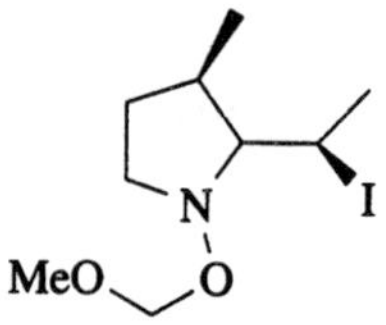

(64) from *cis*-alkene

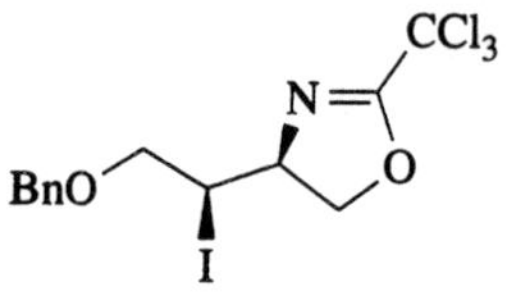

(65)

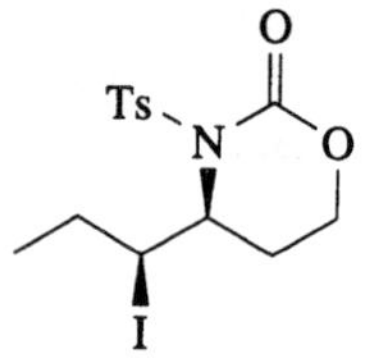

(66)

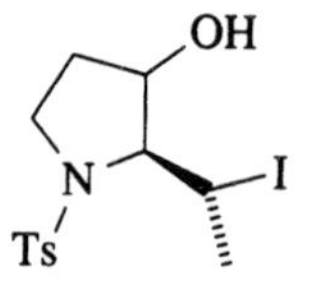

(67) from a *trans*-allylic alcohol

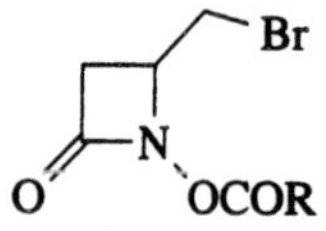

(68)

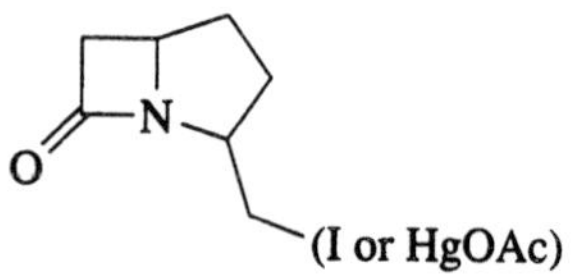

(69) (I or HgOAc)

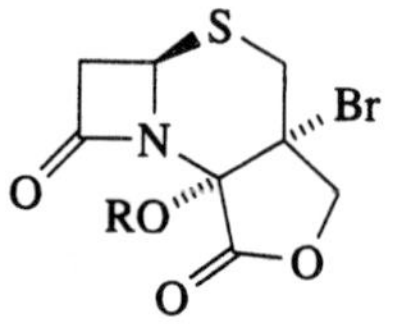

(70)

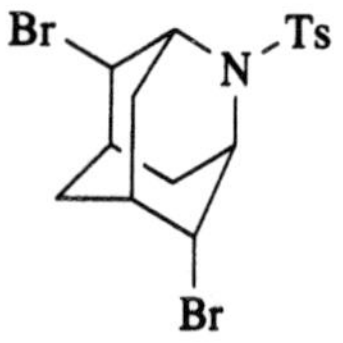

(71) 17%

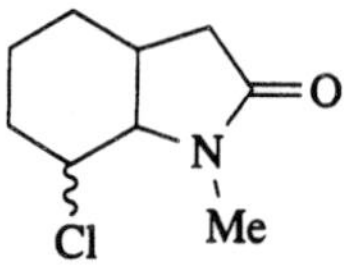

(72) 66%

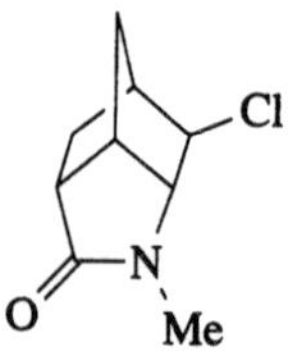

(73)

Induction	*Yield (%)*
hν	70
Bz_2O_2	92
$CrCl_2$	80

3.5.7 ADDITION TO POLYENES

3.5.7.1 Addition to Dienes

The palladium-promoted diamination of alkenes has been modified to yield *cis*-1,4-diamino-2-enes from dienes (Scheme 92).[282]

Lead(IV) azide yields 1,4-diazides from 1,3-dienes.[60,154] Whilst cycloadditions are beyond the scope of this review, the sequence of Scheme 93 is potentially important: 1,3-dienes undergo asymmetric Diels–Alder reactions with an α-chloronitroso derivative of epiandrosterone with high *ee*, and the N—O

i, $(PhCN)_2PdCl_2$/THF, –40 °C; ii, Me_2NH, –40 °C; $PdClNHMe_2$; NMe$_2$; Ph_3P or $AgBF_4$, –40 °C to reflux; Me_2N, NMe$_2$

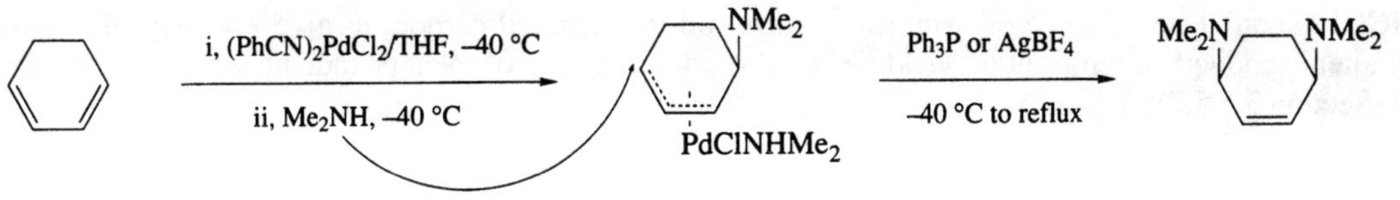

Scheme 92

bond of the adduct can be cleaved to yield a resolved 1-amino-4-hydroxy-2-alkene derivative.[283] In the example shown,[283] the configuration was established as (1*R*,4*S*) by degradation to an L-glutamic acid derivative.[284]

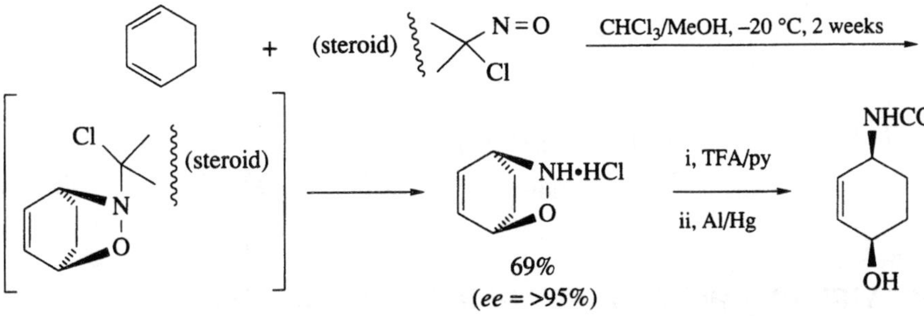

(steroid) N=O, Cl; $CHCl_3$/MeOH, –20 °C, 2 weeks; NH•HCl; i, TFA/py; ii, Al/Hg; $NHCOCF_3$; OH; 69% (*ee* = >95%)

Scheme 93

1,3-Butadiene, in 0.1% water in MeCN, treated with HBF_4 and $I(py)_2BF_4$ in CH_2Cl_2 at –5 °C, 15 min, undergoes a vinylogous Ritter reaction to yield *trans*-$ICH_2CH{=}CHCH_2NHCOMe$ (72%).[285] Conjugated dienes react with arylsulfonyl azides to give enamines rather than either sulfimides or aziridines.[286] 1,3-Dienes react with trifluoroacetyl nitrate in refluxing CH_2Cl_2 and trace HBF_4 to give a mixture of 1,2- and 1,4-nitrotrifluoroacetates; base and NaOAc or NaH then gives 1-nitro-1,3-dienes (35–89%).[287] Benzeneselenenyl iodide (Ph_2Se_2, I_2) adds to dienes in acetonitrile (18 h, 20 °C); the solvent participates and a cyclization, sometimes transannular, takes place, giving acetamido selenides in good yield but moderate stereochemical purity. The precise product depends on the stereochemical opportunity, *e.g.* **(74)** gives **(75)**; **(76)** gives **(77)**.[288]

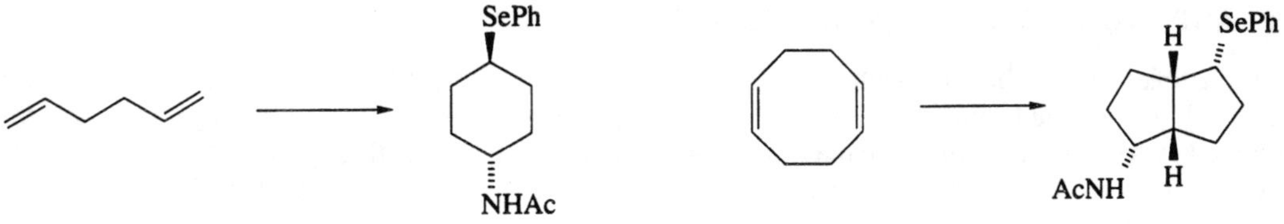

SePh; NHAc; H; SePh; AcNH; H

(74) **(75)** **(76)** **(77)**

Halogenation of **(78)** in excess of tertiary amine solvents yields quaternary adamantane salts **(79)**.[289] Nitryl iodide (N_2O_4, I_2) in ether at –15 °C adds to isoprene and to chloroprene to give good yield of $O_2NCH_2CX{=}CHCH_2I$ (X = Me or Cl).[290] *N*-Chlorodibutylamine adds photolytically to butadiene, in 4M H_2SO_4, 1.5M HOAc aq., giving $Bu_2NCH_2CH{=}CHCH_2Cl$ (60%).[291] *N*-Halocarbamates add (5 °C,

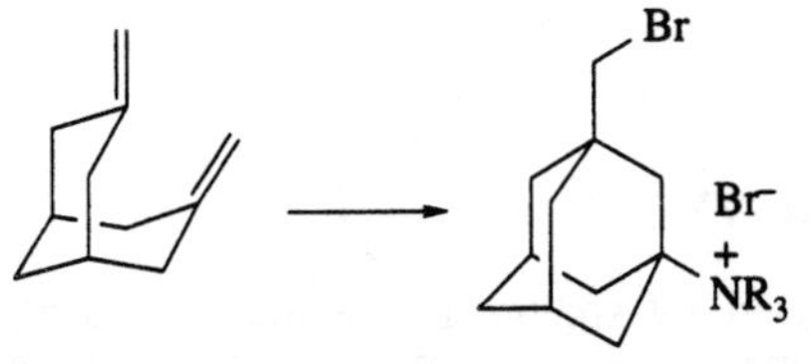

Br; Br^-; $\overset{+}{N}R_3$

(78) **(79)**

$CHCl_3$) to conjugated alkenynes, with the nitrogen on the terminal carbon, to give a variety of alkynic and allenic adducts, often in good yield.[292] A 1,4-azidoselenation of cyclohexadiene was discussed earlier (Section 3.5.5.2).

3.5.7.2 Addition to Cumulenes

Certain congested alkenes add N_2O_4 giving 1,2-dinitro compounds (Et_2O, –10 °C, 27–78%) which can be isolated at 0 °C, but which at room temperature convert to azetines, *e.g.* (**80**; 35–77%).[293] Iodine azide adds to allenes to give RCH=CHICHRN$_3$ or, from allene itself, $(N_3)_2C(CH_2I)_2$.[294] Cyclonona-1,2-diene gives the *cis* adduct; cycloundeca-1,2-diene gives the *trans* adduct.[295]

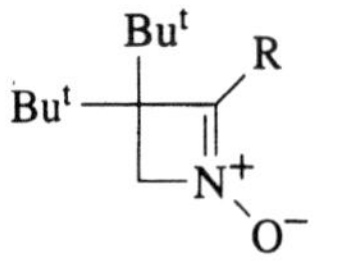

R = H, Me, Cl, $CONH_2$ *etc.*

(80)

3.5.8 ADDITIONS YIELDING MORE HIGHLY OXIDIZED COMPOUNDS

Introduction of oxygen into a Minisci-type reaction mixture leads to formation of amino ketones (Scheme 94).[296] The mixed acetate/azide of lead(IV) with styrene in acetonitrile at –20 °C yields phenacyl azide (60%).[152] One example of azirine formation has already been discussed (Scheme 87). Other related syntheses from vinyl azides are included in a recent review.[297]

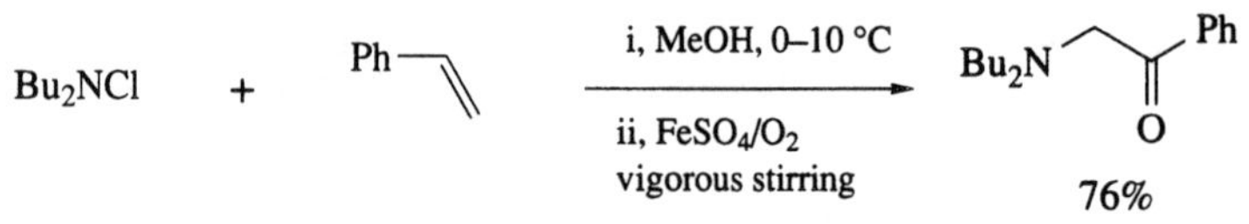

Bu_2NCl + Ph–CH=CH$_2$ → (i, MeOH, 0–10 °C; ii, $FeSO_4/O_2$, vigorous stirring) → $Bu_2NCH_2C(O)CH_2Ph$ 76%

Scheme 94

3.5.9 ADDITIONS LEADING TO REARRANGEMENTS

The examples of aziridination with strongly electronegatively substituted azides cited earlier (Section 3.5.2) were, as indicated, special cases. More typically, imines are formed, and rearrangements are frequent: the key to understanding these reactions is given in Scheme 95, the final product being determined by the chemistry of the ion (**81**). Typical reactions (a–d) are given in Scheme 96 (refs. 298–301 respectively). The C=N bonds of (**82**–**86**) are readily hydrolyzed, yielding lactones, ketones or acids: hydrolysis of (**86**) provides an efficient synthesis of the antiinflammatory drug, naproxen. The diphenyl phosphorazidate procedure has recently been extended to enamines of $ArCH_2COAr$.[302]

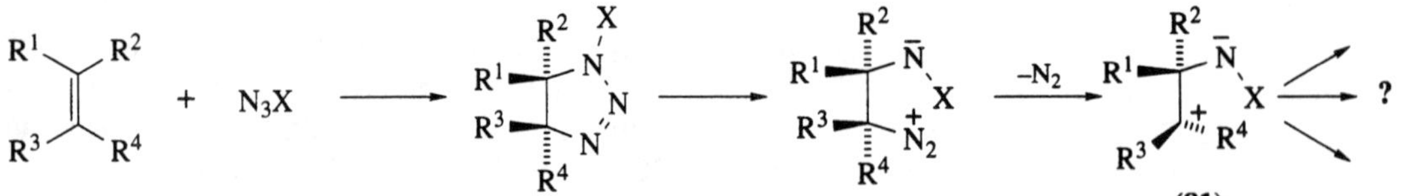

$R^1R^2C=CR^3R^4$ + N_3X → → $-N_2$ → **(81)** → ?

Scheme 95

3.5.10 ADDITIONS CLEAVING THE C=C BOND

Styrenes and stilbenes are cleaved by sodium hydrazide in boiling ether,[303] PhCH=CMe$_2$ giving $Me_2C=NNH_2$ (53%) and toluene (74%) after 6 h reflux. A wide variety of alkenes, dienes and aralkenes are also cleaved by diazonium salts (Scheme 97).[304] Yields and reaction rates vary widely.

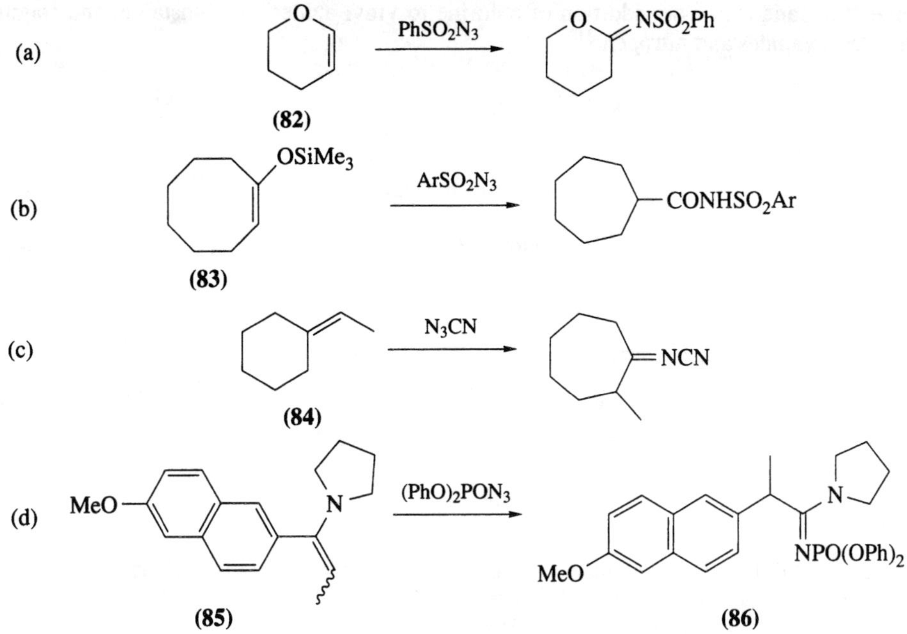

(a) (82) —$PhSO_2N_3$→ NSO$_2$Ph

(b) (83) OSiMe$_3$ —$ArSO_2N_3$→ CONHSO$_2$Ar

(c) (84) —N_3CN→ =NCN

(d) MeO (85) N —$(PhO)_2PON_3$→ MeO N NPO(OPh)$_2$ (86)

Scheme 96

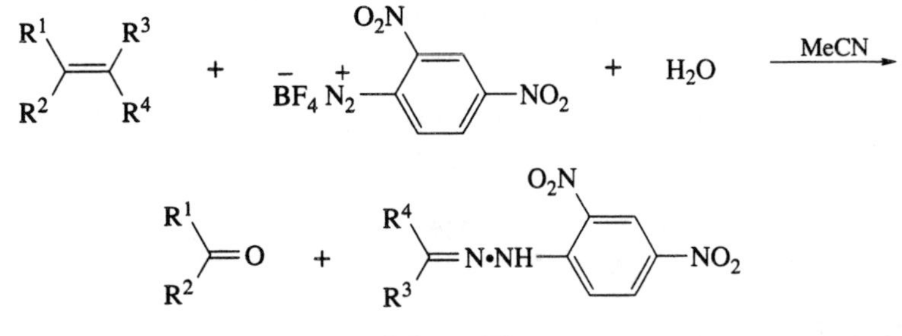

R^1 R^3 R^2 R^4 + $\bar{B}F_4\overset{+}{N}_2$ O$_2$N NO$_2$ + H$_2$O —MeCN→

R^1 R^2 =O + R^4 R^3 =N•NH O$_2$N NO$_2$

Scheme 97

Ozonolysis of cyclic alkenes in aqueous ammonia yields a variety of products, depending on starting material;[305] for example, indene yields isoquinoline (62%). Other less general reactions include conversion of 1,4-cyclohexadiene to *N*-(methoxycarbonyl)azepine in four steps (INCO, NaOMe, Br$_2$, NaOMe);[306] a somewhat similar cleavage has been reported in a [3.8.3] fused ring system.[307] In both cases, the key concept is, in effect, aziridination of the central bond of an (incipient) 1,3,5-triene, and the resulting adduct is then set up to lose the second of the original alkenic bonds in a Cope rearrangement. Enamine aldehydes and ketones are cleaved by reactive azides, *e.g.* Scheme 98, which depicts the first synthesis of a diazoaldehyde.[308]

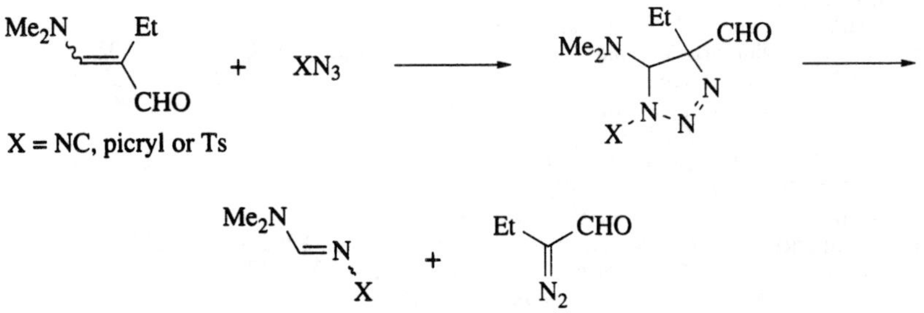

Me$_2$N Et CHO + XN$_3$ → Me$_2$N Et CHO X N N N →

X = NC, picryl or Ts

Me$_2$N =N X + Et CHO N$_2$

Scheme 98

Cyclic enol ethers are cleaved by butyl nitrite in acidic ethanol, providing a useful synthesis of oximinomacrolides (Scheme 99).[309] 1,2-Dichloroazides, from addition of chlorine azide to 1,2-dichloroalkenes (Scheme 100) (and also from addition of chlorine to vinyl azides), are unstable, and fragment to yield alkyl halides, cyanides and nitrogen.[310]

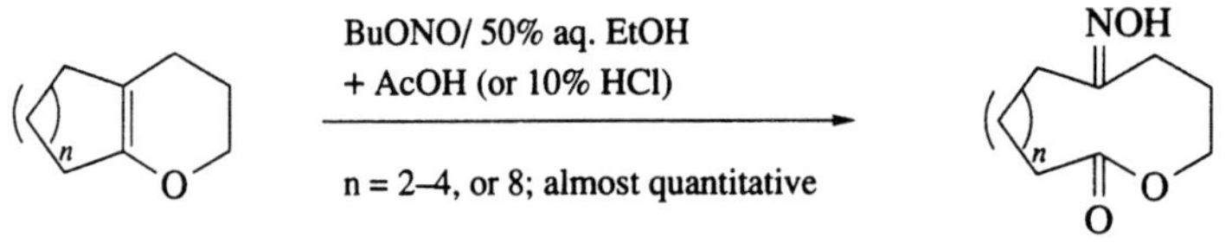

BuONO/ 50% aq. EtOH + AcOH (or 10% HCl)

n = 2–4, or 8; almost quantitative

NOH

Scheme 99

ClN_3/CH_2Cl_2, −10 °C; R = H, Ph; $-N_2$; $PhCCl_3$ + RCN

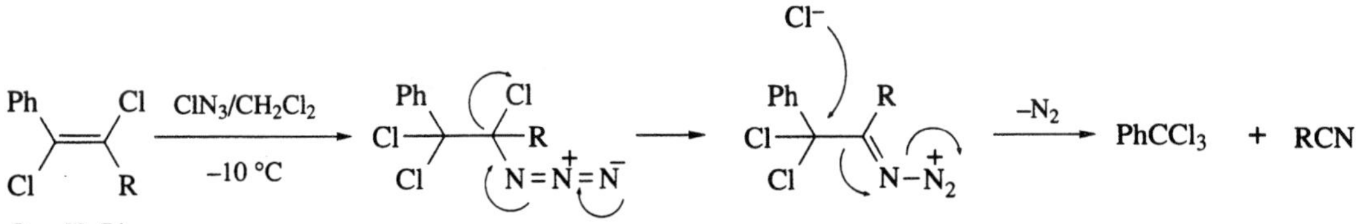

Scheme 100

Steroidal alkenes, with lead(IV) azide/acetate, can yield cleaved derivatives such as (**87**),[311] or its desazido analog, or the uncleaved azido ketone,[154] or the allylic azide,[311] according to conditions, as well as the 1,2-diazides mentioned earlier (Section 3.5.3.2).

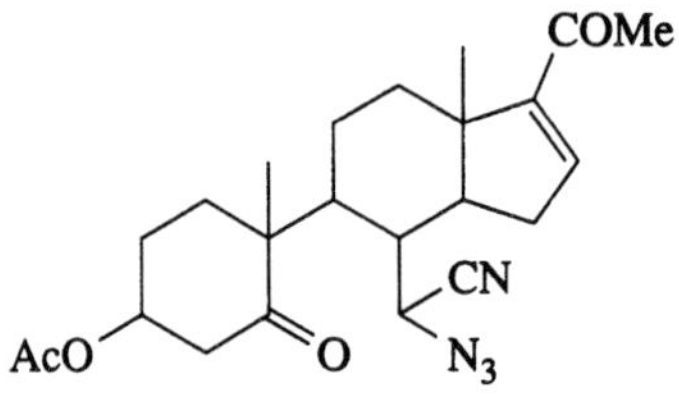

(**87**) 35%

3.5.11 REFERENCES

1. M. B. Gasc, A. Lattes and J. J. Perie, *Tetrahedron*, 1983, **39**, 703.
2. G. H. Whitham, in 'Comprehensive Organic Chemistry', ed. D. H. R. Barton and W. D. Ollis, Pergamon Press, Oxford, 1979, vol. 1, p. 144.
3. L. A. Lazukina and V. P. Kukhar, *J. Gen. Chem. USSR (Engl. Transl.)*, 1988, **58**, 833; *Zh. Obshch. Khim.*, 1988, **58**, 939 (*Chem. Abstr.*, 1989, **110**, 75 640g).
4. A. Padwa and A. D. Woolhouse, in 'Comprehensive Heterocyclic Chemistry', ed. A. R. Katritzky and C. W. Rees, Pergamon Press, Oxford, 1984, vol. 7, p. 47.
5. J. A. Deyrup, *Chem. Heterocycl. Compd.*, 1983, **42** (1), 1.
6. P. E. Fanta, in 'Heterocyclic Compounds with Three- and Four-Membered Rings', ed. A. Weissberger, Wiley, New York, 1964, part 1, p. 524.
7. O. C. Dermer and G. E. Ham, 'Ethyleneimine and Other Aziridines', Academic Press, New York, 1969.
8. A. Laurent, *Bull. Soc. Chim. Belg.*, 1983, **92**, 797 (*Chem. Abstr.*, 1984, **100**, 51 369t).
9. H. Stamm, A. Onistschenko, B. Buchholz and T. Mall, *J. Org. Chem.*, 1989, **54**, 193.
10. N. Furukawa, S. Oae and T. Yoshimura, *Synthesis*, 1976, 30.
11. N. Furukawa, T. Yoshimura, T. Omata and S. Oae, *Chem. Ind. (London)*, 1974, 702.
12. K. Buggle and B. Fallon, *J. Chem. Res. (S)*, 1988, 349; *J. Chem. Res. (M)*, 1988, 2764.
13. N. Farukawa, T. Yoshimura, M. Ohtsu, T. Akasaka and S. Oae, *Tetrahedron*, 1980, **36**, 73.
14. D. O. Spry, *Tetrahedron Lett.*, 1977, 3611.
15. E. Schmitz and K. Jähnisch, *Chem. Heterocycl. Compd. (Engl. Transl.)*, 1974, 1432; *Khim. Geterotsikl. Soedin.*, 1974, 1629 (*Chem. Abstr.*, 1975, **82**, 111 859z).
16. E. Schmitz, *Adv. Heterocycl. Chem.*, 1979, **24**, 63.
17. M. J. Haddadin and J. P. Freeman, *Chem. Heterocycl. Compd.*, 1985, **42** (3), 283.
18. P. Métra and J. Hamelin, *J. Chem. Soc., Chem. Commun.*, 1980, 1038.
19. J. C. Bottaro, *J. Chem. Soc., Chem. Commun.*, 1980, 560.
20. O. Červinka, *Method. Chim.*, 1975, **6**, 591 (*Chem. Abstr.*, 1976, **85**, 20 958b).
21. D. L. Nagel and N. H. Cromwell, *J. Heterocycl. Chem.*, 1974, **11**, 1093.
22. G. R. Harvey, *J. Org. Chem.*, 1968, **33**, 887.
23. E. Schmitz, U. Bicker, S. Schramm and K.-P. Dietz, *J. Prakt. Chem.*, 1978, **320**, 413 (*Chem. Abstr.*, 1978, **89**, 215 130r).
24. I. Ikeda, Y. Machii and M. Okahara, *Synthesis*, 1980, 650.
25. Y. Minoura, M. Takebayashi and C. C. Price, *J. Am. Chem. Soc.*, 1959, **81**, 4689.

26. (a) D. Tanner and P. Somfai, *Tetrahedron*, 1988, **44**, 619; (b) J. Legters, L. Thijs and B. Zwanenburg, *Tetrahedron Lett.*, 1989, **30**, 4881.
27. R. K. Sehgal, B. Almassian, D. E. Rosenbaum, R. Zadrozny and S. K. Sengupta, *J. Med. Chem.*, 1987, **30**, 1626.
28. A. Hassner, G. J. Matthews and F. W. Fowler, *J. Am. Chem. Soc.*, 1969, **91**, 5046.
29. M. M. Campbell, N. Abbas and M. Sainsbury, *Tetrahedron*, 1985, **41**, 5637.
30. D. Van Ende and A. Krief, *Angew. Chem., Int. Ed. Engl.*, 1974, **13**, 279.
31. C. H. Heathcock and A. Hassner, *Org. Synth.*, 1971, **51**, 53.
32. C. H. Heathcock and A. Hassner, *Org. Synth.*, 1971, **51**, 112.
33. R. M. Saito, Y. Kayama, T. Watanabe, H. Fukushima, T. Hara, K. Koyano, A. Takenaka and Y. Sasada, *J. Med. Chem.*, 1980, **23**, 1364.
34. P. Pöchlauer and E. P. Müller, *Helv. Chim. Acta*, 1984, **67**, 1238.
35. G. Lamaty, A. Delbord and W. Werner, *Justus Liebigs Ann. Chem.*, 1969, **726**, 77.
36. G. L. Grunenwald, A. M. Warner and S. L. Hays, *J. Med. Chem.*, 1972, **15**, 747.
37. G. L. Closs and S. L. Brois, *J. Am. Chem. Soc.*, 1960, **82**, 6068.
38. L. M. Kogan, *Russ. Chem. Rev. (Engl. Transl.)*, 1986, **55**, 1164; *Usp. Khim.*, 1986, **55**, 2045 (*Chem. Abstr.*, 1987, **107**, 38 774w).
39. M. E. Hermes and F. D. Marsh, *J. Org. Chem.*, 1972, **37**, 2969.
40. H. C. Brown and A. Tsukamoto, *J. Am. Chem. Soc.*, 1961, **83**, 4549.
41. R. S. Atkinson and B. D. Judkins, *J. Chem. Soc., Perkin Trans. 1*, 1981, 2615.
42. Y. Ito, H. Ida and T. Matsuura, *Tetrahedron Lett.*, 1978, 3119.
43. J.-E. Bäckvall, *J. Chem. Soc., Chem. Commun.*, 1977, 413.
44. B. Åkermark, J.-E. Bäckvall, K. Siirala-Hanséń, K. Sjöberg and K. Zetterberg, *Tetrahedron Lett.*, 1974, 1363.
45. B. Åkermark, J.-E. Bäckvall and K. Zetterberg, *Acta Chem. Scand., Ser. B*, 1982, **36**, 577.
46. K. Harada and I. Nakamura, *Chem. Lett.*, 1978, 1171.
47. K. Harada and I. Nakamura, *J. Chem. Soc., Chem. Commun.*, 1978, 522.
48. A. B. Levy and H. C. Brown, *J. Am. Chem. Soc.*, 1973, **95**, 4067.
49. R. Appel and M. Halstenberg, *Chem. Ber.*, 1976, **104**, 814.
50. I. Shahak, Y. Ittah and J. Blum, *Tetrahedron Lett.*, 1976, 4003.
51. R. Huisgen, G. Szeimies and L. Mobius, *Chem. Ber.*, 1967, **100**, 2494.
52. W. Broeckx, N. Overbergh, C. Samyn, G. Smets and G. L'Abbé, *Tetrahedron*, 1971, **27**, 3527.
53. P. K. Kadaba, B. Stanovnik and M. Tišler, *Adv. Heterocycl. Chem.*, 1984, **37**, 219.
54. P. A. S. Smith, in 'Azides and Nitrenes, Reactivity and Utility', ed. E. F. V. Scriven, Academic Press, New York, 1984, p. 95.
55. K. Burger, J. Fehn and A. Gieren, *Justus Liebigs Ann. Chem.*, 1972, **757**, 9.
56. A. Hassner, in 'Azides and Nitrenes, Reactivity and Utility', ed. E. F. V. Scriven, Academic Press, New York, 1984, p. 35.
57. Y. Nomura, N. Hatanaka and Y. Takeuchi, *Chem. Lett.*, 1976, 901.
58. G. M. Rubottom, in 'Oxidation in Organic Chemistry', ed. W. S. Trahanovsky, Academic Press, New York, 1982, p. 1.
59. T. Kametani and T. Honda, *Adv. Heterocycl. Chem.*, 1986, **39**, 181.
60. E. F. V. Scriven and K. Turnbull, *Chem. Rev.*, 1988, **88**, 297.
61. W. Nagata, S. Hirai, T. Okumura and K. Kawata, *J. Am. Chem. Soc.*, 1968, **90**, 1650.
62. P. G. Gassman and J. G. Schaffhausen, *J. Org. Chem.*, 1978, **43**, 3214.
63. F. Texier and R. Carrie, *Tetrahedron Lett.*, 1969, 823.
64. R. A. Abramovitch, S. R. Challand and Y. Yamada, *J. Org. Chem.*, 1975, **40**, 1541.
65. R. A. Abramovitch and S. R. Challand, *J. Chem. Soc., Chem. Commun.*, 1972, 1160.
66. H. Takeuchi and R. Ihara, *J. Chem. Soc., Chem. Commun.*, 1983, 175.
67. H. Takeuchi, Y. Shiobara, M. Mitani and K. Koyama, *J. Chem. Soc., Chem. Commun.*, 1985, 1251.
68. M. De Poortere and F. C. De Schryver, *Tetrahedron Lett.*, 1970, 3949.
69. P. Spagnolo and P. Zanirato, *J. Chem. Soc., Perkin Trans. 1*, 1988, 3375.
70. J. W. Cornforth, in 'Heterocyclic Compounds', ed. R. C. Elderfield, Wiley, New York, 1957, vol. 5, p. 377.
71. D. H. Boschelli, *Synth. Commun.*, 1988, **18**, 1391.
72. (a) W. Lwowski, in 'Azides and Nitrenes, Reactivity and Utility', ed. E. F. V. Scriven, Academic Press, New York, 1984, p. 205; (b) S. Rhouati and A. Bernou, *J. Chem. Soc., Chem. Commun.*, 1989, 730.
73. R. Huisgen, L. Möbius, G. Müller, H. Stangl, G. Szeimies and J. M. Vernon, *Chem. Ber.*, 1965, **98**, 3992.
74. V. P. Semenov, A. N. Studenikov, A. P. Prosypkina and K. A. Ogloblin, *J. Org. Chem. USSR (Engl. Transl.)*, 1977, **13**, 2056; *Zh. Obshch. Khim.*, 1977, **13**, 2207 (*Chem. Abstr.*, 1978, **88**, 50 560v).
75. D. H. Aue, H. Iwashashi and D. F. Shellhamer, *Tetrahedron Lett.*, 1973, 3719.
76. G. R. Felt and W. Lwowski, *J. Org. Chem.*, 1976, **41**, 96.
77. G. T. Tisue, S. Linke and W. Lwowski, *J. Am. Chem. Soc.*, 1967, **89**, 6303.
78. D. Mansuy, J.-P. Mahy, A. Duréault, G. Bedi and P. Battioni, *J. Chem. Soc., Chem. Commun.*, 1984, 1161.
79. V. V. Dirnens, Yu. Sh. Goldberg and E. Ya. Lukevits, *Dokl. Chem. (Engl. Transl.)*, 1988, **298** (1), 116 (*Chem. Abstr.*, 1989, **110**, 270y).
80. J. S. McConaghy, Jr. and W. Lwowski, *J. Am. Chem. Soc.*, 1967, **89**, 2357.
81. (a) M. Seno, T. Namba and H. Kise, *J. Org. Chem.*, 1978, **43**, 3345; (b) E. Ya. Lukevits, V. V. Dirnens, Yu. Sh. Goldberg and E. Liepins, *J. Organomet. Chem.*, 1986, **316**, 249.
82. P. Scheiner, *J. Am. Chem. Soc.*, 1966, **88**, 4759.
83. A. Mishra, S. N. Rice and W. Lwowski, *J. Org. Chem.*, 1968, **33**, 481.
84. J. S. McConaghy, Jr. and W. Lwowski, *J. Am. Chem. Soc.*, 1967, **89**, 4450.
85. A. C. Oehlschlager, P. Tichman and L. W. Zalkow, *J. Chem. Soc., Chem. Commun.*, 1965, 596.
86. Young Hwan Chang, Fang-Ting Chin and G. Zon, *J. Org. Chem.*, 1981, **46**, 342.
87. E. Kozłowska-Gramsz, *Pol. J. Chem.*, 1980, **54**, 1607 (*Chem. Abstr.*, 1981, **95**, 6326u).

88. P. P. Nicholas, *J. Org. Chem.*, 1975, **40**, 3396.
89. A. V. Kamernitskii, Z. I. Istomina, E. P. Serebryakov and A. M. Turuta, *Izv. Akad. Nauk SSSR, Ser. Khim.*, 1979 (1), 186 (*Chem. Abstr.*, 1979, **91**, 20 886a).
90. L. Pellacani, F. Persia and P. A. Tardella, *Tetrahedron Lett.*, 1980, **21**, 4967.
91. J. Fioravanti, M. A. Loreto, L. Pellacani and P. A. Tardella, *J. Org. Chem.*, 1987, **50**, 5365.
92. R. A. Aitken, I. Gosney, H. Farries, M. H. Palmer, I. Simpson, J. I. G. Cadogan and E. J. Tinley, *Tetrahedron*, 1985, **41**, 1329.
93. Sang-Hun Jung and H. Kohn, *J. Am. Chem. Soc.*, 1985, **107**, 2931.
94. Sang-Hun Jung and H. Kohn, *Tetrahedron Lett.*, 1984, **25**, 399.
95. A. G. Anastassiou, J. N. Shepelavy, H. E. Simmons and F. D. Marsh, in 'Nitrenes', ed. W. Lwowski, Interscience, New York, 1970, chap. 9, p. 305.
96. J. E. G. Kemp, D. Ellis and M. D. Closier, *Tetrahedron Lett.*, 1979, 3781.
97. M. G. Hutchings and D. Swern, *J. Org. Chem.*, 1982, **47**, 4847.
98. M. E. Hermes and F. D. Marsh, *J. Org. Chem.*, 1972, **37**, 2969.
99. K. Ponsold and W. Ihn, *Tetrahedron Lett.*, 1970, 1125.
100. R. S. Atkinson and B. J. Kelly, *J. Chem. Soc., Chem. Commun.*, 1987, 1362.
101. R. S. Atkinson and J. R. Malpass, *J. Chem. Soc., Perkin Trans. 1*, 1977, 2242.
102. P. D. Bartlett, *Rec. Chem. Prog.*, 1950, **11**, 47.
103. (a) R. S. Atkinson, personal communication, 1989; (b) R. S. Atkinson and B. J. Kelly, *J. Chem. Soc., Perkin Trans. 1*, 1989, 1657; (c) R. S. Atkinson and B. J. Kelly, *J. Chem. Soc., Chem. Commun.*, 1989, 836.
104. M. Edwards, T. L. Gilchrist, C. J. Harris and C. W. Rees, *J. Chem. Res. (S)*, 1979, 114; *J. Chem. Res. (M)*, 1979, 1687.
105. L. A. Carpino, R. E. Padykula, Sung-Nung Lee, G. Y. Han and R. K. Kirkley, *J. Org. Chem.*, 1988, **53**, 6047.
106. R. S. Atkinson, M. J. Grimshire and B. J. Kelly, *Tetrahedron*, 1989, **45**, 2875.
107. R. S. Atkinson and B. J. Kelly, *J. Chem. Soc., Perkin Trans. 1*, 1989, 1515.
108. R. S. Atkinson, in 'Azides and Nitrenes, Reactivity and Utility', ed. E. F. V. Scriven, Academic Press, New York, 1984, p. 35.
109. R. W. Butler, in 'Synthetic Reagents', ed. J. S. Pizey, Horwood, Chichester, 1977, vol. 3, p. 277.
110. R. S. Atkinson, J. Fawcett, D. R. Russell and G. Tughan, *J. Chem. Soc., Chem. Commun.*, 1986, 832.
111. R. S. Atkinson and G. Tughan, *J. Chem. Soc., Chem. Commun.*, 1986, 834.
112. S. Shafiullah and J. A. Ansari, *J. Chem. Res. (S)*, 1988, 226.
113. C. D. Campbell and C. W. Rees, *Chem. Commun.*, 1965, 192.
114. F. Schröppel and J. Sauer, *Tetrahedron Lett.*, 1974, 2945.
115. R. S. Atkinson and G. Tughan, *J. Chem. Soc., Perkin Trans. 1*, 1987, 2787.
116. (a) R. S. Atkinson and G. Tughan, *J. Chem. Soc., Chem. Commun.*, 1987, 456; (b) R. S. Atkinson, B. J. Kelly and C. McNicolas, *J. Chem. Soc., Chem. Commun.*, 1989, 562.
117. S. Calet, F. Urso and H. Alper, *J. Am. Chem. Soc.*, 1989, **111**, 931.
118. R. K. Muller, R. Joos, D. Felix, J. Schreiber, C. Wintner and A. Eschenmoser, *Org. Synth.*, 1976, **55**, 114.
119. G. R. Meyer, C. A. Kellert and R. W. Ebert, *J. Heterocycl. Chem.*, 1979, **16**, 461.
120. L. Hoesch, N. Egger and A. S. Dreiding, *Helv. Chim. Acta*, 1978, **61**, 795.
121. A. Padwa and Hao Ku, *Tetrahedron Lett.*, 1980, **21**, 1009.
122. R. S. McDaniel and A. C. Oehlschlager, *Can. J. Chem.*, 1968, **46**, 2318.
123. A. Zwierzak and K. Osowska, *Angew. Chem., Int. Ed. Engl.*, 1976, **15**, 302.
124. B. Davidowitz, T. A. Modro and M. L. Niven, *Phosphorus Sulfur*, 1985, **22**, 255.
125. F. A. L. Anet, R. D. Trepka and D. J. Cram, *J. Am. Chem. Soc.*, 1967, **89**, 357.
126. B. V. Ioffe and E. V. Koroleva, *Tetrahedron Lett.*, 1973, 619.
127. B. V. Ioffe, Yu. P. Artsybasheva and I. G. Zenkovich, *Dokl. Chem. (Engl. Transl.)*, 1976, **231**, 742; *Dokl. Akad. Nauk SSSR*, 1976, **231**, 1130 (*Chem. Abstr.*, 1977, **86**, 105 794r).
128. R. S. Atkinson, B. D. Judkins and N. Khan, *J. Chem. Soc., Perkin Trans. 1*, 1982, 2491.
129. R. S. Atkinson, M. Lee and J. R. Malpass, *J. Chem. Soc., Chem. Commun.*, 1984, 919.
130. W. Bludsuss and R. Mews, *Chem. Ber.*, 1981, **114**, 1539.
131. W. Bludsuss, O. Glemser and G. G. A. Lange, *Isr. J. Chem.*, 1978, **17**, 137 (*Chem. Abstr.*, 1978, **89**, 107 969n).
132. M. S. Kharasch and H. M. Priestley, *J. Am. Chem. Soc.*, 1939, **61**, 3425.
133. J.-P. Mahy, G. Bedi, P. Battioni and D. Mansuy, *J. Chem. Soc., Perkin Trans. 2*, 1988, 1517.
134. J.-P. Mahy, G. Bedi, P. Battioni and D. Mansuy, *Tetrahedron Lett.*, 1988, **29**, 1927.
135. M. Hedayatullah and A. Guy, *J. Heterocycl. Chem.*, 1979, **16**, 201.
136. P. N. Becker and R. G. Bergman, *Organometallics*, 1983, **2**, 787.
137. J.-E. Bäckvall, *Tetrahedron Lett.*, 1978, 163.
138. A. O. Chong, K. Oshima and K. B. Sharpless, *J. Am. Chem. Soc.*, 1977, **99**, 3420.
139. J. P. M. Heude and J. Lambert, *Ger. Pat.* 2 047 995 (*Chem. Abstr.*, 1971, **75**, 5219m).
140. C. J. Michejda and D. H. Campbell, *J. Am. Chem. Soc.*, 1974, **96**, 929.
141. V. G. Aranda, J. Barluenga and F. Aznar, *Synthesis*, 1974, 504.
142. J. Barluenga, L. Alonso-Cires and G. Asensio, *Synthesis*, 1979, 962.
143. H. Natsugari, R. L. Whittle and S. M. Weinreb, *J. Am. Chem. Soc.*, 1984, **106**, 7867.
144. K. B. Sharpless and S. P. Singer, *J. Org. Chem.*, 1976, **41**, 2504.
145. S. Ghomi and D. E. Orr, *Chem. Ind. (London)*, 1983, 928.
146. A. Duréault, L. Tranchepain, C. Greck and J.-C. Depezay, *Tetrahedron Lett.*, 1987, **28**, 3341.
147. M. Egli, L. Hoesch and A. S. Dreiding, *Helv. Chim. Acta*, 1985, **68**, 220.
148. M. J. Fazio, *J. Org. Chem.*, 1984, **49**, 4889.
149. W. E. Fristad, T. A. Brandvold, J. R. Peterson and S. R. Thompson, *J. Org. Chem.*, 1985, **50**, 3647.
150. R. M. Moriarty and J. S. Khosrowshahi, *Tetrahedron Lett.*, 1986, **27**, 2809.
151. F. Minisci, *Acc. Chem. Res.*, 1975, **8**, 165.

152. E. Zbiral, *Synthesis*, 1972, 185.
153. R. Galli and V. Malatesta, *Org. Prep. Proced. Int.*, 1971, **3**, 231.
154. R. W. Draper, *J. Chem. Soc., Perkin Trans. 1*, 1983, 2787.
155. Yuan L. Chow, C. J. Colon, D. W. L. Chang, K. S. Pillay, R. L. Lockhart and T. Tezuka, *Acta Chem. Scand., Ser. B*, 1982, **36**, 623.
156. A. J. Bloom, M. Fleischmann and J. M. Mellor, *J. Chem. Soc., Perkin Trans. 1*, 1984, 2357.
157. J. L. Riebsomer, *Chem. Rev.*, 1945, **36**, 157.
158. O. H. Giamalva, G. B. Kenion, D. F. Church and W. A. Pryor, *J. Am. Chem. Soc.*, 1987, **109**, 7059.
159. M. L. Scheinbaum, *J. Org. Chem.*, 1970, **35**, 2785.
160. A. F. Ellis (Gulf Res. and Dev. Co.), *US Pat.* 3 379 710 (1968) (*Chem. Abstr.*, 1968, **69**, 51 568y).
161. J. F. Brown, *J. Am. Chem. Soc.*, 1957, **79**, 2480.
162. J. M. Birchall, A. J. Bloom, R. N. Haszeldine and C. J. Willis, *J. Chem. Soc.*, 1962, 3021.
163. E. Bagley, J. M. Birchall and R. N. Haszeldine, *J. Chem. Soc. (C)*, 1966, 1232.
164. S. G. Hentges and K. B. Sharpless, *J. Org. Chem.*, 1980, **45**, 2257.
165. (a) J. S. M. Wai, I. Markó, J. S. Svendsen, M. G. Finn, E. N. Jacobsen and K. B. Sharpless, *J. Am. Chem. Soc.*, 1989, **111**, 1123; (b) M. Hirama, T. Oishi and S. Ito, *J. Chem. Soc., Chem. Commun.*, 1989, 665.
166. D. W. Patrick, L. K. Truesdale, S. A. Biller and K. B. Sharpless, *J. Org. Chem.*, 1978, **43**, 2628.
167. E. Herranz and K. B. Sharpless, *Org. Synth.*, 1982, **61**, 85.
168. K. B. Sharpless and E. Herranz, *Org. Synth.*, 1982, **61**, 93.
169. E. Herranz and K. B. Sharpless, *J. Org. Chem.*, 1980, **45**, 2710.
170. S. K. Dubey and E. E. Knaus, *Can. J. Chem.*, 1983, **61**, 565.
171. H. Friege, H. Friege and I. Dyong, *Chem. Ber.*, 1981, **114**, 1822.
172. W. P. Griffith, N. T. McManus and A. D. White, *J. Chem. Soc., Dalton Trans.*, 1986, 1035.
173. K. Kato, T. Saino, R. Nishizawa, T. Takita and H. Umezawa, *J. Chem. Soc., Perkin Trans. 1*, 1980, 1618.
174. J.-E. Bäckvall and E. E. Björkman, *J. Org. Chem.*, 1980, **45**, 2893.
175. J.-E. Bäckvall, E. E. Björkman, S. E. Byström and A. Solladié-Cavallo, *Tetrahedron Lett.*, 1982, **23**, 943.
176. J. Barluenga, L. Alonso-Cires and G. Asencio, *Synthesis*, 1981, 376.
177. R. A. Kretchmer and P. J. Daly, *J. Org. Chem.*, 1976, **41**, 192.
178. K. E. Harding, T. H. Marman and Do-hyun Nam, *Tetrahedron*, 1988, **44**, 5605.
179. J. Beger, C. Pöschmann and I. Thomas, *J. Prakt. Chem.*, 1984, **326**, 519 (*Chem. Abstr.*, 1986, **101**, 173 457m).
180. B. M. Trost and A. R. Sudhakar, *J. Am. Chem. Soc.*, 1987, **109**, 3792.
181. R. S. Garigipati and S. M. Weinreb, *J. Am. Chem. Soc.*, 1983, **105**, 4499.
182. G. Emmer and E. Zbiral, *Justus Liebigs Ann. Chem.*, 1979, 796.
183. N. X. Hu, Y. Aso, T. Otsubo and F. Ogura, *J. Chem. Soc., Chem. Commun.*, 1987, 1447.
184. A. Bongini, G. Cardillo, M. Orena, S. Sandri and C. Tomasini, *J. Chem. Soc., Perkin Trans. 1*, 1985, 935.
185. B. Bernet and A. Vasella, *Tetrahedron Lett.*, 1983, **24**, 5491.
186. S. Knapp and D. V. Patel, *J. Org. Chem.*, 1984, **49**, 5072.
187. P. G. Sammes and D. Thetford, *Tetrahedron Lett.*, 1986, **27**, 2275.
188. (a) J. Das, *Synth. Commun.*, 1989, **18**, 907; (b) S. Karady, E. G. Corley, N. L. Abramson and L. M. Weinstock, *Tetrahedron Lett.*, 1989, **30**, 2191.
189. L. E. Overman and L. A. Flippin, *Tetrahedron Lett.*, 1981, **22**, 195.
190. M. Emziane, K. I. Sutowardogo and D. Sinou, *J. Organomet. Chem.*, 1988, **346**, C7.
191. K. Faber, H. Honig and P. Seuferwasserthal, *Tetrahedron Lett.*, 1989, **29**, 1903.
192. W. Kisan and W. Pritzkow, *J. Prakt. Chem.*, 1978, **320**, 59 (*Chem. Abstr.*, 1978, **88**, 136 092f).
193. Y. Ogata, in 'Oxidation in Organic Chemistry', ed. W. S. Trahanovsky, Academic Press, New York, 1978, Part C, p. 307.
194. D. L. H. Williams, 'Nitrosation', Cambridge University Press, Cambridge, 1988, chap. 2.
195. S. G. Zlotin, M. M. Krayushkin, V. V. Sevost'yanova and S. S. Novikov, *Izv. Akad. Nauk SSSR, Ser. Khim.*, 1977, 2362; *Bull. Acad. Sci. USSR, Div. Chem. Sci. (Engl. Transl.)*, 1977, 2196 (*Chem. Abstr.*, 1978, **88**, 61 948d).
196. A. V. Fokin, Yu. N. Studnev, A. I. Rapkin, V. G. Chilikin and O. V. Verenikin, *Izv. Akad. Nauk SSSR, Ser. Khim.*, 1983, 1437; *Bull. Acad. Sci. USSR, Div. Chem. Sci. (Engl. Transl.)*, 1983, 1306 (*Chem. Abstr.*, 1983, **99**, 157 784a).
197. L. Benati, P. C. Montevecchi and P. Spagnolo, *Tetrahedron*, 1986, **42**, 1145.
198. T. Kamiya, T. Teraji, M. Hashimoto, O. Nakaguchi and T. Oku, *J. Am. Chem. Soc.*, 1975, **97**, 5020.
199. M. Ihara, Y. Haga, M. Yonekura, T. Ohsawa, K. Fukumoto and T. Kametani, *J. Am. Chem. Soc.*, 1983, **105**, 7345.
200. P. Brownbridge, *Tetrahedron Lett.*, 1984, **25**, 3754.
201. B. M. Trost and T. Shibata, *J. Am. Chem. Soc.*, 1982, **104**, 3225.
202. N. S. Zefirov, N. K. Sadovaya, A. M. Magarramov and I. V. Bodrikov, *J. Org. Chem. USSR (Engl. Transl.)*, 1977, **13**, 221; *Zh. Org. Khim.*, 1977, **13**, 245 (*Chem. Abstr.*, 1977, **87**, 22 465d).
203. A. Bewick, D. E. Coe, J. M. Mellor and W. M. Owton, *J. Chem. Soc., Perkin Trans. 1*, 1985, 1033.
204. A. Bewick, J. M. Mellor and W. M. Owton, *J. Chem. Soc., Perkin Trans. 1*, 1985, 1039.
205. T. Mall and H. Stamm, *Chem. Ber.*, 1988, **121**, 1353.
206. D. Wagner, D. Gertner and A. Zilkha, *Tetrahedron Lett.*, 1968, 4875.
207. A. Toshimitsu, T. Aoai, H. Owada, S. Uemura and M. Okano, *J. Org. Chem.*, 1981, **46**, 4727.
208. A. Toshimitsu, G. Hayashi, K. Terao and S. Uemura, *J. Chem. Soc., Perkin Trans. 1*, 1988, 2113.
209. A. Bewick, D. E. Coe, G. B. Fuller and J. M. Mellor, *Tetrahedron Lett.*, 1980, **21**, 3827.
210. C. G. Francisco, E. I. León, J. A. Salazar and E. Suárez, *Tetrahedron Lett.*, 1986, **27**, 2513.
211. D. L. J. Clive, V. Farina, A. Singh, C. K. Wong, W. A. Kiel and S. M. Menchen, *J. Org. Chem.*, 1980, **45**, 2120.
212. A. Toshimitsu, K. Terao and S. Uemura, *J. Org. Chem.*, 1986, **51**, 1724.
213. A. Toshimitsu, K. Terao and S. Uemura, *J. Chem. Soc., Chem. Commun.*, 1986, 530.

214. A. Toshimitsu, S. Uemura, M. Okano and N. Watanabe, *J. Org. Chem.*, 1983, **48**, 5246.
215. J. N. Denis, J. Vicens and A. Krief, *Tetrahedron Lett.*, 1979, 2697.
216. A. Hassner and A. S. Amarasekara, *Tetrahedron Lett.*, 1987, **28**, 5185.
217. T. Hayama, S. Tomoda, Y. Takeuchi and Y. Nomura, *Tetrahedron Lett.*, 1982, **23**, 4733.
218. T. Hayama, S. Tomoda, Y. Takeuchi and Y. Nomura, *J. Org. Chem.*, 1984, **49**, 3235.
219. (a) Nan Xing Hu, Y. Aso, T. Otsubo and F. Ogura, *Chem. Lett.*, 1987, 1327; (b) Nan Xing Hu, Y. Aso, T. Otsubo and F. Ogura, *J. Org. Chem.*, 1989, **54**, 4398; (c) Nan Xing Hu, Y. Aso, T. Otsubo and F. Ogura, *J. Chem. Soc., Perkin Trans. 1*, 1989, 1775; (d) Nan Xing Hu, Y. Aso, T. Otsubo and F. Ogura, *Phosphorus Sulfur*, 1988, **38**, 177; (e) C. Degrand, R. Prest and M. Nour, *Phosphorus Sulfur*, 1988, **38**, 201.
220. A. Bensadat, G. Bodennec, E. Laurent and R. Tardival, *Nouv. J. Chim.*, 1981, **5**, 127 (*Chem. Abstr.*, 1981, **95**, 79 526u).
221. T. N. Wade, *J. Org. Chem.*, 1980, **45**, 5328.
222. G. Alvernhe, S. Lacombe and A. Laurent, *Tetrahedron Lett.*, 1980, **21**, 289.
223. E. Maxa, E. Zbiral, G. Schulz and E. Haslinger, *Justus Liebigs Ann. Chem.*, 1975, 1705.
224. B. Baasner, H. Hagemann and E. Klauke (Bayer A. G.), *Ger. Pat.* 3 305 201 (1984) (*Chem. Abstr.*, 1985, **102**, 5684k).
225. B. Damin, J. Garapon and B. Sillion, *Tetrahedron Lett.*, 1980, **21**, 1709.
226. D. J. Nelson and R. Soundararajan, *Tetrahedron Lett.*, 1988, **29**, 6207.
227. J. R. Whittle and O. W. Rigden (Texaco Inc.), *US Pat.* 4 089 903 (1978) (*Chem. Abstr.*, 1978, **89**, 108 101k).
228. C. B. Colburn, W. E. Hill and R. D. Verma, *J. Fluorine Chem.*, 1981, **17**, 75 (*Chem. Abstr.*, 1981, **94**, 208 022g).
229. T. A. Foglia and D. Swern, *J. Org. Chem.*, 1966, **31**, 3625.
230. (a) F. A. Daniher and P. E. Butler, *J. Org. Chem.*, 1968, **33**, 4336; (b) T. Akasaka, N. Furukawa and S. Oae, *Tetrahedron Lett.*, 1979, 2035.
231. H. Driguez and J. Lessard, *Can. J. Chem.*, 1977, **55**, 720.
232. J. Lessard and J. M. Paton, *Tetrahedron Lett.*, 1970, 4883.
233. (a) L. Stella, B. Raynier and J.-M. Surzur, *Tetrahedron Lett.*, 1977, 2721; (b) J.-L. Stein, L. Stella and J.-M. Surzur, *Tetrahedron Lett.*, 1980, **21**, 287.
234. J. Lessard, M. Mondon and D. Touchard, *Can. J. Chem.*, 1981, **59**, 431.
235. T. P. Seden and R. W. Turner, *J. Chem. Soc. C*, 1968, 876.
236. N. A. Rybakova, V. I. Dostovalova, A. A. Slepushkina, V. I. Robas and R. Kh. Freidlina, *Izv. Akad. Nauk SSSR, Ser. Khim.*, 1973, 359; *Bull. Acad. Sci. USSR, Div. Chem. Sci. (Engl. Transl.)*, 1973, 342 (*Chem. Abstr.*, 1973, **78**, 147 491e).
237. A. Hassner and F. P. Boerwinkle, *Tetrahedron Lett.*, 1969, 3309.
238. A. Hassner and C. H. Heathcock, *J. Org. Chem.*, 1963, **29**, 1350.
239. S. Bresadola, P. Canal, A. Nenz and E. Gallinella, *Chim. Ind. (Milan)*, 1963, **45**, 937 (*Chem. Abstr.*, 1963, **59**, 9771e).
240. H. Schechter, F. Conrad, A. L. Daulton and R. B. Kaplan, *J. Am. Chem. Soc.*, 1952, **74**, 3052.
241. Fu-Lung Lu, Y. M. A. Naguib, M. Kitadani and Yuan L. Chow, *Can. J. Chem.*, 1979, **57**, 1967.
242. G. Peiffer, J. C. Traynard and A. Guillemonat, *Bull. Soc. Chim. Fr.*, 1966, 1910.
243. G. E. Heasley, J. M. Janes, S. R. Stark and B. L. Robinson, *Tetrahedron Lett.*, 1985, **26**, 1811.
244. A. Osowska-Pacewicza and A. Zwierzak, *Tetrahedron*, 1985, **41**, 4717.
245. S. Zawadzki and A. Zwierzak, *Tetrahedron*, 1981, **15**, 2675.
246. Shi-Ching Chang and D. D. DesMarteau, *J. Org. Chem.*, 1983, **48**, 895.
247. A. Hassner and F. P. Boerwinkle, *J. Am. Chem. Soc.*, 1968, **90**, 216.
248. A. Hassner, F. P. Boerwinkle and A. B. Levy, *J. Am. Chem. Soc.*, 1970, **92**, 4979.
249. F. P. Boerwinkle and A. Hassner *Tetrahedron Lett.*, 1968, 3921.
250. E. J. Corey and H. Estreicher, *J. Am. Chem. Soc.*, 1978, **100**, 6294.
251. A. Hassner, R. P. Hoblitt, C. H. Heathcock, J. E. Kropp and M. Lorber, *J. Am. Chem. Soc.*, 1970, **92**, 1326.
252. C. G. Gebelein, S. Rosen and D. Swern, *J. Org. Chem.*, 1969, **34**, 1677.
253. A. Hassner, M. Lorber and C. H. Heathcock, *J. Org. Chem.*, 1967, **32**, 540.
254. A. Hassner and C. H. Heathcock, *Tetrahedron*, 1964, **20**, 1037.
255. (a) C. G. Gebelein, *Chem. Ind. (London)*, 1970, 57; (b) J. E. Baldwin and Y. Yamaguchi, *Tetrahedron Lett.*, 1989, **30**, 3335; (c) J. Barluenga, J. M. Gonzalez, P. J. Campus and G. Asensio, *Angew. Chem., Int. Ed. Engl.*, 1985, **24**, 319.
256. F. W. Fowler, A. Hassner and L. A. Levy, *J. Am. Chem. Soc.*, 1967, **89**, 2077.
257. A. Hassner, *Acc. Chem. Res.*, 1971, **4**, 9.
258. S. Sivasubramanian, S. Aravind, L. T. Kumarasingh and N. Arumugam, *J. Org. Chem.*, 1986, **51**, 1985.
259. A. Hassner, personal communication, 1989.
260. G. Mehta, P. K. Dutta and P. N. Panday, *Tetrahedron Lett.*, 1975, 445.
261. R. C. Cambie, R. C. Hayward, P. S. Rutledge, T. Smith-Palmer, B. E. Swedlund and P. D. Woodgate, *J. Chem. Soc., Perkin Trans. 1*, 1979, 180.
262. L. Ghosez, F. Sainte, M. Rivera, C. Bernard-Henriet and V. Gouverneur, *Recl. Trav. Chim. Pays-Bas*, 1986, **105**, 456.
263. A. Hassner, J. E. Knopp and G. J. Kent, *J. Org. Chem.*, 1969, **34**, 2628.
264. Wing-Wah Sy and A. W. By, *Tetrahedron Lett.*, 1985, **26**, 1193.
265. T. E. Stevens and W. D. Emmons, *J. Am. Chem. Soc.*, 1958, **80**, 338.
266. J. E. McMurry and J. H. Musser, *Org. Synth.*, 1977, **56**, 65.
267. J. W. Patterson and J. E. McMurry, *J. Chem. Soc., Chem. Commun.*, 1971, 488.
268. S. Knapp and A. T. Levorse, *J. Org. Chem.*, 1988, **53**, 4006.
269. A. Bongini, G. Cardillo, M. Orena, S. Sandri and C. Tomasini, *J. Chem. Soc., Perkin Trans. 1*, 1985, 935.
270. M. J. Kurth and S. H. Bloom, *J. Org. Chem.*, 1989, **54**, 411.
271. M. Hirama, M. Iwashita, Y. Yamazaki and S. Itô, *Tetrahedron Lett.*, 1984, **25**, 4963.

272. A. J. Biloski, R. D. Wood and B. Ganem, *J. Am. Chem. Soc.*, 1982, **104**, 3233.
273. (a) Y. Tamaru, S. Kawamura, K. Tanaka and Z. Yoshida, *Tetrahedron Lett.*, 1984, **25**, 1063; (b) H. Takahata, T. Takamatsu and T. Yamazaki, *J. Org. Chem.*, 1989, **54**, 4812; (c) Y. Tamaru, S. Kawamura, T. Bando, K. Tanaka, M. Hojo and Z. Yoshida, *J. Org. Chem.*, 1988, **53**, 5491; (d) T. W. Balko. R. S. Brinkmeyer and N. H. Terando, *Tetrahedron Lett.*, 1989, **30**, 2045.
274. G. Rajendra and M. J. Miller, *Tetrahedron Lett.*, 1985, **26**, 5385.
275. T. Aida, R. Legault, D. Dugat and T. Durst, *Tetrahedron Lett.*, 1979, 4993.
276. (a) D. R. Williams, M. H. Osterhout and J. M. McGill, *Tetrahedron Lett.*, 1989, **30**, 1327; (b) D. R. Williams, M. H. Osterhout and J. M. McGill, *Tetrahedron Lett.*, 1989, **30**, 1331.
277. T. Petrzilka, K. K. Prasad and G. Schmidt, *Helv. Chim. Acta*, 1977, **60**, 2911.
278. J. G. Henkel, W. C. Faith and J. T. Hane, *J. Org. Chem.*, 1981, **46**, 3483.
279. M. E. Kuehne and D. A. Horne, *J. Org. Chem.*, 1975, **40**, 1287.
280. P. Mackiewicz, R. Furstoss and B. Waegell, *J. Org. Chem.*, 1978, **43**, 3746.
281. J. Lessard, R. Cote, P. Mackiewicz, R. Furstoss and B. Waegell, *J. Org. Chem.*, 1978, **43**, 3750.
282. B. Åkermark, J.-E. Bäckvall, A. Löwenborg and K. Zetterberg, *J. Organomet. Chem.*, 1979, **166**, C33.
283. M. Sabuni, G. Kresze and H. Braun, *Tetrahedron Lett.*, 1984, **25**, 5377.
284. H. Braun, R. Charles, G. Kresze, M. Sabuni and J. Winkler, *Liebigs Ann. Chem.*, 1987, 1129.
285. J. Barluenga, J. M. Gonzalez and G. Asensio, *Tetrahedron Lett.*, 1988, **29**, 6497.
286. R. A. Abramovitch, M. Ortiz and S. P. McManus, *J. Org. Chem.*, 1981, **46**, 330.
287. A. J. Bloom and J. M. Mellor, *J. Chem. Soc., Perkin Trans. 1*, 1987, 2737.
288. A. Toshimitsu, S. Uemura and M. Okano, *J. Chem. Soc., Chem. Commun.*, 1982, 87.
289. B. E. Kogai, V. K. Gubernantorov and V. A. Sokolenko, *J. Org. Chem. USSR (Engl. Transl.)*, 1984, **20**, 2324; *Zh. Org. Khim.*, 1984, **20**, 2554 (*Chem. Abstr.*, 1985, **102**, 184 958x).
290. K. B. Rall and A. A. Petrov, *J. Org. Chem. USSR (Engl. Transl.)*, 1966, **2**, 3669; *Zh. Org. Khim.*, 1964, **34**, 3621) (*Chem. Abstr.*, 1965, **62**, 8989f).
291. R. S. Neale and R. L. Hinman, *J. Am. Chem. Soc.*, 1963, **85**, 2666.
292. N. N. Labeish, Yu. I. Porfir'eva and A. A. Petrov, *J. Org. Chem. USSR (Engl. Transl.)* 1984, **20**, 430; *Zh. Org. Khim.*, 1984, **20**, 477 (*Chem. Abstr.*, 1984, **106**, 90 377j).
293. K. Wieser and A. Berndt, *Angew. Chem., Int. Ed. Engl.*, 1975, **14**, 69.
294. A. Hassner and J. Keogh, *J. Org. Chem.*, 1986, **51**, 2767.
295. T. L. Jacobs, in 'The Chemistry of the Allenes', ed. S. R. Landor, Academic Press, New York, 1982, vol. 2, p. 417.
296. F. Minisci, R. Galli and G. Pollina, *Chim. Ind. (Milan)*, 1965, **47**, 736 (*Chem. Abstr.*, 1966, **64**, 691b).
297. V. Nair, *Chem. Heterocycl. Compd.*, 1983, **42** (1), 215.
298. J. E. Franz, H. W. Dietrich, A. Henshall and C. Osuch, *J. Org. Chem.*, 1966, **31**, 2847.
299. R. A. Wohl, *Helv. Chim. Acta*, 1973, **56**, 1826.
300. J. E. McMurry and A. P. Coppolino, *J. Org. Chem.*, 1973, **38**, 2821.
301. T. Shioiri and N. Kawai, *J. Org. Chem.*, 1978, **43**, 2936.
302. N. Kato, Y. Hamada and T. Shioiri, *Chem. Pharm. Bull.*, 1984, **32**, 2496.
303. T. Kauffmann, H. Henkler, C. Kosel, E. Rauch, J. Schulz and R. Weber, *Angew. Chem., Int. Ed. Engl.*, 1962, **1**, 456.
304. H. Marxmeier and E. Pfeil, *Chem. Ber.*, 1964, **97**, 815.
305. M. I. Fremery and E. K. Fields, *J. Org. Chem.*, 1964, **29**, 2240.
306. L. A. Paquette, D. E. Kuhla, J. H. Barrett and R. J. Haluska, *J. Org. Chem.*, 1969, **34**, 2866.
307. A. G. Anastassiou and R. L. Mahaffey, *J. Chem. Soc., Chem. Commun.*, 1978, 915.
308. J. Kučera and Z. Arnold, *Tetrahedron Lett.*, 1966, 1109.
309. J. R. Mahajan, G. A. L. Ferreira and H. C. Araujo, *J. Chem. Soc., Chem. Commun.*, 1972, 1078.
310. R. D. Bowen, J. N. Crabb, C. W. G. Fishwick and R. C. Storr, *J. Chem. Res. (S)*, 1984, 124.
311. H. Hugl and E. Zbiral, *Tetrahedron*, 1973, **29**, 759.

3.6
Addition Reactions with Formation of Carbon–Sulfur or Carbon–Selenium Bonds

KEVIN A. SWISS and DENNIS C. LIOTTA
Emory University, Atlanta, GA, USA

3.6.1 INTRODUCTION

Organosulfur and organoselenium reagents offer many advantages over their first row counterparts for effecting a wide range of important synthetic transformations. These materials can be introduced into substrates either as nucleophiles, radicals or electrophiles. Once introduced, they can facilitate a variety of different processes by stabilizing adjacent positive or negative charges. Finally, after serving their purpose, they can be removed either oxidatively or reductively, thereby introducing either unsaturation or saturation, respectively.

In this chapter addition reactions involving organosulfur and organoselenium reagents are discussed. The examples to be discussed here emphasize those processes which appear to be relatively general and efficient. Although exceptions exist, organoselenium derivatives have generally proven to be more effective than comparable organosulfur derivatives.[1]

3.6.2 THE FORMATION OF CARBON–SULFUR BONDS FROM ADDITIONS TO π-BONDS

3.6.2.1 Formation of Thiiranes

Thiiranes are prepared by two major pathways. The first pathway is the conversion of alkenes to intermediates, such as dihalides, halohydrins or epoxides, followed by nucleophilic attack and subsequent closure of the thiirane sulfur.[2] The other pathway, discussed here, involves the reaction of the alkenes with electrophilic sulfur reagents to produce α-halo-β-sulfur intermediates, which can then be induced to cyclize to thiiranes. There are two methods of introducing sulfur oxidatively to alkenes; these involve the use of either sulfur monochloride or arenethiosulfenyl chloride. In both these methods, the sulfur is introduced stereoselectively (*e.g.* with *trans*-alkenes, *trans*-substituted thiiranes are produced; Scheme 1).

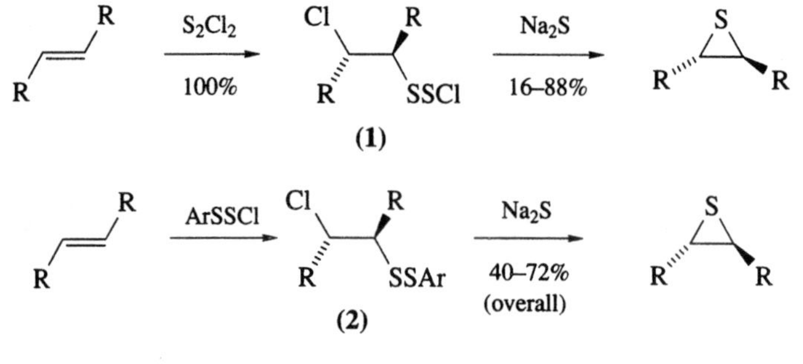

S_2Cl_2 100% (1) SSCl; Na_2S 16–88%

ArSSCl (2) SSAr; Na_2S 40–72% (overall)

Scheme 1

The first method involves sulfur monochloride addition to alkenes forming an α-chloro-β-disulfide (**1**), which can be reduced with sodium sulfide or aluminum amalgam. Unfortunately this procedure suffers from three drawbacks: (i) low yields are realized on reduction of the disulfide; (ii) an excess of alkene (2–3 equiv.) is required to produce good yields; and (iii) the reactions occur poorly with acyclic molecules in general.[3]

A more promising method involves the use of arenethiosulfenyl chlorides. These reagents oxidize alkenes to α-chloro-β-disulfidearenes (*i.e.* **2**) in high yields. The intermediates are then reduced by sodium sulfide and cyclized to thiiranes in good to moderate yield. Unfortunately these arenethiosulfenyl chlorides must be prepared from arenethiols and sulfur dichloride; nevertheless, the yields of acyclic adducts are reliably moderate. In addition, the arene substituents may be changed to modify the nucleophilicity and reducibility of the reagent.[4]

For the formation of thiiranes from alkenes, it seems best to use the arenethiosulfenyl chloride method since this appears to result in the highest yields. While the sulfur monochloride method can be used, the yields are typically poor to moderate over a wide range of substrates. For both methods there are some common problems which include competing group reactivity (other alkenes or electron-rich groups) and, of course, the stench associated with many sulfides.

3.6.2.2 Electrophilic Sulfur Additions

Electrophilic sulfur reagents have been little used in organic synthesis. The reasons include sluggish reactivity, low stability and good alternative methods, such as organoselenium electrophiles. The additions of electrophilic sulfur reagents to alkenes occur in good to excellent yield. Some good examples of electrophile-induced cyclizations are known. Although the electrophilic sulfur reagents are usually divalent, a few reports of hexavalent sulfur electrophiles are also known. These organosulfur reagents offer some synthetic utility, but have seen limited use.

Several types of sulfenyl chlorides are known. Sulfur monochloride and dichloride can add to dienes forming sulfur bridges.[5,6] The most commonly used sulfenyl halides are benzenesulfenyl chlorides and methanesulfenyl chloride. Other sulfenyl halides, such as acetylthiosulfenyl chloride and thiocyanogen chloride, have been added to alkenes, but few subsequent transformations have been carried out with those intermediates.[7,8]

The regiochemistry of additions of alkenes with sulfenyl chlorides exhibits some unusual trends. 2,4-Dinitrobenzenesulfenyl chloride reacts with alkenes, only forming Markovnikov addition products. These results suggest the transition structure possesses little thiiranium character and a large amount of carbonium character.[9] However, methanesulfenyl chloride typically gives anti-Markovnikov addition products (80–95% anti-Markovnikov) with alkyl-substituted terminal alkenes under kinetic control. These initially formed products can be equilibrated to Markovnikov products (all 88% Markovnikov) by trace acid (equation 1). Methanesulfenyl chloride reacts with phenyl-substituted terminal alkenes producing a high preponderance of Markovnikov addition (90–98%). With sterically encumbered terminal alkenes (4-methylbut-1-ene), the degree of anti-Markovnikov addition increases to 95%. Benzenesulfenyl chloride also gives mixtures of regioisomers (63–87%), with anti-Markovnikov products predominating.

These addition products may also be equilibrated to Markovnikov products.[10] The regiochemistry of the addition of sulfenyl chlorides to alkenes is determined by the polarity of the sulfenyl chloride, the relative stability of the carbonium ion intermediates and the steric bulk at or around the alkene.

MeSCl; Cl, SMe 80% + MeS, Cl 20%; AcOH; Cl, SMe 17% + MeS, Cl 83% (1)

Benzenesulfenyl chloride alkene adducts may be transformed to many useful molecules. Intermediates, such as (**3**), can be treated with base to produce vinyl or allyl sulfides (equation 2). Alternatively, the adducts can be oxidized and treated with base to yield vinyl sulfones in high overall yield (equation 3).[11] The thiirane intermediates or adducts, *i.e.* (**3**), may be alkylated with alkyl-titanium and -aluminum reagents which replace the chloride substituent with retention of configuration.[12]

PhSCl, CH_2Cl_2, 100%; SPh, Cl (**3**); DBU, CH_2Cl_2, 73%; SPh (2)

SPh, Cl (**3**); MCPBA, CH_2Cl_2, 99%; SO_2Ph, Cl; DBU, CH_2Cl_2, 83%; SO_2Ph (3)

Cyclizations involving sulfenyl chlorides must be designed carefully, since these reagents can also react with alcohols, amines and other nucleophiles.[13] Benzenesulfenyl chloride has been used to cyclize a 1,4-diene. For example, diene (**4**) cyclizes in good yield to tricyclic heptane product (**5**). This material was then further elaborated to a cyclosativene precursor (**6**; equation 4).[14]

CN (**4**); PhSCl, CH_2Cl_2, 85%; PhS, CN, Cl (**5**); O (**6**) (4)

Another useful cyclization results in the stereoselective synthesis of β-lactams from thiiranium ions derived from α,β-unsaturated amides. Unsaturated amides are treated with benzenesulfenyl chloride and the product is subsequently treated with base under phase transfer conditions. The reaction regenerates a thiiranium ion in the presence of amide anion, which then cyclizes to form β-lactams. The regiochemistry of the alkene addition determines the eventual stereochemical outcome (*e.g. cis*-alkenes produce *cis*-β-lactams; Scheme 2). The yields of the cyclization products are quite sensitive to the amide-protecting group which was employed. With 4-anisyl amide the yield is moderate (73%), but with 4-nitrophenyl amide the yield is excellent (97%), suggesting that the amide must be deprotonated before cyclization can occur.[15]

Benzenesulfenyl chloride can convert unsaturated acids to lactones. The unsaturated acid (**7**) is treated with benzenesulfenyl chloride, followed by Raney nickel, to afford the γ-lactone (**8**) in excellent overall yield (equation 5). Another acid (**9**) has been lactonized to produce the spiro-β-lactone (**10**) which rearranges on silica gel to γ-lactone (**11**; equation 6). δ-Lactones have been also prepared using this approach. An additional advantage of this approach is that the phenyl sulfide moiety can be manipulated into a variety of functional groups to facilitate cyclization.[16] In summary, successful cyclization reactions

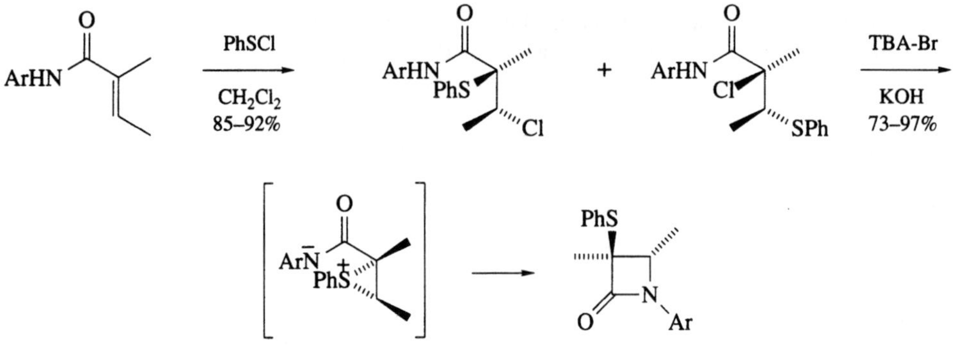

Scheme 2

can be achieved, provided the precursors lack functionality which can react competitively with sulfenyl halides.

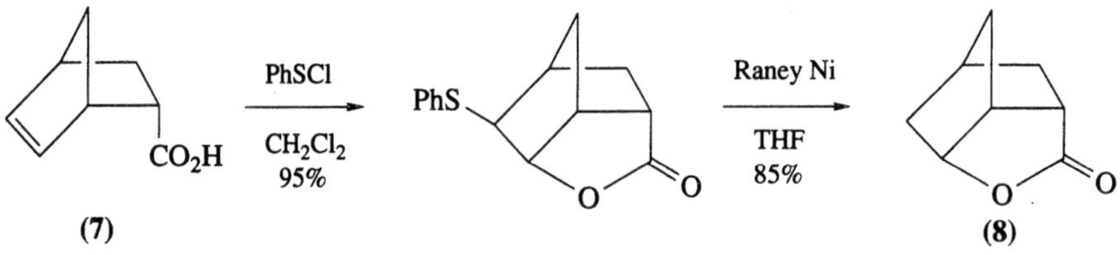

(5)

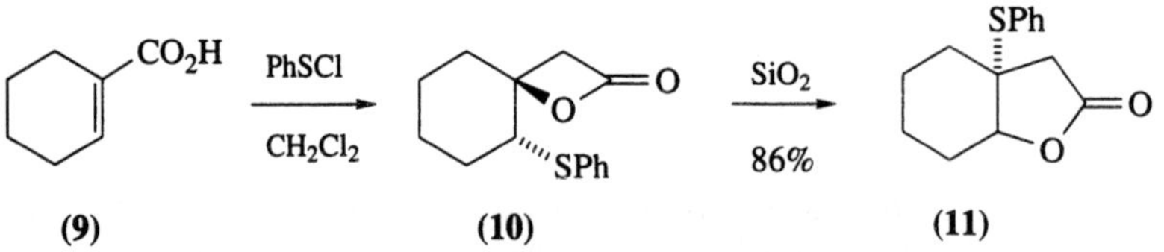

(6)

Arylsulfenyl trifluoroacetates have been generated *in situ* and used for the hydroxysulfenylation of alkenes. These reagents are prepared from diaryl disulfides and LTA in trifluoroacetic acid. Yields for the addition are generally good (42–95%). The hydroxysulfenate intermediate can be oxidatively cleaved by LTA (provided the groups can achieve an antiperiplanar arrangement) to form an α-aldehyde-ω-acetoxy sulfide in good yield (40–88%; Scheme 3).[17]

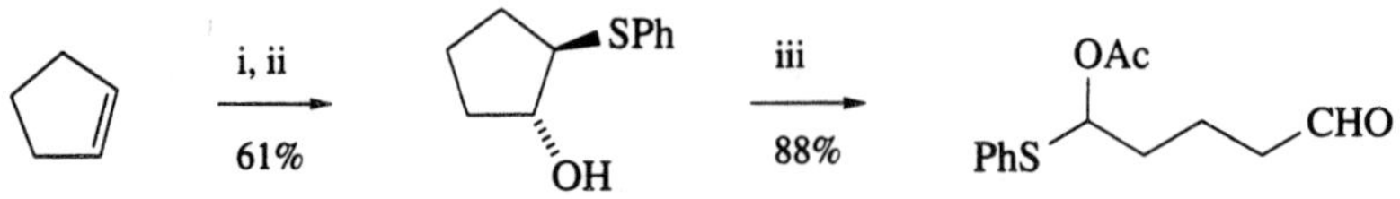

i, LTA, PhSSPh, TFA; ii, H_2O, $NaHCO_3$; iii, LTA, pyridine, AcOH

Scheme 3

Arenesulfonyl halides can add to alkenes with the assistance of a metal catalyst. Several types of catalysts have been employed with varying degrees of success. Substituted phenyl vinyl sulfones, although only the (*E*)-isomers have been prepared from styrenes, benzenesulfonyl chlorides and a ruthenium catalyst in good yield.[18] This same reaction has been tried with chiral ruthenium catalysts with modest success (20–40% *ee*).[19]

3.6.2.3 Radical Sulfur Additions

The number of reported additions of sulfur radicals to π-bonds, while very limited, is growing fast. In general, the major problem which must be overcome in these reactions is the recombination of thiyl radicals to form stable disulfide bonds. This has been partially achieved by the use of dilution techniques and by the use of selenosulfur reagents.

The early work with thiyl radicals primarily involved the intramolecular cyclization of alkenes. The reaction appears to be useful since high yields and stereoselective formation of products are observed (equation 7). Under a variety of conditions the diallyl diester (**12**) produces a 6:1 mixture of stereoisomeric cyclopentanes (**13**) and (**14**) in high yield.[20] When enynes are treated with thiyl radicals, low to moderate yields of addition products are observed (equation 8).[21]

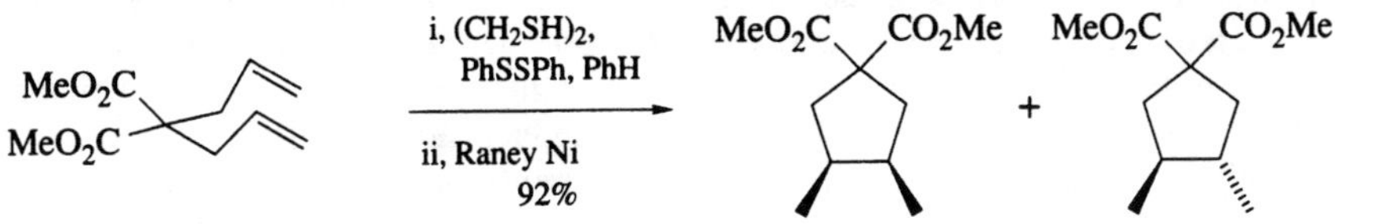

MeO_2C, MeO_2C (**12**) — i, $(CH_2SH)_2$, PhSSPh, PhH; ii, Raney Ni; 92% → MeO_2C CO_2Me (**13**) 86% + MeO_2C CO_2Me (**14**) 14% (7)

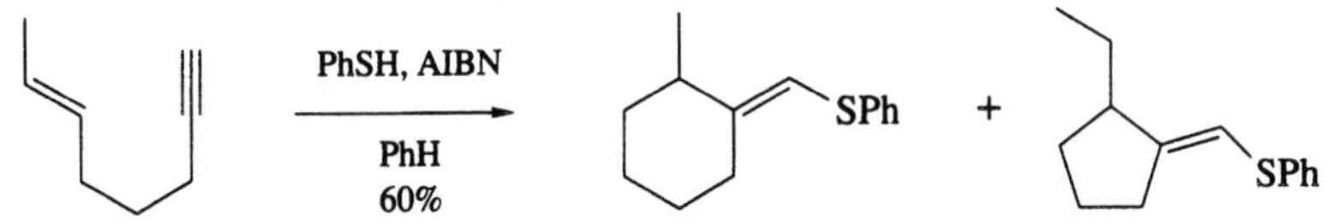

PhSH, AIBN; PhH; 60% → SPh 87% + SPh 13% (8)

A large amount of work has been accomplished using mixed sulfur/selenium reagents, such as selenosulfides and selenosulfones. One example of selenosulfide addition *via* radicals is selenothiolactonization. When selenosulfide (**15**) is treated with AIBN, a mixture of γ-seleno-substituted thiolactones is produced (equation 9). Although these lactonizations result in mixtures of stereoisomers, they usually can be separated.[22] In addition to these reports, selenosulfones have been used to form allenic sulfones, alkynic sulfones and (phenylsulfonyl)dienes (Scheme 4).[23]

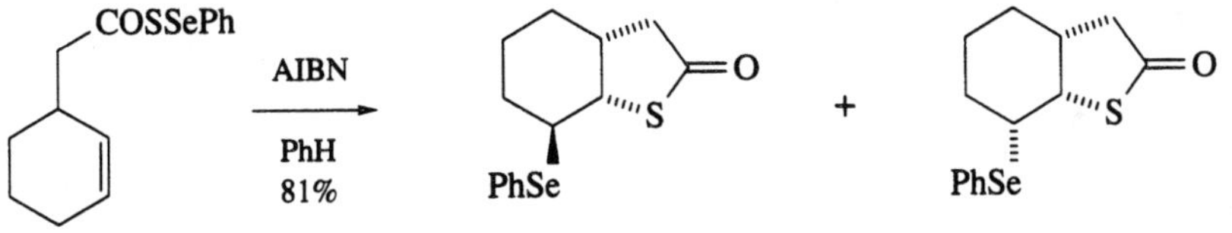

COSSePh (**15**) — AIBN; PhH; 81% → PhSe, S, O 92% + PhSe, S, O 8% (9)

R — $ArSO_2SePh$; PhH, AIBN → R, SO_2Ar, PhSe; [O] 73–98% → R, SO_2Ar; 53–85% → R, SO_2Ar, PhSe, E

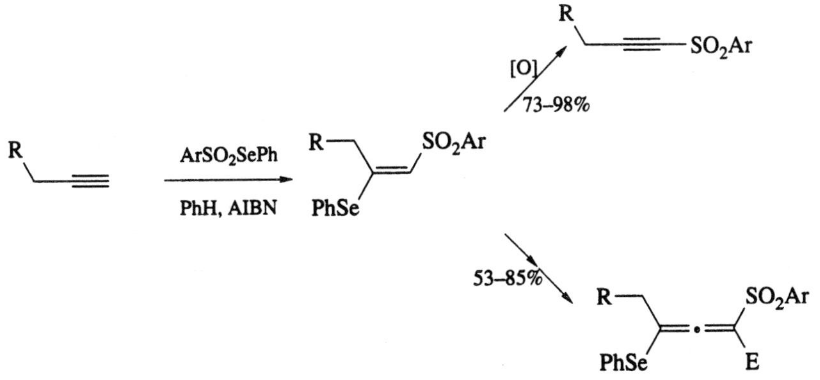

Scheme 4

In summary, much work has been accomplished with selenosulfur radicals, while little has been done with thiyl radicals.

3.6.3 FORMATION OF CARBON–SELENIUM BONDS FROM ADDITIONS TO π-BONDS

3.6.3.1 Regiochemistry of Additions

Selenium electrophiles add to a wide range of π-bonds, usually with good regiochemical and stereochemical results, to form a variety of selenium-containing intermediates.[24] These intermediates can be further elaborated to desired products.

Selenium electrophiles, such as benzeneselenenyl chloride, add to unactivated monosubstituted alkenes, forming Markovnikov or anti-Markovnikov addition intermediates depending on conditions employed. Markovnikov addition products (thermodynamic control) are observed using polar solvents or ambient temperature. Anti-Markovnikov (kinetic control) adducts are initially formed at low temperature and in nonpolar solvents. These adducts can be isolated directly or isomerized to Markovnikov adducts upon warming (equation 10).[25] The isomerization is caused by a reversible seleniranium ion formation and subsequent halide reopening, a process which is accelerated by polar solvents and higher reaction temperatures. Under these conditions 1,2-disubstituted alkenes provide mixtures of regiochemical adducts, while trisubstituted alkenes only give Markovnikov adducts.[24,25]

Bu^n SePh Cl ← PhSeCl, CD_3CN, 25 °C, 94% — Bu^n (alkene) — PhSeCl, CCl_4, –10 °C, 100% → Bu^n Cl PhSe (10)

Benzeneselenenyl chloride adds to allylic alcohols and acetates in a highly regio- and stereo-selective fashion. Substituted cyclohexenyl acetates and benzoates react with benzeneselenenyl chloride, only forming one diastereomer (equation 11). The adduct (**16**) can then be elaborated to an enone. The proposed mechanism for explaining the observed regio- and stereo-selectivity involves the formation of a seleniranium ion *syn* to the ester. The seleniranium ion is then opened by an axial attack of halide ion. With noncyclic allylic acetates, good yields of products possessing the same regio- (>96%) and stereo-chemistry are seen. In additions to allylic alcohols, the *syn* stereochemistry is preserved, but mixtures of regioisomers are observed (70:30).[26]

OAc, R — PhSeCl, CH_2Cl_2, >80% → [O, O, Ph, Se+, H, Cl^-] → OAc, SePh, R, Cl (11)

(**16**)

Selenium electrophiles add to conjugated dienes, only forming 1,2-adducts. Although Markovnikov addition is seen, only a few examples have been reported.[27] With allenes the additions are regiospecific, with the phenylseleno group usually adding to the *sp*-carbon. Unfortunately, all four stereoisomers of halide attack are seen with unsymmetrically substituted allenes.[28]

The addition of selenium electrophiles to activated π-bonds (*i.e.* enol ethers) occurs readily. Enol ethers react with benzeneselenenyl chloride to produce *cis*- and *trans*-α-chloro-β-phenylseleno adducts. These adducts can be transformed into α,β-dichlorides or allylic chlorides.[29] If the reaction is carried out in the presence of alcohols, stereoisomers of β-seleno mixed ketals are isolated (equation 12).[30]

BnO, O — PhSeCl, BnOH, 76% → BnO, SePh, O, OBn (71%) + BnO, SePh, O, OBn (29%) (12)

It has been found that selenium electrophiles add to electron-deficient alkenes to form mixtures of adducts. Benzeneselenenyl chloride adds to chlorocyclohexene (**17**), producing a mixture of adducts (equation 13).[31] 1,1-Difluoroethylene furnishes only one regioisomer (equation 14).[32] Benzeneselenenyl chloride adds to several acrylates (*i.e.* ethyl acrylate, acrolein and others) to produce good yields of regioisomeric adducts. The mixture always favors Markovnikov addition (73–95%), presumably due to electronic effects.[33] However, the pyridine–benzeneselenenyl chloride complex adds to cyclic enones to produce 2-(phenylseleno)enones in good yields. The mechanism is thought to involve a Michael reaction between the enone and pyridine. The enolate is subsequently attacked by the benzeneselenenyl moiety, followed by elimination of pyridine forming the product (equation 15).[34]

(17) —PhSeCl, CH_2Cl_2→ 75% + 25% (13)

—PhSeCl, CH_2Cl_2, 91%→ (14)

—PhSeCl, py, CH_2Cl_2, 71%→ [ ] → (15)

Benzeneselenenyl chloride adds to alkynes to produce mixtures of *trans*-alkene adducts. For example, the addition of benzeneselenenyl chloride to the alkyne (**18**) produces the alkene (**19**), which can be transformed to yield the unusual diene (**20**; equation 16).[35] Alkynic alcohols give anti-Markovnikov addition products under kinetic control. The reaction is thought to proceed through the selenirenium ion (**21**; equation 17).[36] Selenium electrophiles add to α,β-alkynic carbonyl moieties to produce *cis* adducts in good yield (equation 18).[37]

(18) —PhSeCl, CH_2Cl_2, 100%→ (19) —Al(Hg), 76%→ (20) (16)

—PhSeCl, CH_2Cl_2, 100%→ [(21)] → (17)

—PhSeCl, $CHCl_3$, 66–89%→ (18)

In summary, several types of carbon π-bonds react with benzeneselenenyl halides producing usually one or more regioisomers. Several of these regioisomers may be equilibrated *via* their seleniranium ions.

3.6.3.2 Seleno–Heteroatom Additions

Several different functional groups can be introduced by selenenylation. The benzeneselenenyl ion tolerates a wide range of anionic groups. However, with differing anionic moieties the reactivity of the benzeneselenenyl ion varies greatly. All adducts can be subsequently converted to either allylic or vinylic moieties by the *syn* elimination of the corresponding selenoxide (oxidized selenide).

Benzeneselenenyl azide adds to alkenes readily. The addition of the selenenyl azide always occurs with *trans* stereochemistry. The yield of adducts is reliably high with several different alkenes. Unlike benzeneselenenyl chloride, mixtures of regioisomers are found with simple primary alkenes. No addition occurs between benzeneselenenyl azide and ethyl crotonate. The reagent adds to conjugated dienes in a *trans* 1,4-fashion which is thought to be due to an initial *trans* 1,2-addition, followed by a facile 1,3-allylic azide shift (equation 19). Unfortunately, this reagent must be prepared and used *in situ*.[38]

PhSeCl; NaN_3, DMSO; 93%; PhSe; N_3; PhSe; N_3 (19)

Cyanoselenenation of unactivated alkenes occurs only under harsh conditions, such as strong Lewis acid catalysis (*e.g.* tin(IV) chloride). The yields are usually good and only *trans* stereoisomers are observed. Unfortunately, unsymmetrical (terminal and trisubstituted) alkenes yield regioisomeric mixtures. This method provides easy access to unusual trisubstituted carbonitriles starting from relatively simple alkenes (equation 20). Several different cyanoselenates are readily available.[39]

PhSeCN; $SnCl_4$, CH_2Cl_2; 87%; NC; SePh; 83%; + PhSe; CN; 17% (20)

Hydroxyselenenations can be accomplished in good yields by using phenylselenenyl trifluoroacetate. Although the product stereochemistry is consistently *trans*, the observed regioselectivity is poor for unsymmetrical alkenes. The reagent adds to ethyl acrylate, but again the adducts are a mixture of regioisomers. This reagent also adds to alkynes to produce, after hydrolysis, α-phenylseleno ketones (equation 21). This procedure represents an efficient method for the preparation of rearranged allylic alcohols from alkenes similar to allylic oxidations with singlet oxygen but complementary to those obtained from selenium dioxide. The reagent must be prepared *in situ*.[40]

Ph—≡; $PhSeO_2CCF_3$; PhH; 68%; F_3CCO_2; Ph; SePh; KOH; EtOH; O; Ph; SePh (21)

Another method of hydroxyselenenation involves trapping the seleniranium ion by water. The use of *N*-phenylseleno-succinimide (N-PSS) or -phthalimide (N-PSP) as the selenium electrophile facilitates the reaction, since the succinimide or phthalimide anion is not as nucleophilic as water. With dienes, transannular cyclizations can occur, forming bis(phenylseleno) ethers in good yields (equation 22).[41]

N-PSP or N-PSS; CH_2Cl_2, H_2O; 95–97%; PhSe; O; SePh (22)

A method for the conversion of alkenes to α-phenylseleno carbonyl compounds involves the use of benzeneselenenic anhydride. This reagent, which has a relatively short lifetime, is prepared *in situ* from diphenyl diselenide and *t*-butyl hydroperoxide. The alkene is converted to a phenylseleniranium ion which is opened by benzeneselenenic acid. The adduct then loses benzeneselenol, forming the desired product. Unfortunately, high temperatures are required for reaction and all of the adducts from unsymmetrical alkenes are mixtures of regioisomers (equation 23).[42]

'PhSeOSePh', PhH, 56% (23)

Another interesting sequence is the amidoselenenation of alkenes for the synthesis of allylic amides. The seleniranium ion is trapped by a nitrile group which is first converted to an iminium chloride and then hydrolyzed to the amide (similar to the Ritter amide synthesis). Several differing nitriles (*e.g.* methyl to phenyl) have been utilized and all provide good yields of amides. The stereochemistry of addition is always *trans* but mixtures of regioisomers occur with terminal and unsymmetrically substituted olefins (equation 24). The β-seleno amide is easily converted to the allylic amide by oxidation of the phenyl selenide using the standard conditions.[43]

PhSeCl, RCN, 72–98%; H_2O, TfOH (24)

Selenosulfonates (see also Section 3.6.2.3) add to alkenes when catalyzed by Lewis acids. These adducts are transformed *via* oxidative elimination to vinyl sulfones in good overall yields. Similar to other methods only *trans* addition is observed. Mixtures of regioisomers are often produced with unsymmetrical unactivated alkenes. All other alkenes (*i.e.* styrene and acrylonitrile) provide Markovnikov additions. Radical processes yield anti-Markovnikov adducts. Thus, either regioisomeric vinyl sulfone may be prepared from almost any activated alkene by simply varying the mode of addition.[44]

3.6.3.3 Selenium-induced Cyclizations

Several electrophilic selenium-induced cyclizations are known.[1] They include etherifications, lactonizations and lactamidation. Some reasons why selenium electrophiles are used extensively in cyclizations include consistently good yields, few by-products and mild reaction conditions, as well as the ability to further manipulate the seleno group in a variety of straightforward fashions.

Electrophile-induced alkenol cyclizations are well known with many electrophiles.[45] Benzeneselenenyl electrophiles are effective cyclization agents, forming cyclic ethers in high yield. In cyclizations which can involve either a 5-*exo* or a 6-*endo* transition state, the 5-*exo* is highly favored (*i.e.*, tetrahydrofurans, THFs, are produced instead of tetrahydropyrans, THPs). Two examples (equations 25 and 26) illustrate the methodology. When alkenol (**22**; equation 25) is allowed to react with benzeneselenenyl chloride, the allylic ether (**23**) is produced as the only THF.[46] A particularly useful cyclization agent is *N*-phenylselenophthalimide (N-PSP), which possesses a strong selenenylating agent and a very weak nucleophile (phthalimide ion). Using this reagent, ketone alcohol (**24**; equation 26) is transformed to the spiroketal (**25**), a functional group which is often found in ionophore antibiotics.[47] Cyclization can also involve ketones instead of alcohols. For example, ketone (**26**; equation 27) could be cyclized to a mixture of stereoisomeric tetrahydrofurans (**27**) and (**28**). These adducts could be further functionalized *via* both the enol and the phenylseleno group.[48]

Several lactones are ubiquitous in nature. Early workers found that iodine cyclizes unsaturated acids to lactones (iodolactonization). However, since the regiochemistry of selenoxide eliminations is often complementary to the regiochemistry of dehydrohalogenations, selenolactonizations possess obvious synthetic utility. With this methodology several lactones containing five-, six- and seven-membered rings have been prepared in good yield. Also a large macrocyclic lactone (16-membered ring) has been prepared using similar methodology.[41] The first example depicts the conversion of the unsaturated acid (**29**) to the bicyclic lactone (**30**; equation 28).

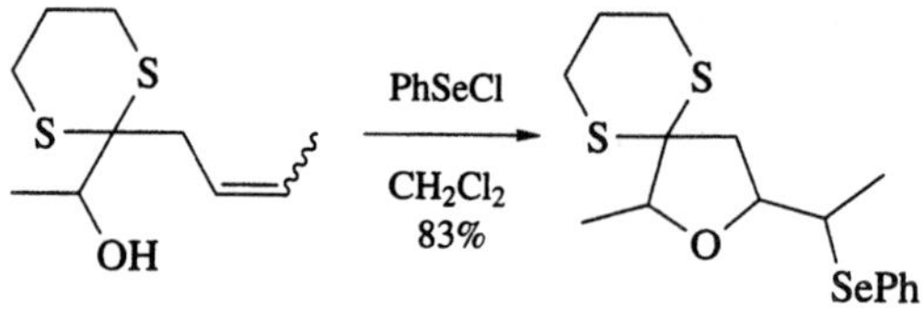

(25)

(22) (23)

(26)

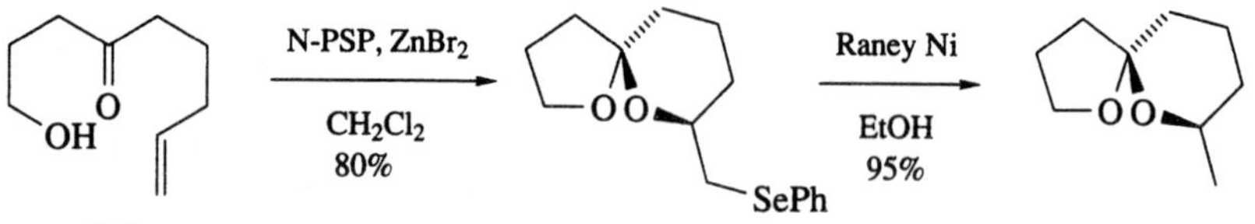

(24) (25)

(27)

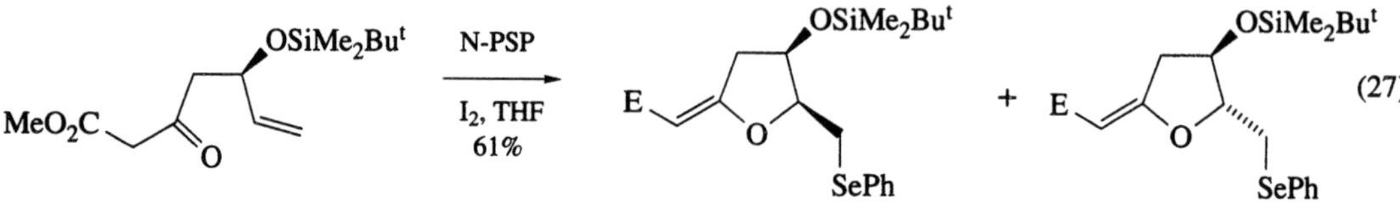

(26) (27) 30% (28) 70%

(28)

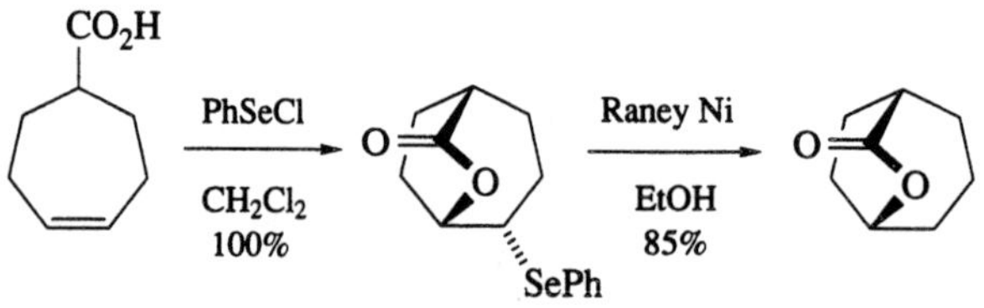

(29) (30)

Several other transannular lactonizations and reductions have been reported to proceed in high overall yields.[49] Also other acid derivatives, such as amides and esters, cyclize to form lactones.[50] Alkynoic acids have been lactonized to γ-alkylidene-γ-lactones in good yield, *e.g.* the conversion of (**31**) to (**32**; equation 29). Unfortunately the vinyl selenide product can isomerize from (*E*) to (*Z*) in a secondary process.[51] Analogous lactam formation is also known. Unsaturated amides, when cyclized with benzeneselenenyl halides, produce good yields of lactams or iminolactones depending upon the alkene utilized.[52] The amide (**33**) cyclizes to the iminolactone (**34**), producing a mixture of stereoisomers (65:35; Scheme 5). The amide (**35**) is cyclized to lactam (**36**) in moderate yield.

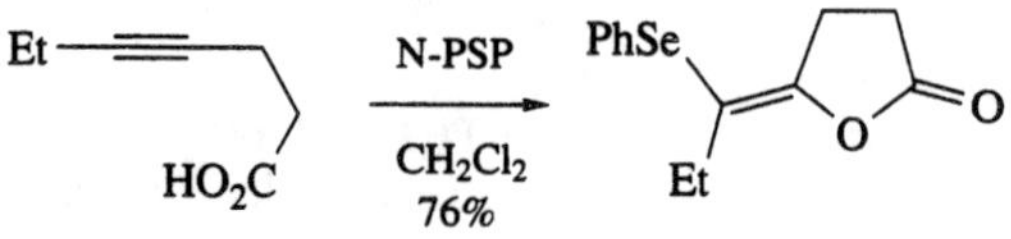

(29)

(31) (32)

Heterocycles can be prepared from selenium-induced cyclizations of urethanes, thioesters and alkenes. These cyclizations are seen infrequently, but have much potential in heterocyclic synthesis. The aniline (**37**) was treated with benzeneselenenyl chloride and silica gel (to facilitate the ring closure) to produce the hexahydrocarbazole (**38**) in good yield (equation 30).[53] Sulfur heterocycles are prepared similarly to nitrogen heterocycles.[54]

Selenium reagents can be used to form carbocycles. A good example of the formation of carbocycles is the elegant synthesis of hirsutene (equation 31). The key step involves the attack of an enol on a seleniranium ion. This type of carbocyclization proceeds very nicely and in high yield.[55]

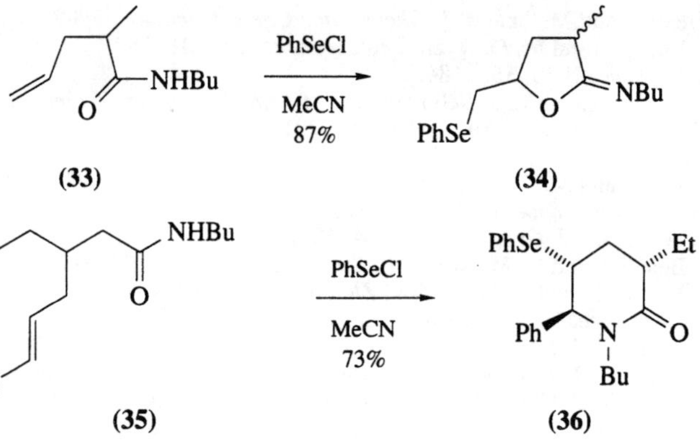

PhSeCl, MeCN, 87%

(33) (34)

PhSeCl, MeCN, 73%

(35) (36)

Scheme 5

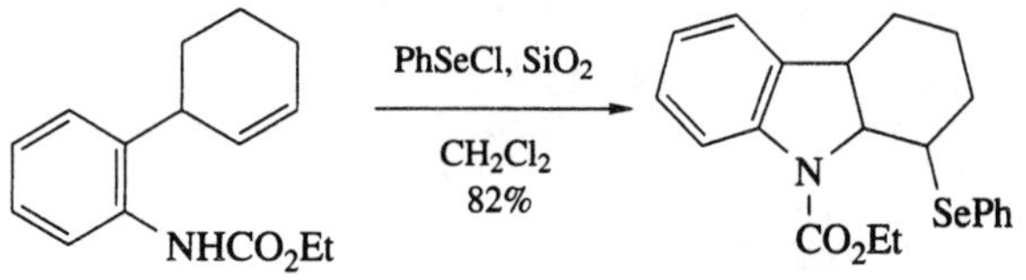

PhSeCl, SiO_2, CH_2Cl_2, 82% (30)

(37) (38)

N-PSP, $SnCl_4$, CH_2Cl_2, 90% (31)

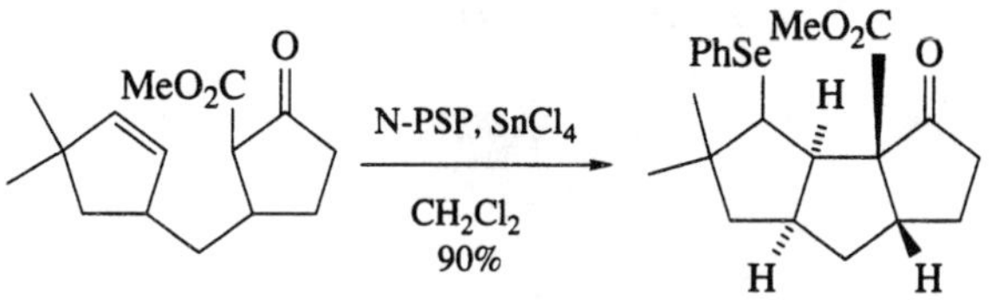

3.6.4 REFERENCES

1. Several organoselenium books are available, these include: D. Liotta (ed.), 'Organoselenium Chemistry', Wiley, New York, 1987; C. Paulmier, 'Selenium Reagents and Intermediates in Organic Synthesis', Pergamon Press, New York, 1986; K. C. Nicolaou and N. A. Petasis, 'Selenium in Natural Products Synthesis', CIS, Philadelphia, 1984; A. Krief and L. Hevesi, 'Organoselenium Chemistry 1', Springer-Verlag, New York, 1988.
2. U. Zoller, in 'The Chemistry of Heterocyclic Compounds', ed. A. Weissberger and E. C. Taylor, Wiley, New York, 1983, vol. 42, part 1, p. 340.
3. F. Lauterschlaeger and N. V. Schwartz, *J. Org. Chem.*, 1969, **34**, 3991.
4. T. Fujisawa and T. Kobori, *Chem. Lett.*, 1972, 935.
5. E. J. Corey and E. Block, *J. Org. Chem.*, 1966, **31**, 1663.
6. D. L. Coffen, D. A. Katonak, N. R. Nelson and F. D. Sancilio, *J. Org. Chem.*, 1977, **42**, 948.
7. W. H. Mueller and P. E. Butler, *J. Org. Chem.*, 1967, **32**, 2925.
8. R. G. Guy and I. Pearson, *J. Chem. Soc., Perkin Trans. 1*, 1973, 281.
9. N. Kharasch and C. M. Buess, *J. Am. Chem. Soc.*, 1949, **71**, 2724.
10. W. H. Mueller and P. E. Butler, *J. Am. Chem. Soc.*, 1968, **90**, 2075.
11. P. B. Hopkins and P. L. Fuchs, *J. Org. Chem.*, 1978, **43**, 1208.
12. M. T. Reetz and T. Seitz, *Angew. Chem., Int. Ed. Engl.*, 1987, **26**, 1028.
13. E. Block, 'Reactions of Organosulfur Compounds', Academic Press, New York, 1978, p. 265.
14. D. Heissler and J.-J. Riehl, *Tetrahedron Lett.*, 1980, **21**, 4707, 4711.
15. M. Ihara and K. Fukumoto, *Heterocycles*, 1982, **19**, 1435.
16. K. C. Nicolaou and Z. Lysenko, *J. Chem. Soc., Chem. Commun.*, 1977, 293.
17. B. M. Trost, M. Ochiai and P. G. McDougal, *J. Am. Chem. Soc.*, 1978, **100**, 7103.
18. N. Kamigata, H. Sawada and M. Kobayashi, *J. Org. Chem.*, 1983, **48**, 3793.
19. M. Kameyama, N. Kamigata and M. Kobayashi, *Chem. Lett.*, 1986, 527.
20. M. E. Kuehne and R. E. Damon, *J. Org. Chem.*, 1977, **42**, 1825.
21. C. A. Broka and D. E. C. Reichert, *Tetrahedron Lett.*, 1987, **28**, 1503.
22. T. Toru, T. Kanefusa and E. Maekawa, *Tetrahedron Lett.*, 1986, **27**, 1583.
23. T. G. Back, M. V. Krishna and K. R. Muralidharan, *Tetrahedron Lett.*, 1987, **28**, 1737; J.-E. Bäckvall, C. Najera and M. Yus, *Tetrahedron Lett.*, 1988, **29**, 1445.
24. D. Liotta and G. Zima, *Tetrahedron Lett.*, 1978, 4977.
25. D. G. Garratt and A. Kabo, *Can. J. Chem.*, 1980, **58**, 1030.
26. D. Liotta, G. Zima and M. Saindane, *J. Org. Chem.*, 1982, **47**, 1258.

27. A. Toshimitsu, S. Uemura and M. Okano, *J. Chem. Soc., Chem. Commun.*, 1982, 965.
28. D. G. Garratt, P. L. Beaulieu and M. D. Ryan, *Tetrahedron*, 1980, **36**, 1507.
29. D. G. Garratt, *Can. J. Chem.*, 1978, **56**, 2184.
30. A. P. Kozikowski, K. L. Sorgi and R. J. Schmiesing, *J. Chem. Soc., Chem. Commun.*, 1980, 477.
31. I. L. Reich and H. J. Reich, *J. Org. Chem.*, 1981, **46**, 3721.
32. A. E. Feiring, *J. Org. Chem.*, 1980, **45**, 1958.
33. A. Toshimitsu, S. Uemura and M. Okano, *J. Chem. Soc., Chem. Commun.*, 1982, 965.
34. G. Zima and D. Liotta, *Synth. Commun.*, 1979, **9**, 697.
35. A. J. Bridges and J. W. Fisher, *J. Org. Chem.*, 1984, **49**, 2954.
36. D. G. Garratt, P. L. Beaulieu and V. M. Morisset, *Can. J. Chem.*, 1981, **59**, 927.
37. E. G. Kataev, T. G. Mannafov and O. O. Saidov, *Zh. Org. Khim.*, 1971, **7**, 2229.
38. A. Hassner and A. S. Amarasekara, *Tetrahedron Lett.*, 1987, **28**, 5185.
39. S. Tomoda, Y. Takeuchi and Y. Nomura, *J. Chem. Soc., Chem. Commun.*, 1982, 871.
40. H. J. Reich, *J. Org. Chem.*, 1974, **39**, 429; other approaches, K. B. Sharpless and R. F. Lauer, *J. Org. Chem.*, 1974, **39**, 430; D. L. J. Clive, *J. Chem. Soc., Chem. Commun.*, 1974, 100.
41. K. C. Nicolaou, D. A. Claremon, W. E. Barnette and S. P. Seitz, *J. Am. Chem. Soc.*, 1979, **101**, 3704.
42. M. Shimizu and I. Kuwajima, *Bull. Chem. Soc. Jpn.*, 1981, **54**, 3100.
43. A. Toshimitsu, T. Aoai, H. Owada, S. Uemura and M. Okano, *J. Org. Chem.*, 1981, **46**, 4727.
44. T. G. Back and S. Collins, *J. Org. Chem.*, 1981, **46**, 3249.
45. A. B. Reitz, S. O. Nortey, B. E. Maryanoff, D. Liotta and R. Monahan, III, *J. Org. Chem.*, 1987, **52**, 4191.
46. K. C. Nicolaou, R. L. Magolda, W. J. Sipio, W. E. Barnette, Z. Lysenko and M. M. Joullie, *J. Am. Chem. Soc.*, 1980, **102**, 3784.
47. S. V. Ley and B. Lygo, *Tetrahedron Lett.*, 1982, **23**, 4625.
48. G. Brussani, S. V. Ley, J. L. Wright and D. J. Williams, *J. Chem. Soc., Perkin Trans. 1*, 1986, 303.
49. K. C. Nicolaou, S. P. Seitz, W. J. Sipio and J. F. Blount, *J. Am. Chem. Soc.*, 1979, **101**, 3884.
50. S. Murata and T. Suzuki, *Chem. Lett.*, 1987, 849.
51. T. Toru, S. Fujita and E. Maekawa, *J. Chem. Soc., Chem. Commun.*, 1985, 1082.
52. A. Toshimitsu, K. Terao and S. Uemura, *J. Org. Chem.*, 1987, **52**, 2018.
53. D. L. J. Clive, V. Farina and A. Singh, *J. Org. Chem.*, 1980, **45**, 2120.
54. K. C. Nicolaou, W. E. Barnette and R. L. Magolda, *J. Am. Chem. Soc.*, 1978, **100**, 2567.
55. S. V. Ley and P. J. Murray, *J. Chem. Soc., Chem. Commun.*, 1982, 1252.

3.7
Addition Reactions with Formation of Carbon–Halogen Bonds

SIGERU TORII and TSUTOMU INOKUCHI
Okayama University, Japan

3.7.1 OXIDATIVE HALOGENATION WITH HALOMETALLIC REAGENTS

Transition metal species play an important role in oxidative halogenation of alkenes, leading to halogenated compounds.[1] Certain oxo-metal compounds, *e.g.* CrO_2Cl_2, $VOCl_3$, $MoCl_5$ and $SbCl_5$, are advantageous for this type of halogenation. Chromyl chloride (CrO_2Cl_2), especially, reacts in unique ways with alkenes, to which its oxygen and halogen atoms transfer. The reaction proceeds *via* initial attack at the metal center. On the other hand, molecular halogens and halide salts, when combined with oxidizing agents such as CrO_3, $Cu(OAc)_2$ and $Pb(OAc)_4$, are capable of converting alkenes to halohydrins and α-halo ketones. Manganese(III) and iron(III) reagents are also beneficial for oxidative halogenation of alkenes through a radical process.

The oxidation of alkenes with CrO_2Cl_2 (1.3 equiv.) in CH_2Cl_2 at –78 °C proceeds in a *cis* stereospecific manner to produce the corresponding chlorohydrin and epoxide. Typically, (*E*)-cyclododecene (**1**; equation 1) gives the corresponding *cis*-chlorohydrin (**2**; 60%), epoxide (**3**; 20%), α-chloro ketone (**4**; 8%) and others (7%).[2] In contrast, the reaction with 2 equiv. of CrO_2Cl_2 in acetone gives preferentially the α-chloro ketones (**5**) and (**6**; equation 2).[4] *cis*-Chlorohydrin acetate (**7**) is formed similarly in a

(1) —[1.3 equiv. CrO_2Cl_2; CH_2Cl_2, –78 °C]→ (2) 60% + (3) 20% + (4) 8% (1)

—[2 equiv. CrO_2Cl_2; acetone, –78 °C]→ (5) 54% + (6) 6% (2)

CH_2Cl_2/AcCl (2:1) system as shown in equation (3).[3] This procedure provides a method which may avoid the over-oxidation of products. Sharpless has proposed a chromyl chloride–alkene π-complex as an intermediate in terms of its *cis* addition. A coordinated alkene (**8**) probably adds to a chlorine–chromium bond to produce an alkylchromium intermediate (**9**; *cis* chlorination), which gives chlorohydrin (**12**) by migration of the alkyl group from chromium to oxygen *via* a chromium derivative (**11**) together with dichloride (**10**) by reductive elimination, as shown in Scheme 1.[2] Migration of the O—Cr bond of the C—Cr complex (**13**) occurs with retention of configuration, giving the epoxide (**15**). *trans*-Chlorohydrin and a ketone are produced by an acid-catalyzed ring opening of the epoxide with either HCl or Lewis acidic chromium species.[5,6]

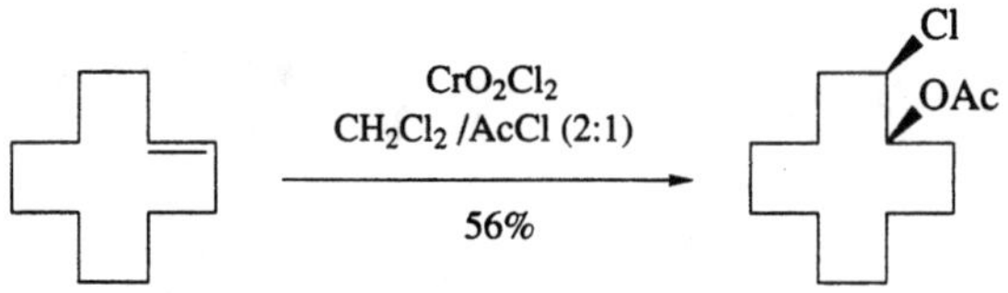

CrO_2Cl_2, CH_2Cl_2/AcCl (2:1), 56% → (**7**) (3)

(**8**) (**9**) (**13**) (**10**) (**11**) (**14**) (**12**) (**15**)

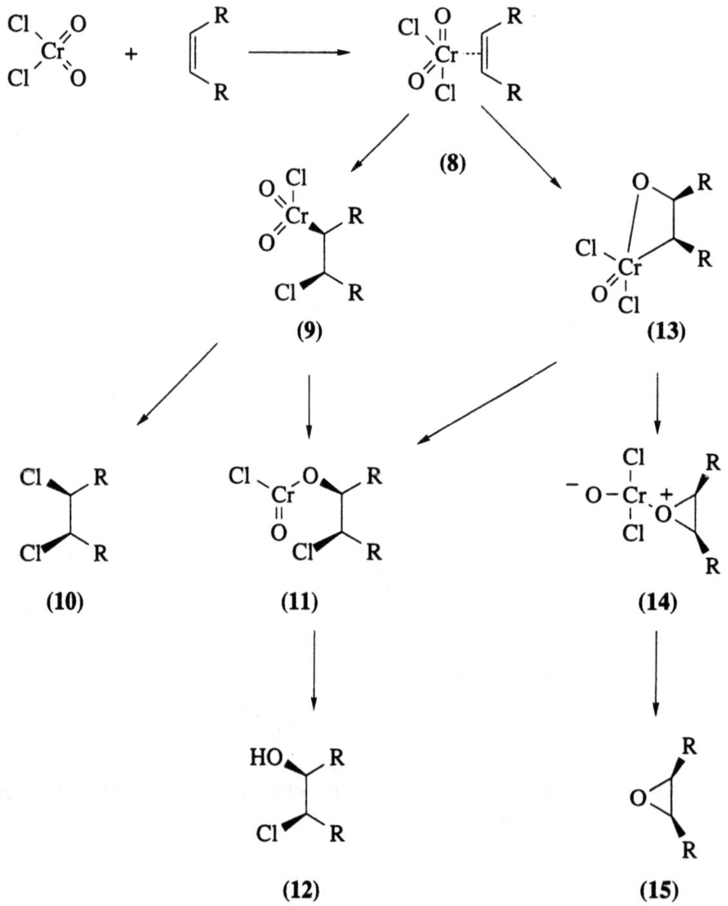

Scheme 1

Norbornadiene can be oxidized with chromyl chloride to give *cis*-1,2-chlorohydrin (**16**; 50%) and rearrangement product (**17**; 37%; equation 4). Thus, chromyl chloride oxidation of norbornadiene generates partially a species of sufficient carbenium ion character to promote the Wagner–Meerwein rearrangement.[7] Similarly, the oxidation of norbornene at –80 °C affords the corresponding *exo-cis*-chlorohydrin (63%), 3-*exo*-chloronorcamphor (11%) and norcamphor (3.1%; equation 5).[8] The chromyl chloride oxidation procedure can be successfully applied to the formation of aldehydes from *exo*-alkenes in monoterpenoid synthesis, *e.g.* the conversion of (**18**) to (**19**) in good yield, by the combination of zinc dust reduction (equation 6).[9] Chromyl fluoride, prepared from CrO_3 and cobalt(III) fluoride at 450 °C,[10] can react with ergosteryl acetate (**20**) to give 5α-hydroxy-6α-fluoro derivatives (**21**) in 43% yield (equation 7).[11] However, the oxidation of other steroid alkenes with CrO_2X_2 (X = Cl, F) is affected by the structure of substrates. For example, cholesteryl acetate (**22**), when treated at –70 to –30 °C, produces 3β-acetoxy-5α-chlorocholestan-6-one (**23**; 12%), a mixture of α- and β-cholesteryl acetate epoxides (**24**; 4%), 3β-acetoxy-6β-chlorocholestan-6α-ol (**25**; 9%) and 3β-acetoxy-5α-chlorocholestan-6β-ol (**26**; 18%; equation 8).[11]

CrO_2Cl_2, CH_2Cl_2, –78 °C → OH, Cl (**16**) 50% + OH, Cl (**17**) 37% (4)

CrO_2Cl_2, –80 °C → OH, Cl 63% + O, Cl 11% + O 3.1% (5)

CO_2Me (**18**) → i, ii, 80% → CO_2Me, CHO (**19**) (6)

i, CrO_2Cl_2; ii, Zn, H_2O

AcO (**20**) → CrO_2F_2, CH_2Cl_2, 43% → AcO, HO, F (**21**) (7)

AcO (**22**) → CrO_2Cl_2, –70 to –30 °C → AcO, Cl, O (**23**) 12% + AcO, O (**24a**) + AcO, O (**24b**) 4% +

AcO, HO, Cl (**25**) 9% + AcO, Cl, OH (**26**) 18% (8)

Perrhenyl chloride (ReO_3Cl) can react with alkenes to produce *cis*-chlorohydrins, while $CrOCl_3$,[2] $MnOCl_3$[2] and $VOCl_3/POCl_3$[12] result in chlorination.

The silver chromate–iodine[13] or pyridinium chlorochromate–iodine[14] reagents can be used for the oxidation of double bonds, leading to the corresponding α-iodo ketones. In the case of the former reagent, the nucleophilic addition of a hypoiodous chromic acid mixed anhydride (**27**) probably produces α-iodo ketones (**28**) in an analogy to a Prevost reaction (Scheme 2). The oxidation of trisubstituted alkenes by use of iodine and pyridinium dichromate (PDC) produces the corresponding iodohydrin (**29**) in a regioselective and stereospecific manner (equation 9).[15] α-Chloro ketones (**30**) are also obtained in 60–90% yields by the oxidation of di- and tri-alkyl-substituted alkenes with a CrO_3/TMS-Cl/CCl_4 system, whose selectivity and yields are superior to those in the chromyl chloride system (equation 10). In this conversion, polyoxochromium dichloride [$Cl(CrO_2)_nCl$] is postulated as an active oxidizing species.[16]

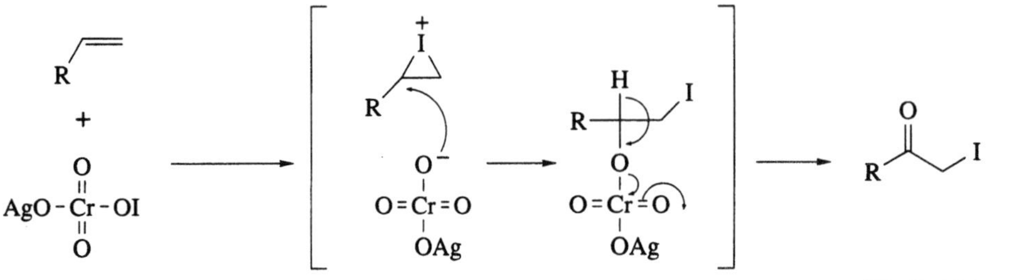

(**27**) (**28**) 60–80%

Scheme 2

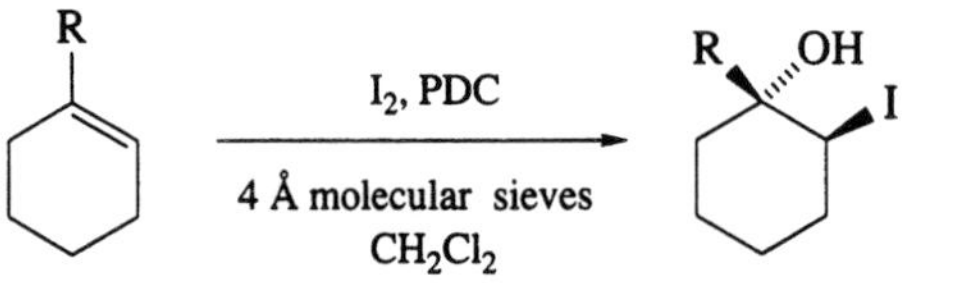

I_2, PDC; 4 Å molecular sieves; CH_2Cl_2 (**29**) 50–60% (9)

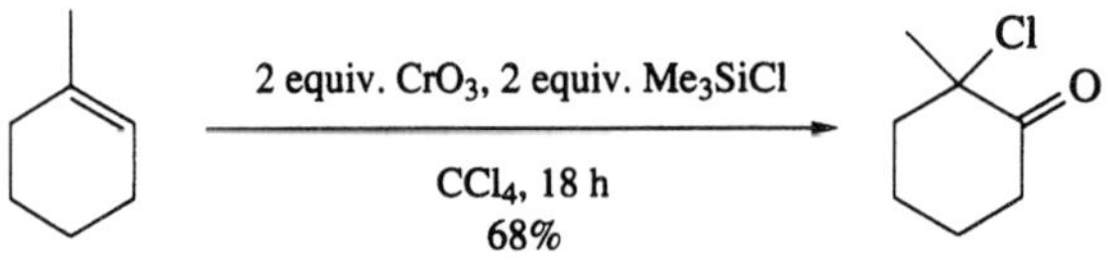

2 equiv. CrO_3, 2 equiv. Me_3SiCl; CCl_4, 18 h; 68% (**30**) (10)

Halogenation of enol ethers and enol esters, leading directly to α-halo ketones is realized by use of molecular halogen or halide salts and metal oxidants. Pyridinium chlorochromate (PCC)/I_2,[17] CrO_3/TMS-Cl/I_2,[18] AgOAc/I_2,[19] TlOAc/I_2,[20] $Pb(OAc)_4$ and metal halides[21] and $Cu(OAc)_2/I_2$[22] are useful classes of reagents for this conversion, and some examples are listed in Table 1.

Antimony(V) chloride ($SbCl_5$)[23,24] and molybdenum(V) chloride ($MoCl_5$)[25,26] can react spontaneously with alkenes to give predominantly the corresponding *cis*-1,2-dichlorides (equation 11). The reaction probably proceeds through a successive insertion and reductive elimination sequence. The chlorination of butadiene with $SbCl_5$[24] and copper(II) chloride[27] results preferentially in the formation of (*Z*)- and (*E*)-1,4-dichloro adducts,[24] while the reaction with chlorine gives an 1:1 mixture of 1,2- and 1,4-adducts, as shown in Table 2 and equation (12).[28] The formation of (*Z*)-1,4-dichloro-2-butene may be ascribed to a transition state as shown in Scheme 3.

Enol silyl ethers can lead to α-chloro ketones on treatment with anhydrous copper(II) chloride in DMF or iron(III) chloride in acetonitrile (equation 13, Table 1).[29] The chlorination of (**36**; equation 14) proceeds through a cation radical intermediate formed by an electron-transfer process with metal halides.

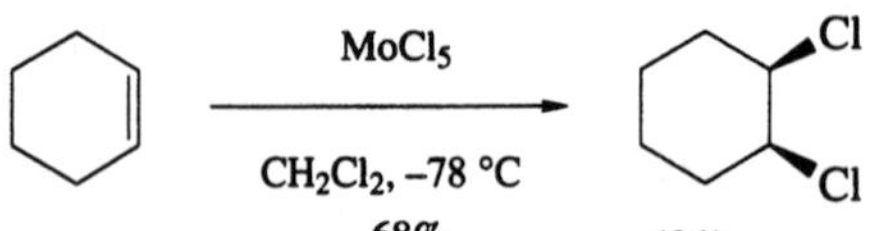

$MoCl_5$; CH_2Cl_2, –78 °C; 68% (**31**) (11)

Table 1 Halogenation of Enol Ethers and Esters with Metallic Reagents

Alkene	*Reagents and conditions*	*Product*	*Yield (%)*	*Ref.*
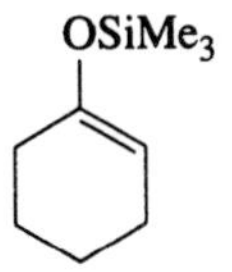 1-(trimethylsiloxy)cyclohexene ($OSiMe_3$)		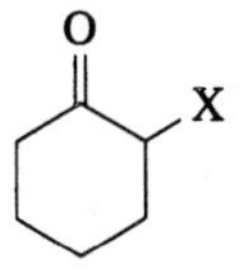 2-X-cyclohexanone		
	2 equiv. PCC, 1 equiv. I_2, CH_2Cl_2	X = I	76	17
	2 equiv. $CrO_3/ClSiMe_3$, 1.5 equiv. I_2	X = I	83	18
	1 equiv. AgOAc, 1 equiv. I_2, CH_2Cl_2	X = I	84	19
	1 equiv. TlOAc, 1 equiv. I_2, AcOH	X = I	46 (11)	20
	3.3 equiv. $Pb(OAc)_4$, 2.2 equiv. NaBr, MeOH	X = Br	47	21
	3.3 equiv. $Pb(OAc)_4$, 2.2 equiv. NaI, MeOH	X = I	87	21
	3 equiv. $CuCl_2$, DMF	X = Cl	58	22
	5 equiv. $FeCl_3$, DMF	X = Cl	52	22
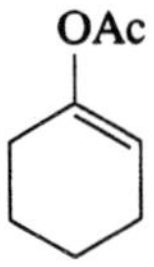 1-acetoxycyclohexene (OAc)		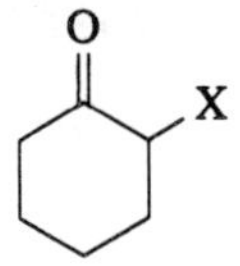 2-X-cyclohexanone		
	3.3 equiv. $Pb(OAc)_4$, 2.2 equiv. NaBr, MeOH	X = Br	82	21
	3.3 equiv. $Pb(OAc)_4$, 2.2 equiv. $CuCl_2$, MeOH	X = Cl	99	21
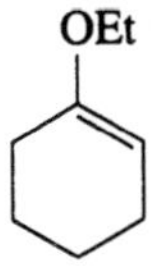 1-ethoxycyclohexene (OEt)		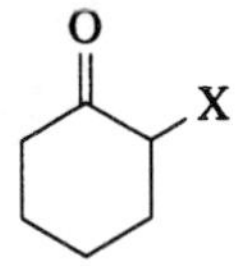 2-X-cyclohexanone		
	3.3 equiv. $Pb(OAc)_4$, 2.2 equiv. NaBr, MeOH	X = Br	90	21
steroidal enol acetate (AcO, C_8H_{17})	1.1 equiv. $Cu(OAc)_2$, 1.1 equiv. I_2, AcOH	2-iodo-3-keto steroid (I, O)	71	22

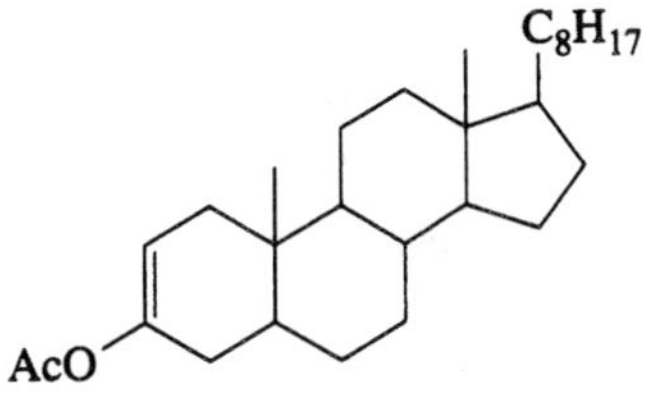

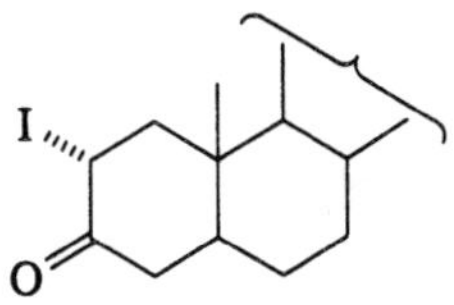

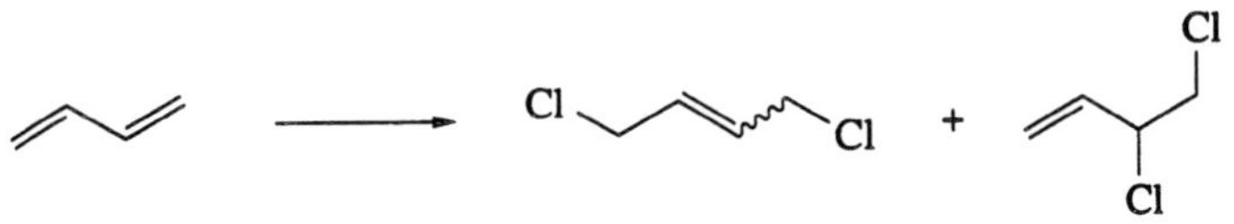

$$\text{CH}_2\text{=CH–CH=CH}_2 \longrightarrow \text{ClCH}_2\text{CH=CHCH}_2\text{Cl } \textbf{(32)} + \text{CH}_2\text{=CH–CHCl–CH}_2\text{Cl } \textbf{(33)} \quad (12)$$

Table 2 Halogenation of 1,3-butadiene

Reagent	*Solvent*	*1,4-Adduct:1,2-Adduct*	*Ref.*
Cl_2	CH_2Cl_2	43.5:56.5	28
$SbCl_5$	CH_2Cl_2	62:38	24
$CuCl_2$	MeCN	85:15	27

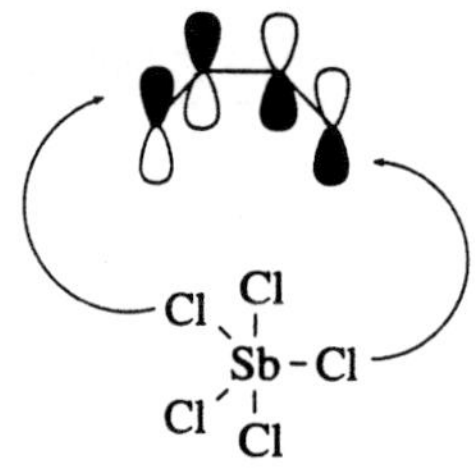

Scheme 3

The reaction of indole (**38**) with copper(II) chloride also proceeds through a cation radical intermediate (**39**) to give mainly 2-chloroindole (**40**), together with a dimeric by-product (**41**), as shown in equation (15).[30]

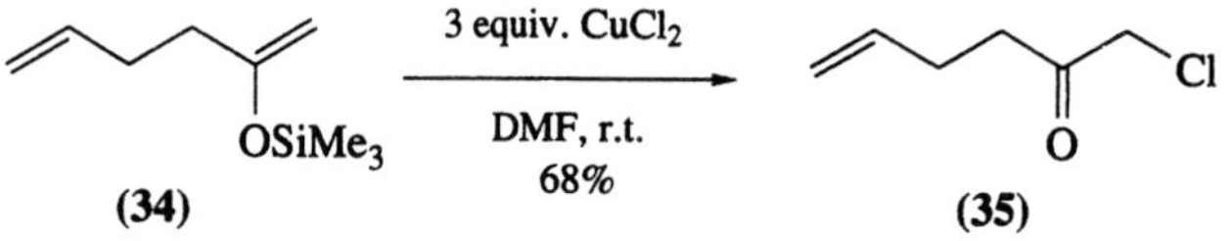

3 equiv. $CuCl_2$, DMF, r.t., 68% (34) → (35) (13)

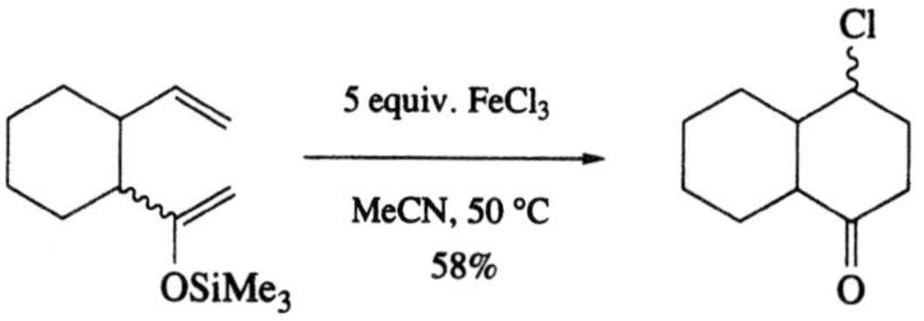

5 equiv. $FeCl_3$, MeCN, 50 °C, 58% (36) → (37) (14)

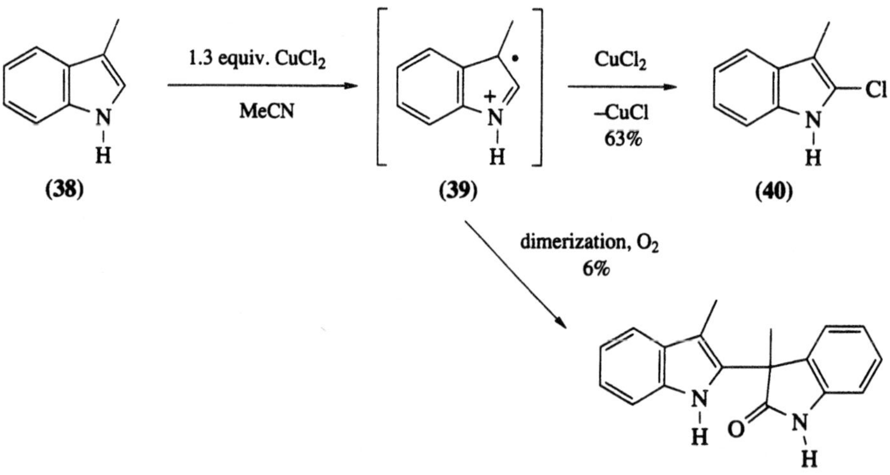

1.3 equiv. $CuCl_2$, MeCN (38) → [(39)] $CuCl_2$, –CuCl, 63% → (40); (39) dimerization, O_2, 6% → (41) (15)

Manganese(III) acetate or chloride salts [$Mn_3O(OAc)_7HOAc$, $MnCl_3$] can react with alkenes to afford 1,2-dichlorides and chlorohydrin acetates (equation 16).[31] The manganese(III) reagent promotes the chlorination of 1,6-heptadiene (**42**) to afford almost equal amount of open chain and cyclized dichlorides

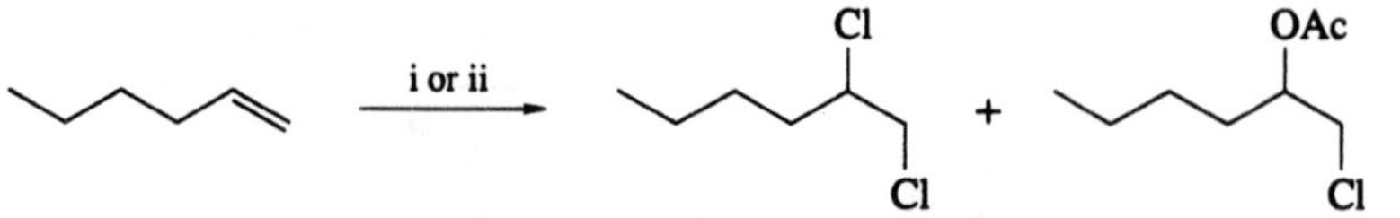

i or ii (16)

i, $[Mn_3O]/CaCl_2$, AcOH, 110 °C	81%	–
ii, $MnCl_3$, AcOH, 25 °C	52%	11%

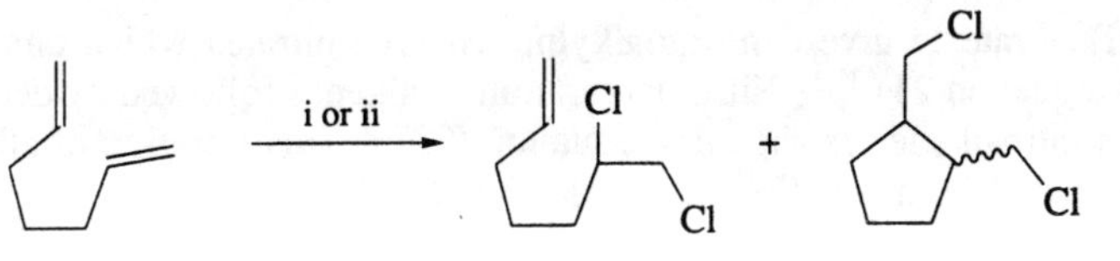

(42) → (43) + (44) (17)

i, $[Mn_3O]/CaCl_2$, AcOH, 110 °C 40.9:1

ii, $MnCl_3$, AcOH, 25 °C 1.12:1

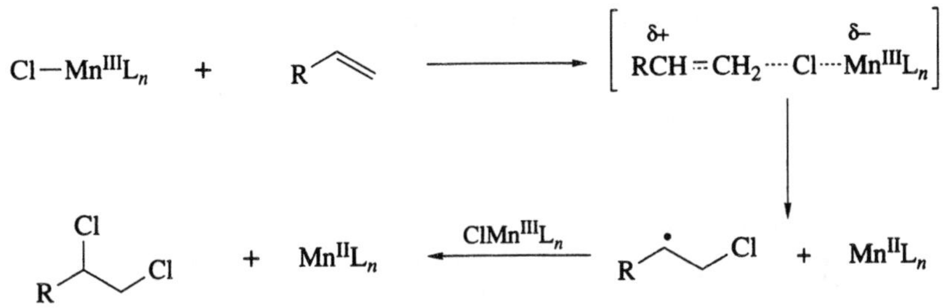

$Cl{-}Mn^{III}L_n$ + R–CH=CH$_2$ → [$\delta+$ RCH=CH$_2$···Cl···$Mn^{III}L_n$ $\delta-$] → R–ĊH–CH$_2$Cl + $Mn^{II}L_n$ —$ClMn^{III}L_n$→ RCHCl–CH$_2$Cl + $Mn^{II}L_n$

Scheme 4

(**43**) and (**44**; equation 17), depending on reaction conditions, which suggests the mechanism involving the radical intermediate as shown in Scheme 4.

Oxidation of pregnenolene (**45**) with Bornstein's reagent $[PbF_2(OAc)_2]$ or $Pb(OAc)_4$–HF (1:4) produces an organolead complex (**46**), and the subsequent treatment with bromine gives a 6β-bromo-5α-fluoro derivative (**47**), a reversed regiochemical bromofluoro isomer which can be obtained by a usual FBr-releasing reagent (equation 18).[32]

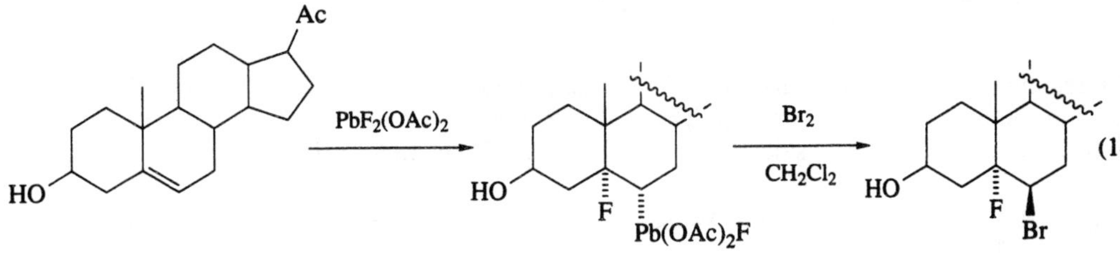

(45) —$PbF_2(OAc)_2$→ (46) —Br_2, CH_2Cl_2→ (47) 95% (18)

Direct synthesis of α-bromo ketones from alkenes is carried out by use of sodium bromite ($NaBrO_2$).[33] The reaction proceeds *via* the bromohydrin (**48**) as intermediate (equation 19).

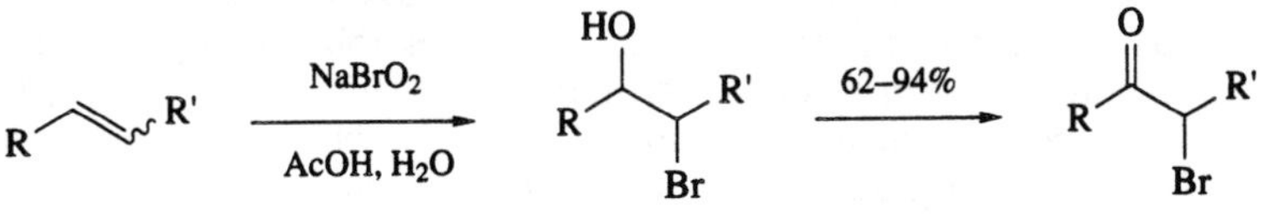

R–CH=CH–R' —$NaBrO_2$, AcOH, H_2O→ (48) —62–94%→ RCO–CHBr–R' (19)

3.7.2 HALOFUNCTIONALIZATIONS WITH Hg^{II}, Tl^{III} AND Te^{IV} REAGENTS

1,2-Bifunctionalization of alkenes has been performed by the reaction of mercury(II) salts (chloride, fluoride, nitrate, *etc.*) and halogens (Br_2 or I_2) with alkenes, through the addition of halogen and mercury(II) salt anion.[34–36] The mercury(II) salt–halogen combination method provides a potential method for the preparation of a variety of 1,2-bifunctionalized organic halides. For instance, the reaction of alkenes with bromine or iodine and different mercury(II) salts (HgX_2; X = F, Cl, Br, HCO_2, AcO, CF_3CO_2, $EtCO_2$, $PhCO_2$, NO_3, $MeSO_3$, 4-$MeC_6H_4SO_3$, SCN or 4-$MeC_6H_4SO_2$) in CH_2Cl_2 affords the corresponding 1,2-bifunctionalized products (**49**) as shown in equation (20). β-Bromoalkyl nitrates (**50**) are formed by treating alkenes with mercury(II) nitrate and bromine, in which the alkenes react rapidly and reversibly with mercury(II) nitrate to give β-nitratoalkylmercury(II) nitrates which undergo brominolysis to give the product (**50**; equation 21).[35,37] Nitromercuration of alkenes followed by demercuration of the resulting (**51**) produces nitroalkenes as shown in equation (22).[38] This nitration of alkenes is improved by using nitryl iodide prepared from $AgNO_2$ and I_2 (equation 23).[39]

HgX_2, Hal_2 (20)

(**49**)

X = F, Cl, Br, RCO_2, NO_3, RSO_3, SCN, *etc.*; Hal = Br, I

R^1–CH=CH–R^2 + Br_2 (or I_2) —$Hg(NO_3)_2$→ O_2NO, X, R^1, R^2 (21)

(**50**) X = Br, I

$HgCl_2$, $NaNO_2$, 80% → NO_2, HgCl (**51**); NaOH, –Hg, –NaCl, 98% → NO_2 (22)

R, NO_2I → I, R, NO_2; base, 70–80% → NO_2, R (23)

The reaction of alkenes with thallium(III) acetate (TTA) forms oxythallium adducts (**52**) in a similar manner to the case of oxymercuration. The thallium moiety of adducts can be replaced by a halogen atom by heating with copper(I) salts (CuX–KX; X = I, Br, Cl) in acetonitrile (equation 24).[40,41]

Ph —$Tl(OAc)_3$, MeOH→ MeO, $Tl(OAc)_2$, Ph (**52**) —CuI–KI, MeCN, 80 °C, 89%→ MeO, I, Ph (24)

Selenium and tellurium reagents have been used for stereoselective halogenations of alkenes. For example, *trans* addition of benzeneselenenyl chloride to alkenes followed by the displacement of the seleno moiety with chloride can lead to *cis*-1,2-dichlorides (equation 25).[42] The addition of 2-naphthyltellurium trichloride proceeds in an *anti* stereospecific manner (equation 26), whereas tellurium tetrachloride gives a mixture of *syn* and *anti* adducts.[43] The reaction of allyl esters with tellurium tetrachloride accompanies acyl migration to give the 1-(trichlorotelluro)-3-chloro adduct (**54**; equation 27).[44]

PhSeCl → Cl, SePh; Cl_2 or PhSeCl → Cl, Y–SePh +; Cl^- → Cl, Cl (25)

Y = Cl, PhSe

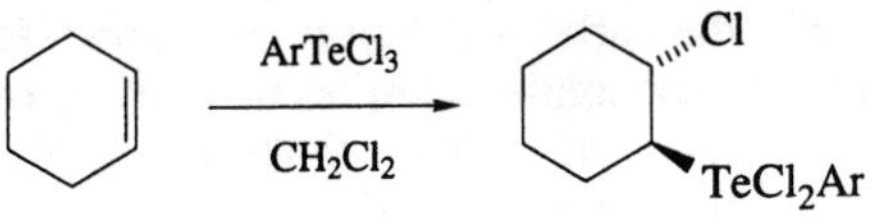

ArTeCl$_3$, CH$_2$Cl$_2$ → (**53**) >97% *anti* (26)

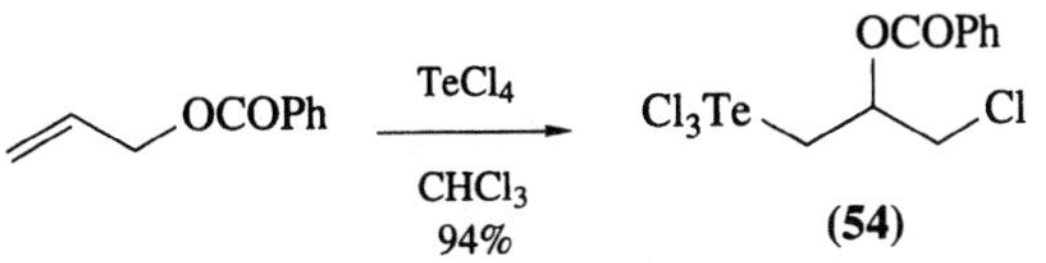

TeCl$_4$, CHCl$_3$, 94% → (**54**) (27)

3.7.3 OXIDATIVE HALOGENATION WITH NONMETALLIC REAGENTS

Generation of positive-like halogen species *in situ* has been realized by the oxidation of halide salts with *m*-chloroperbenzoic acid (MCPBA). The procedure can be used for haloetherification and lactonization (equation 28).[45] Oxidation of potassium bromide with MCPBA in the presence of 18-crown-6 (10 mol %) produces *m*-chlorobenzoylhypobromite, which adds across the double bond to furnish *trans*-1,2-bromocarboxylates (**55**; equation 29).[46]

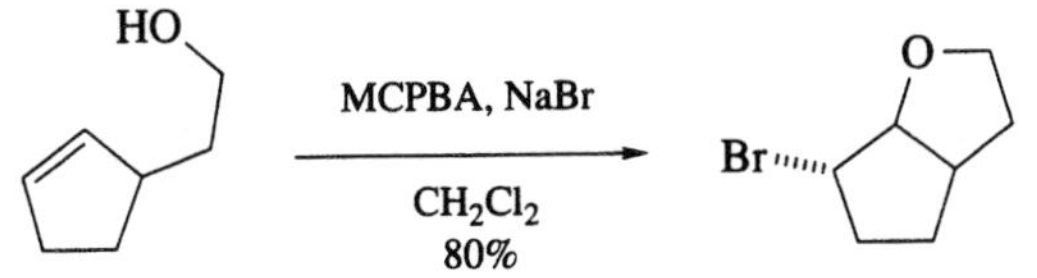

MCPBA, NaBr, CH$_2$Cl$_2$, 80% (28)

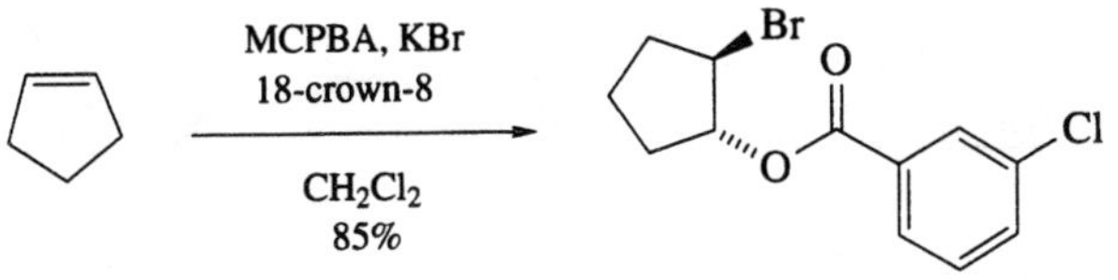

MCPBA, KBr, 18-crown-8, CH$_2$Cl$_2$, 85% → (**55**) (29)

The fluorination of uracil (**56**) and cytosine (**57**) in a F$_2$/AcOF/aq. AcOH system has been performed *via* a radical cation fluoride complex (**58**). This radical cation intermediate (**58**) is probably formed by an electron transfer due to the action of hypofluorite (AcOF) as shown in Scheme 5.[47]

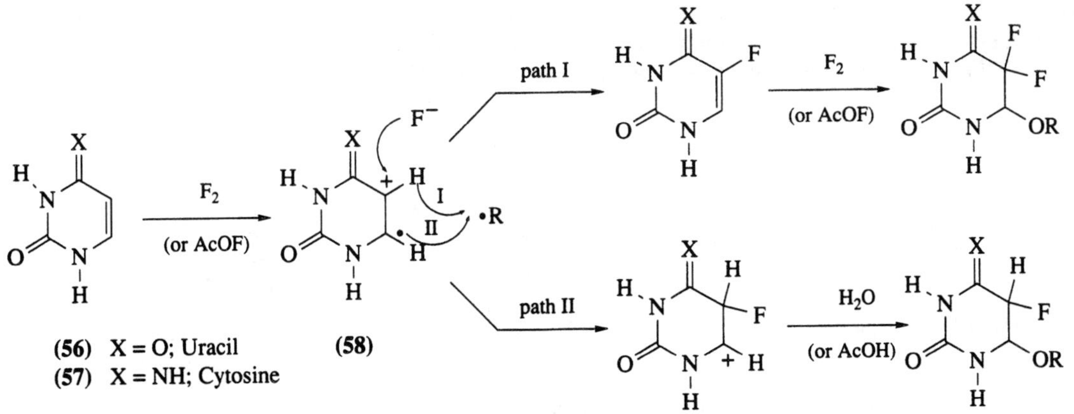

(**56**) X = O; Uracil
(**57**) X = NH; Cytosine

F$_2$ (or AcOF) → (**58**); path I: F$_2$ (or AcOF); path II: H$_2$O (or AcOH)

R = H or OAc

Scheme 5

Positive halogen complexes with pyridine bases are known as versatile halogenating reagents.[48] Bis(*sym*-collidine)iodine(I) tetrafluoroborate (**59**) in dimethyl sulfoxide is a potential reagent for the direct conversion of alkenes to α-iodo carbonyl compounds (Scheme 6).[49] The oxidation involves the transfer of an iodo cation species to the carbon–carbon double bond, forming a three-membered iodonium ring intermediate. Subsequent nucleophilic addition of DMSO, forming a dimethyloxosulfonium salt, followed by proton abstraction with collidine gives the α-iodo carbonyl compound (Scheme 6). Glycals (**60**) can be converted into the corresponding α-iodo-α,β-unsaturated lactones (**61**) with this reagent (equation 30). On the other hand, the reaction of alkenes with this reagent (**59**) in CH_2Cl_2 results in the formation of 1,2-iodofluorides.[50]

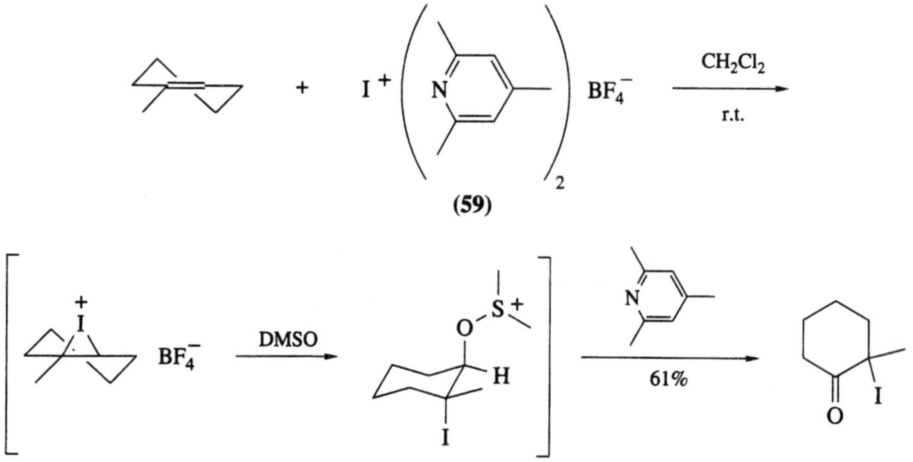

CH_2Cl_2, r.t.; (59); DMSO; 61%

Scheme 6

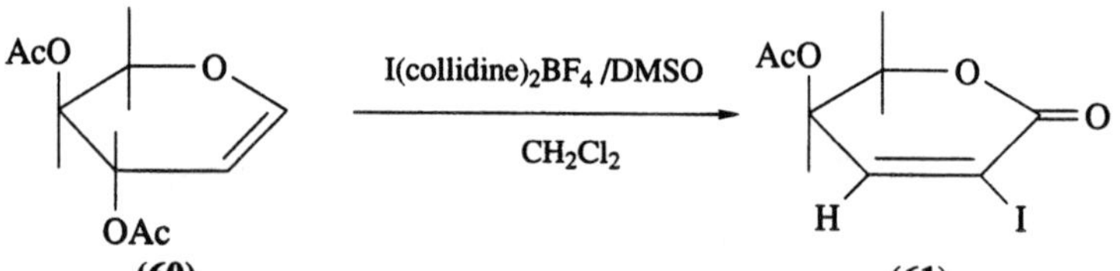

AcO; OAc; (60); $I(collidine)_2BF_4$ /DMSO, CH_2Cl_2; AcO; H; I; (61) (30)

Intramolecular bromoalkylamine addition to alkenes has been performed by using bis(*sym*-collidine)bromine(I) perchlorate (**62**; equation 31).[51] The method plays an important role in the key step of (±)-sporamine synthesis (**63**) → (**65**). The reaction of $I(py)_2BF_4$ and alkenes in the presence of nucleophiles produces 1,2-bifunctionalized iodo compounds (equation 32).[52] The reaction of 1,3-dienes with $I(py)_2BF_4$ allows the regiospecific 1,2-addition of iodine and a nucleophile to terminal dienes to give (**69**; equation 33) and the 1,4-addition to internal 1,3-dienes to give (**70**; equation 34).[53] 1,4-Addition with this reagent is enhanced by the addition of tetrafluoroboric acid. Iodine-induced cyclization of arylalkene system (**71**) occurs with this reagent as shown in equation (35).[54,55]

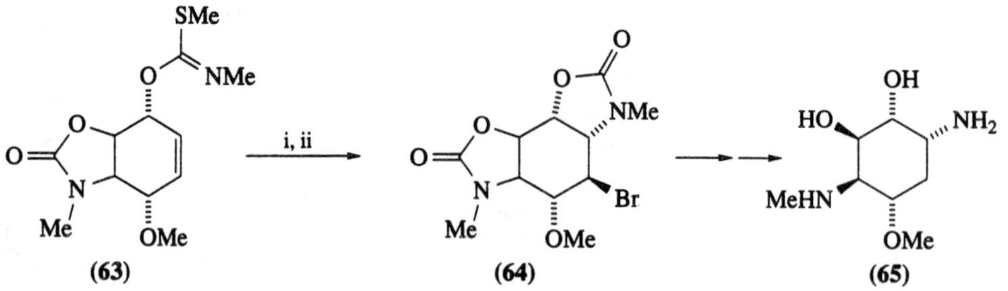

(63) → i, ii → (64) → → (65) (31)

i, $Br(collidine)_2ClO_4$ (**62**); ii, Na_2CO_3, H_2O

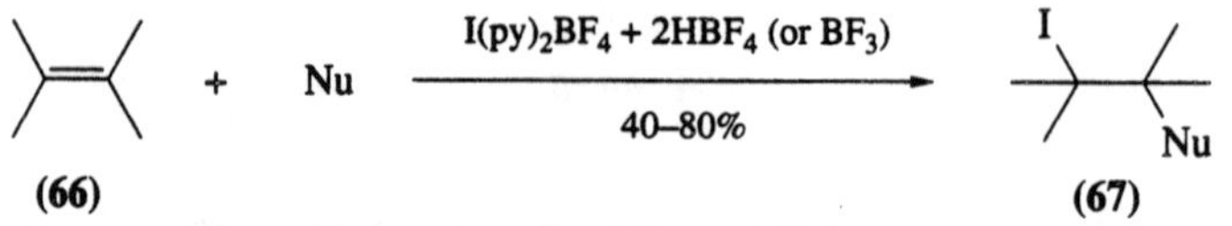

(66) + Nu —$I(py)_2BF_4$ + $2HBF_4$ (or BF_3), 40–80%→ (67) (32)

Nu = F^-, Cl^-, Br^-, NO_2^-, OCN^-, MeOH, AcOH, MeCN

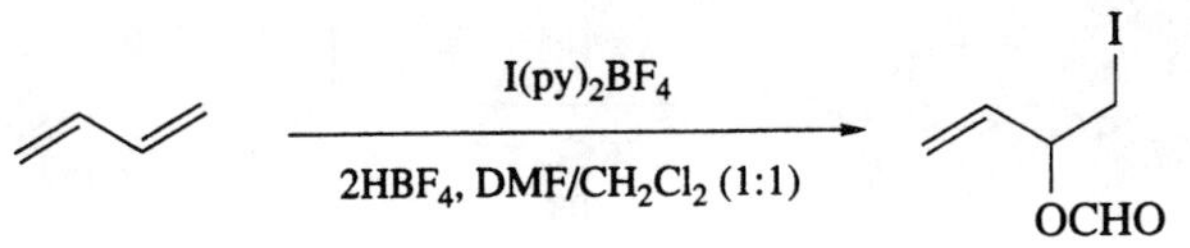

I(py)$_2$BF$_4$; 2HBF$_4$, DMF/CH_2Cl_2 (1:1) — (33)

OCHO

(68) **(69)** 30%

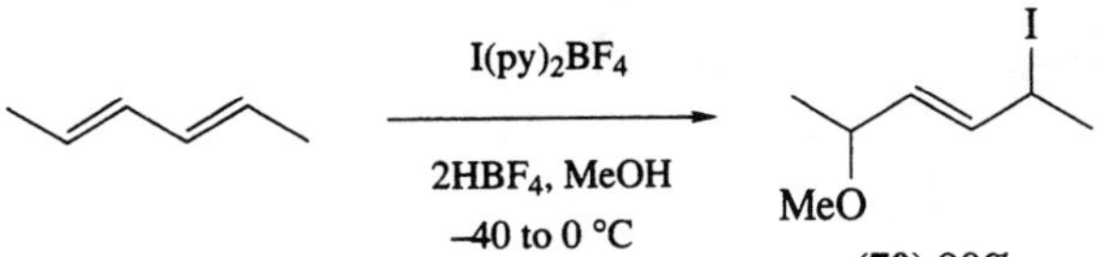

I(py)$_2$BF$_4$; 2HBF$_4$, MeOH; –40 to 0 °C — (34)

MeO

(70) 90%

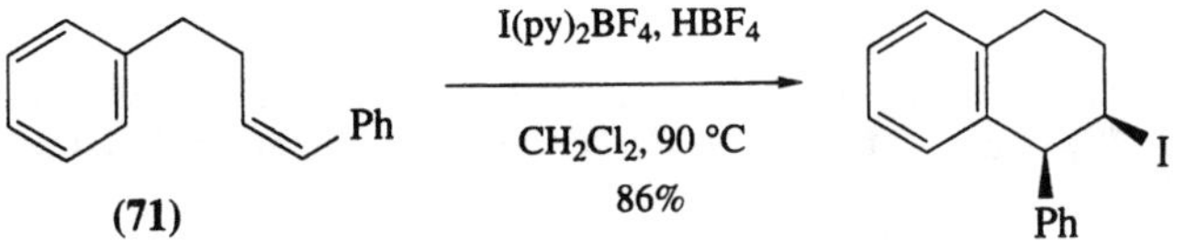

I(py)$_2$BF$_4$, HBF$_4$; CH_2Cl_2, 90 °C; 86% — (35)

Ph **(71)** Ph I **(72)**

Chloramine T (CT) is a powerful positive chlorine releasing reagent toward alkenes in acetic acid, giving mainly a *trans*-chlorohydrin acetate (**73**) as shown in Scheme 7. The similar reaction in an acetone/H_2O(1:1)/H_2SO_4 system under reflux produces chlorohydrins in moderate yields.[56,57]

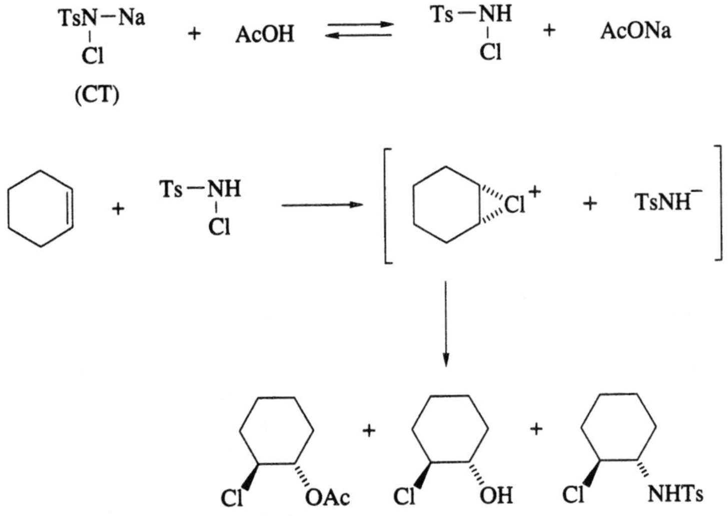

TsN(Cl)–Na (CT) + AcOH ⇌ Ts–NH(Cl) + AcONa

cyclohexene + Ts–NH(Cl) ⟶ [chloronium ion Cl^+ + $TsNH^-$]

↓

Cl, OAc **(73)** 53% + Cl, OH 23% + Cl, NHTs 7%

Scheme 7

The ene-type chlorination, specific with this halogen atom, of alkenes would account for a somewhat different mode of halogen addition, giving useful allylic chlorides. Efficient and convenient reagents and methods developed are dichlorine monoxide (Cl_2O; equation 36),[58] *t*-butyl hypochlorite,[59] and electrochemical reactions with chloride ion (equation 37).[60] The ene-type chlorination proceeds smoothly with 1,1-disubstituted alkenes (**74**) and (**76**) to give (**75**) and (**77**). Especially, dichlorine monoxide is a potential reagent for complex molecules such as penicillin and cephalosporin derivatives.[58]

Electrooxidation of halide salts is quite useful for the generation of reactive species of halogen atoms under mild conditions.[61] Functionalization of alkenes involving the formation of halohydrins, 1,2-halides, α-halo ketones, epoxides, allylic halides and others has been achieved by electrochemical reactions and is well documented in the literature.[62] On the other hand, electrogenerated carbenium ions can be captured by nucleophilic halide anions, providing a new route to halogenated compounds

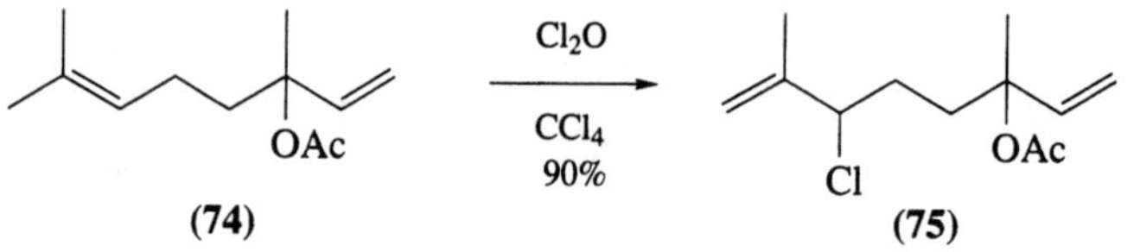

(36)

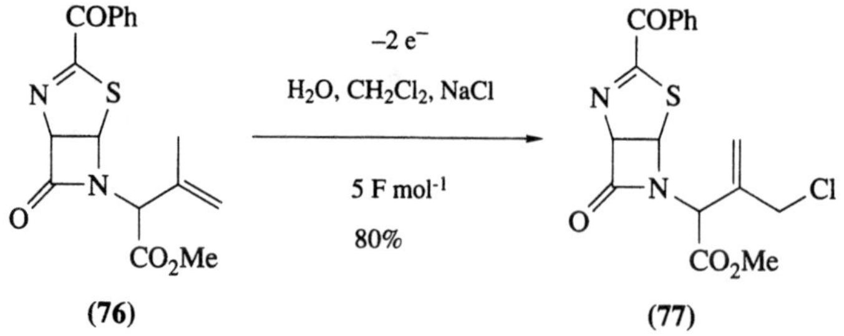

(74) (75)

(37)

(76) (77)

(Scheme 8). For instance, the fluorination at the position α of ketones has been realized by the anodic oxidation of enol acetates (**78**) in an $MeCN/Et_3N{\cdot}3HF/(Pt)$ system under potential control, giving α-fluoro ketone (**79**; equation 38).[63]

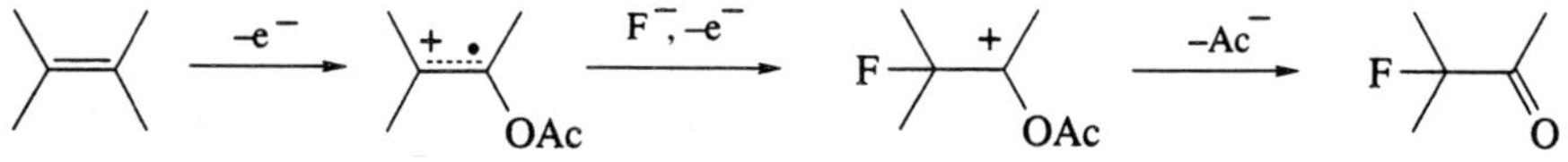

Scheme 8

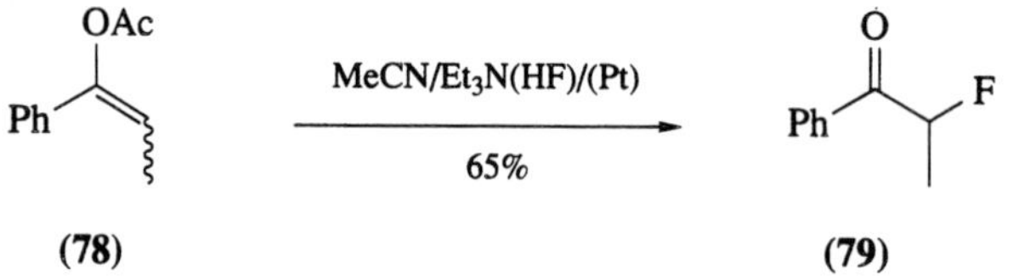

(38)

(78) (79)

3.7.4 MISCELLANEOUS (PHOTOCHEMICAL AND ENZYMATIC REACTIONS)

Photochemically induced oxidative halogenation of alkenes has been carried out in the presence of metal halide/oxygen complexes. The photooxidative halogenation of disubstituted alkenes (**80**; R = H) in a $FeCl_3/O_2$ system can lead to α-chloro ketones (**82**).[64] The initial step of the reaction involves a photoinduced interligand electron transfer from the chlorine ligand to molecular oxygen through the metal ion and alkene molecule. The chlorine radical then adds to the alkene, and successive coupling of the resulting carbon radical with an oxygen radical anion and protonation completes the process (Scheme 9). The

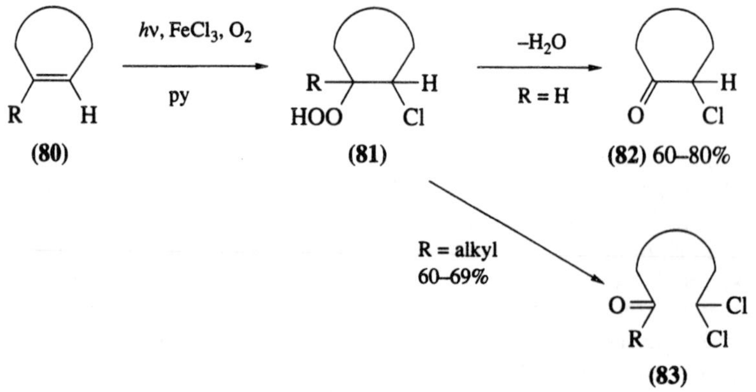

(80) (81) (82) 60–80%

(83)

Scheme 9

hydroperoxides (**81**) further decompose into α-chloro ketones. The metal-catalyzed photooxidation of alkenes is operated with uranyl acetate (1 equiv.) in the presence of bromotrichloromethane (equation 39).[65]

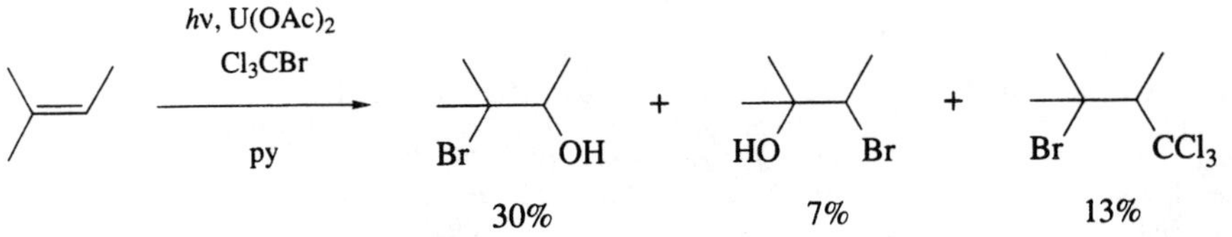

(39)

Semiconductor-mediated photoelectrochemical oxidation of halide salts provides a procedure for the halogenation of alkenes *via* excitated halide species.[66] For example, bromination of cyclohexene has been performed in a TiO_2/Bu_4NBr (or Ph_3PMeBr) $/O_2$ system. The reactive bromine species probably arise from a one-electron oxidation of adsorbed bromide ions on the semiconductor by photoirradiation, which produces surface-bound bromine atoms (equation 40).

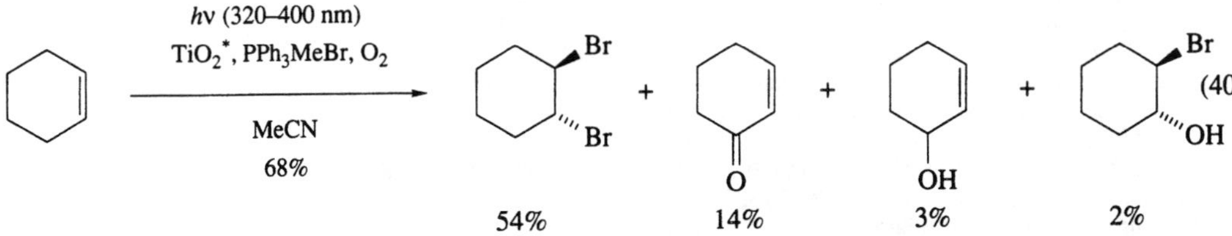

(40)

Enzyme-catalyzed halogenation has been found in biological processes. In the haloperoxidase reaction, halide ions are converted to positive halogen species by hydrogen peroxide. For example, haloperoxidase catalyzes the chlorination, bromination and iodination of cytosine, uracil, *etc.*, to give the corresponding halogenated compounds (**84**; equation 41).[67] The reaction is carried out in a phosphate buffer at pH 3 in the presence of KCl, KBr, or KI by gradually adding H_2O_2.

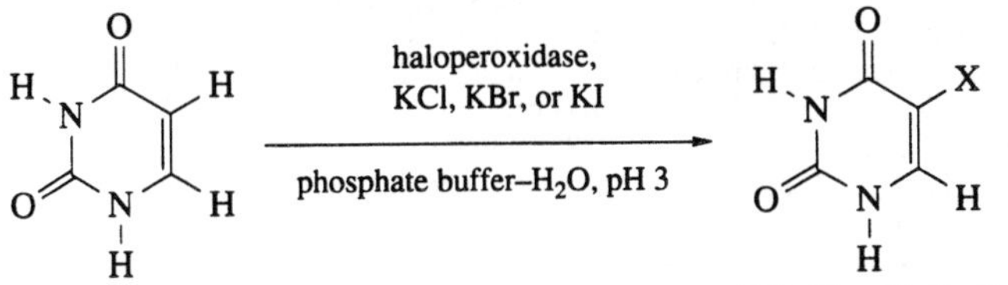

(41)

(**84**) X = Cl, Br, I

3.7.5 REFERENCES

1. (a) R. A. Sheldon and J. K. Kochi, 'Metal-Catalyzed Oxidation of Organic Compounds', Academic Press, New York, 1981; W. J. Mijs and C. R. H. I. de Jonge (eds.), 'Organic Syntheses by Oxidation with Metal Compounds', Plenum Press, New York, 1986.
2. K. B. Sharpless, A. Y. Teranishi and J.-E. Bäckvall, *J. Am. Chem. Soc.*, 1977, **99**, 3120.
3. J.-E. Bäckvall, M. W. Young and K. B. Sharpless, *Tetrahedron Lett.*, 1977, 3523.
4. K. B. Sharpless and A. Y. Teranishi, *J. Org. Chem.*, 1973, **38**, 185.
5. F. Freeman and K. W. Arledge, *J. Org. Chem.*, 1972, **37**, 2656.
6. F. Freeman and N. J. Yamachika, *J. Am. Chem. Soc.*, 1972, **94**, 1214.
7. S.-K. Chung, *Tetrahedron Lett.*, 1978, 3211.
8. F. W. Bachelor and U. O. Cheriyan, *Tetrahedron Lett.*, 1973, 3291.
9. T. H. Jones, M. S. Blum and H. M. Fales, *Tetrahedron Lett.*, 1980, **21**, 1701.
10. G. D. Flesh and H. J. Svec, *J. Am. Chem. Soc.*, 1958, **80**, 3189.
11. A. G. M. Barrett, D. H. R. Barton and T. Tsushima, *J. Chem. Soc., Perkin Trans. 1*, 1980, 639.
12. W. A. Waters and J. S. Littler, in 'Oxidation in Organic Chemistry', ed. K. B. Wilberg, Academic Press, New York, 1965, part A, p. 185.
13. G. Cardillo and M. Shimizu, *J. Org. Chem.*, 1977, **42**, 4268.
14. R. D'Ascoli, M. D'Auria, L. Nucciarelli, G. Piancatelli and A. Scettri, *Tetrahedron Lett.*, 1980, **21**, 4521.
15. R. Antonioletti, M. D'Auria, A. De Mico, G. Piancatelli and A. Scettri, *Tetrahedron*, 1983, **39**, 1765.
16. (a) A. F. Guerrero, H.-J. Kim and M. F. Schlecht, *Tetrahedron Lett.*, 1988, **29**, 6707; (b) J. G. Lee and D. S. Ha, *Tetrahedron Lett.*, 1989, **30**, 193.
17. M. D'Auria, F. D'Onofrio, G. Piancatelli and A. Scettri, *Synth. Commun.*, 1982, **12**, 1127.
18. J. M. Aizpurua, M. Juaristi, B. Lecea and C. Palomo, *Tetrahedron*, 1985, **41**, 2903.

19. G. M. Rubottom and R. C. Mott, *J. Org. Chem.*, 1979, **44**, 1731.
20. R. C. Cambie, R. C. Hayward, J. L. Jurlina, P. S. Rutledge and P. D. Woodgate, *J. Chem. Soc., Perkin Trans. 1*, 1978, 126.
21. S. Motohashi and M. Satomi, *Synthesis*, 1982, 1021.
22. C. A. Horiuchi and J. Y. Satoh, *Synthesis*, 1981, 312.
23. S. Uemura, O. Sasaki and M. Okano, *J. Chem. Soc., Chem. Commun.*, 1971, 1064.
24. R. P. Vignes and J. Hammer, *J. Org. Chem.*, 1974, **39**, 849.
25. S. Uemura, A. Onoe and M. Okano, *Bull. Chem. Soc. Jpn.*, 1974, **47**, 3121.
26. J. S. Filippo, A. F. Sowinski and L. J. Romano, *J. Am. Chem. Soc.*, 1975, **97**, 1599.
27. W. C. Baird, J. H. Surridge and M. Buza, *J. Org. Chem.*, 1971, **36**, 3324.
28. G. E. Heasley, D. A. Smith, J. N. Smith, V. L. Heasley and D. F. Schellhamer, *J. Org. Chem.*, 1980, **45**, 5206.
29. Y. Ito, M. Nakatsuka and T. Saegusa, *J. Org. Chem.*, 1980, **45**, 2022.
30. É. Balogh-Hergovich and G. Speier, *J. Chem. Soc., Perkin Trans. 1*, 1986, 2305.
31. K. D. Donnelly, W. E. Fristad, B. J. Gellerman, J. R. Peterson and B. J. Selle, *Tetrahedron Lett.*, 1984, **25**, 607.
32. M. Ephritikhine and J. Levisalles, *J. Chem. Soc., Chem. Commun.*, 1974, 429.
33. T. Kageyama, Y. Tobito, A. Katoh, Y. Ueno and M. Okawara, *Chem. Lett.*, 1983, 1481.
34. R. C. Larock, *Tetrahedron*, 1982, **38**, 1713.
35. J. Barluenga, J. M. Martinez-Gallo, C. Nájera and M. Yus, *J. Chem. Soc., Chem. Commun.*, 1985, 1422.
36. J. Barluenga, J. M. Martinez-Gallo, C. Nájera and M. Yus, *J. Chem. Res. (S)*, 1986, 274.
37. A. J. Bloodworth and P. N. Cooper, *J. Chem. Soc., Chem. Commun.*, 1986, 709.
38. E. J. Corey and H. Estreicher, *J. Am. Chem. Soc.*, 1978, **100**, 6294.
39. W.-W. Sy and A. W. By, *Tetrahedron Lett.*, 1985, **26**, 1193.
40. S. Uemura, K. Zushi, A. Tabata, A. Toshimitsu and M. Okano, *Bull. Chem. Soc. Jpn.*, 1974, **47**, 920.
41. S. Uemura, H. Miyoshi, A. Toshimitsu and M. Okano, *Bull. Chem. Soc. Jpn.*, 1976, **49**, 3285.
42. A. M. Morella and A. D. Ward, *Tetrahedron Lett.*, 1984, **25**, 1197.
43. J.-E. Bäckvall, J. Bergman and L. Engman, *J. Org. Chem.*, 1983, **48**, 3918.
44. L. Engman, *J. Am. Chem. Soc.*, 1984, **106**, 3977.
45. M. Srebnik and R. Mechoulam, *J. Chem. Soc., Chem. Commun.*, 1984, 1070.
46. M. Srebnik, *Synth. Commun.*, 1989, **19**, 197.
47. G. W. M. Visser, S. Boele, B. W. V. Halteren, G. H. J. N. Knops, J. D. M. Herocheid, G. A. Brinkman and A. Hoekstra, *J. Org. Chem.*, 1986, **51**, 1466.
48. R. U. Lemieux and A. R. Morgan, *Can. J. Chem.*, 1965, **43**, 2190.
49. R. D. Evans and J. H. Schauble, *Synthesis*, 1986, 727.
50. R. D. Evans and J. H. Schauble, *Synthesis*, 1987, 551.
51. S. K. Knapp and D. V. Patel, *J. Am. Chem. Soc.*, 1983, **105**, 6985.
52. J. Barluenga, J. M. González, P. J. Campos and G. Asensio, *Tetrahedron Lett.*, 1986, **27**, 1715.
53. J. Barluenga, J. M. González, P. J. Campos and G. Asensio, *Angew. Chem., Int. Ed. Engl.*, 1985, **24**, 319.
54. J. Barluenga, J. M. González, P. J. Campos and G. Asensio, *Tetrahedron Lett.*, 1988, **29**, 6497.
55. J. Barluenga, J. M. González, P. J. Campos and G. Asensio, *Angew. Chem., Int. Ed. Engl.*, 1988, **27**, 1546.
56. B. Damin, J. Garapon and B. Sillion, *Tetrahedron Lett.*, 1980, **21**, 1709.
57. B. Damin, J. Garapon and B. Sillion, *Synthesis*, 1981, 362.
58. S. Torii, H. Tanaka, N. Tada, S. Nagao and M. Sasaoka, *Chem. Lett.*, 1984, 877.
59. W. Sato, N. Ikeda and H. Yamamoto, *Chem. Lett.*, 1982, 141.
60. S. Torii, H. Tanaka, N. Saito, T. Siroi, M. Sasaoka and J. Nokami, *Tetrahedron Lett.*, 1981, **22**, 3193.
61. S. Torii, K. Uneyama, H. Tanaka, T. Yamanaka, T. Yasuda, M. Ono and Y. Kohmoto, *J. Org. Chem.*, 1981, **46**, 3312.
62. S. Torii, 'Electroorganic Syntheses. Part I: Oxidations Methods and Applications', Kodansha and VCH, Tokyo and Weinheim, 1985, chap. 9.
63. E. Laurent, B. Marquet, R. Tardival and H. Thiebault, *Bull. Soc. Chim. Fr.*, 1986, 955.
64. A. Kohda, K. Ueda and T. Sato, *J. Org. Chem.*, 1981, **46**, 509.
65. E. Murayama, A. Kohda and T. Sato, *J. Chem. Soc., Perkin Trans. 1*, 1980, 947.
66. M. A. Fox and T. L. Pettit, *J. Org. Chem.*, 1985, **50**, 5013.
67. N. Itoh, Y. Izumi and H. Yamada, *Biochemistry*, 1987, **26**, 282.

3.8
Cleavage Reactions

DONALD G. LEE and TAO CHEN
University of Regina, Canada

3.8.1 INTRODUCTION

Oxidative cleavage is a procedure often employed to degrade large compounds or to introduce different functionality into complex molecules. A number of reagents have been used for this purpose with generally good success.[1–4]

The nature of the products obtained is dependent on the choice of oxidant, the structure surrounding the double bond, the reaction conditions, and the work-up procedures. In general, if the double-bonded carbon is tertiary, then ketones or secondary alcohols can be easily obtained. However, if the carbon is secondary, the products will be primary alcohols, aldehydes or, most likely, carboxylic acids. Because they are very susceptible to further oxidation, the most difficult of these products to obtain are the aldehydes. Selective oxidants and mild conditions are required to produce good yields.

Some procedures result in the introduction of nonoxygen functionalities when cleavage occurs. For example, both nitriles and sulfides can be obtained by the use of appropriate reagents and conditions.

Equations (1)–(5) summarize the types of reactions that will be discussed in this chapter.

$R_2C{=}CHR' \xrightarrow[\text{ii, LiAlH}_4\text{ or NaBH}_4]{\text{i, O}_3} R_2CH{-}OH + R'CH_2{-}OH$ (1)

$R_2C{=}CHR' \xrightarrow[\text{or KMnO}_4]{\text{O}_3\text{ or OsO}_4\text{/NaIO}_4} R_2C{=}O + R'CHO$ (2)

$R_2C{=}CHR' \xrightarrow[\text{or Cr}^{VI}]{\text{KMnO}_4\text{ or RuO}_4} R_2C{=}O + R'COOH$ (3)

$PhCH{=}C(NO_2)R \xrightarrow{\text{AlCl}_3\text{/EtSH}} PhCH(SEt)_2$ (4)

$\text{cycloalkene (H, R)} \xrightarrow[\text{or PhI(OAc)}_2\text{/Me}_3\text{SiN}_3]{\text{Pb(OAc)}_4\text{/Me}_3\text{SiN}_3} NC{\sim}C(O)R$ (5)

Transition metal oxidants such as permanganate, ruthenium tetroxide and chromium(VI) oxide are convenient and efficient reagents for routine cleavage reactions. The use of phase transfer catalysts (quaternary ammonium and phosphonium ions, primarily) has made it possible to solubilize transition metal oxides such as permanganate and chromate in nonaqueous solvents, and to thereby increase the scope of these reactions substantially.[5]

Sodium periodate, used along with catalytic amounts of osmium tetroxide, ruthenium dioxide or potassium permanganate, can also be employed to cleave carbon–carbon double bonds. When used with osmium tetroxide, carbonyls are produced; however, the presence of permanganate results in the formation of more highly oxidized products (carboxylic acids) from secondary carbons.

Ozone, while somewhat inconvenient to use, is very specific in its reactions with alkenes.[6–8] It is widely employed for selective synthesis, for qualitative and quantitative analysis of unsaturated compounds, and for studying the position of double bonds in macromolecules. The nature of the products obtained from ozonolysis reactions is determined by the way in which the reaction is carried out. Different work-up procedures (hydrolytic, reductive or oxidative) can be used to produce alcohols, aldehydes, ketones, carboxylic acids or esters.

Oxidative cleavages have been categorized in this chapter according to the products that are produced. Section 3.8.2 describes methods for the cleavage of double bonds to primary or secondary alcohols. Section 3.8.3 describes the formation of carbonyl compounds and Section 3.8.4 those reactions that result in the formation of carboxylic acids, esters, or lactones. Cleavage reactions that give other (nonoxygen containing) functional groups are described in Section 3.8.5. The approach will be to describe sequentially the use of various reagents for these purposes. Each section is followed by a table of representative reactions and a list of references that can be consulted for exact experimental details.

Wherever practical, reaction mechanisms have been used to indicate why the products of a particular reaction can be altered by using different conditions.

3.8.2 CLEAVAGE OF CARBON–CARBON DOUBLE BONDS WITH THE FORMATION OF PRIMARY OR SECONDARY ALCOHOLS

In practice, alcohols can always be obtained from the reduction (in a second step) of the products obtained from oxidative cleavage reactions. However, when ozone is used as the cleavage reagent it is possible to obtain alcohols directly without the need to isolate intermediate products.

3.8.2.1 Ozone

The use of ozone in organic synthesis has been reviewed by Haines,[1] Below,[6] Razumovskii and Zaikov,[7] Bailey,[8] Kuczkowski,[9] Criegee[10] and Carruthers.[11] Although the details of the reaction of ozone with carbon–carbon double bonds are not all completely understood, there is good evidence that the mechanism proposed by Criegee[10] is fundamentally correct.

The first step, a 1,3-dipolar addition, results in the formation of a 'primary' ozonide (**1**; equation 6). This intermediate then opens to give a carbonyl and a zwitterion that can recombine to give the more stable 'normal' ozonide (**2**; equation 7). Reduction of (**2**), without isolation, by lithium aluminum hydride, diborane or sodium borohydride then gives either primary or secondary alcohols, depending on the nature of starting alkene (equation 8).

(**1**) (6)

(7)

(**2**)

LiAlH$_4$ 2 —OH (8)

The cleavage reaction, commonly referred to as 'ozonolysis', is carried out by bubbling ozonized oxygen through a solution of the alkene in various solvents, including methanol,[12] dichloromethane,[13] carbon tetrachloride[14] and ethyl acetate.[15] Other solvents (ethanol, tetrahydrofuran, acetic acid, or a combination of ethyl acetate and hexane) have also been reported for use in individual reactions.[16–20] The reaction is usually performed at low temperatures (about 0 °C), and, since ozonides are potentially explosive compounds, the intermediates are not isolated.

For example, Magari *et al.*[21] have described the preparation of alcohol (**3**) in 90% yield from the corresponding alkene (equation 9). It was found that each mole of ozonide required at least one mole of sodium borohydride for complete reduction to the desired alcohol.

i, O$_3$; ii, NaBH$_4$, aq. EtOH (9)

(**3**) 90%

The reaction can also accommodate other reducible functional groups, such as esters, when sodium borohydride is used as the reducing agent. For example, Dyke *et al.*[22] obtained alcohol (**4**) from the corresponding alkene in 57% yield when using this procedure (equation 10).

Low temperature is required for most reactions, as for example in the preparation of (**5**) recently reported by Boger and Coleman (equation 11).[23]

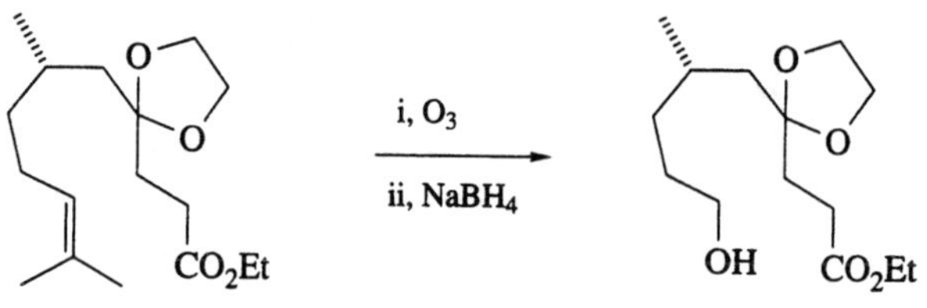

i, O_3 → ii, $NaBH_4$ (10)

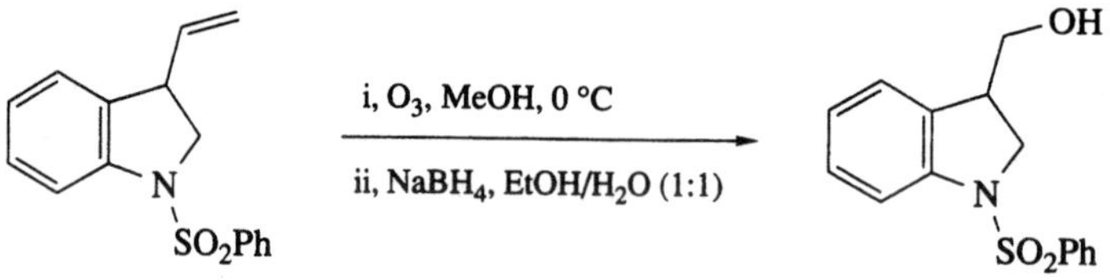

(4) 57%

i, O_3, MeOH, 0 °C → ii, $NaBH_4$, $EtOH/H_2O$ (1:1) (11)

(5)

In a typical procedure,[23] a solution of 3-vinylindoline (65 mg, 0.23 mmol) in 2.0 mL of methanol was cooled to 0 °C and treated with a stream of 3–8% ozone in oxygen (300 mL min^{-1}, 20 min). The reaction mixture was then stirred for an additional 20 min (0 °C) before the excess ozone was removed by passing a stream of nitrogen through the reaction mixture (10 min). Fifty percent aqueous ethanol (1.0 mL) was added at 0 °C, followed by the careful addition of excess sodium borohydride (20 mg, 2.1 mmol). The mixture was allowed to warm up and stirred for 1 h at 23 °C. It was then poured into 10 mL of 10% aqueous HCl and extracted with EtOAc (30 mL). The organic extract was washed with saturated $NaHCO_3$ (10 mL), water (10 mL), saturated aqueous NaCl (10 mL) and dried ($MgSO_4$). Removal of the solvent *in vacuo* and flash chromatography (1 × 15 cm SiO_2, 30–100% Et_2O/hexane gradient elution) afforded the hydroxymethylindoline (**5**) in 59% yield (38.6 mg). Other examples of the formation of alcohols from the cleavage of carbon–carbon double bonds by ozone are summarized in Table 1.

3.8.3 CLEAVAGE TO CARBONYL COMPOUNDS

The conversion of tetrasubstituted double bonds to the corresponding ketones is easily achieved using a number of oxidants. However, if one or more of the alkenic carbons is secondary, the product will be either an aldehyde or a carboxylic acid. Ozone and a combination of osmium tetroxide and sodium metaperiodate are recommended if the desired product is an aldehyde. Under carefully controlled conditions it is also possible to obtain good yields of the aldehyde when permanganate is used as the oxidant. All methods that give aldehydes from secondary carbons can also be used to prepare ketones from tertiary carbons.

3.8.3.1 Ozone

Hydrolysis of ozonides produces carbonyl compounds and hydrogen peroxide, as in equation (12).

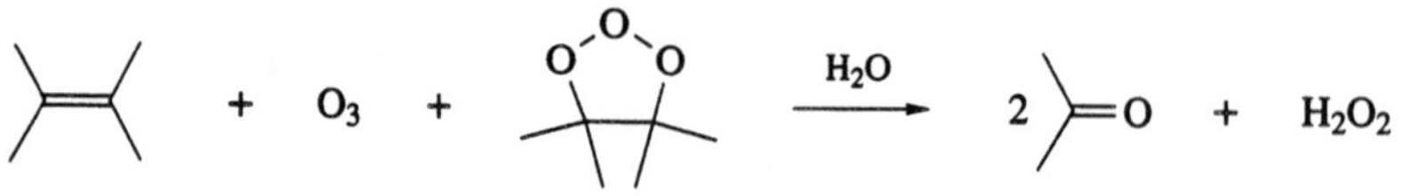

+ O_3 + [ozonide] $\xrightarrow{H_2O}$ 2 >=O + H_2O_2 (12)

Since the formation of peroxides is highly undesirable, a mild reductant is usually added to the reaction medium. Many reducing agents including hydrogen and a catalyst, zinc and acetic acid, potassium iodide and acetic acid, sulfides, disulfites, and phosphenes have been used for this purpose.[34–39] However, the most convenient and efficient reagent is dimethyl sulfide (DMS). It is effective under neutral conditions and highly selective for peroxides, but it does have a low boiling point (37 °C) and a rather obnoxious odor. These disadvantages can be overcome by using thiourea instead of DMS as the reducing agent.[40] Yields of aldehydes and ketones are comparable with both reagents.

In a typical experiment,[40] ozonized oxygen (1.22% w/w O_3 in O_2) was bubbled through a solution of (+)-3-carene (2.74 g, 0.02 mol) in anhydrous methanol (30 mL) at –10 to –15 °C until the required quan-

Table 1 Cleavage of Carbon–Carbon Double Bonds with the Formation of Alcohols

Substrate	*Oxidant and conditions*	*Product*	*Yield (%)*	*Ref.*
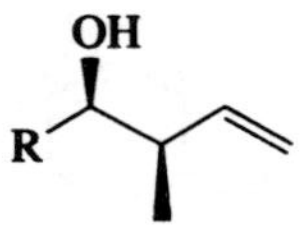	i, O_3, THF, –78 °C, DMS; ii, $LiAlH_4$	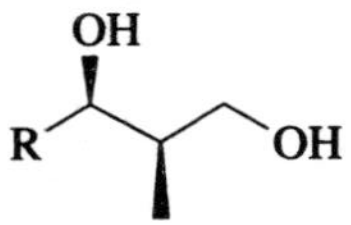	79	24
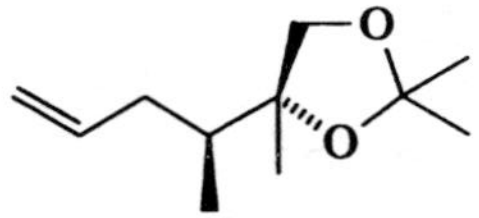	i, O_3, EtOH, –78 °C; ii, DMS, $NaBH_4$, –40 to 25 °C	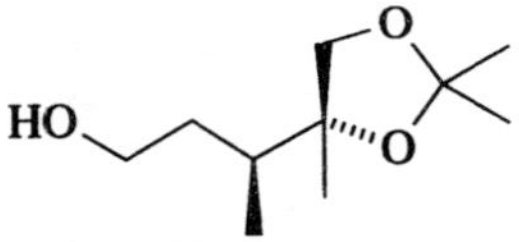	87	25
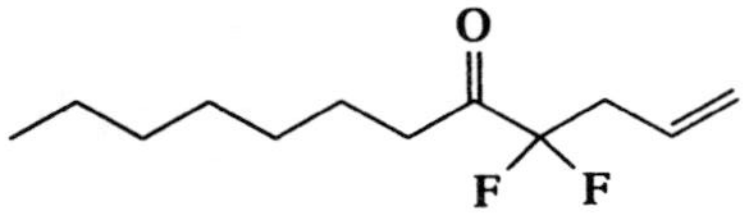	i, O_3, MeOH, –70 °C; ii, $NaBH_4$	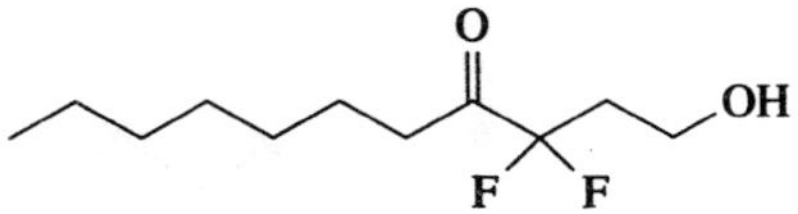	90	26
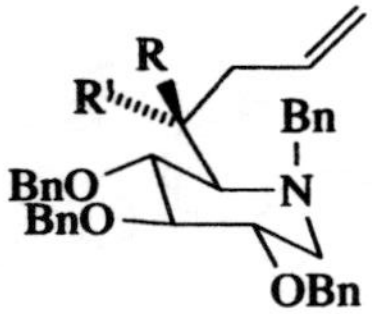	i, O_3, CH_2Cl_2, –78 °C; ii, $NaBH_4$, EtOH	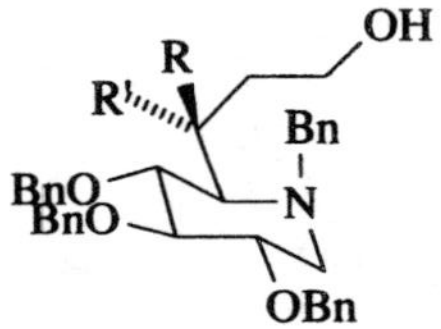	>55	27
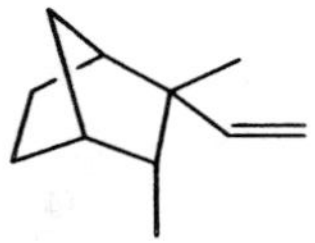	i, O_3; ii, $LiAlH_4$	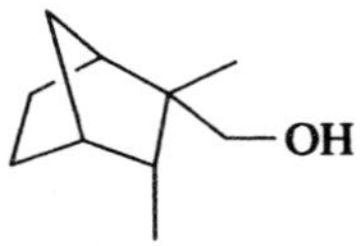		28
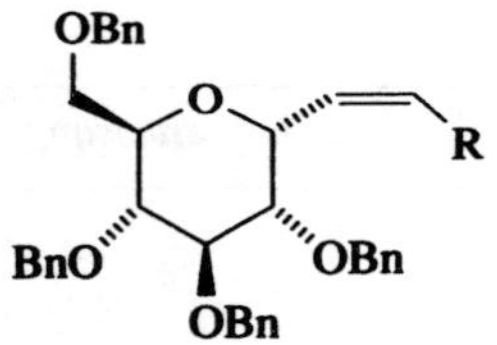	i, O_3, CH_2Cl_2, –78 °C; ii, $NaBH_4$	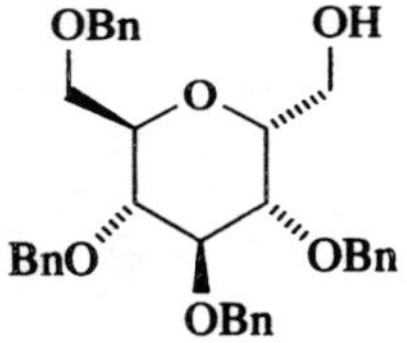		29

Table 1 *(continued)*

Substrate	*Oxidant and conditions*	*Product*	*Yield (%)*	*Ref.*
	i, O_3, MeOH; ii, $NaBH_4$			30
	i, O_3; ii, $NaBH_4$, EtOH, 0 °C		75	31
	i, O_3, MeOH; ii, $NaBH_4$		63	32
	i, O_3, *n*-hexane, –30 °C; ii, $LiAlH_4$		75	19

Table 1 *(continued)*

Substrate	*Oxidant and conditions*	*Product*	*Yield (%)*	*Ref.*
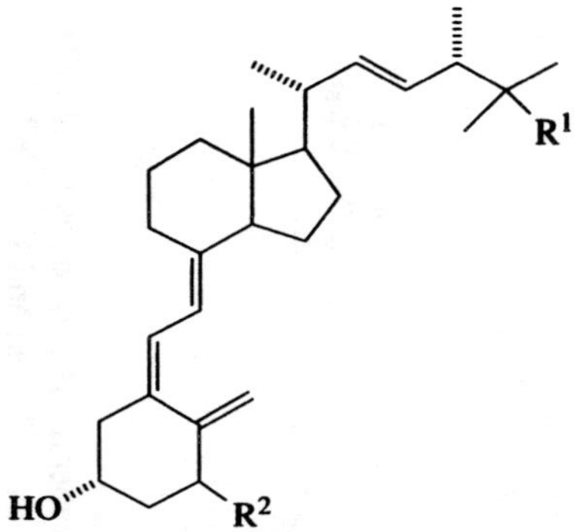 HO, R^1, R^2	i, O_3, MeOH–Py; ii, $NaBH_4$	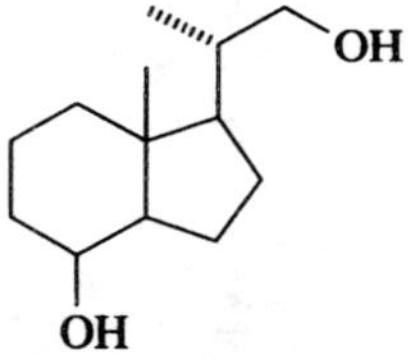 OH, OH	85	33

tity of O_3 had been passed (65 min). Nitrogen was then bubbled through the solution for about 10 min and thiourea (0.767 g. 0.01 mol) in dried methanol (3 mL) was added at 0 °C with stirring. Thiourea *S*-dioxide deposited as white crystals. After continued stirring for another 40 min, the mixture was filtered and the filtrate evaporated under reduced pressure (150 mmHg). The residue was dissolved in light petroleum (b.p. 60–80 °C, 60 mL), washed with 1% sodium bicarbonate (10 mL) and water (3 × 10 mL), and dried (Na_2SO_4). Distillation of the dried solution gave 2,2-dimethyl-3-(2′-oxo)propylcyclopropane-1-acetaldehyde dimethylacetal. Other examples of this reaction are summarized in Table 2.

Selective cleavage of compounds containing two or more sites of unsaturation can also be achieved by the use of ozone. Some examples are given in Table 3. Optimum yields in these selective cleavages requires use of an appropriate amount of oxidant. Addition of too much ozone with subsequent attack on the second site of unsaturation can be avoided by careful monitoring[9,10] or by use of an appropriate dye as an internal standard. Compounds (**6**) and (**7**) have been used for this purpose by Veysoglu *et al.*[68]

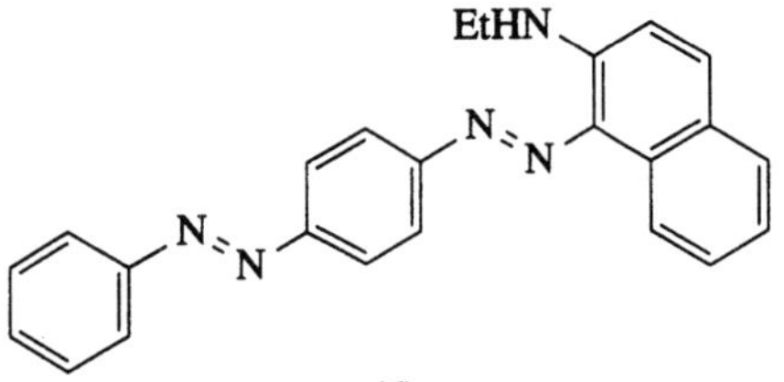

EtHN N=N N=N

(**6**)

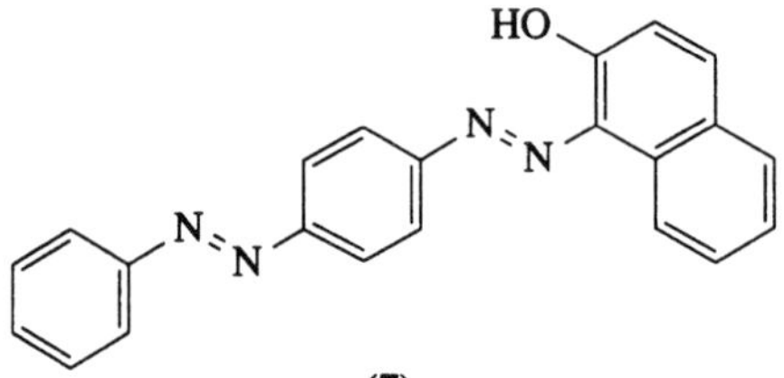

HO N=N N=N

(**7**)

Pyridine has also been used to enhance selectivity.[61,62] When present, reaction occurs at exocyclic rather than at endocyclic double bonds of steroid derivatives, as in equation (13).

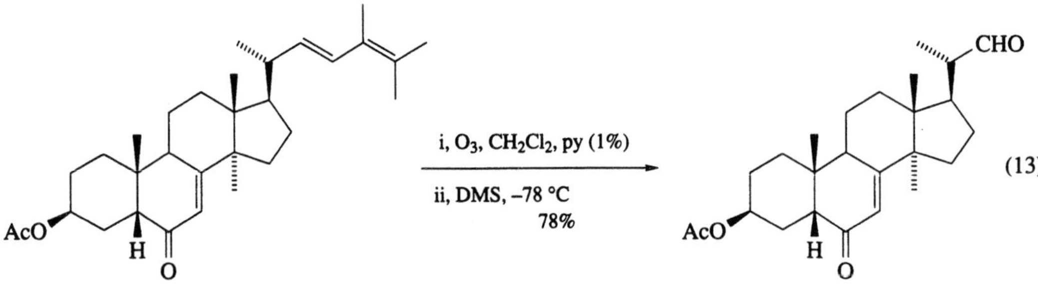

i, O_3, CH_2Cl_2, py (1%)
ii, DMS, –78 °C
78%

AcO H O → AcO H O CHO (13)

These reactions are known to benefit from the addition of phase transfer agents under certain conditions. For example, 3,5,5-trimethylcyclohex-2-enone was cleaved, with loss of one carbon atom as illustrated in equation (14), when Adogen 464 (a quaternary ammonium chloride) was present.

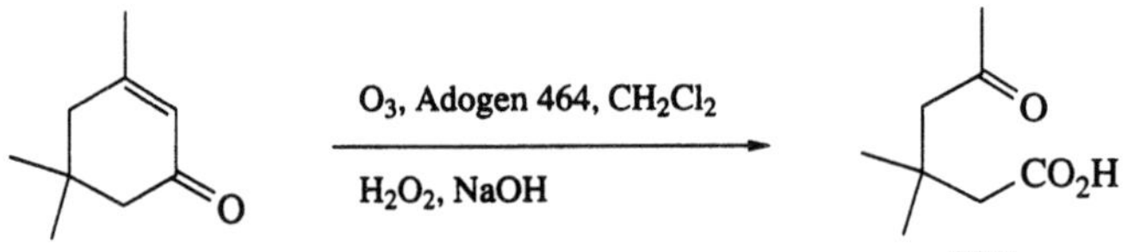

O_3, Adogen 464, CH_2Cl_2
H_2O_2, NaOH

O → O CO_2H 92% (14)

Ozone adsorbed on silica gel has also been found to be an effective cleavage reagent.[56,76] For example, compound (**8**), which is normally very difficult to cleave, underwent the reaction indicated in equation (15) without loss of chirality.

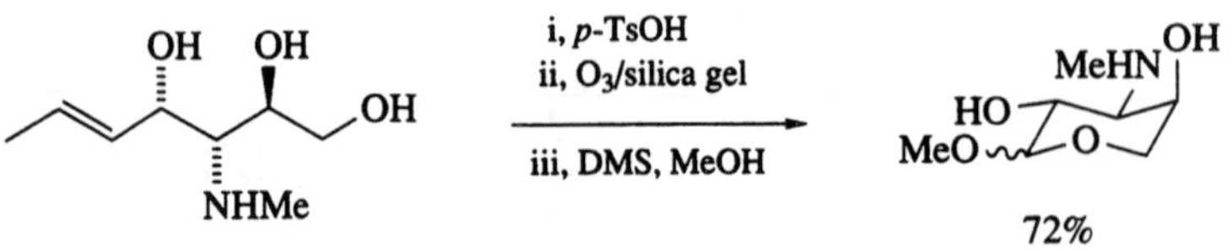

OH OH OH NHMe

(**8**)

i, *p*-TsOH
ii, O_3/silica gel
iii, DMS, MeOH

MeHN OH HO MeO O 72% (15)

The product distribution and mechanism of ozonolysis differ when the oxidant is adsorbed on silica gel. When dried silica gel was used, Besten and Kinstle[76] found that the products were similiar to those

Table 2 Cleavage of Double Bonds with Ozone to give Carbonyl Compounds

Substrate	*Oxidant and conditions*	*Product*	*Yield (%)*	*Ref.*
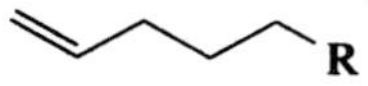 R	i, O_3, CH_2Cl_2, –78 °C; ii, DMS	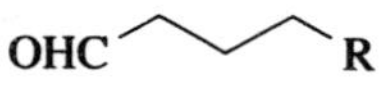 OHC R	80	41
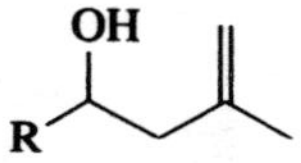 OH R	i, O_3, CH_2Cl_2, –60 °C; ii, DMS	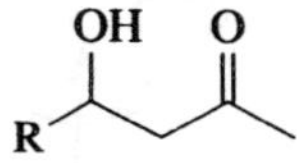 OH O R		42
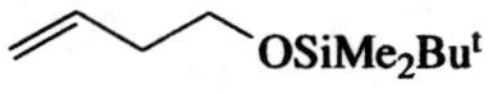 $OSiMe_2Bu^t$	i, O_3, CH_2Cl_2; ii, DMS	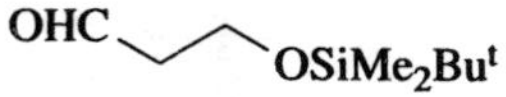 OHC $OSiMe_2Bu^t$	73	43
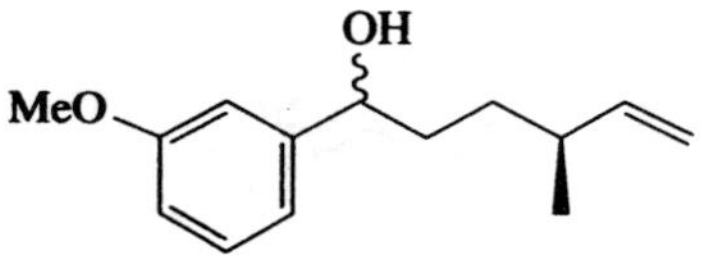 OH MeO	i, O_3, MeOH; ii, DMS	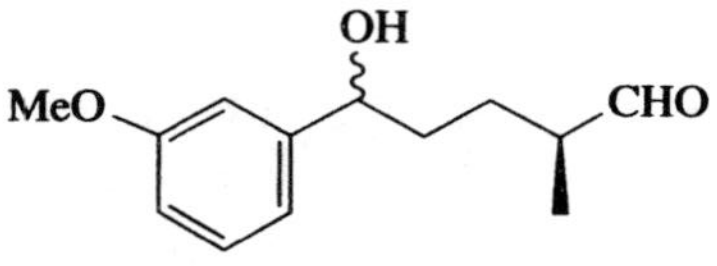 OH MeO CHO	>42	44
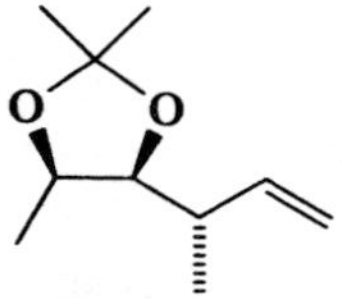 O O	i, O_3, MeOH, –78 °C; ii, PPh_3	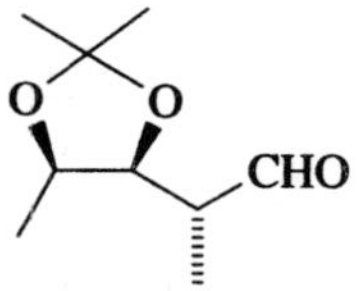 O O CHO	85	45
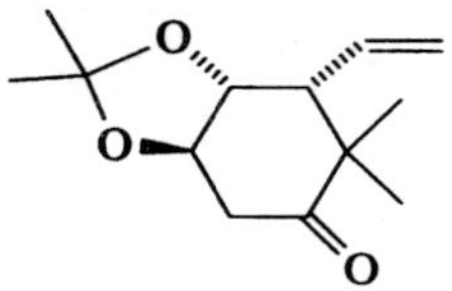 O O O	i, O_3, MeOH; ii, DMS	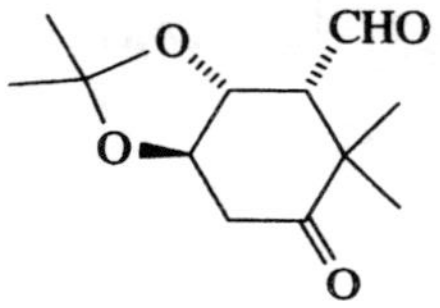 CHO O O O		46
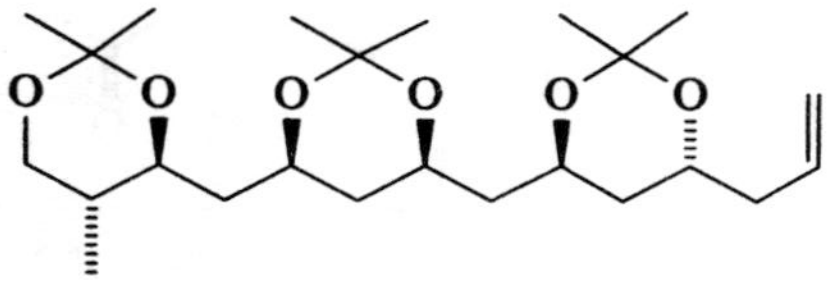 O O O O O O	i, O_3, MeOH, –78 °C; ii, DMS, r.t.	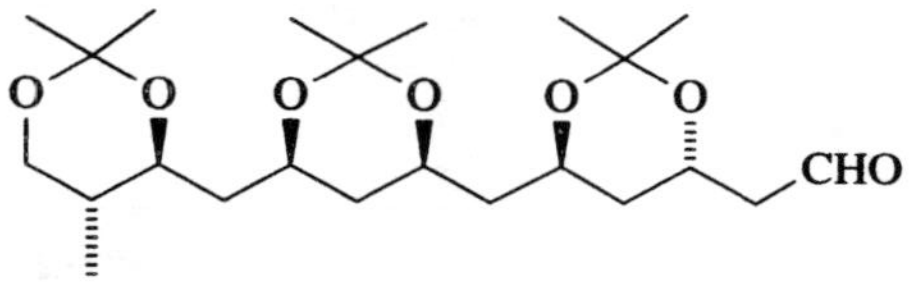 O O O O O O CHO	>84	47

Table 2 *(continued)*

Substrate	*Oxidant and conditions*	*Product*	*Yield (%)*	*Ref.*
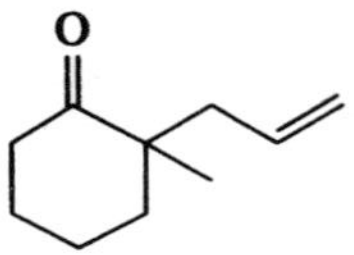 O	i, O_3, CH_2Cl_2; ii, Et_3N	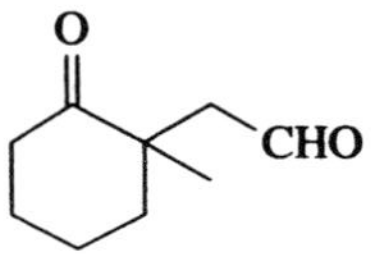 O, CHO		48
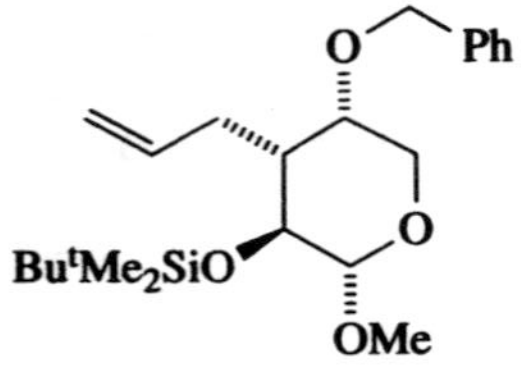 O, Ph, O, Bu^tMe_2SiO, OMe	i, O_3, CH_2Cl_2, –78°C; ii, Et_3N, r.t., 4 h	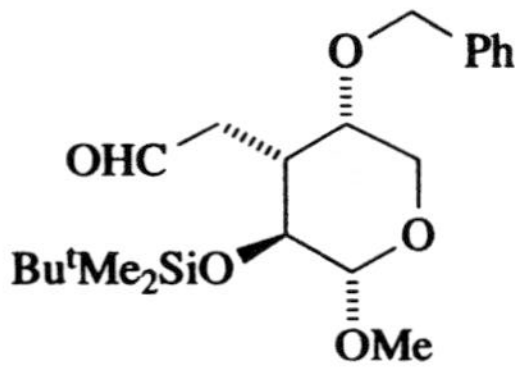 O, Ph, OHC, O, Bu^tMe_2SiO, OMe	93	49
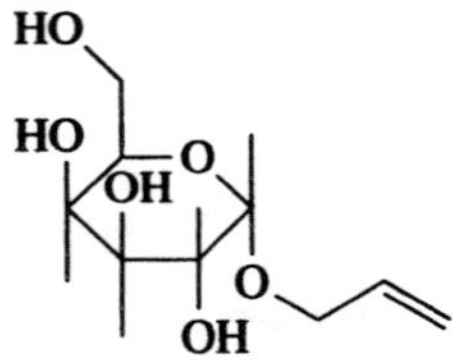 HO, HO, O, OH, O, OH	i, O_3; ii, DMS	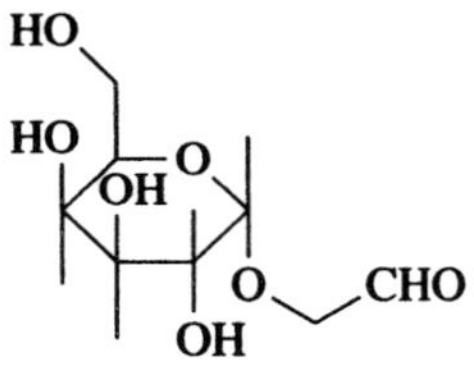 HO, HO, O, OH, O, CHO, OH	70–90	50
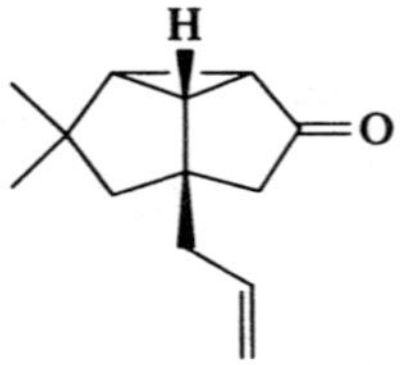 H, O	i, O_3, MeOH, –78 °C; ii, DMS, 25 °C	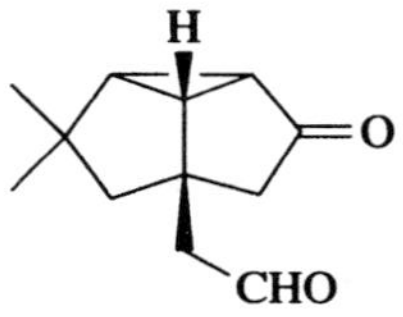 H, O, CHO	>55	51
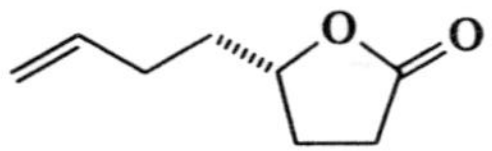 O, O	i, O_3, CH_2Cl_2/MeOH, –78 °C; ii, DMS, –78 °C, 1 h	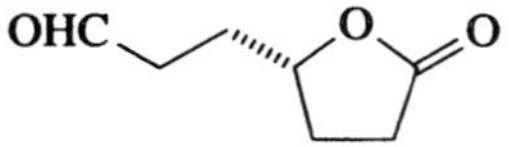 OHC, O, O	>57	52

Table 2 *(continued)*

Substrate	*Oxidant and conditions*	*Product*	*Yield (%)*	*Ref.*
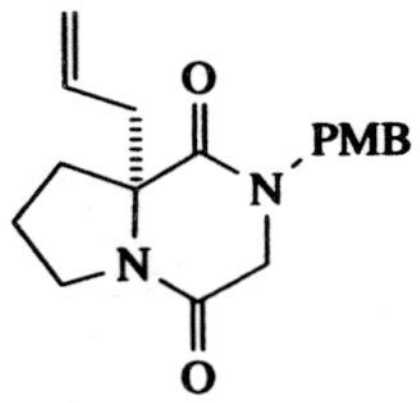 O, PMB, N, N, O	i, O_3, MeOH; ii, DMS	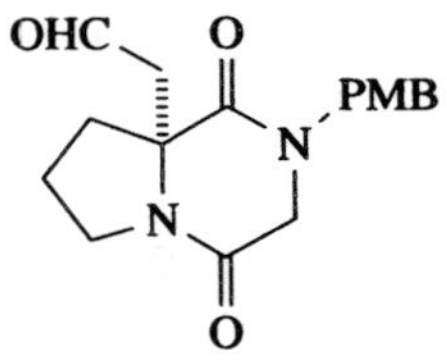 OHC, O, PMB, N, N, O	99	53
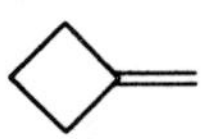	i, O_3, CH_2Cl_2/MeOH, –78 °C; ii, $S{=}C(NH_2)_2$, $NaHCO_3$, CH_2Cl_2, 0 °C	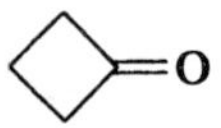 O	71	17
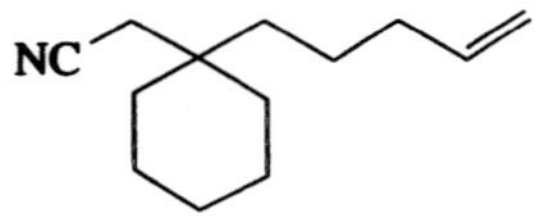 NC	i, O_3, MeOH, –78 °C; ii, DMS, –78 to 25 °C	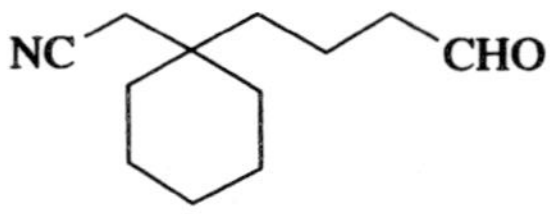 NC, CHO	63	12
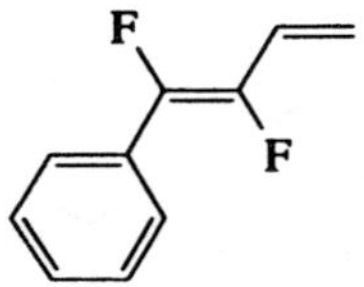 F, F	i, O_3, CH_2Cl_2, –70 °C; ii, DMS, r.t., overnight	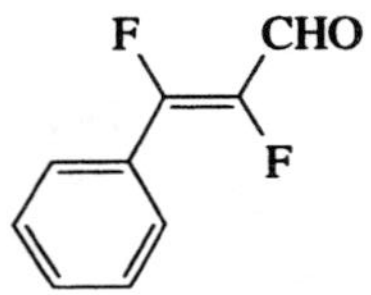 F, CHO, F	64	54
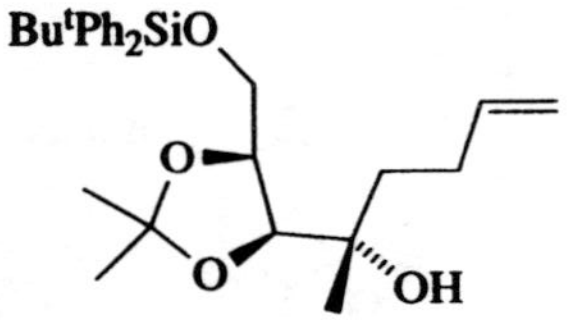 Bu^tPh_2SiO, O, O, OH	i, O_3, CH_2Cl_2, –78 °C; ii, PPh_3	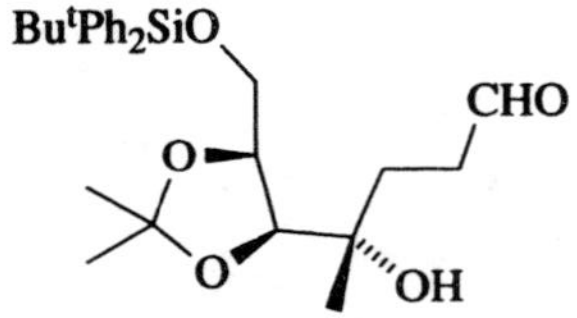 Bu^tPh_2SiO, CHO, O, O, OH	90	38
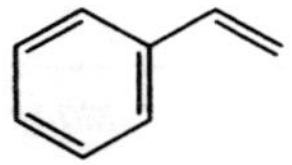	i, O_3, MeOH, –10 to –15 °C; ii, $S{=}C(NH_2)_2$	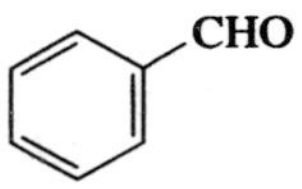 CHO	81	40

Table 2 *(continued)*

Substrate	*Oxidant and conditions*	*Product*	*Yield (%)*	*Ref.*
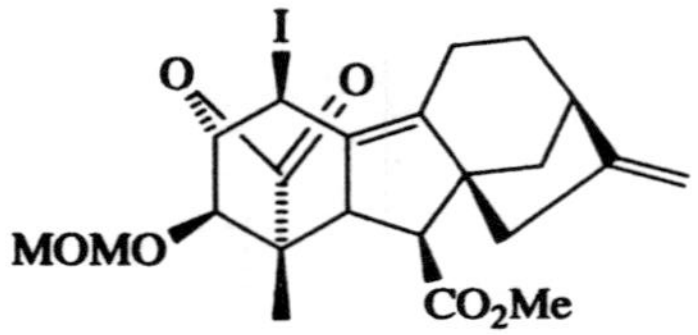	i, O_3, CH_2Cl_2–Py, –78 °C, 5 s; ii, DMS	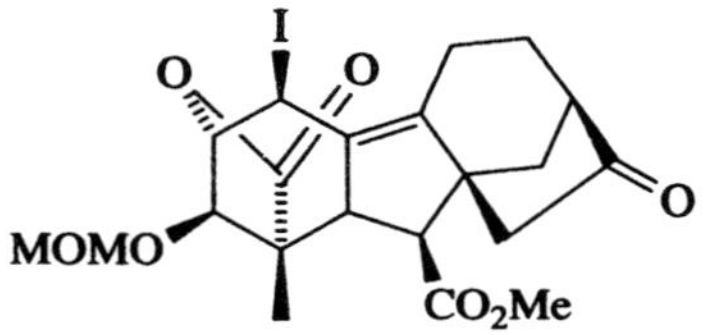	68	55
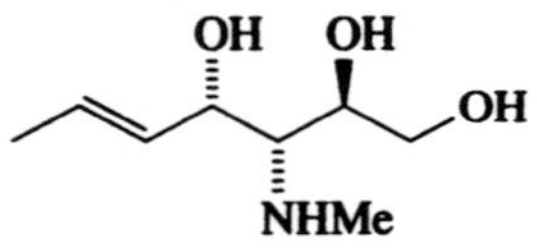	i, *p*-TsOH; ii, O_3/silica gel; iii, DMS, MeOH	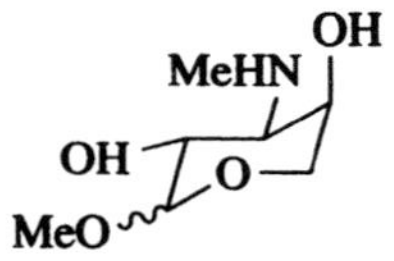	72	56
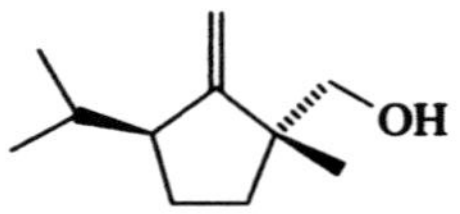	i, O_3, MeOH, –78 °C; ii, DMS, –78 °C, 2 h and r.t., overnight	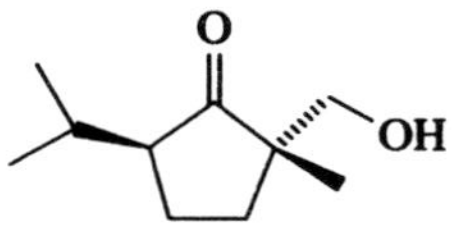	70	57
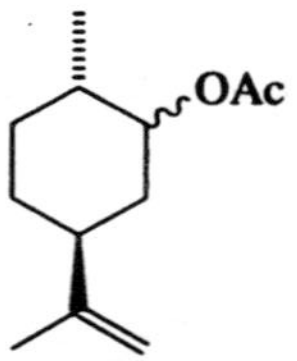	i, O_3, CH_2Cl_2, –78 °C, 1 h; ii, Zn–AcOH, r.t., 30 min	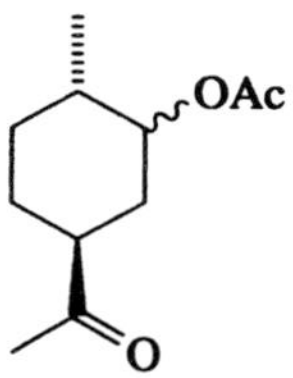	95	58
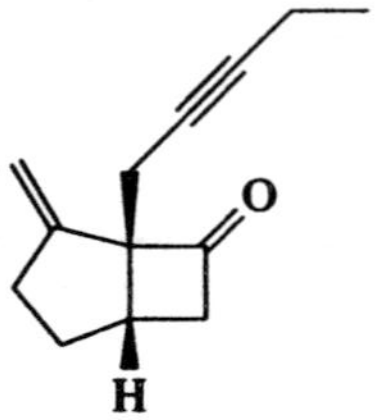	i, O_3, MeOH, –78 °C; ii, DMS	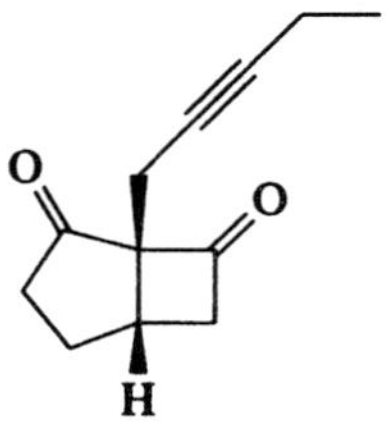		59

Table 2 *(continued)*

Substrate	Oxidant and conditions	Product	Yield (%)	Ref.
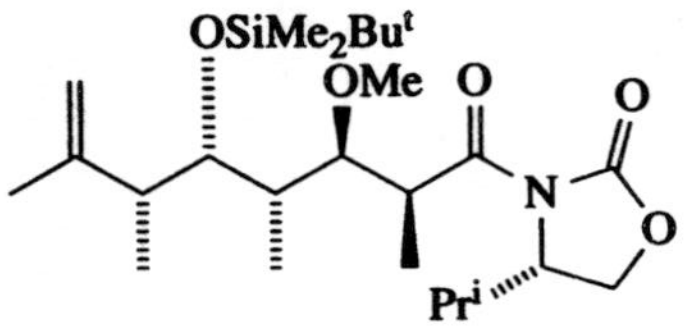 OSiMe₂Bu^t OMe O O N O Pr^i	i, O_3, CH_2Cl_2/MeOH/$NaHCO_3$, –78 °C; ii, DMS, 0 °C to r.t., 10 h	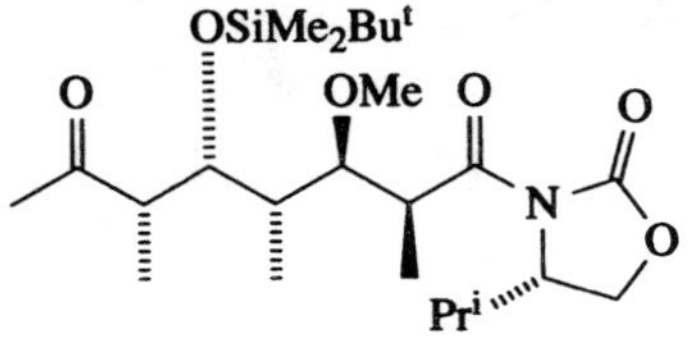 O OSiMe₂Bu^t OMe O O N O Pr^i	92	60
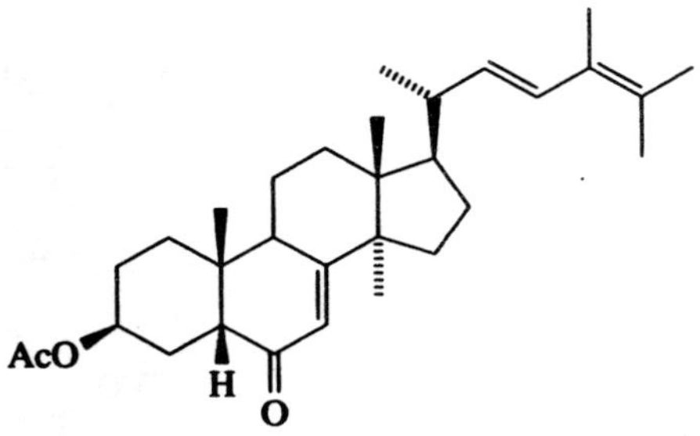 AcO H O	i, O_3, CH_2Cl_2–Py (1%), –78 °C; ii, DMS, –78 °C	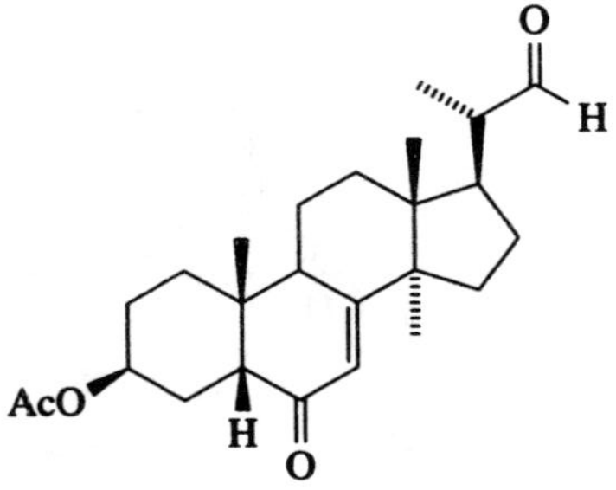 O H AcO H O	78	61
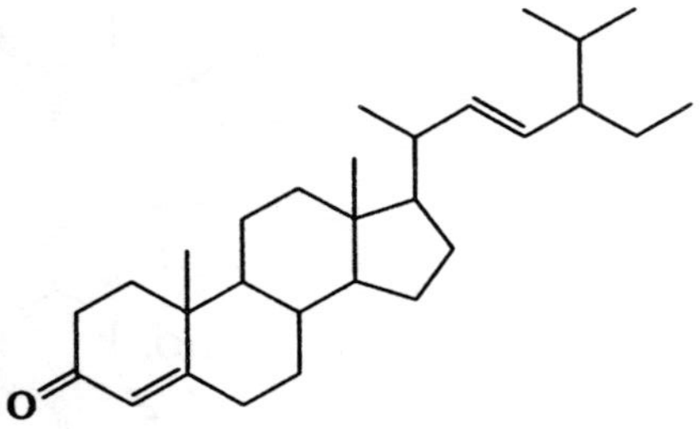 O	i, O_3, CH_2Cl_2–Py, –78 °C	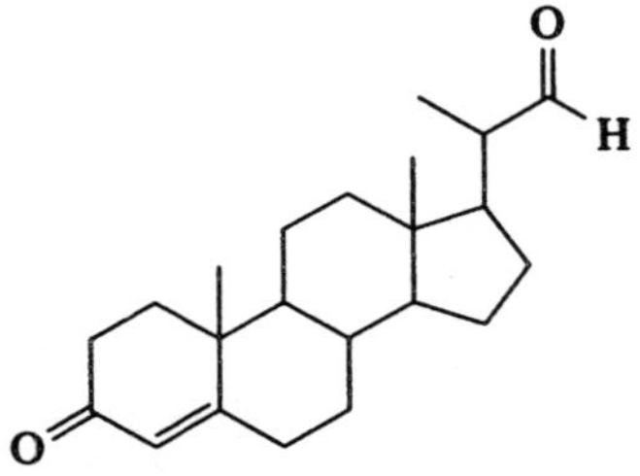 O H O	94	62
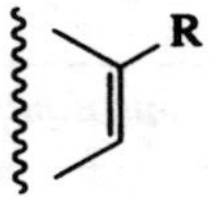 R	i, O_3, EtOAc; ii, S=C(NH_2)$_2$, MeOH	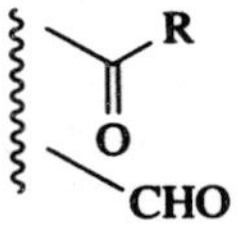 R O CHO		39

Table 2 *(continued)*

Substrate	*Oxidant and conditions*	*Product*	*Yield (%)*	*Ref.*
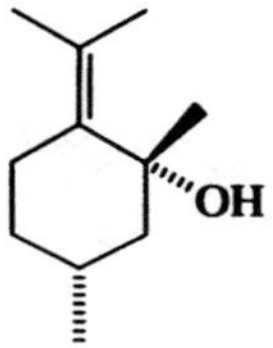 OH	i, O_3, CH_2Cl_2, EtOH, red dye, –60 °C; ii, DMS, r.t., overnight	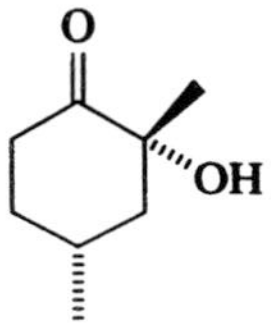 O, OH	80	63
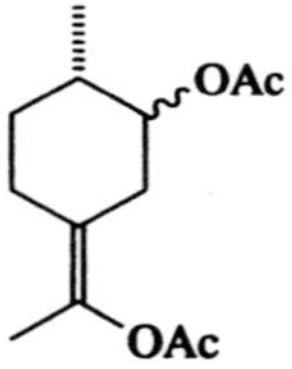 OAc, OAc	i, O_3, CH_2Cl_2/MeOH, –78 °C, 1 h; ii, Zn–AcOH r.t., 30 min	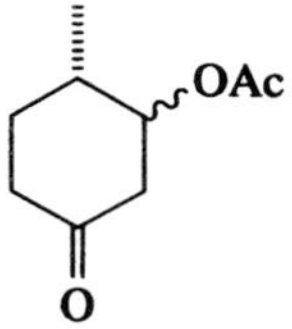 OAc, O	85	58
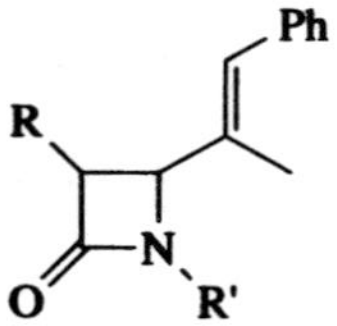 Ph, R, N, R', O	i, O_3, CH_2Cl_2, –78 °C; ii, DMS	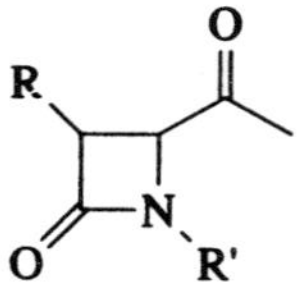 O, R, N, R', O	70	64, 65
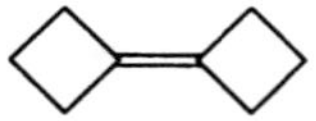	i, O_3, CH_2Cl_2/MeOH, –20 °C; ii, $S{=}C(NH_2)_2$ $NaHCO_3$, CH_2Cl_2, 0 °C	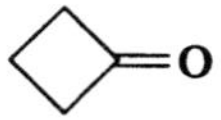 O	67	17
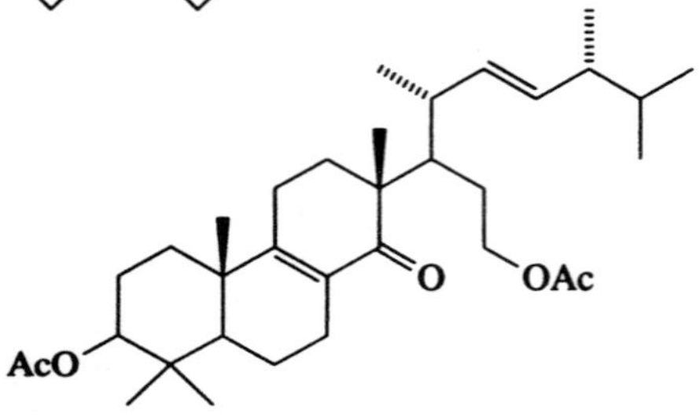 AcO, O, OAc	i, 1.1 equiv O_3, CH_2Cl_2, –78 °C; ii, DMS	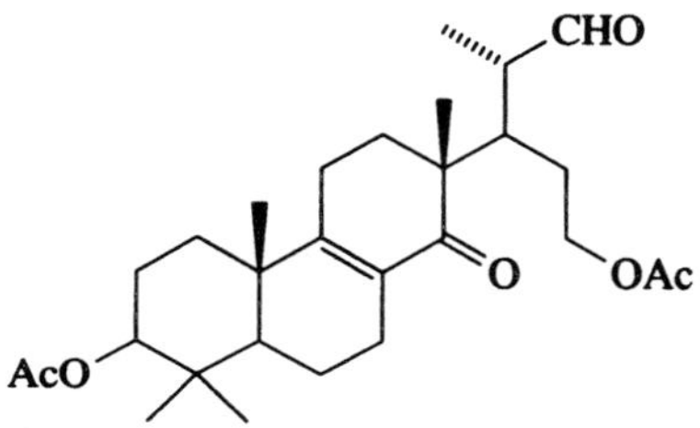 CHO, AcO, O, OAc	76	67

Table 2 *(continued)*

Substrate	*Oxidant and conditions*	*Product*	*Yield (%)*	*Ref.*
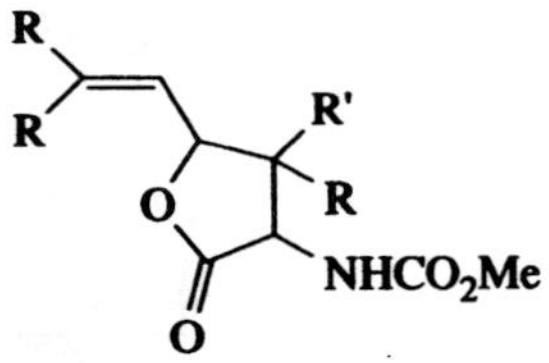 R, R, R', R, O, O, $NHCO_2Me$	i, O_3, MeOH; ii, DMS	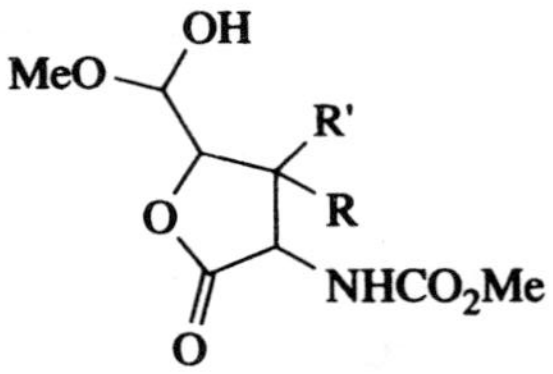 MeO, OH, R', R, O, O, $NHCO_2Me$	93	66
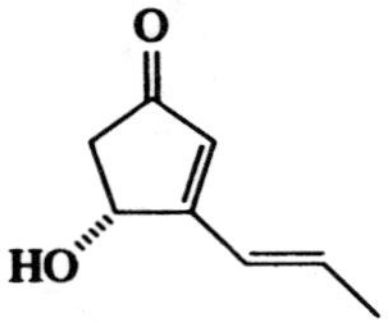 O, HO	i, O_3, EtOH; ii, DMS	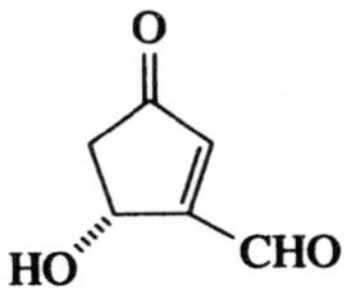 O, HO, CHO	90	68
 $( )_n$ n = 1–3	i, O_3, CH_2Cl_2/MeOH, –78 °C; ii, Et_3N, AcOH/MeOH	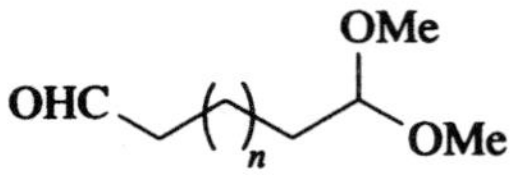 OHC, $( )_n$, OMe, OMe n = 1–3		69
	i, O_3, EtOAc, –20 to –30 °C; ii, $Pd/CaCO_3/H_2$	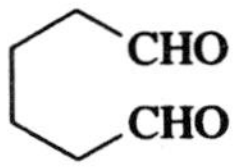 CHO, CHO	61	70
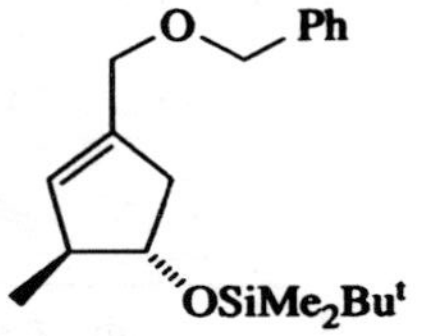 O, Ph, $OSiMe_2Bu^t$	i, O_3; ii, PPh_3	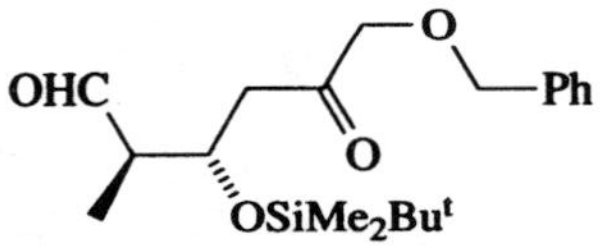 OHC, O, O, Ph, $OSiMe_2Bu^t$	43	71

Table 2 *(continued)*

Substrate	*Oxidant and conditions*	*Product*	*Yield (%)*	*Ref.*
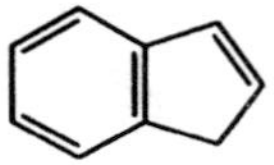	i, O_3, MeOH; ii, $(MeO)_3P$	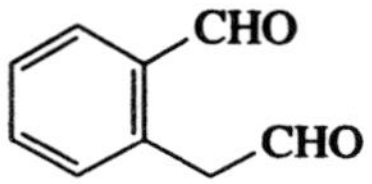	65	72
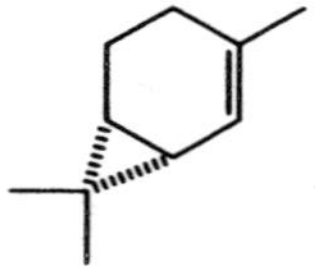	i, O_3, CH_2Cl_2/MeOH, –78 °C, 32 min; ii, DMS, r.t., 3 h	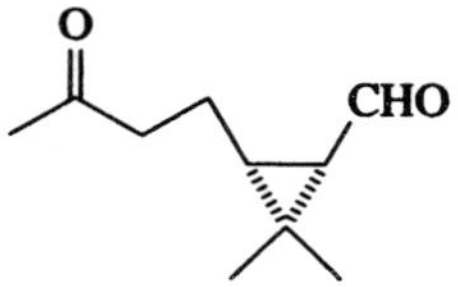	99	73
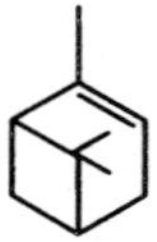	i, O_3, MeOH, –10 to –15 °C; ii, $S{=}C(NH_2)_2$	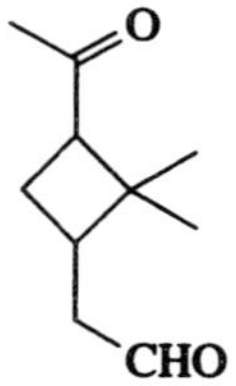	75	40
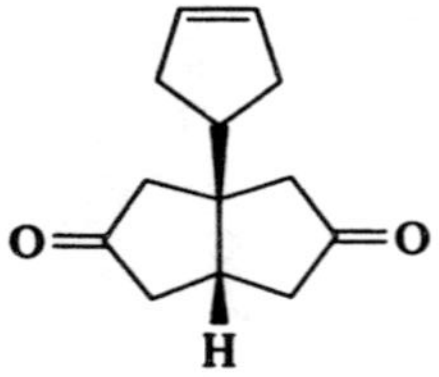	i, O_3, EtOAc, –50 °C; ii, DMS, MeOH	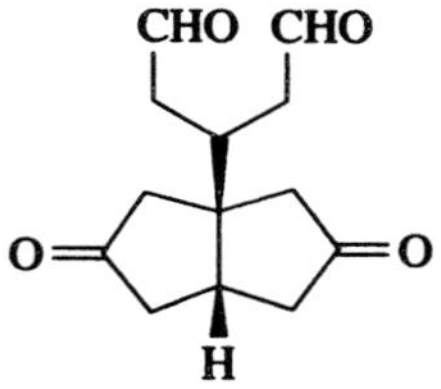		39
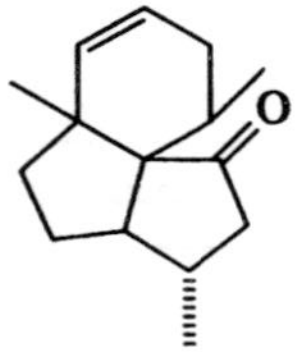	i, O_3, CH_2Cl_2, –78 °C; ii, Zn–AcOH, 0 °C, 5 h	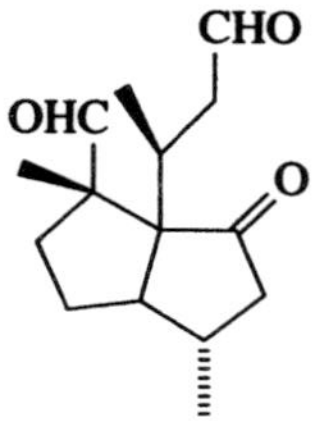	68	35

Table 3 Selective Cleavages by Ozone

Substrate	*Oxidant and conditions*	*Product*	*Yield (%)*	*Ref.*
	i, O_3 (1.5 mol), CH_2Cl_2	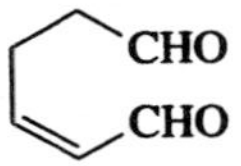 CHO CHO	70	74
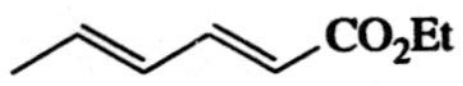 CO_2Et	i, O_3, EtOH; ii, DMS	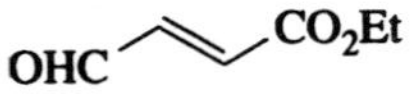 OHC CO_2Et	85	68
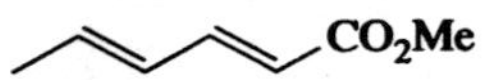 CO_2Me	i, O_3, CH_2Cl_2; ii, DMS	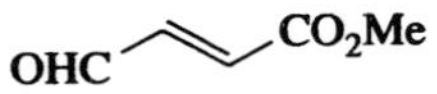 OHC CO_2Me	90	75
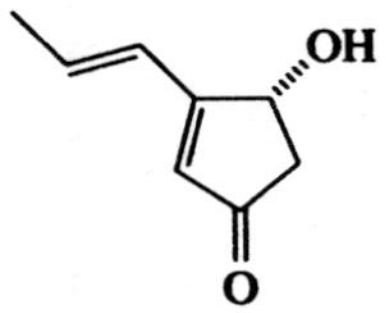 OH O	i, O_3, EtOH; ii, DMS	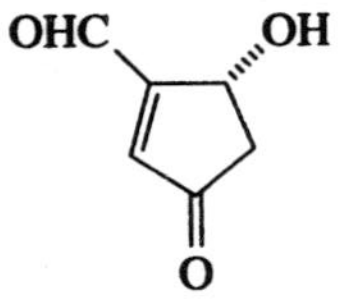 OHC OH O	90	68
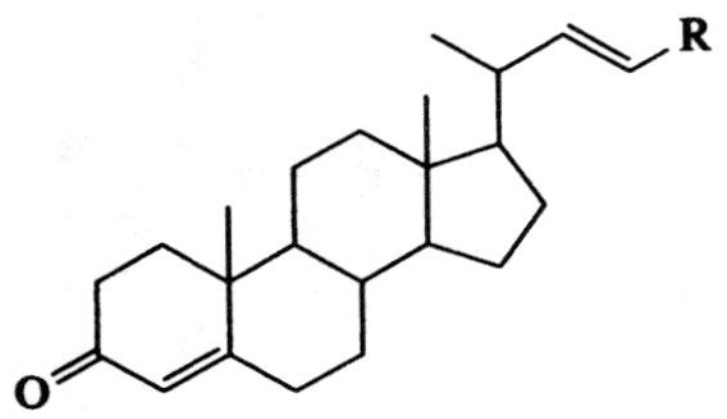 R O	i, O_3, CH_2Cl_2/EtOH (2:1); ii, DMS	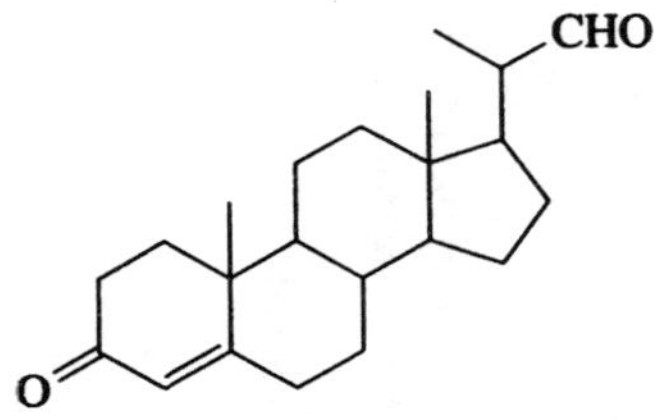 CHO O	85	68
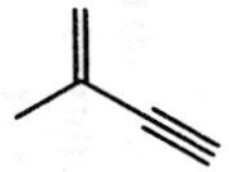	i, O_3, MeOH/Et_2O (1:1); ii, DMS	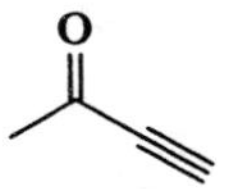 O	64	68

obtained in aprotic and nonparticipating solvents.[9,10] However, when the silica gel was wet, double bond cleavage resulted in the formation of equimolar amounts of aldehyde and carboxylic acid. For example, the oxidative cleavage of cyclopentene by ozone on silica gel containing 5% water gave 5-oxopentanoic acid in 80% yield (equation 16).[76]

$$\text{cyclopentene} \xrightarrow[-78\ ^\circ C]{O_3/SiO_2\ (5\%\ water)} OHC(CH_2)_3CO_2H\ \ 80\% \quad (16)$$

3.8.3.2 Permanganate

3.8.3.2.1 Aqueous potassium permanganate oxidations

It is difficult to prepare aldehydes by the cleavage of carbon–carbon double bonds with permanganate under aqueous conditions. In water, aldehydes exist at least partly as the corresponding hydrates, $RCH(OH)_2$, and are therefore very susceptible to further oxidation by permanganate. Consequently the products obtained are usually carboxylic acids. Aldehydes have been obtained in good yields only when the products are deactivated, as in equation (17).[77]

$$\text{4,4'-dinitrostilbene-2,2'-disulfonic acid} \xrightarrow{MnO_4^-} 2\ O_2N\text{-}C_6H_3(SO_3H)\text{-}CHO\ \ 100\% \quad (17)$$

Wiberg and Saegebarth[78] also obtained fair yields of cyclopentane-1,3-dialdehyde from the oxidation of bicyclo[2.2.1]hept-2-ene under mild conditions (equation 18). However, the oxidation of unsaturated tertiary carbons to the corresponding ketones is much more typical. The reaction depicted in equation (19), where a trisubstituted double bond is cleaved to a ketone and a carboxylic acid, is exemplary of the products that are normally produced when alkenes react with aqueous permanganate.[79]

$$\text{bicyclo[2.2.1]hept-2-ene} \xrightarrow{KMnO_4,\ MgSO_4,\ -15\ ^\circ C} \text{cyclopentane-1,3-(CHO)}_2\ \ 54\text{–}66\% \quad (18)$$

$$\text{1-cyclopentenyl-}(CH_2)_{10}CO_2H \xrightarrow{aq.\ KMnO_4} HO_2C(CH_2)_3C(=O)(CH_2)_{10}CO_2H\ \ 71\% \quad (19)$$

3.8.3.2.2 Mixed solvent systems

Since organic compounds are often not sufficiently soluble in water to permit oxidation in completely aqueous systems, several authors have reported the use of mixed solvent systems (such as acetone and water or alcohol and water) in which the oxidant and reductant are mutually soluble.[3,5] Recent work has shown that the best solvent system to use for the preparation of aldehydes from permanganate cleavages is THF and water. Simandi and coworkers[80] have reported that the treatment of a concentrated aqueous solution of permanganate with a dilute solution of alkene in THF affords the desired aldehydes in good yields. The authors have suggested that the solvent, under these conditions, acts as a quenching reagent that prevents over-oxidation.

In a typical experiment,[80] a solution of 4-formyl-2,2-dimethy-1*H*-1,5-benzodiazepine (0.036 mol) in THF (300 mL) was added in small portions to a solution of potassium permanganate (0.063 mol) in water (100 mL) over a period of 3.5 h. (Addition of solid $KMnO_4$ to neat THF could produce an explosive mixture, and should therefore be avoided.) During the addition, the mixture was allowed to warm to about 40 °C. The mixture was then filtered to remove manganese dioxide and the filtrate was concentrated and extracted with diethyl ether. After drying, the extract was concentrated and the resulting product crystallized from diisopropyl ether. The overall yield was 7 g (79%).

3.8.3.2.3 Phase transfer assisted oxidative cleavages

As depicted in equation (20), the addition of a phase transfer agent, Q^+ (*eg.* quaternary ammonium or phosphonium ions), brings permanganate into solution in nonaqueous solvents.[5] Once solubilized, it reacts with alkenes to produce an intermediate that has been characterized by Ogino *et al.*[81] as a cyclic manganate(V) diester (**9**), that can be decomposed by acidic solutions to yield aldehydes or by mild base to give the corresponding α-diol, as in Scheme 1.

$$Q^+ \text{ (aq.)} \quad + \quad MnO_4^- \text{ (aq.)} \rightleftharpoons Q^+MnO_4^- \text{ (org.)} \qquad (20)$$

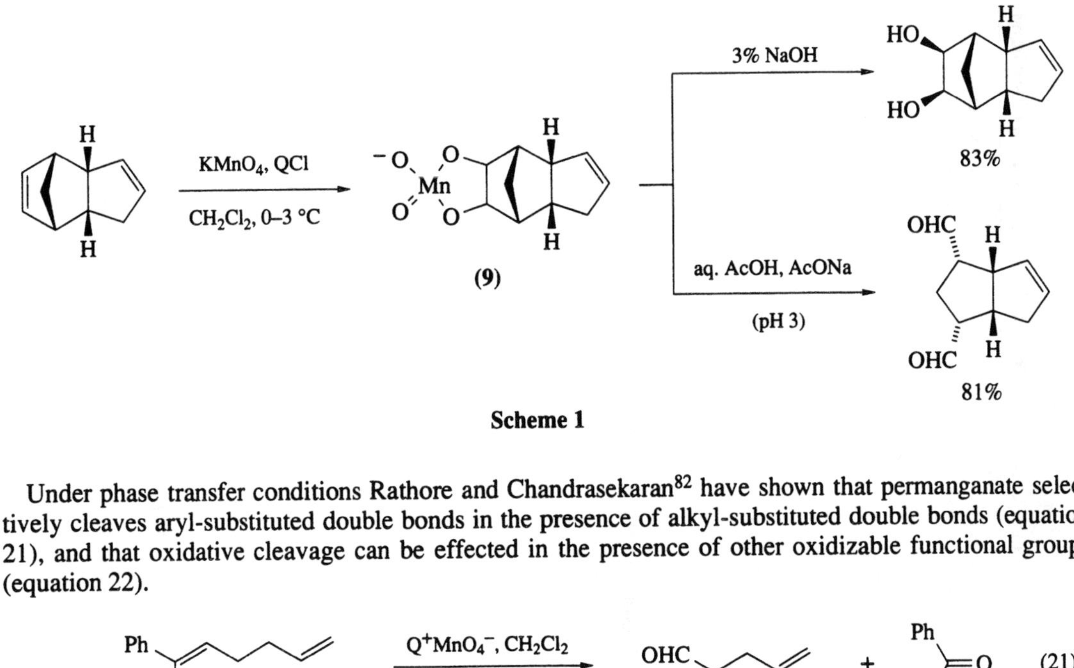

$KMnO_4$, QCl; CH_2Cl_2, 0–3 °C

(**9**)

3% NaOH — 83%

aq. AcOH, AcONa (pH 3) — 81%

Scheme 1

Under phase transfer conditions Rathore and Chandrasekaran[82] have shown that permanganate selectively cleaves aryl-substituted double bonds in the presence of alkyl-substituted double bonds (equation 21), and that oxidative cleavage can be effected in the presence of other oxidizable functional groups (equation 22).

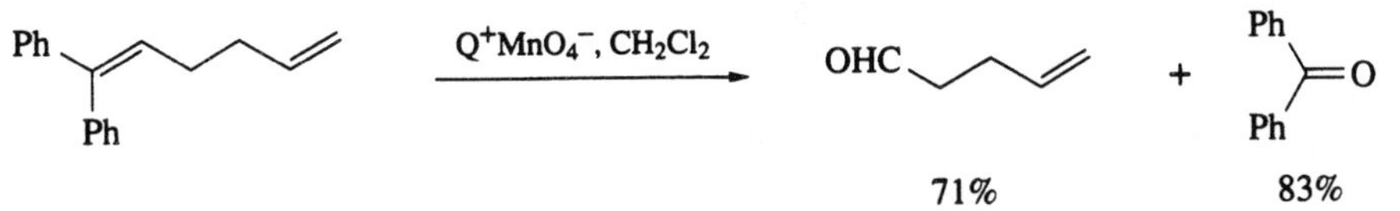

$Q^+MnO_4^-$, CH_2Cl_2 (21)

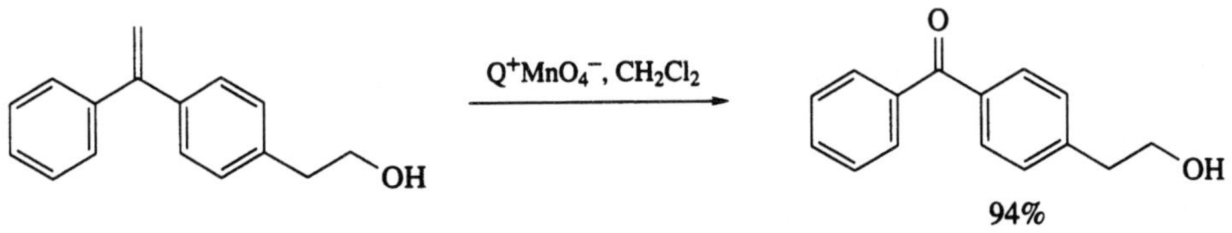

71% 83%

$Q^+MnO_4^-$, CH_2Cl_2 (22)

94%

A typical procedure is provided by the oxidative cleavage of *endo*-dicyclopentadiene to the corresponding dialdehyde (Scheme 1).[81] A solution of potassium permanganate (3.41 mmol) and triethylbenzylammonium chloride (3.41 mmol) in dichloromethane (40 mL) was added dropwise to a solution of *endo*-dicyclopentadiene (2.27 mmol) in 20 mL of the same solvent maintained at 0–3 °C. After the addition, which took 40–50 min, stirring was continued for an additional 30–40 min by which

Table 4 The Oxidative Cleavage of Double Bonds to Carbonyl Compounds by Permanganate

Substrate	*Oxidant and conditions*	*Product*	*Yield (%)*	*Ref.*
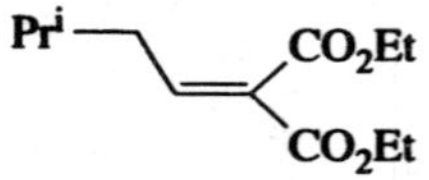 Pr^i, CO_2Et, CO_2Et	$KMnO_4$, THF/H_2O	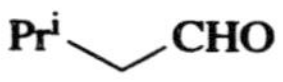 Pr^i CHO	14	80
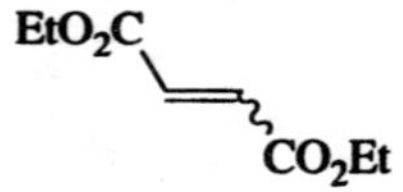 EtO_2C, CO_2Et	$KMnO_4$, THF/H_2O	EtO_2C-CHO	48–51	80
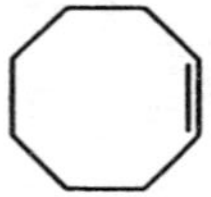	i, $KMnO_4/Et_3\overset{+}{N}CH_2Ph$ Cl^-, CH_2Cl_2 ii, 1 M $HClO_4$	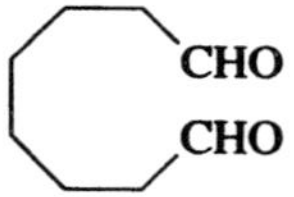 CHO, CHO	74	81
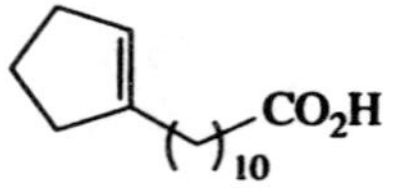 CO_2H, 10	aq. $KMnO_4$	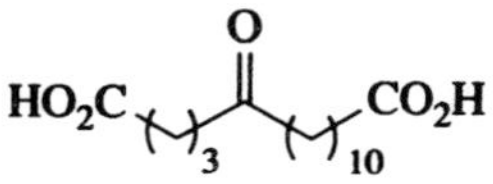 HO_2C, 3, O, 10, CO_2H	71	79
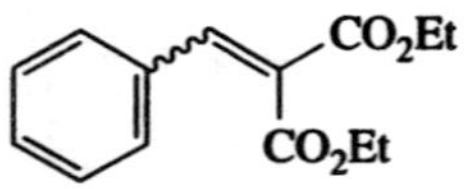 CO_2Et, CO_2Et	$KMnO_4$, THF/H_2O	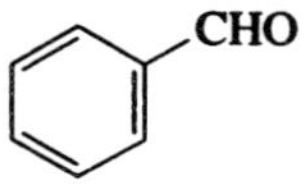 CHO	38	80
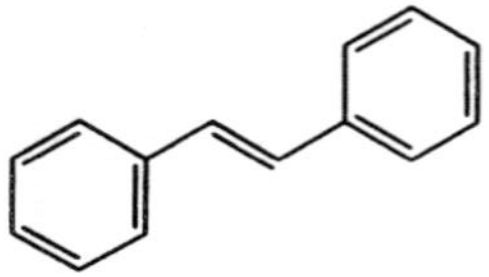	$KMnO_4$, THF/H_2O	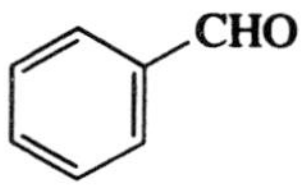 CHO	71	80
	i, $KMnO_4/Et_3\overset{+}{N}CH_2Ph$ Cl^-, CH_2Cl_2; ii, AcOH/NaOAc (pH 5)		74	81
	Cetyltrimethylammonium permanganate, CH_2Cl_2, 25 °C, 2 h		92	82

Table 4 *(continued)*

Substrate	*Oxidant and conditions*	*Product*	*Yield (%)*	*Ref.*
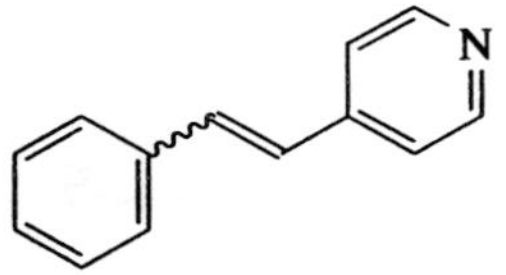	$KMnO_4/MgSO_4$, H_2O/acetone, 0 °C, 2 h or $KMnO_4$, H_2O/NaOH–acetone, 0–5 °C, 2 h	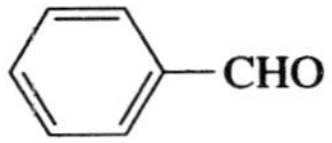 CHO	63–72	84
		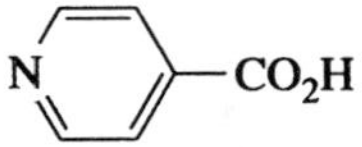 N, CO_2H	≈40–70	84
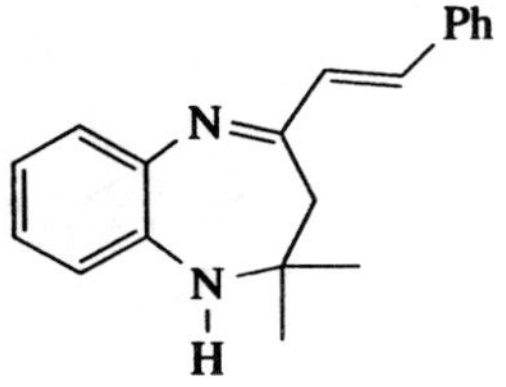 N, Ph, N–H	$KMnO_4$, THF, H_2O	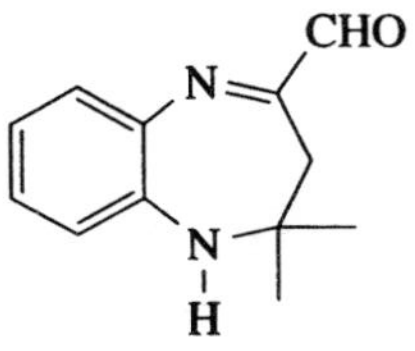 CHO, N, N–H	79	80
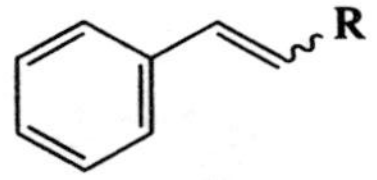 R	Bis(2,2'-bipyridyl)copper(II) permanganate, acetone	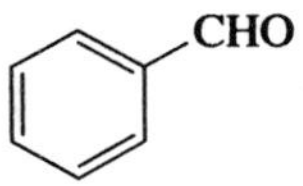 CHO	80–90	85
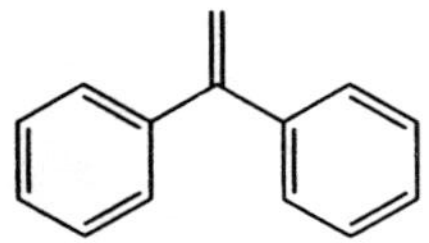	$QMnO_4$, CH_2Cl_2, 25 °C, 2 h	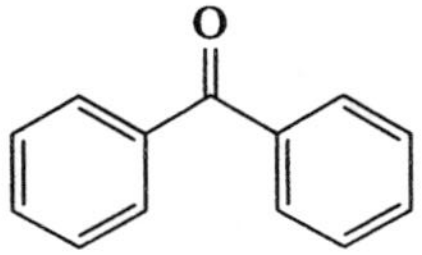 O	94	82
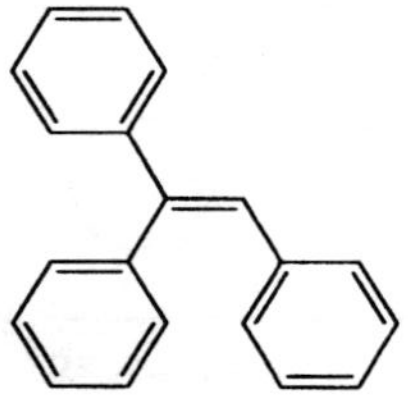	$QMnO_4$, CH_2Cl_2, 25 °C, 2 h	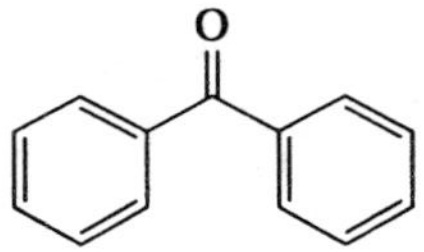 O	96	82
		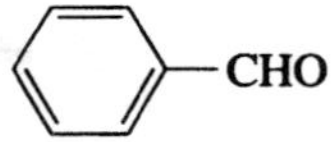 CHO	86	82

Table 4 *(continued)*

Substrate	*Oxidant and conditions*	*Product*	*Yield (%)*	*Ref.*
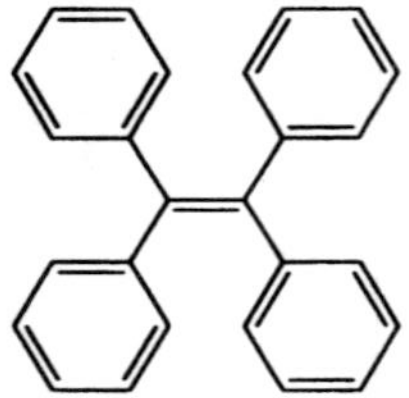	$QMnO_4$, CH_2Cl_2, 25 °C, 2 h	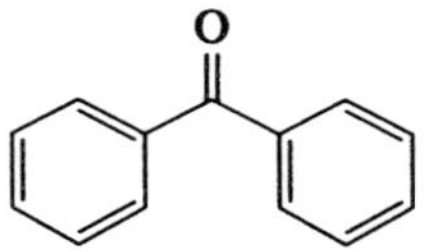 O	90	82
	$KMnO_4$, $MgSO_4$, acetone	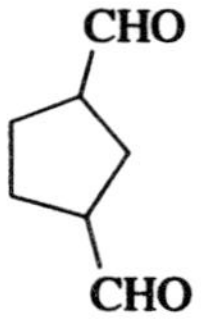 CHO CHO	54–66	78
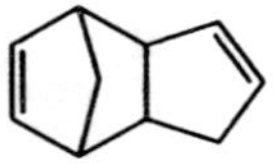	i, $KMnO_4$, $Et_3\overset{+}{N}CH_2Ph\ Cl^-$, CH_2Cl_2 ii, AcOH–NaOAc (pH 3)	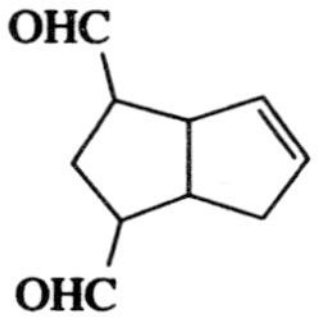 OHC OHC	81	81
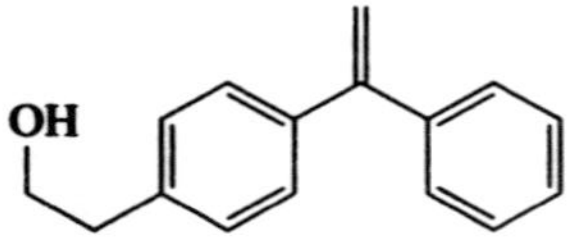 OH	$QMnO_4$, CH_2Cl_2, 25 °C, 2.5 h	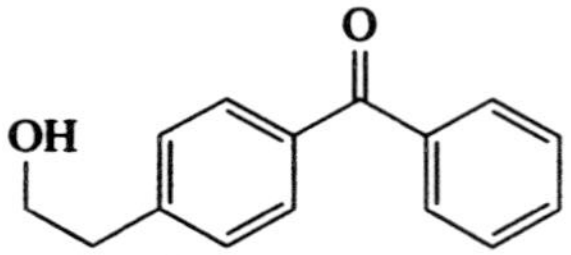 OH O	94	82
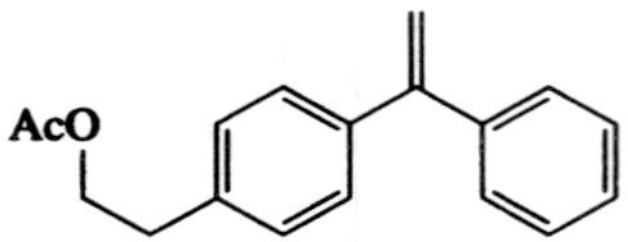 AcO	$QMnO_4$, CH_2Cl_2, 25 °C, 2.5 h	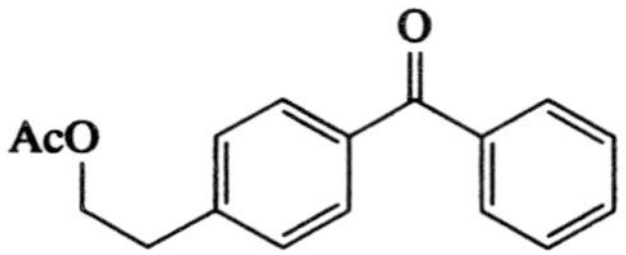 AcO O	92	82

Table 4 *(continued)*

Substrate	*Oxidant and conditions*	*Product*	*Yield (%)*	*Ref.*
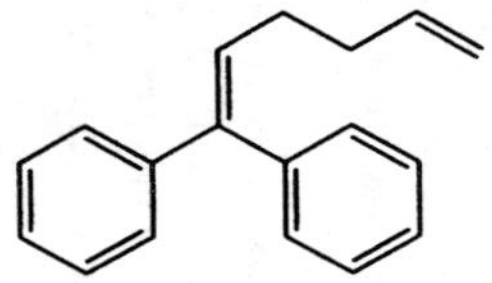	$QMnO_4$, CH_2Cl_2, 25 °C, 5 h	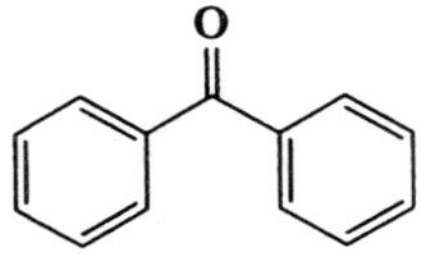 O	83	82
		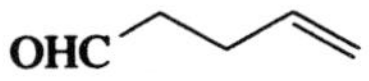 OHC	71	82
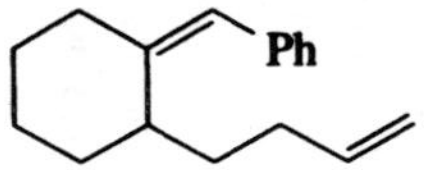 Ph	$QMnO_4$, CH_2Cl_2, 25 °C, 4.5 h	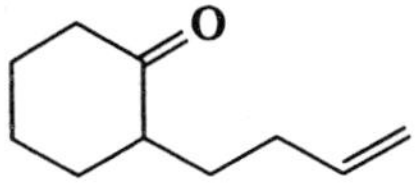 O	86	82
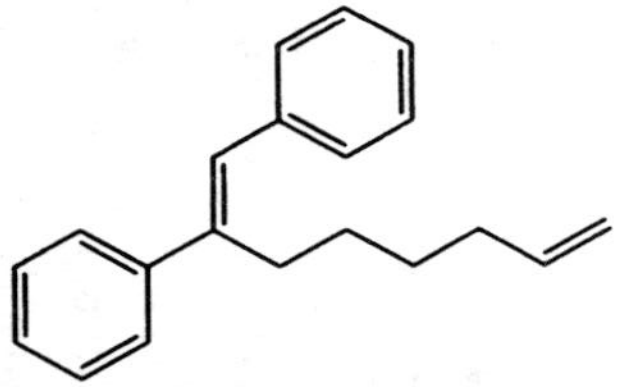	$QMnO_4$, CH_2Cl_2, 25 °C, 4.0 h	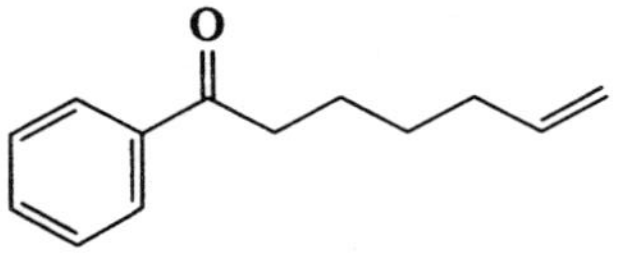 O	86	82

time the permanganate had been completely consumed with the formation of a dark brown solution. Treatment of this solution with 30 mL of water, buffered at pH 3, produced an 81% yield of dialdehyde.

Additional examples of the oxidative cleavage of double bonds by permanganate to produce aldehydes and ketones are summarized in Table 4. A detailed study of the reaction mechanism has also been reported.[83]

3.8.3.3 Osmium Tetroxide and Sodium Periodate

Osmium tetroxide reacts with double bonds to form cyclic osmate(VI) diesters (**10**), which can then be hydrolyzed to provide vicinal diols in good yields.[1,86] If, however, sodium periodate is also present, the diol is cleaved, as in Scheme 2, and carbonyl compounds are the final products. Periodate serves the additional purpose of regenerating osmium tetroxide, thus permitting the use of this expensive and toxic reagent in minimum amounts.

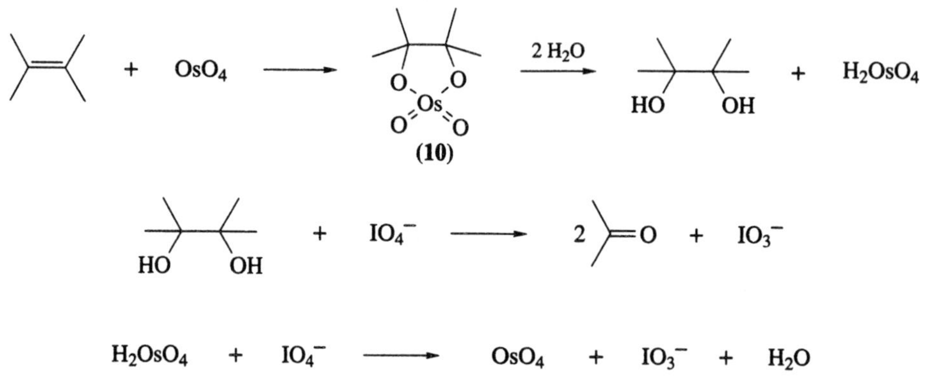

$$\text{R}_2\text{C=CR}_2 + OsO_4 \longrightarrow \textbf{(10)} \xrightarrow{2\,H_2O} \text{HO–CR}_2\text{–CR}_2\text{–OH} + H_2OsO_4$$

$$\text{HO–CR}_2\text{–CR}_2\text{–OH} + IO_4^- \longrightarrow 2\,\text{R}_2\text{C=O} + IO_3^-$$

$$H_2OsO_4 + IO_4^- \longrightarrow OsO_4 + IO_3^- + H_2O$$

Scheme 2

The reaction is usually carried out in a mixed solvent containing water and dioxane, acetone, acetic acid or tetrahydrofuran.[71,87–93] Nonaqueous solvents can also be used if a phase transfer agent is added to bring the periodate ion into solution[94,95] or, alternatively, by use of periodic acid in THF.[96]

In a typical example,[91] sodium periodate (18.2 g, 85 mmol) was added in small portions over a 45 min period to 1,4-dioxa-6-acetyl-6-allylspiro[4.5]decane (8.9 g, 40 mmol) and osmium tetroxide (0.10 g, 0.39 mmol) in a solution of THF (126 mL) and water (42 mL) at room temperature. The mixture was stirred for 2 h at this temperature during which time the black slurry turned brown. Water (600 mL) was introduced, and the mixture was extracted with ether. The extract was dried over anhydrous magnesium sulfate and stripped of solvent to give 7.4 g of crude aldehyde. (Because osmium tetroxide is a toxic and volatile irritant, all preparations should be carried out in a fume hood with use of adequate personal protection, gloves and safety glasses.) Other examples of the use of this reagent have been summarized in Table 5.

3.8.3.4 Ruthenium Tetroxide

The physical properties, preparation and reactions of ruthenium tetroxide have been reviewed by Lee and van den Engh,[109] Rylander,[110] Haines[1] and Henry and Lange.[4] A more vigorous oxidant than osmium tetroxide, its reaction with double bonds produces only cleavage products.[111] Under neutral conditions aldehydes are formed from unsaturated secondary carbons while carboxylic acids are obtained under alkaline or acidic conditions. For example, Shalon and Elliott[112] found that ruthenium tetroxide reacted with compound (**11**) to give the corresponding aldehyde under neutral conditions, but that a carboxylic acid was formed in acidic or alkaline solvents (equation 23).

When used in stoichiometric amounts, ruthenium tetroxide is usually prepared by oxidation of hydrated ruthenium dioxide or trichloride with aqueous periodate or hypochlorite and then extracted into carbon tetrachloride.[86,109] However, since ruthenium compounds are expensive it is more common to use only catalytic amounts of $RuO_2{\cdot}2H_2O$ or $RuCl_3{\cdot}H_2O$ in the presence of a cooxidant that continuously regenerates ruthenium tetroxide.

Table 5 Cleavages to Carbonyl Componds by use of Osmium Tetroxide with Periodate as a Cooxidant

Substrate	*Oxidant and conditions*	*Product*	*Yield (%)*	*Ref.*
	OsO_4/H_5IO_6, THF, r.t.	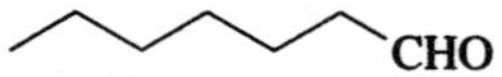	86	96
	$OsO_4/NaIO_4$, dioxane/H_2O, 24–26 °C, 2 h	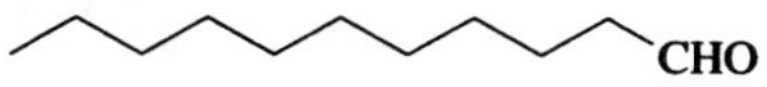	68	93
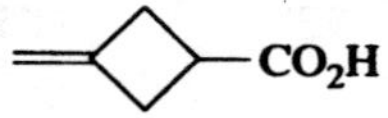	$OsO_4/NaIO_4$, Et_2O, H_2O, 12 h	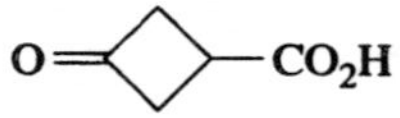	70	90
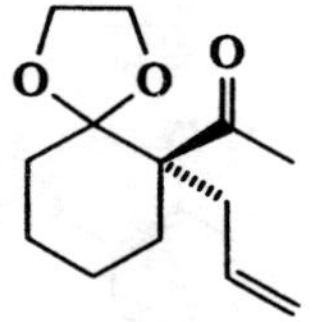	$OsO_4/NaIO_4$, THF/H_2O	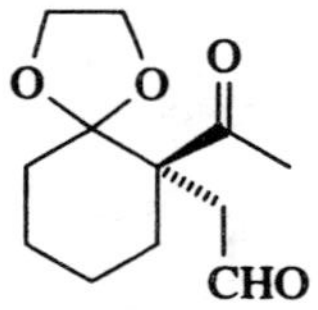	80	91
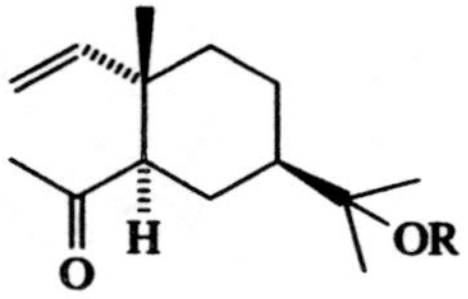	$OsO_4/NaIO_4$, aq. dioxane	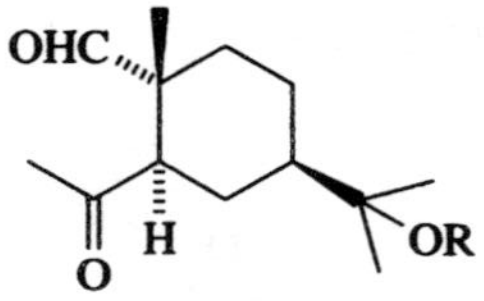	75	87
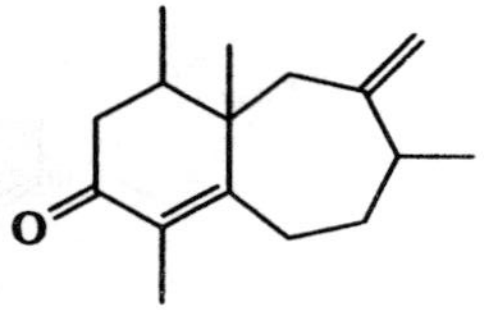	$OsO_4/NaIO_4$, aq. dioxane	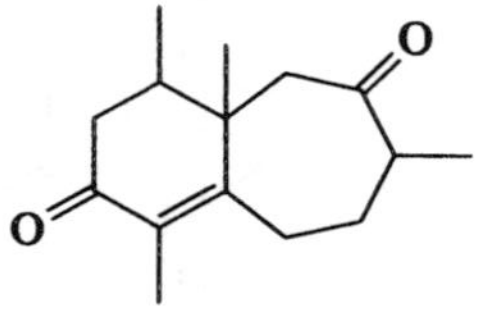	83	97
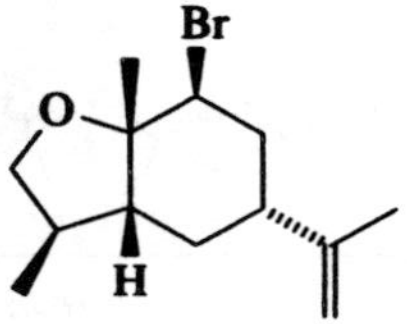	$OsO_4/NaIO_4$, THF/H_2O	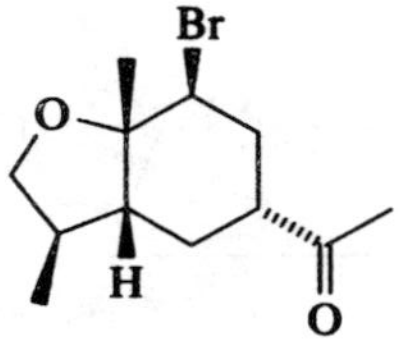	90	98

Table 5 *(continued)*

Substrate	*Oxidant and conditions*	*Product*	*Yield (%)*	*Ref.*
	OsO_4/KIO_4		39–59	99
	$OsO_4/NaIO_4$, dioxane/H_2O/Py		87	35
	OsO_4/KIO_4, dibenzo-18-crown-6		70	94
	$OsO_4/NaIO_4$		>74	100
	$OsO_4/NaIO_4$, dioxane/H_2O/AcOH, r.t. overnight		86	101

Table 5 *(continued)*

Substrate	*Oxidant and conditions*	*Product*	*Yield (%)*	*Ref.*
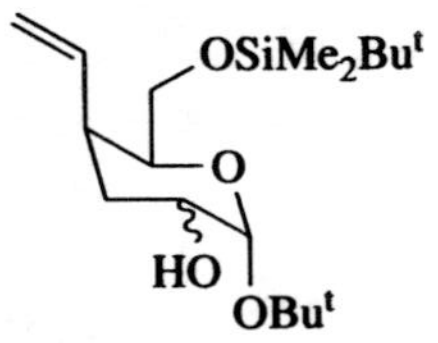 $OSiMe_2Bu^t$, O, HO, OBu^t	$OsO_4/NaIO_4$/NMO, acetone	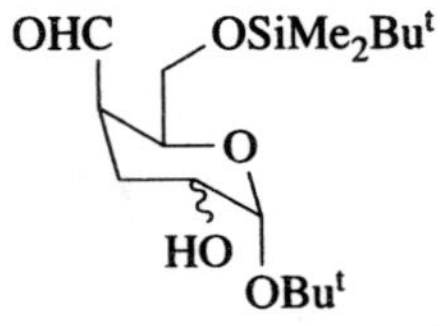 OHC, $OSiMe_2Bu^t$, O, HO, OBu^t	71	102
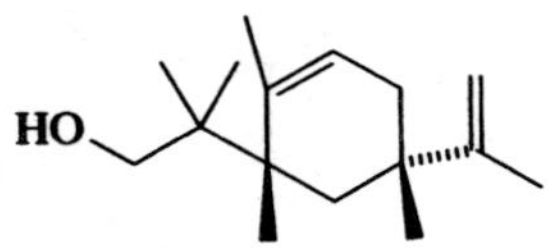 HO	$OsO_4/NaIO_4$, THF/H_2O	 HO, O	90	98
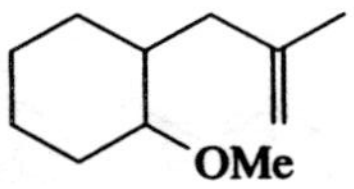 OMe	$OsO_4/NaIO_4$, dioxane, H_2O, 6.5 h	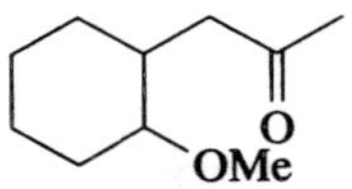 O, OMe	60	103
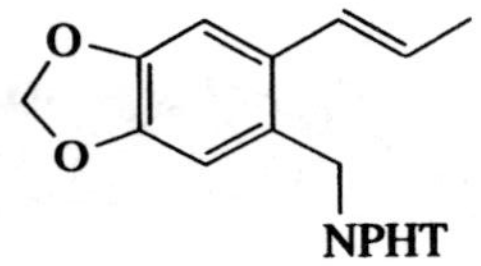 O, O, NPHT	i, OsO_4/NMO, acetone/H_2O (8:5), 0–25 °C, 3 h ii, $NaIO_4$, acetone/H_2O (1:1), r.t., 1 h	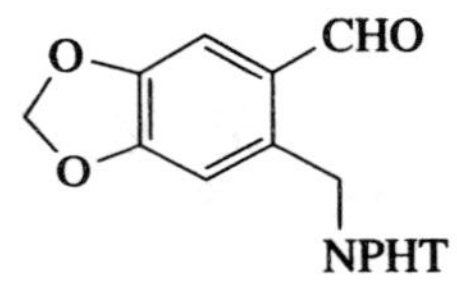 O, O, CHO, NPHT		104
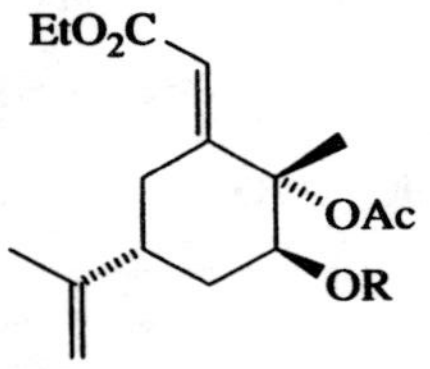 EtO_2C, OAc, OR	OsO_4/KIO_4, THF/H_2O	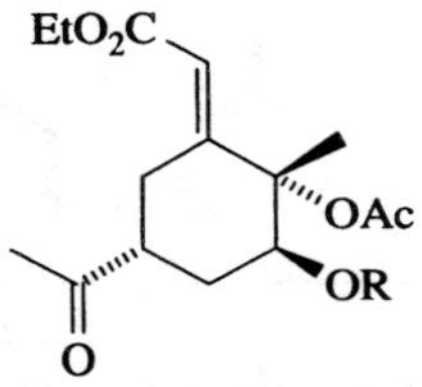 EtO_2C, OAc, OR, O	97	92

Table 5 *(continued)*

Substrate	*Oxidant and conditions*	*Product*	*Yield (%)*	*Ref.*
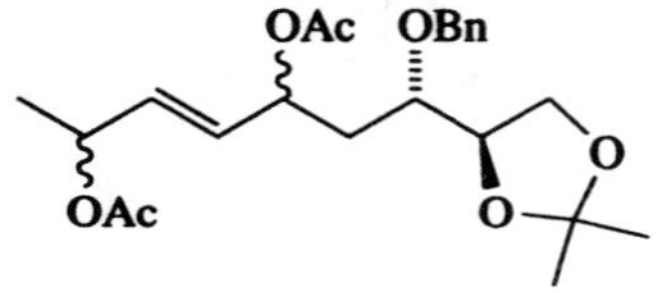 OAc OBn OAc O O	$OsO_4/NaIO_4$, dioxane/H_2O, r.t.	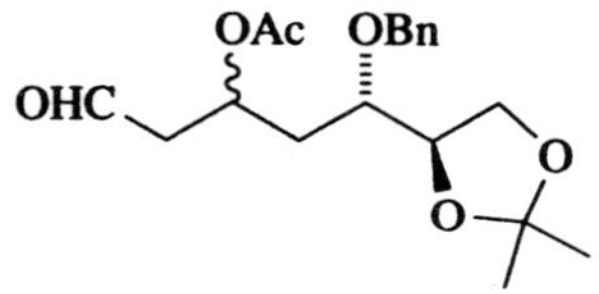 OHC OAc OBn O O		105
 Ph OR O O	$OsO_4/NaIO_4$, MeOH	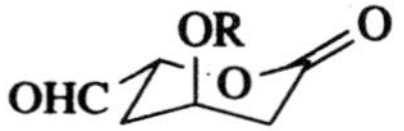 OHC OR O O	53	88
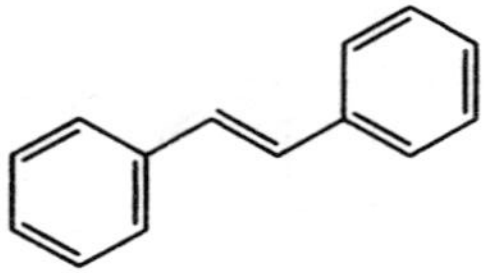	OsO_4/R_4NIO_4	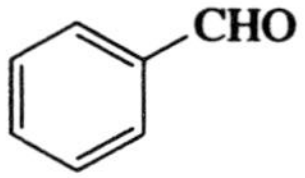 CHO	95	95
	OsO_4/H_5IO_6, THF, r.t.		92	96
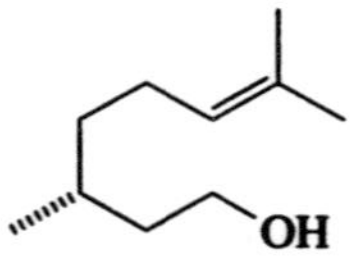 OH	$OsO_4/NaIO_4$, dioxane/H_2O	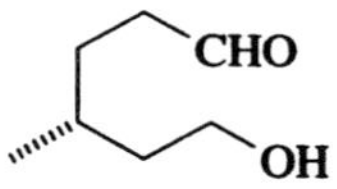 CHO OH	>68	106
	OsO_4/H_5IO_6, THF, r.t.	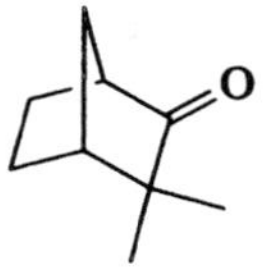 O	91	96
	OsO_4, $NaIO_4$, Et_2O/H_2O, 2 h	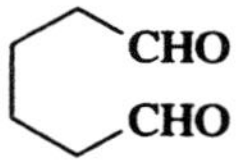 CHO CHO	77	93

Table 5 *(continued)*

Substrate	*Oxidant and conditions*	*Product*	*Yield (%)*	*Ref.*
	$OsO_4/H_5IO_6/THF$, r.t.	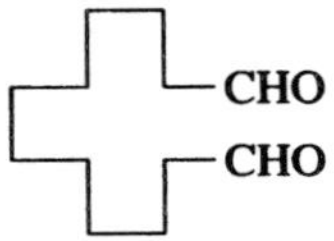 CHO CHO	92	96
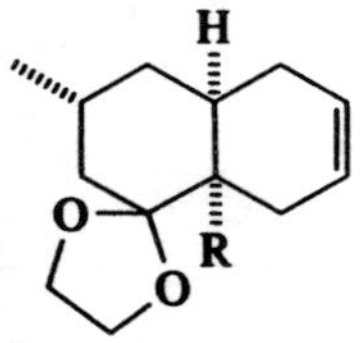 H O R O	OsO_4/H_5IO_6, NMO	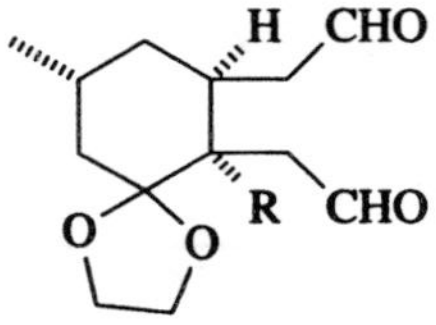 H CHO O O R CHO	92	108
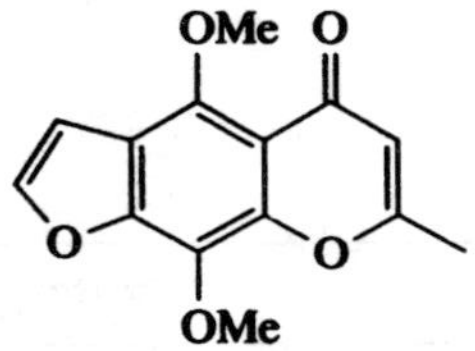 OMe O O O OMe	$OsO_4/NaIO_4$, THF, 50 °C	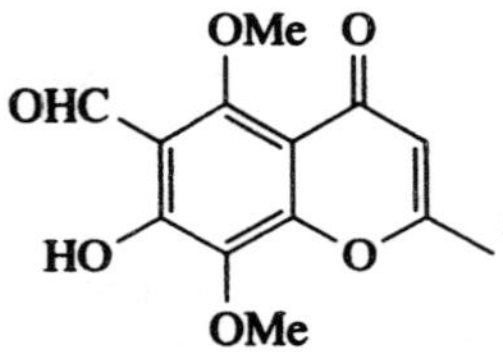 OMe O OHC HO O OMe	73	106
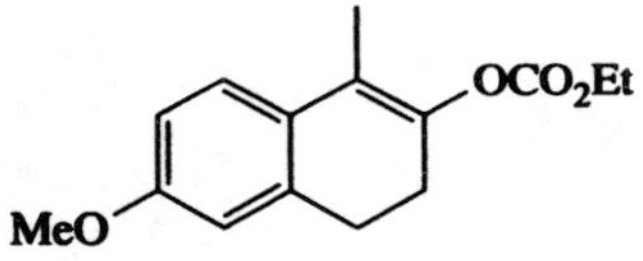 OCO_2Et MeO	$OsO_4/NaIO_4$, Et_2O/H_2O	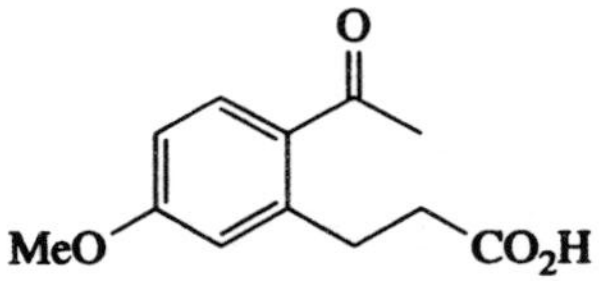 O MeO CO_2H	65	89
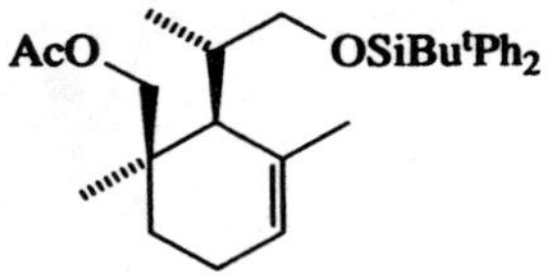 AcO $OSiBu^tPh_2$	i, OsO_4, Py; ii, $NaIO_4$, MeOH/THF; iii, $CH(OMe)_3$ MeOH, $CeCl_3$•xH_2O	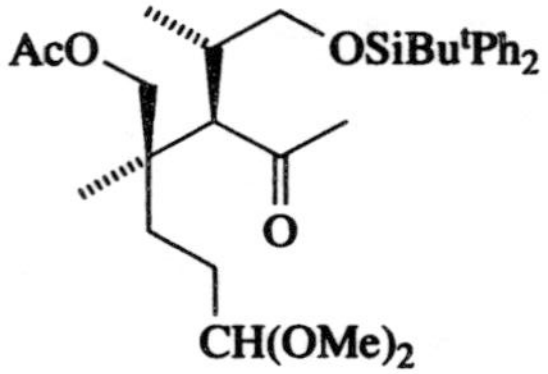 AcO $OSiBu^tPh_2$ O $CH(OMe)_2$	70	107

Table 5 *(continued)*

Substrate	*Oxidant and conditions*	*Product*	*Yield (%)*	*Ref.*
	OsO_4/H_5IO_6/THF, r.t.	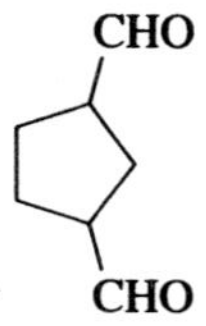 CHO CHO	94	96

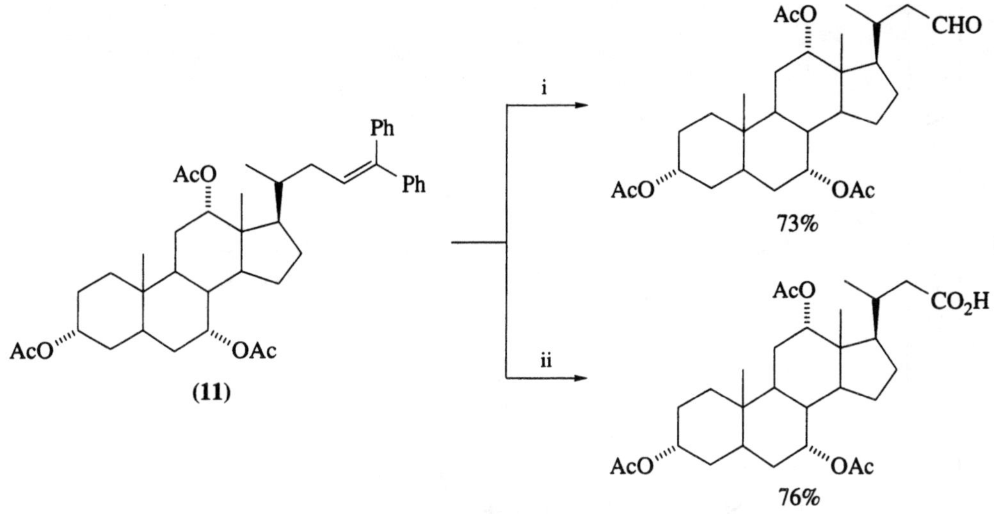

i, RuO_4, CCl_4, acetone/H_2O; ii, RuO_4, CCl_4, AcOH/H_2O

(23)

A two-phase system (carbon tetrachloride and water) is often used for these reactions. It appears that contact between ruthenium tetroxide and the alkene takes place in the organic phase where they are both most soluble. The ruthenium dioxide produced when oxidation occurs is insoluble in all solvents and migrates to the interphase where it contacts the cooxidant (in the aqueous phase) and is reoxidized, as summarized in Scheme 3. Because good contact between all components is essential, best results are obtained when the mixture is shaken or stirred vigorously throughout the course of the reaction. Sharpless and his coworkers[115] have also found that the addition of acetonitrile to the two-phase mixture improves yields.

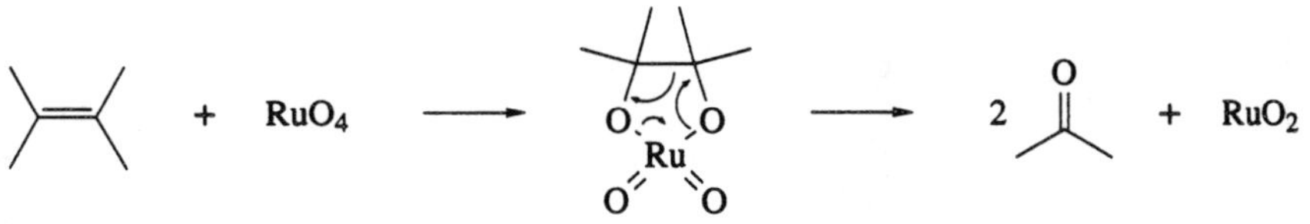

$$RuO_2 + 2\,NaIO_4 \longrightarrow RuO_4 + 2\,NaIO_3$$

Scheme 3

In a typical experiment,[113] a flask was charged with carbon tetrachloride (2 mL), acetonitrile (2 mL), water (3 mL), alkene (1.0 mmol), sodium periodate (877 mg, 4.1 equiv.) and $RuCl_3{\cdot}H_2O$ (5 mg, 2.2 mol %), and the entire mixture was stirred vigorously for 2 h at room temperature. Then dichloromethane (10 mL) was added to assist in the separation of the phases and the aqueous phase was extracted three times with additional volumes of CH_2Cl_2. The combined organic extracts were dried over anhydrous magnesium sulfate and concentrated. The residue was dissolved in 20 mL of ether, filtered through a Celite pad to remove traces of ruthenium dioxide and concentrated again to give the crude carbonyl products. A few typical examples of this reaction have been summarized in Table 6.

3.8.3.5 Hexavalent Chromium Compounds

The reactions of alkenes with chromate or dichromate ions usually leads to an array of products arising from oxidative attack at the double bond and the allylic positions.[3] Only in special cases where the double bond bears one or more phenyl[118] or alkoxy[119] substituents have good yields of the corresponding carbonyl compounds been reported.

Chromium trioxide adsorbed on silica or alumina has been used for the oxidative cleavage[120] of alkenes to aldehydes or ketones with little or no formation of carboxylic acids. A solution of bis(triphenylsilyl) chromate has also been used for the selective cleavage of double bonds to carbonyl compounds.[121]

Table 6 Cleavage of Double Bonds to Carbonyl Compounds by Ruthenium Tetroxide

Substrate	*Oxidant and conditions*	*Product*	*Yield (%)*	*Ref.*
	RuO_4, CCl_4	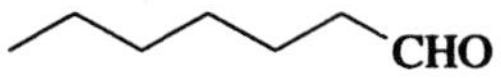	12	114
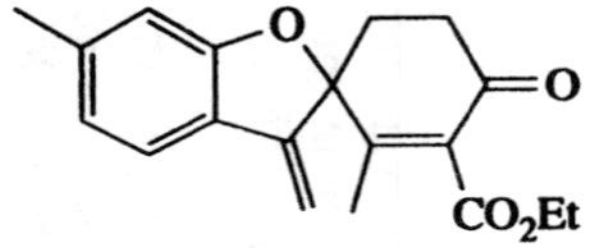	RuO_4, CCl_4	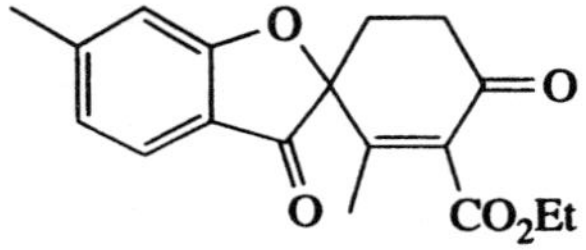	3–32	111
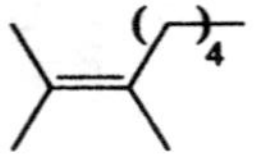	$RuCl_3 \cdot H_2O/NaIO_4$, $CCl_4/MeCN/H_2O$	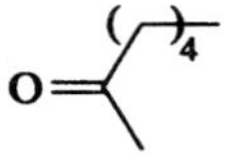	>95	113
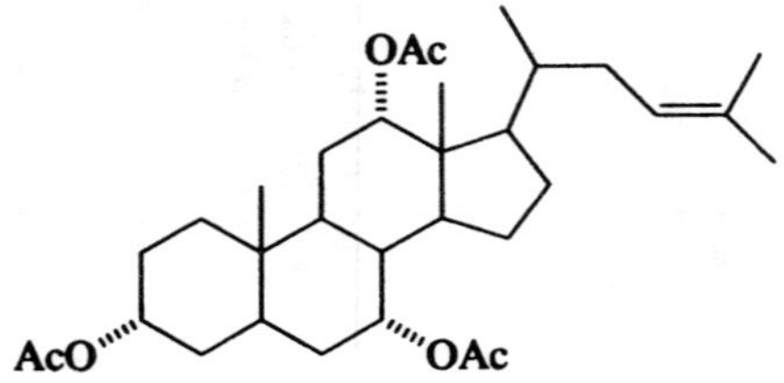	RuO_4, CCl_4/acetone/H_2O	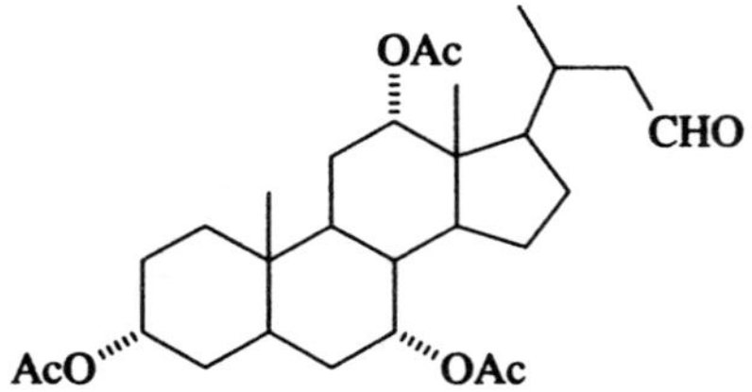	73	112
	RuO_4, CCl_4	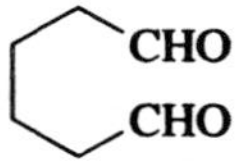	10	114
	$RuCl_3$, 4.0 equiv. $NaClO_4$		25–30	115
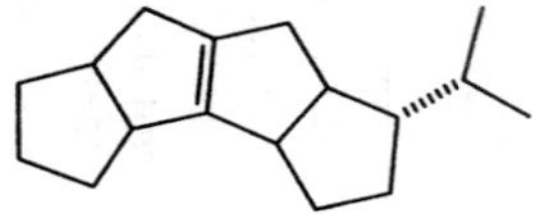	$RuO_2/NaIO_4$, $CCl_4/MeCN/H_2O$	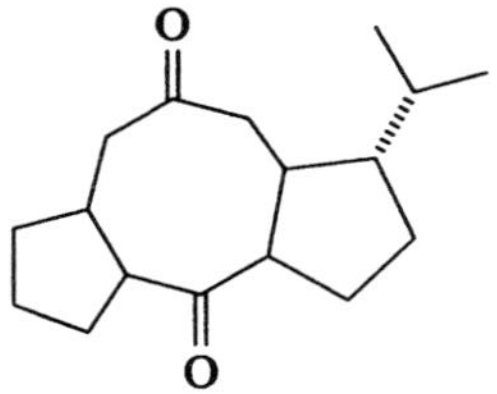	82	116

Table 6 *(continued)*

Substrate	*Oxidant and conditions*	*Product*	*Yield (%)*	*Ref.*
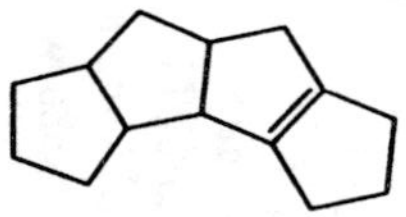	$RuO_2/NaIO_4$, $CCl_4/MeCN/H_2O$	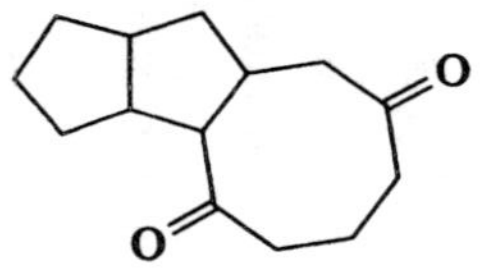	92	116
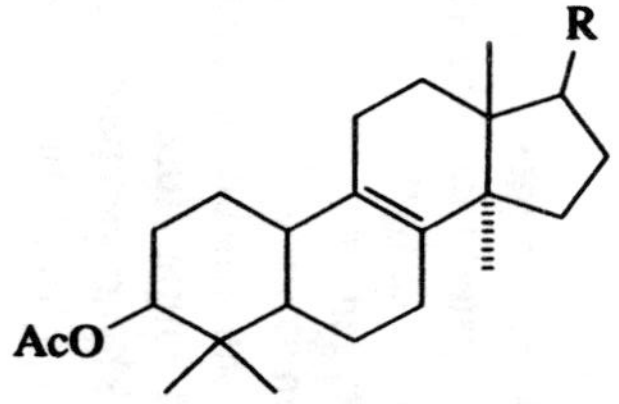	RuO_4, CCl_4	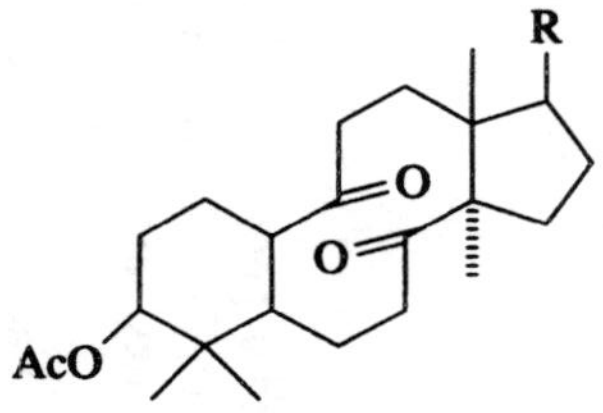	60	117

Finally, a compound formed by dissolving chromium trioxide and 2,2-bipyridyl in glacial acetic acid saturated with dry hydrogen chloride has been reported to cleave double bonds without complicating side reactions.[122] Unfortunately this oxidant, which is reported to have the formula of (bipy)H_2CrOCl_5, is effective only with phenyl-substituted double bonds.

Some examples of the use of hexavalent chromium compounds for oxidative cleavages are given in Table 7.

3.8.4 CLEAVAGE OF DOUBLE BONDS TO YIELD CARBOXYLIC ACIDS, ESTERS OR LACTONES

Oxidative cleavage of a carbon–carbon double bond produces ketones from tertiary carbons with almost all oxidants. If, however, one or both of the carbons are secondary, either aldehydes or, more generally, carboxylic acids are obtained. In some cases these latter products undergo subsequent reactions to form either esters or lactones.

If carboxylic acids are the desired products, the double bonds should be oxidized by potassium permanganate, ruthenium tetroxide, hexavalent chromium, or ozone followed by an oxidative work-up.

3.8.4.1 Ozone Followed by an Oxidative Work-up

Carboxylic acids are produced in good yields if the ozonide, formed when ozone reacts with a double bond as in equation (6), is subjected to oxidative hydrolysis. Although a variety of oxidants (*e.g.* chromic acid, permanganate ion and peroxy acids) have been used for this purpose, hydrogen peroxide is most commonly employed.

Two typical examples are illustrated in equations (24)[125] and (25).[126]

OMe OMe — i, O_3, MeOH; ii, H_2O_2 → HO_2C HO_2C OMe OMe 91% (24)

H D N O O O — i, O_3, CH_2Cl_2; ii, 30% H_2O_2 → H D N HO_2C O O 99% (25)

If the reaction is carried out in an emulsion of sodium hydroxide and hydrogen peroxide, the ozonide intermediates are converted to carboxylic acids directly, with a consequent increase in yields.[127]

Oxidative cleavage of γ-hydroxyalkenes results in the formation of lactones in good yields (equation 26).[128]

OH — i, O_3, acetone; ii, Jones' reagent → O O 65–94% (26)

Ozonolysis of 1,2-dichloroalkenes in methanol affords the corresponding methyl esters in good yield (equation 27).[140] It has been suggested that the intermediates in these reactions must be either the corresponding acid chlorides or α-chloro-α-methoxyalkyl hydroperoxides, as in Scheme 4.

When the alkene bears an oxygen or nitrogen substituent in the allylic position, oxidation often proceeds with the loss of one carbon atom, as in equation (28).[127]

In a typical experiment,[126] ozone was bubbled into a solution of alkene (8.8 mmol) dissolved in dichloromethane (120 mL) at reduced temperatures (–78 °C) until TLC analysis indicated no starting material remained (about 2 h). Then 30% hydrogen peroxide (2 mL) was added and the reaction mixture stirred at room temperature for 18 h. The product mixture was washed with water, dried over anhydrous sodium

Table 7 Cleavage of Double Bonds to Carbonyl Compounds by Hexavalent Chromium Compounds

Substrate	*Oxidant and conditions*	*Product*	*Yield (%)*	*Ref.*
$H_2C{=}CH_2$	$CrO_3/SiO_2/Al_2O_3$, cyclohexane	HCHO		120
	$(Ph_3SiO)_2CrO_2$, heptane or CCl_4	 CHO		121
	(bipy)H_2CrOCl_5 (2 equiv.), CH_2Cl_2, r.t., 0.5 h	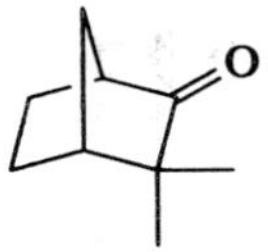 O	80	122
	$(Ph_3SiO)_2CrO_2$, heptane or CCl_4	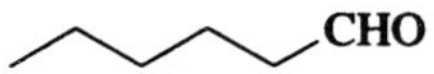 CHO		121
	$CrO_3/SiO_2/Al_2O_3$, cyclohexane	O		120
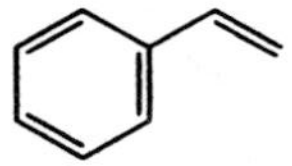	(bipy)H_2CrOCl_5 (4 equiv.), CH_2Cl_2, r.t., 4.5 h	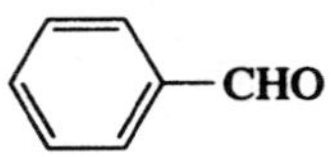 CHO		122
	$CrO_3/SiO_2/Al_2O_3$, cyclohexane			120
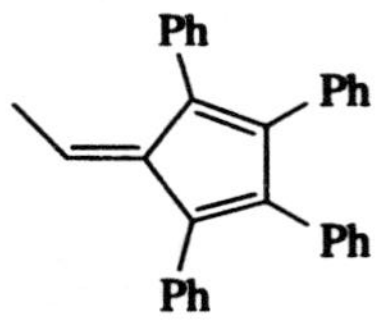 Ph, Ph, Ph, Ph	(bipy)H_2CrOCl_5 (2 equiv.), CH_2Cl_2, r.t., 2 h	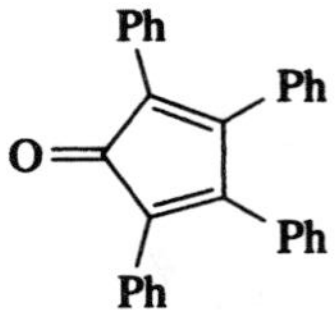 Ph, Ph, Ph, Ph, O	70	122
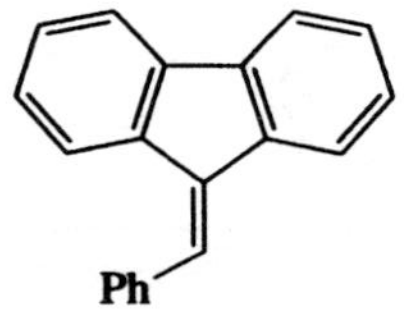 Ph	(bipy)H_2CrOCl_5 (2 equiv.), CH_2Cl_2, r.t., 0.75 h	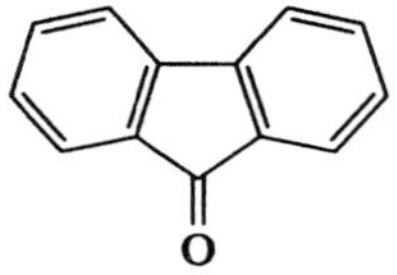 O	90	122

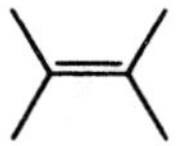

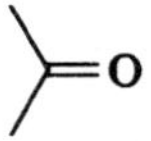

Table 7 *(continued)*

Substrate	*Oxidant and conditions*	*Product*	*Yield (%)*	*Ref.*
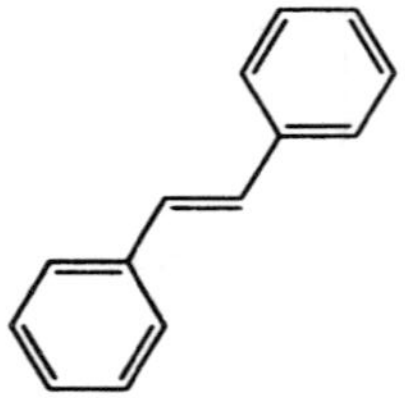	$CrO_3/SiO_2/Al_2O_3$, cyclohexane	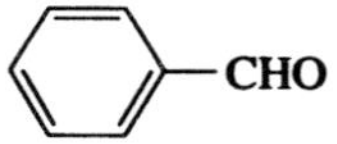 CHO	≈100	120
	(bipy)H_2CrOCl_5 (4 equiv.), CH_2Cl_2, r.t., 4 h		96	122
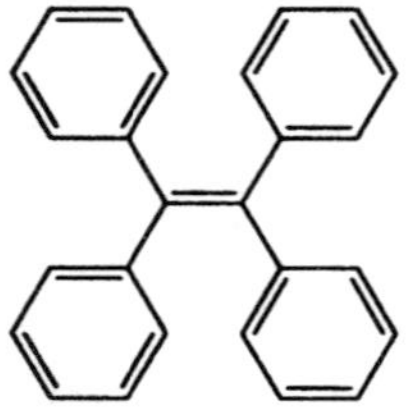	CrO_3, $AcOH/H_2O$, 90–95 °C	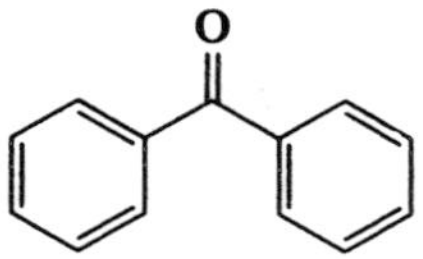 O	26–44	123
	$(Ph_3SiO)_2CrO_2$, heptane or CCl_4	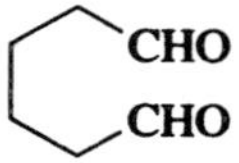 CHO CHO		121
 $(CH_2)_n$ n = 3–6	$CrO_2(OCOCCl_3)_2$, acetone	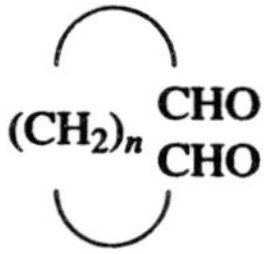 $(CH_2)_n$ CHO CHO n = 3–6	43–70	124
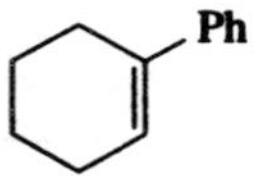 Ph	CrO_3, 30 °C	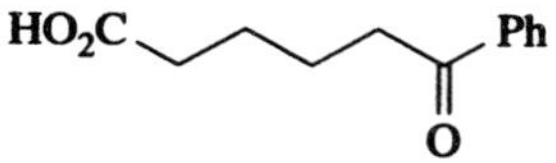 HO_2C Ph O		118

Table 7 *(continued)*

Substrate	*Oxidant and conditions*	*Product*	*Yield (%)*	*Ref.*
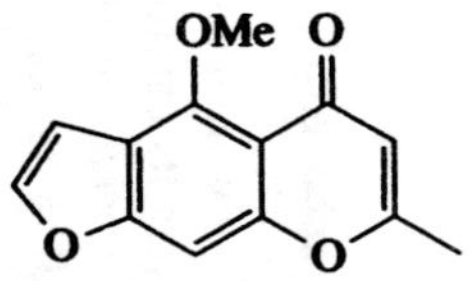 OMe O	CrO_3	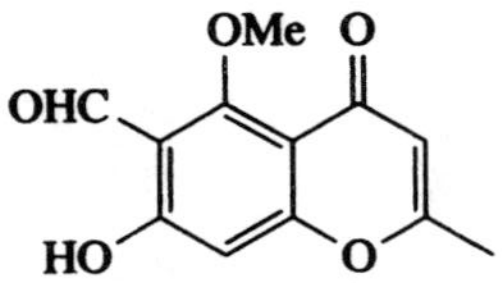 OMe O, OHC, HO		119

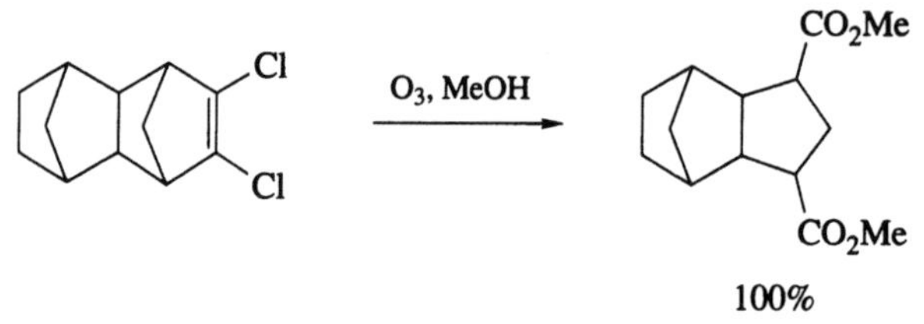

O_3, MeOH (27)

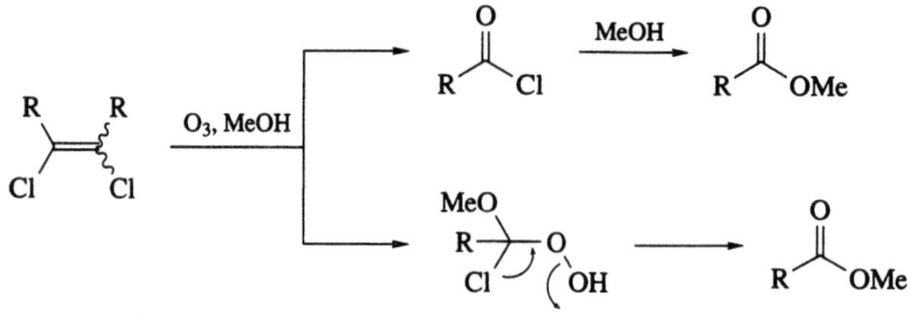

100%

Scheme 4

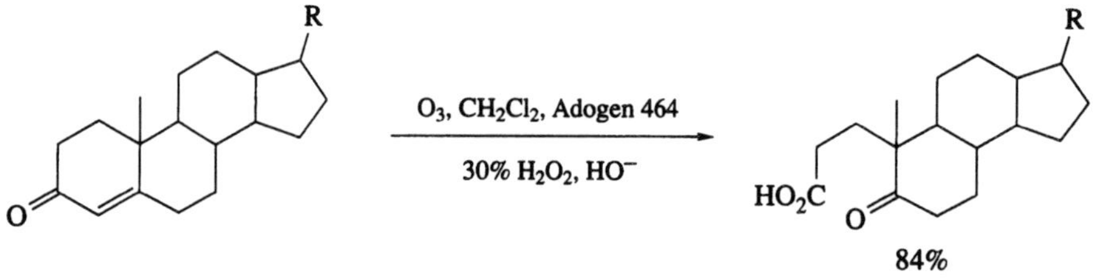

O_3, CH_2Cl_2, Adogen 464; 30% H_2O_2, HO^- (28)

84%

sulfate, and concentrated to give crude product (99% yield). Other examples of this reaction are summarized in Table 8.

3.8.4.2 Permanganate Reactions

3.8.4.2.1 Aqueous potassium permanganate oxidations

The oxidative cleavage of carbon–carbon double bonds has been reviewed by Stewart.[150] In general, carboxylic acids are produced under acidic conditions. However, since many alkenes lack sufficient solubility in water, cosolvents such as pyridine, acetone or acetic acid have often been used to bring the oxidant and reductant into contact. For example, a good yield of 2,6-diphenyl-4-pyridinecarboxylic acid was obtained from the reaction depicted in equation (29).[151] Table 9 contains additional examples.

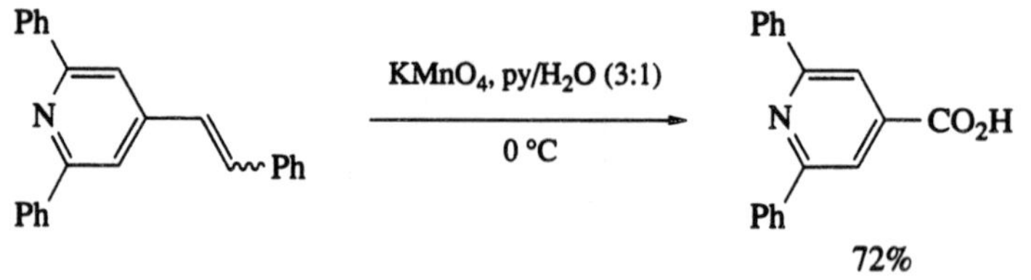

$KMnO_4$, py/H_2O (3:1); 0 °C (29)

72%

3.8.4.2.2 Phase transfer assisted permanganate oxidations

The ability to dissolve permanganate in nonaqueous solvents by use of phase transfer agents (as previously discussed in Section 3.8.2.3.3) extends its use for oxidative cleavages to compounds that are not soluble in aqueous solutions. It has been reported, for example, that 1-eicosene and other long-chain alkenes can be converted into the corresponding carboxylic acids in good yields by use of the following procedure.[154]

Table 8 Cleavage of Double Bonds to give Carboxylic Acids or Esters using Ozone

Substrate	*Oxidant and conditions*	*Product*	*Yield (%)*	*Ref.*
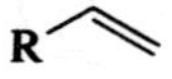 R	i, O_3, THF; ii, H_2/Pd, $CaCO_3$, PbO	R−CO_2H	85	18
 10	i, O_3, $CHCl_3$, −5 °C; ii, Ag_2O, NaOH	 10 CO_2H	94	130
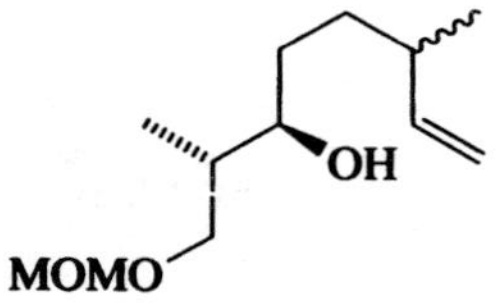 MOMO, OH	i, O_3 (excess), CH_2Cl_2, −78 °C, 15 min; ii, DMS, MeOH −78 to 25 °C, 20 min	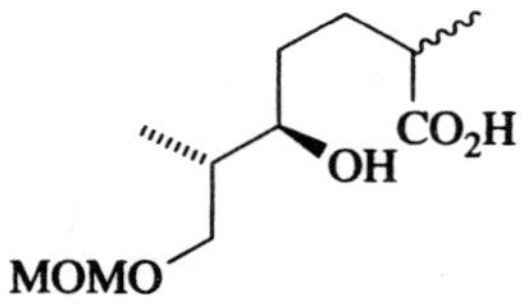 MOMO, OH, CO_2H	>47	131
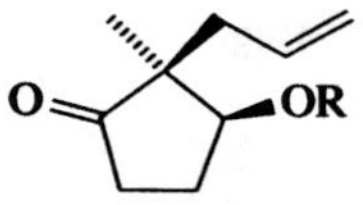 O, OR	O_3, CH_2Cl_2, Py, −78 °C	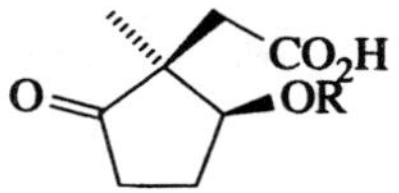 O, OR, CO_2H	>60	132
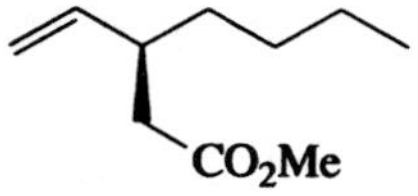 CO_2Me	i, O_3; ii, H_2O_2, 1 M NaOH	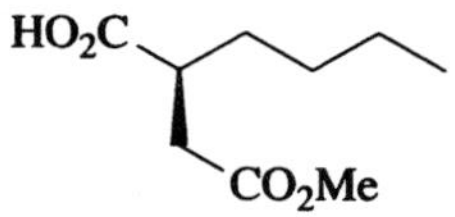 HO_2C, CO_2Me	>95	133
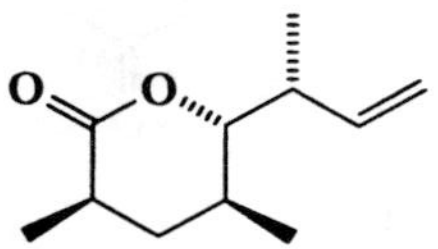 O, O	i, O_3, EtOAc; ii, H_2O_2	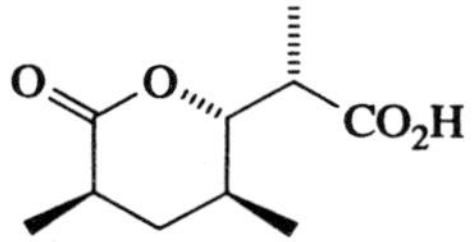 O, O, CO_2H	85	134
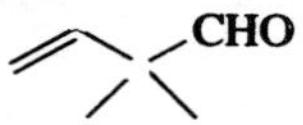 CHO	O_3	 HO_2C, CO_2H		135
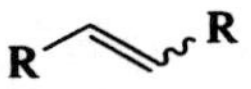 R, R	O_3, HCl, MeOH	R−CO_2Me	62–85	136
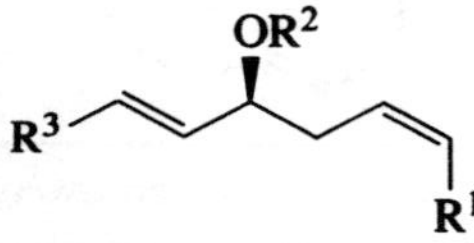 R^3, OR^2, R^1	i, O_3, CH_2Cl_2, −78 °C; ii, 12% H_2O_2, AcOH, 60 °C, 30 min	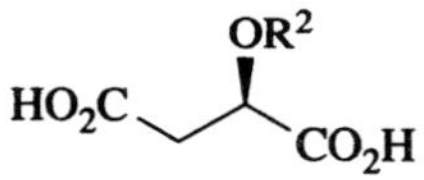 HO_2C, OR^2, CO_2H	>30	137

Table 8 *(continued)*

Substrate	*Oxidant and conditions*	*Product*	*Yield (%)*	*Ref.*
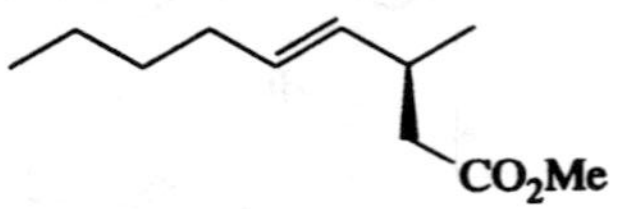 CO_2Me	i, O_3; ii, H_2O_2; iii, CH_2N_2	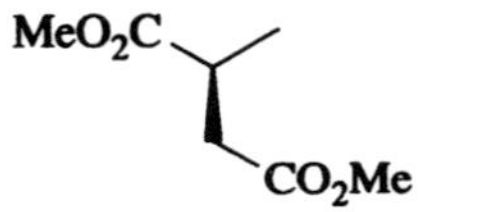 MeO_2C, CO_2Me	>95	133
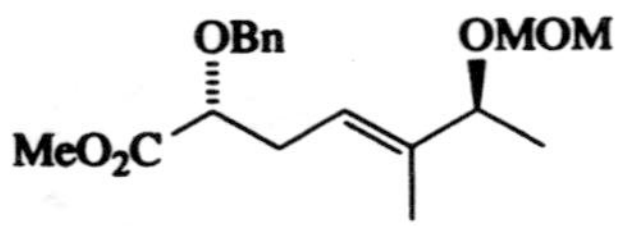 OBn, OMOM, MeO_2C	i, O_3, MeOH; ii, CrO_3, H^+; iii, CH_2N_2	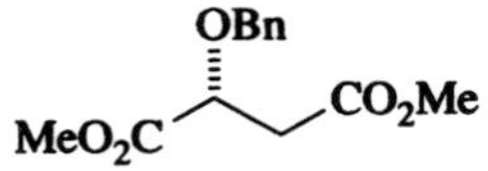 OBn, MeO_2C, CO_2Me		30
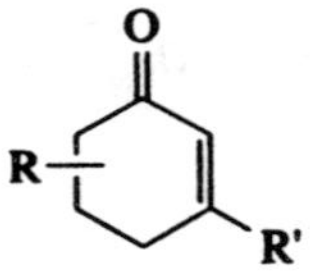 O, R, R'	O_3, 30% H_2O_2–NaOH, Adogen 464	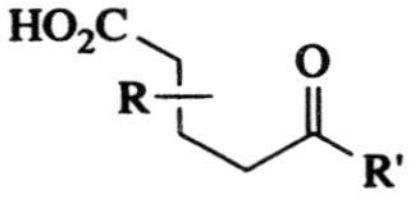 HO_2C, R, O, R'	92	127
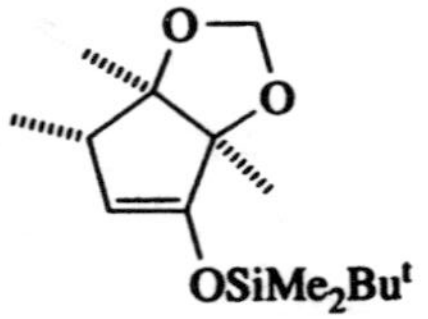 O, O, $OSiMe_2Bu^t$	i, O_3; ii, H_2O_2–HCO_2H	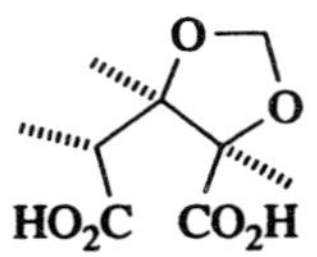 O, O, HO_2C, CO_2H	78	144
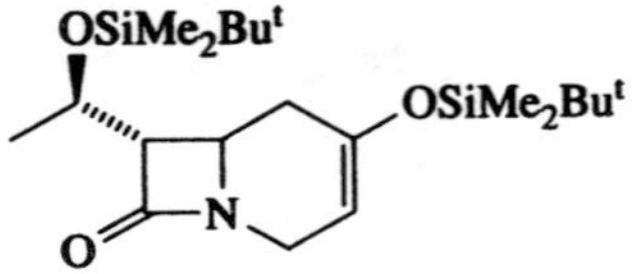 $OSiMe_2Bu^t$, $OSiMe_2Bu^t$, O, N	i, O_3, MeOH, benzyl alcohol; ii, $(PyS)_2$–PPh_3	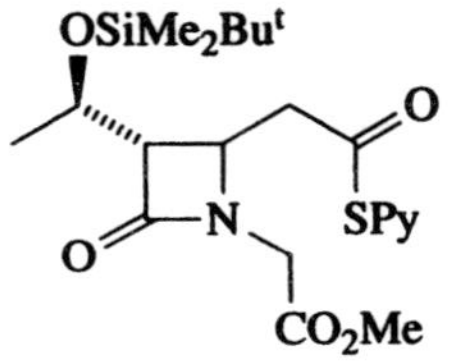 $OSiMe_2Bu^t$, O, SPy, O, N, CO_2Me	85	145
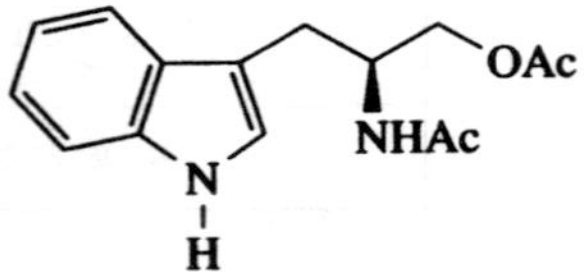 OAc, NHAc, N, H	i, O_3, AcOH; ii, H_2O_2, HCl, Δ	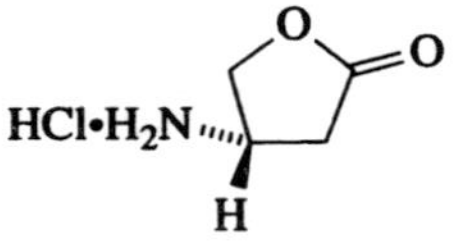 O, O, $HCl{\bullet}H_2N$, H		146

Table 8 *(continued)*

Substrate	*Oxidant and conditions*	*Product*	*Yield (%)*	*Ref.*
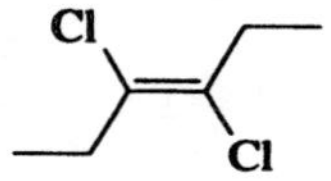 Cl, Cl	O_3, MeOH	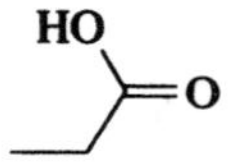 HO, O	92	140
	i, O_3, MeOH, –70 °C; ii, H_2O_2, AcOH	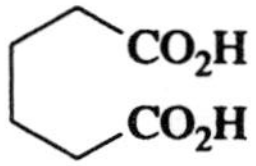 CO_2H, CO_2H	85	141
	i, O_3, AcOH, 0 °C; ii, AcOOH, AcOH		86	142
	i, O_3, AcOH, 0 °C; ii, H_2CrO_4, 50 °C, 12 h		71	142
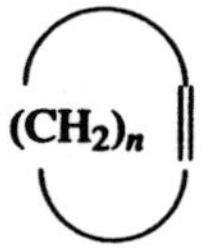 $(CH_2)_n$; $n = 4–10$	i, O_3, THF; ii, H_2/Pd, $CaCO_3$, PbO	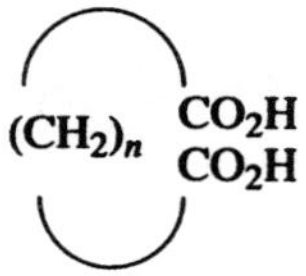 $(CH_2)_n$ CO_2H CO_2H; $n = 4–10$	78–99	18
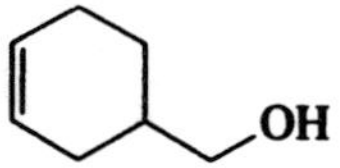 OH	i, O_3, MeOH; ii, H_2O_2, HCO_2H, 90 °C	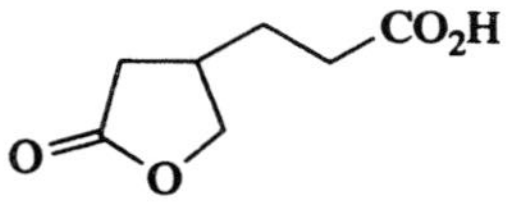 O, O, CO_2H	83	129, 139
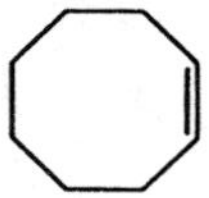	O_3, H_2O, emulsifier, NaOH, H_2O_2, 10 °C	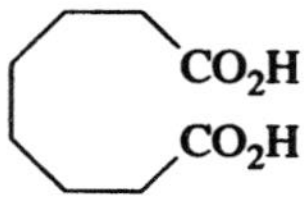 CO_2H, CO_2H	63	143
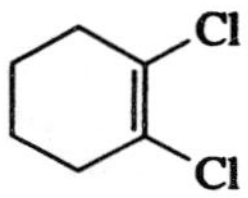 Cl, Cl	O_3, MeOH	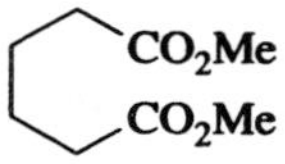 CO_2Me, CO_2Me	84	140

Table 8 *(continued)*

Substrate	*Oxidant and conditions*	*Product*	*Yield (%)*	*Ref.*
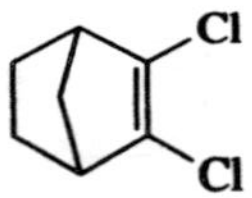 Cl, Cl	O_3, MeOH	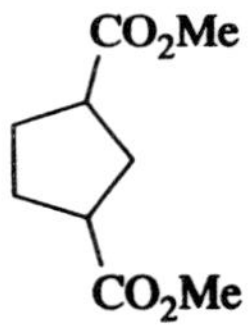 CO_2Me, CO_2Me	≈100	147
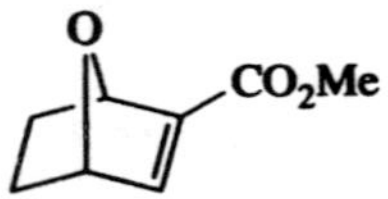 O, CO_2Me	i, O_3, CH_2Cl_2, –78 °C; ii, H_2, Pd–C, MeOH	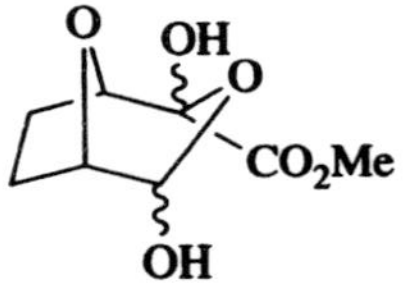 O, OH, O, CO_2Me, OH	>50	148
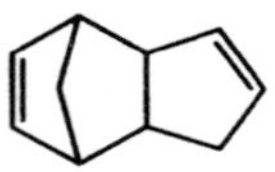	O_3, H_2O, emulsifier, NaOH, H_2O_2, 10 °C	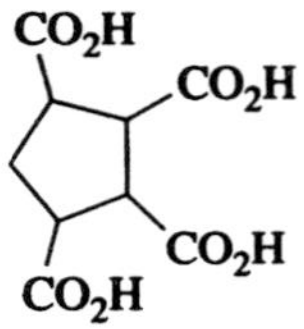 CO_2H, CO_2H, CO_2H, CO_2H	66	149
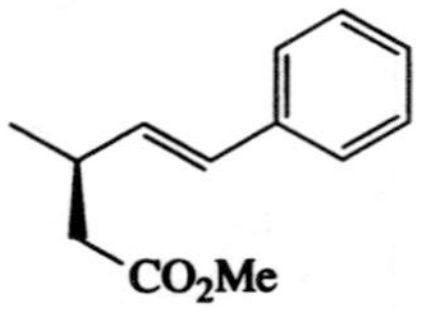 CO_2Me	i, O_3; ii, H_2O_2; iii, CH_2N_2	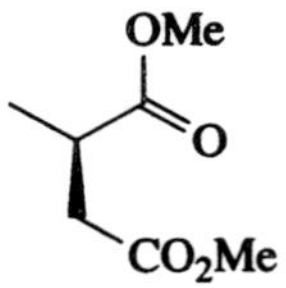 OMe, O, CO_2Me	42	138
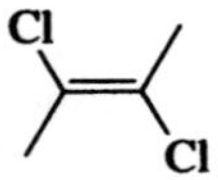 Cl, Cl	O_3, MeOH	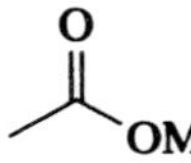 O, OMe	90	140

Table 9 Cleavage of Double Bonds by Permanganate Solutions

Substrate	*Oxidant and Conditions*	*Product*	*Yield (%)*	*Ref.*
[structure: OBn, O-(2-nitrobenzyl) alkene]	$KMnO_4$, $NaIO_4$, Bu^tOH/H_2O	[structure: OBn, CO_2H, O-(2-nitrobenzyl)]	81	44
[structure: vinyl azetidinone, HN, O]	$KMnO_4$, H_2SO_4	[structure: HO_2C azetidinone, HN, O]	54	152
[structure: R, R, R, R polycyclic alkene]	$KMnO_4$, acetone, reflux	[structure: CO_2H, HO_2C, R, R]		153

A 5 L three-necked round-bottomed flask fitted with a mechanical stirrer is placed in an ice bath and charged with 1000 mL of distilled water, 120 mL of 9 M sulfuric acid, 3.0 g of Adogen 464, 20 mL of glacial acetic acid, 1000 mL of dichloromethane, and 0.2 mol of alkene. The solution is rapidly stirred and 80 g (0.544 mol) of potassium permanganate is added in small portions over a 3 h period. Stirring is continued for an additional 18 h at room temperature. The mixture is cooled in an ice bath, and 60 g of sodium hydrogensulfite is added in small portions to reduce any precipitated manganese dioxide. The solution is acidified, if basic, with sulfuric acid and separated. The aqueous layer is extracted with two 400 mL portions of dichloromethane. The organic extracts are combined, washed with two 400 mL portions of water, washed once with brine, and concentrated to 400 mL on a rotary evaporator. The resulting mixture is heated to dissolve any precipitated product, a small amount of amorphous solid is removed by filtration, and the filtrate is cooled to 0 °C. A first crop of white crystals is collected by suction filtration and washed with a minimum amount of ice-cold dichloromethane. Concentration of the mother liquor to 150 mL, and cooling to 0 °C yields a second crop of crystals. The yield is 55–90%.[154,158]

Acetic acid is used in these procedures to neutralize the base that is produced whenever permanganate is reduced (equation 30).

$$MnO_4^- \quad + \quad 2\,H_2O \quad + \quad 3\,e^- \quad \longrightarrow \quad MnO_2 \quad + \quad 4\,OH^- \qquad (30)$$

Over-oxidation occurs if the solution is permitted to become basic. For example, 3-phenylpropene gives approximately equal amounts of phenylacetic acid and benzoic acid when oxidized under phase transfer conditions using a two-phase benzene/water solvent system. However, when acetic acid is added, the yield of phenylacetic acid increases to 80%.[155]

Best results are usually obtained for these reactions when permanganate is transferred into the organic phase from an aqueous solution rather than from a solid ($KMnO_4$) phase. When it is necessary to use solid $KMnO_4$ as the oxidant, care should be taken to add the phase transfer agent to the organic phase before the alkene. When the reverse procedure is followed, the alkene may occasionally form an unreactive complex on the surface of the solid $KMnO_4$.[156] Several examples of preparations using these procedures have been summarized in Table 10.

Table 10 Phase Transfer Assisted Permanganate Oxidations

Substrate	Oxidant and conditions	Product	Yield (%)	Ref.
	$KMnO_4$, Aliquat 336, benzene/H_2O/glacial acetic acid	 CO_2H	80	155
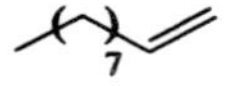 7	$KMnO_4$ (s), dicyclohexano-18-crown-6, benzene	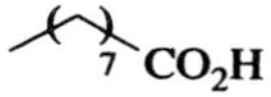 7 CO_2H	86	157
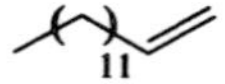 11	$KMnO_4$, Aliquat 336, benzene/H_2O/AcOH	 11 CO_2H	83	155
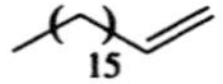 15	$KMnO_4$, Adogen 464, CH_2Cl_2/H_2SO_4/AcOH/H_2O	 15 CO_2H	81	154
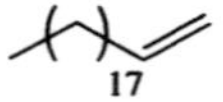 17	$KMnO_4$, Aliquat 336, benzene/H_2O/AcOH	 17 CO_2H	90	155
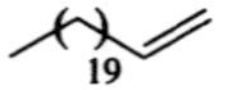 19	$KMnO_4$, Adogen 464, CH_2Cl_2/H_2SO_4/AcOH/H_2O	 19 CO_2H	84	154
 5	$KMnO_4$, Adogen 464, CH_2Cl_2/H_2SO_4/AcOH/H_2O	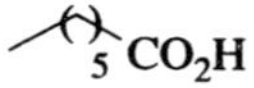 5 CO_2H	88	158
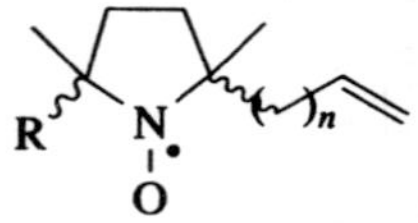 R, N, O, n	$KMnO_4$, 18-crown-6	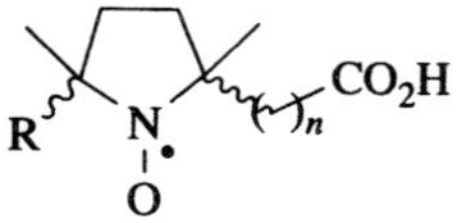 R, N, O, n, CO_2H		102
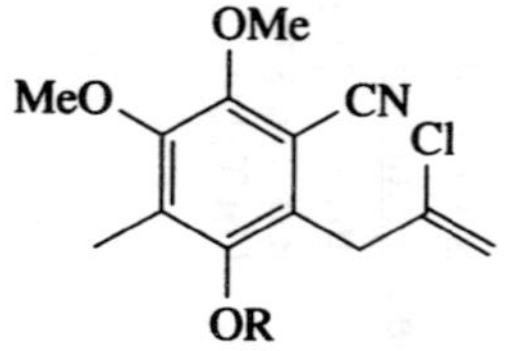 MeO, OMe, CN, Cl, OR	$KMnO_4$, Adogen 464, CH_2Cl_2/H_2SO_4/AcOH/H_2O	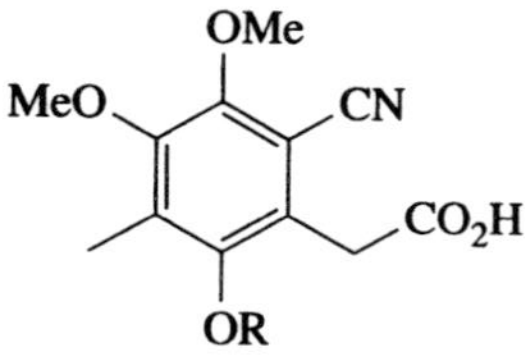 MeO, OMe, CN, CO_2H, OR	>70	159
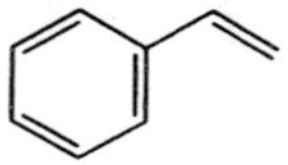	$KMnO_4$, Adogen 464, CH_2Cl_2/H_2SO_4/AcOH/H_2O	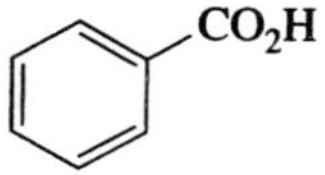 CO_2H	96	154

Table 10 *(continued)*

Substrate	*Oxidant and conditions*	*Product*	*Yield (%)*	*Ref.*
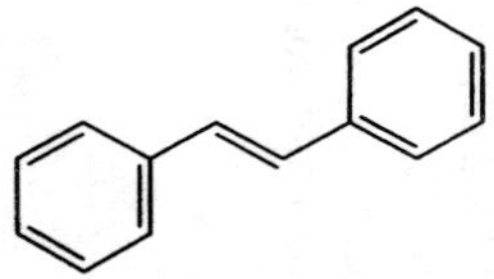	$KMnO_4$, crown ether, benzene	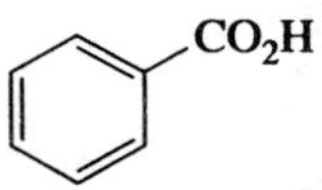 CO_2H	97	160
	$KMnO_4$, crown ether, benzene	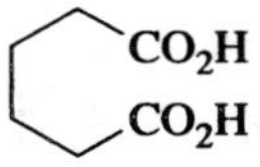 CO_2H CO_2H	≈100	160
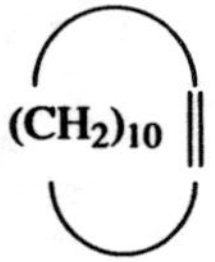 $(CH_2)_{10}$	$KMnO_4$, $MeO(CH_2CH_2O)_nMe/CH_2Cl_2/AcOH/H_2O$	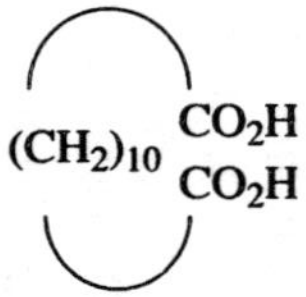 $(CH_2)_{10}$ CO_2H CO_2H	82	161
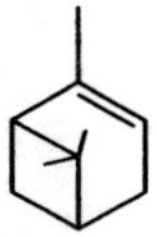	$KMnO_4$, crown ether, benzene	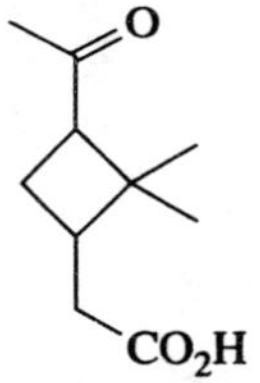 O CO_2H	90	160

3.8.4.2.3 Heterogeneous permanganate oxidations

Potassium permanganate adsorbed on either silica or alumina can also be used to cleave double bonds under mild conditions and in good yields. In one procedure the alkene, dissolved in benzene, is passed through a column packed with $KMnO_4$ on a silica gel support.[162] The reaction occurs rapidly at room temperature and is equally effective for the cleavage of all types of double bonds, even some that are inert to other traditional methods.[162]

It has also been found that it is not necessary to pack the oxidant into a column.[163] The alkene, dissolved in dichloromethane, can be cleaved by adding it to a flask containing $KMnO_4$ and silica gel that have been mixed mechanically. After shaking or stirring the mixture for an appropriate time, the product can be isolated by filtration and evaporation of the solvent. Alumina can also be used equally well as the solid support.[163] Additional examples are summarized in Table 11.

Table 11 Oxidative Cleavages by Permanganate on Solid Supports

Substrate	*Oxidant and Conditions*	*Product*	*Yield (%)*	*Ref.*
CO_2Me	$KMnO_4$, SiO_2 (support), benzene	HO_2C CO_2Me	85	162
CO_2Me	$KMnO_4$, SiO_2 (support), CH_2Cl_2 or $KMnO_4$, Al_2O_3 (support), CH_2Cl_2	CO_2H	50–70	163
	$KMnO_4$, SiO_2 (support), benzene	O CO_2H	84	162
OAc O	$KMnO_4$, SiO_2 (support), benzene	OAc CO_2H O CO_2H	62	162
	$KMnO_4$, SiO_2 (support), benzene	CO_2H CO_2H	74	162

3.8.4.2.4 Permanganate/periodate

A mixture of potassium permanganate and sodium periodate has also been used to cleave double bonds. This procedure, usually referred to as the Lemieux–von Rudloff reaction,[164] can be carried out in several mixed solvent systems such as butanol and water,[144] dioxane and water[165] or acetone and water.[166] It has also been claimed that the addition of phase transfer agents improves yields.[167]

In a typical procedure,[44] a solution of $KMnO_4$ (7 mg), $NaIO_4$ (225 mg) and K_2CO_3 (29 mg) in 29 mL of 7:3 *t*-butyl alcohol/water was added to a solution of alkene (0.213 mmol) in 2 mL of *t*-butyl alcohol. After 2.5 h the reaction mixture was poured into 50 mL of ether and 30 mL of water acidified to pH 2 with 1 M HCl. The aqueous phase was drawn off and extracted with 50 mL of ether. The combined organic layers were washed with 60 mL of 0.1 M HCl, dried (Na_2SO_4) and concentrated under reduced pressure to furnish the expected product.

3.8.4.3 Ruthenium Tetroxide

Although aldehydes are obtained from the cleavage of double bonds by ruthenium tetroxide under neutral conditions (Section 3.8.3.4), carboxylic acids are produced under alkaline or acidic conditions.[112] For example, the oxidation of cyclohexene by RuO_4 under alkaline conditions has been reported to give adipic acid in yields of 86–95%.[168]

Mechanistic studies have indicated that this reaction proceeds as in equations (31) and (32), wiih the initial step being a direct electron transfer that results in the formation of a radical cation–perruthenate complex.[169]

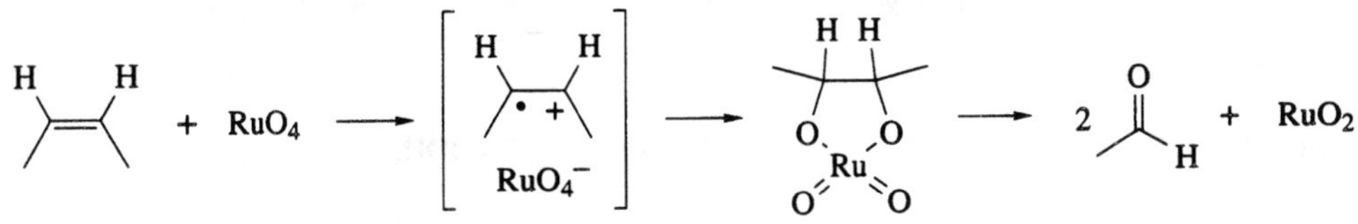

(31)

(32)

When ruthenium dioxide or ruthenium trichloride is used to catalyze periodate cleavages, it is likely that RuO_4 is first formed (equation 33) and then reacts with the double bond as depicted in equations (31) and (32). Sharpless and coworkers[113] have demonstrated that the best solvent system for this reaction is a mixture of carbon tetrachloride, acetonitrile and water, in a volume ratio of 2:2:3.

$$RuO_2 + 2\,NaIO_4 \longrightarrow RuO_4 + 2\,NaIO_3 \qquad (33)$$

3.8.4.4 Chromium Trioxide

Under acidic conditions CrO_3 will cleave double bonds to give the corresponding carboxylic acids. When the alkene also contains a hydroxy group lactones are readily formed, especially when acetic anhydride is used as a cosolvent (equation 34).[170]

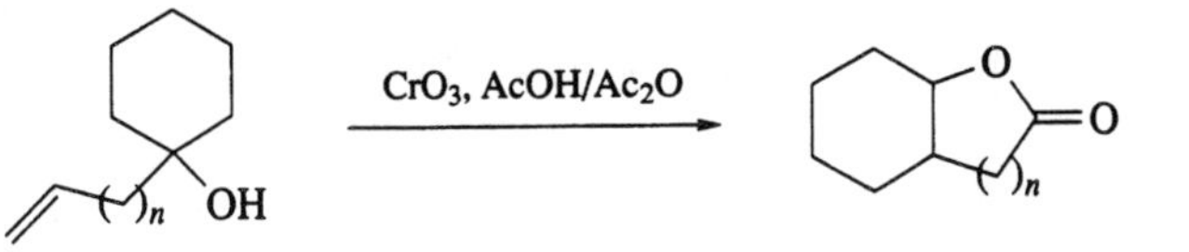

CrO_3, AcOH/Ac_2O

n = 2 or 3

30–80%

(34)

3.8.4.5 *t*-Butyl Peroxide and Molybdenum Dioxide Diacetylacetonate

$MoO_2(acac)_2$ and *t*-butyl peroxide when dissolved in benzene form a reagent that can be used for the specific cleavage of silyl enol ethers.[171] For example, the silyl ether of β-ionine is selectively cleaved as indicated in equation (35), to give β-(2,6,6-trimethylcyclohexyl)acrylic acid.

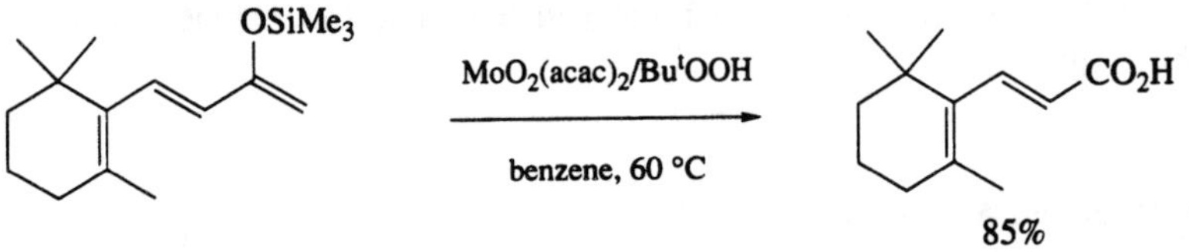

$OSiMe_3$

$MoO_2(acac)_2$/Bu^tOOH

benzene, 60 °C

CO_2H

85%

(35)

Since the formation of silyl enol ethers from the corresponding ketones is subject to either thermodynamic or kinetic control, this reagent can be used (as demonstrated in equation 36) to achieve useful regiospecific cleavages.

i, Me_3SiCl, DMF, Et_3N, Δ; ii, $NaHCO_3$, H_2O; iii, $MoO_2(acac)_2$, Bu^tOOH; iv, LDA, DME; v, Me_3SiCl

(36)

3.8.5 CLEAVAGE WITH THE INTRODUCTION OF NITROGEN AND SULFUR FUNCTIONAL GROUPS

3.8.5.1 Trimethylsilyl Azide and Lead Tetraacetate

Trimethylsilyl azide, ($TMSN_3$) reacts with carbon–carbon double bonds to form a compound which can be cleaved by lead tetraacetate (or phenyliododiacetate) to yield a carbonyl and a nitrile, as in equation (37).[172,173] The reagent has been applied extensively to the cleavage of unsaturated steroids, as illustrated in equation (38).

$$R_2C{=}CHR + Me_3SiN_3 \longrightarrow \left[R_2C{=}C(R)N_3 \right] \xrightarrow{Pb(OAc)_4} R_2C{=}O + R{-}CN \quad (37)$$

Me_3SiN_3, CH_2Cl_2, $Pb(OAc)_4$; –15 °C, 1–2 h; 30% (38)

In a typical procedure,[172] lead tetraacetate (2 mmol) in 50 mL of absolute dichloromethane was slowly added (over a period of 1.5 h) while stirring to a cold (–15 °C) solution of the steroid (2 mmol) and trimethylsilyl azide (8 mmol) in 250 mL of absolute CH_2Cl_2. After cooling for an additional 15 h, the red heterogeneous solution was slowly warmed to room temperature. Water was added and the precipitate removed by filtration through glass wool. The filtrate was washed with saturated $NaHCO_3$ and dried over anhydrous Na_2SO_4. The solvent was removed under vacuum and the residue separated by use of silica gel column chromatography.

3.8.5.2 Ethanethiol and Aluminum Chloride

Double bonds activated by the presence of electron withdrawing groups (NO_2, CO_2Et, COMe, CN) can be cleaved by use of ethanethiol and a hard Lewis acid such as $AlCl_3$, $AlBr_3$, $FeCl_3$ or $ZnCl_2$ to give dithioacetals in good yields.[174,175] For example, dicyanostyrene (**12**) can be converted into the corresponding dithioacetal (**13**) in quantitative yields when treated with aluminum chloride and ethanethiol (equation 39).

The general procedure reported by Fuji *et al.*[175] involves addition of a solution of the alkene (0.5 mmol) in dichloromethane (1 mL) to a mixture of Lewis acid (1.5 mmol) in ethanethiol (1 mL) with ice cooling and under argon. After stirring for an appropriate time, the reaction mixture is poured into

(39)

(12) **(13)** 100%

ice/water and extracted with dichloromethane. The organic layer is washed with brine, dried over Na_2SO_4 and evaporated to give the dithioacetal.

3.8.6 REFERENCES

1. A. H. Haines, in 'Methods for the Oxidation of Organic Compounds', Academic Press, Toronto, 1985, p. 117.
2. D. Arndt, 'Manganese Compounds as Oxidizing Agents in Organic Chemistry', English edn., Open Court, La Salle, Illinois, 1981, p. 241.
3. D. G. Lee, 'The Oxidation of Organic Compounds by Permanganate Ion and Hexavalent Chromium', Open Court, La Salle, Illinois, 1980.
4. P. M. Henry and G. L. Lange, in 'Chemistry of Double-Bonded Functional Groups', ed. S. Patai, Wiley, Chichester, 1977, vol. 2, pp. 1018, 1046.
5. D. G. Lee, in 'Oxidation in Organic Chemistry', ed. W. S. Trahanovsky, Academic Press, New York, 1982, vol. 5, part D, p. 147.
6. J. S. Belew, 'Oxidation Techniques and Applications in Organic Synthesis', ed. R. L. Augustine, Dekker, New York, 1969, vol. 1, p. 259.
7. S. D. Razumovskii and G. E. Zaikov, *Russ. Chem. Rev. (Engl. Transl.)*, 1980, **49**, 1163.
8. P. S. Bailey, 'Ozonation in Organic Chemistry', ed. W. S. Trahanovsky, Academic Press, New York, 1978, vol. 1; P. S. Bailey, *Chem. Rev.*, 1958, **58**, 925.
9. R. L. Kuczkowski, *Acc. Chem. Res.*, 1983, **16**, 42.
10. R. Criegee, *Angew. Chem., Int. Ed. Engl.*, 1975, **14**, 745.
11. W. Carruthers, in 'Some Modern Methods of Organic Synthesis', 2nd edn., Cambridge University Press, New York, 1978, p. 355.
12. J. F. Callahan, K. A. Newlander, H. G. Bryan, W. F. Huffman, M. L. Moore and N. C. F. Yim, *J. Org. Chem.*, 1988, **53**, 1527.
13. R. A. Bartsch, B. R. Cho and M. J. Pugia, *J. Org. Chem.*, 1987, **52**, 5492.
14. N. Nakamura, M. Nojima and S. Kusabayashi, *J. Am. Chem. Soc.*, 1986, **108**, 4671.
15. G. N. Walker, *J. Am. Chem. Soc.*, 1957, **79**, 3508.
16. J. K. Whitesell and D. E. Allen, *J. Am. Chem. Soc.*, 1988, **110**, 3585.
17. L. Fitjer and U. Quabeck, *Synthesis*, 1987, 299.
18. V. N. Odinokov, L. P. Zhemaiduk and G. A. Tolstikov, *J. Org. Chem. USSR (Engl. Transl.)*, 1978, **14**, 48.
19. M. Hinder and M. Stoll, *Helv. Chim. Acta*, 1950, **33**, 1308.
20. R. B. Turner, *J. Am. Chem. Soc.*, 1950, **72**, 579.
21. H. Magari, H. Hirota and T. Takahashi, *J. Chem. Soc., Chem. Commun.*, 1987, 1196.
22. H. Dyke, R. Sauter, P. Steel and E. J. Thomas, *J. Chem. Soc., Chem. Commun.*, 1986, 1447.
23. D. L. Boger and R. S. Coleman, *J. Am. Chem. Soc.*, 1988, **110**, 4796.
24. R. L. Danheiser, D. J. Carini and C. A. Kwasigroch, *J. Org. Chem.*, 1986, **51**, 3870.
25. D. A. Evans, S. L. Bender and J. Morris, *J. Am. Chem. Soc.*, 1988, **110**, 2506.
26. W. Yuan, R. J. Berman and M. H. Gelb, *J. Am. Chem. Soc.*, 1987, **109**, 8071.
27. H. Hamana, N. Ikota and B. Ganem, *J. Org. Chem.*, 1987, **52**, 5492.
28. J. E. Baldwin, T. C. Barden and S. J. Cianciosi, *J. Org. Chem.*, 1986, **51**, 1133.
29. D. Zhai, W. Zhai and R. W. Williams, *J. Am. Chem. Soc.*, 1988, **110**, 2501.
30. M. D. Wittman and J. Kallmerten, *J. Org. Chem.*, 1988, **53**, 4631.
31. B. Witkop and J. B. Patrick, *J. Am. Chem. Soc.*, 1952, **74**, 3855.
32. D. C. Lathbury, P. J. Parsons and I. Pinto, *J. Chem. Soc., Chem. Commun.*, 1988, 81.
33. F. J. Sardina, A. Mourino and L. Castedo, *J. Org. Chem.*, 1986, **51**, 1264.
34. R. Aneja, S. K. Mukerjee and T. R. Seshadri, *Chem. Ber.*, 1960, **93**, 297.
35. C. Iwata, Y. Takemoto, M. Doi and T. Imanishi, *J. Org. Chem.*, 1988, **53**, 1623.
36. P. S. Bailey and R. E. Erickson, *Org. Synth.*, 1961, **41**, 41.
37. R. H. Callighan and M. H. Wilt, *J. Org. Chem.*, 1961, **26**, 4912.
38. D. R. Williams and F. D. Klingler, *J. Org. Chem.*, 1988, **53**, 2134.
39. M. N. Deshpande, S. Wehrli, M. Jawdosiuk, J. T. Guy, Jr., D. W. Bennett, J. M. Cook, M. R. Depp and U. Weiss, *J. Org. Chem.*, 1986, **51**, 2436.
40. D. Gupta, R. Soman and S. Dev, *Tetrahedron*, 1982, **38**, 3013.
41. K.-Y. Ko and E. L. Eliel, *J. Org. Chem.*, 1986, **51**, 5353.
42. G. P. Boldrini, L. Lodi, E. Tagliavini, C. Tarasco, C. Trombini and A. Umani-Ronchi, *J. Org. Chem.*, 1987, **52**, 5447.
43. M. C. Pirrung and N. J. G. Webster, *J. Org. Chem.*, 1987, **52**, 3603.
44. R. E. Ireland, S. Thaisrivongs and P. H. Dussault, *J. Am. Chem. Soc.*, 1988, **110**, 5768.
45. J. Mulzer, T. Schulze, A. Strecker and W. Denzer, *J. Org. Chem.*, 1988, **53**, 4098.
46. J. Lin, M. M. Nikaido and G. Clark, *J. Org. Chem.*, 1987, **52**, 3745.

47. S. L. Schreiber and M. T. Goulet, *J. Am. Chem. Soc.*, 1987, **109**, 8120.
48. E. W. Colvin and S. Cameron, *J. Chem. Soc., Chem. Commun.*, 1986, 1084.
49. H. Hashimoto, K. Furuichi and T. Miwa, *J. Chem. Soc., Chem. Commun.*, 1987, 1002.
50. L. D. Hall and K. B. Holme, *J. Chem. Soc., Chem. Commun.*, 1986, 217.
51. T. Imanishi, M. Matsui, M. Yamashita and C. Iwata, *J. Chem. Soc., Chem. Commun.*, 1987, 1802.
52. H. Mattes and C. Benezra, *J. Org. Chem.*, 1988, **53**, 2732.
53. R. M. Williams, T. Glinka and E. Kwast, *J. Am. Chem. Soc.*, 1988, **110**, 5927.
54. N. Matsuo and A. S. Kende, *J. Org. Chem.*, 1988, **53**, 2304.
55. M. Furber and L. N. Mander, *J. Am. Chem. Soc.*, 1988, **110**, 4084.
56. S. M. Weinreb, *Acc. Chem. Res.*, 1988, **21**, 313.
57. J. Wright, G. J. Drtina, R. A. Roberts and L. A. Paquette, *J. Am. Chem. Soc.*, 1988, **110**, 5806.
58. D. H. Hua and S. Venkataraman, *J. Org. Chem.*, 1988, **53**, 1095.
59. S. Y. Lee, M. Niwa and B. B. Snider, *J. Org. Chem.*, 1988, **53**, 2356.
60. D. M. Walba, W. N. Thurmes and R. C. Haltiwanger, *J. Org. Chem.*, 1988, **53**, 1046.
61. T. Haag and B. Luu, *J. Chem. Soc., Perkin Trans. 1*, 1988, 2353.
62. G. Slomp, Jr. and J. L. Johnson, *J. Am. Chem. Soc.*, 1958, **80**, 915.
63. M. E. Idrissi and M. Santelli, *J. Org. Chem.*, 1988, **53**, 1010.
64. J. M. Odriozola, F. P. Cossio and C. Palomo, *J. Chem. Soc., Chem. Commun.*, 1988, 809.
65. F. P. Cossio, B. Lecea and C. Palomo, *J. Chem. Soc., Chem. Commun.*, 1987, 1743.
66. D. Ben-Ishai and S. Hirsh, *Tetrahedron*, 1988, **44**, 5441.
67. R. E. Dolle and L. I. Kruse, *J. Chem. Soc., Chem. Commun.*, 1988, 133.
68. T. Veysoglu, L. A. Mitscher and J. K. Swayze, *Synthesis*, 1980, 807.
69. P. M. Gannett, D. L. Nagel, P. J. Reilly, T. Lawson, J. Sharpe and B. Toth, *J. Org. Chem.*, 1988, **53**, 1064.
70. B. S. Furniss, A. J. Hannaford, V. Rogers, P. W. G. Smith and A. R. Tatchell, in 'Vogel's Textbook of Practical Organic Chemistry', 4th edn., Longman, London, 1978, p. 420.
71. D. M. Hodgson, P. J. Parsons and P. A. Stones, *J. Chem. Soc., Chem. Commun.*, 1988, 217.
72. W. S. Knowles and Q. E. Thompson, *J. Org. Chem.*, 1960, **25**, 1031.
73. J. E. McMurry and G. K. Bosch, *J. Org. Chem.*, 1987, **52**, 4885.
74. T. Hudlicky, H. Luna, G. Barbieri and L. D. Kwart, *J. Am. Chem. Soc.*, 1988, **110**, 4735.
75. T. Hudlicky, G. Seoane and T. C. Lovelace, *J. Org. Chem.*, 1988, **53**, 2094.
76. I. E. Den Besten and T. H. Kinstle, *J. Am. Chem. Soc.*, 1980, **102**, 5968.
77. A. G. Green and A. R. Wahl, *Ber. Dtsch. Chem. Ges.*, 1897, **30**, 3097.
78. K. B. Wiberg and K. A. Saegebarth, *J. Am. Chem. Soc.*, 1957, **79**, 2822.
79. S. D. Sabnis, H. H. Mathur and S. C. Bhattacharyya, *J. Chem. Soc.*, 1965, 4580.
80. P. Viski, Z. Szeverenyi and L. I. Simandi, *J. Org. Chem.*, 1986, **51**, 3213.
81. T. Ogino and K. Mochizuki, *Chem. Lett.*, 1979, 443.
82. R. Rathore and S. Chandrasekaran, *J. Chem. Res. (S)*, 1986, 458.
83. D. G. Lee, K. C. Brown and H. Karaman, *Can. J. Chem.*, 1986, **64**, 1054.
84. R. Iqbal and F. Malik, *J. Chem. Soc. Pak.*, 1986, **8**, 83.
85. H. Firouzabadi, A. R. Sardarian, M. Naderi and B. Vessal, *Tetrahedron*, 1984, **40**, 5001.
86. D. G. Lee, 'Oxidation Techniques and Applications in Organic Synthesis', ed. R. L. Angustine, Dekker, New York, 1969, vol. 1.
87. F. Kido, H. Kitahara and A. Yoshikoshi, *J. Org. Chem.*, 1986, **51**, 1478.
88. F. Gillard, D. Heissler and J.-J. Riehl, *J. Chem. Soc., Perkin Trans. 1*, 1988, 2291.
89. K. Shishido, K. Hiroya, K. Fukumoto and T. Kametani, *J. Chem. Soc., Chem. Commun.*, 1987, 1360.
90. F. F. Caserio, Jr. and J. D. Roberts, *J. Am. Chem. Soc.*, 1958, **80**, 5837.
91. T. Hayashi, K. Kanehira, T. Hagihara and M. Kumada, *J. Org. Chem.*, 1988, **53**, 113.
92. E. G. Baggiolini, J. A. Iacobelli, B. M. Hennessy, A. D. Batcho, J. F. Sereno and M. R. Uskokovic, *J. Org. Chem.*, 1986, **51**, 3098.
93. R. Pappo, D. S. Allen, Jr., R. U. Lemieux and W. S. Johnson, *J. Org. Chem.*, 1956, **21**, 478.
94. J. P. Marino and J. C. Jaen, *Synth. Commun.*, 1983, **13**, 1057.
95. K. Inomata, Y. Nakayama and H. Kotake, *Bull. Chem. Soc. Jpn.*, 1980, 53, 565.
96. T. Chen, A. Wee and D. G. Lee, University of Regina, unpublished results.
97. G. Majetich, J. Defauw and C. Ringold, *J. Org. Chem.*, 1988, **53**, 50.
98. R. E. Ireland and P. Maienfisch, *J. Org. Chem.*, 1988, **53**, 640.
99. K. Hideg and L. Lex, *J. Chem. Soc., Perkin Trans. 1*, 1987, 1117.
100. E. J. Corey, R. L. Danheiser, S. Chandrasekaran, P. Siret, G. E. Keck and J.-L. Gras, *J. Am. Chem. Soc.*, 1978, **100**, 8031.
101. A. P. Kozikowski, S. H. Jung and J. P. Springer, *J. Chem. Soc., Chem. Commun.*, 1988, 167.
102. U. E. Udodong and B. Fraser-Reid, *J. Org. Chem.*, 1988, **53**, 2131.
103. S. E. Cantor and D. S. Tarbell, *J. Am. Chem. Soc.*, 1964, **86**, 2902.
104. R. K. Boeckman, Jr., J. P. Sabatucci, S. W. Goldstein, D. M. Springer and P. F. Jackson, *J. Org. Chem.*, 1986, **51**, 3740.
105. J. Jurczak and S. Pikul, *Tetrahedron*, 1988, **44**, 4569.
106. T. K. M. Shing, *J. Chem. Soc., Chem. Commun.*, 1986, 49.
107. E. Brandes, P. A. Grieco and P. Garner, *J. Chem. Soc., Chem. Commun.*, 1988, 500.
108. T. Harayama, M. Takatani and Y. Inubushi, *Tetrahedron Lett.*, 1979, 4307.
109. D. G. Lee and M. van den Engh, in 'Oxidation in Organic Chemistry', ed. W. S. Trahanovsky, Academic Press, New York, 1973, vol. 5, part B, p. 177.
110. P. N. Rylander, *Engelhard Ind., Tech. Bull.*, 1969, **9**, 135.
111. F. M. Dean and J. C. Knight, *J. Chem. Soc.*, 1962, 4745.
112. Y. Shalon and W. H. Elliott, *Synth. Commun.*, 1973, **3**, 287.
113. P. H. J. Carlsen, T. Katsuki, V. S. Martin and K. B. Sharpless, *J. Org. Chem.*, 1981, **46**, 3936.

114. L. M. Berkowitz and P. N. Rylander, *J. Am. Chem. Soc.*, 1958, **80**, 6682.
115. S. Wolfe, S. K. Hasan and J. R. Campbell, *J. Chem. Soc. D*, 1970, 1420.
116. G. Mehta and N. Krishnamurthy, *J. Chem. Soc., Chem. Commun.*, 1986, 1319.
117. R. Tschesche, U. Schacht and G. Legler, *Justus Liebigs Ann. Chem.*, 1963, **662**, 113.
118. L. F. Fieser and J. Szmuszkovicz, *J. Am. Chem. Soc.*, 1948, **70**, 3352.
119. R. B. Gammill and S. A. Nash, *J. Org. Chem.*, 1986, **51**, 3116.
120. L. M. Baker and W. L. Carrick, *J. Org. Chem.*, 1968, **33**, 616.
121. L. M. Baker and W. L. Carrick, *J. Org. Chem.*, 1970, **35**, 774.
122. T. K. Chakraborty and S. Chandrasekaran, *Org. Prep. Proced. Int.*, 1982, **14**, 362.
123. W. A. Mosher, F. W. Steffgen and P. T. Lansbury, *J. Org. Chem.*, 1961, **26**, 670.
124. H. Schildknecht and W. Fottinger, *Justus Liebigs Ann. Chem.*, 1962, **659**, 20.
125. F. Plavac and C. H. Heathcock, *Tetrahedron Lett.* 1979, 2115.
126. K. Ramalingam, P. Nanjappan, D. M. Kalvin and R. W. Woodard, *Tetrahedron*, 1988, **44**, 5597.
127. J. A. Cella, *Synth. Commun.*, 1983, **13**, 93.
128. T.-L. Ho, *Synth. Commun.*, 1982, **12**, 53.
129. O. Ceder and B. Hansson, *Acta Chem. Scand., Ser. B*, 1970, **24**, 2693.
130. F. Asinger, *Chem. Ber.*, 1942, **75**, 656.
131. R. K. Boeckman, Jr., E. J. Enholm, D. M. Demko and A. B. Charette, *J. Org. Chem.*, 1986, **51**, 4743.
132. D. W. Brooks, H. Mazdiyasni and P. G. Grothaus, *J. Org. Chem.*, 1987, **52**, 3223.
133. B. Zhou and Y. Xu, *J. Org. Chem.*, 1988, **53**, 4419.
134. K. Maruyama, Y. Ishihara and Y. Yamamoto, *Tetrahedron Lett.*, 1981, **21**, 4235.
135. I. B. Blanshtein, *Soviet Pat.* 859 351 (1981) (*Chem. Abstr.*, 1982, **96**, 6207b).
136. J. Neumeister, H. Keul, M. P. Saxena and K. Griesbaum, *Angew. Chem., Int. Ed. Engl.*, 1978, **17**, 939.
137. A. Guerriero, M. D'Ambrosio and F. Pietra, *Helv. Chim. Acta*, 1988, **71**, 1094.
138. T. Hayashi, A. Yamamoto and T. Hagihara, *J. Org. Chem.*, 1986, **51**, 723.
139. O. Ceder and H. G. Nilsson, *Acta Chem. Scand., Ser. B*, 1977, **31**, 189.
140. T.-S. Huh, J. Neumeister and K. Griesbaum, *Can. J. Chem.*, 1981, **59**, 3188.
141. P. S. Bailey, *J. Org. Chem.*, 1957, **22**, 1548.
142. H. Wilms, *Justus Liebigs Ann. Chem.*, 1950, **567**, 96.
143. M. I. Fremery and E. K. Fields, *J. Org. Chem.*, 1963, **28**, 2537.
144. E. Vedejs, S. Ahmad, S. D. Larsen and S. Westwood, *J. Org. Chem.*, 1987, **52**, 3937.
145. T. J. Sowin and A. I. Meyers, *J. Org. Chem.*, 1988, **53**, 4154.
146. T. Hvidt, W. A. Szarek and D. B. Maclean, *Can. J. Chem.*, 1988, **66**, 779.
147. G. Nagendrappa and K. Griesbaum, *J. Agric. Food. Chem.*, 1978, **26**, 581.
148. D. J. Ager and M. B. East, *J. Chem. Res. (S)*, 1986, 462.
149. P. Bladon, H. B. Henbest, E. R. H. Jones, B. J. Lovell, G. W. Wood, G. F. Woods, J. Elks, R. M. Evans, D. E. Hathway, J. F. Oughton and G. H. Thomas, *J. Chem. Soc.*, 1933, **27**, 2921.
150. R. Stewart, in 'Oxidation in Organic Chemistry', ed. K. B. Wiberg, Academic Press, New York, 1965, vol. 5, part A, p. 1.
151. M. N. Tilichenko, *J. Gen. Chem. USSR (Engl. Transl.)*, 1961, **31**, 1446.
152. Farbwerk Hoechst A G, *Ger. Pat.* 1 670 804, *J. Synth. Method*, **11**, 75 050A.
153. J. A. Akhtar, G. I. Fray and J. M. Yarrow, *J. Chem. Soc. C*, 1968, 812.
154. D. G. Lee, S. E. Lamb and V. S. Chang, *Org. Synth.*, 1981, **60**, 11.
155. A. P. Krapcho, J. R. Larson and J. M. Eldridge, *J. Org. Chem.*, 1977, **42**, 3749.
156. D. G. Lee and Z. Wang, University of Regina, unpublished results.
157. V. S. Chang and D. G. Lee, University of Regina, unpublished results.
158. K. C. Brown, V. S. Chang, F. H. Dar, S. E. Lamb and D. G. Lee, *J. Chem. Educ.*, 1982, **59**, 696.
159. K. A. Parker and D. A. Casteel, *J. Org. Chem.*, 1988, **53**, 2849.
160. D. J. Sam and H. F. Simmons, *J. Am. Chem. Soc.*, 1972, **94**, 4024.
161. D. G. Lee and V. S. Chang, *J. Org. Chem.*, 1978, **43**, 1532.
162. J. T. B. Ferreira, W. O. Cruz, P. C. Vieira and M. Yonashiro, *J. Org. Chem.*, 1987, **52**, 3698.
163. D. G. Lee and W. Tang, University of Regina, unpublished results.
164. R. U. Lemieux and E. von Rudloff, *Can. J. Chem.*, 1955, **33**, 1701, 1710; E. von Rudloff, *Can. J. Chem.*, 1955, **33**, 1714.
165. S. W. Pelletier, K. N. Iyer and C. W. J. Chang, *J. Org. Chem.*, 1970, **35**, 3535.
166. C. G. Overberger and H. Kaye, *J. Am. Chem. Soc.*, 1967, **89**, 5640.
167. S. Schwarz, C. Carl and H. Schick, *Z. Chem.*, 1978, **18**, 401.
168. K. A. Keblys and M. Dubeck (Ethyl Corp.), *US Pat.* 3 409 649 (1968) (*Chem. Abstr.*, 1969, **70**, 114 575).
169. D. G. Lee and U. A. Spitzer, *J. Org. Chem.*, 1976, **41**, 3644.
170. M. F. Schlecht and H.-J. Kim, *Tetrahedron Lett.*, 1985, **26**, 127.
171. K. Kaneda, N. Kii, K. Jitsukawa and S. Teranishi, *Tetrahedron Lett.*, 1981, **22**, 2595.
172. E. Zbiral, G. Nestler and K. Kischa, *Tetrahedron*, 1970, **26**, 1427.
173. E. Zbiral and G. Nestler, *Tetrahedron*, 1970, **26**, 2945.
174. K. Fuji, T. Kawabata, M. Node and E. Fujita, *Tetrahedron Lett.*, 1981, **22**, 875.
175. K. Fuji, T. Kawabata, M. Node and E. Fujita, *J. Org. Chem.*, 1984, **49**, 3214.

4.1
Oxidation of Carbon–Boron Bonds

ANDREW PELTER and KEITH SMITH
University College Swansea, UK

4.1.1 INTRODUCTION

Carbon–boron bonds are generally rather easily oxidized and indeed volatile trialkylboranes such as trimethylborane and triethylborane are spontaneously inflammable when exposed to air. Less volatile organylboranes do not spontaneously inflame but are nevertheless readily oxidized by oxygen and a variety of other reagents. Consequently, it is normally necessary to carry out organoborane reactions in an inert atmosphere.

The high reactivity of organoboranes to oxidizing agents can be ascribed to the availability of a vacant *p*-orbital on boron, which provides the opportunity for a kinetically favorable attack, and to the thermodynamic stability of the B—O bond (111–120 kcal mol^{-1}; 1 cal = 4.18 J) compared with the B—C bond (81–88 kcal mol^{-1}).[1]

Although oxidation of R—B bonds may be defined very generally, for the purposes of this review only processes leading from R—B to R—X, where X is a group bonded *via* an atom more electronegative than carbon (in particular OR, NR_2, SR, SeR, halogen), and processes leading to alkenes [R(–H)], are included.

Oxidations of organoboranes involve numerous reagents and several different general mechanisms, most of which parallel those which occur for other types of organoborane reactions. The mechanisms fall under three broad headings: (i) ionic, with a 1,2-shift from boron to a heteroatom (equations 1–3); (ii) radical; and (iii) electrocyclic.

$$R_2B{-}R + {}^{-}X{-}Y \longrightarrow R{-}\overset{-}{B}R_2{-}X{-}Y \longrightarrow R_2B{-}X{-}R + Y^{-} \quad (1)$$

$$R_2B{-}R + {}^{-}X{-}Y^{+} \longrightarrow R{-}\overset{-}{B}R_2{-}X{-}Y^{+} \longrightarrow R_2B{-}X{-}R + Y \quad (2)$$

$$R_2B{-}R + {:}X{-}Y \longrightarrow R{-}\overset{-}{B}R_2{-}\overset{+}{X}{-}Y \longrightarrow R_2B{-}X(R){-}Y \quad (3)$$

In general, the ionic reactions of equations (1) to (3) proceed with retention of configuration of the migrating organyl group.

There are two common types of radical reactions of organoboranes: bimolecular homolytic substitution (S_H2; equation 4); and α-abstraction processes (equation 5). Both of these can lead to final products by chain processes, as illustrated for the α-abstraction reaction by the continuation shown in equation (6). The radicals produced in these reactions are unlikely to retain complete stereochemical integrity except in special circumstances. The nature of the further reactions, such as equation (6), determines whether or not the products are such as to be included in this section.[2]

$$X^{\bullet} + R_2B{-}R \longrightarrow \left[X\cdots BR_2\cdots R\right]^{\bullet} \longrightarrow R_2B{-}X + R^{\bullet} \quad (4)$$

$$X^{\bullet} + H{-}C(R^1)(R^2){-}BR_2 \longrightarrow {}^{\bullet}C(R^1)(R^2){-}BR_2 + HX \quad (5)$$

$$^{\bullet}C(R^1)(R^2){-}BR_2 + X_2 \longrightarrow X{-}C(R^1)(R^2){-}BR_2 + X^{\bullet} \quad (6)$$

Electrocyclic reactions (equation 7) may also lead to oxidation of organoboranes, usually with retention of configuration of the organyl group.

$$-\overset{|}{B}{-}R \;\; (W{-}X{-}Y{-}Z \text{ cyclic}) \longrightarrow \;\; >B{-}W + X{=}Y + RZ \quad (7)$$

Oxidation reactions of organoboranes have been extensively reviewed as subsections of general accounts of boron chemistry,[3–7] of which reference 3 is the latest and most complete. In the remaining sections of this review, the essential features of the more synthetically useful oxidation reactions are considered in detail.

4.1.2 OXIDATION WITH ALKALINE HYDROGEN PEROXIDE

The reaction of organoboranes with alkaline hydrogen peroxide (equation 8) is one of the oldest[8,9] and most widely used methods for the release of organyl residues from organoboranes. The reaction proceeds with retention of configuration for all three alkyl groups.[10,11] This property is proving invaluable for the production of alcohols of high optical purity. Thus, enantiomerically pure (+)-Ipc_2BH gives (*S*)-(+)-butan-2-ol of 98.4% enantiomeric purity by hydroboration of *cis*-but-2-ene followed by alkaline hydrogen peroxide oxidation of the intermediate organoborane.[12] Similarly, 3-methylbutan-2-ol of 99.6% optical purity has been prepared by the oxidation of diethoxysiamylborane with alkaline hydrogen peroxide.[13,127]

$$R_3B + NaOH + 3\,H_2O_2 \longrightarrow 3\,ROH + NaB(OH)_4 \qquad (8)$$

The mechanism of the reaction[14,15] is believed to be that shown in equation (9), this being a specific example of equation (1), with hydroperoxide anion being the reactive species. Detailed study[14] has led to the proposal of various transition states at different pH values.[16] However, the reaction may be carried out satisfactorily over a wide range of concentrations of base and hydrogen peroxide[17] providing that the reaction mixture is kept alkaline, as otherwise radical reactions may intervene with deleterious stereochemical consequences.

$$R_3B + {}^{-}O{-}OH \longrightarrow \left[R_3\overset{-}{B}{-}O{-}OH \right] \longrightarrow R_2B{-}OR \longrightarrow (RO)_2B{-}OR \qquad (9)$$

In general, oxidation is carried out in the medium used for hydroboration (THF, glyme, diglyme). Immiscibility is generally not a problem, though sometimes ethanol may be added to aid miscibility. Aqueous alkali is added first to the organoborane and then hydrogen peroxide, with caution.[3] Isolation of products is simple, though when 1,2- or 1,3-diols are produced, it is an advantage to add mannitol to liberate them.[18] The reaction mixture is frequently opened to air during the oxidation. However, this may lead to radical oxidation by oxygen competing with alkaline hydrogen peroxide oxidation. Hence, when retention of stereochemistry is important, it is sensible to maintain an inert atmosphere until oxidation is complete.

Oxidation with alkaline hydrogen peroxide is remarkably specific for cleavage of the C—B bond and, in the usual conditions, will tolerate acetal, aldehyde, alkene, alkyne, carboxylic acid, ester, ether, ketone, halogeno, nitrile, silyl and sulfonyl groups.[17] Cyclopropylboranes may need special conditions,[19] though not always.[20,21] In certain cases, such as readily hydrolyzed phenolic esters, the use of acetate or phosphate buffers at pH $\approx$ 8 is advantageous.[22,23] It is even possible to selectively oxidize an organoborane containing a sulfide group as long as 3 mol equiv. or more of sodium hydroxide to 1 of trialkylborane are used.[24,25]

Alkaline hydrogen peroxide easily oxidizes practically all alkyl- and cycloalkyl-boranes in a rapid and quantitative fashion.[17] There is a reactivity trend of $R_3B > R_2BX > RBX_2$ (X = halogen, OH, OR; note that boron halides will anyway be hydrolyzed to hydroxides under the oxidation conditions), which is consistent with reduced acceptor ability of the boron atom when an electron pair of an adjacent group interacts with the vacant boron orbital.[7,8] Increasing the steric hindrance around the boron atom may inhibit the reaction to the point at which it ceases altogether.[26,27]

Although hydroboration of alkenes is the most widely used method for the production of organoboranes, some alkylboranes as well as allyl- and aryl-boranes cannot be made by this method. Another general process that can then be used, and that proceeds in reasonable yields, is the reaction of a boron compound such as diborane[28,29] or, more commonly, an alkoxyborane[30] with an organometallic compound. The organometallic compound may be an organolithium,[5] organomagnesium[28] or organomercury(II) compound.[31] Thus, the overall process of reaction of the organometallic with the boron reagent followed by alkaline hydrogen peroxide allows 'one pot' oxidation of many organometallic compounds (equation 10).

$$ArLi \xrightarrow{B(OMe)_3} Ar{-}\overset{-}{B}(OMe)_3\ Li^+ \xrightarrow{H_2O_2,\ NaOH} ArOH \qquad (10)$$

Many aromatic boron compounds are readily oxidized to phenols, as illustrated in equation (11) for the oxidation of aryldihydroxyboranes (arylboronic acids). This is a useful synthetic pathway as arylboronic acids can resist both alkaline permanganate and nitric acid, so that many derivatives are available.[32]

$$\mathrm{Ar{-}B(OH)_2 + NaOH + H_2O_2 \longrightarrow ArOH + NaB(OH)_4} \quad (11)$$

The process has been used with heteroaromatics, examples being the production of 2- and 3-thiolenones (equation 12),[33a] thienylmethacrylates[33b] and butenolides.[34]

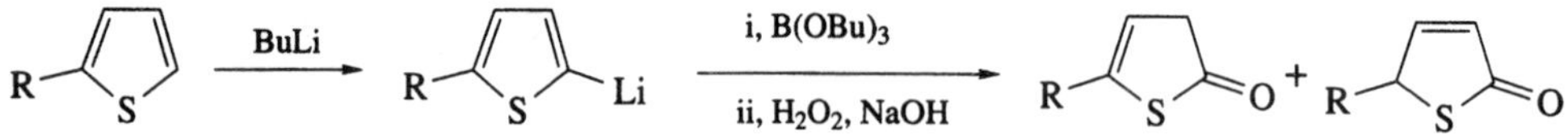

(12)

Oxidation of alkenylboranes with alkaline hydrogen peroxide is an important pathway to aldehydes and ketones (equation 13). Care must be taken to inhibit hydrolysis of the alkenylborane to the corresponding alkene and hence buffered conditions are frequently used.[35–37] The reaction tolerates the same wide range of functionality as the oxidation of alkylboranes.

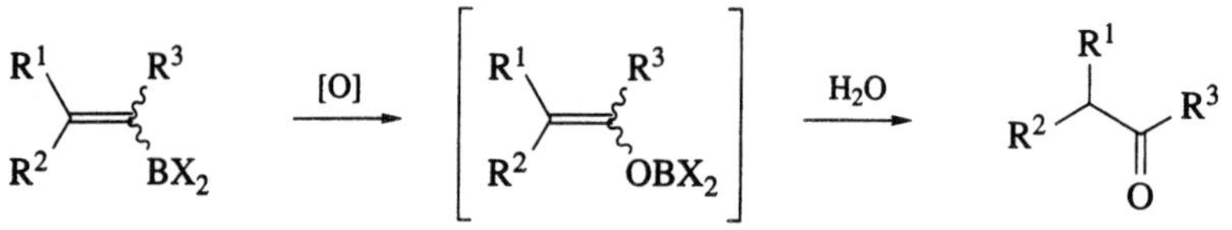

(13)

Allylboranes, which are available from organometallics or by hydroboration, undergo oxidation without allylic transposition and with retention of stereochemistry. This applies both to regio- and stereochemically defined allylic dialkylboryl[38] and allylic dialkoxyboryl[39] compounds (equations 14 and 15).

(14)

(15)

1,1-Diboryl compounds do not yield aldehydes with alkaline hydrogen peroxide. Instead there is rapid hydrolysis, presumably *via* a boron-stabilized carbanion (see Volume 1, Chapter 2.6) which is protonated and then oxidized to the alcohol (equation 16).[35]

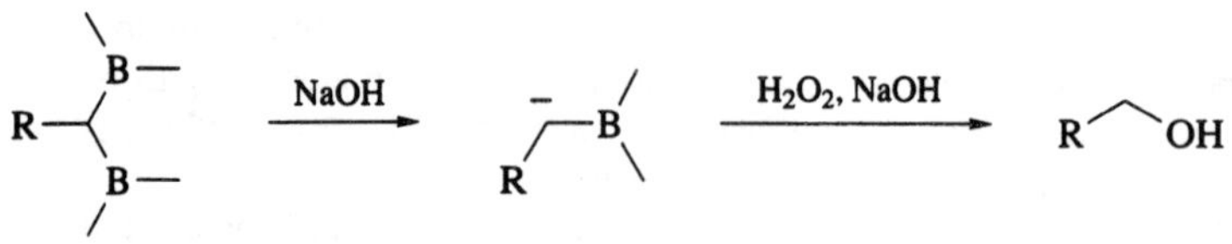

(16)

Boracyclanes are oxidized in the usual fashion to yield either diols or triols (*e.g.* equations 17[40] and 18[41]), as are compounds containing two or more independent boryl groups.

(17)

(18)

4.1.3 OXIDATION WITH HYDROGEN PEROXIDE IN AN ACIDIC MEDIUM

Organoboranes may be oxidized by hydrogen peroxide in an acidic medium. The mechanism of oxidation of alkyldihydroxyboranes in such conditions has been studied and the protonated transition state (**1**) proposed, as compared with transition state (**2**) for the reaction in an alkaline medium.[14]

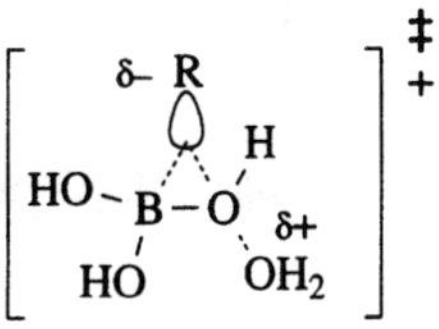

(**1**) 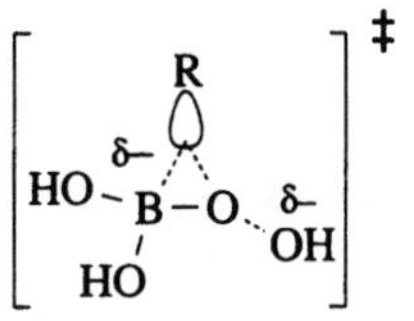 (**2**)

The acid reaction has been used in cases in which complications have been noted with the alkaline hydrogen peroxide reaction. Thus, 1,1-bis(boronates) smoothly yield aldehydes and 1,2-bis(boronates) give 1,2-diols (equations 19 and 20) when oxidized with acidic hydrogen peroxide.[42,43]

(19)

(20)

Vinylboranes react with acidic hydrogen peroxide to give aldehydes,[44] whilst the oxidation of α-haloboronic acids in the presence of DNP/H_2SO_4 to give aldehyde-2,4-DNPs in good yields was used as a proof of structure (equation 21).[45]

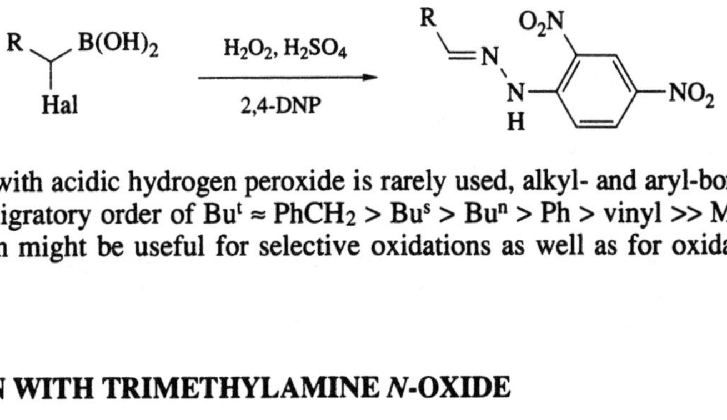

H_2O_2, H_2SO_4; 2,4-DNP (21)

Though oxidation with acidic hydrogen peroxide is rarely used, alkyl- and aryl-boronic acids are readily attacked with a migratory order of $Bu^t \approx PhCH_2 > Bu^s > Bu^n > Ph >$ vinyl $\gg$ Me. It appears, therefore, that the reaction might be useful for selective oxidations as well as for oxidations of base labile organoboranes.[14]

4.1.4 OXIDATION WITH TRIMETHYLAMINE *N*-OXIDE

Trimethylamine *N*-oxide, either anhydrous[46] or as its readily available dihydrate,[47] smoothly oxidizes a wide variety of alkyl, cycloalkyl, aryl and heterocyclic boron derivatives to the corresponding organyloxyboranes (equation 22) which, in the case of the dihydrate, are hydrolyzed in the reaction mixture. Anhydrous trimethylamine *N*-oxide is simply prepared[48] and this reagent must be used for the oxidation of alkenylboranes if prior hydrolysis is not to compete with oxidation.[48,49] Alkynylboranes are not oxidized by trimethylamine *N*-oxide.[46]

$$R_3B + 3\,Me_3NO \longrightarrow 3\,NMe_3\,(g) + (RO)_3B \xrightarrow{H_2O} 3\,ROH \qquad (22)$$

Oxidation with trimethylamine *N*-oxide is not complicated by side reactions and proceeds in a stepwise fashion such that one primary alkyl group is oxidized at 25 °C, the second at 65 °C and the third at 120 °C.[50] The trimethylamine evolved can be estimated and thus the method can be used for the quantitative estimation of organoboranes.[46] The rates of oxidation are not markedly solvent dependent and there is a clear cut ease of oxidation in the order tertiary alkyl > secondary cycloalkyl > secondary alkyl > primary alkyl > branched primary alkyl > vinyl. Some examples of useful differentiation are shown in equations (23)[51] and (24).[52]

$$\text{(R-CH=CH)}_2\text{B-thexyl} \xrightarrow{\text{Me}_3\text{NO}} \text{(R-CH=CH)}_2\text{B-O-thexyl} \quad (23)$$

$$\text{PhCH}_2\text{OCH}_2\text{CH=CH-B(c-C}_6\text{H}_{11})_2 \xrightarrow{\text{2 Me}_3\text{NO}} \text{PhCH}_2\text{OCH}_2\text{CH=CH-B(O-c-C}_6\text{H}_{11})_2 \quad (24)$$

The oxidation tolerates many functional groups, including alkyl sulfide,[25] and has been used for the oxidation of cyclopropylboranes[53] (see references 19 and 54); however, it has also been used to produce acylsilanes (equation 25)[55] which are particularly sensitive to alkaline hydrogen peroxide.

$$\text{RCH=C(SiMe}_3\text{)B(OMe)}_2 \xrightarrow{\text{Me}_3\text{NO}\cdot 2\text{H}_2\text{O}} \text{RCH}_2\text{C(=O)SiMe}_3 \quad (25)$$

The reaction proceeds with retention of configuration, as illustrated for the oxidation of a chiral organoborane in equation (26), which goes in good yield to give product of 98% optical purity.[56] The mechanism is expressed by equation (27) and is an example of general equation (2). Certain betaines (**3**), particularly those derived from tripropynylborane,[57] triphenylborane[58] and BF_3,[59] are reasonably stable and can be characterized.

$$\text{Ph(Me)C}^*\text{H-B(OBu)}_2 \longrightarrow \text{Ph(Me)C}^*\text{H-OH} \quad (26)$$

$$\text{R}_3\text{B} \longrightarrow \left[\text{R}_3\overset{-}{\text{B}}\text{-O-}\overset{+}{\text{N}}\text{Me}_3\right] \longrightarrow \text{R}_2\text{B-OR} + \text{NMe}_3\text{ (g)} \quad (27)$$

(**3**)

The action of *N*-oxides of α-dimethylaminocarboxylic acids on dihydroxyphenylborane produces the internal chelates of the oxidation products.[60] Various pyridine and quinoline *N*-oxides may be used to oxidize organylboranes, but there can be some alkylation of the heteroaromatic rings.[61]

4.1.5 AUTOXIDATION

The autoxidation of organoboranes has been extensively studied and there are many reviews of the topic.[3–7]

Autoxidation of organoboranes generally proceeds to give dialkoxyboranes (equation 28) but by use of pure oxygen and in the presence of THF the reaction may be taken to completion (equation 29).[68]

$$\text{R}_3\text{B} + \text{O}_2 \longrightarrow \text{R-B(OR)}_2 \quad (28)$$

$$\text{R}_3\text{B} + 1.5\ \text{O}_2 \xrightarrow{\text{THF}} \text{(RO)}_2\text{B-OR} \xrightarrow{\text{H}_2\text{O}} 3\ \text{ROH} \quad (29)$$

It is generally agreed,[6,7] on the following grounds, that autoxidation is a radical chain process: (a) reactive radical scavengers (galvinoxyl or iodine) exhibit a measurable inhibiting effect; (b) the reaction can be initiated by radical generators such as Bu^tOCl; and (c) stereodefined alkyl groups lose their stereointegrity in forming the intermediate alkylperoxyboron compounds. The kinetic behavior is reminiscent of hydrocarbon oxidation, but the S_H2 step is up to 10^7 times faster than for the corresponding reaction of hydrocarbons.

The initiation step remains in doubt. It may involve a transient O_2BR_3 complex which rearranges, followed by homolytic cleavage to give a supply of radicals. Alternatively, a displacement reaction (equation 30) might occur. The initiation step is first order in borane and also in oxygen and there is a large steric effect on both its induction period and rate.[62,63] Thus, tri(primary alkyl)boranes generally undergo initiation more readily than tri(secondary alkyl)boranes but an order of tricyclohexylborane > $Bu^s{}_3B$ > $Bu^i{}_3B$ suggests that steric factors cannot be considered alone.[62,63]

For the overall reaction the first B—C bond oxidizes faster than the second and this, in turn, much faster than the third. The observed trends are $R_3B > R_2BX > RBX_2$ (X = OR, OH, Cl), and R^t—B > R^s—B > R^n—B > Me—B > vinyl—B. The overall process has been expressed[7] as in equations (30)–(33).

$$R_3B + O_2 \longrightarrow R^{\bullet} + R_2BO_2^{\bullet} \tag{30}$$

$$R^{\bullet} + O_2 \longrightarrow RO_2^{\bullet} \tag{31}$$

$$RO_2^{\bullet} + R_3B \longrightarrow RO_2BR_2 + R^{\bullet} \tag{32}$$

$$2\,RO_2^{\bullet} \text{ or } 2\,R^{\bullet} \longrightarrow \text{stable products} \tag{33}$$

Although this is the mechanism which is generally written, it has not been rigorously verified and is incomplete. It does not indicate how R_2BO_2R is converted into $RB(OR)_2$ and it may be that this step is not radical in nature. Further evidence for this is that boron peroxides $(RO_2)_3B$ and $(RO)_2BOOR^1$ decompose only above 100 °C,[64] whereas reaction of Pr_3B with $Bu^tO_2B(OBu)_2$ proceeds rapidly at room temperature.[65] Studies on BuO_2BBu_2 and its analog indicate that their decomposition does not proceed by a radical mechanism.[66] Furthermore, although simple chiral boranes autoxidize with racemization,[67,68] this loss of steric integrity need not be total, as shown in equation (34).[69]

$$(\text{2-methylcyclopentyl})_3B \xrightarrow[\text{ii, aq. NaOH}]{\text{i, } O_2\text{, THF}} \textit{trans}\text{-2-methylcyclopentanol (81\%)} + \textit{cis}\text{-2-methylcyclopentanol (19\%)} \tag{34}$$

Whereas the use of 1.5 equiv. of oxygen to 1 of trialkylborane leads to quantitative yield of alcohol, as in equation (29), use of 2 equiv. of oxygen followed by addition of hydrogen peroxide (to prevent internal redox reactions) proceeds according to equation (35) to give 2 mol equiv. of alkyl hydroperoxide per mol of trialkylborane.[70]

$$R_3B \xrightarrow{2\,O_2} R{-}B(OOR)_2 \xrightarrow{H_2O_2} RO{-}B(OOR)_2 \xrightarrow{H_2O} 2\,ROOH + ROH \tag{35}$$

This reaction wastes one of the three alkyl groups, a limitation overcome by use of alkyldichloroborane etherates (equation 36), which are the precursors of choice for the preparation of alkylhydroperoxides.[71]

$$R{-}BCl_2 \xrightarrow[-18\,^{\circ}C]{O_2\text{, }Et_2O} ROO{-}BCl_2 \xrightarrow{H_2O} RO_2H \tag{36}$$

4.1.6 OXIDATION BY PERACIDS

It was early reported[8a] that perbenzoic acid in chloroform at room temperature quantitatively oxidizes tri-*n*-butylborane as in equation (37). The reaction with peroxytrifluoroacetic acid has been used analytically[72] but very little in synthesis.

$$R_3B \;+\; 3\,PhCO_3H \longrightarrow 3\,ROH \;+\; (PhCO_2)_3B \qquad (37)$$

The mechanism of the reaction is not certain, but it appears to have analogies with the protonolysis of organoboranes by carboxylic acids and on that basis could be formulated as in equation (38), which is an example of general equation (7).

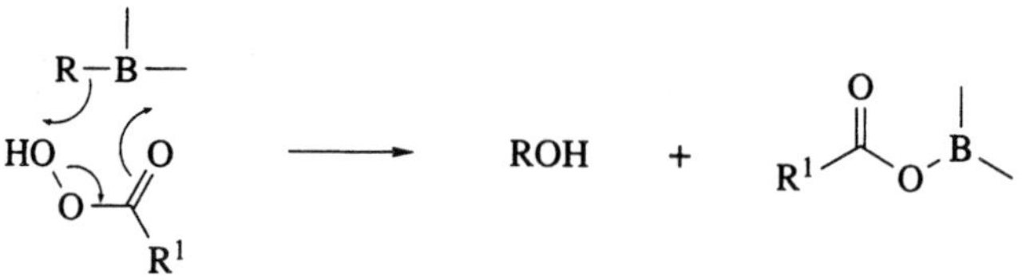

$$\longrightarrow \quad ROH \quad + \quad R^1CO_2B\!< \qquad (38)$$

Although oxidation of 1,1-bis(dialkylboryl) compounds with alkaline hydrogen peroxide proceeds with initial hydrolysis rather than oxidation (equation 16), oxidation with excess MCPBA yields carboxylic acids in good yields. The overall process is an excellent route from 1-alkynes to carboxylic acids (equation 39).[73]

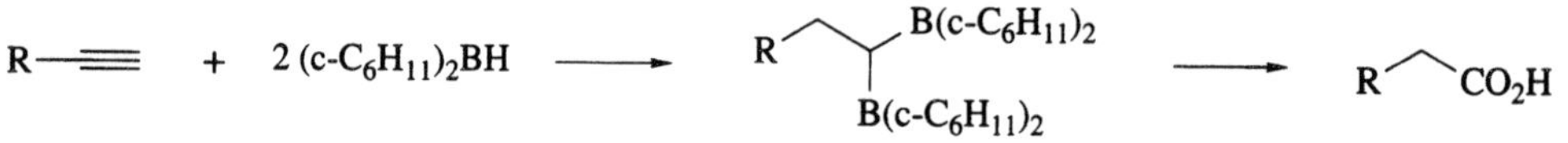

$$R\!-\!\equiv \;+\; 2\,(c\text{-}C_6H_{11})_2BH \longrightarrow RCH_2CH[B(c\text{-}C_6H_{11})_2]_2 \longrightarrow RCH_2CO_2H \qquad (39)$$

Oxidation of 1,1-bis(dialkoxyboryl)alkanes with MCPBA gives aldehydes, and similar oxidation of 2,2-bis(dialkoxyboryl)alkanes yields ketones.[74]

Oxidation by MCPBA offers advantages when other methods give problems. Thus, a series of alkenylboranes containing malonate or acetoacetate units could not be converted directly to the corresponding ketones using alkaline hydrogen peroxide due to complications arising from ester hydrolysis. However, MCPBA oxidation gave the ketones in excellent yields (*e.g.* equation 40).[23]

(Hex)$_2$B, Bu, O, Hex, CO_2Et —MCPBA→ Hex, Bu, O, O, CO_2Et (40)

Similarly, an acetoxyphenyl-substituted boron heterocyclic compound, resulting from a cyanoborate reaction, lost the phenolic acetate group even when buffered hydrogen peroxide was used. However, MCPBA gave the required ketone retaining the labile ester group (equation 41).[75]

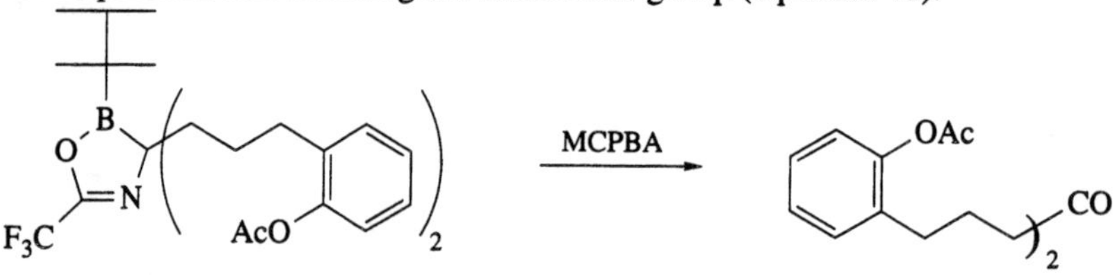

B, O, N, F_3C, AcO, 2 —MCPBA→ OAc, CO, 2 (41)

Peracid oxidation of organylboranes is a smooth and operationally simple procedure that proceeds in mildly acidic conditions in anhydrous solvents. Yields are generally high and the method, though infrequently used at present, should always be considered as an alternative to oxidation with alkaline hydrogen peroxide.

4.1.7 OXIDATION WITH CHROMIUM REAGENTS

4.1.7.1 Chromic Acid

The oxidation of (secondary alkyl)boranes with chromic acid leads to ketones,[76–78] and in combination with hydroboration provides a high-yielding route from alkenes to ketones. The reaction gives ketones at pH < 3, whereas at pH 3–7 secondary alcohols are produced.[78] Therefore, 4M $H_2Cr_2O_7$ in a water–ether two-phase system has been recommended for conversion of cycloalkylboranes to cycloalkanones.[77] The method has been used widely in synthesis,[3,7] an example being given in equation (42) in which a chiral ketone is produced.[79,80]

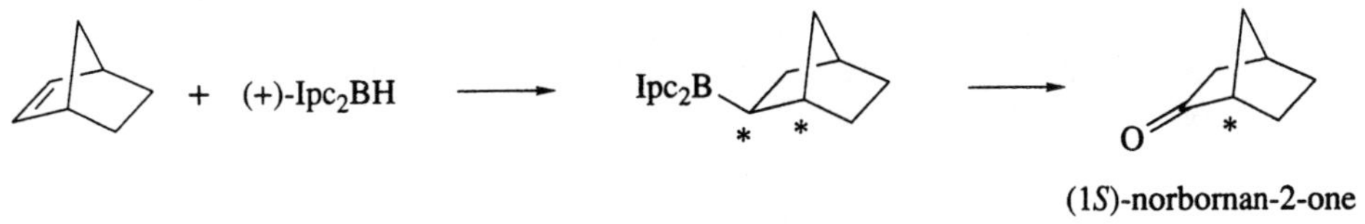

+ (+)-Ipc_2BH ⟶ Ipc_2B ⟶ (42)

(1*S*)-norbornan-2-one

The ketone produced is generally that to be expected, but occasionally the strong acid conditions may lead to unexpected products, as shown in equations (43)[81] and (44).[82]

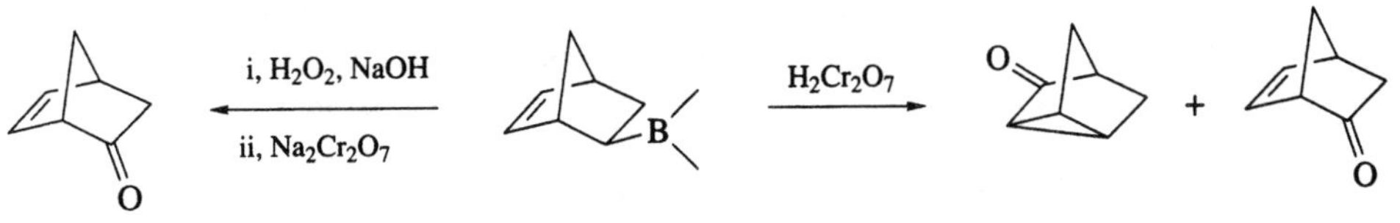

i, H_2O_2, NaOH; ii, $Na_2Cr_2O_7$; $H_2Cr_2O_7$ (43)

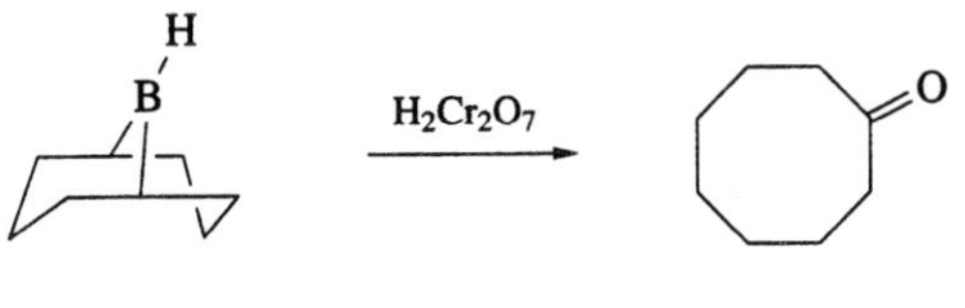

$H_2Cr_2O_7$ (44)

1,2-Diboryl compounds yield alkenes on reaction with chromic acid (equation 45),[83] a reaction also noted in a peracid oxidation[73] and oxidation with pyridinium chlorochromate.[84]

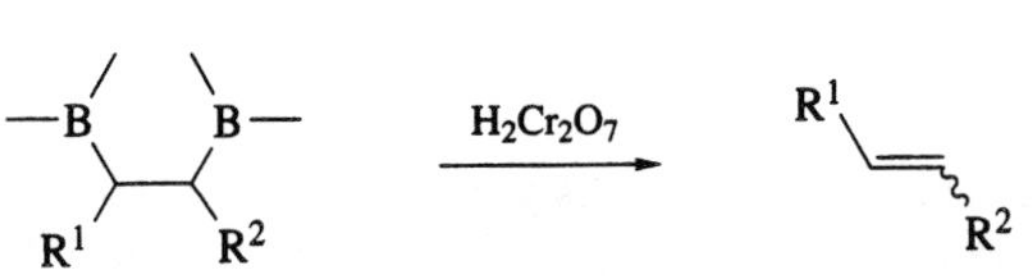

$H_2Cr_2O_7$ (45)

Chromic acid is particularly useful for the oxidation of heterocyclic boranes[85–87] which are resistant to oxidation by the usual oxidants (NaOH/H_2O_2; R_3NO; O_2), *e.g.* equation (46).[85]

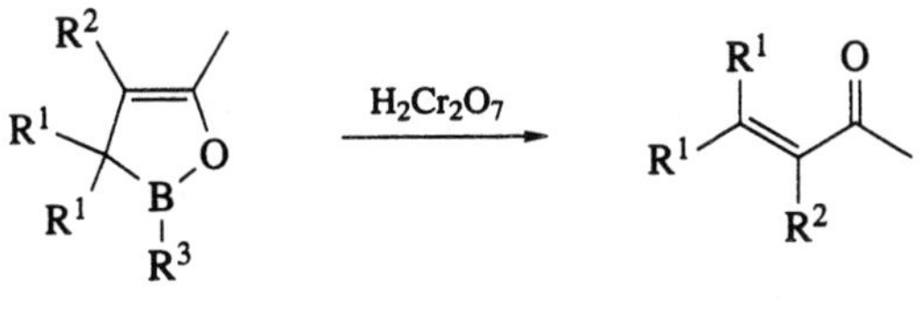

$H_2Cr_2O_7$ (46)

4.1.7.2 Chromyl Trichloroacetate

Some 2,5-diboradihydropyrazines have been successfully oxidized with chromyl trichloroacetate,[88] but the reagent has not been widely used.

4.1.7.3 Pyridinium Chlorochromate

Pyridinium chlorochromate (PCC) is a very useful reagent for the oxidation of organoboranes to carbonyl compounds in mildly alkaline and anhydrous conditions.[89] As well as oxidizing (secondary alkyl)boranes to ketones,[90,91] PCC oxidizes (primary alkyl)boranes to aldehydes in excellent yields (equation 47).[89] The latter transformation cannot be accomplished with chromic acid. The reagent tolerates the presence of alkene, ester and acetal groups.[90]

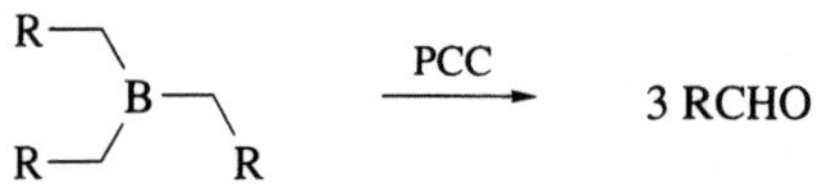

$$R_3B \xrightarrow{\text{PCC}} 3\ \text{RCHO} \quad (47)$$

The observation[92] that trialkylboranes are oxidized to aldehydes as fast as the free alcohols and much faster than acetates prompted the introduction of an efficient one-pot procedure for the conversion of acids to aldehydes (equation 48) in 69–82% isolated overall yields.[93]

PCC oxidizes *gem*-diboryl compounds to ketones.[84]

$$RCO_2H \xrightarrow{H_3B\cdot SMe_2} (RCH_2O)_3B \xrightarrow{\text{PCC, } CH_2Cl_2} 3\ \text{RCHO} \quad (48)$$

4.1.8 MISCELLANEOUS CHEMICAL OXIDIZING AGENTS

Almost any oxidizing agent might be expected to oxidize an organoborane, but few have received systematic investigation. Moreover, aryldihydroxyboranes are relatively resistant to oxidation and nitration of the ring,[94] and oxidation of a methyl group to a carboxy group with potassium permanganate[95] has been achieved without breaking the Ar—B bond.

The reagent MoO_5–pyridine–HMPA (MOPH) was introduced for the anhydrous, ambient temperature oxidation of intermediates in aldol reactions of alkenyloxyboranes.[96] It also oxidizes simple organoboranes, the oxidation proceeding with retention of configuration of the alkyl groups.[97]

A mixture of ruthenium tetroxide, sodium periodate and sodium acetate was effective for the conversion shown in equation (49).[98]

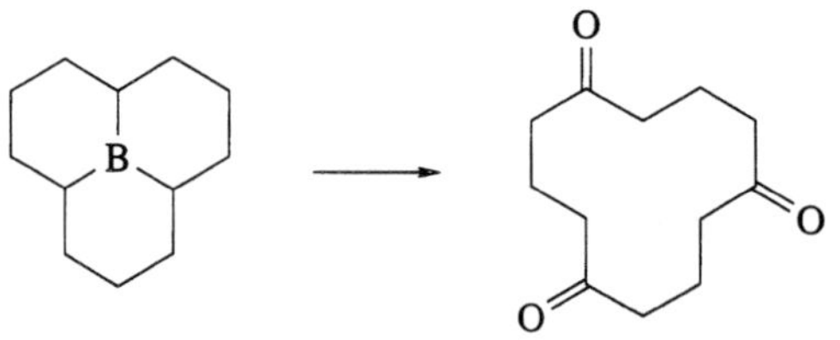

(49)

The action of lead tetraacetate on trialkylboranes produces alkyl acetates.[99]

Although aryldihydroxyboranes appear to resist alkaline permanganate,[94,95] butyldihydroxyborane is oxidized to butanol with this reagent.[100]

Dialkoxy(α-phenylthio)alkylboranes are readily converted to monothioacetals by *N*-chlorosuccinimide in methanol (equation 50).[101]

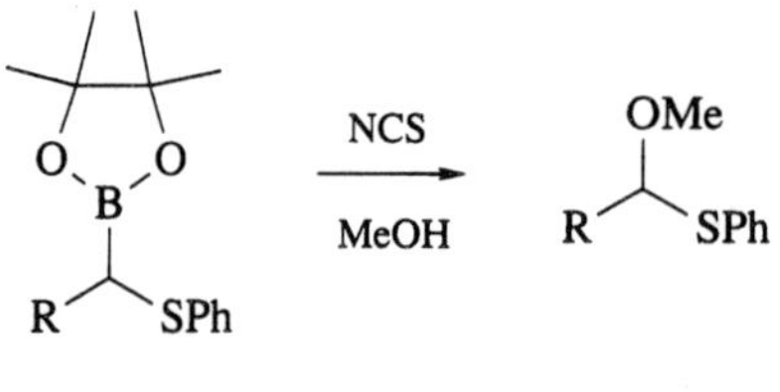

(50)

Sodium hypochlorite oxidizes aryldihydroxyboranes to phenols[102] and trialkylboranes to the corresponding alcohols.[103] Use of sodium hypobromite for the oxidation of phenyldihydroxyborane gave 2,4,6-tribromophenol, regardless of the ratio of the reactants.[102]

Sodium perborate has been recommended for the oxidation of alkenyldialkoxyboranes to aldehydes.[104] In particular, the reagent avoids the formation of possibly hazardous 2:1 aldehyde–hydrogen peroxide adducts as well as C—C bond cleavage.[104a] Sodium percarbonate is an air-stable solid, inexpensive and self-buffered, that appears to be a promising oxidant.[104b]

Alkyl- and phenyl-dihydroxyboranes, their anhydrides and dialkoxy derivatives are oxidized with alkyl hydroperoxides to the corresponding alcohols and phenols.[105–107]

4.1.9 ELECTROCHEMICAL OXIDATION

Electrochemical oxidation of organoboranes can lead to a variety of products, depending on the natures of the electrodes, solvents and electrolytes. It appears that radicals, R·, are generated and that these may couple to give dimeric alkanes, react with other species present or be further oxidized to cations, R^+, which then react. Thus, when trialkylboranes, R_3B, are subjected to electrolysis in methanolic KOH solution between platinum electrodes, the alkyl residues couple to each other to give dimeric alkanes R—R.[108] However, when a graphite anode is used in a methanolic solution containing NaOMe and $NaClO_4$, then trialkylboranes yield alkyl methyl ethers (ROMe) in 84–100% yield.[109] This reaction is thought to pass through R^+, as isomerization occurs to an appreciable extent. Thus the oxidation of tri-*n*-octylborane gives a mixture of 1-, 2-, 3- and 4-methoxyoctanes with a ratio of 8:66:25:1. In the presence of sodium acetate, electrolysis of R_3B gives alkyl acetates.[109] When acetonitrile is used as the medium then RCH_2CN is obtained[110] and if nitromethane is used then RCH_2NO_2 results.[111]

Sodium tetraphenylborate[112] and diphenylhydroxyborane[113] both yield biphenyl on anodic oxidation. Similar results can be obtained using a variety of chemical oxidants.[114]

4.1.10 OXIDATION BY CARBONYL COMPOUNDS TO PRODUCE ALKENES

Trialkylboranes reduce carbonyl compounds by utilizing a β-hydrogen atom, as in equation (51).[115]

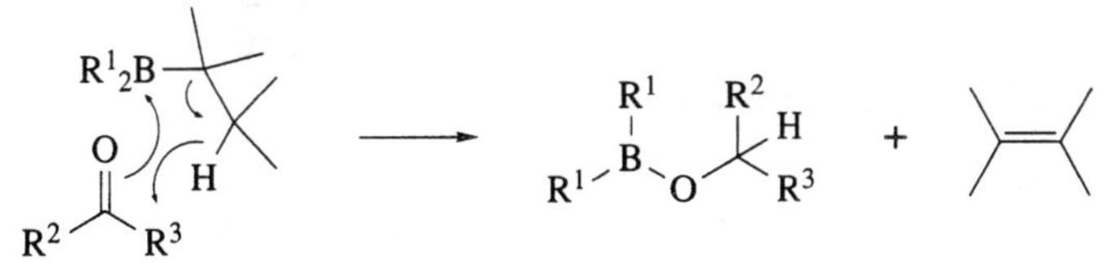

(51)

The reaction was first explored with triethylborane[116] and then extended to higher alkylboranes.[117] Only two unbranched alkyl groups participate and even then prolonged heating at 150–200 °C is required.[117] Only the exocyclic alkyl groups of *B*-alkyl-9-BBN compounds participate in the reaction and hence such compounds may be used as a probe of the intrinsic ease of oxidizability of different alkyl groups.[118] The results show that branching at the β-position strongly enhances the ease of oxidation. The relative rates of oxidation of the *trans*-2-methylcyclopentyl, siamyl, cyclopentyl, isobutyl and ethyl derivatives are 1000:364:267:50:0.72, respectively.[118] Since *B*-alkyl-9-BBN derivatives are mostly prepared from alkenes by hydroboration with 9-BBN-H, the process has little preparative value for the synthesis of alkenes. Instead, the reaction is useful as a very specific method for the reduction of the carbonyl compound, and for this purpose *B*-siamyl-9-BBN has been used for the reduction of aldehydes.[119] α,β-Unsaturated aldehydes yield only allylic alcohols.[119] The rates of reaction of aldehydes vary very little but ketones react at least two orders of magnitude more slowly, thus allowing highly selective reductions to be carried out. The process has been shown to tolerate alkene, amino, aryl, ether, halogen and nitro groups.[119]

3-Pinanyl-9-BBN (Alpine borane; **4**) is a chiral borane that is readily oxidized by aldehydes. Aliphatic deuterioaldehydes undergo chiral reduction to give alcohols with 84–98% *e.e.*[120] The chiral alkene is regenerated in the process, only the hydrogen at the 2-position having been utilized, and can be reused. Equation (52) serves as an illustration of the stereochemistry of the process.

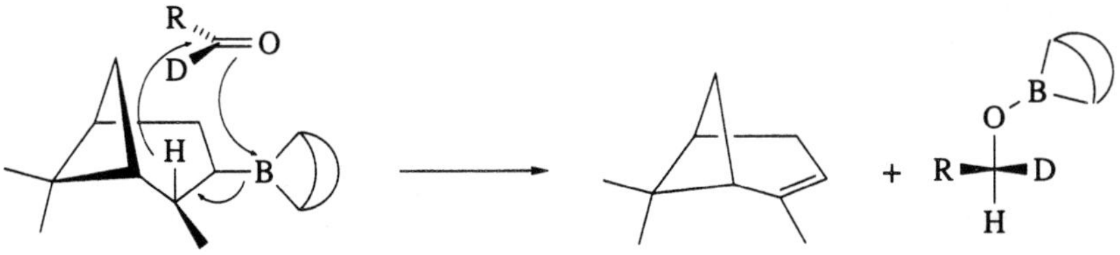

(52)

α,β-Alkynyl ketones are excellent substrates for chiral reduction by Alpine borane.[121] The stereochemistry of the product alcohol is predicted by replacement of the deuterium atom in equation (52) by the alkynyl group.[122] Use of neat reagents[123] and/or pressure[124] allows the reduction of simple ketones, α-halo ketones, α-keto esters and α-ketonitriles.

The nopol derivative (**5**) has advantages in some cases[125] but the most striking advance has been the introduction of commercially available chloroborane (**6**). Compound (**6**) is a stronger Lewis acid than (**4**) or (**5**) and also less subject to steric hindrance.[126] It readily reduces acetophenone (98% *ee*), 2,2-dimethylcyclopentanone (98% *ee*) and pinacolone (95% *ee*).[126]

Another important application of the alkene displacement process is in the synthesis of a whole series of alkyldiethoxyboranes of very high optical purity by displacement of α-pinene from 3-pinanylalkyl-

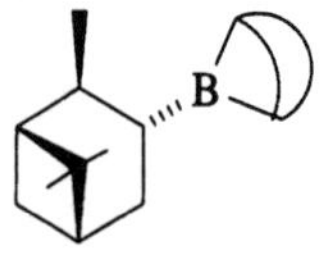

(4) 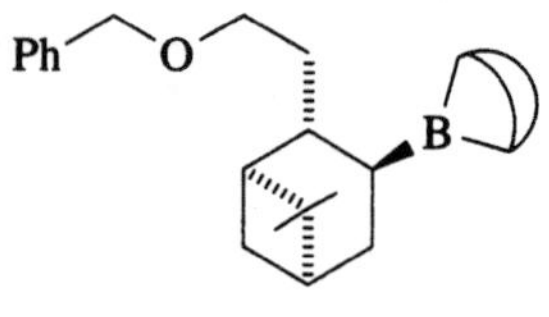 **(5)** 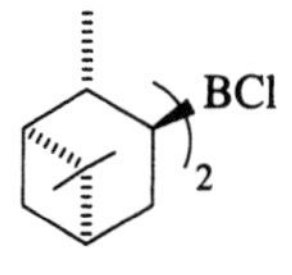 **(6)**

ethoxyboranes, as in equation (53).[127] The chiral alkyldiethoxyborane products are certain to have enormous influence on the field of chiral synthesis.[127]

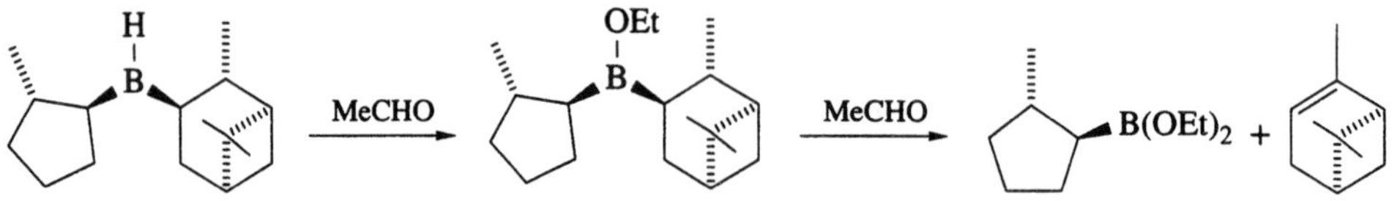

MeCHO → MeCHO → $B(OEt)_2$ + (53)

4.1.11 CHLORINOLYSIS OF C—B BONDS

The chlorination of trialkylboranes has not been well studied. The chlorination of Me_3B at –95 °C yields $ClCH_2BMe_2$.[128] The oxidation of dialkoxy(α-phenylthio)alkylboranes with NCS has been referred to in Section 4.1.8 (equation 50). The reaction of thionyl chloride with $PhSCH_2B(OR)_2$ yields $PhSCH_2Cl$.[101]

Direct chlorination has not been used for the production of organyl chlorides from trialkyl- or triarylboranes. Chlorinolysis has been achieved by the reactions of organoboranes with aqueous copper(II) chloride[129] and iron(III) chloride.[130] However, only two of the three alkyl residues on boron are utilized (equation 54). Arylboranes are also converted to aryl chlorides with copper(II) chloride.[131]

$$R_2B{-}R + 4\,CuCl_2 + 2\,H_2O \longrightarrow 2\,RCl + 2\,Cu_2Cl_2 + 2\,HCl + R{-}B(OH)_2 \qquad (54)$$

Nitrogen trichloride converts all three alkyl groups of R_3B into alkyl chlorides, but 3 equiv. of a hazardous and not readily available reagent must be used.[132] Chlorodimethylamine[133] or dichloramine-T (equation 55)[134] may be used, but here too there is poor utilization of alkyl groups, a problem eased, but not completely overcome, by the use of *B*-alkyl-9-BBN derivatives.[134]

$$R_3B \; + \; Cl_2NSO_2C_6H_4Me \longrightarrow RCl \; + \; R_2BNClSO_2C_6H_4Me \qquad (55)$$

4.1.12 BROMINOLYSIS OF C—B BONDS

All three organyl groups of trialkylboranes[135] or triphenylborane[136] can be cleaved by reaction with bromine in the presence of sodium methoxide. High yields of bromoalkanes (equation 56) or bromobenzene are obtained.[148]

$$R_3B + 3\,Br_2 + 4\,NaOMe \longrightarrow 3\,RBr + 3\,NaBr + NaB(OMe)_4 \qquad (56)$$

The brominolysis of trialkylboranes is unusual among the reactions of organoboranes in that it proceeds mainly with inversion. Thus, tris-*exo*-2-norbornylborane gives 75% of *endo*-2-bromonorbornane and 25% of the *exo* isomer.[137] Over 95% inversion has been observed in the bromination of the *threo* and *erythro* isomers of tris(3,3-dimethyl-1-butyl-1,2-d_2)borane in the presence of NaOMe (equation 57).[138]

erythro —Br_2, NaOMe→ *threo* (57)

A modification involving the use of sodium bromide in the presence of chloramine-T has been devised to allow optimum incorporation of radioisotopic bromine.[139]

In neutral conditions bromine reacts with triorganylboranes to convert one organyl group into the corresponding bromoalkane.[140] The reaction involves a free radical α-bromination (see equations 5 and 6), followed by cleavage with hydrogen bromide, and therefore the product bromide does not retain the configuration of the original organoborane.[140] Although there is normally a strong preference for reaction with secondary rather than primary alkyl groups,[141] *B*-alkyl-9-BBN derivatives react with complete conversion of the primary *B*-alkyl group into the alkyl bromide.[142] The overall process is an efficient anti-Markovnikov addition of HBr to a terminal alkene (equation 58).

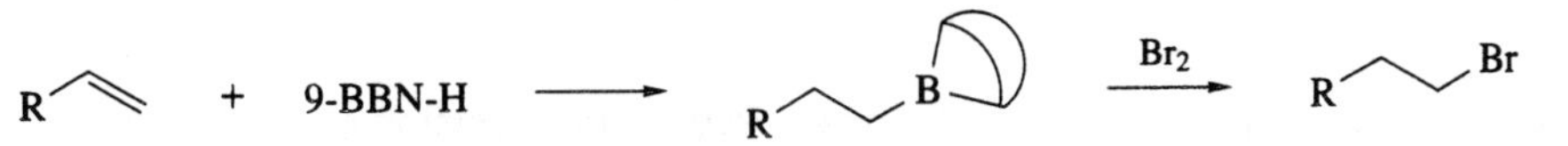

(58)

An alternative indirect but efficient method for the bromination of all three groups of tri(primary alkyl)boranes involves initial reaction with mercury(II) acetate followed by *in situ* bromination.[143]

Alkenyldialkylboranes react with bromine to give bromoalkenes *via* an addition–elimination mechanism.[144] The method of elimination controls the stereochemistry of the product bromoalkenes (Scheme 1). For reasons which are not clear, exactly opposite stereochemical results are obtained from (arylethenyl)dialkylboranes as compared with (alkylethenyl)dialkylboranes (Scheme 1).[144]

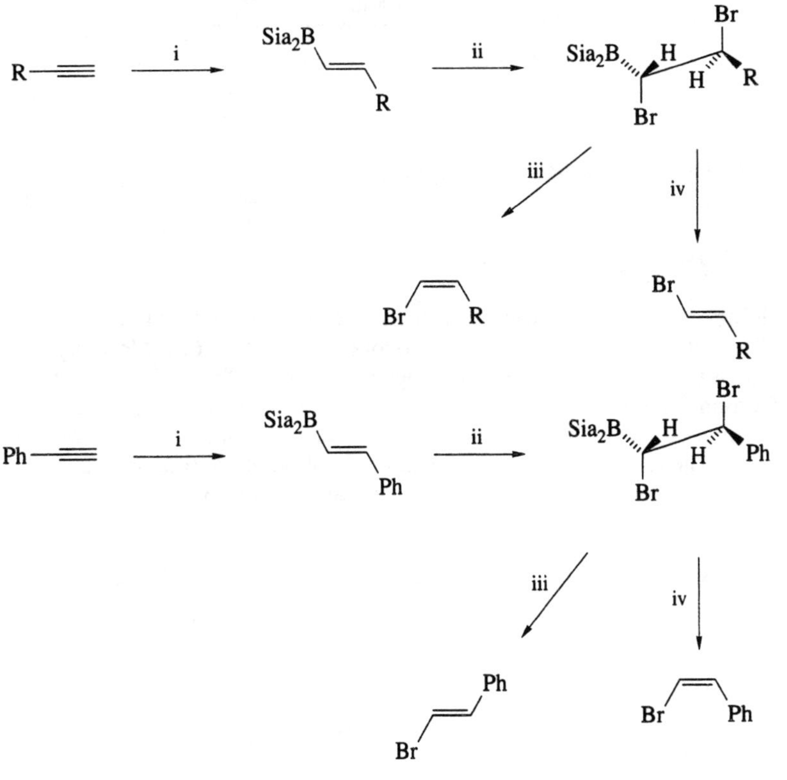

i, Sia_2BH, THF; ii, Br_2, CCl_4, 0 °C; iii, NaOH, 0 °C; iv, 76 °C

Scheme 1

Alkenyldihydroxyboranes react with bromine and base with inversion of configuration irrespective of the nature of the alkenyl group (equation 59).[145] Use of excess bromine allows utilization of catecholborane derivatives.[145]

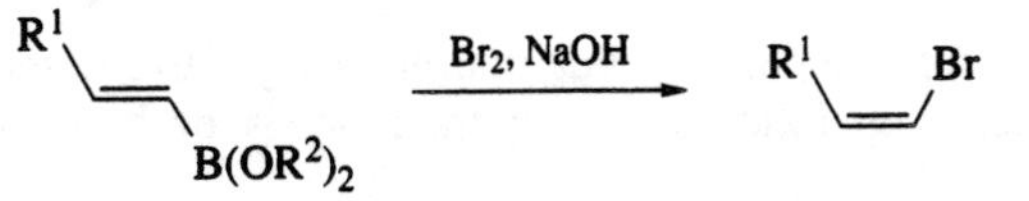

(59)

4.1.13 IODINOLYSIS OF C—B BONDS

In the presence of sodium methoxide, tri(primary alkyl)boranes react smoothly to yield 3 mol equiv. of iodoalkane.[146] In the presence of sodium hydroxide, two C—B bonds of tri(primary alkyl)boranes are cleaved to the corresponding iodides.[147] In the same conditions tri(secondary alkyl)boranes react significantly more slowly and only one C—B bond is broken.[147,148] For anti-Markovnikov addition of HI to terminal alkenes the process shown in equation (60) is therefore applicable.[147]

$$R^1R^2C{=}CH_2 + R^s_2BH \longrightarrow R^1R^2CHCH_2BR^s_2 \xrightarrow{I_2,\ NaOH} R^1R^2CHCH_2I \qquad (60)$$

Like brominolysis, iodinolysis of trialkylboranes occurs with inversion of stereochemistry of the displaced carbon atom.[149,150]

Sodium iodide and chloramine-T can be used for the production of iodides,[150] a process that has particular application to the introduction of radioactive iodine.[151]

The reactions of alkenyldialkylboranes with iodine lead to rearrangements rather than iodinolysis. However, alkenyldihydroxyboranes undergo iodinolysis with *retention* of configuration, in contrast to the corresponding reaction with bromine, to give high yields of the corresponding iodoalkenes (equation 61).[152]

$$RCH{=}CHB(OH)_2 \xrightarrow{I_2,\ NaOH} RCH{=}CHI \qquad (61)$$

4.1.14 REPLACEMENT OF BORON BY NITROGEN

4.1.14.1 Synthesis of Primary Amines

Amines containing good leaving groups, such as hydroxylamine-*O*-sulfonic acid[153–155] and chloramine,[153] yield alkylamines on reaction with organoboranes (equation 62). Mesitylenesulfonylhydroxylamine is more soluble in THF than are the other reagents and reacts faster.[156] Chloramine may be formed *in situ* from ammonium hydroxide and sodium hypochlorite in a reaction designed for the incorporation of ^{15}N.[157] Two of three alkyl groups on boron are displaced and hindered organoboranes may require forcing conditions.[154] The reactions proceed with strict retention of configuration at carbon and probably by an ionic 1,2-migration process,[158] an example of general equation (3).

$$R_3B + NH_2X \longrightarrow \left[R_3\overset{-}{B}{-}\overset{+}{N}H_2X\right] \longrightarrow \left[R_2B{-}\overset{+}{N}H_2R\ X^-\right] \longrightarrow$$

$$RB(\overset{+}{N}H_2R)_2\ 2X^- \xrightarrow{2\ NaOH} 2\ RNH_2 + RB(OH)_2 + 2\ NaX \qquad (62)$$

Primary amines of high optical purity can be obtained with full utilization of the organyl group through the intermediate formation of alkylmethylalkoxyboranes (equation 63).[159]

$$R^*{-}B(O(CH_2)_3O) \xrightarrow[\text{ii, AcCl}]{\text{i, MeLi}} R^*B(Me)O(CH_2)_4OAc \xrightarrow[\text{ii, } H_2O]{\text{i, } NH_2OSO_3H} R^*{-}NH_2 \qquad (63)$$

For the replacement of boron by an amino group in aryl-, styryl- and ferrocenyl-dihydroxyboranes, the reaction of copper(II) phthalimide followed by hydrazinolysis or hydrolysis, has been used (equation 64).[160,161]

$$ArB(OH)_2 + (\text{phthalimido})_2Cu \longrightarrow \text{N-Ar phthalimide} \longrightarrow ArNH_2 \quad (64)$$

4.1.14.2 Synthesis of Secondary Amines

Trialkylboranes and triphenylborane react with alkyl, cycloalkyl and aryl azides with the evolution of nitrogen and the formation of dialkyl(dialkylamino)boranes, which yield secondary amines on solvolysis.[162] As only one of the organyl groups is utilized it is advantageous to use organyldichloroboranes (equation 65), which have the further advantage of requiring milder conditions.[163] The reactions proceed with retention of configuration.[164] 2-Iodoalkyl azides react at room temperature with phenyl- and alkyldichloroboranes to give products that on treatment with base give aziridines (equation 66).[165]

$$R^1{-}BCl_2 + R^2{-}\bar{N}{-}\overset{+}{N}{\equiv}N \longrightarrow N_2(g) + R^1R^2N{-}BCl_2 \longrightarrow R^1R^2NH \quad (65)$$

$$R^2R^1C(N_3){-}CH(R^3)I + R^4BCl_2 \xrightarrow[-N_2]{20\,^\circ C} \xrightarrow{\text{base}} \text{aziridine } (N{-}R^4;\ R^1, R^2, R^3) \quad (66)$$

In one instance a secondary amine has been produced by two migrations from boron to nitrogen using a reagent with two good leaving groups attached to nitrogen.[166]

4.1.14.3 Synthesis of Tertiary Amines and Other Derivatives

N-Chlorodimethylamine reacts with trialkylboranes to give alkyldimethylamines, so long as the alternative radical reaction leading to alkyl chlorides is suppressed by use of the radical scavenger, galvinoxyl (equation 67).[133]

$$R_3B + ClNMe_2 \xrightarrow{\text{galvinoxyl}} R_2BCl + RNMe_2 \quad (67)$$

Trialkylboranes react with chloramine-T to give alkyl toluenesulfonamides in good yields based upon the transfer of one alkyl group (equation 68).[167]

$$R_3B + NaNClSO_2Ar \longrightarrow R_2BNRSO_2Ar \xrightarrow{NaOH} RNHSO_2Ar \quad (68)$$

Trialkylboranes react with sodium azide in the presence of hydrogen peroxide and iron compounds to give azidoalkanes (RN_3) in fair yields based upon the transfer of one alkyl group.[168]

4.1.15 REPLACEMENT OF BORON BY SULFUR OR SELENIUM

Dialkyl or diaryl disulfides react with trialkylboranes in the presence of oxygen or light to give dialkyl or alkyl aryl sulfides in good yields based upon the transfer of one alkyl group (equation 69).[169] If it is

$$R^2_2B{-}R^1 + R^3S{-}SR^3 \xrightarrow{O_2 \text{ or } h\nu} R^2_2B{-}SR^3 + R^1SR^3 \quad (69)$$

necessary to conserve alkyl groups then *B*-alkylborinanes may be used, as only the *B*-alkyl group is utilized in excellent yield.[169]

Alkyl thiocyanates are obtained by the reaction of iron(III) thiocyanate on trialkylboranes in a fashion reminiscent of the reactions of iron(III) chloride (Section 4.1.11). It is not clear whether there is 60% usage of three alkyl groups or whether only two groups are in fact utilized.[170] Equation (70) shows the reaction based on the assumption of usage of three groups.

$$R_3B + 6\,Fe(SCN)_3 + 3\,H_2O \longrightarrow 3\,RSCN + B(OH)_3 + 6\,Fe(SCN)_2 + 3\,HSCN \quad (70)$$

A similar reaction yielding alkyl selenocyanates occurs between trialkylboranes and sodium selenoiso-cyanate in the presence of iron(III) ions.[171] There is a strong preference for the transfer of secondary or tertiary as compared to primary alkyl groups. This allows selective transfer reactions to be carried out in yields of 57–68% (equation 71) and 67–78% (equation 72).[171]

$$R^s_2B{-}H \longrightarrow R^s_2B{-}R^n \longrightarrow 2\,R^sSeCN + R^n{-}B(OH)_2 \quad (71)$$

$$R^t{-}BH_2 \longrightarrow R^t{-}BR^n_2 \longrightarrow R^tSeCN + R^n_2B{-}OH \quad (72)$$

In the presence of base, phenylselenenyl bromide reacts with alkenyldihydroxyboranes to give alkenyl phenyl selenides with retention of configuration at the double bond (equation 73).[172]

$$RCH{=}CH{-}B(OH)_2 \xrightarrow{\text{PhSeBr, NaOH}} RCH{=}CH{-}SePh \quad (73)$$

4.1.16 REFERENCES

1. H. A. Skinner, *Adv. Organomet. Chem.*, 1964, **2**, 49.
2. A. G. Davies, *Pure Appl. Chem.*, 1974, **39**, 497.
3. A. Pelter, K. Smith and H. C. Brown, 'Borane Reagents', Academic Press, New York, 1988.
4. A. Pelter and K. Smith, in 'Comprehensive Organic Chemistry', ed. D. H. R. Barton and W. D. Ollis, Pergamon Press, Oxford, 1979, vol. 3, p. 810.
5. E. Negishi, in 'Comprehensive Organometallic Chemistry', ed. G. Wilkinson, F. G. A. Stone and E. W. Abel, Pergamon Press, Oxford, 1982, vol. 7, pp. 270, 305, 326, 354.
6. T. Onak, 'Organoborane Chemistry', Academic Press, New York, 1975.
7. B. M. Mikhailov and Yu. N. Bubnov, 'Organoboron Compounds in Organic Synthesis', Harwood, Chur, 1984.
8. (a) J. R. Johnson and M. G. van Campen, *J. Am. Chem. Soc.*, 1938, **60**, 121; (b) H. R. Snyder, J. A. Kuck and J. R. Johnson, *J. Am. Chem. Soc.*, 1938, **60**, 105.
9. A. D. Ainley and F. Challenger, *J. Chem. Soc.*, 1930, 2171.
10. W. J. Wechter, *Chem. Ind. (London)*, 1959, 294.
11. H. C. Brown and G. Zweifel, *J. Am. Chem. Soc.*, 1961, **83**, 2544.
12. H. C. Brown, M. C. Desai and P. K. Jadhav, *J. Org. Chem.*, 1982, **47**, 5065.
13. H. C. Brown, B. Singaram and T. E. Cole, *J. Am. Chem. Soc.*, 1985, **107**, 460; H. C. Brown, P. V. Ramachandran and J. Chandrasekharan, *Organometallics*, 1986, **5**, 2138.
14. H. Minato, J. C. Ware and T. G. Traylor, *J. Am. Chem. Soc.*, 1963, **85**, 3024.
15. H. G. Kuivila, *J. Am. Chem. Soc.*, 1954, **76**, 870.
16. H. G. Kuivila, *J. Am. Chem. Soc.*, 1955, **77**, 4014.
17. G. Zweifel and H. C. Brown, *Org. React. (N.Y.)*, 1964, **13**, 1.
18. B. M. Mikhailov, A. Ya. Bezmenov, L. S. Vasil'ev and V. G. Kiselev, *Dokl. Akad. Nauk SSSR*, 1964, **155**, 141; 1966, **167**, 590; B. M. Mikhailov, A. Ya Bezmenov, L. S. Vasil'ev and V. G. Kiselev, *Izv. Akad. Nauk SSSR, Ser. Khim.*, 1965, 2111.
19. H. C. Brown and S. P. Rhodes, *J. Am. Chem. Soc.*, 1969, **91**, 4306.
20. Yu. N. Bubnov, O. A. Nesmeyanova, T. Yu. Rudashevskaya, B. M. Mikhailov and B. A. Kajansky, *Tetrahedron Lett.*, 1971, 2153; *Zh. Obshch. Khim.*, 1973, **43**, 127, 135.
21. K. Utimoto, M. Tamura, M. Tanouti and K. Sisido, *Tetrahedron*, 1972, **28**, 5697.
22. A. Pelter, K. Smith, M. G. Hutchings and K. Rowe, *J. Chem. Soc., Perkin Trans. 1*, 1975, 129.
23. A. Pelter, L. Hughes and J. Madhusudhana Rao, *J. Chem. Soc., Perkin Trans. 1*, 1982, 719.
24. H. C. Brown and A. K. Mandal, *J. Org. Chem.*, 1980, **45**, 916.

25. A. Pelter, P. Rupani and P. Stewart, *J. Chem. Soc., Chem. Commun.*, 1981, 164.
26. H. A. Staab and B. Meissner, *Justus Liebigs Ann. Chem.*, 1971, **753**, 80.
27. A. Pelter, L. Williams and J. W. Wilson, *Tetrahedron Lett.*, 1983, **24**, 627.
28. S. W. Breuer and F. A. Broster, *J. Organomet. Chem.*, 1972, **35**, C5.
29. G. M. Pickles and F. G. Thorpe, *J. Organomet. Chem.*, 1974, **76**, C23.
30. G. W. Kramer and H. C. Brown, *J. Organomet. Chem.*, 1974, **73**, 1.
31. S. W. Breuer, M. J. Leatham and F. G. Thorpe, *J. Chem. Soc., Chem. Commun.*, 1971, 1475.
32. A. N. Nesmeyanov and R. Sokolik, 'Methods in Elemento-Organic Chemistry', ed. A. N. Nesmeyanov and K. A. Kocheskov, North Holland, Amsterdam, 1967, vol. 1. p. 296.
33. (a) A. B. Hornfeldt, *Acta Chem. Scand.*, 1965, **19**, 1249; (b) R. T. Hawkins, *J. Heterocycl. Chem.*, 1974, **11**, 291.
34. A. Pelter and M. Rowlands, *Tetrahedron Lett.*, 1987, **28**, 1203.
35. H. C. Brown and G. Zweifel, *J. Am. Chem. Soc.*, 1961, **83**, 3834.
36. H. C. Brown, D. Basavaiah and S. U. Kulkarni, *J. Organomet. Chem.*, 1982, **225**, 63.
37. H. C. Brown and R. A. Coleman, *J. Am. Chem. Soc.*, 1969, **91**, 4606.
38. E. Negishi, T. Yoshida, A. Silveira, Jr. and B. L. Chiou, *J. Org. Chem.*, 1975, **40**, 814.
39. R. Rauchschwalbe and M. Schlosser, *Helv. Chim. Acta*, 1975, **58**, 1094.
40. B. M. Mikhailov and T. K. Baryshnikova, *Zh. Obschch. Khim.*, 1971, **41**, 1303.
41. R. Köster and G. W. Rotermund, *Angew. Chem.*, 1960, **72**, 138, 563.
42. P. M. Aronovich, V. S. Bogdanov and B. M. Mikhailov, *Izv. Akad. Nauk SSSR, Ser. Khim.*, 1969, 362.
43. B. M. Mikhailov, P. M. Aronovich and V. G. Kiselev, *Izv. Akad. Nauk SSSR, Ser. Khim.*, 1968, 146.
44. Yu. N. Bubnov, S. I. Frolov, V. G. Kiselev, V. S. Bogdanov and B. M. Mikhailov, *Organomet. Chem. Synth.*, 1970, **1**, 37.
45. D. S. Matteson, *J. Am. Chem. Soc.*, 1960, **82**, 4228.
46. R. Köster and Y. Morita, *Justus Liebigs Ann. Chem.*, 1967, **704**, 70.
47. G. W. Kabalka and H. C. Hedgecock, *J. Org. Chem.*, 1975, **40**, 1776.
48. J. A. Soderquist and C. L. Anderson, *Tetrahedron Lett.*, 1986, **27**, 3961.
49. H. C. Brown, N. G. Bhat and J. B. Campbell, *J. Org. Chem.*, 1986, **51**, 3398.
50. G. W. Kabalka and S. W. Slayden, *J. Organomet. Chem.*, 1977, **125**, 273.
51. G. Zweifel, N. L. Polston and C. C. Whitney, *J. Am. Chem. Soc.*, 1968, **90**, 6243.
52. R. W. Hoffman and S. Dresely, *Synthesis*, 1988, 103.
53. R. Köster, S. K. Arora and P. Binger, *Angew. Chem.*, 1969, **81**, 185.
54. R. C. Danheiser and A. C. Savoea, *J. Org. Chem.*, 1985, **50**, 2401.
55. A. Hassner and J. A. Soderquist, *J. Organomet. Chem.*, 1977, **131**, C1.
56. A. G. Davies and B. P. Roberts, *J. Chem. Soc. C*, 1968, 1474.
57. R. Köster, H. J. Horstschäfer and P. Binger, *Justus Liebigs Ann. Chem.*, 1968, **717**, 1.
58. W. Kliegel, *Organomet. Chem. Rev., Sect. A*, 1972, **8**, 153.
59. U. Wanagat and R. Pfeifenschneider, *Z. Anorg. Allg. Chem.*, 1958, **297**, 151.
60. W. Kliegel, *Z. Chem.*, 1969, **9**, 112.
61. T. Kudo, A. Nose and M. Hamana, *Yakugaku Zasshi*, 1975, **95**, 521.
62. H. C. Brown and M. M. Midland, *J. Chem. Soc., Chem. Commun.*, 1971, 699.
63. H. C. Brown and V. H. Dodson, *J. Am. Chem. Soc.*, 1957, **79**, 2302.
64. G. B. Sadikov, V. P. Maslennikov and G. I. Mannin, *Zh. Obshch. Khim.*, 1972, **42**, 1571.
65. G. I. Mannin, V. P. Maslennikov and Y. A. Alexandrov, *Zh. Obshch. Khim.*, 1974, **44**, 1278.
66. P. B. Brindley, J. C. Hodgson and M. J. Scotton, *J. Chem. Soc., Perkin Trans. 2*, 1979, 45.
67. A. G. Davies and B. P. Roberts, *J. Chem. Soc. B*, 1967, 17.
68. A. G. Davies and B. P. Roberts, *J. Chem. Soc., Chem. Commun.*, 1966, 298.
69. H. C. Brown, M. M. Midland and G. W. Kabalka, *J. Am. Chem. Soc.*, 1971, **93**, 1024.
70. H. C. Brown and M. M. Midland, *J. Am. Chem. Soc.*, 1971, **93**, 4078.
71. M. M. Midland and H. C. Brown, *J. Am. Chem. Soc.*, 1973, **95**, 4069.
72. R. D. Strahm and M. F. Hawthorne, *Anal. Chem.*, 1960, **32**, 530.
73. G. Zweifel and H. Arzoumanian, *J. Am. Chem. Soc.*, 1967, **89**, 291.
74. D. J. Pasto, J. Chow and S. K. Arora, *Tetrahedron Lett.*, 1967, 723.
75. A. Pelter, K. Smith, M. G. Hutchings and K. Rowe, *J. Chem. Soc., Perkin Trans. 1*, 1975, 129.
76. R. Pappo, *J. Am. Chem. Soc.*, 1959, **81**, 1010.
77. H. C. Brown and C. P. Garg, *J. Am. Chem. Soc.*, 1961, **83**, 2951, 2952.
78. J. C. Ware and T. G. Traylor, *J. Am. Chem. Soc.*, 1963, **85**, 3036.
79. R. K. Hill and A. G. Edwards, *Tetrahedron*, 1965, **21**, 1501.
80. H. C. Brown, N. R. Ayyangar and G. Zweifel, *J. Am. Chem. Soc.*, 1964, **86**, 397.
81. P. T. Lansbury and E. J. Nienhouse, *J. Chem. Soc., Chem. Commun.*, 1966, 273.
82. M. M. Bhagwat, I. Mehrotra and D. Devaprabhakara, *J. Organomet. Chem.*, 1974, **82**, C27.
83. M. M. Bhagwat, I. Mehrotra and D. Devaprabhakara, *Tetrahedron Lett.*, 1975, 167.
84. C. G. Rao, S. K. Aggarwal, I. Mehrotra and D. Devaprabhakara, *J. Organomet. Chem.*, 1979, **166**, 9.
85. M. Naruse, T. Tomita, K. Utimoto and H. Nozaki, *Tetrahedron Lett.*, 1973, 795.
86. G. Hesse and H. Witte, *Angew. Chem.*, 1963, **75**, 791.
87. J. Casanova and H. B. Kiefer, *J. Org. Chem.*, 1969, **34**, 2579.
88. G. Hesse and H. Witte, *Justus Liebigs Ann. Chem.*, 1965, **687**, 1.
89. C. G. Rao, S. U. Kulkarni and H. C. Brown, *J. Organomet. Chem.*, 1979, **172**, C20.
90. H. C. Brown, S. U. Kulkarni and C. G. Rao, *Synthesis*, 1980, 151.
91. V. V. Ramana Rao, D. Devaprabhakara and S. Chandrasekaran, *J. Organomet. Chem.*, 1978, **162**, C9.
92. H. C. Brown, S. U. Kulkarni and C. G. Rao, *Synthesis*, 1979, 702.
93. H. C. Brown, C. G. Rao and S. U. Kulkarni, *Synthesis*, 1979, 704.
94. F. Bean and J. A. Johnson, *J. Am. Chem. Soc.*, 1932, **54**, 4415.

95. A. Michaelis, *Justus Liebigs Ann. Chem.*, 1901, **315**, 19.
96. D. A. Evans, E. Vogel and J. V. J. Nelson, *J. Am. Chem. Soc.*, 1979, **101**, 6120.
97. M. M. Midland, *J. Org. Chem.*, 1980, **45**, 4214.
98. R. H. Mueller and R. M. Dipardo, *J. Chem. Soc., Chem. Commun.*, 1975, 565.
99. Y. Masuda and A. Arase, *Bull. Chem. Soc. Jpn.*, 1978, **51**, 901.
100. B. Bettman and G. E. K. Branch, *J. Am. Chem. Soc.*, 1934, **56**, 1616.
101. A. Mendoza and D. S. Matteson, *J. Organomet. Chem.*, 1978, **156**, 149.
102. H. G. Kuivila, L. Benjamin, C. Murphy, A. Price and J. Polevy, *J. Org. Chem.*, 1962, **27**, 825.
103. H. C. Brown, *U.S. Pat.* 3 439 046 (1969) (*Chem. Abstr.*, 1969, **71**, 50 273).
104. (a) D. S. Matteson and R. J. Moody, *J. Org. Chem.*, 1980, **45**, 1091; (b) G. W. Kabalka, P. P. Wadgoonkar and T. M. Shoup, *Tetrahedron Lett.*, 1989, **30**, 5103.
105. A. G. Davies, in 'Progress in Boron Chemistry', ed. H. Steinberg and A. L. McCloskey, Pergamon Press, Oxford, 1964, vol. 1, p. 265.
106. G. I. Makin, V. P. Maslennikov and V. A. Shushunov, *Zh. Obshch. Khim.*, 1972, **42**, 834; V. A. Shushunov, V. P. Maslennikov, G. I. Makin and A. V. Gorbunov, *Zh. Org. Khim.*, 1972, **8**, 1409.
107. A. G. Davies and R. B. Moodie, *J. Chem. Soc.*, 1958, 2372; A. G. Davies and R. B. Moodie, *Chem. Ind. (London)*, 1957, 1622.
108. T. Taguchi, M. Itoh and A. Suzuki, *Chem. Lett.*, 1973, 719.
109. T. Taguchi, Y. Takahashi, M. Itoh and A. Suzuki, *Chem. Lett.*, 1974, 1021.
110. Y. Takahashi, M. Tokuda, M. Itoh and A. Suzuki, *Chem. Lett.*, 1975, 523.
111. Y. Takahashi, M. Tokuda, M. Itoh and A. Suzuki, *Synthesis*, 1976, 616.
112. D. H. Geske, *J. Phys. Chem.*, 1959, **63**, 1062; 1962, **66**, 1743; W. R. Turner and P. J. Elving, *J. Phys. Chem.*, 1965, **69**, 1067.
113. W. R. Turner and P. J. Elving, *Anal. Chem.*, 1965, **37**, 207.
114. J. J. Fisch and R. J. Wilcsek, *J. Organomet. Chem.*, 1974, **71**, C21.
115. M. M. Midland, J. E. Petre, S. A. Zderic and A. Kazubski, *J. Am. Chem. Soc.*, 1982, **104**, 528.
116. H. Meerwein, G. Hintz, H. Majert and H. Sönke, *J. Prakt. Chem.*, 1936, **147**, 226.
117. B. M. Mikhailov, V. G. Kiseter and Yu. N. Bubnov, *Izv. Akad. Nauk SSSR, Ser. Khim.*, 1965, 898; B. M. Mikhailov, V. G. Kiseter and Yu. N. Bubnov, *Zh. Obshch. Khim.*, 1966, **35**, 62.
118. M. M. Midland, A. Tramontano and S. A. Zderic, *J. Organomet. Chem.*, 1977, **134**, C17.
119. M. M. Midland and A. Tramontano, *J. Org. Chem.*, 1978, **43**, 1470.
120. M. M. Midland, S. Greer, A. Tramontano and S. A. Zderic, *J. Am. Chem. Soc.*, 1979, **101**, 2352.
121. M. M. Midland and S. Greer, *Synthesis*, 1978, 845.
122. M. M. Midland, D. C. McDowell, R. L. Hatch and A. Tramontano, *J. Am. Chem. Soc.*, 1980, **102**, 867.
123. H. C. Brown and G. C. Pai, *J. Org. Chem.*, 1985, **50**, 1384; 1983, **48**, 1784.
124. M. M. Midland and J. I. McLoughlin, *J. Org. Chem.*, 1984, **49**, 1316.
125. M. M. Midland and A. Kazubski, *J. Org. Chem.*, 1982, **47**, 2814.
126. J. Chandrasekharan, P. V. Ramachandran and H. C. Brown, *J. Org. Chem.*, 1986, **51**, 3394; 1985, **50**, 5446.
127. H. C. Brown, J. V. N. Vara Prasad, A. K. Gupta and R. K. Bakshi, *J. Org. Chem.*, 1987, **52**, 310; H. C. Brown, P. K. Jadhav and M. C. Desai, *Tetrahedron*, 1984, **40**, 1325.
128. J. Rathke and R. Schaeffer, *Inorg. Chem.*, 1972, **11**, 1052.
129. C. F. Lane, *J. Organomet. Chem.*, 1971, **31**, 421.
130. A. Arase, Y. Masuda and A. Suzuki, *Bull. Chem. Soc. Jpn.*, 1974, **47**, 2511; Y. Masuda, M. Hoshi and A. Arase, *Bull. Chem. Soc. Jpn.*, 1979, **52**, 271.
131. R. C. Larock, *Intra-Sci. Chem. Rep.*, 1973, **7**, 95.
132. H. C. Brown and N. R. De Lue, *J. Organomet. Chem.*, 1977, **135**, C57.
133. A. G. Davies, S. C. W. Hook and B. P. Roberts, *J. Organomet. Chem.*, 1970, **23**, C11.
134. V. B. Jigajinni, W. E. Paget and K. Smith, *J. Chem. Res. (S)*, 1981, 376.
135. H. C. Brown and C. F. Lane, *J. Am. Chem. Soc.*, 1970, **92**, 6660.
136. G. W. Kabalka and J. W. Ferrel, *Synth. Commun.*, 1979, **9**, 443.
137. H. C. Brown and C. F. Lane, *J. Chem. Soc., Chem. Commun.*, 1971, 521.
138. D. E. Bergbreiter and D. P. Rainville, *J. Organomet. Chem.*, 1976, **121**, 19.
139. G. W. Kabalka, K. A. R. Sastry, H. C. Hsu and M. D. Hylarides, *J. Org. Chem.*, 1981, **46**, 3113.
140. J. Grotewold, E. A. Lissi and J. C. Sciano, *J. Organomet. Chem.*, 1969, **19**, 431.
141. C. F. Lane and H. C. Brown, *J. Am. Chem. Soc.*, 1970, **92**, 7212.
142. C. F. Lane and H. C. Brown, *J. Organomet. Chem.*, 1971, **26**, C51.
143. J. J. Tufariello and M. M. Hovey, *J. Am. Chem. Soc.*, 1970, **92**, 3221.
144. H. C. Brown, D. H. Bowman, S. Misumi and M. K. Unni, *J. Am. Chem. Soc.*, 1967, **89**, 4531.
145. H. C. Brown, T. Hamaoka and N. Ravindran, *J. Am. Chem. Soc.*, 1973, **95**, 6456.
146. N. R. De Lue and H. C. Brown, *Synthesis*, 1976, 114.
147. H. C. Brown, M. W. Rathke and M. Rogic, *J. Am. Chem. Soc.*, 1968, **90**, 5038.
148. C. F. Lane, *Intra-Sci. Chem. Rep.*, 1973, **7**, 133.
149. H. C. Brown, N. R. De Lue, G. W. Kabalka and H. C. Hedgecock, *J. Am. Chem. Soc.*, 1976, **98**, 1290.
150. G. W. Kabalka and E. E. Gooch, *J. Org. Chem.*, 1981, **46**, 2582.
151. G. W. Kabalka and E. E. Gooch, *J. Chem. Soc., Chem. Commun.*, 1981, 1011.
152. H. C. Brown, T. Hamoaka and R. Ravindran, *J. Am. Chem. Soc.*, 1973, **95**, 5786.
153. H. C. Brown, W. R. Heydkamp, E. Breuer and W. S. Murphy, *J. Am. Chem. Soc.*, 1964, **86**, 3565.
154. M. W. Rathke, N. Inoue, K. R. Varma and H. C. Brown, *J. Am. Chem. Soc.*, 1966, **88**, 2870.
155. N. P. Bullen, K. S. Chiheru and F. G. Thorpe, *J. Organomet. Chem.*, 1980, **195**, 147.
156. Y. Tamura, J. Minamikawa, S. Fujii and M. Ikada, *Synthesis*, 1974, 196.
157. G. W. Kabalka, K. A. R. Sastry, G. W. McCollum and H. Yoshioka, *J. Org. Chem.*, 1981, **46**, 4296; G. W. Kabalka, K. A. R. Sastry, G. W. McCollum and C. A. Lane, *J. Chem. Soc., Chem. Commun.*, 1982, 62.
158. L. Verbit and P. J. Heffron, *J. Org. Chem.*, 1967, **32**, 3199.

159. H. C. Brown, K. W. Kim, T. E. Cole and B. Singaram, *J. Am. Chem. Soc.*, 1986, **108**, 6761.
160. A. N. Nesmeyanov, V. A. Sazonova, A. V. Gerasimenko and V. G. Medvedeva, *Izv. Akad. Nauk SSSR, Otd. Khim. Nauk*, 1962, 2073.
161. A. N. Nesmeyanov, V. A. Sazanova and V. N. Drozd, *Chem. Ber.*, 1960, **93**, 2717.
162. A. Suzuki, S. Sono, M. Itoh, H. C. Brown and M. M. Midland, *J. Am. Chem. Soc.*, 1971, **93**, 4329.
163. H. C. Brown, M. M. Midland and A. B. Levy, *J. Am. Chem. Soc.*, 1972, **94**, 3662.
164. H. C. Brown, M. M. Midland and A. B. Levy, *J. Am. Chem. Soc.*, 1973, **95**, 2394.
165. A. B. Levy and H. C. Brown, *J. Am. Chem. Soc.*, 1973, **95**, 4067.
166. R. H. Mueller, *Tetrahedron Lett.*, 1976, 2925.
167. V. B. Jigajinni, A. Pelter and K. Smith, *Tetrahedron Lett.*, 1978, 181.
168. A. Suzuki, M. Ishidoya and M. Tabata, *Synthesis*, 1976, 687.
169. H. C. Brown and M. M. Midland, *J. Am. Chem. Soc.*, 1971, **93**, 3291.
170. A. Arase, Y. Masuda and A. Suzuki, *Bull. Chem. Soc. Jpn.*, 1974, **47**, 2511.
171. A. Arase and Y. Masuda, *Chem. Lett.*, 1976, 785, 1115.
172. S. Raucher, M. R. Hansen and M. A. Colter, *J. Org. Chem.*, 1978, **43**, 4885.

4.2
Oxidation of Carbon–Metal Bonds

WILLIAM KITCHING
University of Queensland, St. Lucia, Australia

4.2.1 INTRODUCTION

Organometallic compounds based on very electropositive elements such as lithium and magnesium have long served as carbanion sources, but far less frequently in tactical oxidations. Oxidation of Grignard reagents and organolithium compounds to form alcohols has been employed, particularly when the direct halide to alcohol conversion is unattractive, for example because of β-branching.[1] Although organometallics may be formed in a wide variety of ways, their value in synthesis depends on the ease with which the carbon–metal bond may be transformed into some other desired functionality. Thus, in principle there should exist a range of synthetic applications for the oxidation of relatively weak carbon–metal bonds provided some (or most) of the following criteria are met: (a) the metal-containing moiety can be introduced easily and preferably in a number of ways which exhibit high levels of regio- and stereocontrol; (b) the method of introduction shows good functional group tolerance and hence minimizes protection–deprotection sequences; (c) the resulting organometallic is stable enough to withstand standard manipulations if necessary; (d) subsequent oxidation proceeds under suitably mild conditions with reagents that are chemoselective for the carbon–metal system; and (e) the oxidation pathway exhibits a high level of regio- and even stereo-control and is synthetically efficient.

The carbon–metal systems considered appropriate for this section (and not covered elsewhere in this series) are carbon–tin, carbon–mercury and, to a restricted extent, carbon–palladium bonds. Although organomercurials and organostannanes have been studied for over a century, the former continue to attract attention because of their controlled formation by the Markovnikov oxymercuration reaction (OM) and its variants,[2] and the latter because of newer and regiospecific methods for forming C—Sn bonds and oxidatively cleaving them. In both cases, significant examples of synthetically useful oxidations have appeared in recent years and form the substance of this chapter. Organothallium compounds, sometimes usefully compared with the isoelectronic organomercury systems, also exhibit synthetic utility but in general are less stable, less generally acquired and more prone than the mercurials to rearranging (carbocationic), dethallation reactions *etc.* The synthetic aspects of organothallium chemistry have been reviewed recently.[3]

4.2.2 OXIDATION OF CARBON–TIN BONDS

4.2.2.1 Introduction

Although organotin compounds (*i.e.* with at least one carbon–tin bond) date back to 1849, only relatively recently has there been general recognition of the many roles organostannanes have in synthesis. For example, organostannanes find important applications in the free radical based generation of C—H and C—C bonds, transmetallation reactions providing a range of organolithium reagents, and a number of transition metal catalyzed reactions. Much of this chemistry has been nicely organized and discussed elsewhere.[4] A newer and expanding role is related to their entry into oxidation reactions and this aspect is discussed below.

Most organostannanes are relatively stable liquids or solids, which are easily handled in air and are, by and large, insensitive to moisture. The toxicity of organostannanes has been extensively investigated, and as a rule these compounds should be regarded as hazardous and use of gloves and operating in an efficient fume hood are strongly recommended.[5] Methyl- or ethyl-tin derivatives have the highest mammalian toxicity, whereas *n*-butyl derivatives are significantly less toxic. This factor, coupled with the commercial availability of the cheaper *n*-butyl derivatives, indicates that unless methyl derivatives have special advantages, the *n*-butyl derivatives should be preferred. Reactions involving trialkylstannanes (R_3SnR'; R = Me, Bu^n) produce one equivalent of 'trialkyltin salt' (R_3SnX), which is generally removed by thorough washing with aqueous fluoride solution, forming the sparingly soluble R_3SnF derivatives. This procedure works well for Me_3SnX, but less well for $Bu^n{}_3SnX$. Washing the organic phase with aqueous ammonia may be advantageous in those cases.

Two general categories of oxidation may be recognized: (a) 'direct' oxidation of the C—Sn bond to C—O groupings as in alcohols or ketones; and (b) 'oxidative fragmentations' of certain specifically functionalized organostannanes, usually under free radical conditions. With respect to cleavage of the C—Sn bond, the large covalent radius of tin (0.14 nm), the long C—Sn bond (0.22 nm), the relatively low mean C—Sn bond dissociation energy (~50 kcal mol^{-1}; 1 kcal = 4.18 kJ) and the polarizable but essentially covalent C—Sn bond should be noted. Thus it is no surprise that the C—Sn bond exhibits convenient reactivity under both free radical and polar conditions.

4.2.2.2 Unactivated Carbon–Tin Bonds

That unactivated carbon–tin bonds could be oxidized directly in a potentially useful synthetic way was demonstrated in 1964. Symmetrical tetraalkylstannanes reacted with chromic anhydride (CrO_3) in acetic acid to yield aldehydes and acids depending on the conditions (equation 1). Using a large excess of oxidant (12-fold excess) and long reaction times (360 h at 20 °C) a near quantitative yield of *n*-butanoic acid was obtained from $Bu^n{}_4Sn$.[6]

$$Bu^n{}_4Sn \xrightarrow[AcOH]{CrO_3} PrCHO + PrCO_2H + \text{other products} \qquad (1)$$

Development of this approach to a synthetically useful level was reported by Still, who employed large excesses of the CrO_3–pyridine oxidant.[7] The trimethylstannyl group was introduced either by bromide displacement or conjugate addition to enones (using Me_3SnLi) to provide a range of secondary or tertiary stannanes. Oxidation with $CrO_3{\cdot}2py$ (py ≡ pyridine) led to ketones, ketols or alcohols depending on the system (Scheme 1).

In the case of tertiary stannanes, oxidation leads to mixtures of alcohols and alkenes and some allylic oxidation, although 1-adamantyltrimethylstannane, incapable of elimination, provided the tertiary alcohol in good yield (Scheme 2).

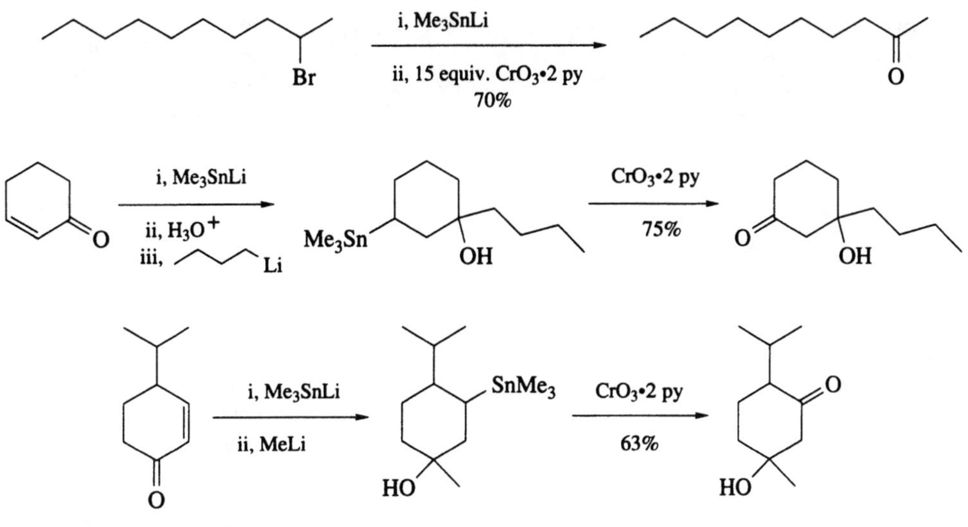

i, Me_3SnLi; ii, 15 equiv. CrO_3•2 py; 70%

i, Me_3SnLi; ii, H_3O^+; iii, Li; Me_3Sn; OH; CrO_3•2 py; 75%; O; OH

i, Me_3SnLi; ii, MeLi; $SnMe_3$; HO; CrO_3•2 py; 63%; O; HO

Scheme 1

$SnMe_3$; [O]; OH; 50%; +; 50%

$SnMe_3$; [O]; + other products

Br; Me_3SnLi; $SnMe_3$; [O]; 74%; OH

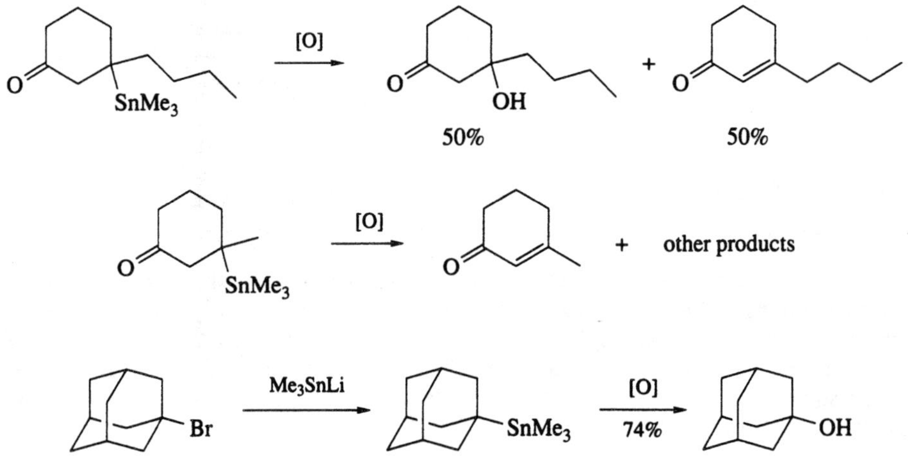

Scheme 2

The examples shown in Scheme 2 illustrate: (a) a method for $R_3C—Br \rightarrow R_2C{=}O$ or $R_3C—OH$; and (b) dialkylative enone transposition as shown in essence in equation (2). The sequence $R_2CH—Cl \rightarrow R_2CH—Sn \rightarrow R_2C{=}O$ was crucial in a synthesis of the troponoid nezukone (**1**), as other methods of functionalizing the $R_2CH—Cl$ system in this ring failed (equation 3).[8] The dialkylative enone transposition (equation 2) was illustrated by a short synthesis of dihydrojasmone (equation 4).[7]

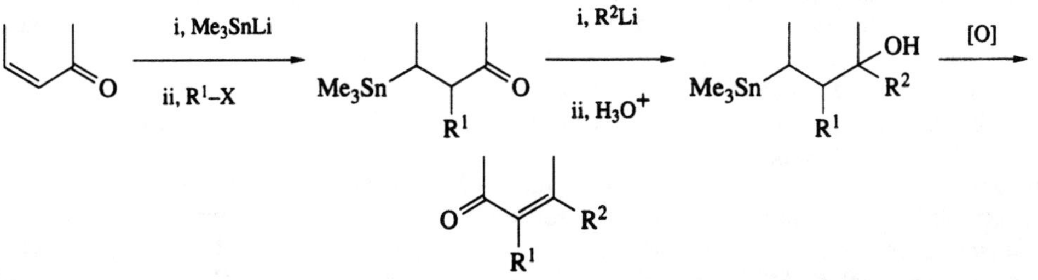

i, Me_3SnLi; ii, R^1–X; Me_3Sn; O; R^1; i, R^2Li; ii, H_3O^+; Me_3Sn; OH; R^2; R^1; [O]; O; R^2; R^1 (2)

Oxidation of dimethylhalotin groups to hydroxy groups with retention of configuration can be achieved under relatively mild conditions using alkaline hydrogen peroxide,[9] a procedure of importance for the oxidation of carbon–silicon bonds (see Volume 7, Chapter 4.3). Thus the iodine(III)-mediated cleavage of Me_3SnR to Me_2SnClR, followed by oxidation, provides an efficient route for $R_2CH—SnMe_3$

(3)

(4)

→ R_2CH—OH with configurational retention (equation 5). This finding may have applications in the synthesis of various hydroxylated natural products based on organotin intermediates.

(5)

4.2.2.3 Allylic Stannanes

In contrast to the oxidation of unactivated stannanes, allylic derivatives are expected to be more reactive, and mild conditions and oxidizing agents can be employed successfully. A particularly useful reaction involves the conversion of an allylstannane to the allylic alcohol, and the commercially available, solid, easily handled *m*-chloroperbenzoic acid (MCPBA) is the reagent of choice for oxidations employing organic solvents such as dichloromethane. Under these conditions epoxystannanes cannot be isolated and allylic alcohols form directly (equation 6).[10,11]

(6)

However, given the periodicity of the Group XIV congeners, and that epoxysilanes can be isolated from reactions of allylsilanes and MCPBA,[12,13] it is reasonable that epoxystannanes are intermediates in the overall conversion. Just as the 'β-silicon effect' may be regarded as dominating the chemistry of allylsilanes with electrophilic reagents,[14] the influence of the SnR_3 group in allylstannanes is similarly dominant. Thus epoxidation of allylstannanes may be viewed as electrophilic addition to the double bond (probably *anti* to the tin moiety), followed by rapid acid-promoted ring opening and destannylation. This yields the allyl alcohol in a regiospecific manner with allylic transposition.

It is possible to write a cyclic destannylation mechanism for allyl alcohol formation as shown in Scheme 3, but this implies hydroxylation *syn* to the departing tin group, and there is evidence that this concerted route (**2**) is not dominant (Scheme 3).[15]

There thus exists a preference for *anti* (or *antara*) hydroxylation in these cyclohexenylstannanes, where electrophilic substitutions are known to proceed faithfully with allylic rearrangement.[16,17] A more likely pathway is shown in Scheme 4, which is supported by results with optically active allylsilanes,[18] which require *anti* attack by MCPBA on the silane conformation maximizing C—Si σ–π interaction.

This chemistry forms the basis of a general method for 1,3-hydroxy transposition in allylic alcohols (equation 7).[19] The starting alcohol is converted by 3,3-sigmatropic rearrangement of the *O*-allyl-*S*-methyldithiocarbonate followed by hydrostannolysis to the allylic stannane, which is oxidized by MCPBA in a completely regiospecific manner. A similar sequence has been reported for allylsilanes.[20]

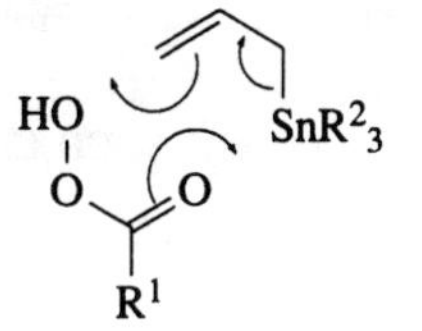

(2)

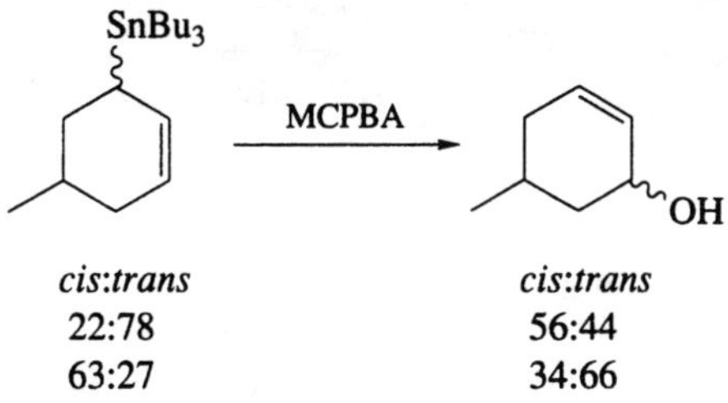

cis:trans 22:78, 63:27 → MCPBA → *cis:trans* 56:44, 34:66

Scheme 3

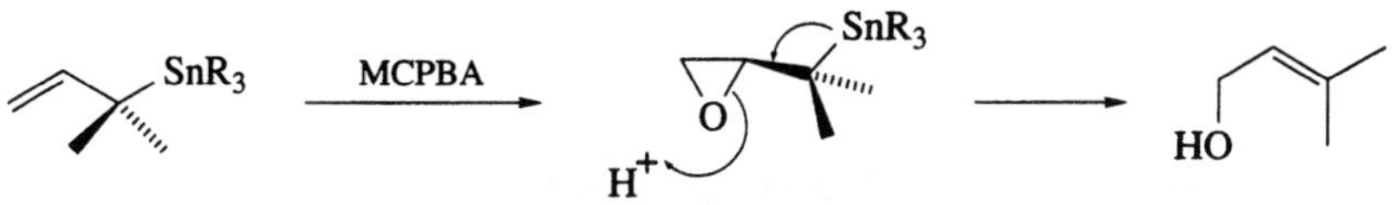

Scheme 4

Some typical results are shown in Scheme 5, and other oxidants such as LTA and bisacetoxyphenyliodine were unsatisfactory.

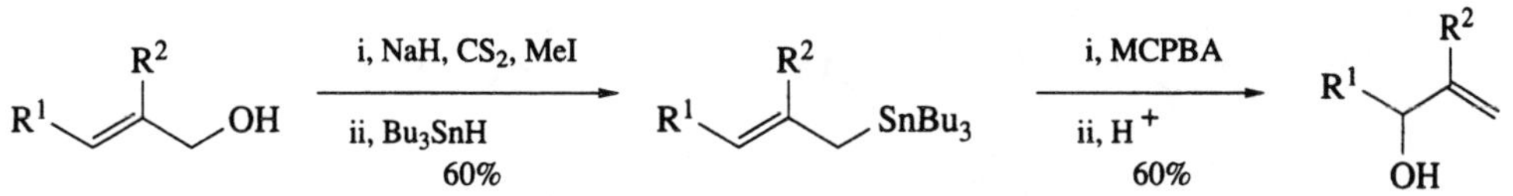

i, NaH, CS_2, MeI; ii, Bu_3SnH 60% — i, MCPBA; ii, H^+ 60% (7)

61%

66%

59%

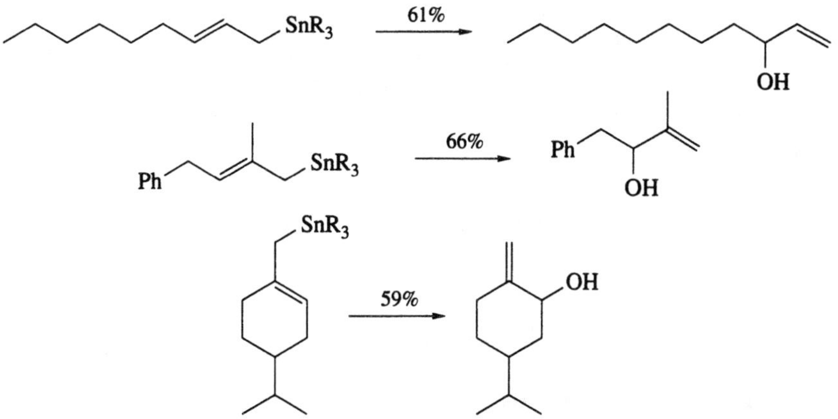

Scheme 5

MCPBA oxidation of an allylic stannane is a key step in the overall conversion of an α,β-unsaturated aldehyde to an (*E*)-β-bromo-α-enone, as shown in equation (8).[21]

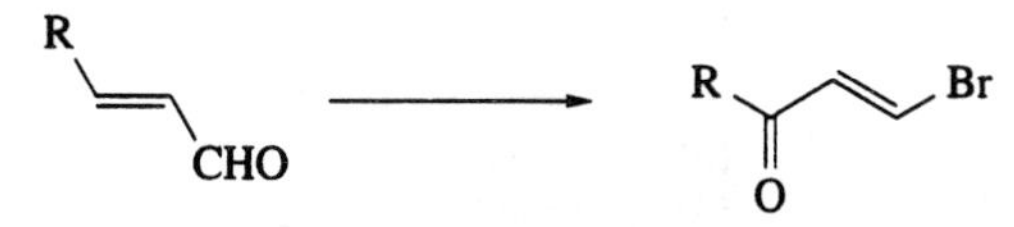

(8)

Addition of $Bu^n{}_3SnLi$ to the α,β-enal provides the sensitive α-hydroxyallylstannane, which is converted directly to the α-bromoallylstannane without allylic rearrangement. Subsequent oxidative destannylation with MCPBA is accompanied by allylic rearrangement, and further oxidation yields the β-bromoenone (Scheme 6).

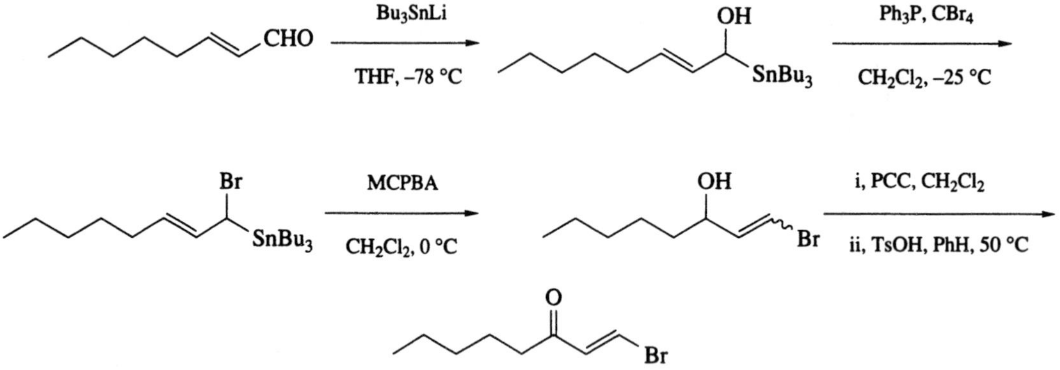

Scheme 6

Replacement of a trialkyltin group with the acetoxy group using LTA in CH_2Cl_2 proceeds with acceptable yields for *O*-activated, allylic and vinylic C—Sn bonds. These sequences may involve intermediate organolead triacetates ($RPbOAc_3$), which demetallate to yield carbonium ions (Scheme 7).[22]

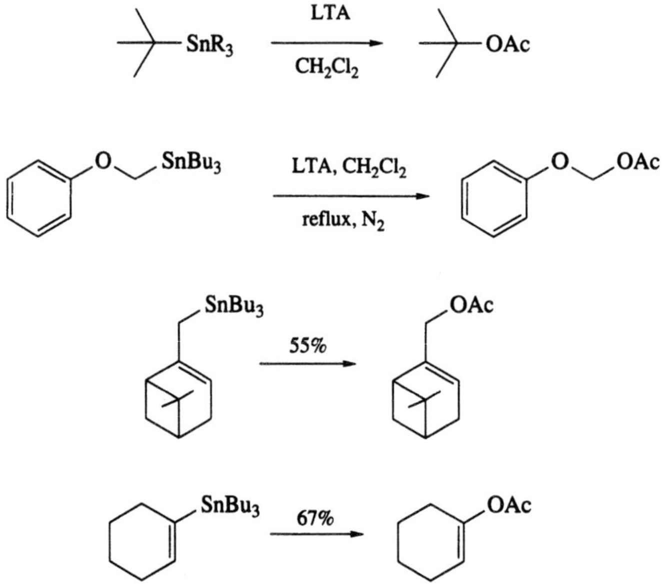

Scheme 7

Heterocyclic stannanes were also employed and yielded 5-acetoxyfuran-2(5*H*)-ones and 5-acetoxy-1-methyl-3-pyrrolin-2-one from 2-stannylfurans and 2-stannyl-*N*-methylpyrrole, respectively (Scheme 8).[22a] In the furan system, it appears initial oxidation of the stannane provides the 2-furyl acetate, which is known to undergo further oxidation with LTA (Scheme 8).[23]

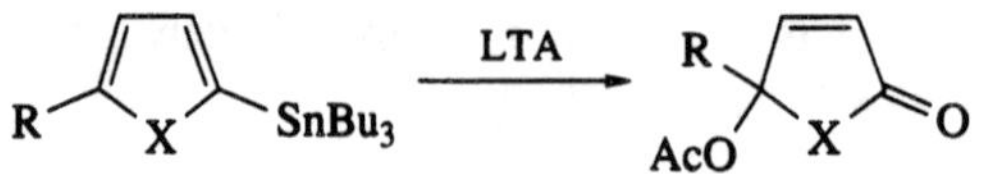

	R = H	X = O	60%
(3)	R = Me	X = O	91%
(4)	R = $PhCH_2CH_2$	X = O	55%
	R = H	X = NMe	20%

Scheme 8

Compound (**3**; Scheme 9) on elimination of AcOH provided (**5**), which dimerized to the natural product anemonin (**6**). This approach provides access to 4-ylidenebutenolides, an arrangement frequently found in biologically active natural products (Scheme 9).

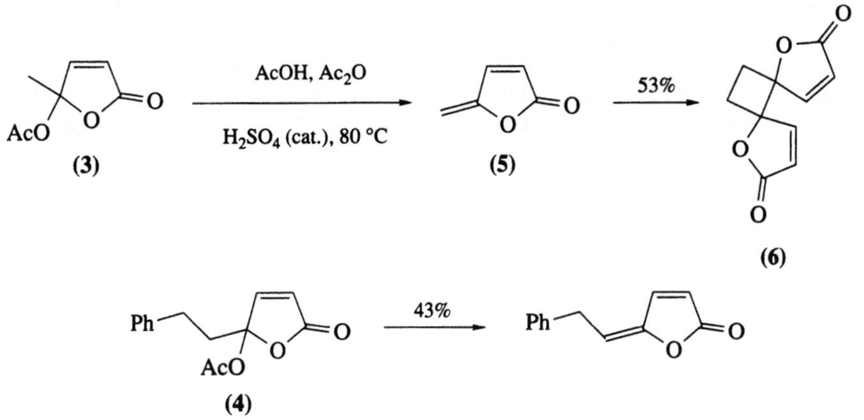
AcOH, Ac_2O
H_2SO_4 (cat.), 80 °C
53%
43%
AcO
Ph
(3)
(5)
(6)
(4)

Scheme 9

The β-stannyl silylenol ether is a useful protection device for α,β-enones, as the ethers are relatively unreactive towards most nucleophiles and are reconverted to the enone on mild oxidation. This form of protection was developed and employed in the acquisition of a crucial disubstituted cyclohex-2-enone (**7**) required in the synthesis of (±)-periplanone-B, a sex pheromone of the American cockroach.[24] Similarly, this sequence was successful in effecting (*Z,E*) to (*E,E*) isomerization of isoacoragermacrone (**8**) to acoragermacrone (**9**), when other methods (*e.g.* photoisomerization) failed (Scheme 10).[25]

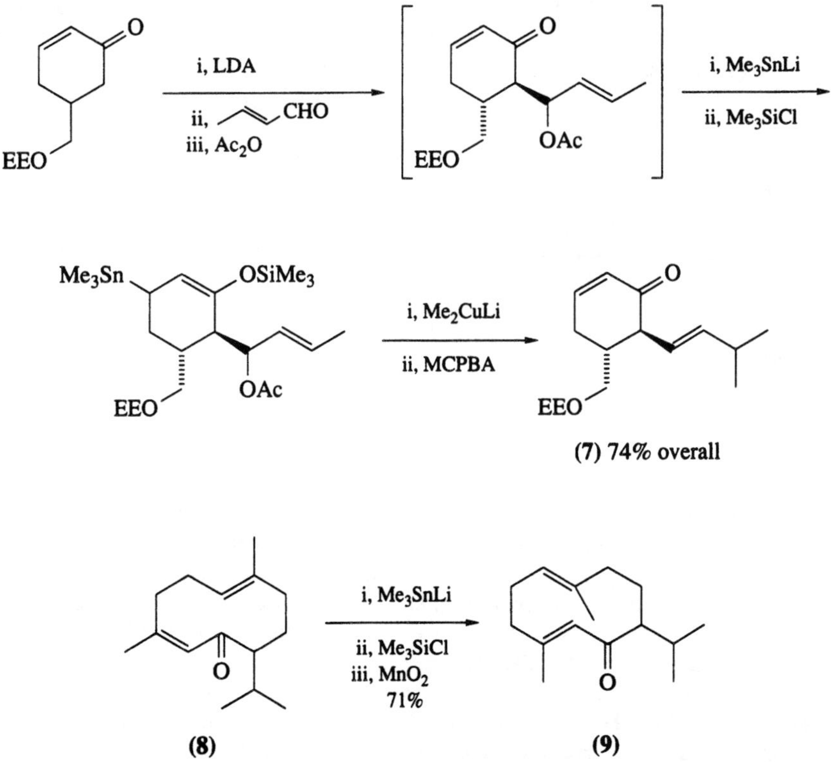
i, LDA
ii, CHO
iii, Ac_2O
EEO
OAc
i, Me_3SnLi
ii, Me_3SiCl
Me_3Sn
$OSiMe_3$
i, Me_2CuLi
ii, MCPBA
(7) 74% overall
i, Me_3SnLi
ii, Me_3SiCl
iii, MnO_2
71%
(8)
(9)

Scheme 10

4.2.2.4 Vinylstannanes

In 1973, it was demonstrated that 1,2-epoxystannanes, produced from vinylstannanes and MCPBA, could be isolated and characterized,[10] in comparison with 2,3-epoxystannanes (from allylstannanes), which are extremely reactive and have not been isolated (see Section 4.2.2.3). Subsequently, useful applications of 1,2-epoxystannanes have been reported, including the internal alkyne → ketone conversion, in the carbapenem and carbacephem (β-lactam antibiotic) skeletons. Ketone (**10**) should be of value in the construction of the biologically interesting 1-carbapen-2-ene ring system. Synthesis of ketoacetates of potential use in the carbacephem system (*e.g.* **11** and **12**) was also achieved by similar sequences shown in Scheme 11.[26]

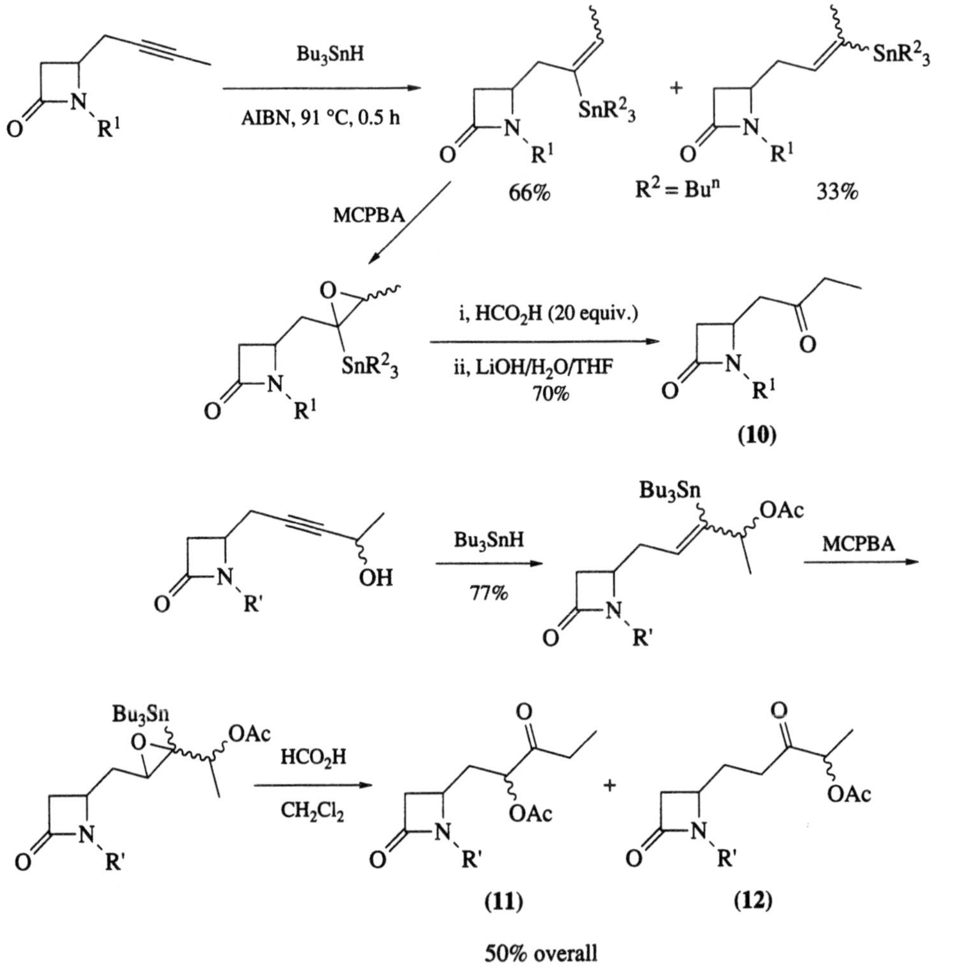

Bu3SnH; AIBN, 91 °C, 0.5 h; SnR23; 66%; R^2 = Bu[n]; 33%; MCPBA; i, HCO2H (20 equiv.); ii, LiOH/H2O/THF; 70%; (10); Bu3SnH; 77%; OAc; MCPBA; HCO2H; CH2Cl2; (11); (12); 50% overall

Scheme 11

Oxidation of vinylstannanes with LTA is a critical aspect of the introduction of angular ethynyl groups, for which the Cu[I]-based methodology, so useful for angular vinyl groups, is precluded because of the efficient binding of ethynyl ligands by copper. Treatment of a range of alkenylstannanes with LTA in acetonitrile resulted in conversion to terminal alkynes,[27] and this novel transformation was rationalized as involving a 'Pb for Sn' substitution to produce (**13**) and (cationic) deplumbation. The approach is shown in Scheme 12 in general terms and then for the bicyclic ketone (**14**).

α-Alkenylation of β-dicarbonyl compounds has been achieved in a similar reaction by generation of what are presumed to be 'alk-1-enyllead triacetates', by treating dialk-1-enylmercurials or alk-1-enyltrialkylstannanes with LTA in $CHCl_3$ in the presence of the β-dicarbonyl compounds.[28]

It should be noted from Schemes 12 and 13 that the mode of decomposition of the presumed 'alk-1-enyllead triacetate' varies depending on the circumstances of its generation, *i.e.* from a mercury or stannane precursor. Similarly, alk-1-ynyltrialkylstannanes (**15**) are oxidized by LTA in $CHCl_3$ to form a species capable of α-alk-1-ynylation of β-dicarbonyl compounds (Scheme 13).[29]

The chemistry outlined in Schemes 12 and 13 has been developed to provide a mild method for the conversion of aldehydes to alkynes *via* vinylstannanes as intermediates, and moderately complex

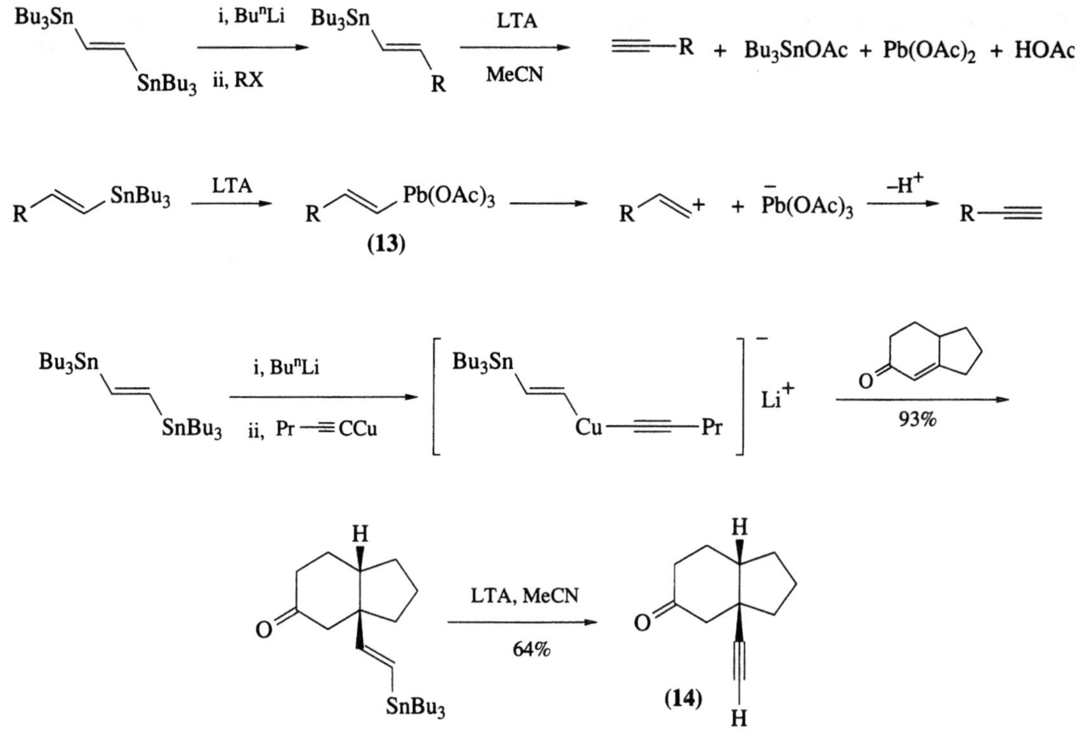

Scheme 12

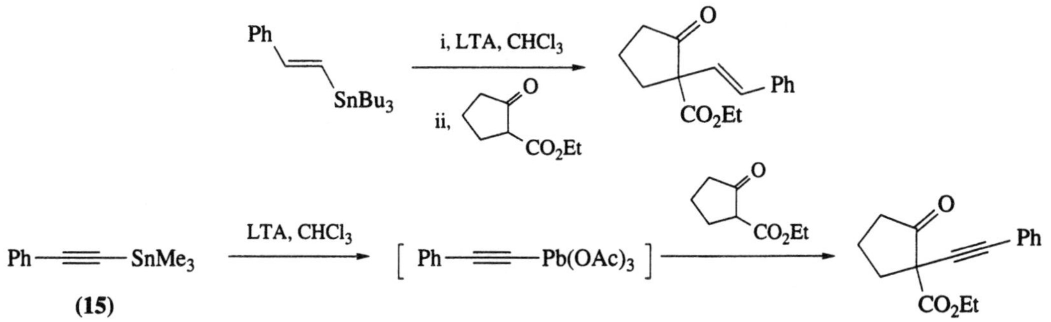

Scheme 13

systems are amenable to the general procedure as shown in Scheme 14.[30] This aldehyde → alkyne transformation was applied in the synthesis of 9(O)-thia-Δ^6-PGI_1 (as shown in Scheme 15)[30] and a PGA_2 ethynyl derivative has been obtained using this method of ethynyl group introduction.[31] It should be mentioned that the intermediate alkenylstannanes provide a route to alkenyllithium reagents from the starting aldehydes.[4]

4.2.2.5 γ-Trialkylstannyl Alcohols; Oxidative 1,4-Fragmentation

γ-Stannyl alcohols display considerable potential for use in organic synthesis, and mention has been made of the chromic anhydride oxidation of certain such alcohols and its role in dialkylative enone transposition (Section 4.2.2.2, equation 2).[7] Two other reactions which have been developed recently into attractive sequences are: (a) 1,3-eliminative cyclization of γ-stannyl alcohols to cyclopropanes;[4,32,33] and (b) 1,4-fragmentation of (cyclic) γ-stannyl alcohols to yield an unsaturated carbonyl compound, which proceeds in a stereospecific manner. 1,4-Fragmentation under Grob conditions utilizes electron-attracting groups, whereas cation or radical induced fragmentations are rarer. Thus oxidation of γ-stannyl alcohols with a hypervalent organoiodine compound (in the presence of DCC)[34] proceeds differently from the chromic anhydride reaction (equation 9).[7]

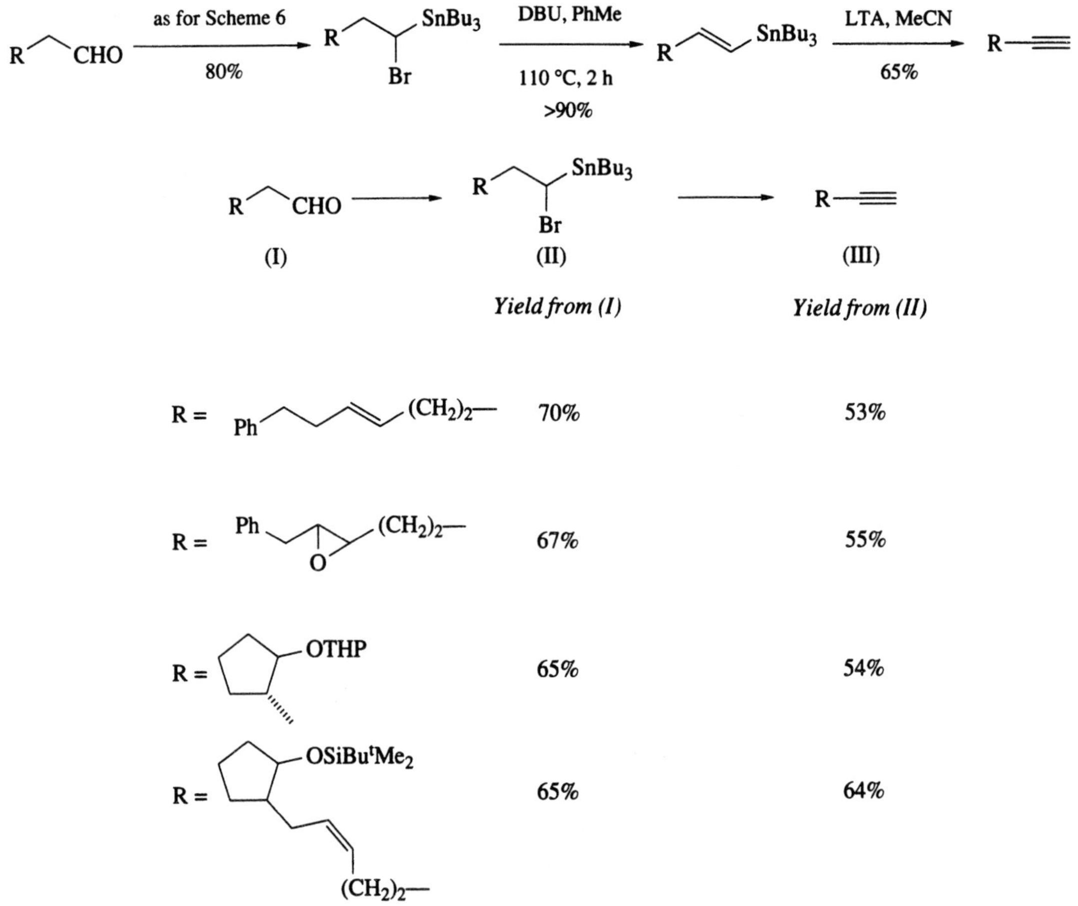

as for Scheme 6 80% DBU, PhMe 110 °C, 2 h >90% LTA, MeCN 65%

R =	Yield from (I)	Yield from (II)
Ph–CH=CH–$(CH_2)_2$—	70%	53%
Ph epoxide $(CH_2)_2$—	67%	55%
cyclopentyl OTHP	65%	54%
cyclopentyl $OSiBu^tMe_2$ $(CH_2)_2$—	65%	64%

Scheme 14

54% 63% 9(O)-Thia-Δ^6-PGI_1

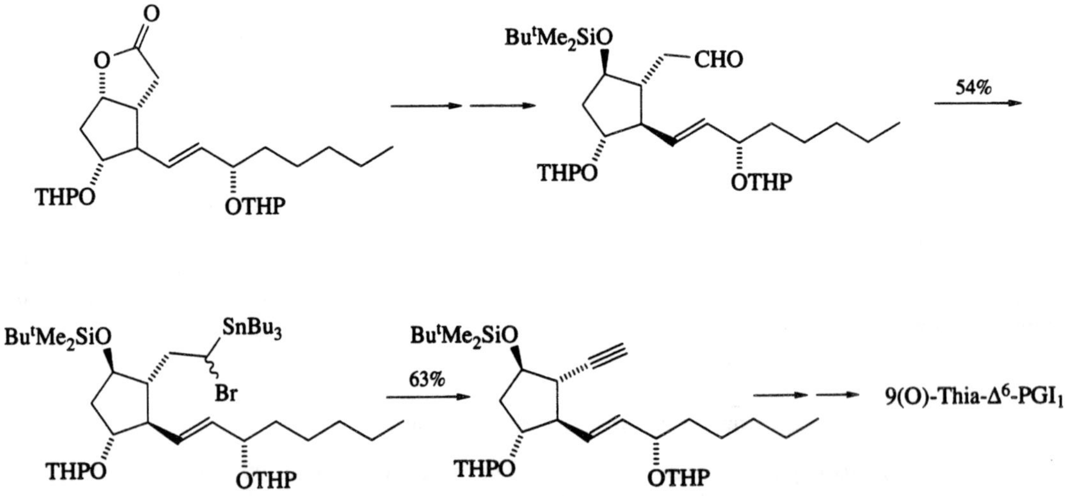

Scheme 15

The required organostannanes are accessible by conjugate addition of trialkyltinlithium reagents to cyclic enones,[7] followed by treatment with the appropriate RLi or RMgX reagent, as shown in Scheme 16.

Fragmentation of the isomeric mixtures of γ-stannyl alcohols was achieved by adding a preformed solution (stir for 1 h at room temperature) of $BF_3{\cdot}Et_2O$ and DCC to the stannane and iodosylbenzene at 0 °C, under a N_2 atmosphere. Fragmentation of five-, six- and seven-membered ring stannanes proceeded

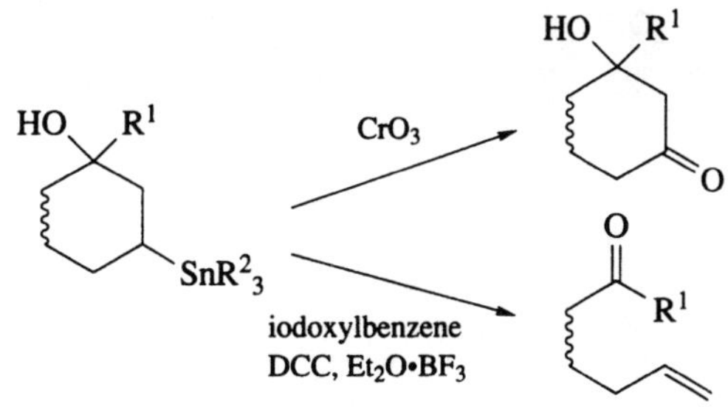

(9)

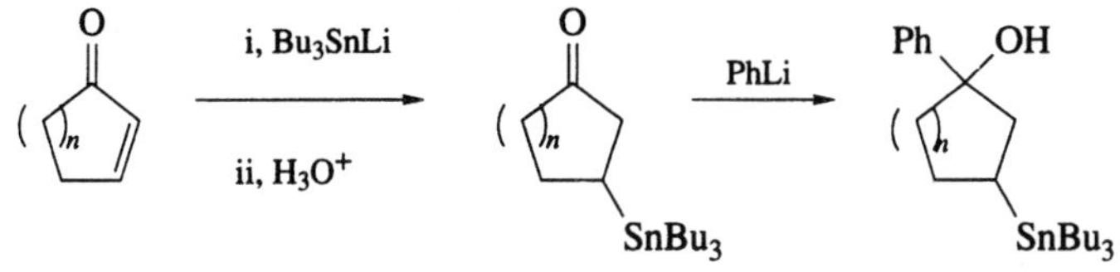

Scheme 16

efficiently, and secondary alcohols (from $NaBH_4$ reduction of the stannyl ketones) provided the unstable enals. These results are listed in Table 1.

Table 1 Iodine(III)-mediated 1,4-Fragmentation of γ-Stannyl Alcohols[a]

System	*Reaction time* (h)	*Product*	*Yield* (%)
n = 1, R = Ph	5	*n* = 1	63
n = 2, R = Ph	4	*n* = 2	81
n = 3, R = Ph	2.5	*n* = 3	86
n = 2, R = H	2	*n* = 2	74
n = 3, R = H	3	*n* = 3	55

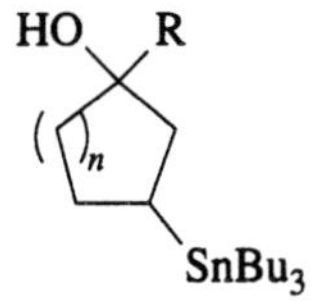

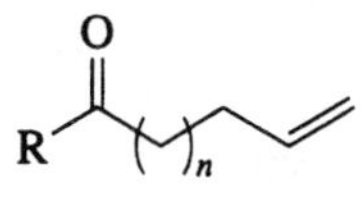

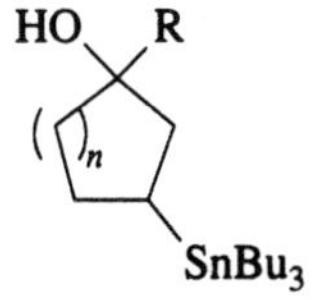

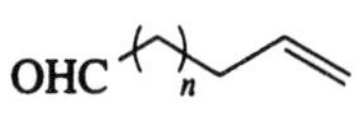

[a] Conditions: add preformed solution (stir for 1 h at room temperature) of $BF_3{\cdot}OEt_2$ and DCC to the stannane and iodosylbenzene at 0 °C with N_2 atmosphere.

Addition of Bu_3SnLi to cyclohex-2-enone followed by enolate trapping with *n*-decyl iodide proceeded with high diastereoselectivity to provide the 2,3-*trans*-stannyl ketone (**16**), which could be equilibrated with the *cis* diastereomer (**17**) upon treatment with base.[35] LAH reduction, followed by separation of the diastereomers, afforded samples of the 2-alkyl-3-stannylcyclohexanols as shown in Scheme 17. Iodine(III)-mediated fragmentation was shown to proceed in a stereospecific *anti* manner, with either of the *trans*-2,3-cyclohexanols affording (*E*)-enal (**18**), and the *cis*-2,3-alcohol the (*Z*)-enal (**19**; Scheme 17). Enal (**18**) was then utilized in a stereoselective synthesis of the mosquito pheromone, *erythro*-6-acetoxyhexadecan-5-olide (**20**; Scheme 18).[35]

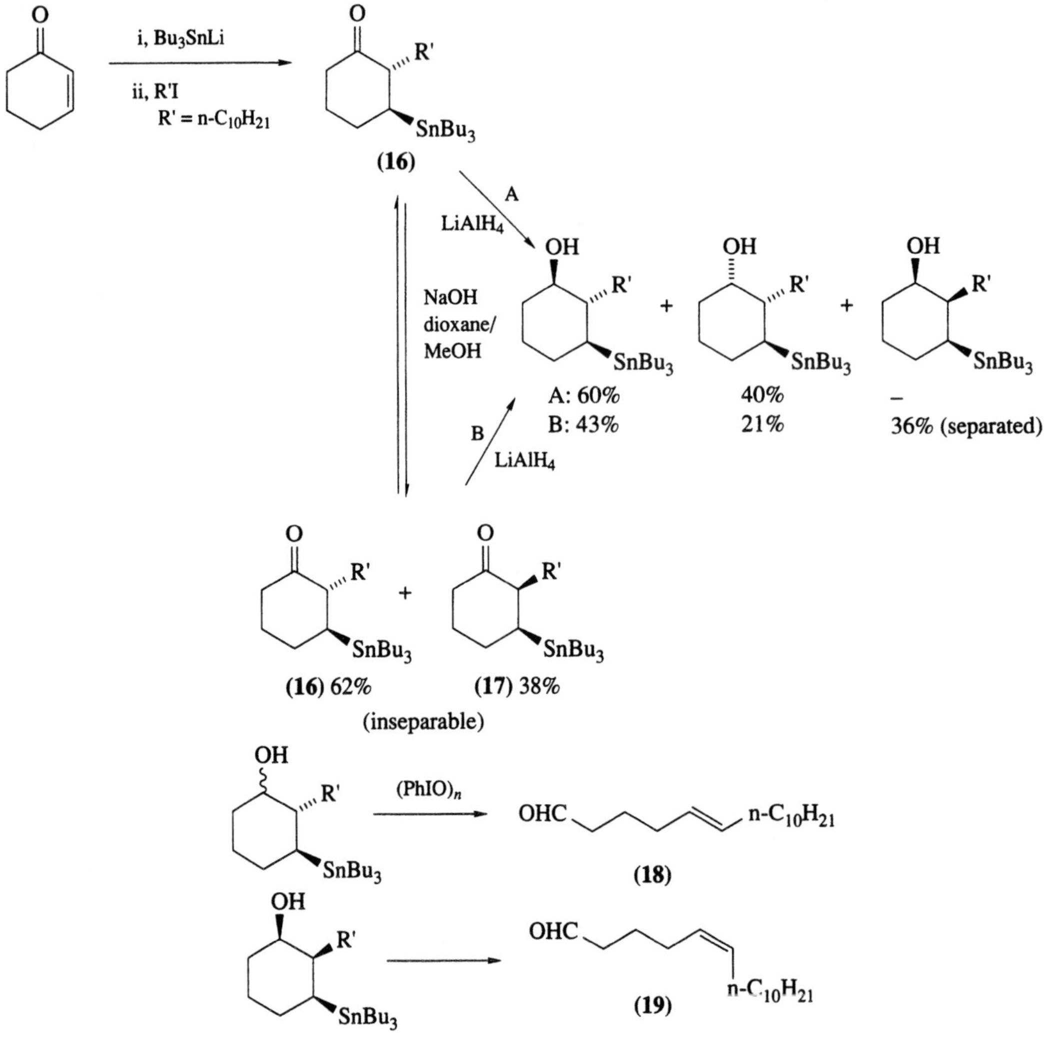

i, Bu_3SnLi
ii, R'I
R' = n-$C_{10}H_{21}$

(16)

A
$LiAlH_4$

NaOH
dioxane/
MeOH

A: 60% 40% –
B: 43% 21% 36% (separated)

B
$LiAlH_4$

(16) 62% (17) 38%
(inseparable)

$(PhIO)_n$

(18)

(19)

Scheme 17

(18) NaOCl, r.t., 17 h, 84%

i, MCPBA
ii, reflux, PhMe, 24 h

Ac_2O, pyridine
r.t., 48 h

(20) 40% from acid

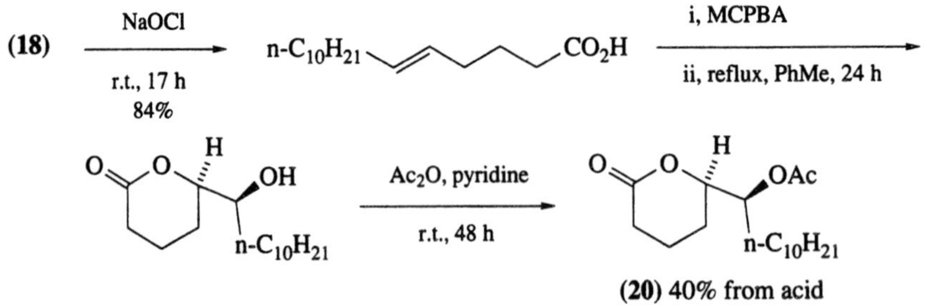

Scheme 18

This iodine(III) oxidative fragmentation has been exploited to generate unsaturated medium ring lactones in good yields and with complete control of double bond stereochemistry as shown in Scheme 19.[36] Use of diacetoxyiodobenzene (DAIB) instead of iodosylbenzene–BF_3–Et_2O–DCC resulted in much enhanced yields (~80%).

LTA in refluxing benzene will also effect fragmentation of γ-stannyl alcohols to (*E*)- and (*Z*)-ketoalkenes in a stereospecific manner depending on the *cis* or *trans* nature of the 2,3-groups, and, for overall synthetic ease, the use of LTA may be more attractive.[37] The fragmentation may be viewed as a trialkyltin-triggered radical cleavage as shown in Scheme 20. Under the conditions for this cleavage of the γ-stannyl alcohols, the corresponding silanes do not react, but there are methods for such silane re-

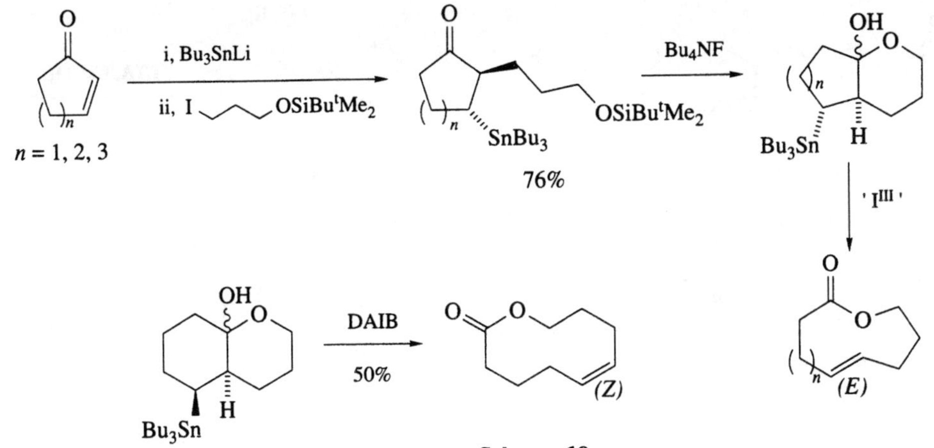

Scheme 19

actions.[38] The LTA-based cleavage has been elegantly applied to part of the sequence leading to brefeldin A seco acid (**21**)[39] and has found application also in the synthesis of dienones[40] capable of ready elaboration to various spiroacetal systems, for example (*E*,*E*)-2-ethyl-8-methyl-1,7-dioxaspiro[5.5]undecane (**22**; Scheme 21).

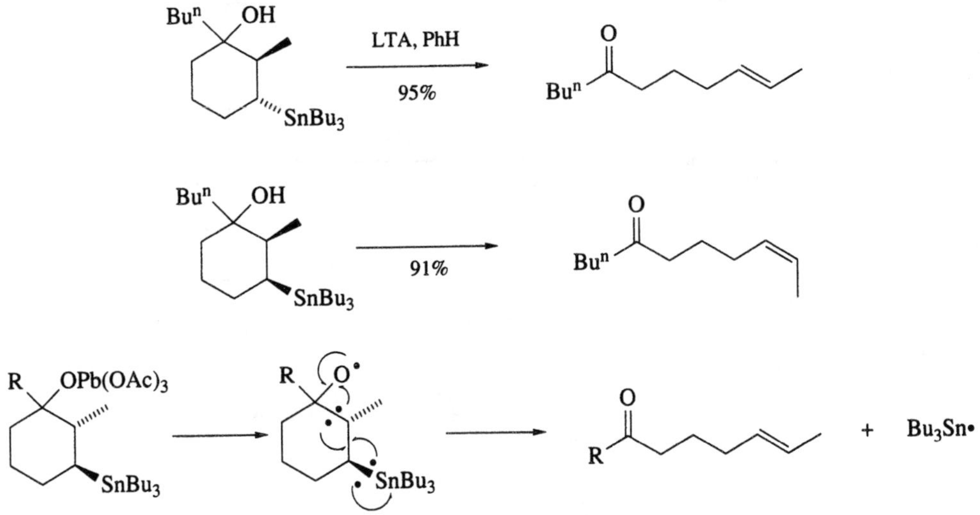

Scheme 20

Oxidative fragmentation (and oxidation) of carbon–tin bonds are part of an overall sequence based on Michael–Michael ring closure reactions (MIMIRC) which represent extraordinarily easy, high yielding, one-pot, four-component coupling reactions, which may lead finally to functionalized cyclohexenes, cyclohexanols, cyclodecenones and aromatic systems, as shown in Scheme 22.[41] Quenching of the trialkyltin enolate with 2 equiv. of methyl acrylate or methyl α-bromoacrylate provides ester enolates which spontaneously execute intramolecular aldol cyclizations to form polyfunctionalized cyclohexanols (**23**) and (**24**), each as a mixture of two diastereomers only. LTA cleavage (of each separated diastereomer) resulted in a stereodefined richly functionalized cyclodecene system (**25**) related to the germacrane class of sesquiterpenes, whereas oxidation–dehydrobromination formed regiospecifically tetrasubstituted aromatics (**26**; Scheme 22).

This general strategy was extended to include five- and seven-membered ring α,β-enones, which also underwent 'one-pot' Michael reactions with Bu_3SnLi, followed by enolate 1,4-addition to vinyl ketones and aldol reactions with aldehydes to provide cyclic hemiketals (MIMIARC).[42] LTA fragmentation results in four-atom enlarged, vicinally disubstituted, regio- and stereo-specifically unsaturated macrolides as summarized in Scheme 23. Yields of chromatographically pure macrolides were in the 30–47% range. Exclusive formation of the (*E*)-alkenolides (Scheme 23) is due to the *trans* nature of bonds a and b in the

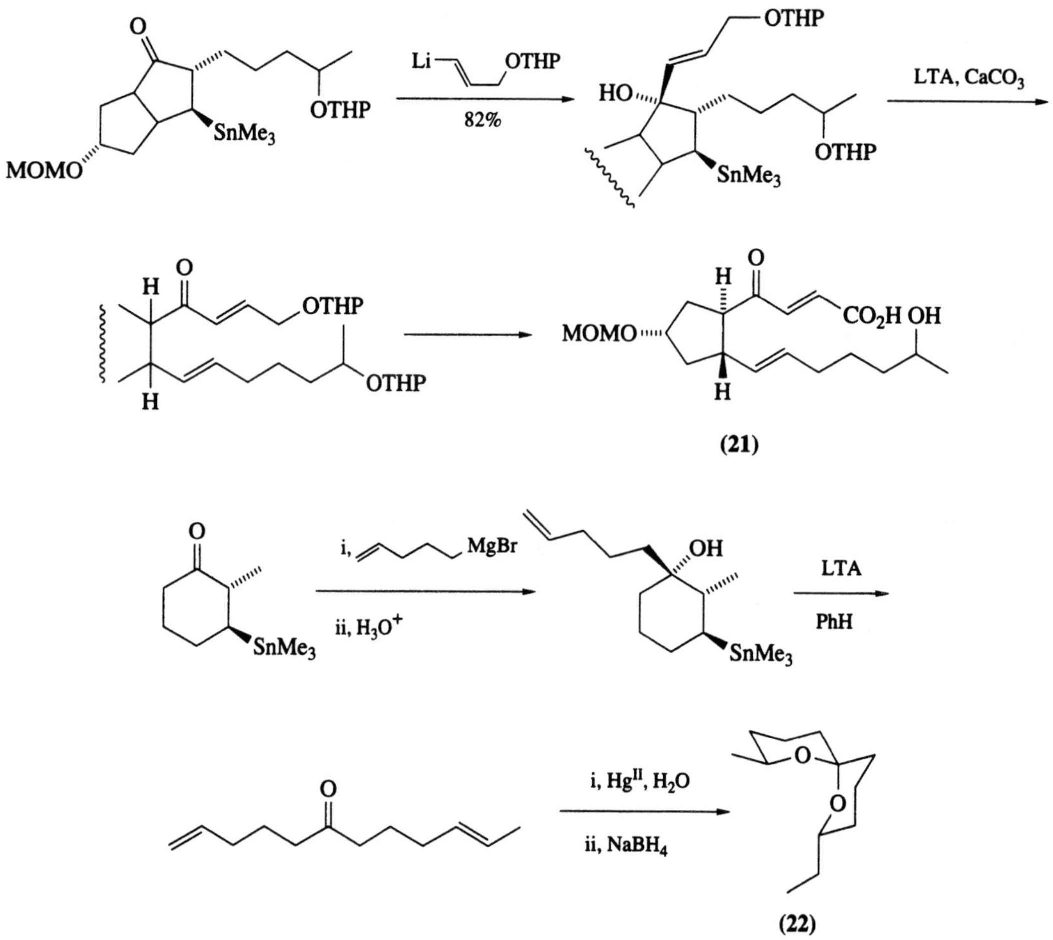

(21)

(22)

Scheme 21

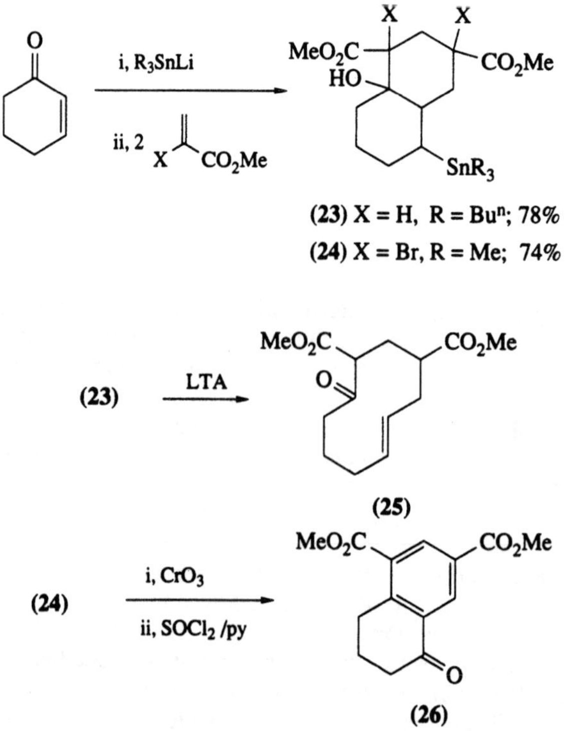

(23) X = H, R = Bu^n; 78%

(24) X = Br, R = Me; 74%

(25)

(26)

Scheme 22

hemiketals, and the *anti* elimination involved in the oxidative cleavage of the γ-stannyl–hydroxy arrangement.

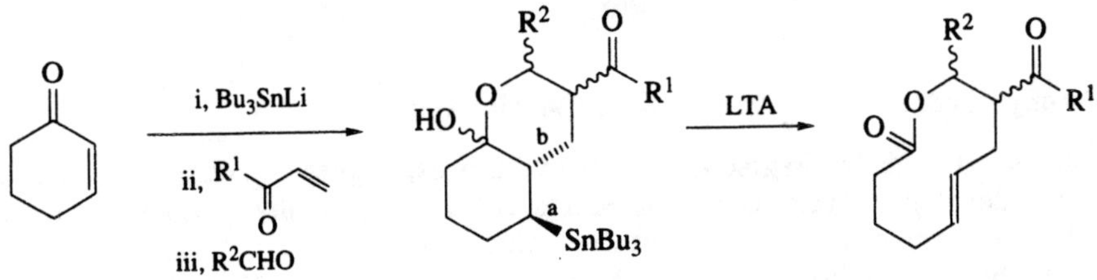

i, Bu_3SnLi
ii, R^1 (vinyl ketone)
iii, R^2CHO

LTA

R^2 = Me	R^1 = Et	47%
R^2 = H_2C=CH	R^1 = Et	39.5%
R^2 = o-IC_6H_4	R^1 = Et	37%

Scheme 23

The general attractiveness and flexibility were illustrated by acquisition of 2,3-disubstituted quinolines (**27**) and also phorocantholide (**28**), a natural 10-membered ring lactone constituent of an insect secretion.[42] The use of α,β-unsaturated ketones as initial receptors lends extra scope to the overall sequence, and permits preparation of substituted 9-, 10- and 11-membered unsaturated macrolides (**29**) bearing double bonds of fixed geometry at specific positions in lactone rings (Scheme 24).[43] As illustrated in Scheme 24, intramolecular cyclizations of the *vic* groups may also lead to regiospecifically substituted

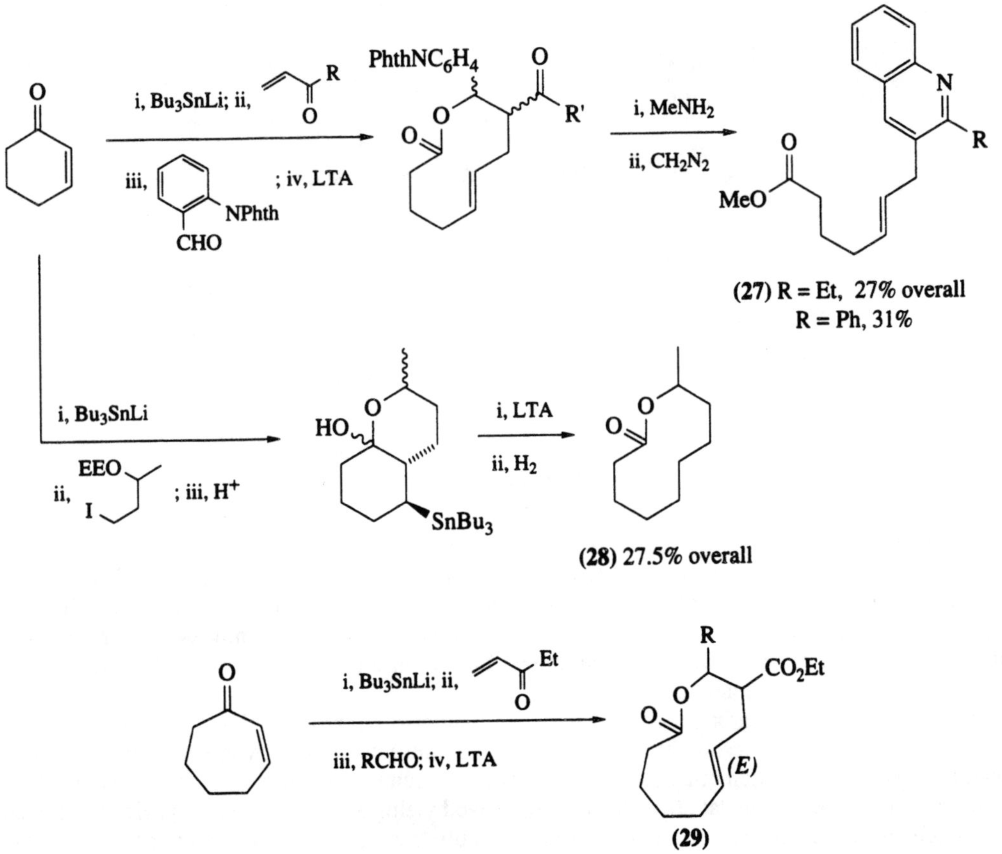

i, Bu_3SnLi; ii, (vinyl ketone, R); iii, (NPhth, CHO aryl aldehyde); iv, LTA

$PhthNC_6H_4$ R'

i, $MeNH_2$
ii, CH_2N_2

MeO N R

(**27**) R = Et, 27% overall
R = Ph, 31%

i, Bu_3SnLi
ii, EEO / I ; iii, H^+

HO $SnBu_3$

i, LTA
ii, H_2

(**28**) 27.5% overall

i, Bu_3SnLi; ii, (Et vinyl ketone)
iii, RCHO; iv, LTA

R CO_2Et (E)

(**29**)

R = c-C_3H_5	44% overall
R = Me	36%
R = 3-$MeOC_6H_4$	42%

Scheme 24

naphthalenes, quinolines, benzofurans and benzothiophenes. This great versatility and flexibility coupled with regio- and stereo-control make the MIMIRC and MIMIARC reactions very attractive, but control is due to the stereochemistry of enolate alkylation and the *anti* fragmentation of the γ-stannyl alcohols.

4.2.2.6 β-Stannyl Hydrazones, Oximes and Carboxylic Acids

Trialkyltin-triggered oxidative fragmentations of hydrazones, oximes and carboxylic acids have been reported, and may develop into very useful procedures. β-Stannyl phenylhydrazones[44] on oxidation with NBS provided the interesting azocyclopropanes (**30**), which could be transformed to the pyrazolines (**31**) by treatment with $SnCl_2$ in refluxing benzene. DDQ and LTA were also efficient oxidants. Homolytic cyclopropanation was suggested (Scheme 25) and shown to proceed with inversion of the C—Sn configuration, eventually yielding pyrazolines, regio- and stereo-specifically (Scheme 25).

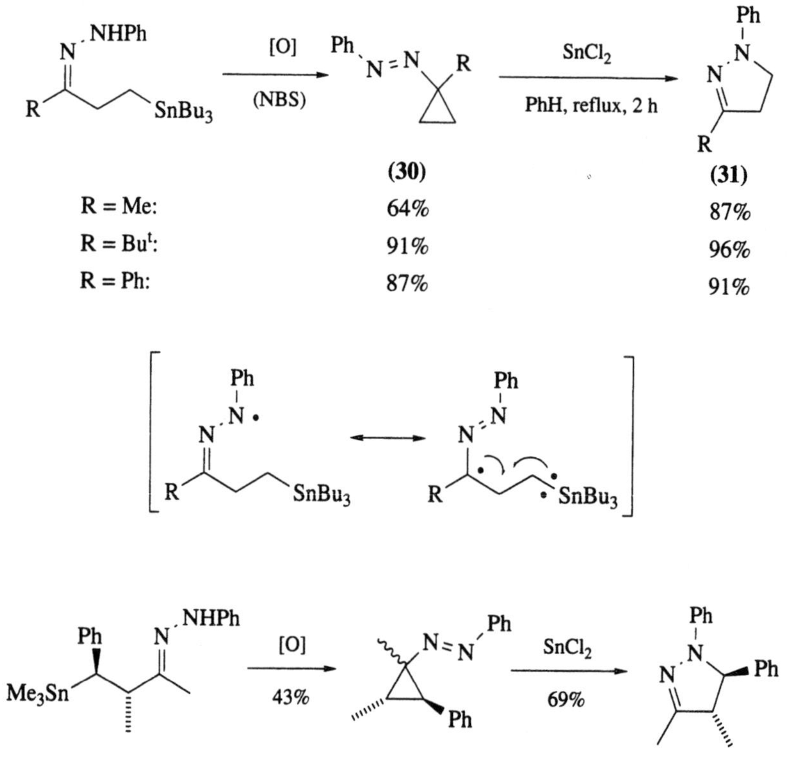

	(30)	(31)
R = Me:	64%	87%
R = Bu^t:	91%	96%
R = Ph:	87%	91%

Scheme 25

Δ^2-Isoxazolines exhibit considerable versatility in synthesis and one-pot oxidative ring contraction of stannyl oximes to Δ^2-isoxazolines with stereocontrol results from the strong directing effect of the tin group, followed by 1,3-dipolar cycloaddition (Scheme 26).[45] In conjunction with other results, it was suggested that oximes with the stannyl group too far removed from the iminoxyl group (3-electron σ-radical) for direct interaction react by fragmentation–recombination, whereas sufficient proximity leads to direct C—O bond formation.

β-Stannyl- (and β-silyl-) carboxylic acids undergo oxidative decarboxylation with LTA under mild conditions to provide the corresponding alkenes.[46] This represents an improvement on the well-known alkene-forming decarboxylation of acids with LTA, which requires thermal or photochemical conditions, for example.[47] The directing metal effect leads to improved yields and regioselectivity. However, stereospecific alkene formation did not occur and this could imply free radical involvement or transmetallation (Pb for Sn) (stereochemistry?) followed by cation formation, see for example Scheme 27.

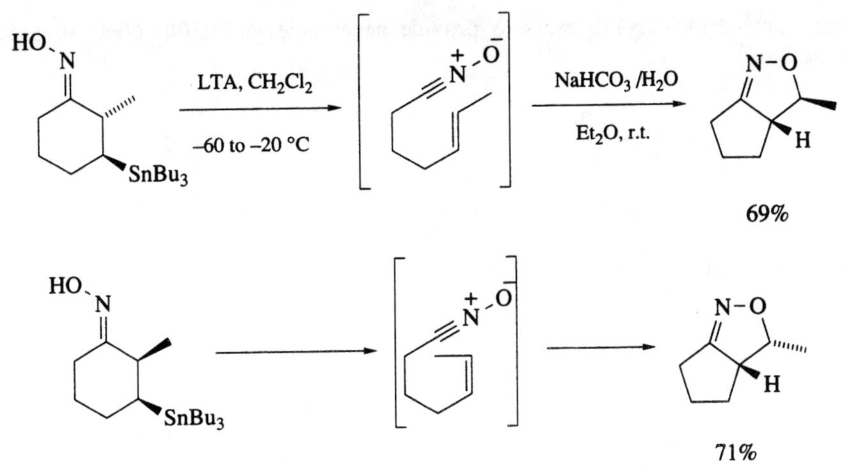

Scheme 26

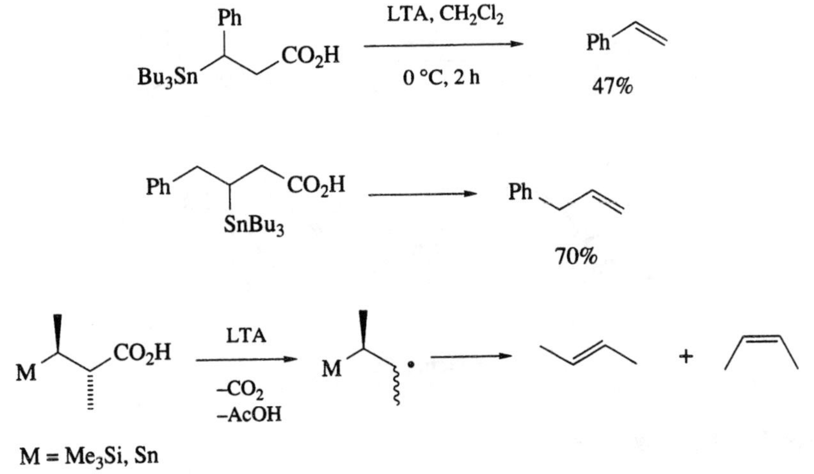

Scheme 27

4.2.3 OXIDATION OF CARBON–PALLADIUM BONDS

The oxidative cleavage of the bond between carbon and palladium is often a key step in Pd-catalyzed reactions, in which transiently formed organopalladium systems are implicated. These latter aspects are covered elsewhere (Volume 7 Chapter 3.4) and the present discussion is restricted to oxidation of preformed, stable organopalladium compounds.

4.2.3.1 π-Allylpalladium Complexes

Despite the existence of a formidable literature on systems in which transitory π-allylpalladium complexes are involved,[48] only a few papers describe the oxidation of preformed π-allylpalladium complexes to alcohols, rather than to carbonyl compounds. Nevertheless, these reports suggest that the procedure has considerable potential, in view of the stereoselectivities reported. In the steroid area, it was established that π-allyl–Pd complexes were oxidized regiospecifically and with high stereoselectivity to allylic alcohols by MCPBA, using pentane–pyridine as solvent.[49] The hydroxy group was delivered preferentially to the same diastereotopic face of the allyl system as that originally occupied by palladium. These oxidations have severe solvent effects, with pyridine appearing to suppress carbonyl compound formation, and enhancing the regio- and stereo-selectivity of the oxidations (Scheme 28). Some studies of the oxidation of organopalladium compounds with Cr^{VI} reagents have also been reported, and only

Collins reagent (Cr^{VI} oxide–2py) appears to provide acceptable yields (40–60%) of aldehydes and ketones (Scheme 28).[50]

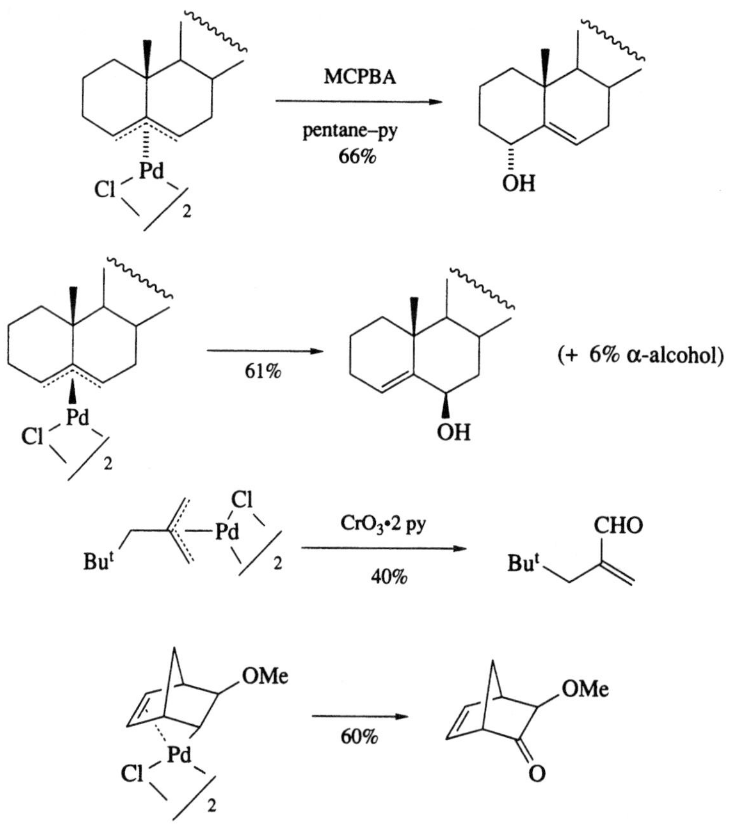

MCPBA

pentane–py
66%

61%

(+ 6% α-alcohol)

CrO_3•2 py
40%

60%

Scheme 28

4.2.3.2 Cyclopalladation–Oxidation

The Shaw cyclopalladation reaction (equation 10), reported in 1978, has been developed sufficiently to conclude that it has considerable potential for functionalization of unactivated methyl groups in the vicinity of a ketone, *via* the oxime.[51] This was demonstrated with lanost-8-en-3-one, which furnished a cyclopalladated derivative (**32**), which allowed functionalization of the 4-Me group to CH_2D ($NaBD_4$) and CH_2I ($I_2/CHCl_3$).[52] Attempted oxidation of the cyclohexanone derivative (**33**) with MCPBA unexpectedly provided the chloromethyl compound (Scheme 29).

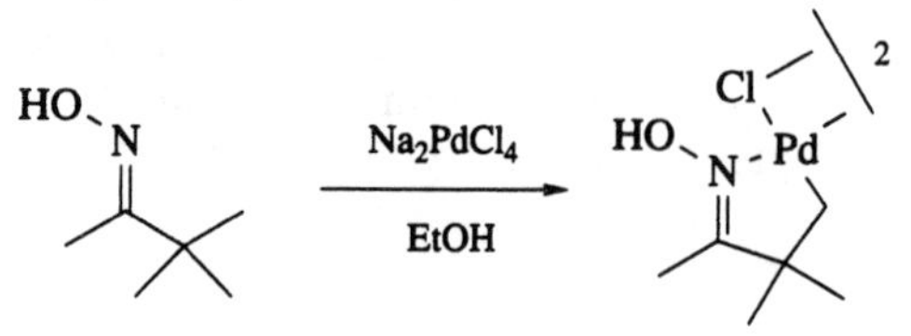

Na_2PdCl_4, EtOH (10)

However, smooth conversion in the sense C—Pd → C—O was achieved when LTA–Py was employed as oxidant, and very high yields (80–100%) of *O*-functionalized methyl compounds were obtained (Scheme 30).[53] Notice that the second palladation to form (**34**) occurs regiospecifically on the second methyl group in 2,2-dimethylcyclohexanone. Hindered ketones, such as 2,2,6,6-tetramethylcyclohexanone also behaved in excellent fashion. An argument based on conformational control has been presented to account for the regiospecificity of the second palladation step.[54] Lupanone oxime (**35**) was successfully converted in high yield to the expected 23-acetoxy derivative (**36**) by this sequence (Scheme 30).[54] Cyclopalladation–oxidation has found interesting applications in the synthesis of carbohydrate-derived, equatorially functionalized *gem*-di-*C*-alkyl derivatives, which are important for certain enantioselective syntheses.[55] Thus (**37**; Scheme 30) provided a single acetoxy oxime (**38**) in nearly quantitative yield, and the stereospecific functionalization is attributed to the necessity for a nearly coplanar arrange-

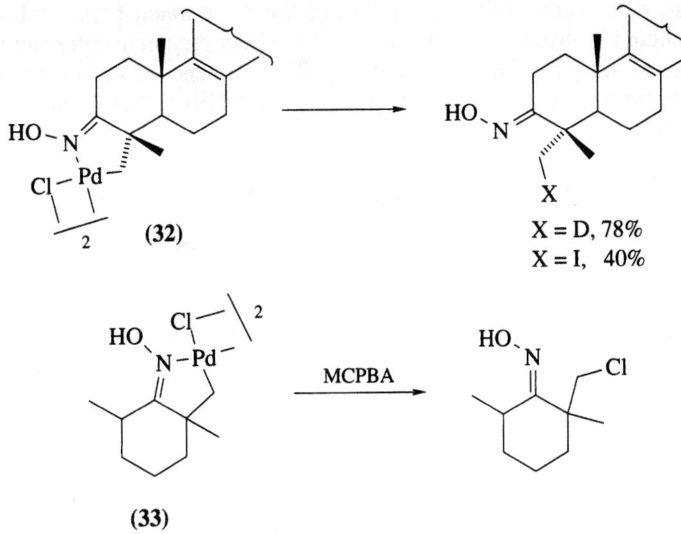

(32)

X = D, 78%
X = I, 40%

(33)

MCPBA

Scheme 29

ment of the oxime carrying Pd and the methyl group. The use of $NaBH_4$ as the final step in the synthetic sequence prevents trapping of Pd^{II} by the oxime, by reduction to Pd^0.

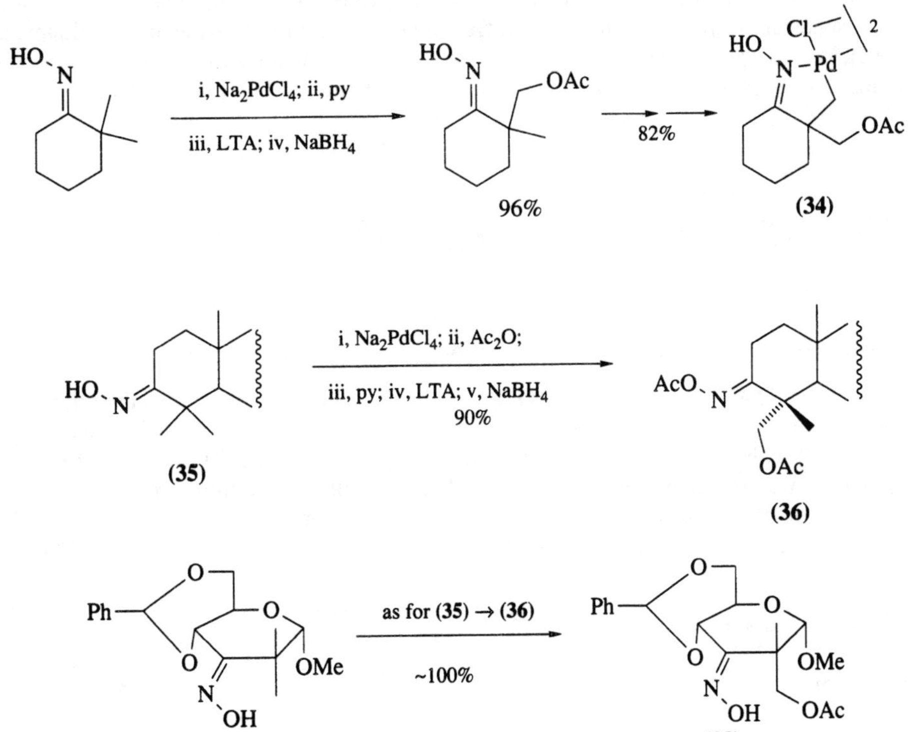

i, Na_2PdCl_4; ii, py
iii, LTA; iv, $NaBH_4$

96%

82%

(34)

(35)

i, Na_2PdCl_4; ii, Ac_2O;
iii, py; iv, LTA; v, $NaBH_4$
90%

(36)

(37)

as for (35) → (36)
~100%

(38)

Scheme 30

4.2.4 OXIDATION OF CARBON–MERCURY BONDS

4.2.4.1 Introduction

Applications of organomercury compounds in synthesis[2,56] overwhelmingly concern Markovnikov conversion of alkenes, allenes and cyclopropanes to alcohols, ethers, amines, peroxides and azides by oxy-, alkoxy-, amino-, peroxy- or azido-mercuration of the unsaturated group, followed by reductive demercuration.[2,57] Sodium borohydride is routinely used to effect reductive demercuration and there has been considerable interest in the mechanism of this step. This sequence is shown in equation (11) for the more common oxymercuration–reductive demercuration. Full details and discussion of these procedures have been presented elsewhere.[2]

$$\text{cyclohexene} \xrightarrow[\text{THF/H}_2\text{O}]{\text{Hg(OAc)}_2} \text{2-(HgX)cyclohexanol} \xrightarrow[\text{NaOH/H}_2\text{O}]{\text{NaBH}_4} \text{cyclohexanol} \qquad (11)$$

There is persuasive evidence that reductive demercuration involves a noncage free radical chain mechanism with a common hydrogen source (RHgH) irrespective of the hydride employed ($NaBH_4$, Bu_3SnH, LAH) (equations 12–14).[58–60]

$$\text{RHgX} \xrightarrow{[\text{H}]} \text{RHgH} \qquad (12)$$

$$\text{RHgH} \longrightarrow \text{R}\bullet \qquad (13)$$

$$\text{R}\bullet + \text{RHgH} \longrightarrow \text{RH} + \text{R}\bullet + \text{Hg} \qquad (14)$$

Hill and Whitesides demonstrated that reduction of organomercurials with $NaBH_4$ in the presence of oxygen gave a spectrum of products in line with free noncaged alkyl radicals as intermediates, and suggested that efficient oxygen-trapping of such radicals would provide a useful method of carbon–oxygen bond formation, and generally of adding functionality to alkene moieties (equations 15 and 16).[58]

$$\text{R}\bullet + \text{O}_2 \longrightarrow \text{ROO}\bullet \qquad (15)$$

$$\text{ROO}\bullet \xrightarrow{\text{HX}} \text{ROOH} \xrightarrow{\text{reduce}} \text{ROH} \qquad (16)$$

4.2.4.2 Oxymercuration–Oxidative Demercuration

Oxymercuration–oxidative demercuration (OM–OD) has considerably broadened the utility of organomercurial applications in synthesis.[56] Typically, a solution of the mercurial in DMF (room temperature) was added to a DMF solution of $NaBH_4$, through which oxygen was rapidly passed.[58] Generally 1.2–1.3 mol of borohydride was required to effect complete demercuration. In addition to the expected alcohols and hydrocarbons, other products presumed to be alkoxyboron compounds were also formed, but these could be hydrolyzed to alcohols. A typical result is shown in equation (17), with yields after the hydrolysis step.

$$\text{2-methoxycyclohexyl-HgBr} \xrightarrow[\text{DMF}]{\text{NaBH}_4,\ \text{O}_2} \underset{52\%}{\text{2-methoxycyclohexanol}} + \underset{35\%}{\text{2-methoxycyclohexanol (diastereomer)}} + \underset{4\%}{\text{methoxycyclohexane}} + \underset{5\%}{\text{cyclohexene}} \qquad (17)$$

Thus good yields of β-alkoxy alcohols can be obtained, albeit as diastereomeric mixtures, but unfortunately hydroxymercurated alkenes under similar conditions do not lead to useful products.[58] Despite this apparent limitation, alkoxymercuration–oxidative demercuration has been very effective in a number of systems described below, and there is no doubt it is a procedure worth consideration for hydroxy group introduction.

The first general application of this procedure was to the synthesis of tetrahydrofurfuryl alcohols, the precursor mercurials of which resulted from an intramolecular reaction of alkenic alcohols (Scheme 31).[61,62] Both mercurials and alcohols were formed as diastereomeric mixtures, the latter in moderate yields (Table 2).

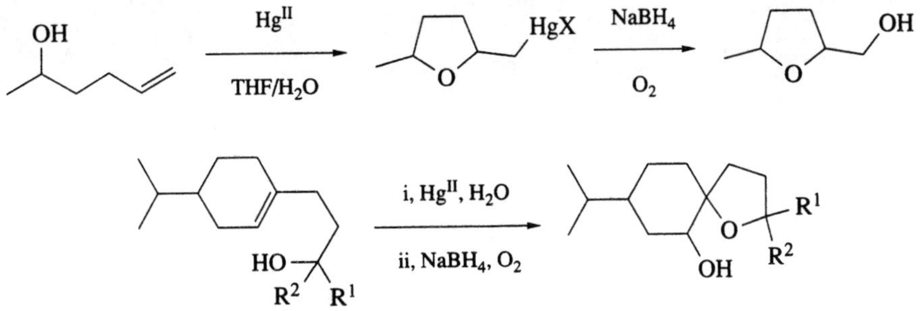

OH → (Hg^{II}, THF/H_2O) → HgX → ($NaBH_4$, O_2) → OH

HO, R^2, R^1 → (i, Hg^{II}, H_2O; ii, $NaBH_4$, O_2) → O, R^1, R^2, OH

Scheme 31

Table 2

Mercurial	*Hydroxy derivative*	*Yield* (%)
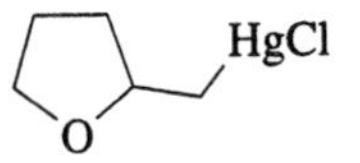 HgCl	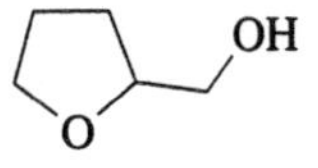 OH	10
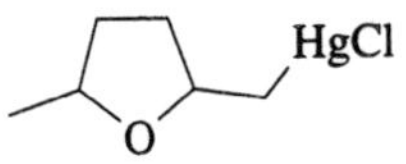 HgCl	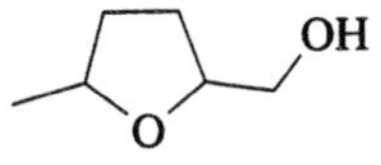 OH	20
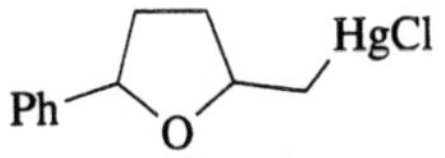 Ph, HgCl	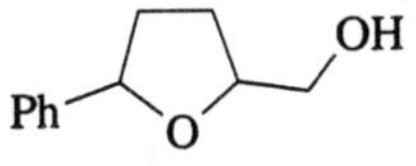 Ph, OH	25
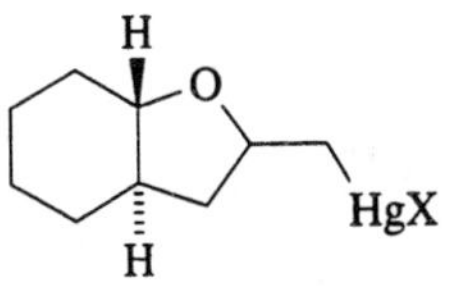 H, O, H, HgX	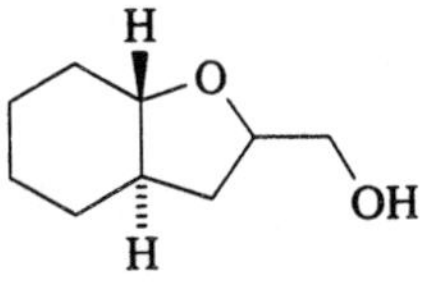 H, O, H, OH	60
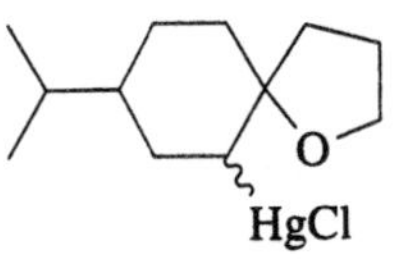 O, HgCl	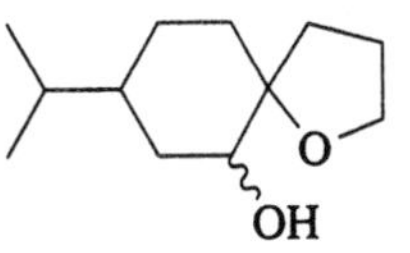 O, OH	45
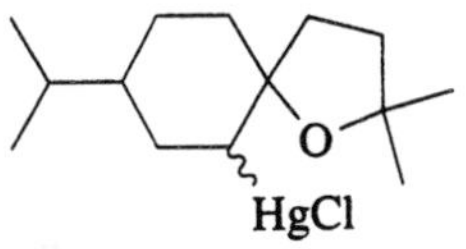 O, HgCl	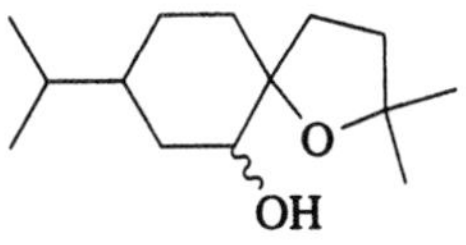 O, OH	60

Oxidative replacement of HgX by hydroxy is straightforward in more complex systems. For example, this conversion was an important step in the total synthesis of the potent antiviral agent aphidicolin (**39**), shown in Scheme 32.[63] The 6,11α-oxygen ring system was introduced by a (cyclization) oxymercuration–oxidative demercuration sequence in the synthesis of 5(*E*)- and 5(*Z*)-11-deoxy-6,11-α-epoxy-Δ^5-prostaglandin $F_{1\alpha}$ in a most efficient manner (Scheme 33).[64] A number of other cyclization methods (I_2, selenium reagents) were tried but were unsuccessful. Thus oxymercuration–cyclization provided the bicyclo[3.2.1] system in good yield, which was then acetylated (for subsequent differentiation of the C-15 and C-5 hydroxy groups) and oxidatively demercurated to provide a separable diastereomeric set of C-5 alcohols. This type of approach has also been employed in the synthesis of 5-hydroxy-PGI_2.[65]

Mercuricyclization–oxidation has been utilized in a short stereoselective approach to the *trans*-fused pyranopyran ring system (**40**), which is found in some squalene-derived tetracyclic ethers such as thyrsiferol and venustatriol.[66] Unsaturated hydroxy nitrile (**41**) on Hg^{II} cyclization (with $Hg(OCOCF_3)_2$ in DMF), followed by metathesis to the chloride afforded the tetrahydropyranylmercurial, which on oxidative demercuration provided an easily separated alcohol mixture, with the desired axial alcohol (**42**)

i, $NaBH_4$, O_2; ii, PCC; iii, Bu_4NF

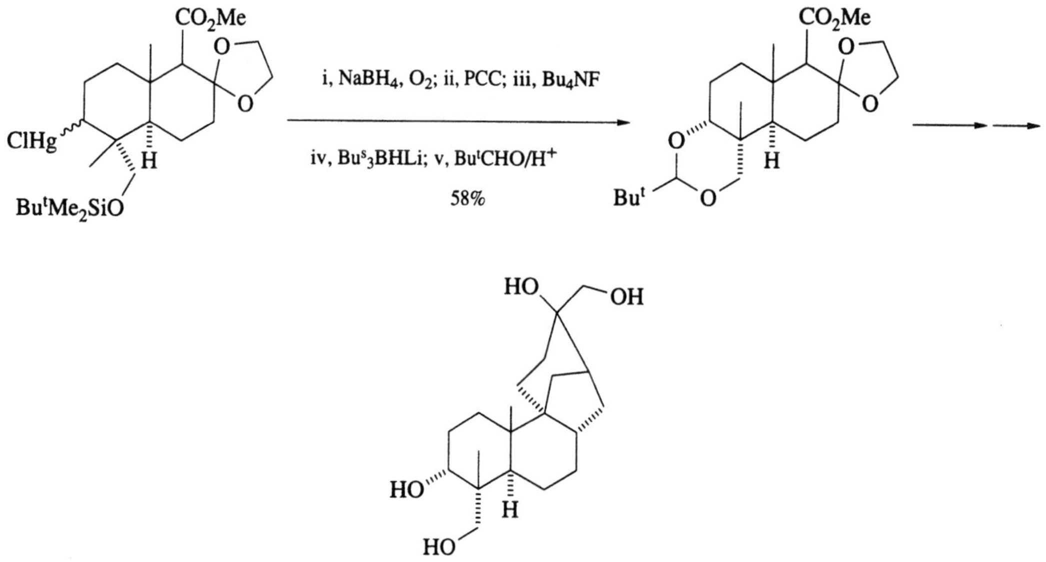

iv, Bu^s_3BHLi; v, Bu^tCHO/H^+

58%

(39)

Scheme 32

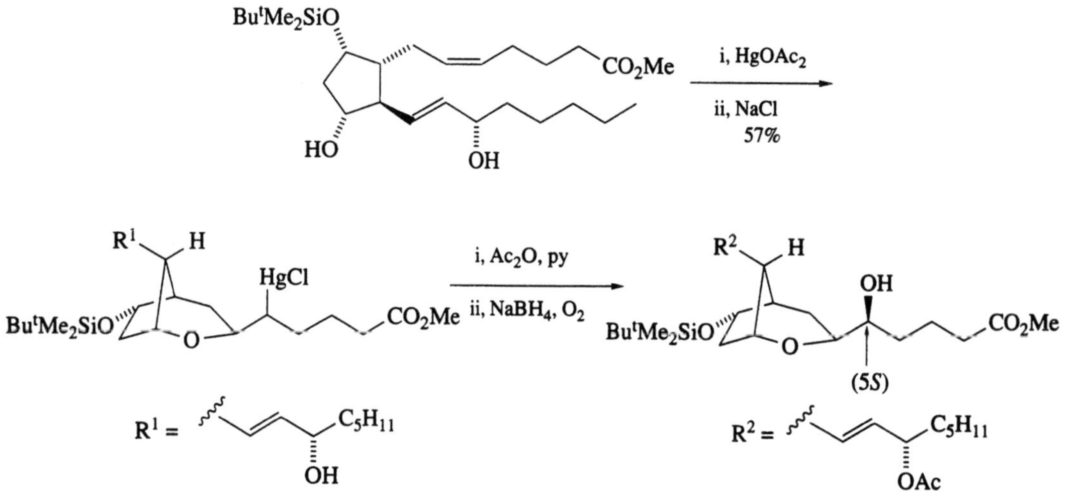

i, $HgOAc_2$

ii, NaCl

57%

i, Ac_2O, py

ii, $NaBH_4$, O_2

Scheme 33

predominating. Alcohol (**42**) was converted to the chair-boat pyrano–pyran system (**43**), again utilizing mercuricyclization–oxidation (Scheme 34).

The synthesis of the limonoid azadiradione, utilizing a Hg^{II} cyclization–oxidative demercuration sequence with an enol phosphate derived from *trans,trans*-farnesol has been reported.[67] Azadiradione, a tetracarbocyclic member of the limonoid group isolated from the neem tree, *Azadirachta indica*, has been converted to other tetracyclic limonoids, and is thus a key intermediate. The sequence is shown in Scheme 35.

Organomercurial intermediates have also been utilized in the biomimetic conversion of communic acids to the pimarane system, during which the radical involved in the $NaBH_4$-demercuration step was captured by oxygen.[68] Treatment of *trans*-communic acid with $Hg(OAc)_2$ (2 equiv.), followed by reduction, led to (**44**) and (**45**) and other products. These results are consistent with the intervention of the radical formed from the dimercurial (**46**) and, indeed, separate reduction in the presence of oxygen provided the peroxy compound (**45**) directly (Scheme 36).

In the area of pheromone synthesis, oxymercuration–oxidative demercuration has also proven valuable. For example, all four stereoisomers of tetrahydro-2,2,6-trimethyl-2*H*-pyran-3-ol, from the elm bark beetle *Pteleobius vittatus* have been acquired by a sequence from (*R*)- and (*S*)-sulcatol, which incorporates this mercury chemistry.[69] The epimeric alcohols (**47**) and (**48**; Scheme 37) were separable (MPLC)

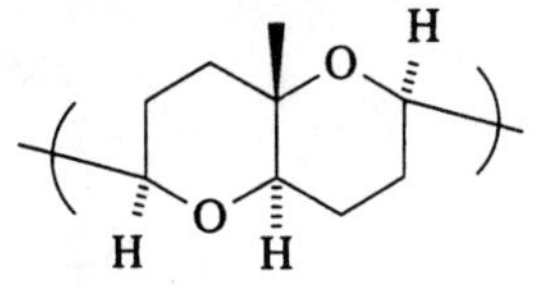

(40)

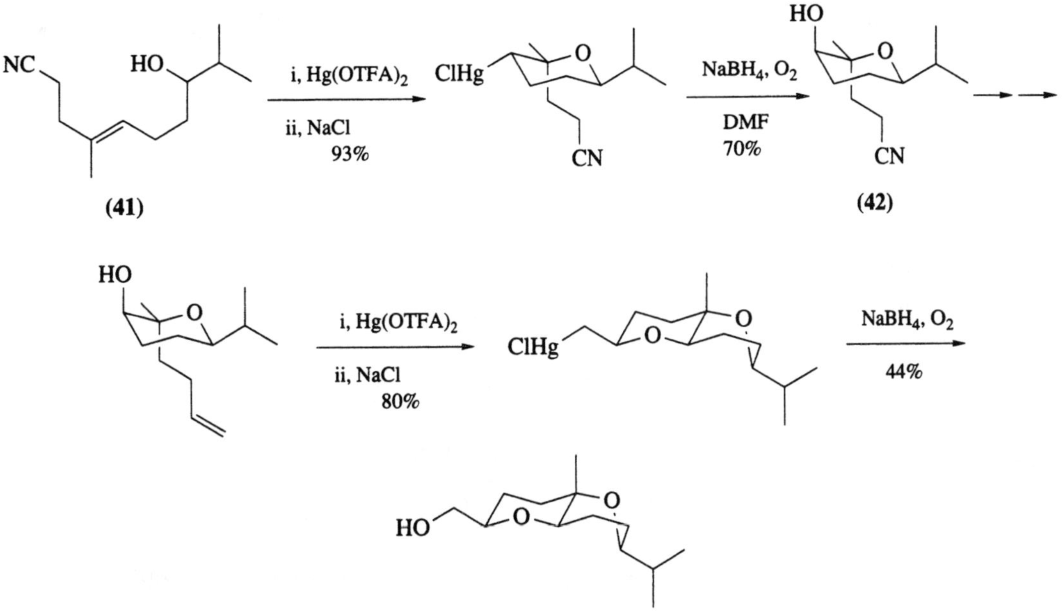

NC HO i, $Hg(OTFA)_2$ ii, NaCl 93% ClHg O CN $NaBH_4$, O_2 DMF 70% HO O CN

(41) (42)

HO O i, $Hg(OTFA)_2$ ii, NaCl 80% ClHg O O $NaBH_4$, O_2 44%

HO O O

(43)

Scheme 34

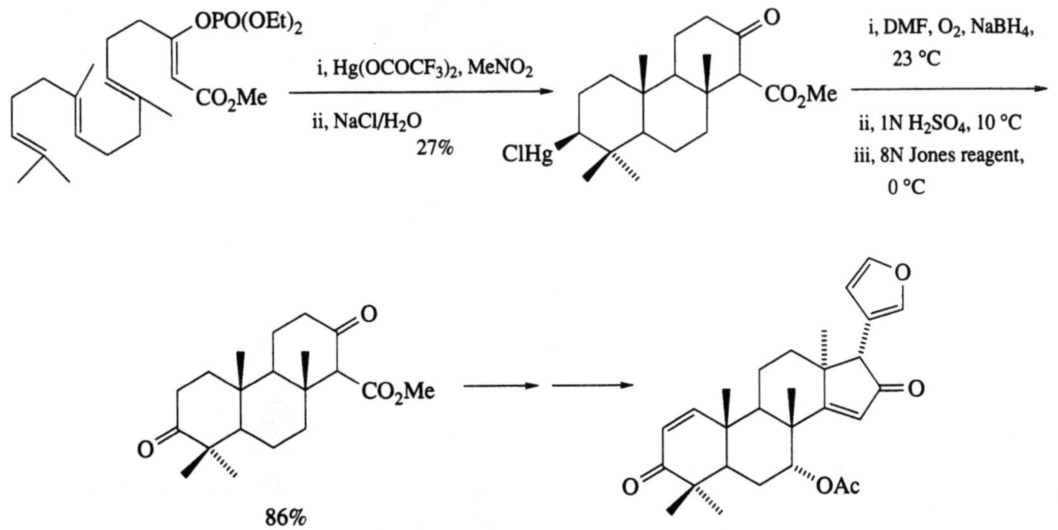

$OPO(OEt)_2$ CO_2Me i, $Hg(OCOCF_3)_2$, $MeNO_2$ ii, NaCl/H_2O 27% ClHg O CO_2Me i, DMF, O_2, $NaBH_4$, 23 °C ii, 1N H_2SO_4, 10 °C iii, 8N Jones reagent, 0 °C

O O CO_2Me 86% O O O OAc Azadiradione

Scheme 35

and shown to possess *ca.* 100% *ee*. In the same manner, the (3*R*,6*S*)- and (3*S*,6*S*)-stereoisomers were obtained. A number of naturally occurring spiroacetals bearing hydroxy substituents have been described and mercury chemistry was employed to obtain (*E*,*E*)-2-hydroxymethyl-8-methyl-1,7-dioxaspiro[5.5]undecane (**49**), a component of the rectal gland secretion of the cucumber fly *D. cucumis* (Scheme 37).[70]

Useful transformations in the carbohydrate field have also been reported.[71] Wittig reaction of 2,3,4,6-tetra-*O*-benzyl-D-glucopyranose provided alkene (**50**), which on Hg^{II}-mediated cyclization provided

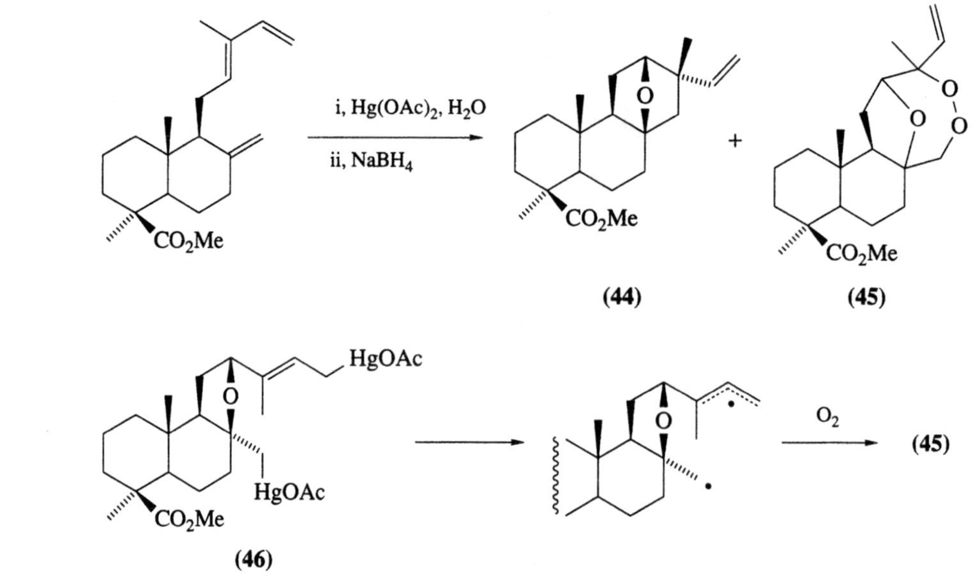

i, $Hg(OAc)_2$, H_2O
ii, $NaBH_4$

CO_2Me

(44)

(45)

HgOAc

HgOAc

CO_2Me

(46)

O_2

(45)

Scheme 36

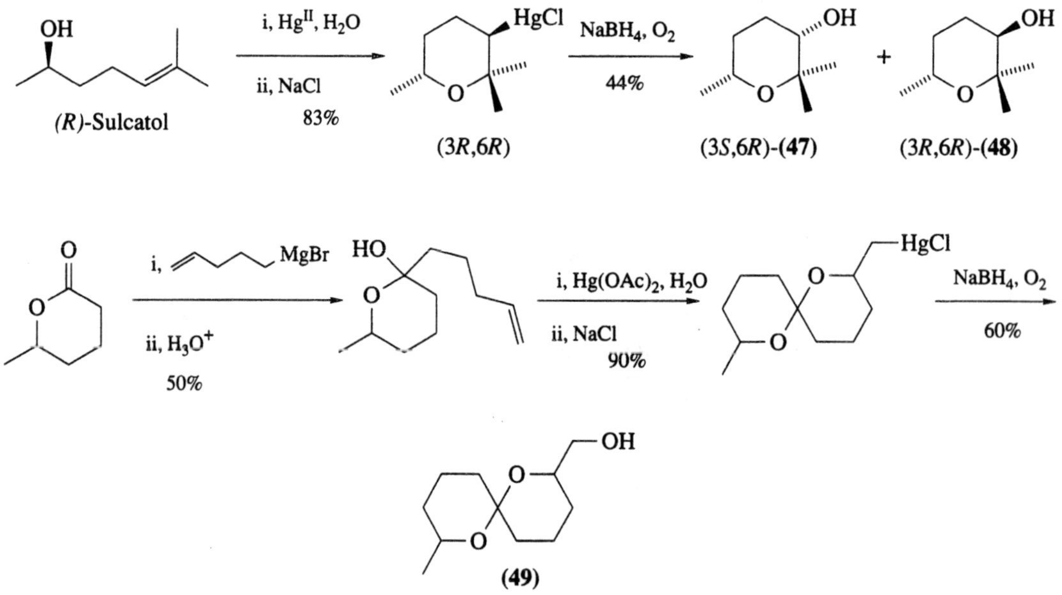

(R)-Sulcatol

i, Hg^{II}, H_2O
ii, NaCl
83%

HgCl

(3R,6R)

$NaBH_4$, O_2
44%

(3S,6R)-(47)

(3R,6R)-(48)

i, MgBr
ii, H_3O^+
50%

i, $Hg(OAc)_2$, H_2O
ii, NaCl
90%

HgCl

$NaBH_4$, O_2
60%

(49)

Scheme 37

essentially pure α-chloromercurial product (**51**) in high yield. Thus the axially oriented mercury group was well equipped for a range of further transformations and this cyclization provides an entry to 1,5-*trans* (*e.g.* α-D) *C*-glucopyranosyl derivatives (Scheme 38).

An especially important example of ring-forming aminomercuration–oxidation has been outlined in the synthesis of 1-deoxynojirimycin (**52**) and 1-deoxymannojirimycin (**53**), which are an interesting class of glycosidase inhibitors.[72] The reported method (Scheme 39) allows the conversion of a natural sugar into an azaalditol possessing the same relative and absolute configuration. Thus aminoalkene (**54**) (from tri-*O*-benzyl-6-bromopyranoside) on cyclizing aminomercuration *etc.* provided bromomercurials (**55**) and (**56**). Oxidation provided (**57**), which on hydrogenolysis *etc.* led to (**52**). The minor mercurial (**56**) after oxidative demercuration could be oxidized (Swern) to the aldehyde, followed by epimerization (DBU) and reduction ($NaBH_4$) to afford additional (**57**). Use of methyl-α-D-mannopyroside provided the epimeric aminoalkene (**58**), which was then transformed as described for (**52**) to (**53**) in 15% overall yield, although the mercuricyclization provided predominantly the unwanted isomer, which was epimerized as outlined (Scheme 39).

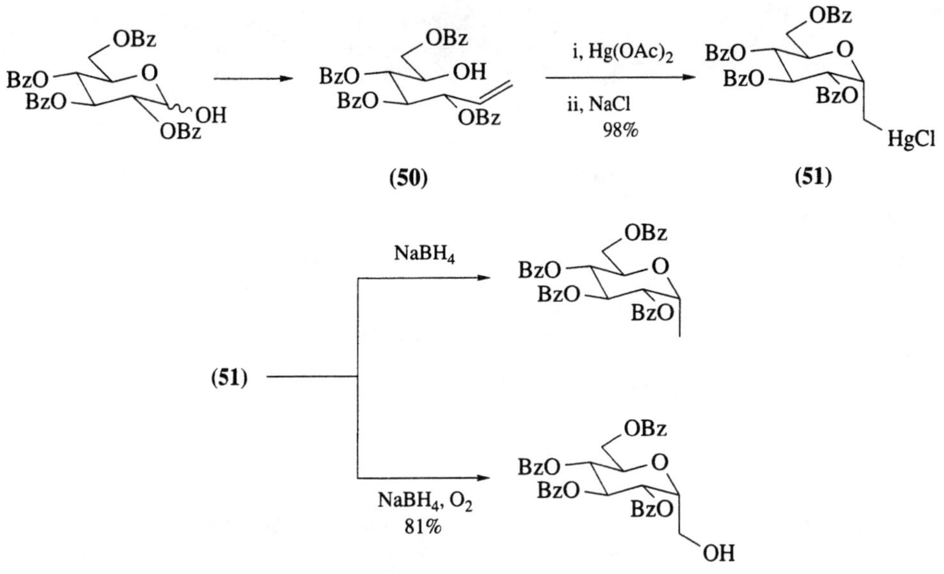

(50) (51)

Scheme 38

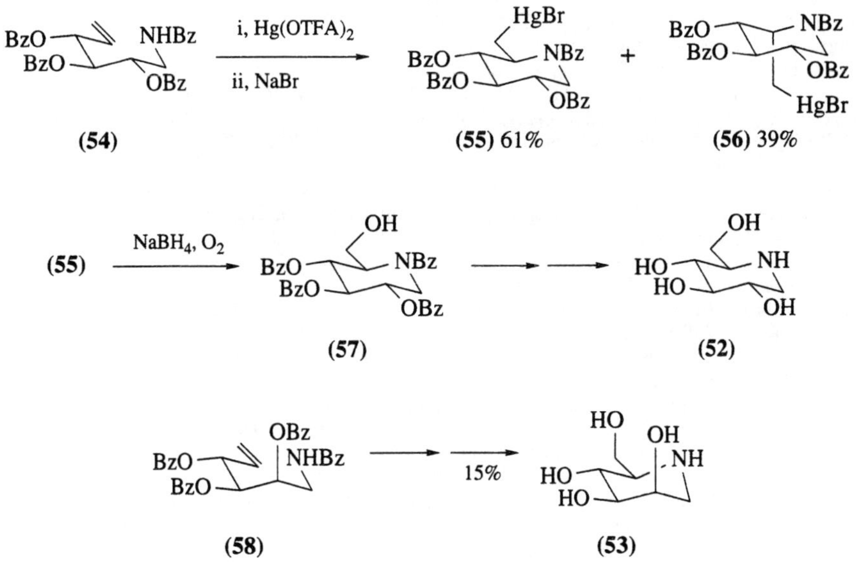

(54) (55) 61% (56) 39%

(57) (52)

(58) (53)

Scheme 39

4.2.4.3 Miscellaneous Oxidations

Some other reactions involving oxidation of the C—Hg bond have been known for some time, but these are either of limited synthetic appeal or have experienced no significant development in recent years. Thus ozonolysis of the C—Hg bond to form carboxylic acids or ketones falls into the first category,[58] whereas allylic acetoxylation of alkenes by $Hg(OAc)_2$ falls into the second category. Nevertheless, this allylic oxidation (Treibe's reaction) has considerable synthetic utility, and has been reviewed quite recently.[56,73]

It is worth noting at this point that palladium(II) salts intervene in a synthetically useful way when alkenes are reacted with Hg^{II} salts in either water or alcohols.[74,75] Thus oxymercuration–palladation–depalladation ensues and results in alkene → ketone conversions (equation 18).[56] The reaction can be catalytic in Pd^{II}, if a reoxidant such as Cu^{II} is employed.

$$R^1CH{=}CHR^2 \xrightarrow[H_2O \text{ or } ROH]{Hg^{II}} \xrightarrow[CuCl_2]{Li_2PdCl_4 \text{ (cat.)}} R^1CH_2C(O)R^2 \quad \text{80–100\%} \tag{18}$$

4.2.5 REFERENCES

1. H. E. Seyfarth, J. Henkal and A. Rieche, *Angew. Chem., Int. Ed. Engl.*, 1965, **4**, 1074; C. Frisell and S.-O. Lawesson, *Org. Synth.*, 1961, **41**, 91; G. Sosnovsky and J. H. Brown, *Chem. Rev.*, 1966, **66**, 529.
2. R. L. Larock, 'Solvomercuration–Demercuration Reactions in Organic Synthesis', Springer, Berlin, 1986.
3. A. McKillop and E. C. Taylor, in 'Comprehensive Organometallic Chemistry', ed. G. Wilkinson, F. G. A. Stone and E. W. Abel, Pergamon Press, Oxford, 1982, vol. 7, p. 465.
4. M. Pereyre, J.-P. Quintard and A. Rahm, 'Tin in Organic Synthesis', Butterworths, London, 1987.
5. P. J. Smith, 'Toxicological Data on Organotin Compounds', International Tin Research Institute, London, 1978, publication 538; see also R. C. Poller, 'The Chemistry of Organotin Compounds', Logos Press, London, 1970, chap. 14.
6. C. Deblandre, M. Gielen and J. Nasielski, *Bull. Soc. Chim. Belg.*, 1964, **73**, 214.
7. W. C. Still, *J. Am. Chem. Soc.*, 1977, **99**, 4836.
8. A. Itoh, T. Saito, K. Oshima and H. Nozaki, *Bull. Chem. Soc. Jpn.*, 1981, **54**, 1456.
9. M. Ochiai, S. Iwaki, T. Ukita, Y. Matsuura, M. Shiro and Y. Nagao, *J. Am. Chem. Soc.*, 1988, **110**, 4606.
10. G. Ayrey, J. R. Parsonage and R. C. Poller, *J. Organomet. Chem.*, 1973, **56**, 193.
11. R. H. Fish and B. M. Broline, *J. Organomet. Chem.*, 1978, **159**, 255.
12. I. Fleming and B. W. Au-Yeung, *Tetrahedron*, 1981, **37** (suppl.), 13.
13. I. Fleming and N. K. Terrett, *Pure Appl. Chem.*, 1983, **55**, 1707.
14. W. P. Weber, 'Silicon Reagents for Organic Synthesis', Springer, New York, 1983.
15. M. Pereyre and J.-P. Quintard, *Pure Appl. Chem.*, 1981, **53**, 2401.
16. J. A. Mangravite, *J. Organomet. Chem. Libr.*, 1979, **7**, 5.
17. G. Wickham, D. Young and W. Kitching, *Organometallics*, 1988, **7**, 1187.
18. T. Hayashi, Y. Okamoto, K. Kabeta, T. Hagihara and M. Kumada, *J. Org. Chem.*, 1984, **49**, 4224.
19. Y. Ueno, H. Sano and M. Okawara, *Synthesis*, 1980, 1011.
20. B. W. Au-Yeung and I. Fleming, *J. Chem. Soc., Chem. Commun.*, 1977, 79.
21. M. Shibasaki, H. Suzuki, Y. Torisawa and S. Ikegami, *Chem. Lett.*, 1983, 1303.
22. (a) M. Yamamoto, H. Izukawa, M. Saiki and K. Yamada, *J. Chem. Soc., Chem. Commun.*, 1988, 560; (b) M. Yamamoto, S. Irie, M. Miyashita, S. Kohmoto and K. Yamada, *Chem. Lett.*, 1989, 221.
23. N. Elming and N. C. Kass, *Acta. Chem. Scand.*, 1952, **6**, 565.
24. W. C. Still, *J. Am. Chem. Soc.*, 1979, **101**, 2493.
25. W. C. Still, *J. Am. Chem. Soc.*, 1977, **99**, 4186.
26. A. Nishida, M. Shibasaki and S. Ikegami, *Tetrahedron Lett.*, 1981, **22**, 4819.
27. E. J. Corey and R. H. Wollenberg, *J. Am. Chem. Soc.*, 1974, **96**, 5581.
28. M. G. Moloney and J. T. Pinhey, *J. Chem. Soc., Chem. Commun.*, 1984, 965.
29. M. G. Moloney, J. T. Pinhey and E. G. Roche, *Tetrahedron Lett.*, 1986, **27**, 5025.
30. M. Shibasaki, Y. Torisawa and S. Ikegami, *Tetrahedron Lett.*, 1982, **23**, 4607.
31. D. P. Strike, *US Pat.* 4 038 308 (1977) (*Chem. Abstr.*, 1977, **87**, 134 059).
32. I. Fleming and C. J. Urch, *Tetrahedron Lett.*, 1983, **24**, 4591.
33. J. F. Kadow and C. R. Johnson, *Tetrahedron Lett.*, 1984, **25**, 5255.
34. M. Ochiai, T. Ukita, Y. Nagao and E. Fujita, *J. Chem. Soc., Chem. Commun.*, 1984, 1007.
35. M. Ochiai, T. Ukita, Y. Nagao and E. Fujita, *J. Chem. Soc., Chem. Commun.*, 1985, 637.
36. M. Ochiai, S. Iwaki, T. Ukita and Y. Nagao, *Chem. Lett.*, 1987, 133.
37. K. Natatini and S. Isoe, *Tetrahedron Lett.*, 1984, **25**, 5335.
38. S. R. Wilson, P. A. Zucker, C. Kim and C. A. Villa, *Tetrahedron Lett.*, 1985, **26**, 1969.
39. K. Nakatini and S. Isoe, *Tetrahedron Lett.*, 1985, **26**, 2209.
40. M. G. O'Shea and W. Kitching, *Tetrahedron*, 1989, **45**, 1177.
41. G. H. Posner and E. Asirvatham, *Tetrahedron Lett.*, 1986, **27**, 663.
42. G. H. Posner, E. Asirvatham, K. S. Webb and S. Jew, *Tetrahedron Lett.*, 1987, **28**, 5071.
43. G. H. Posner, K. S. Webb, E. Asirvatham, S. Jew and A. Degl'Innocenti, *J. Am. Chem. Soc.*, 1988, **110**, 4754.
44. H. Nishiyama, H. Arai, Y. Kanai, H. Kawashima and K. Itoh, *Tetrahedron Lett.*, 1986, **27**, 361.
45. H. Nishiyama, H. Arai, T. Ohki and K. Itoh, *J. Am. Chem. Soc.*, 1985, **107**, 5310.
46. H. Nishiyama, M. Matsumoto, H. Arai, H. Sakaguchi and K. Itoh, *Tetrahedron Lett.*, 1986, **27**, 1599.
47. R. A. Sheldon and J. K. Kochi, *Org. React. (N.Y.)*, 1972, **19**, 279.
48. B. M. Trost and T. R. Verhoeven, in 'Comprehensive Organometallic Chemistry', ed. G. Wilkinson, F. G. A. Stone and E. W. Abel, Pergamon Press, Oxford, 1982, vol. 8, p. 799.
49. D. N. Jones and S. D. Knox, *J. Chem. Soc., Chem. Commun.*, 1975, 166.
50. E. Vedejs, M. F. Salomon and P. D. Weeks, *J. Organomet. Chem.*, 1972, **40**, 221.
51. A. G. Constable, W. S. McDonald, L. C. Sawkins and B. L. Shaw, *J. Chem. Soc., Chem. Commun.*, 1978, 1061.
52. K. Carr and J. K. Sutherland, *J. Chem. Soc., Chem. Commun.*, 1984, 1227.
53. J. E. Baldwin, C. Najera and M. Yus, *J. Chem. Soc., Chem. Commun.*, 1985, 126.
54. J. E. Baldwin, R. H. Jones, C. Najera and M. Yus, *Tetrahedron*, 1985, **41**, 699.
55. U. Rocherolle, J. C. Lopez, A. Olesker and G. Lukacs, *J. Chem. Soc., Chem. Commun.*, 1988, 512.

56. R. C. Larock, 'Organomercury Compounds in Organic Synthesis', Springer, Berlin, 1985.
57. See also: (a) K.-P. Zeller, H. Straub and H. Leditschke, *Methoden Org. Chem. (Houben-Weyl)*, 1974, **13/2b**, 9; (b) A. J. Bloodworth, in 'The Chemistry of Mercury', ed. C. A. McAuliffe, Macmillan, London, 1977, p. 139; (c) L. G. Makarova and A. N. Nesmeyanov, 'Methods of Elemento-Organic Chemistry. The Organic Compounds of Mercury', ed. A. N. Nesmeyanov and K. A. Kocheskov, North-Holland, Amsterdam, 1967, vol. 4.
58. C. L. Hill and G. M. Whitesides, *J. Am. Chem. Soc.*, 1974, **96**, 870.
59. R. P. Quirk and R. E. Lea, *J. Am. Chem. Soc.*, 1976, **98**, 5973.
60. J. Barluenga and M. Yus, *Chem. Rev.*, 1988, **88**, 487.
61. M. C. Benhamou, G. E. Moghadam, V. Speziale and A. Lattes, *J. Heterocycl. Chem.*, 1978, **15**, 1313.
62. V. Speziale and A. Lattes, *J. Heterocycl. Chem.*, 1979, **16**, 465.
63. E. J. Corey, M. A. Tius and J. Das, *J. Am. Chem. Soc.*, 1980, **102**, 1742.
64. J. C. Sih and D. R. Graber, *J. Org. Chem.*, 1982, **47**, 4919.
65. J. C. Sih, R. A. Johnson, E. G. Nidy and D. R. Graber, *Prostaglandins*, 1978, **15**, 409.
66. C. A. Broka, L. Hu, W. J. Lee and T. Shen, *Tetrahedron Lett.*, 1987, **28**, 4993; see also C. A. Broka and Y. T. Lin, *J. Org. Chem.*, 1988, **53**, 5876.
67. E. J. Corey and R. W. Hahl, *Tetrahedron Lett.*, 1989, **30**, 3023.
68. A. F. Barrero, J. F. Sanchez and J. Altarejos, *Tetrahedron Lett.*, 1988, **29**, 3713.
69. K. Mori and P. Puapoomchareon, *Liebigs Ann. Chem.*, 1988, 175.
70. W. Kitching, J. A. Rilatt, M. T. Fletcher, J. J. De Voss, R. A. I. Drew and C. J. Moore, *J. Chem. Soc., Chem. Commun.*, 1986, 853.
71. J. R. Pougny, M. A. M. Nassr and P. Sinay, *J. Chem. Soc., Chem. Commun.*, 1981, 375.
72. R. C. Bernotas and B. Ganem, *Tetrahedron Lett.*, 1985, **26**, 1123.
73. H. Arzoumanian and J. Metzger, *Synthesis*, 1971, 527.
74. G. T. Rodeheaver and D. F. Hunt, *J. Chem. Soc., Chem. Commun.*, 1971, 818.
75. D. F. Hunt and G. T. Rodeheaver, *Tetrahedron Lett.*, 1972, 3595.

4.3
Oxidation of Carbon–Silicon Bonds

ERNEST W. COLVIN
University of Glasgow, UK

4.3.1 INTRODUCTION

This chapter concentrates on those processes in which oxidative cleavage of a carbon–silicon bond results in production of the alkyl/aryl fragment as an alcohol/phenol. Other cleavage processes are dealt with, but more briefly.

The first example of such cleavage was reported in 1958 by Buncel and Davies, in a pioneering study[1] of the rearrangement of triorganosilyl perbenzoates to yield alkoxy- or aryloxy-silanes (Scheme 1). Hydrogen peroxide can also be employed as oxidant, in a sequence which illustrates the migratory preference of the phenyl group. Later kinetic studies[2] confirmed that the rearrangement was intramolecular.

Until recently, little attention has been paid to the synthetic potential of this oxidative cleavage. Due largely to the studies[3] of Tamao and Kumada and, independently, those of Fleming,[4] and their coworkers, such potential has now been revealed. It is the purpose of this chapter to highlight some of its applications. Since the emphasis is on synthetic utility, only high-yielding reactions have been selected for inclusion. Unless otherwise stated, all compounds shown are racemic; only one enantiomer is shown for clarity.

For successful cleavage, the silane must carry at least one electronegative substituent, such as alkoxy or fluorine. This requirement can be fulfilled either at an early stage by hydrosilylation of alkenes or alkynes using suitably functionalized hydridosilanes, or by late-stage electrophilic desilylation of phenyldimethylsilyl or allyldimethylsilyl moieties in the presence of a source of fluoride ion. Either excess hydrogen peroxide or MCPBA may be used as oxidant, and the alcohol is produced with *retention of configuration.* Fluoride ion is normally a mandatory additive in what is believed to be an assisted rearrangement of a silyl peroxide, as shown in Scheme 2.

Anhydrous trimethylamine *N*-oxide[5] has been suggested[6] as an alternative, neutral oxidant, although with dialkoxysilanes relatively high reaction temperatures are required. Alkyltrifluorosilanes, on the other hand, undergo cleavage with this oxidant at room temperature.[7]

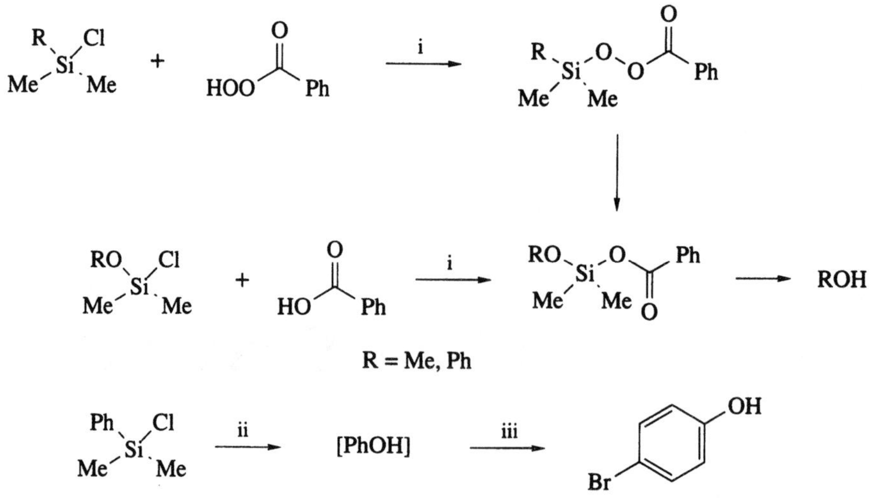

i, NH_3; ii, H_2O_2, AcOH, H_2SO_4 cat.; iii, Br_2

Scheme 1

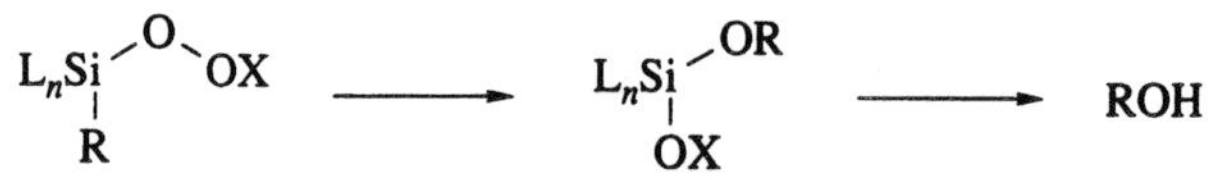

L_n = F, R

Scheme 2

4.3.2 GENERATION OF ORGANOFLUOROSILICATES

4.3.2.1 By Hydrosilylation

4.3.2.1.1 Hydrosilylation of alkenes with Cl_3SiH

Catalyzed addition of trichlorosilane to terminal alkenes and alkynes, followed by the addition of an aqueous solution of potassium fluoride, produces highly reactive organopentafluorosilicates[8] (Scheme 3).

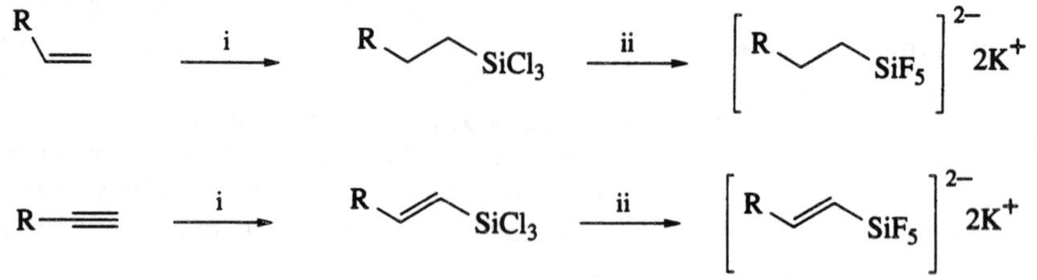

i, Cl_3SiH, H_2PtCl_6 cat.; ii, KF

Scheme 3

Alkylpentafluorosilicates react, in some cases exothermically, with a wide range of electrophilic reagents such as MCPBA,[8] and halogens and halogenoids.[9] Careful stereochemical investigation has shown that oxidative cleavage using MCPBA produces an alcohol with predominant *retention*[8] of stereochemistry, whereas cleavage using NBS gives an alkyl bromide with predominant *inversion* (Scheme 4).[10] Single-electron transfer processes have been implicated.[11] When combined with asymmetric hydrosilylation using a chiral catalyst, optically active alcohols can be generated.[12]

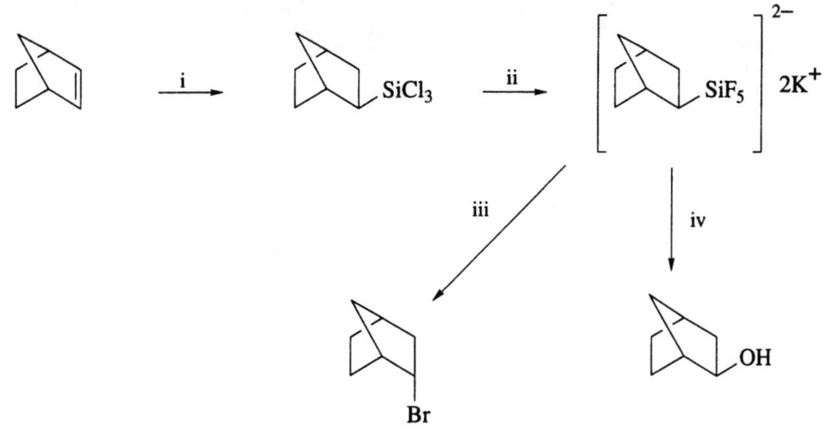

i, Cl_3SiH, chiral Pd^{II} cat.; ii, KF; iii, NBS; iv, MCPBA

Scheme 4

4.3.2.1.2 Hydrosilylation of alkynes with Me_2SiClH and $MeSiCl_2H$

Reaction of $H_2C{=}CHSiMeCl_2$ (obtained from the catalyzed hydrosilylation of acetylene with $MeSiCl_2H$) with an ephedrine-derived lithium dialkylamide produces the chiral vinylsilane (**1**). Addition of BuLi to this vinylsilane followed by treatment with $MgBr_2$ gives the corresponding Grignard reagent (**2**). This latter species can be transformed as shown in Scheme 5 into chiral alcohols in reasonable enantiomeric excess (*ee*).[13]

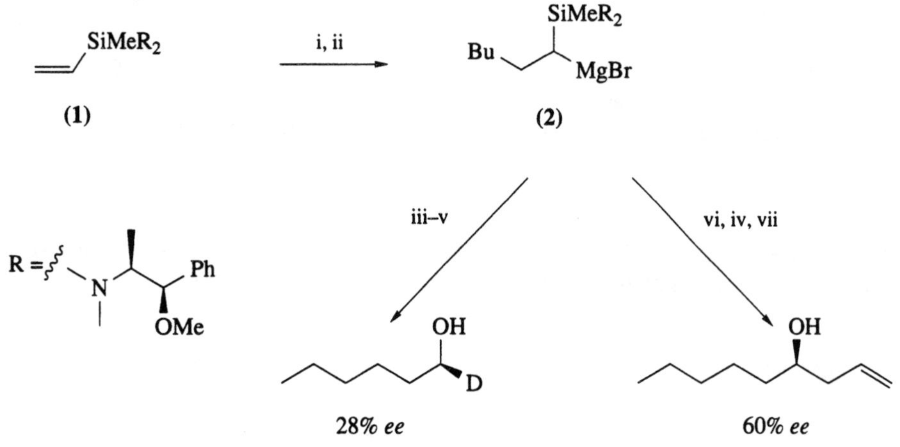

(**1**) (**2**)

28% *ee* 60% *ee*

i, Bu^nLi; ii, $MgBr_2$; iii, D_2O; iv, H_3O^+; v, H_2O_2; vi, $H_2C{=}CHCH_2Br$, CuI; vii, KHF_2, H_2O_2

Scheme 5

The isomeric epoxysilanes (**3**) and (**4**), prepared as shown in Scheme 6, undergo a copper-catalyzed Grignard ring-opening[14] to give β-hydroxysilanes. Oxidative cleavage then completes this selective route to either *syn* (**5**) or *anti* (**6**) 1,2-diols. Such methodology has been utilized in a synthesis of (±)-*exo*-brevicomin.

4.3.2.1.3 Hydrosilylation of Alkynes with $(EtO)_2MeSiH$

Hydrosilylation of terminal alkenes using the air-stable silane $(EtO)_2MeSiH$ in the presence of either H_2PtCl_6 or $(Ph_3P)_3RhCl$ results in the introduction of silicon exclusively at the terminal carbon atom. When coupled with oxidative cleavage, this protocol[15] provides a simple one-pot synthesis of anti-Markovnikov alcohols from terminal alkenes (Scheme 7).

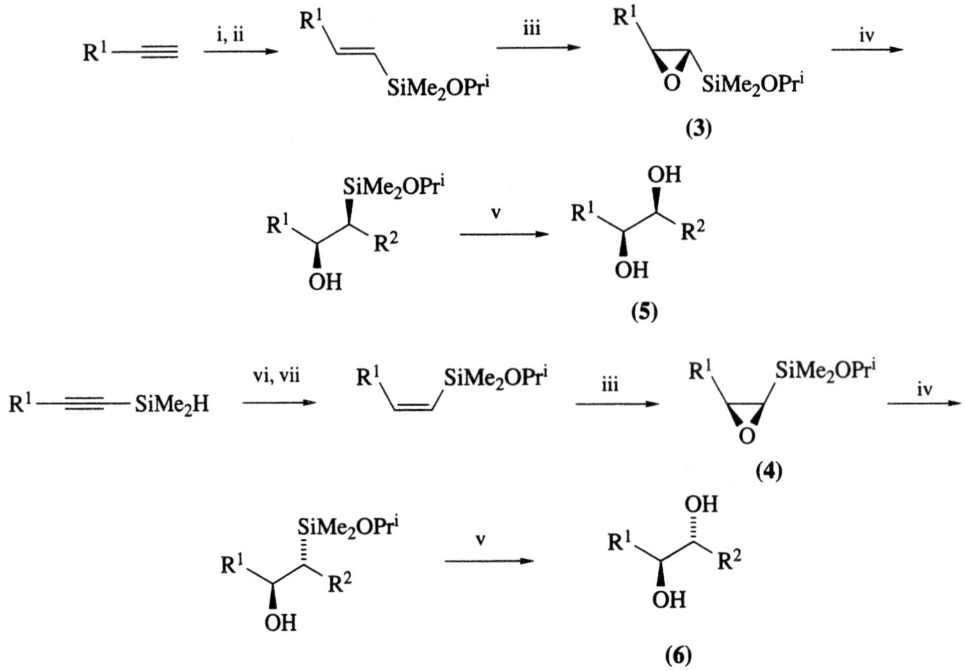

R^1—≡ →(i, ii) R^1CH=CH$SiMe_2OPr^i$ →(iii) (3) →(iv)

→(v) (5)

R^1—≡—$SiMe_2H$ →(vi, vii) →(iii) (4) →(iv)

→(v) (6)

i, Me_2SiClH, H_2PtCl_6 cat.; ii, Pr^iOH, Et_3N; iii, MCPBA; iv, R^2MgX, CuCN; v, H_2O_2, KF, $KHCO_3$, MeOH, THF; vi, hydroalumination then hydrolysis; vii, Pr^iOH, H_2PtCl_6 cat.

Scheme 6

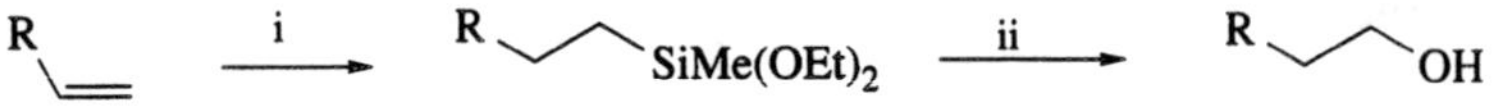

i, $(EtO)_2MeSiH$, cat.; ii, H_2O_2, KHF_2, DMF or MCPBA, KHF_2, DMF

Scheme 7

A related sequence involving alkynes, with the intermediacy of vinyl(alkoxy)silanes, has been described in detail;[16] the various oxidation conditions are summarized in Scheme 8.

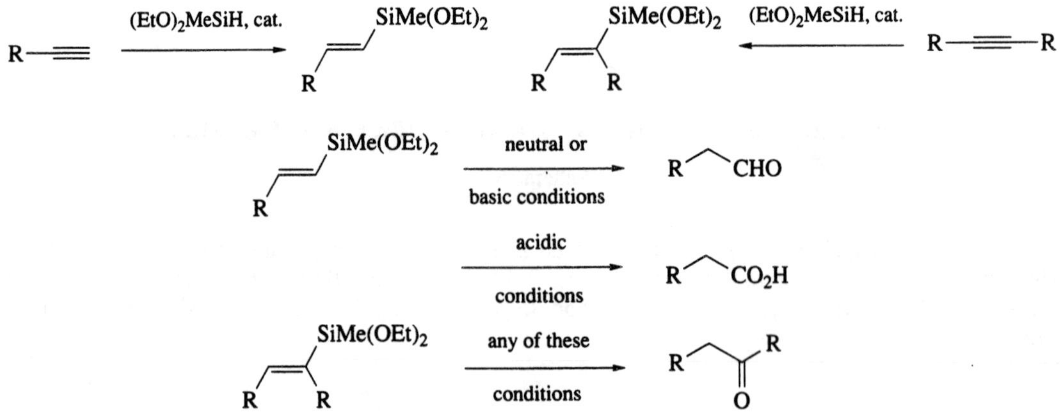

Neutral conditions = 30% H_2O_2, KHF_2, DMF, r.t. to 60 °C
Acidic conditions = 30% H_2O_2, Ac_2O, KHF_2, DMF, r.t.
Basic conditions = 30% H_2O_2, $KHCO_3$, MeOH, THF, 60 °C

Scheme 8

Addition[17] of organolithium and organomagnesium reagents to such vinyl(alkoxy)silanes, followed by catalyzed coupling of the new organometallic with either vinyl or allyl bromide, leads,[18] after oxidative cleavage, to allylic or homoallylic alcohols, respectively (Scheme 9).

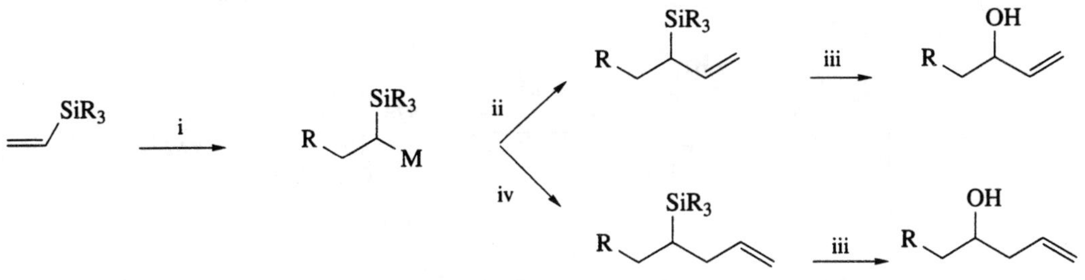

SiR_3 = $SiMe(OEt)_2$, *etc.*

i, RM; ii, $H_2C{=}CHBr$, cat.; iii, MCPBA, KF; iv, $H_2C{=}CHCH_2Br$, cat. [catalyst either $NiCl_2$ or $PdCl_2$(DPPF)]

Scheme 9

4.3.2.1.4 Intramolecular hydrosilylation of allylic and homoallylic alcohols

Intramolecular hydrosilylation of allyl and homoallyl alcohols, with subsequent oxidative cleavage of the resultant C—Si bond, has provided[19] a new approach to the regiocontrolled synthesis of 1,2-and/or 1,3-diols (see also Section 4.3.2.2.3). The example shown[20] (Scheme 10) illustrates nicely the use of *syn* stereoselection in a reiterative manner.

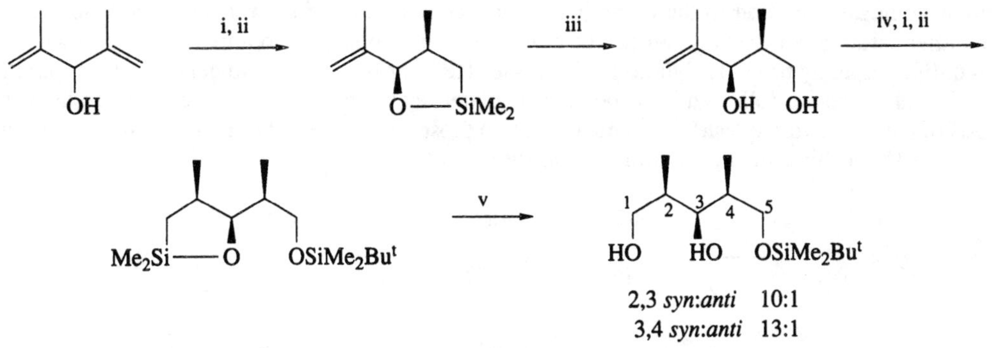

2,3 *syn:anti* 10:1
3,4 *syn:anti* 13:1

i, $(HMe_2Si)_2NH$, r.t. to 60 °C; ii, H_2PtCl_6 cat., 60 °C; iii, H_2O_2, $NaHCO_3$, MeOH, THF; iv, Bu^tMe_2SiCl, Et_3N, DMAP; v, H_2O_2, KF, $KHCO_3$, MeOH, THF

Scheme 10

In an extension of this process, the intramolecular hydrosilylation of α-hydroxy enol ethers has been presented[21] as a new, *syn* selective route to 1,2,3-triols (Scheme 11). With such sensitive substrates, a neutral hydrosilylation catalyst, $Pt\{[(CH_2{=}CH)Me_2Si]_2O\}_2$,[22] must be used. The utility of this method has been demonstrated in a synthesis of the pentitols, D-arabinol and xylitol (as their pentaacetates), in optically pure form.

In a related study[23] of the Lewis acid catalyzed intramolecular hydrosilylation of β-silyloxy ketones, *anti* selective hydrosilylation has been observed.

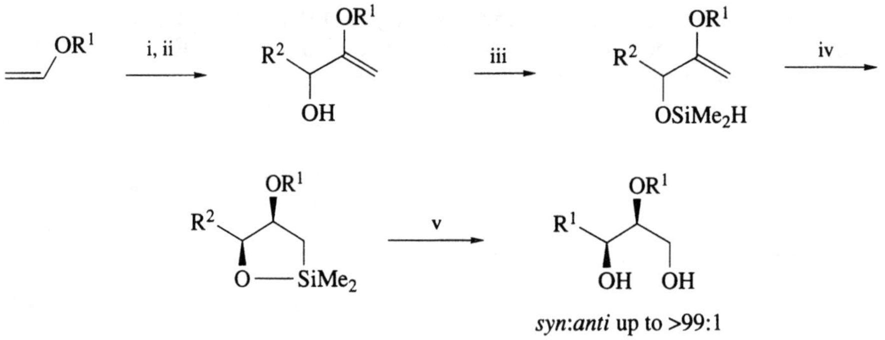

i, BuLi; ii, R^2CHO; iii, $(HMeSi)_2NH$, NH_4Cl cat.; iv, Pt^0 cat.; v, H_2O_2, KOH, MeOH, THF

syn:anti up to >99:1

Scheme 11

4.3.2.2 From Functionalized Silanes

4.3.2.2.1 Use of $PhMe_2Si$ moieties

Extensive studies[4] by Fleming and his group have elegantly demonstrated the utility of lithium bis(phenyldimethylsilyl)cuprate, $(PhMe_2Si)_2CuLi$, for the stereocontrolled nucleophilic introduction of hydroxy groups. Conjugate addition of this reagent to an α,β-unsaturated ester[24] (or an α,β-unsaturated δ-lactone[25]) produces an intermediate β-silyl enolate, which can be trapped with electrophiles such as iodomethane to produce the *anti* diastereoisomer selectively. Access to the *syn* diastereoisomer is provided by similar conjugate addition to the α-methyl-α,β-unsaturated ester followed by protonation (Scheme 12). In either case, the $PhMe_2Si$ group can be converted[26] in two steps into a hydroxy group with retention ofconfiguration, by protiodesilylation using either HBF_4 or $BF_3.2AcOH$ to generate the required fluorosilane (and benzene), followed by a peracid-mediated rearrangement. It can also be applied to the synthesis of chiral β-hydroxy esters[27,28] and amides, in those cases where the α,β-unsaturated carbonyl is functionalized by a chiral alcohol or amine auxiliary.

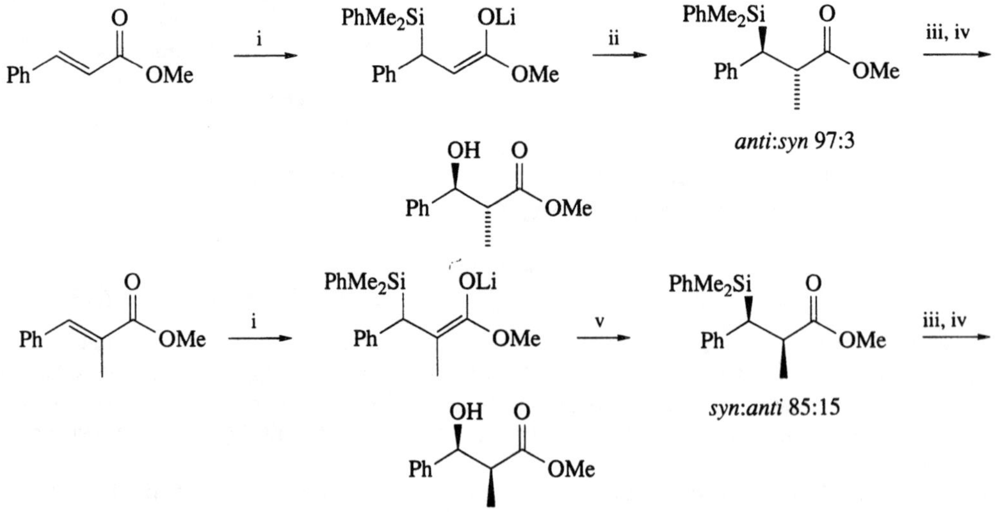

anti:syn 97:3

syn:anti 85:15

i, $(PhMe_2Si)_2CuLi$; ii, MeI; iii, HBF_4; iv, MCPBA, Et_3N; v, NH_4Cl

Scheme 12

Such β-silyl enolate intermediates also react with aldehydes[29] with high diastereoselectivity with respect to both new chiral centers being created, the relative stereochemistry in the aldol reaction being dependent upon the original geometry of the enolate double bond (Scheme 13). This aldol reaction has been applied to a formal synthesis of thienamycin,[30] and, with imines as the electrophilic partners, to a stereocontrolled route[31] to 3-(1-hydroxyethyl)azetidin-2-ones.

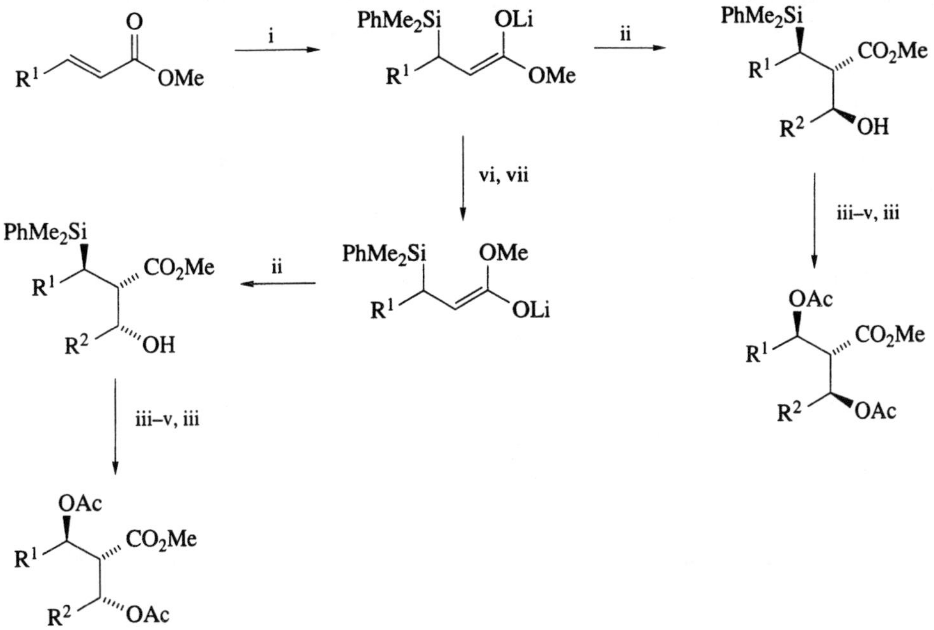

i, $(PhMe_2Si)_2CuLi$; ii, R^2CHO, –78 °C, iii, Ac_2O, Et_3N; iv, $BF_3{\bullet}AcOH$; v, MCPBA, Et_3N; vi, NH_4Cl; vii, LDA, –78 °C

Scheme 13

This two-step protiodesilylation/oxidative cleavage, converting a $PhMe_2Si$ function into a hydroxy group, can be carried out in one pot,[32] using either Br_2 or mercury(II) in an acetic acid solution of peracetic acid. The bromine may be generated by adding the peracid solution to KBr, making it unnecessary to handle bromine itself, and sodium acetate may be used to buffer the sulfuric acid present in commercial peracetic acid. However, the latter device, useful when acid-sensitive groups are present, only works in those reactions using bromine. When using mercury(II), acid is needed to catalyze the mercuration of the benzene ring.

In a synthesis of (–)-reserpine, Stork[33] was unable to effect selective protiodesilylation, using HBF_4, of a $PhMe_2Si$ group in the presence of a benzyl ether. However, alternative use of a 2-furyldimethylsilyl group obviated this difficulty, due to the enhanced reactivity of the furan ring towards fluoride displacement under neutral conditions.

4.3.2.2.2 Nucleophilic hydroxymethylating agents, d^1-methanol synthons

The Grignard reagent $(Pr^iO)_2MeSiCH_2MgCl$ takes part in a metal-catalyzed coupling reaction[34] with alkyl, vinyl (stereochemistry retained), allyl (stereochemistry and regiochemistry retained), aryl and heteroaryl chlorides and bromides. With functionally substituted aryl or heteroaryl halides, conversion into the corresponding organozinc reagent confers increased chemoselectivity. Reaction of the adducts with KF or KHF_2, followed by oxidative cleavage using either H_2O_2 or peracetic acid results in the overall nucleophilic introduction of a hydroxymethyl group, as exemplified in Scheme 14. One application of its use can be seen in a synthesis[35] of (–)-asperdiol, in this case with copper-catalyzed coupling with an alkyl mesylate.

Later studies have shown that a single isopropoxy group suffices for successful oxidative cleavage. The Grignard reagent $Pr^iOMe_2SiCH_2MgCl$ adds[36] to ketones and aldehydes to give adducts which, upon oxidative cleavage, yield 1,2-diols, as shown in Scheme 15. This reagent also effects a copper-catalyzed coupling with allylic chlorides, as demonstrated in a synthesis of (+)-casbene.[37]

The latter reagent undergoes 1,2-addition to α,β-unsaturated aldehydes; 1,4-addition, with copper catalysis, is observed with cyclohexenone alone. A more satisfactory reagent[38] for the conjugate introduction of the hydroxymethyl group is the allyldimethylsilylmethyl Grignard reagent

i, $(Pr^iO)_2MeSiCH_2MgCl$, CuI cat. ; ii, 90% H_2O_2; iii, $(Pr^iO)_2MeSiCH_2MgCl$, $NiCl_2(DPPP)$ cat.; iv, 30% H_2O_2

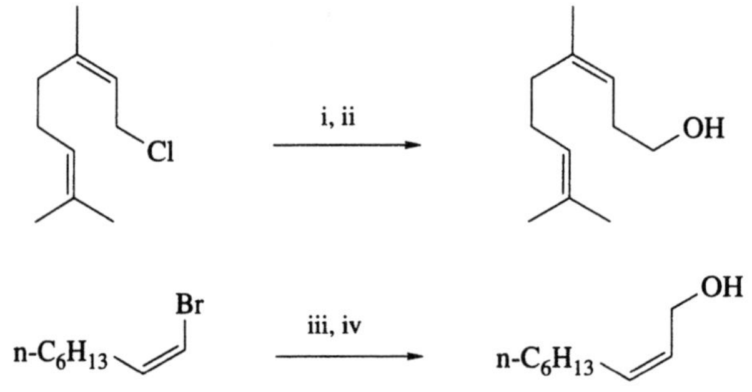

Scheme 14

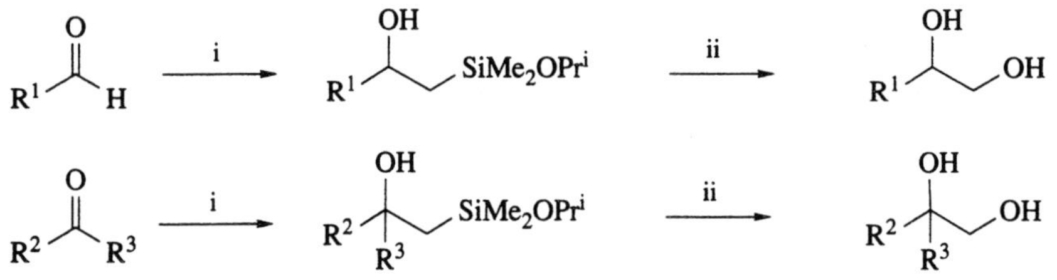

i, $Pr^iOMe_2SiCH_2MgCl$; ii, H_2O_2, $NaHCO_3$, MeOH, THF

Scheme 15

$(CH_2{=}CHCH_2)Me_2SiCH_2MgCl$. Protiodesilylation of the intermediate adduct from isophorone in the presence of fluoride ion generates the fluorosilane (**7**), which then undergoes oxidative cleavage under the normal conditions (Scheme 16). One limitation of this sequence is that it cannot be applied to cyclopentenone nor to $\Delta^{1,9}$-2-octalone systems.

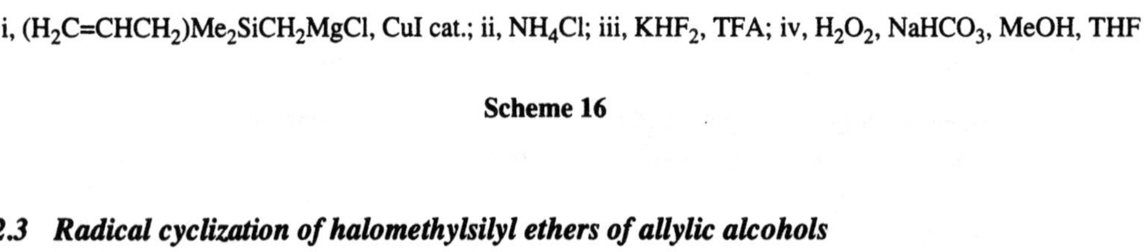

(**7**)

i, $(H_2C{=}CHCH_2)Me_2SiCH_2MgCl$, CuI cat.; ii, NH_4Cl; iii, KHF_2, TFA; iv, H_2O_2, $NaHCO_3$, MeOH, THF

Scheme 16

4.3.2.2.3 Radical cyclization of halomethylsilyl ethers of allylic alcohols

Silylmethyl radicals,[39,40] generated for halomethylsilyl ethers of allylic alcohols, can provide an indirect method of achieving acyclic stereocontrol. Depending on the substrate substitution pattern, either 6-*endo*- or 5-*exo-trig* cyclization can predominate. Stork and coworkers have developed an excellent method[41] for the control of ring junction stereochemistry using such radicals (Scheme 17). In such a 5-*exo-trig* process, transition state geometry dictates a *cis* fusion of the new five-membered ring. The resulting radical, being cup shaped, allows ready access to tributylstannane only from the convex face, resulting, after oxidative cleavage, in the effective overall *anti* addition of a hydroxymethyl group and a hydrogen to the original double bond.

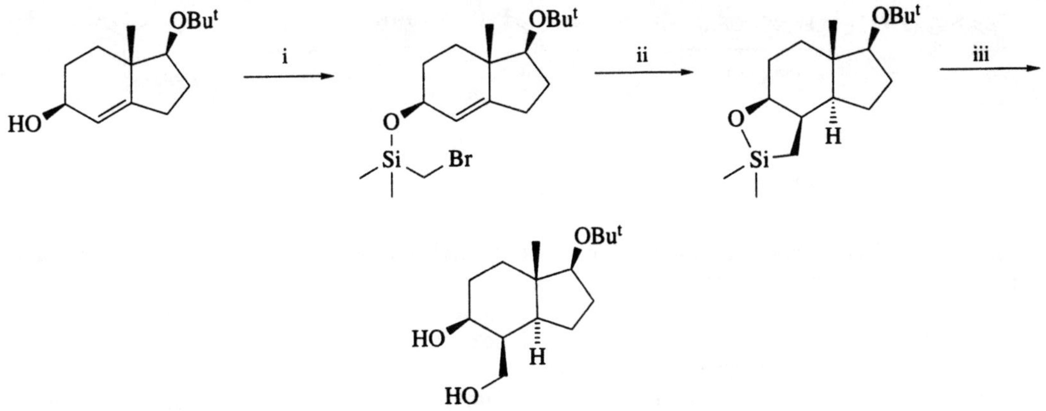

i, $Me_2Si(Cl)CH_2Br$, Et_3N, DMAP; ii, Bu_3SnH, AIBN cat., PhH or $NaCNBH_3$, Bu_3SnCl cat., AIBN cat., Bu^tOH; iii, H_2O_2, KF, DMF

Scheme 17

Similar methodology can be used in acyclic systems[42] for the diastereoselective, and sometimes diastereoselective (Scheme 18), formation of 1,3-diols. With some substrates, the proportion of products arising from 6-*endo-trig* cyclization can be significant.

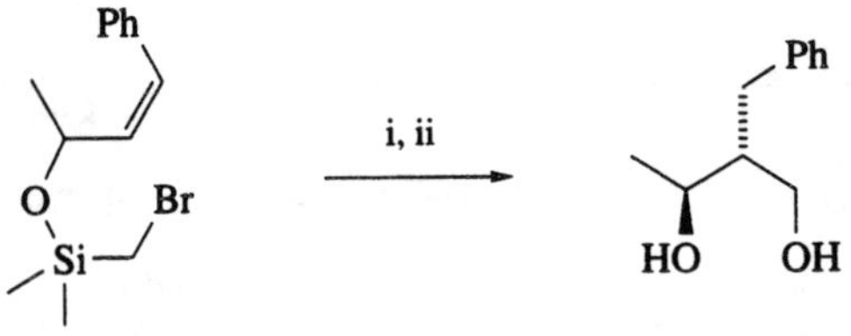

i, Bu_3SnH, AIBN cat., PhH; ii, H_2O_2, KF, DMF

Scheme 18

Indeed, in certain steroidal systems, 6-*endo-trig* cyclization becomes preferred, with ultimate regio- and stereo-controlled production[43] of 1,4-diols (Scheme 19). However, other steroidal systems which react by 5-*exo-trig* cyclization have been described.[44]

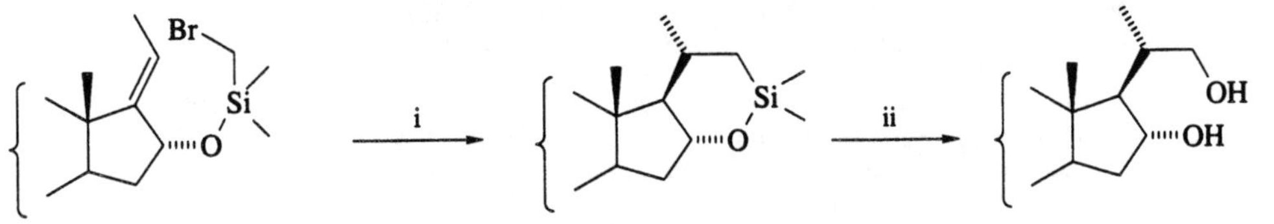

i, Bu_3SnH, AIBN cat., PhH; ii, H_2O_2, $KHCO_3$, MeOH, THF

Scheme 19

4.3.3 MISCELLANEOUS

Aryltrimethylsilanes undergo a facile metal/metal exchange[45] with lead(IV) trifluoroacetate; the intermediate aryllead species then eliminate lead(II) acetate to form the corresponding aryl trifluoroacetates in almost quantitative yield (Scheme 20).

Benzyltrimethylsilanes, on the other hand, give products of C–Si cleavage, *i.e.* benzyl nitrate and acetate, on treatment with cerium(IV) ammonium nitrate in AcOH; based on ring substituent effects, a one-electron transfer mechanism[46] seems to be in operation.

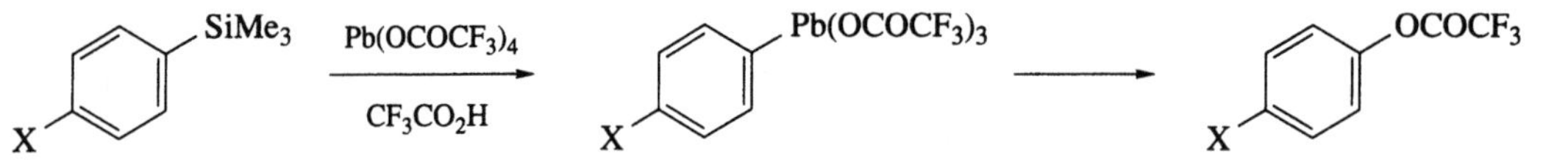

$$p\text{-}X\text{-}C_6H_4\text{-}SiMe_3 \xrightarrow[CF_3CO_2H]{Pb(OCOCF_3)_4} p\text{-}X\text{-}C_6H_4\text{-}Pb(OCOCF_3)_3 \longrightarrow p\text{-}X\text{-}C_6H_4\text{-}OCOCF_3$$

X = F, Cl, Br, Me

Scheme 20

Methoxy(trimethylsilyl)methane and methoxybis(trimethylsilyl)methane have been proposed[47] as new synthons for the formyl anion and the methoxycarbonyl anion, respectively; after alkylation, C–Si cleavage is achieved by anodic oxidation. Similar electrochemical oxidative cleavage[48] of acylsilanes reveals their potential as acyl cation synthons. Anodic oxidation[49] of *N*-silylmethyl carbamates in methanol produces *N*-methoxymethyl carbamates in high yield.

Treatment of tetramethylsilane with TfOH yields[50] trimethylsilyl triflate (Scheme 21). Although not a direct oxidation, this is an excellent, simple method for the preparation of a most useful reagent.[51]

$$Me_4Si + CF_3COOH \longrightarrow Me_3SiOCOCF_3 + CH_4$$

Scheme 21

4.3.4 REFERENCES

1. E. Buncel and A. G. Davies, *J. Chem. Soc.*, 1958, 1550.
2. G. A. Razuvaev, V. A. Yablokov, A. V. Ganyushkin, N. V. Yablokova and G. S. Kalinina, *J. Organomet. Chem.*, 1979, **165**, 281.
3. K. Tamao, in 'Organosilicon and Bioorganosilicon Chemistry', ed. H. Sakurai, Ellis Horwood, Chichester, 1985, p. 231; K. Tamao, *J. Synth. Org. Chem., Jpn.*, 1988, **46**, 861; M. Kumada, K. Tamao and J. Yoshida, *J. Organomet. Chem.*, 1982, **239**, 115.
4. I. Fleming, in 'Organosilicon and Bioorganosilicon Chemistry', ed. H. Sakurai, Ellis Horwood, Chichester, 1985, p. 197.
5. J. A. Soderquist and C. L. Anderson, *Tetrahedron Lett.*, 1986, **27**, 3961.
6. H. Sakurai, M. Ando, N. Kawada, K. Sato and A. Hosomi, *Tetrahedron Lett.*, 1986, **27**, 75.
7. K. Sato, M. Kira and H. Sakurai, *Tetrahedron Lett.*, 1989, **30**, 4375.
8. K. Tamao, T. Kakui and M. Kumada, *J. Am. Chem. Soc.*, 1978, **100**, 2268; K. Tamao, T. Kakui, M. Akita, T. Iwahara, R. Kanatani, J. Yoshida and M. Kumada, *Tetrahedron*, 1983, **39**, 983.
9. K. Tamao, J. Yoshida, H. Yamamoto, T. Kakui, H. Matsumoto, M. Takahashi, A. Kurita, M. Murata, and M. Kumada, *Organometallics*, 1982, **1**, 355.
10. K. Tamao, J. Yoshida, M. Murata and M. Kumada, *J. Am. Chem. Soc.*, 1980, **102**, 3267.
11. J. Yoshida, K. Tamao, M. Kumada and T. Kawamura, *J. Am. Chem. Soc.*, 1980, **102**, 3269.
12. T. Hayashi, K. Tamao, Y. Katsuro, I. Nakae and M. Kumada, *Tetrahedron Lett.*, 1980, 1871.
13. K. Tamao, R. Kanatani and M. Kumada, *Tetrahedron Lett.*, 1984, **25**, 1913.
14. K. Tamao, E. Nakajo and Y. Ito, *J. Org. Chem.*, 1987, **52**, 4412.
15. K. Tamao, N. Ishida, T. Tanaka and M. Kumada, *Organometallics*, 1983, **2**, 1694.
16. K. Tamao, M. Kumada and K. Maeda, *Tetrahedron Lett.*, 1984, **25**, 321.
17. K. Tamao, R. Kanatani and M. Kumada, *Tetrahedron Lett.*, 1984, **25**, 1905.
18. K. Tamao, T. Iwahara, R. Kanatani and M. Kumada, *Tetrahedron Lett.*, 1984, **25**, 1909.
19. K. Tamao, T. Tanaka, T. Nakajima, R. Sumiya, H. Arai and Y. Ito, *Tetrahedron Lett.*, 1986, **27**, 3377.
20. K. Tamao, T. Nakajima, R. Sumiya, H. Arai, N. Higuchi and Y. Ito, *J. Am. Chem. Soc.*, 1986, **108**, 6090.
21. K. Tamao, Y. Nakagawa, H. Arai, N. Higuchi and Y. Ito, *J. Am. Chem. Soc.*, 1988, **110**, 3712.
22. L. N. Lewis and N. Lewis, *J. Am. Chem. Soc.*, 1986, **108**, 7228.
23. S. Anwar and A. P. Davis, *Tetrahedron*, 1988, **44**, 3761.
24. W. Bernhard, I. Fleming and D. Waterson, *J. Chem. Soc., Chem. Commun.*, 1984, 28; W. Bernhard and I. Fleming, *J. Organomet. Chem.*, 1984, **271**, 281.
25. I. Fleming, N. L. Reddy, K. Takaki and A. C. Ware, *J. Chem. Soc., Chem. Commun.*, 1987, 1472.
26. I. Fleming, R. Henning and H. Plaut, *J. Chem. Soc., Chem. Commun.*, 1984, 29.
27. W. Oppolzer, R. J. Mills, W. Pachinger and T. Stevenson, *Helv. Chim. Acta*, 1986, **69**, 1542.
28. I. Fleming and N. D. Kindon, *J. Chem. Soc., Chem. Commun.*, 1987, 1177.
29. I. Fleming and J. D. Kilburn, *J. Chem. Soc., Chem. Commun.*, 1986, 305.
30. I. Fleming and J. D. Kilburn, *J. Chem. Soc., Chem. Commun.*, 1986, 1198.
31. D. A. Burnett, J. C. Gallucci and D. J. Hart, *J. Org. Chem.*, 1985, **50**, 5120.
32. I. Fleming and P. E. J. Sanderson, *Tetrahedron Lett.*, 1987, **28**, 4229.
33. G. Stork, *Pure Applied Chem.*, 1989, **61**, 439.
34. K. Tamao, N. Ishida, and M. Kumada, *J. Org. Chem.*, 1983, **48**, 2120.
35. M. A. Tius and A. Fauq, *J. Am. Chem. Soc.*, 1986, **108**, 6389.
36. K. Tamao and N. Ishida, *Tetrahedron Lett.*, 1984, **25**, 4245.
37. J. E. McMurry and G. K. Bosch, *J. Org. Chem.*, 1987, **52**, 4885.
38. K. Tamao and N. Ishida, *Tetrahedron Lett.*, 1984, **25**, 4249.

39. J. W. Wilt, *J. Am. Chem. Soc.*, 1981, **103**, 5251; *Tetrahedron*, 1985, **41**, 3979.
40. D. P. Curran, *Synthesis*, 1988, 417, 489.
41. G. Stork and M. Kahn, *J. Am. Chem. Soc.*, 1985, **107**, 500; G. Stork and M. J. Sofia, *J. Am. Chem. Soc.*, 1986, **108**, 6826; G. Stork and P. M. Sher, *J. Am. Chem. Soc.*, 1986, **108**, 303; G. Stork and R. Mah, *Tetrahedron Lett.*, 1989, **30**, 3609.
42. H. Nishiyama, T. Kitajima, M. Matsumoto and K. Itoh, *J. Org. Chem.*, 1984, **49**, 2298.
43. M. Koreeda and I. A. George, *J. Am. Chem. Soc.*, 1986, **108**, 8098.
44. A. Kurek-Tyrlik, J. Wicha and G. Snatzke, *Tetrahedron Lett.*, 1988, **29**, 4001.
45. J. R. Kalman, J. T. Pinney and S. Sternhell, *Tetrahedron Lett.*, 1972, 5369; H. C. Bell, J. R. Kalman, J. T. Pinney and S. Sternhell, *Tetrahedron Lett.*, 1974, 853.
46. E. Baciocchi, T. Del Giacco, C. Rol and G. V. Sebastiani, *Tetrahedron Lett.*, 1989, **30**, 3573.
47. J. Yoshida, S. Matsunaga and S. Isoe, *Tetrahedron Lett.*, 1989, **30**, 219.
48. J. Yoshida, S. Matsunaga and S. Isoe, *Tetrahedron Lett.*, 1989, **30**, 5293.
49. J. Yoshida and S. Isoe, *Tetrahedron Lett.*, 1987, **28**, 6621.
50. M. Demuth and G. Mikhail, *Synthesis*, 1982, 827.
51. R. Noyori, S. Murata and M. Suzuki, *Tetrahedron*, 1981, **37**, 3899; H. Emde, D. Domsch, H. Feger, U. Frick, A. Götz, H. H. Hergott, K. Hofmann, W. Kober, K. Krägeloh, T. Oesterle, W. Steppan, W. West and G. Simchen, *Synthesis*, 1982, 1.

4.4
Oxidation of Carbon–Halogen Bonds

S. NICHOLAS KILENYI
Sanofi Research, Brussels, Belgium

4.4.1 INTRODUCTION

Halides are extremely versatile intermediates. One of their more valuable and interesting transformations is their oxidation to carbonyl compounds, thus providing straightforward routes to relatively inaccessible compounds such as 1,2-diketones and heterocyclic aldehydes. As will be seen below, there are many, often complementary, synthetic methods for the oxidation of organic halides. The oxidation of halomethyl compounds has been reviewed recently.[1]

4.4.2 KORNBLUM OXIDATION AND RELATED METHODS

4.4.2.1 Oxidation with Dimethyl Sulfoxide

Known as the Kornblum oxidation, this is perhaps the most widely used and best-known method for the oxidation of halides.[2,3] It works best with activated halides such as benzyl halides, α-halocarbonyl compounds and iodides, though unactivated bromides also work reasonably well. Experimentally, the method consists of heating the halide in DMSO, usually in the presence of a base such as sodium bicarbonate to scavenge the HX. The reaction proceeds by an S_N2 displacement by the oxygen atom of the dimethyl sulfoxide, followed by proton loss and 3,2-sigmatropic rearrangement of the resulting sulfur ylide. The mechanistic analogy with the Swern oxidation is clear, since only the route to the sulfonium ion intermediate differs (Scheme 1).[4]

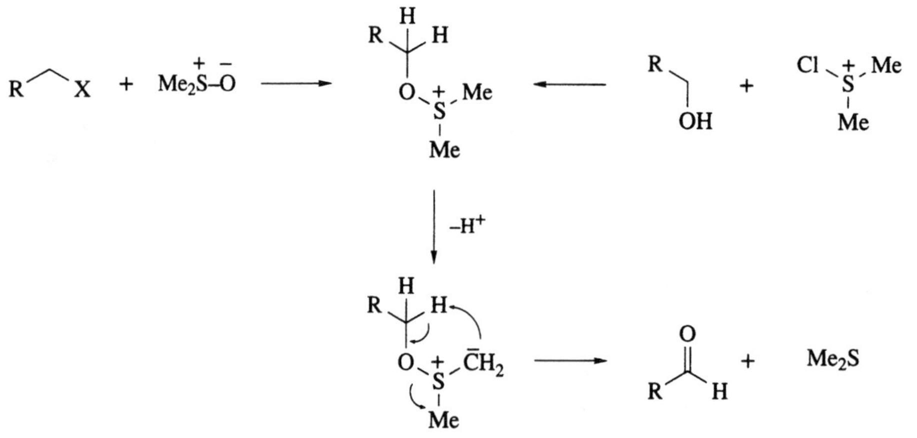

Scheme 1

However, as shown in Scheme 2, a different mechanism intervenes when R is a carbonyl group, since the protons adjacent to it are now more acidic than those of the methyl groups of the sulfonium ion.[5]

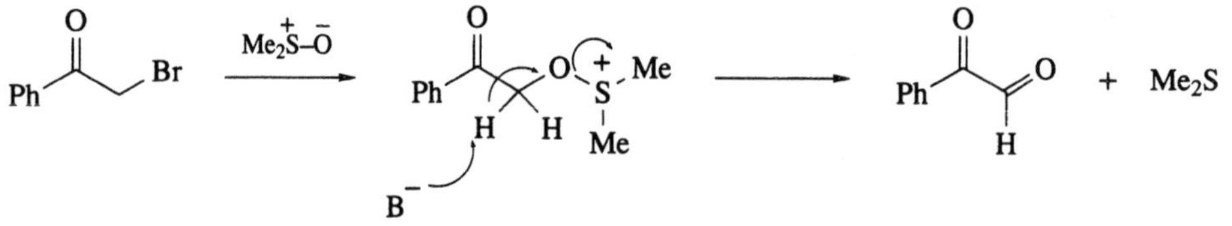

Scheme 2

As can be seen from Table 1 the method is general for a wide variety of activated halides, and provides an excellent alternative to selenium dioxide for the synthesis of 1,2-diketones and keto aldehydes. It constitutes, perhaps, the definitive synthesis of glyoxylates. It is not necessary to heat many α-bromo ketones since the reaction often proceeds smoothly at room temperature.

There are some important limitations on the Kornblum oxidation, however, as would be expected for a reaction relying on an S_N2 displacement as the first step. Secondary halides are prone to elimination under these conditions.[6] Similarly, if the approach of the DMSO is hindered, *e.g.* by a 1,3-diaxial interaction, the oxidation proceeds poorly (equation 1).[7] Overoxidation can also occur, at least in steroidal α-bromo ketones (Scheme 3).[8]

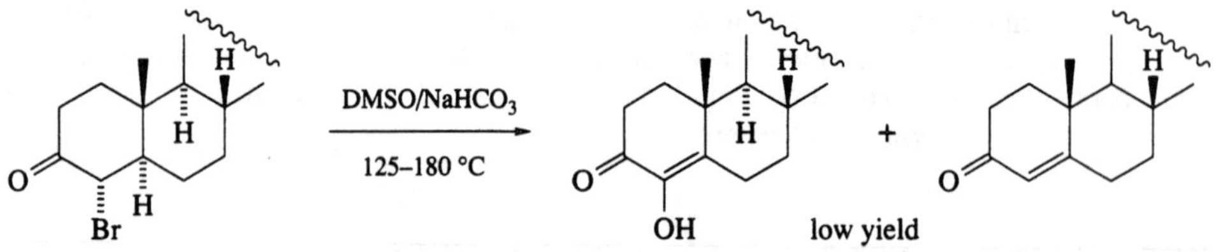

DMSO/$NaHCO_3$, 125–180 °C; low yield (1)

Kornblum *et al.* had asserted[9] that it was necessary to convert primary iodides to tosylates for the oxidation to proceed. However, Johnson and Pelter[10] have claimed that primary iodides can be oxidized directly. They observed that ketonic substrates failed, undergoing aldol condensation under the reaction conditions, but that hydroxy-containing halides reacted normally. For substrates insoluble in DMSO such as 1-bromododecane, they found that DME worked well as a cosolvent. The most interesting example

Table 1

Halide	*Conditions*	*Product*	*Yield (%)*	*Ref.*
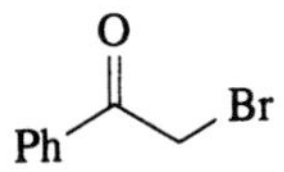 PhCOCH$_2$Br	DMSO, 25 °C	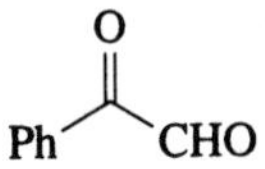 PhCOCHO	71	18
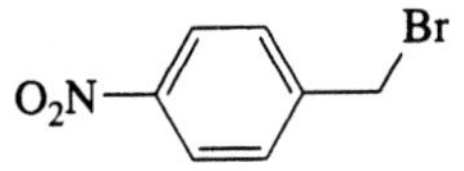 O_2N-C$_6$H$_4$-CH$_2$Br	DMSO, MeCN, 25 °C	 O_2N-C$_6$H$_4$-CHO	48	18
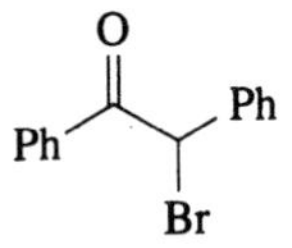 PhCOCH(Br)Ph	DMSO, 45 °C	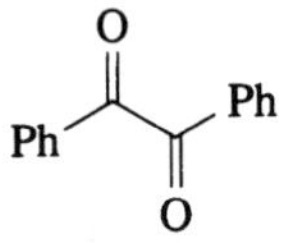 PhCOCOPh	95	18
BrCH$_2$CO$_2$Et	(glycidyl phenyl ether) OPh, DMSO	OHCCO$_2$Et	70	19
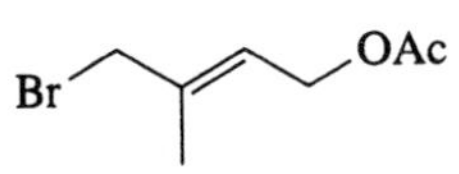 BrCH$_2$C(Me)=CHCH$_2$OAc	DMSO, NaHCO$_3$, 25 °C	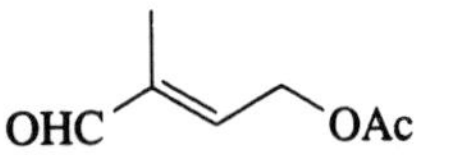 OHCC(Me)=CHCH$_2$OAc	81	20

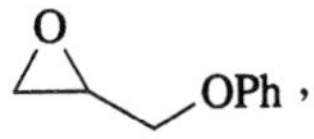

DMSO/NaHCO$_3$, 125 °C

3% + 60–65% + + others

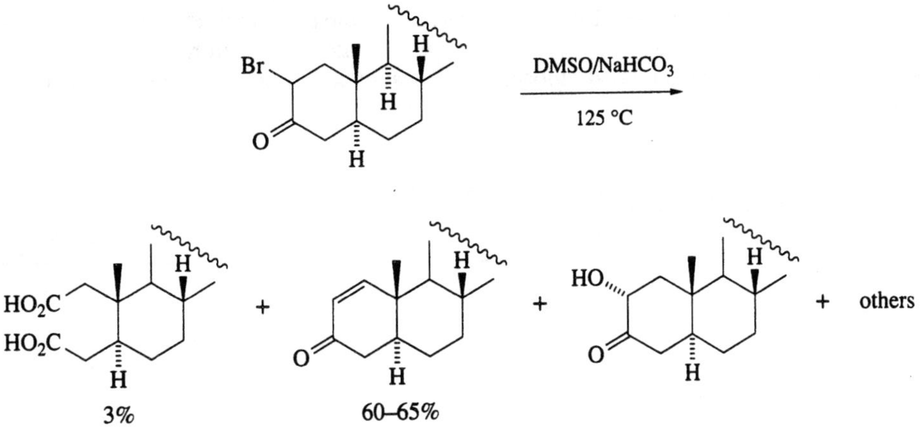

Scheme 3

was the oxidation of 2-iodooctane to the ketone, albeit in a yield of 32%, showing that secondary substrates are sometimes feasible.

Omission of the base in the Kornblum oxidation can have interesting consequences. Bromine, produced by the oxidation of the liberated HBr by DMSO,[11] may lead to bromination of the primary oxidation product in good yield (equation 2).[12] Note the use of epichlorohydrin as an HBr scavenger (equation 3).

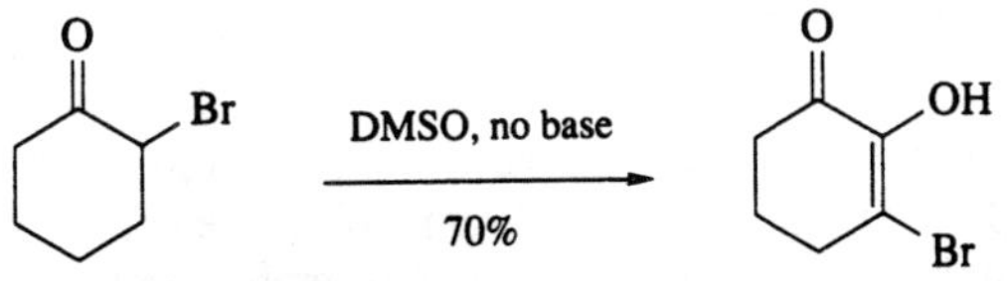

DMSO, no base, 70% (2)

One way to increase the nucleofugacity of halides is to introduce silver ion. Kornblum reported that primary unactivated chlorides, bromides and iodides could be oxidized by prior conversion to the tosylate with silver tosylate, followed by reaction in DMSO (Scheme 4).[9]

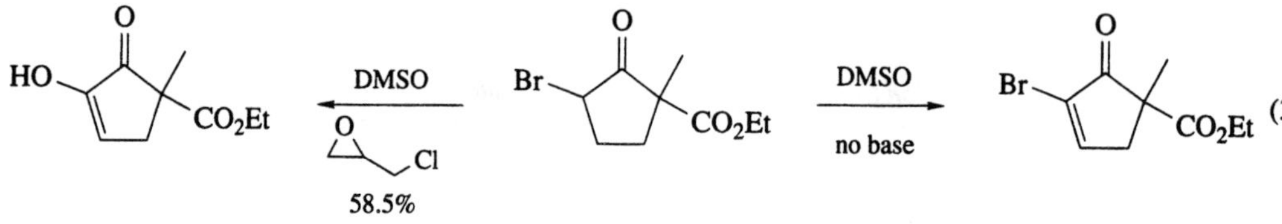

(3)

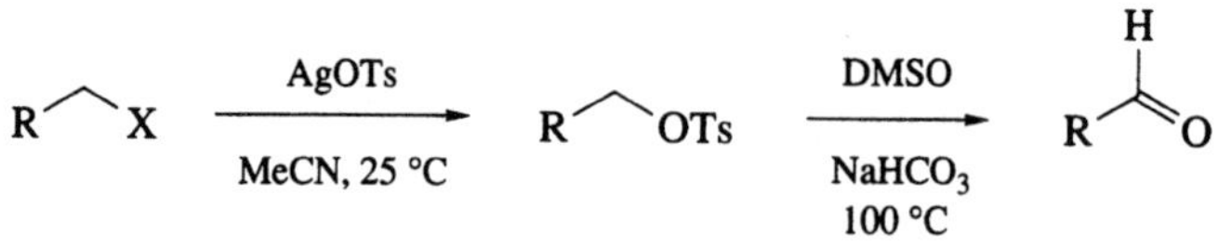

R	X	Yield (%)
n-C_7H_{15}	Cl	71
C_6H_{13}	I	70
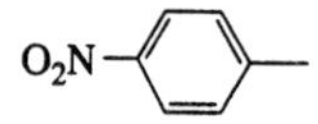 O_2N-C_6H_4-	Br	76

Scheme 4

However, the neopentyl example still failed. Later it was found that addition of a silver salt to the DMSO solution was equally effective (Table 2). In two of the three reports[13,14] triethylamine was added as a base after formation of the sulfonium salt. Silver perchlorate is claimed to be superior to silver nitrate.[15] The method fails for unactivated chlorides, for deactivated bromides, and for substrates where solvolytic reactions are possible.

Table 2

Halide	*Conditions*	*Product*	*Yield (%)*	*Ref.*
	i, $AgBF_4$, DMSO, 25 °C; ii, Et_3N		68.5	14
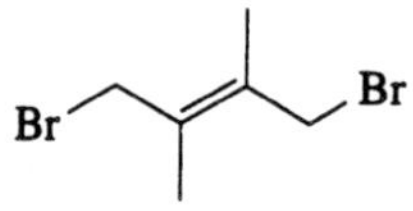	i, $AgBF_4$, DMSO, 25 °C; ii, Et_3N	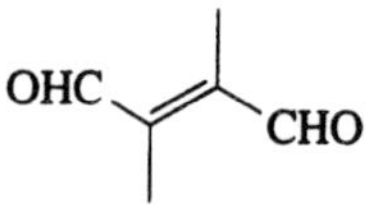	60	13
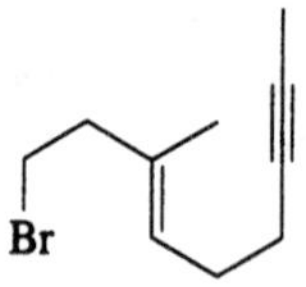	i, $AgBF_4$, DMSO, 25 °C; ii, Et_3N	—	0	13

The Kornblum oxidation can be combined with halogenation to provide a very convenient synthesis of 2- and 4-pyridinecarbaldehydes from the corresponding picolines (equation 4).[16]

The procedure could also be used to synthesize 2,4- and 2,6-pyridinedicarbaldehydes from the lutidines.

Reaction of cyclic 1,3-diones with DMSO containing catalytic bromine gives 1,2,3-triones *via* the 2-bromodione (equation 5).[17] Unfortunately, the reaction gives only poor yields with acyclic diones.

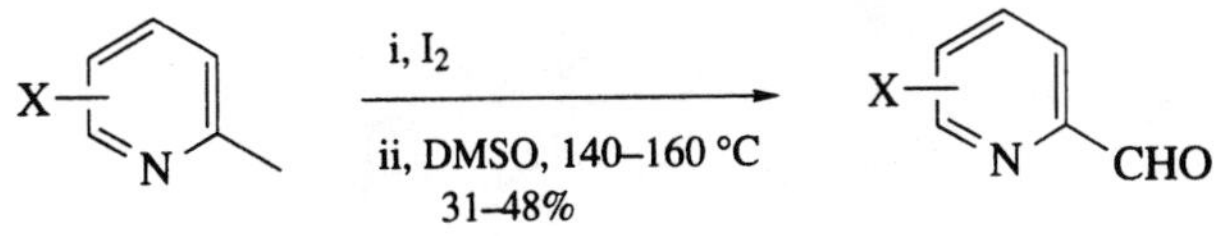

i, I_2; ii, DMSO, 140–160 °C; 31–48% (4)

X = Me, Et, CO_2Me, CHNOH

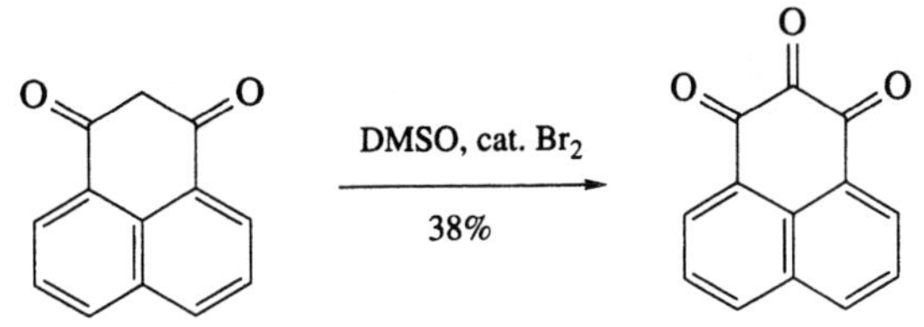

DMSO, cat. Br_2; 38% (5)

4.4.2.2 Oxidation with Selenoxides

Use of dimethyl selenoxide[21] or di(4-anisyl) selenoxide[22] in place of DMSO is claimed to offer some advantages in terms of mildness for polysubstituted or base sensitive benzyl halides. Problems of cost, accessibility and toxicity are not addressed.

4.4.3 THE KRÖHNKE OXIDATION

One of the oldest and most popular methods for oxidizing activated halides is the three-step Kröhnke oxidation.[23] The individual steps are: (i) quaternization of the halide with pyridine; (ii) deprotonation in base and reaction of the resulting pyridinium ylide with *N,N*-dimethyl-4-nitrosoaniline; and (iii) acid hydrolysis of the nitrone to the carbonyl compound (Scheme 5).

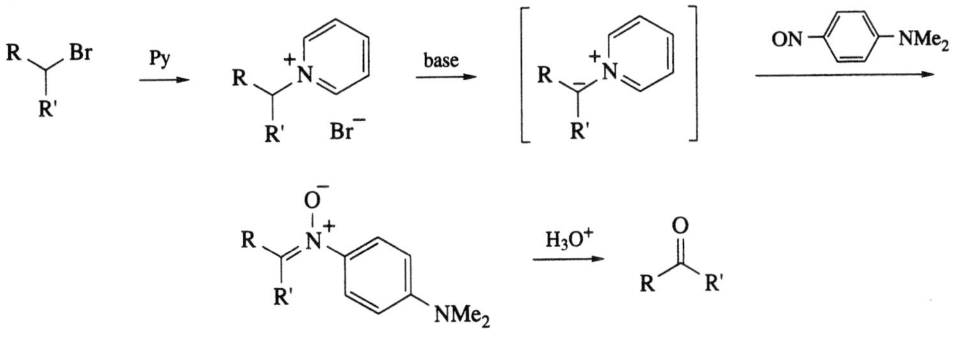

Py; base; ON–C_6H_4–NMe_2; H_3O^+

Scheme 5

The method is limited to activated halides which are stable to alkoxide bases and aqueous acids. According to the discoverer,[23] the Kröhnke oxidation is inferior to the Sommelet reaction for benzyl halides, especially those bearing electron-withdrawing groups, such as nitro. Some representative examples are shown in Scheme 6[27] and equations (6)–(12).[28–34] As can be seen, the Kröhnke oxidation is an excellent method for the preparation of heterocyclic aldehydes which would otherwise be difficult to synthesize. The preparations of a phthalimidoketo aldehyde (equation 11) and benzene-1,3,5-tricarbaldehyde (equation 12) are also noteworthy.

The Kröhnke oxidation often gives good results when other methods fail. In the example given (equation 13)[35] neither the Sommelet or the Hass–Bender method gave any aldehyde.

Similarly, in the preparation of 2-acetoxy-5-nitrobenzaldehyde from the benzyl bromide, the Sommelet reaction affords only 7% of the product, whereas the Kröhnke method worked in good, but unspecified yield.[24]

For acid sensitive substrates, the nitrone can be removed by hydrazinolysis followed by diazotization.[25] In the example given (equation 14),[25] direct acid hydrolysis of the nitrone with 1 M HCl gave only a 4% yield.

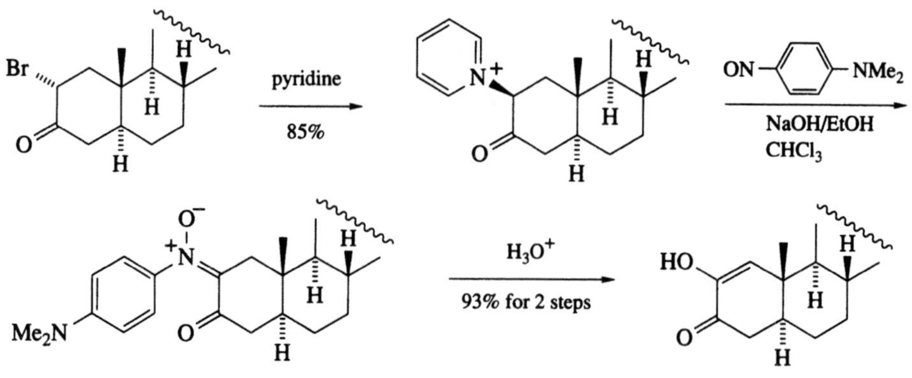

Br
H
pyridine
85%
N+
ON
NMe2
NaOH/EtOH
CHCl3
O−
Me2N
H3O+
93% for 2 steps
HO

Scheme 6

Cl
R
i, pyridine
ii, Me2N NO, NaOH
iii, H3O+
78–88% overall
CHO
(6)

Cl
R
i, pyridine
ii, Me2N NO, NaOH
iii, H3O+
CHO
(7)

Cl
R
i, pyridine, 31–94%
ii, Me2N NO, 56–82%
iii, H3O+, 23–68%
OHC
(8)

NO2
NBS
Br
i, pyridine
ii, Me2N NO
iii, H3O+
47–53%
CHO
(9)

Br
i, pyridine
ii, p-Me2NC6H4NO
iii, H3O+, no yield
CHO
(10)

By combination of the King–Ortoleva reaction and Kröhnke oxidation, activated methyl groups have been transformed directly into carbaldehydes (Scheme 7).[26]

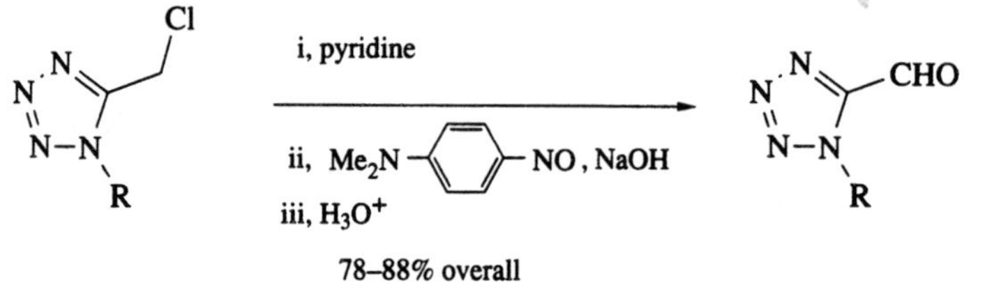

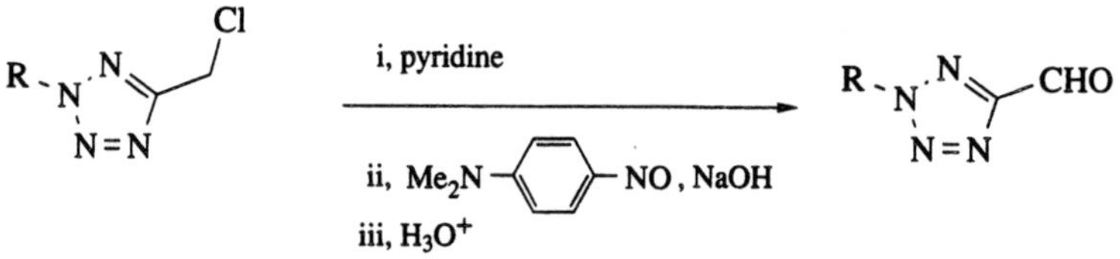

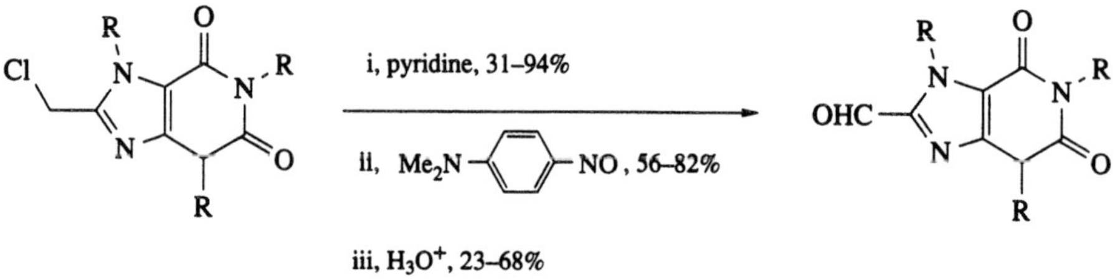

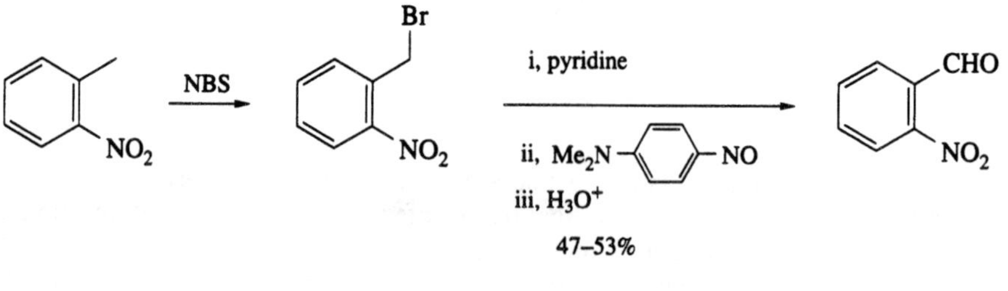

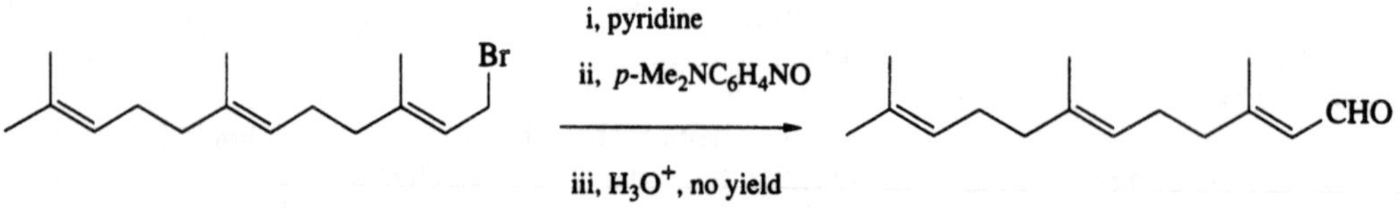

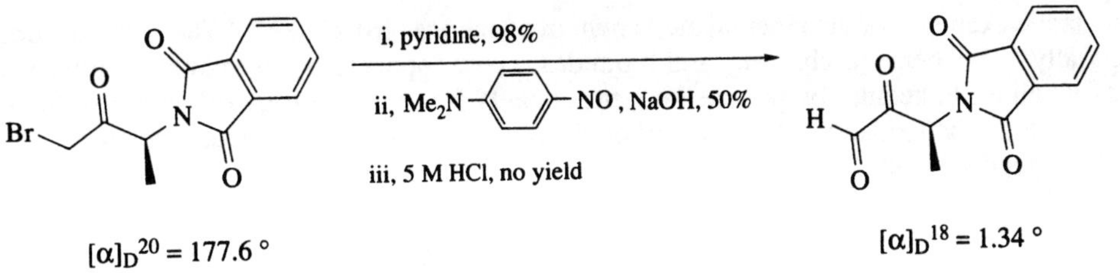

i, pyridine, 98%
ii, Me_2N–C_6H_4–NO, NaOH, 50%
iii, 5 M HCl, no yield

$[\alpha]_D^{20}$ = 177.6 ° $[\alpha]_D^{18}$ = 1.34 ° (11)

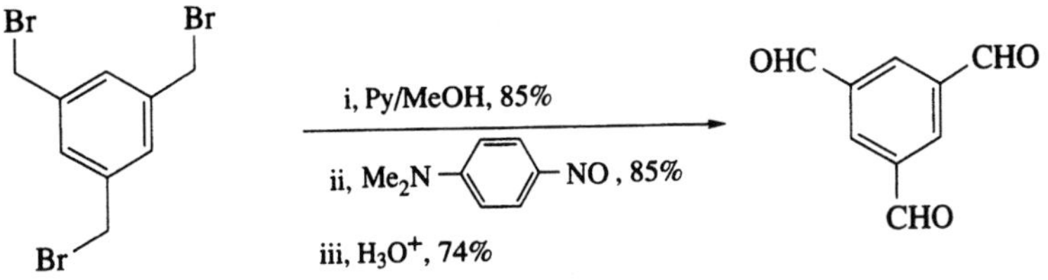

i, Py/MeOH, 85%
ii, Me_2N–C_6H_4–NO, 85%
iii, H_3O^+, 74%

OHC CHO CHO (12)

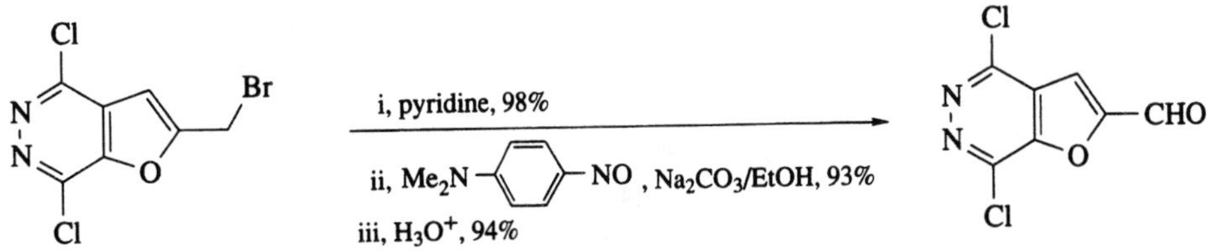

i, pyridine, 98%
ii, Me_2N–C_6H_4–NO, Na_2CO_3/EtOH, 93%
iii, H_3O^+, 94%

CHO (13)

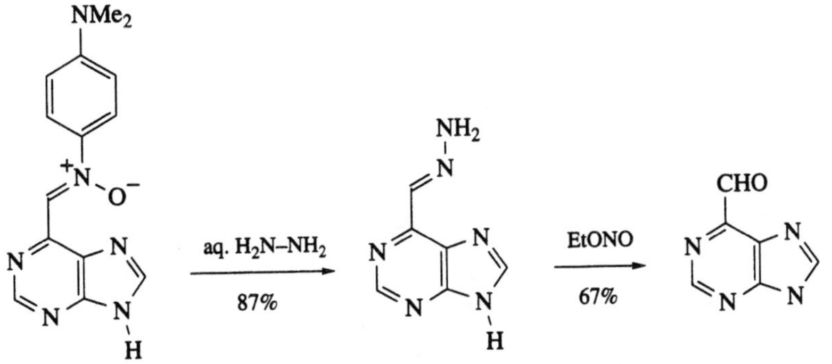

aq. H_2N–NH_2 87% EtONO 67% (14)

I_2, pyridine; as perchlorate salt; Me_2N–C_6H_4–NO, Na_2CO_3; 3 M HCl, 39% overall

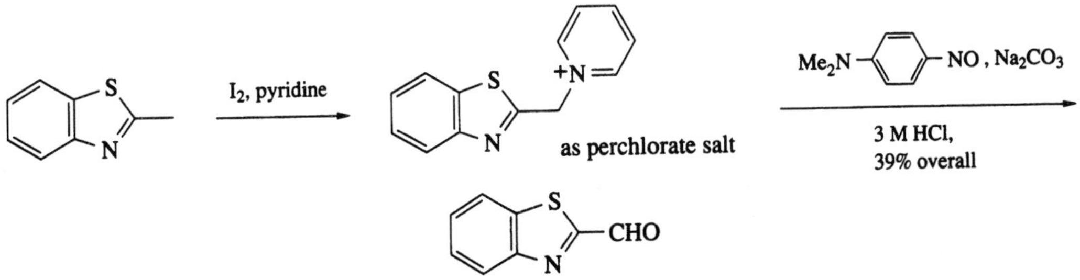

Scheme 7

4.4.4 THE HASS–BENDER REACTION

Reaction of a disubstituted nitronate anion with an allylic or benzylic halide leads not to the expected *C*-alkylated nitro compound, but rather to the carbonyl product.[36] Presumably this reaction, known by the names of its discoverers, proceeds by the displacement of the halide ion by nitronate oxygen followed by loss of the oxime (equation 15).

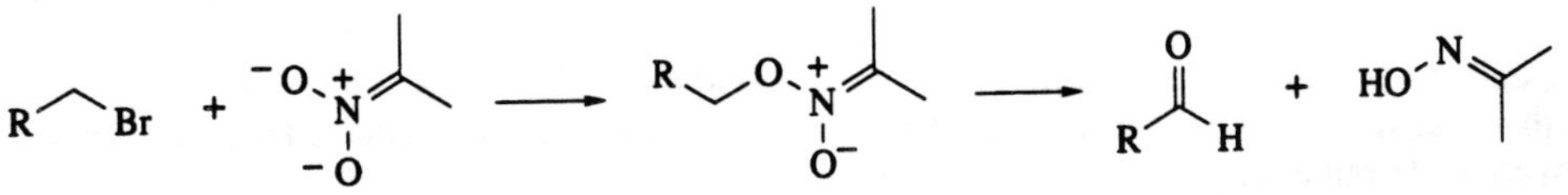

(15)

Representative examples of the reaction are shown in equations (16)–(20).[37–41] The method works well for primary allylic and benzylic chlorides and bromides. There appear to be no examples of the oxidation of secondary halides to ketones by this method, presumably for reasons of lower reactivity. Neither are there any reports of the oxidation of α-halocarbonyl compounds, which is curious since these would be expected to be good substrates.

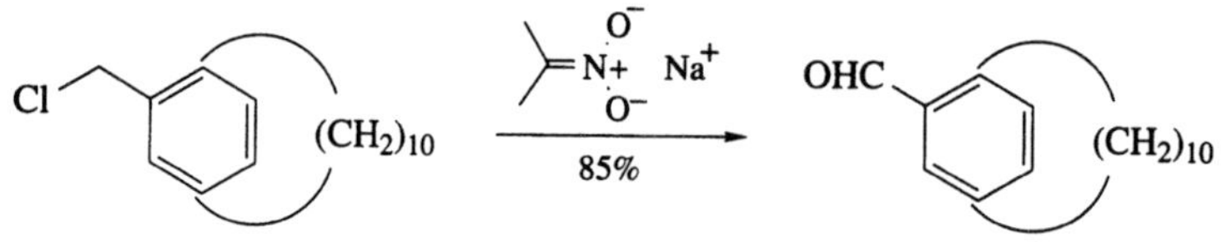

(16)

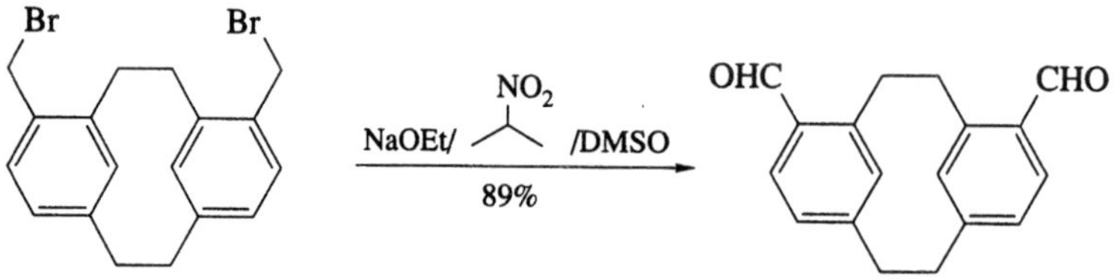

(17)

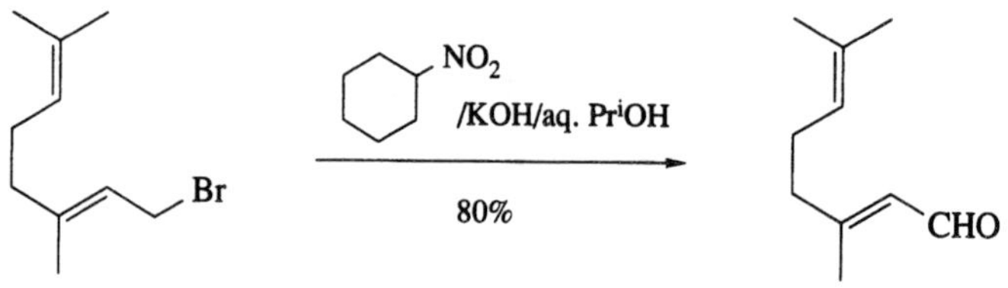

(18)

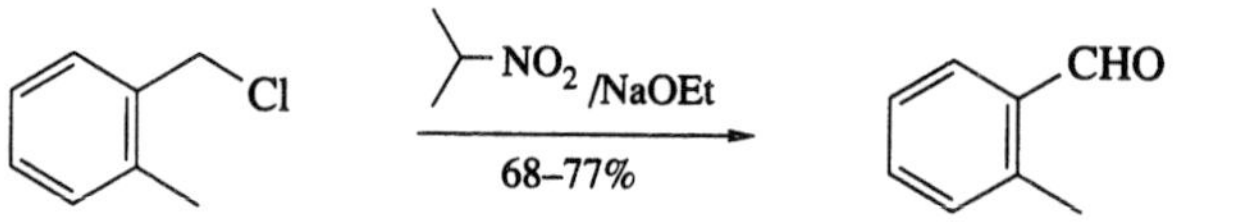

(19)

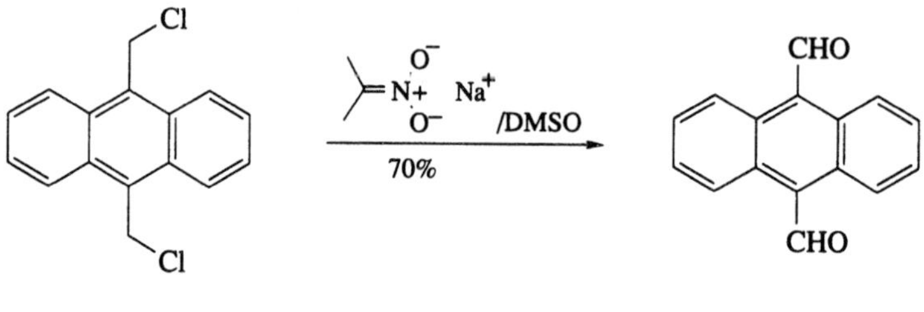

(20)

The Hass–Bender oxidation is often competitive with other methods: in equation (16)[37] the Sommelet reaction gave only 20% of the aldehyde, whereas in equation (20)[41] DMSO and sodium hydrogen carbonate failed to furnish any of the desired product.

There is an interesting limitation to the method: 4-nitrobenzyl chloride gives only 1% of the aldehyde, the major product being the result of single-electron transfer and radical coupling (equation 21).[42]

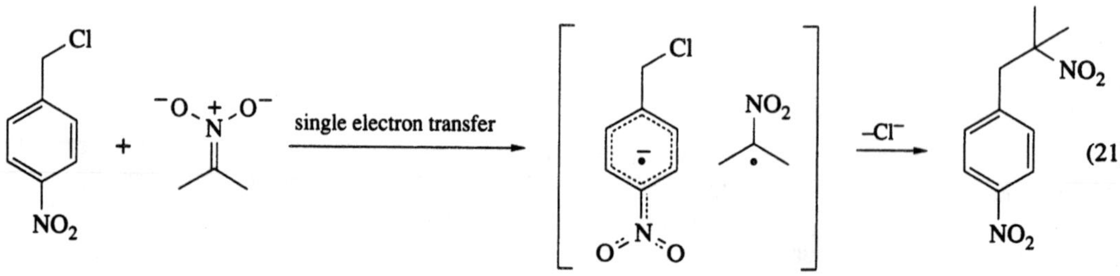

(21)

A Pd^0 catalyst has been used *in situ* to isomerize a secondary allylic halide prior to oxidation (equation 22).[43] In this case it is possible that reaction is occurring by attack of the nitronate anion on the π-allyl–palladium complex rather than on the chloride itself.

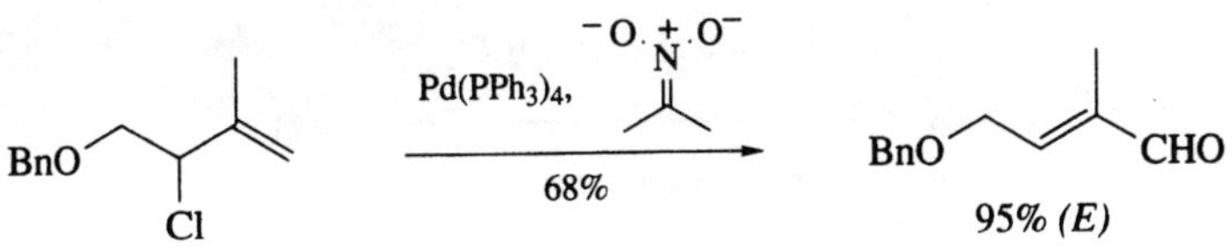

BnO Cl → [Pd(PPh3)4, $^{-}O\overset{+}{N}O^{-}$, 68%] BnO CHO 95% (E) (22)

4.4.5 OXIDATION WITH N-OXIDES, N-HYDROXYPYRIDONES AND HYDROXYLAMINES

4.4.5.1 Pyridine N-Oxides and Derivatives

The oxidation of alkyl halides to carbonyl compounds with pyridine or 2-picoline *N*-oxide is a popular and general method, applicable even to unactivated substrates. The reaction may be performed in two ways. In the first, the halide is heated with the *N*-oxide in the presence of a base such as sodium hydrogen carbonate. In the second, the intermediate *N*-alkoxypyridinium salt is isolated before base treatment. The reaction has been shown by labeling to proceed *via* the pyridinium ylide,[44] or, in the case of picoline *N*-oxide, *via* the anhydrobase (Scheme 8).[45] Some typical examples are shown in equations (23)–(25).[46–48]

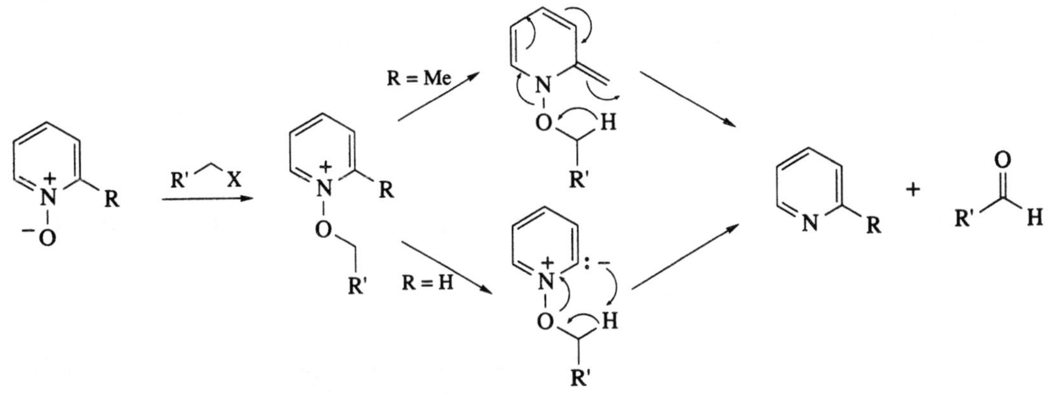

R' X; R = Me; R = H; R'; O; H; N; R

Scheme 8

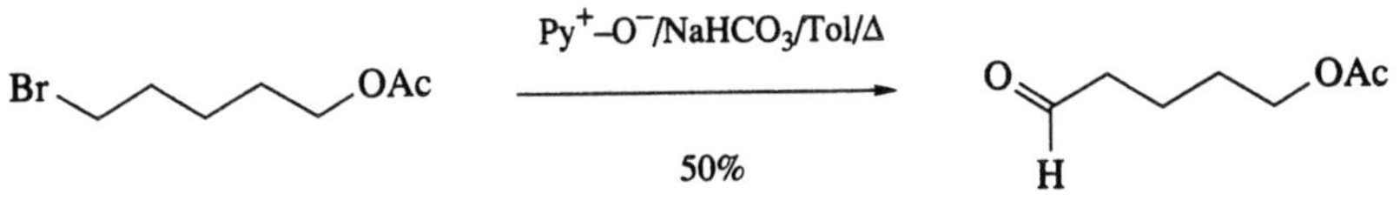

Br OAc → [$Py^{+}–O^{-}$/$NaHCO_3$/Tol/Δ, 50%] O H OAc (23)

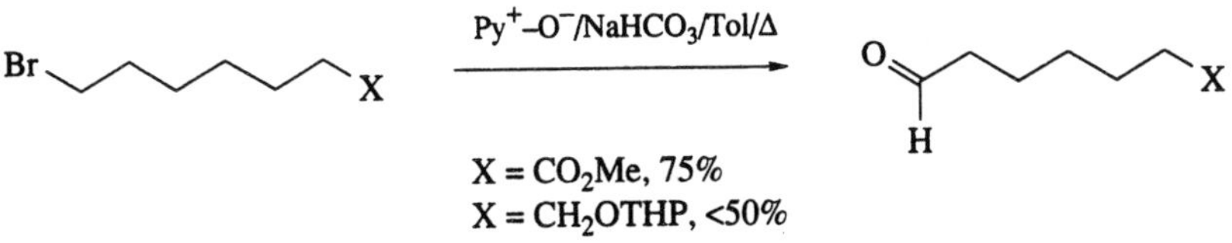

Br X → [$Py^{+}–O^{-}$/$NaHCO_3$/Tol/Δ] O H X (24)

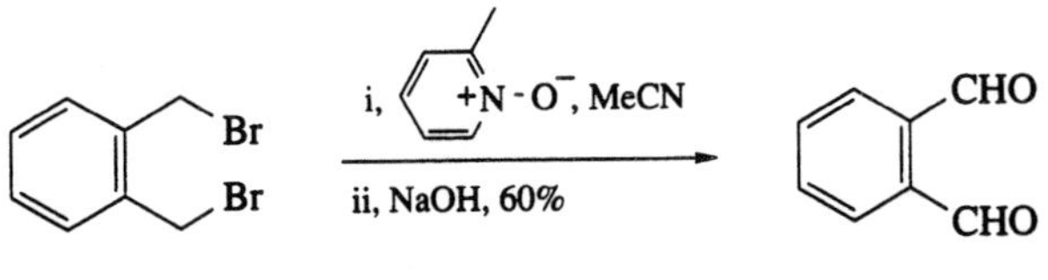

X = CO_2Me, 75%
X = CH_2OTHP, <50%

Br Br → [i, +N - O⁻, MeCN; ii, NaOH, 60%] CHO CHO (25)

The example in equation (24) demonstrates how the yield of these oxidations is sensitive to remote substituents, possibly due to cleavage of the THP ether under the reaction conditions.

The method is excellent for the preparation of α-keto esters and α-keto acids. In the examples given (equations 26–28) the α-bromo ester or acid was treated with pyridine *N*-oxide and silver nitrate at 0 °C. Decomposition of the isolated salt with base gave the dicarbonyl compounds in high yield.[49]

$$\mathrm{BrCH_2CO_2Et} \xrightarrow[\text{ii, Et}_3\text{N/MeOH, 87\%}]{\text{i, Py}^+\text{–O}^-\text{/AgNO}_3\text{/CH}_2\text{Cl}_2} \mathrm{HC(O)CO_2Et} \quad (26)$$

$$\mathrm{BrCH_2CO_2Bu^t} \xrightarrow[\text{ii, Et}_3\text{N/MeOH, 87\%}]{\text{i, Py}^+\text{–O}^-\text{/AgNO}_3\text{/CH}_2\text{Cl}_2} \mathrm{HC(O)CO_2Bu^t} \quad (27)$$

$$\mathrm{RCH(Br)CO_2H} \xrightarrow[\text{ii, Et}_3\text{N/MeOH, 87\%}]{\text{i, Py}^+\text{–O}^-\text{/AgNO}_3\text{/CH}_2\text{Cl}_2} \mathrm{RC(O)CO_2H} \quad (28)$$

The transformation shown in equation (27) is impressive since DMSO had previously been tried and shown to fail for this substrate.[50] The reaction was also performed as a one-pot procedure, with pyridine *N*-oxide and silver nitrate in acetonitrile followed by addition of triethylamine. This is the preferred method of these authors for these substrates.

It is curious that under thermal conditions α-bromo acids preferentially undergo oxidative decarboxylation rather than oxidation to the α-keto acid (equation 29).[51]

$$\mathrm{PhCH(Br)CO_2H} \longrightarrow \mathrm{PhCH(O\text{–}N^+C_5H_5)CO_2H} \longrightarrow \mathrm{PhCHO} + \mathrm{Py} + \mathrm{HBr} + \mathrm{CO_2} \quad (29)$$

The 1-*N*-oxide of 4-dimethylaminopyridine would be expected to be a more powerful nucleophile than the parent compound. This is indeed the case (equations 30–32), and DMAP 1-oxide is able to displace bromide even from secondary unactivated substrates, *e.g.* equation (31). However, the method still fails due to elimination with cyclohexyl bromide.[52]

$$\mathrm{C_7H_{15}CH_2Br} \xrightarrow[\text{ii, DBU}]{\text{i, Me}_2\text{N–C}_5\text{H}_4\text{N}^+\text{–O}^-\text{/MeCN/}\Delta} \mathrm{C_7H_{15}CHO} \quad (30)$$

$$\mathrm{CH_3CH_2CH(Br)CH_3} \xrightarrow[\text{ii, DBU, 98\% (GC)}]{\text{i, Me}_2\text{N–C}_5\text{H}_4\text{N}^+\text{–O}^-\text{/MeCN/}\Delta} \mathrm{CH_3CH_2C(O)CH_3} \quad (31)$$

$$\mathrm{PhCH(Br)CO_2Et} \xrightarrow[\text{ii, DBU}]{\text{i, Me}_2\text{N–C}_5\text{H}_4\text{N}^+\text{–O}^-\text{/MeCN/}\Delta} \mathrm{PhC(O)CO_2Et} \quad (32)$$

4.4.5.2 *N*-Hydroxypyridones

The use of *N*-hydroxypyridone salts offers some advantages over the procedures above in that the second step can be done under nonhydrolytic conditions either thermally or photochemically (equation 33).[53] This could be advantageous for base sensitive substrates. However, these methods are unsuitable for aliphatic halides (alcohols are the major product), and there appear to be no α-halocarbonyl examples. For photochemical cleavage, the *N*-oxides (**1**) and (**2**) offer some advantages.[54,55] However, the need to prepare these rather elaborate and high molecular weight reagents would seem to be a disadvantage.

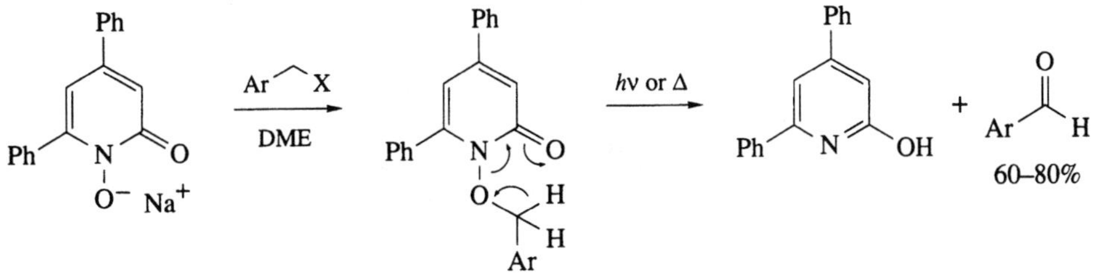

(33)

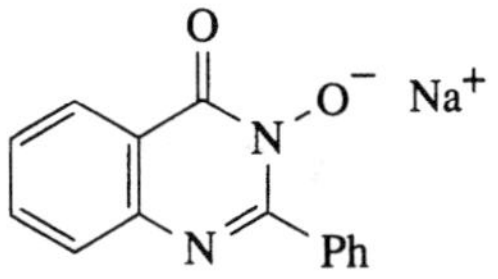

(1)

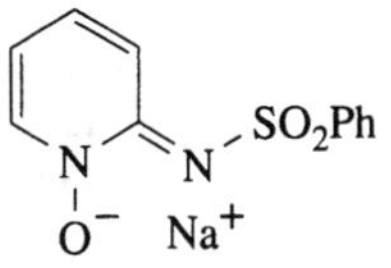

(2)

4.4.5.3 Amine Oxides and Hydroxylamines

Amine oxides can be used in place of pyridine *N*-oxide for the oxidation of activated and unactivated bromides and iodides (equation 34).[56,57] Scope and limitations are similar, though the yields appear to be somewhat lower.

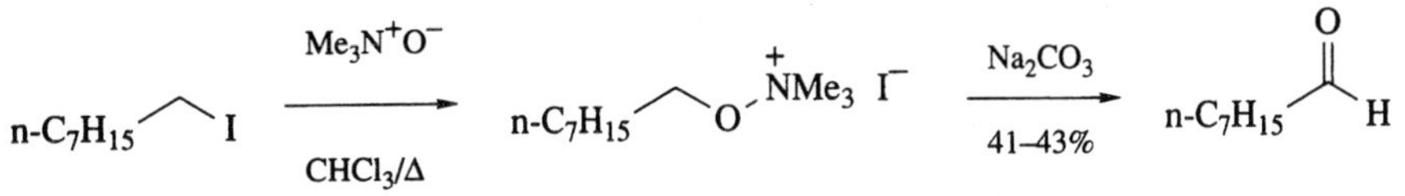

(34)

However, more recently the use of a polymer amine oxide for the oxidation of an alkyl iodide has been reported (equation 35).[58] The yield and experimental simplicity are impressive. The polymer may be regenerated.

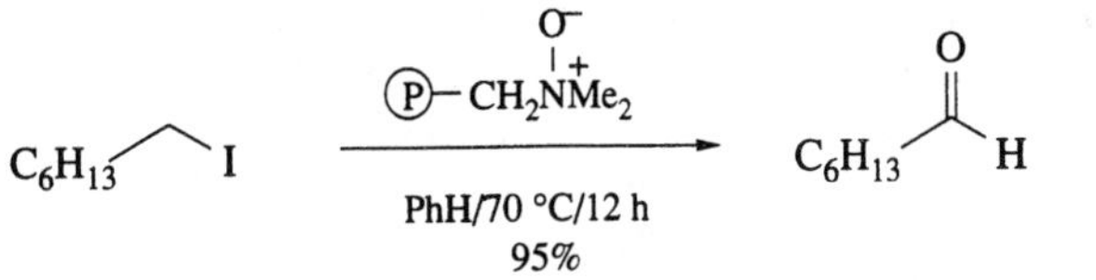

(35)

There is one report of the use of *N,N*-dialkylhydroxylamines to oxidize phenacyl bromides (equation 36).[59]

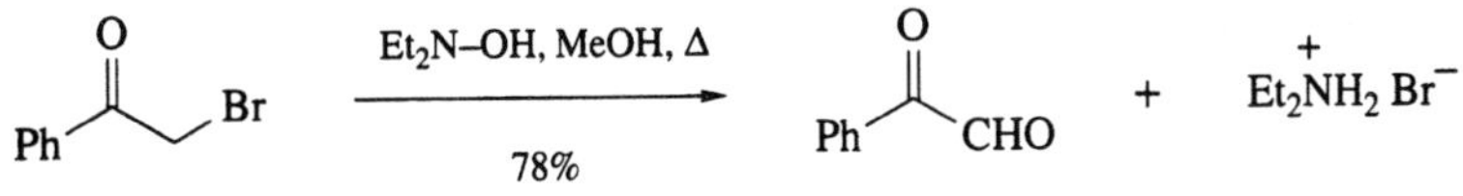

(36)

Advantages over the other methods are not obvious, and the method is not applicable to other α-halocarbonyl compounds such as 2-chlorocyclohexanone and α-bromopropiophenone.

4.4.6 CHROMIUM-BASED METHODS

Chromate and dichromate are capable of displacing halide from benzylic and allylic halides. Oxidation with dichromate has been performed under 'traditional' conditions (equation 37)[60] with the aqueous sodium salt, or with a quaternary ammonium salt in an aprotic solvent (equation 38).[61] Of the two procedures the first would seem to be preferable from the point of view of simplicity and safety.

Two procedures exist for the oxidation of allylic and benzylic halides with chromate ion. In the first,[62] the halide is heated with potassium dichromate in dry HMPA in the presence of 18-crown-6 (equation

$Na_2Cr_2O_7/H_2O/0$ °C, 60–68% (37)

$(Bu_4N)_2Cr_2O_7$, $CHCl_3/\Delta$, 81% (38)

39). Fortunately the same authors report a far more pleasant and less hazardous procedure utilizing a polymer-supported hydrogen chromate ion (equation 40).[63] The yields are all in excess of 95%, and the experimental simplicity is commendable.

K_2CrO_4, HMPA/18–100 °C/2 h, 82% (39)

Ⓟ$-C_6H_4\overset{+}{N}Me_3\ HCrO_4^-$, PhH/Δ, 95% (40)

4.4.7 METAL NITRITES AND NITRATES

4.4.7.1 Silver Nitrate

Reaction of halides with silver nitrate to give nitrate esters has been known for years, but its synthetic application is more recent. Kornblum showed that the nitrate esters derived from α-bromo ketones and esters decompose smoothly with catalytic sodium acetate in DMSO to give the α-dicarbonyl compounds in high yield.[64] It was found unnecessary to isolate the nitrate ester; after reaction of the halide with silver nitrate the solution was filtered to remove AgBr, concentrated, and added to DMSO containing catalytic sodium acetate. The method complements the others for the synthesis of α-dicarbonyl compounds since it employs nonacidic, nonbasic conditions. Unfortunately, the method gave variable results with benzyl halides. The application of the method to bromo esters other than bromoacetates was not reported. Some related oxidations are shown in equations (41) and (42), and Schemes 9 and 10.[65,66] The oxidation of an iminium salt is notable.

$AgNO_3$, MeCN/r.t./24 h; DMSO, cat. $NaOAc{\bullet}3H_2O$, r.t., 82–95% overall (41)

R = Ph, p-C_6H_4Cl, Br, NO_2; R' = H, Me, Ph

i, $AgNO_3$/MeCN/r.t.; ii, cat. $NaOAc{\bullet}3H_2O$/DMSO, 82–95% overall (42)

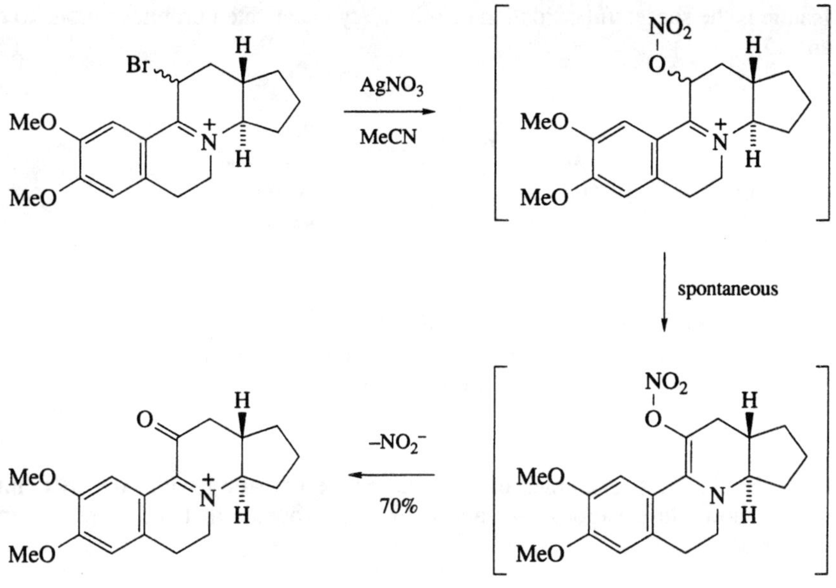

Br MeO MeO H H N+ AgNO$_3$ MeCN NO$_2$ O spontaneous −NO$_2^-$ 70%

Scheme 9

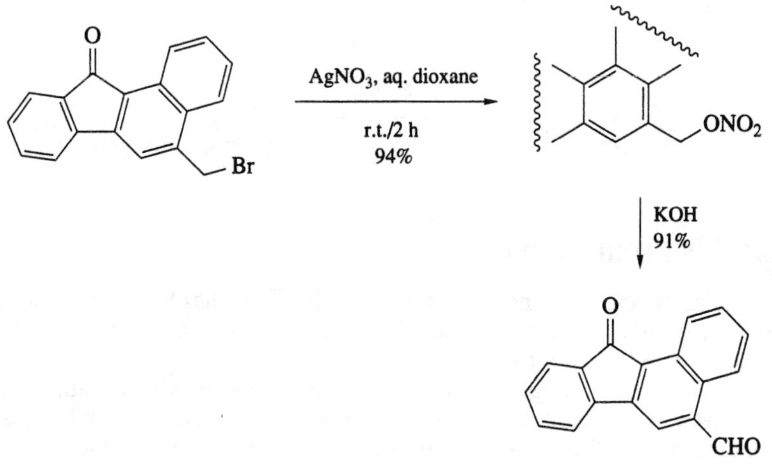

AgNO$_3$, aq. dioxane r.t./2 h 94% ONO$_2$ KOH 91% CHO

Scheme 10

4.4.7.2 Other Metal Nitrates and Nitrites

The difficulty experienced by Kornblum in the oxidation of benzylic halides (*vide supra*) was solved by McKillop and Ford[67] who found that mercury(II) nitrate gave the requisite nitrate esters in high yield with a wide variety of benzylic substrates (equation 43). However, problems were encountered at the hydrolysis stage. If the phenyl ring is substituted at positions 2 or 4 by an alkoxy group, the benzyl alcohol results rather than the aldehyde, due to expulsion of a nitrate ion. No recognizable products resulted when the ring bore a 4-nitro group. The successful oxidation of 2,6-dichlorobenzyl bromide is interesting.

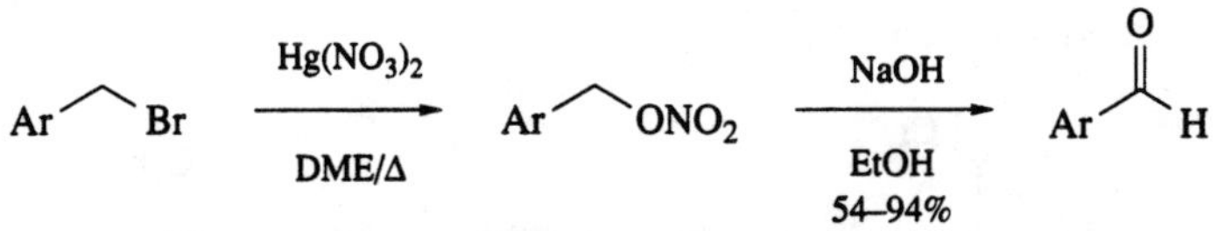

$$\mathrm{ArCH_2Br} \xrightarrow[\mathrm{DME}/\Delta]{\mathrm{Hg(NO_3)_2}} \mathrm{ArCH_2ONO_2} \xrightarrow[\mathrm{EtOH}\ 54\text{–}94\%]{\mathrm{NaOH}} \mathrm{ArCHO} \quad (43)$$

Ar = Ph, 3- and 4-MeC_6H_4, 3-$MeOC_6H_4$, 2- and 4-BrC_6H_4, 2,6-$Cl_2C_6H_3$, 1-naphthyl

Sodium nitrite in DMSO is a powerful nitrogen nucleophile,[68] and this property has been used in a very mild, nonhydrolytic method for the oxidation of activated and unactivated bromides.[69] A particularly impressive feature is the successful oxidation of secondary unactivated bromides prone to elimination (equations 44 and 45).

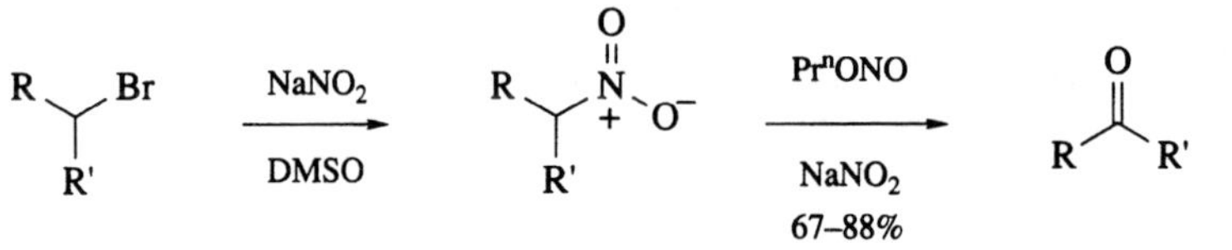

(44)

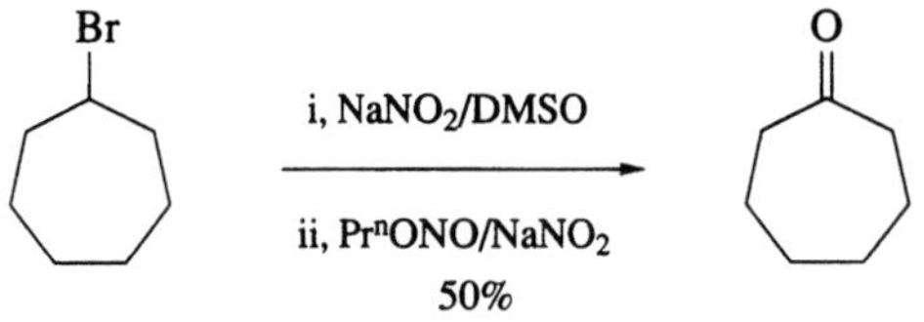

(45)

For the sake of completeness, examples of benzylic halide oxidation with copper(II) nitrate[70] and lead(II) nitrate[71] are shown in equations (46) and (47). It is probable that the more modern methods would give better yields.

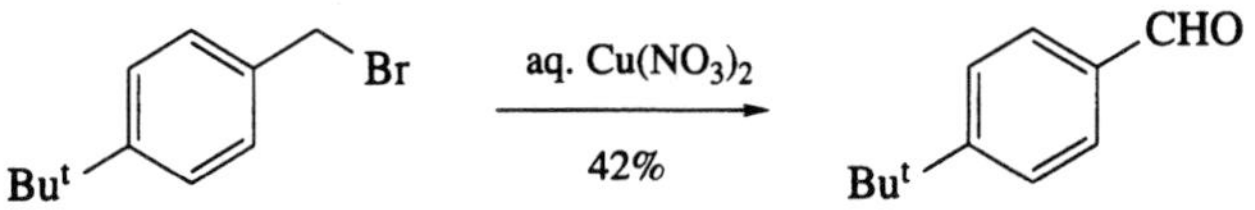

(46)

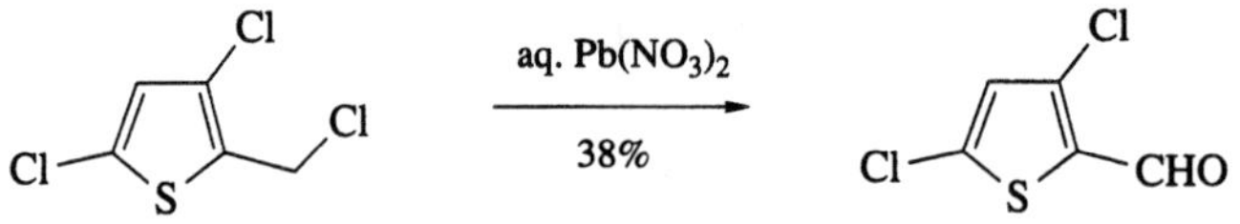

(47)

4.4.8 THE SOMMELET OXIDATION

This is one of the oldest methods for the oxidation of halides,[72] and has been used quite widely for the preparation of benzaldehydes and heteroaromatic aldehydes from the halomethyl compounds. Unactivated aliphatic halides give reduced yields.

The reaction is experimentally simple: either the halide is heated with hexamethylenetetraamine (HMTA) in a polar solvent such as aqueous acetic acid, or, with unreactive halides, the quaternary salt is first prepared in chloroform and then decomposed in a protic medium. The reaction is believed to proceed as shown in Scheme 11.

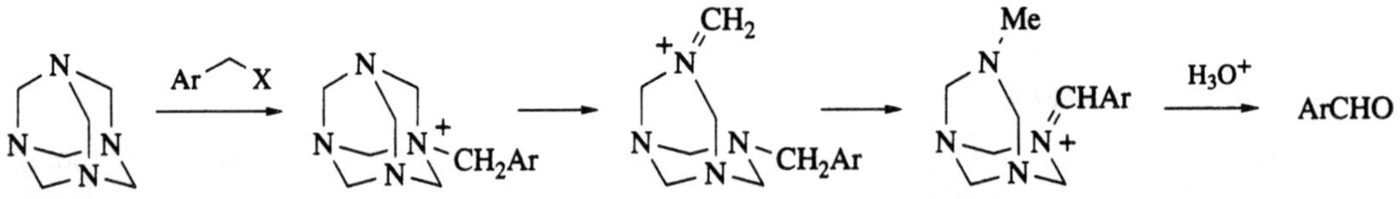

Scheme 11

At the end of the reaction it is normal to add hydrochloric acid to ensure the hydrolysis of the Schiff base. In some cases isolation was performed by steam distillation. As can be seen in equations (48)–(52), the yields are variable but often reasonable.[73–77]

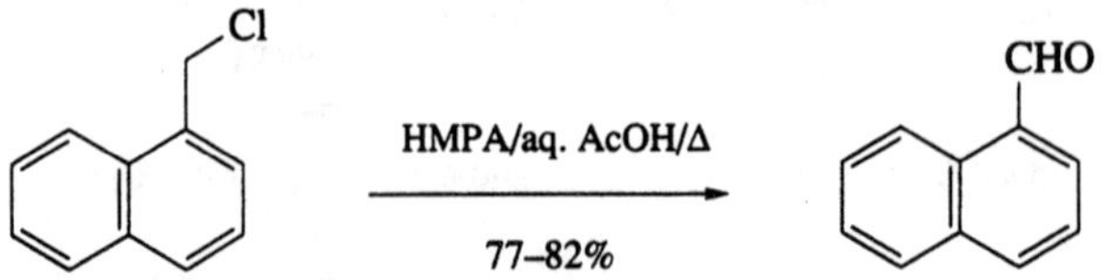

(48)

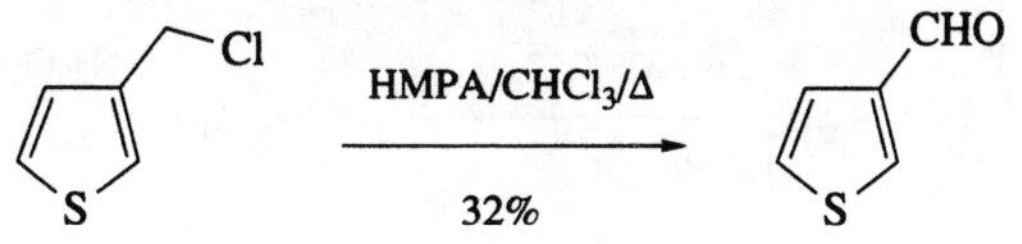

HMPA/CHCl$_3$/Δ, 32% (49)

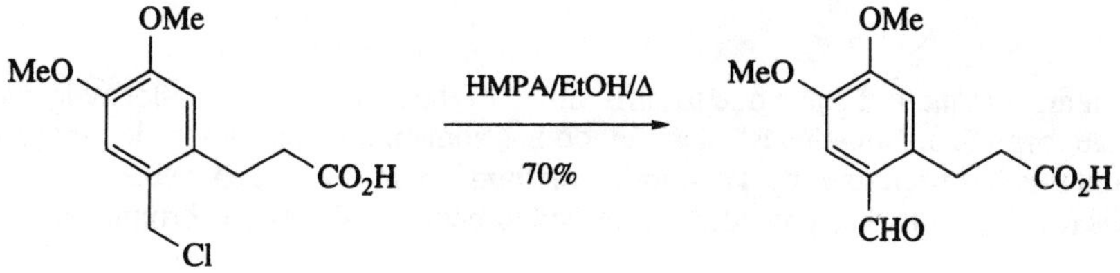

HMPA/EtOH/Δ, 70% (50)

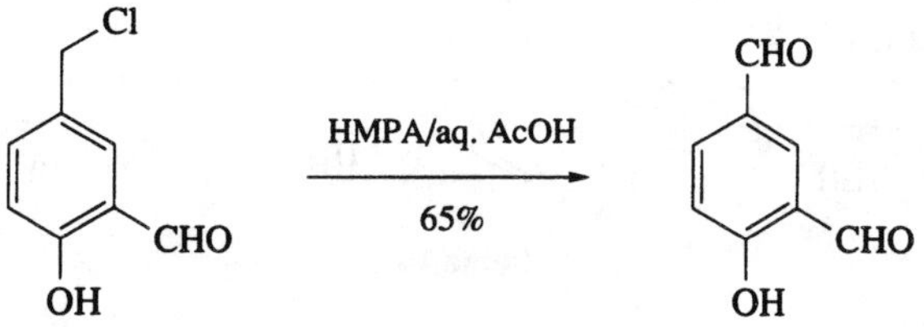

HMPA/aq. AcOH, 65% (51)

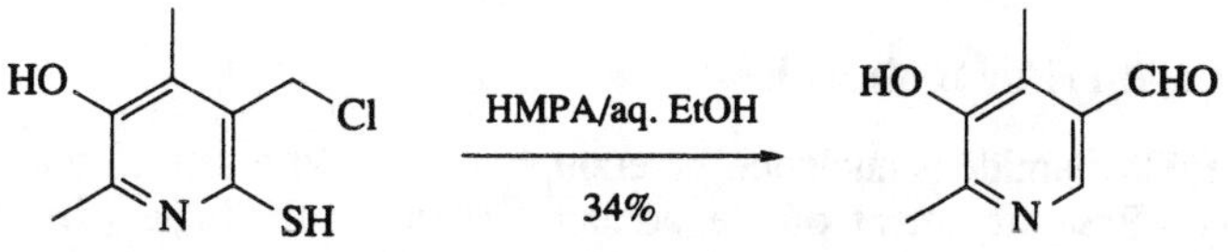

HMPA/aq. EtOH, 34% (52)

The method has an advantage over those involving base since it is applicable to phenolic substrates. The limitations are as one would expect; highly hindered substrates such as 2,6-disubstituted benzyl halides are unreactive, though mononitro-substituted substrates are oxidized.

4.4.9 MISCELLANEOUS METHODS

4.4.9.1 Oxidation *via* the Pummerer Rearrangement

A Japanese group has reported a general oxidation method for activated and unactivated bromides in which the key step is the Pummerer rearrangement of a pyrazinyl sulfoxide (Scheme 12).[78]

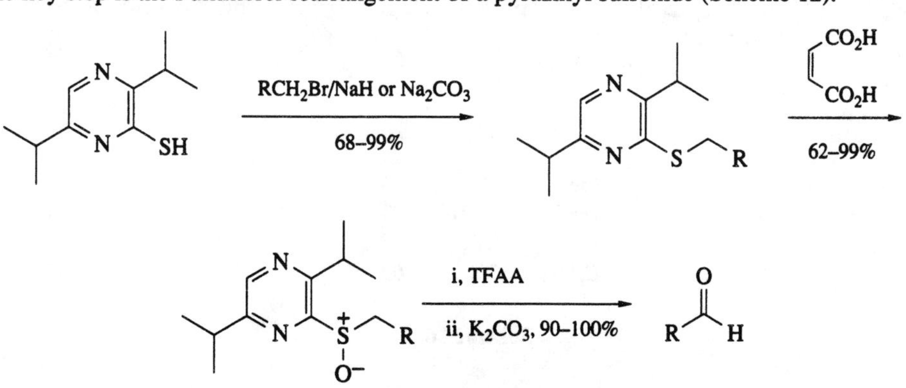

RCH$_2$Br/NaH or Na$_2$CO$_3$, 68–99%; 62–99%; i, TFAA; ii, K$_2$CO$_3$, 90–100%

Scheme 12

This is an improvement of an earlier procedure[79] in which benzenethiol was used as the precursor. The method gives high yields, but several steps are involved and the pyrazinethiol must be prepared separately.

Paquette has reported a related method based on the chlorination of sulfides (Scheme 13).[80]

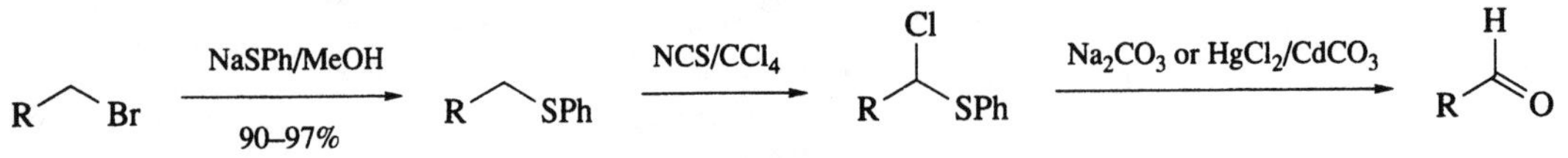

Scheme 13

Problems can arise at the last stage due to difficulties in the isolation of the aldehyde and/or preferential vinyl sulfide formation. Nonetheless, the method has some potential. Sulfoxides are prone to thermal elimination, and this has been used by Trost in his method,[81] which can also be used for the oxidation of primary amines (Scheme 14). The procedure is limited to benzylic and allylic bromides.

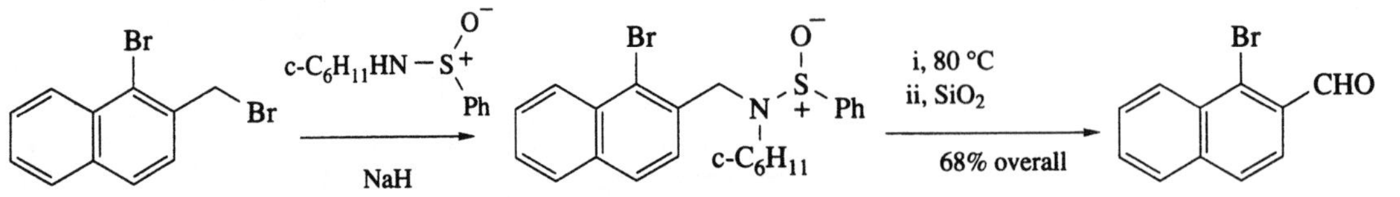

Scheme 14

4.4.9.2 Triflamides and Triflic Hydrazides

The anion of *N*-phenyltriflamide is nucleophilic enough to react with activated and unactivated halides under mild conditions. Base treatment of the adduct eliminates triflinate (trifluoromethanesulfinate, CF_3SO_2–) to give the anil, which is then hydrolyzed in acid to the aldehyde (Scheme 15).[82] The method works quite well with α-bromocarbonyl compounds (Scheme 16).[83]

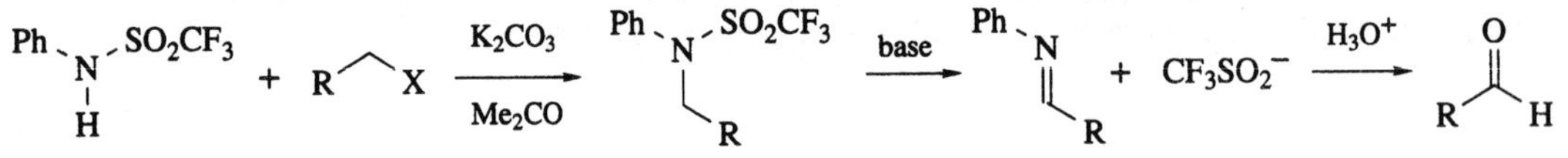

Scheme 15

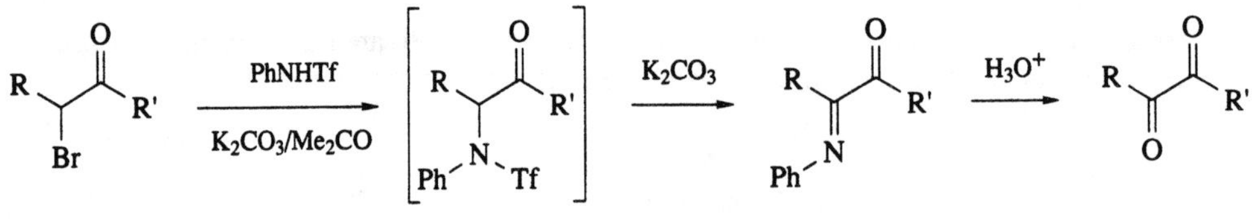

R	*R'*	*Yield* (%)
Me	Ph	73
Me	C_8H_{17}	79
Pr	OEt	67
C_6H_{13}	OEt	62

Scheme 16

With benzylic halides, the elimination requires sodium hydride in hot DMF, whereas aliphatic substrates do not undergo elimination at all. This drawback was surmounted by the use of *N*-4-acetoxyphenyltriflamide.[83] Elimination now occurs *via* the quinoneimines under mild conditions, even with aliphatic substrates (Scheme 17). Despite the ingenious chemistry, there appear to be few advantages over other methods. A procedure exists for the synthesis of hydrazones from halides (Scheme 18).[84]

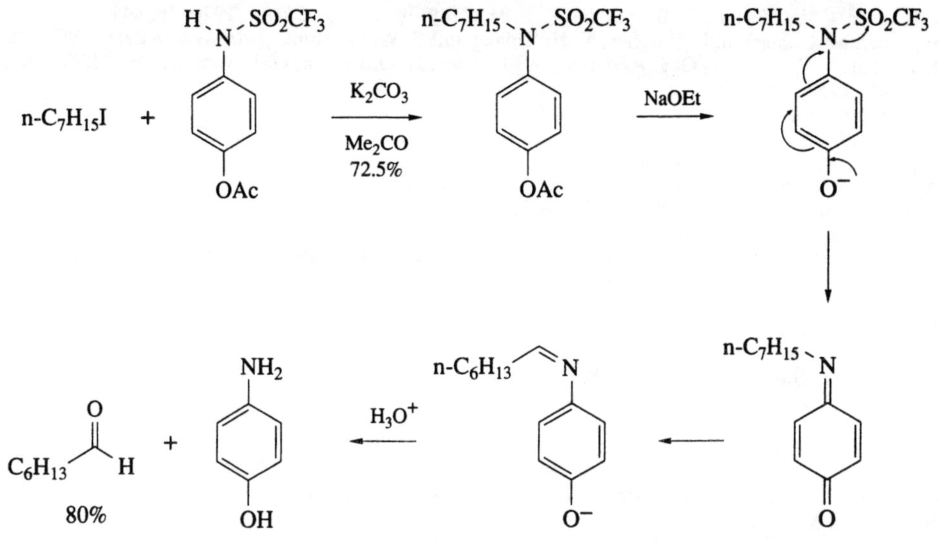

Scheme 17

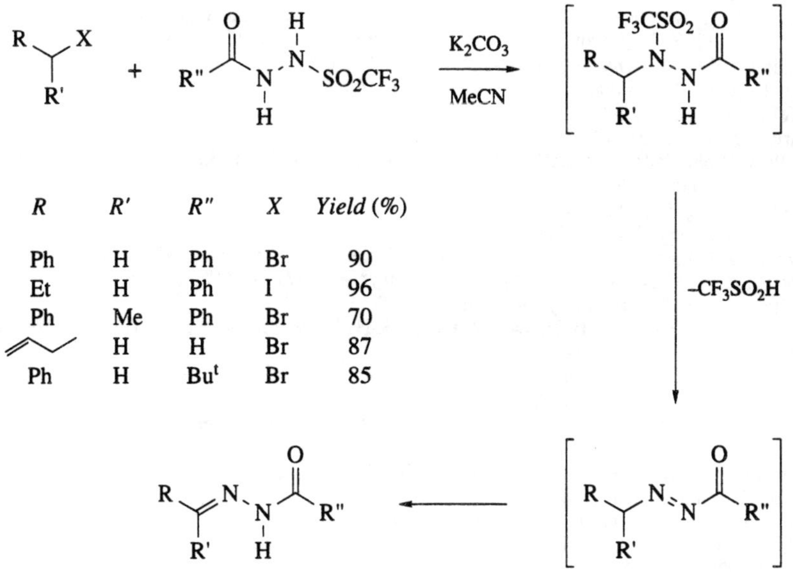

R	*R'*	*R''*	*X*	*Yield* (%)
Ph	H	Ph	Br	90
Et	H	Ph	I	96
Ph	Me	Ph	Br	70
	H	H	Br	87
Ph	H	Bu^t	Br	85

Scheme 18

4.4.10 REFERENCES

1. G. Simchen, *Methoden Org. Chem. (Houben-Weyl)*, 1983, E3.
2. W. W. Epstein and F. W. Sweat, *Chem. Rev.*, 1967, **67**, 247.
3. T. Durst, *Adv. Org. Chem.*, 1969, **6**, 285.
4. K. Torssell, *Acta Chem. Scand.*, 1967, **21**, 1.
5. K. Torssell, *Tetrahedron Lett.*, 1967, 4445.
6. H. R. Nace and J. J. Monagle, *J. Org. Chem.*, 1959, **24**, 1792.
7. R. N. Iacona, A. T. Rowland and H. R. Nace, *J. Org. Chem.*, 1964, **29**, 3495.
8. H. R. Nace and R. N. Iacona, *J. Org. Chem.*, 1964, **29**, 3495.
9. N. Kornblum, W. J. Jones and G. J. Anderson, *J. Am. Chem. Soc.*, 1959, **81**, 4113.
10. A. P. Johnson and A. Pelter, *J. Chem. Soc.*, 1967, 520.
11. T. L. Fletcher and H. Pan, *J. Am. Chem. Soc.*, 1956, **78**, 4812.
12. K. Sato, S. Suzuki and Y. Kojima, *J. Org. Chem.*, 1967, **32**, 339.
13. B. Ganem and R. K. Boeckman, Jr., *Tetrahedron Lett.*, 1974, 917.
14. D. M. Lemal and A. J. Fry, *J. Org. Chem.*, 1964, **29**, 1673.
15. W. W. Epstein and J. Ollinger, *J. Chem. Soc., Chem. Commun.*, 1970, 1338.

16. A. Markovac, C. L. Stevens, A. B. Ash and B. E. Hackley, Jr., *J. Org. Chem.*, 1970, **35**, 841.
17. E. Schipper, M. Cinnamon, L. Rascher, Y. H. Chiang and E. W. Oroshnik, *Tetrahedron Lett.*, 1968, 6201.
18. N. Kornblum, J. W. Powers, G. J. Anderson, W. J. Jones, H. O. Larson, O. Levand and W. M. Weaver, *J. Am. Chem. Soc.*, 1957, **79**, 6562.
19. J. M. Hunsberger and J. M. Tien, *Chem. Ind. (London)*, 1959, 88.
20. J. H. Babler, M. J. Coughlan, M. Feng and P. Fries, *J. Org. Chem.*, 1979, **44**, 1716.
21. L. Syper and J. Mlochowski, *Synthesis*, 1984, 747.
22. K. Ariyoshi, Y. Aso, T. Otsubo and F. Ogura, *Chem. Lett.*, 1984, 891.
23. F. Kröhnke, *Angew. Chem., Int. Ed. Engl.*, 1963, **2**, 380.
24. A. A. Goldberg and H. A. Walter, *J. Chem. Soc.*, 1954, 2540.
25. A. Giner-Sorolla, J. Zimmerman and A. Bendich, *J. Am. Chem. Soc.*, 1959, **81**, 2515.
26. W. Ried and H. Bender, *Chem. Ber.*, 1956, **89**, 1893.
27. H. R. Nace and D. H. Nelander, *J. Org. Chem.*, 1964, **29**, 1677.
28. D. Moderhack, *Justus Liebigs Ann. Chem.*, 1972, **29**, 758.
29. D. Moderhack, *Chem. Ber.*, 1975, **108**, 887.
30. H. Bredereck, E. Siegel and B. Frohlisch, *Chem. Ber.*, 1962, **95**, 403.
31. A. Kalir, *Org. Synth., Coll. Vol.*, 1973, **5**, 825.
32. P. Karrer and A. Epprecht, *Helv. Chim. Acta*, 1941, **24**, 1039.
33. K. Balenovic, D. Cerar and L. Filipovic, *J. Org. Chem.*, 1953, **18**, 868.
34. W. Ried and F.-J. Königstein, *Chem. Ber.*, 1959, **92**, 2532.
35. S. Yoshima, I. Maeda and V. Laohathai, *Chem. Pharm. Bull.*, 1972, **20**, 584.
36. H. B. Hass and M. L. Bender, *J. Am. Chem. Soc.*, 1949, **71**, 1767.
37. A. T. Blomquist, R. E Stahl, J. Meinwald and B. H. Smith, *J. Org. Chem.*, 1961, **26**, 1687.
38. S. Akabori, T. Sato and K. Hata, *J. Org. Chem.*, 1968, **33**, 3277.
39. J. N. Nazarov, S. M. Makin, V. B. Mochalin and D. V. Nazarov, *Zh. Obsch. Khim.*, 1959, **29**, 3965.
40. H. B. Hass and M. L. Bender, *Org. Synth., Coll. Vol.*, 1963, **4**, 932.
41. B. H. Klandermann, *J. Org. Chem.*, 1966, **31**, 2618.
42. N. Kornblum, R. T. Swiger, G. W. Earl, H. W. Pinnick and F. W. Stuchal, *J. Am. Chem. Soc.*, 1970, **92**, 5513.
43. S. Suzuki, T. Onishi, Y. Fujita and J. Otera, *Synth. Commun.*, 1985, **15**, 1123.
44. R. E. Manning and F. M. Schaefer, *Tetrahedron Lett.*, 1975, 213.
45. V. J. Traynelis and J. P. Kimball, *J. Org. Chem.*, 1975, **40**, 2365.
46. J. C. Stowell, *J. Org. Chem.*, 1970, **35**, 244.
47. C. A. Henrick, *Tetrahedron*, 1977, **33**, 1845.
48. W. Feely, W. L. Lehn and V. Boekelheide, *J. Org. Chem.*, 1957, **22**, 1135.
49. H. Sliwa and A. Tartar, *J. Org. Chem.*, 1976, **41**, 160.
50. L. A. Carpino, *J. Org. Chem.*, 1964, **29**, 2820.
51. T. Cohen and I. H. Song, *J. Org. Chem.*, 1966, **31**, 3058.
52. S. Mukaiyama, J. Inanaga and Y. Yamaguchi, *Bull. Chem. Soc. Jpn.*, 1981, **54**, 2221.
53. M. J. Cook, A. R. Katritzky and G. H. Millet, *Heterocycles*, 1977, **7**, 227.
54. A. R. Katritzky and S. B. Brown, *Synthesis*, 1978, 629.
55. A. R. Katritzky, M. J. Cook, S. B. Brown, R. Cruz, G. H. Millet and A. Anani, *J. Chem. Soc., Perkin Trans. 1*, 1979, 2493.
56. V. Franzen and S. Otto, *Chem. Ber.*, 1961, **94**, 1360.
57. V. Franzen, *Org. Synth.*, 1967, **47**, 96.
58. J. M. J. Fréchet, G. Darling and M. J. Farrall, *Polym. Prepr., Am. Chem. Soc., Div. Polym. Chem.*, 1980, **21**, 270.
59. V. E. Gunn and J.-P. Anselme, *J. Org. Chem.*, 1977, **42**, 754.
60. K. Aldar and F. H. Flock, *Chem. Ber.*, 1956, **89**, 1732.
61. D. Landini and F. Rolla, *Chem. Ind., (London)*, 1979, 213.
62. G. Cardillo, M. Orena and S. Sandri, *J. Chem. Soc., Chem. Commun.*, 1976, 190.
63. G. Cardillo, M. Orena and S. Sandri, *Tetrahedron Lett.*, 1976, 3985.
64. N. Kornblum and H. W. Frazier, *J. Am. Chem. Soc.*, 1966, **88**, 865.
65. R. E. Brown, H. V. Hansen, D. M. Lustgarten, R. J. Stanaback and R. I. Meltzer, *J. Org. Chem.*, 1968, **33**, 4180.
66. R. L. Letsinger and J. D. Jamison, *J. Am. Chem. Soc.*, 1961, **83**, 193.
67. A. McKillop and M. E. Ford, *Synth. Commun.*, 1974, 45.
68. N. Kornblum, *Org. React. (N.Y.)*, 1962, **12**, 101.
69. N. Kornblum and P. A. Wade, *J. Org. Chem.*, 1973, **38**, 1418.
70. J. W. Baker, W. S. Nathan and C. W. Shoppee, *J. Chem. Soc.*, 1935, 1847.
71. E. Profft and H. Wolff, *Justus Liebigs Ann. Chem.*, 1959, **96**, 628.
72. S. J. Angyal, *Org. React. (N.Y)*, 1954, **8**, 197.
73. S. J. Angyal, J. R. Tetaz and J. G. Wilson, *Org. Synth., Coll. Vol.*, 1963, **4**, 690.
74. E. Campaigne and W. M. LeSuer, *J. Am. Chem. Soc.*, 1948, **70**, 1555.
75. R. Durand-Dran, M. Lecocq and R. Quelet, *C. R. Hebd. Seances Acad. Sci.*, 1960, **250**, 2727.
76. S. J. Angyal, P. J. Morris, J. R. Tetaz and J. G. Wilson, *J. Chem. Soc.*, 1950, 2141.
77. T. S. Gardner, F. A. Smith, E. Wenis and J. Lee, *J. Org. Chem.*, 1951, **16**, 1121.
78. M. Shimazaki, T. Nakanishi, M. Mechizuki and A. Ohta, *Heterocycles*, 1988, **27**, 1643.
79. H. Sugihara, R. Tanikaga and A. Kagi, *Synthesis*, 1978, 881.
80. L. A. Paquette, L. Klobucar and R. A. Snow, *Synth. Commun.*, 1976, **6**, 575.
81. B. M. Trost and G. Liu, *J. Org. Chem.*, 1981, **46**, 4617.
82. J. B. Hendrickson, R. J. Bergeron, A. Giga and D. Sternbach, *J. Am. Chem. Soc.*, 1973, **95**, 3412.
83. R. J. Bergeron and P. G. Hoffman, *J. Org. Chem.*, 1979, **44**, 1835.
84. J. B. Hendrickson and D. Sternbach, *J. Org. Chem.*, 1975, **40**, 3450.

5.1 The Baeyer–Villiger Reaction

GRANT R. KROW

Temple University, Philadelphia, PA, USA

5.1.1 INTRODUCTION

The oxidation of ketones with organic peroxy acids, hydrogen peroxide or alkyl hydroperoxides to give esters/lactones or derived alcohols and acids is known as the Baeyer–Villiger reaction.[1–15] Similar oxidations of aldehydes to the corresponding formate esters or their hydrolysis products also belong to this class of reactions.[1,2,16] The Baeyer–Villiger reaction is of considerable synthetic use as a component of methods for shortening carbon chains, hydroxylating aromatic rings, converting carbocycles to heterocycles and opening cyclic arrays to prepare functionalized chains and/or rings.

5.1.2 MECHANISM

5.1.2.1 General

The accepted two-step mechanism (Scheme 1) for the Baeyer–Villiger reaction is based upon kinetic, isotopic substitution, stereochemical and substituent effect studies.[2] In step 1 of the reaction, addition of peroxy acid to the ketone carbonyl gives a Criegee intermediate (**1**). In step 2, which is usually rate determining, (**1**) rearranges to product (**2**). Migration of the R_M group from carbon to oxygen in step 2 is normally concerted with O—O bond breaking, although stepwise processes have been noted.[9] Acid catalysts facilitate breaking of the O—O bond by protonation of the carbonyl oxygen. Bases aid the rearrangement by removal of the hydroxy proton. Either electron-donating groups on R_M and R_R or electron-withdrawing groups on R_L facilitate migration.[2]

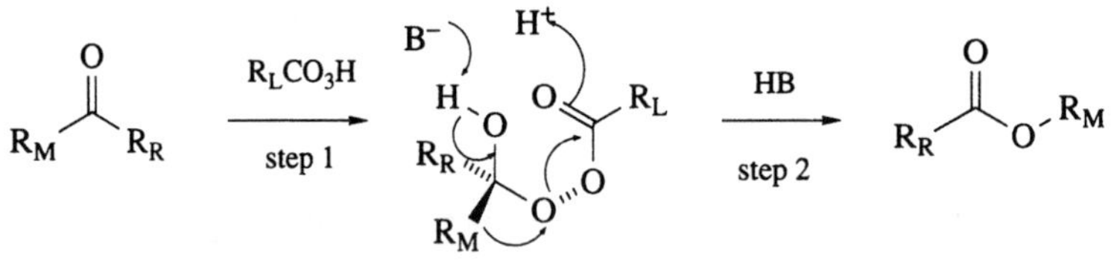

R_LCO_3H, step 1; HB, step 2

(**1**) (**2**)

Scheme 1

The stereoelectronic requirements proposed for the migration step are an antiperiplanar arrangement of the C—R_M bond and the breaking O—O bond. It has been further suggested that one of the hydroxy nonbonding electron pairs must be antiperiplanar to the migrating carbon atom, as in (**3**).[9]

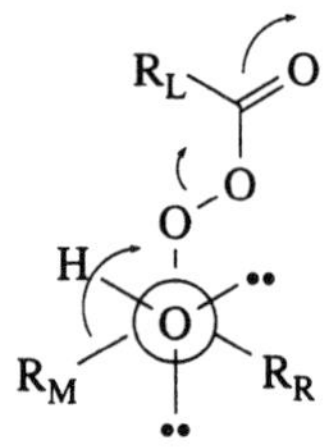

(**3**)

5.1.2.2 Stereochemistry

The Baeyer–Villiger reaction occurs with retention of stereochemistry at the migrating center.[1] This stereoselectivity has been utilized in a practical method for the preparation of isotopically chiral methyl acetic acid (**5**) from [2-^{3}H]cyclohexanone (**4**) prepared by enzyme-catalyzed stereoselective exchange of the *pro-R* α′-proton and enantioconvergent exchange of the α-proton with deuterium (Scheme 2).[17] As a cautionary note, prior epimerization of an acyl group prior to oxidation has been observed.[18]

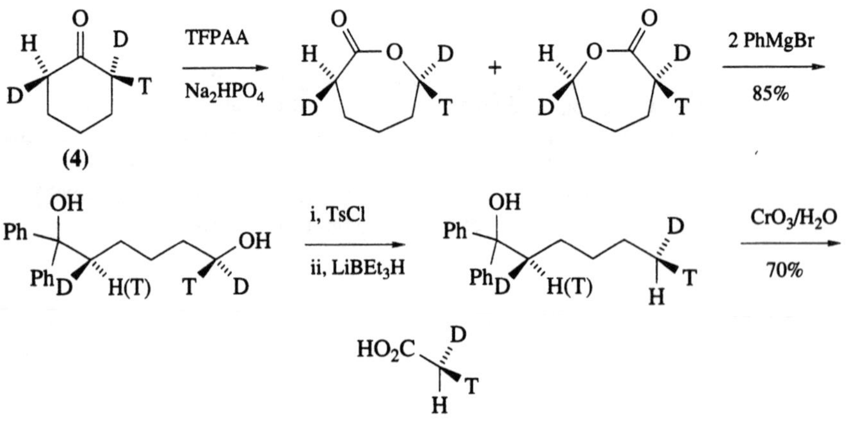

TFPAA, Na_2HPO_4; 2 PhMgBr, 85%; i, TsCl, ii, $LiBEt_3H$; CrO_3/H_2O, 70%

(**4**) (**5**)

Scheme 2

5.1.2.3 Regiochemistry

The usual migratory preference of alkyl groups in the Baeyer–Villiger reaction is tertiary > secondary > primary > methyl.[2,19] This order has been attributed to the greater electron-releasing power[10] of, or to steric acceleration of migration by, the larger group.[19] Conformational, steric and electronic factors can alter the normal migratory preference.[9] Allyl[20] and alkenyl[21] migrations occur in preference to that of a primary alkyl group. Phenyl alkyl ketones with trifluoroperoxyacetic acid (TFPAA) undergo migration of the phenyl group if the alkyl group is primary, but secondary alkyl group rearrangement is favored over phenyl.[19] Aryl substituents may alter this preference (equation 1).[22] Cyclopropyl migration is favored over methyl migration,[2] but not over migration of other primary alkyl substituents.[23]

TFPAA, Na_2HPO_4; CH_2Cl_2; 88% (1)

An α-acyloxy or -ether substituent facilitates migration, even in competition with a secondary carbon (equation 2),[24] while an α-halogen substituent retards migration.[25] Despite its electron-withdrawing capability, an α-ethoxycarbonyl group in (**6**) is not sufficient to deter migration of the attached carbon (equation 3).[26]

$OSiMe_3$ → i, ii; 50–60% → MeO, OMe, OMe (2)

i, MCPBA, ether; ii, MeOH/H^+

COMe, EtO_2C, Pr^i → TFPAA, Na_2HPO_4; CH_2Cl_2; 89% → OAc, EtO_2C, Pr^i (3)

(**6**)

A β-silicon atom directs migration of the proximate carbon (equation 4).[27] Conversely, electron-withdrawing β-substituents retard migration of the carbon near to these groups even, as in (**7**), in competition with a methyl group (equation 5).[28]

p-Tol, $SiMe_3$ → MCPBA, Na_2HPO_4; CH_2Cl_2; 94% → *p*-Tol, $SiMe_3$ (4)

CF_3CONH, COMe, MeO_2C, D (**7**) → TFPAA; CH_2Cl_2; 51% → CF_3CONH, OAc, MeO_2C, D (37%) + CF_3CONH, CO_2Me, MeO_2C, D (63%) (5)

Substituent effects can be catalyst[29] and peroxy acid dependent (equation 6).[30] Occasionally, when regioisomeric lactones are formed, preferential base hydrolysis of one of them may facilitate isolation of a pure product.[9,31]

A reaction sequence which complements the Baeyer–Villiger reaction has been described (equation 7).[32] Regioselective silylenol ether formation allows for introduction of oxygen away from the more highly alkylated side of a ketone.

(6)

i, 38%	100	:	0	
ii, 59%	18	:	82	

i, PAA, NaOAc, AcOH, 72 h; ii, MCPBA, $NaHCO_3$, CH_2Cl_2, 72 h

(7)

i, LDA, Me_3SiCl; ii, O_3, CH_2Cl_2, MeOH; iii, $NaBH_4$

5.1.3 REACTION METHODS

The oxidizing power of a Baeyer–Villiger reagent is related to the strength of the conjugate acid of the leaving group; thus, the reactivity order is Bu^{t}OOH < HOOH << peracetic acid (PAA) < perbenzoic acid (PBA) < MCPBA = HCO_3H[33] < *p*-nitroperbenzoic acid[34] (PNPBA) < mono-*o*-peroxyphthalic acid (MPPA) < monoperoxymaleic acid (PMA)[33] < TFPAA.[3] Among this group, commercial availability, ease of handling or oxidizing power make peracetic acid (40%), *m*-chloroperbenzoic acid (85%) (MCPBA) and TFPAA (90%) the most commonly reported organic peroxy acid reagents used in Baeyer–Villiger oxidations.

PAA solutions of differing strengths can be prepared by adding hydrogen peroxide of varying power to acetic anhydride,[35] glacial acetic acid[36] or aqueous acetic acid[37] containing a catalytic quantity of sulfuric acid. Sodium acetate is the most commonly used buffering agent. Reaction times of one to several days at room temperature are common.

Commercially available 85% MCPBA is generally employed in chlorinated hydrocarbon solvents at room temperature. Reaction times are typically a few hours to several days. Buffers utilized include disodium hydrogen phosphate, sodium acetate and sodium bicarbonate, the catalytic effect of which has been occasionally noted.[38] Acid catalysis with sulfuric acid or Nafion-H are alternatives.[39] Oxidations have been performed at elevated temperature with the aid of radical scavengers.[40]

TFPAA, a remarkably powerful reagent, is prepared prior to use by adding trifluoroacetic anhydride to a suspension of the appropriate strength hydrogen peroxide in dichloromethane at 0 °C.[2,3] Reactions are generally performed in dichloromethane in the presence of dibasic hydrogen phosphate buffer, and are carried out at between 0 °C to reflux temperature for several hours.

Because of safety considerations, 90% hydrogen peroxide, used in most TFPAA oxidations, and 85% MCPBA may soon become commercially unavailable. Possible alternatives are magnesium monoperoxyphthalate for MCPBA[41] and the easily handled acid catalyst Nafion-H which facilitates oxidations with 30% H_2O_2.[39] Weaker strength TFPAA acid solutions are effective,[42] and the strength of MCPBA is easily increased by washing with buffer solution.[43]

Peroxysulfuric acid, PBA and MPPA, common reagents of the past,[1] appear less frequently in the current literature. Basic 30% hydrogen peroxide[44] or *t*-butylhydroperoxide[45] have special utility for oxidation of cyclobutanones and strained, bridged cycloalkanones to lactones. Basic 3–6% hydrogen peroxide is used in the Dakin oxidation of aryl aldehydes to phenols,[1,3] while peroxymonophosphoric acid oxidizes aryl ketones to phenolic acetates.[46]

Rarely used oxidants with potential advantages as chemoselective or regioselective reagents include silylated forms of HOOH and peroxysulfuric acid,[47,48] and the safe and inexpensive weak oxidant sodium perborate.[49] Benzeneseleninic acid/30% hydrogen peroxide has been reported as a polystyrene-bound version,[50] as has peroxyarsenic acid.[51]

5.1.4 CHEMOSELECTIVITY

5.1.4.1 Competitive Reactions with Other Functional Groups

Many of the oxidizing agents used in Baeyer–Villiger oxidations of ketones will also react with alkenes, amines, sulfides and selenides. Reagents have been developed which allow selective oxidations in the presence of some of these functional groups. The reactivities of alkenes and ketones with organic peroxy acids are comparable and are reagent sensitive. Thus, it is possible to either ring expand (equation 8) or epoxidize (equation 9) 2-allylcyclohexanone (**8**).[52] Basic hydrogen peroxide, which does not epoxidize isolated alkenes, will effect Baeyer–Villiger oxidation of strained cyclic ketones (equation 10).[2,53] If alkene epoxidation cannot be avoided, protection as a dibromide prior to oxidation may be necessary.[54] Baeyer–Villiger oxidations with[55] and without[56] *N*-oxide formation have been reported. Only heteroatom oxidations of α-thiophenyl[57] and α-selenenylphenyl[58] ketones have been reported with MCPBA.

PAA, $CHCl_3$, 44% (8)

(**8**)

50% H_2O_2, $KHCO_3$, PhCN, MeOH, 54% (9)

(**8**)

n-C_7H_{15} — 30% H_2O_2, NaOH, MeOH–H_2O, 63% — n-C_7H_{15}, H, O, H (10)

5.1.4.2 Competitive Baeyer–Villiger Reactions

Molecules containing multiple carbonyl groups may be oxidized with group selectivity. For example, cyclobutanones are highly reactive and ring expand even with basic HOOH.[2] Relative reactivities of steroidal ketones depend upon the position of the carbonyl group (equations 11 to 13).[21,59,60] Cyclohexa-

PBA, $HClO_4$, $CHCl_3$, 60% (11)

C_8H_{17} — PBA, TsOH, $CHCl_3$, 63% — C_8H_{17} (12)

nones are normally oxidized faster than acyl side chains;[61] however, steric effects may alter this reactivity pattern (equation 14).[62]

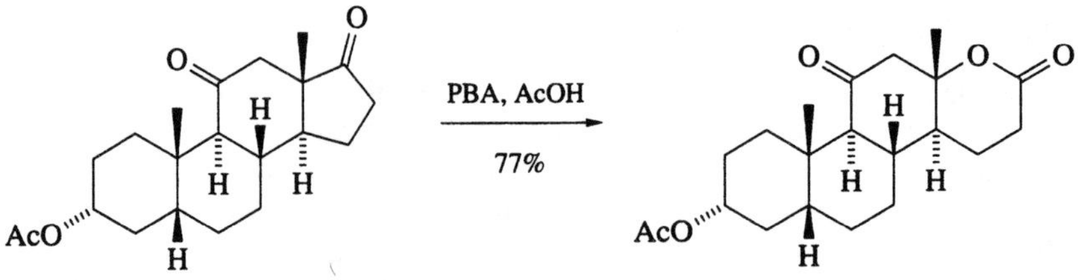

PBA, AcOH; 77% (13)

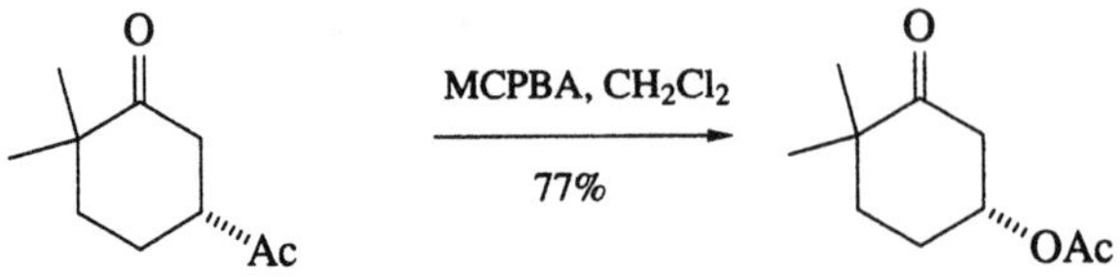

MCPBA, CH_2Cl_2; 77% (14)

5.1.5 REGIOSELECTIVITY

5.1.5.1 Acyclic Aliphatic Ketones

The regioselective Baeyer–Villiger oxidation of acyclic aliphatic ketones normally results in the insertion of oxygen next to the bulkier alkyl chain. Since methyl is a poor migrating group, a common use of this reaction is to reduce the chain length of methyl ketones by two carbons to provide alcohols, after hydrolysis. The ability to synthesize methyl ketones from acids[63] and methyl-substituted alkenes (equation 15)[64] extends the utility of this method. Complementary to this chain cleavage is the formation of carboxylic acids, following migration of the larger alkyl group attached to the carbonyl group (equation 16).[65]

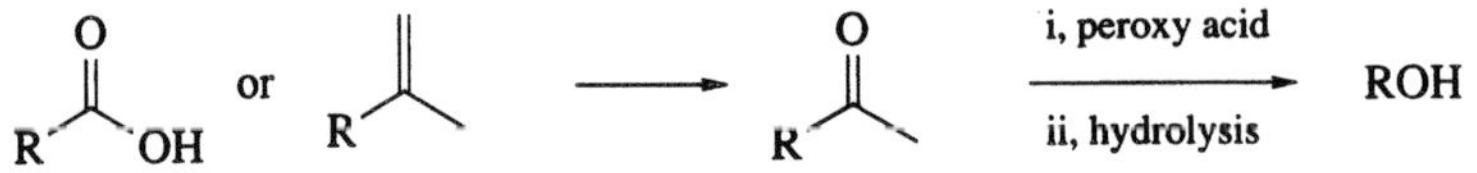

i, peroxy acid; ii, hydrolysis → ROH (15)

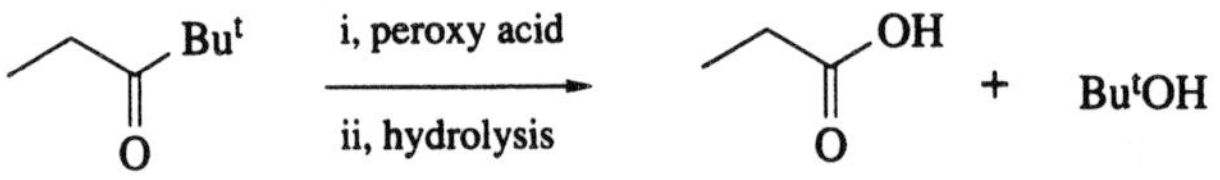

i, peroxy acid; ii, hydrolysis → + Bu^tOH (16)

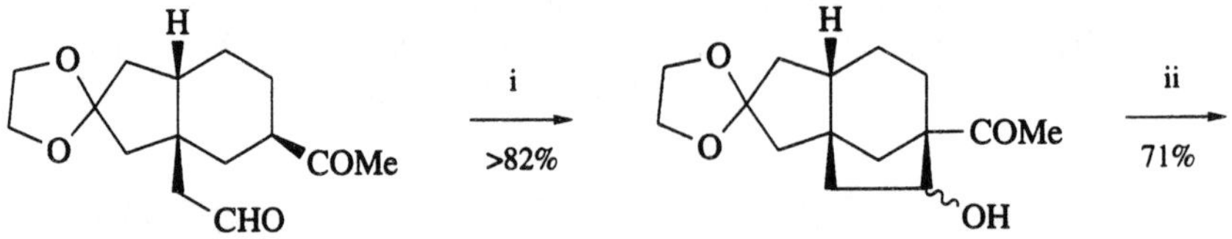

(9) → i, >82% → (10) → ii, 71% → (11) → → (12)

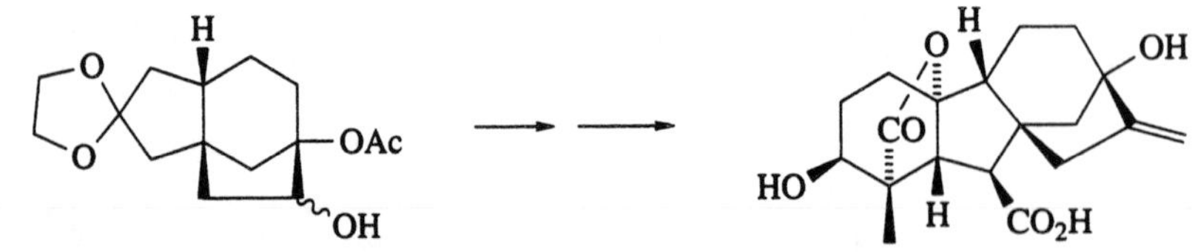

i, NaOH, EtOH; ii, 3,5-$(O_2N)_2$PBA, Na_2CO_3, $ClCH_2CH_2Cl$, 4,4-thiobis(6-*t*-butyl-3-methylphenol), 54 °C

Scheme 3

Corey and Smith,[66] as part of a total synthesis of gibberellic acid (**12**), required the hydroxylated bicyclo[3.2.1]octane (**11**) shown in Scheme 3. An acyl group was first used as the nucleophilic partner in an intramolecular aldol condensation of the keto aldehyde (**9**). The acyl group in (**10**) was subsequently oxidized to become the precursor of the required bridgehead oxygen functionality of (**12**).

During the synthesis of the Woodward reserpine precursor (**17**; Scheme 4), Pearlman[67] used the protected acylacetal (**13**) to control the stereochemistry of an intramolecular photochemical cycloaddition to (**14**). The strategy for opening the cyclobutane ring employed the Baeyer–Villiger reaction to convert the γ-keto ester (**15**) to a β-hydroxy ester (**16**), which underwent retroaldolization to (**17**).

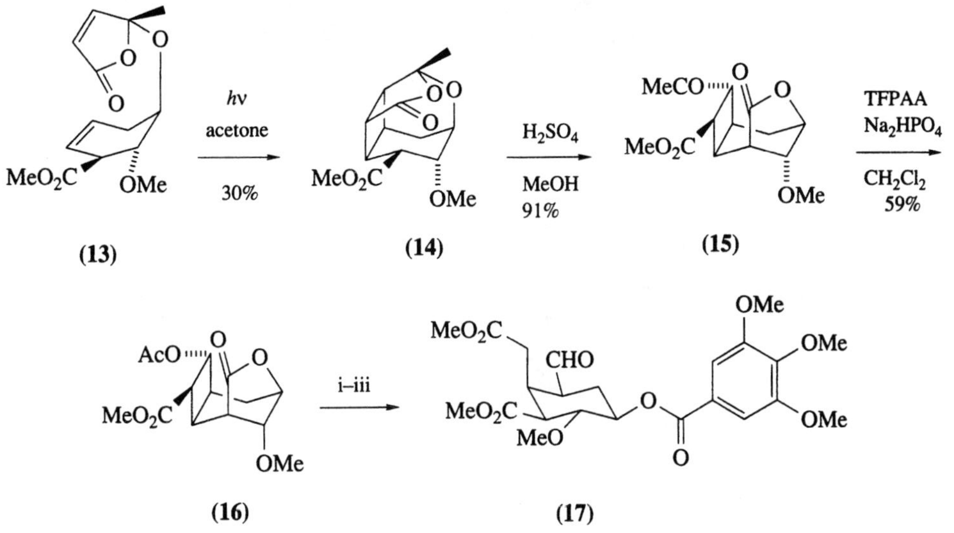

(**13**) —*hv*, acetone, 30%→ (**14**) —H_2SO_4, MeOH, 91%→ (**15**) —TFPAA, Na_2HPO_4, CH_2Cl_2, 59%→ (**16**) —i–iii→ (**17**)

i, MeOH, H_2SO_4; ii, $[3,4,5\text{-}(MeO)_3C_6H_2CO]_2O$, $p\text{-}Me_2NC_6H_4N$; iii, H_3O^+

Scheme 4

Hart and Tsai[68] have found that the allylic acetate (**18**) undergoes directed radical cyclization to afford the pyrrolizidinone (**19**) with good stereoselectivity (Scheme 5). The adduct (**19**) was converted to isoretronecanol (**21**) following conversion of the side chain to methyl ketone (**20**), which was oxidized with TFPAA.

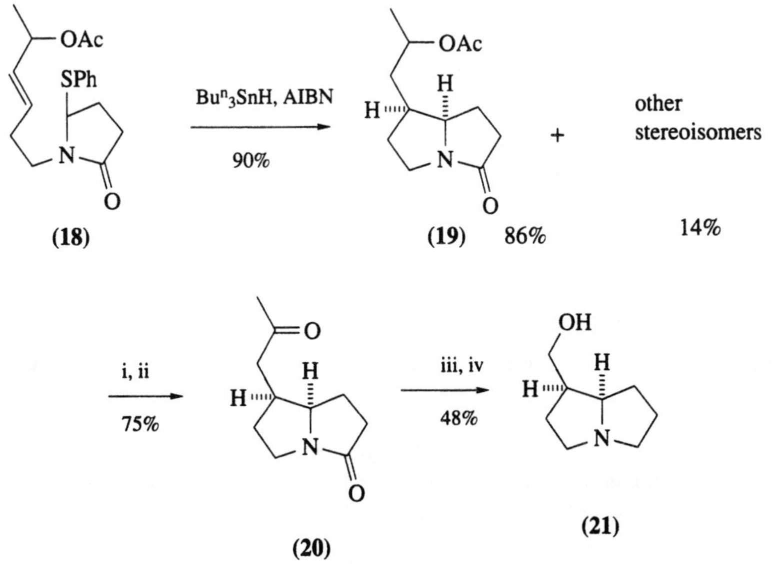

(**18**) —$Bu^n{}_3SnH$, AIBN, 90%→ (**19**) 86% + other stereoisomers 14%

(**19**) —i, ii, 75%→ (**20**) —iii, iv, 48%→ (**21**)

i, NaOH, MeOH, H_2O; ii, $(COCl)_2$, DMSO, Et_3N; iii, TFPAA; iv, $LiAlH_4$

Scheme 5

5.1.5.2 Acyclic Diaryl and Aryl Alkyl Ketones

A two-step procedure consisting of Friedel–Crafts acetylation followed by Baeyer–Villiger oxidation is a useful method for the introduction of oxygen onto an aromatic ring. A conversion of L-tyrosine (**22**) to L-dopa (**23**) utilized this procedure (Scheme 6).[69] An attempt to use this method to introduce a hydroxy group into the 6-position of 1-methylindole-3-carboxylate by oxidation of the 6-acetyl derivative with MCPBA failed.[70] By using a chloro substituent to discourage migration of the attached alkyl carbon atom, the chloroacetylated indole (**24**) was selectively converted to the desired phenol precursor (**25**; equation 17).

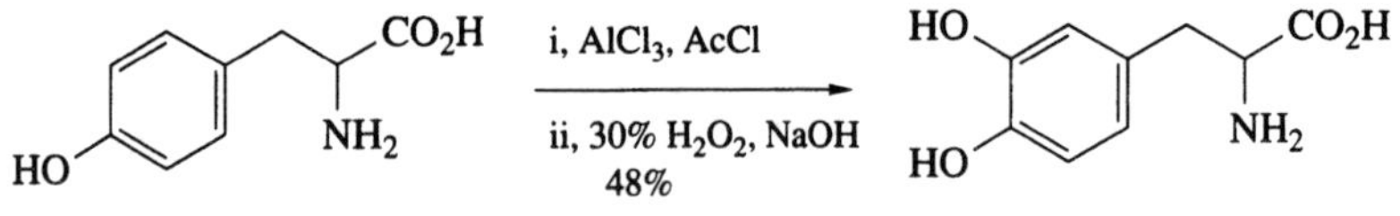

i, $AlCl_3$, AcCl; ii, 30% H_2O_2, NaOH; 48%

(L)-(**22**) (L)-(**23**)

Scheme 6

MCPBA, Na_2HPO_4, $CHCl_3$, 88%

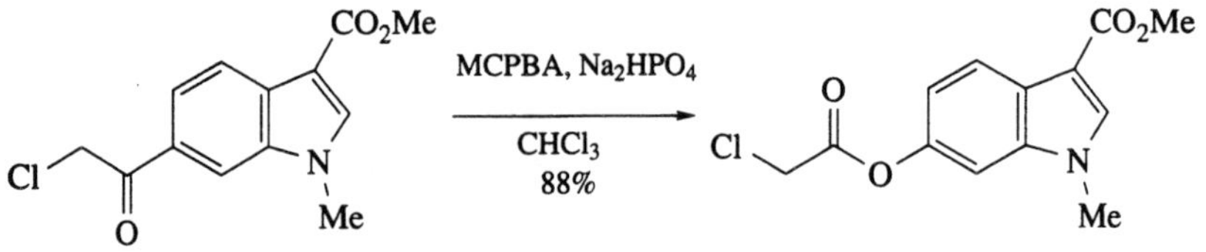

(**24**) (**25**) (17)

A procedure to introduce a C-11 methoxy group into the aryl ring of (**26**; Scheme 7) also used a Baeyer–Villiger reaction.[71] Ring opening and subsequent Friedel–Crafts ring closure at a free aryl position afforded the dimethoxypodocarpic acid derivative (**27**).

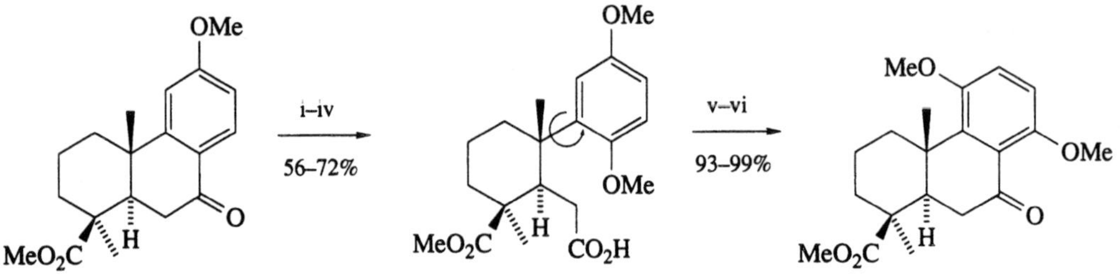

i–iv, 56–72%; v–vi, 93–99%

(**26**) (**27**)

i, MCPBA; ii, MeOH; iii, Me_2SO_4; iv, 10% K_2CO_3; v, TFAA; vi, KOH, MeOH

Scheme 7

5.1.5.3 Monocyclic and Spirocyclic Ketones

Because it is often possible to control the stereochemical orientation of substituents on a cyclic array, Baeyer–Villiger cleavages of substituted cyclic ketones have been used extensively in the stereocontrolled syntheses of substituted carbon chains. An asymmetric synthesis of L-daunosamine intermediate (**30**) from a noncarbohydrate precursor employed the cyclopentenol (**28**), prepared in optically pure form (95% *ee*) from 2-methylcyclopentadiene using asymmetric hydroboration (Scheme 8).[72] Stereoselective epoxidation, conversion to the ketone and regioselective Baeyer–Villiger oxidation afforded lactone (**29**).

The total synthesis of erythronolide B, the biosynthetic progenitor of all the erythromycins, employed a Baeyer–Villiger oxidation of the substituted cyclohexanone (**31**; equation 18). The oxidation was surprisingly slow using customary procedures, but Corey *et al.*[73] found that forcing conditions provided the required lactone (**32**).

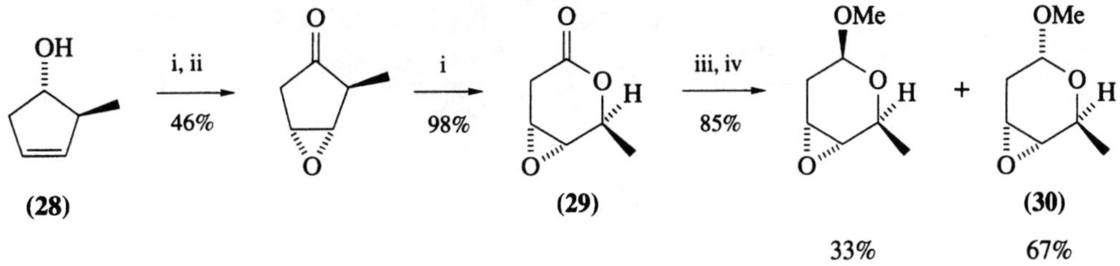
OH
i, ii
46%
O
i
98%
iii, iv
85%
OMe
H
+
(28)
(29)
(30)
33%
67%

i, MCPBA, Na_2CO_3; ii, CrO_3; iii, DIBAL-H; iv, MeOH, BF_3

Scheme 8

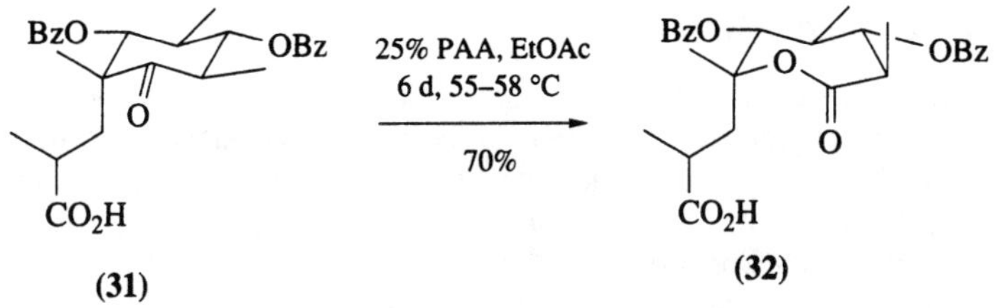
BzO
OBz
25% PAA, EtOAc
6 d, 55–58 °C
70%
CO2H
(31)
(32)

(18)

Integerrinecic acid (**36**), which occurs as the dilactone in the pyrrolizidine alkaloid integerrimine (**33**; Scheme 9), was obtained from ketone (**34**). The observed regioselectivity in the oxidation leading to (**35**) presumably results from steric and dipolar effects.[74]

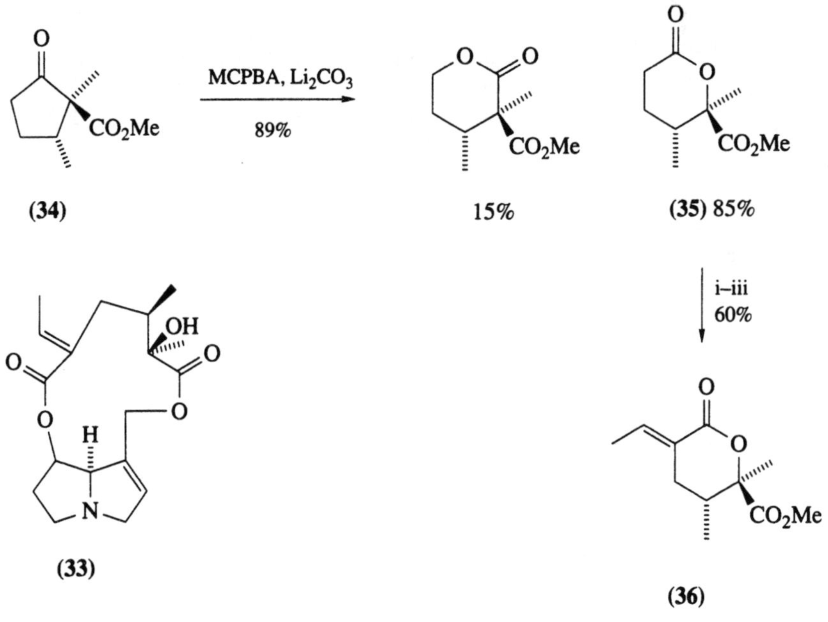
CO2Me
MCPBA, Li2CO3
89%
(34)
15%
(35) 85%
i–iii
60%
OH
H
N
(33)
(36)

i, LDA; ii, MeCHO; iii, 2-F-1-$MeC_5H_3N^+OTos^-$

Scheme 9

The Baeyer–Villiger oxidation can be used to convert large ring cyclic ketones to macrocyclic lactones.[74b] Lactones can be precursors of cyclic ethers. Chiral ketone (**37**) was oxidized to lactone (**38**) and subsequently stereoselectively converted to the *cis*-2,8-disubstituted oxocane (**39**; Scheme 10).[75]

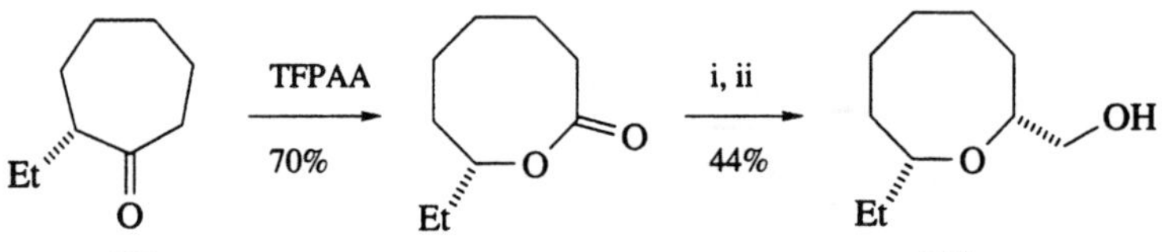

(37) —TFPAA, 70%→ **(38)** —i, ii, 44%→ **(39)**

i, $Cp_2(CH_2)(Cl)AlMe_2$, $Me_2NC_5H_4N$; ii, Sia_2BH, 30% H_2O_2, NaOH

Scheme 10

5.1.5.4 Fused Ring Bicyclic and Polycyclic Ketones

The Baeyer–Villiger reaction has been used to synthesize naturally occurring lactones, such as brassinolide (**41**), a plant growth promoter (equation 19).[76] The usual secondary > primary migratory preference observed during TFPAA oxidation of acyclic ketones is not followed in the oxidation of the B-ring of (**40**) under buffered conditions. Similarly unusual is the completely regioselective oxidation of the cyclobutanone (**42**), observed by Corey *et al.*,[77] during an early stage in the synthesis of the trilactone ginkgolide B (**43**), the platelet-activating factor in ginkgo extract (Scheme 11). Introduction of oxygen into a steroidal A-ring of (**44**), however, follows with the usual regioselectivity, giving an intermediate which leads to a steroidal 3-oxo-4-ene (**45**; Scheme 12).[78]

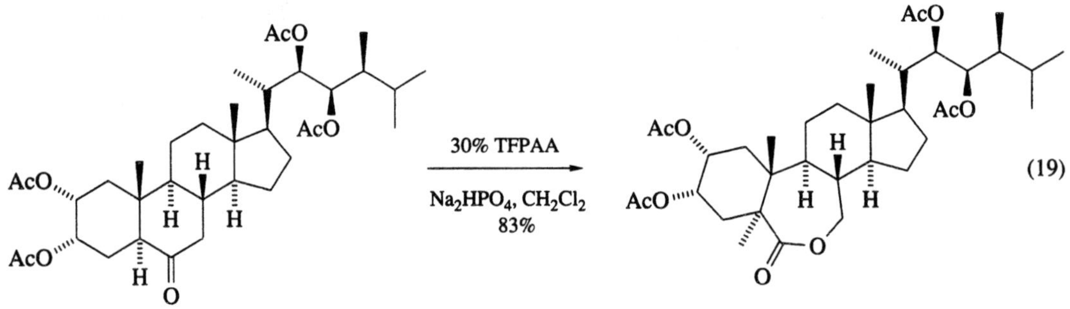

(40) —30% TFPAA; Na_2HPO_4, CH_2Cl_2; 83%→ **(41)** (19)

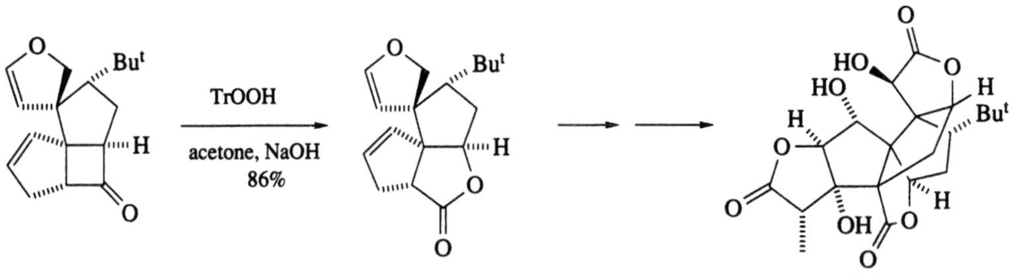

(42) —TrOOH; acetone, NaOH; 86%→ → → **(43)**

Scheme 11

Fused ring ketones have been utilized as templates for stereocontrolled elaboration of substituents fused to smaller rings. Ohno and coworkers[79] have described a regio- and stereo-controlled process for the preparation of the thienamycin intermediate (**48**; Scheme 13). Oxidation of ketone (**46**) provided lactone (**47**), which has three of the required chiral centers of thienamycin.

Murai *et al.*,[80] in their total synthesis of glycinoeclepin A (**53**), exploited the chirality of (*R*)-(–)-carvone (**49**) in the preparation of the functionalized *cis*-decalin (**50**; Scheme 14). Removal of the stereocontrolling isopropenyl side chain from (**50**) involved Baeyer–Villiger methodology. Likewise, a second Baeyer–Villiger oxidation revealed the four contiguous chiral centers of the C–D rings of glycinoeclepin A in (**52**).

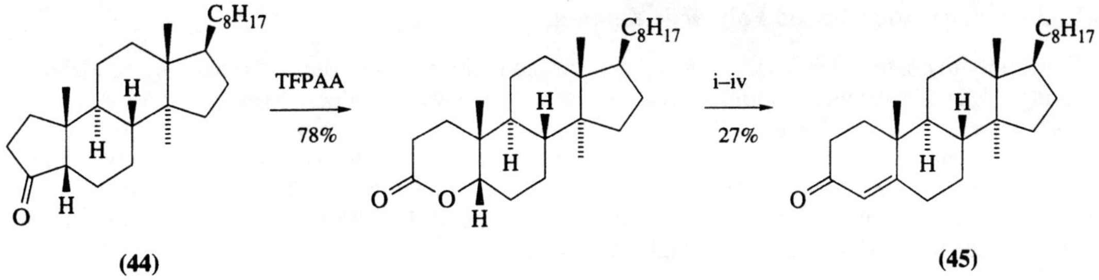

TFPAA 78%

i–iv 27%

(44) (45)

i, CrO_3; ii, Ac_2O, EtOAc; iii, MeMgI; iv, MeOH, NaOH

Scheme 12

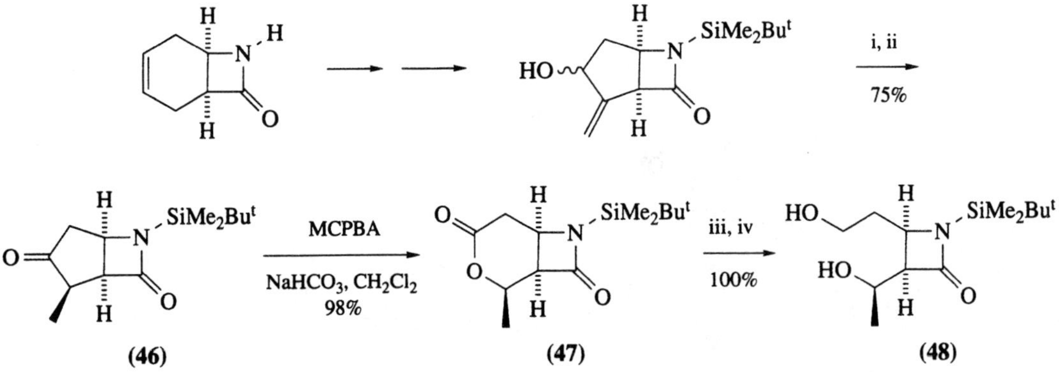

i, ii 75%

MCPBA $NaHCO_3$, CH_2Cl_2 98%

iii, iv 100%

(46) (47) (48)

i, H_2, Pd/C; ii, CrO_3, C_5H_5N; iii, DIBAL-H, Bu^nLi; iv, $NaBH_4$

Scheme 13

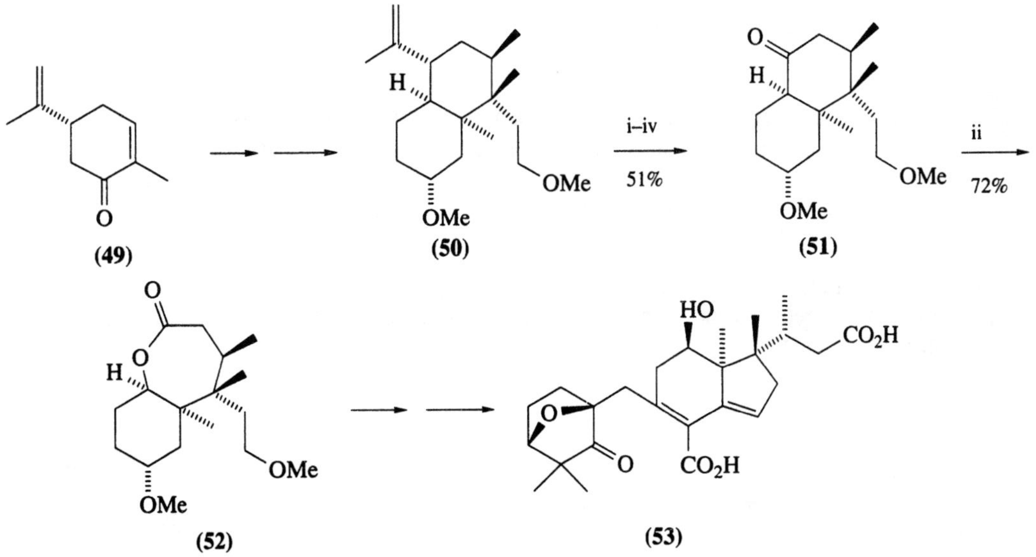

i–iv 51%

ii 72%

(49) (50) (51)

(52) (53)

i, O_3, Me_2S; ii, TFPAA; iii, $LiAlH_4$; iv, CrO_3

Scheme 14

5.1.5.5 Bridged Bicyclic and Polycyclic Ketones

The ability to control substituent stereochemistry and regiochemistry during formation and elaboration of bridged bicyclic ketones combined with regioselective Baeyer–Villiger oxidations provides a useful route to stereoselectively functionalized cycloalkanes. The lactone (**55**), which is available from norcamphor,[81] has been used by Takano *et al.*[82] for the preparation of a number of alkaloids, of which antirhine (**56**; Scheme 15) is one example. Norbornenone (**57**), disubstituted at C-7, has been exploited by Grieco *et al.*[83] in a stereocontrolled synthesis of estrone (**58**; Scheme 16). Both norbornanes and norbornenes have also been used extensively in prostaglandin synthesis.[84,85]

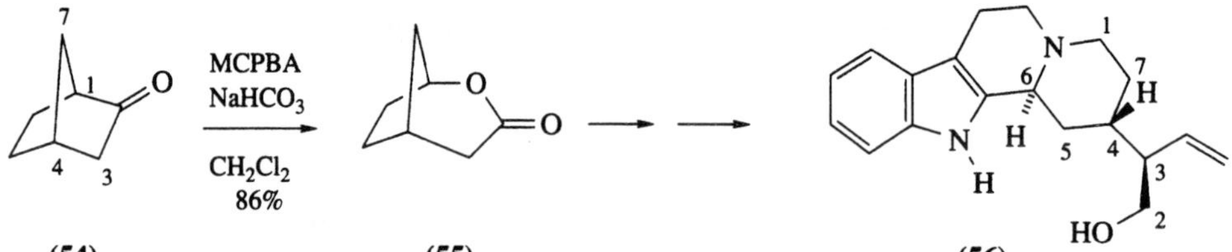

MCPBA, $NaHCO_3$, CH_2Cl_2, 86%

(**54**) (**55**) (**56**)

Scheme 15

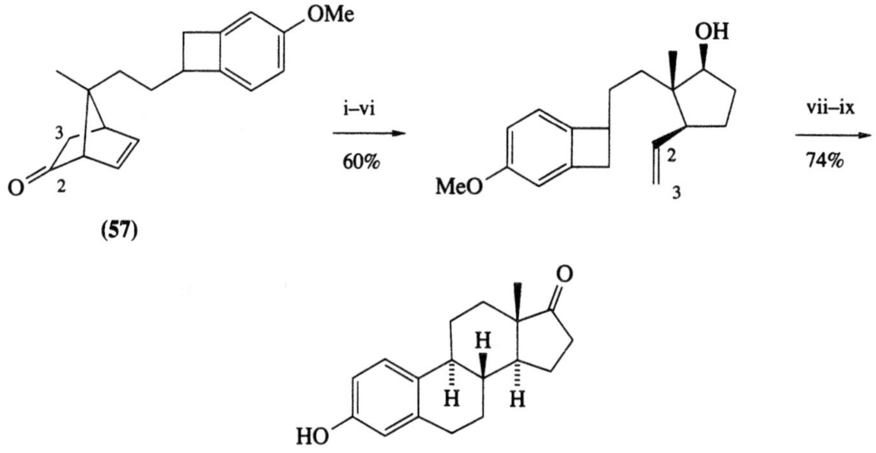

i–vi, 60%; vii–ix, 74%

(**57**) (**58**)

i, 30% H_2O_2, NaOH; ii, CH_2N_2; iii, H_2/PtO_2; iv, $LiAlH_4$; v, p-$NO_2C_6H_4SeCN$, Bu_3P; vi, 50% H_2O_2; vii, o-$Cl_2C_6H_4$, 200 °C; viii, CrO_3; ix, BBr_3

Scheme 16

Stereocontrolled approaches to natural *C*-nucleosides have been based upon Baeyer–Villiger ring opening of bridged oxygen heterocycles. Noyori *et al.*[86] have successfully converted lactone (**59**) to pseudocytidine (**60**; Scheme 17).

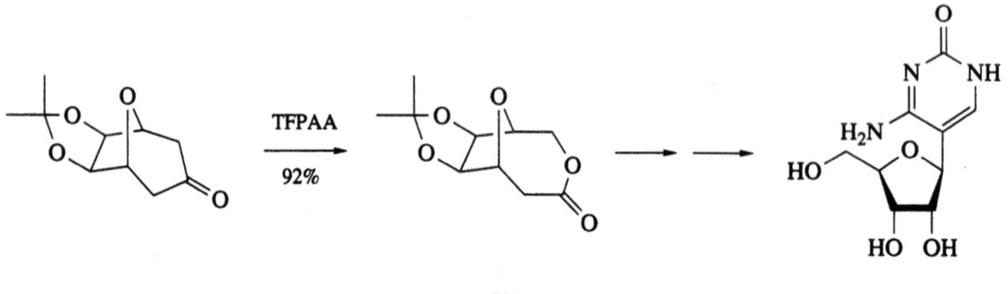

TFPAA, 92%

(**59**) (**60**)

Scheme 17

An example of the utilization of a bridged bicyclic ketone for preparation of an acyclic moiety is the stereoselective synthesis of the C-21 to C-27 segment of rifamycin-S, a member of the ansamycin family of antibiotics (Scheme 18). Rao *et al.*[87] used ketone (**61**), derived from furan, to prepare lactone (**62**). Exhaustive reduction of (**62**) provided the segment (**63**), which contains five chiral centers of rifamycin-S.

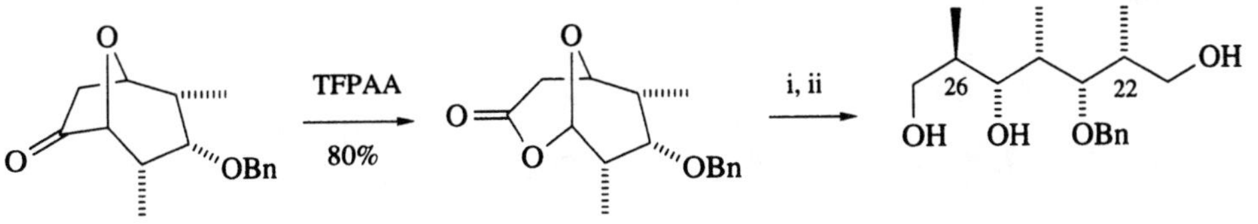

TFPAA, 80%; i, ii

OBn; OBn; OH OH OBn OH; 26; 22

(61) **(62)** **(63)**

i, LDA, MeI; ii, $LiAlH_4$

Scheme 18

The Baeyer–Villiger oxidation has been utilized as an element of several novel functional group manipulations. Suginome and Yamada[88] converted adamantanone (**64**) to 2-thiaadamantane (**66**) *via* the lactone (**65**; Scheme 19). Eaton *et al.*,[89] in the synthesis of pentaprismane (**70**) from homopentaprismanone (**67**; Scheme 20), required that a leaving group be introduced α to the carbonyl group in order to carry out a Favorskii ring contraction. Oxidation of (**67**) afforded lactone (**68**), which was converted in several steps to the requisite hydroxy ketone (**69**).

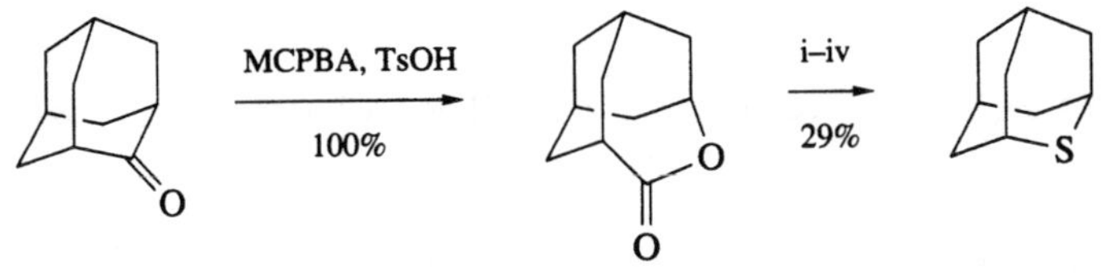

MCPBA, TsOH, 100%; i–iv, 29%

(64) **(65)** **(66)**

i, MeLi; ii, HgO, I_2, *hν*; iii, Me_3SiI ; iv, Na_2S, EtOH

Scheme 19

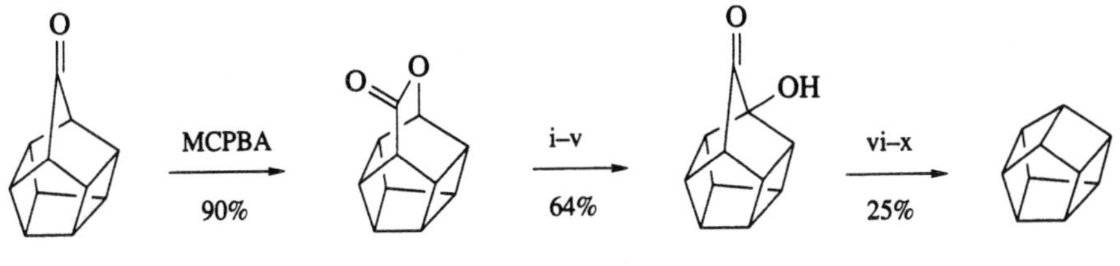

MCPBA, 90%; i–v, 64%; vi–x, 25%

(67) **(68)** **(69)** **(70)**

i, KOH; ii, RuO_2, $NaIO_4$; iii, CH_2N_2; iv, Na/NH_3; v, Cl_2/Me_2S, Et_3N; vi, TsCl; vii, KOH; vii, ClCOCOCl; ix, Bu^tOOH; x, heat

Scheme 20

A 4,4-disubstituted cyclohexenone synthesis has been developed by Holmes and Madge.[90] The procedure is based upon PAA oxidation of anisole-derived bicyclo[2.2.2]oct-5-en-2-ones, followed by acid-catalyzed isomerization of the products (Scheme 21).

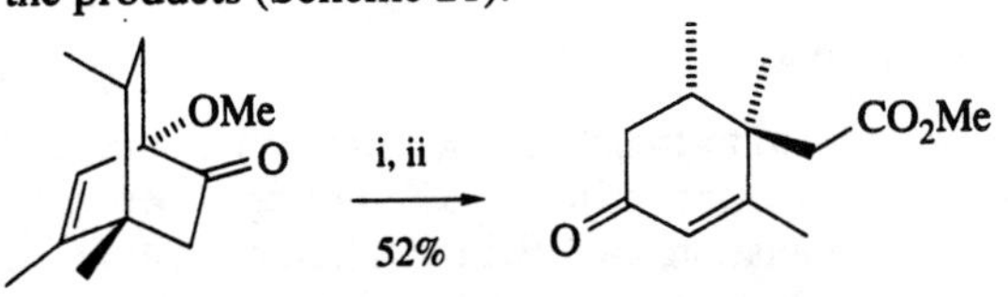

OMe; i, ii, 52%; CO_2Me

i, PAA, NaOAc, AcOH; ii, Me_2SO_4, NaOH

Scheme 21

5.1.5.6 α,β-Unsaturated Ketones

Montury and Gore[91] have developed a 1,2-ketone transposition method (Scheme 22). Baeyer–Villiger oxidation of conjugated ketone (**71**) afforded an enol acetate and subsequent hydrolysis revealed a carbonyl group, one carbon removed from the original position.

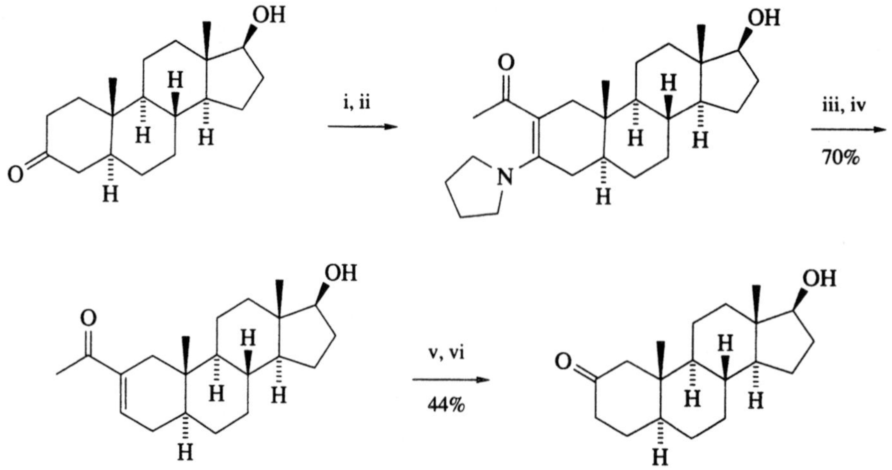

i, C_4H_9N/H^+; ii, AcCl; iii, BH_3; iv, AcOH; v, MCPBA; vi, hydrolysis

Scheme 22

Silverstein and coworkers[92] have used the peroxy acid oxidation of *exo*-alkylidene cycloalkanones as a route to keto acids (Scheme 23). Oxidation of pulegone (**72**) and hydrolysis of the derived enol lactone led to the keto acid (**73**).

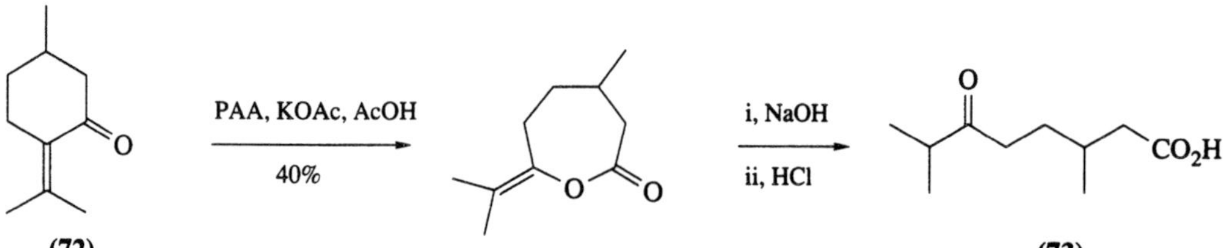

Scheme 23

5.1.5.7 1,2-Dicarbonyl Compounds

Peroxy acids normally form anhydrides when reacted with 1,2-dicarbonyl compounds in inert solvents or acids in alkaline or acidic media.[1] An exception to this generality is the oxidation of the keto ester (**74**; Scheme 24), in which the ring oxygen facilitates tetrahydrofuran ring migration. Ohno and coworkers[93a] have exploited this reaction to develop a chemicoenzymatic approach from furan to the protected L-ribofuranoside (**75**).

5.1.5.8 Aryl- and Alkyl-carbaldehydes

A convenient and inexpensive method to transform electron rich aromatic aldehydes to phenols,[93b] or α,β-unsaturated aldehydes to vinyl formates, utilizes 30% hydrogen peroxide catalyzed by bis(*o*-nitrophenyl) diselenide.[93c] A two-step formylation/MCPBA oxidation procedure (Scheme 25) was utilized by Kishi and coworkers[94] in the 100 g scale conversion of 2,6-dimethoxytoluene to the mitomycin precursor (**76**). An organic peroxy acid was not required for the conversion of 9-formyl-6-methylellipticine (**77**) to 9-hydroxyellipticine (**78**; Scheme 26).[95] Under these conditions, the pyridine nitrogen was not oxidized.

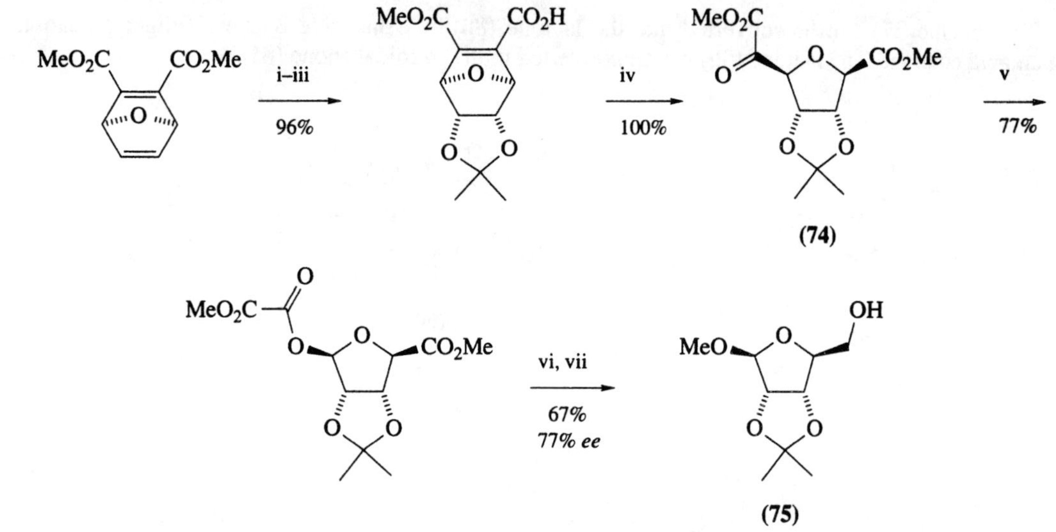

(74)

(75)

i, OsO_4, NMO; ii, 2,2-dimethoxypropane, TsOH; iii, pig liver esterase; iv, O_3, MeOH; v, MCPBA; vi, HCl, MeOH; vii, $LiAlH_4$

Scheme 24

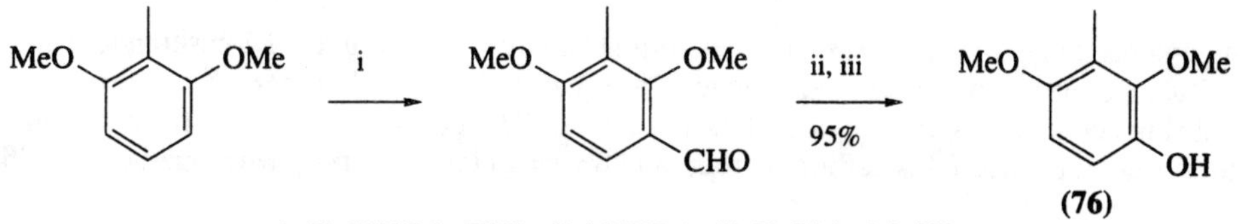

(76)

i, Cl_2CHOMe, $TiCl_4$; ii, MCPBA; iii, NaOMe, MeOH

Scheme 25

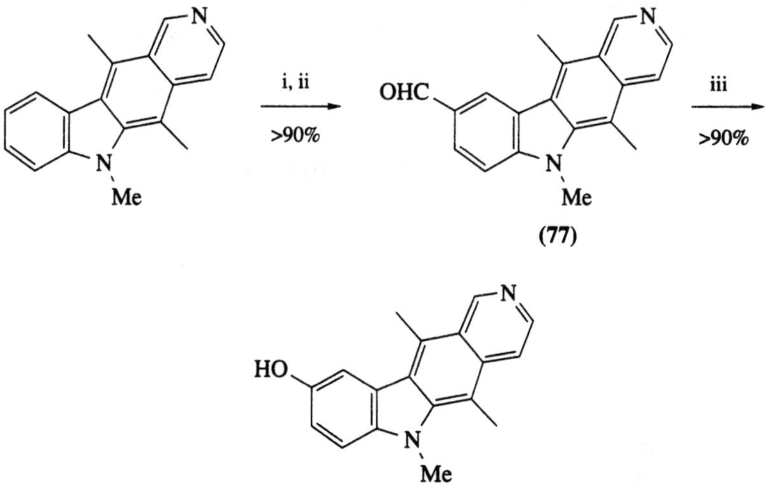

(77)

(78)

i, $ClCH_2OCHCl_2$, $AlCl_3$; ii, hydrolysis; iii, 35% H_2O_2, H_2SO_4, MeOH

Scheme 26

5.1.6 SIDE REACTIONS

While Baeyer–Villiger oxidations of saturated ketones generally occur without skeletal rearrangements, heterolytic cleavages may occur if cations are readily accessible. For example, ring A of triterpenoid (**79**; Scheme 27) can be converted into the lactone (**80**) by exhaustive Baeyer–Villiger oxidation using an acid catalyst. The lactone (**80**) can be converted to the steroidal enone (**81**).[96]

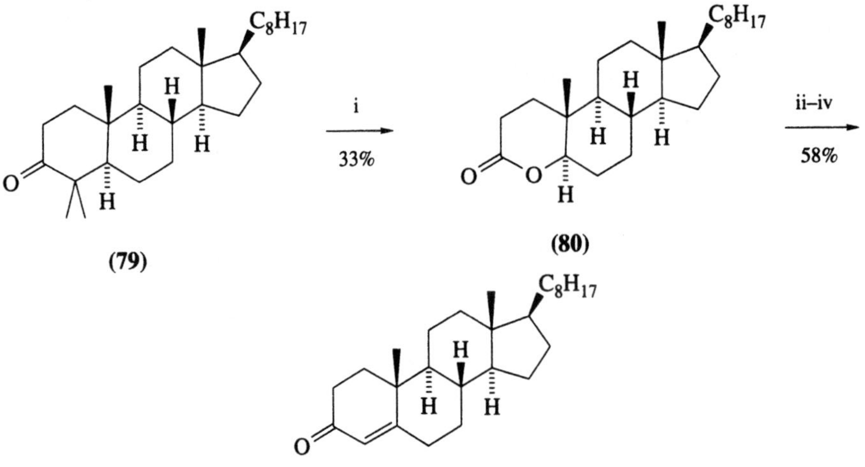

C_8H_{17} H H H H (**79**) i 33% C_8H_{17} H H H H (**80**) ii–iv 58% C_8H_{17} H H H (**81**)

i, 40% PAA, BF_3 etherate; ii, MeLi; iii, PDC; iv, MeOH, NaOH

Scheme 27

Rearrangements can provide access to novel ring structures; The strained 1,3-bishomocubanone (**82**) led to significant amounts of rearranged lactone (**83**; equation 20).[97] Galteri *et al.*[98] have found synthetically useful ring contractions of α-acyldecalones (equation 21). Peroxy acid attacks the exocyclic ketone of (**84**) and the bond between this carbonyl group and the ring cleaves during rearrangement to (**85**).

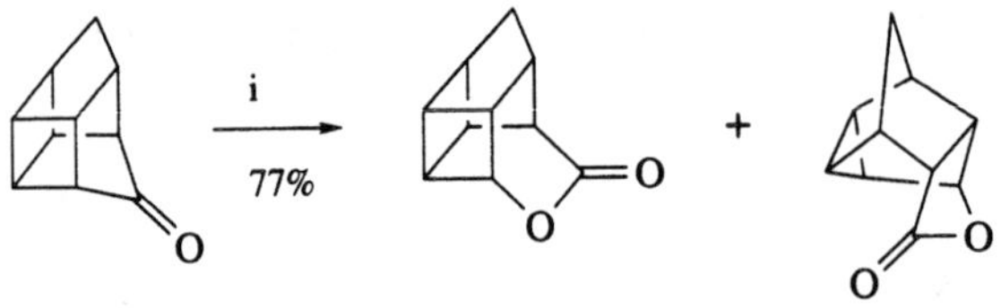

(**82**) i 77% 60% + 40% (**83**) (20)

i, MCPBA, 1:1 dioxane:H_2O, 1 h

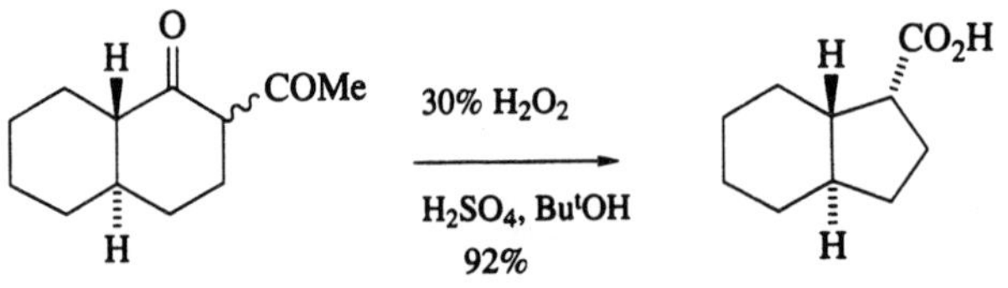

H COMe H (**84**) 30% H_2O_2 H_2SO_4, Bu^tOH 92% H CO_2H H (**85**) (21)

Cationic rearrangement of the bridged bicyclic lactone (**86**) provides a less-strained fused lactone (**87**; equation 22).[99] This rearrangement has proven useful in prostaglandin synthesis.[99,100] Other oxidative rearrangements are discussed in more detail in Chapter 7.2 of this volume.

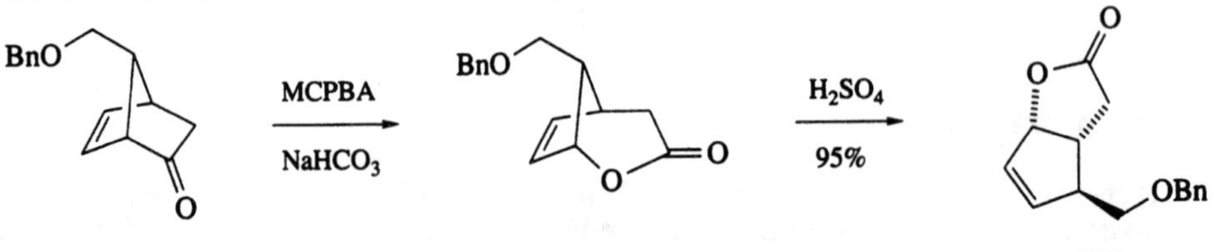

BnO MCPBA $NaHCO_3$ BnO (**86**) H_2SO_4 95% OBn (**87**) (22)

5.1.7 REFERENCES

1. C. H. Hassall, *Org. React. (N.Y.)*, 1957, **9**, 73.
2. B. Plesnicar, in 'Oxidation in Organic Chemistry', ed. W. S. Trahanovsky, Academic Press, New York, 1978, part C, p. 254.
3. S. N. Lewis, in 'Oxidation', ed. R. L. Augustine, Dekker, New York, 1969, vol. 1, p. 213.
4. P. A. S. Smith, 'Molecular Rearrangements', ed. P. de Mayo, Interscience, New York, 1963, vol. I, p. 577.
5. H. O. House, 'Modern Synthetic Reactions', 2nd edn., Benjamin, New York, 1972, p. 327.
6. P. H. Bentley, *Chem. Soc. Rev.*, 1973, **2**, 29.
7. J. B. Lee and B. C. Uff, *Q. Rev. Chem. Soc.*, 1967, **21**, 429.
8. J. E. Leffler, *Chem. Rev.*, 1949, **45**, 385.
9. G. R. Krow, *Tetrahedron*, 1981, **37**, 2697.
10. A. J. Waring, in 'Comprehensive Organic Chemistry', ed. D. H. R. Barton and W. D. Ollis, Pergamon Press, Oxford, 1979, vol. 1, p. 1017.
11. E. W. Colvin, in ref. 10, vol. 2, p. 593.
12. S. M. Roberts, in ref. 10, vol. 2, p. 739.
13. J. M. Brown, in ref. 10, vol. 2, p. 779.
14. I. O. Sutherland, in ref. 10, vol. 2, p. 869.
15. A. F. Hegarty, in ref. 10, vol. 2, p. 1105.
16. T. Laird, in ref. 10, vol. 1, p. 1105.
17. J. D. Rozzell, Jr. and S. A. Benner, *J. Org. Chem.*, 1983, **48**, 1190.
18. H. Suemune, H. Maruoka, S. Saeki and K. Sakai, *Chem. Pharm. Bull.*, 1986, **34**, 4629.
19. M. F. Hawthorne, W. D. Emmons and K. S. McCallum, *J. Am. Chem. Soc.*, 1958, **80**, 6393.
20. R. C. Gadwood, R. Lett and J. E. Wissinger, *J. Am. Chem. Soc.*, 1986, **108**, 6343.
21. E. Caspi, Y. W. Chang and R. I. Dorfman, *J. Med. Pharm. Chem.*, 1962, **5**, 714.
22. P. E. Sonnet and J. E. Oliver, *J. Heterocycl. Chem.*, 1974, **11**, 263.
23. E. S. Olson, *J. Am. Oil Chem. Soc.*, 1977, **54**, 51.
24. G. M. Rubottom, J. M. Gruber, R. K. Boeckman, Jr., M. Ramaiah and J. B. Medwid, *Tetrahedron Lett.*, 1978, 4603.
25. V. Dave, J. B. Stothers and E. W. Warnhoff, *Can. J. Chem.*, 1980, **58**, 2666.
26. H. Wetter, *Helv. Chim. Acta*, 1981, **64**, 761.
27. M. Asaoka, K. Shima and H. Takei, *Tetrahedron Lett.*, 1987, **28**, 5669.
28. D. Gani and D. W. Young, *J. Chem. Soc., Perkin Trans. 1*, 1983, 867.
29. M. J. Thompson, W. J. Meudt, N. B. Mandava, S. R. Dutky, W. R. Lusby and D. W. Spaulding, *Steroids*, 1982, **39**, 89.
30. G. R. Krow, C. A. Johnson, J. P. Guare, D. Kubrak, K. J. Henz, D. A. Shaw, S. W. Szczepanski and J. T. Carey, *J. Org. Chem.*, 1982, **47**, 5239.
31. P. Deslongchamps, *Tetrahedron*, 1975, **31**, 2463.
32. R. D. Clark and C. H. Heathcock, *J. Org. Chem.*, 1976, **41**, 1396.
33. Z. Grudzinski, S. M. Roberts, C. Howard and R. F. Newton, *J. Chem. Soc., Perkin Trans. 1*, 1978, 1182.
34. M. Vikas, *Bull. Soc. Chem. Fr.*, 1959, 1401.
35. A. Gringauz and E. Tosk, *Org. Prep. Proced. Int.*, 1970, **2**, 185.
36. G. Reissenweber and D. Mangold, *Angew. Chem., Int. Ed. Engl.*, 1980, **19**, 222.
37. S. M. Ali, T. V. Lee and S. M. Roberts, *Synthesis*, 1977, 155.
38. J. K. Whitesell, R. S. Matthews and A. M. Helbling, *J. Org. Chem.*, 1978, **43**, 784.
39. G. A. Olah, T. Yamato, P. S. Iyer, N. J. Trivedi, B. P. Singh and G. K. S. Prakash, *Mater. Chem. Phys.*, 1987, **17**, 21.
40. P. A. Grieco, T. Oguri and S. Gilman, *J. Am. Chem. Soc.*, 1980, **102**, 5886.
41. P. Brougham, M. S. Cooper, D. A. Cummerson, H. Heaney and N. Thompson, *Synthesis*, 1987, 1015.
42. M. Anastasia, P. Allevi, P. Ciuffreda, A. Fiecchi and A. Scala, *J. Org. Chem.*, 1985, **50**, 321.
43. N. N. Schwartz and J. H. Blumbergs, *J. Org. Chem.*, 1964, **29**, 1976.
44. P. A. Grieco and T. R. Vedananda, *J. Org. Chem.*, 1983, **48**, 3497.
45. P. A. Grieco, T. Oguri, C.-L. J. Wang and E. Williams, *J. Org. Chem.*, 1977, **42**, 4113.
46. Y. Ogata, K. Tomizawa and T. Ikeda, *J. Org. Chem.*, 1978, **43**, 2417.
47. S. Matsubara, K. Takai and H. Nozaki, *Bull. Chem. Soc. Jpn.*, 1983, **56**, 2029.
48. W. Adam and A. Rodriguez, *J. Org. Chem.*, 1979, **44**, 4969.
49. A. McKillop and J. A. Tarbin, *Tetrahedron*, 1987, **43**, 1753.
50. R. T. Taylor and L. A. Flood, *J. Org. Chem.*, 1983, **48**, 5160.
51. S. E. Jacobson, F. Mares and P. M. Zambri, *J. Am. Chem. Soc.*, 1979, **101**, 6938.
52. G. B. Payne, *Tetrahedron*, 1962, **18**, 763.
53. K. Saki, M. Yamashita and Y. Shibata, *Chem. Lett.*, 1986, 353.
54. N. Koizumi, M. Morisaki, N. Ikekawa, Y. Tanaka and H. F. DeLuca, *J. Steroid Biochem.*, 1979, **10**, 261.
55. P. A. Crooks, L. A. Damani and D. A. Cowan, *Chem. Ind. (London)*, 1981, 335.
56. W. Czuba and C. Walkowicz, *Pol. J. Chem.*, 1983, **57**, 333.
57. B. M. Trost, *Acc. Chem. Res.*, 1978, **11**, 453.
58. H. J. Reich, J. M. Renga and I. L. Reich, *J. Am. Chem. Soc.*, 1975, **97**, 5434.
59. N. L. Wendler, D. Taub and H. L. Slates, *J. Am. Chem. Soc.*, 1955, **77**, 3559.
60. M. S. Ahmad and I. A. Khan, *Acta Chim. Acad. Sci. Hung.*, 1981, **106**, 111 (*Chem. Abstr.*, 1981, **95**, 115 855m).
61. S. L. Schreiber and W.-F. Liew, *Tetrahedron Lett.*, 1983, **24**, 2363.
62. S. F. Lee, M. Edgar, C. S. Pak, G. Barth and C. Djerassi, *J. Am. Chem. Soc.*, 1980, **102**, 4784.
63. P. R. Brook and B. V. Brophy, *J. Chem. Soc., Perkin Trans. 1*, 1985, 2509.
64. R. Sobti and S. Dev, *Tetrahedron*, 1974, **30**, 2927.

65. D. P. Curran, S. A. Scanga and C. J. Fenk, *J. Org. Chem.*, 1984, **49**, 3474.
66. E. J. Corey and J. G. Smith, *J. Am. Chem. Soc.*, 1979, **101**, 1038.
67. B. A. Pearlman, *J. Am. Chem. Soc.*, 1979, **101**, 6404.
68. D. J. Hart and Y.-M. Tsai, *J. Am. Chem. Soc.*, 1984, **106**, 8209.
69. H. Bretschneider, K. Hohenlohe-Oehringen, A. Kaiser and U. Wölcke, *Helv. Chim. Acta*, 1973, **56**, 2857.
70. S.-I. Nakatsuka, K. Ueda, O. Asano and T. Goto, *Heterocycles*, 1987, **26**, 65.
71. S. W. Pelletier and Y. Ohtsuka, *Tetrahedron*, 1977, **33**, 1021.
72. G. Grethe, J. F. Sereno, T. H. Williams and M. R. Uskokovic, *J. Org. Chem.*, 1983, **48**, 5315.
73. E. J. Corey, S. Kim, S. Yoo, K. C. Nicolaou, L. S. Melvin, Jr., D. J. Brunelle, J. R. Falck, E. J. Trybulski, R. Lett and P. W. Sheldrake, *J. Am. Chem. Soc.*, 1978, **100**, 4620.
74. (a) K. Narasaka, T. Sakakura, T. Uchimaru and D. Guedin-Vuong, *J. Am. Chem. Soc.*, 1984, **106**, 2954; (b) A. Prelle and E. Winterfeldt, *Heterocycles*, 1989, **28**, 333.
75. J. S. Clark and A. B. Holmes, *Tetrahedron Lett.*, 1978, **29**, 4333.
76. S. Takatsuto, N. Yazawa, M. Ishiguro, M. Morisaki and N. Ikekawa, *J. Chem. Soc., Perkin Trans. I*, 1984, 139.
77. E. J. Corey, M.-C. Kang, M. C. Desai, A. K. Ghosh and I. N. Houpis, *J. Am. Chem. Soc.*, 1988, **110**, 649.
78. G. R. Pettit and J. R. Dias, *J. Org. Chem.*, 1972, **37**, 973.
79. H. Kaga, S. Kobayashi and M. Ohno, *Tetrahedron Lett.*, 1988, **29**, 1057.
80. A. Murai, N. Tanimoto, N. Sakamoto and T. Masamune, *J. Am. Chem. Soc.*, 1988, **110**, 1985.
81. H. O. House, J. L. Haack, W. C. McDaniel and D. Van Derveer, *J. Org. Chem.*, 1983, **48**, 1643.
82. S. Takano, M. Takahashi and K. Ogasawara, *J. Am. Chem. Soc.*, 1980, **102**, 4282.
83. P. A. Grieco, T. Takigawa and W. J. Schillinger, *J. Org. Chem.*, 1980, **45**, 2247.
84. P. A. Grieco, T. Takigawa and T. R. Vedananda, *J. Org. Chem.*, 1985, **50**, 3111.
85. D. P. Reynolds, R. F. Newton and S. M. Roberts, *J. Chem. Soc., Chem. Commun.*, 1979, 1150.
86. R. Noyori, T. Sato and Y. Hayakawa, *J. Am. Chem. Soc.*, 1978, **100**, 2561.
87. A. V. R. Rao, J. S. Yadav and V. Vidyasagar, *J. Chem. Soc., Chem. Commun.*, 1985, 55.
88. J. Suginome and S. Yamada, *Synthesis*, 1986, 741.
89. P. E. Eaton, Y. S. Or, S. J. Branca and B. K. R. Shankar, *Tetrahedron*, 1986, **42**, 1621.
90. N. C. Madge and A. B. Holmes, *J. Chem. Soc., Chem. Commun.*, 1980, 956.
91. M. Montury and J. Goré, *Tetrahedron*, 1977, **33**, 2819.
92. J. R. Handley, A. A. Swigar and R. M. Silverstein, *J. Org. Chem.*, 1979, **44**, 2954.
93. (a) Y. Ito, T. Shibata, M. Arita, H. Sawai and M. Ohno, *J. Am. Chem. Soc.*, 1981, **103**, 6739; (b) L. Syper, *Synthesis*, 1989, 167; (c) L. Syper, *Tetrahedron*, 1987, **43**, 2853.
94. F. Nakatsubo, A. J. Cocuzza, D. E. Keeley and Y. Kishi, *J. Am. Chem. Soc.*, 1977, **99**, 4835.
95. A. Langendoen, G.-J. Koomen and U. K. Pandit, *Heterocycles*, 1987, **26**, 91.
96. R. Uusvuori and T. A. Hase, *Synth. Commun.*, 1982, **12**, 1081.
97. H. Muira, K.-I. Hirao and O. Yonemitsu, *Tetrahedron*, 1978, **34**, 1805.
98. M. Galteri, P. H. Lewis, S. Middleton and L. E. Stock, *Aust. J. Chem.*, 1980, **33**, 101.
99. J. S. Bindra and A. Grodski, *J. Org. Chem.*, 1978, **43**, 3240.
100. E. J. Corey, S. M. Albonico, U. Koelliker, T. K. Schaaf and R. V. Varma, *J. Am. Chem. Soc.*, 1971, **93**, 1491.

5.2
The Beckmann and Related Reactions

DONALD CRAIG
Imperial College of Science, Technology & Medicine, London, UK

5.2.1 INTRODUCTION

The preeminence of the Beckmann reaction among oxidative synthetic transformations since its discovery[1] over a century ago is borne witness by its frequent review.[2–11] The reaction has found most widespread use in the rearrangement mode, and may be regarded as the nitrogen analog of the Baeyer–Villiger reaction. Thus, nitrogen may be inserted into carbonyl carbon–α-carbon bonds of aldehydes and ketones *via* rearrangement of the derived oximes. The stereoselectivity of the rearrangement allows distinction to be made between the α- and α′-carbon atoms flanking a ketone group (Scheme 1).

Only in relatively recent times has the potential for Beckmann-type reactions other than rearrangements been exploited fully. Reactive Beckmann intermediates have been intercepted successfully both inter- and intra-molecularly, with both carbon and heteroatomic nucleophiles. Reactions in the presence of reducing agents give the products of redox processes insofar as the migrating carbon atom undergoes oxidation whilst the ex-carbonyl carbon is reduced. In certain circumstances fragmentation, rather than rearrangement, becomes dominant, and such reactions have found particular application in ring-cleavage processes.

This chapter is divided into the three broad categories described above, namely rearrangement, addition and fragmentation. Although it is beyond the scope of this account to provide exhaustive coverage of this active area, close attention has been paid to work published since the last major review.[11] Neither

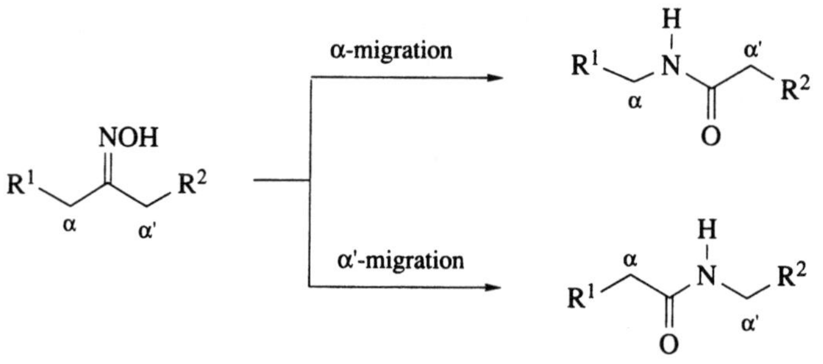

Scheme 1

the photochemically induced Beckmann rearrangement[12] nor the Schmidt reaction of ketones[13] has been covered.

Throughout this chapter, (*E*)- and (*Z*)-nomenclature refers to oxime geometry, and has been indicated whenever specified in the original work.

5.2.2 REARRANGEMENT REACTIONS

5.2.2.1 Mechanism and Stereochemistry

The generally accepted mechanism for the Beckmann rearrangement is presented in Scheme 2. The *N*-hydroxy group of the oxime is rendered more nucleofugal either by protonation or esterification. Migration of hydrogen, alkyl or aryl groups from carbon to nitrogen may be followed by attack of an oxygen nucleophile at the incipiently cationic ex-carbonyl carbon to yield an imidate; this may then undergo Chapman rearrangement[14] to give an *N*-substituted amide. Alternatively, in some instances, free nitrilium species have been implicated. It has also been suggested that nitrenium ions and tetrahedral intermediates may be involved in certain cases.

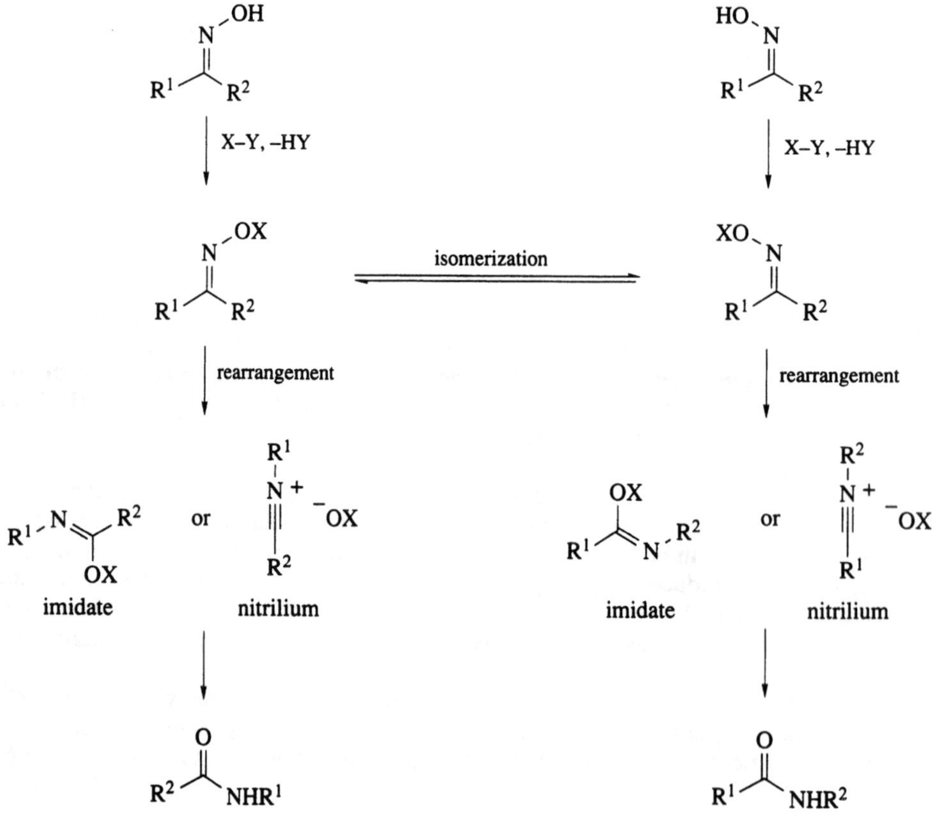

Scheme 2

It has long been considered that migration of the substituent *anti* to the leaving group on nitrogen predominates to the extent that product identity may serve as a reliable indicator of oxime geometry. However, the tendency for oxime isomerization to occur under certain reaction conditions, and the differing migratory aptitudes of oxime substituents are such that this 'rule' should be used with caution. The dependence of product composition upon the rearrangement conditions employed further suggests that varying mechanisms may be operative. Mechanisms are discussed in greater depth in the context of the specific examples presented below.

5.2.2.2 Ketoximes

5.2.2.2.1 O-Unsubstituted ketoximes

Ketoximes may be rearranged directly to amides under a wide variety of conditions. Phosphorus pentachloride, phosphorus oxychloride, thionyl chloride, trimethylsilyl iodide (TMS-I), formic acid, polyphosphoric acid, trimethylsilyl polyphosphate and mineral acids have all been successfully employed. Representative procedures using these and other reagent systems have been documented previously.[5,11]

Compelling evidence for the isomerization of oximes prior to rearrangement was provided by studies on the menthone-derived oxime (**1**) shown in equation (1).[15] The use of strong anhydrous acid in these reactions resulted in amides formed *via* rearrangement of ostensibly the less thermodynamically stable (*Z*)-oxime. This was also observed in the acid-mediated reaction of the (*E*)-polycyclic oxime (**2**),[16] which rearranged with migration of the *syn*-alkyl rather than the *anti*-aryl group (equation 2). It was argued that whilst the (*Z*)-isomer would be thermodynamically disfavored due to steric interactions between the oxime oxygen and the *ortho* hydrogen atom, it would be more reactive for the same reason, and apparent *syn* migration would result.

Under conditions where oxime isomerization is facile, product distribution is clearly the result of a subtle combination of factors, including relative migratory aptitudes of substituents as well as the

(1)

(1)

$POCl_3$, pyridine	98%	2%
$SOCl_2$, pyridine	90%	10%
20% aq. H_2SO_4	43%	57%
HCl/Et_2O	5%	95%

(2)

(2) PPA, 80 °C, 3 h; 76% → **(3)**

thermodynamic stability of oxime isomers. Both geometric isomers of (**3**) gave the lactam (**4**) upon treatment with hot polyphosphoric acid (equations 3 and 4).[17] In contrast, identical processing of tetralone oxime (**5**) gave exclusively ε-lactam (**6**),[18] the product of aryl rather than alkyl migration (equation 5).

The indanone oximes (**7**) gave the aryl- and alkyl-migrated products (**8**) and (**9**) (equation 6), with the proportion of (**9**) increasing with increasing steric demand of arene substituents at the 4- and 7-positions.[18] Exposure of the (*E*)-oxime (**10**) to phosphorus pentoxide and methanesulfonic acid[19] gave the product (**11**) of migration of the *syn*-alkyl group,[20] presumably *via* the (*Z*)-isomer (equation 7).

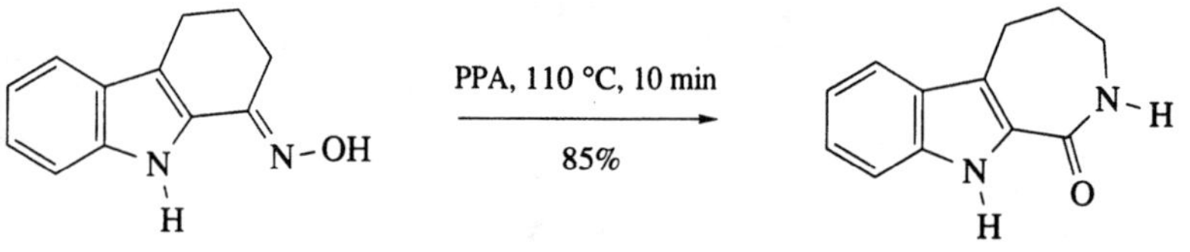

PPA, 110 °C, 10 min; 85%

(E)-(**3**) → (**4**) (3)

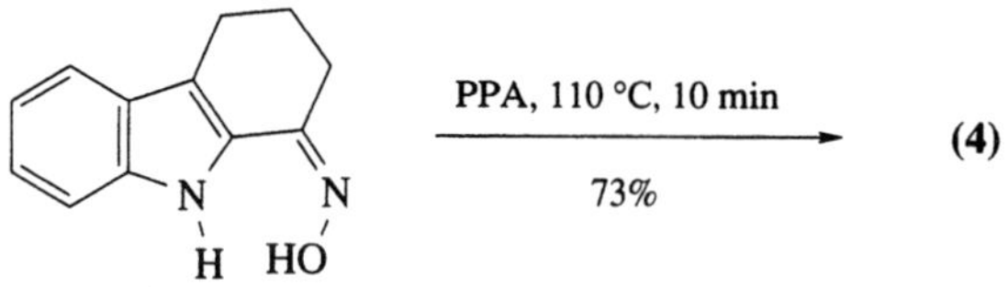

PPA, 110 °C, 10 min; 73%

(Z)-(**3**) → (**4**) (4)

PPA, 110–120 °C, 5–10 min; no yield given

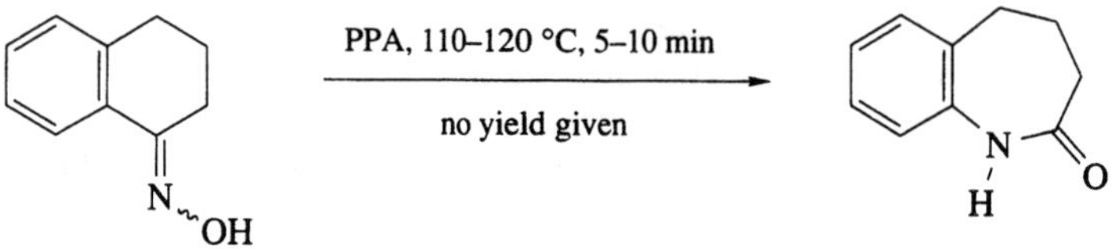

(**5**) → (**6**) (5)

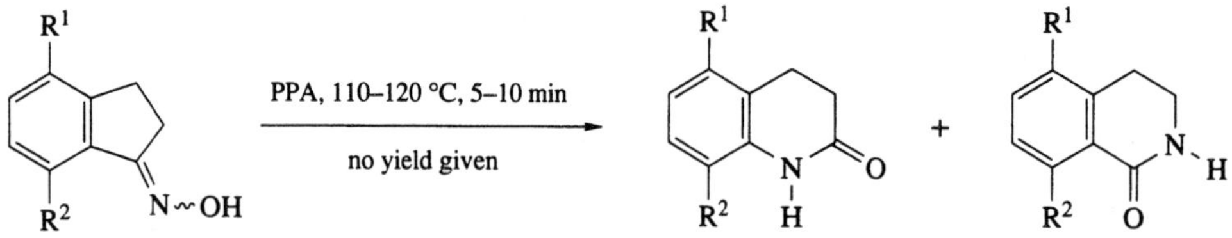

PPA, 110–120 °C, 5–10 min; no yield given

(**7**) → (**8**) + (**9**) (6)

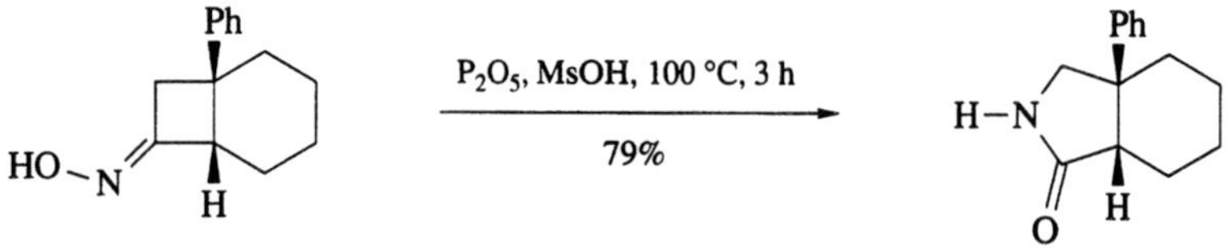

P_2O_5, MsOH, 100 °C, 3 h; 79%

(**10**) → (**11**) (7)

O-Unsubstituted ketoximes undergo rearrangement in nonacidic media also. Thus, both geometric isomers of cholest-4-en-3-one oxime (**12**) rearranged to (**13**) in the presence of triphenylphosphine in refluxing tetrachloromethane (equation 8);[21] the apparent reluctance to migrate of endocyclic unsaturated substituents has also been observed for *O*-substituted oximes (*vide infra*).

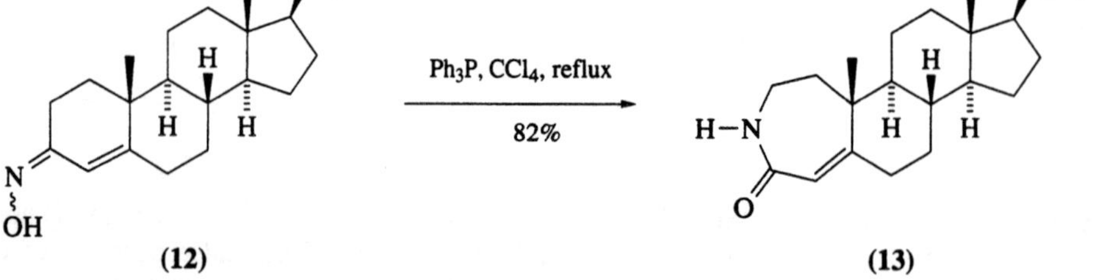

Ph_3P, CCl_4, reflux; 82%

(**12**) → (**13**) (8)

When treated with benzoic acid in the presence of triphenylphosphine and diethyl azodicarboxylate, benzophenone oxime gave a high yield of *N*-benzoylbenzanilide (**14**), formed *via* rearrangement of the intermediate *O*-benzoyloxime (**15**).[22] Reaction of *p*-methoxyacetophenone oxime with 1,1′-carbonyldiimidazole (**16**) gave the *O*-substituted derivative (**17**),[23] whereas in the presence of allyl bromide efficient Beckmann rearrangement was observed (equation 9). Bentonite clay has been effectively employed as a rearrangement promoter.[24]

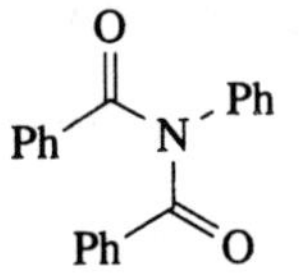
O
Ph
N
Ph
Ph
O

(14)

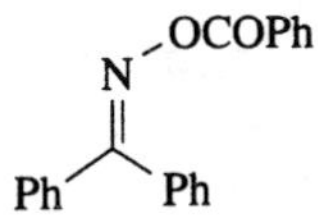
OCOPh
N
Ph
Ph

(15)

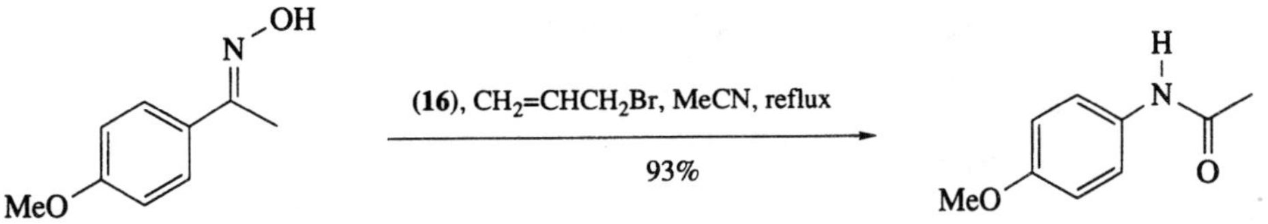
OH
N
MeO
(16), CH2=CHCH2Br, MeCN, reflux
93%
H
N
O
MeO

(9)

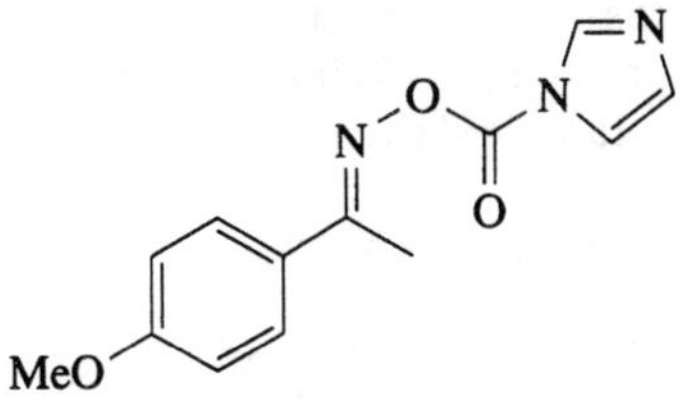
N
O
N
N
O
MeO

(17)

5.2.2.2.2 *O-Substituted ketoximes*

O-Tosyl ketoximes generally rearrange smoothly and with exclusive *anti* migration when dissolved in polar media; a typical example is depicted in equation (10).[25] The tosyl derivatives frequently rearrange under the conditions of their formation (equation 11),[26] and the mildness and specificity of this procedure recommend it for use with acid-sensitive substrates (equation 12).[27]

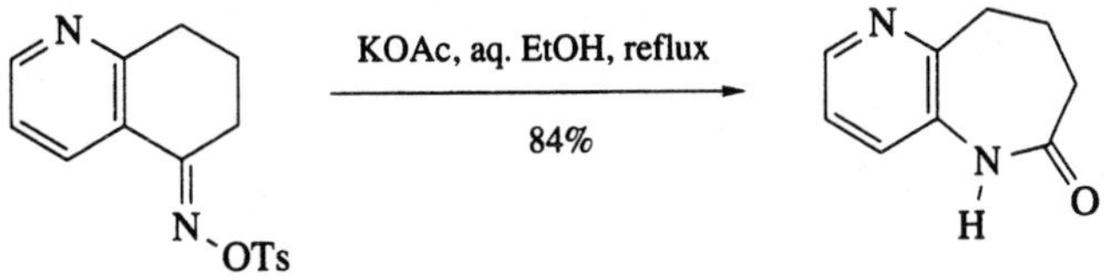
N
N
OTs
KOAc, aq. EtOH, reflux
84%
N
N
H
O

(10)

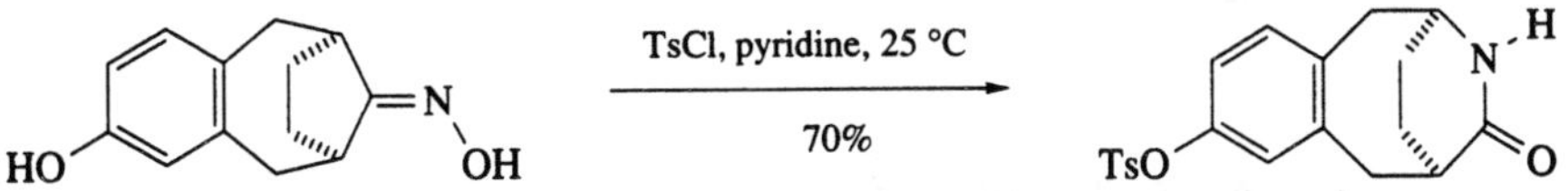
HO
N
OH
TsCl, pyridine, 25 °C
70%
TsO
N
H
O

(11)

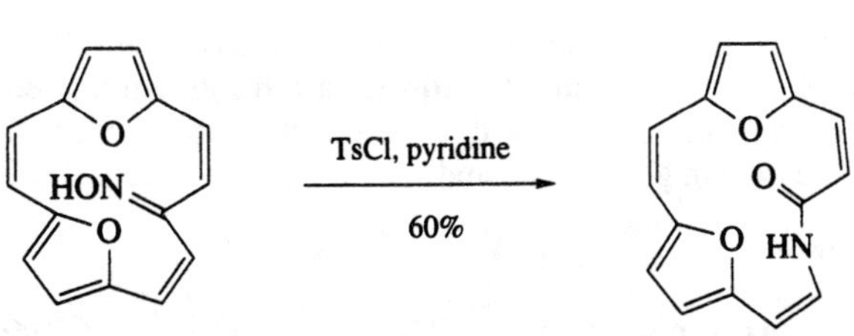
O
HON
O
TsCl, pyridine
60%
O
O
O
HN

(12)

Oxime methanesulfonates are also suitable rearrangement substrates. Thus, the azetidinone oxime methanesulfonate (**18**) underwent smooth rearrangement[28] upon exposure to basic alumina (equation 13);[29] more conventional mineral and organic acid based reagents were ineffective in promoting this transformation.

A striking example of the differing migratory aptitudes of saturated and unsaturated endocyclic groups is represented in equation (14). Under identical conditions, (*E*)-(**19**) was unreactive.[30]

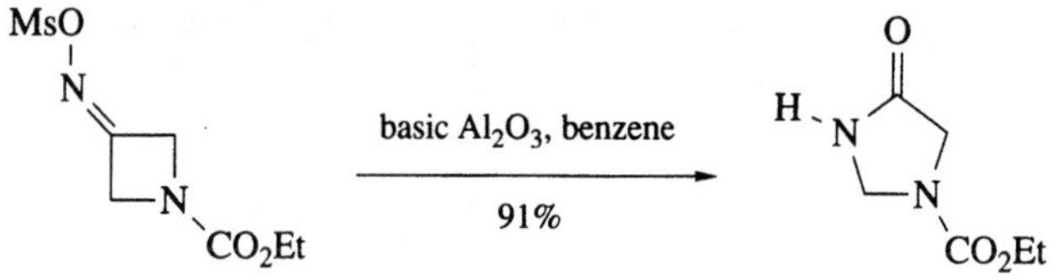
MsO
N
N
CO2Et
basic Al2O3, benzene
91%
O
H
N
N
CO2Et

(18)

(13)

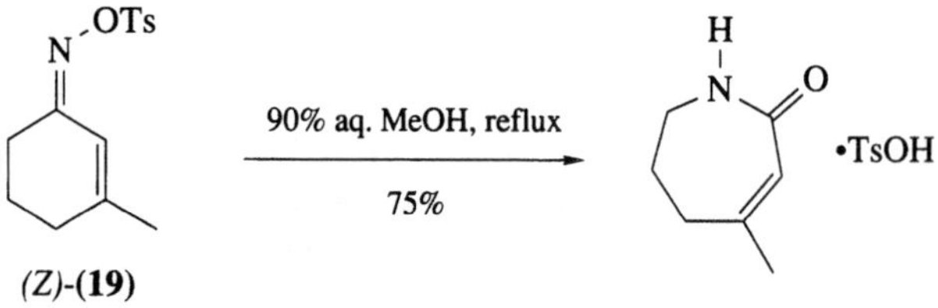
N
OTs
90% aq. MeOH, reflux
75%
H
N
O
•TsOH

(*Z*)-(**19**)

(14)

In contrast, exocyclic vinylic groups migrate readily. (*E*)-α-Benzylidenecyclohexanone oxime (**20**) formed a stable, crystalline addition product (**21**) upon tosylation in pyridine; dilute acidolysis gave the caprolactam derivative (**22**) in high overall yield (Scheme 3).[30]

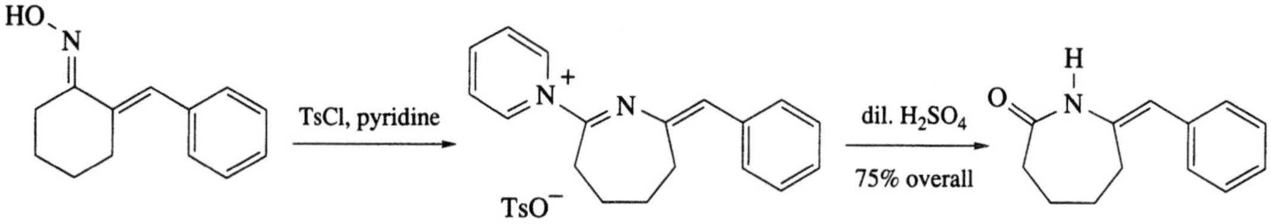
HO
N
TsCl, pyridine
N
+
N
TsO−
dil. H2SO4
75% overall
O
H
N

(**20**) (**21**) (**22**)

Scheme 3

Grob and coworkers have presented evidence for π-participation by migrating acyclic vinylic groups: the 2,4-dinitrophenol derivative (**23**) rearranged more than 2000 times faster than the saturated analog (**24**).[31]

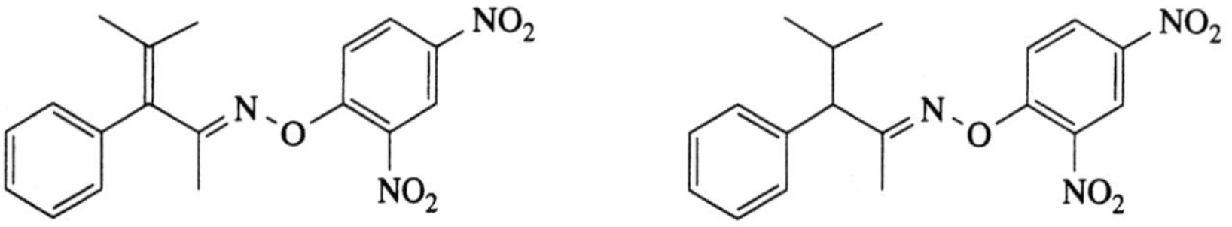
NO2
N
O
NO2
NO2
N
O
NO2

(**23**) (**24**)

5.2.2.2.3 *Direct conversion of ketones to amides*

Under appropriate conditions, ketones may be converted directly to amides. The facility of ketoxime formation from ketones strongly suggests that the former are reactive intermediates in these processes. Treatment of the strained bicyclic ketone (**25**) with *O*-(mesitylsulfonyl)hydroxylamine gave a mixture of the isomeric lactams (**26**) and (**27**) in good yield, and with moderate stereoselectivity (equation 15).[32]

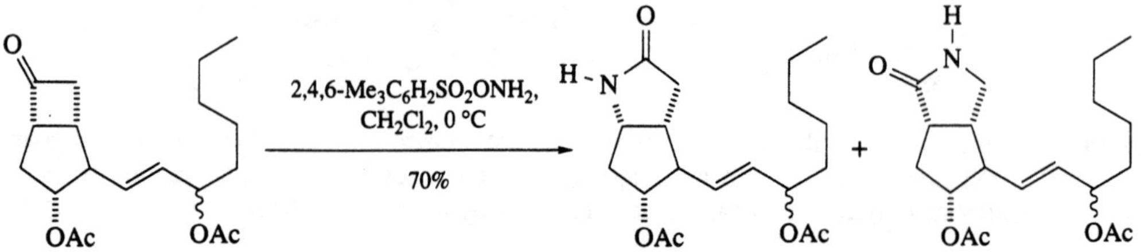
O
OAc
OAc
2,4,6-Me3C6H2SO2ONH2,
CH2Cl2, 0 °C
70%
O
H
N
OAc
OAc
+
H
O
N
OAc
OAc

(15)

(**25**) (**26**) 79% (**27**) 21%

Bicyclo[3.2.1]octan-2-one underwent a highly selective Beckmann rearrangement in the presence of hydroxylamine *O*-hydrogensulfate and formic acid (equation 16).[33] A modification of this procedure using catalytic trifluoromethanesulfonic acid has been reported.[34]

H_2NOSO_3H, HCO_2H; 97%; products 95% + 5% (16)

5.2.2.2.4 *Miscellaneous reactions*

Migration of an acyl substituent was observed during rearrangement of the α-diketone monooxime (**28**),[35] with formation of the tetrasubstituted isoquinoline (**29**) presumably occurring *via* a cyclic imide intermediate (equation 17).

(**28**) —PCl_5, HCl, $POCl_3$; 97%→ (**29**) (17)

A diastereomeric mixture of heterocyclic ketones (**30**) formed a single bicyclic ketolactam (**31**) when heated in refluxing trifluoroacetic acid (equation 18).[36] Ring contraction similarly occurred on thermolysis of (**32**; equation 19).[37]

(**30**) —TFA, 72 °C; 70%→ (**31**) (18)

(**32**) —195 °C; 53%→ (19)

5.2.2.3 Aldoximes

Under a variety of conditions, aldoximes almost invariably rearrange in a nonstereospecific manner to give primary amides. Reagents which have been used to effect this transformation include boron trifluoride, phosphorus pentoxide and methanesulfonic acid, transition metal complexes and silica gel. Certain reagents may promote stereospecific *anti* migration in (*E*)-benzaldoximes to give formanilides. A comprehensive bibliography is provided in ref. 11.

5.2.3 ADDITION REACTIONS

One of the most useful variants of the Beckmann reaction is that in which the intermediate is trapped with a nucleophile other than water. Both carbon and heteroatomic nucleophiles have been used in this context. Intramolecular interception gives cyclized products.

5.2.3.1 Intermolecular Reactions

Treatment of the carbonate derivative (**33**) of acetophenone oxime with TMS-I gave a quantitative yield of the imidoyl iodide (**34**), as determined by ^{1}H NMR (equation 20). The reactive product could be reacted further with a range of nucleophiles giving substituted imines in high yields.[38] Interestingly, exposure of acetophenone oxime itself to similar conditions gave the aryl-migrated Beckmann rearrangement product in moderate yield.[39]

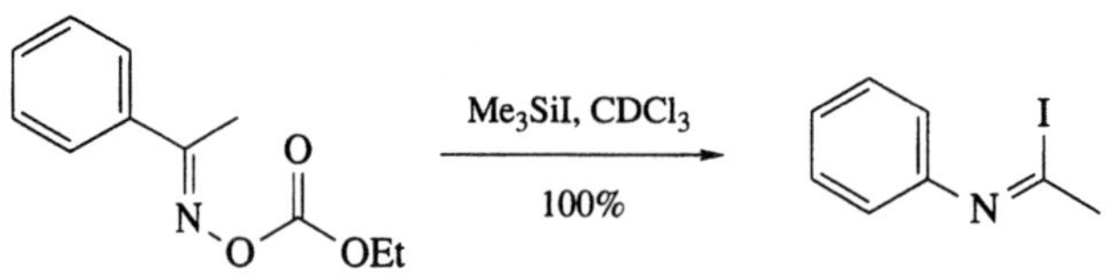

Me$_3$SiI, CDCl$_3$, 100% (**33**) → (**34**) (20)

Intermediates have been trapped by distal attack of halide ion. Reaction of the α,β-unsaturated oxime (**35**) with PCl$_5$ gave a good yield of *N*-phenyl-3-chloropropionamide (equation 21).[40] Similar reactivity was observed with the alkynic oxime (**36**; equation 22).[41]

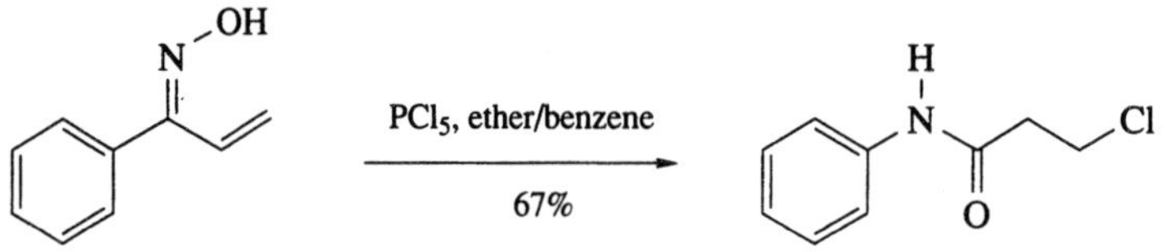

(**35**) → PCl$_5$, ether/benzene, 67% (21)

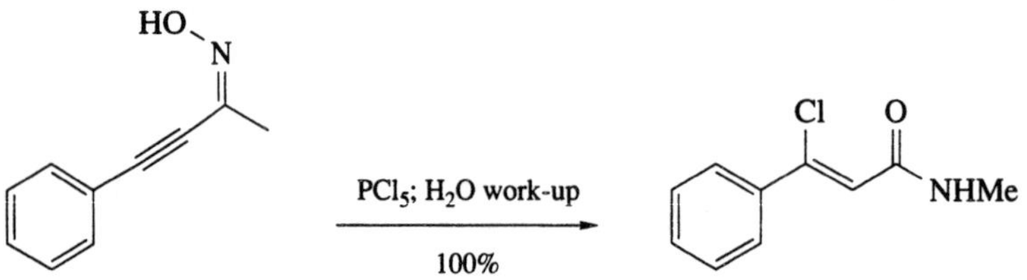

(**36**) → PCl$_5$; H$_2$O work-up, 100% (22)

Azide ion has been shown to be an effective trap for a Beckmann intermediate. Treatment of acetophenone oxime with thionyl cloride in the presence of hydrazoic acid gave tetrazole (**37**), as shown in equation (23).[42] Thiolates participate efficiently in Beckmann reactions, and the sulfur nucleophile may be incorporated into a Lewis acidic reagent, as demonstrated in equation (24).[43]

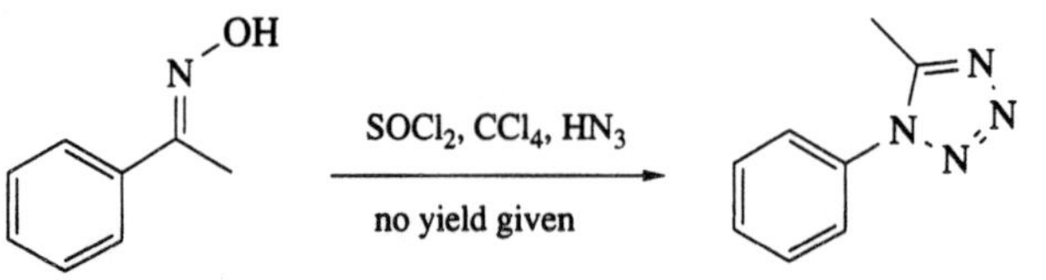

SOCl$_2$, CCl$_4$, HN$_3$, no yield given → (**37**) (23)

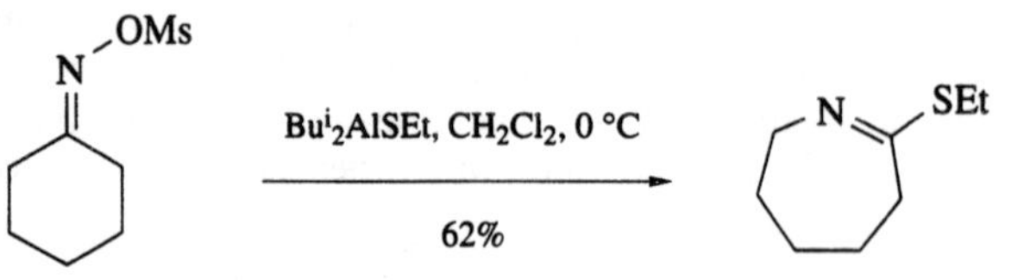

(**38**) → Bu$^i_2$AlSEt, CH$_2$Cl$_2$, 0 °C, 62% (24)

This general strategy has also successfully been exploited for the formation of carbon–carbon bonds. Compound (**38**) underwent ring expansion with concomitant formation of two new carbon–carbon bonds when treated sequentially with Grignard reagents as depicted in equation (25).[44] Similarly, treatment of cyclic ketoxime methanesulfonates with trimethylaluminum in dichloromethane gave cyclic imines, the products of formation of a single new carbon–carbon bond.[43,45] *In situ* treatment of the imines with DIBAL-H gave cyclic amines in good yields, as represented in equation (26).[43,46]

(38) 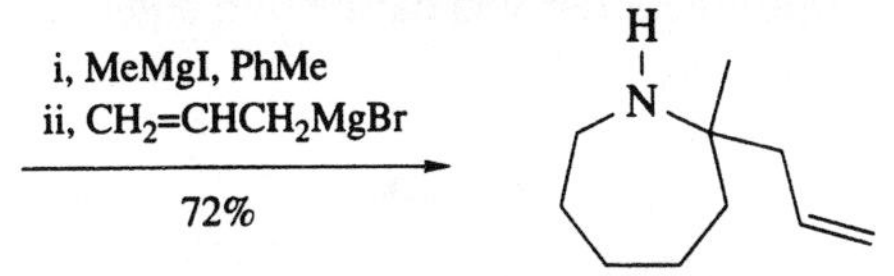 → i, MeMgI, PhMe; ii, CH_2=$CHCH_2MgBr$; 72% (25)

(38) 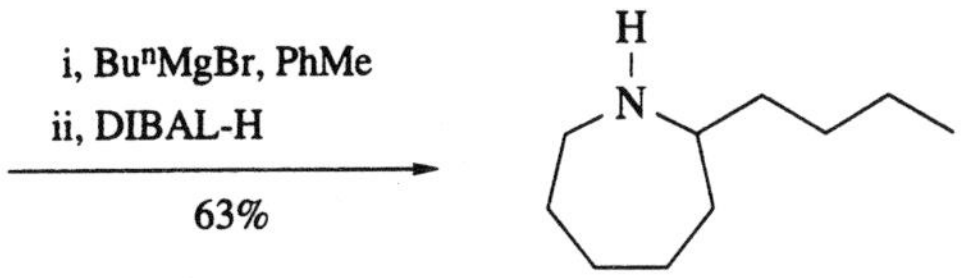 → i, Bu^nMgBr, PhMe; ii, DIBAL-H; 63% (26)

Lewis acid induced Beckmann reaction in the presence of silyl enol ethers has been used to prepare vinylogous amides (equation 27).[47]

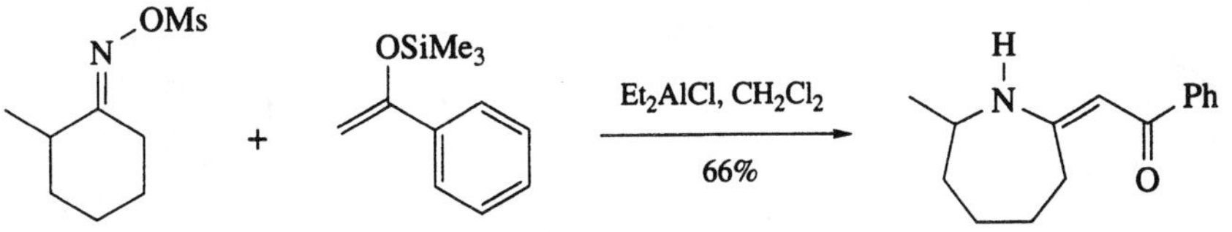

OMs, N + $OSiMe_3$ → Et_2AlCl, CH_2Cl_2; 66%; H, N, Ph, O (27)

5.2.3.2 Intramolecular Reactions

The use of internal nucleophiles in the Beckmann reaction has been demonstrated to be an effective method for the synthesis of N-heterocycles. Substituted pyrrolines are readily available *via* this approach, as exemplified in equation (28).[48]

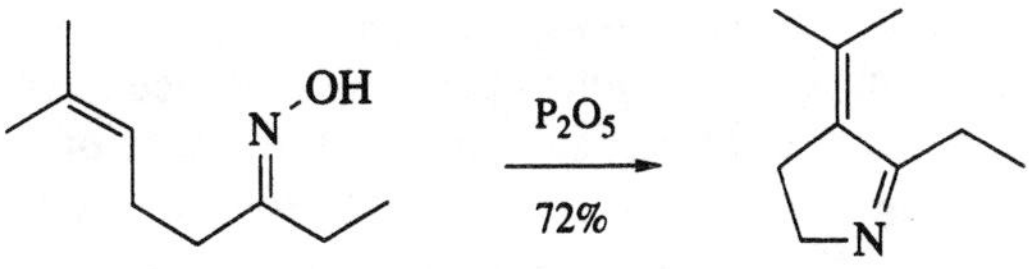

N, OH → P_2O_5; 72%; N (28)

Endocyclic cyclization of double bonds onto Beckmann intermediates gives doubly unsaturated N-heterocycles, which may be oxidized under mild conditions to give aromatic products (Scheme 4).[49]

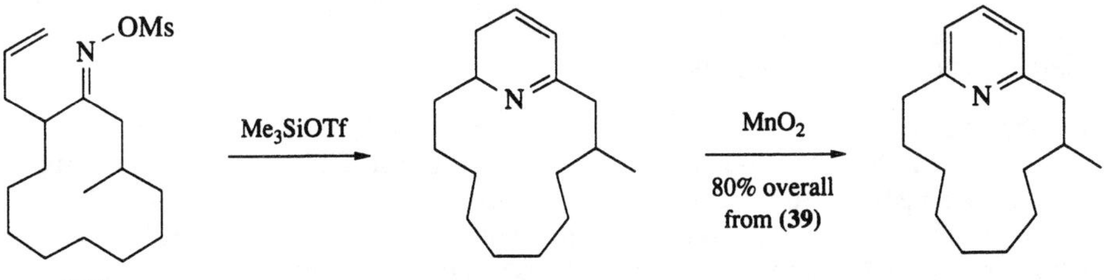

OMs, N (39) → Me_3SiOTf → N → MnO_2; 80% overall from **(39)** → N

Scheme 4

The imines formed in the cyclization reactions may be trapped reductively (*vide supra*).[49] Reaction of the intermediate formed by ring-closure with trimethylaluminum to give a *gem*-dimethyl group as in equation (29)[49] further increases the scope of this transformation.

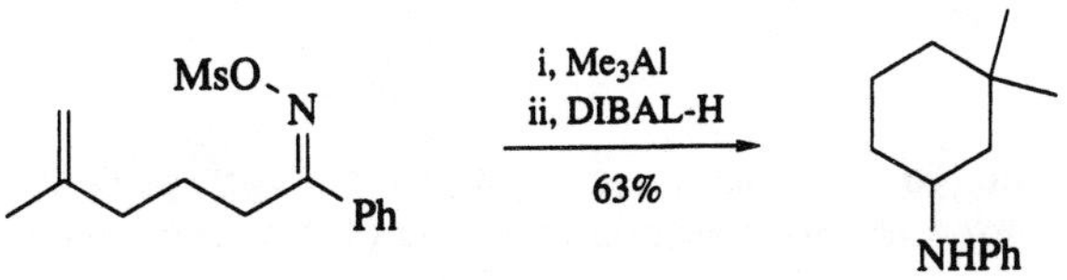

MsO, N, Ph → i, Me_3Al; ii, DIBAL-H; 63% → NHPh (29)

Benzoxazoles have been prepared by intramolecular reaction of a phenolic —OH group with a Beckmann intermediate (equation 30).[50] A cyclic imidate was isolated in high yield from tosylation and *in situ* Beckmann reaction of the erythromycin-derived oxime (**40**; equation 31).[51]

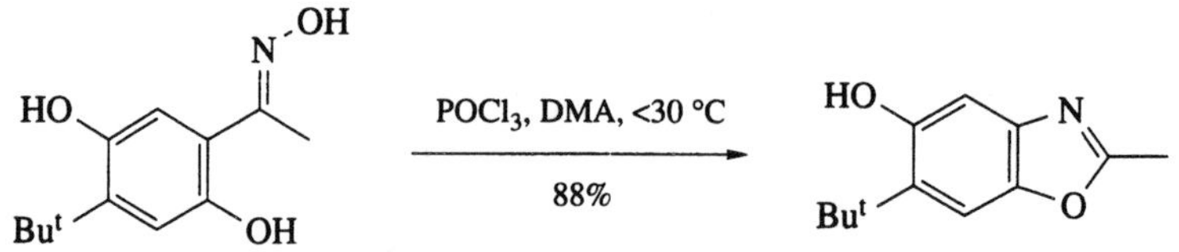

POCl$_3$, DMA, <30 °C

88%

(30)

TsCl, NaHCO$_3$, aq. acetone

87%

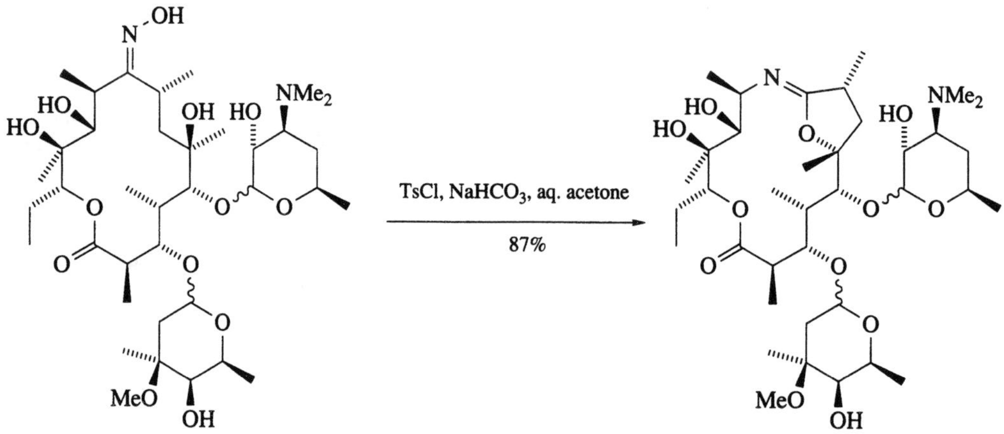

(**40**)

(31)

5.2.4 FRAGMENTATION REACTIONS

Fragmentation reactions, in which the α-carbon–carbon bond breaks, rather than migrates, may compete significantly with rearrangement processes when there is assistance from a neighboring center. Such assistance may be provided in the form of hyperconjugation (in the case of quaternary carbon atoms), or by mesomerically electron-donating heteroatoms. These processes may be stepwise, with the aforementioned assistance taking place during the break-down of the intermediate imidate. Substrates may undergo both rearrangement and fragmentation, depending on the reagents and reaction conditions employed. The fragmentation of aldoximes to give nitriles has been extensively reviewed elsewhere[11] and will not be considered here.

5.2.4.1 Carbon-assisted Fragmentations

Reaction *via* the fragmentation pathway increases with the ability of the α-carbon atom to support positive charge. Fragmentation of *cis*-1-methylbicyclo[4.3.0]nonan-2-one oxime (**41**) competed with normal Beckmann rearrangement to the extent shown in equation (32). The 1-unsubstituted analog gave exclusively the δ-lactam corresponding to (**42**).[52] Analogously, bicyclic oxime (**43**; R = Ph) gave only the fragmentation product (**44**; equation 33), whilst Beckmann rearrangement product (**45**) was the sole compound isolated when (**43**; R = H) was subjected to the same conditions.[53]

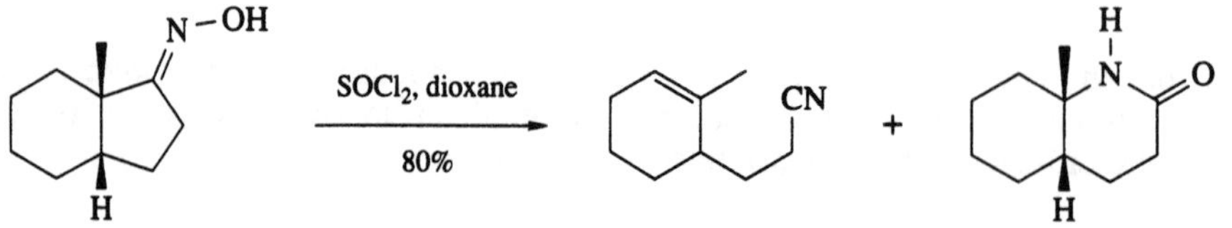

SOCl$_2$, dioxane

80%

(**41**) 50% (**42**) 50%

(32)

The reactivity of spirooxime (**46**) was found to depend on the reagent system employed. In the absence of water, ring cleavage occurred to give unsaturated nitrile (**47**) as the only product. Reaction of (**46**)

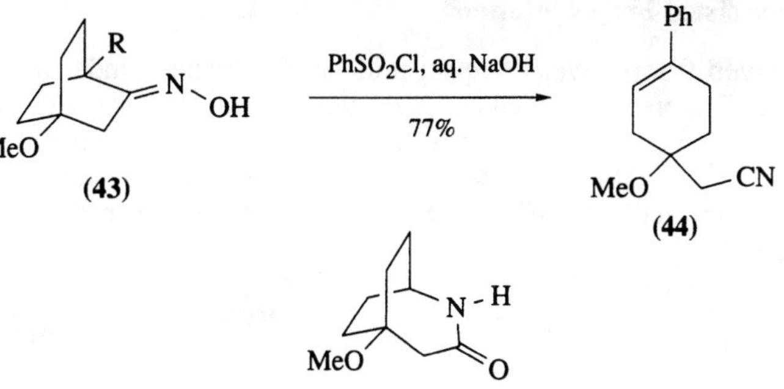

(33)

(43) (44) (45)

with benzenesulfonyl chloride in aqueous acetone gave exclusively the Beckmann rearranged spirolactam (**48**).[54]

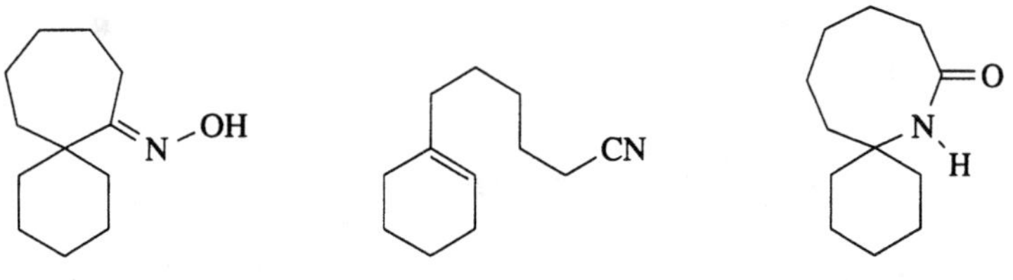

(46) (47) (48)

Under forcing conditions, the presumed intermediate (**49**) of fragmentation of oxime (**50**) underwent recyclization to give the product (**51**) of overall ring contraction, albeit in low yield (Scheme 5).[55]

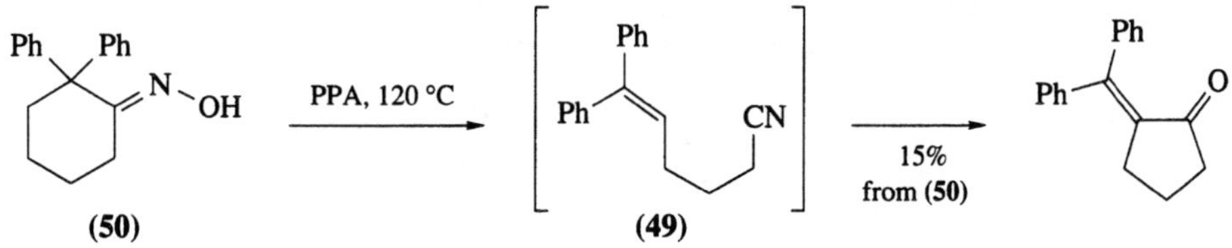

PPA, 120 °C

15% from (**50**)

(50) (49)

Scheme 5

Fragmentations may be governed by stereoelectronic factors.[11] The steroidal oxime (**52**; equation 34) fragmented with loss of a deuteron from the 4α-methyl group, whereas the $\Delta^{5,6}$ analog (**53**; equation 35) fragmented with proton abstraction from the 4β-methyl group.[56,57]

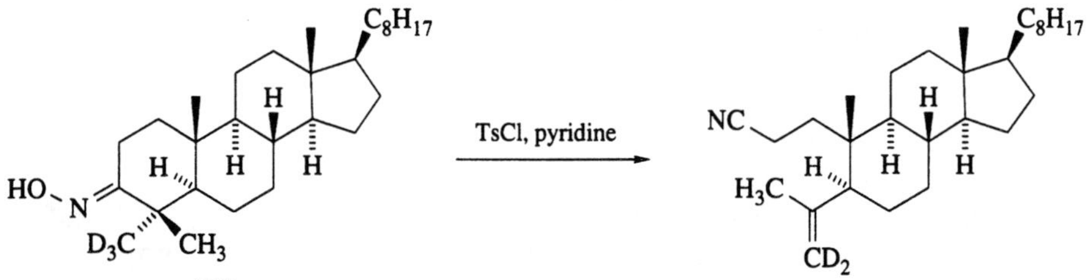

TsCl, pyridine (34)

(52)

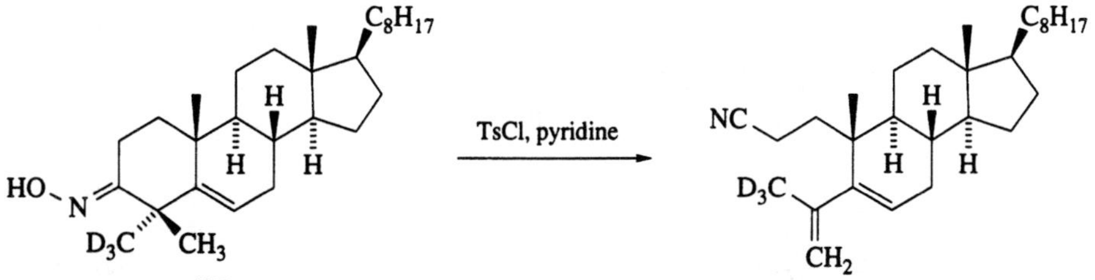

TsCl, pyridine (35)

(53)

5.2.4.2 Heteroatom-assisted Fragmentations

The oxime (**54**), derived from dihydrocamphorquinone, fragmented under phase transfer conditions, with oxime —OH activation presumably taking place *via* nucleophilic attack on dichlorocarbene (equation 36).[58] Carbonyl oxygen assisted ring opening was observed in the fragmentation of the cyclic α-diketone monooxime (**55**),[59] probably *via* a tetrahedral intermediate (equation 37).[11] Fragmentation of oxime orthoester (**56**) using catalytic MsOH gave a quantitative yield of methyl acetate, trimethyl orthoformate and acetonitrile.[60]

(**54**) OH, NOH — $CHCl_3$, EtOAc, 40% aq. NaOH, $BuEt_3N^+Cl^-$; 85% → CHO, CN (36)

(**55**) HO–N, O — $SOCl_2$, NH_3; 96% → NC–$(\ )_{10}$–C(O)NH_2 (37)

(**56**) MeO, OMe, N, O, OMe, OMe

Distal oxygen functionality may assist fragmentation. Basic ethanolysis of the bicyclic ketoxime (**57**) gave the acyclic unsaturated nitrile ester (**58**) as a single geometric isomer (equation 38).[61]

(**57**) O, H, N, TsO — NaOEt, EtOH; 100% → (**58**) O, OEt, CN (38)

Analogously with the reaction depicted in equation (37), α-*N*-morpholino oxime (**59**) underwent oxidative ring cleavage under standard Beckmann conditions (equation 39).[62] Azide nitrogen assisted cleavage of the steroidal azido oxime (**60**) gave a dinitrile (equation 40).[63]

(**59**) HO–N, N, O — PCl_5, Et_2O, 0 °C; 84% → NC–$(\ )_{10}$–CHO (39)

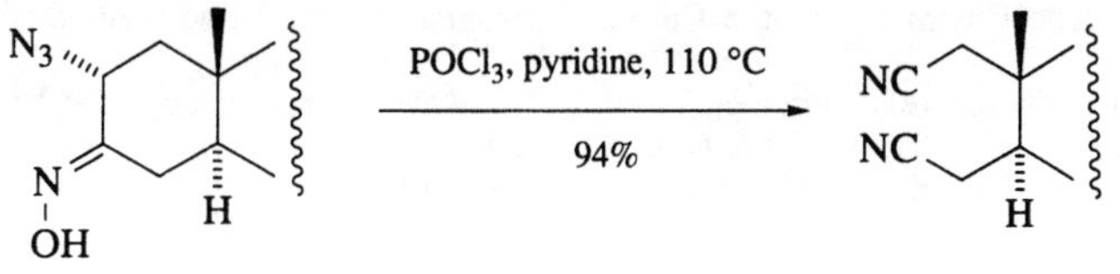

(60)

$$\xrightarrow[94\%]{POCl_3,\ pyridine,\ 110\ °C}\qquad (40)$$

The ability of silicon to stabilize positive charge β to itself has been exploited in silicon-directed fragmentations of β-silyl oximes. Oxime *O*-acetate (**61**) fragmented stereospecifically to the (*E*)-alkenic product in the presence of catalytic TMSOTf (equation 41).[64]

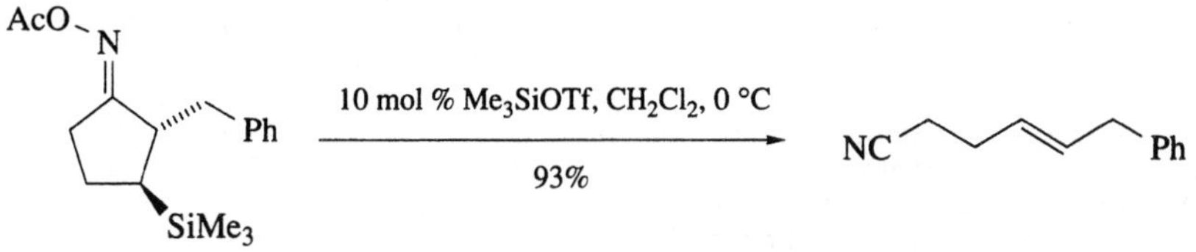

(61)

$$\xrightarrow[93\%]{10\ mol\ \%\ Me_3SiOTf,\ CH_2Cl_2,\ 0\ °C}\qquad (41)$$

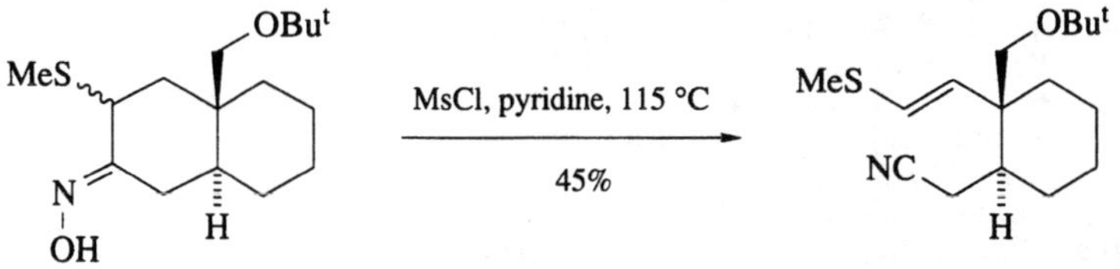

(62)

$$\xrightarrow[45\%]{MsCl,\ pyridine,\ 115\ °C}\qquad (42)$$

Adjacent sulfur substituents have been observed to promote fragmentation of oximes under Beckmann conditions. The bicyclic α-methylthio ketoxime (**62**) gave a thio enol ether on mesylation in pyridine (equation 42).[65] Anchimeric, rather than mesomeric assistance was observed for the fused tetrahydrothiophene derivative (**63**) (Scheme 6);[66] the involvement of sulfur was strongly implicated by the resistance to fragmentation of the corresponding sulfoxide under similar conditions.

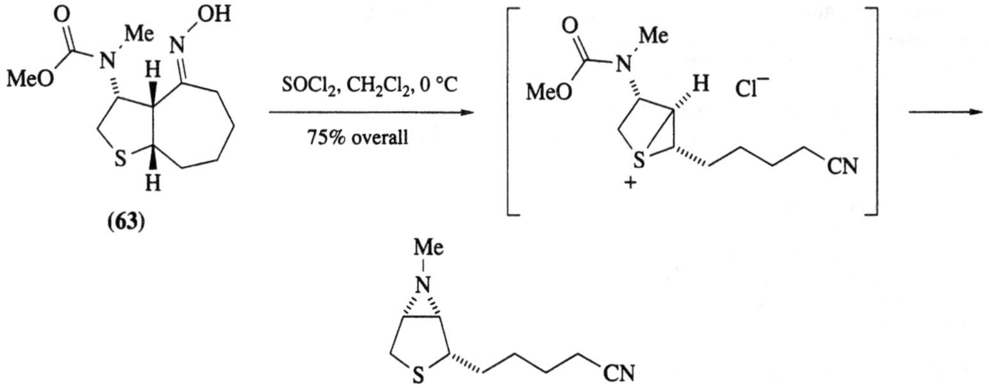

(63)

$$\xrightarrow[75\%\ overall]{SOCl_2,\ CH_2Cl_2,\ 0\ °C}$$

Scheme 6

5.2.5 REFERENCES

1. E. Beckmann, *Ber. Dtsch. Chem. Ges.*, 1886, **89**, 988.
2. A. H. Blatt, *Chem. Rev.*, 1933, **12**, 215.
3. B. Jones, *Chem. Rev.*, 1944, **35**, 335.
4. F. Moller, *Methoden Org. Chem. (Houben-Weyl)*, 1957, **11** (1), 892.
5. L. G. Donaruma and W. Z. Heldt, *Org. React. (N.Y.)*, 1960, **11**, 1.
6. A. L. J. Beckwith, in 'The Chemistry of Amides', ed. J. Zabicky, Interscience, New York, 1970, p. 131.

7. C. G. McCarty, in 'The Chemistry of the Carbon–Nitrogen Double Bond', ed. S. Patai, Interscience, New York, 1970, p. 408.
8. P. A. S. Smith, in 'Molecular Rearrangements', ed. P. De Mayo, Interscience, New York, 1963, p. 457.
9. H. Mukamal, *Nuova Chim.*, 1971, **47**, 79 (*Chem. Abstr.*, 1971, **75**, 75 559).
10. G. Hornke, H. Krauch and W. Kunz., *Chem.-Ztg.*, 1965, **89**, 525.
11. R. E. Gawley, *Org. React. (N.Y.)*, 1988, **35**, 1.
12. A. Padwa, *Chem. Rev.*, 1977, **77**, 37.
13. For reviews, see G. I. Kodolbskii, V. A. Ostrovskii and B. V. Gidaspov, *Russ. Chem. Rev. (Engl. Transl.)*, 1978, **47**, 1084; H. Wolff, *Org. React. (N.Y.)*, 1946, **3**, 307; see also refs. 6 and 8.
14. J. W. Schulenberg and S. Archer, *Org. React. (N.Y.)*, 1965, **14**, 1.
15. F. Fernández and C. Pérez, *J. Chem. Res. (S)*, 1987, 340.
16. P. N. Confalone and E. M. Huie, *J. Org. Chem.*, 1982, **52**, 79.
17. J. B. Hester, Jr., *J. Org. Chem.*, 1970, **35**, 875.
18. P. T. Lansbury and N. R. Mancuso, *Tetrahedron Lett.*, 1965, 2445.
19. P. E. Eaton, G. R. Calson and J. T. Lee, *J. Org. Chem.*, 1973, **38**, 4071.
20. P. W. Jeffs, G. Molina, N. A. Cortese, P. R. Hauck and J. Wolfram, *J. Org. Chem.*, 1982, **47**, 3876.
21. J. C. Kim and S. K. Choi, *Taehan Hwahakhoe Chi*, 1981, **25**, 331 (*Chem. Abstr.*, 1982, **96**, 123 101).
22. S. Bittner and S. Grinberg, *J. Chem. Soc., Perkin Trans. 1*, 1976, 1708.
23. H. Akimoto, A. Kawai, N. Hashimoto and H. Nomura, *Chem. Pharm. Bull.*, 1984, **32**, 2565.
24. C. Alvarez, A. C. Cano, V. Rivera and C. Marquez, *Synth. Commun.*, 1987, **17**, 279.
25. A. Jössung-Yanagida and C. Gansser, *J. Heterocycl. Chem.*, 1978, **15**, 249.
26. P. C. Bélanger, J. Scheigetz and R. N. Young, *Can. J. Chem.*, 1983, **61**, 2177.
27. H. Ogawa, M. Kumemura, T. Imoto, I. Miyamoto, H. Kato and Y. Taniguchi, *J. Chem. Soc., Chem. Commun.*, 1987, 300.
28. Y. Nitta, T. Yamaguchi and T. Tanaka, *Heterocycles*, 1986, **24**, 25.
29. J. C. Craig and A. R. Naik, *J. Am. Chem. Soc.*, 1962, **84**, 3410.
30. T. Sato, H. Wakatsuka and K. Amano, *Tetrahedron*, 1971, **27**, 5381.
31. C. A. Grob and P. Wenk, *Tetrahedron Lett.*, 1976, 4191.
32. W. Bartmann, G. Beck, J. Knolle and R. H. Rupp, *Tetrahedron Lett.*, 1982, **23**, 3647.
33. G. R. Krow and S. W. Szczepanski, *J. Org. Chem.*, 1982, **47**, 1153.
34. I. Ganboa and C. Palomo, *Synth. Commun.*, 1983, **13**, 941.
35. W. Bartmann, E. Konz and W. Rüger, *J. Heterocycl. Chem.*, 1987, **24**, 677.
36. G. M. Iskanderl and V. S. Gulta, *J. Chem. Soc., Perkin Trans. 1*, 1982, 1891.
37. T. S. Sulkowski and S. J. Childress, *J. Org. Chem.*, 1962, **27**, 4424.
38. Y. Ishida, S. Sasatani, K. Maruoka and H. Yamamoto, *Tetrahedron Lett.*, 1983, **24**, 3255.
39. M. E. Jung and Z. Long-Mei, *Tetrahedron Lett.*, 1983, **24**, 4533.
40. F. L. Scott, R. J. MacConaill and J. C. Riordan, *J. Chem. Soc. C*, 1967, 44.
41. Z. Hamlet and M. Rampersad, *J. Chem. Soc., Chem. Commun.*, 1970, 1230.
42. R. N. Butler and D. A. O'Donoghue, *J. Chem. Res. (S)*, 1983, 18.
43. K. Maruoka, T. Miyazaki, M. Ando, Y. Matsumura, S. Sakane, K. Hattori and H. Yamamoto, *J. Am. Chem. Soc.*, 1983, **105**, 2831.
44. K. Hattori, K. Maruoka and H. Yamamoto, *Tetrahedron Lett.*, 1982, **23**, 3395.
45. Y. Matsumura, K. Maruoka and H. Yamamoto, *Tetrahedron Lett.*, 1982, **23**, 1929.
46. K. Hattori, Y. Matsumura, T. Miyazaki, K. Maruoka and H. Yamamoto, *J. Am. Chem. Soc.*, 1981, **103**, 7368.
47. Y. Matsumura, J. Fujiwara, K. Maruoka and H. Yamamoto, *J. Am. Chem. Soc.*, 1983, **105**, 6312.
48. R. E. Gawley and E. J. Termine, *J. Org. Chem.*, 1984, **49**, 1946.
49. S. Sakane, Y. Matsumura, Y. Yamamura, Y. Ishida, K. Maruoka and H. Yamamoto, *J. Am. Chem. Soc.*, 1983, **105**, 672.
50. S. Fujita, K. Koyama and Y. Inagaki, *Synthesis*, 1982, 68.
51. S. Djokic, G. Kobrehel, G. Lazarevski, N. Lopotar, Z. Tamburasev, B. Kamenar, A. Nagl and I. Vickovic, *J. Chem. Soc., Perkin Trans. 1*, 1986, 1881.
52. E. J. Moriconi and M. A. Stemniski, *J. Org. Chem.*, 1972, **37**, 2035.
53. K.-I. Morita and Z. Suzuki, *J. Org. Chem.*, 1966, **31**, 233.
54. R. T. Conley and M. C. Annis, *J. Org. Chem.*, 1962, **27**, 1961.
55. R. T. Conley and B. E. Nowak, *J. Org. Chem.*, 1962, **27**, 1965.
56. G. P. Moss and S. A. Nicolaidis, *J. Chem. Soc. D*, 1969, 1077.
57. A. G. Shavva and T. E. Ryzhkina, *J. Org. Chem. USSR (Engl. Transl.)*, 1974, **10**, 2013.
58. J. N. Shah, Y. P. Mehta and G. M. Shah, *J. Org. Chem.*, 1978, **43**, 2078.
59. M. Kataoka and M. Ohno, *Bull. Chem. Soc. Jpn.*, 1973, **46**, 3474.
60. K. P. Klein, T. R. Demmin, B. C. Oxenrider, M. Rogic and M. T. Tetenbaum, *J. Org. Chem.*, 1979, **44**, 275.
61. W. Eisele, C. A. Grob, E. Renk and H. von Tschammer, *Helv. Chim. Acta*, 1968, **51**, 816; see also C. A. Grob and H. von Tschammer, *Helv. Chim. Acta*, 1968, **51**, 1083.
62. M. Ohno, N. Naruse, S. Torimitsu and I. Teresawa, *J. Am. Chem. Soc.*, 1966, **88**, 3168.
63. T. Takahashi, K. Nomura and J. Y. Satoh, *J. Chem. Soc., Chem. Commun.*, 1983, 1441.
64. H. Nishiyama, K. Sakuta, N. Osaka and K. Itoh, *Tetrahedron Lett.*, 1983, **24**; 4021; see also P. F. Hudrlik, M. A. Waugh and A. M. Hudrlik, *J. Organomet. Chem.*, 1984, **271**, 69; H. Nishiyama, K. Sakuta, N. Osaka, H. Arai, M. Matsumoto and K. Itoh, *Tetrahedron*, 1988, **44**, 2413.
65. P. A. Grieco and K. Hiroi, *Tetrahedron Lett.*, 1973, 1831.
66. P. N. Confalone, G. Pizzolato, D. Lollar Confalone and M. R. Uskokovic, *J. Am. Chem. Soc.*, 1980, **102**, 1954.

5.3
Glycol Cleavage Reactions

TONY K. M. SHING
The Chinese University of Hong Kong, Hong Kong

5.3.1 INTRODUCTION

This chapter discusses the oxidative carbon–carbon bond cleavage of vicinal diols (α- or 1,2-glycols) and related functional groups. There are many oxidants which can effect this transformation; however, the classical oxidants LTA, periodic acid and its salts are still the reagents of choice. The glycol fission reactions by these oxidants are usually very rapid, clean, quantitative and specific. Sodium periodate is particularly popular, attributable to the neutral and mild conditions used which are compatible with a wide range of functionalities. The studies of other reagents have been mainly exploratory or mechanistic. Some have shown promise for a specific application which is indicated in each section, but none has demonstrated the versatility exhibited by LTA and periodate in natural product synthesis. Several reviews of glycol cleavage reactions have appeared.[1–5]

5.3.2 SODIUM BISMUTHATE

Sodium bismuthate ($NaBiO_3$) was first used by Rigby[6,7] as an oxidant for α-glycol cleavage reactions to give carbonyl compounds (equation 1), and was found to be similar in scope to lead tetraacetate and periodic acid. Like lead tetraacetate but unlike periodic acid, it readily cleaves α-hydroxycarboxylic acids and α-hydroxy ketones. Aldehydic products, except formaldehyde,[8] are not oxidized further by the reagent which also oxidizes phenols and alkenes. The rate of alkene oxidation is slow compared with that of glycol cleavage.[9] Sodium bismuthate is generally used in combination with acetic or phosphoric acid in aqueous alcohol and the relatively harsh conditions required limit its synthetic applications. The mechanism of glycol cleavage by sodium bismuthate is still obscure. Since oxidation of *trans*-cyclopentane-1,2-diol with sodium bismuthate gives glutardialdehyde (equation 2),[10] and no difference in reaction rate is observed for the oxidation of *cis*- or *trans*-cyclohexane-1,2-diol, the proposed[7] mechanism involving a bismuth diester intermediate is not justified.

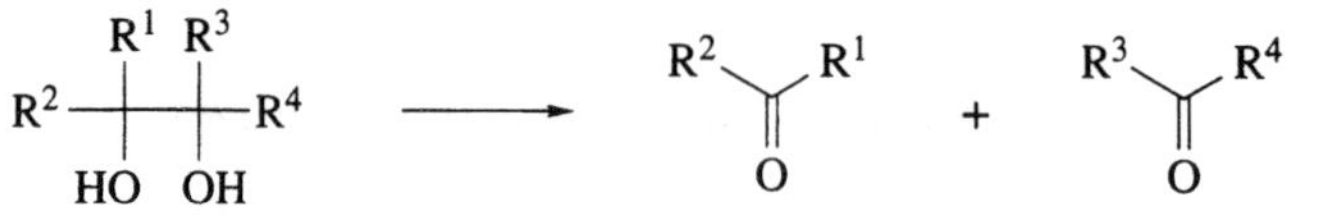

$$R^2R^1C(OH)\text{–}C(OH)R^3R^4 \longrightarrow R^2R^1C{=}O + R^3R^4C{=}O \quad (1)$$

$$\textit{trans}\text{-cyclopentane-1,2-diol} \longrightarrow OHC\text{–}CH_2CH_2CH_2\text{–}CHO \quad (2)$$

Selective oxidative cleavage of the corticosteroid side chain by sodium bismuthate was used as an analytical technique for the determination of urinary corticosteroids.[11] Sodium bismuthate was chosen since it cleaves α-hydroxy acids readily (unlike periodic acid) and tolerates the water in urine samples (in contrast to LTA).

5.3.3 PENTAVALENT ORGANOBISMUTH REAGENTS

Pentavalent derivatives of triphenylbismuth such as μ-oxobis(chlorotriphenylbismuth) (**1**) and triphenylbismuth carbonate (**2**) have been developed[12] recently as oxidizing agents which cleave α-glycols into the corresponding carbonyl derivatives (equation 1). Aldehydic products do not undergo further oxidation. Reagent (**1**) also oxidizes saturated alcohols as well as allylic and benzylic alcohols, whereas (**2**) selectively oxidizes allylic alcohols in the presence of saturated alcohols.[12]

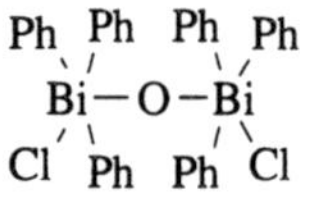

$$Ph_3(Cl)Bi\text{–}O\text{–}Bi(Cl)Ph_3 \qquad Ph_3BiCO_3$$

(1) **(2)**

Triphenylbismuth carbonate (**2**) displays remarkable chemoselectivity, allowing alcohol oxidation in the presence of benzenethiol, pyrrolidine, indole, aniline, dimethyl aniline and 3-pyrolidinocholesta-3,5-diene. The diol moiety in (**3**) is cleaved selectively without oxidizing the dithioacetal function (equation 3).[13] The rate of the stoichiometric oxidative cleavage of *cis*-cyclohexane-1,2-diol to adipic aldehyde with Ph_3BiCO_3 is faster than that of the *trans* isomer, suggesting the formation of a cyclic organobismuth intermediate (**4**; Scheme 1).[12]

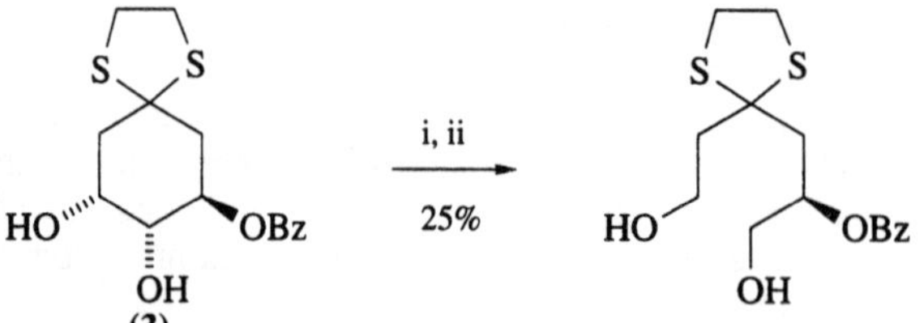

$$(\mathbf{3}) \xrightarrow[25\%]{\text{i, ii}} \text{ring-opened diol (HO}\ldots\text{OH, OBz)} \quad (3)$$

i, PH_3BiCO_3; ii, $NaBH_4$

A catalytic bismuth system (Ph_3Bi–NBS–K_2CO_3–MeCN with 1% water) has been reported[14] to cleave a range of 1,2-glycols efficiently and is shown to have a different mechanism from the cyclic process observed with the stoichiometric bismuth reagent (**2**). The catalytic system cleaves *cis*- and *trans*-decalin-9,10-diols at nearly the same rate, whereas the stoichiometric reagent (**2**) does not cleave the *trans*

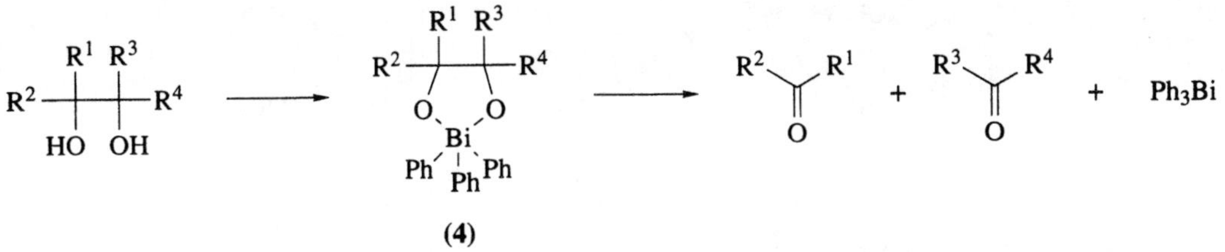

Scheme 1

isomer. The mechanism involves the formation of a hypobromite which oxidizes Ph_3Bi to give a pentavalent alkoxy intermediate (**5**). Base-induced reductive elimination of (**5**) then gives the carbonyl derivatives and triphenylbismuth (Scheme 2).[14]

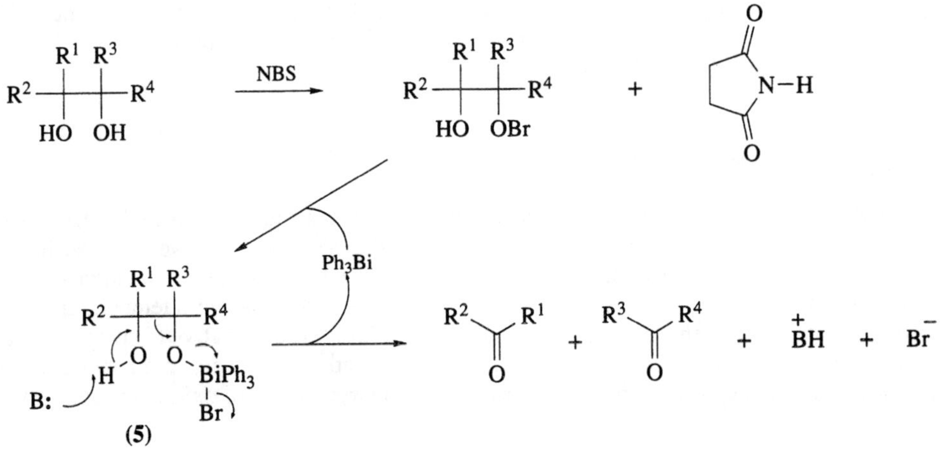

Scheme 2

5.3.4 CERIUM(IV) REAGENTS

Cerium(IV) is an efficient reagent for α-glycol cleavage (equation 1). Vicinal and polyhydric alcohols are quantitatively broken down by the cerium(IV) ion.[15] There are no large differences in the overall oxidation rates of *cis*- and *trans*-cyclohexane-1,2-diols and of *cis*- and *trans*-cyclopentane-1,2-diols. However, the effect of ring size is considerable, cyclopentanediols reacting more rapidly than cyclohexanediols. It is noteworthy that cerium(IV) also cleaves 2-methoxycycloalkanols to give the corresponding dialdehydes.[16] The mechanism for the 1,2-glycol cleavage by cerium(IV) involves the formation of a monodentate complex followed by a one-electron cleavage to give an intermediate radical which is then further oxidized (Scheme 3). The main support for this mechanism comes from the similar rates of oxidation of glycols (**6**) and (**7**) and the monomethyl ether (**8**),[16] radical-trapping experiments[17] and parallel studies of lead(IV) and cerium(IV) glycol cleavage oxidations.[18] However, the synthetic aspects of cerium(IV) oxidation require further investigation.

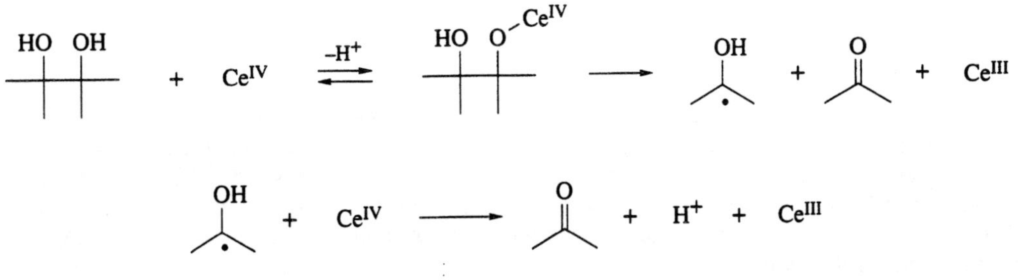

Scheme 3

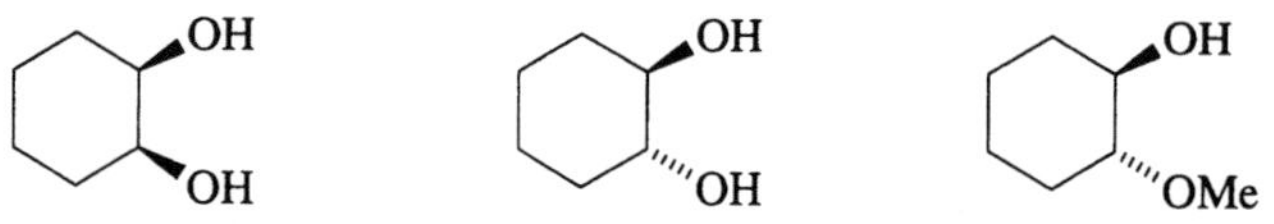

(6) (7) (8)

5.3.5 CALCIUM HYPOCHLORITE

Calcium hypochlorite, $Ca(OCl)_2$, an inexpensive and easily stored oxidant, can be used to cleave α-glycols to the corresponding carbonyl compounds. α-Diones, α-hydroxy ketones and α-hydroxy and α-keto acids are also oxidatively fragmented.[19] Aldehydic products are further oxidized to acids with an excess of the reagent. Consequently, this reaction is more suitable to produce carboxylic acids from 1,2-diols on a preparative scale. α,β-Unsaturated aldehydes and aromatic aldehydes with electron-donating groups undergo a competing nuclear chlorination, whereas aliphatic aldehydes and aromatic aldehydes with electron-withdrawing groups give the expected acids.[20] Reactions are carried out at room temperature in aqueous acetonitrile/acetic acid solution.

5.3.6 CHROMIUM(VI) REAGENTS

Glycol cleavage oxidation by chromic acid usually affords ketones and acids since the first formed aldehydes undergo rapid oxidation.[21] The synthetic aspects of these oxidations have received little attention, probably owing to the relatively harsh conditions used. Recently, PCC has been employed to effect the fission of simple vicinal diols under very mild conditions (a few hours at room temperature in dichloromethane solution) to give aldehydes and ketones in good yields.[22] The oxidative cleavage seems to involve an intermediate chromate ester (**9**). However, the reaction is sensitive to steric crowding: benzpinacol failed to react even after prolonged heating with a large excess of oxidizing agent.[22]

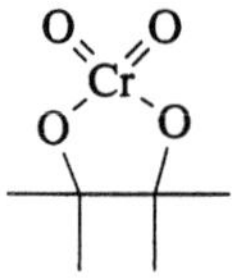

(9)

5.3.7 COBALT(II) REAGENTS

Cobalt(II) salts are effective catalysts for the oxidation of 1,2-glycols with molecular oxygen in aprotic polar solvents such as pyridine, 4-cyanopyridine, benzonitrile, DMF, anisole, chlorobenzene and sulfolane.[23] Water, primary alcohols, fatty acids and nitrobenzene are not suitable as solvents. Aldehydic products are further oxidized under the reaction conditions. Thus, the oxidative fission of *trans*-cyclohexane-1,2-diol gives a mixture of aldehydes and acids. However, the method is of value in the preparation of carboxylic acids from vicinal diols on an industrial scale; for example, decane-1,2-diol is cleaved by oxygen, catalyzed by cobalt(II) laurate, to produce nonanoic acid in 70% yield.[23,24]

5.3.8 IODO REAGENTS

Alicyclic, aromatic, aliphatic, steroidal and triterpenoid 1,2-diols are cleaved by iodine triacetate and iodine(I) acetate to generate carbonyl compounds. Aldehydic products are not further oxidized.[25] Iodine triacetate is prepared from iodine trichloride and silver(I) acetate, whereas iodine(I) acetate is prepared from iodine and silver(I) acetate. Reactions occur in acetic acid at room temperature under nitrogen, and a radical pathway involving a hypoiodite is suggested. The cost and the availability of these reagents are probable reasons for their unpopularity.

Five simple α-diols have been successfully cleaved by *N*-iodosuccinimide (NIS) in THF at ambient temperature.[26] Products from the oxidation are aldehydes, ketones, iodine and succinimide. Irradiation of the reaction increases the cleavage rate, indicating a radical pathway. Its attractiveness is its simplicity of operation and easy availability of the reagent. However, the generation of iodine during the reaction may pose complications with substrates containing sensitive functional groups.

5.3.9 VANADIUM REAGENTS

Although the dioxovanadium cation, VO_2^+, is a useful oxidant for the cleavage of simple ditertiary and secondary–tertiary glycols,[27] the strong acid conditions required for the reaction (H_2SO_4 or $HClO_4$) limit its synthetic application.

Recently, bis(acetylacetonato)oxovanadium [$VO(acac)_2$] has been shown to be a selective oxidant for the quantitative cleavage of ditertiary glycols to ketones under mild conditions.[28] Reactions proceed at room or higher temperatures (up to 70 °C) in aprotic solvents (dichloromethane, benzene or mixtures thereof). The selectivity of the reagent towards ditertiary glycols is demonstrated by the fact that pinacol and benzpinacol are smoothly cleaved to acetone and benzophenone, respectively, whereas butane-2,3-diol and cyclohexane-1,2-diol (*cis–trans* mixture) are inert under the reaction conditions. The reagent also shows excellent chemoselectivity, permitting glycol cleavage of substrates possessing a variety of functional groups (equation 4).

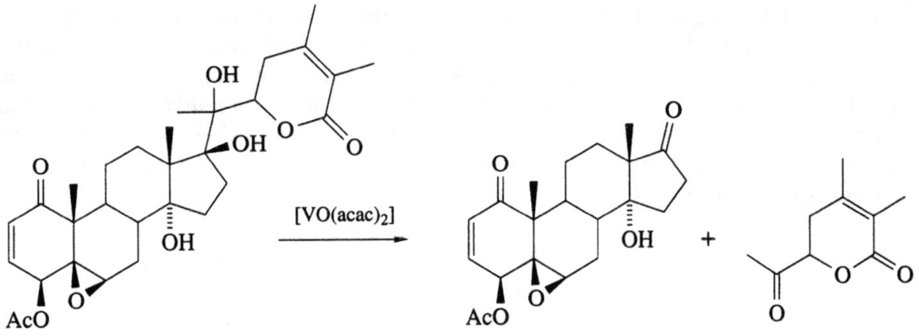

OH O OH O O OH AcO O [$VO(acac)_2$] → O O OH AcO O + O O O (4)

The reaction proceeds also with catalytic amounts of [$VO(acac)_2$], in the presence of *t*-butyl hydroperoxide or *m*-chloroperoxybenzoic acid.[28]

5.3.10 ANODIC OXIDATION

The anodic oxidation of 1,2-diols and their ether derivatives is a simple and clean method for the fission of glycols to the corresponding carbonyl compounds (equation 1). The reaction is carried out using carbon electrodes in methanol containing tetraethylammonium *p*-toluenesulfonate as a supporting electrolyte.[29]

In contrast to other cleavage reagents, this anodic oxidation does not show stereochemical preference. In addition, ether derivatives are oxidized in similar current efficiencies to those of the parent 1,2-diols. Cyclohexene oxide is oxidized *via* the corresponding hydroxy ether.

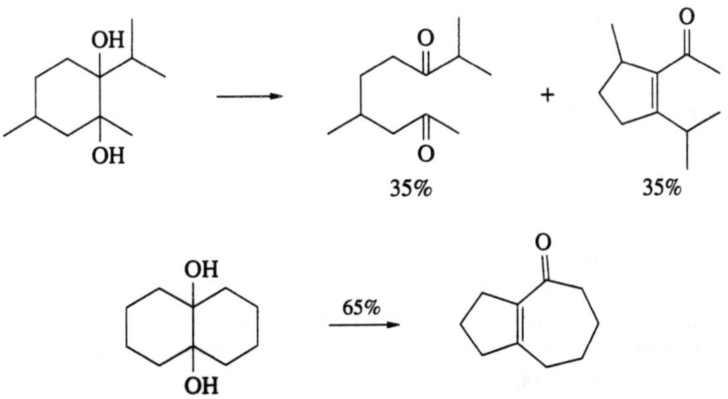

OH OH → O O 35% + O 35%

OH OH —65%→ O

Scheme 4

Aldehydic products are usually isolated as their corresponding acetals. The ketones formed from the oxidation may undergo intramolecular aldol condensation to give enones (Scheme 4).[29]

5.3.11 HYDROGEN PEROXIDE

Aqueous hydrogen peroxide in conjunction with catalytic amounts of tungstate and phosphate (or arsenate) ions, under acidic conditions (pH 2), provides a synthetically useful procedure for the oxidative fission of water soluble 1,2-diols to carboxylic acids.[30] This method, which employs an inexpensive catalyst and a cheap, nonpolluting oxidant, is particularly suitable for large scale operations. Primary–secondary, secondary–secondary and secondary–tertiary 1,2-diols (open chain and cyclic) can be satisfactorily oxidized. Thus, *trans*-1,2-cyclopentanediol, *cis*- and *trans*-1,2-cyclohexanediol, *trans*-1-methyl-1,2-cyclohexanediol and 1,2-hexanediol react with H_2O_2, $Na_2WO_4{\cdot}2H_2O$ and H_3PO_4 at 90 °C for 5 h to give glutaric, adipic, 6-oxoheptanoic and valeric acid, respectively, in 87–96% yield.[30] This method has been applied to water insoluble diols by the use of a phase transfer agent. Thus, 1,2-octanediol gives 78% of heptanoic acid.[30]

A related hydrogen peroxide oxidation of α-glycols catalyzed by tris(cetylpyridinium) 12-tungstophosphate, $[\pi\text{-}C_5H_5N(CH_2)_{15}Me]_3^+(PW_{12}O_{40})_3^-$, (CWP) has also been reported.[31] This catalyst–oxidant system also epoxidizes alkenes and allylic alcohols and converts secondary alcohols to ketones. 1,2-Glycols react with three equivalents of H_2O_2 mediated by CWP in refluxing *t*-butanol to give carboxylic acids in good yields. The oxidation of 4-vinyl-1,2-cyclohexanediol gives 3-vinyladipic acid in 55% yield (equation 5), indicating that the reaction is chemoselective. The cleavage reaction involves the formation of an α-ketol which subsequently undergoes C—C bond fission.[30,31] This CWP–H_2O_2 system, which is also efficient for the oxidative cleavage of carbon–carbon double bonds of alkenes, provides a new way of converting alkenes into carboxylic acids.[31]

OH / OH —(H_2O_2, CWP; Bu^{t}OH; 55%)→ HO_2C ... CO_2H (5)

5.3.12 MANGANESE DIOXIDE

Vicinal diols can be cleaved smoothly under neutral conditions to give carbonyl compounds using an excess of activated manganese dioxide.[32] The reaction proceeds under very mild conditions (stirring in CH_2Cl_2 at room temperature). Only 1,2-*cis*-diols and the analogous *trans* compounds with a flexible arrangement of their hydroxy groups can be oxidized. Even diols subject to extensive steric hindrance undergo oxidative fission (Scheme 5). 9,10-*cis*-Decalindiol is easily cleaved, whereas the 9,10-*trans* isomer remains inert. Reactions work well with glycols containing at least one tertiary hydroxy group. If the hydroxy groups are secondary, ketonization is observed instead of complete oxidative fission. For example, dodecanedial, the oxidation product of 1,2-*cis*-cyclododecanediol, is accompanied by 1,2-cyclododecanedione (14%) and traces of 2-hydroxycyclododecanone.[32] The mild conditions and the ease of work-up of the manganese dioxide oxidation is a valuable method for the fission of 1,2-glycols containing at least one tertiary hydroxy group.

5.3.13 LTA AND PERIODATE

5.3.13.1 General Characteristics

In 1928, Malaprade[33] discovered that periodic acid and its salts cleaved the carbon–carbon bond of 1,2-diols efficiently to give carbonyl compounds. Subsequently, Criegee[34] found that LTA could also effect such transformations. The glycol cleavage reactions of these two reagents are usually very rapid, clean, quantitative and specific. The reactions are usually stoichiometric, with one mole of oxidant being consumed for each carbon–carbon bond cleaved. The carbonyl compounds generated are inert towards further oxidation under the reaction conditions. For many applications, periodate and LTA complement each other. LTA is generally used in acetic acid or aprotic solvents such as benzene, ethyl acetate and

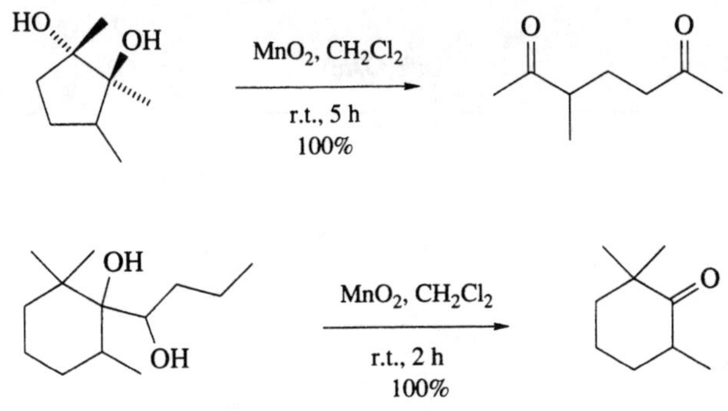

MnO_2, CH_2Cl_2
r.t., 5 h
100%

MnO_2, CH_2Cl_2
r.t., 2 h
100%

Scheme 5

dichloromethane. Sodium and potassium periodate can be used only in water or aqueous organic solvents owing to their solubility properties, whereas periodic acid can be used in water or aprotic solvents (diethyl ether or THF).[35] In order to carry out periodate oxidation in nonaqueous media, sodium periodate supported silica[36] and quaternary ammonium periodates[37] have been developed. Recently, polymer-supported quaternary ammonium periodate, which was used for glycol cleavage reactions in dichloromethane, has been reported as a practically useful alternative.[38]

The fact that periodate fission functions best in aqueous media[39] and LTA in organic solvents makes glycol scission oxidations possible with all types of substrates. These reagents were used extensively for structural elucidation of carbohydrates before the advent of modern spectral instrumentation.[40,41] Now, they are generally used in synthetic work.

5.3.13.2 Reaction Mechanism

The mechanism of cleavage by periodate is consistent with a cyclic, five-membered ring intermediate (**10**) shown in Scheme 6. Support comes from the fact that the *cis* isomers of cyclic diols are more reactive than the *trans* isomers, *threo*-1,2-diols undergo oxidation faster than the *erythro* isomers[42] and the inert behavior of diaxial *trans*-1,2-diols which cannot form a cyclic periodate ester.[43]

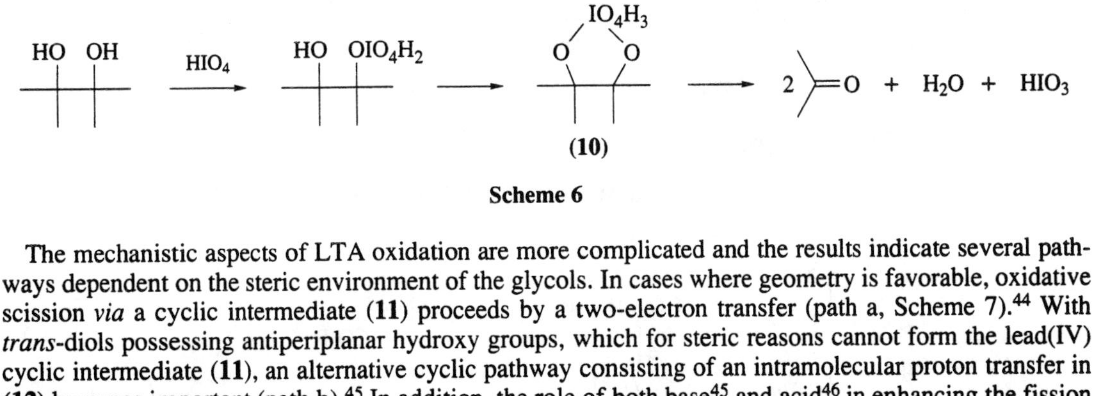

HO OH —HIO_4→ HO OIO_4H_2 → (**10**) → 2 >=O + H_2O + HIO_3

Scheme 6

The mechanistic aspects of LTA oxidation are more complicated and the results indicate several pathways dependent on the steric environment of the glycols. In cases where geometry is favorable, oxidative scission *via* a cyclic intermediate (**11**) proceeds by a two-electron transfer (path a, Scheme 7).[44] With *trans*-diols possessing antiperiplanar hydroxy groups, which for steric reasons cannot form the lead(IV) cyclic intermediate (**11**), an alternative cyclic pathway consisting of an intramolecular proton transfer in (**12**) becomes important (path b).[45] In addition, the role of both base[45] and acid[46] in enhancing the fission of these *trans*-diols has been rationalized by involving two noncyclic transition states, such as (**13**) and (**14**) (paths c and d).

5.3.13.3 Applications in Organic Synthesis

In addition to the oxidative scission of 1,2-diols, the reaction can be extended to related 1,2-bifunctional compounds such as oxiranes,[47] 1,2-dicarbonyl compounds, 2-hydroxy aldehydes, ketones and acids, α-amino alcohols, 1,2-diamines and also to polyols.[1,2] LTA cleaves α-hydroxy acids much more readily than do periodates and both reagents oxidize 2-hydroxy aldehydes and 1,2-dicarbonyl compounds relatively slowly.[1,2] Only periodic acid[47] in water reacts with oxiranes *via* the corresponding diols.

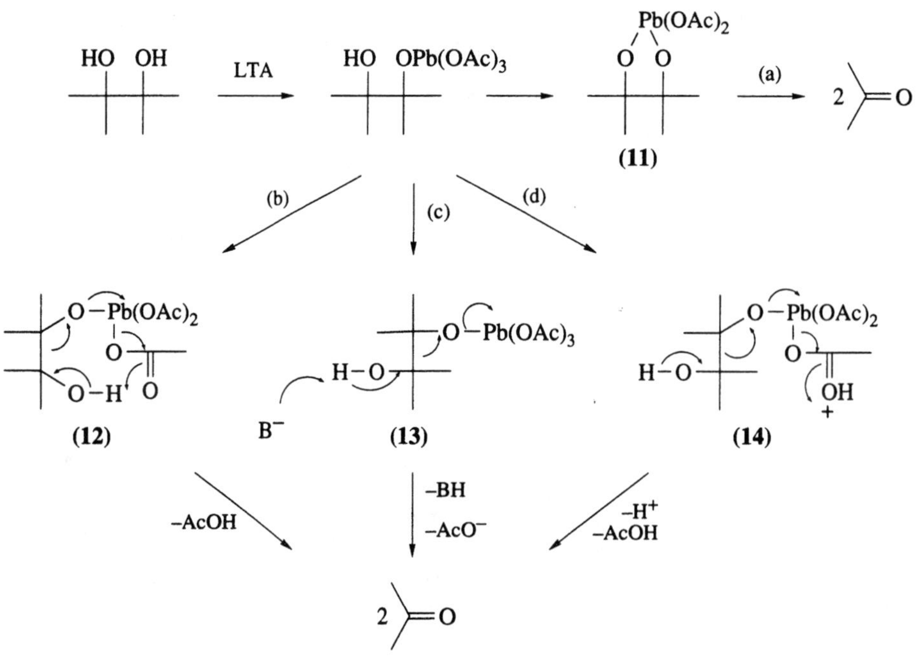

Scheme 7

Although periodates also oxidize polycyclic aromatic hydrocarbons, phenols, hydrazines, active methylene compounds and sulfides,[3] chemoselectivity can usually be achieved and glycol cleavage oxidation takes precedence. For example, the diol moiety in the diethyl dithioacetal derivative of D-glucose can be selectively oxidized in good yield (equation 6).[48] In contrast, LTA is less selective than periodate and oxidizes a far greater variety of organic compounds.[5] Consequently, in order to minimize undesired reactions, it is customary to add LTA slowly to avoid contact of the initially formed products with an excess of the oxidant (equation 7).[49]

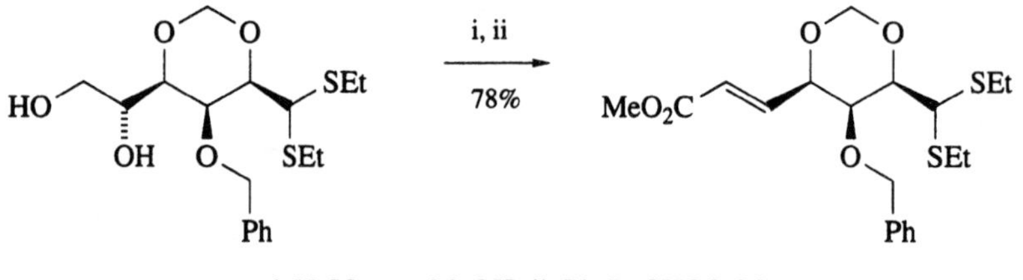

(6)

i, $NaIO_4$, aq. MeOH; ii, $Ph_3P{=}CHCO_2Me$

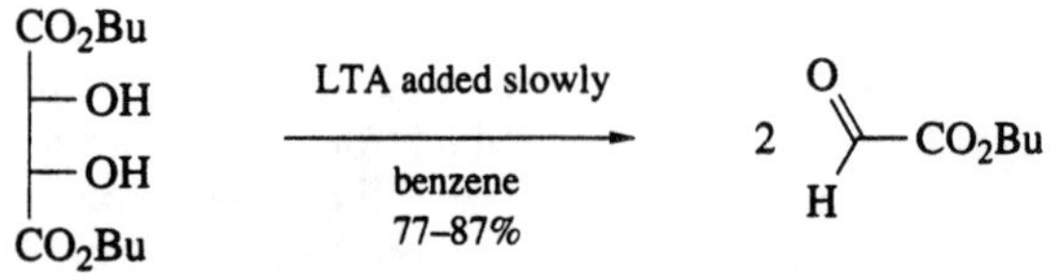

(7)

Periodate oxidation is sensitive to the stereochemistry[42,43] of the substrates: cyclic *trans*-1,2-glycols containing a tertiary hydroxy group and conformationally biased *trans*-diaxial 1,2-diols[1] are generally unreactive. Being a more powerful oxidizing agent, LTA complements periodates in the cleavage of diols which are inert to periodates.[1,2] An interesting application of this type of LTA cleavage is the fission of angular *trans*-diols which allows entry into medium or large rings, illustrated by equations (8)[50] and (9).[51] In cases where the angular diols are *cis*-disposed, oxidative cleavage with periodate occurs readily and thus fragmentation into medium or large rings can be achieved from *trans*-angular alkenes *via* the periodate–ruthenium tetroxide oxidation;[52] an example is provided in the synthesis of the 5-8-5 carbon skeleton of fusicoccins and ophiobolins (equation 10).[53] Trisubstituted alkenes yield keto acids (equation 11).[54] A related reaction, known as the Lemieux–von Rudloff (periodate–permanganate) oxidation,[55] is also used to oxidize alkenes to acids or ketones (equation 12).[56] If aldehydes are required from alkenes, the Lemieux–Johnson (periodate–osmium tetroxide) oxidation[57] is appropriate and equation (13)[58] shows that a free hydroxy group is not oxidized under the reaction conditions. If LTA is used as the oxidant, the dicarbonyl compounds are obtained in two separate steps from cyclic alkenes. Treatment of the dicarbonyl compounds with base causes aldol condensation and hence provides a general method for the preparation of cyclic enones (equation 14)[59] or enals (equation 15).[60]

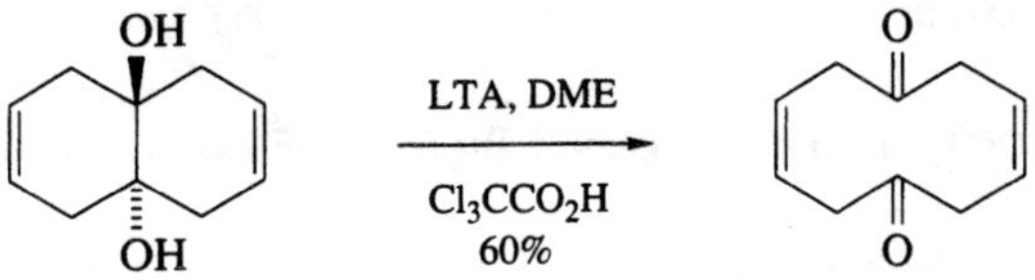

OH / OH —LTA, DME / Cl_3CCO_2H, 60%→ O / O (8)

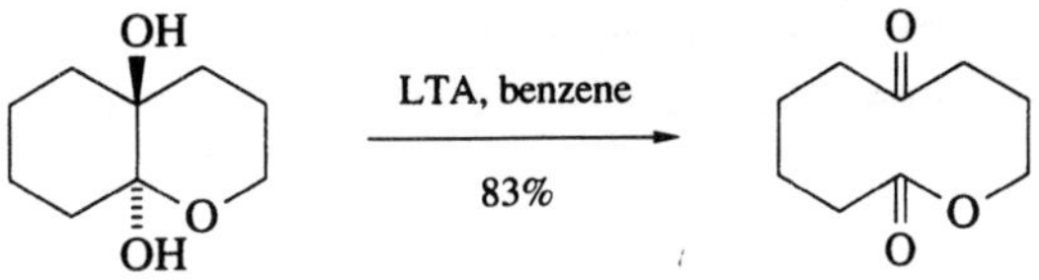

OH / O / OH —LTA, benzene, 83%→ O / O / O (9)

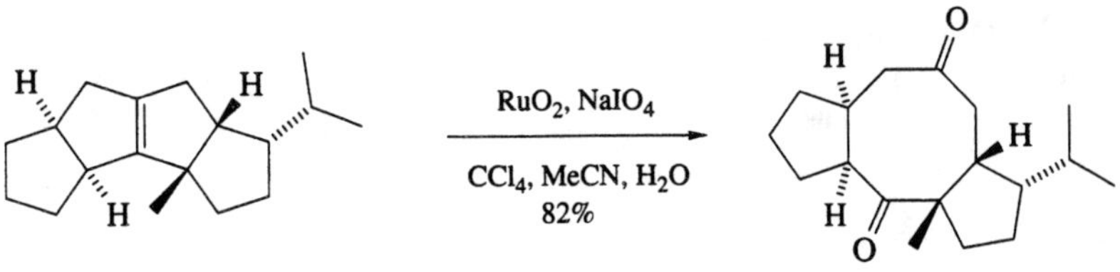

H H H —RuO_2, $NaIO_4$ / CCl_4, MeCN, H_2O, 82%→ O H H H O (10)

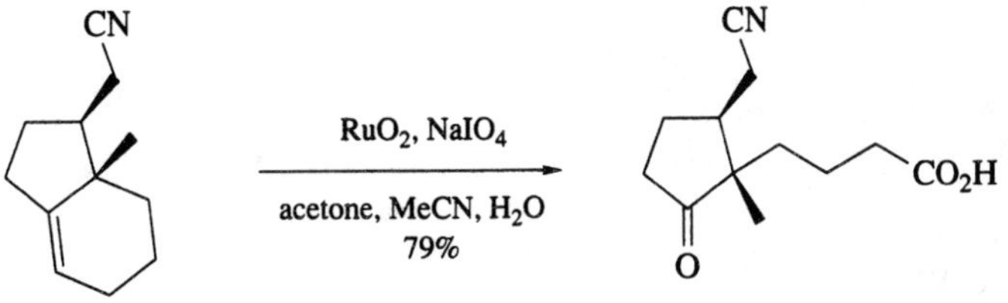

CN —RuO_2, $NaIO_4$ / acetone, MeCN, H_2O, 79%→ CN CO_2H O (11)

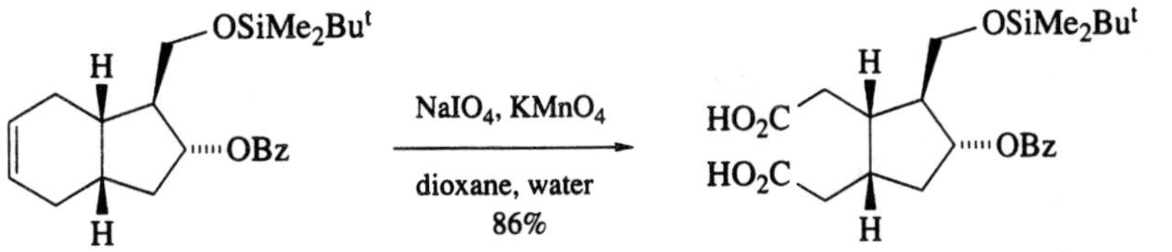

$OSiMe_2Bu^t$ H OBz H —$NaIO_4$, $KMnO_4$ / dioxane, water, 86%→ $OSiMe_2Bu^t$ H HO_2C OBz HO_2C H (12)

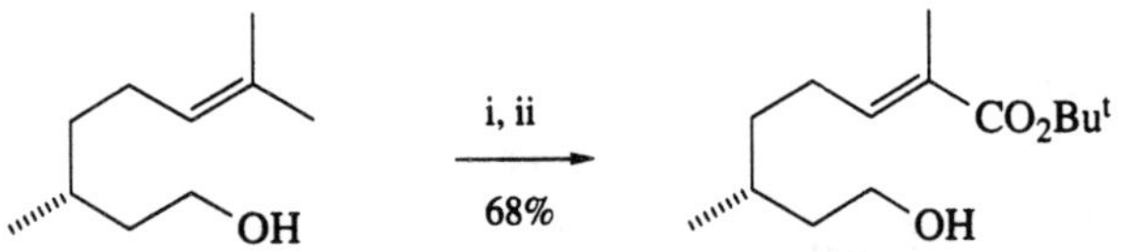

OH —i, ii, 68%→ CO_2Bu^t OH (13)

i, catalytic OsO_4, $NaIO_4$; ii, $Ph_3P{=}CMeCO_2Bu^t$

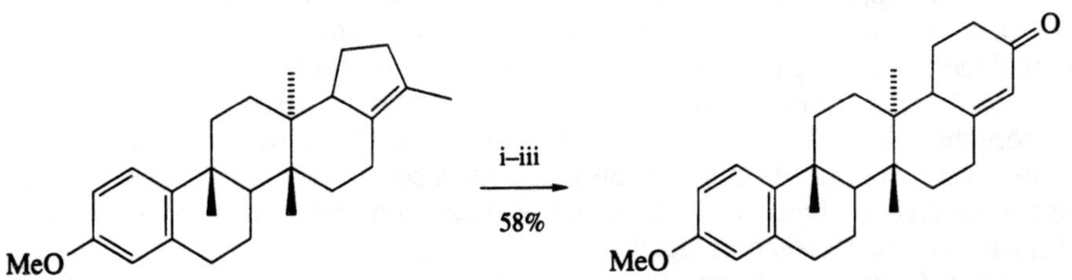

MeO —i–iii, 58%→ O MeO (14)

i, OsO_4; ii, LTA, THF; iii, KOH, H_2O

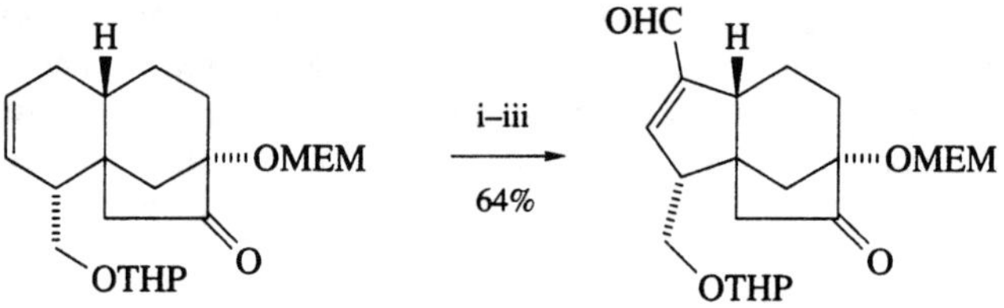

H OHC H

OMEM —i–iii, 64%→ OMEM (15)

OTHP O OTHP O

i, OsO_4; ii, LTA, benzene; iii, $Bn_2NH_2^+CF_3CO_2^-$, benzene

Aldol-type cyclization of dialdehydes with nitroalkanes is a valuable synthetic route to amino sugars, amino cyclitols and nucleosides of amino sugars.[61] Recently, the cyclization of the di- and tetra-aldehydes derived from sucrose (**15**) with nitroalkanes has appeared. It is noteworthy that the oxidative cleavage of sucrose with LTA affords the dialdehyde selectively (Scheme 8).[62]

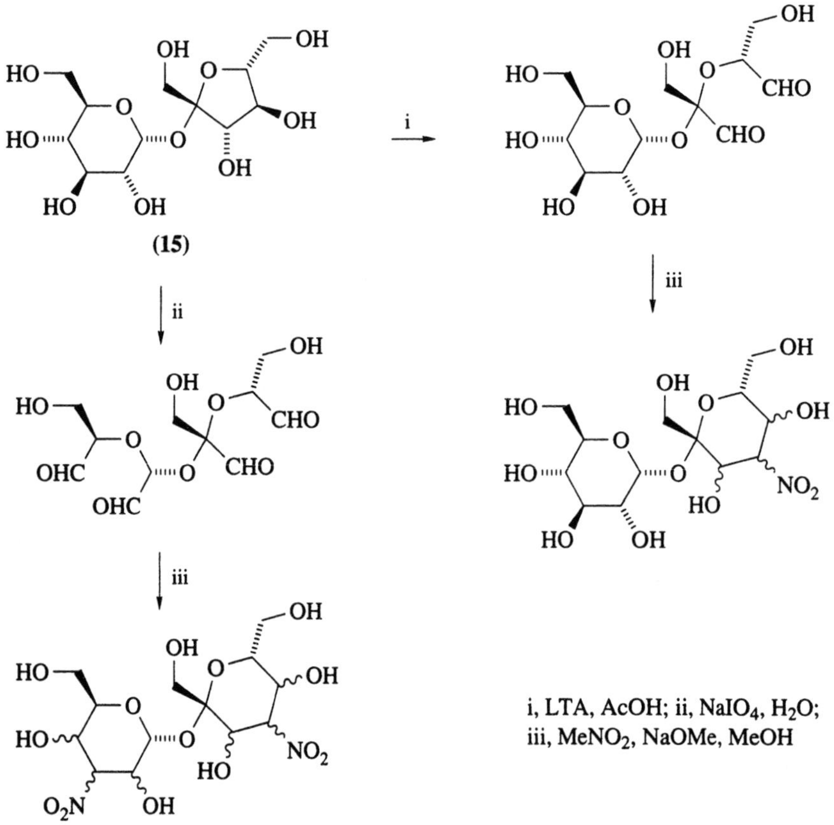

(15)

i, LTA, AcOH; ii, $NaIO_4$, H_2O;
iii, $MeNO_2$, NaOMe, MeOH

Scheme 8

Generation of aldehydes from oxidative fission of diols for further synthetic elaboration is generally more efficient than that from oxidation of the corresponding primary alcohols. The sequence involving a glycol cleavage followed by a Wittig-type homologation, illustrated in equation (16),[63] is particularly attractive and finds wide application in the syntheses of arachidonic acid metabolites. Recently, a series of hydroxylated aldehydes which are useful intermediates in the synthesis of lipoxygenase metabolites of arachidonic acid have been prepared from the corresponding acetonides (equation 17).[64] The use of periodic acid permits the transformation (hydrolysis followed by glycol scission) to proceed in one pot. In a similar way, periodic acid is useful for the hydrolysis of resistant 1,3-dioxolane protecting groups (equation 18).[65] The glycol cleavage–Wittig condensation sequence has also been employed in the total synthesis of other natural products, *e.g.* altholactones (equation 19),[66] and (*S*)-homolaudanosine and (*S*)-2,3,9,10,11-pentamethoxyhomoprotoberberine (equation 20).[67]

The extensive use of periodate, frequently sodium periodate, as oxidants for water insoluble glycols in contemporary organic synthesis contradicts the recommendation made earlier[2] that LTA is the preferred reagent. Hydrophobicity of the substrates does not appear to pose problems since they are readily soluble in aqueous solvents. In addition, periodate cleavage can proceed in a two-phase mixture, indicated in

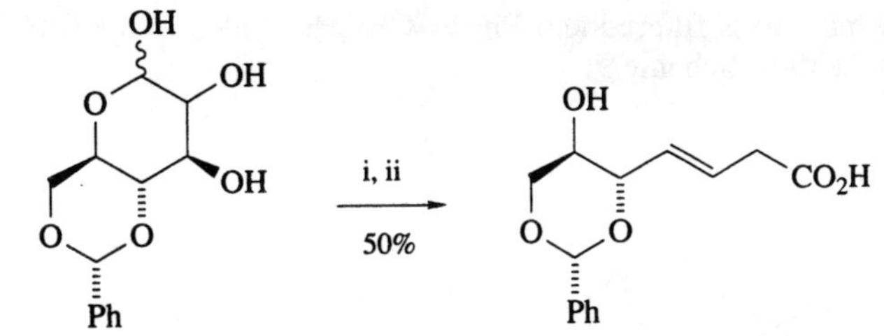

(16)

i, $NaIO_4$, $NaHCO_3$, CH_2Cl_2, H_2O; ii, $HO_2CCH_2CH{=}PPh_3$, THF, DMSO

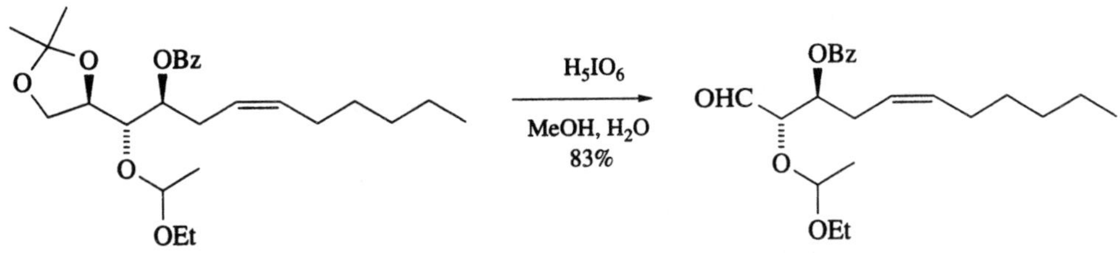

(17)

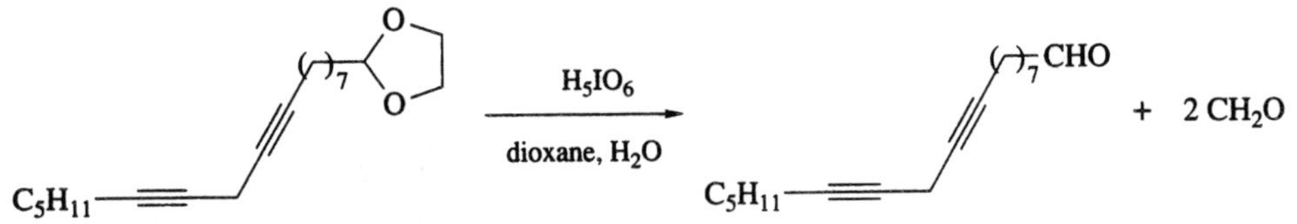

(18)

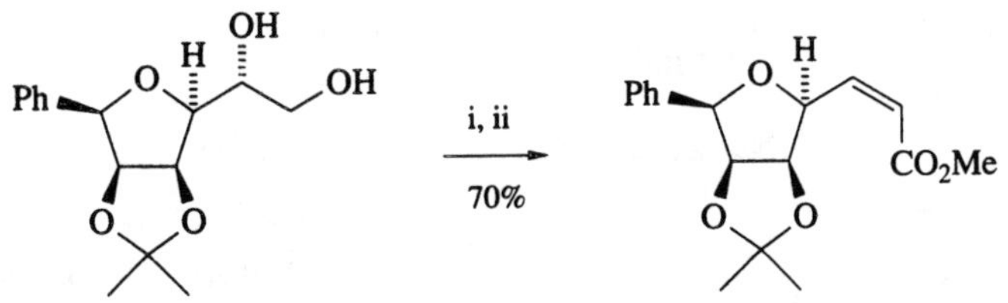

(19)

i, $NaIO_4$, MeOH, H_2O; ii, $Ph_3P{=}CHCO_2Me$, MeOH

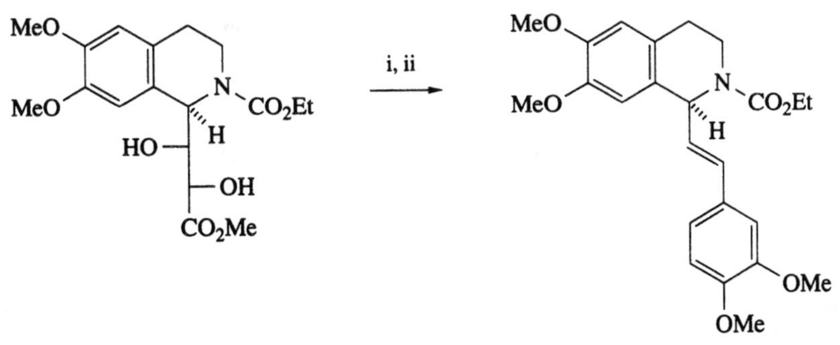

(20)

i, $NaIO_4$, MeOH, H_2O; ii, ylide derived from 3,4-dimethoxybenzyltriphenylphosphonium chloride, THF

equation (16). More examples of the application of sodium periodate mediated glycol fission reactions in synthesis are illustrated in equations (21)–(25).[68–72] The attractiveness of sodium periodate is attributable to its unique features, such as: (i) it has an indefinite shelf life and can be handled easily; (ii) it is used under mild and neutral conditions; (iii) it is highly specific; and (iv) isolation of the reaction products is by extraction into organic solvents. However, if the reaction products are highly water soluble or prone to hydrate formation, LTA is the preferred reagent, as used in the preparation of 2,3-*O*-isopropylidene-D-glyceraldehyde (**16**),[73] a versatile homochiral building block. A convenient synthesis of enal (**17**) results from a simplified work-up procedure for (**16**).[74] The procedure involves filtration of the glycol cleavage reaction mixture in EtOAc through a celite/silica pad followed by neutralization of the filtrate with solid

$NaHCO_3$; the resultant mixture is filtered and $Ph_3P{=}CHCHO$ added to the filtrate to give enal (**17**) in 75% overall yield from the diol (Scheme 9).

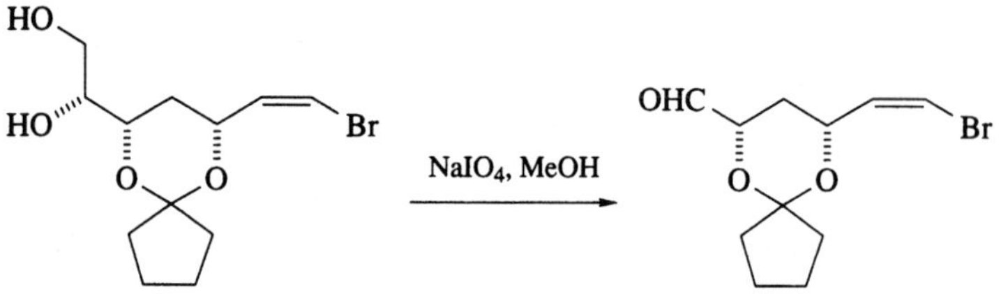

NaIO₄, MeOH (21)

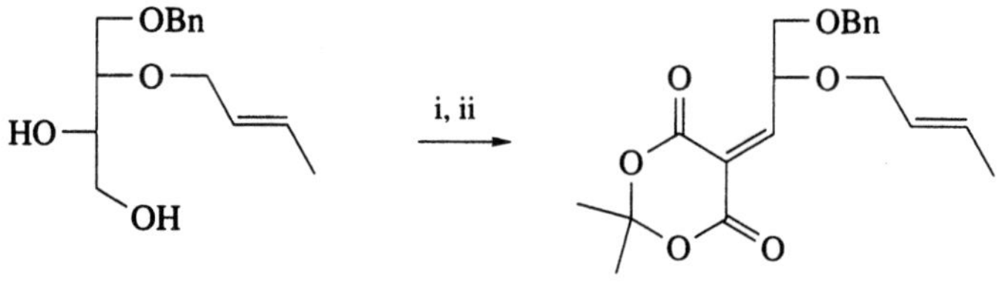

i, ii (22)

i, $NaIO_4$, aq. MeOH, 0 °C to r.t.; ii, Meldrum's acid, $(CH_2NH_3)^+{}_2(AcO^-)_2$, MeOH

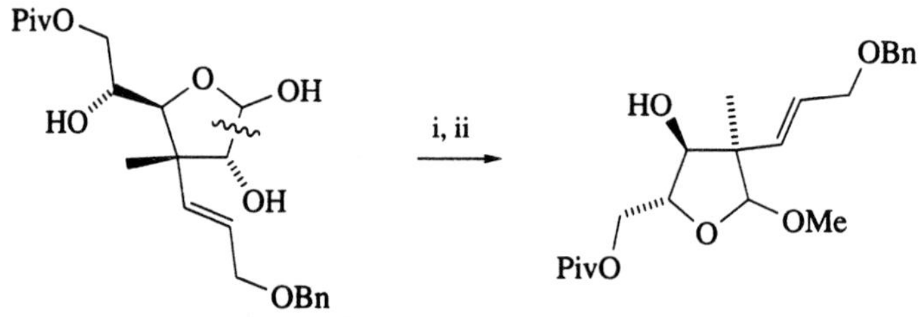

i, ii (23)

i, $NaIO_4$, aq. MeOH; ii, MeOH, acid resin

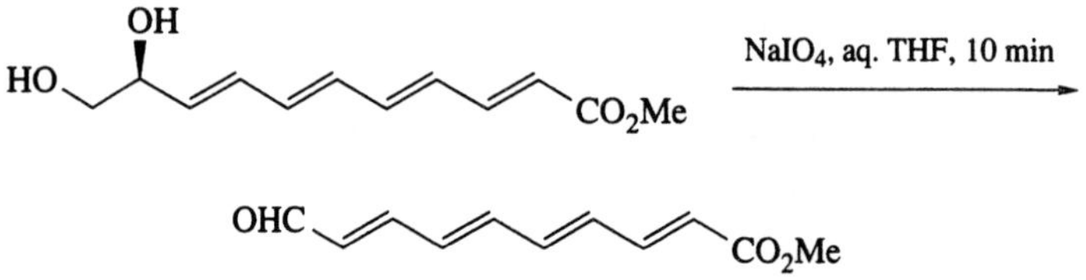

$NaIO_4$, aq. THF, 10 min (24)

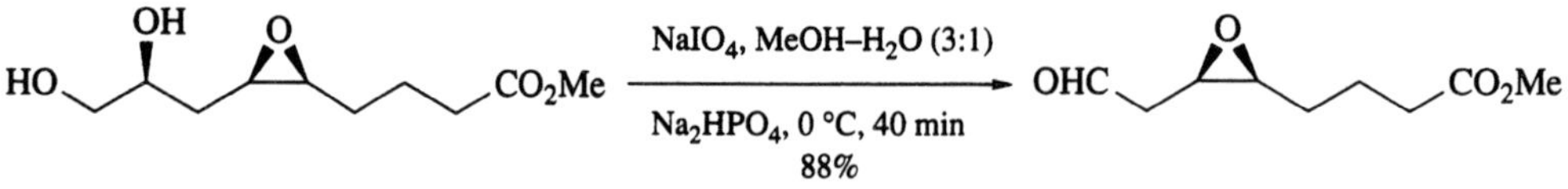

$NaIO_4$, MeOH–H_2O (3:1), Na_2HPO_4, 0 °C, 40 min, 88% (25)

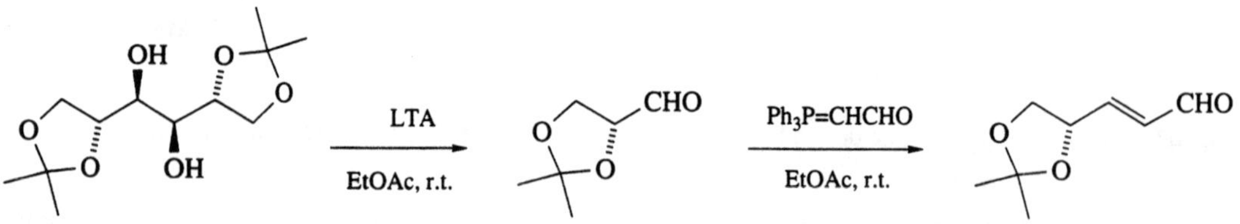

LTA, EtOAc, r.t. (**16**) $Ph_3P{=}CHCHO$, EtOAc, r.t. (**17**) 75% overall

Scheme 9

5.3.14 REFERENCES

1. A. S. Perlin, in 'The Carbohydrates, Chemistry and Biochemistry', ed. W. Pigman, D. Horton and J. D. Wander, Academic Press, London, 1980, vol. 1B, p. 1167.
2. C. A. Bunton, in 'Oxidation in Organic Chemistry', ed. K. E. Wiberg, Academic Press, New York, 1965, vol. 5A, p. 367.
3. A. J. Fatiadi, *Synthesis*, 1974, **4**, 229.
4. B. Skalarz, *Q. Rev. Chem. Soc.*, 1967, **21**, 3.
5. M. L. Mihailovic, Z. Cekovic and L. Lorenc, in 'Organic Syntheses by Oxidation with Metal Compounds', ed. W. J. Mijs and C. R. H. I. de Jonge, Plenum Press, New York, 1986, p. 741.
6. W. Rigby, *J. Chem. Soc.*, 1950, 1907.
7. W. Rigby, *J. Chem. Soc.*, 1951, 793.
8. A. Berka, *Arch. Pharm. (Weinheim, Ger.)*, 1970, **303** (3), 233.
9. L. K. Truesdale and M. E. Reuman, *J. Org. Chem.*, 1980, **45**, 726.
10. A. Stoll, A. Lindenmann and E. Jucker, *Helv. Chim. Acta*, 1953, **36**, 268.
11. G. Copinschi, A. Cornil and J. R. M. Franckson, *Clin. Chim. Acta*, 1962, **7**, 817.
12. D. H. R. Barton, J. P. Kitchin, D. J. Lester, W. B. Motherwell and M. T. B. Papoula, *Tetrahedron*, 1981, **37** (suppl. 1), 73.
13. D. H. R. Barton, S. D. Gero and C. D. Maycock, *J. Chem. Soc., Chem. Commun.*, 1980, 1089.
14. D. H. R. Barton, J.-P. Finet, W. B. Motherwell and C. Pichon, *Tetrahedron*, 1986, **42**, 5627.
15. S. P. Rao, J. N. Gaur and S. K. Sharma, *Naturwissenshaften*, 1961, **48**, 98.
16. H. L. Hintz and D. C. Johnson, *J. Org. Chem.*, 1967, **32**, 556.
17. W. S. Trahanovsky, L. H. Young and M. H. Bierman, *J. Org. Chem.*, 1969, **34**, 869.
18. W. S. Trahanovsky, J. R. Gilmore and P. C. Heaton, *J. Org. Chem.*, 1973, **38**, 760.
19. S. O. Nwaukwa and P. M. Keehn, *Tetrahedron Lett.*, 1982, **23**, 3135.
20. S. O. Nwaukwa and P. M. Keehn, *Tetrahedron Lett.*, 1982, **23**, 3131.
21. H. Kwart, J. A. Ford, Jr. and G. C. Corey, *J. Am. Chem. Soc.*, 1962, **84**, 1252; K. B. Wiberg, in 'Oxidation in Organic Chemistry', ed. K. E. Wiberg, Academic Press, New York, 1965, vol. 5A, p. 69.
22. A. Cisneros, S. Fernandez and J. E. Hernandez, *Synth. Commun.*, 1982, **12** (11), 833.
23. G. de Vries and A. Schors, *Tetrahedron Lett.*, 1968, 5689.
24. U. Zeidler and H. Lepper, *Ger. Pat.* 2 256 888 (1974) (*Chem. Abstr.*, 1974, **81**, 37 251f).
25. R. C. Cambie, D. Chambers, P. S. Rutledge and P. D. Woodgate, *J. Chem. Soc., Perkin Trans. 1*, 1978, 1483.
26. T. R. Beebe, P. Hii and P. Reinking, *J. Org. Chem.*, 1981, **46**, 1927.
27. J. S. Littler, A. I. Mallet and W. A. Waters, *J. Chem. Soc.*, 1960, 2761; J. S. Littler and W. A. Waters, *J. Chem. Soc.*, 1960, 2767.
28. M. Zviely, A. Goldman, I. Kirson and E. Glotter, *J. Chem. Soc., Perkin Trans. 1*, 1986, 229.
29. T. Shono, Y. Matsumura, T. Hashimoto, K. Hibino, H. Hamaguchi and T. Aoki, *J. Am. Chem. Soc.*, 1975, **97**, 2546; T. Shono and H. Hamaguchi, *Kagaku (Tokyo)*, 1982, **37** (4), 302.
30. C. Venturello and M. Ricci, *J. Org. Chem.*, 1986, **51**, 1599.
31. Y. Ishii, K. Yamawaki, T. Ura, H. Yamada, T. Yoshida and M. Ogawa, *J. Org. Chem.*, 1988, **53**, 3587.
32. G. Ohloff and W. Giersch, *Angew. Chem., Int. Ed. Engl.*, 1973, **12**, 401.
33. L. Malaprade, *C. R. Hebd. Seances Acad. Sci.*, 1928, **186**, 382; L. Malaprade, *Bull. Soc. Chim. Fr.*, 1928, **43**, 683.
34. R. Criegee, *Chem. Ber.*, 1931, **64**, 260.
35. L. F. Fieser and M. Fieser, 'Reagents for Organic Synthesis', Wiley, New York, 1967, vol. 1, p. 817.
36. D. N. Gupta, P. Hodge and J. E. Davies, *J. Chem. Soc., Perkin Trans. 1*, 1981, 2835.
37. E. Santaniello, A. Manzocchi and C. Farachi, *Synthesis*, 1980, 563; K. Inomata, Y. Nakayama and H. Kotake, *Bull. Chem. Soc. Jpn.*, 1980, **53**, 565.
38. C. H. Harrison and P. Hodge, *J. Chem. Soc., Perkin Trans. 1*, 1982, 509.
39. R. D. Guthrie, *Chem. Ind. (London)*, 1960, 691.
40. J. M. Bobbitt, *Adv. Carbohydr. Chem.*, 1956, **11**, 1.
41. A. S. Perlin, *Adv. Carbohydr. Chem.*, 1959, **14**, 1.
42. G. J. Buist, C. A. Bunton and J. H. Miles, *J. Chem. Soc.*, 1957, 4567.
43. R. Criegee, E. Buchner and W. Walther, *Chem. Ber.*, 1940, **73**, 571.
44. R. Criegee, C. Kraft and B. Rank, *Justus Liebigs Ann. Chem.*, 1933, **507**, 159.
45. R. Criegee, E. Hoger, G. Huber, P. Kruck, F. Marktscheffel and H. Schellenberger, *Justus Liebigs Ann. Chem.*, 1956, **599**, 81.
46. R. P. Bell, V. G. Rivlin and W. A. Waters, *J. Chem. Soc.*, 1958, 1696.
47. J. P. Nagarkatti and K. R. Ashley, *Tetrahedron Lett.*, 1973, 4599.
48. G. W. J. Fleet and T. K. M. Shing, *J. Chem. Soc., Chem. Commun.*, 1984, 835.
49. F. J. Wolf and J. Weijlard, *Org. Synth., Coll. Vol.*, 1963, **4**, 124.
50. B. W. Roberts, J. J. Vollmer and K. L. Servis, *J. Am. Chem. Soc.*, 1968, **90**, 5264.
51. I. J. Borowitz, G. Gonis, R. Kelsey, R. Rapp and G. J. Williams, *J. Org. Chem.*, 1966, **51**, 3032.
52. P. H. J. Carlson, T. Kalsuki, V. S. Martin and K. B. Sharpless, *J. Org. Chem.*, 1981, **46**, 3936.
53. G. Mehta and N. Krishnamurthy, *J. Chem. Soc., Chem. Commun.*, 1986, 1319.
54. P. T. Lansbury and J. P. Galbo, *Tetrahedron Lett.*, 1988, **29**, 147.
55. R. U. Lemieux and E. von Rudloff, *Can. J. Chem.*, 1955, **33**, 1701.
56. Y. Nagao, T. Nakamura, M. Kume, M. Ochiai, K. Fuji and E. Fujita, *J. Chem. Soc., Chem. Commun.*, 1987, 269, 267.
57. R. Pappo, D. S. Allen, Jr., R. U. Lemieux and W. S. Johnson, *J. Org. Chem.*, 1956, **21**, 478.
58. T. K. M. Shing, *J. Chem. Soc., Chem. Commun.*, 1986, 49.
59. R. E. Ireland, P. Bey, K. F. Cheng, R. J. Czarny, J. F. Moser and R. I. Trust, *J. Org. Chem.*, 1975, **40**, 1000.

60. E. J. Corey, R. L. Danheiser, S. Chandrasekaran, P. Siret, G. E. Keck and J.-L. Gras, *J. Am. Chem. Soc.*, 1978, **100**, 8031.
61. F. W. Lichtenthaler, *Angew. Chem., Int. Ed. Engl.*, 1964, **3**, 211; F. W. Lichtenthaler and H. Zinke, *J. Org. Chem.*, 1972, **37**, 1612.
62. K. J. Hale, L. Hough and A. C. Richardson, *Tetrahedron Lett.*, 1987, **28**, 891.
63. S. R. Baker, D. W. Clissold and A. McKillop, *Tetrahedron Lett.*, 1988, **29**, 991.
64. S. Okamoto, T. Shimazaki, Y. Kitano, Y. Kobayashi and F. Sato, *J. Chem. Soc., Chem. Commun.*, 1986, 1352.
65. H. M. Walborsky, R. H. Davis and D. R. Howton, *J. Am. Chem. Soc.*, 1951, **73**, 2590.
66. J. G. Gillhouley and T. K. M. Shing, *J. Chem. Soc., Chem. Commun.*, 1988, 976.
67. Z. Czarnocki, D. B. Maclean and W. A. Szarek, *J. Chem. Soc., Chem. Commun.*, 1987, 493.
68. G. Just and B. O'Connor, *Tetrahedron Lett.*, 1988, **29**, 753.
69. S. Takano, S. Satoh and K. Ogasawara, *J. Chem. Soc., Chem. Commun.*, 1988, 59.
70. K. Tadano, H. Yamada, Y. Idogaki, S. Ogawa and T. Suami, *Tetrahedron Lett.*, 1988, **29**, 655.
71. S. Hanessian and M. Botta, *Tetrahedron Lett.*, 1987, **28**, 1511.
72. S. Lumin, P. Yadagiri, J. R. Falck, J. Capdevila, P. Mosset and R. Gree, *J. Chem. Soc., Chem. Commun.*, 1987, 389.
73. J. Jurczak, S. Pikul and T. Bauer, *Tetrahedron*, 1986, **42**, 447.
74. T. K. M. Shing and J. R. Wiley, unpublished results.

5.4
The Hunsdiecker and Related Reactions

DAVID CRICH

University of Illinois, Chicago, IL, USA

5.4.1 INTRODUCTION

The main theme of this chapter is the cleavage of alkyl–, aryl– and vinyl–carboxyl single bonds by means of the fragmentation (decarboxylation) of carboxyl radicals (equation 1). The fragmentation of acyl radicals (decarbonylation) (equation 2), although much less efficient, is also considered where appropriate.

$$\mathrm{RC(O)O\bullet} \longrightarrow \mathrm{R\bullet} + CO_2 \quad (1)$$

$$\mathrm{RC(O)\bullet} \longrightarrow \mathrm{R\bullet} + CO \quad (2)$$

Methods for the generation of carboxyl radicals are considered first. The trapping of the ensuing alkyl radicals by various radical-trapping agents allowing the overall transformation of carboxylic acids into a range of diverse functional groups of the first lower homolog is then discussed.

The majority of reactions discussed are readily applicable to simple primary, secondary and tertiary aliphatic acids. The decarboxylation of aryl- and vinyl-carboxyl radicals is a much more difficult process which limits the application of many of the methods described to aliphatic acids. As such, particular attention is drawn in the text to examples of aryl and vinyl decarboxylations.

This chapter does not include electrochemical decarboxylation processes (the Kolbe reaction, Volume 3, Chapter 2.9) and transition metal catalyzed decarbonylation reactions.

5.4.2 GENERATION OF CARBOXYL RADICALS: FUNCTIONAL GROUP COMPATABILITY

The generation of carboxyl radicals requires the preparation of suitable precursors containing a weak carboxyl–X bond susceptible to homolytic cleavage. In the classical Hunsdiecker reaction, the precursor is an acyl hypohalite (X is halogen). More recently, methods have been developed in which X is a lead(IV) salt or a nitrogen atom. These more recent methods tolerate a much wider range of functional groups as they operate under much milder conditions. Throughout the text examples are chosen to illustrate the degrees of functionality and complexity compatible with the various methods. Methods with relatively limited applications, such as the pyrolysis of peroxy esters and the silver-catalyzed oxidation of carboxylate salts by the persulfate anion, are discussed at appropriate points in the text.

5.4.2.1 Acyl Hypohalites

Acyl hypohalites are usually prepared *in situ* by reaction of a metal salt of the carboxylic acid with a halogen (equation 3). Classically the silver salt[1] is used, but problems associated with the preparation of dry silver carboxylates, as well as the more obvious economic factor, have led to the development of methods using mercury[2] and thallium[3] salts. Evidently, those functional groups which react readily with halogens are not compatible with this approach. A major limitation of the acyl hypohalites is the readiness with which they transfer halogen atoms to alkyl radicals; this property essentially limits their use to decarboxylative halogenation reactions.

$$\mathrm{RC(O)OM} \xrightarrow{X_2} \mathrm{RC(O)OX} + \mathrm{MX} \quad (3)$$

5.4.2.2 Lead(IV) Carboxylates

The generation of lead(IV) carboxylates involves exchange of an acetate of LTA for the acid to be decarboxylated (equation 4). The weak bond cleaved to generate the carboxyl radical is a carboxyl–lead(IV) bond. The higher activation energy for decarboxylation of acetoxyl radicals ensures preferential decarboxylation of most other alkanoic acids. On the other hand, attempts at the decarboxylation of aryl acids fail due to competing acetoxyl decarboxylation. Evidently, those functional groups which undergo

$$\mathrm{RC(O)OH} + \mathrm{Pb(OAc)_4} \longrightarrow \mathrm{RC(O)O{-}Pb(OAc)_3} + \mathrm{AcOH} \quad (4)$$

facile oxidation with LTA are not compatible. An excellent review[4] provides further information on functional group compatability and experimental details.

5.4.2.3 *O*-Acyl Oximes and *O*-Acyl Thiohydroxamates

Recently, methods have been devised for carboxyl radical generation by homolytic cleavage of weak carboxyl—N single bonds. Such methods have great potential in organic synthesis as precursor generation requires neither strong oxidants nor strongly electrophilic species. Such precursors are hence compatible with a wide range of functional groups. One such precursor type is composed of the *O*-acyl benzophenone oximes, prepared[5] by the reaction of activated acyl derivatives with benzophenone (equation 5). Carboxyl radical generation is achieved by simple UV photolysis in ordinary pyrex glassware. The inevitable by-product from this method is benzophenone azine.

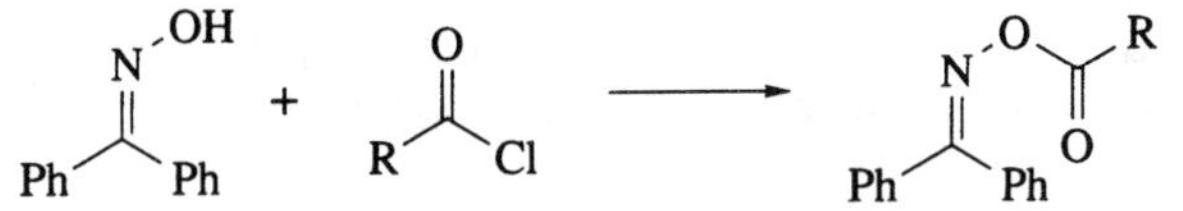

$$Ph_2C{=}N{-}OH + RCOCl \longrightarrow Ph_2C{=}N{-}O{-}C(O)R \qquad (5)$$

A second, more versatile, method involves the *O*-acyl thiohydroxamates.[6] These compounds are generally prepared[7] by reaction of acyl chlorides with the commercial sodium salt (**1**) of 2-mercaptopyridine *N*-oxide (equation 6; X = Cl). Use of mixed anhydrides formed by reaction of the carboxylic acid with isobutyl chloroformate (equation 6; X = $OCO_2CH_2CHMe_2$) renders the procedure compatible with unprotected indoles, phenols, secondary and, presumably, tertiary alcohols. An alternative mode of preparation[7] of the *O*-acyl thiohydroxamates involves the salt (**2**) in reaction with the carboxylic acid (equation 7).

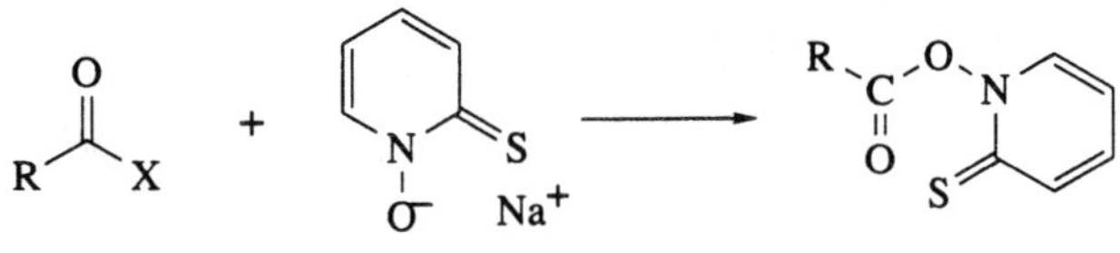

$$RCOX + (\mathbf{1})\ \text{(2-thioxopyridine } N\text{-oxide, } Na^+) \longrightarrow R{-}C(O){-}O{-}N\text{(pyridine-2-thione)} \qquad (6)$$

(1)

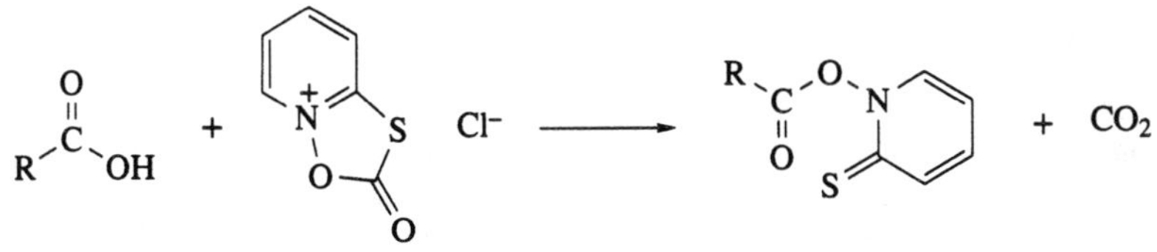

$$RCOOH + (\mathbf{2})\ Cl^- \longrightarrow R{-}C(O){-}O{-}N\text{(pyridine-2-thione)} + CO_2 \qquad (7)$$

(2)

A wide variety of thiophilic radicals induce fragmentation of *O*-acyl thiohydroxamates by addition to the thiocarbonyl group, and in doing so generate carboxyl radicals (Scheme 1). After decarboxylation the

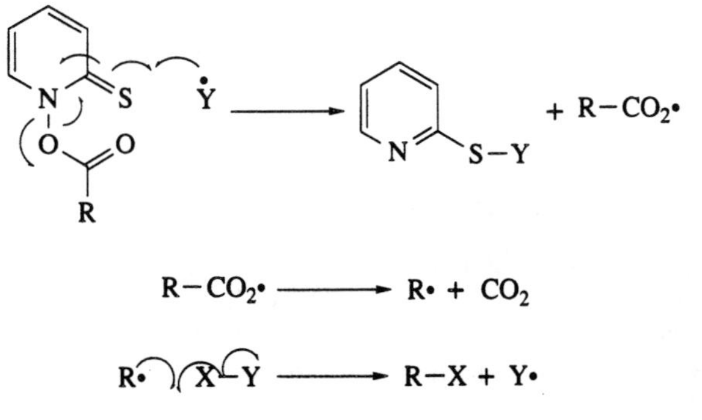

$$\text{RC(O)O-N(pyridine-2-thione)} + Y^\bullet \longrightarrow \text{2-(Y-S)pyridine} + R{-}CO_2^\bullet$$

$$R{-}CO_2^\bullet \longrightarrow R^\bullet + CO_2$$

$$R^\bullet + X{-}Y \longrightarrow R{-}X + Y^\bullet$$

Scheme 1

alkyl radical can be trapped with a variety of radical-trapping reagents. The most efficient reactions occur when the radical trap X—Y (Scheme 1) is designed to release a new thiophilic chain-carrying radical (Y·) on trapping the alkyl radical, so setting up a chain mechanism. The reactions are initiated thermally in benzene or other appropriate solvents, or by simple white light photolysis.

5.4.3 REDUCTIVE DECARBOXYLATION

Reductive decarboxylation requires the generation of the carboxyl and hence the alkyl radical in the presence of a suitable hydrogen donor. Some modifications use the reaction solvent as hydrogen donor (cumene, *p*-cymene, *etc.*), others use added thiols or stannanes. This most basic of reactions has, until recently, been very little used in organic synthesis, owing perhaps to a shortage of suitable methodology for all but simple aliphatic compounds. The discovery of the *O*-acyl thiohydroxamates by the Barton[6] group has made reductive decarboxylation a much more practicable possibility.

Reductive decarboxylation of aryl and vinyl acids, not readily achieved by any of the methods described here, is best brought about by the classical copper/quinoline procedure.[8]

5.4.3.1 Peroxy Esters and Hydrogen Donor Solvents

The pyrolysis of *t*-butyl peroxy esters in suitable hydrogen donor solvents has been reviewed by Rüchardt.[9] The method involves the reaction of an acyl chloride with *t*-butyl hydroperoxide followed by thermolysis of the resulting peroxy ester in cumene or *p*-cymene. Yields are moderate, but the pyrolysis step tolerates a certain degree of functionality as illustrated in equation (8).[10] More recently, the use of ethyl phenylacetate as the pyrolysis solvent and hydrogen donor has been advocated.[11]

BnO … CO_2Me, CO_2OBu^t —(Δ, cumene; 40%)→ … CO_2Me (8)

5.4.3.2 LTA and Hydrogen Donor Solvents

Reductive decarboxylation has been achieved by heating the acid with LTA in chloroform as solvent and hydrogen donor. Only a moderate number of examples are known.[12] The more facile oxidation of secondary and tertiary radicals by LTA effectively limits the method to primary carboxylic acids. It should be noted that stoichiometric quantities of trichloromethyl radicals are generated in the course of this reaction.

5.4.3.3 Persulfate Oxidation in Chloroform

Silver-catalyzed persulfate decarboxylation of carboxylic acids in chloroform provides the noralkane in modest to good yields.[13] Only a limited number of examples with simple aliphatic carboxylic acids are known.

5.4.3.4 *O*-Acyl Benzophenone Oximes and *t*-Butyl Thiol

The photolysis of *O*-acyl benzophenone oximes in isopropyl alcohol in the presence of *t*-butyl thiol leads to overall reductive decarboxylation. Yields for simple aliphatic and amino acids are reported to be good.[14] It is especially noteworthy that reductive decarboxylation of 2- and 4-quinolinecarboxylic acids (although not of the 3-isomer, nor 1- or 2-naphthalenecarboxylic acids) was achieved in moderate yield by this method (equation 9).

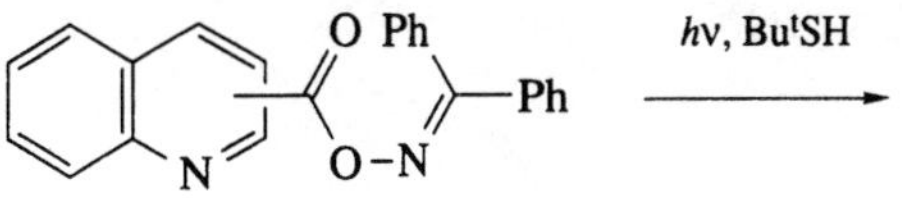

$h\nu$, Bu^tSH (9)

ortho 60%, *para* 40%

5.4.3.5 *O*-Acyl Thiohydroxamates and Tertiary Thiols

The free radical chain reaction of *O*-acyl thiohydroxamates with a tertiary thiol (*t*-butyl, triethylmethyl or more recently *t*-dodecyl) is by far the most wide-ranging reductive decarboxylation method described to date.[7] A wide variety of functional groups, including aldehydes, ketones, esters, amides, isolated and conjugated double bonds, are tolerated. Representative examples are given in equations (10)[7] and (11).[15]

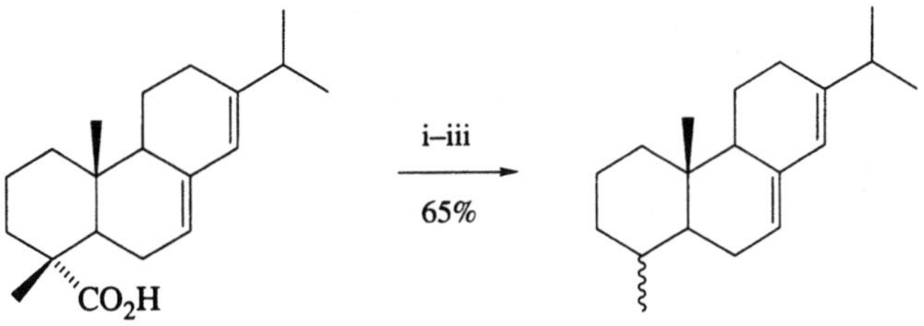

i–iii, 65% (10)

i, $(COCl)_2$; ii, **(1)**; iii, Bu_3SnH, 80 °C

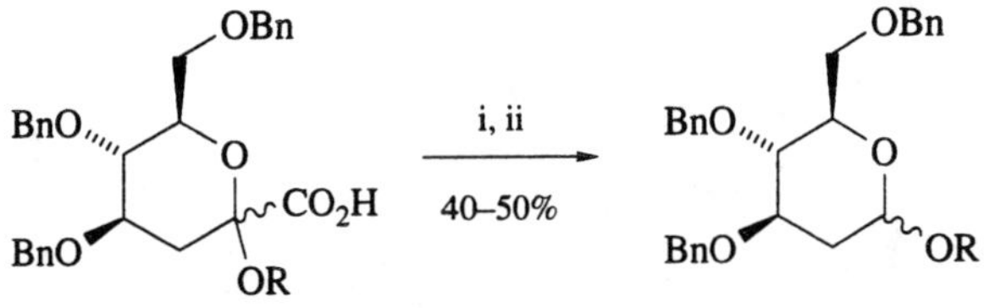

i, ii, 40–50% (11)

i, **(2)**; ii, R^1SH, $h\nu$

5.4.3.6 Decarbonylation Methods

Overall reductive decarboxylation of a carboxylic acid may be achieved[16] by the reaction of the derived acyl chloride with triisopropylsilane (equation 12). Relatively high temperatures are required to bring about efficient decarbonylation of the intermediate acyl radical. A related method[17] involves the reaction of acyl phenyl selenides with tri-*n*-butyltin hydride. Here again relatively high temperatures are required for primary and secondary, although not for tertiary, acids (equation 13).

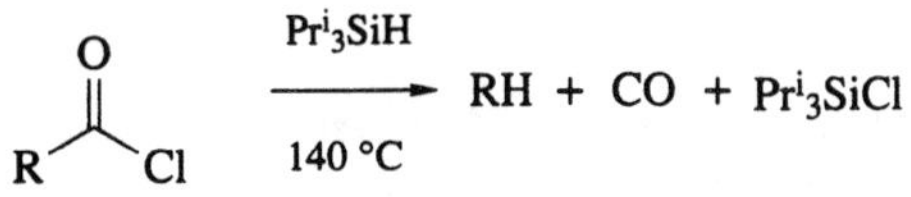

$$RCOCl \xrightarrow[140\ ^\circ C]{Pr^i_3SiH} RH + CO + Pr^i_3SiCl \qquad (12)$$

Bu_3SnH, $h\nu$, 25 °C, 92% (13)

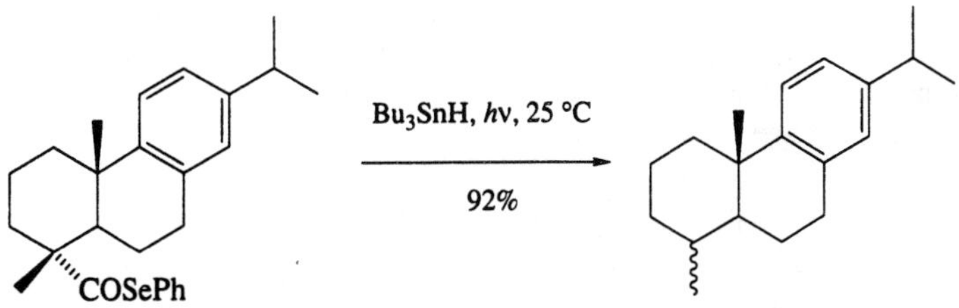

5.4.4 OXIDATIVE DECARBOXYLATION

5.4.4.1 LTA/Copper(II) Acetate and Carboxylic Acids

The oxidative decarboxylation of aliphatic carboxylic acids is best achieved by treatment of the acid with LTA in benzene, in the presence of a catalytic amount of copper(II) acetate. The latter serves to trap the radical intermediate and so bring about elimination, possibly through a six-membered transition state. Primary carboxylic acids lead to terminal alkenes, indicating that carbocations are probably not involved. The reaction has been reviewed.[4] The synthesis[18] of an optically pure derivative of L-vinylglycine from L-aspartic acid (equation 14) is illustrative. The same transformation has also been effected[13] with sodium persulfate and catalytic quantities of silver nitrate and copper(II) sulfate, and with the combination of iodosylbenzene diacetate and copper(II) acetate.[19]

LTA/$Cu(OAc)_2$, 80 °C; 60% (14)

A mild, but indirect, approach to oxidative decarboxylation involves a modification[20] of the *O*-acyl thiohydroxamate decarboxylative rearrangement (Section 5.4.6.1). An *O*-acyl selenohydroxamate is photolyzed to give a noralkyl-2-pyridyl selenide which, after ozonolysis to the selenoxide, undergoes *syn* elimination to the alkene (equation 15).

i, ii; 67% (15)

i, N-hydroxypyridine-2-selone, 80 °C; ii, O_3

5.4.4.2 1,4-Dicarboxylic Acids

The oxidation of 1,4-dicarboxylic acids with LTA in benzene results in double decarboxylation with the formation of a double bond (equation 16).[4] Similarly, the pyrolysis of the di-*t*-butyl peroxy esters of 1,4-dicarboxylic acids in high boiling solvents leads to the formation of double bonds (equation 17).[21] The method is especially useful in so far as 1,4-diacids are readily available from Diels–Alder reactions using derivatives of maleic and fumaric acid as the dienophile. Apparently, application of the *O*-acyl thiohydroxamate method to 1,4-diacids does not result in the formation of double bonds but rather in the product of double decarboxylative rearrangement (Section 5.4.6.1).[22]

C_5H_5N, Δ; 50% (16)

$$\xrightarrow[40\%]{PhBu^t,\ h\nu} \qquad (17)$$

CO$_2$OBu^t CO$_2$OBu^t

5.4.5 DECARBOXYLATIVE HALOGENATION

5.4.5.1 Acyl Hypohalites and Related Species

The classical Hunsdiecker reaction (equation 18), involving the reaction of silver carboxylates with halogens, and the various associated side reactions, has been reviewed several times. Optimum yields are obtained with bromine, followed by chlorine. Iodine gives acceptable yields provided that the correct stoichiometry of 1:1 is used. The reaction is most frequently carried out in tetrachloromethane at reflux. From a practical point of view, one drawback is the difficulty encountered in the preparation of dry silver carboxylates; the reaction of silver oxide on the acyl chloride in tetrachloromethane at reflux has been employed to circumvent this problem.[23] Evidently the use of molecular bromine limits the range of functional groups compatible with the reaction; the different reaction pathways followed by the silver salts of electron poor (equation 19) and electron rich (equation 20) aryl carboxylates illustrate this point well.

$$RCO_2Ag \xrightarrow{X_2} RCO_2X + AgX \xrightarrow{CCl_4,\ \Delta} R{-}X \qquad (18)$$

$$O_2N{-}C_6H_4{-}CO_2Ag \xrightarrow[79\%]{Br_2,\ CCl_4,\ \Delta} O_2N{-}C_6H_4{-}Br \qquad (19)$$

$$MeO{-}C_6H_4{-}CO_2Ag \xrightarrow[92\%]{Br_2,\ CCl_4,\ \Delta} MeO{-}C_6H_3(Br){-}CO_2H + MeO{-}C_6H_3(Br){-}Br \qquad (20)$$

79 : 21

An alternative method for the formation of acyl hypoiodites, developed by Barton,[24] involves the treatment of the acid with *t*-butyl hypoiodite. Subsequent white light photolysis in benzene at room temperature gave good yields of iodides from primary, secondary and tertiary acids (equation 21). The method was not applicable in the presence of alcohols. A more recent technique[19] involving hypervalent iodine is due to Suarez: primary, secondary or tertiary aliphatic acids are heated to reflux in tetrachloromethane with iodosylbenzene diacetate and iodine resulting in good yields of iodides. The method is mild and, with obvious exceptions such as unprotected alcohols, is tolerant of many functional groups, as illustrated in equation (22).[25]

$$\text{CO}_2\text{H} \xrightarrow[62\%]{i,\ ii} \text{I} \qquad (21)$$

i, Bu^{t}OI; ii, C_6H_6, $h\nu$

Decarboxylative fluorination, presumably *via* intermediate acyl hypofluorites, has been achieved[26] in low yield by passing a dilute stream of fluorine in nitrogen into aqueous solutions of sodium carboxylates. A somewhat more promising method, tolerant of aryl groups, ketones and ethers, involves reaction of a dichloromethane solution of the acid with xenon difluoride and hydrogen fluoride (equation 23).[27]

HO O N$_3$ THPO $\xrightarrow[80\%]{PhI(OAc)_2,\ CCl_4,\ \Delta}$ I N$_3$ THPO (22)

$$PhO\frown CO_2H \xrightarrow[64\%]{XeF_2,\ HF,\ CH_2Cl_2,\ 20\ °C} PhO\frown F \qquad (23)$$

5.4.5.2 Mercury(II), Thallium(I) and Lead(IV) Salts

The difficulties and expense associated with the use of silver salts have led to the development of methods using other metal carboxylates. The modification of Cristol[2] involves the action of bromine on a mixture of the acid and red mercury(II) oxide in tetrachloromethane (equation 24). The procedure is easy to carry out and gives good yields with simple aliphatic carboxylic acids. However, concomitant formation of alkyl chlorides, particularly with highly reactive radicals from bridgehead acids, led several groups[28] to prefer bromotrichloromethane as reaction solvent. Iodine has been substituted for bromine leading to overall decarboxylative iodination (equation 25);[29] however, yields are no better than with the corresponding silver salt.

$$2RCO_2H + HgO + 2Br_2 \longrightarrow 2RBr + HgBr_2 + H_2O + 2CO_2 \qquad (24)$$

Cl Cl Cl Cl Cl Cl CO$_2$H $\xrightarrow[70\%]{HgO,\ I_2,\ CCl_4,\ \Delta}$ I (25)

exo:*endo* = 70:30

The use of stable, crystallizable thallium(I) carboxylates with bromine in tetrachloromethane at reflux has also been demonstrated[3] to be effective in bringing about overall decarboxylative halogenation, provided the correct stoichiometry (equation 26) is adhered to.

$$2RCO_2Tl + 3Br_2 \longrightarrow 2RBr + 2CO_2 + Tl_2Br_4 \qquad (26)$$

Two procedures exist for decarboxylative halogenation with LTA. In the first, reported by Barton,[24] the acid is treated with a combination of LTA and iodine in tetrachloromethane, providing good yields of iodides (equation 27). The second, complementary technique, uses a combination of LTA and a lithium halide in a solution of the acid in benzene at reflux.[4] Both lithium chloride and bromide have been employed to good effect leading, respectively, to alkyl chlorides and bromides. Carbocations are not involved; *t*-butylacetic acid gives neopentyl chloride free from *t*-amyl chloride. It is thought that the radical is trapped by halogen abstraction from a lead(IV) halogen complex. LTA can also be used in conjunction with NCS as the chlorine atom donor.[30]

CO$_2$H AcO H $\xrightarrow[85\%]{LTA,\ I_2,\ CCl_4,\ \Delta}$ I (27)

β:α = 50:50

5.4.5.3 *O*-Acyl Thiohydroxamates and Halogen Donor Solvents

The photolytic or thermal decomposition of *O*-acyl thiohydroxamates in halogen donor solvents such as tetrachloromethane or bromotrichloromethane constitutes the most wide-ranging and generally applicable Hunsdiecker system currently available.[7] Decarboxylative iodination by this method uses iodoform as an iodine donor in benzene or, better, cyclohexene. The reactions proceed by chain mechanisms (Scheme 1; X—Y = Cl—CCl_3; Br—CCl_3; I—CHI_2) under very mild conditions; no free halogens or other strongly electrophilic species or oxidants and toxic metal salts are required. A wide variety of primary, secondary and tertiary acids, ranging from steroids[7] through terpenoids[7] to amino acids[19,31] (equation 28), have been subjected successfully to this variant of the Hunsdiecker reaction. The synthesis[32] of an α-chlorooxetane, albeit in low yield, by this method is worthy of particular note (equation 29).

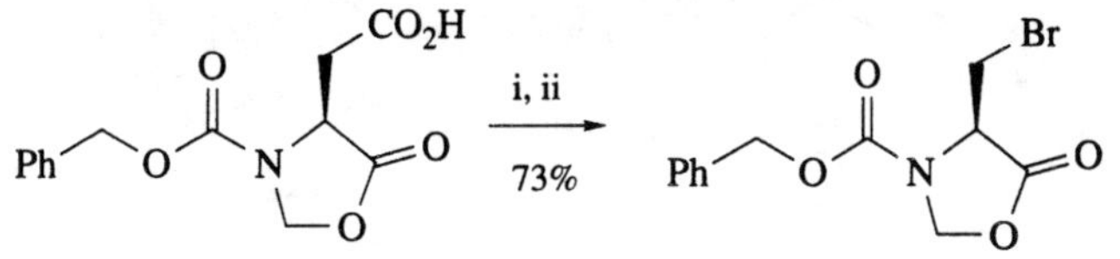

(28)

i, $ClCO_2CH_2CHMe_2$; ii, (**1**), $BrCCl_3$, *h*ν

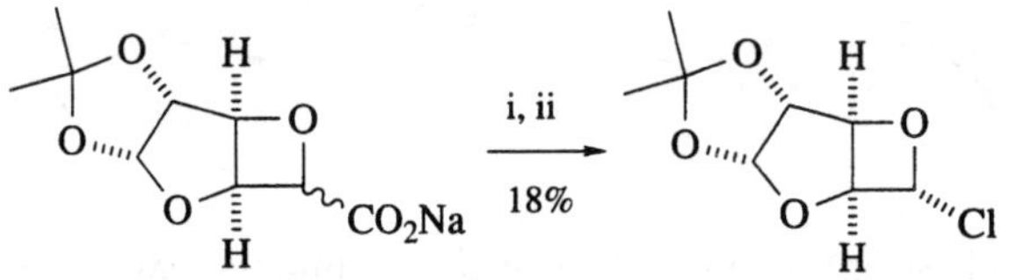

(29)

i, $(COCl)_2$, ii, (**1**), CCl_4, Δ

A variety of aromatic and vinylcarboxylic acids have also been decarboxylated by an adaptation[33] of this method involving the use of AIBN as chain initiator (equations 30 and 31). Unlike the classical Hunsdiecker reaction this variant is applicable to both electron poor and electron rich aryl acids without the risk of electrophilic aromatic halogenation.

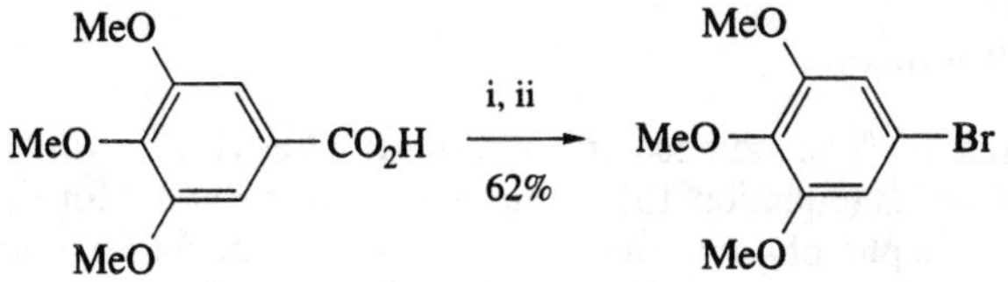

(30)

i, $(COCl)_2$; ii, (**1**), AIBN, $BrCCl_3$, Δ

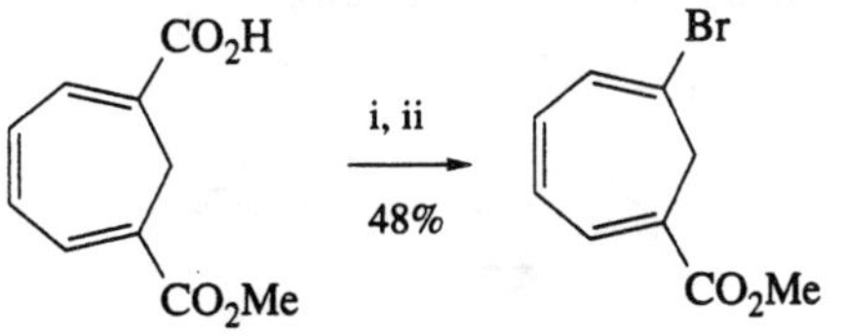

(31)

i, DCC, (**1**); ii, $BrCCl_3$, Δ

5.4.6 DECARBOXYLATIVE CHALCOGENATION AND PHOSPHORYLATION

With the exception of an isolated report[34] on the decomposition of diacyl peroxides in acetonitrile in the presence of copper(II) isothiocyanate and potassium thiocyanate (equation 32), the only preparative methods available for decarboxylative chalcogenation and phosphorylation make use of the *O*-acyl thiohydroxamates.

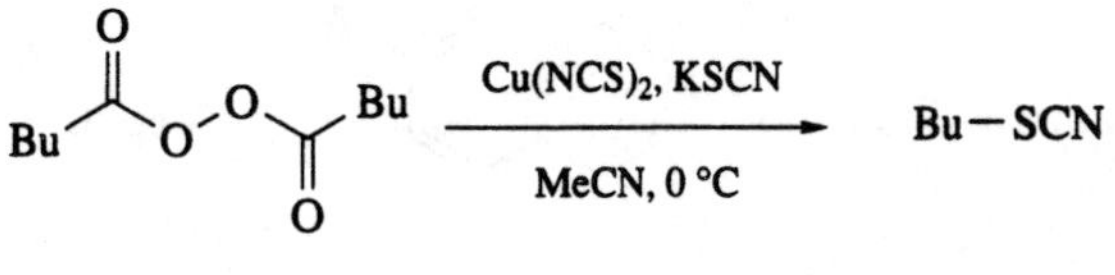

$$\text{BuC(O)OOC(O)Bu} \xrightarrow[\text{MeCN, 0 °C}]{\text{Cu(NCS)}_2\text{, KSCN}} \text{Bu—SCN} \quad (32)$$

5.4.6.1 Sulfuration

Alkyl 2-pyridyl sulfides are formed on simple photolytic or thermal decomposition of *O*-acyl thiohydroxamates in the absence of other radical-trapping agents (equation 33).[7] Other mixed alkyl or alkyl aryl sulfides can be prepared[35] in good yield by irradiation of *O*-acyl thiohydroxamates in the presence of the appropriate dialkyl or diaryl disulfide at low temperature (equation 34).

i, ii; 72% (33)

i, $(COCl)_2$; ii, (**1**), 110 °C

PhSSPh, CH_2Cl_2, $h\nu$, –78 °C; 74% (34)

Thiosulfonates can be synthesized[36] by photolysis of *O*-acyl thiohydroxamates in a 1:1 mixture of dichloromethane and sulfur dioxide at –10 °C (equation 35). For simple primary, secondary and tertiary carboxylic acids yields vary between 30% and 90%.

CH_2Cl_2, SO_2, $h\nu$, –10 °C; 30–90% (35)

5.4.6.2 Selenation and Telluration

O-Acyl selenohydroxamates[6,20] decompose analogously to *O*-acyl thiohydroxamates (Section 5.4.6.1) to give alkyl 2-pyridyl selenides (equation 15). A more general method[35] for decarboxylative selenation or telluration makes use of the photolytic or thermal reaction of *O*-acyl thiohydroxamates with dialkyl or diaryl diselenides or ditellurides (equations 36 and 37). It is reported that under photolytic conditions at low temperature only a slight excess of diselenide or ditelluride is required in order to obtain high yields of mixed diselenides and ditellurides, respectively. A somewhat related method,[37] using dicyanogen triselenide as the radical trap, leads to alkyl selenocyanates.

MeSeSeMe, $h\nu$, –78 °C; 78% (36)

i, ii; 65% (37)

i, $(COCl)_2$; ii, (**1**), $(2\text{-naphthyl-Te})_2$, 50–60 °C

5.4.6.3 Phosphorylation

The reaction of *O*-acyl thiohydroxamates with tris(phenylthio)phosphorus, initiated by adventitious oxygen at room temperature, leads in the first instance to alkylbis(phenylthio)phosphines (Scheme 2) by a chain mechanism. Combination of the latter with the disulfide by-product affords a phosphorus(V) species which on work-up gives alkylbis(phenylthio)phosphonates (Scheme 2 and equation 38) with moderate to good yields.[38] This reaction sequence provides a convenient method for the overall transformation of a carboxylic acid into a readily hydrolyzable ester of the analogous phosphonic acid.

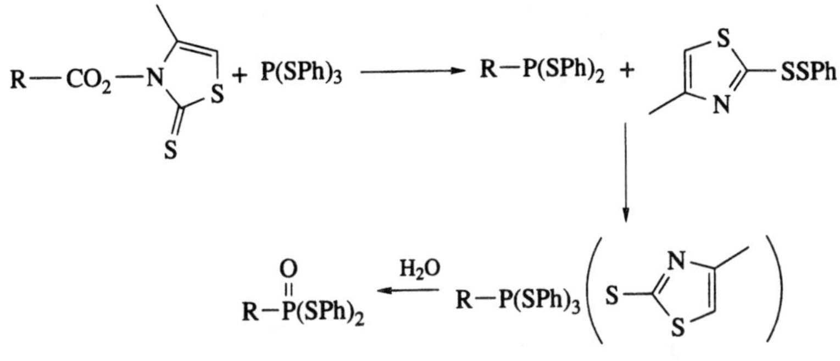

R—CO_2—N + $P(SPh)_3$ ⟶ R—$P(SPh)_2$ + SSPh

R—$P(SPh)_2$ ⟵(H_2O) R—$P(SPh)_3$ (S—)

Scheme 2

AcO H O O—N S S i, ii 60% O= $P(SPh)_2$ 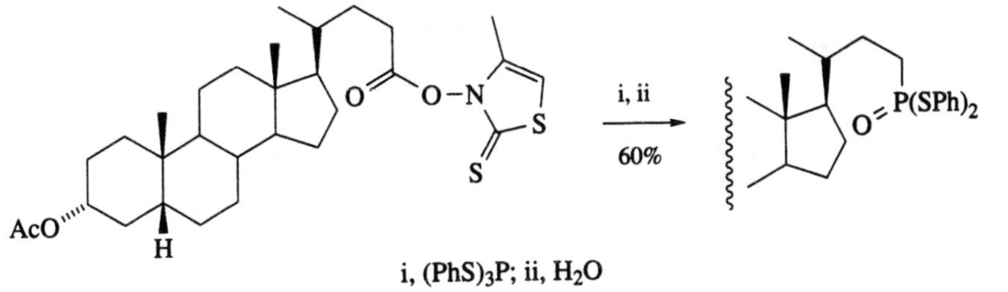 (38)

i, $(PhS)_3P$; ii, H_2O

5.4.7 DECARBOXYLATIVE OXYGENATION

Acetate esters are common by-products of LTA decarboxylation procedures.[4] The yield of these products, derived from further oxidation of the alkyl radical and quenching of the subsequent carbocation by acetate ions, can be improved by working in acetic acid in the presence of potassium acetate. Selective monodecarboxylation of 1,3- and 1,4-dicarboxylic acids leads, *via* an analogous mechanism, to γ- and δ-lactones in moderate to good yields, as illustrated in equation (39).

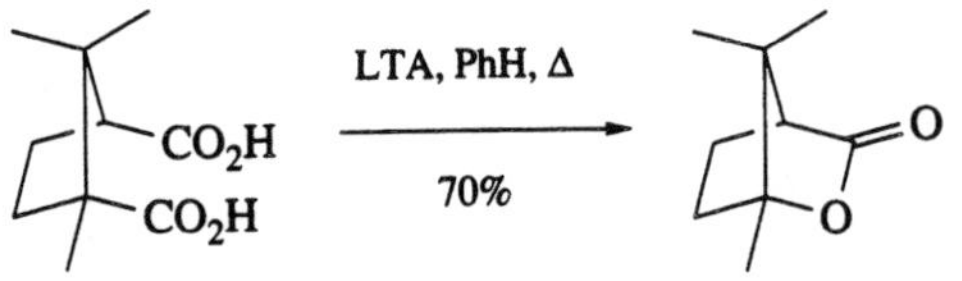

CO_2H CO_2H —(LTA, PhH, Δ; 70%)→ O O (39)

Simple alcohols can be obtained from the decomposition of peroxy acids in cyclohexane or benzene at reflux. This chain reaction, which is efficient for adamantane-1-carboxylic acid (equation 40), is unfortunately usually complicated by side reactions involving hydrogen abstraction from the substrate or solvent.[39]

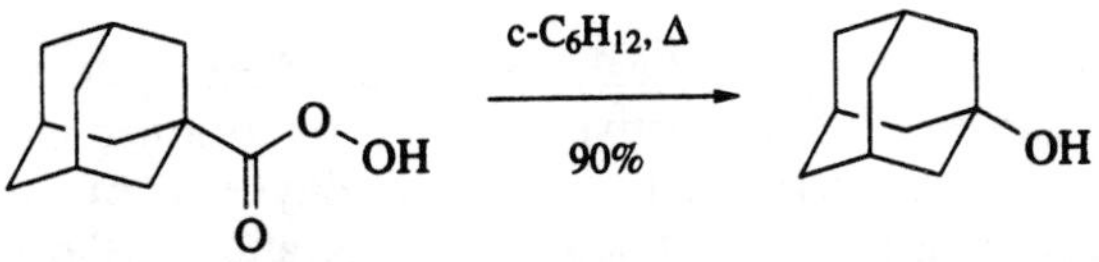

O OH O —(c-C_6H_{12}, Δ; 90%)→ OH (40)

A more efficient and general procedure once again involves the *O*-acyl thiohydroxamates. Decomposition in the presence of triplet oxygen, and *t*-butyl thiol as hydrogen donor, provides noralkyl hydroperoxides. In the original procedure,[7] hydroperoxides obtained in this manner were not normally isolated but immediately reduced *in situ* with trimethyl phosphite to the corresponding alcohols (equation 41). An alternative work-up (equation 42)[7] involves *in situ* treatment of the hydroperoxide with *p*-toluenesulfonyl chloride and pyridine resulting in the isolation of carbonyl compounds. One problem associated with this system is the difficulty encountered in maintaining a suitable concentration of *t*-butyl thiol in the solution during the passage of oxygen. A solution involves the replacement of *t*-butyl thiol with the less volatile triethylmethyl thiol. In this manner,[40] it is possible to prepare and isolate hydroperoxycyclooct-4-ene from cyclooctene-5-carboxylic acid (equation 43).

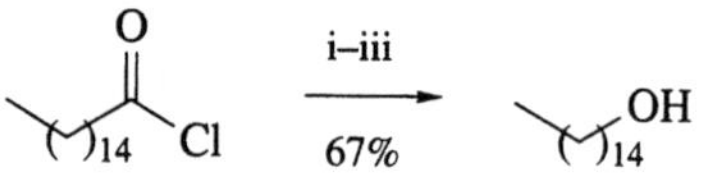

(41)

i, (**1**); ii, O_2, Bu^tSH, $h\nu$; iii, $(MeO)_3P$

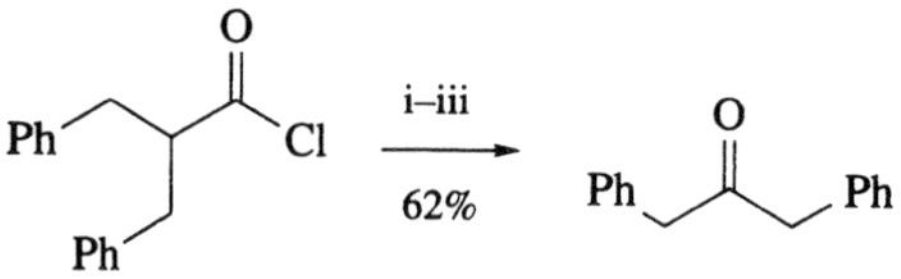

(42)

i, (**1**); ii, O_2, Bu^tSH, $h\nu$; iii, TsCl, C_5H_5N

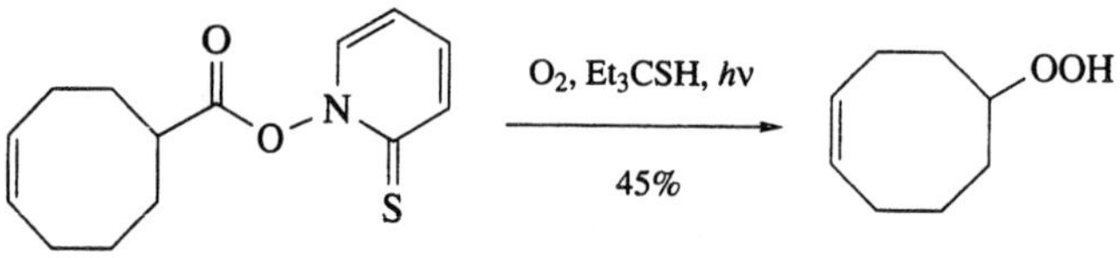

(43)

In an alternative sequence,[41] *O*-acyl thiohydroxamates are reacted with tris(phenylthio)antimony to give the corresponding alkylbis(phenylthio)antimony compound. On admission of air this latter species undergoes oxygen insertion and rearrangement. Finally, hydrolysis provides the alcohol (Scheme 3 and equation 44).

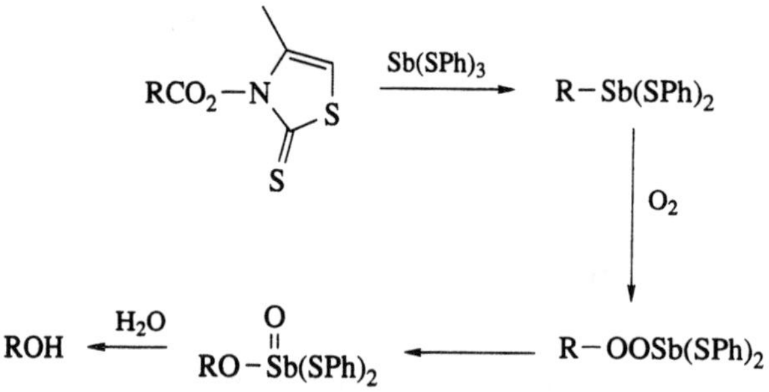

Scheme 3

Finally, a useful, although not strictly a radical, method of effecting decarboxylative oxygenation is the so-called carboxy inversion reaction. The activated acid is transformed into a mixed alkyl aryl diacyl peroxide which suffers decarboxylative rearrangement to the alkyl ester of the aryl acid. This reaction is particularly useful as it takes place with retention of configuration at the migrating center (equation 45).[42]

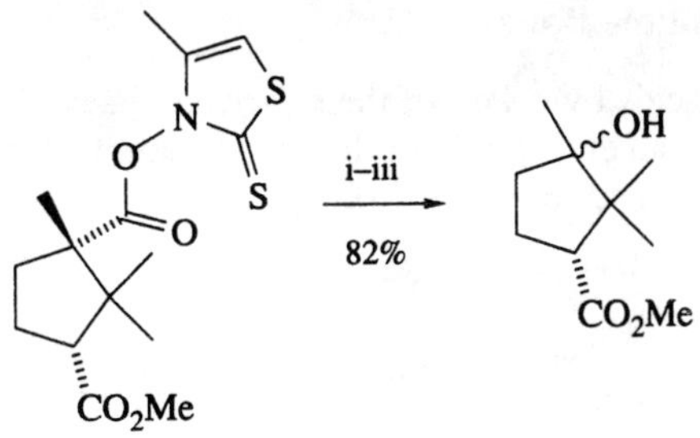

(44)

i, $Sb(SPh)_3$; ii, O_2; iii, H_2O

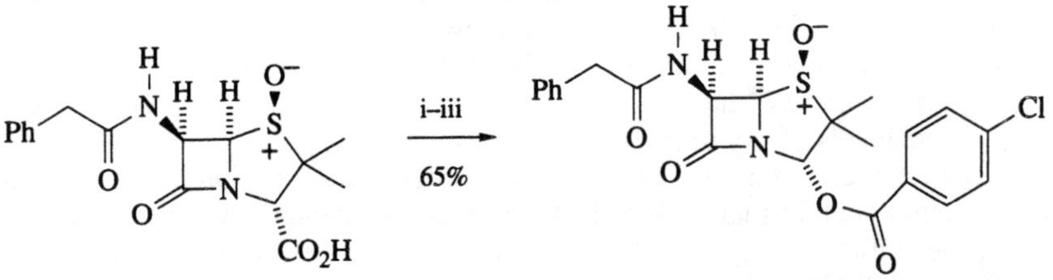

(45)

i, EtOCOCl; ii, MCPBA, –10 °C; iii, –10 °C to 0 °C

5.4.8 DECARBOXYLATIVE AMINATION

The overall transformation of a carboxylic acid into an amide or carbamate or similar nitrogen-containing function is best achieved by one or other of the Hofmann, Curtius, Schmidt, Lössen and related reactions.

Efficient preparative sequences involving radical decarboxylation followed by carbon–nitrogen bond formation are rare. Acyl nitrates decompose at elevated temperatures to give nitroalkanes (equation 46),[43] but are unfortunately explosive and have to be prepared *in situ* and stored in solution. A noteworthy exception[44] is found in the thermal or photochemical decarboxylation of tetrahydro-1,2-oxazine-3,6-diones leading to β-lactams (equation 47). Doubtless a key factor in this reaction, considered to proceed *via* a radical cage mechanism, is the intramolecular nature of the carbon–nitrogen bond formation.

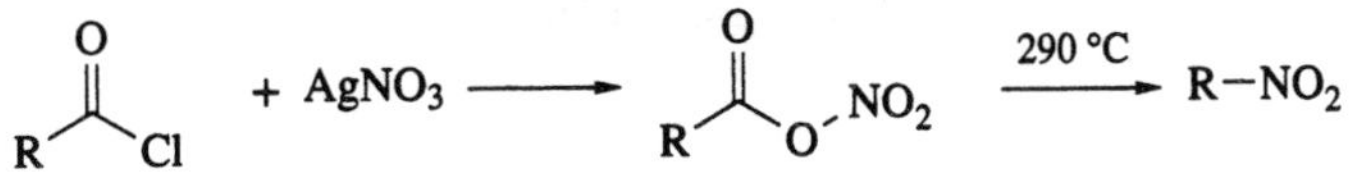

$$RCOCl + AgNO_3 \longrightarrow RCO\text{-}O\text{-}NO_2 \xrightarrow{290\,^\circ C} R\text{-}NO_2 \quad (46)$$

$$\xrightarrow{\Delta \text{ or } h\nu} + CO_2 \quad (47)$$

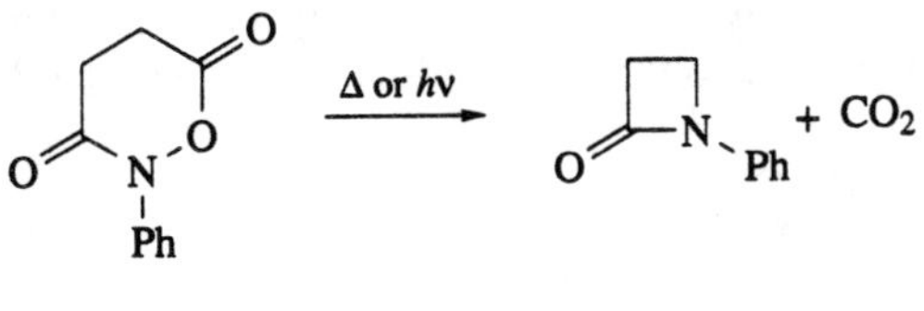

5.4.9 DECARBOXYLATION WITH SUBSEQUENT C—C BOND FORMATION

Alkyl radicals derived by decarboxylation of carboxyl radicals may be added to carbon–carbon multiple bonds resulting in an overall homologation of the starting acid. This reaction type is not strictly a C—C bond oxidation; nevertheless, one of the key steps is C—C bond cleavage by decarboxylation and it is appropriate to briefly consider the scope of such reactions here. A more complete description of inter- and intra-molecular radical C—C bond-forming reactions is given in Volume 4, Chapters 4.1 and 4.2.

5.4.9.1 Addition to C—C Multiple Bonds

Decarboxylation of aliphatic acids by means of their derived *O*-acyl thiohydroxamates in the presence of an electron deficient terminal alkene results in the overall addition of an alkyl radical and a 2-pyridylthiyl radical across the double bond (equation 48).

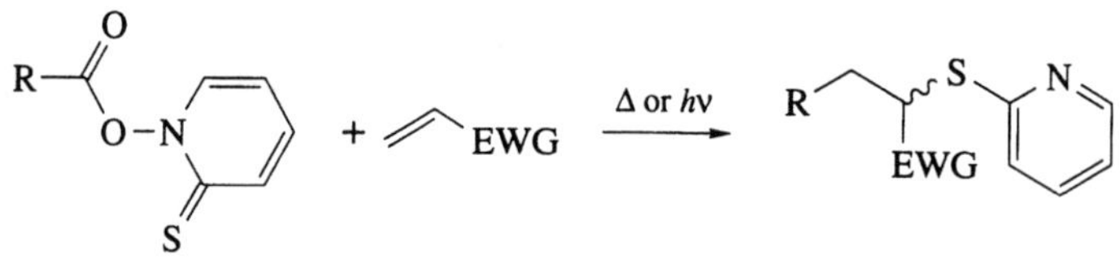

(48)

The alkyl radical, R·, may be any primary, secondary or tertiary radical compatible with the formation of the *O*-acyl thiohydroxamate. The alkene is activated towards radical addition by any strongly electron-withdrawing group, commonly esters, ketones, nitriles and nitro[45,46] groups but also sulfones and phosphonium[47] groups. A recent report[48] has shown that the use of chiral electron-withdrawing groups in this reaction gives moderate levels of asymmetric induction at the newly formed chiral center. With the exception of 1-nitro-1-propene,[46] examples of addition to singly activated internal alkenes are very rare. Doubly activated alkenes such as maleic anhydride undergo the standard addition reaction but suffer *in situ* elimination of the sulfide moiety, thereby providing[45] an excellent route to diversely alkylated maleic anhydrides (Scheme 4). Under thermal conditions, 1,4-benzoquinone behaves in a similar manner to give moderate yields of 2-alkyl-1,4-benzoquinone;[45] however, if the reaction is carried out photochemically at low temperature the main product is the 2-alkyl-3-(2-pyridylthio)-1,4-benzoquinone.[49] Evidently, the primary addition product evolves along different pathways depending on the reaction conditions. Radical alkylation of 1,4-quinones[50] may also be achieved in moderate yield by LTA decarboxylation of alkanoic acids in the presence of the quinone.

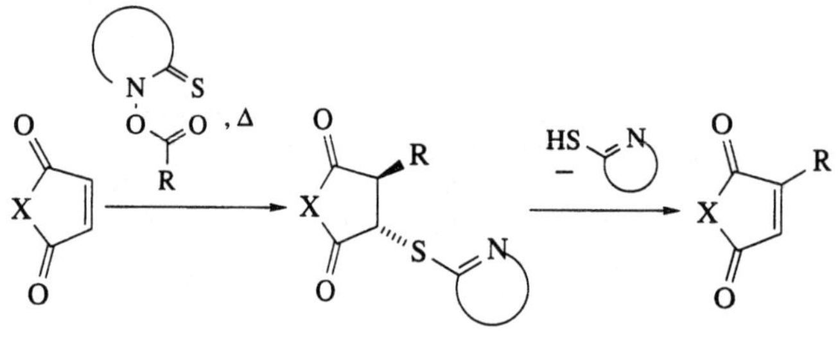

X = O, CMe_2, CH=CH

Scheme 4

Electron deficient terminal alkenes substituted with an alkyl- or aryl-thiomethylene group at the 2-position react efficiently with *O*-acyl thiohydroxamates by a distal addition/elimination sequence,[46,51] as illustrated in equation (49). Alkene polymerization is not a competing side reaction in this process.

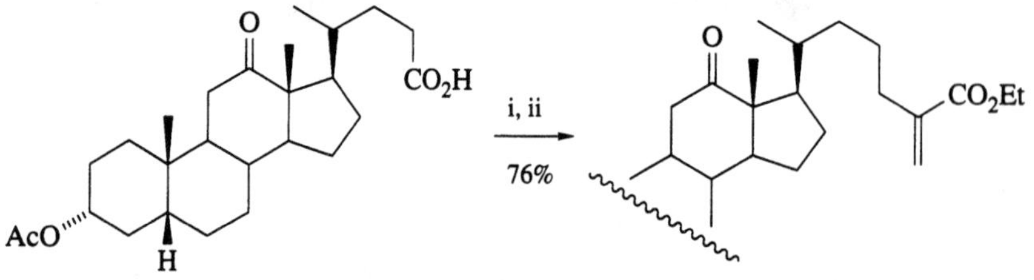

(49)

i, $(COCl)_2$; ii, (**1**), $CH_2=C(CH_2SBu^t)CO_2Et$, 130 °C

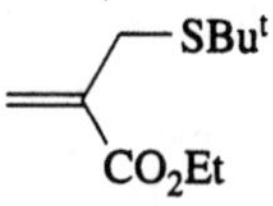

The decarboxylation of perfluoroalkanoic acids may also be achieved *via* the *O*-acyl thiohydroxamates. When carried out in the presence of electron rich alkenes such as ethyl vinyl ether the perfluoroalkyl radical adds to the terminal position of the double bond in moderate to good yields (equation 50).[52] This method provides an attractive alternative to the addition of perfluoroalkyl iodides to alkenes.

$$CF_3(CF_2)_2CO_2{-}N\text{(pyridine-2-thione)} + CH_2{=}CH{-}OEt \xrightarrow[58\%]{h\nu} CF_3(CF_2)_2CH_2CH(OEt){-}S{-}\text{(2-pyridyl)} \quad (50)$$

5-Hexenyl radicals cyclize to cyclopentylmethyl radicals (see Volume 4, Chapter 4.2). Thus radical decarboxylation of 6-heptenoic acids, by whatever means, usually results in the formation of five-membered rings. Although this fact had been appreciated previously[1] it is only recently,[45] with the advent of the *O*-acyl thiohydroxamates, that it has been exploited from a synthetic point of view. An example is provided by the synthesis[53] of bicyclo[4.3.0]proline derivatives from aspartic acid carried out by the Barton group (equation 51). It will be noted that activation of the C—C double bond acting as a radical trap is not necessary in these intramolecular reactions.

(51) [Scheme: N-aroyl-N-(cyclohex-2-enyl) aspartic acid CO_2Bu^t ester → (i–iii, 69%) two diastereomeric 2-pyridylthio-substituted octahydroindole-2-carboxylates (CO_2Bu^t, N-COAr)]

i, Me_2CHCH_2OCOCl; ii, (**1**); iii, *hν*

Radical addition to C—C triple bonds is also possible with the *O*-acyl thiohydroxamate methodology. As with addition to C—C double bonds (*vide supra*), the triple bond must be either terminal and activated with an electron-withdrawing group (equation 52) or doubly activated if internal.[45]

(52) [Scheme: O-acyl thiohydroxamate (thiazolethione, chain $(CH_2)_{14}$) + $HC{\equiv}C{-}CO_2Me$ → (*hν*, 38%) vinyl sulfide bearing CO_2Me and thiazolylthio group]

5.4.9.2 Addition to C–Heteroatom Multiple Bonds

The one-carbon homologation of alkyl radicals by trapping with C–heteroatom multiple bonds is at present extremely rare. Preparative procedures, using *t*-butyl isocyanide and formaldehyde oximes as one-carbon radical traps, have only recently appeared.[54] Application of such procedures to radicals obtained by decarboxylation methods provides a means of reforming the original acid. This is a potentially important reaction sequence as the use of a ^{13}C- or ^{14}C-labeled trap will afford the isotopically labeled acid. The methodology for such a sequence has recently been published by the Barton group.[55] Thus photolysis of *O*-acyl thiohydroxamates in the presence of an isocyanide in which the electron density of carbon is reduced by an electron-withdrawing substituent, as for example in 4-nitrophenyl isocyanide and protonated 3-pyridyl isocyanide, leads after aqueous work-up to the amide of the original acid. Hydrolysis to the acid can then be achieved by various methods. The application of this sequence to arachidonic acid (equation 53) provides a further illustration of the mildness and applicability of the *O*-acyl thiohydroxamate chemistry.

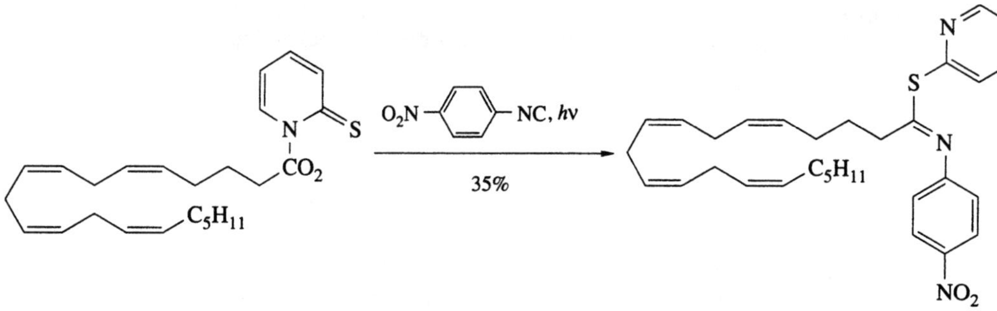

(53)

5.4.9.3 Addition to Aromatic Systems

Decarboxylation of alkanoic acids by means of LTA in benzene as solvent is hindered by the formation of alkylbenzenes as by-products. This side reaction is especially pronounced with radicals derived from primary acids or other acids from which the radical is not easily oxidized by LTA. In some cases, such as that of apocamphane-1-carboxylic acid (equation 54),[56] good yields of alkylbenzene can be obtained. Intramolecular versions of this reaction in which the radical cyclizes onto an aromatic nucleus at the appropriate position in the chain have also been observed.

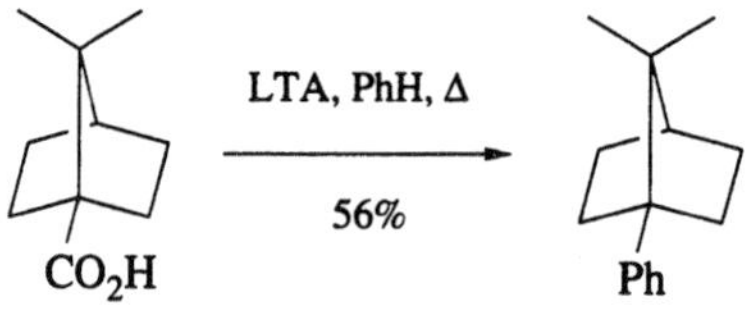

(54)

In an analogous manner, the generation of alkyl radicals in benzene solution by the *O*-acyl oxime method results in the formation of alkylbenzenes with moderate to good yields for simple acids (equation 55).[57] Use of pyridine as solvent leads to the formation of alkylpyridines as mixtures of *ortho, meta* and *para* isomers in which the *para* isomer predominates. The *O*-acyl benzophenone oxime chemistry can also be applied to aryl acids in benzene or pyridine, resulting in the formation of mixed biaryls.[5] A closely related method[58] involves photolysis of mixed anhydrides of arenecarboxylic acids with the hydroxamic acid *N*-hydroxy-2-pyridone in benzene solution (equation 56).

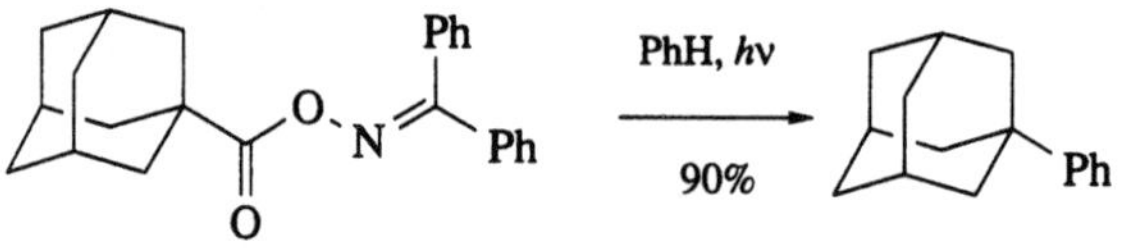

(55)

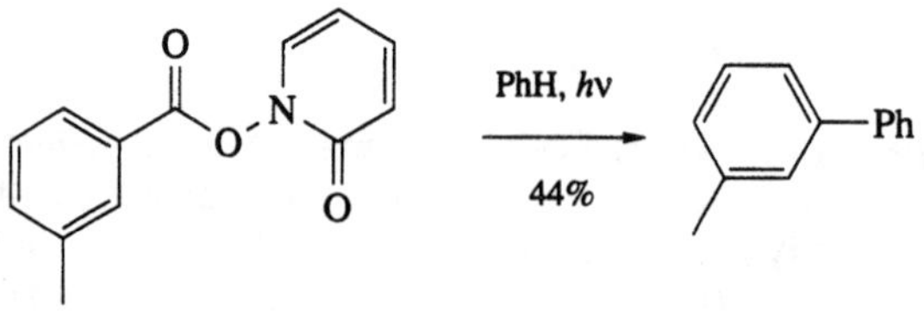

(56)

The simple photolytic or thermal decomposition of *O*-acyl thiohydroxamates in benzene or pyridine as solvent yields the product of decarboxylative rearrangement, and not alkylbenzenes or alkylpyridines. However, photolysis in dichloromethane in the presence of *protonated* heteroaromatic bases results in the formation of alkylated heterocycles in good yield, as illustrated in equation (57).[59] The great advantage of this latter method lies in the fact that the base to be alkylated is not used as the reaction solvent, which evidently permits the use of a much wider range of bases as trapping agents.

CSA, CH_2Cl_2, $h\nu$

66%

(57)

5.4.10 REFERENCES

1. R. G. Johnson and R. K. Ingham, *Chem. Rev.*, 1956, **56**, 219; C. V. Wilson, *Org. React. (N.Y.)*, 1957, **9**, 332.
2. S. J. Cristol and W. C. Firth, *J. Org. Chem.*, 1961, **26**, 280.
3. A. McKillop, D. Bromley and E. C. Taylor, *J. Org. Chem.*, 1969, **34**, 1172.
4. R. A. Sheldon and J. K. Kochi, *Org. React. (N.Y.)*, 1972, **19**, 279; R. N. Butler, *Synth. Reagents*, 1977, **3**, 277.
5. M. Hasebe, K. Kogawa and T. Tsuchiya, *Tetrahedron Lett.*, 1984, **25**, 3887.
6. D. H. R. Barton and S. Z. Zard, *Pure Appl. Chem.*, 1986, **58**, 675; D. Crich, *Aldrichimica Acta*, 1987, **20**, 35.
7. D. H. R. Barton, D. Crich and W. B. Motherwell, *Tetrahedron*, 1985, **41**, 3901.
8. D. M. Burness, *Org. Synth., Coll. Vol.*, 1963, **4**, 628; R. H. Wiley and N. R. Smith, *Org. Synth., Coll. Vol.*, 1963, **4**, 731; R. E. Buckles and N. G. Wheeler, *Org. Synth., Coll. Vol.*, 1963 **4**, 857; T. Cohen and R. A. Schambach, *J. Am. Chem. Soc.*, 1970, **92**, 3189.
9. C. Rüchardt, *Fortschr. Chem. Forsch.*, 1966, **6**, 251.
10. M. A. Beilefeld and R. Oslapas, *J. Med. Chem.*, 1969, **12**, 192.
11. H. Langhals and C. Rüchardt, *Chem. Ber.*, 1975, **108**, 2156.
12. J. K. Kochi and J. D. Bacha, *J. Org. Chem.*, 1968, **33**, 2746.
13. W. E. Fristad, M. A. Fry and J. A. Klang, *J. Org. Chem.*, 1983, **48**, 3575.
14. M. Hasebe and T. Tsuchiya, *Tetrahedron Lett.*, 1987, **28**, 6207.
15. D. Crich and T. J. Ritchie, *J. Chem. Soc., Chem. Commun.*, 1988, 1461.
16. N. C. Billingham, R. A. Jackson and F. Malek, *J. Chem. Soc., Perkin Trans. 1*, 1979, 1137.
17. J. Pfenninger, G. Heuberger and W. Graf, *Helv. Chim. Acta*, 1980, **63**, 2328.
18. S. Hanessian and S. P. Sahoo, *Tetrahedron Lett.*, 1984, **25**, 1425.
19. J. I. Concepcion, C. G. Francisco, R. Freire, R. Hernandez, J. A. Salazar and E. Suárez, *J. Org. Chem.*, 1986, **51**, 402.
20. D. H. R. Barton, D. Crich, Y. Hervé, P. Potier and J. Thierry, *Tetrahedron*, 1985, **41**, 4347.
21. H. H. Wesberg and H. J. Dauben, *Tetrahedron Lett.*, 1968, 5123; E. N. Cain, R. Vukov and S. Masamune, *J. Chem. Soc. D*, 1969, 98.
22. B. Lacher, Ph.D. Thesis, University of Paris XI, 1987.
23. F. A. H. Rice, *J. Am. Chem. Soc.*, 1956, **78**, 3173; F. A. H. Rice and W. Morganroth, *J. Org. Chem.*, 1956, **21**, 1388; but see J. A. Davies, J. Herynk, S. Carroll, J. Bunda and D. Johnson, *J. Org. Chem.*, 1965, **30**, 415.
24. D. H. R. Barton, H. P. Faro, E. P. Serebryakov and N. F. Woolsey, *J. Chem. Soc.*, 1965, 2438.
25. J. Béres, G. Sági, I. Tömösközi, L. Gruber, E. Gulácsi and L. Otvös, *Tetrahedron Lett.*, 1988, **29**, 2681.
26. J. Grakauskas, *J. Org. Chem.*, 1969, **34**, 2446.
27. T. B. Patrick, K. K. Johri and D. H. White, *J. Org. Chem.*, 1983, **48**, 4158.
28. F. W. Baker, H. D. Holtz and L. M. Stock, *J. Org. Chem.*, 1963, **28**, 514; J. A. Davis, J. Herynk, S. Carroll, J. Bunda and D. Johnson, *J. Org. Chem.*, 1965, **30**, 415.
29. S. J. Cristol, L. K. Gaston and T. Tiedeman, *J. Org. Chem.*, 1964, **29**, 1279.
30. K. M. Becker, M. Geisel, C. A. Grob and F. Kuhnen, *Synthesis*, 1973, 493.
31. D. H. R. Barton, Y. Hervé, P. Potier and J. Thierry, *Tetrahedron*, 1988, **44**, 5479.
32. G. W. J. Fleet, J. C. Son, J. M. Peach and T. A. Hamor, *Tetrahedron Lett.*, 1988, **29**, 1449.
33. D. H. R. Barton, B. Lacher and S. Z. Zard, *Tetrahedron*, 1987, **43**, 4321; E. Vogel, T. Schieb, W. H. Schulz, K. Schmidt, H. Schmickler and J. Lex, *Angew. Chem., Int. Ed. Engl.*, 1986, **25**, 723.
34. C. L. Jenkins and J. K. Kochi, *J. Org. Chem.*, 1971, **36**, 3095, 3103.
35. D. H. R. Barton, D. Bridon and S. Z. Zard, *Heterocycles*, 1987, **25**, 449.
36. D. H. R. Barton, B. Lacher, B. Misteriewicz and S. Z. Zard, *Tetrahedron*, 1988, **44**, 1153.
37. D. H. R. Barton, D. Bridon, Y. Hervé, P. Potier, J. Thierry and S. Z. Zard, *Tetrahedron*, 1986, **42**, 4983.
38. D. H. R. Barton, D. Bridon and S. Z. Zard, *Tetrahedron Lett.*, 1986, **27**, 4309.
39. J. Fossey, D. Lefort, M. Massoudi, J.-Y. Nedelec and J. Sorba, *J. Chem. Soc., Perkin Trans. 2*, 1986, 781 and refs. therein.
40. A. J. Bloodworth, D. Crich and T. Melvin, *J. Chem. Soc., Chem. Commun.*, 1987, 786.
41. D. H. R. Barton, D. Bridon and S. Z. Zard, *J. Chem. Soc., Chem. Commun.*, 1985, 1066.
42. D. H. R. Barton, I. H. Coates and P. G. Sammes, *J. Chem. Soc., Perkin Trans. 1*, 1973, 599; F. Kienzle, G. W. Holland, J. L. Jernow, S. Kwoh and P. Rosen, *J. Org. Chem.*, 1973, **38**, 3440 and refs. therein.
43. G. B. Bachman and T. F. Biermann, *J. Org. Chem.*, 1970, **35**, 4229.
44. J. Nally, N. H. R. Ordsmith and G. Proctor, *Tetrahedron Lett.*, 1985, **26**, 4107.
45. D. H. R. Barton, D. Crich and G. Kretzschmar, *J. Chem. Soc., Perkin Trans. 1*, 1986, 39.
46. D. H. R. Barton, H. Togo and S. Z. Zard, *Tetrahedron*, 1985, **41**, 5507.
47. D. H. R. Barton, H. Togo and S. Z. Zard, *Tetrahedron Lett.*, 1985, **26**, 6349.
48. D. Crich and J. W. Davies, *Tetrahedron Lett.*, 1987, **28**, 4205.

49. D. H. R. Barton, D. Bridon and S. Z. Zard, *Tetrahedron*, 1988, **43**, 5307.
50. L. F. Fieser and F. C. Chang, *J. Am. Chem. Soc.*, 1942, **64**, 2043.
51. D. H. R. Barton and D. Crich, *J. Chem. Soc., Perkin Trans. 1*, 1986, 1613.
52. D. H. R. Barton, B. Lacher and S. Z. Zard, *Tetrahedron*, 1986, **42**, 2325.
53. D. H. R. Barton, J. Guilhem, Y. Hervé, P. Potier and J. Thierry, *Tetrahedron Lett.*, 1987, **28**, 1413.
54. D. J. Hart and F. L. Seeley, *J. Am. Chem. Soc.*, 1988, **110**, 1631; G. Stork and P. M. Sher, *J. Am. Chem. Soc.*, 1983, **105**, 6765.
55. D. H. R. Barton, N. Ozbalik and B. Vacher, *Tetrahedron*, 1988, **44**, 3501.
56. D. I. Davies and C. Waring, *Chem. Commun.*, 1965, 263.
57. M. Hasebe and T. Tsuchiya, *Tetrahedron Lett.*, 1986, **27**, 3239.
58. E. C. Taylor, H. W. Altland, F. Kienzle and A. McKillop, *J. Org. Chem.*, 1976, **41**, 24.
59. D. H. R. Barton, B. Garcia, H. Togo and S. Z. Zard, *Tetrahedron Lett.*, 1986, **27**, 1327; E. Castognino, S. Corsano, D. H. R. Barton and S. Z. Zard, *Tetrahedron Lett.*, 1986, **27**, 6337.

6.1
Oxidation of Nitrogen and Phosphorus

THOMAS L. GILCHRIST
University of Liverpool, UK

6.1.1 INTRODUCTION

The aim of this chapter is to describe the main types of functional group interconversion which involve oxidation at nitrogen or phosphorus. 'Oxidation' has been interpreted as including nitrosation, amination, halogenation and other reactions in which attack by an electrophilic heteroatom takes place. The material has been organized on the basis of the nature of the starting materials and the products, rather than on the mechanism of oxidation or the type of oxidant. For example, no attempt has been made to cover comprehensively the chemistry of aminium cation radicals[1] or of aminyl radicals[2] which can occur as intermediates in the oxidation of amines (Scheme 1). Also, oxidation which results in reaction at carbon, such as radical coupling through carbon or reaction at a C—H bond adjacent to nitrogen, is not covered in any detail here, even if the initial oxidation occurs at nitrogen. Reactions of C—H bonds activated by nitrogen are described in Chapter 2.5 of this volume. The biological oxidation of amines and other organic nitrogen compounds has also been reviewed elsewhere.[3]

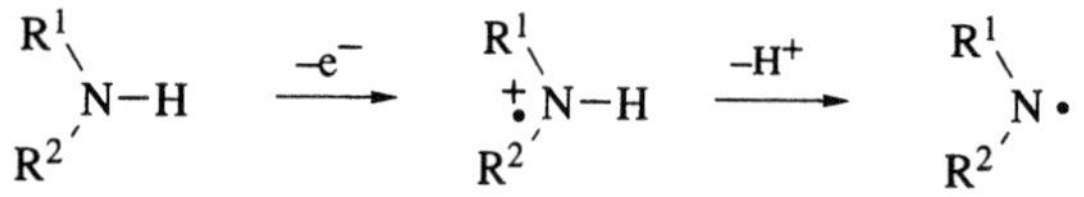

$$R^1R^2N{-}H \xrightarrow{-e^-} R^1R^2\overset{+\bullet}{N}{-}H \xrightarrow{-H^+} R^1R^2N\bullet$$

Scheme 1

Of the many reviews on oxidation which are available, two in particular provide detailed coverage of the literature up to the early 1980s on aspects of oxidation at nitrogen. One of these, by Boyer, is a comprehensive survey of oxidation reactions of nitrogen compounds in which the number of oxygen atoms attached to nitrogen is increased.[4] The other, by Rosenblatt and Burrows, deals with oxidation of amines.[5] References to the primary literature have not always been included here if they are available from these two reviews.

6.1.2 OXIDATION OF THE NH_2 GROUP

6.1.2.1 Primary Amines

6.1.2.1.1 Oxidation to hydroxylamines, nitroso compounds and nitro compounds

Aromatic and aliphatic primary amines can be oxidized to the corresponding nitro compounds by peroxy acids and by a number of other reagents.[5,6] The peroxy acid oxidations probably go by way of intermediate hydroxylamines and nitroso compounds (Scheme 2). Various side reactions can therefore take place, the nature of which depends upon the structure of the starting amine and the reaction conditions. For example, aromatic amines can give azoxy compounds by reaction of nitroso compounds with hydroxylamine intermediates; aliphatic amines can give nitroso dimers or oximes formed by acid-catalyzed rearrangement of the intermediate nitrosoalkanes (Scheme 3).

$$R{-}NH_2 \longrightarrow R{-}N(H)OH \longrightarrow R{-}NO \longrightarrow R{-}NO_2$$

Scheme 2

$$R^1R^2CH{-}NH_2 \longrightarrow R^1R^2CH{-}NO \longrightarrow [R^1R^2CH{-}\overset{+}{N}(O^-){=}\overset{+}{N}(O^-){-}CHR^1R^2]$$

$$R^1R^2CH{-}NO \xrightarrow{H^+} R^1R^2C{=}NOH$$

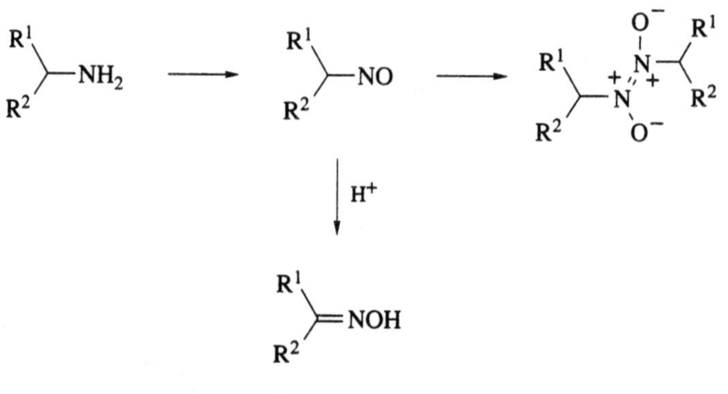

Scheme 3

MCPBA has been regarded as the reagent of choice for the conversion of primary aliphatic amines into the corresponding nitro compounds.[5] The peroxy acid must be used in excess to minimize formation of dimers of the intermediate nitroso compounds. The yield of nitroalkane is also increased if the reaction is carried out at elevated temperature, since this favors the monomeric rather than the dimeric form of the intermediate nitrosoalkane and allows it to be oxidized further.[7] For example, cyclohexylamine gave the dimer of nitrosocyclohexane (43%) when oxidized by MCPBA at 23 °C, but at 83 °C (in boiling 1,2-dichloroethane) the only product was nitrocyclohexane (86%).

Ozone is an alternative oxidant for aliphatic primary amines.[8] The yields are generally not as good as with MCPBA, although a technique of 'dry ozonation' on silica at low temperature has been described which results in good conversions of cyclohexylamine and other aliphatic primary amines into the corresponding nitro compounds.[9] Solid sodium permanganate has been used to prepare 2-methyl-2-nitropropane from *t*-butylamine in good yield.[10] An oxidant which has given excellent yields of both aromatic and aliphatic nitro compounds from the corresponding amines is dimethyldioxirane (**1**).[11] The reagent is readily prepared *in situ* from acetone and a commercial oxidant, Oxone. When used under phase transfer conditions it is a mild, nonacidic oxidant which allows the selective oxidation of aromatic amines in the presence of indoles and furans. This reagent has also been used for the selective oxidation of primary amines to hydroxylamines.[12]

O–O

(1)

Aromatic primary amines have also been converted into the corresponding nitro compounds with peroxy acids: trifluoroperacetic acid, peroxymaleic acid and peracetic acid have all been used. A good alternative reagent for aromatic amines bearing electron-withdrawing substituents is sodium perborate in acetic acid. 4-Nitrobenzonitrile was prepared (91%) from 4-aminobenzonitrile with this reagent.[13] Hindered aromatic amines are oxidized only to the nitroso compounds: MCPBA[14] and perbenzoic acid[15] in stoichiometric amounts give the nitroso compounds in good yield. Thus, 2,6-difluoroaniline is oxidized to 2,6-difluoronitrosobenzene (85%) by perbenzoic acid, and other 2,6-disubstituted anilines react in a similar way. Fremy's salt can also oxidize hindered arylamines to the nitroso compounds.[4] Partial oxidation of *o*-phenylenediamine to 2-nitrosoaniline was achieved by dropwise addition of peracetic acid to the diamine.[4]

Corey and Gross have made use of the peracetic acid oxidation of *t*-butylamine and other tertiary amines to nitrosoalkanes in a synthesis of di-tertiary alkylamines.[16] This is illustrated in Scheme 4 by the synthesis of di-*t*-butylamine. Cyclohexylamine and other alkylamines have been oxidized to the nitroso compounds in high yield by a reagent consisting of sodium percarbonate, sodium hydrogen carbonate and tetraacetylethylenediamine in aqueous dichloromethane.[17] Several nitrosoalkanes have also been obtained from the corresponding amines by oxidation with aqueous hydrogen peroxide and sodium tungstate;[4] various other metal catalysts, including titanium(IV) and vanadium(V) species, have been used in combination with *t*-butyl hydroperoxide.[18] A procedure has been described for the low temperature oxidation of trimethylsilylamines to oximes, by way of the nitroso compounds (Scheme 5): the oxidant is

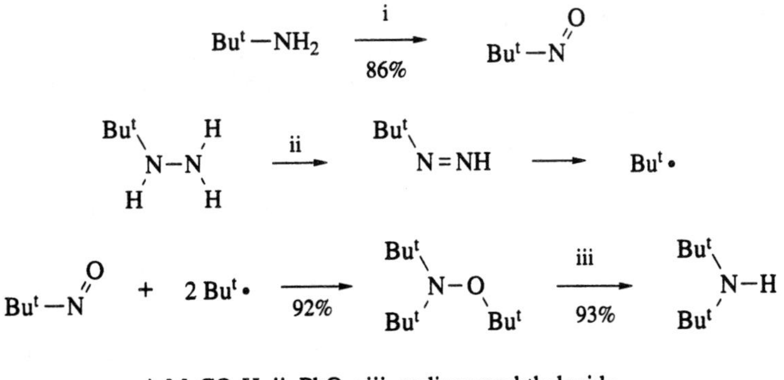

$$Bu^t{-}NH_2 \xrightarrow[86\%]{i} Bu^t{-}N{=}O$$

$$Bu^tNH{-}NH_2 \xrightarrow{ii} Bu^tN{=}NH \longrightarrow Bu^t\bullet$$

$$Bu^t{-}N{=}O + 2\,Bu^t\bullet \xrightarrow{92\%} (Bu^t)_2N{-}OBu^t \xrightarrow[93\%]{iii} (Bu^t)_2N{-}H$$

i, $MeCO_3H$, ii, PbO_2; iii, sodium naphthalenide

Scheme 4

dry air, which reacts with the lithium salts of the amines.[19] This reaction is tolerant of phosphines, sulfides and other functional groups which are susceptible to oxidation.

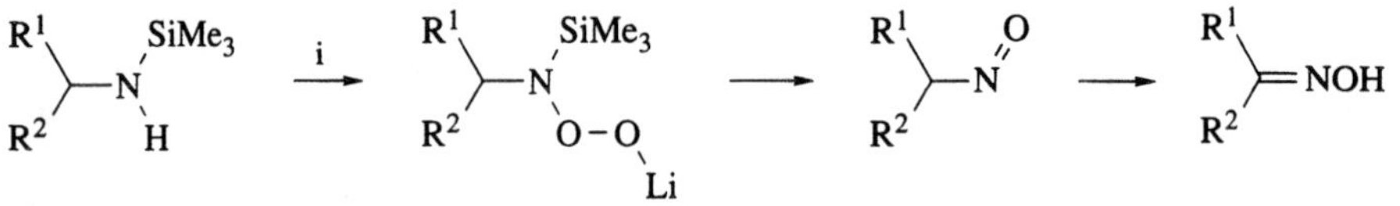

i, BuLi, dry air, –60 to –40 °C

Scheme 5

Hydroxylamines are commonly postulated as intermediates in these oxidations, but they are rarely isolated or detected. Two examples of reactions in which products at the hydroxylamine oxidation level can be isolated both involve peroxides. Oxidation of alkylamines (**2**; R = Bu^t, Me and $PhCH_2$) with arenesulfonyl peroxides bearing an electron-withdrawing group gave the arenesulfonyloxyamines in good yields (Scheme 6).[20] A similar reaction of primary amines with dibenzoyl peroxide gave benzoyloxyamines.[21]

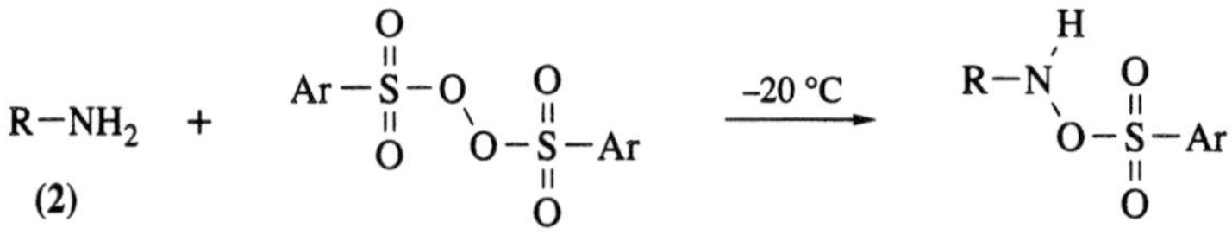

(**2**)

Scheme 6

6.1.2.1.2 Oxidation to azo compounds and related dehydrogenation

Aromatic azo compounds can be obtained by the oxidation of primary arylamines. The reagent most widely used for this purpose is activated manganese dioxide.[22] This converts aniline and substituted anilines into the corresponding azo compounds in moderate to good yield.[23] At room temperature the products are the *cis*-azobenzenes, which are isomerized to the *trans* compounds on heating.[24] It is probable that hydrazobenzenes are intermediates in the reaction, but these are more easily oxidized than the starting anilines (Scheme 7). These oxidations are inhibited by electron-withdrawing substituents and some nitro-substituted anilines fail to react.

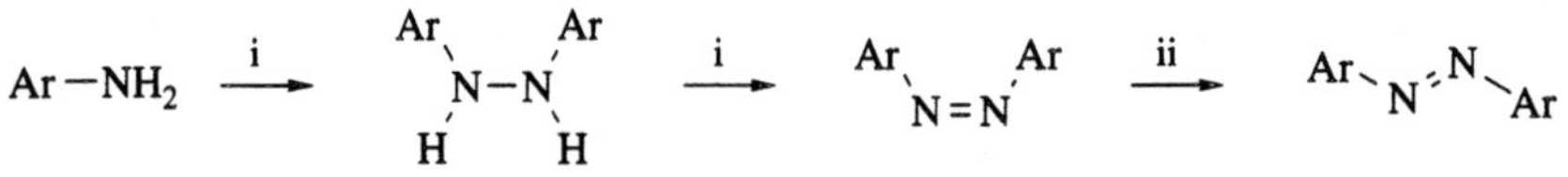

i, MnO_2; ii, heat

Scheme 7

Nickel peroxide also oxidizes anilines to azoarenes.[25] Yields are moderate, although nitroanilines can also be oxidized with this reagent. Other reagents which have been used are silver oxide on Celite,[26] lead(IV) acetate[27] and barium manganate (which is claimed to give azo compounds in higher yield than manganese dioxide).[28] Bispyridinesilver permanganate, $py_2Ag^+MnO_4^-$, which is soluble in polar organic solvents, is also a good alternative to manganese dioxide.[29] Sodium perborate is a convenient oxidant for the conversion of *para*-substituted anilines into azo compounds.[30] Sodium hypochlorite has been used for the oxidation of pentachloroaniline and other chlorinated anilines,[31] while potassium superoxide is capable of selectively oxidizing *ortho*- and *para*-substituted diamines and aminophenols to the corresponding azo compounds in good yield.[32] Thus, 2-aminophenol gave the azo compound (**3**; 70%),

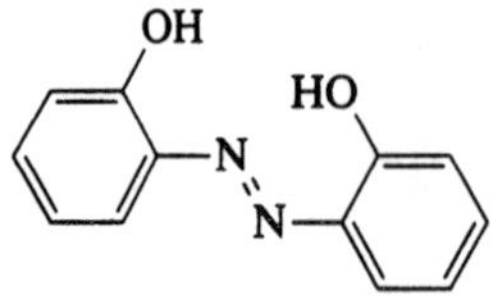

(**3**)

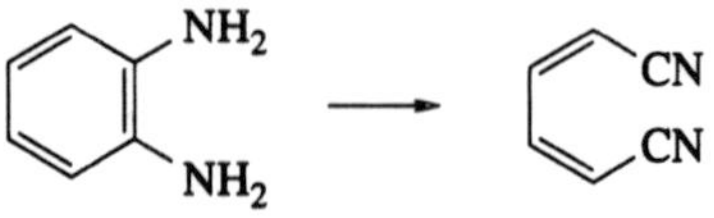

(**4**)

whereas *meta*-substituted anilines failed to react. Oxidation of the anilines to aminyl radicals, followed by their combination to hydrazobenzenes, can account for the observed selectivity.

Manganese dioxide and potassium superoxide both oxidize *o*-phenylenediamine to the azo compound, but the more powerful oxidants nickel peroxide and lead(IV) acetate cause ring cleavage; the product, which can be isolated in low to moderate yield, is the (*Z*,*Z*)-dinitrile (**4**). This oxidative cleavage can be brought about in high yield with oxygen in the presence of copper(I) chloride and pyridine.[33] Some five-membered heteroaromatic amines can be cleaved to nitriles on oxidation: an example is the amine (**5**; Scheme 8) which undergoes ring opening on oxidation by lead(IV) acetate.[34]

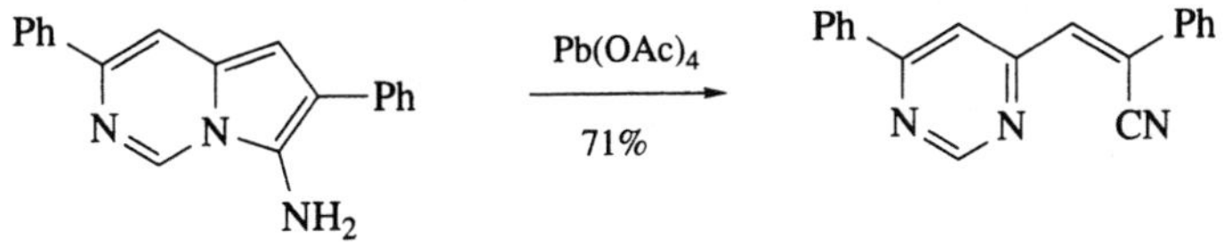

Ph Ph NH$_2$ (**5**) $Pb(OAc)_4$ 71% Ph Ph CN

Scheme 8

Nitriles are also the usual products of oxidation of aliphatic amines RCH_2NH_2 by nickel peroxide and lead(IV) acetate. Aliphatic azo compounds can be prepared from these primary amines by first converting them into sulfamides (**6**), these then being oxidized with sodium hypochlorite or (better) *t*-butyl hypochlorite (Scheme 9).[35] A few aliphatic azo compounds can be formed in good yield by direct oxidation of *t*-alkylamines; for example, AIBN was formed (86%) by oxidation of the amine $Me_2C(CN)NH_2$ with sodium hypochlorite. A special case of azoalkane formation is the synthesis of chlorodiazirines (**7**) from amidines $RC({=}NH)NH_2$ by oxidation with sodium hypochlorite.[36]

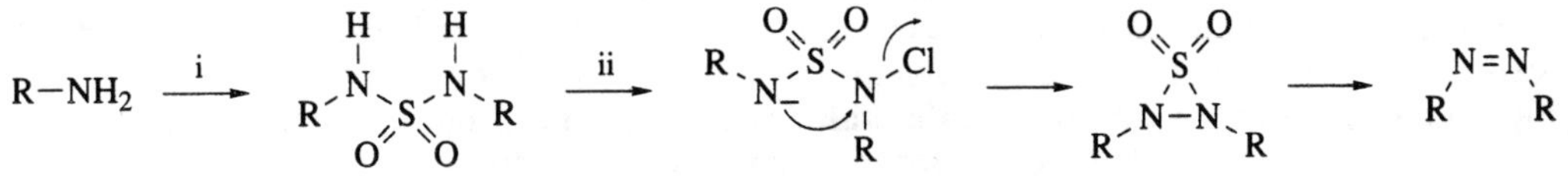

$R{-}NH_2$ →(i) (**6**) →(ii) → → N=N R R

i, SO_2Cl_2, pyridine; ii, Bu^{t}OCl, Bu^{t}OK

Scheme 9

(**7**)

Oxidation of primary aromatic amines bearing a nucleophilic substituent at the *ortho* position can provide a useful route to some heterocyclic compounds. Some examples, shown in Scheme 10, are syntheses of benzofuroxans,[37] benzotriazoles[38] and benzisoxazoles.[39]

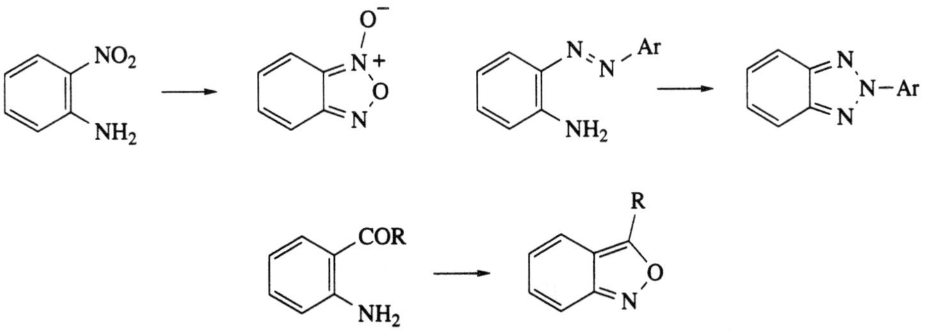

Scheme 10

6.1.2.1.3 Diazotization

The formation of aromatic diazonium salts from aromatic primary amines is one of the oldest synthetic procedures in organic chemistry. Methods based on nitrosation of the amine with nitrous acid in aqueous solution are the best known, but there are variants which are of particular use with weakly basic amines and for the isolation of diazonium salts from nonaqueous media. General reviews include a book by Saunders and Allen[40] and a survey of preparative methods by Schank.[41] There are also reviews on the diazotization of heteroaromatic primary amines[42] and on the diazotization of weakly basic amines in strongly acidic media.[43] The diazotization process (Scheme 11) goes by way of a primary nitrosamine.

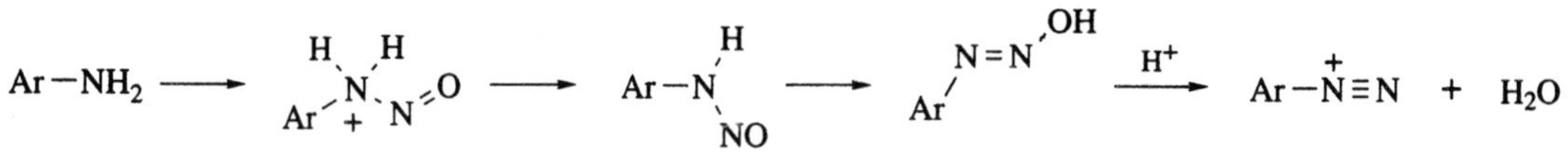

$$Ar-NH_2 \longrightarrow Ar-\overset{+}{N}H_2-N=O \longrightarrow Ar-NH-NO \longrightarrow Ar-N=N-OH \xrightarrow{H^+} Ar-\overset{+}{N}\equiv N + H_2O$$

Scheme 11

If the amine is basic enough to form a salt with dilute mineral acids in aqueous solution, the normal diazotization method of adding sodium nitrite in aqueous solution to a solution or suspension of the amine salt is satisfactory. Variations in the order of addition of the reagents are sometimes used, for example, when there is another functional group present which is sensitive to nitrous acid, or if the amine salt is very insoluble.[40] Weakly basic amines can often be diazotized successfully either in concentrated sulfuric acid or in a mixture of sulfuric acid with acetic or phosphoric acid.[43] It is likely that nitrosylsulfuric acid is formed on addition of sodium nitrite to concentrated sulfuric acid, and that this is the nitrosating agent when the amine is subsequently added.

Diazotization in organic solvents allows solid diazonium salts to be isolated. Diazotization can be carried out using an ester of nitrous acid, such as pentyl nitrite, in a solvent such as acetic acid or methanol. A procedure has also been described for isolating diazonium tetrafluoroborates, in excellent yield, by carrying out the diazotization with boron trifluoride etherate and *t*-butyl nitrite in ether or dichloromethane at low temperature.[44] Another method for the preparation of a variety of diazonium salts in a nonaqueous medium makes use of the chemistry of bis(trimethylsilyl)amines (**8**).[45] These compounds react in dichloromethane with nitrosyl chloride and other nitrosating agents which are generated *in situ*. Thus, benzenediazonium chloride was isolated (96%) from bis(trimethylsilyl)aniline.

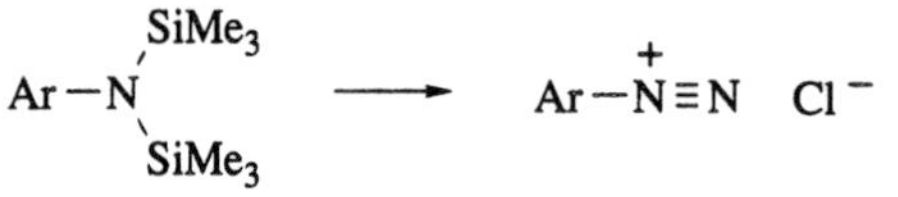

$$Ar-N(SiMe_3)_2 \longrightarrow Ar-\overset{+}{N}\equiv N\ Cl^-$$

(8)

The diazotization of aromatic amines with a nucleophilic substituent at the *ortho* position is a common method of synthesis of benzo-fused heterocyclic compounds with two or more contiguous nitrogen atoms. Benzotriazoles (**9**), benzotriazinones (**10**), and benzothiadiazoles (**11**) are examples of heterocyclic ring systems that can be prepared in this way.

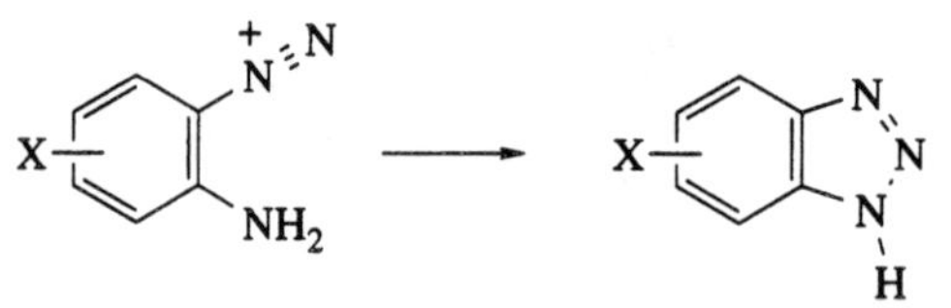

(9)

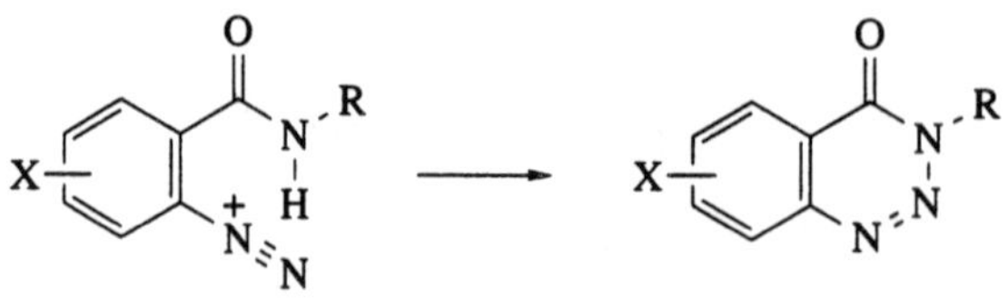

(10)

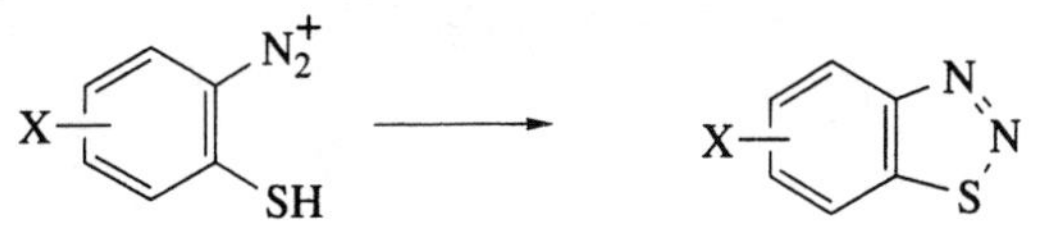

X–C$_6$H$_3$(N$_2^+$)(SH) ⟶ (11)

(11)

6.1.2.1.4 Amination

Simple primary alkylamines can be converted into the corresponding monoalkylhydrazines in moderate yield by amination with chloramine or with hydroxylamine-*O*-sulfonic acid.[46] The method is tolerant of the presence of double bonds: allylhydrazine was prepared (52%) by reaction of chloramine with allylamine. The method is not generally applicable to the preparation of 1,2-disubstituted hydrazines from primary alkylamines and *N*-chloroalkylamines although intramolecular examples are known.[47] Tetrahydropyrazole was prepared in moderate yield in this way (Scheme 12) and piperazine was prepared in low yield by the same type of reaction.

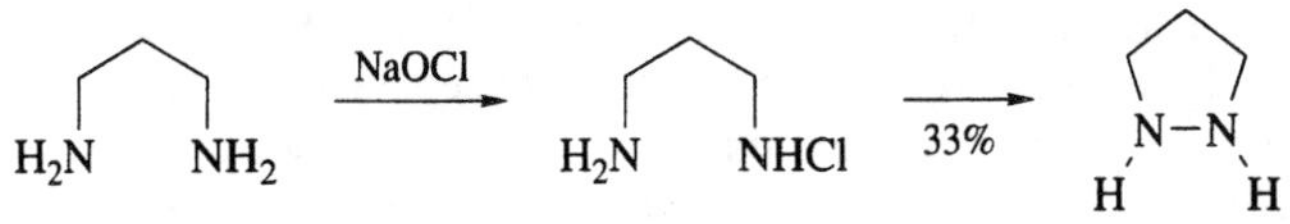

H_2N NH_2 —NaOCl→ H_2N NHCl —33%→ N–N (H, H)

Scheme 12

6.1.2.1.5 Halogenation, sulfenylation and related reactions

Methods for the *N*-chlorination and *N*-bromination of amines have been reviewed.[48] Alkylamines are chlorinated by aqueous sodium hypochlorite, chlorine in aqueous sodium bicarbonate, *N*-chlorosuccinimide, or *t*-butyl hypochlorite at low temperature. By using the appropriate amount of chlorinating agent, selective mono- and di-chlorination can be achieved. Although some of these compounds, especially those derived from *t*-alkylamines, are stable enough to be isolated, the majority are unstable since they can undergo further reaction associated with the loss of a proton from the α-carbon atom. *N*-Chloroanilines are also unstable unless the ring is substituted by an electron-withdrawing group, because of the tendency of the chlorine to migrate to a ring carbon atom. Reagents which have been used for *N*-bromination include bromine and aqueous sodium hypobromite. *N*-Haloamines have also been prepared by the action of the appropriate halogens on *N*-trimethylsilylamines. An indirect method of *N*,*N*-difluorination of *t*-alkylamines is illustrated in Scheme 13.[49]

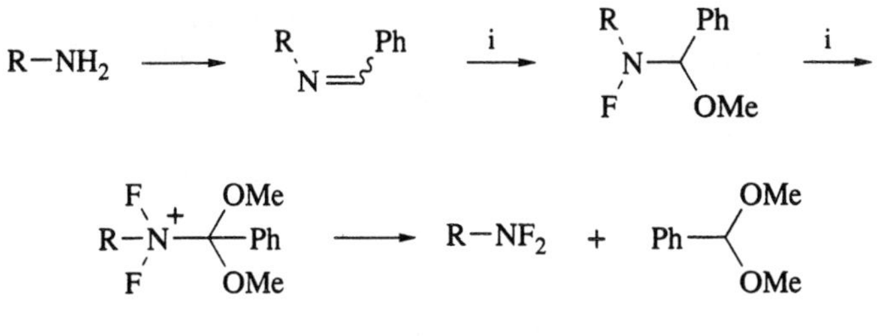

R–NH_2 ⟶ R–N=CHPh —i→ R(F)N–CH(Ph)(OMe) —i→

$R–N^+F_2$–C(Ph)(OMe)$_2$ ⟶ R–NF_2 + Ph–CH(OMe)$_2$

i, CF_3OF, MeOH

Scheme 13

The reaction of primary amines with arenesulfenyl halides leads to the formation of sulfenamides (**12**).[50] These compounds are most stable when the aryl group has electron-withdrawing substituents at the 2- and 4-positions. Selenamides are formed in an analogous manner but are somewhat less stable: aliphatic amines give isolable compounds, but most anilines react with selenyl halides to give products of ring substitution.[51] An example of an isolable selenamide is compound (**13**), which was prepared (79%) from the amine and 2-nitrobenzeneselenyl chloride.[52] This compound was used as an intermediate in the preparation of 7α-methoxycephalosporins.

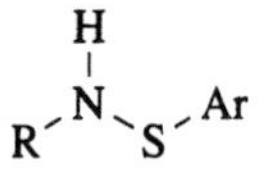

(12)

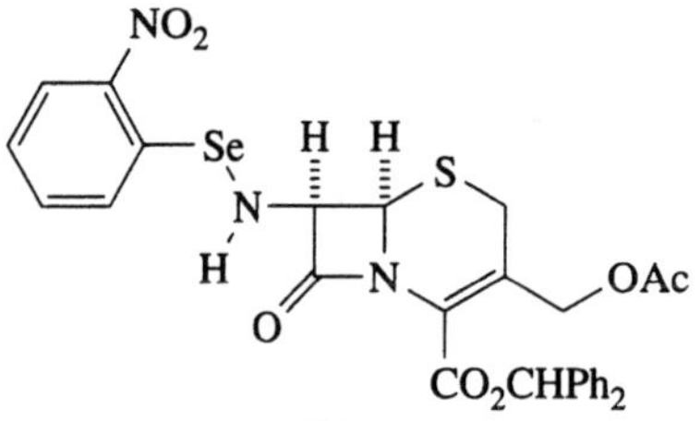

(13)

6.1.2.2 Hydrazones, Hydrazines and Hydroxylamines

6.1.2.2.1 Dehydrogenation

This section includes synthetically useful oxidative reactions of substrates of the general types $R^1R^2C{=}NNH_2$, $R^1R^2NH_2$ and $RONH_2$, where the groups R^1, R^2 and R can be alkyl, aryl or acyl.

The oxidation of many hydrazones provides a method of preparation of the corresponding diazo compounds $R^1R^2C{=}N_2$.[53] The oxidant most commonly used for this purpose is mercury(II) oxide.[54] Fluorenone hydrazone is converted in high yield into diazofluorene and many other diaryldiazomethanes can be prepared in the same way. Diazo ketones of relatively high stability, such as phenylbenzoyldiazomethane, can also be obtained by the oxidation of 1,2-diketone monohydrazones. Silver oxide reacts more rapidly with hydrazones than does mercury(II) oxide, and it is better for the preparation of monoaryldiazomethanes.[53] Activated manganese dioxide has also been used. The usual side products in these oxidations are azines $R^1R^2C{=}NN{=}CR^1R^2$, which are derived from the less stable diazo compounds by decomposition if the contact time with the oxidant is prolonged. Bishydrazones of 1,2-diketones are oxidized by mercury(II) oxide to alkynes (Scheme 14). This reaction provides a good method of synthesis of cycloalkynes such as cyclooctyne.[53]

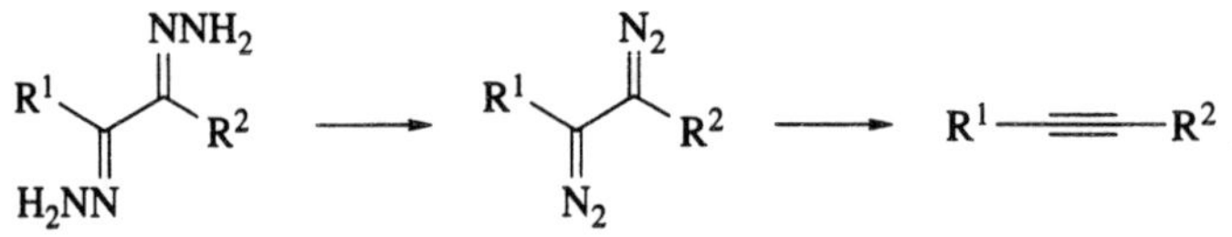

Scheme 14

The oxidation of 1,1-disubstituted hydrazines can be achieved by a wide range of oxidants. The oxidative removal of hydrogen formally leads to the production of aminonitrene, or 1,1-diazene, intermediates and many products of such reactions have been interpreted as being derived from aminonitrene intermediates.[55] Indeed, several of these species, derived from sterically hindered hydrazines by oxidation with nickel peroxide or *t*-butyl hypochlorite, have been detected and characterized in solution at low temperature. Examples include the diazenes **(14)**[56] and **(15)**[57]. The diazene **(16)** is stable enough to persist in solution at room temperature for several days.[58]

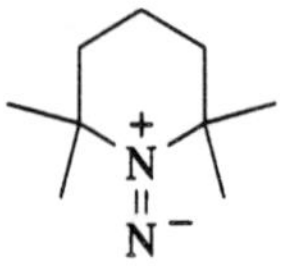

(14) 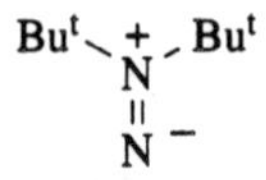 (15) 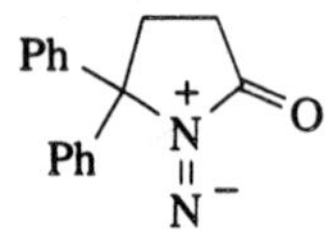 (16)

With many oxidants the most common products derived from 1,1-disubstituted hydrazines are the tetrazenes $R^1R^2NN{=}NNR^1R^2$. Benzeneselenic acid appears to a good reagent for their preparation[59] although several others, including mercury(II) oxide and nickel peroxide, have been widely used. These tetrazenes are formally the dimers of aminonitrenes and indeed are formed from long-lived species such as **(14)** by dimerization. Another possible mode of formation of tetrazenes, which is illustrated in Scheme 15, is the reaction of an aminonitrene, or its precursor, with the starting hydrazine to give a tetrazane, followed by further oxidation. This sequence was established for the oxidation of *N*-aminophthalimide by iodosylbenzene diacetate: the tetrazane (**17**) was isolated and gave the corresponding tetrazene with an excess of the oxidant.[60]

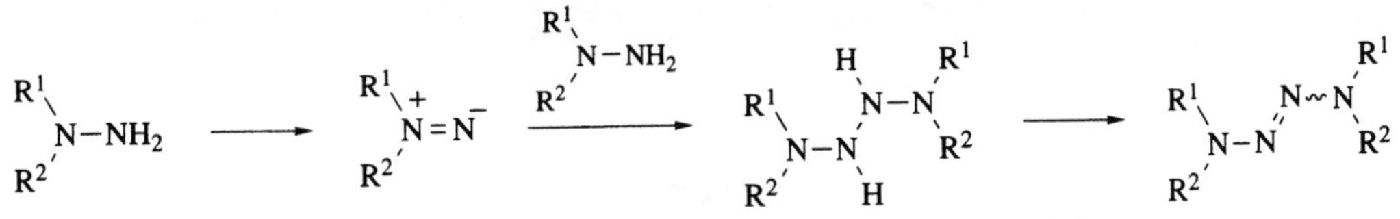

Scheme 15

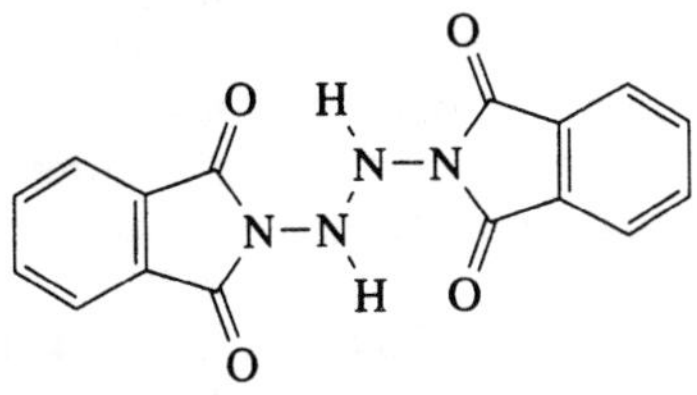

(**17**)

Lead(IV) acetate has proved to be a most efficient oxidant for these hydrazines. The oxidation of *N*-aminoheterocyclic compounds by lead(IV) acetate has, in particular, provided several very useful preparative procedures.[55,61] 1-Aminobenzotriazole (**18**; Scheme 16) is oxidized to 1,2-didehydrobenzene (benzyne) in high yield at low temperature.[62] This reaction has been widely exploited not only for the generation of benzyne but for the formation of cycloalkynes and of other arynes. For example, oxidation of the bisaminotriazole (**19**) gave products derived from 1,2,4,5-tetradehydrobenzene (**20**).[63] Campbell and Rees considered the possibility that benzyne was generated from 1-aminobenzotriazole by way of an aminonitrene and an unstable 1,2,3,4-benzotetrazine (Scheme 16).[62] A similar oxidation, of compound (**21**), did yield an isolable but unstable tetrazine (**22**) as the product.[64] 2-Aminobenzotriazole (**23**) clearly does not give the same intermediates as its isomer (**18**) on oxidation because it is oxidized cleanly to the dinitrile (**4**). On the other hand both 1- and 2-amino-3-phenylindazole give the same product of ring expansion, the benzotriazine (**24**), on oxidation (Scheme 17).[65]

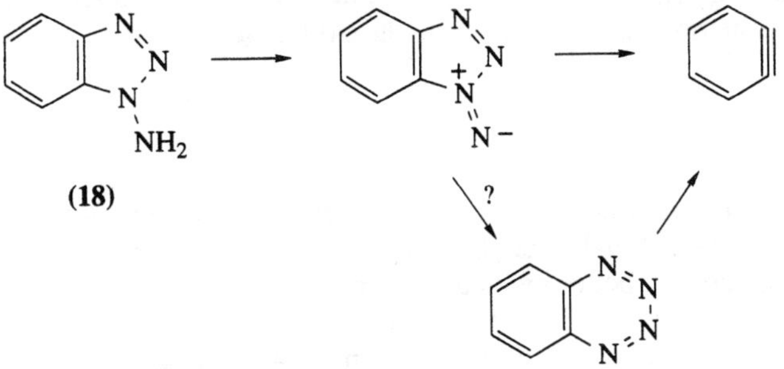

(**18**)

Scheme 16

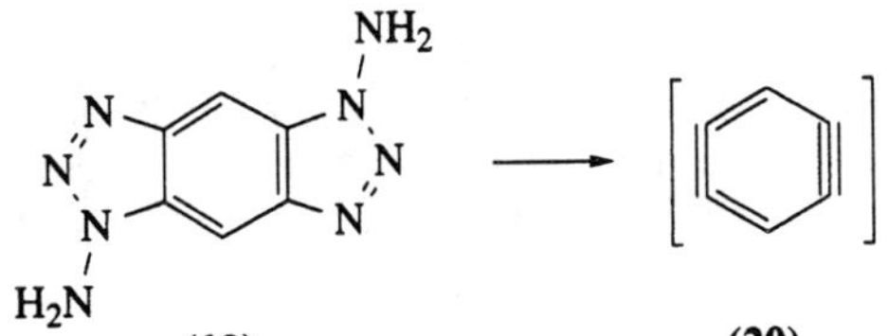

(**19**) (**20**)

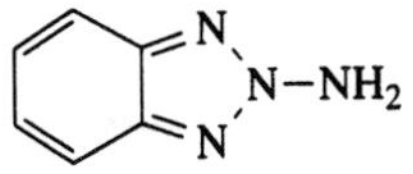

(**21**) (**22**) (**23**)

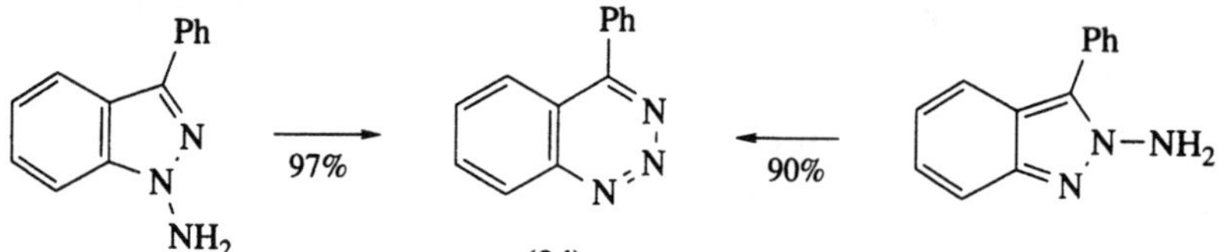

Ph N N NH$_2$ —97%→ (24) ←90%— Ph N–NH$_2$ N

Scheme 17

The oxidation by lead(IV) acetate of *N*-aminophthalimide and of several *N*-aminolactams leads to the formation of intermediates which do not undergo fragmentation or rearrangement, but which can be intercepted by alkenes, alkynes, sulfoxides and other nucleophiles. The reactions have proved particularly useful for the synthesis of aziridines from a variety of alkenes.[55] The mechanism of these reactions has commonly been assumed to require the intermediacy of aminonitrenes, but this is probably not the case. Atkinson and Kelly have shown that oxidation of the aminolactam (**25**) by lead(IV) acetate at –20 °C leads to the formation of an unstable *N*-acetoxy compound.[66] This is the species which can form aziridines with alkenes. The mechanism shown in Scheme 18, which is analogous to that for the epoxidation of alkenes by peroxy acids, has been proposed for the aziridination process.

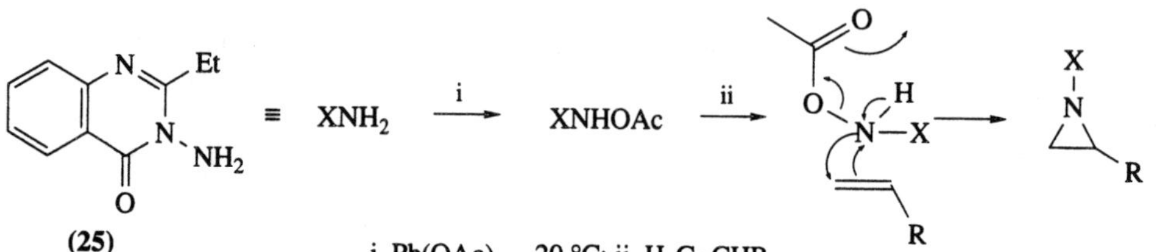

(**25**) ≡ XNH$_2$ —i→ XNHOAc —ii→ → aziridine

i, $Pb(OAc)_4$, –20 °C; ii, $H_2C{=}CHR$

Scheme 18

Oxidation of methoxylamine and some other *O*-substituted hydroxylamines by lead tetraacetate in the presence of alkenes can also lead to the formation of aziridines. The oxidation of 2,4-dinitrobenzenesulfenamide is analogous.[67] In view of the results reported with aminolactams, these reactions do not necessarily establish the intermediacy of nitrenes in the oxidations.

6.1.2.2.2 *Other oxidations*

1,1-Diphenylhydrazine is oxidized to diphenylnitrosamine (50%) by potassium superoxide.[68] The same reagent also oxidizes 1-methyl-1-phenylhydrazine, but here the nitrosamine is a minor product; the major reaction is deamination. A better method of oxidative deamination of some 1,1-disubstituted hydrazines and hydrazinium salts is reaction with nitrous acid. Thus, several hydrazinium salts $Me_2RN^+NH_2\ X^-$ were deaminated to the tertiary amine by treatment with nitrous acid.[69] The method has also been used to deaminate *N*-aminoheterocyclic compounds; for example, some 1,2,3-triazoles are conveniently prepared by deamination of the corresponding 1-aminotriazoles with nitrous acid.[70]

Monosubstituted hydrazines and hydrazides are converted into azides by a variety of nitrosating agents. The mildest reagent appears to be dinitrogen tetroxide, which can be used below 0 °C in acetonitrile to convert benzoylhydrazine, *p*-toluenesulfonylhydrazine and 4-nitrobenzoylhydrazine, among others, into the corresponding azides in high yield.[71] Another mild method involves the use of iron(III) nitrate supported on clay.[72] These reactions probably proceed by way of transient *N*-nitroso compounds (Scheme 19).

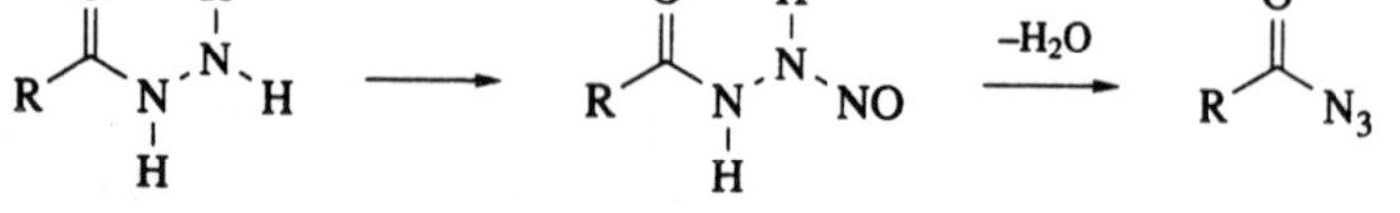

RC(O)NHNH$_2$ → RC(O)NHNHNO —$-H_2O$→ RC(O)N$_3$

Scheme 19

Katritzky and coworkers have nitrated hydrazinium salts and 1,1-disubstituted hydrazines with nitronium tetrafluoroborate and other nitrating agents.[73] Thus, nitrimides $R_3N^+N^-NO_2$ were prepared in good yield from hydrazinium salts derived from tertiary amines such as trimethylamine and 2-methylpyridine; 1-aminobenzotriazole was also nitrated at the amino group.

6.1.3 OXIDATION OF THE NH GROUP

6.1.3.1 Secondary Amines

6.1.3.1.1 Oxidation to nitroxides and hydroxylamines

The conversion of secondary amines R_2NH into nitroxides $R_2NO\cdot$[74] has been carried out using hydrogen peroxide, MCPBA and metal oxides, including silver oxide, mercury(II) oxide and lead(IV) oxide. Secondary amines which have low solubility in water have been oxidized by sodium tungstate and hydrogen peroxide in a mixture of methanol and acetonitrile.[5] Dimethyldioxirane (**1**) is also a good reagent for the oxidation of hindered secondary amines to nitroxides.[75] As illustrated in Scheme 20, such oxidations probably go by way of the hydroxylamines as intermediates. When the secondary amine has a hydrogen atom attached to an α-carbon atom, the hydroxylamine formed by oxidation with sodium tungstate and hydrogen peroxide reacts further by losing this hydrogen atom. For example, tetrahydroisoquinoline was oxidized to the nitrone (**26**; 85%).[76] The reaction has been modified to provide a method of synthesis of *N*-hydroxy amino acids from secondary amines. Thus, *N*-hydroxyproline was prepared from pyrrolidine by oxidation with hydrogen peroxide and sodium tungstate, the intermediate being intercepted by cyanide. This gave the hydroxylamine (**27**) which could then be converted into *N*-hydroxyproline by hydrolysis.[77] A related oxidation has been used for the conversion of tetrahydroquinolines into the corresponding 3,4-dihydro-1-hydroxy-2-quinolones.[78]

$$R_2N{-}H \longrightarrow R_2N{-}OH \longrightarrow R_2N^+(O^-)OH \xrightarrow{-HO\cdot} R_2NO\cdot$$

Scheme 20

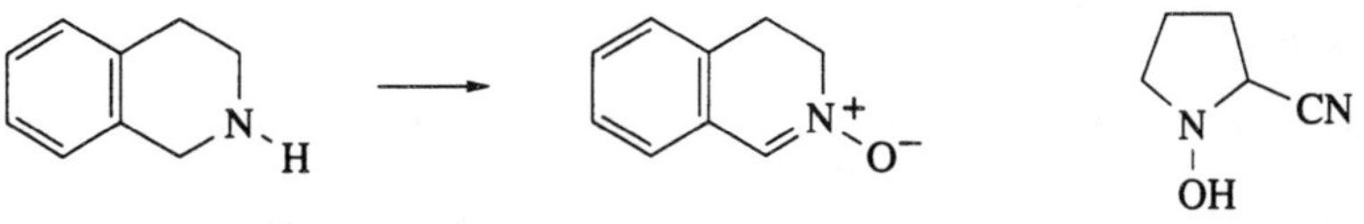

(26) **(27)**

Dibenzoyl peroxide oxidizes morpholine, piperidine and other simple secondary amines in good yield to the corresponding benzoyloxyamines; these compounds can then be hydrolyzed in basic conditions to the free hydroxylamines.[79] An analogous reaction takes place between secondary amines and bis(diphenylphosphinyl) peroxide; for example, diethylamine is converted into the hydroxylamine derivative $Et_2NOPOPh_2$ (97%) by this reagent.[80] The products are easily hydrolyzed to the free hydroxylamines, and they can also be used as aminating agents.

6.1.3.1.2 Oxidation to aminium ions, aminyl radicals and hydrazines

The one-electron oxidation of a secondary amine results in the formation of a secondary aminium ion[1] which on deprotonation gives an aminyl radical (Scheme 1).[2] The nature of the final products derived from these intermediates depends very much on the structure of the substrate and the reaction conditions. If the amine has a hydrogen atom on the α-carbon atom the major products usually result from deprotonation at this α-position. With aromatic secondary amines, products can result from coupling of the delocalized radicals at a ring carbon atom. The formal dimerization of aminyl radicals shown in Scheme 21 is therefore not often a useful method of preparation of hydrazines. Nickel peroxide has been used to oxidize diphenylamine to tetraphenylhydrazine in moderate yield, and other secondary arylamines also give mixtures of products in which the corresponding hydrazines are the major components.[25] The reagent formed by the addition of oxygen to copper(I) chloride in pyridine can also oxidize secondary arylamines. Diphenylamine was oxidized to tetraphenylhydrazine in 83% yield, and *N*-methylaniline gave the corresponding hydrazine in 52% yield.[81] Lithium piperidide and lithium salts of other secondary amines have also been oxidized to the hydrazines with copper(I) chloride and oxygen.[82] Intramolecular coupling of the diamines (**28**) to the pyrazolines (**29**) has also been achieved, using activated manganese dioxide (Scheme 21).[5,83]

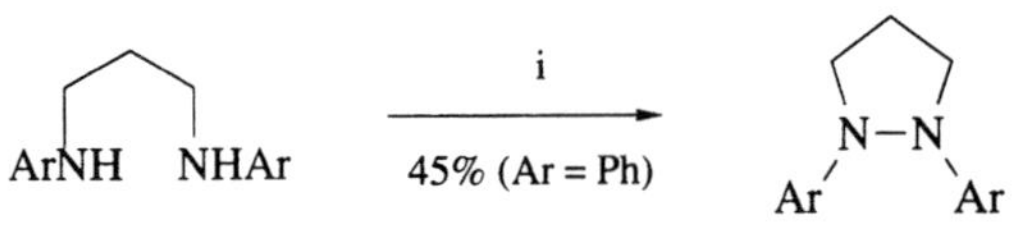

ArNH NHAr (**28**) —i, 45% (Ar = Ph)→ N–N, Ar Ar (**29**)

i, MnO_2, –20 °C

Scheme 21

6.1.3.1.3 Nitrosation and nitration

Methods of formation of *N*-nitrosamines from secondary amines have been reviewed.[84] The most widely used reagent is sodium nitrite in an aqueous acidic medium; others include nitrosyl chloride, nitrogen oxides and nitrite esters. Fremy's salt [$2K^+$ $(SO_3^-)_2NO\cdot$] reacts with hydroxylamine in the presence of secondary amines to give *N*-nitrosamines, or, in the presence of an excess of hydroxylamine, tetrazenes $R_2NN{=}NNR_2$. The nitrosating agent derived from hydroxylamine and Fremy's salt is suggested to be the anion $(SO_3^-)_2NONO$.[85]

Direct *N*-nitration of secondary amines by nitric acid is possible only for weakly basic amines.[86] The more basic amines can be nitrated under neutral conditions with reagents such as dinitrogen pentoxide and nitronium tetrafluoroborate, but nitrosamines are significant by-products. The nitrate ester $CF_3CMe_2ONO_2$ has been recommended as a nonacidic nitrating agent for secondary amines which avoids the problem of contamination of the products by *N*-nitrosamines: piperidine and pyrrolidine were nitrated in yields of 75% and 72%, respectively.[87] Amides and imides are efficiently *N*-nitrated using ammonium nitrate in trifluoroacetic anhydride.[88]

6.1.3.1.4 Amination

The conversion of secondary amines into 1,1-disubstituted hydrazines requires the use of an electrophilic aminating agent. Several such reagents are available, and their use has been reviewed.[46,89–91] Chloramine and hydroxylamine-*O*-sulfonic acid are the commonest reagents of this type. Secondary amines have been aminated by chloramine in moderate yield in aqueous solution, and in good yield by passing gaseous chloramine into methanolic solutions of the amines.[46] Hydroxylamine-*O*-sulfonic acid is a more powerful aminating agent and has been used to aminate heterocyclic compounds such as benzotriazole in good yield. Its disadvantage is its low solubility in nonpolar organic solvents. To overcome this problem several soluble *O*-substituted hydroxylamines have been introduced as aminating agents. These include *O*-mesitylhydroxylamine (**30**), *O*-2,4-dinitrophenylhydroxylamine (**31**), and *O*-mesitylenesulfonylhydroxylamine (MSH; **32**). MSH is a particularly good aminating agent, but it is rather tedious to prepare.[90] A promising, and much more accessible, reagent is *O*-diphenylphosphinylhydroxylamine (**33**).[92] This reagent has been used to aminate imides such as phthalimide in high yield. The oxaziridine (**34**) can act as an aminating agent for amino acids and peptides.[93]

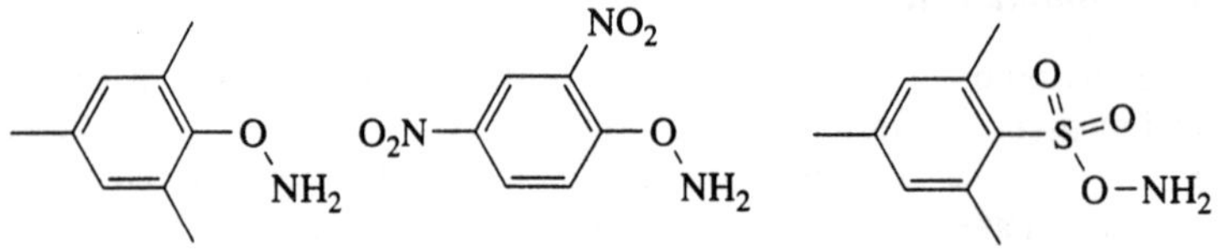

O–NH_2 (**30**) O_2N, NO_2, O–NH_2 (**31**) S(=O)(=O)–O–NH_2 (**32**)

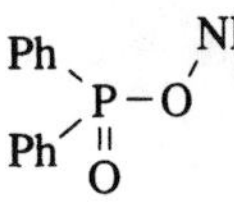

(33) 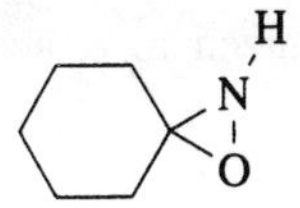 (34)

6.1.3.1.5 Halogenation and other reactions

Methods for the chlorination of secondary amines, secondary amides and imides have been reviewed.[48] Secondary alkylamines can be chlorinated by *t*-butyl hypochlorite at low temperature, or by sodium hypochlorite or *N*-chlorosuccinimide. *N*-Bromination and *N*-iodination can be brought about by using the appropriate halogen, but these *N*-halodialkylamines are unstable and are rarely isolated. *N*-Fluoroamines and *N*-fluoroamides are also rare.[94] The *N*-fluoroimide $(CF_3SO_2)_2NF$, has, however, been prepared by direct fluorination with fluorine. It is a stable liquid which shows promise as a fluorinating agent for aromatic compounds.[95]

N-Chloroaziridines (**35**) are configurationally stable, and there have been attempts to prepare them in optically active form by carrying out the chlorination with *t*-butyl hypochlorite or with *N*-chlorosuccinimide in the presence of the optically active alcohol (*S*)-(+)-PhCH(OH)CF_3.[96] Optical yields were, however, low (less than 10% *ee*).

Secondary amines can be converted into sulfenamides by reaction of the lithium dialkylamides with disulfides.[50,97] Perchlorylamines (**36**) have been prepared in good yields from piperidine and other secondary amines by reaction with chlorine(VII) oxide.[98]

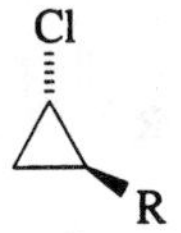

(35) 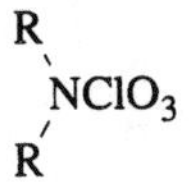 (36)

6.1.3.2 Hydrazines and Hydroxylamines

Methods of oxidation of hydrazo to azo compounds[99] and hydroxylamino to nitroso compounds[100] have been reviewed. Reagents which oxidize aromatic primary amines to azo compounds are also suitable for the oxidation of aromatic hydrazo compounds, since the hydrazo compounds are intermediates in the oxidation of the amines. Thus, manganese dioxide, mercury(II) oxide and lead tetraacetate are all suitable oxidants. Silver carbonate on Celite rapidly oxidizes both diarylhydrazines and acylhydrazines to the corresponding azo compounds in good yield.[26] Another supported oxidant which can convert hydrazobenzene into azobenzene in high yield is sodium periodate on silica gel.[101]

Two-phase oxidation systems are useful for the selective oxidation of the hydrazo to the azo group in the presence of other functional groups. For example, the hydrazotriazine (**37**) was oxidized by chlorine to the azo compound in high yield, without hydrolysis of the delicate ring chloride, in a two-phase system of chloroform and aqueous sodium bicarbonate.[102] Aqueous potassium ferricyanide has also been used in a two-phase system to oxidize hydrazo compounds. A coreagent is required, which is either a hindered phenol[103] or carbon black.[104] In either case the active oxidant is believed to be an aryloxy radical. This type of reagent has been used to produce AIBN and acylazo compounds such as PhN═NCOPh from the corresponding hydrazo compounds in high yield. Oxygen can also oxidize such hydrazo compounds in the presence of a palladium catalyst at room temperature.[105] Benzeneseleninic anhydride has been used to convert arylhydrazones of aldehydes into α-carbonylazo compounds: for example, furfural phenylhydrazone gave the azo compound (**38**) in 92% yield (Scheme 22).[106] Arylhydrazones

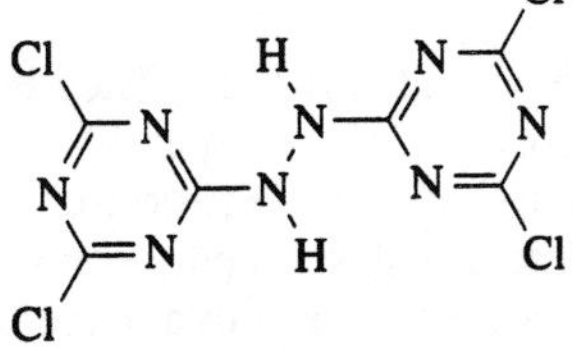

(37)

$R_2C{=}NNHAr$ are also oxidized to α-acetoxyalkylazo compounds $R_2C(OAc)N{=}NAr$ by lead tetra-acetate.[99]

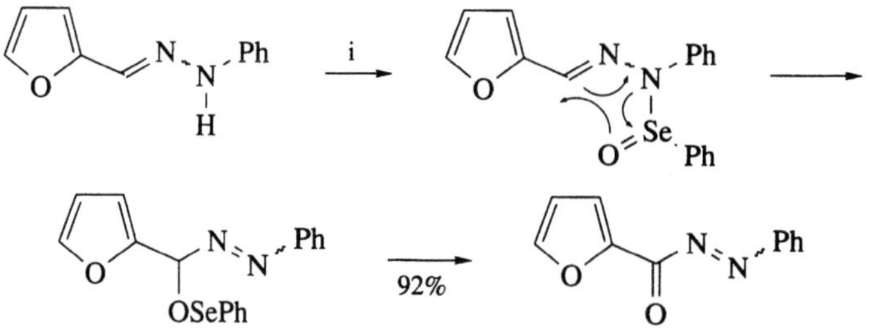

i, $(PhSeO)_2O$

Scheme 22

Several oxidants have been used to produce the highly reactive cyclic azocarbonyl compounds (**39**) from the corresponding hydrazides.[107] These include *t*-butyl hypochlorite, dinitrogen tetroxide and *N*-bromosuccinimide. Anodic oxidation[108] and oxidation by iodosylbenzene diacetate[109] are also effective for preparing these and related azocarbonyl compounds.

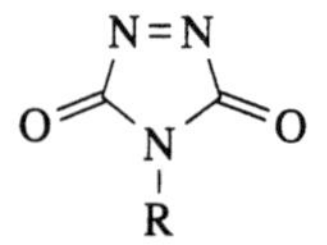

(**39**) 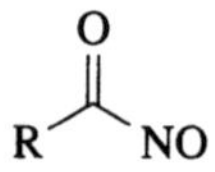 (**40**)

Aliphatic nitroso compounds can be prepared from *N*-alkylhydroxylamines by oxidation with bromine, chlorine or sodium hypochlorite in weakly acidic solution, by reaction with potassium dichromate in acetic or sulfuric acid, and by oxidation with yellow mercury(II) oxide in suspension in an organic solvent.[100] Silver carbonate on Celite has also been used to prepare aliphatic nitroso compounds, such as nitrosocyclohexane, in high yield from the corresponding hydroxylamines.[110] Aqueous sodium periodate and tetraalkylammonium periodates, which are soluble in organic solvents, are the reagents most commonly used for the oxidation of hydroxamic acids and *N*-acylhydroxylamines to acylnitroso compounds (**40**).[111] These compounds are rarely isolated, but are useful as highly reactive dienophiles in the Diels–Alder reaction.[112]

6.1.4 OXIDATION OF TERTIARY (sp^3) NITROGEN

6.1.4.1 Formation of *N*-Oxides

Tertiary alkylamines can be converted into the corresponding *N*-oxides with hydrogen peroxide or with peroxy acids.[113] *t*-Butyl hydroperoxide has also been used in the presence of a catalyst such as $VO(acac)_2$. Sharpless and coworkers have carried out the oxidative kinetic resolution of several β-hydroxy tertiary amines such as (**41**) with *t*-butyl hydroperoxide, titanium(IV) isopropoxide and (+)-diisopropyl tartrate, the titanium(IV):tartrate ratio being about 2:1.[114] After 60% conversion, one enantiomer was selectively oxidized, and the other enantiomer could be recovered in good optical purity (Scheme 23).

Tertiary alkylamines are normally autoxidized and dealkylated by oxygen, but it has been shown that they can slowly be converted into the *N*-oxides by heating at 90–130 °C with oxygen under pressure in a polar solvent.[115] The reaction is thought to involve electron transfer from the amine to oxygen as the rate-determining step. Pyridine does not react with oxygen under these conditions. The cyclic hydrazine *N*-oxide (**42**) has been prepared from the corresponding hydrazine by oxidation with 30% hydrogen peroxide at room temperature.[116]

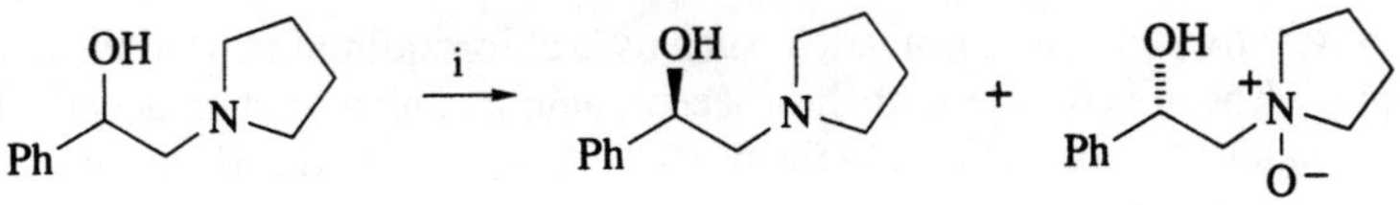

(41) 57%, 95% *ee* 59%, 63% *ee*

i, Bu^tOOH, $Ti(OPr^i)_4$, (+)-diisopropyl tartrate; 60% conversion

Scheme 23

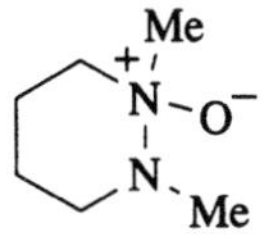

(42)

6.1.4.2 Other Reactions

One-electron oxidants convert tertiary amines into aminium ions.[1] Chlorine dioxide[5] is an example of such a reagent; this converts triethylamine and other tertiary alkylamines into aminium ions, which normally react further by the loss of hydrogen from an α-carbon atom. An example of a synthetic application of this reaction sequence is shown in Scheme 24, the intermediate iminium ion being intercepted intramolecularly.[117] Nitrosonium tetrafluoroborate and dioxygenyl hexafluoroantimonate, $O_2^+SbF_6^-$, are also good one-electron oxidants which can be used at low temperature.[118]

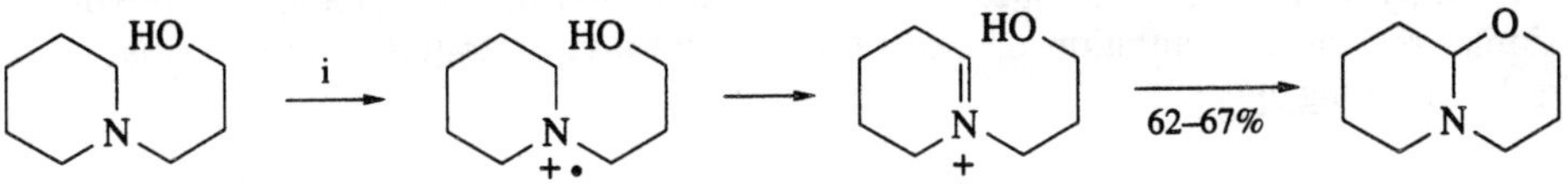

62–67%

i, ClO_2, pH 9–11

Scheme 24

Reagents for amination, nitrosation and nitration of tertiary alkylamines are discussed in the appropriate reviews listed in Sections 6.1.3.1.4 and 6.1.3.1.5. Tertiary amines can be nitrosated with dealkylation by dinitrogen tetroxide: for example, 1-methylpiperidine gave 1-nitrosopiperidine (80%).[119] This reaction probably starts by one-electron oxidation of the amine, the aminium ion then undergoing dealkylation. Other oxidative deacylations and dealkylations include the formation of *N*-nitrosodibenzylamine in high yield from the acid chloride $(PhCH_2)_2NCOCl$ and sodium nitrite[120] and the conversion of the amine **(43)** into the nitramine **(44)** with nitric acid.[121]

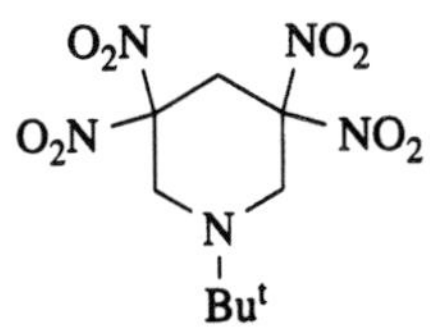

(43) 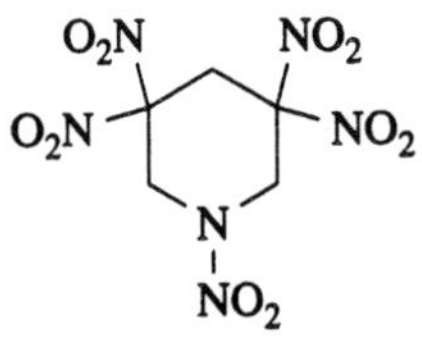 **(44)**

6.1.5 OXIDATION OF TRIGONAL (sp^2) NITROGEN

6.1.5.1 *N*-Oxidation of Heteroaromatic Amines

The standard methods of oxidation of pyridines and other heteroaromatic nitrogen compounds make use of peroxy acids or hydrogen peroxide in carboxylic acid solution.[122] Peracetic acid, peroxymonophthalic acid and MCPBA can all convert simple pyridines to the *N*-oxides. For example, the pyridine **(45)** was oxidized at nitrogen by MCPBA without attack on the vinyl group.[123] Peroxymaleic acid has been used to convert deactivated heteroaromatics, such as 2-chloroquinoline, into the *N*-oxides.[122] Trifluoroperacetic acid, or hydrogen peroxide in trifluoroacetic acid can also oxidize deactivated and hindered heteroaromatic compounds; for example, 2,6-dibromopyridine was oxidized in good yield by 30% hydrogen peroxide in trifluoroacetic acid.

O O N

(45)

N + NO$_2$ BF_4^-

(46)

These and more powerful oxidants are often required for the oxidation of diazines and triazines.[124] Chivers and Suschitzky succeeded in oxidizing pentafluoropyridine, tetrachloropyrazine and other polyhalogenated azines with 90% hydrogen peroxide in acetic and sulfuric acids or in trifluoroacetic and sulfuric acids.[125] The hazards of handling 90% hydrogen peroxide solutions can be avoided if the commercially available solid adduct formed between urea and hydrogen peroxide is used instead.[126] Diazines and diazanaphthalenes which are sensitive to peroxy acids can be oxidized by hydrogen peroxide in the presence of a sodium tungstate catalyst.[127] Sodium perborate in acetic acid is an especially useful oxidant for water-soluble pyrazine *N*-oxides.[128]

Dimethyldioxirane (**1**) is a mild and efficient oxidant for pyridine; pyridines and quinolines have also been oxidized in good yield by *t*-pentylhydroperoxide in the presence of molybdenum(V) chloride.[129]

N-Halogenation and other oxidative reactions of pyridines and related heterocycles have been reviewed.[130] Thus, pyridines and some diazines[124] can be aminated by mesitylenesulfonylhydroxylamine (**32**) and similar reagents. *N*-Nitration of 2-picoline by nitronium tetrafluoroborate gives the salt (**46**) which is itself a nitrating agent.

6.1.5.2 *N*-Oxidation of Imines

Imines are oxidized by peroxy acids to oxaziridines, with the formation of nitrones as a competing process (Scheme 25).[131] Oxaziridines are probably formed by a two-step process involving nucleophilic addition of the peroxy acid to the C═N bond, followed by elimination, as illustrated; nitrones are formed by competing *N*-oxidation.[4] The oxidation of chiral imines to oxaziridines by MCPBA proceeds with good, but not complete, diastereoselectivity.[132]

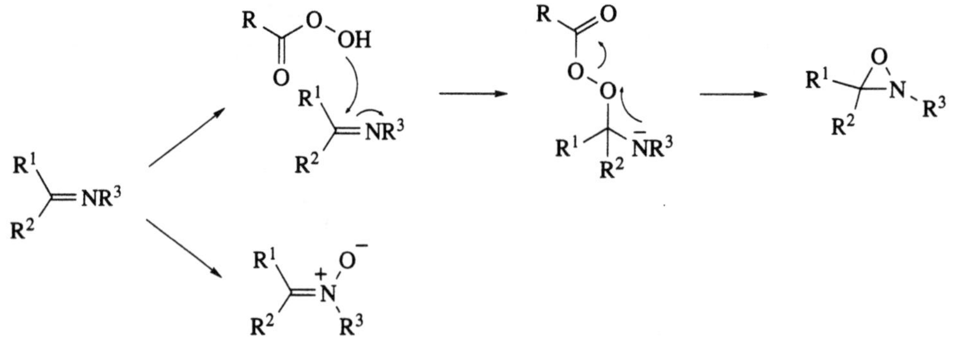

R O OH O R^1 NR^3 R^2 R O O O R^1 NR^3 R^2 R^1 O N R^3 R^2

R^1 NR^3 R^2

R^1 + O$^-$ N R^2 R^3

Scheme 25

6.1.5.3 Oxidation of Azo to Azoxy Compounds

Methods for the oxidation of azo to azoxy compounds have been reviewed.[133] Typical oxidants are MCPBA and other peroxy acids and hydrogen peroxide. *t*-Butyl hydroperoxide is also an effective oxidant for azobenzenes in the presence of molybdenum hexacarbonyl.[134] Trifluoroperacetic acid has been used to oxidize perfluoroazobenzene and other fluorinated azobenzenes to the azoxy compounds in high yield;[135] It is also capable of oxidizing azoxyarenes, such as compound (**47**), to the di-*N*-oxides.[136]

Bridgehead azoalkenes (**48**) can be oxidized to the azoxy compounds and to the di-*N*-oxides with MCPBA.[137]

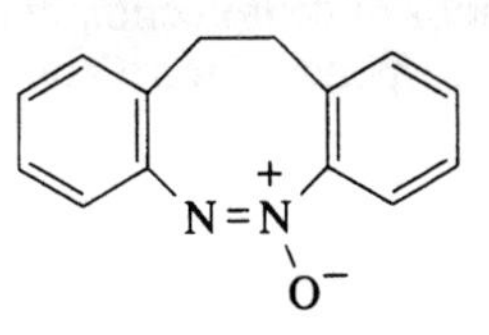

(47)  **(48)**

6.1.5.4 Oxidation of Oximes and Nitroso Compounds

The *N*-oxidation of oximes leads to the formation of *aci*-nitro compounds. In a polar organic solvent such as acetonitrile these compounds are isomerized to nitro compounds (Scheme 26) and are thus protected from further oxidation. The reagent of choice is trifluoroperacetic acid.[138] The oxidation of oximes derived from α-epoxy ketones results in the formation of a nitroalkene with opening of the epoxide, as illustrated in Scheme 27.[139]

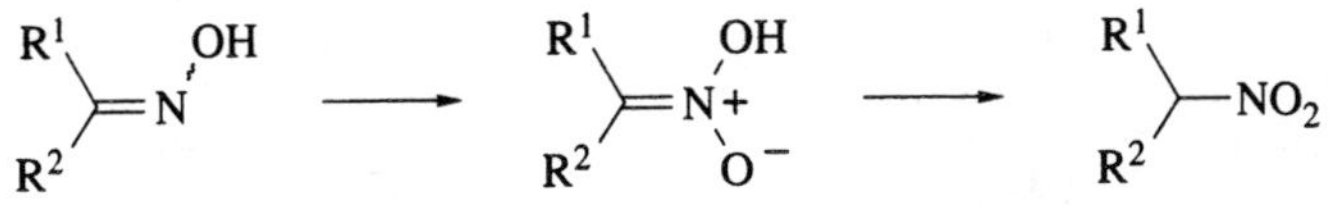

Scheme 26

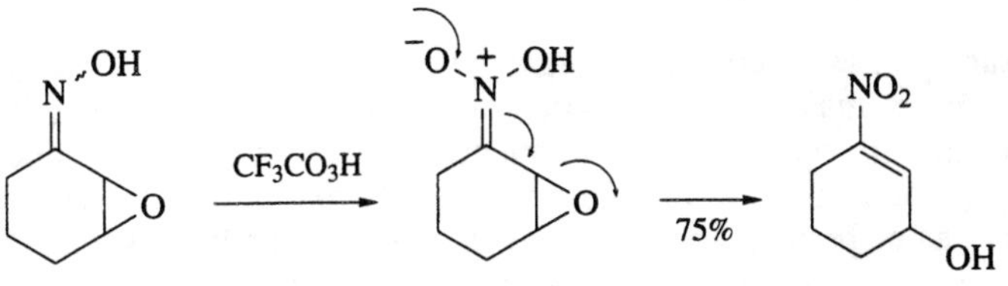

CF_3CO_3H 75%

Scheme 27

N-Nitrosation of oximes by nitrosyl halides or nitrite esters often results in the formation of *N*-nitrimines, $R^1R^2C{=}NNO_2$, these compounds being formed by rearrangement of the initial adducts.[140] *N*-Amination of oximes by chloramine or by hydroxylamine-*O*-sulfonic acid can result in the formation of diazo compounds (Scheme 28). The reaction, known as the Forster reaction,[53] has been used for the preparation of aryldiazoalkanes, although better methods are usually available. Diazo ketones of the general formula (**49**) have also been prepared by this method from the oximes.[141]

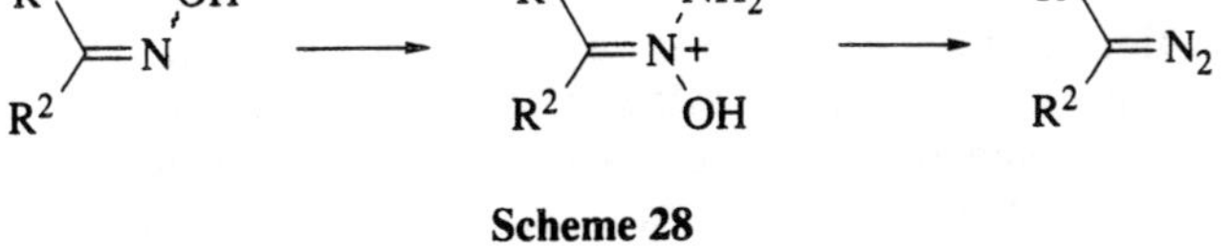

Scheme 28

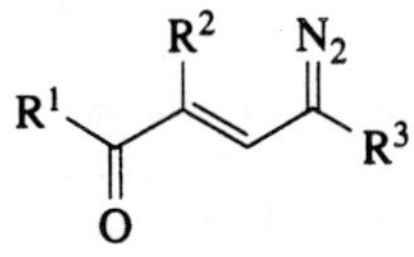

(49)

The oxidation of oximes by lead(IV) acetate, chlorine and other oxidants results in the formation of α-substituted nitroso compounds $R^1R^2C(X)NO$ by attack of the oxidant at carbon. This is also the reaction commonly observed with dinitrogen tetroxide (the Ponzio reaction), the products being *gem*-

dinitro compounds or α-nitronitroso compounds.[142] It has, however, been claimed that many alkyl ketoximes can be converted into the corresponding ketones in good yields by reaction with dinitrogen tetroxide at low temperature.[143] Other oxidative methods of deprotection of oximes exist.[143,144]

Nitroso compounds are readily oxidized to the corresponding nitro compounds by peroxides, peroxy acids, oxygen, ozone and dinitrogen tetroxide.[4]

6.1.5.5 Other Reactions

Reactions which formally involve the oxidation of azides have been reviewed by Boyer.[4] Other oxidations with useful synthetic applications include two which start from nitrogen ylides. Sulfimides (**50**) derived from electron-deficient aromatic and heterocyclic amines are oxidized to the corresponding nitroso compounds by MCPBA.[4,145] This is a very useful method of preparation of some otherwise inaccessible nitroso compounds such as 2-nitrosopyridine and 1-nitrosoisoquinoline. They can be further oxidized, for example by ozone, to the nitro compounds. Phosphimides (**51**) are oxidized directly by ozone to the nitro compounds, although the nitroso compounds are intermediates.[146] Isocyanates can also be oxidized to the corresponding nitro compounds, by dimethyldioxirane (**1**).[147]

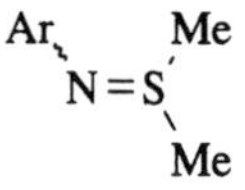

Ar–N=S(Me)Me

(50)

R–N=PPh_3

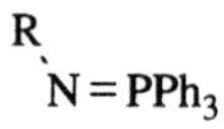

(51)

6.1.6 OXIDATION OF PHOSPHORUS

A comprehensive survey of the chemistry of organophosphorus compounds was published in 1982.[148] This provides descriptions of many of the oxidation methods in phosphorus chemistry, and primary literature references to them. Primary references are not given here if they are available from this source.

Tertiary phosphines (phosphanes) are normally very easily oxidized to the corresponding phosphine oxides.[149] The reaction can be brought about by oxygen alone or with hydrogen peroxide. Bis(trimethylsilyl) peroxide has been recommended as a selective oxygenating agent for phosphines and phosphites.[150] Oxidation of the chiral phosphine (*R*)-(–)-methylpropylphenylphosphine proceeded with 95% retention of configuration and the chlorodioxaphosphorinane (**52**) gave the oxide (**53**) with complete retention of stereochemistry. Diaryl selenoxides[151] and sulfur trioxide[152] have also been used for the oxidation of phosphines. Electrochemical oxidation of phosphines and phosphites proceeds in the presence of diethyl disulfide, which is effectively a catalyst for the reaction.[153]

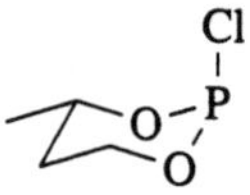

(52) 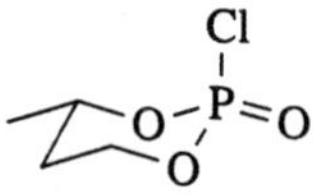 **(53)**

Tertiary phosphine sulfides and selenides are readily obtained from the phosphines and elementary sulfur or selenium. Potassium selenocyanate provides a convenient alternative to selenium for preparing selenides.[149] Phosphinimides (iminophosphoranes), $R_3P{=}NR'$, are prepared by the reaction of phosphines with azides or with *N*-chloroamines, among other methods. Phosphines can be aminated by reaction with *O*-diphenylphosphinylhydroxylamine.[92] The reaction of triphenylphosphine with diethyl azodicarboxylate results in rapid conjugate addition of the phosphine to the N=N bond. The resulting betaine has been used as a method of activation of alcohols (the Mitsunobu reaction), allowing displacement by nucleophiles and conversion of the alcohols into esters and other derivatives (Scheme 29).[154]

Many reagents, including the halogens themselves, are suitable for the conversion of trisubstituted phosphines into dihalides R_3PX_2 (X = F, Cl, Br or I).[155] These compounds can exist either in the pentacovalent form, or in an ionic form R_3PX^+ $R_3PX_3^-$, the covalent structure being favored in the order F > Cl > Br > I and by solvents of low polarity.

Secondary phosphines, R_2PH, can be chlorinated under controlled conditions to give the corresponding chlorophosphines.[156] These compounds can be further oxidized in a number of ways (Scheme 30).[157] The reaction of chlorodiphenylphosphine with dibenzoyl peroxide gives diphenylphosphinic acid

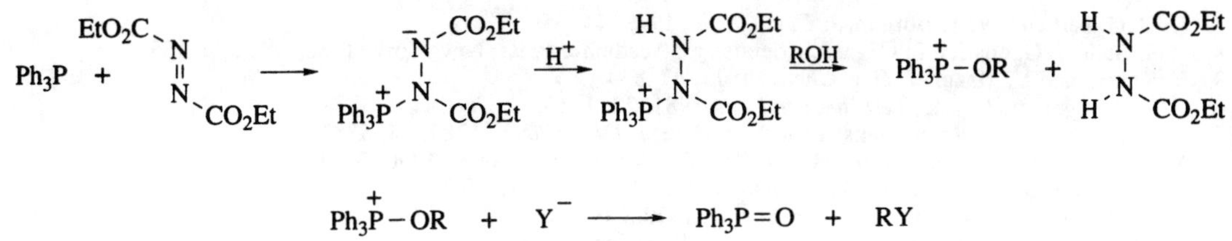

$$Ph_3P + EtO_2C{-}N{=}N{-}CO_2Et \longrightarrow Ph_3\overset{+}{P}{-}N(CO_2Et){-}\overset{-}{N}{-}CO_2Et \xrightarrow{H^+} Ph_3\overset{+}{P}{-}N(CO_2Et){-}NH{-}CO_2Et \xrightarrow{ROH} Ph_3\overset{+}{P}{-}OR + HN(CO_2Et){-}NH(CO_2Et)$$

$$Ph_3\overset{+}{P}{-}OR + Y^- \longrightarrow Ph_3P{=}O + RY$$

Scheme 29

anhydride in 82% yield. With sulfur dioxide, diphenylphosphinyl chloride is formed in high yield. Oxidation with oxygen, or with sodium hydride and oxygen in the presence of a secondary alcohol such as diphenylmethanol, gives diphenylphosphinic acid after hydrolysis. Diphenylphosphinic anhydride can also be obtained by oxidation of the bisoxide $(Ph_2PO)_2$ with perbenzoic acid.

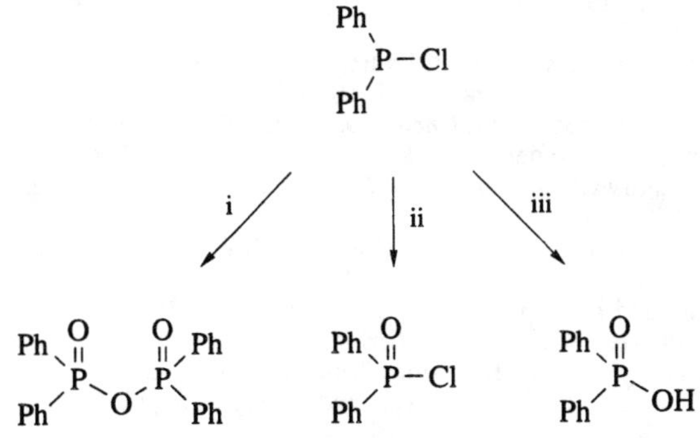

$$Ph_2P{-}Cl \xrightarrow{i} Ph_2P(O){-}O{-}P(O)Ph_2 \qquad Ph_2P{-}Cl \xrightarrow{ii} Ph_2P(O){-}Cl \qquad Ph_2P{-}Cl \xrightarrow{iii} Ph_2P(O){-}OH$$

i, $(PhCO_2)_2$; ii, SO_2; iii, NaH, Ph_2CHOH, O_2

Scheme 30

Derivatives of phosphonic acids, $RP{=}O(OH)_2$, can be prepared by several different oxidative methods.[158] Primary phosphines RPH_2 are oxidized to phosphonic acids by hydrogen peroxide or by sulfur dioxide: thus, phenylphosphine gave benzenephosphonic acid (96%) on reaction with sulfur dioxide at room temperature in a sealed tube. Phosphinic acids, $RP{=}O(OH)H$, can also be oxidized to the corresponding phosphonic acids with hydrogen peroxide. Ozone oxidized the dioxaphosphorane (**54**) to the phosphonic ester in 73% yield. Ozone is also capable of stereospecific oxidation of phosphite esters to phosphates.[159] For example, the cyclic phosphite (**55**) was oxidized to the phosphate (**56**) with retention of configuration. Peroxy acids and selenium dioxide are other common oxidants for phosphite esters.[160]

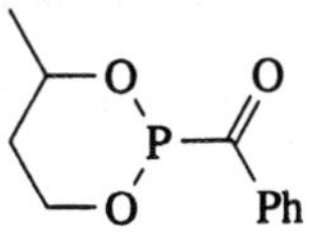

(**54**) 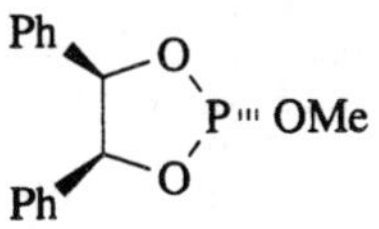 (**55**) 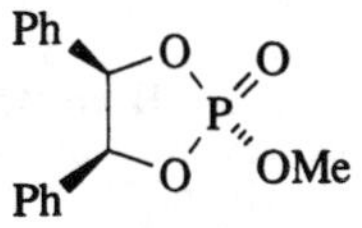 (**56**)

6.1.7 REFERENCES

1. Y. L. Chow, in 'Reactive Intermediates', ed. R. A. Abramovitch, Plenum Press, New York, 1980, vol. 1, p. 151.
2. W. C. Danen and F. A. Neugebauer, *Angew. Chem., Int. Ed. Engl.*, 1975, **14**, 783.
3. J. W. Gorrod and L. A. Damani (eds.), 'Biological Oxidation of Nitrogen in Organic Molecules', Ellis Horwood, Chichester, 1985.
4. J. H. Boyer, *Chem. Rev.*, 1980, **80**, 495.
5. D. H. Rosenblatt and E. P. Burrows, in 'The Chemistry of Amino, Nitroso and Nitro Compounds and Their Derivatives', ed. S. Patai, Wiley, Chichester, 1982, p. 1085.
6. H. O. Larson, in 'The Chemistry of the Nitro and Nitroso Groups', ed. H. Feuer, Interscience, New York, 1969, p. 301.

7. K. E. Gilbert and W. T. Borden, *J. Org. Chem.*, 1979, **44**, 659.
8. P. S. Bailey, 'Ozonation in Organic Chemistry', Academic Press, New York, 1982, vol. 2, p. 155.
9. E. Keinan and Y. Mazur, *J. Org. Chem.*, 1977, **42**, 844.
10. F. M. Menger and C. Lee, *Tetrahedron Lett.*, 1981, **22**, 1655.
11. R. W. Murray, S. N. Rajadhyaksha and L. Mohan, *J. Org. Chem.*, 1989, **54**, 5783.
12. M. D. Wittman, R. L. Halcomb and S. J. Danishefsky, *J. Org. Chem.*, 1990, **55**, 1981.
13. A. McKillop and J. A. Tarbin, *Tetrahedron Lett.*, 1983, **24**, 1505.
14. Y. Yost and H. R. Gutmann, *J. Chem. Soc. C*, 1970, 2497; 1969, 345.
15. L. Di Nunno, S. Florio and P. E. Todesco, *J. Chem. Soc. C*, 1970, 1433.
16. E. J. Corey and A. W. Gross, *Tetrahedron Lett.*, 1984, **25**, 491.
17. W. W. Zajac, Jr., T. R. Walters and J. M. Woods, *Synthesis*, 1988, 808.
18. W. J. Mijs and C. R. H. I. de Jonge (eds.), 'Organic Synthesis by Oxidation with Metal Compounds', Plenum Press, New York, 1986.
19. H. G. Chen and P. Knochel, *Tetrahedron Lett.*, 1988, **29**, 6701.
20. R. V. Hoffman and E. L. Belfoure, *Synthesis*, 1983, 34.
21. M. Psiorz and G. Zinner, *Synthesis*, 1984, 217.
22. A. J. Fatiadi, in ref. 18, p. 119.
23. O. H. Wheeler and D. Gonzalez, *Tetrahedron*, 1964, **20**, 189; I. Bhatnagar and M. V. George, *J. Org. Chem.*, 1968, **33**, 2407.
24. J. A. Hyatt, *Tetrahedron Lett.*, 1977, 141.
25. M. V. George, in ref. 18, p. 373.
26. M. Fetizon, M. Golfier, P. Mourgues and J. M. Louis, in ref. 18, p. 503.
27. M. L. Mihailović, Ž. Čeković and L. Lorenc, in ref. 18, p. 741.
28. H. Firouzabadi and Z. Mostafavipoor, *Bull. Chem. Soc. Jpn.*, 1983, **56**, 914.
29. H. Firouzabadi, B. Vessal and M. Naderi, *Tetrahedron Lett.*, 1982, **23**, 1847.
30. S. M. Mehta and M. V. Vakilwala, *J. Am. Chem. Soc.*, 1952, **74**, 563; L. Huestis, *J. Chem. Educ.*, 1977, **54**, 327.
31. E. T. McBee, G. W. Calundann, C. J. Morton, T. Hodgins and E. P. Wesseler, *J. Org. Chem.*, 1972, **37**, 3140.
32. G. Crank and M. I. H. Makin, *Aust. J. Chem.*, 1984, **37**, 845.
33. T. Kajimoto, H. Takahashi and J. Tsuji, *J. Org. Chem.*, 1976, **41**, 1389.
34. W. J. Irwin and D. G. Wibberley, *J. Chem. Soc. C*, 1971, 3237.
35. J. W. Timberlake, M. L. Hodges and K. Betterton, *Synthesis*, 1972, 632.
36. K. Mackenzie, in 'The Chemistry of Hydrazo, Azo, and Azoxy Groups', ed. S. Patai, Wiley, Chichester, 1975, p. 329.
37. R. M. Paton, in 'Comprehensive Heterocyclic Chemistry', ed. A. R. Katritzky, Pergamon Press, Oxford, 1984, vol. 6, p. 393.
38. H. Wamhoff, in 'Comprehensive Heterocyclic Chemistry', ed. A. R. Katritzky, Pergamon Press, Oxford, 1984, vol. 5, p. 669.
39. S. A. Lang, Jr. and Y.-i. Lin, in 'Comprehensive Heterocyclic Chemistry', ed. A. R. Katritzky, Pergamon Press, Oxford, 1984, vol. 6, p. 1.
40. K. H. Saunders and R. L. M. Allen, 'Aromatic Diazo Compounds', 3rd edn., Arnold, London, 1985.
41. K. Schank, in 'The Chemistry of Diazonium and Diazo Groups', ed. S. Patai, Wiley, Chichester, 1978, p. 645.
42. R. N. Butler, *Chem. Rev.*, 1975, **75**, 241.
43. T. I. Godovikova, O. A. Rakitin and L. I. Khmelnitskii, *Russ. Chem. Rev. (Engl. Transl.)*, 1983, **52**, 440.
44. M. P. Doyle and W. J. Bryker, *J. Org. Chem.*, 1979, **44**, 1572.
45. R. Weiss, K.-G. Wagner and M. Hertel, *Chem. Ber.*, 1984, **117**, 1965.
46. W. Sucrow, in 'Methodicum Chimicum', ed. F. Zymalkowski, Academic Press, New York, 1975, vol. 6, p. 91; S. R. Sandler and W. Karo, 'Organic Functional Group Preparations I', 2nd edn., Academic Press, Orlando, 1983, p. 434; J. W. Timberlake and J. C. Stowell, in 'The Chemistry of the Hydrazo, Azo, and Azoxy Groups', ed. S. Patai, Wiley, Chichester, 1975, p. 69.
47. A. Lüttringhaus, J. Jander and R. Schneider, *Chem. Ber.*, 1959, **92**, 1756.
48. P. Kovacic, M. K. Lowery and K. W. Field, *Chem. Rev.*, 1970, **70**, 639; R. Stroh, *Methoden Org. Chem. (Houben-Weyl)*, 1962, **5** (3), 796; J. Zakrzewski, *Synth. Commun.*, 1988, **18**, 2135.
49. D. H. R. Barton, R. H. Hesse, T. R. Klose and M. M. Pechet, *J. Chem. Soc., Chem. Commun.*, 1975, 97.
50. S. R. Sandler and W. Karo, 'Organic Functional Group Preparations III', Academic Press, New York, 1972, p. 163; F. A. Davis and U. K. Nadir, *Org. Prep. Proced. Int.*, 1979, **11**, 33.
51. T. G. Back, in 'Organoselenium Chemistry', ed. D. Liotta, Wiley, New York, 1987, p. 44.
52. T. Kobayashi and T. Hiraoka, *Bull. Chem. Soc. Jpn.*, 1979, **52**, 3366.
53. M. Regitz, in 'The Chemistry of the Diazonium and Diazo Groups', ed. S. Patai, Wiley, Chichester, 1978, p. 659.
54. A. Friedrich, *Methoden Org. Chem. (Houben-Weyl)*, 1975, **4** (1b), 82.
55. B. V. Ioffe and M. A. Kuznetsov, *Russ. Chem. Rev. (Engl. Transl.)*, 1972, **41**, 131; R. S. Atkinson, in 'Azides and Nitrenes—Reactivity and Utility', ed. E. F. V. Scriven, Academic Press, Orlando, 1984, p. 247.
56. P. B. Dervan, M. A. Squillacote, P. M. Lahti, A. P. Sylvester and J. D. Roberts, *J. Am. Chem. Soc.*, 1981, **103**, 1120.
57. D. K. McIntyre and P. B. Dervan, *J. Am. Chem. Soc.*, 1982, **104**, 6466.
58. R. D. Miller, P. Gölitz, J. Janssen and J. Lemmens, *J. Am. Chem. Soc.*, 1984, **106**, 1508.
59. T. G. Back and R. G. Kerr, *Can. J. Chem.*, 1982, **60**, 2711.
60. D. J. Anderson, T. L. Gilchrist and C. W. Rees, *J. Chem. Soc. D*, 1971, 800.
61. T. L. Gilchrist, C. W. Rees and R. C. Storr, *Proc. Int. Congr. Pure Appl. Chem.*, 1971, **23** (4), 105.
62. C. D. Campbell and C. W. Rees, *J. Chem. Soc. C*, 1969, 742.
63. H. Hart and D. Ok, *J. Org. Chem.*, 1986, **51**, 979.

64. T. Kaiho, T. Itoh, K. Yamaguchi and A. Ohsawa, *J. Chem. Soc., Chem. Commun.*, 1988, 1608; see also T. Nakazawa, M. Kodama, S. Kinoshita and I. Murata, *Tetrahedron Lett.*, 1985, **26**, 335.
65. B. M. Adger, S. Bradbury, M. Keating, C. W. Rees, R. C. Storr and M. T. Williams, *J. Chem. Soc., Perkin Trans. 1*, 1975, 31.
66. R. S. Atkinson and B. J. Kelly, *J. Chem. Soc., Chem. Commun.*, 1987, 1362.
67. R. S. Atkinson and B. D. Judkins, *J. Chem. Soc., Chem. Commun.*, 1979, 832.
68. C.-I. Chern and J. San Filippo, Jr., *J. Org. Chem.*, 1977, **42**, 178.
69. R. F. Smith and K. J. Coffman, *Synth. Commun.*, 1982, **12**, 801.
70. T. L. Gilchrist, *Adv. Heterocycl. Chem.*, 1974, **16**, 33; T. L. Gilchrist, G. E. Gymer and C. W. Rees, *J. Chem. Soc., Perkin Trans. 1*, 1975, 1.
71. Y. H. Kim, K. Kim and S. B. Shim, *Tetrahedron Lett.*, 1986, **27**, 4749.
72. P. Laszlo and E. Polla, *Tetrahedron Lett.*, 1984, **25**, 3701.
73. J. Epsztajn, A. R. Katritzky, E. Lunt, J. W. Mitchell and G. Roch, *J. Chem. Soc., Perkin Trans. 1*, 1973, 2622; A. R. Katritzky and J. W. Mitchell, *J. Chem. Soc., Perkin Trans. 1*, 1973, 2624.
74. H. G. Aurich, in 'The Chemistry of Amino, Nitroso and Nitro Compounds and Their Derivatives', ed. S. Patai, Wiley, Chichester, 1982, p. 565; for further oxidation to nitrosonium cations, see J. M. Bobbitt and M. C. L. Flores, *Heterocycles*, 1988, **27**, 509.
75. R. W. Murray and M. Singh, *Tetrahedron Lett.*, 1988, **29**, 4677; R. W. Murray, *Chem. Rev.*, 1989, **89**, 1187.
76. H. Mitsui, S. Zenki, T. Shiota and S.-I. Murahashi, *J. Chem. Soc., Chem. Commun.*, 1984, 874; S.-I. Murahashi, H. Mitsui, T. Shiota, T. Tsuda and S. Watanabe, *J. Org. Chem.*, 1990, **55**, 1736.
77. S.-I. Murahashi and T. Shiota, *Tetrahedron Lett.*, 1987, **28**, 6469.
78. S.-I. Murahashi, T. Oda, T. Sugahara and Y. Masui, *J. Chem. Soc., Chem. Commun.*, 1987, 1471.
79. A. J. Biloski and B. Ganem, *Synthesis*, 1983, 537.
80. J. J. Yaouanc, G. Masse and G. Sturtz, *Synthesis*, 1985, 807.
81. T. Kajimoto, H. Takahashi and J. Tsuji, *Bull. Chem. Soc. Jpn.*, 1982, **55**, 3673.
82. J. Jackisch, J. Legler and T. Kauffmann, *Chem. Ber.*, 1982, **115**, 659.
83. R. Daniels and B. D. Martin, *J. Org. Chem.*, 1962, **27**, 178.
84. B. C. Challis and J. A. Challis, in 'The Chemistry of Amino, Nitroso and Nitro Compounds and Their Derivatives', ed. S. Patai, Wiley, Chichester, 1982, p. 1151; see also D. L. H. Williams, 'Nitrosation', Cambridge University Press, Cambridge, 1987.
85. M. P. Vázquez Tato, L. Castedo and R. Riguera, *Chem. Lett.*, 1985, 623.
86. G. F. Wright, in 'The Chemistry of the Nitro and Nitroso Groups', ed. H. Feuer, Interscience, New York, 1969, p. 613.
87. J. C. Bottaro, R. J. Schmitt and C. D. Bedford, *J. Org. Chem.*, 1987, **52**, 2292.
88. S. C. Suri and R. D. Chapman, *Synthesis*, 1988, 743.
89. E. Schmitz, *Russ. Chem. Rev. (Engl. Transl.)*, 1976, **45**, 16.
90. Y. Tamura, J. Minamikawa and M. Ikeda, *Synthesis*, 1977, 1.
91. T. Sheradsky, in 'The Chemistry of Amino, Nitroso and Nitro Compounds and Their Derivatives', ed. S. Patai, Wiley, Chichester, 1982, p. 395.
92. M. J. P. Harger, *J. Chem. Soc., Perkin Trans. 1*, 1981, 3284; W. Klötzer, J. Stadlwieser and J. Raneburger, *Org. Synth.*, 1986, **64**, 96; W. Klötzer, H. Baldinger, E. M. Karputschka and J. Knoflach, *Synthesis*, 1982, 592.
93. F. Szurdoki, S. Andreae, E. Baitz-Gács, J. Tamás, K. Valkó, E. Schmitz and Cs. Szántay, *Synthesis*, 1988, 529; F. A. Davis and A. C. Sheppard, *Tetrahedron*, 1989, **45**, 5703.
94. D. H. R. Barton, R. H. Hesse, M. M. Pechet and H. T. Toh, *J. Chem. Soc., Perkin Trans. 1*, 1974, 732.
95. S. Singh, D. D. DesMarteau, S. S. Zuberi, M. Witz and H.-N. Huang, *J. Am. Chem. Soc.*, 1987, **109**, 7194.
96. A. Forni, I. Moretti, A. V. Prosyanik and G. Torre, *J. Chem. Soc., Chem. Commun.*, 1981, 588; M. Bucciarelli, A. Forni, I. Moretti and G. Torre, *J. Org. Chem.*, 1983, **48**, 2640.
97. H. Ikehira and S. Tanimoto, *Synthesis*, 1983, 716.
98. C. D. Beard and K. Baum, *J. Am. Chem. Soc.*, 1974, **96**, 3237.
99. H. D. Padeken, in 'Methodicum Chimicum', ed. F. Zymalkowski, Academic Press, New York, 1975, vol. 6, p. 142; S. R. Sandler and W. Karo, 'Organic Functional Group Preparations', 2nd edn., Academic Press, Orlando, 1986, vol. 2, p. 353.
100. J. H. Boyer, in 'The Chemistry of the Nitro and Nitroso Groups', ed. H. Feuer, Interscience, New York, 1969, p. 215; H. Meier, in 'Methodicum Chimicum', ed. F. Zymalkowski, Academic Press, New York, 1975, vol. 6, p. 23; S. R. Sandler and W. Karo, 'Organic Functional Group Preparations', 2nd edn., Academic Press, Orlando, 1986, vol. 2, p. 457.
101. D. N. Gupta, P. Hodge and J. E. Davies, *J. Chem. Soc., Perkin Trans. 1*, 1981, 2970.
102. P. Loew and C. D. Weis, *J. Heterocycl. Chem.*, 1976, **13**, 829.
103. K. Dimroth and W. Tüncher, *Synthesis*, 1977, 339.
104. N. Tsubokawa, N. Takeda and Y. Sone, *Bull. Chem. Soc. Jpn.*, 1982, **55**, 3541.
105. M. L. Heyman and J. P. Snyder, *Tetrahedron Lett.*, 1973, 2859; G. Gaviraghi, M. Pinza and G. Pifferi, *Synthesis*, 1981, 608.
106. D. H. R. Barton, D. J. Lester and S. V. Ley, *J. Chem. Soc., Chem. Commun.*, 1978, 276.
107. C. J. Moody, *Adv. Heterocycl. Chem.*, 1982, **30**, 1.
108. H. Wamhoff and G. Kunz, *Angew. Chem., Int. Ed. Engl.*, 1981, **20**, 797.
109. R. M. Moriarty, I. Prakash and R. Penmasta, *Synth. Commun.*, 1987, **17**, 409.
110. J. A. Maassen and T. J. DeBoer, *Recl. Trav. Chim. Pays-Bas*, 1971, **90**, 373.
111. G. W. Kirby, *Chem. Soc. Rev.*, 1977, **6**, 1.
112. D. L. Boger and S. M. Weinreb, in 'Hetero Diels–Alder Methodology in Organic Synthesis', Academic Press, San Diego, 1987, p. 82.
113. Y. Ogata and Y. Sawaki, in ref. 18, p. 839; J. Cymerman Craig and K. K. Purushothaman, *J. Org. Chem.*, 1970, **35**, 1721.

114. S. Miyano, L. D.-L. Lu, S. M. Viti and K. B. Sharpless, *J. Org. Chem.*, 1983, **48**, 3608.
115. D. P. Riley and P. E. Correa, *J. Org. Chem.*, 1985, **50**, 1563.
116. R. Askani and I. Alfter, *Tetrahedron Lett.*, 1988, **29**, 1909.
117. C.-K. Chen, A. G. Hortmann and M. R. Marzabadi, *J. Am. Chem. Soc.*, 1988, **110**, 4829.
118. J. P. Dinnocenzo and T. E. Banach, *J. Am. Chem. Soc.*, 1986, **108**, 6063; 1988, **110**, 971.
119. J. H. Boyer, G. Kumar and T. P. Pillai, *J. Chem. Soc., Perkin Trans. 1*, 1986, 1751.
120. M. Nakajima and J.-P. Anselme, *Tetrahedron Lett.*, 1979, 3831.
121. D. A. Cichra and H. G. Adolph, *J. Org. Chem.*, 1982, **47**, 2474.
122. E. Ochiai, 'Aromatic Amine Oxides', Elsevier, Amsterdam, 1967.
123. W. Korytnyk, N. Angelino, B. Lachmann and P. G. G. Potti, *J. Med. Chem.*, 1972, **15**, 1262.
124. M. R. Grimmett and B. R. T. Keene, *Adv. Heterocycl. Chem.*, 1988, **43**, 127.
125. G. E. Chivers and H. Suschitzky, *J. Chem. Soc. C*, 1971, 2867.
126. E. Eichler, C. S. Rooney and H. W. R. Williams, *J. Heterocycl. Chem.*, 1976, **13**, 41.
127. Y. Kobayashi, I. Kumadaki, H. Sato, Y. Sekine and T. Hara, *Chem. Pharm. Bull.*, 1974, **22**, 2097.
128. A. Ohta and M. Ohta, *Synthesis*, 1985, 216.
129. R. W. Murray and R. Jeyaraman, *J. Org. Chem.*, 1985, **50**, 2850; G. A. Tolstikov, U. M. Jemilev, V. P. Jurjev, F. B. Gershanov and S. R. Rafikov, *Tetrahedron Lett.*, 1971, 2807.
130. E. F. V. Scriven, in 'Comprehensive Heterocyclic Chemistry', ed. A. R. Katritzky, Pergamon Press, Oxford, 1984, vol. 2, p. 165.
131. M. J. Haddadin and J. P. Freeman, in 'Small Ring Heterocycles', ed. A. Hassner, Wiley, New York, 1985, part 3, p. 283; D. R. Boyd, P. B. Coulter and N. D. Sharma, *Tetrahedron Lett.*, 1985, **26**, 1673; for oxidation of imines to nitrones see also D. Christensen and K. A. Jorgensen, *J. Org. Chem.*, 1989, **54**, 12b.
132. J. Aubé, *Tetrahedron Lett.*, 1988, **29**, 4509.
133. V. N. Yandovskii, B. V. Gidaspov and I. V. Tselinskii, *Russ. Chem. Rev. (Engl. Transl.)*, 1981, **50**, 164.
134. N. A. Johnson and E. S. Gould, *J. Org. Chem.*, 1974, **39**, 407.
135. J. M. Birchall, R. N. Haszeldine and J. E. G. Kemp, *J. Chem. Soc. C*, 1970, 1519.
136. M. A. Smith, B. Weinstein and F. D. Greene, *J. Org. Chem.*, 1980, **45**, 4597.
137. J. P. Snyder, M. L. Heyman and E. N. Suciu, *J. Org. Chem.*, 1975, **40**, 1395.
138. T. Takamoto, Y. Yokota, R. Sudoh and T. Nakagawa, *Bull. Chem. Soc. Jpn.*, 1973, **46**, 1532.
139. T. Takamoto, Y. Ikeda, Y. Tachimori, A. Seta and R. Sudoh, *J. Chem. Soc., Chem. Commun.*, 1978, 350.
140. J. P. Freeman, *J. Org. Chem.*, 1962, **27**, 1309; G. A. Boswell, Jr., *J. Org. Chem.*, 1968, **33**, 3699; C. Shiue, K. P. Park and L. B. Clapp, *J. Org. Chem.*, 1970, **35**, 2063.
141. T. Severin, P. Adhikary and I. Bräutigam, *Chem. Ber.*, 1976, **109**, 1179.
142. O. von Schickh, H. G. Padeken and A. Segnitz, *Methoden Org. Chem. (Houben-Weyl)*, 1971, **10** (1) 113.
143. S. B. Shim, K. Kim and Y. H. Kim, *Tetrahedron Lett.*, 1987, **28**, 645.
144. J. M. Aizpurua, M. Juaristi, B. Lecea and C. Palomo, *Tetrahedron*, 1985, **41**, 2903.
145. E. C. Taylor, C.-P. Tseng and J. B. Rampal, *J. Org. Chem.*, 1982, **47**, 552.
146. E. J. Corey, B. Samuelsson and F. A. Luzzio, *J. Am. Chem. Soc.*, 1984, **106**, 3682.
147. P. E. Eaton and G. E. Wicks, *J. Org. Chem.*, 1988, **53**, 5353.
148. M. Regitz, *Methoden. Org. Chem. (Houben-Weyl)*, (a) 1982, **E1**; (b) 1982, **E2**.
149. H. Heydt and M. Regitz, in ref. 148b, p. 1.
150. L. Woźniak, J. Kowalski and J. Chojnowski, *Tetrahedron Lett.*, 1985, **26**, 4965; Y. Hayakawa, M. Uchiyama and R. Noyori, *Tetrahedron Lett.*, 1986, **27**, 4191.
151. N. X. Hu, Y. Aso, T. Otsubo and F. Ogura, *Chem. Lett.*, 1985, 603.
152. G. A. Olah, B. G. B. Gupta, A. Garcia-Luna and S. C. Narang, *J. Org. Chem.*, 1983, **48**, 1760.
153. M. Masui, Y. Mizuki, C. Ueda and H. Ohmori, *Chem. Pharm. Bull.*, 1984, **32**, 1236.
154. O. Mitsunobu, *Synthesis*, 1981, 1; D. L. Hughes, R. A. Reamer, J. J. Bergan and E. J. J. Grabowski, *J. Am. Chem. Soc.*, 1988, **110**, 6487.
155. R. Luckenbach, in ref. 148b, p. 873.
156. K. Sasse, *Methoden Org. Chem. (Houben-Weyl)*, 1963, **12** (1), 199.
157. U.-H. Felcht, in ref. 148b, p. 123.
158. B. Gallenkamp, W. Hofer, B.-W. Krüger, F. Maurer and T. Pfister, in ref. 148b, p. 300.
159. B. Gallenkamp, W. Hofer, B.-W. Krüger, F. Maurer and T. Pfister, in ref. 148b, p. 578; M. G. Newton and B. S. Campbell, *J. Am. Chem. Soc.*, 1974, **96**, 7790.
160. For other methods see ref. 150.

6.2
Oxidation of Sulfur, Selenium and Tellurium

SAKAE UEMURA
Kyoto University, Japan

6.2.1 INTRODUCTION

A variety of organo-sulfur, -selenium and -tellurium compounds can be oxidized to their corresponding higher oxidation state species. Thiols are reasonably stable in air and are oxidized by various reagents to disulfides (equation 1), while selenols, although isolable, are unstable in air and are oxidized readily to diselenides (equation 2). Tellurols are very sensitive to air and are normally present as ditellurides (equation 3). These disulfides, diselenides and ditellurides are further oxidized to various higher oxidation products depending on the specific reaction conditions. Sulfides (**1**; M = S), selenides (**1**; M = Se) and tellurides (**1**; M = Te) are generally oxidized to the corresponding sulfoxides (**2**; M = S), selenoxides (**2**; M = Se) and telluroxides (**2**; M = Te), which in turn are further oxidized to sulfones (**3**; M = S), selenones (**3**; M = Se) and tellurones (**3**; M = Te), respectively (Scheme 1). These oxidized species are also written in the alternative formats of (**4**) and (**5**), and syntheses of chiral sulfoxides and selenoxides by direct oxidation are known. A variety of chemical oxidizing agents can be employed for these transformations and some electrochemical, photochemical and enzymatic oxidations are also known.

$$\text{R}-\text{SH} \xrightarrow{[\text{O}]} \text{RS}-\text{SR} \qquad (1)$$

$$\text{R}-\text{SeH} \xrightarrow{[\text{O}]} \text{RSe}-\text{SeR} \qquad (2)$$

$$[\text{R}-\text{TeH}] \xrightarrow{[\text{O}]} \text{RTe}-\text{TeR} \qquad (3)$$

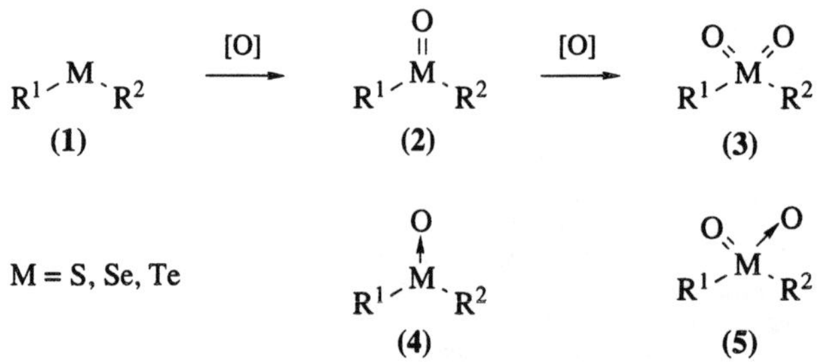

$$\underset{(\mathbf{1})}{R^1{-}M{-}R^2} \xrightarrow{[O]} \underset{(\mathbf{2})}{R^1{-}M({=}O){-}R^2} \xrightarrow{[O]} \underset{(\mathbf{3})}{R^1{-}M({=}O)_2{-}R^2}$$

M = S, Se, Te

$$\underset{(\mathbf{4})}{R^1{-}M(\rightarrow O){-}R^2} \qquad \underset{(\mathbf{5})}{R^1{-}M({=}O)(\rightarrow O){-}R^2}$$

Scheme 1

Disulfides, diselenides and ditellurides are useful compounds in their own right for introducing organic sulfur, selenium and tellurium moieties to other organic molecules. Oxidized species such as $R^1M(O)R^2$ and $R^1M(O)_2R^2$ also find much use in organic synthesis. Many of these applications can be seen in the relevant chapters of this series and the reader is referred to these reviews: (i) stabilization of carbanions by sulfur and selenium moieties [RSO, RSO_2, RSeO and $RSeO_2$] resulting in a favorable carbon–carbon bond formation (Volume 1, Chapters 2.3 and 2.6 and Volume 3, Chapter 1.3); (ii) selenoxide and telluroxide elimination reactions giving alkenes (Volume 6); (iii) 2,3-sigmatropic rearrangement of allylic sulfoxides and selenoxides affording allylic alcohols (Volume 6); (iv) Diels–Alder reaction of heterodienophiles such as alkynyl and alkenyl sulfoxides (Volume 5); and (v) substitution of RMO and RMO_2 moieties by other functional groups (Volume 8). This chapter however deals with oxidation of representative organic sulfur, selenium and tellurium compounds shown in equations (1)–(3) and Scheme 1, concentrating mainly on highly selective methods using chemical oxidizing agents, and covering the literature up to the end of 1989.

6.2.2 OXIDATION OF ORGANIC SULFUR COMPOUNDS

6.2.2.1 Oxidation of Thiols to Disulfides by Molecular Oxygen and Chemical Reagents

The oxidation of thiols to disulfides is a very facile process and many reagents function as oxidants.[1–4] Ease of oxidation usually decreases in the following order: ArSH > *n*-RSH > *s*-RSH > *t*-RSH. The use of vigorous conditions such as strong oxidants, excess oxidants, higher temperatures and longer reaction times *etc.* usually results in further oxidation of initially produced disulfides to give a mixture of several higher oxidation products (Scheme 2). It is normally difficult to synthesize these compounds selectively, although a few methods for obtaining sulfinic acids (RSO_2H) or sulfonic acids (RSO_3H) directly from thiols have been reported. For the selective synthesis of disulfides the recommended oxidants are oxygen or air in the presence of metal catalysts, bromine under phase-transfer conditions, iodine, and various metal salts such as Fe^{III}, Mn^{IV} and Tl^{III}.

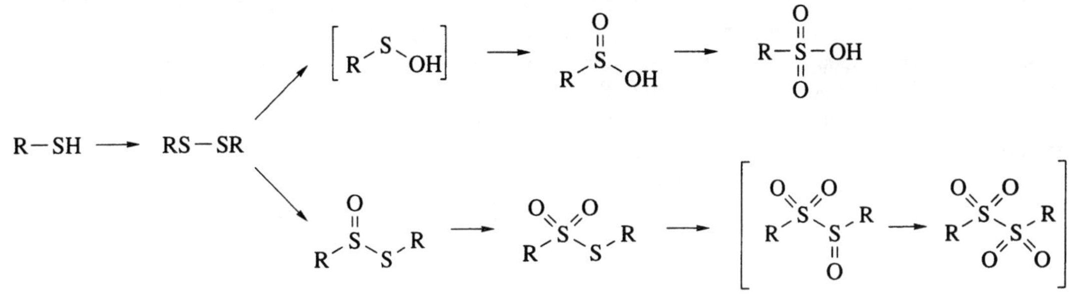

$$R-SH \rightarrow RS-SR; \quad RS-SR \rightarrow [RSOH] \rightarrow RS(O)OH \rightarrow RSO_2OH; \quad RS-SR \rightarrow RS(O)SR \rightarrow RSO_2SR \rightarrow [RSO_2S(O)R \rightarrow RSO_2SO_2R]$$

Scheme 2

6.2.2.1.1 By molecular oxygen

The oxidation of thiols to disulfides by molecular oxygen or air is believed to proceed *via* two routes as shown in Scheme 3.[5] Thus, the reaction may be enhanced under basic conditions but more stable anions react at a slower rate, while aliphatic thiols are oxidized faster than aromatic ones.[6,7]

$$4\ R-SH + O_2 \longrightarrow 2\ RS-SR + 2\ H_2O$$

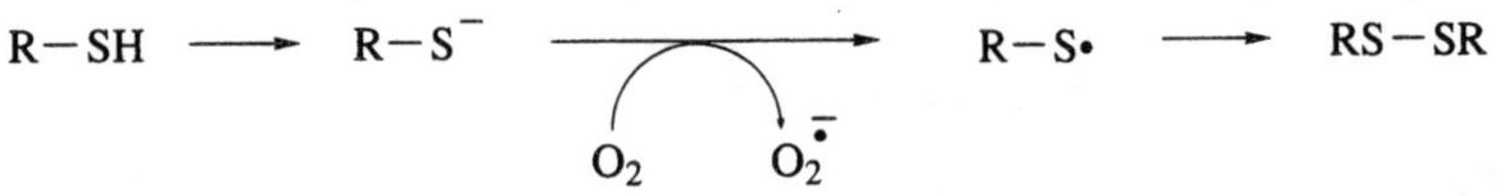

$$R-SH \longrightarrow R-S^- \xrightarrow{O_2 \rightarrow O_2^{\bullet -}} R-S\bullet \longrightarrow RS-SR$$

Scheme 3

The oxidation is catalyzed by various heavy metal ions such as Cu^{II}, Fe^{III} (hemin complex), Ni^{II} and Co^{II} and their complexes,[1,8–10] and more importantly, the addition of these ions leads to the selective formation of disulfides without any overoxidized products. The cluster $(Bu^n{}_4N)_2[Fe_4S_4(SR)_4]$, the analog of the active site of nonheme iron–sulfur proteins, catalyzed extremely smooth oxidation of thiols by oxygen to disulfides in acetonitrile at 0 °C (equation 4), while in the case of $FeCl_2$ or $FeCl_3$ catalysts oxygen uptake was very slow.[11] The catalysis by Al_2O_3 for aerobic oxidation is also common.[12] Thus, by stirring thiols in benzene with exposure to air at room temperature for 4–6 h disulfides were obtained almost quantitatively except in the hindered case of Bu^tSH.

$$4\,R^1-SH + O_2 \xrightarrow[100\%]{[Fe_4S_4(SR^2)_4]^{2-},\ MeCN} 2R^1S-SR^1 + 2\,H_2O \qquad (4)$$

$$R^1 = Ph,\ PhCH_2,\ Et$$

The use of a large excess of base and/or prolonged reaction times results in the absorption of excess oxygen leading to overoxidation.[6,13] The type of product also varies, RSO_3H in the case of KOH/HMPA[13] and a mixture of RSO_2H (major) and RSO_3H (minor) in the case of Bu^tOK/Bu^tOH.[14] More detailed investigations[15] indicated that the main product is RSO_2H which is in accord with the latter result. Similar overoxidation is also known in benzenethiol oxidation by the superoxide anion ($O_2^{\bullet -}$) generated from KO_2 and 18-crown-6.[15] Thus, at room temperature diphenyl disulfide was formed quantitatively, while both $PhSO_2H$ and $PhSO_3H$ were produced by heating with excess $O_2^{\bullet -}$ at 60 °C. In the cases of benzylic thiols desulfurization occurred unexpectedly to give benzoic acids and alkenes by aerobic oxidation in DMSO in the presence of a basic catalyst.[16]

6.2.2.1.2 By metal ions, oxides and carboxylates

Metal ions of higher oxidation states, such as Fe^{III},[17] Ce^{IV}, Co^{III} and V^{V},[18,19] oxidize thiols effectively to disulfides in the absence of oxygen. A large variety of metal oxides such as MnO_2, CrO_3, PbO_2, FeO_3, Co_2O_3 and CuO are also useful for this transformation at low temperatures in chloroform or xylene solutions.[20,21] Various modified chromium compounds such as bis(benzyltriethylammonium) dichromate,[22] bis[trinitratocerium(IV)] chromate,[23] pyridinium chlorochromate[23] and a combination of CrO_3 with chlorosilane[24] are reported to be effective oxidizing agents. Bentonite-supported iron(III) nitrate,[25] CAN[26] and bis(2,2′-bipyridyl)copper(II) permanganate[27] are also selective oxidants to produce disulfides. LTA[28–31] and thallium triacetate (TTA)[32] readily effect the oxidation of thiols to disulfides (equation 5). The use of excess LTA in alcoholic solvents resulted in a selective overoxidation of the initially formed diaryl disulfides to aromatic sulfinic esters ($ArSO_2R$).[30]

$$\text{Bu}^t\text{—SH} \;+\; \text{Tl(OAc)}_3 \xrightarrow[\text{100\%}]{\text{CHCl}_3,\ \text{r.t., 5 h}} \text{Bu}^t\text{S—SBu}^t \;+\; \text{TlOAc} \;+\; 2\ \text{AcOH} \qquad (5)$$

6.2.2.1.3 By organic oxides, hydrogen peroxide and organic peroxy acids

Dialkyl sulfoxides such as DMSO work as oxidants for the selective preparation of disulfides.[33–35] The order of thiol reactivity is ArSH > $ArCH_2SH$ > RSH.[34] Several amine oxides, such as pyridine *N*-oxides[36,37] and triethylamine oxide,[38] also oxidize sulfides to sulfoxides, but the reaction generally requires higher temperatures.

The oxidation of thiols by H_2O_2, dialkyl peroxides and peroxy acids is known to give disulfides as the initial products, which are further oxidized by excess oxidant.[2,39–41] Because of easy overoxidation these reactions are rarely used for preparative purposes of disulfides, but the method is often useful for obtaining some overoxidized compounds. Thus, RSO_3H is produced from tertiary thiols in high yields,[2] and mercaptoimidazole (**6**) was oxidized by alkaline H_2O_2 to the corresponding sulfonic acid (**7**) in a moderate yield (equation 6).[40] Selective and effective oxidation of alkane thiols to the corresponding RSO_2H occurs by treatment with 2 equiv. MCPBA (equation 7).[42]

$$(\mathbf{6}) \xrightarrow[\text{30\%}]{\text{15\% H}_2\text{O}_2\text{/NaOH, 80–90 °C}} (\mathbf{7}) \qquad (6)$$

(**6**): EtO_2C, Ph, N, N, SH (**7**): HO_2C, Ph, N, N, SO_3H

$$\text{R—SH} \xrightarrow[\text{80–85\%}]{\text{2 equiv. MCPBA, CH}_2\text{Cl}_2\text{, –30 °C}} \text{R—SO}_2\text{H} \qquad (7)$$

6.2.2.1.4 By halogens and halogen compounds

In aqueous solvents chlorine and bromine react with thiols to give sulfonyl halides or sulfonic acids (equations 8 and 9), while under anhydrous conditions various reactions occur to give sulfenyl halides (RSX), RSX_3 and/or disulfides.[1,2] On the contrary, oxidation with iodine is prone to give disulfides (equation 10) typically using a solution of I_2 in acetic acid, alcohol, ether or aqueous KI.[2,3,43,44] Under two-phase conditions of aq. $KHCO_3/CH_2Cl_2$ bromine works at room temperature as an excellent oxidant of general utility for the preparation of various disulfides (Scheme 4).[45] 2-Polyvinylpyridine–bromine complex[46] and bromodimethylsulfonium bromide ($Me_2S^+BrBr^-$)[47] are also useful reagents for obtaining disulfides from thiols.

$$\text{R—SH} \;+\; 3\,X_2 \;+\; 2\,H_2O \longrightarrow \text{R—}\overset{\overset{\large O}{\|}}{\underset{\underset{\large O}{\|}}{S}}\text{—X} \;+\; 5\,HX \qquad (8)$$

Several 'positive halogen' compounds such as NBS, NCS, (dichloroiodo)benzene ($PhICl_2$), sulfuryl chloride, thionyl chloride, 2,4,4,6-tetrabromocyclohexa-2,5-dienone[48,49] and diethyl bromomalonate[50] *etc.* react with thiols to produce initially RSX, which then reacts with excess thiol to give disulfides.[2] Tri-

$$R{-}SH + 3\,X_2 + 3\,H_2O \longrightarrow R{-}S(=O)_2{-}OH + 6\,HX \qquad (9)$$

$$2\,R{-}SH + I_2 \longrightarrow RS{-}SR + 2\,HI \qquad (10)$$

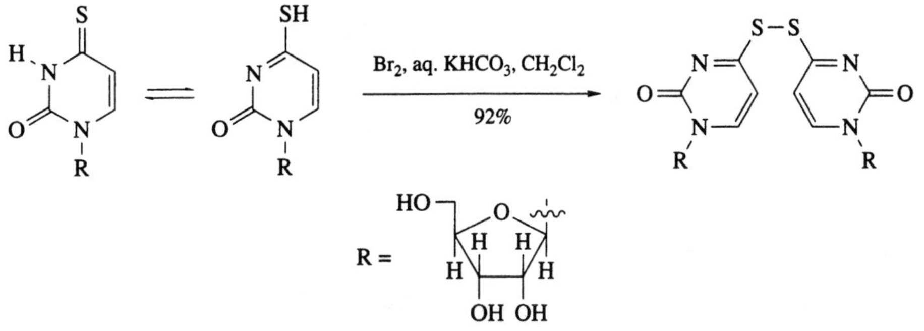

Br$_2$, aq. $KHCO_3$, CH_2Cl_2

92%

Scheme 4

chloromethanesulfonyl chloride is a very good reagent for selective disulfide formation in pyridine (equation 11).[51]

$$R{-}SH \xrightarrow[93\text{–}98\%]{Cl_3CSO_2Cl,\ pyridine} RS{-}SR \qquad (11)$$

R = Bu, Bu^t, Ph, PhCH$_2$

6.2.2.1.5 By other chemical reagents and other methods

A variety of other oxidants are known for thiol oxidation, including diethyl azodicarboxylate,[52] azodicarbonamide,[53] nitrosobenzene and nitrobenzene,[36,37,53] maleic anhydride,[53] nitrogen oxide (NO),[54] iodosylbenzene (PhIO)[55] and nickel peroxide.[56] Excess dinitrogen tetroxide (N_2O_4) reacted with thiols to give either disulfides, thiolsulfonates ($RSSO_2R$) or sulfonic acids selectively by controlling carefully the concentration of N_2O_4 and other reaction conditions (Scheme 5).[57,58] Bis(4-methoxyphenyl) telluroxide[59] and selenoxide[60] as well as their polymer-bound compounds[61] are mild oxidizing agents for preparing disulfides, the ability of the telluroxides being stronger than the selenoxides. The oxidation with powerful oxidants like HNO_3 and $KMnO_4$ has long been known to give overoxidized sulfonic acid products.[2]

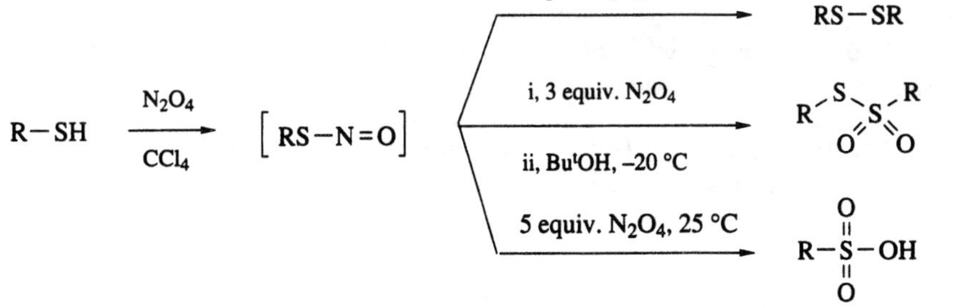

$$R{-}SH \xrightarrow[CCl_4]{N_2O_4} [RS{-}N{=}O]$$

- 1 equiv. N_2O_4, –70 °C → RS–SR
- i, 3 equiv. N_2O_4; ii, Bu^{t}OH, –20 °C → $RS{-}S(=O)_2{-}R$
- 5 equiv. N_2O_4, 25 °C → $R{-}S(=O)_2{-}OH$

Scheme 5

Flavins (**8**),[62,63] 8-azaflavin (**9**)[64] and their analogs are characteristic oxidants toward thiols under anaerobic and other specific conditions (equation 12).[62–64] 5-Arylidene-1,3-dimethylbarbituric acid derivatives (**10**) also work as the oxidant in dioxane at 120–150 °C, and the method was applied to synthesis of unsymmetrical disulfides.[65] Electrochemical oxidation[66] and photolysis[67] of thiols to disulfides are also known.

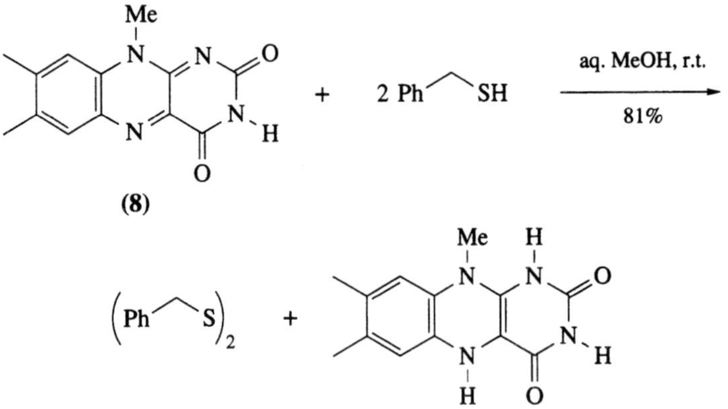

aq. MeOH, r.t.

81%

(8)

(12)

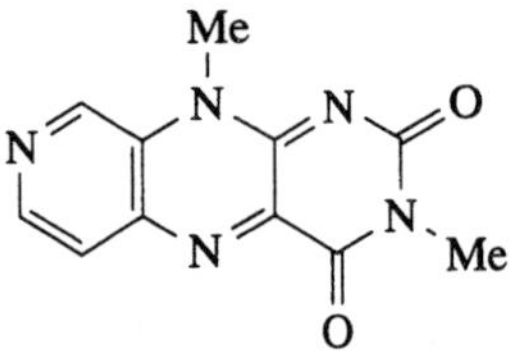

(9)

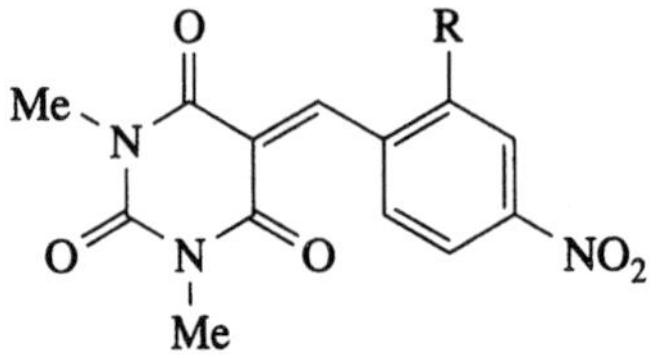

(10) R = NO_2, H

6.2.2.2 Oxidation of Sulfides to Sulfoxides by Chemical Reagents

The oxidation of sulfides (thioethers) gives the corresponding sulfoxides or sulfones or both, depending on the reaction conditions employed (equation 13).[68–70] In order to obtain sulfoxides selectively it is necessary to add equimolar or slightly excess amounts of oxidants to the sulfides under mild conditions. Among a variety of oxidants so far employed for this reaction are: MCPBA, TBHP, $PhICl_2$, PhIO and sodium metaperiodate ($NaIO_4$). All appear to be particularly useful for obtaining sulfoxides selectively in good yields.

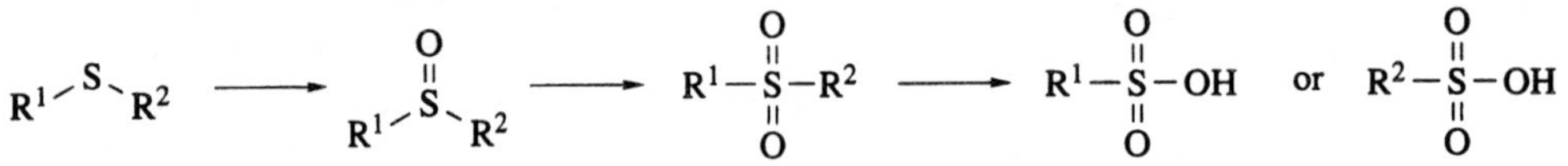

$$R^1-S-R^2 \longrightarrow R^1-S(=O)-R^2 \longrightarrow R^1-S(=O)_2-R^2 \longrightarrow R^1-S(=O)_2-OH \text{ or } R^2-S(=O)_2-OH \quad (13)$$

6.2.2.2.1 *By hydrogen peroxide and organic peroxides and peroxy acids*

One of the simplest methods of oxidation of sulfides to sulfoxides is the use of H_2O_2 in acetone[71–73] or preferably in methanol.[74] The oxidation normally proceeds highly selectively under very mild conditions and can be applied to the preparation of various acid-sensitive sulfoxides such as allylic sulfoxides **(11)**,[73] silyl-substituted vinyl sulfoxides **(12)**[75] and thietane sulfoxides **(13)**.[76] The oxidation is accelerated in the presence of acetic acid although selective oxidation for *S*-monoxides is possible as shown in equations (14),[77] (15)[78,79] and (16).[80] However, the use of excess H_2O_2 often leads to overoxidation to give sulfones. Many metal salts such as SeO_2,[81] V_2O_5,[82] $TiCl_3$[83] and $VO(acac)_2$[84] also function as good catalysts.

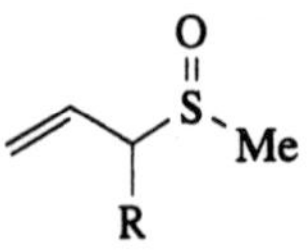

(11) 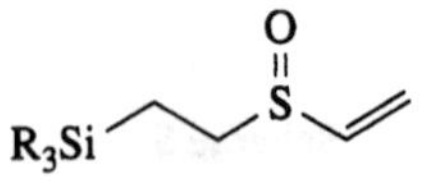 (12) 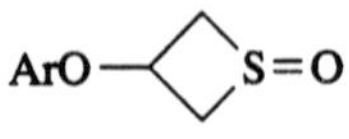 (13)

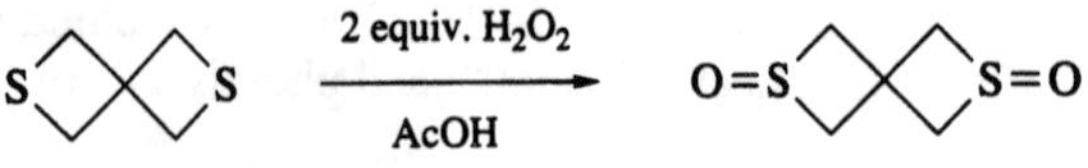

2 equiv. H_2O_2

AcOH

(14)

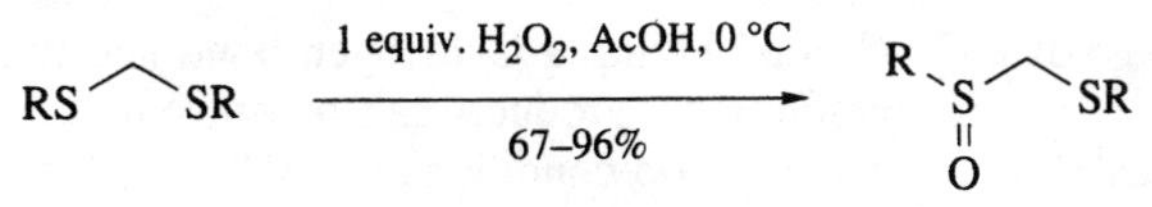

RS⌒SR —(1 equiv. H_2O_2, AcOH, 0 °C; 67–96%)→ R–S(=O)⌒SR (15)

R = Me, Et, Pr^i, Bu^t, Ph

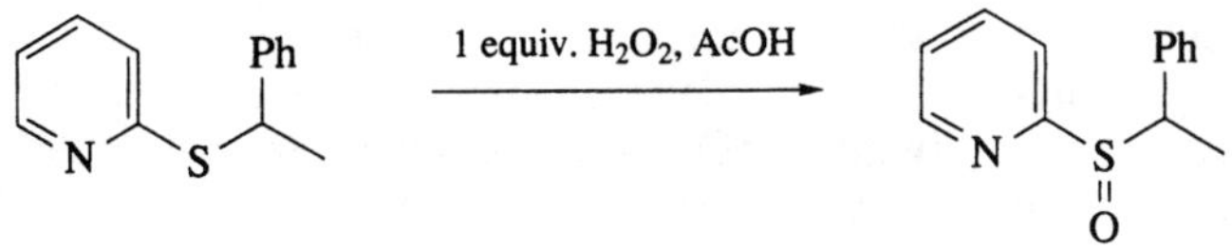

(16)

Organic peroxides such as cyclohexyl or *t*-butyl hydroperoxide[85] and benzoyl peroxides[86] can oxidize various sulfides to the corresponding sulfoxides where oxidation with TBHP in alcohols or benzene appears to be synthetically useful.[85,87,88] 2-Hydroperoxyhexafluoro-2-propanol [$(CF_3)_2C(OH)OOH$], formed *in situ* from hexafluoroacetone and H_2O_2, is also a very effective and convenient reagent for this purpose.[89] 4a-Hydroperoxylumiflavin (**14**) oxidizes sulfides to sulfoxides in organic solvents such as Bu^tOH or dioxane at 30 °C much more effectively than H_2O_2.[90–92] The reactivity of the flavins was shown to be 10^3–10^6 more reactive than TBHP and ~10^3 less reactive than MCPBA.[91]

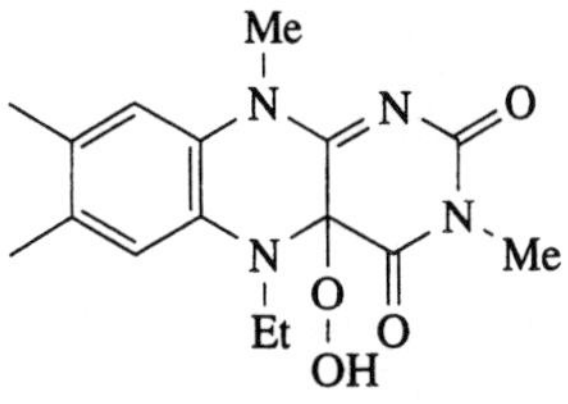

(14)

A variety of organic peroxy acids such as perbenzoic acid,[93,94] MCPBA,[95,96] monoperoxyphthalic acid,[97] peracetic acid[98] and trifluoroperacetic acid[99] are much stronger oxidants than H_2O_2, and oxidize sulfides to sulfoxides under very mild conditions. Usually 1 equiv. of peroxy acid to sulfide is employed, otherwise overoxidation easily occurs to give sulfones.[99] Among these, MCPBA has the advantage of being convenient to use and the oxidation is normally carried out at 0 °C or lower temperatures, in dichloromethane. The preparations of the base-sensitive sulfoxide (**15**),[100] a new dienophile alkynyl sulfoxide (**16**),[101] and thiiraneradialene *S*-oxide (**17**)[102] are typical examples. Selective oxidation of the sulfur atom of penicillins by polymer-supported peroxy acids in DMF or acetone is also known (equation 17).[103]

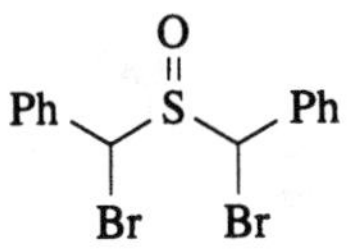

(15) 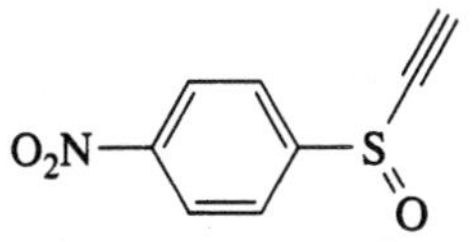 **(16)** 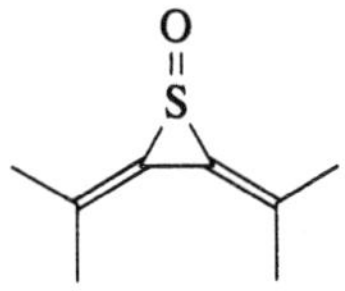 **(17)**

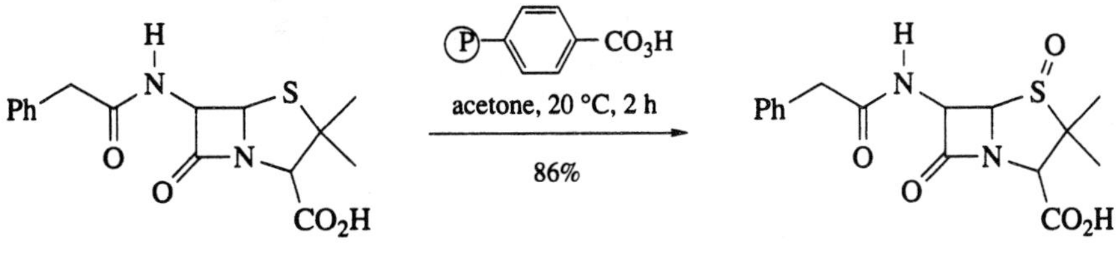

(17)

6.2.2.2.2 By halogens and halogen compounds

Molecular halogens have been known to form some addition compounds (**18**) with organic sulfides which are readily hydrolyzed to sulfoxides (Scheme 6). However, undesired side reactions sometimes occur, such as cleavage of a C—S bond giving a C–halogen bond and halogenation or alkoxylation at various positions.[104–106] The formation of by-products can be prevented by carrying out the reaction in the presence of amines[107] or under two-phase conditions (CH_2Cl_2/H_2O) using $KHCO_3$ as a base.[108] The treatment of sulfides with bromine and then with hexabutyldistannoxane gives sulfoxides in high yields without sulfone by-product contamination (equation 18).[109] The sulfoxide formation by iodine oxidation is relatively slow, but can be accelerated by certain nucleophiles such as phthalate ion[110] or β-cyclodextrin phosphate ion.[111] Sulfides react with the stoichiometric quantity of $PhICl_2$ in aqueous pyridine between –40 and 20 °C to give high yields of the corresponding sulfoxides completely free from sulfones, and use of ^{18}O-enriched water readily gives ^{18}O-labeled sulfoxides (equation 19).[112] PhIO,[55,113] $PhIO_2$ with $VO(acac)_2$ catalyst[114] and (diacetoxyiodo)benzene $[PhI(OAc)_2]$[115] can also be used for oxidation of some sulfides.

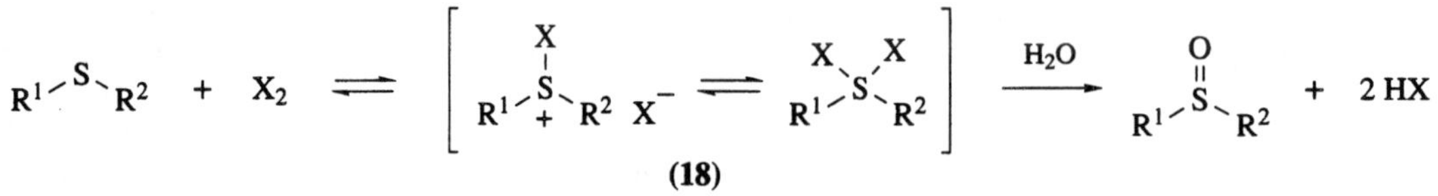

$$R^1SR^2 + X_2 \rightleftharpoons \left[R^1\overset{+}{S}(X)R^2\ X^- \rightleftharpoons R^1SX_2R^2 \right] \xrightarrow{H_2O} R^1S(O)R^2 + 2\,HX$$

(18)

Scheme 6

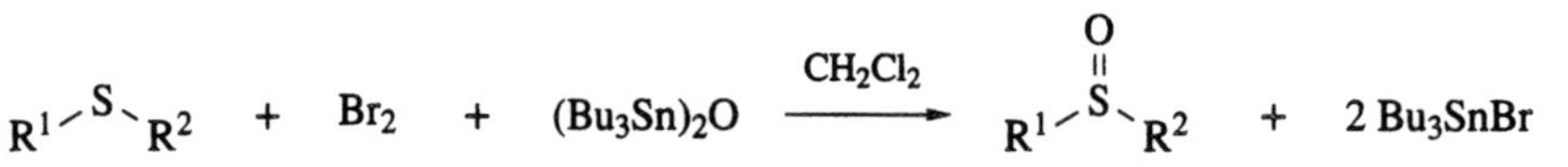

$$R^1SR^2 + Br_2 + (Bu_3Sn)_2O \xrightarrow{CH_2Cl_2} R^1S(O)R^2 + 2\,Bu_3SnBr \tag{18}$$

$$PhSPh \xrightarrow{\text{pyridine, } PhICl_2,\ H_2{}^{18}O,\ 20\ °C} PhS(^{18}O)Ph \tag{19}$$

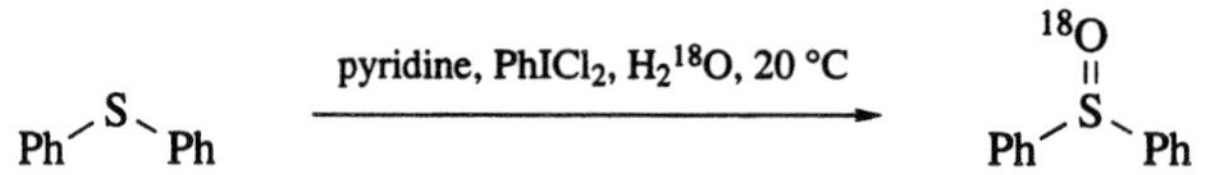

Many 'positive halogen' compounds such as NBS,[116–118] NCS,[118] 1-chlorobenzotriazole,[119] and chloramine-T and bromamine-T[120–122] oxidize sulfides to sulfoxides. Sulfides may also be oxidized efficiently to sulfoxides uncontaminated by sulfones using 2,4,4,6-tetrabromocyclohexa-2,5-dienone in aqueous dioxane or THF (equation 20)[49] and also using sulfuryl chloride at room or lower temperature.[123,124] Hypochlorites such as sodium hypochlorite (NaOCl),[125,126] hypochlorous acid (HOCl)[125] and Bu^{t}OCl[127–129] are sometimes used as stereo- and chemo-selective oxidants for obtaining sulfoxides from sulfides, as shown in equations (21)[126] and (22).[127] However, application to sulfides bearing an ethynyl or a methoxycarbonyl group α to sulfur resulted in sole formation of the corresponding α-alkoxy sulfides, rather than the expected sulfoxides (equation 23).[128] Oxidation of sulfides by $NaIO_4$ in water, aqueous methanol or other organic solvents at 0 °C results in a selective formation of sulfoxides.[130] This reaction was applied to the preparation of various sulfoxides including 1-butadienyl phenyl sulfoxide (**21**),[131] α-phos-

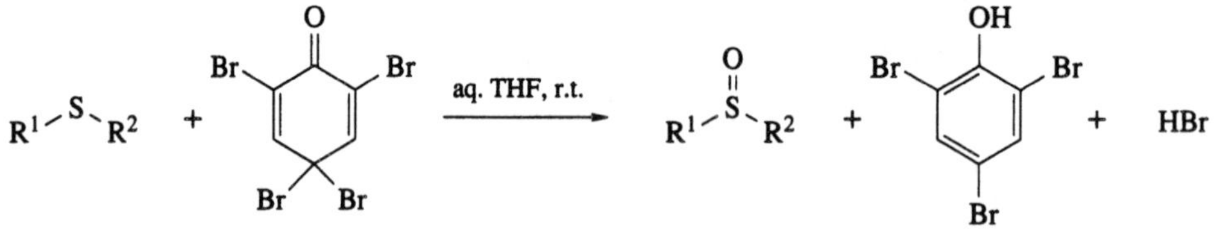

$$R^1SR^2 + \text{2,4,4,6-tetrabromocyclohexa-2,5-dienone} \xrightarrow{\text{aq. THF, r.t.}} R^1S(O)R^2 + \text{2,4,6-tribromophenol} + HBr \tag{20}$$

$$\text{thiacyclooctene} \xrightarrow[88\%]{\text{1 equiv. NaOCl, dioxane, 20 °C}} \text{thiacyclooctene S-oxide} \tag{21}$$

$$\underset{\textbf{(19)}}{\text{4-R-thiane}} \xrightarrow[70\%]{Bu^tOCl,\ MeOH} \underset{\textbf{(20)}\ cis > 89\%}{\text{4-R-thiane S-oxide}} \tag{22}$$

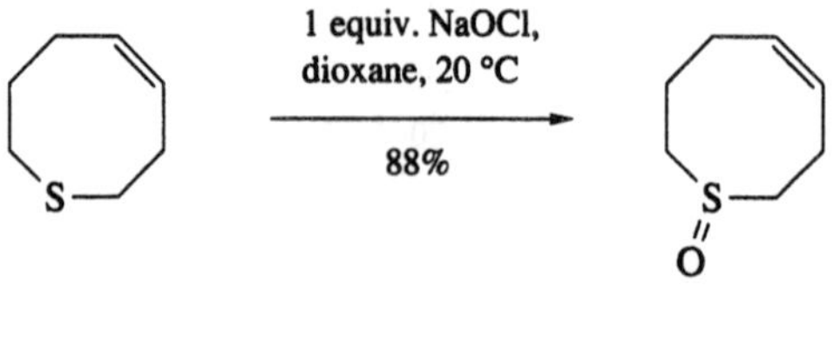

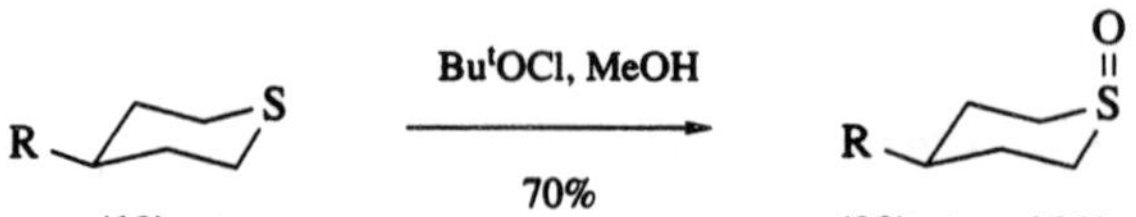

phoryl sulfoxides (**22**),[132] and sulfoxides containing a disulfide moiety (**23**).[133] Various modifications have been made to this method, such as the use of tetrabutylammonium periodate[134] and the use of alumina-[135] and silica gel-supported $NaIO_4$.[136] Sodium bromite ($NaBrO_2$) is also used as an oxidant for sulfoxides in aqueous dioxane.[137]

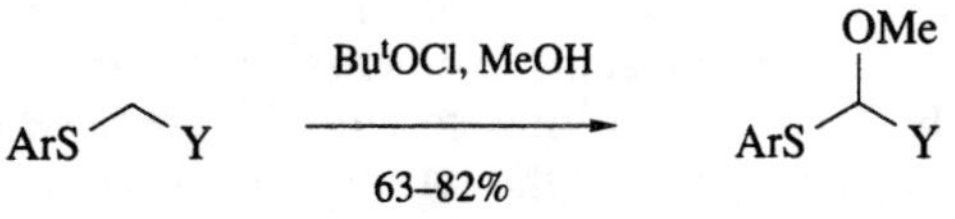

ArS–CH₂–Y → (Bu^{t}OCl, MeOH; 63–82%) ArS–CH(OMe)–Y (23)

Y = C≡CH, CO_2Me

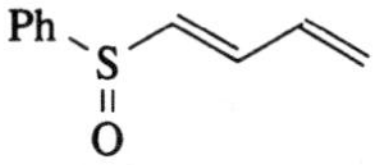

(21) 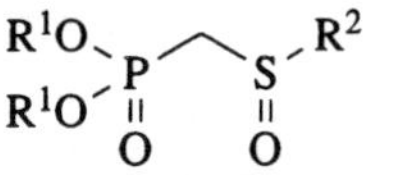 **(22)** 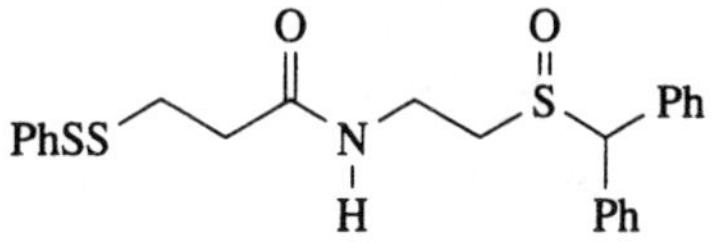 **(23)**

6.2.2.2.3 *By other chemical reagents and other methods*

A variety of other oxidants that produce sulfoxides from sulfides are known, such as nitric acid,[138,139] acyl nitrates,[140] nitronium salts,[141] N_2O_4,[142,143] oxygen with ruthenium complex catalysts,[144,145] pressurized oxygen with CAN as a catalyst,[146] ozone,[147,148] TTN,[149] Ce^{IV} salts,[150,151] LTA[152,153] and potassium peroxodisulfate ($K_2S_2O_8$).[154] Potassium hydrogen persulfate ($KHSO_5$; Oxone) is a very efficient and chemoselective oxidant to produce sulfoxides from sulfides under phase-transfer conditions[155] and also in a catalytic cycle involving *N*-sulfonyloxaziridines (**24**), as shown in Scheme 7.[156] Selenium compounds such as dialkyl[157] or diaryl selenoxides,[60] areneselenonic acids ($ArSeO_3H$)[158,159] and areneseleninic acids ($ArSeO_2H$)[159] are also useful reagents for the sulfoxidation of various sulfides. Sulfides may be oxidized to sulfoxides highly selectively and almost quantitatively by sulfinylperoxy intermediates ($ArSO_3\cdot$ or $ArSO_3^-$; Ar = 2-$NO_2C_6H_4$) generated *in situ* from 2-nitrobenzenesulfinyl chloride (ArSOCl) and superoxide (O_2^-) in acetonitrile at –25 °C.[160]

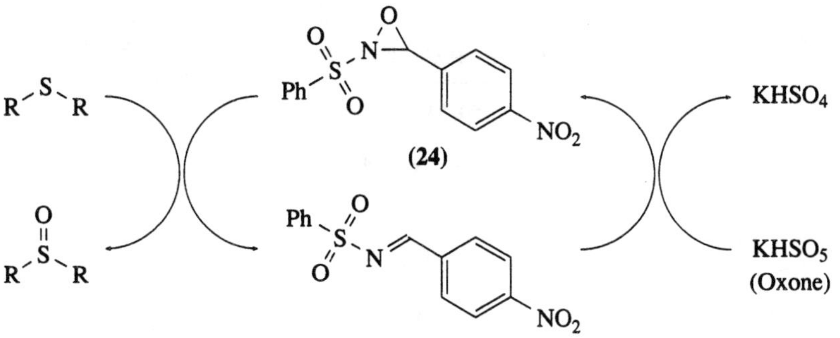

R–S–R, R–S(=O)–R, **(24)**, NO_2, $KHSO_4$, $KHSO_5$ (Oxone)

Scheme 7

Sulfides are also oxidized to sulfoxides under electrochemical[161–165] and photochemical[166–168] conditions. Electrochemical oxidation has remained rather unselective, giving mixtures of sulfoxides and sulfones, but a highly selective electrolytic oxidation is now known where the electric current passes through a bath of 4-polyvinylpyridine hydrobromide (PVP·HBr) and sulfide (equation 24).[163]

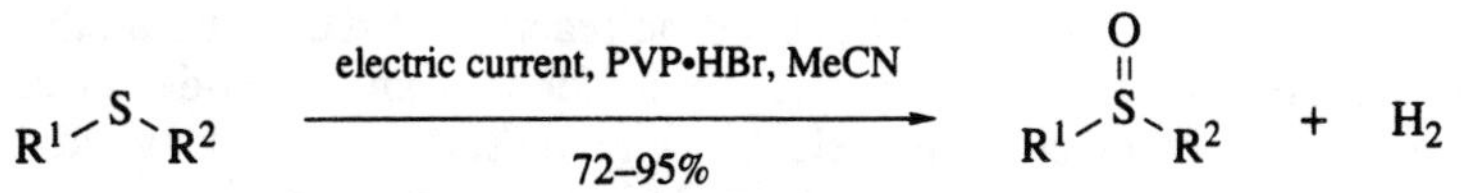

R^1–S–R^2 → (electric current, PVP•HBr, MeCN; 72–95%) R^1–S(=O)–R^2 + H_2 (24)

R^1, R^2 = alkyl, aryl

6.2.2.3 Oxidation of Sulfoxides to Sulfones by Chemical Reagents

Oxidation of sulfoxides usually results in the formation of sulfones, although under extremely vigorous conditions RSO_3H may be produced.[169] Sulfones are also obtained by direct oxidation of sulfides *via* sulfoxides (equation 13). A variety of oxidants have been utilized to effect these transformations, among which H_2O_2 in acetic acid, NaOCl, $PhICl_2$, CrO_3, $KMnO_4$ and $KHSO_5$ (Oxone) appear to be most useful from the synthetic viewpoint. With peroxy acids and peroxides the oxidation of sulfoxides to sulfones generally proceeds more slowly than that of sulfides to sulfoxides because of the reduced nucleophilicity of the sulfoxide sulfur atom compared to that of the sulfide. Completely the reverse is normally observed with oxidants of a nucleophilic nature such as periodates or transition metal salts such as CrO_3 and $KMnO_4$.

6.2.2.3.1 By hydrogen peroxide and organic peroxy acids and peroxides

Hydrogen peroxide has long been known to oxidize sulfoxides to sulfones either alone or in the presence of metal catalysts such as Fe^{III} salts,[170,171] Na_2WO_4,[172,173] $WO_3 \cdot H_2O$[173] and Na_2VO_4.[173] The use of H_2O_2 in acetic acid is also most effective for obtaining sulfones, the reactive species most probably being peracetic acid. The oxidation is catalyzed by $Mn(acac)_3$.[174] The oxidation of diaryl sulfoxides to diaryl sulfones has been shown to proceed several hundred times slower than that of diaryl sulfides to the sulfoxides,[175] in accordance with the result shown (for dialkyl compounds) in equation (14). Probably as an exception, the oxidation of 1,4-dithiadiene monosulfoxide (**25**) is reported to give the corresponding monosulfone instead of the expected bis-sulfoxide (equation 25).[176]

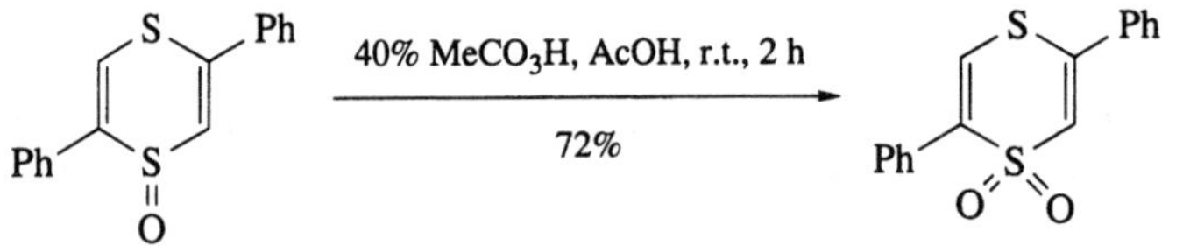

40% $MeCO_3H$, AcOH, r.t., 2 h; 72% (25)

(**25**)

Many other peroxy acids, such as trifluoroperacetic acid (equation 26),[99,177] peroxydodecanoic acid (equation 27)[178] and various perbenzoic acids[100,179–182] are also useful oxidants to give a high yield of sulfones from sulfoxides or directly from sulfides under suitable conditions.

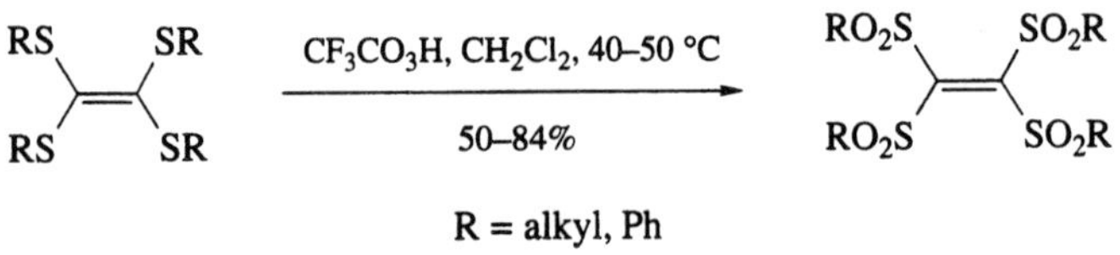

$(RS)_2C{=}C(SR)_2$ → CF_3CO_3H, CH_2Cl_2, 40–50 °C; 50–84% → $(RO_2S)_2C{=}C(SO_2R)_2$ (26)

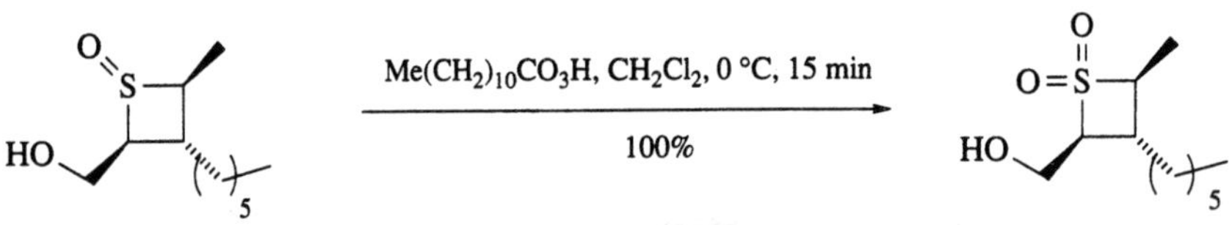

R = alkyl, Ph

$Me(CH_2)_{10}CO_3H$, CH_2Cl_2, 0 °C, 15 min; 100% (27)

Organic hydroperoxides are generally used for the preparation of sulfoxides from sulfides,[85,87,88] while sulfones can be obtained in neutral organic solvents in the presence of metal catalysts such as V, Mo and Ti oxides at 50–70 °C.[183] Two polymer-supported reagents which involve peroxy acid groups[184] and bound hypervalent vanadium(V)[185] and molybdenum(VI)[186] compounds have been developed for facile oxidation of sulfoxides to sulfones.

Solutions of KO_2 and 18-crown-6 in DMSO cause oxidation of the solvent to the sulfone.[187] The phosphorus-containing peroxy anion $(EtO)_2PO_3^-$ obtained on reaction of KO_2 with diethyl chlorophosphate [$(EtO)_2POCl$] in acetonitrile has been used to prepare sulfones from sulfoxides at 20 °C.[188] Similarly, various sulfoxides are readily oxidized chemoselectively to the sulfones in high yields by 2-nitrobenzene peroxysulfur intermediate (**26**) generated *in situ* from 2-nitrobenzenesulfonyl chloride and KO_2 (Scheme 8; Ar = $2\text{-}NO_2C_6H_4$).[189]

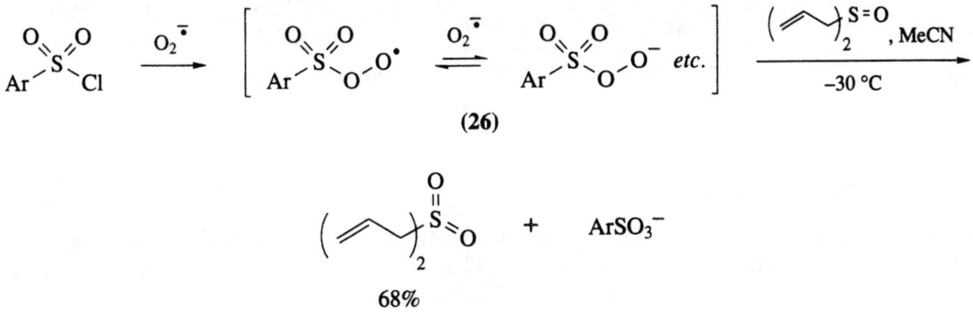

(26)

68%

Scheme 8

6.2.2.3.2 By halogens and halogen compounds

A number of halogen compounds are capable of oxidizing sulfoxides to sulfones, although synthetically useful procedures are rather limited. Chlorine and sulfuryl chloride oxidize sulfoxides to sulfones in aqueous solvents, but the oxidation is often accompanied by many side reactions as shown in equations (28)[190] and (29).[191] NaOCl seems to be a general oxidant to generate sulfones from either sulfides[192] or sulfoxides.[126] At low temperature l-chlorobenzotriazole converts sulfoxides to sulfones in high yields.[119] In aqueous pyridine solution excess $PhICl_2$ oxidizes most diaryl sulfides and sulfoxides to the corresponding sulfones, except those bearing electron-withdrawing substituents on the aryl ring (equations 30 and 31).[112,193] PhIO with ruthenium(II) complex as a catalyst[194] as well as $PhI(OAc)_2$[195] work as oxidants for producing sulfones either from sulfoxides or sulfides, the products depending on the amount of the oxidant used (Scheme 9).[194] Bromine and hypobromite (OBr^-) oxidize sulfoxides to sulfones in alkaline solutions, but multihalogenated products are usually produced (equation 32).[196]

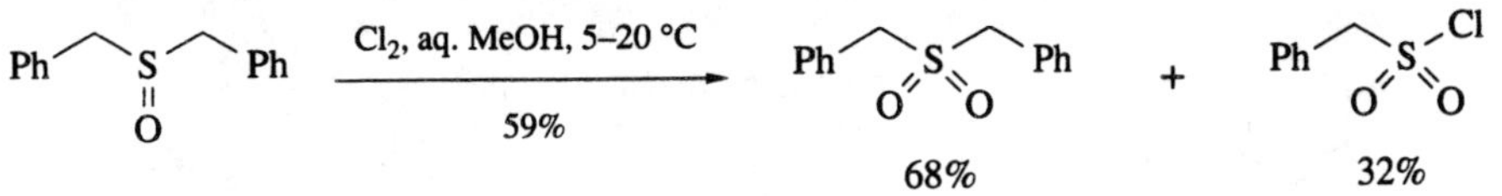

Cl_2, aq. MeOH, 5–20 °C; 59%; 68%; 32% (28)

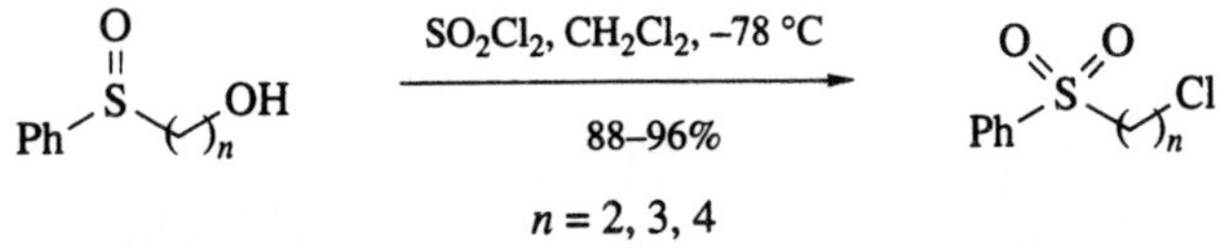

SO_2Cl_2, CH_2Cl_2, –78 °C; 88–96%; $n = 2, 3, 4$ (29)

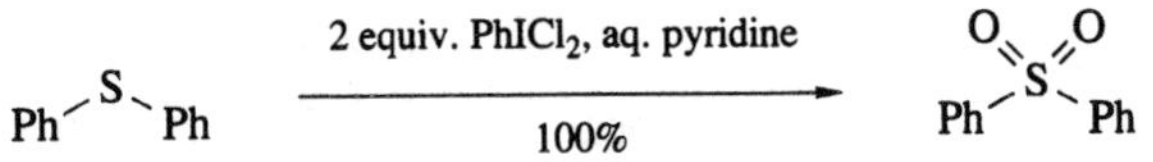

2 equiv. $PhICl_2$, aq. pyridine; 100% (30)

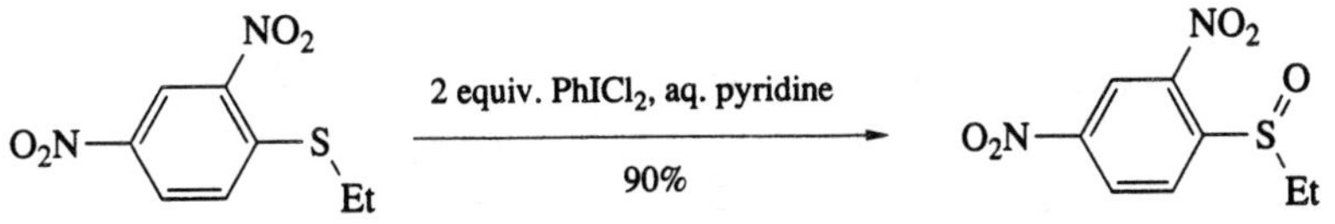

2 equiv. $PhICl_2$, aq. pyridine; 90% (31)

1.2 equiv. PhIO, CH_2Cl_2, $RuCl_2(PPh_3)_3$, r.t., 15 min; 93%; 82%; 18%

2.5 equiv. PhIO, CH_2Cl_2, $RuCl_2(PPh_3)_3$, r.t., 15 min; 100%

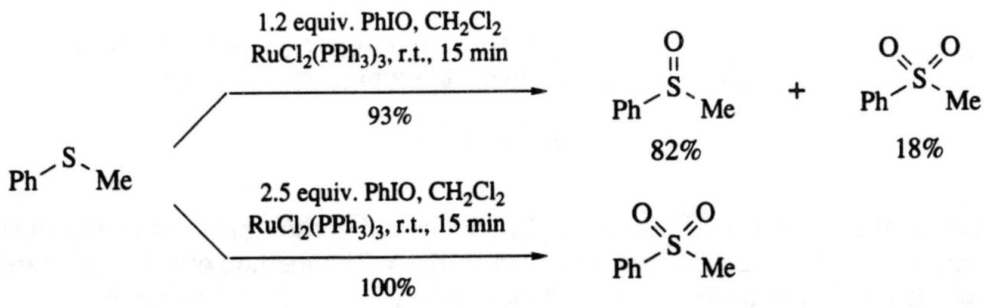

Scheme 9

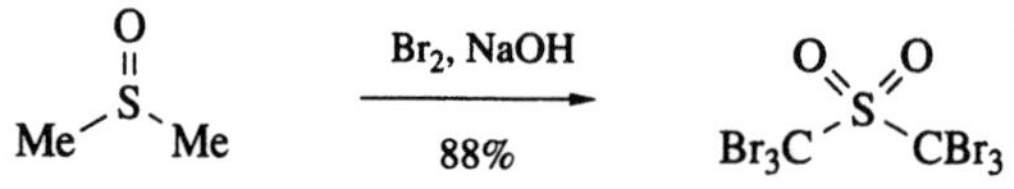

(32)

6.2.2.3.3 *By transition metal salts*

In sharp contrast to peroxy acid oxidation[175] the oxidation of sulfoxides to sulfones with various transition metal salts proceeds much faster than that of sulfides to sulfoxides and consequently sulfoxides may be selectively oxidized to sulfones in the presence of sulfides.

Chromium trioxide is a very effective oxidant in aqueous H_2SO_4,[197] aqueous acetic acid[198] or water[199] where many other functional groups are tolerated (equation 33).[199] Potassium permanganate, though slightly less reactive than CrO_3, can also be used for oxidation of sulfoxides to sulfones in aqueous acid media (equations 34–36).[197,200–202] In the presence of $MgSO_4$ and in acetone, $KMnO_4$ becomes a chemoselective oxidant which reacts faster with sulfoxides than with sulfides, as exemplified in Scheme 10 for the preparation of 1,3-dithietane 1,1-dioxide (**27**).[203] Comparison with other oxidants highlights the selectivity in this reaction.[203]

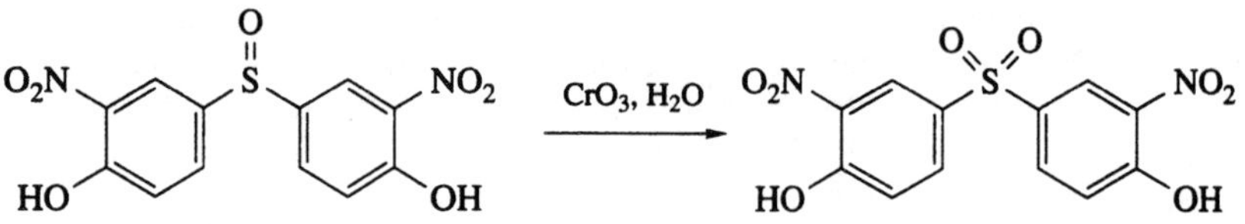

(33)

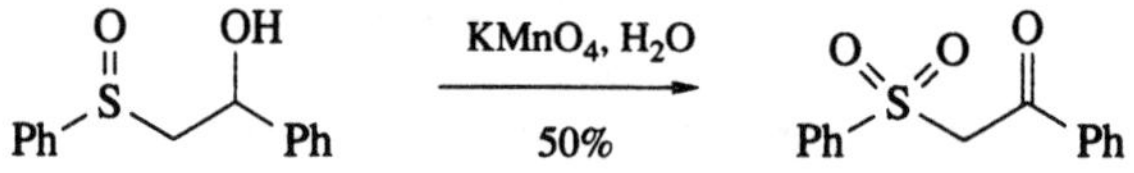

(34)

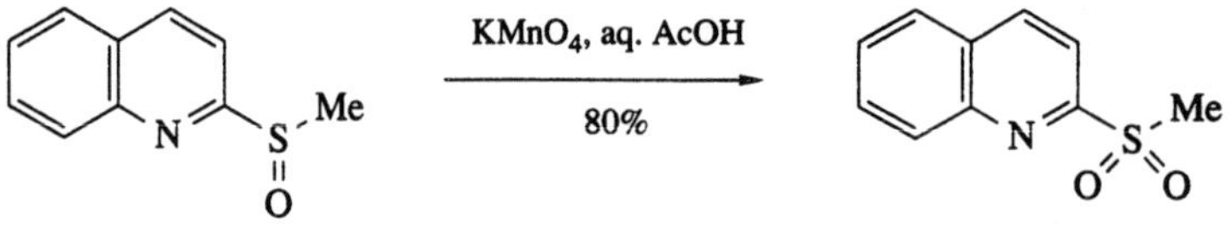

(35)

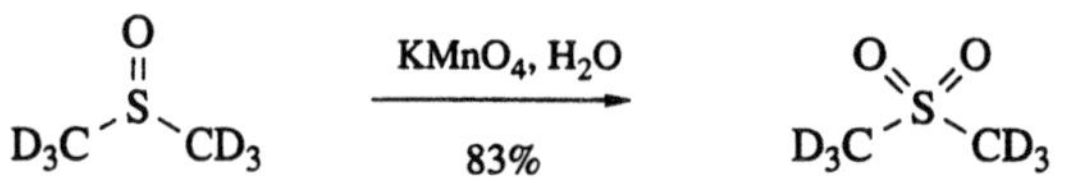

(36)

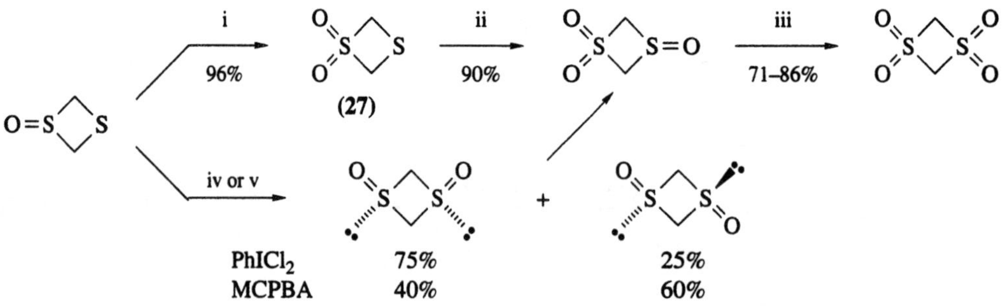

$PhICl_2$	75%	25%
MCPBA	40%	60%

i, $KMnO_4$, $MgSO_4$, acetone, –20 °C; ii, $MeCO_3H$, $CHCl_3$, 0 °C; iii, 30 equiv. $MeCO_3H$, 100 °C, 4 h; iv, $PhICl_2$, aq. pyridine, –30 °C; v, MCPBA, CH_2Cl_2, 0 °C

Scheme 10

Permanganates of zinc,[204] sodium[205] and benzyltriethylammonium[206] have also been shown to be effective and selective oxidants to obtain sulfones from sulfides. The method of sulfoxide oxidation under phase-transfer conditions has also been developed using $KMnO_4$[207] and $Cu(MnO_4)_2$.[208]

Osmium tetroxide is another chemoselective oxidant reported to yield sulfones from sulfoxides. Thus, treatment of a mixture of diphenyl sulfide and sulfoxide with OsO_4 in boiling ether for 48 h affords diphenyl sulfone in 96% yield without any change of the sulfide component.[205] Other transition metal salts such as Ce^{IV}, Ni^{IV} and Ru^{VIII} may be used for sulfoxide oxidation.

6.2.2.3.4 By other chemical reagents and other methods

A number of other oxidants which produce sulfones from sulfoxides are known, such as HNO_3,[209] NO_2BF_4,[210] oxygen with Ir or Rh catalysts,[211] ozone,[147] $KHSO_5$ (Oxone)[212] and $K_2S_2O_8$.[213] Oxone is a highly chemoselective oxidant for the conversion of sulfides to sulfones without affecting hydroxy or alkenic groups (equation 37).[212] Similarly flavin (**14**) oxidizes aryl methyl sulfoxides to sulfones fairly selectively.[91]

OH | $KHSO_5$, aq. MeOH, r.t., 4 h; 94–98% | OH; bicyclic SR → bicyclic $O{=}S{=}O$ (R); R = Ph, Me (37)

It is worthy of note that oxidation of thiolsulfinate (**28**) with sodium periodate in aqueous acetonitrile or dioxane gives thiolsulfonate (**29**) by attack on the sulfur atom of the S=O moiety and not the S atom, as shown in equation (38).[214] These data clearly indicate the nucleophilic nature of the periodate ion.

$$R^1S(O)SR^2 \ (\mathbf{28}) \xrightarrow[\sim 100\%]{\text{1 equiv. NaIO}_4\text{, aq. MeCN, HCl, r.t., 1 h}} R^1S(O)_2SR^2 \ (\mathbf{29}) \quad (38)$$

Diphenyl sulfoxide[215] and dimethyl sulfoxide[216] are oxidized electrochemically to the corresponding sulfones in acetonitrile and 1 M H_2SO_4, respectively. The product yields are enhanced by the presence of transition metal salts or oxides of W, V, Mo or Se.[217] In some cases sulfonic acid salts are formed with C—S cleavage.[218] Sulfoxides react with oxygen under photochemical conditions to give sulfones in good yields.[219–222]

6.2.3 OXIDATION OF ORGANIC SELENIUM COMPOUNDS[223,224]

6.2.3.1 Oxidation of Selenols to Diselenides and Further Oxidized Species by Chemical Reagents

As compared with thiols, selenols are readily oxidized to diselenides (equation 39). Although aryl selenols are more stable than alkyl selenols, both must be stored under inert gas in order to preserve the selenol form.[225] Consequently they are normally prepared *in situ* and used directly as a source of RSe moiety or as a reducing agent. Diselenides can be prepared merely by bubbling air or oxygen through a selenol solution.[226–228] Other oxidizing agents employed are H_2O_2,[229,230] Br_2,[231] R_2NCl[232] and $K_3Fe(CN)_6$.[233]

$$4\ R{-}SeH\ +\ O_2 \longrightarrow 2\ RSe{-}SeR\ +\ 2\ H_2O \quad (39)$$

Oxidation of diselenides with Br_2,[234] H_2O_2,[235–239] HNO_3,[239–241] ozone,[242] TBHP[243,244] and MCPBA[235] affords seleninic acids (**30**) and/or their anhydrides (**31**; equation 40), both of which are known as useful oxidants of various organic compounds.[245] The oxidation has been suggested to proceed as shown in Scheme 11[242] and as evidence intermediate compounds were isolated from the MCPBA oxidation of

$$RSe{-}SeR \xrightarrow{[O]} RSe(O)OH\ (\mathbf{30}) \quad \text{and/or} \quad RSe(O){-}O{-}Se(O)R\ (\mathbf{31}) \quad (40)$$

naphtho[1,8-*cd*]-1,2-diselenole (**33**; Scheme 12).[246] In some cases, however, selenenic acids (RSeOH) are isolated which readily disproportionate into selenols and seleninic acids (**30**).[234] Only a few RSeOH compounds have so far been isolated,[235,247,248] and many compounds previously claimed as RSeOH are in fact anhydrides (**32**).[249,250] The seleninic acids are also synthesized by oxidation of selenocyanates (RSeCN) with HNO_3,[239,251–253] peracetic acid[238] and $KMnO_4$.[234]

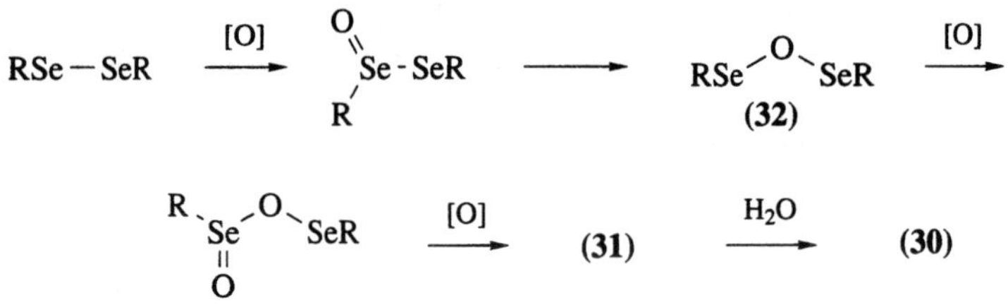

RSe—SeR —[O]→ R–Se(=O)–SeR → RSe–O–SeR (**32**) —[O]→ R–Se(=O)–O–SeR —[O]→ (**31**) —H_2O→ (**30**)

Scheme 11

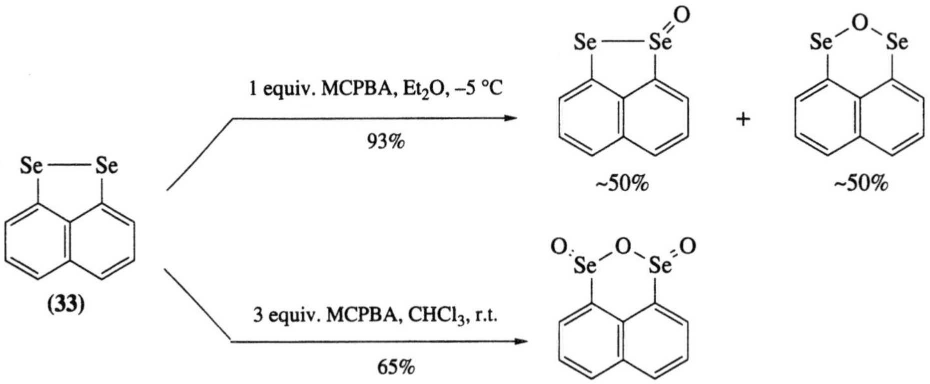

(**33**) —1 equiv. MCPBA, Et_2O, –5 °C, 93%→ Se—Se(=O) naphthalene ~50% + Se–O–Se naphthalene ~50%

(**33**) —3 equiv. MCPBA, $CHCl_3$, r.t., 65%→ O=Se–O–Se=O naphthalene

Scheme 12

Permanganate oxidation of seleninic acids has been reported to give selenonic acids ($RSeO_3H$).[240,254,255] The oxidation product of benzeneseleninic acid seems to be the selenonate (**34**).[256a] Selenonic acids or their K or Na salts may be isolated by ion exchange chromatography[255,256a] and are generally very hygroscopic substances having strong oxidizing properties.

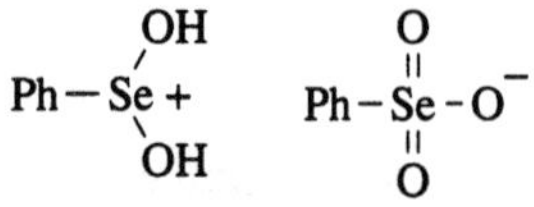

$Ph{-}Se^{+}(OH)_2$ $Ph{-}Se(=O)_2{-}O^{-}$

(**34**)

Oxidation of diphenyl diselenide with $(NH_4)_2S_2O_8$ gives phenylselenenyl cation ($PhSe^+$) which effects the oxyselenenylation of alkenes.[256b] Similar oxidation has also been effected electrochemically.[256c]

6.2.3.2 Oxidation of Selenides to Selenoxides by Chemical Reagents

Selenides may be oxidized by various reagents to selenoxides. When the resulting selenoxides bear a β-hydrogen atom *syn* elimination giving alkenes occurs readily at room temperature with formation of selenenic acid by-products (Scheme 13). For allylic selenides, the oxidation does not lead to conjugated

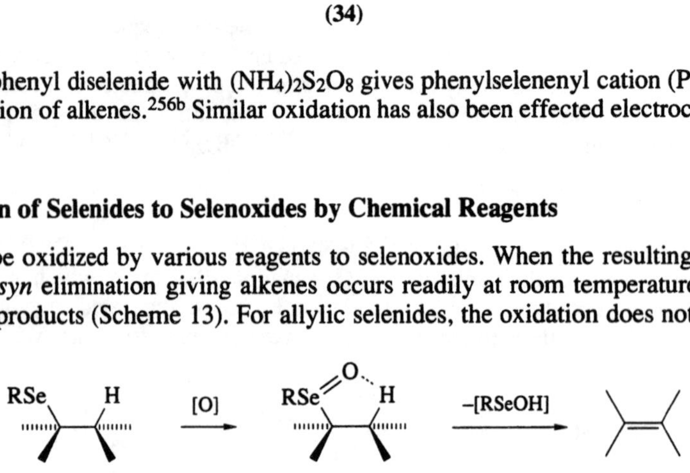

RSe, H —[O]→ RSe(=O)···H —–[RSeOH]→ alkene

Scheme 13

dienes, rather a facile formation of allylic alcohols *via* selenenic esters (**35**) by 2,3-sigmatropic rearrangement occurs (Scheme 14). These two reactions are very useful in synthetic organic chemistry and described in detail in Volume 6 of this series.

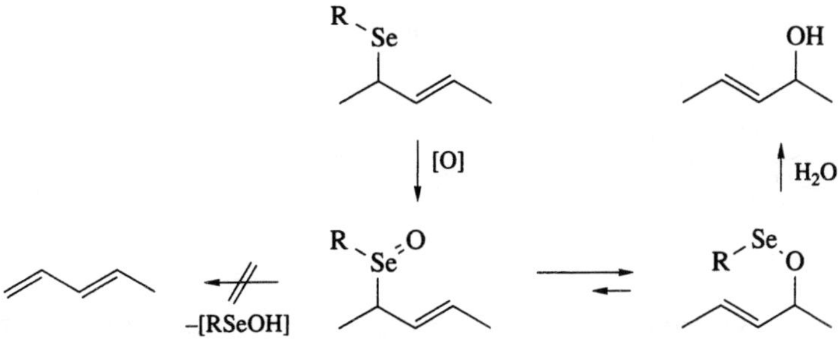

[O]

H_2O

–[RSeOH]

(**35**)

Scheme 14

The isolable and thermally stable selenoxides are, therefore, rather limited. Stable examples are as follows: those derived from selenides which have no hydrogen atoms on the β-carbon, such as dimethyl selenide,[256,257] aryl methyl selenides,[256,257] diaryl selenides[258,259] and benzyl phenyl selenides,[260,261] those with an intramolecular hydrogen bonding, such as (**36**)[262,263] and (**37**),[264] and those leading to an unfavorable double bond such as (**38**).[265] Vinylic selenoxides (**39**)[266] and (**40**)[267] are also generally isolable.

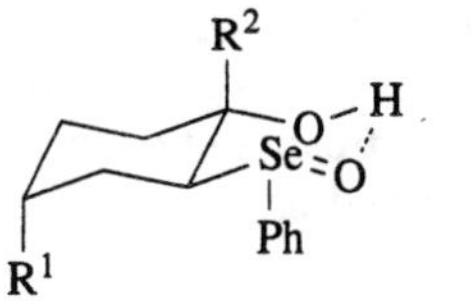

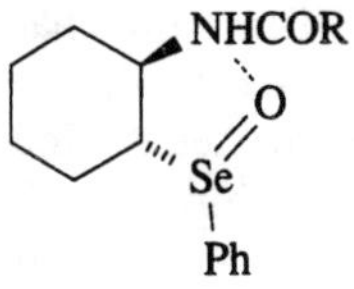

(**36**) R = 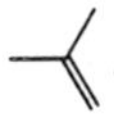 , R^2 = Me; R^1 = R^2 = H

(**37**)

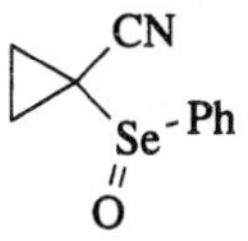

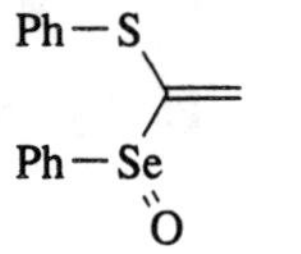

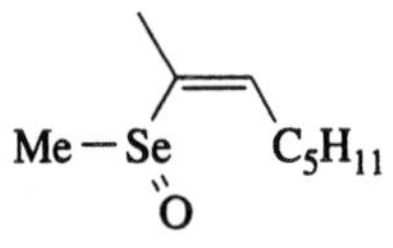

(**38**) (**39**) (**40**)

6.2.3.2.1 *By hydrogen peroxide and organic peroxides and peroxy acids*

The most commonly used oxidant for selenium compounds is H_2O_2.[227,228,268–272] The oxidation procedure normally involves addition of 30% H_2O_2 to a THF or preferably dichloromethane solution of the selenide at 0 °C. The oxidation proceeds chemoselectively and, thus, many potentially oxidizable functional groups such as alkenes, sulfides, amines, sulfoxides, tertiary alcohols, esters, lactones, nitriles and carboxylic acids remain intact. Several selenoxides including (**36**) and (**37**) were isolated by this method.[259,263,264,273]

MCPBA and peracetic acid are also effective oxidants,[272,274] particularly at low temperature (–78 °C) where the selenoxides are stable.[275–277] These reagents may be used in the presence of a double bond, a triple bond[278] or an amino group.[272] Various selenoxides, especially vinylic selenoxides such as (**40**), have been isolated using this route.[267,279–282] *t*-Butyl hydroperoxide is an especially mild oxidant which can be used in excess as a replacement for H_2O_2 without undesirable overoxidation side-reactions taking place.[283]

6.2.3.2.2 *By other chemical reagents*

Ozone is an effective chemoselective oxidant at low temperatures affording selenoxides without affecting alkenic or sulfide groups in the same molecule.[231,242,260,261,276,284,285] For example diastereoisomeric selenoxides (**41**) and (**42**) were prepared by oxidation of the corresponding selenide with ozone.[284] The β-elimination process affording the alkene (**43**) proceeds much faster with (**41**) than with (**42**).

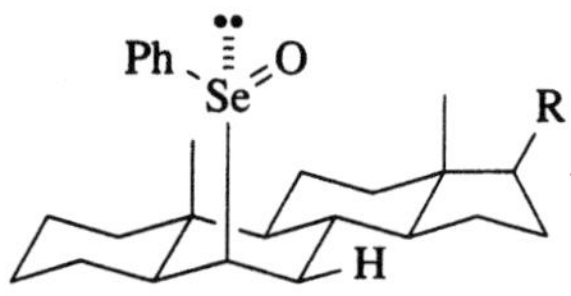

(**41**) 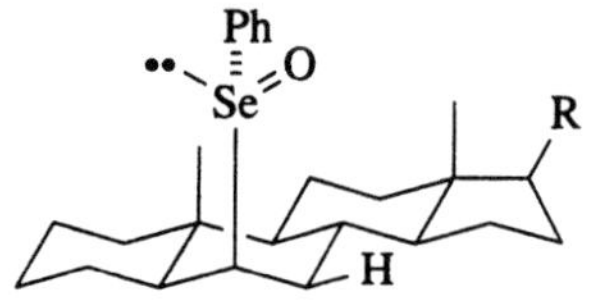 (**42**) 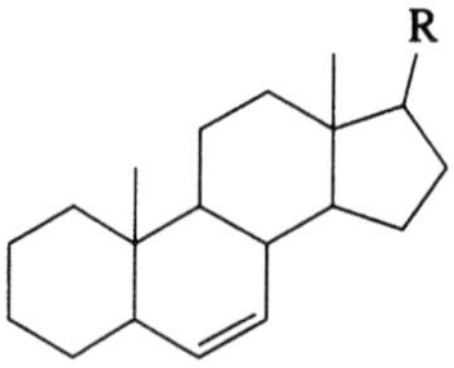 (**43**)

Sodium periodate is also frequently used as an oxidant for selenides, the reaction proceeds slowly in aqueous methanol.[272] Various selenoxides such as methyl phenyl and benzyl phenyl selenoxides,[286] (**38**),[265] (**39**),[266] (**40**)[267] and 2-azidocyclohexyl phenyl selenoxide[287] have been isolated in this way. Other solvents and reaction conditions may also be employed.[231,267,268,272,288–290]

N-Sulfonyloxaziridine (**24**) oxidizes selenides to selenoxides in aprotic solvents at 0–5 °C, sometimes giving quantitative yields.[291] Also halogenation of selenides with Cl_2 or Br_2 followed by alkaline hydrolysis or treatment with silver oxide gives selenoxides in good yields (equation 41).[292–295] The corresponding hydrates (**44**) can be isolated when the aromatic nucleus bears donor groups such as methoxy and ethoxy.[296] A similar hydrate (**45**) was also isolated during H_2O_2 oxidation of the corresponding selenide.[268] It has also been proposed that benzyl phenyl selenoxide is in equilibrium with its hydrate in aqueous DMSO.[260] This halogenation–hydrolysis method has recently been applied to alkyl phenyl selenides for the syntheses of vinylic and allylic chlorides,[297] enones[297] and *cis*-1,2-substituted cyclohexanes.[298] Isolation of the selenoxide (**38**) was achieved by treatment of the selenide with $(Bu^n{}_3Sn)_2O/Br_2$ in dichloromethane.[265] Similarly $PhICl_2$ oxidizes diaryl and aryl benzyl selenides to the corresponding stable selenoxides in aqueous pyridine.[286]

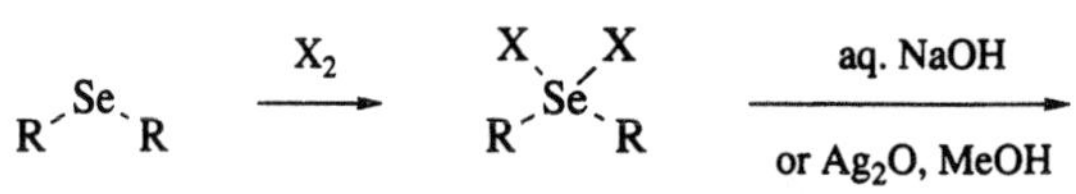

$$R_2Se \xrightarrow{X_2} R_2SeX_2 \xrightarrow[\text{or } Ag_2O,\ MeOH]{\text{aq. NaOH}} R_2Se{=}O \qquad (41)$$

R = Me, $PhCH_2$, aryl

(**44**) 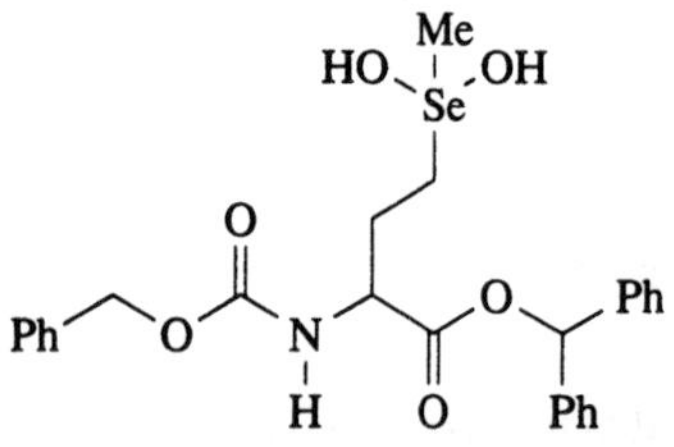 (**45**)

(**46**)

Y = Bu^{t}O or

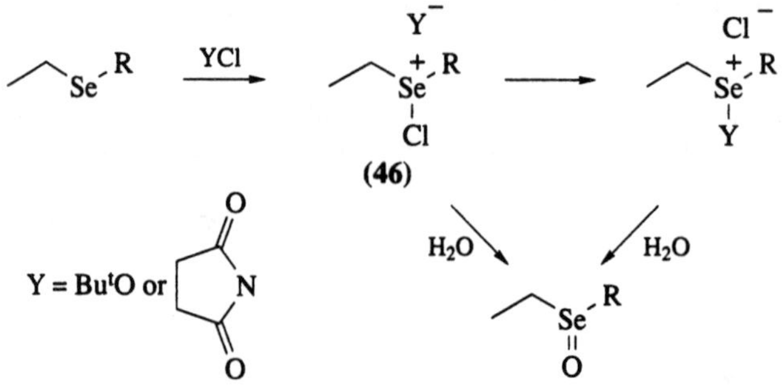

Scheme 15

Treatment of selenides with NBS,[265] NCS[262,265] or Bu^{t}OCl[262,299] followed by hydrolysis gives selenoxides or elimination products in good yields (Scheme 15). The intermediate (**46**) can oxidize alcohols to ketones.[300] Other oxidants such as N_2O_4,[294] TTN[149,301] and CrO_3[302] have been reported to be effective for oxidation of selenides.

6.2.3.3 Oxidation of Selenides and Selenoxides to Selenones

The oxidation of selenoxides to selenones is slow requiring drastic conditions to be used. Diaryl, aryl methyl and dimethyl selenones are prepared by the oxidation of the corresponding selenoxides with prolonged exposure to $KMnO_4$[293,303] or ozone.[295] The direct oxidation of selenides to selenones by PhIO with ruthenium(II) complex catalyst,[194] H_2O_2[304] and $Cu(MnO_4)_2$[208] has also been described. Aryl trifluoromethyl selenides, selenoxides and selenium dichlorides are oxidized with a mixture of trifluoroacetic anhydride and 85% H_2O_2 at low temperature to give the corresponding selenones in good yields (equation 42).[304] However MCPBA appears to be the most effective oxidant of selenides and selenoxides to selenones,[242,279,305] several vinyl selenones being prepared by this method (equation 43)[279] and used for many useful organic transformations.[279,280,306] Recent detailed studies on oxidation of various dialkyl and alkyl aryl selenides to the corresponding selenones revealed that MCPBA, CF_3CO_3H and $KMnO_4$ are the reagents of choice, while oxidation with H_2O_2, TBHP and $NaIO_4$ often stops at the selenoxide stage (Scheme 16).[307]

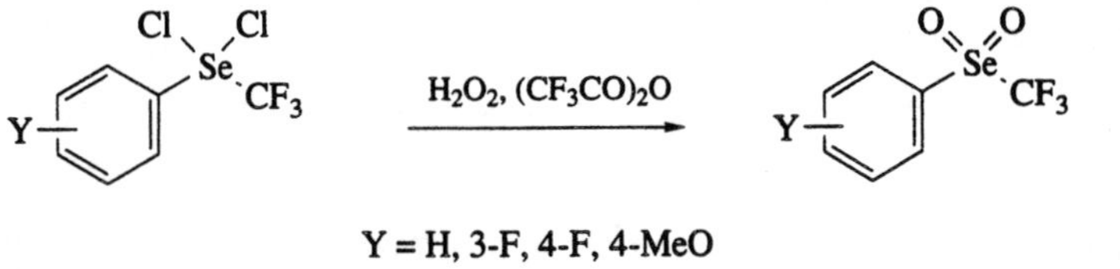

$$Y\text{-}C_6H_4\text{-}SeCl_2CF_3 \xrightarrow{H_2O_2,\ (CF_3CO)_2O} Y\text{-}C_6H_4\text{-}SeO_2CF_3 \qquad (42)$$

Y = H, 3-F, 4-F, 4-MeO

$$PhSe\text{-}CH{=}CH\text{-}Bu \xrightarrow[76\%]{2\ \text{equiv. MCPBA, MeOH, 0 °C–r.t.}} PhSeO_2\text{-}CH{=}CH\text{-}Bu \qquad (43)$$

$$PhSeC_{10}H_{21} \xrightarrow[68\%]{5\ \text{equiv. MCPBA, } CH_2Cl_2,\ 20\ °C,\ 5\ h} PhSeO_2C_{10}H_{21}$$

$$PhSeC_{10}H_{21} \xrightarrow[76\%]{2.5\ \text{equiv. } CF_3CO_3H,\ CH_2Cl_2,\ 20\ °C,\ 1\ h} PhSeO_2C_{10}H_{21}$$

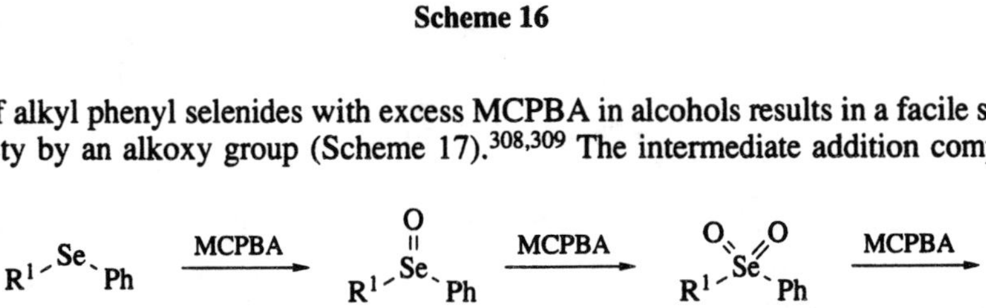

Scheme 16

Oxidation of alkyl phenyl selenides with excess MCPBA in alcohols results in a facile substitution of a selenone moiety by an alkoxy group (Scheme 17).[308,309] The intermediate addition compound (**47**) be-

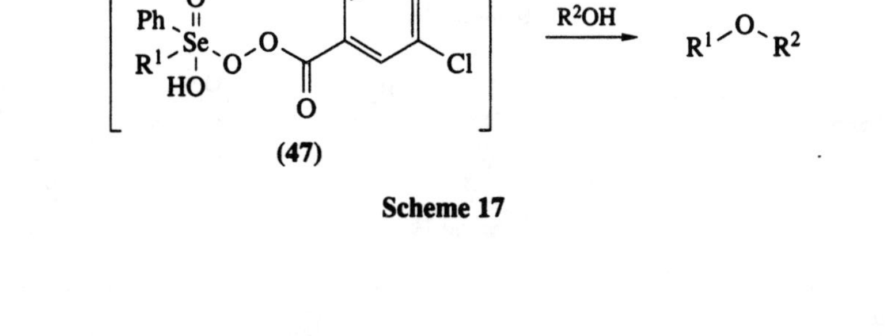

$$R^1SePh \xrightarrow{MCPBA} R^1Se(O)Ph \xrightarrow{MCPBA} R^1SeO_2Ph \xrightarrow{MCPBA} [Ph(R^1)Se(O)(OH)\text{-}O\text{-}O\text{-}C(O)\text{-}C_6H_4\text{-}3\text{-}Cl]\ (\mathbf{47}) \xrightarrow{R^2OH} R^1OR^2$$

Scheme 17

tween the selenones and MCPBA is thought to be the reason why a phenylselenonyl group works as a very good leaving group. MCPBA is the oxidant of choice for this transformation, since the use of other oxidants such as H_2O_2, TBHP and $NaIO_4$ leads to normal selenoxide *syn* elimination providing the corresponding alkenes.

Few examples of photooxidation of selenides are known. Photolysis of dilute aerated solutions of dibenzyl diselenide in benzene resulted in the formation of benzaldehyde and elemental selenium.[310] Without oxygen present during the photolysis only decomposition to dibenzyl selenide and selenium was observed.[310,311] Photolysis in $CDCl_3$ in an NMR tube in the presence of oxygen gave complex mixtures of products derived from benzyl radicals.[312] It appears that monoselenides are more stable than the corresponding sulfides and tellurides to photooxidation.[313]

6.2.4 OXIDATION OF ORGANIC TELLURIUM COMPOUNDS

6.2.4.1 Oxidation of Tellurols to Ditellurides and Further Oxidized Species by Chemical Reagents

Tellurols are extremely air sensitive and are normally not isolated as they are converted to ditellurides (equation 44).[314–319] Nevertheless there are two reports on the isolation of alkanetellurols.[320,321] Benzenetellurol may be prepared *in situ* by treatment of trimethylsilyl phenyl telluride ($PhTeSiMe_3$) with methanol[322] or CF_3CO_2H[323] and also by reduction of diphenyl ditelluride with H_3PO_2 or $NaBH_4$[322,324–326] and used as a reducing agent.[322,323]

$$2\left[\mathrm{R{-}TeH}\right] + \mathrm{O_2} \longrightarrow 2\,\mathrm{RTe{-}TeR} + 2\,\mathrm{H_2O} \qquad (44)$$

Treatment of ditellurides with H_2O_2[327] or air[327–329] gave white products which were not further characterized. Oxidation of diphenyl ditelluride with conc. HNO_3 has been reported to give benzenetellurinic acid nitrate [PhTe(O)ONO_2].[330,331] *t*-Butyl hydroperoxide oxidizes diaryl ditellurides to the corresponding tellurenic esters (**48**) and/or tellurinic esters (**49**) in the presence of carboxylic acids (equation 45).[332] For an exceptional example, the formation of benzotellurophene derivatives (**52**) by SeO_2 oxidation of 1-tellurochromenes (**50**) *via* ditellurides (**51**) was reported (Scheme 18).[333] Halogenation of ditellurides (R_2Te_2) with excess bromine or sulfuryl chloride gives tellurinyl halides ($RTeX_3$).[334,335] With an equimolar amount of bromine an unstable tellurenyl halide (RTeX) was isolated when R is 2-nitrophenyl,[334] while the reaction of di-2-naphthyl ditelluride with iodine produced a stable 2-naphthalenetellurenyl iodide.[336] Oxidation of diphenyl ditelluride with lead tetraacetate gives phenyltellurium triacetate, the hydrolysis of which occurs readily to give a mixture of tellurinic acid and anhydride, (**53**) and (**54**; Scheme 19).[337]

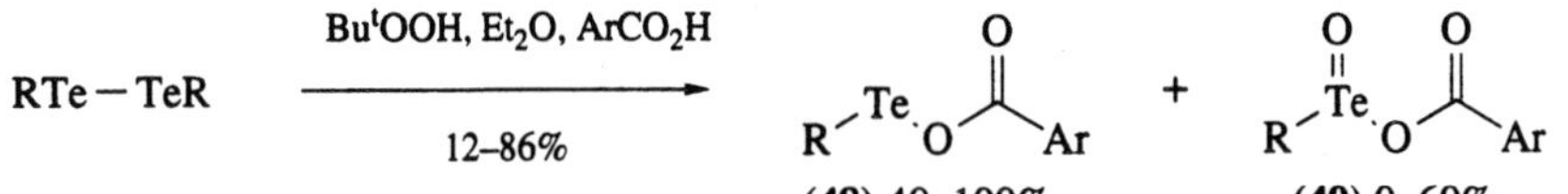

$$\mathrm{RTe{-}TeR} \xrightarrow[\text{12–86\%}]{\mathrm{Bu^tOOH,\ Et_2O,\ ArCO_2H}} \mathrm{RTe{-}O{-}C(O)Ar} + \mathrm{RTe(O){-}O{-}C(O)Ar} \qquad (45)$$

(**48**) 40–100% (**49**) 0–60%

R = Ph, 4-MeC_6H_4, 4-ClC_6H_4

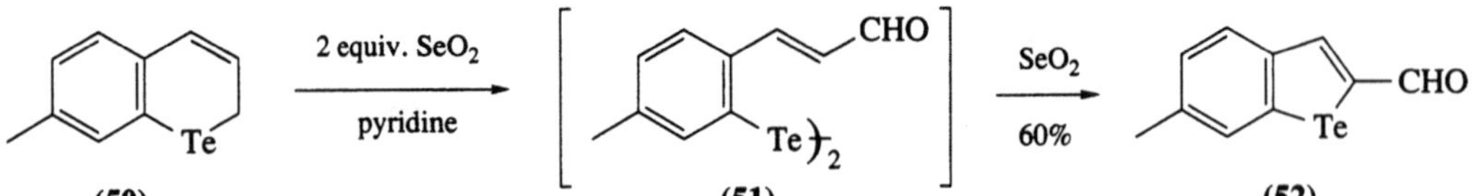

(**50**) $\xrightarrow[\text{pyridine}]{\text{2 equiv. } SeO_2}$ [(**51**)] $\xrightarrow[\text{60\%}]{SeO_2}$ (**52**)

Scheme 18

$$\mathrm{PhTe{-}TePh} \xrightarrow{3\,\mathrm{Pb(OAc)_4}} \mathrm{Ph{-}Te(OAc)_3} \xrightarrow{\mathrm{H_2O}} \mathrm{Ph{-}Te(O)OAc} + \mathrm{PhTe(O){-}O{-}Te(O)Ph}$$

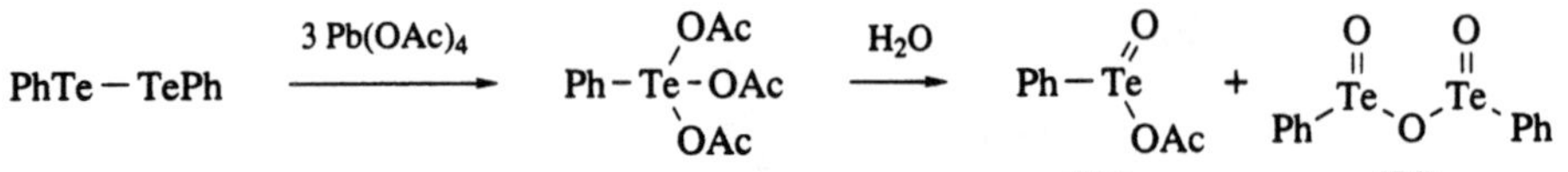

(**53**) (**54**)

Scheme 19

6.2.4.2 Oxidation of Tellurides to Telluroxides

Diaryl and dibenzyl tellurides are oxidized slowly by atmospheric oxygen to the corresponding telluroxides.[338–340] There are also reports of the formation of tellurinic acid ($RTeO_2H$) derivatives by air or alkaline H_2O_2 oxidation of dialkyl tellurides.[341] Treatment of the telluride (**55**) with chloramine-T followed by hydrolysis gives the telluroxide (**57**) *via* *N*-tellurosulfonamide (**56**; Scheme 20).[342] However telluroxides are more generally prepared by alkaline hydrolysis of diorganyltellurium dihalides[343–345] which can be formed by halogenation of tellurides by various halogenating agents such as Br_2, Cl_2, I_2, $SOCl_2$, SO_2Cl_2, $FeCl_3$, $CuCl_2$, vicinal bromides,[345] ICl and IBr.[346] Treatment of tellurides with NCS or Bu^{t}OCl followed by hydrolysis with 10% NaOH or saturated $NaHCO_3$ produces telluroxides in 90–95% yield.[262] The isolable telluroxides are thermally stable, but hygroscopic, white solids, and appear to be often present in their hydrated forms (equation 46).[262,341,347,348] The telluroxide (**58**), however, decomposes in boiling toluene to give alkenes (equation 47).[349]

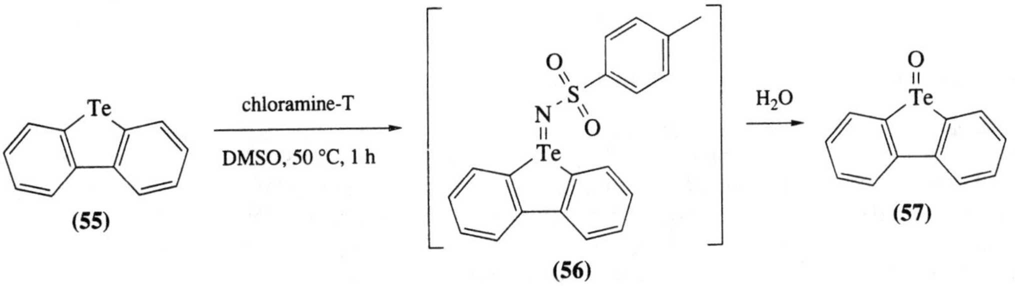

(55) —chloramine-T, DMSO, 50 °C, 1 h→ [(56)] —H_2O→ (57)

Scheme 20

$$R_2Te \xrightarrow{X_2} R_2TeX_2 \xrightarrow{OH^-} R_2Te{=}O{\cdot}H_2O \rightleftharpoons R_2Te(OH)_2 \qquad (46)$$

$$\text{(58)}\ MeOC_6H_4Te(=O{\cdot}H_2O)CH_2CH_2R \xrightarrow[-[PhTeOH]]{\Delta} RCH{=}CH_2 \qquad (47)$$

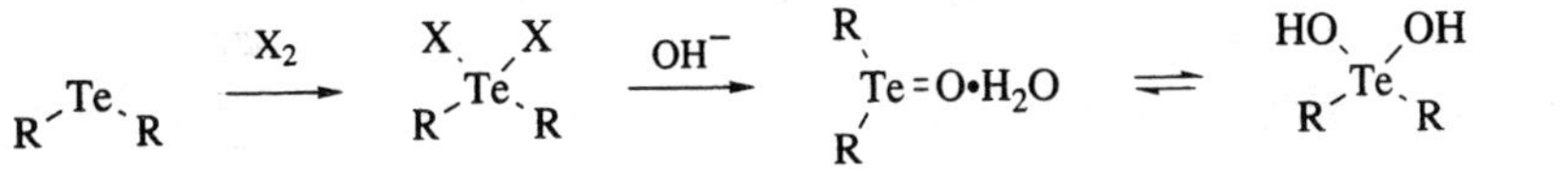

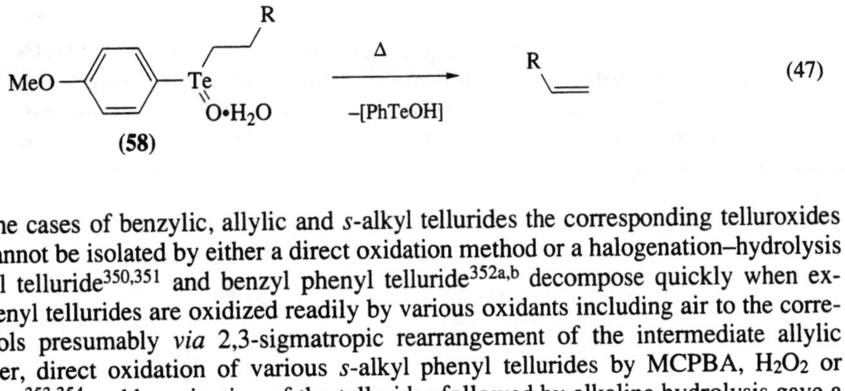

On the contrary, in the cases of benzylic, allylic and *s*-alkyl tellurides the corresponding telluroxides are very unstable and cannot be isolated by either a direct oxidation method or a halogenation–hydrolysis method. Thus, dibenzyl telluride[350,351] and benzyl phenyl telluride[352a,b] decompose quickly when exposed to air. Allylic phenyl tellurides are oxidized readily by various oxidants including air to the corresponding allylic alcohols presumably *via* 2,3-sigmatropic rearrangement of the intermediate allylic telluroxides.[352b,c] Further, direct oxidation of various *s*-alkyl phenyl tellurides by MCPBA, H_2O_2 or TBHP in organic solvents[353,354] and bromination of the tellurides followed by alkaline hydrolysis gave a mixture of alkenes, respectively, by telluroxide elimination (Scheme 21).[355] The double bond geometry

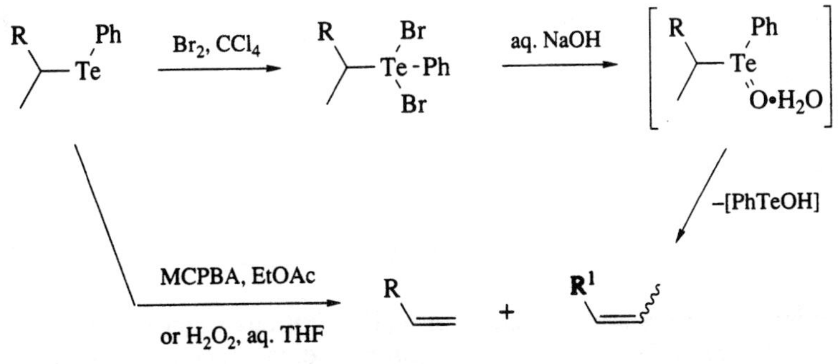

Scheme 21

of the internal alkenes produced by direct oxidation depends upon the amount of oxidant present. Also it should be noticed that the behavior of telluroxides differs from the corresponding selenoxides.[356]

LTA oxidizes diaryl tellurides to diaryltellurium diacetates,[337] while the treatment of diaryl, divinyl, alkyl aryl and dialkyl tellurides with $Pd(OAc)_2$[357] or Li_2PdCl_4[358] results in a new carbon–carbon bond being formed (equations 48 and 49).

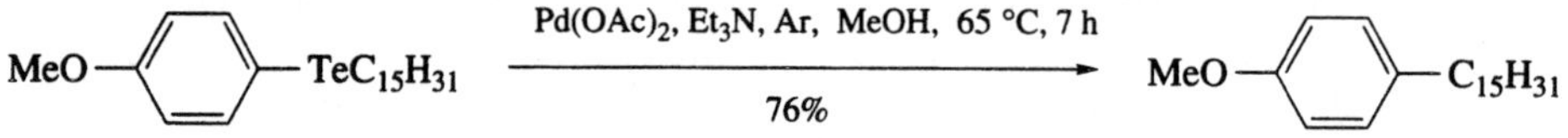

$$\text{MeO–C}_6\text{H}_4\text{–TeC}_{15}\text{H}_{31} \xrightarrow[76\%]{\text{Pd(OAc)}_2,\ \text{Et}_3\text{N, Ar, MeOH, 65 °C, 7 h}} \text{MeO–C}_6\text{H}_4\text{–C}_{15}\text{H}_{31} \quad (48)$$

$$(\text{PhCH=CH})_2\text{Te} \xrightarrow[55\text{–}70\%]{\text{Li}_2\text{PdCl}_4,\ \text{MeCN, 25 °C, 20 h}} (\text{PhCH=CH})_2 \quad (49)$$

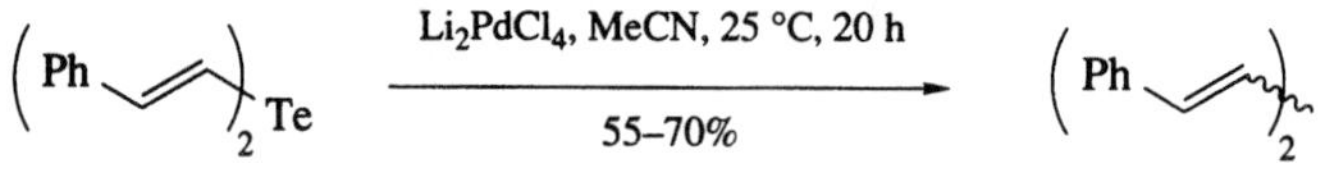

6.2.4.3 Oxidation of Tellurides to Tellurones

Little has so far been reported on the chemistry of tellurones. The preparation of several dialkyl tellurones by H_2O_2 or air oxidation of the corresponding tellurides or telluroxides has been claimed,[341,347,359,360] but it is doubtful whether those compounds were isolated in a pure tellurone form. In 1982 the first definitely characterized tellurone, bis(4-methoxyphenyl) tellurone (**60**), was prepared by periodate oxidation of the corresponding telluroxide (**59**; equation 50).[361] Both (**59**)[59,362] and (**60**)[361] work as mild useful oxidants which show some chemoselectivities and readily oxidize thiols to disulfides. The preparation of dodecyl 4-methoxyphenyl tellurone by a similar method has also been claimed.[349]

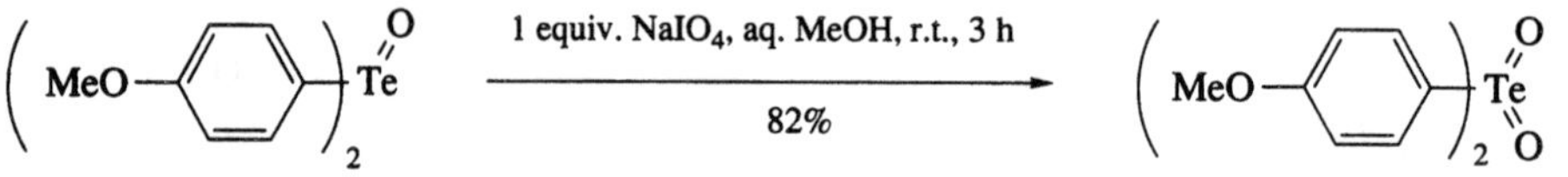

$$(\text{MeO–C}_6\text{H}_4)_2\text{Te=O} \xrightarrow[82\%]{\text{1 equiv. NaIO}_4,\ \text{aq. MeOH, r.t., 3 h}} (\text{MeO–C}_6\text{H}_4)_2\text{Te(=O)}_2 \quad (50)$$

(**59**) (**60**)

As in the selenium case (Scheme 17) the oxidation of alkyl phenyl telluride with excess MCPBA in the presence of alcohols results in a facile substitution of a PhTe moiety by an alkoxy group. The reaction is assumed to proceed *via* a similar tellurone–MCPBA adduct intermediate. Oxidation of cycloalkyl telluride (**61**) was accompanied by ring contraction to produce an acetal (**62**),[308,309] while the bromination–hydrolysis method affords the allylic ether by telluroxide elimination (Scheme 22).[355]

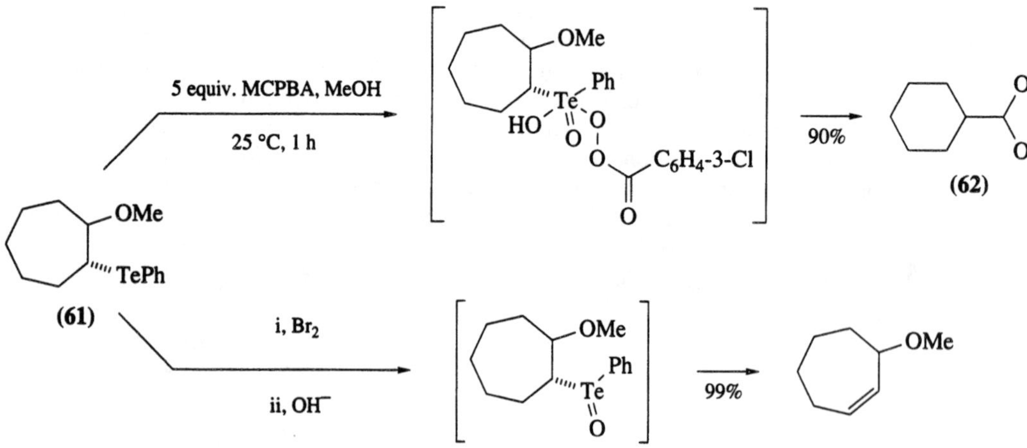

(**61**): 2-methoxycycloheptyl phenyl telluride (OMe, TePh)

Upper route: 5 equiv. MCPBA, MeOH, 25 °C, 1 h → [intermediate: cycloheptyl with OMe, Te(Ph)(=O)(OH)–O–O–C(=O)–C_6H_4-3-Cl] → 90% → cyclohexyl–CH(OMe)(OMe) (**62**)

Lower route: i, Br_2; ii, OH^- → [intermediate: cycloheptyl with OMe, Te(Ph)=O] → 99% → 3-methoxycycloheptene (OMe)

Scheme 22

Oxidative α-elimination occurs with organotellurium(IV) halides when treated with some oxidants, preferably TBHP, in organic solvents to give the corresponding organic halides in good yields with retention of configuration and by *ipso* replacement. A 1,2-halogen shift of the unstable organotellurium(VI) oxyhalide (**63**) was proposed to account for these reactions (Scheme 23).[363]

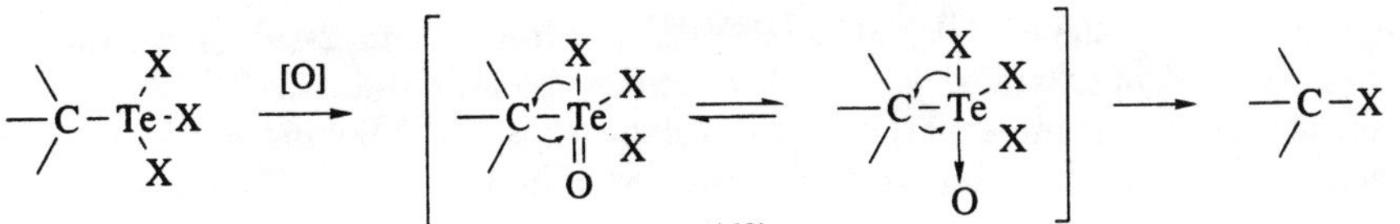

(63)

Scheme 23

6.2.4.4 Photooxidation of Organic Tellurium Compounds

Dibenzyl ditelluride and dibenzyl telluride photodecompose in the presence of oxygen in $CDCl_3$ to give a mixture of oxidation products such as benzaldehyde, benzyl alcohol and diphenylethane thought to be derived from a benzyl radical.[312] Although benzyl tellurocyanate ($PhCH_2TeCN$) is photochemically stable in the absence of oxygen, it rapidly suffers photooxidation in oxygen to give elemental tellurium and a mixture of 60% benzaldehyde and 40% benzyl alcohol.[313]

The tellurophenopyridazine (**64**) decomposed under the influence of light and oxygen to 4,5-dibenzoylpyridazine (**66**) *via* the peroxide (**65**) derived from Diels–Alder reaction of singlet oxygen with (**64**; Scheme 24).[364] Tellurapyrylium dyes such as (**67**) react with singlet oxygen within a few seconds to give dihydroxides (**68**; Scheme 25).[365]

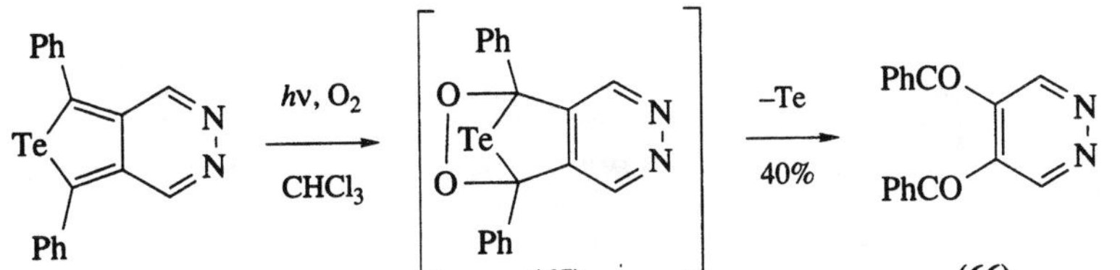

(64) (65) (66)

Scheme 24

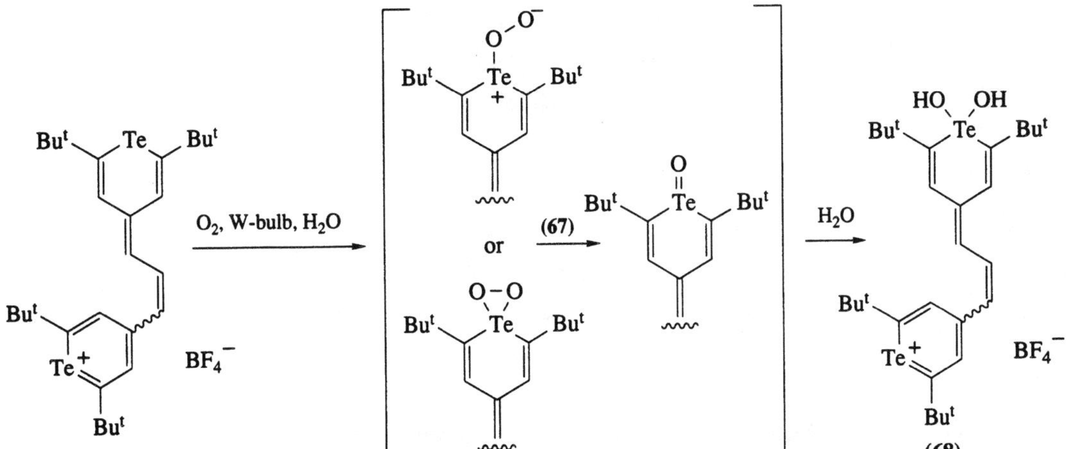

(67) (68)

Scheme 25

6.2.5 SYNTHESIS OF OPTICALLY ACTIVE SULFOXIDES AND SELENOXIDES BY CHEMICAL AND BIOLOGICAL OXIDATION

6.2.5.1 Synthesis of Optically Active Sulfoxides by Chemical Oxidation

Since the first report in 1960,[366,367] many procedures have been reported for the synthesis of optically pure sulfoxides by chemical oxidation of sulfides.[68,69a,b] Typical examples are as follows: (i) the oxidation of achiral sulfides by chiral peroxy acids,[368–375] (ii) oxidation by TBHP in chiral solvents[376] or in the presence of chiral catalysts,[376–382] and (iii) diastereoselective oxidation of sulfides containing another chiral center.[80,383–389] Similar methodologies have been reported using other oxidants such as organic halogen compounds,[390–394] Bu^tOCl,[395] $NaIO_4$,[135,396,397] 2-sulfonyl- and 2-sulfamyl-oxaziridine derivatives,[398–402] organic hydroperoxides[403,404] and electrochemical oxidations.[405,406] Oxidation in the presence of asymmetric 'host' molecules such as cyclodextrins,[407,408] bovine serum albumin[409–411] and chiral clay-chelates[412] has also been reported. As a method of homo-chiral sulfoxide preparation most are less successful than the procedures using chiral sulfenic acid derivatives which are prepared by resolution methods.[413] Some examples of chemical oxidations for obtaining optically active sulfoxides are useful, for example using modified Sharpless reagent [$Ti(OPr^i)_4$/(*R*,*R*)-diethyl tartrate (DET)/TBHP/H_2O (1:2:1:1.1)] was found to oxidize various alkyl aryl sulfides and dialkyl sulfides to the chiral sulfoxides with *ee* in the range of 80–90% (equation 51).[377–379] 2-Aryl-3-sulfamyloxaziridines (**69**)[401a] and (–)-α,α-dichlorocamphorsulfonyloxaziridine (**70**)[401b] were recently described as being equally effective chiral oxidizing agents for nonfunctionalized sulfides to optically active sulfoxides (equation 52).[401]

Bu^tOOH, $Ti(OPr^i)_4$, (*R*,*R*)-DET/H_2O

CH_2Cl_2, –20 °C

SMe

90%

Me S O

90% *ee*

(51)

SMe

(–)-(*S*,*S*)-(**69**)

$CHCl_3$, –78 °C

>90%

O S Me

91% *ee*

(52)

Ph Ph N S O O N O C_6F_5

(–)-(*S*,*S*)-(**69**)

Cl Cl N–SO_2Ph O

(**70**)

6.2.5.2 Synthesis of Optically Active Sulfoxides by Biological Oxidation

Microbiological oxidation of achiral sulfides to homo-chiral sulfoxides has been studied longer than have chemical methods and often gives better results.[68,69,414] It was reported that *Mortierella isabellina* NRRL 1757 converted methyl *p*-tolyl sulfide into (+)-(*R*)-sulfoxide with 100% *ee*, whereas *Helminthosporium sp.* NRRL 4671 oxidized the same sulfide to (–)-(*S*)-sulfoxide with 100% *ee* (Scheme 26).[415] Oxidation of similar sulfides by *Corynebacterium equi* IFO 3730 was also reported to proceed with high enantioselectivities (equation 53).[416]

Similar to biological oxidation, enzymatic oxidation of drugs[417–419] opened a way to studies on the preparation of optically active sulfoxides by enzymatic procedures. Among numerous reports[68,69] a typical example is the oxidation of aryl aminoalkyl sulfides by dopamine β-hydoxylase (DBH) in the presence of some electron donors to afford the sulfoxide with high enantioselectivity (equation 54).[420]

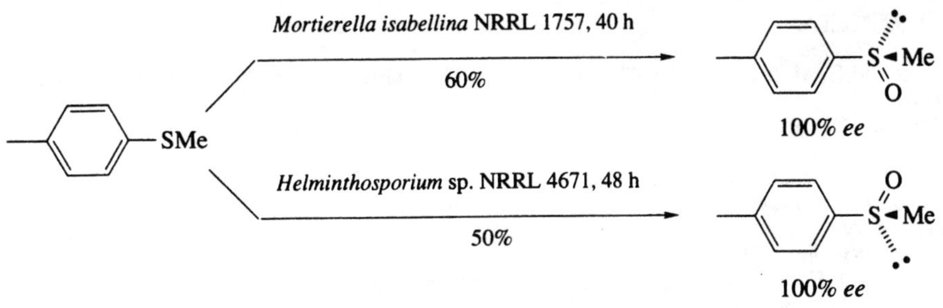

Mortierella isabellina NRRL 1757, 40 h
60%
100% ee
SMe
Helminthosporium sp. NRRL 4671, 48 h
50%
100% ee

Scheme 26

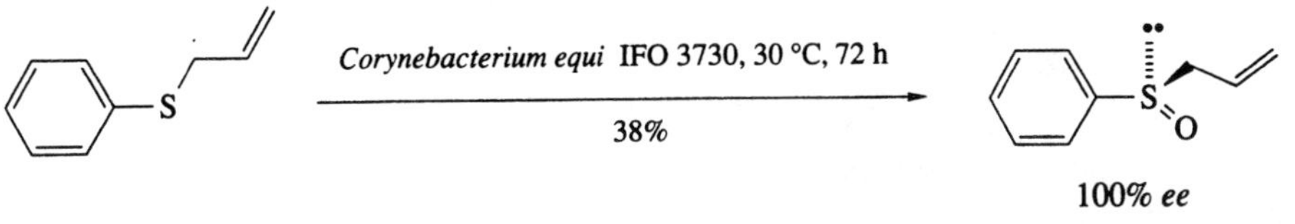

Corynebacterium equi IFO 3730, 30 °C, 72 h
38%
100% ee
(53)

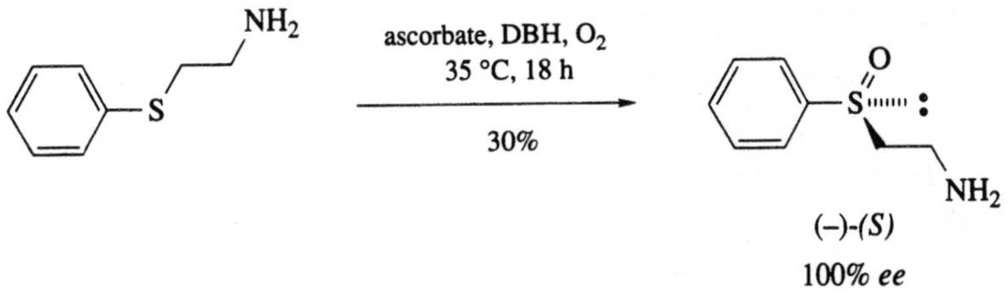

NH_2
ascorbate, DBH, O_2
35 °C, 18 h
30%
NH_2
(−)-(S)
100% ee
(54)

6.2.5.3 Synthesis of Optically Active Selenoxides by Chemical Oxidation

Until quite recently the isolation of optically active selenoxides has been limited to those contained in steroids (isolated as diastereoisomers).[421,422] The difficulty in obtaining these compounds was attributed to the racemization through the achiral hydrated intermediates.[260,269,423–425] Simple optically active selenoxides (5–11% *ee*) were first prepared by kinetic resolution.[425] Direct oxidation of selenides to selenoxides was first reported using optically active oxaziridine derivatives under anhydrous conditions, but the extent of the asymmetric induction was somewhat unsatisfactory with methyl phenyl selenide as substrate (8–9% *ee*).[426] Recently much improved enantiomeric excesses (45–73%) were achieved with new oxaziridine reagents such as (**70**).[401b] An attempt at the asymmetric oxidation of more bulky selenides was independently carried out using Bu^tOCl in the presence of (−)-2-octanol (equation 55),[299] but resulted in unsatisfactory enantioselectivities (*ee* 1%). Much better results were obtained by the oxidation of β-oxyalkyl aryl selenides (*ee* 18–40%; equation 56)[427] and alkyl aryl selenides (*ee* 1–28%)[428] using TBHP in the presence of (+)- or (−)-diisopropyl tartarate (DIPT) and titanium(IV) alkoxide.

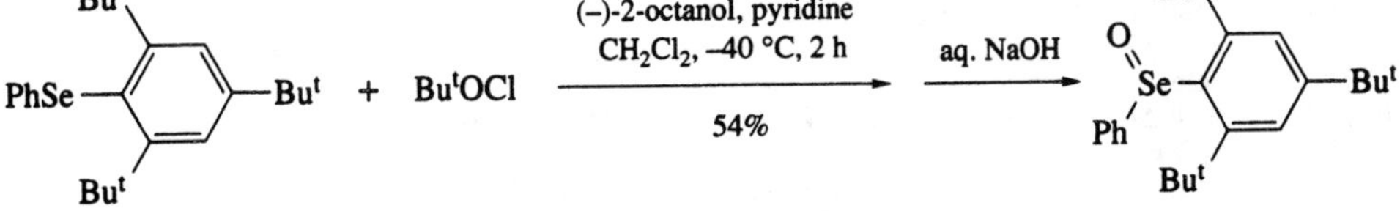

Bu^t PhSe Bu^t Bu^t + Bu^tOCl
(−)-2-octanol, pyridine
CH_2Cl_2, −40 °C, 2 h
54%
aq. NaOH
O Se Ph Bu^t Bu^t Bu^t (55)

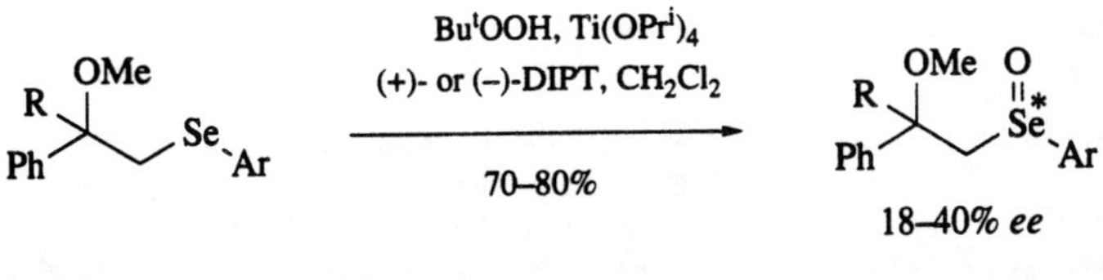

OMe R Ph Se Ar
Bu^tOOH, $Ti(OPr^i)_4$
(+)- or (−)-DIPT, CH_2Cl_2
70–80%
R OMe O Se* Ph Ar
18–40% ee
(56)

An attempt to prepare optically active selenoxides by microbial oxidation of achiral selenides gave poor results.[429] For the preparation of optically pure selenoxides therefore, methodology based on the separation of racemic[430] or diastereoisomeric[431] selenoxides gives the best results, although future developments of asymmetric reagents will undoubtedly become competitive.

6.2.6 REFERENCES

1. G. Capozzi and G. Modena, in 'The Chemistry of the Thiol Group', ed. S. Patai, Wiley, Chichester, 1974, part 2, chap. 17.
2. E. E. Reid, in 'Organic Chemistry of Bivalent Sulfur', Chemical Publishing Co., New York, 1958, vol. 1, p. 118.
3. L. Field, in 'Organic Chemistry of Sulfur', ed. S. Oae, Plenum Press, New York, 1977, chap. 7.
4. K. D. Gundermann and K. Hümke, *Methoden Org. Chem. (Houben-Weyl)*, 1985, **11** (E), 135.
5. T. J. Wallace, A. Schriesheim and W. Bartok, *J. Org. Chem.*, 1963, **28**, 1311.
6. J. Xan, E. A. Wilson, L. D. Roberts and N. H. Norton, *J. Am. Chem. Soc.*, 1941, **63**, 1139.
7. C. F. Cullis, J. D. Hopton and D. L. Trimm, *J. Appl. Chem.*, 1968, **18**, 330.
8. C. F. Cullis, J. D. Hopton, C. J. Swan and D. L. Trimm, *J. Appl. Chem.*, 1968, **18**, 335.
9. I. G. Dance and R. C. Conrad, *Aust. J. Chem.*, 1977, **30**, 305.
10. W. M. Brouwer, A. J. M. S. Robeerst, P. Piet and A. L. German, *Homogeneous Heterogeneous Catal., Proc. Int. Symp. Relat. Homogeneous Heterogeneous Catal., 5th*, 1986, 631 (*Chem. Abstr.*, 1988, **108**, 93 920m).
11. T. Nagano, K. Yoshikawa and M. Hirobe, *Tetrahedron Lett.*, 1980, **21**, 297.
12. K.-T. Liu and Y. C. Tong, *Synthesis*, 1978, 669.
13. T. J. Wallace and A. Schriesheim, *Tetrahedron Lett.*, 1963, 1131; *Tetrahedron*, 1965, **21**, 2271.
14. H. Berger, *Recl. Trav. Chim. Pays-Bas*, 1963, **82**, 773.
15. S. Oae, T. Takata and Y. H. Kim, *Tetrahedron*, 1981, **37**, 37.
16. T. J. Wallace, H. Pobiner and A. Schriesheim, *J. Org. Chem.*, 1964, **29**, 888.
17. T. J. Wallace, *J. Org. Chem.*, 1966, **31**, 3071.
18. J. Hill and A. McAuley, *J. Chem. Soc. A*, 1968, 156, 2405.
19. W. F. Pickering and A. McAuley, *J. Chem. Soc. A*, 1968, 1173.
20. E. P. Papadopulos, A. Jarrar and C. H. Issidorides, *J. Org. Chem.*, 1966, **31**, 615.
21. T. J. Wallace, *J. Org. Chem.*, 1966, **31**, 1217.
22. X. Huang and C.-C. Chan, *Synthesis*, 1982, 1091.
23. H. Firouzabadi, N. Iranpoor, H. Parham, A. Sardarian and J. Toofan, *Synth. Commun.*, 1984, **14**, 717.
24. J. M. Aizpurua and C. Palomo, *Tetrahedron Lett.*, 1983, **24**, 4367; J. M. Aizpurua, M. Juaristi, B. Lecea and C. Palomo, *Tetrahedron*, 1985, **41**, 2903.
25. A. Cornélis, N. Depaye, A. Gerstmans and P. Laszlo, *Tetrahedron Lett.*, 1983, **24**, 3103.
26. D. N. Dhar and A. K. Bag, *Indian J. Chem., Sect. B*, 1984, **23**, 974.
27. H. Firouzabadi, M. Naderi, A. Sardarian and B. Vessal, *Synth. Commun.*, 1983, **13**, 611.
28. E. J. Bourne, W. M. Corbett, M. Stacey and R. Stephens, *Chem. Ind. (London)*, 1954, 106.
29. L. Field and J. E. Lawson, *J. Am. Chem. Soc.*, 1958, **80**, 838.
30. L. Field, C. B. Hoelzel and J. M. Locke, *J. Am. Chem. Soc.*, 1962, **84**, 847.
31. T. Mukaiyama and T. Endo, *Bull. Chem. Soc. Jpn.*, 1967, **40**, 2388.
32. S. Uemura, S. Tanaka and M. Okano, *Bull. Chem. Soc. Jpn.*, 1977, **50**, 220.
33. C. N. Yiannios and J. V. Karabinos, *J. Org. Chem.*, 1963, **28**, 3246.
34. T. J. Wallace, *J. Am. Chem. Soc.*, 1964, **86**, 2108.
35. T. J. Wallace and H. A. Weiss, *Chem. Ind. (London)*, 1966, 1558.
36. T. J. Wallace, J. M. Millor, H. Probner and A. Schriesheim, *Proc. Chem. Soc.*, 1962, 384.
37. D. I. Relya, P. O. Towney and A. R. Williams, *J. Org. Chem.*, 1962, **27**, 477.
38. S. D. Sokolov and N. M. Naidenova, *Zh. Org. Khim.*, 1966, **2**, 1123 (*Chem. Abstr.*, 1966, **65**, 16 849).
39. C. C. Price and G. W. Stacy, *Org. Synth., Coll. Vol.*, 1955, **3**, 86.
40. F. Asinger, A. Saus, H. Offermanns, P. Krings and H. Andree, *Justus Liebigs Ann. Chem.*, 1971, **744**, 51.
41. D. S. Tarbell, in 'Organic Sulfur Compounds', ed. N. Kharasch, Pergamon Press, Oxford, 1961, chap. 10.
42. W. G. Filby, K. Günther and R. D. Penzhorm, *J. Org. Chem.*, 1973, **38**, 4070.
43. J. R. Schaeffer, C. T. Goodhue, H. A. Risley and R. E. Stevens, *J. Org. Chem.*, 1967, **32**, 392.
44. A. Fava, G. Reichenbach and U. Peron, *J. Am. Chem. Soc.*, 1967, **89**, 6696.
45. J. Drabowicz and M. Mikołajczyk, *Synthesis*, 1980, 32.
46. L. W. Christensen and D. J. Heacock, *Synthesis*, 1978, 50.
47. G. A. Olah, M. Arvanaghi and Y. D. Vankar, *Synthesis*, 1979, 721.
48. T.-L. Ho, T. W. Hall and C. M. Wong, *Synthesis*, 1974, 872.
49. V. Caló, F. Ciminale, G. Lopez and P. E. Todesco, *Int. J. Sulfur Chem., Part A*, 1971, **1**, 139 (*Chem. Abstr.*, 1971, **75**, 129 452b).
50. E. Kato, M. Oya, T. Iso and J. Iwao, *Chem. Pharm. Bull.*, 1986, **34**, 486.
51. T.-L. Ho, *Synth. Commun.*, 1977, **7**, 363.
52. F. Yoneda, K. Suzuki and Y. Nitta, *J. Am. Chem. Soc.*, 1966, **88**, 2328; *J. Org. Chem.*, 1967, **32**, 727.
53. F. J. Smentowski, *J. Am. Chem. Soc.*, 1963, **85**, 3036.
54. W. A. Pryor, D. F. Church, C. K. Govindan and G. Crank, *J. Org. Chem.*, 1982, **47**, 156.
55. T. Takaya, H. Enjo and E. Imoto, *Bull. Chem. Soc. Jpn.*, 1968, **41**, 1032.
56. K. Nakayama, S. Shiba, M. Horikawa, K. Sato, H. Nakamura, N. Harada and F. Harada, *Synth. Commun.*, 1980, **10**, 305.
57. Y. H. Kim, K. Shinhama, D. Fukushima and S. Oae, *Tetrahedron Lett.*, 1978, 1211.
58. S. Oae, K. Shinhama, K. Fujimori and Y. H. Kim, *Bull. Chem. Soc. Jpn.*, 1980, **53**, 775.
59. D. H. R. Barton, S. V. Ley and C. A. Meerholz, *J. Chem. Soc., Chem. Commun.*, 1979, 755.
60. F. Ogura, H. Yamaguchi, T. Otsubo and H. Tanaka, *Bull. Chem. Soc. Jpn.*, 1982, **55**, 641.
61. N. X. Hu, Y. Aso, T. Otsubo and F. Ogura, *Bull. Chem. Soc. Jpn.*, 1986, **59**, 879.
62. M. J. Gigian and D. V. Winkelman, *Tetrahedron Lett.*, 1969, 3901.
63. S. Shinkai, R. Ando and F. Yoneda, *Chem. Lett.*, 1977, 147.
64. Y. Yano, I. Yatsu, E. Oya and M. Ohshima, *Chem. Lett.*, 1983, 775; Y. Yano, M. Oshima, I. Yatsu, S. Sutoh, R. E. Vasquez, A. Kitani and K. Sasaki, *J. Chem. Soc., Perkin Trans. 2*, 1985, 753.

65. K. Tanaka, X. Chen, T. Kimura and F. Yoneda, *Tetrahedron Lett.*, 1987, **28**, 4173; K. Tanaka, X. Chen and F. Yoneda, *Tetrahedron*, 1988, **44**, 3241.
66. T. Shono, in 'The Chemistry of Ethers, Crown Ethers, Hydroxyl Groups and Their Sulphur Analogues', ed. S. Patai, Wiley, Chichester, 1980, part 1, chap. 8.
67. A. R. Knight, in 'The Chemistry of the Thiol Group', ed. S. Patai, Wiley, Chichester, 1974, part 1, chap. 10.
68. M. Madesclaire, *Tetrahedron*, 1986, **42**, 5459.
69. (a) J. Drabowicz, P. Kiełbasinski and M. Mikołajczyk, in 'The Chemistry of Sulphones and Sulphoxides', ed. S. Patai, Z. Rappoport and C. J. M. Stirling, Wiley, Chichester, 1988, chap. 8; (b) O. Bortolini, F. Di Furia, G. Licini and G. Modena, *Rev. Heteroatom. Chem.*, 1988, **1**, 66.
70. G. Kresze, *Methoden Org. Chem. (Houben-Weyl)*, 1985, 11 (E), 702.
71. S. Hünig and O. Boes, *Justus Liebigs Ann. Chem.*, 1953, **579**, 23.
72. D. Barnard, J. M. Fabian and H. P. Koch, *J. Chem. Soc.*, 1949, 2442.
73. D. Barnard, L. Bateman, M. E. Cain, T. Colclough and J. I. Cunneen, *J. Chem. Soc.*, 1961, 5339.
74. J. Drabowicz and M. Mikołajczyk, *Synth. Commun.*, 1981, **11**, 1025.
75. M. F. Shostakovskii, N. V. Komarov and N. N. Vlasova, *Int. Symp. Organosilicon Chem., Sci. Commun., Suppl.*, 1965, 21 (*Chem. Abstr.*, 1966, **65**, 8950).
76. M. Sander, *Monatsh. Chem.*, 1965, **96**, 896.
77. A. D. Baker, R. Scharfman and C. A. Stein, *Tetrahedron Lett.*, 1983, **24**, 2957.
78. K. Ogura and G. Tsuchihashi, *Bull. Chem. Soc. Jpn.*, 1972, **45**, 2203.
79. G. Schill and P. R. Jones, *Synthesis*, 1974, 117.
80. S. Oae, T. Kawai and N. Furukawa, *Tetrahedron Lett.*, 1984, **25**, 69.
81. J. Drabowicz and M. Milołajczyk, *Synthesis*, 1978, 758.
82. F. E. Hardy, P. R. H. Speakman and P. Robson, *J. Chem. Soc. C*, 1969, 2334.
83. Y. Watanabe, T. Numata and S. Oae, *Synthesis*, 1981, 204.
84. O. Bartolini, F. Di Furia, P. Serimin and G. Modena, *J. Mol. Catal.*, 1980, **7**, 59.
85. L. Bateman and K. R. Hargrave, *Proc. R. Soc. London, Ser. A*, 1954, **224**, 389; K. R. Hargrave, *Proc. R. Soc. London, Ser. A*, 1956, **235**, 55.
86. L. Horner and E. Jurgens, *Justus Liebigs Ann. Chem.*, 1957, **602**, 135.
87. D. Barnard, *J. Chem. Soc.*, 1956, 489.
88. M. A. P. Dankleff, R. Curci, J. O. Edwards and H. Y. Pynn, *J. Am. Chem. Soc.*, 1968, **90**, 3209.
89. B. Ganem, A. J. Biloski and R. P. Heggs, *Tetrahedron Lett.*, 1980, **21**, 689.
90. C. Kemal, T. W. Chan and T. C. Bruice, *Proc. Natl. Acad. Sci. USA*, 1977, **74**, 405.
91. T. C. Bruice, J. B. Noar, S. S. Ball and U. V. Ventataram, *J. Am. Chem. Soc.*, 1983, **105**, 2452.
92. S. Oae, K. Asada and T. Yoshimura, *Tetrahedron Lett.*, 1983, **24**, 1265.
93. L. N. Levin, *J. Prakt. Chem.*, 1928, **119**, 211; L. N. Levin and I. Chulkov, *J. Prakt. Chem.*, 1930, **128**, 171.
94. C. G. Overberger and R. W. Cummins, *J. Am. Chem. Soc.*, 1953, **75**, 4520.
95. A. L. Ternay, Jr., D. W. Chasar and M. Sax, *J. Org. Chem.*, 1967, **32**, 2465.
96. (a) A. L. Ternay, Jr., L. Ens, J. Herrmann and S. Evans, *J. Org. Chem.*, 1969, **34**, 940; (b) L. A. G. M. van den Broek, P. J. A. In't Veld, J. H. Colstee and H. C. J. Ottenheijm, *Synth. Commun.*, 1989, **19**, 3397.
97. B. Rajanikanth and B. Ravindranath, *Indian J. Chem., Sect. B*, 1984, **23**, 877.
98. E. V. Bell and G. M. Bennett, *J. Chem. Soc.*, 1927, 1798.
99. C. G. Venier, T. G. Squires, Y. Y. Chen, G. P. Hussmann, J. C. Shei and B. F. Smith, *J. Org. Chem.*, 1982, **47**, 3773.
100. L. A. Carpino and H. W. Chen, *J. Am. Chem. Soc.*, 1979, **101**, 390.
101. A. W. M. Lee, W. H. Chan and M. S. Wong, *J. Chem. Soc., Chem. Commun.*, 1988, 1585.
102. W. Ando, Y. Hanyu, Y. Kumamoto and T. Takata, *Tetrahedron*, 1986, **42**, 1989.
103. C. R. Harrison and P. Hedge, *J. Chem. Soc., Perkin Trans. 1*, 1976, 2252.
104. A. H. Schlesinger and D. T. Mowry, *J. Am. Chem. Soc.*, 1951, **73**, 2164.
105. H. Kwart and R. K. Miller, *J. Am. Chem. Soc.*, 1956, **78**, 5008.
106. D. O. Spry, *Tetrahedron Lett.*, 1972, 3717.
107. S. Oae, Y. Onishi, S. Kozuka and W. Tagaki, *Bull. Chem. Soc. Jpn.*, 1966, **39**, 364.
108. J. Drabowicz, W. Midura and M. Mikołajczyk, *Synthesis*, 1979, **39**.
109. Y. Ueno, T. Inoue and M. Okawara, *Tetrahedron Lett.*, 1977, 2413.
110. T. Higuchi and K. H. Gensch, *J. Am. Chem. Soc.*, 1966, **88**, 3874.
111. T. Eiki and W. Tagaki, *Chem. Lett.*, 1980, 1063.
112. G. Barbieri, M. Cinquini, S. Colonna and F. Montanari, *J. Chem. Soc. C*, 1968, 659.
113. A. H. Ford-Moore, *J. Chem. Soc.*, 1949, 2126.
114. R. Barret, F. Pautet, M. Daudon and B. Mathian, *Sulfur Lett.*, 1985, 127 (*Chem. Abstr.*, 1986, **104**, 206 808u).
115. J. R. A. Castrillon and H. H. Szmat, *J. Org. Chem.*, 1967, **32**, 976.
116. W. Tagaki, K. Kikukawa, K. Ando and S. Oae, *Chem. Ind. (London)*, 1964, 1624.
117. K. M. More and J. Wemple, *Synthesis*, 1977, 791.
118. R. Harville and S. F. Reed, Jr., *J. Org. Chem.*, 1968, **33**, 3976.
119. W. D. Kingsbury and C. R. Johnson, *J. Chem. Soc., Chem. Commun.*, 1969, 365.
120. Á. Kucsman, I. Kapovits and B. Tanác, *Tetrahedron*, 1962, **18**, 79.
121. K. Tsujihara, N. Furukawa, K. Oae and S. Oae, *Bull. Chem. Soc. Jpn.*, 1969, **42**, 2631.
122. F. Ruff and Á. Kucsman, *J. Chem. Soc., Perkin Trans. 2*, 1982, 1075.
123. V. J. Traynelis, Y. Yoshikawa, S. M. Tarka and J. R. Livingston, Jr., *J. Org. Chem.*, 1973, **38**, 3986.
124. M. Hojo and R. Masuda, *Tetrahedron Lett.*, 1976, 613; M. Hojo, R. Masuda and K. Hakotani, *Tetrahedron Lett.*, 1978, 1121.
125. L. S. S. Réamonn and W. I. O'Sullivan, *J. Chem. Soc., Chem. Commun.*, 1976, 1012; *J. Chem. Soc., Perkin Trans. 1*, 1980, 1194.
126. V. Sere, A. Guenzi, S. Pollicino, E. Sandri and A. Fava, *J. Org. Chem.*, 1980, **45**, 261.
127. C. R. Johnson and D. McCants, Jr., *J. Am. Chem. Soc.*, 1965, **87**, 1109.

128. L. S. Skattebøl, B. Boulette and S. Solomon, *J. Org. Chem.*, 1967, **32**, 3111.
129. N. K. Sharma, F. de Reinach-Hirtzbach and T. Durst, *Can. J. Chem.*, 1976, **54**, 3012.
130. N. J. Leonard and C. R. Johnson, *J. Org. Chem.*, 1962, **27**, 282; C. R. Johnson and J. E. Keiser, *Org. Synth., Coll. Vol.*, 1973, **5**, 791.
131. D. A. Evans, C. A. Bryan and C. L. Sims, *J. Am. Chem. Soc.*, 1972, **94**, 2891.
132. M. Mikołajczyk and A. Zatorski, *Synthesis*, 1973, 669.
133. R. G. Hiskey and M. A. Harpold, *J. Org. Chem.*, 1967, **32**, 3191.
134. E. Santaniello, A. Manzocchi and C. Farachi, *Synthesis*, 1980, 563.
135. K.-T. Liu and Y. C. Tong, *J. Org. Chem.*, 1978, **43**, 2717.
136. D. N. Gupta, P. Hodge and J. E. Davies, *J. Chem. Soc., Perkin Trans. 1*, 1981, 2970.
137. T. Kageyama, Y. Ueno and M. Okawara, *Synthesis*, 1982, 815.
138. F. G. Bordwell and P. J. Boutan, *J. Am. Chem. Soc.*, 1957, **79**, 717.
139. G. Laurence, *C. R. Hebd. Seances Acad. Sci., Ser. C*, 1969, **269**, 352.
140. R. Louw, H. P. Vermeeren, J. J. A. van Asten and W. J. Ultée, *J. Chem. Soc., Chem. Commun.*, 1976, 496.
141. G. A. Olah, B. G. B. Gupta and S. C. Narang, *J. Am. Chem. Soc.*, 1979, **101**, 5317.
142. L. Horner and F. Hübenett, *Justus Liebigs Ann. Chem.*, 1953, **579**, 193.
143. C. C. Addison and J. C. Sheldon, *J. Chem. Soc.*, 1956, 2705.
144. D. P. Riley, *Inorg. Chem.*, 1983, **22**, 1965.
145. D. P. Riley and R. E. Shumate, *J. Am. Chem. Soc.*, 1984, **106**, 3179.
146. D. P. Riley and P. E. Correa, *J. Chem. Soc., Chem. Commun.*, 1986, 1097; D. P. Riley, M. R. Smith and P. E. Correa, *J. Am. Chem. Soc.*, 1988, **110**, 177.
147. H. Böh and H. Fischer, *Chem. Ber.*, 1942, **75**, 1310.
148. F. G. Bordwell and W. T. Brannen, Jr., *J. Am. Chem. Soc.*, 1964, **86**, 4645.
149. Y. Nagao, M. Ochiai, K. Kaneko, A. Maeda, K. Watanabe and E. Fujita, *Tetrahedron Lett.*, 1977, 1345.
150. T.-L. Ho and C. M. Wong, *Synthesis*, 1972, 561; T.-L. Ho, *Synth. Commun.*, 1979, **9**, 237.
151. E. Baciocchi, A. Piermattei and R. Ruzziconi, *Synth. Commun.*, 1988, **18**, 2167.
152. H. Böhme, H. Fischer and R. Frank, *Justus Liebigs Ann. Chem.*, 1949, **563**, 54.
153. H. E. Barron, G. W. K. Cavill, E. R. Cole, P. T. Gilham and D. H. Solomon, *Chem. Ind. (London)*, 1954, 76.
154. C. Srinivasan, P. Subramanian and S. Radha, *Indian J. Chem., Sect. B*, 1987, **26**, 193.
155. T. L. Evans and M. M. Grade, *Synth. Commun.*, 1986, **16**, 1207.
156. F. A. Davis, S. G. Lal and H. D. Durst, *J. Org. Chem.*, 1988, **53**, 5004.
157. D. Barnard and D. T. Woodbridge, *Chem. Ind. (London)*, 1959, 1603.
158. H. J. Reich, F. Chow and S. L. Peake, *Synthesis*, 1978, 299.
159. L. G. Faehl and J. L. Kice, *J. Org. Chem.*, 1979, **44**, 2357.
160. Y. H. Kim and D. C. Yoon, *Tetrahedron Lett.*, 1988, **29**, 6453.
161. G. S. Wilson, D. D. Swanson, J. T. Klug, R. S. Glass, M. D. Ryan and W. K. Musker, *J. Am. Chem. Soc.*, 1979, **101**, 1040.
162. D. S. Houghton and A. A. Humffray, *Electrochim. Acta*, 1972, **17**, 1421, 2145.
163. J. Yoshida, H. Sofuku and N. Kawabata, *Bull. Chem. Soc. Jpn.*, 1983, **56**, 1243.
164. K. Uneyama and S. Torii, *Tetrahedron Lett.*, 1971, 329.
165. M. Kimura, N. Kuriki, M. Inaishi and Y. Sawaki, *Tetrahedron Lett.*, 1984, **25**, 4665.
166. C. S. Foote and J. W. Peters, *J. Am. Chem. Soc.*, 1971, **93**, 3795.
167. D. Sinnreich, H. Lind and H. Batzer, *Tetrahedron Lett.*, 1976, 3541.
168. T. Tezuka, H. Miyazaki and H. Suzuki, *Tetrahedron Lett.*, 1978, 1959.
169. J. Hoyle, in 'The Chemistry of Sulphones and Sulphoxides,' ed. S. Patai, Z. Rappoport and C. J. M. Stirling, Wiley, Chichester, 1988, chap. 21.
170. H. Sugimoto and D. T. Sawyer, *J. Am. Chem. Soc.*, 1984, **106**, 4283.
171. H. Sugimoto and D. T. Sawyer, *J. Org. Chem.*, 1985, **50**, 1784.
172. Y. Ogata and K. Tanaka, *Can. J. Chem.*, 1981, **59**, 718.
173. H. S. Schultz, H. B. Fryermuth and S. R. Buc, *J. Org. Chem.*, 1963, **28**, 1140.
174. A. R. Doumaux, Jr., J. E. McKeon and D. J. Trecker, *J. Am. Chem. Soc.*, 1969, **91**, 3992.
175. J. Böeseken and E. Arrias, *Recl. Trav. Chim. Pays-Bas*, 1935, **54**, 711 (*Chem. Abstr.*, 1936, **30**, 2832).
176. H. H. Szmat and L. M. Alfonso, *J. Am. Chem. Soc.*, 1956, **78**, 1064; 1957, **79**, 205.
177. N. P. Petukhova, N. E. Dontsova and E. N. Prilezhaeva, *Izv. Akad. Nauk SSSR, Ser. Khim.*, 1982, 2327 (*Chem. Abstr.*, 1983, **98**, 71 447k).
178. D. N. Jones, T. P. Kogan and R. F. Newton, *J. Chem. Soc., Perkin Trans. 1*, 1982, 1333.
179. L. N. Levin, *J. Prakt. Chem.*, 1928, **118**, 282; 1930, **127**, 77.
180. H. H. Szmat, H. F. Harnsberger and F. Krahe, *J. Am. Chem. Soc.*, 1954, **76**, 2185.
181. T. Durst, R. Viau, R. Van Den Elzen and C. H. Nguyen, *J. Chem. Soc., Chem. Commun.*, 1971, 1334.
182. R. Curci and G. Modena, *Tetrahedron Lett.*, 1963, 1749; 1965, 863.
183. L. Kuhnen, *Angew. Chem.*, 1966, **78**, 937.
184. T. Takagi, *Yukagaku*, 1975, **24**, 319 (*Chem. Abstr.*, 1975, **83**, 164 782w).
185. S. Bhaduri, A. Ghosh and H. Khwaja, *J. Chem. Soc., Dalton Trans.*, 1981, 447.
186. S. Bhaduri and H. Khwaja, *J. Chem. Soc., Dalton Trans.*, 1983, 415.
187. H. Gampp and S. J. Lippard, *Inorg. Chem.*, 1983, **22**, 357.
188. M. Miura, M. Nojima and S. Kusabayashi, *J. Chem. Soc., Chem. Commun.*, 1982, 1352.
189. Y. H. Kim and H. K. Lee, *Chem. Lett.*, 1987, 1499.
190. V. I. Dronov, A. U. Baisheva, A. E. Pototskaya and L. M. Soskova, *Khim. Seraorg. Soedin., Soderzh. Neftyakh Nefteprod.*, 1964, **7**, 40 (*Chem. Abstr.*, 1965, **63**, 4235).
191. T. Durst and K. C. Tin, *Can. J. Chem.*, 1971, **49**, 2374.
192. A. E. Wood and E. G. Travis, *J. Am. Chem. Soc.*, 1928, **50**, 1226.
193. R. Annunziata, M. Cinquini and S. Colonna, *J. Chem. Soc., Perkin Trans. 1*, 1972, 2057.
194. P. Müller and J. Godoy, *Helv. Chim. Acta*, 1983, **66**, 1790.

195. A. A. Humffray and H. E. Imberger, *J. Chem. Soc., Perkin Trans. 2*, 1981, 382.
196. L. A. Carpino and J. R. Williams, *J. Org. Chem.*, 1974, **39**, 2320.
197. O. B. Helfrich and E. E. Reid, *J. Am. Chem. Soc.*, 1920, **42**, 1208.
198. A. G. Perkin and W. G. Sewell, *J. Chem. Soc.*, 1923, 3032.
199. T. Matsukawa, B. Ohta and T. Imada, *Yakugaku Zasshi*, 1950, **70**, 77 (*Chem. Abstr.*, 1950, **44**, 5322).
200. H. Bredereck, A. Wagner and A. Kottenhahn, *Chem. Ber.*, 1960, **93**, 2415.
201. G. Malewski and L. Mitzinger, *Monatsber. Dtsch. Akad. Wiss. Berlin*, 1968, **10**, 74 (*Chem. Abstr.*, 1969, **70**, 77 239n).
202. G. B. Barlin and W. V. Brown, *J. Chem. Soc. C*, 1969, 921.
203. E. Block, E. J. Corey, R. E. Penn, T. L. Renken and P. F. Sherwin, *J. Am. Chem. Soc.*, 1976, **98**, 5715.
204. S. Wolfe and C. F. Ingold, *J. Am. Chem. Soc.*, 1983, **105**, 7755.
205. H. B. Henbest and S. A. Khan, *J. Chem. Soc., Chem. Commun.*, 1968, 1036.
206. D. Scholz, *Monatsh. Chem.*, 1981, **112**, 241.
207. D. G. Lee and N. S. Srinivasan, *Sulfur Lett.*, 1982, **1**, 1 (*Chem. Abstr.*, 1983, **98**, 53 128c).
208. N. A. Noureldin, W. B. McConnell and D. G. Lee, *Can. J. Chem.*, 1984, **62**, 2113.
209. D. W. Goheen and C. F. Bennett, *J. Org. Chem.*, 1961, **26**, 1331.
210. G. A. Olah and B. G. B. Gupta, *J. Org. Chem.*, 1983, **48**, 3585.
211. J. Trocha-Grimshaw and H. B. Henbest, *J. Chem. Soc., Chem. Commun.*, 1968, 1035; H. B. Henbest and J. Trocha-Grimshaw, *J. Chem. Soc., Perkin Trans. 1*, 1974, 607.
212. B. M. Trost and D. P. Curran, *Tetrahedron Lett.*, 1981, **22**, 1287.
213. E. Howard, Jr. and L. S. Levitt, *J. Am. Chem. Soc.*, 1953, **75**, 6170.
214. Y. H. Kim, T. Takata and S. Oae, *Tetrahedron Lett.*, 1978, 2305.
215. G. Bontempelli, F. Magno, G. A. Mazzocchin and R. Seeber, *J. Electroanal. Chem. Interfacial Electrochem.*, 1974, **55**, 109.
216. D. Cipris and D. Pouli, *J. Electroanal. Chem. Interfacial Electrochem.*, 1976, **69**, 423.
217. C. F. Bennett and D. W. Goheen, *US Pat.* 3 418 224 (1968) (*Chem. Abstr.*, 1969, **70**, 43 434g).
218. P. T. Cottrell and C. K. Mann, *J. Electrochem. Soc.*, 1969, **116**, 1499.
219. G. O. Schenck and C. H. Kranch, *Chem. Ber.*, 1963, **96**, 517.
220. C. I. Gu, C. S. Foote and M. L. Kacher, *J. Am. Chem. Soc.*, 1981, **103**, 5949.
221. Y. Sawaki and Y. Ogata, *J. Am. Chem. Soc.*, 1981, **103**, 5947.
222. T. Akasaka and W. Ando, *J. Chem. Soc., Chem. Commun.*, 1983, 1203.
223. H. J. Reich, in 'Oxidation in Organic Chemistry', ed. W. S. Trahanovsky, Academic Press, New York, 1978, part C, chap. 1.
224. C. Paulmier, 'Selenium Reagents and Intermediates in Organic Synthesis,' Pergamon Press, Oxford, 1986.
225. See for example, D. G. Foster, *Org. Synth., Coll. Vol.*, 1955, **3**, 771.
226. L. B. Agenas, *Ark. Kemi*, 1965, **24**, 415 (*Chem. Abstr.*, 1965, **63**, 16 247).
227. K. B. Sharpless and R. F. Lauer, *J. Am. Chem. Soc.*, 1973, **95**, 2697.
228. K. B. Sharpless and M. W. Young, *J. Org. Chem.*, 1975, **40**, 947.
229. R. L. N. Rao, *J. Indian Chem. Soc.*, 1941, **18**, 1 (*Chem. Abstr.*, 1941, **35**, 7945).
230. G. H. Schmid and D. G. Garratt, *J. Org. Chem.*, 1983, **48**, 4169.
231. H. J. Reich, J. M. Renga and I. L. Reich, *J. Am. Chem. Soc.*, 1975, **97**, 5434.
232. H. Sisler and N. K. Koita, *J. Org. Chem.*, 1971, **36**, 1700.
233. K. Y. Jen and M. P. Cava, *J. Org. Chem.*, 1983, **48**, 1449.
234. O. Behaghel and H. Seibert, *Chem. Ber.*, 1933, **66**, 708.
235. D. L. Klayman and J. W. Lown, *J. Org. Chem.*, 1966, **31**, 3396.
236. D. L. Klayman, *J. Org. Chem.*, 1965, **30**, 2454.
237. J. D. McCullough and E. S. Gould, *J. Am. Chem. Soc.*, 1949, **71**, 674.
238. E. S. Gould and J. D. McCullough, *J. Am. Chem. Soc.*, 1951, **73**, 1109.
239. E. Rebane, *Ark. Kemi*, 1967, **26**, 345 (*Chem. Abstr.*, 1967, **66**, 94 655e).
240. F. L. Pyman, *J. Chem. Soc.*, 1919, **115**, 166.
241. R. Paetzold and E. Rönsch, *Z. Anorg. Allg. Chem.*, 1965, **338**, 195 (*Chem. Abstr.*, 1965, **63**, 14 692).
242. G. Ayrey, D. Barnard and D. T. Woodbridge, *J. Chem. Soc.*, 1962, 2089.
243. D. T. Woodbridge, *J. Chem. Soc. B*, 1966, 50.
244. R. A. Gancarz and J. L. Kice, *Tetrahedron Lett.*, 1981, **22**, 1661.
245. For example, ref. 224, p. 364.
246. J. L. Kice, Y. H. Kang and M. B. Manek, *J. Org. Chem.*, 1988, **53**, 2435.
247. W. Jenny, *Helv. Chim. Acta*, 1958, **41**, 317.
248. H. Rheinboldt and E. Giesbrecht, *Chem. Ber.*, 1955, **88**, 66, 1937, 1974.
249. H. J. Reich, W. W. Willis, Jr. and S. Wollowitz, *Tetrahedron Lett.*, 1982, **23**, 3319.
250. J. L. Kice, F. McAfee and H. Slebocka-Tilk, *Tetrahedron Lett.*, 1982, **23**, 3323.
251. J. Loevenich, H. Fremdling and M. Föhr, *Chem. Ber.*, 1929, **62**, 2856.
252. V. V. Kozlov and S. E. Suvorova, *Zh. Obshch. Khim.*, 1961, **31**, 3034 (*Chem. Abstr.*, 1962, **57**, 740).
253. A. Cerriani and R. Passerini, *Ann. Chim.*, 1957, **47**, 58 (*Chem. Abstr.*, 1957, **51**, 10 442).
254. M. J. Bird and F. Challenger, *J. Chem. Soc.*, 1942, 570.
255. E. Rebane, *Acta Chem. Scand.*, 1967, **21**, 657.
256. (a) R. Paetzold and D. Lienig, *Z. Chem.*, 1964, **4**, 186 (*Chem. Abstr.*, 1964, **61**, 4249); (b) M. Tiecco, L. Testaferri, M. Tingoli, D. Chianelli and D. Bartoli, *Tetrahedron Lett.*, 1989, **30**, 1417; (c) S. Torii, K. Uneyama, M. Ono and T. Bannou, *J. Am. Chem. Soc.*, 1981, **103**, 4606, and refs. therein.
257. O. K. Edwards, W. R. Gaythwaite, J. Kenyon and H. Phillips, *J. Chem. Soc.*, 1928, 2293.
258. D. G. Foster and S. F. Brown, *J. Am. Chem. Soc.*, 1928, **50**, 1182.
259. W. J. Burlant and E. S. Gould, *J. Am. Chem. Soc.*, 1954, **76**, 5775.
260. M. Ōki and H. Iwamura, *Tetrahedron Lett.*, 1966, 2917.

261. M. Cinquini, S. Colonna and D. Landini, *Boll. Sci. Fac. Chim. Ind. Bologna*, 1969, **27**, 207 (*Chem. Abstr.*, 1970, **72**, 89 955).
262. M. R. Detty, *J. Org. Chem.*, 1980, **45**, 274.
263. R. W. Rickards and W. P. Watson, *Aust. J. Chem.*, 1980, **33**, 451.
264. A. Toshimitsu, H. Owada, T. Aoai, S. Uemura and M. Okano, *J. Chem. Soc., Chem. Commun.*, 1981, 546; A. Toshimitsu, T. Aoai, H. Owada, S. Uemura and M. Okano, *J. Org. Chem.*, 1981, **46**, 3021.
265. Y. Masuyama, Y. Ueno and M. Okawara, *Chem. Lett.*, 1977, 835.
266. B. Harirchian and P. D. Magnus, *J. Chem. Soc., Chem. Commun.*, 1977, 522.
267. M. Servin, W. Dumont and A. Krief, *Tetrahedron Lett.*, 1977, 3835.
268. R. Walter and J. Roy, *J. Org. Chem.*, 1971, **36**, 2561.
269. K. B. Sharpless, M. W. Young and R. F. Lauer, *Tetrahedron Lett.*, 1973, 1979.
270. K. B. Sharpless and R. F. Lauer, *J. Org. Chem.*, 1974, **39**, 429.
271. C. A. Wilson, II and T. A. Bryson, *J. Org. Chem.*, 1975, **40**, 800.
272. K. B. Sharpless, R. F. Lauer and A. Y. Teranishi, *J. Am. Chem. Soc.*, 1973, **95**, 6137.
273. T. Miura and M. Kobayashi, *J. Chem. Soc., Chem. Commun.*, 1982, 438.
274. E. S. Gould and J. D. McCullough, *J. Am. Chem. Soc.*, 1951, **73**, 3196.
275. H. J. Reich and F. Chow, *J. Chem. Soc., Chem. Commun.*, 1975, 790.
276. H. J. Reich and S. K. Shah, *J. Am. Chem. Soc.*, 1975, **97**, 3250.
277. H. J. Reich and J. M. Renga, *J. Org. Chem.*, 1975, **40**, 3313.
278. H. J. Reich and S. K. Shah, *J. Am. Chem. Soc.*, 1977, **99**, 263.
279. R. Ando, T. Sugawara, M. Shimizu and I. Kuwajima, *Bull. Chem. Soc. Jpn.*, 1984, **57**, 2897; M. Shimizu and I. Kuwajima, *J. Org. Chem.*, 1980, **45**, 2921; I. Kuwajima, R. Ando and T. Sugawara, *Tetrahedron Lett.*, 1983, **24**, 4429.
280. T. G. Back, S. Collins, U. Gokhale and K. W. Law, *J. Org. Chem.*, 1983, **48**, 4776; T. G. Back, S. Collins and K. W. Law, *Can. J. Chem.*, 1985, **63**, 2313; T. G. Back and M. V. Krishna, *J. Org. Chem.*, 1987, **52**, 4265.
281. R. H. Mitchell, K. S. Weerawarna and G. W. Bushnell, *Tetrahedron Lett.*, 1987, **28**, 5119.
282. H. J. Reich and W. W. Willis, Jr., *J. Am. Chem. Soc.*, 1980, **102**, 5967.
283. T. Hiroi and K. B. Sharpless, *J. Org. Chem.*, 1978, **43**, 1689.
284. D. N. Jones, D. Mundy and R. D. Whitehouse, *J. Chem. Soc., Chem. Commun.*, 1970, 86.
285. H. J. Reich, J. M. Renga and I. L. Reich, *J. Org. Chem.*, 1974, **39**, 2133.
286. M. Cinquini, S. Colonna and R. Giovini, *Chem. Ind. (London)*, 1969, 1737.
287. A. Hassner and A. S. Amarasekara, *Tetrahedron Lett.*, 1987, **28**, 5185.
288. I. D. Entwistle, R. A. W. Johnstone and J. H. Varley, *J. Chem. Soc., Chem. Commun.*, 1976, 61.
289. D. L. J. Clive, *J. Chem. Soc., Chem. Commun.*, 1973, 695.
290. H. J. Reich, I. L. Reich and J. M. Renga, *J. Am. Chem. Soc.*, 1973, **95**, 5813.
291. F. A. Davis, O. D. Stringer and J. M. Billmers, *Tetrahedron Lett.*, 1983, **24**, 1213.
292. D. G. Foster, *Recl. Trav. Chim. Pays-Bas*, 1935, **54**, 447.
293. H. Rheinboldt and E. Giesbrecht, *J. Am. Chem. Soc.*, 1946, **68**, 2671.
294. R. Paetzold, U. Lindner, G. Boshmann and P. Reich, *Z. Anorg. Allg. Chem.*, 1967, **352**, 295 (*Chem. Abstr.*, 1967, **67**, 81 764k).
295. R. Paetzold and G. Bochmann, *Z. Anorg. Allg. Chem.*, 1968, **360**, 293 (*Chem. Abstr.*, 1969, **70**, 3190t).
296. G. V. Boyd, M. Doughty and J. Kenyon, *J. Chem. Soc.*, 1949, 2196.
297. L. Engman, *J. Org. Chem.*, 1987, **52**, 4086; 1988, **53**, 4031.
298. A. M. Morella and A. D. Ward, *Tetrahedron Lett.*, 1985, **26**, 2899.
299. M. Kobayashi, H. Ohkubo and T. Shimizu, *Bull. Chem. Soc. Jpn.*, 1986, **59**, 503.
300. K. Takaki, M. Yasuhara and K. Negoro, *J. Org. Chem.*, 1983, **48**, 54.
301. E. Fujita, M. Ochiai and K. Watanabe, *J. Indian Chem. Soc.*, 1978, **40**, 1226.
302. K. Yamakawa, K. Nishitani and T. Tominaga, *Tetrahedron Lett.*, 1975, 2829.
303. J. Loevenich, H. Fremdling and M. Föhr, *Chem. Ber.*, 1929, **62**, 2856.
304. L. M. Yagupol'skii, G. P. Syrova, V. G. Voloshchuk and V. R. Bystrov, *Zh. Obshch. Khim.*, 1968, **38**, 2591 (*Chem. Abstr.*, 1969, **70**, 46 677z).
305. H. Hofmeister, K. Annen, G. Cleve, H. Laurent and R. Wiechert, *Liebigs Ann. Chem.*, 1981, 1973.
306. M. Tiecco, D. Chianelli, M. Tingoli, L. Testaferri and D. Bartoli, *Tetrahedron*, 1986, **42**, 4897.
307. A. Krief, W. Dumont, J. N. Denis, G. Evrard and B. Norberg, *J. Chem. Soc., Chem. Commun.*, 1985, 569.
308. S. Uemura, S. Fukuzawa and A. Toshimitsu, *J. Chem. Soc., Chem. Commun.*, 1983, 1501; S. Uemura and S. Fukuzawa, *J. Chem. Soc., Perkin Trans. 1*, 1985, 471.
309. S. Uemura, S. Fukuzawa, T. Yamauchi, K. Hattori, S. Mizutaki and K. Tamaki, *J. Chem. Soc., Chem. Commun.*, 1984, 426; *J. Chem. Soc., Perkin Trans. 1*, 1986, 1983.
310. W. Stanley, M. R. Van De Mark and P. L. Kumler, *J. Chem. Soc., Chem. Commun.*, 1974, 700.
311. J. Y. Chu, D. G. Marsh and W. H. H. Günther, *J. Am. Chem. Soc.*, 1975, **97**, 4905.
312. H. K. Spencer and M. P. Cava, *J. Org. Chem.*, 1977, **42**, 2937.
313. H. K. Spencer, M. V. Lakshmikantham and M. P. Cava, *J. Am. Chem. Soc.*, 1977, **99**, 1470.
314. K. Irgolic, in 'The Organic Chemistry of Tellurium,' Gordon and Breach, New York, 1974, p. 98.
315. M. R. Detty, B. J. Murray, D. L. Smith and Z. Zumbulyadis, *J. Am. Chem. Soc.*, 1983, **105**, 875.
316. D. Seebach and A. K. Beck, *Chem. Ber.*, 1975, **108**, 314.
317. L. Engman and M. P. Cava, *Synth. Commun.*, 1982, **12**, 165.
318. Y. Hirabayashi, H. Ishihara, Y. Ogi, T. Miyake, H. Oida, N. Kondo and H. Goji, *Nippon Kagaku Kaishi*, 1987, 1475 (*Chem. Abstr.*, 1988, **109**, 6166z).
319. M. Herberhold and P. Leitner, *J. Organomet. Chem.*, 1987, **336**, 153.
320. A. Baroni, *Atti Accad. Naz. Lincei, Cl. Sci. Fis., Mat. Nat., Rend.*, 1938, **27**, 238 (*Chem. Abstr.*, 1939, **33**, 163).
321. C. W. Sink and A. B. Harvey, *J. Chem. Soc. D*, 1969, 1023.

322. N. Ohira, Y. Aso, T. Otsubo and F. Ogura, *Chem. Lett.*, 1984, 853; Y. Aso, T. Nishioka, M. Osuka, K. Nakagawa, K. Sakai, T. Otsubo and F. Ogura, *Nippon Kagaku Kaishi*, 1987, 1490 (*Chem. Abstr.*, 1988, **108**, 131 193m).
323. M. Akiba and M. P. Cava, *Synth. Commun.*, 1984, **14**, 1119.
324. J. L. Piette and M. Renson, *Bull. Soc. Chim. Belg.*, 1970, **79**, 353, 367, 383 (*Chem. Abstr.*, 1970, **73**, 66 199k, 66 200d, 66 201e).
325. S. R. Buzilova, I. D. Sadekov, T. V. Lipovich, T. M. Filippova and L. I. Vereshchagin, *Zh. Obshch. Khim.*, 1977, **47**, 1999 (*Chem. Abstr.*, 1978, **88**, 22 289v).
326. S. Uemura, S. Fukuzawa and S. R. Patil, *J. Organomet. Chem.*, 1983, **243**, 9.
327. G. T. Morgan and R. E. Kellett, *J. Chem. Soc.*, 1926, 1080.
328. G. T. Morgan and H. D. K. Drew, *J. Chem. Soc.*, 1925, 2307.
329. H. D. K. Drew, *J. Chem. Soc.*, 1926, 223.
330. M. Giua and F. Cherchi, *Gazz. Chim. Ital.*, 1920, **50**, 362 (*Chem. Abstr.*, 1921, **15**, 521).
331. K. Lederer, *Chem. Ber.*, 1915, **48**, 1345.
332. Y. Hirabayashi, H. Ishihara, H. Manami, S. Matsuda, Y. Morita and T. Fujiwara, *Nippon Kagaku Kaishi*, 1987, 1479 (*Chem. Abstr.*, 1988, **108**, 150 000n).
333. N. Dereu and M. Renson, *J. Organomet. Chem.*, 1983, **258**, 163.
334. P. Wiriyachitra, S. J. Falcone and M. P. Cava, *J. Org. Chem.*, 1979, **44**, 3957.
335. I. D. Sadekov, A. Ya. Bushkov, V. S. Yur'eva and V. I. Minkin, *Zh. Obshch. Khim.*, 1977, **47**, 2541 (*Chem. Abstr.*, 1978, **88**, 61 858z).
336. G. Vicentini, E. Giesbrecht and L. R. M. Pitombo, *Chem. Ber.*, 1959, **92**, 40.
337. B. C. Pant, *Tetrahedron Lett.*, 1972, 4779; *J. Organomet. Chem.*, 1974, **65**, 51.
338. K. Lederer, *Justus Liebigs Ann. Chem.*, 1912, **391**, 326.
339. G. T. Morgan and F. H. Burstall, *J. Chem. Soc.*, 1931, 180.
340. L. Reichel and E. Kirschbaum, *Justus Liebigs Ann. Chem.*, 1936, **523**, 211.
341. M. P. Balfe, C. A. Chaplin and H. Phillips, *J. Chem. Soc.*, 1938, 341; M. P. Balfe and K. N. Nandi, *J. Chem. Soc.*, 1941, 70.
342. D. Hellwinkel and G. Fahrbach, *Justus Liebigs Ann. Chem.*, 1968, **712**, 1.
343. K. Lederer, *Justus Liebigs Ann. Chem.*, 1916, **49**, 334.
344. G. T. Morgan and F. H. Burstall, *J. Chem. Soc.*, 1930, 2599.
345. Ref. 314, pp. 120, 149.
346. T. N. Srivastava, R. C. Srinivastava and M. Singh, *J. Organomet. Chem.*, 1978, **160**, 449.
347. R. H. Vernon, *J. Chem. Soc.*, 1920, 889.
348. G. Kirsch, M. M. Goodman and F. F. Knap, Jr., *Organometallics*, 1983, **2**, 357.
349. H. Lee and M. P. Cava, *J. Chem. Soc., Chem. Commun.*, 1981, 277.
350. R. Sochacka and A. Szuchnik, *Polska Akad. Nauk, Inst. Badan Jad.*, 1960, 149/XIII, 1 (*Chem. Abstr.*, 1960, **54**, 24481).
351. L. Tschugaeff and W. Chlopin, *Chem. Ber.*, 1914, **47**, 1269.
352. (a) J. V. Comasseto, J. T. B. Ferreira and J. A. F. Val, *J. Organomet. Chem.*, 1984, **277**, 261; (b) N. Petragnani and J. V. Comasseto, *Rev. Heteroatom. Chem.*, 1989, **2**, 40; (c) S. Uemura, S. Fukuzawa and K. Ohe, *Tetrahedron Lett.*, 1985, **26**, 921.
353. K. B. Sharpless, K. M. Gordon, R. F. Lauer, D. W. Patrick, S. P. Singer and M. W. Young, *Chem. Scr.*, 1975, **8A**, 9.
354. S. Uemura, K. Ohe and S. Fukuzawa, *Tetrahedron Lett.*, 1985, **26**, 895.
355. S. Uemura and S. Fukuzawa, *J. Am. Chem. Soc.*, 1983, **105**, 2748; K. Ohe, S. Fukuzawa and S. Uemura, *Chem. Express*, 1990, 5, 261.
356. S. Uemura, Y. Hirai, K. Ohe and N. Sugita, *J. Chem. Soc., Chem. Commun.*, 1985, 1037.
357. D. H. R. Barton, N. Ozbalik and R. Ramesh, *Tetrahedron Lett.*, 1988, **29**, 3533.
358. S. Uemura, H. Takahashi and K. Ohe, unpublished data.
359. F. L. Gilbert and T. M. Lowry, *J. Chem. Soc.*, 1928, 3179; T. M. Lowry and F. L. Gilgert, *J. Chem. Soc.*, 1929, 2076.
360. G. T. Morgan and H. Burgess, *J. Chem. Soc.*, 1928, 321.
361. L. Engman and M. P. Cava, *J. Chem. Soc., Chem. Commun.*, 1982, 164.
362. S. V. Ley, C. A. Meerholz and D. H. R. Barton, *Tetrahedron Lett.*, 1980, 1785; *Tetrahedron*, 1981, **37** (suppl. 1), 213.
363. S. Uemura and S. Fukuzawa, *J. Chem. Soc., Chem. Commun.*, 1980, 1033; *J. Organomet. Chem.*, 1984, **268**, 223.
364. E. Luppold, W. Winter and E. Müller, *Chem. Ber.*, 1976, **109**, 3886.
365. M. R. Detty, P. B. Merkel and S. K. Powers, *J. Am. Chem. Soc.*, 1988, **110**, 5920.
366. K. Balenovic, N. Bregant and D. Francetic, *Tetrahedron Lett.*, 1960, 20; K. Balenovic, I. Bregoveć, D. Francetic, I. Monković and V. Tomasić, *Chem. Ind. (London)*, 1961, 469.
367. A. Mayr, F. Montanari and M. Tramontini, *Ric. Sci.*, 1960, **30**, 746 (*Chem. Abstr.*, 1961, **55**, 1499); *Gazz. Chim. Ital.*, 1960, **90**, 739 (*Chem. Abstr.*, 1961, **55**, 16 460).
368. A. Maccioni, F. Montanari, M. Secci and M. Tramontini, *Tetrahedron Lett.*, 1961, 607.
369. K. Mislow, M. M. Green, D. Laur, J. T. Melillo, T. Simmons and A. L. Ternay, Jr., *J. Am. Chem. Soc.*, 1965, **87**, 1958; K. Mislow, M. M. Green and M. Raban, *J. Am. Chem. Soc.*, 1965, **87**, 2761.
370. A. Maccioni, *Boll. Sci. Fac. Chim. Ind. Bologna*, 1965, **23**, 41 (*Chem. Abstr.*, 1965, **63**, 8239).
371. F. Montanari, *Tetrahedron Lett.*, 1965, 3367; U. Folli, D. Iarossi, F. Montanari and G. Torre, *J. Chem. Soc. C*, 1968, 1317.
372. U. Folli, D. Iarossi, F. Montanari and G. Torre, *Boll. Sci. Fac. Chim. Ind. Bologna*, 1967, **25**, 159 (*Chem. Abstr.*, 1968, **69**, 2588w).
373. U. Folli, D. Iarossi and G. Torre, *Ric. Sci.*, 1968, **38**, 914 (*Chem. Abstr.*, 1969, **71**, 3099r); U. Folli and D. Iarossi, *Gazz. Chim. Ital.*, 1969, **99**, 1306 (*Chem. Abstr.*, 1970, **72**, 110 969s).

374. D. Iarossi and A. Pinetti, *Boll. Sci. Fac. Chim. Ind. Bologna*, 1969, **27**, 221 (*Chem. Abstr.*, 1970, **72**, 131 811z).
375. W. H. Pirkle and P. L. Rinaldi, *J. Org. Chem.*, 1977, **42**, 2080.
376. F. Di Furia, G. Modena and R. Curci, *Tetrahedron Lett.*, 1976, 4637; F. Di Furia, G. Modena and R. Seraglia, *Synthesis*, 1984, 325; F. Di Furia, G. Licini, G. Modena and O. De Lucchi, *Tetrahedron Lett.*, 1989, **30**, 2575; V. Conte, F. Di Furia, G. Licini and G. Modena, *Tetrahedron Lett.*, 1989, **30**, 4859.
377. P. Pitchen and H. B. Kagan, *Tetrahedron Lett.*, 1984, **25**, 1049; P. Pitchen, E. Duñach, M. N. Deshmukh and H. B. Kagan, *J. Am. Chem. Soc.*, 1984, **106**, 8188.
378. E. Duñach and H. B. Kagan, *Nouv. J. Chim.*, 1985, **9**, 1; H. B. Kagan, E. Duñach, C. Nemwcek, P. Pitchen, O. Samuel and S. H. Zao, *Pure Appl. Chem.*, 1985, **57**, 1911; O. Samuel, B. Ronan and H. B. Kagan, *J. Organomet. Chem.*, 1989, **370**, 43.
379. H. B. Kagan, *Phosphorus Sulfur*, 1986, **27**, 127.
380. O. Bartolini, F. Di Furia, G. Licini, G. Modena and M. Rossi, *Tetrahedron Lett.*, 1986, **27**, 6257.
381. S. H. Zhao, O. Samuel and H. B. Kagan, *C. R. Hebd. Seances Acad. Sci., Ser. C*, 1987, **304**, 273 (*Chem. Abstr.*, 1987, **107**, 58 584n).
382. S. Colonna, A. Manfredi, M. Spadoni, L. Casella and M. Gullotti, *J. Chem. Soc., Perkin Trans. 1*, 1987, 71.
383. D. J. Cram and S. H. Pine, *J. Am. Chem. Soc.*, 1963, **85**, 1096.
384. G. Barbieri, U. Folli, F. Montanari and G. Torre, *Boll. Sci. Fac. Chim. Ind. Bologna*, 1965, **23**, 47 (*Chem. Abstr.*, 1965, **63**, 8224e); G. Barbieri, V. Davoli, I. Moretti, F. Montanari and G. Torre, *J. Chem. Soc. C*, 1969, 731.
385. B. W. Christensen and A. Kjaer, *J. Chem. Soc., Chem. Commun.*, 1965, 225.
386. E. Bordignon, L. Cattalini, G. Natile and A. Scatturin, *J. Chem. Soc., Chem. Commun.*, 1973, 878.
387. K. Nishihata and M. Nishio, *J. Chem. Soc., Perkin Trans. 2*, 1973, 758.
388. S. Lavielle, S. Bory, B. Moreau, M.-J. Luche and A. Marquet, *J. Am. Chem. Soc.*, 1978, **100**, 1558.
389. F. Ruff, G. Szabó, J. Vajda, J. Kövesdi and Á. Kucsman, *Tetrahedron*, 1980, **36**, 1631.
390. T. Higuchi, I. H. Pitman and K. H. Gensch, *J. Am. Chem. Soc.*, 1966, **88**, 5676.
391. M. Kinoshita, Y. Sato and N. Kunieda, *Chem. Lett.*, 1974, 377; Y. Sato, N. Kunieda and M. Kinoshita, *Bull. Chem. Soc. Jpn.*, 1976, **49**, 3331.
392. Y. Sato, N. Kunieda and M. Kinoshita, *Chem. Lett.*, 1976, 563.
393. Y. Sato, N. Kunieda and M. Kinoshita, *Makromol. Chem.*, 1977, **178**, 683.
394. C. R. Johnson, C. C. Bacon and W. D. Kingsbury, *Tetrahedron Lett.*, 1972, 501.
395. M. Moriyama, S. Oae, T. Numata and N. Furukawa, *Chem. Ind. (London)*, 1976, 163.
396. K.-T. Liu and Y. C. Tong, *J. Chem. Res. (S)*, 1979, 276.
397. R. Herrmann, G. Hübener and I. Ugi, *Tetrahedron*, 1985, **41**, 941.
398. F. A. Davis, R. H. Jenkins, Jr., S. Q. A. Rizvi and T. W. Panunto, *J. Chem. Soc., Chem. Commun.*, 1979, 600; F. A. Davis, R. H. Jenkins, Jr., S. B. Awad, O. D. Stringer, W. H. Watson and J. Galloy, *J. Am. Chem. Soc.*, 1982, **104**, 5412.
399. F. A. Davis, J. C. Towson, M. C. Weismiller, S. G. Lal and P. J. Carroll, *J. Am. Chem. Soc.*, 1988, **110**, 8477.
400. F. A. Davis, J. P. McCauley, Jr. and M. E. Harakal, *J. Org. Chem.*, 1984, **49**, 1465.
401. (a) F. A. Davis, J. P. McCauley, Jr., S. Chattopadhyay, M. E. Harakal, J. C. Towson, W. H. Watson and I. Tavanaiepour, *J. Am. Chem. Soc.*, 1987, **109**, 3370; (b) F. A. Davis, R. Thimma Reddy and M. C. Weismiller, *J. Am. Chem. Soc.*, 1989, **111**, 5964.
402. M. Bucciarelli, A. Forni, S. Marcaccioli, I. Moretti and G. Torre, *Tetrahedron*, 1983, **39**, 187.
403. T. Takata and W. Ando, *Tetrahedron Lett.*, 1986, **27**, 1591.
404. K. Nakajima, M. Kojima and J. Fujita, *Chem. Lett.*, 1986, 1483.
405. B. E. Firth, L. L. Miller, M. Mitani, T. Rogers, J. Lennox and R. W. Murray, *J. Am. Chem. Soc.*, 1976, **98**, 8271; B. E. Firth and L. L. Miller, *J. Am. Chem. Soc.*, 1976, **98**, 8272.
406. T. Komori and T. Nonaka, *J. Am. Chem. Soc.*, 1983, **105**, 5690; 1984, **106**, 2656.
407. M. Mikołajczyk and J. Drebowicz, *J. Am. Chem. Soc.*, 1978, **100**, 2510; J. Drabowicz and M. Mikołajczyk, *Phosphorus Sulfur*, 1984, **19**, 245.
408. A. W. Czarnik, *J. Org. Chem.*, 1984, **49**, 924.
409. T. Sugimoto, T. Kokubo, J. Miyazaki, S. Tanimoto and M. Okano, *J. Chem. Soc., Chem. Commun.*, 1979, 402; T. Sugimoto, T. Kokubo, J. Miyazaki, S. Tanimoto and M. Okano, *Bioorg. Chem.*, 1981, **10**, 311.
410. K. Ogura, M. Fujita and H. Iida, *Tetrahedron Lett.*, 1980, **21**, 2233.
411. S. Colonna, S. Banfi, F. Fontana and H. Sommaruga, *J. Org. Chem.*, 1985, **50**, 769; S. Colonna, S. Banfi, R. Annunziata and L. Casella, *J. Org. Chem.*, 1986, **51**, 891.
412. A. Yamagishi, *J. Chem. Soc., Chem. Commun.*, 1986, 290.
413. G. Solladie, *Synthesis*, 1981, 185.
414. L. D. Wright, E. L. Cresson, J. Valiant, D. E. Wolf and K. Folkers, *J. Am. Chem. Soc.*, 1954, **76**, 4163.
415. E. Abushanab, D. Reed, F. Suzuki and C. J. Sih, *Tetrahedron Lett.*, 1978, 3415.
416. H. Ohta, Y. Okamoto and G. Tsuchihashi, *Chem. Lett.*, 1984, 205; *Agric. Biol. Chem.*, 1985, **49**, 671, 2229.
417. D. J. Cavanaugh, *Science (Washington, D.C.)*, 1957, **125**, 1040.
418. J. R. Gillette and J. J. Kamm, *J. Pharmacol. Exp. Ther.*, 1960, **130**, 262.
419. H. Kexel and H. L. Schmidt, *Biochem. Pharmacol.*, 1972, **21**, 1009.
420. S. W. May and R. S. Phillips, *J. Am. Chem. Soc.*, 1980, **102**, 5981; S. W. May, *Enzyme Eng.*, 1982, **6**, 167.
421. W. G. Salmond, M. A. Barta, A. M. Cain and M. C. Sobala, *Tetrahedron Lett.*, 1977, 1683.
422. T. G. Back, N. Ibrahim and D. J. McPhee, *J. Org. Chem.*, 1982, **47**, 3283.
423. W. R. Graythwaite, J. Kenyon and H. Phillips, *J. Chem. Soc.*, 1928, 2280, 2287.
424. T. Shimizu, M. Yoshida and M. Kobayashi, *Bull. Chem. Soc. Jpn.*, 1987, **60**, 1555.
425. F. A. Davis, J. M. Billmers and O. D. Stringer, *Tetrahedron Lett.*, 1983, **24**, 3191.
426. F. A. Davis, O. D. Stringer and J. P. McCauley, Jr., *Tetrahedron*, 1985, **41**, 4747.
427. M. Tiecco, M. Tingoli, L. Testaferri and D. Bartoli, *Tetrahedron Lett.*, 1987, **28**, 3849.
428. T. Shimizu, M. Kobayashi and N. Kamigata, *Bull. Chem. Soc. Jpn.*, 1989, **62**, 2099.

429. H. L. Holland and I. M. Carter, *Bioorg. Chem.*, 1983, **12**, 1.
430. T. Shimizu and M. Kobayashi, *Bull. Chem. Soc. Jpn.*, 1986, **59**, 2654; *J. Org. Chem.*, 1987, **52**, 3399.
431. T. Shimizu and M. Kobayashi, *Chem. Lett.*, 1986, 161; M. Kobayashi and T. Shimizu, *Nippon Kagaku Kaishi*, 1987, 1437 (*Chem. Abstr.*, 1988, **109**, 22 612x).

7.1
Oxidation by Electrochemical Methods

TATSUYA SHONO
Kyoto University, Japan

7.1.1 INTRODUCTION

The chemistry of reactions promoted by electrochemical oxidation or reduction of organic substrates is called electroorganic chemistry. This term is often used along with organic electrochemistry. The difference between these two terms is not always clearly determined, the latter emphasizing electrochemistry rather than organic chemistry and involving electrochemical and electrotheoretical studies, while the former is more concerned with organic chemistry which is beneficial to synthesis, and it is these methods which are surveyed below.

7.1.1.1 Inversion of Polarity of Substrates

In electroorganic reactions, the active species is generated on the electrode surface by electron transfer between a substrate molecule and the electrode, as shown in equation (1). The substrate molecule is transformed to a cation radical or an anion radical, depending on the direction of electron transfer. When the substrate molecule is a radical or ionic species, the transformation of the substrate is as shown in equation (2).

$$A^{2-} \underset{+e}{\overset{-e}{\rightleftharpoons}} A^{\bullet-} \underset{+e}{\overset{-e}{\rightleftharpoons}} A \underset{+e}{\overset{-e}{\rightleftharpoons}} A^{\bullet+} \underset{+e}{\overset{-e}{\rightleftharpoons}} A^{2+} \quad (1)$$

$$A^{-} \underset{+e}{\overset{-e}{\rightleftharpoons}} A^{\bullet} \underset{+e}{\overset{-e}{\rightleftharpoons}} A^{+} \quad (2)$$

Generally, an organic reaction between two substrate molecules is not achievable when the polarity of the reaction site is the same in both substrates. In other words, the reaction usually takes place between a nucleophile (Nu) and an electrophilic site (E). In organic synthesis, however, it is not uncommon that reaction between two groups of the same polarity is required to synthesize the target compound. As a result, inversion of the polarity of one of the groups is necessary to achieve reaction, although this inversion (Umpolung) is not always easy. As equations (1) and (2) clearly show, in an electroorganic reaction the generation of active species through electron transfer between a substrate and an electrode always involves inversion of polarity of the substrate. Thus, this facile inversion of polarity makes electroorganic chemistry a unique tool in organic synthesis.

7.1.1.2 Interface Reactions

As the active species is formed on the electrode surface at an interface between solid and solution, unique characteristics in reactivity can occur that are useful to the organic chemist.

7.1.1.2.1 Stereochemistry

One of these useful characteristics is that of stereoselectivity. The stereoselectivities observed in the acetoxylation of methylcyclohexenes are good examples of reactions taking place at the interface (see Section 7.1.3.2.1). Further examples of this selectivity have been observed in the anodic acetoxylation of some cyclic dienol acetates carried out in acetic acid containing potassium acetate as the supporting electrolyte (equation 3).

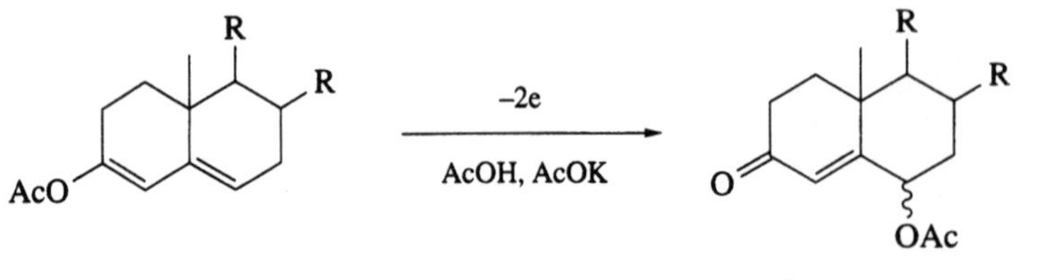

R, R, AcO —(−2e; AcOH, AcOK)→ R, R, O, OAc (3)

α- and β-isomers

The same products are also obtained by oxidation of the dienol acetates with peroxybenzoic acid or by oxidation of the corresponding enones with the liver microsomal oxidation systems.

The configuration of the products obtained by the anodic method ($\beta/\alpha = 13.9$) shows a remarkable similarity with the microsomal oxidation products ($\beta/\alpha = 14.1$), whereas peroxybenzoic acid oxidation exhibits poor stereospecificity ($\beta/\alpha = 3$).[1]

The similarity of anodic and microsomal oxidations may be explained by the fact that both types of oxidation take place at interfaces.

7.1.1.2.2 *Distribution of the active species*

The active species generated on the electrode surface usually reacts with other reagents before it diffuses into the solution, whereas in the usual organic homogeneous reactions the distribution of active species is uniform in solution. Due to this difference the electrogenerated active species displays unique characteristics.

The famous Kolbe electrolysis is a typical example showing the uniqueness of the distribution of the electrogenerated active species. Thus, the free radical species, formed at rather high concentration on the anode surface through anodic oxidation of a carboxylate anion, dimerizes before it is diffused into solution. The same radical species generated in a homogeneous solution by chemical methods forms the dimer as a minor product, the major product being that derived by hydrogen abstraction from the solvent.

7.1.2 APPARATUS AND TECHNIQUES

7.1.2.1 Apparatus

7.1.2.1.1 *Electrolysis cell*

The simplest and most convenient cell is a cylindrical glass cell with a capacity less than several hundred milliliters, although those of approximately one milliliter may be less appropriate for synthetic reaction studies (Figure 1).

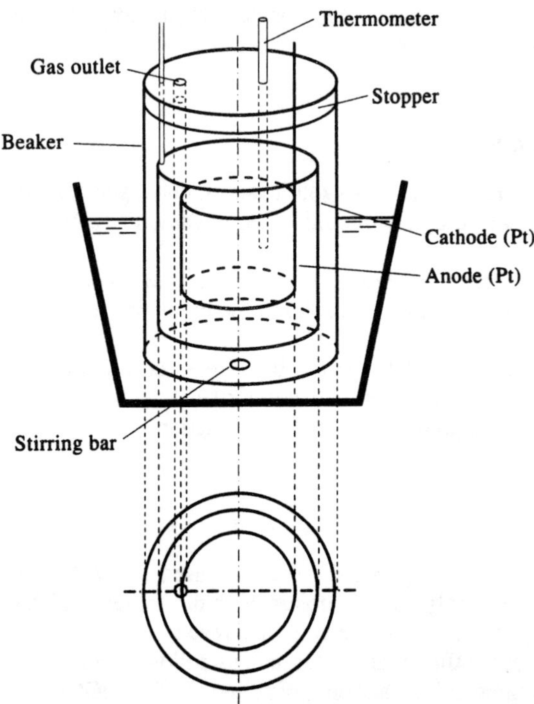

Figure 1 Undivided cell

The cell shown in Figure 1 is usually used for the reaction carried out under constant current conditions. For carrying out the reaction under a constant potential condition a cell equipped with three electrodes is required.

These cells are usually equipped with anode, cathode, stirrer, thermometer and gas inlet and outlet. In the case of a divided cell, a diaphragm is also used.

7.1.2.1.2 Electrode

Generally, the material of the electrode must be stable toward electrochemical oxidation and reduction and also to the chemicals such as substrates, solvents, supporting electrolytes and products.

The materials which are commonly used for the anode are graphite (carbon), Pt, Au, Ti coated by Pt (Pt/Ti), Rh/Ti, TiO_2/Ti, RuO_2/Ti, PbO_2/Pb, some types of alloy of Pb (Pb–Ag, Pb–Sb) and titanium suboxide.

On the other hand, the material of the cathode is generally not limited. Almost any type of metal and graphite may be used as the cathode, though the material of the cathode greatly influences the pattern of the reaction in some cases.

7.1.2.2 Techniques

7.1.2.2.1 Constant current or controlled potential method

In the constant current method, the current is kept constant throughout the reaction and hence the total amount of electricity (F mol^{-1}) is easily calculated by the equation

$$\text{relative amount of electricity (F mol}^{-1}) = (60^2 \times HA)/(96500 \times M)$$

where H = time (h), A = current (A), M = mole (substrate), F = faraday

The correlation between the amount of electricity passed and the extent of the reaction is followed conveniently by this method. The electrode potential is, however, not kept constant in this method and hence it is not always possible to achieve reaction selectively.

On the other hand, the electrode potential is kept constant against a reference electrode in the controlled potential method. The fact that potential is constant throughout the reaction often leads to better regio- and chemo-selectivity than in the constant current method.

7.1.2.2.2 Method with or without a diaphragm

It is not an exceptional case in the electroorganic reaction that a substrate is reactive to both anode and cathode. Also, the product obtained by anodic oxidation or cathodic reduction is often further reduced or oxidized at the counter electrode.

The primary role of the diaphragm, therefore, is to separate the anolyte and catholyte to avoid undesirable side reactions. The diaphragm is also essential in the cases where the anolyte and catholyte are different.

The method using a diaphragm is suitable to the basic research for small-scale reactions, whereas it is not always convenient for synthetic chemistry. Anodic oxidation is often carried out without using the diaphragm, while it is generally necessary for cathodic reduction.

7.1.2.2.3 Selection of solvent

The following characteristics are required for solvents used in electroorganic reactions, namely: (i) good solubility of supporting electrolytes and substrates to the solvent; (ii) high electroconductivity; (iii) high electrochemical stability; and (iv) suitable chemical reactivity.

The commonly used solvents other than water are as follows. For the anolyte: MeOH, $MeCO_2H$, MeCN, CH_2Cl_2, $MeNO_2$, tetramethylene sulfone, pyridine, THF, $MeOCH_2CH_2OMe$ and propylene carbonate. For the catholyte: MeCN, DMF, Me_2 NCOMe, Me_2SO, HMPA, *N*-methylpyrrolidone, THF, dioxane, propylene carbonate, $MeOCH_2CH_2OMe$, MeOH, $MeCO_2H$ and NH_3.

The solvent is used alone or as a mixture with other solvents, including water.

7.1.2.2.4 *Selection of supporting electrolytes*

The supporting electrolyte is essential for the electroorganic reaction. The following points are important for the selection of the supporting electrolyte: (i) solubility to the solvent commonly used for electrolysis; (ii) electrochemical stability; (iii) interaction with reaction intermediate; and (iv) relative difficulty of preparation.

Solvents such as water, methanol, MeCN or DMF dissolve a variety of inorganic supporting electrolytes, while only organic supporting electrolytes are used for organic solvents. The anion part of the commonly used supporting electrolyte is X^- (halide anion), ClO_4^-, BF_4^-, PF_6^-, OTs^- or RO^-, whereas the cation is M^+ (alkali metal cation) or R_4N^+.

7.1.2.2.5 *Amount of electricity passed through the cell*

The amount of electricity needed corresponds to the quantity of the reagents in the chemical reaction. The theoretical amount of the electricity can be calculated on the basis of numbers of electrons which are required to promote the reaction. The unit usually used is the coulomb or the Faraday per mole (F mol^{-1}) as previously shown.

Since the electricity needed corresponds to a reagent, the yield is often calculated on two different bases. Namely, one is the usual material yield and the other is the current yield (or current efficiency), calculated on the basis of the amount of electricity used from the equation

$$\text{current yield (\%)} = (P/T) \times 100$$

where P = amount of product (mol) obtained at the stage where a certain amount of electricity is passed and T = theoretical amount of product (mol) at the stage where the same certain amount of electricity is passed.

In the electroorganic synthesis an excess amount of electricity is often required to achieve the synthesis with a high material yield, and the current yield is often determined at an early stage of reaction.

7.1.3 DIRECT OXIDATION

7.1.3.1 Oxidation of Carbon–Hydrogen and Carbon–Carbon Single Bonds

Direct anodic oxidation of alkanes may be performed if they have ionization potentials lower than about 10 eV.[2] Such oxidations can be classified into two types of reactions, cleavage of C—H bonds (equation 4) and cleavage of C—C bonds (equation 5).

$$RH \xrightarrow{-2e} R^+ + H^+ \quad (4)$$

$$R{-}R \xrightarrow{-2e} 2\,R^+ \quad (5)$$

The high oxidation potentials of alkanes, however, make it difficult to carry out the oxidation in solvents such as acetonitrile since the first intermediates generated in these oxidations are carbonium ions, as illustrated by equations (4) and (5). Their stabilization with strongly acidic solvents like anhydrous fluorosulfonic acid often lowers the oxidation potentials of these hydrocarbons.[3]

7.1.3.1.1 *Cleavage of carbon–hydrogen bonds*

The controlled potential electrolysis of cyclohexane carried out at 1.85 V in fluorosulfonic acid containing 1.15 M acetic acid yields an α,β-unsaturated ketone as a single product in 30% current yield.

In the anodic oxidation, adamantane is a unique compound among alkanes. It has a rather low oxidation potential, and its anodic oxidation in acetonitrile affords acetamidoadamantane (**1**; equation 6) in 90% yield.[4]

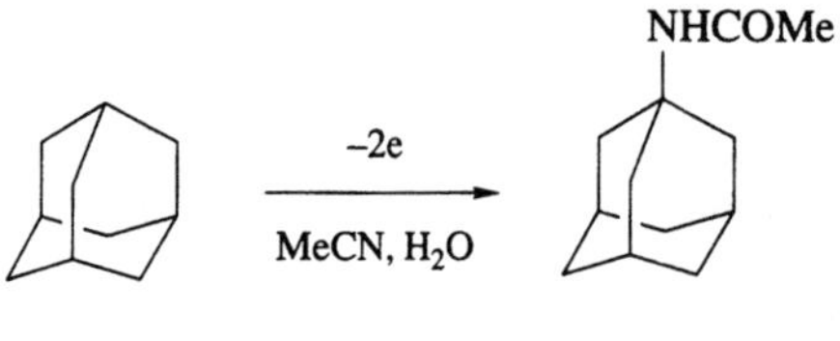

NHCOMe

$$\xrightarrow[\text{MeCN, } H_2O]{-2e} \qquad (6)$$

(1)

7.1.3.1.2 *Cleavage of carbon–carbon single bonds*

The direct anodic cleavage of saturated aliphatic carbon–carbon bonds is only possible if an electron is removed from the highest occupied molecular orbital (HOMO) of the C—C bond, *e.g.* due to the presence of strain in the bond.[5]

Tetramethylcyclopropane (**2**; equation 7) is the simplest strained hydrocarbon which is easily oxidized by the anodic method in methanol to give two products with a total yield of 71%.[6]

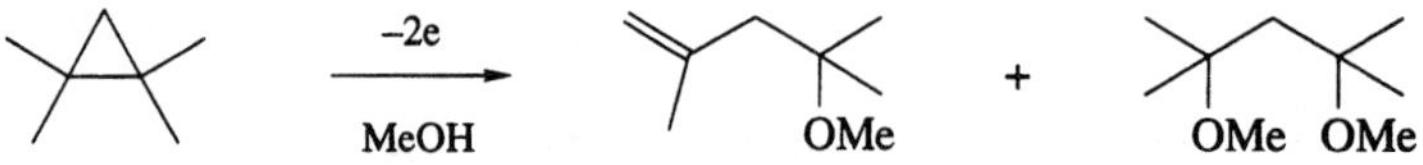

$$\xrightarrow[\text{MeOH}]{-2e} \qquad (7)$$

(2) OMe 47% + OMe OMe 24%

7.1.3.2 Oxidation of Unsaturated Systems

As the oxidation potentials of simple alkenes clearly show, carbon–carbon double bonds are usually anodically oxidized unless electron-withdrawing groups located on the alkene carbon atoms attract electrons from the unsaturated systems to shift the oxidation potentials beyond those accessible by anodic oxidation. On the other hand, electron-donating groups on the unsaturated bonds facilitate oxidation.

The initiation step of the anodic oxidation involves removal of an electron from the double bond leading to a cation radical as the first reactive intermediate. Depending on the structure of the unsaturated compounds, a variety of reactions will take place after the formation of the first intermediate. Thus, typical reactions are addition of nucleophiles (equation 8), allylic substitution (equation 9) and dimerization (equation 10).

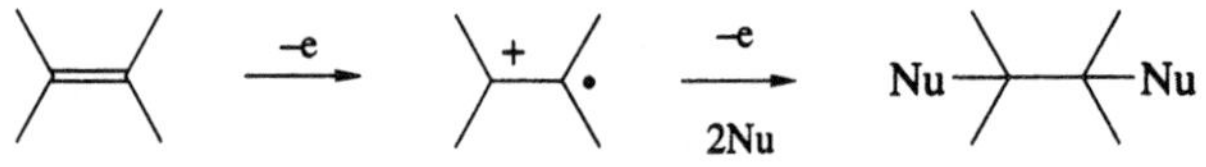

$$\xrightarrow{-e} \; + \bullet \; \xrightarrow[2Nu]{-e} \; \text{Nu} \; \text{Nu} \qquad (8)$$

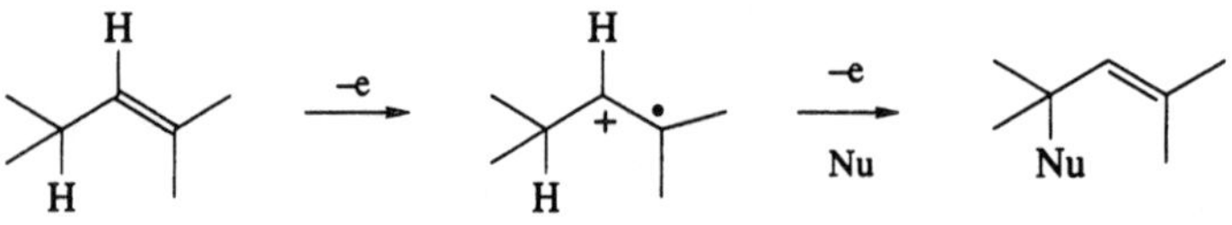

$$\text{H} \; \xrightarrow{-e} \; \text{H} \; + \bullet \; \xrightarrow[\text{Nu}]{-e} \; \text{Nu} \qquad (9)$$

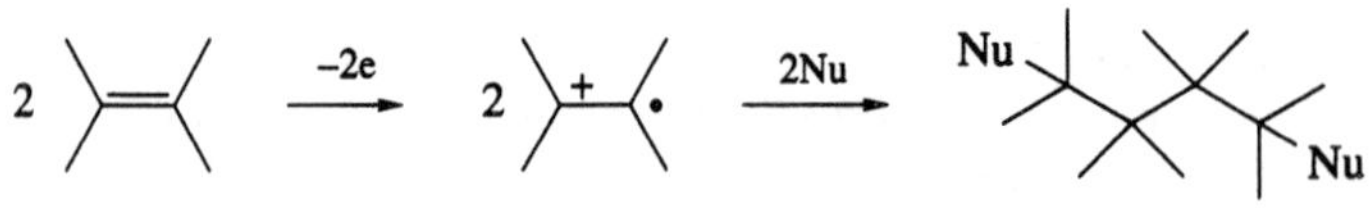

$$2 \; \xrightarrow{-2e} \; 2 \; + \bullet \; \xrightarrow{2Nu} \; \text{Nu} \; \text{Nu} \qquad (10)$$

7.1.3.2.1 *Aliphatic systems*

In general, the anodic oxidation of simple alkenes in nucleophilic solvents yields products resulting from both allylic substitution and oxidative addition of nucleophiles. Cyclohexene has been studied extensively as the starting compound.[7a–7e] The anodic oxidation of cyclohexene in methanol or acetic acid gives three types of products, those from allylic substitution (**3**), oxidative addition (**4**) and rearrangement (**5**; equation 11).

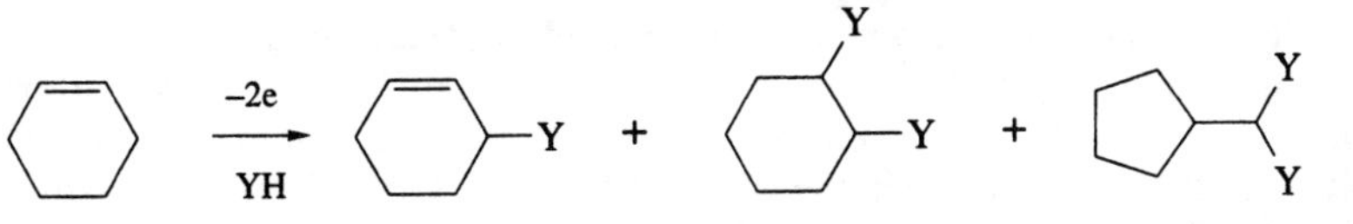

(11)

Y = OAc, OMe

The mechanism of oxidation of cyclohexene has been shown to involve direct removal of one electron from the double bond to generate a cation radical intermediate (**6**; equation 12).[8]

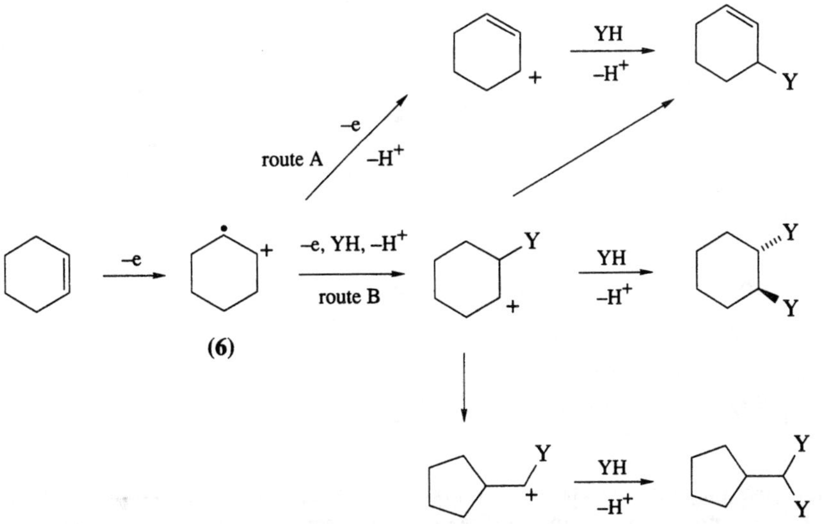

(12)

The relative ratio of routes A and B in equation (12) is controlled by the nucleophilicity of YH. Conjugated dienes are generally more susceptible to oxidation than simple alkenes.[9]

When using a carbon electrode, the anodic oxidation of conjugated dienes (**7**) such as isoprene, piperylene, cyclopentadiene and 1,3-cyclohexadiene in methanol or acetic acid mainly gives oxidative 1,4-addition products (**8**; equation 13). For example, 1,3-cyclohexadiene gives 1,4-dimethoxycyclohex-2-ene (**9**) in 47% yield (equation 14).[10] 1,3-Cyclooctadiene, in a similar experiment, yields a considerable amount of the allylically substituted product.

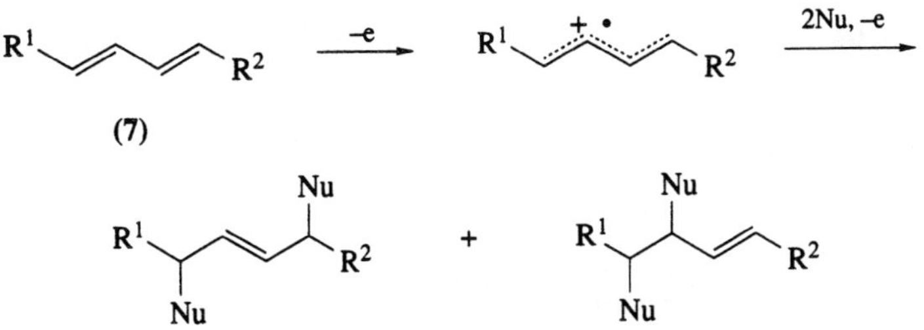

(13)

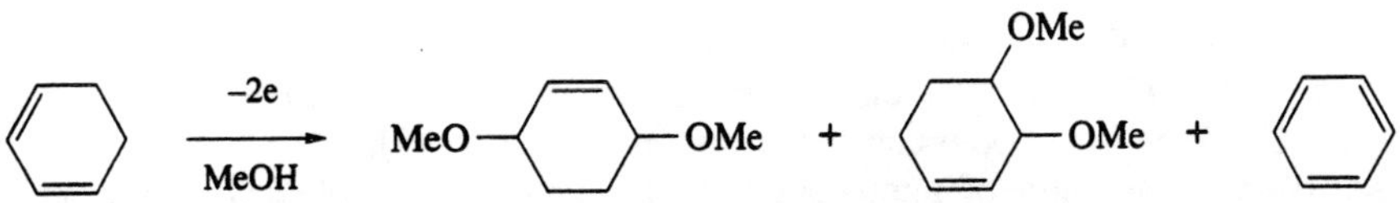

(14)

The oxidation of conjugated dienes has been successfully applied to the synthesis of allethrolone.[11]

Compared with simple aliphatic alkenes and conjugated dienes, the behavior of nonconjugated dienes in anodic oxidation is unique.[12] The possible reaction pathway of the oxidation of nonconjugated dienes (**10**) can be classified into two categories (equation 15).

In route A, one electron is removed from one double bond to generate a cation radical, and subsequent transannular reaction of the cation radical with the other double bond forms a new carbon–carbon bond. On the other hand, in route B, allylic substitution or oxidative addition at one double bond takes place without intramolecular interaction between the double bonds. As exemplified by the anodic oxidation of 4-vinylcyclohexene (**11**) in methanol (equation 16), such dienes as 4-vinylcyclohexene, limonene and 1,5-cyclooctadiene yield only products *via* route B.

(**10**) –e; route A; route B: –e, YH (15)

(**11**) –2e, MeOH, Et_4NOTs, 1.5–1.6 V vs. SCE; 32% + 21% (16)

On the other hand, the electrooxidation of norbornadiene (**12**), in which two double bonds are suitably arranged for the transannular interaction to take place, products *via* route A are seen (equation 17).

(**12**) –2e, YH, Et_4NOTs (17)

Y = OAc	6.2%	14.1%	15.4%
Y = OMe	8.8%	16.5%	18.4%

Cycloheptatriene gives 7-methoxycycloheptatriene (**13**) (7-MCHT) by anodic oxidation in methanol (equation 18).[13]

–2e, MeOH, 71% (**13**) (18)

Although (**13**) gives benzaldehyde dimethyl acetal (63%) upon further anodic oxidation, 3-methoxycycloheptatriene (**14**) (3-MCHT) and 1-methoxycycloheptatriene (**15**) (1-MCHT) afford 7,7-dimethoxycycloheptatriene (**16**), which is a good precursor of tropone (equation 19).

Although the oxidative addition of nucleophiles to the double bond of arylalkenes has long been known,[14] the most interesting reaction from the synthetic point of view is oxidative dimerization. Using a graphite electrode, the anodic oxidation of styrene in methanol containing NaOMe and $NaClO_4$ as supporting electrolytes yields 1,4-dimethoxy-1,4-diphenylbutane (**17**) in 64% yield (equation 20).[15]

OMe –2e MeOH OMe OMe Δ 98% (14) OMe –2e MeOH 83% MeO OMe (16) H_3O^+ 85% O (19)

Δ (15) OMe –2e MeOH 85%

–2e, MeOH MeONa, $NaClO_4$ OMe OMe (17) (20)

Enolic alkenes, *i.e.* alkenes bearing electron-donating substituents such as alkoxy, acyloxy and dialkylamino groups, are easily oxidizable by the anodic method.[16,17]

The addition of methoxy groups to an unsaturated carbon takes place by anodic oxidation of enol ethers in methanol containing sodium methoxide; yields are generally satisfactory (equation 21).[18,19]

OR –2e MeOH MeO OR OMe (21)

The anodic oxidation of enol ethers at a graphite anode in methanol containing 2,6-lutidine and sodium perchlorate results in the dimerization of the enol ethers to acetals of 1,4-dicarbonyl compounds (equation 22).[17] The mechanism of dimerization is thought to involve a tail–tail coupling of the cation radicals generated by the one-electron oxidation of the enol ethers.

2 OEt –2e MeOH EtO MeO OEt OMe current yield 51% (22)

The oxidation of enol acetates in acetic acid containing tetraethylammonium *p*-toluenesulfonate gives four types of compounds (equation 23): conjugated enones (A), α-acetoxycarbonyl compounds (B), geminal diacetoxy compounds (C) and triacetoxy compounds (D).[16] Similar to enol ethers, the first reactive intermediates are cation radicals generated from enol acetates by one-electron oxidation. The yields and the distribution of products A, B, C and D depend on the structure of the starting enol acetates and the reaction conditions.[20]

The formation of α,β-unsaturated enones from enol acetates has been applied to the synthesis of 2,3-disubstituted 2-cyclopentenones, including jasmone homologs. The yields of the anodic oxidation are usually in the range of 80–90%.

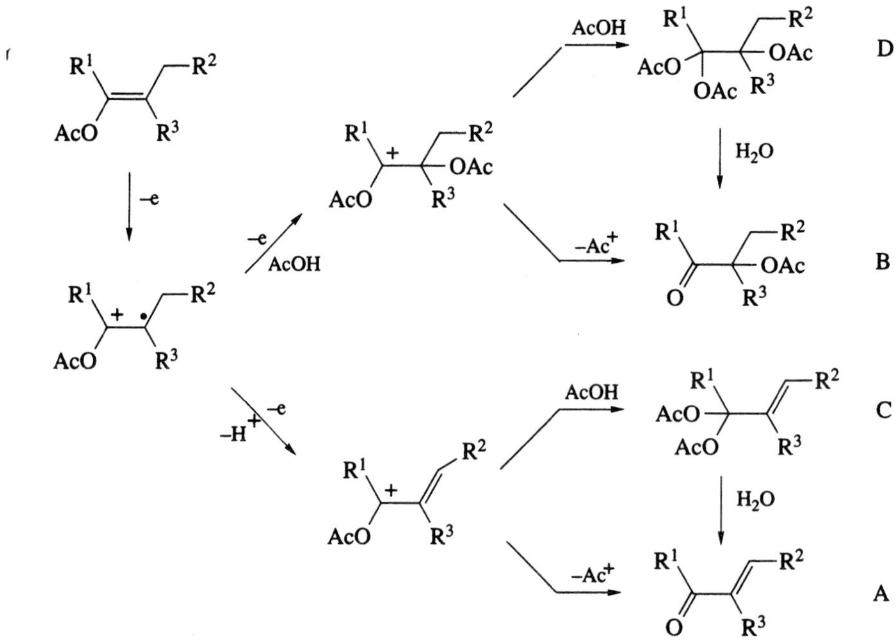

(23)

The anodic α-acetoxylation or α-methoxylation of ketones has been shown to be a powerful tool for the 1,2-transposition of the carbonyl group. The overall process is described by equation (24).[21]

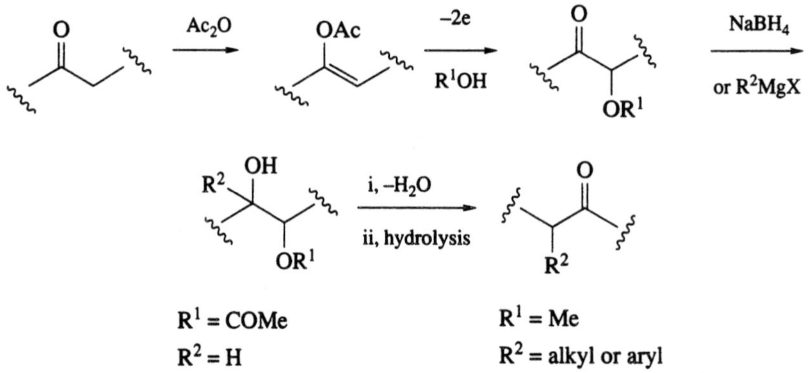

(24)

The concept of this 1,2-transposition can be extended to 1,4-transposition by using enones as the starting compounds. The anodic methoxylation of dienol acetates prepared from enones in a mixed solvent of acetic acid and methanol (1:9) yields γ-methoxylated enones regioselectively (equation 25). Reduction of the γ-methoxylated enones with $NaBH_4$ to the corresponding alcohols followed by solvolysis of the derived tosylates in aqueous acetone gives products in which the carbonyl group is transposed to the γ-position of the starting enones.[22]

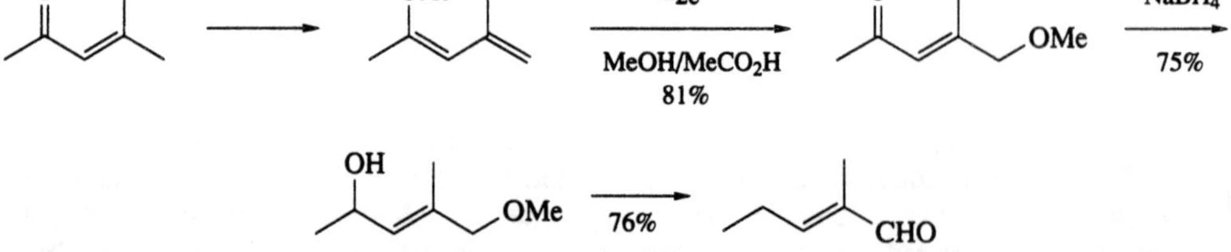

(25)

The anodic oxidation of enamines in methanol containing sodium methoxide as the supporting electrolyte shows a reaction pattern different from that of enol ethers or enol acetates. The main products are mixtures of isomeric methoxylated enamines, (**18**) and (**19**), with yields in the range 74–76% (equation 26).[18b,23]

$$\text{enamine} \xrightarrow[\text{MeOH}]{-2e} \mathbf{(18)} + \mathbf{(19)} \quad n = 3, 4 \tag{26}$$

7.1.3.2.2 Aromatic systems

Removal of electrons from aromatic π-electron systems may be achieved by electrochemical oxidation, and the resulting aromatic cation radical or other aromatic cationic species undergoes interesting and important reactions, such as aromatic substitution (equation 27) and coupling (equation 28).

$$\text{ArH} \xrightarrow[\text{Nu}^-]{-2e} \text{Ar–Nu} + \text{H}^+ \tag{27}$$

$$2\ \text{ArH} \xrightarrow{-2e} \text{Ar–Ar} + 2\ \text{H}^+ \tag{28}$$

(i) Aromatic substitution

(a) Acetoxylation. Although a variety of mechanisms including radical substitution and EE mechanisms have been proposed, an ECEC mechanism (equation 29) is now believed to be most probable for the acetoxylation.[24]

$$\text{ArH} \xrightarrow{-e} \text{ArH}^{+\bullet} \xrightarrow{\text{Nu}^-} \dot{\text{Ar}}(\text{H})(\text{Nu}) \xrightarrow{-e} \overset{+}{\text{Ar}}(\text{H})(\text{Nu}) \xrightarrow{-\text{H}^+} \text{ArNu} \tag{29}$$

When the aromatic substrates contain benzylic hydrogens, anodic benzylic substitution always competes with ring substitution.

The yields of ring acyloxylation products are improved by carrying out the reaction in CF_3CO_2H/CF_3CO_2Na, since the first products, namely trifluoroacetoxylated compounds, are generally stable under the conditions of anodic substitution.[25]

(b) Methoxylation. Ring methoxylation of substrates possessing high oxidation potentials is only achieved with difficulty, although naphthalene[26] and anthracene[27] are readily methoxylated (equation 30).

$$\text{anthracene} \xrightarrow[\text{MeOH–MeONa}\ \ ca.100\%]{-2e} \text{9,10-dimethoxy-9,10-dihydroanthracene}\ (cis{:}trans = 1{:}1) \tag{30}$$

The transformation of easily oxidizable substrates such as 1,4-dimethoxybenzene to the corresponding quinone diacetals occurs in high yields, as shown in equation (31).[28] These quinone diacetals have been used as starting materials in a variety of organic syntheses.

$$\text{MeO–C}_6\text{H}_4\text{–OMe} \xrightarrow[\text{MeOH, KOH}]{-2e} (\text{MeO})_2\text{C}_6\text{H}_4(\text{OMe})_2 \tag{31}$$

Quinone monoacetal, which is obtained by careful hydrolysis of quinone diacetal, can also be prepared directly by anodic oxidation under modified reaction conditions.[29]

(c) Formation of quinones. Quinones can be prepared directly by anodic oxidation of aromatic compounds.[28a,30] An example is shown in equation (32).[30a,30b]

$$\text{PhOH} \xrightarrow[\text{H}_2\text{O, H}_2\text{SO}_4 \;\; 80\%]{-4e} \text{O=C}_6\text{H}_4\text{=O} \quad (32)$$

(d) Oxidation of benzene. Owing to the high oxidation potential of benzene, the direct anodic transformation of benzene to phenol and hydroquinone, and their derivatives, is not always successful. Hence, some modifications are essential to achieve conversion of benzene into phenolic compounds. The anodic oxidation of benzene in trifluoroacetic acid containing sodium trifluoroacetate and subsequent hydrolysis of the reaction product affords phenol in 65% yield.[25a,31]

The hydroxylation of the aromatic nucleus by hydroxyl radicals, generated by decomposition of hydrogen peroxide in the presence of iron(II) ions, may be applied to the electrochemical synthesis of phenol from benzene, since the concentration of the iron(II) ions can be controlled by the cathodic reduction of iron(III) ions formed by oxidation of iron(II) ions with H_2O_2.

(e) Acetamidation and nitration. The anodic oxidation of aromatic compounds in the presence of acetonitrile leads to nuclear acetamidation (equation 33).[32]

$$\text{RCO–C}_6\text{H}_5 \xrightarrow[\text{MeCN, H}_2\text{O}]{-2e} \text{RCO–C}_6\text{H}_4\text{–NHCOMe} \quad (33)$$

	ortho	*para*
R = Me	70%	24%
R = Pr	52%	23%
R = OH	54%	11%

The electrochemical oxidation of aromatic compounds in the presence of ammonium nitrate or N_2O_4 results in the nuclear nitration shown in equation (34).[33]

$$\text{naphthalene} \xrightarrow[\text{MeCN, N}_2\text{O}_4\text{, Bu}_4\text{NPF}_6 \;\; 91\%]{-2e} \text{1-nitronaphthalene (NO}_2\text{)} \quad (34)$$

(f) Halogenation. The mechanism of the halogenation depends on the relative values of the oxidation potentials of the halogen and aromatic substrate. When the oxidation potential of the halogen is lower than that of the aromatic compounds, halogenation is initiated by the oxidation of halogen. Hence, this section is mainly concerned with the fluorination of aromatic compounds. One of the most important points in the anodic fluorination is the choice of the fluoride ion source. The use of anhydrous hydrofluoric acid usually leads to low yields.[34] Higher yields are obtained when a combination of tetraalkylammonium fluoride and hydrofluoric acid are employed, as shown in equation (35).[35]

$$\text{9,10-diphenylanthracene (Ph, Ph)} \xrightarrow[\text{MeCN, Et}_4\text{NF·3HF} \;\; 75\%]{-2e} \text{9,10-difluoro-9,10-diphenyl-9,10-dihydroanthracene (F, Ph; F, Ph)} \quad (35)$$

(g) Cyanation. The direct cyanation of the aromatic nucleus usually affords poor yields of cyanated products. However, considerably higher yields are obtained when one of the alkoxy groups of the starting alkoxyanisole or 4,4′-dialkoxybiphenyl is electrochemically substituted by a cyano group.[36]

(ii) Coupling

(a) Intramolecular coupling. When the structure of the substrate is suitable for coupling, the intramolecular coupling takes place rather easily.[37]

Coupling of the substrates $Ar(CH_2)_nAr'$ (**20**) is controlled by a variety of factors,[38] including solvent and supporting electrolyte, molecular geometry, anode potential and difference in the oxidation potentials of Ar and Ar′ (equation 36). The presence of trifluoroacetic acid or HBF_4 in the reaction system leads to satisfactory yields of coupled products.[39] Intramolecular coupling is readily achieved if n is 1, 2, 3 or 4, whereas with longer chain lengths intermolecular coupling predominates.[40]

Ar–$(CH_2)_n$–Ar′ (**20**) ⟶ Ar–Ar′ bridged by $(CH_2)_n$ (36)

The skeleton of morphine alkaloids has been synthesized by intramolecular coupling (equation 37).[41]

MeO, MeO, MeO, MeO, N–Me, Y substrate —(−2e; MeCN, Me_4NBF_4, Na_2CO_3)⟶ product (OMe, MeO, MeO, N–Me, Y, O) (37)

Y = H, Cl, Br, I, OMe

(b) Intermolecular coupling. When suitable aromatic compounds are oxidized in the absence of nucleophiles, the aromatic compounds themselves behave as nucleophiles to yield dimers. Equation (38) shows a typical reaction.[42]

—(−2e; CH_2Cl_2, Bu_4NBF_4; 53%)⟶ (38)

(iii) Oxidation at the benzylic position

As described in the previous section, substitution at the benzylic position always takes place together with nuclear substitution if the aromatic substrates possess replaceable benzylic hydrogens (equation 39).[43] Benzylic alcohols, esters and ethers can all be oxidized at the benzylic position to yield the corresponding carbonyl compounds.[44]

$$\mathrm{ArCH_2R} \xrightarrow{-e} \mathrm{[ArCH_2R]^{+\bullet}} \xrightarrow{-H^+} \mathrm{Ar\dot{C}HR} \xrightarrow{-e} \mathrm{Ar\overset{+}{C}HR} \xrightarrow{Nu^-} \mathrm{ArCH(Nu)R} \qquad (39)$$

Carbon–carbon bond cleavage can also take place at the benzylic position if the intermediate cationic species is sufficiently stabilized by suitable substituents (equation 40).[45]

$$\mathrm{Ar}\diagdown\!\!\diagup\!\!\diagdown\mathrm{Y} \xrightarrow{-2e} \mathrm{Ar}-\overset{+}{\mathrm{C}}\mathrm{H_2} + \mathrm{H_2}\overset{+}{\mathrm{C}}-\mathrm{Y} \xrightarrow{\mathrm{Nu^-}} \mathrm{Ar}\diagup\!\!\diagdown\mathrm{Nu} + \mathrm{Y}\diagup\!\!\diagdown\mathrm{Nu} \qquad (40)$$

Y = cation-stabilizing group

7.1.3.2.3 Heterocyclic systems

(i) Furans

The anodic oxidation of furans is one of the most extensively studied reactions because electrooxidation of furans in methanol yields 2,5-dimethoxy-2,5-dihydrofurans (**21**; equation 41), which are useful starting materials in organic synthesis.[46]

(41) [Scheme: 2-R^1-5-R^2-furan → (−2e, MeOH, NH_4Br) → 2,5-dimethoxy-2,5-dihydrofuran (**21**) → (H^+/H_2O) → R^1CO–CH=CH–COR^2]

(21)

2,5-Dimethoxy-2,5-dihydrofuran derivatives have been used extensively for the synthesis of aromatic and aliphatic ring systems, as shown in equations (42)[47] and (43).[48]

(42) [Scheme: furan carbamate → (−2e, MeOH) → dimethoxydihydrofuran → (i, H_2O, NaOH; ii, H_2O, HCl) → 3-hydroxypyridine]

R^1, R^2 = H, 67% overall

R^1, R^2 = CH_2OH, 76% overall

(43) [Scheme: furyl carbinol → (−2e, MeOH, 74%) → MeO/OMe dihydrofuran → (H^+, 53%) → methoxypyranone → (H_2O_2, 66%) → epoxide → (H^+, 73%) → 3-hydroxy-2-methyl-4-pyrone]

A variety of cyclopentenone derivatives have been synthesized using the anodic oxidation of furans as a key step.[49]

7.1.3.3 Oxidation of Systems Bearing Lone Pairs of Electrons

7.1.3.3.1 Systems containing oxygen

The direct anodic oxidation of aliphatic saturated alcohols to the corresponding carbonyl compounds is not always effective, because the high oxidation potentials of these alcohols make difficult the direct removal of an electron from the lone pair electrons on the oxygen atom.

The direct electrochemical oxidation of alcohols has been surveyed by Scholl *et al.*,[50] and their conclusions are that this oxidation is best achieved in the neat liquid substrate. Where this is not possible acetonitrile is the best solvent. A fluoroborate as the supporting electrolyte is recommended to obtain higher yields and oxidation with a controlled potential is not effective.

(i) Oxidation of glycols

The anodic oxidation method is highly efficient for the oxidative cleavage of glycols (**22**) and related compounds. The oxidation of glycols and glycol ethers in methanol results in a clean cleavage to the corresponding carbonyl compounds (equation 44).[51]

$$R^1R^2C(OR)\text{–}C(OR)R^3R^4 \;\xrightarrow[\text{MeOH}]{-2e}\; R^1R^2C{=}O + R^3R^4C{=}O \qquad (44)$$

(**22**) R = H, Me

This anodic oxidation does not show any of the stereochemical limitations usually observed in cleavage reactions by chemical oxidizing reagents.[52] Furthermore, 1,2-dimethoxy- and 1-hydroxy-2-methoxyalkanes are also oxidized with similar current efficiencies.

The initiation step of this anodic oxidation of glycols may be the electron transfer from the lone pair electrons of the oxygen atom to the anode. This anodic cleavage of 1,2-glycols has been utilized for a variety of organic syntheses.[53]

The anodic oxidation of enol ethers in methanol yields α-methoxylated carbonyl compounds, which are useful intermediates for the synthesis of carbonyl compounds utilizing the technique of oxidative cleavage of glycols (equation 45).[54]

$$\text{cyclohexanone} \longrightarrow \text{1-methoxycyclohexene} \xrightarrow[\text{MeOH}]{-2e} \text{2-methoxycyclohexanone} \xrightarrow{\text{RMgX or RLi}} \text{1-R-2-methoxycyclohexanol} \xrightarrow[\text{MeOH}]{-2e} RC(O)(CH_2)_4CH(OMe)_2 \qquad (45)$$

R	*Yield (%)*
Me	75
Bu	72
CH_2CN	40
CH_2CO_2Et	44

Saturated aliphatic ethers are oxidized in AcOH/MeOH containing Et_4NOTs or Bu_4NBF_4.[55]

7.1.3.3.2 Systems containing nitrogen

(i) Oxidation of aliphatic amines

The relatively low oxidation potentials of simple aliphatic amines indicate that they should be easily oxidized by the anodic method.[56]

In the presence of an adequate amount of water, aliphatic amines are generally dealkylated by anodic oxidation.[57] Thus, a tertiary amine is successively dealkylated to a secondary amine, a primary amine and finally to ammonia. The mechanism involves initial removal of one electron from the lone pair electrons of nitrogen leading to a cation radical, though a variety of mechanisms have been proposed depending on the structures of the amines and the reaction conditions.

(ii) Oxidation of aromatic amines

In contrast to aliphatic amines, the anodic oxidation of aromatic amines shows a rather complex reaction pattern. Although extensive studies on the electrochemical reaction mechanism have been carried out, there are very few examples for the application of the anodic oxidation of aromatic amines to organic synthesis.

Methoxylation of *N,N*-dimethyl- or *N*-methyl-*N*-alkyl-anilines occurs predominantly at the methyl group (equation 46).[58]

PhN(Me)Et —(−2e; MeOH, KOH; 85%)→ PhN(CH₂OMe)Et (46)

(iii) Oxidation of amides and carbamates

As described in the previous section, the anodic oxidation of aliphatic amines is utilized only rarely in organic synthesis due to the instability of the generated intermediates, whereas amides and carbamates of aliphatic amines yield relatively stable intermediates which are sufficiently promising as starting materials in organic synthesis (equations 47 and 48).[59,60]

$$\text{piperidine-N-CO}_2\text{Me} \xrightarrow[\text{MeOH, 72\%}]{-2e} \text{2-OMe-piperidine-N-CO}_2\text{Me} \quad (47)$$

$$\text{azetidin-2-one (N-H)} \xrightarrow[\text{AcOH, 49\%}]{-2e} \text{4-OAc-azetidin-2-one (N-H)} \quad (48)$$

The reaction mechanism of the α-methoxylation or α-acetoxylation of amides[61] and carbamates[59] has been shown to involve direct one-electron removal from the lone pair electrons of the nitrogen atom in the initial step when inert supporting electrolytes are used.

The anodic oxidation of piperidine derivatives in acetic acid gives α,β-disubstituted products in good yields (equation 49).[62]

$$\text{piperidine-N-CO}_2\text{Me} \xrightarrow[\text{AcOH, AcOK; 88\%}]{-4e} \text{3-OAc-2-OH-piperidine-N-CO}_2\text{Me} \quad (49)$$

The products obtained by the anodic oxidation of amides or carbamates in methanol have the same structures as the compounds which can be synthesized from amides (carbamates), aldehydes and methanol (equation 50). The regeneration of iminium cations from these α-methoxyamides and subsequent reactions of the iminium cations with nucleophiles such as active methylene compounds or nucleophilic aromatic nuclei is well known under the term amidoalkylation (equation 50).[63]

$$R^1C(O)NHR^2 + R^3{-}CHO + MeOH \longrightarrow R^1C(O)N(R^2)CH(R^3)OMe \longrightarrow$$

$$R^1C(O)N^+(R^2){=}CHR^3 \xrightarrow{Nu} R^1C(O)N(R^2)CH(Nu)R^3 \quad (50)$$

In the amidoalkylation, however, the preparation of the starting α-methoxyamides is often difficult since the reaction of aldehydes higher than formaldehyde is not necessarily successful, and even when formaldehyde is employed the yields and purities of the α-methoxyamides are not always satisfactory. On the other hand, the anodic α-methoxylation of amides and carbamates generally allows the synthesis of α-methoxyamides (carbamates) which cannot be prepared by the method described by equation (50).

Since anodically prepared α-formyloxy-*N*,*N*-dimethylformamide has successfully been used as an electrophilic reagent,[64] and it has been found that the α-methoxylation of the carbamates of a variety of higher aliphatic amines and alicyclic amines can be readily performed,[59] extensive studies have been carried out to utilize the anodically synthesized α-methoxy- or α-acyloxy-amides and -carbamates as electrophiles in organic synthesis. One example is shown in equation (51).[65]

$$\text{N-}CO_2Me\text{-pyrrolidine} \xrightarrow[\text{MeOH, 77\%}]{-4e} \text{2,5-(MeO)}_2\text{-N-}CO_2Me\text{-pyrrolidine} \xrightarrow[\text{TiCl}_4\text{, 41\%}]{\text{allyl-SiMe}_3} \text{bicyclic N-}CO_2Me\text{, Cl} \qquad (51)$$

7.1.3.4 Oxidation of Anions

7.1.3.4.1 Carbanions

As the oxidation potential clearly shows, carbanions may easily be oxidized by the anodic method.[66] The most typical process of the anodic oxidation of carbanions is the formation of radical species (equation 52).

$$R^- \xrightarrow{-e} R\bullet \qquad (52)$$

Although generally dimerization is one of the typical reactions of radical species, the yields of the dimers are not always high. Thus the anodic oxidation of anions of monoalkylated malonic esters in acetonitrile gives the corresponding dimers in 20–55% yield.[67] However, Grignard reagents give satisfactory results in anodic dimerization (equation 53).[68]

$$2\ RMgBr \xrightarrow{-2e} R\text{–}R \qquad (53)$$

R	*Yield (%)*
C_5H_{11}	55–60
$C_{18}H_{37}$	54
Ph	55

7.1.3.4.2 Carboxylate anions

Oxidation of carboxylic acids can be classified into two major categories, formation of radical intermediates followed by dimerization and generation of cation intermediates followed by reaction with nucleophiles (equation 54). The reaction is controlled by a variety of factors including anode material, anode potential, current density, solvent, supporting electrolyte, structure of R and temperature.

$$RCO_2^- \xrightarrow{-e} RCO_2\bullet \xrightarrow{-CO_2} R\bullet \begin{cases} \longrightarrow R\text{–}R \\ \xrightarrow{-e} R^+ \xrightarrow{Nu} R\text{–}Nu \end{cases} \qquad (54)$$

(i) Formation of radicals: Kolbe-type reactions

The Kolbe dimerization is believed to be favored by the following reaction conditions: high concentration of carboxylic acid, low pH value, absence of foreign anions, high current density and use of a platinum anode.

Since the Kolbe dimerization has already been reviewed,[69] only a few examples of its application are given in equations (55)[70] and (56).[71]

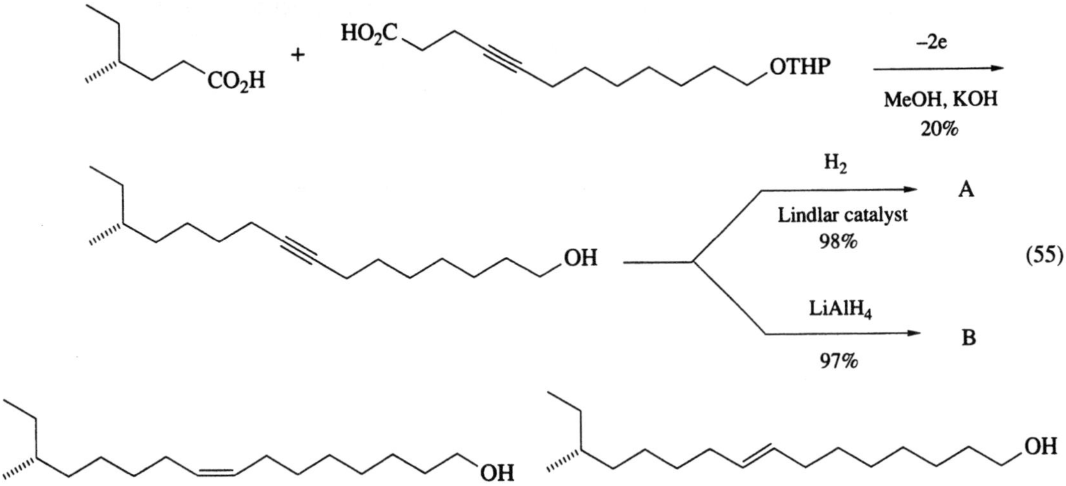

(55)

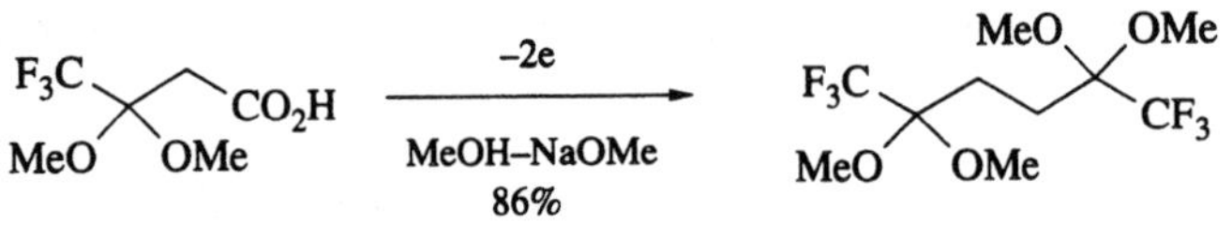

(56)

Intermolecular addition of the radical and mixed coupling with the radical of a coacid gives 3-alkyl-substituted pyrrolidine (**23**; equation 57).[72]

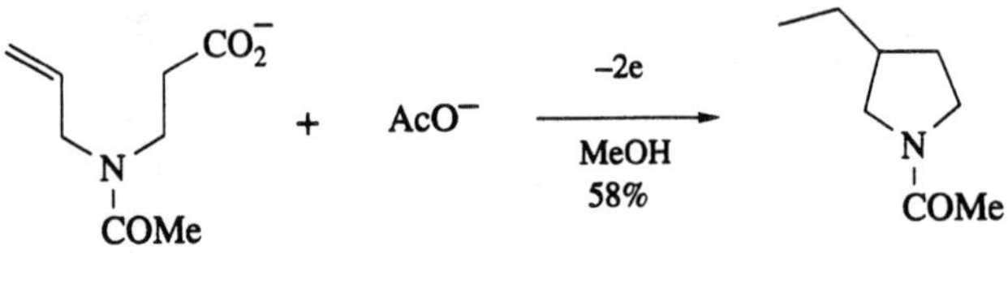

(57)

(ii) Formation of cations

When the cation R^+ is adequately stable and the reaction conditions are favorable for its formation, the radical R· formed from the carboxylic acid RCO_2H is further oxidized to the cation R^+ which is then trapped by a nucleophile, Nu^-.[73] This reaction has been applied to the transformation of a carboxy group to a hydroxy group (equation 58).[74]

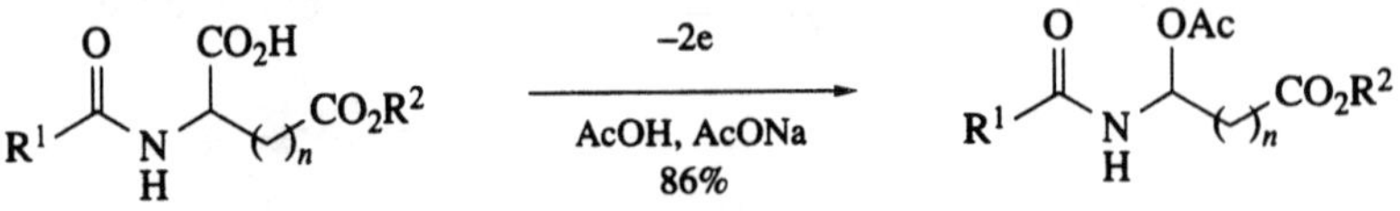

(58)

$n = 2$, R^1 = Ph, R^2 = CH_2Ph

A Wagner–Meerwein-type rearrangement of the cation has been often observed in the oxidation of carboxylic acids (equation 59). The relative migratory aptitude of R^1 and R^2 has been studied, and this type of rearrangement has been applied to the synthesis of (±)-muscone (equation 60).[75]

R^1 OH R^2 CO_2^- R^3 $\xrightarrow{-2e}$ R^1 R^2 O R^3 + R^2 R^1 O R^3 (59)

OH CO_2^- $\xrightarrow[\text{MeOH, KOH}]{-2e}$ O + O (60)

30% 5%

7.1.4 INDIRECT OXIDATIONS

7.1.4.1 Oxidation Using Mediators

7.1.4.1.1 Principles

As described in the previous sections, the active species are generally generated by direct electron transfer between substrate and electrode in the electroorganic reactions. Hence, the formation of active species is highly controlled by the oxidation and reduction potentials of the substrates. When these potentials are beyond the range accessible by the usual electrochemical technique, the direct electron transfer between the substrate and electrode hardly takes place as described in the direct oxidation of aliphatic saturated alcohols. Therefore, it is necessary to devise some other methods to oxidize or reduce the substrates. Also, even if the oxidation and reduction potentials of the substrates are in the accessible range of the electrochemical method, it is more desirable to oxidize or reduce them at much lower potentials than those applied in the direct method. This is achieved by the electroorganic synthesis using mediators. The oxidative reaction system using a mediator is schematically represented in Figure 2.

The oxidation potential of the substrate S in Figure 2 is beyond the range accessible by the electrochemical method so that direct electron transfer from S to the anode hardly occurs, and also the high oxidation potential necessary for the direct oxidation of S causes unexpected side reactions involving oxidation of the solvent or supporting electrolyte. However, when a compound M_{red} (a reduced form of M) which may be oxidized at a sufficiently lower potential than S is added to the reaction system, the oxidation of M_{red} to M_{ox} (an oxidized form of M) will take place prior to the oxidation of S. Provided that M_{ox} is able to oxidize S to product P, the oxidation of S will be achieved at a potential lower than that necessary for its direct oxidation. Oxidation of S with M_{ox} may be effected in two ways, namely by direct electron transfer (homogeneous electron transfer) from S to M_{ox} in solution or by chemical oxidation of S with M_{ox}. The former system is called a homomediatory system and the latter a heteromediatory (or chemomediatory) system. The compound M is called a mediator or an electron carrier, since M mediates electron transfer between S and the anode. When M_{ox} oxidizes S in solution, M_{ox} is reduced to M_{red}

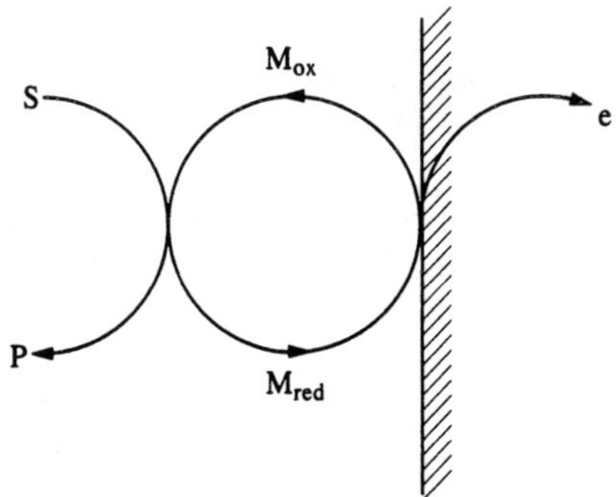

Figure 2 Mediatory system

which is again oxidized at the anode to regenerate M_{ox}. Thus, if the lifetime of the redox system $M_{ox} \leftrightarrow M_{red}$ is sufficiently long, only a catalytic amount of the mediator is required to initiate the entire reaction. As a matter of course, the concept of the mediatory system is not only applicable to oxidations, as illustrated by Figure 2, but also to reductions. Although the term mediator or electron carrier has been introduced rather recently, many types of reaction systems involving a compound which behaves as a mediator were already known.

7.1.4.1.2 Homomediatory systems

The homomediatory system is represented by equations (61) to (63), in which the mediator M is first oxidized to the cation radical $M^{+\cdot}$ at a relatively low oxidation potential. The next step involves a homogeneous electron transfer from S to $M^{+\cdot}$ to form $S^{+\cdot}$; this step is a reversible reaction. In the final step, $S^{+\cdot}$ is transformed to the products P_1^+ and $P_2\cdot$ by an irreversible reaction.

$$\mathrm{M} \xrightarrow{-e} \mathrm{M}^{+\bullet} \qquad (61)$$

$$\mathrm{M}^{+\bullet} + \mathrm{S} \rightleftharpoons \mathrm{M} + \mathrm{S}^{+\bullet} \qquad (62)$$

$$\mathrm{S}^{+\bullet} \longrightarrow \mathrm{P}_1^{+} + \mathrm{P}_2^{\bullet} \qquad (63)$$

Since the oxidation potential of S is more positive than that of M, the equilibrium in equation (62) is largely shifted to the left hand side. Hence, the rate of the whole reaction greatly depends on the rate of the irreversible reaction in equation (63). In fact, the oxidation described by equation (61) proceeds effectively only when $S^{+\cdot}$ is transformed sufficiently fast to products P_1^+ and $P_2\cdot$.

When the oxidation potential of S is much more positive than that of M the oxidation illustrated by equation (61) is almost impossible, even though the irreversible reaction of $S^{+\cdot}$ is fast. In such a case, some further activation of $M^{+\cdot}$ is necessary to make the oxidation possible.[76]

7.1.4.1.3 Heteromediatory systems

In the heteromediatory system, the substrate S is not oxidized by direct electron transfer from S to M_{ox} but by chemical reaction between S and M_{ox}. Many of the mediatory systems which are useful in organic synthesis may be classified into this category. Among a variety of mediators, the redox system consisting of a halide anion and a positive halogen species is one of the most interesting mediators used in organic synthesis.

One of the earliest synthetic reactions in which the halide anion was used as a mediator is the anodic methoxylation of furan in the presence of 0.05 equiv. of ammonium bromide, though the reaction has not been termed a mediated oxidation (equation 64).[77]

$$\text{furan} \xrightarrow[\text{NH}_4\text{Br/MeOH},\ 73\%]{-2e} \text{2,5-dimethoxy-2,5-dihydrofuran (MeO, OMe)} \qquad (64)$$

After this early investigation, a variety of oxidations using the redox system halide anion/positive halogen species as the mediator have been studied. Some of these oxidation systems are shown below in equations (65),[78] (66)[79] and (67).[80]

$$\text{N-hydroxypiperidine (N–OH)} \xrightarrow[\text{NaI, MeOH},\ 85\%]{-2e} \text{2,3,4,5-tetrahydropyridine N-oxide } (\overset{+}{\mathrm{N}}\text{–O}^-) \qquad (65)$$

$$\text{cyclohexanone} \xrightarrow[\text{KI, MeOH, 74\%}]{-2e} \text{2,2-dimethoxycyclohexanol} \quad (66)$$

$$\text{citronellal (CHO)} \xrightarrow[\text{KI, MeOH, 86\%}]{-2e} \text{methyl citronellate } (CO_2Me) \quad (67)$$

Besides halide ions organic sulfides are also efficient mediators. The oxidation of secondary alcohols to ketones has been successfully achieved by using methyl phenyl sulfide (**24**) as the mediator (equation 68).[81]

$$PhSMe\ (\mathbf{24}) \xrightarrow{-e} PhS^{+\bullet}Me \xrightarrow[R^1R^2CHOH]{-e,\ H^+} PhS^+(Me)OCHR^1R^2 \xrightarrow[-H^+]{\text{base}} PhS^+(CH_2^-)OCHR^1R^2 \xrightarrow{70\text{–}99\%} PhSMe + R^1R^2C{=}O \quad (68)$$

It is remarkable that carbon–carbon double bonds are completely inert in these oxidations. Some organic mediators, *e.g.* (**25**),[82] (**26**)[83] and (**27**),[84] have been exploited for the oxidation of alcohols.

(**25**) (**26**) (**27**)

7.1.4.1.4 *Double mediatory systems*

As described above, the mediatory system is an effective tool to oxidize the substrates that cannot be readily oxidized by the direct method. Further development of this concept has led to the combination of two types of mediators (Figure 3). As a result, the oxidation of substrates is achieved at a potential which is far lower than that required when the system contains only one type of mediator.[85]

In this system, the potential (E_p = 1.1 V *versus* SCE) of the oxidation of Br^- to Br^+ is the lowest, and the oxidation of R'_2S to $R'_2S^{+\cdot}$ does not take place at this potential. As described above, alcohols such as R^1R^2CHOH are oxidized by $R'_2S^{+\cdot}$, whereas Br^+ itself is not sufficiently reactive to oxidize alcohols to ketones in satisfactory yields. When both mediators are combined as depicted in Figure 3, however, the oxidation of alcohols may be achieved at a considerably lower potential than that necessary for the oxidation of R'_2S to $R'_2S^{+\cdot}$.

The yields of the obtained ketones are in the range of 80–94%. The mediatory system shown in Figure 3 can be called a double mediatory system. A double mediatory system containing the redox systems $Pd^0 \leftrightarrow Pd^{2+}$ and quinone $\leftrightarrow$ hydroquinone has been reported.[86]

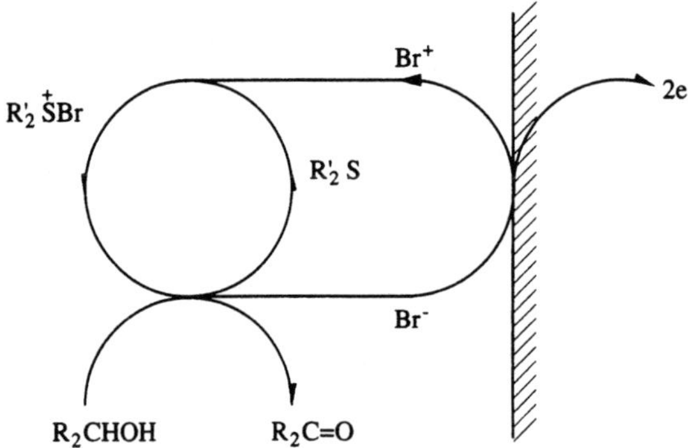

Figure 3 Double mediatory system

7.1.4.2 Formation of Active Species

7.1.4.2.1 Halogenation

The cationic species formed by the anodic oxidation of halide anions add to alkenes in the presence of suitable nucleophiles (equation 69).[87]

$$R_2C{=}CR_2 \;+\; X^- \xrightarrow[\text{Nu}]{-2e} \text{Nu}{-}CR_2{-}CR_2{-}X \qquad (69)$$

Early investigations on the oxidative addition of halogens to alkenes have mainly been focused on the preparation of epoxides from lower alkenes such as ethylene and propene.[88]

Halogenation of aromatic nuclei may also be achieved by halogen or positively charged species of halogen formed in solution by anodic oxidation of halide anions (equation 70).[89]

$$\text{ArH} \;+\; X^- \xrightarrow{-2e} \text{ArX} \;+\; H^+ \qquad (70)$$

The anodic oxidation of a mixture of iodine and aromatic compounds in acetonitrile gives aryl iodides in rather low yields. This iodination is improved by using a stepwise method (equation 71).[90]

$$0.5\,I_2 \xrightarrow[\text{MeCN}]{-e} \text{Me}\overset{+}{C}{=}N{-}I \xrightarrow{\text{ArH}} \text{ArI} \;+\; \text{MeCN} \;+\; H^+ \qquad (71)$$

$$\text{I}_2 \text{ or MeI} \xrightarrow[\text{CH(OMe)}_3\ [\text{TMOF}]]{-2e} \text{'I}^{+}\text{'/TMOF}$$

$$\text{X}{-}C_6H_4{-}CO{-}CH_2R\ \textbf{(28)} \xrightarrow[\text{80–100\%}]{\text{'I}^{+}\text{'/TMOF}} \text{X}{-}C_6H_4{-}CH(R){-}CO_2Me\ \textbf{(29)} \qquad (72)$$

X = H, Bu^i, Cl, Br, Bu^tCO_2; R = alkyl

Anodic oxidation of iodine or MeI in trimethyl orthoformate (TMOF) gives a new positive iodine active species ('I^+'/TMOF), which makes possible a unique rearrangement of aryl alkyl ketones (**28**) to methyl arylalkanoates (**29**; equation 72).[91]

7.1.5 REFERENCES

1. T. Shono, T. Toda and N. Oshino, *Tetrahedron Lett.*, 1984, **25**, 91.
2. M. Fleischmann and D. Pletcher, *Tetrahedron Lett.*, 1968, 6255.
3. J. Bertram, M. Fleischmann and D. Pletcher, *Tetrahedron Lett.*, 1971, 349.
4. V. R. Koch and L. L. Miller, *J. Am. Chem. Soc.*, 1973, **95**, 8631.
5. P. G. Gassman and R. Yamaguchi, *J. Am. Chem. Soc.*, 1979, **101**, 1308.
6. T. Shono and Y. Matsumura, *Bull. Chem. Soc. Jpn.*, 1975, **48**, 2861.
7. (a) T. Shono and T. Kosaka, *Tetrahedron Lett.*, 1968, 207; (b) A. J. Baggaley and R. Brettle, *J. Chem. Soc. C*, 1968, 2055; (c) G. Fiaita, M. Fleischmann and D. Pletcher, *J. Electroanal. Chem.*, 1970, **25**, 455; (d) R. Brettle and J. R. Sutton, *J. Chem. Soc., Perkin Trans. 1*, 1975, 1947; (e) K. Yoshida, T. Kanbe and T. Fueno, *J. Org. Chem.*, 1977, **42**, 2313.
8. T. Shono and A. Ikeda, *J. Am. Chem. Soc.*, 1972, **94**, 7892.
9. H. Baltes, E. Steckhan and H. J. Schäfer, *Chem. Ber.*, 1978, **111**, 1294.
10. T. Shono and A. Ikeda, *Chem. Lett.*, 1976, 311.
11. T. Shono, I. Nishiguchi and M. Okawa, *Chem. Lett.*, 1976, 573.
12. (a) T. Shono, A. Ikeda, J. Hayashi and S. Hakozaki, *J. Am. Chem. Soc.*, 1975, **97**, 4261; (b) T. Shono, A. Ikeda and S. Hakozaki, *Tetrahedron Lett.*, 1972, 4511.
13. T. Shono, T. Nozoe, H. Maekawa and S. Kashimura, *Tetrahedron Lett.*, 1988, **29**, 555.
14. L. Eberson and H. J. Schäfer, in 'Organic Electrochemistry', Springer-Verlag, Berlin, 1971.
15. R. Engels, H. J. Schafer and E. Steckhan, *Justus Liebigs Ann. Chem.*, 1977, 204.
16. T. Shono, Y. Matsumura and Y. Nakagawa, *J. Am. Chem. Soc.*, 1974, **96**, 3532.
17. D. Koch, H. J. Schäfer and E. Steckhan, *Chem. Ber.*, 1974, **107**, 3640.
18. (a) B. Belleau and Y. K. Au-Young, *Can. J. Chem.*, 1969, **47**, 2117; (b) T. Shono, Y. Matsumura and H. Hamaguchi, *Bull. Chem. Soc. Jpn.*, 1978, **51**, 2179.
19. (a) T. Shono, T. Toda and R. Oda, *Nippon Kagaku Zasshi*, 1969, **90**, 1260; (b) R. Couture and B. Belleau, *Can. J. Chem.*, 1972, **50**, 3424; (c) M. A. Michel, P. Martigny and J. Simonet, *Tetrahedron Lett.*, 1975, 3143.
20. T. Shono, M. Okawa and I. Nishiguchi, *J. Am. Chem. Soc.*, 1975, **97**, 6144.
21. T. Shono, I. Nishiguchi and M. Nitta, *Chem. Lett.*, 1976, 1319.
22. T. Shono and S. Kashimura, *J. Org. Chem.*, 1983, **48**, 1939.
23. (a) S. J. Huang and E. T. Hso, *Tetrahedron Lett.*, 1971, 1385; (b) J. M. Fritsch, H. Weingarten and J. D. Wilson, *J. Am. Chem. Soc.*, 1970, **92**, 4038.
24. L. Eberson and K. Nyberg, *Acc. Chem. Res.*, 1973, **6**, 106.
25. (a) Y. H. So and L. L. Miller, *Synthesis*, 1976, 468; (b) Y. H. So, J. Y. Becker and L. L. Miller, *J. Chem. Soc.*, 1975, 262; (c) N. L. Weinberg and C. N. Wu, *Tetrahedron Lett.*, 1975, 262; (d) G. Bockmain, H. P. Fritz and H. Gebauer, *Electrochim. Acta*, 1978, **23**, 21; (e) Z. Blumn, L. Gedheim and K. Nyberg, *Acta Chem. Scand., Ser. B*, 1975, **29**, 715.
26. G. Bockmain and H. P. Fritz, *Electrochim. Acta*, 1976, **21**, 1099.
27. V. D. Parker, J. P. Dirlan and L. Eberson, *Acta Chem. Scand., Ser. B*, 1971, **25**, 341.
28. (a) B. Belleau and N. L. Weinberg, *J. Am. Chem. Soc.*, 1963, **85**, 2525; (b) N. L. Weinberg and B. Belleau, *Tetrahedron*, 1973, **29**, 279.
29. (a) A. Nilsson and A. Ronlán, *Tetrahedron Lett.*, 1975, 1107; (b) A. Nilsson, U. Palmquist, T. Petterson and A. Ronlán, *J. Chem. Soc., Perkin Trans. 1*, 1973, 373; (c) R. F. Stewart and L. L. Miller, *J. Am. Chem. Soc.*, 1980, **102**, 4999.
30. (a) F. H. Cowitz, *US Pat.* 3 509 039 (1970) (*Chem. Abstr.*, 1970, **73**, 115 824); (b) G. C. Jones and O. A. Payne, *US Pat.* 3 994 788 (1976) (*Chem. Abstr.*, 1977, **86**, 35 759); (c) G. A. Bhat, M. Periasamy and M. V. Bhatt, *Tetrahedron Lett.*, 1979, 3097; (d) U. H. Brinker, M. Tyner, III and W. M. Jones, *Synthesis*, 1975, 671; (e) Z. Blumn and K. Nyberg, *Acta Chem. Scand., Ser. B*, 1979, **33**, 73; (f) V. D. Parker, *Acta Chem. Scand., Ser. B*, 1970, **24**, 2775; (g) J. Brachiand and A. Rieker, *Synthesis*, 1977, 708; (h) R. W. Johnson, E. R. Grover and L. J. MacPherson, *Tetrahedron Lett.*, 1981, **22**, 3719.
31. Z. Blumn, L. Gedheim and K. Nyberg, *Acta Chem. Scand., Ser. B*, 1975, **29**, 715.
32. Y. H. So, J. Y. Becker and L. L. Miller, *J. Chem. Soc., Chem. Commun.*, 1975, 262.
33. (a) L. Eberson and F. Radner, *Acta Chem. Scand., Ser. B*, 1980, **34**, 739; (b) C. L. Perrin, *J. Am. Chem. Soc.*, 1977, **99**, 5516.
34. (a) F. Huba, E. B. Yaeger and G. A. Olah, *Electrochim. Acta*, 1979, **24**, 489; (b) V. J. Davis, R. N. Haszeldine and A. E. Tipping, *J. Chem. Soc., Perkin Trans. 1*, 1975, 1263; (c) Y. Inoue, S. Nagase, K. Kodaira, H. Baba and T. Abe, *Bull. Chem. Soc. Jpn.*, 1973, **46**, 2204; (d) M. Yonekura, S. Nagase, H. Baba, K. Kodaira and T. Abe, *Bull. Chem. Soc. Jpn.*, 1976, **49**, 1113.
35. (a) I. N. Rozhkov, N. P. Gambaryan and E. G. Galpern, *Tetrahedron Lett.*, 1976, 4819; (b) C. J. Ludman, E. M. McCarron and R. F. O'Malley, *J. Electrochem. Soc.*, 1972, **119**, 874.
36. (a) S. Andreades and E. W. Zahnow, *J. Am. Chem. Soc.*, 1966, **91**, 4181; (b) N. L. Weinberg, D. H. Marr and C. N. Wu, *J. Am. Chem. Soc.*, 1975, **97**, 1499; (c) K. Yoshida, M. Shigi and T. Fueno, *J. Org. Chem.*, 1975, **40**, 63; (d) L. Eberson and B. Helgée, *Acta Chem. Scand., Ser. B*, 1975, **29**, 451; (e) L. Eberson and B. Helgée, *Acta Chem. Scand., Ser. B*, 1977, **31**, 813.

37. (a) J. D. Stuart and W. E. Ohnesorge, *J. Am. Chem. Soc.*, 1971, **93**, 4531; (b) J. L. Becker, L. L. Miller, V. Boekelheide and T. Morgan, *Tetrahedron Lett.*, 1976, 2939; (c) J. B. Kerr, T. C. Jempty and L. L. Miller, *J. Am. Chem. Soc.*, 1979, **101**, 7338.
38. V. D. Parker and A. Ronlán, *J. Am. Chem. Soc.*, 1975, **97**, 4717.
39. (a) K. Bechgaard, O. Hammerich, N. S. Mope, A. Ronlán, U. Svanholm and V. D. Parker, *Tetrahedron Lett.*, 1972, 2271; (b) E. Kotani and S. Tobinaga, *Tetrahedron Lett.*, 1973, 4759; (c) A. Ronlán, O. Hammerich and V. D. Parker, *J. Am. Chem. Soc.*, 1973, **95**, 7132.
40. (a) A. Ronlán and V. D. Parker, *J. Org. Chem.*, 1974, **39**, 1014; (b) A. Nilsson, U. Palmquist, A. Ronlán and V. D. Parker, *J. Am. Chem. Soc.*, 1975, **97**, 3540.
41. (a) L. L. Miller, F. R. Stermitz and J. R. Falck, *J. Am. Chem. Soc.*, 1971, **93**, 5941; (b) L. L. Miller, F. R. Stermitz and J. R. Falck, *J. Am. Chem. Soc.*, 1973, **95**, 2651; (c) J. R. Falck, L. L. Miller and F. R. Stermitz, *Tetrahedron*, 1974, **30**, 931; (d) L. L. Miller, F. R. Stermitz, J. Y. Becker and V. Ramachandran, *J. Am. Chem. Soc.*, 1975, **97**, 2922; (e) J. M. Bobbitt, J. Noguchi, R. S. Ware, K. N. Ching and S. J. Huang, *J. Org. Chem.*, 1975, **40**, 2924; (f) L. L. Miller and R. F. Stewart, *J. Org. Chem.*, 1978, **43**, 1580.
42. (a) K. Nyberg, *Acta Chem. Scand., Ser. B*, 1970, **24**, 1609; (b) K. Nyberg, *Acta Chem. Scand., Ser. B*, 1971, **25**, 534; (c) E. I. Majeski, J. D. Stuart and W. E. Ohnesorge, *J. Am. Chem. Soc.*, 1968, **90**, 633; (d) K. Nyberg, *Acta Chem. Scand., Ser. B*, 1970, **24**, 2757, 3151, 3162, 3171.
43. L. Eberson and K. Nyberg, *Acc. Chem. Res.*, 1973, **6**, 106.
44. (a) L. L. Miller, V. R. Koch, M. E. Larscheid and J. F. Wolf, *Tetrahedron Lett.*, 1971, 1389; (b) L. L. Miller, J. F. Wolf and E. A. Mayeda, *J. Am. Chem. Soc.*, 1971, **93**, 3306; (c) E. A. Mayeda, L. L. Miller and J. F. Wolf, *J. Am. Chem. Soc.*, 1972, **94**, 6812; (d) R. F. Garwood, N. ud. Dim and B. C. L. Weedon, *J. Chem. Soc., Perkin Trans. 1*, 1975, 2471; (e) S. M. Weinreb, G. A. Epling, R. Comi and M. Reitano, *J. Org. Chem.*, 1975, **40**, 1356; (f) R. Lines and J. H. P. Utley, *J. Chem. Soc., Perkin Trans. 2*, 1977, 803.
45. (a) T. Shono and Y. Matsumura, in 'Japan Chemical Society 3rd Symposium on Oxidation', abstract, 1969, p. 101; (b) T. Sato, K. Torizuka, M. Shimizu, Y. Kurihara and N. Yoda, *Bull. Chem. Soc. Jpn.*, 1969, **52**, 2420.
46. (a) N. C. Kaas, F. Limborg and K. Glens, *Acta Chem. Scand.*, 1952, **6**, 531; (b) N. Elming, *Adv. Org. Chem.*, 1960, **2**, 67 and refs. cited therein; (c) N. L. Weinberg and H. R. Weinberg, *Chem. Rev.*, 1968, 449.
47. (a) N. C. Kaas, N. Elming and Z. Tyle, *Acta Chem. Scand.*, 1952, **9**, 1; (b) N. Elming and N. C. Kaas, *Acta Chem. Scand.*, 1955, **9**, 23; (c) T. Shono, Y. Matsumura, K. Tsubata and J. Takata, *Chem. Lett.*, 1981, 1121.
48. T. Shono and Y. Matsumura, *Tetrahedron Lett.*, 1976, 1363.
49. (a) T. Shono, Y. Matsumura and H. Hamaguchi, *J. Chem. Soc., Chem. Commun.*, 1977, 712; (b) T. Shono, Y. Matsumura, H. Hamaguchi and K. H. Nakamura, *Chem. Lett.*, 1976, 1249; (c) T. Shono, H. Hamaguchi and K. Aoki, *Chem. Lett.*, 1977, 1053; (d) T. Shono, Y. Matsumura, S. Yamane and M. Suzuki, *Chem. Lett.*, 1980, 1619.
50. P. C. Scholl, S. E. Lentsch and M. R. Van De Mark, *Tetrahedron*, 1976, **32**, 303.
51. T. Shono, Y. Matsumura, T. Hashimoto, K. Hibino, H. Hamaguchi and T. Aoki, *J. Am. Chem. Soc.*, 1975, **97**, 2546.
52. S. J. Angyal and R. J. Young, *J. Am. Chem. Soc.*, 1959, **81**, 5467.
53. T. Shono, H. Hamaguchi, Y. Matsumura and K. Yoshida, *Tetrahedron Lett.*, 1977, 3625.
54. T. Shono, Y. Matsumura, T. Imanishi and K. Yoshida, *Bull. Chem. Soc. Jpn.*, 1978, **51**, 2179.
55. T. Shono, Y. Matsumura, O. Onomura and Y. Yamada, *Synthesis*, 1987, 1099.
56. C. K. Mann, *Anal. Chem.*, 1964, **36**, 2425.
57. (a) L. C. Portis, V. V. Bhat and C. K. Mann, *J. Org. Chem.*, 1970, **35**, 2175; (b) S. D. Ross, *Tetrahedron Lett.*, 1973, 1237; (c) L. C. Portis, J. T. Klug and C. K. Mann, *J. Org. Chem.*, 1974, **39**, 3488.
58. T. Shono, Y. Matsumura, K. Inoue, H. Ohmizu and S. Kashimura, *J. Am. Chem. Soc.*, 1982, **104**, 5753.
59. (a) T. Shono, H. Hamaguchi and Y. Matsumura, *J. Am. Chem. Soc.*, 1975, **97**, 4264; (b) Z. Blumn and K. Nyberg, *Acta Chem. Scand., Ser. B*, 1982, **36**, 165.
60. M. Mori, K. Kagechika, K. Tohjima and M. Shibasaki, *Tetrahedron Lett.*, 1988, **29**, 1409.
61. (a) S. D. Ross, M. Finkelstein and R. C. Petersen, *J. Am. Chem. Soc.*, 1966, **88**, 4657; (b) M. Finkelstein and S. D. Ross, *Tetrahedron*, 1972, **28**, 4497; (c) E. J. Rudd, M. Finkelstein and S. D. Ross, *J. Org. Chem.*, 1972, **37**, 1763.
62. T. Shono, Y. Matsumura, O. Onomura, M. Ogaki and T. Kanazawa, *J. Org. Chem.*, 1987, **52**, 536.
63. (a) H. Hellmann, *Angew. Chem.*, 1966, **69**, 463; (b) H. E. Zaugg and W. B. Maartin, *Org. React. (N. Y.)*, 1965, **14**, 52; (c) H. E. Zaugg, *Synthesis*, 1970, 49; (d) J. Mathieu and J. Weill-Raynal, in 'Formation of C—C Bonds', Thieme, Stuttgart, 1977, vol. III, p. 248.
64. S. D. Ross, M. Finkelstein and R. C. Petersen, *J. Org. Chem.*, 1966, **31**, 133.
65. T. Shono, Y. Matsumura, K. Uchida and H. Kobayashi, *J. Org. Chem.*, 1985, **50**, 3243.
66. (a) T. Psarras and R. E. Dessy, *J. Am. Chem. Soc.*, 1966, **88**, 5132; (b) B. Jaun, J. Schwarz and R. Breslow, *J. Am. Chem. Soc.*, 1980, **102**, 5741; (c) J. M. Kern and P. Federlin, *Tetrahedron Lett.*, 1977, 837; (d) J. M. Kern and P. Federlin, *J. Electroanal. Chem.*, 1979, **96**, 209; (e) P. Lochert and P. Federlin, *Tetrahedron Lett.*, 1973, 1109; (f) H. W. Van den Born and D. H. Evans, *J. Am. Chem. Soc.*, 1974, **96**, 4296.
67. (a) H. G. Thomas, M. Streukens and R. Peek, *Tetrahedron Lett.*, 1978, 45; (b) T. D. Binns and R. Brettle, *J. Chem. Soc. C*, 1966, 336; (c) R. Brettle and J. G. Parkin, *J. Chem. Soc. C*, 1967, 1352; (d) R. Brettle and D. Seddon, *J. Chem. Soc. C*, 1970, 1153; (e) R. Brettle, J. G. Parkin and D. Seddon, *J. Chem. Soc. C*, 1970, 1317; (f) R. Brettle and D. Seddon, *J. Chem. Soc. C*, 1970, 2175.
68. J. L. Morgat and R. Pallaud, *C. R. Hebd. Seances Acad. Sci.*, 1965, **260**, 574, 5579.
69. H. J. Schäfer, *Angew. Chem., Int. Ed. Engl.*, 1981, **20**, 911.
70. U. Jensen and H. J. Schäfer, *Chem. Ber.*, 1981, **114**, 292.
71. T. Kubota, T. Ishii, H. Minamikawa, S. Yamaguchi and T. Tanaka, *Chem. Lett.*, 1988, 1987.
72. L. Becking and H. J. Schäfer, *Tetrahedron Lett.*, 1988, **29**, 2729.
73. (a) J. H. P. Utley, in 'Technique of Electroorganic Synthesis', ed. N. L. Weinberg, Wiley, New York, 1974, chap. VI, p. 793; (b) S. Wawzonek, *Synthesis*, 1971, 285.

74. (a) T. Nishitani, T. Iwasaki, Y. Mushika, I. Inoue and M. Miyoshi, *Chem. Pharm. Bull.*, 1980, **28**, 1137; (b) T. Iwasaki, H. Horikawa, K. Matsumoto and M. Miyoshi, *J. Org. Chem.*, 1977, **42**, 2419.
75. T. Shono, J. Hayashi, H. Omoto and Y. Matsumura, *Tetrahedron Lett.*, 1977, 2667.
76. (a) W. Schmidt and E. Steckhan, *J. Electroanal. Chem.*, 1978, **89**, 215; (b) W. Schmidt and E. Steckhan, *Angew. Chem., Int. Ed. Engl.*, 1978, **17**, 673; (c) W. Schmidt and E. Steckhan, *J. Electroanal. Chem.*, 1979, **101**, 123; (d) M. Platem and E. Steckhan, *Tetrahedron Lett.*, 1980, **21**, 511; (e) W. Schmidt and E. Steckhan, *Chem. Ber.*, 1980, **113**, 577; (f) P. Martigny and J. Simonet, *J. Electroanal. Chem.*, 1980, **111**, 133.
77. N. Clauson-Kaas, F. Limborg and K. Glens, *Acta Chem. Scand.*, 1952, **6**, 531.
78. T. Shono, Y. Matsumura and K. Inoue, *J. Org. Chem.*, 1986, **51**, 549.
79. T. Shono, Y. Matsumura, K. Inoue and F. Iwasaki, *J. Chem. Soc., Perkin Trans. 1*, 1986, 73.
80. T. Shono, Y. Matsumura, J. Hayashi, K. Inoue, F. Iwasaki and T. Itoh, *J. Org. Chem.*, 1985, **50**, 4967.
81. T. Shono, Y. Matsumura, M. Mizoguchi and J. Hayashi, *Tetrahedron Lett.*, 1979, 3861.
82. M. F. Semmelhack, C. S. Chou and D. A. Cortes, *J. Am. Chem. Soc.*, 1983, **105**, 4492.
83. A. Deronzier, D. Limosin and J.-C. Moutet, *Electrochim. Acta*, 1987, **32**, 1643.
84. M. Masui, T. Ueshima and S. Ozaki, *J. Chem. Soc., Chem. Commun.*, 1983, 479.
85. T. Shono, Y. Matsumura, J. Hayashi and M. Mizoguchi, *Tetrahedron Lett.*, 1980, **21**, 1867.
86. J. Tsuji and M. Minato, *Tetrahedron Lett.*, 1987, **28**, 3683.
87. (a) V. R. Koch and L. L. Miller, *J. Am. Chem. Soc.*, 1973, **95**, 8631; (b) V. R. Koch and L. L. Miller, *Tetrahedron Lett.*, 1973, 4861; (c) F. Vincent, R. Tradivel and P. Mison, *Tetrahedron Lett.*, 1975, 603; (d) F. Vincent, R. Tradivel and P. Mison, *Tetrahedron*, 1976, **32**, 1681.
88. R. Dietz and H. Lund, in 'Organic Electrochemistry', ed. M. M. Baizer, Dekker, New York, 1973, p. 821.
89. R. Leininger and L. A. Pasiut, *Trans. Electrochem. Soc.*, 1945, **88**, 73.
90. L. L. Miller and B. F. Watkins, *J. Am. Chem. Soc.*, 1976, **92**, 1515.
91. T. Shono, Y. Matsumura, S. Katoh, T. Fujita and T. Kamada, *Tetrahedron Lett.*, 1989, **30**, 371.

7.2
Oxidative Rearrangement Reactions

MATTHEW F. SCHLECHT
Du Pont Agricultural Products, Newark, DE, USA

7.2.1 INTRODUCTION

Oxidative rearrangements comprise a highly diverse group of reactions, some of which enjoy broad usage in synthesis, while others remain curiosities. This chapter necessarily reflects this heterogeneity. The designation 'oxidative rearrangement' is not used uniformly in the literature; the discussion in this chapter is limited to reactions which alter the connectivity in one or more carbon–carbon π- or σ-bonds in the substrate, and in which the molecule undergoes a net oxidation. Most often these changes will occur simultaneously, forming part of a single transformation, and the rearrangement is frequently driven by the oxidation. For the sake of brevity, the scope is further narrowed by excluding oxidative rearrangements of heterocyclic rings such as furans, pyrans, pyrroles and indoles; these reactions in themselves are numerous enough to fill a chapter. The overall emphasis is on selectivity and synthetic utility.

The organization is by type of reacting bond. The first section deals with functional group rearrangements — connectivity changes in carbon–carbon π-bonds and of bonds to heteroatoms which do not alter the carbon skeleton. The second section covers the skeletal rearrangements — connectivity changes in the carbon–carbon σ-bond framework with the concomitant functional group changes. Within each of these sections the discussion is divided according to the functional group undergoing oxidation.[1,2] The strained rings of cyclopropanes[3] and cyclobutanes[4] are treated as functional groups, and the oxidative rearrangements of these small rings which have no counterpart in the chemistry of larger rings are covered separately. Perhaps arbitrarily, oxidative cleavages of small rings are considered with the functional group rearrangements, while other structural reorganizations are covered with the skeletal rearrangements.

Several factors may cause an oxidation to take place with rearrangement. Conformational features of the substrate play an important role; steric crowding at the reaction site may favor strain relief through rearrangement over the normal mode of oxidation, or favorable overlap of the reacting bond with an allylic or an isolated but proximate double bond may cause a rearrangement. The first-formed products or intermediates of oxidation, such as an epoxide[5] or an alkylthallium(III) adduct,[6a] may be unstable and the pathway to a more stable species will involve a rearrangement. Many oxidative rearrangements follow a predictable pattern, and thus constitute reliable synthetic methods. Others are highly substrate dependent, and their utility in synthesis requires a careful conformational analysis of the substrate, or a good measure of luck.[7]

The examples presented here were selected to represent the variety of transformations which have been uncovered, from the heavily used to the seemingly unique cases. The aim is to acquaint practitioners with the more established methods, and to pique the interest in some reactions which could become useful tools with further development.

7.2.2 OXIDATIVE REARRANGEMENT OF FUNCTIONAL GROUPS

The discussion in this section is divided according to the functional group undergoing oxidative rearrangement: alkenes and enols, allylic alcohols, cyclopropanes and cyclobutanes, and miscellaneous functional group rearrangements. For the cyclopropanes and cyclobutanes, the scope is limited to the oxidative cleavages of the small rings which do not have counterparts in the chemistry of larger ring compounds. The major oxidants used commonly for these reactions include chromium(VI),[1,2b,8] lead(IV),[1,2c,9] and singlet oxygen ($^1O_2$).[10] From among the functional group rearrangements, the strongest contributors to synthetic methodology are the allylic oxidations of alkenes (including the singlet oxygen ene reaction), and the 1,3-ketone transposition resulting from the oxidative rearrangement of allylic tertiary carbinols.

7.2.2.1 Alkenes and Enols

The reaction of oxygen with vinyl halides gives acyl derivatives,[11] as shown in equation (1). These reactions proceed in moderate to good yield, and follow a radical chain mechanism. The migratory preference is Br > Cl > F when mixed polyhaloalkenes are used. This reaction has found particular utility in the preparation of functionalized fluorocarbons, as shown in Scheme 1 for the example of perfluoroacrylic acid (**1**). Vinyl sulfides also undergo this oxidative rearrangement to give α-thio acyl derivatives.

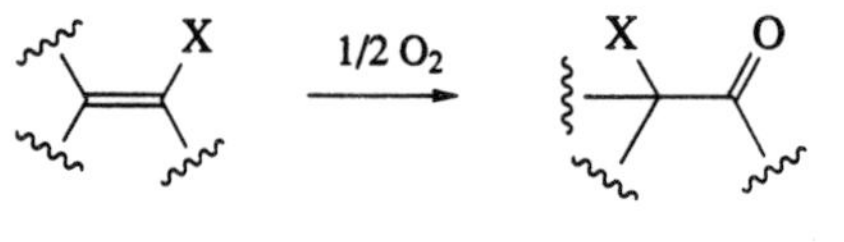

1/2 O_2 (1)

X = halogen

i, O_2, Br_2; ii, aq. NaOH, 42%; Zn, 40%

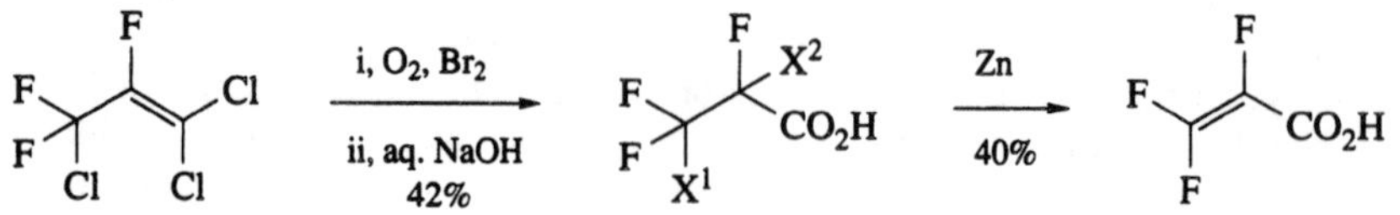

(**1**)

X^1, X^2 = Br, Cl; Cl, Br

Scheme 1

Vinylsilanes follow a similar course on oxidation with peroxy acid or with ozone.[6b,12] Depending upon the conditions of oxidation they can be converted either to a carbonyl compound (**2**) or to the α-hydroxycarbonyl compound (**3**), as in Scheme 2.[12,13] Vinyl silanes are useful synthetic intermediates, and this oxidation rearrangement procedure is an important component in their spectrum of reactivity.

Enol ethers, and in particular silylated enols (see Volume 2, Chapter 2.3), react with peroxy acid reagents to give initially a silyloxy epoxide, which rearranges with silyl migration to yield an α-silyloxy ketone,[12,14] as in Scheme 3. The net result is that a ketone is converted to a protected α-hydroxy ketone, and the stereochemistry is determined by the least hindered approach of the peroxy acid to the enol.

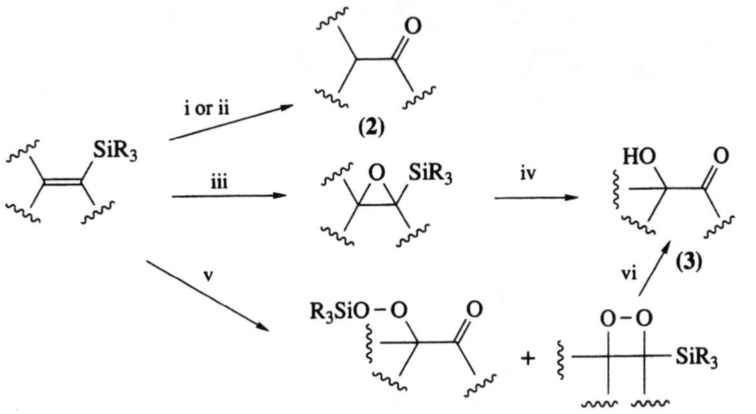

i, MCPBA, KHF_2, DMF; ii, 30% H_2O_2, $NaHCO_3$, MeOH, THF; iii, MCPBA, CH_2Cl_2; iv, 30% H_2O_2, KHF_2, $KHCO_3$, MeOH, THF; v, O_3; vi, [H]

Scheme 2

Although peroxy acid is the reagent of choice, under the proper conditions simple double bonds survive this reaction.

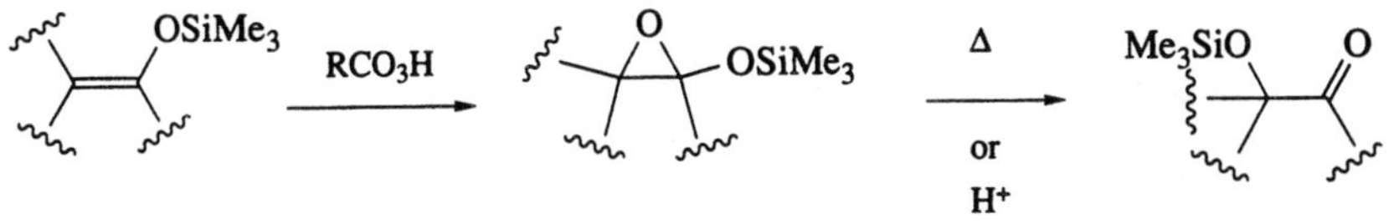

Scheme 3

Allylic oxidation (which is discussed in Chapter 2.1, this volume) takes place with rearrangement in certain substrates, the driving force being either a lower activation energy barrier for the chromate insertion into the C—H bond with rearrangement and/or a greater stability for the transposed enone product. Moderate selectivity is obtained in the case of 4,4-dimethylcyclohexene which serves to demonstrate,[8] as seen in equation (2). The oxidant shows a three-fold preference for hydrogen abstraction at C-3 over C-6 in order to avoid a 1,3-diaxial interaction with one of the methyl groups. The abstraction at C-3 leads mainly to the rearranged enone (**4**) for either steric or stereoelectronic reasons.

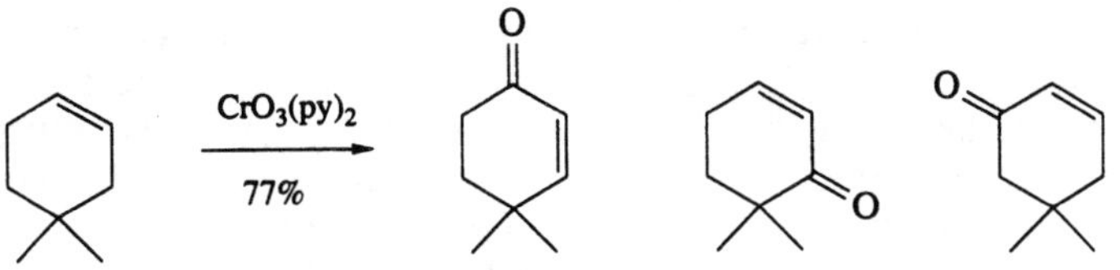

(2)

Two independent syntheses of quadrone employed an allylic oxidation with rearrangement, as shown in equation (3),[8] where the chromium trioxide·(3,5-dimethylpyrazole) reagent (CrO_3·DMP) was used. In some cases, the success of the reaction strongly depends on the nature of the oxidant, as shown in an approach to (–)-upial (equation 4). Here the chromium trioxide–heterocycle reagents, which are weaker oxidants, are quite inferior compared to the Fieser reagent.[15]

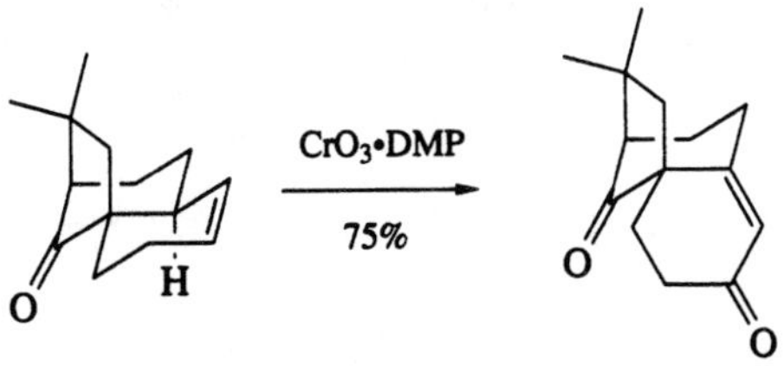

(3)

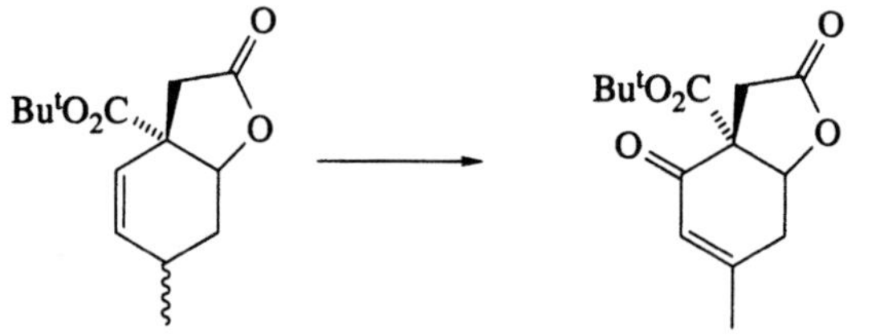

$$\xrightarrow{}$$ (4)

$CrO_3(py)_2$	0%
CrO_3•DMP	18%
CrO_3/HOAc/Ac_2O	67%

The allylic oxidation of alkenes by $^1O_2$ involves an ene reaction, and proceeds with rearrangement[10] as in Scheme 4. The intermediate allylic hydroperoxide (**5**) can be reduced to yield an allylic alcohol (**6**), or be treated with base to give an unsaturated carbonyl compound (**7**). The reaction works best on tri- or tetra-substituted alkenes, and the relative preference for attack is Me ≈ CH_2 >> CH. The $^1O_2$ allylic oxidation has been used in the synthesis of a large number of natural products, including some naturally occurring allylic hydroperoxides. It is possible that $^1O_2$ reactions of this type are involved in biosynthetic processes.

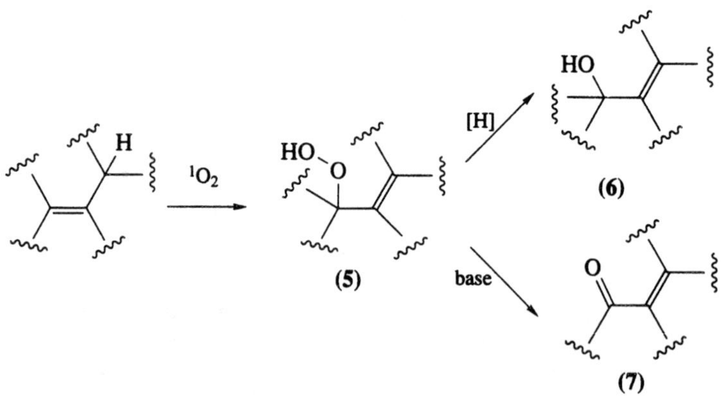

Scheme 4

This oxidation is applicable to a wide variety of both electron-rich and electron-poor alkenes; for example a number of tiglic acid derivatives undergo this reaction in moderate to excellent yield,[16] as in equation (5). In cases where there are several nonequivalent allylic sites the course of this reaction is highly substrate dependent, and the yields and selectivity vary from excellent to mediocre. In trisubstituted alkenes (except for most 1-alkylcyclohexenes), a reactivity pattern has emerged which has been termed a preference for *syn* ene attack (or 'PSEA').[17] This means that $^1O_2$ will preferentially attack one of the two allylic carbons which are *cis* to each other. In practical terms, this still translates into relatively low product selectivity in most cases such as in equation (6), although some notable exceptions are known (equations 7 and 8).

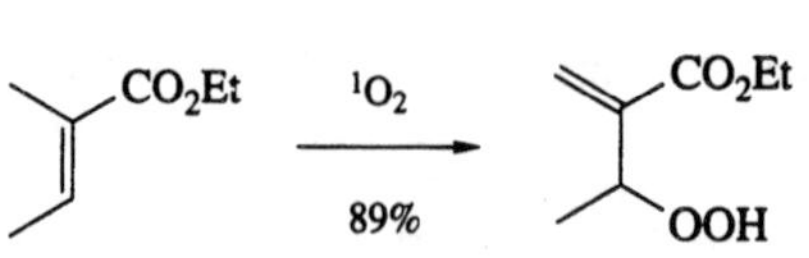

$$\xrightarrow[89\%]{^1O_2}$$ (5)

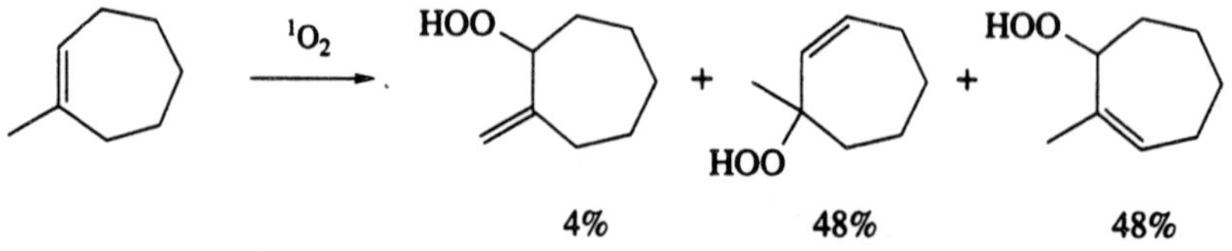

$$\xrightarrow{^1O_2}$$ (6)

4% 48% 48%

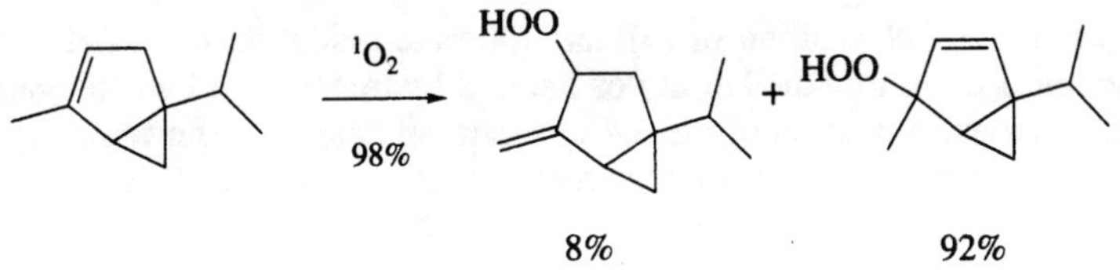

$^1O_2$, 98%; HOO (8%) + HOO (92%) (7)

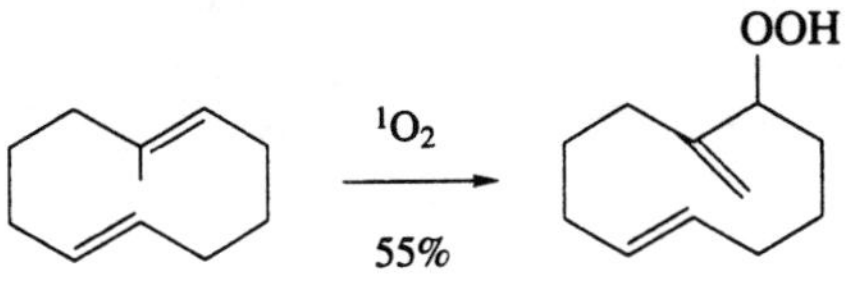

$^1O_2$, 55%; OOH (8)

A procedure for the large-scale conversion of alkenes to unsaturated carbonyl compounds using singlet oxygen has been published,[18] whereby the conversion of cyclopentene to cyclopentenone can be carried out on a molar scale in 60% yield.

Another variant on this chemistry is the use of triphenyl phosphite ozonide as a source of singlet oxygen.[19] This reagent mimics singlet oxygen in many of its reactions, and is easier to quantify.

A further alternative to the singlet oxygen allylic oxidation method involves selenium chemistry.[20] This route involves epoxidation of the alkene, followed by nucleophilic opening of the epoxide with phenyl selenide anion, and finally oxidation to the selenoxide, which eliminates spontaneously to produce an allylic alcohol,[21] as described in Scheme 5. The β-hydroxy phenyl selenide (**8**) need not be isolated but can be oxidized *in situ*. The regioselectivity of this conversion depends on the degree of substitution on the alkene and the conformation of the β-hydrogens. The phenyl selenide anion will attack the least hindered carbon of the corresponding epoxide, and the geometry of the resulting double bond depends on the alignment of the hydrogens allylic to the original double bond; the elimination of the selenoxide is *syn*, and the transition state for elimination may require a rotation. If a ring fusion or conformational restrictions prevent the proper orbital overlap, the elimination may fail, or may give an enol if the hydroxy-bearing carbon is secondary.

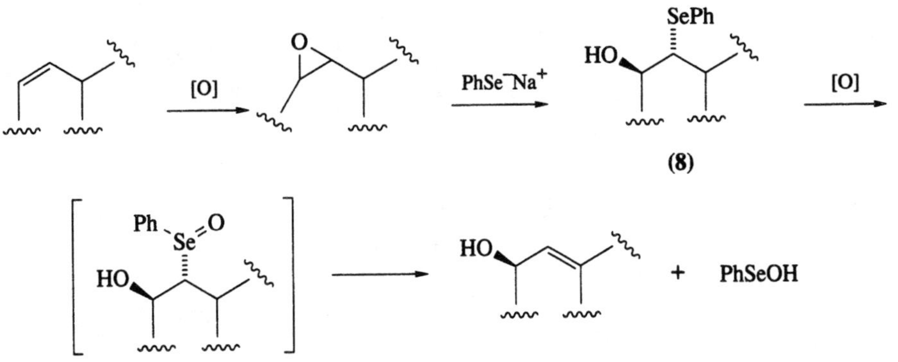

[O]; O; $PhSe^-Na^+$; HO, SePh (**8**); [O]

Ph, Se=O, HO; HO; + PhSeOH

Scheme 5

An electrochemical synthetic process (see Chapter 7.1, this volume) has been reported which requires only a catalytic amount of the selenating agent, and converts an alkene to the allylic alcohol in an aqueous cell, or to the allylic methyl ether if the electrolysis is run in methanol, as in equation (9).[22]

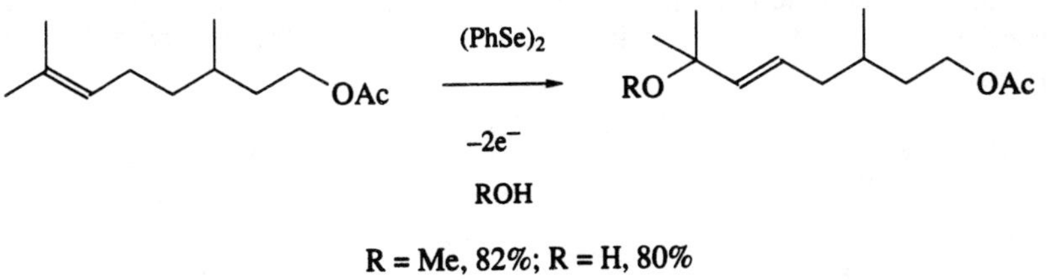

OAc; $(PhSe)_2$, $-2e^-$, ROH; RO, OAc (9)

R = Me, 82%; R = H, 80%

A special case of allylic oxidation with rearrangement occurs in the action of chromium(VI) agents on 3,4-unsaturated ketones (**9**), and this is shown in Scheme 6. Hydrogen abstraction by the oxidant takes place at the doubly activated C-2 position of (**9**), and ultimate oxidation occurs at C-4 to give an ene-dione (**10**). This reaction is driven by the lability of the C-2 hydrogen, and by conjugation of the double bond with the carbonyl group; oxidation of the 3,4-unsaturated ketones is far more rapid than oxidation of their 2,3-unsaturated isomers. This reaction may have a broad scope, but the only examples to date involve six-membered ring-fused polycyclic and acyclic substrates.

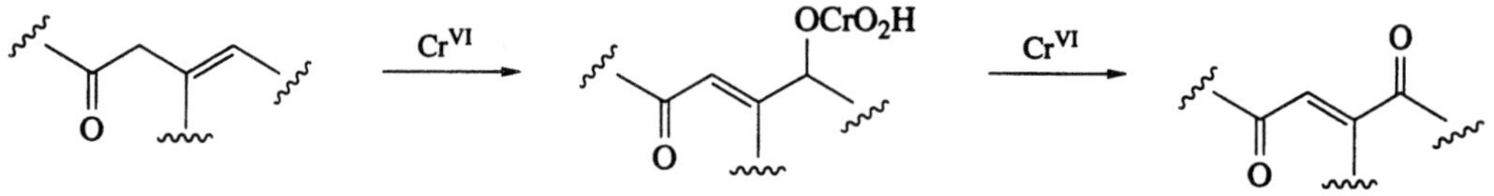

(**9**) $\xrightarrow{Cr^{VI}}$ [$OCrO_2H$ intermediate] $\xrightarrow{Cr^{VI}}$ (**10**)

Scheme 6

The formation of a second carbonyl is not required, and where C-4 is fully substituted a 4-hydroxy-2-en-1-one is obtained.[23] The original carbonyl group may arise by oxidation of an alcohol, the classic example being the oxidation of cholesterol (**11**) to cholest-4-en-3,6-dione (**12**) with chromium(VI) reagents, in which a yield as high as 85% can be obtained, as in equation (10).[24] In degradation work on isopimarenes isolated from the mollusc *Aplysia kurodai*, Jones reagent serves to convert a β-hydroxyalkene (**13**) into the enedione (**14**) in moderate yield, as shown in equation (11).[25]

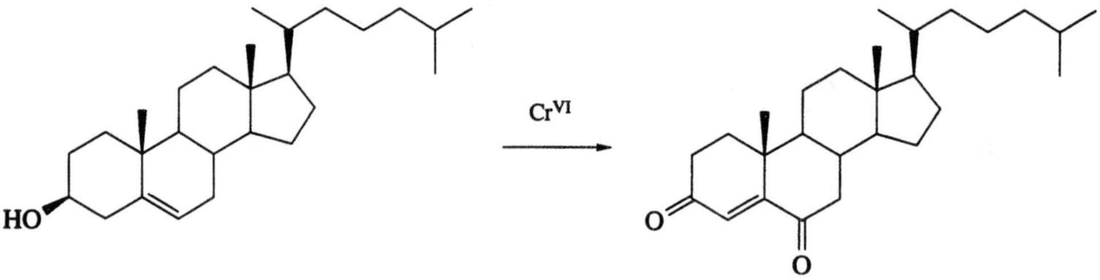

HO (**11**) $\xrightarrow{Cr^{VI}}$ (**12**) (10)

$Na_2Cr_2O_7$, AcOH, 40%; $CrO_3(py)_2$, AcOH, 85%

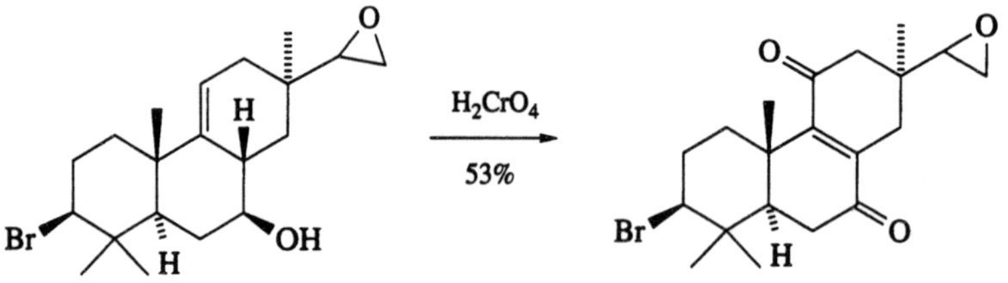

Br, H, OH (**13**) $\xrightarrow[53\%]{H_2CrO_4}$ Br, H, O (**14**) (11)

A related oxidative rearrangement of cephem dioxides has been reported[26] in which an alkene is oxidized stereospecifically with rearrangement to the allylic alcohol in good yield by simple exposure to a palladium/carbon catalyst, as depicted in equation (12). Adventitious oxygen preadsorbed on the catalyst seems the likely oxidant. The reaction fails on the parent cephem or its monoxide, or on the free acid of the dioxide. This reaction would seem to hold some promise for further utility in the cephem field and other related systems.

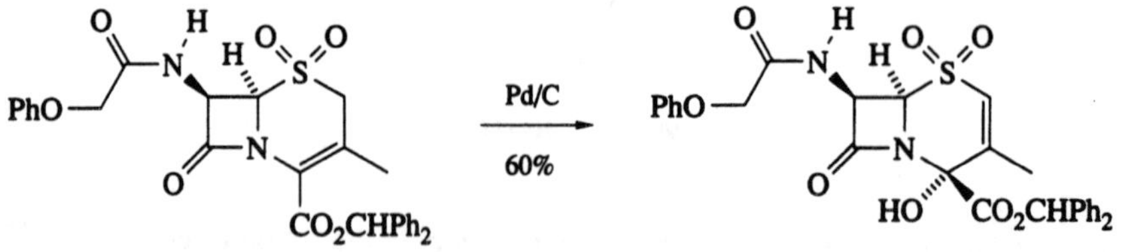

PhO, CO_2CHPh_2 $\xrightarrow[60\%]{Pd/C}$ PhO, HO, CO_2CHPh_2 (12)

7.2.2.2 Allylic Alcohols

The oxidative rearrangement of allylic alcohols to α,β-unsaturated ketones or aldehydes is one of the most widely used synthetic reactions in this group, and forms part of a 1,3-carbonyl transposition sequence.[8] Scheme 7 shows this reaction and the related conversion of the allylic alcohol to an α,β-epoxy carbonyl compound. Chromate reagents induce some allylic alcohol substrates to undergo a directed epoxidation of the alkene without rearrangement, but this reaction is beyond the scope of the present discussion.

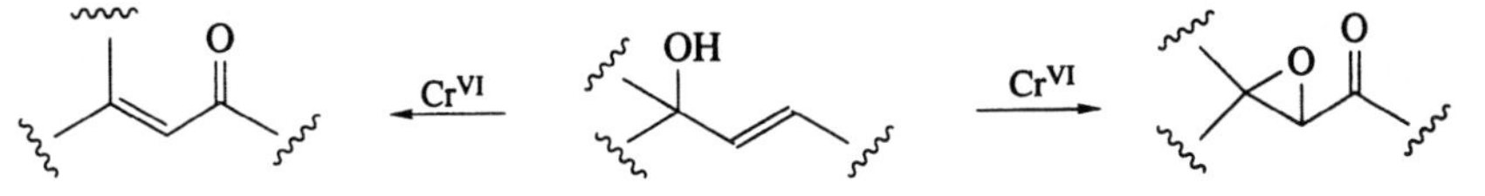

O Cr^{VI} OH Cr^{VI} O O

Scheme 7

The mechanism for this transformation, and the partitioning between unsaturated carbonyl and epoxycarbonyl products has been the subject of several studies.[27] The production of epoxycarbonyl compounds seems to be correlated with the nature of the chromate reagents used, although substrate structure also helps to determine this preference.

However, the conversion to the transposed α,β-unsaturated carbonyl compound is by far the more useful reaction. The full sequence serves both to form carbon–carbon bonds as well as to adjust the functional group array in the synthetic intermediate. Thus, starting with the enone (**15**), organometallic addition generates a tertiary allylic alcohol (**16**) and oxidative rearrangement yields a β-alkyl-α,β-enone (**17**), as shown in Scheme 8.

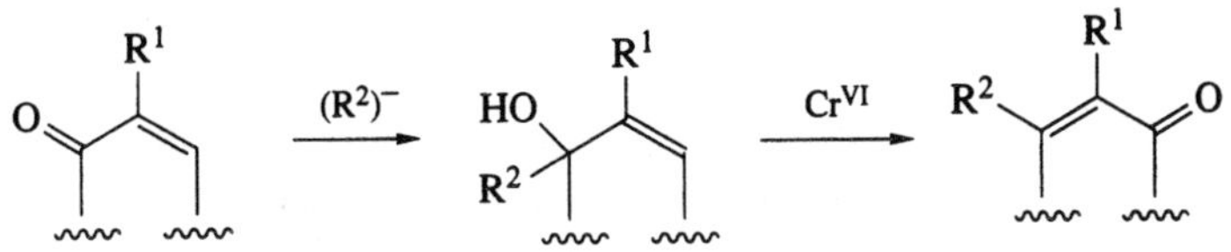

R^1 O $(R^2)^-$ HO R^1 R^2 Cr^{VI} R^1 R^2 O

(15) **(16)** **(17)**

Scheme 8

Applications are found in acyclic as well as five-, six-, seven- and eight-membered ring cyclic substrates, and yields are generally in the range of 50–90%. The best substrate is one in which the C—O bond of the alcohol is (or can easily become) parallel with the *p*-orbitals of the alkene double bond, as the transition state is believed to involve a 3,3-sigmatropic rearrangement of the chromate ester. Isolated double bonds, esters, lactones and silyl ethers, and a number of other functional groups, survive these conditions. Protected carbohydrates undergo degradation which limits the application to such substrates. Examples are found in equations (13)–(15).[28–30]

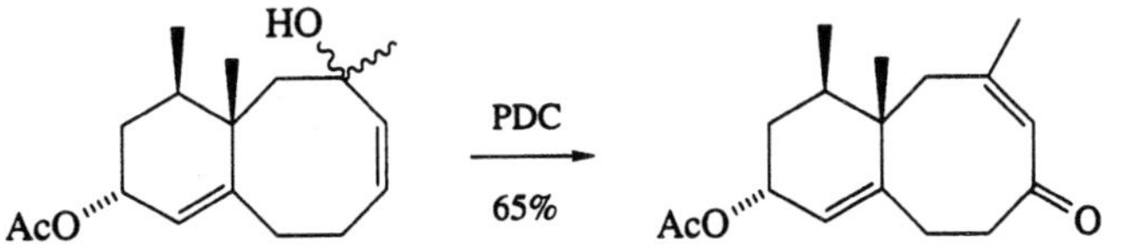

HO AcO PDC 65% AcO O (13)

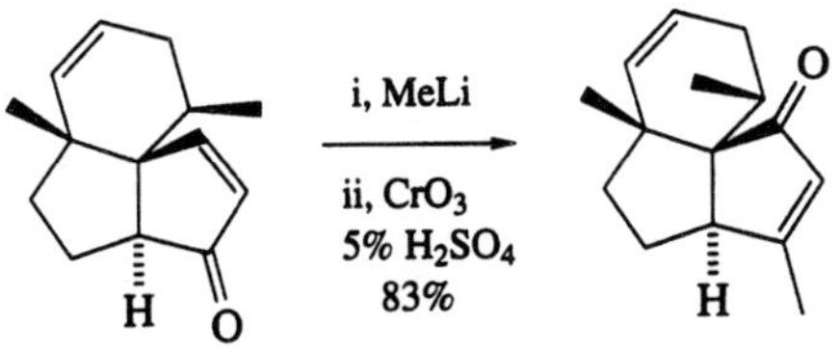

H O i, MeLi ii, CrO_3 5% H_2SO_4 83% O H (14)

Ketones can be homologated to unsaturated ketones or aldehydes by addition of the appropriate vinyl nucleophile followed by oxidative rearrangement, as shown in Scheme 9. The use of this transformation in a synthetic approach to steroids with unsaturated side chains is shown in equation (16).[31]

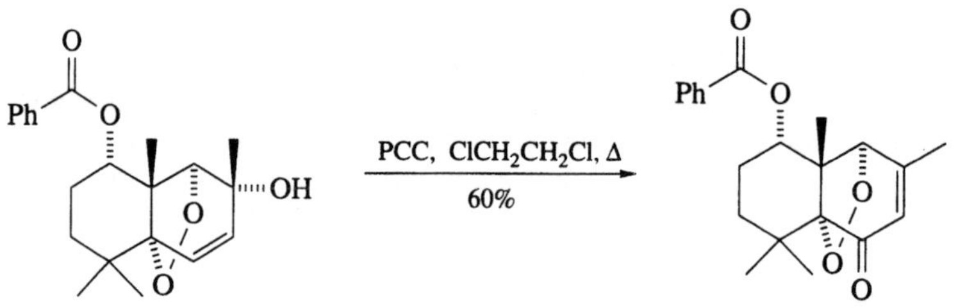

(15)

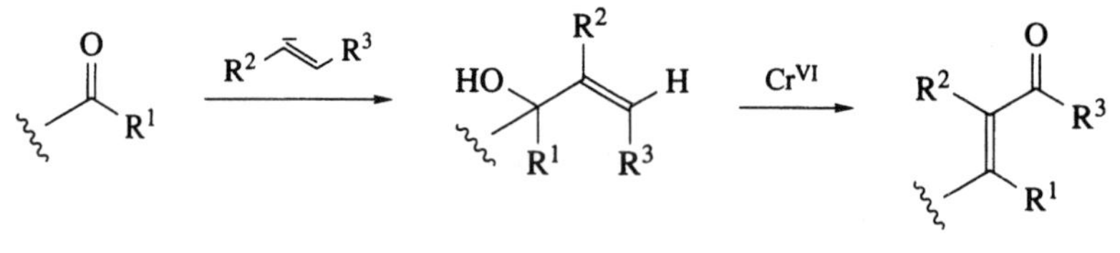

Scheme 9

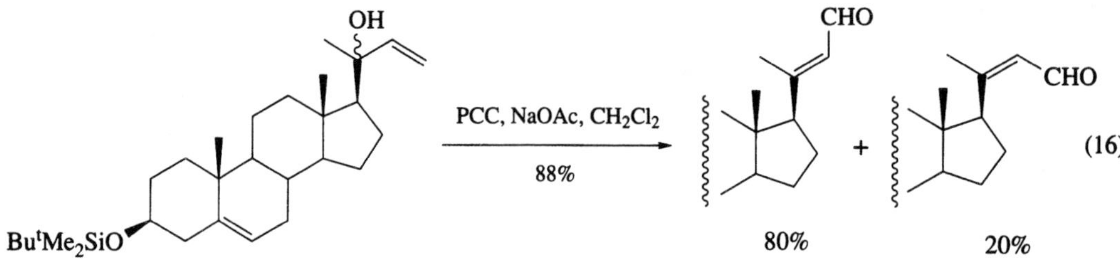

(16)

The regioselectivity of these reactions has been studied in cases where two allylic rearrangements would be possible. In one report tertiary alcohols which were both allylic and propargylic were found to rearrange solely over the allylic system where the alkene is contained in a five- or six-membered ring.[32] In a cyclic system where the alcohol is equatorial, and in acyclic systems, the yield of rearrangement is poor and oxidative cleavage becomes important. In a particularly interesting study a series of bis-allylic alcohols were examined.[33] Vinylic cyclohexenols and cyclopentenols rearrange exclusively within the ring to give 57–80% yields of the β-vinylic cyclohexenone or cyclopentenone, as shown in equation (17). The bis-allylic alcohol (**18**; equation 18), which contains an allylsilane substituent, undergoes clean oxidative rearrangement to the dienone (**19**). This example helps to clarify the mechanism of the rearrangement, since a discreet carbonium ion intermediate would doubtless be trapped through cyclization with the allylsilane moiety; the absence of such cyclization products argues against a cationic intermediate and in favor of the 3,3-sigmatropic mechanism. The fact that the mildly nucleophilic allylsilanes (see Volume 2, Chapter 2.2) survive this reaction is important for its synthetic utility.

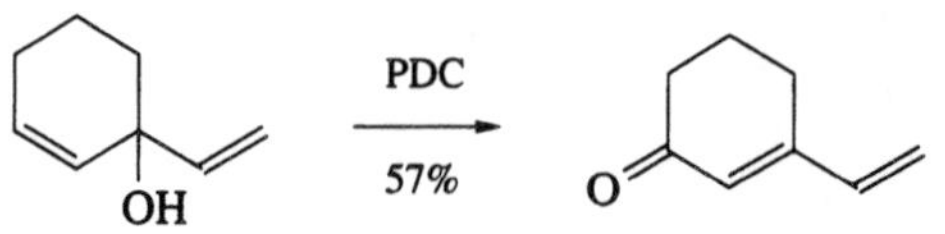

(17)

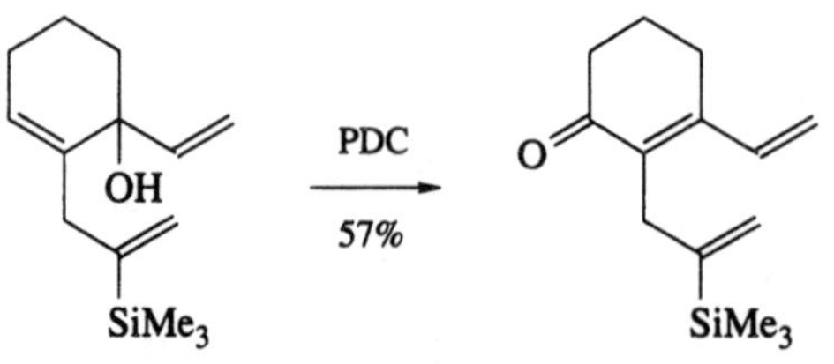

(18)

Where the allylic alcohol is in a secondary position, conformational effects or the character of the oxidant can still favor an oxidative rearrangement over simple oxidation to the ketone (equation 19).[34]

Cr^{VI} 42% (19)

PCC	100%	
PCC, TsOH	17%	83%

An oxidative rearrangement took place during the MCPBA epoxidation of the secondary allylic alcohol auraptenol, leading to the enal shown in equation (20). This reaction has been used in an approach to casegravol and in a synthesis of arnottinin.[35] The reason why the intermediate epoxy alcohol undergoes rearrangement in this case is not known beyond the possibility that the *m*-chlorobenzoic acid by-product could act as an acidic catalyst.

MCPBA, $CHCl_3$, 22% (20)

The peculiar oxidative rearrangement of cycloocta-2,4-dien-1-ol (**20**) shown in equation (21), involves a highly selective *cis* epoxidation (*cis*:*trans* = 20:1) followed by transannular S_N2' attack by the hydroxy group on the allylic epoxide to give the *exo*-β-hydroxy cyclic ether (**21**).[36] This rearrangement is stereospecific in that the *trans*-epoxy alcohol, available by treatment of the cyclooctadienol with MCPBA, does not give a rearranged product. A similar example with a cycloheptadienol was also reported. The pure *cis*-epoxy alcohol rearranges to the bicyclic alcohol at 156 °C, but does so far more rapidly in the presence of the vanadium catalyst at 60 °C.

(**20**) TBHP, $VO(acac)_2$, 60 °C, 78% (**21**) (21)

The cyclodecenol substrate (**22**; equation 22), undergoes a transannular oxidative rearrangement to yield the decalone (**23**) in moderate yield.[37] Although not strictly a simple functional group rearrangement, this reaction can be thought of as the through-space version of the oxidative rearrangement of allylic alcohols. In this case it is quite likely that the reaction proceeds through attack by the alkene on the carbon bearing a preformed chromate ester, which behaves as a leaving group. The intermediate decalinyl carbonium ion is captured by additional chromate and eliminates to the observed product (**23**).

(**22**) PCC, 40% (**23**) (22)

A related example occurs in the adamantane field, as seen in equation (23).[38] It is surprising that a primary alcohol undergoes ring closure instead of the standard oxidation to an aldehyde or acid under the influence of chromate. The chromate ester of this *endo*-oriented alcohol would undoubtedly experience severe crowding, and direct oxidation is probably inhibited for steric reasons.[39] In both of these cases the proximity of the alkene to the carbinol carbon seems to drive the rearrangement, and there is precedent for a related type of oxidative cyclization.[8] In a further example, the norbornadienol (**25**; equation 24) is oxidized by MnO_2 with rearrangement to give the hemiacetal (**26**).[40] This reaction is believed to involve a displacement of the Mn^{4+} ester by the alkene, followed by oxidation of the resulting carbonium ion and cleavage.

OH H_2CrO_4 O (23)

(24)

HO MnO_2, $CHCl_3$ 75% O OH (24)

(25) **(26)**

7.2.2.3 Cyclopropanes and Cyclobutanes

The strained rings of cyclopropanes[3] and cyclobutanes[4] can be considered as functional groups, since these molecules react in ways which are not characteristic of other cycloalkanes. The chemical behavior of cyclopropanes in particular has many analogies to that of alkenes. For the purposes of this discussion the oxidative ring cleavage reactions of cyclopropanes and cyclobutanes which do not have counterparts in the chemistry of larger rings are considered as functional group rearrangements.

Oxidative cleavage of cyclopropanes has been studied mostly with lead(IV),[2c,9] thallium(III)[6] and chromium(VI) reagents.[2b,8] The oxidative cleavage of cyclobutanols has been explored mainly with chromium(VI) reagents,[2b,8] although other oxidants have been studied.[4]

Cyclopropanols undergo oxidative cleavage with a variety of oxidants to give β-functionalized propanal derivatives. Even secondary cyclopropanols give moderate yields of ring-opened products. The activation barrier on the pathway to cyclopropanone is steep, and the alternative pathway of rearrangement is driven by relief of ring strain. The example given in equation (25) shows the use of chromic acid.[8] These oxidations are much faster than the oxidation of a normal secondary alcohol.

HO H_2CrO_4 31% O HO (25)

An approach to a prostaglandin intermediate employed a cyclopropanol oxidation with a mixed chromate/cerate reagent shown in equation (26), but the yield was unacceptably low.[41] Although no information on the selectivity is available, the *trans* stereochemistry of oxidative cleavage in the reported product is of note. In these more complex substrates, side reactions and low yields plague the reaction, which will see only limited use in synthesis unless a better reagent system is developed.

O O H H OH H_2CrO_4, CAN, H_2O, AcOH O O HO CHO (26)

The lead tetraacetate oxidation of tertiary cyclopropanol silyl ethers does show some promise.[42] As shown in Scheme 10, a two-bond oxidative cleavage of the three-membered ring takes place in acetic acid solvent to yield the alkenoic acid (**27**); a carboxy group is produced from the original carbinol carbon and the alkene is derived from the other two ring carbons. The yields for this transformation are from 65–88%, and with a substituent on the methylene bridge the fragmentation is highly stereoselective for alkene geometries. In methylene chloride, a one-bond cleavage takes place to give the acetoxymethylcycloalkanone (**28**). Both types of reaction are useful, and as the starting cyclopropanols are available through Simmons–Smith methylenation of the corresponding silylated enol (see Volume 4, Chapter 4.7), these methods have good synthetic potential.

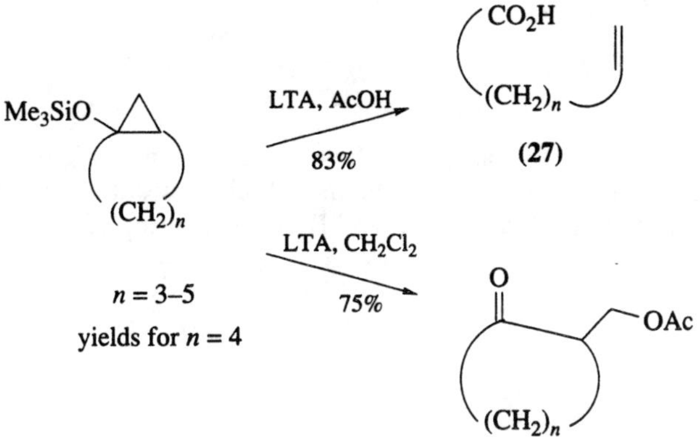

Me_3SiO … $(CH_2)_n$ — LTA, AcOH, 83% → CO_2H … $(CH_2)_n$ (**27**)

$n = 3–5$; yields for $n = 4$ — LTA, CH_2Cl_2, 75% → O … OAc … $(CH_2)_n$ (**28**)

Scheme 10

Cyclopropenes undergo an oxidative cleavage to yield substituted enones, as shown in equation (27).[43] The reaction is believed to proceed through the unstable epoxide. The regioselectivity is generally low if $R^1 \neq R^2$, but if one of the substituents is trimethylsilyl a highly selective conversion to the α-silyl enone takes place. There is one report of a similar oxidative cleavage that takes place with thallium(III).[6]

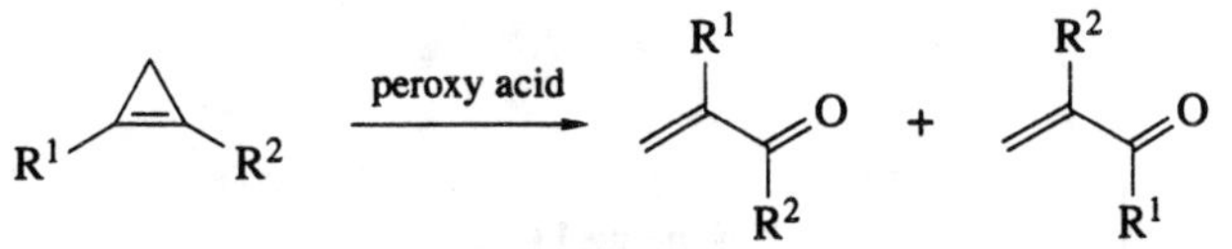

R^1, R^2 — peroxy acid → R^1 … O, R^2 + R^2 … O, R^1 (27)

Methylenecyclopropane undergoes oxidative cleavage and ring expansion with thallium trinitrate in methanol to furnish in quantitative yield a mixture of the ring cleavage product 1-methoxybutan-3-one and cyclobutanone in the ratio of 4:1, as in equation (28).[44]

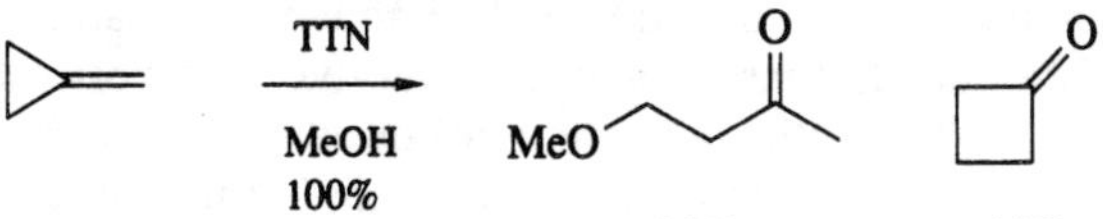

TTN, MeOH, 100% → MeO … O (82%) + O (18%) (28)

The oxidative rearrangement of tertiary cyclopropylcarbinols to 3,4-unsaturated carbonyl compounds is analogous (or homologous) to the reaction of allylic alcohols, and is shown in the example in equation (29).[8] This reaction has been shown to proceed stereospecifically in the conversion of the *cis*-substituted cyclopropylcarbinol (**29**) to the (*Z*)-enynone (**30**) shown in equation (30).[45] The substrates with R = H, Me and TMS all gave comparable yields.

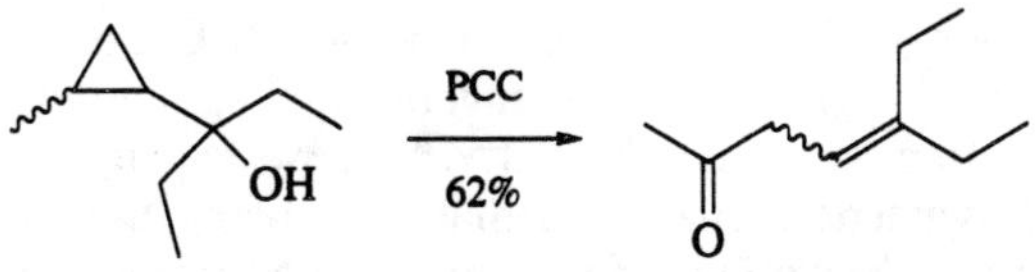

OH — PCC, 62% → O (29)

Cyclobutanols oxidize with ring cleavage to 4-hydroxy ketones, 4-hydroxy acids, or 1,4-diones under the influence of chromium(VI) reagents (Scheme 11).[8] The first formed product is a 4-hydroxycarbonyl compound (**31**) which exists as the five-membered ring hemiacetal (**32**). This form will persist in the absence of excess reagent under nonforcing conditions; otherwise further oxidation takes place to give a

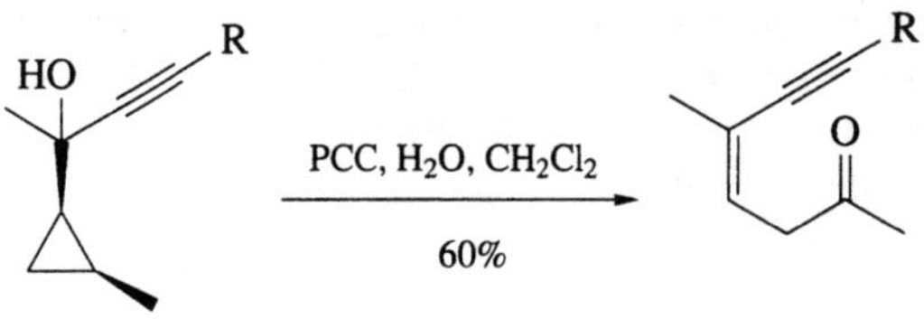

(30)

(29) (30)

4-hydroxy acid (**33**; if R^1 = H) or a 1,4-dione (**34**; if R^3 = H). In unsymmetrical cyclobutanes the bond cleaved is the one between the carbinol carbon and the more highly substituted β-carbon, and the yields are generally good. With a quaternary β-carbon (R^2 and R^3 = alkyl) some of the 3,4- or 4,5-unsaturated carbonyl product arises.

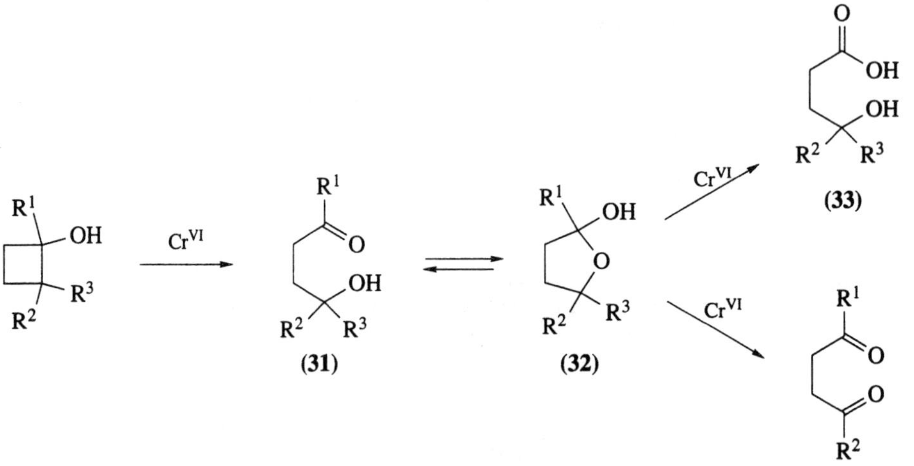

Scheme 11

7.2.2.4 Miscellaneous Functional Group Rearrangements

Treatment of 1,1-disubstituted epoxides of the gibberellin family with sulfuryl chloride results in the formation of the corresponding α,β-enal in good yield, as shown in equation (31).[46] Four examples were reported in which alcohols, esters, lactones and alkenes survive. The postulated mechanism involves an electrophilic opening of the epoxide with elimination, followed by oxidation of the primary chlorosulfate ester. A steroidal 3-spirooxirane also undergoes this reaction, but the yield is poor and several products are obtained, suggesting that the overall scope of this reaction may be limited.

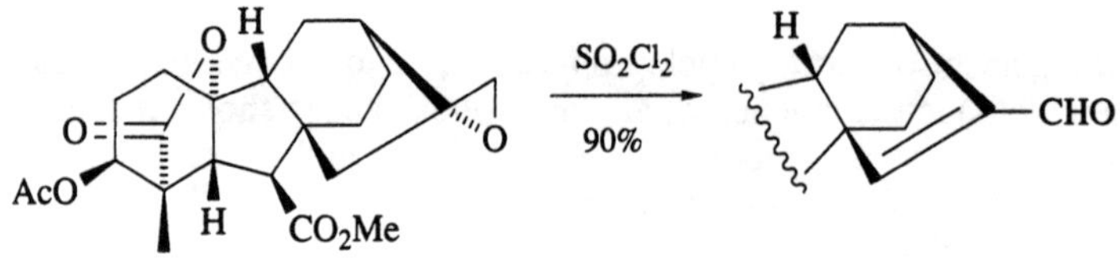

(31)

The Pummerer rearrangement (which is discussed in Volume 6, Chapter 4.7) is a type of oxidative rearrangement, as is the related 2,3-sigmatropic rearrangement of 2,3-unsaturated sulfoxides. Two related examples are presented here from selenium chemistry.[20] These reactions enhance the attractiveness of sulfur- and selenium-based synthetic methods, in that after being used to forge new carbon–carbon bonds, the heteroatom moiety can be exploited for a further functional group interchange.

Propargyl phenyl selenide is a versatile multifunctional acrylate synthon, as shown in Scheme 12.[47] The dianion is prepared and reacted successively with an alkylating agent (R—X) and an electrophile (E^+). The oxidative rearrangement of the propargylic selenoxide (**35**) to an allenic selenenate (**36**), and thence to the α-phenylselenoenone (**37**), forms the keystone of this synthetic method, and overall yields from propargyl phenyl selenide are in the range of 38–68%. Further elaboration of (**37**) is possible through conjugate additions, deselenation or another oxidative rearrangement. This method was used in a synthesis of 7-hydroxymyoporone.[47]

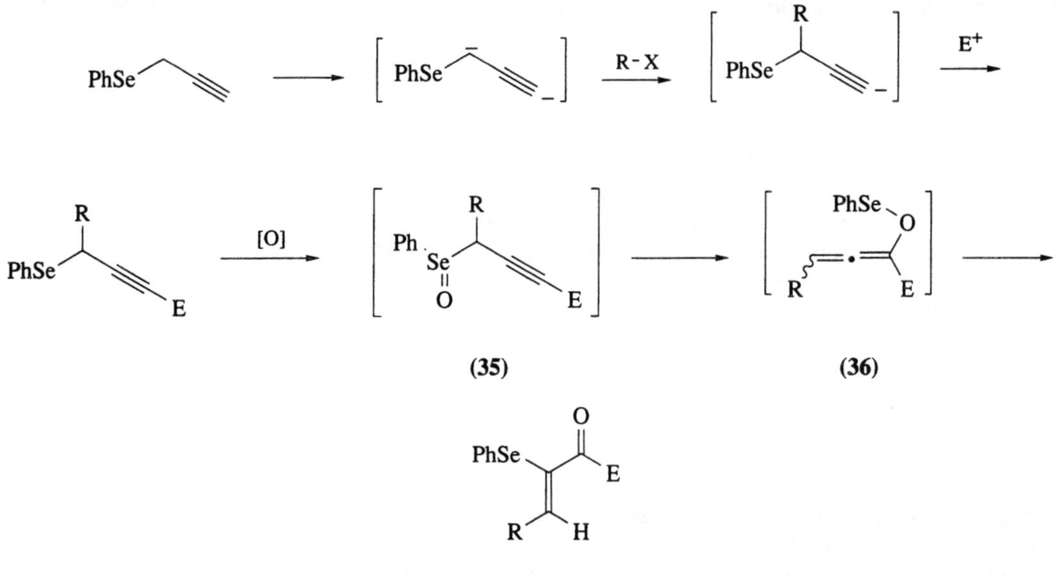

PhSe — [PhSe] R-X [PhSe R] E^+

PhSe R E [O] [Ph Se O R E] — [PhSe O R E] —

(35) **(36)**

PhSe O E R H

(37)

Scheme 12

Equation (32) shows another example of this type of rearrangement, in which a phenylselenoallenic ester is converted to an α-keto alkynic ester in quantitative yield.[48]

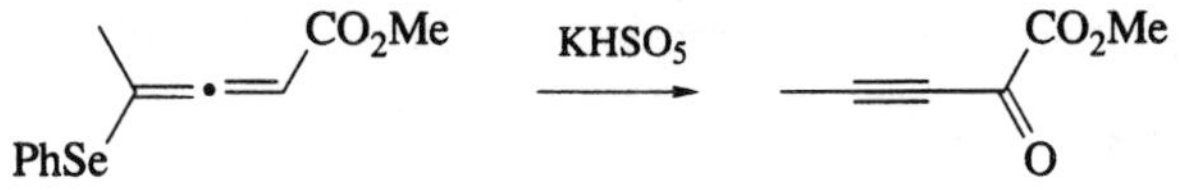

PhSe CO_2Me $KHSO_5$ CO_2Me O (32)

7.2.3 OXIDATIVE SKELETAL REARRANGEMENT

The discussion in this section is divided according to the functional group undergoing oxidative rearrangement: alkenes and enols, dienes, alkynes, cyclopropanes and cyclobutanes, and miscellaneous skeletal rearrangements. In view of the voluminous body of work on alkenes and enols, this area has been subdivided according to particular substrates: arylalkenes, aryl ketones, chalcones and cinnamyl compounds, and cyclic alkenes and ketones. Chalcones and cinnamyl compounds are treated as a special case of arylalkenes because of the extensive synthetic use of the chalcone to isoflavone transformation. In order to focus on the application to ring expansion and ring contraction reactions, cyclic alkenes and ketones are treated as a separate case, and methylenecycloalkanes are discussed with cycloalkenes. For the cyclopropanes and cyclobutanes, the scope is limited to the skeletal rearrangements of the small rings which do not have counterparts in the chemistry of larger ring compounds. The oxidants which have played the largest role in these transformations are thallium(III),[6a] lead(IV)[2c,9] and iodine.[49] The best reagents appear to be lead tetraacetate and thallium trinitrate, although on a larger scale cost is a concern for the latter, and toxicity is a problem for both. The newer hypervalent iodine reagents may prove more amenable in view of these factors.

Three highly useful synthetic transformations are presented in this section: the synthesis of isoflavones from chalcones, the synthesis of α-arylalkanones from arylalkenes, and the synthesis of α-arylalkanoic acids from aryl ketones. Two others are potentially useful methods, but are not as yet widely used: the preparation of α-branched carboxylic acids from alkynes, and the ring expansion and ring contraction of cyclic alkenes and ketones.

7.2.3.1 Alkenes and Enols

The oxidative rearrangement most widely used in synthesis is the oxidative 1,2-shift of an alkene or enol, which is shown in the formal sense in equation (33). The alkene may be electron deficient such as an unsaturated ketone, or electron rich such as an enol, enol ether or enamine.

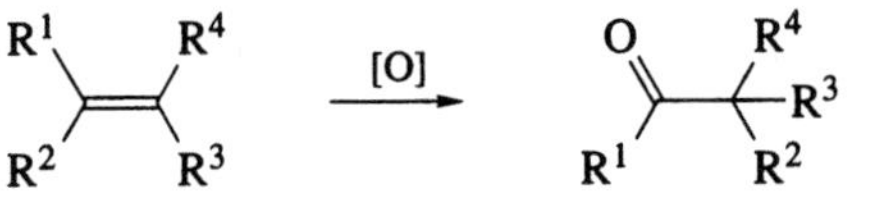

$$R^1R^2C{=}CR^3R^4 \xrightarrow{[O]} R^1C(O)CR^2R^3R^4 \quad (33)$$

If R^2 and R^3 are connected then a ring contraction results. If R^1 and R^2 are connected, a ring expansion takes place. The reagents used to carry out this transformation are strongly electrophilic oxidants, or an oxidant used together with a Lewis acid. Hypervalent main group oxidants such as thallium(III), lead(IV) and iodine(III) have played the largest role in this area. The substrates have been divided into four major groups by compound class: arylalkenes, aryl ketones, chalcones and cinnamyl compounds, and cyclic alkenes and ketones in ring expansion and ring contraction reactions.

7.2.3.1.1 Arylalkenes

Arylalkenes undergo oxidation with 1,2-rearrangement of the aryl group to give α-arylcarbonyl compounds, and this reaction is shown in the formal sense in equation (34). Useful reagents for this transformation include lead(IV),[50] thallium(III),[51] iodine(III),[52] and palladium.[53] The yields for this reaction are moderate to excellent, and there is a reasonable tolerance of functional groups on the aromatic ring (R^1) such as halogen, methyl or methoxy. At least one *ortho* substituent is permissible with no loss in yield. If R^2 = H, the product is an α-aryl aldehyde, and if R^2 = alkyl an α-aryl ketone is obtained. The rearrangement of 2-propenylbenzenes to the 1-arylpropan-2-ones is important due to the interest in the latter compounds as pharmaceutical intermediates, as in equation (35).[50b]

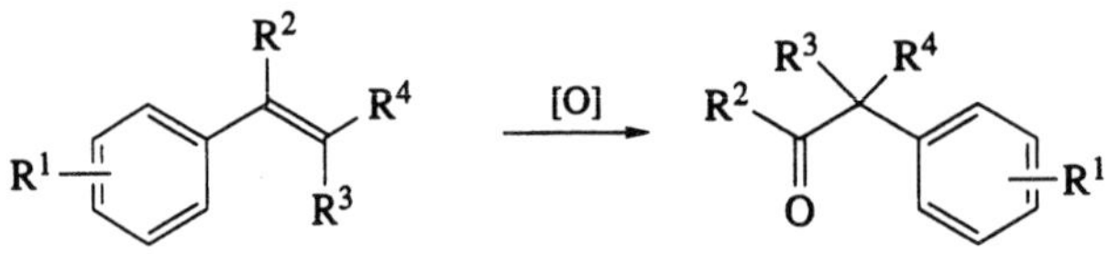

(34): R^1-aryl–C(R^2)=CR^3R^4 —[O]→ R^2C(O)CR^3R^4–aryl-R^1

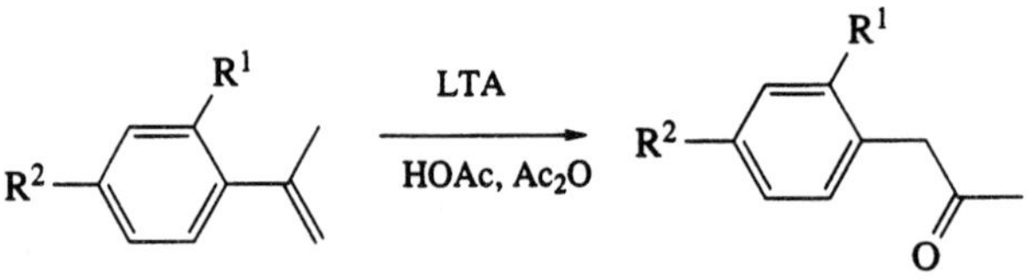

(35): LTA, HOAc, Ac_2O

R^1 = OMe; R^2 = Me; 52%

R^1 = H; R^2 = OMe; 80%

The interesting sequence depicted in Scheme 13 for a sequential oxidative rearrangement and hydroxylation of citral shows some potential for this reaction in nonaromatic alkenes. This transformation affords an elegant, single-step approach to the 6,8-dioxabicyclo[3.2.1]octane skeleton, although the stereoselectivity for the two induced centers is poor.[54]

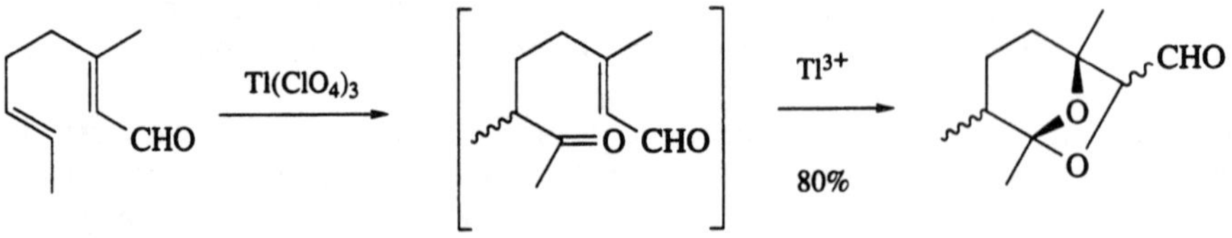

CHO —$Tl(ClO_4)_3$→ [O CHO] —Tl^{3+}, 80%→ CHO

Scheme 13

7.2.3.1.2 Aryl ketones

Aryl ketones undergo oxidative rearrangement with hypervalent main group oxidants such as thallium(III), lead(IV) and iodine(III) to give α-arylalkanoic acids in good to excellent yield,[55] and the general transformation is shown in equation (36).[6a,56] This reaction is related to the Willgerodt–Kindler reaction,[57] and has drawn considerable attention due to the antiinflammatory properties of the product α-arylalkanoic acids. A Friedel–Crafts acylation of an aromatic precursor followed by this oxidative rearrangement forms the synthetic sequence of choice for these compounds. At least a dozen patents have been issued on applications of this method, including one which describes a process catalytic in thallium.

$$R^1\text{-}C_6H_4\text{-}CO\text{-}CH_2R^2 \xrightarrow[\text{MeOH, 50–95\%}]{[O]} R^1\text{-}C_6H_4\text{-}CH(R^2)CO_2Me \qquad (36)$$

The best results are obtained with the above-named oxidants in a mixed solvent of methanol and trimethyl orthoformate in the presence of a strong acid; these conditions presumably ensure rapid acetalization of the carbonyl to prevent α-oxidation. This side reaction is more serious when R^2 is alkyl and the orthoformate is omitted, or if ethyl carbonate or acetonitrile is used as solvent. Preformed enol ethers and enamines give the desired oxidative rearrangement in high yield.

A wide variety of substituents are tolerated. The group R^1 can be alkyl, halogen, alkoxy, *N*-amido, azidomethyl, ester, aryl, aryloxy and aryloyl, and at least one *ortho* substituent is permissible with no loss in yield. The aromatic ring can also be 2-naphthyl, 9,10-dihydro-2-phenanthryl, 3-pyridyl, thiophen-2-yl or pyrrol-3-yl. The group R^2 can be hydrogen, alkyl, acyl or acetic acid. Beyond the antiinflammatory targets, successful reaction substrates include the methyl ketones of a binaphthyl crown ether, a morphinane and a polyaromatic hydrocarbon. The preparation of ibuprofen methyl ester (**38**) is shown in equation (37) as a typical example.[56c]

$$\text{4-}i\text{-BuC}_6H_4COCH_2CH_3 \xrightarrow[87\%]{PhI(OAc)_2,\ HC(OMe)_3,\ H_2SO_4} \text{4-}i\text{-BuC}_6H_4CH(CH_3)CO_2Me \qquad (37)$$

(38)

Dialkyl ketones have been little studied as precursors in this reaction. Selenium dioxide with hydrogen peroxide and *t*-butyl alcohol effects a similar reaction with these substrates to give 35–40% yield of the corresponding carboxylic acid. In methyl alkyl ketones, the regioselectivity is of the order of 5:1 in favor of methyl migration.[58]

7.2.3.1.3 Chalcones and cinnamyl compounds

While following the reactivity patterns of arylalkenes, the extensive use in synthesis that has been made of the chalcone to isoflavone conversion and related sequences warrants a separate treatment. Chalcones (**39**) oxidize with thallium(III)[6a,59] and iodine(III)[52a] under rearrangement of the aromatic B-ring to give 3,3-dimethoxy-1,2-diarylpropan-1-ones (**40**) in yields from 30–90%, as shown in Scheme 14. This intermediate can be hydrolyzed in aqueous acid, and will cyclize to give an isoflavone (**41**) if a hydroxy group is present at C-2′ in the A-ring, or to give a benzoyl benzofuran (**42**) if a hydroxy group is present at C-2 in the B-ring. If the oxidation is carried out in acidic aqueous glyme, deformylation of this intermediate takes place *in situ* to furnish a substituted benzyl phenyl ketone (**43**), which undergoes further oxidation to the benzil (**44**). The isolated dimethyl acetal can be hydrolyzed, deformylated and cyclized to the corresponding phenylbenzofuran (**45**).[60] A preformed chalcone acetal (**46**) undergoes oxidation with migration of the A-ring phenyl to give 3-methoxy-1,2-diarylpropionates (**47**).

By far the most used pathway is that leading to the isoflavones, and literally scores of natural products have been prepared in this way. The yields for these cyclizations vary from 10–90%. Good solubility in methanol is the key to a successful reaction. The oxidation may be carried out with an unprotected C-2′ hydroxy with TTN/methanol, as long as the C-5′ position is unsubstituted; substrates of the latter type

i, TTN, MeOH; ii, HCl (2'-hydroxy); iii, HCl (2-hydroxy); iv, HCl, H_2O; v, heat; vi, heat (2-hydroxy); R^1 and R^2 can be alkyl, alkoxy, single or multiple

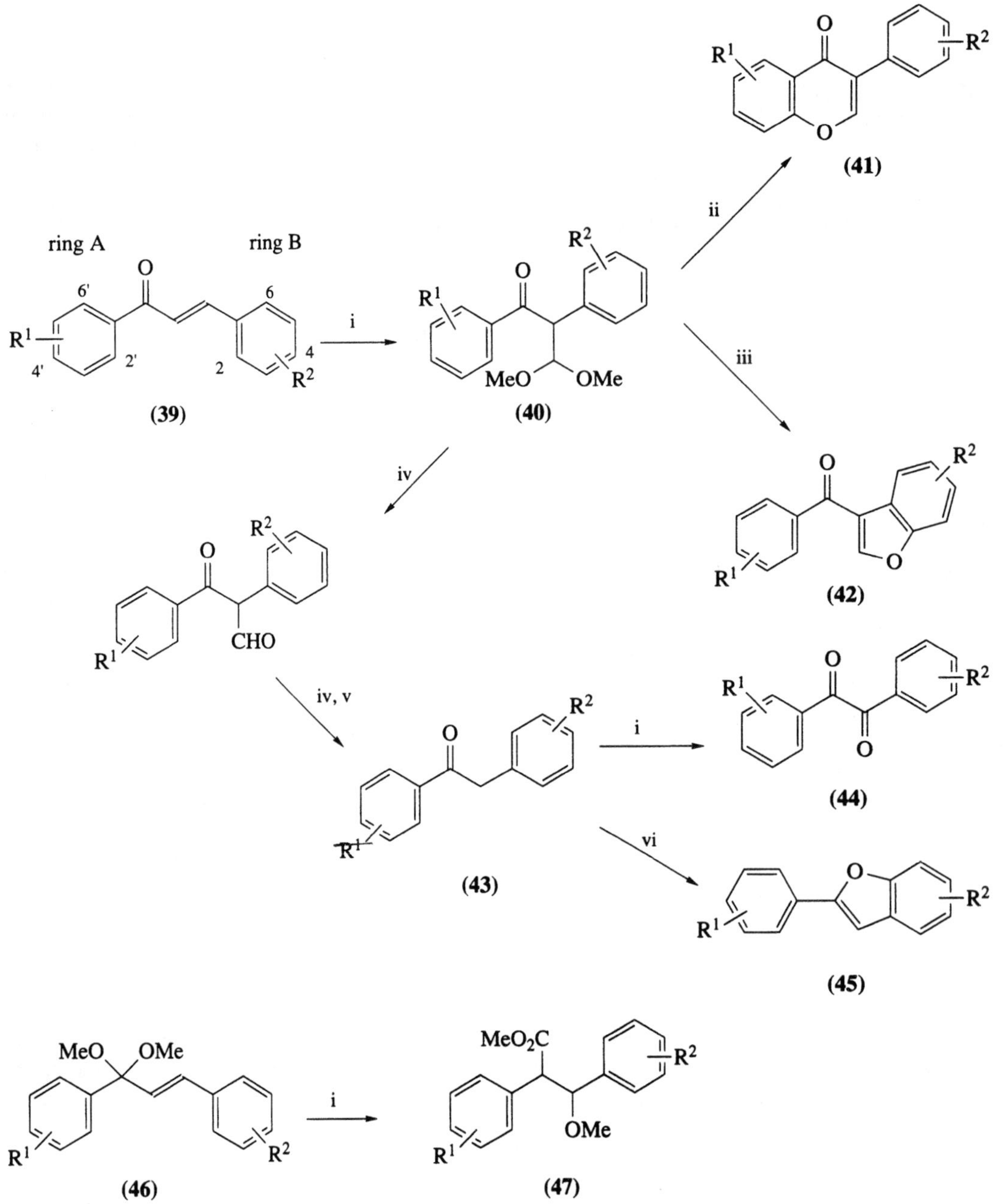

Scheme 14

will oxidize to quinone-type products. Aside from this exception, the tolerance for substitution in both rings is rather broad. R^1 and R^2 can be alkyl, alkoxy, halogen, acetoxy, methylenedioxy and pentaacetyl-β-glycosyl. Attempted oxidative rearrangement of a unprotected C-3′-β-glucosyl chalcone in a synthesis of 7,4′-di-*O*-methylpuerarin gave a low yield, but this was attributed to low solubility and separation problems; the protected glucosyl derivative gave a 90% yield of the target.[61] A dihydropyran ring may be fused to either of the aromatic rings, but some degradation does occur with a similarly fused pyran ring. A free hydroxy group in the B-ring has been used, as in the example found in equation (38), but yields are somewhat better if it is protected as the acetate or the methoxymethyl ether. Ring B can tolerate at least one *ortho* substituent, but ring A can be fully substituted with little loss in yield.

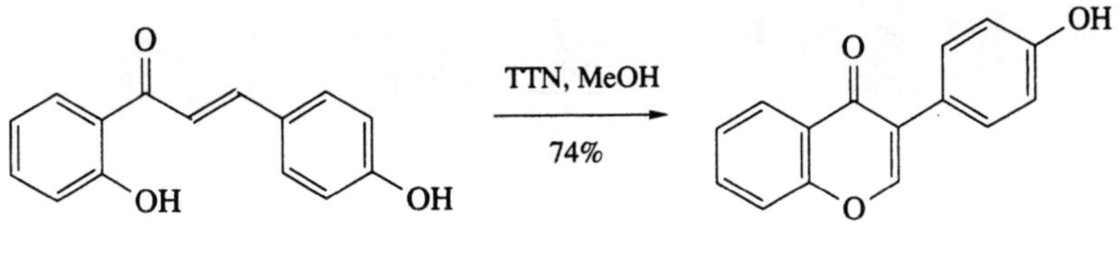

TTN, MeOH; 74% (38)

The ready availability of chalcones, from aldol condensation of acetophenones and benzaldehydes, makes this oxidative rearrangement a useful synthetic entry to isoflavone targets. The isoflavone products may be further elaborated to isoflavanones, isoflavans, pterocarpans and coumestones, broadening the scope of this method.

Cinnamyl compounds rearrange in a similar fashion under the influence of thallium(III), as shown in equation (39); this reaction was used in a synthetic approach to the polystachins.[62] Cinnamaldehydes and cinnamate esters react likewise to give the corresponding α-aryl-substituted malondialdehyde bisacetals and β,β-dimethoxypropionates, respectively.

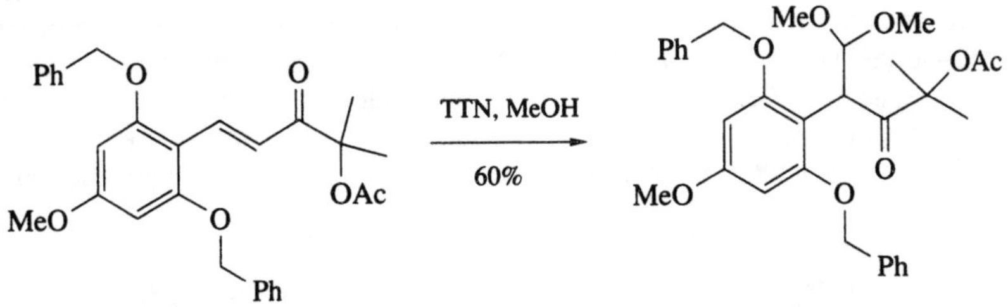

TTN, MeOH; 60% (39)

7.2.3.1.4 *Cyclic alkenes and cyclic ketones: ring expansion and ring contraction*

The oxidative rearrangement of cyclic alkenes and ketones often leads to ring expansion[63] or ring contraction reactions. The reagents generally used for this purpose are hypervalent main group oxidants such as thallium(III), lead(IV), iodine(III) and selenium(IV), although palladium(II) has been used as well.

Methylenecycloalkanes undergo ring expansion to the next higher homologous cycloalkanone, as shown in equation (40). The yields are good to excellent for four- and five-membered rings, and for six-membered rings if fused to an aromatic ring. The example given in equation (41) comes from a synthetic route to dopamine receptor stimulating compounds.[64] Simple methylenecyclohexanes give hydroxylation products, and the reaction does not appear to have been tried in larger rings. Thallium(III),[6a] lead(IV)[65] and palladium(II)[66] reagents have been used for this transformation, which is related to the pinacol rearrangement and the Demjanov rearrangement (Volume 3, Chapters 3.2 and 3.3 respectively).

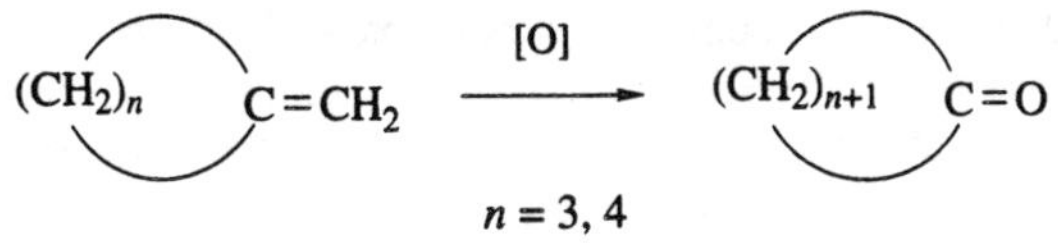

$$(CH_2)_n\ C{=}CH_2 \xrightarrow[n\,=\,3,\,4]{[O]} (CH_2)_{n+1}\ C{=}O \quad (40)$$

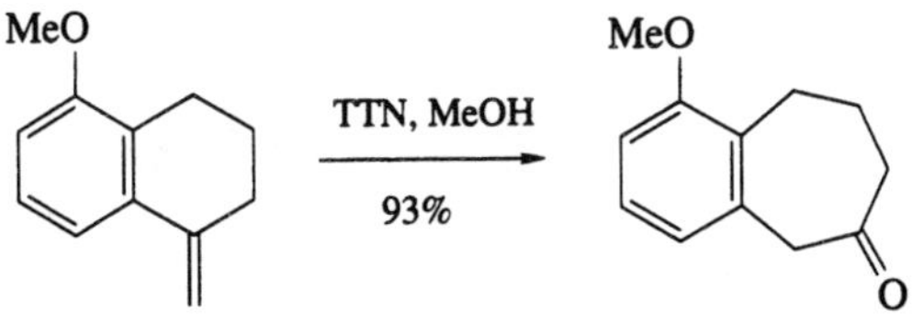

TTN, MeOH; 93% (41)

Cycloalkenes give ring contraction products, as shown in equation (42). This reaction is related to the Favorskii rearrangement and the Wolff rearrangement of ketones (Volume 3, Chapters 3.7 and 3.9, respectively). Moderate to good yields are obtained from four- to seven-membered ring cycloalkene substrates, although cyclopentenes give lower yields in favor of hydroxylation. Dihydropyrans yield the corresponding tetrahydrofuranyl aldehydes. This type of reaction was used in the stereospecific preparation of a key prostaglandin intermediate, as shown in equation (43).[67] Thallium(III)[6a] and lead(IV)/$BF_3{\cdot}Et_2O$[68] are the reagents of choice for this transformation.

Cycloalkanones of ring size from four to six oxidize with ring contraction to give the cycloalkanecarboxylic acid of the next smaller ring size with thallium(III), as shown in equation (44). This reaction

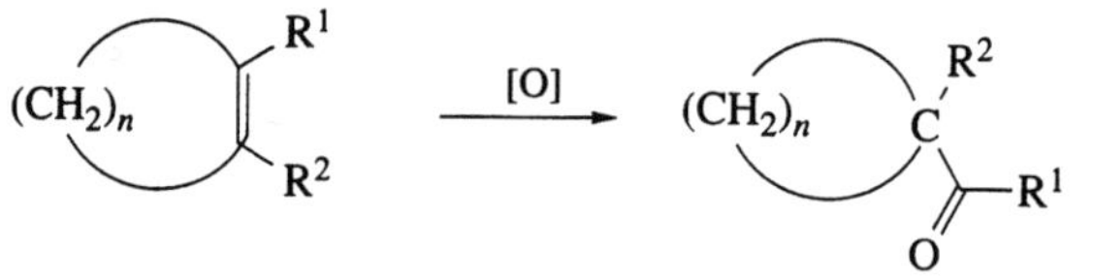

(42)

R^1 = H, alkyl, amino; R^2 = H, alkyl; n = 2–5

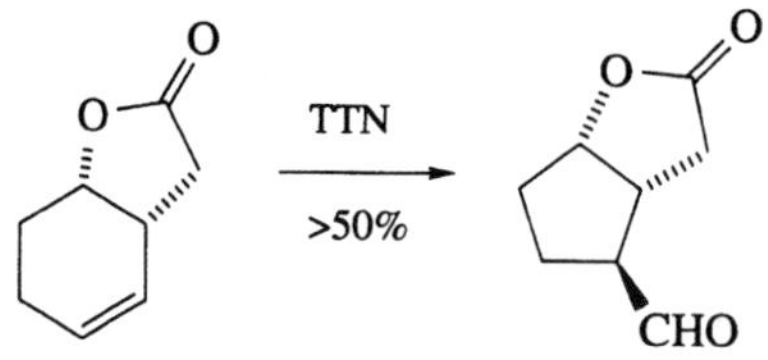

(43)

goes through the enol form, and requires acid, since in base cycloalkanones undergo α-hydroxylation. Cyclohexenones are converted in moderate to good yield to the cyclopentene-3-carboxylic esters by TTN/methanol.[69] Good yields of ring contraction products are obtained from 3-keto steroids, as shown in equation (45),[6a] but ketones at other positions are much less selective in this reaction. Selenium dioxide has been used in this reaction with five-, six-, seven- and twelve-membered ring ketones.[70] This reagent does not tolerate α-branching in the substrate, which leads to Baeyer–Villiger-type reaction.

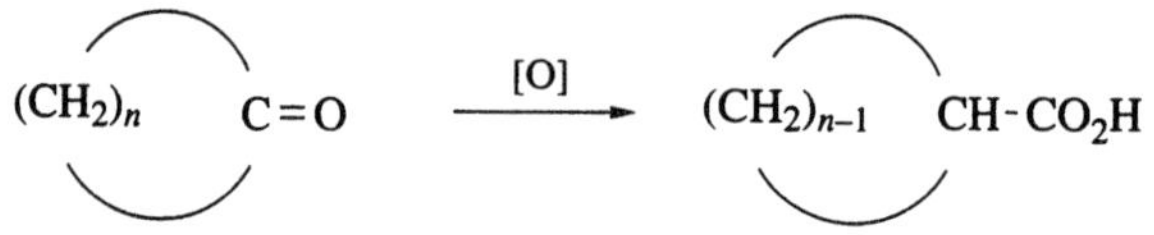

$$(CH_2)_n\ C{=}O \xrightarrow{[O]} (CH_2)_{n-1}\ CH{-}CO_2H \quad (44)$$

n = 3–5

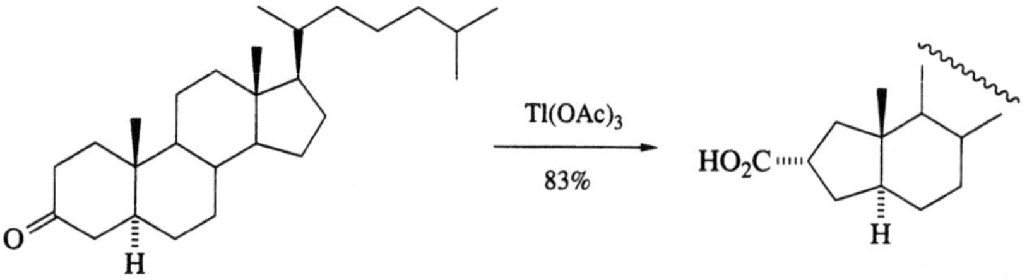

(45)

With some polycyclic substrates, a tandem ring expansion and ring contraction can take place under conditions of oxidative rearrangement. The 11-oxolanostanyl acetate (**48**; equation 46) undergoes such a reaction, in which ring C is contracted and ring D is expanded and aromatized.[71] The yield is poor though, and such a transformation would seem to have limited synthetic potential.

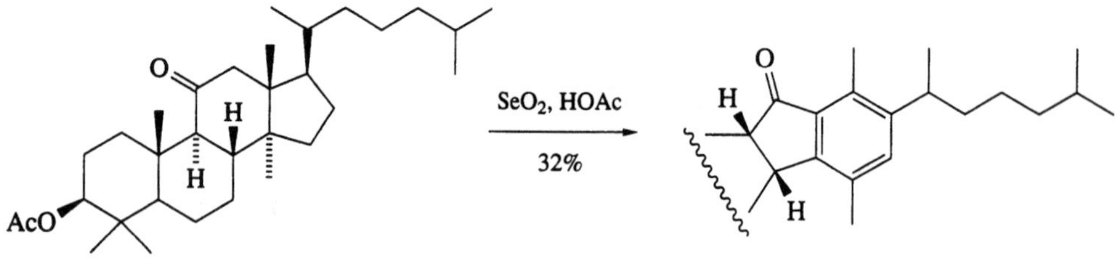

(**48**)

(46)

7.2.3.2 Dienes

Oxidative rearrangements of dienes are related to the dienone/phenol rearrangement, which is discussed in Volume 3, Chapter 3.5. The examples discussed here are limited to cyclohexadienes, and the driving force for the rearrangement is aromatization.

A novel route to the ring B aromatic anthrasteroids (**49**) from 5,7-dienes (**50**) proceeds in two steps and uses 4-phenyl-1,2,4-triazoline-3,5-dione (PTAD) as the oxidant, as is shown in equation (47).[72] Addition of PTAD to the steroidal 5,7-diene gives an adduct which, when treated with boron trifluoride etherate, rearranges to the anthrasteroid in generally greater than 90% yield. This reaction presumably proceeds through a spirocyclohexa-1,4-diene.

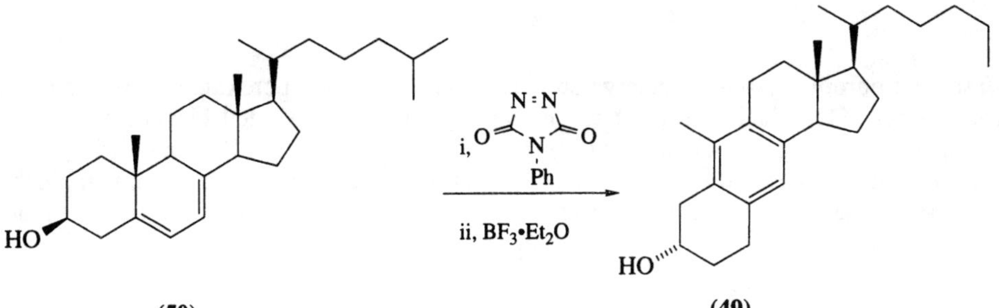

i, PTAD; ii, $BF_3{\cdot}Et_2O$ (47)

(50) **(49)**

Scheme 15 depicts a high yield, general method for specific *ortho* alkylation of polycyclic aromatic hydrocarbons.[73] In this example, biphenyl is subjected to reductive methylation followed by oxidative rearrangement with trityl tetrafluoroborate to give 2-methylbiphenyl. In unsymmetrical substrates the regioselectivity is poor; phenanthrene gives a 3:2 mixture of 4-methyl- and 1-methyl-phenanthrene.

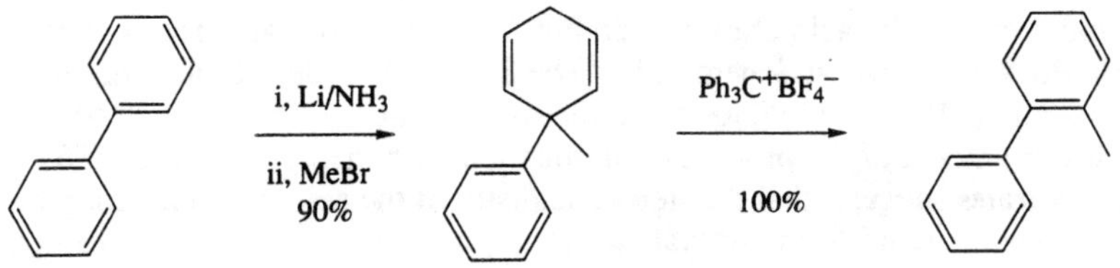

i, Li/NH_3; ii, MeBr; 90%; $Ph_3C^+BF_4^-$; 100%

Scheme 15

7.2.3.3 Alkynes

Disubstituted alkynes will undergo oxidation with a concomitant 1,2-alkyl shift under the right conditions, to yield α-branched carboxylic acid derivatives. A variety of oxidants will effect this transformation, including nitrous oxide,[74] peracetic acid,[75] thallium trinitrate (TTN)[6a] and [hydroxy (tosyloxy)iodo]benzene (HTIB).[76] Yields are moderate to good, as shown in equation (48) for the use of HTIB, where R^1 is alkyl and R^2 may be alkyl or aromatic. The TTN procedure is limited to arylalkylalkynes, as diarylalkynes will oxidize to α-diones and dialkylalkynes yield α-methoxy ketones. The TTN and HTIB reactions proceed through a solvometallation intermediate. The peroxy acid and N_2O reactions are believed to proceed through cycloaddition and rearrangement to a ketene; diphenylacetylene is converted to methyl diphenylacetate by N_2O in methanol.

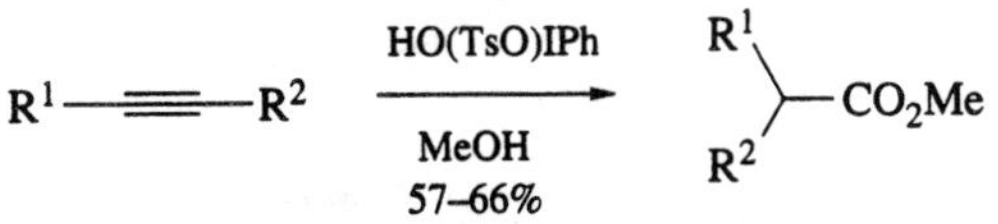

$$R^1{-}{\equiv}{-}R^2 \xrightarrow[\text{MeOH, 57–66\%}]{\text{HO(TsO)IPh}} R^1R^2CH{-}CO_2Me \quad (48)$$

7.2.3.4 Cyclopropanes and Cyclobutanes

As was discussed in Section 7.2.2.3, cyclopropanes[3] and cyclobutanes[4] form a special group, with behavior distinct in many ways from that of other cycloalkanes. Several examples of oxidative skeletal rearrangements of these strained ring compounds are presented here.

Methylenecyclopropanes undergo oxidative ring expansion in a two-step sequence; peroxy acid oxidation to an oxaspiropentane followed by lithium iodide induced rearrangement yields a cyclobutanone in moderate yield, as illustrated in equation (49).[77] Cyclobutanone is a minor product from the reaction of methylenecyclopropane with thallium trinitrate, in contrast to the analogous reaction of the larger methylenecycloalkanes.

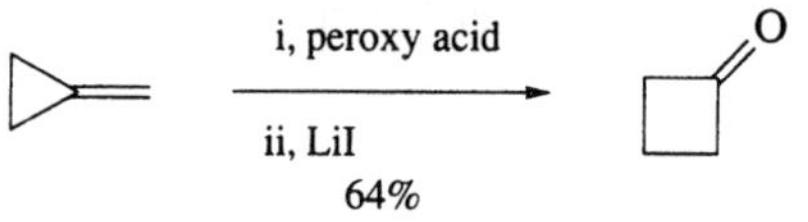

i, peroxy acid; ii, LiI; 64% (49)

Spiro-fused cyclopropyl carbinols undergo solvolysis with hydrogen peroxide to give the corresponding hydroperoxides (**51**), which rearrange to the two carbon ring-expanded bicyclic hydroperoxy hemiacetals (**52**) in good yield, as in equation (50).[78] Yields range from 72–91% for a variety of ring size substrates, and the rearrangement is stereospecific in that the stereochemistry of the initial alcohol is reflected in the stereochemistry of the bridgehead carbon in those rings large enough to accommodate this feature.

$(CH_2)_n$ OOH → $(CH_2)_n$ O OOH (50)

(51) **(52)**

Ring D norsteroidal carboxylic acid chlorides do not follow the normal carboxy inversion reaction on treatment with MCPBA, as shown in Scheme 16.[79] The β-acid chloride (**53**) undergoes rearrangement to the allylic cyclopropane (**54**) in good yield, while the α-acid chloride (**55**) gives mostly the intended alcohol (**57**) and a lesser amount of the product of elimination with methyl migration (**56**). Conformational analysis of these substrates suggests that the stereochemistry of the acid chloride group guides the course of rearrangement, since the bond to the migrating group must be suitably disposed to participate in the decarboxylation.

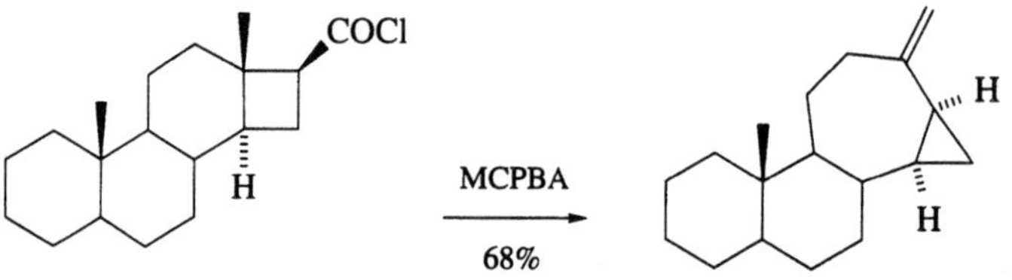

COCl, H → MCPBA, 68% → H, H

(53) **(54)**

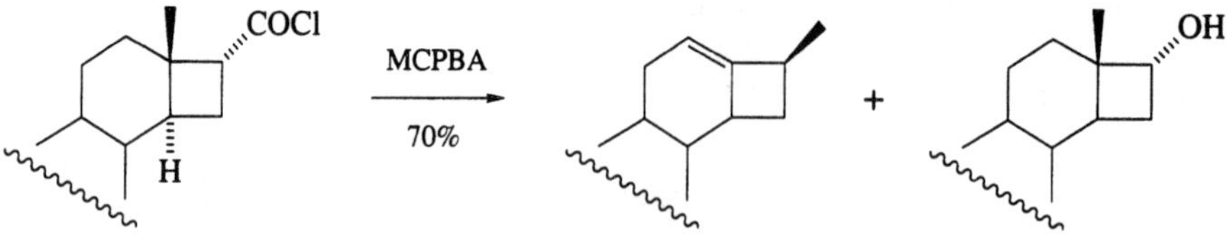

COCl, H → MCPBA, 70% → + OH

(55) **(56)** 37% **(57)** 63%

Scheme 16

Bicyclo[2.2.0]hexan-2-ols oxidize with rearrangement to the isomeric bicyclo[2.1.1]hexan-2-ones. This takes place under Oppenauer oxidation conditions,[80] as well as with chromic acid,[81] and is illustrated for photolevopimarate and chromic acid in equation (51). The yield for this transformation is excellent, although the scope and synthetic potential are probably quite limited. The reaction is highly dependent on the nature of the oxidant, as the chromate/pyridine reagent gave only 15% of the product after several days, and most of the starting alcohol was recovered.

The oxidation of the cyclobutylcarbinol in equation (52) with buffered PCC proceeds with partial rearrangement; a 1:2 ratio of the expected aldehyde (**58**) to the ring-expanded cyclic enol ether (**59**) is ob-

H_2CrO_4, 97% (51)

tained.[82] This latter product is suggested to arise through a 1,3-rearrangement of the first-formed aldehyde, driven by relief of strain; the rearrangement of (**58**) to (**59**) goes to completion on standing.

PCC (52)

(**58**) 33% (**59**) 67%

7.2.3.5 Miscellaneous Skeletal Rearrangements

Into this group fall the named oxidation rearrangement reactions which proceed with carbon–carbon bond cleavage and 1,2-transfer of an alkyl group to a heteroatom, such as the Baeyer–Villiger reaction (discussed in Chapter 5.1, this volume) and the Beckmann reaction (found in Chapter 5.2, this volume) of ketones, as well as the Hofmann reaction/Schmidt reaction/Curtius rearrangement of carboxylic acid derivatives. The two examples discussed here involve related reactions of alcohols.

The oxidative cleavage of the alcohol (**60**; equation 53) by mercury(II) oxide and iodine leads to the iodoacetal (**61**) in good yield.[83] If cholesterol (**11**) is treated with lead tetraacetate and iodine under irradiation, the lactol acetate (**62**) is obtained in moderate yield, as seen in equation (54).[84] These reactions are both believed to go through the hypoiodite, which cleaves heterolytically to an oxygen radical. This intermediate fragments to an aldehyde and an allylic radical, and it is at this point that the mechanisms seem to diverge. In the mercury(II) cyclization the aldehyde adds IO·, and the resulting oxygen radical adds to the terminus of the allyl radical. The second oxygen of the acetal adds to the remaining alkene, and the penultimate intermediate carbon radical is trapped by iodine to give the observed product. In the lead(IV) cyclization the intermediate allyl radical is believed to add to the oxygen of the aldehyde to give an oxepanyl radical, which oxidizes to the lactol acetate. The scope of these reactions seems limited, since other similar substrates give poor selectivity and low yields in these reactions.

$HgO–I_2$, Δ, 66% (53)

(**60**) (**61**)

LTA, I_2, $h\nu$, 47% (54)

(**11**) (**62**)

Sodium periodate is known to oxidize 2-alkylphenols to the corresponding 2-hydroxycyclohexadienones. Phenolic benzhydrol-type compounds (**63**) follow a rearrangement pathway under these conditions, and the results are shown in Scheme 17.[85] The benzylic hydroxy group participates in the periodate oxidation to give an intermediate spiroepoxycyclohexadienone (**64**), which suffers intramolecular attack with carbon–carbon bond cleavage by the ketone carbonyl to yield the benzaldehyde acetal (**65**). Yields for this reaction are in the range of 40–60%.

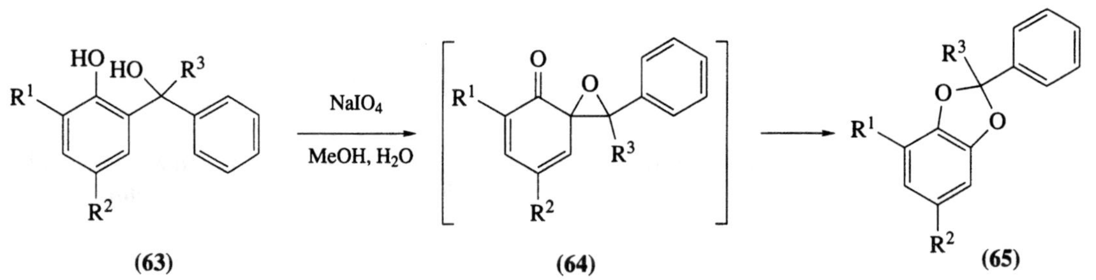

HO HO R³ R¹ R² (**63**) $NaIO_4$ MeOH, H_2O O R¹ O R³ R² (**64**) R³ O R¹ O R² (**65**)

Scheme 17

7.2.4 REFERENCES

1. A. H. Haines, 'Methods for the Oxidation of Organic Compounds: Alkanes, Alkenes, Alkynes and Arenes', Academic Press, London, 1985.
2. (a) W. J. Mijs and C. R. H. I. de Jonge (eds.), 'Organic Synthesis by Oxidation with Metal Compounds', Plenum Press, New York, 1986; (b) F. Freeman, in 'Organic Synthesis by Oxidation with Metal Compounds', ed. W. J. Mijs and C. R. H. I. de Jonge, Plenum Press, New York, 1986, p. 41; (c) M. L. Mihailović, Ž. Čeković and L. Lorenc, in 'Organic Synthesis by Oxidation with Metal Compounds', ed. W. J. Mijs and C. R. H. I. de Jonge, Plenum Press, New York, 1986, p. 741.
3. For a review on cyclopropanes in synthesis see: H. N. C. Wong, M.-Y. Hon, C.-W. Tse, Y.-C. Yip, J. Tanko and T. Hudlicky, *Chem. Rev.*, 1989, **89**, 165.
4. For a review on cyclobutanes in synthesis see: H. N. C. Wong, K.-L. Lau and K.-F. Tam, *Top. Curr. Chem.*, 1986, **133**, 83.
5. N. Miyaura and J. K. Kochi, *J. Am. Chem. Soc.*, 1983, **105**, 2368.
6. (a) A. McKillop and E. C. Taylor, in 'Comprehensive Organometallic Chemistry', ed. G. Wilkinson, F. G. A. Stone and E. W. Abel, Pergamon Press, Oxford, 1982, vol. 7, p. 465; (b) P. D. Magnus, T. Sarkar and S. W. Djuric, in 'Comprehensive Organometallic Chemistry', ed. G. Wilkinson, F. G. A. Stone and E. W. Abel, Pergamon Press, Oxford, 1982, vol. 7, p. 515.
7. While trial and error still furnish the majority of advances in synthetic chemistry, efforts are underway to systematize predictive rationale, even for such a difficult area as oxidation reactions; see: G. D. Paderes and W. L. Jorgensen, *J. Org. Chem.*, 1989, **54**, 2058.
8. G. Cainelli and G. Cardillo, 'Chromium Oxidations in Organic Chemistry', Springer Verlag, New York, 1984.
9. G. M. Rubottom, in 'Oxidation in Organic Chemistry', ed. W. S. Trahanovsky, Academic Press, New York, 1978, vol. 5, part D, p. 1.
10. H. H. Wasserman and J. L. Ives, *Tetrahedron*, 1981, **37**, 1825; L. M. Stephenson, M. J. Grdina and M. Orfanopoulos, *Acc. Chem. Res.*, 1980, **13**, 419; A. A. Frimer, *Chem. Rev.*, 1979, **79**, 359; K. Gollnick and H. J. Kuhn, in 'Singlet Oxygen', ed. H. H. Wasserman, Academic Press, New York, 1979, p. 287; K. Gollnick and H. J. Kuhn, *J. Org. Chem.*, 1979, **40**, 287.
11. R. J. Steltenkamp and W. E. Truice, *Mech. Mol. Migr.*, 1969, **2**, 65.
12. E. Colvin, 'Silicon in Organic Synthesis', Butterworth, London, 1981.
13. K. Tamao and K. Maeda, *Tetrahedron Lett.*, 1986, **27**, 65.
14. For a review on the chemistry of silyl enol ethers see: J. K. Rasmussen, *Synthesis*, 1977, 91.
15. M. J. Taschner, University of Akron, private communication.
16. W. Adam and A. Griesbeck, *Synthesis*, 1986, 1050.
17. K. H. Schulte-Elte and V. Rautenstrauch, *J. Am. Chem. Soc.*, 1980, **102**, 1738, and references cited therein.
18. E. D. Mihelich and D. J. Eickhoff, *J. Org. Chem.*, 1983, **48**, 4135.
19. L. M. Stephenson and M. B. Zielinski, *J. Am. Chem. Soc.*, 1982, **104**, 5819.
20. H. J. Reich, in 'Oxidation in Organic Chemistry', ed. W. S. Trahanovsky, Academic Press, New York, 1978, vol. 5, part C, p. 1; D. L. J. Clive, *Tetrahedron*, 1978, **34**, 1049; D. Liotta, *Acc. Chem. Res.*, 1984, **17**, 28.
21. K. B. Sharpless and R. F. Lauer, *J. Am. Chem. Soc.*, 1973, **95**, 2697.
22. S. Torii, K. Uneyama, M. Ono and T. Bannou, *J. Am. Chem. Soc.*, 1981, **103**, 4606.
23. J. Iriarte, J. N. Shoolery and C. Djerassi, *J. Org. Chem.*, 1962, **27**, 1139.
24. L. F. Fieser, *Org. Synth., Coll. Vol.*, 1963, **4**, 189; K.-E. Stensiö, *Acta Chem. Scand.*, 1971, **25**, 1125.
25. A. G. González, J. F. Ciccio, A. P. Rivera and J. D. Martin, *J. Org. Chem.*, 1985, **50**, 1261.
26. G. D. S. Ananda, A. M. Z. Slawin, R. J. Stoodley and D. J. Williams, *J. Chem. Soc., Chem. Commun.*, 1986, 165.
27. R. N. Warrener, T. S. Lee, R. A. Russell and M. N. Paddon-Row, *Aust. J. Chem.*, 1978, **31**, 1113, and references cited therein; P. Sundararaman and W. Herz, *J. Org. Chem.*, 1977, **42**, 813.
28. G. Majetich, University of Georgia, private communication.
29. C. Iwata, Y. Takemoto, M. Doi and T. Imanishi, *J. Org. Chem.*, 1988, **53**, 1623.
30. E. J. Corey, P. D. S. Jardine and J. C. Rohloff, *J. Am. Chem. Soc.*, 1988, **110**, 3672.

31. J. Drew, G. Gowda, P. Morand, P. Proulx, A. G. Szabo and D. Williamson, *J. Chem. Soc., Chem. Commun.*, 1985, 901.
32. D. Liotta, D. Braun, W. Hoekstra and R. Monahan, III, *Tetrahedron Lett.*, 1987, **28**, 1069.
33. G. Majetich, S. Condon, K. Hull and S. Ahmad, *Tetrahedron Lett.*, 1989, **30**, 1033.
34. P. Baekström, S. Okecha, N. DeSilva, D. Wijekoon and T. Norin, *Acta Chem. Scand., Ser. B*, 1982, **36**, 31.
35. J. Banerji, A. Chatterjee, N. Ghoshal, A. K. Das, S. Surkar, S. Bhattacharya and J. N. Shoolery, *J. Indian Chem. Soc.*, 1982, **59**, 145; J. Banerji, A. K. Das and B. Das, *Chem. Ind. (London)*, 1987, 395.
36. T. Itoh, K. Jitsukawa, K. Kaneda and S. Teranishi, *J. Am. Chem. Soc.*, 1979, **101**, 159, and previous papers in this series.
37. M. F. Schlecht and H.-J. Kim, unpublished results.
38. W. H. Stass and L. A. Spurlock, *J. Org. Chem.*, 1974, **39**, 3822.
39. E. J. Corey and L. S. Melvin, Jr., *Tetrahedron Lett.*, 1975, 929.
40. T. K. Hall and P. R. Storey, *J. Am. Chem. Soc.*, 1967, **89**, 6759.
41. E. J. Corey, Z. Arnold and J. Hutton, *Tetrahedron Lett.*, 1970, 307.
42. G. M. Rubottom, E. C. Beedle, C.-W. Kim and R. C. Mott, *J. Am. Chem. Soc.*, 1985, **107**, 4230.
43. M. S. Baird, *Top. Curr. Chem.*, 1988, **144**, 193.
44. J. Salaun, B. Gardnier and J. M. Conia, *Tetrahedron*, 1974, **30**, 1423.
45. W. Holweger and M. Hannack, *Chem. Ber.*, 1984, **117**, 3004.
46. C. L. Willis, *Tetrahedron Lett.*, 1987, **28**, 2175.
47. H. J. Reich, S. K. Shah, P. M. Gold and R. E. Olson, *J. Am. Chem. Soc.*, 1981, **103**, 3112.
48. R. F. C. Brown, K. J. Coulston, F. W. Eastward and M. P. Hill, *Aust. J. Chem.*, 1988, **41**, 215.
49. R. M. Moriarty and O. Prakash, *Acc. Chem. Res.*, 1986, **19**, 244.
50. (a) A. Lethbridge, R. O. C. Norman and C. B. Thomas, *J. Chem. Soc., Perkin Trans. 1*, 1973, 35; (b) K. J. Divakar and A. S. Rao, *Indian J. Chem., Sect. B*, 1976, **14**, 704.
51. A. McKillop, J. D. Hunt, F. Kienzle, E. Bigham and E. C. Taylor, *J. Am. Chem. Soc.*, 1973, **95**, 3635.
52. (a) R. M. Moriarty, J. S. Khosrowshahi and O. Prakash, *Tetrahedron Lett.*, 1985, **26**, 2961; (b) A. Citterio, M. Gandolfi, C. Giordano and G. Castaldi, *Tetrahedron Lett.*, 1985, **26**, 1665.
53. P. R. Stapp (Phillips Petroleum Co.), *US Pat.* 4 220 604 (1980) (*Chem. Abstr.*, 1980, **93**, 220 381).
54. Y. Tamada, H. Sanjoh and K. Iguchi, *Tetrahedron Lett.*, 1979, 423.
55. For a review on this and other methods to carry out the transformation see: C. Giordano, G. Castaldi and F. Uggeri, *Angew. Chem., Int. Ed. Engl.*, 1984, **23**, 413.
56. (a) T. Yamauchi, K. Nakao and K. Fujii, *J. Chem. Soc., Perkin Trans. 1*, 1987, 1255, and references cited therein; (b) T. Yamauchi, K. Nakao and K. Fujii, *J. Chem. Soc., Perkin Trans. 1*, 1987, 1433, and references cited therein; (c) Y. Tamura, T. Yakura, Y. Shirouchi and J.-I. Haruta, *Chem. Pharm. Bull.*, 1985, **33**, 1097.
57. For a review on this named reaction see: E. V. Brown, *Synthesis*, 1975, 358.
58. N. Sonoda and S. Tsutsumi, *Bull. Chem. Soc. Jpn.*, 1959, **32**, 505.
59. E. C. Taylor, R. A. Conley, D. K. Johnson, A. McKillop and M. E. Ford, *J. Org. Chem.*, 1980, **45**, 3433, and references cited therein.
60. B. Vu, G. Mezey-Vándor and M. Nógrádi, *Liebigs Ann. Chem.*, 1984, 734.
61. R. A. Eade, F. J. McDonald and H.-P. Pham, *Aust. J. Chem.*, 1978, **31**, 2699.
62. S. Antus, F. Boross, M. Katjar-Peredy and M. Nógrádi, *Liebigs Ann. Chem.*, 1984, 1068.
63. H. Heimgartner, *Chimia*, 1980, **34**, 333 (*Chem. Abstr.*, 1981, **94**, 3285); G. R. Krow, *Tetrahedron*, 1987, **43**, 3.
64. U. Hacksell, L.-E. Arvidsson, U. Svensson, J. L. G. Nilsson, H. Wikström, P. Lindberg, D. Sanchez, S. Hjorth, A. Carlsson and L. Paalzow, *J. Med. Chem.*, 1981, **24**, 429.
65. G. Lenz, *J. Org. Chem.*, 1988, **53**, 5791.
66. P. Boontanonda and R. Grigg, *J. Chem. Soc., Chem. Commun.*, 1977, 583.
67. E. J. Corey and T. Ravindranathan, *Tetrahedron Lett.*, 1971, 4753.
68. Ž. Čeković, J. Bosnjak and M. Cvetković, *Tetrahedron Lett.*, 1980, **21**, 2675.
69. E. Mincione, P. Bovicelli, J. B. Gil and M. L. Forcellese, *Gazz. Chim. Ital.*, 1985, **115**, 37 (*Chem. Abstr.*, 1985, **103**, 22 833).
70. W. Dittmann, W. Kirchhof and W. Stumpf, *Justus Liebigs Ann. Chem.*, 1965, **681**, 30, and references cited therein.
71. W. Lawrie, W. Hamilton, J. McLean and J. Meney, *J. Chem. Soc., Perkin Trans. 1*, 1978, 471.
72. N. Bosworth, A. Emke, J. M. Midgley, C. J. Moore, W. B. Whalley, G. Ferguson and W. C. Marsh, *J. Chem. Soc., Perkin Trans. 1*, 1977, 805, and previous papers by this group.
73. D. F. Lindow and R. G. Harvey, *J. Am. Chem. Soc.*, 1971, **93**, 3786.
74. G. D. Buckley and W. J. Levy, *J. Chem. Soc.*, 1951, 3016.
75. V. Franzen, *Chem. Ber.*, 1954, **87**, 1219.
76. R. M. Moriarty, R. K. Vaid, M. P. Duncan and B. K. Vaid, *Tetrahedron Lett.*, 1987, **28**, 2845.
77. J. Salaun, B. Gardnier and J. M. Conia, *Tetrahedron*, 1974, **30**, 1413.
78. T. S. Lillie and R. C. Ronald, *J. Org. Chem.*, 1985, **50**, 5084.
79. H. Suginome and T. Uchida, *J. Chem. Soc., Perkin Trans. 1*, 1980, 943; H. Suginome, Y. Ohue and K. Orito, *J. Chem. Soc., Perkin Trans. 1*, 1987, 1247.
80. R. N. McDonald and C. E. Reineke, *J. Org. Chem.*, 1967, **32**, 1888.
81. W. Herz, M. G. Nair and D. Prakash, *J. Org. Chem.*, 1975, **40**, 1017.
82. W. Kirmse and U. Mrotzeck, *Chem. Ber.*, 1988, **121**, 1013.
83. H. Suginome and N. Maeda, *Bull. Chem. Soc. Jpn.*, 1980, **53**, 2621.
84. H. Suginome, H. Washiyama and S. Yamada, *Bull. Chem. Soc. Jpn.*, 1987, **60**, 1071.
85. H.-D. Becker and T. Bremholt, *Tetrahedron Lett.*, 1973, 197.

7.3
Solid-supported Oxidants

PIERRE LASZLO
Ecole Polytechnique, Palaiseau, France and Université de Liège, Belgium

7.3.1 INTRODUCTION

The benefits of using supported reagents[1] for organic synthesis are considerable, especially as they offer remarkable ease of handling and use; often one can simply weigh the amount of reagent to be used. At the end of the reaction a filtration suffices to remove the contaminating by-products. Solvent evaporation from the filtrate is often sufficient to provide the product in pure form. Likewise, reagent recovery or regeneration can also be very easy. Another advantage is the reduction in product contamination assured by having the reagent fully bound to a solid support. This is very important for oxidation reactions so that overoxidation reactions can be minimized. Solid-supported oxidants are relatively safe to handle owing to full chemisorption of the toxic chemicals. Two examples of toxic oxidants whose contamination is considerably reduced by adsorption to solid supports are thallium(III) salts and chromium salts[2] in all the various oxidation states. Such supported reagents also reduce environmental problems upon work-up. Many inorganic species are powerful oxidants of organic matter and can cause explosions. However they can be tamed by prior adsorption onto the solid supports. These oxidants also enjoy good thermal and mechanical stabilities, allowing higher stirring rates if necessary. High reagent activity and, more importantly, high reagent selectivity are frequent features of supported reagents. The high activity stems from physical factors, such as vastly enhanced collision rates between reactants, due to either the reduced dimensionality on the solid reaction sites or the boost in their local concentration due to trapping in the interstitical spaces of microporous solids.[3] Chemisorption generally activates adsorbates, with respect to bond breaking. The high selectivity is easily understood in terms of shape selectivity (reactant, transition state or product) when, for example, zeolites are involved.

Other more specific advantages of supported reagents exist for oxidants such as chromium(III) anhydride complexes where there are many problems associated with solubility, the ligand and stoichiometry.[4] For example in procedures using pyridine or polar aprotic solvents, the ligand doubles up as solvent, causing complication in work-up with soluble metal species. However, these are avoided by the supported oxidants.

By measurement of aqueous redox potentials an overall impression of the metallic species most often used in the empirical Edinsonian or the rational design of solid-supported oxidants can be obtained (Table 1).

Table 1 Electroreduction Potentials (V)

Metallic species	*Electroreductive potentials (V)*
$MnO_2 + 4\,H^+ + 2\,e^- = Mn^{2+}$	1.23
$IO_3^- + 6\,H^+ + 5\,e^- = I_2$	1.195
$OsO_4 + 8\,H^+ + 8\,e^- = Os$	0.85
$Ag^+ + e^- = Ag$	0.799
$Fe^{3+} + e^- = Fe^{2+}$	0.771
$Cu^+ + e^- = Cu$	0.521
$Cu^{2+} + e^- = Cu$	0.337
$CrO_4^{2-} + 3\,e^- = Cr(OH)_3 + 5\,OH^-$	0.13
$MnO_4^- + 4H^+ + 3\,e^- = MnO_2$	1.695
$Ce^{4+} + e^- = Ce^{3+}$	1.61
$MnO_4^- + 8\,H^+ + 5\,e^- = Mn^{2+}$	1.51
$Mn^{3+} = e^- = Mn^{2+}$	1.51
$Cr_2O_7^{2-} + 14\,H^+ + 6\,e^- = 2Cr^{3+}$	1.33
$Tl^{3+} + 2\,e^- = Tl^+$	1.25

Typically the supports most widely used are alumina, silica and aluminosilicates (clays and zeolites). These inorganic solids all contain surface hydroxy groups and it can be valuable to examine activation of these surfaces by dehydration.

Amorphous SiO_2 contains tetrahedral silicate units whereby each oxygen atom bridges two silicon atoms, together with silanol (Si—OH) groups covering the surface. The vicinal silanol groups may be characterized by lower frequency IR absorptions as compared to isolated silanol groups, owing to their hydrogen-bonding patterns (3550 *versus* 3750 cm^{-1}). Upon dehydration, such vicinal silanol groups give surface Si—O—Si linkages.

Alumina, on the other hand, exists in various forms, of which α-Al_2O_3 is the stable and crystalline form. So-called transition aluminas, η- and γ-Al_2O_3, are defective, metastable solids arising from the heating of aluminum hydroxide gels. Both have spinel structures, with a disordered aluminum lattice interlocked with a cubic close-packed oxygen sublattice. Surface layers are occupied by Al—OH alanol groups. Both the IR[5] spectra and modeling of the alumina surfaces[6] suggest five types of environments for these surface hydroxy groups. Dehydrative activation is performed by heating above 200 °C. The attendant dehydroxylation creates coordinatively unsaturated oxide O^{2-} ions and an adjacent surface anion vacancy. This void in the upper layer exposes, if one considers the case of a close-packed (III) surface plane, either two five-coordinate Al^{3+} ions or one three-coordinate Al^{3+} ion.[7] In a similar manner, silicate and aluminosilicate minerals, such as clays, pillared clays and zeolites, bear surface hydroxy groups whose dehydrative thermal activation creates Lewis and Brønsted acidic and basic sites on the surface. Recent attention has focused[8] on these resulting O^- centers. Such O^- groups with an unpaired electron can dimerize to form peroxo links (Si—O—O—Si), which are relevant to oxidation by these surfaces. Dissociation of peroxy linkages will regenerate two O^- states. Physically, such O^- states are defect electrons or positive holes. They are paramagnetic and they are delocalized over the surface oxygens. For instance, the mineral obsidian suddenly develops positive charge carriers when heated above 450 °C. These O^- states, resulting from dehydrative thermal activation, are also responsible for the yellowing of magnesia, MgO, upon heating. These O^- states are powerful oxidation centers and, accordingly, such minerals serve as supports for solid oxidants.

Historically, the first supported oxidizing reagent, reported by Fétizon and Golfier, was silver carbonate on celite (another diatomaceous earth).[9] This was obtained by precipitation of the reagent onto its support. Ag_2CO_3 on celite smoothly oxidizes primary and secondary alcohols, α,ω-diols, hydroquinones and amines. The main practical asset of the reagent is that it avoids the need to filter off finely divided silver salts after reaction.

7.3.2 ALUMINA-SUPPORTED OXIDANTS

7.3.2.1 Chloral

Chloral, when adsorbed on activated Woelm-200 neutral alumina, is a mild, chemoselective oxidant of secondary alcohols over primary alcohols.[10–12] It oxidizes vinyl alcohols to vinyl ketones without the usual polymerization or oxidation of the ethylenic double bond. Also of importance is that it effects high yield conversion of cyclobutanol into cyclobutanone, without ring cleavage. β-Hydroxy sulfides and selenides are oxidized to the corresponding β-keto sulfides and selenides. Typically, the reaction is run in carbon tetrachloride at room temperature or at reflux with yields being normally in the range of 60–75%. The one drawback of the reagent is the necessity for the vacuum drying at 400 °C of commercial alumina immediately before use. However, the advantage is its selectivity in that other functional groups, such as primary iodides, benzylic chlorides, methyl esters, ethers, lactones and nitro groups,[10] all remain intact.

7.3.2.2 Pyridinium Chlorochromate

Another supported chemoselective alcohol oxidant in common use is pyridinium chlorochromate on alumina[13] which transforms alcohols to carbonyls even in the presence of THP acetals (equation 1).[14] That such acid-sensitive groups resist the reaction conditions derives from the neutralization by the alumina support of the acidity of the PCC. Examples of alcohols which have been oxidized in good yields using this reagent include carveol, 2-ethylhexanol, menthol, tetrahydrogeraniol, citronellol, 2-methylcyclohexanol, cinnamyl alcohol, isopulegol and cholesterol. In typical experiments methylene chloride is used as solvent at room temperature.

$$R^1R^2CH{-}OH \xrightarrow[Al_2O_3]{PCC} R^1R^2C{=}O \qquad (1)$$

7.3.2.3 Periodic Acid

Periodic acid is a versatile oxidant since, depending on pH, the redox potential for the periodate–iodate couple varies from 0.7 V in aqueous basic media to 1.6 V in aqueous acidic media.[15] Based on this observation, Villemin and Ricard devised an oxidative cleavage of glycols,[16] in which *meso*-1,2-diphenyl-1,2-ethanediol was oxidized by periodic acid on alumina to benzaldehyde in 82% yield in aqueous ethanol (90% ethanol) at room temperature in 26 h. The same supported oxidant converted aromatics into quinones. In the presence of transition metal complexes (Mn^{II}), α-arylalkenes suffer oxidative cleavage to aldehydes. For example, *trans*-stilbene gives benzaldehyde at room temperature.

7.3.2.4 Potassium Dichromate and Epoxidizing Agents

Potassium dichromate is another inorganic oxidant that can be supported on alumina and used to convert alcohols into carbonyl compounds.[17] Its chief merit is its selectivity for allylic and benzylic alcohols.[18,19] For instance, 1-phenylpropane-1,3-diol is oxidized selectively to the benzylic oxidation product.

Alumina also serves as the support for reagents effecting the epoxidation of alkenes efficiently under mild conditions. This is the case with triphenylsilyl hydroperoxide; even aryl-substituted alkenes are easily oxidized.[20] The reactions are run in methylene chloride at room temperature and are stereospecific. In a similar manner, alkenes *gem*-disubstituted by two electron-withdrawing groups, dispersed on alumina, are conveniently epoxidized by sodium hypochlorite.[21] Often the reaction, run in acetonitrile at room temperature, is stereospecific and yields are excellent (80–98%). Hydration of the alkene, which is one of the drawbacks of aqueous hypochlorite as an epoxidizing agent is thus avoided.

7.3.2.5 Sodium Metaperiodate

Sodium metaperiodate, $NaIO_4$, is a two-electron chemoselective oxidant and when absorbed on alumina oxidizes alcohols[16] to carbonyl compounds and sulfides to sulfoxides without overoxidation to sulfones.[22] Alkenic double bonds in the substrate remain intact during this oxidation. Typically reactions are performed in 95% ethanol at room temperature for a few hours and with good (85–90%) yields.

7.3.3 SILICA-SUPPORTED OXIDANTS

7.3.3.1 Ozone

Mazur's oxidation method consists of ozone combined with adsorption of the organic substrate on silica gel to effect the clean oxidation of tertiary carbon–hydrogen bonds, in the absence of any solvent (dry medium).[23] Adamantane, for instance, gives a better than 80% yield of 1-adamantanol (equation 2).[24,25] This hydroxylation method has been applied successfully to a number of natural products,[26–28] but does not always work as well with aliphatic substrates[24,29–31] or for the oxidation of secondary carbon–hydrogen bonds.[32] Nevertheless, it converts quantitatively 2-adamantanol to 2-adamantanone (equation 3).[24] This ozonization resembles some biological oxidation processes in its ability to oxidize methylene groups at a distance from other functional groups and is exemplified by transformations of acetates into keto acetates.[32] Dry ozonization is a choice method for oxidizing methylene groups adjacent to cyclopropane rings to carbonyl compounds.[33,34]

O_3, SiO_2; OH (2)

OH; O_3, SiO_2; O (3)

Dry ozonization of alkenes and alkynes has also been explored. Sometimes these results resemble those of the homogeneous ozonization in an aprotic solvent.[35]

Dry ozonization is also an efficient procedure for oxidation of aliphatic primary amines into nitro groups, with yields of about 70%. Arylamines are also oxidized to nitro aromatics, albeit with low yields.[36,37] Ozone on silica gel has also been shown to oxidize arenes in some cases.[38]

7.3.3.2 Sodium Methoxide

Sodium methoxide on silica gel (2 mol equiv. Na per g reagent), is an excellent reagent for effecting the Nef reaction by converting nitro compounds to aldehydes and ketones.[37] The reactions are run either at room temperature or at 80 °C, for a short reaction time. The success of these reactions depends on the neutralization of the normally weakly acidic silica gel by treatment with methoxide in methanol, followed by evaporation to dryness and activation by heating to 400 °C. The nitro substrate is then impregnated onto this methoxide-doped silica gel to effect reaction.

7.3.3.3 Iron(III) Chloride

Two slightly different reagents may be made by adsorption of iron(III) chloride onto silica gel under dehydrating conditions. If anhydrous iron(III) chloride is used, this provides a pale yellow-green powder. This reagent,[39,40] sometimes referred to as the Salaün reagent, dehydrates tertiary alcohols and deprotects THP ethers. Also in dry media Salaün reagent promotes the Wagner–Meerwein ring expansion of tertiary cyclobutanols into cyclopentenes. By similar carbocationic mechanisms, the Salaün reagent will induce the cyclization of alkenic alcohols.[39,40] If iron(III) chloride hexahydrate is used as the starting material, a dry yellow-brown powder results from its deposition onto silica gel under high vacuum (0.1 Torr; 1 Torr ≈ 33 Pa) for 3 h at 60 °C.[41] This reagent performs the dehydration of allylic and tertiary alcohols. In special cases it will dehydrate secondary alcohols in sterically hindered positions. In the steroid series, use of this reagent induces the cholestane–diacholestene and backbone rearrangements under relatively mild conditions.[42] This $FeCl_3$–SiO_2 reagent achieves the oxidative coupling of phenol ethers or the cleavage of the ether function into phenols.[43,44] The latter process is much faster if the solvent is removed.[44] Oxidative desilylation reactions and cleavage of benzyl esters are also possible, all with commendable ease of work-up.

7.3.3.4 Selenium Dioxide

Allylic methyl groups are oxidized to allylic alcohols by the combination of selenium dioxide adsorbed on SiO_2 together with *t*-butyl hydroperoxide (TBHP) in nonpolar solvents such as hexane or methylene chloride. This procedure has been applied to a number of medium-ring sesquiterpenes.[44]

7.3.3.5 Cerium(IV)

Catechols and hydroquinones can be converted (91–98%) into quinones by cerium(IV) salts coated onto silica as free-flowing yellow powder from impregnation with cerium(IV) ammonium nitrate. This reaction is usually performed in the presence of magnesium sulfate.[45] The same $(NH_4)_2Ce(NO_3)_6{\cdot}SiO_2$ reagent in the dry state effects oxidative nitrations of arenes. For example α-naphthol is converted to the *ortho* (42%) and the *para* (38%) nitro compounds, while its methyl or ethyl ethers give exclusively the *para* nitration product (equation 4).[46] In solution, the products are contaminated with the products of dinitration and of oxidation into quinones.[47,48]

$$\text{1-RO-naphthalene} \xrightarrow{(NH_4)_2Ce(NO_3)_6{\bullet}SiO_2} \text{4-NO}_2\text{-1-RO-naphthalene} \qquad (4)$$

7.3.3.6 Periodates

The periodate oxidations (see the above Section 7.3.2)[15] can also occur with dimesoperiodate, $K_4I_2O_9$, supported on silica gel. Poorer loadings of reagent on this support are observed when compared with alumina-based reagents. Nevertheless periodates supported on silica gel, in solvents such as methylene chloride or benzene, at room temperature, are good oxidants of hydroquinones into quinones, or hydrazarenes into azoarenes, and of glycols into dialdehydes.[49] X-Ray and Raman spectroscopy show that the $NaIO_4$ reagent consists of a monomolecular layer of the salt bound to the surface through the silanol groups.

$$R^1{-}S{-}R^2 \xrightarrow[SiO_2]{SO_2Cl_2} R^1{-}S({=}O){-}R^2 \qquad (5)$$

This IO_4^-–SiO_2 reagent is not very effective in the oxidation of sulfides into sulfoxides.[49] However, sulfuryl chloride adsorbed on wet silica gel is an excellent reagent for this transformation (equation 5).[50] High yields of methyl aryl, diaryl, allylic, benzylic and dialkyl sulfoxides are thus obtainable. The procedure commends itself by its simplicity and its extension to thioacetals provides a good, quantitative regeneration of the carbonyl groups by oxidative cleavage.[51] This supported sulfuryl chloride oxidation has been used as a method for partial ^{18}O enrichment of sulfoxides and of carbonyl compounds from ^{18}O-enriched water.[52]

7.3.3.7 Permanganates

The seminal observation that a number of solid supports, such as aluminosilicates (clays and zeolites) and silica gel, activate potassium permanganate[53] led to the use of this reagent for synthesis. Oxidation of an alcohol in benzene normally does not proceed, due to the insolubility of the oxidant.[54] However, using $KMnO_4$ adsorbed onto silica gel at 70 °C, the quantitative conversion of benzyl alcohol to benzaldehyde is effected without overoxidation to benzoic acid.[54] The optimized reagent was also applied successfully to the Nef reaction, converting nitro groups into carbonyls.[55] Thus 1-nitro 4-ketones are turned into 1,4-diketones in boiling benzene, in variable yields (4–55%). The poor yields are probably due to the difficulty in product recovery arising from adsorption of the 1,4-diketones to the surface covered with the silanol hydrogen-bond donors.

Potassium permanganate impregnated on silica gel is the reagent of choice for the cleavage of ethylenic double bonds.[56] The reaction requires only mild conditions, such as room temperature, for 20–30 min. The process can be applied to terminal, secondary, tertiary and also electron withdrawing substituted double bonds and provides good to excellent yields of cleavage products.

Another very active permanganate is the zinc salt, which functions as an oxidant *via* a three-electron conversion of Mn^{VII} to Mn^{IV}. The advantage of $Zn(MnO_4)_2$ is that it is a neutral oxidant. However, it strongly complexes organic substrates.[57] The oxidizing power is reported to be 13% that of TNT, *i.e.* about twice that of potassium permanganate. Use of a silica gel support permits the safe handling of this strong oxidant.[57] Halogenated solvents (methylene chloride, chloroform) are best for the oxidation reactions, which are best conducted at room temperature or at reflux. For instance, thioanisole is oxidized in 92% isolated yield into the corresponding methyl phenyl sulfone by 1.2 equiv. of $Zn(MnO_4)_2$ in methylene chloride at 20 °C for 2.5 h. Alkynes are oxidized to diketones, ethers lactones, cyclic ketones to diacids, cyclic ketals to ketones and acylated amines into acylimides.[57]

It has also been found that addition of a catalytic amount of calcium hydride and silica gel to the Sharpless reagent can greatly reduce the reaction time for asymmetric epoxidation of an allylic alcohol. The time saving is often a factor of 10 or 15, and always at least a factor of 3.[58]

7.3.3.8 Chromium(VI)

The general scheme for Cr^{VI} oxidation of organic compounds[59] makes use of the Cr^{V}, Cr^{IV} and Cr^{III} oxidation states.

$$H_2A + Cr^{VI} \rightleftharpoons A + Cr^{IV} \quad \text{slow}$$

$$Cr^{IV} + Cr^{VI} \rightleftharpoons 2\,Cr^{V}$$

$$Cr^{V} + H_2A \rightleftharpoons Cr^{III} + A$$

Chromic acid, H_2CrO_4, is a well-known powerful oxidant, with an electroreduction potential determined by the process:

$$CrO_4^{2-} + 4\,H_2O + 3\,e^- \rightleftharpoons Cr(OH)_3\,(s) + 5\,OH^- \qquad E° = -0.13\text{ V}$$

Chromic acid deposited on silica gel from the anhydride CrO_3 in aqueous (or in aqueous acidified) solutions[60] affords a useful oxidant.

Other preparations of this reagent are known[61] but the reagent has only a limited shelf life of less than a week. Oxidations with the reagent are conducted in diethyl ether, using 3 g of reagent per mmol of alcohol substrate. Generally they are rapid at ambient temperature, giving good yields (60–98%) of ketone products.[60,61] It is interesting to note that impregnation of chromium(IV) anhydride on alumina provides only an inactive reagent.[62] Conversely, chromium(IV) anhydride can be adsorbed on resins, such as polyvinylpyridines, polyacrylates, *etc.* (typically 6 mmol CrO_3 per g resin).[4,63] These reagents oxidize primary and secondary alcohols and may be used in nonpolar solvents. The ratio Cr:alcohol is usually in the range 1–4. Yields are, however, very variable (10–80%).[4] The procedure is improved by the use of catalytic amounts of quaternary ammonium salts, such as $(NBu^n{}_4)^+$, and by recourse to trifluoroacetato chromate as the Cr^{VI} impregnated species.[63]

The procedure is commendable for its simplicity, reduced toxicity (chromium in all its oxidation states is carcinogenic) and achieves good yields of ketones from alcohol, for example, octan-2-ol is oxidized into octan-2-one (92%), cyclohexanol into cyclohexanone (90%) and menthol into menthone (98%).[63] Pyridinium chromate is also a well-known oxidant for allylic oxidations.[64] As a silica gel supported reagent, this is turned into an efficient alcohol oxidant that will leave acid-labile functions unscathed.[61] Another advantage of the reagent is the long shelf-life of more than a year. These solid-supported oxidants also greatly facilitate product work-up, when compared with their solution counterparts.

Oxidation of alcohols typically proceeds in 4–12 h giving excellent yields of carbonyl products.[61]

Chromyl chloride, CrO_2Cl_2, like other Cr^{VI} species, is a vigorous oxidant of organic compounds. It may be, however, tamed as a silica gel adsorbate.[65] This reagent combination is also a good oxidant of alcohols, alkenes and alkynes, that will tolerate halides, esters, lactones, ethers and nitriles. The shelf life of the reagent appears to be indefinite in the dry state.

Potassium dichromate, $K_2Cr_2O_7$, adsorbed on silica gel (or on alumina, magnesia or Florisil) is a selective oxidant of allylic and benzylic alcohols and halides in neutral media.[66]

Similarly, supported ammonium dichromate is an effective reagent for hydroquinone to quinone oxidation.[67]

7.3.4 CLAY-SUPPORTED OXIDANTS

Firstly it should be recognized that clays will oxidize organic matter. One of the most effective means for detoxification of 1,4-dioxin is oxidation into the radical cation and subsequent polymerization on Cu^{II}–smectites.[68] Transition metal centers in phyllosilicate clays, such as Fe^{III}, are known to oxidize aromatic molecules by such single-electron transfers.[69–76] Some radical-coupling products have also been obtained.[73,75] A study of the oxidation of hydrocortisone by the two fibrous clay minerals sepiolite and palygorskite (a commercial variety of attalpugite)[77] has been reported. The former clay has the lower Fe^{III} content and accordingly has lower oxidizing power. The latter oxidizes the steroid due to surface-adsorbed iron oxides and to octahedral Fe^{III}, present in 2–3% amounts.

7.3.4.1 Permanganate

Clay-supported potassium permanganate oxidizes secondary alcohols into ketones.[78] A great asset is the ability of this reagent to selectively oxidize an allylic alcohol into the α,β-unsaturated ketone without any double bond oxidation. The procedure is extremely simple, involving grinding of the inorganic salt with a bentonite clay, prior to heating the alcohol and reagent in methylene chloride. Although reaction times are rather long (days), good yields are obtained (80–100%). Note that with a very large excess of oxidant and prolonged reaction times, alkenic bonds may be cleaved. The advantage of this method, as compared with the use of manganese dioxide,[79] is that prior activation of the reagent is not necessary.

7.3.4.2 Thallium Trinitrate

The oxidative potential of thallium(III) trinitrate was discovered by Taylor and McKillop.[80] The reagent was supported on the K10 acidic montmorillonite ($H_0 = -6$ to -8) by stirring the clay suspended in a methanol–trimethyl orthoformate solution of the thallium salt, followed by evaporation to dryness. In this way a colorless, free-flowing solid is obtained. Oxidations are very easily performed in inert solvents such as toluene, heptane, methylene chloride or carbon tetrachloride. A first application of this reagent was to the oxidative rearrangement of alkyl aryl ketones, leading to alkyl arylcarboxylates in *ca.* 90% yields.[80–82] Acetophenones are converted into methyl arylacetates. Likewise, propiophenone and butyrophenone are cleanly oxidized into methyl α-methyl- and α-ethyl phenylacetate, respectively. The thallium trinitrate–K10 system effects the rapid and effective oxidative rearrangement of alkenes into acetals. Cyclohexene is converted in less than a minute into the dimethylacetal of cyclopentanecarbaldehyde, and styrene and 1-phenyl-1-propene also give clean rearrangement into the corresponding aryl acetal propionaldehyde dimethylacetals, all in 85–92% yield. Likewise, cinnamaldehydes and cinnamic esters undergo such oxidative rearrangements in 85–90% yields. The Princeton–East Anglia coworkers also demonstrated the superiority of lamellar K10 montmorillonite support to microporous supports for thallium trinitrate.[80] Another good application of this system was the oxidative rearrangement of 3-acetylpyrroles into the corresponding methoxycarbonylpyrroles.[83]

7.3.4.3 Iron(III) and Copper(II) Nitrates

An asset of lamellar clays as supports for oxidants is their effective surface dimensionality. This leads to fast diffusional kinetics on the clay surfaces, which translate through the Smoluchowsky–Debye equation into high collision rates and, in turn, through the preexponential term, into high kinetic rates.[3,84]

Dehydrative activation of surface O^- centers has already been mentioned (Section 7.3.1) and produces powerful oxidation centers. Doping by transition metal ions is best performed under strong dehydration conditions such that the metal atoms are associated with resulting new anionic sites.

Addison in the 1960s prepared covalent metallic nitrates under anhydrous conditions and gave vibrational spectroscopic criteria to ascertain if the nitrato group is coordinated as a unidentate, a bidentate or a bridging ligand. This group showed that metallic nitrates were powerful oxidants if: (i) there is covalent bonding of the nitrato group, as a bidentate ligand; and (ii) the metal can fall back on lower oxidation states.[85] The oxidizing power of such covalent metallic nitrates is such that copper(II) nitrate effects both these coupled transformations, at liquid nitrogen temperature (equations 6 and 7).[86]

$$Et_2O + Cu(NO_3)_2 \longrightarrow EtONO + MeCHO + Cu(NO_3)OH \qquad (6)$$

$$MeCHO + Cu(NO_3)_2 \rightarrow Cu(NO_3)OAc + HONO \qquad (7)$$

Although it is possible to obtain the acetone solvate of anhydrous iron(III) nitrate, this oil decomposes in a vigorous exothermic reaction. However, it is possible to stabilize the oil by impregnation on the K10 acidic montmorillonite. The name 'clayfen' has been given to the resulting reagent.[87–89]

This reagent was first applied to the oxidation of alcohols where use of this 'clayfen' system gave satisfactory yields (65–90%), avoided overoxidation of aromatic aldehydes and required only inexpensive reagents.[87] A great practical advantage of this and other 'clayfen' oxidations is the opportunity for visual monitoring of the reactions. The start of the reaction is signalled by the evolution of reddish nitrous fumes and their cessation indicates completion. 'Clayfen' smoothly oxidizes benzoins into benzils,[90] with better isolated yields (85–95%) than the ytterbium(III) nitrate catalyzed Kagan procedure.[91] The alcohol oxidation proceeds through intermediate nitrous esters.[88] Likewise, oxidative coupling of thiols into disulfides is effected by 'clayfen' *via* thionitrite intermediates.[92] 'Clayfen' also converts *N*,*N*-dimethylhydrazones into the parent carbonyl compounds in 67–91% isolated yield.[93]

Regeneration of the carbonyl group from various protecting groups may be achieved by 'clayfen', for example from imine-protecting groups, tosylhydrazones, phenylhydrazones, 2,4-dinitrophenylhydrazones and semicarbazones.[94] 'Clayfen', because it is an inexpensive and mild source of nitrosonium ions, NO^+,[89] can be used to convert hydrazines into azides which in turn are transformed into iminophosphoranes.[95]

'Claycop' is a related reagent to 'clayfen'. This reagent is based on anhydrous copper(II) nitrate and is somewhat less reactive but enjoys much greater stability than 'clayfen'.[86,96,97] Clay-supported copper(II) nitrate ('claycop') is prepared in a process similar to the preparation of 'clayfen', by adding K10 clay to a solution of copper(II) nitrate trihydrate in acetone. This reagent has been applied to the aromatization of dihydropyridines, with consistently better isolated yields (40–93%) than with 'clayfen'.[97] It is also the reagent of choice for quantitative regeneration of carbonyl groups from protective bisthioacetals.[98] 'Claycop' also regenerates carbonyls from selenoacetals[99] and from thiocarbonyls[100] with very good isolated yields, although 'clayfen' is superior to 'claycop' for this last application.

7.3.5 CONCLUSIONS

Owing to the limited format, this review has not discussed the use of enzyme-supported materials or catalytic oxidants with industrial applications, even though both these are of current interest. The illustrative examples of solid oxidants that have been discussed display considerable differences in reactivity and selectivity from solution chemistry using the same or similar reagents. The choice of which support to use is still very much a matter of trial and error, although progress is being made rapidly. The single most useful asset of supported oxidants is their ease of use, offering cleanliness, safety and simplicity. These reagents now have a secure future in organic synthesis.

7.3.6 REFERENCES

1. P. Laszlo (ed.), 'Preparative Chemistry Using Supported Reagents', Academic Press, San Diego, CA, 1987.
2. P. H. Connet and K. E. Wetterham, *Struct. and Bonding (Berlin)*, 1983, **54**, 94.
3. P. Pfeifer, in 'Preparative Chemistry Using Supported Reagents', ed. P. Laszlo, Academic Press, San Diego, CA, 1987, chap. 2, p. 13.
4. T. Brunelet and G. Gelbard, *Nouv. J. Chim.*, 1983, **7**, 483.
5. H. Knözinger, *Adv. Catal.*, 1976, **25**, 184.
6. H. Knözinger and P. Ratnasamy, *Catal. Rev.-Sci. Eng.*, 1978, **17**, 31.
7. H. H. Lamb, B. C. Gates and H. Knözinger, *Angew. Chem., Int. Ed. Engl.*, 1988, **27**, 1127.
8. F. Freund, in 'Abstracts of the 196th ACS National Meeting, Los Angeles, CA., Sept. 25–30 1988'; 'Characterization of Active Sites of Minerals', ed. L. M. Coyne, ACS Symposium Series, Washington, DC, 1989, vol. 145.
9. M. Fétizon and M. Golfier, *C. R. Hebd. Seances Acad. Sci.*, 1968, **267**, 900.
10. G. H. Posner, R. B. Perfetti and A. W. Runquist, *Tetrahedron Lett.*, 1976, 3499.
11. G. H. Posner and M. J. Chapdelaine, *Tetrahedron Lett.*, 1977, 3227.
12. G. H. Posner, *Angew. Chem., Int. Ed. Engl.*, 1978, **17**, 487.
13. Y.-S. Cheng, W.-L. Liu and S. Chen, *Synthesis*, 1980, 223.
14. D. Savoia, C. Trombini and A. Umani-Ronchi, *J. Org. Chem.*, 1982, **47**, 564.
15. A. J. Fatiadi, *Synthesis*, 1974, 229.
16. D. Villemin and M. Ricard, *Nouv. J. Chim.*, 1982, **6**, 605.
17. E. Santaniello and P. Ferraboschi, *Nouv. J. Chim.*, 1980, **4**, 279.
18. R. O. Hutchins, N. R. Natale, W. J. Cook and J. Ohrr, *Tetrahedron Lett.*, 1977, 4176.
19. D. G. Lee and U. A. Spitz, *J. Org. Chem.*, 1970, **35**, 3589.
20. J. Rebek and R. McCready, *Tetrahedron Lett.*, 1979, 4337.
21. A. Foucaud and M. Bakovetila, *Synthesis*, 1987, 854.
22. W. T. Liu and Y. C. Tong, *J. Org. Chem.*, 1978, **43**, 2717.
23. Y. Mazur, *Pure Appl. Chem.*, 1975, **41**, 145.
24. Z. Cohen, E. Keinan, Y. Mazur and T. H. Varkony, *J. Org. Chem.*, 1975, **40**, 2141.
25. Z. Cohen, M. Varkony, E. Keinan and Y. Mazur, *Org. Synth.*, 1980, **59**, 176.
26. Z. Cohen and Y. Mazur, *J. Org. Chem.*, 1979, **44**, 2318.
27. E. Trifilieff, L. Bang and G. Ourisson, *Tetrahedron Lett.*, 1977, 2991.
28. E. Trifilieff, L. Bang, A. S. Naruta and G. Ourisson, *J. Chem. Res. (S)*, 1978, 601.
29. A. L. J. Beckwith, C. L. Bodkin and Thach Duong, *Aust. J. Chem.*, 1977, **30**, 2177.
30. A. L. J. Beckwith, C. L. Bodkin and Thach Duong, *Chem. Lett.*, 1977, 425.
31. D. Tal, E. Keinan and Y. Mazur, *J. Am. Chem. Soc.*, 1979, **101**, 501.
32. A. L. J. Beckwith and Thach Duong, *J. Chem. Soc., Chem. Comm.*, 1978, 413.
33. E. Proksch and A. de Meijere, *Angew Chem., Int. Ed. Engl.*, 1976, **15**, 761.
34. E. Proksch and A. de Meijere, *Tetrahedron Lett.*, 1976, 4851.
35. E. Keinan and T. H. Varkony, in 'The Chemistry of Functional Groups, Peroxides', ed. S. Patai, Wiley, New York, 1983, p. 649.
36. E. Keinan and Y. Mazur, *J. Org. Chem.*, 1977, **42**, 844.
37. E. Keinan and Y. Mazur, *J. Am. Chem. Soc.*, 1977, **99**, 3888.
38. H. Klein and A. Steinmetz, *Tetrahedron Lett.*, 1975, 4249.
39. A. Fadel and J. Salaun, *Tetrahedron*, 1981, **41**, 413.
40. A. Fadel and J. Salaun, *Tetrahedron*, 1981, **41**, 1267.
41. E. Keinan and Y. Mazur, *J. Org. Chem.*, 1977, **43**, 1020.
42. D. Tal, E. Keinan and Y. Mazur, *Tetrahedron*, 1981, **37**, 4327.
43. T. C. Jempty, L. L. Miller and Y. Mazur, *J. Org. Chem.*, 1980, **45**, 749.
44. T. C. Jempty, K. A. Z. Gogins, Y. Mazur and L. L. Miller, *J. Org. Chem.*, 1981, **46**, 4545; M. A. Umbreit and K. B. Sharpless, *J. Am. Chem. Soc.*, 1977, **99**, 5526.
45. A. Fisher and G. N. Henderson, *Synthesis*, 1985, 641.
46. H. M. Chawla and R. S. Mittal, *Synthesis*, 1985, 70.
47. H. M. Chawla and R. S. Mittal, *Indian J. Chem., Sect. B*, 1983, **22**, 753.
48. Y. Hatenaka, T. Imamoto and M. Yokoyama, *Tetrahedron Lett.*, 1983, **24**, 2399.
49. D. N. Gupta, P. Hodge and J. E. Davies, *J. Chem. Soc., Perkin Trans. 1*, 1981, 2970.
50. M. Hojo and R. Masuda, *Tetrahedron Lett.*, 1976, 613.
51. M. Hojo and R. Masuda, *Synthesis*, 1976, 678.
52. M. Hojo, R. Masuda and K. Hakotani, *Tetrahedron Lett.*, 1978, 1121.
53. S. L. Regen and C. Koteel, *J. Am. Chem. Soc.*, 1977, **99**, 3837.
54. A. Al Jazzaa, J. H. Clark and M. Robertson, *Chem. Lett.*, 1982, 405.
55. J. H. Clark and D. G. Cork, *J. Chem. Soc., Perkin Trans. 1*, 1983, 2253.
56. J. T. B. Ferreira, W. O. Cruz, P. C. Vieira and M. Yonashiro, *J. Org. Chem.*, 1987, **52**, 3698.
57. S. Wolfe and C. F. Ingold, *J. Am. Chem. Soc.*, 1983, **105**, 7755.
58. W. Zhi-min, Z. Wei-shan and L. Guo-giang, *Tetrahedron Lett.*, 1985, **26**, 6221.
59. F. A. Cotton and G. Wilkinson, 'Advanced Inorganic Chemistry', 5th ed., Wiley-Interscience, New York, 1988, 683.

60. E. Santaniello, F. Ponti and A. Manzocchi, *Synthesis*, 1978, 534.
61. R. P. Singh, H. N. Subbarao and S. Dev, *Terahedron*, 1979, **35**, 1789.
62. E. Santaniello, in '*Preparative Chemistry Using Supported Reagents*', ed. P. Laszlo, Academic Press, San Diego, CA, 1987, chap. 18, p. 345.
63. T. Brunelet, C. Jouitteau and G. Gelbard, *J. Org. Chem.*, 1986, **51**, 4016.
64. W. G. Dauben, M. Lorber and D. S. Fullerton, *J. Org. Chem.*, 1969, **34**, 3587.
65. J. San Filippo, Jr. and C.-I. Chern, *J. Org. Chem.*, 1977, **42**, 2182.
66. E. Santaniello and P. Ferraboschi, *Nouv. J. Chim.*, 1980, **4**, 279.
67. A. Fischer and G. N. Henderson, *Synthesis*, 1985, 641.
68. S. A. Boyd and M. M. Mortland, *Nature (London)*, 1985, **316**, 532.
69. H. E. Doner and M. M. Mortland, *Science (Washington, DC)*, 1969, **166**, 1406.
70. M. M. Mortland and T. J. Pinnavaia, *Nature (London)*, 1971, **229**, 75.
71. T. J. Pinnavaia and M. M. Mortland, *J. Phys. Chem.*, 1971, **75**, 3957.
72. J. P. Rupert, *J. Phys. Chem.*, 1973, **77**, 784.
73. D. Fenn, M. M. Mortland and T. J. Pinnavaia, *Clays Clay Miner.*, 1973, **21**, 315.
74. T. J. Pinnavaia, P. L. Hall, S. S. Cady and M. M. Mortland, *J. Phys. Chem.*, 1974, **78**, 994.
75. M. M. Mortland and L. J. Halloran, *Proc. Soil Sci. Soc. Am.*, 1976, **40**, 367.
76. P. J. Isaacson and B. L. Sawhney, *Clay Miner.*, 1983, **18**, 253.
77. J. Cornejo, M. C. Hermosin, J. L. White, J. R. Barnes and S. L. Hem, *Clays Clay Miner.*, 1983, **31**, 109.
78. D. G. Lee and N. A. Noureldin, *Tetrahedron Lett.*, 1981, **22**, 4889.
79. A. J. Fatiadi, *Synthesis*, 1976, 65.
80. C.-S. Chiang, A. McKillop and E. C. Taylor, *J. Am. Chem. Soc.*, 1976, **98**, 6750.
81. A. McKillop and D. W. Young, *Synthesis*, 1979, 401.
82. A. McKillop and D. W. Young, *Synthesis*, 1979, 481.
83. E. Adelakun, A. H. Jackson, N. S. Ooi and K. R. N. Rao, *Tetrahedron Lett.*, 1984, **25**, 6049.
84. P. Laszlo, *Acc. Chem. Res.*, 1986, **19**, 121.
85. C. C. Addison, *Coord. Chem. Rev.*, 1966, **1**, 59.
86. L. C. Coard and R. E. Powell, *J. Chem. Soc. A*, 1967, 296.
87. A. Cornélis and P. Laszlo, *Synthesis*, 1980, 849.
88. A. Cornélis, P. Y. Herzé and P. Laszlo, *Tetrahedron Lett.*, 1982, **23**, 5035.
89. A. Cornélis and P. Laszlo, *Synthesis*, 1985, 909.
90. M. Besemann, A. Cornélis and P. Laszlo, *C. R. Hebd. Seances Acad. Sci.*, 1984, **299**, 427.
91. P. Girard and H. Kagan, *Tetrahedron Lett.*, 1975, 4513.
92. A. Cornélis, N. Depaye, A. Gerstmans and P. Laszlo, *Tetrahedron Lett.*, 1983, **24**, 3103.
93. P. Laszlo and E. Polla, *Tetrahedron Lett.*, 1984, **25**, 3309.
94. P. Laszlo and E. Polla, *Synthesis*, 1985, 439.
95. P. Laszlo and E. Polla, *Tetrahedron Lett.*, 1984, **25**, 3701.
96. A. Cornélis and P. Laszlo, *Aldrichimica Acta*, 1988, 97.
97. M. Balogh, I. Hermecz, Z. Mészaros and P. Laszlo, *Helv. Chim. Acta*, 1984, **67**, 2270.
98. M. Balogh, A. Cornélis and P. Laszlo, *Tetrahedron Lett.*, 1984, **25**, 3313.
99. P. Laszlo, P. Pennetreau and A. Krief, *Tetrahedron Lett.*, 1986, **27**, 3153.
100. S. Chalais, A. Cornélis, P. Laszlo and A. Mathy, *Tetrahedron Lett.*, 1985, **26**, 2327.

7.4
Electron-transfer Oxidation

JAY K. KOCHI
University of Houston, TX, USA

7.4.1 SCOPE AND DEFINITIONS

Electron-transfer oxidation of the vast majority of organic compounds involves multiple steps with transient radicals as key reactive intermediates. Since stable organic compounds are mostly diamagnetic donors with even numbers of electrons, the electron transfer must perforce generate an odd-electron species. In the case of a neutral organic donor (generically represented hereafter as RH), electron-transfer oxidation produces a radical cation (equation 1a),[1–3] which is constrained to undergo a second electron transfer before it ultimately yields the diamagnetic product. It is the unique properties of organic radical cations that lead to the rich menu of organic transformations exploitable for organic synthesis.

$$\mathrm{RH} \xrightarrow{-e^-} \mathrm{RH}^{\overset{+}{\bullet}} \tag{1a}$$

$$\mathrm{RH} \xrightarrow{+e^-} \mathrm{RH}^{\overset{-}{\bullet}} \tag{1b}$$

Oxidation is the microscopic reverse of reduction, and electron transfer in equation (1a) has its counterpart in reduction, *i.e.* equation (1b). Accordingly, for every organic oxidation there is a conjugate process involving reduction, as simply illustrated by the electron-transfer equilibria among carbenium ions, free radicals and carbanions (equation 2).[4,5]

$$\mathrm{R}^+ \underset{}{\overset{e^-}{\rightleftharpoons}} \mathrm{R}\bullet \underset{}{\overset{e^-}{\rightleftharpoons}} \mathrm{R}\overset{-}{:} \tag{2}$$

Indeed the combination of the reactive intermediates in equations (1a) and (2) forms the chemical basis of electron-transfer oxidation (reduction) of organic compounds in both stoichiometric and catalytic processes.[6]

The energetic basis for electron-transfer oxidation includes the thermodynamic potential E°_{ox} for the initial act of electron transfer from RH in equation (1a).[7,8] Such an electron detachment is commonly effected: (a) at an electrode; (b) by an oxidant; or (c) with light. Thus the organic oxidation in equation (1a) is driven electrochemically by the anodic electrode potential (*E*) to match the value of E°_{ox},[9] *i.e.* equation (3a). Likewise, the driving force in the chemical oxidation of RH is provided by the redox potential E°_{red} of the electron acceptor or oxidant (hereinafter referred to as A; equation 3b).[10]

$$\mathrm{RH} \xrightarrow{[E]} \mathrm{RH}^{\overset{+}{\bullet}} \tag{3a}$$

$$\mathrm{RH} + \mathrm{A} \longrightarrow \mathrm{RH}^{\overset{+}{\bullet}} + \mathrm{A}^{\overset{-}{\bullet}} \tag{3b}$$

Photochemical electron transfer proceeds either by the prior actinic activation of the organic donor RH followed by quenching by the electron acceptor (equation 4), or by the reverse sequence involving the prior acceptor activation and quenching with donor.[11] Photochemical electron transfer can also be effected by the irradiation of the charge-transfer (CT) absorption band of the precursor electron donor–acceptor (EDA) complex (equation 5).[12]

$$\mathrm{RH} \xrightarrow{h\nu} \mathrm{RH^*} \xrightarrow{\mathrm{A}} \mathrm{RH}^{\overset{+}{\bullet}} + \mathrm{A}^{\overset{-}{\bullet}} \tag{4}$$

$$\mathrm{RH} + \mathrm{A} \rightleftharpoons [\mathrm{RH,A}] \xrightarrow{h\nu_{CT}} \mathrm{RH}^{\overset{+}{\bullet}} + \mathrm{A}^{\overset{-}{\bullet}} \tag{5}$$

The actinic irradiation of the charge-transfer band of the EDA complex in equation (5) is the most direct method for the photoactivation of electron-transfer oxidation, since the absorbed energy $h\nu_{CT}$ is directly applied to the conversion of a bonding electron in the HOMO of the donor RH to an antibonding electron in the LUMO of the acceptor A. Such a spontaneous generation of $[\mathrm{RH}^{+\bullet},\mathrm{A}^{-\bullet}]$ represents the contact ion pair (CIP)[13] in Figure 1 with an interionic separation that is essentially that originally present in the EDA precursor [RH,A].[14] However in the alternative mode of photoactivation (equation 4) the excitation of only RH (see $h\nu_{RH}$ in Figure 1) is followed by electron transfer to A in a subsequent step. Since the latter takes place by a diffusional process,[15] $[\mathrm{RH}^{+\bullet},\mathrm{A}^{-\bullet}]$ is not necessarily the same contact ion pair formed by the direct charge-transfer activation. Indeed, there are examples of diffusional quenching by electron transfer over long distances to form initially a less intimate, solvent-separated ion pair (SSIP).[16] The same situation pertains to the photoinduced electron-transfer oxidation by the prior excitation of the acceptor (see $h\nu_A$ in Figure 1). The modulating effect of varying ion-pair structures lies at the core of electron-transfer oxidation, as will be elaborated in the following sections. Finally, in the electron-transfer oxidation of a particular organic donor RH, the thermal process in equation (3b) invariably requires a stronger oxidant (*i.e.* A with more positive E°_{red}) than its photochemical counterparts in equations (4) and (5) owing to the contribution from the actinic input (see Figure 1).

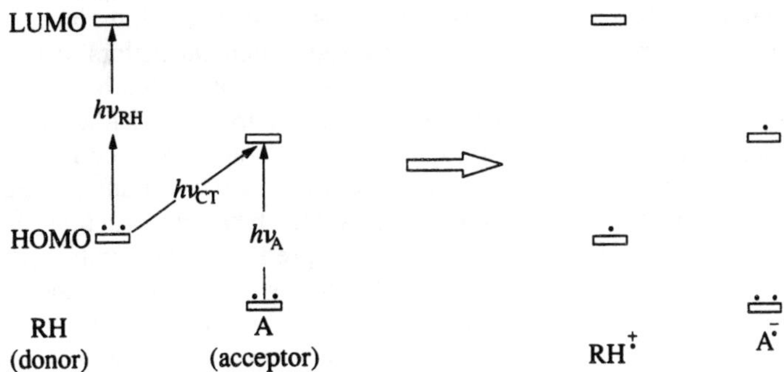

Figure 1 Energy level diagram (qualitative) for the charge-transfer excitation ($h\nu_{CT}$) of the electron donor-acceptor complex (RH, A) in comparison with that for the excitation of either the donor ($h\nu_{RH}$) followed by quenching of RH* with the acceptor or the acceptor ($h\nu_A$) followed by quenching of A* with the donor

Since electrochemical methods are described in Volume 7, Chapter 7.1, emphasis will be placed on the thermal and photochemical activation of electron-transfer oxidation. Even with this restriction the scope of electron-transfer oxidation is too extensive to be covered completely in a single chapter. Therefore the approach here is to present those fundamental aspects that allow electron-transfer oxidations to be developed for synthetic transformations. Hopefully this format will encourage the creative chemist to devise myriad oxidative syntheses from a limited number of principles. Fortunately, there are already available a variety of recent monographs with each presenting a restricted coverage to permit the inclusion of detailed and useful examples. For the convenience of the reader these articles are listed as references 17 to 32, with the chapter titles included where appropriate. Taken all together they offer the reader an interesting panoply of electron-transfer oxidations that are intertwined by the principles outlined herein.

It is important to emphasize the anodic, chemical and actinic activations of electron-transfer oxidation to be complementary methods that all commonly involve the reactive intermediates like those presented in equations (1a) and (2). As such, cognizance must always be taken of the subtle differences of concentration, temperature, solvent polarity, *etc.* that affect the behavior of the transient radicals and ion radicals sufficient to alter the unique complexion of products obtainable in the course of electrochemical, chemical or photochemical oxidation of a given organic substrate. For this reason it is helpful to define first the features that are critical to electron-transfer oxidation, independent of the methodology to be employed. These include the consideration of: (a) the driving force for electron transfer in terms of the oxidation potential E°_{ox} of the organic substrate and the reduction potential E°_{red} of the oxidant; and (b) the chemical properties of the oxidized donor ($RH^{+\bullet}$) as well as those of the reduced acceptor ($A^{-\bullet}$).

7.4.2 THE FORMULATION OF ELECTRON-TRANSFER OXIDATION

Electron-transfer oxidation in equation (3b) can be considered to consist of a series of preequilibria, in the limit where the radical cation of the organic donor and radical anion of the acceptor are both persistent species (equation 6a).[33] The first set of brackets encloses the electron donor–acceptor or EDA precursor complex, and the second set the contact ion pair or CIP successor complex that is constrained by the solvent cage.[34] Intermolecular reactions of $RH^{+\bullet}$ that lead to the oxidation products largely occur subsequent to cage escape (k_3).

$$RH + A \rightleftharpoons [RH, A] \underset{k_2}{\overset{k_1}{\rightleftharpoons}} [RH^{+\bullet}, A^{-\bullet}] \overset{k_3}{\rightleftharpoons} RH^{+\bullet} + A^{-\bullet} \qquad (6a)$$

Electron-transfer oxidation of an organic substrate in equation (6a) derives from a driving force given as $-\Delta G = F(E^\circ_{ox} + E^\circ_{red})$, where F is the Faraday constant and the conversion factor is 1 V ≈ 23 kcal mol^{-1} (1 cal = 4.18 J).[8] In the simplest cases of anionic donors (*e.g.* carbanions) reacting with cationic oxidants, the ion-pair annihilation will proceed by electron transfer even when the driving force is endergonic by as much as 0.7 V (*i.e.* uphill by ~15 kcal mol^{-1}).[35] Moreover electron transfer between uncharged donors and acceptors may occur with as little as –0.4 V of driving force. These qualitative estimates are of course strongly tempered by solvent effects and inherent factors that are intrinsic to the donor and acceptor. The latter in outer-sphere electron transfer is represented by the reorganization energy (λ) as described by Marcus.[36] (For an excellent account of the use of Marcus theory in organic chemistry the reader is referred to the recent monograph by Eberson.[37]) Suffice it to mention here that most organic reactions proceed *via* inner-sphere electron transfer in which the intermolecular interactions in the first-formed contact ion pair $[RH^{+\bullet}, A^{-\bullet}]$ must be explicitly taken into account.[38] As such, the facility with which electron transfer occurs is not so readily predicted from only a knowledge of the measurable quantities E°_{ox}, E°_{red}, λ(RH), λ(A) and ΔG_s. In other words, simply a knowledge of E°_{ox} and E°_{red} alone is insufficient to predict whether electron transfer will or will not be a viable process in the oxidation of the organic donor RH. This caveat must be underscored, since the thermodynamic driving force is often and erroneously taken as a predictor of electron-transfer oxidation. At best, the driving force ($E^\circ_{ox} + E^\circ_{red}$) relates only to the electron-transfer equilibrium. The critical element in oxidation efficiency is the behavior of the contact ion pair, as determined by the competition between its formation (k_1) and further reaction (k_3) relative to the energy-wasting, back electron transfer (k_2) in equation (6a).[32]

7.4.2.1 Oxidation Potentials of Organic Compounds

Except for very electron-rich organic donors that yield stable, persistent radical cations, the values of the one-electron potential E°_{ox} for equation (1a) are not generally available for organic compounds. Thus the radical cations $RH^{+\bullet}$ that are derived from most organic donors are too reactive to allow the measurement of their reversible potentials E°_{ox} in either aqueous or organic solvents by the standard techniques. This problem is partially alleviated by the measurement of the irreversible anodic peak potentials E_a that are readily obtained from the linear-sweep or cyclic voltammograms (CV) of RH.[39] Since the values of E_a contain contributions from kinetic terms, a comparison with the values of the thermodynamic E°_{ox} is restricted to a series of related donors, *i.e.* $E^\circ_{ox} = \beta E_a$ + constant, where $\beta \approx 1.0$, as illustrated in Figure 2(a).[40] It is important to emphasize this limitation when values of E_a (at a constant CV sweep rate) are employed as measures of the electron-donor properties of various organic donors, as in Table 1.[41,42] Alternatively, the energetics of electron detachment from RH are obtained in the gas-phase measurements of the ionization potential *IP*. The ionization potentials of many organic donors have been determined experimentally, most conveniently from the photoelectron spectra (PES) obtained by the photoionization of RH.[43] The values of *IP* measured in the gas phase differ from the values E°_{ox} in solution largely by solvation, *i.e.* $E^\circ_{ox} = IP + \Delta G_s$ + constant, where ΔG_s is the solvation energy of the radical cation, owing to a negligible contribution from the solvation of the uncharged donor RH. Since the variations in ΔG_s are usually minor, the values of *IP* such as those listed in Table 1 can be adequate measures of the electron-donor abilities of organic compounds applicable to a particular solvent. This generalization is especially tenable for a series of related compounds, as illustrated in Figure 2(b). Independently of whether the electron-donor properties are evaluated by such indirect measures as E_p and *IP*, note must always be taken of the approximations that relate them to the thermodynamic values of $E°_{ox}$.

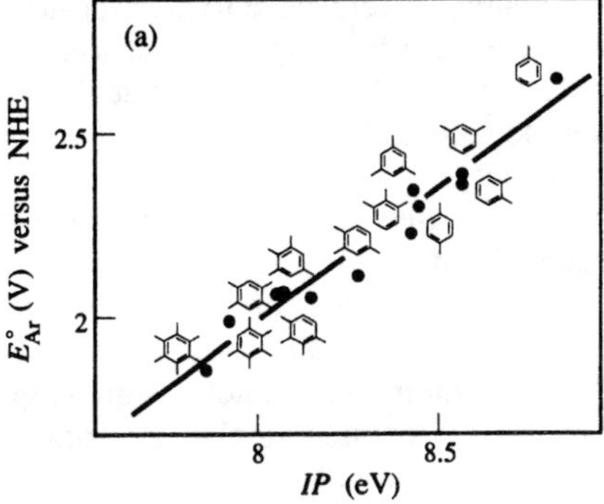

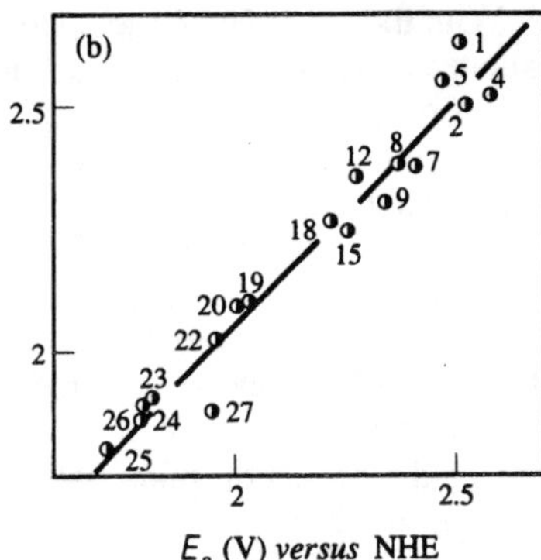

Figure 2 (a) Correlation of the reversible oxidation potentials and the vertical ionization potentials of methylarenes; (b) the correlation of the standard oxidation potentials E^o_{Ar} of various alkylbenzenes with the irreversible CV peak potentials E_a. Numbers refer to the aromatic hydrocarbons identified in ref. 40

Table 1 Some Representative Values of the Oxidation Potentials and Ionization Potentials of Organic Electron Donors[a]

Donor (RH)	*IP*	$E°_{ox}$	*Donor (RH)*	*IP*	$E°_{ox}$
2-Methylpentane	10.11	3.01	Benzene	9.24	2.04
2,2-Dimethylbutane	10.05	3.28	Toluene	8.82	1.96
Ethylene	10.51	2.90	*o*-Xylene	8.56	1.58
1-Butene	9.58	2.78	*m*-Xylene	8.56	1.60
1-Octene	9.52	2.70	*p*-Xylene	8.45	1.54
2-Methylpropene	9.23	2.65	*p*-Bromotoluene	8.67	1.72
2-Butene	9.13	2.21	Iodobenzene	8.73	1.77
1,4-Cyclohexadiene	8.40	1.60	Anisaldehyde	8.86	1.64
1,3-Butadiene	9.07	2.03	*p*-Chlorotoluene	8.69	1.76
Cyclohexene	8.95	1.98	Chlorobenzene	9.07	2.07
2-Methyl-1-butene	9.12	1.97	Bromobenzene	8.89	1.98
2,3-Dimethyl-1,3-butadiene	8.27	1.84	Biphenyl	8.27	1.48
2-Iodopropane	9.17	2.04	1-Propylbenzene	8.72	1.97
Methyl iodide	9.54	2.12	2-Propylbenzene	8.69	1.88
n-Butanethiol	9.14	1.34	Pentamethylbenzene	7.92	1.28
Dimethyl sulfide	8.69	1.26	1,2,4,5-Tetramethylbenzene	8.03	1.29
Diethyl sulfide	8.43	1.35	1,2,3-Trimethylbenzene	8.48	1.58
Dimethyl sulfoxide	8.84	1.73	1,2,4-Trimethylbenzene	8.27	1.41
Diphenylamine	7.40	0.53	Mesitylene	8.39	1.53
1-Naphthylamine	7.30	0.34	Indene	8.81	1.25
2-Naphthylamine	7.25	0.44	Hexamethylbenzene	7.85	1.20
Dimethylaniline	7.14	0.45	1,4-Dimethoxybenzene	7.90	1.04
Triethylamine	7.50	0.79	Naphthalene	8.12	1.34
Trimethylamine	7.82	0.82	1-Methylnaphthalene	7.96	1.24
Aniline	7.70	0.70	2-Methylnaphthalene	7.96	1.22
n-Butylamine	8.71	1.87	Phenanthrene	7.80	1.23
N,N-Dimethylacetamide	8.81	1.82	Anthracene	7.23	0.84
Pyridine	9.27	1.82	Tetracene	6.88	0.53
Quinoline	8.30	1.73	Fluorene	8.63	1.25
Phenol	8.50	1.04	Triphenylene	7.80	1.35
1,4-Dioxane	9.13	1.97	Coronene	7.60	0.93
Anisole	8.22	1.40	Perylene	7.15	0.55
Thiophene	8.86	1.70	Azulene	7.43	0.61
t-Butyl alcohol	9.71	2.94	Chrysene	7.75	1.22

[a] *IP* in eV; $E°_{ox}$ in V *versus* $Ag^+/AgNO_3$, in MeCN (consult the text in ref. 41 for the reliability).

7.4.2.2 Reduction Potentials of Oxidants (Electron Acceptors)

The electron-acceptor properties of oxidants are most readily evaluated by the reversible potentials E°_{red} for the one-electron reduction, *i.e.* equation (6b). Values of E°_{red} for many types of oxidants, particularly those based on transition metal cations, have been tabulated, and some of the more common ones in water are listed in Table 2.[44] However there are a number of useful oxidants that undergo a multiple electron change, *e.g.* $Tl^{3+} + 2e^- \rightarrow Tl^+$, $O_2Cr^{2+} + 3e^- \rightarrow Cr^{3+}$, *etc.*, and E°_{red} is known only for the overall change. With these oxidants, the one-electron potential of relevance to electron-transfer oxidation must be evaluated separately by such transient electrochemical techniques as linear-sweep microvoltammetry.[45] Reduction potentials are also highly dependent on the solvent, particularly in those oxidants undergoing a pronounced change in charge. Since the values of E°_{red} are generally unattainable in organic solvents, an alternative measure of the electron-acceptor properties of A can be evaluated from the irreversible cathodic peak potential E_c. For a series of related compounds the values of E_c can parallel the gas-phase electron affinities (E_A).[46] (Note the same limitations apply to the use of E_c as those described above for the anodic counterpart.) Moreover, there are a number of stable organic and nonmetallic radicals that are useful in electron-transfer oxidations. Table 3 also includes several varieties of organic acceptors that afford persistent radical anions. Owing to their use as photochemical quenchers, the enhanced values of the reduction potentials E_S and E_T for the excited singlet and triplet acceptor species, respectively (see $h\nu_A$ in Figure 1), are also included in Table 3.[47–59]

$$A + e^- \overset{E^\circ_{red}}{\rightleftharpoons} A^{\bullet -} \qquad (6b)$$

Table 2 Reduction Potentials of Some Common Metal Oxidants[a]

Oxidant (A)	E°_{red}	*Oxidant (A)*	E°_{red}
Ag^{II}	2.00	Ru^{IV}	0.86
Co^{III}	1.81	Ag^{I}	0.80
$O_4Bi_2^{IV}$	1.59[b]	Tl^{III}	1.26[b]
Ce^{IV}	1.61	O_4Re^{VII}	0.77
$(5\text{-NOphen})Fe^{III}$	1.53[c]	Fe^{III}	0.77
ORh^{IV}	1.43	$(phen)_2Fe^{III}$	1.33[c]
Hg^{II}	0.91[b]	O_4Ru^{VII}	0.59
Ru^{IV}	1.01	O_4Mn^{VII}	0.57
Au^{III}	1.4[b]	$(NC)_6Fe^{III}$	0.55
O_2V^{V}	1.00	Cu^{I}	0.52
O_4Ru^{VIII}	1.00	$(NC)_8W^{V}$	0.46
Pb^{IV}	1.65[b]	W^{VI}	0.26
Pu^{VI}	0.92	O_4Os^{VIII}	0.18[c]
Cl_6Ir^{IV}	0.87	Cu^{II}	0.17
XeF_2	2.2[b]	OTi^{IV}	0.1
Pd^{II}	0.92[b]	O_2Cl^{IV}	0.06

[a]One-electron potential (NHE) in water with oxo and aquo ligands, unless indicated otherwise.[44] [b]Two-electron potential. [c]MeCN.

7.4.3 GENERIC BEHAVIOR OF RADICAL IONS AS REACTIVE INTERMEDIATES IN ELECTRON-TRANSFER OXIDATION

The fate of the contact ion pair $[RH^{\bullet +}, A^{\bullet -}]$ is critical to electron-transfer oxidation. Oxidative efficiency is the highest with those organic donors that yield unstable radical cations, such as hexamethyl(Dewar benzene), which undergoes spontaneous rearrangement (equation 7).[60,61]

$$\xrightarrow{\sim 10^{-10}\ \text{s}} \qquad (7)$$

When such a unimolecular process occurs faster than back electron transfer (k_2 in equation 6a), electron-transfer oxidation in Scheme 1 proceeds rapidly despite an unfavorable driving force ($E^\circ_{ox} + E^\circ_{red} \approx$ 30 kcal mol^{-1}) for electron transfer.[62]

Table 3 Reduction Potentials of Organic Electron Acceptors[a]

Acceptor (A)	E°_{red}	*Ref.*
Thianthrenium ClO_4^-	1.28	47
2,4,6-Triphenylpyrilium ClO_4^-	–0.29 (2.8)	48
Tropylium BF_4^-	–0.18	49
Nitrosonium BF_4^-	1.28	50
Tris-*p*-bromophenylaminum BF_4^- ($BA^{+\cdot}$)	0.80	51
2-Phenylpyrrolinium ClO_4^-	— (2.9)	52
Nitronium BF_4^-	1.27	50
1,2,4,5-Tetracyanobenzene (TCB)	–0.65 (3.83)	53
9,10-Dicyancyanoanthracene (DCA)	–0.98 (2.88)	53
2,6,9,10-Tetracyanoanthracene (TCA)	–0.45 (2.82)	53
1,4-Dicyanonaphthalene (DCN)	–1.28 (3.45)	53
9-Cyanoanthracene	–1.39 (2.96)	53
1-Cyanonaphthalene (CN)	–1.98 (3.75)	54
1,4-Dicyanobenzene (DCB)	–1.60 (4.2)	53
Chloranil (CA)	0.02 [2.70]	55
Dioxygenyl ($O_2^{+\cdot}$) SbF_6^-	5.3	56
2,4,4,6-Tetrabromocyclohexa-2,5-dienone	0.29	57
Tetracyanoethylene (TCNE)	0.24	58
Tetracyanoquinodimethane (ICNQ)	0.19	53
Tetranitromethane (TNM)	~0.0	59
Dioxygen (O_2)	–0.78 (0.98)	55
1,2-Benzoquinone	0.12 [2.3]	53
Dichlorodicyano-1,4-benzoquinone (DDQ)	0.52	55
1,4-Dinitrobenzene	–0.69 [2.6]	55
Nitrobenzene	–1.15	55
N,N′-Dimethyl-4-bipyridinium (MV^{2+}) ClO_4^-	–0.45 [3.1]	53

[a]In V *versus* SCE in MeCN solution; E_S (parentheses) and E_T [brackets] in eV.

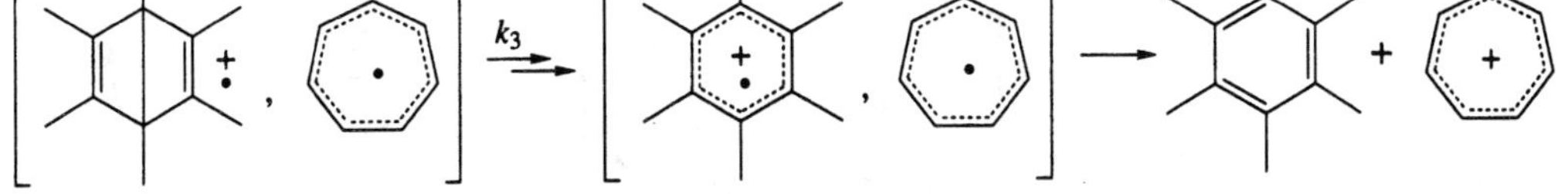

Scheme 1

Analogously, those oxidants that produce unstable radical anions, *e.g.* tetranitromethane, which suffers spontaneous fragmentation (equation 10),[63] similarly facilitate electron-transfer oxidation by pulling the redox equilibria in equation (6a) to the right. As a result, the alkene addition of tetranitromethane (Scheme 2) occurs despite an unfavorable redox equilibrium.[64]

$$C(NO_2)_4^{\overline{\cdot}} \xrightarrow{\sim 10^{-12}\,s} C(NO_2)_3^{-} + NO_2^{\bullet} \quad (10)$$

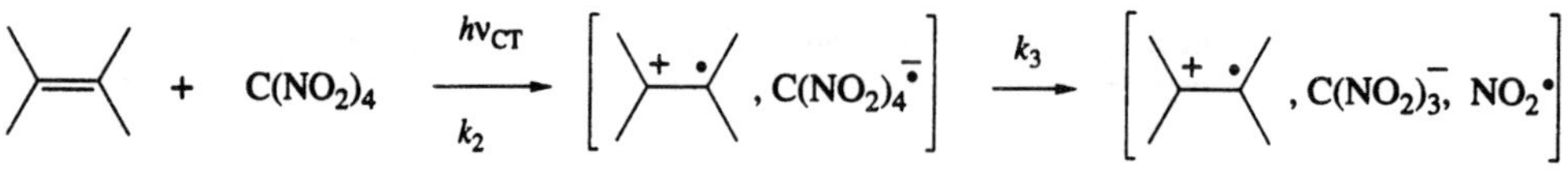

$$\text{alkene} + C(NO_2)_4 \underset{k_2}{\overset{h\nu_{CT}}{\longrightarrow}} [\text{alkene}^{+\bullet}, C(NO_2)_4^{\overline{\cdot}}] \xrightarrow{k_3} [\text{alkene}^{+\bullet}, C(NO_2)_3^{-}, NO_2^{\bullet}] \quad (11)$$

$$[\text{alkene}^{+\bullet}, C(NO_2)_3^{-}, NO_2^{\bullet}] \xrightarrow{\text{fast}} O_2N\text{–}C(CH_3)_2\text{–}C(CH_3)_2\text{–}C(NO_2)_3 \quad (12)$$

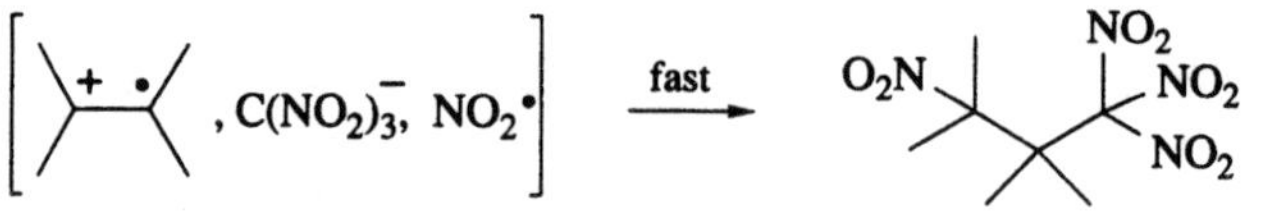

Scheme 2

Moreover, the facile bimolecular reactions of the cationic donor RH^{+} and/or the anionic acceptor $A^{\overline{\cdot}}$, especially with additives that are present during oxidation, can accomplish the same displacement of the redox equilibria in measure with the competition from back electron transfer. For example, the arene activation with nitrosonium ion merely reaches a low steady-state concentration of the radical pair, which persists indefinitely in equation (13). However, oxygen rapidly traps even small amounts of nitric oxide to render back electron transfer ineffective, and successfully effects aromatic nitration (Scheme 3).[65]

$$ArH + NO^{+} \underset{k_2}{\rightleftharpoons} [ArH^{+\bullet}, NO] \quad (13)$$

$$[ArH^{+\bullet}, NO] + 1/2\,O_2 \xrightarrow{\text{fast}} [ArH^{+\bullet}, NO_2] \longrightarrow ArNO_2 + H^{+} \quad (14)$$

Scheme 3

In the related photochemical context, the EDA complex of hexamethylbenzene and maleic anhydride merely reaches a photostationary state (equation 15; Scheme 4) with no productive photochemistry, except when acid is present to trap the acceptor anion on its way to the photoadduct in equation (16).[66]

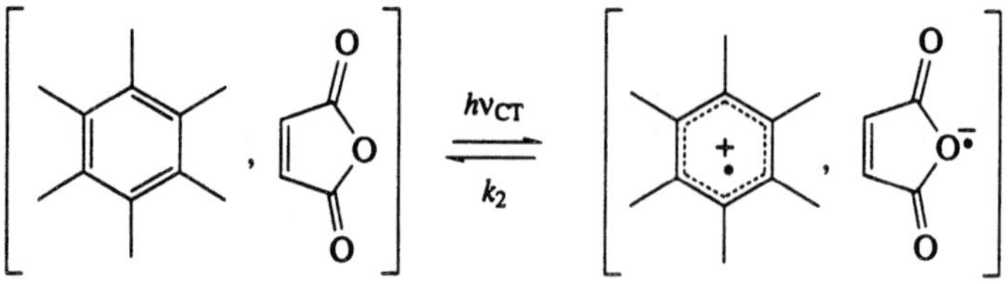

$$[\text{hexamethylbenzene}, \text{maleic anhydride}] \underset{k_2}{\overset{h\nu_{CT}}{\rightleftharpoons}} [\text{hexamethylbenzene}^{+\bullet}, \text{maleic anhydride}^{\overline{\cdot}}] \quad (15)$$

$$[\text{hexamethylbenzene}^{+\bullet}, \text{maleic anhydride}^{\overline{\cdot}}] \xrightarrow{H^{+}} [\text{hexamethylbenzene}^{+\bullet}, \text{hydroxyfuranone radical (OH)}] \xrightarrow{-H^{+}} \text{photoadduct} \quad (16)$$

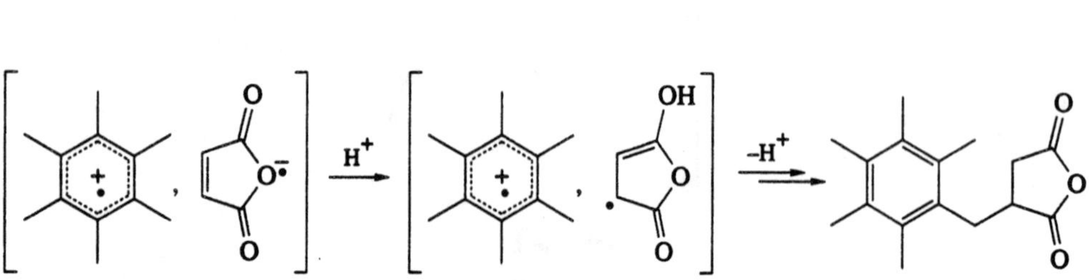

Scheme 4

Owing to the central role of radical cations and radical anions, any general description of electron-transfer oxidation must rely on their individual behavior, as described in the next section.[67,68]

7.4.3.1 Chemistry of Organic Radical Cations

With few exceptions the removal of a bonding electron from the HOMO generates a radical cation of greatly enhanced reactivity in both fragmentation and rearrangement as well as homolytic and electrophilic activity. For purposes of organization the reactions of organic radical cations can be broadly classified according to their kinetic behavior, unimolecular reactions occurring optimally within the contact ion pair $[RH^{+\cdot},A^{-\cdot}]$ and the bimolecular reactions of $RH^{+\cdot}$ taking place largely after diffusive separation as in equation (6a).

Unimolecular reactions of organic radical cations are fragmentation, rearrangement and cyclization, as illustrated by the following generic examples. The specific details of each of these transformations are included in Section 7.4.6. (Note Ar and R represent aryl and alkyl groups, respectively.)

7.4.3.1.1 *α-Fragmentation*

$$ArCH_2CO_2H^{+\cdot} \longrightarrow Ar{-}CH_2^{\cdot} + CO_2 + H^+ \quad (a)$$

$$RCH_2MR_n^{+\cdot} \longrightarrow R{-}CH_2^{\cdot} + R_nM^+ \quad (b)$$

$$RCH_2MR_n^{+\cdot} \longrightarrow R{-}CH_2^+ + R_nM^{\cdot} \quad (c)$$

M = Si, Sn, Pb, Hg, Mg, B, *etc.*

$$RS{-}SR^{+\cdot} \longrightarrow RS^{\cdot} + RS^+ \quad (d)$$

(e) bicyclic azoalkane radical cation ⟶ cyclopentane-1,3-diyl radical cation + N_2

7.4.3.1.2 *β-Fragmentation*

$$Ar{-}CMe_2{-}CMe_2{-}Ar^{+\cdot} \longrightarrow Ar{-}CMe_2^+ + Ar{-}CMe_2^{\cdot} \quad (a)$$

$$Ar_2CH{-}CH_2OR^{+\cdot} \longrightarrow Ar_2CH^+ + RO{-}CH_2^{\cdot} \quad (b)$$

(c) 9,9'-dimethyl dimer of anthracene radical cation ⟶ 9-methylanthracene radical cation + 9-methylanthracene

$$Ph_2CH{-}N_2^{+\cdot} \longrightarrow Ph_2C^{+\cdot} + N_2 \quad (d)$$

7.4.3.1.3 Rearrangement

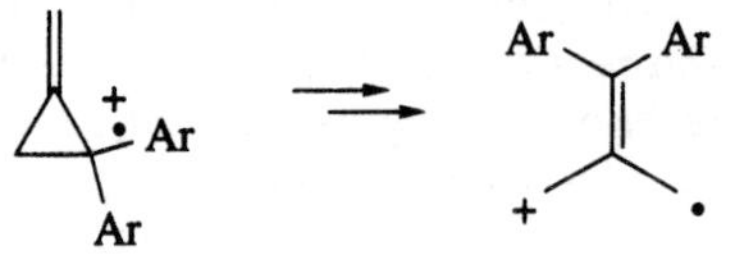

(a)

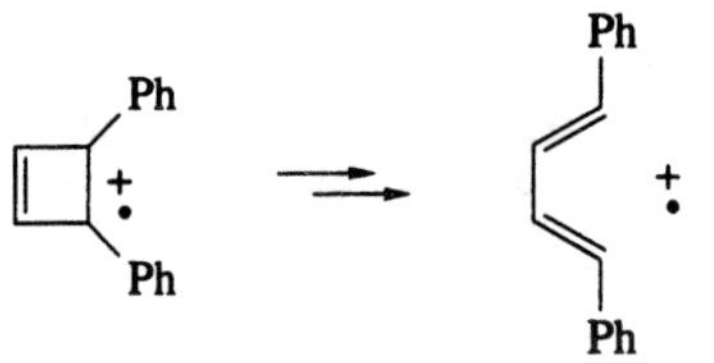

(b)

(c)

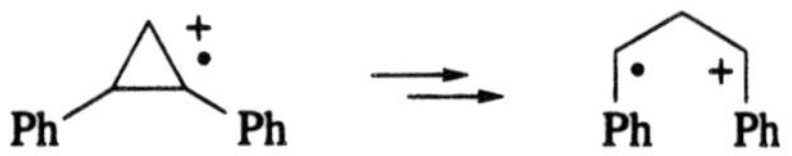

(d)

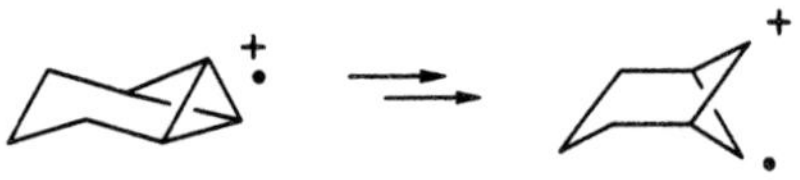

(e)

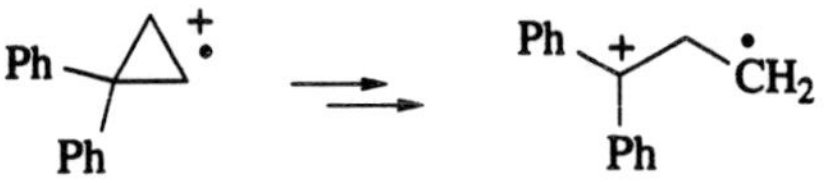

(f)

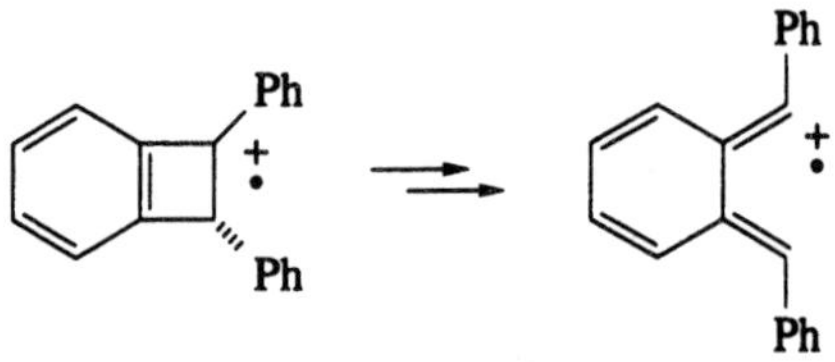

(g)

7.4.3.1.4 Cyclization

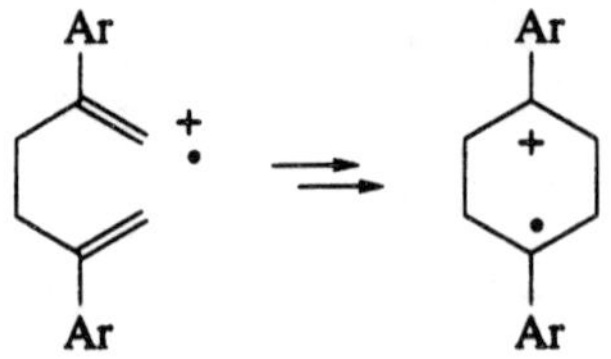

(a)

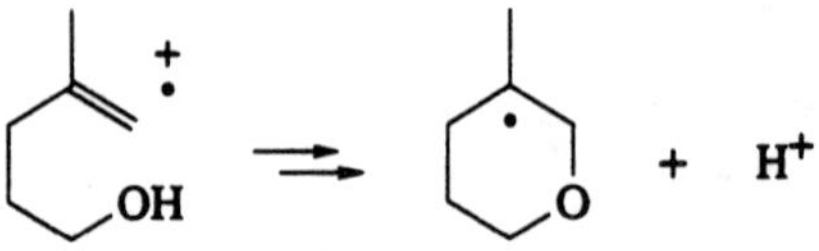

(b)

Bimolecular reactions of organic radical cations relating to their ambivalent character involve the reactions with bases, nucleophiles and radicals.

7.4.3.1.5 *Deprotonation*

$$ArCH_3^{+\bullet} + Py \longrightarrow ArCH_2\bullet + PyH^+ \quad \text{(a)}$$

Py = pyridine

$$Me_3CH^{+\bullet} + H_2O \longrightarrow Me_3C\bullet + H_3O^+ \quad \text{(b)}$$

(Note hydrocarbon radical cations are conjugate acids of the hydrocarbyl radical.)

$$Et_3N^{+\bullet} + MeOH \longrightarrow Et_2N\dot{C}HMe + MeOH_2^+ \quad \text{(c)}$$

7.4.3.1.6 *Nucleophilic addition*

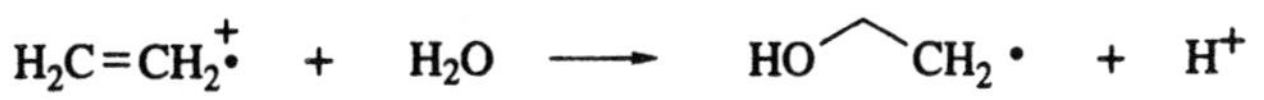

$$H_2C{=}CH_2^{+\bullet} + H_2O \longrightarrow HOCH_2CH_2\bullet + H^+ \quad \text{(a)}$$

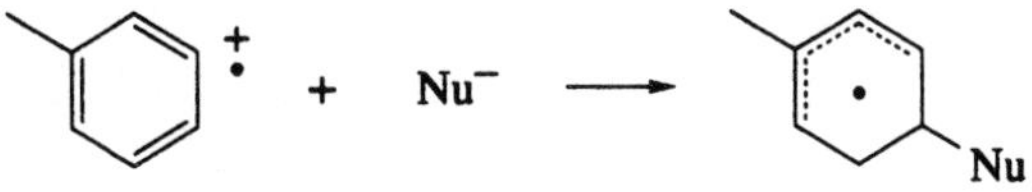

(b)

Nu = amine, OAc^-, OH^-, CN^-, MeOH, *etc.*

7.4.3.1.7 *Dimerization*

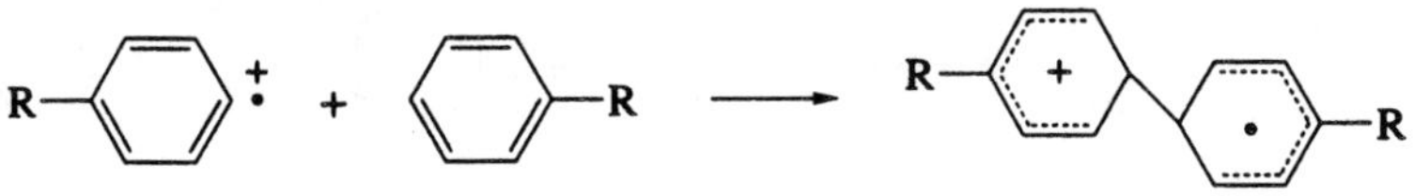

(a)

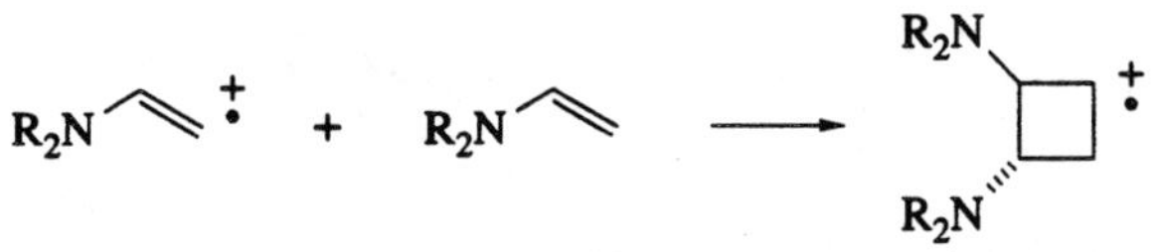

(b)

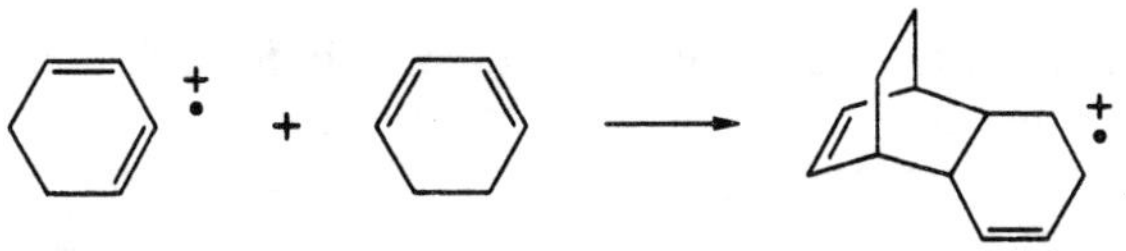

(c)

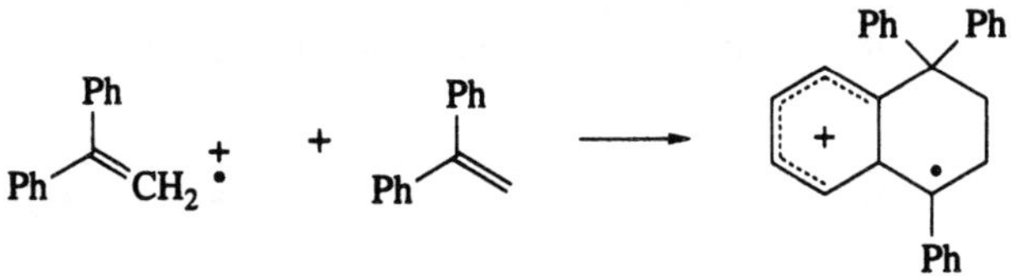

(d)

7.4.3.1.8 *Cycloaddition*

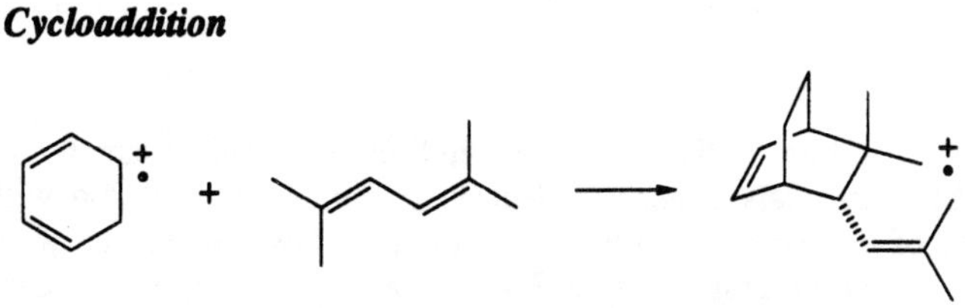

(a)

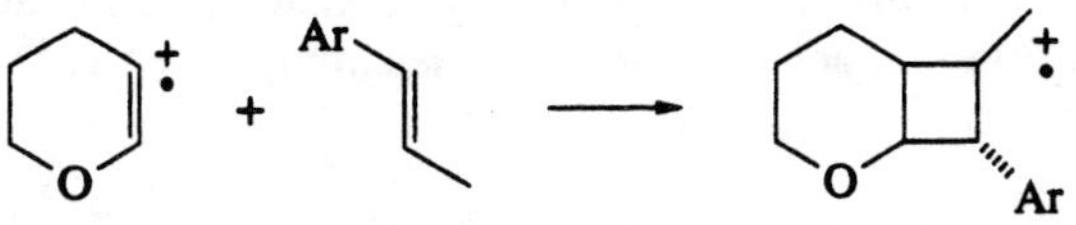

(b)

7.4.3.1.9 Homolytic addition

(a)

NO_2 $NO_2\bullet$

(b)

O_2 O–O•

(c)

S S R S $\overset{+}{S}$ R

7.4.3.1.10 Electron transfer

$$Ar_3N^{+\bullet} + \text{butadiene} \longrightarrow Ar_3N + \text{butadiene}^{+\bullet} \quad (a)$$

(b)

$$O_2^{+\bullet} + R_3N \longrightarrow O_2 + R_3N^{+\bullet} \quad (c)$$

7.4.3.2 The Follow-up Reactions of Organic Radical Cations

Each of the generic reactions of organic radical cations (as presented in Sections 7.4.3.1.1–7.4.3.1.10) generates a new radical and cation center. In the dissociative processes (such as α-fragmentation and deprotonation) the radical and cation centers become separated; the further follow-up oxidations of the uncharged organic radical by electron transfer and ligand transfer are already well described,[69] and the reader is referred to several monographs on free radical chemistry.[70,71] Suffice it to mention two examples here to illustrate this point with the electron-transfer oxidation of acids by LTA (equations 17 and 18).[72]

$$\text{C}_4\text{H}_7\text{–CO}_2\text{H} + Pb(OAc)_4 \xrightarrow{Cu^{II}} \text{C}_4\text{H}_6 + CO_2 + Pb(OAc)_2 \quad (17)$$

$$\text{C}_4\text{H}_7\text{–CO}_2\text{H} + Pb(OAc)_4 \xrightarrow{Cl^-} \text{C}_4\text{H}_7\text{–Cl} + CO_2 + Pb(OAc)_2 \quad (18)$$

Thus the first electron transfer to Pb^{IV} relates to the reaction (a) in Section 7.4.3.1.1, and the second involves the oxidation of the cyclobutyl radicals either by electron transfer/deprotonation with Cu^{II} in equation (17) or by ligand transfer of chlorine with $Pb^{IV}Cl$ in equation (18). When the product of a generic reaction is itself a radical cation (such as in Sections 7.4.3.1.8 and 7.4.3.1.9), an electron-transfer chain or ETC process[73] can ensue, as in the hole-catalyzed cycloadditions and autoxidations of dienes.[74,75] The electron-transfer propagation sequence for the latter is simply given as in equations (19) and (20).

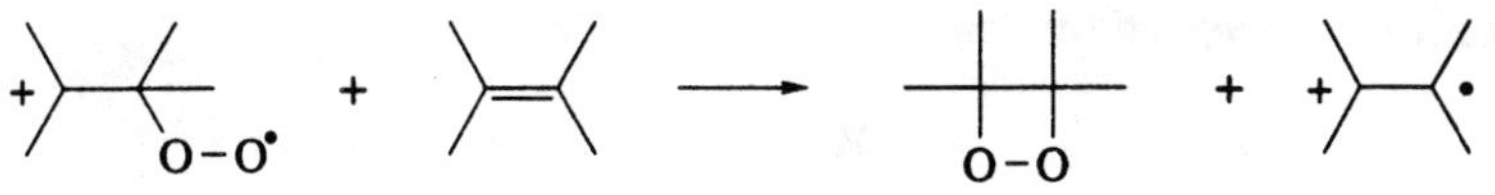

(19)

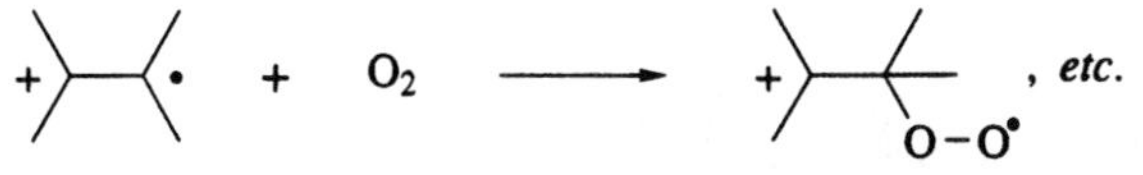

(20)

7.4.3.3 Chemistry of Organic Radical Anions

Electron attachment to the LUMO of an organic acceptor can produce a radical anion that is subject to ready unimolecular decomposition, as given by the following generic examples.[76,77]

7.4.3.3.1 *Fragmentation*

$$\mathrm{ArX}^{\overline{\bullet}} \longrightarrow \mathrm{Ar}\bullet + \mathrm{X}^- \qquad \text{(a)}$$

$$\mathrm{X = I > Br > Cl}$$

$$\mathrm{RX}^{\overline{\bullet}} \longrightarrow \mathrm{R}\bullet + \mathrm{X}^- \qquad \text{(b)}$$

$$\mathrm{Ar_2O}^{\overline{\bullet}} \longrightarrow \mathrm{Ar}\bullet + \mathrm{ArO}^- \qquad \text{(c)}$$

$$\mathrm{RO{-}OR}^{\overline{\bullet}} \longrightarrow \mathrm{RO}^- + \mathrm{RO}\bullet \qquad \text{(d)}$$

$$\mathrm{R = alkyl, H}$$

$$\mathrm{RS{-}SR}^{\overline{\bullet}} \longrightarrow \mathrm{RS}\bullet + \mathrm{RS}^- \qquad \text{(e)}$$

Bimolecular reactions of radical anions are largely restricted to arene acceptors owing to their generally more persistent character. The ambivalence of arene radical anions generally relates to the reactivity towards acids, electrophiles and electron acceptors.

7.4.3.3.2 *Protonation*

$$\mathrm{ArX}^{\overline{\bullet}} + \mathrm{H^+B^-} \longrightarrow \overset{\bullet}{\mathrm{Ar}}(\mathrm{H})(\mathrm{X}) + \mathrm{B}^- \qquad \mathrm{B^- = base} \qquad \text{(a)}$$

7.4.3.3.3 *Electrophilic addition*

$$\mathrm{ArH}^{\overline{\bullet}} + \mathrm{Ac_2O} \longrightarrow \overset{\bullet}{\mathrm{Ar}}(\mathrm{H})(\mathrm{Ac}) + \mathrm{AcO}^- \qquad \text{(a)}$$

7.4.3.3.4 *Electron transfer*

$$\mathrm{ArH}^{\overline{\bullet}} + \mathrm{ArX} \longrightarrow \mathrm{ArH} + \mathrm{ArX}^{\overline{\bullet}} \qquad \text{(a)}$$

$$\mathrm{ArH}^{\overline{\bullet}} + \mathrm{RX} \longrightarrow \mathrm{ArH} + \mathrm{R}\bullet + \mathrm{X}^- \qquad \text{(b)}$$

(See nucleophilic aromatic substitution (S_{RN}) in Volume 4, Chapter 2.1)

$$\mathrm{ArH}^{\overline{\bullet}} + \mathrm{O_2} \longrightarrow \mathrm{ArH} + \mathrm{O_2}^{\overline{\bullet}} \qquad \text{(c)}$$

7.4.3.3.5 Dimerization/disproportionation

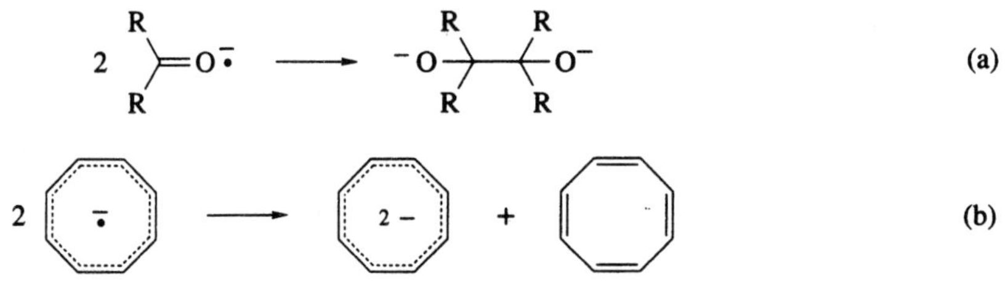

(a)

(b)

Although it may appear that the collapse of the contact ion pair [$RH^{+\cdot}$,$A^{-\cdot}$] with bond formation would frequently be the most favored pathway for its annihilation, only a few examples are presently available. These include the osmylation of arenes to be described in Section 7.4.4.8 (equation 24), as well as the following examples.

7.4.3.3.6 C—C bond formation[78,79]

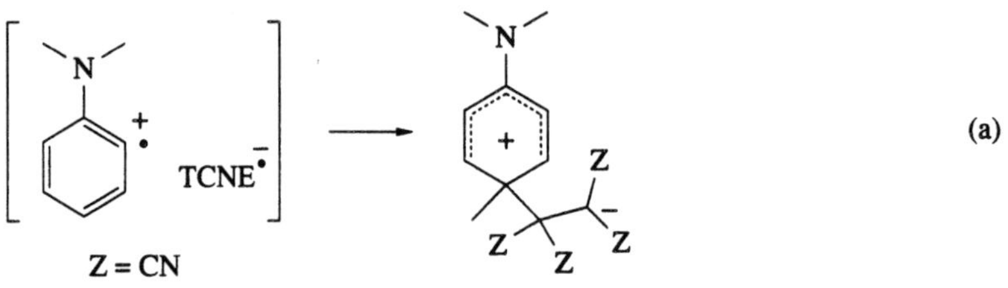

TCNE$^{-\cdot}$

Z = CN

(a)

7.4.3.3.7 [3 + 2] cycloaddition[80]

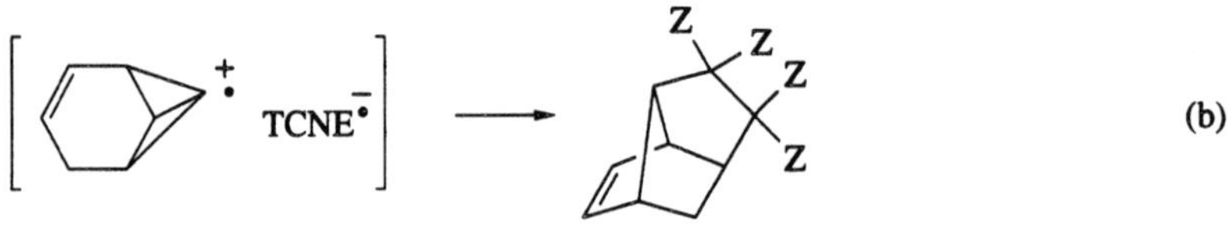

TCNE$^{-\cdot}$

(b)

7.4.3.3.8 [4 + 2] cycloaddition[81]

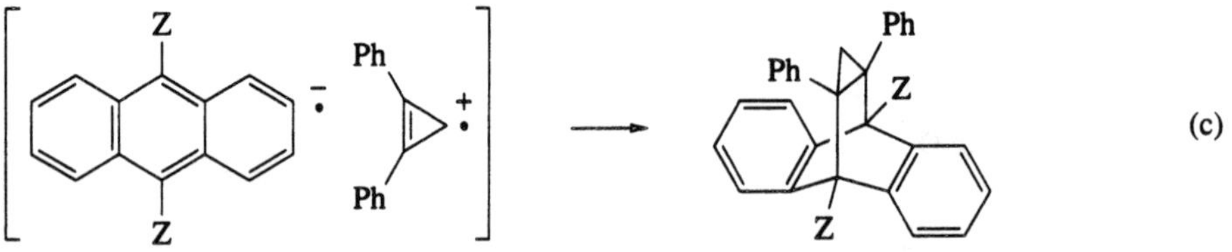

(c)

In each case the formation of the σ-bond(s) between $RH^{+\cdot}$ and $A^{-\cdot}$ must compete with back electron transfer.

7.4.4 THERMAL AND PHOTOCHEMICAL ACTIVATION OF ELECTRON-TRANSFER OXIDATION

When the oxidation–reduction equilibria in equation (6a) are included, the thermal activation of electron-transfer oxidation in equation (3b) follows a course that is akin to the charge-transfer activation in equation (5). In both, the EDA complex [RH,A] is the important precursor which is directly converted into the critical contact ion pair [$RH^{+\cdot}$,$A^{-\cdot}$]. Such an involvement of reactive intermediates in common does widen the scope of electron-transfer oxidations to include both thermal and photochemical processes in related contexts. The latter is especially useful in organic synthesis since a much wider range of organic acceptors become employable as oxidants that are otherwise too weak to effect the thermal oxidation of many organic donors. Accordingly, it is necessary to delineate the intimate relationship between charge-transfer activation and thermal activation, especially with regard to the reactive intermediates.

The osmylation of arenes (Ar) with osmium tetroxide is a particularly informative system with which to illustrate the close interrelationship between the thermal and photochemical activation of electron-transfer oxidation. For example, a colorless solution of osmium tetroxide in *n*-hexane or dichloromethane upon exposure to benzene turns yellow instantaneously.[82] With durene an orange coloration develops and a clear bright red solution results from hexamethylbenzene. The quantitative effects of the dramatic color changes are illustrated in Figure 3 by the spectral shifts of the electronic absorption bands that accompany the variations in aromatic conjugation and substituents. The progressive bathochromic shift parallels the decrease in the arene ionization potentials (*IP*) in the order: benzene 9.23 eV; naphthalene 8.12 eV; anthracene 7.55 eV. Such spectral behaviors are diagnostic of electron donor–acceptor complexes [Ar,OsO_4^-]. According to Mulliken,[83] the new absorption bands derive from charge-transfer excitation with the energetics defined by[84] $h\nu_{CT} = IP - E_A - \omega$, where E_A is the electron affinity of the OsO_4 acceptor and ω is the dissociation energy of the CT excited ion-pair state [$Ar^{+\cdot}$,OsO_4^-].

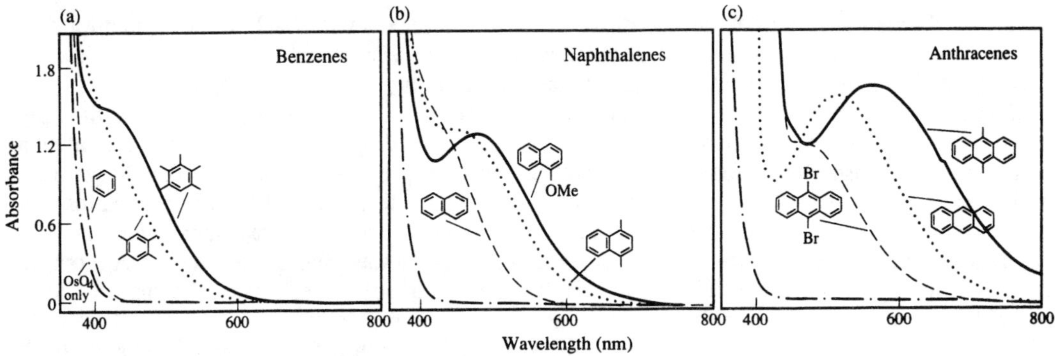

Figure 3 Charge-transfer absorption bands from OsO_4 and: (a) benzene, durene and pentamethylbenzene; (b) naphthalene, 1,4-dimethylnaphthalene, and 1-methoxynaphthalene; (c) 9,10-dibromoanthracene, anthracene and 9,10-dimethylanthracene. Solution of OsO_4 only (—·—)

7.4.4.1 Thermal Osmylation of Naphthalene, Anthracene and Phenanthrene

Benzene shows no signs of osmylation in the absence of light, as indicated by the persistence of the yellow color of the [C_6H_6,OsO_4] complex in *n*-hexane even upon prolonged standing. On the other hand, the orange CT color of the phenanthrene complex [$C_{14}H_{10}$,OsO_4] slowly diminishes over a period of weeks, accompanied by the formation of a dark brown precipitate of composition $C_{14}H_{10}OsO_4$. Dissolution of the solid in pyridine yields the 1:1 adduct (**1**; $C_{14}H_{10}OsO_4Py_2$) as the sole product in very low conversion. Anthracene behaves similarly to afford the 2:1 adduct in 10% conversion only after two months. The thermal osmylation can be expedited in a purple solution of refluxing *n*-heptane (100 °C) to effect a 68% conversion in 30 h. However even at these relatively elevated temperatures naphthalene is converted to the corresponding 2:1 adduct to only a limited extent. In every case the dark brown primary adducts are easily collected from the reaction mixture as insoluble solids, and then immediately ligated with pyridine for structural characterization. Indeed the characteristic IR and ^1H NMR spectra of the anthracene, phenanthrene and naphthalene adducts (**2**), (**1**) and (**3**) respectively, allow the ready analysis of the osmylated adducts. Since these adducts are derived from the arenes with only OsO_4 present, the chemical transformation is hereinafter designated as the direct thermal or DT osmylation. For comparison, the same polynuclear arenes can be osmylated in the presence of promoter bases, typically pyridine. Under these conditions the adducts (**2**), (**1**) and (**3**) are formed directly in the reaction mixture and at substantially increased rates of reaction, as previously established with the related family of alkene substrates.[85] Such a procedure differs visually from the DT osmylation described above in that the charge-transfer colors are not observed as transients, owing to the preferential complexation of OsO_4 with pyridine. Accordingly, this promoted thermal or PT osmylation is to be distinguished by the enhanced reactivity of the pyridine complex relative to the free OsO_4 in the DT osmylation. The corresponding increase in the yields of adducts such as (**2**), (**1**) and (**3**) within a shorter span of reaction times is apparent from the comparison of the results of DT and PT osmylations.

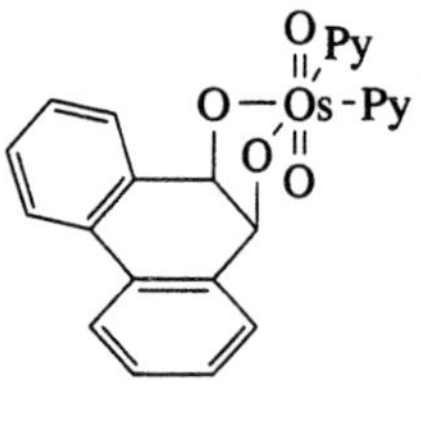

(**1**) 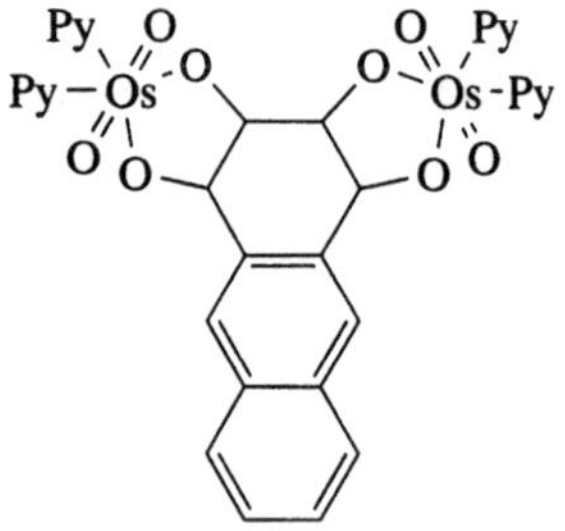 (**2**) 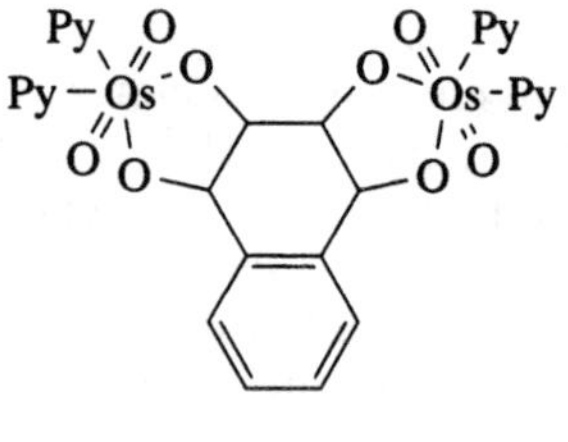 (**3**)

7.4.4.2 Charge-transfer Osmylation of Benzene, Naphthalene and Anthracene

The various charge-transfer colors for the different arene complexes with OsO_4 are persistent for days. However when the colored solutions are deliberately exposed to visible light with energy sufficient to excite only the charge-transfer band, they always deposit a highly insoluble, dark brown solid of the OsO_4 adducts obtained from the direct thermal osmylation of arenes (*vide supra*). Since this actinic process must have arisen *via* the electronic excitation of the EDA complex, it is referred to hereafter as charge-transfer or CT osmylation for individual arenes. For example, the irradiation of the charge-transfer bands (see Figure 3) of the OsO_4 complexes with various benzenes, naphthalenes and phenanthracene yields the same osmylated adducts such as (**3**) and (**2**) described above.[86] Anthracene is unique in that it affords two entirely different types of products upon the photoexcitation of the EDA complex $[C_{14}H_{10},OsO_4]$ in dichloromethane and hexane, despite only minor solvent effects on the charge-transfer bands. Irradiation of the purple solution of anthracene and OsO_4 in dichloromethane at $\lambda > 480$ nm yields the 2:1 adduct (**2**) together with its *syn* isomer as the sole products. On the other hand, irradiation of the same purple-colored solution in *n*-hexane under otherwise identical conditions leads to a small amount of polymeric osmium dioxide $(OsO_2)_x$. Work-up of the hexane solution yields anthraquinone as the major product contaminated with only traces (<1%) of the 2:1 adduct (**2**). Interestingly, even higher yields of anthraquinone are obtained from 9-bromo-, 9-nitro- and 9,10-dibromo-anthracene when the CT osmylation is carried out in *n*-hexane. Such an accompanying loss of the electronegative substituents (X = Br, NO_2) probably occurs *via* osmylation at the *meso* (9,10) positions followed by oxidative decomposition of the unstable adduct with the stoichiometry shown in equation (21).

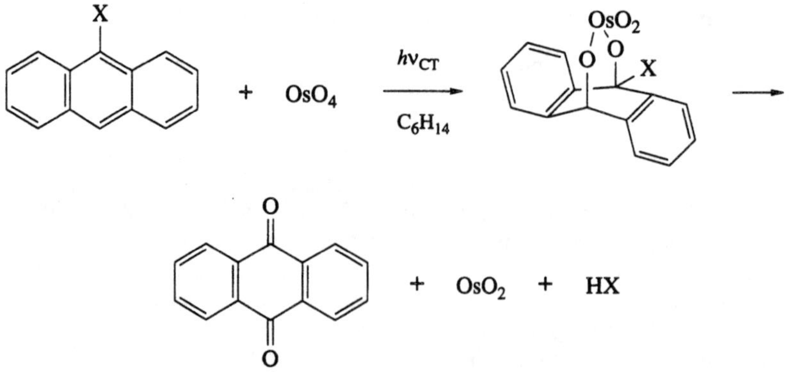

$$\text{9-X-anthracene} + OsO_4 \xrightarrow[C_6H_{14}]{h\nu_{CT}} \text{9,10-}(O_2OsO_2)\text{ adduct} \longrightarrow \text{anthraquinone} + OsO_2 + HX \qquad (21)$$

7.4.4.3 Time-resolved Spectra of Arene Radical Cations in Charge-transfer Osmylation

In order to identify the reactive intermediates in the charge-transfer excitation of arene–OsO_4 complexes, the time-resolved spectra are measured immediately following the application of a 30 ps pulse consisting of the second harmonic at 532 nm of a mode-locked Nd:YAG laser. The wavelength of this excitation source corresponds to the maxima (or near maxima) of the charge-transfer absorption bands of the series of anthracene complexes with osmium tetroxide illustrated in Figure 3(c). Accordingly, the time-resolved spectra from the anthracene–OsO_4 system relate directly to the CT osmylation since there is no ambiguity about either the adventitious local excitation[83] of complexed (or uncomplexed) chromophores, or the photogeneration of intermediates that did not arise from the photoexcitation of the EDA complex. Indeed, intense absorptions are observed in the visible region between 700 and 800 nm from the excitation of the anthracene–OsO_4 complex, as shown in Figure 4(a). This time-resolved absorption spectrum from anthracene is obtained in the time interval of ~30 ps following the application of the 532 nm laser pulse. Comparison with the steady-state absorption spectrum of the anthracene radical cation (see inset Figure 4(a) generated by the spectroelectrochemical technique,[87] thus establishes the identity of the charge-transfer transient. Similar time-resolved spectra of arene radical cations are obtained from various anthracene and naphthalene EDA complexes despite the excitation of only the low-energy tails of the CT bands in Figure 3 with the 532 nm laser pulse. The evolution of the anthracene radical cation is followed by measuring the absorbance change at λ_{max} = 742 nm upon the charge-transfer excitation of the EDA complex with a single laser shot of ~10 mJ. The time evolution of the absorbance shown in Figure 4(b) includes the initial onset for ~20 ps owing to the rise time of the 30 ps (fwhm) laser pulse. The first-order plot of the decay portion is shown in the inset to the figure. Decay curves similar to those shown in Figure 4(b) are also observed for the disappearances of the radical cations derived from all of the other arene–OsO_4 complexes. In each case the highest concentration can be obtained of the arene radical cation, the decays of which are all first-order processes. The magnitudes of the rate constant k_1 are applicable to the complete disappearance of $Ar^{+\cdot}$, as indicated by the return of the radical cation absorbances to the baseline.

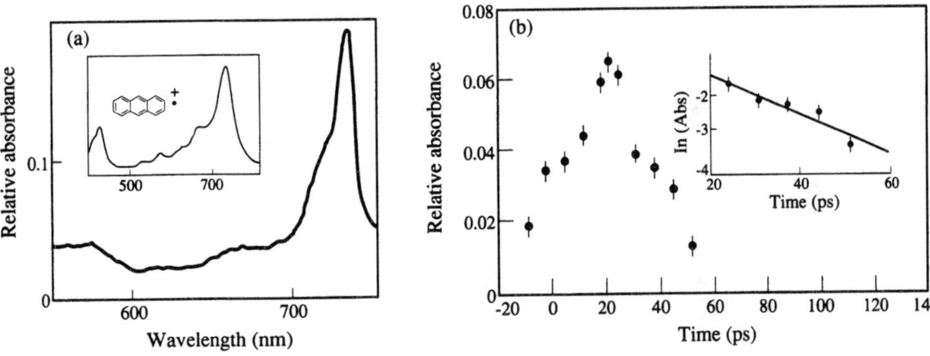

Figure 4 (a) Transient absorption spectrum of the cation radical from anthracene at ~40 ps following the 532 nm CT excitation of the OsO_4 complex with a 30 ps (fwhm) laser pulse. The inset is the steady-state spectrum of $Ar^{+\cdot}$ obtained by spectroelectrochemical generation. (b) Appearance and decay of the radical cation from anthracene by following the change of the absorbance at λ_{max} = 742 nm. The inset shows the first-order plot of the absorbance decay subsequent to the maximum at ~20 ps

7.4.4.4 Common Features in Thermal and Charge-transfer Osmylations

The [Ar,OsO_4] complexes are involved as the common precursors in the oxidative addition of osmium tetroxide to various arenes by the three independent procedures designated as direct thermal (DT), promoted thermal (PT) and charge-transfer (CT) osmylation. For example, the anthracenes react rather slowly with osmium tetroxide *via* the EDA complex to effect DT osmylation in nonpolar solvents and afford 2:1 adducts that are then converted to the more tractable pyridine derivatives such as (**2**). Alternatively, the same ternary product (**2**) is directly formed at a significantly enhanced rate by the PT osmylation of anthracene with a mixture of OsO_4 and pyridine. Finally, the OsO_4 adduct to anthracene is instantly produced by CT osmylation involving actinic excitation of the [Ar,OsO_4] precursor complex. As such, the three procedures represent different activation mechanisms for arene oxidation. Thus DT and PT osmylations are adiabatic processes in which the transition states are attained *via* the collapse of an arene donor with the OsO_4 and the base-coordinated OsO_4(Py) electrophile, respectively. On the other hand, CT osmylation is a nonadiabatic process resulting from the vertical excitation of the [Ar,OsO_4] complex. For the latter, time-resolved picosecond spectroscopy can define the relevant photophysical and photochemical events associated with the charge-transfer excitation of an arene EDA complex, as previously established with arene complexes involving other electron acceptors. Accordingly, the CT osmylation is delineated first and then related to DT and PT osmylation. Before proceeding, however, it is important to emphasize that the DT, PT and CT osmylations all share in common the formation of the 1:1 osmium(VI) cycloadduct $ArOsO_4$ in the initial rate-limiting step, since the concomitant loss of aromaticity produces a reactive alicyclic diene that is highly susceptible to the further thermal osmylation.[88] The universal adherence to the 2:1 adduct $Ar(OsO_4)_2$ (except phenanthrene), irrespective of the molar ratios of arene/OsO_4 and the particular procedure employed, accords with the rapid addition of a second mole of OsO_4 in DT, PT and CT osmylations. This allows the focus on the formation of a single intermediate $ArOsO_4$ in order to delineate the unifying activation processes for DT, PT and CT osmylations.

7.4.4.5 Electron Transfer in the Charge-transfer Osmylation of Arenes

The direct observation of the reactive intermediates by the use of time-resolved picosecond spectroscopy and fast kinetics (Figure 4) enables the course of CT osmylation to be charted in some detail. The analysis proceeds from the mechanistic context involving the evolution and metamorphosis of the CT ion pair, as summarized in Scheme 5 (the brackets denote solvent-caged pairs) for the critical initial step (equation 24) to form the 1:1 adduct to a benzene donor.

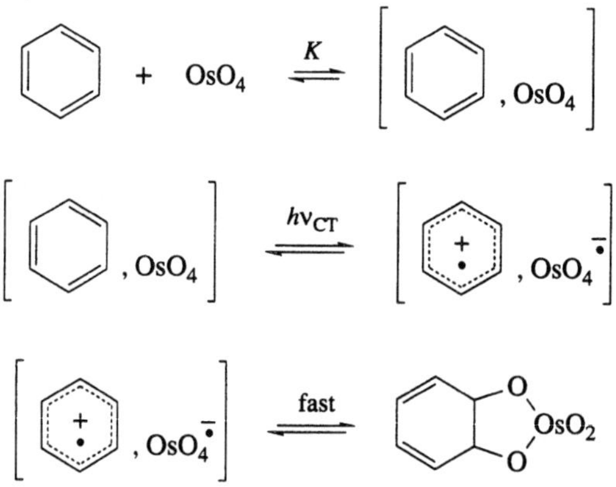

$$C_6H_6 + OsO_4 \overset{K}{\rightleftharpoons} [C_6H_6, OsO_4] \quad (22)$$

$$[C_6H_6, OsO_4] \overset{h\nu_{CT}}{\rightleftharpoons} [C_6H_6^{+\bullet}, OsO_4^{-\bullet}] \quad (23)$$

$$[C_6H_6^{+\bullet}, OsO_4^{-\bullet}] \xrightarrow{\text{fast}} C_6H_6O_2OsO_2 \quad (24)$$

Scheme 5

All the experimental observations on CT osmylation indeed coincide with the formulation in Scheme 5. Thus the exposure of arene to osmium tetroxide leads immediately to new absorption bands (Figure 3) that are readily associated with the formation of the EDA complex in equation (22). These binary complexes are always present in low steady-state concentrations owing to the limited magnitudes of K determined by the Benesi–Hildebrand method. The complexes are so weak that every attempt at isolation, including the freezing of various mixtures of OsO_4 in neat aromatic donors, merely leads to phase separation. The absorption bands are thus properly ascribed to contact charge transfer, as formulated by Orgel and Mulliken,[89] who predicted the CT absorption bands in these EDA complexes to be associated with the electronic excitation to the ion-pair state (equation 23). As such, the time-resolved spectrum in Figure 4(a) indicates that the formation of the arene radical cation occurs within the rise time of the 30ps laser pulse. (The accompanying presence of the perosmate(VII) (OsO_4^-) counteranion is obscured by the arene absorptions.) The electron transfer from the arene donor to the OsO_4 acceptor in the EDA complex in equation (23) effectively occurs with the absorption of the excitation photon ($h\nu_{CT}$), in accord with Mulliken's theory. Furthermore the appearance at <30 ps demands that $Ar^{+\bullet}$ and $OsO_4^{-\bullet}$ are born as an intimate ion pair with a mean separation essentially that of the precursor complex [Ar,OsO_4] since this timescale obviates significant competition from diffusional processes. The seminal role of the ion pair [Ar^+,OsO_4^-] as the obligatory intermediate from the photoexcitation of the EDA complex must be included in any formulation of CT osmylation, by taking particular note of how it decays. The spontaneous collapse of the CT ion pair in equation (24) represents the most direct pathway to arene cycloaddition—the measured half-life of $\tau \approx 35$ ps for the disappearance of the anthracene radical cations in Figure 4(b) largely precluding diffusive separation of such ion pairs. However the magnitudes of the product quantum yield, $\phi_p \approx 10^{-2}$, indicate that the primary route for ion-pair decay is the back electron transfer (k_2) as the reverse step of equation (23). Such an energy-wasting process with an estimated rate constant of $k_2 \approx 10^{11}$ s^{-1}, derives from a highly exergonic driving force that is estimated to be $\Delta G \approx -30$ kcal mol^{-1}, based solely on the standard redox potentials of $E° = +1.30$ and -0.06 V for anthracene and the perosmate(VII) anion, respectively. More relevant to this issue is an estimated first-order rate constant for cycloaddition of $k_c = 10^9$ s^{-1} for the ion-pair collapse to the arene cycloadduct in Scheme 5. Such a relatively large rate constant also points to a highly exergonic (bond-making) process for the cycloaddition in equation (24). Therefore the selectivity in adduct formation can be considered for various polynuclear arenes in which the initial addition of OsO_4 is possible at several sites. The regiospecificity observed in the CT osmylation of phenanthrene and 1,4-dimethylnaphthalene to produce only one isomeric adduct, (**1**) and (**3**), respectively, accords with the reactive site centered on the arene HOMO.[90] However in the extended polynuclear anthracenes the separation of the HOMO and subjacent SHOMO (*i.e.* HOMO-1) is not so well delineated,[91] and the regiospecificity is strikingly modulated by solvent polarity. Ion-pair annihilation is known to occur with the greatest ease in highly nonpolar alkanes.[92] Accordingly in *n*-hexane as solvent, the immediate collapse of the first-formed ion pair (**4**) centered at the anthracene HOMO is expected to occur at the *meso* (9,10) positions. Such an ion-pair collapse would produce anthraquinone in a manner similar to that presented in equation (21). On the other hand, the formation of only adduct (**2**) from the initial addition of OsO_4 to the terminal ring (**5**) represents a very unusual regiospecificity insofar as other addition (and substitution) reactions of anthracene are concerned. It suggests that the initially formed HOMO ion pair (HIP) has time to relax in the more polar dichloromethane medium to the isomeric SHOMO ion pair (SIP) that rapidly leads to adduct (**2**). This proposal receives support from the observation of adducts related to (**2**) from the CT osmylation of both 9-methyl- and 9,10-dimethyl-anthracene in hexane. The enhanced stability of the radical cations from these relatively electron-rich anthracenes will optimize the opportunity to convert the HIP to the more reactive SIP even in the nonpolar hexane medium, particularly if the collapse of the former is reversible.

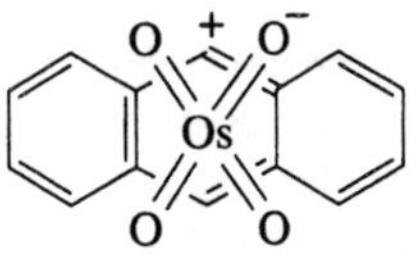

(**4**) 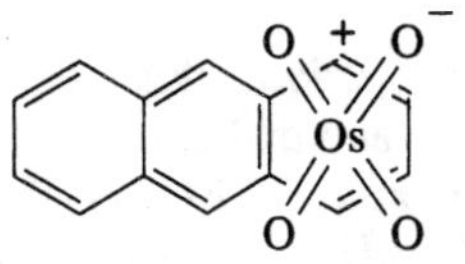 (**5**)

7.4.4.6 Electron Transfer as the Common Theme in Arene Osmylation

The wide-ranging reactivity of various aromatic hydrocarbons to OsO_4 offers the unique opportunity to probe the activation process for oxidative osmylation, especially with regard to the role of the EDA complex and the reactive intermediates. In particular, the deliberate photoexcitation ($h\nu_{CT}$) of the EDA complex in hexane or dichloromethane effectively activates various arenes including benzenes, naphthalenes and anthracenes to CT osmylation. This photoactivated process is readily associated with the charge-transfer ion pair, *i.e.* equation (25), as established by the growth and decay of arene radical cations with the aid of time-resolved picosecond spectroscopy. When kept in the dark, the same solutions of the EDA complexes slowly afford arene–OsO_4 adducts that are identical to those derived by CT osmylation. Indeed the close kinship between the thermal and charge-transfer activation of osmylation is underscored by the unique adduct (**2**) in which OsO_4 addition occurs exclusively to the terminal ring and not to the usual *meso* (9,10) positions of anthracene. The activation process to form the kindred adiabatic ion pair $[Ar^+,OsO_4^-]_s$ in the thermal osmylation provides the unifying theme in arene oxidation. Furthermore the promoted thermal osmylation of arenes *via* the five-coordinate pyridine analog OsO_4Py is related to the widely used procedure for alkene bishydroxylation[88] and the same regiochemistry is observed, especially with anthracene donors, indicating that the activated complex for PT osmylation is strongly related to that for DT osmylation.

$$[Ar, OsO_4] \xrightarrow{h\nu_{CT}} [Ar^{+\bullet}, OsO_4^{-\bullet}] \xrightarrow{fast} ArOsO_4, \textit{etc.} \tag{25}$$

The variable regiochemistry observed in the collapse of $[Ar^{+\bullet},OsO_4^{-\bullet}]$ to the cycloadduct $ArOsO_4$ underscores the importance of CIP structures in determining the course of electron-transfer oxidation. Since CIP structures are not readily determined as yet, the structural effects induced by qualitative changes in solvent polarity, salts, additives and temperature are reaction variables that must always be optimized in the synthetic utilization of electron-transfer oxidation by either thermal or photochemical activation.

7.4.5 ELECTRON-TRANSFER OXIDATION *VERSUS* ELECTROPHILIC OXIDATION

With oxidants such as Mn^{VII}, Cr^{VI}, Bi^{V}, Pb^{IV}, Tl^{III}, Pd^{II}, *etc.* (as well as most organic electron acceptors) that are capable of several electron changes, the multistep electron-transfer oxidation of an organic donor must be distinguished from the one-step electrophilic process. Thus a series of one-electron transformations will always have a concerted counterpart. This dichotomy can be considered in the oxidative thallation and mercuration of arenes with thallium(III) and mercury(II) trifluoroacetates, in which kinetic studies establish the principal active forms of the electrophile to be the cationic $Tl(O_2CCF_3)_2^+$ and the neutral $Hg(O_2CCF_3)_2$, respectively.[93] Thus the ionic dissociation of mercury(II) trifluoroacetate to the cation is not important, even in the polar trifluoroacetic acid with a dielectric constant of $\varepsilon_{CF_3CO_2H} = 42.1$. On the other hand, the uncharged form of thallium(III) trifluoroacetate is an inactive electrophile. As such, thallation requires prior dissociation primarily to the monocation for activation even in the nonpolar dichloromethane ($\varepsilon_{CH_2Cl_2} = 9.08$). The second dissociation to the dication (equation 26) can also be relevant (*vide infra*). Although the active electrophiles $Hg(O_2CCF_3)_3$ and $Tl(O_2CCF_3)_2^+$ are isoelectronic (and probably isostructural) species, they basically differ in the charge they bear. Accordingly, the activation processes for mercuration and thallation show strong similarities, although at the same time they exhibit some striking differences.

$$Tl(O_2CCF_3)_3 \rightleftharpoons Tl(O_2CCF_3)_2^+ + CF_3CO_2^- \rightleftharpoons TlO_2CCF_3^{2+} + 2\,CF_3CO_2^- \qquad (26)$$

7.4.5.1 EDA Complexes as Intermediates in Mercuration and Thallation. Comparison of Their Ground and CT Excited States

Quantitative spectrophotometric analysis establishes the transient charge-transfer absorption spectra observed during mercuration and thallation to derive from the same electrophilic species involved in the kinetics, *viz.* $Hg(O_2CCF_3)_2$ and $Tl(O_2CCF_3)_2^+$, respectively. Indeed these species form two series of arene–EDA complexes, $[ArH,Hg(O_2CCF_3)_2]$ and $[ArH,Tl(O_2CCF_3)_2^+]$, which bear strong resemblances to each other, both in the ground state and in the CT excited state (CIP). Ground state similarities of the mercury(II) and thallium(III) EDA complexes are reflected in the linear relationship of the association constants (log *K*) in Figure 5(a), which indicates that the stabilization of both series of EDA complexes is affected in the same way with changes in the arene structure. The correlation with a slope of 1.4 indicates that the cationic complexes $[ArH,Tl(O_2CCF_3)_2^+]$ are stabilized about 40% more than their neutral counterparts $[ArH,Hg(O_2CCF_3)_2]$. CT excited state similarities of the arene complexes of $Hg(O_2CCF_3)_2$ and $Tl(O_2CCF_3)_2^+$ are revealed in the parallel trend in their absorption bands in Figure 5(b). Thus for weak electron donor–acceptor complexes of the type described as EDA complexes, the spectral transition $h\nu_{CT}$ represents an electronic excitation of the arene moiety from the neutral ground state to the contact ion pair, *i.e.* its photoionization to an electrophilic acceptor.[83] For the mercury(II) complexes the relevant CT transition corresponds to equation (27), and for the thallium(III) complexes it is as shown in equation (28).

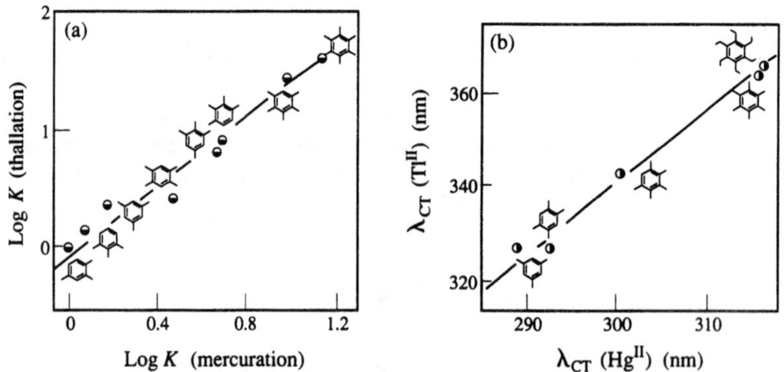

Figure 5 Ground-state and excited-state similarities of the arene π-complex of mercury(II) and thallium(III), as shown by (a) association constants (*K*) and (b) CT absorption spectra (λ_{CT})

$$[ArH,\ Hg^{II}(O_2CCF_3)_2] \xrightarrow{h\nu_{CT}} [ArH^{+\bullet},\ Hg^{I}(O_2CCF_3)_2^{-\bullet}\ ] \qquad (27)$$

$$[\mathrm{ArH}, \mathrm{Tl^{III}(O_2CCF_3)_2^+}] \xrightarrow{h\nu_{CT}} [\mathrm{ArH^{\overset{+}{\bullet}}}, \mathrm{Tl^{II}(O_2CCF_3)_2^{\bullet}}] \quad (28)$$

Both series of EDA complexes share the arene cation in the form of the radical pair shown in the brackets in equations (27) and (28), as the CT excited state according to expectations of Mulliken theory. The slope of 1.3 in the linear correlation (Figure 5b) indicates that the energy of the CT excitation of the cationic EDA complex from $Tl(O_2CCF_3)_2^+$ is merely ~30% more sensitive to changes in arene structure compared to that derived from $Hg(O_2CCF_3)_2$. Otherwise both series of EDA complexes show parallel behavior in the transformation to the CT excited state. The similarity in the EDA complexes of $Hg(O_2CCF_3)_2$ and $Tl(O_2CCF_3)_2^+$ with the series of sterically crowded 1,3,5-trialkylbenzenes indicates that the CT interaction occurs at relatively long range. The latter is confirmed by the pair of long Hg—C bond distances in the η^2-bonding of the electrophilic mercury(II) to the hexamethylbenzene donor as determined by X-ray crystallography.[94] As a result, any minor difference which may exist in the steric properties of $Hg(O_2CCF_3)_2$ and $Tl(O_2CCF_3)_2^+$ is expected to be obscured in the EDA complexes, both in the ground state and in the CT excited state.

7.4.5.2 Comparison of the Activation Barriers for Mercuration and Thallation

The kinetic studies also establish the neutral $Hg(O_2CCF_3)_2$ and the cationic $Tl(O_2CCF_3)_2^+$ to be the principal electrophiles in mercuration and thallation,[95] respectively. The reactivity trends in the two types of metallations are quantitatively compared with a graded series of arene ranging from the electron-rich mesitylene at one end to the least reactive arene chlorobenzene at the other extreme. The relative reactivity of an arene to electrophilic metallation is represented by the activation free energy difference: $\Delta G_r^{\ddagger} = -2.3RT \log k_{rel}$, where $k_{rel} = k/k_0$ represents the second-order rate constant relative to that of benzene (k_0) arbitrarily chosen as the reference arene. The direct comparison between mercuration and thallation is shown in Figure 6, in which the logarithms of the ratio of second-order rate constants for mercuration are plotted against those for thallation in trifluoroacetic acid. The striking linear free energy correlation spans more than six orders of magnitude in rate with a 1:1 relationship, as shown by the fit of the data to the line drawn with a slope of unity. In other words, those factors relevant to surmounting the activation barrier for mercuration are mirrored in exactly the same way during thallation as a consequence of systematic changes in the arene donor.

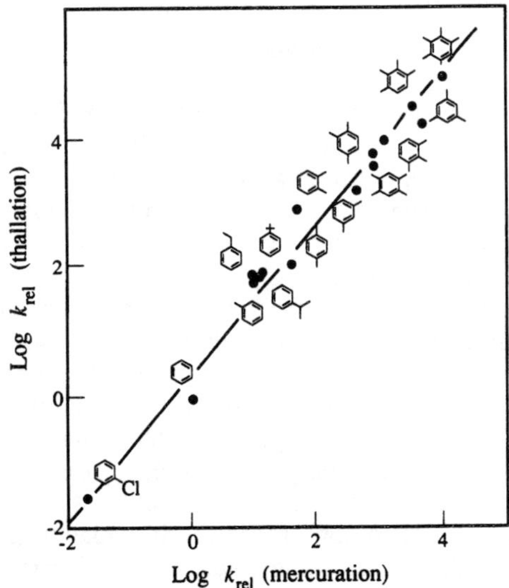

Figure 6 Direct relationship of the relative reactivities (k_{rel}) of arenes in mercuration and thallation

7.4.5.3 Correlation of the Rates of Mercuration and Thallation with the CT Excitation Energies of the EDA Complexes

The relative reactivities of arenes to metallation are represented by the activation free energy difference $\Delta G_r^\ddagger$ (*vide supra*). In the same way, the transition energy $h\nu_{CT}$ associated with the charge-transfer excitation of the EDA complex can be evaluated from the absorption spectrum (λ_{CT}) by a similar comparative method, *i.e.* $\Delta h\nu_{CT} = h\nu_{CT} - h\nu^0_{CT}$, where $h\nu^0_{CT}$ is the CT transition energy of the benzene–EDA complex. In the comparative method the values of $\Delta h\nu_{CT}$ focus primarily on the contribution from the arene moiety, since the electrophile component largely cancels out in the difference procedure. In the correlation of the activation barriers with the CT excitation energies, a linear plot is observed for the CT transition energy with a slope of close to unity. In other words these mercuration rates relate to the CT excited state of the EDA complex with a free energy relationship described as: $\log k/k_0 = -\Delta h\nu_{CT}/2.3\,RT + \text{constant}$.

7.4.5.4 The Relevance of Arene Radical Cations in Electrophilic Aromatic Substitution

The linear free energy relationship observed for arene donors relates the activation barrier $\Delta G^\ddagger$ for aromatic substitution directly to the CT transition energy $h\nu_{CT}$ of the EDA complex. Since $h\nu_{CT}$ pertains to the energetics of the photoionizations in equations (27) and (28), the correlation suggests that these arene contact ion pairs are reasonable approximations to the transition states for both mercuration and thallation, *e.g.* Scheme 6.

$$[\mathrm{ArH}, \mathrm{Hg(O_2CCF_3)_2}] \underset{k_{-e}}{\overset{k_e}{\rightleftharpoons}} [\mathrm{ArH}^{+\bullet}, \mathrm{Hg(O_2CCF_3)_2}^{-\bullet}] \quad (29)$$

$$[\mathrm{ArH}^{+\bullet}, \mathrm{Hg(O_2CCF_3)_2}^{-\bullet}] \xrightarrow{k_c} \overset{+}{\mathrm{Ar}}(\mathrm{H})\mathrm{Hg(O_2CCF_3)_2^-} \quad (30)$$

$$[\mathrm{ArH}^{+\bullet}, \mathrm{Hg(O_2CCF_3)_2}^{-\bullet}] \xrightarrow{k_d} \mathrm{ArH}^{+\bullet} + \mathrm{Hg(O_2CCF_3)_2}^{-\bullet} \quad (31)$$

$$\overset{+}{\mathrm{Ar}}(\mathrm{H})\mathrm{Hg(O_2CCF_3)_2^-} \xrightarrow{k_s} \mathrm{ArHgO_2CCF_3} + \mathrm{CF_3CO_2H} \quad (32)$$

Scheme 6

In Scheme 6 the activation step for electrophilic substitution proceeds by electron transfer (k_e) to the arene ion pair, which is to be likened to the photoactivation of the EDA complex in equations (27) and (28). The substitution product determining process (k_s) is then dependent on the rate constants k_c and k_d for collapse to the Wheland intermediate and diffusive separation to free ions, respectively. Since the rate constant k_d is likely to be invariant between a given electrophile and a series of structurally related arenes, the rate of aromatic substitution will be strongly mediated by the value of k_c. As the rate of arene ion pair collapse is retarded, the competition from back electron transfer (k_{-e}) will become increasingly important. Under these circumstances the reaction rate will no longer follow the linear free energy relationship (*vide supra*). Indeed at some point back electron transfer will dominate (*i.e.* $k_{-e} >> k_c$), and it is conceivable that little or no thermal reaction will take place. The experimental variables pertaining to electronic effects, steric effects, solvent effects, and product studies can be reconciled with Scheme 6 in terms of the facility with which such an arene ion pair collapses. For example, electronic effects in the collapse of the arene ion pair can be viewed as the influence of substituents on the spin density in the singly occupied orbital (SOMO) requisite to bond formation in equation (30). With mesitylene as the donor, the pair collapse at any of the three unsubstituted nuclear positions is unimpeded since the SOMO is degenerate. Similarly the SOMOs in the cations of *p*-xylene, *m*-xylene, and pseudocumene are conducive to pair collapse at the free positions owing to the available spin densities. By contrast, the SOMOs of the cations of durene and pentamethylbenzene have nodes at the free 3- and 6-positions and collapse therefore is not favored at these positions. Indeed the lack of spin density at these positions of durene and pentamethylbenzene cations accords with the observed magnitudes of the ^{1}H hyperfine splittings in the ESR spectra,[96] *i.e.* (**6**) and (**7**) respectively.

Thus the pair collapse of durene and pentamethylbenzene cations is only favored at the already substituted *ipso* positions, and is probably a reversible process. Product studies provide a divergent view of

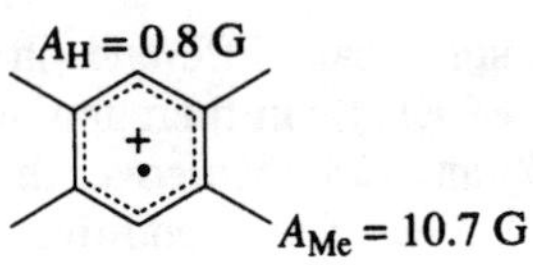

$A_H = 0.8$ G

$A_{Me} = 10.7$ G

(6)

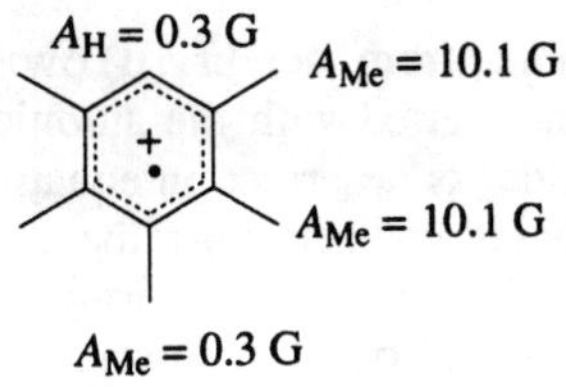

$A_H = 0.3$ G

$A_{Me} = 10.1$ G

$A_{Me} = 10.1$ G

$A_{Me} = 0.3$ G

(7)

the arene ion pair collapse during mercuration and thallation. The dichotomy is most pronounced in pentamethylbenzene. Thus the treatment of pentamethylbenzene with $Hg(O_2CCF_3)_2$ affords the substitution product pentamethylphenylmercury trifluoroacetate in high yields (equation 33).

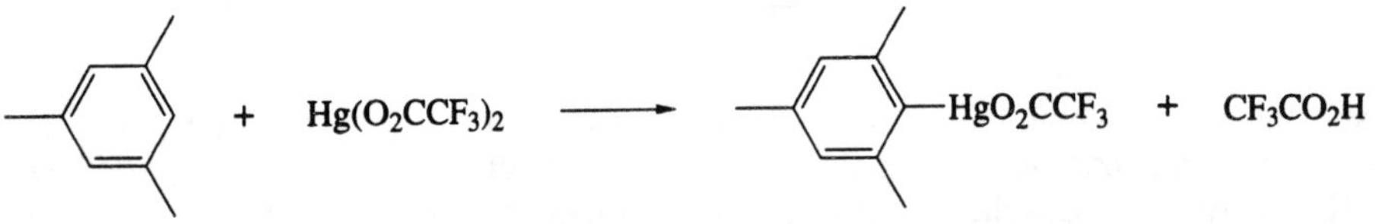

$$+ \quad Hg(O_2CCF_3)_2 \longrightarrow \quad -HgO_2CCF_3 \quad + \quad CF_3CO_2H \qquad (33)$$

No transient intermediates other than the π-complex are observed in either the UV–visible or ESR spectra. On the other hand, the thallation of pentamethylbenzene proceeds only 26% to nuclear substitution, the remainder being accounted for by side products resulting from side chain substitution and dimer formation (equation 34).

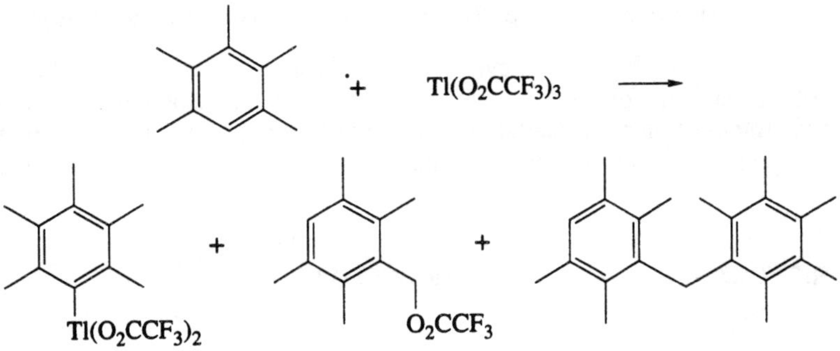

$$+ \quad Tl(O_2CCF_3)_3 \longrightarrow \qquad (34)$$

$Tl(O_2CCF_3)_2$ + O_2CCF_3 +

Such side products are known to derive from the radical cation of pentamethylbenzene,[9] *i.e.* equations (35) and (36).

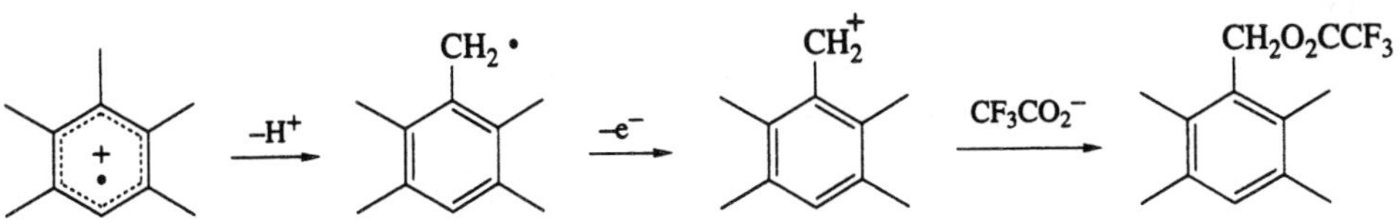

$$\xrightarrow{-H^+} CH_2\bullet \xrightarrow{-e^-} CH_2^+ \xrightarrow{CF_3CO_2^-} CH_2O_2CCF_3 \qquad (35)$$

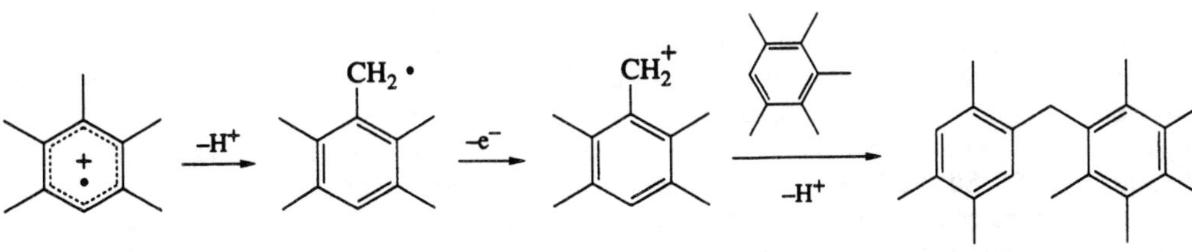

$$\xrightarrow{-H^+} CH_2\bullet \xrightarrow{-e^-} CH_2^+ \xrightarrow[-H^+]{} \qquad (36)$$

Indeed the diversion to side products during thallation coincides with the direct observation of the arene radical cation as a transient intermediate both by UV–visible and ESR spectroscopy. A similar dichotomy between the products of mercuration and thallation exists with durene, albeit to a lesser degree. Finally no discrepancy is observed with mesitylene, nuclear substitution occurring exclusively in both mercuration and thallation. Such a divergence between mercuration and thallation can be reconciled by the formulation in Scheme 6 if they differ by the extent to which diffusive separation (k_d) occurs in equation (31). All factors being the same, diffusive separation of the radical pair from thallium(III) should occur more readily than that from mercury(II) owing to a significant difference in the coulombic interactions, the arene ion being paired with the anionic $Hg^{I}(O_2CCF_3)_2^-$ in mercuration and with the neutral $Tl^{II}(O_2CCF_3)_2$ in thallation, as described in equations (27) and (28). Moreover the same electrostatic argument provides a ready rationalization for the ability of Lewis acids as additives to promote arene radical cation formation (leading to biaryls) during both mercuration and thallation of even unexceptional arenes.[97] Thus the addition of boron trifluoride (as the etherate) will foster ionic dissociation of mercury(II) trifluoroacetate, *i.e.* equation (37), as well as thallium(II) trifluoroacetate cation, *i.e.* equation (38).

$$Hg(O_2CCF_3)_2 \;+\; BF_3 \;\rightleftharpoons\; HgO_2CCF_3^+ \;+\; CF_3CO_2BF_3^- \qquad (37)$$

$$Tl(O_2CCF_3)_2^+ \;+\; BF_3 \;\rightleftharpoons\; TlO_2CCF_3^{2+} \;+\; CF_3CO_2BF_3^- \qquad (38)$$

The results lead to a diminution in the coulombic interaction in the radical pair $[ArH^{+\cdot}, Hg^{I}O_2CCF_3]$ during mercuration, and enhancement of the coulombic repulsion in the radical pair $[ArH^{+\cdot}, Tl^{II}O_2CCF_3^+]$ during thallation. In both cases the increased amounts of cage escape will lead to a higher component of electron transfer derived products (such as biaryls). Attractive as such a simple electrostatic explanation may seem, cognizance must also be taken of the attendant change in the driving force for back electron transfer k_{-e} in equation (29).

7.4.5.5 Electron Transfer *versus* Electrophilic Pathways for Aromatic Substitution

The study of mercuration and thallation provides a sharp focus on the experimental delineation of stepwise and concerted mechanisms for arene activation. Thus the unequivocal demonstration of arene radical cations as key intermediates in thallation, particularly of durene and pentamethylbenzene, is consistent with a stepwise (electron-transfer) mechanism for arene activation (compare Scheme 6[98] and equation 39).

$$[ArH, Tl(O_2CCF_3)_2^+] \longrightarrow [ArH^{+\cdot}, Tl(O_2CCF_3)_2^{\cdot}] \longrightarrow \overset{+}{Ar}\genfrac{}{}{0pt}{}{H}{Tl(O_2CCF_3)_2}, \textit{etc.} \qquad (39)$$

By the same token, the singular absence of any experimental evidence for such intermediates during mercuration is directly accommodated by a concerted (electrophilic) mechanism for arene activation, *i.e.* equation (40).

$$[ArH, Hg(O_2CCF_3)_2] \longrightarrow \overset{+}{Ar}\genfrac{}{}{0pt}{}{H}{Hg(O_2CCF_3)_2^-}, \textit{etc.} \qquad (40)$$

The difficulty with two separate mechanisms for arene activation by mercury(II) and thallium(III) is underscored by the striking correlation in Figure 6 which establishes the activation barriers to follow identical trends. In other words the rate-determining processes for mercuration and thallation are similar yet they distinctly differ in the products derived for the electron-rich arenes (durene and pentamethylbenzene in equation 35). Such a kinetics situation commonly demands that there exists at least one intermediate which separates the activation process from the products, as in Scheme 6. This paradox can be resolved in one of two ways. Firstly, the formulation in Scheme 6 merges stepwise and concerted processes by the modulation of a pair of rate constants. According to Scheme 6, the two principal pathways are differentiated during the competition between cage collapse (k_c) and diffusive separation (k_d) of the arene ion pair. The inability to observe the arene cation (*e.g.* from mesitylene) could be attributed to a rate of cage collapse to the Wheland intermediate occurring substantially faster than diffusive separation (*i.e.* $k_c >> k_d$), which is tantamount to a concerted process. Likewise the ESR observation of the arene cation would derive from a diffusive process occurring faster than collapse (*i.e.* $k_c << k_d$), which could appear as a stepwise process. Some of the structural and environmental factors which influence both rate constants have been presented in the discussion above. However this mechanism is not without its problems since it does not as yet address the observed kinetic isotope effects in a quantitative way (*vide supra*). Secondly, the stepwise and concerted processes for arene activation may simply represent competing pathways, namely equations (41) and (42), where the superscript ‡ represents the activated complex. If so those factors related to the donor properties of arenes, such as EDA complex formation, CT excitation, activation barriers, *etc.*, are too common to both to allow any distinction between these pathways. Moreover the common dependence on isoelectronic electrophiles, independent of charge, further obscures any difference in their transition states. This formulation thus recognizes two such dissimilar mechanisms as electron transfer and electrophilic processes in mercuration and thallation to be remarkably alike. Crucial to the resolution of this dilemma is the understanding of the stepwise process for arene activation. Particularly germane are the microdynamics of the contact ion pairs in Scheme 6, especially as they collapse or evolve to 'loose' (solvent-separated) ion pairs and finally to 'free' (separate) pairs of radicals, as described in Section 7.4.4.

$$[\mathrm{ArH}, \mathrm{Hg(O_2CCF_3)_2}] \longrightarrow \left[\overset{+}{\mathrm{Ar}}\begin{matrix} \diagup \mathrm{H} \\ \diagdown \mathrm{Hg(O_2CCF_3)_2^-} \end{matrix} \right]^{\ddagger} \quad (41)$$

$$[\mathrm{ArH}, \mathrm{Tl(O_2CCF_3)_2^{+}}] \longrightarrow [\mathrm{ArH}^{+\bullet}, \mathrm{Tl(O_2CCF_3)_2}]^{\ddagger} \quad (42)$$

7.4.6 SYNTHETIC TRANSFORMATIONS *VIA* ELECTRON-TRANSFER OXIDATION

The foregoing elements of electron-transfer oxidation presented in terms of the energetics (Section 7.4.2) of contact ion-pair $[RH^{+\bullet},A^{-\bullet}]$ formation (Section 7.4.1) and the generic reactions of the individual organic radical cations (Section 7.4.3.1) and acceptor radical anions (Section 7.4.3.3), allow a variety of thermal and photochemical processes (Section 7.4.4) to be predicted in terms of CIP behavior (Section 7.4.5). The following examples are from the chemical literature and only the reactants, principal products and predominant methodologies are included. They are deliberately chosen to illustrate how electron-transfer oxidations can be utilized in various types of synthetic transformations. Representative citations to the original literature are included to facilitate the delineation of the roles of radical cations and anions in sometimes complex pathways. Accordingly, the examples in Section 7.4.6.1 and Section 7.4.6.2 are presented in the same sequence as the generic reactions are listed in Section 7.4.3.1 for organic radical cations ($RH^{+\bullet}$) and in Section 7.4.3.3 for acceptor radical anions ($A^{-\bullet}$), respectively. For conciseness and generality the complete structures are avoided, and the following abbreviations employed: Ar (aromatic moiety); R (aliphatic moiety); Z (CN); E (CO_2Et); M (CO_2Me); Py (pyridine); $BA^{+\bullet}$ [tris(*p*-bromophenyl)aminium]; $h\nu_{CT}$ (charge-transfer activation); $h\nu_A$ (acceptor (sensitizer not specified) activation); $h\nu$ (general photochemical activation unspecified). A blank space over the arrow represents thermal activation, and only the principal products are given.

7.4.6.1 Donor Radical Cations

7.4.6.1.1 α-Fragmentation

(a) 3-pyridiniomethyl carboxylate (CO_2^-, N–H⁺) $\xrightarrow{h\nu}$ 3-methylpyridine + CO_2 (ref. 99)

(b) $ArCH_2CO_2H$ + 1-cyanonaphthalene $\xrightarrow{h\nu_A}$ ArMe + 1-cyano-1,2-dihydronaphthalene (ref. 100)

(c) 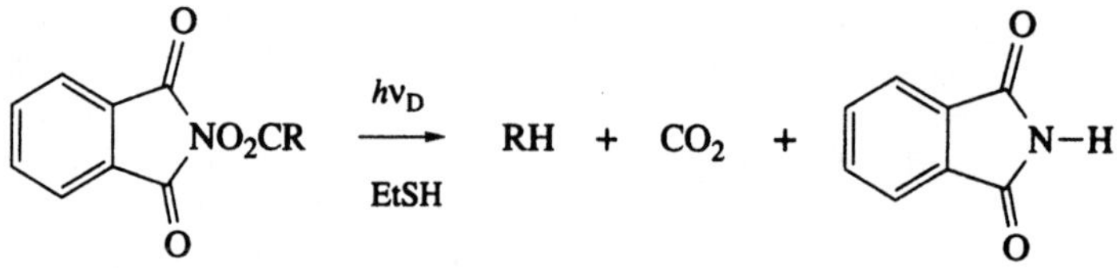 (ref. 101)

NO2CR
hνD
EtSH
RH + CO2 +
N−H

(d) 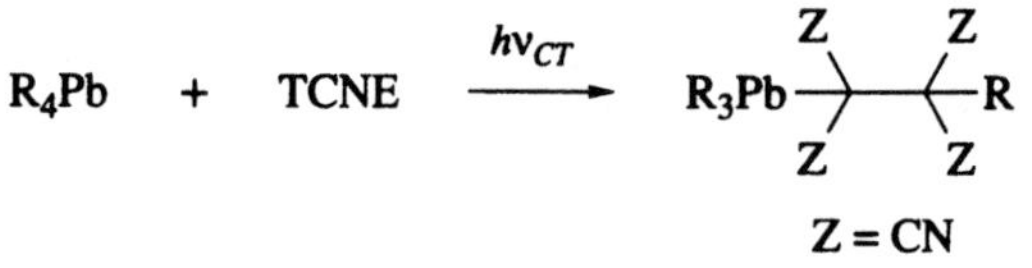 (ref. 102)

R4Pb + TCNE
hνCT
R3Pb
R
Z = CN

(e) 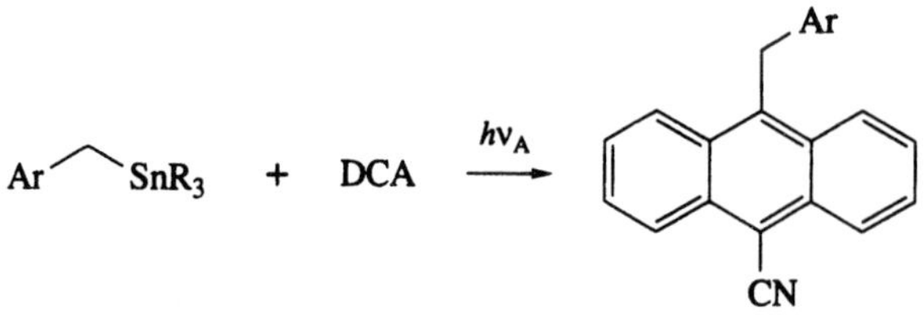 (ref. 103)

Ar SnR3 + DCA
hνA
Ar
CN

(f) 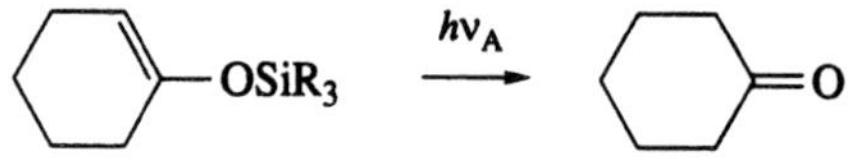 (ref. 104)

OSiR3
hνA

(g) 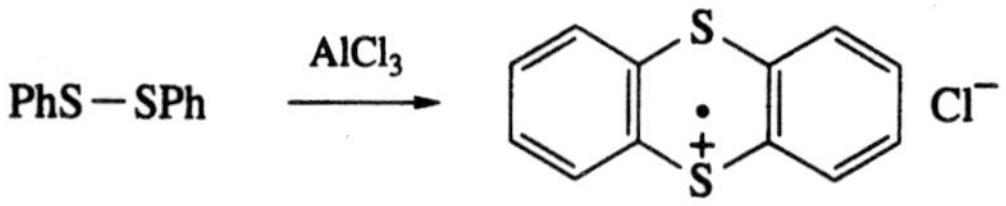 (ref. 105)

PhS − SPh
AlCl3
Cl−

(h) 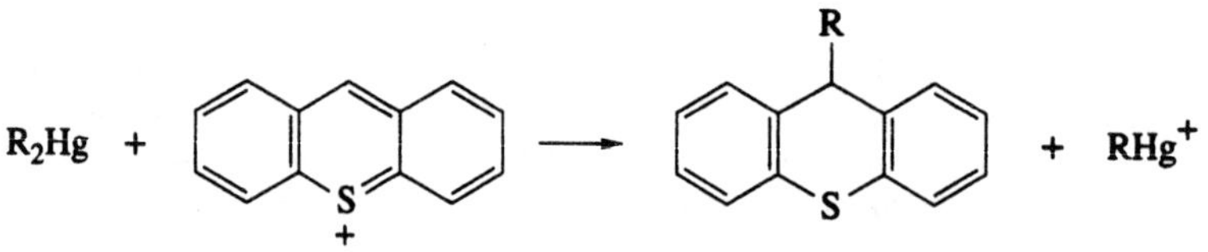 (ref. 106)

R2Hg +
+ RHg+

7.4.6.1.2 *β-Fragmentation*

(a) 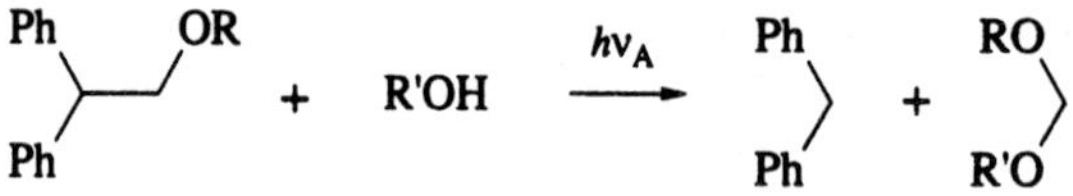 (ref. 107)

+ R'OH
hνA

(b) 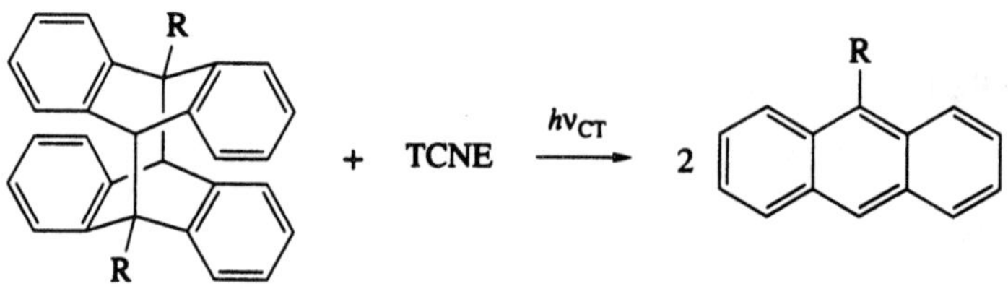 (ref. 108)

+ TCNE
hνCT
2

(c) 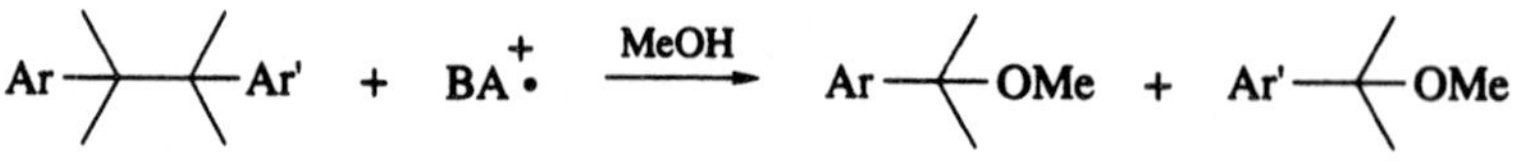 (ref. 109)

Ar Ar' + BA+•
MeOH
Ar OMe + Ar' OMe

(d) 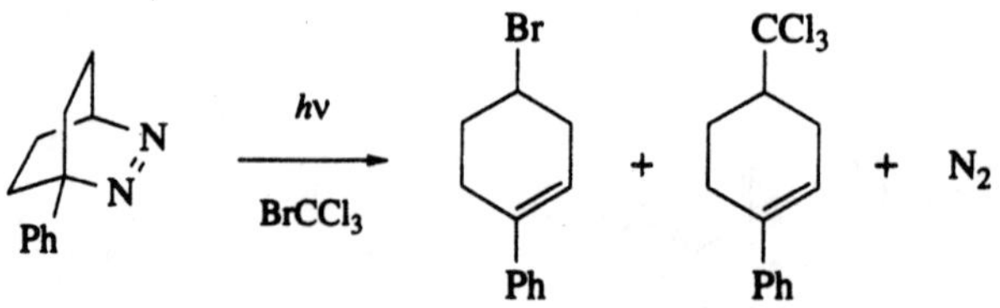 (ref. 110)

hν
BrCCl3
+ N2

7.4.6.1.3 Rearrangement

(a) Ph, Ph; $h\nu_A$ → Ph, Ph (ref. 111)

(b) Ar, Ar; $h\nu_A$, O_2 → Ar, Ar, O—O (ref. 112)

(c) Ar, Ar + CA; $h\nu_A$ → Ar, Ar, O, O, Cl_4 (ref. 113)

(d) Ph, Ph + TCNE; $h\nu_{CT}$ → Ph, Z, Z, Z, Z, Ph; Z = CN (ref. 114)

(e) Ar, Ar; $h\nu_A$, O_2 → Ar, Ar, O, O (ref. 115)

(f) Ar, O, Ar + TCNE; $h\nu_{CT}$, O_2 → Ar, O, Ar, O–O (ref. 116)

(g) Ph, Ph; $h\nu_A$, MeOH → Ph, OMe, Ph (ref. 117)

(h) Ph, Ph; $h\nu_A$ → Ph, Ph (ref. 118)

(i) + I_2 → I, I (ref. 119)

7.4.6.1.4 Cyclization

(a) Ar, Ar; $h\nu_A$, O_2 → Ar, O, O, Ar (ref. 120)

(b) N; $h\nu_A$, MeOH → N, OMe (ref. 121)

(c) CN, O; $h\nu$ → CN, O (ref. 122)

(d) Ph, $\overset{+}{N}$, $SiMe_3$; $h\nu$ → Ph, N (ref. 123)

(e) $\overset{+}{N}$, O, $SiMe_3$; $h\nu$ → N, O (ref. 124)

(f) Ar, O, N, CO_2H + DDQ → Ar, O, O, N, O, H (2*S*,5*S*) (ref. 125)

R*, Y + Nu; DDQ → H, R*, Nu, Y; R* = chiral (ref. 125)

(g) HO, O, O, OH; *[E]* → O, O, O, O (ref. 126)

7.4.6.1.5 Deprotonation

(a) + PhMe $\xrightarrow{h\nu_A}$ (ref. 127)

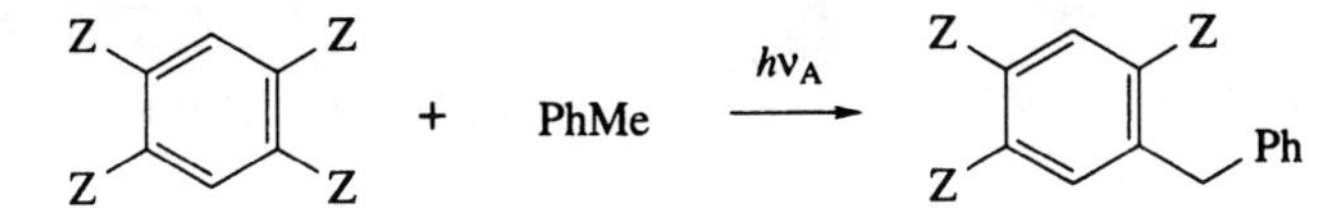

(b) + PhMe $\xrightarrow{h\nu}$ (ref. 128)

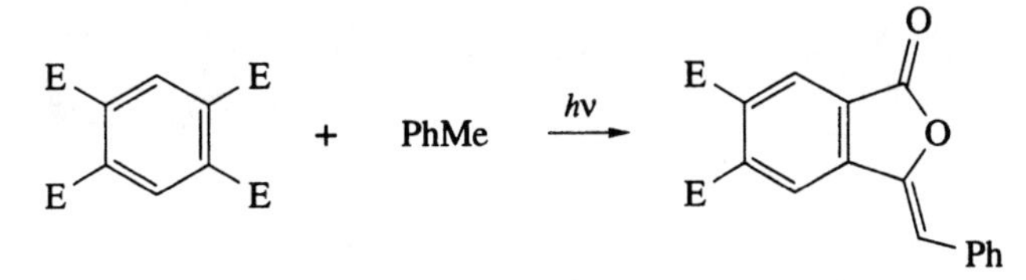

(c) 2 + $S_2O_8^{2-}$ $\longrightarrow$ (ref. 129)

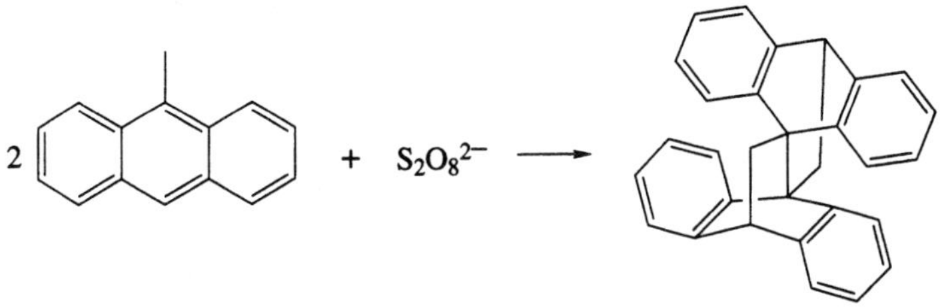

(d) + Et_2NH $\xrightarrow{h\nu}$ (ref. 130)

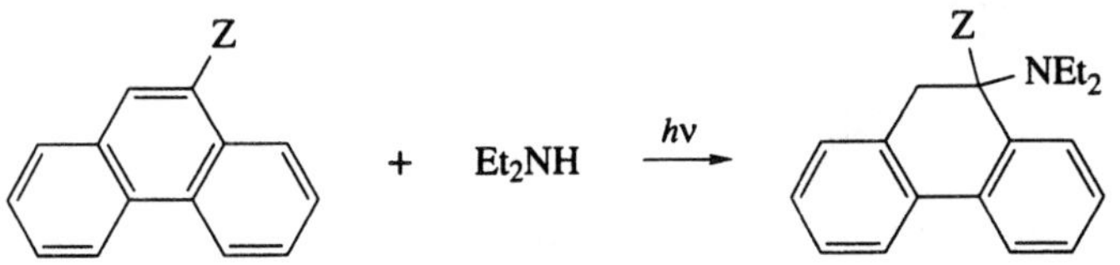

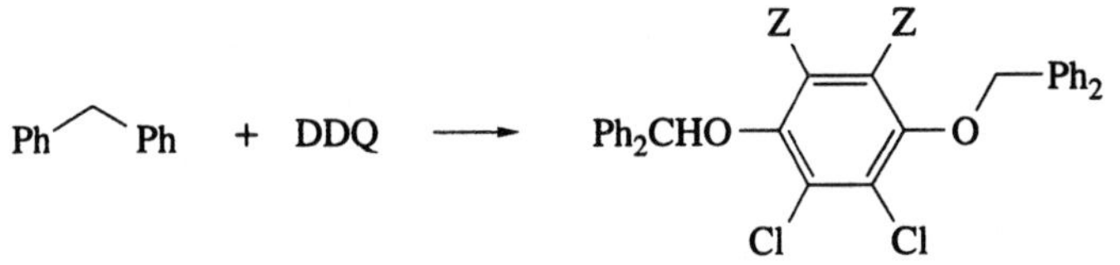

(e) Ph⁀Ph + DDQ $\longrightarrow$ (ref. 131)

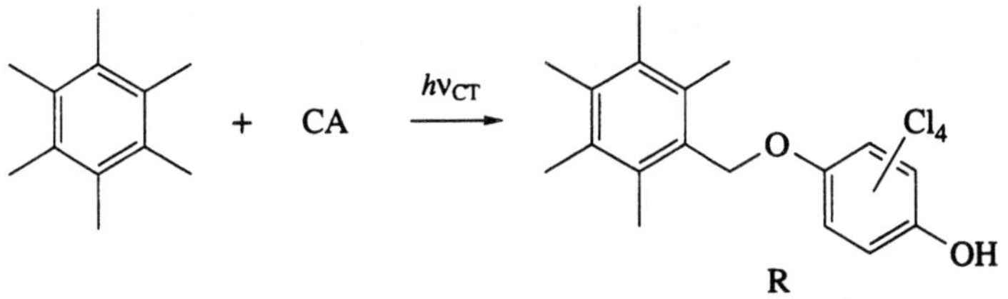

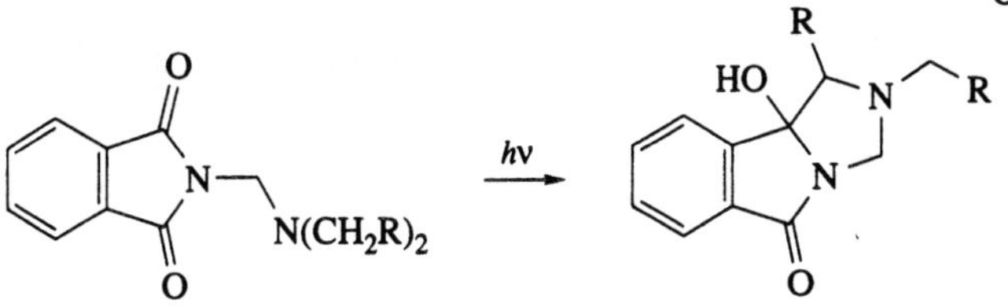

(f) + CA $\xrightarrow{h\nu_{CT}}$ (ref. 132)

(g) $\xrightarrow{h\nu}$ (ref. 133)

(h) $\xrightarrow{h\nu}$ (ref. 134)

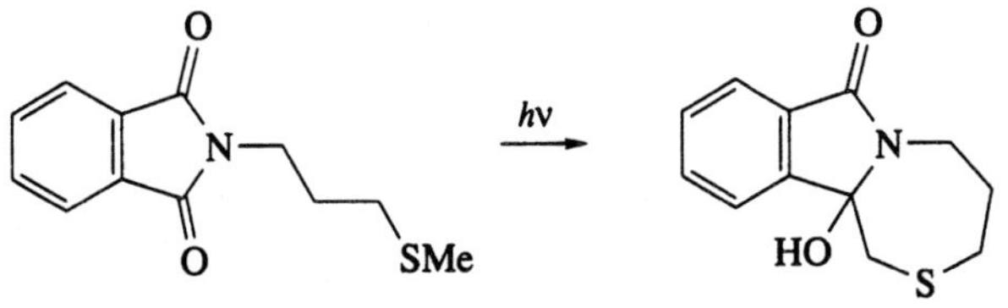

(i) $PhNMe_2$ + PyO $\xrightarrow{h\nu_{CT}}$ PhNHMe + H_2CO + Py (ref. 135)

7.4.6.1.6 Nucleophilic addition

(a) $ArH + C(NO_2)_4 \xrightarrow{h\nu_{CT}} ArC(NO_2)_3 + HONO$ (ref. 136)

(b) 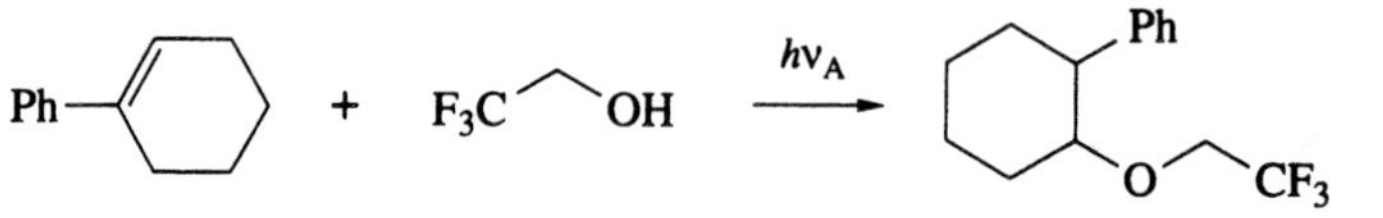 (ref. 137)

(c) 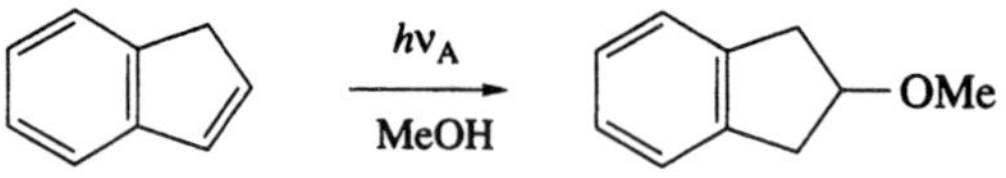 $\xrightarrow[MeOH]{h\nu_A}$ OMe (ref. 138)

(d) $ArF + AcO^- \xrightarrow{[E]} ArOAc + F^-$ (ref. 139)

(e) 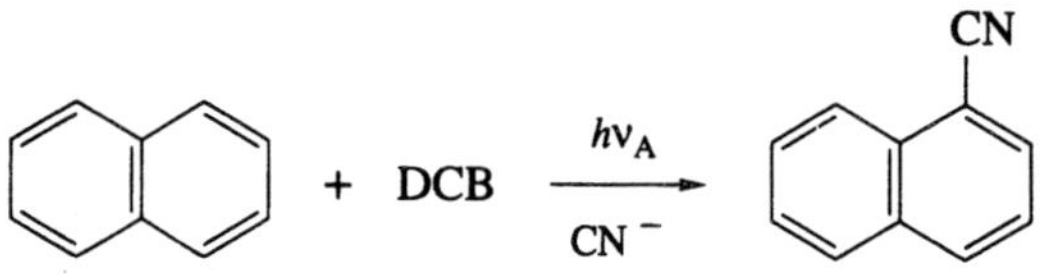 + DCB $\xrightarrow[CN^-]{h\nu_A}$ (ref. 140)

(f) $ArH + XeF_2 \longrightarrow ArF + HF + Xe$ (ref. 141)

(g) $ArH + WCl_6 \longrightarrow ArCl$ (ref. 142)

(h) 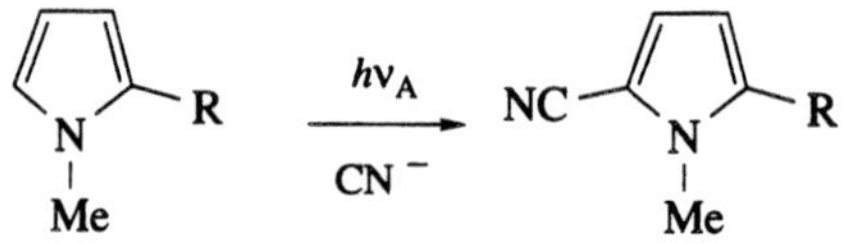 $\xrightarrow[CN^-]{h\nu_A}$ (ref. 143)

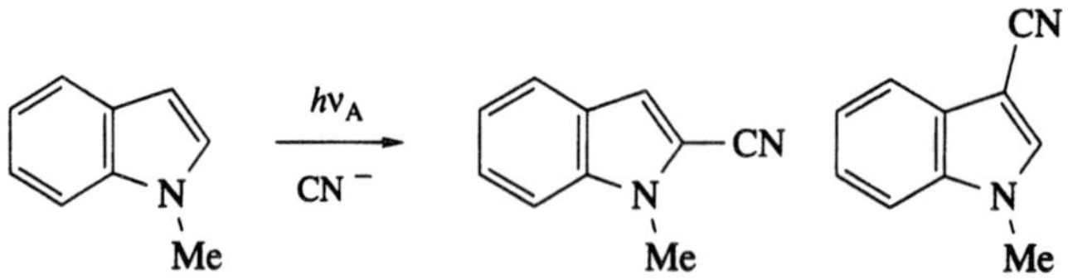

$\xrightarrow[CN^-]{h\nu_A}$ (ref. 143)

(i) 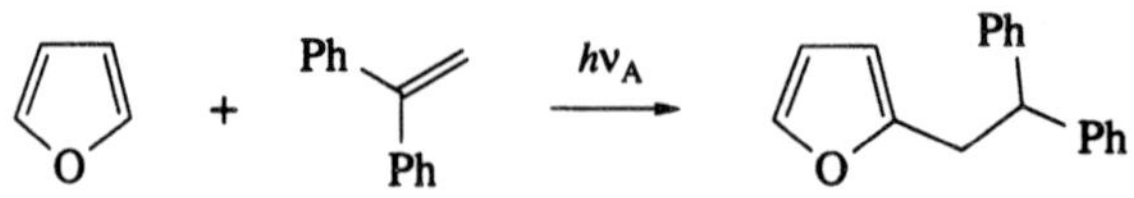 $\xrightarrow{h\nu_A}$ (ref. 144)

(j) 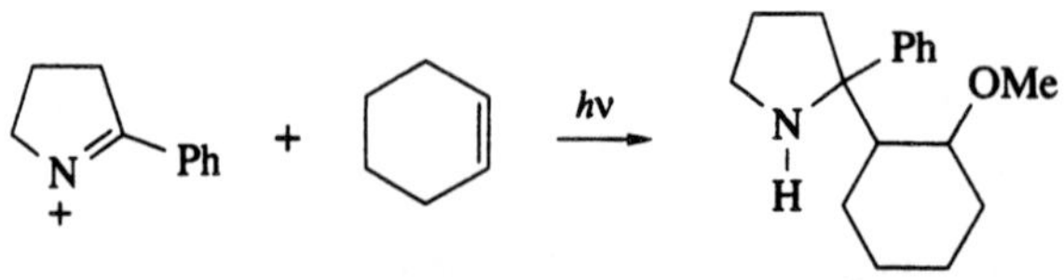 $\xrightarrow{h\nu}$ (ref. 121)

(k) 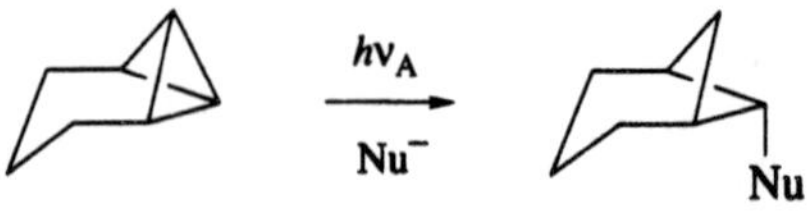 $\xrightarrow[Nu^-]{h\nu_A}$ (ref. 145)

7.4.6.1.7 Dimerization

(a) Ar_2N-CH=CH$_2$ + Fe^{3+} $\xrightarrow{MeOH}$ [cyclobutane, 1,2-bis(Ar_2N)] + [Ar_2N–CH(OMe)–CH$_2$–]$_2$ (ref. 146)

$Ar_2C=CH_2$ + Fe^{3+} ⟶ [1,1,3,3-tetraarylcyclobutane] + $Ar_2C=CH-CH=CAr_2$ (ref. 146)

(b) Ar_2N-CH=CH$_2$ $\xrightarrow[O_2]{h\nu_A}$ [cyclobutane, 1,2-bis(Ar_2N)] (ref. 147)

$Ar_2C=CH_2$ $\xrightarrow[O_2]{h\nu_A}$ [1,1,3,3-tetraarylcyclobutane] (ref. 147)

(c) [1,3-cyclohexadiene] + $BA^{+\bullet}$ ⟶ [Diels–Alder dimer] (ref. 148)

(d) Ar-CH=CH-CH$_3$ + $BA^{+\bullet}$ ⟶ [1,2-diaryl-3,4-dimethylcyclobutane] + [stereoisomeric 1,2-diaryl-3,4-dimethylcyclobutane] (ref. 149)

(e) $Ph_2C=CH_2$ $\xrightarrow{h\nu_A}$ [1,1,4-triphenyl-1,2,3,4-tetrahydronaphthalene] (ref. 150)

(f) ArH + NO_2^+ ⟶ Ar−Ar + NO_2 (ref. 151)

7.4.6.1.8 Cycloaddition

(a) + $\longrightarrow$ (ref. 152)

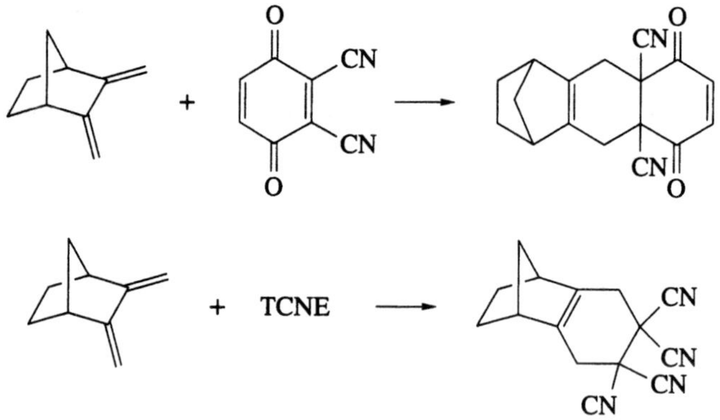

CN, CN, O, O

+ TCNE $\longrightarrow$ (ref. 152)

(b) NMe_2, Me_2N + TCNE $\longrightarrow$ (ref. 153)

(c) + X $\xrightarrow{h\nu_A}$ (ref. 154)

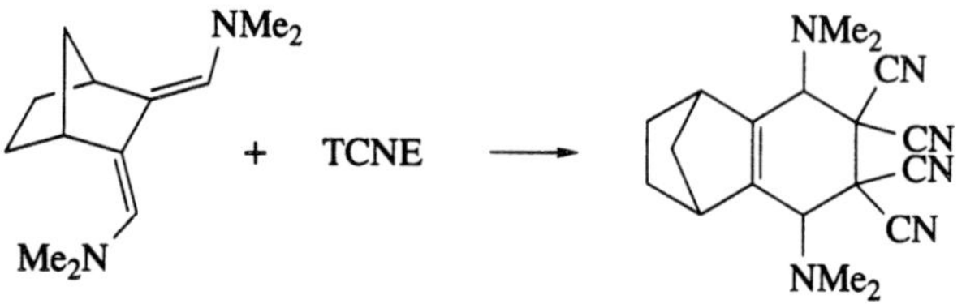

X = OEt, OPh, SPh, OCH_2CH_2Cl

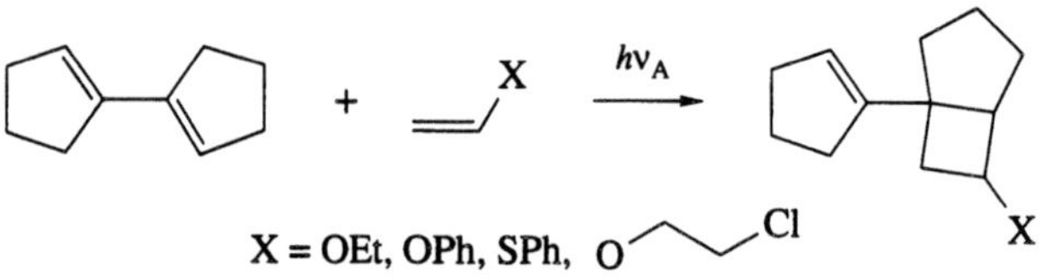

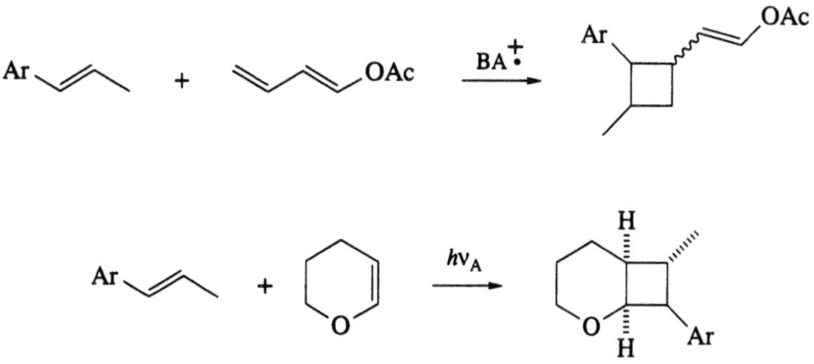

(d) Ar + OAc $\xrightarrow{BA^{+\bullet}}$ (ref. 154)

Ar + $\xrightarrow{h\nu_A}$ (ref. 149)

(e) + $\xrightarrow{BA^{+\bullet}}$ (ref. 148)

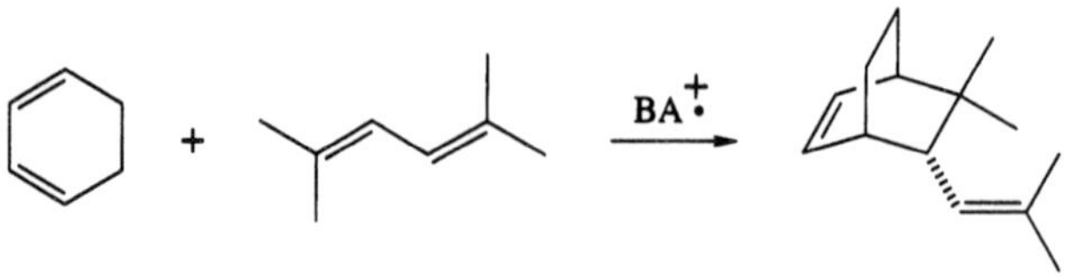

(f) M, Ph, Ph + DCA $\xrightarrow{h\nu_A}$ (ref. 155)

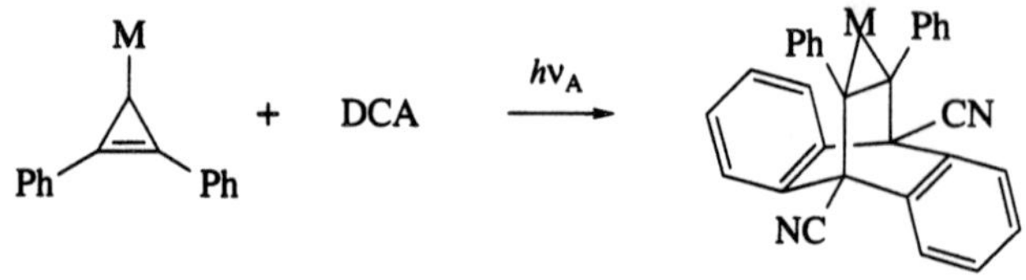

7.4.6.1.9 Homolytic addition

(a) $R_2C{=}CR_2 \xrightarrow[O_2]{h\nu_A}$ 1,2-dioxetane (O–O, R, R, R, R) (ref. 156)

(b) norbornene $\xrightarrow[Ag^+,\ MeCN]{h\nu}$ norbornyl–CH_2CN (ref. 157)

(c) $Ph_2C{=}CPh_2 \xrightarrow[O_2]{h\nu_A} Ph_2C{=}O$ + tetraphenyloxirane (ref. 158)

(d) PhCH=CHPh $\xrightarrow[O_2]{h\nu_A}$ PhCH=O + stilbene oxide (Ph, O, Ph) + PhCOCOPh (ref. 159)

(e) p-xylene $\xrightarrow[O_2]{h\nu_A}$ $CH_3C_6H_4$CHO (ref. 160)

(f) MeO–C_6H_4–CH=CHCH_3 $\xrightarrow[O_2]{h\nu_A}$ AnCHO An-epoxide (O, An) (ref. 161)

(g) ArH + $C(NO_2)_4 \xrightarrow{h\nu_{CT}} ArNO_2 + HC(NO_2)_3$ (ref. 162)

(h) Ar_2N–CH=CH_2 + $C(NO_2)_4 \longrightarrow Ar_2N$–CH($CH_2NO_2$)–$C(NO_2)_3$ (ref. 163)

$Ph_2C{=}CH_2$ + $C(NO_2)_4 \xrightarrow{h\nu_{CT}} Ph_2C{=}CHNO_2$ (ref. 163)

(i) ArH + thianthrene radical cation (S, S, $\overset{+}{\bullet}$) $\longrightarrow$ S-aryl thianthrenium ($\overset{+}{S}$–Ar) (ref. 164)

7.4.6.1.10 Electron transfer

(a) hexamethylbenzene + $C(NO_2)_4$ $\xrightarrow[\text{DPA}]{h\nu_{CT}}$ 9,10-diphenylanthracene radical cation (ref. 165)

(b) 1-methoxycyclohexa-1,4-diene + $BA^{+\bullet}$ ⟶ radical cation + BA (ref. 166)

(c) 1,5-dithiacyclooctane + $O_2^{+\bullet}$ ⟶ $S^{+\bullet}$ radical cation + O_2 (ref. 167)

diazabicyclooctane + $O_2^{+\bullet}$ ⟶ $N^{+\bullet}$ radical cation + O_2 (ref. 167)

(d) C_6F_6 + $O_2^{+\bullet}$ ⟶ $C_6F_6^{+\bullet}$ + O_2 (ref. 168)

7.4.6.2 Acceptor Radical Anions

7.4.6.2.1 Fragmentation

(a) Ar⌒OAc + R_3N $\xrightarrow{h\nu}$ Ar—Me + Ar⌒⌒Ar (ref. 169)

(b) nitrodibenzo-p-dioxin (NO₂) $\xrightarrow[\text{RNH}_2]{h\nu}$ OH, O, NO_2, HN—R product (ref. 170)

(c) Ar—C(Me)₂—Cl + PhS^- ⟶ Ar—C(Me)₂—SPh + Cl^- (ref. 171)

(d) Ar—C(Me)₂—NO_2 $\xrightarrow{\text{Me}_2\bar{C}\text{—NO}_2,\ N_3^-}$ Ar—C(Me)₂—N_3 (ref. 172)

(e) Ar—CH_2Cl + Me_2C—NO_2^- ⟶ Ar—CH_2—C(Me)₂—NO_2 (ref. 173)

(f) $RNO_2 + Bu_3SnH \longrightarrow RH + NO_2^-$ (ref. 174)

$Ar-C(CH_3)_2-NO_2 + Bu_3SnH \longrightarrow Ar-CH(CH_3)_2$ (ref. 175)

(g) o-(allyloxy)benzenediazonium (N_2^+) + $Bu_3SnH \longrightarrow$ 3-methyl-2,3-dihydrobenzofuran (ref. 176)

(h) $ArCH_2CH_2X + HO^- \longrightarrow$ ArCH=CHAr $X = Cl, \overset{+}{S}Me_2$ (ref. 177)

(i) 1-benzyl-1,4-dihydronicotinamide (NH_2, Ph) + RS–SR ⟶ 1-benzylnicotinamide pyridinium (N^+) RS^- (ref. 178)

(j) 3,5-E-1,4-dihydropyridine (N–H) + $(AcO)_2$ ⟶ 3,5-E-pyridinium (N^+–H) AcO^- + CH_4 + CO_2 (ref. 179)

(k) $ArCH_2Br$ + 1-benzyl-1,4-dihydronicotinamide $\xrightarrow{h\nu}$ ArMe + 1-benzylnicotinamide pyridinium (N^+) Br^- (ref. 180)

O_2N–C(CN)(R)–CH_2CH_2–Ac + 1-benzyl-1,4-dihydronicotinamide $\xrightarrow{h\nu}$ R–CH(CN)–CH_2CH_2–Ac + 1-benzylnicotinamide pyridinium (N^+) NO_2^-

(l) 6-bromohept-1-ene + Me_3SnNa ⟶ (2-methylcyclopentyl)methyl-$SnMe_3$ (ref. 181)

7.4.6.2.2 *Protonation*

(a) naphthalene + Et_3N $\xrightarrow[ROH]{h\nu}$ 1,4-dihydronaphthalene (ref. 182)

(b) anisole (OMe) + Li $\xrightarrow{ROH}$ 1-methoxycyclohexa-1,4-diene (OMe) (ref. 183)

7.4.6.2.3 *Electrophilic addition*

$Ar_2CO^{\bar{\cdot}}$ + Ac_2O ⟶ $Ar_2CH(OAc)$ + $AcO-C(Ar)_2-C(Ar)_2-OAc$ (ref. 184)

7.4.6.2.4 *Electron transfer*

(a) Ph_3C^+ + Bu^tO^- ⟶ Ph_3C-OBu^t (ref. 185)

(b) ArN_2^+ + PhO^- ⟶ Ar–O–Ph + N_2 (ref. 186)

(c) O_2 $\xrightarrow[D]{h\nu_A}$ $O_2^{\bar{\cdot}}$ (ref. 187)

7.4.6.2.5 *Dimerization/disproportionation*

(a) $Ar_2CO^{\bar{\cdot}}$ ⟶ $^-O-C(Ar)_2-C(Ar)_2-O^-$ (ref. 188)

(b) 4-nitrotoluene $\xrightarrow{K}$ $O_2N-C_6H_4-CH_2CH_2-C_6H_4-NO_2$ (ref. 189)

(c) ArN_2^+ + I^- ⟶ Ar–Ar + N_2 (ref. 190)

(d) 2 $C_8H_8^{\bar{\cdot}}$ ⟶ $C_8H_8^{2-}$ + cyclooctatetraene (ref. 191)

ACKNOWLEDGMENTS

I thank S. Sankararaman and T. M. Bockman for their invaluable help and illuminating discussions in the preparation of this chapter.

7.4.7 REFERENCES

1. (a) E. Weitz and H. W. Schwechten, *Chem. Ber.*, 1926, **59**, 2307; 1927, **60**, 545; (b) J. Weiss, *Trans. Faraday Soc.*, 1946, **42**, 116; (c) E. T. Kaiser and L. Kevan, 'Radical Ions', Wiley, New York, 1968.
2. A. Ledwith, *Acc. Chem. Res.*, 1972, **5**, 133.
3. A. J. Bard, A. Ledwith and H. J. Shine, *Adv. Phys. Org. Chem.*, 1976, **13**, 156.
4. D. J. Mackinnon and W. A. Waters, *J. Chem. Soc.*, 1953, 323.
5. I. V. Khudyakov and V. A. Kuz'min, *Russ. Chem. Rev. (Engl. Transl.)*, 1978, **47**, 22.
6. J. K. Kochi, 'Free Radicals', Wiley, New York, 1973, vol. 1, p. 591.
7. By convention, the electrode potentials for redox are referred to only in terms of reduction (E°_{red}). For electron-transfer oxidation, however, the values of E°_{ox} are employed for RH donors here with the understanding that $E^\circ_{ox} = -E^\circ_{red}$ of the radical cation.
8. A. J. Bard and L. R. Faulkner, 'Electrochemical Methods', Wiley, New York, 1980.
9. K. Yoshida, 'Electrooxidation in Organic Chemistry. The Role of Cation Radicals as Synthetic Intermediates', Wiley, New York, 1984.
10. (a) C. J. Schlesener, C. Amatore and J. K. Kochi, *J. Am. Chem. Soc.*, 1984, **106**, 3567; (b) J. Grodkowski, P. Neta, C. J. Schlesener and J. K. Kochi, *J. Phys. Chem.*, 1985, **89**, 4373.
11. R. S. Davidson, *Adv. Phys. Org. Chem.*, 1983, **19**, 1; *Molecular Association*, 1975, **1**, 216.
12. G. Jones, II, 'Photochemistry and Photophysics of Organic Charge-Transfer Complexes,' in 'Photoinduced Electron Transfer', ed. M. A. Fox and M. Chanon, Elsevier, Amsterdam, 1988, p. 245.
13. M. Szwarc (ed.), 'Ions and Ion Pairs in Organic Reactions', Wiley, New York, 1972 and 1974, Vols. 1 and 2.
14. E. F. Hilinski, J. M. Masnovi, J. K. Kochi and P. M. Rentzepis, *J. Am. Chem. Soc.*, 1984, **106**, 8021.
15. G. J. Kavarnos and N. J. Turro, *Chem. Rev.*, 1986, **86**, 401.
16. J. D. Simons and K. S. Peters, *Acc. Chem. Res.*, 1984, **17**, 277.
17. R. A. Sheldon and J. K. Kochi, 'Metal-Catalyzed Oxidations of Organic Compounds. Mechanistic Principles and Synthetic Methodology Including Biochemical Processes', Academic Press, New York, 1981.
18. W. J. Mijs and C. R. H. I. de Jonge (eds.), 'Organic Synthesis by Oxidation with Metal Compounds', Plenum Press, New York, 1986. Includes chapters on: (a) Oxidation by Vanadium(V), by F. Freeman; (b) Oxidation by Oxochromium(VI), by F. Freeman; (c) Oxidation by Active Manganese Dioxide, by A. J. Fatiadi; (d) Oxidation with Manganese(III) Acetate, by W. J. de Klein; (e) Oxidations by Cobalt(III), by F. Freeman; (f) Oxidation with Nickel Peroxide, by M. V. George; (g) Catalytic Oxidations with Copper and Cobalt–Amine Complexes, by C. R. H. I. de Jonge; (h) Ruthenium(VIII) Tetroxide Oxidations, by J. L. Courtney; (i) Oxidations with Palladium Compounds, by S. F. Davison and P. M. Maitlis; (j) Silver Carbonate on Celite Oxidations, by M. Fetizon, M. Golfier, P. Mourges and J. M. Louis; (k) Oxidation with Cerium(IV), by T. L. Ho; (l) Oxidations with Osmium(VIII) Tetroxide, by H. S. Singh; (m) Oxidation with Thallium(III), by A. McKillop and E. C. Taylor; (n) Oxidation with Lead(IV) Tetraacetate, by M. Lj. Mihailović, Ž. Čeković and L. Lorenc; (o) Bismuth(V, III) Oxidations, by J. P. Kitchin; (p) Metal Catalysis of Peroxide Oxidations, by Y. Ogata and Y. Sawicki.
19. K. B. Wiberg (ed.), 'Oxidation in Organic Chemistry', Academic Press, New York, 1965, vol. 5A. Includes chapters on: (a) Oxidation by Permanganate, by R. Stewart; (b) Oxidation by Chromic Acid and Chromyl Compounds, by K. B. Wiberg; (c) Oxidation by Vanadium(V), Cobalt(III) and Manganese(III), by W. A. Waters and J. S. Littler; (d) Cerium(V) Oxidation, by W. H. Richardson; (e) Oxidations with Lead(IV) Tetraacetate, by R. Criegee; (f) Iodine(V,III) and Bismuth(V) Oxidations, by C. A. Bunton.
20. W. S. Trahanovsky (ed.), 'Oxidation in Organic Chemistry', Academic Press, New York, 1973, vol. 5B. Includes chapters on: (a) Oxidation by Copper(II), by W. G. Nigh; (b) Oxidative Coupling of Phenols, by P. D. McDonald and G. A. Hamilton; (c) Oxidation by Thallium(III), by R. J. Ouellette; (d) Ruthenium(VIII) Tetroxide Oxidation, by D. G. Lee and M. van den Engh.
21. W. S. Trahanovsky (ed.), 'Oxidation in Organic Chemistry', Academic Press, New York, 1982, vol. 5D. Includes chapters on: (a) Oxidations with Lead(IV) Tetraacetate, by G. M. Rubottom; (b) Phase-Transfer Assisted Manganese(VII) Oxidation, by D. G. Lee.
22. D. Arndt, 'Manganese(VIII, VI, V, IV, III) Compounds as Oxidizing Agents in Organic Chemistry', Open Court Publishing, LaSalle, IL, 1981.
23. A. E. Martell and D. T. Sawyer (eds.), 'Oxygen Complexes and Oxygen Activation by Transition Metals', Plenum Press, New York, 1988. Includes chapters on: (a) Metal Oxo Complexes and Oxygen Activation, by T. J. Meyer; (b) Radical Cation ($R_2S^{+\cdot}$) Pathways for Selective Catalytic Oxidation, by D. P. Riley and M. R. Smith, among many other contributions.
24. W. Ando and Y. Moro-oka, 'Role of Oxygen in Chemistry and Biochemistry', Elsevier, Amsterdam, 1988. Includes chapters on: (a) Cation-Radical Chain Catalyzed Oxygenation of Alkenes, by S. F. Nelsen, *et al.*; (b) Photooxygenation of Organic Compounds *via* Their Radical Ions, by K. Mizono and Y. Otsuji; (c) Triphenylpyrylium Sensitized Oxygenations, by K. Tokumaru *et al.*; (d) Catalytic Oxidation of Methylarenes to Benzaldehydes, by R. A. Sheldon and N. de Heij; (e) Oxidation of Arylamines with Horseradish Peroxidase, by Fujimori *et al.*; (f) Arene Hydroxylation by Cytochrome P-450, by M. Tsuda *et al.*, and many others.
25. E. S. Huyser (ed.), 'Methods in Free Radical Chemistry', Dekker, New York. The series includes chapters on: (a) Free-Radical Chlorination *via* Nitrogen Cation Radicals, by N. C. Deno, 1971, vol. 3, p. 93; (b)

One-Electron Reductions of Aromatic Cations, by M. Siskind, 1971, vol. 3, p. 152; (c) 'Metal-Ion Oxidative Cleavage of Alcohols and Glycols', by W. S. Trahanovsky, 1973, vol. 4, p. 133.
26. J. Mattay, *Angew. Chem., Int. Ed. Engl.*, 1987, **26**, 825.
27. K. Katsumi, *Rev. Chem., Intermed.*, 1979, **2**, 321.
28. (a) D. Creed and R. A. Caldwell, *Photochem. Photobiol.*, 1985, **41**, 715; (b) H. D. Roth, *Tetrahedron*, 1986, **28**, 6097; see also *Tetrahedron*, 1982, **38**, 1027; (c) Y. Shirota and H. Mikawa, *J. Macromol. Sci., Rev. Macromol. Chem.*, 1978, **162**, 129.
29. R. Foster and M. I. Foreman, in 'Chemistry of Quinonoid Compounds, Part I', ed. S. Patai, Wiley, New York, 1974, p. 257.
30. (a) L. M. Tolbert, *Org. Photochem.*, 1983, **6**, 177; (b) R. J. Sundberg, *Org. Photochem.*, 1983, **6**, 121.
31. S. L. Mattes and S. Farid, *Org. Photochem.*, 1983, **6**, 233.
32. M. A. Fox, *Adv. Photochem.*, 1986, **13**, 237.
33. R. D. Cannon, 'Electron Transfer Reactions', Butterworths, London, 1980.
34. (a) T. Koenig and H. Fischer, *Free Radical*, 1973, **1**, 157; (b) J. E. Gordon, 'Organic Chemistry of Electrolyte Solutions', Wiley, New York, 1975.
35. T. M. Bockman and J. K. Kochi, submitted for publication.
36. (a) R. A. Marcus, *J. Chem. Phys.*, 1956, **24**, 966; (b) R. A. Marcus, B. Zwolinski and H. Eyring, *J. Phys. Chem.*, 1954, **58**, 432; (c) N. Sutin, in 'Inorganic Biochemistry', ed. G. L. Einhorn, Elsevier, Amsterdam, 1973, vol. 2, p. 611.
37. L. Eberson, 'Electron Transfer in Organic Chemistry', Springer Verlag, New York, 1987; see also C. J. Schlesener, C. Amatore and J. K. Kochi, *J. Phys. Chem.*, 1986, **90**, 3747; S. Fukuzumi, C. L. Wong and J. K. Kochi, *J. Am. Chem. Soc.*, 1980, **102**, 2928.
38. J. K. Kochi, *Angew. Chem., Int. Ed. Engl.*, 1988, **27**, 1227.
39. J. Heinze, *Angew. Chem., Int. Ed. Engl.*, 1984, **23**, 831.
40. J. O. Howell, J. M. Goncalves, C. Amatore, L. Klasinc, R. M. Wightman and J. K. Kochi, *J. Am. Chem. Soc.*, 1984, **106**, 3968.
41. L. L. Miller, G. D. Nordblom and E. A. Mayeda, *J. Org. Chem.*, 1972, **37**, 916.
42. (a) V. I. Vedeneyev, L. V. Gurvich, V. N. Kondrat'yev, V. A. Medvedev and Ye. L. Frankevich, 'Bond Energies, Ionization Potentials and Electron Affinities', St. Martins Press, New York, 1966; (b) J. B. Birks, 'Photophysics of Aromatic Molecules', Wiley, New York, 1970.
43. C. R. Brundle and A. D. Baker (eds.), 'Electron Spectroscopy', Academic Press, New York, 1977.
44. J. A. Dean, 'Lange's Handbook of Chemistry', XIIIth edn., McGraw-Hill, New York, 1985.
45. (a) J. O. Howell and R. M. Wightman, *Anal. Chem.*, 1984, **56**, 524; (b) J. O. Howell and R. M. Wightman, *J. Phys. Chem.*, 1984, **88**, 3915; (c) D. O. Wipf, E. W. Kristensen, M. R. Deakin and R. M. Wightman, *Anal. Chem.*, 1988, **60**, 306; (d) C. Amatore, A. Jutand and F. Pflüger, *J. Electroanal. Chem. Interfacial*, 1987, **218**, 361.
46. (a) S. Chowdhury and P. Kebarle, *J. Am. Chem. Soc.*, 1986, **108**, 5453; (b) S. Chowdhury, T. Heinis and P. Kebarle, *J. Am. Chem. Soc.*, 1986, **108**, 4662; (c) E. K. Fukuda and R. T. McIver, Jr., *J. Chem. Phys.*, 1982, **77**, 4942; (d) For the electron affinities of some common acceptors see J. K. Kochi in ref. 69b, p. 503.
47. (a) U. Svanholm, O. Hammerich and V. D. Parker, *J. Am. Chem. Soc.*, 1975, **97**, 101; (b) A. Zweig and A. K. Hoffmann, *J. Org. Chem.*, 1965, **30**, 3997.
48. F. D. Saeva and G. R. Olin, *J. Am. Chem. Soc.*, 1980, **102**, 299.
49. M. R. Wasielweski and R. Breslow, *J. Am. Chem. Soc.*, 1976, **98**, 4222.
50. K. Y. Lee, D. J. Kuchynka and J. K. Kochi, to be published.
51. A. Ledwith, *Acc. Chem. Res.*, 1972, **5**, 133.
52. P. S. Mariano, in 'Organic Photochemistry', ed. A. Padwa, Dekker, New York, 1987, vol. 9, p. 10.
53. S. L. Mattes and S. Farid in 'Organic Photochemistry', ed. A. Padwa, Dekker, New York, 1983, vol. 6, p. 238.
54. J. J. McCullough, R. C. Miller, D. Fung and W.-S. Wu, *J. Am. Chem. Soc.*, 1975, **97**, 5942.
55. C. K. Mann and K. K. Barnes, 'Electrochemical Reactions in Nonaqueous Systems', Dekker, New York, 1970.
56. T. E. Banach and J. P. Dinnocenzo, *J. Am. Chem. Soc.*, 1986, **108**, 6063.
57. L. Lopez, V. Calo and R. Aurora, *J. Photochem.*, 1986, **32**, 95.
58. D. Rehm and A. Weller, *Isr. J. Chem.*, 1970, **8**, 259.
59. J. M. Masnovi and J. K. Kochi, unpublished results.
60. (a) C. J. Rhodes, *J. Am. Chem. Soc.*, 1988, **110**, 4446; (b) H. D. Roth, *Acc. Chem. Res.*, 1987, **20**, 343.
61. (a) G. Jones, II and W. G. Becker, *J. Am. Chem. Soc.*, 1983, **105**, 1276; (b) N. J. Peacock and G. B. Schuster, *J. Am. Chem. Soc.*, 1983, **105**, 3632.
62. S. Sankararaman, Y. Takahashi and J. K. Kochi, *J. Am. Chem. Soc.*, 1989, **111**, 2954.
63. (a) S. A. Chaudhuri and K. D. Asmus, *J. Phys. Chem.*, 1972, **76**, 26; (b) J. M. Masnovi, E. F. Hilinski, P. M. Rentzepis and J. K. Kochi, *J. Am. Chem. Soc.*, 1986, **108**, 1126.
64. (a) J. M. Masnovi and J. K. Kochi, *Recl. Trav. Chim. Pay-Bas*, 1986, **105**, 286; (b) K. V. Altukhov and V. V. Perekalin, *Russ. Chem. Rev. (Engl. Transl.)*, 1976, **45**, 1052.
65. E. K. Kim and J. K. Kochi, *J. Org. Chem.*, 1989, **54**, 1692. Compare S. Brownstein, E. Gabe, F. Lee and A. Piotrowski, *Can. J. Chem.*, 1986, **64**, 1661.
66. (a) Z. Rappoport, *J. Chem. Soc.*, 1963, 4498; (b) Z. Rappoport and A. Horowitz, *J. Chem. Soc.*, 1964, 1348; (c) Z. Raciszewski, *J. Chem. Soc. (B)*, 1966, 1147; (d) H. Kobashi, M. Funabashi, T. Kondo, T. Morita, T. Okada and N. Mataga, *Bull. Chem. Soc. Jpn.*, 1984, **57**, 3557; (e) E. F. Hilinski, S. V. Milton and P. M. Rentzepis, *J. Am. Chem. Soc.*, 1983, **105**, 5193.
67. O. Hammerich and V. D. Parker, *Adv. Phys. Org. Chem.*, 1984, **20**, 56.
68. M. Schlosser, *Angew. Chem., Int. Ed. Engl.*, 1964, **3**, 287.
69. (a) J. K. Kochi, *Pure Appl. Chem.*, 1971, 377; (b) J. K. Kochi, 'Organometallic Mechanisms and Catalysis', Academic Press, New York, 1978.

70. (a) C. Walling, 'Free Radicals in Solution', Wiley, New York, 1957; (b) W. A. Pryor, 'Free Radicals', McGraw-Hill, New York, 1966.
71. (a) D. C. Nonhebel and J. C. Walton, 'Free-Radical Chemistry', Cambridge University Press, London, 1974; (b) B. Giese, 'Radicals in Organic Synthesis. Formation of Carbon–Carbon Bonds', Pergamon Press, Oxford, 1986.
72. R. A. Sheldon and J. K. Kochi, *Org. React (N.Y.)*, 1972, **19**, 279.
73. (a) M. Julliard and M. Chanon, *Bull. Soc. Chim. Fr.*, 1982, 197; (b) M. Chanon, *Chem. Rev.*, 1983, **83**, 425; (c) M. Chanon and M. L. Tobe, *Angew. Chem., Int. Ed. Engl.*, 1982, **21**, 1.
74. N. L. Bauld, *Tetrahedron*, 1989, **45**, 5307.
75. S. F. Nelsen, *Acc. Chem. Res.*, 1987, **20**, 276.
76. N. L. Holy, *Chem. Rev.*, 1974, **74**, 243.
77. (a) B. J. McClelland, *Chem. Rev.*, 1964, **64**, 301; (b) K. D. Jordan and P. D. Burrow *Chem. Rev.*, 1987, **87**, 557.
78. (a) A. Yoshino, K. Yamasaki and T. Yonezawa, *J. Chem. Soc., Perkin Trans. 1*, 1975, 735; (b) K. Yamasaki, T. Yonezawa and M. Ohashi, *J. Chem. Soc., Perkin Trans. 1*, 1975, 94.
79. (a) M. Ohashi, K. Tsujimoto and Y. Furukawa, *J. Chem. Soc. Perkin Trans. 1*, 1979, 1148; (b) M. Ohashi, S. Suwa, Y. Osawa and K. Tsujimoto, *J. Chem. Soc, Perkin Trans. 1*, 1979, 2220.
80. E. K. Kim and J. K. Kochi, to be submitted for publication.
81. S. Farid and K. A. Brown, *J. Chem. Soc., Chem. Commun.*, 1976, 564; see also D. P. Kjell and R. S. Sheridan, *J. Photochem.*, 1985, **28**, 205.
82. J. M. Wallis and J. K. Kochi, *J. Am. Chem. Soc.*, 1988, **110**, 8207.
83. (a) R. S. Mulliken, *J. Am. Chem. Soc.*, 1952, **74**, 811; (b) R. S. Mulliken and W. B. Person, 'Molecular Complexes', Wiley, New York, 1969.
84. R. Foster, 'Organic Charge-Transfer Complexes', Academic Press, New York, 1969.
85. R. L. Clark and E. J. Behrman, *Inorg. Chem.*, 1975, **14**, 1425.
86. J. M. Wallis and J. K. Kochi, *J. Org. Chem.*, 1988, **53**, 1679.
87. J. M. Masnovi, J. C. Huffman, J. K. Kochi, E. F. Hilinski and P. M. Rentzepis, *Chem. Phys. Lett.*, 1984, **106**, 20.
88. M. Schröder, *Chem. Rev.*, 1980, **80**, 187.
89. L. E. Orgel and R. S. Mulliken, *J. Am. Chem. Soc.*, 1957, **79**, 4839.
90. F. Brogli and E. Heilbronner, *Theor. Chim. Acta*, 1972, **26**, 289.
91. L. Klasinc, B. Kovac and H. Gusten, *Pure Appl Chem.*, 1983, **55**, 289.
92. J. M. Masnovi and J. K. Kochi, *J. Am. Chem. Soc.*, 1985, **107**, 7880.
93. W. Lau and J. K. Kochi, *J. Am. Chem. Soc.*, 1986, **108**, 6720.
94. W. Lau and J. K. Kochi, *J. Org. Chem.*, 1986, **51**, 1801.
95. W. Lau and J. K. Kochi, *J. Am. Chem. Soc.*, 1984, **106**, 7100.
96. R. M. Dessau, S. Shih and E. I. Heiba, *J. Am. Chem. Soc.*, 1970, **92**, 412.
97. A. McKillop, A. G. Turrell, D. W. Young and E. C. Taylor, *J. Am. Chem. Soc.*, 1980, **102**, 6504.
98. (a) P. J. Andrulis, M. J. S. Dewar, R. Dietz and R. L. Hunt, *J. Am. Chem. Soc.*, 1966, **88**, 5483; (b) J. K. Kochi, R. T. Tang and J. Bernath, *J. Am. Chem. Soc.*, 1973, **95**, 7114.
99. F. R. Stermitz and W. H. Huang, *J. Am. Chem. Soc.*, 1970, **92**, 1446.
100. (a) J. Libman, *J. Am. Chem. Soc.*, 1975, **97**, 4139; (b) J. Libman, *Tetrahedron Lett.*, 1975, 2507.
101. K. Okada, K. Okamoto and M. Oda, *J. Am. Chem. Soc.* 1988, **110**, 8736.
102. H. C. Gardner and J. K. Kochi, *J. Am. Chem. Soc.*, 1976, **98**, 2460.
103. (a) D. F. Eaton, *Pure Appl. Chem.*, 1984, **56**, 1191; (b) D. F. Eaton, *J. Am. Chem. Soc.*, 1980, **102**, 3280; (c) D. F. Eaton, *J. Am. Chem. Soc.*, 1981, **103**, 7235.
104. P. G. Gassman and K. J. Bottorff, *J. Org. Chem.*, 1988, **53**, 1097.
105. (a) H. Bock, U. Stein and P. Rittmeyer, *Angew. Chem.*, 1982, **94**, 540; (b) J. Giordau and H. Bock, *Chem. Ber.*, 1982, **115**, 2548.
106. D. M. Shin and K. Kim, *Bull. Korean Chem. Soc.*, 1981, **2**, 114.
107. D. R. Arnold and A. J. Maroulis, *J. Am. Chem. Soc.*, 1976, **98**, 5931.
108. (a) J. M. Masnovi and J. K. Kochi, *J. Am. Chem. Soc.*, 1985, **107**, 6781; (b) J. M. Wallis and J. K. Kochi in ref. 82; (c) S. Sankararaman, Y. Takahashi and J. K. Kochi in ref. 62; (d) see also A. Albini and E. Fasani, *J. Am. Chem. Soc.*, 1988, **110**, 7760.
109. P. Maslak and S. L. Asel, *J. Am. Chem. Soc.*, 1988, **110**, 8260.
110. (a) P. S. Engel, A. Kitamura and D. E. Keys, *J. Org. Chem.*, 1987, **52**, 5015; (b) P. S. Engel, D. E. Keys and A. Kitamura, *J. Am. Chem. Soc.*, 1985, **107**, 4964; (c) see also S. C. Blackstock and J. K. Kochi, *J. Am. Chem. Soc.*, 1987, **109**, 2484; D. Bethell and V. D. Parker, *Acc. Chem. Res.*, 1988, **21**, 400.
111. P. C. Wong and D. R. Arnold, *Tetrahedron Lett.*, 1979, 2101. See also P. C. Du, D. A. Hrovat and W. T. Borden, *J. Am. Chem. Soc.*, 1988, **110**, 3405.
112. K. Mizuno, N. Kamiyama and Y. Otsuji, *Chem. Lett.*, 1983, 477.
113. T. Miyashi, Y. Takahashi, T. Mukai, H. D. Roth and L. M. Schilling, *J. Am. Chem. Soc.*, 1985, **107**, 1079.
114. Y. Takahashi and J. K. Kochi, *Chem. Ber.*, 1988, **121**, 253.
115. Y. Takahashi, T. Miyashi and T. Mukai, *J. Am. Chem. Soc.*, 1983, **105**, 6511.
116. T. Miyashi, M. Kamata and T. Mukai, *J. Chem. Soc., Chem. Commun.*, 1986, 1577.
117. V. R. Rao and S. S. Hixon, *J. Am. Chem. Soc.*, 1979, **101**, 6458.
118. A. Padwa, C. S. Chou and W. F. Rieker, *J. Org. Chem.*, 1980, **45**, 4555.
119. S. Hoz, M. Livneh and D. Cohen, *J. Am. Chem. Soc.*, 1987, **109**, 5149.
120. T. Miyashi, A. Konno and Y. Takahashi, *J. Am. Chem. Soc.*, 1988, **110**, 3676.
121. (a) P. S. Mariano, J. L. Stavinoha, G. Pepe and E. Meyer, *J. Am. Chem. Soc.*, 1978, **100**, 7114; (b) J. L. Stavinoha, E. Bay, A. Leone and P. S. Mariano, *Tetrahedron Lett.*, 1980, **21**, 3455. For a review see P. S. Mariano in ref. 52.
122. J. J. McCullough, W. K. MacInnis, C. J. L. Lock and R. Faggiani, *J. Am. Chem. Soc.*, 1982, **104**, 4644.

123. (a) A. J. Y. Lan, R. O. Heuckeroth and P. S. Mariano, *J. Am. Chem. Soc.*, 1987, **109**, 2738; (b) R. M. Borg, R. O. Heuckeroth, A. J. Y. Lan, S. L. Quillen and P. S. Mariano, *J. Am. Chem. Soc.*, 1987, **109**, 2728.
124. M. A. Brumfield, S. L. Quillen, U. C. Yoon and P. S. Mariano, *J. Am. Chem. Soc.*, 1984, **106**, 6855.
125. M. Lemaire, A. Guy, D. Imbert and J.-P. Guette, *J. Chem. Soc., Chem. Commun.*, 1986, 741.
126. P. Margaretha and P. Tissot, *Helv. Chim. Acta*, 1975, **58**, 933.
127. (a) A. Yoshino, M. Ohashi and T. Yonezawa, *J. Chem. Soc., Chem. Commun.*, 1971, 97; (b) K. Yamasaki, T. Yonezawa and M. Ohashi, *J. Chem. Soc., Perkin Trans. 1*, 1975, 93. For similar adduct from TCNE see M. Ohashi, S. Suwa, Y. Osawa and K. Tsujimoto, *J. Chem. Soc., Perkin Trans. 1*, 1979, 2219.
128. K. Fukui and Y. Odaira, *Tetrahedron Lett.*, 1969, 5255.
129. L. A. Deardurff, M. S. Alnajjar and D. M. Camaioni, *J. Org. Chem.*, 1986, **51**, 3686, 3693.
130. F. D. Lewis and P. E. Correa, *J. Am. Chem. Soc.*, 1981, **103**, 7347; F. D. Lewis and P. E. Correa, *J. Am. Chem. Soc.*, 1984, **106**, 194. For a review see F. D. Lewis, *Acc. Chem. Res.*, 1986, **19**, 401.
131. H.-D. Becker, *J. Org. Chem.*, 1969, **34**, 1203.
132. G. Jones, W. A. Haney and X. T. Phan, *J. Am. Chem. Soc.*, 1988, **110**, 1922.
133. (a) Y. Kanaoka, K. Sakai, R. Murata and Y. Hatanaka, *Heterocycles*, 1975, **3**, 719; (b) J. D. Coyle, G. L. Newport and A. Harriman, *J. Chem. Soc., Perkin Trans. 1*, 1978, 133; (c) M. Machida, H. Takechi and Y. Kanaoka, *Heterocycles*, 1977, **1**, 273.
134. (a) Y. Sato, H. Nokai, H. Ogiwara, T. Mizoguchi, Y. Migita and Y. Kanaoka, *Tetrahedron Lett.*, 1973, 4565; (b) Y. Sato, H. Nokai, T. Mizoguchi and Y. Kanaoka, *Tetrahedron Lett.*, 1976, 1889; (c) Y. Sato, H. Nokai, T. Mizoguchi and Y. Kanoaka, *J. Am. Chem. Soc.*, 1979, **98**, 2349.
135. M. Sako, K. Shimada, K. Hirota and Y. Maki, *J. Am. Chem. Soc.*, 1986, **108**, 6039; S. Sankararaman, S. Perrier and J. K. Kochi, *J. Am. Chem. Soc.*, 1989, **111**, 6448; A. Bhattacharya, L. M. DiMechele, V.-H. Dolling, A. W. Douglas and E. J. J. Grabowski, *J. Am. Chem. Soc.*, 1988, **110**, 3318.
136. S. Sankararaman, W. A. Haney and J. K. Kochi, *J. Am. Chem. Soc.*, 1987, **109**, 7824.
137. D. R. Arnold and A. J. Maroulis, *Synthesis*, 1979, **39**, 819.
138. (a) C. Pac, A. Nakasone and H. Sakurai, *J. Am. Chem. Soc.*, 1977, **99**, 5806; (b) J. Majima, C. Pac, A. Nakasone and H. Sakurai, *J. Am. Chem. Soc.*, 1981, **103**, 4499.
139. (a) L. Eberson and L. Jonsson, *J. Chem. Soc., Chem. Commun.*, 1980, 1187; (b) R. W. Alder, *J. Chem. Soc., Chem. Commun.*, 1980, 1184.
140. (a) R. Beugelmans, M. T. LeGoff, J. Pusset and G. Roussi, *Tetrahedron Lett.*, 1976, 2305; (b) K. Mizuno, C. Pac and H. Sakurai, *J. Chem. Soc., Chem. Commun.*, 1975, 553; (c) R. Beugelmans, H. Ginsburg, A. Lecas, M. T. LeGoff and G. Roussi, *Tetrahedron Lett.*, 1978, 3271.
141. R. Filler, *Isr. J. Chem.*, 1978, **17**, 71.
142. L. Eberson, L. Jonsson and O. Sanneskog, *Acta Chem. Scand., Ser. B*, 1985, **39**, 113.
143. (a) K. Yoshida, *J. Am. Chem. Soc.*, 1979, **101**, 2116; (b) C. Pac, A. Nakasone and H. Sakurai in ref. 138a.
144. K. Mizuno, M. Ishii and Y. Otsuji, *J. Am. Chem. Soc.*, 1981, **103**, 5570.
145. (a) P. G. Gassman, K. D. Olson, L. Walter and R. Yamaguchi, *J. Am. Chem. Soc.*, 1981, **103**, 4977; (b) P. G. Gassman and J. L. Smith, *J. Org. Chem.*, 1983, **48**, 4438; (c) P. G. Gassman and K. D. Olson, *J. Am. Chem. Soc.*, 1982, **104**, 3740.
146. (a) F. A. Bell, R. A. Crellin, H. Fujii and A. Ledwith, *J. Chem. Soc., Chem. Commun.*, 1969, 251; (b) A. Ledwith in ref. 2.
147. R. A. Carruthers, R. A. Crellin and A. Ledwith, *J. Chem. Soc., Chem. Commun.*, 1969, 252.
148. D. J. Bellville, D. D. Wirth and N. L. Bauld, *J. Am. Chem. Soc.*, 1981, **103**, 718.
149. (a) N. L. Bauld and R. Pabon, *J. Am. Chem. Soc.*, 1983, **105**, 633; (b) K. T. Lorenz and N. L. Bauld, *J. Am. Chem. Soc.*, 1987, **109**, 1157.
150. (a) R. A. Neunteufel and D. R. Arnold, *J. Am. Chem. Soc.*, 1973, **95**, 4080; for reviews see (b) R. A. Caldwell and D. Creed, *Acc. Chem. Res.*, 1980, **13**, 45; (c) S. L. Mattes and S. Farid, *Acc. Chem. Res.*, 1982, **15**, 80.
151. F. Radner, *J. Org. Chem.*, 1988, **53**, 702; see also A. McKillop *et al.* in ref. 97.
152. R. Sustmann, M. Dern, R. Kasten and W. Sicking, *Chem. Ber.*, 1987, **120**, 1315.
153. D. Dern, H.-G. Korth, G. Kopp and R. Sustmann, *Angew. Chem., Int. Ed. Engl.*, 1985, **24**, 337.
154. R. A. Pabon, D. J. Bellville and N. L. Bauld, *J. Am. Chem. Soc.*, 1984, **106**, 2730.
155. K. A. Brown, S. L. Mattes and S. Farid, *J. Am. Chem. Soc.*, 1978, **100**, 4162; see also Farid and Brown in ref. 81.
156. R. Akaba, S. Aihara, H. Sakuragi and K. Tokumaru, *J. Chem. Soc., Chem. Commun.*, 1987, 1262.
157. J. W. Bruno, T. J. Marks and F. D. Lewis, *J. Am. Chem. Soc.*, 1981, **103**, 3608.
158. J. Eriksen and C. S. Foote, *J. Am. Chem. Soc.*, 1980, **102**, 6083.
159. L. E. Manring, J. Eriksen and C. S. Foote, *J. Am. Chem. Soc.*, 1980, **102**, 4275.
160. I. Saito, K. Tamoto and T. Matsura, *Tetrahedron Lett.*, 1979, 2889.
161. F. D. Lewis and M. Kojima, *J. Am. Chem. Soc* 1988, **110**, 8664.
162. (a) S. Sankararaman and J. K. Kochi, *Recl. Trav. Chim. Pays-Bas*, 1986, **105**, 278; (b) S. Sankararaman, W. A. Haney and J. K. Kochi, *J. Am. Chem. Soc.*, 1987, **109**, 5235.
163. J. M. Masnovi and J. K. Kochi, *Recl. Trav. Chim. Pays-Bas*, 1986, **105**, 289.
164. (a) K. Kim, V. J. Hull and H. J. Shine, *J. Org. Chem.*, 1974, **39**, 2534; (b) J. J. Silber and H. J. Shine, *J. Org. Chem.*, 1971, **36**, 2923.
165. J. M. Masnovi, S. Sankararaman and J. K. Kochi, *J. Am. Chem. Soc.*, 1989, **111**, 2263.
166. R. A. Pabon, D. J. Bellville and N. L. Bauld, *J. Am. Chem. Soc.*, 1983, **105**, 5158.
167. J. P. Dinnocenzo and T. E. Banach, *J. Am. Chem. Soc.*, 1986, **108**, 6063.
168. T. J. Richardson and N. Bartlett, *J. Chem. Soc., Chem. Commun.*, 1974, 427.
169. M. Ohashi, K. Tsujimoto and Y. Kurakawa, *Chem. Lett.*, 1977, 543.
170. M. A. Leoni, G. F. Betlinetti, G. Minoli and A. Albini, *J. Org. Chem.*, 1980, **45**, 2331.
171. N. Kornblum, T. M. Davies, G. W. Earl, N. L. Holy, R. C. Kerber, M. T. Musser and D. H. Snow, *J. Am. Chem. Soc.*, 1967, **89**, 725.
172. N. Kornblum, R. T. Swiger, G. W. Earl, H. W. Pinnick and F. W. Stuchal, *J. Am. Chem. Soc.*, 1970, **92**, 5513.

173. (a) R. C. Kerber, G. W. Urry and N. Kornblum, *J. Am. Chem. Soc.*, 1964, **86**, 3904; (b) R. C. Kerber, G. W. Urry and N. Kornblum, *J. Am. Chem. Soc.*, 1965, **87**, 4520; (c) N. Kornblum, R. E. Michel and R. C. Kerber, *J. Am. Chem. Soc.*, 1966, **88**, 5660, 5662; (d) G. A. Russell and W. C. Danen, *J. Am. Chem. Soc.*, 1966, **88**, 5663.
174. N. Ono, H. Miyake, R. Tamura and A. Kaji, *Tetrahedron Lett.*, 1981, **22**, 1705.
175. (a) D. D. Tanner, E. V. Blackburn and G. E. Diaz, *J. Am. Chem. Soc.*, 1981, **103**, 1557. (b) J. Dupuis, B. Giese, J. Hartung, M. Leisung, H. G. Korth and R. Sustmann, *J. Am. Chem. Soc.*, 1985, **107**, 4332.
176. A. L. J. Beckwith and G. F. Meijs, *J. Chem. Soc., Chem. Commun.*, 1981, 136.
177. G. L. Closs and S. H. Goh, *J. Chem. Soc., Perkin Trans. 2*, 1972, 1473.
178. (a) C. H. Wang, S. M. Linnell and W. Wang, *J. Org. Chem.*, 1971, **36**, 525; (b) J. A. K. Harmony, 'Methods in Free Radical Chemistry', ed. E. Huyser, Dekker, New York, 1974, vol. 5, p. 101.
179. E. S. Huyser, J. A. K. Harmony and F. L. McMillan, *J. Am. Chem. Soc.*, 1972, **94**, 3176.
180. S. Fukuzumi, K. Hironaka and T. Tanaka, *J. Am. Chem. Soc.*, 1983, **105**, 4722.
181. (a) E. C. Ashby and R. DePriest, *J. Am. Chem. Soc.*, 1982, **104**, 6144; *Acc. Chem. Res.*, 1988, **21**, 415; (b) H. G. Kuivila and M. S. Alnajjar, *J. Am. Chem. Soc.*, 1982, **104**, 6146.
182. (a) J. A. Barltrop and A. J. Owers, *J. Chem. Soc., Chem. Commun.*, 1970, 1462; (b) J. A. Barltrop, *Pure Appl. Chem.*, 1973, **33**, 179.
183. (a) A. J. Birch, *Q. Rev., Chem. Soc.*, 1950, **4**, 69; (b) R. G. Harvey, *Synthesis*, 1970, 161.
184. J. Honzl and J. Lovy, *Tetrahedron*, 1984, **40**, 1885.
185. K. A. Bilevitch, N. N. Bubnov and O. Yu. Okhlobystin, *Tetrahedron Lett.*, 1968, 3465.
186. N. N. Bubnov, K. A. Bilevitch, L. A. Poljakova and O. Yu. Okhlobystin, *J. Chem. Soc., Chem. Commun.*, 1972, 1058.
187. L. E. Marning, C. I. Gu and C. S. Foote, *J. Phys. Chem.*, 1983, **87**, 40.
188. R. Adams and E. W. Adams, *Org. Synth.*, 1932, **1**, 448.
189. G. A. Russell and E. G. Janzen, *J. Am. Chem. Soc.*, 1962, **84**, 4153.
190. P. R. Singh and R. Kumar, *Aust. J. Chem.*, 1972, **25**, 2133.
191. T. J. Katz, *J. Am. Chem. Soc.*, 1960, **82**, 3785.

Author Index

This Author Index comprises an alphabetical listing of the names of over 7000 authors cited in the references listed in the bibliographies which appear at the end of each chapter in this volume.

Each entry consists of the author's name, followed by a list of numbers, each of which is associated with a superscript number. For example

Abbott, D. E., 6[12,12c], 10[40], 573[53,54]

The numbers indicate the text pages on which references by the author in question are cited; the superscript numbers refer to the reference number in the chapter bibliography. Citations occurring in the text, tables and chemical schemes and equations have all been included.

Although much effort has gone into eliminating inaccuracies resulting from the use of different combinations of initials by the same author, the use by some journals of only one initial, and different spellings of the same name as a result of transliteration processes, the accuracy of some entries may have been affected by these factors.

Subject Index